ANIMAL PHYSIOLOGY
FOURTH EDITION

Theme	An Example of the Theme in Action	See Pages
The Dynamic State of Body Constituents: Great quantities of many of the key constituents of the body are added and subtracted every day in many animals under many conditions. Thus the constituents of the body—far from being static—are continuously in a dynamic state of flux. This is true even though additions and subtractions are often relatively balanced, resulting in relatively constant concentrations (a phenomenon termed *homeostasis*).	Averaged over the course of an ordinary 24-h day, an adult person is likely to process more than 2 kg of adenosine triphosphate (ATP) each hour, synthesizing that amount of ATP from adenosine diphosphate (ADP) and, with only a short delay, breaking it back down to ADP. To synthesize the ATP, the person—during each hour—will use about 20 liters of oxygen (O_2) that he or she takes up from the atmosphere. During a 24-h day, the oxygen used will combine with almost 100 g (a fifth of a pound) of hydrogen atoms that have been removed from food molecules, forming about 800 milliliters of water. This water is added to the body fluids.	12–13 *189–190* 385 723–724 769 (Fig. 28.21)
Multiple Forms of Key Molecules: Animals have often evolved multiple molecular forms (sometimes called *isoforms*) of particular proteins or other sorts of molecules. Physiologists hypothesize that when two species or two tissues exhibit different molecular forms of a molecule, the forms are often specialized to function in the specific settings in which the animals live or the tissues function.	The cell membranes of all animals are composed principally of lipid molecules. Physiologists have found, however, that the membranes of all animals are not composed of chemically identical lipid molecules. Instead, multiple molecular forms of lipids are employed by different animals living under different circumstances. Cold-water fish species, for instance, construct their cell membranes using molecular forms of lipids that are less likely to harden at low temperatures than the molecular forms synthesized by warm-water species.	*38–39* *(Fig. 2.3)* 250–251 (Fig. 10.19) 552–553 639 (Fig. 24.2) 656–657 (Fig. 24.20)
Phenotypic Plasticity: An individual animal is often able to change its phenotype in response to changes in the particular circumstances under which it is living (e.g., its particular environment). This ability of an individual animal to adopt two or more phenotypes despite having a fixed genotype is termed *phenotypic plasticity*.	Animals that eat only occasionally, such as pythons, often alternate between two intestinal phenotypes. When they have not had a meal for weeks, their intestinal tract is physically small, and it has poorly developed molecular mechanisms for absorbing food. After a meal, the tissues of the intestinal tract enlarge greatly, and the intestinal tract expresses well-developed absorption mechanisms.	16–18 (Fig. 1.10) 82 (Table 3.1) 94–97 *161–162 (Fig. 6.23)* 273 (Fig. 10.40) 570 (Fig. 21.5)
Interdependency of Function and Form: The *function* of a biological system typically cannot be understood without knowledge of its *structure*, and vice versa.	The kidney tubules of mammals not only produce the most concentrated urine observed in vertebrates but also differ from other vertebrate kidney tubules in that they all have distinctive hairpin shapes. Physiologists have shown that the *functional* ability to produce highly concentrated urine depends on the hairpin *structure*, which guides the urine (as it is being formed) to flow first in one direction and then in the opposite direction.	144–145 (Fig. 6.13) 268–269 (Fig. 10.35) 380 (Fig. 14.10) 670 (Fig. 25.1) *794 (Fig. 29.12)*
Applicability of the Laws of Chemistry and Physics: Animals must adhere to the laws of chemistry and physics. Sometimes chemistry and physics act as constraints, but sometimes animals gain advantages by evolving systems that capitalize on particular chemical or physical principles.	Heat transfer through air follows different physical laws when the air is still rather than moving; heat tends to move much more slowly through still air than moving air. Animals cannot change such laws of physics. They sometimes can affect which law applies to them, however, as when the ancestors of mammals evolved fur. The hairs of a furred mammal keep the layer of air next to the body relatively motionless. Heat transfer through that air is therefore slow, helping mammals retain internal heat when living in cold environments.	*238–239* 505 (Fig. 18.8) 592 (Box 22.2) 716–717
The Interdependency of Levels of Organization: An animal's *overall* functional properties depend on how its *tissues* and *organs* function, and the function of its tissues and organs depends on how its *cells* and *molecular systems* function. All these levels of organization are interdependent. An important corollary is that properties at one level of organization often cannot be fully understood without exploring other levels of organization.	When your physician strikes a tendon near your knee with a mallet, your leg straightens. For this response, electrical signals must travel along nerve cells to the spinal cord and back. The rate of travel depends in part on the *molecular* properties of ion-transporting proteins in the cell membranes of the nerve cells. It also depends in part on key *cellular* properties, such as the spacing between the sections of each nerve cell membrane that are fully exposed to the fluids bathing the cell. Molecular and cellular properties of these sorts determine the *overall* properties of the process. For instance, they determine the length of time that passes between the moment the mallet strikes and the moment your leg muscles contract.	5–6 (Figs. 1.2, 1.3) 120 (Box 5.2) 206 (Fig. 8.12) *316–318* 521 (Fig. 19.4)
The Crucial Importance of Control Mechanisms: In addition to mechanisms for reproducing, breathing, moving, and carrying out other overt functions, animals require *control* mechanisms that orchestrate the other mechanisms. The control mechanisms—so diverse that they include controls of gene expression as well as those exerted by the nervous and endocrine systems—determine the relations between inputs and outputs in physiological systems. They thereby crucially affect the functional properties of animals.	Although sheep and reindeer are born at cold times of year, newborns receive no heat from their parents and must keep warm on their own or die. They possess a process for rapid heat production. Proper control of this process requires that it be activated at birth, but not before birth when it would tend needlessly to exhaust fetal energy supplies. The control mechanism has two key properties: It activates heat production when neural thermal sensors detect cold, but its capacity to activate heat production is turned off by chemical factors secreted by the placenta. The control mechanism remains in a turned-off state until a newborn is separated from the placenta at birth. The cold environment is then able to stimulate rapid heat production.	54 (Fig. 2.19) 260–261 (Box 10.3) *293 (Fig. 11.10)* 492 (Fig. 17.18)

Animal Physiology

FOURTH EDITION

Richard W. Hill
Michigan State University

Gordon A. Wyse
University of Massachusetts, Amherst

Margaret Anderson
Smith College

Sinauer Associates, Inc. Publishers
Sunderland, Massachusetts

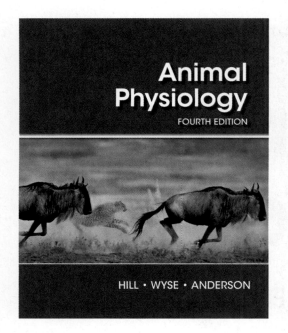

About the Cover

"Physiology in action" aptly describes this scene, photographed in the Serengeti eco-system in Tanzania, East Africa. The cheetahs have just sprinted to make a close ap-proach to the wildebeest antelopes they are hunting. Now exercise physiology will de-termine the outcome. The running speeds of the two species depend on how fast their muscle cells can produce ATP (adenosine triphosphate), the required energy source for muscular work. As soon as the cheetahs started sprinting, they started making ATP anaerobically. This source of ATP is fast and abundant, but short-lived. Thus the chee-tahs will tire quickly. The wildebeests need evade them for only another minute or two to survive. Although cheetahs usually hunt by themselves, family groups like this one sometimes hunt together. Photo © Fred Ward/Corbis.

Address editorial correspondence and orders to:
Sinauer Associates, 23 Plumtree Road, Sunderland, MA 01375 U.S.A.
FAX: 413-549-1118
Email: publish@sinauer.com
Internet: www.sinauer.com

Library of Congress Cataloging-in-Publication Data

Names: Hill, Richard W. | Wyse, Gordon A. | Anderson, Margaret, 1941-
Title: Animal physiology / Richard W. Hill, Michigan State University, Gordon
  A. Wyse, University of Massachusetts, Amherst, Margaret Anderson, Smith
  College.
Description: Fourth edition. | Sunderland, Massachusetts : Sinauer
  Associates, Inc., [2016] | Includes bibliographical references and index.
Identifiers: LCCN 2016005528 | ISBN 9781605354712 (casebound)
Subjects: LCSH: Physiology, Comparative.
Classification: LCC QP33 .H54 2016 | DDC 571.1--dc23
LC record available at http://lccn.loc.gov/2016005528

Printed in U. S.A
8 7 6 5 4 3 2 1

Preface

Our most important goal in this book is to articulate the themes and overarching hypotheses of modern animal physiology. In the contemporary information era, individual facts are easier to find than ever. Themes—which collect facts into useful, coherent stories and worldviews—have always been the key challenge in textbook writing. Today that challenge is greater than ever because of the ceaseless escalation in the numbers of available facts. While recognizing this, our central goal remains the same as it was in earlier editions. We have not sought to produce an encyclopedia. Instead, we emphasize themes, worldviews, and overarching hypotheses.

Another of our key goals is to use graphics, drawings, and photographs to maximum advantage to communicate knowledge of animal physiology. Thus we have an ambitious art program. For example, as often as possible, we include drawings and photographs of animals being discussed, to ensure that readers are familiar with the subjects under discussion.

Clear, rigorous explanations of physiological concepts are a third key goal. Toward this end, we work with experienced reviewers, editors, and artists to address complex material with lucid text and revealing conceptual diagrams.

In these pages, we consistently and deliberately address animal physiology as a discipline integrated with other disciplines in biology—especially molecular biology, genetics, evolutionary biology, and ecology. We also consistently emphasize the roles of physiology throughout the life cycles of animals by discussing physiological development and by examining animal function during such important life-cycle processes as exercise, long-distance migration, seasonal rhythms, and accommodation to severe conditions (we generally omit pathology and parasitism, however). Although we give particular attention to mammals, we make a point of recognizing the other vertebrate groups and at least the arthropods and molluscs among invertebrates. We address all levels of organization that are germane, from the genome to the ecology of the animals involved.

We want to mention three specific strategies we have adopted to add interest and breadth to the book. First, we start every chapter with a vivid example of the application of the chapter's material to the lives of animals in their natural habitats. Second, we devote five entire chapters (our "At Work" chapters) to in-depth explorations of how physiologists do their work; in these chapters we break out of the usual textbook mold to discuss exciting topics—such as the diving physiology of seals and whales—with emphasis on experiments, theory maturation, integration of physiological systems, and prospects for future research. Third, we invite specialists to contribute expert Guest Boxes on emerging topics that expand the book's scope.

With our aspirations being as numerous as we have described, we have put a great deal of effort into balancing competing demands for space. The product is a complete physiology textbook that in one volume will meet the requirements of a diversity of one- or two-semester courses in animal function. Our intended audience is sophomores through beginning graduate students. To make the book accessible to as wide an audience as possible, we include both a glossary of more than 1500 terms and 11 appendices on important background concepts.

Our approach to the writing has been to work from the original scientific literature and obtain extensive peer review. Another aspect of our approach is that we have opted for the pedagogical consistency of a book written by just three principal authors. Margaret Anderson wrote Chapters 16, 20, and 21, and Gordon Wyse wrote Chapters 12–15, 18, and 19. Richard Hill wrote Chapters 1–11, 17, and 22–30. David S. Garbe, Scott A. Huettel, Matthew S. Kayser, Kenneth J. Lohmann, and Margaret McFall-Ngai wrote Guest Boxes. Matthew S. Kayser and Gordon Fain assisted with topic development in certain parts of the principal text.

The book is organized in modular fashion with the express purpose of providing instructors and students with flexibility in choosing the order in which they move through the book. The first of the six parts (modules) consists of Chapters 1 to 5, which are background chapters for the book as a whole. Most instructors will want to assign those chapters at the beginning of the course of study. Each of the subsequent five parts of the book is written to be free-standing and self-contained, so that students who have mastered the material in Part I will be well prepared to work through any of the other five parts. Two of the final five parts begin with explicitly introductory chapters that present fundamentals. All five of these parts end with "At Work" chapters. Within a part, although chapters will probably be best read in order, most chapters are themselves written to be relatively self-contained, meaning that the order of reading chapters within a part is flexible. Three additional features promote flexibility in the order of reading: the glossary, the new index, and page cross-references. Text is extensively cross-referenced both forward and backward, so that instructors and students can link material across chapters.

Throughout the book we strive to take advantage of all the assets of traditional book production to achieve a text that integrates the full range of relevant pedagogical elements to be a first-rate learning tool. Many sorts of professionals have important contributions to make for a book to be excellent. Thus many sorts of professionals have traditionally found personal fulfillment by engaging in the cooperative, synergistic production of books. The authors listed on the cover are just the tip of the iceberg. A book's art program depends on scientific illustrators. Coordination between the art and the text—a key to the success of any textbook—depends on the editorial expertise of the book's editors. An attractive science text needs to be designed and physically executed by talented people who combine scientific acumen with artistic sensibility. This book is the creative product of a team of at least a dozen people playing diverse, mutually reinforcing roles—all dedicated to providing students with a superior textbook.

We have tried to keep animals front and center. At the end of our production, as the orchestra goes silent and the klieg lights dim, we hope that animals leading their lives in their natural habitats will be the enduring image and memory left by this book—animals now better understood, but still with much to attract the curiosity of upcoming generations of biologists.

New to this Edition

We have added genomics material to most chapters to reflect the rapidly growing role of genomics and related disciplines in the modern study of animal physiology. As in all new editions, two of our central goals were to update content and enhance pedagogical effectiveness throughout. To these ends, we have reconsidered every sentence and every element of the art program. Now the total of figures and tables exceeds 750, of which over 130 are revised or new. Chapters that have received exceptional attention are: Chapter 4 (development and epigenetics), Chapter 6 (nutrition, feeding, and digestion), Chapter 10 (thermal relations), Chapter 14 (sensory processes), Chapter 17 (reproduction), Chapter 18 (animal navigation), Chapter 19 (control of movement), and Chapter 29 (kidney physiology, edited more completely to emphasize plasma regulation). The index is new and upgraded. In addition to the genomics content already mentioned, many topics have been added or substantially upgraded, including (but not limited to): bioluminescence, cardiac muscle of vertebrates, CRISPR, evolution of electric fish, gills of ram-ventilating fish, human individual variation in maximal rate of O_2 consumption, human thermoregulation, hunger and satiation, insect juvenile hormone, metabotropic sensory transduction, monarch butterfly migration, neuron population coding, ocean acidification, opah endothermy, owl auditory localization, PGC-1α roles in exercise training, physiological "personalities," python gut and cardiac genomics, voltage-gated channels, and TRP channels.

Acknowledgments

In the preparation of this edition, we were assisted in particularly memorable ways by Arnoldus Blix, Matthew Kayser, Zachary Nuttall, and David Hillis. Caroline R. Cartwright, Christina Cheng, and Nigel D. Meeks provided new photographs from their research. Special thanks also to Ewald R. Weibel, who in a sense has been a lifelong friend of this project; he contributed scanning EMs to the very first published version of this book, and for this edition he has provided a marvelous new image of human lung morphology. We also want to thank Anapaula S. Vinagre, Luciano S. de Fraga, Luiz C. R. Kucharski, and the rest of their team in Brazil, who have honored us with their enthusiasm for our book, which they have translated into Portuguese.

Our peer reviewers are particularly important to the quality of the book, although we of course accept full responsibility and at times have followed our own inclinations rather than theirs. We are thus happy to acknowledge our current peer reviewers as well as individuals who acted as reviewers for earlier editions and whose influence remains evident: Doris Audet, Brian Bagatto, Michael Baltzley, Jason Blank, Robert Bonasio, Charles E. Booth, Eldon Braun, Warren Burggren, Heather Caldwell, John Cameron, Jeffrey C. Carrier, David Coughlin, Sheldon Cooper, Daniel Costa, Emma Creaser, David Crews, Elissa Derrickson, Stephanie Gardner, Stephen Gehnrich, Steve George, Joseph Goy, Bernd Heinrich, Raymond Henry, James Hicks, Carl S. Hoegler, Richard Hoffman, Mark A. Holbrook, Jason Irwin, Steven H. Jury, William Karasov, Fred J. Karsch, Leonard Kirschner, Andor Kiss, Courtney Kurtz, Mark Kyle, Sharon Lynn, Megan M. Mahoney, Robert Malchow, Duane McPherson, Jennifer Marcinkiewicz, Ulrike Muller, Barbara Musolf, Randy Nelson, Scott Parker, Gilbert Pitts, Fernando Quintana, Matthew Rand, Beverly Roeder, Susan Safford, Malcolm Shick, Bruce Sidell, Mark Slivkoff, Paul Small, George Somero, Frank van Breukelen, John VandenBrooks, Itzick Vatnick, Curtis Walker, Zachary Weil, Alexander Werth, Eric Widmaier, and Heather Zimbler-DeLorenzo.

Another group to whom we offer special thanks are the many scientists who in the past have provided us with photographs, drawings, or unpublished data for direct inclusion in this book: Jonathan Ashmore, William J. Baker, Lise Bankir, Jody M. Beers, Rudolf Billeter-Clark, Walter Bollenbacher, Richard T. Briggs, Klaus Bron, Marco Brugnoli, Jay Burnett, Daniel Costa, Matthew Dalva, Hans-Ranier Duncker, Aaron M. Florn, Jamie Foster, Peter Gillespie, Greg Goss, Bernd Heinrich, Dave Hinds, Michael Hlastala, Hans Hoppeler, José Jalife, Kjell Johansen, Toyoji Kaneko, Matthew S. Kayser, Mary B. Kennedy, Andor Kiss, Daniel Luchtel, David Mayntz, Margaret McFall-Ngai, Nathan Miller, Eric Montie, Michael Moore, Mikko Nikinmaa, Sami Noujaim, Dan Otte, Thomas Pannabecker, R. J. Paul, Steve Perry, Bob Robbins, Ralph Russell, Jr., Josh Sanes, Klaus Schulten, Stylianos Scordilis, Bruce Sidell, Helén Nilsson Sköld, Jake Socha, Kenneth Storey, Karel Svoboda, Emad Tajkhorshid, Irene Tieleman, Christian Tipsmark, Shinichi Tokishita, Walter S. Tyler, Tom Valente, Tobias Wang, Rüdiger Wehner, Judith Wopereis, and Eva Ziegelhoffer.

Thanks are due too for encouragement, feedback, and other help with writing that we have gratefully received from Brian Barnes, Richard T. Briggs, Alex Champagne, John Dacey, Giles Duffield, Aaron M. Florn, Fritz Geiser, Loren Hayes, Gerhard Heldmaier, Richard Marsh, Steve Perry, George Somero, Mark Vermeij, Tobias Wang, and Joseph Williams.

Of course, no book of this scope emerges fully formed in a single edition. Thus, we also thank the following who played important roles in earlier versions of this work: Simon Alford, Kellar Autumn, Robert Barlow, Al Bennett, Eric Bittman, Jeff Blaustein, Batrice Boily, Beth Brainerd, Richard C. Brusca, Gary Burness, Bruce Byers, John Cameron, Donald Christian, Barbara Christie-Pope, Corey Cleland, Randal Cohen, Joseph Crivello, Peter Daniel, Bill Dawson, Gregory Demas, Linda Farmer, Jane Feng, Milton Fingerman, Dale Forsyth, Christopher Gillen, Kathleen Gilmour, Judy Goodenough, Edward Griff, Jacob Gunn, James Harding, Jean Hardwick, John Harley, Ian Henderson, Kay Holekamp, Charles Holliday, Henry John-Alder, Kelly Johnson, Alexander Kaiser, Reuben Kaufman, M. A. Q. Khan, William Kier, Peter King, Rosemary Knapp, Heather Koopman,

Richard Lee, John Lepri, Robert Linsenmeier, Stephen Loomis, William Lutterschmidt, Steffen Madsen, Don Maynard, Grant McClelland, Kip McGilliard, Stephen McMann, Allen Mensinger, Tim Moerland, Thomas Moon, Thomas Moylan, Richard Nyhof, David O'Drobinak, Linda Ogren, Sanford Ostroy, Christine Oswald, Linda Peck, Sandra Petersen, Chuck Peterson, Richard Petriello, Nathan Pfost, Robert Rawding, Heinrich Reichert, Larry Renfro, David Richard, R. M. Robertson, Robert Roer, William Seddon, Brent Sinclair, Laura Smale, Amanda Southwood, Tony Stea, Philip Stephens, Georg Striedter, Rebekah Thomas, Heather Thompson, Irene Tieleman, Lars Tomanek, Terry Trier, Kay Ueno, Joshua Urio, Mark Wales, Winsor Watson, Leonard E. White, Susan Whittemore, Steve Wickler, Robert Winn, and Tom Zoeller.

Of the many colleagues who have made contributions, Richard Hill would like in particular to thank Kjell Johansen, one of the greats, who way back at the beginning said without a moment's hesitation, "This is good." Energy still emanates from those words decades later.

Thanks to our students, who have challenged us, encouraged us, taught us, and—if nothing else—listened to us over our many years of classroom teaching. Our classes with our students have been our proving ground for teaching physiology and our most fundamental source of reinforcement to take on a project of this magnitude. We are grateful to work and teach at institutions—Michigan State University, the University of Massachusetts, and Smith College—at which efforts of this sort are possible.

Special thanks to Andy Sinauer, who has helped us to think big and provided the resources to realize ambitious goals for four editions. We have all worked with many editors and publishers in our careers, and Andy is tops: an entrepreneur dedicated to putting the life of ideas on the printed page. We also extend special thanks to our editor, Laura Green, who has brought expertise and sound judgment to our work on every aspect of the book, including text, art, and pedagogy. Warm thanks, too, to Joanne Delphia, production specialist, who has completely redesigned the book for this edition; David McIntyre, photo editor; head of production Chris Small; and the others at Sinauer Associates whose talents and dedication have been indispensable. We feel privileged to have had Elizabeth Morales execute the art, which makes such a contribution to our pages. Elizabeth Pierson once again has helped with her exceptional talent as a copy editor.

We each have particular thanks to offer to the people in our personal lives whose support and patience have been indispensable. Richard Hill thanks Sue, Dave, and Chrissie, who have always been there even though the hours of writing have often meant long waits between sightings of their husband and father. Sue in particular has been a major contributor by repeatedly offering the benefits of her knowledge and judgment as a biologist. Gordon Wyse thanks Mary for her editorial talents, support, and willingness to keep planning around this long project, and Jeff, Karen, and Nancy for inspiration. Likewise, Margaret Anderson expresses gratitude to her family, especially Andy and Anita, and to her friends and students, whose boundless enthusiasm and idealism provide great inspiration.

While acknowledging the many ways others have helped, we of course accept full responsibility for the finished product and invite readers' opinions on how we could do better. Please contact us with your observations.

One of the gratifications of writing a book like this is the opportunity to participate in the raw enthusiasm of scientists for science. On countless occasions, many colleagues have performed great favors on short notice without the slightest hint of wanting pay for their professional expertise. Pure science must be one of the last redoubts of this ethic in today's professional world. We are honored to play the role of synthesizing and communicating the insights and questions that arise from the exciting search for knowledge.

Richard W. Hill
East Lansing, Michigan

Gordon A. Wyse
Amherst, Massachusetts

Margaret Anderson
Northampton, Massachusetts

March 2016

To Our Readers

If you've ever been to a show and one of the producers stepped out on stage before the curtain went up to offer remarks about the upcoming event, you will understand the nature of these two pages. We, your authors, want to say a few words about the way we approached writing this book. We would also like to mention how we have handled several challenging issues.

One of our primary goals has been to create a book in which you will find the fascination of physiology as well as its content. Thus we start each of the 30 chapters with an intriguing example that illustrates the application of the chapter to understanding the lives of animals. Collectively, these examples highlight the many ways in which the study of physiology relates to biology at large.

Besides our desire to emphasize the fascination of physiology, we have also wanted to stress the importance of integrating knowledge across physiological disciplines—and the importance of integrating physiology with ecology, behavior, molecular biology, and other fields. We have wanted, in addition, to discuss how concepts are tested and revised during research in physiology and to focus on the cutting edges in physiological research today. To help meet these goals, we include five "At Work" chapters, which appear at the ends of five of the book's six parts. You will find that the "At Work" chapters are written in a somewhat different style than the other chapters because they give extra emphasis to the *process* of discovery. The topics of the "At Work" chapters are deliberately chosen to be intriguing and important: diving by seals and whales, animal navigation, muscle in states of use (e.g., athletic training) and disuse, mammals in the Arctic, and desert animals. Each "At Work" chapter uses concepts introduced in the chapters preceding it. We hope you will find these chapters to be something to look forward to: enjoyable to read and informative.

In the information age, with easy access to facts online, you might wonder, why should I take the course in which I am enrolled and why should I read this book? The answer in a few words is that the extraordinary quantities of information now available create extraordinary challenges for synthesis.

To explain, the more information each of us can locate, the more we need frameworks for organizing knowledge. Scientists, philosophers, and historians who comment on the practice of science are almost never of one mind. Nonetheless, they nearly universally reach one particular conclusion: In science, the mere accumulation of facts leads quite literally nowhere. The successful pursuit of scientific knowledge requires testable concepts that organize facts. Scientists create concepts that organize raw information. Then, in science, it is these concepts that we test for their accuracy and utility. A good course taught with a good textbook provides a conceptual framework into which raw information can be fitted so that it becomes part of the essential life of ideas and concepts.

We hope we have provided you not simply with a conceptual framework, but one that is "good for the future." By this we mean we have not tried merely to organize the knowledge already available. We have tried in equal measure to articulate a conceptual framework that is poised to grow and mature as new knowledge becomes available.

Just briefly we want to comment on three particular topics. First, our Box design. Boxes that start on the pages of this book often continue online. To find the web content, go to the book's website that is mentioned prominently at the end of each chapter. The part of a Box that you will read online is called a Box Extension. All the Box Extensions are fully integrated with the rest of the book in terms of concepts, terminology, and artistic conventions. Moreover, many of the Box Extensions are extensive and include informative figures. Thus, we urge that you keep reading when a Box directs you to a Box Extension.

Second, units of measure. For 40 years there has been a revolution underway focused on bringing all human endeavor into line with a single system of units called the Système International (SI). Different countries have responded differently, as have different fields of activity. Thus, if you purchase a box of cereal in much of the world, the cereal's energy value will be quoted on the box in kilojoules, but elsewhere it will be reported in kilocalories. If you go to a physician in the United States and have your blood pressure measured, you will have it reported in millimeters of mercury, but if you read a recent scientific paper on blood pressures, the pressures will be in kilopascals. The current state of transition in units of measure presents challenges for authors just as it does for you. We have tried, in our treatment of each physiological discipline, to familiarize you with the pertinent units of measure you are *most likely to encounter* (SI or not). Moreover, you will find in Appendix A an extensive discussion of the Système International and its relations to other systems of units.

The third and final specific matter on our minds is to mention our referencing system. For each chapter, there are three reference lists: (1) a brief list of particularly important or thought-provoking references at the end of the chapter, (2) a longer list of references in the section titled Additional References at the back of the book, and (3) a list of all the references cited as sources of information for figures or tables in the chapter. The final list appears in the Figure and Table Citations at the back of the book.

All three of us who wrote this book have been dedicated teachers throughout our careers. In addition, we have been fortunate to develop professional relationships and friendships with many of our students. This book is a product of that two-way interaction. In the big universities today, there are many forces at work that encourage passivity and anonymity. We urge the opposite. We encourage you to talk science as much as possible with each other and with your instructors, whether in classroom discussions, study groups, office hours, or other contexts. Active learning of this sort will contribute in a unique way to your enjoyment and mastery of the subjects you study. We have tried, deliberately, to write a book that will give you a lot to talk about.

Richard W. Hill

Gordon A. Wyse

Margaret Anderson

Media and Supplements
to accompany
Animal Physiology, Fourth Edition

For the Student
Companion Website

(sites.sinauer.com/animalphys4e)
The *Animal Physiology* Companion Website includes content that expands on the coverage in the textbook plus convenient study and review tools. The site includes the following resources:

- *Chapter Outlines* include all of the headings and sub-headings from each chapter.
- *Chapter Summaries* provide quick chapter overviews.
- *Box Extensions* expand on topics introduced in the textbook and cover important additional conceptual material.
- *Online Quizzes* cover all the key material in each chapter (instructor registration required).
- *Flashcards* allow students to learn and review the many new terms introduced in the textbook.
- A complete *Glossary*.

For the Instructor
(Available to qualified adopters)
Instructor's Resource Library

The *Animal Physiology* Instructor's Resource Library includes a range of resources to help instructors in planning the course, preparing lectures and other course documents, and assessing students. The IRL includes the following resources:

Presentation Resources

- *Textbook Figures & Tables*: All of the textbook's figures (both line art and photographs) are provided as JPEG files at two sizes: high-resolution (excellent for use in PowerPoint) and low-resolution (ideal for web pages and other uses). All the artwork has been reformatted and optimized for exceptional image quality when projected in class.
- *Unlabeled Figures*: Unlabeled versions of all figures.
- PowerPoint Presentations:
  - *Figures & Tables*: Includes all the figures and tables from the chapter, making it easy to insert any figure into an existing presentation.
  - *Layered Art*: Selected key figures throughout the textbook are prepared as step-by-step and animated presentations that build the figure one piece at a time.

Answers to End-of-Chapter Questions Answers to all of the end-of-chapter Study Questions are provided as Word documents.

Test Bank Revised and updated for the Fourth Edition, the Test Bank consists of a broad range of questions covering key facts and concepts in each chapter. Both multiple-choice and short-answer questions are provided. The Test Bank also includes the Companion Website online quiz questions. All questions are ranked according to Bloom's Taxonomy and referenced to specific textbook sections.

Computerized Test Bank The entire Test Bank, including all of the online quiz questions, is provided in Blackboard's Diploma format (software included). Diploma makes it easy to assemble quizzes and exams from any combination of publisher-provided questions and instructor-created questions. In addition, quizzes and exams can be exported for import into many different course management systems, such as Blackboard and Moodle.

Online Quizzing

The Companion Website features pre-built chapter quizzes (see above) that report into an online gradebook. Adopting instructors have access to these quizzes and can choose to either assign them or let students use them for review. (Instructors must register in order for their students to be able to take the quizzes.) Instructors can add questions and create their own quizzes.

Course Management System Support

The entire Test Bank is provided in Blackboard format, for easy import into campus Blackboard systems. In addition, using the Computerized Test Bank provided in the Instructor's Resource Library, instructors can easily create and export quizzes and exams (or the entire test bank) for import into other course management systems, including Moodle, Canvas, and Desire2Learn/Brightspace.

Value Options
eBook

Animal Physiology is available as an eBook, in several different formats, including VitalSource, RedShelf, and BryteWave. The eBook can be purchased as either a 180-day rental or a permanent (non-expiring) subscription. All major mobile devices are supported. For details on the eBook platforms offered, please visit www.sinauer.com/ebooks.

Looseleaf Textbook

(ISBN 978-1-60535-594-8)
Animal Physiology is also available in a three-hole punched, looseleaf format. Students can take just the sections they need to class and can easily integrate instructor material with the text.

Brief Contents

Contents

CHAPTER 3
Genomics, Proteomics, and Related Approaches to Physiology   71

CHAPTER 4
Physiological Development and Epigenetics   89

**CHAPTER 9
The Energetics of Aerobic Activity   215**

**CHAPTER 10
Thermal Relations   233**

CHAPTER 11
Food, Energy, and Temperature AT WORK: The Lives of Mammals in Frigid Places 287

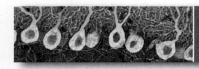

PART III Integrating Systems 303

CHAPTER 12
Neurons 305

CHAPTER 13
Synapses 337

CHAPTER 14
Sensory Processes   369

CHAPTER 15
Nervous System Organization and Biological Clocks   407

CHAPTER 16
Endocrine and Neuroendocrine Physiology 429

CHAPTER 17
Reproduction 465

CHAPTER 18
Integrating Systems
AT WORK: Animal
Navigation 497

PART IV   Movement and Muscle   513

CHAPTER 19
Control of Movement
The Motor Bases of Animal Behavior 515

CHAPTER 20
Muscle 537

CHAPTER 21
Movement and Muscle AT WORK: Plasticity in Response to Use and Disuse 565

CHAPTER 26
Oxygen, Carbon Dioxide, and Internal Transport AT WORK: Diving by Marine Mammals   701

PART VI   Water, Salts, and Excretion   721

CHAPTER 27
Water and Salt Physiology
Introduction and
Mechanisms   **723**

CHAPTER 28
Water and Salt Physiology
of Animals in Their
Environments   741

CHAPTER 29
Kidneys and Excretion
(with Notes on Nitrogen Excretion) 779

Basic Mechanisms of Kidney Function 780

Urine Formation in Amphibians 783

Urine Formation in Mammals 788

Urine Formation in Other Vertebrates 802

Urine Formation in Decapod Crustaceans 804

Urine Formation in Molluscs 805

Urine Formation in Insects 805

Nitrogen Disposition and Excretion 809

PART I
Fundamentals of Physiology

Previous page
Bioluminescent fireflies, when contemplated, are among the animals that most vividly exemplify how the extreme endpoints of the hierarchy of life—molecules and ecology—interact and are interdependent. Having evolved unusual molecular mechanisms that permit light production, fireflies are able to communicate ecologically in unique ways. Shown is a long exposure of fireflies in early summer in Japan.

Animals and Environments

Function on the Ecological Stage

Animal physiology is the study of animal function—the study of "how animals work." Physiologists—the scientists who carry out this study—bring a special perspective to scenes such as birds migrating. They wonder how much energy the birds must expend to fly, where and when the birds obtain the energy, and how the birds stay oriented toward their destination so as to arrive efficiently. More broadly stated, physiologists seek to identify the functional challenges that migrating birds face, and they seek to understand the functional properties of the birds that allow them to meet the challenges.

Billions of animals migrate over the face of the planet every year, making the functional properties of migrating animals a subject of paramount importance. By definition, however, migrating animals are on the move; they do not stay in one place where they might be investigated with ease. Consequently, researchers have had to be inventive to study these animals.

Physiologists now have high-quality methods for measuring the cost of flight. Sandpipers such as those shown here are trained to fly in a wind tunnel, where their speed of flight can be controlled. While the birds fly, their rates of energy use are measured by techniques designed not to disturb them. One such technique makes use of unusual, benign isotopes of oxygen

Long-distance migrants

Some populations of these sandpipers, which are known as red knots (*Calidris canutus*), breed in the high Arctic every summer but overwinter in southern Argentina. The birds thus migrate almost halfway around the globe twice a year. They use energy at relatively high rates not only while migrating but also during several other phases of their annual life cycle, such as their period of nesting on the cold, exposed Arctic tundra.

and hydrogen. These isotopes are injected into the sandpipers before they start flying, and then the rates of loss of the isotopes from the birds are measured as they fly unencumbered in the wind tunnel. From the isotope data, the birds' rates of energy use can be calculated as they fly at speeds typical of migratory flight. These rates turn out to be very high: about seven or eight times the birds' resting rates of energy use.[1] Physiologists have then combined this information with field observations on food ingestion and processing to learn how the birds obtain sufficient energy and how they manage their energy supplies to meet their flight needs during migration. One population of the sandpipers is famous for migrating every spring from southern South America to the Arctic—a distance of 15,000 km (9300 miles [mi]). In common with other populations of the same species, when these birds migrate, they alternate between extended *stopover periods*, during which they "refuel" by feeding, and *flight periods*, during which they fly nonstop for long distances—sometimes more than 5000 km (3100 mi). Based on the information available, the sandpipers fuel each long, uninterrupted flight by eating lots of clams, snails, and other food during the stopover period—often lasting 3–4 weeks—that immediately precedes the flight. By the time they take off, the birds must have enough stored energy to fuel the entire next leg of their journey because they do not eat as they fly.

During a stopover period, as the birds eat day after day, they store a great amount of energy as fat, and their body weight can increase by 50%. Physiologists have discovered, however, that the birds' adjustments during a stopover period are far more complex than simply storing fat. For part of the time, the birds' stomach and intestines are large, aiding them in processing food at a high rate. During the week before they take flight, however, several organs that they will not use during flight, including their stomach and intestines, decrease significantly in size. Other organs, such as their heart, grow larger. Overall, during that week, the body of each bird is reproportioned in ways that poise the bird to fly strongly, while reducing the amount of unnecessary weight to be carried. By investigating these phenomena, physiologists have revealed that the fascinating migrations of these birds are, in truth, far more fascinating than anyone could have imagined prior to the detailed study of function.

As you start your study of physiology, we—your authors—believe you are at the beginning of a great adventure. We feel privileged to have spent our professional lives learning how animals work, and we are eager to be your guides. If we could hop with you into a fantastic travel machine and tour Earth in the realms we are about to travel in this book, we would point out sperm whales diving an hour or more to depths of a mile or more, electric fish using modified muscles to generate 500-V shocks, just-born reindeer calves standing wet with amniotic fluid in the frigid Arctic wind, reef corals growing prolifically because algae within their tissues permit internal photosynthesis, and moths flying through cool nights with bodies as warm as those of mammals. Each of these scenes draws the interest of physiologists and continues to spark new physiological research.

---

[1] The method of measuring rate of energy consumption discussed here, known as the *doubly labeled water method*, is explained in greater detail on page 217.

FIGURE 1.1 Pacific salmon migrating upriver to their spawning grounds Having spent several years feeding and growing in the Pacific Ocean, these fish have once again found the river in which they were conceived. Now they must power their way back to their birthplace to spawn, even though they ate their last meal at sea and will starve throughout their upriver journey. Shown are sockeye salmon (*Oncorhynchus nerka*).

The Importance of Physiology

Why is the study of animal physiology important to you and to people in general? Not the least of the reasons is the one we have already emphasized—namely, that a full understanding and appreciation of all the marvels and other phenomena of the animal world depend on an analysis of how animals work. The study of physiology draws us beyond surface impressions into the inner workings of animals, and nearly always this venture is not only a voyage of discovery, but also one of revelation.

The study of physiology also has enormous practical applications because physiology is a principal discipline in the understanding of health and disease. The analysis of many human diseases—ranging from aching joints to heart failure—depends on understanding how the "human machine" works. A physician who studies heart disease, for instance, needs to know the forces that make blood flow into the heart chambers between one heartbeat and the next. The physician also needs to know how pressures are developed to eject blood into the arteries, how the cells of the heart muscle coordinate their contractions, and how the nutrient and O_2 needs of all parts of the heart muscle are met. We discuss these and other aspects of mammalian physiology extensively in this book. Even when we turn our attention to other types of animals, our study will often have application to human questions. One reason is that nonhuman animals are often used as "models" for research that advances understanding of human physiology. Research on squids, for instance, has been indispensable for advancing knowledge of human neurophysiology because some of the nerve cells of squids are particularly large and therefore easily studied.

Physiology is as important for understanding the health and disease of nonhuman animals as it is for understanding health and disease in humans. An example is provided by studies of another group of migrating animals, the Pacific salmon—which swim up rivers to reach their spawning grounds (**FIGURE 1.1**). Physiologists have measured the costs these fish incur to swim upstream and leap waterfalls. This research has enabled better understanding of threats to their health. For instance, although each individual dam along a river might be designed to let salmon pass, a series of dams might so increase the overall cost of migration that the fish—which don't eat and live just on their stores of energy when migrating—could run out of energy before reaching their spawning grounds. With knowledge of the energetics of swimming and leaping, managers can make rational predictions of the cumulative effects of dams, rather than simply altering rivers and waiting to see what happens. The effects of water pollutants, such as heavy metals and pesticides, are other important topics in salmon physiology. Examples in other animals include studies of stress and nutrition. Careful physiological studies have solved mysterious cases of population decline by revealing that the animals were under stress or unable to find adequate amounts of acceptable foods.

In brief, physiology is one of the key disciplines for understanding

- The fundamental biology of all animals
- Human health and disease

- The health and disease of nonhuman animals of importance in human affairs

The Highly Integrative Nature of Physiology

Physiology is also important in the study of biology because *it is one of biology's most integrative disciplines*. Physiologists study all the levels of organization of the animal body. In this respect, they are much like detectives who follow leads wherever the leads take them. To understand how an organ works, for instance, information about the nervous and hormonal controls of the organ might be required, plus information about enzyme function in the organ, which might lead to studies of the activation of genes that code for enzyme synthesis. Physiology not only pursues all these levels of biological organization within individual animals but also relates this knowledge to the ecology and evolutionary biology of the animals. Students often especially enjoy their study of physiology because the discipline is so integrative, bringing together and synthesizing many concepts that otherwise can seem independent.

Suppose we observe a pheasant pursued by a fox. This scene has obvious evolutionary and ecological importance (**FIGURE 1.2A**); if the pheasant is to continue passing its genes to future generations, it must win the contest, but if it loses, the fox gets to eat. In other words, the outcome is significant at the evolutionary and ecological levels of organization. But what determines the outcome? The

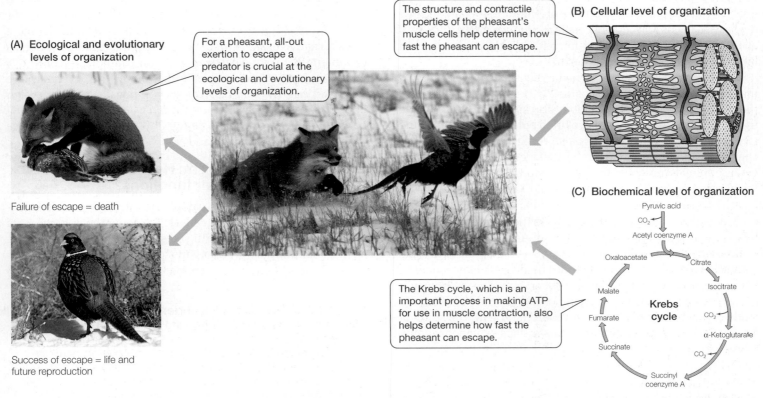

(A) Ecological and evolutionary levels of organization

For a pheasant, all-out exertion to escape a predator is crucial at the ecological and evolutionary levels of organization.

Failure of escape = death

Success of escape = life and future reproduction

The structure and contractile properties of the pheasant's muscle cells help determine how fast the pheasant can escape.

(B) Cellular level of organization

(C) Biochemical level of organization

Pyruvic acid
CO_2
Acetyl coenzyme A
Oxaloacetate
Citrate
Malate
Isocitrate
Krebs cycle
Fumarate
CO_2
Succinate
α-Ketoglutarate
CO_2
Succinyl coenzyme A

The Krebs cycle, which is an important process in making ATP for use in muscle contraction, also helps determine how fast the pheasant can escape.

FIGURE 1.2 The study of physiology integrates knowledge at all levels of organization (A) When we see a pheasant or other animal escaping from a predator, we tend to think of the process as being entirely a matter of behavior. Actually, however, internal physiological processes such as (B) muscle contraction and (C) ATP production play key roles, because the outward behavior of the animal can be only as effective as these internal processes permit. The study of physiology considers all the relevant levels of organization, from the cellular and biochemical processes involved in running and flying to the ecological and evolutionary consequences. See Chapter 8 for more detail on ATP production and Chapter 20 for more on muscle structure and force production.

physiological properties of many of the pheasant's internal tissues play major roles. The speed at which the pheasant can escape, for example, depends on the precise ways that the pheasant's muscles work (**FIGURE 1.2B**) as they contract to produce the mechanical forces that propel the pheasant forward. The pheasant's speed also depends on its biochemical mechanisms of producing adenosine triphosphate (ATP) (**FIGURE 1.2C**), because the muscles use ATP as their source of energy for producing mechanical forces, and consequently they can work only as fast as their ATP supply permits. In a very real sense, the pheasant's success at the ecological and evolutionary levels of organization is a direct consequence of the pheasant's physiological capabilities at the cellular and biochemical levels of organization in its tissues.

To explore the highly integrative nature of the study of physiology in greater detail, let's again consider the Pacific salmon. As juveniles, these fish migrate from rivers to the open ocean. Years later, they return to the very rivers of their conception to procreate the next generation. Before a returning salmon enters freshwater, it maintains its blood more dilute than the seawater in which it swims. After it enters freshwater, however, it must maintain its blood more concentrated than the dilute freshwater now surrounding it. Another challenge the salmon faces is meeting the energy costs of its migration. Once in its natal river, a salmon no longer eats. Yet it may swim for many weeks before it reaches its spawning grounds—sometimes traveling against the river current as far as 1100 km (680 mi) and, in mountainous regions, climbing 1.2 km (0.75 mi) in altitude. During this trip, because the fish is starving, it gradually breaks down the substance of its body to supply its energy needs (e.g., needs for metabolic fuels to produce ATP); 50–70% of all tissues that can supply energy for metabolism are typically used by the time the fish reaches its destination.

As physiologists study salmon, they consider all levels of organization (**FIGURE 1.3**). As part of their background of knowledge, they recognize that the populations and species of salmon alive today not only are products of evolution but also are still evolving (see Figure 1.3A). Physiologists also recognize that the laws of chemistry and physics need to be considered (see Figure 1.3B), because animals must obey those laws—and sometimes they *exploit* them. For understanding swimming, multiple levels of organization must be considered (see Figure 1.3C). The nervous system generates coordinated nerve impulses that travel to the swimming muscles, which contract using energy drawn from ATP that is synthesized from organic food molecules that serve as metabolic fuels. The contraction of the swimming muscles then exerts biomechanical forces on the water that propel the fish forward. The investigation of swimming illustrates, too, the important general point that the study of *function* typically goes hand in hand with the study of *form*; knowledge of anatomy often sets the stage for understanding physiology, and as function is clarified, it typically helps account for anatomy. Often, the ultimate goal of a physiological study is to understand how an animal functions in its natural environment. Thus an ecological perspective is vital as well. As seen in Figure 1.3D, when an individual salmon's fluid environment changes from saltwater to freshwater, the fish alters the set of ion-transporting proteins expressed in its gills, permitting inward ion pumping in freshwater whereas ions were pumped outward in saltwater. The distance a fish swims is another important ecological consideration.

Different populations of salmon travel vastly different distances. Going far upriver can provide advantages of certain kinds, such as providing pristine spawning grounds. However, this ecological factor has other consequences as well. Females that exert great effort to reach their spawning grounds, such as by swimming great distances, spawn fewer eggs because swimming diverts energy away from use in reproduction (see Figure 1.3D).

Mechanism and Origin: Physiology's Two Central Questions

Physiology seeks to answer two central questions about how animals work: (1) What is the mechanism by which a function is accomplished, and (2) how did that mechanism come to be? To understand why there are *two* questions, consider the analogous problem of how a car works. In particular, how is an engine-driven wheel made to turn?

To understand this function, you could disassemble a car and experiment on its parts. You could study how the pistons inside the cylinders of the engine are made to oscillate by forces released from exploding gasoline, how the pistons and connecting rods turn the drive shaft, and so forth. From studies like these, you would learn how the car works.

At the conclusion of such studies, however, you would have only half the answer to the question of how the car works. Presuming that you have investigated a routine design of modern car, your experiments will have revealed how a routine internal combustion engine turns a wheel by way of a routine transmission. Let your mind run free, however, and you may quickly realize that there are alternative designs for a car. The engine could have been a steam engine or an electric engine, for example. Accordingly, when you ponder how a wheel turns, you see that you really face two questions: the *immediate* question of how a particular design of car makes a wheel turn, and the *ultimate* question of how that particular design came into being. Physiologists also face these two questions of *mechanism* and *origin*.

The study of mechanism: How do modern-day animals carry out their functions?

If you examine a particular car and its interacting parts to understand how it works, you are learning about the *mechanisms* of function of the car. Likewise, if you study the interacting parts of a particular animal—from organs to enzymes—to learn how it works, you are studying the animal's *mechanisms*. In physiology,

FIGURE 1.3 To develop a full understanding of salmon swimming, physiologists integrate studies at multiple levels of organization To understand the physiology of fish, physiologists consider (A) evolutionary biology, (B) the laws of chemistry and physics, and (D) ecological relations—as well as (C) body function at all levels of organization. All elements shown are for fish in a single genus, *Oncorhynchus*, the Pacific salmonid fish. In (C) the drawing in "Systems physiology" is a cross section of the body; the salmon in "Morphology" is a chinook salmon. (Graph in A—which pertains to populations of chum salmon—after Hendry et al. 2004; cross section, salmon, and biomechanics illustration in C after Videler 1993; graph in D—which pertains to sockeye salmon—after Crossin et al. 2004.)

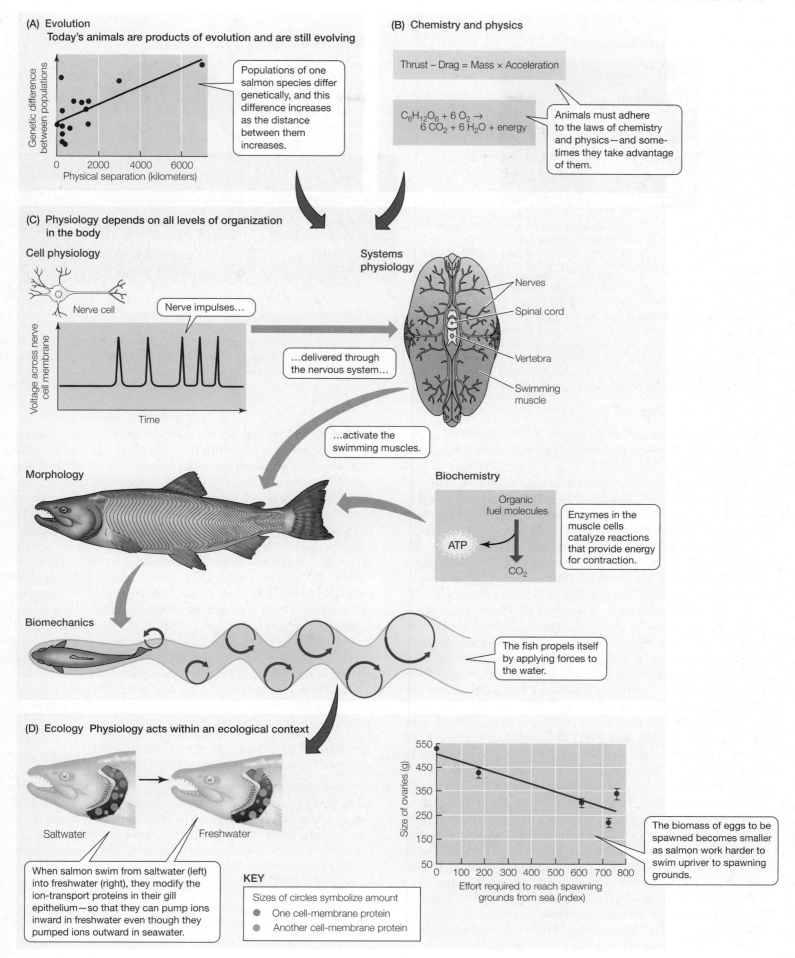

(A) Evolution
Today's animals are products of evolution and are still evolving

Genetic difference between populations / Physical separation (kilometers)

0 2000 4000 6000

Populations of one salmon species differ genetically, and this difference increases as the distance between them increases.

(B) Chemistry and physics

Thrust − Drag = Mass × Acceleration

$C_6H_{12}O_6 + 6 O_2 \rightarrow$
$6 CO_2 + 6 H_2O + energy$

Animals must adhere to the laws of chemistry and physics—and sometimes they take advantage of them.

(C) Physiology depends on all levels of organization in the body

Cell physiology

Nerve cell

Voltage across nerve cell membrane / Time

Nerve impulses…

…delivered through the nervous system…

Systems physiology

Nerves
Spinal cord
Vertebra
Swimming muscle

…activate the swimming muscles.

Morphology

Biochemistry

Organic fuel molecules

ATP

CO_2

Enzymes in the muscle cells catalyze reactions that provide energy for contraction.

Biomechanics

The fish propels itself by applying forces to the water.

(D) Ecology Physiology acts within an ecological context

Saltwater → Freshwater

When salmon swim from saltwater (left) into freshwater (right), they modify the ion-transport proteins in their gill epithelium—so that they can pump ions inward in freshwater even though they pumped ions outward in seawater.

KEY
Sizes of circles symbolize amount
● One cell-membrane protein
● Another cell-membrane protein

Size of ovaries (g) / Effort required to reach spawning grounds from sea (index)

550 450 350 250 150 50
0 100 200 300 400 500 600 700 800

The biomass of eggs to be spawned becomes smaller as salmon work harder to swim upriver to spawning grounds.

(A) A firefly producing light

(B) Anatomy of the light organ

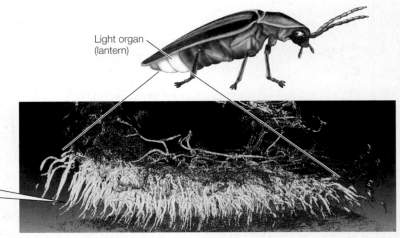

Light organ (lantern)

A gas-filled tubule (trachea or tracheole) brings O_2 to each light cell in the light organ. The final tubules in this branching system have diameters so small (< 1 μm) that they are invisible to the human eye.

FIGURE 1.4 Light production by fireflies Fireflies are not in fact flies. Instead they are beetles. (A) Light is produced by a light organ at the posterior end of the abdomen. *Photinus pyralis*, found in the eastern and central United States, is shown here. Insects employ a system of branching gas-filled tubules to deliver oxygen (O_2) to all their cells. O_2 enters this system at pores on the surface of the insect's body.

(B) A detailed image of the exceedingly fine gas-transport tubules in the light organ was finally obtained in Asian fireflies (*Luciola*) in 2014 by use of advanced X-ray methods. Each light cell in the light organ is supplied by a tiny gas-transport tubule. (Tubule image in B courtesy of École Polytechnique Fédérale de Lausanne, after Tsai et al. 2014.)

mechanism refers to the components of actual, living animals and the interactions among those components that enable the animals to perform as they do.

Curiosity about mechanism is what inspires most physiologists to study animals, and studies of mechanism dominate physiological research. Physiology, in fact, is most clearly distinguished from other biological disciplines with which it is related, such as morphology or ecology, by its central focus on the study of mechanism. A physiologist typically begins an investigation by observing a particular capability that excites curiosity or needs to be understood for practical purposes. The capability of the human visual system to distinguish red and blue is an example. Another example is the ability of certain types of nerve cells to conduct nerve impulses at speeds of over 100 meters (m) per second. Whatever the capability of interest, the typical goal of physiological research is to discover its mechanistic basis. What cells, enzymes, and other parts of the body are employed, and how are they employed, to enable the animal to perform as it does?

For a detailed example of a mechanism, let's consider light production by fireflies (**FIGURE 1.4**). Each light-producing cell in a firefly's light organ receives molecular oxygen (O_2) by way of a branching system of gas-transport tubules. O_2 enters this system from the atmosphere and is carried by the tubules to all the tissues of the insect's body (see Chapter 23). To understand how a light-producing cell emits flashes of light, let's start with the biochemistry of light production (**FIGURE 1.5A**). A chemical compound (a benzothiazol) named *firefly luciferin* first reacts with ATP to form luciferyl-AMP (*AMP*, adenosine monophosphate). Then, if O_2 can reach the luciferyl-AMP, the two react to form a chemical product in which electrons are boosted to an excited state, and when this electron-excited product returns to its ground state, it emits photons. This sequence of reactions requires a protein

catalyst, an enzyme called *firefly luciferase*. A question only recently answered is how cells within the light organ are controlled so that they flash at certain times but remain dark at others. When a firefly is not producing light (**FIGURE 1.5B**), any O_2 that reaches the insect's light cells via its gas-transport tubules is intercepted (and thereby prevented from reacting with luciferyl-AMP) by mitochondria that are positioned (in each cell) between the gas-transport tubules and the sites of the luciferin reactions. The light cells produce light (**FIGURE 1.5C**) when, because of stimulation by the nervous system, the mitochondria become bathed with nitric oxide (NO).[2] The NO blocks mitochondrial use of O_2, allowing O_2 through to react with luciferyl-AMP. Facts like these form a description of the *mechanism* by which fireflies produce light.

The study of a mechanism may become so intricate that decades or centuries are required for a mechanism to be fully understood. By definition, however, the complete mechanism of any given function is present for study in the here and now. A scientist can, in principle, fully describe the mechanism of a process merely by studying existing animals in ever-finer detail.

The study of origin: Why do modern-day animals possess the mechanisms they do?

Suppose a youngster observes a firefly produce a flash of light and asks you to explain what he has seen. One way you could interpret the request is as a question about mechanism. Thus you could answer that the brain of the insect sends nerve impulses that cause the light cells to become bathed with nitric oxide, resulting in the production of excited electrons through the reaction of O_2 with luciferyl-AMP. However, the youngster who asks you to explain

---

[2] Interestingly, NO—the signaling molecule that controls firefly flashing—is the same molecule that initiates erection of the penis in human sexual activity (see page 490).

(A) Light-emitting chemical reactions

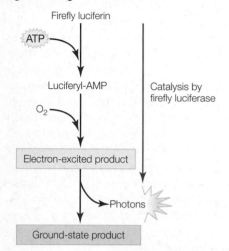

(B) Light cell in dark state

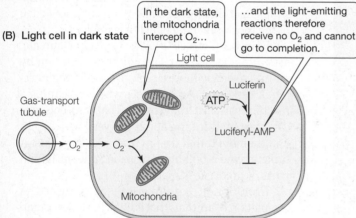

(C) Light cell in flashing state

In the flashing state, nitric oxide is produced under nervous control and bathes the mitochondria, preventing them from intercepting O_2.

Pulses of O_2 reach the luciferin reactions, resulting in pulses of light.

FIGURE 1.5 The mechanism of light production by fireflies
(A) The chemistry of light production. (B,C) In the light cells—the cells that compose the light organ—the luciferin reactions are spatially separated from mitochondria. When a light cell is not flashing (B), the mitochondria intercept O_2. However, when a cell *is* flashing (C), O_2 gets through to the luciferin reactions. AMP = adenosine monophosphate; ATP = adenosine triphosphate.

the flashing of a firefly is probably interested in something else. The *reason* the firefly makes light is probably what is on your young friend's mind, rather than the mechanism. That is, the youngster is probably wondering *why* the firefly possesses a mechanism to make light.

For biologists, the answer lies in *evolutionary origins*. The mechanisms of modern-day animals are products of evolution, and thus the reasons for the existence of mechanisms lie in evolutionary processes. The study of evolutionary origins is a central aim of modern physiology because it promises to reveal the *significance* of mechanisms. If we can learn why evolution produced a mechanism, we will better understand what (if anything) animals gain by having the mechanism.

Because modern-day mechanisms evolved in the past, the question of origins is fundamentally historical. The origins of a mechanism, unlike the mechanism itself, cannot usually be observed directly in the here and now. Instead, origins must usually be studied indirectly, by means of inferences about the past derived from observations we can make in the present. The reliance on indirect reasoning means that evolutionary origins are rarely understood with the same certainty as mechanisms.

Natural selection is a key process of evolutionary origin

Natural selection is just one of several processes by which animals acquire traits during evolution, as we discuss later in this chapter. Natural selection, however, holds a place of special importance for biologists because, of all the modes of evolutionary change, natural selection is believed to be the principal process by which animals become fit to live in their environments.

Natural selection is the increase in frequency of genes that produce phenotypes that raise the likelihood that animals will survive and reproduce. During evolution by natural selection, such genes increase in frequency—over the course of generations—because animals with the genes are differentially successful relative to other members of their species. If we find that a physiological mechanism originated by natural selection within the prevailing environment, we can conclude that the mechanism is an asset; that is, it improves an animal's chances of survival and reproduction within the environment the animal occupies.

Adaptation is an important sister concept to natural selection. Because we discuss adaptation at length later, here we simply state that an adaptation is a physiological mechanism or other trait that is a product of evolution by natural selection. Adaptations are assets; because of the way they originated, they aid the survival and reproduction of animals living in the environment where they evolved. When we speak of the **adaptive significance** of a trait evolved by natural selection, we refer to the reason *why* the trait is an asset: that is, the reason *why* natural selection favored the evolution of the trait.

The light flashes of fireflies usually function to attract mates. The males of each species of firefly emit light flashes in a distinctive, species-specific pattern as they fly, thereby signaling their species identity to females (**FIGURE 1.6**). Using various sorts of evidence, students of fireflies infer that the firefly light-producing mechanism evolved by natural selection because light flashes can be used to

1 Soon after it has cleared the tree, the uppermost firefly emits three flashes in close sequence.

2 Then it flies awhile without flashing.

3 Again it emits a set of three flashes and goes dark.

4 Finally it emits three more flashes just before it disappears from your field of vision to the right.

FIGURE 1.6 Male fireflies employ their mechanism of light production for an adaptive function: mate attraction The drawing shows representative flashing patterns and flight paths of males of nine different species of fireflies of the genus *Photinus* from the eastern and central United States. Each line of flight represents a different species. For instance, the uppermost line represents *Photinus consimilis*, a species that flies high above the ground. To understand the format, imagine that you are watching the uppermost firefly as it leaves the tree at the left and follows the numbered sequence. The differences in flashing and flight patterns among species allow males to signal their species to females. (From a drawing by Dan Otte in Lloyd 1966.)

bring the sexes together. Thus the mechanism of light production is an adaptation, and its adaptive significance is mate attraction.

Mechanism and adaptive significance are distinct concepts that do not imply each other

Why have we stressed that physiology faces *two* central questions? We have emphasized both that physiology studies mechanism and that it studies evolutionary origins to understand adaptive significance. Why *both*? Physiologists must seek answers to both questions because *mechanism and adaptive significance do not imply each other*. If you know the mechanism of a process, you do not necessarily know anything about its adaptive significance. If you

know the adaptive significance, you do not necessarily know anything about the mechanism. Thus, to understand both mechanism and adaptive significance, you must study both.

As an example, consider light production by fireflies once again. Physiologists know of many mechanisms by which organisms can produce light.[3] Thus, even if fireflies were *required* to attract their mates with light, their mechanism of making light would not be limited theoretically to just the mechanism they use. The mechanism of light production by fireflies cannot be deduced from simple knowledge of the purpose for which the mechanism is used. Conversely, light flashes could be used for purposes other than mate attraction, such as luring prey, distracting predators, or synchronizing biorhythms. The significance of light production cannot be deduced from the simple fact that light is made or from knowledge of the mechanism by which it is made.

François Jacob (1920–2013), a Nobel laureate, asked in a famous article whether evolution by natural selection more closely resembles engineering or tinkering. An engineer who is designing a machine can start from scratch. That is, an engineer can start by thinking about the very best design and then build that design from raw materials. A tinkerer who is building a new machine starts with parts of preexisting machines.

Evolution is like tinkering, Jacob argued: A population of animals that is evolving a new organ or process rarely starts from scratch; instead, it starts with structures or processes that it already has on hand for other reasons. The lungs of mammals, for example, originated as outpocketings of a food-transport tube, the *esophagus*, in the ancient fish that gave rise to the tetrapods living on land today. Those fish, moreover, were not the only fish to evolve air-breathing organs. Today, as we will discuss in Chapter 23 (see page 611), there are various different groups of fish that use the *stomach, intestines, mouth cavity,* or outpocketings of the *gill chambers* as air-breathing organs. This diversity reminds one of a tinkerer who, in the course of assembling a garden cart, might try to use axles and wheels taken from a discarded bicycle, an outdated trailer, or an old children's wagon.

Throughout evolution, animals in a sense have had to remain capable of living in their old ways even as they have developed new ways. Thus design from first principles—the engineering approach—has not been possible.

[3] The number of known mechanisms is partly obscured by the fact that *luciferin* and *luciferase* are generic terms, each used to refer to many distinctly different chemical compounds. For example, more than 30 distinctly different compounds are called *luciferin*. Accordingly, although many bioluminescent organisms are said to use a "luciferin–luciferase system" to make light, all such organisms are not by any means employing the same chemistry.

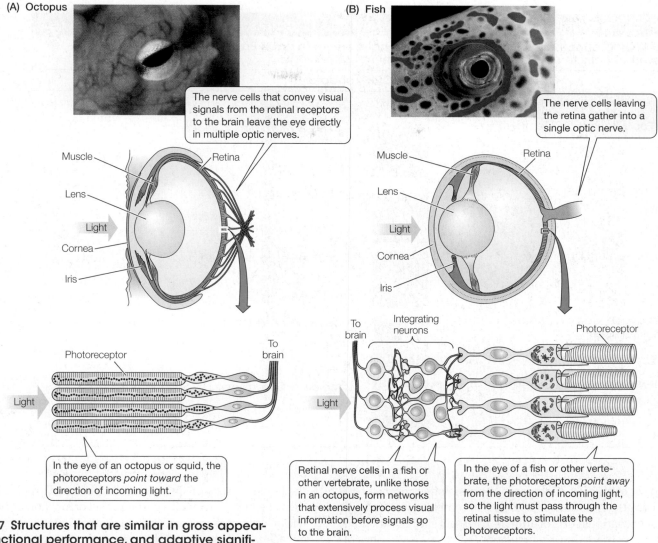

(A) Octopus

The nerve cells that convey visual signals from the retinal receptors to the brain leave the eye directly in multiple optic nerves.

Muscle
Retina
Lens
Light
Cornea
Iris

Photoreceptor
To brain
Light

In the eye of an octopus or squid, the photoreceptors *point toward* the direction of incoming light.

(B) Fish

The nerve cells leaving the retina gather into a single optic nerve.

Muscle
Retina
Lens
Light
Cornea
Iris

To brain
Integrating neurons
Photoreceptor
Light

Retinal nerve cells in a fish or other vertebrate, unlike those in an octopus, form networks that extensively process visual information before signals go to the brain.

In the eye of a fish or other vertebrate, the photoreceptors *point away* from the direction of incoming light, so the light must pass through the retinal tissue to stimulate the photoreceptors.

FIGURE 1.7 Structures that are similar in gross appearance, functional performance, and adaptive significance can differ dramatically in details of how they are assembled and work Both cephalopod molluscs and fish have evolved excellent vision, but they see using different detailed mechanisms. (A after Wells 1966 and Young 1971; B after Walls 1942.)

The tinkering aspect of evolution is a key reason why mechanism and adaptive significance do not imply each other. The mechanism employed to perform a particular function is not an abstraction but instead bears an imprint of the structures and processes that came before in any particular evolutionary line. Consider, for instance, the eyes of two groups of active aquatic animals: the cephalopod molluscs (squids and octopuses) and the fish. Both groups have evolved sophisticated eyes that permit lifestyles based on excellent vision. However, these eyes are built on very different retinal designs (**FIGURE 1.7**). The photoreceptors in the retinas of fish point *away* from the light; those of squids and octopuses point *toward* the light. Moreover, whereas visual signals from the fish photoreceptors are extensively processed by networks of integrating nerve cells *within* the retina before visual information is sent to the brain in a single optic nerve, in squids and octopuses the photoreceptors send their visual signals more directly to the brain in multiple optic nerves. The adaptive significance of excellent vision in the two groups of animals is similar, but the detailed mechanisms are not. A major reason for the differences in mechanisms is that the two groups built their eyes from different preexisting structures.

This Book's Approach to Physiology

Mechanistic physiology, which emphasizes the study of mechanism, and **evolutionary physiology**, which emphasizes the study of evolutionary origins, have become recognized as major approaches to the study of animal physiology in recent years. The two approaches share the same overall subject matter: They both address the understanding of animal function. They differ, however, in the particular aspects of physiology they emphasize. The viewpoint of this book, as stressed already, is that both approaches are essential for physiology to be fully understood.

Comparative physiology and **environmental physiology** are additional approaches to the study of animal physiology. These approaches overlap mechanistic and evolutionary physiology, and they overlap each other. Comparative physiology is the synthetic study of the function of all animals. It contrasts, for example, with *human physiology* or *avian physiology*, each of which addresses only

TABLE 1.1 Major Parts of this book and "AT WORK" chapters

The "AT WORK" chapters exemplify how the material covered in each Part of the book can be used synthetically to understand a problem in animal physiology.

PART	"AT WORK" chapter
PART I: Fundamentals of Physiology	
PART II: Food, Energy, and Temperature	*The Lives of Mammals in Frigid Places* (Chapter 11)
PART III: Integrating Systems	*Animal Navigation* (Chapter 18)
PART IV: Movement and Muscle	*Plasticity in Response to Use and Disuse* (Chapter 21)
PART V: Oxygen, Carbon Dioxide, and Internal Transport	*Diving by Marine Mammals* (Chapter 26)
PART VI: Water, Salts, and Excretion	*Mammals of Deserts and Dry Savannas* (Chapter 30)

a limited set of animals. Comparative physiology is termed *comparative* because one of its major goals is to compare systematically the ways that various sorts of animals carry out similar functions, such as vision, breathing, or circulation. Environmental physiology (also called *physiological ecology*) is the study of how animals respond physiologically to environmental conditions and challenges, or—more briefly—"ecologically relevant physiology." **Integrative physiology** is a relatively new term referring to investigations with a deliberate emphasis on *synthesis across levels of biological organization*, such as research that probes the relations between molecular and anatomical features of organs.

Our viewpoint in this book is mechanistic, evolutionary, comparative, environmental, and integrative. In other words, we stress:

- The mechanisms by which animals perform their life-sustaining functions
- The evolution and adaptive significance of physiological traits
- The ways in which diverse phylogenetic groups of animals both resemble each other and differ
- The ways in which physiology and ecology interact, in the present and during evolutionary time
- The importance of all levels of organization for the full understanding of physiological systems

Overlapping with the classifications already discussed, physiology is divided also into various branches or disciplines based on the *types of functions* that are performed by animals. The organization of this book is based on the types of function. As **TABLE 1.1** shows,

the book consists of six major subdivisions, Parts I through VI, each of which focuses on a particular set of functions. The chapters within each part are listed in the Table of Contents and at the start of each part. Here we want to stress that each part (except Part I) ends with a unique type of chapter that we call an "At Work" chapter. Each "At Work" chapter takes a synthetic approach to a prominent, curiosity-provoking topic in the part. Table 1.1 lists these topics. *The principal goal of the "At Work" chapters is to show how the material in each of the parts can be used in an integrated way to understand animal function.*

Now, as they say in theater, "Let the play begin." As we consider the principal subject of this chapter—function on the ecological stage—the three major players are animals, environments, and evolutionary processes (see Figure 1.3). We now address each.

Animals

The features of animals that deserve mention in an initial overview are the features that are of overriding importance. These include that (1) animals are *structurally dynamic*, (2) animals are *organized* systems that *require energy* to maintain their organization, and (3) both *time* and *body size* are of fundamental significance in the lives of all animals.

One of the most profoundly important properties of animals is that the atoms of their bodies—their material building blocks—undergo continuous, dynamic exchange with the atoms in their environments *throughout life*. This structural dynamism—memorably termed "the dynamic state of body constituents" by Rudolf

Schoenheimer, who discovered it[4]—is a fundamental and crucially important way in which animals differ from inanimate objects such as telephones. After a telephone is manufactured, the particular carbon and iron atoms that are built into its substance remain as long as the telephone exists. One might think by casual observation that the composition of a person, lion, or crab is similarly static. This illusion was abruptly dispelled, however, when Schoenheimer and others began using chemical isotopes as research tools.

Isotopes proved to be revealing because they permit atoms to be labeled and therefore tracked. Consider iron (Fe) as an example. Because most iron atoms in the natural world are of atomic weight 56 (^{56}Fe), an investigator can distinctively label a particular set of iron atoms by substituting the unusual (but stable and nonradioactive) alternative isotope of iron having an atomic weight of 58 (^{58}Fe). Suppose that we make a telephone in which all the iron atoms are of the unusual ^{58}Fe isotope, so that we can distinguish those iron atoms from the ones generally available. Years later, all the iron atoms in the telephone will still be of the unusual ^{58}Fe type. Suppose, however, that we create a ^{58}Fe-labeled person by feeding the person over the course of a year the unusual ^{58}Fe isotope, so that isotopically distinctive iron atoms are built into hemoglobin molecules and other iron-containing molecules throughout the person's body. Suppose we then stop providing the unusual iron isotope in the person's diet. Thereafter—as time goes by—the isotopically distinctive ^{58}Fe atoms in the body will leave and will be replaced with atoms of the ordinary isotope, ^{56}Fe, from the environment. Years later, all the unusual iron atoms will be gone. We see, therefore, that although the person may outwardly appear to be structurally constant like a telephone, the iron atoms in the substance of the person's body at one time differ from those at another time.

The mechanistic reason for the turnover of iron atoms in an animal is that the molecular constituents of an individual's body break down and are rebuilt. A human red blood cell, for example, typically lives for only 4 months. When a red blood cell is discarded and replaced, some of the iron atoms from the hemoglobin molecules of the old cell are excreted into the environment, and some of the iron atoms built into the new cell are acquired from food. In this way, even though the number of red blood cells remains relatively constant, the iron atoms of the cells are in dynamic exchange with iron atoms in the environment.

Essentially all the atoms in the substance of an animal's body undergo similar dynamic exchanges. Calcium atoms enter an animal's skeleton and later are withdrawn; some of the withdrawn atoms are replaced with newly ingested calcium atoms from the environment. Proteins and fats throughout an animal's body are continually broken down at substantial rates,[5] and their resynthesis is carried out in part with molecules newly acquired from the environment, such as amino acids and fatty acids from foods. Adult people typically resynthesize 2–3% of their body

protein *each day*, and about 10% of the amino acids used to build the new protein molecules are acquired from food and therefore new to the body.

Have you ever wondered why you need to worry *every week* about whether you are eating enough calcium, iron, magnesium, and protein? The explanation is provided by the principles we are discussing. If you were an inanimate object, enough of each necessary element or compound could be built into your body at the start, and you would then have enough forever. Instead, because you are alive and dynamic—rather than inanimate and static—you lose elements and compounds every day and must replace them.

As this discussion has illustrated, the *material boundaries* between an animal and its environment are blurred, not crisp. Atoms cross the boundaries throughout life, so that an atom that is part of an animal's tissues on one day may be lying on the forest floor or drifting in the atmosphere the next day, and vice versa. Possibly the most profound implication of these facts is that *an animal is not a discrete material object*.

The structural property of an animal that persists through time is its organization

If the atomic building blocks of an animal are transient and constantly changing, by what structural property is an animal *defined*? The answer comes from imagining that we can see the individual molecules in an adult animal's body. If we could, we would observe that the molecular structures and the spatial relations of molecules in tissues are relatively constant over time, even though the particular atoms constructing the molecules change from time to time. A rough analogy would be a brick wall that retains a given size and shape but in which the bricks are constantly being replaced, so that the particular bricks present during one month are different from those present a month earlier.

The structural property of an animal that persists through time is the *organization* of its atomic building blocks, not the building blocks themselves. *Thus an animal is defined by its organization*. This characteristic of animals provides the most fundamental reason why animals require inputs of energy throughout life. As we will discuss in detail in Chapter 7, the second law of thermodynamics says that for organization to be maintained in a dynamic system, use of energy is essential.

Most cells of an animal are exposed to the internal environment, not the external environment

Shifting our focus now to the cells of an animal's body, it is important first to stress that the conditions experienced by most of an animal's cells are the conditions *inside* the body, not those outside. Most cells are bathed by the animal's tissue fluids or blood. Thus the *environment* of most cells consists of the set of conditions prevailing in the tissue fluids or blood. Claude Bernard (1813–1878), a Frenchman who was one of the most influential physiologists of the nineteenth century, was the first to codify this concept. He coined the term **internal environment** (*milieu intérieur*) to refer to the set of conditions—temperature, pH, sodium (Na$^+$) concentration, and so forth—experienced by cells within an animal's body. The conditions outside the body represent the **external environment**.

---

[4] As chemists learned about and started to synthesize unusual isotopes in the 1930s, Rudolf Schoenheimer (1898–1941) was one of the first to apply the newfound isotopes to the study of animal metabolism. His classic book on the subject, published posthumously as World War II raged, is titled *The Dynamic State of Body Constituents*.

[5] See Chapter 2 (page 59) for a discussion of the ubiquitin–proteasome system that tags proteins for breakdown and disassembles them.

The internal environment may be permitted to change when the external environment changes, or it may be kept constant

Animals have evolved various types of relations between their internal environment and the external environment. If we think of the organization of the body as being hierarchically arranged, the relations between the internal and external environments represent one of the potential hierarchical levels at which animals may exhibit organization. *All* animals consistently exhibit structural organization of their atoms and molecules. When we think about the relation between the internal and external environments, we are considering another level at which organization may exist. At this level, animals sometimes—but only sometimes—exhibit further organization by keeping their internal environment distinct from their external environment.

Animals display two principal types of relation between their internal and external environments. On the one hand, when the conditions outside an animal's body change, the animal may permit its internal environment to match the external conditions and thus change along with the outside changes. On the other hand, the animal may maintain constancy in its internal environment. These alternatives are illustrated with temperature in **FIGURE 1.8**. If the temperature of an animal's external environment changes, one option is for the animal to let its internal temperature change to match the external temperature (see Figure 1.8A). Another option is for the animal to maintain a constant internal temperature (see Figure 1.8B). If an animal permits internal and external conditions to be equal, it is said to show **conformity**. If the animal maintains internal constancy in the face of external variability, it shows **regulation**. Conformity and regulation are extremes; intermediate responses are common.

Animals frequently show conformity with respect to some characteristics of their internal environment while showing regulation with respect to others. Consider a salmon, for example (**FIGURE 1.9**). Like most fish, salmon are *temperature conformers*; they let their

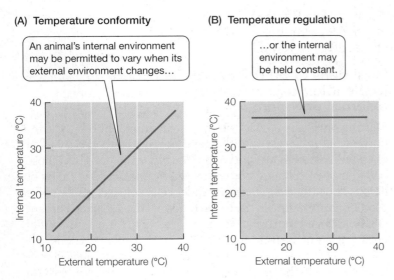

FIGURE 1.8 Conformity and regulation These examples from the study of temperature illustrate the general principles of conformity (A) and regulation (B).

(A) **Temperature conformity**

When a salmon enters a river from the sea, its body temperature (including blood temperature) changes if the river water is warmer or cooler than the ocean water…

(B) **Chloride regulation**

…but its blood Cl⁻ concentration remains almost constant, even though river water is very dilute in Cl⁻ and seawater is very concentrated in Cl⁻.

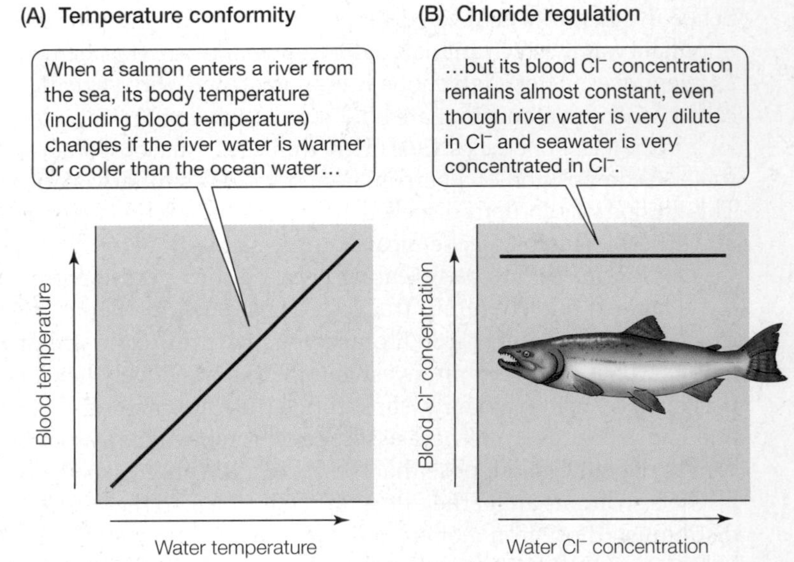

FIGURE 1.9 Mixed conformity and regulation in a single species Salmon are temperature conformers but chloride regulators. The presentation of Cl⁻ regulation is diagrammatic; the blood Cl⁻ concentration is not in fact absolutely constant but is a little higher when the fish are in seawater than when they are in freshwater.

internal temperature match the surrounding water temperature (see Figure 1.9A). Simultaneously, salmon are excellent *chloride regulators*; they maintain a nearly constant concentration of Cl⁻ ions in their blood, regardless of how salty their environmental water is—that is, regardless of how high or low the outside Cl⁻ concentration is (see Figure 1.9B).

Regulation demands more energy than conformity because regulation represents a form of organization. How does regulation represent organization? The answer is that during regulation, an animal maintains *constancy* inside its body, and it maintains *distinctions* between inside and outside conditions. Both the constancy and the distinctions are types of organization. A familiar analogy for the energy costs of regulation in animals is provided by home heating. Considerable energy is required to keep the inside of a house at 22°C (72°F) during the cold of winter. This energy cost is entirely avoided if the inside temperature is simply allowed to match the outside temperature.

Homeostasis in the lives of animals: Internal constancy is often critical for proper function

Homeostasis is an important concept regarding the *nature* and *significance* of internal constancy. Soon we will define homeostasis using the words of Walter Cannon (1871–1945), who coined the term. To fully appreciate the concept, however, we must first recognize its historical roots in medicine. The two men who contributed the most toward developing the concept of homeostasis, Claude Bernard and Walter Cannon, were physicians and medical researchers, concerned primarily with human physiology. Healthy humans maintain remarkable constancy of conditions in their blood and tissue fluids. The concept of homeostasis was thus conceived during studies of a species that exhibits exceptional internal constancy, and later the concept was extrapolated to other animals.

BOX 1.1 Negative Feedback

The type of control that Claude Bernard discovered in his studies of blood glucose is what today we call *negative feedback*. In any control system, the **controlled variable** is the property that is being kept constant or relatively constant by the system's activities. The **set point** is the level at which the controlled variable is to be kept. **Feedback** occurs if the system uses information on the controlled variable itself to govern its actions. In **negative feedback**, the system responds to changes in the controlled variable by bringing the variable back toward its set point; that is, the system *opposes deviations* of the controlled variable from the set point. There are many detailed mechanisms by which negative feedback can be brought about in physiological systems. Negative feedback, however, is virtually synonymous with homeostasis and occurs in all homeostatic systems.

In the case of the blood glucose level that so intrigued Claude Bernard,

the control system adds glucose to the blood if the blood glucose concentration—the controlled variable—falls below its set-point concentration, thereby opposing the deviation of the blood concentration from the set point. The control system removes glucose from the blood if the glucose concentration rises too high, thereby again opposing the deviation of the concentration from its set point. Biologists and engineers who study control systems have established that no control system can maintain perfect constancy in a controlled variable; putting the case roughly, a controlled variable must be a moving target for a control system to act on it. Thus the blood glucose concentration is not kept perfectly constant by the glucose control system, but during normal health it is kept from varying outside a narrow range. You will find a more thorough discussion of control systems based on negative feedback in Box 10.3.

In **positive feedback**, a control system *reinforces deviations* of a controlled variable from its set point. Positive feedback is much less common in physiological systems than negative feedback. It is more common during normal function than is usually recognized, however. For example, positive feedback occurs when action potentials (nerve impulses) develop in nerve cells (see Figure 12.16), and it also occurs during the birth process in mammals (see Figure 17.18). In the first case, a relatively small change in the voltage across the cell membrane of a nerve cell modulates the properties of the membrane in ways that make the voltage change become greater. In the second, muscular contractions acting to expel the fetus from the uterus induce hormonal signals that stimulate ever-more-intense contractions.

Claude Bernard was the first to recognize the impressive stability of conditions that humans maintain in their blood and tissue fluids. One of Bernard's principal areas of study was blood glucose in mammals. He observed that the liver takes up and releases glucose as necessary to maintain a relatively constant glucose concentration in the blood. If blood glucose rises, the liver removes glucose from the blood. If blood glucose falls, the liver releases glucose into the blood. Bernard stressed that, as a consequence, most cells in the body of a mammal experience a relatively constant environment with respect to glucose concentration (**BOX 1.1**). Bernard's research and that of later investigators also revealed that most cells in a mammal's body experience a relatively constant temperature—and a relatively constant O_2 level, osmotic pressure, pH, Na^+ concentration, Cl^- concentration, and so on—because various organs and tissues regulate these properties at consistent levels in the body fluids bathing the cells.

Claude Bernard devoted much thought to the *significance* of internal constancy in humans and other mammals. He was greatly impressed with how freely mammals are able to conduct their lives regardless of outside conditions. Mammals, for example, can wander about outdoors in the dead of winter, seeking food and mates, whereas fish and insects—in sharp contrast—are often driven into a sort of paralysis by winter's cold. Bernard reasoned that mammals are able to function in a consistent way regardless of varying outside conditions because the cells inside their bodies enjoy constant conditions. He thus stated a hypothesis that remains probably the most famous in the history of animal physiology: *"Constancy of the internal environment is the condition for free life."*

A modern translation might go like this: Animals are able to lead lives of greater freedom and independence to the extent that they maintain a stable internal environment, sheltering their *cells* from the variability of the outside world.

Walter Cannon, a prominent American physiologist who was born in the same decade that Claude Bernard died, introduced the word *homeostasis* to refer to internal constancy in animals. In certain ways, Bernard's and Cannon's views were so similar that Bernard might have invented the homeostasis concept, but the implications of internal constancy were clearer by Cannon's time. Because animals dynamically interact with their environments, the temperature, pH, ion concentrations, and other properties of their bodies are incessantly being drawn away from stability. Cannon emphasized that for an animal to be internally stable, vigilant physiological mechanisms must be present to correct deviations from stability. Thus when Cannon introduced and defined the term **homeostasis**, he intended it to mean not just internal constancy, but also the existence of regulatory systems that automatically make adjustments to maintain internal constancy. In his own words, Cannon described homeostasis as "the coordinated physiological processes which maintain most of the [constant] states in the organism."

An essential aspect of Cannon's perspective was his conviction that homeostasis is good. Cannon argued, in fact, that homeostasis is a signature of highly evolved life. He believed that animal species could be ranked according to their degree of homeostasis; in his view, for example, mammals were superior to frogs because mammals maintain a greater degree of homeostasis.

HOMEOSTASIS IN THE MODERN STUDY OF ANIMAL PHYSIOLOGY
The concept of ranking animals using degrees of homeostasis seems misguided to most biologists today. Bernard and Cannon, having focused on mammals, articulated ideas that are truly indispensable for understanding mammalian biology and medicine. However, the mere fact that mammals exhibit a high degree of homeostasis does not mean that other animals should be held to mammalian standards. Animals that exhibit less-complete homeostasis than mammals coexist in the biosphere with mammals. Indeed, the vast majority of animals thriving today do not achieve "mammalian standards" of homeostasis. Thus most biologists today would argue that a high degree of homeostasis is merely one of several ways to achieve evolutionary and ecological success. In this view, Bernard and Cannon did not articulate universal requirements for success, but instead they clarified the properties and significance of *one particular road* to success.

Recent research has clarified, in fact, that organisms sometimes achieve success in the biosphere precisely by letting their internal environment *vary* with the external environment: the antithesis of homeostasis. Consider, for example, insects that overwinter within plant stems in Alaska. They survive by ceasing to be active, allowing their internal temperatures to fall below −40°C, and tolerating such low tissue temperatures. Any attempt by such small animals to maintain an internally constant temperature from summer to winter would be so energetically costly that it would surely end in death; thus the tolerance of the insects to the *change* of their internal temperature in winter is a key to their survival. Even some mammals—the hibernators—survive winter by *abandoning constancy* of internal temperature; hibernating mammals allow their body temperatures to decline and sometimes match air temperature. For lizards in deserts, tolerance of profound dehydration is often a key to success.

Both constancy and inconstancy of the internal environment—regulation and conformity—have disadvantages and advantages:

- *Regulation.* The chief disadvantage of regulation is that it costs energy. The great legacy of Bernard and Cannon is that they clarified the advantage that animals enjoy by paying the cost: Regulation permits cells to function in steady conditions, independent of variations in outside conditions.
- *Conformity.* The principal disadvantage of conformity is that cells within the body are subject to changes in their conditions when outside conditions change. The chief advantage of conformity is that it is energetically cheap. Conformity avoids the energy costs of keeping the internal environment different from the external environment.

Neither regulation nor conformity is categorically a defect or an asset. One cannot understand mammals or medical physiology without understanding homeostasis, but one cannot understand the full sweep of animal life without recognizing that physiological flexibility is sometimes advantageous.

Time in the lives of animals: Physiology changes in five time frames

Time is a critical dimension for understanding the physiology of all animals because the physiology of animals invariably changes from time to time. Even animals that exhibit homeostasis undergo change. Details of their internal environment may change. Moreover, the regulatory processes that *maintain* homeostasis must change from time to time so that homeostasis can prevail, much as day-to-day adjustments in the fuel consumption of a home furnace are required to maintain a constant air temperature inside the home during winter.

An important organizing principle for understanding the role of time in the lives of animals is to recognize *five major time frames* within which the physiology of an animal can change. The time frames fall into two categories: (1) responses of physiology to changes in the external environment and (2) internally programmed changes of physiology. **TABLE 1.2** lists the five time frames classified in this way. We will recognize these five time frames throughout this book as we discuss various physiological systems.

The concept of the five time frames overlies other ways of organizing knowledge about animal function. For example, the concept of time frames overlies the concepts of regulation, conformity, and homeostasis that we have just discussed. When we speak of regulation, conformity, and homeostasis, we refer to *types* of responses that animals show in relation to variations in their external environments. When we speak of the time frames, we address *when* those responses occur.

PHYSIOLOGY RESPONDS TO CHANGES IN THE EXTERNAL ENVIRONMENT IN THREE TIME FRAMES Individual animals subjected to a change in their external environment exhibit *acute* and *chronic* responses to the environmental change. **Acute responses**, by definition, are responses exhibited during the first minutes or hours after an environmental change. **Chronic responses** are expressed following prolonged exposure to new environmental conditions. You might wonder why an individual's immediate responses to an environmental change differ from its long-term responses. The answer is that the passage of time permits biochemical or anatomical restructuring of an animal's body. When an animal suddenly experiences a change in its environment, its immediate responses must be based on the "old," preexisting properties of its body because the animal has no time to restructure. A morphological example is provided by a person who suddenly is required to lift weights after months of totally sedentary existence. The sedentary person is likely to have small arm muscles, and her immediate, *acute* response to her new weight-lifting environment will likely be that she can lift only light weights. However, if the person lifts weights repeatedly as time goes by, restructuring will occur; her muscles will increase in size. Thus her *chronic* response to the weight-lifting environment will likely be that she can lift heavy weights as well as light ones.

A familiar *physiological* example of acute and chronic responses is provided by human reactions to work in hot weather. We all know that when we are first exposed to hot weather after a period of living in cool conditions, we often feel quickly exhausted; we say the heat is "draining." We also know that this is not a permanent state: If we experience heat day after day, we feel more and more able to work in the heat.

FIGURE 1.10 shows that these impressions are not merely subjective illusions. Twenty-four physically fit young men who lacked recent experience with hot weather were asked to walk at a fixed pace in hot, relatively dry air. Their endurance was measured as a

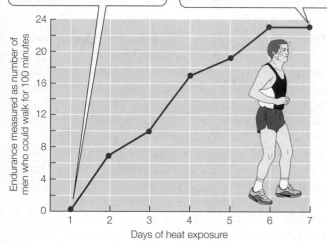

The **acute response**, displayed when the men were first exposed to the hot environment on day 1, was low endurance; none could continue walking for 100 minutes.

The **chronic response**, displayed after a week of experience with the hot environment, was dramatically increased endurance; 23 of the 24 men could continue walking for 100 minutes.

FIGURE 1.10 Heat acclimation in humans as measured by exercise endurance Twenty-four fit young men without recent heat experience were asked to walk at 3.5 miles per hour in hot, dry air (49°C, 20% relative humidity). Their endurance was used as a measure of their physiological capability to engage in moderate work under hot conditions. The acclimation illustrated by the chronic response is reversible; if heat-acclimated men return to a life of no heat exposure, they gradually revert to the level of endurance evident on day 1. (After Pandolf and Young 1992.)

From research on the physiology of human work under hot conditions, physiologists know that endurance under hot conditions changes because as people gain increased experience with heat, their rates of sweat secretion increase, their sweat glands are able to maintain high rates of sweat secretion for dramatically lengthened periods of time, their sweat becomes more dilute (so they lose less salt), the blood flow to their skin becomes more vigorous (improving delivery of internal heat to the body surface), and their heart rates during exercise in the heat become lower. Thus human physiology is restructured in many ways by repeated exposure to heat. For a person who has been living in cool conditions, the *acute* physiological responses to heat exposure are low exercise endurance, a low rate of sweat production, and so forth. Heat training poises a person to express *chronic* physiological responses to heat, such as high exercise endurance and a high capacity to sweat.

The acute and chronic responses are, by definition, phenotypic responses of *individual animals* to environmental change. *Populations* may exhibit a third category of response to environmental change: **evolutionary responses** involving changes of genotypes. Collectively, therefore, animals display responses to environmental change in three time frames:

- Individuals exhibit immediate, *acute* responses.
- Individuals exhibit long-term, *chronic* responses. The length of time that an individual must be exposed to a new environment for chronic responses to be fully expressed is usually a few days to a few weeks.
- Populations exhibit *evolutionary* responses.

Chronic responses by individual animals to environmental change are so common, diverse, and important that their study forms a special discipline with its own terminology. For many physiologists, the concepts of *acclimation* and *acclimatization* provide an important way to classify the chronic responses of individuals to environmental change. A chronic response to a changed environ-

way of quantifying their physiological ability to sustain moderate exercise under the hot conditions. None of the men had sufficient endurance to walk for 100 minutes (min) on the first day. However, as the days passed and the men had more and more experience with hot conditions, their endurance increased, as indicated by a steady increase in the number of men who could keep walking for 100 min.

TABLE 1.2 The five time frames in which physiology changes	
Type of change	**Description**
Changes in physiology that are responses to changes in the external environment	
1. Acute changes	Short-term changes in the physiology of individual animals: changes that individuals exhibit soon after their environments have changed. Acute changes are reversible.
2. Chronic changes (termed *acclimation* and *acclimatization*; also termed *phenotypic plasticity* or *phenotypic flexibility*)	Long-term changes in the physiology of individual animals: changes that individuals display after they have been in new environments for days, weeks, or months. Chronic changes are reversible.
3. Evolutionary changes	Changes that occur by alteration of gene frequencies over the course of multiple generations in populations exposed to new environments.
Changes in physiology that are internally programmed to occur whether or not the external environment changes	
4. Developmental changes	Changes in the physiology of individual animals that occur in a programmed way as the animals mature from conception to adulthood and then to senescence (see Chapter 4).
5. Changes controlled by periodic biological clocks	Changes in the physiology of individual animals that occur in repeating patterns (e.g., each day) under control of the animals' internal biological clocks (see Chapter 15).

The Evolution of Phenotypic Plasticity

When animals express different genetically controlled phenotypes in different environments—when they acclimate and acclimatize—they require controls that determine which particular phenotypes are expressed in which particular environments. As an illustration, suppose that an individual animal has four possible phenotypes, *P1* through *P4*, and that there are four environments, *E1* through *E4*. One option is that the individual could express phenotype *P1* in environment *E3*, *P2* in *E4*, *P3* in *E1*, and *P4* in *E2*. This set of *correspondences* between phenotypes and environments constitutes the individual's **norm of reaction**; that is, if we think of the phenotypes as one list and the environments as a second list in a matching game, the *norm of reaction* is like the set of lines that we would draw between items on the two lists to indicate which item on one matches which on the other.

The norm of reaction is genetically coded in the genome. Because of this, *the norm of reaction itself can evolve and is subject to natural selection.* To see this, suppose that an individual other than the one just discussed expresses phenotype *P1* in environment *E1*, *P2* in *E2*, *P3* in *E3*, and *P4* in *E4*. In this case, the two individuals would differ in their norms of reaction. Suppose, now, that there is a population—living in a variable environment—that is composed half of individuals with the first reaction norm and half of individuals with the second. If individuals of the first sort were to survive and reproduce more successfully as the environment varied, natural selection for the first reaction norm would occur. In this way the reaction norm itself would evolve in ways that would better adapt the animals to the variable environment in which they live.

A simple example is provided by tanning in people with light complexions.

During tanning, the skin changes in its content of a dark-colored pigment called melanin. Suppose that there are two possible cutaneous phenotypes: *high melanin* and *low melanin*. Suppose also that there are two environments: *high sun* and *low sun*. One possible norm of reaction would be to express high melanin in low sun and low melanin in high sun. Another norm of reaction would be to express high melanin in high sun and low melanin in low sun. Over the course of evolution, if both of these reaction norms once existed, it is easy to understand why individuals with the second reaction norm would have left more progeny than those with the first. That is, it is easy to understand the evolution of the sort of reaction norm we see today among people with light complexions.

Phenotypic plasticity *itself* can evolve, and norms of reactions can themselves be adaptations.

---

ment is called **acclimation** if the new environment differs from the preceding environment in just a few highly defined ways.[6] Acclimation is thus a laboratory phenomenon. **Acclimatization** is a chronic response of individuals to a changed environment when the new and old environments are different *natural* environments that can differ in numerous ways, such as winter and summer, or low and high altitudes. Thus animals are said to *acclimatize* to winter, but they *acclimate* to different defined temperatures in a laboratory experiment.

Acclimation and acclimatization are types of **phenotypic plasticity**: the ability of an individual animal (a single genotype) to express two or more genetically controlled phenotypes. Phenotypic plasticity is possible because the genotype of an individual can code for multiple phenotypes (**BOX 1.2**). Growth of the biceps muscle during weight training provides a simple example of a change in phenotype under control of genetically coded mechanisms. Another example is that the particular suite of enzymes active in an adult person may change from one time to another because the genes for one suite of enzymes are expressed under certain environmental conditions, whereas the genes for another suite are expressed under different conditions.[7] We will discuss phenotypic plasticity in more detail—with several additional examples—in Chapter 4.

[6] Some authors restrict use of the word *acclimation* to cases in which just one property differs between environments.

[7] Enzymes that vary in amount as a result of changes in environmental conditions are termed *inducible* enzymes. An excellent illustration is provided by the P450 enzymes discussed at length in Chapter 2 (see page 52).

PHYSIOLOGY UNDERGOES INTERNALLY PROGRAMMED CHANGES IN TWO TIME FRAMES   The physiological properties of individuals sometimes change even if their external environment stays constant. For instance, the type of hemoglobin in your blood today is different from the type you produced as a newborn. This change in hemoglobin is internally programmed: It occurs even if your external environment stays constant. Sometimes internally programmed changes interact with environmental changes. For instance, an internally programmed change might occur sooner, or to a greater amplitude, in one environment than in another. However, the internally programmed changes do not require any sort of environmental activation. There are two principal types of internally programmed change: developmental changes and changes controlled by periodic biological clocks.

Development is the progression of life stages from conception to senescence in an individual. Different genes are internally programmed to be expressed at different stages of development, giving rise to **developmental changes in an animal's phenotype**. Puberty is a particularly dramatic example of internally programmed developmental change in humans. The environment may change the timing of puberty—as when the advent of sexual maturity is delayed by malnutrition—but puberty always occurs, no matter what the environment, illustrating that internally programmed changes do not require environmental activation. We will discuss physiological development in much greater depth in Chapter 4.

Biological clocks are mechanisms that give organisms an internal capability to keep track of the passage of time. Most biological clocks resemble wristwatches in being periodic; that is, after they complete one timing cycle, they start another, just as a wristwatch starts to time a new day after it has completed timing of the previous day. These sorts of biological clocks emit signals that cause cells and organs to undergo internally programmed, repeating cycles in their physiological states, thereby giving rise to **periodic, clock-controlled changes in an animal's phenotype**. An enzyme under control of a biological clock, for instance, might increase in concentration each morning and decrease each evening, not because the animal is responding to changes in its outside environment, but because of the action of the clock. The changes in enzyme concentration might mean that an animal is inherently better able to digest a certain type of food at one time of day than another, or is better able to destroy a certain type of toxin in the morning than in the evening. Biological clocks typically synchronize themselves with the external environment, but they go through their timing cycles inherently, and they can time physiological changes for days on end without environmental input. We will discuss them in greater detail in Chapter 15.

Size in the lives of animals: Body size is one of an animal's most important traits

How big is it? is one of the most consequential questions you can ask about any animal. This is true because within sets of related species, many traits vary in regular ways with their body sizes. The length of gestation, for example, is a regular function of body size in mammals (**FIGURE 1.11**). Brain size, heart rate, the rate of energy use, the age of sexual maturity, and hundreds of other physiological and morphological traits are also known to vary in regular, predictable ways with body size in mammals and other phylogenetically related sets of animal species. The study of these regular relations is known as the study of **scaling** because related species of large and small size can be viewed as *scaled-up* and *scaled-down* versions of their type.

Knowledge of the statistical relationship between a trait and body size is essential for identifying specializations and adaptations of particular species. To illustrate, let's ask if two particular African antelopes, the bushbuck and mountain reedbuck, have specialized or ordinary lengths of gestation. Answering this question is complicated precisely because there is no single norm of mammalian gestation length to use to decide. Instead, because the length of gestation varies in a regular way with body size, a biologist needs to consider the body sizes of the species to know what is average or ordinary.

Statistical methods can be used to derive a line that best fits a set of data. In the study of scaling, the statistical method that has traditionally been considered most appropriate is **ordinary least squares regression** (see Appendix D). The line in Figure 1.11 was calculated by this procedure. This line shows the average trend in the relationship between gestation length and body size. The line is considered to show the length of gestation *expected* of an ordinary species at each body size.

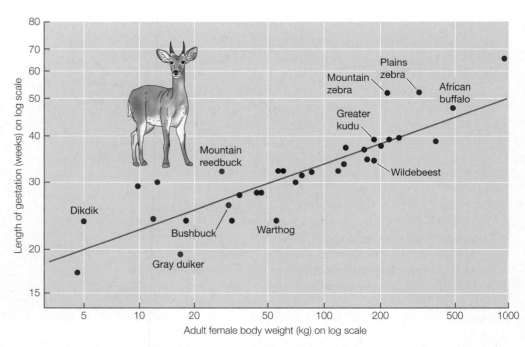

FIGURE 1.11 Length of gestation scales as a regular function of body size in mammals The data points—each representing a different species—are for African herbivorous mammals weighing 5–1000 kg as adults. The line (fitted by ordinary least squares regression; see Appendix D) provides a statistical description of the overall trend and thus depicts the gestation length that is statistically expected of an average or ordinary animal at each body size. Both axes use logarithmic scales, explaining why the numbers along the axes are not evenly spaced (see Appendix E). The red-colored data points are for animals discussed in the text. (After Owen-Smith 1988.)

With this information on *expected* gestation lengths, now we can address the question asked earlier: Are the bushbuck and mountain reedbuck specialized or ordinary? Notice that the length of gestation in the bushbuck is very close to what the line in Figure 1.11 predicts for an animal of its size. The bushbuck, therefore, adheres to what is expected for its size: It has an ordinary gestation length when its size is taken into account. The mountain reedbuck, however, is far off the line. According to the line, as shown in **TABLE 1.3**, an animal of the reedbuck's size is expected to have a gestation lasting 26.5 weeks, but actually the reedbuck's gestation lasts 32 weeks. Thus the reedbuck seems to have evolved a specialized, exceptionally

TABLE 1.3 Predicted and actual gestation lengths for two African antelopes of about the same body size

Species	Predicted gestation length (weeks)[a]	Actual gestation length (weeks)
Bushbuck (*Tragelaphus scriptus*)	27	26
Mountain reedbuck (*Redunca fulvorufula*)	26.5	32

[a]Predicted lengths are from the statistically fitted line shown in Figure 1.11.

long gestation. Similarly, the gray duiker seems to have evolved an exceptionally short length of gestation for *its* size (see Figure 1.11).

In the last 20 years, physiologists have recognized that ordinary least squares regression may not always be the best procedure for fitting lines to scaling data because the ordinary least squares procedure does not take into account the family tree of the species studied; it simply treats each data point as being fully independent of all the other data points (see Appendix D). Increasingly, therefore, physiologists have fitted lines not only by the ordinary least squares procedure but also by an alternative procedure based on **phylogenetically independent contrasts**, a method that takes the family tree into account (see Appendix G).[8] Although these two approaches sometimes yield distinctly different results, they most often yield similar results, and in this book, the lines we present for scaling studies will be derived from the method of traditional, ordinary least squares regression.

Body-size relations are important for analyzing almost all sorts of questions in the study of physiology, ecology, and evolutionary biology. If all one knows about an animal species is its body size, one can usually make useful predictions about many of the species' physiological and morphological traits by consulting known statistical relationships between the traits and size. Conversely, there is always the chance that a species is specialized in certain ways, and as soon as one has actual data on the species, one can identify potential specializations by the type of scaling analysis we have discussed.

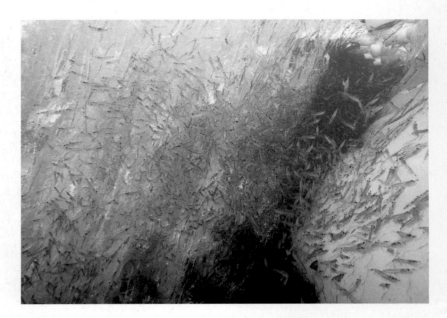

FIGURE 1.12 Krill and fish in the sea around Antarctica spend their entire lives at body temperatures near –1.9°C Despite the low temperatures, these shrimplike krill (*Euphausia superba*) occur in huge, gregarious populations that blue whales greatly favor for food. The krill—which grow to lengths of 3–6 cm—eat algal cells in the water and also graze on algae growing on ice surfaces. They hatch, grow, feed, and mate at body temperatures near –1.9°C.

Environments

What is an environment? An important starting point in answering this question is to recognize that an *animal* and its *environment* are interrelated, not independent, entities. They are in fact defined in terms of each other: The environment in any particular case cannot be specified until the animal is specified. A dog, for instance, is an animal from our usual perspective, but if the animal of interest is a tapeworm in the dog's gut, then the dog is the environment. All animals, in fact, are parts of the environments of other animals. The birds in the trees around your home are part of your environment, and you are part of theirs. The interdependence of animal and environment is reflected in standard dictionary definitions. A dictionary defines an animal to be a living organism. An **environment** is defined to be all the chemical, physical, and biotic components of *an organism's* surroundings.

Earth's major physical and chemical environments

The physical and chemical environments on our planet are remarkably diverse in their features, providing life with countless challenges and opportunities for environmental specialization. Temperature, oxygen, and water are the "big three" in the set of physical and chemical conditions that set the stage for life. Here we discuss the ranges of variation of temperature, oxygen, and water across the face of the globe. We also discuss highlights of how animals relate to these features. In later chapters, we will return to these topics in greater detail.

TEMPERATURE The **temperature** of the air, water, or any other material is a measure of the intensity of the random motions that the atoms and molecules in the material undergo. All atoms and molecules ceaselessly move at random on an atomic-molecular scale. A high temperature signifies that the intensity of this atomic-molecular agitation is high. Most animals are temperature conformers, and as we discuss temperature here, we will focus on them. The conformers are our principal interest because the level of atomic-molecular agitation in their tissues matches the level in the environments where they live.

The lowest temperature inhabited by active communities of relatively large, temperature-conforming animals is –1.9°C, in the polar seas.[9] The open waters of the polar oceans remain perpetually at about –1.9°C, the lowest temperature at which seawater is liquid. Thus the shrimplike krill (**FIGURE 1.12**), the fish, the sea stars and sea urchins, and the other invertebrates of these oceans have tissue temperatures near –1.9°C from the moment they are conceived until they die. They do not freeze. Whereas some, such as the krill, do not freeze because their ordinary freezing points are similar to the

---

[8] Appendix G explains the reasons why the family tree should ideally be taken into account, as well as providing a conceptual introduction to phylogenetically independent contrasts.

[9] The very lowest temperature at which *any* active communities of temperature-conforming animals live occurs *within* the sea ice near the poles. Minute nematodes and crustaceans, as well as algae, live and reproduce within the sea ice at temperatures that, in some places, are a few degrees colder than the temperature of –1.9°C that prevails in the surrounding polar water.

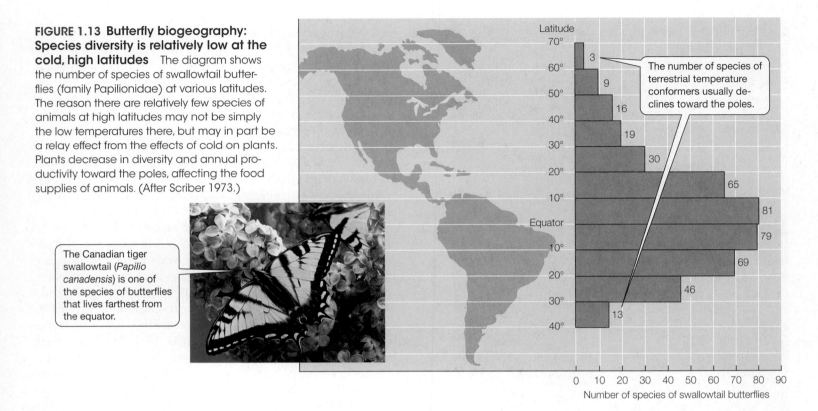

FIGURE 1.13 Butterfly biogeography: Species diversity is relatively low at the cold, high latitudes The diagram shows the number of species of swallowtail butterflies (family Papilionidae) at various latitudes. The reason there are relatively few species of animals at high latitudes may not be simply the low temperatures there, but may in part be a relay effect from the effects of cold on plants. Plants decrease in diversity and annual productivity toward the poles, affecting the food supplies of animals. (After Scriber 1973.)

The number of species of terrestrial temperature conformers usually declines toward the poles.

The Canadian tiger swallowtail (*Papilio canadensis*) is one of the species of butterflies that lives farthest from the equator.

freezing point of seawater,[10] others, such as many fish, metabolically synthesize antifreeze compounds that keep them from freezing. Because the tissues of these animals are very cold, one might imagine that the animals live in a sort of suspended animation. Actually, however, the communities of temperature-conforming animals in the polar seas are active and thriving. In the ocean around Antarctica, for example, a sure sign of the vigor of the populations of krill and fish is that they reproduce and grow prolifically enough to meet the food needs of the famously huge populations of Antarctic whales, seals, and penguins.

Are the low tissue temperatures of polar fish and invertebrates actually challenging for them, or do these low tissue temperatures only seem challenging? One way to obtain an answer is to compare polar species with related nonpolar ones. Tropical species of fish clearly find low temperatures to be challenging. Many tropical fish, in fact, die if cooled to +6°C, even if they are cooled very gradually. Such observations emphasize that success at –1.9°C is not "automatic," and that the polar species of fish have had to evolve special adaptations to thrive with their tissues perpetually at –1.9°C. The polar species themselves often die if they are warmed to +6°C, indicating that the tropical species also have special adaptations—adaptations that poise them to live at tropical temperatures. The evolutionary divergence of these fish is dramatized by the fact that a *single* temperature can be lethally *cold* for tropical species and yet be lethally *warm* for polar species!

Far greater extremes of cold are found on land than in aquatic environments. In Antarctica, the air temperature sometimes drops

to –90°C (–130°F); in the Arctic, it descends to –70°C (–90°F). The extremes of animal adaptation to low tissue temperature are represented by certain extraordinary species of Arctic insects that spend winters inside exposed plant stems or on the surface of pack ice. These insects are in a state of suspended animation at these times; they are quiescent, not active. Nonetheless, it is impressive that some endure tissue temperatures of –60°C to –70°C, either in a frozen state (which they have adaptations to tolerate) or in an unfrozen supercooled state.

The diversity of terrestrial temperature-conforming animals typically becomes lower and lower as latitude increases from the temperate zone toward the poles, as exemplified in **FIGURE 1.13**. This decline in diversity toward the poles indicates that the very cold terrestrial environments are demanding places for animals to occupy, despite evolutionary adaptability.

At the high end of the temperature scale, the temperature of the air or water on Earth usually does not go higher than +50°C (+120°F). Animals on land may experience even higher heat loads, however, by being exposed simultaneously to hot air and the sun's radiation. Some temperature-conforming animals from hot environments—such as certain desert insects and lizards—can function at *tissue* temperatures of 45°C–55°C (**FIGURE 1.14**).[11] These are the highest tissue temperatures known for animal life, suggesting that the high levels of molecular agitation at such temperatures pose the greatest challenge that can be met by evolutionary adaptation in animal systems.

The hottest places in the biosphere are the waters of geothermally heated hot springs and underwater hot vents. These waters are often far above the boiling point when they exit Earth's crust. Although aquatic animals typically stay where the waters have

---

[10] Dissolved salts and other dissolved compounds lower the freezing points of solutions. Krill and most other marine invertebrates have total concentrations of dissolved matter in their blood similar to the concentration in seawater. Consequently, their blood freezing points are about the same as the freezing point of seawater, and they do not freeze, provided that the seawater remains unfrozen.

[11] Normal human body temperature is 37°C.

FIGURE 1.14 A thermophilic ("heat-loving") lizard common in North American deserts The desert iguana (*Dipsosaurus dorsalis*) can often be seen abroad as the sun beats down on hot days. Although it does not usually expose itself to body temperatures higher than 42°C, it can survive 48.5°C, one of the highest body temperatures tolerated by any vertebrate animal.

cooled to 35°C–45°C or lower, many prokaryotic microbes—bacteria and archaea—thrive at much higher temperatures than animals can. Some prokaryotes even reproduce at temperatures above 100°C.

OXYGEN The need of most animals for oxygen (O_2) is a consequence of their need for metabolic energy. The chemical reactions that animals use to release energy from organic compounds remove some of the hydrogen atoms from the compounds. Each adult person, for example, liberates about one-fifth of a pound of hydrogen every day in the process of breaking down food molecules to obtain energy. Hydrogen liberated in this way cannot be allowed to accumulate in an animal's cells. Thus an animal must possess biochemical mechanisms for combining the hydrogen with something, and O_2 is the usual recipient. O_2 obtained from the environment is delivered to each cell, where it reacts with the free hydrogen produced in the cell, yielding water (see Figure 8.2).

The suitability of an environment for animals often depends on the availability of O_2. In terrestrial environments at low and moderate altitudes, the open air is a rich source of O_2. Air consists of 21% O_2, and at low or moderate altitudes it is relatively dense because it is at relatively high pressure. Thus animals living in the open air have a plentiful O_2 resource. Even within burrows or other

secluded places on land, O_2 is often freely available because—as counterintuitive as it may sound—O_2 diffuses fairly readily from the open atmosphere through soil to reach burrow cavities, provided the soil structure includes gas-filled spaces surrounding the soil particles.

As altitude increases, the O_2 concentration of the air declines because the air pressure decreases and the air therefore becomes more and more rarified. Air at the top of Mt. Everest—8848 m above sea level—is 21% O_2, like that at sea level; but the total air pressure is only about one-third as high as at sea level, and gas molecules within the air are therefore so widely spaced that each liter of air contains only about one-third as much O_2 as at sea level.

The high altitudes are among Earth's most challenging places, reflected in the fact that the numbers of animal species are sharply reduced. At such altitudes, the maximum rate at which animals can acquire O_2 is often much lower than at sea level, and functions are consequently limited. At elevations above 6500 m (21,000 ft), for example, people breathing from the atmosphere find that simply walking uphill is a major challenge because their level of exertion is limited by the low availability of O_2 (**FIGURE 1.15**). Some animal species have evolved adaptations to succeed in the dilute O_2 of rarefied air in ways that humans cannot. One of the most remarkable species is the bar-headed goose (*Anser indicus*), which—in ways that physiologists still do not fully comprehend—is able to fly (without an oxygen mask!) over the crests of the Himalayas at 9000 m.

Water-breathing animals typically face a substantially greater challenge to obtain O_2 than air-breathing animals do because the supply of O_2 for water breathers is the O_2 *dissolved* in water, and the solubility of O_2 in water is not high. Because of the low solubility of O_2, water contains much less O_2 per liter than air does, even when the water is fully aerated. For example, aerated stream or river water at sea level contains only 3–5% as much O_2 per liter as air at sea level does.

A common problem for animals living in slow-moving bodies of water such as lakes, ponds, or marshes is that the O_2 concentration in the water may be even lower than in aerated water because dissolved O_2 may become locally depleted by the metabolic activities of animals or microbes. *Density layering* of water—which prevents the water from circulating freely—is a common contributing factor to O_2 depletion in the deep waters of lakes and ponds. Density layering occurs when low-density water floats on top of high-density water, causing distinct water layers to form. When this happens, there is often almost no mixing of oxygenated water from the low-density surface layer (where photosynthesis and aeration occur) into the high-density bottom layer. Thus O_2 in the bottom layer is not readily replaced when it is used, and as microbes and animals in the bottom layer consume O_2, the O_2 concentration may fall to very low levels.

In lakes during summer, density layering occurs because of temperature effects: Sun-heated warm water tends to float on top of colder and denser bottom water.[12] The lake studied by a group of university students in **FIGURE 1.16** provides an example of this sort of density layering. The bottom waters of this lake contained

---

[12] In estuarine bodies of water along seacoasts—where freshwater and seawater mix—layering can occur because of salinity effects as well as temperature effects. Low-salinity water is less dense than—and tends to float on top of—high-salinity water.

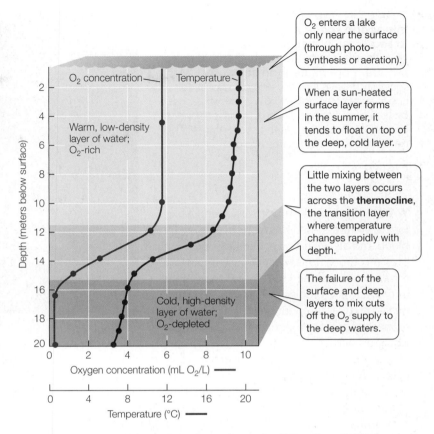

FIGURE 1.15 Performance in an O₂-poor environment
Because of the difficulty of acquiring O₂ from rarefied air, the rate at which energy can be released from food molecules for use in work by humans is reduced at high altitudes, and the simple act of walking uphill becomes extremely arduous. Well-conditioned mountaineers are slowed to a walking rate of 100–200 m per hour near the tops of the world's highest mountains if they are breathing from the air rather than from oxygen tanks. Shown here is Chantal Mauduit (1964–1998) during an unsuccessful attempt to reach the summit of Mt. Everest while breathing only atmospheric air. On an earlier expedition she had been the fourth woman to climb to the peak of K2 (8611 m), second highest mountain on Earth, without supplemental oxygen.

FIGURE 1.16 Density layering can cut off the O₂ supply to the deep waters of a lake Different densities of water do not mix readily. The O₂ concentration in the deep waters of a lake may fall to near zero because the animals and microbes living there consume O₂ that is not replaced. (From data gathered by a group of animal physiology students on a lake in northern Michigan in July.)

essentially *no* dissolved O₂ on the July day when the data were collected. Deep-water O₂ depletion has become more common in recent decades in lakes, ponds, and estuaries as human populations have enriched waters with organic matter. The organic matter supports the growth of microbes that deplete dissolved O₂. For animals in deep waters to survive, they must be able to tolerate low O₂ levels, or they must temporarily move to other places where O₂ is more available.

In certain sorts of water bodies, animals have faced the challenge of low O₂ concentrations for millennia. Unlike animals confronted with new, human-induced O₂ depletion, the animals living in primordially O₂-poor waters have been able to undergo long-term evolutionary adaptation to low-O₂ conditions. Tropical rivers that are naturally very rich in organic matter, as in the Amazon basin, provide examples of waters that have been O₂-poor for millennia. The warmth of these rivers not only lowers the solubility of O₂ in the water but also promotes rapid multiplication of microbes that use O₂. In addition, thick forest canopies may create deep shade over the rivers, impeding algal

photosynthesis that otherwise could replenish O₂. Among the fish living in such waters, the evolution of air breathing is one of the most remarkable features. Hundreds of species of fish in these waters are air breathers. Some take up inhaled O₂ across well-vascularized mouth linings or lunglike structures. Others swallow air and absorb O₂ in their stomachs or intestines, as mentioned previously. For animals confronted with short-term or long-term O₂ deficiency, whether in O₂-depleted freshwater environments or elsewhere, a potential solution over evolutionary time is to adopt a biochemistry that can attach hydrogen to molecules other than O₂. Many species—both air breathers and water breathers—have temporary options of this sort. Certain tissues in our own bodies, for example, can live without O₂ for 10 min at a time by attaching hydrogen to pyruvic acid (making lactic acid). Suppose, however, that an animal's entire body must live with little or no O₂ for many hours, days, weeks, or months. Doing so is possible for some animals, but as the period without O₂ lengthens, ever-fewer species have evolved biochemical specializations that enable them to survive. Some exceptional animals are able to meet the most extreme challenge of living indefinitely in O₂-free environments. Most that are currently known to science are parasites (e.g., nematodes and tapeworms) that live in the O₂-free environment of the vertebrate gut cavity.

FIGURE 1.17 The sea star and the corals in this ocean scene have body fluids similar to seawater in their total concentration of salts, although the body fluids of the fish are more dilute than seawater Most scientists believe that the difference between marine invertebrates and marine bony fish is based on their evolutionary histories. The invertebrates' ancestors always lived in the sea, but the fish's ancestors once lived in freshwater.

WATER Water is the universal solvent in biological systems—and therefore is required for blood and all other body fluids to have their proper compositions. Water is also important for animals because H_2O bound to proteins and other macromolecules as *water of hydration* is often required for the macromolecules to maintain their proper chemical and functional properties.

Animal life began in Earth's greatest watery environment, the oceans. Most invertebrates that live in the open oceans today—sea stars, corals, clams, lobsters, and so forth—are thought to trace a continuously marine ancestry. That is, their ancestors never left the seas from the time that animal life began, and thus the salinity of the oceans has been a perennial feature of their environment. The blood of these invertebrates (**FIGURE 1.17**), although differing a bit from seawater in composition, is similar to seawater in its total salt concentration. These animals therefore do not tend to gain much H_2O from their environment by osmosis, nor do they tend to lose H_2O from their blood to the seawater. Because this situation is almost universal among these animals, we believe it is the primordial condition of animal life. Thus we believe that for much of its evolutionary history, animal life lived in a setting where (1) H_2O was abundant in the environment and (2) little danger existed for an animal to be either dehydrated or overhydrated.

This benign situation was left behind by the animals that migrated from the oceans into rivers during their evolution. Freshwater has a very low salinity compared with seawater. When animals from the oceans, with their salty blood, started to colonize freshwater, they experienced a severe challenge: H_2O tended to seep osmotically into their bodies and flood their tissues because osmosis transports H_2O relentlessly from a dilute solution into a more concentrated one. Today, lakes and rivers are populated by fish, clams, crayfish, sponges, hydras, and so forth, all descended from ocean ancestors. Over evolutionary time, freshwater animals reduced their tendency to gain H_2O osmotically from their environment, but they have not eliminated it. A 100-g goldfish, for example, osmotically gains enough water to equal 30% or more of its body weight every day.

Vertebrates and several groups of invertebrates invaded the land from freshwater. In so doing they came face to face with the most severe of all the water challenges on Earth: On land, evaporation of water into the atmosphere tends to dehydrate animals rapidly. Some terrestrial habitats are so dry, in fact, that replacing lost water borders on impossible. When animals first invaded terrestrial habitats, they probably possessed integuments (body coverings), inherited from aquatic ancestors, that provided little or no barrier to evaporative water loss. This problem ultimately had to be solved for animals to be able to live entirely freely in the open air.

Some of today's land animals have integuments that resemble the primordial types. Leopard frogs, earthworms, and wood lice, for example, have integuments that lack significant evaporation barriers and permit virtually free evaporation. In some cases, water evaporates across these sorts of integuments as fast as it evaporates from an open dish of water of the same surface area! Animals with such integuments dehydrate so rapidly that they cannot possibly live steadily in the open air. Instead, they must stay in protected places where the humidity of the air is high, or if they venture into the open air, they must return often to places where they can rehydrate. The danger of dehydration severely constrains their freedom of action.

For a terrestrial animal to be liberated from these constraints and lead a fully exposed existence in the open air, it must have evolved highly effective barriers to water loss across its integument. Only a few major groups of animals possess such novel water barriers: mammals, birds, other reptiles (see Figure 1.14), insects, and spiders. In each of these groups, thin layers of lipids are deposited in the integument and act as barriers to evaporative water loss. The evolution of these lipid layers liberated animals to occupy the open air and was a prerequisite for animals to invade the driest places on Earth, the deserts. In hyperarid deserts, a year or two can pass without rain, yet there are populations of insects, lizards, birds, and mammals that succeed there.

Some terrestrial animals have adapted to land in part by evolving exceptional tolerance of dehydration. Although most terrestrial animals die if they lose half or more of their body water without replacing it, the exceptional types can dehydrate more. The most extreme cases are certain invertebrates that can lose essentially all their body water and survive in a dormant, crystalline state until water returns. Certain tardigrades ("moss animals" or "water bears"), for example, dry completely when deprived of water and then can blow about like dust, ready to spring back to active life if water becomes available.

Contrary to what intuition might suggest, even some aquatic animals are threatened with dehydration. The bony fish of the oceans, such as the reef fish seen in Figure 1.17, are the most important example. These fish have blood that is only one-third to one-half as salty as seawater, probably because they are descended from freshwater ancestors rather than from ancestors that always lived in the sea. The ocean is a desiccating environment for animals with dilute blood because osmosis transports H_2O steadily from the blood to the more concentrated seawater. These desiccating fish have an advantage over terrestrial animals desiccating in a desert—namely, that H_2O to replace their losses is abundant in their watery environment. To incorporate H_2O from seawater into their dilute bodies, however, they must in essence possess mechanisms to "distill" the seawater: They must be able to separate H_2O from the salty seawater solution.

The environment an animal occupies is often a microenvironment or microclimate

In a forest, lake, or any other large system, small places inevitably exist where physical or chemical conditions are significantly different from the average in the system. For instance, when the average temperature of the open air in a woodland is 30°C, the temperature under a pile of brush on the forest floor might be 24°C. Although large-bodied animals are often, by necessity, exposed to the statistically average conditions where they live, small-bodied animals may enter the small places—the nooks and crannies—where they may find conditions that are far from average. Places within an environment that potentially differ from the environment at large in their physical or chemical conditions are called **microenvironments**. A related concept is that of **microclimates**. A microclimate is the set of climatic conditions (temperature, humidity, wind speed, and so forth) prevailing in a subpart of a system.

Because we humans are large organisms, our perception of the prevailing conditions in a place may bear little relation to the microclimates that smaller creatures can find by entering distinctive subparts of the place. George Bartholomew (1919–2006), one of the founders of environmental physiology, expressed this important point well:

> Most vertebrates are much less than a hundredth of the size of man…, and the universe of these small creatures is one of cracks and crevices, holes in logs, dense underbrush, tunnels and nests—a world where distances are measured in yards rather than miles and where the difference between sunshine and shadow may be the difference between life and death. Climate in the usual sense of the word is, therefore, little more than a crude index to the physical conditions in which most terrestrial animals live.[13]

Desert environments nicely illustrate the point that Bartholomew makes (**FIGURE 1.18**). At head level (about 2 m aboveground), a human or a horse standing in the Arizona desert may experience

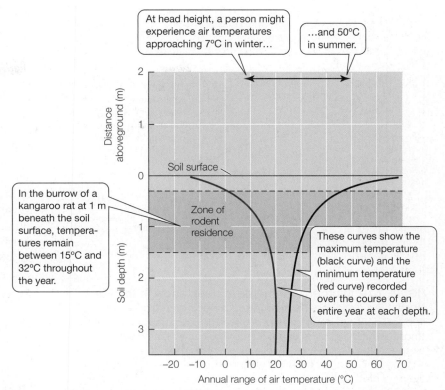

FIGURE 1.18 Microenvironments in the Arizona desert near Tucson The plot shows the annual range of temperatures in the soil and air and at the soil surface. (After Misonne 1959.)

daytime air temperatures that reach almost 50°C during the summer, combined with intense solar radiation. Humans and horses typically have no choice but to cope physiologically with these conditions because they are too large to escape by going underground or squeezing into patches of shade cast by cacti or desert bushes. Small desert rodents such as kangaroo rats and pocket mice are in a very different situation, however, because they can burrow deep into the soil, where thermal conditions are far different from those that humans associate with deserts. On the surface of the desert soil, the annual temperature range is actually greater than that in the air above (see Figure 1.18); the soil surface becomes hotter than the air during the day as it absorbs solar radiation, and it becomes cooler than the air at night because it radiates infrared energy to the cold nighttime sky (see page 241). Beneath the soil surface, however, the annual range of temperature decreases dramatically as depth increases. At a depth of 1 m, the temperature remains well below the maximum aboveground air temperature during summer and well above the minimum air temperature during winter. In fact, in certain desert regions, such as that shown in Figure 1.18, the rodents never face significant heat or cold stress throughout the year when they are in their burrows![14]

Microenvironments must be considered in the study of virtually all the physical and chemical features of the places where animals live (e.g., see Figure 1.16). In tall grass, for example, the *wind speed* is likely to be lower than in adjacent open areas, and because the

---

[13] From G. A. Bartholomew. 1964. The roles of physiology and behavior in the maintenance of homeostasis in the desert environment. *Symp. Soc. Exp. Biol.* 18: 7–29.

[14] There are hotter desert regions where even the burrow environment presents thermal challenges in some seasons, but the burrow environment is still far more moderate than the environment aboveground (see Chapter 28).

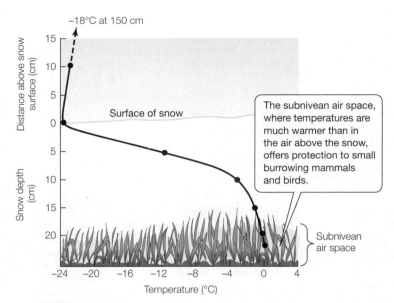

FIGURE 1.19 Microenvironments in deep snow in the Far North An air space—termed the subnivean air space—is often trapped beneath deep snow. When lemmings, ptarmigans, or other small mammals or birds burrow under the snow into the subnivean air space, they enter a windless environment where the temperature may be 20°C warmer than in the air above the snow, and where they are protected by the overlying snow from losing heat radiatively to the frigid nighttime sky. The temperatures shown were measured on a March night in Sweden. (After Coulianos and Johnels 1963.)

weak wind permits moisture evaporating from the soil and grass to accumulate in the air rather than being blown away, the *humidity* also tends to be higher than in adjacent open air. Animals that enter tall grass are thus likely to find a less desiccating microenvironment than in open fields nearby. In winter, spaces under deep snow in the Far North form distinctive thermal microenvironments. The temperature there is often higher by 20°C (or more) than in the open air above (**FIGURE 1.19**). Thus a lemming burrowing under the snow experiences a very different environment from a reindeer standing above.

Animals often modify their own environments

An important way in which *animal* and *environment* are interdependent is that animals modify their own environments. In the simplest case, animals behaviorally select the actual environments that they experience from the many that they *could* experience. A frog stressed by desiccation on open terrain, for example, can raise the humidity and lower the wind speed of its environment by hopping into tall grass. The environment of an animal is the animal's surroundings, and the surroundings depend on where the animal places itself.

A more subtle but equally important point is that the very presence of an animal in a place often alters the physical and chemical features of the place. The environmental alteration caused by the animal may then change that selfsame animal's behavior or physiology. Consider, for instance, a squirrel in a small cavity in a tree. In winter, the squirrel will warm the cavity to a higher temperature than would prevail in the squirrel's absence. The squirrel will then respond physiologically to the elevated temperature in the cavity, by cutting back its rate of metabolic heat production. In a like manner,

a school of fish can deplete water of dissolved O_2 and then must cope with low "environmental" O_2 levels.

Because of phenomena like these, the analysis of an animal–environment interaction often requires dynamic calculations that take into account that the interaction is of a two-way, back-and-forth sort. After an animal has initially altered an environment, the animal may function differently because it is in a changed environment, and thus the animal's future effect on the environment may be different from its original effect.

Global warming represents a planet-sized example of this phenomenon. Most scientists believe that the use of fossil fuels by the global human population is changing Earth's atmosphere toward a composition that increases planetary heat retention. Human activities are therefore raising the global temperature. The warming environment then will alter the ways in which human societies function in the future.

Evolutionary Processes

The evolutionary origins of physiological traits—and the continuing evolution of physiological traits in today's world—form the subject matter of *evolutionary physiology*, one of the two most important branches of the modern study of physiology, as stressed earlier. Physiologists have long recognized that the traits of species are often well matched to the environments they occupy. For example, polar bears are well suited to deal with cold, and dromedary camels with heat. Evolution by natural selection is believed by modern biologists to be the primary process that produces this match between species and the environments they inhabit.

Carefully defined, **evolution** is *a change of gene frequencies over time* in a population of organisms. Suppose a population of animals contains a gene that codes for the oxygen affinity of hemoglobin (the ease with which hemoglobin combines with O_2). The gene has two alleles (alternative forms), one coding for high oxygen affinity (H allele) and one for low oxygen affinity (L allele). At one time in the history of the population, suppose 30% of all copies of this gene were of the H allele and 70% were of the L allele. After 1000 generations have passed, however, suppose 60% of all copies are of the H allele and 40% are of the L allele. In this case, gene frequencies have changed. Therefore evolution has occurred.

A more complex question by far is whether *adaptation* has occurred. There are several known processes by which gene frequencies can change. Only one, natural selection, leads to adaptation.

Some processes of evolution are adaptive, others are not

Returning to the hypothetical example just discussed, suppose that in a population of animals occupying a particular environment, individuals with hemoglobin of high oxygen affinity are more likely to survive and reproduce than those with hemoglobin of low affinity. By this very fact, an allele that codes for high affinity will tend to increase in frequency in the population from one generation to the next (and an allele that codes for low affinity will tend to decrease). After many generations, the H allele might become so common that essentially all individuals born into the population have it. You will recognize this as the process of *natural selection*. Natural selection creates a better match between animals and their environments.

The concept of *adaptation*, which is intimately related to that of natural selection, has a specific meaning in the study of evolutionary biology. By definition, a trait is an **adaptation** if it has come to be present at high frequency in a population because it confers a greater probability of survival and successful reproduction in the prevailing environment than available alternative traits. Thus adaptations are products of the process of natural selection. An adaptation is not necessarily an *optimum* or *ideal* state, because constraints on the freedom of natural selection may have precluded the optimum state from being an option (the optimum state, for instance, might never have arisen through mutation). As the definition stresses, an adaptation is the trait favored by natural selection from among the *available* alternative traits.

Now let's repeat our thought exercise but substitute different assumptions. Consider a population, in the same environment as we have just been analyzing, in which the H and L alleles are both common. Suppose that the population experiences a drop in size, so that it contains fewer than 100 individuals. Suppose also that during this low point, a catastrophe strikes, killing individuals *at random*, regardless of whether they possess the H or L allele. In a small population of 100 or fewer animals, deaths at random could *by sheer chance* eliminate all individuals possessing one of the alleles. All copies of the H allele might, in fact, be eliminated. In a population subjected to this process, when the population later regrows in size, it will have only the L allele, the *less adaptive* allele. In this case, the process of gene frequency change we have described is a process of **nonadaptive evolution**. Because of chance, an allele that provides a *lower* probability of survival and reproduction than an available alternative comes to be the predominant allele in the population.

Processes in which chance assumes a preeminent role in altering gene frequencies are termed **genetic drift**. We have described, in the last paragraph, one scenario for genetic drift: Gene frequencies may shift in chance directions because of random deaths (or other random blocks to individual reproduction) in populations transiently reduced to small size. Another scenario for genetic drift is that when a species enters a new area and founds a new population there, the new population may exhibit changed gene frequencies, relative to the parent population, simply because of chance—because the founding individuals may by chance be genetically nonrepresentative of the population from which they came (a so-called *founder effect*).

Researchers who study allele frequencies in natural populations believe that they often observe evidence of genetic drift. For example, two populations of mice living 10 km (6 mi) apart in seemingly identical woodlots usually exhibit many differences in allele frequencies thought to be produced by drift of one sort or another. Often the genes affected by drift seem to be ones that have little or no fitness effect; drift, in other words, seems to have its greatest influence on genes not subject to strong natural selection. That is not always the case, however.

Additional processes are known by which evolution may lead to nonadaptive outcomes. These include:

- A trait may be common in a population simply because it is closely correlated with *another* trait that is favored by natural selection. For instance, a trait that itself diminishes the fitness of animals (their ability to survive and reproduce)

may occur because it is coded by a gene that is subject to positive selection because of other, fitness-enhancing effects. The control by an allele of a single gene of two or more distinct and seemingly unrelated traits is called **pleiotropy**. An example is provided by a recently discovered allele of an enzyme-coding gene that simultaneously has two effects in the mosquito *Culex pipiens*. The allele both (1) increases the resistance of the mosquitoes to organophosphate insecticides and (2) decreases the physiological tolerance of the mosquitoes to the cold of winter. When a population of mosquitoes is sprayed with insecticides, the population *may evolve toward a diminished physiological ability to survive winter* because of this pleiotropy. Selection will favor the allele we've been discussing because it confers insecticide resistance, but the allele will also diminish the odds of winter survival.[15]

- A trait may have evolved as an adaptation to an ancient environment, yet persist even though the environment has changed. In the new environment, the trait may no longer be beneficial and thus may not be an adaptation. Traits of this sort are thought by some evolutionary biologists to be common, because animals often move to new places, and even if they stay in one place, climates often change radically over relatively short periods of geological time.[16] The need of many desert amphibians for pools of water to breed, and the possession of eyes by numerous species of arthropods that live in caves, are two examples of traits that seem to exist today because they are carryovers from the past, not because they are adaptations to the animals' present environments. Similarly, the dilute blood of the bony fish of the oceans is probably a legacy of life in a different environment, not an adaptation to life in seawater.

A trait is not an adaptation merely because it exists

Prior to about 1980, many physiologists referred to *all* traits of organisms as adaptations. Traits were called *adaptations* merely because they existed, and stories (now sometimes recalled as "just-so" stories) were concocted to explain how the traits were beneficial. This habit ignored the possibility of genetic drift and other forms of nonadaptive evolution. The habit, in fact, reduced adaptation to a nonscientific concept because no empirical evidence was required for a trait to be considered an adaptation.

A major shift in the use of the concept of adaptation was precipitated by Stephen J. Gould and Richard C. Lewontin with the publication of a critique in 1979. They stressed that natural selection in the present environment is just one of several processes by which a species may come to exhibit a trait. A trait, therefore, is not an adaptation merely because it exists. Instead, when physiologists call

---

[15] In addition to causation by pleiotropy, traits may also evolve in tandem because of *linkage disequilibrium*, in which alleles of two or more genes on the same chromosome—*because* of being on a single chromosome—tend to be inherited together to a nonrandomly great extent.

[16] Just 18,000 years ago, the arid, warm deserts of Arizona and New Mexico were far more moist than today, and they were on average about 6°C cooler, because of the last ice age. About 10,000 years ago, large expanses of the Sahara Desert experienced far more rain than they do today and were savannas (prairielike landscapes) rather than desert.

a trait an *adaptation*, they are really making a *hypothesis* that natural selection has occurred.

Data must be gathered to assess whether a hypothesis of adaptation is true or false, just as we gather data to assess the truth of other types of hypotheses. With this objective in mind, the study of adaptation has been maturing gradually into an empirical (i.e., data-based) science.

Adaptation is studied as an empirical science

Biologists today are giving a great deal of attention to the question of how to obtain *data* that will guide a decision on whether or not a trait is an adaptation. Sometimes the biosphere presents a "natural experiment" that permits scientists to *observe* evolution taking place over multiple generations in a natural setting. Scientists cannot depend entirely on such natural experiments to study adaptation, because the natural experiments are uncommon and may not speak to questions of greatest interest. Nonetheless, a natural experiment may provide particularly useful insights into adaptation because it may allow the adaptiveness of a trait to be judged from all the angles that matter.

Industrial melanism is a famous phenomenon—with which you are likely familiar from your study of general biology—that exemplifies a natural experiment for adaptation. *Melanism* refers to a genetically coded dark body coloration. *Industrial melanism* is an evolutionary increase in the frequency of melanism in a population of animals living in an environment modified by human industries. A species of moth in the industrial regions of England has two genetically determined color states: light and dark. The moths were predominantly light-colored prior to the industrial era, when light-colored lichens covered the tree trunks on which they rested during the day. With increasing industrialization, the lichens on the trees were killed by pollutants, and soot from factories darkened the tree trunks. Within 50 years of the beginning of industrialization, the moth populations in the industrial areas became predominantly dark-colored because, from generation to generation, genes for dark coloration increased in frequency. Studies demonstrated that on dark tree trunks, the dark-colored moths were less likely than light-colored ones to be seen by avian predators.

From the *direct observation* of this natural experiment, we can say the following: In an environment affected by industrial pollution, dark coloration became common in the moth populations by way of natural selection because it increased an individual's likelihood of survival in comparison with the available alternative coloration. Dark coloration thus met all the standards of our formal definition of adaptation and could be judged, based on evidence, to be an adaptation to the sooty environment.

Usually biologists are not able to observe evolution in action in this way. Thus, to study adaptation empirically, they must adopt other approaches. Several techniques have been developed—or are being developed—to study the question of adaptation when nature fails to provide an ideal natural experiment:

■ *The comparative method.* The **comparative method** seeks to identify adaptive traits by comparing how a particular function is carried out by related and unrelated species in

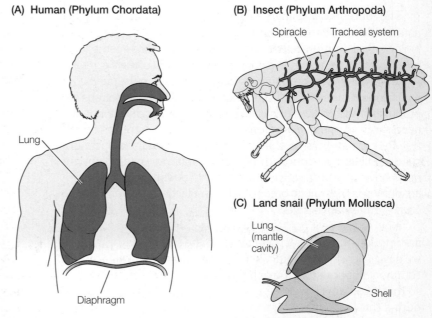

(A) Human (Phylum Chordata)

Lung

Diaphragm

(B) Insect (Phylum Arthropoda)

Spiracle   Tracheal system

(C) Land snail (Phylum Mollusca)

Lung (mantle cavity)

Shell

FIGURE 1.20 The comparative method   Terrestrial vertebrates (A), insects (B), and land snails (C)—representing three phyla that separately colonized the land—have independently evolved breathing organs that are invaginated into the body. This convergence in the type of breathing organ suggests that invaginated breathing organs are adaptive for living on land.

similar and dissimilar environments. *This method is based on the premise that although we cannot see evolution occurring in the past, the many kinds of animals alive today provide us with many examples of outcomes of evolution, and patterns we identify in these outcomes may provide insights into processes that occurred long ago.* **FIGURE 1.20** presents a simple example of the use of the comparative method. Terrestrial vertebrates have lungs for breathing. If we were to look *only* at terrestrial vertebrates, we would have just that single isolated bit of knowledge about breathing mechanisms. However, if we also examine other unrelated terrestrial organisms, we discover a pattern: In insects, in land snails, *and* in terrestrial vertebrates, the breathing surfaces are parts of *invaginated* structures that hold the air rather than projecting into the air. This pattern is striking because *evaginated* breathing surfaces, which project into the water, are nearly universal among aquatic animals (note the gills of fish and crayfish). The occurrence of invaginated structures in multiple independent lines of modern terrestrial animals suggests that if we could see into the distant evolutionary past, we would witness individuals with invaginated breathing organs outcompete ones with alternative breathing structures on land. The pattern suggests that natural selection was at work, and that the invaginated breathing organs are adaptations to life on land.

■ *Studies of laboratory populations over many generations.* Changes in gene frequencies can be observed over multiple generations in laboratory populations of fast-breeding animals such as fruit flies. By exposing such populations to specific, controlled conditions (e.g., high or low desiccation stress), physiologists may observe which alleles are favored by selection when a particular condition prevails. An illustration is provided by studies of fruit fly populations exposed for many generations to high desiccation stress; in such populations, the genetically coded blood volume of flies increases dramatically, and the flies become able to tolerate desiccation for greatly enhanced lengths of time (see Box 28.6). The selection that occurs in cases like this is usually considered to be *laboratory selection* or *artificial selection* because humans are manipulating the circumstances. A concern, therefore, is to assess whether outcomes of *natural selection* in the wild would be likely to be similar.

■ *Single-generation studies of individual variation.* Individuals in a natural population of a single species typically vary in their physiological properties, as discussed in the final section of this chapter. Such natural variation among individuals of a species can be exploited to carry out single-generation experiments to determine which traits are most advantageous. To illustrate, suppose we trap several hundred mice in a wild population and measure the maximum rate of O_2 consumption of each, and then we release all the mice back into their natural population, where we monitor them until they die. If we find that individuals with particular O_2-consumption capabilities produce more young before dying than individuals with other O_2-consumption capabilities, we will have evidence of which capabilities are adaptive.

■ *Creation of variation for study.* Biologists may be able to *create* variation in a trait that shows little or no natural variation among individuals of a species. Then competitive outcomes in natural or laboratory settings may be observed. Years ago, the principal application of this approach was morphological; for instance, the size of the ear cavities of desert rats was morphologically altered to assess which ear-cavity dimensions allowed the surest detection of predators. One newer approach is to employ genetic manipulations. Suppose that the vast majority of individuals of a species have a certain allele for a digestive enzyme but that an unusual mutant allele is found that produces a different molecular form of the enzyme. By controlled breeding, one could create a population rich in both alleles and then observe the relative advantages of the two enzyme forms. Another genetic approach is to employ genetic engineering methods to silence genes. As we will discuss in Chapter 3 (see page 83), **knockout animals** that lack functional copies of a gene of interest can be produced, or **RNA interference (RNAi)** can be employed to block transcription of a gene. Individuals manipulated in these ways are unable to synthesize the protein coded by the affected gene and thus can be used to evaluate the functional significance of the protein. Other forms of "engineering" are available for creating individual diversity that can be tested for effects. These include "allometric engineering," in which the body sizes of individuals are artificially manipulated during development to create variation, and "hormonal engineering," in which hormone injections are used.

■ *Studies of the genetic structures of natural populations.* Natural populations are sometimes genetically structured in revealing ways. Genetic **clines** provide excellent examples. A genetic cline is a progressive change in allele frequencies or gene-controlled phenotype frequencies along an environmental gradient. Investigators find, for instance, that within certain species of fish of the East Coast of the United States, alleles that are common in warm-water Georgia individuals become progressively less common toward the north and are almost absent in cold-water New England individuals (see Figure 2.22). Genetic patterns of this sort often point to ways in which natural selection differs in its effects from place to place.

■ *Phylogenetic reconstruction.* The goal of phylogenetic reconstruction is to determine the structure of the *family tree* (the ancestry) of groups of related species, often using molecular genetic data. The family tree is useful in two major ways. First, a family tree often facilitates the estimation of exactly *when* in evolutionary history each trait evolved; thus, for example, we might learn from a family tree whether the evolution of one trait preceded or followed the evolution of another—knowledge that can help us understand the context of the evolution of each trait. Second, a family tree clarifies whether a trait evolved independently more than once; several independent origins in one environment suggest that a trait is adaptive to the environment. In this book we discuss several analyses of adaptation based on family trees. Chapter 3, for example, starts with a family-tree analysis of the icefish of Antarctic seas, fish that lack red blood cells and sometimes lack myoglobin, a key O_2-transport compound, in their heart muscle (see Figures 3.3 and 3.4).

Evolutionary potential can be high or low, depending on available genetic variation

A key determinant of the course of evolution of a trait in an animal population is the amount of genetic diversity for the trait in the population. If there is no genetic diversity for a trait—that is, if all the individuals in a population are homozygous for a single allele—then evolutionary change in the trait is essentially impossible. As an example, imagine a population of mammals in which all individuals are homozygous for an eye-color allele that codes for brown eyes. In this population as a whole, there would be no genetic diversity for eye color. Thus natural selection of alleles could not possibly alter eye color. By contrast, if the individuals in a population collectively have several different alleles of the gene for eye color—some alleles coding for brown, others for blue or olive—then the frequencies of the various alleles can be modified by natural selection, and eye color can evolve.

Physiologists are just beginning to take into full account the importance of genetic diversity in understanding evolutionary potential. **FIGURE 1.21** provides a model example of the sorts of insights that can be obtained from considering genetic diversity. House mice were collected from five locations in eastern North America, locations chosen to represent a progression in winter severity, from mild winters in the south to severe winters in the north. The mice collected were from wild populations that had

FIGURE 1.21 The effects of genetic diversity on evolutionary potential Tested in a constant laboratory setting, house mice (*Mus domesticus*) from the more northern locations attain larger body sizes and build bigger nests than mice from the more southern populations. In contrast, despite having evolved in different climates, mice from the five locations show no differences in body temperature or in the amount of brown fat they have relative to body size. *Independent* studies reveal that in house mouse populations, there is high diversity in the genes that control the body sizes of the mice and the sizes of the nests they build, whereas the diversity of genes controlling body temperature and the amount of brown fat is low. The results support the hypothesis that the effectiveness of natural selection depends on how much genetic diversity exists in populations. The dashed lines on the map show average winter temperatures. (After Lynch 1992.)

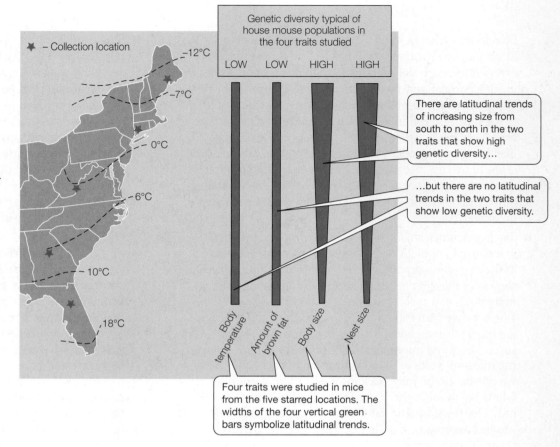

presumably reproduced at the five locations for many generations and were subjected to natural selection. Mice from the five locations had litters in the laboratory, and their offspring were studied. The reason for studying the offspring rather than the wild-caught animals was to gain as clear insight as possible into *genetic* differences among the populations; because all the offspring were born and reared in one environment, their differences were likely to be mostly or entirely caused by differences in genetics, rather than differences in early upbringing.

Four traits of the offspring were measured: their body temperatures, the sizes of the nests they constructed, their body weights, and the fractions of their bodies occupied by *brown fat*, a tissue capable of intense heat production (see page 266). The offspring of the mice from the five locations differed significantly in body size and nest size; both the body size and the nest size were higher in the colder, northern populations than in the southern ones, as one might expect (see Figure 1.21). However, the offspring from all five sets of mice had the same average body temperature and the same average amount of brown fat. In the abstract, one might expect animals in a cold climate to evolve a lower body temperature and a larger quantity of heat-producing tissue than ones in a warm climate, but neither of these expectations is fulfilled in reality. Why has adaptation occurred in only two of the four traits studied?

Genetic diversity provides an important part of the answer. House mouse populations exhibit relatively high genetic diversity in the genes that control body size and nest size; these two traits have responded to natural selection. However, house mouse populations exhibit little diversity in the genes that control body temperature and the amount of brown fat; these two traits have failed to respond to

natural selection in the very same mice. We do not know why genetic diversity is high for some traits and low for others. Examples such as this show, however, that evolution by natural selection depends on the underlying genetic structure of populations. It can be only as effective as genetic diversity permits it to be.

Individual Variation and the Question of "Personalities" within a Population

When biologists make measurements on many individuals within a single interbreeding population of animals, they usually discover that the individuals vary somewhat in their physiological properties. For example, 35 randomly selected, adult deer mice (*Peromyscus maniculatus*) were captured from a single wild population living in a forest. Using a standardized procedure, each individual was tested to measure the maximum rate at which it could take in and use O_2—a rate that is a key determinant of how long and vigorously an individual can sustain metabolic effort. Although all the mice looked similar to each other and were of similar body size, they varied significantly in this important physiological property, as seen in **FIGURE 1.22**. Evidence indicates that the differences among individuals in this trait are consistent: Two individuals that differ at one point in time are likely to differ in about the same way if tested again at a later time.

Individual variation such as seen in this example—if it is heritable—provides the raw material for evolutionary change. As we have already discussed, if a particular genetically coded O_2-uptake capacity endows animals with increased chances of surviving and reproducing, then the frequencies of the genes responsible will

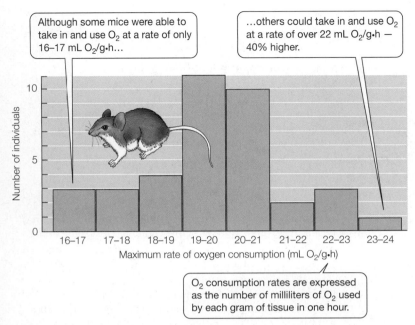

FIGURE 1.22 **Physiological variation among individuals of a species** This histogram summarizes the measured maximum rates of O_2 consumption of 35 deer mice (*Peromyscus maniculatus*) from a single natural population. The individual variation observed was attributable to both genetic and nongenetic causes, although the proportions of each were not measured.

increase in the gene pool of the population over time through the action of natural selection.

Similar individual variation often occurs in human populations. For example, to better understand individual variation in the maximum rate of O_2 consumption among people, investigators took advantage of the fact that in 1983 all 18-year-old Norwegian men were required to report for a medical exam. Of the 33,000 men examined, each exercised on a bicycle-like device at an intensity that permitted estimation of his maximum rate of O_2 consumption. **FIGURE 1.23** summarizes the results. Collectively, the men showed

a wide range of individual variation, with some men able to take in and use O_2 at rates twice as high as others, even though they were all the same age and from a fairly homogeneous society.

Within the world of competitive sports, people tend to sort themselves out in ways that correlate with their maximum rates of O_2 consumption.[17] For example, a high maximum rate is an advantage during long-distance races, and when individual athletes who excel in long races are tested, they often prove to have relatively high maximum rates of O_2 consumption. By contrast, a high maximum rate is not of central importance for excelling in sports such as weight lifting or gymnastics, and when individual athletes who excel in those sports are tested, they often prove to be only average (or less) in their maximum rates of O_2 consumption.

In the past one or two decades, certain physiologists have postulated that there may be a similar sorting out in nonhuman animals. For example, let's consider again the deer mice living in a single population in a forest (see Figure 1.22). The individuals with high maximum rates of O_2 consumption would be far more able to run at high, sustained speeds than those with low maximum rates of O_2 consumption. Possibly, therefore, if a fox or cat starts to hunt mice in the population, the individuals with high maximum rates of O_2 consumption would most often escape by running, whereas those with low maximum rates would instead emphasize hiding. Looking back at human athletes who excel in long-distance races versus weight lifting or wrestling, we might describe the athletes as having different athletic "personalities." With this in mind, some physiologists have hypothesized that the various individuals in a population of mice might have different functional "personalities" (consistent differences in their physiological and associated behavioral traits) in the ways they evade predators. Similarly, if we consider a population of predatory fish that use sight to find prey, we might hypothesize that there is

---

[17] This phenomenon will be discussed at greater length in Chapter 9.

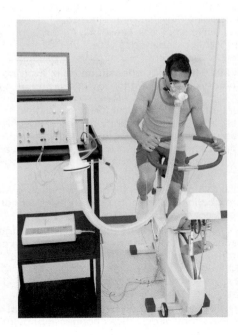

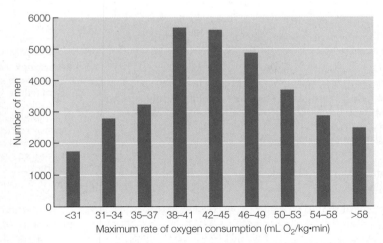

FIGURE 1.23 **Individual variation in the maximum rate of O_2 consumption among 33,000 18-year-old men who reported for compulsory medical exams in Norway in 1983** The individual variation observed was attributable to both genetic and nongenetic causes, although the proportions of each were not measured. (After Dyrstad et al. 2005.)

individual variation in how well the fish see. Some individuals, for example, may have greater visual acuity in dim light than others. If so, various individuals might differ in the details of how they hunt.

Traditionally, biologists have emphasized averages. When measurements have been made on large numbers of individuals in a population, the measurements have been averaged, and only the averages have been used for further analysis. By contrast, the biologists who focus on "personalities" are hypothesizing that when the individuals in a population exhibit consistent individual differences, this individual variation may be important in its own right.

STUDY QUESTIONS

1. There is a chance that a calcium atom or carbon atom that was once part of Caesar's or Cleopatra's body is now part of your body. Part of the reason is that most calcium and carbon atoms that were parts of these rulers' bodies did not go to their graves with them. Explain both statements. (If you enjoy quantifying processes, also see Question 11.)

2. Animals do not keep all their detoxification enzymes in a constant state of readiness. Thus they depend on phenotypic plasticity to adapt to changing hazards. An example is provided by the enzyme *alcohol dehydrogenase*, which breaks down ethyl alcohol. People who do not drink alcoholic beverages have little alcohol dehydrogenase. Expression of the enzyme increases when people drink alcohol, but full expression requires many days, meaning that people are incompletely defended against alcohol's effects when they first start drinking after a period of not drinking. Consider, also, that muscles atrophy when not used, rather than being maintained always in a fully developed state. Propose reasons why animals depend on phenotypic plasticity, instead of maintaining all their systems in a maximum state of readiness at all times.

3. Whereas the larvae of a particular species of marine crab are bright orange, the adults of the species are white. An expert on the crabs was asked, "Why are the two different life stages different in color?" She replied, "The larvae accumulate orange-colored carotenoid pigments, but the adults do not." Did she recognize all the significant meanings in the question asked? Explain.

4. Referring to Figure 1.11, do plains zebras, warthogs, and greater kudus have normal or exceptional gestation lengths? Justify your position in each case.

5. At least three hemoglobin alleles in human populations alter hemoglobin structure in such a way as to impair the transport of O_2 by the blood but enhance resistance of red blood cells to parasitization by malaria parasites. Explain how such alleles exemplify pleiotropy, and discuss whether such alleles could lead to nonadaptive evolution of blood O_2 transport in certain situations.

6. What are some of the microclimates that a mouse might find in your professor's home?

7. Figure 1.16 seems at first to be simply a description of the physical and chemical properties of a lake. Outline how living organisms participate in determining the physical and chemical (i.e., temperature and O_2) patterns. Consider organisms living both in the lake and on the land surrounding the lake. Consider also a research report that shows that dense populations of algae sometimes change the temperature structure of lakes by raising the thermocline and thereby increasing the thickness of the deep, cold layer; how could algal populations do this, and what could be the consequences for deep-water animals?

8. Do you agree with François Jacob that evolution is more like tinkering than engineering? Explain.

9. Explain how the comparative method, knockout animals, and geographical patterns of gene frequencies might be used to assess whether a trait is adaptive. As much as possible, mention pros and cons of each approach.

10. Certain species of animals tolerate body temperatures of 50°C, but the vast majority do not. Some species can go through their life cycles at very high altitudes, but most cannot. What are the potential reasons that certain exceptional species have evolved to live in environments that are so physically or chemically extreme as to be lethal for most animals? How could you test some of the ideas you propose?

11. Using the set of data that follows, calculate how many of the molecules of O_2 that were used in aerobic catabolism by Julius Caesar are in each liter of atmospheric air today. All values given are expressed at Standard Conditions of Temperature and Pressure (see Appendix C) and therefore can be legitimately compared. Average rate of O_2 consumption of a human male during ordinary daily activities: 25 L/h. Number of years after his birth when Caesar was mortally stabbed near the Roman Forum: 56 years. Number of liters of O_2 per mole: 22.4 L/mol. Number of moles of O_2 in Earth's atmosphere: 3.7×10^{19} mol. Number of molecules per mole: 6×10^{23} molecules/mol. Amount of O_2 per liter of air at sea level (20°C): 195 mL/L. Be prepared to be surprised! Of course, criticize the calculations if you feel they deserve criticism.

Go to **sites.sinauer.com/animalphys4e**
for box extensions, quizzes, flashcards,
and other resources.

REFERENCES

Alexander, R. M. 1985. The ideal and the feasible: Physical constraints on evolution. *Biol. J. Linn. Soc.* 26: 345–358. A thought-provoking discussion of the relevance of physics for animals.

Bennett, A. F. 1997. Adaptation and the evolution of physiological characters. In W. H. Dantzler (ed.), *Comparative Physiology*, vol. 1 (Handbook of Physiology [Bethesda, MD], section 13), pp. 3–16. Oxford University Press, New York. A thorough, compact review of adaptation and modes of evolution as seen through the eyes of a physiologist.

Berenbrink, M. 2007. Historical reconstructions of evolving physiological complexity: O_2 secretion in the eye and swimbladder of fishes. *J. Exp. Biol.* 209: 1641–1652. A challenging but worthwhile paper that employs phylogenetic reconstruction to analyze the evolution, in fish, of the astounding capability to inflate the swim bladder with almost pure O_2. The author aspires to present a model approach for understanding the evolution of complex systems in which multiple physiological and morphological traits interact.

Biro, P. A., and J. A. Stamps. 2010. Do consistent individual differences in metabolic rate promote consistent individual differences in behavior? *Trends Ecol. Evol.* 25: 653–659.

Buehler, D. M., and T. Piersma. 2008. Travelling on a budget: predictions and ecological evidence for bottlenecks in the annual cycle of long-distance migrants. *Philos. Trans. R. Soc., B* 363: 247–266. A thought-provoking, comprehensive discussion of the life cycle in populations of red knots, which are among the most thoroughly studied long-distance migrants.

Chevillon, C., D. Bourguet, F. Rousset, and two additional authors. 1997. Pleiotropy of adaptive changes in populations:

Comparisons among insecticide resistance genes in *Culex pipiens*. *Genet. Res.* 70: 195–204.

Costa, D. P., and B. Sinervo. 2004. Field physiology: physiological insights from animals in nature. *Annu. Rev. Physiol.* 66: 209–238.

Eliason, E. J., T. D. Clark, M. J. Hague, and seven additional authors. 2011. Differences in thermal tolerance among sockeye salmon populations. *Science* 332: 109–112.

Feder, M. E., A. F. Bennett, and R. B. Huey. 2000. Evolutionary physiology. *Annu. Rev. Ecol. Syst.* 31: 315–341. A review of the evolutionary approach to physiology, emphasizing empirical methods to assess adaptation.

Gibbs, A. G. 1999. Laboratory selection for the comparative physiologist. *J. Exp. Biol.* 202: 2709–2718.

Gould, S. J., and R. C. Lewontin. 1979. The spandrels of San Marco and the Panglossian paradigm: A critique of the adaptationist programme. *Proc. R. Soc. London, B* 205: 581–598. The article that launched the modern reconsideration of adaptationist thinking. Deliberately provocative, it excites critical thought about key biological concepts that are often treated as truisms.

Hughes, L. 2000. Biological consequences of global warming: is the signal already apparent? *Trends Ecol. Evol.* 15: 56–61.

Jacob, F. 1977. Evolution and tinkering. *Science* 196: 1161–1166.

Laland, K. N., K. Sterelny, J. Odling-Smee, and two additional authors. 2011. Cause and effect in biology revisited: Is Mayr's proximate–ultimate dichotomy still useful? *Science* 334: 1512–1516. A probing discussion of modern strengths and weaknesses of the proximate–ultimate distinction in studies of causation.

Lewis, S. M., and C. K. Cratsley. 2008. Flash signal evolution, mate choice, and predation in fireflies. *Annu. Rev. Entomol.* 53: 293–321.

Lynch, C. B. 1992. Clinal variation in cold adaptation in *Mus domesticus:* Verification of predictions from laboratory populations. *Am. Nat.* 139: 1219–1236. A fine example of the modern application of quantitative genetics to the study of physiological and behavioral evolution.

Mangum, C. P., and P. W. Hochachka. 1998. New directions in comparative physiology and biochemistry: Mechanisms, adaptations, and evolution. *Physiol. Zool.* 71: 471–484. An effort by two leaders to project the future of animal physiology, emphasizing the integration of mechanistic and evolutionary approaches.

Somero, G. N. 2000. Unity in diversity: A perspective on the methods, contributions, and future of comparative physiology. *Annu. Rev. Physiol.* 62: 927–937. A brief, modern, and stimulating review of why comparative physiology has been important in the past and will continue to be so in the future, written by a prominent comparative biochemist.

Widder, E. A. 2010. Bioluminescence in the ocean: origins of biological, chemical, and ecological diversity. *Science* 328: 704–708. A thought-provoking, beautifully illustrated review of bioluminescence in marine animals and other organisms.

Wikelski, M., and S. J. Cooke. 2006. Conservation physiology. *Trends Ecol. Evol.* 21: 38–46. A survey of the ways in which physiological knowledge can be applied to animal conservation.

See also **Additional References** *and* **Figure and Table Citations**.

Molecules and Cells in Animal Physiology

The two slow-moving animals pictured here are able to consume fast-moving prey because they have evolved ways to defeat the function of essential molecules and cellular structures in their prey. The puff adder is one of the slowest moving of snakes. It feeds on fast-moving rats, however, because it needs only a split second of contact with its prey to set in motion processes that will destroy key molecular–cellular properties on which a rat depends for life. Like rattlesnakes and other adders, the puff adder sits and waits for an unsuspecting animal to come close enough for a strike. It then lunges at its victim, sinks its fangs in, and in less than 1–2 s, injects a complex mix of compounds that attack critical molecules and cells. The snake then immediately releases the rat and tracks it as the rat's molecular–cellular mechanisms fall apart. Some of the injected compounds strip the outer membranes from the rat's muscle cells, whereas other compounds make tiny holes in the rat's blood capillaries, permitting widespread internal hemorrhaging. When, finally, the molecular–cellular damage is so great that the victim can no longer function, the slow-moving adder moves in to eat.

The second example of a slow-moving animal that consumes fast-moving prey—the cone snail—feeds on fish. The snail lures fish to its vicinity by waving a wormlike body part that deceives the fish into coming close to eat. The snail then harpoons the fish with a hollow barbed

Two slow-moving predators that use molecular weapons to capture fast-moving prey A cone snail sits virtually motionless in the coral reef ecosystems it occupies, yet feeds routinely on fish. The puff adder is a notoriously sluggish, but deadly, African snake that resembles rattlesnakes in its strategy of hunting small mammals. Both predators produce venoms that disrupt vital molecular structures or mechanisms.

tooth. The fish could easily tear itself loose if permitted just a moment's time to do so. The snail preempts such escape by injecting the fish through the tooth with compounds that almost instantly disrupt the function of proteins that are essential for the function of the fish's nerve and muscle cells. In this way the fish's most promising defense, its ability to swim rapidly away, is immediately defeated. With the cells in its nervous system in disarray and its muscles paralyzed, the fish is ingested by the sedentary snail.

The actions of venoms and poisons remind us that all the higher functions of animals depend on molecules and on the *organization* of molecules into cellular structures and cells. An animal as spectacular as a racehorse or a mind as great as that of Socrates can be brought down in a moment if the function of key molecules is blocked or the normal organization of cells is disrupted.

A case can be made that the study of molecules and the cellular organization of molecules is the most fundamental study of biology, because molecules and cells are the building blocks of tissues and other higher levels of organization. Some scientists believe that all the properties of tissues, organs, and whole animals will eventually be fully predictable from knowledge of molecules and cells alone. Other scientists, however, believe that animals have **emergent properties**: properties of tissues, organs, or whole animals that will never, in principle, be predictable from mere knowledge of molecules and cells because the properties *emerge* only when cells are assembled into interactively functioning sets. Regardless of the resolution of this important debate, molecules and cells are critically important.

The goal of this chapter is to discuss fundamental structural and functional properties of molecules and cells. Many of the properties discussed here will come up in more specific ways throughout the book. Four topics receive greatest attention:

- Cell membranes and intracellular membranes
- Epithelia—the sheets of tissue that line body cavities and form the outer surfaces of organs
- Enzyme function, diversity, and evolution
- Mechanisms by which cells receive and act on signals

In addition we will discuss fundamental properties of proteins, the ways that proteins are repaired or destroyed, and the abilities of some cells to produce light or modify an animal's external color.

Cell Membranes and Intracellular Membranes

Each animal cell is enclosed in a **cell membrane** (*plasma membrane*). Each cell also includes many sorts of **intracellular membranes** (*subcellular membranes*), such as the endoplasmic reticulum, the inner

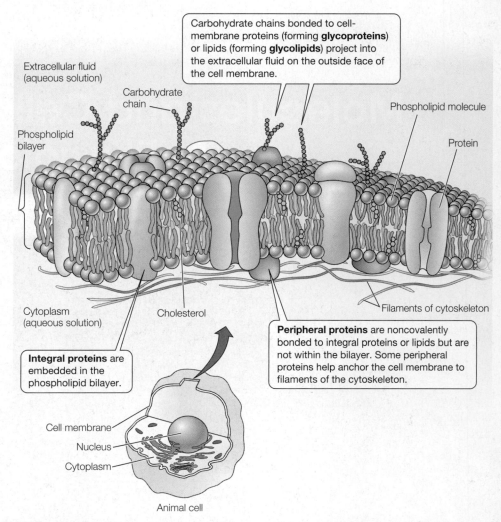

FIGURE 2.1 The structure of a cell membrane The cell membrane consists primarily of two layers of phospholipid molecules with protein molecules embedded and attached. Intracellular membranes also have a structure based on proteins embedded in a phospholipid bilayer.

and outer membranes of each mitochondrion, and the two closely associated membranes that form the nuclear envelope. These membranes are exceedingly thin, measuring 6–8 nanometers (nm) from one side to the other. They play vitally essential roles nonetheless. They physically compartmentalize systems in functionally essential ways; the cell membrane, for instance, separates the inside of a cell from the cell's surroundings, permitting the inside to have different properties from the outside. In addition, far from being inert barriers, the membranes are dynamic systems that *participate* in cellular and subcellular functions. For example, the cell membrane acts to receive and transmit signals that arrive at the cell surface.

The cell membrane is ordinarily composed primarily of a bilayer (double layer) of phospholipid molecules in which protein molecules are embedded (**FIGURE 2.1**). Similarly, the fundamental structure of the intracellular membranes is also a bilayer of phospholipid molecules with protein molecules embedded in it. Recognizing the ubiquity and importance of phospholipids, it is not surprising that they are targets of venoms. A principal weapon in the complex venom of a puff adder or a rattlesnake is a set of enzymes known as *phospholipases*, which break up phospholipids. Among other effects, these enzymes destroy the phospholipid matrix in the cell membranes of a victim's skeletal muscle cells, thereby exposing the insides of the cells, setting membrane proteins adrift, and wreaking other havoc.

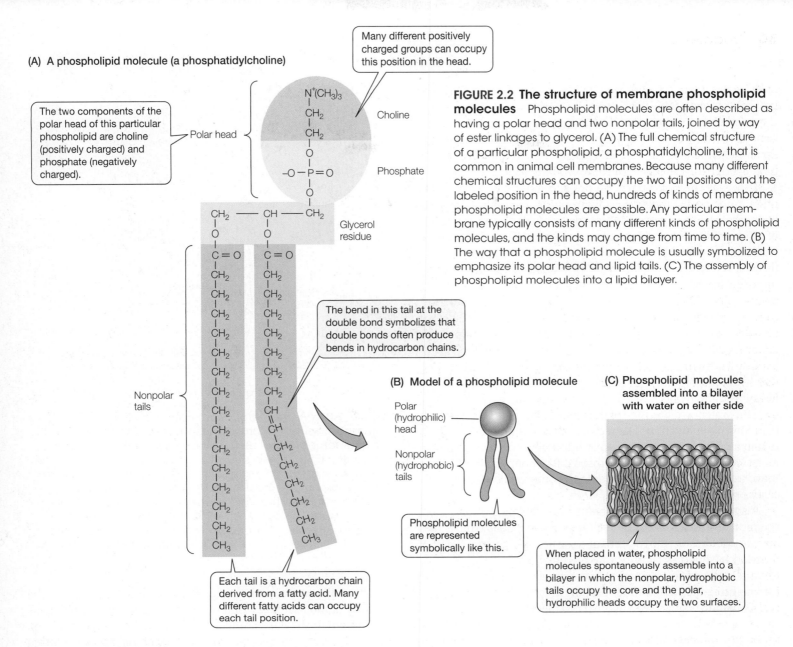

(A) A phospholipid molecule (a phosphatidylcholine)

Many different positively charged groups can occupy this position in the head.

The two components of the polar head of this particular phospholipid are choline (positively charged) and phosphate (negatively charged).

Polar head

N⁺(CH₃)₃ — Choline

Phosphate

Glycerol residue

The bend in this tail at the double bond symbolizes that double bonds often produce bends in hydrocarbon chains.

Nonpolar tails

Each tail is a hydrocarbon chain derived from a fatty acid. Many different fatty acids can occupy each tail position.

FIGURE 2.2 The structure of membrane phospholipid molecules Phospholipid molecules are often described as having a polar head and two nonpolar tails, joined by way of ester linkages to glycerol. (A) The full chemical structure of a particular phospholipid, a phosphatidylcholine, that is common in animal cell membranes. Because many different chemical structures can occupy the two tail positions and the labeled position in the head, hundreds of kinds of membrane phospholipid molecules are possible. Any particular membrane typically consists of many different kinds of phospholipid molecules, and the kinds may change from time to time. (B) The way that a phospholipid molecule is usually symbolized to emphasize its polar head and lipid tails. (C) The assembly of phospholipid molecules into a lipid bilayer.

(B) Model of a phospholipid molecule

Polar (hydrophilic) head

Nonpolar (hydrophobic) tails

Phospholipid molecules are represented symbolically like this.

(C) Phospholipid molecules assembled into a bilayer with water on either side

When placed in water, phospholipid molecules spontaneously assemble into a bilayer in which the nonpolar, hydrophobic tails occupy the core and the polar, hydrophilic heads occupy the two surfaces.

To understand the molecular logic of the structure of cell membranes and intracellular membranes, it is necessary to consider the *polarity* of molecules and the attendant attributes of *hydrophilic* and *hydrophobic* interactions. Consider vinegar-and-oil salad dressing as an everyday example of the effects of molecular polarity. Vinegar consists of acetic acid and water. Thus the dressing has three principal components: oil, acetic acid, and water. If the dressing sits still for a while, the acetic acid remains in solution in the water, but the oil forms a separate layer. This outcome occurs because the acetic acid is **hydrophilic** ("water-loving"), whereas the oil is **hydrophobic** ("water-hating"). Why do the two substances behave in these different ways? A principal reason is the polarity of the molecules. Acetic acid is polar and because of its polar nature is attracted to water. Oil is nonpolar and therefore repelled from water.

The distribution of electrons in a molecule is the property that determines whether the molecule is polar or nonpolar. Within a **polar molecule**, electrons are unevenly distributed; thus some regions of a polar molecule are relatively negative, whereas others are relatively positive. Water is a polar molecule. Other polar molecules, such as acetic acid—and ions—intermingle freely with polar water molecules by charge interaction, forming solutions. Within a **nonpolar molecule**, electrons are evenly distributed

and there are no charge imbalances between different molecular regions. Nonpolar molecules, such as the oil in salad dressing, do not freely intermingle with polar water molecules. Because of this, after oil is dispersed into water by violent shaking, the oil-and-water system can move toward a more thermodynamically stable state if the water molecules assemble *with other water molecules* into arrays that surround the dispersed droplets of nonpolar oil molecules. These arrays of water molecules then join with each other until there is a complete separation of the water and nonpolar molecules. As we will shortly see, these principles help explain the structure of the phospholipid bilayer in cell membranes and intracellular membranes, and they also help explain the positioning of other chemical constituents within the bilayer.

The lipids of membranes are structured, diverse, fluid, and responsive to some environmental factors

Phospholipids are lipids that contain phosphate groups (**FIGURE 2.2A**). They are the principal constituents of the matrix in which proteins are embedded in cell membranes and intracellular membranes. They are **amphipathic**, meaning that each molecule consists of a polar part (within which there are regional differences of charge) and a nonpolar part (which lacks regional differences of charge). A membrane phospholipid consists of a *polar head* and two *nonpolar tails* (**FIGURE 2.2B**). The polar head is composed of the phosphate group, which forms a region of negative charge, bonded to another group that forms a region of positive charge, such as choline (see Figure 2.2A). Each nonpolar tail consists of a long-chain hydrocarbon derived from a fatty acid.

Whereas the polar part of a phospholipid molecule or any other amphipathic molecule is hydrophilic, the nonpolar part is hydrophobic. When phospholipid molecules are placed in a system of oil layered on water, they collect at the interface of the oil and water in a predictable way, with their polar, hydrophilic heads in the water and their nonpolar, hydrophobic tails in the oil. Of greater importance for understanding living cells is the fact that when phospholipid molecules are placed simply in an aqueous solution, they spontaneously assemble into bilayers, adopting the same bilayer conformation they take in cell membranes and intracellular membranes (**FIGURE 2.2C**). This bilayer conformation forms because it is thermodynamically stable. All the hydrophobic regions (the hydrocarbon tails) get together in the interior of the bilayer (away from the water), whereas the hydrophilic heads associate with the water on either side of the membrane. The energy barrier to mixing polar and nonpolar regions in the membrane is so great that in a cell membrane, it is nearly impossible for a phospholipid molecule to "flip" its polar head through the nonpolar interior and move from one side of the bilayer to the other (unless specifically catalyzed to do so).

A striking attribute of membrane phospholipids is their great chemical diversity. Many different types of phospholipid molecules are possible because the two tails and the positively charged part of the head, as shown in Figure 2.2A, can differ widely in their specific chemical composition. The cell membranes of human red blood cells contain more than 150 different chemical forms of phospholipids, and similar diversity is seen in other cell membranes. The two layers of phospholipid molecules in any particular membrane, known as the two **leaflets** of the membrane, typically are composed of different mixes of phospholipid molecules.

The phospholipids in a cell membrane or intracellular membrane are **fluid**. Individual phospholipid molecules are not covalently bound to one another. Therefore they move relative to each other. They are able to move about rather freely by diffusion *within each membrane leaflet*. The rate of this diffusion is great enough that a particular phospholipid molecule is able to travel, by diffusion, around the entire circumference of a cell in a matter of minutes. The ease of motion of the phospholipid molecules in a membrane leaflet is termed their **fluidity**.

Fluidity depends in part on the degree of *chemical saturation* of the hydrocarbons that make up the phospholipid tails. What do we mean by chemical saturation? A hydrocarbon is **saturated** if it contains no double bonds. It is **unsaturated** if it includes one or more double bonds; different *degrees* of unsaturation are possible because the number of double bonds can be high or low. As shown in Figure 2.2A, a double bond often imparts a bend to a hydrocarbon chain. Bent tails of membrane phospholipids prevent tight, crystal-like packing of the tails in the hydrophobic interior of the membrane. This disruption of tight packing helps keep the phospholipid molecules free to move. Accordingly, a greater proportion of unsaturated fatty acids in the tails of phospholipids results in a membrane with more fluidity.

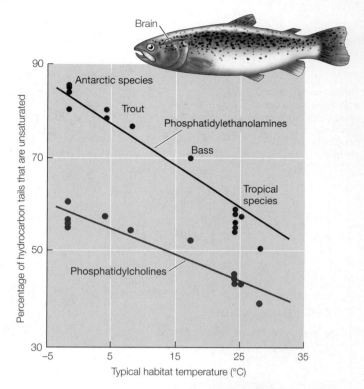

FIGURE 2.3 The degree of chemical unsaturation of the hydrocarbon tails of brain phospholipids in fish varies with habitat temperature   Brain synaptic membranes of 17 species of teleost (bony) fish were studied. Measurements were made of the composition of the hydrocarbon tails of two categories of cell-membrane phospholipids, the phosphatidylcholines and the phosphatidylethanolamines, which differ in whether the group at the top of the head in Figure 2.2A is choline [$-CH_2-CH_2-N(CH_3)_3$] or ethanolamine ($-CH_2-CH_2-NH_2$), respectively. Each plotted symbol corresponds to the average value for one species. (After Logue et al. 2000.)

In addition to chemical composition, temperature affects the fluidity of membranes; just as butter and other household lipids stiffen when they are chilled, the phospholipid leaflets in cell membranes tend to become stiffer at lower temperatures. During evolution, one important way in which cells have become adapted to different temperatures is alteration of the number of double bonds (the degree of unsaturation) in their membrane phospholipids. This is evident in fish of polar seas, for instance. The fish experience tissue temperatures so low that their cell membranes could be overly stiff. This problem is avoided, however, because these fish have cell membranes constructed of phospholipids that are particularly rich in double bonds; the highly unsaturated phospholipids are inherently quite fluid and thus less likely than other phospholipids to become detrimentally stiff at low temperatures. Recent research on the cell membranes of brain neurons (nerve cells) in fish demonstrates that the degree of phospholipid unsaturation depends in a regular way on the environmental temperatures to which various species are adapted (**FIGURE 2.3**). Tropical species of fish, which face little risk of having their membranes rendered too stiff by low temperatures, have evolved relatively saturated phospholipids, but as the temperature of the habitat falls, the degree of unsaturation of the phospholipids increases.

Evidence is accumulating that individual animals sometimes restructure their membrane phospholipids in response to environmental factors. For example, lab mice alter the mix of membrane phospholipids in their heart muscle cells after just 4–12 h of fasting and reverse the changes when fed. At least some hibernating species of mammals substantially alter the mix of phospholipids in their cell and mitochondrial membranes as they approach hibernation, in ways thought to promote the hibernating physiological state (e.g., suppression of metabolism).

Besides phospholipids, cell membranes and intracellular membranes contain other classes of lipids, one of which is **sterols**. The principal membrane sterols are **cholesterol** and **cholesterol esters**. In cell membranes, which are typically much richer in sterols than intracellular membranes are, sterols collectively occur in ratios of 1 molecule per 10 phospholipid molecules, up to 8 per 10. Cholesterol is mildly amphipathic and positioned within the phospholipid leaflets (see Figure 2.1), where it exerts complex effects on membrane fluidity.

Proteins endow membranes with numerous functional capacities

Proteins are the second major constituents of cell membranes and intracellular membranes. According to the **fluid mosaic model** of membranes, a membrane consists of a mosaic of protein and lipid molecules, all of which move about in directions parallel to the membrane faces because of the fluid state of the lipid matrix. As we start to discuss proteins, an important fact to recall from the study of organic chemistry is that—in terms of their chemical makeup—proteins are considered to have primary, secondary, tertiary, and sometimes quaternary structure. **BOX 2.1** reviews this aspect of protein structure.

Membrane proteins are structurally of two principal kinds: integral and peripheral. **Integral membrane proteins** are part of the membrane and cannot be removed without taking the membrane apart. Most integral proteins (see Figure 2.1) span the membrane and thus are called *transmembrane* proteins. These molecules have both hydrophobic and hydrophilic regions. As we will see in detail shortly, each hydrophobic region typically has an amino acid composition and a molecular geometry that allow it to associate with the hydrophobic hydrocarbon tails of the membrane interior. The hydrophilic regions of transmembrane protein molecules, by contrast, typically protrude into the aqueous solutions bathing the two sides of the membrane.

BOX 2.1 Protein Structure and the Bonds That Maintain It

All protein molecules have *primary*, *secondary*, and *tertiary* structure (and some have *quaternary* structure). Primary structure refers to the string of covalently bonded amino acids. As essential as primary structure is, protein *function* depends most directly on secondary and tertiary structure—the three-dimensional conformation of the protein molecule. Because secondary and tertiary structure are stabilized by *weak* chemical bonds rather than covalent bonds, the three-dimensional conformation of a protein can *change* and *flex*—a process essential for protein function. *Denaturation* is a disruption of the correct tertiary structure; because primary structure is not altered, denaturation may be reversible (reparable). This box continues on the web at **Box Extension 2.1**. There you will find detailed infformation on—and illustrations of—all levels of protein structure, strong versus weak bonds, the types of weak bonds, denaturation, and potential repair of denaturation.

Peripheral membrane proteins are associated with the membrane but can be removed without destroying the membrane. They are bonded noncovalently (i.e., by weak bonds) to membrane components (e.g., integral proteins) and are positioned on one side of the membrane or the other (see Figure 2.1). Their positioning means that *the two leaflets of a membrane differ in protein composition, as well as phospholipid composition.*

The proteins of cell membranes and intracellular membranes endow the membranes with capabilities to *do* many things. Five *functional* types of membrane proteins are recognized: **channels, transporters (carriers), enzymes, receptors,** and **structural proteins**. Because these types are classified by function, the *actions* listed in **TABLE 2.1** *define* the five types. The categories are not

---

TABLE 2.1 The five functional types of membrane proteins and the functions they perform

Functional type	Function performed (defining property)
Channel	A channel protein permits simple or quasi-simple *diffusion* of solutes in aqueous solution (see page 108)—or *osmosis* of water (see page 126)—through a membrane. A simplified view of a channel is that it creates a direct water path from one side to the other of a membrane (i.e., an aqueous pore) through which solutes in aqueous solution may diffuse or water may undergo osmosis.
Transporter (carrier)	A transporter protein binds noncovalently and reversibly with specific molecules or ions to move them across a membrane intact. The transport through the membrane is *active transport* (see page 113) if it employs metabolic energy; it is *facilitated diffusion* (see page 113) if metabolic energy is not employed.
Enzyme	An enzyme protein catalyzes a chemical reaction in which covalent bonds are made or broken (see page 45).
Receptor	A receptor protein binds noncovalently with specific molecules and, as a consequence of this binding, initiates a change in membrane permeability or cell metabolism. Receptor proteins mediate the responses of a cell to chemical messages (signals) arriving at the outside face of the cell membrane (see page 62).
Structural protein	A structural protein attaches to other molecules (e.g., other proteins) to anchor intracellular elements (e.g., cytoskeleton filaments) to the cell membrane, creates junctions between cells (see Figure 2.7), or establishes other structural relations.

FIGURE 2.4 The structure of a transmembrane protein—a voltage-gated Na⁺ channel—illustrating several modes of presentation

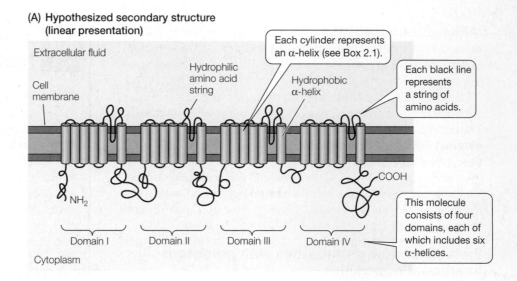

(A) Hypothesized secondary structure (linear presentation)

Extracellular fluid

Cell membrane

Hydrophilic amino acid string

Each cylinder represents an α-helix (see Box 2.1).

Hydrophobic α-helix

Each black line represents a string of amino acids.

NH₂

COOH

This molecule consists of four domains, each of which includes six α-helices.

Domain I   Domain II   Domain III   Domain IV

Cytoplasm

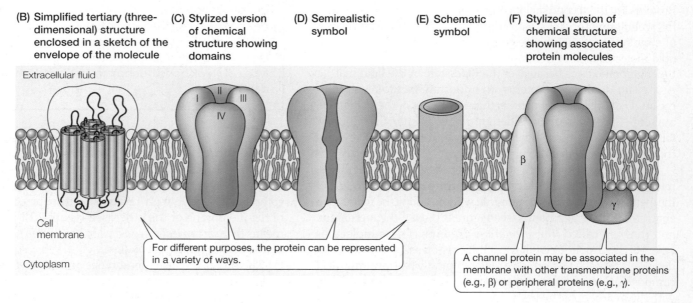

(B) Simplified tertiary (three-dimensional) structure enclosed in a sketch of the envelope of the molecule

(C) Stylized version of chemical structure showing domains

(D) Semirealistic symbol

(E) Schematic symbol

(F) Stylized version of chemical structure showing associated protein molecules

Extracellular fluid

Cell membrane

Cytoplasm

For different purposes, the protein can be represented in a variety of ways.

A channel protein may be associated in the membrane with other transmembrane proteins (e.g., β) or peripheral proteins (e.g., γ).

mutually exclusive: A membrane protein can be both a receptor and a channel, or a transporter and an enzyme, for example.

The molecular structures of membrane proteins are complex and are diagrammed in several ways, depending on the degree of chemical detail to be shown. To illustrate, let's focus on a channel, which is a type of membrane-spanning integral protein. Channels provide paths for ions or other materials in aqueous solution to pass through membranes. In our example the channel is formed by a single protein molecule, the secondary structure of which is shown in **FIGURE 2.4A** (see Box 2.1 for an explanation of secondary structure).

Each cylinder in Figure 2.4A represents a sequence of amino acids that forms a helix-shaped subunit, called an *α-helix* (see Box 2.1), within the protein structure. The whole protein molecule exemplifies a common property of membrane proteins, in that it consists of repeating structural patterns known as **domains**. To identify the domains, review the molecule from left to right. You will note five α-helices linked closely together, and then a sixth helix separated from the others by a longer string of amino acids; then you will note that this pattern of five closely spaced helices followed by a sixth more-separated helix is repeated three more times. On

the basis of this repeating pattern, this molecule is said to show four domains, numbered I to IV, as illustrated in Figure 2.4A. The α-helices are predominantly hydrophobic and span the membrane by associating with the hydrophobic interior of the phospholipid bilayer. The strings of amino acids that connect successive helices are hydrophilic and protrude from the membrane into the aqueous solutions on either side. In its natural state in a membrane, this protein is believed to be shaped into a closed ring in which the four domains form cylinder-like structures surrounding a central pore, as diagrammed in **FIGURE 2.4B**.

The three additional representations of the membrane protein shown in Figure 2.4 are progressively simpler. The sort of representation in **FIGURE 2.4C**, which still shows that there are four domains, is a simplified way to represent the chemical structure of the molecule. The diagrammatic, semirealistic representation in **FIGURE 2.4D**, which leaves one guessing about the number of domains, is more simplified yet, and in **FIGURE 2.4E** the channel is represented schematically (without any intention of resembling the actual molecule).

The interrelations of the presentations in Figure 2.4A–E are important to note because all of these sorts of presentations are commonly used in biological literature. An important additional detail is that the major subunits of membrane proteins are not always parts of one molecule (**FIGURE 2.4F**).

Carbohydrates play important roles in membranes

Cell membranes and intracellular membranes also contain carbohydrates, which occur mostly in covalently bonded combination with lipids or proteins, or both (see Figure 2.1). *Glycolipids* (e.g., gangliosides), *glycoproteins*, and *proteoglycans* are some of the major categories of carbohydrate-containing membrane compounds.[1] Carbohydrate groups are hydrophilic and thus are associated with the membrane surface and adjacent aqueous solution. Carbohydrates reinforce the point, stressed earlier, that the two leaflets of a membrane are typically different. In cell membranes, for example, the carbohydrate groups always project from the outer, extracellular face, not the inner, cytoplasmic face (see Figure 2.1). These carbohydrate groups serve as attachment sites for extracellular proteins and as cell-recognition sites.

SUMMARY
Cell Membranes and Intracellular Membranes

- The matrix of a cell membrane or intracellular membrane consists of a bilayer of phospholipid molecules. The phospholipids are chemically very diverse, even within a single membrane, and in a particular cell the phospholipid composition can undergo change in response to environmental or other factors. The phospholipids are fluid, meaning that individual molecules move about relatively freely by diffusion within each membrane leaflet.

- Animals exhibit adaptive trends in the phospholipid compositions of their cell membranes. Cells that function routinely at low temperatures tend to have a phospholipid composition that permits membranes to remain fluid under cold conditions (e.g., they have high proportions of double bonds in the hydrocarbon tails).

- Five functional categories of proteins occur in cell and intracellular membranes: channels, transporters (carriers), enzymes, receptors, and structural proteins. A single protein may engage in more than one function.

- In addition to phospholipids and proteins, which are the principal components, membranes often have other components such as cholesterol (a lipid) and glycoproteins (composed of covalently bonded carbohydrate and protein subunits).

Epithelia

An **epithelium** (plural *epithelia*) is a sheet of cells that covers a body surface or organ, or lines a cavity. Although epithelia are radically different from cell membranes and intracellular membranes, they—to some degree—perform parallel functions on a larger structural scale. Epithelia compartmentalize the body by forming boundaries between body regions. They also form a boundary

[1] The word fragment *glyco* refers to carbohydrates (after the Greek *glykeros*, "sweet").

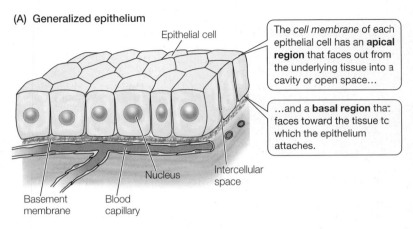

(A) Generalized epithelium

Epithelial cell

The *cell membrane* of each epithelial cell has an **apical region** that faces out from the underlying tissue into a cavity or open space…

…and a **basal region** that faces toward the tissue to which the epithelium attaches.

Nucleus
Intercellular space
Basement membrane
Blood capillary

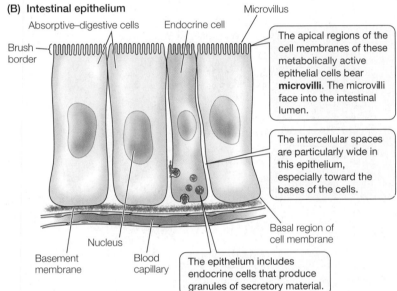

(B) Intestinal epithelium

Microvillus
Absorptive–digestive cells
Endocrine cell
Brush border

The apical regions of the cell membranes of these metabolically active epithelial cells bear **microvilli**. The microvilli face into the intestinal lumen.

The intercellular spaces are particularly wide in this epithelium, especially toward the bases of the cells.

Basal region of cell membrane
Nucleus
Basement membrane
Blood capillary

The epithelium includes endocrine cells that produce granules of secretory material.

FIGURE 2.5 Simple epithelia (A) A generalized simple epithelium covering a free surface of a tissue. (B) The specific simple epithelium lining the mammalian small intestine (midgut). This epithelium consists of several cell types. Most cells are the absorptive–digestive cells emphasized here. Scattered among these cells are mucin-secreting cells (not shown) and at least ten types of endocrine or endocrine-like cells. Each endocrine cell produces granules of secretory material; the granules move to juxtapose themselves with the basal or near-basal regions of the cell membrane and then release their secretions into the spaces outside the cell, after which the secretions enter the blood for transport elsewhere. Endocrine-like cells termed *paracrine cells* (not shown) are also present. Paracrine cells produce secretions that affect nearby cells rather than acting on distant cells by way of the circulation (see Figure 16.1).

between an animal and its external environment. Moreover, like cell membranes, epithelia have numerous functional capacities and play major functional roles in animal physiology.

A **simple epithelium** consists of a single layer of cells (**FIGURE 2.5**). Simple epithelia are exceedingly common; in the human body, for instance, the intestines, kidney tubules, blood vessels, and sweat glands are all lined with a simple epithelium. Each cell in a simple epithelium has an **apical surface** (*mucosal surface*) facing into a cavity or open space, and a **basal surface** (*serosal surface*) facing toward the underlying tissue to which the epithelium is attached. An epithelium typically rests on a thin, permeable, noncellular, and nonliving sheet of matrix material, positioned beneath the basal

(A) Epithelial cells can form tubules and follicles

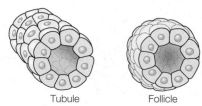

Tubule Follicle

FIGURE 2.6 Tubules and follicles formed by simple epithelia (A) Both tubules and follicles are formed by the wrapping of simple epithelia into closed curves. Cross sections of two important tubular structures are shown in (B) and (C); in each case the basal cell surfaces and basement membrane of the epithelium are on the outside. For historical reasons, the cells of blood capillaries are usually called *endothelial* cells, but they are a form of epithelium.

(B) Proximal part of a mammalian nephron (kidney tubule) in cross section

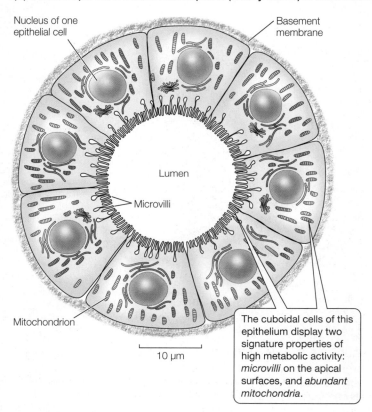

Nucleus of one epithelial cell

Basement membrane

Lumen

Microvilli

Mitochondrion

10 μm

The cuboidal cells of this epithelium display two signature properties of high metabolic activity: *microvilli* on the apical surfaces, and *abundant mitochondria*.

(C) Mammalian blood capillary in cross section

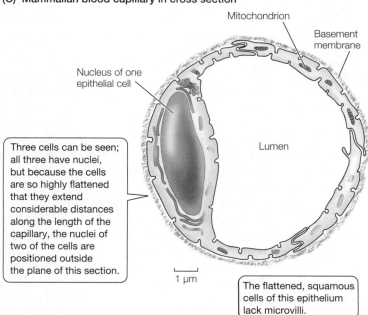

Mitochondrion

Basement membrane

Nucleus of one epithelial cell

Lumen

Three cells can be seen; all three have nuclei, but because the cells are so highly flattened that they extend considerable distances along the length of the capillary, the nuclei of two of the cells are positioned outside the plane of this section.

1 μm

The flattened, squamous cells of this epithelium lack microvilli.

cell surfaces. This sheet is called the **basement membrane** (*basal lamina*) and is composed of glycoproteins and particular types of collagen. It is secreted mostly by the epithelial cells, although the underlying cells also contribute. Simple epithelia are classified as *squamous, cuboidal,* or *columnar,* depending on how tall the cells are. The cells in a squamous epithelium are low and flat, whereas those in a columnar epithelium are high relative to their basal dimensions; the epithelium in Figure 2.5A is classified as cuboidal because the cells are about as tall as they are wide. Blood vessels usually do not enter epithelia. Instead, epithelial cells exchange O_2, CO_2, and other materials through the underlying basement membrane with blood capillaries located on the opposite side of the basement membrane (see Figure 2.5A).

The columnar epithelium that lines the small intestine (midgut) of a mammal (see Figure 2.5B) is a type of simple epithelium that will be featured prominently in this book (e.g., in Chapters 5 and 6) and that introduces additional aspects of epithelial morphology and function. The apical surfaces of the cells in this epithelium face into the lumen (open central cavity) of the intestine. As digestion occurs, liberating food molecules from foods, the molecules pass through the epithelium and basement membrane to reach blood vessels and lymph passages that transport them to the rest of the body.

The intestinal epithelium illustrates that a simple epithelium can consist of two or more cell types. Whereas the epithelium is composed mostly of absorptive–digestive cells, it also includes endocrine cells (see Figure 2.5B) and additional cell types.

The intestinal epithelium also illustrates **microvilli** (singular *microvillus*), which are a common (but not universal) feature of epithelial cells. Microvilli are exceedingly fine, fingerlike projections of the apical cell membrane (see Figure 2.5B). In the intestinal epithelium, the microvilli greatly increase the area of contact between the epithelial cells and the contents of the gut. Microvilli are most often found in epithelia that are active in secreting or absorbing materials, such as the epithelia of certain kidney tubules and the pancreatic ducts, as well as the intestinal epithelium. Microvilli are often described collectively as a **brush border** because they look like the bristles on a brush when viewed microscopically.

Another significant aspect of diversity in simple epithelia arises from the geometric arrangement of the cells. *Tubules* or *follicles* (hollow globes) are often formed by the wrapping of a simple epithelium into a closed curve (**FIGURE 2.6A**) supported by the basement membrane on the outside. A tubule formed by cuboidal epithelium bearing microvilli forms the proximal region of each mammalian nephron (kidney tubule), for example (**FIGURE 2.6B**). Vertebrate blood capillaries are an especially important example. Each blood capillary consists of a single layer of highly flattened epithelial cells (lacking microvilli) supported by the epithelial basement membrane (**FIGURE 2.6C**). The basement membranes of capillaries are one of the important biochemical targets of the venoms of puff adders and

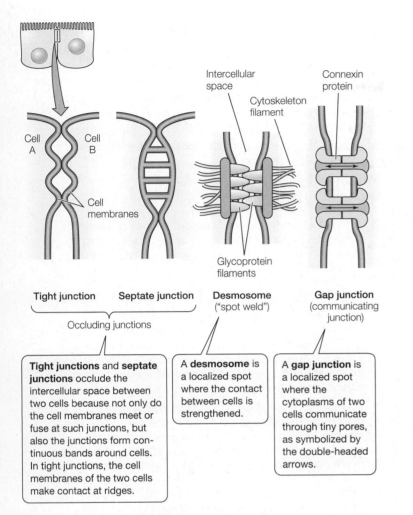

Tight junction   Septate junction   Desmosome ("spot weld")   Gap junction (communicating junction)

Occluding junctions

Tight junctions and **septate junctions** occlude the intercellular space between two cells because not only do the cell membranes meet or fuse at such junctions, but also the junctions form continuous bands around cells. In tight junctions, the cell membranes of the two cells make contact at ridges.

A **desmosome** is a localized spot where the contact between cells is strengthened.

A **gap junction** is a localized spot where the cytoplasms of two cells communicate through tiny pores, as symbolized by the double-headed arrows.

FIGURE 2.7 Types of junctions between cells   At a pore in a gap junction, each cell has a ring of six connexin proteins that together form the pore, and the rings of the two cells line up to create continuity between cells. Each ring of connexin proteins is called a *connexon*.

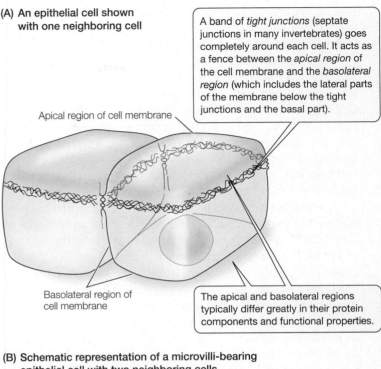

(A) An epithelial cell shown with one neighboring cell

A band of *tight junctions* (septate junctions in many invertebrates) goes completely around each cell. It acts as a fence between the *apical region* of the cell membrane and the *basolateral region* (which includes the lateral parts of the membrane below the tight junctions and the basal part).

Apical region of cell membrane

Basolateral region of cell membrane

The apical and basolateral regions typically differ greatly in their protein components and functional properties.

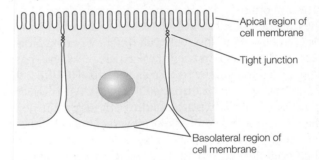

(B) Schematic representation of a microvilli-bearing epithelial cell with two neighboring cells

Apical region of cell membrane

Tight junction

Basolateral region of cell membrane

FIGURE 2.8 The organization of epithelial cells into apical and basolateral regions   (A) The cell membrane of an epithelial cell is divided into apical (pink) and basolateral (orange) regions—which differ in their protein components and functions—by a band of tight junctions formed with adjacent epithelial cells. In this book we will often use the schematic format shown in (B) to represent an epithelium. Microvilli do not always occur, but when they do, they are on the apical side only.

rattlesnakes. The venoms contain enzymes (metalloproteases) that break down the basement membranes, destroying the integrity of blood capillaries. In this way the venoms cause widespread internal hemorrhaging.

Adjacent cells in an epithelium are physically joined by cell-membrane junctions of several sorts; the four most important are *tight junctions, septate junctions, desmosomes,* and *gap junctions* (**FIGURE 2.7**). In the paragraphs that follow, we look at each of these junctions in turn.

A **tight junction** is a place where the cell membranes of adjacent cells are tightly joined so that there is no intercellular space between the cells; adjacent cells are perhaps 10–20 nm apart for the most part, but at tight junctions the cell membranes meet or fuse. Tight junctions typically occur between the sides of adjacent cells, just a short distance away from their apical surfaces (**FIGURE 2.8**). Any given epithelial cell has tight junctions with adjacent epithelial cells in a continuous ring around its entire perimeter. *This ring of tight junctions demarcates the apical surface of the cell from its lateral and basal surfaces, giving rise to one of the most important distinctions in the physiological study of epithelia, the distinction between the **apical region** and the **basolateral region** of each cell membrane* (see Figure

2.8). Many invertebrate groups have **septate junctions** instead of tight junctions. Septate junctions differ from tight junctions in their fine structure (see Figure 2.7), but they resemble tight junctions in their position and in the fact that they fully encircle each cell. Tight and septate junctions are sometimes aptly called **occluding junctions** because they block, or occlude, the spaces between adjacent epithelial cells, preventing open passage between the fluids on either side of an epithelium.

A **desmosome** (see Figure 2.7) is a junction at which mutually adhering glycoprotein filaments from two adjacent cells intermingle across the space between the cells. Desmosomes are often likened to rivets or spot welds because they occur as tiny isolated spots, not continuous bands, and their principal function is believed to be to strengthen and stabilize contacts between adjacent cells.

Gap junctions (see Figure 2.7) are like desmosomes in that they occur at discrete spots, but otherwise they are very different from all the other junctions we have discussed because within a gap junction there are open pores between cells. At these pores, which are formed by **connexin proteins** (see Figure 2.7), the two adjacent cells lack cell-membrane boundaries, and there is continuity between the cytoplasms of the cells. Molecules and ions smaller than 1000–1500 daltons (Da) in molecular mass are able to pass between cells at gap junctions, although large solutes such as proteins cannot. Gap junctions are important in cell–cell communication—including passage of intracellular signaling agents in some tissues and direct electrophysiological interactions between cells in nerve or muscle (gap junctions are treated in detail in Chapter 13; see Figure 13.2).

A central feature of epithelia is that *each epithelial cell is functionally asymmetric*. The proteins in the cell membrane of an epithelial cell are unable (for reasons only poorly known) to diffuse through tight junctions. Thus the ring of tight junctions around each epithelial cell *acts as a fence* that keeps proteins from crossing between the apical and basolateral regions of the cell membrane. *The two regions of the cell membrane therefore have different sets of channels, transporters, membrane enzymes, and other classes of membrane proteins, and they are functionally different in many ways.* Differences also exist between the apical and basolateral regions in the membrane phospholipids composing the outer (but not inner) leaflet of the cell membrane.

One of the important functions of an epithelium is to control and mediate the transport of substances between the apical and basal sides of the epithelium and thus between different body regions. Substances—such as ions, nutrient molecules, or water—pass through a simple epithelium by two types of paths (**FIGURE 2.9**). They may pass *through* cells by **transcellular** paths. Alternatively, they may pass *between* cells, in **paracellular** paths. Tight junctions interfere with or block the paracellular movement of substances across an epithelium. In some epithelia the tight junctions prevent almost all paracellular movement. In others, however, the tight junctions permit extensive paracellular movement of certain sorts of molecules or ions, and the epithelia are described as **leaky**.

A substance that crosses an epithelium by a transcellular path must pass through two *cell* membranes. One of the most important principles in the study of epithelia is that for scientists to understand the physiology of transcellular transport, they must understand the membrane proteins and functions of both the apical cell membranes and the basolateral cell membranes of the epithelial cells.

SUMMARY
Epithelia

- An epithelium is a sheet of cells that lines a cavity or covers an organ or body surface, thereby forming a boundary between functionally different regions of the body or between the animal and the external environment.

- In a simple epithelium, each cell is fully encircled by a ring of tight or septate junctions formed with adjacent epithelial cells. These occluding-type junctions seal the spaces between adjacent cells. Moreover, the ring of junctions around each cell divides the cell membrane into chemically and functionally distinct apical and basolateral regions.

- An epithelium rests on a nonliving, permeable basement membrane secreted by the epithelial cells and underlying tissue. The apical membranes of metabolically active epithelial cells often bear a brush border of microvilli, greatly enhancing their surface area. In addition to the occluding junctions, adjacent epithelial cells are joined by structurally reinforcing "spot welds," called desmosomes, and sometimes by gap junctions at which continuity is established between the cytoplasms of the cells.

- Materials pass through epithelia by paracellular paths between adjacent cells and by transcellular paths through cells. Materials traveling through a cell must pass through both the apical and the basolateral cell membranes of the cell.

Elements of Metabolism

At this point in the chapter, we shift to focus principally on processes, rather than morphology. The basics of metabolism constitute a good beginning for this new emphasis.

Metabolism is the set of processes by which cells and organisms acquire, rearrange, and void commodities in ways that sustain life. Metabolism involves myriad chemical and physical processes. To give order to their research, animal physiologists subdivide the study of metabolism. One way of doing this is according to specific commodities. For example, **nitrogen metabolism** is the set of processes by which nitrogen is acquired, employed in synthetic reactions to create proteins and other functional nitrogenous compounds, and ultimately transferred to elimination compounds such as urea or ammonia. **Energy metabolism** consists of the processes by which energy is acquired, transformed, channeled into useful functions, and dissipated.

Metabolism also may be subdivided according to the type of transformation that occurs. **Catabolism** is the set of processes by

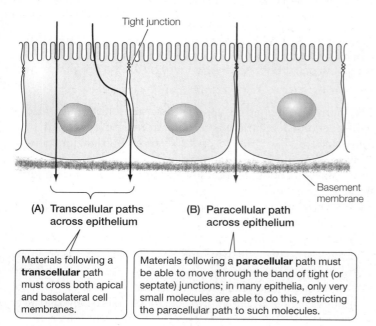

Tight junction

Basement membrane

(A) Transcellular paths across epithelium

(B) Paracellular path across epithelium

Materials following a **transcellular** path must cross both apical and basolateral cell membranes.

Materials following a **paracellular** path must be able to move through the band of tight (or septate) junctions; in many epithelia, only very small molecules are able to do this, restricting the paracellular path to such molecules.

FIGURE 2.9 Transcellular and paracellular paths across an epithelium

which complex chemical compounds are broken down to release energy, create smaller chemical building blocks, or prepare chemical constituents for elimination. **Anabolism**, by contrast, consists of the processes that synthesize larger or more complex chemical compounds from smaller chemical building blocks, using energy. Whereas catabolism is destructive, anabolism is constructive.

Metabolism depends on *sets* of biochemical reactions, such as the 30 or so linked reactions that cells employ to oxidize glucose into CO_2 and H_2O. The prominence of biochemistry in metabolism can give the impression that cells are just like test tubes: merely places where chemicals react. There is a massive distinction between cells and test tubes, however. Whereas test tubes are simply places where chemical reactions occur, cells orchestrate their own chemistry. The cellular orchestration of metabolism is directed by genes and mediated, in major part, by enzymes.

Enzyme Fundamentals

In his story "The Celebrated Jumping Frog of Calaveras County," Mark Twain appealed to the imagination of his readers by extolling the awesome jumping abilities of a frog, probably a common leopard frog (*Rana pipiens*) (**FIGURE 2.10A**), named Dan'l Webster. Anyone who has ever tried to catch leopard frogs knows that when first disturbed, they hop away at lightning speed. Thus it is hard not to smile in knowing admiration as Twain describes Dan'l Webster's celebrated jumping feats. Muscles can work only as fast as they are supplied with adenosine triphosphate (ATP). Amphibians, however, have only modest abilities to make ATP using oxygen (O_2), because they have relatively simple lungs and can supply their cells with O_2 only relatively slowly. For leopard frogs to hop along as fast as they do when fleeing danger, they need to make ATP faster than the O_2 supply to their muscles permits. That is, they must make ATP by *anaerobic* mechanisms not requiring O_2. A crucial reason they can do this is that their leg muscles are well endowed with the enzyme *lactate dehydrogenase*.

Compared with leopard frogs, toads such as the common western toad of North America (*Anaxyrus boreas*, formerly *Bufo boreas*) (**FIGURE 2.10B**) are not nearly as well endowed with lactate

dehydrogenase. Thus they cannot make ATP to a great extent without O_2, and the slow rate of O_2 delivery to their muscle cells means a slow rate of ATP production, explaining why they cannot hop along as fast as frogs. Mark Twain could not have known this, because the study of enzymes was just beginning during his life, but when he searched his mind for an amphibian that could inspire his readers as a "celebrated" jumper, he chose a frog rather than a toad in major part (we now know) because frogs have more of the enzyme lactate dehydrogenase.

Enzymes are protein catalysts that play two principal roles: They *speed* chemical reactions, and they often *regulate* reactions.[2] To appreciate the extreme importance of enzymes, it is crucial to recognize that the vast majority of the biochemical reactions that occur in animals *do not take place on their own at significant rates* under physiological conditions. Cells are biochemically complex enough that, in principle, tens of thousands of reactions might occur in them. However, because reactions in general require catalysis to occur at significant rates, the particular reactions that do take place in a cell—out of all those that *could* take place—depend on the cell's own biosynthesis of enzyme proteins. *Enzymes represent one of the foremost means by which cells take charge of their own biochemistry.*

When we say that an enzyme is a **catalyst**, we mean it is a molecule that accelerates a reaction without, in the end, being altered itself. The reaction catalyzed by lactate dehydrogenase (LDH) that is important for escape by frogs is the reduction of pyruvic acid to form lactic acid, a reaction in which each pyruvic acid molecule is combined with two hydrogen atoms (**FIGURE 2.11A**). Although the presence of LDH speeds this reaction, LDH is not itself altered by the reaction. Thus a molecule of LDH persists as it catalyzes the reduction of many pyruvic acid molecules, one after another.

Enzymes are described as having *substrates* and *products*, and often there are two or more of each. To be exact about the substrates and products of LDH, a chemically complete presentation of the LDH-catalyzed reaction is needed (**FIGURE 2.11B**). The hydrogen atoms that reduce pyruvic acid are taken from a molecule we

[2] Research over the past 35 years has shown that protein catalysts—enzymes—are not the only catalysts. Some types of RNA molecules also serve as catalysts.

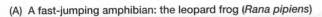

(A) A fast-jumping amphibian: the leopard frog (*Rana pipiens*)

(B) A slow-jumping amphibian: the western toad, *Anaxyrus* (*Bufo*) *boreas*

FIGURE 2.10 Two amphibians with different jumping capabilities based in part on different levels of a key enzyme, lactate dehydrogenase

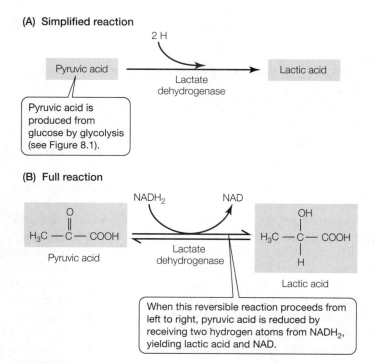

(A) Simplified reaction

2 H

Pyruvic acid → Lactic acid
Lactate dehydrogenase

Pyruvic acid is produced from glucose by glycolysis (see Figure 8.1).

(B) Full reaction

NADH₂ → NAD

Pyruvic acid ⇌ Lactic acid
Lactate dehydrogenase

When this reversible reaction proceeds from left to right, pyruvic acid is reduced by receiving two hydrogen atoms from NADH₂, yielding lactic acid and NAD.

FIGURE 2.11 The reaction catalyzed by lactate dehydrogenase (LDH) The enzyme cofactor nicotinamide adenine dinucleotide (NAD) acts as an electron (or hydrogen) shuttle by undergoing reversible reduction (forming NADH₂) and oxidation (forming NAD). As (B) shows, when the reaction catalyzed by LDH proceeds from left to right, NADH₂ produced elsewhere is converted to NAD, renewing the supply of NAD. The reaction catalyzed by LDH is reversible, but the NAD reaction involved in the reverse direction is not shown. Chapter 8 discusses the important role of the LDH-catalyzed reaction in ATP production.

symbolize as NADH₂. NAD is an enzyme cofactor (nicotinamide adenine dinucleotide) found in all animal cells; and NADH₂ symbolizes the reduced form of this cofactor, the form that is combined with hydrogen. The **substrates** of an enzyme are the initial reactants of the reaction that the enzyme catalyzes; the **products** of the enzyme are the compounds produced by the reaction. Thus in the reaction we are discussing—the left-to-right reaction in Figure 2.11B—the substrates of LDH are pyruvic acid and NADH₂, and the products are lactic acid and NAD. Chapter 8 discusses how this reaction aids not only rapid jumping by frogs, but also other forms of sudden, intense vertebrate exercise, such as sprinting by people. Put simply, the way the reaction helps is precisely that it produces NAD, an essential compound for ATP synthesis by glycolysis.

There are many kinds of enzymes. Mammalian cells, for instance, collectively synthesize several thousand kinds. Usually, the names of enzymes end in *-ase*. Thus when you see a biochemical term that ends in *-ase*, it usually refers to an enzyme. Later we will see that a single enzyme may exist in multiple molecular forms in different tissues or different animal species. The name of an enzyme typically refers to the *reaction catalyzed*. Lactate dehydrogenase, for example, is defined to be an enzyme that catalyzes the reaction in Figure 2.11B. All molecular forms that catalyze this reaction are considered to be forms of lactate dehydrogenase, even though they vary in their exact molecular structures and detailed functional properties.

Enzyme-catalyzed reactions exhibit hyperbolic or sigmoid kinetics

For an enzyme molecule to catalyze a reaction, it must first combine with a molecule of substrate to form an **enzyme–substrate com-**

plex. (Here, for simplicity, we assume there is only one substrate.) This complexing of enzyme and substrate, which usually is stabilized by *noncovalent* bonds, is *essential* for catalysis because the enzyme can alter the readiness of the substrate to react only if the two are bonded together. Substrate is converted to product *while united with the enzyme*, forming an **enzyme–product complex**, also usually held together by *noncovalent* bonds. The enzyme–product complex then dissociates to yield free product and free enzyme. Symbolically, if E, S, and P represent molecules of enzyme, substrate, and product, the major steps in enzyme catalysis are:

$$E + S \rightleftharpoons E\text{–}S \text{ complex} \rightleftharpoons E\text{–}P \text{ complex} \rightleftharpoons E + P \qquad (2.1)$$

Note that, as stressed earlier, the enzyme emerges unaltered.

An enzyme-catalyzed reaction occurs at a rate that is affected by the relationship between the available number of enzyme molecules and the concentration of substrate. The **reaction velocity** (*reaction rate*) is the amount of substrate that is converted to product per unit of time. At relatively low substrate concentrations, the reaction velocity increases as the substrate concentration increases. However, this process does not go on indefinitely: As the substrate concentration is raised, the reaction velocity eventually reaches a maximum. The reason for this overall behavior is precisely that substrate must combine with enzyme molecules to form product. As shown in **FIGURE 2.12A**, when the substrate concentration is low (as at ❶), all of the available enzyme molecules are not occupied by substrate at any given time and the amount of substrate available is therefore the limiting factor in determining the reaction velocity. Raising the substrate concentration (as from ❶ to ❷) increases the reaction velocity by using more of the available enzyme molecules. At high substrate concentrations (as at ❸), however, the amount of enzyme is the limiting factor in determining the reaction velocity. When the substrate concentration is high, the population of available enzyme molecules becomes **saturated**, meaning that each enzyme molecule is occupied by a substrate molecule nearly all of the time. Increasing the substrate concentration, therefore, cannot increase the reaction velocity further.

Because of the principals just discussed, enzyme-catalyzed reactions are one of the types of reactions that exhibit *saturation kinetics*. **Kinetics** refers to the velocity properties of reactions. A reaction exhibits **saturation kinetics** if it is limited to a maximum velocity because there is a limited supply of a molecule (the enzyme in the case of enzyme-catalyzed reactions) with which other molecules must reversibly combine for the reaction to take place.

Two types of saturation kinetics are exhibited by various enzyme-catalyzed reactions. One is **hyperbolic kinetics** (*Michaelis–Menten kinetics*), illustrated by the reaction we have been discussing in Figure 2.12A. The second is **sigmoid kinetics**, seen in **FIGURE 2.12B**. Whether the kinetics are hyperbolic or sigmoid depends in major part on the chemical properties of the enzyme. Hyperbolic kinetics occur when each enzyme molecule has just one substrate-binding site for the particular substrate of interest, or alternatively, such kinetics can occur when there are multiple sites but the sites behave independently. Sigmoid kinetics occur when each enzyme molecule has multiple substrate-binding sites and the multiple sites influence each other by way of ripple effects within the enzyme molecule (discussed later) so that catalytic activity at any one site depends on whether binding has occurred at other sites.

(A) Hyperbolic kinetics

The reaction velocity increases from **1** to **2** because the increased availability of substrate allows a greater fraction of enzyme molecules to engage in catalysis at any given time.

At **3**, however, the reaction velocity cannot increase further, because substrate is so abundant that all the enzyme molecules are engaged to the fullest extent possible.

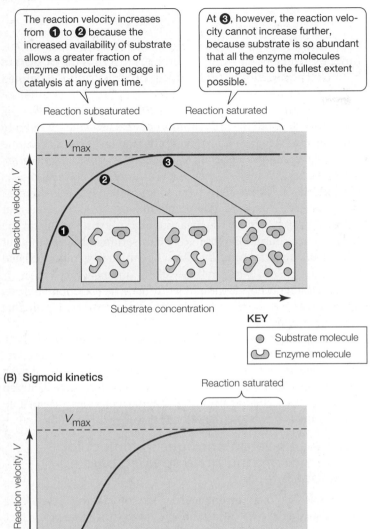

Reaction subsaturated

Reaction saturated

KEY

- ○ Substrate molecule
- ◡ Enzyme molecule

(B) Sigmoid kinetics

Reaction saturated

FIGURE 2.12 Reaction velocity as a function of substrate concentration (A) Some enzymes exhibit hyperbolic kinetics, in which the reaction velocity increases as shown, asymptotically approaching a maximum velocity, called V_{max}. (B) Some enzymes exhibit sigmoid kinetics, in which the approach to V_{max} follows an S-shaped (sigmoid) trajectory.

The catalytic effectiveness of an enzyme molecule is expressed as its **turnover number (k_{cat})**, the number of substrate molecules converted to product per second by each enzyme molecule when saturated. Different enzymes vary enormously in turnover number. Indeed, even the molecular variants of a single enzyme can vary substantially in this crucial property. Some enzymes are so catalytically effective that when they are saturated, each enzyme molecule converts 10,000 substrate molecules to product each second, whereas others convert only 1 substrate molecule to product per enzyme molecule per second.

The catalytic effectiveness of an enzyme depends partly on the *activation energy of the enzyme-catalyzed reaction*. To understand the implications of activation energy, it is necessary to recognize that a substrate molecule must pass through an intermediate chemical state termed a **transition state** to form a product molecule. Thus one can think of any reaction, whether or not it is enzyme catalyzed, as involving first the conversion of the substrate to a transition state, and second the conversion of the transition state to the product. For a substrate molecule to enter the transition state, its content of energy must increase. The amount by which it must increase is the **activation energy** of the reaction. Molecules gain the energy they need by random collisions with other molecules. Any particular substrate molecule has a continuously fluctuating energy content as it gains and loses energy through intermolecular collisions; as its energy content rises and falls, it undergoes reaction when its energy content is boosted by an amount at least equal to the activation energy. An enzyme accelerates a reaction by lowering the activation energy (**FIGURE 2.13**). The *extent* to which it lowers the activation energy is one factor that determines the enzyme's catalytic effectiveness.

A mathematical description of hyperbolic kinetics was first provided by Leonor Michaelis and Maude Menten in 1913. Their equation, after being revised by other chemists about a decade later, is called the **Michaelis–Menten equation**:

$$V = \frac{V_{max}[S]}{[S] + K_m} \tag{2.2}$$

where V is the reaction velocity at any given substrate concentration [S], V_{max} is the maximum reaction velocity (assuming a certain fixed amount of enzyme to be present), and K_m is a constant that is usually termed the **Michaelis constant**.[3] This equation describes the curve plotted in Figure 2.12A.

Maximum reaction velocity is determined by the amount and catalytic effectiveness of an enzyme

Two properties determine the **maximum velocity (V_{max})** at which a saturated enzyme-catalyzed reaction converts substrate to product (see Figure 2.12). One is the number of active enzyme molecules present. The second is the catalytic effectiveness of each enzyme molecule.

[3] Square brackets signify concentration. Thus [S] is the concentration of compound S.

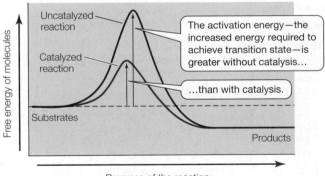

The activation energy—the increased energy required to achieve transition state—is greater without catalysis...

...than with catalysis.

FIGURE 2.13 Enzymes accelerate reactions by lowering the needed activation energy Starting at their average energy level, substrate molecules must gain sufficient energy to enter a transition state before they can react to form product. The amount of energy required, the activation energy, is lowered by enzyme catalysts. Catalysts do not, however, alter the average free energy of either substrates or products; nor do they affect the relative concentrations of substrates and products at equilibrium. Vertical arrows indicate the activation energy.

According to modern theories of how enzymes work, catalytic effectiveness also depends critically on the rates at which enzyme molecules can go through molecular conformational changes required for catalysis. As we discuss below, enzyme molecules change shape when they bind with substrate and again when they release product. There is reason to believe that different enzymes vary in the rates at which they can go through these necessary conformational changes, and differences in these rates may be as important as differences in activation energy in determining the relative turnover numbers of different enzymes.

Enzyme–substrate affinity affects reaction velocity at the substrate concentrations that are usual in cells

In a cell, a collision between an enzyme molecule and substrate molecule does not necessarily result in the formation of an enzyme–substrate complex. The two molecules may instead collide and "bounce apart" (i.e., separate). The outcome of a collision depends on a property of the enzyme called **enzyme–substrate affinity**, which refers to the proclivity of the enzyme to form a complex with the substrate when the enzyme and substrate meet. An enzyme that is highly likely to form complexes with substrate molecules it contacts has a *high* enzyme–substrate affinity. Conversely, an enzyme that is unlikely to form complexes has a *low* enzyme–substrate affinity.

The affinity of an enzyme for its substrate affects the shape of the velocity–concentration relation at *subsaturating* concentrations of substrate (concentrations too low to saturate the reaction), as illustrated in **FIGURE 2.14A** by three enzymes with hyperbolic kinetics. Curve *x* in the figure represents an enzyme having a high affinity for its substrate; curve *z* represents one having a low affinity. All three enzymes in the figure have the same maximum velocity. The key difference among them is that, at any given subsaturating substrate concentration, the reaction velocity more closely approaches V_{max} if the enzyme has high substrate affinity (*x*) rather than low substrate affinity (*z*).

A convenient numerical expression of enzyme–substrate affinity for reactions showing hyperbolic kinetics is the **apparent Michaelis constant** or **half-saturation constant**, K_m, defined to be the substrate concentration required to attain one-half of the maximum reaction velocity. **FIGURE 2.14B** shows how K_m is determined for both the high-affinity enzyme *x* and the low-affinity enzyme *z* from Figure 2.14A. Note that the low-affinity enzyme has the greater K_m value. Thus K_m *and enzyme–substrate affinity are related inversely*: A high K_m means low affinity, and a low K_m means high affinity. K_m is one of the parameters in the Michaelis–Menten equation (see Equation 2.2). For enzyme-catalyzed reactions that follow sigmoid kinetics, the measure of affinity is once again the substrate concentration required to half-saturate the enzyme, but it is calculated in technically different ways and symbolized like this: $(S_{0.5})_{substrate}$.

Substrate concentrations in cells are usually subsaturating. Thus the affinities of enzymes for substrates are important determinants of reaction velocities in cells.

In sum, therefore, reaction velocities in cells depend on all three of the enzyme properties we have discussed: (1) the number of active enzyme molecules present (which affects V_{max}), (2) the catalytic effectiveness of each enzyme molecule when saturated (which also affects V_{max}), and (3) the affinity of enzyme molecules for substrate (which affects how close the velocity is to V_{max}).

Enzymes undergo changes in molecular conformation and have specific binding sites that interact

Like any other protein, an enzyme depends on its three-dimensional molecular shape—its conformation—for its functional properties. One of the single most important attributes of enzymes and other proteins is that *their three-dimensional structure is stabilized mostly by weak, noncovalent bonds*, such as hydrogen bonds, van der Waals interactions, and hydrophobic interactions (see Box 2.1). Weak bonds create flexible links between molecular regions that allow an enzyme's three-dimensional structure to change its detailed

(A) Three enzymes with high, intermediate, and low affinity for substrate

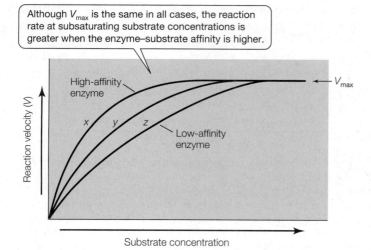

Although V_{max} is the same in all cases, the reaction rate at subsaturating substrate concentrations is greater when the enzyme–substrate affinity is higher.

High-affinity enzyme

V_{max}

x *y* *z*

Low-affinity enzyme

Reaction velocity (V)

Substrate concentration

FIGURE 2.14 The approach to saturation depends on enzyme–substrate affinity

(B) Determination of K_m for two of the enzymes from (A)

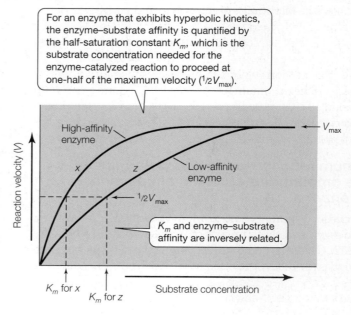

For an enzyme that exhibits hyperbolic kinetics, the enzyme–substrate affinity is quantified by the half-saturation constant K_m, which is the substrate concentration needed for the enzyme-catalyzed reaction to proceed at one-half of the maximum velocity ($^1/_2 V_{max}$).

High-affinity enzyme

V_{max}

x *z*

Low-affinity enzyme

$^1/_2 V_{max}$

K_m and enzyme–substrate affinity are inversely related.

Reaction velocity (V)

K_m for *x*

K_m for *z*

Substrate concentration

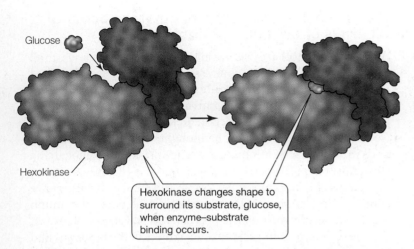

Glucose

Hexokinase

Hexokinase changes shape to surround its substrate, glucose, when enzyme–substrate binding occurs.

FIGURE 2.15 Molecular flexibility is important for enzyme function as exemplified here by hexokinase Hexokinase is the enzyme that ushers glucose into glycolysis by catalyzing glucose phosphorylation using a phosphate group from ATP (see Figure 2.19). The enzyme molecule is a single protein; the different shading of the two parts of the molecule is for visual clarity only.

shape while retaining its overall organization. Such shape changes, as already suggested, are crucial for proper enzyme function. For example, an enzyme changes shape when it binds with its substrate (**FIGURE 2.15**).

A substrate molecule binds with an enzyme molecule at a particular molecular region, at or near the surface of the enzyme, called the **active site** or **substrate-binding site**. The three-dimensional shape of the active site and the peculiarities of its chemical constituents complement a particular three-dimensional part of the substrate molecule (and its chemical constituents) such that the substrate molecule and the enzyme molecule match up and fit together. The binding of the substrate and enzyme molecules is typically stabilized entirely by weak bonds, not covalent bonds. If an enzyme requires two or more substrates, the enzyme molecule has an active site specific for each. Enzyme–substrate binding is sometimes said to resemble a lock and key fitting together, but this analogy is flawed in two important respects. First, the binding between the substrate and the corresponding active site on an enzyme is principally chemical and electrochemical in nature, not mechanical. Second, the lock-and-key analogy erroneously suggests mechanical rigidity. In fact, as we have seen, the active site and other regions of an enzyme molecule are flexible and change conformation when enzyme–substrate binding occurs (see Figure 2.15). They also change conformation when product is released.

Many enzyme molecules consist of two or more noncovalently bonded proteins, and these often interact in important ways to determine enzyme properties. Enzyme molecules composed of two, three, or four protein subunits are called *dimeric, trimeric,* or *tetrameric,* respectively. All the subunits in a multisubunit enzyme may be chemically identical, or they may consist of two or more types. Multisubunit enzymes typically have multiple binding sites. The simplest version of this property is that a multisubunit enzyme may have an active site on each subunit. However, multisubunit enzymes often also have specific binding sites for molecules other than the substrate. These nonsubstrate-binding sites have important similarities to the active (substrate-binding) sites: They are at or near

the surface of the enzyme molecule; they bind *noncovalently* and *reversibly* with specific molecules; and their specificity arises because they are complementary in three-dimensional shape and chemistry to parts of the molecules they bind. The substrates of enzymes and the molecules that bind to specific nonsubstrate-binding sites are collectively known as *enzyme ligands*. A **ligand** is any molecule that selectively binds by noncovalent bonds to a structurally and chemically complementary site on a specific protein; not just enzymes but also certain other sorts of proteins (e.g., transporters and receptors) are said to bind or combine with ligands, as we will see.

When an enzyme molecule has multiple binding sites, the binding of any one site to its ligand may facilitate or inhibit the binding of other sites to their ligands. Such interactions between the binding behaviors of different sites are termed **cooperativity**, whether they are facilitating or inhibiting. In **positive cooperativity**, ligand binding at one site facilitates binding of other sites on the same molecule to their ligands; in **negative cooperativity**, binding at one site inhibits binding at other sites on the same molecule. In addition to being classified as positive or negative, cooperativity is also categorized as *homotropic* or *heterotropic*. In **homotropic cooperativity** the binding of a particular type of ligand facilitates or inhibits the binding of other molecules of the *same ligand* to the same enzyme molecule; homotropic cooperativity occurs, for example, when the binding of a substrate molecule to one of the active sites on a multisubunit enzyme molecule facilitates or inhibits the binding of other substrate molecules to other active sites (this is the phenomenon that causes the kinetics to be sigmoid). In **heterotropic cooperativity** the binding of one type of ligand to an enzyme molecule influences the binding of other types of ligands.

An important point is that when cooperativity occurs, the interactions between binding sites on a molecule are interactions *at a distance.* The various binding sites on a multisubunit enzyme— whether they are sites for substrates or nonsubstrates—are usually not immediately next to each other. Instead, they are found at separate locations in the multisubunit molecular structure. Cooperativity occurs because the binding of a ligand to its particular binding site causes the detailed conformation of the enzyme molecule to change in a way that ripples through the whole molecule, affecting the shapes and binding characteristics of all its other binding sites. A type of cooperativity that has great importance in the *control* of multisubunit enzymes is **allosteric modulation** (*allosteric modification*). Allosteric modulation means modulation of the *catalytic* properties of an enzyme by the binding of *nonsubstrate ligands* to specific nonsubstrate-binding sites, which are called **regulatory sites** or **allosteric sites**. The nonsubstrate ligands that participate in this sort of modulation are called **allosteric modulators**. In **allosteric activation** the binding of an allosteric modulator to its binding site on an enzyme molecule increases the affinity of the molecule's active sites for the substrate or otherwise increases the catalytic activity of the enzyme. In **allosteric inhibition** the binding of an allosteric modulator impairs the catalytic activity of an enzyme, such as by decreasing its affinity for substrate. Allosteric modulation, as we will discuss, opens up vast regulatory possibilities.[4]

---

[4] Although the term *allosteric* was originally used only in the context of allosteric modulation, its meaning has evolved. Today *allosteric* is often used to refer to *any* form of enzyme conformational change that results from the noncovalent bonding of ligands to ligand-specific sites, not just allosteric modulation.

Enzymes catalyze reversible reactions in both directions

Like all other catalysts, enzymes accelerate reversible reactions in both directions. LDH, for example, can accelerate either the reduction of pyruvic acid (in Figure 2.11B, the reaction going from left to right) or the oxidation of lactic acid (in Figure 2.11B, the reaction going from right to left). Although all the reactions that take place within animals are reversible in principle, only some are reversible in practice. This is true because some reactions—for reasons unrelated to the enzymes that catalyze them—always proceed significantly in just one direction under the conditions that prevail in the body.

The direction of a reversible enzyme-catalyzed reaction is determined by the principles of **mass action**. Consider the following reversible reaction (where A, B, C, and D are compounds):

$$A + B \rightleftharpoons C + D \qquad (2.3)$$

If the four compounds A, B, C, and D are mixed and then left alone, they will react until they reach equilibrium. The reaction equilibrium is characterized by a particular *ratio of concentrations* of the four compounds. This ratio—[C][D]/[A][B]—always has the same value at equilibrium. The principles of mass action state that if compounds are out of equilibrium, the reaction will proceed in the direction of equilibrium as dictated by the ratios of concentrations. For example, if the reactants on the left, A and B, are collectively too concentrated relative to C and D for the equilibrium state to exist, the reaction will proceed to the right, thereby lowering the concentrations of A and B and raising those of C and D.

An enzyme does not alter the principles of mass action. The catalytic effect of an enzyme on a reversible reaction is to increase the rate of approach to equilibrium from either direction. To see an important aspect of this action, consider that the substrate or substrates are different from the two directions. For instance, when LDH catalyzes the reaction in Figure 2.11 going from left to right, its substrates are pyruvic acid and $NADH_2$; when it catalyzes the reaction going from right to left, its substrates are lactic acid and NAD. The enzyme–substrate affinity of an enzyme and its other kinetic properties are typically different for the substrates of the reaction going in the left-to-right direction than for the substrates of the reaction going in the right-to-left direction. Thus, although an enzyme always catalyzes a reversible reaction in both directions, its catalytic behavior in the two directions may be very different.

In cells reversible reactions are typically directional at any given time because reactions operate dynamically in a state that remains far from equilibrium. In a test tube, if A and B in Equation 2.3 are initially at high enough concentrations for the reaction to proceed to the right, the reaction itself will draw down the concentrations of A and B and create an equilibrium state. In a cell, however, the substrates of any one reaction are typically being produced by other reactions. Thus, in a cell, A and B are likely to be replaced as they are converted to C and D—meaning that their concentrations are not drawn down and a condition of disequilibrium in Equation 2.3 is maintained, driving the reaction steadily to the right (an example of a steady state). In this way, the enzyme-catalyzed reaction in a cell can display directionality, even though the enzyme itself catalyzes both directions of reaction.

Multiple molecular forms of enzymes occur at all levels of animal organization

A single enzyme often exists in multiple molecular forms, which all catalyze the same reaction, as stressed earlier. Dozens of described forms of lactate dehydrogenase are known, for example, in the animal kingdom. All the enzyme forms are called *lactate dehydrogenase* because they catalyze one reaction. In terms of primary structure (see Box 2.1), an enzyme can be thought of as a string of amino acids in which each amino acid occupies a specific position in the string; an enzyme composed of 300 amino acids has 300 positions, for example. Multiple molecular forms of an enzyme typically have similar string lengths and are identical in the particular amino acids that occupy most of the positions on the string. However, they differ in the amino acids at one or more of the positions, and these differences in their primary structures alter the details of their tertiary structures and function.

You might guess from what we have said that multiple molecular forms of enzymes are often related by evolutionary descent—that is, that certain forms evolved from others by mutations that caused changes in the amino acid sequence. As we will see, biochemists in fact know enough about the exact structures of many different LDH molecules to be almost certain that the various forms of LDH are related by evolutionary descent. It is probably a general rule that the multiple molecular forms of enzymes are families of evolutionarily related molecules.

LDH provides a good example for understanding multiple molecular forms of enzymes in greater detail. Among vertebrates, individuals have two or three different gene loci that code for LDH proteins. Thus two or three different forms of LDH protein are synthesized in any one individual; these are called LDH-A, LDH-B, and—if a third form is present—LDH-C. The various gene loci are not, however, expressed equally in all tissues. An additional complexity is that each "finished" LDH molecule is a tetramer, consisting of four LDH protein molecules that are independently synthesized but linked together as subunits of the mature enzyme. The A and B forms are produced in all or nearly all vertebrates. Usually, skeletal muscle cells express the A genetic locus strongly and the B locus weakly. Thus, although some of the LDH tetramers produced in skeletal muscle consist of mixed A and B subunits, the principal type of LDH tetramer in the skeletal muscles consists of all A subunits, symbolized LDH-A_4. In contrast, the cells of heart muscle express the B genetic locus strongly, and their principal type of LDH tetramer is composed entirely of B subunits: LDH-B_4. In mammals, LDH-C is expressed in just a single organ, the mature testis; mammalian sperm LDH is mostly LDH-C_4.

Both finer-scale and larger-scale variations occur in the forms of LDH. At a finer scale than we have already described, two or more alleles may exist at each genetic locus within a species. Thus, for instance, a species might have two alleles for the A locus, meaning that two different types of the A protein can be synthesized; the skeletal muscles of the species would then exhibit multiple molecular forms of the finished LDH-A_4 enzyme (i.e., LDH-A_4 in which all four A subunits are of the sort coded by one allele, LDH-A_4 in which all four A subunits are coded by the other allele, and LDH-A_4 in which some of the A subunits are coded by one allele and some by the other allele). On a larger scale, different species typically differ in the A, B, and C proteins. For example, although the A proteins

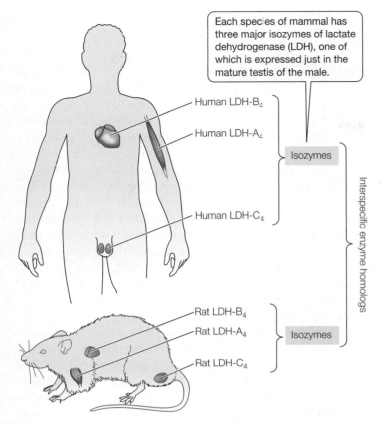

Each species of mammal has three major isozymes of lactate dehydrogenase (LDH), one of which is expressed just in the mature testis of the male.

Human LDH-B₄

Human LDH-A₄

Isozymes

Human LDH-C₄

Rat LDH-B₄

Rat LDH-A₄ — Isozymes

Rat LDH-C₄

Interspecific enzyme homologs

FIGURE 2.16 Isozymes and interspecific enzyme homologs The forms of LDH in rats are interspecific enzyme homologs of the forms in humans.

synthesized by laboratory rats and by humans are similar, they are not identical, so rat LDH-A₄ differs from human LDH-A₄.

Considering multiple molecular forms in general, researchers have developed a complex terminology to describe all the possibilities. For the purposes of an introduction to animal physiology, a simple dichotomy between *isozymes* (*isoenzymes*) and *interspecific enzyme homologs*, illustrated in **FIGURE 2.16**, is sufficient. **Isozymes** are the different molecular forms of an enzyme produced by a single species; an example is that the LDH-A₄, LDH-B₄, and LDH-C₄ produced in humans are three isozymes. **Interspecific enzyme homologs** are the different molecular forms of an enzyme coded by homologous gene loci in different species; an example is that human LDH-A₄ and rat LDH-A₄ are interspecific enzyme homologs. Functionally, isozymes and interspecific enzyme homologs often differ not only in their *catalytic* properties but also in their *regulatory* properties.

When functional differences exist between isozymes or interspecific enzyme homologs, they often seem to be adaptive differences; that is, they often seem to *assist* the proper functioning of the animal. For an example that pertains to isozymes, consider LDH-A₄ and LDH-B₄, the two isozymes of LDH usually found in the skeletal and heart muscles, respectively, of a vertebrate. Of these two isozymes, LDH-A₄ is much more effective in using pyruvic acid as a substrate. As noted earlier, the reduction of pyruvic acid to lactic acid is an essential part of the mechanism by which skeletal muscles can at times make more ATP than their O_2 supply permits. Whenever people sprint for sport, cheetahs sprint for food, or frogs sprint for prizes in the Calaveras County races, the ability of their skeletal muscles to make ATP without O_2 allows the muscles to work exceptionally vigorously. Because this ATP production depends on the reduction of pyruvic acid, the particular isozyme of LDH found

in the skeletal muscles—LDH-A₄, the isozyme that is superior in reducing pyruvic acid—plays a key role in the performance of intense exercise. The heart muscle, in contrast, seldom makes ATP without O_2, and its isozyme, LDH-B₄, is more suited to other functions than to the rapid reduction of pyruvic acid.

Interspecific enzyme homologs are often instrumental in the adaptation of species to different habitats. For example, two closely related species of thornyhead rockfish (*Sebastolobus*)—both found in the ocean along the West Coast of the United States—possess different genetically coded homologs of LDH-A₄ in their skeletal muscles. One of the species is found in waters shallower than 500 m; the other lives at depths of 500–1500 m. A problem with the forms of LDH-A₄ found in shallow-water species of marine fish, including the shallow-water rockfish, is that their affinity for substrate is highly sensitive to water pressure; these enzyme forms lose their affinity for substrate as pressure increases—so much so that at the high pressures of the deep oceans, the forms become ineffective as enzymes. The homolog of LDH-A₄ synthesized by the deep-water species of rockfish is distinctive. It is relatively insensitive to water pressure and retains a suitably high affinity for substrate even when the pressure is high, helping to account for the adaptation of the species to live in the deep ocean.

SUMMARY
Enzyme Fundamentals

■ Enzymes are protein catalysts that accelerate reactions by lowering the activation energy required for reactants to reach transition state. For most reactions to occur in cells, they must be catalyzed by enzymes. Thus a cell controls which reactions occur within it by the enzymes it synthesizes.

■ An enzyme must bind with its substrate to catalyze the reaction of substrate to form product. This binding, which is usually stabilized entirely by noncovalent bonds, occurs at a specific active site on the enzyme molecule, a site that is complementary in its three-dimensional chemical and electrochemical configuration to a portion of the substrate molecule. Enzyme molecules change shape when they bind to substrate or release product. These changes are permitted because the tertiary structure of a protein is stabilized by weak bonds.

■ Enzyme properties that determine the velocity of an enzyme-catalyzed reaction in a cell are: (1) the number of active enzyme molecules present in the cell, (2) the catalytic effectiveness of each enzyme molecule when saturated, and (3) the enzyme–substrate affinity. Enzyme-catalyzed reactions exhibit saturation kinetics because the reaction velocity is limited by the availability of enzyme molecules at high substrate concentrations. The maximum reaction velocity (V_{max}) that prevails at saturation depends on properties 1 and 2: the amount and catalytic effectiveness of the enzyme. Property 3, the enzyme–substrate affinity, determines how closely the reaction velocity approaches V_{max} when (as is typical in cells) substrate concentrations are subsaturating. The enzyme–substrate affinity is measured by the half-saturation constant (i.e., the Michaelis constant, K_m, for enzymes displaying hyperbolic kinetics).

(Continued)

- Multisubunit enzymes often exhibit cooperativity, a phenomenon in which the binding of certain binding sites to their ligands affects (positively or negatively) the binding of other binding sites to their ligands. An important type of cooperativity is allosteric modulation, in which a nonsubstrate ligand called an allosteric modulator affects the catalytic activity of an enzyme by binding noncovalently with a specific regulatory (allosteric) binding site. Both allosteric activation and allosteric inhibition are possible.

- Enzymes catalyze reversible reactions in both directions because their action is to accelerate the approach toward reaction equilibrium (determined by principles of mass action), regardless of the direction of approach.

- Multiple molecular forms of enzymes occur at all levels of biological organization. Isozymes are multiple molecular forms within a single species. Interspecific enzyme homologs are homologous forms of an enzyme in different species. Functional differences between isozymes and interspecific enzyme homologs often prove to be adaptive to different circumstances.

Regulation of Cell Function by Enzymes

The catalytic nature of enzymes often receives such exclusive attention that enzymes are viewed merely as molecules that speed things up. At least as important, however, is the role that cellular enzymes play as agents of *regulation* of cell function. The biochemical tasks in a cell are typically accomplished by sequences of enzyme-catalyzed reactions called **metabolic pathways**. Enzymes participate in the regulation of cell function in two principal ways. First, the types and amounts of enzymes synthesized by a cell determine which metabolic pathways are functional in the cell; any particular pathway is functional only if the cell synthesizes (through gene expression) all the enzymes the pathway requires. Second, the catalytic activities of the enzyme molecules that actually exist in a cell at any given time can be modulated as a way of controlling the rates at which the functional metabolic pathways operate.

The types and amounts of enzymes present depend on gene expression and enzyme degradation

Essentially all cells in an animal's body have the same genome, and the genome includes the genetic code for all enzymes that the animal can produce. Cells of different tissues differ, however, in their suites of enzymes. Moreover, any one cell typically differs from time to time in the types and amounts of enzymes it contains. A gene that codes for an enzyme is said to be **expressed** in a cell if the cell actually synthesizes the enzyme. The reason that cells of various tissues differ in their enzymes—and that one cell can differ from time to time—is that only some genes are expressed in each cell at any given time. Gene expression is not all-or-none. Thus, for enzymes that are being synthesized by a cell, the rate of synthesis can be varied by modulation of the degree of gene expression.

The amount of a particular enzyme in a cell depends not just on the rate of enzyme synthesis but also on the rate of degradation of the enzyme. All enzymes are broken down in specific and regulated ways by various pathways, of which the ubiquitin–proteasome

system discussed later in this chapter is best understood. Because of degradation, unless an enzyme is synthesized in an ongoing manner, the enzyme will disappear from a cell. The amount of an enzyme present in a cell depends, then, on an interplay of synthesis and degradation; the amount can be increased, for example, by either accelerated synthesis or decelerated degradation.

Variation in the rate of enzyme synthesis is the best-understood way that animal cells modify the amounts of their enzymes. The synthesis of an enzyme molecule requires several sequential steps: *transcription* of the stretch of DNA coding for the enzyme protein to form pre-messenger RNA, *posttranscriptional processing* to form mature mRNA, *exit* of the mature mRNA from the nucleus to associate with ribosomes in the cytoplasm, *translation* of the mature RNA into the amino acid sequence of the protein, and *posttranslational processing* that transforms the immature polypeptide chain into a mature protein. Each of these steps is potentially modulated by a cell to control the rate at which the enzyme is synthesized.

The first step, the transcription of DNA, for instance, is typically modulated by two types of specific regulatory regions of the DNA molecule that control whether, and how fast, transcription occurs in the relevant coding region of DNA. One type of regulatory region is the **promoter**, a DNA sequence located just upstream (toward the 5′ end) from the site where transcription starts. The second type of regulatory region consists of one or more **enhancers**, which are DNA sequences that may occur at various locations, even thousands of nucleotide bases away from the promoter. Proteins called **transcription factors** bind with the promoter and enhancer regions of DNA by way of DNA-matching subparts of their molecular structures, and this binding controls the extent to which RNA polymerase attaches to and transcribes the DNA-coding region responsible for a given enzyme. Transcription factors are highly specific and often work in sets, permitting different genes to be independently and finely controlled.

The processes that control the rates of synthesis of enzymes act on a variety of timescales to determine which metabolic pathways are functional in a cell. A useful distinction for discussing timescales is that between *constitutive* and *inducible* enzymes. **Constitutive enzymes** are present in a tissue in relatively high and steady amounts regardless of conditions. **Inducible enzymes**, however, are present at low levels (or not at all) in a tissue, unless their synthesis is activated by specific *inducing agents*.

The differentiation of tissues in an animal's body exemplifies the control of constitutive enzymes on a long timescale. Tissues become different in their sets of functional metabolic pathways during development, and they remain different throughout life, because of long-term (often epigenetic) controls on gene expression. For example, the bone marrow cells and skin cells of mammals differ in whether they express the genes that code for the enzymes required for hemoglobin synthesis. All the genes are relatively steadily expressed—and the enzymes are therefore constitutive—in marrow cells but not skin cells. Accordingly, the marrow cells have a functional metabolic pathway for hemoglobin synthesis at all times throughout life, whereas skin cells do not.

Inducible enzymes that undergo up- and downregulation on relatively short timescales are excellently illustrated by the **cytochrome P450 enzymes** found in the liver, kidneys, and gastrointestinal tract of vertebrates (and also found in most or all invertebrates). The P450 enzymes are a complex family of enzymes;

more than 30 kinds occur in humans, for example. Their function is to help detoxify foreign compounds by oxidizing them. The foreign compounds themselves often serve as inducing agents for the enzymes.

Low levels of P450 enzymes are found in an individual animal that has not been exposed in the immediately preceding weeks or months to suitable inducing agents, because little or no enzyme synthesis occurs in such an individual. However, even a single exposure to an inducing agent will strongly induce increased synthesis of certain P450 enzymes. A mammalian example is provided by barbiturate anesthetics. If a person or other mammal is administered an identical dose of barbiturate on two occasions that are a few days or weeks apart, the second administration will have much less effect than the first. The reason is that P450 enzymes that break down barbiturates are induced by the first administration, and the levels of the enzymes are therefore higher when the second dose is given. Another example that is particularly well understood is induction by halogenated aromatic hydrocarbons (HAHs)—a class of modern pollutants. When an animal is exposed to HAHs, the HAHs enter cells and activate an *intracellular receptor*, which acts as a *transcription factor*, causing expression of P450-coding genes. Levels of P450 enzymes then rise, poising the animal to better detoxify HAHs if a second exposure occurs. Another control system for inducible enzymes that is well understood is the system for response to hypoxia (low O_2 levels) described in Chapter 23 (see page 604).

In Chapter 1 (see pages 17–18) we discussed *acclimation* and *acclimatization*—the chronic (i.e., long-term) modifications of phenotype that individual animals commonly exhibit in response to changes in their environments. These important phenomena are often dependent at the cellular level on changes in the amounts of key enzymes.

Modulation of existing enzyme molecules permits fast regulation of cell function

Cells require speedier mechanisms of regulating their functions than are provided by even the fastest inducible enzyme systems. They achieve speedier regulation by modulating the catalytic activity of their *existing* enzyme molecules. A control mechanism that depends on changing the amounts of enzymes in cells usually requires many hours to be even minimally effective because increasing the synthesis or the degradation of enzymes cannot alter enzyme amounts more rapidly; if such mechanisms were the only controls, a cell would be like a car that could be accelerated or decelerated only once every several hours. Moment-by-moment acceleration and deceleration of cell functions, however, can often be achieved by changing the catalytic properties of already existing enzyme molecules.

Some enzymes are better positioned than others in metabolic pathways to serve the requirements of rapid metabolic regulation. One category of well-positioned enzymes consists of those that catalyze **rate-limiting reactions**. In a linear metabolic pathway, it is possible in principle for the rate of one of the reactions to set the rate of the entire pathway. Suppose, for example, that of all the reactions in the pathway in **FIGURE 2.17A**, the conversion from B to C is inherently the slowest. Because the rate of the *entire pathway* would then be limited by the rate of that reaction, the conversion of B to C would be the *rate-limiting reaction* of the pathway. The catalytic effectiveness of enzyme enz_2 would then be crucial.

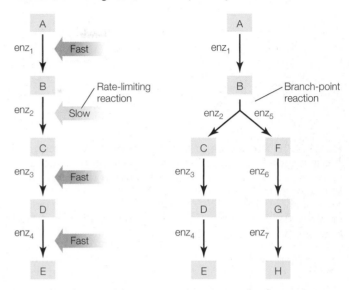

FIGURE 2.17 Enzymes that catalyze rate-limiting reactions and branch-point reactions are well positioned to exert control over metabolism The two reaction sequences in (A) and (B) are independent. A–H are reacting compounds; enz_1–enz_7 are enzymes. For example, B is the substrate of enz_2, and C is its product.

Enzymes that catalyze **branch-point reactions** in metabolic pathways are another category of enzymes that are well positioned to effect rapid metabolic regulation. A branching metabolic pathway permits two or more final products to be made from a single initial reactant. For instance, in **FIGURE 2.17B** either E or H can be made from A. The relative activities of the enzymes at the branch point—enz_2 and enz_5—determine which product is favored.

ALLOSTERIC MODULATION OF EXISTING ENZYMES Although binding sites for allosteric modulators do not occur in all enzymes, they are a common feature of enzymes that play regulatory roles. Allosteric modulation is a principal mechanism by which cell function is regulated. Recall that allosteric modulators are nonsubstrate molecules that bind noncovalently with specific sites, termed *allosteric sites* or *regulatory sites*, on enzyme molecules and that thereby affect the catalytic activities of the enzymes. The binding of an allosteric modulator with a regulatory site is *reversible* and follows the principles of *mass action*. To illustrate, suppose that enz_2 in Figure 2.17A, the rate-limiting enzyme in the reaction sequence, is allosterically modulated by a compound M. The reaction between M and the regulatory site on enz_2 would then be

$$M + enz_2 \rightleftharpoons M\text{–}enz_2 \text{ complex} \qquad (2.4)$$

Increasing the concentration of M shifts the reaction to the right by mass action, causing more enz_2 molecules to form M–enz_2 complexes (**FIGURE 2.18A**). Decreasing the concentration of M shifts the reaction to the left, causing fewer enz_2 molecules to be in complexes with the allosteric modulator (**FIGURE 2.18B**). These adjustments, being driven by mass action, occur almost instantly, and they almost instantly affect the catalytic activity of the enzyme. Thus allosteric modulation can occur *very rapidly*.

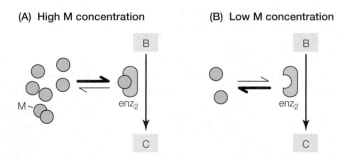

(A) High M concentration (B) Low M concentration

FIGURE 2.18 An allosteric modulator follows the principles of mass action in binding with the enzyme it modulates This figure shows how enz$_2$ in Figure 2.17A could (A) associate with an allosteric modulator, M, and (B) dissociate from the modulator. B and C are reacting compounds. Combination of enz$_2$ with M might promote or inhibit the action of the enzyme, depending on the exact nature of M, as exemplified in Figure 2.19.

As stressed previously, when an allosteric modulator binds with (or dissociates from) an enzyme, it alters the enzyme's ability to catalyze the conversion of *substrate* to *product*. This outcome occurs because the binding of the modulator to its *regulatory site* induces changes in the conformation of the enzyme molecule that ripple through the enzyme's molecular structure, affecting the *catalytically important properties* of the molecule, such as molecular flexibility or the conformation of the active site. An enzyme molecule that has its catalytic activity increased by a modulator is said to be **upregulated**; conversely, one that has its catalytic activity decreased is said to be **downregulated**.[5] A single enzyme molecule may have two or more regulatory sites, each specific for a different allosteric modulator. In this case, the individual modulators can exert reinforcing or canceling effects on the catalytic activity of the enzyme, offering elaborate regulatory possibilities.

When an allosterically modulated enzyme is the rate-limiting enzyme in a metabolic pathway, the *entire pathway* may be upregulated or downregulated by allosteric modulation. The downregulation of an entire pathway occurs, for example, during the phenomenon known as **feedback inhibition (end-product inhibition)**, a common process in which a product of a metabolic pathway decreases the catalytic activity of a rate-limiting enzyme earlier in the pathway. Feedback inhibition would occur in the pathway in Figure 2.17A, for example, if enz$_2$, the rate-limiting enzyme, were downregulated by allosteric combination with compound E, the final product of the pathway. In this case, an abundance of E in the cell would diminish the further formation of E by slowing the entire pathway. Conversely, if E were scarce, the rapid dissociation of E–enz$_2$ complexes by mass action would accelerate the reaction sequence that produces E. A metabolic pathway of this sort would act to stabilize levels of E in the cell by negative feedback (see Box 1.1).

<hr/>

[5] The concepts of *upregulation* and *downregulation* are used in additional contexts as well. Another application in the study of enzymes, for example, is to the amounts of inducible enzymes in cells. When the cellular concentration of an inducible enzyme is increased, the enzyme is said to be upregulated; when the enzyme is permitted to fall to low concentration, it is said to be downregulated. Common usage also refers today to the upregulation and downregulation of *processes* and *receptors*.

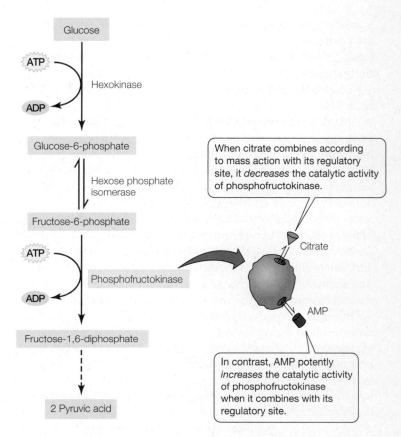

FIGURE 2.19 Phosphofructokinase, an allosterically modulated enzyme, is a key regulatory enzyme for glycolysis The first three reactions of the glycolytic metabolic pathway are shown. The enzymes catalyzing the three reactions are in red print. Citrate and adenosine monophosphate (AMP) are allosteric modulators of the mammalian phosphofructokinase known as PFK-2. The first and third reactions are not reversible under physiological conditions and thus are symbolized with single, rather than double, arrows. Although the reactions shown consume ATP, subsequent reactions in glycolysis produce more ATP than is consumed, so the catabolism of glucose brings about a net production of ATP. ADP = adenosine diphosphate; ATP = adenosine triphosphate.

The potential complexity of allosteric modulation is illustrated by the reactions at the start of glycolysis, the metabolic pathway that converts glucose into pyruvic acid (see Figure 8.1). As shown in **FIGURE 2.19**, the third reaction in glycolysis is catalyzed by *phosphofructokinase*, an enzyme of pivotal regulatory significance. The form of phosphofructokinase that occurs in most mammalian tissues (PFK-2) is allosterically modulated by more than six different substances, of which citrate and adenosine monophosphate (AMP) are particularly influential. Binding of citrate to a citrate-specific regulatory site on the phosphofructokinase molecule inhibits catalysis. This modulation by citrate is essentially a case of *feedback inhibition* because in a cell with plenty of O_2, the pyruvic acid produced by glycolysis forms citric acid in the tricarboxylic acid cycle; if the citrate concentration in a cell is high, allosteric downregulation of phosphofructokinase tends to restrain further entry of glucose into the glycolytic pathway that would produce more citrate. AMP potently upregulates phosphofructokinase. A high concentration of AMP in a cell signals that the cell has depleted its ATP (because AMP is formed from the use of ATP). Under such circumstances, the allosteric modulation of phosphofructokinase by AMP can increase the catalytic activity of the enzyme 100-fold, accelerating the use of glucose to make more ATP.

COVALENT MODULATION OF EXISTING ENZYMES   In addition to allosteric modulation, **covalent modulation** (also called *covalent modification*) is the second major way that the function of cells is regulated by changes in the catalytic activity of existing enzymes. Covalent modulation occurs by way of chemical reactions that make or break **covalent bonds** (strong bonds) between modulators and enzymes. Although allosteric modulators are chemically very diverse, just a few principal chemical entities are employed in covalent modulation. Of these, the most common is phosphate.

The most important processes of covalent modulation are **phosphorylation** and **dephosphorylation**—the covalent attachment and removal of orthophosphate groups (HPO_4^{2-}). In discussions of these processes, the orthophosphate groups are usually called simply "phosphate" groups and symbolized P_i or PO_4^{2-}. The phosphate groups are added to and removed from specific parts of modulated enzyme molecules, usually bonding with units of serine, threonine, or tyrosine in the protein structure. When a phosphate group forms a covalent bond with an enzyme that is covalently modulated, the enzyme's activity is modulated because the shape of the protein changes, leading to changes in the catalytically important properties of the molecule. Often phosphorylation and dephosphorylation act as a very rapid type of *on–off* switch. That is, for example, an enzyme molecule might be completely inactive ("turned off") when it lacks a phosphate group and become activated ("turned on") when it bonds with a phosphate group. The transition between the downregulated "off" form and the upregulated "on" form can occur almost instantaneously.

A crucial property of covalent modulation is that, unlike allosteric modulation, it *requires the action of enzymes* to catalyze the making and breaking of covalent bonds. The enzymes that catalyze phosphorylation belong to a large class called **protein kinases**, which are enzymes that covalently bond phosphate to proteins using ATP as the phosphate donor (see the inset of Figure 2.20). The enzymes that catalyze dephosphorylation are *protein phosphatases*, which break covalent bonds between proteins and phosphate, liberating phosphate in the simple form of inorganic phosphate ions. Here we emphasize the protein kinases.

A significant question with regard to covalent modulation that you may have already wondered about is this: If phosphate is nearly always the modulator in covalent modulation, how does a cell prevent the simultaneous modulation of all of its covalently modulated enzymes? A key part of the answer is that the protein kinases required for phosphorylation are *specific to the enzymes being modulated*. Hundreds of major types of protein kinases are known. Two different enzymes that are modulated by phosphorylation require the action of two different protein kinases to bind with phosphate, meaning that each can be controlled independently of the other. Some protein kinases phosphorylate proteins other than enzymes, as we will see later.

Protein kinases often act in *multiple-enzyme sequences* in carrying out their control functions. That is, one protein kinase often activates another protein kinase! Then the second protein kinase may activate a different sort of enzyme or possibly even a third protein kinase. The principal advantage of such sequences is **amplification** of the final effect. Amplification occurs because each molecule of an activated protein kinase can catalyze the activation of *many* molecules of the enzyme following it.

FIGURE 2.20 presents a simple example of an amplifying sequence of enzymes, consisting of two protein kinases and a final target enzyme that controls a critical metabolic process. The sequence is set in motion by an initial activating agent that activates a single molecule of the first protein kinase. That one protein kinase molecule then catalyzes the phosphorylation—and thus the activation—of four molecules of the second protein kinase. Each of the four activated molecules of the second protein kinase then catalyzes the phosphorylation—and activation—of four molecules of the final target enzyme. In total, therefore, 16 target-enzyme molecules are activated. The initial activating agents of such sequences are often signaling compounds that arrive at cells in amounts so minute they could not by themselves exert large effects. A multi-enzyme sequence like that in Figure 2.20 allows a tiny quantity of a signaling compound to have a 16-fold greater effect than otherwise would occur.

SUMMARY
Regulation of Cell Function by Enzymes

■ The metabolic pathways active in a cell depend on which enzymes are present in the cell, as determined by the processes of enzyme synthesis (dependent on gene expression) and enzyme degradation. The presence or absence of enzymes in a cell is regulated on long and short timescales. During individual development (an example of a long timescale), tissues acquire tissue-specific patterns of gene expression that establish tissue-specific suites of enzymes and metabolic pathways. Inducible enzymes, such as the cytochrome P450 enzymes, exemplify shorter-term regulation of the presence or absence of enzymes and metabolic pathways.

■ Very fast regulation of enzyme-catalyzed metabolic pathways is achieved by the modulation (upregulation or downregulation) of the catalytic activity of enzyme molecules already existing in a cell. Enzymes that catalyze rate-limiting or branch-point reactions are well positioned to mediate the rapid regulation of entire metabolic pathways in this way.

■ Allosteric modulation and covalent modulation are the two principal types of modulation of existing enzyme molecules. Allosteric modulation occurs by way of the noncovalent binding of allosteric modulators to regulatory sites, governed by the principles of mass action. Covalent modulation requires the enzyme-catalyzed making and breaking of covalent bonds—most commonly with phosphate. Phosphorylation is catalyzed by enzyme-specific protein kinases, which usually are the principal controlling agents in covalent modulation.

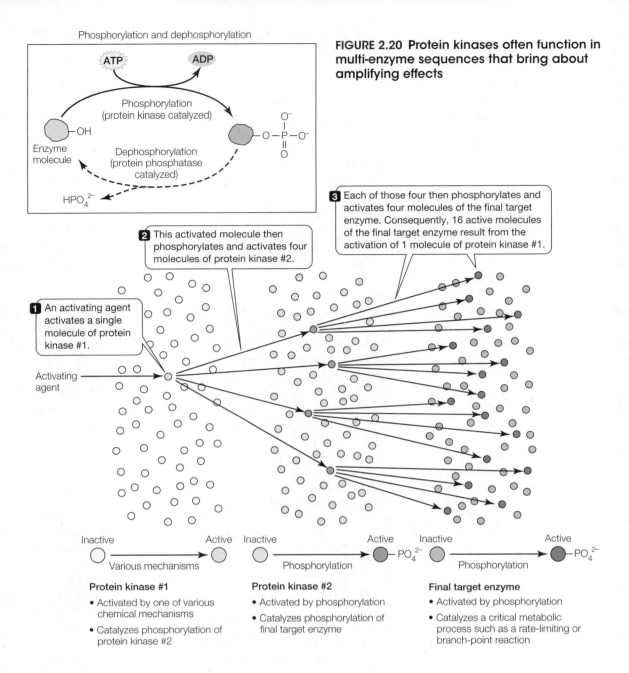

FIGURE 2.20 Protein kinases often function in multi-enzyme sequences that bring about amplifying effects

Phosphorylation and dephosphorylation

1 An activating agent activates a single molecule of protein kinase #1.

2 This activated molecule then phosphorylates and activates four molecules of protein kinase #2.

3 Each of those four then phosphorylates and activates four molecules of the final target enzyme. Consequently, 16 active molecules of the final target enzyme result from the activation of 1 molecule of protein kinase #1.

Activating agent

Inactive — Various mechanisms → Active

Inactive — Phosphorylation → Active $-PO_4^{2-}$

Inactive — Phosphorylation → Active $-PO_4^{2-}$

Protein kinase #1
- Activated by one of various chemical mechanisms
- Catalyzes phosphorylation of protein kinase #2

Protein kinase #2
- Activated by phosphorylation
- Catalyzes phosphorylation of final target enzyme

Final target enzyme
- Activated by phosphorylation
- Catalyzes a critical metabolic process such as a rate-limiting or branch-point reaction

Evolution of Enzymes

A great achievement of modern molecular biology is that protein evolution can now be studied at the biochemical level. Here we discuss studies in which proteins themselves are examined to elucidate protein evolution. Our focus will be on enzyme proteins. In Chapter 3 we will discuss studies of protein evolution that are based on examination of the genes that code for proteins.

In the study of enzyme protein evolution, investigators typically adopt one of two perspectives. One perspective is to focus on the *evolutionary relationships of the multiple enzyme forms that are found in sets of related species*. The goal of research on these relationships is to reconstruct the *family tree* of the enzyme forms, so as to clarify enzyme evolution over geological timescales. In research directed at this goal, scientists extract homologous enzymes from all the species of interest and determine the sequence of amino acids in each enzyme. They then employ the amino acid sequences to estimate the evolutionary relationships among the enzymes by drawing logical conclusions from similarities and differences in the sequences. **FIGURE 2.21A** illustrates this approach using a set of five simplified enzymes; note, for example, that it is logical to

conclude that the enzyme forms with red + green and red + blue mutations are descended from a form with only the red mutation because all the red mutations are identical.

FIGURE 2.21B presents a far more elaborate family tree of 24 vertebrate lactate dehydrogenases (LDHs). The evidence available indicates that there was just a single LDH gene when vertebrates first arose. All the modern, vertebrate LDH enzyme forms are coded by genes descended from that original gene and belong to a single family tree. According to amino acid sequence data, all the A forms of LDH in modern vertebrates are relatively closely related to each other (and thus all fall on one major branch—the upper branch—of the family tree in Figure 2.21B). Moreover, all the B forms are also relatively closely related to each other (and fall on a second major branch of the tree). This pattern indicates that the single original LDH gene duplicated (i.e., gave rise to two genes) early in vertebrate evolution at the point marked *. After that early duplication, each individual vertebrate animal had two LDH genes. The two diverged evolutionarily to give rise to two lineages (two evolutionary families) of genes: one coding for the A forms of LDH, and the other for the B forms. The analysis also

(A) Five molecular forms of an enzyme protein (at the top) and their arrangement (below) into a logical family tree based on similarities and differences in their amino acid sequences

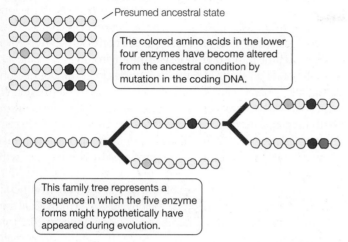

(B) Twenty-four vertebrate LDH proteins arranged into a family tree based on their amino acid sequences

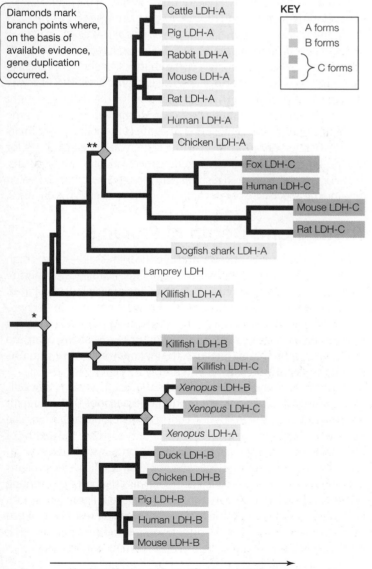

Number of amino acids modified by mutation

FIGURE 2.21 Enzymes and other proteins can be arranged into family trees based on their amino acid sequences
(A) Five simplified enzymes, each consisting of a string of eight amino acids, and a proposal for how they could logically be arranged into a family tree. (B) The most logically cogent arrangement of 24 vertebrate LDH proteins into a family tree. The tree was determined by a computer program that took into account the primary chemical structures of the 24 proteins (300+ amino acids per protein) but was provided with no information on the presumed relationships of the animals sampled. Each red horizontal line segment is proportional in length to the number of amino acids altered in that segment of the tree. Asterisks are referred to in the text. At the time this analysis was carried out, 24 LDH proteins had been sequenced, and all were included in this tree. (B after Stock et al. 1997.)

indicates that the A gene itself duplicated prior to the appearance of mammals at the point marked **. Following that duplication, one copy of the gene continued to code for the A form, whereas the other diverged to produce the C form. According to the amino acid sequence data, the C forms of LDH in fish and amphibians are only distantly related to the C forms in mammals (these were all named "C" forms long ago, before their relationships were known). In sum, the study of the 24 LDHs illustrates that important features of the family tree of enzymes can be elucidated by the study of the amino acid sequences of the enzyme forms in modern animals.

The second perspective that investigators adopt in studying enzyme protein evolution is to focus on the *evolution of allele frequencies within single species*. An important goal of this type of research is to study *evolution in action*. Changes in allele frequencies within species can be highly dynamic and occur on relatively short time-scales. A case can often be made, therefore, that when differences in allele frequencies are observed—from place to place, or time to time—within a living species, the differences reflect the *present-day action* of natural selection or other evolutionary mechanisms.

A famous example of this type of research focuses on the killifish *Fundulus heteroclitus*, a small fish (5–10 cm long) found commonly in estuaries along the Atlantic seaboard. The waters along the coast from Georgia to Maine represent one of the sharpest marine temperature gradients in the world; killifish living in Georgia experience body temperatures that, averaged over the year, are about 15°C higher than those of their relatives in Maine. In killifish, there are two major alleles of the B form of LDH (the form that occurs in the heart, red blood cells, liver, and red swimming muscles of fish). As shown in **FIGURE 2.22**, killifish in the coastal waters of Georgia have mostly the *a* allele (symbolized B^a), whereas those in Maine have only the *b* allele (B^b). Moreover, the *a* allele becomes *progressively* less frequent from Georgia to Maine.

Several sorts of studies indicate that *modern-day natural selection maintains this geographical gradient of allele frequencies* in killifish. Individual killifish, for instance, have been demonstrated to travel substantial distances. Because of these long travels, interbreeding would rapidly even out the frequencies of the *a* and *b* alleles along the entire Atlantic seaboard if simply left to its own devices. The fact that different allele frequencies persist from place to place indicates that fish with different alleles undergo differential survival and reproduction: Those with the *b* allele, for example, survive and reproduce better than those with the *a* allele in Maine. Evidently, in the study of killifish, we are witnessing natural selection right before our eyes because otherwise there would be no differences in allele frequencies!

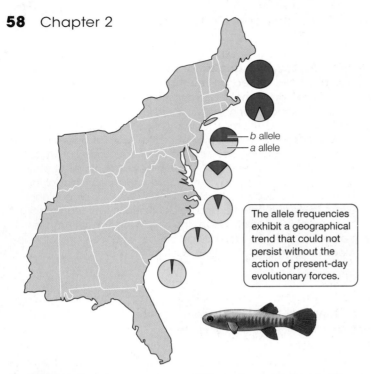

The allele frequencies exhibit a geographical trend that could not persist without the action of present-day evolutionary forces.

FIGURE 2.22 Contemporary evolution of an enzyme in the killifish *Fundulus heteroclitus* The pie diagrams show the frequencies of the *a* and *b* alleles of the gene for the B form of lactate dehydrogenase in fish of this one species collected at seven latitudes from Georgia to Maine. The green region in each diagram is the frequency of the *b* allele; the yellow region is the frequency of the *a* allele. (After Powers et al. 1993.)

To try to understand this natural selection, physiologists have explored how the two isozymes of the B protein, coded by the *a* and *b* alleles, differ in their functional properties as enzymes. They have found that the isozyme coded by the *b* allele has several functional advantages at low temperatures, and the one coded by the *a* allele has advantages at high temperatures. Thus mutation has given rise to two forms of this one enzyme protein, and both forms are retained because each is superior to the other in certain environments where the fish live.

Enzymes Are Instruments of Change in All Time Frames

Enzymes are primary instruments of physiological change in all five of the time frames identified in Chapter 1 (see Table 1.2). Three of the time frames, you will recall, refer to changes in animal physiology that are *responses to changes in the external environment*. The first of these three is *acute physiological responses by individuals*, the responses that occur rapidly after the environment changes. Allosteric modulation and covalent modulation of existing enzymes are major mechanisms of acute enzyme responses. For instance, if an animal is frightened by a predator and runs rapidly away, allosteric upregulation of phosphofructokinase by accumulation of adenosine monophosphate (AMP) in its muscle cells will immediately increase the rate at which glucose is processed to manufacture more ATP to sustain muscular work (see Figure 2.19).

The second major time frame of response to the environment, the *chronic (long-term) physiological responses of individuals*, depends on reconstructions of physiological systems requiring hours, days, or longer periods to complete. Environmentally induced changes in the expression of enzyme-coding genes constitute a major mechanism

of chronic responses. For an example, consider a fish acclimated to toxin-free water. If the fish encounters toxins, it will be unable to defend itself immediately using P450 enzymes, because the enzymes must be synthesized, a process requiring many hours or days. In the long term, however, the fish will assume a new phenotype—characterized by superior toxin defenses—because of induction of its P450 enzymes.

The third time frame of response to the environment, *evolutionary change*, depends on shifts of gene frequencies in entire populations over multiple generations. Genes that code for enzymes are frequently known to evolve by mutation, natural selection, and other mechanisms on both long and short scales of evolutionary time, as we have seen in Figures 2.21B and 2.22. In this way, populations of animals take on new catalytic and regulatory phenotypes by comparison with their ancestors.

In addition to the responses of animals to their environments, there are two time frames in which the physiology of individual animals is internally programmed to change, and enzymes are primary participants in these time frames as well. One time frame of internally programmed change consists of *developmental (ontogenetic) changes in an animal's physiology*, the changes that occur in a programmed way as an animal matures from conception to adulthood. The expression of particular enzymes is often programmed to start at particular stages of development, as we will discuss in Chapter 4 (see Figure 4.5).

Individual animals also undergo *periodic physiological changes— such as changes between day and night—under control of internal biological clocks*.[6] Enzymes often mediate these changes, as shown by the fact that—in the tissues of animals—the catalytic activities of many enzymes rise and fall in rhythms that parallel the daily day–night cycle even when the animals have no external information on the prevailing time. Some of these enzymes affect the abilities of animals to metabolize particular foodstuffs. Others affect capabilities for detoxifying foreign chemicals, including medications as well as toxins. Thus food metabolism and responses to foreign agents vary between day and night because of internally programmed enzyme changes.

The Life and Death of Proteins

In the lives of animals, physical and chemical stresses—such as high tissue temperatures or exposure to toxic chemicals—can denature enzymes and other proteins. When we say a protein is *denatured*, we mean that its three-dimensional conformation—its tertiary structure—is altered in a way that disrupts its ability to function (see Box 2.1). Usually, the primary structure—the string of amino acids—remains intact. Because the primary structure remains intact, the denatured state is potentially reversible.

One of the stunning discoveries of the last 25 years is that cells synthesize proteins termed **molecular chaperones** that can repair damage to other proteins by correcting reversible denaturation. The molecular chaperones use ATP-bond energy to guide the folding of other proteins. They are often active when proteins are first synthesized, and they are active in the repair of "old" but damaged proteins as emphasized here. Molecular chaperones assist repair by preventing protein molecules that are in unfolded states from aggregating with each other and by promoting folding patterns that restore damaged proteins to their correct three-dimensional structures. Because ATP is used, repair by molecular chaperones has a metabolic cost.

---

[6] Biological clocks are discussed at length in Chapter 15.

Heat-shock proteins are the most famous and best understood molecular chaperones. They are called *heat-shock proteins* because they were initially discovered in cells of organisms that had been exposed to stressfully high tissue temperatures. We realize now, however, that "heat-shock" proteins often function as molecular chaperones following many other types of cell stress, sometimes even including cold stress! An alternative name is **stress proteins**. The heat-shock proteins belong to several protein families of characteristic molecular weights (especially 70 and 90 kilodaltons, kDa) and display highly conserved amino acid sequences, indicating that they are evolutionarily related in most or all animals. Often they are named by combining the prefix *hsp* with their molecular weight; thus *hsp*70 and *hsp*90 refer to heat-shock proteins with molecular weights of 70 and 90 kDa. Although some are constitutive proteins, heat-shock proteins are principally inducible proteins: Most are absent except during times when a stress has elicited expression of the genes that encode them.

Rocky shores along seacoasts are known from recent research to be one of the ecological settings in which heat-shock proteins routinely play critical roles. Mussels, snails, and other attached or slow-moving animals living on the rocks can experience heat stress on clear, hot days when the tide goes out and they are exposed to the sun. During or soon after such events, these animals express heat-shock proteins.

Of course, repair is not always possible, or proteins once needed may become unnecessary, or regulatory processes may require that proteins be lowered in concentration. That is—speaking metaphorically—enzymes and other proteins in cells often die. Biochemists have discovered in the last 25 years that cells possess metabolic processes that specifically target enzymes and other proteins for destruction.

The most important known protein-degradation mechanism is the **ubiquitin–proteasome system**. There are three major players or sets of players in this complex system. One is a small protein called **ubiquitin**. Another is a multiprotein complex, termed a **proteasome**, which functions as an enzyme. The third is a suite of additional enzymes that catalyze steps in the process. These include *E1*, an enzyme that activates ubiquitin; *E2*, an enzyme that conjugates activated ubiquitin to a lysine unit within the protein that is destined to be broken down (sometimes aided by another enzyme, *E3*); and *cytoplasmic peptidases*.

A stunning (and almost scary) attribute of the ubiquitin–proteasome system is that it *tags* proteins prior to destroying them. Tagging occurs by the attachment of ubiquitin molecules to a protein molecule that is targeted to be degraded—a process termed **ubiquitination**. After the ubiquitin molecules are attached, the ultimate destruction of the targeted protein is inevitable: No reprieve or reversal is possible. Ubiquitination is the kiss of death.[7]

As shown in **FIGURE 2.23**, after a protein is tagged, it is recognized by a proteasome, which breaks the protein into peptides (short strings of amino acids). The cytosolic peptidases then break up the peptides into amino acids, which replenish a cell's amino acid supplies, or can be oxidized. Ubiquitin is released unaltered and can be reused.

---

[7] The one exception is that histone molecules in the chromosomes are routinely combined with ubiquitin and yet not degraded. Histones in the nucleus are the only proteins that are tagged with ubiquitin and live to tell about it (paraphrasing a lecture by Michael S. Brown).

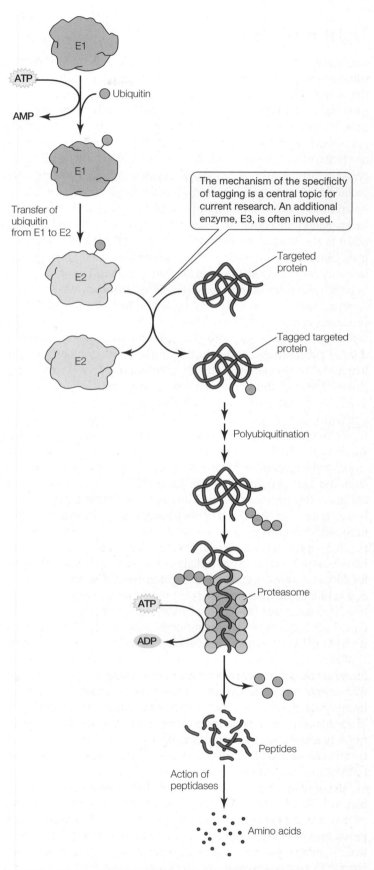

The mechanism of the specificity of tagging is a central topic for current research. An additional enzyme, E3, is often involved.

FIGURE 2.23 The ubiquitin–proteasome system tags proteins and then inevitably destroys those that are tagged   Note that ATP is used by both the initial reactions leading up to tagging and the proteasome reactions that break up a protein into peptides. Tagged proteins are repeatedly tagged, a process termed *polyubiquitination*.

Light and Color

The ability of cells to produce light biochemically—called **bioluminescence**—is widespread in animals. It is most common in the ocean, where bioluminescent species are found in most of the principal animal groups, including many invertebrate groups and fish—as well as bacteria and protists. On land, the more than 2000 species of beetles in the family Lampyridae—known as fireflies—are bioluminescent (see Figures 1.4 and 1.5), as are some other types of beetles, and some flies, millipedes, and earthworms. Few bioluminescent animals are known in freshwater, however. Based on analysis of chemical mechanisms, bioluminescence has evolved independently more than 40 times, indicating that it confers functional advantages. Animals employ bioluminescence to attract mates, lure prey, camouflage themselves where there is ambient light of similar intensity, frighten predators, and in other functions. The animal cells in which bioluminescence occurs are called **photocytes**.

Bioluminescence must be distinguished from *fluorescence*. Both processes can occur within a photocyte. In bioluminescence, light is produced de novo. In fluorescence, light is not produced de novo; instead, preexisting light is absorbed and re-emitted at longer wavelengths (although, as soon noted, actual events may not exactly follow this dictionary definition). *Green fluorescent protein* (GFP), discovered in a bioluminescent species of jellyfish (genus *Aequorea*), provides a prominent example of fluorescence. Photocytes in the jellyfish have a biochemical pathway that, in isolation, generates light at blue wavelengths. In the intact cells, however, the pathway is intimately associated with GFP, and in the final step of light production, energy from the light-producing pathway is transferred by resonance—a radiation-less process—to the GFP. The GFP emits light at green wavelengths, so the clusters of photocytes on the margin of the bell of the jellyfish glow green (**FIGURE 2.24**).

In a very general sense, light is often said to result from a *luciferin–luciferase reaction*. That is, light is said to be generated when a **luciferin**—a compound capable of light emission—is oxidized by the action of a **luciferase**—an enzyme that catalyzes luciferin oxidation. This statement is valid, but it hides the fact that there are many chemically different luciferins and luciferases, and thus an enormous diversity of light-producing reactions exists. The luciferases are particularly diverse. Typically, the luciferase catalyzes combination of the luciferin with O_2 to form a peroxide intermediate compound, which then spontaneously decomposes to generate a singlet electronically excited product, which in turn decays, emitting a photon of visible light. The color of the light depends on the particular luciferin–luciferase reaction and on fluorescent proteins if present. Marine animals most commonly emit at blue wavelengths—the wavelengths that travel farthest in clear seawater.

In some marine animals—notably the *Aequorea* jellyfish (see Figure 2.24)—luciferin, O_2, and an inactive form of the catalyzing protein are assembled into a complex called a **photoprotein**. Light production in these cases is initiated by exposure of the photoprotein to Ca^{2+} or Mg^{2+} (or another agent), which induces a conformational change that activates catalysis.

Bioluminescent animals probably, in most cases, synthesize their own luciferin and luciferase—and produce their own light. However, many departures from this straightforward scenario are known in marine animals. Some obtain their luciferin in their diet. A more common variant is that some animals depend on symbi-

FIGURE 2.24 Bioluminescence and fluorescence in *Aequorea victoria*, a hydromedusan jellyfish found along the West Coast of the United States   Clusters of light-producing cells are found on the margin of the bell. Although the light-producing mechanism generates blue light, the light emitted is green because of the presence of green fluorescent protein (GFP), which converts the emission wavelengths from blue to green. The light-producing photoprotein is Ca^{2+}-activated. Both the photoprotein and the GFP, discovered in this jellyfish, have revolutionized biology because of their widespread use as molecular probes.

otic bacteria for light production, rather than having endogenous photocytes. This phenomenon is best understood in the Hawaiian bobtail squid (*Euprymna scolopes*), in which each generation must acquire specific light-emitting bacteria (*Vibrio fischeri*) from the water in its ocean environment. As discussed in **BOX 2.2**, the squid and its bacteria provide probably the greatest insight available today on the mechanisms by which animal–microbial symbioses are established—as well as being a striking example of how an animal can achieve bioluminescence by the use of microbial light.

When biologists discuss *animal color*, they usually refer to colors that result from pigments in the skin (or other outer integument) and the wavelengths that those pigments absorb or reflect when illuminated by solar light. If an animal's skin is rich in a pigment that absorbs wavelengths other than green, for example, the animal looks green when viewed in solar light because only the green wavelengths are reflected into our eyes.

Speaking of animal color in this sense, a process of great physiological interest and ecological importance is *rapid color change* (*physiological color change*)—the ability of individuals to change color (or color pattern) in seconds, minutes, or at most a few hours. For example, in many species of frogs, flatfish, and crayfish, individuals darken rapidly when placed on a dark substrate, and lighten on a light substrate. In these animal groups, this color change depends on

BOX 2.2 Squid and Bioluminescent Bacteria, a Study in Cross-Phylum Coordination: The *Euprymna scolopes–Vibrio fischeri* Symbiosis   MARGARET MCFALL-NGAI

The Hawaiian bobtail squid (*Euprymna scolopes*) forms a lifelong symbiotic relationship with the Gram-negative bioluminescent bacterium *Vibrio fischeri*. The animal houses populations of the bacterium in a bi-lobed *light organ* in the center of its mantle (body) cavity (**Figure A**). This squid is nocturnally active and uses the light produced by the bacterial symbiont as an antipredatory mechanism. Specifically, bacterial light is emitted from the ventral surface of the squid at an intensity that matches the intensity of moonlight and starlight shining down through the water (a phenomenon termed *counterillumination*), so that the animal does not cast a shadow that can be perceived by a predator looking up from below. Each squid acquires its own bacteria from its environment early in life: A juvenile squid recruits *V. fischeri* cells from the seawater in which it develops within hours of hatching from its egg. Careful studies have revealed that this recruitment—the formation of the symbiosis—entails an intimate interaction between the squid and the bacteria (**Figure B**). A young squid presents specialized epithelia to its seawater environment to acquire the specific bacterial symbionts, which populate deep crypts within the squid's light organ. Once acquired, the symbionts initiate the lifelong loss of those very epithelia, making further acquisition impossible! For more on this fascinating story, see **Box Extension 2.2**.

> Bacteria in the light organ, in the center of the mantle cavity of the squid, emit bioluminescent light downward so that at night, the squid blends in with moon- or starlight when viewed from below.

FIGURE A The ecological function of the symbiosis for the squid (Photo courtesy of Margaret McFall-Ngai.)

> Soon after a squid hatches, *Vibrio fischeri* bacteria specifically attach to key epithelial surfaces that the bacteria, after they are acquired, promptly induce to be lost. Simultaneously, …

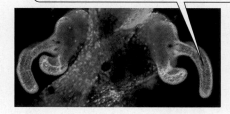

> … the bacteria enter ciliated pores to populate the squid's light organ.

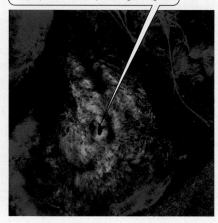

FIGURE B Acquisition of bacterial symbionts Both images were obtained by use of immunocytochemistry and confocal microscopy. (Images courtesy of J. Foster [upper] and E. Ziegelhoffer [lower].)

---

the function of **chromatophores**—flattened pigment-containing cells—in the skin or other integument. An individual may have several types of chromatophores that differ in their pigment colors. Thus chromatophores containing brown-black pigments, ones containing red pigments, and still others containing yellow or white pigments may be present. The pigment in a chromatophore is in the form of **pigment granules** (pigment-containing organelles), each about 0.3–1.0 micrometer (μm) in diameter. As a first approximation, each chromatophore *cell* has a *fixed size* in these animals. Color change is achieved by *dispersing* or *aggregating* the pigment granules within the cell. When the granules are dispersed throughout the cell, the cell as a whole takes on the color of the granules and imparts that color to the skin. When the granules within a cell are aggregated tightly together at a tiny spot in the center of the cell, the color of the granules may be essentially invisible and exert hardly any effect on the color of the skin.

The process of fully dispersing or fully aggregating pigment granules in a chromatophore takes as little as 2–8 s in some flatfish but as long as a few hours in some frogs. Several mechanisms—still being elucidated even in vertebrates, where they are best understood—are involved. One is that the pigment granules within a chromatophore move along microtubules (part of the cytoskeleton) that radiate out to the cell periphery from the cell center in complex geometries (**FIGURE 2.25A**). Movement of the granules toward the periphery disperses them, whereas movement toward the cell center aggregates them (**FIGURE 2.25B**). Movement is driven by ATP-using intracellular motor proteins such as kinesin and dynein. Chromatophores are signaled to disperse or aggregate their pigment granules by hormones, such as (1) melanocyte-stimulating hormone in amphibians and fish and (2) several well-defined peptide hormones (e.g., red-pigment-concentrating hormone) in crustaceans. Chromatophores in some fish are also directly innervated, poising them for relatively fast neuronally-stimulated responses.

The most rapid color change in the animal kingdom is displayed by squids, cuttlefish, and octopuses—the cephalopod molluscs. Their color change is based on an entirely different principle than that in amphibians, fish, and crustaceans. It occurs so rapidly in some species that an individual can switch from a fully dark to a fully light coloration in less than 1 s! Color change

(A) Microtubules

Microtubules are seen here as fibrous white material. Note that they radiate out from the cell center.

Pigment granules are tightly aggregated at the center in this cell but can travel outward along microtubules to disperse.

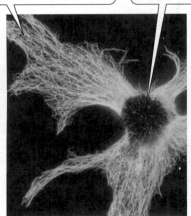

10 μm

(B) Aggregation and dispersal

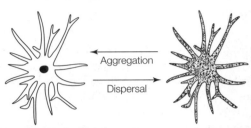

Aggregation

Dispersal

FIGURE 2.25 Pigment aggregation and dispersal within black-pigmented chromatophores (melanophores) from the skin of a codfish (*Gadus morhua*) (A) Photomicrograph of a cell treated so that the microtubules are visible. Pigment granules are transported along the microtubules during dispersal and aggregation. In this cell, the black pigment granules are aggregated at the cell center. (B) Diagram of a cell in aggregated and dispersed states. Note that the branched shape of the cell ensures that pigment granules will be widely spread out when in the dispersed state. (A, photo courtesy of Helén Nilsson Sköld; from Nilsson and Wallin 1998.)

in cephalopod molluscs is mediated by tiny *color-change organs* (**FIGURE 2.26A**).[8] Each of these organs consists of a pigment cell of *variable size* that is surrounded (in three dimensions) by dozens of radially arranged muscles that are innervated directly from the brain. Relaxation of the muscle cells allows the pigment cell to contract to minimal size (**FIGURE 2.26B**). Alternatively, contraction of the muscles—which can be very fast (as is typical of muscles)—expands the pigment cell so the pigment inside is spread out and easily visible (**FIGURE 2.26C**), imparting its color to the integument.

---

[8] These organs are often called chromatophores, although this usage of "chromatophores" is entirely different from the usage we have been discussing in the preceding paragraphs.

(A) A young squid

Pigment cell of a color-change organ

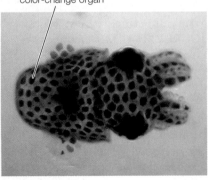

(B) Color-change organ with muscles relaxed

Pigment cell   Muscle

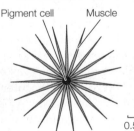

(C) Color-change organ with muscles contracted

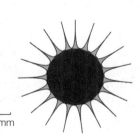

0.5 mm

FIGURE 2.26 Color-change organs in a squid (A) A juvenile Hawaiian bobtail squid, *Euprymna scolopes*. The entire squid is only 2 mm long, accounting for the large size of the color-change organs relative to the size of the animal. (B,C) Diagrams showing a color-change organ with muscles relaxed and contracted. (B) Relaxation of the muscles allows the pigment cell to contract to a small, barely visible size. (C) Contraction of the muscles expands the pigment cell so it has a prominent effect on integumentary color. (A, photo courtesy of Margaret McFall-Ngai; B,C after Bozler 1928.)

Reception and Use of Signals by Cells

Cells send signals to each other that serve to coordinate cell functions throughout the body. Nerve cells, for example, employ neurotransmitter molecules to signal muscle cells to contract, and endocrine cells secrete hormones that affect the functions of other cells. When a signal arrives at a target cell, the cell must have mechanisms of **signal reception** to detect the signal. It must also have mechanisms of **signal transduction**—meaning mechanisms by which it modifies its intracellular activities in appropriate ways in response to the extracellular signal. Here we address signal reception and transduction.

Extracellular signals initiate their effects by binding to receptor proteins

Extracellular signaling molecules such as neurotransmitters or hormones initiate their actions on a cell by binding with certain protein molecules of the cell, called **receptors**. A molecule that binds specifically and noncovalently to a receptor protein is considered a *ligand* of the receptor. Ligand binding occurs at a specific *receptor site* (or sites) on the receptor protein and results in a change in the molecular conformation of the receptor protein, a process that sets in motion a further response by the cell.

Receptors may be categorized into four functional classes: (1) ligand-gated channels, (2) G protein–coupled receptors, (3) enzyme/enzyme-linked receptors, and (4) intracellular receptors (**FIGURE 2.27**). Receptors in the first three categories reside in the cell membrane. This prevalence of receptors at the cell surface reflects the fact that *most signaling molecules cannot enter cells*. For the most part, signaling

(A) Ligand-gated channel

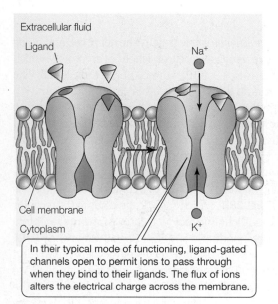

Extracellular fluid

Ligand

Na+

Cell membrane

Cytoplasm

K+

In their typical mode of functioning, ligand-gated channels open to permit ions to pass through when they bind to their ligands. The flux of ions alters the electrical charge across the membrane.

(B) G protein–coupled receptor and associated G protein system

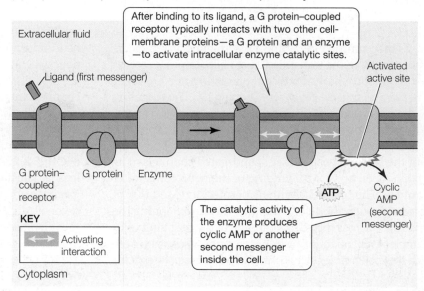

Extracellular fluid

After binding to its ligand, a G protein–coupled receptor typically interacts with two other cell-membrane proteins—a G protein and an enzyme—to activate intracellular enzyme catalytic sites.

Ligand (first messenger)

Activated active site

G protein–coupled receptor

G protein

Enzyme

KEY

⟷ Activating interaction

Cytoplasm

ATP

Cyclic AMP (second messenger)

The catalytic activity of the enzyme produces cyclic AMP or another second messenger inside the cell.

(C) Enzyme/enzyme-linked receptor

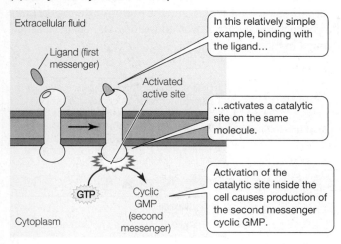

Extracellular fluid

Ligand (first messenger)

In this relatively simple example, binding with the ligand…

Activated active site

…activates a catalytic site on the same molecule.

GTP

Cyclic GMP (second messenger)

Cytoplasm

Activation of the catalytic site inside the cell causes production of the second messenger cyclic GMP.

(D) Intracellular receptor

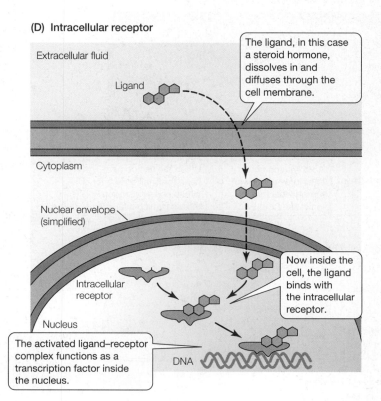

Extracellular fluid

Ligand

The ligand, in this case a steroid hormone, dissolves in and diffuses through the cell membrane.

Cytoplasm

Nuclear envelope (simplified)

Intracellular receptor

Now inside the cell, the ligand binds with the intracellular receptor.

Nucleus

DNA

The activated ligand–receptor complex functions as a transcription factor inside the nucleus.

FIGURE 2.27 The four types of receptor proteins involved in cell signaling (A) A ligand-gated channel. The particular example shown, an acetylcholine receptor of a muscle cell, must bind a ligand molecule at two sites for the channel to open. (B) A G protein–coupled receptor. Details of the molecular interactions symbolized by the yellow, double-headed arrows are discussed later in this chapter. (C) Enzyme/enzyme-linked receptors are themselves enzymes or, when activated, interact directly with other membrane proteins that are enzymes. Either way, binding with the ligand activates an enzyme catalytic site inside the cell. The example shown is the atrial natriuretic peptide receptor, which is particularly simple because it consists of just a single protein with both a ligand-binding site and a catalytic site. (D) Intracellular receptors are effective only for ligands that can dissolve in and diffuse through the lipid bilayer of the cell membrane. After a ligand enters the cell, it forms a complex with the receptor to initiate cellular responses. The example shown is a steroid hormone receptor, a type of receptor protein that is composed of a hormone-binding region and a region capable of binding with DNA. Binding with the hormone activates the receptor, and the activated hormone–receptor complex functions as a transcription factor. ATP = adenosine triphosphate; cyclic AMP = cyclic adenosine monophosphate; cyclic GMP = cyclic guanosine monophosphate; GTP = guanosine triphosphate.

molecules are proteins or other *hydrophilic* molecules that are unable to pass through the hydrophobic interior of the cell membrane. Instead of entering cells, these signaling molecules bind to receptors on the cell-membrane surface, and the receptors then initiate their intracellular effects. Only hydrophobic or very small signaling molecules can enter a cell at meaningful rates through the cell membrane; once inside, such molecules bind to intracellular receptors. Now let's discuss the properties of the four principal classes of receptors.

LIGAND-GATED CHANNELS A **ligand-gated channel** is a cell-membrane protein that acts as both a receptor and a channel.[9] This sort of channel opens to create a passageway for specific solutes, typically inorganic ions, through the cell membrane when the receptor site or sites on the protein bind to specific signaling

---

[9] Table 2.1 defines the functional categories of membrane proteins.

molecules, as diagrammed in Figure 2.27A. Ligand-gated channels function mostly in the transmission of nerve impulses across synapses, the narrow spaces between interacting nerve cells or between nerve and muscle cells (see Chapter 13). The signaling molecules that carry signals across synapses are called *neurotransmitters*. When a neurotransmitter is released by one cell into a synaptic gap, it diffuses across the gap to the receiving cell. The initial response of the receiving cell is that ligand-gated channels in its cell membrane open because of binding of the neurotransmitter to the channels. The opened channels permit increased flux of inorganic ions through the cell membrane, thereby changing the voltage difference across the cell membrane of the receiving cell.

An example of synaptic transmission is provided when a nerve cell stimulates a skeletal muscle cell to contract. In this case, the specific neurotransmitter *acetylcholine* is released by the nerve cell and binds noncovalently to the receptor sites of acetylcholine receptors—which are ligand-gated channels—on the surface of the muscle cell. The channels then open and allow sodium (Na^+) and potassium (K^+) ions to flow through the cell membrane of the muscle cell, initiating a change in voltage across the membrane and a series of subsequent changes culminating in muscle contraction. Fish-eating cone snails (**FIGURE 2.28A**), which we introduced at the opening of this chapter, incapacitate their prey in part by using toxins that block these ligand-gated channels. One of the most potent of a cone snail's *conotoxins* is α-conotoxin (**FIGURE 2.28B**), which specifically binds to the receptor sites on muscle cell acetylcholine receptors, preventing the receptors from binding with or responding to acetylcholine, as shown in **FIGURE 2.28C**. Because α-conotoxin binds to the receptors very rapidly and tightly, the swimming muscles of a fish attacked by a cone snail are promptly blocked from responding to nervous stimulation, and the fish becomes paralyzed. Thus the fish is condemned to death by the incapacitation of a crucial ligand-gated channel protein in its body. Certain Asian krait snakes and the poison-dart frogs of Latin America have independently evolved toxins that also incapacitate this channel in their prey.

(A) A fish-hunting cone snail in action

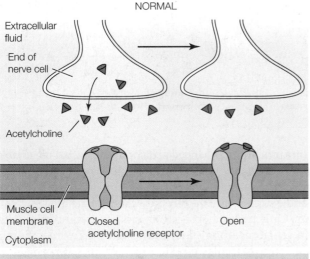

A fish-eating cone snail lures a fish with a proboscis that looks like food.

When the fish comes near, the snail spears it with a hollow harpoon through which it injects a potent mix of toxins.

(B) An example of an α-conotoxin

Different species of cone snails synthesize different forms of α-conotoxin, but all the conotoxins are very small molecules. This one, for example, consists of a string of only 13 amino acids (symbolized by purple ovals).

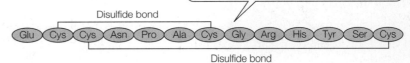

Disulfide bond

Glu — Cys — Cys — Asn — Pro — Ala — Cys — Gly — Arg — His — Tyr — Ser — Cys

Disulfide bond

(C) Block of receptor action by α-conotoxin

NORMAL

Extracellular fluid

End of nerve cell

Acetylcholine

Muscle cell membrane

Closed acetylcholine receptor

Open

Cytoplasm

In a normal fish, when a nerve cell releases acetylcholine, the muscle cell receptors bind the acetylcholine, causing the receptors, which are ligand-gated channels, to open, thereby stimulating the muscle cell to contract.

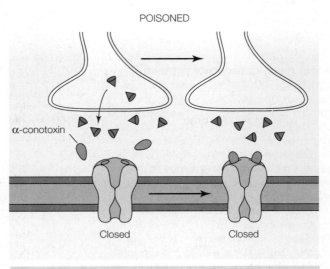

POISONED

α-conotoxin

Closed

Closed

In a poisoned fish, the muscle cell receptors are unable to bind the acetylcholine because the receptor sites are blocked by the α-conotoxin. The receptors thus fail to open in the normal way. Consequently, the muscle cell is not stimulated and does not contract—the fish is paralyzed.

FIGURE 2.28 The defeat of a vital molecule by a venom
(A) The speed of action in this cone snail's capture of a fish is imperative because the slow-moving snail could not pursue a fish that had even seconds to swim away. (To see an electron micrograph of the harpoon the snail uses, see Figure 6.11D.) (B) One of the most important toxins for the quick immobilization of the fish is α-conotoxin, a small polypeptide. (See Appendix I for three-letter codes for amino acids.)
(C) α-Conotoxin binds quickly and tightly to the receptor sites on the acetylcholine receptors of the fish's swimming muscles. Consequently, as shown in the "Poisoned" side of the diagram, the receptors become incapable of binding acetylcholine.

G PROTEIN–COUPLED RECEPTORS **G protein–coupled receptors** in cell membranes mediate cellular responses to many hormones and neurotransmitters. They also mediate many responses of sensory neurons. When a G protein–coupled receptor in the cell membrane of a cell is activated by binding its ligand, it activates a separate cell-membrane protein termed a **G protein**. The activated G protein may then directly exert an intracellular effect, or more commonly, it interacts with still another cell-membrane protein, usually an enzyme, and activates it so that a distinctive intracellular signaling compound is synthesized in the cytoplasm of the cell by the catalytic activity of the enzyme (see Figure 2.27B).

A major difference between reception based on G protein–coupled receptors and reception based on ligand-gated channels is that in general, *no sort of chemical passes through the cell membrane in the case of G protein–mediated reception*. In the most common type of such reception, one chemical brings the cell-signaling message to the extracellular side of the cell membrane, and a second, different, chemical is produced on the intracellular side to carry the signal onward to the interior of the cell. The molecules that bring signals to the cell membrane from the outside—such as hormones or neurotransmitters—are called **first messengers**, whereas the *intracellular* signaling molecules that carry the signals to the interior of the cell are called **second messengers**. The action of the G protein–mediated mechanism in the cell membrane is analogous to a relay race in which the first messenger brings the message to a certain point but then can go no farther and must activate a second messenger for the message to go on. An example is provided by the action of epinephrine (adrenaline) on a liver cell. Epinephrine, the first messenger, binds to a G protein–coupled receptor in the cell membrane, which initiates steps resulting in intracellular synthesis of the second messenger **3′-5′-cyclic adenosine monophosphate** (**cyclic AMP**, or **cAMP**). Cyclic AMP then activates the intracellular responses to the epinephrine signal. Shortly we will return to this and other second-messenger systems in greater detail.

ENZYME/ENZYME-LINKED RECEPTORS **Enzyme/enzyme-linked receptors** are cell-membrane proteins that either are enzymes themselves or that interact directly with enzyme proteins when activated. They are a more structurally and functionally diverse class of receptors than the two types of cell-membrane receptors we have already discussed. As in the case of G protein–mediated reception, molecules or ions do not pass through the cell membrane in this sort of reception, and enzyme/enzyme-linked receptors often activate the formation of second messengers. The simplest sort of enzyme/enzyme-linked receptor is a receptor protein that is itself an enzyme; such a protein is composed of an extracellular receptor region, a membrane-spanning region, and an intracellular catalytic region (see Figure 2.27C). Binding of the extracellular signaling molecule to the receptor site activates the catalytic site at the other end of the molecule. The hormone atrial natriuretic peptide (ANP) acts on target cells in the human kidney to increase Na^+ excretion by way of this sort of receptor. When ANP binds to the receptor region on the outside of a cell, the receptor molecule catalyzes the formation of a second messenger, **3′-5′-cyclic guanosine monophosphate** (**cyclic GMP**, or **cGMP**), inside the cell.

INTRACELLULAR RECEPTORS **Intracellular receptors** are the only class of receptors not localized at the cell surface. As noted earlier, most signaling molecules cannot enter cells. Those that do are typically relatively small, hydrophobic molecules that can dissolve in and diffuse through the core of the lipid bilayer of the cell membrane. These signaling molecules include steroid hormones, thyroid hormones, retinoic acid, vitamin D, and the gas nitric oxide (NO). The receptors for these substances are located intracellularly, in the cytoplasm or nucleus. The usual pattern for intracellular receptors is that, after they are activated by binding with their ligands, they interact with DNA (see Figure 2.27D) to activate specific primary-response genes, the products of which may secondarily activate other genes.

When the steroid hormone estrogen arrives at a cell, for example, it passes through the cell membrane and binds to an estrogen-specific intracellular receptor protein, forming a hormone–receptor complex. The complex is itself a transcription factor that activates specific promoter and enhancer regions of the nuclear DNA, causing the expression of specific genes. The resulting effects can alter much of a target cell's metabolism, often promoting female cellular phenotypes.

RECEPTORS OCCUR AS MULTIPLE MOLECULAR FORMS RELATED BY EVOLUTIONARY DESCENT In terms of molecular diversity, receptor proteins of any given type follow the same general principles as enzyme proteins: Each type of receptor exists in multiple molecular forms that typically are related by evolutionary descent. Most of the ligand-gated channels in modern-day animals, for example—although they vary in molecular details—have very similar chemical structures and are coded by a single lineage of genes that diversified over evolutionary time to give rise to the channels seen today. Similarly, all the G protein–coupled receptors belong to a single family tree, as do the intracellular steroid receptors.

Cell signal transduction often entails sequences of amplifying effects

When signaling molecules bind to cell-membrane receptors, *sequences of amplifying effects*—analogous to a chain reaction—are often involved between the moment that signal reception occurs and the moment that the final intracellular response occurs. For a classic example of this widespread pattern, let's look at the process by which epinephrine leads to the activation of glycogen breakdown to produce glucose in vertebrate liver cells, shown in **FIGURE 2.29**.

When a human or other vertebrate experiences stress, such as the stress that occurs in anticipation of physical conflict, the adrenal glands secrete epinephrine into the blood. The circulation carries the epinephrine to the liver, where the hormone bathes liver cells, which contain abundant supplies of glycogen, a glucose-storage compound. The epinephrine itself cannot cross the cell membranes of the liver cells. "News" of its arrival reaches the inside of each cell, instead, by way of a G protein–coupled receptor system.

The receptor system itself has important amplifying properties. To set the stage for discussing these, we need to note some details of G-protein function. Recall that a G protein–coupled receptor activates a cell-membrane G protein. G proteins get their name from the fact that they are modulated by binding with guanine nucleotides. A G protein bonded with **guanosine diphosphate** (**GDP**) is inactive. A G protein is activated when it is induced to change from being

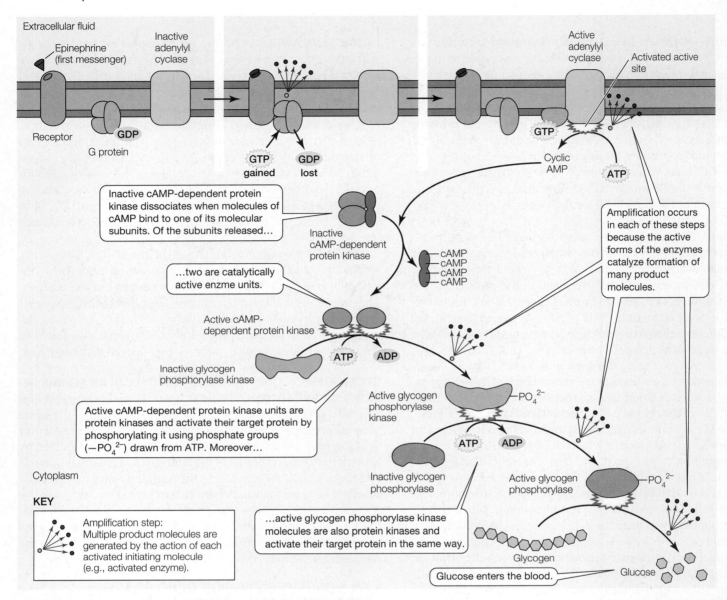

Extracellular fluid

Epinephrine (first messenger)

Inactive adenylyl cyclase

Active adenylyl cyclase

Activated active site

Receptor

GDP

G protein

GTP gained

GDP lost

Inactive cAMP-dependent protein kinase dissociates when molecules of cAMP bind to one of its molecular subunits. Of the subunits released...

Inactive cAMP-dependent protein kinase

...two are catalytically active enzme units.

Active cAMP-dependent protein kinase

GTP

Cyclic AMP

ATP

cAMP
cAMP
cAMP
cAMP

Amplification occurs in each of these steps because the active forms of the enzymes catalyze formation of many product molecules.

ATP   ADP

Inactive glycogen phosphorylase kinase

Active cAMP-dependent protein kinase units are protein kinases and activate their target protein by phosphorylating it using phosphate groups ($-PO_4^{2-}$) drawn from ATP. Moreover...

Active glycogen phosphorylase kinase $-PO_4^{2-}$

ATP   ADP

Inactive glycogen phosphorylase

Active glycogen phosphorylase $-PO_4^{2-}$

Cytoplasm

KEY

Amplification step: Multiple product molecules are generated by the action of each activated initiating molecule (e.g., activated enzyme).

...active glycogen phosphorylase kinase molecules are also protein kinases and activate their target protein in the same way.

Glycogen

Glucose enters the blood.

Glucose

FIGURE 2.29 In a liver cell, amplifying signal transduction of an extracellular epinephrine signal results in enzymatic release of glucose Because five steps in the epinephrine signal transduction pathway are amplifying, a very low epinephrine concentration can trigger a very large increase in glucose concentration. The cyclic AMP signal is ultimately terminated by the action of a cytoplasmic enzyme, phosphodiesterase. ATP = adenosine triphosphate; cyclic AMP = cAMP = cyclic adenosine monophosphate; GDP = guanosine diphosphate; GTP = guanosine triphosphate.

bonded with GDP to being bonded with **guanosine triphosphate (GTP)**. However, G proteins exhibit intrinsic GTP-destructive activity: When bonded with GTP, they tend to break down the GTP to GDP by hydrolysis. In this way, a G protein that has been activated by binding with GTP *tends to inactivate itself* by reverting to the inactive GDP-bonded form. The membrane G proteins, which are our focus here, are trimers in their inactive state. They dissociate into two parts when activated by GTP binding.

When epinephrine binds to its specific G protein–coupled receptor in the cell membrane of a liver cell, what first occurs is a series of amplifying reactions *within the cell membrane*, diagrammed across the top of Figure 2.29. The activated receptor first interacts with molecules of G protein in the membrane to activate them by promoting loss of GDP in exchange for GTP. The G protein–coupled receptor and the G protein are separate, however, and both diffuse freely and independently in the fluid mosaic of the cell membrane. Accordingly, as an activated receptor diffuses about in the membrane,

it must randomly bump into a G-protein molecule to activate it, a situation that sounds inefficient until one realizes that it *makes amplification possible*. During its active life, a *single* activated receptor can bump into and activate *many* (perhaps 100) G-protein molecules. Each activated G-protein molecule then remains active for a period of time, the duration of which depends on how long it takes to inactivate itself (tens of seconds to several minutes), and while it is active, it can activate a cell-membrane enzyme, **adenylyl cyclase** (also called *adenylate cyclase*), which it bumps into by diffusion in the membrane; probably each activated G-protein molecule activates just one adenylyl cyclase molecule because the activation requires steady linkage of the two proteins. Adenylyl cyclase has an active site on the cytoplasmic side of the cell membrane, and when it is activated, it catalyzes the formation of the second messenger cyclic AMP (cAMP) from ATP inside the cell. Further amplification occurs at this step because a *single* activated molecule of adenylyl cyclase can catalyze the formation of *many* molecules of cAMP during its active life.

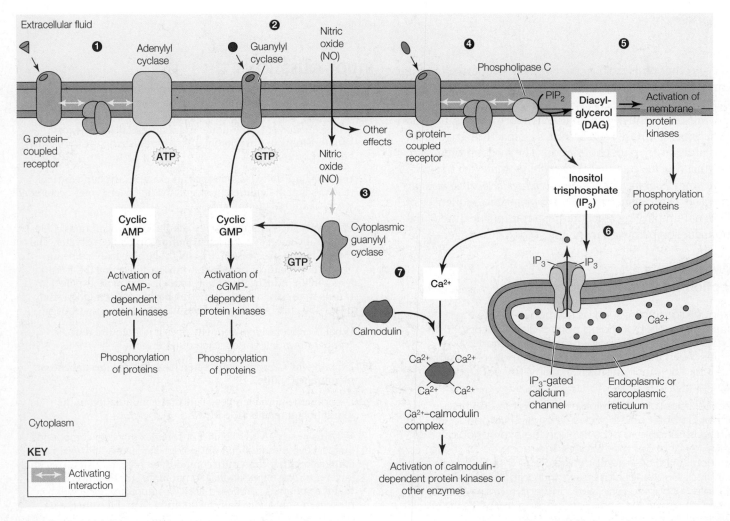

FIGURE 2.30 Second messengers in overview The production and the actions of five important second messengers are shown: cyclic AMP, cyclic GMP, diacylglycerol, inositol trisphosphate, and calcium ion. This figure includes only some of the major cell signal transduction pathways that employ second messengers. ❶ Some G protein–coupled receptor systems employ cyclic AMP as a second messenger, as seen previously in Figure 2.29. ❷ Receptor systems based on guanylyl cyclase enzymes employ cyclic GMP as a second messenger. When activated, a guanylyl cyclase produces cyclic GMP from guanosine triphosphate (GTP). In some cases, as in the atrial natriuretic peptide receptor system (see Figure 2.27C), the guanylyl cyclase is a cell-membrane enzyme. ❸ Some guanylyl cyclase enzymes are cytoplasmic. Many of the actions of nitric oxide (NO) are mediated by cyclic GMP produced by the activation of a cytoplasmic guanylyl cyclase. ❹ Some G protein–coupled receptor systems employ diacylglycerol and inositol trisphosphate as second messengers. When such receptor systems are activated, the two second messengers are synthesized simultaneously from a cell-membrane phospholipid, PIP_2 (phosphatidylinositol 4,5-bisphosphate), by the catalytic action of activated phospholipase C, a membrane-associated enzyme. ❺ Diacylglycerol stays in the cell membrane because it is hydrophobic. ❻ Inositol trisphosphate, which is hydrophilic, enters the cytoplasm, where its major action is to open ligand-gated channels that release Ca^{2+} from intracellular stores such as the endoplasmic reticulum. ❼ Ca^{2+} ions sometimes act as second messengers, as when Ca^{2+} released by action of inositol trisphosphate activates the cytoplasmic protein calmodulin, which then can activate protein kinases or other enzymes.

The cAMP signal inside a liver cell triggers the activation of a series of intracellular enzymes (see Figure 2.29). *Multiple amplifying steps occur in this series.* The series starts with two protein kinases and is a classic example of the type of amplification sequence shown in Figure 2.20. The cAMP second messenger activates a protein kinase named *cAMP-dependent protein kinase* (*cAPK*) by causing it to dissociate, forming two active enzyme units. The activated cAPK units phosphorylate, and thereby activate, a second protein kinase called *glycogen phosphorylase kinase* (*GPK*). Finally, the active GPK molecules phosphorylate and activate the ultimate target enzyme, *glycogen phosphorylase* (*GP*). Great numbers of activated GP molecules are produced. Each of them catalyzes the removal of glucose units from the glycogen polymers stored in the cell, and the glucose is then released into the blood for distribution throughout the body.

Because of the cumulative effect of all the amplifications that occur in this cell signal transduction pathway, a *minute quantity* of epinephrine can cause a *flood* of blood glucose. A cumulative amplification of about 10,000-fold can occur between the binding of an epinephrine molecule to a G protein–coupled receptor molecule and the formation of cAMP. Accordingly, a blood epinephrine concentration of 10^{-10} molar (M) can result in an intracellular concentration of cAMP of 10^{-6} M. Then the protein-kinase cascade within the cell can result in a further amplification of about 1000, so that the concentration of activated GP—which directly catalyzes release of glucose —is 10 million times the blood concentration of epinephrine that initiated the signal transduction process!

Several second-messenger systems participate in cell signal transduction

Several compounds—not just cyclic AMP—act as second messengers. The common second messengers, in addition to cyclic AMP, are cyclic GMP (cGMP); inositol 1,4,5-trisphosphate (IP_3); 1,2-diacylglycerol (DAG); and Ca^{2+} ions. **FIGURE 2.30** provides an overview of some prominent second-messenger systems in which these compounds participate. For the most part, all of the second messengers share with cyclic AMP the property that their immediate intracellular effect is to activate a protein kinase that is already present in the cell in an inactive form, as Figure 2.30 shows. The protein kinase then activates or inactivates its target protein or proteins by phosphorylation. The target proteins are often enzymes, but *they may be cell-membrane channels or receptors, channels in intracellular membranes, transcription factors that regulate gene expression, or virtually any other sort of protein*. Sequences of multiple signal-amplifying reactions are a common feature of the signal transduction pathways involving second messengers.

SUMMARY
Reception and Use of Signals by Cells

■ Extracellular signals such as hormones initiate their actions on cells by binding noncovalently with specific receptor proteins. Receptor proteins activated by binding with their signal ligands set in motion cell signal transduction mechanisms that ultimately cause cell function to be altered.

■ Most extracellular signaling molecules are chemically unable to enter cells because they are hydrophilic, or otherwise unable to pass through the hydrophobic, lipid interior of cell membranes. The receptors for these molecules are cell-membrane proteins that fall into three principal functional classes: ligand-gated channels, G protein–coupled receptors, and enzyme/enzyme-linked receptors. Extracellular signaling molecules that readily pass through cell membranes, such as steroid hormones, thyroid hormones, and nitric oxide (NO), bind with receptors that belong to a fourth functional class: intracellular receptors.

■ Activation of ligand-gated channels by their ligands most commonly results in changed fluxes of inorganic ions, such as Na^+ and K^+, across cell membranes, thereby altering voltage differences across the membranes. The altered voltage differences may then trigger other effects.

■ Activation of G protein–coupled receptors and enzyme/enzyme-linked receptors by their extracellular signaling ligands typically initiates the formation of second messengers, such as cyclic AMP or cyclic GMP, on the inside of the cell membrane. The second messengers, in turn, often trigger sequences of additional intracellular effects in which preexisting enzymes are modulated, most notably protein kinases. A function of these sequences is dramatic amplification of the ultimate effect.

■ Intracellular receptors, when activated by their ligands, usually bind with nuclear DNA and directly activate specific primary-response genes.

STUDY QUESTIONS

1. It is becoming possible for molecular biologists to synthesize almost any protein desired. Suppose you use a phylogenetic tree of modern-day enzymes (e.g., Figure 2.21B) to predict the amino acid sequence of a now-nonexistent ancestral enzyme form. What insights might you obtain by synthesizing the ancestral enzyme protein?

2. Using lactate dehydrogenase as an example, explain why it is true to say that "multiple molecular forms of enzymes occur at all levels of animal organization."

3. Pollutants such as halogenated aromatic hydrocarbons (HAHs) are usually spotty in their distributions in bodies of water. Thus even if HAHs are present, fish might be able to avoid being exposed to them. Suppose you want to determine if the fish living in an industrialized harbor are in fact more exposed to HAHs than fish in a more pristine harbor. Why might a study of liver P450 enzymes be particularly useful for your purposes?

4. What is cooperativity, and why does it not require that "cooperating" sites affect each other directly?

5. Explain why G protein–mediated receptor systems depend on membrane fluidity.

6. Describe the possible roles of allosteric modulation in the regulation of metabolic pathways.

7. Venoms nearly always consist of complex mixes of compounds. Suggest evolutionary and physiological reasons why *mixes* are employed rather than pure compounds. Assume that mixes imply lower amounts of individual components; for instance, assume that if a venom is composed of two compounds, each will be present in only about half the quantity than if it were the only component.

8. What are your views on the two sides of the debate over whether emergent properties exist? Explain and justify.

9. Outline the functional roles of conformational changes in proteins, being sure to consider the various categories of proteins such as enzymes, channels, and receptors.

10. Present additional plausible family trees for the enzyme forms in Figure 2.21A, and explain which tree you judge to be most likely.

11. Cone snails, krait snakes, and poison-dart frogs (dendrobatid frogs) have independently evolved venoms that block the muscle acetylcholine receptor. Why do you suppose this receptor has so often become a target of venoms? Explain your answer in terms of the cellular mechanisms involved.

Go to **sites.sinauer.com/animalphys4e**
for box extensions, quizzes, flashcards,
and other resources.

REFERENCES

Alberts, B., A. Johnson, J. Lewis, and four additional authors. 2014. *Molecular Biology of the Cell,* 6th ed. Garland, New York. A peerless treatment of its subject, including advanced discussion of many of the topics in this chapter.

Aspengren, S., D. Hedberg, H. N. Sköld, and M. Wallin. 2008. New insights into melanosome transport in vertebrate pigment cells. *Int. Rev. Cell. Mol. Biol.* 272: 245–302. A fine, general introduction to the subject followed by details on mechanisms of pigment transport within chromatophores.

Dill, K. A., and J. L. MacCallum. 2012. The protein-folding problem, 50 years on. *Science* 338: 1042–1046.

Feder, M. E., and G. E. Hofmann. 1999. Heat-shock proteins, molecular chaperones, and the stress response: Evolutionary and ecological physiology. *Annu. Rev. Physiol.* 61: 243–282.

Glickman, M. H., and A. Ciechanover. 2002. The ubiquitin-proteasome pathway: destruction for the sake of construction. *Physiol. Rev.* 82: 373–428.

Golding, G. B., and A. M. Dean. 1998. The structural basis of molecular adaptation. *Mol. Biol. Evol.* 15: 355–369. A daring and thought-provoking introduction to the new field of paleomolecular biochemistry.

Hardison, R. 1999. The evolution of hemoglobin. *Am. Sci.* 87: 126–137. An accessible treatment of research on the evolution of a protein, including studies of the evolution of the relevant promoter and enhancer regions of DNA, as well as coding regions.

Hochachka, P. W., and G. N. Somero. 2002. *Biochemical Adaptation: Mechanism and Process in Physiological Evolution.* Oxford University Press, New York. A peerless review of the modern study of biochemical adaptation written by two of the foremost scientists in the field. Pedagogically exceptional.

Hofmann, G. E. 2005. Patterns of Hsp gene expression in ectothermic marine organisms on small to large biogeographic scales. *Integr. Comp. Biol.* 45: 247–255.

King, J., C. Haase-Pettingell, and D. Gossard. 2002. Protein folding and misfolding. *Am. Sci.* 90: 445–453.

Lodish, H., A. Berk, C. A. Kaiser, and five additional authors. 2012. *Molecular Cell Biology,* 7th ed. W. H. Freeman, New York.

Olivera, B. M. 1997. E. E. Just Lecture, 1996. Conus venom peptides, receptor and ion channel targets, and drug design: 50 million years of neuropharmacology. *Mol. Biol. Cell* 8: 2101–2109.

Powers, D. A., and P. M. Schulte. 1998. Evolutionary adaptations of gene structure and expression in natural populations in relation to a changing environment: A multidisciplinary approach to address the million-year saga of a small fish. *J. Exp. Zool.* 282: 71–94. This challenging paper records one of the most successful efforts to understand the molecular physiology and evolutionary biology of the ecological relationships of a species.

Somero, G. N. 2002. Thermal physiology and vertical zonation of intertidal animals: Optima, limits, and costs of living. *Integr. Comp. Biol.* 42: 780–789.

Viviani, V. R. 2002. The origin, diversity, and structure function relationships of insect luciferases. *Cell. Mol. Life Sci.* 59: 1833–1850.

Widder, E. A. 2010. Bioluminescence in the ocean: origins of biological, chemical, and ecological diversity. *Science* 328: 704–708. An articulate, compact review—with many informative images—of light production by marine animals.

Wirgin, I., N. K. Roy, M. Loftus, and three additional authors. 2011. Mechanistic basis of resistance to PCBs in Atlantic tomcod from the Hudson River. *Science* 331: 1322–1325. This paper argues that rapid evolutionary increase in frequency of a mutated intracellular receptor protein provides resistance to toxic effects of polychlorinated biphenyls.

NOTE: A truly marvelous historical account of the ubiquitin–proteasome system was authored by Michael S. Brown and can be read with the year 2000 awards at the Lasker Foundation website.

See also **Additional References** *and* **Figure and Table Citations***.*

Genomics, Proteomics, and Related Approaches to Physiology

Old-time Antarctic whalers believed that some of the fish in the polar seas had no blood because when they lifted the opercular flaps of the fish to see their gills, the gills were white, and when they cut the fish, only a whitish fluid ran out. A young Norwegian named Johan Ruud, recently graduated from college, was introduced to these fish by whalers during an Antarctic voyage in the late 1920s. His curiosity whetted, he remembered the unique fish throughout the middle years of his life and, 20 years later, seized an opportunity to investigate them. Convinced of their distinctive properties, he then brought the fish to the attention of biologists worldwide. That was in the 1950s, and by now the fish have become the focus of one of the most startling and instructive efforts to fuse studies of physiology and genetics.

Although Johan Ruud originally referred to the fish using the whalers' term *bloodless fish*, his studies revealed that they in fact have blood. Their blood lacks hemoglobin, however, and is virtually devoid of red blood cells. Thus the blood is translucent and whitish, rather than dense red like most vertebrate bloods (**FIGURE 3.1**). Today the fish are usually called *icefish*, a reference to their clear blood and the icy seas they inhabit. There are more than 30,000 species of fish alive today—and more than 60,000 species of vertebrates of all kinds—yet the icefish are the only vertebrates that do not have red blood as adults.

This Antarctic fish differs from most fish in that it has no hemoglobin in its blood, giving it an almost ghostlike appearance The fish is one of 16 species known as *icefish* because their blood is clear like ice and they live in the icy polar seas around Antarctica. The icefish differ dramatically from most fish in the proteins they synthesize: They fail to synthesize blood hemoglobin, explaining why their blood is clear instead of red, but they produce antifreeze glycoproteins that are not made by the great majority of fish. (The photo is of *Chaenocephalus aceratus*.)

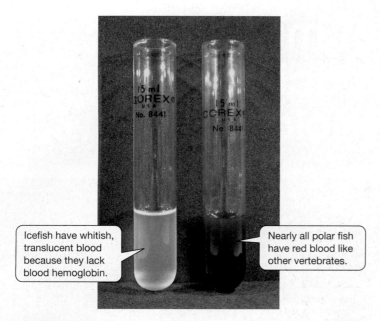

Icefish have whitish, translucent blood because they lack blood hemoglobin.

Nearly all polar fish have red blood like other vertebrates.

FIGURE 3.1 Freshly drawn blood from two species of Antarctic fish Both species belong to a single suborder, the Notothenioidei, the dominant group of fish in Antarctic waters. Most species in this group, such as the yellowbelly rockcod (*Notothenia coriiceps*) represented on the right, are red-blooded. The icefish, represented here by the blackfin icefish (*Chaenocephalus aceratus*) on the left, lost blood hemoglobin during their evolution. (Photo courtesy of Jody M. Beers.)

The icefish function, overall, as quite ordinary fish even though one might imagine that their lack of blood hemoglobin would be a crippling defect. They are neither rare nor small. Some species have been sufficiently common at times in the past to form commercially valuable fisheries, and several species grow to be 0.5–0.6 m long. Some are active swimmers that move between deep and shallow waters each day.

If icefish have any sort of obvious limitation, it is that they are restricted to the Antarctic seas,[1] where the waters are persistently very cold (often –1.9°C) and saturated with dissolved oxygen (O_2). The coldness plays a key role in their survival because it tends to depress their metabolic needs for O_2, and it tends to make O_2 particularly soluble in both the seawater and their body fluids. The icefish evolved after the time—about 30 million years ago—when the Antarctic seas became functionally isolated from most of the world's oceans because of dramatically altered patterns of ocean circulation. The seas became much colder than they had been before, and the icefish evolved in that frigid context.

A question that immediately arises in considering icefish is how they came to lose their ability to synthesize blood hemoglobin. Physiologists recognize that studies of genetics can often provide insight into such questions. Do the icefish still have the genes that code for hemoglobin and not transcribe those genes? Or have the genes become nonfunctional, or

possibly entirely lost? Modern molecular genetic studies can answer these sorts of key questions.

Studies of genetics from an evolutionary perspective also may help to clarify the adaptive significance of the loss of blood hemoglobin. Fewer than 20 species of icefish exist today; most authorities say 16. If the fish in this small group turn out to lack functional genes for synthesis of blood hemoglobin, can we trace the loss of the genes back to a single common ancestor of all members of the group, meaning the genes were lost a single time? Or did certain species lose the genes independently of others during evolution? The answer, as discussed later, might help us think more confidently about whether the loss of blood hemoglobin was a disadvantageous accident or an advantageous change favored by natural selection.

The protein portion of the blood hemoglobin of vertebrates consists of alpha (α) and beta (β) globin units. Specifically, each hemoglobin molecule is composed of two α-globin units and two β-globin units (see Figure 24.1C). The genes that code for the α- and β-globin units are members of an evolutionarily ancient *gene family*. Biologists know the family is ancient, in part, because genes with a clear structural similarity are found in bacteria and yeasts, indicating that genes of this basic type existed before the time that animals branched off from other life forms. The ancestral genes became duplicated during evolution. Because of this duplication, each individual vertebrate animal today has multiple copies. Over the course of millions of years of animal evolution, copies of the genes in different species diversified by accumulating changes, and the multiple copies within a single species also underwent diversification. All the genes retained their family resemblance nonetheless. In modern birds and mammals, the two distinct—but structurally similar—genes that code for the α- and β-globin units are located on different chromosomes. In fish, by contrast, the two genes are found on a single chromosome, relatively close to each other.

Physiologists reasoned that if they could look at the exact DNA structure of the α- and β-globin genes in icefish—and compare those genes with ordinary fish globin genes—they might be able to determine how the DNA of icefish became modified during evolution to produce the hemoglobin-free condition (**FIGURE 3.2**). The physiologists employed knowledge of the basic DNA structure

[1] One species, of undoubted polar ancestry, occurs in adjacent cold waters near the southern tip of South America.

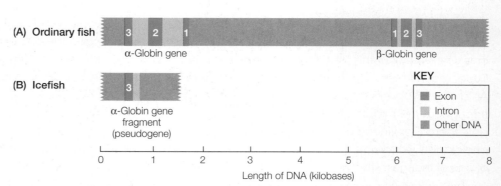

(A) Ordinary fish

α-Globin gene   β-Globin gene

KEY
■ Exon
■ Intron
■ Other DNA

(B) Icefish

α-Globin gene fragment (pseudogene)

Length of DNA (kilobases)

FIGURE 3.2 Genes and pseudogenes for blood hemoglobin The diagrams depict homologous stretches of DNA in (A) ordinary fish and (B) icefish. In ordinary fish (exemplified here by *Notothenia coriiceps*, an Antarctic red-blooded fish), functional genes for α- and β-globin are found near each other on a single chromosome; each globin gene consists of three exons (coded green) and two introns (coded yellow). In nearly all icefish, the entire β-globin gene, most of the α-globin gene, and the DNA between the original globin genes have been deleted. The icefish retain only a nonfunctional pseudogene, a fragment of the α-globin gene consisting of exon 3 and a part of the adjacent intron. (After Near et al. 2006.)

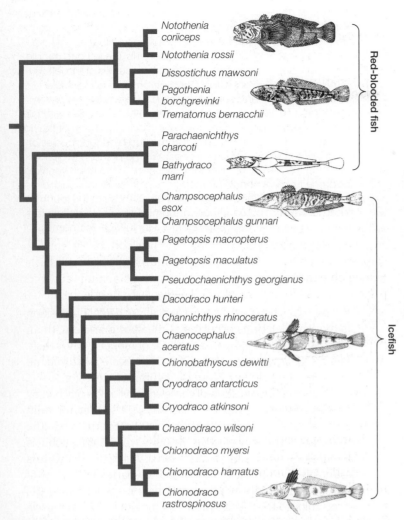

FIGURE 3.3 The evolution of icefish The diagram shows the evolutionary tree of 22 species of related Antarctic fish belonging to the suborder Notothenioidei. The tree is the product of the most recent research on phylogenetic reconstruction of this suborder. Information on blood hemoglobin was *not* used to construct the tree, which is primarily based on studies of mitochondrial DNA. (After Near et al. 2004.)

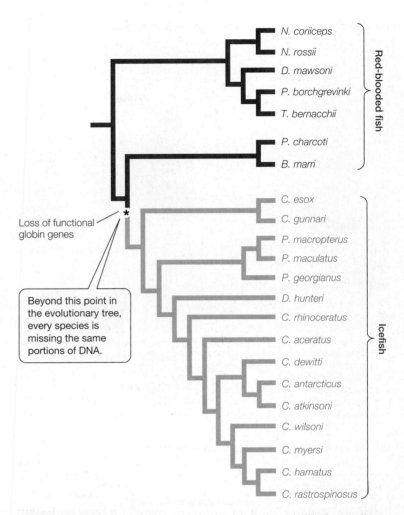

FIGURE 3.4 The most likely point at which the ability to synthesize blood hemoglobin was lost during the evolution of icefish Red and gray are used to symbolize which species and lines of evolution are characterized by the presence of functional globin genes (red) and which are not (gray). All of the 15 icefish species shown exhibit the condition diagrammed in Figure 3.2B: They lack the β-globin gene and possess only a nonfunctional fragment of the α-globin gene. One species of icefish (*Neopagetopsis ionah*) is omitted because, although it does not contradict the conclusions described here, it represents a special case. See Figure 3.3 for full species names.

of fish globin genes to find the relevant stretches of DNA in icefish. Then they used the polymerase chain reaction (PCR) to make enough copies of the icefish DNA so that they could determine the sequences of nucleotide bases in the DNA. Research of this type carried out in just the last 20 years showed that in 15 of the 16 species of icefish, the DNA is modified in exactly the same way![2]

The relevant DNA in ordinary fish consists of a complete α-globin gene separated by an intermediate stretch of DNA from a complete β-globin gene (see Figure 3.2A). In the icefish, however, the β-globin gene is completely gone, and the α-globin gene is missing parts, rendering it nonfunctional (see Figure 3.2B). A substantial stretch of DNA was deleted during the evolution of the icefish from their red-blooded ancestors.

To visualize *when* an event occurred during the evolution of a set of species, biologists often plot the event at the most logical position on an evolutionary tree of the species concerned. **FIGURE 3.3** shows the most likely evolutionary tree of the icefish and some of their closest red-blooded relatives. This tree is itself based

principally on genetic information. To construct the tree, biologists determined the nucleotide base sequences of mitochondrial DNA in all the fish involved. Then they used the sequences of the various species to identify logical relationships among the species, based on the same principles we discussed in Chapter 2 (see Figure 2.21) for interpreting amino acid sequences. No information on hemoglobin or the α- or β-globin gene was used in constructing the evolutionary tree in Figure 3.3.[3] The tree is therefore *completely independent of our knowledge of the globin genes.*

The most logical spot on the evolutionary tree to plot the loss of the globin genes is shown in **FIGURE 3.4**. All the lines of evolution drawn in red in Figure 3.4 end in species that have functional genes for both α- and β-globin and that synthesize blood hemoglobin. All the lines of evolution drawn in gray end in species that lack functional globin genes because of DNA deletions. Moreover, *all the species with deletions exhibit the same deletions*, those evident in

---

[2] The 16th species, although it does not contradict the conclusions described herein, is a special case. If you are interested in more detail, see Near et al. 2006, listed in the Additional References.

[3] The globin genes are found in chromosomal DNA in the cell nucleus, not in the mitochondrial DNA used to construct the tree.

(A) Hearts of two species of icefish and a red-blooded Antarctic fish

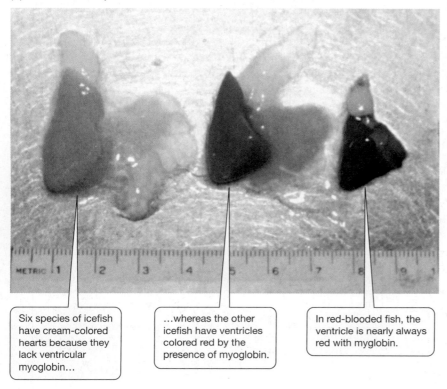

Six species of icefish have cream-colored hearts because they lack ventricular myoglobin…

…whereas the other icefish have ventricles colored red by the presence of myoglobin.

In red-blooded fish, the ventricle is nearly always red with myglobin.

(B) The points at which the myoglobin genes became nonfunctional

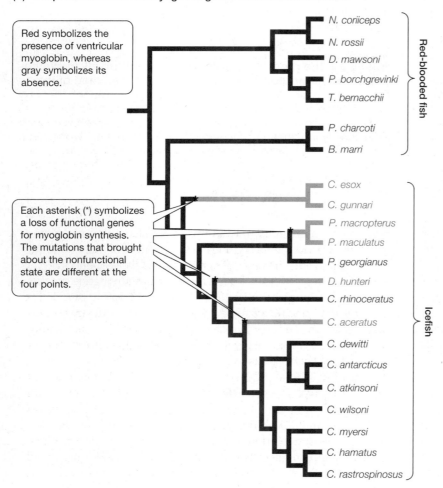

Red symbolizes the presence of ventricular myoglobin, whereas gray symbolizes its absence.

Each asterisk (*) symbolizes a loss of functional genes for myoglobin synthesis. The mutations that brought about the nonfunctional state are different at the four points.

Red-blooded fish
- N. coriiceps
- N. rossii
- D. mawsoni
- P. borchgrevinki
- T. bernacchii
- P. charcoti
- B. marri

Icefish
- C. esox
- C. gunnari
- P. macropterus
- P. maculatus
- P. georgianus
- D. hunteri
- C. rhinoceratus
- C. aceratus
- C. dewitti
- C. antarcticus
- C. atkinsoni
- C. wilsoni
- C. myersi
- C. hamatus
- C. rastrospinosus

FIGURE 3.5 Presence and absence of myoglobin in the ventricular heart muscle (A) Three hearts—representing three species of Antarctic fish—removed from individuals of approximately the same body size. Because blood has been drained from the hearts, the color of the tissues depends on whether or not myoglobin is present in the cardiac muscle cells. The left and middle hearts are from two species of icefish; the species at the left (*Chaenocephalus aceratus*) has a cream-colored ventricle because it lacks ventricular myoglobin, whereas the species in the middle (*Chionodraco rastrospinosus*) has a reddish ventricle because it synthesizes ventricular myoglobin. The heart at the right, red with ventricular myoglobin, is from a red-blooded species of Antarctic fish (*Notothenia coriiceps*). (B) The points in evolution at which the genes for myoglobin synthesis in the heart ventricle became nonfunctional. The mutations at three of the four points are known to be different from each other; those at the fourth (the icefish *Dacodraco hunteri*) are presumed also to be distinct but remain to be described. See Figure 3.3 for full species' names. (A from Moylan and Sidell 2000; B after Sidell and O'Brien 2006.)

Figure 3.2B. Therefore the most logically coherent proposition is that the deletions occurred at the spot marked with an asterisk in Figure 3.4, in an ancestor of modern-day icefish. Later, as the various existing species of icefish evolved, all inherited the deletions from their common ancestor.

A deeper appreciation of these conclusions is reached by looking at another property of icefish that is similar in certain respects but dissimilar in others. In most vertebrates, the blood is not the only place in the body where hemoglobin is found (see Chapter 24). Hemoglobin of distinctive structure is found also within the cells of muscles—particularly many of the skeletal muscles and the heart muscle—where it imparts a red color to the muscle tissue. Hemoglobin within muscle cells helps increase the rate at which O_2 diffuses into the cells, and it sometimes acts as an important internal store of O_2 for the cells. Muscle hemoglobin is known as *myoglobin* (*myo-*, "muscle").

Six of the 16 species of icefish lack myoglobin in the cells of their ventricular heart muscle. The ventricle in these species is cream-colored, in contrast to the ventricle in the other 10 species of icefish, which have ventricular myoglobin.[4] **FIGURE 3.5A** shows examples of these two sorts of icefish (left and middle images). To increase understanding of the evolution of the myoglobin-free condition, physiologists took the same approach as in the study of blood hemoglobin. They examined the DNA sequences of the genes for myoglobin and asked what had happened to render the genes nonfunctional in the icefish that lack ventricular myoglobin.

The physiologists found that the myoglobin genes in some species of myoglobin-free icefish are altered in distinctly different ways from the genes in other myoglobin-free species. This discovery indicates that the myoglobin-free condition evolved independently more than once. In fact, based on the evidence currently available, there were four independent occasions when the myoglobin genes became nonfunctional. These plot on the evolutionary tree as shown in **FIGURE 3.5B**.

Comparing Figures 3.4 and 3.5B, you can see that during the evolution of the icefish, the loss of blood hemoglobin and that of muscle myoglobin followed very different paths.

[4] Fish with red blood almost always have ventricular myoglobin.

Blood hemoglobin was lost once: A large stretch of DNA was deleted, eliminating one globin gene and rendering the other irretrievably nonfunctional, and this deletion was passed on to all the species of icefish alive today.[5] By contrast, most icefish synthesize myoglobin in their heart muscle, just like most of their red-blooded relatives do; myoglobin was not lost when blood hemoglobin was. After the icefish first appeared, however, mutations that eliminated myoglobin occurred independently in four of the lines of icefish evolution, and today six of the species exhibit one of those mutations and have myoglobin-free ventricular muscle.

Was the loss of blood hemoglobin and myoglobin advantageous or disadvantageous? Answering this question is complicated, and surely biologists will debate some aspects long into the future. Here let's look just at blood hemoglobin.

Most researchers conclude that the initial loss of blood hemoglobin was probably a disadvantage for the ancestors of icefish. The genetic discovery that the loss occurred once—rather than multiple times—enhances the plausibility of this conclusion. Evolution by natural selection tends to weed out deleterious mutations. It is easy to imagine that a single deleterious loss might survive the immediate selection against it; multiple surviving losses would be much less plausible. Dramatic morphological and physiological specializations of icefish provide the most compelling evidence that the loss of blood hemoglobin was a disadvantage. Compared with related red-blooded fish of the same body size, existing icefish have very large hearts, and they circulate their blood at rates that are far higher than usual.[6] These specializations strongly suggest that the original loss of blood hemoglobin was a defect that significantly decreased the ability of the circulatory system to transport O_2, and the circulatory system thereafter became modified to make up for the defect by evolving a capacity to circulate the blood exceptionally rapidly. Looking at the evidence from both genetics and physiology, a persuasive case can be made that the loss of blood hemoglobin initially decreased the fitness of icefish and thereby favored the subsequent evolution of other attributes that compensated for the shortcoming.[7]

The *antifreeze glycoproteins* are a final set of proteins of icefish that raise intriguing genetic and evolutionary questions. All species of ocean fish with bony skeletons, including icefish, have body fluids that are more dilute in total dissolved matter than seawater is (see Chapter 28). Because of this, the body fluids of ocean fish tend to freeze at a higher temperature than

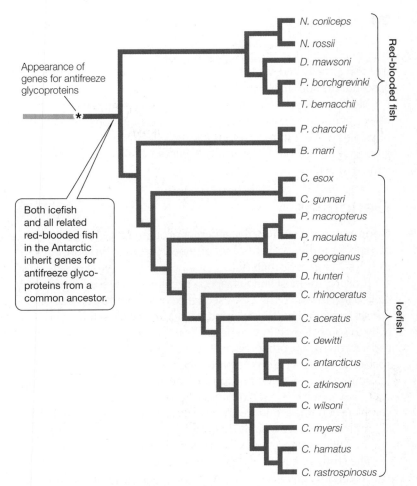

Appearance of genes for antifreeze glycoproteins

Both icefish and all related red-blooded fish in the Antarctic inherit genes for antifreeze glycoproteins from a common ancestor.

FIGURE 3.6 Evolution of antifreeze glycoproteins Blue symbolizes the species and lines of evolution characterized by antifreeze glycoproteins. Genes coding for the glycoproteins appeared prior to the evolution of icefish. See Figure 3.3 for full species' names. (After Cheng et al. 2003.)

seawater freezes.[8] The seawater in the frigid polar seas, in fact, is often cold enough to freeze fish even though the seawater itself remains unfrozen. Antarctic species of fish typically differ from the great majority of fish species in that they synthesize specialized antifreeze compounds, the antifreeze glycoproteins. These compounds bind to any minute ice crystals that appear in the body and arrest crystal growth, thereby preventing freezing of the blood and other body fluids. An individual fish synthesizes a suite of chemically similar antifreeze glycoproteins, coded by a suite of evolutionarily related genes.

The antifreeze glycoproteins present still another evolutionary scenario, compared with blood hemoglobin and ventricular myoglobin. According to available chemical and genetic evidence, the glycoproteins and the genes that code for them are essentially the same not only in all icefish but also in all the Antarctic red-blooded fish related to icefish. The most logical conclusion, therefore, is that the genes for antifreeze synthesis evolved before icefish appeared, as shown in **FIGURE 3.6**. When blood hemoglobin became deleted, giving rise to the first ancestors of icefish, those early hemoglobin-

---

[5] The overall change from normal to altered DNA, diagrammed in Figure 3.2, probably involved several sequential mutations, not just a single large and simultaneous deletion. The change is described as occurring once because from the viewpoint of the species existing today, there was a single net alteration in the DNA, regardless of the intermediate steps involved in its occurrence.

[6] The distinctive cardiovascular physiology of icefish is discussed further in Chapter 24 (see page 657). You can see that the icefish hearts are enlarged in Figure 3.5A.

[7] An obvious question is how the fish that initially exhibited hemoglobin loss were able to survive. Researchers working on the subject argue that survival occurred in part because of the particular conditions existing in the Antarctic seas at the time, including relatively little overall competition and the existence of ecological refuges from competition. The arguments rest on studies of the ancient oceanography and plate tectonics of the region, as well as studies of biology.

[8] Typically, in aqueous solutions, the freezing point depends on the concentration of dissolved entities: The higher the concentration, the lower the freezing point. More specifically, doubling the concentration of dissolved entities (e.g., Na^+ ions or glucose molecules) approximately doubles the extent to which the freezing point of a solution is lowered below 0°C (see page 122). Antifreeze proteins and glycoproteins are unusual, compared with other dissolved materials, in that they exert a far greater effect on freezing point than can be accounted for by just their concentrations (see page 256).

free fish already had the types of antifreeze compounds that their descendants still display today.

Modern research on icefish dramatically illustrates the fruitful use of genome science to help understand questions in animal physiology. We will return to the icefish examples several times as we now look more systemically at genomics and the disciplines related to genomics.

Genomics

Genomics is the study of the genomes of organisms. The **genome** of a species is the species' full set of genes, or—more broadly put—its full set of genetic material.

Probably the most famous aspect of genomics at present is *genome sequencing*, in which the DNA sequence of the entire genome of a species is determined. Because individuals of a species differ genetically to some degree (e.g., whereas some people have genetically coded brown eyes, others have blue eyes), the DNA sequence for a species depends a bit on the particular individual from which the DNA for sequencing was acquired. The sequence is enormously useful nonetheless, even if based on just one individual. At present, complete genome sequences have been determined for more than 100 animal species. Probably thousands of animal species' genomes will be fully sequenced within the next decade.

The study of genomics is not limited to species for which the entire genome has been sequenced. Sequencing of just a limited set of individual genes—a subpart of the genome—can sometimes set the stage for major new insights into a physiological system. An example is provided by research on the evolutionary origins of the genes that code for the antifreeze glycoproteins of Antarctic fish. Sequence comparisons of those genes with a limited but relevant set of other genes reveal that the antifreeze genes are derived from genes that in ordinary fish code for pancreatic proteins similar to trypsinogen, the precursor of the digestive enzyme trypsin (see Box 10.2). That is, copies of genes that originally coded for pancreatic digestive proteins evolved to code for the antifreezes!

Genomics is inextricably linked with advanced methods of information processing

Genomics, especially when entire genome sequences are studied, involves the processing of massive quantities of information. The genome of a single species, for example, may consist of a string of more than 1 billion nucleotide bases. To compare the genomes of two species, researchers often need to search for stretches of similar and dissimilar DNA in two sequences that each exceed 1 billion bases in length. Modern genomics is defined in part by massive information processing.

Progress in genomics relies typically on the use of computer programs and robots that carry out great numbers of steps without direct human attention. The computer programming is itself sufficiently demanding that it is carried out by scientists in new specialties termed *computational biology* and *bioinformatics*. One key task for these specialties is the organization of data: The great masses of information gathered in genomic studies need to be recorded in ways that permit reliable retrieval by multiple users, many of whom were not involved in the original data collection. A second key task is to articulate the operational meaning of similarity and difference among base sequences within stretches of

DNA and write algorithms that efficiently identify similarities and differences. Algorithms of this sort are used, for example, to locate apparently homologous strings of nucleotides in DNA—similar DNA sequences—in two or more genomes.

Much of genomic research is carried out by what are called **high-throughput methods**. The term refers to methods of the sort we have been discussing, in which computer programs and robots—after being designed to be as effective and error-free as possible—are "turned loose" to carry out procedures and generate results without moment-to-moment human attention or detailed human quality control. The process of adding direct human interpretation is known as **annotation**. To illustrate the interplay between high-throughput methods and annotation, consider a genome composed of 20,000 genes. When the genome is first fully sequenced, both the sequencing itself and the initial identification of individual genes will be carried out largely by high-throughput methods. Thereafter, experts on various genes in the global scientific community will directly or indirectly (e.g., by use of data catalogs) assess and add annotations to the information on genes of interest, but this time-consuming process is never complete. Possibly, therefore, the information on just 6000 of the genes will be annotated within the first few years. Knowledge of the other 14,000 genes would then consist, for the moment, only of the decisions of computer programs.

The World Wide Web is the primary vehicle by which the vast quantities of genomic information are shared among scientists worldwide. Gene and genome sequences are posted at dedicated websites (e.g., National Center for Biotechnology Information, GenBank, FlyBase, and WormBase). Web-based tools—notably BLAST programs—are available online to search for similarities among nucleotide base sequences in two or more stretches of DNA of interest.[9] Web-based tools, such as the Gene Ontology, are also available to facilitate and standardize annotation.

One overarching goal of genomics is to elucidate the evolution of genes and genomes

Genomics, whether based on entire or partial genome sequences, can be said to have two overarching goals:

■ First, elucidate the evolution of genes and genomes.
■ Second, elucidate the current functioning of genes and genomes.

Genomics does not proceed in isolation in the pursuit of either of these goals. Instead, progress is most effective when genomics is integrated with physiology, biochemistry, and other disciplines.

In the study of the evolution of genes and genomes, one central topic is the elucidation of *mechanisms of gene modification*. The genome of each species is descended from the genomes of ancestral species, and genomes become modified as they evolve. Genomic studies help clarify the mechanisms by which genes and genomes become modified. One mechanism documented by genomic research is that genes sometimes become modified during evolution by the accumulation of beneficial base substitutions

---

[9] BLAST stands for *basic local alignment search tool*. BLAST programs are available to search for similarities among amino acid sequences in different proteins, as well as similarities among base sequences in samples of DNA. See "Learn" at the National Center for Biotechnology Information website (www.ncbi.nlm.nih.gov) for useful tutorials on BLAST programs.

or other mutations favored by natural selection. This process can occur to such a significant extent that certain genes come to code for new proteins, as we have seen in the evolution of the genes for antifreeze glycoproteins in Antarctic fish from genes for pancreatic proteins such as trypsinogen. Another mechanism of modification is that genes sometimes become duplicated during evolution, and following duplication, the multiple copies within a single species often come to code for different proteins (illustrated by the evolution of the α- and β-globin genes in red-blooded vertebrates). Genes also sometimes become nonfunctional as they evolve, by partial or full deletion or by mutations that block transcription (illustrated by the icefish hemoglobin and myoglobin genes). The *coding* regions of genes are not the only parts that change by the operation of these and other mechanisms. *Regulatory* regions also sometimes change, thereby affecting the circumstances in which genes are transcribed.

Besides analyzing mechanisms of gene change, another central topic in the study of the evolution of genomes is the *reconstruction of paths of evolution in the past*. In research of this sort, the evolutionary tree of species of interest is first determined as accurately as possible from genomic, morphological, and biochemical evidence. Then, as illustrated by Figures 3.4–3.6, the most likely points of occurrence of particular evolutionary developments are located on the tree. Methods exist to add a time scale, so that the amount of time between events can be approximated. Reconstructions of the sort described are useful in several ways. One is that they clarify the order of evolutionary events. For example, as physiologists reason about why certain icefish lost ventricular myoglobin, they can be virtually certain that blood hemoglobin was already gone (compare Figures 3.4 and 3.5B): To hope to explain accurately the evolution of myoglobin-free ventricular muscle in icefish, one needs to think of the loss of myoglobin as having occurred, not in fish with ordinary vertebrate blood, but in ones with hemoglobin-free blood.

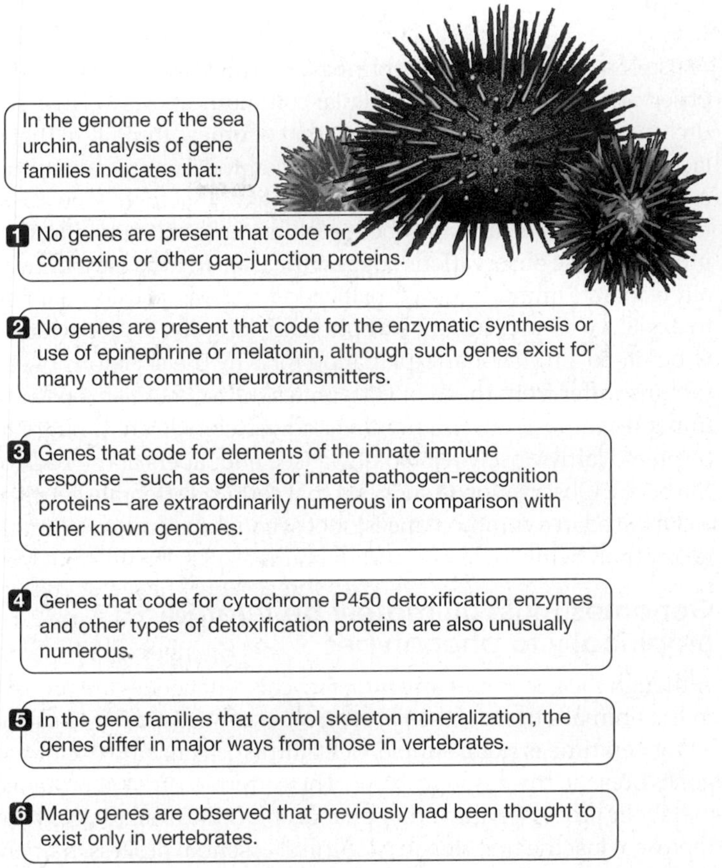

A second overarching goal of genomics is to elucidate the current functioning of genes and genomes

Because of the evolutionary continuity of life, when a new genome is sequenced, many of the genes found in the new genome are likely to be similar to genes observed already in genomes that were sequenced at earlier times. One may therefore be able to predict the function of genes in the new genome by extrapolating from preexisting knowledge of homologous genes. Suppose, to illustrate, that a particular gene of physiological importance is known to occur in the laboratory mouse (one of the most intensively studied mammals), and suppose that the mouse gene has been extensively annotated with information on how it is regulated and the physiological roles it plays. Suppose also that when researchers sequence the genome of a new mammal species, they find a gene that has a DNA sequence closely similar to the mouse gene. It would then be reasonable for the researchers to predict that the gene in the newly sequenced species has a function like that of the gene already known. Not all such predictions prove to be accurate when tested, because genes evolve and can take on new properties. Nonetheless, homologous genes in different species often have similar functional properties.

The existence of *gene families* creates opportunities for large-scale interpretations based on the same sort of logic as just described. We

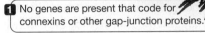

In the genome of the sea urchin, analysis of gene families indicates that:

1 No genes are present that code for connexins or other gap-junction proteins.

2 No genes are present that code for the enzymatic synthesis or use of epinephrine or melatonin, although such genes exist for many other common neurotransmitters.

3 Genes that code for elements of the innate immune response—such as genes for innate pathogen-recognition proteins—are extraordinarily numerous in comparison with other known genomes.

4 Genes that code for cytochrome P450 detoxification enzymes and other types of detoxification proteins are also unusually numerous.

5 In the gene families that control skeleton mineralization, the genes differ in major ways from those in vertebrates.

6 Many genes are observed that previously had been thought to exist only in vertebrates.

FIGURE 3.7 Characteristics of the genome of a sea urchin The six observations listed here arise from a close study of the genome of the purple sea urchin (*Strongylocentrotus purpuratus*), a common echinoderm in shallow coastal waters and tidal pools along the West Coast of the United States. Sequencing of the urchin's genome was completed in 2006. (After Sea Urchin Genome Sequencing Consortium 2006.)

have already seen an example of a gene family, namely the set of genes that code for vertebrate globin proteins. You will recall that the various globin genes within a single vertebrate species bear a family resemblance, and the globin genes in different species also do—all being related through evolutionary descent. All these genes are considered to belong to one gene family. The property that defines a **gene family** is that all the genes in a family share distinctive DNA base sequences. All the genes in a family also *tend* to code for functionally similar proteins; for example, just as the genes in the globin family code for hemoglobin proteins, the genes in another family might code for a particular type of enzyme, and those in still another family might code for a particular type of voltage-gated channel protein. With these concepts in mind, it is clear that when the genome of a species is initially sequenced, the simple process of scanning the genome for distinctive DNA sequences—the signature sequences of gene families—can be instructive. The process can provide a great deal of tentative insight into the functions of the genes present and the types of proteins likely to be synthesized.

An illustration is provided by the genome of the purple sea urchin. **FIGURE 3.7** lists just six of the many major insights that have been gained into urchin biology by surveying this genome. Consider, first, observations **1** and **2**, which state that when the urchin genome is surveyed, no genes are found that would be expected to code for gap-junction proteins or mediate the synthesis or use of melatonin and epinephrine (also called adrenaline). These observations suggest that the cellular communication system of sea urchins is unusual, compared with that of other animals, in that it lacks gap junctions and certain of the common neurotransmitter compounds. Observations ❸ and ❹ highlight that there seem to be lots of genes in the urchin genome for immune and detoxification proteins; these observations suggest that sea urchins have unusually elaborate immune and detoxification systems, possibly helping to explain why urchins are exceptionally long-lived. Observation ❺ points to a functional explanation for why the skeletons of sea urchins differ from those of vertebrates in the chemistry of their mineral composition. Sea urchins and vertebrates are thought to belong to fairly closely related phyla (see endpapers at the back of the book). Observation ❻ suggests that certain vertebrate genes in fact evolved in a common ancestor of sea urchins and vertebrates, rather than being exclusive to vertebrates as previously thought.

Genomes must ultimately be related empirically to phenotypes

Although the genome of an animal reveals what genes are present in the animal's tissues, the *phenotype*[10] of any particular tissue at any given time is not a simple, deterministic consequence of the genes present, the tissue *genotype*. Thus efforts to predict proteins, metabolic processes, and other phenotypic traits from the genome represent just the first step in a long and essential process, namely that the genome must be related by empirical studies to the phenotype. When predictions are made from just the genome, they may ultimately prove to be wrong for several reasons. These include that (1) the actual functions of newly discovered genes may in fact not match the functions predicted by extrapolation from already-known genes and (2) even if the true function of a new gene is known, the gene may not be expressed when and where predicted. Predictions from the genome are in fact *hypotheses*, and they must be tested before they can be accepted or rejected.

The process of testing genomic predictions entails, in part, the study of *which genes are transcribed and expressed* under various circumstances. It also entails the study of the *actual proteins synthesized* as a consequence of gene expression, and the *metabolites*[11] that are synthesized, used, and otherwise processed by the proteins. We will return to these sorts of studies—transcriptomics, proteomics, and metabolomics—after looking at important issues in research strategy in the next two sections of this chapter.

The study of a species is said to enter a **postgenomic era** after the genome of the species has been sequenced. *Postgenomic* does not mean that the genome can be relegated to history. Quite the contrary, it emphasizes that in the era "after the genome is known," the study of a species' biology is forever altered. In the postgenomic era, the sequence of a species' genetic material is entirely known. The monumental task of empirically evaluating the full *significance* of this knowledge remains, however.

---

[10] The *phenotype* of a tissue consists of its outward characteristics—its structure, activities (such as contraction or secretion), biochemical constituents, and metabolic pathways—as opposed to its genetic material. Its *genotype* is its genetic material, its genome.

[11] A *metabolite* is an organic compound of modest to low molecular weight that is currently being processed by metabolism. An example would be glucose that is being processed by glycolysis.

SUMMARY
Genomics

■ Genomics is the study of the genomes—the full sets of genes—of organisms. Because of the large numbers of genes, genomics depends on high-throughput methods to collect data and on advanced information processing to catalog and use data.

■ One of the two major goals of genomics is to elucidate the evolution of genes and genomes. In pursuit of this goal, students of genomics seek to understand the mechanisms of evolutionary modification of genes and genomes (e.g., deletion and duplication). They also seek to reconstruct the paths followed by evolution in the past so that, for example, the order of evolutionary events is better defined.

■ The second major goal of genomics is to elucidate the current functioning of genes and genomes. In pursuit of this goal, genomics uses information on already-known genes and gene families to predict the likely functions of newly identified genes and the likely ranges of action and competence of newly sequenced genomes.

■ Although knowledge of an animal's genome permits many useful predictions to be made about the animal's biochemical phenotype, these predictions must ultimately be tested empirically. For example, although the suite of proteins synthesized in an animal's tissues can be predicted from the genome, the proteins present must ultimately be studied directly, as by proteomic methods.

Top-down versus Bottom-up Approaches to the Study of Physiology

The traditional approach to the study of the multiple levels of organization in animal physiology can be described as *top-down*. To see this, consider the diagram of levels of organization and the chain of causation in **FIGURE 3.8A**. To *study* a phenomenon using the traditional approach, as shown at the left in **FIGURE 3.8B**, the order of study proceeds from the top of the diagram to the bottom. Physiologists first recognize an attribute of animal function of interest; a human example would be the *exercise training effect*: the improved ability of previously sedentary people to engage in exercise when they participate in a program of athletic training. After the attribute of animal function is specified, physiologists seek to identify the aspects of tissue function that are involved. In the case of the exercise training effect, a key aspect of tissue function is that skeletal muscles increase their capacity for physical work when they are trained. Physiologists then look for the specific proteins—and the properties of the proteins—that are responsible for the tissue functions they have identified. Finally, physiologists identify the genes coding for the proteins, and they study how the expression of the genes is controlled, and the evolution of the genes. In the traditional, **top-down order of study**, investigation proceeds from animal function to tissue function, then to tissue biochemistry, and finally to genes.

Genomics sets the stage for physiologists to adopt a new *bottom-up* approach to the study of physiological phenomena, shown at the

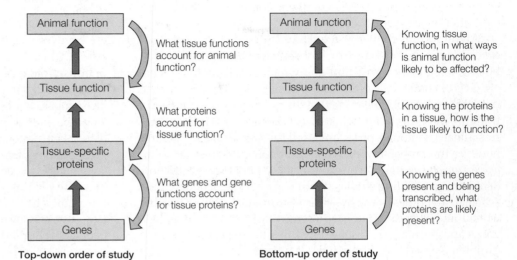

(A) The levels of organization and chain of causation in animal physiology

(B) The order of questions posed in top-down and bottom-up studies

FIGURE 3.8 Top-down versus bottom-up studies (A) Four of the principal levels of organization that must be taken into account in the study of animal physiology. Arrows show cause–effect relationships among the levels of organization: the order of causation. Arrows are labeled with just a few of the processes that are instrumental in the cause–effect relationships. (B) The order in which investigators study the levels of organization (yellow arrows) in top-down versus bottom-up studies.

right in Figure 3.8B. In this approach, physiologists first sequence the genome of a species, or they approximate the genome by extrapolating from other, related species. Actually, the entire genome need not be known or approximated; physiologists need only have information on the portion of the genome that is relevant to them. After the genomic information is available, physiologists study the transcription of the genes in key tissues. For example, in research on athletic training, physiologists would study gene transcription in the skeletal muscles to determine which genes are transcribed to a greater extent after training than before (i.e., genes upregulated by training), and which genes are transcribed less than before training (downregulated). After changes in gene transcription are known, physiologists predict changes in tissue proteins. They also look directly at extensive suites of tissue proteins to see which are increased and decreased in amount. Then physiologists seek to understand how the changes in proteins alter tissue function, and how the changes in tissue function are likely to affect animal function. In the new, **bottom-up order of study**, investigation proceeds from genes to gene expression, then to tissue biochemistry, and finally to tissue and animal function.

What would be the potential advantage of the bottom-up approach? The answer depends on recognizing the methods used. Researchers who employ the bottom-up approach have developed *high-throughput* methods to study all the genetic and biochemical steps. Thus not only the genes, but also the gene transcripts, proteins, and metabolites, are surveyed and monitored in great numbers simultaneously.

The bottom-up approach carried out with high-throughput methods has two distinctive advantages: (1) It can be extremely thorough in searching for the genes, proteins, and metabolites that are instrumental in a physiological process; and (2) precisely because it is thorough, it can proceed without preexisting biases regarding

which genes or proteins are likely be involved. Advocates of the bottom-up approach point to examples in which the traditional top-down approach missed important genes or proteins because investigators looked only at the small subsets that they imagined would be important. With the high-throughput methods employed in the bottom-up approach, investigators can, in principle, look at everything and therefore miss nothing.

Defenders of the top-down approach emphasize that looking at everything is not always an advantage. For example, when dozens or even hundreds of genes prove to be up- or downregulated during a physiological process—as is often the case—sorting through the implications for tissue function and animal function can be mind-boggling. Advocates of the top-down approach stress that it brings a needed focus to research because it starts with a defined phenomenon of known importance to whole organisms.

The top-down and bottom-up approaches seem certain to coexist for the foreseeable future. Each has advantages. Thus the two approaches can work together synergistically.

Screening or Profiling as a Research Strategy

At each stage of a bottom-up study, the most common strategy to-day—termed **screening** or **profiling**—is for investigators to look as comprehensively as possible at the class of compounds of interest, whether the compounds are messenger RNAs (produced by transcribed genes), proteins, or metabolites. To study the compounds comprehensively, high-throughput methods are used.

In screening or profiling studies, the most common type of research design is to compare a tissue of interest before and after a change of interest. For example, investigators might screen skeletal muscle before and after exercise training. Or they might screen a tissue before and after stress, or at different times of day, or in young individuals versus old. In all cases, the full suite of compounds present before a change is compared with the full suite after the change.

An important challenge for screening studies is the *statistical* challenge of deciding which observed changes are likely to be physiologically significant and which are likely to be mere artifacts of chance. We often read that statistical tests are carried out "at the 5% level," meaning that the probability of error (i.e., of thinking we see a change when in fact there is none) is 5% or less. In standard computer programs used for statistical calculations, however, this probability level is calculated on the assumption of an *a priori hypothesis*, that is, a hypothesis stated prior to data collection. Screening or profiling studies typically gather the data first, then articulate hypotheses; the hypotheses from such studies, in other words, are not a priori. Suppose a screening study examines 1000 genes and—among the 1000—identifies 100 genes that seem to exhibit increased transcription when a tissue is stressed in some particular way. Of those 100, if an investigator has simply used a standard statistical program and an error probability of 5%, 50 of the instances of increased transcription (5% of 1000) will doubtless be false positives. They will not be real and repeatable changes, but instead will be chance events. Specialized statistical methods already exist and are used to reduce false positives. Moreover, research is ongoing to develop improved methods that will deal in superior ways with the problem that when genes, proteins, or metabolites are surveyed in great numbers, some of the effects observed are bound to be artifacts of chance.

The Study of Gene Transcription: Transcriptomics

The study of gene transcription—that is, the study of which genes are being transcribed to make messenger RNA (mRNA) and the rates at which they are transcribed—is known as **transcriptomics** or **transcription profiling**.[12] Because physiologists have recognized for decades that the genes that matter are those that are transcribed, physiologists have studied transcription and mRNA synthesis for a long time. *Transcriptomics* and *transcription profiling* are new terms. One of their key connotations is that they imply the simultaneous study of great numbers of mRNAs, often by use of highly automated methods.

Changes in gene transcription during and after exercise nicely illustrate the transcriptomic approach. Investigators have found that numerous genes are upregulated—transcribed at an increased rate—in exercising muscles each time a person engages in an extended period of endurance exercise. In one study (**FIGURE 3.9A**), subjects performed 90 min of leg extension (kicking) exercise with one leg while the other leg remained at rest. For 24 h following the exercise, biopsy methods (see Figure 21.3) were used to remove tiny samples of muscle tissue from thigh muscles periodically, and the levels of dozens of mRNAs in the exercised leg were compared with those in the unexercised leg. Three groups of genes were identified that

---

[12] **Expression profiling** is another commonly used term. However, *expression* is sometimes used to mean protein synthesis. *Transcription profiling* is a more exacting term to use when *transcription* per se is the process under study. See Nikinmaa and Schlenk 2009 in the Additional References for a discussion.

(A) The type of exercise performed

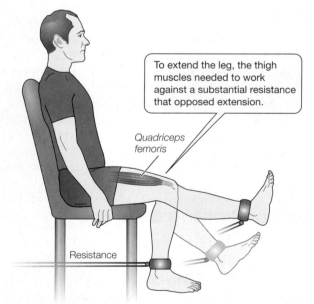

To extend the leg, the thigh muscles needed to work against a substantial resistance that opposed extension.

Quadriceps femoris

Resistance

(B) Changes in three categories of mRNAs in the exercised leg

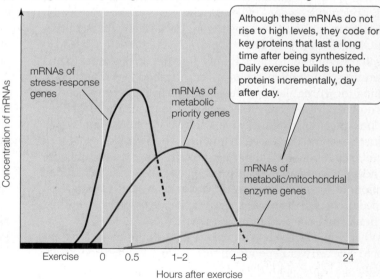

Although these mRNAs do not rise to high levels, they code for key proteins that last a long time after being synthesized. Daily exercise builds up the proteins incrementally, day after day.

mRNAs of stress-response genes

mRNAs of metabolic priority genes

mRNAs of metabolic/mitochondrial enzyme genes

Concentration of mRNAs

Exercise   0   0.5   1–2   4–8   24

Hours after exercise

FIGURE 3.9 Exercise is followed by increased transcription of groups of genes   (A) Subjects performed a 90-min bout of leg-extension exercise once each day for 5 days. Only one leg was exercised; the other remained at rest. On the fifth day, biopsy samples were taken repeatedly from the *quadriceps femoris* muscles of the two legs, and concentrations of many messenger RNAs were measured in the samples. (B) Based on the mRNA data—and using the unexercised leg as a control for the exercised one—investigators identified three groups of genes that exercise causes to be upregulated. One group of

genes—termed the *stress-response genes*—quickly exhibits highly enhanced transcription (and therefore increased mRNA synthesis) during and after exercise of the sort performed. Another group—dubbed the *metabolic priority genes*—is slower to be upregulated but also exhibits a dramatic increase in transcription. The third group—the *metabolic/mitochondrial enzyme genes*—is the slowest to be upregulated and exhibits only modest (but long-duration) increases in transcription. Note that times are not evenly spaced on the *x* axis. (B after Booth and Neufer 2005.)

(A) A fluorescent image of a DNA microarray, greatly magnified

~1 mm

underwent upregulation in the muscles of the exercised leg during and following exercise (**FIGURE 3.9B**).

One of the groups of upregulated genes—termed the *metabolic/ mitochondrial enzyme genes*—is especially interesting. These genes code for mitochondrial proteins. As seen in Figure 3.9B, the metabolic/ mitochondrial enzyme genes undergo just a small degree of upregulation after a single bout of exercise and thus boost mitochondrial protein synthesis just a bit. The mitochondrial proteins, however, have long half-lives; once synthesized, they last for a long time. The investigators believe they have found a key mechanistic reason for why muscles keep increasing their exercise capability, day after day, over many weeks of daily endurance training. Each day's training produces just a modest increase in transcription of the mitochondrial genes and a modest increase in synthesis of mitochondrial proteins, but these small effects cumulate when training is repeated day after day.

DNA microarrays (also called *DNA microchips* or *gene chips*) are the basis for a set of particularly important, high-throughput methods for the study of gene transcription. A single microarray can permit investigation of thousands or tens of thousands of genes at a time.

As a physical object, a microarray consists of a grid of spots of DNA placed on a glass plate or other solid substrate by a robot (**FIGURE 3.10A**). Each spot might, for instance, consist of a stretch of DNA that represents a single gene or presumptive gene.[13] Given the minute physical size of each bit of DNA and the technology used to apply the spots, a grid of 10,000 spots—representing 10,000 different genes—will fit within an area of only 1 cm^2, or less.

In one common type of experiment using a DNA microarray, a single array is used to carry out a direct comparison of the mRNAs produced by a tissue under two different conditions.[14] For illustration, let's assume that we are doing a microarray study of muscle before and after exercise. The mixes of mRNA molecules present under the two conditions are extracted (❶ in **FIGURE 3.10B**), and all the mRNA molecules in each extract are then labeled with a distinctive fluor, a compound that has the potential to emit light by fluorescence.[15] A common approach (❷ in the figure) is to label one extract (e.g., that from muscle before exercise) with a fluor

---

[13] The nature of the DNA spots depends on the method used to prepare them, and several methods are in common use. One method is to reverse-transcribe mRNA molecules to produce DNA sequences (cDNA) that code for the mRNA molecules; this method, you will note, does not require a sequenced genome. Another method is to start with the entire genome and essentially cut it up into many, often overlapping, pieces of DNA (producing a *tiling* array). Still another method is to synthesize DNA sequences from scratch (essentially from raw nucleotides) using knowledge of the genome to select the sequences made.

[14] This experimental design is said to employ a *two-color spotted microarray*.

[15] For conceptual simplicity, we here describe the procedure as if the raw mRNA itself were used, although in reality, technical steps must be taken to stabilize the mRNA.

(B) A study to compare gene transcription under two different conditions

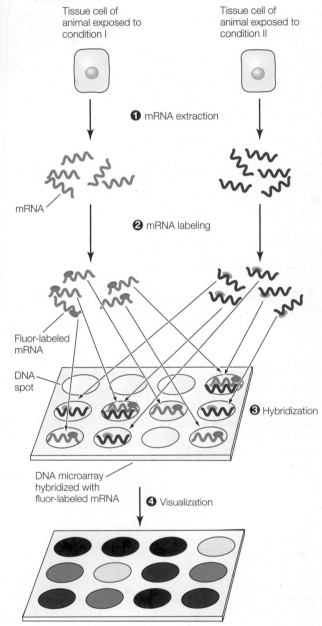

FIGURE 3.10 With DNA microarrays, the transcription of thousands of genes can be studied at once (A) A DNA microarray. Each spot represents one gene. The color of a spot after processing indicates whether the gene was transcribed under either or both of two conditions being investigated. (B) An outline of the procedure followed to compare directly the mRNAs produced by a tissue under two different conditions, I and II. The mRNAs extracted from the tissue of an animal exposed to the two conditions are labeled with two different fluors (green and red) prior to hybridization with spots of DNA on the microarray. Each mRNA hybridizes with the specific DNA that represents the specific gene that codes for the mRNA. When the fluors are visualized, spots emit either green or red fluorescence if they represent genes that were transcribed under just one of the conditions. Spots that represent genes that were transcribed about equally under both conditions emit yellow (the combination of green and red). Spots emit nothing and appear black if they represent genes that were transcribed under neither condition.

TABLE 3.1 Numbers of genes studied and discovered to be up- or downregulated in transcriptomic research on seven phenomena using DNA microarrays

Phenomenon and tissue studied (reference in parentheses)[a]	Number of genes studied	Number of up- or downregulated genes[b]	Some of the functions controlled by affected genes
Endurance exercise in humans: Thigh muscle 3 h following lengthy (~75 min), high-intensity bicycling compared with before bicycling (Mahoney et al. 2005)	8432	126	Mitochondrial biogenesis, tolerance of oxidative stress, membrane ion transport, nuclear receptor function (all categories mostly upregulated)
Tissue freezing in freeze-tolerant wood frogs: Heart muscle after frogs experienced extracellular freezing compared with before (Storey 2004)	>19,000	>200	Glucose metabolism, antioxidant defense, membrane ion transport, ischemia-related signaling (all categories mostly upregulated)
High water temperature in killifish: Liver after fish were exposed to high temperature compared with ordinary temperature (Podrabsky and Somero 2004)	4992	540	Heat-shock protein synthesis, cell membrane synthesis, nitrogen metabolism, protein biosynthesis
Continuous swimming for 20 days in rainbow trout: Ovaries in fish that swam compared with resting fish (Palstra et al. 2010)	1818	235	Protein biosynthesis, energy provision, ribosome functionality, anion transport (all categories mostly downregulated)
Transfer from freshwater to seawater in eels: Several tissues in eels after transfer to seawater compared with before (Kalujnaia et al. 2007)	6144	229	Transport across membranes, cell protection, signal transduction, synthesis of structural proteins
Exposure to juvenile hormone in developing honeybees: Animals treated with a juvenile hormone analog compared with ones not treated (Whitfield et al. 2006)	5559	894	Foraging behaviors, RNA processing, protein metabolism, morphogenesis
Ocean acidification in sea urchin larvae: Larvae exposed to high CO_2 and low pH compared with ones not exposed (Todgham and Hofmann 2009)	1057	178	Biomineralization, energy metabolism, cellular defense responses, apoptosis

[a]Full citations are listed in the Figure and Table Citations.
[b]The numbers of affected genes are quoted from the references cited; different criteria might have been used in different studies.

that will emit green light after it is fully processed, and the other extract (e.g., that from muscle after exercise) with a different fluor that will emit red light. The two fluor-labeled mRNA preparations are then permitted to hybridize with the microarray. Each mRNA hybridizes with the DNA spot representing its specific gene (❸ in the figure). Thus, at this step, each spot in the microarray potentially becomes labeled. A spot is labeled with mRNA from both preparations if the gene it represents was being transcribed under both conditions. A spot is labeled with mRNA from only one preparation if the gene it represents was being transcribed under only one of the two conditions. A spot that represents a gene that was not transcribed under either condition is not labeled at all. Finally, the fluors bonded to the microarray are visualized by laser scanning (❹ in the figure) so they emit their distinctive fluorescent wavelengths: green or red. If the gene represented by a spot on the microarray was not being transcribed under either condition, the spot emits nothing and appears black. If the gene was being transcribed approximately equally under both conditions, the spot emits both green and red and thus appears yellow. The most interesting spots are those that emit only green or only red, because those represent the genes that were being transcribed under only one of the two conditions studied.

Transcription profiling often identifies large numbers of genes that exhibit altered transcription in response to environmental or other conditions

TABLE 3.1 illustrates that animals routinely modify the transcription of hundreds of genes—by up- or downregulating them—in response to environmental changes of many sorts, or in response to other changes such as exercise or hormone exposure. Physiologists never imagined until recently that such large numbers of genes would be involved. Determining the *significance* of gene transcription changes numbering in the hundreds will obviously be a challenging and revelatory phase in the advance of physiological knowledge. One of the "At Work" chapters in this book, Chapter 21, discusses exercise responses in far greater detail.

Transcription profiling reveals that many genes routinely undergo daily cycles of transcription

Periodic cycles of gene transcription—*transcription rhythms*—are common, based on recent research. Most known cycles in gene transcription are daily cycles in which the pattern of falling and rising transcription repeats approximately every 24 h. Based on studies on a

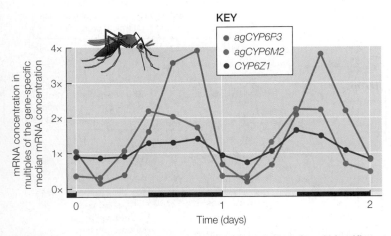

KEY
- agCYP6P3
- agCYP6M2
- CYP6Z1

FIGURE 3.11 Daily cycles in three mRNAs coding for detoxification enzymes in the malaria mosquito _Anopheles gambiae_ The mRNAs produced by transcription of three genes—agCYP6P3, agCYP6M2, and CYP6Z1—were measured over two day–night cycles (see scale at bottom: yellow = day, black = night). The genes and mRNAs code for three detoxification enzymes in the P450 system (see page 50). Because the enzymes help the mosquitoes resist pyrethrin-based insecticides, the cycles in gene transcription suggest that vulnerability to such insecticides is likely cyclic. mRNA levels are expressed on the _y_ axis as multiples of the gene-specific median mRNA concentration (e.g., 2× for a particular gene indicates a concentration twice as high as that gene's median mRNA concentration over the course of the study). Because the mosquitoes studied here could not actually see day and night, biological clocks probably controlled the transcription cycles shown. (After Rund et al. 2011.)

wide variety of animals, 2%–40% of genes exhibit such daily cycles. For example, in a recent study of gene transcription in the lung tissue of laboratory rats, more than 600 genes exhibited clear 24-h cycles in their transcription. These genes participate in just about every aspect of lung function. Some, for example, play roles in tissue maintenance, others in defense against airborne foreign materials, and still others in the genesis of asthma. Because the genes are transcribed more rapidly at some times in the 24-h daily cycle than at others, lung repair and defense are likely to occur more effectively at some times of day than others, and treatments for asthma may be more effective at certain times. All the genes are not synchronous; some reach their peaks and troughs of transcription at different times than others. Some of these transcription cycles are undoubtedly under control of biological clocks (see page 420). Other cycles, however, are probably direct responses to day–night changes in the outside environment.

Malaria mosquitos (_Anopheles gambiae_) provide another thought-provoking example. Hundreds of genes in these mosquitos exhibit daily cycles (**FIGURE 3.11**). Inasmuch as some of these genes play roles in defending the mosquitoes against toxic environmental agents, the mosquitoes may be inherently more vulnerable at some times of day than others to agents (e.g., insecticides) used to combat them (see Figure 3.11).

Manipulations of protein synthesis can be used to clarify gene function

We are concerned in this chapter mostly with the unmanipulated chain of events by which genes are naturally transcribed and translated, leading to changes in proteins and other aspects of the biochemical phenotype. A significant aspect of gene expression, nonetheless, is that it can be manipulated experimentally as a way of gaining insight into gene function.

One strategy of this sort is **gene deletion** or **gene knockout**, in which a gene is manipulated so that experimental animals lack functional copies of the gene. The animals therefore cannot synthesize the mRNA ordinarily associated with transcription of the gene, and consequently they do not synthesize the protein (or proteins) coded by the gene. In principle, such animals will be deficient or inferior in one or more phenotypic traits, and their deficiencies will reveal the function of the missing gene. A converse strategy is **forced overexpression**, in which tissues are subjected to experimentally increased synthesis of the mRNA associated with a gene of interest.[16]

Several considerations can cloud the interpretation of gene knockout or overexpression studies. _Compensation_ is one of the most important: When animals artificially lack a certain protein, for example, they often exhibit other phenotypic alterations that tend to make up for the loss of function they would otherwise exhibit. Mice engineered to lack functional genes for myoglobin synthesis in their heart muscle,[17] for example, have more blood capillaries in their heart muscle than ordinary mice, and they circulate blood faster through the muscle. These compensations, and others, prove that the lack of myoglobin is physiologically significant. Because of the compensations, however, the lack of myoglobin does not have a simple deterministic effect on the overall phenotype of the animals, and the mice are actually quite normal in their overall vigor and appearance.

RNA interference (RNAi) is a cellular process that silences protein synthesis and, among other things, can be manipulated to gain insight into gene function. When certain types of short RNA molecules are introduced into cells and operate in tandem with so-called Argonaute proteins in the RNAi pathway, specific mRNA molecules that are naturally produced by the cells are destroyed or blocked from being translated. In effect, the genes that produce the targeted mRNAs are silenced because the mRNAs produced by transcription of the genes are rendered inoperative before they lead to protein synthesis. The consequence is in many ways similar to that of gene deletion: Certain proteins are not synthesized, providing an opportunity to learn what the proteins normally do. Unlike the case when gene deletion is used, however, animals with _normal genomes_ can be manipulated with RNAi because gene action is blocked following gene transcription.

The **CRISPR/Cas system** is a recently discovered system that evolved in prokaryotes to protect against foreign genetic material. Introduced into an animal cell, this system can be used to "edit" nuclear DNA, meaning it can insert specific deletions into the DNA (a type of gene deletion), or it can be used to introduce specific mutations into the DNA. Small RNA molecules introduced into a cell guide the CRISPR/Cas system to remove a highly defined short, double-stranded stretch of DNA. Effectively, if this is all that happens, a deletion results. However, the system can also be used to replace the removed stretch of DNA with an alternative stretch that can be engineered to have a specific, desired nucleotide sequence. When this inserted sequence is transcribed and translated, the cell will produce the protein coded by the inserted DNA, providing a means of identifying the function of this protein.

---

[16] This might be achieved, for example, by genetically engineering a tissue to have an unusually large number of copies of the gene.

[17] Myoglobin aids intracellular O_2 transport and storage, as discussed earlier in this chapter.

SUMMARY
The Study of Gene Transcription: Transcriptomics

■ Transcriptomics or transcription profiling (also sometimes called expression profiling) is the study of which genes are transcribed in a tissue and the degrees to which they are transcribed. Transcription is evaluated by measuring the messenger RNAs (mRNAs) produced in the tissue.

■ DNA microarrays are a major tool in modern transcriptomic research. A microarray consists of a grid of thousands of DNA spots, each representing a particular gene. Each mRNA produced by a tissue binds to the DNA that corresponds to the gene that produced the mRNA. The DNA spots that thus become labeled with mRNA when exposed to the mix of mRNAs produced by a tissue collectively mirror the genes being transcribed in the tissue.

■ DNA microarrays are often used to carry out a direct comparison of the mRNAs produced by a tissue under two different conditions. The mRNAs produced under the two conditions are labeled with green and red fluors. Gene-specific DNA spots in the microarray then glow with green light, red light, or yellow light (the result of green + red), depending on whether the corresponding genes were transcribed under just one of the two conditions or both.

■ Studies using microarrays often identify hundreds of genes that undergo changes of transcription when an animal is exposed to a change in conditions. Many genes also routinely exhibit daily cycles of transcription.

■ One way to study gene function is to manipulate gene transcription or translation and observe the consequences. Gene knockout, gene overexpression, RNA interference, and the CRISPR/Cas system are four of the major methods used. All modify the capacities of cells to produce specific proteins.

Proteomics

Proteomics is the study of the proteins being synthesized by cells and tissues. The term implies the simultaneous study of large numbers of proteins, even to the extent of screening all proteins that can be detected. One reason for the study of proteomics as a separate discipline is that the proteins coded by many genes are unknown, and therefore biologists cannot *predict* the full list of proteins in a cell from knowledge of gene transcription alone. A second reason for studying proteomics as a separate discipline is that, even if the proteins coded by mRNAs are known and the mRNAs have been quantified, protein concentrations cannot necessarily be predicted because protein synthesis is sometimes only loosely correlated with mRNA synthesis. Cell proteins constitute a part of the cell phenotype. Proteomics is thus a branch of the study of the *biochemical phenotype*.

The set of proteins assayed in a proteomic study can be narrowed down by a variety of methods. Thus a particular proteomic study might examine just the set of proteins that bind ATP—such as ATPase enzymes—or just the set of phosphorylated protein kinases (types of regulatory proteins; see Chapter 2).

(A) A gel employed for protein identification

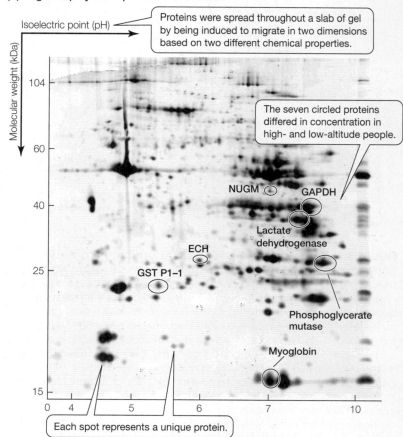

Isoelectric point (pH)

Proteins were spread throughout a slab of gel by being induced to migrate in two dimensions based on two different chemical properties.

Molecular weight (kDa)

The seven circled proteins differed in concentration in high- and low-altitude people.

NUGM GAPDH
Lactate dehydrogenase
ECH
GST P1–1
Phosphoglycerate mutase
Myoglobin

Each spot represents a unique protein.

(B) A modern-day Sherpa carrying a heavy load at high altitude

FIGURE 3.12 Proteomics: A study of tissue protein response to life at high altitude (A) For a proteomic study of muscle proteins in high-altitude human populations, tiny (15-mg) samples of muscle tissue were cut by biopsy from the *vastus lateralis* muscles in the thighs of volunteers and subjected to two-dimensional gel electrophoresis. The gel shown here resulted from the processing of one sample. The proteins in the sample were induced to migrate both from left to right and from top to bottom. When the proteins migrated from left to right, they were separated by isoelectric point (measured in pH units). When they migrated from top to bottom, they were separated by molecular weight (measured in kilodaltons, kDa). The proteins were in solution and invisible during migration. Afterward, however, they were visualized in the gel by staining, as seen here. The seven circled proteins were found to be present in significantly different concentrations in high- and low-altitude people. (B) An individual typical of the high-altitude populations studied. ECH = Δ²-enoyl-CoA-hydratase; GAPDH = glyceraldehyde-3-phosphate dehydrogenase; GST P1-1 = glutathione S-transferase; NUGM = NADH-ubiquinone oxidoreductase. (A after Gelfi et al. 2004.)

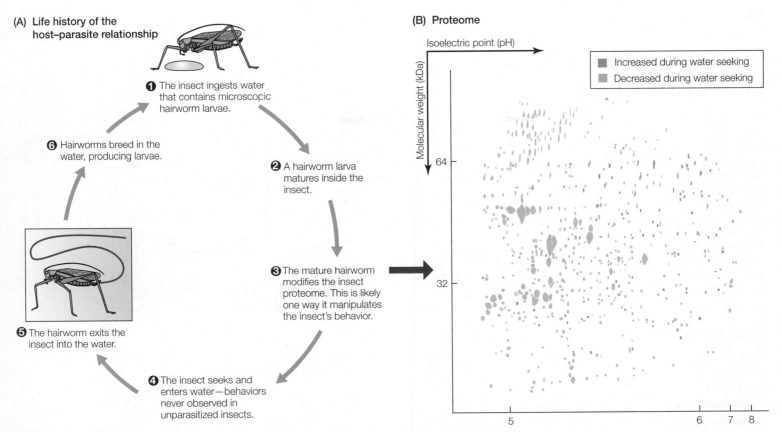

(A) Life history of the host–parasite relationship

❶ The insect ingests water that contains microscopic hairworm larvae.

❷ A hairworm larva matures inside the insect.

❸ The mature hairworm modifies the insect proteome. This is likely one way it manipulates the insect's behavior.

❹ The insect seeks and enters water—behaviors never observed in unparasitized insects.

❺ The hairworm exits the insect into the water.

❻ Hairworms breed in the water, producing larvae.

(B) Proteome

Isoelectric point (pH)

Molecular weight (kDa)

64

32

5   6   7   8

■ Increased during water seeking
■ Decreased during water seeking

FIGURE 3.13 Proteomics: Parasites sometimes alter the protein profile of the central nervous system in their hosts (A) The life history of the parasite–host relationship. (B) Proteins in the brains of katydids (*Meconema thalassinum*) were separated in two dimensions by the methods described in the caption of Figure 3.12A. Unlike in Figure 3.12A, however, the image seen here is not just that of a single gel. Instead, this image is a computer-generated synthesis of evidence from multiple gels, some from control katydids and some from katydids that were in the act of seeking water. Orange-colored proteins were at elevated concentrations during water-seeking behavior. Green-colored proteins were at reduced concentrations. Water-seeking occurs only when katydids are parasitized by hairworms (*Spinochordodes tellinii*), which manipulate the katydids' behavior in ways advantageous to the parasites. Although the insects are called katydids in the United States, they are more often called grasshoppers in Europe, where this study was done. (B after Biron et al. 2005.)

Two-dimensional (2D) gel electrophoresis is a common proteomics method. In this method, the proteins in a tissue sample, after being extracted, are first forced to migrate linearly through a gel material that separates them according to their isoelectric points, and thereafter the proteins are forced to migrate at a right angle to the first migration through a second gel material that separates them according to their molecular weights. In this way, the proteins in the original mixture are spread out in two dimensions—based on two different chemical properties—as seen in **FIGURE 3.12**. The proteins are then chemically identified by, first, excising spots of interest from the gel and, second, analyzing the spots by mass spectrometry or another analytical method.

Figure 3.12A is a gel from a proteomic study carried out to understand how Tibetan Sherpas (see Figure 3.12B) are able to climb with great endurance—often carrying heavy loads—at altitudes above 5500 m (18,000 ft) in the Himalaya Mountains. Tissue was taken by biopsy from a thigh muscle of six Tibetans who had spent their entire lives at altitudes of at least 3500–4500 m and from six Nepal natives who had lived at 1300 m. Proteins in the muscle samples were then subjected to 2D electrophoresis. Seven of the proteins—those circled in Figure 3.12A—proved to be different in concentration in the high- and low-altitude people. One of these, myoglobin, was hardly a surprise; the high-altitude people had relatively high levels of myoglobin, which as we earlier discussed, helps O_2 diffuse into muscle cells. The discovery of very high levels of another protein, the enzyme glutathione S-transferase (GST P1-1), in the high-altitude people was a surprise, however. Investigators now must learn what role this protein plays at altitude; possibly it helps defend against oxidative damage that can cause muscle deterioration at high elevations. When biologists first started to study high-altitude physiology, they focused on breathing and blood circulation. Later they realized that tissue biochemistry is also of great importance, as discussed in Box 23.2. Proteomics now promises to expand rapidly our understanding of which tissue proteins are important.

As a second example of the use of proteomics in research, let's consider a study of a parasite's ability to modify the behavior of its host. A hairworm that parasitizes certain katydids is known to manipulate the behavior of the insects, so that—entirely contrary to their normal behavior—the katydids jump into bodies of water at night. This strange behavior is induced when the hairworm within a katydid has grown to full size, and the behavior enables the hairworm to exit its host into water (**FIGURE 3.13A**). Investigators reasoned that the parasite might exert its effect on the behavior of its host by increasing or decreasing the levels of brain proteins. They also reasoned that the host might upregulate defensive proteins. A proteomic study confirmed that many proteins are increased or decreased in the brain of a parasitized katydid at the time it bizarrely seeks water and jumps in. All the proteins colored orange or green in **FIGURE 3.13B** are significantly altered. The next step will be to determine the roles the protein changes are playing.

FIGURE 3.14 Metabolomics of heat stress Fruit flies (*Drosophila melanogaster*) were exposed to 38°C—a stressfully high temperature—for 1 h and compared with control flies that lived always at 25°C. Whole flies from both groups were analyzed immediately after the heat-stress period. (A) The NMR spectrum obtained from a single fly. The signatures of multiple metabolites are readily seen, and the amplitudes of the signatures can be used to calculate metabolite concentrations. The portion of the spectrum at 5.5 and higher was recorded at greater amplification than the portion at 5.4 and lower. (B) The average concentrations in control and heat-stressed flies of 10 metabolites that exhibited statistically significant changes in heat-stressed flies as compared with control flies, based on the NMR data. The concentrations are in arbitrary units and should be compared only *within* metabolites because different metabolites are scaled differently. The metabolite labeled "Fatty acid" was identified only in general terms, as being a fatty acid–like compound. (After Malmendal et al. 2006.)

Metabolomics

Metabolomics parallels and complements proteomics. Like proteomics, metabolomics seeks to describe the biochemical phenotype. Unlike proteomics, metabolomics does not focus on gene products.

Metabolomics, in a few words, is the study of all the organic compounds in cells and tissues other than macromolecules coded by the genome. The compounds encompassed by metabolomics are generally of relatively low molecular weight (roughly <1500 daltons). They include many types of small molecules found in cells and tissues, such as sugars, amino acids, and fatty acids. Most of these molecules are *metabolites*, that is, compounds currently being processed by metabolism. A central goal of metabolomics is to clarify the metabolic pathways operative in cells and the ways the pathways are modulated. To illustrate metabolomics, let's consider a study of fruit flies subjected to heat stress (**FIGURE 3.14**). Some flies were exposed for 1 h to a heat-stress temperature of 38°C. Their tissues and the tissues of control flies were then extracted with solvents and analyzed by nuclear magnetic resonance (NMR) spectroscopy, a method capable of detecting and quantifying a great diversity of compounds. Numerous metabolites could be observed—because of their distinctive signatures—in the NMR spectrum obtained from each fly (see Figure 3.14A). Researchers therefore could paint a very broad picture of metabolite changes in the cells of the flies (see Figure 3.14B). The results obtained have been used to generate new hypotheses that will help direct future research. For instance, the rise in alanine (see Figure 3.14B) in the heat-stressed flies suggests that flies might depend to an increased extent on anaerobic catabolism when they are under heat stress, because alanine is a product of the anaerobic pathways in flies. From the rise in tyrosine, one might hypothesize that synthesis of several hormones is accelerated during heat stress, because tyrosine is involved in hormone synthetic pathways.

The metabolomic approach differs from traditional metabolite studies in that traditionally just one, two, or three compounds would typically have been studied and the others disregarded. In a metabolomics study, however, as many compounds as possible are measured, so effects are studied without preconceived limits. A metabolomics study often leads to hypotheses for future research rather than to definitive conclusions. In part, this is true

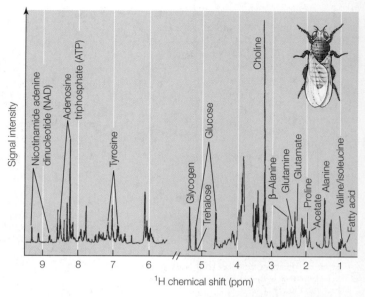

(A) The NMR spectrum obtained from analysis of a single fruit fly

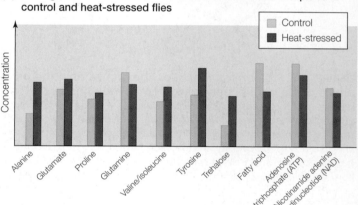

(B) Average concentrations of 10 metabolites that were compared in control and heat-stressed flies

precisely because metabolomics measures any and all effects that are detectable, including ones for which there may be no existing background knowledge. All the areas of *-omics* biology discussed in this chapter are characterized by similar assets and limitations.

STUDY QUESTIONS

1. A case can be made that the sedentary way of life characteristic of many people today is a very recent development in human history, dating from at most a century or so ago. According to this line of argument, physical exertion was a regular part of daily life for most people during most of human evolution. From this point of view, in what way do recent gene-transcription data suggest that physical exertion is in fact probably essential for full health? (For reading on this question, see the article by Booth and Neufer in the References.)

2. Geneticists can synthesize any gene—any stretch of DNA— desired. Suppose you use a tree of genome evolution to predict the structure of a now nonexistent, ancient gene. What insights might you obtain by synthesizing the ancient gene and inserting it into a living animal?

3. The solubility of O_2 in aqueous solutions—such as seawater or blood plasma—increases as temperature decreases. Moreover, the rates at which tissues metabolically consume O_2 tend to decrease as temperature decreases. With these facts in mind, explain why—in the world's oceans—polar seas are the most likely place where fish could survive a mutation that deletes their ability to synthesize blood hemoglobin.

4. Suppose you carry out genomic, transcriptomic, and proteomic studies on a single tissue. What type of information would you obtain from each sort of study? For understanding physiology, what are the uses and shortcomings of each type of information?

5. Compare and contrast the top-down and bottom-up approaches to the study of physiology.

6. One of the most famous quotes in the history of inquiry is Francis Bacon's statement: "The truth emerges more readily from error than confusion." What did he mean? High-throughput methods are sometimes accused of producing such an information overload that they generate confusion. Carefully consider the pros and cons of high-throughput methods as viewed from Bacon's perspective.

7. We tend to think that each expressed gene has a particular effect on an animal's phenotype. Why, therefore, if a gene is knocked out by genetic engineering methods, is the consequent change in an animal's phenotype not necessarily a direct reflection of the gene's effect? For example, how is it possible to knock out a gene that in fact has a phenotypic effect and yet not, in some cases, see any measurable change in an animal's overall ability to function? How might your answer to Study Question 2 be affected by the considerations raised here?

8. The icefish alive today that lack ventricular myoglobin are often cited as examples of "natural knockout animals" because they lack functional genes for myoglobin synthesis. Suppose geneticists engineer some ordinary, modern-day fish to lack functional genes for myoglobin synthesis. Compare the icefish—the "natural knockout animals"—with the first generation of engineered knockout animals. Explain how these two groups of animals would be likely to differ in the mechanisms available for compensating for their lack of myoglobin.

9. Suppose you want to determine if excessive lipid ingestion alters gene transcription. Describe and explain all the steps you would carry out to do a microarray-based study of the question. Include your choice of subjects and tissues to be investigated. (The statement of the problem to be studied—effects of "excessive lipid ingestion"—is deliberately vague, and you should refine it by specifying and explaining an operational procedure.)

10. Suppose you want to determine if dehydration changes the types or amounts of enzymes. Describe and explain all the steps you would carry out to do a two-dimensional gel study of the question. As in Study Question 9, refine the experimental objectives, and include your choice of subjects and tissues.

Go to **sites.sinauer.com/animalphys4e**
for box extensions, quizzes, flashcards,
and other resources.

REFERENCES

Booth, F. W., and P. D. Neufer. 2005. Exercise controls gene expression. *Am. Sci.* 93: 28–35.

Cheng, C.-H. C., and H. W. Detrich III. 2007. Molecular ecophysiology of Antarctic notothenioid fishes. *Philos. Trans. R. Soc., B* 362: 2215–2232.

Gibson, G., and S. V. Muse. 2009. *A Primer of Genome Science*, 3rd ed. Sinauer Associates, Sunderland, MA. A widely applauded introduction to not just genomics but also related topics such as proteomics and bioinformatics. This book emphasizes practical techniques (computer-based and lab-based) that are used in these fields as well as the theoretical perspectives that drive research in the fields. The book is a good place to seek information on methods.

Hardison, R. 1999. The evolution of hemoglobin. *Am. Sci.* 87: 126–137. An accessible treatment of research on the evolution of a protein, emphasizing a genomic perspective.

Miller, K. M., S. Li, K. H. Kaukinen, and 14 additional authors. 2011. Genomic signatures predict migration and spawning failure in wild Canadian salmon. *Science* 331: 214–216. A truly intriguing effort to use genomic methods to find out why a salmon population is failing.

Palstra, A. P., D. Crespo, G. E. E. J. M. van den Thillart, and J. V. Planas. 2010. Saving energy to fuel exercise: swimming suppresses oocyte development and downregulates ovarian transcriptomic response of rainbow trout *Oncorhynchus mykiss*. *Am. J. Physiol.* 299: R486–R499. A wide-ranging study of the potential conflict between sexual maturation and long-distance swimming in migratory fish.

Peck, L. S., M. S. Clark, A. Clark, and 11 additional authors. 2005. Genomics: Applications to Antarctic ecosystems. *Polar Biol.* 28: 351–365. A wide-ranging discussion of genomic concepts and applications, which gains interest from its focus on a single natural system, the Antarctic.

Schlieper, G., J.-H. Kim, A. Molojavyi, and five additional authors. 2004. Adaptation of the myoglobin knockout mouse to hypoxic stress. *Am. J. Physiol.* 286: R786–R792.

Sidell, B. D., and K. M. O'Brien. 2006. When bad things happen to good fish: The loss of hemoglobin and myoglobin expression in Antarctic icefishes. *J. Exp. Biol.* 209: 1791–1802.

Storey, K. B. 2006. Genomic and proteomic approaches in comparative biochemistry and physiology. *Physiol. Biochem. Zool.* 79: 324–332. A thought-provoking—sometimes opinionated—overview of what can be expected from the brave new world of gene/protein studies and how the new agenda will interrelate with (and sometimes commandeer) earlier agendas in the study of animal physiology. Although the paper seems awkward at times because it is focused on a particular symposium, it has the virtue that the writing is readily accessible.

Thornton, J. W. 2004. Resurrecting ancient genes: Experimental analysis of extinct molecules. *Nat. Rev. Genet.* 5: 366–375. A discussion of the startling but realistic proposition that genome science can be used to reconstruct the DNA base sequences of extinct genes, permitting the genes to be resurrected.

See also **Additional References** *and* **Figure and Table Citations**.

Physiological Development and Epigenetics

Hooded seals (*Cystophora cristata*) spend most of their adult lives in the open ocean, where they undertake deep dives to find and capture food, principally fish and squids. The diving behavior of free-living seals has been studied by outfitting the animals with dive recorders that collect information on each dive and radio the data to satellites. From these studies, we know that adults in the open ocean spend about 90% of their time underwater as they dive three to four times per hour, reaching depths as great as 1000 m: more than half a mile. Hooded seals are native to the far-northern reaches of the Atlantic Ocean. The females, like the females of other seals, require a solid substrate to give birth to their young, and they use pack ice (sheets of ice floating on the ocean surface) for this purpose. Each March they haul out on the pack ice and give birth. The newborn hooded seals start life with one of the most astounding bouts of suckling known. Born at an average weight of about 20 kg, they nurse for only 3 days: the shortest nursing period of any mammal. Milk transfer from the mother to the youngster during these 3 days is prodigious, however, and enables each newborn to essentially double its weight to about 40 kg. Afterward the newborns remain on the pack ice without eating for about a month. They are on their own at that point, and when the time comes to eat again, they must venture into the ocean and dive to catch food for themselves.

Hooded seals (*Cystophora cristata*) on pack ice, where the mothers come to give birth After a newborn seal takes millk for a few days and spends additional weeks on the ice, the youngster will be on its own to dive in the ocean to obtain food. Young seals are more limited than adults in their physiological capacities for diving. They gradually acquire adult competence as they mature.

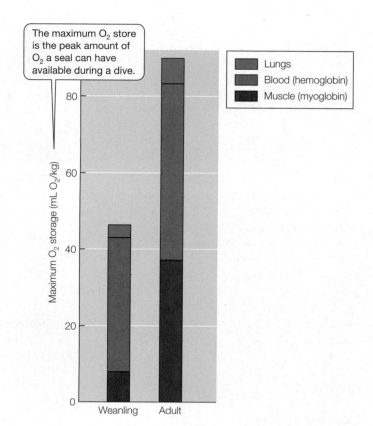

The maximum O$_2$ store is the peak amount of O$_2$ a seal can have available during a dive.

Legend:
- Lungs
- Blood (hemoglobin)
- Muscle (myoglobin)

y-axis: Maximum O$_2$ storage (mL O$_2$/kg)

x-axis: Weanling, Adult

FIGURE 4.1 The maximum O$_2$ stores of hooded seals at two stages of development The total O$_2$ available per kilogram of body weight at each stage of development (weanling and adult) is estimated by adding (from bottom to top) the O$_2$ bound to myoglobin in muscles, the O$_2$ in blood (mostly bound to hemoglobin), and the O$_2$ in lung air. Muscle O$_2$ and blood O$_2$ are higher in adults than weanlings to a statistically significant extent. (After Burns et al. 2007.)

Each individual animal that is born must pass successfully through all the stages of maturation, from infancy to adulthood, if it is ever to reproduce. This is a key principle in the study of development, and it brings into focus the challenge faced by a newborn seal. Each seal must function successfully with the physiological capabilities of a newborn, then the physiological capabilities of a 1-month-old, then those of a 6-month-old, and so forth. For a young seal, capabilities for diving are among the most important.

Oxygen stores play a crucial role in diving. During many dives—although not all (see Chapter 26)—the tissues of a diving mammal maintain their metabolism by using O$_2$ that the animal stored in its body while it was at the surface, breathing air. Oxygen is stored in three major ways. First, during diving, an animal's lung air contains O$_2$, which can potentially be transported from the lungs to other tissues. Second, O$_2$ is combined with hemoglobin in the blood; as a dive progresses, this O$_2$ is released from the blood hemoglobin and taken up by tissues. Third, within the cells of the skeletal muscles and heart, O$_2$ is combined with myoglobin, a red, iron-containing protein—a specialized form of hemoglobin—that gives red muscles their reddish color. The O$_2$ combined with myoglobin is released for use in muscle function—such as swimming in pursuit of prey—during a dive.

Biologists studying the individual development of hooded seals have discovered that the capacity of a young seal to store O$_2$ undergoes a dramatic developmental change as the animal matures to adulthood. This can be seen in **FIGURE 4.1**. Note that the amount of O$_2$ that can be stored at the start of a dive—plotted on the y axis—is expressed per unit of body weight. As you can see, a weanling hooded seal, which is just starting to dive for food, is able to store only about half as much O$_2$ as it will be able to store when it reaches adulthood. This is not because the animal is smaller. Instead, what the data show is that the amount of O$_2$ that can be stored *per unit of tissue* is half as great in a weanling as an adult.

Recognizing that the O$_2$ storage capacities of immature seals are relatively low, we might well expect that when a seal is young, its diving capabilities are limited, compared with an adult. This expectation is supported by the available evidence (**TABLE 4.1**). Young weanlings—like adults—spend about 90% of their time underwater when foraging in the open ocean. However, most dives of weanlings are only 2–5 min long, whereas most dives of adults are 5–25 min long—so weanlings undertake, on average, about 14 short dives per hour in contrast to about 4 long dives per hour in adults. Moreover, weanlings dive to much shallower depths than adults (see Table 4.1). It seems inevitable that the hunting abilities of weanlings are limited by their shorter, shallower diving. Despite such limitations, each animal must find ways to succeed as a youngster, or it will never be a reproductive adult. In this way, the physiology of all the successive stages of individual development—**developmental physiology**—is a life-and-death matter for every individual.[1]

Focusing on the storage and use of O$_2$, what properties of weanlings are most likely to account for their short dives? One factor is that small individuals of a species tend to have higher weight-specific rates of O$_2$ consumption than large individuals of the same species (see Chapter 7). That is, a small weanling is likely to use O$_2$ faster per unit of body weight than a large adult. Thus, if the weanling and adult were to have the same O$_2$ stores per unit of body weight, the

TABLE 4.1 Dive durations and depths observed in free-living weanling and adult hooded seals (*Cystophora cristata*) in the Atlantic Ocean between Greenland and Norway[a]

Property measured	Weanlings	Adults
Durations of most dives (i.e., ~80% of dives)	2–5 min	5–25 min
Longest dives[b]	10–15 min	25–52 min
Depths of most dives (i.e., ~70% of dives)	< 20 m	100–600 m
Deepest dives[b]	100–250 m	970–1020 m

Sources: Folkow and Blix 1999; Folkow, et al. 2010.

[a]Over 170,000 dives were recorded using satellite-linked dive recorders.

[b]Outliers that represented < 0.1% of the data set are omitted.

[1] Another, commonly used term for individual development is *ontogeny*.

weanling would tend to run out of O_2 faster during a dive simply because of its smaller body size. Actually, as we have seen (see Figure 4.1), a weanling has a far smaller O_2 store per unit of weight than an adult, compounding the problem.

The most important factor that limits the O_2 stores of weanlings, relative to adults, is that weanlings have very small capacities to store O_2 in combination with myoglobin in their muscles, as Figure 4.1 shows. Whereas adults on average have about 37 mL of O_2 bound to muscle myoglobin per unit of body weight, weanlings have only about 8 mL: less than a quarter as much. There are two reasons for this. First, weanlings have less muscle (in proportion to their total size) than adults. Second—and more important—weanling muscles have far lower concentrations of myoglobin than adult muscles. Myoglobin per gram of muscle rises during individual development. As this happens, a seal's muscles become redder, and its ability to store O_2 in its body increases dramatically, poising the animal to dive longer and deeper.

The Physiology of Immature Animals Always Differs from That of Adults

Universally, the physiology of youngsters differs in important ways from that of the adults of their species, gradually changing in age-specific ways as postnatal development takes place. For one way to see these points, consider the maturation of three categories of tissues in humans, shown in **FIGURE 4.2**. The brain matures rapidly. Already by 7 years of age, a person's brain has reached its full adult size, whereas general body tissues are less than half developed at that age and the reproductive organs have hardly started to grow. Lab rats are very different, emphasizing that there is not simply one pattern of growth that applies to all animals. In rats, the reproductive organs reach full maturation before the brain does.

One consequence of the developmental program evolved by humans is that young children are in a dramatically different position from adults in the challenge they face to meet their brain's energy needs. Brain tissue in mammals has a particularly high metabolic rate per gram—and therefore a particularly high need for fuel per gram—compared with most tissues; the metabolic rate per gram of brain is more than ten times that of skeletal muscle, for example. The high per-gram metabolic rate of brain tissue explains why, in human adults, the brain accounts for about 20% of total body metabolism (see page 186) even though brain weight is only 2% of body weight.

In young children, the brain accounts for far more than 20% of total metabolism. Because of the rapid, early development of the brain in humans, a youngster who is 4–5 years old has a brain that is almost full size (see Figure 4.2) even though the youngster's body is small. Biologists have calculated that in children of that age, the brain accounts for approximately 50% of total body metabolism; that is, half of the food metabolized by a 4- or 5-year-old is for his or her brain! One implication is that childhood starvation—a common phenomenon around the world—is a particular threat to the most crucial of all human tissues.

Beyond brain energetics, countless cases are known in which brain performance undergoes maturation during the normal development of young animals, with long-term consequences for each individual. Striking examples come from the study of birds that employ the stars to help guide them during nocturnal migration, such as indigo buntings (*Passerina cyanea*), which breed in North America, and garden warblers (*Sylvia borin*), which breed in Europe. In both of these species, youngsters hatch and undergo their early development in the Northern Hemisphere. As adults, these birds migrate to warmer latitudes each winter. They migrate at night, and in the dark they are able to determine compass directions, so they know whether they are flying north or south. How is this possible? From careful research we know that these birds can determine compass directions by looking at the stars! Obviously, the stars, *in and of themselves*, do not provide compass information. When a person determines north and south from the stars, the person does so by having learned that particular stars serve this purpose. Birds must do much the same thing. Experiments have shown that young indigo buntings and garden warblers, during their first summer of life, observe the rotation of the stars in the sky (**FIGURE 4.3**) and undergo a maturational

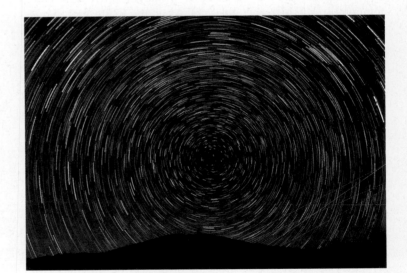

FIGURE 4.3 The apparent rotation of the stars when looking north in the Northern Hemisphere The stars seem to make circles around the northern end of Earth's axis because of Earth's rotation. This apparent movement of the stars is evident in this photograph, which was made by pointing the camera toward the northern sky and keeping the shutter open continuously for over an hour. As newly hatched indigo buntings and garden warblers undergo their early development, they use the apparent rotation of the stars to identify north (the center of the apparent circles), and they undergo a maturational change whereby they thereafter possess a brain record of the stars positioned at the north.

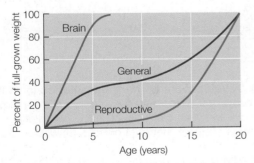

FIGURE 4.2 Growth of the brain, reproductive organs, and other ("general") tissues in humans Each tissue type is plotted, at each age, as a percentage of its full-grown weight. (After Bogin 1999.)

change whereby thereafter they possess a brain record of the stars at the center of the rotation. They then associate those stars with north. Without this maturation during their early weeks of life, they cannot determine compass directions from the stars. Normally, of course, they are able to observe the stars during their early development, and consequently, when they head south for the first time in the autumn of their first year of life, they possess a fully functional star compass to guide them. The necessity for star observations during early development was discovered by manipulative experiments in a planetarium, where the stars can be made artificially to rotate around any spot in the sky. If birds undergo their early development in a planetarium where the stars rotate around a spot in the eastern sky, they later treat the eastern stars as if those stars are in the northern sky, so when they travel north based on their internal calculation of direction, they actually travel east.

Thermoregulation is another aspect of physiology that undergoes postnatal development. Adult mammals and birds are noted for physiological regulation of body temperature, termed **homeothermy**. Placental mammals, for example, typically maintain a deep body temperature of about 37°C as adults (see page 260). Universally, however, newborn mammals are not as effective in maintaining homeothermy as adults of their species. Individuals develop their full capacity for homeothermy as they mature from birth to adulthood.

An example is provided by the white-footed mouse (*Peromyscus leucopus*), one of the most common native rodents in North America (see Chapter 11 for additional examples). Individual adult white-footed mice can maintain a deep body temperature near 37°C when the air temperature is below freezing. Individual newborns, however, quickly cool to the environmental temperature; in a near-freezing environment, their tissues cool to near-freezing temperatures (which they tolerate). As young white-footed mice grow and mature during their 3-week-long nestling period, the peak rate per gram at which they can metabolically generate heat increases dramatically (**FIGURE 4.4A**). In addition, they develop fur, and their resistance to heat loss (their whole-body insulation) increases (**FIGURE 4.4B**). The mice therefore become increasingly able to thermoregulate as isolated individuals, away from their nest and siblings (**FIGURE 4.4C**). By 18 days of age, a lone youngster can thermoregulate for several hours even in freezing-cold air, and it capitalizes on this newfound ability by making its first excursions outside its nest (**FIGURE 4.4D**).

Physiological development occurs at all scales: all levels of organization. The scale of individual proteins is particularly fundamental. This is true because the proteins expressed in a tissue determine the tissue phenotype: the overt structural and metabolic properties of the tissue.

One class of proteins of great importance is the enzymes. During the development of a tissue, as the tissue matures, a common pattern is for enzymes in the tissue to be upregulated—increasing their amounts and activities—in an almost stepwise fashion, under genetic control. For example, **FIGURE 4.5** shows such stepwise upregulation of three enzymes in the liver of developing rats. As new enzymes appear in cells (and old ones disappear), their collective catalytic and regulatory properties alter cell metabolism. For instance, although early fetal rats cannot synthesize liver glycogen because they lack glycogen synthetase, newborn rats *can* because

4 days old

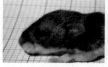
10 days old

14 days old

(A) Peak rate of metabolic heat production (cal/g•h)

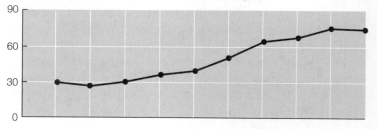

(B) Whole-body insulation below thermoneutrality (°C•g•h/cal)

(C) Lowest ambient temperature at which isolated individuals can thermoregulate (°C)

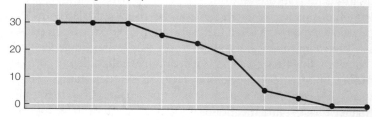

(D) Time per day spent out of nest (h)

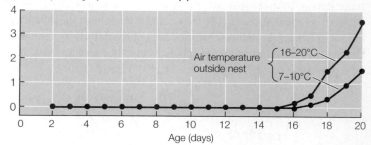

FIGURE 4.4 Development of thermoregulation in white-footed mice (*Peromyscus leucopus*) Four properties are shown as functions of age between birth (age 0) and 3 weeks of age, when weaning occurs. (A) Peak thermogenic rate. (B) Whole-body insulation (*I*) (see Equation 10.10). (C) The lowest air temperature at which an isolated individual can thermoregulate for 2.5–3.0 hours. (D) The number of hours per day that nestlings spend out of the nest when the air temperature outside is 7°C–10°C or 16°C–20°C. In all parts, the data are means for all individuals studied. (After research by Richard W. Hill; photos courtesy of Robert J. Robbins.)

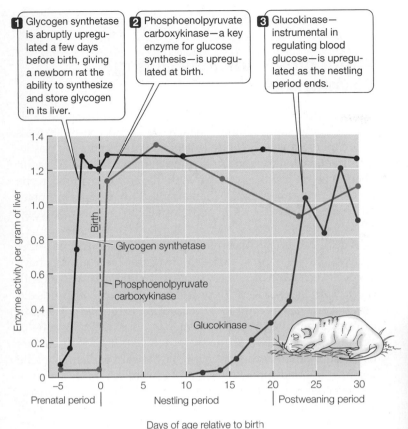

1. Glycogen synthetase is abruptly upregulated a few days before birth, giving a newborn rat the ability to synthesize and store glycogen in its liver.

2. Phosphoenolpyruvate carboxykinase—a key enzyme for glucose synthesis—is upregulated at birth.

3. Glucokinase—instrumental in regulating blood glucose—is upregulated as the nestling period ends.

FIGURE 4.5 Sequential upregulation of three liver enzymes during development in laboratory rats Activities of three enzymes in the developing liver are shown. Day 0 on the *x* axis is the day of birth; negative times are days before birth. (After Walker 1983.)

the required enzyme is suddenly expressed during the 5 days before birth (see Figure 4.5).

The tissues in which a protein is highly expressed may also undergo developmental change. To illustrate this phenomenon, we will use a fish example.

Fish living in seawater need to transport Cl^- ions out of their bodies to prevent their body fluids from becoming overly salty. In adults, this transport is carried out by cells in the gills. However, many species of fish complete a substantial portion of their early development before their gills form, raising the question of where Cl^- transport occurs in their early life stages. The cells that transport Cl^- ions out of the body are known as **chloride cells** (see Box 28.1) and can be identified because they express the transporter protein Na^+–K^+-ATPase (see page 115) in abundance (the protein is crucial for the cells' ion-transport function). Because these cells profusely express this specific protein, they can be located using immunocytochemistry. In this technique, an antibody is made against a protein of interest—Na^+–K^+-ATPase in this case—and when tissues are exposed to this antibody, the antibody binds wherever its target protein occurs. By visualizing the binding sites, one can learn the locations where the target protein is expressed (and how abundant it is). This method has revealed the fascinating developmental patterns depicted in **FIGURE 4.6**. In the species shown, a killifish, young that are 8 days old or younger do not have gills. They do, however, have chloride cells, which—at these early ages—are found in the yolk sac membrane and scattered over the surface of the skin (see Figure 4.6A). By 15 days of age, the young

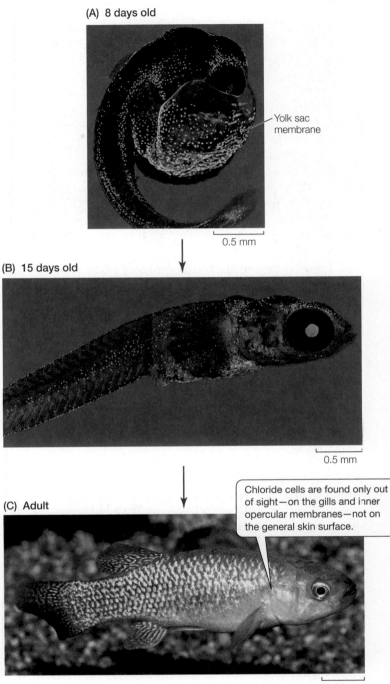

(A) 8 days old

Yolk sac membrane

0.5 mm

(B) 15 days old

0.5 mm

(C) Adult

Chloride cells are found only out of sight—on the gills and inner opercular membranes—not on the general skin surface.

1 cm

FIGURE 4.6 The locations of chloride cells at three stages of development in killifish (*Fundulus heteroclitus*) living in seawater The chloride cells are responsible for transporting Cl^- ions out of the body so the tissue fluids do not become overly salty. In (A) and (B) the chloride cells were labeled with an antibody that glowed green in the light used to produce the images. Ages are measured from conception. (A and B after Katoh et al. 2000; images courtesy of Toyoji Kaneko.)

have emerged from the egg membrane (typically at 11 days), and the yolk sac has been reabsorbed. At this stage, chloride cells are still found widely on the skin surface (see Figure 4.6B), and they are beginning to appear in the gills, which have started to develop. Later, as killifish progress to adulthood, chloride cells cease to be found on the general skin surface and become mostly localized in

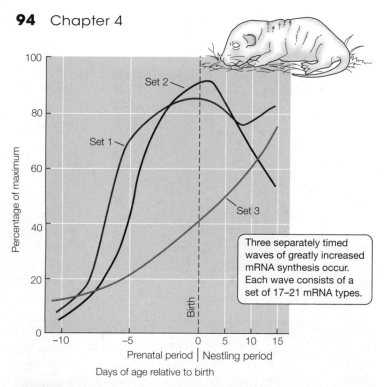

FIGURE 4.7 Developmental programming of gene transcription in fetal and newborn rats The graph shows quantities of three sets of mRNAs in the developing spinal cord. Almost 120 mRNAs were monitored in this study. Some (not shown here) were steady or declined. Day 0 on the *x* axis is the day of birth; negative times are days before birth. Time is scaled differently before and after birth. (After Wen et al. 1998.)

Text inside figure:
- y-axis: Percentage of maximum
- Set 2, Set 1, Set 3
- Birth
- Three separately timed waves of greatly increased mRNA synthesis occur. Each wave consists of a set of 17–21 mRNA types.
- x-axis: Prenatal period | Nestling period; Days of age relative to birth

the gills (and inner opercular membranes). In all, a major shift occurs in the location of outward Cl⁻ transport as a killifish develops. For the vast majority of individuals conceived, the site of Cl⁻ transport is the yolk sac and skin because early death rates are very high, and most individuals die before they develop gills.

The development of an animal from conception to adulthood is, in part, under genetic control. Lately, investigators have become able to study patterns of genetic control directly by using DNA microarrays and other transcriptomic methods to monitor gene expression (see Chapter 3). From studies of this kind on laboratory rats and other species, we now know that different sets of mRNAs are upregulated sequentially during development (**FIGURE 4.7**), confirming that different sets of genes are transcribed at different times in a developmental program. Some of the mRNAs in the studied sets code for enzymes. Programmed transcription of the sort observed in Figure 4.7 can thereby give rise to sequences of enzyme upregulation such as those seen in Figure 4.5.

Introduction to Phenotypic Plasticity and Epigenetics

For the rest of this chapter, we will focus on phenotypic plasticity and epigenetics, both of which have intimate connections with development. These phenomena are essential for understanding numerous topics in addition to development: Their importance does not arise exclusively from their roles in development. Nonetheless, their roles in development are so vital and significant that they are logical subjects to which to turn in this chapter.

Phenotypic Plasticity during Development

Phenotypic plasticity, as discussed in Chapter 1 (see page 18), is the ability of a single animal—with a fixed genotype—to express two or more genetically controlled phenotypes. Phenotypic plasticity is frequently observed during development. An animal's adult phenotype, for instance, often differs if the animal matured in one type of environment rather than another. Two human examples will provide a good starting point for discussing this highly significant phenomenon.

The first example concerns the age at which **menarche**—first menstruation—occurs in girls. Many European countries have historical records on menarche dating back to at least the nineteenth century. Analyses of these records show that, on average, girls did not attain menarche until 16–17 years of age in the mid-nineteenth century, whereas they attained menarche at about 13 years of age in the second half of the twentieth century. Most scholars have concluded that this dramatic shift toward earlier menarche between the mid-nineteenth and late twentieth centuries was a consequence of improved nutrition, public health, and medical care. That is, population genotypes did not evolve in ways that caused the recorded shifts in menarche. Instead, changes in the environment were responsible. Girls in the late twentieth century reached reproductive maturity when they were several years younger than girls in the mid-nineteenth century because they enjoyed a more benign childhood environment.

The second human example concerns the restriction of growth that is often observed in people who live in a relative state of nature. Based on recent, detailed studies, 4- to 12-year-old children in populations of Maya people in Guatemala are, on average, 5.5 cm (2 inches) shorter at each age than children in populations of the *same* ethnic group who recently settled in the United States. From every perspective—public health, sufficiency of the food supply, access to medical care, and political security—Maya populations in Guatemala face greater environmental challenges than do those in the United States. Evidently, the environmental stresses in Guatemala impair growth, and when those stresses are relaxed, children attain greater stature at each age.

Several historical comparisons—although not as well documented as the case of the Maya—also point to a strong environmental effect on human growth. **FIGURE 4.8**, for example, compares the stature of youngsters in two early populations in the British Isles with the stature of recent British youngsters. The factory children studied in 1833 were far shorter than modern children at each age—16–20 cm (6–8 inches) shorter in their teenage years. A 14-year-old factory child in 1833 was about as tall as a 10- or 11-year-old today! The factory children were, in fact, dramatically shorter than aristocratic children living at the same time; Francis Galton (1822–1911) commented that working-class children and aristocratic children of that time could be told apart by their heights, and the fact that the elite grew faster helped reinforce their social status. Youngsters in a medieval English rural population were even a bit shorter than the nineteenth-century factory children, as seen in Figure 4.8. The rude conditions of life that existed for medieval people and nineteenth-century factory children are believed to have been responsible for their short statures.

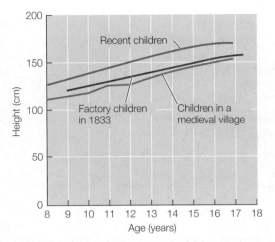

FIGURE 4.8 Height as a function of age in three groups of children in England The "Recent children" and "Factory children in 1833" were living subjects when measured. The "Children in the medieval village" (Wharram Percy) had died in childhood, and their skeletons were later excavated; height was calculated from bone lengths, and age was estimated from dental features. (After Mays 1999.)

Examples such as these emphasize that the developmental environment can exert strong effects on an individual's phenotype. The developmental program does not unfold as a strictly genetic process, and the phenotype is not simply a fixed property set by genes (as some popularizations of genetic determinism argue). The phenotype, instead, is a product of interaction between genes and environment—with the developmental environment often being particularly important.

Phenotypic plasticity is known today to occur in such a wide range of contexts that the question arises of whether a single term—*phenotypic plasticity*—is appropriate. Possibly several distinct types of phenomena are being lumped under that one rubric. Here, nonetheless, we use the global term *phenotypic plasticity* because a consensus has not yet emerged on an alternative terminology.

Environmental effects during development may arise from programmed responses to the environment or may be forced by chemical or physical necessity

Now we need to focus on a significant question that arises in all studies of phenotypic plasticity—the question of whether envi-

ronmental effects are programmed responses or are forced on the animal. Two further examples—deliberately chosen to illustrate extremes—will set the stage.

The flat periwinkle snail (*Littorina obtusata*) and blue mussel (*Mytilus edulis*) (**FIGURE 4.9**) are marine molluscs that are preyed on by crabs. Recent research has shown that when these molluscs grow in the presence of crabs over a period of weeks, they upregulate shell deposition and develop thicker shells than members of the same species that grow without predators being present. To eat the flesh of a snail or mussel, a crab must first crack the mollusc's shell with its claws. A thicker shell therefore protects against crab predation. No direct contact is required between the crabs and molluscs to trigger shell thickening. For example, when snails share the same water with crabs, the snails develop shells that are more than 10% thicker, even when the snails and crabs are kept separate; the snails detect that crabs are nearby by use of chemical cues in the water.

A different effect of the environment on development is seen in studies of brain function following malnutrition. Particularly rigorous studies can now be conducted on *spatial learning*—the ability to learn and remember locations—in rats and mice, by using two devices, the *Morris water maze* and *radial-arm maze* (see page 510). In the Morris water maze, a small platform is positioned beneath the water surface in a pool in which a test animal swims (rats and mice swim readily). An opaque dye is added to the water so that the animal cannot see the platform. When an animal is first introduced into the water maze, it swims at random until it finds the platform and can crawl out of the water. Later, as the procedure is repeated over and over, the animal will exhibit learning with each repetition. The rat or mouse will arrive at the platform sooner and travel less distance to find it. Many studies have shown that when recently born rats or mice are subjected to a low-protein diet during their postnatal development, they exhibit impaired performance in the water maze, even long after being returned to

(B) Blue muscle (*Mytilus edulis*)

(A) Flat periwinkle (*Littorina obtusata*)

FIGURE 4.9 Some marine molluscs synthesize thicker shells when in the presence of predators These two species of shelled molluscs, both of which are common seashore inhabitants, have been demonstrated to produce thicker shells when crabs that eat them are nearby.

a fully adequate diet. Thus deficiencies in an animal's nutritional environment during development cause an easily demonstrated, long-term deficit in capacity to learn.

A persistent question in studies of phenotypic plasticity is whether adaptive evolution in the past is responsible for the pattern of phenotypic plasticity observed. In the case of snails having their development altered by the presence or absence of predators, the pattern of phenotypic plasticity is summarized as follows: When predators are absent, a snail grows an ordinary shell, but when predators are present, a snail grows a thickened shell.[2] This pattern of plasticity could readily be advantageous in all respects, in that a snail does not expend extra energy to synthesize and carry a thickened shell when such a shell is unneeded, but it develops an augmented shell in the presence of danger. Accordingly, natural selection could well have favored the evolution of this pattern of plasticity. That is, the plasticity might well be genetically programmed because it has become inbuilt over evolutionary time by positive selection.

In contrast, in the case of young rats having their brain development altered by availability of dietary protein, the pattern of plasticity is summarized as follows: When dietary protein is adequate, a rat develops a normal capacity for learning, but when dietary protein is inadequate, a rat grows up to be deficient in its ability to learn. Probably, this pattern of plasticity, instead of being a product of adaptive evolution, simply represents pathology. In a protein-poor environment, the relevant part of a rat's brain (the hippocampus) may simply be unable to develop normally, and in that case the observed learning deficiency would be *forced* on the rat by chemical or physical necessity.

In many cases of developmental phenotypic plasticity, no means presently exist to resolve the question of causation. Accordingly, different biologists sometimes take opposite stances. Recall, for example, the low growth rates and delay of menarche in human populations living under stressful environmental conditions. Most biologists tend to think that these effects are simply pathologies; according to this reasoning, poor public health and uncertain nutrition prevent normal growth and maturation. Some biologists, however, argue that when environmental conditions are harsh, growth and maturation are programmed to be slowed because there are advantages to being small-bodied when food is scarce, and there are advantages to delaying reproduction when daily life is difficult.

Insect polyphenic development underlies some of the most dramatic cases of phenotypic plasticity

Many species of insects exhibit **polyphenic development**, a phenomenon in which *genetically identical* individuals can assume two or more distinct body forms, induced by differences in the environment. The body forms of a polyphenic species can be so dramatically different that anyone but a specialist would think they represent different species. **FIGURE 4.10**, for example, shows a dramatic case of polyphenic development. The two butterflies shown do not differ genetically. They belong to one species, and indeed either of the individuals could have developed either of the patterns of wing coloration. Polyphenic development is a particularly striking form of phenotypic plasticity.

The species shown in Figure 4.10 exemplifies a particular type of polyphenic development. Many species of insects, including the

[2] In technical terms, this sentence is a statement of the norm of reaction discussed in Box 1.2.

Developed in spring   Developed in summer

FIGURE 4.10 An example of polyphenic development: Butterflies of a single species differ dramatically in phenotype depending on the season when they develop In the western white butterfly (*Pontia* [*Pieris*] *occidentalis*), any one individual can express either of the phenotypes shown; the difference between phenotypes represents phenotypic plasticity and is not caused by differences in genotype. This species goes through more than one generation each year, and the phenotype changes with the season in which an individual develops—a case of seasonal polyphenism. (Courtesy of Tom Valente.)

one shown in the figure, go through two or more generations each year, and individuals that develop in one season differ in body form from those that develop in another season—a phenomenon termed **seasonal polyphenism**. Day length and temperature during development often serve as cues that help determine which body form an individual will assume. Two hormones, juvenile hormone and ecdysone (see Table 16.5)—and the timing of their secretion relative to endogenously timed sensitive periods—are frequently implicated in the control of seasonal polyphenism.

Extensive studies have established that in the western white butterfly shown in Figure 4.10—and in several other butterfly species —seasonal polyphenism aids thermoregulation. In these butterflies, an individual requires a high thoracic temperature to fly; if the thorax is too cool, the flight muscles cannot develop sufficient power. The required high body temperature is achieved by complex sun-basking behaviors. Owing to seasonal polyphenism, adults have different amounts and patterns of dark pigment (melanin) in their wings during different seasons. Careful studies have shown that seasonal timing of the different phenotypes aids thermoregulation in that (1) during cool seasons, the prevailing wing phenotypes aid absorption of solar radiation for warming (as in the spring butterfly in Figure 4.10, which has dark wings), whereas (2) during hot seasons, the prevailing wing phenotypes offer more options for avoiding solar overheating.

Probably the most stunning examples of phenotypic plasticity in the animal kingdom are provided by migratory locusts—notorious since ancient times for their capacity to descend from the sky in incomprehensible numbers and ravage crops. Swarms are large and fast-moving—capable of decimating vegetation over large swaths of land (**FIGURE 4.11**). In some cases, a swarm may extend at any single time over a total land area of 100 km[2] or more, and swarms may include tens of millions of locusts per square kilometer.

Both *Locusta migratoria* and *Schistocerca gregaria*, the species of migratory locusts that have been best studied, exhibit polyphenic development. They have two behavioral phenotypes—*solitary* and *gregarious*. These two phenotypes, besides differing in behavior, also differ strikingly in their morphology (e.g., coloration; see 4.11 inset). No genetic difference exists between individuals that show the solitary and gregarious phenotypes. Instead, any individual can

FIGURE 4.11 Migratory locusts in their gregarious pheno-type move across the landscape in voracious, fast-moving swarms Both major species are shown. The swarm consists of *Locusta migratoria*, photographed in Madagascar. The inset shows the gregarious (left) and solitary (right) phenotypes of *Schistocerca gregaria*.

exhibit either phenotype. When an individual exhibits the solitary phenotype, it avoids other individuals. Solitary individuals are colored in ways that make them inconspicuous, and they are difficult to find in their natural setting. However, when circumstances bring individuals displaying the solitary phenotype into contact with each other for a few hours, they transform to the gregarious phenotype and begin to associate avidly with other locusts. That transformation is the starting point for swarm formation. The adaptive advantage is postulated to be that the fast-moving, voracious behavior of swarms enables locusts in a swarm to collect large quantities of food. When locusts in the solitary phenotype become crowded together—signaling that numbers in their population may be outstripping their local food supply—the transformation to swarming poises them to collect food from a large area at an enhanced rate.

Recent research has pinpointed two biogenic amines (see Table 13.2) as key agents that physiologically mediate the transformation from solitary to gregarious. In *Schistocerca gregaria*, serotonin is both necessary and sufficient to induce transformation to the gregarious phenotype. In *Locusta migratoria*, dopamine is the principal agent of phenotype change, although serotonin also plays a role. We will have more to say about migratory locusts in the next section.

Other animals besides insects also sometimes exhibit polyphenic development

Some species of fish, lizards, and turtles exhibit *temperature-dependent sex determination*. In these cases, an individual can develop into a male or a female, depending on the body temperatures it experiences at certain stages of its development. The phenomenon is an example of polyphenic development. Other vertebrate examples occur in certain species of amphibians, in which an egg can develop into either of two very different tadpole types, depending on the environment. In Cope's gray treefrog (*Hyla chrysoscelis*), for example, eggs develop into one tadpole type if no predators are present in the pond where they live, but they develop into another tadpole type—differing in behavior and tail anatomy—if predators are present. Similarly, the presence of predators can influence the development of several species of water fleas—small freshwater

crustaceans in the genus *Daphnia*. The water fleas develop into adults of a different body form if predators are present than if their pond is predator-free. They are useful for studying polyphenic development because they can reproduce parthenogenetically (asexually): A mother can produce sets of offspring that are genetic clones of herself and therefore genetically identical to each other. Among such offspring, those that develop in the presence of predators form stronger, larger exoskeletons than those not exposed to predators.

Examples such as these emphasize that insects are not the only animals to exhibit polyphenic development. Nonetheless, it is in insects that polyphenic development is most widely found and takes its most dramatic forms.

SUMMARY
Phenotypic Plasticity during Development

■ Phenotypic plasticity is the capacity for an individual of fixed genotype to exhibit two or more genetically controlled phenotypes. Because the phenotype expressed is often dependent on the prevailing environment, phenotypic plasticity is a process by which genotype and environment interact to determine the phenotypic characteristics of an individual.

■ Phenotypic plasticity is often programmed by a genetically coded, physiological control system that determines which specific phenotypes are expressed under which specific environmental conditions. Such control systems are subject to natural selection over evolutionary time and may, therefore, in themselves represent adaptations. At the opposite extreme, changes of phenotype in different environments may simply be forced by chemical or physical necessity. A major challenge for biologists is to develop empirical ways to determine whether instances of phenotypic plasticity represent adaptations or products of chemical or physical forcing.

■ Polyphenic development in insects is perhaps the most striking form of phenotypic plasticity. In a species with polyphenic development, an individual with a fixed genotype can express two or more highly distinct phenotypes. Solitary and gregarious forms of migratory locusts provide an example.

Epigenetics

To start our discussion of epigenetics, we will adopt the formal definition that is most widely adopted by specialists today. Later we will say a bit more about definitions.

Epigenetics refers to modifications of gene expression—*with no change in DNA sequence*—that are transmitted when genes replicate. Although we usually think of *genes* being transmitted during gene replication, *gene expression* differences are transmitted along with the genes when epigenetic effects are at work. To illustrate, let's take a brief look at one of the mechanisms involved (later we will say more about mechanisms). As shown in the transition from cell ❶ to cell ❷ in **FIGURE 4.12**, an environmental agent might lead methylation mechanisms in a cell to attach methyl ($-CH_3$) groups to the DNA in a cell's nucleus at specific locations. These methyl groups will affect expression of certain genes in cell ❷. When cell ❷ divides to produce two new cells, ❸ and ❹, the DNA is replicated

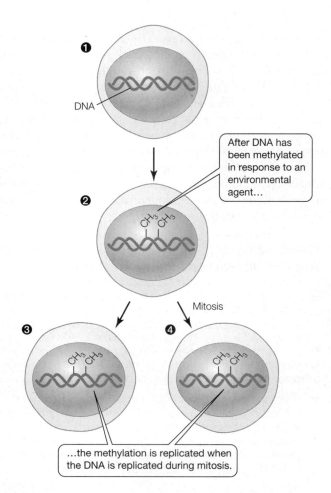

FIGURE 4.12 Methylation of DNA caused by epigenetic mechanisms is perpetuated during DNA replication in mitosis   In this way, the gene expression modification caused by methylation is transmitted to daughter cells while the DNA sequence remains unaltered.

to produce two identical copies, one for each daughter cell. Consequently, the DNA sequence (sequence of nucleotide bases) remains the same in all cells. Cell ❷, however, differs from cell ❶ in having methyl groups attached to the DNA at specific locations, and when an epigenetic mechanism is at work, this methylation pattern is passed on to cells ❸ and ❹ when cell ❷ replicates, as shown. Because the methyl groups affect expression of certain genes in cell ❷ and because cells ❸ and ❹ have the same methyl groups, cells ❷, ❸, and ❹ will have the same pattern of gene expression, which will differ from that of cell ❶. Thus we see, as said earlier, that the *gene expression* difference caused by the attached methyl groups in cell ❷ is transmitted during cell replication to cells ❸ and ❹ without any change in the *DNA sequence*.

As already suggested, some epigenetic modifications of gene expression are environmentally induced. Some, having been environmentally induced in embryonic cells during the early development of an animal, are passed to all the cells in the adult that arise, during development, from the cell lineages initially affected. Some are passed, many researchers argue, from parents to offspring—meaning that epigenetics provides a mechanism whereby environmental effects can be transmitted from one generation to the next. The existence of these stunning phenomena

discredits the old dogma that, in the transmission of hereditary material, the genes are isolated from environmental influences. Interest in epigenetics has soared in the past 20 years, as evidenced by a tenfold increase in the number of pertinent scientific papers published each year. Much of this flurry of interest is focused on human disease processes and the effects of toxic chemicals. However, biologists are rapidly discovering the operation of epigenetic mechanisms in the ordinary lives of a diversity of animals, and ever-increasing attention can be expected from physiologists in the near future.

Epigenetics provides an entirely new dimension in which genes and environment interact. To see this, let's first reflect again on programmed phenotypic plasticity, which—as we have seen earlier in this chapter—is another process that mediates interaction between genes and environment in the determination of phenotype. In cases of programmed phenotypic plasticity, although genes control the program, the genes themselves are not modified in any way. For example, consider the case of snails and crabs discussed earlier (see Figure 4.9A). Snails develop an ordinary shell when crabs are absent from their environment, but they develop a thickened shell when crabs are present. A genetic program orchestrates the control sequence by which snails use sensory information from their environment to modulate shell formation. However, the genes responsible for this program—speaking loosely—*use* the environmental information *without being modified* by it.

By contrast, in some forms of epigenetic modification, the expression of genes is semi-permanently altered by interaction with the environment. A gene affected in this manner is said to be **marked** or **tagged** in such a way that its expression is modified, and when the gene replicates, its marking also replicates so that the resulting gene copies also have their expression modified in the same way (see Figure 4.12). The environmental modification of gene expression is thereby passed along as gene replication takes place. All the while, the DNA sequence of the gene remains unaltered. Phenotype is modified by the sort of gene–environment interaction seen in epigenetics, but it is modified in a different way than in programmed phenotypic plasticity. In the case of epigenetics, phenotype is altered because genes are directly modified in a transmissible way by marking, and the marking causes their expression to be different than it was before the environmental effect.

Phenotypes arising from epigenetic modification (similarly to ones arising from phenotypic plasticity) may, in principle, be adaptive for the organism or not. If a particular epigenetic modification has been subject to natural selection over evolutionary time, it may prove to be a product of positive selection and advantageous. However, when environmental agents chemically force epigenetic marking (as is known to occur), the epigenetic modifications may have negative effects. Adaptive effects cannot be assumed.

Two major mechanisms of epigenetic marking are DNA methylation and covalent modification of histone proteins

As already implied, one of the most widely observed and best-understood mechanisms of epigenetic marking is DNA methylation. In this process methyl ($-CH_3$) groups are attached by covalent bonds to specific cytosine residues in the structure of DNA. In the most common version of this process, methylation is at a promoter region of DNA, and when it occurs there, expression of the gene regulated by the promoter is reduced because the gene is repressed or silenced. When a cell with DNA methylation replicates, an enzyme (usually a DNA methyltransferase in the DNMT1 family) acts to perpetuate its pattern of methylation by methylating the same cytosine residues in each daughter cell as are methylated in the starting cell (see Figure 4.12).

DNA methylation is not employed equally by all animal groups. It is an important mechanism in vertebrates and some types of insects. However, the mechanism seems to be absent (or of low importance) in many types of invertebrates, including some insect groups.

Another important mechanism of epigenetic marking is covalent modification of histone proteins in a cell's nucleus. The histone proteins, which occur in the nucleosomes of the chromatin, assemble into discoid structures around which DNA is wrapped. The histone molecules in these structures can be modified by methylation, acetylation, phosphorylation, or other covalent binding of chemical groups at specific sites. These chemical changes are often termed *posttranslational modifications* of the histones, referring to the fact that they occur after the histone proteins have been synthesized by translation. Because of the intimate relationship between DNA and histones in the cell nucleus, histone modifications can alter gene expression. As in DNA methylation, mechanisms exist to perpetuate the modifications during cell replication. Additional mechanisms of epigenetic marking also exist. Small RNA molecules play roles, for example.

For general conversation about epigenetics, all the chemical modifications responsible for epigenetic effects—such as DNA methylation and histone modification—are often referred to simply as epigenetic *marks* or *tags*, as already suggested. This shorthand simplifies discussion. When all the epigenetic marks in a cell or tissue are described together as a set, this global summary of marks is called the **epigenome** of the cell or tissue (in parallel to the *genome*, which is the set of all genes). The set of all methylated DNA sites is called the **methylome**.

Epigenetic inheritance can be within an individual or transgenerational

Let's focus briefly now on patterns of *inheritance* of epigenetic marks, disregarding the chemical nature of the marks. There are two major types of epigenetic inheritance. One of these entails the perpetuation of marks during the process of cell division by mitosis within an individual animal. This is the type of inheritance illustrated in Figure 4.12, and it is termed *mitotic* or *somatic* inheritance.

The second major type of epigenetic inheritance entails passing marks from one generation to the next. In principle, marks in an individual animal, if present in the germ cells, can be incorporated during meiosis into sperm or eggs, which can then pass the marks to the individual's offspring. This type of inheritance is termed *meiotic* or *transgenerational* inheritance. Transgenerational epigenetic inheritance is well documented in plants and in model animals (e.g., laboratory mice) subjected to toxic agents. Biologists are less certain, however, that such inheritance occurs in animals living in their natural environments, although suggestive cases exist.

A challenge at present is that the word *epigenetics* has been in use for over 70 years and has not always been used in the same way. Some authorities today would limit discussion of epigenetics just to cases that exhibit the types of inheritance we have specified here. Other authorities, however, use the word in additional ways.[3]

Epigenetic marking plays a key role in tissue differentiation during ordinary development

All the cells in an individual animal's body typically have the same genes. How, then, can liver cells be consistently different from muscle cells, and muscle cells consistently different from fat cells? The immediate answer is that each tissue is characterized by a consistent, tissue-specific pattern of gene *expression*. Liver cells reliably *express* only a subset of the genes present in each cell, muscle cells reliably express a different subset, and fat cells express still a different subset.

Early in an individual animal's development, the developmental program establishes many distinctive cell lineages differing in their patterns of gene expression. For example, one lineage may ultimately produce liver cells, whereas another may ultimately produce muscle cells. Epigenetic marking that takes place early in development plays a key role in defining, establishing, and maintaining the cell lineages. Within each lineage, epigenetic marks help determine the suite of genes expressed and thus the cell phenotype (e.g., liver or muscle). As cells in a lineage divide during development, these marks are copied into their daughter cells (see Figure 4.12), ensuring that the new cells remain true to their lineage and perpetuating the lineage-specific phenotype.

Evidence increasingly points to epigenetic control of polyphenic development

In colonies of social insects such as bees and ants, individuals that are *identical or very similar in genotype* may belong to different castes, in which they have *dramatically different phenotypes*. For example, in honeybees (*Apis mellifera*), females that are genetically identical can become either workers or queens. An individual becomes a queen instead of a worker if she eats a particular food termed royal jelly. The development of females into these two differing castes—workers and queens—is a type of polyphenic development.

In recent years, cutting-edge research using manipulations of epigenetic marking has provided strong evidence that such marking plays a major role in caste determination. In honeybees, for example, researchers have shown that larvae destined to become workers can be redirected to become queens by manipulating DNA methylation in their cells. In this case, the change of epigenetic marking radically alters the physiology of an individual; queens have a functioning reproductive system, whereas workers do not, for example. The change of epigenetic marking also alters body size and longevity; queens are larger than workers and live far longer. More than 20% of the genes in the brain of a developing female are expressed differently if she becomes a queen rather than a worker, depending on epigenetic marks.

In the Florida carpenter ant (*Camponotus floridanus*), another social insect, there are two castes of female workers, termed *minor* and *major*. As shown in **FIGURE 4.13**, minors (red line) start to

---

[3] For example, it is fairly common for biologists to use the word "epigenetic" whenever the biochemical marking mechanisms we have discussed are used to control gene expression. Fully mature cells that are no longer dividing may be marked by these biochemical mechanisms, for example. In such cases, DNA is not being replicated, yet the process may be called "epigenetic."

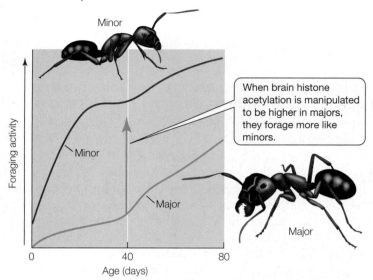

FIGURE 4.13 Foraging activity of the major caste in the Florida carpenter ant (*Camponotus floridanus*) can be elevated to resemble that in the minor caste by manipulating epigenetic marks The red line shows average intensity of foraging as a function of postlarval age in minors, which carry out most of the foraging for a colony. The blue line shows similar data for majors. The blue arrow depicts the change in foraging of majors caused by brain injections of agents that alter epigenetic marking. (After Simola et al. 2016.)

forage when very young and, at each age, forage far more actively than majors (blue line). When chemical agents that promote histone acetylation (a type of histone modification) were injected into the brains of majors, the majors started to forage far more actively (at a level resembling that in minors) and continued to do so for weeks. However, if other agents that had opposite effects on histone acetylation were injected at the same time as the acetylation-promoting agents, the switch of majors to minor-like foraging was suppressed. These results point strongly to epigenetic control of caste-specific foraging activity. In earlier experiments, researchers also found that they could manipulate head size (far larger in majors than minors; see Figure 4.13) by modifying DNA methylation during larval development.

In water fleas (*Daphnia*) and migratory locusts—both of which we have seen exhibit polyphenic development—there is evidence for transgenerational epigenetic inheritance. Epigenetic control of polyphenic development has not been definitively demonstrated as yet in these cases. However, circumstantial evidence—including the evidence of transgenerational inheritance—points to epigenetic control.

When solitary locusts first experience crowding, they change within a few hours to the gregarious phenotype (see Figure 4.11). During this process, their appearance changes, they become attracted to other locusts, they start to fly during the day instead of night, and they are far more predisposed to flying. However, this transformation to the gregarious phenotype is incomplete, in the sense that the transformation becomes even more dramatic in succeeding generations, when the changes already listed intensify and additional, radical changes—such as remodeling of the brain—take place. One prominent hypothesis is that the initial transformation from solitary to gregarious is controlled epigenetically, and then—when the newly gregarious individuals reproduce—their epigenetic marks for gregariousness are passed to their offspring

by transgenerational epigenetic inheritance. The offspring would then not only experience close contact with other locusts—an experience that promotes the gregarious phenotype—but also possess inherited epigenetic marks that promote the gregarious phenotype. These mutually reinforcing drives would then account for intensification of the transformation.

Epigenetic marking may account for lifelong effects of early-life stress

Evidence has accumulated in the last few decades that early-life stress can have lifelong physiological effects. Epigenetic marking is implicated. Early-life stress can change the epigenetic marking of cells in a young animal. Then, as the animal grows and develops, these marks can be passed from cells to daughter cells by mitotic inheritance (see Figure 4.12), so that—years later in adulthood—the marks from early development persist. The marks from early stress may then affect the physiology of the adult.

A fascinating recent study illustrates the application of these concepts to humans. During the winter of 1944–1945, Nazi Germany imposed a food embargo on part of The Netherlands as one of the final spasms of World War II. The Dutch citizens were subjected to severe famine in what has come to be called the Dutch Hunger Winter. Studies of pregnant laboratory rats subjected to malnutrition throughout conception and pregnancy have shown that certain genes in their offspring have fewer methylation marks than in ordinary rats. To see if this same phenomenon occurs in humans, researchers tracked down 60 people, all of whom were over 60 years old at the time of the study, who were conceived during the Dutch famine and compared them with same-sex control siblings who were not conceived during the famine. DNA was extracted from the whole blood of each person and analyzed for methylation marks in a gene of particular relevance. As **FIGURE 4.14** shows, the researchers discovered that there is a statistically strong tendency for the individuals who were conceived during the famine to exhibit a lower percentage of marked DNA sites than their sibling controls. As noted already, methylation marks are generally associated with reduced gene expression. Accordingly, reason exists to expect that the reductions of methylation marks induced by famine are allowing greater expression of some genes in the adult people whose lives began during the Hunger Winter.

A stunning consideration to recognize in this study of Dutch people conceived in famine is the timescale. The original environmental effect was exerted on their cells when they were embryos. More than 60 years later, that effect was still evident in their blood cells, probably having been transmitted through a great many cell divisions (i.e., throughout their lives) in the cells that multiply to produce blood cells. Shortly we will come back to this fascinating case.

Let's first look more generally at lifelong effects of early-life stress. Much of the research on this subject has been focused on chronic diseases of adulthood. A central hypothesis is that early-life famine or malnutrition is associated with an increased likelihood of cardiovascular and metabolic disease later in life. Researchers have found a lot of evidence for associations of this type. The associations have been studied by two methods: epidemiological research on people and experimental research on laboratory animals, principally laboratory rats. Studies of both types indicate that famine or malnutrition during preg-

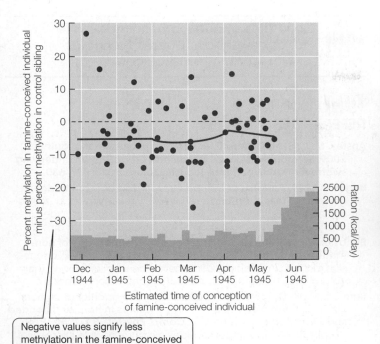

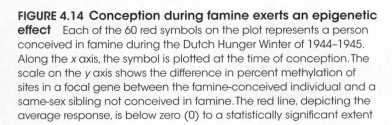

Negative values signify less methylation in the famine-conceived individual than in the control sibling.

FIGURE 4.14 Conception during famine exerts an epigenetic effect Each of the 60 red symbols on the plot represents a person conceived in famine during the Dutch Hunger Winter of 1944–1945. Along the *x* axis, the symbol is plotted at the time of conception. The scale on the *y* axis shows the difference in percent methylation of sites in a focal gene between the famine-conceived individual and a same-sex sibling not conceived in famine. The red line, depicting the average response, is below zero (0) to a statistically significant extent ($P < 0.00002$), indicating that famine-conceived individuals tend to exhibit a reduced extent of DNA methylation. The daily ration of food per person in the food-deprived population was tightly regulated and recorded, and is shown at the bottom; note that during the famine (December 1944 to May 1945), the daily ration per person was far below 2000–2500 kilocalories (kcal), the range of values considered ordinary. The two people in the photograph were victims of the food embargo in the winter of 1944–1945. (After Heijmans et al. 2008.)

nancy—especially early pregnancy—predisposes offspring to a variety of disease states that often develop relatively late in life, including cardiovascular diseases and metabolic diseases such as diabetes and obesity.

Although the mechanisms of these associations are not fully understood, strong evidence indicates that epigenetic marks play roles in the associations. In fact, the associations are among the best documented examples of epigenetic control of individual phenotype. Possibly, therefore, the alteration of marks observed in the 60-plus-year-old Dutch people is not merely of abstract, academic interest. The research on those people may have directly revealed epigenetic changes—wrought when they were embryos—that now, late in their lives, affect their odds of developing coronary artery disease, hypertension, and other ailments. In fact, separate studies have shown that these people show an unusually high prevalence of obesity and coronary artery disease.

Effects of maternal care of offspring have also been studied and represent another of the best-documented examples of epigenetic control of individual phenotype. After genetically homogeneous female lab rats have given birth, they exhibit two patterns of maternal care during their offsprings' first week of life. One pattern—termed *low-LG*—is characterized by low amounts of licking and grooming of the young; the other—*high-LG*—is characterized by high amounts of licking and grooming. Although the young of these two types of mothers are homogeneous in their DNA sequences, they develop into adults that display multiple, consistent differences.

In adulthood, the low-LG offspring are more behaviorally fearful and show greater endocrine responses to stress than the high-LG offspring. Direct chemical measurements on genes in hippocampal tissue from the brain show that the low-LG offspring differ significantly in epigenetic marking from the high-LG offspring. The results of extensive analysis show that offspring differences are epigenetically programmed by the care provided by their mothers and ripple through their lives to affect fearfulness and endocrine responsiveness to stress in adulthood!

SUMMARY
Epigenetics

- Epigenetics refers to changes in gene expression that are transmitted during gene replication despite the lack of any alteration in the DNA sequence. The mechanism of epigenetic control is that genes are marked in ways (e.g., DNA methylation or histone modification) that modify their expression and the marks are replicated when the genes are replicated.

- Epigenetic marking is often initiated by environmental conditions because of the action of programmed control systems or of chemical or physical forcing. Accordingly, epigenetic control is a specific mechanism by which genotype and environment interact to determine phenotype.

- Because epigenetic marks are replicated from cell to cell as cells divide in an individual, epigenetic marks induced by the environment early in development can be perpetuated throughout life and affect the adult phenotype.

- Increasing evidence indicates that polyphenic development, such as differentiation into castes in social insects, is controlled by epigenetic marking.

STUDY QUESTIONS

1. You are asked to give a 20-minute talk to high-school students on why nutrition matters in a child's first years of life. Your assignment is to present a biologist's perspective. Write your talk.

2. Genotype and environment interact to produce the phenotype of an animal. Explain comprehensively the types of mechanisms by which this interaction takes place.

3. Rigorously define phenotypic plasticity, polyphenic development, and epigenetics. Outline how they relate to each other.

4. Explain why the physiological properties of individual animals *at each stage of their development* are crucially relevant for their ecological and evolutionary success.

5. As seen in Figure 24.20, freshwater water fleas (*Daphnia*) develop without hemoglobin (Hb) and are whitish when living in well-aerated water but produce abundant Hb and are red in O_2-depleted water. Explain how each of these phenotypes could be advantageous (relative to the other) in the environment where it is expressed.

6. Regarding the water fleas in Question 5, describe experiments you could carry out to test empirically whether (a) the low-Hb phenotype is more advantageous than the high-Hb one in well-aerated water and (b) the high-Hb phenotype is superior to the low-Hb one in O_2-poor water.

7. After studying Box 1.2, state in your own words the meaning of norm of reaction. Give an example.

8. Write a short explanatory essay on the following statement: "In cases of genetically programmed phenotypic plasticity, natural selection acts on the norm of reaction. Thus the norm of reaction as a whole—rather than any one phenotype—determines if selection is positive or negative."

9. In northern latitudes, a variety of birds and mammals create storage depots of food, termed *caches*, in the autumn and later can find the caches in winter to obtain food. What experiments could be done to determine if the ability to find caches is fully innate at birth or must undergo postnatal development?

10. Some molecules found in foods are known to be capable of driving DNA methylation or DNA demethylation. For example, folic acid (a vitamin), vitamin B_{12}, and choline are thought to increase methylation. Based on this consideration, why would it be important to avoid eating *excessive* quantities of such molecules (while being certain to eat sufficient quantities for health)?

11. In principle, what consequences might be expected to arise from an environmentally induced shift in the age of reproductive maturation? Think as broadly as possible. Answer the question for the human populations, discussed in this chapter, in which menarche shifted to occur more than 3 years earlier between the mid-nineteenth and late twentieth centuries. Also answer for a nonhuman mammal such as a species of squirrel or antelope.

Go to **sites.sinauer.com/animalphys4e** for box extensions, quizzes, flashcards, and other resources.

REFERENCES

Anstey, M. L., S. M. Rogers, S. R. Ott, and two additional authors. 2009. Serotonin mediates behavioral gregarization underlying swarm formation in desert locusts. *Science* 323: 627–630.

Bogin, B. 1999. *Patterns of Human Growth*, 2nd ed. Cambridge University Press, Cambridge. This highly recommended book deliberately adopts a biological, rather than medical, perspective to analyze the full range of knowledge regarding human growth.

Bonasio, R. 2015. The expanding epigenetic landscape of non-model organisms. *J. Exp. Biol.* 218: 114–122. A peerless discussion of the roles of epigenetic inheritance in animals living their natural lives in their natural habitats. Highly recommended.

Burns, J. M., K. C. Lestyk, L. P. Folkow, and two additional authors. 2007. Size and distribution of oxygen stores in harp and hooded seals from birth to maturity. *J. Comp. Physiol., B* 177: 687–700. Despite its title, this paper includes a commendable overview of modern knowledge of diving throughout the life cycle in seals in general.

Chittka, A., and L. Chittka. 2010. Epigenetics of royalty. *PLoS Biol.* 8(11): e1000532. doi:10.1371/journal.pbio.1000532. A thought-provoking overview of caste determination in honeybees that starts in a memorably entertaining way.

Fraga, M. F., E. Ballestar, M. F. Paz, and 18 additional authors. 2005. Epigenetic differences arise during the lifetime of monozygotic twins. *Proc. Natl. Acad. Sci. U.S.A.* 102: 10604–10609. A fascinating paper, providing evidence that epigenetic mechanisms play a large role in the common phenomenon of identical twins developing distinctive phenotypes.

Ho, D. H., and W. W. Burggren. 2010. Epigenetics and transgenerational transfer: a physiological perspective. *J. Exp. Biol.* 213: 3–16. This intellectual tour de force is the best available treatment of epigenetics as a process in animal physiology. However, a much broader definition of epigenetics is adopted than in this chapter. Readers should recognize that several very disparate concepts are under discussion in the paper.

Kingsolver, J. G., and R. B. Huey. 1998. Evolutionary analyses of morphological and physiological plasticity in thermally variable environments. *Am. Zool.* 38: 545–560. A particularly cogent effort to assess the evolutionary adaptive value of phenotypic plasticity, with emphasis on empirical tests of relevant hypotheses.

Piersma, T., and J. Drent. 2003. Phenotypic flexibility and the evolution of organismal design. *Trends Ecol. Evol.* 18: 228–233. This provocative article is a noteworthy attempt to provide a modern theory of phenotypic plasticity.

Raichle, M. E., and M. A. Mintun. 2006. Brain work and brain imaging. *Annu. Rev. Neurosci.* 29: 449–476. This fascinating paper discusses cutting-edge knowledge of brain energetics, with a focus on insights provided by—and questions raised by—modern neuroimaging.

Trussell, G. C. 1996. Phenotypic plasticity in an intertidal snail: The role of a common crab predator. *Evolution* 50: 448–454.

West-Eberhard, M. J. 2003. *Developmental Plasticity and Evolution*. Oxford University Press, New York.

See also **Additional References** *and* **Figure and Table Citations**.

Transport of Solutes and Water

Hummingbirds are dietary specialists that feed almost exclusively on flower nectars, which are principally solutions of sucrose (table sugar), plus other sugars, notably glucose and fructose. When sucrose itself is digested, it yields glucose and fructose. Thus a hummingbird has an abundance of glucose and fructose in its intestines after digesting a meal, and these simple sugars represent its immediate reward for feeding. The sugars, however, are in solution in the lumen—the hollow central core—of the intestines, and for them to do the bird any good, they must move into the bloodstream to be carried throughout the body.

You might at first imagine that glucose and fructose in the lumen of a hummingbird's intestines could simply diffuse across the intestinal epithelium into blood vessels. However, the hydrophilic or hydrophobic nature of the molecules is a factor (see Chapter 2). Glucose and fructose are hydrophilic, and because hydrophilic molecules tend not to mingle with hydrophobic ones, the sugars can be expected to have difficulty dissolving in the highly hydrophobic interior of the apical and basolateral cell membranes of the epithelial cells lining the intestine. In fact, the glucose and fructose from a meal cannot diffuse into a hummingbird's bloodstream at significant rates. How then do the sugars move into the blood? Questions of this sort are among the most common and important faced in the study of animal physiology.

Mechanisms must exist to transport sugars from the open central cavity of a hummingbird's intestines into its blood following a meal
Nectars are unusually rich in sugars. We may well wonder if sugar-transport mechanisms are exceptionally developed in hummingbirds because of the birds' nectar diets.

(A) A single animal cell: relative ion concentrations inside and outside

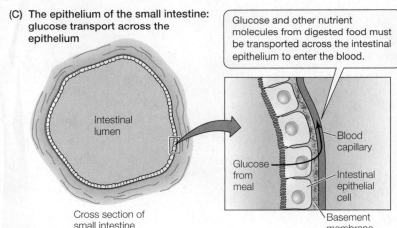

(B) The gill epithelium of a freshwater fish: relative ion concentrations on the two sides

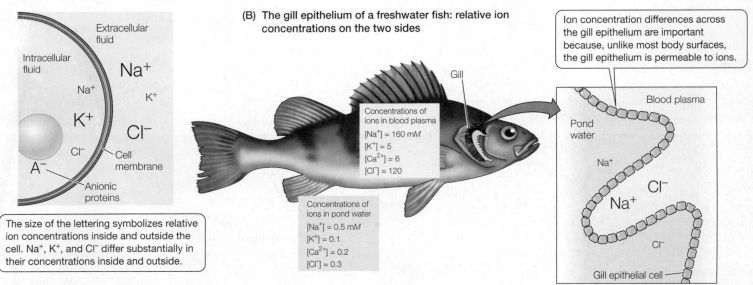

Ion concentration differences across the gill epithelium are important because, unlike most body surfaces, the gill epithelium is permeable to ions.

Concentrations of ions in blood plasma
$[Na^+] = 160$ mM
$[K^+] = 5$
$[Ca^{2+}] = 6$
$[Cl^-] = 120$

Concentrations of ions in pond water
$[Na^+] = 0.5$ mM
$[K^+] = 0.1$
$[Ca^{2+}] = 0.2$
$[Cl^-] = 0.3$

The size of the lettering symbolizes relative ion concentrations inside and outside the cell. Na$^+$, K$^+$, and Cl$^-$ differ substantially in their concentrations inside and outside.

(C) The epithelium of the small intestine: glucose transport across the epithelium

Glucose and other nutrient molecules from digested food must be transported across the intestinal epithelium to enter the blood.

Cross section of small intestine

FIGURE 5.1 Three focal examples in which transport occurs Study of the three examples shown in this figure will bring to light most of the basic principles of solute transport. (A) A typical cell in an animal's body, surrounded by extracellular fluid. A$^-$ represents anionic (negatively charged) proteins and other solutes that cannot cross the cell membrane and thus are trapped inside the cell. (B) A freshwater fish, showing representative ion concentrations in the blood plasma and surrounding pond water. A thin epithelium is all that separates the blood plasma and water in the gills of the fish. (C) A diagram of the small intestine of a bird or mammal, emphasizing the intestinal epithelium lining the lumen. In reality the epithelium is microscopically thin.

In the life of any animal, many materials must move across cell membranes and epithelia in substantial quantities to resupply cells or tissues with needed raw materials, to void wastes, to maintain the proper composition of body fluids, and to otherwise maintain the animal's integrity. The **solutes**—that is, the dissolved materials—that move across cell membranes and epithelia are chemically very diverse, and the mechanisms by which solutes and water cross are numerous. Thus a question that quickly arises is this: What term should be used to describe the movements globally? In this book we use the term **transport** to refer, in an entirely general way, to any and all movements of solutes or water across cell membranes or epithelia, regardless of the mechanisms of movement.

Three examples in which transport is important to animals will help define the subject of this chapter and provide a focus for our study of transport. Here we describe the examples. Later, as the chapter unfolds, we will return to all three to elucidate the specific transport processes involved.

1. One of our focal examples is *a single cell inside an animal's body*. A typical animal cell is bathed by extracellular fluid. Thus we can think of the cell membrane as separating an *intracellular fluid* on the inside from an *extracellular fluid* on the outside (**FIGURE 5.1A**). The solutions on the inside and outside of a cell are normally similar in their *sum-total*

concentrations of dissolved matter; speaking more exactly, they have similar osmotic pressures.[1] Consequently the solutions inside and outside a typical cell do not tend to exchange water in net fashion by osmosis to a great degree. However, the *solute compositions* of the two solutions differ dramatically. The concentration of potassium ions (K+) is much higher inside an animal cell than outside, for instance, whereas sodium (Na+) and chloride (Cl–) ions are typically less concentrated inside than outside. These differences in solute composition suggest that the intracellular fluid of a cell is not at equilibrium with the extracellular fluid bathing the cell. What are the transport mechanisms by which ions such as Na+ and K+ move toward equilibrium across a cell membrane, and what are the transport processes that keep the intracellular fluid different in composition from the extracellular fluid?

2. Our second focal example is *the outer gill membrane of a freshwater fish living in a pond or stream*. The fluid portion of the blood of a freshwater fish—known as the fish's *blood plasma*—is much more concentrated in Na+, Cl–, and other inorganic ions than the surrounding pond or stream water is (**FIGURE 5.1B**). In the fish's gills, the relatively

---

[1] Osmotic pressures and osmosis are discussed later in this chapter.

concentrated blood plasma is separated from the dilute pond water by just a thin epithelium. Because the gill epithelium is a permeable rather than impenetrable barrier, the blood plasma tends to lose ions such as Na+ and Cl– to the pond water across the epithelium, and H2O tends to enter the blood plasma from the pond water across the epithelium. What are the transport mechanisms by which inorganic ions and H2O cross the gill epithelium? Are there transport mechanisms in the epithelium that help maintain the difference in composition between the fish's blood plasma and the environmental water?

3. Our third focal example is *the intestinal epithelium of the small intestine of a bird or mammal.* As we saw in Chapter 2 (see Figure 2.5B), this epithelium consists of a single layer of cells bearing microvilli on their apical cell membranes (**FIGURE 5.1C**). As stressed already in our discussion of hummingbirds, dissolved sugar molecules such as glucose and fructose must cross this epithelium from the intestinal lumen to the blood after a meal if they are to be of use to an animal. The same is true of dissolved amino acids and other compounds. What are the transport mechanisms that move each of these materials across the epithelium?

As we start our study of transport mechanisms, an important organizing principle is the distinction between *passive* transport mechanisms and *active* ones. To define *passive* and *active* in this context, we first need to define **equilibrium**. Be cautious of the term, because it is commonly used in nonscientific ways, such as when people say that an angry child has "recovered his equilibrium" or that a city has reached "equilibrium size." These sorts of uses tend to confuse rather than clarify the scientific meaning of the term.

An accurate definition of *equilibrium* is based on the second law of thermodynamics, which describes the behavior of systems, termed **isolated systems**, that have no inputs or outputs of energy or matter.[2] The second law of thermodynamics states that when an isolated system undergoes change, it is not able to change in all conceivable ways. Instead, an isolated system can change in only certain, limited ways. Equilibrium is, by definition, the state toward which an isolated system changes; that is, it is the state toward which a system moves—internally—when it has no inputs or outputs of energy or matter. A system is *at equilibrium* when internal changes have brought it to an internally stable state from which further *net* change is impossible without system inputs or outputs. *The state of equilibrium is a state of minimum capacity to do work under locally prevailing conditions.* Thus a change toward equilibrium is always in the direction of decreasing work potential.[3]

Passive-transport mechanisms, by definition, are capable of carrying material only in the direction of equilibrium. **Active-transport mechanisms**, by contrast, *can* carry material in the direction opposing equilibrium. The word *can* is emphasized for an important reason: An active-transport mechanism does not *necessarily* carry materials in the direction opposing equilibrium, but it is *capable* of doing so.

---

[2] We say much more about the second law of thermodynamics in Chapter 7.

[3] Appendix J provides background on work and other physics concepts used in this chapter.

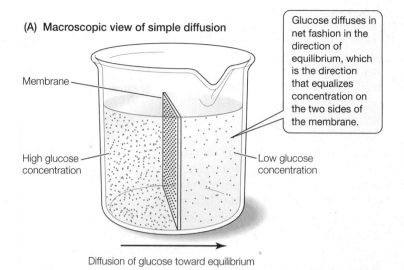

(A) Macroscopic view of simple diffusion

Membrane

High glucose concentration

Low glucose concentration

Glucose diffuses in net fashion in the direction of equilibrium, which is the direction that equalizes concentration on the two sides of the membrane.

Diffusion of glucose toward equilibrium

(B) Microscopic view of simple diffusion

As a glucose molecule on the left side of the membrane collides with other molecules, its random motions, such as depicted here, can line it up to pass through a pore to the right side of the membrane. More molecules pass through the membrane from left to right by this chance process than from right to left, simply because the density of glucose molecules is higher on the side with higher glucose concentration.

High glucose concentration

Low glucose concentration

Glucose

FIGURE 5.2 Simple diffusion viewed macroscopically and microscopically (A) A beaker is divided in half by a porous artificial membrane that permits passage of glucose molecules, but nothing else. The solutions on either side of the membrane differ in glucose concentration (symbolized by different dot densities), and glucose diffuses through the membrane. (B) Looking at the same system microscopically reveals that random motions are more likely to carry glucose molecules from left to right than from right to left. A point to stress is that glucose pores of the sort in this example occur only in artificial membranes, not in living biological membranes.

Passive Solute Transport by Simple Diffusion

Simple diffusion is the most straightforward form of passive solute transport. To understand simple diffusion, consider a beaker divided by a nonliving membrane that is penetrated by many microscopic pores through which the sugar glucose, but nothing else, can pass. A solution of glucose in water is placed on each side of the membrane, with the solution on the left having a higher glucose concentration than that on the right. Viewing this beaker macroscopically (**FIGURE 5.2A**), we know from everyday experience what will happen: Glucose will move in net fashion from left to right across the membrane until ultimately the concentration of glucose is equal on both sides. However, what is the *mechanism of motion* that carries glucose from left to right? In addition, why do the concentrations of glucose on the two sides *stay equal* once they have become equal?

Answering these questions will reveal the nature of simple diffusion. To answer them, we need to look at the beaker *microscopi-*

cally. At temperatures above absolute zero, all atoms and molecules undergo ceaseless random motions. Imagine that you're taking a microscopic look at the membrane and the solutions just on either side, and you fix your eye on a particular glucose molecule to the left of the membrane (**FIGURE 5.2B**). This molecule, like all the others, will be moving constantly, sometimes colliding with and bouncing off other molecules in random directions. There is a chance that one of these random motions will carry the molecule through a pore in the membrane to the right-hand side, as the figure shows. Over a period of time, more glucose molecules will move in this way from left to right than from right to left, merely because there are more glucose molecules per unit of volume on the left. Macroscopically, these individual molecular events will result in net glucose transport from left to right. Once the concentrations on the two sides have become equal, glucose molecules will continue to move at random from left to right and from right to left. The numbers of molecules going in the two directions will then be equal, however, because of the equal numbers of glucose molecules per unit of volume on the two sides. Thus, at a macroscopic level, the concentrations on the two sides will remain equal once they have become equal.

The mechanism of glucose transport that we have just described is **simple solute diffusion**. Such diffusion is transport that arises from the molecular agitation that exists in all systems above absolute zero and from the simple statistical tendency for such agitation to carry more molecules out of regions of relatively high concentration than into such regions. Note that simple solute diffusion could not possibly increase the difference in glucose concentration between the two sides of a membrane. The equilibrium state for the beaker in Figure 5.2 is a state of equal glucose concentration on both sides, and simple solute diffusion can change the glucose concentrations only toward equilibrium. Thus simple solute diffusion is a passive transport mechanism.

Simple diffusion is not limited just to solutes in aqueous solutions. Because molecular agitation is universal, simple diffusion is universal: Gases diffuse in air; water diffuses; even heat diffuses. As we now discuss the quantitative principles of the simple diffusion of solutes in aqueous solutions, it is worth noting that all simple-diffusion phenomena follow similar principles. Thus, later when we discuss the quantitative laws of gas, water, and heat diffusion,[4] you will notice close similarities to the principles we now address. Finally, note that simple diffusion is often called just *diffusion*.

Concentration gradients give rise to the most elementary form of simple solute diffusion

Solutes diffuse between any two regions of a solution, whether or not a membrane is present. To understand the quantitative laws of solute diffusion, we start with diffusion in an open solution, which is the most general case. Consider a still, open solution in which a solute, such as glucose, is at a relatively high concentration (C_1) in one region and at a relatively low concentration (C_2) in another. Imagine a representative 1-cm^2 cross-sectional area in the low-concentration region, facing toward the high-concentration region. As solute molecules diffuse into the low-concentration region from

the high-concentration region, we can count the numbers of molecules passing through this representative cross-sectional area per second. Specifically, we can count the *net* numbers that diffuse into the low-concentration region per second (the numbers diffusing in, minus the numbers diffusing out). Let J be the net rate of diffusion into the low-concentration region; that is, J is the net number of solute molecules passing into the low-concentration region per second through each unit of cross-sectional area. Then

$$J = D\frac{C_1 - C_2}{X} \tag{5.1}$$

where X is the distance separating the region of high concentration (C_1) from the region of low concentration (C_2), and D is a proportionality factor termed the **diffusion coefficient**. This formulation is often called the **Fick diffusion equation**, after Adolf Fick (1829–1901), who devised it.

The terms on the right-hand side of the Fick equation reveal important aspects of diffusion. Note that the rate (J) at which solute molecules diffuse into the low-concentration region is directly proportional to the difference in concentration ($C_1 - C_2$). Note also that the rate increases as the distance separating the two concentrations (X) decreases: Diffusion is a notoriously slow process for transporting substances from one place to another in the macroscopic world, but when only a tiny distance separates regions of differing concentration, diffusion can transport substances very rapidly. **TABLE 5.1** illustrates this crucially important point. The table focuses on rates of diffusion when there is a concentration difference of a small solute between two regions of a water solution. The table addresses the question: How much time is required for enough solute molecules to cross into the low-concentration region to reduce a concentration difference to half of its starting magnitude? The time required is *32 years* if the distance between regions is 1 meter (m), but it is only *100 nanoseconds* if the distance between regions is 10 nanometers (nm)—the approximate thickness of a cell membrane! The high rate of diffusion across minute distances helps explain why diffusion is a crucially important process in the lives of cells.

In the Fick equation, the ratio ($C_1 - C_2$)/X is called the **concentration gradient**. It expresses how much the concentration changes per unit of distance. Note that J, the diffusion rate, is proportional to it. Sometimes the term *concentration gradient* is used in a looser sense to refer only to the difference of concentration ($C_1 - C_2$).

For analyzing complex solutions, an aspect of the concentration effects on diffusion that is worth emphasizing is this: The rate of diffusion of any particular solute is determined by the concentrations of *that* solute, not the concentrations of other solutes that might be present. *Each solute diffuses according to its own concentration gradient.*

What determines the diffusion coefficient D? One important factor is the ease with which the solute of interest moves through the material separating the two concentrations; when diffusion is occurring through a cell membrane or an epithelium, this factor is termed the *permeability* of the membrane or epithelium to the solute (we discuss permeability later in this chapter). Another factor that determines D is temperature. As temperature rises, random molecular motions become more vigorous (see Chapter 10). Thus the rate of diffusion increases.

If a fluid current flows over the surfaces of an animal or cell, that flow can also affect the rate of diffusion. Suppose the con-

---

[4] Diffusion of gases is discussed in Chapter 22 (see page 588). Diffusion of liquid water is discussed later in this chapter, and that of water vapor in Chapter 27 (see page 729). Diffusion of heat (heat conduction) is covered in Chapter 10 (see page 238).

TABLE 5.1 The time required for diffusion through water to halve a concentration difference

Values are calculated for small solutes such as O_2 or Na^+. For each tabulated distance between solutions, the time listed is the time required for diffusion to transport half the solute molecules that must move to reach concentration equilibrium. It is assumed that no electrical effects exist, and thus only diffusion based on concentration effects is occurring.

A biological dimension that exemplifies the distance specified	Distance between solutions	Time required to halve a concentration difference by diffusion
Thickness of a cell membrane	10 nanometers	100 nanoseconds
Radius of a small mammalian cell	10 micrometers	100 milliseconds
Half-thickness of a frog sartorius muscle	1 millimeter	17 minutes
Half-thickness of a human eye lens	2 millimeters	1.1 hours
Thickness of the human heart muscle	2 centimeters	4.6 days
Length of a long human nerve cell	1 meter	32 years

Source: After Weiss 1996.

centration of a solute is 100 millimolar (mM) in the body fluids of an aquatic animal and 2 mM in the ambient water. The outward diffusion of the solute will then tend to create a *boundary layer* of elevated solute concentration next to the animal's body surface (**FIGURE 5.3**). The presence of the boundary layer decreases the rate of diffusion. One way to understand this effect in terms of Equation 5.1 is to consider the effect of the boundary layer on X, which is the distance separating the two concentrations, 100 mM and 2 mM; without the boundary layer, these two concentrations would be separated just by the thickness of the outer epithelium of the animal, but with the boundary layer, the two concentrations

are also separated by the thickness of the boundary layer itself, which increases X. Increasing the flow of water over an animal or cell tends to carry solute away from areas of accumulation at the animal or cell surface, thereby decreasing boundary-layer thickness. This decreases X and therefore increases the rate of diffusive solute loss from the animal or cell.

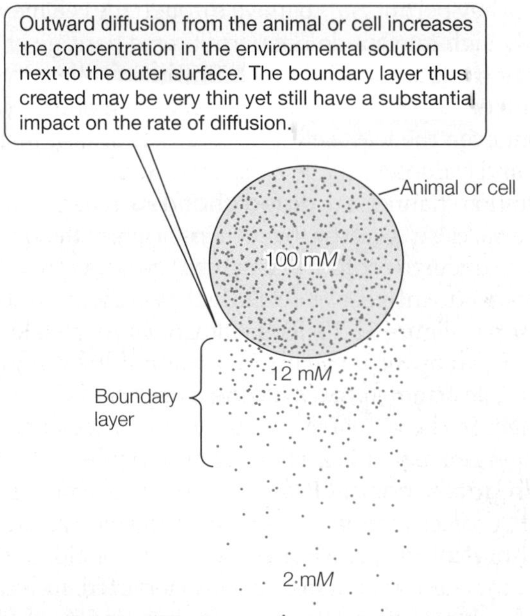

Electrical gradients often influence the diffusion of charged solutes at membranes

Many solutes of biological importance—such as Na^+, Cl^-, and other inorganic ions—bear an electrical charge. The diffusion of charged solutes—in addition to being affected by concentration gradients—is influenced by forces of *electrical attraction* or *repulsion*, forces that do not affect the movement of uncharged solutes such as glucose. As individual ions or other charged solutes move ceaselessly because of atomic-molecular agitation, their paths of motion are affected by the attraction of positive charges toward negative ones, and by the tendency of like charges (e.g., two positive ones) to repel.

To understand *where* such electrostatic effects need to be considered in the analysis of diffusion, a key principle is that *solutions at large* are electrically neutral. In **bulk solution**—the solution that is away from contact with a membrane—the concentrations of positive and negative charges are always equal (**FIGURE 5.4**). The *net* charge in any region of bulk solution is therefore zero. Moreover, because all regions of bulk solution have zero net charge, regions do not differ in net charge. This means that in bulk solution, the charges on solutes do not affect their diffusion, and diffusion simply follows the *concentration*-based principles we have already discussed.

In contrast to what happens in bulk solution, electrical attraction and repulsion can play large roles in diffusion *along or across cell membranes or epithelia*. This is true because the lipid bilayers in cell membranes can maintain separation of oppositely charged ions—meaning that different regions of solution can differ in net electrical charge—by acting like capacitors in electrical circuits.[5]

> Outward diffusion from the animal or cell increases the concentration in the environmental solution next to the outer surface. The boundary layer thus created may be very thin yet still have a substantial impact on the rate of diffusion.

Animal or cell

100 mM

12 mM

Boundary layer

2 mM

FIGURE 5.3 Diffusional concentration of a solute in a boundary layer next to an animal or cell In this example, the concentration of the solute is 100 mM inside the animal or cell and 2 mM in the open environmental solution.

[5] Appendix J discusses capacitance and other physics concepts used in this chapter.

FIGURE 5.4 Ionic charge separation occurs only within nanometers of membranes Thus only in the vicinity of membranes do conditions exist for the diffusion of charged solutes to be affected by electrical attraction and repulsion.

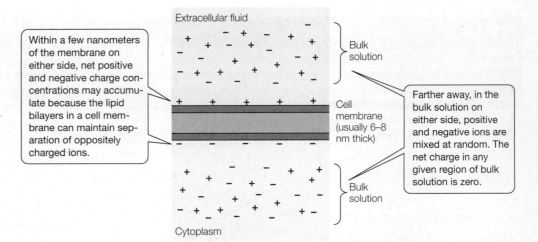

Within a few nanometers of the membrane on either side, net positive and negative charge concentrations may accumulate because the lipid bilayers in a cell membrane can maintain separation of oppositely charged ions.

Extracellular fluid

Bulk solution

Cell membrane (usually 6–8 nm thick)

Farther away, in the bulk solution on either side, positive and negative ions are mixed at random. The net charge in any given region of bulk solution is zero.

Bulk solution

Cytoplasm

Within a few *nanometers* of a cell membrane or epithelium, the solution on one side may have a net positive charge because positive ions outnumber negative ones. Conversely, within a few nanometers of the membrane or epithelium, the solution on the other side may have a net negative charge because negative ions outnumber positive ones (see Figure 5.4). Under these circumstances the diffusion of electrically charged solutes may be greatly affected by electrical attraction and repulsion.

Biological aspects of diffusion across membranes: Some solutes dissolve in the membrane; others require channels

Principles we discussed at the start of Chapter 2 return to center stage when we consider the chemical mechanisms of the diffusion of solutes through cell membranes or epithelia. The interior of a cell membrane is hydrophobic because it consists principally of the hydrocarbon tails of phospholipid molecules composing the lipid bilayer (see Figure 2.1). The ease with which a solute can diffuse directly through the lipid interior of a cell membrane thus depends on the hydrophobic or hydrophilic nature of the solute.

Let's start by considering *lipid solutes*, such as steroid hormones and fatty acids, which are hydrophobic. Molecules of these solutes *dissolve* in the lipid interior of a cell membrane. In this dissolved state, they make their way from the side of the membrane where they are more concentrated to the side where they are less concentrated because of molecular agitation and the other principles of simple diffusion we have been discussing. *Molecular oxygen* (O_2), which is small and nonpolar, is also generally believed to make its way through cell membranes primarily or exclusively by simple diffusion through the lipid layers, without the need of channels or other proteins. Certain signaling molecules besides steroid hormones, such as *thyroid hormones* and *nitric oxide*, are also believed to cross cell membranes in this way.

Inorganic ions present a very different picture because they are hydrophilic and therefore have very low solubilities in membrane lipids. Studies using experimental *all-phospholipid* membranes demonstrate that the rates of simple diffusion of ions directly through membrane lipids are exceedingly low. However, ions of physiological importance—such as Na^+, K^+, Ca^{2+} (calcium), and Cl^-—can sometimes move passively through *actual* cell membranes at very rapid rates. They are able to do this because their passive transport through cell membranes takes place through integral membrane proteins termed *ion channels*.

The defining characteristic of **ion channels** is that they permit the *passive* transport of inorganic ions by diffusion through a membrane. This means that the directions and eventual equilibrium states of the ion movements through channels are determined not by the channels themselves, but by the chemical-concentration and electrical-charge gradients that exist across a membrane. Another key characteristic of channels is that the ions that pass through them do not bind to the channel proteins. Structurally, an ion channel consists of one or more protein molecules that extend across the full thickness of a cell membrane or intracellular membrane and that encircle a lipid-free central passageway through the membrane (see Figures 2.4 and 13.16). From this fact and the fact that ions do not bind to channels, one might think that channels are simple holes through the membrane. Actually, they are intricate protein structures that are *selective* in determining which ions can pass readily through them.[6] Some ion channels, for instance, are specific for Na^+. They permit Na^+ to move through the cell membrane at a relatively high rate but do not permit rapid passage of other ions. Other channels are specific for K^+. Still others allow Na^+ and K^+ to pass with equal ease. Some channels allow several sorts of ions to pass, but even the least selective channels discriminate between anions and cations.[7]

Many ion channels are **gated channels**, meaning that they can "open" and "close" because the proteins of which they are composed are able to undergo conformational changes that cause their central passageways to increase or decrease the ease with which ions pass through (see Figure 12.20). Four categories of gated ion channels, distinguished by the mechanisms of gating, are recognized (**FIGURE 5.5**). **Voltage-gated channels** open and close in response to changes in the voltage difference across a membrane; they are very important in the generation of nerve impulses. **Stretch-gated (tension-gated) channels** open or close in response to stretching or pulling forces that alter the physical tension on a membrane. **Phosphorylation-gated channels** open or close according to whether the channel proteins are phosphorylated; such channels are under the control of protein kinases (see page 55), which are often

---

[6] The ion selectivity of channels is discussed in detail in Chapter 12.

[7] *Anions* are negatively charged ions; *cations* are positively charged ions.

(A) Voltage-gated channel

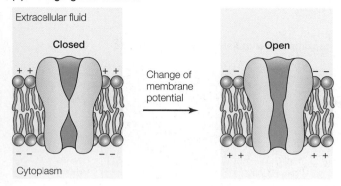

Extracellular fluid

Closed

Change of membrane potential

Open

Cytoplasm

(B) Stretch-gated (tension-gated) channel

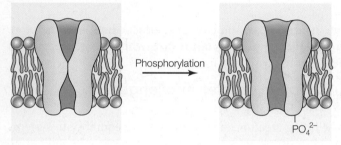

Stretch

Cytoskeleton

(C) Phosphorylation-gated channel

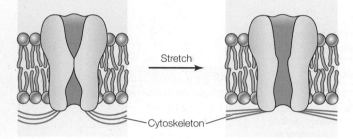

Phosphorylation

PO_4^{2-}

(D) Ligand-gated channel

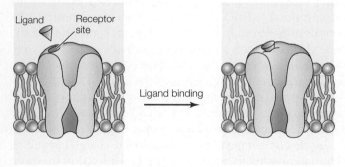

Ligand

Receptor site

Ligand binding

FIGURE 5.5 Gated ion channels Gated ion channels are classified into four functional categories based on which property must change for them to open and close: (A) transmembrane voltage, (B) stretch or tension, (C) phosphorylation, or (D) noncovalent binding with ligands at receptor sites.

themselves controlled by second messengers (see pages 66–67). **Ligand-gated channels**, discussed in Chapter 2 (see page 63), act both as *receptors* of extracellular signals and as ion channels. They are closed to the passage of ions in the absence of a signal, but they open when their receptor sites bind to receptor-specific ligands (e.g., neurotransmitter molecules), thereby allowing the passage of selected ions.

The **permeability of a cell membrane to a solute** is defined to be the ease with which the solute can move through the membrane by diffusion. The concept of permeability applies both to solutes that dissolve in the membrane interior and to solutes that pass through channels. When Equation 5.1 is applied to solute diffusion through a membrane, the membrane permeability to the solute of interest helps determine D, the diffusion coefficient. For solutes, such as lipids and O_2, that dissolve in the lipid bilayer to pass through a membrane, the permeability of a cell membrane depends on the factors that affect solute diffusion through lipid, such as the molecular size of the solute. For inorganic ions, the permeability of a cell membrane depends on the number of channels per unit of membrane area and on the proportion of the channels that are open. Cell membranes are often described as **selectively permeable** because they permit some solutes to pass through by diffusion with greater ease than others. One explanation for selective permeability is that different sorts of ion channels are often unequally present. For example, if a membrane has a high density of Na^+ channels and a low density of Ca^{2+} channels, it is more permeable to Na^+ than to Ca^{2+}.

Diffusion of ions across cell membranes is determined by simultaneous concentration and electrical effects

The final aspect of the mechanism of simple diffusion that we need to consider is the manner in which concentration effects and electrical effects *interact* in affecting the diffusion of ions and other charged solutes. We have already seen that, in animal cells, there is a *concentration gradient* across the cell membrane for each major solute (see Figure 5.1A). The diffusion of each solute is affected by its own concentration gradient but not by the concentration gradients of other solutes. All cell membranes also exhibit charge separation. That is, there is always an **electrical gradient** across a cell membrane, defined to be the difference in voltage (electrical potential) between the two sides.[8] The electrical gradient affects the diffusion of all charged solutes.

It is obvious that the diffusion of an ion or other charged solute must depend on *both* that solute's concentration gradient *and* the electrical gradient. A problem that soon arises in trying to *quantify* this dual effect is that concentration gradients and electrical gradients are expressed in different units: molarity and volts, respectively. Walther Nernst (1864–1941), an early winner of the Nobel Prize, goes down in history for solving this problem. Using his *Nernst equation* (discussed in Chapter 12), investigators can directly compare concentration and electrical gradients to determine which is stronger, and thus predict the diffusion of an ion.

Here we take a more informal approach to understanding the dual effects of concentration and electrical gradients on an ion's diffusion across a cell membrane. We term the influence of the concentration gradient across the membrane on the ion's diffusion the **concentration effect** on the ion, and we term the influence of the electrical gradient on the ion's diffusion the **electrical effect** on the ion.

Sometimes the electrical charge difference across a membrane pulls an ion in the same direction as the ion is moving because of

[8] A truly formal definition of the *electrical gradient* is that it is the difference in voltage divided by the distance between the high and low voltage levels. Less formally, the *electrical gradient* is simply the difference in voltage.

(A) Reinforcing concentration and electrical effects

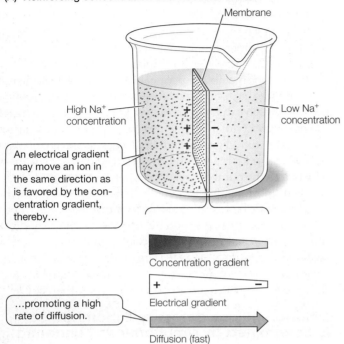

(B) Opposing concentration and electrical effects

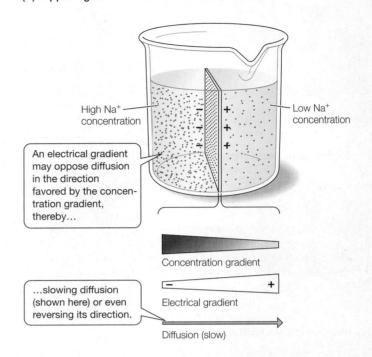

FIGURE 5.6 Ion diffusion depends on the dual effects of concentration and electrical gradients   (A) Fast diffusion resulting from reinforcing concentration and electrical effects. (B) Slow diffusion resulting from opposing concentration and electrical effects. In both parts the different dot densities on the two sides of the membrane in the beaker symbolize the different Na$^+$ concentrations.

its concentration gradient. **FIGURE 5.6A** shows this situation in a model system. In such cases the electrical and concentration effects reinforce each other, and the presence of the electrical gradient *accelerates* diffusion compared with the rate at which it would proceed if only the concentration gradient were present.

However, the electrical charge difference and the concentration gradient sometimes act in opposite directions, as in **FIGURE 5.6B**. If the electrical effect on Na$^+$ diffusion in Figure 5.6B is weaker than the concentration effect, Na$^+$ ions diffuse, as shown, in net fashion from left to right—the direction favored by the concentration gradient—but *more slowly* than if the opposing electrical gradient were absent. If the electrical effect in Figure 5.6B is stronger than the concentration effect, Na$^+$ ions diffuse in net fashion from right to left—the direction favored by the electrical gradient—even though that direction of diffusion is opposite to what the concentration gradient alone would cause. A large voltage difference can cause an ion or other charged solute to diffuse in a direction that *increases* its concentration gradient!

A charged solute is *at equilibrium* across a membrane when the concentration effect on its diffusion and the electrical effect are equal but opposite. Under such conditions the solute will not move in net fashion in either direction. Such an equilibrium is called an **electrochemical equilibrium** to stress that both electrical and "chemical" (concentration) effects are involved. The concept of an electrochemical equilibrium is, in fact, an entirely general concept that can be applied whether a solute is charged or uncharged (if

a solute is uncharged, no electrical effect exists, and therefore electrochemical equilibrium prevails when the concentration gradient is zero).

Diffusion often creates challenges for cells and animals

Because living creatures are inherently nonequilibrium systems and because simple diffusion moves solutes toward equilibrium, diffusion often creates challenges that must be met by energy expenditure. This point is illustrated by two of our focal examples, the individual animal cell and the freshwater fish (see Figure 5.1A,B).

Considering the cell first, let's examine the concentration and electrical effects on each of the three principal ions: Cl$^-$, Na$^+$, and K$^+$ (**FIGURE 5.7**). A typical cell membrane is positively charged on the outside and negatively charged on the inside. The concentration and electrical effects on the diffusion of Cl$^-$ across the cell membrane are therefore opposite to each other, because Cl$^-$ is more concentrated outside a cell than inside (favoring inward diffusion) but the inside of a cell membrane is negatively charged (repelling Cl$^-$). The magnitudes of the concentration and electrical effects on Cl$^-$ are typically about equal, meaning that Cl$^-$ is typically at or near electrochemical equilibrium across a cell membrane.

Na$^+$ presents an extremely different picture, however. The concentration and electrical effects on Na$^+$ diffusion are in the same direction; Na$^+$ is more concentrated on the outside of a cell than the inside, favoring inward diffusion, and the negative charges at the inside of the cell membrane also tend to draw Na$^+$ in by electrical attraction. Thus *Na$^+$ is very far from electrochemical equilibrium across a cell membrane and has a great tendency to diffuse from the extracellular fluid into a cell.* Of course, as stressed previously, Na$^+$ cannot cross the cell membrane unless open channels are present. However, there are sets of channels—called *resting channels*—that are open all or most of the time. These permit Na$^+$ to leak across the cell membrane. A cell therefore experiences a steady inward diffusion of

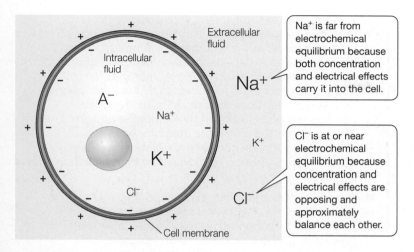

FIGURE 5.7 An electrochemical view of a typical animal cell The size of the lettering symbolizes relative ion concentrations inside and outside the cell. A^- represents anionic proteins that cannot cross the cell membrane and thus are trapped inside the cell.

Na^+ is far from electrochemical equilibrium because both concentration and electrical effects carry it into the cell.

Cl^- is at or near electrochemical equilibrium because concentration and electrical effects are opposing and approximately balance each other.

Na^+. This diffusion presents a challenge because it would do away with the Na^+ concentration gradient if it were unopposed. For a cell to maintain the normal Na^+ concentration gradient across its cell membrane, it must steadily get rid of Na^+. That is, it must steadily use some of its energy resources to do work to oppose the effects of inward Na^+ diffusion.

The third major ion, K^+, is not at electrochemical equilibrium across the cell membrane, but it is not as far from equilibrium as Na^+. The concentration and electrical effects on K^+ tend to cancel rather than add (see Figure 5.7). The concentration effect on K^+, however, exceeds the electrical effect. Animal cells therefore tend steadily to lose K^+ by diffusion through resting channels. For cells to keep their inside K^+ concentration high, they must use energy to take in K^+ and counteract this diffusion. When we address active transport later in this chapter, we will discuss the mechanisms that cells use to move Na^+ out and K^+ in.

The major blood ions of a freshwater fish—Na^+ and Cl^-—are much more concentrated in the blood plasma of the fish than in surrounding freshwater—such as pond water—suggesting that the ions tend to diffuse outward across a fish's gills (see Figure 5.1B). At this point in our discussion, we recognize that electrical effects also need to be taken into account, not just concentration effects. When investigators take both types of effects into consideration, however, they come to the same conclusion: For both Na^+ and Cl^-, the direction of simple diffusion is from the blood into the pond water. That is, the ceaseless and unstoppable random motions of ions present a challenge to a freshwater fish by causing the fish to lose these important ions steadily from its blood. A fish could theoretically block the losses by making its whole body impermeable to the ions. However, fish need to breathe, and gills that are permeable to O_2 are inevitably also permeable to Na^+ and Cl^-. Freshwater fish lose Na^+ and Cl^- by diffusion, and

they must replace what they lose. This task is accomplished by active transport, and it costs energy, as we will discuss soon.

Concentration gradients can create electrical gradients that alter concentration gradients

Thus far we have discussed the diffusion of ions as if the process were purely the *consequence* of concentration and electrical gradients. Ion diffusion is actually more dynamic, however. Because cell membranes and epithelia are selectively permeable—excluding some solutes while allowing the passage of others—the concentration gradients of ions across cell membranes or epithelia can *create* electrical gradients that may then affect ion diffusion and ion concentration gradients.

To see how this process takes place, let's study an experiment in which the diffusion of K^+ along its concentration gradient produces a counteracting electrical gradient. For this purpose we will use the model system in **FIGURE 5.8**, in which a membrane that is *permeable only to K^+* separates two solutions. This membrane represents an example of selective permeability. At the start, shown in Figure 5.8A, the solutions on the two sides are equimolar, with one composed of K^+ and A_1^- (an anionic solute) and the other composed of Na^+ and A_2^- (another anionic solute). After the apparatus is set up in this way, K^+ will start to diffuse from left to right across the membrane because of its concentration gradient. This diffusion will *create* a voltage difference across the membrane because K^+ is charged; specifically, the diffusion of K^+ will cause more positive charges—K^+ ions—to accumulate along the right-hand surface of the membrane than along the left-hand surface. As the K^+ diffusion continues, this voltage difference will grow steadily larger, and it will keep growing larger until the electrical effect on K^+ diffusion (repelling K^+ from the right) fully counteracts the concentration effect (favoring diffusion to the right). At that point, illustrated

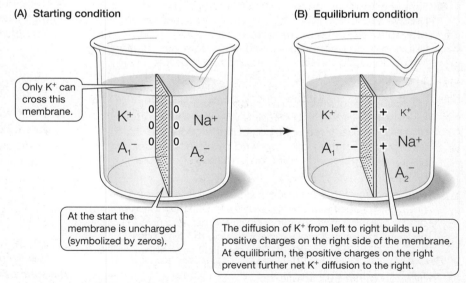

(A) Starting condition

Only K^+ can cross this membrane.

At the start the membrane is uncharged (symbolized by zeros).

(B) Equilibrium condition

The diffusion of K^+ from left to right builds up positive charges on the right side of the membrane. At equilibrium, the positive charges on the right prevent further net K^+ diffusion to the right.

FIGURE 5.8 The development of an equilibrium in which a voltage difference generated by diffusion exactly opposes the remaining concentration gradient The size of the lettering symbolizes relative ion concentrations. A_1^- and A_2^- represent anionic solutes that cannot cross the membrane.

by Figure 5.8B, K$^+$ will be at electrochemical equilibrium, and no further net diffusion of K$^+$ will occur.

Upon reviewing this experiment, you will see that when a membrane is selectively permeable and blocks some solutes from crossing, the diffusion of a permeating ion along its concentration gradient can *create* an electrical gradient. This is the sort of process that produces the voltage difference, termed the *membrane potential*, that typically exists across the cell membrane of an animal cell (see Figure 5.7).[9]

When a membrane is permeable to more than one sort of ion, an electrical gradient produced by diffusion of any one ion will affect the diffusion of the others because of electrical attraction and repulsion. A complex type of electrochemical equilibrium involving diffusion of multiple ions (and osmosis of water) then tends to develop. The **Donnan equilibrium**, which you will see mentioned in advanced literature, is this type of equilibrium. Specifically, a Donnan equilibrium occurs across a membrane when several ions (anions and cations) can cross the membrane but there is a set of nonpermeating ions (ions that cannot cross the membrane) that are more abundant on one side of the membrane than the other. Animal cells tend toward Donnan equilibrium because, relative to the extracellular fluids bathing them, they have high concentrations of nonpermeating anionic materials (e.g., proteins and nucleic acids) within them.

SUMMARY
Passive Solute Transport by Simple Diffusion

■ A mechanism of solute transport is *passive* if it transports only *toward* electrochemical equilibrium. For an uncharged solute, the electrochemical equilibrium is the same as concentration equilibrium. For a charged solute, the electrochemical equilibrium is achieved when the voltage difference exactly counterbalances the solute's concentration gradient across a cell membrane or epithelium.

■ Simple diffusion is the most straightforward type of passive solute transport. The fundamental mechanism of simple diffusion is exemplified most clearly by the diffusion of an uncharged solute in an aqueous solution. Molecular agitation tends by simple statistics to carry more molecules of such a solute out of regions of high concentration than into such regions, thereby tending to produce equality of concentration everywhere. If a solute is charged and diffusing where a voltage difference exists, forces of electrical attraction and repulsion affect the paths followed by molecules or ions during molecular agitation and thus contribute to diffusion.

■ Voltage differences can occur across cell membranes or epithelia because the lipid layers in cell membranes act as capacitors, permitting positive and negative charges to be unequally distributed on the two sides. Bulk solutions are electrically neutral, however, and voltage differences are therefore not a factor in simple diffusion within bulk solutions.

■ When lipid solutes such as steroid hormones and fatty acids cross cell membranes by simple diffusion, they typically dissolve—because of their hydrophobic nature—in the lipid layers of the membranes. Ions, however, cannot cross membranes by this mechanism because of their poor solubility in lipids; thus their diffusion requires mediation by membrane proteins. Inorganic ions typically cross membranes by simple diffusion through ion channels. Channels are often gated—that is, opened and closed by factors such as ligand binding.

■ The permeability of a membrane to a lipid solute depends on how readily the solute dissolves in and moves through the membrane lipid layer. The permeability of a membrane to an inorganic ion depends on the number of channels for the ion per unit of membrane area, and can be changed by the opening and closing of the channels.

Passive Solute Transport by Facilitated Diffusion

You may have noticed that when we discussed the passive movements of solutes across cell membranes in the last section, we did not mention *polar organic solutes*—such as glucose and amino acids. The reason we did not mention them is that when these solutes move passively across cell membranes, they typically do not do so by simple diffusion. Polar organic solutes cannot diffuse through the lipid interior of cell membranes because they are hydrophilic. Moreover, they cannot diffuse through channels because there are no channels for them. Passive transport of polar organic solutes across cell membranes occurs typically by the noncovalent and reversible *binding* of the solutes to solute-specific *transporter (carrier)* proteins[10] in the membranes, a process called *facilitated diffusion*. As we discuss facilitated diffusion, it will be important to keep in mind that although polar organic solutes sometimes cross cell membranes by active transport, our focus now is exclusively on their *passive* transport.

Facilitated diffusion has three defining properties:

1. It always occurs in the direction of electrochemical equilibrium. This is why it is a form of passive transport and is called *diffusion*.

2. Solutes transported by this mechanism move across membranes much faster than they could if they did not associate with transporter proteins. This is why the transport is called *facilitated*.

3. The mechanism requires solutes to bind reversibly with binding sites on the transporter proteins. This is why it is not a type of simple diffusion.

An important example of facilitated diffusion is the transport of glucose from the blood plasma into cells throughout an animal's body. Blood glucose typically crosses cell membranes to enter cells

---

[9] The generation of the *membrane potential* is discussed in detail in Chapter 12.

[10] Recall that Table 2.1 provides basic definitions of the functional categories of membrane proteins. Soon, in this chapter, we will provide an expanded definition of transporter proteins.

passively, aided by glucose transporters.[11] One of the most important effects of the hormone insulin in vertebrates is that it stimulates certain types of cells (notably muscle and adipose cells) to increase the numbers of transporter molecules in their cell membranes, thereby poising the cells to be able to take up glucose by facilitated diffusion at an accelerated rate (see pages 446–447).

Active Transport

Animals are able to transport many solutes across their cell membranes and epithelia in directions away from electrochemical equilibrium by using ATP or some other input of energy from metabolism. Mechanisms capable of such transport are called **active-transport mechanisms** or **pumps**. Active-transport mechanisms do not *necessarily* move solutes away from electrochemical equilibrium, but they are *capable* of doing so. Historically, the secretion of stomach acid was the first case in which biologists realized that cellular mechanisms can generate nonequilibrium distributions of solutes. During the production of stomach acid by humans or dogs, cells with an internal pH of 7.1–7.2 (near neutrality) secrete dramatically acidic fluids with a pH as low as 0.8. The concentration of H^+ in the secretions is more than 2 million times greater than the H^+ concentration in the intracellular fluids! This huge difference in concentration is generated by active-transport pumps that draw the required energy (in this specific case) from ATP.

Active transport plays critical roles in all three of our focal examples. To see this, let's look first at the individual animal cell. We have seen that K^+ is more concentrated inside a cell than in the extracellular fluids bathing the cell, and Na^+ is less concentrated inside than out (see Figure 5.1A). We have also seen that diffusion tends to obliterate these concentration differences by carrying K^+ outward and Na^+ inward. Cells keep the inside K^+ concentration high and the inside Na^+ concentration low by using an active-transport mechanism that transports K^+ in and Na^+ out *at a cost,* using metabolic energy.

Active transport also plays critical roles in our other two focal examples—the freshwater fish and the vertebrate small intestine. In the case of the freshwater fish (see Figure 5.1B), Na^+ and Cl^- are so much more concentrated in the fish's blood than in pond water that the two ions tend to diffuse out of the fish. A major mechanism by which a freshwater fish replaces lost ions is active transport of Na^+ and Cl^- from the dilute pond water into its concentrated blood. By means of active-transport mechanisms in their gills, many fish can pump Cl^- from freshwater, where it is as dilute as 0.02 mM, into their blood plasma, where the concentration exceeds 100 mM. In the small intestines of vertebrates (see Figure 5.1C), active-transport mechanisms help transport sugars and amino acids from digested food into the blood. Active transport also plays countless other vital roles in the lives of animals.

An instructive alternative name for *active transport* is **uphill transport**. For a rock on a hillside, the approach to equilibrium is to roll downhill. Uphill motion is away from equilibrium and therefore requires an input of energy. Active-transport mechanisms are analogous to uphill motion in that they can make solutes move away from electrochemical equilibrium and require energy to do so.

---

[11] Because cells continuously use glucose, the concentration gradient for glucose is *from* the blood *into* the cells.

Active transport and facilitated diffusion are types of carrier-mediated transport

For a solute to undergo active transport across a cell membrane, it must combine noncovalently and reversibly with a solute-specific *transporter* or *carrier* protein in the cell membrane. Active transport is therefore a form of **carrier-mediated transport**. Facilitated diffusion is also a form of carrier-mediated transport because, as we have noted already, it too requires binding of the transported solute to a transporter (carrier) protein. The two forms of carrier-mediated transport are distinguished by whether transport is driven by metabolic energy. In *facilitated diffusion*, no mechanism exists for metabolic energy to drive the process; thus transport is always toward electrochemical equilibrium. In *active transport*, metabolic energy is used to drive the process, so transport can be away from electrochemical equilibrium.

We have now reached a point where we can fully define *transporters* (*carriers*), one of the five principal functional classes of membrane proteins (see Table 2.1). A **transporter**, or **carrier**, is a membrane protein that mediates *active* or *passive* transport across the membrane by binding noncovalently and reversibly with solute. *Carrier* is the older term and is the basis for the expression *carrier-mediated transport*. Physiologists originally called these proteins *carriers* because they believed the proteins linked up with solutes on one side of a cell membrane and then moved to the other side, carrying the solutes with them. By now it is clear that the proteins do not move through the cell membrane but instead stay in one place while they undergo *intramolecular conformational changes* that usher solutes through. The solutes that bind noncovalently and reversibly with a transporter protein are the **ligands** of the transporter.

An important attribute of carrier-mediated transport, whether active or passive, is that it exhibits saturation kinetics (see page 46). Saturation occurs because solute molecules must bind with transporter molecules. When the concentration of solute is so high that all transporter molecules are engaged in transport (i.e., bonded with solute) at all times, the transport rate is maximal and cannot increase if more solute is made available.

Basic properties of active-transport mechanisms

Active-transport mechanisms in animal cells draw their energy directly or indirectly from ATP made by the catabolism of foodstuffs (see page 190). Because of the requirement for ATP, active transport is sensitive to factors that affect a cell's ATP supply, such as poisons or O_2 deficiency. In some sorts of cells that are very involved with transport, such as certain kidney-tubule cells, active-transport mechanisms are the single biggest consumers of ATP, using as much as 40% of the ATP a cell makes.

A great diversity of active-transport mechanisms is known. Thus although any particular active-transport mechanism is specific for certain solutes, a great variety of solutes can be actively transported in one situation or another, including inorganic ions, amino acids, and sugars.

However, active-transport mechanisms seem *not* to have evolved in animals for two of the most important chemical substances in physiology: H_2O and O_2. Current evidence indicates that *H_2O and O_2 always move by passive transport in animals.*

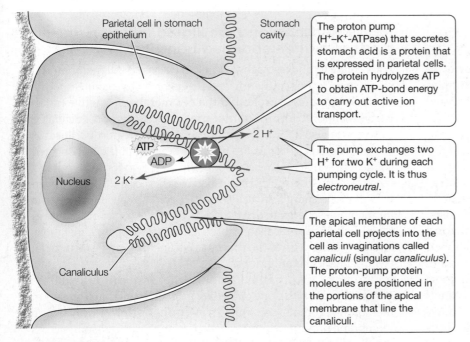

The proton pump (H⁺–K⁺-ATPase) that secretes stomach acid is a protein that is expressed in parietal cells. The protein hydrolyzes ATP to obtain ATP-bond energy to carry out active ion transport.

The pump exchanges two H⁺ for two K⁺ during each pumping cycle. It is thus *electroneutral*.

The apical membrane of each parietal cell projects into the cell as invaginations called *canaliculi* (singular *canaliculus*). The proton-pump protein molecules are positioned in the portions of the apical membrane that line the canaliculi.

FIGURE 5.9 Electroneutral active transport is responsible for secretion of stomach acid in the vertebrate stomach lining This acid-secreting *parietal cell* is bounded by other types of cells in the epithelium lining the stomach.

An active-transport mechanism that transports ions is potentially able to create a voltage difference across a membrane because the mechanism moves charges. Active ion-transport mechanisms do not *necessarily* generate voltage differences, however, because they often simultaneously transport two ions, and the transport of two ions may be electrically neutral. For example, consider the pump—informally known as the **proton pump**—that in vertebrates is responsible for the extreme acidification of stomach contents. This active-transport mechanism (**FIGURE 5.9**) simultaneously transports H⁺ (protons) into the stomach cavity and K⁺ in the opposite direction *in a 1:1 ratio*. Mechanisms such as this, which do not generate an imbalance of electrical charge, are termed **electroneutral (nonelectrogenic)**. In contrast, an active-transport mechanism is **electrogenic** if the actions of the mechanism create a charge imbalance across a membrane.

Recognition of active transport completes our overview of a single animal cell

At this point we can complete our discussion of the first of our three focal examples, the single animal cell (see Figure 5.1A). We have seen that, in terms of its Na⁺ and K⁺ concentrations, a cell is not at equilibrium with its surroundings. This nonequilibrium state is maintained by a ubiquitous, exceedingly important pump, the **Na⁺–K⁺ pump**. During each cycle of pumping, this active-transport mechanism transports three Na⁺ ions out of a cell and two K⁺ ions in (**FIGURE 5.10A**). Among other things, the pump is therefore electrogenic: It helps create a charge difference across the cell membrane (outside positive) because it pumps more positive charges in one direction than the other.[12] The Na⁺–K⁺ pump is found in the basolateral membranes of all epithelial cells. It is found in all other sorts of animal cells as well.

---

[12] The Na⁺–K⁺ pump, being electrogenic, contributes a portion of the normal charge difference across the cell membrane (outside positive). However, much of the charge difference originates from ion *diffusion* processes of the type exemplified in Figure 5.8. The generation of the membrane potential is discussed in detail in Chapter 12.

FIGURE 5.10 Summary of active and passive ion transport in a typical animal cell (A) The action of the Na⁺-K⁺ pump in the cell membrane. (B) Active and passive ion transport processes, and related phenomena, in a cell. Again, the size of the lettering in the solutions symbolizes relative ion concentrations inside and outside the cell. The reason that Cl⁻ is approximately at electrochemical equilibrium is that its concentration gradient across the cell membrane is almost exactly opposed by the voltage difference across the membrane.

(A) The action of the Na⁺–K⁺ pump

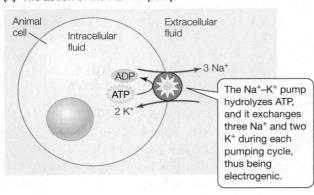

The Na⁺–K⁺ pump hydrolyzes ATP, and it exchanges three Na⁺ and two K⁺ during each pumping cycle, thus being electrogenic.

(B) An animal cell in summary

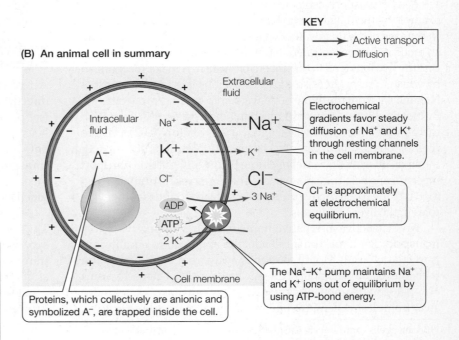

KEY
→ Active transport
----→ Diffusion

Electrochemical gradients favor steady diffusion of Na⁺ and K⁺ through resting channels in the cell membrane.

Cl⁻ is approximately at electrochemical equilibrium.

The Na⁺–K⁺ pump maintains Na⁺ and K⁺ ions out of equilibrium by using ATP-bond energy.

Proteins, which collectively are anionic and symbolized A⁻, are trapped inside the cell.

Earlier (in discussing Figure 5.7) we concluded that a cell must transport Na$^+$ outward and K$^+$ inward, against electrochemical gradients, if it is to maintain observed differences of ion composition across the cell membrane (between the intracellular and extracellular fluids). The Na$^+$–K$^+$ pump carries out this transport using ATP-bond energy. **FIGURE 5.10B** summarizes the processes of both active and passive ion transport between an animal cell and the extracellular fluids that bathe it.

Primary and secondary active transport differ in their cellular-molecular mechanisms

Active-transport mechanisms are classified as *primary* or *secondary* based on whether they use ATP directly or indirectly. We will discuss primary mechanisms first. Doing so will set the stage for drawing a rigorous distinction between the primary and secondary types.

One of the best-known transporter proteins engaged in active transport is **Na$^+$–K$^+$-ATPase**. This protein *is* the Na$^+$–K$^+$ pump discussed a moment ago. The molecule is called an **ATPase** because it is an *enzyme* that catalyzes the hydrolysis of ATP as well as being a *transporter*. This hydrolysis provides the energy that the protein uses for active Na$^+$–K$^+$ transport. Na$^+$–K$^+$-ATPase belongs to a category of ATPases called **P-type ATPases**. This name refers to the fact that in these ATPases, the protein becomes phosphorylated and dephosphorylated during each pumping cycle.[13]

The mechanism by which ATP-bond energy is transduced into ion motive energy by Na$^+$–K$^+$-ATPase—and by other P-type ATPases—has been dramatically clarified in the last 20 years and seems now to be largely understood. One key property is that these ATPases exhibit *strict coupling* between their molecular conformation and ATP hydrolysis: During each pumping cycle, the conformation of the transporter protein switches back and forth between two or more states, depending on whether the enzymatic hydrolysis of an ATP molecule has occurred. These conformation changes have two major effects on protein function. To understand these, we need to recognize that each molecule of a P-type ATPase has *cation-binding sites* deep within its molecular structure. Transported ions bind to these sites as they pass through the transporter protein on their way from one side of the cell membrane to the other side. One major effect of the conformation changes that occur during a pumping cycle is to *open passages between the cation-binding sites and the intra- and extracellular fluids*. However, unlike a channel protein, there is never an open passage all the way from the intracellular fluid to the extracellular fluid. Instead, as the molecular conformation undergoes change during a pumping cycle, at certain times a half-channel opens between the cation-binding sites and the *intracellular* fluid (while the sites remain closed off from the extracellular fluid); and conversely, at other times a half-channel opens between the cation-binding sites and the *extracellular* fluid (while the sites remain closed off from the intracellular fluid). The second major effect of the conformation changes that occur during a pumping cycle is that they *dramatically modify the affinity of the cation-binding sites for specific ions*. These affinity changes are

synchronized with the opening and closing of the half-channels to control which ions are bound and which are expelled when each half-channel is open.

Let's look at the action of Na$^+$–K$^+$-ATPase to clarify these points. **FIGURE 5.11** presents a model of one pumping cycle. At step ❶, the half-channel to the intracellular fluid is open, and the cation-binding sites have high affinity for Na$^+$ while having low affinity for K$^+$. Accordingly, Na$^+$ is taken up from the intracellular fluid, and K$^+$ is expelled into it. A molecule of ATP is bound at the ATPase catalytic site. Enzymatic hydrolysis of the ATP energizes the protein molecule and helps induce a conformation change to state ❷ —in which three Na$^+$ ions are *occluded* (out of communication with both intra- and extracellular fluid) within the deep interior of the protein. The energized protein then changes conformation to state ❸, in which the half-channel to the extracellular fluid opens and the cation-binding sites undergo a dramatic reduction in their affinity for Na$^+$ and an increase in their affinity for K$^+$, thereby expelling Na$^+$ into the extracellular fluid and taking up K$^+$. Also in state ❸, an ATP molecule binds to the protein at an ATP regulatory site, where it is not poised for hydrolysis but instead exerts modulatory effects. The state of the protein in ❸ poises it to dephosphorylate, which leads to state ❹, in which two K$^+$ ions are now occluded. The dephosporylation then leads to a return to state ❶. As this occurs, the ATP molecule moves from the ATP regulatory site to the catalytic site and thereby becomes poised to be hydrolyzed, energizing the protein once again.

The P-type ATPases include not only Na$^+$–K$^+$-ATPase but also **Ca^{2+}-ATPase**, which is responsible for critical Ca^{2+} pumping in the sarcoplasmic reticulum of muscle (see Chapter 20), and **H$^+$–K$^+$-ATPase**, the proton pump that acidifies stomach contents (see Figure 5.9). All the P-type ATPases are thought at present to function in closely similar ways. Besides the P-type ATPases, there are three other principal categories of ATPases, known as F-type, V-type, and ABC-type ATPases.

Primary active transport is defined to be active transport that draws energy immediately from the hydrolysis of ATP. That is, the transporter protein in primary active transport is an ATPase. As you can see, the active transport of Na$^+$ and K$^+$ by Na$^+$–K$^+$-ATPase is primary active transport. So also is the transport of Ca^{2+} by Ca^{2+}-ATPase, and that of protons by H$^+$–K$^+$-ATPase.

Secondary active transport draws energy, in an immediate sense, not from ATP but from an electrochemical gradient of a solute. ATP is required for secondary active transport. However, ATP is not the immediate source of energy tapped by the transporter protein. Instead, ATP is employed to create an electrochemical gradient, which is employed by the transporter protein to drive transport.[14]

Glucose absorption in the small intestine of a hummingbird or other bird, or of a mammal, provides a classic example of secondary active transport. Thus, to explore the principles of secondary active transport and how energy is provided for it, we may now return to the focal example with which this chapter opened, the case of a hummingbird with abundant glucose in its

---

[13] As discussed on page 55, phosphorylation is covalent bonding to an orthophosphate group, $-PO_4{}^{2-}$.

[14] Secondary active transport is sometimes called *indirect* active transport, whereas primary active transport is sometimes called *direct* active transport.

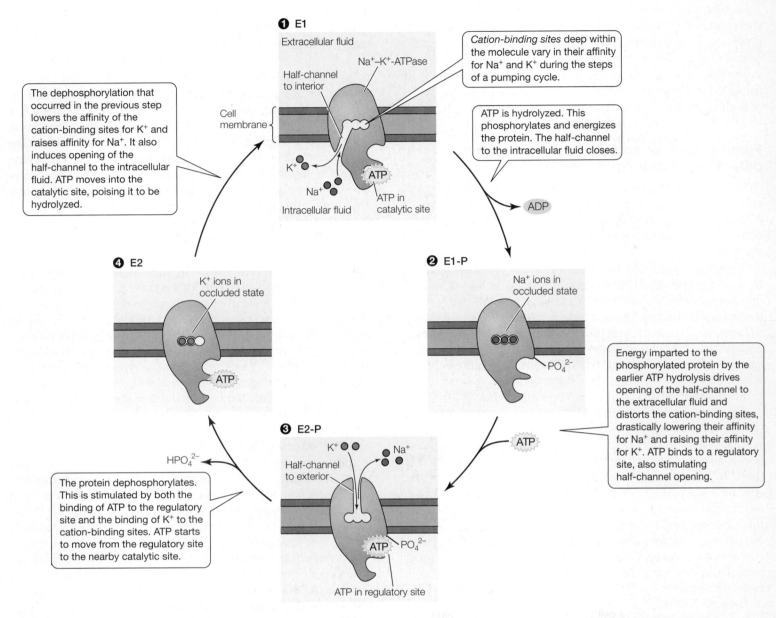

① E1

Extracellular fluid

Na⁺–K⁺-ATPase

Half-channel
to interior

Cation-binding sites deep within
the molecule vary in their affinity
for Na⁺ and K⁺ during the steps
of a pumping cycle.

Cell
membrane

ATP is hydrolyzed. This
phosphorylates and energizes
the protein. The half-channel
to the intracellular fluid closes.

K⁺

Na⁺

ATP

Intracellular fluid

ATP in
catalytic site

ADP

The dephosphorylation that
occurred in the previous step
lowers the affinity of the
cation-binding sites for K⁺ and
raises affinity for Na⁺. It also
induces opening of the
half-channel to the intracellular
fluid. ATP moves into the
catalytic site, poising it to be
hydrolyzed.

④ E2

K⁺ ions in
occluded state

ATP

② E1-P

Na⁺ ions in
occluded state

PO_4^{2-}

Energy imparted to the
phosphorylated protein by the
earlier ATP hydrolysis drives
opening of the half-channel to
the extracellular fluid and
distorts the cation-binding sites,
drastically lowering their affinity
for Na⁺ and raising their affinity
for K⁺. ATP binds to a regulatory
site, also stimulating
half-channel opening.

HPO_4^{2-}

ATP

③ E2-P

K⁺

Na⁺

Half-channel
to exterior

The protein dephosphorylates.
This is stimulated by both the
binding of ATP to the regulatory
site and the binding of K⁺ to the
cation-binding sites. ATP starts
to move from the regulatory site
to the nearby catalytic site.

ATP PO_4^{2-}

ATP in regulatory site

FIGURE 5.11 Na⁺–K⁺-ATPase transduces ATP-bond energy into ion motive energy   Shown here is a current model of the steps in a pumping cycle. Control of the cycle is intrinsic, in the sense that the molecular changes occurring at each step induce conformation changes that lead to the next step. Note that there is never an open channel all the way between the extra- and intracellular fluids; such a channel would permit ions to diffuse "backward" and reduce the gradients the pump is working to create. Instead, half-channels to the intra- and extracellular fluids open alternately (states ① and ③), with occlu-

sion periods in between (states ② and ④). The protein has two binding sites for ATP. ATP first binds to a regulatory site where it helps stimulate conformation changes, then moves into a catalytic site where it is hydrolyzed by the enzymatic activity of the protein. Because of the energy provided by ATP hydrolysis, ion movement can be against the electrochemical gradient. The protein is considered to have two major states, E1 and E2. In turn, each of those can be phosphorylated (E1-P and E2-P) or not. (After Bublitz et al. 2011.)

intestinal lumen after feeding on nectar. Specifically, let's consider how glucose is transported into a single intestinal epithelial cell from the intestinal lumen across the apical membrane of the cell (**FIGURE 5.12**). Recall that Na⁺–K⁺-ATPase (the Na⁺–K⁺ pump) is found in the basolateral membranes of all epithelial cells. This Na⁺–K⁺-ATPase steadily transports Na⁺ out of the intestinal epithelial cell using ATP-bond energy. This transport lowers the concentration of Na⁺ inside the cell and thus creates a Na⁺

electrochemical gradient across the cell's apical membrane.[15] Because of the action of Na⁺–K⁺-ATPase, Na⁺ is less concentrated on the inside of the apical membrane than on the outside, and in addition, the inside of the apical membrane is negative relative to the outside. As a result of *both* of these properties (see Figure

[15] The electrochemical gradient is created across all parts of the cell membrane, but we focus here on the apical part because it is where glucose is transported into the cell from the intestinal lumen.

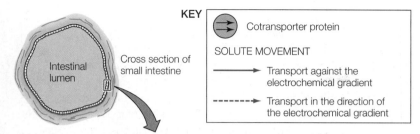

KEY

⬤➜ Cotransporter protein

SOLUTE MOVEMENT

➝ Transport against the electrochemical gradient

----➤ Transport in the direction of the electrochemical gradient

Intestinal lumen

Cross section of small intestine

(A) Na⁺–K⁺-ATPase in the basolateral membrane and the Na⁺ electrochemical gradient it generates across the apical membrane

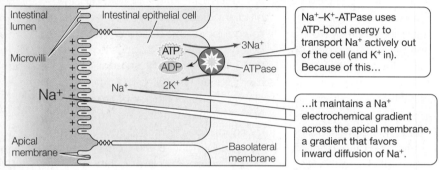

Na⁺–K⁺-ATPase uses ATP-bond energy to transport Na⁺ actively out of the cell (and K⁺ in). Because of this…

…it maintains a Na⁺ electrochemical gradient across the apical membrane, a gradient that favors inward diffusion of Na⁺.

(B) The Na⁺–glucose cotransporter in the apical membrane; glucose transport into the cell

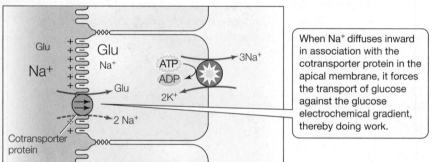

When Na⁺ diffuses inward in association with the cotransporter protein in the apical membrane, it forces the transport of glucose against the glucose electrochemical gradient, thereby doing work.

FIGURE 5.12 The secondary active transport of glucose into an epithelial cell of the vertebrate small intestine Throughout this figure, the size of the lettering used for solutes in the intra- and extracellular solutions symbolizes relative solute concentrations inside and outside the cell. Glu = glucose. (A) An intestinal epithelial cell emphasizing the actions of the Na⁺–K⁺-ATPase. (B) The same cell, now also showing the Na⁺–glucose cotransporter protein in the apical membrane and glucose transport. After glucose is inside the cell, it crosses the basolateral membrane into the blood by mechanisms discussed in Chapter 6 (see Figure 6.20).

5.12A), a strong electrochemical gradient exists for Na⁺ across the apical membrane.

From the perspective of energetics, the Na⁺ electrochemical gradient across the apical membrane is an *energy store* that can be used to drive transport processes across the membrane. **BOX 5.1** explains this critically important concept. The hydrolysis of ATP by the Na⁺–K⁺-ATPase adds energy to the cell system. This energy—instead of being used immediately for work—is partially stored as potential energy by the creation of the Na⁺ electrochemical gradient. In the presence of this gradient, Na⁺ tends to diffuse inward across the apical membrane. That is, the potential energy inherent in the electrochemical gradient tends to set Na⁺ in motion across the apical membrane. This Na⁺ movement can be used to do work. It can be used to pump glucose into the cell.

To understand the work performed, the properties of a remarkable, dual transporter (*cotransporter*) protein must be considered.

This protein facilitates the *obligatorily linked* transport of Na⁺ and glucose. The cotransporter protein does not use ATP and is not an ATPase. The protein, however, has the following essential attribute: For every two Na⁺ ions it carries across the membrane, it *must*, because of its particular chemistry, carry a molecule of glucose in the same direction (see Figure 5.12B). Thus, as Na⁺ diffuses into the cell from the intestinal lumen across the cell's apical membrane, glucose obligatorily enters with it. In fact, as Figure 5.12B shows, the cotransporter protein can move glucose into the cell *even when the concentration of glucose inside the cell exceeds the concentration in the intestinal lumen*.

In total, you can see, the transport of glucose into an intestinal epithelial cell from the intestinal lumen occurs by way of two energy-driven processes (see Box 5.1). First, Na⁺ is transported out of the cell by the Na⁺–K⁺ pump using ATP-bond energy, a process that creates and maintains a Na⁺ electrochemical gradient (representing potential energy) across the apical membrane. Second, glucose is transported into the cell across the apical membrane in linked fashion with Na⁺, driven by the energy stored in the Na⁺ electrochemical gradient.

Most investigators today consider this type of transport of glucose to be a form of *active* transport because it meets three critical criteria: (1) The mechanism can move glucose away from glucose equilibrium, from a solution of low concentration into one of high concentration; (2) the glucose transport is carrier-mediated; and (3) the energy source for the uphill transport of glucose is metabolism. The active transport of glucose deserves, nonetheless, to be distinguished from forms of active transport that use ATP-bond energy directly. This is why *secondary active transport* (transport that draws energy from an electrochemical gradient of a solute) is distinguished from primary active transport (transport that draws energy from ATP).

Available evidence indicates that when *organic solutes* are *actively* transported by animal cells, secondary active transport is the usual mechanism. In the vertebrate intestine, for example, secondary-active-transport mechanisms are responsible for the active uptake not only of glucose but also of certain other sugars, amino acids, and water-soluble vitamins (see Chapter 6).[16] A variety of apical-membrane transporter proteins are involved. All move two solutes simultaneously in an obligatorily linked fashion. When a transporter protein moves two solutes in linked fashion in *one* direction, the transport is called **cotransport**, and the protein is a **cotransporter**. When a transporter moves two solutes in obligatorily linked fashion in *opposite* directions, the transport is **countertransport**, and the protein is a **countertransporter**.[17] By

---

[16] All intestinal uptake of organic solutes is not necessarily active, however. Recent research indicates, for example, that in several species of birds, following ingestion of glucose-rich meals, glucose is taken up both by secondary active transport and by paracellular diffusion (see Figure 2.9).

[17] Cotransporters and countertransporters are sometimes called *symporters* and *antiporters*, respectively.

Energy Coupling via the Potential Energy of Electrochemical Gradients

In 1961, Peter Mitchell revolutionized knowledge of energy metabolism in cells by proposing an entirely novel idea for a mechanism by which cells can couple energy sources to work-performing processes. He won the Nobel Prize for this idea, termed the *chemiosmotic hypothesis* in certain contexts. Mitchell proposed that *physical compartmentalization of cells by membranes* is critical. That is—contrary to the assumption prior to his research—chemistry alone is not sufficient for certain biochemical processes to take place. Instead, the chemistry occurs in a physical context—compartmentalization by membranes—that imparts *spatial directionality*, which is itself an essential functional attribute of the processes.

To see the underlying concept, consider a type of mechanism that engineers sometimes employ for energy coupling (e.g., in pumped-storage plants). At the left in Part 1 of the figure is a power plant; at the right is an apparatus that does work by use of energy. The question to be addressed is how energy from the power plant can be coupled to the work-performing apparatus. The approach shown in Part 1 is to store energy from the power plant in the form of potential energy, then allow the work-performing apparatus to draw energy from the potential energy. Specifically, energy from the power plant is delivered to a pump that lifts water into an elevated tank (**❶**). The water, by being moved to a higher elevation, acquires potential energy of position in Earth's gravitational field, thereby storing a fraction of the energy from the plant (**❷**). Then, an apparatus is designed that can use energy from falling water to perform the type of work desired. When water from the elevated tank is allowed to fall back to its original elevation under force of gravity, a fraction of the potential energy present at **❷** acts to run the apparatus, and the work desired is accomplished (**❸**).

Living organisms started to employ similar energy-coupling systems almost from the start of their evolution. These energy-coupling systems do not use the elevation of water to store potential energy. Instead, they store potential energy in the form of ionic electrochemical gradients between solutions separated by cell membranes or intracellular membranes. As seen in Part 2 of the figure, energy from an energy source—usually ATP or bonds of food molecules—is used to pump ions across a membrane (**❹**), producing an electrochemical gradient that is far from equilibrium (**❺**). The gradient, created by use of energy from the source, represents a store of that energy as potential energy, analogous to the store in Part 1. Ions tend to diffuse back across the membrane in the direction dictated by the electrochemical gradient. Membrane-spanning proteins have evolved that obligatorily couple ion diffusion to specific types of work performance. When ions diffuse across the membrane by passing through these proteins, work is performed (**❻**). In this way, energy from the source—ATP or food molecules—is coupled to the performance of a specific type of work.

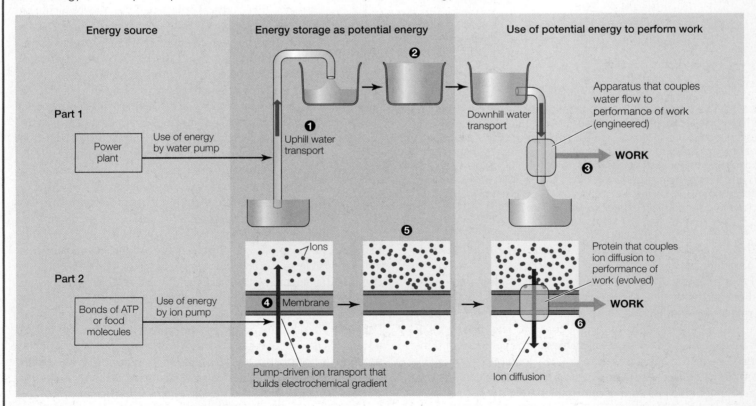

Two examples of energy coupling via an intermediate step in which energy is temporarily stored as potential energy Part 1 is an example from human engineering, whereas Part 2 shows a process that commonly occurs in cells. Both depend on an exacting energy-transducing mechanism (colored green) that performs a specific type of work by use of the type of potential energy available. In Part 1, this mechanism is a product of engineering (e.g., a turbine that turns an axle used to do work). In Part 2, the energy-transducing mechanism is a membrane protein that is a product of evolutionary natural selection. When energy is stored—**❶** to **❷** in Part 1, **❹** to **❺** in Part 2—the process is less than 100% efficient. Similarly, when energy is used—**❸** in Part 1, **❻** in Part 2—the process is also less than 100% efficient.

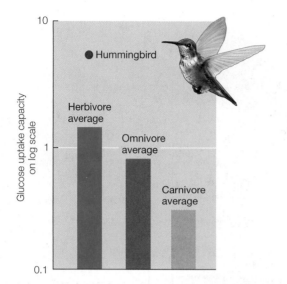

FIGURE 5.13 A species of hummingbird exhibited the highest capacity for intestinal glucose absorption of 42 vertebrate species measured Glucose uptake by secondary active transport across the intestinal epithelium was measured. On the y axis, the measured glucose uptake is expressed as a ratio of proline uptake as a way of providing a common reference point for the glucose measures. Note that the data on the y axis are plotted on a log scale; the numerical values are therefore more divergent than a quick visual impression might suggest. (After Karasov and Diamond 1988.)

these definitions, in our focal example the transporter for glucose is a cotransporter, as we have already implied. All vertebrates have this cotransporter in the apical membranes of their intestinal epithelial cells. Hummingbirds are distinguished, however, by exhibiting extraordinarily high activity of this cotransporter, far higher than observed in most vertebrates (**FIGURE 5.13**). This exceptional glucose cotransporter activity in hummingbirds seems almost surely to be an adaptation to the exceptional glucose yield of their nectar diet.

Active transport across an epithelium does not imply a specific transport mechanism

Scientists who study *epithelia* often know that active transport occurs across a particular epithelium without knowing much, if anything, about the specific molecular mechanisms that carry out the transport. There are, in fact, two different perspectives in the study of active transport across epithelia. The simpler *whole-epithelium* perspective—exemplified by **FIGURE 5.14A**—regards the epithelium as a "black box": The behavior of the epithelium itself is described without the underlying cell-membrane mechanisms being identified. The more-complex *cell-membrane* perspective—exemplified by the options in **FIGURE 5.14B**—seeks to identify the cell-membrane mechanisms

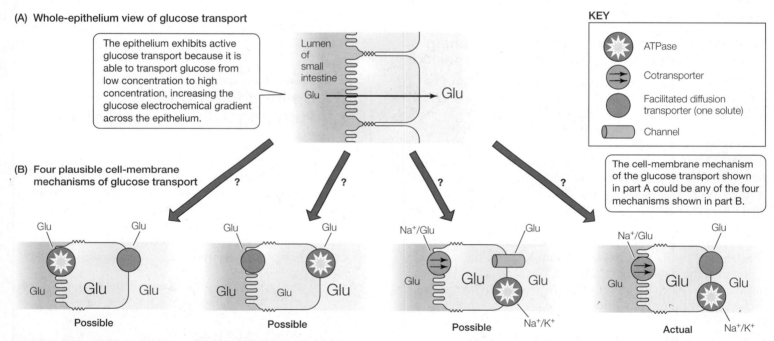

FIGURE 5.14 Two perspectives on epithelial active transport (A) Whole-epithelium and (B) cell-membrane views of glucose transport across the epithelium of the vertebrate small intestine. Each diagram in (B) depicts a potential mechanism by which transporter proteins and channel proteins in the apical and basolateral membranes could endow the whole epithelium with the property shown in (A). In each diagram, proteins are labeled with the solutes they transport. Free-standing Glu symbols show relative glucose concentrations in the intra- and extracellular fluids. Explanations of the four plausible cell-membrane mechanisms in (B): *Far left:* Glucose could cross the apical membrane by means of **primary active transport** mediated by an ATPase and powered with energy derived directly from ATP. The high glucose concentration thereby created within the cell would favor diffusion of glucose out of the cell, so transport across the basolateral membrane could occur by facilitated diffusion. *Middle left:* Glucose could enter the

cell across the apical membrane by **facilitated diffusion**. This could occur only if the cell concentration of glucose were low enough to be below the intestinal-lumen concentration, but the problem of then making glucose move out across the basolateral membrane could be solved by the presence of a glucose-transporting ATPase that uses ATP-bond energy to make glucose move against its concentration gradient. *Middle right:* Glucose could enter the cell across the apical membrane by **secondary active transport**, involving the Na^+–K^+-ATPase in the basolateral membrane and a Na^+–glucose cotransporter in the apical membrane (see Figure 5.12B). Then glucose could cross the basolateral membrane by simple diffusion through a glucose channel. *Far right:* This is what actually happens. Glucose enters the epithelial cell across the apical membrane by **secondary active transport** and leaves across the basolateral membrane by facilitated diffusion mediated by a glucose-transporter protein (see Figure 6.20 for full detail).

of transport, including all the particular proteins that are involved (e.g., ATPases, cotransporters, and channels) in both the apical and basolateral membranes of the epithelial cells.

To make this clearer, suppose that a researcher carries out the following measurements on an epithelium: First the researcher measures the concentration of the solute of interest on the two sides of the epithelium, and if the solute is charged, the researcher measures the voltage difference across the epithelium. From these data, the electrochemical gradient across the epithelium is calculated. Next the researcher studies whether the epithelium can move the solute of interest from one side to the other in a way that *increases* the electrochemical gradient across the epithelium. If so, it is common practice to say that "the epithelium actively transports the solute." By this standard, it is easy to show that the epithelium of the small intestine of a vertebrate actively transports glucose (see Figure 5.14A).

It is always the case, however, that many different cell-membrane mechanisms could permit an epithelium to function as it does. In the case of the intestinal epithelium, for example, *any* of the four mechanisms in Figure 5.14B (plus others not shown) would permit the epithelium to transport glucose as it does. Research at the level of cell membranes is required to identify the actual mechanism. As we have already seen, the actual mechanism in the intestinal epithelium is known; it is the one shown at the far right of Figure 5.14B. Often, however, physiologists have not fully resolved the molecular mechanisms of epithelial transport.

Two epithelial ion-pumping mechanisms help freshwater fish maintain their blood composition

Now we can complete our analysis of the movement of Na^+ and Cl^- across the gill epithelial membranes of freshwater fish (see Figure 5.1B), concluding our discussion of our focal examples. We have seen that freshwater fish lose Na^+ and Cl^- from their blood into pond water by diffusion, and thus they face a constant threat of having their blood become too dilute. Food is one means by which they replace lost Na^+ and Cl^-. A more important means of ion replacement in most fish is that they actively take up Na^+ and Cl^- across their gill epithelial membranes. What are the mechanisms of this active transport?

Studies of whole gill epithelia demonstrate that the mechanisms of active Na^+ uptake and active Cl^- uptake are, for the most part, different and independent (**FIGURE 5.15**). Each active-transport mechanism is electroneutral (or approximately so) because each exchanges two ions of identical charge across the epithelium in opposite directions and in a 1:1 ratio. When the Na^+-pumping mechanism actively transports Na^+ against its electrochemical gradient from dilute pond water into concentrated blood, it simultaneously secretes H^+ into the water in exchange for the Na^+. The Cl^--pumping mechanism secretes HCO_3^- (bicarbonate ion) into the water in exchange for Cl^-. The H^+ and HCO_3^- that are exchanged for Na^+ and Cl^- are, in essence, metabolic wastes because both are derived from the CO_2 produced by metabolism. Thus freshwater fish employ waste ions to keep the active uptake mechanisms for Na^+ and Cl^- from generating large charge differences across the gill epithelium. You will recognize that our presentation of the active-transport mechanisms in Figure 5.15 is a whole-epithelium, black-box view. **BOX 5.2** discusses the cell-membrane view.

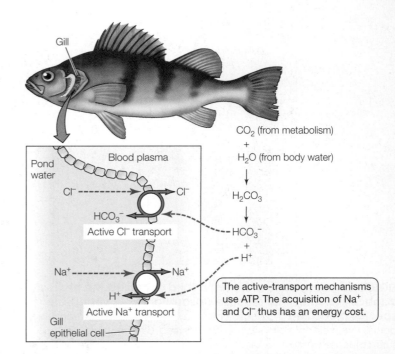

FIGURE 5.15 A whole-epithelium view of active ion transport across the gill epithelium of a typical freshwater fish   The Na^+ and Cl^- active-transport mechanisms are different and independent, for the most part. The pumping mechanisms exist within single epithelial cells. Some cells take up Na^+ principally or solely, whereas other cells take up Cl^-. Carbon dioxide (CO_2) from metabolism reacts with body water to form bicarbonate ions (HCO_3^-) and protons (H^+). These products are exchanged for Cl^- and Na^+ by the active-transport mechanisms.

BOX 5.2 Cellular Mechanisms of Ion Pumping in Freshwater Fish Gills

What are the cell-membrane proteins and other cellular mechanisms that enable the gill epithelium of a freshwater fish to take up Na^+ and Cl^- ions from pond or stream water? This important question is still under active investigation. Nonetheless, some of the principal attributes of the ion-pumping mechanisms are well understood, as you will learn by reading **Box Extension 5.2** on the web.

Bluegill sunfish (*Lepomis macrochirus*)

SUMMARY
Active Transport

■ Active-transport mechanisms—also called uphill-transport mechanisms or pumps—are able to convert energy obtained from the catabolism of foodstuffs into solute motive energy and thus can transport solutes away from electrochemical equilibrium. Such mechanisms are known for many solutes but not for O_2 or H_2O.

■ Solutes must bind noncovalently to a transporter protein for active transport to occur. Thus active transport is a type of carrier-mediated transport. A second type is facilitated diffusion, which differs in that it cannot tap metabolic energy and is therefore strictly toward equilibrium.

■ Active transport is primary if the transporter protein is an ATPase and thus draws energy directly from ATP bonds. One of the best-known transporters of this sort is Na^+-K^+-ATPase (a P-type ATPase), which is found in all animal cells, including the basolateral membranes of all epithelial cells. Primary-active-transport mechanisms pump ions. Active transport is secondary if the immediate source of energy for transport is a solute electrochemical gradient, rather than ATP. Secondary-active-transport mechanisms depend on transporter proteins—cotransporters or countertransporters—that obligatorily transport two solutes simultaneously. Secondary-active-transport mechanisms pump organic solutes and ions.

■ Active transport of ions is electrogenic if it produces a voltage difference but electroneutral if it does not.

Diversity and Modulation of Channels and Transporters

Like enzyme proteins, channel and transporter proteins exist in multiple molecular forms and can be modulated in numerous ways, providing many opportunities for the regulation of cell and tissue function. A detailed treatment of this topic would be at least as long and full of possibilities as our discussion of enzyme diversity and modulation in Chapter 2. Instead, the goal in this case will be simply to outline known phenomena.

■ *Multiple molecular forms.* Multiple molecular forms of channel and transporter proteins are common. For example, just as the genome of an individual vertebrate animal codes for the synthesis of two or more LDH (lactate dehydrogenase) enzyme proteins, it codes for the synthesis of two or more Na+–K+-ATPase proteins, and the various forms of Na+–K+-ATPase are differentially expressed in different tissues. Multiple forms of Ca2+-ATPase and H+–K+-ATPase, expressed in different tissues, also occur. In addition, different species often have evolved different molecular forms of channels and transporters. Because different molecular forms can exhibit distinct transport, catalytic, and modulation characteristics, the presence of different forms in different tissues of a species or in different species provides opportunities for adaptation.

■ *Modulation by gene expression.* In the life of an individual, it is common for channels and transporters to be modulated by means of gene expression. All cells of a mammal, for example, have the gene that codes for the form of H+–K+-ATPase responsible for stomach acidification. The gene is vigorously expressed, however, only in certain cells (the acid-secreting cells of the stomach), giving those cells their unique ability to secrete strong acid. Gene expression is also often modulated in response to an individual's circumstances, often under hormonal control. The adrenal hormone aldosterone, for instance, increases Na+ reabsorption in the kidney tubules, thus increasing retention of Na+ in the body. It exerts this effect in part by increasing expression of the gene that codes for Na+–K+-ATPase in kidney-tubule cells.

■ *Noncovalent and covalent modulation.* Existing channel and transporter molecules often are subject to modulation. One form of modulation is analogous to the allosteric modulation of enzymes in that it is mediated by noncovalent and reversible bonding with modulating agents. Ligand gating of channels provides an illustration (see Figure 5.5D). Covalent modulation (e.g., by phosphorylation) also occurs. Both noncovalent and covalent modulation provide for very rapid adjustment of a channel or transporter molecule's function.

■ *Insertion-and-retrieval modulation.* Because channel and transporter proteins perform their functions only when in the cell membrane (or in a function-specific intracellular membrane), the *location* of the proteins in a cell provides a means of controlling their activity. Cells often have reserves of cell-membrane channel or transporter proteins held in locations other than the cell membrane. Protein molecules from these reserves can be **inserted** into the cell membrane—becoming functional—or molecules in the cell membrane can be **retrieved**. An excellent example is provided by H+–K+-ATPase in the acid-secreting cells of the stomach. Resting acid-secreting cells (see Figure 5.9) have relatively few molecules of H+–K+-ATPase in their apical cell membranes, but they have large reserves of ATPase molecules bound to intracellular membranes (where the molecules are nonfunctional). When a meal is ingested, the cells are hormonally stimulated to become active. This stimulation causes *insertion* of intracellular H+–K+-ATPase molecules into the apical membranes of the cells. The microvilli lining the canaliculi of the apical membranes (see Figure 5.9) grow dramatically (e.g., from six- to tenfold), and the number of ATPase molecules positioned to function by secreting protons into the stomach cavity is vastly increased. After digestion, the ATPase molecules are *retrieved* from the apical membranes. Insertion mechanisms are much faster than gene-expression mechanisms because the time-requiring steps of transcription and translation are carried out prior to the need for the channel or transporter proteins. The only step required to activate the proteins at a time of need is to insert them in the membranes where they are active, a process that can be completed in minutes.

Osmotic Pressure and Other Colligative Properties of Aqueous Solutions

Having discussed the mechanisms of *solute* transport, we now turn to the transport of the *solvent* in aqueous solutions: water. A logical starting point is to discuss the **colligative properties** of aqueous solutions, because osmosis of water, discussed in the next section, is fundamentally a colligative phenomenon. The colligative properties of aqueous solutions are the properties that depend simply on the *number* of dissolved entities per unit of volume rather than the *chemical nature* of the dissolved entities.

The three colligative properties of greatest significance for the study of animal physiology are the osmotic pressure, freezing point, and water vapor pressure of a solution.[18] A useful way to define **osmotic pressure** is that it is the property of a solution that allows one to predict whether the solution will gain or lose water by osmosis when it undergoes exchange with another solution; in other words, if you know the osmotic pressures of two solutions, you can *predict* the direction of osmosis between them. The **freezing point** of a solution is the highest temperature capable of inducing freezing. For a formal definition of **water vapor pressure**, we defer to Chapter 27 (see page 729), but here we can say that the water vapor pressure measures the tendency of a solution to evaporate.

The study of colligative properties requires a shift in view, compared with earlier sections of this chapter, because the colligative properties—unlike solute-transport mechanisms—do not depend on the particular chemical nature of solutes. In **FIGURE 5.16**, the two upper drawings are microscopic views of two complex solutions. The same two solutions are shown again in the two lower drawings, except that in the lower drawings, the chemical identity of each dissolved entity is not recognized; instead, each dissolved entity is represented just with a dot. For quantifying colligative properties, the view in the lower drawings is more straightforward because the colligative properties are independent (or approximately independent) of the chemical nature of dissolved entities. They depend instead on the simple matter of *how many dissolved entities* (of any kind) *are present per unit of volume of solution*. Whether a dissolved entity is a glucose molecule, a large protein molecule such as albumin, or a Na^+ ion, each makes an approximately equal contribution to the quantitative magnitudes of the colligative properties of a solution. Thus the two solutions in Figure 5.16 are identical or nearly identical in their osmotic pressures, freezing points, and water vapor pressures, even though they are very different in their chemical mixes of solutes.

All the colligative properties of a solution change in magnitude as the concentration of dissolved entities changes. Raising the concentration of dissolved entities in a solution increases the osmotic pressure of the solution. It also lowers the solution's freezing point and lowers its water vapor pressure. Speaking quantitatively, the osmotic pressure is approximately *proportional* to the concentration of dissolved entities; doubling the concentration approximately doubles the osmotic pressure. The other colligative properties also exhibit simple proportionalities to concentration, but these proportionalities can be a little less obvious because of the way the properties are usually expressed. Consider the freezing point and

[18] A fourth colligative property of occasional interest is the **boiling point**, the lowest temperature that will cause boiling.

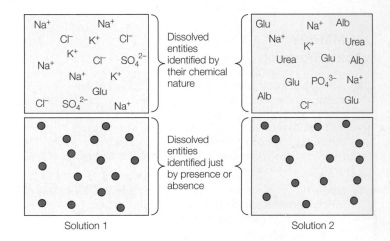

FIGURE 5.16 Magnified views of two solutions that are similar in colligative properties All the drawings depict identical volumes. The upper drawings of Solutions 1 and 2 show the chemical nature of each dissolved entity in the solutions. The lower drawings of the same two solutions show merely the numbers of dissolved entities within the volumes shown, disregarding chemical differences. Because the two solutions have identical numbers of dissolved entities per unit of volume, they are virtually the same in their osmotic pressures, freezing points, and other colligative properties. Alb = albumin; Glu = glucose.

water vapor pressure: When the concentration of dissolved entities in a solution is doubled, the factor that approximately doubles is the *difference* (sign ignored) between the actual freezing point or the actual water vapor pressure of the solution and the freezing point or the water vapor pressure of pure water. To see this more clearly, consider the **freezing-point depression**, defined to be the difference (sign ignored) between the actual freezing point of a solution and the freezing point of pure water. The freezing-point depression of a solution is approximately proportional to the concentration of dissolved entities. For example, if the freezing-point depression is 0.4°C before the concentration of dissolved entities is doubled, it is about 0.8°C after. This means that the freezing point itself is lowered from −0.4°C to −0.8°C. The **water-vapor-pressure depression** is the difference (sign ignored) between a solution's water vapor pressure and the water vapor pressure of pure water under the same conditions. It is approximately proportional to the concentration of dissolved entities.

A useful corollary of the points already made is that the osmotic pressure, the freezing-point depression, and the water-vapor-pressure depression of a solution *are all proportional to each other*. This means that if you have measured any one of these properties in a solution, you can calculate the others, a principle that we will see shortly has considerable practical significance.

Only dissolved materials affect the colligative properties of solutions. Suspended materials, such as clay particles in river water or red blood cells in blood plasma, do not affect the osmotic pressures or the other colligative properties of solutions because they are not dissolved.

Solutions of *nonelectrolytes* that are equal in their molar chemical concentrations typically are identical in their osmotic pressures and other colligative properties. This is easy to understand when you consider two principles: (1) a mole of solute always contains a fixed number of molecules (Avogadro's number, 6×10^{23}); and (2) when nonelectrolytes go into solution, each individual molecule remains intact in the solvent, meaning that each constitutes a single dissolved entity. These two principles dictate that if several solutions of nonelectrolytes have equal molarities, they are also equal in their numbers of dissolved entities per unit of volume. A 0.1-*M* solution of glucose, for example, has the same number of dissolved entities per unit of volume as a 0.1-*M* solution of urea. For the determination of colligative properties, it does not matter that the dissolved entities are glucose molecules in one solution and urea molecules in the other. Thus both solutions are virtually identical in their osmotic pressures and other colligative properties. Similarly, the colligative properties of a 0.1-*M* solution of a large, nondissociating protein such as serum albumin (molecular mass ~66,000 daltons [Da]) are essentially the same as those of the equimolar solutions of the lower-mass solutes: 0.1-*M* glucose (180 Da) and 0.1-*M* urea (60 Da).[19]

Solutions of *electrolytes* present additional complexities for predicting colligative properties because individual molecules of electrolytes dissociate when placed in solution, giving rise to more than one dissolved entity. To take the simplest case, consider *strong electrolytes*, which dissociate fully when they are dissolved. When NaCl, a strong electrolyte, is dissolved, each molecule dissociates fully into two dissolved entities, a Na^+ ion and a Cl^- ion. This means that a 0.1-*M* solution of NaCl has twice as many dissolved entities per unit of volume as a 0.1-*M* glucose solution. Similarly, a 0.1-*M* solution of Na_2SO_4 (sodium sulfate) has three times as many dissolved entities as a 0.1-*M* glucose solution. The colligative properties of salt solutions exhibit nonideal behavior because such solutions do not behave exactly as if each dissociated ion is an independent dissolved entity. In most situations relevant to animal physiology, however, a relatively close approximation to ideal behavior is observed. From the viewpoint of ideal behavior, for example, we would expect a 0.1-*M* solution of NaCl to have an osmotic pressure and freezing-point depression that are 2 times higher than those of a 0.1-*M* glucose solution. The actual osmotic pressure and freezing-point depression are about 1.9 times higher in the NaCl solution.

Physiologists usually express osmotic pressure in osmolar units

The most commonly used system of units for osmotic pressure in biology today is the **osmolarity** system. A **1-osmolar (*Osm*)** solution is defined to be one that behaves osmotically as if it has 1 Avogadro's number of independent dissolved entities per liter. Saying the same thing in another way, a 1-*Osm* solution has the same osmotic pressure as is exhibited by a 1-*M* solution of ideal nonelectrolyte. Some solutions relevant to animal physiology, such as seawater and the bloods of most marine invertebrates, are concentrated enough to be about 1 *Osm*. Many solutions in animal physiology, however, are more dilute, and their osmotic pressures are typically expressed in units of **milliosmolarity**; a **1-milliosmolar (*mOsm*)** solution behaves osmotically as if it has 0.001 Avogadro's number of independent dissolved entities per liter.

---

[19] To be strictly accurate, solutions of various nonelectrolytes have the same ratio of dissolved entities to water molecules when they are of the same *molality*, not molarity. Thus it is solutions of identical molality that exhibit identical colligative properties. Biologists usually measure and discuss molarities, and for most purposes this habit does not introduce consequential errors.

Osmotic pressures can be measured in several ways

For day-to-day practical purposes in physiology and medicine today, people measure osmotic pressures by employing instruments that actually measure either the freezing-point depression or the water-vapor-pressure depression. These are called *freezing-point osmometers* and *vapor-pressure osmometers*. The instruments take advantage of the point, stressed earlier, that all the colligative properties of a solution are typically proportional to each other. Thus the osmotic pressure can be calculated after the freezing-point depression or the water-vapor-pressure depression has been measured.[20]

Physical chemists have devised methods of measuring osmotic pressure itself, rather than calculating it from other colligative properties. These methods are the gold standard of osmotic-pressure measurement, but usually they are not employed for day-to-day practical work by biologists because they are cumbersome to carry out. **FIGURE 5.17** illustrates the basic approach to the direct measurement of the osmotic pressure of a solution. The solution is

---

[20] Prior to about 1980, it was common for biologists who measured freezing-point depression to leave their results in temperature units rather than converting them to osmolarity. Thus in the literature of that era you will find osmotic pressures expressed in degrees Celsius! The equivalency is that a freezing-point depression of 1.86°C corresponds to 1 *Osm*.

(A) A piston device for direct measurement of osmotic pressure

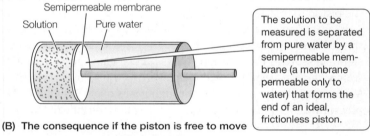

(B) The consequence if the piston is free to move

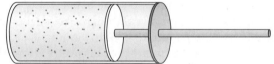

(C) The consequence if the solution is placed under hydrostatic pressure that exactly opposes osmosis

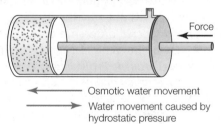

FIGURE 5.17 How to measure the osmotic pressure of a solution directly   (A) A piston osmometer. (B) A later state of the system if the piston is free to move. Water moves into the solution by osmosis. (C) The stable state of the system if the solution is subjected to increased hydrostatic pressure by a force applied to the piston and the difference of hydrostatic pressure across the membrane is equal to the osmotic pressure of the solution. The open sidearm on the pure-water compartment in (C) permits the hydrostatic pressure in that compartment to remain constant as the hydrostatic pressure in the solution compartment is increased.

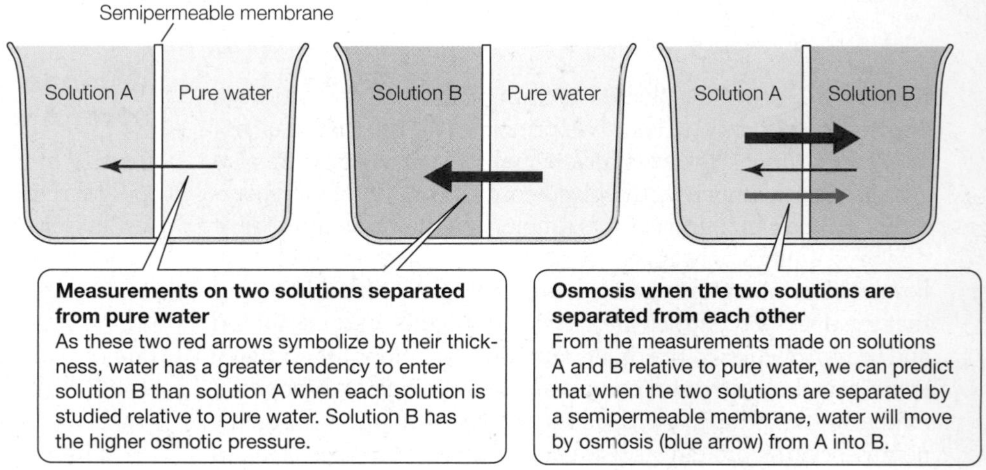

FIGURE 5.18 Predicting the direction of osmosis between two solutions from measurements made independently on each The red arrows in the two diagrams on the left symbolize the tendency for water to move by osmosis into each of two solutions, A and B, when the solutions are separated by a semipermeable membrane from pure water. The thickness of each arrow is proportional to the hydrostatic pressure difference that is required between each solution and the pure water to block any net transfer of water by osmosis; thus the thicknesses of the red arrows are proportional to the osmotic pressures of the two solutions. In the right-hand diagram, the red arrows are repeated from the diagrams on the left, and the blue arrow shows the difference between them.

Semipermeable membrane

Solution A | Pure water | Solution B | Pure water | Solution A | Solution B

Measurements on two solutions separated from pure water
As these two red arrows symbolize by their thickness, water has a greater tendency to enter solution B than solution A when each solution is studied relative to pure water. Solution B has the higher osmotic pressure.

Osmosis when the two solutions are separated from each other
From the measurements made on solutions A and B relative to pure water, we can predict that when the two solutions are separated by a semipermeable membrane, water will move by osmosis (blue arrow) from A into B.

placed in a piston device as shown in Figure 5.17A. The end of the piston consists of a **semipermeable membrane**, defined to be a membrane permeable only to water. The reason for using such a membrane is that it ensures the solution will not be corrupted by losing some of its solute during the measurement process. A semipermeable membrane is a laboratory device; no animal or plant membranes are semipermeable. Note in Figure 5.17A that whereas the solution of interest is on one side of the semipermeable membrane, *pure water* is on the other side. If the piston is simply allowed to move freely, water molecules will travel by osmosis from the side filled with pure water to the side filled with the solution, and the changes in volume of the two fluid compartments will push the piston to the right (see Figure 5.17B).

This piston movement can be prevented by exerting a force on the piston as shown in Figure 5.17C. The force will raise the hydrostatic pressure in the solution so it is higher than that in the pure water. The elevated hydrostatic pressure in the solution will then tend to force water molecules through pores of the membrane by pressure-driven, streaming bulk flow (termed **ultrafiltration**) from the solution into the pure water. For any given solution, there is a particular difference of hydrostatic pressure between the solution and pure water that will exactly prevent the piston from moving in either direction from its initial position, as in Figure 5.17C. When this difference of hydrostatic pressure prevails, water is forced out of the solution by the hydrostatic-pressure difference at the same rate that osmosis transports water into the solution. This difference of hydrostatic pressure therefore serves as a measure of the degree to which the solution tends to take on water by osmosis.

To a physical chemist, the specified difference of hydrostatic pressure *is* the osmotic pressure of the solution. Stated succinctly, *the osmotic pressure of a solution is the difference of hydrostatic pressure that must be created between the solution and pure water to prevent exactly any net osmotic movement of water when the solution and the pure water are separated by a semipermeable membrane.* Historically, osmotic pressures came to be called *pressures* precisely because they can be measured as hydrostatic pressures; and to this day, in physical chemistry they are usually expressed in pressure units, such as pascals (Pa) or atmospheres (atm).[21]

To see the logic of the physical chemist's approach, suppose you place two solutions, called A and B, in the piston device at different

times so that you can measure each solution's osmotic pressure, and suppose that the hydrostatic pressure required to negate osmosis into B is greater than that required to negate osmosis into A. From these results, you would know that when the solutions are studied in relation to pure water, B has a greater tendency than A to take on water by osmosis, as shown in the two images to the left in **FIGURE 5.18**. Knowing this, you could then accurately predict that if the two solutions, A and B, are placed on opposite sides of a water-permeable membrane, as at the right in Figure 5.18, the direction of water transport by osmosis will be from A into B. Recall that one way to define osmotic pressure is that it is the property of a solution that allows one to predict whether the solution will gain or lose water by osmosis when it undergoes osmotic exchange with another solution. The physical chemist's approach permits measurement of this property and is in fact the most fundamental of all ways to measure osmotic pressure (again, the gold standard). The measurements of osmotic pressure that biologists obtain using freezing-point osmometers or vapor-pressure osmometers typically provide the same information by different means.[22]

---

[22] The physical chemist's approach is the *direct* method of measuring osmotic pressure. All other approaches are *indirect*. See Box 7.3 for a discussion of the fundamental distinctions between direct and indirect measurements.

SUMMARY
Osmotic Pressure and Other Colligative Properties of Aqueous Solutions

■ Colligative properties depend on the number of dissolved entities per unit of volume of solution rather than the chemical nature of the dissolved entities. The three colligative properties of greatest importance in biology are the osmotic pressure, freezing point, and water vapor pressure. Because the three are quantitatively related, any one can be calculated from the others.

■ Osmotic pressures are generally expressed in osmolar units by biologists today. A 1-osmolar (*Osm*) solution behaves osmotically as if it has 1 Avogadro's number (6×10^{23}) of independent dissolved entities per liter.

■ Physical chemists measure osmotic pressures by using hydrostatic pressure, explaining why osmotic pressures are called *pressures*.

---

[21] The equivalency to convert to osmolar units is that a 1-*Osm* solution has an osmotic pressure in pressure units of 22.4 atm (technically, at 0°C, because temperature matters to a small degree in the conversion).

Osmosis

Osmosis is the passive transport of water across a membrane, which can be a cell membrane, an epithelium, or an artificial membrane. Sometimes osmosis is referred to as "diffusion of water." This is misleading as applied to biological membranes because strong evidence indicates that the rate of osmosis across some sorts of biological membranes is greater than can be explained by simple water diffusion. Because osmosis is a form of passive transport, it is strictly toward equilibrium, regardless of its rate or mechanistic details. Active water transport is not generally believed to occur.

When two solutions exchange water by osmosis, *water always moves from the one with the lower osmotic pressure into the one with the higher osmotic pressure.* This attribute of osmosis tends to be confusing at first because we are accustomed to thinking of things moving passively from high to low; heat moves from high temperature to low, glucose diffuses from high concentration to low, and so forth. Water undergoing osmosis, however, moves from low osmotic pressure to high. One way to remember this crucial point is to note that water itself is more abundant per unit of volume where dissolved matter is less abundant. Thus during osmosis water moves from where *it* is more abundant per unit of volume to where *it* is less abundant.

Osmosis is important in the lives of animals in many ways. Striking examples are provided by the osmotic exchanges between freshwater animals and the environmental waters in which they live. Freshwater animals ceaselessly take on water by osmosis from the pond or stream waters they inhabit because their blood plasma and other body fluids have far higher concentrations of dissolved entities—higher osmotic pressures—than freshwater has (**FIGURE 5.19**). A 100-g goldfish (*Carassius auratus*) living in freshwater, for example, gains about 30 g of water per day by osmosis—a third of its body weight! Osmotic influxes like this tend to bloat freshwater fish and dilute their blood. The fish must continually expend energy to rid themselves of the excess water.

As you study transport, never forget that osmotic pressures have just one major claim to consideration: They govern the movements of *water* across membranes. It is easy to slip into a habit of thinking that osmotic pressures affect solute movements. They do not do so directly. When solutes diffuse, they are affected by gradients of their specific chemical concentrations. Only water follows the gradient of osmotic pressure.

Quantification and terminology

The *rate* at which water crosses a membrane by osmosis follows the equation

$$\text{Rate of osmotic water transport per unit of membrane area} = K\frac{\Pi_1 - \Pi_2}{X} \qquad (5.2)$$

where Π_1 and Π_2 are the osmotic pressures of the solutions on the two sides of the membrane, and X is the distance separating Π_1 and Π_2. The ratio $(\Pi_1 - \Pi_2)/X$ is called the **osmotic-pressure gradient** or **osmotic gradient**, although sometimes the term *osmotic gradient* is used less formally to refer to just the difference in osmotic pressure. The proportionality coefficient K depends in part on the temperature, and in part on the permeability of the membrane to osmotic water transport, termed the **osmotic permeability**.

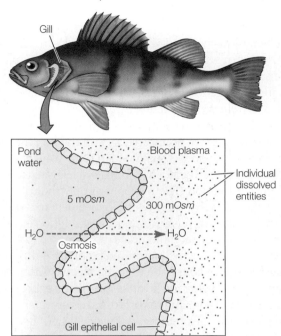

FIGURE 5.19 Osmotic uptake of water by a freshwater animal Because freshwater animals have osmotic pressures in their blood that greatly exceed the osmotic pressure of freshwater, they gain water by osmosis, especially across their gills or other highly permeable body surfaces. Each dot represents a single osmotically effective dissolved entity.

If two solutions, A and B, have the same osmotic pressure, they are said to be **isosmotic**. If solution A has a lower osmotic pressure than solution B, A is said to be **hyposmotic** to B, and B is said to be **hyperosmotic** to A.[23] Note that the terms *isosmotic, hyposmotic,* and *hyperosmotic* are *relative*; the two solutions being compared must be specified for the terms to have meaning. In this system of terminology, the *direction* of net water movement by osmosis is from the hyposmotic solution into the hyperosmotic one. Osmotic equilibrium is reached when two solutions become isosmotic.

Hydrostatic pressures develop from osmotic pressures only when two or more solutions interact

For most people, the most confusing aspect of studying osmosis is that osmotic pressures are called *pressures*. To avoid confusion, it is important to be clear that *a solution by itself does not exert a hydrostatic pressure because of its osmotic pressure.* In pressure units, a 1-*M* glucose solution has an osmotic pressure of about 22 atm. Thus this simple sugar solution would explode glass bottles if it exerted a hydrostatic pressure equal to its osmotic pressure! The glucose solution can be stored in exactly the same sorts of containers as pure water precisely because the osmotic pressure of an isolated solution is not a hydrostatic pressure. Osmotic pressures are named *pressures* because of a historical accident—namely, that the earliest scientists who studied

---

[23] The *tonicity* system of nomenclature, often inappropriately used, is not equivalent to the *osmoticity* system mentioned here. The tonicity system refers to effects on cellular volume. An **isotonic** solution is one into which cells of a specified kind can be placed without the volumes of the cells being affected. If cells are placed in a **hypotonic** solution, they swell because of an osmotic influx of water. In a **hypertonic** solution, they shrink because of osmotic water loss. In many cases, solutions isosmotic to cells are not isotonic in the long run, emphasizing that the two systems of nomenclature are not equivalent.

them were physical chemists, who (as we have seen) used hydrostatic pressure as *a measurement device* to quantify osmotic pressures.

There are circumstances relevant to animal physiology when osmotic pressures generate hydrostatic pressures. These circumstances *always involve interaction between two or more solutions* across a membrane. If two solutions of differing osmotic pressure are separated by a water-permeable membrane and the solution of higher osmotic pressure is contained in some way that limits its freedom to expand, then osmosis will create an elevated hydrostatic pressure in that solution. The reason is simply that transport of water by osmosis adds matter (water molecules) to the solution with the higher osmotic pressure, and if that solution is prevented from freely expanding to accommodate the increased matter, it will be pressurized. This is why human red blood cells dropped into freshwater enlarge and burst. When a red blood cell first falls into freshwater, its cell membrane is intact and, like a rubber balloon, tends to resist expansion. Thus as osmosis carries water molecules into the cell from the surrounding freshwater, an elevated pressure develops inside the cell, as in a balloon being blown up. Eventually the pressure becomes so great that the cell explodes.

Water may dissolve in membranes or pass through aquaporin water channels during osmosis

Experiments with artificial all-phospholipid membranes show that water molecules can dissolve in phospholipid bilayers and move through them by the random molecular motions of simple diffusion. This result seems odd at first because H_2O is a polar molecule. Evidently H_2O is small enough (0.3 nm in diameter) for the dissolve-and-diffuse mechanism to operate. Osmosis across many real cell membranes is thought to occur entirely by simple diffusion of water through the lipid bilayers.

However, in many sorts of cells, channel proteins called **water channels**—found in the cell membrane—provide avenues for water molecules to move by osmosis through the cell membrane without encountering the lipid bilayers. The most important of these water-channel proteins, which speed osmosis, are the *aquaporins*.

Aquaporins

The **aquaporins** ("water pores") are a large family of water-channel proteins found in all groups of organisms. Although extremely important, they were discovered only in 1992. Prior to that date, mystery shrouded many of the processes and illnesses in which the aquaporins play roles.

The aquaporins are in general remarkably specific to H_2O. For example, although they allow H_2O to pass through, they do not allow hydronium ions (H_3O^+) to pass. The study of aquaporins is an exploding area of research. Thirteen molecular forms of aquaporins—identified as AQP-1, AQP-2, and so forth—are known in the placental mammals, and hundreds are known among organisms taken as a whole. As the chemical diversity of aquaporins has become better understood in recent years, so has their functional diversity. Some aquaporins, we now realize, are sufficiently permeable to certain non-water solutes—such as glycerol or CO_2—that they likely play additional functional roles besides their water-channel roles. Some cells and tissues noted for having aquaporins in critical cell membranes are red blood cells, the brain, and certain kidney-tubule epithelia (see Chapter 29) in

vertebrates in general, and the bladder and skin of amphibians. Aquaporins play essential roles in urine formation, production of the aqueous humor of the eyes, secretion of tears and sweat, and in the blood–brain barrier (they are involved in brain swelling after trauma). Cell membranes endowed with aquaporins are said to exhibit **channel-mediated water transport**.

Molecules of H_2O move through an aquaporin in single file (**FIGURE 5.20**). Each aquaporin molecule is a tetramer, and—very unlike the structure of ion channels—each of the four monomers in a molecule has a pore through which H_2O moves. Electrochemical and conformation factors in the pores are responsible for H_2O selectivity.

Water transport through aquaporins is strictly passive. It does not employ metabolic energy, and it is always in the direction of equalizing osmotic pressure on the two sides. In a cell membrane without aquaporins, osmosis across the membrane occurs only by diffusion of H_2O molecules through the lipid bilayer. In a cell membrane with aquaporins, osmosis across the membrane occurs both by diffusion through the lipid bilayer and by channel-mediated H_2O transport through the aquaporin proteins. Membranes endowed with aquaporins exhibit high—sometimes very high—H_2O permeability. Osmosis occurs through such cell membranes at rates that are typically 5–50 times higher than through membranes lacking aquaporins! The mechanism by which aquaporins accelerate osmosis is not fully understood. Interestingly, a single aquaporin molecule has only a small effect on H_2O transport. Because of this, for a membrane to have high H_2O permeability it must be densely occupied by aquaporin molecules. Mammalian red blood cells, for example, are a classic example of cells with high H_2O permeability. Each red blood cell has about 200,000 copies of the aquaporin AQP-1 in its cell membrane.

Aquaporins are subject to modulation like other membrane proteins. Some aquaporins, for instance, have phosphorylation-modulation sites and are modulated covalently by protein kinases. Certain (but not all) aquaporins are controlled by *insertion* and *retrieval*. Insertion increases the number of aquaporin molecules in a membrane and thus increases membrane water permeability. Retrieval exerts opposite effects. The cells in some kidney-tubule epithelia, for example, undergo rapid changes in water permeability by the insertion and retrieval of AQP-2 in the cell membranes, under control of the hormone vasopressin (antidiuretic hormone, ADH), as seen in Figure 29.17. Responses of this sort are important in the ability of people and other vertebrates to raise and lower their urine concentration (see Chapter 29).

Osmosis and solute physiology often interact

Because the osmotic pressures of solutions depend on solute concentrations, osmotic water movements are interrelated with solute physiology. The three concepts we now discuss demonstrate the often-complex interrelations between water and solutes.

NONPERMEATING SOLUTES OFTEN CREATE PERSISTENT OSMOTIC-GRADIENT COMPONENTS ACROSS CELL MEMBRANES OR EPITHELIA If a cell membrane or epithelium is impermeable to a solute and the solute is more concentrated on one side than the other, the solute creates a persistent difference of osmotic pressure across the cell membrane or epithelium. The osmotic-gradient component attributable to the nonpermeating solute is persistent because the solute cannot diffuse to concentration equilibrium across the membrane or epithelium.

FIGURE 5.20 Water molecules moving through a single aquaporin pore, one at a time Pictured is an aquaporin monomer within a cell membrane, modeled by means of computational biophysics. Blue represents H_2O molecules. The layers of H_2O molecules at the top and bottom represent the fluids outside and inside a cell. The H_2O molecules actually passing through the aquaporin pore are represented at a larger size than the other H_2O molecules. The aquaporin monomer protein—represented by the gold, purple, and red cylinders, and by the other gold structures—permits H_2O molecules to move readily through the cell membrane by channel-mediated water transport. The finely divided red molecules at left and right represent the lipid matrix of the cell membrane. (Courtesy of Drs. Emad Tajkhorshid and Klaus Schulten, Theoretical and Computational Biophysics Group, University of Illinois at Urbana-Champaign; see also *Am. Sci.* 92: 554–555, 2004.)

This principle is critically important for understanding water movement across the walls of blood capillaries, for example. Blood plasma typically contains relatively high concentrations of dissolved proteins that do not cross blood-capillary walls. These proteins create a persistent tendency for the blood inside the capillaries to take up water osmotically across the capillary walls from the tissue fluids surrounding the capillaries. However, the hydrostatic pressure of the blood (the *blood pressure*) tends to force water out of the capillaries. The net effect of these processes (see Figure 25.13) is for blood to lose water to the tissue fluids as it flows through the capillaries, but without the osmotic effect of the blood-plasma proteins trapped inside the capillaries, this loss would be vastly greater than it is. The portion of the blood-plasma osmotic pressure that is caused by nonpermeating dissolved proteins is termed the **colloid osmotic pressure** or **oncotic pressure** of the blood.

PASSIVE SOLUTE TRANSPORT AND OSMOSIS INTERACT When permeating solutes diffuse across a cell membrane or epithelium because they are not at electrochemical equilibrium, their movement tends to alter the osmotic-pressure gradient across the membrane by removing osmotically effective solute from one side and transferring it to the other. Conversely, osmosis of water across a cell membrane or epithelium tends to alter the electrochemical gradients of solutes because solutes tend to become concentrated on the side losing water and diluted on the opposite side. In addition to these interactive effects of solute diffusion and osmosis on osmotic and electrochemical gradients, the rates of passive solute transport and passive water transport frequently interact by direct mechanical, chemical, or electrochemical coupling of solute and water movements. When water is crossing a membrane in one direction, for example, solute may tend to move physically with the water in the same direction by **solvent drag**, a phenomenon roughly analogous to minute bubbles being carried along by a water current.

ACTIVE SOLUTE TRANSPORT PROVIDES A MEANS TO CONTROL PASSIVE WATER TRANSPORT Animals are generally believed to lack active water transport. However, active solute transport can create a gradient of osmotic pressure, thereby setting up conditions favorable for osmosis to occur in a particular direction. In this way, active solute transport provides a means by which animals can use metabolic energy to exert control over the direction of water transport. Suppose, for example, that it would be adaptive for an animal to force water to cross a membrane from left to right. The animal could employ an active solute pump to drive such water movement by concentrating a solute on the right side of the membrane, so that the solution on the right is hyperosmotic to that on the left. Then water would cross the membrane by osmosis in the adaptive direction. Indirectly, in this way ATP-bond energy can be used to provide the motive force for water transport. This principle is believed to be used by animals in many situations; for example, many insects use the principle to concentrate their urine (see Chapter 29), and vertebrates are believed to use it to absorb water into their blood from the materials passing through their intestines.

SUMMARY
Osmosis

■ Osmosis is the passive transport of water across a membrane, such as a cell membrane or an epithelium. Osmosis always occurs from the solution of lower osmotic pressure (more abundant water) to the solution of higher osmotic pressure (less abundant water). Osmosis is always toward equilibrium, tending to bring the solutions on the two sides of a membrane to equal osmotic pressure.

■ Two solutions are isosmotic if they have the same osmotic pressure. If two solutions have different osmotic pressures, they are described as being hyposmotic and hyperosmotic to each other, the one with the lower osmotic pressure being the hyposmotic one.

■ A single solution exerts no increase in hydrostatic pressure because of its osmotic pressure. However, hydrostatic pressures can be generated by osmosis between two solutions interacting across a membrane.

■ In some cell membranes, the only mechanism by which osmosis occurs is that H_2O molecules dissolve in the membrane and move through by molecular agitation (simple diffusion). Other cell membranes, such as the membranes of red blood cells and certain kidney tubules, however, have aquaporins, providing a second path for osmosis to occur and increasing the rate of osmosis 5- to 50-fold in comparison with membranes without aquaporins. Aquaporins permit highly specific, channel-mediated H_2O transport.

STUDY QUESTIONS

1. People often say things like "A city has reached equilibrium size" or "A person has reached equilibrium between needs and wants." Discuss whether these uses of *equilibrium* are compatible with the word's thermodynamic meaning.

2. What are the similarities and differences between the mathematical equation for the rate of simple solute diffusion and the equation for the rate of osmosis?

3. Consider three groups of solutes: (i) steroid hormones, fatty acids, and other lipids; (ii) inorganic ions; and (iii) polar organic solutes such as glucose and amino acids. What is the principal mechanism by which each group crosses cell membranes passively? Why do members of the first group cross in a fundamentally different way from solutes belonging to the other two groups?

4. (a) Life-threatening diarrhea is a shockingly common problem in the developing world. People with life-threatening diarrhea are often Na^+-depleted, and to save their lives, replacing Na^+ is essential. However, "raw" Na^+ in the intestines is not absorbed. Drinking a solution of NaCl does not, therefore, replenish body Na^+. In fact, drinking such a solution can actually worsen a person's situation by osmotically dehydrating the blood and other body fluids. Explain how drinking an NaCl solution could have this effect. (b) One of the greatest physiological discoveries of the twentieth century was that drinking a solution of mixed glucose and NaCl can promote restoration of the body's Na^+. With the glucose concentration high enough in the solution, glucose "drives" the glucose–Na^+ cotransporter in the apical membranes of intestinal epithelial cells, promoting Na^+ uptake in sick people. Explain the concept behind this manipulation of the cotransporter for therapeutic ends. The approach has saved millions of lives.

5. Explain why active transport of an ion shows saturation kinetics, whereas transport of an ion through an ion channel does not.

6. Whereas electrical currents are carried by electrons in copper wire, they are carried by ions in aqueous solutions. Explain how an active-transport mechanism can create an electrical *current* across a membrane.

7. One way to produce freshwater from seawater is "reverse osmosis," in which high hydrostatic pressures are used to force water to move against its osmotic gradient, from seawater to freshwater, across a membrane. How would you calculate the minimum hydrostatic pressures required? Why might it be preferable to use salty water from a coastal bay diluted with river water rather than full-strength seawater as the water source?

8. Outline all the principal transport processes at work in the three focal examples shown in Figure 5.1.

9. When we discussed the microscopic mechanism of simple diffusion, we made the following point: After the concentrations of glucose on the two sides of a membrane have become equal, glucose molecules continue to move at random from left to right and from right to left; the numbers of glucose molecules going in the two directions are equal, however, explaining why the two concentrations stay equal once they have become equal. Taking advantage of the options provided by multiple isotopes of elements, how could you do an experiment, in an actual physical system, to determine whether the point we have made here is true?

10. The cell membranes of mammalian red blood cells are permeable to urea. If red blood cells are dropped into a solution of urea that is identical in osmotic pressure (isosmotic) to the cytoplasm of the cells, although the cells do not swell and burst as quickly as when they are dropped simply into pure water, they eventually swell and burst. Explain. Also discuss how you would design a solution into which red cells could be placed without ever swelling. (Hint: Think about whether urea will stay on the outside of the cells and the implications for osmotic pressures.)

11. Amphibian eggs laid in freshwater exhibit low water permeabilities and thus do not swell and burst osmotically. When investigators first believed they had identified an aquaporin, they manipulated amphibian eggs so the eggs expressed the purported aquaporin protein. When the investigators observed those eggs swell and burst, they knew they had made a monumental discovery: They had found the first aquaporin. Recalling what we have discussed in the text of this chapter regarding red blood cells, explain why this experiment provided convincing evidence for channel-mediated water transport.

Go to **sites.sinauer.com/animalphys4e**
for box extensions, quizzes, flashcards,
and other resources.

REFERENCES

Agre, P. 2004. Aquaporin water channels (Nobel Lecture). *Angew. Chem.* 43: 4278–4290. *doi*: 10.1002/anie.200460804. A delightful, easy-to-understand essay on aquaporins, presented by Peter Agre on the occasion of his receiving the Nobel Prize for the discovery of aquaporins.

Alberts, B., A. Johnson, J. Lewis, and four additional authors. 2014. *Molecular Biology of the Cell*, 6th ed. Garland, New York.

Bublitz, M., J. P. Morth, and P. Nissen. 2011. P-type ATPases at a glance. *J. Cell Sci.* 124: 2515–2519.

Evans, D. H. 2011. Freshwater fish gill ion transport: August Krogh to morpholinos and microprobes. *Acta Physiol.* 202: 349–359.

Finn, R. N., and J. Cerdà. 2015. Evolution and functional diversity of aquaporins. *Biol. Bull.* 229: 6–23.

Harold, F. M. 2001. Cleanings of a chemiosmotic eye. *BioEssays* 23: 848–855. An intellectually stimulating essay on the discovery of the chemiosmotic principle and the novel implications it still has today, decades later.

Kirschner, L. B. 2004. The mechanism of sodium chloride uptake in hyperregulating aquatic animals. *J. Exp. Biol.* 207: 1439–1452.

Levin, R. 2011. The ancient patterns of work in living systems. In R. Levin, S. Laughlin, C. de la Rocha, and A. Blackwell (eds.), *Work Meets Life: Exploring the Integrative Study of Work in Living Systems*, pp. 11–38. MIT Press, Cambridge, MA. Focuses on chemiosmosis.

McWhorter, T. J., B. H. Bakken, W. H. Karasov, and C. Martínez del Rio. 2006. Hummingbirds rely on both paracellular and carrier-mediated intestinal glucose absorption to fuel high metabolism. *Biol. Lett.* 2: 131–134.

Perry, S. F. 1997. The chloride cell: Structure and function in the gills of freshwater fishes. *Annu. Rev. Physiol.* 59: 325–347.

Verkman, A. S. 2011. Aquaporins at a glance. *J. Cell Sci.* 124: 2107–2112.

Voigt, C. C., and J. R. Speakman. 2007. Nectar-feeding bats fuel their high metabolism directly with exogenous carbohydrates. *Func. Ecol.* 21: 913–921. In parallel with hummingbirds, some bats have become nectar-feeding specialists that feed while flying. This paper provides an entry point to the large literature on these bats.

Weiss, T. F. 1996. *Cellular Biophysics*. MIT Press, Cambridge, MA.

Wilson, J. M., P. Laurent, B. L. Tufts, and four additional authors. 2000. NaCl uptake by the branchial epithelium in freshwater teleost fish: an immunological approach to ion-transport protein localization. *J. Exp. Biol.* 203: 2279–2296.

See also **Additional References** *and* **Figure and Table Citations**.

PART II
Food, Energy, and Temperature

Previous page
All growing animals require food as a source of matter for constructing their tissues. They require food also as a source of energy that enables them to mature into highly organized, functionally potent organisms despite the second law of thermodynamics. Adult mammals donate the essential resources of matter and energy to each new generation.

Nutrition, Feeding, and Digestion

The Maasai and other cattle-raising peoples of East Africa have long presented a mystery with respect to how they cope with milk as a major part of their diet in adulthood. Contrary to superficial impressions, most adult people on Earth—60%–70% according to current estimates—are not able to digest lactose, the principal sugar in the milk of humans and many other mammals. Newborn babies and infants are able to digest the sugar. They do this by producing a digestive enzyme, lactase-phlorizin hydrolase—*lactase* for short—in their small intestine. This enzyme splits each lactose molecule into two smaller sugar molecules (glucose and galactose) that are readily absorbed into the bloodstream and distributed to cells throughout the body. In most species of mammals, however, individuals cease producing lactase after they stop nursing, and most human beings are no different. Humans usually cease production of lactase by the age of 3–4 years (an example of genetically controlled developmental programming, similar to that shown in Figure 4.5). People who follow this pattern can suffer from lactose intolerance disease in adulthood because lactose itself cannot be absorbed from the digestive tract and they are unable to break it down into simpler sugars that can be absorbed. If people lacking lactase ingest milk, the lactose in the milk simply passes the entire length of the digestive tract and is voided in the feces. Along the way, its presence can cause multiple problems.

Adults of the cattle-raising tribes of Africa continue to synthesize a key enzyme required for milk digestion Most African people can digest milk sugar (lactose) only when they are young children because they cease in childhood to produce the required digestive enzyme, lactase. Cattle-raising was adopted by certain cultural groups only a few thousand years ago. Despite the short time since then, these cultural groups have evolved distinctive mutations that permit continued production of lactase into adulthood. Thus substantial fractions of the people are able to digest milk sugar throughout their lives. This is a Maasai man with his herd.

TABLE 6.1 Percentages of people who synthesize lactase in adulthood[a]

Group	Percentage synthesizing lactase
Dutch	100
Danes	96
French (northern)	78
White Americans	76
Greeks	55
Indigenous people of Australia	45
Mexican Americans	44
African Americans	19
Ibo and Yoruba peoples, Nigeria	18
Zulu people, South Africa	15
Native Americans (Oklahoma)	5
!Kung and Herero peoples, Namibia	3
Thai	2
Asians in the United States	0

Sources: Data for Old World and Australia from Molecular and Cultural Evolution Lab, University College London (http://www.ucl.ac.uk/mace-lab/resources/glad); data for New World and Netherlands from Buller and Grand 1990.

[a]The data presented in this table are based on clinical tests in which a person ingests a standardized test dose of lactose and investigators look for evidence of digestion of the sugar. Outside a clinical setting, the experience of "lactose intolerance disease" is partly subjective (depending, e.g., on a person's sensitivity to gut distress). Thus there is not necessarily a one-to-one relation between the proportion of a population that is clinically classified as failing to synthesize lactase and the proportion complaining of lactose intolerance disease.

Northern Europeans such as the Dutch and Danes have long been known to be exceptions to the usual human pattern, as illustrated in **TABLE 6.1**. High proportions of people in northern Europe—and in populations descended from them around the world—continue to produce lactase throughout their lives, a condition known as **lactase persistence**, enabling them to digest lactose in adulthood. This explains why they can drink milk and eat dairy products in large quantities throughout their lives without ill effects.

In many parts of Africa, however, most adults do not produce lactase and are unable to digest lactose (see Table 6.1), thus the mystery mentioned in the first paragraph. In the cattle-raising African groups, how are adults able to consume lots of milk without suffering from lactose intolerance disease?

Cattle-raising began in East Africa only a few thousand years ago. Following the start of the practice, the Maasai and other cattle-raising peoples developed extremely close cultural associations with their herds. Many of these peoples were nomadic prior to the European colonization of Africa; they moved across the landscape to find suitable grazing lands and watering places for their cattle, all the while depending on their animals for food and hides. Milk became a major food and a crucial source of liquid in times of drought.

Some years ago, genetic studies of northern Europeans revealed that they differ from most people in the *control of expression* of the gene for lactase synthesis. A specific nucleotide mutation at a chromosome position near the gene for lactase synthesis acts on the lactase gene promoter to upregulate transcription of the lactase gene in adulthood, resulting in lactase persistence. This mutation is inherited as a dominant Mendelian trait and is so common in the gene pool of northern Europeans that almost all individuals inherit at least a single copy. The mutation is entirely absent, however, in East Africa.

Recent studies in East Africa have now not only clarified how the cattle-raising peoples gain food value from milk but also provided biologists with startling new insight into the *rate* at which evolution can proceed in human populations. The studies have revealed that large percentages (60%–80%) of individuals in the populations of cattle-raising peoples exhibit lactase persistence—continuing to synthesize lactase as adults—and they evidently do so in the same fundamental way as northern Europeans, namely that they have mutations that affect control of the lactase gene, causing the gene to be expressed in adulthood. These mutations are entirely novel, however, and evidently arose in Africa in just the few thousand years since cattle-raising began there. One can reason in the abstract that mutations favoring lactase persistence would be of considerable advantage for cattle-raising peoples. Based on the actual evidence recently revealed, we now know that in fact such mutations are able to spread in human populations so rapidly as to become common within a few thousand years! One especially common mutation in East African cattle-raising peoples is estimated to have first appeared only 5000 years ago.

A key general principle is implicit in the story of lactose: Eating a food substance is usually not sufficient for the substance to have nutritional value. In addition, for an ingested material to have nutritional value, an animal must be able to digest the material (or break it down by other processes), and the animal must be able to absorb the products of digestion. *Feeding, nutrition, digestion,* and *absorption,* we see, are intimately interrelated. They evolve interactively in relation to the foods available in a population's environment. **FIGURE 6.1** provides a map of the basic processes involved in meeting an animal's nutritional needs through the ingestion of food. To meet its nutritional needs, an animal feeds on potentially useful food substances in its environment. The exact substances it ingests depend on the nature of its feeding apparatus, and also on its feeding behavior, which often is adjusted to match nutritional needs. After organic compounds are ingested, they are broken down by the action of digestive enzymes and other digestive mechanisms. These mechanisms, however, are not omnipotent and in fact are often highly specific. Thus whereas some substances in a meal may be broken down rapidly and completely, others may not be digested at all. A key factor is the suite of digestive enzymes an animal synthesizes. Fermentation by symbiotic microbes may also be important in breaking down food. *Absorption* refers to the transfer of molecules from the open central cavity of an animal's gastrointestinal tract into the living tissues of the animal. Absorption of many substances requires transporter proteins that are specific in the molecules they transport. Thus the successful use of ingested compounds often depends, in the end, on whether digestion can break up the compounds into molecules that can be transported

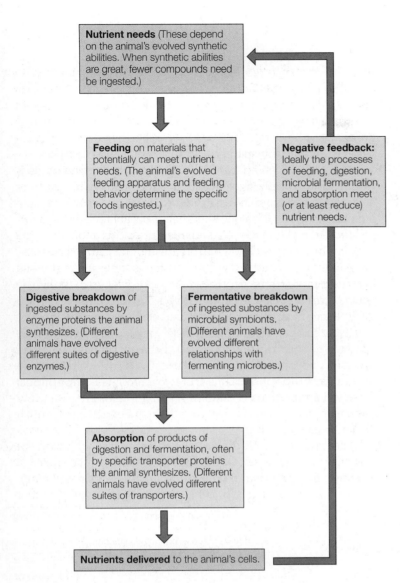

FIGURE 6.1 Relations of nutrition, feeding, digestion, and absorption in vertebrates and most types of invertebrates
"Feeding" includes the processes of selecting, acquiring, and ingesting foods. Although evolved properties are emphasized in this presentation, phenotypic plasticity is also important; the particular digestive enzymes or absorption transporters expressed by a given animal at a given time may depend on the amounts and types of foods it has recently been eating. Fermentative breakdown by symbiotic microbes is of conspicuous quantitative importance in some animals (e.g., cows), although not in others (e.g., humans). Certain types of invertebrates have novel mechanisms of meeting nutrient needs not represented by this diagram; reef-building corals, for example, are photosynthetic because of algal symbionts.

by the specific absorption transporters an animal synthesizes. All the processes discussed here, taken together, determine which nutrient molecules are actually delivered to an animal's tissues when the animal eats. Ideally, these molecules meet the animal's nutrient needs.

A notable property of animals is that all must eat regularly throughout their lives. The frequency of eating varies: American people, for instance, eat three meals a day, whereas some pythons might eat three meals a *year*. Regardless of such differences in timing, all must eat repeatedly. Why is this so? We of course readily understand why growing animals need to eat over and over: They

must accumulate all the diverse atomic building blocks required for their increasingly large bodies. But why do adults need to eat repeatedly for as long as they live? This question is especially compelling because feeding is often dangerous for animals in a state of nature; they must emerge from safe refuges to eat, for example—as when a squirrel leaves its nest cavity to search in the open for nuts. Why can't a fully grown adult simply make do with the chemical building blocks and energy it already has? One of the principal reasons that animals must feed throughout their lives is the *dynamic state of body constituents* emphasized in Chapter 1 (see page 13). Individual cells and molecules in an animal's body age or become damaged throughout the animal's life, and they are discarded and replaced. During this turnover, some of the chemical building blocks of the old cells and molecules are lost from the body and must be replaced with new chemical building blocks from foods. Another key consideration, stressed in Chapter 7, is that *the chemical-bond energy used in metabolism cannot be reused.* Instead, new chemical-bond energy must be acquired by food ingestion.

Nutrition

Nutrition is the study of the chemical compounds that compose the bodies of animals and how animals are able to synthesize the chemical components of their bodies from the chemical materials they collect from their environments. Nutrition also includes the study of the energy available from foods.

An instructive starting point for the study of nutrition is to look at the chemical composition of a familiar example, the human body (**FIGURE 6.2**). Excluding water, the human body is composed principally of proteins and lipids. Minerals are third in overall importance, followed by two additional categories of organic molecules: nucleic acids and carbohydrates. The most abundant types of atoms in the human body are carbon, hydrogen, and oxygen—the ubiquitous building blocks of organic compounds. Nitrogen, calcium, and phosphorus are additional, particularly abundant types of atoms— nitrogen because it appears in all proteins and nucleic acids, calcium because of its importance in the skeleton, and phosphorus because

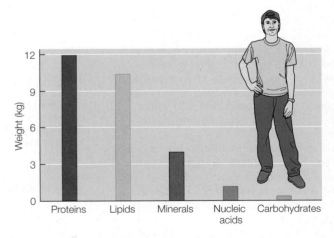

FIGURE 6.2 The composition of the adult human body
The weights in kilograms of the five principal body components, other than water, in an average, slender adult weighing 70 kg. About 42 kg of the total body weight is water (not shown).

it occurs not only in the skeleton but also in nucleic acids and the phospholipids of all cell membranes.

A useful way to understand animal nutrition is to consider the major components of the body—such as proteins or lipids—and ask several questions about each. Why is each sort of body component needed? How do animals acquire the chemical building blocks that are required to synthesize each sort? What peculiar challenges are likely to be faced for an animal to synthesize adequate amounts of each sort of body component? These are the questions that dominate this section.

Proteins are "foremost"

Among the major components of the body, proteins deserve to be discussed first because they occupy a position that is both uniquely important and uniquely tenuous in animal nutrition. The very word *protein* is derived from the Greek for "foremost," referring to the preeminent position that proteins occupy in the structure and function of all animals. One way to see the importance of proteins is to look at their abundance in the bodies of animals. About half the organic matter in a mammal is protein (see Figure 6.2), and proteins are similarly abundant in other types of animals. A second way to see the importance of proteins is to consider the many vital roles they play. Enzyme proteins *speed and regulate biochemical reactions*. Muscle proteins are responsible for *locomotion*, one of the most characteristic features of animals. Structural proteins, such as collagen in connective tissue and keratin in skin, *determine the struc-*

tural properties of tissues. Other proteins are so diverse in function as to defy simple classification. They include the receptor, channel, and transporter proteins in cell membranes; proteins dissolved in blood plasma (e.g., albumin); antibodies; proteins that function as hormones; oxygen-transport proteins such as hemoglobin; protein venoms; and the transparent crystallin proteins of eye lenses.

A set of 20–22 amino acids, called the **standard amino acids**, is required for the synthesis of proteins in all organisms. The structures of five amino acids are shown in **FIGURE 6.3A**, and all the standard amino acids are listed (along with their usual abbreviations) in Appendix I. In addition to the standard amino acids, more than 200 other amino acids are known in organisms and play diverse roles, although they are not involved in protein synthesis. As stressed in Box 2.1, protein structure is complex. However, from the simple perspective of *primary* structure (see Box 2.1), **proteins** are defined to be strings of amino acids composed of many amino acid units. Relatively short strings are often distinguished as **polypeptides**, although no strict dividing line exists between proteins and polypeptides, and for simplicity we will usually use the word *protein* to refer to both. **Dipeptides** consist of two amino acids; **tripeptides**, three. All amino acids contain nitrogen. Proteins are about 16% nitrogen by weight, in contrast to carbohydrates and lipids, which do not contain nitrogen.

A fact that colors every aspect of the nutritional biology of proteins is that nitrogen is a limiting element in many ecosystems, including many terrestrial ecosystems and more than half of the

FIGURE 6.3 Amino acid chemistry
(A) Structures of five of the standard amino acids required for protein synthesis. All amino acids contain nitrogen even if lacking an α-amino group. Amino acids with α-amino groups may contain additional nitrogen, as exemplified by lysine. Box 2.1 discusses the relation between amino acids and protein structure. (B) The fate of amino acids that are ingested but that an animal either does not need or is unable to use for protein synthesis. The reaction shown here, although it can literally occur as drawn (catalyzed by an oxidase), is a conceptual version of what usually happens. Typically, the unneeded amino acid undergoes a transamination reaction (catalyzed by an aminotransferase) that removes the amino group from the unneeded amino acid and attaches it to another carbon chain, usually synthesizing another amino acid (only nonessential amino acids can be made in this way). A common product of this process is the amino acid glutamate, which then may be deaminated to produce NH_3. Amino acids are typically not stored for future use. After a meal, all the amino acids from the meal that cannot be promptly used (including ones formed by transamination) are deaminated directly or indirectly.

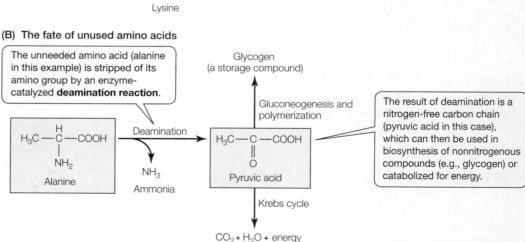

(A) Five of the standard amino acids

The H_2N- group on the second carbon from the top in each molecule shown is an α-amino group. This type of amino group occurs in most, but not all, standard amino acids.

Alanine

Phenylalanine

Serine

Valine

Lysine

(B) The fate of unused amino acids

The unneeded amino acid (alanine in this example) is stripped of its amino group by an enzyme-catalyzed **deamination reaction**.

Alanine

Deamination

NH_3
Ammonia

Glycogen (a storage compound)

Gluconeogenesis and polymerization

Pyruvic acid

The result of deamination is a nitrogen-free carbon chain (pyruvic acid in this case), which can then be used in biosynthesis of nonnitrogenous compounds (e.g., glycogen) or catabolized for energy.

Krebs cycle

$CO_2 + H_2O$ + energy

oceans. Nitrogen limitation may at first seem strange because about 78% of the gas in the atmosphere is molecular nitrogen, N_2. However, animals, plants, and most microbes cannot use N_2 as a source of nitrogen for building proteins or other nitrogen-containing body components. Instead, as their source of nitrogen they require the element in chemically combined, nongaseous ("fixed") forms. Plants and algae most commonly use nitrate (NO_3^-), ammonium (NH_4^+), and other inorganic nitrogen-containing compounds as their nitrogen sources. The nitrogen available in these chemical forms in an ecosystem is often low enough that the growth and reproduction of plants or algae are held back by nitrogen insufficiency. Herbivorous animals most commonly acquire nitrogen from proteins and other nitrogen-containing organic compounds in the tissues of the plants or algae they eat. One consequence of this fact is that nitrogen limitation can be relayed up food chains. If the production of plants and algae in an ecosystem is nitrogen-limited, growth and reproduction in herbivore populations may be nitrogen-limited, and fewer carnivores may be able to live there than if herbivores were more successful.

Of all the nutritional vulnerabilities of animals, one of the greatest is that they are biochemically unable to synthesize about ten of the standard amino acids at rates sufficient to meet their requirements (even if nitrogen is freely available). The amino acids that cannot be adequately synthesized are called **essential amino acids** because there is an "essential" requirement that they be acquired *fully formed* from food or another *outside* source.[1] Experiments to determine the list of essential amino acids in a species are challenging. Thus ambiguity often exists about the list. The set of essential amino acids also varies a bit from one animal group to another. **TABLE 6.2** lists the ten amino acids that are essential for the best-studied animal model, the growing laboratory rat. As the table notes, eight of the ten are required in the diet of adult humans, and nine are essential in children. The ten essential amino acids of the growing rat are also essential in insects, fish, and birds. For some insects and birds, there are additional amino acids that are essential.

The inability of animals to synthesize certain amino acids might not be a problem if animals were to keep stores of amino acids or proteins in their bodies in the same way they store fat. If stores of "extra" amino acids were maintained, presumably there would always be some of each kind available. However, animals do not generally store amino acids for future use, either as free amino acids or as storage proteins. Instead, when an animal eats amino acids in excess of those it needs for the synthesis of functioning proteins at the time, it promptly strips the nitrogen-containing amino groups ($-NH_2$) from the carbon chains of the excess amino acids (**FIGURE 6.3B**). The stripped nitrogen is excreted, and the nitrogen-free carbon chains that are formed are used for biosynthesis or catabolized for energy. Consequently, the amino acids in an animal's body at any one time are generally constituents of *functioning* protein molecules.

---

[1] The word *essential* has a special meaning in the study of nutrients. A nutrient is essential to an animal if the animal—in and of itself—cannot synthesize it in amounts sufficient for its requirements and thus must get it from other organisms or from the nonliving environment. Vitamins, minerals, and certain fatty acids are essential in this sense, in addition to the essential amino acids. Essential amino acids are sometimes called *indispensable amino acids.*

TABLE 6.2 Essential amino acids in the most thoroughly studied model, the growing laboratory rat, and in people

Essential amino acids in the growing rat	Ways that people are similar and different
Arginine	Arginine not essential in humans
Histidine	Histidine essential for children, probably not for healthy adults
Isoleucine	
Leucine	
Lysine	
Methionine	All essential in children and adults
Phenylalanine	
Threonine	
Tryptophan	
Valine	

Sources: After Burton and Foster 1988; Morris 1991.

Another important fact about the amino acid physiology of animals concerns the source of nitrogen for synthesis of **nonessential amino acids**—that is, those amino acids that an animal is capable of synthesizing. A property of the biochemical pathways used by animals for such synthesis is that generally the nitrogen-containing amino groups are derived from other amino acids. Thus, for a nonessential amino acid to be synthesized, other amino acids must be available to serve as amino-group donors.

With these background points in mind, you will see that animals employ a "just in time" strategy for acquiring the raw materials they need for protein synthesis. In automobile-assembly plants, the standard practice throughout most of the twentieth century was to maintain large stores of all the necessary parts. The approach used nowadays in many assembly plants, however, is the "just in time" strategy. Large stores of parts for future use are not maintained. Instead, all parts are brought to the assembly plant just in time to be used. The "just in time" approach eliminates storage costs, but it makes car assembly more vulnerable to parts shortages; if shipments of *any* required parts are interrupted, assembly of cars must stop. Animals use a "just in time" strategy in building their proteins. The amino acid building blocks that an animal needs for protein synthesis must arrive in the body at the same time that the proteins are actually being synthesized. What is more, the essential amino acids must arrive fully formed. The nonessential amino acids either must arrive preformed or must be synthesized on the spot; synthesis requires the ingestion of alternative amino acids that can serve as nitrogen donors for making the amino acids that are synthesized.

To illustrate the weighty consequences of this strategy for protein synthesis, assume that lysine is an essential amino acid and consider an animal that eats too little lysine in a particular week. The animal will be unable to synthesize lysine-containing proteins in the full amounts needed that week. If too little lysine is obtained during the following week, the shortages of proteins will worsen.

The animal's health will deteriorate because the functions of the unmade proteins will not be carried out. Lysine deficiency, in fact, is a serious health problem for millions of people in impoverished parts of the world.

The failure of animals to store amino acids for future use leads to wastage of available essential amino acids when other essential amino acids are in too-short supply for protein synthesis. Returning to the animal we were just discussing, suppose, to illustrate, that a particular meal provides all the essential amino acids in abundance, except that lysine is in short supply. Needed proteins will go unmade because of the lysine shortage. In addition, as the meal is metabolized, the essential amino acids, besides lysine, *that are abundant in the meal and that could have been used to make needed proteins* will not be fully used and will be broken down, as discussed earlier. These other essential amino acids will therefore be wasted.

For the amino acids in a meal to be used optimally for protein synthesis, the meal typically must supply the full required amounts of all essential amino acids. Mixing foods is a mechanism to achieve this objective and thereby reduce or eliminate amino acid wastage. To illustrate, consider an animal for which both lysine and methionine are essential. If food A contains little lysine but much methionine and food B contains lots of lysine but little methionine, eating the two together will allow the animal to use the amino acids from the foods more fully than if either food were eaten alone. Foods must be eaten nearly simultaneously to complement each other in this way. In the human diet, combinations such as corn and beans, or milk and cereal, eaten approximately simultaneously, increase the use of amino acids for protein synthesis because the foods tend to make up for each other's shortcomings in essential amino acids.

Where do essential amino acids originate in ecosystems? Plants and algae are able to synthesize from scratch all 20–22 of the standard amino acids required for protein synthesis. For most animals, plants and algae are the principal ultimate sources of essential amino acids.[2] Herbivores eat the essential amino acids in their plant or algal foods, and then the molecules are relayed up food chains when carnivores eat the herbivores.

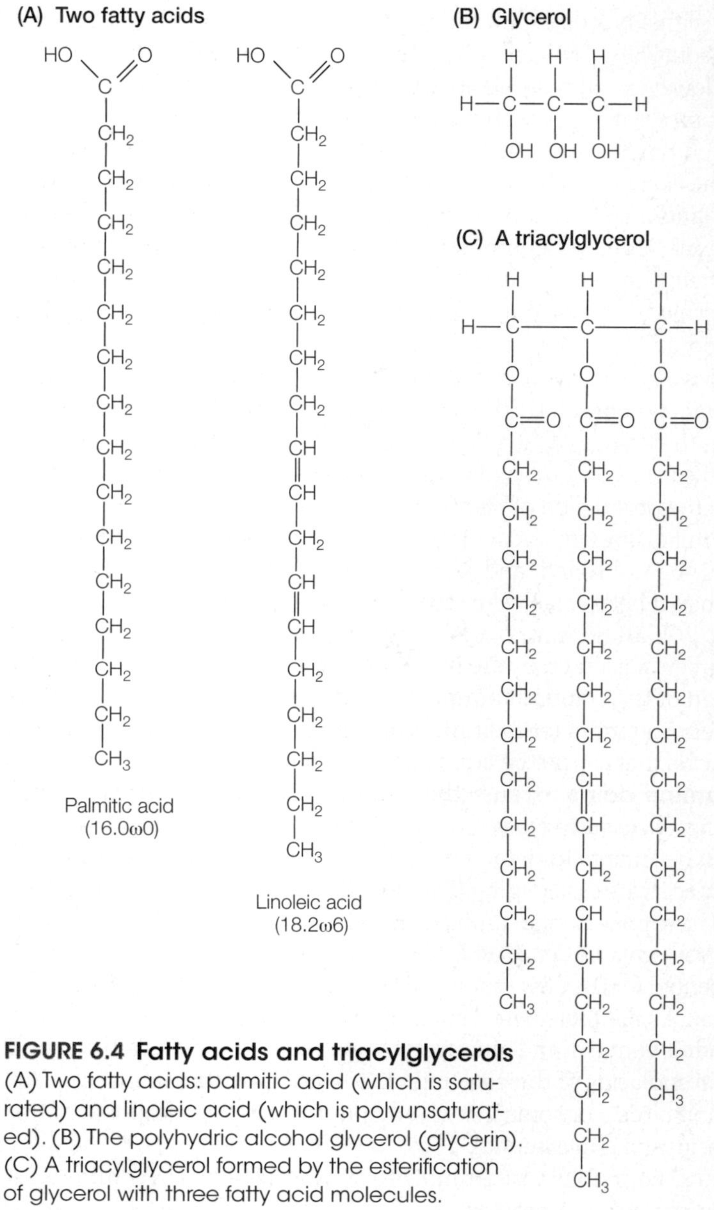

Lipids are required for all membranes and are the principal storage compounds of animals

Lipids are often about as abundant as proteins in the structure of animal bodies (see Figure 6.2). They are structurally very diverse and thus problematic to define. Lipids are usually defined as organic molecules composed principally of carbon and hydrogen that are predominantly nonpolar and therefore hydrophobic. The chemical components that most lipids have in common are **fatty acids**, which are hydrocarbons built on a backbone consisting of a chain of carbon atoms (**FIGURE 6.4A**). One important chemical category of lipids is the **triacylglycerols (triglycerides)**, also known as **fats** and **oils**; a triacylglycerol consists of three fatty acid molecules combined (esterified) with a molecule of glycerol (**FIGURE 6.4B,C**). Waxes, phospholipids (see Figure 2.2), and sterols (see Figure 16.2) are additional categories of lipids.

Fatty acids are chemically diverse; more than 50 different fatty acids are known in organisms. This diversity sets the stage for lipids

[2] Bacteria or other heterotrophic microbes may also act as ultimate sources, and some animals maintain symbioses of major nutritional importance with such microbes, as discussed later in this chapter.

FIGURE 6.4 Fatty acids and triacylglycerols (A) Two fatty acids: palmitic acid (which is saturated) and linoleic acid (which is polyunsaturated). (B) The polyhydric alcohol glycerol (glycerin). (C) A triacylglycerol formed by the esterification of glycerol with three fatty acid molecules.

to be very diverse in their chemical structures and other properties; many different triacylglycerols, for example, can be made by employing different fatty acids in the three fatty-acid positions (see Figure 6.4C). Most fatty acids in organisms have an even number of carbon atoms, between 8 and 24, in their carbon-chain backbone, and they are unbranched. Besides the number of carbon atoms, additional features that impart diversity to fatty acids—and help define their properties—are their degree of saturation and the positions of double bonds. A fatty acid is **saturated** if all the bonds between carbon atoms are single bonds, as exemplified by palmitic acid (see Figure 6.4A). A fatty acid is **unsaturated** if one or more of the bonds between carbon atoms are double. **Polyunsaturated** fatty acids are the subset of unsaturated fatty acids that contain multiple (two to six) double bonds, exemplified by linoleic acid (see Figure 6.4A).

A common system for symbolizing fatty acids specifies three integer numbers, as in the example "18.2ω6." The first integer, preceding the decimal point, is the number of carbons in the fatty

TABLE 6.3 The energy values of mixed lipids, carbohydrates, and proteins in aerobic catabolism[a]		
	Energy value in kilocalories (kcal) per gram	Energy value in kilojoules (kJ) per gram
Mixed lipids	9.3–9.5	39–40
Mixed carbohydrates	4.0–4.1	17
Hydrated glycogen	1.1	4.6
Mixed proteins	4.3–4.8	18–20

[a]Data are for mammals and were measured by bomb calorimetry (see Chapter 7). The values for proteins depend on the chemical form of nitrogen excreted; therefore moderately different values apply to groups of animals that differ from mammals in their nitrogenous excretory compounds. All values are for dry foodstuff material except those for hydrated glycogen; the glycogen values are for glycogen combined with its usual water of hydration.

acid. The second, written between the decimal and the Greek letter omega (ω), is the number of double bonds. The third, which follows ω, is the position of the first double bond encountered when the molecule is scanned from its methyl (—CH_3) end. The example "18.2ω6" refers to linoleic acid (see Figure 6.4A), which has 18 carbons and 2 double bonds, with the first double bond at the sixth position from the methyl end. When people speak of "omega-3" or "omega-6" fatty acids, they are referring to unsaturated fatty acids with the first double bond in the third (ω3) or sixth (ω6) position.

Lipids play several functional roles. One role of lipids is that phospholipids and cholesterol (a lipid) are principal *components of cell and intracellular membranes* (see Chapter 2). Lipids are also *storage compounds* in both animals and plants. Lipids serve especially well as energy stores because they far exceed proteins and carbohydrates in their energy value per unit of weight (**TABLE 6.3**). The high energy density of lipids means that the weight of material an animal must carry around to store a given amount of chemical-bond energy is far lower if lipids are the storage compounds than if other compounds are used. A third functional role played by lipids is that in fully terrestrial animals such as mammals and insects, lipids in or on the integument *greatly reduce the permeability of the integument to water*; by thus slowing the evaporative loss of body water, they permit the animals to live in the open air. Lipids play many other roles. Steroid lipids, for example, serve as hormones.

Animals must synthesize a great variety of lipids to construct their cells and tissues and meet other requirements for lipids in their bodies. They are able to meet these biosynthetic requirements using a variety of organic carbon chains that they obtain in their diets. Indeed, carbon chains used to synthesize lipids may be derived from dietary carbohydrates and proteins, as well as dietary lipids (as is well known, we can add body fat by eating sugar). This biochemical flexibility is one reason that lipid nutrition is less likely than protein nutrition to be fraught with problems. Another reason is the fact that animals maintain stores of lipids; contrary to the case with proteins, lipids obtained in excess of needs at one time are often held over in the body for use in the future.

Problems can arise in lipid nutrition, however, because although animals have extensive capabilities to synthesize and structurally modify fatty acids, many types of animals—including mammals—lack the enzymes needed to create double bonds at the omega-3 and omega-6 positions. The omega-3 and omega-6 fatty acids therefore cannot be synthesized and must be obtained from foods or other *outside* sources, at least during critical life stages. The omega-3 and omega-6 fatty acids thus are **essential fatty acids**. Essential amino acids, you will recall, are specific compounds. The situation with essential fatty acids is usually a little different because when animals eat foods containing fatty acid carbon chains with double bonds in the ω3 or ω6 position, they often can modify those chains to obtain others that they need. For example, relatively long fatty acids with a double bond at the ω3 or ω6 position can meet the dietary requirement for shorter omega-3 or omega-6 fatty acids because the long fatty acids can be shortened. The simplest fatty acids that will meet omega-3 requirements and omega-6 requirements are α-linolenic acid and linoleic acid (see Figure 6.4A), respectively. Exceptions to this general picture occur. Cats and certain fish, for example, have dietary fatty acid requirements that are more chemically specific than the requirements of most vertebrates. These requirements must be taken into account in the design of foods for pets and aquaculture.

Carbohydrates are low in abundance in many animals but highly abundant when they play structural roles

The simplest carbohydrates are **monosaccharides** such as glucose and fructose (**FIGURE 6.5A**). Other types of carbohydrates are composed of two or more monosaccharides bonded together. Those consisting of two monosaccharides are called **disaccharides**; examples are sucrose (glucose + fructose) (**FIGURE 6.5B**), trehalose (glucose + glucose), and the sugar with which we started this chapter, lactose (glucose + galactose). **Polysaccharides** have more than ten monosaccharides per molecule and often consist of hundreds or thousands of polymerized monosaccharide units.

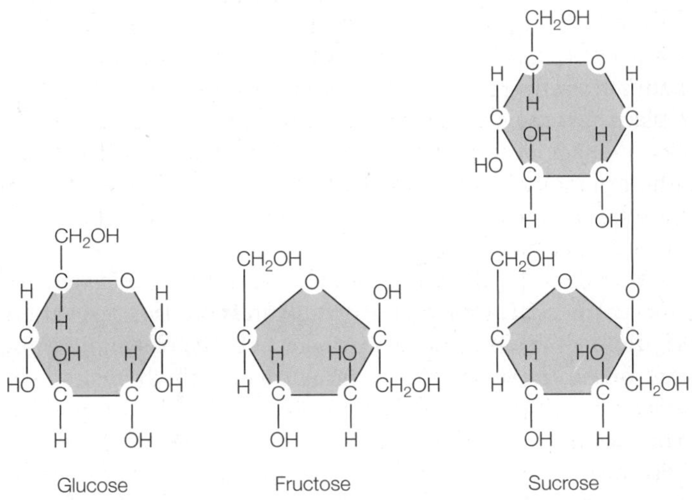

(A) **Two monosaccharides**   (B) **A disaccharide**

Glucose   Fructose   Sucrose

FIGURE 6.5 Carbohydrate chemistry Sucrose, a disaccharide, is composed of glucose and fructose units.

Carbohydrates play three principal functional roles in animals and plants. First, large polysaccharides often *provide structural support and shape* to cells and other parts of organisms. The most important structural carbohydrate in animals is the polysaccharide *chitin*, which is the principal component of the exoskeletons of insects and many other arthropods. Often, more than half the dry weight of an arthropod consists of chitin. The structural polysaccharides of plants and algae, which include *cellulose* and *hemicelluloses*, are enormously abundant in ecosystems and thus are potentially major food sources for animals. Chitin and cellulose are the two most abundant organic compounds in the biosphere!

In their second principal role, carbohydrates function as *storage* compounds. Like the structural carbohydrates, the storage carbohydrates are polysaccharides. They are accumulated and broken down far more dynamically than structural polysaccharides, however. *Starch*, a form of polymerized glucose, is one of the principal storage carbohydrates in plants. *Glycogen*, also a form of polymerized glucose, is the principal storage carbohydrate of animals, stored primarily in the liver and muscles in vertebrates. In a simplistic sense, the "value" of stored glycogen is that it is a source of energy. Glycogen, however, is far inferior to lipid as a general energy-storage compound for the following reason: Molecules of glycogen are highly hydrated in the tissues of living animals, and because of the added weight of their water of hydration, they yield a low amount of energy per unit of total weight (see Table 6.3). This attribute of glycogen probably explains why the amount of glycogen stored by animals is typically *far* less than the amount of lipid stored (see Figure 6.2). Glycogen serves mainly as a store of *glucose* which in turn serves as a *chemically specific energy source* for certain tissues and metabolic processes that require glucose as their energy source. The central nervous system in vertebrates requires glucose as its principal fuel; moreover, the fastest mechanism of ATP production in vertebrate skeletal muscles (anaerobic glycolysis) also requires glucose as its fuel. Stores of glycogen supply glucose energy to the brain and skeletal muscles in times of need.

The third principal functional role of carbohydrates is as *transport* compounds. The transport carbohydrates are small molecules—monosaccharides or disaccharides—found dissolved in the blood or other moving body fluids. When the body fluids travel from one place to another, the dissolved carbohydrates move with them—thereby, for example, transporting energy from place to place. Glucose is the principal blood transport carbohydrate ("blood sugar") in vertebrates and most other groups of animals, although many insects employ the disaccharide trehalose in this role. Lactose is transported from mother to offspring in the milks of mammals.

Animals typically have a great deal of latitude in the biochemical mechanisms by which they can synthesize the variety of carbohydrates they need to construct their cells and tissues. The carbon chains used to synthesize carbohydrates may be derived from dietary proteins or from the glycerol components of dietary lipids, as well as from dietary carbohydrates. Moreover, there are no essential carbohydrates; animals can synthesize all the carbohydrates they need. The one noteworthy "nutritional problem" that animals have with carbohydrates is that many animals are unable to digest cellulose, chitin, or some of the other structural polysaccharides. Such animals are potentially unable to use some of the most abundant organic materials on Earth. People, for example, cannot digest cellulose; consequently, cellulose in our food—for the most part—simply travels the entire length of our digestive tract and is eliminated in our feces. Cellulose may have been favored as a structural material in plant evolution precisely because it is indigestible and therefore potentially unusable by many animals.

Vitamins are essential organic compounds required in small amounts

Vitamins are organic compounds that animals must obtain in small quantities from food or another *outside* source because the compounds cannot be synthesized by the animals and yet are required in small amounts. Possibly this definition will seem strange because of its emphasis on small quantities. For historical reasons, however, compounds that must be ingested in relatively large amounts, such as essential amino acids, are not called vitamins. The existence of vitamins initially came to light when, in about 1880, scientists tried to maintain normal growth in mice by feeding them purified diets consisting only of carbohydrates, lipids, nutritionally complete proteins, and inorganic salts. The mice could not maintain normal growth. Decades of research then gradually revealed the need for small (often minute) amounts of additional organic compounds, which were called *vitamins*.

The need for vitamins in modern-day animals is often a consequence of "opportunism" during evolution. Vitamin A illustrates this point. The term *vitamin A* refers to *retinol* (**FIGURE 6.6**) and several closely related compounds. Animals cannot synthesize the retinol structure from small building blocks. Instead, they are dependent on plants or algae to make the structure. In plants and algae, the structure occurs as a subpart of photon-collecting *carotenoid pigments* such as β-carotene. An animal can obtain vitamin A by eating carotenoids and liberating the vitamin-A structure from these large molecules. The animal can alternatively obtain vitamin A by eating another animal that obtained the vitamin A from carotenoids. Either way, the source of the structure is plant or algal biosynthesis. In animals, the vitamin-A structure is incorporated into crucial visual pigments called *rhodopsins*.[3] *Remarkably,*

---

[3] The functions and synthesis of rhodopsins are discussed further in Chapter 14 (see Figure 14.23).

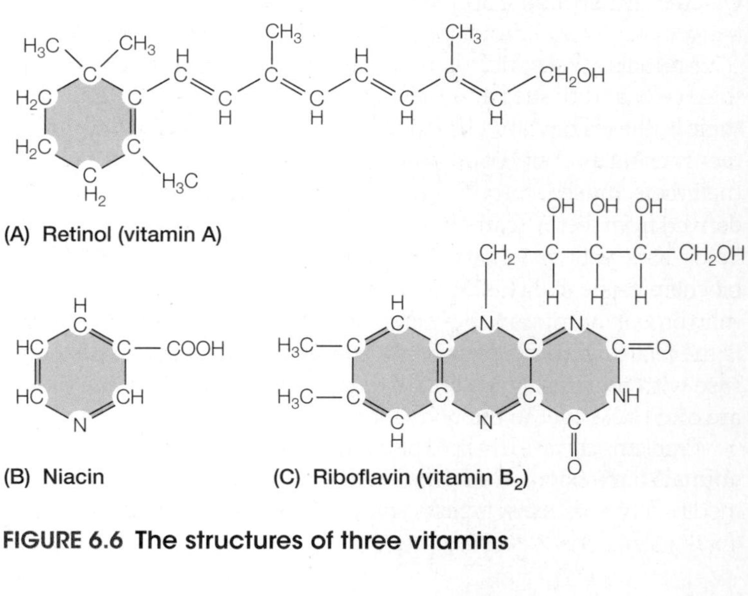

(A) Retinol (vitamin A)

(B) Niacin

(C) Riboflavin (vitamin B$_2$)

FIGURE 6.6 The structures of three vitamins

TABLE 6.4 Water-soluble vitamins and their physiological roles

Of the vitamins listed, all except ascorbic acid (vitamin C) are classified as B vitamins, although some are not usually referred to using "B" symbols. The B vitamins are required for the reactions specified because they are incorporated into relatively small organic molecules that—binding to enzymes—are required for enzyme catalytic function.

Vitamin	Physiological role in animals
Thiamin (B$_1$)	Required for oxidative decarboxylation reactions.
Riboflavin (B$_2$)	Required for oxidation–reduction reactions (required for synthesis of FAD, flavin adenine dinucleotide).
Niacin	Required for oxidation–reduction reactions (required for synthesis of NAD, nicotinamide adenine dinucleotide).
Pyridoxine (B$_6$)	Required for reactions of amino acid metabolism.
Pantothenate	Required for acetyl-transfer reactions (required for synthesis of coenzyme A).
Folate	Required for single-carbon-transfer reactions (as in nucleic acid synthesis).
Cobalamin (B$_{12}$)	Required for single-carbon-transfer reactions (as in nucleic acid synthesis).
Biotin	Required for carboxylation reactions.
Ascorbic acid (C)	Protects cells against damage by reactive oxygen compounds (antioxidant; see Box 8.1). Plays many additional roles in reactions involving oxygen.

TABLE 6.5 Lipid-soluble vitamins and their physiological roles

Unlike the water-soluble vitamins, lipid-soluble vitamins are usually referred to by their vitamin designations rather than by their chemical names. The reason is that multiple related compounds can often meet the vitamin requirement for lipid-soluble vitamins.

Vitamin	Physiological role in animals
Vitamin A	Light-activated component of visual pigments. Also needed for normal bone growth, reproductive function (e.g., sperm production), cell membrane integrity, and other functions, but exact biochemical mechanisms are not always known. Vitamin A is a regulator of gene transcription and can cause deranged development of a fetus if ingested in artificially large amounts by the mother during pregnancy.
Vitamin D	Activator of pathways of calcium and phosphorus metabolism; acts by binding, like a hormone, to specific receptors.
Vitamin E	Protects cells against damage by reactive oxygen compounds (antioxidant), preserving integrity of critical molecules, especially membrane phospholipids. See Box 8.1.
Vitamin K	Required for production of blood-clotting factors.

the light-absorbing portion of a rhodopsin is the subpart of the molecule derived from vitamin A! In plants and algae, the carotenoids serve as photon-collecting molecules for photosynthesis. Evidently, as vision evolved, animals that ingested the plant or algal photon-absorbing molecules in their foods "opportunistically" used the vitamin-A structure from those molecules to serve as the photon-absorbing portion of their visual pigments, rather than evolving an ability to synthesize their own photon-absorbing structure from scratch. This strategy means that animals need not be able to synthesize the photon-absorbing substructure of their rhodopsins. It also means, however, that animals are vulnerable to shortages if they are unable to get enough in their foods.

Because vitamins are consequences of opportunism during evolutionary history, it is not surprising to find that they are extremely diverse in their chemical structures (see Figure 6.6). They are also very diverse in their functions, meaning that there is no simple way to summarize what they do. The use made of most vitamins is that they are incorporated as *key molecular subsystems* into vital molecules, notably molecules that enzymes must bind with in order for the enzymes to function as catalysts. Vitamins are usefully subdivided into a *water-soluble* set and *lipid-soluble* set. The water-soluble vitamins include the B vitamins and vitamin C (**TABLE 6.4**). Virtually all animals have a mandatory requirement for the B vitamins because the vitamins are required for biochemical

reactions of universal importance. The lipid-soluble set of vitamins includes vitamins A, D, E, and K (**TABLE 6.5**). All four are required by vertebrates. Invertebrates, however, may or may not require them. The physiological uses of the lipid-soluble vitamins are more specialized than those of the B vitamins (see Tables 6.4 and 6.5), helping to explain why the need for them is not nearly so universal as that for B vitamins.

Elemental nutrition: Many minerals are essential nutrients

Many chemical elements—referred to as *minerals* in the study of nutrition—are required by animals, in addition to the carbon, hydrogen, oxygen, and nitrogen that predominate in organic molecules. One reason a wide diversity of elements is essential is that many enzymes and other proteins include in their structures trace amounts of atoms that otherwise are uncommon in protoplasm. About 40% of all proteins[4] are in fact metal-containing *metalloproteins*; the metal atoms in the structures of such proteins include iron, copper, molybdenum, zinc, manganese, vanadium, selenium, and cobalt, among others. Humans and other mammals must obtain all of these metals in their diets for proper nutrition. They also require iodine for thyroid hormone synthesis. In addition, turning to some of the elements that are more abundant in the body, phosphorus

---

[4] Including those of plants and microbes as well as animals.

FIGURE 6.7 The great Serengeti migration: A quest for minerals? Wildebeests and zebras, in herds sometimes numbering a million or more, travel from the far north of the Serengeti ecosystem to the southeast at the start of the rainy season and return to the north for the dry season each year. According to one prominent hypothesis, the trek to the southeast is a quest for minerals. See Chapter 30 for more on the Serengeti migration.

is required for the synthesis of phospholipids, nucleic acids, and bone; and sodium, chlorine, and potassium are essential solutes in the body fluids. One striking way to see the elemental needs of humans is to look at a stoichiometrically correct chemical formula of the human body (published by Sterner and Elser in their book, which is listed in the References).

A person as a chemical formula:

$$H_{375000000}O_{132000000}C_{85700000}N_{6430000}Ca_{1500000}P_{1020000}S_{206000}$$

$$Na_{183000}K_{177000}Cl_{127000}Mg_{40000}Si_{38600}Fe_{2680}Zn_{2110}$$

$$Cu_{76}I_{14}Mn_{13}F_{13}Cr_7Se_4Mo_3Co_1$$

Other kinds of animals have lists of elemental needs similar to those of humans and other mammals.

Mineral deficiencies are a potential problem for essentially all animals. Considering just terrestrial animals, for example, mineral deficiencies are common in a great many regions of the world because of mineral shortages in soils or because of soil-chemistry processes that render minerals unavailable. Sodium and phosphorus deficiencies are common threats to livestock worldwide; moreover, in much of tropical Africa, soil concentrations of calcium, magnesium, and copper are too low for plants to meet readily the needs of lactating mothers and the growing young of ungulate species. Iodine deficiencies provide another illustration; terrestrial regions so expansive that they are occupied by more than 1.5 billion people have soils that provide too little iodine for naturally growing foods to meet human iodine needs. Iodine deficiency is prevented in developed nations by the use of dietary iodine supplements, notably iodized salt. In impoverished regions, however, insufficient iodine ingestion is a major cause of mental retardation and abnormal neurological development in children. (Recent calculations indicate that average IQ could be raised in these regions by 10–15 points at a very low cost by alleviating iodine deficiency.)

The migration of wildebeests and zebras in the Serengeti ecosystem (**FIGURE 6.7**), justly called one of the wonders of the living world, may well be a quest for essential minerals. At the start of each rainy season, a herd of a million or more migrates 200 km (more than 100 miles) from the far north of the Serengeti—where rivers flow year-round—to the dry southeast. As the rains fall in the southeast, parched grasses spring to life, and dry streambeds fill with running water. The animals thrive there during the rains, giving birth to most of their young. When the rains end months later, however, the thin soils of the southeast dry out quickly, and the animals are forced by lack of water into a mass exodus that ultimately takes them back to their starting place in the north. Many die along the way. No one is certain why the wildebeests and zebras undertake their migration. Why don't they simply stay year-round in the north, where water is available in all seasons, rather than migrating a great distance into an area—the southeast—that cannot possibly be a permanent home because it is too dry for half the year?

The most compelling modern hypothesis is that the wildebeests and zebras migrate to the southeast to acquire minerals. In the Serengeti ecosystem, the soils of the southeast—because they are of relatively recent volcanic origin—are especially rich in needed minerals. This leads to plants rich in minerals. For example, grasses in the southeast contain 1.5 times as much calcium per gram of tissue weight as grasses in the north. Lactating mothers and their growing young require especially large amounts of calcium. According to the mineral–nutrition hypothesis, the reason the fabled herds migrate to the southeast, and calve there, is that the migration ensures their young of adequate minerals for development.

SUMMARY
Nutrition

■ Essential organic nutrients are organic compounds that animals must obtain from other organisms because the animals are biochemically unable to synthesize them. About ten of the standard amino acids required for protein synthesis are essential in most sorts of animals. Omega-3 and omega-6 fatty acids are essential in many animals, and vitamins are essential.

- Proteins often present particular nutritional problems because nitrogen can be in short supply in ecosystems and essential amino acids may be unavailable in the amounts needed. Because animals employ a "just in time" strategy for protein synthesis (rather than storing amino acids for future use), the essential amino acids must be eaten together in the amounts they are needed if they are to be used. A shortage of any one essential amino acid can cause wastage of available supplies of others.

- Vitamins are very diverse in their chemical structures. What they have in common is not their chemical nature but the fact that they are organic compounds required in small (often minute) amounts. Most are used as substructures in molecules of critical importance for animal function. The water-soluble vitamins, notably the B vitamins, are essential for most or all animals. The lipid-soluble vitamins, such as vitamins A and K, are more specialized in their functions and not as universally required.

- Minerals are also essential nutrients. Metal atoms occur in approximately 40% of proteins. Minerals are also required as constituents of body fluids and skeletons. More than 20 chemical elements are required for the construction of animal bodies.

- Structural carbohydrates such as chitin, cellulose, and hemicelluloses are the most abundant organic compounds on Earth, but many animals lack enzymes required to digest them and thus are unable (on their own) to tap those carbohydrates for nutritional value.

Feeding

Feeding—by which we mean the process of obtaining and ingesting foods—needs to be targeted at food items that will meet an animal's nutritional needs. One factor of importance in the success of feeding is the structure of an animal's feeding apparatus. Behavioral modulation of the use of the feeding apparatus is often equally important. For example, individuals with particular nutrient needs often behaviorally select foods that will meet those specific needs. Mountain gorillas, for example, are attracted to eating rotting wood because it is a particularly effective source of the sodium (Na^+) they require. A more intricate example comes from recent research on wolf spiders. The spiders are predatory, and they were provided with fruit flies to eat. Some of the flies—all of which were of one species and looked the same—had been manipulated to have a body composition high in protein, whereas others were high in lipid. As seen in **FIGURE 6.8**, the spiders chose prey that helped balance their needs for protein and lipid. For example, if a spider had ingested lots of protein-rich flies during the 24 h before testing, its appetite for lipid-rich flies was heightened and its appetite for protein-rich flies was diminished.

The feeding mechanisms employed by animals are stunningly diverse. Consider, for example, three animals that live in the oceans: orcas, blue whales, and reef-building corals. Orcas (killer whales) feed in a way that is very familiar to us. They have large teeth and strong jaw muscles, and they feed by attacking other animals that are modestly smaller than they are, such as fish and seals (**FIGURE 6.9**). Blue whales (see Figure 6.13) lack teeth and do not attack individual prey. Instead, they are *suspension feeders*, meaning they ingest organisms suspended in the water that are minute by com-

FIGURE 6.8 Behavioral food selection: Wolf spiders select the flies they eat in ways that balance their intake of protein and lipid Two groups of spiders (*Pardosa prativaga*; 22–24 spiders in each group) were, respectively, given only protein-rich or only lipid-rich fruit flies (*Drosophila melanogaster*) to eat for 24 h prior to testing. Then they were tested by being permitted to feed on lipid-rich (left) or protein-rich (right) flies for 72 h. As seen on the left, when the two groups of spiders were presented with lipid-rich flies to eat, the spiders that had been eating protein-rich flies before testing caught and ate far more than those that had been feeding on lipid-rich flies before testing. By contrast, as seen on the right, when the two groups were presented with protein-rich flies to eat, the spiders that had been eating lipid-rich flies ingested far more than those that had been feeding on protein-rich flies. The differences between groups are statistically significant. Error bars show the standard error. (After Mayntz et al. 2005. Photograph courtesy of David Mayntz.)

parison with their own size. The whales capture their tiny prey by sieving them from the water. Reef-building corals (see Figure 6.15) are photosynthetic; they get much of their nutrition from algae with which they live in symbiotic associations. All three of the feeding modes exemplified by the orca, blue whale, and corals—attack of individual prey items, suspension feeding, and association with symbiotic microbes—are widespread in the animal kingdom.

Many animals feed on organisms that are individually attacked and ingested

Many animals, both terrestrial and aquatic, are like orcas in that their feeding is directed at individually targeted food items. This feeding mode requires that the individual food organisms be located, identified, subdued, and ingested. In contrast to the simplicity of these basic imperatives, the array of actual mechanisms employed is wonderful in its diversity. Carnivorous mammals that

FIGURE 6.9 Some species feed by targeting and subduing individual food items Orcas (*Orcinus orca*) eat fish, sea otters, and seals by attacking them with sharp teeth. Groups (pods) of orcas tend to be behavioral specialists in the prey they attack. Some groups, for example, eat mostly fish, whereas others eat mostly marine mammals.

feed in this way grasp prey with teeth, and usually they use their teeth to reduce the prey to smaller pieces by tearing or chewing before swallowing. Carnivorous fish grasp prey with teeth as well, but they usually swallow their prey whole. Some sea stars eat by enveloping prey in their evaginated stomachs. Fragile butterflies harvest the nectar of flowers with long proboscises. Peregrine falcons violently strike flying birds at high speed in midair, then circle back to grasp their victims as the latter—stunned—fall toward ground in the grip of gravity.

Related animals that share a particular mechanism of feeding usually exhibit specialized variations in the mechanism. Bird bills provide a familiar example. Virtually all birds collect food items using their bills. The bills, however, exhibit a wide range of specializations. Some bird species have long, strong, sharp bills that they use as chisels to reach prey within tree trunks (**FIGURE 6.10A**); other species have stout bills that help them crush seeds (**FIGURE 6.10B**); still others have slender bills that they insert, stiletto-like, deep into mud to collect burrowing prey (**FIGURE 6.10C**).

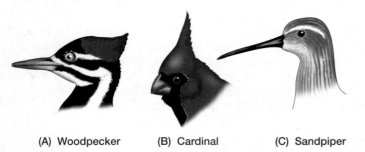

(A) Woodpecker   (B) Cardinal   (C) Sandpiper

FIGURE 6.10 Specialization of a vertebrate feeding apparatus Bill specialization in a woodpecker, a cardinal, and a sandpiper (a type of shorebird).

The food-collection devices of grazing mammals—their teeth and lips, plus their prioritization mechanisms—provide another vertebrate example of variations on a theme. Grazing can at first look as nonselective as lawn mowing. In fact, however, species of grazers that are ostensibly similar sometimes feed on very different sets of the plant species available. The teeth, jaws, and lips of zebras, for example, enable them to feed more readily on tall grasses than wildebeests can. Wildebeests, in contrast, are more effective at ingesting short grasses in large quantities, in part because they have an exceptionally blunt muzzle and a wide row of incisors. Differences such as these are believed by most biologists to help account for the diversity of coexisting species.

Snails provide a dramatic invertebrate example of diversification in the use of a single basic feeding mechanism. The primitive feeding device of snails is a scraping organ, the **radular apparatus**. It consists of a ribbonlike band of connective tissue (the radular ribbon), studded with chitinous teeth and stretched over a cartilaginous rod (**FIGURE 6.11A**). Most snails use this apparatus to rasp organic matter off objects, such as algae off rocks. The cartilaginous rod is pressed against a surface, and muscles at the two ends of the radular ribbon pull the ribbon back and forth over the rod to produce a rasping action. The radular teeth, which are elaborate (**FIGURE 6.11B**), vary widely in shape from species to species. In some species, the teeth are hardened with iron or silica compounds, enabling the snails to rasp tissue from armored food sources.

The most exotic specializations of the radular feeding mechanism occur in snails that attack and feed on living animals. One specialization for carnivory is the use of the radular apparatus as a *drill* to cut a hole through the shell of a clam or barnacle to gain entry to the soft tissues inside (**FIGURE 6.11C**). The most extreme specialization of the radular apparatus occurs in the carnivorous cone snails we discussed in the opening of Chapter 2. In cone snails, the radular teeth have become detachable, elongated harpoons, complete with barbs (**FIGURE 6.11D**). Cone snails synthesize some of the most toxic venoms known (conotoxins), and they use their highly modified teeth to inject the venoms into fish or other prey (see Figure 2.28).

TOXIC COMPOUNDS Toxic compounds incorporated into venoms or employed in other ways are involved in the feeding biology of a wide diversity of animals that attack individual food items. Predators that employ poisons to attack prey include not just cone snails but also scorpions, spiders, many snakes, and many coelenterates, such as jellyfish. The poisons used by all these predators are proteins. The most common actions of the poisons, as we saw in Chapter 2, are to inflict structural damage on cell or basement membranes, or to interfere with nerve and muscle function by combining with receptor or channel proteins.

Another way that toxic compounds play roles in predation is that they are used as defenses by potential prey. Bees, wasps, and some skates, for instance, have defensive venoms. Far more common are toxic or repellent compounds *within the tissues* of sessile prey. Sessile organisms such as sponges or plants may seem at first to be easy meals—sitting in plain view, waiting to be eaten. Many, we now realize from advances in analytical chemistry, are far from defenseless, however, because they synthesize and

(A) The mouth cavity of a snail showing the radular apparatus in its up and down positions

Apparatus up

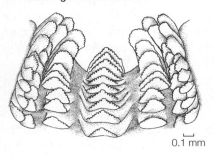

Radular ribbon with teeth

Cartilaginous rod

Retractor muscle

Algae

Protractor muscle

Apparatus down

When the apparatus is pushed down, food is scraped or rasped off the substrate by back-and-forth motions of the ribbon and teeth.

An herbivorous snail feeding by rasping algal growth off a rock

(B) A radular ribbon viewed side-to-side, showing rows of teeth

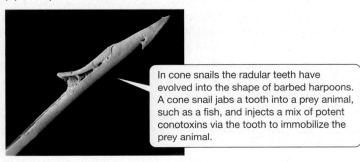

0.1 mm

(C) The shell of a clam that was killed by a carnivorous drill snail

Some carnivorous snails, called drills, have a specialized radular apparatus that cuts a neat hole through the shell of a clam, barnacle, or other armored prey item, permitting the snail to attack the living tissues inside.

(D) A harpoonlike radular tooth of a cone snail

In cone snails the radular teeth have evolved into the shape of barbed harpoons. A cone snail jabs a tooth into a prey animal, such as a fish, and injects a mix of potent conotoxins via the tooth to immobilize the prey animal.

FIGURE 6.11 Specialization of an invertebrate feeding apparatus The primitive form of the radular apparatus in snails is a scraping or rasping organ used to scrape algae off rocks, rasp flesh off dead animals, and so forth. The radular apparatus is greatly specialized in carnivorous snails that prey on living animals. (A) Longitudinal sections of the primitive radular apparatus in the mouth cavity of a snail. (B) A magnified drawing of a radular ribbon of the primitive sort seen in (A), viewed side-to-side and showing the intricate shapes and multiple rows of teeth; these details are species-specific. (C) The telltale sign of specialized radular feeding by a carnivorous snail called a drill. (D) A highly specialized radular tooth of a fish-eating cone snail, shown at great magnification. (B from Hyman 1967.)

deposit within their tissues chemicals that discourage attack. In addition to sponges and plants, other groups sharing this property include seaweeds, soft corals, and tunicates. The compounds *within the tissues* of these organisms that deter consumption are called **secondary compounds**, **secondary metabolites**, or **allelochemicals**. Some are merely distasteful; others disturb the digestion or metabolism of herbivores or predators. Chemically, the compounds are extremely diverse and include alkaloids, terpenes, polymerized phenolics (e.g., tannins), steroids, nonprotein amino acids, polyethers, and quinones.

Some herbivores and predators have become specialists at defeating defenses based on secondary compounds. They may, for instance, have a specialized biochemistry that detoxifies defensive chemicals, or they may be able to tolerate the chemicals or rapidly excrete them. Therefore the battle to eat and avoid being eaten, on both sides, is often waged biochemically as well as mechanically.

Suspension feeding is common in aquatic animals

Suspension feeding, as noted earlier, is feeding on objects, suspended in water, that are very small by comparison with the

feeding animal.[5] Clams and oysters—which are often many *centimeters* long—typically feed on suspended particles that are 5–50 *micrometers* in size. They thus qualify as suspension feeders. Blue whales that are 20–30 *meters* long feed on shrimplike krill that are 2–3 *centimeters* long. Blue whales thus also qualify as suspension feeders. A requirement for suspension feeders is to collect large numbers of food items because each food item is so small. Although suspension feeders often display evidence of selectivity, they typically do not single out and attack food items individually. Instead, they collect food items in numbers.

Suspension feeding is very common in both freshwater and the ocean. A factor that helps explain this is that small, suspended food items are abundant in aquatic systems.

Besides food abundance, additional factors probably help explain why suspension feeding has evolved as often as it has in water. One of the most important factors is that suspension feeding permits

---

[5] Suspension feeding can in principle occur in air; in fact, some terrestrial animals, such as certain spiders, function as suspension feeders. However, little if any primary production occurs in the atmosphere, and potential food items for suspension feeders are sparse in air. Thus suspension feeding is far more common in aquatic animals than terrestrial ones.

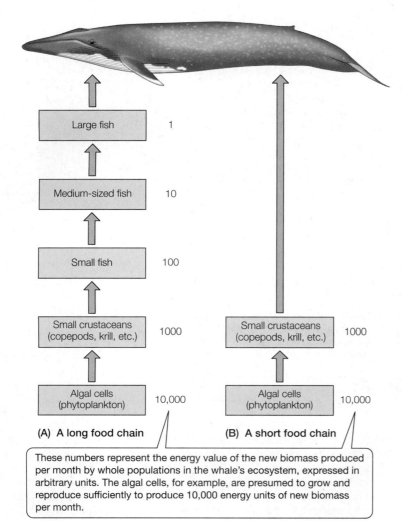

These numbers represent the energy value of the new biomass produced per month by whole populations in the whale's ecosystem, expressed in arbitrary units. The algal cells, for example, are presumed to grow and reproduce sufficiently to produce 10,000 energy units of new biomass per month.

FIGURE 6.12 Short food chains deplete energy less than long food chains do In this thought exercise, we imagine a whale with two possible food chains. In (A), the whale eats fish just a bit smaller than itself, and the food chain leading to the whale's prey has four steps. In (B), the whale eats animals that are minute by comparison with its own size, and the food chain leading to the whale's prey has one step. The short food chain (B) depletes far less of the energy, prior to consumption by the whale, than the long food chain (A) does.

animals to feed lower on food chains, where food productivity tends to be particularly great. Ecologists find that as food is passed through the steps of a food chain, only about 10% of the energy value of the food makes the transition through each step. Feeding lower on a food chain reduces the number of steps and thereby increases the energy available.

To see these principles more clearly, let's focus on a particular example (even though the principles apply to most or all suspension feeders). Let's contrast two possible mechanisms by which a large whale can obtain food. One possibility is for the whale to eat fish just somewhat smaller than itself. Such fish might live by eating other fish only somewhat smaller than they are, and so forth. The entire food chain leading to the whale might thus look like that in **FIGURE 6.12A**. Note that there are four steps in this food chain between the photosynthetic organisms (the algal cells) and the prey of the whale (the large fish). Suppose that the population of algal cells grows and reproduces sufficiently over a period of a month to

produce 10,000 units of food energy. Using the value 10%, we can then estimate that the population of small crustaceans will grow and reproduce enough to produce 1000 units of food energy during the month (10% of 10,000). Next, the food energy produced by the population of small fish during the month will be 100 units. That produced by the population of medium-sized fish will be 10 units, and finally, that produced by the population of large fish will be 1 unit. This 1 unit is the food available for a whale population at the top of the food chain. Thus a long food chain with many steps, as in Figure 6.12A, depletes most of the energy value of algal production before the energy reaches a form that the whales eat. In contrast, as presented in **FIGURE 6.12B**, a whale can function as a suspension feeder and eat organisms much smaller than it is, such as the small crustaceans. In this case, there is just one step between the photosynthetic algal cells and the prey of the whale. The energy *available to the whale population* in a month is thus 1000 units rather than 1. Shortening the food chain in this example increases the food available to the whales by a factor of 1000!

Many of the largest animals on Earth, and some of the most productive animal populations on Earth, are suspension feeders. The large individual size and high productivity observed in suspension feeders are believed to be consequences of (and evidence for) the energy advantages of suspension feeding. Blue whales, the suspension feeders we have already highlighted, are by far the largest animals that have ever lived. The two largest species of fish alive today are both suspension feeders: the whale shark (12 tons) and the basking shark (5 tons). About 30% of all the biomass of fish caught commercially in the world each year consists of herrings, sardines, menhadens, anchovies, and other species that belong to the clupeid group of fish—a group noted for suspension feeding. Crabeater seals, which are by far the most abundant and productive of all seals, are not actually "crab eaters" but use highly modified teeth to eat small crustaceans (notably krill) by suspension feeding. In the Antarctic Ocean, the shrimplike krill are so productive that they are among the most important foods for many animals. The krill are suspension feeders.

For an animal to have a shortened food chain and reap the energy advantages of suspension feeding, the *mechanistic* challenge it must meet is that it must be able to gather food items far smaller than itself in great numbers. A blue whale feeding on krill, for example, must have a mechanism of collecting from the water great numbers of objects that are individually one-thousandth of its size. A rough analogy would be for a person to live by collecting objects the size of individual tomato seeds from the air.

The food-collection mechanism employed by suspension-feeding whales is based on sets of **baleen plates**, constructed mostly of keratin, that hang from the upper jaw on the two sides of the head, each plate oriented approximately perpendicular to the longitudinal axis of the whale's body (**FIGURE 6.13**). In an adult blue whale, there are hundreds of these plates on each side of the mouth cavity; most are a meter or more long, and they hang like pages of a book from front to rear. The medial (inside) edge of each plate frays into strands, and the frayed strands of adjacent plates become tangled with each other. In this way, the entire inner face of the baleen array on each side of the whale's mouth consists of a mat of tangled keratin fibers. A suspension-feeding whale typically feeds by taking a huge mass of water into its mouth cavity and forcing the water out laterally on each side through the baleen

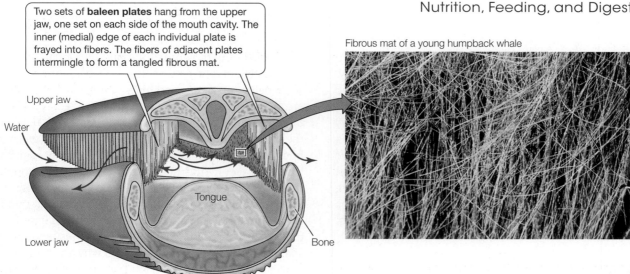

Two sets of **baleen plates** hang from the upper jaw, one set on each side of the mouth cavity. The inner (medial) edge of each individual plate is frayed into fibers. The fibers of adjacent plates intermingle to form a tangled fibrous mat.

Upper jaw

Water

Tongue

Lower jaw

Bone

Fibrous mat of a young humpback whale

FIGURE 6.13 The feeding apparatus of a baleen whale In these toothless whales, water enters at the mouth, flows through an array of baleen plates on each side of the mouth cavity from the inside, and exits laterally. A thick fibrous mat on the inside face of each array of baleen plates forms a sieve. Small food items accumulate on the inside face of each fibrous mat because of sieving. To be swallowed, the food items are freed from the fibrous mats by some mix of backflushing, tongue licking, and vibrational movements that shake them free. (Drawing after Slijper 1979.)

arrays. Krill, copepods, or tiny fish in the water cannot fit through the mats of fibers. Thus these small food items are sieved out in huge numbers and swallowed.

Although the baleen whales employ mechanical sieving (analogous to filtering) to collect food and therefore might be termed *filter feeders*, most suspension feeders do not use simple mechanical sieving or filtering. Suspension-feeding devices usually collect food items from the water by employing hydrodynamic forces, electrostatic attractions, chemical attractions, targeted grasping by tiny appendages, and other processes distinct from sieving.

In the types of fish that are suspension feeders (which are often of great commercial importance, as we earlier mentioned), specialized gill rakers are used to collect food items (**FIGURE 6.14A**). People used to think water was rammed perpendicularly through the arrays of gill rakers, similarly to forcing water through a kitchen sieve. Actually, according to recent research employing tiny cameras positioned to observe the process, water flows past, not through, the arrays of gill rakers toward the throat, and as it does so, pure water is bled off, as seen in **FIGURE 6.14B**. This process concentrates food particles prior to swallowing.

Symbioses with microbes often play key roles in animal feeding and nutrition

Various types of animals maintain symbioses of nutritional importance with three different categories of microbes. One important distinction in categorizing the microbes is whether they

FIGURE 6.14 Suspension feeding in fish (A) A gill arch of a suspension-feeding clupeid fish. The gill rakers are extraordinarily long, numerous, and closely spaced compared with those of fish that do not suspension feed. Gill rakers are skeletal elements covered with ordinary epidermis and are not involved in gas exchange. (B) Based on research on several species of clupeid fish, after water is taken up through the mouth, it flows past (not through) the elaborate arrays of gill rakers in the buccal cavity. The arrays are employed hydrodynamically to concentrate particulates in the stream of water prior to swallowing. (B after Brainerd 2001.)

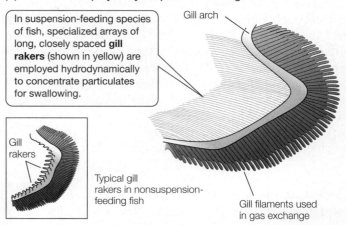

(A) Gill rakers employed by suspension-feeding fish

In suspension-feeding species of fish, specialized arrays of long, closely spaced **gill rakers** (shown in yellow) are employed hydrodynamically to concentrate particulates for swallowing.

Gill arch

Gill rakers

Typical gill rakers in nonsuspension-feeding fish

Gill filaments used in gas exchange

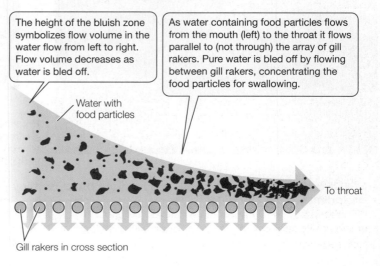

(B) Mechanism of separating suspended food particles from water in clupeid fish with specialized gill rakers

The height of the bluish zone symbolizes flow volume in the water flow from left to right. Flow volume decreases as water is bled off.

As water containing food particles flows from the mouth (left) to the throat it flows parallel to (not through) the array of gill rakers. Pure water is bled off by flowing between gill rakers, concentrating the food particles for swallowing.

Water with food particles

To throat

Gill rakers in cross section

are *heterotrophs* or *autotrophs*. Heterotrophic microbes require organic compounds of external origin as their sources of energy. Autotrophic microbes, in contrast, are able to synthesize organic molecules from inorganic precursors (e.g., CO_2 and NO_3^-) using nonorganic energy sources. Some autotrophs are *photosynthetic*; their source of energy for the synthesis of organic compounds is photon energy from the sun. Other autotrophs are *chemosynthetic* and obtain their energy from energy-releasing inorganic chemical reactions (oxidation reactions). In summary, therefore, the three categories of microbes with which animals maintain symbioses are **photosynthetic autotrophs**, **chemosynthetic autotrophs**, and **heterotrophs**.

SYMBIOSES WITH PHOTOSYNTHETIC AUTOTROPHS (PHOTO-AUTOTROPHS)   Several sorts of aquatic animals obtain organic food molecules from internal populations of algae—photosynthetic autotrophs—with which they maintain symbiotic associations. After the algae synthesize organic molecules from inorganic precursors employing sunlight, the algae export some of the organic molecules to the tissues of their animal host, where the animal's cells use the organic molecules of algal origin as food molecules. All the animals characterized by these symbioses also have other modes of feeding. Nonetheless, the algal symbionts are in some cases so important that their photosynthetic products are able to meet 100% of their host's energy needs.

The reef-building corals are the most famous of the animals that feed by obtaining organic compounds from endogenous algal symbionts (**FIGURE 6.15**). More than 600 species of warm-water, reef-building corals are known, and all have algal symbionts. A topic of great current importance in coral-reef biology is the *stability of the symbiotic association*. Stresses can destabilize the association so that algal symbionts leave the coral polyps, an unhealthy condition that can lead to polyp death and reef disintegration. Corals that have lost their symbionts are called *bleached* because they no longer have chlorophyll or other algal pigments.

Considering the entire animal kingdom, in most cases when there are symbiotic associations between animals and algae, the algae belong to the taxonomic group known as *dinoflagellates* and are called **zooxanthellae** (pronounced as if the *x* were a *z*; singular *zooxanthella*).[6] The reef-building corals have dinoflagellate zooxanthellae, as do most other ocean animals with algal symbionts. In freshwater animals that have algal symbionts—such as some of the freshwater hydras, sponges, and flatworms—the algae belong to taxonomic groups other than the dinoflagellates.

All animals that depend on algal symbionts for nutrition must ensure that their symbionts receive adequate light for photosynthesis. This is a major reason why reef corals occur only in clear or shallow waters. It also helps explain why any human activity that makes water cloudy (turbid) tends to destroy coral reefs; cloudiness in the water can *starve* corals!

SYMBIOSES WITH CHEMOSYNTHETIC AUTOTROPHS (CHEMO-AUTOTROPHS)   Animals that depend on chemosynthetic autotrophs were virtually unknown to science until their sudden, unexpected discovery in the deep sea in 1977. The bottom of the deep sea—being totally dark—is generally an energy-poor habitat. With very little food available, the animals living there tend to be small and sparse. In 1977, however, a group of geologists exploring deep-sea tectonic spreading centers made the stunning discovery that crowded communities of large animals—including clams as big as quart bottles and worms as big as baseball bats—often live near the spreading centers. The communities are now named **hydrothermal-vent communities** because they occur at places where warm water rises out of underwater cracks (vents) in Earth's crust.

Chemoautotrophic **sulfur-oxidizing bacteria** are the key to the profusion of animal life in the hydrothermal-vent communities.

---

[6] The term *zooxanthellae* refers rather loosely to "golden-colored" algal symbionts. Most are dinoflagellates, but some belong to other taxa of algae, such as the diatoms or cryptophytes.

(A) A tropical coral-reef ecosystem

(B)

About a dozen living, coral polyps

Populations of symbiotic algae live in the **gastrodermis**. Photosynthetic products from the algae pass directly to the animal cells in each polyp.

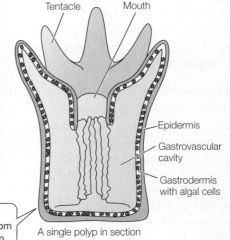

Tentacle　　Mouth

Epidermis

Gastrovascular cavity

Gastrodermis with algal cells

A single polyp in section

FIGURE 6.15 Reef-building corals require light because they are animal–algal symbioses   The reef-building corals (A) are colonies of polyps (B) that secrete skeletal material. The polyps of warm-water species maintain a symbiosis with dinoflagellate algae (zooxanthellae), which live inside coral cells. In addition to gaining nutrition from algal photosynthesis, the polyps have stinging cells with nematocysts and use them to capture small animals, which are taken into the gastrovascular cavity for absorption and digestion.

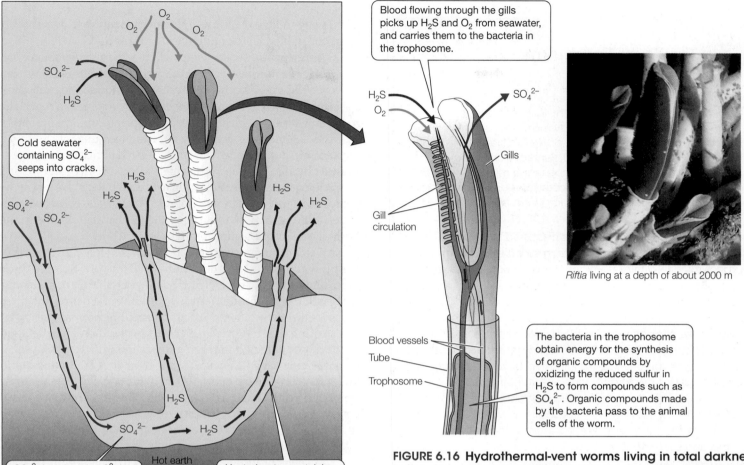

Blood flowing through the gills picks up H_2S and O_2 from seawater, and carries them to the bacteria in the trophosome.

The bacteria in the trophosome obtain energy for the synthesis of organic compounds by oxidizing the reduced sulfur in H_2S to form compounds such as SO_4^{2-}. Organic compounds made by the bacteria pass to the animal cells of the worm.

Cold seawater containing SO_4^{2-} seeps into cracks.

SO_4^{2-} is reduced to S^{2-} by complex reactions under heat and pressure.

Heated water containing H_2S rises to be spewed out in plumes.

Riftia living at a depth of about 2000 m

FIGURE 6.16 Hydrothermal-vent worms living in total darkness in the deep sea are symbiotic with chemoautotrophic bacteria The hydrothermal-vent worm *Riftia pachyptila* has no mouth, gut, or anus. Its food comes from sulfur-oxidizing chemoautotrophic bacteria, living in its trophosome, that require H_2S. The worms must live near hydrothermal vents because the vents are their source of H_2S. SO_4^{2-} made in the trophosome is voided across the gills. (After Arp et al. 1985; photograph courtesy of Woods Hole Oceanographic Institution.)

These bacteria oxidize inorganic sulfide (S^{2-}) to obtain energy for the synthesis of organic molecules. The source of S^{2-} for hydrothermal-vent communities is in part *local* reduction of sulfate (SO_4^{2-}) by Earth's heat. Seawater is rich in SO_4^{2-}. At tectonic spreading centers, seawater seeps into deep cracks in Earth's crust, where it comes into contact with hot rock of Earth's core and is heated to temperatures as high as 400°C. Because of inorganic chemical reactions between the seawater and rock at the high temperatures and pressures that prevail, a portion of the sulfur in the heated water is chemically reduced from SO_4^{2-} to S^{2-}. In addition, S^{2-} is sometimes added by being dissolved out of the rock. Heated water has a reduced density because of its elevated temperature. Thus water heated in the cracks of Earth's crust rises back up into the open sea, carrying S^{2-} with it in the form of hydrogen sulfide (H_2S) (**FIGURE 6.16**). Chemoautotrophic bacteria in the vent communities take up the H_2S, oxidize the S^{2-} to chemical forms such as SO_4^{2-}, and use the energy released to synthesize organic molecules.

Sulfur, you will note, acts as an *energy shuttle* in these communities. As seawater passes down into cracks in the seafloor and contacts hot Earth, sulfur acquires energy by undergoing reduction. The sulfur later yields this energy when it is reoxidized by sulfur-oxidizing bacteria. In this way, the bacteria are able to draw on Earth's heat as an energy source for primary production.

Many animals in the hydrothermal-vent communities obtain organic nutrient molecules by maintaining symbiotic associations with sulfur-oxidizing bacteria. The most dramatic example is provided by the highly specialized annelid worm *Riftia* (see Figure 6.16), a large worm (up to 1.5 m long) named after the undersea rifts (cracks in Earth's crust) near which it lives. *Riftia* has no mouth, no gastrointestinal tract, and no anus. It cannot ingest food! However, about one-fifth of its body is filled with a tissue, the *trophosome*, where abundant populations of sulfur-oxidizing bacteria live. The blood hemoglobin of the worm transports not only O_2 but also H_2S. As the blood flows through the worm's gills, it picks up H_2S from the vent water. The symbiotic bacteria obtain H_2S from the blood, oxidize the S^{2-}, and use the energy released to synthesize organic compounds, some of which enter the worm's tissues and meet the worm's food needs. Clams and snails in the hydrothermal-vent communities also have sulfur-oxidizing bacterial symbionts in their tissues (e.g., gills), although they are also able to ingest and digest foods.

Animal symbioses with chemosynthetic autotrophs are now known to occur in many locations in the sea. They are common near oil seeps, for example, not just in hydrothermal-vent communities.

HETEROTROPHIC MICROBIAL POPULATIONS IN ANIMAL GUTS Humans—and probably all other animals—have popula-

BOX 6.1 **A Genomic Discovery: The Gut Microbiome**

Until recently, the only way to study the function of gut microbes was to culture the microbes in test tubes or other laboratory containers. This meant, in fact, that scientists were completely unable to study most of these microbes because most of the microbes could not be cultured. The microbes require specific—and generally unknown—conditions in the gut to survive. Because the microbes could not be cultured, they could not be studied, and microbial life in the gut lumen of most animals remained almost entirely a mystery.

The advent of genomics has changed all this and opened up a new field of research that is expanding spectacularly. With genomic methods, a nucleotide sequence can be acquired for an animal's entire microbial community—a type of *metagenome* (a mixed-species genome). Such a sequence is then searched for the signatures of particular microbial species to determine species composition. Moreover, when genes are identified, researchers can often make informed predictions about their functions by extrapolating from functional studies of homologous genes in culturable organisms. The use of genomic methods is permitting biologists to obtain detailed knowledge about the species present in the gut microbiome and their functional abilities.

within one individual animal. However, it can vary dramatically from one individual to another.

Evidence exists that the composition of the gut microbiome can potently affect the physiology of the animal host. For the most part, however, the effects known at present relate to pathological states. For example, chronic changes in the gut microbiome can contribute to chronic diseases such as obesity, diabetes, and inflammatory diseases—sometimes acting by affecting immune function. In a recent study, gut microbes were transplanted from lab mice exhibiting a syndrome of metabolic disorders into germ-free lab mice, and the recipient mice thereafter exhibited the metabolic syndrome. By comparison with evidence of gut microbial effects in pathological states, little is known about the roles of the microbiome in the normal physiology of most animals.

In a recent study, investigators discovered that most people in the human populations they studied fell into one of three types (termed *enterotypes*) in the composition of their gut microbiome. Although the composition is variable within each type, the three types are distinctly different. For understanding normal physiology, it is striking that the three types vary—based on interpretation of the genes detected in them—in several important ways. For example, one type is especially effective at synthesizing biotin and riboflavin—two B vitamins (see Table 6.4)—whereas another type is especially effective at producing thiamin and folate—two other B vitamins. The types also differ in how effective they are in breaking down complex carbohydrates in the gut contents, potentially making breakdown products available to the host for use. Conceivably, individual people vary in their normal physiology in ways that depend on functions being performed by their gut communities.

tions of many species of heterotrophic microbes living in their *gut lumen*—the hollow central core of the gut. These populations are collectively known as the **gut microbiome**. The home of the gut microbiome is unlike any other place in the bodies of most animals. For example, the gut lumen is entirely devoid of oxygen (O_2) in vertebrates and probably most invertebrates. Detailed knowledge of the gut microbiome was almost impossible to acquire until recently, but that situation changed with the advent of genomic methods (**BOX 6.1**).

A person or other mammal lacks gut microbes prior to birth. Colonization of the gut begins during the birth process, however (e.g., microbes are acquired from the mother's vagina), and thereafter colonization rapidly escalates as a youngster interacts with potential microbial sources such as other individuals or environmental surfaces. In some species of animals, including horses and certain lizards, juveniles consume the feces of adults. Conceivably, this behavior has evolved to establish a suitable gut microbiome in the youngsters. By the time the microbiome of a mammal reaches a steady state in its size and species composition, the number of gut-microbe cells is about ten times the number of mammalian cells in the animal's body (a circumstance that is possible because microbial cells are far smaller than mammalian cells). In humans, most of the species of gut microbes are bacteria; species of viruses, archaea, and microbial eukaryotes account for less than 10% of the total. According to current evidence, after the composition of the gut microbiome reaches steady state, it tends to be relatively stable

SPECIALIZED SYMBIOSES WITH HETEROTROPHIC MICROBES It has been recognized for more than a century that some animals have dramatic, specialized gut structures in which microbes live. A cow's rumen is a striking example. Until recently, studies of gut microbes were largely limited to the microbes in these specialized gut structures. Scientists didn't realize that the gut microbiome is probably important in all animals, not just ones with rumens or other obvious structural specializations. How does the "old" knowledge of microbes in specialized gut structures relate to the "new," genomic knowledge of gut microbiomes? Only time will tell.

In this section we discuss the symbioses with gut microbes in animals with specialized gut structures. The microbes living in rumens or other such specialized structures are heterotrophic and typically occur *as mixed communities* of multiple microbial types (e.g., bacteria, protists, yeasts, fungi) living in the gut lumen. These mixed microbial communities carry out important functions in the processing of food and other nutrients. The microbes are anaerobes, given that the gut lumen is typically anaerobic (devoid of O_2). They are often called **fermenting microbes** to emphasize that they carry out fermentation. **Fermentation** refers to several sorts of enzyme-catalyzed reactions that occur without O_2, such as reactions that break down organic compounds anaerobically to liberate energy substrates for metabolic use. Animals that maintain symbiotic associations with fermenting microbes in specialized gut structures are often described as **fermenters**.

Types of Meal Processing Systems

Physiologists have looked to industrial models to gain insight into the types of meal processing systems that are possible and their properties. When chemical or microbial processes are used to make commercial products in industry, the apparatus used to carry out a process is termed a **reactor**. Three types of reactors are recognized, as shown in the figure. You will immediately recognize parallels in the meal processing systems of animals. Our intestines, for example, function like the type of reactor shown in Part 2 of the figure, and the rumen of a cow functions like the type of reactor in Part 3 of the figure. **Box Extension 6.2** discusses these reactor types in greater depth. It focuses on the type of meal processing that occurs in ruminants, which takes place in two very different stages: a first stage in which self-replicating microbes act on the meal, and a second stage in which non-self-replicating digestive enzymes produced by the animal act on it.

Reactor models applied to meal processing in animals   Three meals (coded red, blue, and yellow) are ingested at three successive times.

(1) **Batch reactor:**
Each meal is processed before the next meal enters.

(2) **Continuous-flow reactor without mixing:**
Meals line up in a tubular reaction vessel so they do not mix, but each can be processed for far longer than in a batch reactor.

(3) **Continuous-flow reactor with mixing:**
New meals mix with meals that have already undergone some processing.

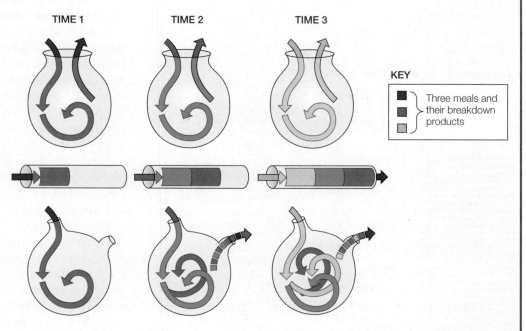

TIME 1   TIME 2   TIME 3

KEY

Three meals and their breakdown products

The specialized gut structures in which the microbes are housed are typically enlarged or dilated chambers, and the fluids in them are suitable for microbes to thrive (e.g., not strongly acidic). In many cases the specialized gut chambers of vertebrate fermenters are part of the foregut, which is the part of the gut composed of the esophagus and stomach. Those animals are called **foregut fermenters**. The most famous of them are the *ruminant mammals*. Other than the specialized fermentation chamber, the rest of the gut of a vertebrate foregut fermenter typically consists of a relatively ordinary, acid-secreting stomach chamber, and intestines.

Mathematical models used to understand gut function indicate that *breakdown of foods by microbial fermentation* is generally most effective when fresh, newly ingested food material is mixed with older material already colonized by the microbes. Conversely, *enzymatic digestion of foods*—carried out by enzymes the animal produces—is generally most effective when the enzymes are added to new material that is not mixed with material already subjected to enzymatic digestion. In foregut fermenters, these considerations are believed to explain (**BOX 6.2**) why the specialized chambers for microbial fermentation are enlarged and vatlike, whereas the intestine remains tubular as in other mammals. Foods from suc-

cessive meals mix in the vatlike part of the gut [see reactor type (3) in Box 6.2], creating optimal conditions for microbial fermentation. However, after material enters the tubular intestine, it does not mix with material that has already been in the intestine for a substantial time [see reactor type (2)], an optimal condition for enzymatic digestion.

VERTEBRATE FOREGUT FERMENTERS   Ruminant mammals—all of which are herbivores—provide the most famous example of foregut fermenters, as already noted. The stomach of a ruminant has a complex structure, consisting of several compartments (**FIGURE 6.17**). The first and largest compartment, to which the esophagus connects, is the **rumen**, where mixed communities of heterotrophic, fermenting microbes thrive in a nonacidified setting. When a ruminant swallows grasses, leaves, twigs, or other plant parts, the first step in processing is that the microbes colonize the materials and ferment them. Ruminants include sheep, antelopes, cattle, goats, camelids,[7] moose, deer, giraffes, and buffaloes. Not all vertebrate foregut fermenters are ruminants. Others—all of

---

[7] In some classifications, camelids (e.g., camels and llamas) are termed *pseudoruminants*.

(A) An African buffalo showing the size and position of the rumen

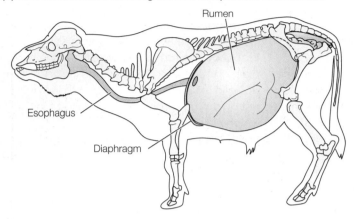

(B) The digestive tract of a domestic sheep

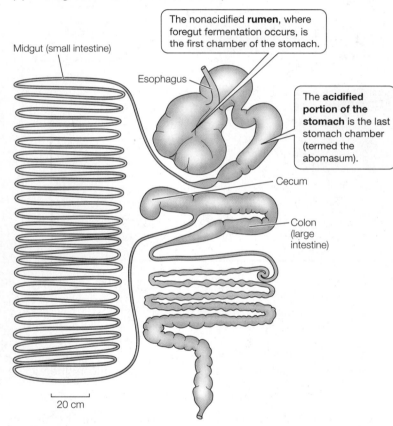

FIGURE 6.17 The digestive tract of ruminants (A) An African buffalo, showing the position and impressively great size of the rumen. (B) The digestive tract of a domestic sheep. Ruminants periodically regurgitate fermenting material from the rumen into the mouth, break the material up by further grinding using their teeth, and reswallow it. In cows and sheep, the stomach chambers are, in order, the rumen, reticulum, omasum, and abomasum. The rumen and reticulum—the two largest chambers—are connected by a broad opening and function more or less together; the omasum, filled with leaflike folds, acts as a filter, allowing only liquids and fine particulates to move through to the abomasum; and the abomasum is the true stomach, secreting acid and digestive enzymes. There is an enormous range of variation in stomach morphology and function among other ruminant species. (A after Hofmann 1989; B after Stevens 1977.)

which are herbivores like ruminants—include kangaroos, hippos, colobus monkeys, sloths, and a rainforest bird called the hoatzin.

In ruminants and other foregut fermenters, three categories of function are carried out by the microbes in the mixed microbial community in the rumen. The first function is the *fermentative breakdown of compounds that the animal cannot digest*, notably cellulose and some of the other structural carbohydrates of plant cell walls, such as hemicelluloses. As noted earlier in this chapter, cellulose is one of the most abundant organic compounds in the biosphere and thus a major potential source of energy. Vertebrates, however, are unable to digest it on their own, and therefore—left to their own devices—allow it to pass into their feces. Some of the microbes in the rumen microbial community produce a complex of enzymes called **cellulase** that breaks down cellulose into compounds that their animal host can absorb and metabolize. **Short-chain fatty acids (SCFAs)**, sometimes called **volatile fatty acids**, are the principal useful products of the microbial breakdown of plant structural carbohydrates. SCFAs are fatty acids, structurally similar to those in Figure 6.4A, that consist of relatively few carbon atoms; they include acetic acid (two carbons), propionic acid (three), and butyric acid (four). SCFAs are readily absorbed and metabolized by animals. Other products of the fermentation of structural carbohydrates include CO_2 and methane (CH_4), which in foregut fermenters are voided to the atmosphere by belching (retrograde passage of gas up the esophagus). Methane is a potent greenhouse gas, and as absurd as it may seem, the collective burps of proliferating populations of cattle around the world are a factor in the worrisome global heat budget.

The second major function carried out by the microbes is *synthetic*. Rumen microbes are able to *synthesize B vitamins*, which vertebrate animals themselves cannot synthesize. The microbes of sheep and cattle, for instance, synthesize sufficient amounts of all the B vitamins to meet the animals' needs fully. The rumen microbes also *synthesize essential amino acids*. As the microbial cells grow and divide in an animal's rumen, they synthesize their own cellular proteins from nitrogenous and nonnitrogenous compounds in the animal's food. Some of the amino acids they synthesize to make their cellular proteins are essential amino acids for the animal. When the microbial cells later pass down the animal's digestive tract to the acid stomach and intestines, their protein constituents are digested by the animal's digestive enzymes, and the amino acid products of digestion are absorbed for use in the animal's metabolism. The essential amino acids obtained by the animal in this way reduce its requirement for essential amino acids in its ingested food.

The third noteworthy function of the microbial community is related to the second. The microbes *permit waste nitrogen from animal metabolism to be recycled into new animal protein rather than being excreted*. To see how this occurs, consider first that mammals incorporate nitrogen from their metabolic breakdown of proteins mostly into a waste compound called *urea* (see Figure 29.24). We ourselves and many other species excrete the urea in our urine. In ruminants, however, the urea can diffuse from the blood into the rumen, where certain of the microbes break down the urea to make ammonia (NH_3). Neighboring microbes are then able to use the NH_3 as a nitrogen source for the *synthesis of proteins*. When the microbial proteins are later enzymatically digested, the animal

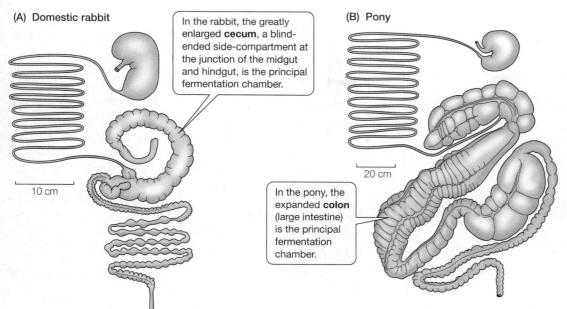

(A) Domestic rabbit

In the rabbit, the greatly enlarged **cecum**, a blind-ended side-compartment at the junction of the midgut and hindgut, is the principal fermentation chamber.

10 cm

(B) Pony

20 cm

In the pony, the expanded **colon** (large intestine) is the principal fermentation chamber.

FIGURE 6.18 The digestive tracts of two hindgut fermenters (A) A rabbit. (B) A pony. Both species have simple, one-chambered stomachs. Their fermentation chambers are parts of the hindgut. See Figure 6.17 for labeling of other parts of the digestive tract. Rabbits produce hard and soft feces. The hard feces are the familiar "rabbit pellets." The soft feces, which are less familiar, consist of material from the cecum that is enclosed in a mucous membrane and that passes through the full length of the colon without being altered. Soft feces are eaten as they emerge from the anus. (After Stevens 1977.)

obtains amino acids, which it uses to synthesize animal proteins. In this way, nitrogen—which is potentially a very valuable asset—is saved from excretion and recycled into the animal's tissues.

VERTEBRATE HINDGUT AND MIDGUT FERMENTERS Some herbivorous vertebrates, called **hindgut fermenters**, have specialized gut chambers to support communities of mixed, fermenting microbes in their hindgut, the part of the gut corresponding to the mammalian large intestine and cecum. Many species of mammals, for example, are hindgut fermenters. They have an enlarged cecum or enlarged colon—or both—where thriving mixed microbial communities reside (**FIGURE 6.18**). The mammalian hindgut fermenters include rabbits, horses, zebras, rhinos, apes, elephants, koalas, and some rodents. Hindgut fermentation in the cecum or colon—or both—occurs also in many ground-dwelling herbivorous birds—geese, grouses, chickens, and ostriches—and some herbivorous lizards and turtles. Many herbivorous fish are classified as **midgut fermenters** because part of the midgut (the gut between the foregut and hindgut) serves as the principal fermentation chamber. These fish include tilapias, carps, and catfish—all of exceptional importance in aquaculture.

The functions of microbial communities in hindgut fermenters differ from those in foregut fermenters because the hindgut is posterior to the acid stomach and the small intestine, meaning (among other things) that microbes from hindgut communities are not automatically digested. One function occurs in hindgut fermenters in much the same way as in foregut fermenters—namely, the microbial breakdown of cellulose and other structural carbohydrates to form short-chain fatty acids (SCFAs). SCFAs readily diffuse across the gut wall to enter an animal's blood wherever they are produced in the gut (see page 158). Thus the SCFAs produced by hindgut or midgut fermentation of structural carbohydrates are readily available to the animal.

However, the roles of microbes in supplying B vitamins and essential amino acids, and recycling nitrogen, are not fulfilled in the same way in hindgut fermenters as in foregut fermenters—if they occur at all. Let's first consider the B vitamins. To be absorbed, they require specialized absorption mechanisms (e.g., transporter proteins) present mainly in the small intestine. The B vitamins synthesized by hindgut microbes do not pass through the small intestine and thus tend mostly to be lost in an animal's feces. Similarly, for an animal to gain from microbial synthesis of essential amino acids (or nitrogen recycling), the cellular proteins of the microbes must be digested; yet when microbial growth occurs in the cecum or colon, the microbes are eliminated in the feces, undigested. As odd as it may seem at first, a *common* solution to the problems just described is for hindgut fermenters to eat defecated material. Some eat ordinary feces; others (such as rabbits) eat special soft feces derived from the contents of the cecum. In either case, the animals ingest microbes and microbial products in defecated material just like other foods and thus can digest and absorb them.

INVERTEBRATE SYMBIOSES WITH HETEROTROPHIC MICROBES A variety of invertebrate animals also maintain nutritionally important symbiotic associations with heterotrophic microbes in their digestive tracts. The "lower" termites (consisting of four of the six termite families) are the most famous example. Because termites are wood eaters, much of the nutritional value of their food lies in cellulose. Some invertebrates, unlike vertebrates, synthesize cellulose-digesting enzymes; certain insects, for example, are entirely self-sufficient in their capacity to digest and metabolize cellulose. More commonly, however, both insects and other invertebrates rely on microbial symbionts to aid cellulose breakdown. In their hindguts, the lower termites maintain communities of anaerobic flagellated protists and bacteria, which ferment cellulose to acetic acid the termites use. Similarly, scarabid beetles—which are sometimes of immense ecological importance—depend on gut bacteria to ferment cellulose to SCFAs. Microbial symbionts of invertebrates may also provide B vitamins, essential amino acids, and nitrogen recycling.

Some roles of microbial symbionts in invertebrates are novel by comparison with their roles in vertebrates. Sterol synthesis in insects provides an important example. Unlike most animals, insects cannot synthesize required precursors for making sterols.

Microbial symbionts commonly supply these precursors. A second distinctive role for symbionts in invertebrates occurs in those that feed on vertebrate blood—such as leeches, tsetse flies, ticks, and sucking lice. Bloodsucking animals seem to have a universal requirement for symbiotic bacteria; the symbionts help them digest blood and may produce antibiotics that prevent the decay of blood during processing.[8]

SUMMARY
Feeding

- One mode of feeding is for animals to target and ingest individual food items, such as when orcas or cone snails catch fish. Toxic compounds, such as venoms and secondary compounds, often play roles as weapons or defenses in feeding of this sort.

- Suspension feeding is a second major mode of feeding. In this mode, an animal feeds on living or nonliving food items, suspended in water, that are individually tiny in comparison with the animal and that typically are collected in numbers rather than being individually targeted. Suspension feeding permits animals to feed lower on food chains and thus gain access to higher food productivity. Many of the largest and most productive animals on Earth are suspension feeders.

- Symbiotic associations with microbes are a third major mode of feeding in animals. Some animals maintain symbiotic associations with algae; warm-water reef corals, which are symbiotically associated with dinoflagellate zooxanthellae, are prominent examples. Other animals, notably species in hydrothermal-vent communities, are symbiotically associated with chemosynthetic autotrophs. Many herbivorous animals, both vertebrate (e.g., ruminants) and invertebrate (e.g., lower termites), maintain specialized symbiotic associations with heterotrophic, anaerobic fermenting microbes.

- Fermenting microbes most commonly provide three sorts of nutritional advantages to their animal hosts. They synthesize vitamins and essential amino acids. They break down structural polysaccharides such as cellulose so that the host can gain food value from them. They permit waste nitrogen to be recycled for use in animal protein synthesis.

- Apart from specialized symbioses with heterotrophic microbes, biologists now recognize that large assemblages of bacteria—termed the gut microbiome—are found in the guts of all mammals and probably most other animals. The functions of these assemblages are only starting to be understood.

Digestion and Absorption

Of all the topics in physiology, digestion and absorption are often the first to be studied in detail. Many students learn about the parts of the digestive tract and the major digestive enzymes on several occasions before studying physiology at the university level. These early treatments rightfully emphasize humans and other vertebrates. A problem

can arise in the *comparative* study of animal digestion and absorption, however, because vertebrate anatomy and physiology do not provide effective models for understanding the anatomy and physiology of many other animal groups. The comparative study of digestion and absorption therefore calls for a particular effort to be open-minded.

Digestion is defined as the breakdown of food molecules by enzyme action (or other animal-mediated processes) into smaller chemical components that an animal is capable of distributing to the tissues of its body. In the case of food proteins, for example, digestion splits up the proteins—which typically cannot be absorbed and distributed throughout the body—into amino acids that can be taken up from the digestive tract and distributed. **Extracellular digestion** is digestion in an extracellular body cavity, such as the lumen of the stomach or intestines. It is the principal mode of digestion in vertebrates, arthropods, and many other animals. In **intracellular digestion**, food particles are taken into specialized cells prior to digestion, and digestion occurs within the cells. Digestion is principally intracellular in sponges, coelenterates, flatworms, and some molluscs.

In human physiology, *absorption* is defined to be the transfer of products of digestion from the lumen of the gastrointestinal tract to the blood or lymph. Because digestion is almost entirely extracellular in humans, digestion and absorption occur in sequence: digestion first, absorption second. These organizing concepts, however, are inadequate for the animal kingdom as a whole. When digestion is predominantly intracellular, *absorption* is defined to be the transfer of food particles from the gut lumen into the cells that digest the food particles intracellularly, and only later are digestive products passed to the blood. *Absorption* of small organic molecules (e.g., amino acids and B vitamins) even occurs directly from seawater solution into the bloods of some marine animals, without being preceded by ingestion or digestion. Defined generally, **absorption**—also called **assimilation**—is the entry of molecules into the living tissues of an animal from outside those tissues.[9]

To understand digestion and absorption more thoroughly, our first step will be to examine the major features of the digestive–absorptive systems in some of the principal groups of animals. Terminology will continue to be a problem. The reason is that biologists have often tried to apply a single set of human anatomical terms to digestive systems throughout the animal kingdom—meaning that many, very diverse digestive systems are described as consisting of the parts we ourselves have: esophagus, stomach, small intestine, pancreas, and so forth. This use of human terms can be extremely misleading because the *functional* properties of organs in invertebrates sometimes differ dramatically from those of the like-named organs in humans. Therefore one must be cautious about drawing any conclusions about function from names alone.

Vertebrates, arthropods, and molluscs represent three important digestive–absorptive plans

Modern morphologists recognize four sequential segments of the tubular digestive tract in vertebrate animals: **headgut**, **foregut**, **midgut**, and **hindgut** (see Figures 6.17 and 6.18). The *headgut* of a vertebrate consists of the parts of the digestive tract in the head

---

[8] Similar symbionts are also required by vampire bats.

[9] The lumen of an animal's gut is continuous with the ambient air or water. Thus the gut lumen is outside the animal, and materials pass into the animal's living tissues when they are taken up from the gut lumen.

and neck, such as the lips, buccal cavity, tongue, and pharynx. The principal functions of the headgut are to capture and engulf food and to prepare the food for digestion. The preparation of food by the headgut is minimal in some vertebrates, such as many predatory fish, which swallow food quickly and whole. In other cases, preparation is extensive and involves chewing or grinding, the addition of lubricating agents, and the addition of digestive enzymes.

The *foregut* of a vertebrate consists of the parts of the digestive tract between the headgut and the intestines. Therefore the foregut consists of the esophagus and stomach, and sometimes a crop (storage chamber) or gizzard (grinding chamber). The function of the esophagus is to move food from the headgut to the stomach. The functions of the stomach are to store ingested food, initiate protein digestion, and break up food by a combination of muscular, acid, and digestive-enzyme effects.[10] With few exceptions, the stomach of a vertebrate secretes acid (HCl)—often at very high chemical activity, such as pH 0.8 in humans and dogs (see page 113). The stomach also secretes pepsins, a set of protein-digesting enzymes.

The *midgut* and *hindgut* are the first and second segments of the intestines. In humans, they are distinguished by diameter: The midgut is of smaller diameter and thus termed the *small intestine*, whereas the hindgut is called the *large intestine*. The distinction between "small" and "large" does not apply in many vertebrates because the two parts are of similar diameter. The midgut of a vertebrate is the principal site of digestion of proteins, carbohydrates, and lipids. It is also typically the principal site of absorption of the products of digestion of all three categories of foodstuffs, as well as vitamins, minerals, and water. The main functions of the hindgut are to store wastes between defecations and complete the absorption of needed water and minerals from the gut contents prior to elimination. The absorption of water and minerals by the hindgut may sound like a humble function, but it has life-and-death importance. In humans, for example, about 7 L (almost 2 gallons) of a watery solution containing Na^+, Cl^-, and other ions are secreted from the blood into the gut per day to facilitate processing of food. Reabsorption of this water and electrolytes is essential. When people suffer from untreated cholera or dysentery, they can die in just a few days because these diseases block water reabsorption in the hindgut, and profuse water loss in feces causes lethal dehydration. In addition to the headgut, foregut, midgut, and hindgut, the vertebrate digestive system includes two additional components that secrete important materials into the midgut: the **pancreas** and the **biliary system**. The pancreas secretes many digestive enzymes into the midgut. The biliary system, which is a part of the liver, secretes bile, which plays a crucial emulsifying role in the digestion of lipids.

Muscular action is the principal mechanism by which food and other materials are moved along as they pass through the digestive systems of vertebrates. The muscles of the gut wall, which are composed mostly of smooth muscle cells (see page 557), are arrayed in two primary layers: an outer layer of *longitudinal* muscles that *shorten* the gut when they contract, and an inner layer of *circular* muscles that *constrict* the gut when they contract. **Peristalsis** is one of the gut's principal modes of muscular activity. It is a highly coordinated pattern of contraction in which constriction of the gut at one point on its length initiates constriction at a neighboring point farther along, producing a "wave" of constriction that moves progressively along the gut, propelling food material before it. **Segmentation** is a second important mode of muscular activity, in which constrictions of circular muscles appear and disappear in patterns that push the gut contents *back and forth*. The control of peristalsis and segmentation depends principally on the *enteric nervous system* (see page 420), a network of nerve cells within the gut walls. **Gut motility** is a general term that refers to peristalsis, segmentation, or other muscular activity by the gut in vertebrates or invertebrates.

The digestive systems of arthropods and bivalve molluscs are discussed and illustrated in **BOX 6.3**. The arthropods resemble vertebrates in that digestion is chiefly extracellular, and food is moved through the digestive tract by muscular contraction. However, the bivalve molluscs—clams, mussels, oysters, and their relatives—provide an instructive contrast. In them, digestion is mostly intracellular, and cilia play a major role in moving food through the digestive tract.

Digestion is carried out by specific enzymes operating in three spatial contexts

In all animals, digestion is carried out for the most part by **hydrolytic enzymes**, which catalyze the breakup of large molecules into smaller parts by bond-splitting reactions with H_2O, termed *hydrolytic reactions*. The dynamics of the interaction between food molecules and digestive enzymes depend in part on the gross morphology of the part of the gut where digestion occurs, plus patterns of filling and emptying, as discussed in Box 6.2. Here our principal focus will be on the digestive enzymes themselves: their chemistry and the fine-scale spatial contexts in which they operate.

Each digestive enzyme is specific (more or less) in the type of chemical bond it can break. Because of the specificity of digestive enzymes, a species can digest only those ingested molecules for which it has enzymes that hydrolyze the specific types of bonds in the molecules. The enzyme *chitinase*, for example, is required for a species to digest chitin;[11] the enzyme initiates chitin breakdown by attacking key bonds in the large chitin polysaccharide to divide it into disaccharide (chitobiose) subunits. Whereas many species of rodents and bats synthesize chitinase, humans and rabbits do not. Thus humans and rabbits cannot digest chitin even though some other mammals can. We are reminded by this example of the crucial point stressed at the start of this chapter (see Figure 6.1): The nutritional value of a food depends in part on the digestive enzymes an animal possesses, not just the chemical composition of the food.

Digestive enzymes, like other enzymes, typically occur in multiple molecular forms (see Chapter 2). When two species both have a particular enzyme, they often synthesize different molecular forms of the enzyme. If a digestive enzyme is synthesized by two tissues in one species, the molecular forms are often different in the two tissues.

Digestive enzymes act in *three spatial contexts* in animals. An enzyme that acts in one context in one group of animals may act in another context in a different group. Thus the three

---

[10] In this general overview, we cannot address the full range of vertebrate variation. Thus, for example, the special stomach features of ruminants—discussed already—are not repeated here. Nor are the special hindgut features of hindgut fermenters.

[11] Chitin (see page 138) is a structural polysaccharide of great importance in the exoskeletons of insects and other arthropods.

BOX 6.3 The Digestive Systems of Arthropods and Bivalve Molluscs

The insects and the crustaceans (e.g., crayfish, crabs, and shrimp) nicely illustrate the digestive system of arthropods (see the figure). In them, digestion is principally extracellular. Another similarity to vertebrates is that food is moved through the insect or crustacean digestive tract by muscular contraction.

Morphologists describe the digestive tract of an insect or crustacean as consisting of a **foregut**, **midgut**, and **hindgut**. For defining these, we need first to mention that in these animals the early and late parts of the digestive tract have an ectodermal developmental origin. Because of this they synthesize exoskeleton and are *lined with a thin layer of chitinous exoskeleton material, called cuticle*. The foregut and hindgut are defined to be the parts of the digestive tract at the anterior and posterior ends that have this cuticular lining (see the figure). The midgut is of different (endodermal) developmental origin and lacks a lining of cuticle.

Another point deserving early mention is that in insects (but not crustaceans) the urine-producing tubules, called *Malpighian tubules*, empty into the gut at the junction of the midgut

The digestive systems of two types of arthropods: insects and crustaceans (1) An insect. (2) A crustacean. Although the excretory system of crustaceans is separate from the digestive system, in insects the Malpighian tubules—which constitute the initial part of the excretory system—connect to the digestive system at the junction of the midgut and hindgut.

and hindgut (see Part 1 of the figure). Thus urine joins other gut material at that point and travels through the hindgut to be excreted by way of the anus. Because the hindgut plays a crucial role in adjusting the final composition and quantity of the urine, it is part of the excretory system as well as the digestive system of an insect.

Our discussion of arthropods continues in **Box Extension 6.3**, where you will also find an illustration of the bivalve mollusc digestive system and a discussion of how it works.

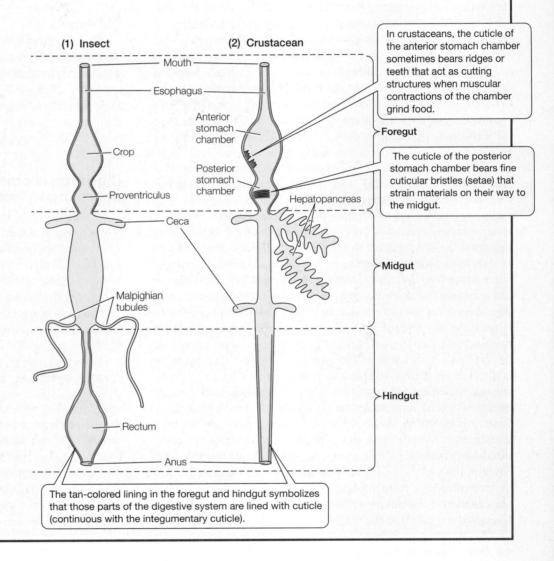

In crustaceans, the cuticle of the anterior stomach chamber sometimes bears ridges or teeth that act as cutting structures when muscular contractions of the chamber grind food.

The cuticle of the posterior stomach chamber bears fine cuticular bristles (setae) that strain materials on their way to the midgut.

The tan-colored lining in the foregut and hindgut symbolizes that those parts of the digestive system are lined with cuticle (continuous with the integumentary cuticle).

spatial categories of enzyme action do not bear a one-to-one correspondence with the chemical categories of enzymes. Many examples of each spatial category will be provided as we move on shortly to discuss carbohydrate, protein, and lipid digestion. For the present, we simply list and define the three spatial contexts.

1. **Intraluminal enzymes**. Sometimes digestive enzymes are secreted into the lumen of a body cavity, such as the lumen of the stomach or the midgut, where the enzyme molecules mix freely with food materials and catalyze digestive

reactions. When digestive enzymes operate in this spatial context, their action is termed *intraluminal*, meaning "in the lumen." Intraluminal digestion is a type of extracellular digestion because it occurs outside cells.

2. **Membrane-associated enzymes**. Sometimes digestive enzymes are cell-membrane proteins that are synthesized by epithelial cells lining the gut and that remain associated with the epithelial cells while digesting food molecules in the gut lumen. Such *membrane-associated enzymes* are

located in the apical membranes of the epithelial cells and positioned so that their catalytic sites are exposed to the gut lumen. Food molecules must contact the epithelial cells to be digested by the membrane-associated enzymes. Digestion by these enzymes is extracellular because the food molecules affected are in the lumen, not inside cells.

3. **Intracellular enzymes**. Digestive enzymes may also be positioned inside cells, where they carry out intracellular digestion. For food particles or molecules to be affected by these enzymes, the particles or molecules must be taken into the cells.

Individual animals commonly have digestive enzymes functioning in all three spatial contexts. However, the emphasis on the three varies from one group of animals to another, as we have already mentioned. For example, in some clams (see Box 6.3) the great majority of digestive reactions are carried out by intracellular enzymes. In vertebrates, by contrast, the majority of digestive reactions are extracellular, catalyzed by intraluminal or membrane-associated enzymes; only some of the final reactions of protein digestion occur intracellularly.

CARBOHYDRATE DIGESTION   Disaccharides are sometimes ingested as foods, as when hummingbirds feed on sucrose-rich flower nectars, infant mammals ingest lactose in their mother's milk, or people eat candies rich in sucrose. Disaccharides are also produced in the digestive tract by the digestion of more-complex ingested carbohydrates. Enzymes that hydrolyze disaccharides into their component monosaccharides are called **disaccharidases**. They include *sucrase*—which breaks sucrose into two monosaccharides—*lactase, trehalase*, and others. In vertebrate animals, the disaccharidases are membrane-associated enzymes found in the apical membranes of the midgut epithelium.

When a polysaccharide is digested, it is typical for two enzymes to carry out the process, acting in sequence. The first enzyme splits the polysaccharide into disaccharides or oligosaccharides (short chains of three or more monosaccharides). Then the second enzyme splits the products of the first into monosaccharides.

As stressed earlier, the enzymes that carry out the initial breakup of *structural* polysaccharides are far from universally distributed in the animal kingdom. *Cellulase*, required for the initial breakup of cellulose, is not synthesized by any vertebrates, for example, meaning that vertebrates cannot tap cellulose for its food value unless they have symbiotic associations with microbes that ferment cellulose. *Chitinase*, required for the initial breakup of chitin, is synthesized by some species of vertebrates and invertebrates, but not others; therefore only some animals can obtain food value from the exoskeletons of insects and other arthropods.

Although the ability to digest cellulose or chitin is spotty, most animals are able to digest starch and glycogen, two of the principal *storage* polysaccharides. The first enzyme to act on starch or glycogen is **amylase**. It hydrolyzes the molecules to yield the disaccharides maltose and isomaltose, plus oligosaccharides such as maltotriose. In a mammal, molecular forms of amylase are present in both the saliva and the pancreatic juices. The disaccharides and oligosaccharides produced by amylase digestion are hydrolyzed to yield free glucose by *maltase*, a disaccharidase, and by other enzymes.

PROTEIN DIGESTION   Protein digestion involves a much larger array of enzymes than carbohydrate (or lipid) digestion. This is true because proteins include, by far, the greatest diversity of types of chemical bonds that must be hydrolyzed for digestion to occur. Three or more enzymes are often required for the complete breakdown of a protein. The enzymes that digest proteins and polypeptides are categorized as *endopeptidases* and *exopeptidases*. Both types break bonds between amino acids (i.e., peptide bonds). **Endopeptidases** create breaks *within* chains of amino acids, whereas **exopeptidases** split off terminal amino acids from amino acid chains.

An interesting problem associated with protein digestion is that digestive enzymes targeted at proteins have the potential to attack an animal's own body substance! When protein-digesting enzymes function intraluminally, they are typically synthesized in inactive forms called **proenzymes** or **zymogens**. The enzymes are activated only after they have arrived at the location where they are to act in food digestion.

Protein digestion in vertebrates usually begins in the stomach with the action of a set of intraluminal endopeptidases named *pepsins*, which require acid conditions to be enzymatically effective. Pepsins are secreted by stomach cells—and in some vertebrates by esophageal cells—as proenzymes called *pepsinogens*. Exposure to acid in the stomach lumen activates the pepsins by cleaving the pepsinogens.

When proteins and protein fragments arrive in the midgut of a vertebrate, they are subjected to further intraluminal digestion by enzymes synthesized by the pancreas. The midgut intraluminal enzymes are secreted by the pancreas as proenzymes, which are carried in the pancreatic juice to the midgut; the proenzymes are activated in the midgut by enzyme-catalyzed cleavage reactions that release the active peptidases. Some of the pancreatic enzymes are endopeptidases, including *trypsin, chymotrypsin, elastase*, and *collagenase* (not all are found in all species). The pancreatic enzymes also include exopeptidases, notably *carboxypeptidases A and B*. **FIGURE 6.19** shows how these enzymes act together to break up a protein.

After the intraluminal protein-digesting enzymes of the vertebrate stomach and midgut have had their effect, the resulting product is a mix of both free amino acids and short amino acid chains called *oligopeptides*. The next step in protein digestion is for the oligopeptides to be further hydrolyzed. They are digested by a great diversity of membrane-associated endo- and exopeptidases (about 20 in humans) located in the apical membranes of the midgut epithelium. The amino acids from the original proteins are then in the form of free amino acids, dipeptides, and tripeptides. These products are transported into the digestive–absorptive cells of the midgut epithelium (see Figure 2.5). There, protein digestion is completed by intracellular peptidases, which hydrolyze the di- and tripeptides. In the end, therefore, the products passed to the blood are mostly free amino acids.

LIPID DIGESTION   Lipid digestion is both simpler and more complex than protein digestion. It is simpler because fewer types of chemical bonds need to be hydrolyzed, and thus fewer enzymes are required. It is more complex because although digestive enzymes are in the aqueous phase, lipids are insoluble in water. Successful

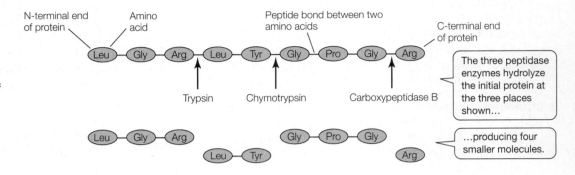

FIGURE 6.19 The digestion of a short protein by three pancreatic peptidases, each specific for particular types of peptide bonds Arg = arginine; Gly = glycine; Leu = leucine; Pro = proline; Tyr = tyrosine.

lipid digestion depends not just on enzymes but also on nonenzymatic, emulsifying processes that break up lipids into small droplets having lots of surface area relative to their lipid content. The principal digestive enzymes of lipids are **lipases**. They act at lipid–water interfaces, and therefore their effectiveness depends on the surface area produced by emulsifying processes. Emulsification can be brought about by mechanical agitation or chemical action.

The midgut is the principal site of lipid digestion in vertebrates. One reason for the vigor of lipid digestion in the midgut is the presence there of intraluminal *pancreatic lipases*, secreted by the pancreas as active enzymes. A second important factor promoting effective lipid digestion in the midgut is that chemical emulsifying agents—namely **bile salts** and other bile constituents—are secreted into the midgut by the biliary system of the liver. Bile salts are amphipathic molecules (see page 37) that act as detergents, dispersing lipids into droplets having a large collective surface area at which lipases can act. Interestingly, bile salts are not destroyed in the gut. Instead, they are reabsorbed downstream by active transport in terminal portions of the midgut and returned to the liver biliary system, which again secretes them upstream. Any given bile-salt molecule may be recycled in this way several times during the digestion of a single meal!

The products of the digestion of fats and oils (triacylglycerols) by lipases are free fatty acids, glycerol, and 2-monoacylglycerols (glycerol esterified at the middle position with a single fatty acid) (see Figure 6.4). Lipids such as phospholipids and cholesterol esters (containing ring structures) require other enzymes in addition to lipases for digestion (e.g., *phospholipases* and *esterases*).

Absorption occurs by different mechanisms for hydrophilic and hydrophobic molecules

Absorption can mean different things in different sorts of animals, as stressed earlier. Here we discuss just the type of absorption that occurs in vertebrates, arthropods, and other groups in which digestion is principally extracellular. Absorption in these animals consists primarily of the transport of chemically simple compounds—such as monosaccharides, free amino acids, and free fatty acids—across the epithelial cells lining the digestive tract, from the gut lumen into blood or lymph.

Surface area is important for absorption. Often the lining of the gut is configured in ways that greatly increase the surface area of the epithelium across which absorption occurs. (This expansion of surface area also aids digestion.) The inset drawings in **FIGURE 6.20** show how the surface area of the midgut digestive–absorp-

tive epithelium is expanded by the presence of folds and minute fingerlike projections in humans and many other vertebrates.

Three mechanisms of transport are involved in absorption: simple diffusion, facilitated diffusion, and active transport.[12] Active transport is always of the secondary type in this context, meaning that the transporter proteins are cotransporters or countertransporters that are able to transport nutrient molecules uphill by use of energy from metabolically generated electrochemical gradients of inorganic ions such as Na^+ (see Box 5.1 and Figure 5.12).

Monosaccharides, amino acids, and *water-soluble vitamins* are aptly considered together in the study of absorption because all three are hydrophilic and thus do not pass readily by simple diffusion through cell membranes (see page 112). Transporter proteins in cell membranes are required for these compounds to be absorbed. Because of the necessity of transporters, these compounds are absorbed only in the parts of the gut where the epithelial cells of the gut lining synthesize the transporters, notably the midgut in vertebrates.

The absorption of glucose in the vertebrate midgut is the best understood of all transporter-mediated absorptive systems. For glucose to be absorbed into the bloodstream across a midgut digestive–absorptive epithelial cell, the glucose must first enter the cell from the gut lumen across the apical, brush-border cell membrane, and then it must exit the cell into the blood across the basolateral cell membrane. The transport of glucose across both cell membranes is transporter-mediated (carrier-mediated). The best-known transporter in the apical membrane is a cotransporter, *sodium–glucose transporter 1 (SGLT1)*, that mediates the secondary active transport of glucose using energy from the Na^+ electrochemical gradient across the apical membrane (see Box 5.1). Figure 6.20 shows this transporter, and Figure 5.12 explains in detail how a transporter of this type mediates the active transport of glucose into an epithelial cell. After glucose has entered a midgut epithelial cell by *active* transport, it leaves the cell to enter the blood across the cell's basolateral membrane by transporter-mediated *passive* transport—that is, by facilitated diffusion. The best-known transporter protein that mediates this facilitated diffusion is GLUT2 (see Figure 6.20).

The absorption of monosaccharides other than glucose by vertebrates exhibits similarities and differences relative to glucose absorption. Galactose—a monosaccharide released by the diges-

---

[12] See Chapter 5 for a detailed discussion of these transport mechanisms. A fourth mechanism that may be important in certain situations is *solvent drag*, in which solutes are carried along by osmotic water movement.

FIGURE 6.20 The structure of the vertebrate midgut and the mechanism of absorption of monosaccharides (glucose, fructose, and galactose) The insets show how the surface area of the midgut epithelium—composed principally of digestive–absorptive epithelial cells—is enlarged by the presence of folds and projections. The drawings are of the human midgut. There are huge numbers of minute fingerlike structures (each less than 1 mm long), called intestinal *villi* (singular *villus*). Collectively they greatly increase the total surface area of the epithelium and thus the numbers of digestive–absorptive epithelial cells. Four membrane transporter proteins of those cells—SGLT1, Na⁺–K⁺-ATPase, GLUT2, and GLUT5—play key roles in the absorption of monosaccharides, as shown. GLUT2, shown here only in the basolateral membrane, sometimes also is present in the apical membrane. Cells turn over rapidly in the midgut epithelium; for example, the average life span of an epithelial cell in humans is about 3–4 days. This rate of cell turnover means equally rapid turnover of the transporter proteins involved in absorption. See Figure 2.5 for more cytological detail.

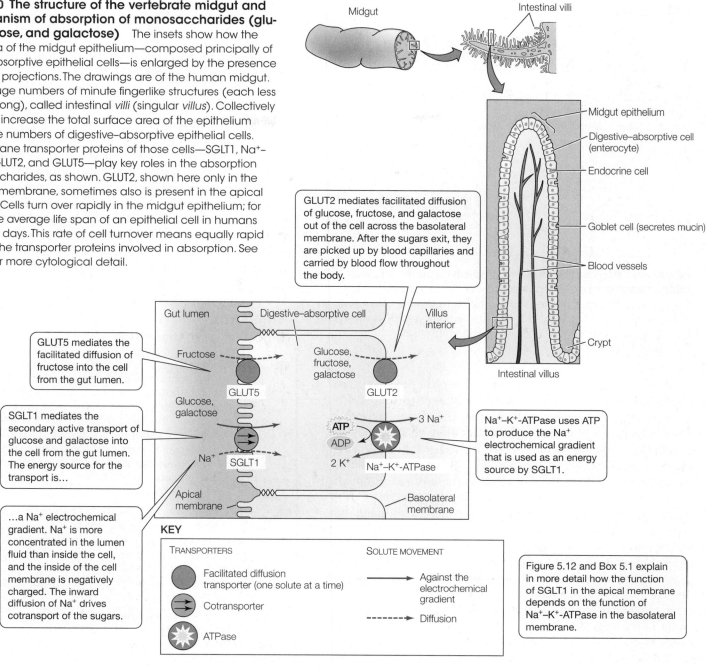

GLUT2 mediates facilitated diffusion of glucose, fructose, and galactose out of the cell across the basolateral membrane. After the sugars exit, they are picked up by blood capillaries and carried by blood flow throughout the body.

GLUT5 mediates the facilitated diffusion of fructose into the cell from the gut lumen.

SGLT1 mediates the secondary active transport of glucose and galactose into the cell from the gut lumen. The energy source for the transport is…

…a Na⁺ electrochemical gradient. Na⁺ is more concentrated in the lumen fluid than inside the cell, and the inside of the cell membrane is negatively charged. The inward diffusion of Na⁺ drives cotransport of the sugars.

Na⁺–K⁺-ATPase uses ATP to produce the Na⁺ electrochemical gradient that is used as an energy source by SGLT1.

Figure 5.12 and Box 5.1 explain in more detail how the function of SGLT1 in the apical membrane depends on the function of Na⁺–K⁺-ATPase in the basolateral membrane.

KEY

TRANSPORTERS

- Facilitated diffusion transporter (one solute at a time)
- Cotransporter
- ATPase

SOLUTE MOVEMENT

- → Against the electrochemical gradient
- ⤍ Diffusion

---

tion of lactose—is transported into vertebrate midgut epithelial cells actively by SGLT1. But fructose—a monosaccharide product of sucrose digestion—enters the cells principally by facilitated diffusion mediated by an apical-membrane transporter, GLUT5 (see Figure 6.20). GLUT2 transports fructose and galactose, as well as glucose, out of the cells.

In the last decade, evidence has developed that GLUT2 can occur in the apical membrane as well as in the basolateral membrane. When monosaccharides are more concentrated in the gut lumen than they are within the midgut epithelial cells, GLUT2 in the apical membrane facilitates inward diffusion. Under some conditions, more glucose enters this way than via SGLT1, a discovery that may well be pertinent to understanding diseases affecting sugar absorption.

Amino acid absorption in the vertebrate midgut is complicated (partly because of the diversity of amino acids) and not thoroughly understood. As many as seven distinct transporter proteins for amino acids are found in the apical, brush-border membranes of digestive–absorptive epithelial cells. Each transporter is specialized

to transport a distinct *set* of amino acids (e.g., the set of cationic amino acids) into the epithelial cells. Most of the amino acid transporters carry out secondary active transport using energy drawn from electrochemical gradients of Na⁺ or other ions. Additional transporters are involved in moving dipeptides into epithelial cells and in moving amino acids out of the cells into the blood across the basolateral membranes.

Transporters are also known in the apical membranes of vertebrate midgut epithelial cells for many of the water-soluble vitamins. Active transporters driven by the Na⁺ electrochemical gradient are known for some of the B vitamins, for example. Many primates (including humans), and some other vertebrates, also possess a Na⁺-driven active transporter for vitamin C.[13]

---

[13] The acquisition mechanisms for some vitamins are particularly complex. Looking at vitamin B₁₂ in mammals, for example, at least five animal-synthesized proteins are involved in protecting the vitamin from destruction during digestion and absorbing it. Knowledge of such systems is clinically important, in part because defects at any step along the way can lead to vitamin deficiency.

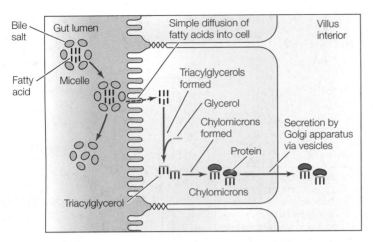

FIGURE 6.21 Major steps in fatty acid absorption across a midgut epithelial cell Figure 6.20 shows the location of such cells. Chylomicrons are packed into vesicles by the Golgi apparatus. The vesicles then move to the basolateral membrane and, fusing with the membrane, release their contents to the lymphatic circulation in the villus interior.

The principles that govern the absorption of the products of *lipid* digestion typically are different from those that govern the absorption of hydrophilic molecules. Fatty acids and monoacylglycerols (monoglycerides) produced by lipid digestion, because they are hydrophobic, are able to dissolve in the lipid-rich interior of cell membranes. Thus they can move through cell membranes readily by simple diffusion (see page 108). In fact, simple diffusion—unmediated by transporter proteins—often seems to account for much or all of the transport of fatty acids and monoacylglycerols into gut epithelial cells. However, transporters for these compounds—some active—are known.

To fully understand the absorption of fatty acids and monoacylglycerols, one must take into account that most are poorly soluble in the aqueous solutions inside and outside epithelial cells. Vertebrates have mechanisms to emulsify or solubilize the lipids in both solutions. In the midgut lumen, as fatty acids and monoacylglycerols are produced by lipid digestion, they are dramatically solubilized by combining with bile salts to produce minute disc- or sphere-shaped molecular aggregates called *micelles*; a micelle is smaller than 10 nanometers (nm) and maintains its lipid constituents in solution by means of the emulsifying effects of the amphipathic bile salts. Lipid-soluble vitamins participate in micelle formation as well. The solubilization of fatty acids, monoacylglycerols, and lipid-soluble vitamins by micelle formation greatly aids the absorption of the molecules. Micelles themselves are not absorbed, however. Instead, the fatty acids and other molecules dissociate from micelles next to the apical membranes of gut epithelial cells, and then move through the membranes by simple diffusion.

After the products of lipid digestion are inside gut epithelial cells (**FIGURE 6.21**), enzyme-catalyzed metabolic pathways synthesize intracellular triacylglycerols, phospholipids, and cholesterol esters from them. These complex lipids are then combined inside the cells with proteins to form **lipoprotein molecular aggregates**, especially ones called **chylomicrons**. The formation of chylomicrons sets the stage for the lipids to be carried throughout the body by way of the lymphatic and circulatory systems, because the protein constituents of the chylomicrons are amphipathic and emulsify the lipids in the aqueous solution of the lymph or blood.

The chylomicrons produced inside a midgut epithelial cell are assembled into secretory vesicles by the cell's Golgi apparatus. The vesicles then fuse with the cell's basolateral membrane to release the chylomicrons into the midgut lymphatic circulation, which carries the chylomicrons into the bloodstream.

The absorption of *short-chain fatty acids* (*SCFAs*) produced by microbial fermentation is a special case because these particular fatty acids (e.g., acetic acid) are both water-soluble and lipid-soluble. The SCFAs do not require emulsification, because of their water solubility. Moreover, they tend to be absorbed wherever they are produced in the gut. The least complicated model for their absorption is that they simply diffuse through cell membranes and cell interiors by being lipid- and water-soluble. Detailed studies show, however, that their absorption is sometimes transporter-mediated in ways that remain poorly understood.

SUMMARY
Digestion and Absorption

- Digestion is the process of splitting food molecules into smaller parts that an animal can take into its living tissues and distribute throughout its body. Absorption (assimilation) is the process of taking organic compounds into the living tissues of an animal from the gut lumen or from other places outside those tissues. In vertebrates, arthropods, and some other groups, digestion precedes absorption. However, absorption precedes digestion in certain other animals, such as certain bivalve molluscs.

- Digestion is carried out mostly by hydrolytic enzymes, each of which catalyzes the splitting of specific types of chemical bonds. The enzymes may be intraluminal, membrane-associated, or intracellular. Extracellular digestion is carried out by intraluminal and membrane-associated enzymes, whereas intracellular digestion is carried out by intracellular enzymes.

- The absorption of relatively simple hydrophilic compounds—such as monosaccharides, amino acids, and water-soluble vitamins—usually requires transporter proteins in the cell membranes involved, and it occurs by either facilitated diffusion or secondary active transport. The absorption of the hydrophobic fatty acids and monoacylglycerols produced by lipid digestion can occur to a large extent by simple diffusion across cell membranes. However, complexities arise in the absorption of these compounds because (being hydrophobic) they need to be emulsified in the aqueous solutions outside and inside cells while being absorbed. SCFAs are a special case because they are water- and lipid-soluble.

- In vertebrates, the midgut is typically the most important site of digestion and absorption. This is true because (1) the apical membranes of midgut epithelial cells are richly populated by membrane-associated digestive enzyme proteins and by transporter proteins, and (2) pancreatic and biliary secretions enter the midgut.

- An animal's digestive and absorptive capabilities are major determinants of the nutritional value of foods because ingested organic compounds can be used only to the extent that they can be digested and absorbed.

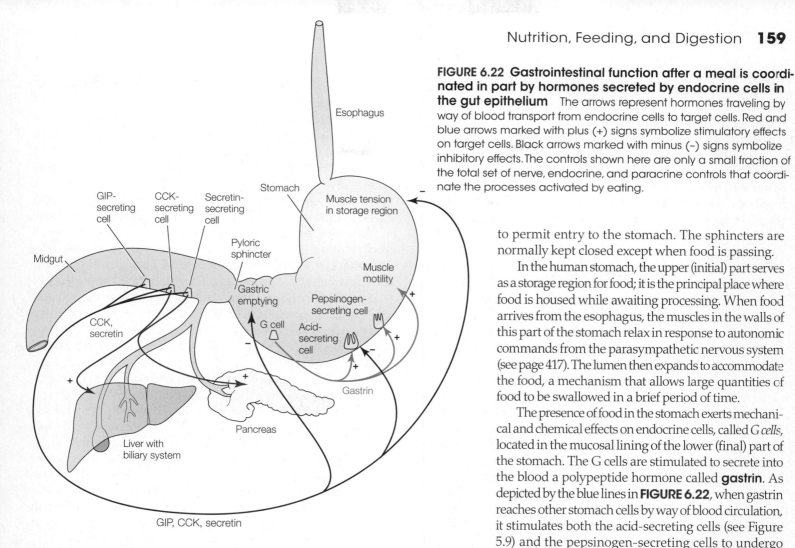

FIGURE 6.22 Gastrointestinal function after a meal is coordinated in part by hormones secreted by endocrine cells in the gut epithelium The arrows represent hormones traveling by way of blood transport from endocrine cells to target cells. Red and blue arrows marked with plus (+) signs symbolize stimulatory effects on target cells. Black arrows marked with minus (−) signs symbolize inhibitory effects. The controls shown here are only a small fraction of the total set of nerve, endocrine, and paracrine controls that coordinate the processes activated by eating.

to permit entry to the stomach. The sphincters are normally kept closed except when food is passing.

In the human stomach, the upper (initial) part serves as a storage region for food; it is the principal place where food is housed while awaiting processing. When food arrives from the esophagus, the muscles in the walls of this part of the stomach relax in response to autonomic commands from the parasympathetic nervous system (see page 417). The lumen then expands to accommodate the food, a mechanism that allows large quantities of food to be swallowed in a brief period of time.

The presence of food in the stomach exerts mechanical and chemical effects on endocrine cells, called *G cells*, located in the mucosal lining of the lower (final) part of the stomach. The G cells are stimulated to secrete into the blood a polypeptide hormone called **gastrin**. As depicted by the blue lines in **FIGURE 6.22**, when gastrin reaches other stomach cells by way of blood circulation, it stimulates both the acid-secreting cells (see Figure 5.9) and the pepsinogen-secreting cells to undergo secretory activity, and it stimulates the muscle layers of the nonstorage part of the stomach to begin and sustain peristaltic contractions. This motility of the stomach physically mixes and breaks up food at the same time that acid and pepsins initiate digestion of the meal. Additional controls are involved in modulating stomach function. For example, in addition to having gastrin receptors and being stimulated by gastrin, the acid-secreting cells have receptors for acetylcholine—released by the parasympathetic nervous system—and for histamine. The smell and taste of food cause stimulatory signals to be passed from the brain to the acid-secreting cells by way of the parasympathetic system. Histamine—which acts as a paracrine agent—is released by cells neighboring the acid-secreting cells (partly in response to gastrin and parasympathetic stimulation) and stimulates the acid-secreting cells.

The midgut (small intestine) can accommodate only a relatively small amount of material at a time. A sphincter between the stomach and midgut, the *pyloric sphincter* (see Figure 6.22), meters material into the midgut by opening only enough to let an appropriate amount of the most liquefied stomach contents through. At the end of each peristaltic wave passing along the stomach toward the sphincter, most of the material being processed in the stomach is refluxed back toward the main body of the stomach. Only a metered amount passes into the midgut. The midgut, as we have already seen, is the principal site of both digestion and absorption in the human gastrointestinal tract.

One of the most important principles of gut coordination is that *endocrine cells of many sorts are found in the epithelium lining much of the gut, and the types of endocrine cells differ from one region of the gut to another, meaning that each region can send out distinctive hormonal*

Responses to Eating

When you eat a meal, your gastrointestinal system responds to the arrival of food in intricate ways that enable the system to carry out its function of making the constituents of the food available to your body. These responses belong to the category of *acute responses* defined in Chapter 1 (see Table 1.2). Cellular activities that are modified to process a meal include the secretion of digestive enzymes, the secretion of acid, and the contraction of muscles. Nerve, endocrine, and paracrine cells control and coordinate the responses.[14]

After chewing and secretion of saliva, the first step to consider is the transit of food from the mouth to the stomach. Each mouthful of food that we force by *voluntary* swallowing to the pharynx in the back of our mouth initiates an *involuntary* wave of peristaltic muscular contraction that passes from the pharynx along the length of the esophagus. Complexities arise because of the presence of sphincters at the two ends of the esophagus. A **sphincter** is a circular muscle that can contract tightly and steadily (tonically) for long periods, thereby preventing exchange between one segment of the gut and another. When a peristaltic wave initiated by the presence of food in the pharynx reaches the upper esophagus, the upper esophageal sphincter transiently relaxes, allowing the food to pass. The food is then propelled by peristalsis down the esophagus, and when it reaches the lower esophageal sphincter, the latter opens transiently

---

[14] As discussed in Chapter 16 (see Figure 16.1), *endocrine cells* secrete hormones into the blood, which carries them to often-distant target cells where they exert their effects. *Paracrine cells* secrete communicatory chemicals into the fluids between cells, where the chemicals affect other neighboring cells.

messages. An illustration is provided by three classes of endocrine cells that occur in the epithelium of the upper midgut. These three classes of cells secrete three polypeptide hormones: **secretin,**[15] **cholecystokinin (CCK),** and **gastric inhibitory polypeptide (GIP).**

The endocrine cells in the upper midgut are stimulated to secrete their respective hormones by acidity and by exposure to nutrient molecules. Thus when acidified and partly digested food is metered into the upper midgut from the stomach, it causes release of the three hormones we have mentioned into the blood. As shown by the red lines in Figure 6.22, CCK and secretin synergistically stimulate the pancreas and the biliary system of the liver to secrete into the midgut the array of critical digestive agents we have already discussed, notably pancreatic digestive enzymes and bile. An additional critical function stimulated by secretin is the secretion by the pancreas of bicarbonate (HCO_3^-), which neutralizes acid from the stomach. The midgut hormones also affect the stomach. As shown by the black lines in Figure 6.22, all three inhibit gastric-acid secretion, gastric emptying, and muscle contraction in the storage part of the stomach, for example; in this way the hormones promote retention of food in the stomach when acidity and nutrient concentrations in the midgut are high, signifying that the midgut—for the moment—already has adequate input from the stomach. Stomach function is also affected by neuron-mediated reflexes from the midgut; excessive acidity or accumulation of unabsorbed digestive products in the midgut, for instance, can inhibit stomach motility by this means.

Motility of the midgut is under control of a complex array of hormonal and neuronal influences, both stimulatory and inhibitory. Segmentation is the principal type of midgut motility. It pushes the midgut contents back and forth. This action mixes food products with intraluminal digestive enzymes and bile, and it ensures contact of all digesting material with the walls of the midgut, where the membrane-associated digestive enzymes and absorption transporters are found. In addition to segmentation, the midgut exhibits progressive, peristaltic contractions that move material toward the hindgut.

Just as humans exhibit responses to the arrival of food, so also do all other animals. In oysters, mussels, and clams, for example, the principal cells lining the digestive diverticula (the cells responsible for intracellular digestion) *grow dramatically* as they take in food particles. They rapidly synthesize new cell membrane at this time to meet the membrane requirements of phagocytosis and pinocytosis. Some cells grow flagella when food is present, permitting them to circulate fluid and food particles within the diverticula by flagellar action. All these changes reverse after food has been processed.

The Control of Hunger and Satiation

Hunger and its converse, satiation, are subjects of active research, especially in mammals, the group of animals we will briefly consider here. Years ago researchers studying the regulation of hunger and satiation expected to find negative feedback controls in which a few agents would play overwhelmingly dominant roles. By now it is clear that the regulation of hunger and satiation is much more diffuse and complex than expected. Two brain regions, the hypothalamus and hindbrain, are the preeminent control centers. They receive relevant sensory information from many sources. All parts of the gastroin-

testinal tract, for example, send neuronal or endocrine signals to the control centers in the brain (see Appendix K).

Ghrelin, a peptide hormone secreted by endocrine cells in the mucosal lining of the stomach and upper intestines, often plays a role in stimulating hunger. Ghrelin is typically secreted before a meal; then, after eating has occurred, its blood concentration rapidly declines. There are also other controls of the hunger felt at a meal. For example, loss of body weight affects the magnitude of the ghrelin signal, leading to heightened ghrelin secretion in advance of a meal.

As a meal is eaten and processed, the controls of satiation and meal termination come into play. These controls are complex, involving all parts of the gastrointestinal tract and other organs such as white adipose tissue (ordinary white fat). The presence of food materials in the gut causes signals to be sent from the stomach, intestines, and pancreas to the control centers in the brain. These signals promote satiation and meal termination. A variety of peptides serve as signaling molecules. Some act as hormones carried by blood circulation to the control centers. Others initiate neuronal signaling, often transmitted to the control centers via the vagus nerve. Signals sent from the stomach exert effects that in general are not dependent on the composition of the food eaten. Instead, the quantity (volume) of material accumulated in the stomach is the most important stimulus; as the quantity increases, the signals from the stomach promote a sense of satiation and act to terminate eating. In contrast, the signals sent from the intestines and pancreas as a meal is processed depend on the composition of the meal and can exert complex composition-dependent effects on satiation and meal termination. Five peptides that directly or indirectly promote satiation and meal termination are secreted after a meal: CCK, secreted by endocrine cells in the upper midgut epithelium (see Figure 6.22); peptide YY and glucagon-like peptide-1, secreted by endocrine cells in the lower midgut epithlium and hindgut; and glucagon and amylin, secreted by the pancreas.

White adipose tissue, which secretes the polypeptide hormone **leptin,** is also involved in the regulation of hunger and satiation. Leptin is secreted by the white adipose tissue in proportion to the amount of stored fat in the body and thus serves as an "adiposity signal" that can provide information on the magnitude of an animal's existing fat stores. Studies on laboratory rodents have shown that leptin tends to inhibit feeding. Some of the gastrointestinal peptides that promote satiation after a meal—such as peptide YY—are secreted for long enough periods after one meal to suppress hunger for the next meal. Leptin works in tandem with these gastrointestinal peptides to suppress meal-to-meal hunger and food intake. Whether leptin acts in exactly this way in humans remains uncertain because of differences between humans and the rodents used in most research.

Nutritional Physiology in Longer Frames of Time

For the most part in this chapter we have focused on the immediate present in the lives of animals as they acquire foods and process meals. We have asked how animals gather their food, how food is digested and absorbed, and other short-term questions. Rarely have we addressed processes requiring more than a single day to take place. Nonetheless, fascinating and important processes occur on longer time scales. Here we consider just a few of those, starting with long-term natural fasting.

---

[15] Secretin, discovered in about 1900, was the first hormone known to science.

Surprising numbers of animals undergo periods of *natural fasting* that last for days, weeks, or months. **BOX 6.4** reviews several examples. Such animals undergo dramatic shifts in their nutritional physiology, from periods in which they have abundant inputs of nutrients to periods in which they have no inputs at all.

Of all the animals that undergo natural fasting, pythons are among the most dramatic and the best studied physiologically. They get food by sitting and waiting for edible animals to wander by, which means weeks often go by between one meal and the next. One adaptation of the pythons to this pattern of feeding is that they are equipped to ingest large animals when the opportunity arises. Fully grown snakes can weigh more than a person and can easily eat animals that weigh 70% as much as they do. Thus stories of goats, antelopes, and teenagers being consumed are true. With this type of lifestyle a meal can be huge and demand a massive effort of digestion and absorption to be processed. Then, however, there may be no food—and no need at all for digestion or absorption—for a long time.

Unlike humans and many other animals—which maintain their digestive systems in a continuous state of readiness—pythons deconstruct their gastrointestinal systems between meals! Burmese pythons (*Python bivittatus*, also called *P. molurus*), for example, undergo extensive deconstruction of their digestive apparatus if they are deprived of food for a month. Then, when they obtain food, they rapidly reconstruct their digestive apparatus. In the first 24 h after feeding, these pythons *double* the mass of their midgut, largely by the growth of new gut epithelium (critical for both digestion and absorption). Simultaneously, they undergo massive synthesis of midgut transporter proteins; the total numbers of molecules of both glucose transporters and amino acid transporters may increase 20-fold or more.

These pythons provide unique opportunities to learn about the role of gene-expression changes in the control of tissue growth

and remodeling. In one recent study, investigators fasted Burmese pythons for a month, knowing that when they fed the snakes again, there would immediately be a period of exceedingly rapid midgut growth. Using genomic methods, they sampled midgut tissue every 6 h after the snakes ate to determine the numbers and types of genes that responded to the meal by being upregulated or downregulated. The results were even more dramatic than expected. In only 6 h after the pythons ingested food—after 720 h of fasting—the expression of almost 2500 midgut genes was modified (**FIGURE 6.23**). Some genes were upregulated, including the gene named *Frizzled-4*, which codes for a key receptor protein in a signaling system (the Wnt system) for tissue growth and differentiation. Some genes were downregulated, an example being *TCF7L2*, which codes for a protein that can inhibit targets of the same signaling system. Collectively the midgut genes that underwent expression changes are involved not only in signaling of tissue growth but also in the control of apoptosis and the synthesis of proteins involved in microvillus growth, digestion, and absorption. This research indicates that the massive growth of the midgut and its functional response after a meal are orchestrated and controlled by dramatic, genome-wide regulation of gene expression.

As you might guess, the midgut response to feeding in pythons does not occur in isolation. In extreme cases, the metabolic rates of fasted pythons increase 40-fold after eating. Part of the reason for this steep increase in energy use is undoubtedly that the pythons need to expend energy to rebuild their digestive systems. Pythons also increase the size of their heart muscle by 40% within 48–72 h after eating a large meal, as discussed in Chapter 21 (see Figure 21.10).

Let's now look at some of the other ways that the nutritional physiology of animals undergoes long-term changes. In fact,

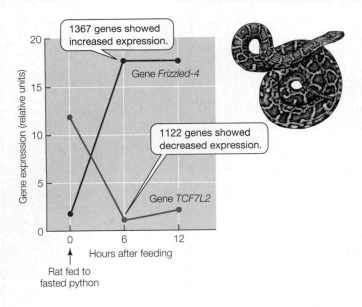

FIGURE 6.23 Midgut genomics: Massive upregulation and downregulation of midgut genes after food consumption in fasted pythons After a group of Burmese pythons (*Python bivittatus*, also called *P. molurus*) was fasted for 30 days, each python was fed a rat at time 0. Gene expression was measured in midgut tissue before feeding and at 6 h and 12 h after feeding. The midgut tissue responded to feeding by upregulating or downregulating almost 2500 genes. The graph shows the results for two genes (discussed in the text) as examples, one that was upregulated and another that was downregulated. (After Andrew et al. 2015.)

- 1367 genes showed increased expression.
- Gene *Frizzled-4*
- 1122 genes showed decreased expression.
- Gene *TCF7L2*
- Gene expression (relative units)
- Hours after feeding
- Rat fed to fasted python

nutritional physiology is commonly adjusted—often in adaptive ways—in *all five time frames* stressed in Chapter 1 (see Table 1.2).

Nutritional physiology responds to long-term environmental change

Individual animals and populations commonly undergo changes in their nutritional physiology in response to their environments. Three aspects of this theme deserve emphasis here: responses of the nutritional physiology of individuals to long-term (chronic) environmental changes, alterations of gene expression in individuals in response to the composition of their diet, and evolutionary responses of populations to changed nutritional conditions.

INDIVIDUALS RESTRUCTURE THEIR DIGESTIVE–ABSORPTIVE SYSTEMS WHEN THEY EXPERIENCE NEW SETS OF ENVIRONMENTAL CONDITIONS FOR EXTENDED PERIODS Often, when individual animals chronically alter their diet—so that they eat new types of foods compared with what they were previously eating—they upregulate digestive enzymes and absorption transporters targeted at the new foods. These changes, which typically require a few days to a few weeks to be completed, quite literally *increase the nutritional value* of the new foods because, as earlier stressed (see Figure 6.1), the nutritional value of a food depends in part on how readily an animal is able to digest and absorb it.

The sugar-transporter proteins in the midgut epithelial cells of a person (see Figure 6.20) provide an excellent example. The numbers of molecules of SGLT1, GLUT2, and GLUT5 *per cell* are not static in an adult human. Instead, the numbers of these molecules per cell become relatively low if an individual eats little carbohydrate for many days. Then, if the person changes his or her diet so that a lot of carbohydrate is ingested for several days, the numbers of transporter molecules per cell increase. Therefore the capacity for carbohydrate absorption changes chronically, depending on the amount of carbohydrate in the diet. Similarly, acclimation of midgut sugar transporters has been observed in experiments on many other vertebrates, from fish to mammals. Midgut amino acid transporters are also often upregulated or downregulated when dietary protein is chronically increased or decreased.

Sustained changes in diet commonly also bring about chronic adjustments in the quantities of digestive enzymes produced. The intraluminal digestive enzymes secreted by the pancreas in vertebrates provide dramatic examples. In a variety of species, a sustained increase in the amount of carbohydrate in the diet results in chronically increased pancreatic secretion of amylase. Similarly, sustained increases in the amounts of dietary proteins or lipids lead to chronically heightened secretion of pancreatic peptidases or lipases. Laboratory rats provide the best-studied model of these responses. In a rat, changes in pancreatic enzyme secretion start within 24 h after a change in diet and require about a week to be fully expressed. At that point a rat may exhibit a fivefold or greater increase in pancreatic secretion of digestive enzymes targeted at the new foods.

The size of the gastrointestinal tract is also subject to chronic adjustments. When a mouse or a small bird is placed chronically at subfreezing air temperatures, its intestines grow. House wrens, for example, gradually increase their intestinal length by 20% when placed at –9°C after living at +24°C. For small birds or mammals to stay warm in cold environments, they must metabolize more food energy per day than when they are living in warm environments. Growth of the midgut in cold environments helps them increase their rates of food digestion and absorption. Similarly, female mice increase their midgut length when they are nursing young. Pythons, which we've already discussed, provide the most stunning known display of chronic changes in gut size.

DIETARY CONSTITUENTS SOMETIMES ALTER GENE TRANSCRIPTION One of the cutting-edge stories in nutritional physiology today is the increasing realization from genomic research that food constituents or their immediate metabolic products sometimes induce upregulation or downregulation of genes. That is, some genes are *regulated by diet composition*. The exact implications of this discovery will probably not be fully understood for many years. Perhaps, for example, we will find that the effects discussed in the last section are mediated by relatively straightforward effects of food constituents on the control of genes that code for enzyme or transporter proteins. Already, researchers have found that dietary fatty acids often bind with receptor molecules, in the nuclei of cells, that control genes that are pivotal in fatty acid metabolism! Thus, when a person makes a long-term change in the lipids—the suites of fatty acids—he or she eats, that change can alter the manner in which the person metabolizes lipids. This phenomenon is one way in which changes in the composition of a person's diet can affect long-term health for better or for worse.

IN POPULATIONS, NUTRITIONAL PHYSIOLOGY UNDERGOES EVOLUTIONARY CHANGE IN RESPONSE TO THE NUTRITIONAL ENVIRONMENT A good starting point to discuss this theme is to recall the intriguing recent studies of lactase persistence (adult lactase expression) in human populations discussed at the start of this chapter. When populations of people alter their nutritional environment over the course of many generations by adding milk and milk products to their adult diet, it seems clear that—at least sometimes—mutations that permit lactose digestion in adulthood are favored by natural selection and come to be shared by large proportions of individuals. This phenomenon has long been recognized in northern

European populations (e.g., the Dutch and Danes in Table 6.1). The recent research in Africa reveals the same phenomenon in East African cattle-raising peoples. The studies of African peoples emphasize that different populations of a single species can differ in nutritional physiology, and when selective pressures are great, evolutionary change can take place in human populations in only a few thousand years.

When groups of related species are compared, they are often found to differ in their digestive physiology in ways that are well correlated with the types of foods they eat, providing interspecific comparative evidence that digestive physiology evolves in parallel with diet. The distributions of disaccharidases in relation to disaccharides in the diet provide well-documented illustrations. For example, sucrase tends to be synthesized in abundance just by species that consume sucrose-rich foods. Although sucrose is common in our kitchens in the form of "table sugar," it is actually not a common sugar in nature; flower nectars that are rich in sucrose are relatively distinctive foods. Species of birds, bats, and insects that specialize in feeding on sucrose-rich nectars synthesize abundant sucrase. However, related birds, bats, and insects that do not eat such nectars usually do not synthesize sucrase in particular abundance. The distributions of trehalase and dietary trehalose parallel each other in an equally striking way. The disaccharide trehalose has a very limited distribution in the biosphere. One of the few places it is found in abundance is in the blood of insects. Mammal species that do not eat insects typically synthesize little or no trehalase. By contrast, insectivorous mammals, such as insectivorous bats, exhibit high trehalase activity.

The midgut glucose transporters of vertebrates provide another example. Starches from plants yield glucose when digested. Vertebrate species that are principally plant eaters (herbivores), from fish to mammals, tend to show higher activities of midgut glucose transporters than related carnivorous vertebrates, which get less glucose in their diets (see Figure 5.13). Some direct evidence exists (from studies of herbivores and carnivores subsisting on similar foods) that these differences are at least partly genetic, not merely consequences of acclimation to different diets.

The nutritional physiology of individuals is often endogenously programmed to change over time: Developmental and clock-driven changes

Nutritional physiology often undergoes *developmentally programmed changes* as individuals mature from birth to adulthood. Developmental changes in lactase synthesis in mammals provide an example we've already discussed; individuals usually stop synthesizing lactase when they become old enough to have stopped taking in milk by nursing. Other developmentally programmed changes sometimes occur simultaneously. In some mammal species, for example, disaccharidases other than lactase, such as sucrase and maltase, increase as nursing comes to an end. The substrates of these alternative disaccharidases appear in the diet only as young mammals start to eat solid foods.

Programmed developmental change is particularly dramatic in amphibians. Whereas the tadpoles of frogs and toads are herbivores or omnivores, the adults are carnivores. The gut of a frog or toad is rapidly shortened and restructured during metamorphosis.

In addition to developmental changes, animals also often exhibit *periodically repeating programmed changes in nutritional physiology under the control of biological clocks*. The disaccharidases in the midgut of laboratory rats illustrate this phenomenon. In the apical membranes of epithelial cells, these enzymes are upregulated at night and downregulated during the day under the control of an endogenous biological (circadian) clock. This rhythm ensures greatest digestive capability at the time that the rats—being nocturnal—are most likely to ingest food.

On another time scale, the annual fattening cycles of some mammals and birds are among the most dramatic of the clock-controlled cycles in nutritional physiology. Some species of hibernating ground squirrels, to illustrate, increase their ingestion and absorption of food energy—and fatten dramatically—in the autumn of each year even when they are kept under constant environmental conditions in a laboratory. **FIGURE 6.24** shows this phenomenon over a period of more than 3 years in golden-mantled ground squirrels. The cycles in eating behavior, fat storage, and hibernation in such animals are under control of endogenous timekeeping mechanisms called *circannual biological clocks*.[16] Although Figure 6.24 shows sustained cycling that lasted about 3 years, some species of ground squirrels have now been shown to maintain circannual rhythms in nutritional physiology for a decade under constant laboratory conditions. Circannual clocks in fact were originally discovered by research on annual cycles of fat storage in ground squirrels.

[16] The reason for the name *circannual*—meaning "approximately annual"—is that when animals are in a constant environment and receive no exogenous information about the time of year, these clocks maintain *approximately* 365-day rhythms, not *exactly* 365-day rhythms.

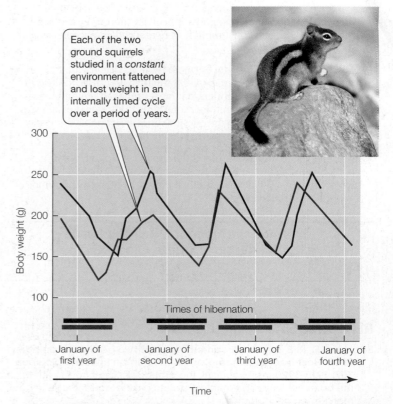

Each of the two ground squirrels studied in a *constant* environment fattened and lost weight in an internally timed cycle over a period of years.

Times of hibernation

January of first year | January of second year | January of third year | January of fourth year

Time

FIGURE 6.24 Endogenously timed annual rhythms in nutritional physiology Two golden-mantled ground squirrels (*Callospermophilus lateralis*)—one coded red, the other black—were kept under *constant* environmental conditions in a laboratory for more than 3 years, during which they received no information about the seasons or the time of year. The plotted lines show how their body weights changed; each ground squirrel lost a great deal of weight when it hibernated and gained weight by fattening between periods of hibernation. The bars near the bottom mark the time periods when each animal hibernated (bars are color coded to match the plots of body weight). (After Pengelley and Asmundson 1969.)

STUDY QUESTIONS

1. Malnutrition of children is a profound global problem, and protein deficiencies are the most common form of childhood malnutrition. Why are people so vulnerable to protein deficiencies?

2. Explain how and why the absorption of hydrophobic and hydrophilic organic molecules differs.

3. List three major roles played by each class of foodstuffs: proteins, carbohydrates, and lipids.

4. Many of the most important vertebrate species in agriculture and aquaculture are foregut, midgut, or hindgut fermenters. Name an example of each, and discuss why species of this sort have advantages in agriculture and aquaculture.

5. Explain why wastage of essential amino acids can occur and how eating two foods simultaneously can prevent wastage.

6. Many species of hindgut fermenters, including horses, eat their feces to a greater extent when they have protein insufficiency than when they have plenty of protein. How might increased fecal consumption help?

7. It can be difficult or impossible to find a single diet that will serve the needs of both an herbivore and a carnivore. If you were trying to determine if two related species, one an herbivore and the other a carnivore, differ *genetically* in their digestive enzymes or absorption transporters, why would it be important to try to find a common diet for your test subjects of both species?

8. On the basis of our discussion of food-chain energetics, explain how the global human population's need for food could be met more readily if meat consumption were reduced and people lived more on plant foods.

9. Discuss hypotheses for why Burmese pythons deconstruct their digestive systems between meals. Similarly, discuss hypotheses for why small birds and mammals, having lengthened their intestines during cold seasons, permit their intestines to shorten during warm seasons. In other words, why might natural selection favor plastic, rather than static, gastrointestinal morphology?

10. Most people tend to think that if the sun ever stopped emitting light, all animal life on Earth would soon end. Populations of hydrothermal-vent animals might continue to exist for thousands of years, however. Explain.

Go to **sites.sinauer.com/animalphys4e**
for box extensions, quizzes, flashcards,
and other resources.

REFERENCES

Bäckhed, F., R. E. Ley, J. L. Sonnenburg, and two additional authors. 2005. Host-bacterial mutualism in the human intestine. *Science* 307: 1915–1920. An especially thought-provoking introduction to the potential for the gut microbiome to play important roles in normal human physiology.

Cummings, D. E., and J. Overduin. 2007. Gastrointestinal regulation of food intake. *J. Clin. Invest.* 117: 13–23.

Delange, F. 2001. Iodine deficiency as a cause of brain damage. *Postgrad. Med. J.* 77: 217–220. This article discusses the global toll of inadequate mineral nutrition in childhood and the large benefits that would result from inexpensive interventions.

Gerbault, P., A. Liebert, Y. Itan, and five additional authors. 2011. Evolution of lactase persistence: an example of human niche construction. *Phil. Trans. R. Soc., B* 366: 863–877.

Goldbogen, J. A. 2010. The ultimate mouthful: lunge feeding in rorqual whales. *Am. Sci.* 98: 124–131. Fascinating new information, and thoughts, on feeding by blue whales and other rorquals based on use of high technology.

Hess, M., A. Sczyrba, R. Egan, and 14 additional authors. 2011. Metagenomic discovery of biomass-degrading genes and genomes from cow rumen. *Science* 331: 463–467. Use of genomic methods to explore ruminant microbes for enzymes that can break down cellulose and other plant polysaccharides.

Jackson, S., and J. M. Diamond. 1996. Metabolic and digestive responses to artificial selection in chickens. *Evolution* 50: 1638–1650.

Karasov, W. H., and C. Martinez del Rio. 2007. *Physiological Ecology: How Animals Process Energy, Nutrients, and Toxins.* Princeton University Press, Princeton, NJ.

Lankford, T. E., Jr., J. M. Billerbeck, and D. O. Conover. 2001. Evolution of intrinsic growth and energy acquisition rates. II. Trade-offs with vulnerability to predation in *Menidia menidia. Evolution* 55: 1873–1881. The intrinsic growth rates of Atlantic silversides become lower, the farther south the fish live. This research explores the reasons why intrinsic growth rate is not always maximized, and is an important contribution to the modern exploration of the crucial role of trade-offs in the evolution of animal features.

McCue, M. D. (ed.) 2012. *Comparative Physiology of Fasting, Starvation, and Food Limitation.* Springer-Verlag, Berlin. A valuable collection of chapters written by specialists on many aspects of fasting and starvation in animals as diverse as fish, spiders, and birds.

Nicholson, J. K., E. Holmes, J. Kinross, and four additional authors. 2012. Host-gut microbiota metabolic interactions. *Science* 336: 1262–1267. A concise, up-to-date review of roles of the gut microbiome, with a thought-provoking summary of potential bacterial roles.

Piersma, T., and J. Drent. 2003. Phenotypic flexibility and the evolution of organismal design. *Trends Ecol. Evol.* 18: 228–233. A thought-provoking review of phenotypic plasticity that emphasizes morphological adjustments, many of them involved in food acquisition or processing.

Secor, S. M. 2005. Evolutionary and cellular mechanisms regulating intestinal performance of amphibians and reptiles. *Integr. Comp. Biol.* 45: 282–294.

Sterner, R. W., and J. J. Elser. 2002. *Ecological Stoichiometry: The Biology of Elements from Molecules to the Biosphere.* Princeton University Press, Princeton, NJ. A challenging book that synthesizes information on elemental nutrition in a groundbreaking, almost revolutionary, way. The chapter on biological chemistry is particularly relevant to this chapter.

Stevens, C. E., and I. D. Hume. 2004. *Comparative Physiology of the Vertebrate Digestive System*, 2nd ed. Cambridge University Press, New York. The best available book-length treatment of digestion and digestive systems throughout the vertebrates.

Tishkoff, S. A., F. A. Reed, A. Ranciaro, and 16 additional authors. 2007. Convergent adaptation of human lactase persistence in Africa and Europe. *Nat. Genet.* 39: 31–40. A stunning application of modern molecular genetics to the study of digestion and recent human evolution.

Wang, T., C. C. Y. Hung, and D. J. Randall. 2006. The comparative physiology of food deprivation: from feast to famine. *Annu. Rev. Physiol.* 68: 223–251.

See also **Additional References** *and* **Figure and Table Citations**.

Energy Metabolism

In 1986 explorers set out to reach the North Pole by dogsled from Ellesmere Island at latitude 83° north. Their objective was to make the journey without resupply along the way and thus to reenact the famous expedition by Admiral Robert Peary in 1909. Each of the modern explorers' sleds was huge—5 m (16 ft) long, and loaded with 630 kg (1400 pounds) of material. Of the two-thirds of a ton on each sled, most was food for people and dogs. Most of the weight of the food, moreover, was required to meet energy needs; if the only food materials that had needed to be hauled were vitamins, minerals, amino acids, and other sources of chemical building blocks for biosynthesis, the pile of food on each sled would have been much smaller. The human food on each sled was sufficient to meet the energy needs of just two people. A team of sled dogs for each sled also had to be fed so they could do most of the hauling, including pulling the sled over numerous ice ridges 6–18 m (20–60 ft) high on the way to the pole. A trek to the North Pole by dogsled would be immeasurably easier if there were no need for food energy for people and dogs. The need for energy is equally consequential in the natural world. Animals regularly risk their lives to obtain energy, or die because they did not obtain enough.

The burden of food A trek to the North Pole from the nearest land requires that hundreds of pounds of food be hauled per explorer, to supply energy for the explorers and their dogs.

Why do animals need energy? Even if the answer to that question seems obvious, why do animals need new inputs of energy on a regular basis? Why do people typically need new food energy every day, for instance? Why not take in an adequate amount of energy early in life and simply reuse it, over and over, for the rest of life? These are some of the key questions we address in this chapter.

Other questions to be discussed are more practical. The food on the sleds of the North Pole explorers was rich in fats and oils; the meat they carried, for instance, was a 50:50 mix of ground beef and lard (pure fat). Why did they emphasize lipids? If you were planning an Arctic expedition, how would you use physiological principles to predict your energy needs and calculate the amount of food to pack on your sleds?

Energy metabolism, the subject of this chapter, is the sum of the processes by which animals acquire energy, channel energy into useful functions, and dissipate energy from their bodies. Energy metabolism consists of two subsets of processes mentioned in Chapter 2: *catabolic processes* that break down organic molecules to release energy, and *anabolic processes* that use energy to construct molecules.

Why Animals Need Energy: The Second Law of Thermodynamics

Animals are *organized* or *ordered* systems. As we saw in discussing the dynamic state of body constituents in Chapter 1, the atoms composing an animal's body are routinely exchanged with atoms in the environment. The *organization* of atoms in the body persists, however, even as particular atoms come and go. As this observation suggests, organization is a far more fundamental feature of animals than their material construction is (see page 13).

The **second law of thermodynamics**, one of the greatest achievements of intellectual history, provides fundamental insight into the nature of organized systems. The law applies to **isolated systems**. An isolated system is a part of the material universe that exchanges nothing—neither matter nor energy—with its surroundings. Animals, you will recognize, are not isolated systems. In fact, true isolated systems are difficult to create even in the highly controlled setting of a physics laboratory. Thus the concept of an isolated system is largely an abstraction. Nonetheless, we can gain insight into animal energetics by analyzing isolated systems. The second law of thermodynamics states that *if an isolated system undergoes internal change, the direction of the change is always toward greater disorder.*

A corollary of the second law is that order can be maintained or increased within a system only if the system is *not* isolated. If "energy" is permitted to enter a system from the outside, order may be maintained or increased within the system. **Energy** is defined in the field of mechanics to be the capacity to do mechanical work, measured as the product of force and distance. A broader definition that is often more useful for biologists is that *energy is the capacity to increase order.*

To illustrate the second law of thermodynamics, let's examine an isolated system that consists of a closed loop of copper pipe filled with water, with the water initially flowing around and around in the loop. We need not know how the water started moving; it is enough to know that the loop of pipe has water flowing in it and

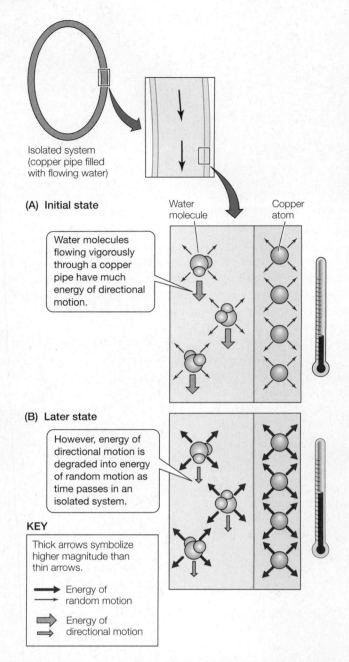

(A) Initial state

Water molecules flowing vigorously through a copper pipe have much energy of directional motion.

(B) Later state

However, energy of directional motion is degraded into energy of random motion as time passes in an isolated system.

KEY

Thick arrows symbolize higher magnitude than thin arrows.

→ Energy of random motion

⇒ Energy of directional motion

FIGURE 7.1 The second law of thermodynamics in action Energy of directional motion is converted to energy of random motion in this isolated system. This transformation of energy increases the intensity of random motions of both the water molecules and the copper atoms, causing the system temperature to rise.

that the loop, being isolated, exchanges no energy or matter with its surroundings. Let's focus on the motions of atoms and molecules *on an atomic-molecular scale*. Initially, the distribution of atomic-molecular motions in this system is highly nonrandom, because each water molecule is moving in an ordered way in its direction of travel around the loop of pipe (**FIGURE 7.1A**). In addition to this directional motion, all of the water molecules—and all of the copper atoms in the walls of the pipe—are also undergoing ceaseless random motions on an atomic-molecular scale.[1]

---

[1] Random atomic-molecular motions are a universal property of matter that is at a temperature above absolute zero. In fact, the *temperature* of an object is a measure of the intensity of these motions within the material substance of the object.

As time passes in this system, the energy of *directional* motion of the water molecules is gradually transformed into energy of *random* motion (**FIGURE 7.1B**). This transformation occurs because each time water molecules collide with other water molecules or with copper atoms, some of their energy of directional motion is transferred in such a way as to increase the intensity of random motions of the atoms and molecules with which they collide. Over time, the rate of flow of water around the loop of pipe gradually decreases as energy of directional motion is lost in this way. Simultaneously, a gradual increase occurs in the energy of random molecular agitation (heat), and the temperature of the system rises. Eventually all the energy of directional motion is lost, and the flow of water comes to a halt. At that point the water molecules and copper atoms in the system display only random motions, and the original order in the system (the directional motion of water molecules) is entirely degraded to disorder. *This inevitable outcome represents the second law of thermodynamics in action.*

The only way to keep the water flowing—and thus maintain the original order in the water-filled loop of pipe—would be to convert the system into an **open system**, a system that is *not* isolated. If there were a pump in the loop of pipe, and if electricity were provided to the pump from outside, the initial nonrandom state could be sustained indefinitely. That is, an energy input to the system could create order in the system as rapidly as processes within the system tend to diminish order.

In terms of their thermodynamics, animals must function as open systems. Without an energy input, the blood coursing through an animal's circulatory system will slow to a halt, just as the water in a loop of pipe does when there is no energy supplied from outside. Without an energy input, vital molecules in an animal's tissues will become more disorganized by spontaneously breaking down or by other processes; eventually, therefore, many vital molecules will lose their critical structural and functional properties. Without an energy input, positive and negative ions will distribute themselves randomly across an animal's cell membranes; this randomization of electrical charges, among other things, will make nerve impulses impossible. The second law of thermodynamics dictates that if an animal were required to function as an isolated system, all forms of order within its body would decay. This loss of order would eventually kill the animal because order is essential for life. *Animals require energy from the outside because energy is necessary to create and maintain their essential internal organization.*

Fundamentals of Animal Energetics

To understand animal energetics more thoroughly, the first step is to recognize that energy exists in different *forms*, and the various forms differ in their significance for animals. We will focus here on four forms of energy of particular importance: chemical energy, electrical energy, mechanical energy, and heat. **Chemical energy (chemical-bond energy)** is energy liberated or required when atoms are rearranged into new configurations. Animals obtain the energy they need to stay alive by reconfiguring atoms in food molecules, thereby liberating chemical energy. **Electrical energy** is energy that a system possesses by virtue of separation of positive and negative electrical charges. All cell membranes possess electrical energy because there is charge separation across them

(see Figure 5.10B). There are two forms of energy of motion (kinetic energy) that are important for animals. One, **mechanical energy**, is energy of *organized* motion in which many molecules move simultaneously in the same direction. The motion of a moving arm, or that of circulating blood, provides an example. **Heat**, often called **molecular kinetic energy**, is the energy of *random atomic-molecular* motion. Heat is the energy that matter possesses by virtue of the ceaseless, random motions of all the atoms and molecules of which it is composed.[2]

The forms of energy vary in their capacity for physiological work

Although, by definition, all forms of energy are capable of doing work in one context or another, all forms of energy are *not equally* capable of doing physiological work in animals. **Physiological work** is any process carried out by an animal that increases order. For example, an animal does physiological work when it synthesizes macromolecules such as proteins—or when it generates electrical or chemical gradients by actively transporting solutes across cell membranes. An animal also does physiological work when it contracts its muscles to move materials inside or outside its body (or set its whole body in motion).

How do the forms of energy that we have mentioned—chemical energy, electrical energy, mechanical energy, and heat—differ in their ability to do physiological work? Animals can use chemical energy (directly or indirectly) to do *all* forms of physiological work; hence, for animals, chemical energy is *totipotent* (*toti*, "all"; *potent*, "powerful"). Animals use electrical and mechanical energy to accomplish some forms of physiological work, but neither form of energy is totipotent. For example, animals use electrical energy to set ions in motion and mechanical energy to pump blood, but they cannot use either form of energy to synthesize proteins. Finally, *animals cannot use heat to do any form of physiological work.*

This last point is important. According to thermodynamics, a system can convert heat to work only if there is a temperature difference between one part of the system and another. Temperature is a measure of the intensity of random atomic-molecular motions. If the intensity of random motions differs from one place to another within a system, this difference represents a form of order (nonrandomness), and the system can be used as a machine to convert heat to work; the high temperatures in the cylinders of an internal combustion engine, for example, permit heat from the burning of fuel to be converted into mechanical energy that propels a car. Consider, however, a physical system that has a uniform temperature. The purely random molecular motions that exist throughout such a system cannot do work. Within cells—the relevant functional systems of organisms—temperature differences from place to place are very small and transient, when they exist at all. Hence cells cannot in theory use heat to do physiological work, and biological experiments confirm that they cannot. Heat is hardly unimportant to animals; as we discuss later in this chapter and in Chapter 10, inputs of heat influence animal metabolic rates and affect the abilities of macromolecules such as proteins to carry out their

---

[2] The study of forms of energy is part of thermodynamics. Some specialists in thermodynamics emphasize energy transfer. For them, heat and mechanical work are not forms of energy, but rather *heating* and *working* are processes that *transfer* energy.

functions. However, heat has no importance as a source of energy for physiological *work* because heat cannot do work in organisms.

The forms of energy are placed into two categories based on their ability to do physiological work. **High-grade energy** can do physiological work; chemical, electrical, and mechanical energy are included in this category. **Low-grade energy**—heat—cannot do physiological work. When we say that animals **degrade** energy, we mean that they transform it from a high-grade form to heat.

Transformations of high-grade energy are always inefficient

When organisms transform energy from one high-grade form to another high-grade form, the transformation is always incomplete, and some energy is degraded to heat. The **efficiency of energy transformation** is defined as follows:

$$\text{Efficiency of energy} \atop \text{transformation} = \frac{\text{output of high-grade energy}}{\text{input of high-grade energy}} \quad (7.1)$$

The efficiency of energy transformation is typically much less than 1. For example, when a cell converts chemical-bond energy of glucose into chemical-bond energy of adenosine triphosphate (ATP), at most only about 70% of the energy released from glucose is incorporated into bonds of ATP; the other 30%—which started as high-grade energy—becomes low-grade energy: heat. When, in turn, a muscle cell uses the chemical-bond energy of ATP to contract, typically a maximum of only 25%–30% of the energy liberated from the ATP appears as energy of muscular motion; again, the remainder is lost as heat. The contraction efficiency of muscles in fact depends on the type of work they are doing. If you plant one of your feet on a wall and use your leg muscles to push your foot steadily against the wall (a type of isometric exercise), no motion occurs, and the muscles' efficiency in producing motion is zero. If you ride an exercise bike, however, the efficiency of your leg muscles in producing external motion might be as high as 25%–30%. Even then, 70%–75% of the energy liberated from ATP in the process would become heat inside your body, rather than producing external motion.

You may be familiar with the Hollywood image of a jungle explorer caught in quicksand. The victim sinks deeper every time he moves. In certain ways an animal's use of food energy is analogous to this image. To make use of the chemical energy from a meal, an animal *must* transform the energy, usually in multiple steps. Each step, however, robs the energy of part of its value because energy transformations are always inefficient and degrade energy to heat. Thus with each step an animal takes to use the high-grade energy in its food, the

resource shrinks, just as each move of the hapless jungle explorer lowers his body farther into the quicksand.

Animals use energy to perform three major functions

It can be helpful to visualize the energy used for physiological work as a resource that "flows" through an animal during its lifetime (**FIGURE 7.2**). From the viewpoint of energetics, each time an animal eats, it gains chemical energy from its external environment. Chemical energy, therefore, enters an animal repeatedly throughout life. This energy, the energy in the chemical bonds of food, is known as the animal's **ingested chemical energy** or **ingested energy**. Forms of energy derived from the ingested energy later pass back into the external environment.

If we follow ingested energy after it is first taken into an animal's body, we find that although some of the ingested energy is absorbed, some is not. This distinction parallels the principle—emphasized in Chapter 6—that some chemical compounds in foods can be successfully digested (or fermented) and absorbed into the bloodstream, whereas others cannot. Ingested compounds that an animal is unable to absorb are egested in the feces. The chemical-bond energy in these compounds is known as the animal's **fecal chemical energy** or **fecal energy** (see Figure 7.2). By contrast, the chemical-bond energy of the organic compounds that are absorbed (assimilated) is known as the animal's **absorbed chemical energy** or **absorbed energy** (or **assimilated chemical energy**).[3] The absorbed energy

---

[3] The term *digestible energy* is widely used as a synonym, but it is not used in this book because the absorbed chemical energy depends on not only digestive but also absorptive processes.

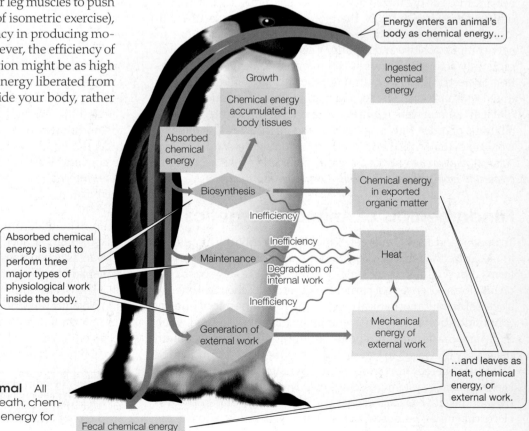

FIGURE 7.2 The uses of energy by an animal All the green boxes refer to chemical energy. At death, chemical energy in body tissues becomes ingested energy for other organisms.

from ingested food is the energy that is distributed to the animal's living tissues and that therefore is made available to the animal's cells for use in physiological work.

An animal uses its absorbed chemical energy to carry out three major types of physiological work. As we discuss these, note that Figure 7.2 diagrams many of the points made.

1. *Biosynthesis.* An animal synthesizes its body constituents, such as its proteins and lipids, by use of absorbed energy. As this process—called **biosynthesis**—takes place, some of the absorbed energy that is used *remains in chemical form* because the products of biosynthesis are organic molecules with significant chemical-energy content. During *growth*, chemical energy accumulates in the body in the form of biosynthesized products that are used to assemble new cells and tissues. Some of the chemical energy accumulated in body tissues through growth (e.g., the chemical energy of fat) may be used by an animal as food energy during times of fasting or starvation; ultimately, all of the chemical energy accumulated in body tissues becomes food for predators or decay organisms when the animal dies. In addition to contributing to tissues during growth, biosynthesis also produces *organic compounds that are exported from the body during an animal's life*, taking chemical-bond energy with them. Gametes, milk, externally secreted mucus, sloughed skin or hair, and shed exoskeletons are just a few of the organic products that animals synthesize and lose to the environment during their lives. Whether biosynthesis yields growth or exported organic products, this form of physiological work typically involves elaborate biochemical pathways requiring many steps. Each step is inevitably inefficient. Biosynthesis therefore produces heat, as well as organic products, because of inefficiency.

2. *Maintenance.* An animal's **maintenance** functions are all the processes that *maintain the integrity* of its body; examples include circulation, respiration, nervous coordination, gut motility, and tissue repair. With only trivial exceptions, *the energy used for maintenance is degraded entirely to heat within the body*. To see why, consider the circulation of the blood as an example. First, the chemical-bond energy of the absorbed food molecules that are used as fuel must be converted into chemical-bond energy of ATP, and energy is lost as heat in the process because of the inefficiency of ATP synthesis discussed previously. Additional energy is then lost as heat when the energy of ATP is used to drive contraction of the heart muscle. Finally, a small fraction of the chemical-bond energy originally obtained from food molecules appears as mechanical energy in the motion of the blood ejected from the heart. Even that mechanical energy is degraded to heat within the body, however, in overcoming the viscous resistances that oppose the motion of the blood through the blood vessels. Mechanical work that takes place *inside an animal's body* is termed **internal work**. Several maintenance functions, such as the circulation of the blood and gut motility, are types of internal work. The energy of internal work is degraded to heat within the body.

FIGURE 7.3 In this type of external work, some of the energy driving locomotion is converted to potential energy of position As this bicyclist goes uphill, although much of his mechanical energy of external work is transformed into heat, a fraction is stored as potential energy because he is propelling the mass of his body and bicycle higher in Earth's gravitational field.

3. *Generation of external work.* Animals perform **external work** when they apply mechanical forces to objects outside their bodies. A mouse running across a field and a bicyclist ascending a hill, for example, are performing external work using their leg muscles. Much of the absorbed chemical energy used to fuel external work is degraded to heat within the body (e.g., in using ATP to drive muscle contraction). However, when external work is performed, some energy leaves the body as mechanical energy transmitted to the environment. The fate of that energy depends on whether it is stored. *Energy of external work is stored if it is converted into increased potential energy of position.* When a bicyclist ascends to the top of a hill, as in **FIGURE 7.3**, part of his energy of external work is stored as increased potential energy of position because his body and bike move to a higher position in Earth's gravitational field. (When the bicyclist later descends, this potential energy of position is converted into mechanical energy [motion downhill] and then to heat.) By contrast, consider horizontal motion. If a mouse scurries a *horizontal* distance across a field—or a person bicycles along a *horizontal* road—no energy is stored in repositioning the body relative to gravity, and the mechanical energy transmitted to the environment is entirely, and quickly, degraded to heat in overcoming resistances to motion.

Reviewing the flow of energy through an animal's body (see Figure 7.2), we see that *all uses of energy by animals generate heat*. All living animals, therefore, produce heat. Because frogs, fish, clams, and other poikilotherms are often cool to the touch, one can get the erroneous impression that they do not produce heat.

Views on Animal Heat Production

Heat is an inevitable by-product of the use of high-grade, chemical-bond energy to create and maintain the vital organization of living organisms. Interestingly, from the time of Aristotle until the nineteenth century, the significance of heat was generally viewed in a completely opposite way. Far from being a by-product, heat was usually seen as a primary source of life, a vital force that endowed many parts of organisms with their living attributes. This "vital heat" was thought to differ from the heat of a fire. It was believed to originate exclusively in the heart, lungs, or blood and to suffuse the rest of the body. When William Harvey first described the circulation of the blood in the early seventeenth century, one of the principal roles attributed to the newfound circulation was transport of "vital heat" from tissues where it was produced to other tissues, which it animated.

The old view of animal heat began to change at about the time of the American Revolution, when Antoine Lavoisier, in France, showed that the ratio of heat production to CO_2 production was about the same for a guinea pig as for burning charcoal. From this and other evidence, Lavoisier and the Englishman Adair Crawford argued that animal respiration is a slow form of combustion, and that animal heat is the same as the heat produced by fire. Still, for several more decades, all animal heat was believed to originate in the lungs, and the lungs were thought to be the exclusive site of O_2 use. Not until 1837 did Heinrich Gustav Magnus show that the blood takes O_2 from the lungs to the rest of the body and returns CO_2. Evidence for the all-important concept that tissues throughout the body make heat came a decade later when Hermann von Helmholtz demonstrated that muscular contraction liberates heat. In 1872, Eduard Pflüger presented evidence that all tissues consume O_2.

The discovery that all tissues use O_2 and produce heat was one of several lines of thought and investigation that came together in the nineteenth century to give birth to our modern understanding of animal energetics. Other important developments were the flowering of the science of thermodynamics (sparked by the Industrial Revolution) and profound changes in the understanding of energy. In the 1840s, Julius Robert von Mayer in Germany and James Joule in England developed the seminal concept that heat, motion, electricity, and so on are all forms of one thing: energy. Mayer, a physician, was probably the first person to conceptualize the true nature of animal energy transformations, as described in this chapter.

However, such animals are cool not because they fail to produce heat, but because their rates of heat production are so low and their bodies are so slightly insulated that they are not warmed by the heat they produce (see page 265). Animal heat, which is universal, has been studied for centuries—far longer than most physiological phenomena—and these studies have led to fundamental discoveries about the nature of life (**BOX 7.1**).

Another point to stress as we conclude our discussion of energy flow through animals is that *the conversion of chemical-bond energy to heat is one-way*: No animal or other living creature is able to convert heat back to chemical-bond energy or to any other form of high-grade energy. Thus *energy is not recycled* within individual animals or within the biosphere as a whole. This principle provides the answer to a key question we asked at the start of this chapter—namely, why do animals need to obtain food energy regularly throughout their lives? When an animal ingests and uses totipotent chemical-bond energy, it converts much of the energy in a one-way, irreversible fashion to heat, which is useless for physiological work. Accordingly, as the animal uses the chemical-bond energy from a meal, it inevitably develops a need to eat again to acquire new chemical-bond energy. The biosphere as a whole requires a continuing input of high-grade photon energy from the sun for much the same reason. The photon energy captured in bonds of organic compounds by photosynthesis is converted progressively to heat by plants as well as animals, meaning that new photon energy is required if organic compounds are to continue to be available. The heat that all organisms collectively make is radiated from Earth into outer space.

SUMMARY
Fundamentals of Animal Energetics

■ Forms of energy vary in their capacity to do physiological work. Chemical-bond energy is totipotent for animals. Electrical and mechanical energy can do certain types of physiological work but are not totipotent. Heat cannot perform physiological work of any kind.

■ Animals use their absorbed chemical energy for three major functions: biosynthesis, maintenance, and generation of external work. Biosynthesis, which preserves some of the absorbed energy in the form of chemical energy, includes both growth and the synthesis of organic materials that are exported from the body during an individual's life.

■ Some energy is degraded to heat (low-grade energy) whenever one high-grade form of energy is transformed to another. Energy transformations are always inefficient.

■ Animals take in chemical-bond energy and put out heat, chemical-bond energy, and external work.

Metabolic Rate: Meaning and Measurement

We have seen that an animal takes in chemical energy in its food, and in the process of living it releases chemical energy, heat, and external work to its environment (see Figure 7.2). The energy that an animal converts to heat and external work is defined by physi-

BOX 7.2 Units of Measure for Energy and Metabolic Rates

The traditional unit of measure for energy is the **calorie** (**cal**), which is the amount of heat needed to raise the temperature of 1 g of water by 1°C. Although the calorie is defined as a quantity of heat, it can be used as a unit of measure for all other forms of energy, because the forms of energy bear strict equivalencies to each other. A **kilocalorie** (**kcal**) is 1000 cal. Sometimes the kilocalorie is written *Calorie*, with a capital *C*, an unfortunate approach that often produces confusion. In the United States, the "calories" listed in formal nutrition labels for foods (see the figure) are kilocalories. To illustrate the sort of confusion that arises, note in the figure that "calorie" is capitalized at the top of the label but not at the bottom, which would be correct only if the meaning at the top were a unit 1000 times greater than at the bottom. Popular books and periodicals sometimes write "calorie" with a lowercase *c* when they mean kilocalorie, a practice that creates three orders

of magnitude of ambiguity. If energy is expressed in calories or kilocalories, then *rates* of energy exchange or transformation—such as metabolic rates—are expressed in calories or kilocalories *per unit of time*.

The fundamental unit of measure for energy in the SI system of units is the **joule** (**J**), named in honor of James Joule. Appendix A discusses the derivation of the joule from the SI base units. A **watt** (**W**), which is equivalent to 1 joule/second (J/s), is the fundamental SI unit for rates of energy exchange or transformation.

One calorie is equivalent to 4.186 J, a relation that permits the interconversion of units in the calorie and SI systems. For example, as you sit quietly reading this page, your body is likely producing heat at a rate near 23 cal/s, equivalent to 23 × 4.186 = 96 J/s, or 96 W. That is, you are producing heat about as rapidly as a 100-W incandescent light bulb.

Nutrition Facts

Serving Size 1 Package (283g)
Servings Per Container 1

Amount Per Serving	
Calories 270 Calories from Fat 20	
	% Daily Value*
Total Fat 2g	**3%**
Cholesterol less than 5mg	**1%**
Sodium 790mg	**33%**
Total Carbohydrate 52g	**17%**
Dietary Fiber 2g	**10%**
Sugars 5g	
Protein 11g	

* Percent Daily Values are based on a 2,000 calorie diet.

A nutrition label for frozen macaroni and cheese sold in the United States In most parts of the world, energy values are given in unambiguous SI units.

---

ologists to be **consumed** because that energy is "spent" or "exhausted": The heat cannot be used at all to do physiological work, and the energy of external work—far from being totipotent—(1) can be used for only a narrowly defined function (the specific external work performed) and (2) soon will itself become heat in most cases.

The *rate* at which an animal consumes energy is its **metabolic rate**. That is, an animal's metabolic rate is the rate at which it converts chemical energy to heat and external work. *Heat is always the dominant component of the metabolic rate.* Accordingly, for simplicity we will sometimes speak of metabolic rates as rates of heat production. Energy is measured in *calories* or *joules*. Metabolic rates, therefore, are expressed in *calories per unit of time* or *watts* (**BOX 7.2**).

At the start of this chapter, we raised the question of how one could predict the food needs of people and sled dogs during a polar expedition. The *metabolic rates* of the people and dogs are the basis for making this prediction. Knowing the average metabolic rates of the people and dogs, one can calculate how much chemical energy they will need per day and then calculate the total food energy they will require for all the days of their trek.

Speaking broadly, metabolic rates are significant for three principal reasons:

1. An animal's metabolic rate is one of the most important determinants of *how much food it needs*. For an adult, food needs depend almost entirely on metabolic rate.

2. Because every energy-using process that takes place in an animal produces heat, an animal's metabolic rate—its *total* rate of heat production—provides a *quantitative measure*

of the total activity of all its physiological mechanisms. An animal's metabolic rate, roughly speaking, represents the animal's *intensity* of living.

3. Ecologically, an animal's metabolic rate measures the *drain the animal places on the physiologically useful energy supplies of its ecosystem*, because the metabolic rate is the pace at which the animal degrades the chemical energy of organic compounds in its ecosystem.

Direct calorimetry: The metabolic rate of an animal can be measured directly

Physiologists sometimes measure metabolic rates directly using a **direct calorimeter**, a device that measures the rate at which heat leaves an animal's body (**BOX 7.3**). Although modern direct calorimeters are technically complex instruments, the basic operation of a direct calorimeter is illustrated nicely by the device that Antoine Lavoisier used in the very first measurements of animal heat production (**FIGURE 7.4**). The heat produced by the test animal melted ice in an ice-filled jacket surrounding the animal, and Lavoisier collected the melt water over measured periods of time. By knowing the amount of heat required to melt each gram of ice, he was able to calculate the animal's rate of heat output, and thus its metabolic rate.

For measures of metabolic rate by direct calorimetry to be fully accurate, external work—not just heat—sometimes must be considered. If the animal under study is at rest, it is not performing external work; a measurement of heat production alone is then fully sufficient for measuring the animal's metabolic rate. If the

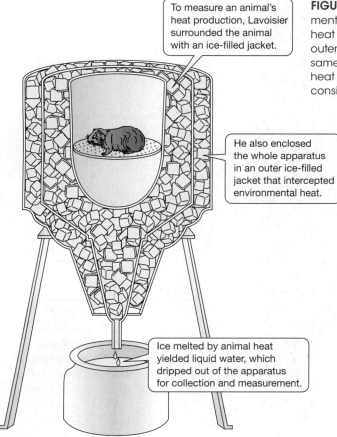

To measure an animal's heat production, Lavoisier surrounded the animal with an ice-filled jacket.

He also enclosed the whole apparatus in an outer ice-filled jacket that intercepted environmental heat.

Ice melted by animal heat yielded liquid water, which dripped out of the apparatus for collection and measurement.

FIGURE 7.4 Lavoisier's direct calorimeter Heat from the general environment must be excluded from measurement of animal heat. In Lavoisier's device, heat entering from the air surrounding the calorimeter was intercepted by an outer ice-filled jacket, which prevented the environmental heat from melting the same ice as the animal heat. Modern direct calorimeters, although they measure heat in a different way and more precisely, still reflect the fundamental design considerations that Lavoisier introduced. (After Lavoisier 1862.)

animal *is* performing external work, the energy of the external work is often rapidly degraded to heat; this is true, for example, if the animal merely moves around horizontally. In such cases, a measure of heat production includes the energy of external work; again, therefore, a measurement of heat production alone is fully sufficient for measuring the metabolic rate. However, if the test animal is performing external work and some of the energy of its external work fails to be converted to heat because it is stored, that energy must be measured independently and added to heat production to quantify the animal's metabolic rate accurately by direct calorimetry.

Indirect calorimetry: Animal metabolic rates are usually measured indirectly

Indirect calorimetry measures an animal's metabolic rate by means other than quantifying heat and work (see Box 7.3). Biologists today usually measure metabolic rates indirectly because the methods of indirect calorimetry are cheaper and easier than those of modern, sophisticated direct calorimetry. Here we consider two methods of

---

BOX 7.3 **Direct Measurement versus Indirect Measurement**

The distinction between *direct* and *indirect* methods of measurement is one of the most fundamental and important concepts in the science of measurement. Direct and indirect methods give results that can be expressed in the same units; a metabolic rate, for example, can be written down in watts whether it is measured directly or indirectly. The distinction between direct and indirect methods is in the property that is *actually measured*.

Rigorously speaking, any physiological trait is *defined* by specific properties. A **direct method of measurement** actually measures the *very properties that are specified by the definition of a trait.* An animal's metabolic rate, for example, is defined to be its rate of production of heat and external work. Thus a direct method of measuring metabolic rate actually measures heat and external work. Direct methods are the *gold standards* of measurement because their results relate unambiguously to the trait being studied.

By contrast, an **indirect method of measurement** actually measures *properties that are different from those specified by the definition of a trait.* The properties that are actually measured by an indirect method serve as "stand-ins" or "proxies" for the properties that define the trait. For instance, some indirect methods for measuring metabolic rate record O_2 consumption. Researchers use such methods because O_2 consumption is relatively easy to measure and often closely reflects an animal's rate of production of heat and external work. After measuring the amount of O_2 consumed per unit of time, researchers often convert the amount of O_2 to calories or joules, units of energy. It is crucial to understand that *the act of converting the measurement results does not change the nature of the measurement method.* A measure of metabolic rate obtained by recording O_2 consumption is indirect regardless of the units in which it is ultimately expressed.

Why is the distinction between direct and indirect methods important? A direct method, if carried out carefully with suitable instruments, *must provide information of the sort desired* because it measures exactly what the definition specifies. An indirect method, however, may introduce errors regardless of how carefully it is carried out, because it measures something different from what is stated by the definition. The accuracy of an indirect method is assessed by determining how well its results agree with a direct method. Whereas indirect methods often have *practical* advantages, they are usually less accurate than direct ones, at least under certain circumstances. Thus the choice of method involves a compromise. In the measurement of metabolic rate, the indirect methods commonly introduce uncertainties of ±1%–5% in the results, but they might reduce costs of time and money by tenfold, making them attractive.

indirect calorimetry: (1) measuring an animal's rate of respiratory gas exchange with its environment (termed *respirometry*) and (2) measuring the chemical-energy content of the organic matter that enters and leaves an animal's body (the *material-balance method*).[4]

INDIRECT CALORIMETRY BASED ON RESPIRATORY GAS EXCHANGE An animal's rate of oxygen (O_2) consumption provides a convenient and readily measued estimate of its metabolic rate. To understand the use of O_2 consumption for this purpose, consider first an oxidation reaction occurring in a test tube. If a mole of glucose ($C_6H_{12}O_6$) is burned completely, chemical stoichiometry dictates that 6 moles of O_2 will be used, and 6 moles of CO_2 will be produced. The reaction will also release heat, known as *heat of combustion*. For glucose, the heat of combustion during complete oxidation is about 2820 kilojoules per mole (kJ/mol), or 673,000 calories per mole (cal/mol). The following equation therefore applies:

$$C_6H_{12}O_6 + 6\,O_2 \rightarrow 6\,CO_2 + 6\,H_2O + 2820\ kJ/mol \quad (7.2)$$

Note that when glucose is oxidized, a fixed proportional relation exists between the amount of heat produced and the amount of O_2 used: 2820 kJ of heat per 6 mol of O_2. Similarly, a fixed proportional relation exists between heat production and CO_2 production: 2820 kJ per 6 mol of CO_2. Knowing these relations, if you oxidize an *unknown* quantity of glucose in a test tube and you measure only the amount of O_2 used, or only the amount of CO_2 produced, you can *calculate* the exact amount of heat produced.

When an animal metabolically oxidizes glucose (or any other chemical substance), if the chemical end products are the same in the animal as in a test tube, then the stoichiometric relations that prevail in the animal are the same as those in the test tube. This important principle, established by Max Rubner and Wilbur Atwater in the 1890s, is true even though the *intermediate* steps of the metabolic oxidation reactions in an animal differ from the intermediate steps of test-tube oxidation. Because of this principle, if an animal oxidizes glucose to CO_2 and H_2O, the stoichiometry in Equation 7.2 applies to the animal. Accordingly, if you measure either the animal's O_2 consumption or its CO_2 production, you can calculate the animal's heat production. This is the rationale for estimating animal metabolic rates from rates of respiratory exchange of O_2 and CO_2.

TABLE 7.1 lists conversion factors for calculating the amount of heat generated when a milliliter of O_2 is consumed or a milliliter of CO_2 is produced.[5] To understand how to use the table, imagine that an animal consumes O_2 at a rate of 10 mL/min, and suppose you know that the animal's cells are oxidizing only glucose (a carbohydrate). The animal's metabolic rate would then be 10 mL O_2/min × 21.1 J/mL O_2 = 211 J/min.

As you can see from Table 7.1, problems can arise in the use of respiratory gas exchange to measure metabolic rates because the correct conversion factor for calculating heat production from O_2 consumption (or from CO_2 production) is not a simple, fixed number.

---

[4] Additional methods of indirect calorimetry that are used for active or free-living animals are discussed in Chapter 9.

[5] According to universal convention, gas volumes in metabolic studies are expressed at Standard Conditions of Temperature and Pressure (STP). That is the way they are expressed in this chapter and throughout this book. For more information, see Appendix C.

TABLE 7.1 Ratios of heat production to O_2 consumption, and heat production to CO_2 production, during the aerobic catabolism of carbohydrates, lipids, and proteins

Values given are for representative mixtures of each of the three foodstuffs. Gas volumes are at Standard Conditions of Temperature and Pressure, STP (see Appendix C).

Foodstuff	Heat produced per unit of O_2 consumed (J/mL O_2)	Heat produced per unit of CO_2 produced (J/mL CO_2)
Carbohydrates	21.1	21.1
Lipids	19.8	27.9
Proteins[a]	18.7	23.3

Source: After Brown and Brengelmann 1965.

[a] For proteins, values depend on the metabolic disposition of nitrogen; the values tabulated here apply to mammals and other animals in which urea is the dominant nitrogenous end product.

Instead, the conversion factor for calculating heat production varies depending on the foodstuffs being oxidized. If animals were to oxidize only glucose, calculating their metabolic rates by measuring their rates of O_2 consumption would be unambiguous, as we have already seen. However, animals oxidize a *variety* of foodstuffs, which yield *different* quantities of heat per unit of volume of O_2 consumed (or CO_2 produced; see Table 7.1). This consideration introduces the possibility of ambiguity or inaccuracy. Returning to our previous example, we saw that if an animal consumes O_2 at a rate of 10 mL O_2/min, its metabolic rate is 211 J/min (10 mL/min × 21.1 J/mL) if its cells are oxidizing carbohydrates. Its metabolic rate is only 198 J/min, however, if its cells are oxidizing lipids (10 mL/min × 19.8 J/mL). Hence an investigator cannot calculate metabolic rate (the rate of heat production) exactly from measurements of O_2 consumption (or CO_2 production) unless the investigator knows the exact mixture of foodstuffs that the study animal's cells are oxidizing.

An animal's recent diet often does not provide accurate insight into the foodstuffs its cells are oxidizing, because animals store and interconvert foodstuffs. We can determine the foodstuffs that *cells* are oxidizing only by looking at indices of *cellular function*. One useful index of this sort is obtained by simultaneously measuring both CO_2 production and O_2 consumption and taking their ratio:

$$\frac{\text{moles of } CO_2 \text{ produced per unit time}}{\text{moles of } O_2 \text{ consumed per unit time}}$$

This ratio is called the **respiratory quotient (RQ)** (or sometimes, depending on details, the *respiratory exchange ratio* [R]).

The value of *RQ* is, in essence, a metabolic signature that reveals the particular sorts of foodstuffs being oxidized by an animal's cells (**TABLE 7.2**). If an animal exhibits a value of *RQ* near 1.0, for example, its cells are likely oxidizing mostly carbohydrates. However, if the animal's *RQ* value is near 0.7, its cells are likely catabolizing mostly lipids. An *RQ* value near 1.0 or 0.7 strongly suggests which conver-

TABLE 7.2 Respiratory quotients (*RQ* values) during the aerobic catabolism of carbohydrates, lipids, and proteins

Foodstuff	Respiratory quotient
Carbohydrates	1.0
Lipids	0.71
Proteins	0.83[a]

Source: After Kleiber 1975.

[a]The value listed for proteins is for animals such as mammals in which urea is the dominant nitrogenous end product. Different values apply to animals that produce ammonia or uric acid as their principal nitrogenous end product.

BOX 7.4 Respirometry

Respirometry is the process of measuring an animal's gas exchange with its environment. The devices used are called *respirometers*. For studies of metabolic rate, the most common type of respirometry is the measurement of an animal's rate of O_2 consumption. In **Box Extension 7.4** you will find illustrations and explanations of the two basic types of respirometry configurations that are used to measure O_2 consumption: (1) *closed* configurations, in which an animal is housed in a fully sealed chamber with a relative fixed volume of nonmoving air and (2) *open* configurations, in which the animal draws its O_2 from an air stream flowing by during measurement.

sion factor from Table 7.1 should be used to calculate an animal's metabolic rate from its O_2 consumption. Specifically, an *RQ* value near 1.0 suggests use of the carbohydrate factor, 21.1 J/mL O_2, whereas an *RQ* value near 0.7 suggests use of the lipid factor, 19.8 J/mL O_2. Unfortunately, *RQ* values that differ substantially from 1.0 or 0.7 are often difficult to interpret. Investigators often sidestep the uncertainty concerning foodstuffs by using a "representative" or "standard" conversion factor of about 20.2 J/mL O_2 (4.8 cal/mL O_2) to calculate an animal's metabolic rate from its O_2 consumption. This conversion factor approximates the heat produced by an animal that is assumed to be catabolizing a representative mixture of carbohydrates, lipids, and proteins. Using the representative conversion factor to calculate metabolic rate does not do away with the uncertainty we have been discussing. To illustrate, suppose that an investigator uses the representative conversion factor of 20.2 J/mL O_2 but the animal is actually oxidizing only carbohydrates. Because the true conversion factor (for carbohydrates) is 21.1 J/mL O_2, the investigator will underestimate the metabolic rate by 4.5% by using the approximate factor of 20.2 J/mL O_2. If the animal is oxidizing only proteins, the true conversion factor (see Table 7.1) is 18.7 J/mL O_2; thus the investigator will overestimate the metabolic rate by 8% by using the approximate factor. The use of the representative conversion factor is a "good news, bad news" situation. The bad news is that the metabolic rate can be misestimated by ±5%–8% if one ignores the foodstuff question and simply multiplies the O_2 consumption by the representative conversion factor to calculate metabolic rate. The good news is that the use of this conversion factor is convenient, and for many purposes an error of ±5%–8% may not be much of a worry.[6]

Among all the methods available to measure metabolic rate, by far the most common approach used today for routine studies is to measure the rate of O_2 consumption—nothing more—and "live with" the relatively small potential errors that are inherent in the method. Metabolic rates, in fact, are often expressed simply as rates of O_2 consumption. The O_2 consumption method has four notable advantages: Two of these, already mentioned, are its technical ease of accomplishment (**BOX 7.4**) and its relatively small inherent uncertainty under many conditions. A third advantage

is that external work does not, in most cases, have to be measured independently when the O_2 consumption method is used. Because the O_2 consumption of an animal is proportional to the ultimate yield of heat from the foodstuffs it aerobically catabolizes, the heat equivalent of any aerobic external work performed by the animal is included in the metabolic rate calculated from its O_2 consumption. The fourth advantage of the O_2 consumption method is that it excludes the metabolism of microbes in the gut (the gut microbiome) because those microbes are anaerobic and do not consume O_2.

Related to the point just made, a limitation of the O_2 consumption method—and a vital point to keep in mind—is that an animal's metabolic rate will not be measured accurately by O_2 consumption if some or all of the animal's tissues are employing anaerobic mechanisms of catabolism to release energy. As discussed in Chapter 8, for example, anaerobic ATP production is used often by skeletal muscles during sudden, highly intense exercise in people and many other sorts of animals. To measure an animal's metabolic rate during anaerobically fueled exercise, one must turn to more elaborate methods than just measuring O_2 consumption.

INDIRECT CALORIMETRY BASED ON MATERIAL BALANCE   In addition to methods based on respiratory gas exchange, the second most commonly used approach for measuring animal metabolic rates today is a type of indirect calorimetry in which one measures the chemical-energy content of *organic materials* entering and leaving an animal's body. This approach, which is widely used in agricultural production research, is described as the study of **material balance**. To apply the method, one measures the chemical-energy content of all the food an animal eats over a period of time, as well as the chemical-energy content of the feces and urine eliminated over the same period.[7] Subtracting the energy content of the excreta from that of the food then gives an estimate of the animal's metabolic rate. The logic of the method is straightforward: Any energy that an animal ingests as chemical energy, but does not void as chemical energy, must be consumed.

Complications arise if the animal under study is increasing or decreasing its biomass. If, for example, an animal is growing and

---

[6] If one measures only CO_2 production and calculates metabolic rate with a representative conversion factor, the potential error is much greater, ±15%–20%, because the relationship between CO_2 production and heat production (see Table 7.1) depends very strongly on foodstuff.

[7] An instrument known as a *bomb calorimeter* is used to measure the energy values of organic materials. It does this by burning them explosively in pressurized, pure O_2 and measuring the heat evolved.

thus increasing the chemical-energy content of its body, some of the chemical energy ingested but not voided is nonetheless not consumed; an estimate of this quantity must enter the calculation of metabolic rate. Another type of complication is that chemical energy may enter or leave an animal's body in ways other than in food, feces, and urine. For instance, an animal could lose chemical energy by shedding feathers or secreting mucus. For the material-balance method to be applied, the chemical-energy content of *all* significant inputs and outputs of organic material must be measured.

To use the material-balance method, measurements of ingestion, egestion, and other relevant processes must extend over a substantial period—typically 24 h or more—so that average, steady-state rates of input and output of chemical energy are quantified. The metabolic rate calculated from the method is the animal's average rate over the entire study period. Thus the material-balance method is suited only for *long-term measurements of average metabolic rates.* To measure minute-by-minute variations in metabolic rate, the methods of choice are those based on respiratory gas exchange or direct calorimetry.

SUMMARY
Metabolic Rate: Meaning and Measurement

■ An animal's metabolic rate is the rate at which it converts chemical energy into heat and external work.

■ Metabolic rate is important because it helps determine the amount of food an animal needs, and therefore the food energy that the animal removes from its ecosystem. An animal's metabolic rate also provides a quantitative measure of the total activity of all its physiological mechanisms.

■ An animal's rate of O_2 consumption is the most common measure of metabolic rate. Metabolic rates can also be measured by direct calorimetry or studies of material balance.

Factors That Affect Metabolic Rates

Now that we have discussed how metabolic rates are defined and measured, we can turn our attention to the experiences of animals and the processes within them that influence their metabolic rates. The two factors that typically exert the greatest effects on an animal's metabolic rate are the intensity of its physical activity (e.g., speed of running) and the temperature of its environment. Other factors that commonly influence animal metabolic rates include the ingestion of food, age, gender, time of day, body size, reproductive condition, hormonal state, psychological stress, and for aquatic animals, the salinity of the ambient water. **TABLE 7.3** provides an overview of many of these factors and identifies where they are discussed in this book.

TABLE 7.3 Some factors that affect the metabolic rates of individual animals

Factor	Response of metabolic rate	Chapter(s) where discussed in this book
Factors that exert particularly large effects		
Physical activity level (e.g., running speed)	↑ with rising activity level	8, 9
Environmental temperature	*Mammals and other homeotherms:*	10
	Lowest in thermoneutral zone	
	↑ below thermoneutral zone	
	↑ above thermoneutral zone	
	Fish and other poikilotherms:	
	↑ with increasing temperature	
	↓ with decreasing temperature	
Factors that exert smaller effects		
Ingestion of a meal (particularly protein-rich)	↑ for several hours to many hours following ingestion	7
Body size	Weight-specific rate ↑ as size ↓	7
Age	Variable; in humans, weight-specific rate ↑ to puberty, then ↓	—
Gender	Variable; in humans, ↑ in male	—
Environmental O_2 level	Often ↓ as O_2 ↓ below a threshold; not affected above threshold	8, 23
Hormonal status	Variable; example: ↑ by excessive thyroid secretions in mammals	16
Time of day	Variable; in humans, ↑ in daytime	15
Salinity of water (aquatic animals)	Variable; in osmoregulating marine crabs, ↑ in dilute water	28

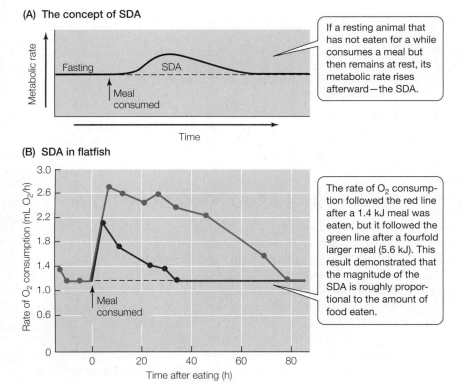

(A) The concept of SDA

If a resting animal that has not eaten for a while consumes a meal but then remains at rest, its metabolic rate rises afterward—the SDA.

(B) SDA in flatfish

The rate of O_2 consumption followed the red line after a 1.4 kJ meal was eaten, but it followed the green line after a fourfold larger meal (5.6 kJ). This result demonstrated that the magnitude of the SDA is roughly proportional to the amount of food eaten.

FIGURE 7.5 Specific dynamic action (SDA) (A) Following a meal, the SDA begins after a delay, which may be up to 1 h. The solid line shows the animal's actual metabolic rate. The dashed line depicts what the rate would have been, had the meal not been eaten. The area shaded blue is the *magnitude* of the SDA. The timing of the process varies enormously; the SDA might be over in a few hours in a mouse, in 12 h in a cow, and in 1–3 days in a fish. *Fasting* in such studies means the animal has not eaten for long enough that the SDA of its last meal is over. (B) Actual data for predatory flatfish (*Pleuronectes*) fed two different-sized meals of fish meat. The apparent absence of a delay in the beginning of the SDA is an artifact of the sampling schedule: Data were not gathered in the immediate aftermath of feeding, when the delay would have been evident. (B after Jobling 1993.)

Ingestion of food causes metabolic rate to rise

Among the factors that affect metabolic rate, the ingestion of food deserves some extended consideration—even though it is not the most influential factor quantitatively—because it must be taken into account in almost all metabolic studies. Under many circumstances, if an animal has not eaten for a while and then consumes food, its metabolic rate temporarily increases following the meal *even though all other conditions are kept constant*. This increase in metabolic rate caused by food ingestion is known as **specific dynamic action (SDA)**, the **calorigenic effect of ingested food**, or the **heat increment of feeding** (**FIGURE 7.5**). Although we may often not notice this process in our day-to-day lives, it is very apparent at certain times. Think back, for example, to a festive holiday dinner when everyone ate lots of high-protein food such as turkey or other meat. After such a meal, people may feel so warm that they remove sweaters and loosen neckties or scarves. The reason for the sense of excessive warmth is the SDA of the ingested protein. The occurrence of SDA means that a certain portion of the energy available from a meal is degraded to heat in processing the meal; only the remaining portion of the energy is available for subsequent physiological uses.

The *magnitude* of the SDA following a meal is the *total excess* metabolic heat production induced by the meal, integrated from the time metabolism first rises to the time that it falls back to the background level. Thus the blue area in Figure 7.5A, showing the integrated difference between the actual metabolic rate after a meal and the metabolic rate that would have prevailed without eating, represents the magnitude of the SDA. The magnitude of the SDA that occurs after a meal of a particular type of food tends to be roughly proportional to the amount of food eaten, for given animals under given conditions (see Figure 7.5B): Doubling the amount of food eaten tends approximately to double the SDA. Protein foods exhibit

much higher SDAs, in proportion to the amount eaten, than do lipids or carbohydrates. Traditionally the SDA of a protein meal has been considered to be equivalent to 25%–30% of the total energy value of the meal. Recent research indicates, however, that the percentage (while virtually always high) can vary considerably with prevailing conditions.

The mechanism of SDA remains uncertain. Although digestive processes make a contribution, strong evidence exists that the SDA in most animals arises principally *after the absorption* of digestive products from the gastrointestinal tract, as a consequence of cellular processing of the absorbed organic compounds. An important cause of the SDA associated with protein meals, for example, is believed to be the energy expenditure required to synthesize nitrogenous waste products (e.g., urea in mammals) to dispose of nitrogen from excess amino acids.

The SDA is a relatively short-term phenomenon, but sometimes an animal's diet induces a *semipermanent*, or *chronic*, change in its metabolic rate. Experiments have revealed that if laboratory rats are enticed to eat unusually large amounts of food day after day (as by adding sweets to their food), some individuals do not fatten, because their metabolic rates chronically rise, turning the excess food energy into heat. The long-term increase in metabolic rate induced by persistent overeating is dubbed **diet-induced thermogenesis (DIT)**. Research on DIT has been intense ever since its discovery, because DIT is an anti-obesity process of potential human importance. The relation between DIT and SDA is confused at present. We will say more about DIT in Chapters 8 and 10.

Basal Metabolic Rate and Standard Metabolic Rate

Physiologists, ecologists, and other biologists often wish to compare metabolic rates. A physician, for example, might want to know how the metabolic rate of a particular patient compares with the average metabolic rate of all people of similar age, because some diseases are distinguished by abnormal rates of energy consumption. An ecologist might want to compare the metabolic rates of two species in an attempt to learn which species is more likely to place high demands for food on an ecosystem.

In making comparisons, it is often important to standardize factors that could confound results. For example, physicians typically standardize food ingestion during diagnostic metabolic studies by having patients fast for at least 12 h prior to measurements of their metabolic rates; otherwise, some patients would have their measured metabolic rates elevated by SDA, whereas others would not, creating confounding variation. Similarly, biologists who want to know if species have inherently different metabolic rates typically standardize physical activity, so that results are not confounded by having the individuals of one species walking around while those of the other rest during measurement. Several standardized measures of metabolic rate have been defined to facilitate valid comparisons. The two most commonly used standardized measures are the *basal metabolic rate* and the *standard metabolic rate*.

The **basal metabolic rate** (**BMR**) is a standardized measure of metabolic rate that applies to *homeotherms*, animals that physiologically regulate their body temperatures, such as mammals and birds. For each homeothermic species, there is a range of environmental temperatures within which the metabolic rate is minimal. This range, called the *thermoneutral zone*, is illustrated in Figure 10.28. The basal metabolic rate of a homeotherm is the animal's metabolic rate while it is (1) in its thermoneutral zone, (2) fasting, and (3) resting. The term *fasting* here has a different meaning than in some other contexts. In studies of metabolic rate, **fasting** (or **postabsorptive**) means that a subject's last meal took place sufficiently long ago for the SDA of the meal to be over.

The concept of **standard metabolic rate** (**SMR**) applies to *poikilotherms* (*ectotherms*), animals that allow their body temperatures to fluctuate freely with variations in environmental temperature, such as amphibians, molluscs, and most fish. The standard metabolic rate is the metabolic rate of a poikilothermic animal while it is (1) fasting and (2) resting. Again, *fasting* signifies that the SDA of the last meal is over. An animal's standard metabolic rate is specific for its prevailing body temperature; thus for a given animal there are as many SMRs as there are different body temperatures.

Besides specifying a fasting state, both of the standardized measures of metabolic rate mentioned here call for subjects to be resting. The term *resting* can have somewhat different meanings in different studies because inducing animals to rest is often not simple. Specific levels of rest are formally recognized in some subdisciplines of physiology. Fish physiologists, for instance, often use the term **routine metabolic rate** to refer to metabolic rates of reasonably quiet fish exhibiting only small, spontaneous movements; when they speak of *standard metabolic rate*, they refer to fish that have been coaxed to truly minimal levels of activity. In human medicine, *resting* means lying down but awake.

Metabolic Scaling: The Relation between Metabolic Rate and Body Size

How does metabolic rate vary with body size within a set of phylogenetically related species? This simple question turns out to have a profoundly important, intricate answer. The study of the relation between metabolic rate and body size is known as the study of **metabolic scaling** or of the **metabolism–size relation**.

A comparison between meadow voles and white rhinos provides a revealing starting point for understanding metabolic scaling (**FIG-**

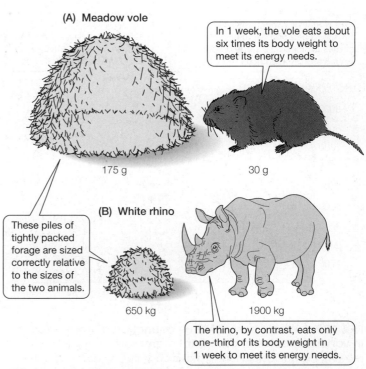

FIGURE 7.6 The effect of body size on weekly food requirements Both species—(A) the 30-g meadow vole (*Microtus pennsylvanicus*) and (B) the 1900-kg white rhino (*Ceratotherium simum*)—are grazers. (Calculated from Golley 1960 and Owen-Smith 1988, assuming 70% moisture content in the forage.)

URE 7.6). Both of these species are mammals, and both eat similar foods, being "pure grazers" that eat little else besides grassland plants. They are very different in body size, however. An instructive way to gain insight into the effect of their different body sizes on their metabolic rates is to compare how much food they must eat to meet their metabolic needs. If we pile up all the grass that a vole and a rhino must eat in a week under similar measurement conditions, we find, not surprisingly, that the rhino requires more food than the vole. However, a week's pile of food for the vole is far larger than the vole itself, whereas the pile for the rhino is much smaller than the rhino. This disparity reveals that *the energy needs of the species are not proportional to their respective body sizes*.

Resting metabolic rate is an allometric function of body weight in related species

To fully understand the relation between metabolic rate and body size, it is important to compare large numbers of related species, not just two. The BMRs of more than 600 species of placental mammals have been measured. When all of these BMRs are plotted against the body weights of the species, statistics can be used to draw the best-fitted line through the data.[8] From this sort of

---

[8] As discussed in Chapter 1 (see page 19), the most common statistical approach for analyzing these sorts of data is to apply ordinary least squares regression (see Appendix D) to all the available data. A case can be made, however, that methods based on phylogenetically independent contrasts (see Appendix G) should often be used. In the literature today, investigators often analyze data by both approaches, recognizing that a consensus does not exist. The fitted equations in this book are from ordinary least squares regression.

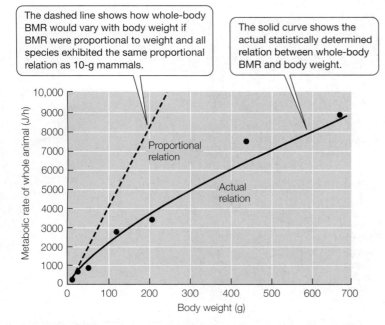

The dashed line shows how whole-body BMR would vary with body weight if BMR were proportional to weight and all species exhibited the same proportional relation as 10-g mammals.

The solid curve shows the actual statistically determined relation between whole-body BMR and body weight.

FIGURE 7.7 Whole-body BMR as a function of body weight in various species of placental mammals The solid curve—showing the actual relation—is statistically fitted to data for all sizes of mammals, although this plot includes body weights up to only 700 g. The points are data for seven North American species (see Figure 7.8 for identifications), illustrating that although the statistical line runs through the data, individual species do not necessarily fall right on the line. (After Hayssen and Lacy 1985.)

analysis, physiologists have discovered that the pattern we have observed in comparing voles and rhinos is in fact a general pattern that applies across the full range of mammalian body sizes. Let's first discuss *whole-body BMR*, meaning the basal metabolic rate of the whole animal. Although the whole-body BMR of species of placental mammals increases with body weight, it does not increase in proportion to weight. Instead, the whole-body BMR increases less than proportionally with body weight (**FIGURE 7.7**). Consider, for example, that an average 10-g species of placental mammal exhibits a whole-body BMR of about 400 J/h. If BMR increased in proportion to body size, a 100-g species would have a whole-body BMR of 4000 J/h. In actuality, the average whole-body BMR of a 100-g species is much less, about 2200 J/h. This quantitative trend persists throughout the entire range of mammalian weights. For instance, the

average whole-body BMR of a 400-g species of placental mammal is only about 2.7 times higher than that of a 100-g species, not 4 times higher.

An alternative way to examine the relation between metabolic rate and body weight is to calculate the metabolic rate *per unit of body weight*, termed the **weight-specific metabolic rate**, and plot it as a function of body weight. **FIGURE 7.8** presents the data points and curve from Figure 7.7 in this new way. This representation illustrates that the weight-specific BMR of mammals decreases as weight increases (Figure 7.6 also shows this). Under basal conditions, a 670-g desert cottontail rabbit produces only about 40% as much metabolic heat per gram as a 21-g white-footed mouse. These differences become even more dramatic if we examine mammals still larger than those represented in Figure 7.8. A 70-kg human produces about 10% as much heat per gram as the mouse, and a 4000-kg elephant produces about 5% as much. The rate of basal energy expenditure of a gram of mammalian tissue is far lower if it is a gram of elephant than a gram of mouse!

Suppose that instead of focusing on mammals, we expand our investigation of metabolic scaling by looking at many species of fish of different body weights—or many species of crustaceans of different sizes. Do poikilothermic animals (ones having variable body temperatures) exhibit the same sorts of relations between metabolic rate and body size as mammals do? Yes, they do. Within a set of phylogenetically related poikilothermic species of different body sizes, the SMR at a particular body temperature usually varies with body size in the same basic pattern as seen in mammals. The whole-body SMR increases with body weight, but it increases less than proportionally. Thus the weight-specific SMR decreases as body weight increases.

Metabolic scaling is investigated in detail by use of equations. With total (whole-body) metabolic rate symbolized M and body weight symbolized W, the equation that has been applied for more than half a century is

$$M = aW^b \tag{7.3}$$

FIGURE 7.8 Weight-specific BMR as a function of body weight in various species of placental mammals Although the *x* axis (body weight) and animals are the same as in Figure 7.7, here the BMRs (*y* axis) are expressed per gram of body weight. (After Hayssen and Lacy 1985.)

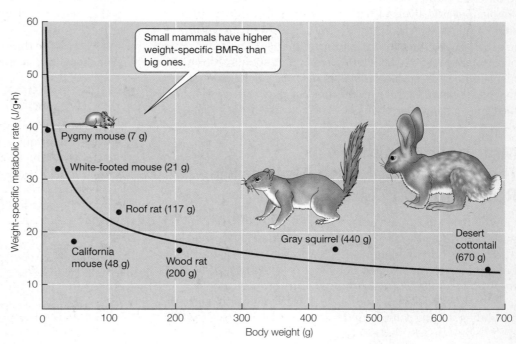

Small mammals have higher weight-specific BMRs than big ones.

Pygmy mouse (7 g)

White-footed mouse (21 g)

Roof rat (117 g)

California mouse (48 g)

Wood rat (200 g)

Gray squirrel (440 g)

Desert cottontail (670 g)

where *a* and *b* are constants. This type of equation describes the relation between *M* and *W* for species of various body sizes within a defined set of related species, such as the bony fish or, alternatively, the placental mammals. The constants in the equation are determined—that is, fitted—statistically. To determine *a* and *b* for placental mammals, for example, researchers begin by tabulating *M* and *W* for as many species as possible. The raw information used for the process, in other words, is a series of points on *M*, *W* coordinates, such as those plotted on the graph in Figure 7.7. The researchers then use a statistical algorithm to fit an equation of the form $M = aW^b$ as closely as possible to the data points. Such a statistical procedure calculates the values of *a* and *b* that make the equation match the data points as closely as possible.[9]

Note that the values of *a* and *b* depend on the *particular data used* as well as the animal group being studied. This is important. If two researchers independently investigate a *single* group of animals but use data for different sets of species, they will obtain somewhat different values of *a* and *b*.

In the past 15 years, several teams of researchers have presented strong evidence that Equation 7.3—despite its long history—is sometimes too simple. Sometimes a more complex equation, with more terms and fitted constants, is required to describe a metabolism–weight relation in the full detail required for advanced scaling research. Nonetheless, for an introduction to animal physiology, Equation 7.3 is appropriate.[10]

If *b* were to equal 1.0, Equation 7.3 would become $M = aW$, a proportional relation. However, in studies of metabolic scaling, *b* is almost always less than 1.0, meaning the relation is not proportional. When *b* is not 1.0, Equation 7.3 is known as an **allometric equation**.[11] Often, therefore, the metabolic rates of animals are said to be *allometric functions of body size*.

Let's now continue with our discussion of the ways in which BMR and SMR vary with body weight. Because both BMR and SMR are measured when animals are at rest, we can use *resting metabolic rate* to refer to them both, rather than saying *BMR and SMR* every time we wish to refer to them. When biologists have analyzed the allometric relation between whole-body resting metabolic rate (*M*) and body weight (*W*), they have discovered that the exponent *b* in Equation 7.3 exhibits an impressive consistency in its value from one phylogenetic group to another.[12] *The value of* b *for the resting metabolic rates of diverse groups of animals tends to be about 0.7.* This is true of mammals, fish, decapod crustaceans, snails, echinoderms, and most other animal groups. Most commonly, *b* is 0.6–0.8; it is nearly always between 0.5 and 0.9. Universal biological principles seem to be at work in determining *b*.

Unlike the value of *b*, the value of *a* in studies of resting metabolic rate is not at all consistent from one phylogenetic group to another. From Equation 7.3, you can see that $M = a$ when $W = 1$.

[9] The considerations discussed in footnote 8 are again relevant.

[10] Equation 7.3 is useful because it provides a good approximation to the metabolism–weight relation in great numbers of animal groups, and as a consequence of its long history, thousands of published scaling relations are expressed in this form. Moreover, it has obvious conceptual meaning (as discussed in the next paragraph), whereas the more complex equations often require specialized knowledge to be conceptually meaningful.

[11] An allometric equation is defined to be an equation of form $Y = aX^b$ with $b \neq 1$. Allometric equations are discussed in Appendix F.

[12] The exponent *b* is dimensionless and thus has no units of measure.

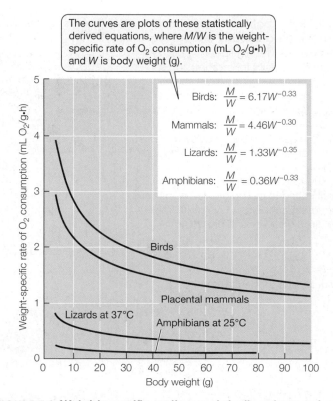

The curves are plots of these statistically derived equations, where *M/W* is the weight-specific rate of O_2 consumption (mL O_2/g·h) and *W* is body weight (g).

Birds: $\dfrac{M}{W} = 6.17W^{-0.33}$

Mammals: $\dfrac{M}{W} = 4.46W^{-0.30}$

Lizards: $\dfrac{M}{W} = 1.33W^{-0.35}$

Amphibians: $\dfrac{M}{W} = 0.36W^{-0.33}$

FIGURE 7.9 Weight-specific resting metabolic rate as a function of body weight in four groups of vertebrates The lines for birds and mammals show BMRs. The line for lizards shows the SMR when the lizards are at the same body temperature as placental mammals, 37°C. The line for amphibians shows the SMR in temperate-zone anurans and salamanders at a body temperature of 25°C. (Sources for equations: birds, McKechnie and Wolf 2004; mammals, Hayssen and Lacy 1985; lizards, Templeton 1970; amphibians, Whitford 1973.)

Thus *a* is the metabolic rate of a 1-g animal (real or theoretical) in the phylogenetic group under consideration. Some phylogenetic groups (e.g., mammals) have metabolic rates that are intrinsically far higher than the metabolic rates of other groups (e.g., fish) and thus also have much higher values of *a*. A high value of *a* means high metabolic intensity.

What is the mathematical relation between *weight-specific* resting metabolic rate and body weight? This relation is easily derived if both sides of Equation 7.3 are divided by *W*, yielding

$$M/W = aW^{(b-1)} \tag{7.4}$$

The expression *M/W* is the weight-specific BMR or SMR, and you can see that it is an allometric function of *W*. The value of *a* is the same as in Equation 7.3, but the exponent in Equation 7.4 is $(b-1)$. Because *b* is usually about 0.7, the exponent here is usually about −0.3. The negative value of $(b-1)$ in Equation 7.4 signifies what we have already said; namely, weight-specific resting metabolic rate *decreases* with increasing body weight (see Figure 7.8). **FIGURE 7.9** presents four examples of Equation 7.4, fitted to four different groups of vertebrates. Note the similarity of the exponents but the differences in *a*, signifying different metabolic intensities in the four types of animals.

A useful property of Equations 7.3 and 7.4 is that they are *linear* equations when plotted on log–log coordinates (see Appendix F).

FIGURE 7.10 Metabolic rate and body weight are related linearly on log–log coordinates (A) Weight-specific BMR as a function of body weight for mammalian species that eat primarily vertebrate flesh, plotted on log–log coordinates. The points represent individual species; the line is statistically fitted to them. (B) A log–log plot of weight-specific metabolic rate as a function of body weight in a common Pacific shore crab (*Pachygrapsus crassipes*) at a body temperature of 16°C. Each point represents a particular individual. The line is statistically fitted to the points. See Appendix E for information on the axis layouts. (A after McNab 1986; B after Roberts 1957.)

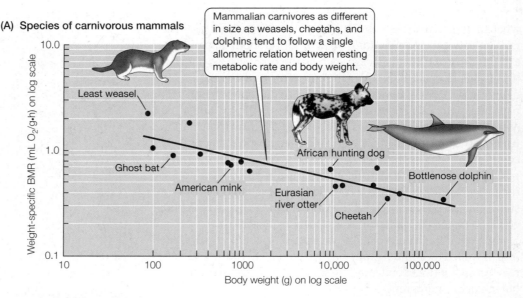

(A) Species of carnivorous mammals

Mammalian carnivores as different in size as weasels, cheetahs, and dolphins tend to follow a single allometric relation between resting metabolic rate and body weight.

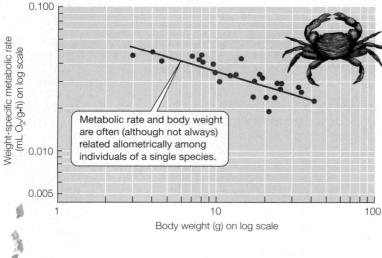

(B) Individuals of a species of crab

Metabolic rate and body weight are often (although not always) related allometrically among individuals of a single species.

For example, if we take the logarithm of both sides of Equation 7.3, we get

$$\log M = \log a + b \log W \qquad (7.5)$$

Note that the dependent variable in Equation 7.5 ($\log M$) equals a constant ($\log a$) plus the independent variable ($\log W$) multiplied by another constant (b). This means that Equation 7.5 describes a straight line. Accordingly, a plot of $\log M$ as a function of $\log W$ is linear, and similarly, a plot of $\log M/W$ against $\log W$ is linear. Data relating metabolic rate to weight are nearly always graphed on a log–log plot, as exemplified by **FIGURE 7.10**. An advantage of logarithmic axes is that they can accommodate very wide ranges of values. The log–log format of Figure 7.10A, for instance, permits species of mammals ranging in weight from 80 g to almost 200,000 g to be analyzed together on one graph.

The metabolic rate of active animals is often also an allometric function of body weight

Sustained vigorous physical activity causes an animal's rate of aerobic metabolism to reach a maximum. How does the maximum rate of aerobic metabolism, which we will symbolize M_{max}, compare with the resting metabolic rate, and how does it vary with body weight within a set of phylogenetically related species? A useful rule of thumb in vertebrates and some groups of invertebrates is that the

exercise-induced maximum aerobic metabolic rate tends to be about ten times higher than the resting metabolic rate (BMR or SMR). Usually the whole-body M_{max} is an allometric function of body weight: $M_{max} = a'W^{b'}$, where a' and b' are the constants that apply in the case of the maximum metabolic rate. The value of a' tends to be roughly ten times the value of a, which applies to the resting metabolic rate (corresponding to the tenfold difference we noted). The exponent b' for M_{max} is usually similar to b, the exponent for resting metabolic rate. However, b' and b clearly differ in mammals, flying insects, and some other groups that can be analyzed in close detail using available data; in placental mammals, for example, whereas the exponent for resting metabolic rate is about 0.7, that for M_{max} is about 0.85. This difference is noteworthy because it has theoretical significance, as we will soon discuss.

What about the *average* metabolic rates of animals *living in nature*—a state in which they are active at some times but resting at other times? Recognizing that both M_{max} and resting metabolic rate are typically allometric functions of body weight, we might expect that *average daily* metabolic rate is also an allometric function of weight within sets of phylogenetically related animals. It is.[13]

The metabolism–size relation has important physiological and ecological implications

The metabolism–weight relation pervades almost every aspect of an animal's physiology. This is illustrated by the breathing and circulatory systems. Because these systems are responsible for delivering the O_2 consumed by an animal's tissues, you might predict that key features of the physiology of breathing and circulation are also allometrically related to body size within a set of related species, and this is the case.

The resting heart rate in placental mammals, to cite one example, varies with body weight in almost exactly the same functional relation as weight-specific BMR: Small species have far higher heart rates than do large ones (**TABLE 7.4**). This pattern makes sense in view of the fact that, statistically, all mammals have about the same size of heart in relation to their body size (**FIGURE 7.11**; see Table

---

[13] Chapter 9 discusses the methods that are used to measure the average daily metabolic rates of free-living animals, and it discusses the results in greater detail.

TABLE 7.4 Resting heart rate, and heart size relative to body weight, in seven species of mammals
Species are listed in order of decreasing individual size.

Species and average body weight	Resting heart rate (beats/min)[a]	Heart weight per unit of body weight (g/kg)[b]
African elephant (4100 kg)	40	5.5
Horse (420 kg)	47	7.5
Human (69 kg)	70	5.2
Domestic dog (19 kg)	105	9.2
Domestic cat (3 kg)	179	4.1
Roof rat (0.34 kg)	340	2.9
Lab mouse (0.03 kg)	580	4.0

Source: After Seymour and Blaylock 2000.

[a]According to the source of these data, the statistical relation between resting heart rate (*RHR*) and body weight (*W*) in mammals is RHR (beats/min) $= 227W^{-0.23}$, where *W* is in kilograms.

[b]Although heart weight per unit of body weight varies, it shows little or no consistent relation to body size.

FIGURE 7.11 Hearts of a horse, cat, and mouse: Heart size in mammals is roughly proportional to body size As a corollary, the relatively high weight-specific O_2 demands of small species are *not* met by pumping more blood, relative to body size, *per heartbeat*. Instead, the number of heartbeats per minute is far greater in small species than in large ones. (From Noujaim et al. 2004; photograph courtesy of Sami Noujaim and José Jalife.)

7.4). Small mammals require more O_2 per gram of body weight than large ones, but their hearts are no bigger relative to body size than are the hearts of large mammals. Thus the hearts of small mammals must beat faster than those of large mammals for O_2 to be delivered at a greater rate per unit of body weight. How is the function of the mammalian heart scaled to permit different rates of contraction and recovery in small and large species? Investigators are just now exploring the answers to this fascinating question (**BOX 7.5**). Similar to the case with the heart, various species of mammals also have lungs of about the same size in relation to

body size. Thus by much the same logic as we have just outlined for the heart, small mammals must take breaths at higher rates than large mammals. Whereas humans, for example, breathe about 12 times per minute when resting, mice breathe about 100 times per minute. The allometric relation between metabolic rate and body weight (a relation seen during both rest and activity) suggests that different-sized, but related, species are likely also to differ in how well their cells and tissues are endowed with mitochondria and

BOX 7.5 **Scaling of Heart Function**

Each contraction of the heart muscle entails a complex sequence of coordinated electrical and mechanical events (see Chapter 25). In outline, at the time of each heartbeat, the heart's pacemaker initiates a wave of electrical activity that sweeps through the heart muscle. This electrical wave first causes the atrial chambers to contract, then pauses as it passes from the atrial to the ventricular chambers, and finally causes the ventricular chambers to contract. For the hearts of two species to contract at different rates, adjustments are required in all of these individual processes. The sweep of electrical activity in two hearts of very different size is shown in the figure and discussed in **Box Extension 7.5**.

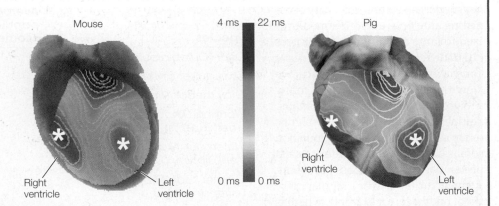

Views of the ventricles of a mouse heart and pig heart showing the sweep of activation of the superficial ventricular muscle layer during a single heartbeat The wave of electrical activity responsible for activating contraction first appears in the superficial muscle layer at the points marked by the white asterisks. It then sweeps through the superficial muscle layer in the color-coded order shown in the center, taking a total of 4 ms in the mouse heart but 22 ms in the pig heart. (From Noujaim et al. 2007; images courtesy of Sami Noujaim and José Jalife.)

FIGURE 7.12 Herbivores of different body sizes co-existing on an African grassland When extensive landscapes are analyzed statistically, species are found to vary in population biomass in a way that depends in part on body size and metabolic allometry.

other components of the aerobic catabolic apparatus. Cellular properties indeed often vary allometrically with body size. In animals as diverse as mammals and fish, for example, the skeletal muscles of small species have more mitochondria per unit of tissue than those of related, large species. The density of mitochondria varies allometrically with body weight, paralleling the relation between weight-specific metabolic rate and weight.

Numerous ecological and practical implications also arise from the allometric relation between metabolic rate and body weight. We have already seen in our initial comparison of voles and rhinos, for example, that within a set of phylogenetically related animals, small-bodied species typically require food at a greater rate per unit of body weight than large-bodied species (see Figure 7.6). Ecologically, the uninitiated might expect that—in terms of basal food requirements—3500 mice, each weighing 20 g (total weight 70,000 g), would place about the same demand on a woodland ecosystem as a single 70,000-g deer would. Because the weight-specific BMR of a 20-g mouse is about eight times greater than that of a deer, however, the total basal food requirement of only about 440 mice is equivalent to that of a deer.

When entire ecosystems are analyzed, ecologists sometimes observe that the allometric relation between metabolic rate and body weight has a significant structuring effect. Consider, for example, the savannas and woodlands of eastern and southern Africa. These ecosystems are among the marvels of life on Earth, in part because they support an extreme diversity of co-existing antelopes and other medium-sized to large mammalian herbivores (**FIGURE 7.12**). From aerial surveys of major national parks in Africa, we know the average numbers of many herbivore species per square kilometer. Multiplying numbers by body weights, we can calculate the average population biomass per square kilometer of each species. Population biomass per square kilometer turns out to be a regular function of body size; for example, whereas all the warthogs (a relatively small species) living per square kilometer together weigh about 95 kg, the zebras in a square kilometer collectively weigh 460 kg, and the elephants weigh 1250 kg (**TABLE 7.5**). Metabolic allometry,

while emphatically not the only factor at work, helps explain this trend, because each kilogram of a large-bodied species requires less food than each kilogram of a small-bodied species.

The allometric relation between metabolic rate and body size also means that related small- and large-bodied species often process foreign chemicals differently. Because small-bodied species eat food and breathe air at greater rates per unit of body weight than their larger counterparts, they tend to receive greater weight-specific doses of food-borne and airborne toxins such as pesticides; this consideration, in itself, creates a tendency for toxins to accumulate more readily to high concentrations in the tissues of the small-bodied

TABLE 7.5 Biomasses of populations of selected herbivores living in mixed communities in African national parks

Species are listed in order of increasing individual size. These species were chosen for listing because they are statistically about average in population biomass for their body sizes.

Species	Average biomass of entire population per square kilometer (kg/km^2)	Average individual body weight (kg)
Oribi (*Ourebia ourebi*)	44	13
Gray duiker (*Sylvicapra grimmia*)	62	16
Gray rhebok (*Pelea capreolus*)	105	25
Warthog (*Phacochoerus aethiopicus*)	95	69
Waterbuck (*Kobus ellipsiprymnus*)	155	210
Greater kudu (*Tragelaphus strepsiceros*)	200	215
Plains zebra (*Equus burchelli*)	460	275
White rhino (*Ceratotherium simum*)	2400	1900
African elephant (*Loxodonta africana*)	1250	3900

Source: After Owen-Smith 1988.

species. However—correlated with (not necessarily caused by) their high weight-specific metabolic rates—small-bodied animals tend to catabolize or excrete some substances faster per unit of weight than do related larger animals; a practical consequence is that small-bodied species may require relatively high doses of a veterinary drug per unit of weight to achieve and sustain the drug's intended effect. Overall, the dynamics of accumulation and dissipation of foreign chemicals often differ between related large- and small-bodied species.

The explanation for allometric metabolism–size relations remains unknown

The fact that b, the allometric exponent, tends to be near 0.7 in widely diverse groups of animals is profoundly intriguing. For a century, some of the greatest minds in biology have grappled with the questions of *why* metabolic rate and body size are related allometrically and *why* the allometric exponent is sometimes impressively consistent. Great minds have been drawn to these questions because of a conviction that the allometries are manifestations of fundamental organizing principles of life. As yet, however, no consensus exists about how to explain the allometries.

A century ago, the problem seemed solved! Physiologists thought then that they understood the reasons for not only the allometric metabolism–size relation, but also the particular value of b. The theory offered at that time has been reinvented by every generation of biologists because it seems so "obvious." Thus an understanding of the theory's flaws remains important even today. At the time the theory first appeared in the early twentieth century, all the data on metabolism–size relations were data gathered on mammals, and mammals therefore dominated thinking about the subject. During that period, Max Rubner articulated an explanatory theory that is still known as *Rubner's surface "law."*

Euclidean geometry provides the starting point for understanding this "law" that is not a law. Recall from your study of geometry that the surface area s of a sphere is proportional to the square of r, the sphere's radius: $s \propto r^2$. The volume v of a sphere, however, is proportional to the cube of the radius: $v \propto r^3$. From the rules of exponents, we can write $r^2 = (r^3)^{2/3}$. Thus $s \propto (r^3)^{2/3}$; and substituting v for r^3, we get

$$s \propto v^{2/3} \tag{7.6}$$

In words, as spheres increase in size, their surface area increases only as the two-thirds power of their volume. Big spheres, therefore, have less surface area per unit of volume (or of weight) than little spheres. Similar relationships hold true for all sets of geometrically similar objects. Whether you consider cubes, cylinders, hearts, or whole animals, as the objects within a geometrically similar set become larger, the area of the outside surface is expected to increase approximately as the two-thirds power of volume, and the ratio of outside-surface area to volume declines.

Rubner's surface "law" stated that the BMR of a mammal is proportional to its body-surface area[14] and that the allometric

relation between BMR and body weight is a corollary of this proportionality. Rubner's explanation of the allometric relation rested on four logical steps:

1. Placental mammals maintain high, relatively constant body temperatures (near 37°C) and thus tend to lose heat to the environment when studied at thermoneutral environmental temperatures.

2. Because heat is lost across an animal's outer body surfaces, the rate of heat loss from a mammal is approximately proportional to the animal's body-surface area.[15]

3. Small mammals have more surface area per unit of weight than do **large** mammals and thus lose heat more rapidly per unit of weight.

4. Heat lost must be replaced metabolically for a mammal to stay warm. Accordingly, small mammals must produce heat at a greater rate per unit of weight than large ones.

The surface "law" as just outlined can hardly be faulted as a thought exercise. Why, then, do most physiologists today believe that it is not the correct mechanistic explanation of the allometric relation between BMR and body weight? The answer is that data contradict the "law" in two respects. First, although the surface "law" predicts an exponent b equal to about 0.67 (2/3), most physiologists who have estimated values of b for mammals have concluded that b is statistically higher than 0.67 to a significant degree. Second, by now we realize, as emphasized already, that poikilothermic animals—such as fish, frogs, and crabs—display allometric metabolism–size relations (see Figures 7.9 and 7.10). Rubner's "law" cannot possibly explain these relations in poikilotherms because the reasoning behind the "law" applies only to animals that warm their bodies to elevated, regulated temperatures using metabolic heat production. The consistency of the metabolism–size relation across many animal groups suggests a single mechanistic explanation. Because Rubner's "law" is irrelevant for most types of animals (most are poikilotherms), it cannot be that explanation.

Since the time in the mid-twentieth century when the surface "law" started to be rejected by most physiologists, several alternative hypotheses have been put forward to explain allometric metabolism–size relations. Physiologists, however, have not reached a consensus in supporting any of the hypotheses. Until recently, debate tended to center on whether the true value of b is 2/3 or 3/4. Physiologists assumed that a single universal exponent existed and that, if it could be nailed down, the underlying mechanistic basis for allometry would be revealed. Why did difficulty exist in nailing down the exponent? For a long time, the amount of relevant data was not great, and everyone could assume, therefore, that when abundant data became available, the correct value for b would become obvious. What has actually happened is quite different.

Now that great quantities of carefully scrutinized data are available, many specialists have concluded that in fact there is no universal exponent. Recent research reports have concluded that although the exponent b generally tends to be *similar* in many animal groups, it is not *identical*, and the *exponent* b *is not a constant*. Among

---

[14] A modern holdover of the early emphasis on body-surface area is that surface areas are employed to calculate certain sorts of critical variables in the contemporary practice of medicine. In breast cancer chemotherapy, for example, the doses of chemotherapeutic agents administered to a woman are still sometimes calculated from her body-surface area.

[15] During the era when Rubner's surface law was accepted, this concept seemed too obvious to be questioned. In fact, it is not exactly true because of details in the physics of heat transfer.

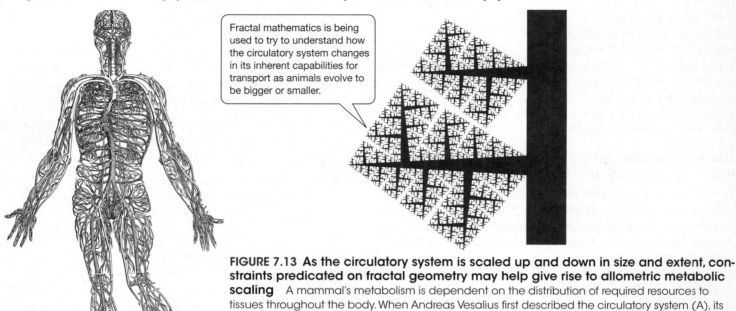

(A) Vesalius 1543: One of the first anatomically accurate images of the human circulatory system

(B) Mandelbrot 1983: A fractal model of a branching system such as the circulatory system

> Fractal mathematics is being used to try to understand how the circulatory system changes in its inherent capabilities for transport as animals evolve to be bigger or smaller.

FIGURE 7.13 As the circulatory system is scaled up and down in size and extent, constraints predicated on fractal geometry may help give rise to allometric metabolic scaling A mammal's metabolism is dependent on the distribution of required resources to tissues throughout the body. When Andreas Vesalius first described the circulatory system (A), its function was a mystery. Even oxygen transport could not be considered because oxygen had not yet been discovered! By now, with many old questions answered, new questions about circulatory function have come to the fore. The invention of fractal mathematics by Benoit Mandelbrot in about 1980 may help biologists gain a better understanding of the evolution of circulatory structure and function. Fractal systems, as seen in (B), are "self-similar" at multiple scales, meaning that the patterns of branching of fine elements are miniatures of the patterns of branching of large elements. (A from *De Humani Corporis Fabrica*, produced by Andreas Vesalius in 1543, as reproduced in Saunders and O'Malley 1950; B after Mandelbrot 1983.)

placental mammals, for example, several meticulous efforts have concluded that *b* is different in some mammalian orders than in others. In addition, as already noted, *b* is greater when mammals are exercising than when they are at rest. Moreover, *b* is higher when only large-bodied species are analyzed than when only small-bodied species are.[16]

As physiologists have searched for the mechanistic basis of metabolism–size relations, a key question has been, what attributes of animals are so *common* and so *fundamental* that they could explain the way in which metabolism varies with size? One attribute in particular has attracted a great deal of attention: internal transport. For metabolism to occur, internal transport of metabolic resources—notably O_2 and metabolic fuels—is critical. In mammals and many other types of animals, this transport is carried out by the circulatory system. Physiologists therefore realized that they had to understand how the circulatory system—first accurately described in 1543 by Andreas Vesalius (1514–1564) (**FIGURE 7.13A**)—changes in its inherent capabilities for transport as animals evolve to be bigger or smaller. The new mathematics of *fractal geometry*—invented more than 400 years later to describe the properties of branching systems (**FIGURE 7.13B**)—was marshaled to analyze this question. From this fractal research, a hypothesis was propounded, that allometric metabolism–size relations occur in part because of geometrically imposed constraints. This hypothesis stresses that in fractally structured transport systems, *rates* of transport—and thus rates of supply of resources required for metabolism—are geometrically constrained in distinctive ways as body size is scaled up or scaled

down over the course of evolution. Researchers are continuing to explore hypotheses based on the properties of circulatory systems and other transport systems. Several other types of fascinating hypotheses are also being investigated at present.

SUMMARY
Metabolic Scaling: The Relation between Metabolic Rate and Body Size

■ BMR, SMR, and other measures of resting metabolic rate are allometric functions of body weight within phylogenetically related groups of animals ($M = aW^b$, where *b* is usually in the vicinity of 0.7). Small-bodied species tend to have higher weight-specific metabolic rates than do related large-bodied species, an effect so great that the weight-specific BMR is 20 times higher in mice than in elephants.

■ Maximum aerobic metabolic rate also tends to be an allometric function of body weight in sets of related species. In many cases studied thus far, the allometric exponent for maximum metabolic rate differs from that for resting metabolic rate.

■ The allometric relation between metabolic rate and weight exerts important effects on the organization and structure of both individual animals and ecosystems. Heart rates, breathing rates, mitochondrial densities, and dozens of other features of individual animals are allometric functions of body weight within sets of phylogenetically related species. In ecosystems, population biomasses and other features of community organization may vary allometrically with individual body size.

[16] Accordingly, the log–log plot of metabolism–size data exhibits a bit of curvature and requires a more complex equation than Equation 7.3 to be described in detail.

■ Physiologists are not agreed on the explanation for the allometric relations between metabolic rate and body weight. Rubner's surface "law," based on heat loss from homeothermic animals, does not provide a satisfactory explanation.

Energetics of Food and Growth

Food and growth are important topics in animal energetics, aptly discussed together because one animal's growth is another's food. A consequential attribute of foods as energy sources is that lipids are at least twice as high in energy density—energy value per unit of weight (see Table 6.3)—as proteins and carbohydrates are. We asked at the start of this chapter why polar explorers carry lipid-rich foods, such as meat mixed with pure lard. If they are going to pull, push, and lift their food for many miles before they eat it, the explorers should choose food that provides a lot of energy per kilogram transported. Similarly, migrating animals often capitalize on the high energy density of lipids by carrying their fuel as body fat.

A key question about any food in relation to an animal's physiology is how efficiently the animal can digest (or ferment) the food and absorb the products of digestion. The **energy absorption efficiency** is defined to be the fraction of ingested energy that is absorbed for use:[17]

$$\text{Energy absorption efficiency} = \frac{\text{absorbed energy}}{\text{ingested energy}} \quad (7.7)$$

This efficiency matters because the absorbed energy is the energy actually available to an animal for use in its metabolism. To illustrate the importance of absorption efficiency, consider the processing of ingested cellulose by humans and ruminants. Because people cannot digest cellulose, they cannot absorb it, and their absorption efficiency for cellulose is essentially 0%; if they eat only cellulose, they starve. Ruminants such as cows, in contrast, commonly achieve about 50% absorption efficiency for cellulose because their rumen microbes ferment cellulose into compounds that the animals can absorb; thus ruminants are able to use about half of the energy available from cellulose in their metabolism. This example illustrates how the physiology of digestion and absorption, discussed in Chapter 6, bears on the physiology of energy.

Growing animals accumulate chemical-bond energy in their bodies by adding tissue consisting of organic molecules. An important question in many contexts is how efficiently they are able to use their available food energy to build tissue. Two types of growth efficiency, termed **gross growth efficiency** and **net growth efficiency**, are defined on the basis of whether the food energy is expressed as the ingested energy or the absorbed energy:[18]

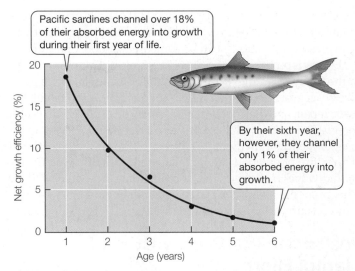

FIGURE 7.14 Net growth efficiency during each year of life in Pacific sardines (*Sardinops sagax*) When their populations are thriving, these fish are a major food source for seals, predatory fish, birds, and humans. (After Lasker 1970.)

> Pacific sardines channel over 18% of their absorbed energy into growth during their first year of life.

> By their sixth year, however, they channel only 1% of their absorbed energy into growth.

$$\text{Gross growth efficiency} = \frac{\substack{\text{chemical-bond energy of tissue} \\ \text{added in net fashion by growth}}}{\text{ingested energy}} \quad (7.8)$$

$$\text{Net growth efficiency} = \frac{\substack{\text{chemical-bond energy of tissue} \\ \text{added in net fashion by growth}}}{\text{absorbed energy}} \quad (7.9)$$

The growth efficiency of animals (gross and net) typically declines with age (**FIGURE 7.14**). This pattern is important in the analysis of energy flow in ecological communities. It is also important in agriculture and aquaculture because as growth efficiency declines as animals age, a decline occurs in the amount of product (e.g., meat) that is obtained in return for a farmer's or aquaculturalist's investment in feed. In the production of broiler chickens, for example, the birds are slaughtered at just 2–3 months of age because at that point they are large enough to be meaty but their growth efficiency—their growth in return for feed provided—is declining.[19]

Conclusion: Energy as the Common Currency of Life

Energy features in virtually every biological process and in many inanimate processes as well. It is a factor in animal growth, body maintenance, migration, photosynthesis, automobile operation, building construction, ecosystem degradation, and war.

When scientists attempt to analyze complex systems—from individual animals to entire ecosystems or even the entire planet—they inevitably come up with long lists of processes that they must take into account. Although the isolated study of individual processes may be straightforward, the *integration* of multiple processes is usually not. One of the greatest challenges for the integrated study

---

[17] Recall that the energetic efficiency of a process is the output of high-grade energy expressed as a ratio of input (see Equation 7.1). When digestion, fermentation, and absorption are the functions of interest, the output of high-grade energy is the absorbed energy, whereas the input is the ingested energy.

[18] When growth is analyzed in terms of Equation 7.1, the output of high-grade energy is the chemical-bond energy of added tissue, whereas the input is food energy.

[19] Feed accounts for 60%–75% of a farmer's costs.

of complex systems is to find a common set of units of measure—a "common currency"—in which all the operative processes can be expressed so that they can be compared, added, or multiplied.

Energy is probably the single most promising common currency. In the study of an individual animal, for example, processes as diverse as growth, running, nerve conduction, blood circulation, tissue repair, and thermoregulation can all be expressed in units of energy. The costs of all these processes can therefore be summed to estimate the individual's total cost of life, and the cost of life of an entire population can be calculated by multiplying the cost per individual by the number of individuals present. Few, if any, other properties come close to *energy* in their potential to serve as common currencies in this way.

Postscript: The Energy Cost of Mental Effort

Not the least of the energy costs of analyzing complex systems is the cost of operating our brain. This cost has some fascinating and unexpected properties. From studies of tissue metabolic rates, we know that in adult humans, the brain accounts for about 20% of resting metabolic rate (although it accounts for a much higher percentage in young children; see page 91); loosely put, in adulthood one-fifth of our food is for our brain when we are at rest. This cost resembles an "idling" cost; the energy is expended whether we subjectively feel we are doing hard mental labor or not. Decades ago, the prominent physiologist Francis Benedict (1870–1957) wanted to estimate how much the brain's energy needs increase with mental "effort." So, of course, he recruited a group of college students to find out. He told the students on one occasion to sit for tens of minutes keeping their minds as blank as possible. Then he had them spend an equal amount of time working mental arithmetic problems at a fevered pace. Measures of their metabolic rates under the two conditions indicated that the increase in energy consumption caused by an hour of hard mental effort is slight, equivalent to the energy of half a peanut! Benedict's methods were crude by today's standards. Nonetheless, recent calculations from modern neuroimaging methods confirm his conclusions. Thus the brain's high costs are largely steady costs, and thinking hard is not a way to stay slim.

STUDY QUESTIONS

1. Assuming that ten people plan to trek 500 miles to the North Pole, outline the steps you would take to calculate the amount of food they should pack, taking into account the number of sled dogs needed and the food needed for the dogs.

2. Suppose you use a tire pump to inflate a tire on a bicycle. The elevated pressure created in the tire represents a form of potential energy because the release of the pressure can do mechanical work (such as making a pinwheel turn). The potential energy in the tire is derived from chemical-bond energy in your food. Trace the energy from the time it enters your mouth at a meal until it ends up in the tire, identifying losses of energy as heat along the way.

3. Define *absorbed energy* (*assimilated energy*). Then list the major categories of use of absorbed energy, and specify the fate of

energy used in each category. Explain rigorously why heat is always made, regardless of the way energy is used.

4. Small animals tend to expire sooner than related large ones if forced to live on stored supplies. For instance, suppose you have a mouse and a dog that both start with body stores of fat equal to 20% of body weight. Explain why the mouse would be likely to die sooner if these animals could not find any food and thus had to live on their fat reserves. Which one would die sooner if they were trapped underwater and had only their stores of O_2 to live on while trying to escape?

5. Suppose that over a period of 4 h a dog was observed to consume 20 L of O_2 and produce 14 L of CO_2. Using Tables 7.1 and 7.2, estimate the dog's total heat production over the 4 h. Explain why Table 7.2 is essential for your calculation.

6. Poultry scientists are doing research on the design of diets that are nutritionally complete for chickens but minimize the SDA. These scientists believe that such diets would be particularly helpful to the poultry industry in southern states during the heat of summer. Why might this be true?

7. Before Mayer and Joule came along (see Box 7.1), people were well aware that if a person cranked a drill, heat appeared. For instance, the drilling of the bores of cannons was legendary for the heat produced. However, heat per se was believed to be neither created nor destroyed, and thus no one thought that the motion associated with drilling *turned into* heat. Mayer and Joule go down in history in part because they demonstrated the real relation between motion and heat. Imagine that you were alive in the early nineteenth century, and like Mayer and Joule, you hypothesized that animal motion could turn into heat. Design an experiment that would provide a rigorous test of your hypothesis.

8. Suppose you are measuring the metabolic rate of a young, growing cow by using the material-balance method. What procedures could you use to take account of the cow's growth, so that you measure a correct metabolic rate?

9. Suppose you have measured the average rate of O_2 consumption of two groups of laboratory rats that are identical, except that one group was injected with a hormone that is being tested to see if it affects metabolic rate. If the hormone-treated group has a rate of O_2 consumption 5% higher than the other, there are physiological reasons why you cannot conclude that the hormone has changed the metabolic rate. Explain, referring to Table 7.1. According to the table, what might the hormone have done to change the rate of O_2 consumption without changing the metabolic rate?

10. Only nine species of existing land mammals grow to adult body weights over 1000 kg (1 megagram). All are herbivores that employ fermentative digestion. These "megaherbivores" are the two species of elephants, the five species of rhinos, the common hippo, and the giraffe. What are the metabolic pros and cons of such large size? Can you suggest why no terrestrial carnivores achieve such large size?

11. If there are many species of herbivores in a grassland ecosystem, and if the species as populations are equally competitive in acquiring food, predict b in the following allometric equation: population biomass per square kilometer = aW^b, where W is individual body weight. Do the data in Table 7.5 follow your equation? What hypotheses are suggested by the comparison?

Go to **sites.sinauer.com/animalphys4e**
for box extensions, quizzes, flashcards,
and other resources.

REFERENCES

Atkins, P. 2007. *Four Laws That Drive the Universe.* Oxford University Press, Oxford. Another gem from Peter Atkins, providing succinct explanations of the principles of thermodynamics, accessible to general readers.

Atkins, P. W. 1984. *The Second Law.* Scientific American Library, New York. A serious book on the second law of thermodynamics that is accessible to general readers, containing many useful diagrams.

Glazier, D. S. 2005. Beyond the '3/4-power law': variation in the intra- and interspecific scaling of metabolic rate in animals. *Biol. Rev.* 80: 611–662. A review of metabolic scaling in most groups of animals and in plants, including discussion of hypotheses of causation.

Kolokotrones, T., V. Savage, E. J. Deeds, and W. Fontana. 2010. Curvature in metabolic scaling. *Nature* 464: 753–756. A breakthrough paper, limited to mammals, showing that the time-honored allometric equation for metabolic scaling does not in fact fit available data in detail because the data exhibit curvilinearity on a log–log plot, with implications.

Lankford, T. E., Jr., J. M. Billerbeck, and D. O. Conover. 2001. Evolution of intrinsic growth and energy acquisition rates. II. Trade-offs with vulnerability to predation in *Menidia menidia*. *Evolution* 55: 1873–1881. A thought-provoking study of why growth rates are not always maximized, focusing on evolutionary trade-offs in a fish of great ecological importance.

Levine, J. A., N. L. Eberhardt, and M. D. Jensen. 1999. Role of non-exercise activity thermogenesis in resistance to fat gain in humans. *Science* 283: 212–214. An application of the concepts of energy metabolism to a significant human problem.

Nagy, K. A. 2005. Field metabolic rate and body size. *J. Exp. Biol.* 208: 1621–1625. This paper discusses in five lucid pages almost all the issues currently in the forefront of vertebrate metabolic allometry.

Owen-Smith, R. N. 1988. *Megaherbivores: The Influence of Very Large Body Size on Ecology.* Cambridge University Press, New York. A searching discussion of extremely large body size in terrestrial mammals, providing an intriguing way to see the application of many principles of animal energetics.

Raichle, M. E., and M. A. Mintun. 2006. Brain work and brain imaging. *Annu. Rev. Neurosci.* 29: 449–476. This fascinating paper discusses cutting-edge knowledge of brain energetics, with a focus on insights provided by—and questions raised by—modern neuroimaging.

Sinervo, B., and R. B. Huey. 1990. Allometric engineering: An experimental test of the causes of interpopulational differences in performance. *Science* 248: 1106–1109.

Speakman, J. R., and E. Król. 2010. Maximal heat dissipation capacity and hyperthermia risk: neglected key factors in the ecology of endotherms. *J. Anim. Ecol.* 79: 726–746. Exposition of an interesting and novel idea for explaining metabolic scaling in endotherms.

See also **Additional References** *and* **Figure and Table Citations**.

Aerobic and Anaerobic Forms of Metabolism

A startled crayfish swims backward at jetlike speed by flipping its tail in a series of powerful contractions. In this way, a crayfish discovered under a rock in a stream can move to another rock and disappear almost before its presence is appreciated. The tail muscles, which account for 25% of a crayfish's weight in some species, are a powerful tool for escape and survival. To contract, however, the crayfish's muscles—like all muscles—require adenosine triphosphate (ATP) as their immediate source of energy for contraction. When a crayfish is startled, its first tail flip occurs instantly and is followed in short order by multiple additional flips. Each of these massive contractions requires a substantial amount of ATP. How is the ATP made available so promptly? The same question can be asked about the ATP that a human sprinter requires to run 100 meters (m) in 10 seconds (s).

Burst exercise is a general term that refers to sudden, intense exercise. Besides escaping crayfish and sprinting people, burst exercise is illustrated by salmon leaping waterfalls during their upstream journey, cheetahs racing toward antelopes, and scallops jetting away from danger by clapping their shells. In most animals, burst exercise cannot be continued for long periods. Crayfish escaping by flipping their tails become exhausted after 15–30 flips, just as human sprinters are exhausted or nearly exhausted at the ends of their brief races.

The biochemistry of survival
Crayfish can escape rapidly from danger by tail flipping, provided that the cells in their tail muscles can generate adenosine triphosphate (ATP) at high enough rates. The contractile apparatus in each muscle cell requires ATP at a high rate to have energy for rapid, powerful contractions.

A second general form of physical activity is **sustained exercise**, defined to be exercise that can be continued at a steady rate for a long period. Jogging by humans, migratory flight by birds, and steady cruising by fish or crayfish are examples. During sustained exercise, the locomotory muscles must be supplied with ATP minute after minute for long, uninterrupted periods. What are the mechanisms that supply ATP in this way?

Comparing sustained exercise and burst exercise, do the mechanisms that supply ATP during sustained exercise differ from the higher-intensity, shorter-duration mechanisms used during burst exercise? Do different types of animals differ in their ATP-producing mechanisms? Is exercise ever forced to stop because of limitations in ATP-generating mechanisms? These are some of the questions we discuss in this chapter.

ATP is not transported from one cell to another. *Each cell, therefore, must make its own ATP.* This is one of the most critically important properties of ATP physiology. Another key property is that *ATP is not stored by cells to any substantial extent.* Because of these two properties, the *rate* at which a cell can do muscular work (or carry out other forms of ATP-requiring physiological work) at any given moment depends strictly on the rate at which that *very cell* is able to produce ATP at the moment.

The following two complementary reactions, taken together, serve as a crucial *energy shuttle* and *energy transduction* mechanism in cells:

$$\text{ADP} + P_i + \text{energy from foodstuff molecules} \rightarrow \text{ATP} \qquad \textbf{(8.1)}$$

$$\text{ATP} \rightarrow \text{ADP} + P_i + \text{energy usable by cell processes} \qquad \textbf{(8.2)}$$

Each cell uses energy from carbohydrates, lipids, or other foodstuff molecules to drive Equation 8.1, causing ATP to be formed from adenosine diphosphate (ADP) and inorganic phosphate ions (HPO_4^{2-}), symbolized P_i. In this way, the energy from the bonds of the foodstuff molecules is moved into bonds of ATP in the cell and becomes poised for use in physiological work. Energy-demanding processes in the cell—such as muscle contraction or ion pumping—then split the ATP, as shown in Equation 8.2, thereby releasing the energy they need. Cellular energy-demanding processes are not able to draw energy directly from the bonds of foodstuff molecules. The energy-demanding processes, therefore, (1) are utterly dependent on the cellular mechanisms that drive Equation 8.1 and (2) can take place only as fast as those mechanisms supply ATP.

Mechanisms of ATP Production and Their Implications

There are two major categories of catabolic, biochemical pathways by which animal cells release energy from foodstuff molecules to synthesize ATP and thus make energy available for the performance of physiological work. Some pathways, termed **aerobic**, require O_2; others, termed **anaerobic**, can function without O_2. In this section we examine the mechanistic features of the aerobic and anaerobic catabolic pathways. In later sections of this chapter we will look at the interplay between the aerobic and anaerobic modes of energy release when animals engage in burst or sustained exercise, or when they face environmental stresses such as O_2 deficiency. There is a compelling reason to start with study of the mechanistic features of the pathways of ATP production: The overall performance of animals during exercise and in other situations often depends on the mechanistic peculiarities of the particular biochemical pathways by which energy is extracted for use from foodstuff molecules.

Aerobic catabolism consists of four major sets of reactions

Each cell in most animals possesses **aerobic catabolic pathways**: pathways that, by use of O_2, completely oxidize foodstuff molecules to CO_2 and H_2O and capture in ATP bonds much of the chemical energy thereby released. These aerobic pathways typically can oxidize *all* the major classes of foodstuffs. Here, for simplicity, we emphasize just the catabolism of carbohydrates. Moreover, our aim is to provide an overview, not to duplicate detailed treatments available in texts of cellular physiology or biochemistry. The principal aerobic catabolic pathway can be subdivided into four major sets of reactions: (1) *glycolysis*, (2) the *Krebs cycle* (*citric acid cycle*), (3) the *electron-transport chain*, and (4) *oxidative phosphorylation*.

GLYCOLYSIS Glycolysis is the series of enzymatically catalyzed reactions shown in **FIGURE 8.1**, in which glucose (or glycogen) is converted to pyruvic acid. The enzymes and reactions of glycolysis occur in the cytosol. The first step in glycolysis is that glucose is phosphorylated at the *cost* of an ATP molecule to form glucose-6-phosphate. Glucose-6-phosphate is then converted to fructose-6-phosphate, and the latter is phosphorylated—also at the *cost* of an ATP molecule—to form fructose-1,6-diphosphate. The latter is cleaved to form two three-carbon molecules: dihydroxyacetone phosphate and glyceraldehyde-3-phosphate. These compounds are interconvertible, and when glucose is being catabolized for release of energy, the former is converted to the latter, yielding two molecules of glyceraldehyde-3-phosphate. The reactions subsequent to glyceraldehyde-3-phosphate in Figure 8.1 are all multiplied by 2 to emphasize that two molecules follow these pathways for each glucose molecule catabolized.

The reaction that uses glyceraldehyde-3-phosphate *is the only oxidation reaction* in glycolysis and is particularly significant for understanding not only aerobic but also anaerobic catabolism. Each molecule of glyceraldehyde-3-phosphate is oxidized, with the addition of inorganic phosphate (P_i), to a three-carbon diphosphate: 1,3-diphosphoglyceric acid. Although this reaction is an oxidation reaction, it does not itself require O_2. Instead, it occurs by the simultaneous reduction of one molecule of nicotinamide adenine dinucleotide (NAD) per molecule of glyceraldehyde-3-phosphate. NAD—which a cell synthesizes from the vitamin niacin—is a relatively small, nonprotein molecule that undergoes reversible oxidation and reduction. When a molecule of glyceraldehyde-3-phosphate is oxidized, two hydrogen atoms are removed from it and transferred to NAD, reducing the NAD to form $NADH_2$.[1] The two hydrogens remain bound to $NADH_2$ only temporarily because they are soon passed to another compound, regenerating NAD.

An alternative way to think about an oxidation–reduction reaction—such as the oxidation of glyceraldehyde-3-phosphate—is that electrons are transferred. In the reaction under discussion, NAD

---

[1] The reduction of NAD is symbolized "NAD + 2 H → $NADH_2$" or "NAD → $NADH_2$" in this book because these simplified expressions compactly emphasize the features of relevance for us. The true reaction is $NAD^+ + 2H$ → $NADH + H^+$.

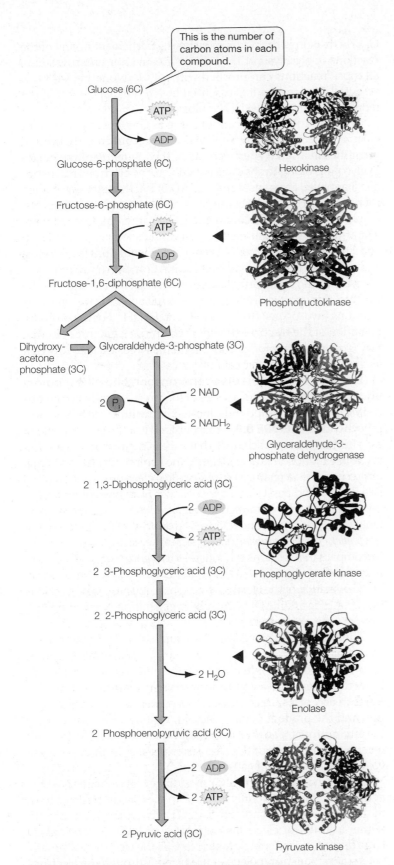

FIGURE 8.1 The major reactions of glycolysis Each reaction requires catalysis by an enzyme protein. The three-dimensional structures of six of these enzyme proteins are shown. Subunits of an enzyme are shown in different colors. The expression for reduction of NAD, NAD → $NADH_2$, is shorthand; the actual reaction is $NAD^+ + 2 H → NADH + H^+$. P_i = inorganic phosphate. (Protein structures from Erlandsen et al. 2000.)

serves as the *immediate electron acceptor* because it combines with electrons (hydrogens) removed from glyceraldehyde-3-phosphate.

The 1,3-diphosphoglyceric acid formed from the oxidation of glyceraldehyde-3-phosphate is next converted to a monophosphate, 3-phosphoglyceric acid, with the formation of one ATP per molecule. The 3-phosphoglyceric acid is then converted in two steps to phosphoenolpyruvic acid, and the latter reacts to form pyruvic acid, again with the formation of one ATP per molecule.

Three important consequences of glycolysis deserve note:

- Each molecule of glucose is converted into two molecules of pyruvic acid.
- Two molecules of NAD are reduced to $NADH_2$ per molecule of glucose catabolized.
- Two molecules of ATP are used and four are formed for each glucose processed, providing a *net yield of two ATP molecules per glucose molecule.*

THE KREBS CYCLE (CITRIC ACID CYCLE) During aerobic catabolism, the pyruvic acid formed by glycolysis enters the mitochondria by facilitated diffusion (mediated by a carrier protein just recently defined). The pyruvic acid is then oxidized in the mitochondria by a cyclic series of enzymatically catalyzed reactions called the **Krebs cycle** (**citric acid cycle**). This set of reactions, named after Hans Krebs, who in 1937 was the first to envision its features, is diagrammed in **FIGURE 8.2**. For simplicity the reactions involved in the oxidation of just one molecule of pyruvic acid are shown, although two molecules are in fact processed for each molecule of glucose.

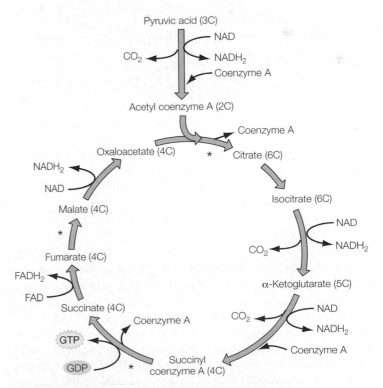

FIGURE 8.2 The major reactions of the Krebs cycle (citric acid cycle) A molecule of H_2O enters the reactions at each asterisk (*). The expression for reduction of NAD, NAD → $NADH_2$, is shorthand; the actual reaction is $NAD^+ + 2 H → NADH_+ H^+$.

Pyruvic acid enters the Krebs cycle by participating in a complex set of reactions in which it is oxidatively decarboxylated, forming CO_2 and a two-carbon acetyl group that is combined with coenzyme A in the form of acetyl coenzyme A.[2] In the process, a molecule of NAD is reduced. Acetyl coenzyme A then reacts with oxaloacetate, with the end result that coenzyme A is released and the acetyl group is condensed with oxaloacetate (four-carbon) to form citrate (six-carbon). In the ensuing series of reactions, oxaloacetate is ultimately regenerated and then again can combine with acetyl coenzyme A. For our purposes, there is no need to review the reactions stepwise. It is more important to emphasize the overall outcomes of the reactions:

- *The six carbons of each glucose molecule catabolized emerge in the form of six molecules of CO_2 as the pyruvic acid molecules produced by glycolysis are processed by the Krebs cycle.* The CO_2 is formed by decarboxylation reactions. Such reactions occur at two points in the Krebs cycle: in the conversion of isocitrate to α-ketoglutarate and in the conversion of α-ketoglutarate to succinyl coenzyme A. These two decarboxylations, plus the one in the reaction of pyruvic acid to form acetyl coenzyme A, account for the formation of three molecules of CO_2 for every molecule of pyruvic acid processed (thus six molecules of CO_2 formed per glucose molecule).

- *For each glucose molecule catabolized, the Krebs cycle produces eight molecules of $NADH_2$ and two molecules of $FADH_2$.* Oxidation reactions occur at four points in the Krebs cycle. At three of these, NAD is reduced, forming $NADH_2$. At one (the oxidation of succinate to fumarate), another small, nonprotein molecule that undergoes reversible oxidation and reduction—flavin adenine dinucleotide (FAD)—is reduced, forming $FADH_2$.[3] Recall also that one $NADH_2$ is formed in the reaction of pyruvic acid to form acetyl coenzyme A. Considering all the oxidation reactions, the processing of each pyruvic acid molecule results in four $NADH_2$ and one $FADH_2$ (thus eight $NADH_2$ and two $FADH_2$ per glucose molecule).

- *Two molecules of ATP are produced in the Krebs cycle for each molecule of glucose catabolized.* Guanosine triphosphate (GTP) is formed from guanosine diphosphate (GDP) when succinyl coenzyme A reacts to form succinate in the Krebs cycle. GTP donates its terminal phosphate group to ADP, resulting in GDP and ATP. Thus one molecule of ATP is generated for each molecule of pyruvic acid processed (two ATPs per glucose).

ELECTRON TRANSPORT, OXIDATIVE PHOSPHORYLATION, AND THE ROLE OF O_2 The final two of the four sets of reactions in the aerobic catabolic pathway are the *electron-transport chain* and *oxidative phosphorylation*. We discuss them together because they are often tightly linked.

A paradox you may have noticed is that thus far, in discussing *aerobic* catabolism, we have not mentioned the involvement of

O_2. The reason is that O_2 is in fact not a participant in any of the reactions of glycolysis or the Krebs cycle; in a very narrow sense, all those reactions can proceed without O_2. Nonetheless, O_2 is essential. The reason it is essential lies in the disposition of the reduced $NADH_2$ and $FADH_2$ molecules.

$NADH_2$ and $FADH_2$ cannot serve as final resting places for electrons because NAD and FAD are present in only limited quantities in a cell. Whenever one of the mainstream molecules in glycolysis or the Krebs cycle is oxidized, the electrons (or hydrogens) removed are transferred to NAD or FAD. As stressed earlier, therefore, NAD and FAD are the *immediate* electron acceptors. However, if the $NADH_2$ and the $FADH_2$ thereby formed were simply allowed to accumulate, a cell would soon run out of NAD and FAD. Running out of NAD and FAD would bring glycolysis and the Krebs cycle to a halt because NAD and FAD are required. Thus NAD and FAD cannot serve as *final* electron acceptors. The electron-transport chain regenerates NAD and FAD by removing electrons (hydrogens) from $NADH_2$ and $FADH_2$. For the ordinary operation of the electron-transport chain, O_2 is required. In that way O_2 is necessary for glycolysis and the Krebs cycle to function as they do during aerobic catabolism.

Let's now focus on the **electron-transport chain (respiratory chain)** itself. It consists of a series of four major protein complexes (I–IV), plus other compounds, located in the inner membranes of mitochondria (**FIGURE 8.3**). A key property of the constituents of the electron-transport chain is that each is capable of undergoing reversible reduction and oxidation. The compounds function in a discrete order—a chain. The chain takes electrons from $NADH_2$ and $FADH_2$ and passes them in sequence from one compound to the next in a series of reductions and oxidations (see Figure 8.3). Finally, the last compound in the electron-transport chain, known as *complex IV* or *cytochrome oxidase*, passes the electrons—along with "accompanying" protons (H^+ ions)—to oxygen, reducing the O_2 to water.[4] In this way, *O_2 acts as the final electron acceptor*. The net effect of the operation of the electron-transport chain is to take electrons from $NADH_2$ and $FADH_2$ (thereby regenerating NAD and FAD) and pass the electrons to O_2.

The special role played by O_2 is important. The constituents of the electron-transport chain (e.g., cytochrome b-c_1 and cytochrome oxidase), just like NAD and FAD, are present in limited quantities in a cell and therefore cannot act as terminal electron acceptors. In contrast, O_2 is continuously supplied to a cell, and the product of its reduction, water, can be voided into the environment, *thereby carrying electrons out of the cell*. An adult person produces about 0.8 L of water per day in the process of voiding electrons from cells!

The electron-transport chain, in addition to reoxidizing $NADH_2$ and $FADH_2$, is also pivotally involved in the *transfer of energy* from the bonds of foodstuff molecules to ATP. Molecular O_2 has a much higher affinity for electrons than NAD or FAD. Accordingly, a large decline in free energy takes place as the electrons originally taken from foodstuff molecules are passed through the electron-transport chain. Much of this energy is captured in bonds of ATP. The process of forming ATP from ADP by use of energy released

---

[2] Coenzyme A, an essential compound for aerobic metabolism, is synthesized from pantothenic acid, a B vitamin.

[3] Flavin adenine dinucleotide is synthesized from riboflavin (vitamin B_2).

[4] Two electrons and two protons combine with one oxygen *atom*. A molecule of oxygen (O_2), therefore, can react with four electrons and four protons.

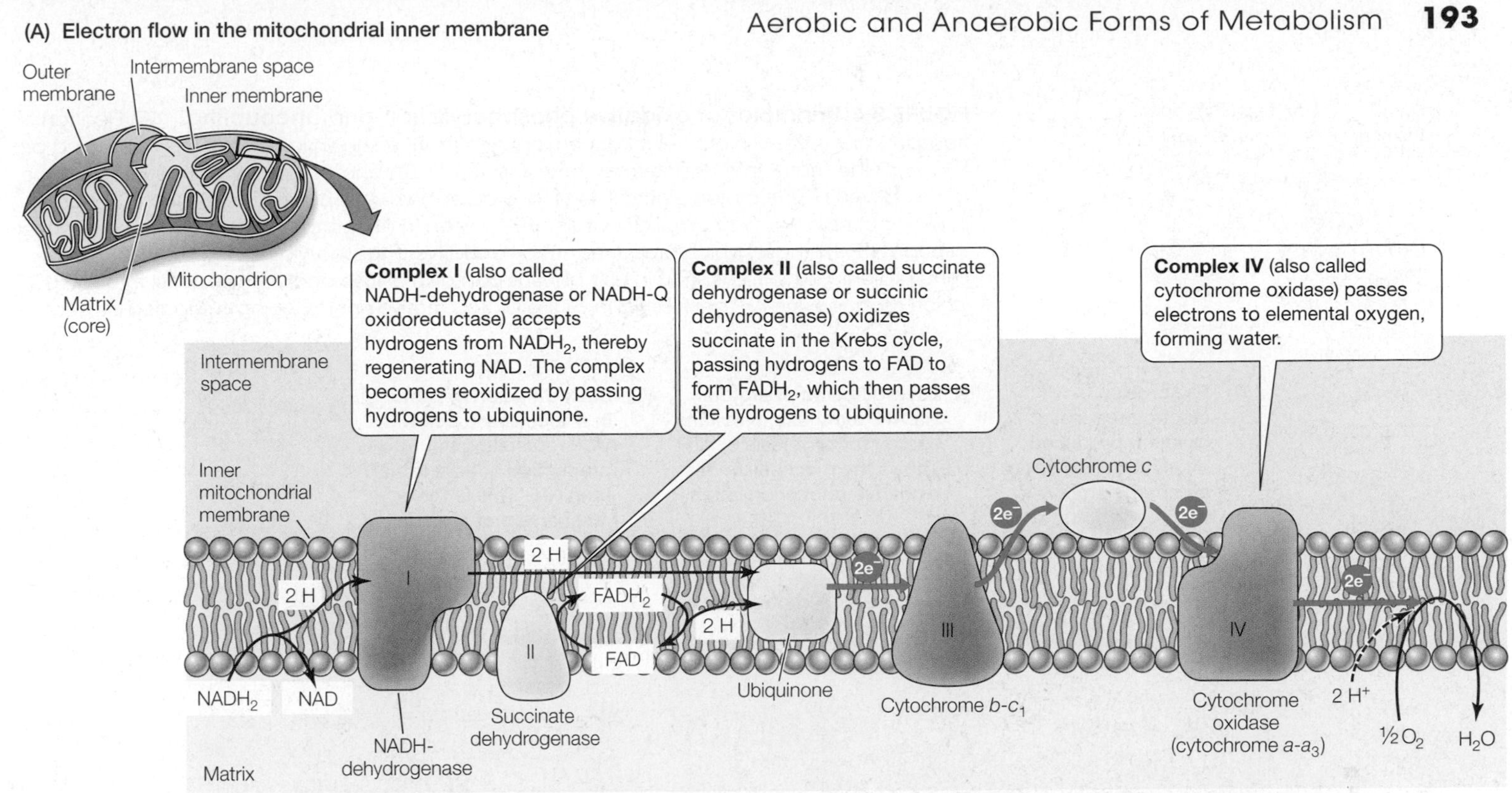

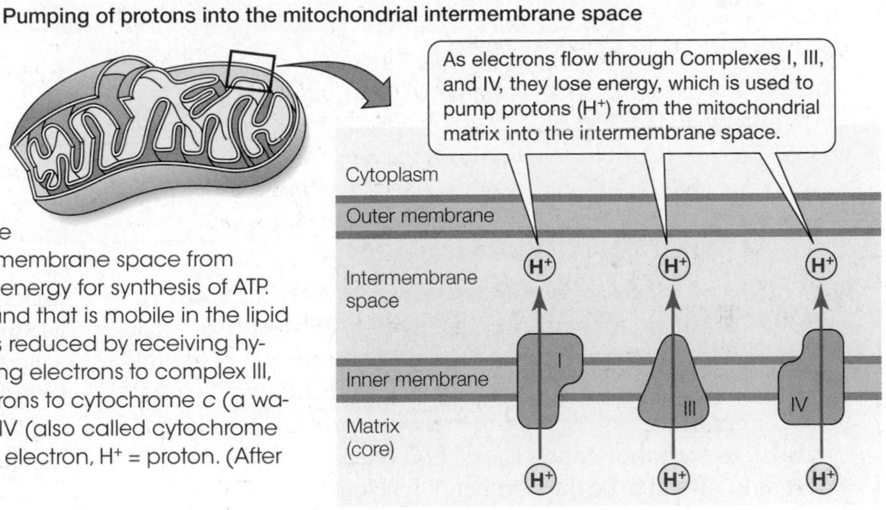

(A) Electron flow in the mitochondrial inner membrane

Outer membrane

Intermembrane space

Inner membrane

Mitochondrion

Matrix (core)

Intermembrane space

Inner mitochondrial membrane

Complex I (also called NADH-dehydrogenase or NADH-Q oxidoreductase) accepts hydrogens from NADH₂, thereby regenerating NAD. The complex becomes reoxidized by passing hydrogens to ubiquinone.

Complex II (also called succinate dehydrogenase or succinic dehydrogenase) oxidizes succinate in the Krebs cycle, passing hydrogens to FAD to form FADH₂, which then passes the hydrogens to ubiquinone.

Complex IV (also called cytochrome oxidase) passes electrons to elemental oxygen, forming water.

Cytochrome c

2e⁻ 2e⁻

2 H

FADH₂

2e⁻

2 H

FAD

I

II

III

IV

2e⁻

NADH₂ NAD

Ubiquinone

Cytochrome b-c₁

Cytochrome oxidase (cytochrome a-a₃)

2 H⁺

½O₂ H₂O

NADH-dehydrogenase

Succinate dehydrogenase

Matrix

(B) Pumping of protons into the mitochondrial intermembrane space

As electrons flow through Complexes I, III, and IV, they lose energy, which is used to pump protons (H⁺) from the mitochondrial matrix into the intermembrane space.

Cytoplasm

Outer membrane

Intermembrane space

Inner membrane

Matrix (core)

H⁺ H⁺ H⁺

I III IV

H⁺ H⁺ H⁺

FIGURE 8.3 The electron-transport chain As shown in (A), the electron-transport chain consists of four enzyme protein complexes, called complexes I through IV, embedded in the inner mitochondrial membrane, plus ubiquinone and cytochrome c. The three purple complexes (I, III, and IV) capture energy from the flux of electrons to pump protons into the intermembrane space from the matrix across the inner membrane, as shown in (B), storing energy for synthesis of ATP. Ubiquinone, also called coenzyme Q, is a lipid-soluble compound that is mobile in the lipid core of the membrane. As also shown in (A), ubiquinone that is reduced by receiving hydrogens from complex I or FADH₂ becomes reoxidized by passing electrons to complex III, also called cytochrome b-c₁. Complex III then passes the electrons to cytochrome c (a water-soluble compound), which in turn passes them to complex IV (also called cytochrome oxidase, cytochrome a-a₃, COX, or cytochrome c oxidase). e⁻ = electron, H⁺ = proton. (After Saraste 1999.)

in the transport of electrons through the electron-transport chain is called **oxidative phosphorylation**.

How is the energy released in electron transport used by oxidative phosphorylation? How, in other words, are the two processes coupled? According to the **chemiosmotic hypothesis** currently favored by most biochemists, the coupling is achieved indirectly by a two-step process of the sort discussed in Box 5.1.[5] As energy is released in electron transport, it is used to pump protons across the inner mitochondrial membrane, creating—across that membrane—a proton electrochemical gradient that is far from equilibrium. This gradient represents an energy store (see Box 5.1). Driven by the electrochemical gradient, protons diffuse back across the membrane toward equilibrium by way of molecules of a membrane-spanning, ATP-synthesizing protein. In this process, energy from the proton

[5] The term *chemiosmotic hypothesis* refers to the entire process outlined in this paragraph. Box 5.1 is essential reading for fully understanding the hypothesis.

electrochemical gradient becomes energy of proton movement, and this proton motive energy is used by the ATP-synthesizing protein to synthesize ATP from ADP.

Let's now look at further details of electron transport and oxidative phosphorylation. Of the four protein complexes in the electron-transport chain, three (complexes I, III, and IV) use energy released from electron transport to pump protons across the inner mitochondrial membrane, as shown in Figure 8.3B. As already indicated, this proton pumping is away from equilibrium and *builds* an electrochemical gradient, seen in **FIGURE 8.4A**. Collectively, according to current estimates, the three complexes pump ten protons into the intermembrane space for each pair of electrons (or hydrogens) that flow the full length of the electron-transport chain from NADH₂ to oxygen. Although protons in the intermembrane space (see Figure 8.4A) have a strong tendency to diffuse back across the inner mitochondrial membrane, they are largely unable to pass through the membrane itself. Their

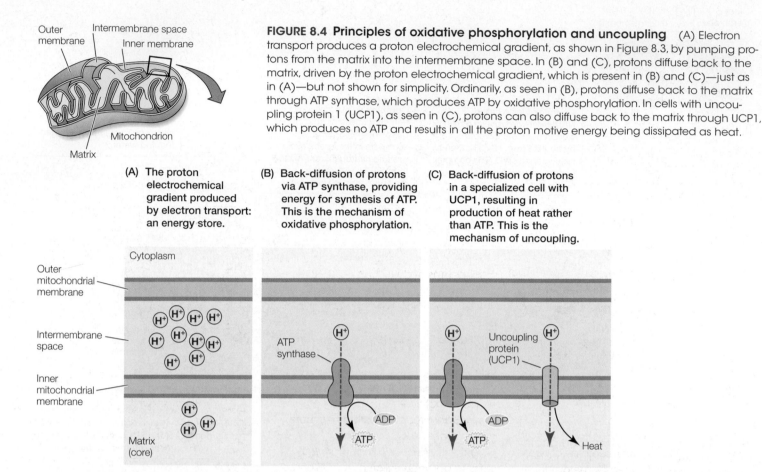

Outer membrane
Intermembrane space
Inner membrane
Mitochondrion
Matrix

FIGURE 8.4 Principles of oxidative phosphorylation and uncoupling (A) Electron transport produces a proton electrochemical gradient, as shown in Figure 8.3, by pumping protons from the matrix into the intermembrane space. In (B) and (C), protons diffuse back to the matrix, driven by the proton electrochemical gradient, which is present in (B) and (C)—just as in (A)—but not shown for simplicity. Ordinarily, as seen in (B), protons diffuse back to the matrix through ATP synthase, which produces ATP by oxidative phosphorylation. In cells with uncoupling protein 1 (UCP1), as seen in (C), protons can also diffuse back to the matrix through UCP1, which produces no ATP and results in all the proton motive energy being dissipated as heat.

(A) The proton electrochemical gradient produced by electron transport: an energy store.

(B) Back-diffusion of protons via ATP synthase, providing energy for synthesis of ATP. This is the mechanism of oxidative phosphorylation.

(C) Back-diffusion of protons in a specialized cell with UCP1, resulting in production of heat rather than ATP. This is the mechanism of uncoupling.

Cytoplasm

Outer mitochondrial membrane

Intermembrane space

Inner mitochondrial membrane

Matrix (core)

H^+

ATP synthase
ADP
ATP

Uncoupling protein (UCP1)
ADP
ATP
Heat

back-diffusion therefore occurs, as shown in **FIGURE 8.4B**, via molecules of a membrane-spanning protein—**ATP synthase**—that is permeable to protons and that couples the motive energy of the diffusing protons to the synthesis of ATP from ADP.[6]

A common mode of expressing the *efficiency* of ATP production by oxidative phosphorylation is as a **P/O ratio** (P = phosphate, O = oxygen), defined to be the number of ATP molecules formed per *atom* of oxygen reduced to water. According to current estimates, in mammalian mitochondria, ATP synthase makes 1 ATP for every 4.3 protons that cross through it. As earlier noted, electron transport pumps 10 protons for each pair of electrons (or hydrogens) that pass through the electron-transport chain. Inasmuch as a pair of electrons (or hydrogens) combines with 1 atom of oxygen at the end of the chain, that's 10 protons per O atom reduced. Those 10 protons, when they pass through ATP synthase, result in production of 2.3 ATP molecules (1 ATP per 4.3 protons). Thus the P/O ratio is 2.3. *This value, P/O = 2.3, is currently considered to be the maximum P/O ratio for electrons that pass the entire length of the electron-transport chain from $NADH_2$ to oxygen.*

Linkage of electron transport with oxidative phosphorylation is called **coupling**. An important aspect of coupling is that it is capable of being *graded* in its completeness (efficiency). Because of this graded nature, the P/O ratio can be lower than the maximum possible. In fact, a P/O ratio of 0 (zero) is observed at times in some types of cells, meaning that the P/O ratio can be as high as 2.3 or as low as 0. Electron transport and oxidative phosphorylation are said to be *fully ("tightly") coupled* when the P/O ratio is high and *completely uncoupled* when P/O = 0. Intermediate states of coupling are possible. Usually, coupling is relatively tight.

The mitochondria of *certain types of specialized cells* express in their inner membrane—in addition to ATP synthase—a second type of membrane-spanning protein, **uncoupling protein 1** (**UCP1**). This protein forms the basis for the most thoroughly understood mechanism of uncoupling. As shown in **FIGURE 8.4C**, UCP1 provides a path for protons to diffuse from the intermembrane space across the inner membrane to the mitochondrial matrix without ATP synthesis.[7]

Uncoupling opposes ATP production: In the fully uncoupled state (P/O = 0), even though electron transport continues to convert $NADH_2$ and $FADH_2$ to NAD and FAD—thereby permitting continued oxidation of foodstuff molecules by glycolysis and the Krebs cycle—none of the energy released by electron transport is captured as ATP. The energy instead is immediately converted to heat. Uncoupling by UCP1 therefore is a specialized biochemical state that is ordinarily employed only when there is an advantage to converting food energy to heat rather than ATP.

---

[6] Spectacular simulations of the operation of this "splendid molecular machine" are available on the web.

[7] There are at least four different hypotheses regarding how exactly UCP1 functions to achieve this outcome. The concept that UCP1 provides a straightforward back-diffusion path for protons is the simplest of the models.

BOX 8.1 Reactive Oxygen Species (ROS)

The ordinary metabolism of aerobic cells produces molecules—chemical species—that because of the status of their oxygen atoms have a high potential to react with (e.g., oxidize) cell lipids, proteins, and nucleic acids. These molecules are called **reactive oxygen species (ROS)**. ROS can have benign functions; they play signaling roles, for example. ROS are more famous, however, for their ability to damage macromolecules on which health and life depend. ROS, in fact, are believed by some biologists to be the primary causative agents of aging. According to this hypothesis, we age because of incessant damage done to vital macromolecules by metabolically produced ROS; this damage is repaired, but not entirely, causing defects to accumulate over time. ROS are also implicated in muscle fatigue and many other phenomena.

The superoxide radical (anion)—an O_2 molecule with an extra electron—is often the initial ROS produced by metabolism (see figure). Mitochondria routinely generate superoxide as a by-product of electron transport, and in turn, superoxide and its ROS products can injure mitochondria by reacting in destructive ways with mitochondrial macromolecules. Superoxide is produced also by several extramitochondrial reactions. Superoxide is converted to three other ROS: hydrogen peroxide, hydroxyl radicals, and peroxinitrite (see figure). All ROS are short-lived—sometimes exceedingly so—precisely because they are extraordinarily reactive. Hydroxyl radicals, for example, exist for less than 1-billionth of a second after they are formed.

Cells possess **antioxidant mechanisms**—enzymatic and non-enzymatic mechanisms of detoxifying ROS. Enzyme antioxidants catalyze the transformation of ROS to less-reactive chemical forms. *Superoxide dismutase*, for example, converts superoxide to hydrogen peroxide, and *catalase* converts hydrogen peroxide to water and O_2 (see figure). Non-enzymatic antioxidants often function in a sacrificial way: By reacting with ROS, they prevent the ROS from reacting with other, more critical molecules.

Oxidative stress (oxidant stress) refers to damage inflicted by ROS on cellular integrity. It occurs when the rate of ROS production exceeds the rate at which antioxidant mechanisms can dispose of ROS. **Box Extension 8.1** discusses antioxidants and the roles of ROS in more detail, including references.

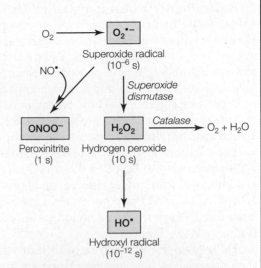

An outline of the chemical relationships of O_2 and four reactive oxygen species (ROS) The ROS are shown in boxes. A rough estimate of the average lifespan of each ROS is shown below the name of the ROS; for example, superoxide radicals typically react—and therefore cease to be superoxide radicals—within 1-millionth of a second (10^{-6} s) after they are formed. Superoxide dismutase (SOD) and catalase are enzymes. A superscript dot next to a compound symbolizes an unpaired electron. The reactions are shown in outline only, not as fully balanced reactions. (After Allen et al. 2008.)

For example, uncoupling of oxidative phosphorylation is a major mechanism of increasing metabolic heat production in many mammals in cold seasons (see Chapter 10). It also sometimes serves as a mechanism of body-weight regulation by "burning off" the energy of excess food molecules.[8] In a tissue with UCP1, the degree of coupling is controlled by modulation of the extent to which protons back-diffuse by way of the two possible paths: ATP synthase and uncoupling protein (see Figure 8.4C). This control is mediated by a variety of agents, including fatty acids, hormones, and Ca^{2+} ions.

Uncoupling by UCP1 is not the only mechanism that causes the P/O ratio to be less than maximum. A mechanism that is observed in a wider variety of cells than those with UCP1 is *leakage* of protons across the inner mitochondrial membrane. In this case, protons diffuse across the membrane, from the intermembrane space to the matrix, by way of largely unknown, nonspecific pathways. Leakage reduces the amount of ATP synthesized (it reduces the P/O ratio) because protons that leak do not pass through ATP synthase. In some tissues, 20% or more of the proton electrochemical gradient is dissipated by leakage, according to the evidence now available. This value for leakage is so high that it suggests there may be advantages achieved by allowing leakage to occur.

An important incidental effect of the processing of O_2 by mitochondria is that highly reactive oxygen-containing compounds are produced as the processes we have described take place. These highly reactive entities—known as *reactive oxygen species (ROS)*—can react with and damage mitochondrial constituents (**BOX 8.1**). Moreover, some can exit the mitochondria and possibly react with and damage extramitochondrial cell constituents.

THE YIELD OF ATP FROM AEROBIC CATABOLISM Now let's look at the total yield of ATP from the aerobic catabolism of glucose when coupling is tight. For each molecule of glucose that is catabolized, 10 molecules of $NADH_2$ and 2 molecules of $FADH_2$ are formed in glycolysis and the Krebs cycle. When these molecules are processed via the electron-transport chain, the maximum yield of ATP by oxidative phosphorylation is 25 ATP mol-

---

[8] Recall, for example, the rats mentioned in Chapter 7 that remain slim despite chronic overeating by undergoing diet-induced thermogenesis (DIT). Uncoupling is the principal mechanism of DIT in those animals.

ecules per glucose.[9] Even in tightly coupled mitochondria, this theoretical yield is probably not fully achieved in reality because of leakage. Nonetheless, for simplicity, we will use this value in this book without further qualification.

Besides the ATP molecules made by oxidative phosphorylation, additional ATP molecules are generated by phosphorylation reactions associated with several steps in glycolysis and the Krebs cycle: the reaction of 1,3-diphosphoglyceric acid to form 3-phosphoglyceric acid (see Figure 8.1), that of phosphoenolpyruvic acid to form pyruvic acid (see Figure 8.1), and that of succinyl coenzyme A to form succinate (see Figure 8.2). These phosphorylations, in contrast to oxidative phosphorylations, are known collectively as **substrate-level phosphorylations** because they occur immediately in the reactions of substrates of glycolysis and the Krebs cycle. Altogether, 6 ATP molecules are formed by substrate-level phosphorylation per molecule of glucose under aerobic conditions. In total, therefore, 31 molecules of ATP are generated for each glucose molecule catabolized. However, 2 ATP molecules are consumed at the start of glycolysis (see Figure 8.1).

The *net* yield of ATP by aerobic catabolism is therefore 29 molecules of ATP per glucose. The *energetic efficiency* of this ATP production—defined by Equation 7.1—is not altogether certain and can vary even in a tightly coupled cell, depending on metabolite concentrations and other conditions.[10] In glucose catabolism, assuming tight coupling, the efficiency is usually considered to be 60%–70%. That is, 60%–70% of the chemical energy released by oxidizing glucose to CO_2 and H_2O is captured as chemical-bond energy of ATP.

O_2 deficiency poses two biochemical challenges: Impaired ATP synthesis and potential redox imbalance

What are the biochemical implications if cells are denied O_2 or supplied with O_2 at an inadequate rate? To explore this question, let's for simplicity consider cells deprived entirely of O_2, even though for many animals this would be an unrealistically extreme state.

Without O_2, electrons entering the electron-transport chain cannot be discharged at the end by the reduction of O_2. Thus the electron-transport chain becomes a *dead end* for electrons. The consequences for most animals are easily summarized. Soon after the supply of O_2 is cut off, all the cytochromes (see Figure 8.3A) and other constituents of the electron-transport chain become fully reduced because electrons entering the chain accumulate. The electron-transport chain then can no longer accept further electrons. Two major consequences follow. First, oxidative phosphorylation cannot take place, eliminat-

---

[9] This value is calculated using the latest estimates of P/O ratios: P/O = 2.3 for most $NADH_2$ oxidation, P/O = 1.4 for $FADH_2$ oxidation. The paper by Martin Brand in the References provides a full, accessible explanation. By way of background, note that although both $NADH_2$ and $FADH_2$ are oxidized by the electron-transport chain, their electrons enter the chain at different points, giving the compounds different P/O ratios; also, a factor discussed in biochemistry texts is that $NADH_2$ produced in glycolysis has a lower effective P/O ratio than most $NADH_2$ because the former is produced outside the mitochondria and its electrons must be shuttled in. The P/O ratios used here are lower than those traditionally used (P/O = 3 for most $NADH_2$ oxidation, P/O = 2 for $FADH_2$ oxidation) because of modern revisions in the estimates.

[10] For example, ATP does not have a fixed energy value, as often implied by introductory biology texts. Instead, the "energy content" of ATP depends on how close, or far away, the ATP–ADP interconversion reaction is to chemical equilibrium, which in turn depends on relative concentrations of the reactants in a particular cell.

ing a cell's ability to produce 25 of the 29 ATP molecules that can be produced in net fashion by the aerobic catabolism of glucose. Second, the electron-transport chain can no longer serve as a mechanism for reoxidizing the reduced molecules of $NADH_2$ and $FADH_2$ that are produced by glycolysis and the Krebs cycle. The cell's supply of NAD and FAD is thus threatened. A complete failure to regenerate NAD and FAD would mean that a cell could produce no ATP at all, because NAD is needed for glycolysis, and both NAD and FAD are required for operation of the Krebs cycle.

The inability of the electron-transport chain to regenerate NAD and FAD from $NADH_2$ and $FADH_2$ in the absence of O_2 can be described as a problem of *redox balance*, meaning *reduction–oxidation balance*. **Redox balance** is a key concept in the study of all cellular compounds that undergo alternating reduction and oxidation. By definition, a cell is in redox balance for such a compound if the cell possesses the means to remove electrons from the compound as fast as electrons are added to it.

Applying the concept specifically to NAD, a cell is in redox balance for NAD if it can convert $NADH_2$ back to NAD (a process that removes electrons) as fast as NAD is being converted to $NADH_2$ (a process that adds electrons). During aerobic catabolism, O_2 and the electron-transport chain together provide the means for maintaining redox balance for NAD and FAD, thereby ensuring that NAD and FAD are steadily available to perform their essential roles in glycolysis and the Krebs cycle. *To make ATP without O_2, a cell must possess alternative mechanisms that will permit redox balance to be maintained while at least some ATP-generating reactions continue to take place.*

Certain tissues possess anaerobic catabolic pathways that synthesize ATP

Only some tissues have evolved an ability to make ATP at significant rates without O_2. The mammalian brain is a prime example of a tissue that lacks this ability. The brain succumbs rapidly when denied O_2 (as during cardiac arrest) because when O_2 is unavailable, the brain is unable to synthesize ATP at the rate it needs it. Many other vertebrate tissues possess an anaerobic catabolic pathway that can produce ATP at a substantial rate without O_2. These other tissues are therefore less rigidly dependent on an O_2 supply than the mammalian brain is.

Anaerobic glycolysis is the principal anaerobic catabolic pathway of vertebrates

Vertebrate skeletal muscle and many other vertebrate tissues are able to use the substrate-level phosphorylations of glycolysis to transfer energy from the bonds of glucose to ATP at substantial rates in the absence of O_2. For a tissue to have this ability, it must have a way of maintaining redox balance for NAD without O_2. Glycolysis requires NAD to serve as the immediate electron acceptor for the oxidation of glyceraldehyde-3-phosphate, and in using NAD in this way, it converts the NAD to $NADH_2$ (see Figure 8.1). Under anaerobic conditions in skeletal muscles and other such tissues, the $NADH_2$ molecules are reoxidized to NAD by passing their electrons (hydrogens) to pyruvic acid, reducing the latter to lactic acid (**FIGURE 8.5**). Pyruvic acid, in essence, is used as the final electron acceptor by the anaerobic cell. The entire sequence of reactions that converts glucose to lactic acid is called **anaerobic glycolysis**.

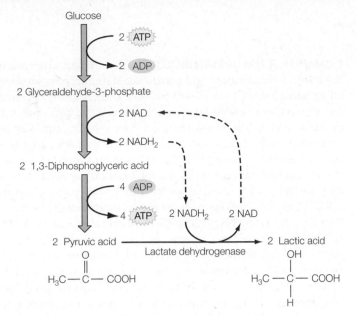

FIGURE 8.5 Anaerobic glycolysis The reduction of pyruvic acid is catalyzed by the enzyme lactate dehydrogenase (LDH) and permits redox balance to be maintained without O_2. The net yield of ATP is two ATP molecules per glucose molecule if the initial fuel is glucose itself, as shown here. If glycogen is used as fuel, there is a net yield of three ATPs per glucose unit catabolized.

The principal determinant of whether a cell can carry out anaerobic glycolysis at a substantial rate is its expression of the enzyme lactate dehydrogenase (LDH), the enzyme that catalyzes reduction of pyruvic acid. LDH occurs in multiple molecular forms, as discussed in detail in Chapter 2 (see page 50). *For a cell to carry out anaerobic glycolysis, it must synthesize significant amounts of one or more LDH forms that are catalytically effective in converting pyruvic acid to lactic acid under the intracellular conditions that prevail during O_2 deprivation.* In cells that meet these criteria, a true redox balance exists during anaerobic glycolysis. As Figure 8.5 reveals, one molecule of pyruvic acid is produced for each molecule of NAD that is reduced to $NADH_2$ during glycolysis. Thus the supply of pyruvic acid keeps exact pace with the need for it as an electron acceptor. A cell with appropriate amounts and types of LDH is thus able to convert every $NADH_2$ that it forms back to NAD without O_2. This means that from a strictly biochemical perspective, the cell can maintain a steady production of ATP without O_2.

What fuels can anaerobic glycolysis use, and how much ATP can it make? Glucose and glycogen are the only foodstuff molecules that can serve as fuels for anaerobic glycolysis under most conditions. The amount of ATP synthesized in net fashion by anaerobic glycolysis is two molecules of ATP per glucose molecule catabolized (see Figure 8.5). This yield of ATP is vastly lower than that from the aerobic catabolism of glucose. However, as we will see, the anaerobic pathway can nonetheless be an impressively important source of ATP. The major reason for the comparatively low yield of ATP per glucose molecule during anaerobic glycolysis is that only a small fraction (about 7%) of the free energy available from glucose is released by the conversion of glucose to lactic acid. Lactic acid is itself an energy-rich molecule.

Anaerobic glycolysis is not the only mechanism of anaerobic catabolism in animals. However, we will take up other anaerobic pathways later and now address the important matter of the disposition of end products.

What happens to catabolic end products?

For an animal to make use of any ATP-producing pathway, it must possess satisfactory means of disposing of the chemical end products generated. The principal products of *aerobic* catabolism, CO_2 and H_2O, are fully oxidized and not capable of being tapped for further energy. Animals typically dispose of CO_2 and H_2O by voiding them into the environment, as when we exhale CO_2 by breathing.[11]

The end products of anaerobic catabolic pathways, in contrast to CO_2 and H_2O, are always organic compounds that are far from fully oxidized and possess considerable further potential to yield energy. The high energy value of these products places a premium on retaining them in the body for future use as energy sources; excreting lactic acid, for instance, would be equivalent to voiding more than 90% of the energy value of catabolized carbohydrates. However, unlimited retention of the organic end products of anaerobic catabolism is usually not possible, because the end products typically exert harmful effects if allowed to accumulate to high concentrations. These two competing considerations seem to have been major operative factors in the evolution of animal strategies for disposing of anaerobic end products. A third factor of importance is time. An animal that makes only short-term use of anaerobic catabolic pathways is more likely to be able to retain all the end-product molecules without adverse effects than an animal that uses anaerobic catabolism for days or weeks on end.

THE DISPOSITION OF LACTIC ACID IN VERTEBRATES When vertebrates use anaerobic glycolysis, they universally retain the lactic acid they produce in their bodies, and they ultimately get rid of the lactic acid *metabolically*. The metabolism of lactic acid requires O_2 and thus can take place only when and where O_2 becomes available. Lactic acid itself is a metabolic cul-de-sac (**FIGURE 8.6**), meaning that for lactic acid to be metabolized, it must first be converted back to pyruvic acid by a reversal of the very reaction that formed it. The

---

[11] In the study of water balance, the H_2O produced by aerobic catabolism is known as *metabolic water* or *oxidation water*. This water is sometimes an important part of the water budget (see Chapter 27).

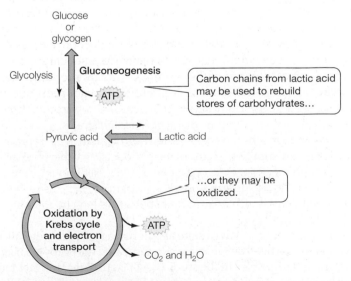

FIGURE 8.6 Major paths by which lactic acid is metabolized when O_2 is available The green arrows show the paths followed by the carbon chains of lactic acid. The carbon chains are first converted to pyruvic acid—and then can be converted to glucose or oxidized.

conversion of lactic acid to pyruvic acid is an oxidation reaction, with NAD acting as the immediate electron acceptor. After conversion to pyruvic acid, the carbon chains of lactic acid are metabolized by one of two major paths, both of which require O_2 (see Figure 8.6):

1. One path is for the carbon chains to be used to form glucose or glycogen, thereby replenishing a cell's carbohydrate stores. The conversion of lactic acid or pyruvic acid to glucose or glycogen is a form of **gluconeogenesis** ("new formation of glucose"). The process *uses ATP* (six ATP molecules per glucose molecule formed). It requires O_2 because the ATP must be made aerobically.

2. The second path is for the carbon chains to be fully oxidized by way of the Krebs cycle and electron-transport chain. This process *makes ATP* (27 ATP molecules per pair of lactic acid molecules). It requires O_2 because the electron-transport chain must be functional.

To determine the relative importance of the two paths followed by lactic acid, investigators inject animals with lactic acid containing the unusual carbon isotope ^{14}C. They then monitor the fractions of the ^{14}C that show up in CO_2 and carbohydrate. These fractions vary, even within single species, in ways that are still poorly understood.

Lactic acid may be metabolized by the same tissues that produce it, or it may be carried by the blood to other tissues that metabolize it. Quite often, tissues that are well supplied with O_2 metabolize lactic acid made by other tissues that are deprived of O_2; this can occur even while lactic acid production continues.

The functional roles of ATP-producing mechanisms depend on whether they operate in steady state or nonsteady state

One of the most consequential properties of any ATP-producing mechanism is whether or not it can operate in **steady state**. Ideally, a mechanism is in steady state if (1) it produces ATP as fast as ATP is used, (2) it uses raw materials (e.g., foodstuff molecules) no faster than they are replenished, (3) its chemical by-products (besides ATP) are voided (or metabolically destroyed) as fast as they are made, and (4) it does not cause other changes in cell function that progress to the point of disrupting cell function. When a cell uses a **steady-state mechanism** of ATP production, even as ATP is made and used in the cell, the cell remains essentially constant in its levels of ATP and of the precursors and by-products of ATP production, and it also remains essentially homeostatic in other ways as well. At least in principle, therefore, *a steady-state mechanism can go on and on indefinitely, free of intrinsic limitations*.

In contrast, a **nonsteady-state mechanism** of ATP production depletes supplies, accumulates products, or otherwise alters the conditions of its own operation at rates fast enough that the mechanism is *self*-limiting or *self*-terminating. For this reason, a nonsteady-state mechanism cannot persist for long by comparison with steady-state ones.

The concepts of steady-state and nonsteady-state mechanisms are relative rather than absolute. Nonetheless, they provide useful organizing principles in the study of ATP synthesis. In vertebrates and many invertebrates, aerobic ATP production is usually a steady-state process, whereas anaerobic ATP production is usually a nonsteady-state process.

EXAMPLES FROM HUMAN BIOLOGY To understand the practical meaning of steady-state and nonsteady-state mechanisms better, let's examine some examples in our own lives. Routine daily living provides a good illustration of a circumstance in which ATP is produced by aerobic catabolism functioning in steady state. Consider what you are doing now, for instance. As you read this page, unless you have bizarre reading habits, you are in an aerobic steady state. You are making all your ATP by aerobic catabolism; your needs for ATP are fully met; you are taking in O_2 at the rate you use it; you are voiding CO_2 and H_2O approximately as you produce them; and your cells are not undergoing any progressive alterations that might disrupt their function. The one way in which you may not be in steady state is in regard to the foodstuff molecules you are using as fuels. If a person is not eating while reading, fuel levels are obviously falling. However, the rate of decline during reading is very low relative to fuel stores, and intakes of fuels probably closely match uses when averaged over a day (including meals). Thus, even with respect to fuels, the departure from steady state is merely technical.

The capability of the aerobic catabolic pathways to operate in steady state is of great significance when we consider the ATP requirements of daily living. Because aerobic catabolism can supply ATP at the required rates while operating in steady state, it can go on and on without self-limitation and meet our ATP needs for a lifetime!

A mile race is a good example of an activity that is fueled substantially by anaerobic ATP production functioning in nonsteady state. As we will see in detail shortly, the rate at which ATP is used in a competitive mile race is far greater than the maximum rate at which aerobic catabolism can make ATP. Thus additional ATP must be provided by anaerobic glycolysis throughout such a race. Anaerobic glycolysis is a nonsteady-state mechanism of ATP synthesis during a mile race because its use progressively induces alterations in the cells of the working muscles—alterations that will ultimately render continued running *physiologically impossible*.[12] The process is self-limiting. The exact alterations that are mechanistically responsible for this self-limitation have yet to be nailed down. As we will discuss later in greater detail, the alterations include (1) accumulation of lactic acid (probably not responsible for fatigue), (2) acidification (probably partly responsible for fatigue but not the major player), and (3) other changes that are probably the principal causative agents.

For now, two principal points regarding the mile race are important to see. First, the mile race contrasts with routine daily living in that it cannot go on indefinitely. Second, the key reason for this difference lies in the mechanisms of ATP production: nonsteady-state during the mile race, steady-state during daily living.

Phosphagens provide an additional mechanism of ATP production without O_2

Phosphagens are compounds that serve as temporary stores of high-energy phosphate bonds. They occur in the skeletal muscles of vertebrates and the muscles of many invertebrates. *Creatine phosphate* (**FIGURE 8.7A**) is the phosphagen of vertebrate muscle and also occurs in some groups of invertebrates. The most widespread phosphagen of invertebrates (including the crayfish with which

---

[12] Mile runners collapse near the finish line fairly often in championship races.

(A)

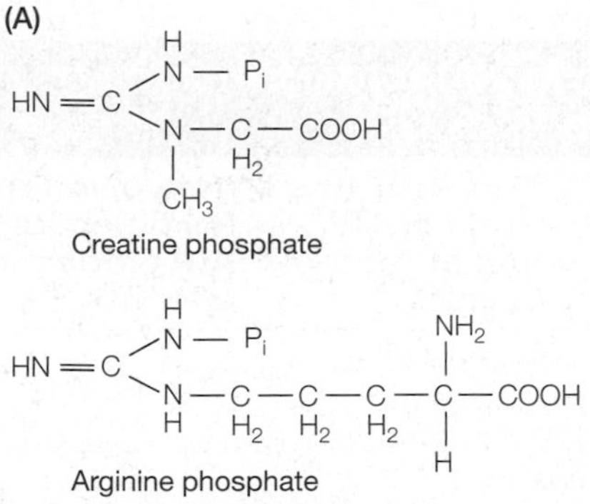

Creatine phosphate

Arginine phosphate

(B)

Creatine phosphate + ADP ⇌ creatine + ATP
Creatine kinase

Arginine phosphate + ADP ⇌ arginine + ATP
Arginine kinase

FIGURE 8.7 Two important phosphagens and the reactions they undergo (A) Creatine phosphate and arginine phosphate. (P_i = inorganic phosphate). (B) The reversible reactions by which these phosphagens are formed from ATP or can donate their phosphate groups to ADP to make ATP. The reactions are catalyzed by creatine kinase and arginine kinase.

we started this chapter) is *arginine phosphate* (see Figure 8.7A). Other phosphagens are known, especially among annelids. Phosphagens are synthesized by use of high-energy phosphate bonds taken from ATP, and later they can donate the phosphate bonds to ADP to form ATP. The reversible reactions for creatine phosphate and arginine phosphate, catalyzed by *phosphagen kinases* (*creatine kinase* and *arginine kinase*), are shown in **FIGURE 8.7B**.

Chemical mass action determines the direction of a phosphagen reaction.[13] For example, consider the reaction involving creatine phosphate and creatine in Figure 8.7B. During times when a vertebrate muscle cell is at rest, when ATP supplies in the cell are relatively untaxed and ATP concentrations are high, the reaction is shifted to the left: ATP reacts with molecules of creatine, and consequently most of the creatine in the resting cell becomes phosphorylated. The molar concentration of creatine phosphate in a vertebrate muscle cell might then be three to six times the concentration of ATP. Later, during times when ATP supplies are taxed and ATP concentrations fall, the reaction is shifted to the right, *forming ATP without a simultaneous need for O_2*. Each ADP molecule in a vertebrate muscle cell can be rephosphorylated three to six times by this mechanism, assuming the concentration ratios we just mentioned. Anaerobic production of ATP in this way is a nonsteady-state mechanism because it exhausts the supply of creatine phosphate and causes buildup of the end product creatine.

Internal O_2 stores may be used to make ATP

For certain purposes, it is useful to recognize that aerobic catabolism can function in two different modes, steady-state and nonsteady-state, depending on its source of O_2. Up to now, whenever we have

discussed aerobic catabolism, we have assumed that O_2 is taken into the body from the environment and distributed to cells as it is used. Aerobic catabolism is a steady-state mechanism under such conditions. Aerobic catabolism, however, can also draw its O_2 from preexisting stores of O_2 in a cell or tissue. Aerobic catabolism using O_2 stores is a nonsteady-state process because it depletes the O_2 stores on which it depends. Muscle cells often have enhanced O_2 stores because they contain *myoglobin*, a red O_2-binding pigment that is actually a type of hemoglobin (see Chapter 24). The myoglobin binds with O_2 when O_2 is abundant. Aerobic catabolism then can draw on these O_2 stores at other times.

SUMMARY
Mechanisms of ATP Production and Their Implications

■ Aerobic catabolism occurs in four steps: (1) glycolysis, (2) the Krebs cycle, (3) electron transport, and (4) oxidative phosphorylation. It can oxidize carbohydrates, lipids, or proteins to produce ATP. The maximum net yield of ATP in carbohydrate oxidation is currently estimated to be 29 ATP molecules per glucose molecule.

■ Anaerobic glycolysis is a redox-balanced process by which ATP can be made without O_2 in certain tissues. Whether a tissue can employ anaerobic glycolysis depends on the amount and type of its lactate dehydrogenase (LDH). Anaerobic glycolysis typically can use only carbohydrate fuel. It releases only about 7% of the energy of glucose. Thus it produces just two ATP molecules per glucose molecule, and the lactic acid produced is itself an energy-rich compound.

■ Vertebrates retain the lactic acid that they produce. The metabolism of lactic acid requires O_2, either to oxidize the lactic acid via the Krebs cycle—thereby producing ATP—or to convert lactic acid to glycogen or glucose (gluconeogenesis)—a process that uses ATP.

■ Phosphagens in muscle cells serve as temporary stores of high-energy phosphate bonds, which they can transfer to ADP to make ATP anaerobically.

Comparative Properties of Mechanisms of ATP Production

We have now identified four mechanisms of ATP production: (1) aerobic catabolism by use of O_2 acquired simultaneously from the environment, (2) anaerobic catabolism, (3) anaerobic ATP production by use of phosphagen, and (4) aerobic ATP production by use of internal O_2 stores. To compare the functional properties of these four, we now discuss several key questions. **TABLE 8.1** can be used as a summary of this discussion.

Question 1: What is each mechanism's total possible ATP yield per episode of use?

If an animal uses a particular mechanism to make ATP and continues using the mechanism until no more ATP can be made, what is the total amount of ATP that can be produced? Because aerobic catabolism using environmental O_2 operates in steady state, it is capable of supplying an indefinite amount of ATP (a lifetime's amount).

---

[13] See Chapter 2 (page 50) for a discussion of the meaning of mass action.

TABLE 8.1 Comparative properties of mechanisms of ATP production in vertebrates, including numerical estimates for some of the properties in human beings

Mechanism of ATP production	Mode of operation (mandatory or assumed)	Total possible ATP yield per episode of use[a] (moles)	Rate of acceleration of ATP production at onset of use	Peak rate of ATP production[b] (μmol ATP/g·min)	Rate of return to full potential for ATP production after use
Aerobic catabolism using O_2 from environment	Steady state	Very large (~200 in a marathon, $> 4 \times 10^6$ in a lifetime)	Slow	Moderate (30 with glycogen fuel, 20 with fatty acid fuel)	—
Anaerobic glycolysis	Nonsteady state	Moderate (1.5)	Fast	High (60)	Slow
Phosphagen use	Nonsteady state	Small (0.4)	Fast	Very high (96–360)	Fast
Aerobic catabolism using O_2 preexisting in body	Nonsteady state	Small (0.2)	Fast	High	Fast

[a]Numerical estimates of total yields are computed from information in Åstrand and Rodahl 1986; a 75-kg person living 70 years is assumed.
[b]Peak rates of production are from Hochachka and Somero 2002.

In contrast, the three mechanisms that operate in nonsteady state are limited in the amount of ATP they can produce during any one episode of use because of their mechanistic self-limitations. For example, the amount of ATP that can be made from phosphagen is limited by the quantity of phosphagen available; after the phosphagen originally present has been used, no more ATP can be made until (at some later time) the phosphagen is regenerated. Table 8.1 shows representative yields of ATP in humans. Phosphagen and O_2 stores permit only relatively small quantities of ATP to be made per episode of use. The yield from anaerobic glycolysis is larger but modest.

Question 2: How rapidly can ATP production be accelerated?

In vertebrates and many types of invertebrates (although not insects), aerobic catabolism using environmental O_2 requires a relatively long time (minutes) to accelerate its rate of ATP production fully to a new high level. Acceleration is slow because this mechanism of ATP production is *not self-contained in cells*. The mechanism requires inputs of O_2, and the processes of breathing and blood circulation—which supply the needed O_2 by transporting it from the environment to the cells—increase their rate of O_2 delivery *gradually, not instantly*.

By contrast, the other three mechanisms of ATP production can accelerate very rapidly because all *are self-contained in the cells*. At the start of a bout of exercise, for example, a vertebrate muscle cell contains not only all the enzymes of anaerobic glycolysis but also the glycogen fuel required. Nothing needs to be brought to the cell for anaerobic glycolysis to take place, and in fact a cell is capable of nearly instantly stepping up its glycolytic rate of ATP production to a high level. ATP production by use of phosphagen and O_2 stores can also accelerate very rapidly because the phosphagen and O_2 stores are already present in cells when the mechanisms are called into play.

In both vertebrates and crustaceans such as crayfish, the first stages of burst exercise receive their ATP supply principally from anaerobic glycolysis, phosphagen use (**BOX 8.2**), and use of O_2 stores. This is true because these are the three mechanisms of ATP

production that can accelerate their rates of ATP production very rapidly. Vertebrates would be *incapable* of burst exercise without these mechanisms. As bizarre as it might be to imagine, if aerobic catabolism using O_2 from the environment were the only means of making ATP, vertebrates would be unable to start running suddenly at top speed; instead, they would have to accelerate their running gradually over a period of minutes!

Question 3: What is each mechanism's peak rate of ATP production (peak power)?

After a mechanism has accelerated its ATP production to the fastest rate possible, how fast can ATP be made? As shown in Table 8.1, the peak rate at which anaerobic glycolysis can make ATP is much greater than the peak rate of ATP production by aerobic catabolism using environmental O_2, and the rate of ATP production by use of phosphagen is greater yet. Although the phosphagen mechanism cannot make a lot of ATP, it can make its contribution very rapidly; thus it can briefly support very intense exertion. Anaerobic glycolysis can make a modest amount of ATP at a high rate, and aerobic catabolism using environmental O_2 can make an indefinite amount at a relatively low rate.

Question 4: How rapidly can each mechanism be reinitialized?

Whenever ATP has been made by a nonsteady-state mechanism, cells are left in an altered state and must be returned to their original state before the mechanism can be used again to full effect. When internal O_2 stores have been used, the stores must later be recharged. When phosphagen has been used, it must be remade. When lactic acid has accumulated, it must be metabolized. In a word, the nonsteady-state mechanisms must be *reinitialized*.

The reinitialization of anaerobic glycolysis requires much more time than the reinitialization of the phosphagen or O_2-store mechanisms. This is true because the length of time required to rid tissues of lactic acid is much greater than the time required to rebuild phosphagen or O_2 stores. In people, for example, a substantial accumulation of lactic acid may require 15–20 min for

BOX 8.2 Genetic Engineering as a Tool to Test Hypotheses of Muscle Function and Fatigue

Experiments based on genetic engineering are increasingly being used to test physiological hypotheses. Genetic engineering methods, for example, are being used extensively to study muscle function and fatigue. One example is provided by research on the role of the phosphagen *creatine phosphate*. Because creatine phosphate in mammalian muscle can be mobilized extremely rapidly to make ATP at a high rate (see Table 8.1), physiologists have long hypothesized that a cell's phosphagen serves as a principal source of ATP during the first seconds of burst exercise. ATP synthesis from creatine phosphate depends on the enzyme *creatine kinase* (*CK*) (see Figure 8.7B). One way to test the hypothesis of phosphagen function is to lower or raise the levels of CK in muscle cells by genetic engineering

methods. If the hypothesized role of creatine phosphate in burst exercise is correct, lowering or raising the levels of CK should interfere with or facilitate burst exercise.

Mutant mice *deficient* in CK have been reared by genetic engineering methods. The muscles of these mice clearly exhibit compensatory adjustments that tend to make up for the effects of their CK deficiency (see page 83). Even with such compensations, however, the muscles exhibit a subnormal ability to perform burst activity, and this performance deficit increases with the extent of their CK deficiency. Mice have also been engineered to produce unusually *high* amounts of CK. Their muscles contract faster than normal muscles in the first moments of isometric twitches. These experiments

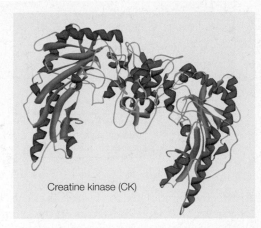

Creatine kinase (CK)

support the hypothesized role of creatine phosphate in burst exercise.

Box Extension 8.2 discusses another example of the application of genetic engineering methods to the study of muscle, focusing on fatigue.

---

half dissipation and even 1–2 h for full dissipation, whereas the half-time for reconstituting phosphagen and O_2 stores is just 30 s. Fish, frogs, lizards, and other poikilothermic vertebrates require even more time than mammals to dissipate lactic acid; 1–10 h may be required for half dissipation.

Conclusion: All mechanisms have pros and cons

A review of Table 8.1 and the preceding discussion reveals that each mode of ATP production has pros and cons; none is superior in all respects. Let's consider, for example, the relative pros and cons of the two major sources of ATP in vertebrates: aerobic catabolism using environmental O_2 and anaerobic glycolysis.

When one looks simply at biochemical maps, aerobic catabolism tends to seem far superior to anaerobic glycolysis. Aerobic catabolism fully releases the energy of food molecules (producing 29 ATPs per glucose), whereas anaerobic glycolysis unlocks only a small fraction of the energy value of food molecules (producing 2 ATPs per glucose). Moreover, anaerobic glycolysis typically is able to use only carbohydrate fuels, produces only a limited total amount of ATP in any one episode, and creates a product, lactic acid, that requires a long time to be cleared from the body.

Anaerobic glycolysis also has great advantages in comparison with aerobic catabolism, however. Because anaerobic glycolysis does not require O_2, it can provide ATP when O_2 is unavailable, or it can supplement aerobic ATP production when O_2 is insufficient. Of equal importance, anaerobic glycolysis can accelerate very rapidly and reach an exceptionally high rate of ATP production almost instantly. These properties make anaerobic glycolysis indispensible for meeting the ATP needs of burst exercise.

Two Themes in Exercise Physiology: Fatigue and Muscle Fiber Types

During muscular exercise, performance is affected by fatigue and by the fact that a muscle often consists of two or more types of fibers (cells) that differ in their contractile and metabolic properties. Here we discuss these important themes.

Fatigue has many, context-dependent causes

Muscle fatigue is an exercise-induced reduction in a muscle's ability to generate peak forces and maintain power output. It is a fascinating and critical physiological phenomenon—having life-and-death consequences in natural settings—that has defied full understanding. Fatigue clearly has multiple causes, depending on the type and duration of exercise, and on the physiological status of the individual performing the exercise.

The fatigue associated with lactic acid accumulation is one of the types of fatigue that is best understood phenomenologically. Humans or other vertebrates undergoing intense exercise that involves sustained net lactate production characteristically become profoundly overcome with fatigue when lactic acid accumulates to a certain level; virtual paralysis often sets in. Even in the case of this very regular and predictable type of fatigue, however, the cellular and molecular mechanisms are not well understood. In the 1960s and earlier, the lactate ion was considered to be a specific "fatigue factor." By 1990, however, investigators had decided that the acid–base disturbance associated with lactic acid accumulation—acidification of muscle cells—was the real cause of this fatigue. By now, acidification—although it plays a role—seems unlikely to be the principal cause, based on studies in which muscle cell pH has

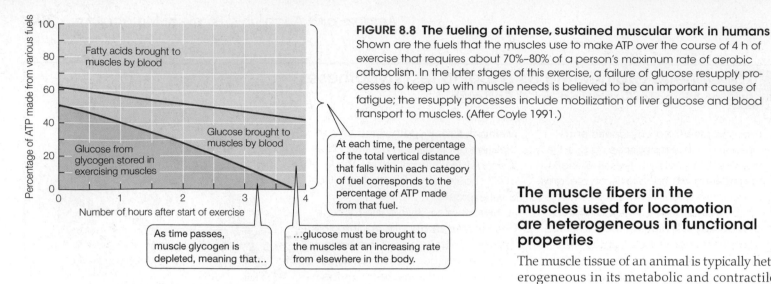

FIGURE 8.8 The fueling of intense, sustained muscular work in humans Shown are the fuels that the muscles use to make ATP over the course of 4 h of exercise that requires about 70%–80% of a person's maximum rate of aerobic catabolism. In the later stages of this exercise, a failure of glucose resupply processes to keep up with muscle needs is believed to be an important cause of fatigue; the resupply processes include mobilization of liver glucose and blood transport to muscles. (After Coyle 1991.)

been directly manipulated to observe its effects. Investigators have not, however, reached a consensus on the actual mechanism. Accumulation of lactic acid clearly serves as an indicator—a proxy—for the development of this sort of fatigue without lactic acid being the principal causative agent.[14]

Some types of fatigue seem to be partly caused by changes in organ-level systems. For example, a type of fatigue that meets this description and is fairly well understood is the fatigue that develops after a long time during intense, *sustained* (aerobically fueled) exercise (**FIGURE 8.8**). When people engage in this sort of exercise, fatigue is associated with (and probably partly caused by) inadequate muscle glucose. In the first tens of minutes of such exercise, glucose from muscle glycogen fuels the portion of muscle ATP production that is based on glucose oxidation. As time passes, however, muscle glycogen becomes depleted,[15] and the muscles become dependent on glucose brought to them from other organs (notably the liver) by blood flow (see Figure 8.8). When the rate of this glucose resupply is inadequate, fatigue sets in. Hyperthermia (unusually high body temperature) is another organismal factor that sometimes causes fatigue.

Usually, however, fatigue is caused by changes in cell function and molecular function in the exercising muscles, explaining why the overwhelming focus of modern fatigue research is on cellular-molecular mechanisms (see Box 8.2). As discussed in Chapter 20, both the excitation of muscle and the contraction of muscle entail ion fluxes across cellular or intracellular membranes mediated by ion transport proteins. Contraction, moreover, entails interactions among the contractile proteins, ATP, and other cellular molecules. Disruption of any of these constituents has the potential to diminish a muscle cell's ability to develop force and maintain power output. For instance, during the sort of sustained exercise discussed in the last paragraph, part of the cause of fatigue seems to be a gradual accumulation of critical ions (e.g., Ca^{2+}) in incorrect cellular locations as a consequence of the repetitive ion fluxes involved in long-duration muscular effort.

The muscle fibers in the muscles used for locomotion are heterogeneous in functional properties

The muscle tissue of an animal is typically heterogeneous in its metabolic and contractile properties. In vertebrates, for example, several different types of muscle fibers (muscle cells)[16] occur in skeletal muscle. Two of the principal types in mammals are termed **slow oxidative (SO)** fibers and **fast glycolytic (FG)** fibers. The SO fibers are called *slow* because they contract and develop tension relatively slowly, in contrast to the FG fibers, which are *fast* in contracting and developing tension (see Table 20.2). The SO fibers are called *oxidative*, whereas the FG fibers are called *glycolytic*, because of differences in the physiology of their ATP production. The SO fibers are poised principally to make ATP by aerobic catabolism; they have high levels of key enzymes specific to aerobic catabolism, such as succinic dehydrogenase (a Krebs-cycle enzyme), and are well endowed with mitochondria. The FG fibers, by contrast, are poised principally to make ATP anaerobically; they are high in enzymes of anaerobic glycolysis, such as lactate dehydrogenase, and are relatively sparse in mitochondria.

Correlated with the differences in the catabolic pathways they use, the SO and FG fibers differ in how readily they can take up and store O_2. SO fibers are relatively rich in myoglobin, which not only helps store O_2 within the fibers but also aids diffusion of fresh O_2 into the fibers. FG fibers are low in myoglobin. Because of the difference in myoglobin content, SO fibers are reddish in color, whereas FG fibers are whitish, explaining why the SO fibers are sometimes called *red* fibers, and the FG fibers *white*.

SO and FG fibers also differ in their power-generation and fatigue properties in ways that correlate with the features already discussed. The properties of SO fibers are predictable from the fact that they mainly use steady-state aerobic catabolism to make ATP: Although SO fibers have relatively low peak mechanical-power outputs (see Table 8.1), they are relatively resistant to fatigue and readily sustain work over long periods. FG fibers represent the opposite extreme. They can generate a high power output, but they rely strongly on nonsteady-state mechanisms of ATP production, accumulate lactic acid, and fatigue quickly.

In mammals, major skeletal muscles are typically built of mixes of intermingled SO fibers, FG fibers, and other fiber types (see Chapter 20). Fish, by contrast, often have large muscle masses composed principally of a single type of muscle fiber similar to the mammalian SO or FG fibers; thus entire blocks of muscle in fish are red or white and exhibit the performance properties of red or white fibers.

---

[14] Some investigators even argue that lactic acid *aids* muscular work, such as through effects on cell-membrane ion channels that help prevent deleterious changes in intracellular and extracellular ion concentrations.

[15] Recall (see Figure 6.2) that carbohydrate stores are small in comparison with fat stores. Even when a person's glycogen stores are fully loaded, the human body contains only about 450 g (1 pound) of glycogen. The small size of the glycogen stores helps explain why glycogen depletion can occur during prolonged exercise.

[16] A muscle fiber is a muscle cell. "Fiber" and "cell" are synonyms in the study of muscle.

The Interplay of Aerobic and Anaerobic Catabolism during Exercise

When animals engage in exercise, their performance typically reflects the underlying mechanisms that they are using to produce the ATP required. Fish that are cruising about at relatively leisurely speeds, for example, do so with their red swimming muscles, employing steady-state aerobic catabolism to make ATP. Cruising, therefore, can be sustained for long periods. However, fish that engage in sudden, intense exertion—such as cod avoiding a trawling net or salmon leaping waterfalls—use their white swimming muscles and anaerobic glycolysis to generate the high power they require.[17] They thus accumulate lactic acid, and if they must perform repeatedly in a short time, they are in danger of the sort of fatigue associated with lactic acid.

Crustaceans similarly illustrate that exercise performance reflects the underlying mechanisms of ATP production. When crayfish, lobsters, and crabs walk or cruise about at modest, sustainable speeds, they produce ATP by steady-state aerobic catabolism. When a crayfish or lobster employs tail flipping to power itself rapidly away from danger, however, the tail muscles require ATP at a greater rate than aerobic catabolism can provide. Thus the muscles turn to mechanisms that can produce ATP at exceptionally high rates. In species of crayfish studied in detail, the tail muscles exhibit high levels of arginine kinase and lactate dehydrogenase—enzymes instrumental in anaerobic ATP production—and low levels of aerobic catabolic enzymes. The primary mechanism of ATP production at the start of escape swimming by tail flipping is use of the phosphagen arginine phosphate. Later, if tail flipping continues, anaerobic glycolysis—producing lactic acid—is brought into ever-greater play to meet the ATP demands of the tail muscles. Because both of these mechanisms are nonsteady-state and self-limiting, tail flipping cannot be sustained for long.

A huge amount of information exists on lactic acid concentrations in relation to fatigue. Although lactic acid is not the *cause* of fatigue, it is nonetheless often a useful, readily measured *index* of cell conditions causing fatigue. As a practical index, animals are often considered to have an upper limit of lactic acid accumulation. According to this view, after one bout of exercise has caused a buildup of lactic acid, a closely following second bout is limited in the degree to which anaerobic glycolysis can be used because newly made lactic acid adds to the preexisting amount. A human example is provided by athletes who compete in two burst-type races in a single track meet; performance in the second race is often impaired if lactic acid accumulated in the first race has not yet been fully metabolized. Clearing lactic acid from the body takes tens of minutes in mammals and can take hours in poikilotherms. For animals in nature, the slow rate of removal of lactic acid may well mean that an animal is impaired for a substantial time in its ability to pursue prey or escape danger. These important behavioral consequences are additional manifestations of the high degree of relevance of the biochemistry of ATP production.

Some types of animals follow patterns that are different from those of vertebrates and crustaceans. Insects are the most dramatic

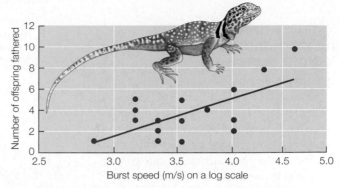

FIGURE 8.9 Modern genetic paternity testing reveals that the number of offspring fathered by male collared lizards (*Crotaphytus collaris*) depends directly on how fast they can run during burst running Each symbol represents one male in a free-living population. Males defend territories where they mate with females. The defense of a territory depends in part on sudden, high-intensity running to repel male invaders or meet other threats. (After Husak et al. 2006.)

and important example. Insect flight muscles typically have little or no ability to make ATP anaerobically and remain fully aerobic even when they suddenly increase their power output, as at the onset of flight. Correlated with the fact that virtually all flight-muscle work is aerobic, insect flight muscles contain very high levels of aerobic catabolic enzymes, and in some species, half the tissue volume of these muscles is mitochondria! The flight muscles of certain insects, in fact, are the most aerobically competent of all animal tissues as judged by the rate per gram at which they can synthesize ATP by aerobic catabolism. A characteristic of insects that helps explain their aerobic fueling of burst exercise is their tracheal breathing system, which provides O_2 directly to each flight-muscle cell by way of gas-filled tubes (see page 629).

When squids and other molluscs use anaerobic catabolism for burst swimming, they exhibit a curious difference from vertebrates. Their swimming muscles reduce pyruvic acid to *octopine* rather than lactic acid. The octopine-generating anaerobic pathway provides a high rate of ATP synthesis, as in vertebrates, but has different detailed biochemical implications.

Underlying much of scientific thinking about exercise in the wild is an assumption that a high capacity for exercise performance is advantageous for survival and reproduction—meaning that a high capacity for performance is favored by natural selection. Although some tests of this hypothesis have not obtained positive results, direct evidence is in fact accumulating that, in a state of nature, individual animals of a single species vary in exercise performance and this variation affects their survival and reproductive success as hypothesized. An example is provided by male collared lizards (*Crotaphytus collaris*) that are defending territories where they mate with females. To defend these territories the males employ sudden, high-speed running to exclude intruders and meet other threats. Their burst running almost surely depends on ATP produced by anaerobic glycolysis because (see Table 8.1), in such running, they accelerate almost instantly and achieve high work intensities. Individual males vary substantially in how fast they can cover distance during burst running, from about 3 to 5 m/s. Investigators measured the burst running speeds of 16 males in a free-living population. Then they used genetic paternity testing to determine the number of offspring fathered by each. They found a statistically significant trend for the faster males to father more offspring (**FIGURE 8.9**).

---

[17] When we speak of red and white muscles in this context, we are referring to the distinction between the myoglobin-rich, SO-like muscles and the myoglobin-poor, FG-like muscles discussed in the previous section. This is an entirely different matter from the overall hue of a fish's muscle tissue, whether off-white in a cod or orange in a salmon.

Metabolic transitions occur at the start and end of vertebrate exercise

Multiple metabolic processes are involved in the provision of ATP for vertebrate exercise of all types. Details depend on the intensity of exercise, which is usefully indexed by comparison with an individual's maximum rate of aerobic catabolism. A given individual in a particular state of training is capable of a certain maximum rate of O_2 consumption.[18] Exercise that requires exactly this maximum is classified as **maximal exercise**. Exercise that requires less than the maximum rate of O_2 consumption is called **submaximal exercise**, and exercise that requires more than an individual's maximum rate of O_2 consumption is called **supramaximal exercise**.

Consider a bout of submaximal exercise that starts and ends abruptly, requires about 80% of an individual's maximum rate of O_2 consumption, and lasts 30 min or so. Let's assume we are talking about a person who is running, although the principles we will develop apply to all or most vertebrates. **FIGURE 8.10A** shows how the person's rate of O_2 consumption by breathing would change during this bout of exercise if *all* ATP were made on a moment-by-moment basis by aerobic catabolism using O_2 taken up from the environment by breathing. The person's rate of O_2 consumption would increase stepwise at the start of running and decrease stepwise at the end.

Actually, however, during this type of exercise, a person's rate of O_2 uptake by breathing changes as shown by the red line in **FIGURE 8.10B**. In the middle of the bout of exercise, the person's actual rate of O_2 uptake matches the theoretical O_2 demand of the exercise. However, there is a transition phase at the start of the exercise, when the person's actual rate of O_2 uptake is lower than the theoretical O_2 demand, and there is another transition phase at the end when the person's actual rate of O_2 uptake exceeds the theoretical O_2 demand.

THE TRANSITION PHASE AT THE START: THE OXYGEN DEFICIT
The reason for the transition phase at the start of the exercise we are considering is the fact, already mentioned, that the breathing and circulatory systems in vertebrates do not instantly increase the rate at which they deliver O_2 to the body. Instead, even if *exercise* starts *abruptly*, O_2 *delivery* to the tissues increases *gradually*. In people, 1–4 min are required for the breathing and circulatory systems to accelerate fully. During the period when the breathing and circulatory systems are accelerating their actual rate of O_2 delivery at the start of a bout of exercise, the body's supply of O_2 from the environment (actual O_2 uptake) is less than its theoretical O_2 demand for the exercise. This difference is termed an **oxygen deficit** (see Figure 8.10B).

[18] See Chapter 9 for an extensive discussion of the maximum rate of O_2 consumption and its determinants.

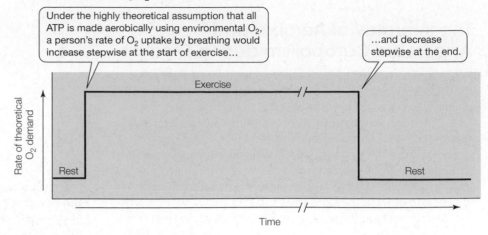

(A) Theoretical rate of O_2 consumption assuming that, moment-by-moment, all ATP is made using O_2 from the atmosphere

> Under the highly theoretical assumption that all ATP is made aerobically using environmental O_2, a person's rate of O_2 uptake by breathing would increase stepwise at the start of exercise...

> ...and decrease stepwise at the end.

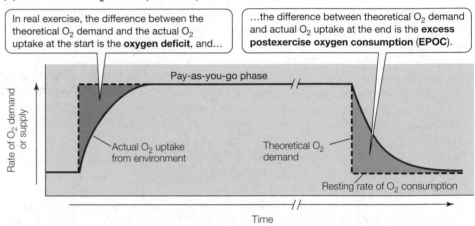

(B) Actual rate of O_2 consumption compared with theoretical

> In real exercise, the difference between the theoretical O_2 demand and the actual O_2 uptake at the start is the **oxygen deficit**, and...

> ...the difference between theoretical O_2 demand and actual O_2 uptake at the end is the **excess postexercise oxygen consumption (EPOC)**.

FIGURE 8.10 The concepts of oxygen deficit and excess postexercise oxygen consumption (A) The rate of O_2 demand of a person who is initially at rest, suddenly starts vigorous submaximal exercise, continues for 30 min or so, and then suddenly stops, assuming unrealistically that all ATP is produced aerobically, using atmospheric O_2, at all times. (B) The actual rate of O_2 uptake from the environment (red line) of the person in (A), showing that there is an initial transition period during which the full ATP demand is not met by O_2 uptake, then a period when O_2 uptake matches the full O_2 demand, and finally a transition period when the person's actual rate of O_2 uptake exceeds the resting rate even though the person is at rest.

During the period of the oxygen deficit, the full ATP demand of exercise is *not* met by aerobic catabolism based on environmental O_2. Where, then, does the other ATP come from? In the sort of exercise we are discussing, the answer is that ATP is contributed during the period of oxygen deficit by anaerobic glycolysis, use of phosphagen, and use of O_2 stores.[19] These three mechanisms, in fact, are *essential* for exercise to start in a *stepwise* way. They make up for the slow acceleration of ATP production by aerobic catabolism based on environmental O_2. They thereby permit the *overall* rate of ATP production to increase abruptly to a high level when exercise begins.

THE PAY-AS-YOU-GO PHASE During the exercise we are discussing or any other sort of submaximal exercise, the breathing and circulatory systems ultimately accelerate their rate of O_2 delivery sufficiently to meet the full O_2 demand of the exercise. The exercise is then said to enter a **pay-as-you-go phase** (see Figure 8.10B)

[19] Anaerobic glycolysis, use of phosphagen, and use of O_2 stores are sometimes termed the *mechanisms of oxygen deficit*.

(A) Light submaximal exercise

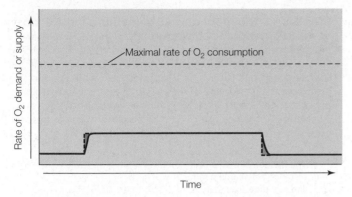

(B) Heavy submaximal exercise

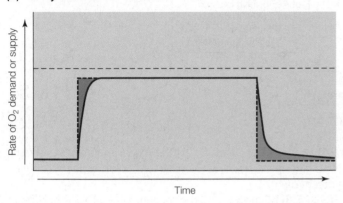

(C) Supramaximal exercise

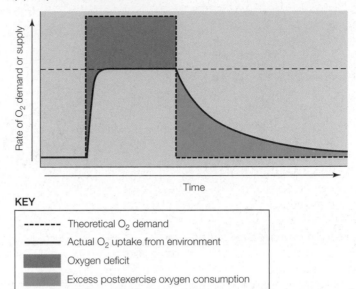

KEY

- - - - - Theoretical O_2 demand

——— Actual O_2 uptake from environment

▓▓ Oxygen deficit

▒▒ Excess postexercise oxygen consumption

FIGURE 8.11 Stylized O_2 supply–demand diagrams for light submaximal, heavy submaximal, and supramaximal exercise   The format is as in Figure 8.10. The patterns illustrated are those observed in mammals. Two important determinants of exercise performance are an individual's maximum rate of O_2 consumption and maximum oxygen deficit. Both are increased by training (repeated exercise), thereby increasing performance. A person, for example, might increase his or her maximum rate of O_2 consumption by 10%–30% through appropriate training.

because thereafter its full O_2 cost is met on a moment-to-moment basis by use of O_2 taken up from the environment by breathing. With all ATP being made by steady-state aerobic catabolism, the exercise can in principle be sustained indefinitely.

The pay-as-you-go phase in submaximal exercise begins soon enough, in general, to prevent biochemical self-termination of anaerobic glycolysis, use of phosphagen, and use of O_2 stores. Although those three mechanisms are susceptible to self-termination, they are not called upon for long enough in this sort of exercise to reach their inherent limits.

THE TRANSITION PHASE AT THE END: EXCESS POSTEXERCISE OXYGEN CONSUMPTION   At the end of the exercise we are describing, the exercising person suddenly stops running, but thereafter his or her actual rate of O_2 uptake does not suddenly drop in stepwise fashion. Instead, it declines gradually, remaining above the resting rate of O_2 consumption for many minutes (see Figure 8.10B). This elevation of the actual rate of O_2 uptake above the resting rate of O_2 uptake—even though the person is behaviorally at rest—is termed **excess postexercise oxygen consumption (EPOC)**. In everyday language, people describe the EPOC as "breathing hard" after exercise.[20] The EPOC has multiple causes, and the reasons for it are not fully understood at present.

---

[20] For several decades in the mid-twentieth century, the EPOC was called *oxygen debt*. That term has been discredited, however, because it was based on an assumption that the cause of the EPOC is simply the need to metabolize any lactic acid that was accumulated at the start of the exercise. Actually, there is often a nearly total lack of correspondence between the time course of the EPOC and the time course of lactic acid metabolism in vertebrates. The term *EPOC* is preferred over *oxygen debt* because it is functionally neutral and implies no particular mechanistic explanation.

THE TRANSITION PHASES VARY IN THEIR NATURE DEPENDING ON THE INTENSITY OF EXERCISE   Different intensities of exercise in vertebrates produce different sorts of transition phases. For example, if a vertebrate undertakes *light submaximal* exercise that requires less than 50%–60% of the individual's maximum rate of O_2 consumption, transition phases occur at the start and end, as shown in **FIGURE 8.11A**, but lactic acid does not accumulate at these exercise intensities. Without a lactic acid accumulation, the only processes required to "reinitialize" the body at the end of exercise are the replenishing of O_2 stores and phosphagen stores, both of which occur very rapidly (see Table 8.1).

In *heavy submaximal* exercise that requires more than 50%–60% of maximum O_2 consumption (**FIGURE 8.11B**), net accumulation of lactic acid occurs at the start. Accordingly, lactic acid must be metabolized at the end—a long process. The EPOC therefore lasts longer than in light submaximal exercise.

The most dramatic transition phases occur during *supramaximal* exercise (**FIGURE 8.11C**). Such exercise demands ATP at a greater rate than can *ever* be supplied by steady-state aerobic catabolism. Thus a pay-as-you-go phase is never reached, and anaerobic glycolysis must continue to be tapped for ATP for as long as the exercise continues—causing a steadily increasing oxygen deficit and buildup of lactic acid. *Supramaximal exercise is the principal form of exertion in which lactic acid accumulates to such high levels that the profound, debilitating, "lactic acid type" of fatigue can occur.* In fact, unless an individual electively stops exercising, supramaximal exercise typically undergoes metabolic self-termination within minutes. Then, associated with the large accumulation of lactic acid, a long time is required for the body to recover, during which the individual's capability for further supramaximal exertion is impaired.

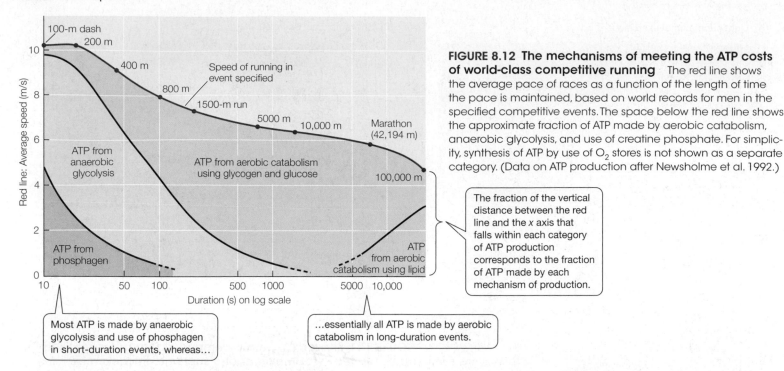

FIGURE 8.12 The mechanisms of meeting the ATP costs of world-class competitive running The red line shows the average pace of races as a function of the length of time the pace is maintained, based on world records for men in the specified competitive events. The space below the red line shows the approximate fraction of ATP made by aerobic catabolism, anaerobic glycolysis, and use of creatine phosphate. For simplicity, synthesis of ATP by use of O_2 stores is not shown as a separate category. (Data on ATP production after Newsholme et al. 1992.)

The fraction of the vertical distance between the red line and the x axis that falls within each category of ATP production corresponds to the fraction of ATP made by each mechanism of production.

Most ATP is made by anaerobic glycolysis and use of phosphagen in short-duration events, whereas…

…essentially all ATP is made by aerobic catabolism in long-duration events.

The ATP source for all-out exercise varies in a regular manner with exercise duration

If you consider human competitive running and reflect on the progression from a sprint to a mile race to a marathon, you will notice that as the duration of all-out exertion increases, the pace slows. Mile races are run slower than sprints, and marathons are run slower than mile runs. This trend in performance, which is a fairly general property of animal exercise, is a *direct reflection of the biochemistry of ATP production.*

In a person or other vertebrate, when all-out exertion lasts 10 s or so, the total amount of ATP needed, from start to finish, is relatively small (simply because ATP must be provided for only a short time). This means that anaerobic glycolysis, phosphagen, and O_2 stores can in principle meet the full ATP requirement. In actual practice, when people run the 100-yard dash (or 100-m dash), these three mechanisms of ATP production meet at least 90% of the ATP cost; some champions hold their breath from start to finish! Anaerobic glycolysis and the use of phosphagen and O_2 stores—while unable to make a great quantity of ATP—are able to produce ATP at exceptionally high *rates* (see Table 8.1). Thus, when these mechanisms are sufficient to meet most of the cost of running, the pace of running can be very fast, as it is in the 100-yard dash.

In a mile race (about the same as a 1500-m race), the total ATP requirement is much greater than in a 100-yard dash. Even if anaerobic glycolysis, phosphagen, and O_2 stores are fully exploited to make as much ATP as they possibly can during a mile race, they can meet no more than 25%–50% of the total ATP need. The rest of the ATP must be made by steady-state aerobic catabolism using O_2 from the environment, a mechanism that cannot produce ATP as fast as the other mechanisms (see Table 8.1). Thus, taking into account the rates of all the processes that are required to contribute ATP during a mile race, the overall rate of ATP production is lower than that during a 100-yard dash. The pace of the mile race must therefore be slower.

The pace of a marathon must be slower yet because the total ATP requirement of a marathon is so great that only 2%–3% of it can be met by use of anaerobic glycolysis, phosphagen, and O_2

stores. In a marathon, *almost all* of the ATP is made by steady-state aerobic catabolism, limiting the pace to that which is permitted by that mechanism.

FIGURE 8.12 shows how the mechanisms of ATP production vary with the duration of exertion in world-class competitive runs. The red line depicts the speed of running as a function of duration. The subdivision of the space below the red line shows how ATP is made. With increasing duration, ATP production shifts from being principally anaerobic (based on phosphagen and anaerobic glycolysis) to being chiefly aerobic. Moreover, at marathon and ultramarathon distances, aerobic catabolism shifts from exclusive use of carbohydrate fuels toward substantial use of lipid fuels, which permit only a lower rate of ATP synthesis than carbohydrate fuels (see Table 8.1).

Trends similar to those in Figure 8.12 apply to vertebrates living in the wild. Thus ecologically relevant performance depends in a regular way on the biochemical mechanisms by which ATP is made. As the primary mechanism of ATP synthesis shifts from phosphagen use to anaerobic glycolysis and then to aerobic catabolism based on carbohydrate and lipid fuels, the pace slows.

Related species and individuals within one species are often poised very differently for use of aerobic and anaerobic catabolism

Related species sometimes have evolved very different emphasis on aerobic and anaerobic ATP production during intense exertion, and these differences can have important life-history consequences. Thus the biochemistry of ATP production is one of the ways that species become specialized to live as they do.

The terrestrial amphibians we briefly noted in Chapter 2 (see Figure 2.10) provide a classic illustration. Some species, such as the common leopard frog (*Rana pipiens*) and many other ranid and hylid frogs, have a muscle biochemistry that emphasizes anaerobic glycolysis as the principal mechanism of ATP production during all-out exertion. If you chase a leopard frog, at first the frog flees by jumping away very rapidly, but within a few minutes it collapses in fatigue. Both the high speed of the initial jumping and

FIGURE 8.13 Two top athletes who differ in the fiber composition of their thigh muscles (A) The swimmer competes in 50-m sprints. (B) The cyclist competes in long-distance races. Shown to the right of each man is a microscopic section of his *vastus lateralis*, a thigh muscle, stained to make slow oxidative (SO) fibers dark. The sections are labeled with a different nomenclatural system from the one we use in this chapter: I = SO fibers; II = either FG fibers or other fast-contracting fibers. (From Billeter and Hoppeler 1992; courtesy of Rudolf Billeter-Clark and Hans Hoppeler.)

(A)

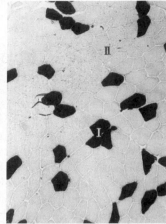

(B)

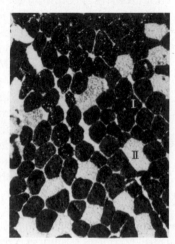

the quick fatigue reflect the emphasis on anaerobic glycolysis to make ATP. Many of the bufonid toads, such as the western toad (*Anaxyrus boreas*, formerly *Bufo boreas*), exemplify the opposite extreme. The toads have a muscle biochemistry that emphasizes aerobic production of ATP during all-out exertion. When chased, they do not flee as quickly as leopard frogs, but they can continue jumping at a steady pace for a long time. Both their slow speed and their resistance to fatigue reflect the aerobic fueling of their exercise. Lactate dehydrogenase (LDH) represents a key respect in which the frogs and toads differ; the leg muscles of the frogs express LDH at higher concentration than those of the toads.

A parallel, and very dramatic, example is provided by species of fish. The skipjack tuna (*Katsuwonus pelamis*) is a supremely vigorous species of fish. The enzymatic activity of LDH in its white swimming muscles exceeds the activity of LDH in sluggish fish species by more than 1000 times.[21] The tuna is thereby able to engage in exceptionally intense burst exercise.

At a different scale, individuals within a single species are often poised differently for use of aerobic and anaerobic catabolic pathways. A particularly intriguing example of this sort of variation is that humans vary widely in the fiber composition of their muscles, as illustrated in **FIGURE 8.13**. The swimmer at the top competes in sprints that require intense exertion for short periods. The cyclist, however, competes in sustained cycling races. Small pieces of tissue were removed from identical thigh muscles of the two men and subjected to a histochemical procedure that darkly stains slow oxidative (SO) muscle fibers. The images to the right in Figure 8.13 show that few of the muscle fibers of the swimmer are SO fibers, whereas most of the fibers of the cyclist are SO fibers. Most researchers believe that a difference of this specific sort (percentage of SO fibers) is relatively fixed. That is, people like the cyclist, it is thought, have principally SO fibers from early childhood. Such people discover through experience that their muscles, although not extremely powerful, are resistant to fatigue; and if such people are inclined toward athletic competition, they choose sports such as long-distance cycling that are well served by SO muscle fibers. In contrast, people like the swimmer, whose muscles are low in SO fibers and presumably high in fast glycolytic (FG) fibers, gravitate toward sports such as sprint swimming in which high power—rather than long-term endurance—is a key to success.

---

[21] Enzyme **activity** is measured as the maximum rate at which substrate can be converted to product by a unit weight of tissue.

SUMMARY
The Interplay of Aerobic and Anaerobic Catabolism during Exercise

- Behavior and biochemistry are linked during physical activity because attributes of performance depend on how the ATP for muscular effort is synthesized.

- Submaximal forms of exercise can be supported entirely (except during transition phases) by aerobic catabolism using O_2 taken in from the environment by breathing. From the viewpoint of ATP supply and demand, submaximal forms of exercise can thus be sustained indefinitely.

- Supramaximal forms of exercise in vertebrates, crustaceans, and some other animals require a continuing input of ATP from anaerobic glycolysis. The steady use of anaerobic glycolysis—manifested by a steady accumulation of lactic acid—eventually causes metabolic self-termination of the exercise.

- In vertebrates, metabolic transitions occur at the start and the end of even light submaximal exercise. An oxygen deficit occurs at the start, and excess postexercise oxygen consumption (EPOC) occurs at the end. The oxygen deficit is a consequence of the fact that the breathing and circulatory systems increase O_2 delivery gradually, not stepwise, at the start of exercise.

- As the duration of all-out exertion increases, ATP must increasingly be supplied by steady-state aerobic catabolism, rather than by nonsteady-state mechanisms that can produce ATP exceptionally rapidly but cannot produce a great deal of it. The pace of all-out exertion therefore declines as duration increases.

- Closely related species, and even individuals within one species, often differ greatly in their emphasis on aerobic and anaerobic mechanisms of producing ATP for exercise. These metabolic differences help explain differences in exercise performance.

Responses to Impaired O₂ Influx from the Environment

In addition to vigorous exercise, impaired O_2 influx from the environment is the second major reason that animals turn to anaerobic catabolic pathways to make ATP. Many animals experience reduced O_2 influx from their environments during parts of their lives. The situation can arise in two ways. First, O_2 influx may be reduced because the concentration of O_2 in the environment is low. Alternatively, animals may enter environments in which they cannot breathe, as when seals or whales dive. Under either set of conditions, it is common for at least some tissues to experience **hypoxia** or **anoxia**, defined respectively to be an especially low level of O_2 in tissues or an absence of O_2 in tissues.

Some animals display metabolic depression when they are faced with reduced O_2 influx. **Metabolic depression** is a regulated reduction in the ATP needs of the animal (or certain of its tissues) to levels below the needs ordinarily associated with rest in a way that does not present an immediate physiological threat to life. When deprived of O_2, for example, brine shrimp embryos switch to a greatly reduced rate of metabolism (signifying a reduced rate of ATP use), and they are able to sustain this state for long periods and later recover without ill effect (**FIGURE 8.14**). Study of the mechanisms of metabolic depression is in its infancy. In some cases, key, rate-limiting enzymes are downregulated, or numbers of mitochondria decline. The hypoxia-inducible factor 1 (HIF-1) control system is often involved (see Figure 23.6), as are other control systems.

To make ATP during anoxia, animals must turn to anaerobic catabolic pathways, and they also often use such pathways to help make ATP during hypoxia. When vertebrates turn to anaerobic catabolism, they almost universally produce lactic acid,[22] which they never excrete. In sharp contrast, invertebrates adapted to life without influx of O_2 rarely produce lactic acid. Instead, they usually employ anaerobic catabolic pathways that are more elaborate than anaerobic glycolysis and that yield a variety of different products, which often are excreted. All anaerobic catabolic pathways produce far less ATP per food molecule than aerobic catabolism does. Thus, when animals turn from aerobic to anaerobic catabolic pathways as their means for long-term ATP production, they are less able to make ATP. Metabolic depression helps with this situation because it reduces the rate at which animals *need* ATP.

Air-breathing vertebrates during diving: Preserving the brain presents special challenges

Most vertebrate brains are obligatorily aerobic; they must have O_2. When air-breathing vertebrates dive for extended periods, therefore, the status of their brain is of special interest and concern.

In diving species of mammals and birds, most dives are kept short enough that the ATP needs of all tissues can be met aerobically using stored O_2. Dives of exceptionally long duration, however, cause these vertebrates to resort to anaerobic glycolysis. The animals then metabolically subdivide their bodies (see Chapter 26 for details). They ensure continuing O_2 delivery to their brain by reserving certain of their O_2 stores for the brain, while simultaneously they deprive large portions of their bodies of O_2 delivery; the latter parts run out of O_2 and become dependent on anaerobic glycolysis. Diving crocodilians, sea turtles, lizards, and most freshwater and terrestrial turtles must also keep their brains aerobic.

Certain species of freshwater and terrestrial turtles, however, are dramatic exceptions in that they can tolerate total-body anoxia—*full O_2 depletion of all tissues, including the brain*—during protracted dives. These anoxia-tolerant turtles are able to dive for exceptionally long periods. **FIGURE 8.15** shows the lengths of time that various vertebrates can survive anoxia as a function of their body (tissue) temperatures when they are tested. The anoxia-tolerant turtles (in which the brain can survive without O_2), such as those in the genera *Trachemys* and *Chrysemys*, are able to survive about 1000 times longer at any given body temperature than most other vertebrates.

An informative way to understand the implications of brain anoxia in diving turtles is to look briefly at the threat anoxia poses to vertebrate central nervous system (CNS) tissue. In a person or other mammal, catastrophe strikes when the brain is deprived of O_2 because the ATP requirement of mammalian brain cells per unit of time far exceeds the rate at which those cells can make ATP by anaerobic means. Within *seconds* after O_2 influx to the mammalian brain is cut off, the concentration of ATP in brain cells starts to fall precipitously. Soon, ATP-dependent ion pumps (e.g., Na^+–K^+-ATPase) are unable to pump ions across the cell membranes rapidly enough to maintain normal membrane polarization. The cell membranes thus depolarize, with numerous cataclysmic consequences; for instance, nerve impulses (action potentials) become impossible, and voltage-gated Ca^{2+} channels are inappropriately triggered to open, allowing Ca^{2+} to flood into the cells. For this reason and others, the Ca^{2+} concentration in the

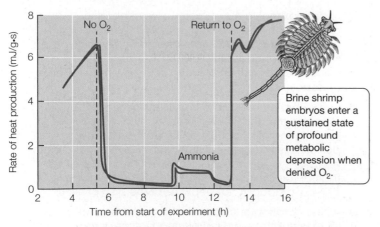

FIGURE 8.14 Metabolic depression in an invertebrate faced with anoxia Metabolic rate was quantified directly, by measurement of heat production. Two groups of embryos of brine shrimps (*Artemia*)—represented by the two lines—were studied. They were living in water near equilibrium with the atmosphere at the start of the experiment. Then, at the time marked "No O_2," they were switched into O_2-free water. The embryos were exposed to ammonia while in O_2-free water, as a way of raising their pH. Their response to ammonia supports the hypothesis that metabolic depression in the embryos is partly dependent on a low pH. The drawing shows an adult brine shrimp. (After Hand and Gnaiger 1988.)

---

[22] A few vertebrate species, notably two species of fish discussed later, are exceptions.

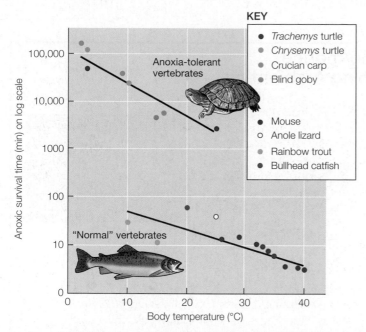

FIGURE 8.15 **Survival times during total-body anoxia in vertebrates as a function of body temperature** The survival times are plotted on a logarithmic scale. (After Nilsson and Lutz 2004.)

cytoplasm rises to levels that trigger a variety of inappropriate and disastrous Ca^{2+}-mediated responses.

The turtles that tolerate brain anoxia employ metabolic depression of the brain as a key mechanism of maintaining the integrity of their brain tissue during anoxia. Synaptic transmission between brain cells is suppressed in the absence of O_2, and ion-mediated bioelectrical activity of cells is reduced so that the brain becomes electrically relatively silent. This response has a significant cost: The turtles cease to be behaviorally alert. They become comatose. Because of this response, however, the brain ion pumps have much less work to do to maintain normal ion distributions and cell-membrane polarization. The ATP requirement of maintaining tissue integrity is lowered, anaerobic glycolysis is able to meet the ATP requirement, and brain ATP concentrations do not fall. With its entire body anoxic, a turtle accumulates lactic acid, which can reach extraordinary concentrations during prolonged anoxia. The shell and the bones of a turtle play key roles in preventing lethal acidification under these circumstances, by buffering the acid.[23]

Animals faced with reduced O_2 availability in their usual environments may show conformity or regulation of aerobic ATP synthesis

When animals are living in the environments in which they can breathe, such as fish in water or mammals in air, how do their rates of aerobic ATP production change when they are confronted with changes of O_2 concentration in the environmental water or air? As the environmental O_2 level is lowered, the usual pattern is for an animal's rate of O_2 consumption to be unaffected over a certain range of O_2 levels (**FIGURE 8.16A**). This maintenance of a steady

[23] A parallel phenomenon has recently been discovered in certain crayfish, in which the shell buffers lactic acid produced when the animals are placed in air, where they cannot breathe.

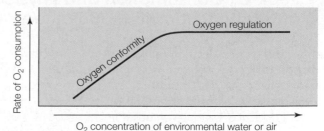

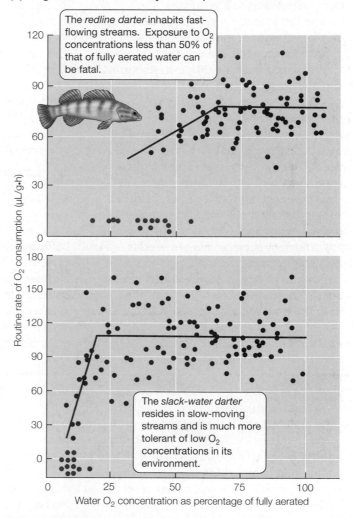

FIGURE 8.16 **Oxygen regulation and conformity** (A) The general concepts of oxygen regulation and conformity. (B) Rates of O_2 consumption of two species of fish in the genus *Etheostoma*—the redline darter (*E. rufilineatum*) and the slack-water darter (*E. boschungi*)—during routine activity at 20°C. Red circles mark ambient O_2 levels at which deaths occurred. On the x axis, 100 is the O_2 concentration of water that is at equilibrium with the atmosphere and therefore fully aerated; values higher than 100 were created in some studies by bubbling water with pure O_2. (After Ultsch et al. 1978.)

rate of O_2 consumption (and aerobic ATP production) regardless of the level of O_2 in the environment is termed **oxygen regulation** (see Figure 1.8). It often involves active responses, such as an increase in breathing rate as the O_2 level in the water or air declines. Ultimately, if the environmental O_2 level is lowered further and

BOX 8.3 Peak O₂ Consumption and Physical Performance at High Altitudes in Mountaineers Breathing Ambient Air

When people (and other mammals) are exposed to decreased atmospheric concentrations of O_2, they marshal vigorous physiological defenses, discussed in Boxes 23.2 and 24.5. Because of these responses, *resting* or *moderately active* people show a substantial degree of oxygen regulation as the ambient O_2 level falls with increasing altitude. Here, however, we consider the most demanding of circumstances: the capability of mountaineers for *all-out physical effort* at the highest altitudes on Earth.

As the figure shows, when people are asked to work hard enough that they take in O_2 at their peak rate, their maximum rate of O_2 consumption becomes a smaller and smaller fraction of their rate at sea level as altitude increases. The cost of any particular form of physical exertion remains the same regardless of altitude, however. Accordingly, a rate of climbing that is distinctly submaximal at low altitudes can become maximal, or even supramaximal, at high altitudes. At altitudes near the top of Mt. Everest (8848 m), the maximum rate of O_2 consumption is so low that even *minimum* rates of climbing require about the *maximum* possible rate

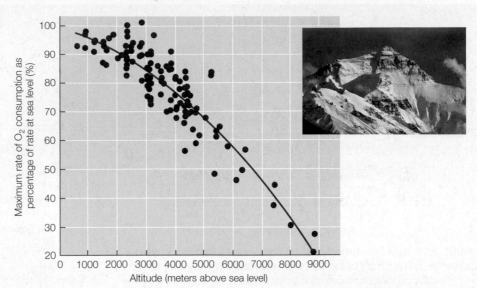

Maximum rates of O₂ consumption of human mountaineers at increasing altitudes Values are expressed as percentages of the values at sea level. Mountaineers were breathing from the ambient air, not from O_2 tanks. (After Fulco et al. 1998.)

of O_2 consumption. Work that requires 100% of a person's maximum rate of O_2 consumption is always extremely taxing: barely possible. When, in 1978, Reinhold Messner and Peter Habeler became the first to reach the summit of Mt. Everest without supplemental O_2, they reported

climbing so slowly near the top that, even though they felt they were working at their limits, they required an hour to cover the final 100 meters!

Box Extension 8.3 provides more information on climbing Mt. Everest.

---

further, oxygen regulation can no longer continue. Instead, the rate of O_2 consumption starts to fall as the environmental O_2 level falls. This condition is termed **oxygen conformity** (see Figure 8.16A). High altitudes are an intriguing situation in which these concepts apply to human beings (**BOX 8.3**).

When related species are compared under similar test conditions, their abilities for oxygen regulation and conformity often correlate with the types of habitats in which they live. These abilities therefore seem to have evolved in parallel with habitat selection. An illustration is provided by two species of related freshwater fish living in a single Alabama watershed (**FIGURE 8.16B**). One species, the redline darter, occurs in fast-flowing streams where O_2 levels tend to be consistently high because turbulence promotes aeration. The other species, the slack-water darter, is found in slow-moving streams where the O_2 concentration may be only one-third as high as in fully aerated water. As Figure 8.16B shows, the slack-water darter, which often experiences low-O_2 waters, exhibits a much broader range of oxygen regulation than the redline darter, which rarely must cope with low O_2 levels. Indeed, lowering the concentration of O_2 to 40%–50% of the fully aerated level causes a sharp depression of O_2 uptake and deaths in redline darters, whereas it does not affect the O_2 uptake of slack-water darters at all.

Water-breathing anaerobes: Some aquatic animals are capable of protracted life in water devoid of O₂

O_2-free microenvironments are far more common in bodies of water than in terrestrial environments (see Chapter 22), and biologists have discovered quite a few cases of water-breathing aquatic animals that can live in O_2-free settings because they can function as **anaerobes**, meaning they are able to survive whole-body anoxia for long periods. Some of the best-studied examples among invertebrates are certain species of clams, mussels, and other bivalve molluscs. Ribbed mussels (*Geukensia demissa*), for example, live in the mud of salt marshes, where they can become buried; they are able to survive in an atmosphere of pure N_2 for 5 days. Certain marine clams that live in seas prone to O_2 depletion, such as the ocean quahog (*Arctica islandica*), can live for 1–2 months in O_2-free water. Worms and other invertebrates that live at the bottoms of lakes or ponds are other animals that may experience severe and prolonged O_2 deprivation (see Figure 1.16), and some are among the most tolerant to anoxia of all free-living animals. Certain annelid worms (*Tubifex*) that burrow in O_2-free sediments in pond or lake bottoms, for example, have been shown not only to survive but to feed, grow, and *reproduce* while deprived of O_2 for 7 months!

Anaerobes are rare among aquatic vertebrates. Nonetheless, two species of related cyprinid fish are known to have extraordinary abilities to live without O_2. One of these is the common goldfish (*Carassius auratus*), which is reported to survive in O_2-free water for 11–24 h at 20°C and for 1–6 days at 10°C. No wonder goldfish survive the tender loving care of 5-year-olds! The second species, which is even more tolerant of anoxia than the goldfish, is the crucian carp (*Carassius carassius*), a common inhabitant of northern European ponds. It can survive without O_2 for several months at temperatures below 10°C (see Figure 8.15)! The ability of crucian carp to live without O_2 permits them to evade predators by living in ponds that become O_2 depleted, where the predators die. Their physiology, therefore, is a key to their ecological success.

For both invertebrate and vertebrate aquatic anaerobes, metabolic depression (see Figure 8.14) is typically a key strategy used to survive anoxia. **TABLE 8.2** presents data on metabolic depression in goldfish. Although their O_2 consumption falls to zero during anoxia, direct measurements of heat production show that metabolism continues, but it does so at a highly depressed rate. Metabolic depression lowers ATP requirements and thus relaxes demands on the ATP-producing mechanisms that are available to make ATP in the absence of O_2.

MECHANISMS OF INVERTEBRATE ANAEROBIOSIS The most common principal products of the anaerobic biochemical pathways used for ATP synthesis by invertebrate aquatic anaerobes are acetic acid, succinic acid, propionic acid, and the amino acid alanine. These products obviously signal that the biochemical pathways of anaerobic catabolism in these animals differ from simple anaerobic glycolysis. The pathways are elaborate. They often permit the animals to catabolize anaerobically not only carbohydrates but also other classes of food molecules, notably amino acids, and they typically yield more ATP per food molecule than anaerobic glycolysis does. Invertebrate anaerobes often excrete their anaerobic end products or derivatives of them. Excretion wastes the energy value of the carbon compounds excreted, but it helps limit acidification of the body fluids and helps prevent self-limitation of the ATP-generating mechanisms, allowing the animals to sustain all their vital functions in the absence of O_2 for protracted periods.

MECHANISMS OF ANAEROBIOSIS IN GOLDFISH AND CRUCIAN CARP The goldfish and crucian carp provide a possibly amusing end to this chapter because they raise the question of whether certain animals are constantly drunk. Of all vertebrate animals, these fish are the most proficient known anaerobes because not only do they survive without O_2 for long periods (see Figure 8.15), but also, unlike the turtles we discussed earlier, they remain conscious and capable of responding behaviorally to their environments.

TABLE 8.2 Average rates of O_2 consumption, heat production, and carbohydrate use in goldfish (*Carassius auratus*) before and during exposure to anoxia at 20°C

Note that although metabolic rate decreases in anoxia, the rate of use of carbohydrate stores increases because production of ATP is far less efficient by anaerobic glycolysis than by aerobic catabolism.

Property measured[a]	At normal O_2 levels	After 3 h of anoxia	After 8 h of anoxia
Oxygen consumption (mmol/kg·h)	1.51	0	0
Heat production (J/ kg·h)	709	203	206
Carbohydrate catabolism (mg/kg·h)	43	221	234

Source: van Waversveld et al. 1989.

[a]All rates are expressed per kilogram of adjusted body weight. Adjusted weights were calculated in a way intended to remove allometric effects of different body sizes.

The *swimming muscles* of these fish possess biochemical specializations, including an unusual form of the enzyme alcohol dehydrogenase (ADH). This form of ADH strongly favors the formation of ethanol (ethyl alcohol) under prevailing tissue conditions. During anoxia all the tissues of the fish, including the brain, synthesize ATP by anaerobic glycolysis and produce lactic acid. The swimming muscles convert this lactic acid (some of it brought to the muscles from the other tissues) to ethanol and CO_2. The production of ethanol does not increase ATP yield. Rather, its principal advantage is believed to be that it makes possible the *excretion* of the carbon chains produced by anaerobic glycolysis. Unlike lactic acid, ethanol is lost across the gills into the water that the fish inhabit (**FIGURE 8.17**). This excretion limits end-product accumulation and

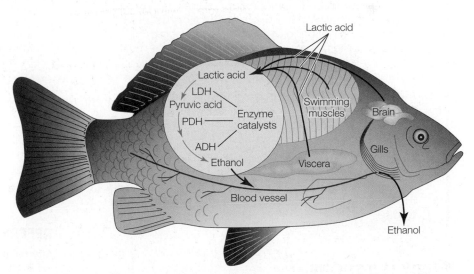

FIGURE 8.17 Excretion of the end product of anaerobic glycolysis as ethanol When crucian carp (*Carassius carassius*, shown here) and goldfish are living in O_2-free water, all tissues—including the brain—make ATP by anaerobic glycolysis from stored glycogen. In most fish, when lactic acid is made, it accumulates in the body. In the crucian carp and goldfish, however, the swimming muscles have an unusual ability to convert lactic acid to ethanol, which then is readily lost across the gills. In the swimming muscles, lactic acid is first converted to pyruvic acid, a reaction catalyzed by lactate dehydrogenase (LDH). Pyruvic acid is then converted to acetaldehyde—catalyzed by pyruvate dehydrogenase (PDH)—and acetaldehyde is converted to ethanol—catalyzed by alcohol dehydrogenase (ADH). (After a concept by Bickler and Buck 2007.)

body-fluid acidification, thereby helping to prevent self-limitation of the ATP-generating mechanisms.

The brains of goldfish and crucian carp living in O_2-free water exhibit far less metabolic depression than those of turtles. The more-limited metabolic depression permits the fish to remain conscious and responsive. It also increases demands for anaerobic ATP synthesis by brain cells. Crucian carp build up stores of brain glycogen with the approach of winter—the season when they are most likely to face anoxia. These stores provide fuel for making the ATP that keeps the brain functional.

The fish do not get drunk! Their rate of excretion of ethanol is great enough to keep their tissue concentrations of ethanol below the drunken level.

SUMMARY
Responses to Impaired O_2 Influx from the Environment

■ Many of the animals that are adapted to living without O_2 undergo metabolic depression when deprived of O_2. Metabolic depression can be so profound as to lower an animal's metabolic rate to less than 5% of the usual rate, thereby greatly reducing the rate at which ATP must be supplied by catabolic mechanisms.

■ Invertebrate anaerobes deprived of O_2 produce ATP by means of a diversity of complex anaerobic catabolic pathways that generate end products such as acetic acid, succinic acid, and propionic acid. The invertebrates commonly excrete these organic products during anoxia as a way of avoiding end-product accumulation in their bodies.

■ Virtually all vertebrates use simple anaerobic glycolysis to produce ATP in tissues deprived of O_2, and vertebrates invariably retain lactic acid in their bodies, setting the stage for potential metabolic self-limitation. Usually when vertebrates experience anoxia, it is strictly regional; whereas some tissues become anoxic, others—most notably the CNS—retain an O_2 supply. Only a few vertebrates can tolerate total-body anoxia.

■ Turtles capable of total-body anoxia employ anaerobic glycolysis to make ATP. A key part of their strategy for survival is a metabolic depression of the CNS sufficiently profound to produce a comatose state.

■ Goldfish and crucian carp undergoing total-body anoxia remain alert. They have the unusual ability to convert lactic acid to ethanol, which they can excrete, thereby preventing end-product accumulation in their bodies.

STUDY QUESTIONS

1. One approach to conservation of fish populations is to release unwanted fish accidentally caught in trawling nets. Such fish often have very high concentrations of lactic acid in their bodies. Why do you think they have these high concentrations of lactic acid, and how might their survival after release be affected by their condition?

2. Explain how the reactions that produce and use ATP serve, together, as an *energy shuttle* mechanism in cells.

3. Explain the concept of redox balance. What conditions must exist, for example, for cytochrome oxidase to be in redox balance?

4. How does the reduction of pyruvic acid create a state of redox balance in anaerobic glycolysis?

5. Outline the chemiosmotic hypothesis for the mechanism by which oxidative phosphorylation is coupled with electron transport. How does uncoupling occur in tissues with uncoupling protein 1 (UCP1)? Under what circumstances would uncoupling be disadvantageous, and under what circumstances might it be advantageous?

6. Using two or three carefully chosen examples, illustrate the point that during physical activity, behavior and biochemistry are intimately linked, such that an animal's exercise performance depends on the mechanisms that are making ATP for the exercise.

7. Assuming that an animal uses a catabolic pathway that produces organic products, such as lactic acid or propionic acid, compare the pros and cons of retaining or excreting the organic molecules.

8. Why is it important to distinguish *temporary* electron (hydrogen) acceptors in cells from *final* electron acceptors? What are the unique advantages of O_2 as an electron acceptor?

9. Why does an oxygen deficit occur at the start of submaximal exercise in vertebrates? What are the mechanisms of ATP production during the oxygen deficit phase, and how is ATP made in the ensuing pay-as-you-go phase?

10. A single individual can differ from time to time in his or her maximum rate of O_2 consumption. For example, athletic training in people can raise the maximum rate of O_2 consumption by 10%–30%, whereas going to high altitudes can lower it (see Box 8.3). Explain how these sorts of changes in the maximum rate of O_2 consumption can make a single type of exercise (such as jogging at 6 miles per hour) shift from being submaximal to supramaximal, or vice versa. What are the physiological implications of such shifts?

11. There has been a great deal of debate over whether the ratio of SO to FG fibers in the muscles of individual people or other animals is fixed genetically. Researchers have asked whether the ratio of fiber types can be altered during an individual's lifetime by various sorts of training or other experiences. Why would a change in the ratio of fiber types be of interest and importance? Design experiments or other sorts of studies that would help elucidate whether the ratio of fiber types can undergo change.

Go to **sites.sinauer.com/animalphys4e**
for box extensions, quizzes, flashcards,
and other resources.

REFERENCES

Alberts, B., A. Johnson, J. Lewis, and four additional authors. 2014. *Molecular Biology of the Cell*, 6th ed. Garland, New York. A superior and time-proven presentation of all aspects of cellular physiology, including the catabolic mechanisms, emphasizing mammals. Excellent illustrations for understanding difficult concepts.

Åstrand, P.-O., K. Rodahl, H. A. Dahl, and S. B. Strømme. 2003. *Textbook of Work Physiology: Physiological Bases of Exercise*, 4th ed. Human Kinetics, Champaign, IL. A classic treatment of the physiology of all forms of human work, from housework to Nordic skiing.

Brand, M. D. 2005. The efficiency and plasticity of mitochondrial energy transduction. *Biochem. Soc. Trans.* 33: 897–904. A superb, accessible review of oxidative phosphorylation and other aspects of mitochondrial function.

Brooks, G. A., T. D. Fahey, and K. M. Baldwin. 2005. *Exercise Physiology: Human Bioenergetics and Its Applications*, 4th ed. McGraw-Hill, Boston.

Divakaruni, A. S., and M. D. Brand. 2011. The regulation and physiology of mitochondrial proton leak. *Physiology* 26: 192–205.

Enola, R. M., and J. Duchateau. 2008. Muscle fatigue: what, why and how it influences muscle function. *J. Physiol.* 586: 11–23. Although this article takes a limited approach, it is one of the more accessible reviews of the complex question of muscle fatigue.

Hand, S. C., and I. Hardewig. 1996. Downregulation of cellular metabolism during environmental stress: Mechanisms and implications. *Annu. Rev. Physiol.* 58: 539–563. A compact but fascinating and thought-provoking review of the physiology of metabolic dormancy. See also S. C. Hand's 1991 paper listed in the Additional References for a great deal more on the natural history of metabolic dormancy.

Hochachka, P. W., and G. N. Somero. 2002. *Biochemical Adaptation*. Oxford University Press, New York. A definitive synthesis of the themes discussed in this chapter.

Nilsson, G. E., and P. L. Lutz. 2004. Anoxia tolerant brains. *J. Cerebr. Blood Flow Metab.* 24: 475–486. A review of a fascinating phenomenon written by two of the leaders in research on fish and turtles.

Milton, S. L., and H. M. Prentice. 2007. Beyond anoxia: The physiology of metabolic downregulation and recovery in the anoxia-tolerant turtle. *Comp. Biochem. Physiol., A* 147: 277–290. A progress report on numerous molecular aspects of brain tolerance to episodes of anoxia in an anoxia-tolerant turtle. Several other relevant research reports can be found in the same issue of Comparative Biochemistry and Physiology.

Shephard, R. J., and P.-O. Åstrand (eds.). 2000. *Endurance in Sport*, 2nd ed. Blackwell, Oxford, UK. Excellent chapters by many different authorities on most of the major topics in human exercise performance. Chapters 22 and 23 on metabolism in muscle are especially relevant to our discussion here.

Weibel, E. W., and H. Hoppeler. 2004. Modeling design and functional integration in the oxygen and fuel pathways to working muscle. *Cardiovasc. Eng.* 4: 5–18.

See also **Additional References** *and* **Figure and Table Citations**.

Breeding colonies of seabirds are memorable for the amount of activity taking place, as great numbers of birds come and go all day and sometimes into the night. Adults congregate in large numbers on islands and beaches in these colonies. All their food, however—and all the food for their offspring—comes from the ocean. More than 100,000 masked boobies (*Sula dactylatra*) congregate for breeding each year on Clipperton Island, a tiny atoll far out in the Pacific Ocean off the coast of Costa Rica. Both sexes typically forage at sea each day (**FIGURE 9.1**). Some foraging trips are relatively short: only about 1 h. Others last most of a day: 18 h. The average is 9 h. During an average 9-h trip, a bird spends about 80% of its time—7 h—flying out and back to fishing areas that, on average, are 103 km (64 miles [mi]) from land. In the fishing areas, the birds feed by use of short (3-s), spectacular plunge dives, during which they undertake aerodynamically controlled, high-speed free falls into the water to capture fish—mostly species of flying fish. Each adult booby (weighing 1.5–2 kg) must catch and digest about 0.5 kg of fish each day to meet its own metabolic needs, especially its costs of flight. Parents must also provide each offspring with about 0.3 kg of fish per day. All told, the breeding colony on Clipperton Island takes about 70 metric tons of fish from the surrounding Pacific Ocean each day.[1] According to a recent study, the seabirds of the world—summing up all species—remove about 190,000 metric tons of fish from the oceans daily, making the global fish consumption by seabirds approximately equal to global harvesting by humans.

The sustained foraging activity exhibited each day by masked boobies and other seabirds must be aerobic.

---

[1] A metric ton is 1000 kg, or about 2200 pounds.

Masked boobies (*Sula dactylatra*) nest on islands but get all their food from the sea Their foraging flights to distant fishing areas far out over the ocean—in common with sustained activities of other animals—are fueled by aerobic ATP production and elevate the costs of daily life.

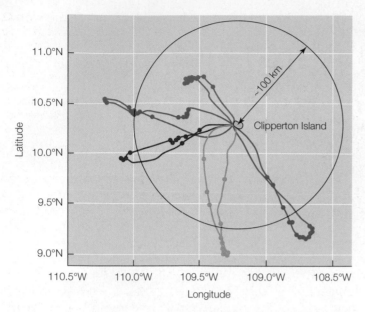

FIGURE 9.1 **The routine daily flights of masked boobies (***Sula dactylatra***) to and from feeding areas over the ocean depend on aerobic catabolism to supply energy for muscular work** The map shows five daily foraging flights by birds that were nesting on Clipperton Island. Each color represents a different individual; for one individual, two flights are shown (blue plots). Dots along the flight paths show places where the birds engaged in fishing. The circle represents a flight distance from the island of 100 km (62 mi). (After Weimerskirch et al. 2008.)

That is, the ATP for the activity must be produced by aerobic catabolism, using O_2 delivered steadily to the exercising muscles from the atmosphere. Anaerobic mechanisms of ATP production could not possibly supply the ATP because—in birds, as in mammals—the anaerobic mechanisms can produce only very limited total amounts of ATP (see Table 8.1). Throughout the animal kingdom, as discussed in Chapter 8, the ATP requirements of routine daily living are mostly met by aerobic catabolism because the mechanism of aerobic ATP production can operate in steady state, meaning aerobic ATP synthesis can go on and on without self-limitation.

In this chapter we examine the energy costs of routine daily living and of sustained, aerobically fueled exercise. This focus is justified because although highly anaerobic forms of exertion (e.g., sprints) sometimes spell the difference between life and death (see Chapter 8), aerobic types of activity predominate in the lives of most animals in regard to time spent and energy required.

As a first example of the energy costs of sustained, aerobically fueled activity, the metabolic rates of humans engaged in various physical activities are listed in **TABLE 9.1**. Walking increases a person's metabolic rate—and therefore the rate at which ATP must be made—by two- to fourfold compared with rest. More intense types of sustained exercise can increase metabolic rate by tenfold or more.

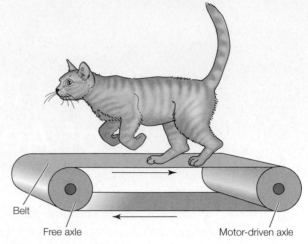

FIGURE 9.2 **A treadmill provides a way to control the running or walking speed of an animal during study** When the motor-driven axle turns, the belt moves as indicated by the arrows. After being trained, animals run at the same speed as the belt to keep their position relative to the surrounding room.

How Active Animals Are Studied

Studies of actively moving animals present challenging methodological problems, whether carried out in the laboratory or in the wild. One important question that is usually studied under laboratory conditions is the relation between the speed of locomotion and the metabolic rate. The greatest challenge for such studies is controlling and measuring an animal's speed as it runs, flies, or swims.

For running or walking animals, the device most commonly used to control the speed of a moving individual is a motor-driven treadmill (**FIGURE 9.2**). The animal stands on a belt, which is driven round and round by a motor. For the animal to keep its position, it must run or walk at the same speed as the belt passing beneath its feet. The treadmill can be tilted at an angle relative to the horizontal to simulate uphill or downhill running. Animals as diverse as cockroaches, land crabs, turkeys, and cheetahs have been

TABLE 9.1	Representative metabolic rates of young adult people of average build during sustained, aerobically fueled forms of exercise
Type of activity[a]	Metabolic rate (kJ/minute)[b]
Lying down	6.3
Sitting	7.1
Standing	8.8
Walking at 2 miles per hour (mph)	12
Walking at 4 mph	21
Bicycling at 13 mph	32
Jogging at 7 mph	59
Crawl swimming at 2 mph	59
Running at 10 mph	84

Source: After Åstrand and Rodahl 1986.
[a]All forms of locomotion are assumed to be on level ground.
[b]In aerobic catabolism, 1 kJ = 49.5 mL O_2.

FIGURE 9.3 Modern telemetry devices permit information on O₂ consumption to be radioed from a freely moving subject In the cases shown, the person's or horse's rate of O₂ consumption is measured using the principle of open-flow respirometry (see Box 7.4). Valves in the mask allow the subject's breathing to cause a measured stream of air to flow through the mask. An O₂ probe measures how much O₂ is removed from each unit of volume of the flowing air. The device has been used to study racehorse performance and the energetics of running in Kenyan athletes (inset). (Courtesy of Marco Brugnoli and Cosmed S.r.l.)

trained to run and walk on treadmills! Lobsters and crayfish have even been trained to use underwater treadmills. As the speed of a treadmill is varied, the subject's rate of O₂ consumption is measured, permitting investigators to describe the relation between the speed and the metabolic cost of locomotion.

Wind tunnels are used to control the speeds of flying animals. A bird, insect, or bat in a wind tunnel must fly into the forced air current produced by the tunnel at the same speed as the current if it is to maintain its position. For swimming animals such as fish, a device analogous to a wind tunnel—but filled with water instead of air—is commonly used to control speed.

For carrying out field studies of the energetics of activity,[2] the greatest challenge is often that of measuring or estimating the metabolic rate, because standard laboratory metabolic techniques usually cannot be used. In some cases, O₂ consumption can be measured in the field by use of a mask and valve system, so that the subject breathes from a defined air stream rather than from the atmosphere at large (**FIGURE 9.3**). This sort of approach has been used in a wide variety of studies (**BOX 9.1**). Typically, however, alternatives to the measurement of O₂ consumption must be used in field studies.

One of the most important innovations for studying the energetics of air-breathing animals in the wild is the **doubly labeled water method**. With this technique, researchers can measure the metabolic rate of an individual animal that is entirely free to engage in its normal behaviors in its natural habitat. The method takes its name from the fact that the individual under study is injected with water labeled with unusual isotopes of both hydrogen and oxygen. Deuterium (a hydrogen isotope) and oxygen-18 are the most commonly used isotopes.[3] The method is then called the

[2] *Field studies* are studies carried out in natural or seminatural environments outside the laboratory.

[3] The ordinary isotope of oxygen is oxygen-16.

BOX 9.1 **The Cost of Carrying Massive Loads**

Nepalese porters carry massive loads up steep mountain trails at high altitudes. Investigators recently interviewed more than 500 male porters in a remote village at an elevation of 3500 m (11,500 ft) near Mt. Everest. The porters had just arrived in the village after a trek of 100 km (62 mi). The average load they carried was 93% of body weight. One-fifth of the porters carried loads greater than 125% of their body weight! The porters employ a unique head-supported system to carry these enormous loads.

The investigators took a device similar to that in the inset of Figure 9.3 to the Himalayan village and used it to measure the O₂ consumption of eight porters.

Nepalese porters: Their cost of carrying a load as a function of the size of the load The black line shows average results from a previous study of Europeans carrying heavy loads in standard backpacks. Error bars show standard deviations. (After Bastien et al. 2005.)

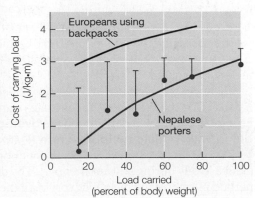

They discovered that for these Nepalese porters, the cost of carrying heavy loads was only 50%–60% as great as the cost incurred by Europeans carrying identical loads in backpacks, as shown by the data in the figure. African women carrying loads on their heads also exhibit relatively low carriage costs. The biomechanical explanation remains a mystery.

D$_2$^{18}O method. We explain the method here with reference to these particular isotopes, although the rationale is the same regardless of the isotopes used.

For an animal to be studied using the D$_2$^{18}O method, it is injected at the start of the study period with measured amounts of both D$_2$^{16}O and H$_2$^{18}O, which mix with its body water. Later, at the end of the study period, a blood sample is taken to determine how rapidly the deuterium and oxygen-18 were eliminated from the animal's body during the study period. In the interval between the initial injection of the isotopes and the collection of the final blood sample, the subject is released and is free to live normally in its natural habitat. The method measures the individual's *average* rate of CO$_2$ production during the study period. Average metabolic rate is then computed from CO$_2$ production.[4]

The reason the doubly labeled water method works is that the oxygen atoms in expired CO$_2$ are in isotopic equilibrium with the oxygen atoms in body water; this means that if the body H$_2$O consists of given proportions of H$_2$^{16}O (ordinary water) and H$_2$^{18}O, expired CO$_2$ contains oxygen-16 and oxygen-18 in the same proportions.[5] After the concentration of H$_2$^{18}O is elevated in an animal's body at the start of a study, the excess atoms of oxygen-18 are gradually lost by way of expired CO$_2$. The *rate* of loss of oxygen-18 thus depends on the subject's rate of CO$_2$ production. A problem is that atoms of oxygen-18 are lost from the body in the form of H$_2$O as well as CO$_2$ because H$_2$O that leaves the animal's body by evaporation, urination, or other mechanisms consists in part of H$_2$^{18}O. Accordingly, after an animal has been injected with oxygen-18, its *total* rate of elimination of the oxygen-18 is in fact a function of *both* its rate of CO$_2$ expiration *and* its rate of H$_2$O loss. The hydrogen isotope—deuterium—is injected in the doubly labeled water method to obtain an independent measure of the rate of H$_2$O loss; because CO$_2$ contains no hydrogen, after an animal has been injected with deuterium, its rate of loss of the deuterium from its body depends just on its rate of H$_2$O elimination. Knowing the animal's rate of H$_2$O loss from the hydrogen-isotope data, one can calculate how much oxygen-18 is lost in H$_2$O during the study period. Knowing this value, one can then subdivide the total rate of oxygen-18 loss into two components: the rate of loss in H$_2$O and the rate of loss in CO$_2$. In this way, the rate of CO$_2$ expiration—and the metabolic rate—is calculated.

The most common use of the doubly labeled water method is to determine the average metabolic rate of an animal in the wild during the entire 24-h day, termed the **average daily metabolic rate** (**ADMR**) or **field metabolic rate** (**FMR**). The animal might sleep for part of the day, stand guard for another part, and move about actively during still another part. The doubly labeled water method measures the metabolic cost of all these behavioral states lumped together.

---

[4] As discussed in Chapter 7 (see Table 7.1), uncertainties are involved in calculating metabolic rate from the rate of CO$_2$ production. These uncertainties need to be kept in mind in interpreting results of the doubly labeled water method.

[5] The reason for the isotopic equilibrium between oxygen in H$_2$O and oxygen in CO$_2$ is that atoms of oxygen are freely exchanged between molecules of H$_2$O and CO$_2$ during the reactions of the Krebs cycle (citric acid cycle) (see Figure 8.2).

TABLE 9.2 A time-energy budget for an adult African penguin (*Spheniscus demersus*) during the breeding season

Behavior	A Hours per day devoted to behavior[a]	B Hourly cost of behavior[b] (kJ/h)	C Daily cost of behavior: A × B (kJ)
Maintenance on land	19.5	49	956
Swimming underwater	2.0	358	716
Resting at water's surface	2.5	83	208
Total daily energy expenditure (sum of column C):			1880 kJ/day

Source: After Nagy et al. 1984.
[a]Time devoted to each behavior was determined from field studies of African penguins.
[b]The hourly energetic cost of each behavior was estimated from comparative laboratory studies of multiple bird species.

Time–energy budgets represent an older, but still useful, approach for estimating the ADMRs of animals free in their natural habitats. To construct a time–energy budget, an investigator first categorizes all of an animal's behaviors into a few categories. For instance, the categories might be sleeping, standing guard, and actively moving. By using laboratory studies or other sources of information, the investigator then estimates the animal's rate of energy expenditure (metabolic rate) while it engages in each category of behavior. Finally, field observations are used to estimate the amount of time the animal spends in each sort of behavior during a day. To calculate the total energy cost of each behavior, the time spent in the behavior is multiplied by the rate of energy expenditure during the behavior. The energy costs of all behaviors are then summed to obtain the total daily energy expenditure.

TABLE 9.2 presents an example of a time–energy budget for African penguins (*Spheniscus demersus*). These penguins live on southern African shores and forage for fish by underwater swimming. In the time–energy budget, three categories of behavior are recognized: maintenance on land (which includes costs of thermoregulation), swimming underwater, and resting at the water's surface while at sea. The values in column *C* show the estimated daily costs of the three behaviors separately. Summation of the daily costs of all three yields the total daily energy expenditure of an African penguin, about 1900 kJ.

A third technique useful for studying the physiology of activity in the wild is the use of miniaturized electronic monitors, which are placed on or in a study animal. Two basic types of monitors are in use: **telemetric devices** that transmit data as the data are acquired, and **data loggers** that accumulate data in onboard digital memory for later retrieval. Miniaturized electronic monitors can be used to record heart rates, behavioral activity, and other useful indices in many sorts of animals. Heart-rate data can sometimes be used to estimate metabolic rates, and activity data may be useful in constructing time–energy budgets.

The Energy Costs of Defined Exercise

If we consider each of the major forms of animal locomotion—swimming, running, and flying—we find that there is often a characteristic relation between the metabolic cost per unit of time

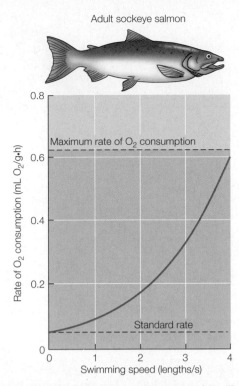

Adult sockeye salmon

FIGURE 9.4 Rate of O₂ consumption as a function of swimming speed in yearling sockeye salmon (*Oncorhynchus nerka*) The fish were studied in a water tunnel. (After Brett 1964.)

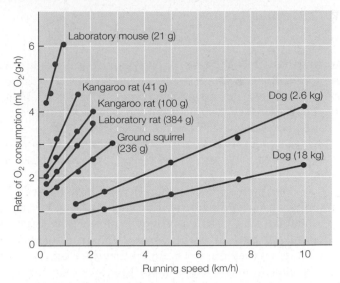

FIGURE 9.5 Rate of O₂ consumption as a function of running speed in six species of mammals of different body sizes studied on treadmills The species are laboratory mouse (*Mus domesticus*, average weight 21 g), Merriam's kangaroo rat (*Dipodomys merriami*, 41 g), banner-tailed kangaroo rat (*D. spectabilis*, 100 g), laboratory rat (*Rattus norvegicus*, 384 g), round-tailed ground squirrel (*Xerospermophilus tereticaudus*, 236 g), and domestic dog (mongrels weighing 2.6 kg and Walker foxhounds weighing 18 kg). (After Taylor et al. 1970.)

and the speed of locomotion. Consider swimming by fish, for example. In this form of exercise, metabolic rate typically increases in a *J-shaped* power function as speed increases (**FIGURE 9.4**).[6] This relation occurs because the drag that a fish must overcome to move through water increases approximately in proportion to the square of its speed of swimming.

When a mammal runs, its metabolic rate usually increases as a *linear* function of its speed, as is shown for six species in **FIGURE 9.5**. Other running or walking animals, such as running insects and terrestrial crabs, also usually exhibit linear relations between metabolic rate and speed.

For birds flying by flapping their wings,[7] aerodynamic theory predicts a *U-shaped* relation between metabolic rate and speed (**FIGURE 9.6A**). The actual existence of this sort of relation has been difficult to test in practice, in part because of difficulties in

[6] If M is metabolic rate and u is speed, the expected relation between M and u for swimming fish is described by a power function: $M = a + bu^c$, where a, b, and c are constants for a particular fish species. The characteristics of power functions are discussed in Appendix F.

[7] Flapping flight is distinguished from soaring flight and from gliding and diving.

(A) Predicted relation from aerodynamic theory

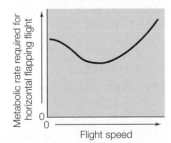

(B) Empirical data for five species of birds

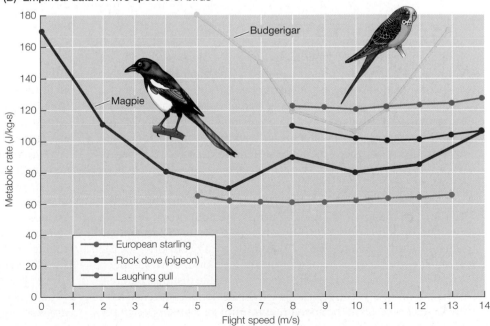

FIGURE 9.6 The metabolic rate of birds moving by flapping flight as a function of flight speed (A) The predicted relation between energy output and airspeed (speed relative to the air) based on aerodynamic theory. (B) The empirically measured rate of output of metabolic energy as a function of flight speed in five species of birds studied in wind tunnels. The results for Eurasian magpies (*Pica pica*) were obtained by analysis of data on wing movements and forces. The results for the other four species were gathered by the measurement of O₂ consumption and have been calculated to be in the same units as the magpie data. (A after Alerstam and Hedenström 1998; B after Dial et al. 1997.)

measuring the metabolic rates of birds flying at low speeds in wind tunnels. **FIGURE 9.6B** shows results for five species of birds. Two species, the magpie and budgerigar (a type of parakeet), exhibited U-shaped relations when studied, but the data for the other three species are incomplete because of no information at low speeds. A recent review of more than 20 studies reached the conclusion that most studies support, or are compatible with, the hypothesis of a U-shaped relation.

For all three forms of locomotion that we have discussed—swimming by fish, running by terrestrial animals, and flying by birds—small-bodied species tend to require greater weight-specific metabolic rates to move at any particular speed than large-bodied species. Note, for example, in Figure 9.5 that the weight-specific metabolic rate required to run at about 1 km/h is far higher for a mouse than for a rat. Similarly, small birds such as budgerigars have higher weight-specific metabolic rates than large ones (e.g., gulls) when flying at the same speed (see Figure 9.6B).

The most advantageous speed depends on the function of exercise

We are well aware that we can judge the ideal speed of a vehicle in more than one way, depending on what we are attempting to accomplish by travel in the vehicle. For instance, we might want to get to our destination as fast as possible, or we might want to travel at a speed that promotes energy efficiency. Similarly, there are several ways to judge the most advantageous speed for an animal to swim, run, or fly.

To illustrate this point, the study of flapping flight by birds is particularly revealing. In the analysis of animal locomotion, two measures of cost are useful: *energy cost per unit of time* (metabolic rate) and *energy cost per unit of distance traveled*—termed the **cost of transport**. **FIGURE 9.7A** shows the weight-specific *energy cost per unit of time* (i.e., weight-specific metabolic rate) of flying budgerigars as a function of their flight speed. **FIGURE 9.7B** shows the weight-specific *energy cost per unit of distance traveled* (i.e., weight-specific cost of transport) as a function of speed. The following equation explains how the two plots are related:

$$\frac{energy}{time} \div \frac{distance}{time} = \frac{energy}{distance} \qquad (9.1)$$

On the left side of this equation, the first expression is the metabolic rate, and the second expression is speed. The equation shows that if an animal is traveling at any given speed, its energy cost per unit of distance is calculated by dividing its metabolic rate at that speed by the speed. Each of the six data points in Figure 9.7A depicts the metabolic rate at a particular speed. Dividing each metabolic rate by the corresponding speed produces the six data points in Figure 9.7B. *The two plots, therefore, are merely different ways of looking at the exact same information.*

To see the significance of the two ways of analyzing budgerigar flight costs in Figure 9.7, consider these two questions: First, if a budgerigar takes off with certain fuel reserves and does not eat during flight, at what speed should it fly to stay airborne for the longest time before it runs out of fuel? Second, at what speed should the bird fly to cover the greatest possible distance before it runs out of fuel? Clearly, to stay aloft for the longest possible *time*, the bird should fly at the speed that minimizes its energy cost per

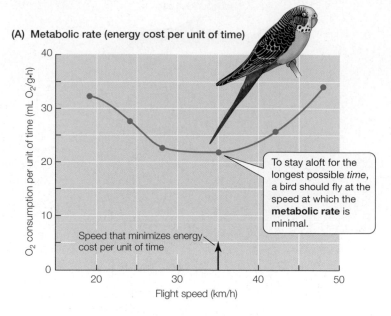

(A) Metabolic rate (energy cost per unit of time)

y-axis: O$_2$ consumption per unit of time (mL O$_2$/g·h)

x-axis: Flight speed (km/h)

Speed that minimizes energy cost per unit of time

To stay aloft for the longest possible *time*, a bird should fly at the speed at which the **metabolic rate** is minimal.

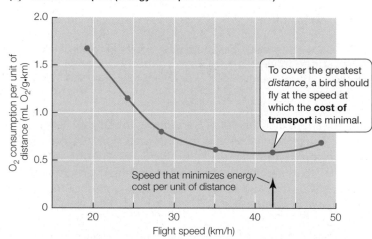

(B) Cost of transport (energy cost per unit of distance)

y-axis: O$_2$ consumption per unit of distance (mL O$_2$/g·km)

x-axis: Flight speed (km/h)

Speed that minimizes energy cost per unit of distance

To cover the greatest *distance*, a bird should fly at the speed at which the **cost of transport** is minimal.

FIGURE 9.7 Two ways to view the energetics of flapping flight by budgerigars (*Melopsittacus undulatus*) (A) Metabolic rate as a function of speed. (B) Cost of transport as a function of speed. To minimize *metabolic rate*, the flight speed must be significantly lower than the speed that minimizes *cost of transport*. Both graphs express energy expenditure in weight-specific terms. Figure 9.6B presents the data in (A) in transformed units. (After Tucker 1968, 1969.)

unit of time—the speed at which the metabolic rate in Figure 9.7A is minimal. By contrast, to cover the greatest *distance*, the bird should fly at the speed that minimizes its energy cost per unit of distance—the speed at which the cost of transport in Figure 9.7B is minimal. The two speeds are different.

On the basis of these theoretical considerations, therefore, the most adaptively advantageous speed depends on the function being performed by flight. If the function of flight is best served by staying airborne as long as possible, the speed marked in Figure 9.7A would be most advantageous. However, if the function of flight is to cover as much distance as possible—as when a bird migrates—the speed marked in Figure 9.7B would be most beneficial.

Data collected on the behavior of birds in the wild over the last 25 years strongly suggest that birds actually tend to follow these principles in selecting their flight speeds. When skylarks (*Alauda arvensis*), for example, engage in song flights that act as displays to attract mates, they fly at about the speed that minimizes their energy cost per unit of time in the air. When they undertake migratory flights, however, their average speed increases more than twofold and more closely approximates the speed at which they require the least energy to cover a kilometer of distance.

Besides the perspectives already stressed, there are additional ways in which the adaptive advantage of speed can be assessed, depending on circumstances. When an animal must run a long distance to escape a pursuing predator, for example, *maximization of sustained speed* is likely to be paramount, regardless of the efficiency of travel at that speed. Another perspective arises when we consider animals undertaking long-distance migrations during which they feed along the way. If forward travel will likely bring an animal into a habitat where food is more abundant than in the habitat it is leaving, a substantial body of theory indicates that—energetically speaking—the optimal speed of travel is a bit higher than the speed at which the animal's cost of transport is minimized; this is true because faster travel allows the animal to arrive sooner in places where food is easier to obtain.

You can see from this discussion that travel at a speed that minimizes cost of transport is relevant only when certain functions are being performed by exercise; it is not a universal standard of performance. Nonetheless, the study of travel at the minimum cost of transport (minimum energy cost per distance covered) has provided powerful insights into many types of animal performance—insights we now discuss.

The minimum cost of transport depends in regular ways on mode of locomotion and body size

Suppose that for each species of animal we study, we identify the *minimum* weight-specific cost of transport. Specifically, we identify the minimum energy cost to move 1 kg of body weight 1 m during horizontal locomotion, regardless of the speed at which the minimum occurs. Then we plot the minimum cost of transport as a function of body weight, as in **FIGURE 9.8**. This sort of plot reveals one of the most striking patterns ever discovered in exercise physiology: Namely, species that engage in a particular primary mode of locomotion—running, flying, or swimming—tend to exhibit a regular relation between minimum cost of transport and body size. Furthermore, the relation for runners differs from that for fliers, and both relations differ from that exhibited by swimming fish. To interpret the lines showing these three relations in Figure 9.8, note that the graph employs log–log coordinates. Minimum cost of transport is an allometric function of body weight for animals that engage in each primary mode of

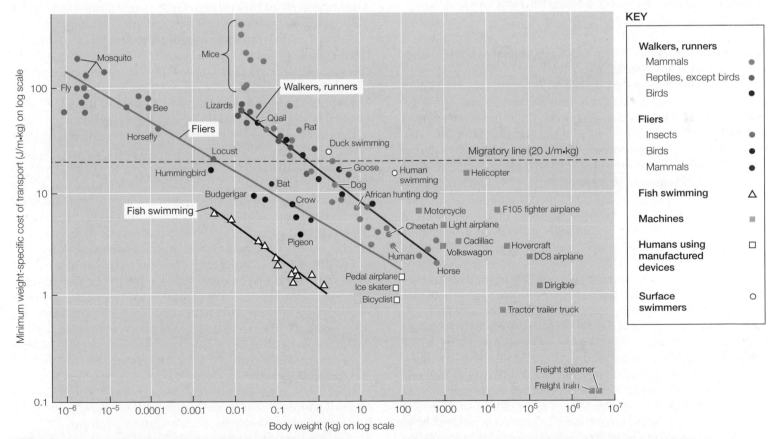

FIGURE 9.8 Minimum weight-specific cost of transport in relation to body weight for running, flying, and swimming animals and for machines Representative species and machines are identified at random. The three solid lines show the relation between cost of transport and body weight for animals that employ walking or running, flapping flight, or swimming as their *primary* modes of locomotion. Only fish are included in the swimming line. By now, huge numbers of data points have been added to this sort of analysis. However, this older plot—partly because it is relatively uncluttered—nicely brings out the basic principles. (After Tucker 1975.)

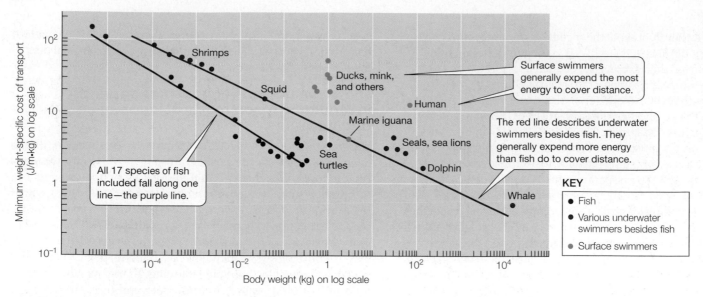

FIGURE 9.9 Minimum weight-specific cost of transport during swimming The animals besides fish that swim underwater include shrimps (*Palaemon*), a species of squid (*Loligo*), the green sea turtle (*Chelonia*), the harbor seal (*Phoca*), the California sea lion (*Zalophus*), the bottlenose dolphin (*Tursiops*), and the gray whale (*Eschrichtius*). These various species roughly fall on a single line. Animals swimming at the water's surface include humans, several species of ducks and geese, a mink (*Mustela*), a pengiun (*Eudyptula*), and the marine iguana (*Amblyrhynchus cristatus*); the iguana plots very close to the underwater-swimming sea turtles. (After Videler 1993.)

locomotion, and for this reason the relations plot as straight lines when both axes are logarithmic.[8]

To appreciate the full significance of the regular relations shown in Figure 9.8, consider first the fliers—the species for which the primary mode of locomotion is flapping flight. These species are taxonomically very diverse—including insects, bats, and birds—yet they all fall approximately on a single line in terms of their relation between minimum cost of transport and body weight. Similarly, the walkers and runners plotted in Figure 9.8—which consist of lizards and running birds as well as mammals—fall on a single line. In fact, data are now available for more than 150 species of animals that move primarily by running—including running insects, semiterrestrial crabs, additional running birds (e.g., roadrunners and ostriches), centipedes, and millipedes, as well as additional lizards and mammals—and with a few exceptions, all the species fall statistically along a single line. The line for swimming in Figure 9.8 is just for fish. Nonetheless, data are available now for more than 20 species of fish, some of which are very different from one another in body form and swimming style, and essentially all fall along one line. The overall picture that emerges from analyses such as the one shown in Figure 9.8 is remarkable: Among animals engaged in their primary form of locomotion, the minimum cost of transport displayed by a species of given body size typically *depends principally on the species' mode of locomotion rather than its clade (phylogenetic position) or the details of its locomotor mechanisms.*

For animals of a particular body size engaging in their primary form of locomotion, running is the most costly way to cover distance, whereas swimming—as practiced by fish—is the least costly. The differences in cost among the three forms of locomotion are more substantial than they might appear from a simple visual inspection of Figure 9.8 because the logarithmic scale used for cost of transport tends to make differences look smaller than they are. For a 100-g animal, the cost of running a unit of distance is about 4 times higher than the cost of flying the same distance and about 14 times higher than the cost of swimming!

Among animals that share a single primary mode of locomotion, large-bodied species cover distance at considerably lower weight-specific cost than small-bodied species do. This means, among other things, that if two animals of different sizes set off on travels with equal proportions of body fat (which they use as fuel), the larger one will be able to cover more distance before exhausting its fat.

Swimming has proved to be a particularly interesting and revealing mode of locomotion for study. As we have noted, even though different species of fish employ a diversity of swimming styles, fish species usually fall along a single line in their relation between minimum cost of transport and body size. What about other types of animals that—like fish—travel by *underwater swimming* as their primary mode of locomotion, such as shrimps, sea turtles, and marine mammals? As shown in **FIGURE 9.9**, such animals (shown in red) typically exhibit higher costs of transport than fish (shown in purple), possibly because they often are less streamlined than fish. Costs of transport are in general even higher in animals that *swim at the water's surface* rather than underwater, such as humans, ducks, and other primarily terrestrial animals (see Figure 9.9). The high costs observed in most surface swimmers are consequences of two factors: (1) swimming is not the primary mode of locomotion for the surface-swimming animals, and (2) according to hydrodynamic principles, swimming at the surface tends to be intrinsically more costly than underwater swimming.

For us humans, an interesting corollary of our high cost of swimming is this: Our own, human experience of the effort required to swim provides not the slightest insight into the effort required of a fish! In fact, if you want to get a sense of how strenuous it would be for a human-sized fish to cover distance by underwater swimming, climb on a bicycle. Bicycling ranks as one of the least costly

---

[8] For each mode of locomotion, minimum weight-specific cost of transport $= aW^b$, where W is body weight, a and b are constants, and b is typically about −0.2 to −0.4. See Chapter 7 (pages 178–179) and Appendix F for discussions of allometric functions and their shapes on various sorts of plots.

of all animal-powered forms of locomotion, and the cost to cover distance by bicycling approaches the cost of swimming expected for fish of human size (see Figure 9.8).

Let's now briefly consider the various sorts of animals that undertake long-distance migrations. When we do this, we encounter a remarkable fact: Although certain small or medium-sized fish and flying animals (insects, bats, and birds) undertake long-distance migrations, such migrations are rare among small or medium-sized running animals. Some *large* running animals, such as reindeer (caribou), are noted for their long migrations, but as Figure 9.8 shows, their cost of covering distance is likely to be similar to that of relatively small fish! Vance Tucker pointed out years ago that if a line (which he called the *migratory line*) is drawn across Figure 9.8 at about 20 J/m·kg, most migratory species are found below the line, and species above the line are unlikely to be migrants. Apparently, long-distance migration has a high chance of being favored by natural selection only if the cost of covering distance is relatively low, and the body size that permits an adequately low cost depends on the mode of locomotion. We discuss long-distance migration at more length shortly.

SUMMARY
The Energy Costs of Defined Exercise

- When a mammal or other animal runs, its metabolic rate typically increases linearly with its speed. Metabolic rate and speed are related by a J-shaped power function in swimming fish. For animals engaged in flapping flight, metabolic rate is expected to exhibit a U-shaped relation to speed, but this theoretical expectation is not always observed in real animals.

- Cost of transport is the energy cost of covering a unit of distance. The speed that minimizes the cost of transport is the speed that maximizes the distance that can be traveled with a given amount of energy.

- Running animals, animals flying by flapping flight, and swimming fish exhibit three distinctive and coherent allometric relations between minimum cost of transport and body weight. For animals of any particular body size, running is the most expensive way to cover distance, flying is intermediate, and swimming by fish is the least expensive. Within any one locomotory group, the minimum weight-specific cost of transport decreases as body size increases.

The Maximum Rate of Oxygen Consumption

An animal's **maximum rate of oxygen consumption**, symbolized $\dot{V}_{O_2max}$ and sometimes called **aerobic capacity** or **maximum aerobic power**, is a key property for the study of aerobic activity.[9]

[9] A symbolic convention used in some branches of physiology is to place a dot over a symbol to indicate *rate*. V_{O_2} symbolizes volume of O_2. Placing a dot over the V symbolizes the rate at which the volume of O_2 is used (i.e., rate of O_2 consumption in volume units). In this chapter we assume that to measure $\dot{V}_{O_2max}$, investigators induce animals to exercise intensely, thereby raising their rates of O_2 consumption to peak levels. Another way to raise the rates of O_2 consumption of some animals, such as mammals, is to expose them to low environmental temperatures. Maximum rates of O_2 consumption induced by exercise and cold are often not the same because different organs participate in exercise and cold defense.

One reason an animal's $\dot{V}_{O_2max}$ is of interest is that it determines the peak rate at which the animal can synthesize ATP by aerobic catabolism. Thus it determines how intensely the animal can exercise in a pay-as-you-go mode (see Figure 8.10B). Suppose two species, A and B, exhibit the same relation between metabolic rate and speed, but species A has a higher $\dot{V}_{O_2max}$ than species B. In that case, although A and B would require the same rate of energy expenditure to travel *at any given speed*, species A would be able to cover long distances *faster* because it would be able to make ATP more rapidly by aerobic catabolism and thus reach higher sustained speeds.

A second reason for interest in $\dot{V}_{O_2max}$ is more subtle and remarkable: In studies of humans and other vertebrates, the maximum rate of O_2 consumption provides a benchmark by which to judge the *strenuousness* of all *aerobic* physical activity. Specifically, the strenuousness to an individual of any particular form of aerobic exercise depends roughly on how high a proportion of that individual's $\dot{V}_{O_2max}$ is required by the exercise. In people on average, if a fully aerobic activity demands no more than 35% of an individual's $\dot{V}_{O_2max}$, the activity can likely be carried out for 8–10 continuous hours. As the percentage of $\dot{V}_{O_2max}$ required by aerobic activity increases, the activity becomes more strenuous. An activity that requires 75% of $\dot{V}_{O_2max}$, for instance, will probably, for most people, be exhausting in 1–2 h.

The use of $\dot{V}_{O_2max}$ as a benchmark is informative in several ways. One is the analysis of human-powered flight (**BOX 9.2**). A more practical application for most people is the analysis of the relation between exercise and aging. As people age beyond young adulthood, their $\dot{V}_{O_2max}$ tends to decline; after age 30, the decline is about 9% per decade for sedentary people, although it is less than 5% per decade for people who stay active. Because of the decline in $\dot{V}_{O_2max}$, an activity that requires any particular *absolute* rate of O_2 consumption tends to demand an ever-greater proportion of $\dot{V}_{O_2max}$ as people age. Thus the activity becomes more strenuous. A form of exercise that demands 35% of $\dot{V}_{O_2max}$ in youth might require 50% in old age; the exercise would therefore shift from being sustainable for 8–10 h in youth to being sustainable for perhaps half that time in old age. This process helps explain why jobs involving physical labor become more difficult for people to perform for a full workday as aging occurs.

Work that requires O_2 consumption at 100% of $\dot{V}_{O_2max}$ is strenuous in the extreme; people can ordinarily continue it for only a few minutes. We saw in Box 8.3 that mountaineers who are breathing the ambient air exhibit a lower and lower $\dot{V}_{O_2max}$ as they go to higher altitudes. Accordingly, for such mountaineers, although slow uphill walking requires just a small percentage of $\dot{V}_{O_2max}$ at low altitudes, it demands approximately 100% of $\dot{V}_{O_2max}$ at the highest altitudes on Earth. Slow uphill walking therefore shifts from being simple at low altitudes to being barely possible, or impossible, in high mountains. As mountaineers breathing the ambient air approach the top of Mt. Everest, they may require an hour to walk the final 100 m (see Box 8.3)!

The physiological causes of the limits on maximum O_2 consumption are hotly debated. Some physiologists argue that particular organ systems set the limits on $\dot{V}_{O_2max}$. For instance, some point to the circulatory system as being the "weak link" in mammals, arguing that all other organ systems could transport and use O_2 at a greater rate if it were not for more-restrictive limits on how fast the circulatory system can transport O_2. An alternative position is the hypothesis of **symmorphosis**, which states that all organ systems that serve a single function in an animal are interactively

BOX 9.2 Finding Power for Human-Powered Aircraft

The world was electrified in 1977 when a human-powered aircraft named *Gossamer Condor* first flew a mile. Two years later, Bryan Allen powered another such aircraft, the *Gossamer Albatross*, across the English Channel in a flight of 36 km (22 mi) that required almost 3 h. Human-powered aircraft fly at an average altitude of about 6 m. At these heights, even a few *seconds* of loss of power are out of the question, because without continuous power the aircraft will quickly land or crash. Thus the feat achieved by Bryan Allen required 3 h of truly *uninterrupted* effort at a very high work intensity.

Bryan Allen powering the *Gossamer Albatross* over the English Channel on June 12, 1979

The physiological properties of human exercise discussed in this chapter dictate that the rate of O_2 consumption of a pilot must be no higher than 65%–70% of his or her $\dot{V}_{O_2max}$ for there to be a reason- able chance that a very motivated pilot will be able to maintain uninterrupted effort for 3 h. This biological fact set the engineering goals for the design of the *Gossamer Albatross*: The aircraft could not demand more power from its "power plant" than 65%–70% of $\dot{V}_{O_2max}$. **Box Extension 9.2** discusses pilot selection and the next stage in human-powered aircraft following the *Gossamer Albatross*.

---

adapted to have approximately equal limits because it would make no sense for any one system to have evolved capabilities that could never be used because of more-restrictive limits in other systems. For biologists who subscribe to the concept of symmorphosis, the reason an animal cannot attain a rate of O_2 consumption higher than its $\dot{V}_{O_2max}$ is that multiple organ systems simultaneously reach their performance limits at $\dot{V}_{O_2max}$.

Aerobic scope for activity and *aerobic expansibility* are two concepts that are sometimes employed in the study of $\dot{V}_{O_2max}$. An animal's **aerobic scope for activity** at a particular temperature is usually defined to be the *difference* between its $\dot{V}_{O_2max}$ at that temperature and its resting rate of O_2 consumption at the same temperature. Its **aerobic expansibility** is the *ratio* of its $\dot{V}_{O_2max}$ over its resting rate of O_2 consumption.[10] To illustrate, suppose that at a particular body temperature, a fish has a resting rate of O_2 consumption of 0.05 mL/g·h and a $\dot{V}_{O_2max}$ of 0.30 mL/g·h. Its aerobic scope for activity would be $(0.30 - 0.05) = 0.25$ mL/g·h. Its aerobic expansibility would be $0.30/0.05 = 6$.

$\dot{V}_{O_2max}$ differs among phyletic groups and often from species to species within a phyletic group

A rough but useful rule of thumb for vertebrates is that $\dot{V}_{O_2max}$ is about ten times higher than the resting rate of O_2 consumption.[11]

---

[10] There has never been a successful effort to achieve universal consensus on the use of these terms. Thus, for example, aerobic *expansibility* values are sometimes called aerobic *scope* values. In reading the scientific literature, it is important to ascertain the meaning in each context.

[11] In the case of mammals and birds, the $\dot{V}_{O_2max}$ is about ten times the *basal* metabolic rate. Keep in mind that in this chapter we use $\dot{V}_{O_2max}$ to refer to the highest rate of O_2 consumption elicited by exercise.

That is, the aerobic expansibility of vertebrates tends to be about 10. The consistency of the aerobic expansibility in vertebrates has some remarkable implications. For example, consider that the standard metabolic rates of fish, amphibians, and nonavian reptiles are typically—at most—only one-tenth to one-fourth as high as the basal metabolic rates of mammals and birds of similar body size (see page 265). Given that aerobic expansibility averages about 10 in all vertebrate groups, you can see that the *peak* rates of O_2 consumption of fish, amphibians, and nonavian reptiles are of the same order of magnitude as the *basal* (resting) rates of mammals and birds.

TABLE 9.3 illustrates this important relation with data on two pairs of vertebrates: (1) a salmon and a rat of similar size and (2) a monitor lizard and a guinea pig of similar size. Salmon and monitor lizards are among the most aerobically competent of all fish and nonavian reptiles. Yet their *peak* rates of O_2 consumption ($\dot{V}_{O_2max}$) resemble the *basal* rates of O_2 consumption of mammals. $\dot{V}_{O_2max}$ in the mammals is thus far above $\dot{V}_{O_2max}$ in the salmon and lizards. Mammals and lizards of the same body weight typically resemble each other in the ATP cost to cover a unit of distance when they run at speeds that minimize their cost of transport (see Figure 9.8). Mammals, however, can make ATP aerobically far faster than lizards of the same size can. Thus mammals can achieve far higher sustained speeds than lizards can. One of the principal hypotheses offered for the evolution of homeothermy is that it permitted animals to run faster in a sustained manner.

Within any one vertebrate phyletic group, $\dot{V}_{O_2max}$ per gram of body weight tends to be an allometric function of body weight, with small species having a higher $\dot{V}_{O_2max}$ per gram than large species (see Chapter 7). The line in **FIGURE 9.10** shows the average relation between $\dot{V}_{O_2max}$ and size in mammals. The line, in other words,

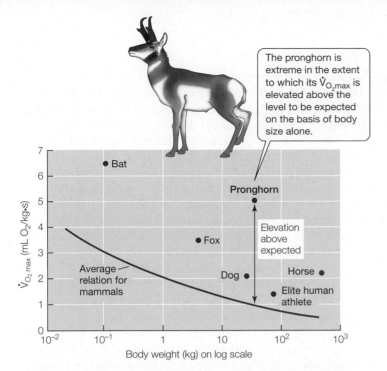

The pronghorn is extreme in the extent to which its $\dot{V}_{O_2max}$ is elevated above the level to be expected on the basis of body size alone.

FIGURE 9.10 The pronghorn (*Antilocapra americana*) represents an extreme case of evolutionary specialization for high $\dot{V}_{O_2max}$ The line shows the average statistical relation between weight-specific $\dot{V}_{O_2max}$ and body weight in mammals; it is curved rather than straight in this plot because although the x axis is logarithmic, the y axis is not. Five nonhuman species that are strong athletic performers and that have $\dot{V}_{O_2max}$ values higher than average (for their body sizes) are plotted individually. The high $\dot{V}_{O_2max}$ of the pronghorn enables it to be the fastest known sustained runner on Earth. (After Lindstedt et al. 1991.)

TABLE 9.3 Rates of aerobic catabolism (mL O_2/g·h) during rest and during sustained exercise of peak intensity in two pairs of vertebrates

One comparison is between a fish and mammal of similar body weight. The other is between a lizard and mammal of similar weight. The mammals were studied at 30°C. The other animals were studied at the temperatures that maximized their aerobic scopes (lizard, 40°C; salmon, 15°C).

Species	Basal or standard rate of O_2 consumption	$\dot{V}_{O_2max}$
Animals weighing 230 g		
Rat (*Rattus*)	0.9	4.6
Salmon (*Oncorhynchus*)	0.05	0.49
Animals weighing 700–900 g		
Guinea pig (*Cavia*)	0.6	3.7
Monitor lizard (*Varanus*)	0.11	1.0

Source: From Hill and Wyse 1989.

shows what is statistically "expected" for mammals of each size. Physiologists have been interested in species that exhibit a $\dot{V}_{O_2max}$ that is above the "expected" value for their size, suggesting evolution of an unusually high capacity for sustained, aerobic exercise (see page 22). The pronghorn, found in grasslands of the American West, is the most extreme of these species (see Figure 9.10). Its $\dot{V}_{O_2max}$ is more than four times higher than the value that would be average or "expected" for a mammal of its size. Pronghorns are grazers that are legendary for the speeds at which they flee danger. They are not quite as fast as cheetahs. However, unlike cheetahs—which produce ATP anaerobically when running fast—pronghorns produce the ATP required for fast running *aerobically*. Cheetahs fatigue within a minute or two when they run fast. Pronghorns, however, can maintain speeds of at least 65 km/h (40 mph) for long periods of time. Pronghorns are the fastest known sustained runners on Earth. They have enormous lungs for their size and exceptional abilities to maintain high rates of blood circulation. Compared with goats and dogs of the same total body weight, pronghorns have 1.2–1.7 times more muscle, and—*per gram* of muscle—they have 1.2–2.6 times more mitochondria.

Considering the entire animal kingdom, the highest known levels of weight-specific $\dot{V}_{O_2max}$ are found in certain types of flying animals. Among invertebrates, certain strong-flying insects achieve extreme $\dot{V}_{O_2max}$ levels. Among vertebrates, certain small-bodied fliers—hummingbirds and bats (see Figure 9.10)—do. In the flight muscles of such aerobic heroes, mitochondria often occupy 35%–45% of tissue volume—very high values. With mitochondria packed this densely in muscle cells, a sort of competition arises between mitochondria and contractile elements for cell space; as mitochondria occupy more space, contractile elements have less. Animals seem never to go higher than about 45% mitochondria in working muscles, suggesting that the evolution of higher values would be pointless because the contractile apparatus would be too diminished to use the ATP that the mitochondria could make. The peak in mitochondrial packing helps set an ultimate peak on $\dot{V}_{O_2max}$ per gram of muscle tissue.

$\dot{V}_{O_2max}$ varies among individuals within a species

Individuals of a species vary in $\dot{V}_{O_2max}$, as we saw in Figures 1.22 and 1.23. The phenomenon is readily documented in humans (see Figure 1.23). For example, when groups of young men entering the military are tested for $\dot{V}_{O_2max}$, they typically exhibit a normal statistical distribution (bell curve) in $\dot{V}_{O_2max}$ (expressed per unit of body weight), with the $\dot{V}_{O_2max}$ of the highest-$\dot{V}_{O_2max}$ men being twice as high as that of the lowest-$\dot{V}_{O_2max}$ men. As we discuss in greater detail shortly, some of this variation is a consequence of life experience, whereas some (about half, according to recent research) is genetic.

Recall that the strenuousness of any given form of sustained exercise depends on the fraction of $\dot{V}_{O_2max}$ that the exercise requires. This principle helps explain why a single form of sustained exercise can be differently taxing to different individuals. Suppose, to illustrate, that for a person to run at a particular speed, a rate of O_2 consumption of 30 mL O_2/kg·min is required. Suppose also that two otherwise identical people have $\dot{V}_{O_2max}$ values of 40 and 70 mL O_2/kg·min, respectively. For the former individual the running would require 75% of $\dot{V}_{O_2max}$ and would be very taxing; for the latter, it would require just 43% and would be sustainable for hours.

TABLE 9.4	Average $\dot{V}_{O_2max}$ in male Swedish athletes who compete in various events at the world-class level
Event (or other category)	**Average $\dot{V}_{O_2max}$ (mL O_2/kg·min)**
Highest values ever recorded	90–95
Cross-country skiing	84
Long-distance running	83
Canoeing	67
Ice hockey	63
Soccer	58
Weight lifting	53
Gymnastics	52
General population (young adult Swedish men)	44

Source: After Åstrand and Rodahl 1986.

Investigators consistently find that, among successful human athletes, individuals who excel in different types of competition tend to exhibit striking differences in $\dot{V}_{O_2max}$. As **TABLE 9.4** shows, for example, men who compete at the world-class level in cross-country skiing or long-distance running have far higher average $\dot{V}_{O_2max}$ values than men who compete in weight lifting or gymnastics. Based on careful analysis, the differences in Table 9.4—taken as a whole—cannot be attributed primarily to differences in training (except for the "general population"). The differences in $\dot{V}_{O_2max}$ among successful athletes in various events are believed to result to a substantial extent from the athletes' choosing to compete in events in which they have inherent abilities to succeed. A high $\dot{V}_{O_2max}$, meaning a high peak rate of aerobic ATP production, is an asset in long-sustained events, especially ones like cross-country skiing in which most major muscle masses are employed. Athletes who are endowed by heredity or by early developmental influences with a high $\dot{V}_{O_2max}$ choose sports such as cross-country skiing and other sports in which aerobic ATP synthesis is particularly important, and vice versa. Although almost nothing is known about nonhuman animals in this regard, it seems reasonable to hypothesize that in them also, individuals with different aerobic competence elect different lifestyles (see page 32).

$\dot{V}_{O_2max}$ responds to training and selection

Training that emphasizes aerobic ATP production—known as *endurance training*—often increases an individual's $\dot{V}_{O_2max}$. When sedentary people, for example, participate in endurance training, they typically increase their $\dot{V}_{O_2max}$ by 10%–30%, although individuals vary widely and some show no change. The physiological reasons for the increase in $\dot{V}_{O_2max}$ are starting to be well understood, at least in humans and lab animals. Training activates widespread changes in gene expression in exercising muscles (see Figure 3.9). One of several major consequences is an increase in numbers of mitochondria in

muscle cells and upregulation of mitochondrial enzymes of aerobic catabolism, such as citrate synthase (a Krebs cycle enzyme) and cytochrome oxidase (an electron-transport enzyme), as exemplified in **TABLE 9.5**. Other changes include increases in the heart's capacity to pump blood, the density of blood capillaries in muscles, and muscle glucose transporters. Training is discussed in detail in Chapter 21.

On a different scale of time, there are several ongoing projects focused on *evolutionary selection*—over many generations—for high or low aerobic competence. In one study, investigators started with random-bred lab rats that were individually tested for their ability to run long distances during aerobic exercise. Rats with both high and low aerobic endurance were selected to be parents of future generations. As **FIGURE 9.11** shows, ten generations of this process resulted in two populations of rats (shown in blue and red) that differed dramatically in aerobic endurance, emphasizing that there is a partial genetic basis for exercise performance; on average, rats in the population selected for high running capacity could run 905 m prior to exhaustion, whereas rats in the population selected for low running capacity could run only 217 m. $\dot{V}_{O_2max}$ diverged in the two populations; by the 15th generation, it was 1.5 times higher, on average, in the high-endurance population. The capacity of the heart to pump blood was far higher in that population, blood-capillary density in the muscles was higher, and enzymes of aerobic catabolism were elevated. The research also brought to light a discovery that is disturbing in light of today's sedentary lifestyles: Rats in

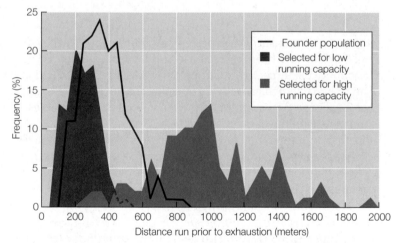

FIGURE 9.11 Consequences of selection on endurance running in rats Hundreds of rats were tested on a treadmill tilted so they had to run uphill. The distance each rat could run prior to exhaustion was measured. Rats led a sedentary existence except during testing. Plotted here are frequency distributions: the percentages of rats that ran various distances prior to exhaustion. The study started with a "founder" population. Animals with exceptionally great endurance were chosen as parents for the next generation, and this process was repeated for a total of ten generations to give rise to the population shown here (blue) as "selected for high running capacity." Similarly, animals in the founder population that displayed exceptionally low endurance were mated, and the process was repeated for ten generations to obtain the population (red) "selected for low running capacity." Both the high- and low-running-capacity populations exhibited shifted distributions in comparison with the founder population. (After Koch and Britton 2005.)

TABLE 9.5 Average physiological measures before and after 7 weeks of endurance training in young adult people

The training consisted of 60 min of cycling, requiring 60% of $\dot{V}_{O_2max}$, 5 days a week. All before–after differences are statistically highly significant, except for the testosterone values. Muscle measures are for samples from the *vastus lateralis*, a thigh muscle.

Property	Gender	Before	After
$\dot{V}_{O_2max}$ (mL O_2/kg·min)	Male	41.5	48.7
	Female	31.9	41.5
Muscle citrate synthase activity (μmol/mg·min)	Male	7.3	10.2
	Female	8.4	12.0
Muscle cytochrome oxidase activity (μmol/mg·min)	Male	4.8	5.9
	Female	4.1	5.3
Blood testosterone (nmol/L)	Male	19.3	21.5
	Female	1.7	1.3

Source: Carter et al. 2001.

the population selected for low aerobic exercise capacity tended to exhibit symptoms of chronic cardiovascular and metabolic diseases, such as high blood pressure and diabetes.

SUMMARY
The Maximum Rate of Oxygen Consumption

■ An animal's maximum rate of O_2 consumption ($\dot{V}_{O_2max}$) is significant for two principal reasons. First, it determines the maximum rate at which sustained, aerobic exercise can be performed. Second, it serves as a benchmark by which the strenuousness of submaximal aerobic work can be assessed. Sustained work becomes more strenuous—and more quickly fatiguing—for an individual as it demands a higher proportion of the individual's $\dot{V}_{O_2max}$.

■ Major phyletic groups sometimes exhibit consistent differences in $\dot{V}_{O_2max}$. Among vertebrates—as a rough but important rule of thumb—$\dot{V}_{O_2max}$ in mammals and birds is about an order of magnitude higher than $\dot{V}_{O_2max}$ in fish, amphibians, and nonavian reptiles of similar body size, assuming that the latter groups are at body temperatures near mammalian and avian levels. Within a single phyletic group, $\dot{V}_{O_2max}$ typically tends to vary with body size in an allometric fashion, with small-bodied species having higher $\dot{V}_{O_2max}$ per gram of body weight than large-bodied species.

■ Individuals of a species that are similar in age and gender typically vary considerably in $\dot{V}_{O_2max}$. Some of this variation can be attributed to differences in training. Usually a significant proportion of the variation can also be attributed to inheritance and/or early developmental effects.

The Energetics of Routine and Extreme Daily Life

From studies using the doubly labeled water method, physiologists now know a great deal about the *average daily metabolic rates* (ADMRs; also called *field metabolic rates*) of terrestrial animals as they lead their routine lives in their natural habitats. Within sets of related species, the ADMR is an allometric function of body size: To meet the costs of their ordinary daily lives in the wild, small-bodied species expend more energy (and must therefore gather more food) per unit of body weight than large-bodied species.

An extremely important insight from the ADMR studies is quantification of the vast difference in energy costs between poikilothermic ("cold-blooded") and homeothermic ("warm-blooded") animals. In comparison with lizards, snakes, and other nonavian reptiles living in the wild, mammals and birds of the same body size living in the wild have ADMRs that are 12–20 times higher! Physiologists have long known that homeothermy is expensive, but now we can put realistic numbers on the phenomenon (discussed further in Chapter 10).

Let's focus a bit on how much the ADMR of a species differs from its resting metabolic rate. Among wild mammals and birds, ADMR is typically about 2.5–3.5 times higher than basal metabolic rate (BMR). Routine daily existence, in other words, requires about 2.5–3.5 times more energy than resting all the time in a thermoneutral environment.

How do people compare? In both developed and developing societies, the ADMRs of hundreds of people going about their ordinary daily lives have been measured by use of the doubly labeled water method. ADMR typically varies from about 1.2 to 2.5 times BMR in the general population. For the most part, only dedicated athletes have ADMRs that exceed BMR by more than 2.5.

A question that has drawn the interest of physiologists is how high the ADMR can possibly be. That is, what is an animal's maximum possible metabolic rate that can be *sustained day after day*? Note that this is a very different question from asking how high the $\dot{V}_{O_2max}$ can be, because $\dot{V}_{O_2max}$ reflects a rate of metabolism that can be maintained for only tens of minutes, at most.

To gain insight into how high the human sustained metabolic rate can possibly be, investigators studied cyclists in the Tour de France, a long-distance bicycle race. In 1984 the race covered nearly 4000 km and lasted 22 days, during which the cyclists fought for the lead while going up and down 34 mountains. The athletes maintained stable body weights over the 3 weeks of the race by eating large amounts of food during their nighttime breaks. The metabolic rates of four of the cyclists were measured with the doubly labeled water method. The ADMRs of these four averaged about 4.5 times higher than their BMRs. These data indicate that when well-conditioned people in a competitive situation expend the maximum effort that they can possibly sustain for periods of many days, their time-averaged metabolic rates are about 4.5 times BMR (even though $\dot{V}_{O_2max}$ is at least 10 times BMR).

TABLE 9.6 presents analogous information for nonhuman mammals and birds. In each study listed, the animals were required to sustain a highly demanding, all-out effort day after day, analogous to the effort put forth by competitors in the Tour de France. Whereas

TABLE 9.6 Particularly high sustained metabolic rates measured in nonhuman mammals and birds	
Animals and circumstances	Average daily metabolic rate as ratio of resting metabolic rate
Female laboratory mice nursing litters of 14 young	6.5
Females of three other species of mice nursing litters	3.7–6.7
Four species of mice living at –10°C	3.7–6.1
Grey seals nursing pups	7.4
Three species of perching birds rearing young	2.4–3.9
Six species of seabirds rearing young	3.1–6.6
Migrating birds:	
Bar-tailed godwits migrating over Pacific Ocean	8.0–10.0
Several species flying in wind tunnels	9.0

Sources: After Gill et al. 2009; Hammond and Diamond 1997; Mellish et al. 2000; Piersma 2011.

some of the animals in the table were in cold environments requiring high energy costs to keep warm, others were rearing young. The data indicate that usually, at the extreme, the sustained ADMR can be as high as 6–7 times resting metabolic rate. Migrating birds are exceptions. In some cases, they sustain metabolic rates 8–10 times resting metabolic rate.

Clearly, if animals could achieve higher peak sustained metabolic rates than they do, they could raise more young, migrate faster, or survive in more-demanding environments. Why then are the peak sustained metabolic rates of mammals and birds not usually higher than 6–7 times their resting metabolic rates? What constraints prevent sustained metabolic rates from being higher than they are? These are current questions for research.

Long-Distance Migration

When we see an osprey (*Pandion haliaetus*)—also called a fish hawk—at the seashore, going about its life, we might easily assume that the bird is always at that area where we see it. Applications of satellite and global-positioning technology in the past 20 years, however, have provided an entirely different view. **FIGURE 9.12** shows the directly observed locations of a single osprey over a 3-year period of its life. The osprey traveled back and forth between Sweden and Ivory Coast—a one-way distance of 7000 km (4300 mi)—three times. Each autumn, having spent the summer in Scandinavia, the osprey traveled a route that must have started to seem familiar—given that the route was almost the same each year. Stopping over on repeated occasions to rest and potentially

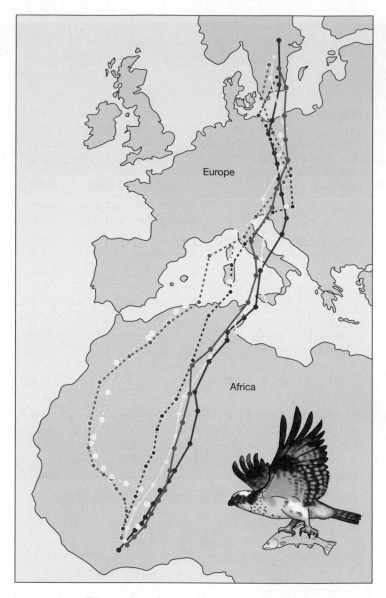

FIGURE 9.12 Three years in the life of a male osprey (*Pandion haliaetus*) Ospreys migrate singly rather than in groups. As this male flew between Scandinavia and Africa, a small radiotransmitter on him, mounted externally, radioed his position, which was detected by satellite. Three colors are used to distinguish the three successive years in the order yellow, red, and blue. Symbols along the lines represent the bird's locations at intervals of 1 day or more. Solid lines mark his southbound migrations in autumn. Dashed lines mark his northbound migrations in spring. (After Alerstam et al. 2006.)

refuel, the osprey took 34–55 days to reach its destination in Africa. There, it resumed the life it had left behind a half-year before: living in its individual wintering locality. Then, each spring, it returned in 21–33 days to Scandinavia to its individual breeding location. Migratory patterns of this sort already existed in the deep past of Earth's geologic history and continue today despite burgeoning human influences all along the routes traveled.

Some of the questions raised by this natural phenomenon are put into greater relief, perhaps, by studies recently completed on bar-tailed godwits (*Limosa lapponica*)—shorebirds weighing

Eel Migration and Energetics: A 2300-Year Detective Story

European eels (*Anguilla anguilla*) are currently believed to breed in the western Atlantic Ocean near Bermuda, although the distance between Europe and this breeding area is about 5500 km (3400 mi). Skeptics have questioned whether adult eels—which do not eat in the ocean—could swim that far and still have enough stored energy for reproduction. Recently, investigators placed eels in a water tunnel for a simulated 5500-km migration. The eels had to swim for almost 6 months to cover the equivalent of 5500 km. Not only did they succeed, but they did so at a remarkably low cost. As seen in the figure, the metabolic rates of the swimming eels tended to be only about twice as great as those of resting eels. The migrating eels achieved a cost of transport that was only about one-fifth as great as expected for their body size, pointing to exceptional swimming efficiency. As a result, they lost only about 20% of their body weight and would have had reasonably large energy stores remaining on arrival in Bermuda.

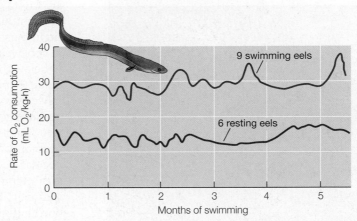

Metabolic rate measured on each day of a 6-month, 5500-km simulated migration   Nine eels (*Anguilla anguilla*), 0.7 m in average body length, swam for 173 days in a water tunnel. They were compared with 6 resting eels. The eels were not fed. The temperature was 19°C. (After van Ginneken et al. 2005.)

European eels have been a focal mystery in marine biology for more than 2300 years—ever since the time of the ancient Greeks, when Aristotle, in his writing on natural history, highlighted the literal statement that no one in Europe had ever seen sperm or eggs in a European eel or seen the eels breed! **Box Ex-**

tension 9.3 outlines the long story of how scientists ultimately concluded that the eels do in fact breed like other fish, but do so far from Europe. The box extension also discusses the latest efforts to confirm this hypothesis by direct observation.

roughly 0.5 kg—during their southward migration from Alaska each autumn. These cutting-edge studies have revealed that the godwits fly nonstop to New Zealand over the open water of the central Pacific Ocean—a trip of approximately 10,000 km (6200 mi) that takes them 6–9 days. Over those 6–9 days, they starve (they do not feed at sea), and they have no freshwater to drink. Moreover, they presumably do not sleep and thus suffer sleep deprivation; and their flight muscles must engage in powering or controlling flight continually day and night—not to mention the necessity of navigating accurately over endless expanses of water. Although the godwits take advantage of tailwinds, this flight seems to challenge some of our most basic assumptions about the very nature of animal life.

Physiologists have long recognized that birds principally use stored fat as fuel for long-distance migrations. Fat is particularly suited for this function because of its high energy density (see Table 6.3). The *mobilization* and *distribution* of fatty acids from fat during migration are significant concerns, however. This point is highlighted by the fact that mammals that have been studied cannot mobilize fat rapidly enough for it to be the principal fuel for high-intensity, long-term exertion. In migrating birds, by contrast, roughly 90% of the fuel is fat. The birds evidently have highly developed mechanisms for fatty acid transport across cell membranes.

Moreover, because the fatty acids are not water-soluble, the birds evidently also have highly developed mechanisms for solubilizing fatty acids, as by formation of lipoprotein complexes, so the fatty acids can be distributed by the circulatory system to the working muscles from storage depots.

Birds accumulate fat prior to migration (they often increase 50%–100% in body weight) and use their stored fat during migration—giving rise to a long-held view that the process is much like fueling an aircraft. In the last 20 years, however, evidence has accumulated that—at least in some long-distance migrants—the nature of the aircraft changes along the way! For example, some of the organs that are involved in building up fat stores (e.g., the intestines) shrink substantially after the fuel-storage phase and during migration, thereby lightening the load that must be transported (see the discussion of red knot sandpipers at the start of Chapter 1). Other long-distance migrations also have properties that draw the interest of physiologists. Studies of migrating sea turtles, for example, have revealed that the turtles are able to navigate by use of Earth's magnetic fields (see Chapter 18). Certain eels are believed to undertake such long migrations that physiologists have wondered if their energy resources are adequate. **BOX 9.3** discusses studies of energetics during the spawning migration of European eels to the Sargasso Sea near Bermuda.

FIGURE 9.13 A case study in ecological energetics   A breeding colony of seabirds, such as this colony of terns, is ecologically dependent on the populations of fish in its vicinity to obtain energy for life and reproduction. With modern methods, physiologists can quantify the rate at which a colony harvests energy relative to the rates of production of prey populations.

Ecological Energetics

Ecological energetics is the study of energy needs, acquisition, and use in ecologically realistic settings. An example is provided by research on the energy needs and acquisition of breeding colonies of seabirds (**FIGURE 9.13**). Using the doubly labeled water method or time–energy budgets, investigators estimate the daily energy demand of a colony, including costs of growth in the young birds and costs of foraging flight in the adults, as discussed at the start of this chapter. The energy demand of the colony can then be compared with the energy available from the fish populations on which the birds feed. Studies of this sort have revealed that seabird colonies sometimes consume one-fourth to one-third of all the productivity of prey fish in their foraging areas. This sort of ecological energetic analysis has helped biologists better understand seabird population dynamics. For example, the high energy needs of some colonies help explain why some have been devastated by competition from human fishing.

A more elaborate illustration of the power of ecological energetic research is provided by Bernd Heinrich's analysis of costs and rewards in bumblebee foraging—an example of what Heinrich terms *bumblebee economics*. The starting point of his analysis is the recognition that in ecologically realistic settings, the acquisition of food has energy costs as well as energy rewards.

When bumblebees (*Bombus*) forage, they fly from one flower (or flower cluster) to another, landing on each long enough to collect available nectar. Two major *costs* of bumblebee foraging must be considered:

1. Flight is itself very costly. It can easily elevate the metabolic rate of a bumblebee to 20–100 times its resting rate. The cost of flight per unit of time is essentially independent of air temperature.

2. Temperature regulation may also be costly. To fly, bumblebees require the temperature of their flight muscles to be about 30°C or higher (see page 281). When the bees are flying, temperatures that high are maintained by the heat produced by the wing-flapping contractions of their flight muscles. When bees land on flowers (and stop flying) in cool weather, however, they are at risk of quickly cooling to below the necessary flight temperature—which would make them unable to take off again. To keep their flight muscles warm while they are alighted on flowers, bees produce heat by a process analogous to human shivering (see Chapter 10). The intensity and *energetic cost* of this form of shivering become greater as the air temperature decreases. Although shivering may be unnecessary at an air temperature of 25°C, shivering at 5°C may raise a stationary bee's metabolic rate to as high a level as prevails during flight.

Considering the costs of *both* flying and shivering, the *average* metabolic expenditure per unit of time for a bee to forage tends to increase as the air becomes cooler. If the air is warm enough that no shivering is needed, a bee has a high metabolic rate when it is flying but a low rate when it is not. If the air is cold, the bee has a high metabolic rate all the time, whether flying or stationary.

Now let's turn to the energy *rewards* of foraging. The energy reward that can be obtained per unit of time from any particular species of flowering plant depends on (1) the volume of nectar obtained from each flower, (2) the sugar concentration of the nectar, and (3) the number of flowers from which a bee can extract nectar per unit of time. The third property depends on the spacing of the flowers and the difficulty of penetrating flowers to obtain their nectar.

Some species of plants yield sufficient sugar per flower that bumblebees can realize a *net* energy profit when foraging from them regardless of the air temperature. For example, the rhododendron *Rhododendron canadense*, a plant with large flowers, typically yields sugar equivalent to about 1.7 J/flower. At 0°C, a large bee expends energy at a time-averaged rate of about 12.5 J/min while foraging. Accordingly, the bee could break even energetically by taking the nectar from about 7–8 flowers per minute. In fact, bees can tap almost 20 rhododendron flowers per minute. Thus, even at 0°C, bees foraging on the rhododendron are able to meet their costs of foraging *plus* accumulate a surplus of nectar to contribute to the hive.

In contrast, some plants yield so little sugar per flower that they are profitable sources of nectar only when air temperatures are relatively high (and the bees' costs of foraging are thereby reduced). For example, bees typically visit flowers of wild cherry (*Prunus*) only when the air is warm. The flowers yield sugar equivalent to only about 0.21 J/flower. At 0°C, a bee would therefore have to tap about 60 flowers per minute just to meet its costs of foraging. Tapping so many flowers is impossible, meaning the bees cannot profitably forage on cherry flowers when the air is cold.

The study of bumblebee foraging exemplifies how an ecologically realistic accounting of energy *costs* and *gains* can help ecologists better understand the foraging choices that animals make. Bumblebees are observed to forage on rhododendron in all weather, but they forage on wild cherry only in warm weather. From the study of ecological energetics we now understand that energetic considerations place constraints on foraging choices by helping to dictate which flowers the bees can profitably exploit.

STUDY QUESTIONS

1. How does the doubly labeled water method depend on the existence of isotopic equilibrium between the oxygen in H_2O and that in CO_2?

2. From a list of your friends, select one (theoretically) for study to determine his or her average daily metabolic rate. How would you carry out research to create a time–energy budget for your friend?

3. In your own words, explain why foraging on wild cherry flowers is beneficial for bumblebees in warm weather but not in cold weather.

4. As noted in this chapter, the $\dot{V}_{O_2max}$ of people tends to decline after age 30 by about 9% per decade for sedentary individuals, but it declines less than 5% per decade for people who stay active. The average $\dot{V}_{O_2max}$ in healthy 30-year-olds is about 3.1 L/min. Using the information given here, what would the average $\dot{V}_{O_2max}$ be in 60-year-olds who have been sedentary throughout their lives and in 60-year-olds who have stayed active (keep in mind that the decline is exponential)? Consider the activities in Table 9.1, and recall from Chapter 7 that 1 kJ is equivalent to about 0.05 L of O_2 in aerobic catabolism. How would you expect sedentary and active people to differ in their capacities for each of those activities in old age? Explain.

5. For an animal engaging in sustained exercise, why is there not one single ideal speed?

6. List the possible reasons why two individuals of a certain species might differ in $\dot{V}_{O_2max}$.

7. Suppose that a bird's metabolic rate while flying at 30 km/h is 8 kJ/h. What is the bird's cost of transport when flying at 30 km/h?

8. Looking at Figure 9.8, how would you say animals and machines compare in their efficiencies in covering distance?

9. African hunting dogs depend on sustained chases by groups of cooperating individuals to capture antelopes for food. If the members of two groups differ in their average $\dot{V}_{O_2max}$, how might the two groups differ in the strategies they use during hunting?

10. In mammals of all species, the peak rate of O_2 consumption of each mitochondrion is roughly the same. On the basis of patterns of how $\dot{V}_{O_2max}$ varies with body size in species of mammals, how would you expect the muscle cells of mammals of various body sizes to vary in how densely they are packed with mitochondria? Explain your answer.

11. What is the hypothesis of symmorphosis? How might you evaluate or test the hypothesis?

12. Explain the concept that in high-performance muscle cells, mitochondria and contractile elements compete for space over scales of evolutionary time.

Go to **sites.sinauer.com/animalphys4e**
for box extensions, quizzes, flashcards,
and other resources.

REFERENCES

Alexander, R. M. 2003. *Principles of Animal Locomotion.* Princeton University Press, Princeton, NJ.

Åstrand, P.-O., K. Rodahl, H. A. Dahl, and S. B. Strømme. 2003. *Textbook of Work Physiology: Physiological Bases of Exercise,* 4th ed. Human Kinetics, Champaign, IL. A definitive textbook treatment of work in the full range of human endeavor, from daily life to sport.

Biro, P. A., and J. A. Stamps. 2010. Do consistent individual differences in metabolic rate promote consistent individual differences in behavior? *Trends Ecol. Evol.* 25: 653–659.

Dickinson, M. H., C. T. Farley, R. J. Full, and three additional authors. 2000. How animals move: An integrative view. *Science* 288: 100–106. A short but broadly conceived introduction to the biomechanics and functional morphology of animal locomotion. Although the present chapter does not include these topics, they are intimately related to the energetic themes that the chapter stresses.

Gill, R. E., Jr., T. L. Tibbits, D. C. Douglas, and seven additional authors. 2009. Extreme endurance flights by landbirds crossing the Pacific Ocean: ecological corridor rather than barrier? *Proc. R. Soc. London, Ser. B* 276: 447–457.

Hammond, K. A., and J. Diamond. 1997. Maximal sustained energy budgets in humans and animals. *Nature* 386: 457–462.

Hannah, J. B., D. Schmitt, and T. M. Griffin. 2008. The energetic cost of climbing in primates. *Science* 320: 898.

Heinrich, B. 1979. *Bumblebee Economics.* Harvard University Press, Cambridge, MA. One of the great essays on ecological energetics of the twentieth century. A rewarding and thoroughly enjoyable book.

Nagy, K. A. 2005. Field metabolic rate and body size. *J. Exp. Biol.* 208: 1621–1625. A brief but thought-provoking paper on the costs of existence in free-living vertebrates.

Piersma, T. 2011. Why marathon migrants get away with high metabolic ceilings: towards an ecology of physiological restraint. *J. Exp. Biol.* 214: 295–302. An enticing paper on maximum sustained rates of energy expenditure in people and other endotherms—and why the rates of energy expenditure are what they are. Provides lots to think about.

Suarez, R. K., L. G. Herrera M., and K. C. Welch, Jr. 2011. The sugar oxidation cascade: aerial refueling in hummingbirds and nectar bats. *J. Exp. Biol.* 214: 172–178. Hummingbirds and nectar bats have converged in that they hover at flowers to collect nectar. This research shows that they reach a steady state in which they use the sugars they are collecting immediately for the muscular work of hovering.

Weibel, E. R. 2000. *Symmorphosis. On Form and Function in Shaping Life.* Harvard University Press, Cambridge, MA. Even if one is unconvinced by the theory of symmorphosis, this book provides a compact and lucid introduction to the suite of systems responsible for O_2 delivery in sustained exercise.

See also **Additional References** *and* **Figure and Table Citations**.

As this bumblebee flies from one flower cluster to another to collect nectar and pollen, temperature matters for the bee in two crucial ways. First, the temperature of the bumblebee's flight muscles determines how much power they can generate. The flight muscles must be at a tissue temperature of about 30–35°C to produce enough power to keep the bee airborne; if the muscles are cooler, the bee cannot fly. The second principal way in which temperature matters is that for a bumblebee to maintain its flight muscles at a high enough temperature to fly, the bee must expend food energy to generate heat to warm the muscles. In a warm environment, all the heat required may be produced simply as a by-product of flight. In a cool environment, however, as a bumblebee moves from flower cluster to flower cluster—stopping at each to feed—it must expend energy at an elevated rate even during the intervals when it is not flying, either to keep its flight muscles continually warm enough to fly or to rewarm the flight muscles to flight temperature if they cool while feeding. Assuming that the flight muscles must be at 35°C for flight, they must be warmed to 10°C above air temperature if the air is at 25°C, but to 30°C above air temperature if the air is at 5°C. Thus, as the air becomes cooler, a bee must expend food energy at a higher and higher rate to generate heat to warm its flight muscles to flight temperature, meaning it must collect food at a higher and higher rate.

For a foraging bumblebee, warming the thorax to a high temperature is a critical requirement The process adds to the bee's energy costs and food needs on cool days. However, the flight muscles in the thorax require high temperatures to produce sufficient power for flight.

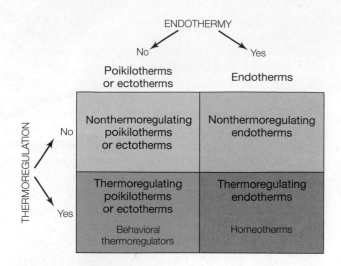

FIGURE 10.1 Animals fall into four categories of thermal relations based on whether they display endothermy and whether they display thermoregulation

Overall, tissue temperatures have a two-fold significance in many animals, including ourselves. The temperature of a tissue helps determine how the tissue performs. Tissue temperature also helps determine an animal's rate of energy expenditure. Bumblebees illustrate both of these points. The temperature of a bumblebee's flight muscles determines how intensely the muscles are able to perform their function of generating lift, and it determines how much food energy the bee must employ for heat production.

Physiologists now realize that animals are very diverse in the types of thermal relations they maintain with their environments. For categorizing the thermal relations of animals, one key concept is **endothermy**; if an animal's tissues are *warmed by its metabolic production of heat*, the animal is said to exhibit endothermy. A second key concept is **thermoregulation**, which refers to the maintenance of a relatively constant tissue temperature.[1] Suppose we classify animals according to whether or not they exhibit endothermy and whether or not they display thermoregulation. Doing so results in the matrix in **FIGURE 10.1**, which identifies the four most fundamental types of thermal relations that animals have with their environments.

Most animals are incapable of endothermy and thus fall on the left side of the matrix in Figure 10.1.[2] Animals of this sort are termed **ectotherms** because their body temperatures (not being elevated by their metabolism) are determined by the thermal conditions outside their bodies (*ecto*, "outside"). These animals are also called **poikilotherms** because they have variable body temperatures (*poikilo*, "variable"); their body temperatures are high in warm environments but low in cool ones. Most fish are excellent examples of ectotherms or poikilotherms; their tissues are

not warmed metabolically and therefore are at essentially the same temperature as the environmental water in which the fish swim.

A poikilotherm or ectotherm may or may not exhibit thermoregulation (see the vertical dimension of Figure 10.1). When a poikilotherm displays thermoregulation and thus falls into the lower left category of our matrix, it does so by *behavior*: It keeps its tissues at a certain "preferred" temperature by behaviorally positioning itself in environments that will warm or cool its tissues as needed.

Animals that exhibit endothermy—that is, animals that warm their tissues by their production of metabolic heat—are termed **endotherms** and fall on the right side of the matrix in Figure 10.1. Although endotherms may or may not be thermoregulators, most in fact exhibit thermoregulation (placing them in the lower right category of the matrix). Mammals and birds are outstanding examples of animals that exhibit both endothermy and thermoregulation. Additional examples include many species of medium-sized and large insects, such as the bumblebees we have already discussed, which exhibit both endothermy and thermoregulation in their flight muscles when they are flying. A **homeotherm** is an animal that thermoregulates by *physiological* means (rather than just by behavior). Humans provide an excellent example of homeothermy. Under many circumstances, the principal way we thermoregulate is by adjusting how rapidly we produce and retain metabolic heat: We thermoregulate by modulating endothermy! Other mammals and birds do the same under many circumstances, as do insects such as bumblebees.

As we attempt to categorize animal thermal relations, *temporal and spatial variation* often add complexity. Let's focus first on *temporal variation*: An individual animal may exhibit different thermal relations to its environment at different times. Species of mammals that hibernate illustrate this phenomenon; in such species, individuals are homeotherms during the seasons of the year when they are not hibernating, but often they exhibit neither endothermy nor thermoregulation when they are hibernating. Thermal relations may also exhibit *spatial variation*, differing from one region of an animal's body to another. The abdomens of bumblebees and other active insects, for example, are typically neither endothermic nor thermoregulated, even in individuals that exhibit endothermy and thermoregulation in their thoracic flight muscles. **Heterothermy** refers to a difference in thermal relations from one time to another, or one body region to another, within a single individual. Hibernating species of mammals exemplify **temporal heterothermy**. Flying bumblebees illustrate **regional** (i.e., spatial) **heterothermy**.

Temperature is *always* a major factor in the lives of individual animals, regardless of the particular thermal relations the animals exhibit. Whether animals are poikilotherms or homeotherms, for example, temperature is universally important in at least two ways, as already illustrated in our opening discussion of bumblebees:

■ The environmental temperature—also known as *ambient temperature*—universally is a principal determinant of an animal's metabolic rate and therefore the rate at which the animal must acquire food.

■ The temperature of an animal's tissues universally plays a principal role in determining the functional properties of the tissues. For example, tissue temperature affects whether protein molecules in a tissue are in high-performance

---

[1] *Thermoregulation* is a specific type of *regulation* as defined in Chapter 1 (see Figure 1.8).

[2] As stressed in Chapter 7, metabolic heat production is a universal feature of living organisms. When we say "most animals are incapable of endothermy," we do not mean they fail to produce heat metabolically. Instead, keep in mind that endothermy is *warming of the tissues* by metabolic heat production. Most animals, although they produce heat, do not make heat fast enough or retain heat well enough for their tissues to be warmed by their metabolic heat production.

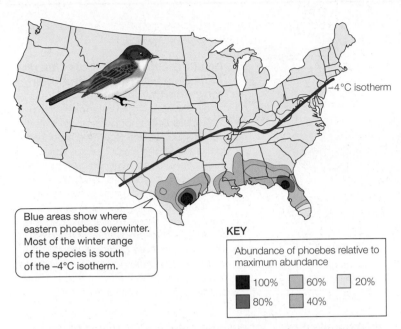

Blue areas show where eastern phoebes overwinter. Most of the winter range of the species is south of the –4°C isotherm.

KEY

Abundance of phoebes relative to maximum abundance

■ 100%	■ 60%	□ 20%
■ 80%	■ 40%	

FIGURE 10.2 Eastern phoebes (*Sayornis phoebe*) overwinter where the average minimum air temperature in January is –4°C or warmer The data shown were compiled in the 1980s. The average minimum air temperature in January was –4°C or warmer below the red line (the "–4°C isotherm") and colder than –4°C above the line. (After Root 1988.)

or low-performance molecular conformations. Tissue temperature also affects the rates of biophysical processes (e.g., diffusion and osmosis) and the rates of biochemical reactions in a tissue.

Temperature also exerts major effects on the properties of entire ecological communities. You will see this vividly if you walk through a temperate woodland during the various seasons of the year. On a walk in summer, you will be aware of vigorous photosynthesis by the plants, and you will witness sustained activity by mammals, birds, insects, turtles, snakes, amphibians, and other animals. In the winter, however, plants and most animals become cold and quiescent; activity in the woodland becomes restricted largely to the mammals and birds that keep their tissues warm. We cannot always say with certainty whether temperature is the primary determinant of the seasonal changes that we observe in a particular species, because in an entire community of this sort, the responses of any one species may be ripple effects of impacts on others. We cannot doubt, nonetheless, that much of the change in the animal life of a temperate woodland from summer to winter is a direct consequence of the seasonal change of temperature.

Beyond the obvious effects of temperature in a local ecological community, temperature also helps determine where each animal species can live. That is, temperature affects a species' geographical range. In North America, for example, if we consider the geographical ranges of resident birds in winter, we often find that the northern limits of these ranges correlate well with particular winter temperatures. Eastern phoebes illustrate this pattern. The northern limit of their geographical range in winter corresponds closely with a line that connects all the places where the average minimum air temperature is –4°C (**FIGURE 10.2**). Eastern

phoebes in winter do not extend northward to a fixed latitude, mountain range, river, or other geographical limit. Instead, they extend northward to a relatively fixed severity of winter cold stress. Where winter nights average warmer than about –4°C, these birds are to be found. Where winter nights average colder than –4°C, they do not occur.

Temperature is a particularly prominent focus for biologists today because of the threat of global warming (**BOX 10.1**). Society needs accurate predictions of the potential effects of global warming. The need to make such predictions constitutes a major reason for the study of animal thermal relations in today's world.

Temperature and Heat

The distinction between *temperature* and *heat* is tricky, and it is important for understanding the thermal relations of animals. To elucidate the distinction, consider a simple inanimate system: two blocks of copper—one of which is ten times more massive than the other, and both of which have been sitting in a room at 20°C long enough that they are at temperature equilibrium with the room. If you measure the *temperature* of each block, you will find that it is 20°C, even though one block is small and the other is large. Suppose, however, that you remove and measure the *heat* from each block; suppose, for instance, that you place each block at absolute zero and measure the amount of heat liberated as the block temperature falls from 20°C to absolute zero. You will find that the large block yields ten times more heat than the small one. Thus, as the two blocks sit in the room at 20°C, their *temperatures* are the same and independent of the amount of matter in each block, but their contents of *heat* are different and directly proportional to the amount of matter in each block.

To understand in greater depth these contrasting attributes of temperature and heat, recall from Chapter 5 (see page 106) that the atoms and molecules within any substance undergo constant random motions on an atomic-molecular scale. The **temperature** of a substance is a measure of the *speed*—or *intensity*—of these incessant random motions.[3] In the two copper blocks sitting in the room at 20°C, the average speed of atoms during the random atomic-molecular motions is identical; thus, even though the blocks differ in size, they are the same in temperature. **Heat**, unlike temperature, is a form of energy; it is the energy that a substance possesses by virtue of the random motions of its atomic-molecular constituents (see page 167). The amount of heat in a piece of matter thus depends on the *number* of atoms and molecules in the piece, as well as the *speed* of each atom and molecule. A copper block with many copper atoms moving at a given average speed contains proportionally more heat energy than one with fewer atoms moving at the same speed.

A key property of temperature is that it dictates the *direction of heat transfer*. Heat always moves by conduction or convection from a region of high temperature to one of low temperature. To refine this concept, suppose you have a large copper block at

---

[3] Temperature, more specifically, is proportional to the product of molecular mass and the mean square speed of random molecular motions. The speeds of the motions are astounding. In a gas, molecules collide with each other, bounce apart, and then fly through free space until they collide with other molecules. At 20°C, the average speed during each period of free flight is about 500 m/s! The speed is lower at lower temperatures, and higher at higher temperatures.

Global Warming

The great majority of scientists who have assessed the evidence on global climate change agree that effects of global warming are already right before our eyes or can be predicted with confidence.

Species are tending to shift their ranges poleward. Surveys of large sets of animal species find that there is a strong statistical bias for species in both hemispheres to be shifting their ranges toward the poles. For example, of 36 fish species studied in the North Sea over a recent 25-year period (1977–2001), 15 species changed their latitudinal center of distribution, and of those, 13 (87%) shifted

FIGURE B Little brown bats (*Myotis lucifugus*) will likely extend their range northward by 2080 Considering bats in hibernation, the map shows predicted northern range limits in eastern Canada, based on a bioenergetic model. The model makes various predictions, depending on assumptions made (e.g., cave types used). Blue shows model predictions of the present northern range limit for hibernation, whereas red shows model predictions for 2080. Model predictions of the present range limit are compatible with actual known hibernation sites, bolstering confidence in the predictions for the future. (After Humphries et al. 2002.)

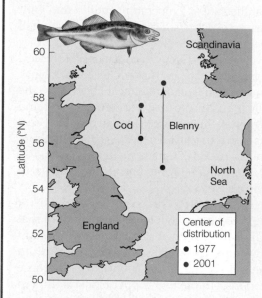

northward (**Figure A**). This is the pattern expected as a response to global warming: Faced with a warming environment, many species will shift to more-polar latitudes. A synthetic study of almost 900 animal and plant species that were monitored over a median observation period of 66 years found that 434 of the species shifted their

FIGURE A Shifts in the centers of distribution of two fish species in the North Sea from 1977 to 2001 Arrows symbolize the latitudinal shifts of cod (*Gadus morhua*) and blenny (*Lumpenus lampretaeformis*). Average sea temperature increased 1.1°C over the period. Both shifts of species distribution have been monitored annually and have strong statistical support. (After Perry et al. 2005.)

ranges, and of those, 80% shifted as expected in response to a warming world.

Physiological principles enable researchers to predict that some species will be required to shift their ranges poleward in the future. Little brown bats provide an example. When small mammals hibernate, they allow their body temperature to fall to environmental temperature. The fall of body temperature is critical because it helps inhibit metabolism and thereby save energy. However, temperature cannot safely decline without limit. When a hibernator's body temperature reaches the lowest tolerable level, the hibernating animal increases its metabolic rate to keep its body temperature from falling further. Consequently, energy costs during hiber-

---

20°C in contact with a tiny copper block at 30°C; although the large block contains more heat than the small one, heat will move from the small block into the large one because temperature, not energy content, dictates the direction of energy transfer. The net addition of heat to any object causes an increase in the temperature of the object. All in all, therefore, *temperature* and *heat* have intimate interactions:

- Heat moves by conduction or convection from high temperature to low.

- The transfer of heat raises the temperature of the object receiving heat and lowers the temperature of the object losing heat.

- In a simple physical system such as two solid objects in contact with each other, objects are at thermal equilibrium when their temperatures are the same because then heat does not tend to move in net fashion between them.

Heat Transfer between Animals and Their Environments

A living animal positioned in an environment, besides making heat internally because of its metabolism, exchanges heat with its surroundings by four distinct heat-transfer mechanisms: *conduction, convection, evaporation,* and *thermal radiation* (**FIGURE 10.3**). The animal may well gain heat by one mechanism of heat transfer while it simultaneously loses heat by another. A familiar illustration of this important point is that on a hot day in summer, people may simultaneously gain heat from the sun by *thermal radiation* while they lose heat by the *evaporation* of sweat. Because the four mechanisms of heat transfer follow distinct laws and can operate simultaneously in opposite directions, they cannot simply be lumped together. Instead, each mechanism needs to be analyzed in its own right, and then the effects of all four can be summed to determine an animal's overall heat exchange with its environment.

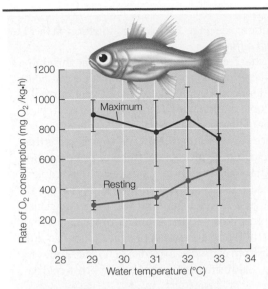

FIGURE C Rising water temperatures on the Great Barrier Reef could impair the ability of cardinalfish (*Ostorhinchus doederleini*) to engage in aerobic exercise When water temperature is raised from 29°C to 31°, 32°, or 33°C, the difference between maximum and resting O_2 consumption becomes significantly smaller with each step. Symbols are means; error bars show ± 1 standard deviation. (After Nilsson et al. 2009.)

for overwinter survival. The northern limits of latitude at which hibernating little brown bats are likely to find suitable temperatures are predicted to shift poleward as global warming proceeds (**Figure B**).

Experiments on physiological effects of anticipated temperatures sometimes point to severe future challenges. Some species of fish on the Great Barrier Reef, for example, are in danger of losing much of their capability to be active, as exemplified by cardinalfish (**Figure C**). Experiments show that the difference between their maximum rate of O_2 consumption and their resting rate of O_2 consumption (i.e., their aerobic scope) becomes dramatically smaller if the water in which they live is warmed from its current temperature of 29°C to temperatures 2°–4°C higher. A rise in water temperature on the reef could reduce the ability of the fish to increase their O_2 consumption, limiting their ability to engage in aerobic exercise (see Chapter 9).

Effects of environmental warming on some animals can pose ecological challenges for others. Because animals live in interconnected ecological communities, effects on one species affect others. At a study site in the Netherlands, for example, peak caterpillar abundance in the spring has been occurring progressively earlier from year to year because spring temperatures have been rising, speeding caterpillar development and causing trees (food for the caterpillars) to leaf out earlier. For great tits (chickadee-like birds), caterpillars for feeding their nestlings are a key to repro-

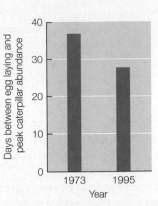

FIGURE D Days between egg laying by great tits (*Parus major*) and peak caterpillar abundance in a Netherlands woodland When young birds hatch out of the eggs and then undergo their nestling development, the abundance of caterpillars in the woodland affects how well parents can feed them. (After Visser et al. 1998.)

ductive success. However, the dates when the tits lay their eggs have hardly changed at all from year to year. Accordingly, although the time between egg laying and peak caterpillar abundance was about ideal in 1973, it had shortened—and become too short to be ideal—by 1995 and remains so today (**Figure D**). Food for nestlings has become detectably inadequate because of the mismatch of ever-earlier caterpillar abundance—caused by warming temperatures—while bird reproduction has not shifted to be equally earlier.

nation are elevated if the environmental temperature is too high *and* if it is too low. Many hibernators live entirely on fat stores. To survive, they must not exhaust their fat stores before winter's end. Environmental temperatures determine their metabolic rates and therefore the rates at which they use up their fat stores. Accordingly, certain environmental temperatures are required

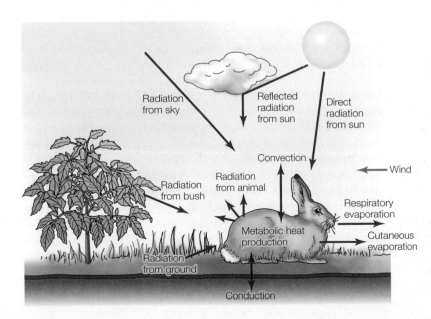

FIGURE 10.3 An animal exchanges heat with its environment by conduction, convection, evaporation, and thermal radiation The animal exchanges heat by conduction with the ground and by convection with the wind. It loses heat by evaporation (both respiratory and cutaneous). It receives heat by thermal radiation from all objects in its surroundings and also emits heat by thermal radiation toward all objects. Finally, it gains heat from its own metabolism. (Because rabbits lack sweat glands, the cutaneous evaporation from a rabbit is entirely of a nonsweating sort.)

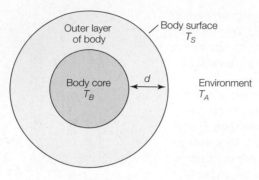

FIGURE 10.4 A model of an animal's body showing key temperatures The body core is at body temperature T_B, while the surrounding environment is at ambient temperature T_A. The temperature of the body surface is T_S. The outer layer of the body, separating the body core from the surface, has thickness d.

For the body temperature of an animal to be constant, the sum total of its heat gains by all mechanisms taken together must equal the sum total of all its heat losses. For instance, suppose that an animal is gaining heat from its environment by conduction and thermal radiation, as well as from metabolism, while losing heat by convection and evaporation. Its body temperature will be constant if and only if the sum of its heat gains by conduction, radiation, and metabolism per unit of time is exactly matched by the sum of its heat losses through convection and evaporation per unit of time.

FIGURE 10.4 presents a simple model of an animal that will be useful as we discuss the individual mechanisms of heat exchange. The core of an animal's body is considered to be at a uniform **body temperature**, symbolized T_B. The temperature of the environment is called **ambient temperature**, T_A. The temperature of the body surface often differs from T_B and T_A and thus is distinguished as **surface temperature**, T_S. Separating the body core from the body surface is the outer layer of the body, where temperature gradually changes from T_B on the inside to T_S on the outside.

Conduction and convection: Convection is intrinsically faster

Conduction and convection are usefully discussed together because, in a sense, these two mechanisms of heat transfer define each other. What they have in common is that when heat moves through a material substance by either mechanism, the atoms and molecules of the substance *participate* in the transfer of heat. **Conduction** is the transfer of heat through a material substance that is macroscopically motionless. A familiar example of conduction is the transfer of heat through a block of copper. We know that if the temperature of one side of a copper block is raised, heat will move through the block and appear on the other side even though the copper undergoes no macroscopic motion. The way heat makes its way through such a macroscopically motionless substance is strictly by *atomic-molecular interactions*; if atoms (or molecules) on one side are especially agitated, they increase the agitation of atoms farther into the substance by interatomic collisions, and by repetition of this process, successive layers of atoms relay the increased agitation through the entire thickness of the substance. Conduction mechanistically has much in common with simple solute diffusion (the movement of solute through a macroscopi-

cally motionless solution; see page 106), and conduction in fact is sometimes called *heat diffusion*.

Convection, in sharp contrast, is transfer of heat through a material substance *by means of macroscopic motion of the substance*. Fluid flow is required for convection. If a wind or water current is present, the macroscopic motion of matter carries heat from place to place. This transfer of heat is convection.

A critical difference between conduction and convection is that, for a given difference of temperature, heat transfer by convection is *much* faster than that by conduction. Consider, for example, a horizontal surface that is 10°C warmer than the surrounding air. If the air is moving at just 10 miles per hour (4.5 m/s), convection will carry heat away from the surface about 70 times faster than if the air is perfectly still! The acceleration of heat transfer by fluid movement is familiar from everyday experience. We all know, for instance, that a wind greatly increases the thermal stress of a cold day.

THE LAWS OF CONDUCTION We can better understand conduction if we focus on a specific object, such as a sheet of material of thickness d. If the temperature on one side of the sheet is T_1, that on the other is T_2, and heat is moving through the sheet by conduction, then the rate of heat transfer H from one side to the other per unit of cross-sectional area is

$$H_{\text{conduction}} = k \frac{T_1 - T_2}{d} \tag{10.1}$$

where k is a constant. The ratio $(T_1 - T_2)/d$ is called the **thermal gradient**.[4] You can see from the equation that the rate of heat transfer by conduction through a sheet of material increases as the temperature difference between the two sides increases. In addition, the rate at which heat moves from one side of the sheet to the other decreases as the thickness of the sheet (d) increases. The coefficient k depends in part on the type of material through which conduction is occurring. Some biologically important materials, such as air, conduct heat poorly; they are said to exhibit low *thermal conductivity* and have low values of k. Other materials, such as water, exhibit higher thermal conductivity and higher k values (water's conductivity is about 20 times that of air).

Heat transfer through the fur of a furred mammal, or through a winter jacket worn by a person, is typically analyzed as a case of conduction because fur traps a layer of relatively motionless air around the body of a furred mammal, and a winter jacket envelops a person's body in a shell of relatively still air. Motionless air is one of the most highly insulating materials in the natural world, and the stillness of the air layer trapped by fur or a jacket is the key to the insulative value of the fur or jacket. To the extent that the air is motionless, heat must move through it by conduction; thus heat moves much more slowly than if convection were at work. Indeed, from the viewpoint of physics, the benefit of fur or a jacket in a cold environment is that it favors an intrinsically slow mechanism of heat loss from the body, conduction, over an intrinsically faster mechanism, convection. In Figure 10.4, the "outer layer" of the body might be taken to represent the fur or jacket. Equation 10.1 shows that increasing the thickness (d) of the motionless air layer trapped

---

[4] Although the *thermal gradient* is technically defined to be $(T_1 - T_2)/d$ (i.e., temperature difference per unit of distance), the expression *thermal gradient* is sometimes used to refer simply to a temperature difference, $(T_1 - T_2)$.

by the fur or jacket will tend to slow heat loss from an animal or person to a cold environment.

THE LAWS OF CONVECTION   When air or water flows over an object, the rate of heat transfer by convection between the object and the moving fluid depends directly on the difference in temperature between the *surface* of the object and the fluid. Suppose, for instance, that the model animal in Figure 10.4 is exposed to a wind. Then the rate of convective heat transfer between the animal and the air per unit of surface area is calculated as follows:

$$H_{\text{convection}} = h_c(T_S - T_A) \qquad \textbf{(10.2)}$$

The animal will lose heat by convection if its surface temperature (T_S) exceeds the ambient air temperature (T_A); however, it will gain heat by convection if T_A is higher than T_S.

The coefficient h_c, called the **convection coefficient**, depends on many factors, including the wind speed, the shapes of the body parts of the animal, and orientation to the wind. If the shape of a body part is approximately cylindrical (as is often true of the limbs) and the wind is blowing perpendicularly to the cylinder's long axis, then

$$h_c \propto \frac{\sqrt{V}}{\sqrt{D}} \qquad \textbf{(10.3)}$$

where V is the wind speed and D is the diameter of the cylinder. This equation shows that the rate of heat transfer per unit of surface area by convection tends to increase with the square root of the wind speed. The rate of heat transfer per unit of surface area also tends to increase as the square root of the diameter of a cylindrically shaped body part is decreased; this physical law helps explain why body parts of small diameter (e.g., fingers) are particularly susceptible to being cooled in cold environments.

Evaporation: The change of water from liquid to gas carries much heat away

Evaporation of body water from the respiratory passages or skin of an animal takes heat away from the animal's body because water absorbs a substantial amount of heat whenever its physical state changes from a liquid to a gas. The amount of heat required to vaporize water, called the **latent heat of vaporization**, depends on the prevailing temperature. It is 2385–2490 joules (J) (570–595 calories [cal]) per gram of H_2O at physiological temperatures. These *large* values mean that evaporation can be a highly effective cooling mechanism for an animal. The heat is absorbed from the body surface where the vaporization occurs, and it is carried away with the water vapor.[5]

Thermal radiation permits widely spaced objects to exchange heat at the speed of light

For terrestrial animals, including people, thermal-radiation heat transfer often ranks as one of the quantitatively dominant mechanisms of heat exchange with the environment, yet it tends to be the least understood of all the mechanisms. Although we are all familiar with radiant heating by the sun, such heating is only a special case of a sort of heat transfer that is in fact ubiquitous.

[5] See Chapter 27 (page 730) for a detailed discussion of the physical laws of evaporation.

FIGURE 10.5 An antelope jackrabbit (*Lepus alleni*)   This species of jackrabbit is found principally in the low-altitude desert plains of southern Arizona and northern Mexico.

The first fact to recognize in the study of thermal-radiation heat transfer is that all objects *emit* electromagnetic radiation. That is, all objects are *original sources* of electromagnetic radiation. If you look at a wall, your eyes see electromagnetic radiation (light) coming from the wall, but that radiation is merely reflected; it originated from a lamp or the sun and reflected off the wall to enter your eyes. As a completely separate matter, the wall also is the original source of additional electromagnetic radiation. The radiation *emitted* by the wall is at infrared wavelengths and thus invisible. It travels at the speed of light, essentially unimpeded by the intervening air, until it strikes a solid surface (such as your body), where it is absorbed. Simultaneously, your body emits electromagnetic radiation, some of which strikes the wall. In this way the wall and your body can exchange heat even though they are not touching and in fact may be far apart. Any two objects that are separated only by air undergo exchange of heat at the speed of light by thermal-radiation heat transfer.[6]

An interesting application of the principles of thermal-radiation heat transfer is to the huge ear pinnae of jackrabbits (**FIGURE 10.5**). In some species, such as the one pictured, the ear pinnae constitute 25% of the total body surface area. Despite decades of interest, physiologists still do not definitely know the function of these pinnae. The most likely function is that they act as radiators. Jackrabbits modulate blood flow to the pinnae. When blood flow is brisk and the pinna blood vessels are engorged (as in Figure 10.5), the pinnae are warmed, and they thereby increase the intensity at which they emit electromagnetic radiation. When heat is lost in this

[6] Water, being far more opaque to infrared radiation than air, largely blocks this sort of heat transfer in aquatic environments.

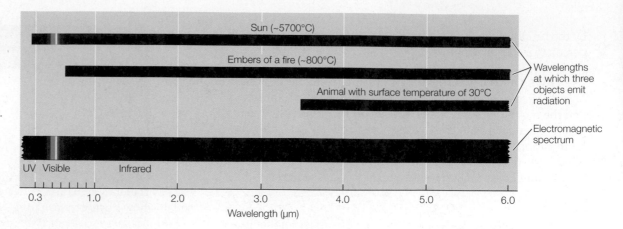

FIGURE 10.6 As objects reach higher surface temperatures, the ranges of wavelengths at which they emit thermal radiation extend to shorter wavelengths Temperatures specified are surface temperatures. All three of the objects shown also emit energy at wavelengths longer than 6 μm (not shown).

way, it need not be lost by panting or other forms of evaporation—a water-saving benefit for animals that live in deserts or semideserts.[7]

When objects emit electromagnetic radiation, they do so over a *range* of wavelengths. A key principle of thermal-radiation physics, illustrated in **FIGURE 10.6**, is that the range of wavelengths emitted by an object depends on the *surface* temperature of the object (T_S) and shifts toward shorter wavelengths as the surface temperature increases. The lowest thin black bar in Figure 10.6 shows the wavelengths emitted by an animal or other object with a surface temperature of about 30°C. Note that the shortest wavelengths emitted by a surface at this temperature are between 3 and 4 micrometers (μm); energy is also emitted over a broad range of longer wavelengths. All the emitted wavelengths are in the infrared range and thus invisible. The embers of a fire (middle thin black bar in the figure) emit at shorter wavelengths because they are hotter. They are in fact hot enough that the shortest wavelengths they emit are within the visible range. Because we *see* those wavelengths, we see the coals *glow*. The visible wavelengths emitted by the coals are limited to the red-orange end of the visible spectrum; thus the glow of the coals is red-orange. The sun is so hot that it emits electromagnetic energy (upper thin black bar in the figure) at all wavelengths of the visible spectrum and therefore glows with a nearly white light. The most important concept illustrated by Figure 10.6 is that the radiative emissions from organisms are of the same basic nature as those from a fire or the sun. The only reason we do not see organisms glow is that the wavelengths they emit are out of our visible range.

An important principle of thermal-radiation physics is that the *total intensity* of radiation emitted by an object—summing the radiation emitted at all wavelengths—increases as surface temperature increases:

$$H_{\text{radiative emission}} = \varepsilon \sigma T_S^4 \qquad (10.4)$$

In this equation, which is known as the **Stefan–Boltzmann equation**, H is the rate of emission per unit of surface area at all wavelengths combined, ε is a surface property called *emissivity* (*emittance*), σ is a constant called the Stefan–Boltzmann constant, and the surface temperature T_S must be expressed in absolute degrees (K).

Another important principle of thermal-radiation physics is that when electromagnetic radiation strikes an object, the radiant energy may be *absorbed* or *reflected*, or it may *pass through*. The fractions of the energy absorbed, reflected, and transmitted depend on the surface properties of the object and are wavelength-specific. Energy that is absorbed is converted into heat at the surface of the absorbing object, as illustrated in everyday experience by the fact that our skin is warmed by radiant energy from the sun or from the embers of a fire.

RADIANT EXCHANGES IN THE BIOSPHERE THAT DO NOT IN-VOLVE THE SUN In natural biological communities, the sun is usually the only object that is hot enough to emit energy at wavelengths shorter than 3–4 μm. The surface temperatures of animals, plants, rocks, and all other objects besides the sun are typically between −50°C and 50°C, and surfaces at such temperatures emit only wavelengths of 3–4 μm or longer (see Figure 10.6). Thus, if we exclude the sun from consideration, all radiant exchanges among objects in the biosphere are at such wavelengths: Various organisms and objects emit at 3–4 μm and longer, and the emitted radiation that they receive from other organisms and objects is at 3–4 μm and longer. This fact massively simplifies the analysis of radiant exchanges because although organisms and objects in the biosphere commonly differ from one another in surface temperature, all are essentially identical in their other radiative properties at wavelengths of 3–4 μm and longer. Specifically, all exhibit about the same value for ε in the Stefan–Boltzmann equation (Equation 10.4) at these wavelengths; and all are highly absorptive at these wavelengths, meaning that they absorb (rather than reflect or transmit) most energy that strikes them. Put loosely, organisms and objects in the biosphere do not differ in color at these wavelengths. If this idea sounds strange, recognize that the color you see with your eyes is a property at *visible* wavelengths of 0.4–0.72 μm. Whether the visible color of an organism or object is brown, green, or even white, the color at wavelengths of 3–4 μm and longer is, in all cases, nearly black.

Because all organisms and objects in the biosphere are virtually identical in ε and in their absorptive properties at wavelengths of 3–4 μm and longer, surface temperature (T_S) is the sole major determinant of radiative heat exchange when the sun is excluded from consideration. If two organisms or objects are exchanging heat radiatively, each can be considered to emit a beam of energy toward the other. Whereas the warmer of the two emits a relatively strong beam (see Equation 10.4), the cooler emits a relatively weak beam. Each absorbs most of the energy that it receives from the other. For these reasons, energy is passed in *net* fashion from the

---

[7] When the pinnae are warmer than the air, heat will also be carried away from them by convection if a breeze or wind is present. Like heat loss by thermal-radiation heat transfer, loss by convection also occurs without making demands on body water.

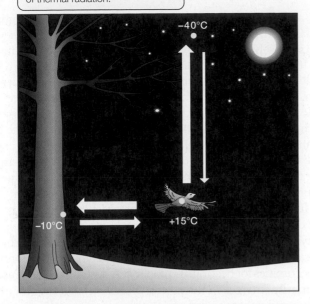

The widths of the arrows symbolize the relative intensities of the beams of thermal radiation.

−40°C

−10°C

+15°C

FIGURE 10.7 A bird loses heat in net fashion to tree trunks by thermal radiation as it flies past them on a cold winter night The bird also loses heat in net fashion to the night sky. More than half of a bird's total heat loss may be by thermal-radiation heat transfer. The temperatures shown for the tree and bird are their surface temperatures; that shown for the sky is the radiant sky temperature expected on a night when the air temperature near the ground is −10°C. Quantitatively, thermal-radiation heat transfer depends on temperature on the absolute (Kelvin) scale. On that scale, +15°C = 288 K; −10°C = 263 K; and −40°C = 233 K.

warmer object to the cooler one. Quantitatively, if the surface temperatures of the two objects (on the Kelvin scale) are T_1 and T_2, the net rate of heat transfer between them is proportional to $(T_1^4 − T_2^4)$, and the direction of net heat transfer is from the one with the higher T_S to the one with the lower T_S.

As examples, consider first a relatively cool lizard standing in the early nighttime hours near a rock that remains hot from the preceding day. The rock emits a relatively strong beam of radiant energy toward the lizard, and the lizard absorbs most of this radiant energy; simultaneously, the lizard emits a weaker beam of energy toward the rock, and the rock absorbs that energy. The net effect is that the lizard is warmed by standing near the rock. A less familiar example is provided by a bird flying past cold trees on a frigid winter night (**FIGURE 10.7**). The surface temperature of the bird (+15°C in Figure 10.7) is higher than that of the tree trunks (−10°C). In this case the beam of energy carrying heat away from the bird is more intense than the beam striking the bird from each tree, and the net effect of thermal-radiation heat transfer is to cause a loss of heat from the bird to the trees.

THE NIGHT SKY AS A RADIANT OBJECT The sky is one of the objects in the biosphere that deserves special note. Here we consider just the night sky; because the sun is absent at night, the discussion in this section is a special case of the last section's discussion. In the atmosphere above us at night, each gas molecule—whether positioned just above Earth's surface or at the limits of outer space—emits radiation as a function of its temperature. In this way, the surface of Earth steadily receives a beam of radiation emitted from the sky above. One way to express the intensity of this

radiation is to pretend that the sky is a solid surface and ask what the temperature of that surface would have to be for it to emit at the intensity observed (assuming ε = 1.0). This temperature is called the **radiant temperature of the sky** (or the *black-body sky temperature*). A characteristic of the radiant temperature of the clear night sky is that it is far lower than the simultaneous air temperature at ground level. For example, during a particular summer night in the Arizona desert when the air temperature near the ground was +30°C, the radiant temperature of the clear sky was simultaneously −3°C; that is, the sky on that warm night behaved like a subfreezing object! The low radiant temperature characteristic of the clear night sky explains how frosts can form on nights when the air temperature at ground level stays above freezing.

When animals are exposed to the clear night sky, they emit a beam of radiation toward the sky. In return, they receive only a weak beam of radiation from the sky (see Figure 10.7). Accordingly, animals tend to lose energy in net fashion to the clear night sky, which is often, therefore, said to act as a "radiant heat sink." The radiative loss of heat to the clear sky can be of substantial importance. Because of this, animals confronted with cold stress may benefit considerably by avoiding exposure to the clear sky.

SOLAR RADIATION The sun is the one object in the biosphere that routinely emits radiation at wavelengths shorter than 3–4 μm (see Figure 10.6). Most of the solar radiant energy is at visible or near-visible wavelengths. Accordingly, when we consider objects exposed to solar radiation, the visible colors of the objects matter; visible color affects the fraction of the energy that is absorbed. If an animal's body surfaces are opaque (nontransparent), the analysis of the effects of the animal's visible color on the absorption of the visible and near-visible solar radiation is straightforward: Dark surfaces absorb more of this solar radiation—and are heated more by it—than light ones. Black beetles, for instance, absorb the visible and near-visible wavelengths relatively well, whereas light-colored beetles tend more to reflect these wavelengths and absorb them relatively poorly. Animals that can change their skin color, such as many species of lizards, can increase and decrease the solar heating of their bodies by darkening and lightening, respectively.

SUMMARY
Heat Transfer between Animals and Their Environments

- In addition to making heat metabolically, animals exchange heat with their environments by conduction, convection, evaporation, and thermal radiation. An animal's body temperature depends on heat gains and losses; it is constant only if the sum total of gains equals the sum total of losses.

- Conduction and convection have in common the property that when heat moves through a material substance by either mechanism, the atoms and molecules of the substance participate in the transfer of heat. Conduction occurs when a material substance is macroscopically motionless. Convection is heat transfer brought about by flow of a material substance (e.g., by wind). Convection is much faster than conduction.

(Continued)

- Evaporation is a potentially potent mechanism for heat transfer because the change of state of water from a liquid to a gas absorbs a great deal of heat per gram of water. The heat is absorbed from the surface where evaporation occurs and is carried away with the water vapor.

- Thermal-radiation heat transfer occurs by means of beams of radiant energy that all objects emit and that travel between objects at the speed of light. Because of thermal-radiation heat transfer, objects can exchange heat at a distance. In most instances of thermal-radiation heat transfer in the biosphere, the heat transfer occurs at invisible infrared wavelengths; because all objects are nearly black at such wavelengths, visible color plays little role, and the net transfer of heat is from the object with higher surface temperature to the one with lower surface temperature. Visible color, however, is a major factor in how well objects absorb the visible and near-visible wavelengths of solar radiation.

Poikilothermy (Ectothermy)

Poikilothermy is by far the most common type of thermal relation exhibited by animals. Amphibians, most fish, most nonavian reptiles, all aquatic invertebrates, and most terrestrial invertebrates are poikilotherms. The defining characteristic of poikilothermy is that the animal's body temperature is determined by equilibration with the thermal conditions of the environment and varies as environmental conditions vary. *Poikilothermy* and *ectothermy* are the same thing. The two terms simply emphasize different aspects of one phenomenon; whereas *poikilothermy* emphasizes the variability of body temperature, *ectothermy* emphasizes that outside conditions determine the body temperature (see page 234).

Poikilothermy manifests itself differently depending on whether an animal is aquatic or terrestrial. Aquatic poikilotherms typically have body temperatures that are essentially the same as water temperature. Terrestrial poikilotherms, however, do not necessarily have body temperatures that equal "air" temperature, because thermal-radiation heat transfer or evaporation on land can tend to draw the body temperature away from air temperature. For instance, if a frog or snail on land basks in the sun, its body temperature may be much higher than the air temperature. Such animals nonetheless still meet the definition of poikilothermy or ectothermy, because their body temperatures are determined simply by equilibration with the sum total of thermal conditions in their environments.

Poikilothermic or ectothermic animals are often called *cold-blooded* in nonscientific writing, in reference to their coolness to the touch under certain conditions. Many species, however, may have high body temperatures when in warm environments. For example, desert lizards and insects that are perfectly fine poikilotherms often have body temperatures that substantially exceed human body temperature! *Cold-blooded* is therefore not a suitable general term to describe poikilotherms or ectotherms.

Poikilotherms often exert behavioral control over their body temperatures

The natural environments of poikilotherms typically vary from place to place in thermal conditions. In a forest, for example, the temperature on the exposed forest floor might be higher than that under a log, and the temperature in a spot of sunlight might be higher yet. Poikilotherms in the wild can behaviorally choose where they position themselves and, in this way, control their body temperatures.

If a poikilotherm behaviorally maintains a *relatively constant* body temperature, it is said to exhibit **behavioral thermoregulation**. Sometimes behavioral thermoregulation is rather simple. In a lake, for instance, various large water masses (such as those at the surface and at greater depth) often differ in temperature (see Figure 1.16). Fish that elect to stay in one water mass, rather than another, take on the temperature of the water they occupy and remain at that temperature for extended periods. The behavior of the fish is accordingly a simple form of behavioral thermoregulation.

In other cases, behavioral thermoregulation is far more complex and dynamic. Many lizards, for example, maintain relatively stable body temperatures during daylight hours, and they do so by complex, moment-to-moment behavioral exploitation of environmental opportunities for heating and cooling. A desert lizard, for instance, ordinarily emerges in the morning and basks in the sun until its body temperature rises to be within a "preferred" range that it maintains during its daily activity. Thereafter the lizard keeps its body temperature within that range until nightfall by a variety of mechanisms. One common strategy is to shuttle back and forth between sun and shade; when its body temperature starts to drop too low, the lizard moves into sunlight, and then later, when its body temperature starts to rise too high, it enters shade. The lizard might also modify the amount of its body surface exposed to the direct rays of the sun by changing its posture and orientation to the sun. It might flatten itself against the substrate to lose or gain heat (depending on substrate temperature), and when the substrate has become very hot during midday, the lizard might minimize contact by elevating its body off the ground or even climbing on bushes. By thus exploiting the numerous opportunities for heating and cooling in its thermally heterogeneous environment, a lizard may well maintain a body temperature that varies only modestly for long periods. The desert iguana illustrated in Figure 1.14, for instance, typically maintains an average abdominal temperature of 38°–42°C during daylight hours, and it often keeps its temperature within 2°–3°C of the mean for hours on end.

Investigators have worried a lot about the question of documenting true behavioral thermoregulation. They thus have compared living animals with inanimate model animals. In one study, living lizards in a natural setting on a Mediterranean island were found to exhibit far less variable body temperatures than lizard models placed widely in the same environment (**FIGURE 10.8**). Such evidence documents that real lizards do not simply position themselves at random, but *behave* in ways that keep their body temperatures within a relatively narrow preferred range.

Poikilotherms must be able to function over a range of body temperatures

A limitation of behavioral thermoregulation is that it is dependent on the thermal opportunities available in the environment, and thus it may be thwarted by changes of weather or other conditions outside an animal's control. A desert iguana, for example, may never reach a body temperature that is even close to its "preferred" level of 38°–42°C on a day that happens to be cloudy and cool. Similarly, a fish that would select a cool water mass if it could, cannot do so if all the water in its lake or pond is warm.

(A) Temperatures of actual lizards

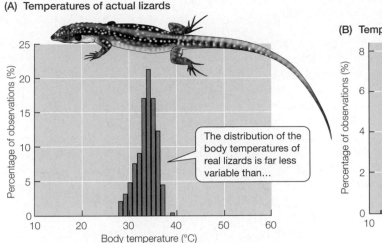

(B) Temperatures of lizard models

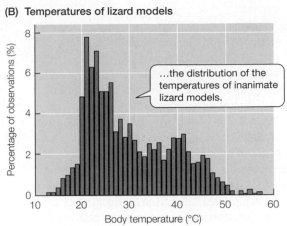

...the distribution of the temperatures of inanimate lizard models.

The distribution of the body temperatures of real lizards is far less variable than...

FIGURE 10.8 Behavioral thermoregulation documented by comparison of real lizards with inanimate lizard models Multiple daytime measurements of the body temperatures of real lizards (A) and inanimate lizard models (B) were made. The lizards (*Podarcis hispanica*) were living freely on a Mediterranean island. The lizard models were placed as comprehensively as possible in all the various microhabitats available to real lizards during their daytime activities on the same island. Data on the *y* axes are the percentages of all observations in various 1°C intervals of temperature. (After Bauwens et al. 1996.)

For these and other reasons, poikilotherms must typically be thermal generalists: They must be capable of functioning at a variety of different body temperatures. Species differ in how wide a range of body temperatures is acceptable. Some species, termed **eurythermal**, can function over wide ranges of body temperature; goldfish, for instance, maintain normal body orientation, feed, and swim at body temperatures of 5°–30°C. Other poikilotherms, termed **stenothermal**, have comparatively narrow ranges of body temperature over which they can function.

Poikilotherms respond physiologically to their environments in all three major time frames

The three major time frames of physiological response to the environment identified in Chapter 1 (see Table 1.2) provide a useful way to organize knowledge of the relations of poikilotherms to their thermal environments. In three of the next four sections, we discuss poikilotherms in each of the three time frames. First, in the next section, we address the *acute responses* of poikilotherms to changes in their body temperatures. The acute responses are those that individual animals exhibit *promptly* after their body temperatures are altered. After that we address the *chronic responses* of poikilotherms, termed *acclimation* and *acclimatization*:[8] What changes do individual animals undergo when they live in an altered thermal environment (and have altered body temperatures) for a prolonged period? Finally, after discussing temperature limits, we discuss *evolutionary changes*—the ways in which the physiology of poikilotherms may be modified by changes in the frequencies of genes when populations live in different environments over many generations.

Acute responses: Metabolic rate is an approximately exponential function of body temperature

When the body temperature of an individual poikilotherm is raised in a series of steps and its metabolic rate is measured promptly after each upward step, the usual pattern is that the resting metabolic rate increases approximately exponentially with the animal's body temperature (**FIGURE 10.9A**).[9] An *exponential* relation signifies that the metabolic rate increases by a particular *multiplicative factor* each time the body temperature is stepped up by a particular *additive increment* (see Appendix F). For example, the metabolic

---

[8] *Acclimation* and *acclimatization* are forms of *phenotypic plasticity*. The distinction between them is discussed on pages 17–18.

[9] There are limits to this process: An exponential increase is seen only within a particular range of body temperatures, a range that depends on the species and individual. We discuss the limits later in the chapter.

(A) Plot on linear coordinates

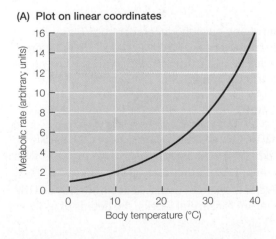

(B) Plot on semilogarithmic coordinates

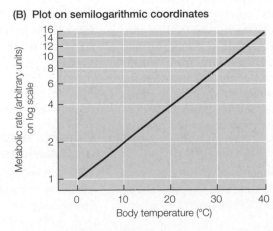

FIGURE 10.9 An exponential relation between metabolic rate and body temperature plotted in two ways (A) A plot employing linear scales for both variables. (B) A semilogarithmic plot of the same data as in part (A); metabolic rate is plotted on a logarithmic scale, whereas body temperature is plotted on a linear scale. Metabolic rate is expressed in the same arbitrary units in both parts. See Appendix E for background on logarithmic scales.

rate might increase by a factor of 2 for each increment of 10°C in body temperature. Then, if the metabolic rate were 1 J/min at 0°C, it would be 2 J/min at 10°C, 4 J/min at 20°C, and 8 J/min at 30°C (see Figure 10.9A). The acute relation between metabolic rate and body temperature is usually, in fact, only *approximately* exponential. That is, the factor by which the metabolic rate increases for a given increment in temperature is usually not precisely constant from one temperature range to the next but might, for example, be 2.5 between 0°C and 10°C but only 1.8 between 20°C and 30°C.

The reason that the metabolic rate of a poikilotherm increases as its body temperature goes up relates back to the concept of activation energy discussed in Chapter 2. Each biochemical reaction involved in metabolism is characterized by a particular activation energy, a certain minimum energy level that a reacting molecule must attain in order to undergo the reaction (see Figure 2.13). As the temperature of a cell increases, all molecules in the cell tend to become more agitated and have higher energy levels. Svante Arrhenius (1859–1927) demonstrated in the late nineteenth century that if one specifies any particular activation energy, the fraction of molecules that have that level of energy—or more—at any moment increases approximately exponentially as temperature increases. Reactions tend, therefore, to speed up approximately exponentially as cellular temperature rises. In this context, it is vital to recall that most metabolic reactions are enzyme catalyzed, and the enzymes determine the activation energies. Thus the detailed, quantitative relations between biochemical reaction rates and cellular temperature depend on the particular enzyme proteins that cells synthesize.

If the resting metabolic rate of a poikilotherm, symbolized M, were a true exponential function of its body temperature (T_B), the relation would be described by an exponential equation (see Appendix F):

$$M = a \cdot 10^{n \cdot T_B} \qquad \textbf{(10.5)}$$

where a and n are constants. If one takes the common logarithm of both sides of Equation 10.5, one gets

$$\log M = \log a + n \cdot T_B \qquad \textbf{(10.6)}$$

According to this second equation, $\log M$ is a *linear* function of T_B ($\log a$ and n are constants).

Thus, if M is an exponential function of T_B as in Equation 10.5, $\log M$ is a linear function of T_B (Equation 10.6). This result represents the basic reason why physiologists usually plot metabolism–temperature data for poikilotherms on semilogarithmic coordinates. The logarithm of the animal's metabolic rate is plotted on the y axis, and the animal's body temperature itself is plotted on the x axis. The curve of Figure 10.9A is replotted on semilogarithmic coordinates in **FIGURE 10.9B**, illustrating the "linearizing" effect of semilogarithmic coordinates. A similar comparison is seen in **FIGURE 10.10** using data on actual animals. As we have emphasized, metabolic rate in fact is usually an *approximately* exponential function of body temperature, not a truly exponential one. Thus the semilogarithmic plot for actual animals is typically not precisely linear, as exemplified in Figure 10.10B.

One simple way to describe an exponential relation between metabolic rate (or any other physiological rate) and temperature is to specify the multiplicative factor by which the rate increases when

(A) Plot on linear coordinates

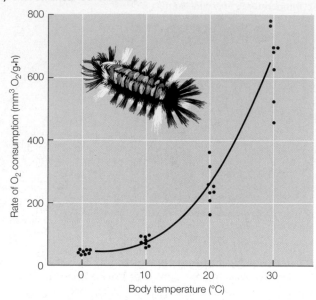

(B) Plot on semilogarithmic coordinates

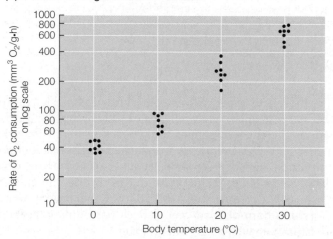

FIGURE 10.10 The relation between metabolic rate and body temperature in tiger moth caterpillars (family Arctiidae), plotted in two ways The metabolic rate was measured as the rate of O_2 consumption. (A) A plot employing linear scales for both variables. (B) A semilogarithmic plot. (After Scholander et al. 1953.)

the body temperature is increased by a standardized increment of 10°C. This factor is called the **temperature coefficient, Q_{10}**:

$$Q_{10} = \frac{R_T}{R_{(T-10)}} \qquad \textbf{(10.7)}$$

where R_T is the rate at any given body temperature T, and $R_{(T-10)}$ is the rate at a body temperature 10°C lower than T. To illustrate, if the resting metabolic rate of an animal is 2.2 J/min at a body temperature of 25°C and 1.0 J/min at 15°C, the Q_{10} is 2.2. As a rough rule of thumb, the Q_{10} for the metabolic rates of poikilotherms is usually between 2 and 3.

Chronic responses: Acclimation often blunts metabolic responses to temperature

When an individual poikilotherm is kept chronically at one body temperature for several weeks and then is kept chronically at a different body temperature for several weeks, the details of its acute

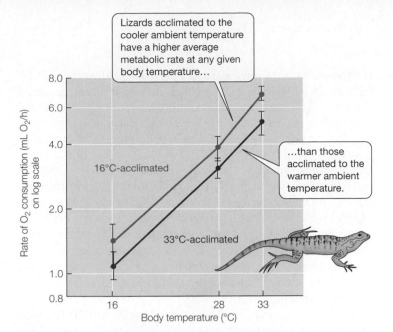

FIGURE 10.11 Acclimation of the metabolism–temperature relation to two different chronic temperatures in a poikilotherm One group of fence lizards (*Sceloporus occidentalis*) was acclimated for 5 weeks to 33°C prior to testing. A second, otherwise identical group was acclimated to 16°C for 5 weeks prior to testing. After the 5 weeks, the animals in each group were placed briefly at body temperatures of 16°C, 28°C, and 33°C, and their standard metabolic rates were measured at all three temperatures as rates of O_2 consumption. The circles show the average metabolic rates; error bars indicate ±2 standard deviations of the mean. (After Dawson and Bartholomew 1956.)

metabolism–temperature relation usually change. Such a change is an example of *acclimation* (see page 18). Understanding this sort of acclimation and its implications can be tricky. The best way to gain clear insight is to start with the actual procedures that are followed to study acclimation. To this end, let's discuss the acclimation study in **FIGURE 10.11**.

In the experiment represented by Figure 10.11, a group of lizards, named the "33°C-acclimated" group, was maintained for 5 weeks at 33°C. At the end of this chronic exposure to 33°C, the lizards were exposed acutely (i.e., briefly) to three different body temperatures—16°C, 28°C, and 33°C—and their resting metabolic rates were measured at each of the three. The line labeled "33°C-acclimated" shows the results. It represents the *acute* relation between resting metabolic rate and body temperature for lizards that were living *chronically* at 33°C during the weeks before the measurements were made.

Another group of lizards, called the "16°C-acclimated" group, was maintained for 5 weeks at 16°C. These 16°C-acclimated lizards were a closely matched but different set of individuals from the 33°C-acclimated group; however, physiologists know from other research that if the individuals that had been acclimated to 33°C were themselves later acclimated to 16°C, the results for the 16°C-acclimated group would be the same as shown. After 5 weeks at 16°C, the 16°C-acclimated lizards were exposed acutely to the same three study temperatures employed for the 33°C-acclimated group, and their metabolic rates were measured. The line labeled "16°C-acclimated" in Figure 10.11 shows the results and thus represents the *acute* relation between resting metabolic rate and body temperature for lizards that were living *chronically* at 16°C.

As Figure 10.11 shows, the acute metabolism–temperature relation is altered when lizards have been living chronically at 16°C rather than 33°C. Lizards acclimated to the cooler ambient temperature, 16°C, have a higher average metabolic rate at any given body temperature than those acclimated to the warmer ambient temperature, 33°C. Although this specific sort of change during temperature acclimation is not universal, it is the most common type of acclimation response in poikilotherms and has been observed in well over half the species studied.

What is the *significance* of this acclimation response? One way to understand the significance is provided by **FIGURE 10.12**. As a thought exercise, imagine that we have some lizards that have been living at 33°C for 5 weeks. The average metabolic rate of these lizards—that is, the metabolic rate of 33°C-acclimated lizards at 33°C—is marked *x* in Figure 10.12. Imagine now that we suddenly lower the temperature of these lizards to 16°C and leave the lizards at 16°C for 5 weeks. The key question we need to address is: How will their average metabolic rate change from the moment their temperature is lowered? Let's begin by considering the *first hour*. In other words, what is the *acute* (prompt) response of the lizards to the change of their temperature? As the animals cool from a body temperature of 33°C to 16°C during the first hour, their average metabolic rate will decline along the acute-response line for 33°C-acclimated animals, following the thin arrows from *x* to *y*. Immediately after the lizards have cooled fully to 16°C, their average metabolic rate will be *y*, the metabolic rate of 33°C-acclimated lizards at 16°C. Note that the drop of body temperature causes a profound

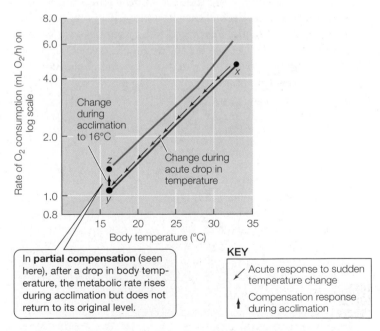

FIGURE 10.12 Compensation through acclimation This diagram shows one way to interpret the data on acclimation by fence lizards in Figure 10.11 (the blue and red lines in the diagram are carried over from Figure 10.11). If lizards that initially are 33°C-acclimated and living at 33°C are switched abruptly to 16°C and left at 16°C, their metabolic rate changes from *x* to *y* during the first hour as they cool acutely from 33°C to 16°C. Then their metabolic rate rises from *y* to *z* over the next 5 weeks as they become acclimated to 16°C: an example of partial compensation.

fall in metabolic rate. Now we come to the most critical question to answer for understanding acclimation: *What will happen to the average metabolic rate of the lizards during the following 5 weeks at 16°C?* The answer is that the metabolic rate will *rise* from *y* to *z* because during those 5 weeks the lizards will become 16°C-acclimated animals! At the end of the 5 weeks, they will have the metabolic rate of 16°C-acclimated animals at 16°C (*z*). Acclimation in these lizards thus reduces—blunts—the effect of the change of their body temperature. Although cooling to 16°C initially lowers the lizards' average metabolic rate by a profound amount, the metabolic rate is lowered to a lesser extent after acclimation has occurred. Put another way, acclimation tends to return the metabolic rate toward its level prior to the drop in body temperature (see Figure 10.12).

After a physiological rate has been raised or lowered by an abrupt change in body temperature, any subsequent, long-term tendency for the rate to return toward its original level even though the new temperature continues is called **compensation**. The rise from *y* to *z* in Figure 10.12 illustrates compensation. Compensation is **partial** if the rate returns only partially to its original level, as in Figure 10.12. When compensation occurs, it is nearly always partial.

An alternative way to understand the *significance* of the sort of acclimation response we have been discussing is presented in **FIGURE 10.13**. Fish of a particular species were acclimated to 10°C, 20°C, and 30°C by being kept at the three temperatures for several weeks. The 30°C-acclimated fish were then tested acutely at all three temperatures (purple symbols [circles]), resulting in the lowermost solid line in Figure 10.13. Similarly, the 20°C- and 10°C-acclimated fish were tested acutely at the three temperatures (red and green symbols, respectively). Note that each of the three solid lines is an *acute-response line*: Each shows how the metabolic rate of fish varies when it is measured promptly after changes in their body temperature. Now let's construct the *chronic-response line* for these fish: The chronic-response line will show how metabolic rate varies

with temperature when the fish are permitted to live at each temperature for several weeks before their metabolic rate is measured. The three bold, black symbols are the metabolic rates of the fish when living chronically at the three temperatures. For instance, the black symbol at the left is the metabolic rate at 10°C of fish that have been living at 10°C for several weeks (10°C-acclimated fish), and the black symbol at the right is the metabolic rate at 30°C of fish that have been living at 30°C. We obtain the chronic-response line by connecting the three black symbols. The chronic-response line has a shallower slope than *any* of the acute-response lines. This means that *if the fish are allowed to acclimate to each temperature before their metabolic rate is measured, their metabolic rate is less affected by changes of body temperature than if they are shifted rapidly from one temperature to another.* Again we see: Acclimation blunts the response to changes of temperature.

What are the mechanisms of metabolic acclimation? The best-understood mechanism involves changes in the *amounts* of key, rate-limiting enzymes, notably enzymes of the Krebs cycle and the electron-transport chain. When poikilotherms acclimate to cold temperatures, their cells synthesize greater amounts of these enzymes.[10] For example, in the red swimming muscles of fish, the number of mitochondria per unit of tissue increases dramatically during cold acclimation in some species (**FIGURE 10.14A**); this increase in mitochondria points to increased amounts of enzymes because the mitochondria are the sites where the enzymes of the Krebs cycle and electron-transport chain reside and operate. In other species of fish, although the number of mitochondria changes little, if at all, the amounts of key enzymes per mitochondrion are increased during cold acclimation (**FIGURE 10.14B**). Responses of these sorts require time; this is one reason why the acclimation response is not observed immediately after a drop in temperature

---

[10] Chapter 2 reviews the effects of enzyme concentration and the processes by which cells modify it.

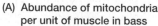

FIGURE 10.13 Because of acclimation, the chronic metabolism–temperature curve is relatively flat compared with the acute metabolism–temperature curves The three solid lines show the *acute* relations between metabolic rate and body temperature for hypothetical fish when 10°C-, 20°C-, and 30°C-acclimated. The dashed line shows the relation between metabolic rate and body temperature when the fish live *chronically* at each temperature.

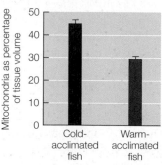
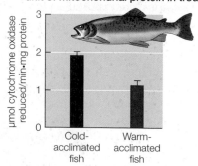

FIGURE 10.14 Mitochondrial and biochemical mechanisms of cold acclimation in the red swimming muscles of fish (A) Striped bass (*Morone saxatilis*) increase the abundance of mitochondria per unit of muscle tissue when acclimated to 5°C (cold-acclimated) rather than 25°C (warm-acclimated). (B) Rainbow trout (*Oncorhynchus mykiss*) increase the activity per unit of mitochondrial protein of the key electron-transport enzyme cytochrome oxidase when acclimated to 5°C (cold-acclimated) rather than 15°C (warm-acclimated). Error bars show ± 1 standard error. (A after Egginton and Sidell 1989; B after Kraffe et al. 2007.)

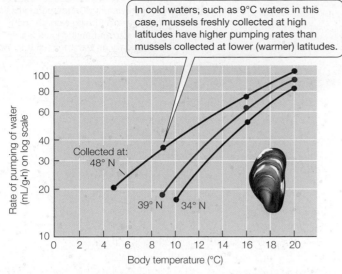

In cold waters, such as 9°C waters in this case, mussels freshly collected at high latitudes have higher pumping rates than mussels collected at lower (warmer) latitudes.

FIGURE 10.15 Acclimatization in mussels Mussels (bivalve molluscs) are extremely important members of intertidal and subtidal marine communities. They pump water through their bodies to obtain both O₂ and food. Mussels of the most abundant West Coast species (*Mytilus californianus*) were collected from nature at three latitudes: 48° north near Seattle, 39° north near San Francisco, and 34° north near Los Angeles. The three lines are acute-response lines for the three groups of mussels, which were acclimatized to the three different latitudes; symbols along the lines represent actual data, to which the lines were fitted. (After Bullock 1955.)

but requires a more extended length of time to be expressed. As the amounts of key, rate-limiting enzymes increase in cells, the presence of the increased enzymes tends to speed metabolic reactions, helping to account for the compensation observed (see Figure 10.12). During acclimation to warm temperatures, enzyme amounts are reduced. Thus, although a shift to a higher body temperature initially speeds an animal's metabolism dramatically, the metabolic rate tends to slow as acclimation occurs (another manifestation of compensation).

Such acclimation responses illustrate in an outstanding way that animals can modify their own cell composition and biochemistry in potentially adaptive ways. In studying biochemistry, it can be easy to get the impression that cells are simply miniature reaction vessels in which test-tube reactions take place. Actually, because most reactions must be catalyzed by enzymes to occur and the cells make the enzymes, cells in fact orchestrate their own biochemistry.

A classic study of *acclimatization* by poikilotherms living in their natural habitats (see page 18) was conducted on three groups of mussels (believed to be genetically similar) of a single species that were collected at three latitudes along the West Coast of the United States. Each group was acutely exposed to several test temperatures, and the rate at which the animals pumped water across their gills was measured. Because of acclimatization, as seen in **FIGURE 10.15**, the populations of mussels living in relatively cold, high-latitude waters and warm, low-latitude waters were more similar to each other in pumping rates than they otherwise would have been.

As a consequence of acclimation and acclimatization, the physiology of an individual animal often depends significantly on its recent *individual history*. This point is important in many ways. For example, when doing research on poikilotherms, investigators need to recognize that the recent histories of the individuals studied may affect the results obtained.

The rate–temperature relations and thermal limits of individuals: Ecological decline occurs at milder temperatures than the temperatures that are lethal

Animals need to *perform* in a variety of ways to succeed. They need to move, grow, raise their rate of O₂ delivery so they can be active, and so forth. With these points in mind, we can ask how the performance of an individual animal varies with its body temperature. From research on this question, physiologists have developed the concept of a generalized, asymmetrical *performance curve*, seen in **FIGURE 10.16A**. Many types of performance roughly follow a curve of this shape. The rate of performance is low at low body temperature. It increases gradually as body temperature rises, over a relatively wide range of temperatures, up to a certain body temperature where the rate of performance peaks. Then, however, if body temperature goes still higher, *performance limitations* set in: The rate of performance declines relatively rapidly, over a relatively narrow range of temperatures, to a low level. In discussing the performance curve, we will focus here on the high-temperature end, because doing so simplifies discussion while still illuminating the most important basic concepts. The high-temperature end is also the end most relevant to understanding the effects of global warming.

To clarify the significance of performance limitations, let's consider points ❶ to ❹ on the generalized performance curve (see Figure 10.16A). When body temperature is at ❶, the rate of performance is at its peak. This means that if the type of performance we are studying is elevation of O₂ delivery, the rate of O₂ delivery is highest at ❶; if the performance we are studying is growth, growth is fastest at ❶. If body temperature rises above ❶, performance will shift to the range labeled ❷. European researchers have created a new term—**pejus temperatures**—to refer to the range of body temperatures at ❷. *Pejus* is from Latin and means "turning worse." If we assume that the highest possible rate of performance is best—that is, if we assume that an animal's fitness is highest when its capacity to perform is highest—then a rise in body temperature from ❶ to ❷ will place the animal in a weakened ("turning worse") condition. If we are interested in O₂ delivery, the animal will not be able to deliver O₂ at the rate that is best for its fitness; if we are interested in growth, the animal will not be able to grow at the rate that is best. If body temperature rises still further to ❸, the animal is still alive, but it is unable to *do* much. Point ❸ marks the body temperature at which an animal's maximum rate of O₂ consumption is little higher than its resting rate of O₂ consumption. At ❸ the animal is passive, and its survival—if it cannot lower its body temperature—is time-limited. Point ❹ is the temperature at which elevated body temperature is itself directly lethal.[11]

The most important message of this analysis is that, as body temperature rises beyond the point of peak performance, an animal's circumstances probably usually "turn worse" in subtle ways before the body temperature becomes high enough to render the animal passive or kill it outright. This distinction is believed to explain why animals *living in natural ecological communities* can

---

[11] When an animal dies because of too high a body temperature, people often say "it died because of protein denaturation." Actually, as shown in Figure 10.16A, irreversible protein denaturation typically occurs only at body temperatures significantly higher than the temperature that kills. What kills the animal, then? *Performance limitations*, as discussed in this section, are thought often to be the answer.

(A) Generalized performance curve

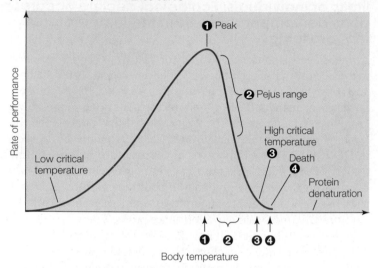

(B) Actual performance curve for aerobic scope in sockeye salmon

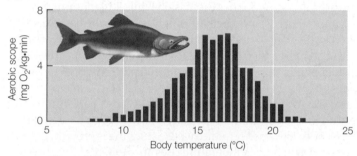

FIGURE 10.16 Performance curves (A) A generalized performance curve, showing key benchmarks discussed in the text. Numbers along the *x* axis show body temperatures that cause performance to be at the numbered spots on the performance curve. In the study of performance curves, the term *critical temperature* is the temperature at which the rate of O_2 consumption can barely be raised above the resting rate (although *critical temperature* has other meanings in other contexts). (B) An example of an actual performance curve. Plotted on the *y* axis is the aerobic scope of salmon, where aerobic scope refers to maximum ability to increase the rate of O_2 consumption above the resting rate. The data are for a population of sockeye salmon (*Oncorhynchus nerka*) during migration (see Figure 17.5). The ability to increase O_2 consumption is highly relevant for these fish because they must generate swimming power to swim up the Fraser River (British Columbia) to reach spawning areas that are more than 1000 km from the sea. (B after Eliason et al. 2011.)

be weakened—and their populations may even go extinct—at temperatures that are distinctly lower, and therefore milder, than lethal temperatures measured *in laboratories*. In a natural ecological community, a capacity for mere survival is often far from adequate. "Turning worse" may reduce an animal's competitive ability so that in a natural community the animal is eliminated by superior competitors, even though it would live if isolated from the competitors. Or "turning worse" may impair the animal's ability to swim or run so that, even though it is not killed outright by temperature, it fails because it cannot catch sufficient prey when living in a natural community.

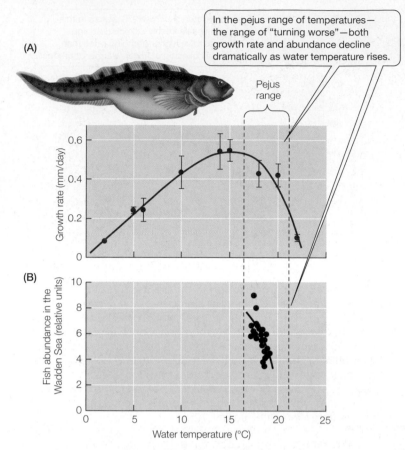

FIGURE 10.17 Nonlethal water temperatures that suppress growth are also associated with ecological decline in populations of common eelpouts (*Zoarces viviparus*) The fish were studied in the Wadden Sea, where water temperature has risen at least 1.1°C over the past 45 years. The upper plot is the performance curve for growth: rate of growth as a function of water (= body) temperature. The lower curve shows fish abundance in the wild. As water temperature becomes higher in the pejus range for growth, fish abundance plummets. (After Pörtner and Knust 2007.)

Where is the performance curve positioned on the scale of body temperature? Various species differ greatly in this regard. For a terrestrial species that evolved at temperate latitudes, the low and high critical temperatures (see Figure 10.16A)[12] might be –10°C and +33°C. For a terrestrial species that evolved in tropical rainforests, the critical temperatures might be +10°C and +35°C. As a specific example, **FIGURE 10.16B** shows the range for an aquatic species, the sockeye salmon. The performance curve is a general concept of how the rate of performance varies within each species' range of body temperatures compatible with life.

A recent study of the common eelpout—a nonmigratory fish—in the Wadden Sea in northern Europe illustrates the sorts of insights that can be gained by interpretation of performance curves. **FIGURE 10.17A** shows the eelpouts' performance curve for growth. You can see that as water temperature rises, the pejus range of "turning worse" starts at 17°C, a temperature 6°C *lower than temperatures the fish can tolerate in a laboratory setting*! Eelpouts

---

[12] Later, when we discuss homeotherms, we will again encounter critical temperatures. Because the term *critical* is used in many different contexts and its meaning sometimes varies, the "critical" temperatures of ectotherms are entirely unrelated to those of homeotherms. Be sure, therefore, to apply the analysis here only to ectotherms.

are abundant in the wild at water temperatures cooler than 17°C. However, as shown in **FIGURE 10.17B**, their abundance in the wild declines sharply as the water temperature increases within the pejus range—indicating that "turning worse" has severe consequences. Temperatures in the pejus range—although not high enough to kill the fish outright—are associated with *ecological* demise of the fish.

Why are animals impaired at temperatures in the pejus range? In fish and other aquatic poikilotherms, O_2 limitation seems to be the most likely general answer. As water warms, its ability to dissolve O_2 declines (see page 596). In natural habitats, therefore, as water temperature rises, the availability of O_2 in the water tends to decline, yet the metabolic needs of animals for O_2 tend to rise (see Figure 10.9). These clashing trends evidently impair function in subtle ways at temperatures (the pejus temperatures) that are distinctly lower than those that directly bring about death. The concept that rising temperatures cause O_2 limitation (see Figure 10.16B), which in turn limits other critical functions, is termed the *theory of oxygen- and capacity-limited thermal tolerance.*

Evolutionary changes: Species are often specialized to live at their respective body temperatures

Related species of poikilotherms often spend much of their time at different body temperatures. Dramatic examples are provided by animals that live in different geographical regions. For example, species of fish, sea stars, and shrimp living on coral reefs in the tropical oceans (see Figure 1.17) live at tissue temperatures that are 25°–30°C higher than those of related species of fish, sea stars, and shrimplike krill that live in polar seas (see Figure 1.12). As another example, among species of lizards that live in the American West, some differ substantially from others in the behaviorally regulated "preferred" body temperatures they maintain during the daylight hours of each day. Whereas one species might employ behavior to thermoregulate at an average body temperature of 34°C, another coexisting species might thermoregulate at a body temperature of 40°C. A key question for physiologists is whether related species that live and reproduce at different body temperatures have evolved adaptations to their respective temperatures.

Some physiological differences among species living at different body temperatures are so dramatic that there can be no doubt about the existence of evolved, adaptive specializations. For example, certain Antarctic species of molluscs promptly die if their body temperature rises above +2°C, even though other species of molluscs live with great success in tropical oceans. Many Antarctic species of fish thrive at temperatures near freezing and die of *heat* stress when warmed to 4°–6°C. And many tropical species of fish thrive at tropical temperatures and die of *cold* stress if cooled to 4°–6°C. One can hardly doubt that these Antarctic and tropical species have evolved adaptive specializations to their respective body temperatures. Most differences among species, however, are not so categorical and are more difficult to interpret.

Physiologists face challenges when they try to understand genetic, evolutionary adaptation to temperature. One challenge is that most animal species cannot be bred in captivity, meaning that individuals must be collected from nature for study. When biologists try to interpret data gathered on wild-caught adults, they must always worry that differences may exist among species not because the species differ genetically, but because the study animals underwent their early development under different conditions in their respective natural habitats. Another challenge is that species from thermally different environments are often unable to live successfully at a single temperature; in such cases, biologists cannot do the "obvious" experiment of comparing species in a single laboratory environment.

LIZARD SPECIES WITH DIFFERENT PREFERRED BODY TEMPERATURES Of what advantage is thermoregulation? A plausible hypothesis is that when a species thermoregulates, its tissues and cells can improve their performance by becoming thermally specialized to function at the body temperatures maintained. In the complete absence of thermoregulation, tissues are equally likely to be at almost any temperature; accordingly, specialization to function at particular temperatures might be disadvantageous. However, if thermoregulation occurs and tissue temperatures are thereby maintained for substantial periods of time in a narrow range, a tissue might profit by becoming specialized (over evolutionary time) to function at temperatures in that range.

Species of lizards with different preferred body temperatures provide excellent models for testing the hypothesis that tissues become specialized to function at the body temperatures maintained by thermoregulation. If the hypothesis is correct, species with relatively high preferred body temperatures should have tissues specialized to function at relatively high temperatures, whereas species with lower preferred temperatures should exhibit tissue specializations to lower temperatures.

Many tissue functions of lizards, when tested, seem in fact to be carried out best in various species when the species are at their respective preferred body temperatures. For example, in species that have preferred body temperatures near 40°C, testicular development at the onset of the breeding season is often most rapid and complete at such high temperatures; in other species that prefer body temperatures near 30°C, the testicles develop optimally near 30°C and are damaged by 40°C. For another example, consider the optimum body temperature for sprint running in various species of lizards. This temperature is well correlated in certain groups of related lizards (but not in all groups) with the respective preferred body temperatures of the species (**FIGURE 10.18**). Hearing, digestion, and the response of the immune system to bacterial invasion are just some of the other processes known to take place optimally, in at least certain sets of related species, when body temperatures are at preferred levels. There are exceptions to these patterns, and there are traits that seem in general not to be optimized at preferred temperatures. Nonetheless, the data on lizards indicate that thermoregulation and tissue thermal specialization have often evolved in tandem.

FISH AND INVERTEBRATES OF POLAR SEAS Many decades ago, investigators hypothesized that the species of fish and invertebrates in polar seas maintain higher resting and average metabolic rates in cold waters than related temperate-zone or tropical species could maintain in the same waters. Today, most specialists conclude that the hypothesis is correct, at least for certain groups of polar poikilotherms. This conclusion, however, follows 60 years of contentious debate, which continues today.

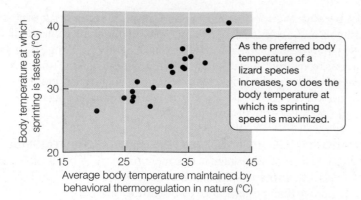

As the preferred body temperature of a lizard species increases, so does the body temperature at which its sprinting speed is maximized.

FIGURE 10.18 The body temperatures at which 19 species of iguanid lizards are able to sprint fastest correlate well with the behaviorally regulated preferred body temperatures of the species In each species of lizard, as the body temperature is raised, sprinting speed increases to a certain point, but then if the body temperature is raised further, sprinting speed starts to decline (a typical performance curve). The body temperature at which the sprinting speed is maximized is plotted on the *y* axis for each of the 19 species. Preferred body temperatures maintained by behavioral thermoregulation in nature are on the *x* axis. (After Huey and Kingsolver 1993.)

Studies of isolated tissues provide more-certain evidence for evolutionary specialization in polar poikilotherms. For instance, investigators have studied the rate of protein synthesis in isolated fish livers. At near-freezing tissue temperatures, protein synthesis is much more rapid in livers taken from polar species than in those taken from temperate-zone species. Similarly, the skeletal muscles of polar fish are able to generate more mechanical power at polar temperatures than are the muscles of temperate-zone fish. These sorts of evidence point to the evolution in polar fish of distinctive physiological properties that permit them to function more vigorously at low body temperatures than unspecialized fish can.

Temperature and heat matter because they affect the functional states of molecules, as well as the rates of processes

One of the most important reasons to study poikilotherms is that they clarify the fundamental ways in which temperature and heat are significant factors for the tissues of animals. Recall from Chapter 7 that heat energy cannot be used to do work by organisms. If heat cannot do work, why does it even matter?

There are two principal reasons why temperature and heat are important for animal tissues. The first we have already discussed: The temperatures of tissues (which are determined by heat inputs and outputs) *affect the rates of tissue processes.*[13]

Now we turn to the second reason: The temperatures of tissues *affect the molecular conformations and therefore the functional states of molecules.*

The exact three-dimensional conformation (the "molecular shape") of a protein molecule depends on prevailing temperature because three-dimensional conformation is stabilized by weak,

noncovalent bonds—not by strong, covalent bonds (see Box 2.1). The various weak bonds in a molecule change in their relative strengths when the temperature is modified, and thus the molecule assumes a different conformation at each temperature. The *functional properties* of a protein molecule depend on its molecular conformation.[14] With this background in mind, we can understand in principle why the functional properties of protein molecules often vary with the prevailing temperature.

One of the most significant discoveries of the last few decades in the study of comparative physiology is the realization that animals living in different temperature regimes often have evolved *different molecular forms* of proteins: forms that are differentially suited to function in the divergent temperature regimes. **FIGURE 10.19** provides a dramatic visual illustration of this point. At the left in Figure 10.19B are the freshly removed eye lenses of three vertebrates that live in different temperature regimes. The eye lens of the cow normally functions at 37°C. The two fish are from coral-reef ecosystems (the soldierfish) and the Antarctic Ocean (the toothfish), and their lenses normally function at 25°C and –2°C, respectively. Ostensibly the three lenses at the left are all the same: All are composed of a type of protein—called crystallin protein—that is perfectly clear. Testing the lenses revealed, however, that they are not the same. When the cow lens and tropical-fish lens were placed at 0°C, they underwent denaturation: a type of protein-conformation change that disrupts normal protein function. As a consequence of the denaturation, instead of being clear, the lenses became opaque (a phenomenon called cold cataract). This sort of change would have blinded the animals! The lens of the Antarctic toothfish, however, exists for a lifetime (up to 30 years) at –2°C without undergoing denaturation; and tests showed that it could be cooled to –12°C without denaturing. In brief, all these vertebrates have lenses made of crystallin proteins, but they have *different molecular forms* of the proteins: forms differentially suited to the distinct temperatures at which their eye lenses function. This is a theme that is repeated throughout the study of proteins and other macromolecules.

The enzyme–substrate affinity of an enzyme molecule is one of the molecule's most important functional properties because it determines how readily the molecule is able to form an enzyme–substrate complex (see page 48). The enzyme–substrate affinity, however, is not a fixed property of an enzyme molecule. Instead, it changes as the prevailing temperature is raised and lowered. Biochemists believe that a certain intermediate level of enzyme–substrate affinity is ordinarily ideal. Whereas too low an affinity can render an enzyme molecule incapable of forming complexes with substrate molecules, too high an affinity can make the enzyme molecule so prone to forming complexes with substrate that it becomes uncontrollable by regulatory processes. **FIGURE 10.20A** illustrates how the enzyme–substrate affinity of one particular enzyme molecule—lactate dehydrogenase (LDH) isolated from the muscles of a goby fish—varies with the prevailing temperature because of reversible, temperature-induced conformational changes in the protein.

---

[13] These rates include metabolic rates, rates of particular biochemical reactions, and rates of biophysical processes such as diffusion and osmosis.

[14] This is probably true for several reasons. One important reason is that (as discussed in Chapter 2) a protein molecule often must *flex* (change shape) to carry out its functions, and conformation affects how readily various molecular subregions are able to flex.

(A) An Antarctic toothfish

(B) Eye lenses of a cow, a coral-reef soldierfish, and an Antarctic toothfish

Cow at 25°C

> The cow lens looks like this after 1.5 h at 0°C.

⊢ ⊣ 0.5 cm

Soldierfish at 15°C

> The soldierfish lens looks like this after 48 h at 0°C. A cold cataract takes longer to form than in the cow, but forms.

⊢ ⊣ 0.5 cm

Antarctic toothfish at –2°C

> The lens of the Antarctic toothfish looks like this after a lifetime at –2°C.

⊢ ⊣ 0.5 cm

FIGURE 10.19 Seeing at –2°C requires specialized eye-lens crystallin proteins (A) An Antarctic toothfish (*Dissostichus mawsoni*) living at –2°C in the ocean near Antarctica. Toothfish sometimes live for 30 years, and their eye lenses remain crystal clear throughout. (B) At the left are normal eye lenses taken from three species: a cow, a coral-reef fish called the blackbar soldierfish (*Myripristis jacobus*), and the Antarctic toothfish. In life, the lenses of these three species function at about 37°C, 25°C, and –2°C, respectively. Note at the right in (B) that the lenses of the cow and soldierfish develop cold cataracts—which would blind the animals—with only short-term exposure to 0°C. (Photographs in B courtesy of Andor Kiss and C.-H. Christina Cheng; photographs from Kiss et al. 2004.)

Because the functional properties of enzymes depend on the prevailing temperature, any particular enzyme protein can be highly functional at certain tissue temperatures while being only marginally functional (or even nonfunctional) at other tissue temperatures. How, then, can animals living in different thermal regimes all have suitably functional enzymes?

An important part of the answer is that during evolution, species that have different body temperatures have often evolved different molecular forms of enzyme proteins. Not all species of vertebrates, for instance, have the same molecular form of LDH that the goby fish in Figure 10.20A has. If they did, species that ordinarily have low body temperatures would routinely have far higher enzyme–substrate affinities than species that have high body temperatures. Instead, as **FIGURE 10.20B** shows, different species have evolved different molecular forms of LDH. The six species of poikilotherms shown in Figure 10.20B, some of which ordinarily live at very different body temperatures than others, have six different (although homologous) LDH proteins.[15] Although all six LDH proteins catalyze the same reaction, they differ in their detailed structures and functional properties, so the six exhibit different relations between enzyme–substrate affinity and temperature. The line for each species in Figure 10.20B is thickened and colored blue at the temperatures that correspond to the usual body temperatures of the species. For example, the line for the warm-water goby is thickened and colored blue at temperatures between 25°C and 40°C because that species of fish ordinarily has

body temperatures of 25°–40°C, and the line for the Antarctic fish is thickened and colored blue at temperatures near –1°C because that species ordinarily has a body temperature near –1°C. All the blue, thickened segments are at about the same height on the *y* axis. Specifically, all fall within the vertical distance marked by the shaded band that runs across the figure near the top. What this shows is that *all six species have about the same enzyme–substrate affinity when they are at their respective body temperatures.* The way they have achieved this remarkable condition, even though they live at body temperatures as much as 40°C apart, is by having evolved different molecular forms (homologs) of the enzyme.

The conservation of enzyme–substrate affinity by the evolution of enzyme homologs that are adapted to function best at different temperatures is very common. One of the most striking and instructive examples is provided by the four species of barracudas in **FIGURE 10.21**. These four species, all closely related evolutionarily, behaviorally elect to live in waters that are just modestly different in temperature. For example, the waters occupied by *Sphyraena lucasana* average just 3°–4°C warmer than those occupied by *S. argentea*, and those occupied by *S. ensis* average just 3°–4°C warmer yet. Even these relatively small differences in habitat temperature (and body temperature) have led to the evolution of different molecular forms of the LDH protein. Consequently, the four species all have similar enzyme–substrate affinities when living at their respective (and different) temperatures.

Earlier we noted that there are two major ways in which temperature and heat matter for animals. The second of those ways should now be clear enough that we can state it succinctly: *Particular enzyme*

---

[15] These are *interspecific enzyme homologs*. See page 51 in Chapter 2 for an extensive discussion of both LDH and the concepts of protein homology.

(A) Enzyme–substrate affinity as a function of temperature in a goby

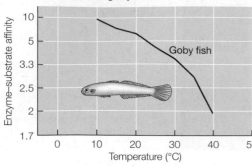

(B) Enzyme–substrate affinity as a function of temperature in six species of poikilotherms

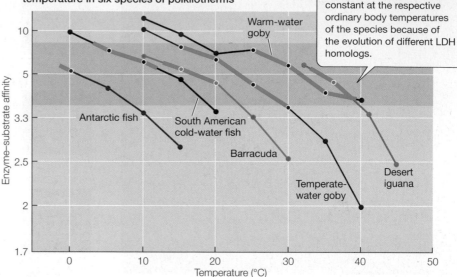

All the blue line segments, which identify the ordinary body temperatures of the species, fall within the narrow vertical distance marked by the shaded band. Thus affinity for substrate is kept relatively constant at the respective ordinary body temperatures of the species because of the evolution of different LDH homologs.

FIGURE 10.20 The affinity of the enzyme lactate dehydrogenase (LDH) for substrate as a function of temperature This relation is shown in (A) for a goby fish (*Gillichthys mirabilis*) and in (B) for six species of poikilotherms—five fish and a desert lizard—that ordinarily live at different body temperatures. The blue, thickened portion of each line identifies the range of body temperatures ordinarily experienced by the species. The enzyme–substrate affinity shown in both plots is the affinity of muscle LDH (LDH-A$_4$) for pyruvic acid. Affinity is expressed as the inverse of the apparent Michaelis constant (m*M* pyruvate); see Chapter 2 (page 47) for background. The Antarctic fish is an Antarctic notothenioid; the South American fish is also a notothenioid; the barracuda is *Sphyraena idiastes*; the temperate-zone goby is *Gillichthys mirabilis*; the warm-water goby is *Gillichthys seta*; and the desert iguana is *Dipsosaurus dorsalis* (see Figure 1.14). (After Hochachka and Somero 2002.)

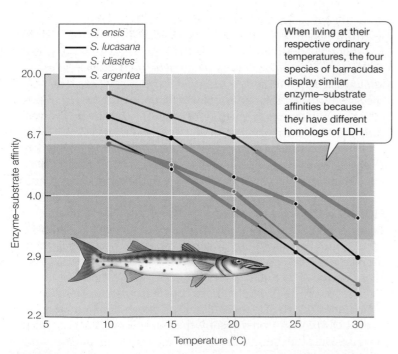

When living at their respective ordinary temperatures, the four species of barracudas display similar enzyme–substrate affinities because they have different homologs of LDH.

FIGURE 10.21 Enzyme adaptation in four species of barracudas The four species, all in the genus *Sphyraena*, live at somewhat different temperatures. The enzyme studied is LDH. All details are the same as in Figure 10.20B. (After Somero 1997.)

molecules (*and other sorts of protein molecules*) *are typically specialized to function best within certain temperature ranges. The protein molecules therefore* require *certain temperatures to function optimally.* With few known exceptions, the tissues of the adults of any particular species are fixed in the homolog of each enzyme they synthesize; although a tissue may change the *amount* of the enzyme it synthesizes (as often occurs during acclimation or acclimatization), it cannot change the *type* of enzyme.[16] Thus individuals of a species of fish (or other aquatic poikilotherm) ordinarily found in warm waters typically *require* warm tissue temperatures for their enzyme molecules to have ideal functional forms. Conversely, individuals of a cold-water species of fish typically *require* cold tissue temperatures for their particular types of enzyme molecules to have ideal functional forms. The same principles apply to homeotherms. For instance, the LDH of cows needs to be at about 37°C to have an appropriate enzyme–substrate affinity, just as the crystallin proteins of cows need to be warm to be clear. Certain tissue temperatures, in brief, are crucial because the conformations and functional properties of proteins are not deterministically set by the chemical compositions of the proteins but depend as well on the prevailing temperature.

IMPLICATIONS FOR GLOBAL WARMING A key question in the study of global warming is *how much* the tissue temperatures of poikilotherms must change for the changes to have significant consequences. Data such as those on the barracudas (see Figure

---

[16] As discussed later in the chapter, this statement does not necessarily apply to other types of proteins besides enzymes.

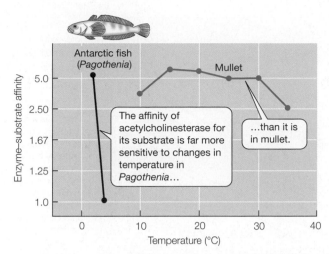

FIGURE 10.22 An enzyme that shows extreme sensitivity to temperature change   The plot shows the affinity of brain acetylcholinesterase for acetylcholine in a stenothermal, polar fish (*Pagothenia borchgrevinki*, pictured) and a eurythermal, warm-water species of fish, a mullet. Because acetylcholinesterase is a lipoprotein enzyme, lipid moieties may be involved in interspecific differences. Affinity is expressed as the inverse of the apparent Michaelis constant (m*M* acetylcholine). (After Somero 1997.)

10.21) suggest that the answer is sometimes "not very much." The species of barracudas, which live in waters differing by 3°–4°C and have evolved different enzyme homologs, appear to be telling us that a change in body temperature of 3°–4°C is sufficiently consequential that natural selection favors the evolution of new molecular variants of key enzymes. A worry about human-induced global warming is that it may occur so rapidly that evolution will not immediately "keep up," and many poikilotherms may be forced to function for years with nonoptimized molecular systems.

In situations in which tissue temperatures are different from ideal, an important consideration is the steepness of the relation between molecular functional properties and temperature. A classic example of a very steep relation is provided by the acetylcholinesterase homolog found in the brain of *Pagothenia borchgrevinki*, a red-blooded Antarctic fish (**FIGURE 10.22**). Acetylcholinesterase is essential for brain function because it keeps the neurotransmitter acetylcholine from building up excessively at synapses (see page 353). The enzyme–substrate affinity of *Pagothenia*'s acetylcholinesterase is exceptionally sensitive to changes of tissue temperature, so much so that the enzyme undergoes functional collapse—it loses almost all affinity for its substrate—when warmed to 5°–10°C. An enzyme form of this sort would cause any species possessing it to be unusually vulnerable to climate change. It also probably helps explain why *Pagothenia* is one of the most stenothermal fish known; it ordinarily lives its entire life at temperatures near –2°C and dies of heat stress at +4°–6°C.

ADDITIONAL ASPECTS OF ENZYME ADAPTATION TO TEMPERATURE   Besides enzyme–substrate affinity, another critical functional property of enzymes is the *catalytic rate constant*, k_{cat}, which measures the number of substrate molecules that an enzyme molecule is capable of converting to product per unit of time. If the k_{cat} of a particular type of enzyme, such as LDH, is measured

under fixed conditions, the usual pattern is that homologs of the enzyme from cold-water species tend to exhibit higher k_{cat} values than homologs from related warm-water species. Thus the enzyme homologs of the species living in cold waters have a greater intrinsic ability to speed reactions, an attribute that in nature helps offset the reaction-slowing effects of low temperatures. An example is provided by the four barracuda species shown in Figure 10.21. The k_{cat} of their LDH enzymes, measured at a fixed study temperature, increases as the temperature of their habitat decreases.[17] The k_{cat}s of LDH forms in Antarctic fish are four to five times higher than the k_{cat}s of LDH forms in mammals.

An important question from the viewpoint of evolutionary biochemistry is how much the amino acid composition of an enzyme must change for the enzyme to take on new functional properties. One of the most interesting studies on this question also involved the barracudas. Using modern sequencing techniques, researchers found that in the LDH protein—which consists of about 330 amino acid units—four amino acids at most are changed from one barracuda species to another; only one amino acid is different between some of the species. Of equally great interest, none of the changes in amino acid composition in these LDH homologs is at the substrate-binding site; the changes therefore affect the function of the enzyme by altering properties such as molecular flexibility, not the properties of the catalytic site per se. These characteristics, exemplified by the barracudas, are emerging as important generalizations: (1) Homologous enzyme molecules often differ in only a relatively few amino acid positions—helping explain how species can readily evolve adaptively different enzyme homologs; and (2) the altered amino acid positions are located outside the substrate-binding site, so the substrate-binding site is constant or near-constant, explaining why all homologs catalyze the same reaction.

MYOSIN ISOFORMS EXEMPLIFY THAT PROTEIN ISOFORMS SOMETIMES CHANGE DURING ACCLIMATION AND ACCLIMATIZATION   Muscle contractile function depends on a variety of muscle-specific proteins, such as myosin, troponin, and actin (see Chapter 20). An individual of any particular species is typically capable of synthesizing multiple molecular forms of each of these proteins, providing the basis for a wide range of phenotypic plasticity in muscle function. The various molecular forms of a particular protein that can be synthesized by a species are termed *isoforms* of the protein—a term with similar meaning to *isozymes* (see page 51) but preferred in this context because not all muscle proteins have enzymatic activity. How is it possible for an individual to synthesize multiple forms of one protein? Often the answer is that an individual can possess multiple genes in the gene family coding for the protein.

The isoforms of the *myosin heavy-chain* protein in fish fast muscle provide an elegant and instructive example of the importance of the properties just outlined. This protein plays a central role in muscle contraction (see Chapter 20) and thus is critical for swimming—one of the most important of all activities of a fish. After individual carp

---

[17] Enzyme–substrate affinity and k_{cat} tend to coevolve because of molecular structural reasons that are only starting to become clear. Thus the evolution of particular interspecific patterns in k_{cat} is not entirely independent of the evolution of particular patterns in enzyme–substrate affinity.

(*Cyprinus carpio*) or goldfish (*Carassius auratus*) that have been living in warm water are switched to cold water, they initially cannot swim particularly fast, but over several weeks, they exhibit increasing swimming performance in the cold water. A change in isoforms of the myosin heavy-chain protein is a key part of this acclimation. That is, according to available evidence, the muscle cells of the fish synthesize isoforms in altered proportions, and they switch out old isoforms for new ones in the contractile apparatus! Although the proteins in thoroughly cold-acclimated individuals are relatively unstable if subjected to warm temperatures, at cold temperatures they have molecular properties that enhance contractile performance. The change in isoforms is thus a key reason that the swimming performance of the fish increases as cold acclimation takes place. Whereas the enzymes of glycolysis and the Krebs cycle that have been so thoroughly studied (see the preceding sections of this chapter) typically are adjusted only in *amount,* not in *type,* during acclimation, the myosin isoforms illustrate that some proteins undergo isoform changes.

LIPIDS AND HOMEOVISCOUS ADAPTATION   As is true of proteins, the functional properties of lipids depend on the prevailing temperature as well as the chemical compositions of the molecules. One of the most important functional properties in the study of lipids is the fluidity of the phospholipids in cell membranes and intracellular membranes. As stressed in Chapter 2 (see page 38), individual phospholipid molecules—and protein molecules embedded in the phospholipid matrix—diffuse from place to place within the leaflets of cell membranes and intracellular membranes, and this mobility is exceedingly important for membrane function. **Membrane fluidity** is a measure of how readily the phospholipid molecules in a membrane move.

FIGURE 10.23 depicts membrane fluidity as a function of temperature for membrane lipids extracted from the brains of nine vertebrate species—seven fish from a broad range of habitats, a mammal, and a bird. If you focus on any *particular* species, you will note that fluidity is a regular function of the prevailing temperature. Fluidity increases as temperature increases, much as any particular household lipid, such as butter, becomes more fluid as it is warmed.

When different species of animals are taken from their natural habitats and analyzed, they typically differ in the compositions of their membrane phospholipids. Consequently, as can be seen when all nine species in Figure 10.23 are compared, species differ in the details of their relations between membrane fluidity and temperature. The line for each species is thickened and colored blue at temperatures that correspond to the usual body temperatures of the species. As in Figures 10.20B and 10.21, the blue, thickened line segments all fall within a narrow range on the *y* axis, marked by the shaded band in the figure. The meaning of this result is that *all nine species have about the same membrane fluidity when they are living at their respective, normal body temperatures.* Such maintenance of a relatively constant membrane fluidity regardless of tissue temperature is called **homeoviscous adaptation** (*homeoviscous,* "steady viscosity").

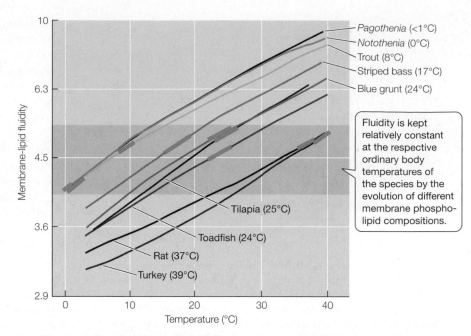

FIGURE 10.23 The fluidity of lipid-bilayer membranes from brain tissue as a function of temperature   The relation between fluidity and temperature is shown for each of nine species of vertebrates—seven fish that ordinarily live at different body temperatures (see temperatures listed after each species name), a mammal, and a bird. The blue, thickened portion of each line marks the body temperatures ordinarily experienced by the species. Fluidity is measured in terms of the mobility of a molecular probe, to which the units of measure refer. (After Hochachka and Somero 2002.)

Homeoviscous adaptation is possible because the chemical composition of membrane phospholipids is not fixed but instead can differ among species. If all animal species had the same membrane phospholipid composition, the species with high body temperatures would have very fluid membranes, whereas those with low body temperatures would have stiff membranes. In reality, all have about the same membrane fluidity because species that have evolved to operate at different body temperatures have also evolved systematically different phospholipid compositions. As we saw in studying enzymes, again this means that tissue temperature is critical because it must be "matched" to the particular molecules present: *A tissue in which cell membranes are built of particular phospholipids will have the "correct" membrane fluidity only if its temperature is correct.*

The best-understood chemical basis for homeoviscous adaptation is modification of the number of double bonds in the fatty acid tails of the membrane phospholipids. Double bonds create bends in the fatty acid tails (see Figure 2.2A), and these bends interfere with close packing of the tails in a membrane. Thus membrane fluidity tends to increase as the number of double bonds increases—that is, as the lipids become more chemically unsaturated. We saw in Chapter 2 (see Figure 2.3) that among fish species native to different thermal environments, the degree of unsaturation of brain phospholipids increases as habitat temperature decreases: Whereas polar species have highly unsaturated lipids that, because of their chemical structure, remain reasonably fluid at polar temperatures, tropical species have much more saturated lipids that, because of their chemical structure, resist becoming too fluid at tropical temperatures.

Individual animals are able to alter the membrane phospholipids that they synthesize: Phospholipid composition is phenotypically plastic (often greatly so). The phospholipid composition of cell membranes and intracellular membranes is commonly restructured during acclimation and acclimatization in ways that promote homeoviscous adaptation. The restructuring of membrane phospholipids by an individual exposed to a chronically changed temperature typically requires many days or more. However, some fish in desert ponds undergo substantial phospholipid restructuring on a day–night cycle, thereby keeping membrane fluidity relatively constant even as the ponds heat up during the day and cool at night.

Poikilotherms threatened with freezing: They may survive by preventing freezing or by tolerating it

If poikilotherms are exposed to temperatures even slightly colder than those necessary to freeze water, they face a threat of freezing. A classic example of this threat is provided by barnacles, mussels, and other animals attached to rocks along the seacoast. In places such as Labrador in the winter, when the tide is out, such animals may be exposed to extremely cold air and become visibly encased in ice (**FIGURE 10.24**). In this way, they themselves may freeze. Animal body fluids have lower freezing points than pure water because the freezing point is ordinarily a colligative property and becomes lower as the total concentration of dissolved matter increases.[18] Although animal body fluids have lowered freezing points because of their solute content, they nonetheless typically freeze at –0.1°C to –1.9°C (depending on the animal group) unless they are specially protected.

THE FREEZING PROCESS IN SOLUTIONS AND TISSUES To understand the threat of freezing and the possible strategies that animals might use to avoid freezing damage, the first step is to examine the freezing process. An important and seemingly strange point to mention at the outset is that when aqueous solutions are progressively cooled, they commonly remain unfrozen even when their temperatures have fallen below their freezing points, a phenomenon called **supercooling**. Supercooling is an intrinsically unstable state, and a supercooled solution can spontaneously freeze at any moment. Nonetheless, solutions in the supercooled state sometimes remain supercooled for great lengths of time.

An important determinant of a supercooled solution's likelihood of freezing is its *extent* of supercooling; as a solution's temperature drops further below its freezing point, freezing becomes more likely. If the temperature of a supercooled solution is gradually lowered while the solution is not otherwise perturbed, a temperature is reached at which the likelihood of freezing becomes so great that the solution spontaneously freezes within a short time. This temperature is called the **supercooling point** of the solution.

Exposure to ice *induces* freezing in a supercooled solution. This fact has two important implications. First, an unfrozen but supercooled solution immediately freezes if it is seeded with even just a tiny ice crystal, regardless of its extent of supercooling. Second, if a solution, initially at 0°C, is gradually cooled in the presence of an ice crystal, the crystal will prevent supercooling.

[18] See Chapter 5 (see page 122) for a discussion of the colligative properties of solutions, including the freezing point.

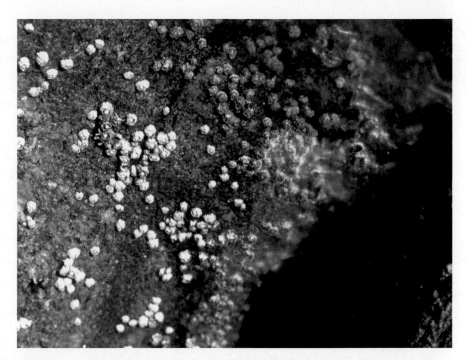

FIGURE 10.24 Barnacles encased in ice during low tide along a northern seacoast The animals—glued to rocks and unable to flee when exposed to frigid air at low tide—face a threat of freezing. They have met the threat, not by preventing freezing, but by evolving an ability to tolerate—and thereby survive—extensive freezing of their body fluids.

An important application of this second point is that the cooling of a solution in the presence of an ice crystal permits determination of the solution's **freezing point**. The freezing point is the temperature above which a solution cannot freeze and below which it deterministically freezes in the presence of preexisting ice. A solution's freezing point is typically a colligative property, depending on the concentration of dissolved entities (see page 122). If a frozen solution is gradually warmed, its **melting point** is the lowest temperature at which melting occurs. The freezing point and the melting point are usually equal.

In tissues, additional complexity is involved in understanding freezing because the *location* of freezing is an important factor. Under natural conditions, freezing *within cells* (*intracellular freezing*) almost always kills the cells in which it occurs. Intracellular freezing is thus fatal for animals unless they can survive without the cells that are frozen. However, many animals are remarkably tolerant of widespread ice formation in their *extracellular* body fluids. This tolerance of extracellular freezing is significant because, for reasons only partly understood, when freezing occurs in an animal, the formation of ice often begins in the extracellular fluids and thereafter tends to remain limited to the extracellular fluids.

To understand the implications and dangers of extracellular freezing more thoroughly, we need to look at the *process* of extracellular ice formation (**FIGURE 10.25**). An important attribute of the slow freezing of a solution is that water tends to freeze out of the solution in relatively pure form. Thus, when ice crystals form in extracellular fluid, solutes (excluded from the ice crystals) tend to accumulate in the portion of the extracellular fluid that remains unfrozen, raising the total solute concentration of the unfrozen fluid (see Figure 10.25B). The freezing point of the unfrozen fluid is lowered by the increase in its solute concentration. Thus, at a

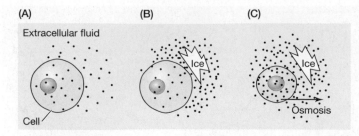

FIGURE 10.25 The process of extracellular freezing in a tissue Dots represent dissolved entities, and therefore the density of dots represents osmotic pressure. (A) The unfrozen condition. The intracellular fluid inside the cell and the surrounding extracellular fluid have the same osmotic pressure. (B) Slow extracellular freezing produces ice that consists of virtually pure water. Solutes excluded from the ice elevate the osmotic pressure of the unfrozen extracellular fluid. (C) Because of the difference in osmotic pressure created in part (B), water leaves the cell by osmosis, shrinking the cell and raising the osmotic pressure of the intracellular fluid.

fixed temperature, the formation of ice in extracellular fluid is a *self-limiting* process: Water freezes out of the extracellular fluid only until the freezing point of the unfrozen fluid becomes low enough to equal the prevailing temperature.

The intracellular and extracellular fluids have similar osmotic pressures in an unfrozen animal, meaning that water has little or no tendency to enter or leave cells by osmosis. This benign state is disrupted by freezing in an animal's extracellular fluids. Immediately after extracellular ice forms, the unfrozen extracellular fluids are osmotically more concentrated than the intracellular fluids (see Figure 10.25B). Thus the ice formation leads to the osmotic loss of water from cells (see Figure 10.25C). This loss of intracellular water is itself self-limiting; it stops after the intracellular osmotic pressure has risen to equal the extracellular osmotic pressure. Within limits, the osmotic loss of water from cells is protective: By concentrating the intracellular fluids and thus lowering the intracellular freezing point, the loss of water from cells helps prevent intracellular freezing, which usually is fatal.

THE ADAPTIVE RESPONSES OF ANIMALS TO FREEZING CONDITIONS: INTRODUCTION
Many poikilotherms behaviorally avoid environments where freezing conditions prevail. For example, many species of frogs, turtles, and crayfish move to the bottoms of lakes and ponds during winter. This location is a safe microhabitat because lakes and ponds do not normally freeze to the bottom.

Many poikilotherms, in contrast, are actually exposed to freezing conditions and must cope physiologically. The mechanisms by which they do so are classified into three types: (1) production of antifreeze compounds, (2) supercooling, and (3) tolerance of freezing. Antifreeze production and supercooling are mechanisms of *preventing* freezing. Usually, species that employ antifreezes and supercooling are **freezing-intolerant**; they die if they freeze and thus are absolutely dependent on successful prevention. By contrast, some species are **freezing-tolerant**; they have evolved the ability to survive extensive freezing of extracellular body water and typically respond to freezing conditions by freezing. It remains largely a mystery why some species have evolved along lines of freezing intolerance, whereas others, sometimes closely related, have evolved toward freezing tolerance.

PRODUCTION OF ANTIFREEZE COMPOUNDS Many poikilotherms gain protection against freezing by synthesizing **antifreeze compounds**, defined to be dissolved substances that are added to the body fluids specifically to lower the freezing point of the body fluids. Two types of physiologically produced antifreezes are recognized.

1. *Colligative antifreezes.* Some antifreezes lower the freezing point of the body fluids strictly by colligative principles: They affect the freezing point by increasing the total concentration of solutes in the body fluids, not by virtue of their particular chemical properties. The most common of these colligative antifreezes are polyhydric alcohols, especially glycerol, sorbitol, and mannitol.

2. *Noncolligative antifreezes.* Some antifreezes lower the freezing point of the body fluids because of specialized chemical properties. Certain proteins and glycoproteins produced by a variety of insects and marine fish (see Figure 3.6) are the best-understood antifreezes of this sort. They are believed to act by binding (through weak bonds such as hydrogen bonds) to nascent ice crystals in geometrically specific ways, thereby suppressing growth of ice by preventing water molecules from freely joining any ice crystals that start to form. The noncolligative antifreezes can be quite dilute and yet highly effective because they depress the freezing point hundreds of times more than can be accounted for by simple colligative principles. The noncolligative antifreezes, however, do not depress the melting point any more than colligative principles explain. Thus solutions containing these antifreezes exhibit the unusual property—termed **thermal hysteresis**—that their freezing points are substantially lower than their melting points. The noncolligative antifreezes are usually called **thermal hysteresis proteins (THPs)** or **antifreeze proteins**. Fascinating insights into protein evolution have been gained by study of the evolutionary origins of these antifreezes (**BOX 10.2**).

Antifreezes are synthesized principally by certain species in two sets of animals: the marine teleost fish (bony fish) and the insects. The marine teleost fish, in comparison with most other aquatic animals, face unique problems of freezing because their body fluids are osmotically more dilute than seawater (see page 750).[19] Specifically, marine teleosts have blood and other body fluids that—without special protection—freeze at temperatures of −0.6°C to −1.1°C. Seawater, being more concentrated, has a lower freezing point: −1.9°C. Marine teleost fish therefore can potentially freeze even when they are swimming about in unfrozen seas!

A great many of the marine teleost species that live at polar and subpolar latitudes prevent freezing by synthesizing antifreeze proteins. These proteins are found in the blood and most other extracellular fluids of the fish. Although some polar species main-

---

[19] Marine invertebrates generally have body fluids that are as concentrated as seawater. Their freezing points thus match the freezing point of seawater, and—when they are immersed in seawater—they are not threatened with freezing unless the seawater itself freezes. Freshwater animals of all kinds have body fluids that are more concentrated than freshwater. Thus their freezing points are below the freezing point of freshwater, and—when they are immersed in freshwater—they also do not freeze unless the water in which they are living freezes.

BOX 10.2 Evolutionary Genomics: The Genes for Antifreeze Proteins Are Descended from Genes for Other Functional Proteins

Among the antifreeze proteins (AFPs) in polar fish, there are several known major types that are independently evolved. How did they evolve? What were their evolutionary precursors?

One type of AFP is indirectly evolved from a digestive enzyme protein! This astounding insight comes from genomic studies of the Antarctic toothfish (see Figure 10.19A). By means of gene sequencing, researchers recognized similarities in the gene that codes for the AFP and the gene that codes for a trypsin-like digestive enzyme secreted by the pancreas (see pages 155–156). Then, exploring the genome of the toothfish, they found an unusual gene: a single gene that encompasses the genetic code for both the digestive protein and the AFP. This gene demonstrates and illustrates how the original gene for the trypsin protein could have evolved into the gene that codes for the AFP. In fish that produce this particular type of AFP, most of the AFP is synthesized in the stomach and pancreas, and secreted into the gut lumen. Only later does it enter the blood. This strange path to the blood reinforces the conclusion that the gene for the AFP is descended from the digestive trypsin gene.

In 2010, the evolutionary origin of a second type of AFP was discovered, and this evolutionary scenario turns out to be amazingly parallel. Sialic acid is an important cytosolic compound. Through studies based on gene sequencing and controlled gene expression, researchers

A fish that provides insight into gene evolution A gene found in the genome of the Antarctic eelpout (*Lycodichthys dearborni*) probably evolved to have a rudimentary antifreeze function in addition to its traditional function. After duplication of the gene, one copy evolved to code for one of the major types of AFPs in Antarctic fish. (Courtesy of Christina Cheng.)

discovered in an Antarctic fish (see the figure) a sialic-acid synthesis gene that also includes the code for a protein with rudimentary antifreeze properties. They then established that this gene duplicated during evolution. Whereas one copy continued its prior role, the other ceased to be involved in that role and instead evolved into a gene coding for a highly effective AFP.

Based on these cases, it appears that when polar fish first confronted the threat of freezing, variants of old protein-synthesizing genes underwent evolution to produce new genes specialized for syn-

thesis of proteins with antifreeze properties. This discovery reminds us of François Jacob's "tinkering" model for evolution (see page 10). Jacob emphasized that evolution makes new things from old, preexisting things, rather than starting from scratch. In the case of the AFPs discussed here, the gene for one was made from a gene for a digestive enzyme, and the gene for the other was made from a gene for cytosolic sialic-acid synthesis. **Box Extension 10.2** provides references that will enable you to learn more about these fascinating genomic insights.

---

tain high antifreeze-protein concentrations in all months, most species synthesize antifreeze proteins just in the cold seasons. The winter flounder (*Pleuronectes americanus*) is one of the best-known antifreeze-producing species. It has an unusually large number of copies of the gene for antifreeze-protein synthesis (about 100), and these gene copies are transcribed and translated in an anticipatory way as winter approaches (**FIGURE 10.26**), under photoperiodic control.

Antifreezes are also found in the body fluids of many species of insects during winter. Colligative antifreezes such as glycerol, sorbitol, and mannitol are common and sometimes accumulate to impressive levels; in extreme cases, 15%–25% of an insect's over-wintering body weight consists of glycerol and other polyhydric alcohols. Antifreeze proteins—noncolligative antifreezes—are also

known in many insects, probably having evolved independently several times. When insects are devoid of antifreezes, they have freezing points higher than –1°C. With high levels of antifreeze solutes, however, insects may exhibit freezing points as low as –10°C or, in one known case, –19°C. Even freezing points this low, however, may be well above temperatures that insects encounter in many terrestrial environments. Prevention of freezing in freezing-intolerant insects is thus often a result of both antifreezes and supercooling working in concert. Both polyhydric alcohols and antifreeze proteins promote supercooling and may have other favorable effects in addition to their antifreeze effects.

SUPERCOOLING Supercooling is a perfectly ordinary, commonplace phenomenon in both the inanimate and animate worlds;

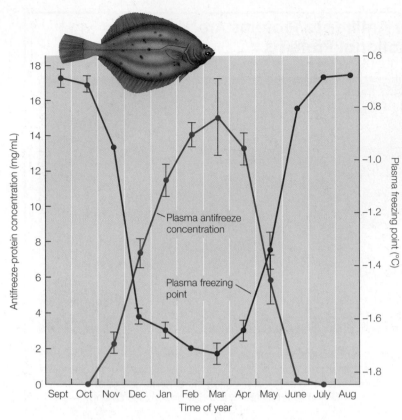

FIGURE 10.26 Seasonal changes in antifreeze protection in winter flounder (*Pleuronectes americanus*) The concentration of antifreeze protein in the blood plasma (blue line) rises as winter approaches, because of increased expression of the genes coding for antifreeze protein. The freezing point of the plasma is synchronously lowered (red line) and in winter is below the lowest winter temperatures the fish experience, ensuring protection against freezing. The winter flounder—an important commercial species—is named for the fact that it spawns in frigid waters in late winter or early spring. (After Fletcher et al. 1998.)

animals do not *cause* themselves to supercool. However, animals can modify their probabilities of spontaneous freezing during supercooling. Many animals, in fact, undergo adjustments whereby they exhibit low probabilities of spontaneous freezing even when they are supercooled to temperatures far below their freezing points. At one level of understanding, animals achieve this result by altering the quality or quantity of **ice-nucleating agents** in their bodies. Ice-nucleating agents are dissolved or undissolved substances that act as foci for the initiation of freezing. An animal containing an abundance of ice-nucleating agents may freeze when its body temperature is barely below the freezing point of its body fluids. In contrast, an animal that has substantially cleansed its body of ice-nucleating agents may have a supercooling point that is far below its freezing point.

Extensive supercooling is the principal means by which the overwintering life stages of many freezing-intolerant species of insects survive, and it is among these insects that the greatest known capacities to supercool are found. An ability to supercool to 20°–25°C below the freezing point of the body fluids—and remain unfrozen because of supercooling for prolonged periods of time—is about average for such insects, and prolonged supercooling to 30°–35°C

below the freezing point is not uncommon. At the extreme, there are now several known examples of insects that remain unfrozen at –50°C to –65°C by virtue of extensive supercooling, combined with antifreeze depression of their freezing points. These species can overwinter, unfrozen, in exposed microhabitats, such as plant stems, in some of the most severe climates on Earth.[20]

Less extreme supercooling is employed by a variety of other types of animals to avoid freezing. For example, some species of *deep-water* marine teleost fish found in polar seas have been shown to have freezing points of about –1.0°C, yet they swim about unfrozen in waters that have a temperature of about –1.9°C. Supercooled fish in deep waters are unlikely to encounter floating ice crystals that might induce them to freeze.

TOLERANCE OF FREEZING An ability to survive extracellular freezing is far more widespread than was appreciated even 35 years ago. In the intertidal zone along ocean shores at high latitudes, sessile or slow-moving invertebrates clinging to rocks frequently experience freezing conditions when exposed to the air during winter low tides (see Figure 10.24). Many of these animals—including certain barnacles, mussels, and snails—actually freeze and survive; some tolerate solidification of 60%–80% of their body water as ice. Increasing numbers of insect species are also known to tolerate freezing of their blood; tolerance of freezing is probably the most common overwintering strategy of Arctic insects, and some survive temperatures lower than –50°C in their frozen state. One of the extreme examples is a larval insect (*Gynaephora*)—one of the type called woolly bears—that lives in places such as Ellesmere Island in the Arctic. These woolly bears live for many years as larvae and thus must survive many winters before they can metamorphose into adults. They overwinter, frozen, in relatively exposed sites, tolerating body temperatures as low as –70°C! Certain amphibians that overwinter on land, notably wood frogs (*Rana sylvatica*) and spring peepers (*Hyla crucifer*), survive freezing at body temperatures of –2°C to –9°C (**FIGURE 10.27**)—or at even lower temperatures (e.g., –16°C) in Alaskan populations.

For freezing-tolerant animals, whereas intracellular freezing is destructive, extracellular freezing is safe and helps prevent intracellular freezing (see Figure 10.25). These animals commonly undergo physiological changes in winter that *limit the degree of supercooling* that is possible in their extracellular fluids—thereby promoting freezing in the extracellular fluids, where the freezing is safe. Some synthesize ice-nucleating agents (e.g., proteins or lipoproteins) and add them to their extracellular fluids. In some cases, the animals expose themselves to environmental ice and have body surfaces that permit external ice to induce freezing (inoculative freezing) of their extracellular fluids.

The ability of animals to *tolerate* freezing depends in part on the addition of certain organic solutes to their body fluids. Polyhydric alcohols (principally glycerol) are the primary organic solutes promoting tolerance of freezing in insects. Glucose and glycerol

[20] Disruption of supercooling is a potential tool for insect control. Some bacteria and other microbes are known to act as highly effective ice-nucleating agents. Such microbes are being investigated as biological control agents against insect pests that depend on supercooling for winter survival.

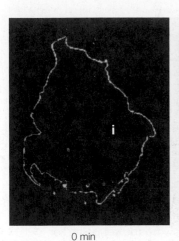 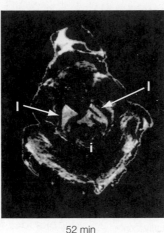 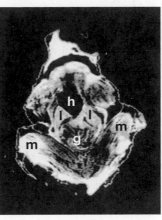

0 min   52 min   87 min

FIGURE 10.27 Frozen wood frogs (*Rana sylvatica*) thaw approximately synchronously throughout the body Magnetic resonance imaging (MRI) was used to detect the state of water in the body of a thawing wood frog. In the images, ice is dark; liquid water is light. The images were taken at specified times after the frozen frog was placed at +4°C. Contrary to what might be expected, wood frogs do not thaw from outside to inside. Instead, deep and superficial regions of a frog's body thaw approximately simultaneously, probably because deep regions have lower melting points than superficial ones have. Synchronous thawing may ensure that blood flow to thawed tissues can start promptly. g = gut; h = heart; i = ice; l = liver; m = leg muscle. (Photographs used with gratitude; from Rubinsky et al. 1994.)

are the solutes of primary importance in most freezing-tolerant amphibians. These organic solutes enter both the intracellular and the extracellular fluids of the animals, thereby increasing the amount of solute in both places. The increased solute in the extracellular fluids limits the amount of extracellular ice formation that occurs before the concentration of the unfrozen extracellular fluids rises high enough to prevent further freezing (see Figure 10.25B). The increased solute in the intracellular fluids limits the amount of water that must be lost from the cells for cells to come to osmotic equilibrium with freeze-concentrated extracellular fluids (see Figure 10.25C), thus limiting cell shrinkage. These are thought to be some of the principal ways by which the organic solutes aid the tolerance of freezing by poikilotherms.

SUMMARY
Poikilothermy (Ectothermy)

■ Poikilotherms, also called ectotherms, are animals in which body temperature (T_B) is determined by equilibration with external thermal conditions. They often thermoregulate. Their mechanism of thermoregulation is behavioral; a poikilotherm controls its T_B by positioning its body in environments that will bring its T_B to the set-point ("preferred") level.

■ The resting metabolic rate of a poikilotherm is usually an approximately exponential function of its T_B. The Q_{10} is typically 2–3. The metabolism–temperature curves of poikilotherms are often plotted on semilogarithmic coordinates because exponential functions are straight on such coordinates.

■ From the viewpoint of metabolic rate, when poikilotherms acclimate to cold or acclimatize to low-temperature environments in nature, their most common response is partial compensation. Partial compensation returns an animal's metabolic rate toward the level that prevailed prior to the change in environment, and thus it blunts the effect of environmental change. The most common known mechanism of partial compensation is for cells to change their concentrations of key, rate-limiting enzymes.

■ Different species of poikilotherms that have long evolutionary histories of living at different body temperatures frequently display evolved physiological differences that suit them to function best at their respective body temperatures. Species of lizards sprint fastest at their respective preferred body temperatures, and polar species of fish function at higher rates in frigid waters than temperate-zone species can. The important mechanisms of evolutionary adaptation to different body temperatures include molecular specialization: Species with evolutionary histories in different environments often synthesize different molecular forms of protein molecules and different cell-membrane phospholipids. The evolution of *structurally* distinct protein forms and phospholipids conserves *functional* properties of the molecules; as a consequence, species living in different thermal environments are similar to each other in their enzyme–substrate affinities and membrane-lipid fluidities.

■ When exposed to threat of freezing, some poikilotherms actually freeze and are freezing-tolerant; freezing must be limited to the extracellular body fluids, however. Other poikilotherms are freezing-intolerant and exploit one of three strategies—behavioral avoidance, antifreeze production, or supercooling—to avoid freezing. Antifreezes lower the freezing point. Stabilization of supercooling permits animals to remain unfrozen while at temperatures below their freezing points.

Homeothermy in Mammals and Birds

Homeothermy, the regulation of body temperature by physiological means, gives mammals and birds a great deal more independence from external thermal conditions than is observed in lizards, frogs, or other poikilotherms. On a cool, cloudy day, a lizard or other behaviorally thermoregulating poikilotherm may be unable to reach its preferred body temperature, because warming in such animals depends on a source of outside heat. A mammal or bird, however, produces its own heat for thermoregulation and thus can maintain its usual body temperature whether the environment is warm, moderately cold, or subfreezing.

BOX 10.3 Thermoregulatory Control, Fever, and Behavioral Fever

Of all the physiological control systems, the system for thermoregulation is the one that usually seems the most straightforward conceptually. Virtually every introductory treatment of control theory in physiology uses the thermoregulatory control system as its central example. This is undoubtedly true because analogies can so readily be drawn with engineered thermal control systems, which are common in our everyday lives.

In a house with a furnace and air conditioner, the thermostat controls heat production by the furnace and heat removal by the air conditioner to maintain a stable air temperature. Using the terminology of control theory to describe this system, the air temperature is the **controlled variable** (see Box 1.1), and the furnace and air conditioner are **effectors**, instruments that are capable of changing the controlled variable. The thermostat itself actually includes three separate elements that are essential for a control system:

1. A **sensor**, a device that can measure the controlled variable so that the control system knows its current level (the current air temperature).

2. A **set point** or **reference signal**. The set point is a type of information that remains constant in a control system even as the controlled variable goes up and down, and that tells the system the desired level of the controlled variable. We usually call the set point of a home thermostat its "setting." If, for example, we "set" the thermostat to 20°C, the device is able to retain that set-point information in an invariant form, so that the air temperature detected by the sensor can be compared with it. An important point to recognize is that a thermostat does not remember its set point by having inside it an object that is kept literally at the set-point temperature. Instead, the set-point temperature is *represented* in the thermostat by a physical system that is not a temperature, but corresponds to a temperature.

3. A **controller**, a mechanism that compares the set point with the current level of the controlled variable to decide whether the controlled variable is too high or low.

The control system in a house, considered as a whole, operates as a **negative feedback** system (see Box 1.1). It controls the effectors to bring the controlled variable back toward the set point. For example, if the air temperature goes below the set point, the furnace is commanded to add heat to the house.

By analogy, it is easy to describe the thermoregulatory control system of a lizard or mammal (or any other thermoregulator) in terms of the same basic concepts. The principal *effectors* in a lizard (a behavioral thermoregulator) are the skeletal muscles that move the limbs and control posture. Effectors in a mammal include muscle cells that can produce heat by shivering, sweat glands that can promote evaporative cooling, hair-erector muscles that determine how fluffed the pelage is,

---

Mammals and birds *independently* evolved the full-fledged forms of homeothermy they exhibit today. Although the extent of convergence in their physiology of homeothermy is remarkable, they also exhibit consistent differences, one being in their average body temperatures.

Placental mammals typically maintain deep-body temperatures averaging about 37°C when they are at rest and not under heat or cold stress.[21] Birds maintain higher temperatures under similar conditions: about 39°C. One of the most remarkable attributes of mammals and birds is that in both groups, the average body temperatures of thermally unstressed animals do not vary much with climate. One might expect, for instance, that species of mammals living in the Arctic would have lower average body temperatures than related species living in the tropics. Actually, however, differences of this sort are slight, if present at all.

Deep-body temperature is not absolutely constant. Daily cycles occur; the body temperatures of mammals and birds are typically 1°–2°C higher during their active phases each day than during their resting phases. Moreover, in some species the body temperature is permitted to rise when individuals are under heat stress, or it is permitted to fall in winter.

Regardless of the variations that occur, the body temperatures of mammals and birds are among the most stable in the animal kingdom. Thus one consequence of homeothermy is that cellular functions are able to be specialized to take place *especially reliably* at certain temperatures. However, as we will see, homeothermy has a very high energy cost and greatly increases the food requirements of mammals and birds in comparison with like-sized nonavian reptiles or fish.

Thermoregulation by a mammal or bird requires neurons (nerve cells) that sense the current body temperature and also requires thermoregulatory control centers in the brain that, by processing thermal sensory information, properly orchestrate the use of heat-producing and heat-voiding mechanisms in ways that stabilize the body temperature (**BOX 10.3**). The detection of body temperature in a mammal or bird occurs in multiple parts of the body; thermosensitive neurons of importance are found in the skin, spinal cord, and brain, and sometimes also in specialized locations such as the scrotum. The principal control centers—which process the multiple sensory inputs and command the thermoregulatory mechanisms—are located in the hypothalamus and the associated preoptic regions of the brain.

A behavioral thermoregulator like a fish or a lizard must also have thermoregulatory control centers that receive and process thermosensory information and that orchestrate the processes of thermoregulation (see Box 10.3). Physiologists generally hypothesize that during the course of vertebrate evolution, there has been continuity in the control centers. According to this hypothesis, the

---

[21] Marsupials, some of the primitive placental mammals, and especially monotremes have lower body temperatures; the platypus, for example, exhibits a deep-body temperature of 30°–33°C.

and so forth. A lizard or mammal has multiple *sensors*: temperature-sensitive neurons that measure the current temperatures of the skin, spinal cord, and brain. These sensors send their temperature data to a *controller* in the brain that compares the current temperatures with a *set point* to decide what to do. The exact natures of the controller and set point remain far from fully understood because they consist of many tiny neurons in the depths of the brain. As in the case of the home thermostat, however, we recognize that the set point is not literally a temperature in the brain but is *represented* in some way by neurons.

If we disregard the uncertainties that exist about the nature of the set point and simply use the terminology of control theory, the set point of a lizard or mammal can be adjusted to different "settings" at different times, just as the setting of a home thermostat can be adjusted. **Fever** in a mammal provides an elegant example of resetting of the thermostat (see the figure). Fish, amphibians, and nonavian reptiles sometimes develop

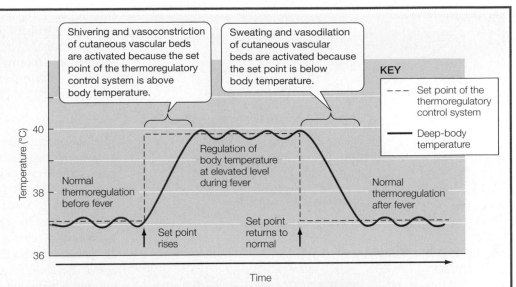

A bout of fever in a placental mammal: The relation between the set point of the thermoregulatory control system and body temperature When the set point jumps up at the start of a bout of fever and falls back down at the end, the thermoregulatory control system detects the mismatch between the set point and the body temperature and commands vigorous effector responses to correct the mismatches. These responses include shivering at the start and sweating at the end.

fevers. These are called **behavioral fevers** because the effectors causing elevated body temperatures are skeletal muscles that modify behaviors. The animals move to warmer environments so their body

temperature is higher. **Box Extension 10.3** discusses thermoregulatory control and especially fever in more detail.

---

control centers were already present in a rudimentary way when the only vertebrates were fish and the only thermoregulatory mechanisms to be controlled were behaviors. Recognizing that some modern lizards pant, change color to aid their thermoregulation, or employ other physiological mechanisms, physiologists usually conclude that the early reptilian ancestors of mammals and birds probably had some physiological thermoregulatory mechanisms that supplemented their dominant behavioral mechanisms. The control centers of those early reptiles would therefore have had both physiological and behavioral mechanisms to control. Then, as mammals and birds appeared, the control centers assumed control of predominantly physiological mechanisms. There is some evidence for this sort of scenario—with the evidence from the study of fever being particularly intriguing (see Box 10.3).

A comparison of modern nonavian reptiles with mammals and birds suggests that the single most revolutionary step that occurred in the evolution of mammalian and avian homeothermy was the evolution of endothermy. Modern lizards, turtles, crocodilians, and snakes (with isolated exceptions) cannot warm their bodies by metabolic heat production. Mammals and birds, in dramatic contrast, have an endogenous ability to stay warm in cold environments because of endothermy. With endothermy plus their physiological mechanisms of keeping cool in hot environments, mammals and birds are able to maintain relatively constant tissue temperatures over exceedingly wide ranges of environmental conditions.

Metabolic rate rises in cold and hot environments because of the costs of homeothermy

The resting metabolic rate of a mammal or bird typically varies with ambient temperature, as shown in **FIGURE 10.28**. Within a certain range of ambient temperatures known as the **thermoneutral zone (TNZ)**, an animal's resting metabolic rate is independent of ambient temperature and constant. The lowest ambient temperature in the TNZ is termed the **lower-critical temperature**; the highest is the **upper-critical temperature**.[22] The lower-critical and upper-critical temperatures depend on the species, and they can also be affected by acclimation or acclimatization. An animal's **basal metabolic rate (BMR)** is its metabolic rate when resting and fasting[23] in its thermoneutral zone.

The resting metabolic rate of a mammal or bird increases as the ambient temperature falls below the animal's lower-critical temperature or rises above its upper-critical temperature (see Figure 10.28A). These increases in metabolic rate in both cold and warm environments arise from the animal's need to perform physiologi-

---

[22] The meaning of *critical temperature* in the study of homeotherms is very different from the meaning of the same term in the study of poikilotherms (see Figure 10.16).

[23] In this context, *fasting* means that the animal has not eaten for long enough that the specific dynamic action of its last meal has ended. *Postabsorptive* is a synonym.

(A) The general relation

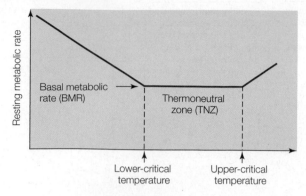

(B) An example

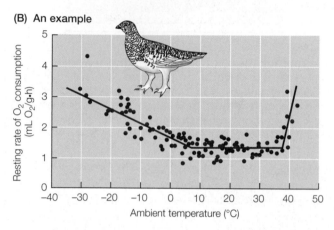

FIGURE 10.28 The relation between resting metabolic rate and ambient temperature in mammals and birds (A) The general relation and the terminology used to describe it. (B) An example, the metabolism–temperature relation of the white-tailed ptarmigan (*Lagopus leucurus*). (B after Johnson 1968.)

cal work to keep its deep-body temperature constant regardless of whether the ambient temperature is low or high.

The shape of the metabolism–temperature curve depends on fundamental heat-exchange principles

Before we study the physiological mechanisms used by mammals and birds to thermoregulate, it is important to analyze why the metabolism–temperature curve has the specific shape it does. A useful first step for this analysis is to recognize the concept of **dry heat transfer**, defined to be heat transfer that does not involve the evaporation (or condensation) of water. Dry heat transfer occurs by conduction, convection, and thermal radiation. As stressed earlier, these three mechanisms of heat transfer must be analyzed separately in heterogeneous thermal environments. However, in a *uniform thermal environment* such as a laboratory cage or test chamber, where the radiant temperatures of all environmental objects are typically similar to air temperature, these three mechanisms of heat transfer can be meaningfully lumped together. They can be lumped in this circumstance because, in all three cases, the rate of heat transfer between an animal and its environment tends to increase approximately in proportion to the difference in tempera-

ture between the animal's body and the environment ($T_B - T_A$) (see Figure 10.4). By lumping the three together, we can study dry heat transfer as a whole (because dry heat transfer = conduction + convection + thermal-radiation heat transfer).

In a uniform thermal environment, if factors other than the environmental temperature (T_A) are held constant,

$$\text{Rate of dry heat transfer} \propto T_B - T_A \qquad (10.8)$$

Heat moves out of an animal's body by dry heat transfer when T_B exceeds T_A; conversely, heat moves into the body when T_B is less than T_A. The rate of dry heat transfer is proportional to $(T_B - T_A)$ in either case, and thus $(T_B - T_A)$ can be thought of as being the "driving force" for dry heat transfer.

To analyze the shape of the metabolism–temperature curve, Equation 10.8 *taken by itself* can be used at ambient temperatures that are *within* and *below* the thermoneutral zone (TNZ). At temperatures above the TNZ, evaporative heat transfer is too important to be ignored. The body temperature of a mammal or bird is typically higher than the animal's upper-critical temperature. Thus, when the ambient temperature is within or below the TNZ, $(T_B - T_A)$ is always positive, and dry heat transfer carries heat out of the body at a rate predicted by Equation 10.8.

Under such conditions, in which an animal is losing heat to its environment, the only way the animal can maintain a constant body temperature is to make heat *metabolically* at a rate that matches its rate of heat loss. Thus, if we use M to symbolize the animal's metabolic rate, M must equal the animal's rate of heat loss. Based on Equation 10.8, therefore, at ambient temperatures within and below the TNZ, $M \propto (T_B - T_A)$. We can rewrite this expression as an equation by introducing a proportionality coefficient (C):

$$M = C\,(T_B - T_A) \qquad (10.9)$$

This equation, which is a famous equation for analyzing a mammal's or bird's thermal relations, is called the **linear heat-transfer equation**, also described sometimes as **Newton's law of cooling** or **Fourier's law of heat flow**. The coefficient C, which is termed the animal's **thermal conductance**, is a measure of *how readily* heat can move by dry heat transfer from an animal's body into its environment.

To see the significance of C, suppose that two placental mammals are in the same environment and therefore have the same driving force for dry heat loss ($T_B - T_A$), but one has a higher thermal conductance than the other. The one with the higher C will lose heat faster because heat can move out of its body more readily than heat can move out of the body of the other. Therefore the one with the higher C will require a higher metabolic rate to stay warm.

An animal with a *high C* can be thought of as having a *low resistance* to dry heat loss. Conversely, an animal with a *low C* can be thought of as having a *high resistance* to dry heat loss. Physiologists, accordingly, define an animal's **resistance to dry heat loss** to be the inverse of C: $1/C$. The resistance to dry heat loss is often called **insulation** (*I*). Thus $I = 1/C$. The linear heat-transfer equation can therefore also be written as:[24]

---

[24] In this form, the linear heat-transfer equation bears a close similarity to Ohm's Law. ($T_B - T_A$), the driving force, is analogous to potential difference (voltage); I is analogous to electrical resistance; and M is analogous to current flow. Current = voltage/resistance.

$$M = \frac{1}{I}(T_B - T_A) \qquad \textbf{(10.10)}$$

An important point to note about the concept of insulation (I) introduced here is that it is not simply a measure of the heat-retaining properties of the fur or feathers. Instead, insulation (I) is a measure of an animal's *overall* resistance to dry heat loss. For instance, because both posture and fur affect a mammal's resistance to dry heat loss, the value of I for a mammal depends on its posture as well as its fur (and also on additional factors).

THE THERMONEUTRAL ZONE: INSULATION IS MODULATED TO KEEP THE RATE OF HEAT LOSS CONSTANT Let's now use the concepts we have developed to understand why the metabolism–temperature curve of a mammal or bird has the shape it does, starting with the thermoneutral zone. The defining property of the TNZ is that an animal's metabolic rate (M) remains constant at all the different ambient temperatures in the TNZ. This property probably seems impossible or paradoxical at first. After all, if T_A changes, then ($T_B - T_A$) changes, and Equation 10.10 suggests that M would have to change. The answer to this paradox is that in its TNZ, a mammal or bird *varies* its insulation. *Modulation of insulation against a background of constant metabolic heat production is the principal means by which a mammal or bird thermoregulates in its thermoneutral zone.*

Let's discuss this key concept in more detail. As the ambient temperature is lowered in the TNZ and ($T_B - T_A$) accordingly becomes greater, a mammal or bird responds by increasing its insulation, I.[25] This increase in the animal's resistance to heat loss *counterbalances* the increase in the *driving force* for heat loss, ($T_B - T_A$), so that the animal's *actual rate of heat loss* remains constant (or nearly so). The animal's rate of metabolic heat production, therefore, can also remain constant. These points are mathematically apparent in Equation 10.10. In the TNZ, as T_A decreases and ($T_B - T_A$) therefore increases, I is increased in a precisely counterbalancing way so that the ratio ($T_B - T_A$)/I remains constant. The metabolic rate of the animal, M, can therefore be constant.

The *width* of the TNZ varies enormously from species to species, depending in part on the extent to which various species are able to modulate their insulation. Small-bodied species tend to have narrower TNZs than large-bodied species do. Species of mice, for instance, often have TNZs extending only from about 30°C to 35°C. At another extreme, Eskimo dogs have a TNZ extending from –25°C to +30°C—a range of 55°C!

TEMPERATURES BELOW THERMONEUTRALITY Unlike the case within the thermoneutral zone, *the principal means by which a mammal or bird thermoregulates at ambient temperatures below thermoneutrality is modulation of its rate of metabolic heat production.* Specifically, below the TNZ, as the environment becomes colder, a mammal or bird must raise its rate of metabolic heat production to higher and higher levels if it is to stay warm. In this way, mammals and birds closely resemble a furnace-heated house in which the furnace must increase the rate at which it produces heat as the air outside becomes colder.

What determines the lower-critical temperature? To see the answer, consider an animal, initially in its TNZ, that is subjected to a steadily declining ambient temperature. As T_A declines while remaining in the TNZ, the rate at which the animal loses heat to its environment stays constant because the animal increases its insulation, I. Insulation cannot be increased without limit, however. An animal's lower-critical temperature represents the T_A below which its insulatory adjustments become inadequate to counterbalance fully the increase in the driving force favoring heat loss. As T_A falls below the lower-critical temperature, the rate at which an animal loses heat increases, and the animal must therefore increase its rate of heat production to match the increased rate of heat loss.

The insulation of a mammal or bird sometimes becomes maximized at the lower-critical temperature. Cases like this are particularly straightforward to understand in terms of the linear heat-transfer equation (Equation 10.9 or 10.10).

If an animal maximizes its insulation at the lower-critical temperature, then its value of I at ambient temperatures below the TNZ is a *constant* (equaling its maximum value of I). In addition, because $C = 1/I$, the animal's value of C below the TNZ is a *constant* (equaling its minimum value of C). For such an animal, therefore, T_B, I, and C in the linear heat-transfer equation are all constants below the TNZ. Accordingly, the linear heat-transfer equation—whether written as Equation 10.9 or 10.10—is a simple linear equation (accounting for its name) that has two variables: M and T_A. If we plot M as a function of T_A for this linear equation—using Equation 10.9—we obtain a straight line having two particular properties, illustrated in **FIGURE 10.29A**: First, the slope of the line is $-C$. Second, the line intersects the x axis at the ambient temperature that is equal to T_B.

As a model of an animal's metabolism–temperature curve, the plot in Figure 10.29A is flawed because it ignores the fact that an animal's metabolic rate (M) does not truly fall below the basal level. **FIGURE 10.29B** is thus more realistic. By comparing Figures 10.29A and B, you can see that *the portion of an animal's metabolism–temperature curve below the TNZ is simply a plot of the linear heat-transfer equation.*

The fact that the slope of the metabolism–temperature curve below the TNZ is equal to $-C$ (for animals that maintain a constant C) provides a useful tool for the visual interpretation of metabolism–temperature curves. As shown in **FIGURE 10.29C**, if two otherwise similar animals differ in thermal conductance (C) below the TNZ, the relative slopes of their metabolism–temperature curves mirror their differences in C: The animal with a high value of C (low insulation) has a steeper slope than the animal with low C (high insulation). Using this principle, one can look at Figure 10.40B (see page 273), for example, and tell at a glance that the winter fox has lower conductance and higher insulation than the summer fox. Figure 10.29C also highlights the energy advantages of high insulation. Note that the animal with relatively high insulation (low C)—analogous to a well-insulated house—has a relatively low requirement for metabolic heat production and a low metabolic rate at any given ambient temperature below the TNZ.[26]

---

[25] Starting on page 265, we discuss the actual mechanisms of increasing insulation.

[26] Although the slopes of metabolism–temperature curves were used to calculate C quantitatively some years ago, better approaches for the calculation of C have been developed. Thus the use of slopes today should be reserved for just qualitative, visual interpretation.

(A) $M = C(T_B - T_A)$ with C and T_B constant ($T_B = 37°C$)

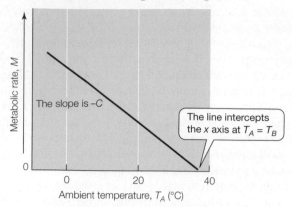

The slope is $-C$

The line intercepts the x axis at $T_A = T_B$

Ambient temperature, T_A (°C)

(B) The plot from **(A)**, recognizing that M actually falls only to the basal level

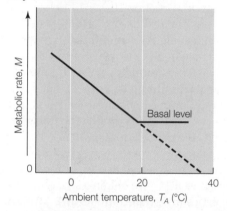

Basal level

Ambient temperature, T_A (°C)

(C) Comparison of two animals that differ in C below thermoneutrality

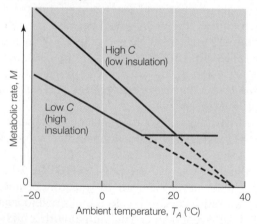

High C (low insulation)

Low C (high insulation)

Ambient temperature, T_A (°C)

FIGURE 10.29 A model of the relation between metabolic rate and ambient temperature in and below the thermoneutral zone

TEMPERATURES ABOVE THERMONEUTRALITY Mammals and birds employ two principal processes to respond to ambient temperatures above their thermoneutral zones:

■ Most mammals and birds actively increase the rate at which water evaporates from certain of their body surfaces, a process called **active evaporative cooling**. Sweating, panting, and gular fluttering (**FIGURE 10.30**) are the most common mechanisms of active evaporative cooling.

FIGURE 10.30 Gular fluttering is one means of actively increasing the rate of evaporative cooling During gular fluttering, which occurs in birds but not mammals, an animal holds its mouth open and vibrates the floor of the mouth, termed the gular area (arrow). In this way, airflow is increased across the moist, vascular mouth membranes, promoting a high rate of evaporation. The birds seen here are young great egrets (*Ardea albus*) on a hot day in Florida. In some species the gular area vibrates up and down at 800–1000 cycles/min during gular fluttering.

■ Some mammals and most birds allow their body temperatures to rise to unusually high levels, a phenomenon called **hyperthermia**.

Both active evaporative cooling and hyperthermia can cause an animal's metabolic rate to rise at temperatures above thermoneutrality. Active evaporative cooling causes a rise in metabolic rate because physiological work must be done to enhance water evaporation (panting, for example, requires an increase in the rate of breathing). Hyperthermia can also cause a rise in metabolic rate because tissues tend to accelerate their metabolism when they are warmed; according to recent research, hyperthermia does not always cause metabolic acceleration in mammals and birds, but in some cases it does.

To appreciate more fully the processes at work above the TNZ, it is informative to take a dynamic approach by considering an animal that is initially within its TNZ and subjected to a steadily increasing ambient temperature. As T_A rises, the driving force for dry heat loss ($T_B - T_A$) decreases, meaning that the animal faces a greater and greater challenge to get rid of its *basal* metabolic heat production. While T_A remains in the TNZ, the animal responds to the rising T_A by decreasing its resistance to dry heat loss, its insulation. Consequently, even high in the TNZ, metabolic heat is carried away as fast as it is produced by a combination of dry heat transfer and *passive* evaporation. This handy state of affairs comes to an end when T_A reaches the upper-critical temperature and goes higher. Near the upper-critical temperature, insulation either reaches its minimum or at least becomes incapable of sufficient further reduction to offset additional decreases in $(T_B - T_A)$. Thus as T_A rises above the upper-critical temperature, the rate of dry heat loss tends to fall too low for the combination of dry heat loss and passive evaporation to void metabolic heat. Both of the principal responses of mammals and birds—active evaporative cooling and hyperthermia—serve to promote heat loss so that animals are not overheated by their metabolic heat production. Hyperthermia does this because a rise in T_B increases the driving force for dry heat loss $(T_B - T_A)$.

If T_A keeps rising and becomes so high that it exceeds T_B, heat stress becomes extraordinary because—when T_A is above T_B—dry heat transfer carries environmental heat *into* the body![27] Then active evaporative cooling must assume the entire burden of removing heat from the body.

From a quick glance at the metabolism–temperature curve above the TNZ (see Figure 10.28), it may seem extremely paradoxical that a mammal or bird increases its metabolic rate—its rate of internal heat production—when it is under heat stress. To understand this paradox, it is important to recall the very large amount of heat carried away by the evaporation of each gram of water (see page 239). Although an animal must increase its metabolic rate to pant, gular flutter, or otherwise actively increase its rate of evaporation, the amount of heat carried away by the evaporation of each gram of water far exceeds the heat produced per gram by the physiological processes that accelerate evaporation.

Homeothermy is metabolically expensive

One of the most important attributes of homeothermy in mammals and birds is that it is metabolically expensive in comparison with vertebrate poikilothermy. Homeothermy in mammals and birds in fact provides an outstanding example of a point stressed in Chapter 1: When physiological *regulation* and *conformity* are compared, the greatest downside of regulation is that its energy costs are high.

To quantify the cost of homeothermy, physiologists have compared the metabolic rates of vertebrate homeotherms and poikilotherms at similar tissue temperatures. Specifically, they have compared the basal metabolic rates of mammals and birds with the resting metabolic rates of like-sized poikilotherms held at the same body temperatures as the mammals and birds. A typical experiment would be to obtain a 100-g placental mammal and place it in its thermoneutral zone, and simultaneously obtain a 100-g lizard and place it in a chamber at 37°C so that its body temperature matches that of the mammal. If both animals are at rest and fasting and you measure their metabolic rates, you will obtain (1) the basal metabolic rate (BMR) of the mammal, and (2) the standard metabolic rate (SMR) of the lizard *at mammalian body temperature*. Typically what you will find is that the metabolic rate of the mammal is four to ten times higher than that of the lizard, even though the cells of the two animals are at one temperature and the mammal's metabolic rate under these conditions is its *minimum* rate! Many studies of this sort have been carried out on a variety of species, and they have confirmed repeatedly that the BMRs of mammals and birds are four to ten times the SMRs of poikilothermic vertebrates at mammalian or avian body temperatures (see Figure 7.9). *Metabolic intensity stepped up dramatically when vertebrates evolved homeothermy.*

If mammals, birds, and poikilothermic vertebrates studied as we have just described are transferred to cold ambient temperatures, the metabolic rates of the mammals and birds *rise* (see Figure 10.28), whereas the metabolic rates of the poikilotherms *fall* (see Figure 10.9A). At cold ambient temperatures, therefore, the difference in metabolic intensity between homeotherms and poikilotherms is far greater than just four- to tenfold.

Animals living in the wild experience both high and low ambient temperatures at various times. Their *average* metabolic rates thereby integrate the effects of different temperatures. As discussed in Chapter 9, field metabolic rates have now been measured in many free-living terrestrial vertebrates by use of the doubly labeled water method. Those measures reveal that the average field metabolic rate is typically 12–20 times higher in mammals and birds than in lizards or other nonavian reptiles of the same body size! The mammals and birds must therefore acquire food at a much higher rate.

Insulation is modulated by adjustments of the pelage or plumage, blood flow, and posture

Now we turn (in this section and several that follow) to the *mechanisms* that mammals and birds employ to thermoregulate physiologically. First we discuss the mechanisms by which mammals and birds modulate their resistance to dry heat transfer, their insulation. As we have seen, these are the predominant mechanisms of thermoregulation within the thermoneutral zone.

One means of varying insulation is erection or compression of the hairs or feathers. Each hair or feather can be held upright or allowed to lie flat against the skin by the contraction or relaxation of a tiny muscle at its base, under control of the sympathetic nervous system. These responses are termed **pilomotor responses** in mammals and **ptilomotor responses** in birds. If the ambient temperature declines within the TNZ, the hairs or feathers are erected to an increased degree. In this way the pelage or plumage is fluffed out and traps a thicker layer of relatively motionless air around the animal, thereby increasing the resistance to heat transfer through the pelage or plumage (see Equation 10.1).

Another mechanism of modulating insulation is the use of **vasomotor responses** in blood vessels (see page 680)—responses that alter the rate of blood flow to the skin surface and other superficial parts of the body. Arterioles supplying superficial vascular beds are constricted at cool ambient temperatures because of stimulation by the sympathetic nervous system. This response retards transport of heat to the body surfaces by blood flow. Conversely, vasodilation at warm ambient temperatures enhances blood transport of heat to body surfaces where the heat is readily lost.

Insulation may also be modified by **postural responses** that alter the amount of body surface area directly exposed to ambient conditions. At low ambient temperatures, for example, mammals often curl up, and some birds tuck their heads under their body feathers or squat so as to enclose their legs in their ventral plumage. Many birds hold their wings away from their bodies when ambient temperatures are high.

---

[27] For T_A to exceed T_B, it must rise substantially above the upper-critical temperature. When T_A is just moderately above the upper-critical temperature, it is typically below T_B.

In addition to the insulative properties that can be modulated by an individual animal, there are also properties that affect insulation but are more or less fixed for any given individual. Outstanding among these is body size. At temperatures *below thermoneutrality*, small size tends to increase the weight-specific rate at which animals lose heat—and thus the weight-specific cost of thermoregulation—because relatively small animals have more body surface per unit of weight than large ones have.[28] Another reason that small size tends to enhance heat loss is that small animals cannot have as thick pelage or plumage as large ones. Whereas large mammals commonly have pelage that is at least 5–6 cm thick, mice could not conceivably have such thick pelage; a mouse with 5–6 cm of pelage would be trapped inside its own hair!

Heat production is increased below thermoneutrality by shivering and nonshivering thermogenesis

When a mammal or bird is below its lower-critical temperature, it must increase its rate of heat production as the ambient temperature declines. Although all metabolic processes produce heat as a by-product, mammals and birds have evolved mechanisms, termed **thermogenic mechanisms**, that are specialized to generate heat for thermoregulation. One of these, shivering, is universal in adult mammals and birds.

SHIVERING   **Shivering** is unsynchronized contraction and relaxation of skeletal-muscle motor units in high-frequency rhythms, mediated by motor neurons (nerve cells) of the somatic nervous system. Skeletal muscles can basically contract in two patterns. When muscles are being employed in locomotion to move a limb, all the motor units in each muscle contract synchronously, and antagonistic muscles contract in ways that they do not work against each other. When the same muscles are employed in shivering, various motor units within each muscle contract more or less at random relative to each other, antagonistic muscles are activated simultaneously, and the muscles quiver. Either mode of contraction uses ATP and liberates heat. When a muscle shivers, the conversion of ATP-bond energy to heat becomes the primary function of contraction because no useful mechanical work is accomplished.

NONSHIVERING THERMOGENESIS   The concept of **nonshivering thermogenesis (NST)** is most readily understood by taking a look at the classic studies on laboratory rats that originally led to the discovery of NST. If lab rats that have been living at warm temperatures are transferred to a room at 6°C, they shiver violently during their first days there. If one observes them visually over the next few weeks as they acclimate to 6°C, however, they appear gradually to stop shivering even though they continue to maintain elevated metabolic rates. This visual observation suggests that during acclimation to cold, the rats develop mechanisms of thermogenesis that do not involve shivering. To test if this is in fact the case, cold-acclimated rats can be injected with curare, a plant extract that blocks the contraction of skeletal muscle and therefore prevents shivering. Curare-injected, cold-acclimated rats continue to have

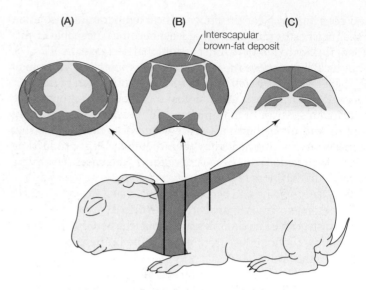

FIGURE 10.31 The deposits of brown adipose tissue in a newborn rabbit (A), (B), and (C) are cross sections of the body at the positions indicated on the side view. Brown adipose tissue also occurs typically in discrete deposits in adult mammals that have the tissue. (After Dawkins and Hull 1964.)

elevated metabolic rates and thermoregulate, confirming that they have well-developed nonshivering thermogenic mechanisms.

Whereas shivering is universal in mammals and birds, NST is not. NST is best known, and very common, in placental mammals. It has been reported in the young of a few species of birds (e.g., ducklings), but its occurrence in adult birds remains controversial.

Of all the possible sites of NST in placental mammals, the one that is best understood and dominant is **brown adipose tissue (BAT)**, also called **brown fat**.[29] Although, like white fat, this is considered a type of "adipose tissue" or "fat," brown fat does not develop from the same type of precursor cells as white fat during the embryonic development of mammals. Instead brown fat develops from a type of precursor cell that also gives rise to skeletal muscle! Correlated with these different developmental origins, brown fat and white fat differ greatly in both structure and function. Brown fat—named for the fact that it is often reddish brown—tends to occur in discrete masses, located in such parts of the body as the interscapular region, neck, axillae, and abdomen (**FIGURE 10.31**). Deposits of BAT receive a rich supply of blood vessels and are well innervated by the sympathetic nervous system. The cells of BAT are distinguished by having great numbers of relatively large mitochondria (just as skeletal muscle, similar in its developmental origin, is rich in mitochondria). The rich, red blood supply of BAT and the abundant, yellow cytochrome pigments in its mitochondria impart to the tissue its distinctive color.[30]

---

[28] Be certain that you do not extend this argument to the thermoneutral zone. The argument is valid below thermoneutrality but probably does not apply in the thermoneutral zone (see page 183).

[29] BAT does not occur in birds. Nor does uncoupling protein 1, which we soon discuss. In young birds that show NST, skeletal muscles are apparently the NST site.

[30] Evidence has accumulated in just the past 5 years that BAT-like cells can appear within deposits of white fat in human adults. The discovery is revolutionary, because although human infants have dramatic BAT deposits (see Figure 11.9), BAT has often been considered absent in human adults. One reason the new discovery is important is that the BAT-like cells could be involved in body-weight control by oxidizing excess body fat.

When the sympathetic nervous system releases norepinephrine in BAT, the BAT responds by greatly increasing its rate of oxidation of its stored lipids, resulting in a high rate of heat production. BAT is biochemically specialized to undergo uncoupling of oxidative phosphorylation from electron transport (see Figure 8.4C) and uses this mechanism to produce heat rapidly. Uncoupling does two things that result in rapid heat production: (1) It suspends the ordinary controls on the rate of aerobic catabolism, permitting unbridled rates of lipid oxidation; and (2) it causes the chemical-bond energy of oxidized lipid molecules to be released immediately as heat (rather than being stored in ATP). The property that gives BAT its specialized ability to undergo uncoupling is that BAT expresses a distinctive proton-transport protein, **uncoupling protein 1** (**UCP1**; **thermogenin**), in the inner membranes of its mitochondria (see Figure 8.4C).

Norepinephrine released in BAT binds to β-adrenergic receptors (and other receptors) in the cell membranes of the BAT cells. These receptors are G protein–coupled receptors; as discussed in Chapter 2 (see Figure 2.27), the binding of norepinephrine to the receptors activates G proteins in the cell membranes and leads to the intracellular production of the second messenger cyclic AMP. Cyclic AMP then activates (by phosphorylation) an intracellular lipase enzyme that rapidly hydrolyzes triacylglycerols stored in the cells to release free-fatty-acid fuels for mitochondrial oxidation. Simultaneously, by a mechanism that remains ambiguous, existing molecules of the uncoupling protein UCP1 are activated, and thus the mitochondria carry out the lipid oxidation in an uncoupled state. In addition, if norepinephrine stimulation continues for tens of minutes or longer, increased amounts of UCP1 are synthesized because β-adrenergic activation stimulates increased transcription of the gene that encodes UCP1. Still another effect of stimulation of BAT is that a fatty acid transport protein (FATP) in cell membranes is upregulated, permitting the cells to carry out rapid uptake of fatty acids brought from elsewhere in the body.

Brown fat—like NST in general—is particularly prominent in three types of placental mammals: (1) cold-acclimated or winter-acclimatized adults (particularly in species of small to moderate body size), (2) hibernators, and (3) newborn individuals (see Figures 10.31 and 11.9) In mice, rats, and other small- to medium-sized species, when adults acclimate to cold or acclimatize to winter, their BAT often markedly increases its potential to produce heat; deposits of the tissue grow, BAT cells increase their numbers of mitochondria, and the mitochondria become richer in uncoupling protein. In part, this development of BAT probably serves to free the skeletal muscles to perform exercise. A muscle cannot shiver and exercise at the same time; during acclimation to cold or acclimatization to winter, as BAT develops, muscles are less likely to need to be employed in shivering, leaving them free to be used in exercise. BAT in newborns (including human babies) and hibernators is discussed further in Chapter 11.

Regional heterothermy: In cold environments, allowing some tissues to cool can have advantages

Appendages such as legs, tails, and ear pinnae present particular thermal challenges when mammals and birds are below thermoneutrality. The appendages are potentially major sites of heat loss because they have a great deal of surface area relative to their sizes,

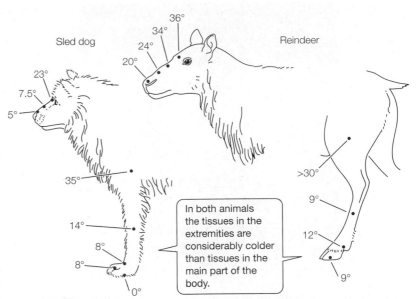

FIGURE 10.32 Regional heterothermy in Alaskan mammals The air temperature was –30°C when these data were gathered. The temperatures shown are subcutaneous temperatures (°C) at various locations on the body. Both animals had deep-body temperatures in the typical mammalian range: 37°–38°C. (After Irving and Krog 1955.)

are often thinly covered with fur or feathers, and exhibit (because of their dimensions) intrinsically high rates of convective heat exchange (see Equation 10.3). If a mammal or bird in a cold environment were to keep its appendages at the same temperature as its body core, the appendages would contribute disproportionally to the animal's overall weight-specific metabolic cost of homeothermy.

A mammal or bird can limit heat losses across its appendages in cool environments by allowing the appendage tissues to cool. The difference between the temperature of an appendage and the ambient temperature is the driving force for heat loss from the appendage. Allowing the appendage to cool toward ambient temperature reduces this driving force, in effect compensating for the appendage's relatively low resistance to heat loss. Cooling of the appendages, a type of *regional heterothermy*, is in fact very common. When the ambient temperature is low, the tissues of appendages—especially their distal parts—are often 10°–35°C cooler than tissues in the core parts of an animal's thorax, abdomen, and head (**FIGURES 10.32** and **10.33**).[31]

The usual mechanism by which appendages are allowed to cool is by curtailing circulatory delivery of heat to them. Appendages (or parts of appendages) often consist in large part of bone, tendon, cartilage, skin, and other tissues that metabolically are relatively inactive. Such appendages typically do not have sufficient endogenous heat production to keep themselves warm in cold environments. Their temperatures depend, therefore, on how rapidly heat is brought to them from the thorax, abdomen, or head by the circulating blood. Accordingly, curtailing circulatory heat delivery

---

[31] Because regional heterothermy reduces the total metabolic cost of maintaining a given core body temperature, it effectively increases the animal's overall insulation (I) in the linear heat-transfer equation (Equation 10.10).

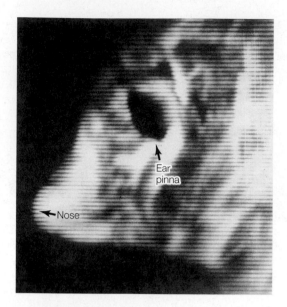

FIGURE 10.33 A thermal map of an opossum showing regional heterothermy in the pinna of the ear In this image, which was produced by infrared radiography, shades of gray represent the temperatures on the animal's body surface, ranging from low (black) to high (white). The animal, a Virginia opossum (*Didelphis marsupialis*), was resting at an ambient temperature of 10°C. The surface temperature of its ear pinna was the same as ambient temperature.

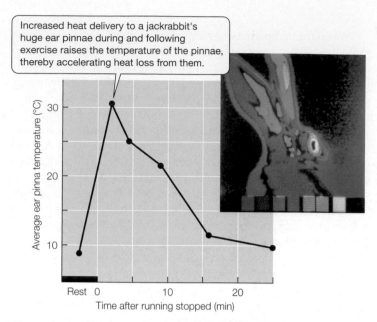

Increased heat delivery to a jackrabbit's huge ear pinnae during and following exercise raises the temperature of the pinnae, thereby accelerating heat loss from them.

FIGURE 10.34 Heat loss across appendages is sometimes modulated in ways that aid thermoregulation The average surface temperature of the ear pinnae of this black-tailed jackrabbit (*Lepus californicus*) was near ambient temperature (8°C) when the rabbit was resting but increased to more than 30°C following running. The inset is an infrared radiograph of the jackrabbit when it had an elevated ear-pinna temperature after exercise. In this presentation, temperature is color-coded. The color blocks at the bottom symbolize increasing temperatures from left to right. The environment fell into the range of temperatures coded by green. Part of the right ear pinna was warm enough to fall within the much higher temperature range coded by yellow. (After Hill et al. 1980.)

to the appendages lets them cool. Heat delivery to an appendage may be curtailed simply by the restriction of blood flow to the appendage, but as we will see in the next section, more elaborate mechanisms of restricting heat delivery are usually employed.

Species that have long evolutionary histories in frigid climates often display exquisite control over the extent of appendage cooling. For example, in a variety of Arctic canids—including foxes and wolves, as well as sled dogs—the tissues of the footpads are routinely allowed to cool to near 0°C in winter (see Figure 10.32), but even when the feet are in contact with much colder substrates (e.g., −30°C to −50°C), the footpads are not allowed to cool further. The footpads, therefore, are thermoregulated at the lowest temperature that does not subject them to a risk of frostbite!

Appendages also often play special roles in the dissipation of excess metabolic heat. If a high rate of circulatory heat delivery is provided to an appendage, the heat is lost readily to the environment because of the ease of heat loss from appendages. Accordingly, whereas animals curtail circulatory heat delivery to their appendages when heat conservation is advantageous, they often augment heat delivery to their appendages when they need to get rid of heat. In a cool environment, for example, when jackrabbits are at rest, they limit blood flow to their huge ear pinnae (see Figure 10.5)—so much so that the pinnae become as cool as the air. However, when the jackrabbits run, they increase blood flow and pinna temperature considerably (**FIGURE 10.34**). Running evidently produces an excess of metabolic heat, and the pinnae are used to void the excess heat. Opossums, rats, and muskrats sometimes warm their tails when they exercise; seals heat up their flippers; and goats warm their horns.

Countercurrent heat exchange permits selective restriction of heat flow to appendages

To reduce circulatory heat flow into an appendage, one option would be simply to reduce the rate of blood flow to the appendage. However, this mechanism would have the disadvantage of being highly nonspecific. Reducing the rate of blood flow would not only limit heat flow into an appendage but also reduce O_2 delivery to the appendage and all other circulatory transport.

Heat flow to an appendage can be *selectively* curtailed by *countercurrent heat exchange*, a process that depends on a specialized morphological arrangement of the blood vessels that carry blood to and from the appendage. To understand countercurrent heat exchange, let's examine the two different arrangements of the arteries and veins in a limb diagrammed in **FIGURE 10.35**. The arteries (red) are located deep within the appendage. In Figure 10.35A the veins (blue) are superficial, but in Figure 10.35B the veins are closely juxtaposed to the arteries. The vascular arrangement in Figure 10.35A does nothing to conserve heat; as blood flows into the appendage through the arteries and then flows back through the veins, it loses heat all along the way, without any opportunity to regain it. In contrast, the vascular arrangement in Figure 10.35B promotes heat conservation because it encourages a transfer of heat from the arterial blood to the venous blood, which then can carry that heat back to the body core, keeping the heat in the body. If the area of contact between the veins and arteries in Figure 10.35B is

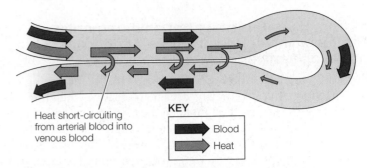

Heat short-circuiting
from arterial blood into
venous blood

KEY

| | Blood |
| | Heat |

FIGURE 10.36 Countercurrent heat exchange short-circuits the flow of heat in an appendage In a vascular countercurrent exchanger, commodities that can pass through the walls of the blood vessels short-circuit from one fluid stream to the other while the blood travels all the way out in the appendage and all the way back. This illustration shows heat being short-circuited. The widths of the arrows symbolize the relative magnitudes of heat flow and blood flow from place to place.

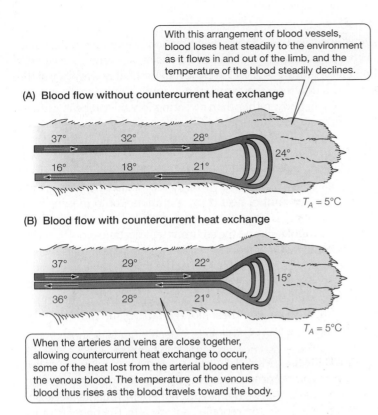

With this arrangement of blood vessels, blood loses heat steadily to the environment as it flows in and out of the limb, and the temperature of the blood steadily declines.

(A) Blood flow without countercurrent heat exchange

37°   32°   28°
16°   18°   21°
24°
$T_A = 5°C$

(B) Blood flow with countercurrent heat exchange

37°   29°   22°
36°   28°   21°
15°
$T_A = 5°C$

When the arteries and veins are close together, allowing countercurrent heat exchange to occur, some of the heat lost from the arterial blood enters the venous blood. The temperature of the venous blood thus rises as the blood travels toward the body.

FIGURE 10.35 Blood flow with and without countercurrent heat exchange Arrows show blood flow. All temperatures are in degrees Celsius (°C). (A) In this arrangement, which does not permit countercurrent heat exchange, the veins (blue) returning blood from the limb are just under the skin and separate from the arteries (red) that carry blood into the limb. (B) In this case, countercurrent heat exchange can occur because the veins returning blood from the limb are closely juxtaposed to the arteries carrying blood into the limb. In part (B) the arterial blood is cooled more than in part (A) because of the close proximity of cool venous blood. Furthermore, in (B) more heat is returned to the body than in (A) because heat that enters the venous blood is carried back to the body rather than being lost to the environment.

sufficiently extensive, blood may be little cooler when it reenters the body core in veins than it was when it flowed out into the appendage in arteries. The heat exchange in Figure 10.35B is **countercurrent heat exchange**. By definition, such heat exchange depends on the transfer of heat between two closely juxtaposed fluid streams flowing in opposite directions (*counter* = "opposite"; *current* = "flow").

A particularly useful way to conceive of the effect of countercurrent heat exchange in an appendage is to think of it as *short-circuiting the flow of heat into the appendage*. **FIGURE 10.36** illustrates that in the presence of a suitable vascular arrangement, although *blood* flows all the way to the end of an appendage before returning to the body core, *heat* tends to flow only part of the length of the appendage before it short-circuits from the arteries to the veins and starts its return to the body core. This short-circuiting impedes the access of heat to the outer extremities of the appendage. The outer extremities are therefore cooler than they otherwise would be, limiting heat loss to the environment.

A vascular countercurrent exchange system short-circuits the flow of only those commodities that are able to pass through the walls of the blood vessels involved. Heat is short-circuited by the vascular

systems we have been discussing precisely because heat can pass through the walls of arteries and veins. If O_2, nutrients, or wastes could pass through the walls of arteries and veins, they too would be short-circuited. However, they cannot pass through the walls of such thick-walled vessels as those we are discussing, and thus they travel with the blood all the way to the outer limits of an appendage and back. This is how selectivity is achieved: This is how a vascular system can conserve heat while not affecting the flow of other commodities in and out of an appendage.

Vascular arrangements that meet the prerequisites for countercurrent heat exchange (close juxtaposition of arteries and veins) are commonly found in appendages that display regional heterothermy. Such vascular arrangements are known, for example, in the arms of humans, the legs of many mammals and birds, the flippers and flukes (tail fins) of whales, the tails of numerous rodents, and the ear pinnae of rabbits and hares. Anatomically the vascular arrangements vary from relatively simple to highly complex. The vessels in some cases are simply ordinary veins and arteries touching each other; this is the case in the human arm. Alternatively, the main arteries and veins in a limb may split up to form a great many fine vessels that intermingle. A complex network of tiny vessels like this is termed a **rete mirabile** ("wonderful net") or simply a **rete**.[32]

A common way for countercurrent heat exchange to be controlled is for an appendage to have *two* sets of veins, only one of which is juxtaposed to the arteries. Countercurrent exchange can then be activated or deactivated by control of the set of veins in use. In the arm of a person, for example, one set of veins is deep in the arm and closely juxtaposed to the arteries, whereas a second set is just under the skin. Under control of the autonomic nervous system, the deep set of veins is used when there is a premium on heat conservation, but the superficial set is used when heat loss is advantageous. These controls explain why the superficial veins of our arms seem to disappear on cold days whereas they bulge with blood on warm days.

---

[32] The word *rete* is pronounced with both syllables rhyming with *sea*: "reetee." *Rete mirabile* and *rete* are general terms used to refer to intricately complex systems of small-diameter arterial and venous vessels wherever they occur. We will encounter many additional examples in this book.

Mammals and birds in hot environments: Their first lines of defense are often not evaporative

Sweating, panting, and other modes of actively increasing the rate of evaporative cooling are so easy to observe when they occur that they are often thought to be the principal or only means by which mammals and birds cope with high environmental or metabolic heat loads. Evaporation, however, has a potentially lethal price: It carries body water away. Although evaporative cooling may *solve* problems of temperature regulation, it may *create* problems of water regulation. For many mammals and birds, especially species that have long evolutionary histories in hot, arid climates, active evaporative cooling is in fact a *last* line of defense against heat loading. Other defenses are marshaled preferentially, and only when these other defenses have done as much as they can is body water used actively to void heat. In this section we discuss the nonevaporative defenses. When these defenses are employed as the preferential or first-line defenses, they act as water-conservation mechanisms.

Behavioral defenses are one set of commonly employed nonevaporative defenses. Desert rodents, for instance, construct burrows, which they occupy during the day (see Figure 1.18), and most emerge on the desert surface only at night. They thus evade the extremes of heat loading that could occur in deserts. Mammals and birds that are active during daylight hours often rest during the heat of the day, thereby minimizing their metabolic heat loads. Resting camels shift the positions of their bodies to present a minimum of surface area to the sun throughout hot days.

Insulatory defenses are also important nonevaporative defenses in some cases. For example, some species of large, diurnal mammals and birds native to hot, arid regions have evolved strikingly thick pelages and plumages. The dorsal pelage of dromedary camels in summer can be at least 5–6 cm thick, and when ostriches erect their plumage, it can be 10 cm thick. Such thick pelages and plumages probably evolved because in very hot environments they can act as heat shields, increasing body insulation and thereby acting as barriers to heat influx from the environment. The outer surface of the dorsal pelage of camels and sheep has been measured to get as hot as 50°–80°C when exposed to solar radiation on hot days! The pelage shields the living tissues of the animals from these enormous heat loads.

Body temperature is a third nonevaporative attribute of mammals and birds that can be used in the first line of defense against the challenges of hot environments. Both high-amplitude cycling of body temperature and profound hyperthermia can act as defenses and in fact are commonly employed as water-conservation mechanisms by species adapted to hot environments.

CYCLING OF BODY TEMPERATURE   Dromedary camels provide a classic and instructive example of how animals can employ high-amplitude cycling of body temperature as a nonevaporative defense and water-conservation mechanism in hot environments (see also Figure 30.11). A dehydrated dromedary in summer permits its deep-body temperature to fall to 34°–35°C overnight and then increase to more than 40°C during each day. Its body temperature therefore cycles up and down by about 6°C. The advantage of such cycling is that it permits some of the heat that enters the body during the intensely hot part of each day to be temporarily

stored in the body and later voided by nonevaporative rather than evaporative means. When dawn breaks on a given day, a camel's body temperature is at its lowest level. As the day warms and the sun beats down on the camel, the animal simply lets heat accumulate in its body, rather than sweating to void the heat, until its body temperature has risen by 6°C. Physiologists have measured that about 3.3 J (0.8 cal) is required to warm 1 g of camel flesh by 1°C. From this figure, one can calculate that a 400-kg camel will accumulate about 7920 kilojoules (kJ) (1900 kilocalories [kcal]) of heat in its body by allowing its body temperature to rise 6°C; to remove this amount of heat by evaporation would require more than 3 L of water, but the camel simply stores the heat in its body. Later, after night falls and the environment becomes cooler, conditions become favorable for convection and radiation to carry heat out of the camel's body. At that point the camel is able to get rid of the heat stored during the day by nonevaporative means. Its body temperature falls overnight to its minimum, poising the animal to take full advantage of heat storage during the following day, thereby again saving several liters of water.

HYPERTHERMIA   Many mammals and birds employ controlled, profound hyperthermia as a principal nonevaporative, water-conserving mechanism for coping with hot environments. Because a rise in body temperature entails heat storage, the benefits of hyperthermia are to some extent the very ones we have just noted in discussing cycling. In addition, however, a high body temperature *in and of itself* holds advantages for water conservation. As mentioned already, under conditions when dry heat *loss* occurs, a high T_B promotes such nonevaporative heat loss by elevating the driving force $(T_B - T_A)$ that favors it. A high T_B also aids water conservation under conditions when an animal has stored as much heat as it can and yet the environment is so hot $(T_A > T_B)$ that dry heat *gain* occurs. Under such conditions, evaporation of water must be used to get rid of all the heat that enters an animal's body. A high T_B impedes heat gain from the environment by decreasing the driving force $(T_A - T_B)$ that favors heat influx, and thus the high T_B reduces the rate at which body water must be evaporated to void the incoming heat.

Birds commonly permit their body temperatures to rise to profoundly high levels when in hot environments; whereas resting birds typically have body temperatures near 39°C in the absence of heat stress, they commonly have body temperatures as high as 43°–46°C in hot environments. Among mammals, profound hyperthermia typically occurs only in species with long evolutionary histories in hot, arid climates, but among such species it is common. Certain antelopes native to the deserts and dry savannas of Africa provide the extreme examples. Two such species, the beisa oryx (*Oryx beisa*) and Grant's gazelle (*Gazella granti*), sometimes permit their rectal temperatures to reach 45.5°–47°C (114°–116°F) without ill effect!

KEEPING A COOL BRAIN   Considerable evidence indicates that the brain is kept cooler than the thorax and abdomen in many species of mammals and birds when the animals are in warm or hot environments, especially during exercise. To cite an extreme example, when a Thomson's gazelle (*Gazella thomsonii*) runs vigorously in a warm environment, its brain is kept as much as 2.7°C cooler than its thorax. Camels, dogs, pronghorns, sheep, and harp seals are other animals known to exhibit brain cooling.

The advantage of brain cooling is believed to be that it permits an animal to take enhanced advantage of the benefits of high-amplitude body-temperature cycling and hyperthermia. The brain tolerates less elevation of temperature than most organs. Thus the bulk of an animal's body can cycle to a higher temperature, and become more hyperthermic, if the brain can be prevented from becoming as hot as most of the body.

What is the mechanism of brain cooling? In many cases, the key process is cooling of the arterial blood supplying the brain by countercurrent heat exchange (**FIGURE 10.37**). The arteries carrying blood toward the brain from the heart come into intimate contact with veins or venous blood draining the nasal passages and other upper respiratory passages. The site of this contact in many of the mammals involved is the cavernous sinus located at the base of the skull; there the arteries divide into a plexus of small vessels (the *carotid rete mirabile*) that is immersed in a lake of venous blood. As noted, the venous blood juxtaposed to the arteries is traveling back toward the heart from the upper respiratory passages. Blood in the upper respiratory passages is cooled by the inevitable evaporation of water from the walls of the respiratory passages into breathed air. As the cooled venous blood traveling back to the heart flows by the arteries, it cools the arterial blood traveling toward the brain.

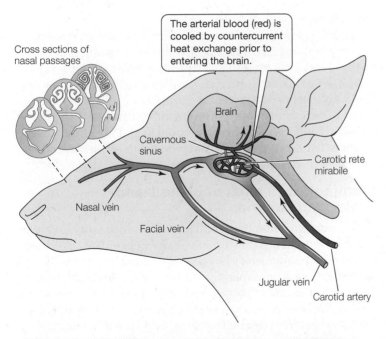

> The arterial blood (red) is cooled by countercurrent heat exchange prior to entering the brain.

Cross sections of nasal passages

Brain

Cavernous sinus

Carotid rete mirabile

Nasal vein

Facial vein

Jugular vein

Carotid artery

FIGURE 10.37 Structures hypothesized to be responsible for cooling the brain in sheep and other artiodactyls The carotid artery branches and anastomoses in the cavernous sinus, forming the carotid rete mirabile. Venous blood from the upper respiratory passages (e.g., nasal passages) flows around the vessels of the carotid rete. The insets above the snout show representative cross sections of the nasal passages of an artiodactyl (specifically, reindeer), illustrating that surface area in many species is greatly elaborated by folds and scrolls of tissue termed the *nasal turbinates*. The high surface area facilitates evaporation—and thus evaporative cooling of blood—in the nasal passages, as well as having other effects. (Principal drawing after Maloney and Mitchell 1997; turbinates after Johnsen 1988.)

Active evaporative cooling is the ultimate line of defense against overheating

Active facilitation of evaporation is the ultimate line of defense for mammals and birds faced with high environmental or metabolic (e.g., exercise-induced) heat loads. If heat is accumulating in the body to excessive levels and all the other means we have already discussed fail to stop the accumulation, active evaporative cooling becomes the only mechanism available to reestablish a balance between heat gain and heat loss. As stressed earlier, the loss of water during evaporative cooling can dehydrate an animal if replacement water is not readily available; this probably explains why species native to arid habitats employ other defenses against overheating before turning to evaporative cooling. Three major mechanisms of active evaporative cooling are known: sweating, panting, and gular fluttering.[33]

SWEATING During **sweating**, a fluid called *sweat* is secreted, by way of the ducts of sweat glands, through the epidermis of the skin onto the skin surface. Even when an animal is not sweating, water loss occurs through the substance of the skin—but at a low rate.[34] Sweating increases the rate of cutaneous evaporation by a factor of 50 or more by wetting the outer surface of the skin. Sweat is not pure water but instead is a saline solution. Concentrations of Na^+ and Cl^- in sweat are lower than in the blood plasma, and during acclimation to hot conditions the salinity of sweat becomes reduced. Nonetheless, prolonged sweating can cause a significant depletion of the body's pool of Na^+ and Cl^-. Secretion by the sweat glands is activated by the sympathetic nervous system.

A capability to sweat vigorously is found in a variety of mammals, including humans, horses, camels, and some kangaroos. Sweat production can be profuse. Humans working strenuously in the desert, for example, can attain sweating rates of 2 L/h! Many types of mammals, however, do not sweat. Rodents, rabbits, and hares lack integumentary sweat glands. Although dogs and pigs have sweat glands, the secretion rates of the glands are so low that sweating appears to play little or no role in thermoregulation. Birds do not sweat.

PANTING **Panting** is an increase in the rate of breathing in response to heat stress. It is common in both birds and mammals. Panting increases the rate of evaporative cooling because water evaporates from the warm, moist membranes lining the respiratory tract into the air that is breathed in and out.

In some species, the respiratory frequency (number of breaths per minute) during panting increases progressively as the extent of heat stress increases. In others, the respiratory frequency changes abruptly at the onset of panting, and within a wide range of thermal stress, the rate of breathing during panting is independent of the

---

[33] A fourth mechanism is **saliva spreading**, seen in many rodents and marsupials, which spread saliva on their limbs, tail, chest, or other body surfaces when under heat stress. Spreading of saliva on furred regions of the body is a relatively inefficient use of body water for cooling because the evaporative surface created—on the outer surface of the fur—is insulated from the living tissues of the animal's body by the pelage. For many rodents, however, saliva spreading is the only means available to increase evaporative cooling, and the animals use it in heat-stress emergencies.

[34] Water lost through the skin in the absence of sweating is termed *transpirational water loss* or *insensible* ("unperceived") *water loss*.

degree of heat stress. Dogs exemplify this second pattern; whereas in cool environments they breathe 10–40 times per minute, their respiratory frequency jumps abruptly to 200 or more breaths per minute when panting begins. Analysis indicates that animals with such a stepwise change in respiratory frequency often pant at the *resonant frequency* of their thoracic respiratory structures. At the resonant frequency, the thorax has an intrinsic tendency to "vibrate" between its inspiratory and expiratory positions. Thus less muscular work needs to be done—and less heat is produced by the muscular work—than at other frequencies.

By comparison with sweating, panting holds certain advantages. One is that no salts are lost during panting because evaporation occurs within the body and only pure water vapor leaves the body in the exhalant air. A second advantage of panting is that it *forcibly* drives air saturated with water vapor away from the evaporative surfaces.

Panting also has liabilities in comparison with sweating. Because of the muscular effort required for panting, evaporation of a given quantity of water is likely to require more energy—and entail more heat production—when panting is employed than when sweating is. Another potential liability of panting is that it can induce *respiratory alkalosis*, an elevation of the pH of the body fluids caused by excessive removal of carbon dioxide (see page 664). Ordinarily, when animals are not panting, ventilation of the respiratory-exchange membranes deep in the lungs (e.g., the alveolar membranes of mammals) is closely regulated so that the rate at which CO_2 is voided is equal to the rate of metabolic production of CO_2. During panting, the potential exists for breathing to carry CO_2 away faster than it is produced, because the rate of breathing is increased for thermoregulation rather than being governed only by metabolic needs. If CO_2 is carried away by breathing faster than it is produced by metabolism, the concentration of CO_2 in the blood will fall, causing the following reactions in the blood to shift to the left:

$$CO_2 + H_2O \rightleftharpoons H_2CO_3 \rightleftharpoons H^+ + HCO_3^- \qquad \textbf{(10.11)}$$

Consequently, the concentration of H^+ in the blood will fall, and the pH of the blood will rise. Such excessive alkalinity—*alkalosis*—can have major deleterious effects because many enzymes and cellular processes are acutely sensitive to pH. (In middle school, we probably all witnessed friends make themselves get dizzy and fall down by deliberately breathing too rapidly.)

From extensive research, physiologists now know that little or no alkalosis develops during panting in many species of mammals and birds when the heat stress to which they are exposed is light to moderate. These animals avoid alkalosis by restricting the increased air movement during panting to just their *upper* airways,[35] where no exchange of CO_2 occurs between the air and blood (**FIGURE 10.38**); the respiratory-exchange membranes *deep* in the lungs receive about the same rate of airflow during panting as they usually do. By contrast, when heat stress becomes extreme, resting but panting animals often develop severe alkalosis. Some panting species have evolved superior tolerance to alkalosis.

GULAR FLUTTERING Many birds (but not mammals) augment evaporative cooling by rapidly vibrating their gular area (the floor

---

[35] In birds, both the upper airways and air sacs may be involved.

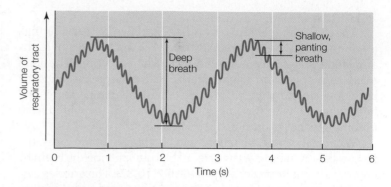

FIGURE 10.38 A breathing pattern that limits hyperventilation of the respiratory-exchange membranes during panting Shown here is one of the known breathing patterns whereby the upper airways receive a greatly increased flow of air during panting while simultaneously the respiratory-exchange membranes deep in the lungs are ventilated about as much as usual. In this pattern, sometimes called compound breathing, shallow breaths are superimposed on deep breaths.

of the buccal cavity) while holding their mouth open, a process termed **gular fluttering** (see Figure 10.30). The process is driven by flexing of the hyoid apparatus and promotes evaporation by increasing the flow of air over the moist, highly vascular oral membranes. Gular fluttering usually occurs at a consistent frequency, which apparently matches the resonant frequency of the structures involved. Birds commonly use gular fluttering simultaneously with panting.

Gular fluttering shares certain positive attributes with panting: It creates a vigorous, forced flow of air across evaporative surfaces and does not entail salt losses. Unlike panting, gular fluttering cannot induce severe alkalosis, because it enhances only *oral* airflow, and CO_2 is not exchanged between air and blood across oral membranes. Gular fluttering involves the movement of structures that are less massive than those that must be moved in panting; thus it entails less muscular work—and less heat production—to achieve a given increment in evaporation.

Mammals and birds acclimatize to winter and summer

When individual mammals and birds live chronically in cold or warm environments, they usually undergo long-term alterations in their thermoregulatory physiology. During acclimatization to winter, for example, a mammal or bird typically exhibits one or more of three sorts of chronic responses that we discuss in this section. Because the change of seasons is complex, these responses are not necessarily triggered solely (or even primarily) by the drop in temperature as winter approaches, but may be triggered by photoperiod (shortening day length) or other seasonal cues.[36]

One possible chronic response to the approach of winter is **acclimatization of peak metabolic rate**. When a mammal or bird exhibits this response, it increases the máximum rate at which it can produce heat by sustained, aerobic catabolism. If an

---

[36] *Acclimation* of mammals or birds to cold in a laboratory sometimes has dramatically different effects than *acclimatization* to winter (see page 18 for the distinction). This is true because laboratory acclimation usually entails only exposure to cold, whereas during acclimatization to winter, photoperiod and other environmental factors are modified as well as temperature.

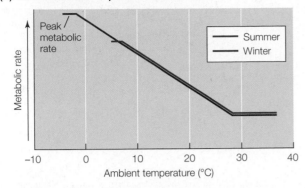

(A) Acclimatization of peak metabolic rate

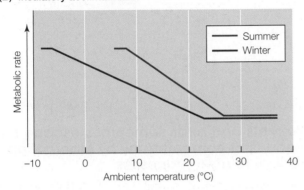

(B) Insulatory acclimatization

FIGURE 10.39 Two types of seasonal acclimatization
(A) Acclimatization of peak metabolic rate without insulatory acclimatization. (B) Insulatory acclimatization without acclimatization of
peak metabolic rate. The plateau at the left of each curve indicates
where metabolic rate has peaked.

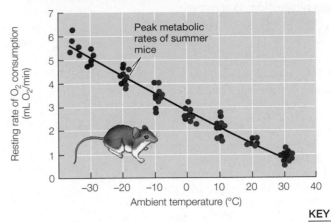

(A) Acclimatization of peak metabolic rate without insulatory
acclimatization in deer mice

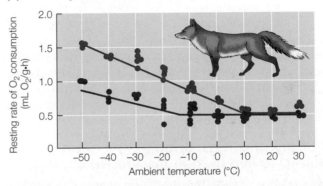

(B) Insulatory acclimatization in a red fox

**FIGURE 10.40 Seasonal acclimatization in two species of
mammals** (A) The deer mice (*Peromyscus maniculatus*) studied
had the same insulation in winter and summer, but their peak metabolic rates rose in winter, meaning they could thermoregulate at lower
ambient temperatures. (B) A single red fox (*Vulpes vulpes*) individual,
studied in both seasons, had far greater insulation in winter than in
summer. (After Hart 1957.)

animal displays *only* this sort of acclimatization, the metabolic rate
it requires to thermoregulate at any given ambient temperature
remains unchanged, but it can thermoregulate in colder environments than it could before, as shown by **FIGURE 10.39A**. The
development in winter of enlarged brown adipose tissues in which
cells are biochemically especially poised for heat production is a
common mechanism by which small and medium-sized mammals
increase the rate at which they can produce heat and thus undergo
acclimatization of peak metabolic rate.

A second possible chronic response to the approach of winter is
acclimatization of metabolic endurance, meaning an increase in
the length of time that a high rate of metabolic heat production can
be maintained. Although current evidence indicates that this sort of
acclimatization is common, little is known about its mechanisms.

The third major sort of chronic response that a mammal or
bird might exhibit in winter is **insulatory acclimatization**, an
increase in the animal's maximum resistance to dry heat loss
(maximum insulation). If this sort of acclimatization occurs, the
metabolic rate required to thermoregulate at any particular ambient temperature (below thermoneutrality) is reduced. Accordingly,
even if an animal's peak metabolic rate remains unchanged, the
animal is able to thermoregulate in colder environments than it
could before (**FIGURE 10.39B**). The most obvious way for insulatory acclimatization to occur is for an animal to molt into a more
protective pelage or plumage in winter, but other determinants of

insulation (such as peripheral blood flow) can also change. Of the
three chronic responses to winter we have described, two—or all
three—can occur together.

Acclimatization of peak metabolic rate is the norm in small
and medium-sized mammals, and occurs also in perhaps half
the species of small birds. As for insulatory acclimatization, some
small-bodied species of mammals and birds fail to exhibit it and
thus undergo only metabolic forms of acclimatization (**FIGURE
10.40A**). Among the mammals that undergo insulatory acclimatization, medium-sized and large species tend to exhibit greater
changes in insulation between summer and winter than do small
species. Red foxes (**FIGURE 10.40B**), collared lemmings, and
varying hares in northern Alaska all exhibit substantial increases
in insulation in winter. The air temperature in northern Alaska
averages −30°C in winter and +5°C in summer. For the foxes,
lemmings, and hares, the metabolic cost of thermoregulating at
−30°C in winter is little higher than the cost of thermoregulating
at +5°C in summer, because of their winter increase in insulation
(see Figure 10.40B).

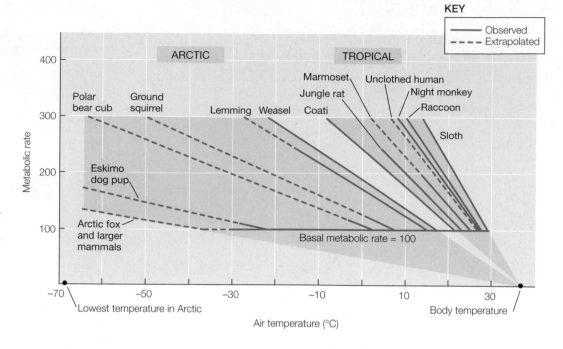

FIGURE 10.41 Mammalian physiological specialization to different climates Species found in the Arctic (Alaska) expend less energy to thermoregulate at cold ambient temperatures, and they can thermoregulate at lower temperatures, than species found in the tropics (Panama) can. In this presentation, each species' basal metabolic rate is set equal to 100, and metabolic rates outside the thermoneutral zone are expressed relative to basal; this convention facilitates comparison in certain ways but means that the slopes of the metabolism–temperature curves below thermoneutrality can be used in only a qualitative way to compare insulation. (After Scholander et al. 1950.)

Mammals and birds commonly acclimatize to heat stress as well as cold stress, as seen in Chapter 1 (see Figure 1.10). Among humans, acclimatization to heat stress occurs much more rapidly than that to cold stress. Partly for that reason, we tend to notice our own acclimatization to heat more than our acclimatization to cold.

Evolutionary changes: Species are often specialized to live in their respective climates

Abundant evidence indicates that the thermoregulatory physiology of mammals and birds has undergone evolutionary adaptation to different climates. One sort of evidence is shown in **FIGURE 10.41**, which is one of the classic sets of data in animal physiology. As the figure shows, species of mammals native to the Arctic and ones native to the tropics differ dramatically in their thermal relations; Arctic species—compared with tropical species—have lower-critical temperatures that are lower (i.e., they have broader TNZs), and they increase their metabolic rates proportionally less above basal levels at ambient temperatures below thermoneutrality. Direct studies of pelage insulation demonstrate that the Arctic species typically have thicker and better insulating pelages than do similarly sized tropical species. As a consequence of all these differences, Arctic species are in a far better position to thermoregulate under Arctic conditions than tropical species are.

In hot climates, a major pattern that has emerged with ever-increasing clarity in recent decades is that species of both mammals and birds native to such climates often have lower basal metabolic rates than are observed in related species native to temperate or cold climates. The evolution of an exceptionally low BMR has probably been favored in hot climates because, with a low BMR, an animal has a particularly low internal heat load.

As mentioned earlier, body temperature is basically a *conserved* character; within any taxonomic group of mammals or birds, the core body temperature maintained in the absence of heat or cold stress tends to be the same in species from various climates. Adaptation of body temperature to climate is clearly evident, however, in one specific respect among mammals exposed to heat stress: Mammal species native to hot climates typically tolerate greater degrees of hyperthermia than species native to temperate or cold climates do.

Mammals and birds sometimes escape the demands of homeothermy by hibernation, torpor, or related processes

Many species of mammals and birds allow their body temperatures to fall in a controlled manner under certain circumstances. **Controlled hypothermia** is a general term for this sort of phenomenon; **hypothermia** is the state of having an unusually low body temperature, and in the cases we are discussing, it is "controlled" because the animals orchestrate their entry into and exit from hypothermia rather than being forced.

The most well known and profound forms of controlled hypothermia are **hibernation**, **estivation**, and **daily torpor**. According to definitions that have been in place for several decades, these are all states in which an animal *allows its body temperature to approximate ambient temperature within a species-specific range of ambient temperatures*. Hibernation, estivation, and daily torpor are generally viewed as being different manifestations of a *single* physiological process. They are distinguished by differences in their durations and seasons of occurrence. When an animal allows its body temperature to fall close to ambient temperature for periods of several days or longer during winter, the process is termed *hibernation*. When this occurs during summer, it is called *estivation*. When an animal permits its body temperature to fall close to ambient temperature for only part of each day (generally on many consecutive days), the process is termed *daily torpor* in any season. **FIGURES 10.42** and **10.43** illustrate the sorts of changes in body temperature and metabolic rate that occur in episodes of controlled hypothermia.

Hibernation, estivation, and daily torpor permit mammals and birds to escape the energy demands of homeothermy. As stressed earlier, homeothermy is energetically costly. A hamster, for example, needs to acquire and consume a great deal of food energy to keep its body temperature at 37°C when the temperature of its environment is near freezing. If the hamster abandons homeothermy and temporarily allows its body temperature to fall close to ambient temperature, it is temporarily freed of homeothermy's energy costs. Animals capable of hibernation, estivation, or daily torpor are in essence able to switch back and forth between two very

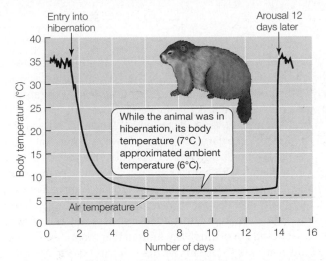

FIGURE 10.42 Changes in body temperature during hibernation A woodchuck (groundhog) (*Marmota monax*) was implanted with a small temperature transmitter that broadcast its body temperature continuously, and after it healed from the surgery, it was studied at an air temperature of 6°C. The record shows its body temperature during a 12-day episode of hibernation. (After Armitage et al. 2000.)

different thermal worlds. They are *temporal heterotherms*. When they function as ordinary homeotherms do, they reap the benefits of homeothermy, such as physiological independence of external thermal conditions; but they pay the high energy cost. When the animals suspend homeothermy, they take on many of the attributes of poikilotherms: Their tissues are subjected to varying tissue temperatures, but the animals have low energy needs.

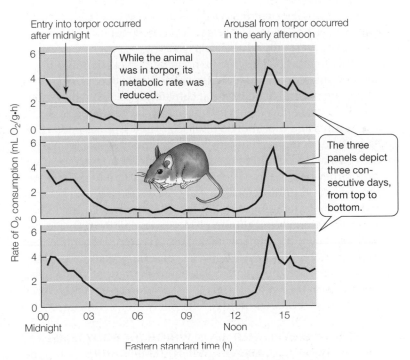

FIGURE 10.43 Changes in metabolic rate during daily torpor The rate of O_2 consumption of a white-footed mouse (*Peromyscus leucopus*) studied at an air temperature of about 14°C is shown for 3 consecutive days. The x axis shows time of day on a 24-h scale (e.g., 15 = 3:00 PM). The animal required a resting metabolic rate of about 3.0 mL O_2/g·h to be able to maintain high body temperatures. It underwent a prolonged episode of daily torpor on each day, as indicated by the drop in its metabolic rate. Its body temperature measured during an episode of torpor was 17°C. (After Hill 1975.)

Quantitatively, the amount of energy saved by controlled hypothermia depends on the ambient temperature at which hypothermia occurs and the duration of the hypothermia. To elucidate the importance of ambient temperature, **FIGURE 10.44** shows the different metabolism–temperature relations that exist in a single species when the animals are homeothermic and when they are in controlled hypothermia. At any given ambient temperature, the difference between the two curves (double-headed arrow) shows the degree to which animals can reduce their energy costs per unit of time by entering hypothermia; the amount of energy saved per unit of time becomes greater as the ambient temperature falls. If a hibernating animal remains in hibernation at low ambient temperatures for long periods of time, its total energy savings can be enormous. For example, free-living ground squirrels of at least two species, living in cold climates, have been measured to expend only 10%–20% as much energy per month by hibernating as they would if they failed to hibernate, and they reap these monthly savings throughout their 7- to 8-month hibernating seasons.

Controlled hypothermia also permits mammals and birds to escape the high water demands of homeothermy. This point is not as widely significant as the escape from energy demands, because the escape from water demands matters only for animals that face water shortages. Sometimes, nonetheless, the escape from water demands can be the most important consequence of entering controlled hypothermia; this is especially true for animals that enter estivation or daily torpor in hot, dry environments. As we will discuss in detail in Chapter 28, homeotherms have relatively high rates of water loss. One reason is simply that they must breathe rapidly to acquire the amounts of O_2 they need for their high metabolic rates. Another is that the air they exhale tends to be relatively warm, and warm air

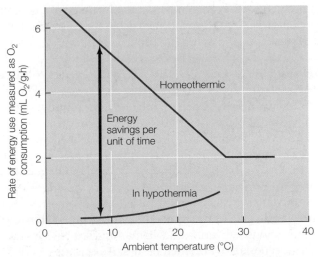

FIGURE 10.44 Energy savings depend on temperature In common with other species that undergo hibernation, estivation, or daily torpor, kangaroo mice (*Microdipodops pallidus*) alternate between two metabolism–temperature relations, shown here. The double-headed arrow shows how much a kangaroo mouse's rate of energy use is reduced when the animal is in hypothermia instead of being homeothermic. The amount of energy saved by being in hypothermia is greater at low ambient temperatures than at higher ones because the metabolic cost of homeothermy is particularly high at low ambient temperatures, whereas the cost of hypothermia is particularly low at low ambient temperatures. (After Brown and Bartholomew 1969.)

holds more water vapor (which is exhaled with the air) than cooler air. Entry into controlled hypothermia reduces an animal's rate of water loss by reducing both (1) its breathing rate and (2) the temperature, and therefore the water vapor content, of its exhaled air.

WHAT ARE THE MECHANISMS BY WHICH METABOLIC RATE IS LOWERED DURING CONTROLLED HYPOTHERMIA? Recent research has established that—in at least some mammalian hibernators—*biochemical downregulation of metabolism* takes place during hibernation.

Until about 25 years ago, the almost universal view was that animals initiate their entry into controlled hypothermia simply by turning off thermoregulation. According to this view, the sequence of events during entry into hypothermia is that thermoregulation is deactivated, body temperature falls because of cooling by the environment in the absence of thermoregulatory responses, and tissue metabolic rates then decline because the tissues cool. This sort of lowering of metabolic rate—driven by tissue cooling and therefore following the Q_{10} principle (see Equation 10.7)—is often described as a "Q_{10} effect."

The newer view is that the first step in the sequence of events during entry into hypothermia is biochemical downregulation of tissue metabolism. Body temperature then falls as a consequence of the reduced metabolic rate. In this sequence of events, after biochemical downregulation initiates the fall of body temperature, the declining body temperature can potentially exert a Q_{10} effect that reinforces the biochemical downregulation in depressing metabolism.

The evidence currently available indicates that both of the sequences of events discussed are observed during controlled hypothermia in mammals. One recent analysis identifies a divergence between species that undergo only daily torpor (which follow the first sequence described) and those that hibernate (which follow the second sequence). In some hibernators, the metabolic rate during hibernation is determined by biochemical controls in a way that body temperature, over wide ranges, does not matter.

IN WHAT RESPECTS IS "CONTROLLED" HYPOTHERMIA CONTROLLED? Mammals and birds that display controlled hypothermia orchestrate their entry into and exit from hypothermia, and they exhibit control over their situation in other respects as well. The most dramatic evidence of the controlled nature of hibernation, estivation, and daily torpor is the fact that animals are able to arouse from these conditions. **Arousal** is the process of rewarming the body by metabolic heat production. The animals do not require outside warming to return to homeothermy. Instead, they are in control: They return to homeothermy on their own by employing intense shivering and, in mammals, intense nonshivering thermogenesis to warm their tissues. All episodes of controlled hypothermia *end* with arousal. In addition, hibernating mammals universally undergo periodic, short arousals during the period of time they are hibernating; for instance, an animal that hibernates for 6 months might arouse for a few hours every 14 days or so. The possible functions of periodic arousals are discussed in Chapter 11.

A second, particularly fascinating sort of control exhibited by animals in controlled hypothermia is the control they display when their body temperatures start to fall too low. Each species that undergoes hibernation, estivation, or daily torpor has a species-specific range of body temperatures that it can tolerate, and for an animal to

survive hypothermia, it must respond if its body temperature starts to go below the tolerable range. *Within* the tolerable range, animals typically let their body temperatures drift up and down as the ambient temperature rises and falls. For instance, if an animal can tolerate a T_B as low as 3°C and T_A varies between 5°C and 15°C, the animal typically allows its T_B to vary as T_A varies (always being a bit higher than T_A). What happens, however, if the ambient temperature falls below 3°C? Frequently (although not always), the animal exerts control in one of two life-preserving ways. It may arouse. Alternatively and more remarkably, it may start to *thermoregulate* at a reduced body temperature, its thermoregulatory control system functioning with a lowered set point. For example, an animal that must stay at a body temperature of at least 3°C to survive may keep its body temperature at 3°C even if the ambient temperature drops to –10°C or –20°C, increasing its metabolic rate as the ambient temperature falls so as to offset the increasing cooling effect of the air (see Figure 11.11).[37]

DISTRIBUTION AND NATURAL HISTORY Hibernation is known to occur in at least six different orders of mammals. Species that hibernate include certain hamsters, ground squirrels, dormice, jumping mice, marmots, woodchucks, bats, marsupials, and monotremes. Because of its seasonal nature, hibernation is often preceded by long-term preparation. Hibernating mammals, for instance, typically store considerable quantities of body fat during the months before their entry into hibernation (see Figure 6.25). Hibernation is rare in birds; it is known to occur in only a single species, the common poorwill (*Phalaenoptilus nuttallii*). We discuss mammalian hibernation at considerably more length in Chapter 11.

Estivation is not nearly as well understood as hibernation, partly because it is not as easy to detect. It has been reported mostly in species of desert ground squirrels.

Daily torpor is widespread among both mammals and birds, and it occurs not only in species facing cold stress but also in species occupying tropical or subtropical climates. It occurs in numerous species of bats and rodents and in certain hummingbirds, swallows, swifts, and caprimulgid birds (e.g., nightjars and poorwills). Animals undergoing daily torpor are homeothermic for part of each day, and feed at that time. When bats are undergoing daily torpor, they become hypothermic during daylight hours and forage at night; hummingbirds, in contrast, become torpid at night and feed in daylight. In some species, the proclivity to enter daily torpor is seasonally programmed. However, daily torpor seems to be employed most commonly, regardless of season, as an immediate response to hardship. Many species, for example, undergo daily torpor only when they are suffering food shortage; in some cases they increase the length of time they spend in torpor each day as food shortage becomes more severe.

CONTROLLED HYPOTHERMIA IN WHICH THE BODY TEMPERATURE REMAINS WELL ABOVE AMBIENT TEMPERATURE Over the last 35 years, there has been an escalating realization that many species of small birds undergo hypothermia without ever allowing their body temperatures to approximate ambient temperature. Black-capped chickadees (*Poecile atricapillus*) provide an excellent example. They sometimes allow their core body temperature to fall by roughly 7°C while sleeping overnight in freezing-cold winter

---

[37] This phenomenon features prominently in the prediction of bat hibernation ranges discussed in Box 10.1.

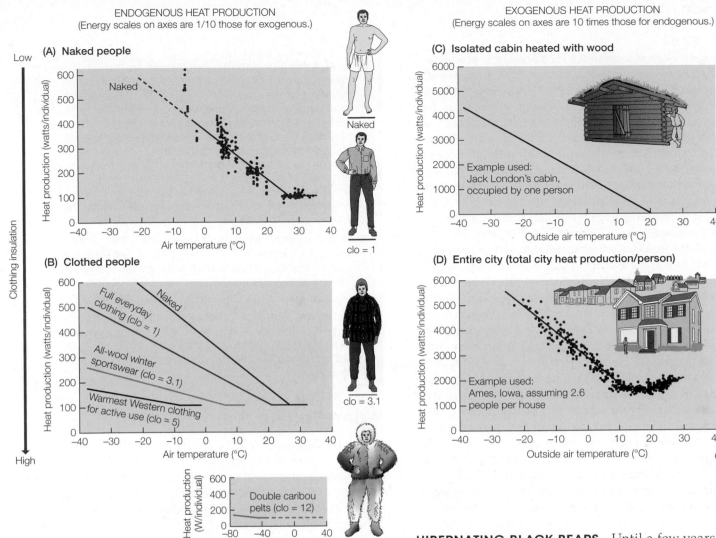

ENDOGENOUS HEAT PRODUCTION
(Energy scales on axes are 1/10 those for exogenous.)

(A) Naked people

Low

Clothing insulation

High

Heat production (watts/individual)

Naked

Air temperature (°C)

Naked

clo = 1

(B) Clothed people

Heat production (watts/individual)

Full everyday clothing (clo = 1)

Naked

All-wool winter sportswear (clo = 3.1)

Warmest Western clothing for active use (clo ≈ 5)

Air temperature (°C)

clo = 3.1

Heat production (W/individual)

Double caribou pelts (clo = 12)

Air temperature (°C)

clo = 12

EXOGENOUS HEAT PRODUCTION
(Energy scales on axes are 10 times those for endogenous.) Local fuel; small, dispersed dwellings

(C) Isolated cabin heated with wood

Heat production (watts/individual)

Example used: Jack London's cabin, occupied by one person

Outside air temperature (°C)

(D) Entire city (total city heat production/person)

Heat production (watts/individual)

Example used: Ames, Iowa, assuming 2.6 people per house

Outside air temperature (°C)

Fuel source, dwelling size, and landscape

Outsourced fuel; large, clustered dwellings

FIGURE 10.45 Per capita energy demands of human thermoregulation by endogenous and exogenous heat production For individual people thermoregulating by means of metabolic heat production using endogenous fuels, the metabolism–temperature relation is shown for (A) naked people (dots are measured data points) and (B) people wearing various types of clothing (line colors match lines in sketches). For people keeping warm by occupying dwellings heated by use of exogenous fuels, the rate of heat production per person is shown as a function of outside air temperature in (C) a log cabin and (D) houses in a modern cirty. The cabin is modeled on the one in the Yukon where Jack London spent the winter of 1897–1898. (From Hill et al. 2013.)

weather. They then have body temperatures (31°–34°C) that are distinctly hypothermic but nonetheless far above ambient temperature. This hypothermia does not eliminate their need to expend metabolic energy to stay warm. However, because the hypothermia reduces the difference between body temperature and ambient temperature, the birds lose heat more slowly—and have lower costs for thermoregulation—than if they maintained higher body temperatures. Chickadees are so small (11 g) that they may virtually exhaust their body fat in a single night of frigid weather; their hypothermia helps them survive until they can feed the next day.

A variety of mammals also exhibit subtle forms of hypothermia in which core body temperature falls to only a relatively small extent. Bears of some species are the most famous examples of mammals exhibiting moderate core hypothermia.

HIBERNATING BLACK BEARS Until a few years ago, many physiologists doubted that black bears (*Ursus americanus*) should be considered hibernators because, although they allow their core body temperature to fall for 5–7 months in winter, their body temperature stays at 30°C or higher and therefore is far above ambient temperature, violating the traditional criterion for hibernation. Bears stand out because of their size. Nearly all hibernating species of mammals weigh 5 kg or less, whereas black bears weigh 30–150 kg. Recent research has revealed that black bears, during their winter hypothermia, exhibit a dramatic degree of biochemical downregulation of metabolism, so much so that their metabolic rates are only one-quarter as high as their basal rates; metabolic downregulation is the principal control of their metabolic rates, with body temperature (Q_{10} effect) playing little role. Under these circumstances, despite their body temperatures being only mildly hypothermic, the bears have weight-specific metabolic rates similar to those of most hibernators. Hibernation physiologists now, therefore, rank the black bear as a specialized hibernator. The reason its body temperature fails to fall to ambient temperature (even though the species metabolically resembles other hibernators) may be its large body size.

Human thermoregulation

As we finish our exploration of homeothermy in mammals and birds, let's take a look at thermoregulation by people to see how the principles we've developed apply to our own species (**FIGURE 10.45**). People are unique in that they often burn external fuels such as wood, coal, or natural gas to keep warm in cold environments.

People also keep warm by means of metabolic heat production, using metabolic fuels, as during shivering. Thus in considering human thermoregulation we need to recognize the distinction between *exogenous* and *endogenous* heat production: that is, heat production by use of external fuels versus internal, metabolic fuels.

When people are studied in their birthday suits—that is, naked—they exhibit a typical mammalian relation between metabolic rate and air temperature (see Figure 10.45A). The slope below thermoneutrality is steep (see Figure 10.41), reflecting the fact that, not having fur, we are relatively poorly insulated. When we add clothing, we change the slope of the metabolism–temperature curve below thermoneutrality. The insulative value of clothing is measured in *clo* units. As we attire ourselves in better- and better-insulating clothing (higher and higher clo), our cost of endogenous thermoregulation at any given cold air temperature becomes lower and lower (see Figure 10.45B). Astoundingly, traditional, indigenous peoples in the subarctic all around the world devised clothing, made from caribou (reindeer) pelts (see Figure 11.2), that insulated them so well that they could keep warm with only their basal rate of metabolic heat production even when the air was as cold as −50°C.

Cabins and houses heated by burning exogenous fuels follow the same laws of physics as apply to individual people or other mammals. Consequently, the rate of heat production in a cabin or house varies with the outside air temperature in a way that parallels the metabolism–temperature curves of mammals (see Figure 10.45C,D). However, the per-person cost of thermoregulation in such dwellings exceeds the cost of individual, endogenous thermoregulation by a large measure: 10- to 30-fold. The principal reason is that, when we live in a cabin or house, we not only keep ourselves warm but we also heat a large living space around us.

SUMMARY
Homeothermy in Mammals and Birds

- Homeothermy—thermoregulation by physiological means—is energetically expensive.

- The principal way that a mammal or bird thermoregulates in its thermoneutral zone is that it varies its body insulation to offset changes in the driving force for dry heat loss ($T_B - T_A$). Insulation can be modulated by changes in posture, cutaneous blood flow, the thickness of the relatively motionless air layer trapped by the pelage or plumage, and regional heterothermy.

- Below thermoneutrality, variation in the rate of metabolic heat production (thermogenesis) is the principal mechanism of thermoregulation. The two most prominent mechanisms of increasing heat production are shivering—found in both mammals and birds—and nonshivering thermogenesis (NST)—found mainly in placental mammals. The principal site of NST in mammals is brown adipose tissue, which, by expressing uncoupling protein 1, is able to employ uncoupling of oxidative phosphorylation to achieve very high rates of lipid oxidation with immediate heat release.

- Regional heterothermy, which is often exhibited when animals are at ambient temperatures below thermoneutrality, usually depends on countercurrent heat exchange. Close juxtaposition of arteries and veins short-circuits the flow of heat into appendages.

- Above thermoneutrality, species with long evolutionary histories in hot, dry environments typically use nonevaporative mechanisms—notably hyperthermia and cycling of body temperature—as first lines of defense. When active evaporative cooling occurs, the specific mechanisms usually employed to increase the rate of evaporation are sweating (only in certain mammals), panting (mammals and birds), and gular fluttering (only birds). Both hyperthermia and the effort involved in active evaporative cooling can cause metabolic rate to rise at ambient temperatures above thermoneutrality.

- Acclimatization to changing seasons is the norm and may involve one or more of three mechanisms: acclimatization of peak metabolic rate, acclimatization of metabolic endurance, and insulatory acclimatization.

- Controlled hypothermia permits animals to evade temporarily the high energy costs and water costs of homeothermy. During hibernation, estivation, and daily torpor, T_B is generally allowed to fall close to T_A within a species-specific range of T_A. Forms of shallow hypothermia also occur.

Warm-Bodied Fish

The body temperatures of 99% of all species of fish closely approximate water temperature. However, in tunas, lamnid sharks, and billfish, temperatures within *certain body regions* exceed water temperature, sometimes substantially. All these warm-bodied fish are large, streamlined, fast-swimming predators that lead wide-ranging lives and feed on such speedy prey as squids and herring. The lamnid sharks include the great white shark. The billfish include the marlins and swordfish. Besides the tunas, lamnids, and billfish—which are similar enough in their thermal physiology that we can meaningfully discuss them as a set—another, unrelated fish, called the opah, was discovered in 2015 to be warm-bodied in a different way. We will discuss the opah separately at the end of this section.

In tunas and lamnid sharks, the red (dark) swimming muscles are warmed above water temperature.[38] These muscles provide the power for steady swimming in these vigorously active animals, and the contractile activity of the muscles produces the heat that warms the muscles. A critical principle to recognize, however, is that *a high rate of heat production is never in itself adequate to elevate tissue temperature in water-breathing animals*. If metabolic heat is carried freely to the gills by the circulation of the blood, the heat is lost so readily to the surrounding water across the gills that no significant elevation of body temperature can occur. Thus for a region of the body to be warmed, transport of heat away from that body region by the circulation must be impeded. Not only in the red swimming muscles of tunas and lamnid sharks, but universally in warm-bodied fish, the mechanism of impeding heat loss is countercurrent heat exchange.

The vasculature of the red swimming muscles in tunas and lamnid sharks is diagrammed in **FIGURE 10.46**. Note that the red muscles (which are usually located superficially in fish) are found deep in the body near the spinal column in these fish, an unusual

---

[38] The roles of the red and white muscles in powering swimming in fish are discussed in Chapter 8 (see page 203).

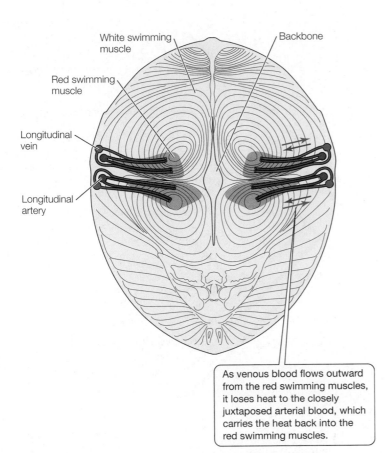

FIGURE 10.46 A cross section of a tuna showing the nature of the blood supply to the red swimming muscles   The longitudinal arteries, which carry blood along the length of the body, give off small arteries that penetrate (toward the backbone) into the muscles. Small veins running in close juxtaposition to the small arteries return blood peripherally to the longitudinal veins, which lead back to the heart. Red vessels and arrows refer to arterial blood flow; blue vessels and arrows refer to venous flow.

As venous blood flows outward from the red swimming muscles, it loses heat to the closely juxtaposed arterial blood, which carries the heat back into the red swimming muscles.

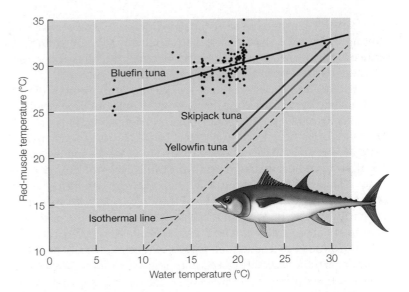

FIGURE 10.47 Red-muscle temperatures of tunas at various ambient water temperatures   The upper line and data points are for wild bluefin tunas (*Thunnus thynnus*) captured in waters of various temperatures. The other two solid lines show the average relation between red-muscle temperature and water temperature in small, captive skipjack tunas (*Katsuwonus pelamis*) and yellowfin tunas (*Thunnus albacares*) swimming in an aquarium; larger, wild fish of these species are sometimes observed to exhibit greater temperature differentials between muscles and water (e.g., 5°–10°C in skipjacks). The isothermal line shows how tissue temperature would vary with water temperature if there were no endothermy and tissue temperature simply equaled water temperature. The fish shown is a bluefin tuna. (After Carey and Teal 1969; Dizon and Brill 1979.)

placement. The major longitudinal arteries and veins that carry blood along the length of the body, to and from the swimming muscles, run just under the skin on each side of the body (again, an unusual placement). Small arteries branch off from the longitudinal arteries and penetrate inward to the red muscles. In turn, blood is brought outward from the muscles in veins that discharge into the longitudinal veins leading back to the heart. The arteries carrying blood inward to the red swimming muscles and the veins carrying blood outward from those muscles are closely juxtaposed, forming countercurrent exchange networks. Figure 10.46 is highly simplified in the way it presents these networks. In actuality, the arteries and veins going to and from the red muscles branch profusely, forming thick layers of vascular tissue in which huge numbers of minute arterial and venous vessels, each only about 0.1 mm in diameter, closely intermingle—a true rete mirabile (see page 269). Because of the countercurrent-exchange arrangement, much of the heat picked up by the venous blood in the red muscles is transferred to the ingoing arterial blood rather than being carried by the venous flow to the periphery of the body and the gills, where it would readily be lost to the water. Thus heat produced by the red swimming muscles tends to be retained within them.

Bluefin tunas, which reach body weights of 700 kg and are the largest of all tunas, maintain fairly constant red-muscle temperatures over a wide range of water temperatures (**FIGURE 10.47**). In most other tunas, such as the yellowfin and skipjack tunas, red-muscle temperature is elevated over water temperature by a relatively constant amount regardless of the water temperature (see Figure 10.47). Referring back to our scheme for classifying animal thermal relations (see Figure 10.1), all the tunas are endotherms, but species differ in whether they also thermoregulate. Whereas yellowfin and skipjack tunas are endotherms without being thermoregulators,[39] bluefin tunas are endothermic thermoregulators (homeotherms).

The warming of the red swimming muscles in tunas and lamnid sharks is generally thought to aid power development and locomotory performance, although exactly how is debated. Any aid to the performance of the swimming muscles would be significant for fish that are so dependent on high-intensity exertion for their livelihood.

The swimming muscles are not the only tissues kept warm in tunas and lamnid sharks. In certain species, the stomach and other viscera are warmed when food is being digested. The brain and eyes are also warmed in some species. Each warmed organ is served by arteries and veins that form a rete mirabile, which short-circuits the outflow of heat produced in the organ, thereby favoring heat accumulation in the organ.

[39] There is some evidence for active thermoregulation in these fish. For example, they decrease heat retention when they are highly active in warm water, thus preventing their activity from driving their muscle temperature too high.

Now let's turn to the billfish. They differ in two ways from the tunas and lamnid sharks. First, in the billfish, only the brain and the retinas of the eyes are warmed. Second, the billfish possess "heater tissues" specialized for exceptional heat output.[40]

The heater tissues of billfish are derived from portions of the extraocular eye muscles (the muscles on the outside of each eyeball that serve to turn the eyeball to look in various directions). These portions of the muscles have lost most of their contractile apparatus and are very rich in mitochondria. Current evidence suggests that they produce heat at a high rate by a "futile cycle" of Ca^{2+} pumping: ATP is used to transport Ca^{2+} actively from one intracellular compartment to another, and then the Ca^{2+} leaks back to where it started, where once again ATP is used to pump it; the principal net result is breakdown of ATP at a high rate to release heat. The heat produced by the heater tissues is retained in the head by countercurrent vasculature and in that way warms the brain and retinas. If warming of the *brain* by specialized *eye* muscles sounds impossible, remember that in a fish, the eyes and eye muscles are far larger than the brain! Warming of the brain and the retinas is hypothesized to aid marlins, swordfish, and other billfish in their pursuit of prey because the tissues are kept from becoming cold when the fish swim through cold water.

A family tree (phylogeny) has been developed for the warm-bodied teleost fish and their close relatives (**FIGURE 10.48**), to provide a basis for better understanding the evolution of the warm-bodied condition.[41] In a manner similar to the family trees discussed in Chapter 3, this family tree is based entirely on information other than physiology and thus is independent of physiological knowledge of the fish.

One of the physiological features mapped onto the tree is endothermy in the red swimming muscles. Specifically, all the little red boxes represent fish with red-muscle endothermy. When endothermy is mapped in this way onto the independently derived family tree, we can see that red-muscle endothermy probably appeared just once in the evolutionary history of these fish, at the spot marked A. The letter B marks the spot in evolutionary history where the red swimming muscles started to shift to an unusual, deep location near the spinal column. Evidently, the new red-muscle position evolved first, and then tunas (but not bonitos) capitalized on it to evolve red-muscle endothermy. The concept that the new red-muscle position set the stage for the evolution of endothermy in the red swimming muscles is bolstered by information on the lamnid sharks; they and the tunas exhibit a remarkable convergence in the mechanics of how they swim, and part of that convergence is that the lamnids as well as the tunas have red swimming muscles positioned in an unusual position near the spinal column.

If we reflect on the warm-bodied fish discussed up to now, we see from Figure 10.48 that the warm-bodied condition (endothermy) has evolved independently four times in fish: at spots A, C, and D in the figure, plus at least one additional time in the lamnid sharks. Now, based on research published in 2015, we know of a fifth independent origin in a large (40 kg), poorly known fish, the opah

---

[40] In tunas and lamnids, the rate of heat production in each warmed organ or tissue is believed to be simply the ordinary rate, based on available evidence.

[41] The sharks, which are not teleosts, were not included in the study to produce the family tree.

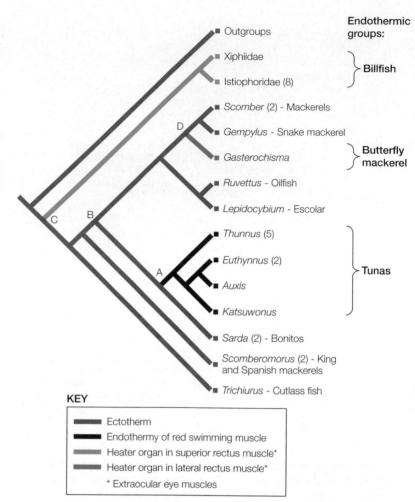

FIGURE 10.48 A family tree (phylogeny) of teleost fish belonging to the suborder Scrombroidei Physiological features (see key) are mapped onto the family tree. The tree, however, was derived entirely independently of physiological information, being based on an analysis of DNA nucleotide sequences in the gene for mitochondrial cytochrome *b* in the species included. The colors in the little boxes on the right side indicate the actual, known physiological nature of the various modern-day fish specified. The coloring of the lines of the family tree represents an interpretation of past history. Endothermy occurs only in the groups so identified at the right side of the diagram. Numbers are numbers of species studied if greater than one. "Outgroups" are other species of fish used to establish a base of comparison. The text explains the letters at branch points. (After Block and Finnerty 1994.)

(*Lampris guttatus*). The opah is distinctive in that its entire body is maintained at a temperature above ambient water temperature. Tissue temperatures in the opah reach levels 5°C higher than ambient temperature—significant warming, although far short of that sometimes seen in bluefin tunas (see Figure 10.47). As in all the other warm-bodied fish, the "secret" to endothermy in the opah is the presence of vascular countercurrent heat exchangers that prevent metabolically produced heat from being carried freely to the gills, where it would be readily lost. In the opah, the countercurrent heat exchangers are located *in the gills*, thereby enabling the entire body—other than the gills—to be warmed.

SUMMARY
Warm-Bodied Fish

■ Tunas, lamnid sharks, and billfish are distinguished from other fish by exhibiting endothermy in certain body regions, and the opah displays whole-body endothermy. The tissues that are endothermic in tunas and lamnids are (1) the red swimming muscles and (2) sometimes the stomach, other viscera, brain, and retinas. In billfish, only the brain and retinas are endothermic.

■ A countercurrent vascular array that short-circuits outflow of heat from a tissue—thereby preventing the heat from reaching the gills—is required for the tissue to be endothermic.

■ Ordinary metabolic heat production is the source of heat for endothermy in all cases except the billfish, which have specialized "heater" tissues derived from extraocular eye muscles.

Endothermy and Homeothermy in Insects

A solitary insect at rest metabolizes at a sufficiently low rate that no part of its body is warmed by its metabolic heat production. Insects in flight, however, often exhibit very high metabolic rates; species that are strong fliers in fact release heat more rapidly per gram than active mammals or birds. This high rate of heat production is localized in the flight muscles in the thorax. Given that insects do not have the profound problems of retaining heat that characterize water-breathers, it is quite possible for the thorax to be warmed by the high metabolism of the flight muscles during flight, and thus, as we saw at the beginning of this chapter, the thorax may be endothermic.

Some insects that display thoracic endothermy during flight do not thermoregulate; examples are provided by certain species of small geometrid moths, which maintain a thoracic temperature that is about 6°C above air temperature regardless of what the air temperature is. Other sorts of insects physiologically thermoregulate during flight and thus exhibit *thoracic homeothermy*. The thermal relations of endothermic insects are particularly complex because they exhibit both *temporal* and *spatial* heterothermy. The insects exhibit endothermy only when they are active, not when they are resting. Moreover, even when they exhibit endothermy, they usually do so just in their thorax, not their abdomen.

Historically, sphinx moths were the first group of insects discovered to display thoracic physiological thermoregulation during flight, and to this day they are model examples of the phenomenon. Sphinx moths are strong fliers and often (for insects) are particularly large; some species weigh as much as several grams and thus are similar in weight to some of the smallest mammals and birds. Flying sphinx moths closely regulate their thoracic temperatures. The species shown in **FIGURE 10.49**, for example, maintains its thoracic temperature within a narrow range, 38°–43°C, over a wide range of air temperatures. Thermoregulation is not limited just to insects of such large body size. Worker bumblebees (*Bombus vagans*), averaging 0.12 g in body weight, for instance, maintain thoracic temperatures near 32°–33°C whether the air temperature is 9°C

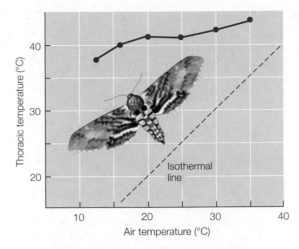

FIGURE 10.49 The average thoracic temperature of freely flying sphinx moths (*Manduca sexta*) as a function of air temperature The isothermal line shows how thoracic temperature would vary with air temperature if there were no endothermy or thermoregulation and the temperature of the thorax simply equaled air temperature. (After Heinrich 1971.)

or 24°C when they are foraging. Honeybees, averaging 0.09 g in body weight, exhibit impressive thoracic thermoregulation over a somewhat narrower range of air temperature, and also illustrate the usual insect pattern that—at moderate to cool air temperatures—the abdominal temperature tends approximately to match air temperature (**FIGURE 10.50**). The list of insects known today to exhibit thoracic homeothermy during flight also includes many other lepidopterans and bees, some dragonflies, and some beetles.

Although endothermy and physiological thermoregulation occur principally during flight in insects, a few types of insects display the phenomena during solitary terrestrial activities. In nearly all such cases, the primary source of heat is the flight muscles, which instead of being used to fly, are activated to "shiver" (as we discuss shortly). Dung beetles—which transport energy-rich elephant dung or other dung to preferred locations by forming the dung into balls—sometimes become markedly endothermic while working in dung piles and rolling their dung balls. Some crickets and katydids thermoregulate while they sing.

The insects that thermoregulate during flight require certain flight-muscle temperatures to fly

The flight muscles of an insect must be able to generate mechanical power at a certain minimum rate (which is species-specific) for the insect to be able to fly. Within a broad range of temperatures, the power output that flight muscles can attain increases as their temperature increases. Thus the temperature of an insect's flight muscles is potentially an important determinant of whether the insect can fly.

Tiny insects such as fruit flies, mosquitoes, and midges have such high surface-to-volume ratios that the activity of their flight muscles cannot warm the thorax significantly. Correlated with their inability to be endothermic, the tiny insects commonly can fly with very broad ranges of thoracic temperatures, including, in some species, thoracic temperatures as low as 0°–5°C. An important property of the flight physiology of these tiny, poikilothermic

(A) Temperatures of thorax and abdomen

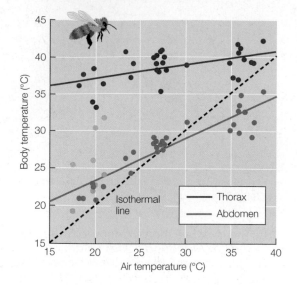

(B) Metabolic rates

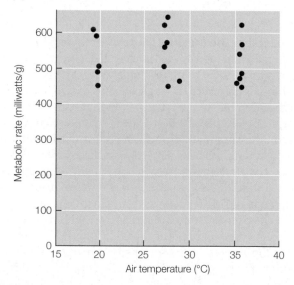

FIGURE 10.50 Temperature and metabolism in steadily flying honeybees (*Apis mellifera*) Honeybees vary considerably in how well they fly under controlled circumstances, and the data presented here are specifically for individuals that flew without prodding. (A) Temperatures in the thorax and abdomen; lighter-colored symbols are for four bees that showed low wing-beat frequency while flying. (B) Metabolic rates during flight. (After Woods et al. 2005.)

fliers is that they apparently require only a modest fraction of their maximum power output to stay aloft; thus they can fly at relatively low thoracic temperatures, at which their power output is substantially submaximal.

In sharp contrast, many medium-sized and large insects, including the species known to thermoregulate, require a near-maximum power output from their flight muscles to take off and remain airborne. They therefore require that their flight muscles be at high temperatures to fly. The sphinx moth *Manduca sexta*, for example, cannot fly unless its thorax is at least as warm as 35°–38°C, and worker bumblebees (*Bombus vagans*) require about 30°C.

The need for high flight-muscle temperatures for flight raises the question of how resting insects are able to get warm enough to take off. Because insects typically cool to environmental temperature when they are fully at rest, an insect that requires a high flight-muscle temperature to fly will often be too cold to take off after it has been resting for a while. Diurnal species may be able to warm their flight muscles to flight temperature by basking in the sun. Most species, however, have an endogenous ability to warm their flight muscles to flight temperature, a phenomenon known as **physiological preflight warm-up**.

Physiological preflight warm-up is accomplished by contraction of the flight muscles in a nonflying mode, a process often called **shivering** (not homologous to vertebrate shivering). Several forms of shivering are known. In many types of insects, including moths and butterflies, what happens during shivering is that the muscles responsible for the upstroke and downstroke of the wings contract simultaneously (rather than alternately as they do in flight), thus working against each other. The wings merely vibrate during shivering, rather than flapping, but heat is evolved by the muscular contraction, warming the flight muscles. When a sphinx moth warms from a low temperature, its flight muscles shiver in this manner at an ever-higher intensity as its thoracic temperature increases to the flight level. Then suddenly the pattern of muscular contraction changes, the wings are driven through the flapping motions of flight, and the moth takes to the air.

Solitary insects employ diverse mechanisms of thermoregulation

Innovative investigators continue to progress in understanding the mechanisms that insects employ to thermoregulate, despite the obstacles of working on such small animals.

As the ambient temperature drops, one mechanism of maintaining a constant thoracic temperature is for an insect to increase its rate of heat production, much as mammals and birds do below thermoneutrality. Many insects do this when they are *not flying*. Heat is generated in these circumstances by shivering of the flight muscles, and because the muscles can engage in various intensities of shivering, they can modulate their rate of heat production to serve thermoregulatory needs. Honeybees and bumblebees working in the hive, for example, often maintain high and stable body temperatures for long periods by increasing and decreasing their rates of shivering heat production as the air temperature falls and rises. An intriguing example is also provided by the brood incubation of queen bumblebees (**FIGURE 10.51**). A queen, which overwinters alone and thus is solitary when she rears her first brood in the spring, incubates her brood by keeping her abdomen at an elevated temperature and pressing it against the brood. Heat is brought to her abdomen from her thorax, where it is produced by her flight muscles. As the ambient air temperature falls, the queen thermoregulates by increasing her rate of heat production (see Figure 10.51).

Modulation of shivering can also be used to thermoregulate during *intermittent flight*. Bumblebees are known to do this, for instance. As a bumblebee, such as that pictured at the start of this chapter, flies from flower to flower during foraging, it can shiver or not shiver while it is clinging to each flower. More shivering of this sort occurs as the air temperature falls, and thus the bumblebee's overall, time-averaged metabolic rate increases as air temperature decreases.

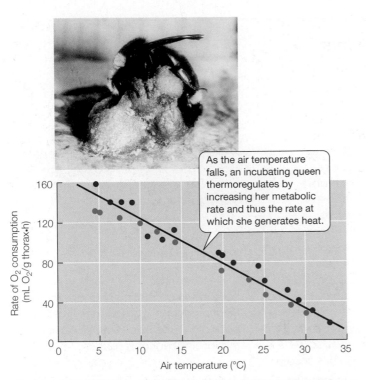

FIGURE 10.51 The rate of O₂ consumption by queen bumble-bees as a function of air temperature when they are incubating their broods In the species studied, *Bombus vosnesenskii*, a queen incubates her brood by pressing her abdomen against it as seen in the photograph. In the plot, the two colors of symbols refer to two different individuals. (After Heinrich 1974; photograph courtesy of Bernd Heinrich.)

When insects *fly continuously*, their flight muscles are employed in flight movements all the time and cannot shiver. Investigators hypothesized years ago that under these circumstances, the rate of heat production by the flight muscles would be determined by the requirements of flight and not modulated to serve thermoregulation. Early, seminal experiments on sphinx moths supported the truth of this hypothesis, because when the moths flew at a certain speed, their metabolic rates were essentially constant whether the air temperature was 15°C or 30°C. Honeybees in flight often (although not always) are similar (see Figure 10.50B). If insects in continuous flight do not modulate their rates of heat production as a means of thermoregulating, how do they thermoregulate?

Studies of sphinx moths, bumblebees, and some other insects reveal that their primary mechanism of thermoregulation during continuous flight is much akin to that used by mammals and birds in the thermoneutral zone; namely, they vary their insulation—in this case their *thoracic insulation*. A flying insect keeps its thorax at a steady temperature by modulating how readily heat can exit the thorax. This modulation is accomplished in some moths, dragonflies, and bumblebees by control of the rate of blood flow between the thorax and abdomen. In a continuously flying sphinx moth, for example, when the air temperature is low, the heart beats weakly and blood circulates slowly between the thorax and abdomen; thus heat produced by the flight muscles tends to remain in the thorax, which retains the heat effectively because it is densely covered with furlike scales. As the air temperature is raised, the heart beats more vigorously and circulates blood to the abdomen

more rapidly; in this way, heat is transported at an increased rate out of the thorax. Honeybees sometimes carry out an analogous process in which they modulate blood transport of thoracic heat to the head; at elevated air temperatures, heat is transported at an increased rate into the head, where it is lost in part by evaporation of fluid regurgitated out of the mouth.[42]

Colonies of social bees and wasps often display sophisticated thermoregulation

Physiological regulation of colony temperature is widespread within colonies of social bees and wasps. Honeybee (*Apis mellifera*) hives provide the best-studied example. Thermoregulation by honeybee hives is so dramatic that it was recognized for almost two centuries before thermoregulation by solitary insects was first demonstrated.

A honeybee hive that is rearing a brood maintains the temperature of its brood combs within a narrow range, about 32°–36°C, even if the air temperature outside the hive falls to −30°C or rises to +50°C. When the air outside the hive is cold, worker bees cluster together within the hive and shiver. When the air outside becomes warm enough that the hive is threatened with metabolic overheating, workers disperse within the hive and fan with their wings in a cooperative pattern that moves fresh air from outside the hive across the brood combs. At very high outside air temperatures, workers also collect water and spread it within the nest, where it evaporates into the airstream produced by fanning. Honeybees provide an outstanding example of coevolution between thermal requirements and thermoregulation. Their broods of young must have temperatures of about 32°–36°C for proper development. Thus sophisticated thermoregulation of the hive by the workers is essential for the hive's reproductive success.

SUMMARY
Endothermy and Homeothermy in Insects

- Many solitary insects, especially those of medium to large size, display thoracic endothermy or homeothermy during flight or certain other sorts of activity. Warming of the flight muscles increases their power output. Often in these insects, a certain minimum flight-muscle temperature is required for flight.

- When insects are not flying, activation of the flight muscles in a nonflight mode—termed *shivering*—is the mechanism they employ to warm the thorax. Shivering is used for preflight warm-up. Nonflying insects also sometimes thermoregulate by modulation of shivering, as observed in bees working in their hives.

- When insects are flying, the best-known mechanism of thermoregulation is modulation of thoracic insulation, brought about by raising and lowering circulatory transport of heat out of the thorax.

- Colonies of social bees and wasps sometimes employ group efforts to maintain exquisitely stable hive temperatures.

[42] Over the last 20 years, the original paradigm of thermoregulation during continuous flight—which holds that all flying insects thermoregulate by modulating thoracic heat loss but not heat production—has been challenged by studies showing that occasionally some species modulate their metabolic rate.

Coda

Endothermy can provide organisms with distinct advantages. Accordingly, despite the fact that endothermy usually has a high energy cost, it has evolved independently in animals multiple times. It even occurs in a few plants (**BOX 10.4**).

STUDY QUESTIONS

1. As discussed in Chapter 1 (see page 13), Claude Bernard, a nineteenth-century French physiologist often considered the father of modern animal physiology, is still remembered today for his famous dictum: "Constancy of the internal environment is the condition for free life." Does the study of thermal relations lend support to his dictum? Explain.

2. There is currently a worldwide movement to create protected marine parks. If the parks have an Achilles heel, it is that they have fixed geographical positions, just in the way that Yellowstone National Park is at a fixed geographical location. Suppose that a certain endangered species of fish exists only in a marine park. If the ocean temperature rises in the park because of global warming, explain what physiological problems the species of fish might confront. How might the species face a brighter future if parks could have moveable boundaries rather than fixed ones?

3. Referring to Figure 10.11, suppose you have some lizards that are at 16°C and have been living at that temperature for 5 weeks. What is their resting metabolic rate? If the lizards are suddenly shifted to a room at 33°C, trace on the graph how their metabolic rate will change from the moment they are placed in the new room until 5 weeks have passed. According to the graph, will they exhibit compensation?

4. Discuss ways that the cryobiology ("freezing biology") of insects could be manipulated to control insect pests. One factor to consider is that certain bacteria and fungi act as highly effective ice nucleators.

5. In the animal kingdom today, poikilotherms outnumber homeotherms by a great margin. Why is poikilothermy a successful way of life even though poikilotherms sometimes must compete successfully with homeotherms to survive?

6. Suppose you travel to a tropical place such as the Bahamas and watch the coastal poikilotherms, such as fish, crabs, and sea stars, swim and crawl about in the warm waters. Suppose then that you travel to northern Maine and watch the related species of poikilotherms in the cold waters there. In the abstract, it would not be unreasonable to expect to see the animals in Maine moving about in slow motion compared with those in the Bahamas. In fact, however, rates of locomotion are likely to look to your eye to be more similar than different in the two places. Design experiments to assess whether the Maine animals are especially able to be active in cold waters. If you find that they are, how might their high ability for activity in cold waters be explained? For each hypothesis you present, design an experiment to test the hypothesis.

BOX 10.4 Warm Flowers

In the early spring when snow is still on the ground, the flower structures of the arum lily called eastern skunk cabbage (*Symplocarpus foetidus*) melt their way to the snow surface by being as much as 30°C warmer than the ambient temperature. In this way, this species dramatically announces that plants have evolved endothermy! The eastern skunk cabbage in fact displays thermoregulatory properties, in that its flower structures increase their rate of metabolic thermogenesis—responsible for endothermy—as the ambient temperature becomes colder. The function of endothermy in this case is believed to be that it enhances the volatilization of odor compounds that attract pollinators. **Box Extension 10.4** discusses this fascinating topic further.

Eastern skunk cabbage (*Symplocarpus foetidus*)

---

7. During winter, when people are in a well-insulated house, they usually feel comfortable if the air temperature is near 22°C (72°F). If you have ever spent a night in a poorly insulated cabin in winter, however, you will recognize that paradoxically, when people are in poorly insulated buildings, they often feel chilly even when the air inside is heated to 22°C or higher. One important reason for the difference in how warm people feel in the two sorts of buildings is that even if a well-insulated and poorly insulated building are identical in the air temperature inside, they differ in thermal-radiation heat transfer. Specifically, a person standing in the two types of buildings experiences different heat exchange by thermal radiation in the two. Explain how thermal-radiation heat transfer accounts for the sense of chill in the poorly insulated building. (Hint: Think of the outer walls of the two types of buildings, and think specifically of the temperatures of the interior surfaces of those walls.)

8. What is homeoviscous adaptation? Although we discussed it in our study of poikilotherms, the phenomenon was actually first discovered about a century ago in studies of pigs in Sweden. Some pigs were dressed in blankets during winter while others were allowed to roam about stark naked. When their subcutaneous fat was analyzed, the two sets of pigs turned out to have laid down lipids of differing chemical composition. How could different lipids give the two groups similar lipid fluidities?[43]

9. Suppose you are trying to choose between two winter jackets. Suppose also that you have a heat-producing mannequin available for your use and you are able to adjust the mannequin's rate of heat production. According to Equation 10.10, insulation is equal to $(T_B - T_A)/M$ (this is in fact a general equation for insulation). How would you make a quantitative comparison of the insulation provided by the two jackets?

---

[43] To avoid introducing any confusion, it may be important to mention that the subcutaneous lipid deposits of mammals (which are known as *depot fats*) consist of ordinary triacylglycerols, not phospholipids like membrane lipids, but the basic concepts of homeoviscous adaptation remain the same.

10. In the rete mirabile serving the red swimming muscles of tunas, some key enzymes of catabolism show gradients of concentration: They are more concentrated at the cold end of the rete, and less concentrated at the warm end. These variations parallel variations that are often seen in the thermal acclimation of poikilotherms, when enzyme concentrations rise during acclimation to cold and fall during acclimation to heat. What do you think could be some of the reasons for these spatial and temporal variations in enzyme concentration? Why not have the highest observed enzyme concentrations everywhere at all times?

11. Humphries, Thomas, and Speakman presented a bioenergetic model to predict how global warming might cause insectivorous bats to alter the northernmost latitudes at which they hibernate. The investigators stress that not only is the model fairly simple, but also it allows the *existing* distribution of hibernation sites to be predicted reasonably well. Study their model, and assess its pros and cons. Why is the little brown bat a particularly suitable species for the application of this method? See M. M. Humphries, D. W. Thomas, and J. R. Speakman. 2002. Climate-mediated energetic constraints on the distribution of hibernating mammals. *Nature* 418: 313–316.

Go to **sites.sinauer.com/animalphys4e**
for box extensions, quizzes, flashcards,
and other resources.

REFERENCES

Cannon, B., and J. Nedergaard. 2004. Brown adipose tissue: function and physiological significance. *Physiol. Rev.* 84: 277–359. A definitive and detailed review of the biology of brown adipose tissue by two of the leaders in the field.

Cossins, A., J. Fraser, M. Hughes, and A. Gracey. 2006. Post-genomic approaches to understanding the mechanisms of environmentally induced phenotypic plasticity. *J. Exp. Biol.* 209: 2328–2336. A nicely articulated and broadly conceived discussion of the use of genomic methods to understand responses to cold in fish.

Deutsch, C., A. Ferrel, B. Seibel, and two additional authors. 2015. Climate change tightens a metabolic constraint on marine habitats. *Science* 348: 1132–1135. A challenging but rewarding paper on ways that the interaction between O₂ availability and temperature may affect marine poikilotherms in a warming world, based on the theory of oxygen- and capacity-limited thermal tolerance.

Eliason, E. J., T. D. Clark, M. J. Hague, and seven additional authors. 2011. Differences in thermal tolerance among sockeye salmon populations. *Science* 332: 109–111. A dense but particularly worthwhile paper on temperature-dependent performance and thermal tolerance in seven distinct populations of migratory salmon.

Geiser, F. 2004. Metabolic rate and body temperature reduction during hibernation and daily torpor. *Annu. Rev. Physiol.* 66: 239–274. A detailed exploration of several key themes in modern research on the comparative biology of hibernation and other hypothermic states by a leader in the field.

Heinrich, B. 1993. *The Hot-Blooded Insects: Strategies and Mechanisms of Thermoregulation.* Harvard University Press, Cambridge, MA. A searching and enlightening treatment (although becoming dated) of all aspects of insect thermal relations written by a pioneer in the field. This book does a wonderful job of putting the physiological information into the larger context of insect life histories and ecology.

Heldmaier, G., and T. Ruf. 1992. Body temperature and metabolic rate during natural hypothermia in endotherms. *J. Comp. Physiol., B* 162: 696–706. Also: Ortmann, S., and G. Heldmaier. 2000. Regulation of body temperature and energy requirements of hibernating Alpine marmots (*Marmota marmota*). *Am. J. Physiol.* 278: R698–R704. These companion papers grapple in an illuminating way with the question of what governs the metabolic rates of hibernators: their tissue temperatures or some sort of biochemical metabolic depression. The first paper takes a very broad approach. The second focuses on just one species and is simpler to follow. Geiser 2004 takes the analysis one step further.

Hill, R. W., T. E. Muhich, and M. M. Humphries. 2013. City-scale expansion of human thermoregulatory costs. *PLoS ONE* 8(10): e76238. A far-ranging paper that compares the energy costs of human thermoregulation in cold environments when people are exposed as isolated individuals to the cold and when they live in heated dwellings. The paper reveals that the energy usage of entire cities varies with ambient temperature in the same way as mammalian metabolic rate varies with ambient temperature.

Hochachka, P. W., and G. N. Somero. 2002. *Biochemical Adaptation: Mechanism and Process in Physiological Evolution.* Oxford University Press, New York. A definitive, stimulating treatment of the biochemistry and molecular biology of temperature, written by two leaders in the field (Hochachka now deceased).

Ortmann, S., and G. Heldmaier. 2000. *See Heldmaier and Ruf 1992 earlier in this list.*

Pörtner, H. O., L. Peck, and G. Somero. 2007. Thermal limits and adaptation in marine Antarctic ectotherms: an integrative view. *Philos. Trans. R. Soc., B* 362: 2233–2258. A model of broadly conceived integrative thinking, focused on Antarctic ectotherms, written by a multidisciplinary team of leaders in the field.

Rommel, S. A., D. A. Pabst, and W. A. McLellan. 1998. Reproductive thermoregulation in marine mammals. *Am. Sci.* 86: 440–448. The testicles of marine mammals are internal—a placement that helps create a hydrodynamically efficient body shape—but how are they kept cool?

Somero, G. N. 2011. Comparative physiology: a "crystal ball" for predicting consequences of global change. *Am. J. Physiol.* 301: R1–R14. A thought-provoking paper on biochemistry and global change written by a master.

Tøien, Ø., J. Blake, D. M. Edgar, and three additional authors. 2011. Hibernation in black bears: Independence of metabolic suppression from body temperature. *Science* 331: 906–909.

See also **Additional References** *and* **Figure and Table Citations**, *and* **References in Chapter 11**.

Food, Energy, and Temperature
AT WORK
The Lives of Mammals in Frigid Places

Reindeer (*Rangifer tarandus*), which occur in Siberia and other far-northern regions, typically give birth in May, when the ground remains snow-covered and the air often cools to below 0°C overnight. At the moment of its birth, a reindeer calf experiences a drop in its environmental temperature from 37°C inside its mother's uterus to the prevailing air temperature. If the air temperature is –3°C, the calf's environmental temperature plummets by 40°C at birth. The air sometimes is much colder, and some calves experience a drop of 50°–60°C when they are born. Newborns are wet—covered with amniotic fluid—and a strong wind may blow.

Reindeer calves must thermoregulate on their own from the moment they are born, or die, because they do not huddle with each other or with adults, and they have no nest to protect them. They are perhaps the most precocial (adultlike) of all the newborns of land mammals. They stand the moment they are born. When 2 days old, they can run faster than a person. Because herds typically move from place to place incessantly to find food, the newborn calf has no luxury of resting to gather strength. It must keep up. Within a week, a reindeer calf can swim across broad rivers.

As stressed in Chapter 4, the physiology of young animals is as important as the physiology of adults, because each individual animal must survive first as a youngster if it is ever to have the chance of surviving (and reproducing) as an adult. We will return to the young of reindeer shortly. First, however, let's focus on adult reindeer and the environment in which they live—matters that set the context for fully understanding the young.

A newborn reindeer calf must thermoregulate on its own
Although mother reindeer provide milk to their young, they do not keep them warm, and the Arctic environment where birth occurs may be very cold.

Food, Nutrition, Energy Metabolism, and Thermoregulation in the Lives of Adult Reindeer

Reindeer, as a species, are probably the most adapted of all inland mammals to cold exposure; some herds live year-round in places where the average annual temperature is below –5°C. An intriguing aspect of animal species that are extremely well suited physiologically to live in stressful habitats is that their very presence can permit *other* species to exist there as well. In deserts, the existence of rodents that have evolved extremely low requirements for ingested water permits many water-dependent predators to exist as well; the rodents provide the watery food the predators need. Similarly, in the Far North, the existence of reindeer permits the existence there of wolves and other predators—including indigenous human cultures. During preindustrial times, people could not possibly have survived in the interior of the Far North on their own. The existence of humankind there was made possible, all around the Arctic, by herds of reindeer, which served as sources of food and of highly insulating pelts that people employed for necessary shelter and clothing. The same points still apply today to many indigenous human cultures that live apart from modern comforts, such as communities of Sami people who live in traditional ways in Scandinavia.

Because of the intimate relations between reindeer and people over long periods in the past, reindeer were domesticated to some degree in many regions. Today, therefore, the species consists of both wild populations and numerous domesticated strains or races. Caribou are considered by most mammalogists to be the same species. As adults, reindeer in some populations are roughly the same size as humans.[1]

Reindeer have many characteristics that help them prosper in the Far North. Although adult reindeer weigh, on average, about one-third as much as adult moose (another northern species), the feet of reindeer are so unusually broad that the contact area of a reindeer with the ground is about the same as that of a moose. Accordingly, the downward gravitational force on each unit of area of a reindeer's feet is very low compared with that of a moose, explaining how reindeer can readily negotiate snowfields that stop moose.

Adult reindeer in their winter pelage have lower-critical temperatures of –30°C or lower in still air. The air temperature in the Arctic rarely goes below –50°C. Thus the air temperature does not drop far below their lower-critical temperature, and reindeer in their winter pelage in still air never have to increase their metabolic rates by much above basal (**FIGURE 11.1**). As an amusing comparison, Laplanders and Norwegians equipped just with the insulation "nature gave them" (i.e., naked) have lower-critical temperatures of +24°–27°C! Only by dressing in something like a reindeer pelt can a person have a lower-critical temperature that is compatible with Arctic life.

One reason for the low energy costs of adult reindeer in the frigid air of winter is their pelage. A dense underfur of fine hairs thickly covers their skin, and the longer "guard" hairs of the pelage, which protrude beyond the underfur and hang over its outer

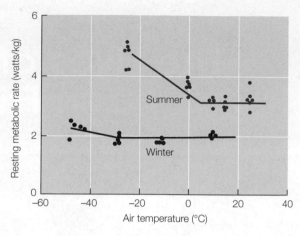

FIGURE 11.1 Resting metabolic rate as a function of air temperature down to –50°C in adult reindeer The same Norwegian animals, weighing about 70 kg, were studied in winter and summer. The air was still. Lines are drawn by eye to approximate the trends in the data. (After Nilssen et al. 1984.)

surface, are filled with tiny air spaces (like microscopic bubble wrap) (**FIGURE 11.2**). The air inside these hairs is motionless, meaning that each hair in itself has a high insulative value (see page 238). In common with other large species of mammals, reindeer undergo dramatic seasonal molts. In winter, a reindeer's fur is 3–4 cm thick over much of its body (**FIGURE 11.3**). All body surfaces of a reindeer, even the nose, have a hair covering.

Besides the pelage, another reason for the low energy costs of adult reindeer in frigid air is that they employ regional heterothermy. Reindeer do not keep the tissues of their legs and other exposed body parts as warm as their body core (see Figure 10.32).

Reflecting on what we learned about lipid fluidity in our study of fish and other poikilotherms in Chapter 10, an interesting question about the heterothermic legs of reindeer is whether the lipids in their legs are hard and stiff where the limbs are cold, much as butter is hard when cold. Actually, it is a matter of ancient knowledge that

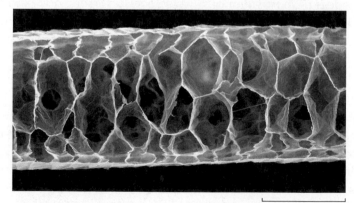

0.1 mm

FIGURE 11.2 A reindeer hair in longitudinal section This image, obtained by scanning electron microscopy, shows the open cellular ("bubblelike") structure of the interior of the hair. (With thanks to Nigel D. Meeks and Caroline R. Cartwright, Department of Scientific Research at the British Museum © The British Museum; after Meeks and Cartwright 2005.)

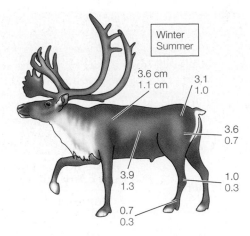

FIGURE 11.3 Fur thickness of adult reindeer in winter and summer The thickness was measured perpendicular to the skin surface and is expressed in centimeters. (After Johnsen et al. 1985.)

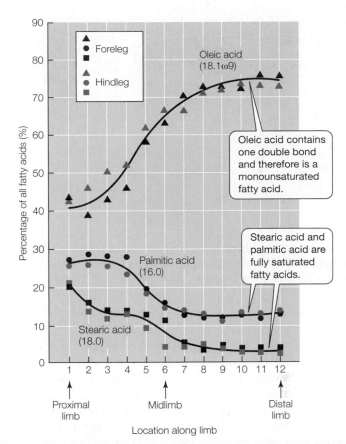

FIGURE 11.4 Fatty acid composition of bone marrow lipids in the legs of reindeer Marrow lipids were sampled at 12 locations from the proximal (upper) end of each limb to the distal (lower) end. The proximal locations were the proximal ends of the humerus and femur; the midlimb locations were the distal ends of the radius and tibia; the distal locations were the third phalanges. In the numbering system for the fatty acids (discussed in Chapter 6, see pages 136–137), the number before the decimal is the number of carbon atoms; that following the decimal is the number of double bonds; and that following ω (omega) designates the position of a key double bond (ω is not written when there are no double bonds). (After Meng et al. 1969.)

lipids from the outer extremities of reindeer legs—and also lipids from the hooves of cattle—are particularly fluid, compared with body-core lipids. People have long used lipids (oils) from the outer extremities to soften boot leather and give flexibility to leather bow strings in cold climates.

Homeoviscous adaptation exists from one end of a heterothermic leg to the other: Whereas the lipids in the upper leg of a reindeer or cow have chemical structures that give them a gel-like texture at 37°C and cause them to become hard at 0°C, the lipids from the outer extremities have different chemical structures that give them a gel-like or oily texture near 0°C. One way to examine the spatial diversity of chemical structure is to look at the abundances of key, diagnostic fatty acids in the marrow lipids of the limb bones. As seen in **FIGURE 11.4**, oleic acid—an unsaturated fatty acid—becomes dramatically more abundant as a constituent of the marrow lipids as one moves out along a leg, whereas palmitic acid and stearic acid—both saturated fatty acids—become less abundant. In this way, the bone marrow lipids in the legs of reindeer are reminiscent of the brain phospholipids of fish from various climates (see Figure 2.3). The same basic trends exist even in the legs of many tropical mammals. Thus, although the trends are significant for reindeer, they do not seem to be specific adaptations to a truly frigid climate.

Food and nutrition represent great challenges in the environments where reindeer live. Like other deer, reindeer are ruminants. Thus rumen processes, as well as digestive and absorptive processes, play pivotal roles in their physiology of food and nutrition. A key to the survival of reindeer in the Far North is that they eat a great diversity of plants (37 genera were found in the rumens of one herd). Moreover, they obtain an exceptional degree of nutritional benefit from eating species of lichens ("reindeer moss") and a variety of other species of far-northern plants that are not much eaten by other mammals.[2] The exact mechanisms by which reindeer are able to exploit their unusual foods remain poorly known, although studies of ingestion and egestion show, for example, that reindeer obtain twice the nutrient value from lichens as sheep or cows do. Some lichens produce and accumulate toxic phenolic compounds such

as usnic acid, raising the question of how reindeer handle these compounds. Recent research points to degradation by specialized rumen microbes. Usnic acid in ingested lichens is evidently fully broken down during rumen processing and never enters the tissues of the reindeer.

The seasonal cycle of plant growth in the Far North is dramatic—not just because winters are cold, but also because of winter darkness (in some places that reindeer occupy, the sun does not rise above the horizon for 3 months in winter). The summer foods of reindeer in most places are collectively high in protein and mineral nutrients, and they are high in the proportion of total carbohydrate that is in readily digestible forms (rather than cellulose or hemicelluloses). Winter foods are a different story.

Reindeer often depend heavily on lichens for food in the winter. This has been known for years from visual observation of their feeding and has recently been confirmed using genomic methods. By analyzing the DNA in reindeer fecal pellets and applying DNA bar-coding methods, investigators have obtained impressively detailed information on the foods wild reindeer consume.

---

[2] For simplicity of language, we use the term *plants* in a loose, comprehensive sense to refer to all the photosynthetic organisms that reindeer eat, including lichens and mosses, as well as true plants.

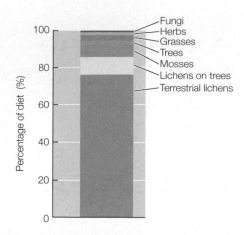

FIGURE 11.5 Diet composition of winter reindeer as determined by DNA bar coding DNA was extracted from feces and analyzed to determine the foods eaten by 125 individuals. (After Newmaster et al. 2013.)

TABLE 11.1 Responses of the rumen microbes of reindeer to seasonal changes in diet

The mixed communities of fermenting microbes in rumen fluid were classified using standard microbiological methods. Compared with the summer microbial community, the winter microbial community was more able to ferment plant fiber, including particular structural carbohydrates such as cellulose, and less able to carry out hydrolysis of proteins.

Food breakdown activity	Percentage of microbes that could carry out each activity (%)	
	Summer	Winter
Fiber digestion	31	74
Cellulose digestion	15	35
Hemicellulose (xylan) digestion	30	58
Proteolysis (protein hydrolysis)	51	28
Starch digestion	68	63

Source: After Orpin et al. 1985.

In winter, lichens may account for over 80% of the diet (**FIGURE 11.5**). Lichens are often the salvation of reindeer in winter *in terms of supplying energy* because they are abundant, and as we have already said, reindeer can tap a high proportion of their nutrient value. However, lichens are far from being nutritionally complete because they are low in protein and low in minerals such as Na⁺. **FIGURE 11.6**, which summarizes the composition of the foods available to reindeer throughout the year in Finland, is well worth close study, because it illustrates dramatically that animals in the wild—away from veterinarians and manufactured feeds—often face substantial nutritional stresses. During winter the lichens and senescent vascular plants that dominate the diets of reindeer are collectively low in protein, low in multiple minerals, low in highly digestible carbohydrates, and high in cellulose and hemicelluloses.

Reindeer and the microbial symbionts in their rumens make adjustments as the seasons change. For instance, certain strains of reindeer fatten dramatically as winter approaches, thereby reducing their need for winter food. In addition, the community of fermenting microbes in the rumen changes in composition in ways that respond to the shifts in the types of foods eaten. **TABLE 11.1** presents one

example, indicating that microbes that digest woody, fibrous plant material (including cellulose and hemicellulose) increase in winter when reindeer ingest considerable amounts of such material. A recent study showed that the mixed rumen microbial community in Norwegian reindeer became four to six times more capable of breaking down lichens when lichens were chronically in the diet than when lichens were chronically not eaten.

Despite such adjustments, by the time spring arrives after a long winter, reindeer have lost body weight, often are somewhat emaciated, often exhibit other signs of having been in negative nitrogen balance for months, and may exhibit blood mineral levels diagnostic of mineral deficiency. In some places, they are renowned for having high "mineral appetites" in spring.

In addition to the physiological and anatomical features we have already discussed, the distinctive feeding behaviors and migratory behaviors of reindeer herds play key roles in their success in the Far North. Food is thinly distributed there, even for animals that feed

FIGURE 11.6 Seasonal changes in the protein and mineral content of the foods eaten by Finnish reindeer The foods available to—and eaten by—reindeer vary from place to place. The particular seasonal changes seen here are not, therefore, observed everywhere. (After Nieminen 1980.)

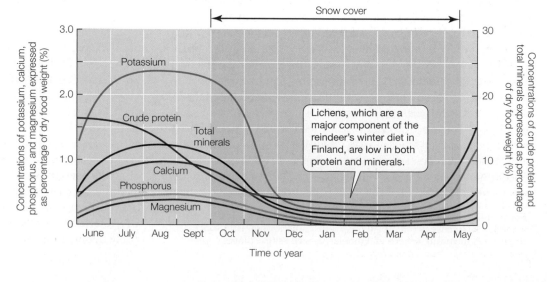

on a diversity of plants. Thus herds must range widely to obtain enough food. Reindeer herds are legendary, in general, for being incessantly on the move during daylight hours, covering large areas of ground every day, nibbling as they go. The extraordinary annual migrations of reindeer herds are additional behavioral adaptations, helping them to find not only food but also shelter from wind during winter. Although some herds do not migrate, most do. As winter approaches, the norm is for a herd to migrate about 1000 km from its summer site to its winter site. Some travel farther: Recent satellite-based tracking studies in Alaska and the Yukon revealed that some herds make a round-trip of 5000 km per year. Of all animals that travel by walking or running, reindeer migrate the greatest distances!

The movements of the adults in a herd are a part of reality for all calves born into the herd. Calves must join in the wide-ranging daily movements of their herd soon after birth, as we have already mentioned, and by autumn of their first year of life, they must be ready for the annual migration.

Newborn Reindeer

When reindeer are born, they already have a well-developed pelage, consisting of woolly, hollow hairs. Their fur provides substantial insulation as soon as the uterine fluids have evaporated away and the hairs are dry. From the moment of birth, reindeer also exhibit a typical homeothermic relation between their metabolic rate and the air temperature, as shown by the red line in **FIGURE 11.7**. Newborns are able to raise their rate of metabolic heat production to at least twice their resting rate. By virtue of the combined effects of their pelage insulation and this thermogenic ability, newborns are able to keep their body temperatures at 39°–40°C when the air is –20°C to –25°C (a difference of 60°C or more) for at least a few hours in still air. This performance probably represents the pinnacle or near-pinnacle of thermoregulatory ability among all the terrestrial newborns on Earth.[3] That said, it remains true that reindeer are born into an environment that can be very harsh, and the thermoregulatory abilities of newborns are far inferior to those of adults. Many newborns die if they get wet from precipitation, or if the wind blows briskly or the air temperature remains very low for a day or more.

Reindeer calves grow rapidly compared with other deer. Partly as a consequence of their increasing body size, their metabolism–temperature relation becomes noticeably more favorable with each passing week (see Figure 11.7). When calves are 2 weeks old, the weight-specific rate of heat production that they require to stay warm at an air temperature of –20°C is already reduced to only 70% of that required at birth. For understanding the rapid growth of reindeer calves, it is undoubtedly significant that the milk produced by their mothers is about 20% lipid, compared with about 4% lipid in cow's milk. Reindeer milk is accordingly very energy dense (see Table 6.3), having about three times the energy value per liter as cow's milk has. It is also particularly rich in protein. The nutrient-rich milk of reindeer aids the rapid growth of the calves. In addition, calves may start eating vegetation within 2–3 days of being born,

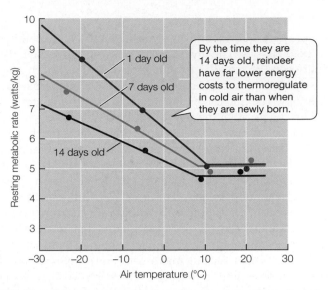

FIGURE 11.7 Resting metabolic rate as a function of air temperature in newborn and growing reindeer The air was still when these measurements were made. Dots are averages for the sets of animals studied at various conditions. (After Markussen et al. 1985.)

and by 2 weeks after birth they eat considerable amounts of plants, while continuing to nurse. By late autumn of their first year, when they are 5–6 months old and fully stop nursing, they have reached 50%–60% of their adult weight. This extent of growth in the first season of life far exceeds the average for other, related mammals and is believed to be important in enabling young reindeer to participate successfully in their herd's long migration to its wintering area.

A great deal of interest has focused on the mechanisms by which newborn reindeer and other young mammals increase their metabolic heat production for thermoregulation. Nonshivering thermogenesis (NST) by brown fat plays an extraordinary role in this regard. Brown fat is nearly always the principal thermogenic tissue in newborn placental mammals, and NST is thus the principal mechanism of thermogenesis in newborns (**BOX 11.1**). *Why* brown fat and NST should be of paramount importance in newborns is largely an unresolved mystery, as we discuss later.

The ways in which brown adipose tissue and NST have been identified in reindeer calves illustrate well the steps that physiologists typically take to identify the tissues and mechanisms responsible for physiological processes. The first study indicating that brown

---

[3] Newborn muskox (*Ovibos moschatus*) may be superior, partly because of greater body size. At birth, reindeer calves weigh about 4–5 kg, whereas muskox calves weigh about 8 kg (on average, adult muskox are about twice the size of adult reindeer). Newborn muskox have a highly insulating pelage and large deposits of brown fat.

| **BOX 11.1** | **Knockout Mice Document the Importance of Brown Fat** |

Molecular genetic tools have been used to produce laboratory mice that cannot synthesize the type of mitochondrial protein, *uncoupling protein 1* (*UCP1*), that mediates nonshivering thermogenesis (NST). The gene coding for UCP1 is inactivated. As explained in **Box Extension 11.1**, research using these knockout mice has provided strong support for two critical concepts: (1) brown fat is the sole tissue in which NST occurs in mice, and (2) UCP1 is the only molecular form of UCP that mediates NST.

fat occurs in newborn reindeer involved microscopic examination of adipose tissues. The investigators observed that the fatty tissue between the scapulae (shoulder blades) of newborn calves is reddish brown and, suspecting it to be brown fat, prepared it for microscopy. The microscopic approach to identifying brown fat is based on the fact that the tissue usually differs from white fat in several structural ways. For example, brown fat is much richer in mitochondria than white fat is, and its intracellular fat deposits are multilocular (meaning there are many small lipid droplets per cell) rather than unilocular (one droplet per cell) as in white fat. Microscopic studies performed in about 1980 indicated that *most of the major deposits of adipose tissue in the body of a newborn reindeer are brown fat*. Nonetheless, physiologists assume that function, not just morphology, must ultimately be directly studied for function to be understood. Thus, soon after microscopy had established the likely presence of brown fat in newborn reindeer, investigators tested calves to determine their responses to injected norepinephrine.

As discussed in Chapter 10, thermogenesis by brown fat is ordinarily activated by the sympathetic nervous system, secreting norepinephrine. On the basis of this fact, an animal's response to a norepinephrine injection has often been used as an informative (although relatively crude) test for brown-fat function. When newborn reindeer are injected subcutaneously with a standard dose of norepinephrine, they respond with a large increase in their rate of O_2 consumption (**FIGURE 11.8**). This response to norepinephrine, considered together with the microscopic observations discussed earlier, convinced physiologists that brown fat and NST are important for heat production in newborn reindeer.

Always, however, thoughtful scientists are wondering if their standards of evidence are adequate. Anyone who has spent months of his or her life gathering data becomes aware that there is a risk of being fooled. Maybe, one fears at times, the hard-won data do not say exactly what we have imagined they say. Because of these concerns, people using the microscopic approach kept reassessing whether the microscopic criteria used to identify brown fat were infallible. By the last decade of the twentieth century, these inves-

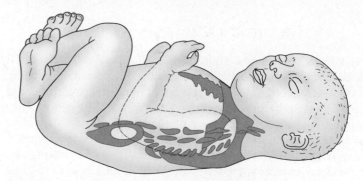

FIGURE 11.9 Brown adipose tissue in a human infant The tissue occurs in discrete masses in many parts of the body. (After Hull 1973.)

tigators had concluded—from hundreds of studies on brown fat in various mammals—that in fact the traditional microscopic criteria are not 100% reliable. On unusual occasions, when the traditional criteria are used, white fat can seem to be brown fat, or vice versa.

A third test for brown fat was therefore devised. It employed a distinctive molecular marker: the mitochondrial protein *uncoupling protein 1* (*UCP1*; see Figure 8.4), which is believed to occur exclusively (or virtually exclusively) in brown fat, not white fat or other tissues. Antibodies to UCP1 can be produced in a laboratory. Then, UCP1 in a newborn reindeer or other animal can be unambiguously identified by its antigen–antibody reaction (immunocytochemistry). Studies have shown that all of the major adipose-tissue deposits in newborn reindeer react with UCP1 antibodies, demonstrating more convincingly than ever that the tissues are brown fat.

Reindeer newborns are not alone in being well endowed with brown fat. In fact, most placental mammals, including humans (**FIGURE 11.9**), have extensive masses of brown fat at birth (pigs are exceptions, as discussed in **BOX 11.2**). In large-bodied species, such as reindeer and humans, the brown fat of newborns typically declines rapidly with age. This decline is particularly rapid in ruminants. In reindeer, sheep, goats, and other species of ruminants that have been studied, all the brown fat present in newborns undergoes a genetically programmed decline fast enough that there is no brown fat by 1 month of age; this is demonstrated by the fact that in 1-month-old animals, the gene for UCP1 is no longer expressed and no tissue reaction with UCP1-specific antibodies occurs. Simultaneously, the metabolic response of the young animals to a norepinephrine injection declines substantially (see Figure 11.8).

In the sorts of species we are discussing, as the capacity of a young mammal for *nonshivering* thermogenesis declines, *shivering* becomes more important as a source of heat production; ultimately, shivering becomes the sole mechanism of increasing heat production for thermoregulation. This transition is quite obvious in people; whereas we rely principally on nonshivering thermogenesis when we are newly born, we become dependent principally on shivering thermogenesis in youth and adulthood. In reindeer, as already said, the transition occurs relatively rapidly. Shivering becomes their only substantial mechanism of increasing metabolic heat production for thermoregulation by the time they are 1 month old, as far as is now known. This is not to say that young reindeer shiver a lot. By the time reindeer are 1 month old, the warm air temperatures of summer have started to prevail, and the need for any sort of thermoregulatory thermogenesis is reduced. When winter arrives, the young reindeer are 6 months old and—similar to adults—have lower-critical temperatures of about −30°C (see Figure 11.1).

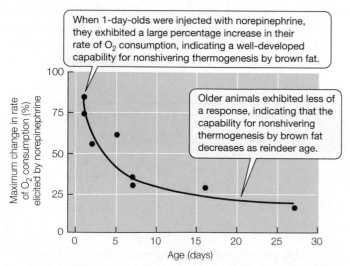

When 1-day-olds were injected with norepinephrine, they exhibited a large percentage increase in their rate of O_2 consumption, indicating a well-developed capability for nonshivering thermogenesis by brown fat.

Older animals exhibited less of a response, indicating that the capability for nonshivering thermogenesis by brown fat decreases as reindeer age.

FIGURE 11.8 A test for brown-fat thermogenesis in newborn and growing reindeer Reindeer of various ages were injected subcutaneously with a standard dose of norepinephrine per kilogram of body weight. (After Soppela et al. 1986.)

Genomics Confirms That Piglets Lack Brown Fat

The completion of a partial sequence of the genome of domestic pigs in 2005 set the stage for a remarkable discovery. Investigators located the gene for uncoupling protein 1 (UCP1) in this genome, but when they looked in detail, they found the gene to be disrupted by several mutations and deletions, including complete deletion of two exons. The gene is nonfunctional, a fact that greatly strengthens a conclusion reached through earlier morphological studies, that piglets lack brown fat.

Studies of additional types of pigs quickly revealed that the UCP1 gene is disrupted in all of them. European wild boars, warthogs, red river hogs, and Bornean bearded pigs display the same exon deletions that are observed in domestic pigs.

Phylogenetic analysis indicates that the UCP1 gene became disrupted in the pig lineage about 20 million years ago.

A European wild boar, the only wild pig that lives in cold climates

As yet, no one can be certain why this occurred. Nor can anyone be certain whether the gene became nonfunctional first and then brown fat disappeared, or vice versa. In any case, members of the pig family today are unusual among placental mammals in that the newborns lack nonshivering thermogenesis and depend entirely on shivering for physiological heat production. Piglets tend to be especially vulnerable to cold stress.

The primary role of brown fat in newborn ruminants, such as reindeer and sheep, seems to be to act as a transition source of heat production: the dominant site of thermogenesis when the animals first make the transition from life in the uterus to life outside their mother. For brown fat to play this role, it must develop extensively prior to birth, a fact that raises numerous interesting questions.

Do fetuses, for example, develop more brown fat when the environment into which they will be born is cold rather than warm? This question has been studied in sheep, in an indirect but fascinating way, by shearing off the fur of some mothers during their final month of pregnancy and comparing their newborns with the newborns of unshorn mothers (shearing serves in these experiments to mimic colder weather for the mothers by lowering their insulation). Lambs born to shorn mothers have more brown fat at birth than lambs born to unshorn mothers have. Moreover, when newborns are exposed to an air temperature that is chilly for lambs (14°C), the newborns of shorn mothers never shiver, whereas many of the newborns of unshorn mothers shiver. These results indicate that the intrauterine development of brown fat by fetuses is modified, depending on the cold stress experienced by their mothers, in ways that help ensure that NST will be sufficient for thermoregulation in the newly born.

Another interesting question is how the large masses of brown fat in unborn, near-term young are regulated so that they do not catabolize large amounts of the young's foodstuffs. Logic suggests that, prior to birth, a young animal will profit most from the foodstuffs it has by using them to grow and mature; brown fat, if uncontrolled, could turn large quantities of the foodstuffs into heat and chemical wastes. Experiments reveal that if near-term sheep fetuses are cooled inside the uteruses of their mothers (**FIGURE 11.10**), they do not activate their brown fat or engage in NST. However, if blood flow from the placenta of a near-term fetus

is blocked by pinching shut the umbilical cord, the fetus rapidly exhibits substantial NST. These results and others indicate that a near-term sheep fetus in the uterus receives from its placenta one or more types of signaling molecules (probably prostaglandins) that inhibit its activation of brown fat. These signaling molecules

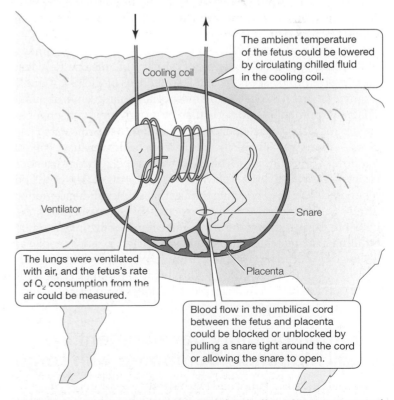

The ambient temperature of the fetus could be lowered by circulating chilled fluid in the cooling coil.

Cooling coil

Ventilator

Snare

The lungs were ventilated with air, and the fetus's rate of O₂ consumption from the air could be measured.

Placenta

Blood flow in the umbilical cord between the fetus and placenta could be blocked or unblocked by pulling a snare tight around the cord or allowing the snare to open.

FIGURE 11.10 A near-term sheep fetus in the uterus prepared for study of "simulated birth" (After Power et al. 1987.)

are cut off at birth by the severing of the umbilical cord, and the brown fat is then able to respond vigorously to the chilly outside environment by producing heat! A similar process may occur in reindeer. Future research will tell.

The Future of Reindeer: Timing and Ice

Demographers have sounded an alarm that wild reindeer may be in decline throughout their circumpolar distribution. Since the 1980s, 80% of populations studied have exhibited declines. If we presume that these data represent a long-term trend, the causes are of great interest. Some of the causes may be intimately related to the environmental physiology of the animals in a changing world.

Migratory ungulates commonly have evolved life histories closely synchronized with cycles of plant growth. This phenomenon is very evident in reindeer. As spring unfolds, herds of reindeer migrate back to the areas where they will spend the summer, and after arriving there, the pregnant females in a herd give birth almost synchronously with each other. Historically, births have tended to occur at a time when mothers and young can make maximum use of the peak period of nutritious spring plant growth on the summering grounds. For several reasons—one being that when reindeer start their migration, they are at a distant place from their summering grounds—the question arises, in a changing world, of whether birthing will continue to coincide with peak spring plant growth. A recent study on West Greenland herds indicates that it will not. Over the past 25 years, the peak of spring plant growth has occurred earlier and earlier because of warming spring temperatures. The reindeer, however, have birthed always at about the same time each spring. A growing mismatch of 5–10 days has consequently developed between the time when young are born and the time when nutritious food is most readily available. As this mismatch has developed, death rates of offspring have risen.

The consequences of this sort of mismatch highlight, among other things, the precariousness of reindeer life history. Reindeer have evolved to be specialists in leading lives of lifelong ("cradle to grave"), direct exposure to an especially harsh environment. That environment is also particularly unforgiving: small changes can impose large consequences. In addition to the mismatch just discussed, another "small" change that may be driving the decline of reindeer populations is the change of snow type. Arctic indigenous peoples often have dozens of words for the forms of ice, snow, and other precipitation—reflecting the fact that subtle differences (often unrecognized by non-Arctic people) may be of large importance. Evidence exists that in some areas where reindeer live, precipitation is falling to a greater extent as rain (and a lesser extent as snow) as years go by because of global warming. When precipitation falls as rain, it can freeze on the ground into a hard, thick crust. This crust (unlike fluffy snow) can block reindeer from reaching food plants, leading to hunger and starvation.

Thermoregulatory Development: Small Mammals Compared with Large

Mammals the size of mice and rats have only very modest capabilities to thermoregulate when they are first born. The white-footed mouse (*Peromyscus leucopus*), one of the most abundant native small mammals in North America, provides a typical example of the course of development in such animals. The species occurs in northern states such as Michigan and Wisconsin, as well as into Canada. In these areas, the mice give birth to their first litters of young each year in March and April, when the cold of late winter still prevails. Their litters consist of four to six young born without fur.

In the days immediately following birth, young white-footed mice can respond to cold exposure by increasing their metabolic rates to a small extent, and if all the young in a litter huddle together within the nest their mother provides, they are able collectively to thermoregulate reasonably well for a few hours even when the air outside the nest is near freezing. A *newborn litter*, therefore, can stay relatively warm for a while when its mother is away foraging. However, if a *single newborn mouse* is removed from the nest and studied by itself, it cannot marshal a high enough metabolic rate to stay warm even when the air temperature is +25°C—a temperature higher than "room temperature" in American buildings. At an ambient temperature of 25°C, the body temperature of a solitary 2-day-old soon drops from 37°C to about 28°C—not because the animal is in some sort of controlled hypothermia, but because its thermoregulatory abilities are overwhelmed. The young of lemmings and other small mammals characteristic of the Far North are not much different; to thermoregulate in their first days of life, they require a protective microhabitat, including the nest their mother provides and their siblings with which they huddle.

The length of the nestling period of white-footed mice is typical of that of most mice and rats: about 3 weeks. During those 3 weeks, young mice must become physiologically capable of setting off on their own. They must, for example, become able to thermoregulate as isolated individuals in the environment outside the nest, however cold that environment may be. As Figure 4.4 shows, individuals dramatically increase their body insulation and the rate at which they can produce heat during their 3 weeks of nestling life. Consequently they become capable of thermoregulating by themselves.

Regarding the *mechanisms* of thermogenesis in developing mice and rats, the evidence available indicates that shivering is not functional in most species for roughly the first week of postnatal life. Brown fat, in contrast, is present at birth. The brown fat grows as the young grow. In fact, studies of laboratory rats have shown that the maturation of brown fat as a thermogenic tissue outpaces body growth for a period during nestling development, so that the brown fat is able to produce ever more heat per unit of body weight as time passes.

Let's focus now on *small* mammals as compared with *large* mammals. The trajectory of brown-fat development in small mammals is extremely different from the trajectory in large mammals. In reindeer and sheep, brown fat is maximally or near maximally developed at birth, and it starts to wane soon after birth, so it is gone or approximately gone by about a month of age. In small mammals such as mice and rats, by contrast, brown fat is far from fully developed at birth, and during much of the first month of life, it develops an ever-greater thermogenic ability. Correlated with these differences in developmental trajectory, the adults of large and small mammals tend to exhibit consistent differences in brown fat. Brown fat is not present to any great extent in the adults of most large-bodied species of mammals.[4] By contrast, brown fat is conspicuous in the adults of most species of placental mammals smaller than about 5 kg. It becomes a particularly prominent tissue in these adults when they are acclimated to cold or acclimatized to winter, as we saw in Chapter 10; and it is a prominent tissue in hibernators, as we will discuss later in this chapter.

Looking back over our discussion of the development of thermoregulation in large and small species, it is striking to observe that in *both* large and small placental mammals, brown fat and nonshivering thermogenesis are the favored means of producing heat for thermoregulation at birth and during the period immediately following birth. Why is this true? One hypothesis is that skeletal muscles are inherently too immature at birth for shivering to be a viable primary mechanism of thermogenesis in newborns. There are other hypotheses as well, but only further research will answer the question. The fact that we humans rely on brown fat for thermogenesis when we are newly born adds interest to finding an answer.

The Effect of Body Size on Mammals' Lives in Cold Environments: An Overview

A retrospective look at the topics we have discussed so far in this chapter makes clear that body size is a principal determinant of the options available to mammals for thermoregulation. Although both reindeer and white-footed mice are warm in the days following their birth, they are warm for different reasons. Newborn reindeer are so large that they have few options for using protective microhabitats; they cannot burrow underground, and on the tundra they cannot readily find other refuges. Thus newborn reindeer must confront the full harshness of their environment, and they must *physiologically* meet the full challenges of their environment if they are to stay warm. Fortunately, large size—while limiting behavioral options—has physiological advantages; because of their size, newborn reindeer have a surface-to-volume ratio that is favorable for retaining heat in cold environments, and they can have a thick pelage. For small-bodied newborns such as white-footed mice, the interplay between behavior and physiology is almost opposite. Such newborns are smaller than the little fingers on our hands, and just as naked. Thus physiologically there is no chance that newborn mice could evolve mechanisms that would allow them to thermoregulate while fully exposed to a cold external environment. Being small, however, they can exist in highly protective microhabitats; their mothers can place them in secluded, benign locations such as underground burrows and can readily ensconce them in an insulating nest. The reasons newborn white-footed mice are warm are principally behavioral: When their mother is present, the newborns are warmed by her; and when she is absent, they benefit from the behavioral provisions she has made for them.

The same trade-offs between physiology and behavior are equally evident in the lives of adult mammals. During winter in Earth's frigid places, small nonhibernating mammals such as lemmings in the Arctic and pikas in the high mountains are able to escape the biting cold and howling winds of the larger environment by living under the snow (see Figure 1.19) or in other protective hideaways. Their ability to escape in this way is fortunate because, physiologically, a mouse- or rat-sized adult could not survive full exposure to the cold of winter in such places. Large mammals, in contrast, are in a far more advantageous position in terms of their physiology of thermoregulation; the body size of an adult reindeer is one of the major reasons it can have a lower-critical temperature below −30°C (see Figure 11.1). The physiological advantage of large size is itself fortunate, because large size limits behavioral options. A large mammal, such as an adult reindeer in the Arctic or a bighorn sheep in the high mountains, cannot escape the severity of the cold season by burrowing under the snow.[5]

The single greatest behavioral option available to large-bodied species is migration. In fact, as we saw in Chapter 9 (see Figure 9.8), migration is energetically more feasible for large species than for small ones. Often, therefore, with the approach of winter, as the small mammals in a place go underground or under the snow, the large ones get out. Bighorn sheep trek to the lowlands, and reindeer often migrate into more-forested areas where they can find windbreaks, as well as better winter feeding grounds.

Hibernation as a Winter Strategy: New Directions and Discoveries

The hibernating species of placental mammals are noted for retaining brown fat as adults. Indeed, when Conrad Gessner first described brown fat in 1551, he was studying the adults of a hibernating species, the European marmot. For four centuries thereafter, brown fat was observed in the adults of one hibernating species after another. The association of brown fat with hibernation was so consistent, in fact, that—although the function of brown fat was unknown—the tissue was dubbed "the hibernation gland." The actual function of brown fat—not a gland but a thermogenic tissue—was not discovered until the mid-twentieth century.

Today we know a great deal about the function of brown fat in adult hibernators. First and foremost, brown fat is the *thermogenic tissue that takes the lead in rewarming during arousal* (emergence from hibernation). Intense thermogenesis by brown fat is activated at the very beginning of the arousal process. Because of the placement of brown fat in the body (some deposits actually surround major arteries), and because of blood flow patterns, heat produced by the activation of brown fat is believed to be delivered particularly to the vital organs such as the heart, lungs, and brain. This focused warming of the vital organs, starting early in arousal, may be important in poising those organs to play their roles (e.g., coordination mediated by the brain) in the overall sequence of events by which homeothermy is gradually restored throughout the body.

As we discuss hibernators further, two species will receive special attention, the Arctic ground squirrel (*Urocitellus parryii*) and the alpine marmot (*Marmota marmota*). The ground squirrel, which weighs 500–800 g, is found widely in the Far North—Alaska, the Yukon, and neighboring parts of Canada. The marmot, which

---

[4] See Chapter 10 (pages 266–267) for a fuller discussion.

[5] The basic theme—that large body size tends, overall, to be a physiological advantage, whereas small body size tends overall to be a behavioral advantage—is also evident in desert mammals, as discussed in Chapters 28 and 30 (e.g., see Figure 30.1).

FIGURE 11.11 The allometry of energy savings by hibernators Weight-specific resting metabolic rate is shown as a function of body weight in nonhibernating and hibernating mammals, on log–log coordinates. The blue line shows the relation in nonhibernating mammals; this metabolic rate decreases substantially as body weight increases. The red line shows the relation in hibernating mammals; this metabolic rate is statistically the same for animals of all body weights. The nonhibernating metabolic rates are basal rates. Hibernating metabolic rates were measured at ambient temperatures of 2°–7°C. In the equations describing the two lines, M is weight-specific metabolic rate (mL O_2/g·h); W is body weight (g). (After Heldmaier et al. 2004.)

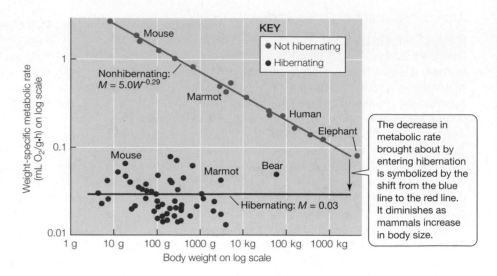

is larger and weighs about 4–5 kg, occurs high in the European Alps. Both species employ hibernation as a strategy to cope with the stresses of winter in some of Earth's most frigid places.

The great majority of hibernators have adult body sizes of 5 kg or less. This means that alpine marmots are near the upper limit of the size range that typifies most hibernators. Typically, species of mammals larger than 5 kg that occupy frigid environments, such as reindeer, remain active and fully homeothermic through all seasons (although some exhibit nonhibernation hypothermic or hypometabolic states at times). Physiologists have pondered why most large-bodied species have not evolved hibernation. One reason is probably that the energy savings afforded by hibernation decline with body size. As shown in **FIGURE 11.11**, the weight-specific metabolic rates of hibernators in hibernation (red line) are statistically the same regardless of their body size. However, when the animals are not hibernating, small-bodied species have higher weight-specific metabolic rates than large-bodied species. Thus the energy savings achieved by hibernation decrease as body size increases. In the face of this allometric trend, natural selection seems not often to have favored the evolution of hibernation in large-bodied species. Bears stand out as a dramatic exception.

Among species of mammals that weigh 5 kg or less, hibernation is not at all universal. In the Arctic, for example, whereas ground squirrels hibernate, lemmings and flying squirrels do not. We do not understand why species have diverged in these ways. At our present level of understanding, we can only conclude that the two modes of dealing with the cold of winter—hibernating and staying active—must each have pros and cons for small and medium-sized mammals. Neither strategy is so obviously "better" that it has become universal.

As we discuss Arctic ground squirrels, alpine marmots, and other hibernators, it is worth reflecting on the fact that much of what we know about them today is a consequence of a technological revolution that has occurred in the last 50 years in which biologists, engineers, and computer scientists have teamed up to create ever-better technologies for the remote monitoring of physiological traits. At the beginning of this revolution, small radio transmitters were designed that could be implanted in animals and report

body temperatures by *radiotelemetry*. Today's transmitters are able to operate on their built-in battery power for many continuous months, and the data they transmit are recorded automatically by computers. A more recent advance is the introduction of *data loggers* to physiological research. These are dedicated microcomputers that have large amounts of internal memory and that periodically log the temperatures they measure into memory.

With these advances in technology, records like that in **FIGURE 11.12** have become possible—and have revolutionized knowledge of the physiology of hibernation. The record in the figure shows the body temperature of a *free-living* and *undisturbed* Arctic ground squirrel for *8 continuous months*—the full duration of its hibernation season!

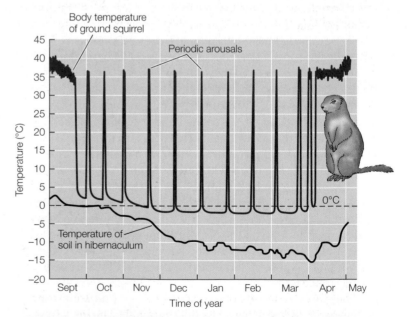

FIGURE 11.12 The body temperature of an Arctic ground squirrel during its hibernation season in Alaska The body temperature was recorded with a data logger. The soil temperature in the ground squirrel's hibernaculum is also shown. (From Boyer and Barnes 1999.)

Arctic ground squirrels supercool during hibernation and arouse periodically throughout their hibernation season

Although most hibernating species spend the winter in microhabitats where the temperature always stays above freezing, Arctic ground squirrels cannot do so over much of the range where the species occurs. The reason is *permafrost*—soil that never melts throughout the year. At the Arctic Circle, permafrost starts at 0.8–1.0 m below the ground surface. The animals, therefore, cannot burrow deeper than 0.8–1 m. At that depth the ambient temperature in the winter can drop to far below freezing, even sometimes to –25°C, within the **hibernacula** (singular *hibernaculum*) of the squirrels—the chambers or cavities where they hibernate.

As mentioned in Chapter 10, essentially all hibernating species periodically undergo temporary arousals during their hibernation season. The animals typically remain in their hibernacula during these arousals. This behavior explains why early naturalists believed that hibernation simply started in the early winter and ended in the spring; hibernating animals disappeared into their hibernacula for the entire period. When automated long-term records of the body temperatures of hibernators were obtained, however, the records revealed that periodic arousals almost universally occur. The Arctic ground squirrel in Figure 11.12 aroused 11 times between October and April before it aroused for the final time and emerged from hibernation. Continuous periods of hypothermia lasted 1–3 weeks. Between those bouts of hibernation, the ground squirrel raised its body temperature to 36°–37°C for 12–24 h during each arousal.

A striking aspect of the record in Figure 11.12 is that, by early December, the ground squirrel's body temperature fell below 0°C during periods of hypothermia! The soil temperature fell as winter progressed, and as the soil temperature fell, the ground squirrel maintained a larger difference between its body temperature during hypothermia and the soil temperature. Nonetheless, the body temperature fell below 0°C, a very unusual state for hibernators. The freezing point of the body fluids of Arctic ground squirrels remains the same during winter as it is in summer: –0.6°C (the typical value for mammals). Yet Arctic ground squirrels sometimes cool during hibernation to body temperatures of –2°C to –3°C. They do not freeze at these times. Instead, they supercool, just as some poikilotherms do in winter (see page 258).

Although supercooling is important, the *principal* way that Arctic ground squirrels cope with the threat of freezing during their hibernation bouts is by *thermoregulating* at body temperatures of –2°C to –3°C, thereby maintaining their tissues well above the soil temperature when the soil becomes profoundly cold. By midwinter, for example, the ground squirrel in Figure 11.12 kept its body temperature 10°C above the soil temperature during its bouts of hibernation. To be endothermic in this way, hibernating Arctic ground squirrels must increase their rates of heat production in midwinter to rates above those seen at other times—such as September and October—when the soil is warmer and they can simply let their body temperature approximate soil temperature. The rate of heat production of a hibernating ground squirrel is increased, as needed, by brown-fat thermogenesis (**FIGURE 11.13**). As noted in Chapter 10, it is common for mammals and birds that are in states of controlled hypothermia to prevent their body temperatures from falling lower than certain levels by elevating metabolic heat

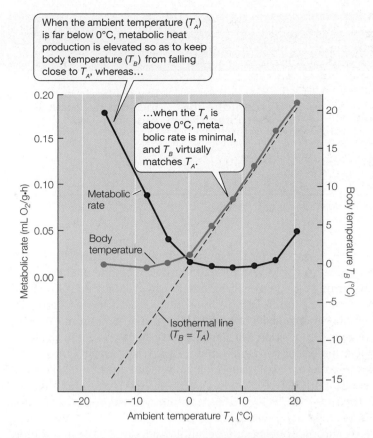

FIGURE 11.13 Excessively low body temperatures are prevented during hibernation bouts by elevated metabolic heat production The plot—which has a scale for metabolic rate on the left and a scale for body temperature on the right—shows the relations among metabolic rate, body temperature, and ambient temperature in Arctic ground squirrels *during bouts of hibernation*. As the ambient temperature drops from 0°C to –20°C, the body temperature is held higher and higher above ambient because of ever-accelerating metabolic heat production. The isothermal line is a line of equality between body temperature (T_B) and ambient temperature (T_A). (After Barnes and Buck 2000.)

production. The increased energy expenditure lowers the energy savings of hibernation, however.

The composition of the lipids consumed before hibernation affects the dynamics of hibernation

As stressed in Chapter 6, the composition of the foods that animals eat is often as significant as the amount of food. A dramatic illustration of the interaction between food composition and physiology has emerged in recent years from studies of hibernators. In addition to brown fat, hibernators often accumulate large stores of white fat with the approach of winter. The white fat is a storage tissue; lipids are deposited in the tissue as fattening occurs and later are mobilized from the tissue to meet metabolic needs (including the need to replenish lipids oxidized in brown fat). The lipids stored are triacylglycerols (see Figure 6.4C) and thus meet the chemist's definition of fats and oils. They are accumulated in droplets within specialized cells called *adipocytes*.

Fattening requires time—often many months. Accordingly, it must be set in motion by cues or processes that occur well in advance of winter stresses. In some species, the primary controls of fatten-

TABLE 11.2 Hibernation performance in chipmunks (*Eutamias amoenus*) fed three diets

All values are means. All differences between the group on a PUFA-rich diet and the group on a SFA-rich diet are statistically significant.

Diet	Percentage that hibernated (%)	Lowest body temperature that did not provoke arousal (°C)	Rate of O_2 consumption at an ambient temperature of 2°C (mL O_2/g·h)	Length of each continuous hibernation bout at an ambient temperature of 5°C (hours)
PUFA-rich	100	0.6	0.034	138
Intermediate	100	1.2	0.047	110
SFA-rich	75	2.2	0.064	92

Source: After Geiser and Kenagy 1987.

ing are photoperiodic; the shortening of day length in autumn, for example, may serve as a cue for fattening to begin. In other species, the timing of fattening is endogenously programmed (see Figure 6.24) under control of a circannual biological clock (see Chapter 15).

The storage lipids of hibernators (and other mammals) reflect in their composition the suites of fatty acids present in the foods the animals eat during fattening. Each triacylglycerol molecule is built from three fatty acid molecules (see Figure 6.4C). The fatty acids fall into three chemical categories (see page 136). *Saturated fatty acids*, also termed *SFAs*, contain no carbon–carbon double bonds. *Monounsaturated fatty acids—MUFAs—*contain one carbon–carbon double bond per molecule. Finally, *polyunsaturated fatty acids—PUFAs—*contain two or more such double bonds per molecule. As we discussed in Chapter 6, mammals are incapable of synthesizing most PUFAs from scratch. However, if they eat plants rich in PUFAs, mammals can use those PUFAs directly or employ them as substrates for the synthesis of other PUFAs. Because of this relationship, animals that eat foods rich in PUFAs typically deposit fats that are richer in PUFAs than are the fats deposited by animals eating PUFA-poor foods. Similarly, individuals that eat MUFA-rich diets tend to deposit MUFA-rich fats.

Biochemists postulated years ago that storage fats must be in a physically fluid state to be capable of being mobilized and metabolized. If this is true, fats composed primarily of SFAs could become useless during hibernation because the body temperatures of hibernators are often low enough to cause SFA-rich fats to solidify (for a mental image, recall the hardness of the SFA-rich fats of beefsteaks at refrigerator temperature). Reasoning from these thoughts, researchers hypothesized that hibernators might hibernate in a more effective manner if they deposit fats rich in PUFAs and MUFAs during their fattening periods. Such fats tend to remain fluid at far colder temperatures than fats composed strictly of SFAs (see page 254).

A great deal of evidence now exists indicating that the lipid composition of the diet of hibernators affects the dynamics of their hibernation! Relatively high levels of PUFAs (and sometimes MUFAs) in the diet, as predicted, improve the effectiveness of hibernation. Laboratory studies using defined diets provide one sort of evidence supporting this conclusion. For example, chipmunks in one study were fed three diets: a particularly PUFA-rich diet, an intermediate diet, and a particularly SFA-rich diet. Those on the PUFA-rich diet, compared with those on the SFA-rich diet, were more likely to hibernate, tolerated lower body temperatures, had lower metabolic rates, and had longer bouts of hibernation (meaning they aroused less frequently) (**TABLE 11.2**). For evaluating the predicted relation between diet and

hibernation, another sort of approach has been to look at correlations between white-fat composition and hibernation performance in free-living animals in the wild, as we will discuss shortly.

Whereas there are many studies by now that demonstrate a relation between hibernation performance and diet (e.g., see Table 11.2), these studies do not in fact demonstrate that lipid fluidity is the *reason* for the relation, as originally postulated. At the present time, a cutting edge in this area of research is that investigators are starting to do detailed studies to determine the *mechanisms* by which diet—and the fatty-acid composition of an animal's fat—affect hibernation. These studies suggest that, whereas lipid fluidity is a factor, other factors are also important.

There are two major types of PUFAs: the omega-3 and omega-6 fatty acids (see page 137). These differ considerably in structure and conformation, as seen in **FIGURE 11.14**. Recent research indicates that the type of PUFA deposited in white fat sometimes matters, not just the total amount of all PUFAs.

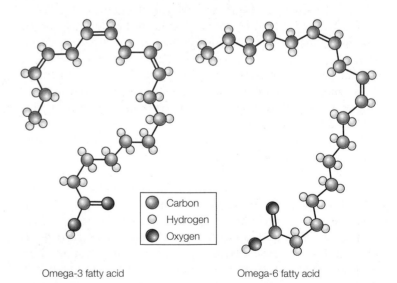

Omega-3 fatty acid   Omega-6 fatty acid

FIGURE 11.14 Structure and conformation of representative omega-3 and omega-6 polyunsaturated fatty acids The particular fatty acids shown are α-linolenic acid (omega-3) and linoleic acid (omega-6); these are the simplest precursors of the two types of PUFAs. Because their structures and conformations are different, the two types of PUFAs, when incorporated into membrane phospholipids, have different effects on the structure of the lipid matrix and create a different context for the function of membrane proteins. (After Ruf and Arnold 2008.)

In free-living populations of alpine marmots, because individuals eat different foods, they differ in the amounts and types of PUFAs in their white fat when they start the hibernation season. Individuals with high percentages of PUFAs tend to be more effective hibernators; they have lower body temperatures during hibernation and lose less weight during the winter than do individuals with lower percentages of PUFAs—confirming the same sort of conclusion as Table 11.2. Closer analysis reveals, however, that this relation is in fact chemically specific. It depends on the percentages of omega-6 PUFAs rather than omega-3 PUFAs.

To explain the relation between omega-6 PUFAs and hibernation success, a current working hypothesis is that when dietary omega-6 PUFAs are incorporated into membrane phospholipids, they provide a superior membrane lipid matrix for the function at low temperatures of key membrane proteins. One of these, a Ca^{2+}-pumping protein, is critical for heart contraction. According to this hypothesis, therefore, adequate omega-6 PUFAs from an animal's food help ensure stable heart action when the heart is cold.

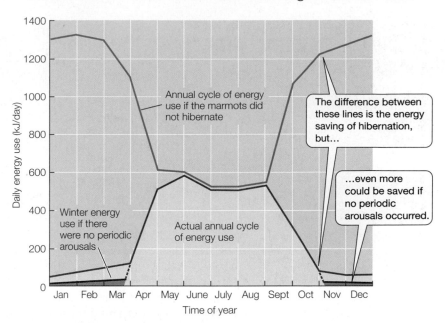

FIGURE 11.15 The annual cycle of energy use in alpine marmots (red line) Also shown are the cycle of energy use that would exist if the marmots did not hibernate (green line), and the energy costs in winter if the marmots did not undergo periodic arousals during hibernation (black lines). (After Heldmaier 1993.)

Although periodic arousals detract from the energy savings of hibernation, their function is unknown

Periodic arousals significantly reduce the energy savings of hibernation. To illustrate, let's look at calculations done by Lawrence Wang for Richardson's ground squirrels (*Urocitellus richardsonii*) hibernating in Alberta, Canada. Suppose that these ground squirrels would use 100 units of energy over the course of their winter hibernation season if they did not hibernate. They actually use 12 units of energy. However, of the 12 units they use, 10 are expended for their periodic arousals. Thus, if they did not periodically arouse and simply remained in continuous hibernation, their expenditure of energy would be 2 units. Presenting these numbers slightly differently: Of all the energy these ground squirrels use during the hibernation season, more than 80% is used for periodic arousals!

A qualitatively similar pattern is exhibited by other species. Alpine marmots, for example, reap great energy rewards by entering hibernation each winter (**FIGURE 11.15**). However, about two-thirds of their energy expenditure during their hibernation period is for arousals, and thus their winter energy expenditure would be only one-third as great if they did not undergo periodic arousals.

Given that animals pay an energy price for periodic arousals, it seems that the arousals must have important functions. In the early days of research on this topic, a lively hypothesis for periodic arousals was that they allow hibernators to void wastes. Several lines of evidence contradict this hypothesis, however. In the last 25 years, multidisciplinary attention has been focused on the question of periodic arousals, and several innovative new hypotheses have been put forward. One postulate that has some experimental support is that brain dendrites tend to be lost—and synapses tend to deteriorate—during hypothermia, and arousal is required to restore dendrites and synapses. An entirely different line of thinking—again with some experimental support—is that normal immune responses may be downregulated or blocked during hypothermia. Periodic arousals, according to this hypothesis, allow periodic function of the immune system and thereby enable hibernators to combat pathogenic organisms during the hibernation season. As yet, no consensus exists on an explanation for the periodic arousals.

The intersection of sociobiology and hibernation physiology

Recent analyses of data from many species have established that hibernation is correlated with increased survival in the wild. For example, one question of interest is whether individuals capable of hibernation enjoy lower odds of death during months when they are hibernating than when they are not. In this regard, the available data provide an emphatic answer. On average, in species of hibernating small mammals, an individual is about five times more likely to die in a month when it is active than when it is hibernating. Hibernation probably has this effect mostly because of avoidance of predators; hibernating animals remain always in their hibernacula, instead of moving around in the larger environment where predators can spot them. Avoidance of predation may, in fact, have been a significant factor in the evolution of hibernation. Because of its effect on odds of survival, hibernation can affect an animal's odds of reproduction.

Many additional, interesting questions arise at the interface of hibernation biology and reproductive biology. For example, are hibernation and pregnancy compatible? Do they occur synchronously in the life of an animal? In 2006, researchers reported for the first time that they had observed full-blown hibernation during pregnancy in a placental mammal. Specifically, they had observed three insectivorous bats (hoary bats, *Lasiurus cinereus*) in advanced pregnancy enter deep hibernation for 3–6 days during cold late-spring weather. The bats later aroused normally and gave birth. Prior to this observation on the bats, scattered pieces of less conclusive evidence—pointing to occasional hypothermia during pregnancy—had accumulated despite the traditional assumption of biologists that pregnant mothers would not enter hibernation. It is clear that hibernation and other forms of hypothermia are less likely

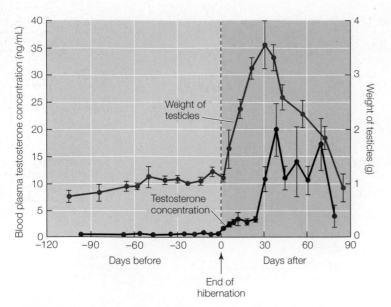

FIGURE 11.16 A constraint of hibernation: Testicular development in ground squirrels requires that hibernation end Shown are testicular size (weight) and blood testosterone as functions of time before and after the end of hibernation in adult male golden-mantled ground squirrels (*Callospermophilus lateralis*). Error bars show the standard errors. (After Barnes 1996.)

to occur during pregnancy than otherwise. Equally clearly, however, we now know that pregnant mothers do sometimes hibernate.

Male ground squirrels illustrate another consideration that arises in some cases at the interface of hibernation biology and reproductive biology. In ground squirrel species that have been studied, the testicles are regressed during hibernation; and from what is known, no regrowth is possible while hypothermia continues. **FIGURE 11.16** shows this phenomenon in male golden-mantled ground squirrels. Testicular development starts only after high body temperatures have been restored following the end of hibernation, and because full development requires several weeks, males are unable to breed until long after they have terminated hibernation. Why the testicles face these limits is unknown. However, the limits have very real consequences for reproduction. In wild populations of ground squirrels, the males emerge from hibernation many weeks before the females. This early emergence of the males seems necessary for them to prepare to be successful mates.

Social hibernation is another theme of considerable interest. Although ground squirrels hibernate as isolated individuals, many of the world's 14 species of marmots undergo social hibernation—meaning that individuals hibernate together in social groups. Alpine marmots, for example, live throughout the year in social groups. A typical social group of alpine marmots consists of a dominant male and female and an assemblage of related, younger animals of various ages (up to several years old). All these animals hibernate together, in groups that number up to 20 individuals.

Social hibernation in alpine marmots has been shown to increase each individual's probability of overwinter survival, particularly among the young-of-the-year, termed *juveniles*. One reason that survival is aided relates to the fact that these marmots do not allow their body temperatures to go below 3°–4°C in winter, yet temperatures in their hibernacula are often lower, meaning that hibernating animals must elevate their metabolic rates to keep their body temperatures from going too low. Animals huddled together in a large cluster are better insulated and require less of an increase in metabolic rate than those hibernating alone or in small groups.

One of the most remarkable revelations in the recent annals of the radiotelemetric study of hibernation is the discovery that adults and subadults in a hibernating group of alpine marmots usually undergo their periodic arousals in close synchrony (**FIGURE 11.17**). This synchrony (which has recently been observed also in Alaskan marmots, *Marmota browerii*) lowers their mutual energy costs to arouse. If a single adult were to arouse alone within a hibernating group, it would experience heightened energy costs by being in the group, because the cold tissues of the other animals in the group, pressed closely against it, would increase the heat production it would require to raise its own body temperature. What actually happens is diametrically opposite. Adults and subadults in a group arouse simultaneously, thereby lowering each other's energy costs to arouse by mutually warming each other. The simultaneity of this arousal is particularly intriguing because all the animals are hibernating and therefore are seemingly comatose when they initiate the process!

Juveniles in a hibernating group of alpine marmots often do not initiate their arousals simultaneously with the adults and subadults, but instead lag behind, letting the older animals warm them for several hours (or even days). Only after receiving this benefit do the juveniles invest their own energy in the arousal process.

By statistical calculations, an *index of synchrony* can be calculated for arousals in groups of marmots. A high index for a hibernating

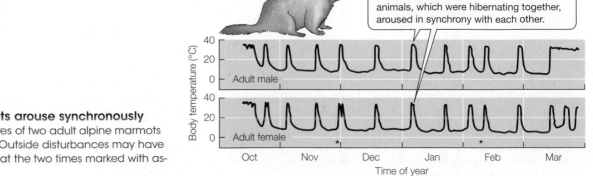

FIGURE 11.17 Alpine marmots arouse synchronously Shown are the body temperatures of two adult alpine marmots that were hibernating together. Outside disturbances may have affected arousal of the animals at the two times marked with asterisks (*). (After Arnold 1988.)

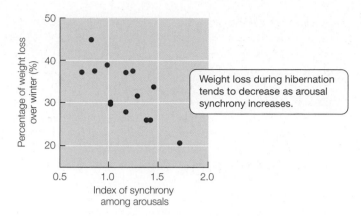

FIGURE 11.18 Weight loss during hibernation as a function of arousal synchrony in hibernating groups of alpine marmots Each point refers to a distinct group of hibernating marmots and shows both the group's average weight loss during winter and the group's degree of arousal synchrony. High values of the index of synchrony signify high degrees of arousal synchronization among group members. (After Ruf and Arnold 2000.)

group means that the individuals monitored with radiotelemeters in the group aroused highly synchronously. Based on data for 14 hibernating groups of alpine marmots, **FIGURE 11.18** shows the relation between loss of body weight and synchrony. As you can see, individuals in highly synchronized groups lost approximately 20%–25% of their body weight during the winter hibernation season, but individuals in some poorly synchronized groups lost about 40%–45%—roughly twice as much. Because relatively great loss of body weight lowers animals' likelihood of survival, this analysis reveals that synchrony during periodic arousals has important survival consequences.

Further analysis of the data available on alpine marmots also reveals that the presence of juveniles is very important in determining the dynamics of group hibernation. When juveniles are present in a group, they themselves have greater chances of survival than if they lived alone. However, because they tend to diminish arousal synchrony by delaying their own arousals relative to others in their group, they increase the amount of weight lost by adults and subadults in their group, and therefore decrease the odds of survival of the older animals. The interplay between sociobiology and hibernation biology is particularly vivid. For an individual adult to survive the winter, avoidance of all contact with juveniles during hibernation would be an advantage. Without immortality, however, adults require juveniles if they are to pass genes to future generations, even though when they associate with juveniles during hibernation, their own odds of individual survival go down.

STUDY QUESTIONS

1. What are the relative advantages and disadvantages of large and small body size in frigid places? In your explanation, try to go beyond the points mentioned in this chapter.

2. When the composition of the community of rumen microbes in a ruminant changes, there are several possible causes. What are some of them? Could you do experiments to decide what the actual causes are? Explain.

3. Long migrations such as those of reindeer are rare in animals that walk across the land, despite being common in flying birds. In what relevant way are reindeer similar to birds in their energetics of covering distance? (Review Chapter 9 if you are stumped.)

4. List possible hypotheses for the function of periodic arousal, and design experiments to test your hypotheses as rigorously as possible.

Go to **sites.sinauer.com/animalphys4e**
for box extensions, quizzes, flashcards,
and other resources.

REFERENCES

Geiser, F. 2004. Metabolic rate and body temperature reduction during hibernation and daily torpor. *Annu. Rev. Physiol.* 66: 239–274. A detailed exploration of several key themes in modern research on the comparative biology of hibernation and other hypothermic states by a leader in the field.

Heldmaier, G., S. Ortmann, and R. Elvert. 2004. Natural hypometabolism during hibernation and daily torpor in mammals. *Respir. Physiol. Neurobiol.* 141: 317–329. A sweeping look at the phylogeny of mammalian hypothermic states, their adaptive significance, and the comparative physiology of hypometabolism.

Heldmaier, G., and T. Ruf. 1992. Body temperature and metabolic rate during natural hypothermia in endotherms. *J. Comp. Physiol., B* 162: 696–706. Although dated, an insightful treatment by two leaders in hibernation research. The paper by Heldmaier et al. 2004 follows up on themes raised in this paper.

Nedergaard, J., and B. Cannon. 1990. Mammalian hibernation. *Philos. Trans. R. Soc. London, B* 326: 669–686. Probably the most delightful, scientifically serious article ever written on hibernation.

Ruf, T., and W. Arnold. 2008. Effects of polyunsaturated fatty acids on hibernation and torpor: a review and hypothesis. *Am. J. Physiol.* 294: R1044–R1052.

Ruf, T., C. Bieber, W. Arnold, and E. Millesi (eds.). 2012. *Living in a Seasonal World: Thermoregulatory and Metabolic Adaptations.* Springer, New York. This is the symposium volume for the 14th International Hibernation Symposium. As such, it provides up-to-date articles, as well as links to the wider literature, in most areas of hibernation research.

Stokkan, K.-A., B. E. H. van Oort, N. J. C. Tyler, and A. S. I. Loudon. 2007. Adaptations for life in the Arctic: evidence that melatonin rhythms in reindeer are not driven by a circadian oscillator but remain acutely sensitive to environmental photoperiod. *J. Pineal Res.* 43: 289–293.

Turbill, C., C. Bieber, and T. Ruf. 2011. Hibernation is associated with increased survival and the evolution of slow life histories among mammals. *Proc. R. Soc. London, Ser. B* 278: 3355–3363.

Vors, L. S., and M. S. Boyce. 2009. Global declines of caribou and reindeer. *Glob. Change Biol.* 15: 2626–2633.

See also **Additional References** *and* **Figure and Table Citations**.

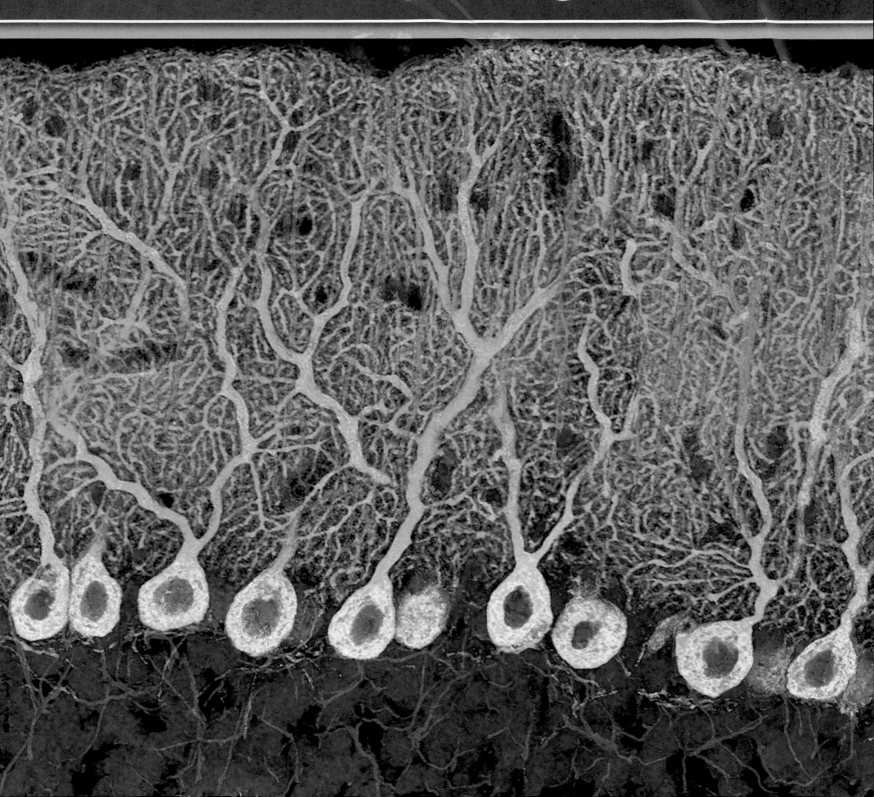

PART III
Integrating Systems

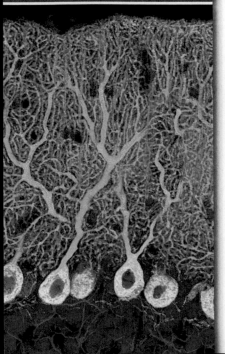

Previous page
Purkinje neurons (green) send branching dendrites toward the surface of the cerebellum (top); they receive synaptic input from axons of other neurons (blue). Nuclei of
neurons and glial cells are red in this fluorescence micrograph. The cerebellum plays
roles in movement coordination and motor learning. A major challenge in physiology is
to explain behavioral activities in terms of neural functions and interactions.

Locomotion in a squid, whether for capturing a meal or to avoid becoming one, depends on jet propulsion: The contraction of muscles in the squid's outer mantle expels seawater through a moveable siphon, propelling the animal in the opposite direction. As is true in all animals, feeding, escape, and similar behaviors in the squid are controlled by nervous system signals, which travel rapidly in a point-to-point manner, from one specific cell to another. These signals arise from properties of nerve cells—termed *neurons*—which have long cablelike processes—termed *axons*—that convey electrical signals rapidly and faithfully from place to place in the body, even over long distances. In the squid, sensory neurons such as those in the eyes encode information about the squid's environment and convey signals to the brain. There, the signals are integrated into a decision to attack or retreat. The brain then sends commands to the mantle muscles, in part through a set of large neurons with large ("giant"), rapidly conducting axons.

As you will discover in this chapter, squid giant axons have played an important role in our understanding of neuronal functions. The diameter of these giant axons can be as large as 1 mm (1000 micrometers [μm]), and for more than half a century investigators have taken advantage of this prodigious cellular size to perform noteworthy experiments that have

Squid axons are important to physiologists—and to the squid.

revealed the mechanisms of neuronal signaling. Sir Alan Hodgkin (1914–1998), who received the Nobel Prize in 1963 for his work on squid axons, recalled that a colleague had remarked (not, he thought, with the greatest tact) that it was the squid that really ought to be awarded the prize!

This chapter describes the electrical basis of neuronal function—the ability of neurons to generate electrical signals and propagate them over relatively large distances. The cellular mechanisms of neuronal signaling are similar in all animals, whether we examine neurons of squid, cockroaches, jellyfish, or humans. Before we turn to neuronal function, however, it is important to take a broader look at the challenges of integration and control. Doing so will clarify the range of physiological control processes and the contrasting functions of neuronal and hormonal modes of integration.

The Physiology of Control: Neurons and Endocrine Cells Compared

An animal needs to function like a coherent organism, not like a loose collection of cells and intracellular mechanisms. **Integration** is a general term that refers to processes—such as summation and coordination—that produce coherency and result in harmonious function. *Cellular integration* refers to processes within cells. *Whole-animal integration* refers to the selective combination and processing of sensory, endocrine, and central nervous system (CNS) information in ways that promote the harmonious functioning of the whole organism—including all its cells, tissues, and organs—within its environment. Just as some cells are specialized to produce movements, secrete acid, or carry oxygen, nerve cells and endocrine cells are specialized for control and coordination. Whole-animal integration is carried out by the nerve and endocrine cells. The integrative functions carried out by those cells ensure that an animal's responses are smooth and coordinated, rather than clashing or disjointed.

Control systems, initially described in Boxes 1.1 and 10.3, occupy a central place in the achievement of integration. In the abstract, a **control system** is a system that sets the level of a particular variable (temperature, blood pressure, muscle force, and so on) that is being controlled. To do so, it uses information from sensors to determine signals it sends to effectors that can modify the controlled variable. Control systems often (but not always) operate on negative feedback principles (see Box 1.1) and are stabilizing: When the controlled variable deviates from a desired level, the control system activates effectors to reverse the deviation.

The nervous and endocrine systems are also often described as *control systems* because nerve cells and endocrine cells control the ways in which other cells function. This use of the concept of control systems is complementary to the use discussed in the previous paragraph. To see the relations, consider that a control system of the sort discussed in the previous paragraph is present in inanimate objects such as cars and computers, where the physical entities that implement control functions are made of materials such as copper and silicon. In animals, control functions are mostly carried out by nerve cells and endocrine cells.

The nervous system and the endocrine system work in systematically different ways to control and coordinate the cells of an animal. As **FIGURE 12.1A** shows, a signal in a neuron travels electrically along a cell process all the way to its target cell; transmission along

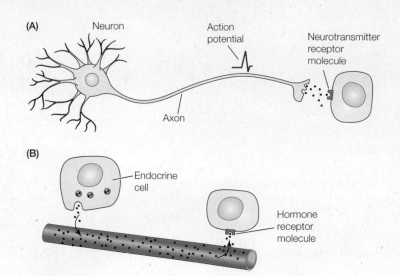

FIGURE 12.1 Neuronal and hormonal signaling both convey information over long distances Red dots are signaling molecules: neurotransmitter molecules in (A) and hormone molecules in (B). (A) Neurons have long axons that rapidly propagate action potentials, and also use short-distance chemical neurotransmitter signaling to communicate from cell to cell. (B) Endocrine cells release chemical hormones into circulatory fluids that carry the hormonal message over long distances to activate hormone receptors on other cells.

the cell process is very fast and spatially highly defined (a signal travels only along the cell process in which it was initiated). When the electrical signal arrives at the end of the neuron process, it causes the release of a chemical substance—a neurotransmitter—that diffuses quickly across the minute gap between the neuron process and the target cell. When this chemical substance arrives at the target cell, it binds (noncovalently) with specific receptor molecules on the cell, activating target-cell responses. In contrast, as **FIGURE 12.1B** shows, when an endocrine cell emits a signal, it does so by secreting a chemical substance—termed a **hormone**—into the general blood circulation. The signal travels more slowly than a neuronal signal because it is carried by blood flow, but instead of being spatially highly circumscribed, the signal is transmitted to all cells in the body. The target cells—the cells that respond—are the subset of cells that have receptor proteins for the hormone in their cell membranes. In the ensuing paragraphs, we will discuss the broad features of neural and endocrine control at greater length. Then, in the remainder of this chapter and in Chapters 13–15, we will consider aspects of neural control in detail (neurons, synapses, sensory functions, and the organization of whole systems of neurons). We will discuss endocrine control in detail in Chapter 16.

Neurons transmit electrical signals to target cells

Because neurons are commonly likened to the wires in a telephone or computer network, most people have an intuitive understanding of what these cells do. A **neuron** is a cell that is specially adapted to generate an electrical signal—most often in the form of a brief, self-propagating impulse called an *action potential*—that travels from place to place in the cell. As **FIGURE 12.2** reveals, a neuron has four parts—dendrites, cell body, axon, and presynaptic terminals—that generally correspond to its four functions—input, integration, conduction, and output—as a controller cell within an animal's body.

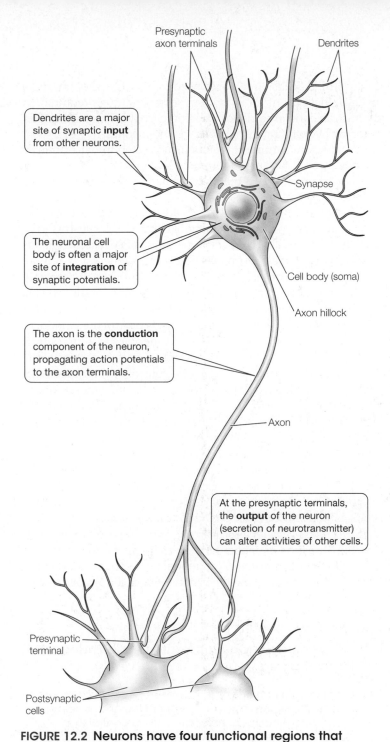

Presynaptic axon terminals

Dendrites

Dendrites are a major site of synaptic input from other neurons.

Synapse

The neuronal cell body is often a major site of integration of synaptic potentials.

Cell body (soma)

Axon hillock

The axon is the conduction component of the neuron, propagating action potentials to the axon terminals.

Axon

At the presynaptic terminals, the output of the neuron (secretion of neurotransmitter) can alter activities of other cells.

Presynaptic terminal

Postsynaptic cells

FIGURE 12.2 Neurons have four functional regions that typically correspond to their four major structural regions The descriptions in the figure provide a functional model of a neuron, showing typical functional properties that it mediates. The labels identify the structural parts of a neuron that are associated with these functions. The correlation between structures and functional properties is imperfect: Synaptic input often occurs at the cell body as well as the dendrites, for example, and some dendrites can generate action potentials. In contrast, some local neurons generate no action potentials at all, and thus lack a separate function of active conduction.

A neuron receives input—signals from other neurons or sensory cells—at specialized cell–cell contact points called **synapses**. Usually the synaptic input occurs along branching *processes* known as **dendrites**, although synapses may occur on the cell body as well. Impulses arriving at a synapse from a *presynaptic* cell cause the release of a chemical substance called a **neurotransmitter** into the synaptic cleft, or space between the cells. The chemical neurotrans-

mitter exerts specific physiological effects on the postsynaptic cell by binding to neurotransmitter receptors. These changes can result in a new electrical impulse in the target neuron. Thus a synapse allows for transmission of information between neurons through conversion of a signal from electrical to chemical to electrical.

The **cell body** (also called the *soma*) is commonly the part of a neuron where signal integration and impulse generation occur. A single neuron may receive thousands of synaptic contacts from other neurons. The neurotransmitters released across some synapses excite the neuron; those released across other synapses inhibit it. From moment to moment, the cell membrane of the cell body combines the inhibitory and excitatory synaptic inputs, and if excitatory inputs surpass inhibitory inputs, the neuron may respond by generating one or more action potentials.

The long slender **axon** is the conduction component of a neuron, serving to propagate action potentials along its length. The axon typically arises from the soma via a conical **axon hillock**, which leads to the **axon initial segment**, a specialized area that is commonly the site of action potential initiation. The microscopic axons from individual neurons sometimes collect together in long macroscopically visible bundles that are called *tracts* in the CNS and *nerves* in the peripheral nervous system (PNS).

Where an axon ends, it usually divides into several **presynaptic terminals**, which constitute the places where neuronal output occurs. The presynaptic terminals form synapses with other neurons or other types of cells, such as muscle fibers (muscle cells). An action potential arriving at the presynaptic terminals triggers the release of molecules of neurotransmitter across the synapses to exert a specific physiological effect—excitatory or inhibitory—on the target cell. Neurons that form synaptic endings on a cell are said to **innervate** that cell.

The extended networks of neurons in an animal's body (along with supporting cells, described later) constitute its *nervous system*. Neurons perform various roles in the nervous system. Some neurons perform sensory functions by initiating signals in response to physical or chemical stimuli. As we have just described, other neurons integrate signals arriving from other neurons, generate nerve impulses of their own, and transmit these signals over distances that can be very long, at least on a cellular scale. As we will discuss in Chapter 15, animals have a CNS (brain and spinal cord in vertebrates) and a PNS. Neurons that relay sensory signals to integrative centers of the CNS are called **afferent neurons** (*afferent,* "to bring toward"). Other neurons, called **efferent neurons** (*efferent,* "to carry off"), relay control signals (instructions) from the CNS to target cells that are under nervous control, such as muscle cells or secretory cells. Neurons that are entirely within the CNS are called **interneurons**.

Neural control has two essential features: It is fast and addressed. Neuronal signals are *fast* in that they travel very rapidly and begin and end abruptly. A mammalian neuronal axon, for example, might conduct impulses along its length at 20–100 m/s, and it might be capable of transmitting 100 or more impulses in a second. The connections of neurons are said to be *addressed* because they provide highly discrete lines of communication (like a letter or a telephone call). A neuron normally must make synaptic contact with another cell to exert control, and it typically innervates multiple, but relatively few, cells that are its potential targets. Neuronal lines of communication therefore provide opportunities for fine control of other cells both *temporally* and *spatially*, sending fast, rapidly changing signals to some potential targets and not to others.

Endocrine cells broadcast hormones

In contrast to the signals of the neurons in nervous systems—which are precisely targeted—the signals produced by the *endocrine system* are broadly distributed throughout the animal's body. Endocrine cells release *hormones* into the blood (or sometimes just into other extracellular fluids). These chemicals are carried throughout the body by the blood, bathing the tissues and organs at large. For a hormone to elicit a specific response from a cell, the cell must possess *receptor proteins* for that hormone (see Chapter 2, page 62). Thus cells of only certain tissues or organs respond to a hormone and are called *target cells*. The responsiveness of target cells is under control of gene expression; that is, the tissues that respond to a hormone are tissues that express the genes encoding its receptor proteins.

Endocrine control has two essential features: It is slow and broadcast. Individual hormonal signals are relatively *slow* because they operate on much longer timescales than individual neuronal signals. Initiation of hormonal effects requires at least several seconds or minutes because a hormone, once released into the blood, must circulate to target tissues and diffuse to effective concentrations within the tissues before it can elicit responses. After a hormone has entered the blood, it may act on targets for a substantial amount of time before metabolic destruction and excretion decrease its concentration to ineffective levels. In the human bloodstream, for example, the hormones vasopressin, cortisol, and thyroxine display half-lives of about 15 min, 1 h, and nearly 1 week, respectively. Thus a single release of hormone may have protracted effects on target tissues.

Unlike addressed neural control, endocrine control is said to be *broadcast*. Once a hormone is released into the blood, all cells in the body are potentially bathed by it. The specificity of hormone action depends on which cells have receptor molecules for the hormone. Many types of cells may respond to the hormone, perhaps with different types responding in different ways. Alternatively, a hormone may affect only one type of target cell, because only those target cells have the kind of receptor to which the hormone attaches. Although in principle hormones may exert either limited or widespread effects, in practice they commonly affect at least a whole tissue, and often multiple tissues.

Nervous systems and endocrine systems tend to control different processes

Neural lines of communication are capable of much finer control—both temporal and spatial—than is possible for endocrine systems. Not surprisingly, the two systems tend to be used to control different functions in the body. *Whereas the nervous system controls predominantly the fine, rapid movements of discrete muscles, the endocrine system typically controls more widespread, prolonged activities* such as metabolic changes.

Consider, for example, running to catch a fly ball in baseball. It requires rapid computation and very specific control of discrete muscles in split-second time, functions that can be mediated only by the nervous system. In contrast, the control of metabolism or growth requires the modulation of many tissues over a protracted period. In principle, an animal's nervous system *could* carry out a coordination task of this sort. To do so, however, the nervous system would need thousands of discrete axons between integrating centers

and controlled cells, and would need to send trains of impulses along all these axons for as long as the modulation is required. In contrast, an endocrine gland can accomplish this task with greater economy, by secreting a single long-lasting chemical into the blood. For this reason, control of metabolism is often under primarily hormonal control, as are other processes (growth, development, reproductive cycles, etc.) that involve many tissues and occur on timescales of days, months, or years.

Most tissues in an animal's body are under the dual control of the nervous and endocrine systems. Skeletal muscle illustrates the relationship of this dual control. A typical vertebrate muscle contains thousands of muscle cells (muscle fibers) and is innervated by more than 100 motor neurons. Each motor neuron innervates a separate set of muscle fibers, controlling the contraction of just these fibers. The nervous system can selectively activate a few, many, or all of the motor neurons, to rapidly and precisely control the amount of force the muscle generates. Whereas the nervous system controls the contractile activity of the muscle cells over the short term, the hormone insulin provides endocrine control of their metabolic activity over the long term. Insulin facilitates resting muscle fibers' uptake of glucose from the blood and their rate of glycogen synthesis. This example emphasizes the spatial and temporal distinctions between the two types of control: The nervous system controls moment-to-moment, differential contractile actions of the muscle cells in a muscle, whereas the endocrine system provides long-term metabolic control of all the muscle cells en masse.

Nervous and endocrine systems can exert control over each other, as well as over other targets. *Interaction between the nervous and endocrine systems occurs in both directions.* Nervous systems can affect the function of endocrine cells, as in innervated endocrine glands. Likewise, hormones can modulate nervous system function; for example, sex steroid hormones affect certain neurons in mammalian brains.

SUMMARY
The Physiology of Control: Neurons and Endocrine Cells Compared

■ Control by a nervous system involves neurons that send axons to discrete postsynaptic cells. Neurons generate rapidly conducting action potentials to control the specific targets on which they end. They exert fast, specific control by releasing neurotransmitters at synapses.

■ Endocrine cells release hormones into the bloodstream to mediate endocrine control. All body cells are potential targets of a hormone, but only those with specific receptors for the hormone actually respond. Hormonal control is slower, longer lasting, and less specific than neural control.

Neurons Are Organized into Functional Circuits in Nervous Systems

The functions of a nervous system depend on "wiring"—the anatomical organization by which neurons are connected into circuits. Any behavioral activity (such as swimming, in the squid

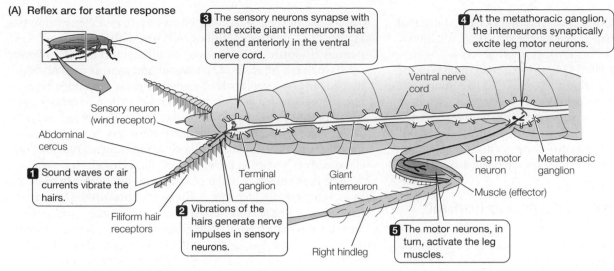

(A) Reflex arc for startle response

3 The sensory neurons synapse with and excite giant interneurons that extend anteriorly in the ventral nerve cord.

4 At the metathoracic ganglion, the interneurons synaptically excite leg motor neurons.

Ventral nerve cord

Sensory neuron (wind receptor)

Abdominal cercus

1 Sound waves or air currents vibrate the hairs.

Terminal ganglion

Giant interneuron

Leg motor neuron

Metathoracic ganglion

Muscle (effector)

Filiform hair receptors

2 Vibrations of the hairs generate nerve impulses in sensory neurons.

Right hindleg

5 The motor neurons, in turn, activate the leg muscles.

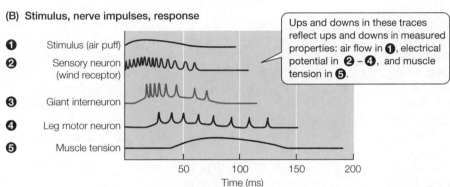

(B) Stimulus, nerve impulses, response

Ups and downs in these traces reflect ups and downs in measured properties: air flow in **1**, electrical potential in **2** – **4**, and muscle tension in **5**.

1 Stimulus (air puff)

2 Sensory neuron (wind receptor)

3 Giant interneuron

4 Leg motor neuron

5 Muscle tension

50   100   150   200
Time (ms)

FIGURE 12.3 The neural circuit mediating the startle response in the cockroach *Periplaneta americana* (A) Hairlike wind receptors located on an abdominal cercus trigger this reflex. (B) Nerve and muscle cells in the reflex circuit respond to a controlled puff of air lasting 50 ms. The action potentials in successive neurons in the circuit lead to contraction (tension) in the muscle of the leg. (After Camhi 1984.)

with which we opened the chapter) is a property of the neural circuit that mediates it. We will discuss nervous system organization in Chapter 15, but here we provide a simple illustrative example. Suppose you walk into the kitchen and surprise a cockroach. The cockroach jumps, exhibiting a *startle response* in which it turns away from the disturbance and prepares to run. This simple behavioral act is mediated by electrical signals and chemical synapses within the cockroach's nervous system.

The cockroach's jump is a **reflex**, a simple, stereotyped behavioral response to a distinct stimulus. Air currents or airborne sound waves vibrate filiform hairs that act as wind receptors at the cockroach's posterior end (**FIGURE 12.3, 1**), providing the stimulus that evokes the reflex. This stimulus initiates a brief series of action potentials in sensory neurons **2** located at the bases of the hairs. The action potentials travel along the conducting afferent processes (axons) of the sensory neurons toward the CNS, where the sensory neurons contact other neurons in the CNS. In the cockroach, the sensory axons make synaptic contacts with a few large *interneurons* (neurons that do not extend outside the CNS). These synapses are excitatory, so the barrage of action potentials from the sensory neurons excites the interneurons **3**, which generate their own action potentials.

The interneuron axons extend anteriorly in the ventral nerve cord (part of the CNS). They in turn make synaptic contact with efferent **motor neurons**, whose outgoing axons exit the CNS and innervate a muscle. The interneurons synaptically excite the motor neurons **4**, which in turn excite the extensor muscles of the legs **5** that produce the jump. At the same time, the interneurons inhibit motor neurons that excite the antagonist flexor muscles of the cockroach's legs.

As the barrage of action potentials in Figure 12.3 indicates, this startle response happens very quickly: It is less than 150 milliseconds (ms) from stimulus to jump! This rapid and selective activation of particular muscles to generate a behavioral response is the essential element of neural control.

The Cellular Organization of Neural Tissue

Nervous systems are composed primarily of neural tissue, which in turn is composed of discrete cells: neurons and glial cells (see page 311), as well as connective tissue cells and cells of the circulatory system. The cellular organization of nervous systems is a corollary of the **cell theory**, which states that organisms are composed of cells, that these cells are the structural and functional units of organization of the organism, and that all cells come from preexisting cells as a result of cell division. Matthias Schleiden (1804–1881) and Theodor Schwann (1810–1882) formulated the cell theory in 1839.

The cell theory gained widespread and rather rapid acceptance—except as applied to nervous systems. Instead, the dominant view of the organization of nervous systems in the latter half of the nineteenth century was the **reticular theory**, most strongly argued by Joseph von Gerlach (1820–1896) and Camillo Golgi (1843–1926). The reticular theory held that nervous systems were composed of complex, continuous meshworks of cells and processes in protoplasmic continuity with each other (i.e., the cells ran together without any boundaries).

The reticular theory was supplanted only gradually, over the first third of the twentieth century, by an outgrowth of the cell theory known as the **neuron doctrine**. The neuron doctrine states that neurons are anatomically distinct and are the structural, functional, and developmental units of organization of nervous systems. Santiago Ramón y Cajal (1852–1934), the main champion of the neuron doctrine, used special staining techniques to demonstrate convincingly that neurons are contiguous (in contact with each other) but are not continuous (connected without interruption). However, the debate on contiguity versus continuity persisted until the 1950s, when electron microscopy permitted resolution of cell membranes and rigorously demonstrated the discontinuity of neurons in contact.

Neurons are structurally adapted to transmit action potentials

Neurons, as seen earlier, are cells that are specialized for generating electrical impulses and transmitting those impulses from place to place within the body, sometimes over considerable distances. They have long processes, which relate to their functions—acting, for example, as the conduits for long-distance transmission. As you will recall, a neuron consists of a *cell body*, or *soma* (plural *somata*) (also called the *perikaryon*), which is the region that contains the nucleus, and one or more *processes* arising from it (**FIGURE 12.4**).

The cytology of a neuronal soma is broadly similar to that of nonneuronal cells. A neuronal soma contains a nucleus and most of the organelles and cytoskeletal elements familiar to cytologists: mitochondria, Golgi apparatus, smooth endoplasmic reticulum (ER), rough ER, microtubules, neurofilaments, and actin microfilaments. Neurons are very active in protein synthesis and thus have extensive, well-developed rough ER, aggregates of which can be stained to appear in light microscopy as *Nissl substance*.

Neurons can be classified according to the number of processes emanating from the soma. Neurons may be unipolar (having one process), bipolar (two processes), or multipolar (three or more processes). Unipolar neurons predominate in the CNS of most invertebrates, multipolar neurons predominate in the vertebrate CNS, and many sensory neurons are bipolar in various taxa. The neuronal processes themselves exhibit a bewildering geometric variety and complexity. Early anatomists attempted to bring order to this variety by classifying processes as *axons* and *dendrites*. Their

classifications were usually based on vertebrate CNS neurons (see Figure 12.4) and are useful for cells resembling vertebrate CNS neurons in form. Definitions of dendrites and axons, however, are based on a mixture of functional and morphological criteria that do not always coincide in a single neuron. Functionally (as we noted previously) a *dendrite* is considered to be a *receptive element* of a neuron that conveys information toward the soma (see Figure 12.2). An *axon*, by contrast, is the *output element* of a neuron, carrying information away from the cell body to other cells. This functional classification applies to most, but not all, neurons.

The dendrites of spinal motor neurons are relatively short and branch repeatedly (*dendrite* is Greek for "branch"). Dendrites of most neurons have continuously varying diameters and lack myelin sheaths (which we'll discuss shortly). In general, the broader

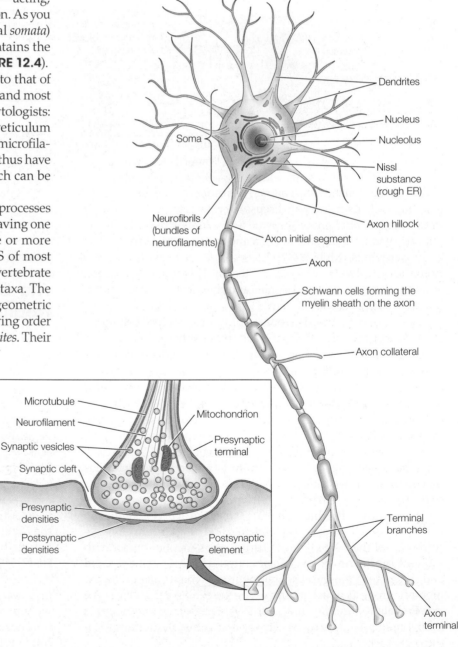

FIGURE 12.4 The cellular structure of a vertebrate neuron Every neuron has a cell body (soma, or perikaryon) and processes usually classified as axons and dendrites. The inset shows the structure of the very end of the axon, the axon terminal. The soma contains organelles, including rough endoplasmic reticulum (ER), Golgi apparatuses (not shown), and mitochondria. Stained aggregates of rough ER appear in light microscopy as Nissl substance. Cytoskeletal elements—microtubules and neurofilaments (see inset)—are present in the soma, dendrites, and axon. This neuron is myelinated, with periodic thickenings of myelin insulation around its axon. (The importance of the myelin sheath for the rate of propagation of nerve impulses is discussed later in this chapter; see page 333.) The axon ends in terminals, where synaptic vesicles (see inset) store molecules of neurotransmitter for synaptic transmission.

dendritic trunks resemble the soma in fine structure; they contain rough ER, mitochondria, microtubules, neurofilaments, and an occasional Golgi apparatus. Thinner dendritic branches may lack Golgi apparatuses and rough ER. The dendrites of many vertebrate neurons bear numerous short, thin protrusions termed *dendritic spines* that, when present, are important sites of synaptic input.

The axon of a neuron is classically single and long, with a relatively constant diameter and few collateral branches. The larger vertebrate axons are surrounded by **myelin** sheaths—multiple wrappings of insulating glial cell membranes (see below) that increase the speed of impulse transmission. Not all axons are myelinated; the smaller axons of vertebrate neurons and nearly all invertebrate axons lack myelin and are termed *unmyelinated*. At the fine structural level, axons contain microtubules, neurofilaments, elongated mitochondria, and sparse smooth endoplasmic reticulum (see Figure 12.4). Axons generally lack rough ER and Golgi apparatuses. Functionally, the axon is usually the portion of the neuron that supports action potentials, which *propagate* or conduct along the axon without decrement, carrying information away from the cell body to the axon terminals.

Glial cells support neurons physically and metabolically

Cells that are referred to collectively as **glial cells** or **neuroglial** ("nerve glue") **cells** surround the neurons (**FIGURE 12.5**). Rudolf Virchow (1821–1902) discovered and named the neuroglial cells in 1846 and thought that their primary function was to bind the neurons together and maintain the form and structural organization of the nervous system. The ratio of glial cells to neurons increases with increasing evolutionary complexity, from brains of fish to mammals. Glial cells are estimated to make up half the volume of the mammalian brain and to outnumber neurons by ten to one. These measures suggest that glial cells are important in nervous system function, perhaps in ways that are not yet fully understood.

Different types of glial cells play diverse functional roles in nervous systems. Vertebrate nervous systems have two kinds of ensheathing glial cells, called **Schwann cells** (in the PNS) and **oligodendrocytes** (in the CNS). Ensheathing glial cells envelop the axons of neurons (see Figure 12.5). The glial sheath can be a simple encircling of an *unmyelinated* axon or a group of axons, or a *myelin sheath* consisting of multiple concentrically wrapped layers of glial membrane that insulate the axon and increase the velocity of nerve-impulse propagation (discussed in detail later in this chapter). Other glial cells called **astrocytes** line the outside surfaces of capillaries in the vertebrate CNS and act as metabolic intermediaries between the capillaries and neurons. Astrocytes take up neurotransmitters from extracellular space and help supply metabolic substrates to neurons. They also regulate extracellular ion concentrations and play important roles in nervous system development. *Microglial cells* mediate immune responses in neural tissue and may act as phagocytes, consuming pathogens and cell debris in brain injury.

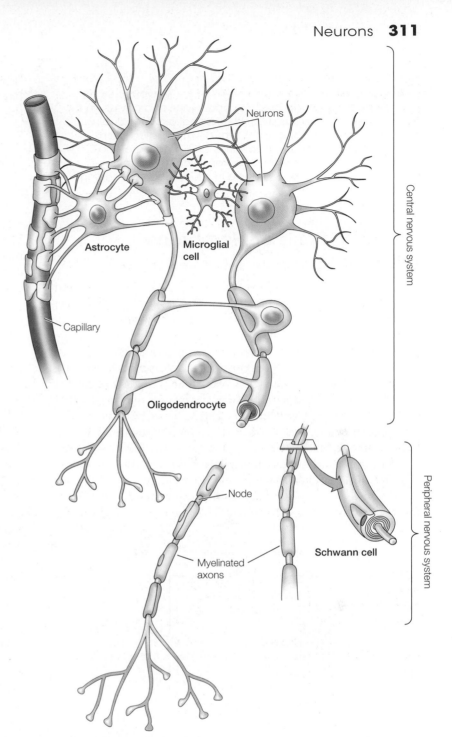

FIGURE 12.5 Glial cells There are four types of glial cells in vertebrate nervous systems. Schwann cells ensheathe axons (myelinated are shown; unmyelinated are not shown) in the PNS. Oligodendrocytes ensheathe axons in the CNS. Astrocytes are metabolic support cells in the CNS. Microglial cells are phagocytes related to cells of the immune system.

SUMMARY
The Cellular Organization of Neural Tissue

■ Neurons are the principal cells of nervous systems. They have long processes (dendrites and axons) that are specialized to receive signals from other neurons (via dendrites) and to generate and propagate action potentials (via axons).

(Continued)

■ Glial cells are the support cells of the nervous system. Schwann cells (in the PNS) and oligodendrocytes (in the CNS) form sheaths around neuronal axons, including insulating myelin sheaths around myelinated axons. Astrocytes surround capillaries and act as metabolic intermediaries between neurons and their circulatory supply. Microglial cells serve immune and scavenging functions.

The Ionic Basis of Membrane Potentials

What are the properties of the electrical signals of neurons, and how are these signals generated? Let's begin with a brief review of basic electrical concepts. Protons and electrons have *electrical charge*, and **ions** are atoms or molecules that bear a net charge because they have unequal numbers of protons and electrons. The net movement of charges constitutes an **electric current** (*I*), which is analogous to the hydraulic current of fluids flowing in a system of pipes. The separation of positive and negative electrical charges constitutes a **voltage**, or electrical *potential difference* (*V*). This potential difference can do work when charges are allowed to flow as current. Voltage is analogous to a height difference or head of pressure in a hydraulic system, allowing water to flow downhill.

FIGURE 12.6 shows a simple electrical circuit, that of a flashlight. A battery provides voltage; closing the switch allows current to flow through the electrical circuit. The electric current in the

(A) Flashlight circuit elements

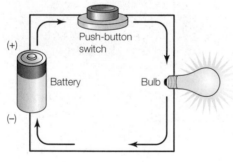

(B) Circuit diagram for the flashlight

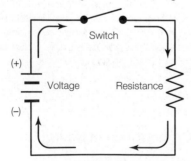

FIGURE 12.6 The simple electrical circuit of a flashlight
(A) Elements of the electrical circuit. (B) Circuit diagram. When the switch is closed, electric current flows through the resistance of the lightbulb filament, causing it to glow. Current must flow through the complete circuit. (Note that, by convention, current always flows from plus to minus.)

flashlight is the flow of free electrons along metal wires. Current flows through the lightbulb filament, which acts as **resistance** (*R*) that limits the current flow. Consequently, the filament heats and glows, emitting light.

Electrical circuits in cells are similar to the circuit in a flashlight, but they differ in some important ways. In cells, both the inside and outside media are *aqueous* solutions in which the electrical charges are ions rather than free electrons. Furthermore, all currents in cells are carried by ions, and any voltage or potential difference results from local imbalances of ion charges. Recall from Chapter 5 (see Figure 5.4) that fluids farther than a few nanometers from a membrane are electrically neutral, with equal numbers of positive and negative charges.

Because of this *charge neutrality of bulk solutions*, the only portion of a cell that *directly* determines its electrical properties is its outer-limiting cell membrane. Any electrical activity of a nerve cell is a property of the cell membrane, and the electrical potentials observed are called *transmembrane* potentials. The only immediately important attribute of the rest of the cell is the concentration of ions in solution in the intracellular fluid.

Cell membranes have passive electrical properties: Resistance and capacitance

All cells respond to electric currents, but not all cells generate action potentials (nerve impulses). The universal responses are *passive* responses (meaning that the cell's electrical properties do not change), but action potentials are *active* responses in which the properties do change. A cell's passive responses depend on the **passive electrical properties** of the membrane, principally its resistance and capacitance. A membrane exhibits *resistance* (measured in ohms, Ω) resulting from the fact that ions must flow through restrictive ion channels because the membrane's lipid bilayer is impermeable to ions. A membrane exhibits **capacitance** (measured in farads, F) because of the insulating properties of the bilayer. In electrical circuits, a capacitor has two conducting plates separated by an insulating layer; in cells, the conducting fluids on either side of the membrane act as plates, and the lipid bilayer separates and stores oppositely charged ions.

The resistance and capacitance of a cell's membrane depend on membrane area; specific membrane resistance and capacitance are measured per unit of area (e.g., $R_m = 1000\ \Omega \times cm^2$; $C_m = 1\ \mu F/cm^2$).[1] Whereas the specific membrane capacitance does not change, resistance may or may not change (depending on the behavior of specific populations of ion channels). *When we speak of a cell's passive electrical properties, we mean those conditions in which membrane resistance does not change.* A cell's passive electrical properties govern how voltages change over space and time along neuronal axons. Passive electrical properties do not explain the generation of action potentials (in which resistances change), but they are important for understanding how neurons generate and propagate action potentials.

We can use a squid giant axon to demonstrate the passive electrical properties of cells. The largest axons of a common squid may be 2 cm long and 700–1000 μm in diameter. Because these axons are so

[1] The resistance of a whole cell's membrane is usually called its *input resistance*.

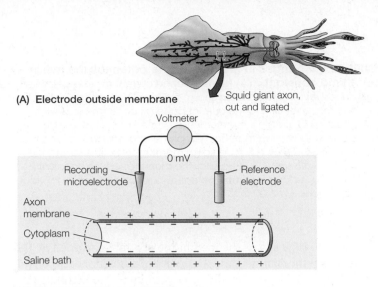

(A) Electrode outside membrane

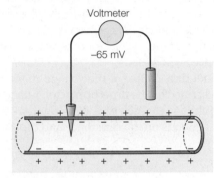

(B) Electrode inside membrane

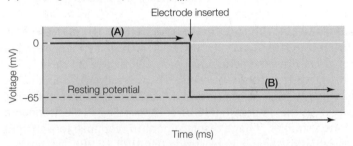

(C) Resting membrane potential (V_m)

FIGURE 12.7 Recording the resting membrane potential of a squid giant axon A section of the squid axon is removed and its ends ligated to seal the axon segment. (A) A voltmeter measures the potential difference between a glass capillary microelectrode (the recording electrode) and a reference electrode in the saline bath around the axon. When the microelectrode is outside the axon, there is no potential difference between the two electrodes. (B) The recording microelectrode has been advanced through the axon membrane, and the resting membrane potential (V_m) is recorded. (C) The output of the voltmeter, recorded on a chart writer or an oscilloscope, demonstrates that the resting membrane potential is inside-negative, a condition true for all cells. (By convention, negative is down for intracellular recording.)

large, it is relatively easy to cut out a length of the axon, ligate (close off) the ends, and penetrate the isolated axon with a microelectrode (**FIGURE 12.7**). The microelectrode consists of a glass capillary that has been heated and pulled to a fine tip (<1 μm in diameter) so that it can penetrate the cell membrane without causing damage. The capillary is filled with a solution of strong electrolyte, such as 3 molar (*M*) KCl, to minimize its electrical resistance.

When the tip of the microelectrode is outside the axon (see Figure 12.7A), a voltmeter records no potential difference (voltage) between the recording microelectrode and a neutral (reference) electrode suspended in the surrounding saline bath. Both electrodes are electrically neutral because of the charge neutrality of bulk solutions.

When the recording microelectrode is advanced just past the axon membrane into the cytoplasm (see Figure 12.7B), the voltmeter records a potential difference (see Figure 12.7C). This potential difference across the axon membrane is the **resting membrane potential (V_m)**. For all known cells, the polarity of the resting membrane potential is *inside-negative*; that is, the inner membrane surface is negative with respect to the outer membrane surface.[2] This example also reveals the axon's **membrane resistance (R_m)** to current flow, because without such resistance, ions would freely diffuse across the membrane, and it would not be able to maintain a potential difference.

What would happen to the resting membrane potential if we inserted a second microelectrode into the squid axon and generated a pulse of electric current between it and an extracellular electrode (**FIGURE 12.8**)? The current pulse would either depolarize or hyperpolarize the membrane, depending on the direction of the current. **Depolarization** is a *decrease* in the absolute value of the membrane potential toward zero (becoming less negative inside the cell). **Hyperpolarization** is an *increase* in the absolute value of the membrane potential away from zero (becoming more negative inside the cell).

In the example of Figure 12.8, we apply a current that flows outward across the membrane to cause a depolarization of the membrane potential, perhaps from −65 to −55 mV. (Remember that we are not considering action potentials yet, so assume that this depolarization is too small to trigger an action potential.) According to **Ohm's law**, the current should change the membrane potential by an amount proportional to the resistance to current flow:

$$\Delta V = IR \qquad (12.1)$$

where ΔV is the change in potential (termed a **graded potential**), I is the current (in amperes), and R is the resistance (in ohms). If the membrane exhibited only resistance, the change in membrane potential would occur instantaneously, as shown by the "theoretical" line in Figure 12.8B. However, the actual change in membrane potential occurs more gradually, reaching a plateau after a short delay, as shown by the "observed" line in Figure 12.8B.

PASSIVE ELECTRICAL PROPERTIES RETARD MEMBRANE VOLTAGE CHANGES The delay in depolarizing (or hyperpolarizing) a cell membrane occurs because the membrane behaves electrically like a resistor and a capacitor in parallel (see Figure 12.8C). On the one hand, the lipid bilayer of a cell membrane behaves like a *capacitor*: The bilayer blocks the exchange of ions between the extracellular fluid and the intracellular fluid, and its insulative properties enable oppositely charged ions to accumulate along the inner and outer surfaces of the membrane. On the other hand, the membrane-spanning ion channels behave like *resistors*: They al-

---

[2] By convention, the membrane potential is given as the inside value, with the outside considered zero.

(A) A current pulse changes membrane potential

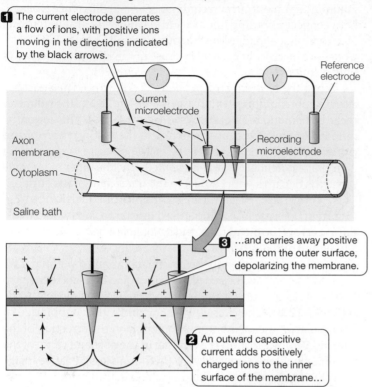

1 The current electrode generates a flow of ions, with positive ions moving in the directions indicated by the black arrows.

Reference electrode

Current microelectrode

Axon membrane

Cytoplasm

Recording microelectrode

Saline bath

3 ...and carries away positive ions from the outer surface, depolarizing the membrane.

2 An outward capacitive current adds positively charged ions to the inner surface of the membrane...

(B) Time course of voltage and current changes

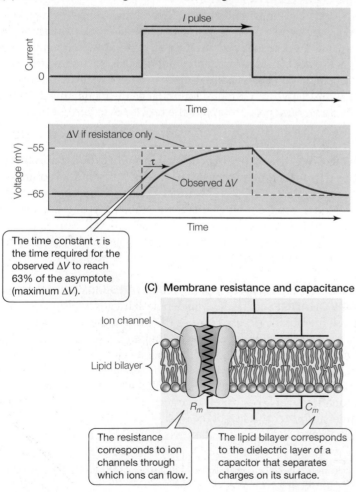

I pulse

Current

0

Time

ΔV if resistance only

−55

Voltage (mV)

τ

Observed ΔV

−65

Time

The time constant τ is the time required for the observed ΔV to reach 63% of the asymptote (maximum ΔV).

(C) Membrane resistance and capacitance

Ion channel

Lipid bilayer

R_m

C_m

The resistance corresponds to ion channels through which ions can flow.

The lipid bilayer corresponds to the dielectric layer of a capacitor that separates charges on its surface.

FIGURE 12.8 Changes in the membrane potential: the membrane time constant (A) Here a second current microelectrode is advanced into the squid axon shown in Figure 12.7. (B) The redistribution of charges during the current pulse (*I*) occurs more slowly than would be expected if the membrane acted as a pure resistance (dashed line), indicating that the membrane also has capacitive properties (see inset in [A]). (C) A neuronal membrane exhibits resistance (*R*) and capacitance (*C*) in parallel. The membrane *time constant* (τ) is the time it takes for membrane potential to reach 63% of its final value. In many neurons, τ is 2–20 ms.

low ions to flow across the membrane at a rate governed by the structure of the channels and the potential difference between the inside and outside of the membrane. Current first redistributes the charges on the membrane capacitance (capacitive current) and then flows through the membrane resistance (resistive or ionic current). This redistribution of charges slows (retards) the change in voltage on the membrane, by a factor that increases if the resistance or capacitance is increased.[3]

The exponential time course of the voltage change shown in Figure 12.8B is described by the **time constant**, τ (tau), the time it takes the voltage change to reach 63% of its final value. The time constant of a cell depends on the resistance and capacitance of its membrane:

$$\tau = RC \qquad (12.2)$$

where *R* is the cell's resistance (input resistance) and *C* is capacitance. For many cells, τ is in the range of 2–20 ms.

PASSIVE ELECTRICAL PROPERTIES LIMIT THE SPREAD OF GRADED POTENTIALS How does a change in voltage spread over *distance* in a membrane? (Remember that we are considering only the *passive* electrical properties of a membrane.) Suppose we insert yet another electrode into the squid axon, this one farther away from the other two (**FIGURE 12.9A**). The V_2 electrode will record a smaller voltage displacement in response to a current pulse than the nearer V_1 electrode (**FIGURE 12.9B**). The voltage change (ΔV) will decrease exponentially with distance from the source producing it, a property called **passive spread (decremental spread)** or **electrotonic conduction**. The steepness of this decrease with distance is described by the membrane **length constant**, λ (lambda), which represents the distance at which the decaying voltage change (ΔV_m) is 37% of its value at the origin (**FIGURE 12.9C**).

The reason for this decrease with distance is that as current flows along the inside of the axon, some of it leaks out through ion channels. For simplicity we ignore the slowing effects of membrane capacitance, and we lump the resistive pathways as R_m (the resistance to current flow out across the membrane ion channels in a segment) and R_i (the resistance to current continuing down the axon to the next segment). An axon (or dendrite) with a high R_m value and a low R_i value will have a large λ value. We

---

[3] To understand this, it is helpful to realize that *capacitance* (expressed in farads) is the amount of charge stored per unit of voltage ($C = Q/V$). The greater the capacitance, the more ions the membrane can separate and store for a given potential difference, and therefore the more time it takes for ions to be redistributed in response to a pulse in current.

(A) Recording the spread of a potential

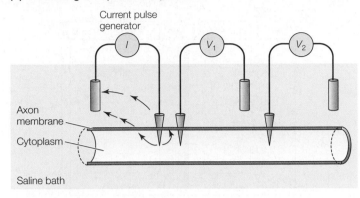

(B) Passive potentials spread decrementally

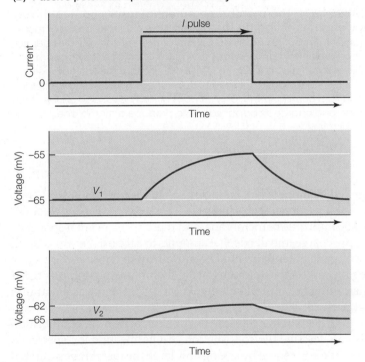

(C) The membrane length constant describes the exponential decrement

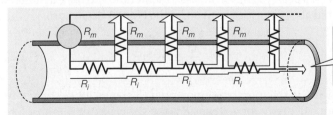

Because most of the current flows through shorter paths of lower resistance, less and less current is available to change the membrane potential at greater distances from the source.

will consider the membrane length constant again later, when we discuss action-potential propagation.

The electrical properties of membrane resistance (R_m), membrane capacitance (C_m), and resting membrane potential (V_m), and the related time constant (τ) and length constant (λ), adequately describe the passive electrical properties of a neuron or any other cell. Because the passive electrical properties of axons are similar to those of underwater telephone cables, these passive electrical properties are often called *cable properties*. For the same reason, the equations describing the length constant and time constant of neuronal membranes are called *cable equations*.

Resting membrane potentials depend on selective permeability to ions: The Nernst equation

Our model of the passive electrical properties of cells describes the inside-negative resting membrane potential maintained by all living cells. A cell membrane's permeability to different kinds of ions is the mechanism that establishes and maintains this voltage. Dissolved ions have charges and attract polar water molecules around them. The charged ions cannot mix with the nonpolar tails of the lipid molecules in the center of the membrane bilayer, so the ions cannot pass through the bilayer (see Chapter 5). Instead, ions must pass (if they pass at all) by way of the protein ion channels that span the bilayer. There are many kinds of ion channels, each kind *selectively permeable* to specific ions. Moreover, some ion channels can open and close, which means that *membrane permeability to specific ions is a controlled condition*.

How does a membrane's selective permeability to ions produce a membrane potential? Consider a simplified cell (**FIGURE 12.10**) that contains a solution of potassium ions (K^+) and nonpermeating anions (represented by A^-), such as charged proteins. (The identity of A^- is unimportant; all we're concerned with is its charge.) The cell is in a bath of two nonpermeating ions (Na^+ and A^-); we stipulate that the cell membrane is permeable only to K^+. When the cell is placed in the bath, K^+ tends to diffuse out of the cell, down its concentration gradient. Because charge neutrality is always maintained in bulk solution, A^- would tend to follow the K^+ across the membrane; however, the membrane is not permeable to A^-. Therefore A^- tends

FIGURE 12.9 Graded potentials decrease exponentially with distance The amplitude of a voltage change decreases with distance along the axon. (A) A third electrode (V_2) is added to the setup diagrammed in Figure 12.8 and is used to measure the potential change at some distance from the source of current (I). (B) The voltage change measured at V_2 is smaller than that at V_1, which is closer to the source of current. This decremental spread of graded potentials is referred to as *electrotonic conduction*. (C) The decrement in voltage change is exponential with distance from I. Arrows show the local paths of current flow that depolarize the membrane. (Capacitances are ignored.) The membrane *length constant* (λ) describes the exponential decay of a change in voltage (ΔV) with distance.

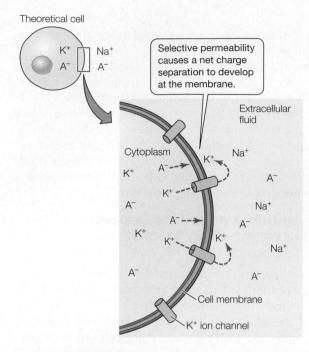

Theoretical cell

Selective permeability causes a net charge separation to develop at the membrane.

FIGURE 12.10 Selective permeability of a membrane gives rise to a membrane potential A simplified, theoretical cell containing a solution of K⁺ and A⁻ is bathed by a solution of Na⁺ and A⁻. The cell membrane is permeable only to K⁺. At the membrane, K⁺ ions tend to diffuse out, down their concentration gradient. A⁻ ions attempt to follow (to maintain charge neutrality), but the membrane is not permeable to them, so they cannot pass through it. The resulting charge separation produces a membrane potential.

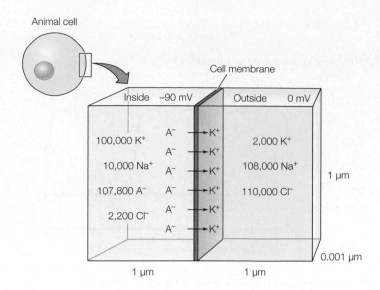

FIGURE 12.11 The membrane potential results from relatively few charges sitting on the membrane A small patch of membrane (1 μm × 0.001 μm in area) from a mammalian muscle fiber, with a small volume (1 μm × 1 μm × 0.001 μm) of adjacent cytoplasm and extracellular fluid. As in Figure 12.10, assume that the membrane is most permeable to K⁺. Of the 110,000 cations and 110,000 anions in each fluid compartment, only *six* pairs of ions need to sit on the membrane and charge its capacitance to produce a membrane potential of –90 mV. (After Schmidt 1985.)

to accumulate at the inner surface of the membrane, while K⁺ tends to accumulate at the outer surface (because of charge attraction). Thus a net charge separation develops, but only at the membrane.

The net negative charge on the inner surface of the membrane and net positive charge outside tend to move K⁺ ions back into the cell, by forces of charge attraction and repulsion. Eventually this system reaches *electrochemical equilibrium* (see Chapter 5, page 110), in which there is no *net* movement of ions and no work is done. In Figure 12.10, equilibrium is reached when the tendency for K⁺ ions to diffuse out of the cell (down the concentration gradient) is exactly balanced by their tendency to move in (down the electrical gradient of the membrane potential). That is, when the concentration–diffusion force just equals the opposing electrical force, there is no *net* flow of K⁺ ions. *For this system to come to equilibrium, there must be an electrical force (i.e., a membrane potential) across the membrane.* This simple example shows that any cell that has a transmembrane concentration difference of a permeating ion tends to generate a membrane potential. It is a good starting point for visualizing a cell's resting potential because cells at rest are more permeable to K⁺ than to other ions.

It is important to grasp that the resting membrane potential results from relatively few ion charges sitting on the membrane, and to remember as well that the charge separation producing the membrane potential is an extraordinarily local phenomenon. Charge neutrality always prevails in the bulk solutions that make up the intracellular and extracellular fluids. For example, there are approximately 110,000 cations and 110,000 anions in a 1 μm × 1 μm × 0.001 μm "slice" of the fluid compartments on either side of

the cell membrane of a mammalian muscle fiber (**FIGURE 12.11**). Of these portions, approximately six pairs of ions sit along the 1 μm × 0.001 μm area of membrane, and these six pairs are responsible for the charge imbalance that amounts to a robust –90-mV resting membrane potential! Thus the movement of only a few ions in a region can establish (or change) a membrane potential without disrupting the overall charge neutrality of the intracellular and extracellular fluids.

Although living cells are not at electrochemical equilibrium, we can describe the contribution of a permeating ion species to membrane potential by asking how large the membrane potential *would be* at equilibrium. The relation between the concentration difference of a permeating ion across a membrane and the membrane potential at equilibrium is given by the **Nernst equation**:

$$E = \frac{RT}{zF} \ln \frac{C_{out}}{C_{in}} \tag{12.3}$$

in which E is the membrane potential (E stands for *electromotive force*, an older term for voltage), R is the gas constant, T is absolute temperature, z is the valence of the ion species (charge for the kind of ion), F is Faraday's constant (charge per mole of ions), and C_{out} and C_{in} are the ion concentrations on the two sides of the membrane. Notice that *the larger the concentration difference across the membrane, the larger the membrane potential* at which the ion species is in equilibrium. The reason for this relation is that increasing the concentration difference increases the concentration gradient of the ion species and thus increases the electrical force necessary to oppose it.

(A) Ion concentrations

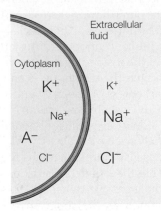

(B) Pump maintains Na⁺, K⁺ concentrations

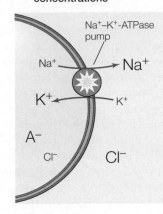

(C) Steady state

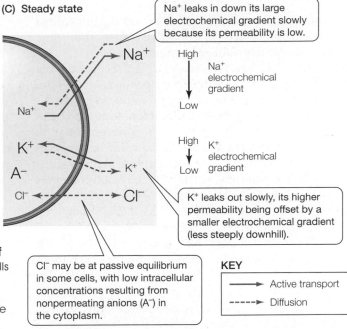

Na⁺ leaks in down its large electrochemical gradient slowly because its permeability is low.

Na⁺ electrochemical gradient

K⁺ electrochemical gradient

K⁺ leaks out slowly, its higher permeability being offset by a smaller electrochemical gradient (less steeply downhill).

Cl⁻ may be at passive equilibrium in some cells, with low intracellular concentrations resulting from nonpermeating anions (A⁻) in the cytoplasm.

KEY

→ Active transport

‐‐‐‐‐▶ Diffusion

FIGURE 12.12 Ion pumps help maintain the concentration of major ions in intracellular and extracellular fluids (A) All cells maintain low intracellular concentrations of Na⁺ and Cl⁻ and high concentrations of K⁺ and nonpermeating anions (A⁻) within intracellular fluids, relative to extracellular fluid. (Symbol sizes represent relative concentrations.) (B) An active sodium–potassium exchange pump transports Na⁺ out and K⁺ in, counteracting the tendency of Na⁺ to diffuse in and K⁺ to diffuse out. (C) Here the Na⁺ and K⁺ concentrations are maintained in a steady state across the membrane. The slow, passive leaks of Na⁺ and K⁺ (dashed arrows) are counteracted by active transport by the Na⁺–K⁺ exchange pump (solid arrows). Chloride, in contrast, can be at passive equilibrium in some cells.

We can simplify the Nernst equation by calculating R/F, converting to $\log_{10}$, and considering an ion of a given valence at a given temperature. For K⁺, a monovalent cation, at 18°C (British room temperature),

$$E\left(\text{in mV}\right) = 58 \log_{10} \frac{C_{\text{out}}}{C_{\text{in}}} \qquad (12.4)$$

For a mammal at 37°C,

$$E\left(\text{in mV}\right) = 61 \log_{10} \frac{C_{\text{out}}}{C_{\text{in}}} \qquad (12.5)$$

Thus, for our simplified cell permeable only to K⁺, if the internal K⁺ concentration is, say, 100 mM and the external concentration is 10 mM, at 18°C, $E = 58 \log_{10}(0.1) = -58$ mV. (By convention, the minus sign means that the inside of the membrane is negative relative to the outside.)

The value of −58 mV is the **equilibrium potential** for potassium (E_K) in our system as we have defined it—that is, the value of the membrane potential at which K⁺ ions are at electrochemical equilibrium and the internal K⁺ concentration is ten times the external concentration. In other words, the electrical force holding K⁺ inside the cell is just balanced by the chemical (concentration) force for K⁺ diffusion out of the cell. There is an equilibrium potential for each ion species (E_{Na}, E_{Cl}, etc.).

The Nernst equation relates membrane potential to the concentration ratio of only one ion species. For this and other reasons, the generation of membrane potentials in real cells is considerably more complex than in this simplified model.

Ion concentration differences result from active ion transport and from passive diffusion

All cells maintain higher concentrations of potassium and lower concentrations of sodium and chloride in the intracellular fluids than are present in the surrounding extracellular fluid (**FIGURE 12.12A**). The concentrations of these ions differ from organism to organism, as shown in **TABLE 12.1**. Despite quantitative differences, the *concentration ratios* of ions in all cells are similar to those represented in the table. The difference in ion concentrations

TABLE 12.1	Concentrations of major ions in intracellular fluid (cytoplasm) and extracellular fluid					
	Squid axon			**Mammalian muscle**		
Ion type	Out (mM)	In (mM)	Out/In	Out (mM)	In (mM)	Out/In
Na⁺	440	50	8.8	145	12	12.1
K⁺	20	400	0.05	4	155	0.03
Cl⁻	560	60	9.3	120	3.8	31.6
A⁻ (organic anions)	—	270	—	—	—	—

between the intracellular and extracellular fluids results from a combination of two processes: (1) the active transport of some ions and (2) the passive distribution of other ions.

Examine Figure 12.12A and consider the Nernst equation. It is impossible for both Na^+ and K^+ ions to be in passive equilibrium, because the ratios of their concentrations differ. Whereas K^+ would require an inside-negative membrane potential to be in equilibrium, Na^+ would require an inside-positive potential (to counteract inward Na^+ diffusion). In fact, *neither* ion species is in passive equilibrium in cells, because both ions are maintained at nonequilibrium levels by *pumps*—that is, by active ion transport that requires the input of energy from hydrolysis of ATP.

The most important pump is the *Na^+–K^+-ATPase pump* (see Chapter 5, page 115), which actively transports Na^+ out of the cell and K^+ into it. For most cells, Na^+ ions slowly leak into the cell and K^+ ions slowly leak out. (The channels through which these leaks occur are normally open and are called *leakage channels*.) The Na^+–K^+-ATPase pump counteracts these leaks, using ATP energy to pump Na^+ out as fast as it leaks in, and to pump K^+ in as fast as it leaks out (**FIGURE 12.12B**). The function of the Na^+–K^+-ATPase pump is analogous to a bilge pump in a boat, bailing water out as fast as it leaks in. Another good analogy for the pump is a battery charger, which can work in the background to prevent the "batteries" of Na^+ and K^+ concentration distributions from running down. For known Na^+–K^+ exchange pumps, the ratio of Na^+ and K^+ pumped is 3:2. (We will consider a secondary consequence of the 3:2 ratio shortly.)

However, a permeating ion species may have very different concentrations inside and outside a cell *without* being pumped. Consider the Cl^- ions in Figure 12.12A. Because cells have large intracellular concentrations of nonpermeating anions, permeating ions such as Cl^- must become distributed unequally across the membrane. The A^- ions that limit the tendency of K^+ ions to diffuse out of the cell (see Figure 12.10) also limit the tendency of Cl^- ions to diffuse in. As described by the Nernst equation, the membrane potential counteracts the tendency of Cl^- ions to diffuse from their high extracellular concentration to the lower intracellular concentration (see Figure 12.12C). In some cells, Cl^- ions are in passive equilibrium despite strikingly unequal concentrations inside and outside the cell. This equilibrium, sometimes called a *Donnan equilibrium* or *Gibbs–Donnan equilibrium*, explains how nonpermeating anions inside the cell can lead to unequal concentrations across the membrane of permeating ions such as Cl^-.

Donnan-type phenomena are one way in which inequalities of ions such as Cl^- are maintained in living cells. The nonpermeating anions (A^-) in cells are a mixture of relatively small organic anions such as amino acids, along with proteins and other large molecules that bear net negative charges. Chloride ions appear to be passively distributed across the membranes of many cells, even though the ratio of outside to inside Cl^- concentrations may be quite large, as it is, for example, in mammalian muscle (see Table 12.1). Neurons of the vertebrate CNS, however, actively transport Cl^- out, so the ratio of $[Cl^-]_{out}$ to $[Cl^-]_{in}$ is larger than that predicted by Donnan equilibrium, and E_{Cl} is more negative than V_m.

Now we are in a position to explain ion concentration distributions of living cells. Neither Na^+ nor K^+ is at equilibrium in Figure 12.12, so passive diffusion alone produces net movement of Na^+ and K^+. (This disequilibrium results from the pump having changed Na^+ and K^+ concentrations from their equilibrium levels.) Na^+ especially is far out of equilibrium because both the concentration–diffusion gradient and the electrical gradient of the inside-negative membrane potential drive Na^+ inward. Because the membrane is only slightly permeable to Na^+ at rest, Na^+ enters only slowly and is pumped out as fast as it diffuses in. K^+ is closer to equilibrium but not at it. The cell loses K^+ passively at a slow rate because although permeability to K^+ is large, the driving force for K^+ flow ($V_m - E_K$) is small. The slow passive loss of K^+ is also counteracted by the sodium–potassium exchange pump. Thus the cation concentrations are maintained in a *steady state* in which passive-transport leaks are counteracted by active-transport pumps. Metabolic energy is required to maintain ions at concentrations different from the concentrations at equilibrium.

FIGURE 12.12C summarizes the roles of active and passive transport in maintaining the steady-state concentrations of Na^+ and K^+ ions in intracellular and extracellular fluids: Passive leaks of Na^+ into the cell and of K^+ out of the cell are counteracted by active ion transport. Chloride ions are passively distributed in this cell. The balance of active and passive transport that controls concentrations of different ions is important because, as we will see, the steady-state concentrations of K^+, Na^+, and Cl^- ions all contribute to membrane potentials in animal cells.

Membrane potentials depend on the permeabilities to and concentration gradients of several ion species: The Goldman equation

The ion concentrations of living cells are in a steady state in which the ions are unequally distributed across the cell membrane and many are out of equilibrium. *The resting membrane potential is largely determined by K^+ concentrations because the cell membrane is more permeable to K^+ than to other ions.* If the membrane were permeable only to K^+, then the membrane potential would be exactly equal to the K^+ equilibrium potential (i.e., $V_m = E_K$), as predicted by the Nernst equation employing the K^+ concentrations across the membrane. Because the membrane is somewhat permeable to other ions, however, they also contribute to the membrane potential.

The contribution of each ion is weighted by its ability to permeate the membrane, with the more-permeating ions having more effect. The value of the membrane potential (V_m) produced by the contributions of several permeating ion species can be determined by the **Goldman equation**:

$$V_m = \frac{RT}{F} \ln \frac{P_K [K^+]_o + P_{Na} [Na^+]_o + P_{Cl} [Cl^-]_i}{P_K [K^+]_i + P_{Na} [Na^+]_i + P_{Cl} [Cl^-]_o} \quad (12.6)$$

in which P_K, P_{Na}, and P_{Cl} are relative permeability values for potassium, sodium, and chloride ions, respectively. (The chloride term in the equation is inverted to reflect its negative charge.) In principle, it is necessary to add a term in the Goldman equation for every permeating ion species, but in practice it is necessary to include terms only for Na^+, K^+, and Cl^-. The contributions of other ion species can be neglected, by reason of either low permeability

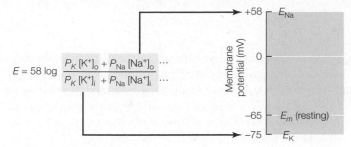

$$E = 58 \log \frac{P_K\,[K^+]_o + P_{Na}\,[Na^+]_o \,\cdots}{P_K\,[K^+]_i + P_{Na}\,[Na^+]_i \,\cdots}$$

FIGURE 12.13 The Goldman equation and the "voltage thermometer" A simplified Goldman equation describes membrane potential in terms of relative permeabilities (P) of the membrane to K^+ and Na^+. The voltage scale graphs the membrane potential determined by these permeabilities. For the resting membrane, $P_K \gg P_{Na}$, so E_m is close to E_K. If P_{Na} increases to become greater than P_K, E_m will approach E_{Na}. Each arrow relates the dominant term in the equation to the value of E_m toward which it drives the membrane.

of the membrane to those ions (e.g., HCO_3^-) or low concentrations of those ions (e.g., $[H^+] = 10^{-7}\,M$).[4]

In fact, for some purposes it is useful to consider sodium and potassium only, ignoring chloride. In such a simplification of the Goldman equation (**FIGURE 12.13**), we can view the membrane potential as a result of the membrane's relative permeabilities to sodium and potassium ions, visualized with a sliding voltage scale, rather like a thermometer but in units of voltage. Consider a squid axon with the following ion concentrations:

$$\left[K^+\right]_{out} = 20\ mM \qquad \left[Na^+\right]_{out} = 440\ mM$$

$$\left[K^+\right]_{in} = 400\ mM \qquad \left[Na^+\right]_{in} = 44\ mM$$

$$\left[K^+\right]_{out} : \left[K^+\right]_{in} = 0.05 \quad \left[Na^+\right]_{out} : \left[Na^+\right]_{in} = 10$$

$$E_K = 58 \log\ 0.05 = -75\ mV\ (\text{inside-negative})$$

$$E_{Na} = 58 \log\ 10 = +58\ mV\ (\text{inside-positive})$$

If the cell were permeable only to K^+, then V_m would equal E_K, or -75 mV; and if it were permeable only to Na^+, then V_m would equal E_{Na}, or $+58$ mV. The actual membrane potential can be anywhere between these values and is governed by the ratio of membrane permeabilities to Na^+ and K^+. The "voltage thermometer" shows that when permeability to potassium is much greater than permeability to sodium, the membrane potential approaches E_K (see Figure 12.13):

$$\text{If } P_K = 10 \times P_{Na}$$

$$\text{then } V_m = 58 \log_{10} \frac{10[20] + 1[440]}{10[400] + 1[44]} = 58 \log\,(644/4044) = -46.3\text{mV}$$

In contrast, when permeability to sodium is much higher than permeability to potassium, the membrane potential approaches E_{Na}. This visualization of membrane potential in terms of the Goldman equation will be important for our consideration of action potentials later in this chapter.

---

[4] Strictly speaking, the ion concentration values in the Nernst and Goldman equations should be those for *free* (unbound) ions rather than total concentrations. The **activity** of an ion is the concentration of the ion in its dissociated, freely diffusible form. Monovalent ions dissociate relatively completely in cytoplasm and extracellular fluids, so not much correction is needed for ion activities. Divalent ions such as Ca^{2+}, however, are predominantly bound in cytoplasm, so corrections for their activity are important.

Electrogenic pumps also have a small direct effect on V_m

Our explanation to this point of the generation of membrane potentials has been termed the ionic hypothesis. The **ionic hypothesis** argues that the concentrations of ions inside and outside a cell are maintained in a *steady state* by a mixture of active-transport processes (ATPase pumps) and passive-transport processes (diffusion and Donnan effects). The ionic hypothesis further asserts that the concentrations of ions inside and outside the cell, and the permeability of the cell membrane to these ions, determine the resting membrane potential (V_m) as described by the Goldman equation. The ionic hypothesis is substantially accurate, and provides a useful description of the factors giving rise to membrane potentials in living cells. A more complete explanation of the causes of membrane potentials, however, must include the fact that some ion pumps are electrogenic.

There are two kinds of active ion-transport mechanisms: electroneutral pumps and electrogenic pumps. An **electroneutral pump** transports equal quantities of charge inward and outward across a membrane and thus changes ion concentrations without generating an electric current. An **electrogenic pump** transports unequal quantities of charges inward and outward across the membrane. As noted already, the Na^+–K^+ exchange pump has a 3:2 ratio, transporting 3 Na^+ ions out for each 2 K^+ ions transported into the cell.

Any ion pump that is not 1:1 generates a net current (net movement of charge) across the membrane. This current, acting across the cell's membrane resistance, directly generates a potential, via Ohm's law. The potential resulting from pump current changes V_m from the value predicted by the Goldman equation. Thus an electrogenic pump has two functional properties: It changes concentrations to offset passive leaks (its major function), and it alters V_m directly via the pump current (a smaller, secondary function).

The 3:2 sodium–potassium pump generates an outward ionic current (outward movement of positive charge) that hyperpolarizes the cell to a level more inside-negative than is predicted by the Goldman equation. Because sodium–potassium exchange pumps can be selectively poisoned with toxins such as *ouabain*, their electrogenic contribution to resting membrane potentials can be measured as the initial change in V_m before concentrations change. (Inactivation of the pump will also [more slowly] lead to changes in ion concentrations, and thus have an additional, indirect effect on membrane potential.) In many neurons the direct contribution of an electrogenic pump accounts for only a few millivolts of the resting membrane potential, although electrogenicity can make a larger contribution in small axons and in some invertebrate neurons.

SUMMARY
The Ionic Basis of Membrane Potentials

- Cell membranes have properties of electrical resistance and capacitance, which allow them to maintain a voltage (membrane potential) and regulate current flow across the membrane. Cells have inside-negative resting membrane potentials. The passive electrical properties of membranes determine how membrane potentials change with time (the time constant, τ) and with distance (the length constant, λ).

(Continued)

■ Membrane potentials depend on selective permeability to ions. Any ion species to which the membrane is permeable will tend to drive the membrane potential toward the equilibrium potential for that ion. The Nernst equation calculates the equilibrium potential of a single ion species in terms of its concentrations on both sides of the membrane.

■ All cells have higher concentrations of K^+ inside than outside, higher concentrations of Na^+ outside than inside, and higher concentrations of Cl^- outside than inside. Ion concentrations inside and outside cells are maintained by active ion pumps, as well as by passive Donnan-equilibrium effects.

■ Membrane potentials depend on the permeabilities to and concentration gradients of several ion species: The resting membrane is dominated by permeability to K^+, so the resting membrane potential is near E_K. The Goldman equation describes how changing the membrane permeability of an ion species changes the membrane potential.

■ In addition to their major role of maintaining the nonequilibrium concentrations of ions, electrogenic ion pumps generate a current that makes a small, direct contribution to V_m. In addition, only those ions that are freely diffusible contribute to V_m, so corrections for bound ions may be necessary.

The Action Potential

Excitable cells such as neurons, muscle fibers, and a few others have the ability to generate electrical signals. The hallmark electrical signal of an excitable cell is the *action potential*. Action potentials (which in neurons may also be called *nerve impulses*) are one of the most important kinds of electrical signals underlying the integrative activity of nervous systems. Some kinds of neurons do not generate action potentials, however, so the association of neurons with action potentials is not universal.

Action potentials are voltage-dependent, all-or-none electrical signals

Action potentials result from *voltage-dependent* changes in membrane permeabilities to ions because the ion channels that produce action potentials are *voltage-gated*—that is, their opening depends on the membrane potential (see Figure 5.5). An action potential is initiated by a change in the resting membrane potential, specifically by a depolarization sufficiently strong to open the voltage-gated channels. The voltage dependence of ion permeabilities is a critical feature of action potentials, and it makes action potentials fundamentally different from resting potentials or from graded potentials.

Action potentials have characteristic features. An **action potential** is a momentary reversal of membrane potential from about –65 mV (inside-negative) to about +40 mV (inside-positive)—a voltage change of about 100 mV, lasting about 1 ms, followed by restoration of the original membrane potential (**FIGURE 12.14A**). The action potential is triggered by any depolarization of the membrane that reaches a critical value of depolarization, the **voltage threshold**.

After the suprathreshold (above threshold) depolarization, the action potential has a rapid *rising phase* that reaches a peak more positive than zero potential (*overshoot*) followed by a rapid repolarization (the *falling phase*). In the squid axon and in many other neurons, the action potential is followed by an *undershoot*, a transient after-hyperpolarization lasting a few milliseconds.

To illustrate the voltage-dependent properties of action potentials, let's perform a hypothetical experiment using a squid giant axon (**FIGURE 12.14B**). As we did in Figure 12.8, we penetrate the axon with two glass capillary microelectrodes—one to apply current pulses and one to record voltage. The first three inward-flowing current pulses hyperpolarize the membrane in the vicinity of V_1; the amount of hyperpolarization is proportional to the strength of each current pulse (**FIGURE 12.14C**). This relation follows from Ohm's law and (ignoring the time constant) indicates that with hyperpolarization the membrane resistance does not change. Thus hyperpolarization can't induce action potentials, because it doesn't change the permeabilities of membrane ion channels. Weak, outward-flowing current pulses in the opposite direction (pulses 4 and 5 in Figure 12.14C) elicit small depolarizations that approximately mirror the preceding hyperpolarizations, again indicating no significant change in membrane resistance.

Stronger depolarizing currents (pulses 6–8 in Figure 12.14C) that exceed the voltage threshold produce action potentials. However, a stronger depolarizing current (beyond threshold) does not produce a larger action potential (compare the responses to pulses 6 and 7). Instead, action potentials are **all-or-none** phenomena; that is, a depolarization below threshold elicits no impulse, but all suprathreshold depolarizations produce complete impulses substantially alike in amplitude and duration.

Immediately following an action potential, another action potential cannot be generated for at least 1 ms (the **absolute refractory period**) and is harder to generate for a few milliseconds longer (the **relative refractory period**). We will discuss the membrane properties that impose these refractory periods later in this chapter. Because of the all-or-none property of the action potential and the succeeding refractory period, impulses cannot summate. Instead, a prolonged suprathreshold depolarizing current (pulse 8 in Figure 12.14C) can elicit a *train* of discrete action potentials. For many neurons, the frequency of impulses in a train increases with increasing strength of depolarizing current (within limits).

An action potential, once initiated, *propagates* along the axon without a decrease in amplitude and at a constant velocity that depends on the diameter of the axon (among other factors). If in Figures 12.14B and C a remote electrode measured voltage at the end of the axon (not shown), it would record each action potential that the local electrode (V_1) records, with no decrease in amplitude. Each impulse recorded remotely follows the impulse at V_1 by a short latency that represents the time required for the impulse to propagate along the axon between the two electrodes. The distant V_2 electrode would not record the subthreshold depolarizations and hyperpolarizations, because they are not propagated; instead they spread decrementally and so are weakened before reaching V_2 (compare Figure 12.9).

In summary, action potentials are all-or-none electrical signals in excitable cells that propagate rapidly and without degradation over long distances. This ability to send signals over long distances

(A) An action potential

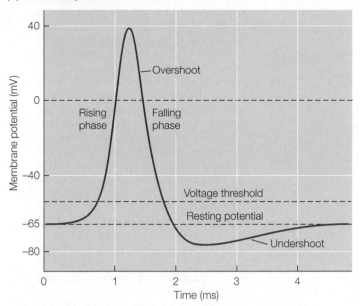

(B) Stimulating and recording action potentials

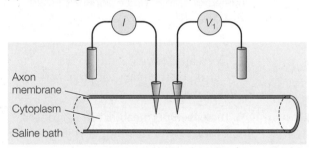

(C) Subthreshold responses and action potentials

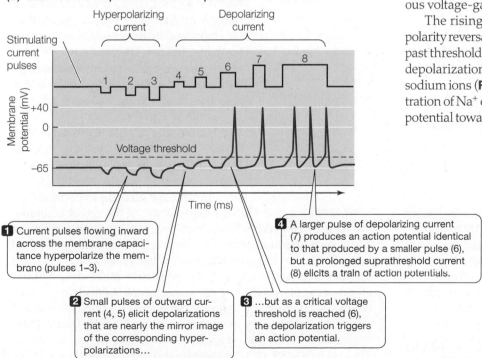

1 Current pulses flowing inward across the membrane capacitance hyperpolarize the membrane (pulses 1–3).

2 Small pulses of outward current (4, 5) elicit depolarizations that are nearly the mirror image of the corresponding hyperpolarizations...

3 ...but as a critical voltage threshold is reached (6), the depolarization triggers an action potential.

4 A larger pulse of depolarizing current (7) produces an action potential identical to that produced by a smaller pulse (6), but a prolonged suprathreshold current (8) elicits a train of action potentials.

FIGURE 12.14 General features of action potentials (A) An action potential is a brief voltage change characterized by a rising phase that overshoots zero and a falling phase (repolarization) that may be followed by an after-hyperpolarization, or undershoot. (B) Recording action potentials in a squid giant axon, using a stimulating electrode (*I*) and a recording electrode (V_1). (C) Responses of the axon to stimulating current pulses.

rapidly and without distortion was presumably an important factor allowing the evolution of large animals whose complex physiology and behavior require extensive neural coordination.

Action potentials result from changes in membrane permeabilities to ions

The permeability terms in the Goldman equation (Equation 12.6) show that any factor that changes the permeability of the membrane to one or more ion species will change the value of the membrane potential. *An action potential results from intense, localized increases in permeabilities to specific ions*—increases that are both voltage- and time-dependent. What's more, the permeability increases are selective for specific ions: first sodium and then potassium.

PERMEABILITIES AND ION CHANNELS Let's follow the rise and fall of one action potential to see when and how these changes in the membrane's permeability to sodium and potassium ions occur. At the resting membrane potential of –65 mV, the membrane is most permeable to K⁺ ions (**FIGURE 12.15A**). Neurons contain some K⁺ channels that are normally open and are not voltage-gated. These *leakage channels* allow K⁺ to diffuse across the membrane following the electrochemical gradient. The K⁺ leakage channels remain open throughout an action potential, but the more numerous voltage-gated channels swamp their effects.

The rising phase of the action potential (depolarization and polarity reversal) begins when a stimulus depolarizes the membrane past threshold. Voltage-gated Na⁺ channels open in response to the depolarization, vastly increasing the membrane's permeability to sodium ions (**FIGURE 12.15B**). Because of the much higher concentration of Na⁺ outside the cell, Na⁺ rushes in, driving the membrane potential toward E_{Na} (which is inside-positive). The inward-rushing sodium current is the cause of depolarization and polarity reversal at the rising phase of the action potential. Just as a dominant permeability to K⁺ at rest makes the resting membrane potential inside-negative, the inflow of Na⁺ during the rising phase of the action potential makes the membrane momentarily inside-positive.

The falling phase of the action potential results from two changes in the membrane's permeability to ions (**FIGURE 12.15C**). First, the opening of the voltage-gated sodium channels is rapidly terminated by a process called Na⁺ channel **inactivation**, which abruptly decreases permeability to Na⁺. Second, after a slight delay, voltage-gated potassium channels open, greatly increasing permeability to K⁺. Potassium ions flow out and drive the membrane toward E_K.

(A) Resting membrane potential

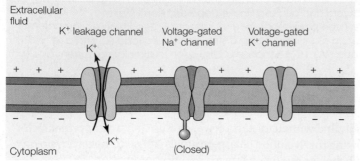

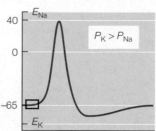

(B) Rising phase

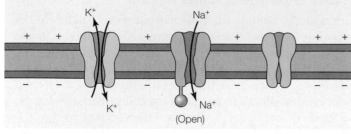

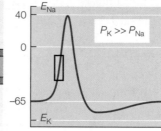

(C) Falling phase

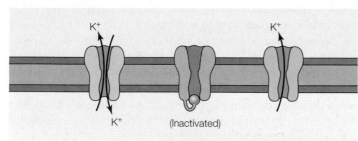

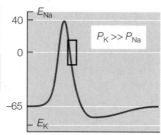

(D) Recovery

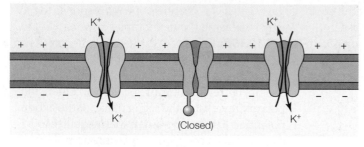

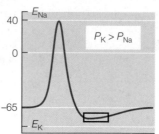

FIGURE 12.15 Membrane permeability changes that produce an action potential (A) At rest the membrane is most permeable to K⁺, as represented here by leakage channels that are always open. The box in the graph on the right indicates the membrane potential at this stage, described by the "voltage thermometer" (see Figure 12.13). (B) During the rising phase of the action potential, voltage-gated Na⁺ channels open, and the high permeability to Na⁺ dominates, driving the membrane potential toward E_{Na}. (C) Na⁺ channels are inactivated soon after they open, and voltage-gated K⁺ channels begin to open. Thus, during the falling phase permeability to K⁺ again dominates, driving the membrane toward E_K. (D) K⁺ channels remain open for a short time after an action potential, producing an undershoot in some cells. Na⁺ channels recover from inactivation and again become ready to be opened by depolarization. (Leakage channels remain open throughout, but their effects are swamped by those of the more numerous voltage-gated channels.) (After Bear et al. 2001.)

2. Decreased permeability to Na⁺, caused by the inactivation of Na⁺ channels

3. Increased permeability to K⁺, caused by the slower opening of voltage-gated K⁺ channels

All three permeability changes are initiated by depolarization of the membrane and thus are characterized as *voltage-dependent* permeability changes. The voltage dependence of neuronal membrane permeabilities permits action potentials and gives the action potentials their unique all-or-none property.

THE HODGKIN CYCLE EXPLAINS THE RISING PHASE OF THE ACTION POTENTIAL To see how voltage dependence makes an action potential all-or-none, let's examine the increase in permeability to sodium that underlies the rising phase of the action potential. We have discussed how increased permeability to and inflow of Na⁺ depolarizes the membrane. *The critical feature of action-potential generation is that the permeability to Na⁺ that produces depolarization itself depends on depolarization.* The **Hodgkin cycle** describes the effects of depolarizing an excitable membrane in which the permeability to sodium (P_{Na}) is voltage-dependent. (The cycle is named after Sir Alan Hodgkin, who was a corecipient of the Nobel Prize for his work clarifying the ionic mechanism of action potentials.) The cycle (**FIGURE 12.16**) consists of three processes that feed back on each other in a cyclic manner.

At the conclusion of an action potential, the membrane remains highly permeable to K⁺ for a brief period (**FIGURE 12.15D**). Voltage-gated potassium channels remain open for a few milliseconds, producing a characteristic undershoot (after-hyperpolarization) in many neurons. The voltage-gated sodium channels recover from inactivation and again become ready to be opened by depolarization.

In summary, the action potential results from three overlapping permeability changes:

1. Increased permeability to Na⁺, caused by the rapid opening of voltage-gated Na⁺ channels

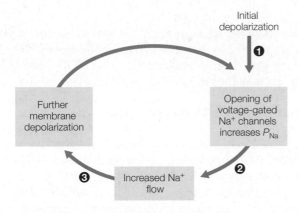

FIGURE 12.16 The Hodgkin cycle produces the rising phase of the action potential The critical feature of the cycle is that permeability to Na⁺ is *voltage-dependent*. ❶ An initial depolarization increases P_{Na} by opening voltage-gated Na⁺ channels. ❷ The increased permeability to Na⁺ allows inflow of Na⁺ down its electrochemical gradient, which further depolarizes the membrane. ❸ The cycle intensifies as each depolarization step opens additional Na⁺ channels.

The Hodgkin cycle describes a positive feedback loop that starts with depolarization: Changing V_m changes P_{Na}, and (as predicted by the Goldman equation) changing P_{Na} changes V_m. At rest, the membrane is 20–50 times as permeable to K⁺ as to Na⁺, so the resting V_m is near E_K. Subthreshold depolarizations open some voltage-gated Na⁺ channels, but not enough to overcome the effects of the higher resting permeability to K⁺. At threshold, the current carried by Na⁺ inflow just equals the K⁺ current, and at any depolarization above threshold the Hodgkin cycle "wins." The regenerative increase in P_{Na} in the Hodgkin cycle makes the membrane transiently much more permeable to Na⁺ than to K⁺, so V_m approaches E_{Na} (+40 to +55 mV inside-positive).

The Hodgkin cycle explains only the rising phase of the action potential, because if the cycle alone were operating, the membrane potential would remain near E_{Na} indefinitely. Instead, the polarity reversal lasts only about 1 ms because the sodium channels become inactivated and voltage-gated potassium channels open, causing the membrane to repolarize rapidly.

SINGLE-CHANNEL CURRENT RECORDING FROM ION CHANNELS The changes in membrane permeability that cause action potentials can be visualized as the actions of individual ion channels. (They can also be seen as whole-cell ionic currents, which we discuss next.) Evidence at the level of single ion channels comes from **single-channel current recording**, also termed **patch-clamp recording**.

In this procedure, a patch of membrane containing (with a little luck) a single Na⁺ ion channel is sealed by suction onto the smoothed tip of a fine glass micropipette electrode, so that any current must flow through a channel in the isolated patch (**FIGURE 12.17A**). The electrode records the opening and closing of the membrane ion channel by recording the ionic current that flows through the single channel when it is open. In this configuration, the experimenter has access only to the outside of the patch in the extracellular medium inside the micropipette. It is also possible, however, to rapidly pull

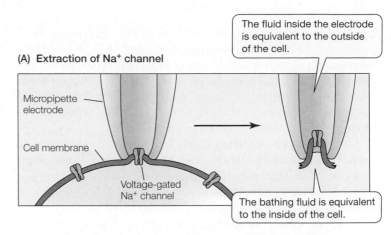

(A) Extraction of Na⁺ channel

The fluid inside the electrode is equivalent to the outside of the cell.

Micropipette electrode

Cell membrane

Voltage-gated Na⁺ channel

The bathing fluid is equivalent to the inside of the cell.

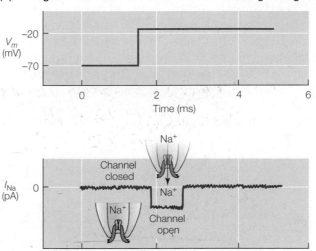

(B) Voltage across a Na⁺ channel and current flowing through it

V_m (mV)

Na⁺

Channel closed

I_{Na} (pA)

Na⁺

Na⁺

Channel open

FIGURE 12.17 Patch-clamp recording of single-channel currents (A) A fine, fire-polished glass electrode is fused to the membrane with suction, making what is known as a *gigaohm seal*, and the patch of membrane is pulled away from the cell. The electrode will then record current flowing through the channel when it opens. (B) A voltage-gated Na⁺ channel is closed at resting potential (–70 mV), and no current flows through it. When the membrane patch is depolarized, the channel opens transiently, allowing an inward current carried by Na⁺ ions. (Note that "inward" is toward the cytoplasmic side, not relative to the pipette, and that ion concentrations of the solutions in the bath and inside the electrode on either side of the patch are similar to the respective concentrations inside and outside the cell before detachment.) pA = picoampere (a measure of electric current).

a patch away from a cell and maintain the tight seal. In the inside-out arrangement of this detached patch, the inside of the electrode is the equivalent of the outside of the cell and the bathing fluid is equivalent to the inside of the cell. In response to a depolarization (caused by setting the voltage across the membrane patch to a less negative value) (**FIGURE 12.17B**), the channel opens, allowing Na⁺ ions to flow out of the electrode and into the bathing medium (remember, the bathing medium is now acting as the *inside* of the cell). This inward-flowing current lasts for about 1 ms before the channel closes again (conventionally, an inward-flowing current is shown downward and an outward-flowing current is upward). By providing data about the opening and closing of single channels, patch-clamp recording allows direct visualization of the permeability changes underlying action potentials. This technique has resulted in such major advances in our understanding of single channel function in neurons (as well as other cell types) that its developers were awarded the Nobel Prize, in 1991.

To show a more complete picture of the ionic currents flowing in and out of a neuron during an action potential, let's consider three responses of voltage-gated Na⁺ channels and three responses of voltage-gated K⁺ channels to a depolarization that is similar to the depolarization in an action potential (**FIGURE 12.18**). Recordings of the voltage-gated Na⁺ channels and voltage-gated K⁺ channels reveal conspicuous differences in latency and action of each type of channel. The channels are normally closed at resting potential, and depolarization increases the *probability* that they will open.

The voltage-gated Na⁺ channels have a short latency and open first, but rapidly become inactivated and remain so until membrane potential returns near baseline (see Figure 12.18A). The K⁺ channels open with a slightly longer latency but do not become inactivated, and tend to stay open until the depolarization ends (see Figure 12.18B). The six individual single-channel currents illustrate the three effects of depolarization on Na⁺ and K⁺ channels: (1) Na⁺ channels open first in response to the depolarization, (2) they are then inactivated during depolarization, and (3) K⁺ channels open slightly later than the Na⁺ channels but do not become inactivated.

VOLTAGE-CLAMP EXPERIMENTS SHOW WHOLE-CELL IONIC CURRENTS Before the development of single-channel current recording, researchers used a whole-cell current-measuring technique called a *voltage clamp* in experiments to study action-potential generation. These experiments remain a cornerstone of the physiological investigation of action potentials. A **voltage clamp** is an electronic device that allows the experimenter to measure whole-cell ionic currents, by setting membrane potential very rapidly to a predetermined value, delivering whatever current is necessary to keep it there, and measuring the imposed current.

Recall that we described the Hodgkin cycle as a positive feedback loop in which a change in membrane potential changes the permeability to sodium ions, and vice versa. As the Hodgkin cycle exemplifies, any ion flow through the membrane constitutes an ionic current that tends to change the membrane potential. Clamping the membrane potential uncouples the feedback loop of the Hodgkin cycle. To keep the potential constant, the clamp circuit must generate an opposing (bucking) current that is exactly opposite to the net ionic current (current carried by ion flows through ion channels). *By measuring the bucking current, the experimenter has an accurate measure of the amplitude and time course of the net ionic current, because the two must be equal and opposite to each other.* Hence a voltage clamp uncouples the feedback loop of the Hodgkin cycle (at step ❸ in Figure 12.16) so that ionic currents resulting from permeability changes are prevented from changing the membrane potential.

In 1952 Alan Hodgkin and Andrew Huxley published a series of landmark papers in which they used voltage-clamped squid axons to demonstrate and quantify the voltage-dependent permeability changes underlying the action potential. **FIGURE 12.19** shows the most fundamental result of such a voltage-clamp experiment. When the membrane potential is clamped to a hyperpolarized value (see Figure 12.19A), the current-measuring circuit shows only a brief blip of capacitive current required to set the membrane potential to a new level (changing

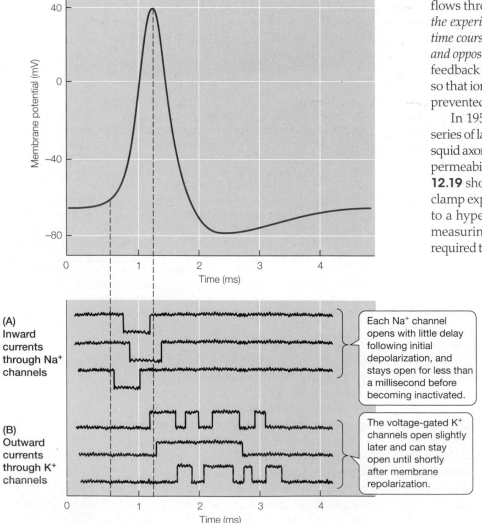

(A) Inward currents through Na⁺ channels

Each Na⁺ channel opens with little delay following initial depolarization, and stays open for less than a millisecond before becoming inactivated.

(B) Outward currents through K⁺ channels

The voltage-gated K⁺ channels open slightly later and can stay open until shortly after membrane repolarization.

FIGURE 12.18 Patch-clamp recording of single-channel currents underlying an action potential These diagrams illustrate simulated patch-clamp recordings of inward currents through three representative voltage-gated Na⁺ channels (A), and outward currents through three representative voltage-gated K⁺ channels (B) of the hundreds that produce the action potential. Note that the voltage-gated Na⁺ channels open in a narrow time window that corresponds to the rising phase of the action potential. The extended permeability to K⁺ can lead to an after-hyperpolarization of the membrane. (See Figure 12.14A for a description of the different phases of the action potential diagrammed at the top of this illustration.) (After Bear et al. 2001.)

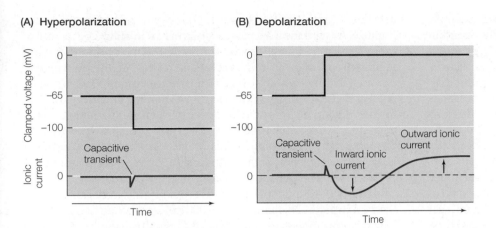

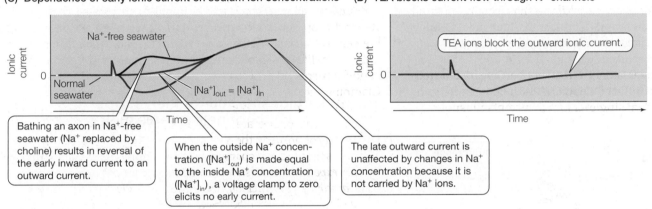

FIGURE 12.19 A voltage-clamp experiment reveals ionic currents during the action potential (A) The membrane potential is clamped at a hyperpolarized level (–100 mV) relative to the resting potential (–65 mV). After a brief capacitive transient, this hyperpolarization results in no significant ionic current. (B) The membrane is clamped at a depolarized level (0 mV). Depolarization induces an early *inward ionic current*, followed by a later *outward ionic current* which persists as long as the depolarization is maintained. (C) A voltage-clamp demonstration that the early inward current is carried by sodium ions. (D) Tetraethylammonium (TEA) ions block K+ channels, leaving only the early inward (Na+) current.

the charge stored by the membrane capacitance). The capacitive transient is not ionic current (i.e., not current flowing through ion channels), but just a shift in accumulated charges on either side of the membrane. Following the capacitive transient there is only a slight leakage current resulting from holding the membrane at a hyperpolarized level (too small to appear within the scale of the current record shown in Figure 12.19A). Hyperpolarization thus leads to no significant flow of ionic current because it doesn't increase permeability to any ions.

In contrast, clamping the membrane potential to a value more *depolarized* than the resting potential produces quite different effects (see Figure 12.19B). Following the initial capacitive transient, bucking current is required to hold the membrane at the set value. The bucking current (which is not shown) flows first outward and then inward. Because the bucking current is equal and opposite to the net ionic current, this pattern shows that there is an *early inward ionic current* that is reversed in 1–2 ms to a *later outward ionic current*. Depolarization of the membrane thus induces permeability changes that (if the currents are carried by cations) result in first an inward movement of cations and then an outward movement of cations. If the membrane were not clamped, these ionic currents would produce first a depolarization and then a repolarization of the membrane, as in an action potential.

Hodgkin and Huxley proposed that the early inward ionic current (which generates the rising phase of the action potential in unclamped axons) is an influx of Na+ ions. How could this prediction be tested? Hodgkin and Huxley replaced the Na+ in the seawater

with which they bathed the axon with choline, a nonpermeating cation. In the absence of extracellular Na+, the early inward current was replaced by an early outward current (see Figure 12.19C). That is, depolarization induced an increase in permeability to Na+, which, in the absence of extracellular Na+, resulted in Na+ diffusion *outward* down its concentration gradient.

This interpretation predicts that if the Na+ concentration is equal on both sides of the membrane, there will be no Na+ concentration gradient and no early Na+ current in either direction. Hodgkin and Huxley replaced about 90% of the extracellular Na+ with nonpermeating ions so that $[Na^+]_{in} = [Na^+]_{out}$. When the membrane was clamped to 0 mV (so that there was no voltage gradient), there was no early current (see Figure 12.19C).

Further evidence that the early inward current is carried by Na+ was provided by experiments in which a squid axon in normal artificial seawater was clamped to the sodium equilibrium potential (E_{Na} = +50 mV). There was no resultant early current because there was no driving force on Na+ ions at E_{Na}. Clamping the membrane at a level beyond E_{Na} (more inside-positive than E_{Na}) resulted in an early outward current, representing Na+ efflux toward E_{Na}. These experiments demonstrate that Na+ ions carry the early inward current during a voltage clamp, but not the later outward current, which is unchanged by changing Na+ concentrations. Other experiments demonstrate that the late current is a K+ efflux.

Pharmacological agents used in conjunction with a voltage clamp confirm that sodium and potassium currents flow through separate ion channels. Certain drugs can selectively block Na+ and

K+ channels when they are applied to the membrane. For example, tetrodotoxin (TTX), an extremely poisonous substance found in puffer fish, selectively blocks voltage-dependent Na+ channels. If a squid axon is bathed in seawater containing TTX and is voltage-clamped to a depolarized level such as 0 mV, the early inward Na+ current is blocked. The delayed outward (K+) current, however, is completely unaffected. However, tetraethylammonium (TEA) ions selectively block the delayed outward current flowing through K+ channels (see Figure 12.19D). TEA ions have no effect on the early inward current flowing through Na+ channels.

From their voltage-clamp experiments, Hodgkin and Huxley were able to quantify the voltage dependence and time course of the changes in permeability to Na+ and K+.[5] They developed a set of equations by which they showed that the magnitudes and time courses of these three voltage-dependent processes are sufficient to describe the behavior of action potentials in unclamped squid giant axons. These studies remain critical for our understanding of the physiology of excitable membranes.

ION MOVEMENTS IN ACTION POTENTIALS DO NOT SIGNIFI-CANTLY CHANGE BULK ION CONCENTRATIONS In the generation of an action potential, a neuron gains a small amount of Na+ and loses a small amount of K+. These amounts have been

---

[5] Permeabilities are often measured in electrical units of **conductance**, the inverse of resistance ($g = 1/R$). *Conductance* and *permeability* are not synonymous, because increasing ion concentrations increases conductance but not permeability.

calculated to be 3×10^{-12} to 4×10^{-12} mol/cm^2 of membrane per impulse. As with the slow passive leaks of Na+ in and K+ out across the resting membrane, the ions crossing the membrane during an impulse must be pumped back again by the Na+–K+ exchange pump. It is important to realize that the pumping process is *slow* relative to the time course of the action potential, and serves only to keep the ion concentrations constant over minutes, hours, and days. The Na+–K+ exchange pump does not contribute directly to the generation of action potentials, and the ion movements underlying impulse generation are very small relative to the quantities of ions inside and outside the axon.

If the Na+–K+ exchange pump of a squid giant axon is poisoned, the axon can still generate about 100,000 impulses before the internal Na+ concentration is increased by 10%! Smaller axons, however, have a greater ratio of membrane surface to internal volume, so the concentration changes produced by impulses are greater. Therefore the smallest axons (0.1 μm in diameter) presumably cannot generate impulses at a rate that greatly exceeds the moment-to-moment ability of the Na+–K+ exchange pump to maintain normal ion concentrations.

The molecular structure of the voltage-dependent ion channels reveals their functional properties

The voltage-gated Na+ channel protein changes its tertiary structure in response to membrane depolarization, to achieve three conformations: closed, open, and inactivated. Molecular analysis of

---

FIGURE 12.20 The molecular structure of voltage-gated Na+ channels (A) The predicted structure of the principal (α) subunit of the voltage-gated Na+ channel. This subunit makes up the pore-forming channel itself and is a single polypeptide chain, with four homologous domains labeled I, II, III, and IV. Each domain has six membrane-spanning α-helical segments (labeled 1 through 6). The P (for pore) loops between segments 5 and 6 of each domain are thought to line the pore and determine ion selectivity. (B) A surface view of a voltage-gated K+ channel, which is structurally closely related to the voltage-gated Na+ channel. The numbered α-helical segments of the red portion correspond to those of domain II in part (A). Segments 5 and 6 of each domain are near the central pore, whereas segments 1 to 4 are further away. Other segments (not shown) mediate channel inactivation, intracellular regulation, and association with other channel subunits. (C) A hypothetical three-dimensional structure of the Na+ channel, showing closed and open conformations. The four domains surround the central pore; domains I and III are visible in this cross-section. The segment-4 voltage sensors are thought to rotate and slide upward in response to depolarization, leading to channel opening. (B from Long et al. 2005.)

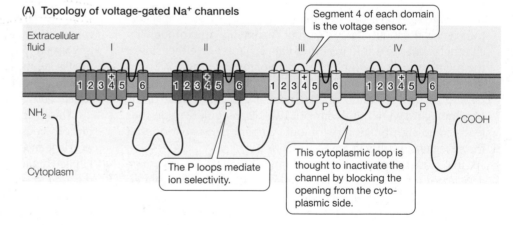

(A) Topology of voltage-gated Na+ channels

Segment 4 of each domain is the voltage sensor.

Extracellular fluid

NH₂

Cytoplasm

The P loops mediate ion selectivity.

This cytoplasmic loop is thought to inactivate the channel by blocking the opening from the cytoplasmic side.

COOH

(B) Surface view of a K+ channel

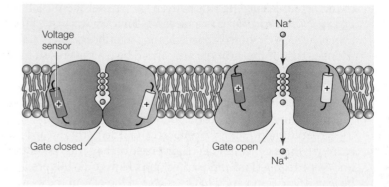

(C) Voltage-dependent conformational change

Voltage sensor

Gate closed

Gate open

Na+

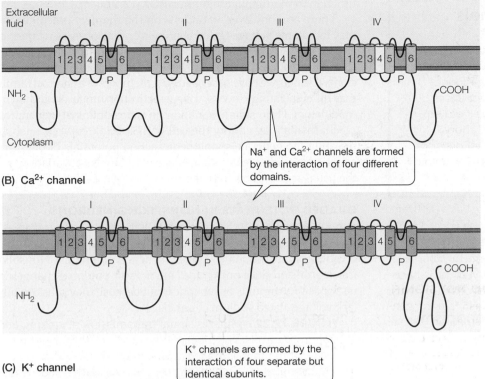

(A) Na⁺ channel

Extracellular fluid

I II III IV

1 2 3 4 5 6   1 2 3 4 5 6   1 2 3 4 5 6   1 2 3 4 5 6

P   P   P   P

NH₂

COOH

Cytoplasm

Na⁺ and Ca²⁺ channels are formed by the interaction of four different domains.

(B) Ca²⁺ channel

I II III IV

1 2 3 4 5 6   1 2 3 4 5 6   1 2 3 4 5 6   1 2 3 4 5 6

P   P   P   P

NH₂

COOH

K⁺ channels are formed by the interaction of four separate but identical subunits.

(C) K⁺ channel

1 2 3 4 5 6

P

NH₂

COOH

FIGURE 12.21 The voltage-gated channel superfamily All the voltage-gated channels have principal subunits with extensive sequence homology and thus are evolutionarily related. Voltage-gated Na⁺ channels (A) and Ca²⁺ channels (B) have four domains, each with six α-helical membrane-spanning segments and a P loop (P). The voltage-sensing segment of each domain is highlighted in yellow. (C) The voltage-gated K⁺ channel, in contrast, has only a single domain of six α-helices and the P loop, homologous to one domain of a Na⁺ channel. Four separate subunits interact to form a complete K⁺ channel (see Figure 12.20B).

the voltage-gated Na⁺ channel suggests which parts of the protein mediate specific aspects of its function.

The major (α) protein of the voltage-gated Na⁺ channel forms the channel itself. It consists of a single polypeptide chain (**FIGURE 12.20**; see also Figure 2.4). (There are two smaller, variable β-peptide subunits that interact with and modify the single α protein. However, the α protein alone is sufficient to produce voltage-gated Na⁺ currents, and we consider it only.) The Na⁺ channel α polypeptide chain has four *domains*, or regions, each domain consisting of amino acid sequences very similar to the other three. Thus there is said to be extensive *sequence homology* among the four domains, suggesting that they evolved from a common ancestral peptide. Each of the four domains contains six *membrane-spanning segments*, regions of the polypeptide that contain predominantly hydrophobic amino acid side chains that can form α-helices and cross the lipid bilayer of the membrane. The four domains of the voltage-gated Na⁺ channel α protein surround an aqueous channel pore, through which Na⁺ ions can diffuse in response to depolarization. The voltage-gated K⁺ channel is very similar in overall structure (see Figure 12.20B).

Particular structural regions of the channel protein impart to it particular functional properties. For example, the region of the

protein that responds to voltage is membrane-spanning segment 4 of each domain (see Figure 12.20). The voltage-sensor region of the channel must be charged, but it must also be in or close to the membrane in order to detect changes in the transmembrane electric field, and segment 4 has a collection of positively charged amino acids appropriate to act as the voltage-sensor region. Moreover, mutations affecting segment 4 selectively alter the voltage sensitivity of the channel. The channel's voltage-sensor segments are thought to move outward in response to depolarization, leading to an overall conformational change in the channel from closed to open (see Figure 12.20C).

Another structural correlate of a critical channel function is the P loop connecting segments 5 and 6 of each domain. This loop lines the pore of the ion channel and helps mediate ion *selectivity*. Mutations in the P-loop region alter ion selectivity in ways consistent with this idea. Finally, the cytoplasmic loop between domains III and IV appears to mediate inactivation of the Na⁺ channel; it is thought to act like a "ball on a string" that can block the (open) channel from the cytoplasmic side (see Figure 12.15C).

Other voltage-gated channels are structurally similar to the voltage-gated Na⁺ channel (**FIGURE 12.21**). Channels showing such similarity include the K⁺ channels that repolarize the membrane in an action potential, as well as Ca²⁺ channels involved in neurotransmitter release. (Like the Na⁺ channels, these channels have modulatory polypeptides, but only the principal α protein of each is considered here.) Na⁺, Ca²⁺, and K⁺ channels are similar in overall structure and have extensive homology in amino acid sequence. The sequence homology is greatest in certain regions (conserved regions), such as the voltage-sensor region of membrane-spanning segment 4. Their homology suggests that the various ion channels are evolutionarily related, and therefore they are referred to as the **voltage-gated channel superfamily** of membrane proteins.

The Na⁺ and Ca²⁺ channels are most similar in structure (see Figure 12.21A,B). The K⁺ channels (see Figure 12.21C) consist of four identical subunits that resemble one of the four domains of a Na⁺ channel. Four K⁺ channel proteins interact as subunits to form

| BOX 12.1 | Evolution and Molecular Function of Voltage-Gated Channels |

Voltage-gated channels are amazing molecular mechanisms that make possible the functions of nervous systems. Recent molecular studies have suggested a sequence of steps in the evolution of voltage-gated channels, and have largely clarified the structural basis of their action. **Box Extension 12.1** shows how voltage-gated channels are thought to have evolved, and how their critical features—ion selectivity and voltage gating—work at the molecular level.

a channel, aligning like the four domains of a Na⁺ channel protein around a central pore (see Figure 12.20B). There are several subtypes of each kind of ion channel, with an especially large number of K⁺ channel subtypes. **BOX 12.1** discusses the evolution of the voltage-gated channel superfamily and the molecular bases of their actions.

There are variations in the ionic mechanisms of excitable cells

How universal are the ionic mechanisms of action potentials? Studies have shown that the basic aspects of impulse generation elucidated in squid axons apply to most excitable cells. The action potentials of vertebrate and invertebrate unmyelinated axons, amphibian myelinated axons, and vertebrate skeletal twitch-muscle fibers have ionic mechanisms qualitatively similar to those of squid

axons. In fact, even ion channels from algae have been shown to function in mammalian neurons (**BOX 12.2**).

There are, however, variations on the theme: Some neurons may lack some voltage-gated channels or may possess additional channels. Genomic and other studies reveal entire families of ion channels related to the voltage-gated sodium, potassium, and calcium channels we have described. The members of a channel family may differ in gating kinetics, voltage-activation range, or binding by modulators. The resultant variations in action potential–generating mechanisms may endow the cell with special features, such as the ability to generate spontaneous action potentials or bursts of action potentials. Here we consider two examples of variations on the usual action-potential theme.

GRADED POTENTIALS IN NONSPIKING NEURONS Not all nerve cells generate action potentials. Researchers report increasingly numerous examples of **nonspiking neurons** (neurons that do not generate the sharp "spikes" of action potentials). Nonspiking neurons produce only *graded* membrane-potential changes in response to a stimulus or synaptic input because they substantially lack voltage-gated sodium channels.

FIGURE 12.22 compares signal transmission in spiking and nonspiking neurons. The input of both neurons (sensory stimulation or synaptic input) is graded in amplitude, and so is their output (neurotransmitter release). Spiking neurons encode the graded input signals into trains of action potentials for long-range transmission; the action potentials are recoded at the terminal to control graded

FIGURE 12.22 Nonspiking neurons do not generate action potentials Information transmission in an ordinary, spiking neuron (A) is contrasted with that in a nonspiking neuron (B). (A) Input to the spiking neuron is graded in amplitude and evokes graded potentials in the dendrites and soma of the neuron. (The input could be an external stimulus for a sensory neuron, or synaptic input for a nonsensory neuron.) (B) Nonspiking neurons do not encode graded potentials into action potentials; instead the graded potentials spread electrotonically to the axon terminal. If the axon is short relative to its membrane length constant, the amplitude of the graded potential will be sufficient to evoke graded neurotransmitter release. Action-potential generation, then, may be an adaptation for signal transmission in long-axon neurons that is unnecessary in short-axon neurons.

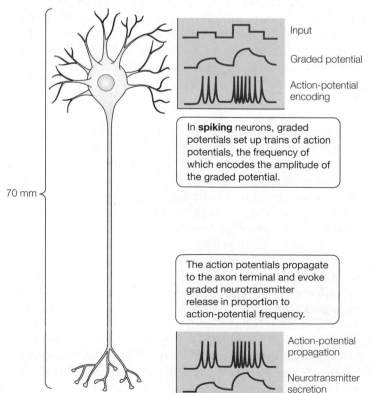

(A) Spiking neuron

Input

Graded potential

Action-potential encoding

In **spiking** neurons, graded potentials set up trains of action potentials, the frequency of which encodes the amplitude of the graded potential.

70 mm

The action potentials propagate to the axon terminal and evoke graded neurotransmitter release in proportion to action-potential frequency.

Action-potential propagation

Neurotransmitter secretion

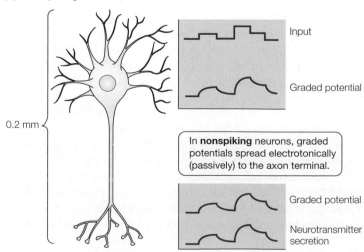

(B) Nonspiking neuron

Input

Graded potential

0.2 mm

In **nonspiking** neurons, graded potentials spread electrotonically (passively) to the axon terminal.

Graded potential

Neurotransmitter secretion

Optogenetics:
Controlling Cells with Light   MATTHEW S. KAYSER

The human brain is a remarkably complex organ, with billions of neurons and trillions of synapses communicating through precisely timed electrical signals. A major limitation in understanding how our brains work is that researchers have been unable to manipulate the system on the same millisecond timescale on which it normally operates. To learn how every movement, thought, and experience we have results from groups of neurons talking to each other, don't we need a way to speak to neurons on the same timescale they use when communicating with each other? The field of optogenetics has begun to accomplish just this by combining optics (the use of light) with manipulation of genes (thus *opto* + *genetics*). Specifically, scientists have figured out how to put genes into cells that make those cells responsive to pulses of light. Unexpectedly, this technol-

ogy is possible because of light-sensitive transporter proteins and ion-channel proteins first discovered in microorganisms a few decades ago. **Optogenetics** involves taking the genes encoding these light-sensitive transporter and channel proteins, inserting them into target cells, and then delivering light to those cells as a way of controlling their functions. For example, neurons in the mammalian brain can be targeted to express the light-sensitive channels. Then, by delivering light to those neurons (see the figure), investigators are able to exert millisecond control over neuronal firing

A mouse prepared for an optogenetic experiment
(Courtesy of Mike J. F. Robinson.)

patterns, shedding light—literally—on mysteries of neuroscience in the process. **Box Extension 12.2** describes how optogenetics was developed and its many potential applications.

---

release of neurotransmitter. Nonspiking neurons, in contrast, are typically compact cells with short axons or no axons, so a graded potential change at one part of the cell can spread passively (electrotonically) to the terminal without major decrement.

The inputs and outputs of spiking and nonspiking neurons are the same, but the short-axon nonspiking neuron does not require spike encoding to carry the signal over large distances. Examples of nonspiking neurons include the photoreceptors, bipolar cells, and horizontal cells of the vertebrate retina (see Chapter 14), granule cells of the olfactory bulb, and many arthropod interneurons.

PACEMAKER POTENTIALS OF SPONTANEOUSLY ACTIVE CELLS   Many neurons are spontaneously active, generating action potentials at rather regular intervals without an external source of depolarization. The somata of some molluscan neurons, for example, generate action potentials in regular trains, or even in repetitive bursts, in the absence of synaptic input. Vertebrate cardiac muscle fibers and some other excitable cells are also spontaneously active.

The membrane potential of a spontaneously active cell, instead of maintaining a fixed resting value, undergoes a continuous upslope of depolarization between action potentials, until it reaches threshold for the generation of the next action potential. The repolarizing phase of an action potential restores the membrane to a relatively hyperpolarized level, from which the next ramp of depolarization begins. These ramp depolarizations are termed **pacemaker potentials** because they determine the rate of impulse generation by the cell. For example, in a cardiac muscle cell in the pacemaker region of a vertebrate heart, the greater the rate of depolarization during the ramp phase, the sooner the cell reaches threshold for

the next action potential and, thus, the faster the heart rate. For vertebrate cardiac muscle fibers, norepinephrine increases the rate of depolarization during the ramp phase, whereas acetylcholine decreases it. (See Chapter 25 for discussion of heart rate control.)

The ionic basis of pacemaker potentials can be complex and may vary somewhat among cells. Spontaneously active cardiac muscle fibers have a slow inward cationic current (termed I_h) that is activated by hyperpolarization rather than by depolarization. The channels that open to produce this current are not very selective among cations, so both Na^+ and K^+ ions contribute to the current. I_h is activated at potentials more negative than −40 mV. The unique features of this current—that it is inward and activated by hyperpolarization—give it pacemaker properties. I_h produces a ramp depolarization that triggers an action potential and then is inactivated at −40 mV. After the action potential, the repolarizing potassium current (I_K, flowing through voltage-gated K^+ channels) restores the membrane to a value near E_K. The hyperpolarization turns off I_K and turns on I_h for the next ramp depolarization.

CARDIAC MUSCLE ACTION POTENTIALS   The action potentials of heart muscle fibers differ from action-potential generation in squid axons in another way—their duration. In contrast to most action potentials, which last about 0.4–3 ms, vertebrate cardiac muscle fibers have action potentials with typical durations of 100–500 ms (**FIGURE 12.23A**). The long duration of cardiac muscle action potentials is functionally important because action-potential depolarization is the necessary stimulus for myocardial contraction, and because cardiac muscle must contract for about 100 ms to pump blood effectively. A cardiac muscle fiber action potential has a rapid upstroke and a rapid initial recovery to near 0 mV, but

(A) Cardiac action potential

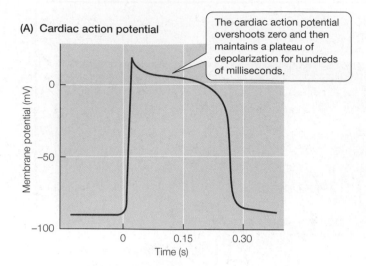

> The cardiac action potential overshoots zero and then maintains a plateau of depolarization for hundreds of milliseconds.

(B) Permeability changes

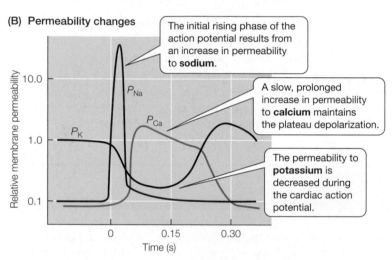

> The initial rising phase of the action potential results from an increase in permeability to **sodium**.

> A slow, prolonged increase in permeability to **calcium** maintains the plateau depolarization.

> The permeability to **potassium** is decreased during the cardiac action potential.

FIGURE 12.23 The cardiac muscle fiber action potential (A) An action potential in a vertebrate cardiac muscle fiber has a long duration. (B) Permeability changes underlie the cardiac action potential. Eventually, increased [Ca^{2+}] opens Ca^{2+}-activated K^+ channels, leading to repolarization.

it remains depolarized near zero for many milliseconds. This prolonged depolarization, the *plateau* of the action potential, gradually decreases and is followed by a relatively slow repolarization.

Two separate inward currents underlie the cardiac action potential. The first is a fast current resulting from increased permeability to Na^+ (P_{Na}) (**FIGURE 12.23B**), very similar to the squid axon's permeability to Na^+. The fast Na^+ current produces the rapid upstroke of the cardiac action potential and is inactivated within a few milliseconds. A second, slow inward current results mainly from increased permeability to Ca^{2+} (P_{Ca}) and helps produce the plateau. The Ca^{2+} channels take at least 20 ms to open, and their slow inward current is much weaker than the fast Na^+ current.

The other factor sustaining the plateau is a *decrease* during the plateau in permeability to K^+ (P_K) from the resting level (in contrast to the increased permeability to K^+ that occurs in axons). Thus the plateau represents a balance between two small currents, a slow inward Ca^{2+} current and a diminishing outward K^+ current. Repolarization depends on two factors: (1) the Ca^{2+} channels gradually become inactivated during the plateau depolarization, and (2) permeability to K^+ gradually increases. Some of the K^+ channels are

activated by intracellular Ca^{2+} ions (see Box Extension 12.1), which accumulate during the plateau and eventually open the K^+ channels.

The complex ionic basis of cardiac action potentials increases their energetic efficiency. Recall that in a 1-ms action potential of a typical neuron, only the in-rushing fast sodium current sustains the brief depolarization. A cardiac muscle cell would be flooded with Na^+ if the fast current alone had to sustain a 100-ms action potential, and metabolically expensive ion-exchange pumps would be needed to clear the Na^+ ions from the cytoplasm. During a cardiac action potential, however, the duration of the fast sodium current is about the same as that of a standard action potential. Instead, *depolarization is sustained by the slow Ca^{2+} current and by the decreased membrane permeability to K^+*. Thus a cardiac muscle fiber can generate a protracted action potential with only a relatively modest exchange of ions across its membrane.

SUMMARY
The Action Potential

■ An action potential is a voltage change—a brief, transient reversal of membrane potential from inside-negative to inside-positive. Action potentials are all-or-none responses to any depolarization beyond a voltage threshold and are each followed by a brief refractory period.

■ Action potentials result from voltage-dependent changes in membrane permeability to ions. Depolarization first opens voltage-gated Na^+ channels, allowing Na^+ ions to flow in and further depolarize the membrane toward E_{Na}. The voltage-gated Na^+ channels rapidly become inactivated to terminate the rising phase of the action potential; then voltage-gated K^+ channels open to repolarize the membrane.

■ The effects of depolarization on membrane permeability to ions can be studied at the level of single channels by patch clamp, and at the whole-cell level by voltage clamp.

■ Ongoing investigations are clarifying the molecular structures of voltage-gated channels. The principal protein subunit of a K^+ channel is a single chain with six transmembrane regions; a K^+ channel consists of four of these protein subunits around a central pore. Na^+ and Ca^{2+} channels consist of a single polypeptide chain with four similar domains; each domain corresponds to one of the four subunits of the K^+ channel. Functional attributes of the channels can be localized to particular regions of the proteins.

■ Nonspiking neurons do not generate action potentials, and the ionic mechanisms of action potentials in excitable cells can vary. Calcium ions can make substantial contributions to action potentials in cardiac muscle cells and in some neurons. Other varieties of voltage-gated channels modify the excitable properties of neurons.

The Propagation of Action Potentials

Consider that in large animals, single axons—such as those that control wiggling of your toes—can be at least 1 m long. Now recall that an electrotonic voltage change at one point on a membrane decreases exponentially with distance. Indeed, with passive or electrotonic spread, the amplitude of a voltage change typically decreases to a third of its starting value in a fraction of a millimeter! Because of this decrement, passive electrotonic spread of a voltage change cannot serve as an electrical signal over long distances. Instead, there must be a mechanism to amplify or refresh the electrical signal.

We have seen that an action potential is all-or-none because the voltage-dependent, regenerative permeability increases bring the membrane potential toward a limiting value, which is the sodium equilibrium potential (E_{Na}). *Nondecremental propagation of the action potential is possible because the action potential at one location on an axon can itself initiate an action potential at a neighboring location, and the induced action potential will have the same all-or-none amplitude as the original action potential.* By repeating this process, a signal can travel 1 m along an axon without any decrease in amplitude.

Local circuits of current propagate an action potential

An action potential at one locus on an axon depolarizes an adjacent locus by setting up local circuits of current flow (**FIGURE 12.24**). A complete local circuit is composed of a series of ionic and capacitive currents, which can be broken down into four components, as shown in Figure 12.24B:

❶ At the locus of the action potential, an ionic current begins with the inflow of sodium ions through open Na⁺ channels into the intracellular fluid (cytosol).

❷ Ions flow in intracellular fluid, carrying the current to more distant parts of the membrane (see Figure 12.9).

❸ At the membrane, the ion movements change the distribution of charges on the membrane capacitance (see Figure 12.24C): Cations accumulate along the membrane interior, displacing negative charges and repelling an equivalent number of cations from the membrane exterior. Although ions do not physically cross the membrane at this point, the movements of ions onto and off the membrane surface constitute a capacitive current.

❹ An (extracellular) ionic current completes the local circuit as cations move toward the locus of the action potential and anions move away.

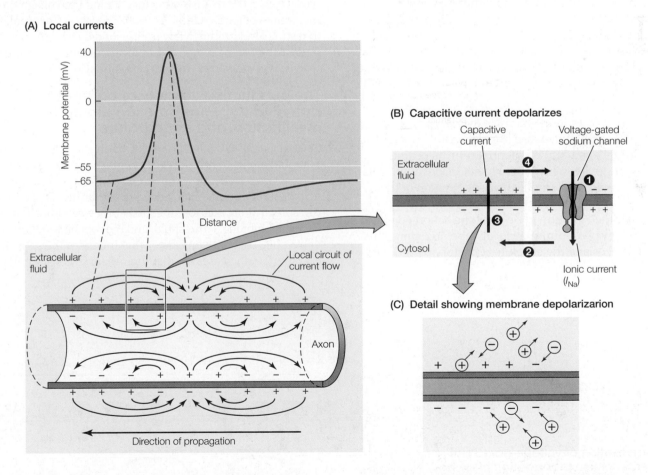

(A) Local currents

(B) Capacitive current depolarizes

(C) Detail showing membrane depolarizarion

FIGURE 12.24 Propagation of an action potential An action potential is diagrammed here at an instant in time, while propagating from right to left. (A) The action potential induces local circuits of current flow along the axon, ahead of the advancing action potential and behind it. (B) The local circuit ahead of the action potential can be divided into four components, as shown. These components are described in the text. (C) This diagram illustrates how the capacitive current on the left in (B) depolarizes the membrane ahead of the advancing action potential, by adding cations and removing anions at the inner side of the membrane, and adding anions and removing cations on the extracellular side. These changes in the distribution of charges on the membrane depolarize it, without any ions crossing the membrane. The depolarization resulting from the local currents opens voltage-gated sodium channels, leading (via the Hodgkin cycle) to an action potential at the new location.

During an action potential, local circuits of current such as the one described here spread the depolarization passively (electrotonically) along the surface of the membrane. *The action potential propagates to an adjacent portion of the axon because the capacitive depolarization produced by the local current lowers the membrane potential to threshold.* When this happens, the Hodgkin cycle takes over; the region of the membrane that reaches threshold undergoes a regenerative increase in permeability to Na$^+$ and generates its own action potential. In this way the nerve impulse passes along the entire length of the axon without any decrease in amplitude.

Membrane refractory periods prevent bidirectional propagation

An axon can conduct impulses equally well in either direction. For example, an action potential that is triggered via an electrode placed in the middle of an axon will be propagated in both directions from that point. Normally, however, impulses start at or near one end of an axon and travel along the axon in one direction. As an impulse is propagated, its local currents depolarize the membrane behind it, as well as the membrane ahead of it. Why don't these local currents initiate reverse-traveling impulses going the other way? The membrane behind a traveling impulse is not reexcited by the local currents because the membrane is still in its refractory period (**FIGURE 12.25**).

Three aspects of the ionic mechanisms of action potentials produce the absolute and relative refractory periods following an impulse, and thereby prevent reexcitation and bidirectional propagation:

1. The inactivation of sodium channels (which turns off the voltage-dependent increase in permeability to Na$^+$) persists until the membrane potential returns near its negative resting state, which means that inactivation lasts for at least 1 ms after an impulse passes a region of the membrane. Na$^+$ channel inactivation prevents the channels from entering the Hodgkin cycle until the action potential is far enough away to minimize local depolarization. This is the primary basis of the absolute refractory period.

2. The increased permeability to potassium (the slowest of the three voltage-dependent processes in onset) does not decrease to resting levels until after repolarization. The lingering P_K increase after an impulse may hyperpolarize the membrane toward E_K for a few milliseconds after the impulse. Thus (for those neurons whose action potentials have an undershoot) a region of membrane that has just generated an impulse is hyperpolarized *away* from its voltage threshold for new impulse generation.

3. The increase in P_K also renders a membrane refractory because it represents a decreased membrane resistance. The decreased resistance means that by Ohm's law ($V = IR$), local currents will cause a smaller voltage change, so more current is needed to depolarize the membrane to threshold.

Sodium-channel inactivation causes the absolute refractory period, and the effects of residual P_K increase are largely responsible for the relative refractory period. During the absolute refractory period (>1 ms), the membrane's voltage threshold is infinite because no amount of depolarization can open the inactivated Na$^+$ channels. During the relative refractory period the voltage threshold, membrane potential, and membrane resistance gradually return to resting levels within a few milliseconds. The refractory periods outlast the backward spread of local currents, thus preventing reverse propagation of the action potential.

The conduction velocity of an action potential depends on axon diameter, myelination, and temperature

Several factors can affect the velocity of propagation of an action potential along an axon. In general, these factors affect either or both of two conduction parameters: (1) the spatial parameter and (2) the temporal parameter. Spatially, the farther that local currents can spread along an axon, the farther they can (directly) depolarize the membrane to threshold, and the sooner an action potential will result. Therefore any factor that increases the spread of local currents (i.e., increases the membrane length constant, λ) tends to

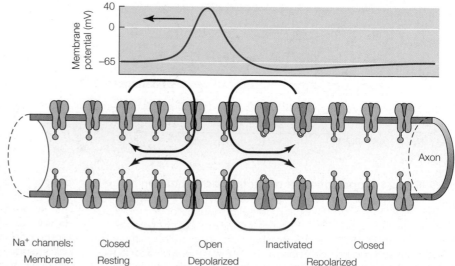

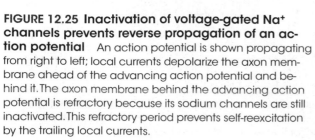

FIGURE 12.25 Inactivation of voltage-gated Na$^+$ channels prevents reverse propagation of an action potential An action potential is shown propagating from right to left; local currents depolarize the axon membrane ahead of the advancing action potential and behind it. The axon membrane behind the advancing action potential is refractory because its sodium channels are still inactivated. This refractory period prevents self-reexcitation by the trailing local currents.

increase the conduction velocity of an action potential. Both large axon diameter and myelination increase the spatial spread of local currents.

With respect to the temporal parameter, the less time it takes the membrane to reach threshold, the faster the conduction velocity. Intrinsic membrane properties such as differences in the density of sodium channels may have minor effects on conduction velocity by influencing the spatial and temporal parameters. The three major evolutionary variables that influence conduction velocities are axon diameter, myelination, and temperature.

AXON DIAMETER AND CONDUCTION VELOCITY Large-diameter axons tend to conduct action potentials more rapidly than small-diameter axons. Many animal groups have evolved rapidly conducting neuronal giant axons (**BOX 12.3**). Conduction velocity increases with axon diameter because larger-diameter axons have longer length constants and thus more distant spread of local currents. The length constant depends principally on two types of resistance (see Figure 12.9C): the resistance across the membrane (R_m) and the axoplasmic resistance (R_i) to current flow along the length of the axon (the cytoplasm inside the axon is called *axoplasm*). (The external longitudinal resistance is usually small and is ignored.) A slightly simplified equation for the length constant (λ) is

$$\lambda = K \left[\frac{R_m}{R_i} \right]^{1/2} \qquad (12.7)$$

where K is a constant.

The membrane surface area increases proportionally with increasing axon diameter, which lowers R_m by adding resistances in parallel. However, R_i decreases in proportion to an increase in cross-sectional area of the axoplasm—that is, in proportion to the square of the diameter. The net effect is that the ratio R_m/R_i increases linearly with increasing diameter. If other factors are equal, the length constant and the conduction velocity should increase with the square root of the diameter. This would make the conduction velocity an allometric function of diameter: $V \propto D^b$, with $b = 0.5$ (see Appendix F).

FIGURE 12.26 shows that conduction velocity empirically increases with increasing axon diameters. The data are plotted on log–log coordinates. On these coordinates, the allometric relation $V \propto D^b$ will plot as a straight line regardless of the value of b (see Appendix F). The two lines forming a V at the center show the slopes for $V \propto D^1$ (simple proportionality) and $V \propto D^{0.5}$. As you can see, the empirical plots (red lines) have slopes between these extremes. For some unmyelinated axons, V and D follow the square-root relationship ($V \propto D^{0.5}$), but many myelinated axons have a more nearly proportional relationship of velocity to diameter ($V \propto D^1$). That the relation of velocity

Because an increase in axon diameter increases the conduction velocity of an action potential and because animals often face circumstances in which a rapid response is advantageous for survival, **giant axons** have evolved in several animal groups. No particular diameter qualifies an axon as *giant*. Rather, the term is relative: A giant axon is of exceptional diameter compared with other axons in the same animal. Some axons are truly giant in cellular dimensions, such as the third-order giant axons in the squid, which may be 1 mm (1000 μm) in diameter. At the other extreme, the giant axons in the fruit fly *Drosophila* are only about 4 μm in diameter, but they are still an order of magnitude larger than other nearby axons. **Box Extension 12.3** describes the structure and function of giant axons in squid and in some other invertebrates.

and diameter differs for different kinds of axons implies that other factors, including intrinsic membrane differences, are also involved.

MYELINATION INCREASES CONDUCTION VELOCITY Myelinated axons of vertebrates represent a tremendous evolutionary advance because they allow very high conduction velocities with relatively small axon diameters. A **myelinated axon** (**FIGURE 12.27**) is wrapped with 200 or more concentric layers of glial membrane (the membrane of Schwann cells in PNSs and of oligodendrocytes in CNSs; see Figure 12.5). The glial cytoplasm is extruded from be-

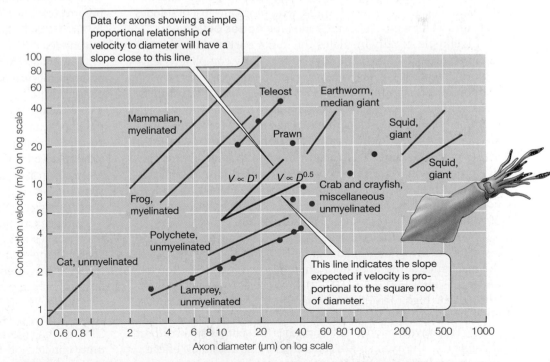

FIGURE 12.26 The velocity of nerve-impulse conduction increases with increasing axon diameter in both myelinated and unmyelinated axons Points not connected by lines are axons of different types. (After Bullock and Horridge 1965.)

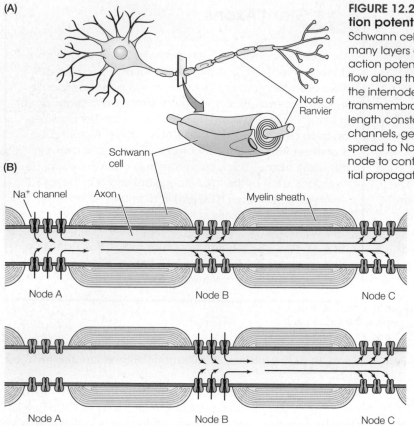

(A)

Node of Ranvier

Schwann cell

(B)

Na⁺ channel Axon Myelin sheath

Node A Node B Node C

Node A Node B Node C

FIGURE 12.27 Myelinated axons speed the propagation of an action potential (A) Each segment of an axon in the PNS is myelinated by a Schwann cell. The Schwann cell wraps around the axon segment, providing many layers of Schwann cell membrane without intervening cytoplasm. (B) An action potential at one node of Ranvier (Node A) sets up local currents that flow along the axon, as in Figure 12.25. Currents cannot cross the membrane in the internode, which has very high resistance and low capacitance. Therefore transmembrane current flow is restricted to the nodes, and the membrane length constant is much longer. Below, depolarization of Node B opens Na⁺ channels, generating the action potential and setting up local currents that spread to Node C. Note that some local current can flow past the nearest node to contribute to the depolarization of more distant nodes. Action-potential propagation in myelinated axons is saltatory, jumping from node to node.

tween the glial membrane layers so that the whole wrapping serves as an insulating layer. This multiply-wrapped insulating layer, termed *myelin*, stops at intervals of 1 mm or so along the length of the axon. The gaps at which the glial wrappings are absent (Nodes A, B, and C in Figure 12.27) are called the **nodes of Ranvier**.

Myelin electrically insulates the major part of the axon (the regions between nodes, or **internodes**) nearly completely, leaving only the nodes of Ranvier as loci of ion flow across the axon membrane (see Figure 12.27). *In myelinated axons, action potentials occur only at the nodes of Ranvier*, in contrast to the continuous sweep of action potentials over an unmyelinated axon. Myelinated axons are therefore said to exhibit **saltatory conduction**, in which the action potential jumps (saltates) from node to node without active propagation in the internode. Action potentials are typically initiated at the axon initial segment, and then saltate from node to node because these are the only regions with high concentrations of voltage-gated Na⁺ channels (**FIGURE 12.28**).

The principal effect of myelin is to increase the membrane resistance of myelinated axon regions by 1000- to 10,000-fold over the resistance at the nodes of Ranvier. When a node of Ranvier undergoes an action potential, the local currents cannot leak out through the high membrane resistance of the adjacent internode, but instead must flow farther to the next node of Ranvier (see Figure 12.27). Thus myelination greatly increases the spatial spread of local currents (i.e., the axon length constant) by this resistance effect.

A second, equally important function of myelin is to decrease membrane capacitance. If myelin only increased membrane resistance (without decreasing capacitance), its effect on the axon

length constant would be largely offset by an increase in the membrane time constant, τ. (Recall that the time constant is equal to the product of membrane resistance and membrane capacitance, $R_m C_m$.) An increase in the time constant would tend to *slow* conduction velocity because it would take more time for a current to depolarize a patch of membrane to threshold. Capacitance is inversely proportional to the distance separating the charges on the "plates" of a capacitor, which in this case is the distance between the axoplasm and the extracellular fluid. Myelin increases this distance in proportion to the number of glial membrane wrappings, so that capacitance is decreased about 1000-fold. Thus the increase in R_m of myelinated regions is offset by a decrease in C_m, and the membrane time constant is nearly unaffected.

Myelination, then, greatly increases conduction velocity by increasing the axon length constant without increasing the time constant. Currents from an action potential at one node must travel to the next node before crossing the membrane, and the currents are not slowed by having to displace much charge in the intervening myelinated internode.

Myelinated axons permit vertebrates to have neural coordinating and control systems with small-diameter axons that nonetheless conduct rapidly. A frog myelinated axon 12 µm in diameter has a conduction velocity of 25 m/s at 20°C. An unmyelinated squid giant axon must be about 500 µm in diameter to achieve the same 25-m/s velocity at 20°C! Thus myelination allows the same velocity to be achieved with a 40-fold reduction in diameter and a 1600-fold reduction in axon cross-sectional area and volume. *With a reduction in axon dimensions, many more axons can be incorporated into a nervous system of a given size.*

Although myelin is usually considered to have evolved exclusively in vertebrates, some crustaceans have axons with analogous sheaths of glial wrappings that exhibit increased conduction velocities similar to those for vertebrate compact myelin. The 30-µm myelinated axons of a shrimp conduct at 20 m/s at 17°C, a velocity comparable to that of a 350-µm squid axon. Other myelinated shrimp axons 100–120 µm in diameter conduct at velocities exceeding 90 m/s at 20°C, rivaling the fastest mammalian myelinated axons. In contrast, unmyelinated lobster axons of the same 100- to 120-µm diameter have conduction velocities of only 8 m/s. Vertebrates, however, remain the only group with substantial numbers of myelinated neurons. Crustacean myelinated neurons are rare, specialized

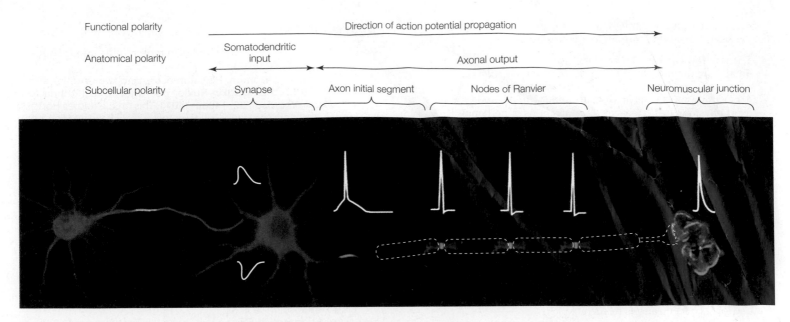

FIGURE 12.28 Spatial distribution of voltage-gated channels at the surface of a myelinated neuron This composite fluorescence micrograph depicts the locations of high concentrations of voltage-gated Na⁺ channels (green, yellow) and K⁺ channels (blue). The Na⁺ channels are localized at the axon initial segment (site of action-potential initiation) and at the nodes of Ranvier. K⁺ channels are localized at paranodal regions just next to the nodes. (From Rasband 2010.)

adaptations in high-velocity escape systems, but myelinated neurons are ubiquitous features of vertebrate nervous systems.

TEMPERATURE The gating kinetics of voltage-dependent ion channels are temperature-dependent. Thus the time course of membrane depolarization to threshold shortens with increased temperature. For both myelinated and unmyelinated axons, an increase of 10°C tends to nearly double conduction velocity (the value of Q_{10} is approximately 1.8; see Chapter 10, page 244). A frog myelinated axon 12–14 μm in diameter conducts at 25 m/s at 20°C, but a cat myelinated axon of only 3.5–4 μm conducts at the same 25 m/s at 37°C. Thus the evolution of homeothermy in birds and mammals—along with myelination—has allowed further axon miniaturization and higher conduction velocities.

Finally, the propagation of action potentials merely ensures that a neuron's electrical signals reach the end of the axon undiminished. For a neuron's signals to affect other cells, another process—synaptic transmission (the topic of Chapter 13)—is required.

SUMMARY
The Propagation of Action Potentials

■ Action potentials propagate because the membrane's underlying permeabilities to ions are voltage-dependent. Local circuits of current flow spread the depolarization along the axon, depolarizing a new region to threshold. Behind an advancing action potential, Na⁺ channels remain inactivated long enough to prevent reexcitation by the local currents.

■ The conduction velocity of an action potential depends on axon diameter, myelination, and temperature. Larger-diameter axons have higher conduction velocities because their length constants are longer, so local currents spread farther along the axon. Myelin greatly increases conduction velocity by increasing R_m (increasing the length constant) while decreasing C_m (preventing an increase in the time constant). Increasing temperature speeds the gating of channels so that the membrane responds faster to the local currents.

STUDY QUESTIONS

1. Suppose that the cell shown in Figure 12.12A is permeable to Na⁺, K⁺, and Cl⁻ but not to A⁻. In the absence of ion pumps, Na⁺ and Cl⁻ will diffuse in, the cell will become out of osmotic balance, and water will enter. Explain why a cell cannot be in thermodynamic equilibrium and also be in osmotic balance, unless there is a nonpermeating ion in the extracellular fluid (as would be the case if Na⁺ were nonpermeating). What does this suggest about the evolutionary origin of the Na⁺–K⁺ exchange pump?

2. Unmyelinated axons conduct action potentials without decrement, but when myelinated axons lose myelin in demyelinating diseases such as multiple sclerosis, conduction of action potentials is blocked. Why?

3. Using Figure 12.26, compare and explain the difference (a) between the velocity of action-potential propagation in lamprey axons and frog myelinated axons of equal diameter, (b) between

squid axons and frog myelinated axons of equal velocity, and (c) between frog and mammalian myelinated axons.

4. The ion flows across neuronal membranes at rest and during an action potential do not significantly change bulk ion concentrations, except for that of Ca^{2+} ions. Resting Ca^{2+} ion concentrations in cells are usually about 10^{-7} M, and Ca^{2+} ions exert physiological effects at concentrations of perhaps 10^{-5} M. Explain why relative changes of intracellular $[Ca^{2+}]$ are much greater than for, say, $[Na^+]$ (12–50 mM).

5. For the ion concentrations in Table 12.1, calculate the equilibrium potentials for each ion species in squid axons and in mammalian muscle fibers.

6. Using the values in question 5, determine what ions are in passive equilibrium at a membrane potential of –60 mV in the squid at 18°C, and at –91 mV in mammalian muscle fibers at 37°C.

7. Suppose you voltage-clamp a squid axon from a resting membrane potential of –60 mV to a clamped value of +55 mV. Describe the early ionic current (say, at 0.5 ms after clamping). Use ionic concentrations from Table 12.1.

8. Suppose a squid axon at rest is 20 times as permeable to K^+ as to Na^+. Using the simplified version of the Goldman equation in Figure 12.13, calculate the resting membrane potential. If during the rising phase of an action potential the permeability to Na^+ increases to 100 times the permeability to K^+, at what value will the action potential peak?

9. Suppose you stimulate an axon so that you generate an action potential at both ends at the same instant. Describe the propagation of these action potentials. What happens when they meet?

10. With increased neuronal electrical activity in a brain area, the rates of glucose uptake, O_2 consumption, and blood flow increase. (These changes are the basis of activity imaging such as functional magnetic resonance imaging [fMRI], which is illustrated in Figure 15.7.) Why does neuronal activity increase local metabolic rate?

11. Toxins such as tetrodotoxin (from puffer fish and newts) and saxitoxin (from red-tide dinoflagellates) block voltage-gated Na^+ channels selectively. What effect would they have on currents in a voltage clamp to 0 mV? How do you suppose such toxins evolved?

12. Part of the evidence for a "ball-and-string" model of Na^+ channel inactivation is that the proteolytic enzyme pronase can selectively disable channel inactivation when perfused into the axoplasm of a squid axon. Where would you expect pronase to cleave the Na^+ channel protein?

Go to **sites.sinauer.com/animalphys4e**
for box extensions, quizzes, flashcards,
and other resources.

REFERENCES

Bean, B. P. 2007. The action potential in mammalian central neurons. *Nat. Rev. Neurosci.* 8: 451–465.

Catterall W. A., I. M. Raman, H. P. Robinson, and two additional authors. 2012. The Hodgkin-Huxley heritage: from channels to circuits. *J. Neurosci.* 32: 14064–14073. This article relates Hodgkin and Huxley's original formulation of the currents underlying action potentials to contemporary findings.

Catterall, W. A., and N. Zheng. 2015. Deciphering voltage-gated Na^+ and Ca^{2+} channels by studying prokaryotic ancestors. *Trends Biochem. Sci.* 40: 526–534.

Debanne, D., E. Campanac, A. Bialowas, and two additional authors. 2011. Axon physiology. *Physiol. Rev.* 91: 555–602.

Fain, G. L. 2014. *Molecular and Cellular Physiology of Neurons,* 2nd ed. Harvard Univ. Press, Cambridge, MA.

Gadsby, D. C. 2009. Ion channels versus ion pumps: the principal difference, in principle. *Nat. Rev. Molec. Cell Biol.* 10: 344–352.

Kandel, E. R., J. H. Schwartz, T. M. Jessell, S. A. Siegelbaum, and A. J. Hudspeth (eds.). 2013. *Principles of Neural Science,* 5th ed. McGraw-Hill, New York. An excellent but heavy, medically oriented neuroscience text.

Moran, Y., M. G. Barzilai, B. J. Liebeskind, and H. H. Zakon. 2015. Evolution of voltage-gated ion channels at the emergence of Metazoa. *J. Exp. Biol.* 218: 515–525.

Nicholls, J. G., A. R. Martin, P. A. Fuchs, and three additional authors. 2012. *From Neuron to Brain,* 5th ed. Sinauer, Sunderland, MA. Update of a classic neurobiology text, with narrower coverage than Kandel et al. but with full explanations and excellent experimental focus.

Tye, K. M., and K. Deisseroth. 2012. Optogenetic investigation of neural circuits underlying brain disease in animal models. *Nat. Rev. Neurosci.* 13: 251–266.

See also **Additional References** *and* **Figure and Table Citations**.

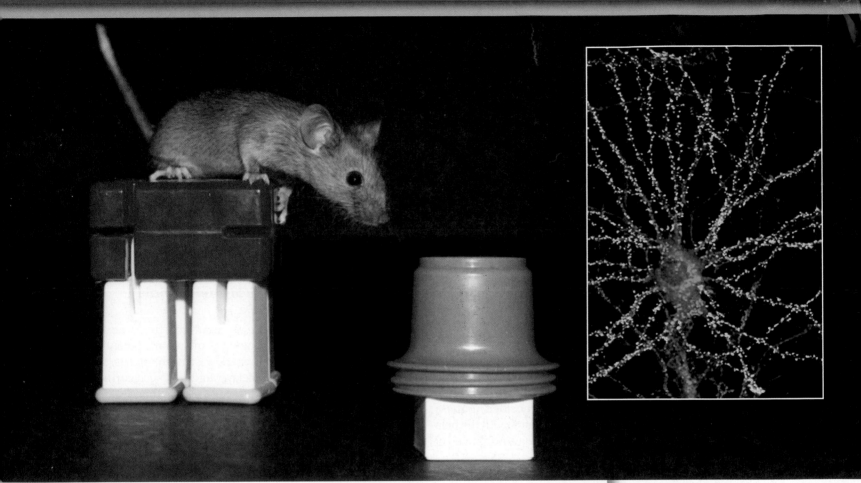

A **synapse**, as we have noted, is a specialized site of contact of a neuron with another neuron or with an effector. It is the locus where one cell (the *presynaptic* neuron) influences the function of another cell (the *postsynaptic* neuron or effector). Neurons—except in the special case of those having electrical synapses—are discontinuous with each other; even at synapses, they are typically separated by a 20- to 30-nanometer (nm) space called the *synaptic cleft*. Essentially all influences that neurons have on each other and on effectors are exerted at synapses. The green neuron shown here receives synaptic input from axon terminals of other neurons, which exhibit red fluorescence. Each yellow spot is a synapse at which the red presynaptic terminal overlaps with the bright green postsynaptic membrane, producing a combined yellow fluorescence.

You often hear the analogy that the brain is like a computer. Like all analogies, this one is imperfect, but it is useful here. Inasmuch as the brain is like a computer, the neuronal axons (which transmit action potentials long distances without degradation) are analogous to the wires in the computer, and the synapses are analogous to logical junctions by which signals in one element affect others. It is the connections and properties of the logical junctions (the synapses) that determine the performance—and improvement in performance—of the brain computer, more than the properties of the wires.

The *Doogie* mouse
This genetically engineered mouse is synaptically smarter than the average mouse. Mouse brains contain neurons that receive synaptic input; the neuron shown here has a postsynaptic protein labeled with a green fluorescent dye, with the presynaptic terminals of other neurons labeled with a red fluorescent dye. Each synapse is a yellow spot of overlap of the red presynaptic ending and the green postsynaptic membrane in a yellow mix of both red and green fluorescence. Changes in synaptic strength underlie learning and memory. (Neuron photograph courtesy of Mary B. Kennedy, Division of Biology, Caltech.)

The *Doogie* mouse introduces two sorts of synaptic functions that illustrate the range of ways in which a synapse can work: *synaptic transmission* and *synaptic plasticity*. In **synaptic transmission**, a presynaptic signal—usually an action potential—has an effect on a postsynaptic cell. The effect is rapid and transient, and it can be excitatory or inhibitory. This transmission of a signal across the synapse is the simplest kind of synaptic action. We introduce this fast-transmission function of synapses by examining a neuromuscular junction: the synapse between a neuron and a muscle fiber. Neuromuscular synaptic transmission causes muscle fibers to contract, allowing, for example, a mouse to jump. The second sort of synaptic function is **synaptic plasticity**: the ability to change the functional properties of synapses. Synaptic actions can change the synapse itself to make it stronger or weaker, and they can also produce long-lasting changes in the postsynaptic cell. This modulation of synaptic processes is the basis of much functional change in the nervous system, both in development and in learning and memory.

Synapses have diverse actions, as befits their functional importance. A synapse can work chemically or electrically. It can excite the postsynaptic cell or inhibit it. A synaptic potential can result from a permeability increase or a permeability decrease. Synaptic action can be fast or slow, and it can *mediate* transmission or *modulate* the plastic properties of the synapse itself. Synapses can act immediately and directly on the membrane potential of the postsynaptic cell, or they can have indirect and longer-lasting effects. Some of these diverse actions are functionally linked, but others are not. In this chapter, we first consider electrical transmission and then give an overview of the more widespread chemical transmission. Later in the chapter we examine the presynaptic and postsynaptic mechanisms of chemical synapses and study the distinction between *ionotropic* synaptic action (which is fast and produces direct changes in ion permeability and thus membrane potential) and *metabotropic* synaptic action (which is slow and produces chemical signal transduction changes in the postsynaptic cell). Only after considering these kinds of synaptic mechanisms can we try to explain (with the help of the *Doogie* mouse) the synaptic changes that may form much of the basis of learning and memory.

Synaptic Transmission Is Usually Chemical but Can Be Electrical

At the most fundamental level, there are two kinds of synapses: electrical and chemical (**FIGURE 13.1**). Both kinds can rapidly change the membrane potential of a postsynaptic cell. Electrical and chemical synapses play different functional roles, so both kinds are adaptive in nervous systems. We discuss electrical transmission first, setting the stage for treatment of the more common and more complex chemical transmission process. **TABLE 13.1** classifies the different types of synapses.

Fast synaptic transmission—in contrast to slow synaptic effects, which we will discuss later—works in milliseconds: An action potential in a presynaptic neuron leads to a rapid postsynaptic voltage change, with a typical delay of less than a millisecond. The mechanism of this fast synaptic action was a subject of much debate in the first half of the twentieth century. One group (the "sparks") argued that synaptic transmission was by direct electrical means. The other group (the "soups") postulated that transmission was by a chemical mechanism: Depolarization of the presynaptic terminal caused it to release a chemical neurotransmitter, which diffused across the synaptic cleft to affect the postsynaptic cell. It is now clear that both kinds of synaptic transmission can occur: Most synaptic transmission is chemical, whereas direct electrical transmission occurs uncommonly.

Electrical synapses transmit signals instantaneously

In an **electrical synapse**, electric currents from one cell flow directly into the next cell, changing its membrane potential (see Figure 13.1A). Because this current flow is instantaneous, an electrical synapse has essentially no delay. Moreover, such current flow can usually be in either direction, so electrical synapses are often not polarized. Because of their instantaneous transmission of signals and their synchronizing ability, electrical synapses are found in nervous systems where speed is most important, and where synchronous activity of several cells is an advantage.

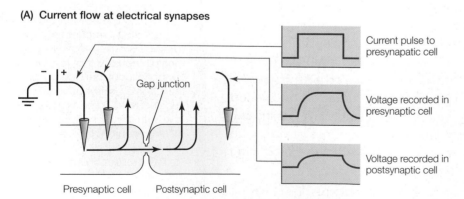

(A) Current flow at electrical synapses

Gap junction

Current pulse to presynapatic cell

Voltage recorded in presynaptic cell

Voltage recorded in postsynaptic cell

Presynaptic cell   Postsynaptic cell

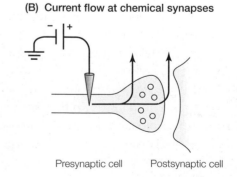

(B) Current flow at chemical synapses

Presynaptic cell   Postsynaptic cell

FIGURE 13.1 Electrical and chemical synapses Both kinds of synapses play roles in neural communication, but they work in fundamentally different ways. (A) Electrical synapses have low-resistance pathways that allow currents to flow directly between neurons. Current flow through the low-resistance pathways of gap junctions electrically couples the neurons. If current is injected into one neuron, the depolarization (or hyperpolarization) spreads to another electrically coupled neuron, and the depolarization of one cell produces a smaller depolarization of the adjacent cell. (B) In chemical synapses, the currents escape between neurons and do not enter the postsynaptic neuron. Instead, released molecules of neurotransmitter (not shown) carry the signal across the synaptic cleft.

TABLE 13.1 Kinds of synapses

Characteristic	Chemical synapse		Electrical synapse
	Ionotropic	Metabotropic	
Mechanism and time course	Fast	Slow	Instantaneous current flow
Function	Signal transmission	Signal transmission, neuronal modulation	Electrical transmission
Effect	Excitation (fast EPSP), inhibition (fast IPSP)[a]	Excitation (slow EPSP), inhibition (slow IPSP), other (cytoplasmic and genetic effects)	Electrical coupling

[a]EPSP = excitatory postsynaptic potential; IPSP = inhibitory postsynaptic potential.

The major structural specialization for electrical transmission is the **gap junction**. A gap junction (as introduced in Chapter 2; see page 43) is a specialized locus where protein channels bridge the gap between two cells, directly connecting their cytoplasm. Gap junctions provide a low-resistance path for current flow, electrically coupling the cells that they join. Thus any electrical change in one cell is recorded in the other, with some weakening but with negligible delay.

Figure 13.1A shows the typical effects of this electrical coupling. Depolarization or hyperpolarization of one cell produces a weaker corresponding change in the other cell. The voltage change is always accompanied by some decrease because every gap junction has some resistance, but the amount of the decrease varies. Thus electrical synapses differ in strength: The larger the gap-junction area, the lower the resistance between cells and the stronger the coupling.

Electrical synapses can act as short-latency synaptic relays, in which each presynaptic action potential triggers a postsynaptic action potential. In other cases, with weaker electrical coupling or high-impulse thresholds, a presynaptic action potential may produce only subthreshold depolarization of the postsynaptic cell. Most electrical synapses are bidirectional, transmitting voltage changes roughly symmetrically in both directions. However, in the electrical synapse between the lateral giant axon and the giant motor neuron of crayfish, the electrical synapse is rectifying; that is, it allows current flow preferentially in one direction. Thus an action potential in the lateral giant axon excites the giant motor neuron, but excitation of the motor neuron by other pathways cannot "backfire" to the lateral giant axon.

The structure of vertebrate gap junctions has been examined by electron microscopy and X-ray diffraction studies. Gap junctions narrow the space between adjacent cells, holding the pre- and postsynaptic membranes only about 3.5 nm apart (instead of the 20–30 nm that usually separates neighboring cells). In the region of close membrane apposition, the narrow gap separating the membranes is bridged by a regular array of channel structures termed **connexons** (**FIGURE 13.2**).

Each connexon is composed of hexamers of the protein *connexin* surrounding a 2-nm pore. Connexons are hemichannels; each connexon of one cell pairs with a connexon of the adjacent cell to form a channel that connects the cytoplasm of the cells. The channels of all the connexons are the low-resistance pathways that allow electric current to flow between the cells. They are large enough to allow the passage of most ions, as well as dye and tracer molecules smaller than about 1000 daltons. Connexon channels are thought to be open normally, but they can close in response to electrical or chemical changes in the cells. Gap junctions in invertebrate protostomes have similar structures, called *innexons*, which are composed of subunits of the protein *innexin*. Innexin proteins are analogous but not homologous to the connexin proteins of vertebrates.

Where in nervous systems has natural selection favored the evolution of electrical synapses? In most cases it is where speed or synchronization is paramount. We noted that electrical synapses can act as fast intercellular relays, in which action potentials are instantaneously transmitted from cell to cell. Such relay electrical synapses are found in escape systems—for example, in crayfish (**FIGURE 13.3**) and fish. In such fast escape responses, every millisecond saved by an electrical synapse can produce a selective advantage of not being eaten. Electrical synapses are also advantageous for groups of neurons that normally fire synchronously, such as the neurons controlling electric-organ discharge in electric fish.

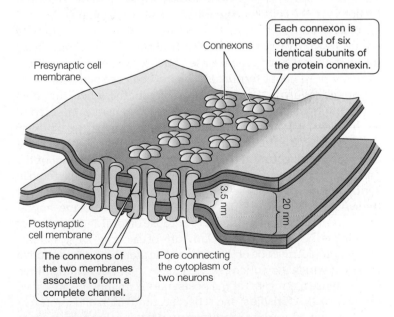

Connexons

Presynaptic cell membrane

Each connexon is composed of six identical subunits of the protein connexin.

Postsynaptic cell membrane

3.5 nm

20 nm

The connexons of the two membranes associate to form a complete channel.

Pore connecting the cytoplasm of two neurons

FIGURE 13.2 The molecular structure of gap junctions
A gap junction is a localized patch of close membrane apposition where protein hexamers called connexons provide channels that electrically connect the cytoplasm of the cells.

(A)

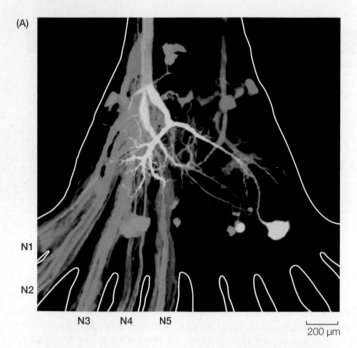

(B)

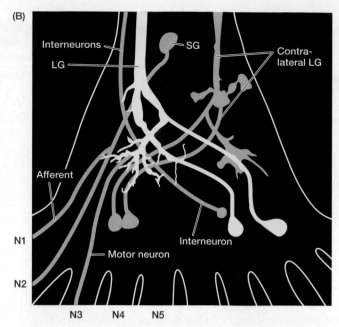

200 μm

FIGURE 13.3 Electrical synapses in the crayfish escape circuit   Neurons electrically connected by gap junctions are said to be dye-coupled, because low-molecular-weight dyes pass through the gap junctions to the coupled cells. (A) A fluorescence micrograph and (B) diagram show the coupling of cells in the last abdominal ganglion of a crayfish. The lateral giant neuron (LG, yellow) has been injected with two fluorescent dyes: a small dye molecule that fluoresces green and a larger dye molecule that fluoresces red. The smaller, green-fluorescing molecule passes through gap junctions, so the green images are all neurons that are dye-coupled, and electrically coupled, to the LG. The red-fluorescing molecule is too large to pass through gap junctions and thus stays in the LG. The LG is yellow because it contains both the red- and green-fluorescing dyes. Neurons that are electrically coupled to the LG include sensory neurons from tactile hairs on the crayfish's tail (diffuse green in nerves N1–N5), sensory interneurons, motor neurons to tail muscles (bright green cell bodies and axons in N1–N5), the segmental giant neuron (SG), and the contralateral LG. Electrically mediated synaptic transmission through many elements of this neural circuit allows for a very rapid, short-latency escape tail flip in response to threatening stimuli. (A from Antonsen and Edwards 2003.)

Chemical synapses can modify and amplify signals

Unlike electrical synapses, **chemical synapses** have a discontinuity between the cells because the 20- to 30-nm synaptic cleft of a chemical synapse is a barrier to direct electrical communication (see Figure 13.1B). The presynaptic electrical signal is first transduced into a chemical signal: the release of neurotransmitter molecules from the presynaptic terminals. The molecules of neurotransmitter rapidly diffuse to the postsynaptic membrane, where they bind to receptor molecules that are specialized to generate an electrical or chemical change in response to the neurotransmitter binding.

Viewed with an electron microscope, chemical synapses exhibit a distinctive structure (**FIGURE 13.4**). The axon terminal of the presynaptic neuron contains neurotransmitter molecules stored in **synaptic vesicles**, with several thousand molecules per vesicle. At the synaptic cleft, both the pre- and postsynaptic membranes appear denser and thicker than elsewhere because of local aggregations of proteins at these membranes. Dark tufts of electron-dense material on the cytoplasmic side of the presynaptic membrane mark **active zones** at which synaptic vesicles release their neurotransmitter. Other dense aggregates at the postsynaptic membrane are called **postsynaptic densities**, and reflect accumulation of neurotransmitter receptors as well as **scaffolding proteins** that help organize these receptors and other proteins at the synapse.

The vast majority of excitatory synaptic inputs in the vertebrate central nervous system (CNS) occur on neuronal dendrites, and

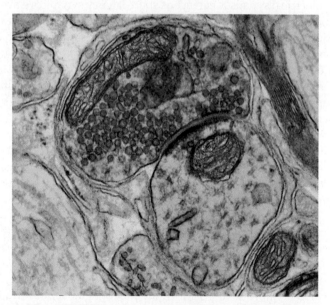

FIGURE 13.4 The structure of a chemical synapse   Electron microscopy reveals distinctive features of chemical synapses. A presynaptic terminal (above) contains mitochondria (here colorized purple) and numerous synaptic vesicles (red), some of which are clustered at dense release sites called active zones (dark presynaptic densities). Opposite the active zones across the synaptic cleft (pink), a dark postsynaptic density underlies the postsynaptic membrane. The densities represent proteins involved in neurotransmitter release (presynaptically) and in clustering of neurotransmitter receptors (postsynaptically).

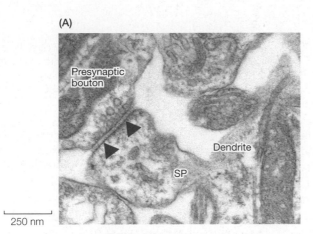

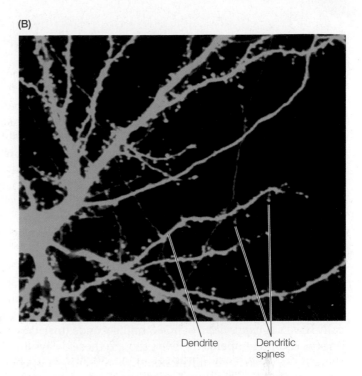

FIGURE 13.5 The majority of excitatory synapses in a mammalian brain occur on dendritic spines (A) Electron microscopy of a synapse shows a mushroom-shaped dendritic spine (SP) extending from the postsynaptic dendrite. Also visible are the presynaptic axon ending ("presynaptic bouton") containing docked vesicles and the synaptic junction (blue arrowheads). (B) A cortical neuron grown in culture for 21 days expresses green fluorescent protein (GFP) which fills the cell, demonstrating that dendrites are covered in small mushroom-shaped protrusions (spines), most of which are sites of synaptic contact. (A from Waites et al. 2005; B courtesy of Matthew Kayser and Matthew Dalva.)

specifically on small mushroom-shaped protrusions known as dendritic spines (**FIGURE 13.5**). As we will discuss later, these dendritic spines remain malleable even in the mature CNS: Their size and shape appear to change in the intact brain of living mice in response to alterations of external stimuli!

A presynaptic neuron releases neurotransmitter molecules in response to an arriving action potential (**FIGURE 13.6**). Neu-

rotransmitter is synthesized in the presynaptic neuron and stored in synaptic vesicles until release. Neurotransmitter is released by calcium-dependent exocytosis—fusion of the synaptic vesicles to the presynaptic membrane (see Figure 13.6B). The released neurotransmitter molecules bind to receptor proteins embedded in the postsynaptic membrane. Neurotransmitter receptors are transmembrane proteins that are effectors for change in the

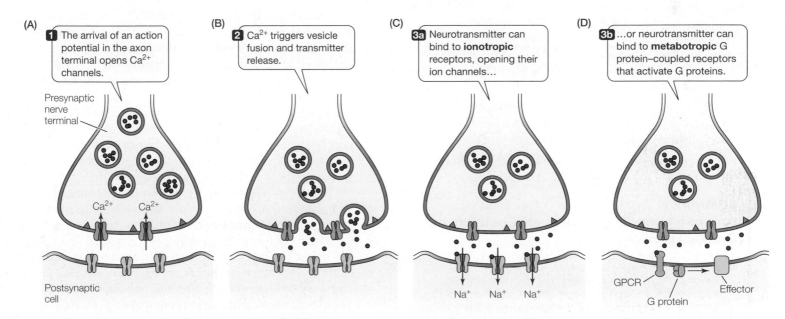

FIGURE 13.6 The function of a chemical synapse (A) A presynaptic action potential depolarizes the axon terminal and opens voltage-gated Ca^{2+} channels located near active zones. (B) Ca^{2+} ions enter the terminal and induce vesicles at the active zones (indicated by triangles on the membrane) to fuse with the presynaptic membrane and release neurotransmitter by exocytosis. Neurotransmitter molecules diffuse across the synaptic cleft and bind to postsynaptic receptor proteins. (C) Ionotropic neurotransmitter receptors open ion channels, creating an ionic current that changes the postsynaptic cell's membrane potential. In this example, Na^+ influx would cause depolarization. (D) Metabotropic receptors activate a metabolic cascade in the postsynaptic cell. Metabotropic receptors are G protein–coupled receptors (GPCRs) that activate G proteins to produce a second messenger such as cyclic AMP (cAMP).

postsynaptic cell, usually producing a change in postsynaptic membrane potential.

Neurotransmitter receptors work in one of two ways. They can produce *fast* changes in membrane potential (depolarization or hyperpolarization) by directly increasing permeability to ions (see Figure 13.6C). When such receptors bind their neurotransmitter, they open to allow ion flow. In this case, a single molecule constitutes both the receptor and the ion channel. Receptors working in this way are termed **ionotropic receptors** because they directly alter permeability to ions. Other receptors (see Figure 13.6D) trigger a signaling cascade of second messengers in the postsynaptic cell and are called **metabotropic receptors**. Metabotropic receptors often have relatively *slow*, long-lasting modulatory effects on synaptic processes, as we will discuss later in this chapter.

Transmission at chemical synapses is necessarily slower than transmission at electrical synapses because the steps of transmitter release and receptor action take more time than the instantaneous spread of electric current. Fast ionotropic chemical synapses typically have a *synaptic delay* of 0.3–3 milliseconds (ms), depending on species and temperature. Even though these delays are short, there must be countervailing adaptive advantages of chemical synapses that explain their prevalence. Several advantages can be recognized:

- Chemical synapses can amplify current flow. A presynaptic action potential can lead to the release of a few or many synaptic vesicles. Each vesicle contains a few thousand molecules of transmitter. Depending on the number of active zones and the size of the presynaptic terminal, the amount of transmitter released can open many channels and amplify the postsynaptic current.

- Chemical synapses can be either excitatory or inhibitory, unlike electrical synapses, which are nearly always excitatory.

- Chemical synapses are one-way; a presynaptic neuron excites or inhibits a postsynaptic cell, but not vice versa. Most electrical synapses are two-way.

- Chemical synapses are much more modifiable in their properties than electrical synapses are. Use and circumstance can make them stronger, and disuse can make them weaker. This plasticity is important for nervous system development and for learning.

SUMMARY
Synaptic Transmission Is Usually Chemical but Can Be Electrical

- Most synapses are chemical; some are electrical. Electrical synapses are very fast and usually are bidirectional. Gap junctions are the anatomical basis of electrical synapses; they contain connexons that allow current to flow directly between the cells, electrically coupling them.

- Chemical synapses are unidirectional, with a presynaptic neuron that releases neurotransmitter when stimulated, and a postsynaptic neuron (or effector) that bears receptor molecules to which the neurotransmitter binds.

- Neurotransmitter receptors may directly open their own ion channels; or they may act indirectly through a signal transduction cascade that involves second messengers, to open, close, or change ion channels that are separate molecules.

- Electrical synapses mediate fast, synchronizing actions of neurons. Chemical synapses integrate neuronal functions, by fast (ionotropic) excitation and inhibition, or by slow (metabotropic) modulation of neuronal and synaptic properties.

Synaptic Potentials Control Neuronal Excitability

The primary function of synaptic transmission is to control the excitability of the postsynaptic cell. When a burst of neurotransmitter diffuses across a synapse, it generates a **synaptic potential**—a transitory, graded change in the resting membrane potential—in the postsynaptic cell. A synaptic potential that tends to *depolarize* the cell membrane is excitatory, and one that tends to *hyperpolarize* the cell membrane is inhibitory. **Excitation** is an increase in the probability that a cell will generate an impulse (an action potential), or if the cell is already generating impulses, excitation causes an increase in the impulse frequency. **Inhibition**, by contrast, is a decrease in the probability of impulse generation or a decrease in impulse frequency. Excitatory and inhibitory synapses summate their voltage effects to control action-potential generation of the postsynaptic cell.

Synapses onto a spinal motor neuron exemplify functions of fast synaptic potentials

We can witness the excitatory and inhibitory effects of synaptic potentials by stimulating the peripheral nerves containing axons that synapse on a cat spinal motor neuron, while recording the changes in the motor neuron's membrane potential (**FIGURE 13.7**). In Figure 13.7B, stimulation of certain peripheral nerves (sensory pathway A or B) produces synaptic potentials that are graded depolarizations in the motor neuron, each with a brief rising phase and an exponential decay over a time course of 10–20 ms. These depolarizations are **excitatory postsynaptic potentials** (EPSPs). Each excitatory synapse usually produces a very small EPSP, one that depolarizes the membrane by less than 1 mV.

EPSPs can be combined in two ways. If peripheral nerve A is stimulated repeatedly and rapidly, the resultant EPSPs combine in a process called **temporal summation**. Simultaneously occurring EPSPs produced by different nerves (Stimulus A + B in Figure 13.7B) also combine, in a process called **spatial summation**. With sufficient presynaptic stimulation, EPSPs can summate temporally and spatially to the voltage threshold and cause the motor neuron to generate one or more action potentials.

Inhibitory synapses produce similar, but countervailing, effects on the membrane potential of a motor neuron. In the example in Figure 13.7C, stimulating peripheral nerve D or E evokes synaptic potentials that briefly hyperpolarize the motor-neuron cell membrane. Because they drive the membrane potential away from threshold, these synaptic potentials are called **inhibitory postsynaptic potentials** (IPSPs).

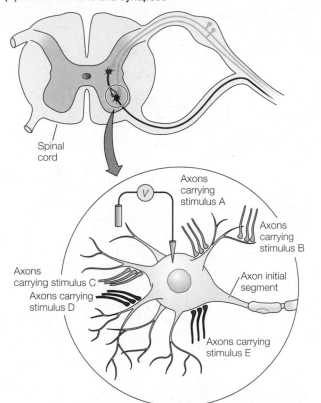

(A) Motor neurons and synapses

Spinal cord

Axons carrying stimulus A

Axons carrying stimulus B

Axons carrying stimulus C

Axons carrying stimulus D

Axon initial segment

Axons carrying stimulus E

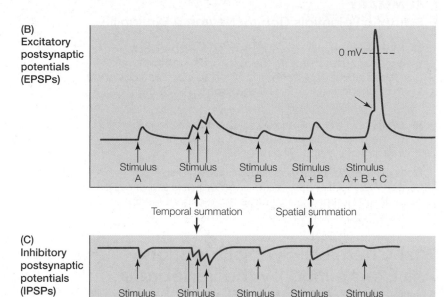

(B) Excitatory postsynaptic potentials (EPSPs)

0 mV

Stimulus A

Stimulus A

Stimulus B

Stimulus A + B

Stimulus A + B + C

Temporal summation

Spatial summation

(C) Inhibitory postsynaptic potentials (IPSPs)

Stimulus D

Stimulus D

Stimulus E

Stimulus D + E

Stimulus A + D

FIGURE 13.7 Excitatory and inhibitory postsynaptic potentials (A) Clusters of synapses onto a spinal motor neuron of a cat. Yellow presynaptic terminals are excitatory; red presynaptic terminals are inhibitory. The axon initial segment is the site of action-potential initiation. (B) Stimulation of excitatory presynaptic neurons elicits excitatory postsynaptic potentials (EPSPs), graded depolarizations that depolarize V_m (membrane potential) toward threshold (indicated by the red arrow). (C) Inhibitory postsynaptic potentials (IPSPs) are hyperpolarizing, moving V_m away from threshold. Both EPSPs and IPSPs can summate with other postsynaptic potentials of the same pathway (temporal summation) or a different pathway (spatial summation).

As Figure 13.7C demonstrates, IPSPs also produce temporal and spatial summation effects. What's more, EPSPs and IPSPs that occur simultaneously can summate spatially, reducing or canceling their respective excitatory and inhibitory effects on the postsynaptic neuron's membrane potential. A typical cat motor neuron receives input from about 10,000 synaptic terminals, and the moment-to-moment balance of EPSPs and IPSPs determines whether the motor neuron generates impulses or remains quiescent.

Synapses excite or inhibit a neuron by depolarization or hyperpolarization at the site of impulse initiation

Because most neurons receive thousands of synaptic endings, the output of a neuron—its temporal sequence of action potentials—is a complex function of its synaptic input. Thus a neuron's output is not the same as its input, but is instead an integral function of that input, a property called **neuronal integration**. The major process underlying neuronal integration is the spatial and temporal summation of EPSPs and IPSPs. However, this process is more complex than the simple algebraic summation of postsynaptic potentials. Neuronal integration also is a function of the spatial relationships of excitatory and inhibitory synapses to one another, as well as their relative proximity to the site of impulse initiation. For spinal

motor neurons, this site is the initial segment of the axon, adjacent to the axon hillock.

Because postsynaptic potentials are *graded potentials*, their spread is governed by the cable properties of the postsynaptic cell membrane (see Chapter 12; page 315). Consequently, the functional effect of a synapse depends in part on where it is located on a neuron. A synapse on the soma or cell body (called an **axosomatic synapse**) is only a short electrotonic distance from the axon initial segment; that is, the amplitudes of postsynaptic potentials are decreased only a small amount in their passive spread from the soma to the impulse initiation site.

A synapse onto a dendrite (**axodendritic synapse**), however, may be as much as 200 micrometers (μm) away from the axon hillock. The electrotonic length of motor neuron dendrites is estimated to be 1–2 λ (one or two times the membrane length constant [symbolized by the Greek letter lambda]; see page 314), so a synaptic potential at the dendrite tip is decreased to 14–37% of its initial amplitude in its spread to the axon hillock. Although other factors may partially compensate for this decrease with distance, synapses that are closer to the axon hillock usually have more effect on the output of the postsynaptic cell than do synapses on the distal ends of dendrites. Thus the summation of synaptic input is weighted by the electrotonic distance of the synapses from the axon hillock.

SUMMARY
Synaptic Potentials Control Neuronal Excitability

■ Most synapses in nervous systems are chemical synapses that mediate fast excitation and inhibition. Neurotransmitters act at receptors to open ion channels, to depolarize (EPSP) or hyperpolarize (IPSP) the postsynaptic neuron.

■ EPSPs and IPSPs summate, so the membrane potential of the postsynaptic neuron is a moment-to-moment integral of synaptic input.

■ Postsynaptic potentials are graded and spread passively to the axon initial segment (the site of action-potential initiation). Therefore more-distant synapses may have smaller effects on the neuron's output.

Fast Chemical Synaptic Actions Are Exemplified by the Vertebrate Neuromuscular Junction

We turn now to the mechanisms of action of fast chemical synapses, the sort that gave rise to the synaptic potentials in the preceding section. Such synapses are called *fast* because they produce postsynaptic potentials within a few milliseconds of a presynaptic action potential. Fast synaptic action is the conceptual model that has dominated scientific thinking about how nervous systems work. In fact, until about 30 years ago, fast synaptic transmission was the only kind known, or even suspected. All synapses were thought to work by producing fast EPSPs and IPSPs such as those shown in Figure 13.7. Although investigators now know that synapses can also work in other ways, fast chemical transmission is still considered the "workhorse" of synaptic mechanisms, the basis of most neuronal interaction in nervous systems.

To show the *function* of fast chemical synapses, in the previous section we considered synapses onto motor neurons in the CNS. Much of what we know about synaptic *mechanisms*, however, has been learned through the study of *peripheral* synapses. For technical reasons it is harder to study synaptic mechanisms in the CNS. Synapses in the vertebrate CNS are buried inside a large mass of tissue. They usually cannot be seen through a microscope, nor can their immediate environment be changed readily. Moreover, a single vertebrate CNS neuron may have tens of thousands of synapses from other neurons ending on it, each neuron producing its own synaptic effects. Hence researchers have turned to simpler *model systems*, in which detailed investigations of synaptic transmission are easier, and have later applied their findings to the more complex CNS. This strategy is analogous to studying squid giant axons to clarify the mechanisms by which action potentials are generated (see Chapter 12).

The vertebrate skeletal **neuromuscular junction** (also called the *motor end-plate*) is the model system for chemical synaptic transmission from which much of our basic knowledge of synaptic physiology is derived (**FIGURE 13.8**). For most vertebrate skeletal muscles, each muscle fiber is innervated by only one motor neuron. (Hundreds of motor neurons may innervate the whole muscle, but only one innervates a given fiber.) The neuromuscular junction functions as a *relay* synapse: Each action potential in the motor

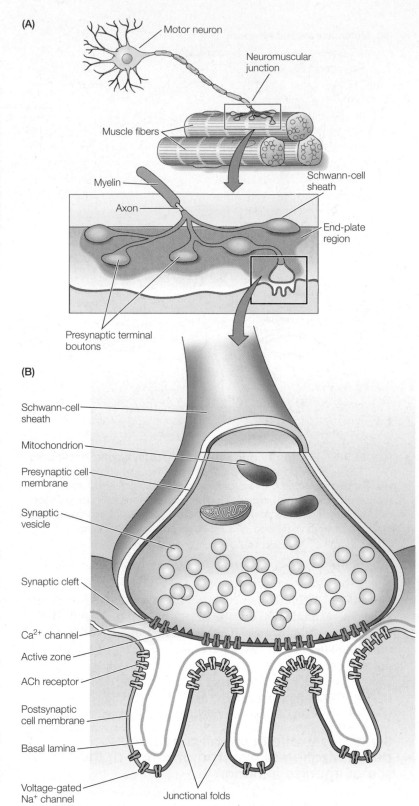

FIGURE 13.8 A vertebrate neuromuscular junction (A) The neuromuscular junction consists of a single presynaptic motor neuron contacting a single postsynaptic skeletal-muscle fiber. (B) Presynaptic vesicles fill part of the terminal and cluster near active zones. The postsynaptic muscle fiber membrane is deeply infolded, and these junctional folds contain ligand-gated acetylcholine (ACh) receptors at the upper parts of the fold. Voltage-gated channels are absent at the tops of the folds but are present deep in the folds and at the nonjunctional membrane. A basal lamina of extracellular, secreted proteins covers the muscle fiber. (After Kandel et al. 2000.)

neuron evokes a large EPSP in the muscle fiber, which reaches threshold and elicits an action potential in the muscle fiber. The muscle fiber action potential propagates to all parts of the muscle fiber, depolarizing its membrane and triggering its contraction (we will discuss muscle contraction in Chapter 20).

The vertebrate neuromuscular junction has several advantages as a model synapse. In addition to its anatomical simplicity and large synaptic response, the neuromuscular junction is accessible and microscopically visible at the muscle surface. A researcher can impale a muscle fiber with an electrode under visual control and can readily change the fluid bathing the junction.

Chemical synapses work by releasing and responding to neurotransmitters

The sequence of events of synaptic transmission at the vertebrate skeletal neuromuscular junction is summarized in **FIGURE 13.9**. Many of these steps and their sequence are characteristic of most synapses, as previously introduced in Figure 13.6. At the neuromuscular junction the neurotransmitter is acetylcholine (ACh), so when an action potential depolarizes the presynaptic terminal and opens voltage-gated Ca^{2+} channels (step ❶ in Figure 13.9), the Ca^{2+} entry (step ❷) triggers vesicle fusion to release ACh into the synaptic cleft (step ❸). Diffusion of neurotransmitter molecules across the synaptic cleft to the postsynaptic membrane is rapid (20–50 microseconds [μs]) because the diffusion path is so short (50 nm at the neuromuscular junction, 20–30 nm at CNS synapses).

At the postsynaptic membrane, neurotransmitter molecules bind to specific receptor proteins (step ❹). At the neuromuscular junction, ACh receptor molecules are ligand-gated channels that are opened by the binding of transmitter. The channel opening allows ion flow, producing an *EPSP that spreads decrementally* (step ❺).

At the neuromuscular junction (unlike most synapses) the EPSP is large enough to depolarize the muscle fiber membrane to its threshold and initiate a muscle fiber action potential (step ❻). The muscle fiber action potential is generated by voltage-gated channels and propagates to the ends of the fiber, depolarizing the entire membrane and thereby initiating contraction of the fiber (see Chapter 20).

The action of the neurotransmitter at most synapses is terminated by enzymatic degradation or by reuptake. At the neuromuscular junction, ACh is destroyed by acetylcholinesterase (step ❼), an enzyme synthesized by the postsynaptic muscle fibers and located in the extracellular matrix (basal lamina) within the synaptic cleft. A choline transporter (step ❽) retrieves choline into the presynaptic terminal for ACh resynthesis.

Postsynaptic potentials result from permeability changes that are neurotransmitter-dependent and voltage-independent

In Chapter 12 we learned that the permeability changes that produce an action potential are voltage-dependent, resulting from depolarization of the membrane. In contrast, the permeability (P) changes that produce a synaptic potential depend on neurotransmitter and not on voltage. As noted above, the molecules that control permeability changes at fast synapses are ligand-gated channels (see Chapter 2, page 63, for review). Ligand-gated channels open as a result of

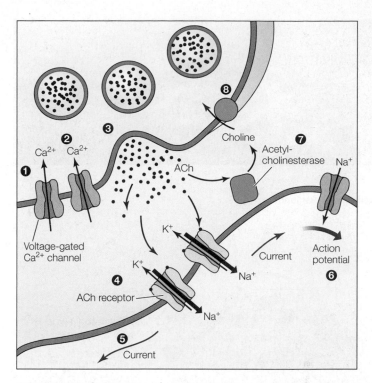

FIGURE 13.9 Summary of events in chemical synaptic transmission at the vertebrate neuromuscular junction ❶ An action potential depolarizes the axon terminal of the motor neuron. ❷ The depolarization opens voltage-gated Ca^{2+} channels. ❸ Depolarization of the terminal triggers vesicle exocytosis at an active zone, releasing acetylcholine (ACh). ❹ ACh diffuses rapidly across the synaptic cleft and binds to acetylcholine receptors at the postsynaptic membrane. ❺ The receptor channel opens to allow Na^+ and K^+ ion flow, producing a depolarizing excitatory postsynaptic potential (EPSP). The EPSP spreads to depolarize nearby regions to threshold and triggers a muscle fiber action potential. ❻ The action potential propagates to all parts of the muscle fiber, eliciting contraction. ❼ Acetylcholinesterase hydrolyzes the acetylcholine into acetate and choline. ❽ Choline is actively transported back into the motor axon terminal to be resynthesized into acetylcholine.

binding neurotransmitter, not in response to depolarization. (We will consider ligand-gated channel structure in a later section.)

AN EPSP RESULTS FROM A SIMULTANEOUS INCREASE IN THE POSTSYNAPTIC MEMBRANE'S PERMEABILITY TO Na^+ AND K^+

When the channel of an acetylcholine receptor opens, it becomes permeable to both Na^+ and K^+ ions. Therefore, during an EPSP, P_{Na} and P_K increase *simultaneously*, rather than sequentially as in generation of the action potential. The ion flows through all the channels that open in response to release of neurotransmitter constitute the **synaptic current**, which produces the depolarization that is the rising phase of an EPSP. Although the channels' permeabilities to Na^+ and K^+ are similar, most of the synaptic current underlying an EPSP is carried by Na^+ entry, because the two ions have different driving forces. The membrane potential of the muscle fiber is very far from the sodium equilibrium potential (E_{Na}; see Chapter 12), so there is a large driving force for sodium entry ($E_m - E_{Na}$). In contrast, the resting membrane potential is closer to the potassium equilibrium potential (E_K), so the driving force for potassium entry ($E_m - E_K$) is smaller.

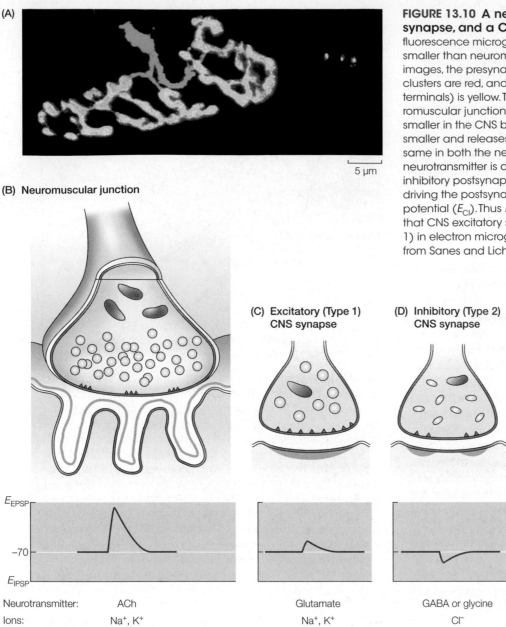

(A)

5 μm

(B) Neuromuscular junction

(C) Excitatory (Type 1) CNS synapse

(D) Inhibitory (Type 2) CNS synapse

E_{EPSP}

−70

E_{IPSP}

| Neurotransmitter: | ACh | Glutamate | GABA or glycine |
| Ions: | Na^+, K^+ | Na^+, K^+ | Cl^- |

FIGURE 13.10 A neuromuscular synapse, a CNS excitatory synapse, and a CNS inhibitory synapse (A) As shown in these fluorescence micrographs, CNS synapses (three on the right) are far smaller than neuromuscular junction synapses (on the left). In both images, the presynaptic axon terminal is green, postsynaptic receptor clusters are red, and the synapse (overlap of pre- and postsynaptic terminals) is yellow. The mechanism of an excitatory synapse at a neuromuscular junction (B) and in the CNS (C) is similar, but the EPSP is smaller in the CNS because the presynaptic neuron terminal is much smaller and releases fewer vesicles. The ionic basis of the EPSP is the same in both the neuromuscular and the CNS synapses, although the neurotransmitter is different. (D) A CNS inhibitory synapse produces an inhibitory postsynaptic potential (IPSP) by opening chloride channels, driving the postsynaptic membrane toward the chloride equilibrium potential (E_{Cl}). Thus E_{Cl} is the reversal potential for the IPSP, or E_{IPSP}. Note that CNS excitatory synapses often have a different appearance (Type 1) in electron micrographs from CNS inhibitory synapses (Type 2). (A from Sanes and Lichtman 2001, courtesy of Josh Sanes.)

The underlying ionic mechanisms of these postsynaptic potentials are also similar. At both synapses, neurotransmitter molecules bind to postsynaptic receptors to increase the permeability of the membrane to both Na^+ and K^+ ions. Neuronal EPSPs have reversal potentials near zero, similar to the reversal potentials of neuromuscular EPSPs. Because of their similarities in mechanism and ionic basis, we use the same term, *EPSP*, to refer to both neuronal EPSPs and neuromuscular EPSPs.[1]

However, there are two major differences between neuronal and neuromuscular EPSPs. First, in the CNS of vertebrates, the common neurotransmitter mediating fast EPSPs is *glutamate*, rather than acetylcholine (see Figure 13.10C). Thus some glutamate receptors have the same effect as the ACh receptors at the neuromuscular junction: rapidly increasing permeability to Na^+ and K^+ ions when they bind their neurotransmitter.

The second major difference between neuromuscular and neuronal EPSPs is their size. At a neuromuscular junction, a single presynaptic action potential liberates enough neurotransmitter to produce an EPSP that depolarizes the muscle fiber by 25–60 mV, more than enough to exceed the voltage threshold for a muscle fiber action potential. In the CNS, the EPSP resulting from a single presynaptic action potential is typically a small fraction of a millivolt in amplitude. The major cause of this difference is that a neuromuscular junction is large (see Figure 13.10A,B) and contains many active zones, so the axon terminal releases many vesicles of neurotransmitter per action potential. In contrast, at synapses in the CNS, each presynaptic action potential leads to the release of only one or a few vesicles. The smaller the quantity of neurotransmitter released, the fewer the receptors activated and the smaller the synaptic current generated.

The difference in size and amount of neurotransmitter release at neuromuscular and CNS neuronal synapses is an adaptation

Because both Na^+ and K^+ contribute to the synaptic current, they drive the membrane toward a potential near zero. Suppose for a minute that neurotransmitter molecules opened ion channels that allowed only Na^+ ions to flow into the cell. If this were so, then the EPSP would always displace V_m (membrane potential) toward E_{Na}. This is not the case, however. If a muscle fiber is voltage clamped to different values and the synaptic current is measured, the current always drives the membrane toward a value near zero, intermediate between E_{Na} and E_K (rather than toward E_{Na}). This value is the **reversal potential** of the EPSP (E_{EPSP}), the value of V_m beyond which an EPSP would reverse its polarity. Typical values of E_{EPSP} are between 0 and −15 mV in different cells.

EPSPs between neurons resemble neuromuscular EPSPs but are smaller

Fast EPSPs between neurons in the CNS are generally similar to those at the neuromuscular junction, but they also exhibit some differences (**FIGURE 13.10**). They are similar in waveform, with a fast rise of 1–2 ms and an exponential return typically lasting 10–20 ms.

[1] Neuromuscular EPSPs are sometimes called end-plate potentials (EPPs), named after the motor end-plate.

that largely determines the functional properties of the synapses. Recall that the vertebrate skeletal neuromuscular junction is a relay synapse, at which each presynaptic action potential produces a postsynaptic action potential. In contrast, CNS synapses are usually integrating synapses, in which each of many presynaptic neurons has only a rather small effect on the postsynaptic neuron, and excitatory input from 20–50 neurons may have to be summated to depolarize the integrating neuron past its voltage threshold (see Figure 13.7).

Fast IPSPs can result from an increase in permeability to chloride

What about IPSPs? IPSPs are the major mechanism of synaptic inhibition in CNSs. The waveform of a fast IPSP resembles that of an EPSP but is typically hyperpolarizing rather than depolarizing. Most fast IPSPs result from an increase in permeability to Cl⁻ ions and are mediated by one of two neurotransmitters: gamma-aminobutyric acid (GABA) or glycine. We will consider IPSPs mediated by GABA as our example because most inhibitory synapses in mammalian brains use GABA as their neurotransmitter (see Figure 13.10D).

The common mechanism for GABA-mediated synaptic inhibition is as follows: As with other synapses, an action potential in the presynaptic, GABA-containing neuron causes vesicle fusion and GABA release. GABA diffuses to the postsynaptic membrane and binds to GABA receptors that open to allow selective permeability to Cl⁻ ions. In most mammalian neurons, Cl⁻ is pumped out of the cell, so E_{Cl} is at a hyperpolarized value relative to E_m. Then when the permeability to Cl⁻ increases, Cl⁻ diffuses in, down its concentration gradient, hyperpolarizing the cell by driving E_m toward E_{Cl} (see Figure 13.10D). In other neurons, Cl⁻ is not pumped, so E_{Cl} and E_m are equal. Nevertheless, inhibitory synapses that increase permeability to Cl⁻ are still effective in inhibition because they "lock" the membrane potential at a value more hyperpolarized than the threshold voltage.

Excitatory and inhibitory synapses in the mammalian CNS have different characteristic appearances in electron micrographs (as shown schematically in Figure 13.10C,D). Type 1 synapses are usually excitatory and have a relatively wide synaptic cleft with a relatively large area of prominent density under the postsynaptic membrane. Synaptic vesicles are round. Type 2 synapses are usually inhibitory and have a narrower cleft with smaller areas of membrane density that are more symmetrically distributed at the pre- and postsynaptic membranes. The vesicles appear flattened after aldehyde fixation for electron microscopy. This classification can be valuable in identifying which synapses in a micrograph are inhibitory and which are excitatory. The classification is not completely reliable, however, because the correlation of structural type with function is imperfect, and not all synapses fit the two types.

Overall, then, we see that at fast chemical synapses, neurotransmitters bind receptors and increase permeability to ions, leading to a postsynaptic potential. Whether the postsynaptic potential is excitatory or inhibitory depends on what kinds of ions flow through the ion channels when the channels open. Different neurotransmitters act at different receptors; they may produce different effects (e.g., glutamate → EPSP, GABA → IPSP) or similar effects (e.g., glutamate → EPSP, ACh → EPSP).

Note that, in addition, a single neurotransmitter substance can have several different functions because it can act on different kinds of transmitter receptors. For example, ACh produces fast EPSPs at neuromuscular junctions (by acting on one kind of ACh receptor), but the same neurotransmitter produces IPSPs to inhibit the heart, by acting on a different kind of ACh receptor. We will consider this important principle in more depth later.

SUMMARY
Fast Chemical Synaptic Actions Are Exemplified by the Vertebrate Neuromuscular Junction

- At the vertebrate skeletal neuromuscular junction, the neurotransmitter is acetylcholine. When stimulated, the presynaptic axon terminal releases acetylcholine, which diffuses to postsynaptic receptors.

- Acetylcholine binding to its receptors opens ion channels to increase permeability to both Na⁺ and K⁺ ions. The resulting Na⁺ and K⁺ currents drive the membrane toward a value (E_{EPSP}) that is more depolarized than the threshold of the muscle fiber. At the neuromuscular junction, the amplitude of the EPSP is sufficient to exceed threshold and triggers a muscle fiber action potential.

- The EPSP itself is a nonregenerative, nonpropagated local response because the neurotransmitter-dependent permeability changes are not voltage-dependent.

- Fast excitatory synapses in CNSs work by mechanisms similar to those at neuromuscular junctions. Neurotransmitter-gated ion channels increase membrane permeability to Na⁺ and K⁺ ions to produce depolarizing EPSPs.

- Neuronal EPSPs are much smaller than neuromuscular EPSPs because at neural synapses the postsynaptic membrane encompasses a small area that has a small number of receptor molecules, and the presynaptic axon releases less neurotransmitter, activating fewer postsynaptic receptors.

- Fast synaptic inhibition results from the opening of ion channels to increase permeability to chloride. E_{Cl} is commonly at a hyperpolarized value relative to the resting potential, leading to a hyperpolarizing IPSP.

- CNS excitatory and inhibitory synapses often have characteristic differences in their appearance in electron micrographs of the vertebrate CNS.

Presynaptic Neurons Release Neurotransmitter Molecules in Quantal Packets

Neurotransmitter molecules are synthesized and stored in the presynaptic terminal and are released by presynaptic impulses. Much of our knowledge of the mechanisms of neurotransmitter release comes from the vertebrate neuromuscular junction, and because acetylcholine is the neurotransmitter at this synapse, we will consider primarily cholinergic (acetylcholine-mediated) transmission here. However, because mechanisms of neurotransmitter release are thought to be similar for different neurotransmitters, much of this information is general.

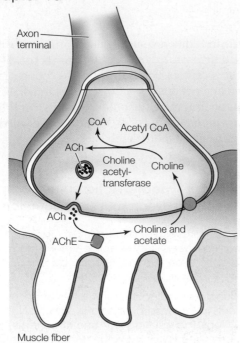

FIGURE 13.11 Acetylcholine synthesis and breakdown Acetylcholine is synthesized in the cytoplasm of the presynaptic terminal and then taken up into synaptic vesicles. Many vesicles are stored away from active-zone release sites; vesicles move from the storage compartment to release sites and dock to the presynaptic membrane at active zones. Acetylcholine is released by exocytosis of vesicles; the enzyme acetylcholinesterase (AChE) in the synaptic cleft terminates its action, hydrolyzing it to choline and acetate. A high-affinity transporter reabsorbs choline into the presynaptic terminal.

Acetylcholine is synthesized and stored in the presynaptic terminal

Acetylcholine (ACh) is synthesized from choline and acetyl coenzyme A in the cytoplasm of the presynaptic terminal (**FIGURE 13.11**). The reaction is catalyzed by the specific enzyme choline acetyltransferase. After synthesis in the cytoplasm, ACh is taken up into vesicles by a specific vesicular transporter molecule. Like all other proteins, choline acetyltransferase is synthesized in the cell body and must be transported the length of the axon. Nevertheless, the availability of choline is the limiting factor in the rate of ACh synthesis. Choline is supplied via the blood and by choline uptake transporters at the presynaptic membrane, but as the available choline in the terminal is converted to ACh, synthesis slows and eventually ceases.

Acetylcholine in an axon terminal is stored in various compartments. Some vesicles are "docked" at release sites and constitute the rapidly releasable compartment of ACh. Other vesicles are anchored to the cytoskeleton distant from release sites in the interior of the terminal; these constitute the storage compartment of ACh.

Neurotransmitter release requires voltage-dependent Ca²⁺ influx

Under normal circumstances, neurotransmitter release requires both presynaptic depolarization and Ca^{2+} ions. The normal stimulus for neurotransmitter release is the depolarization of the presynaptic terminal by an action potential. Experiments indicate that the amount of neurotransmitter release increases with increasing depolarization over the physiological range. The release of ACh into the synaptic cleft depends on an influx of Ca^{2+} ions into the presynaptic terminal. If a neuromuscular junction is bathed in a saline solution free of Ca^{2+} ions, then a depolarization of the presynaptic membrane does not elicit an EPSP, because exocytosis does not occur.

All neurotransmitter release appears to be Ca^{2+}-dependent. Depolarization of the presynaptic membrane opens voltage-gated Ca^{2+} channels, allowing Ca^{2+} to enter and trigger neurotransmitter release. Because the intracellular concentration of Ca^{2+} is much lower than the extracellular concentration, there is a strong inward driving force on Ca^{2+} ions; Ca^{2+} entry locally increases the Ca^{2+} concentration near the channels. In preparations of the squid giant synapse (between second-order and third-order giant axons in the stellate ganglion; see Box 12.3), the presynaptic terminal is large enough to impale with a micropipette. Experimenters have injected the presynaptic terminal with aequorin, a protein that emits light in the presence of Ca^{2+}. Depolarization of the presynaptic terminal induces aequorin luminescence, experimentally confirming that presynaptic depolarization leads to local Ca^{2+} entry. In fact, *the only role of presynaptic depolarization is to trigger Ca^{2+} entry*. In the squid giant synapse, investigators can inject Ca^{2+} directly into the presynaptic terminal with a microelectrode. Pulsed Ca^{2+} injection triggers postsynaptic EPSPs, indicating that Ca^{2+} entry is sufficient to release neurotransmitter.

Neurotransmitter release is quantal and vesicular

Acetylcholine is not released from the presynaptic membrane of a neuromuscular junction a molecule at a time; rather, it is released in multimolecular "packets" called **quanta** (singular *quantum*), units of about 5000 molecules each. Experimental evidence has demonstrated that each quantum is the equivalent of the contents of one synaptic vesicle.

The original evidence for quantal release comes from the vertebrate neuromuscular junction. If we impale a muscle fiber near the neuromuscular junction with an intracellular microelectrode (**FIGURE 13.12**), we can record a series of small depolarizations in the absence of any stimulation. These depolarizations have the same shape as a neuromuscular EPSP but are about 1/50th the amplitude; thus they are termed spontaneous **miniature EPSPs** (**mEPSPs**).

Each mEPSP is the postsynaptic response to a quantum. Miniature EPSPs typically have amplitudes of about 0.4 mV and occur nearly randomly over time (see Figure 13.12B). They do not represent responses to individual molecules of acetylcholine, because several thousand ACh molecules are required to produce a depolarization of 0.4 mV. The presence of spontaneous mEPSPs indicates that in the absence of presynaptic stimulation, there is a spontaneous, low-frequency release of ACh quanta—that is, about 5000 molecules at a time.

Can we show that the release of ACh evoked by presynaptic depolarization is also quantal? The neuromuscular EPSP evoked by a presynaptic impulse has an amplitude of 20–40 mV and would require the nearly simultaneous discharge of 100–300 quanta. This number is too large to determine whether the release is quantal or not, because the difference between the response to, say, 200 and 201 quanta is not detectable.

(A) Experimental setup

Axon of
motor neuron

Recording
electrode

V

(B) Spontaneous miniature EPSP

mEPSP

0.5 mV [

300 ms

Occurrences

40

30

20

10

0

0   0.2   0.4   0.6   0.8
Average amplitude (mV) of
spontaneous mEPSPs

FIGURE 13.12 Spontaneous mEPSPs (A) An electrode is positioned close to a neuromuscular synapse for recording in the absence of any stimulation. (B) The recording contains small (miniature) EPSPs at an average frequency of 1–2 per second. The average amplitude of spontaneous mEPSPs is 0.4 mV (histogram). Spontaneous mEPSPs show that the nerve terminal has a background low level of transmitter release in the absence of stimulation. The transmitter is released in multimolecular quantal packets, each quantum eliciting a mEPSP at the postsynaptic membrane.

Researchers solve this problem by *decreasing the number of quanta* released with an applied stimulus, so that only 1 or 2 quanta are released at a time into the synaptic cleft, rather than hundreds. To achieve this, they lower the Ca^{2+} concentration in the bath surrounding the muscle fiber, and they raise the bath concentration of Mg^{2+}, an ion that competitively inhibits the action of the Ca^{2+}. Are the resultant *evoked mEPSPs* quantal? As **FIGURE 13.13** shows, evoked mEPSPs fall into amplitude classes that are multiples of the amplitudes of spontaneous mEPSPs. That is, each presynaptic stimulus evokes the release of 0, 1, 2, or 3 quanta (but never 1.5 quanta).

The experiment in Figure 13.13 illustrates that neurotransmitter release evoked by a presynaptic impulse is quantal. Similar results occur with other types of synapses, and in some of these cases quantal release can be demonstrated without manipulating extracellular ion concentrations. It is thought, then, that quantal release is the general rule for all chemical synapses.

Research has shown that cholinergic vesicles contain 10^3–10^4 molecules of acetylcholine, and that neurotransmitter is released in quanta of 10^3–10^4 molecules; thus the conclusion seems inescapable

that a quantum corresponds to a synaptic vesicle. The synaptic vesicles fuse with the presynaptic membrane and discharge their content of neurotransmitter by exocytosis. The fusion of vesicles with the presynaptic membrane is sporadic in the resting terminal, producing spontaneous mEPSPs. Depolarization of the terminal by the presynaptic action potential greatly increases the *probability* of release of each of many vesicles, so that at a neuromuscular junction 150–300 synaptic vesicles discharge in a millisecond and produce a neuromuscular EPSP.

Synaptic vesicles are cycled at nerve terminals in distinct steps

The idea that quantal release of neurotransmitter corresponds to vesicular exocytosis is termed the *vesicular release hypothesis*, which is now widely accepted as explaining how neurotransmitter is released at all chemical synapses. The vesicular release hypothesis suggests that fusing exocytotic vesicles should be seen in electron micrographs, and that vesicular membranes must be recycled (to prevent vesicle depletion and expansion of the postsynaptic mem-

(A) Experimental setup

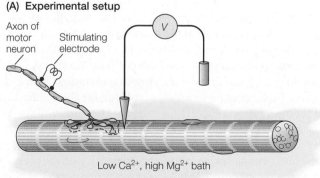

Axon of
motor
neuron

Stimulating
electrode

V

Low Ca^{2+}, high Mg^{2+} bath

(B)

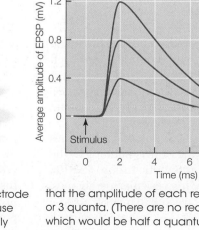

Average amplitude of EPSP (mV)

1.2

0.8

0.4

0

Stimulus

Quanta released per stimulus

3

2

1

0

0   2   4   6   8   10
Time (ms)

FIGURE 13.13 Evoked miniature EPSPs (A) A stimulating electrode evokes presynaptic action potentials in the motor axon, but because of the low Ca^{2+}, high Mg^{2+} bath, each action potential releases only a few quanta of neurotransmitter. (B) The amplitudes of the resultant EPSPs cluster around the size classes shown in the graph, indicating

that the amplitude of each response results from the release of 0, 1, 2, or 3 quanta. (There are no recorded responses of 0.2 mV, for example, which would be half a quantum.) Therefore stimulus-evoked release of neurotransmitter, like spontaneous release, is quantal.

(A) Overview of vesicle recycling

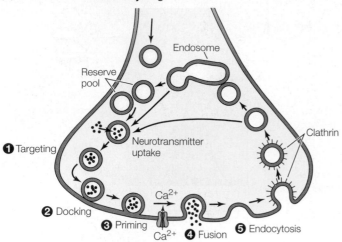

❶ Targeting

❷ Docking

❸ Priming

❹ Fusion

❺ Endocytosis

(B) Retrieval of the vesicular membrane

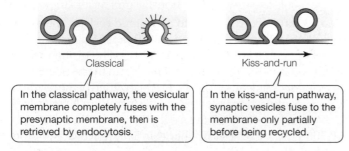

Classical

Kiss-and-run

In the classical pathway, the vesicular membrane completely fuses with the presynaptic membrane, then is retrieved by endocytosis.

In the kiss-and-run pathway, synaptic vesicles fuse to the membrane only partially before being recycled.

FIGURE 13.14 Synaptic vesicle recycling at the neuromuscular junction (A) Synaptic vesicles cycle though several steps: ❶ mobilization (also called targeting); ❷ docking; ❸ priming; ❹ exocytotic fusion; and ❺ endocytosis. Endocytotic vesicles may have to fuse with an inner membrane compartment, the endosome, before budding off as synaptic vesicles again. (B) After synaptic vesicles fuse with the presynaptic membrane and undergo exocytosis, at least two different sorts of pathways can retrieve the vesicular membrane. In the *classical* pathway, after complete fusion with the presynaptic membrane, the vesicular membrane is retrieved by endocytosis of clathrin-coated pits into endocytotic vesicles. This process selectively aggregates specific vesicle components and occurs away from the active zone. In the *kiss-and-run* pathway, synaptic vesicles fuse transiently to the presynaptic membrane. They discharge their transmitter at a fusion pore and then are retrieved without having been fully integrated into the presynaptic membrane.

brane with release). Exocytosis occurs rapidly, and it is difficult to show it directly in routinely fixed electron micrographs. However, electron micrographs of neuromuscular junctions stimulated at modest rates can show a temporary depletion of synaptic vesicles, but within minutes the vesicular membranes are pinched off and returned to the inside of the terminal by endocytosis.

The recycling of vesicular membranes allows synaptic vesicles to re-form and prevents the cell membrane from expanding by the addition of vesicular membrane from exocytotic fusion. **FIGURE 13.14** summarizes the processes of recycling vesicular membranes. Synaptic vesicles first are mobilized or targeted to move to release sites (active zones; step ❶ in Figure 13.14A), where they dock and are primed until Ca^{2+} ions trigger fusion of the vesicular and presynaptic membranes (steps ❷–❹).

There appear to be at least two modes of fusion and subsequent retrieval of vesicular membranes: classical and kiss-and-run (see Figure 13.14B). In *classical* exocytosis, the vesicular membrane merges with the terminal membrane and a new vesicle is later retrieved by endocytosis, a pinching off from the terminal membrane (step ❺ in Figure 13.14A). Classical endocytosis selectively retrieves vesicular membranes with the aid of two proteins: *clathrin* and *dynamin*. In *kiss-and-run* fusion, the docked vesicle opens a fusion pore to release transmitter into the synaptic cleft, without completely becoming integrated into the terminal membrane; the vesicle is then reinternalized without requiring clathrin-mediated pinching off. The kiss-and-run pathway is faster and (paradoxically) may predominate at lower rates of neurotransmitter release, whereas the classical exocytotic–endocytotic pathway is slower and may predominate at higher rates of release. Controversy remains, however, about how widespread the kiss-and-run mechanism is.

Several proteins play roles in vesicular release and recycling

The molecular mechanism of the exocytotic release of neurotransmitters is a topic of active investigation. Researchers have identified the proteins of vesicular membranes and other proteins with which they interact. Interestingly, these proteins are similar to those of exocytotic release in other cells, including yeast cells. These similarities suggest that a common mechanism of exocytotic secretion evolved in early eukaryotic cells, and although neurotransmitter release is much faster than other exocytotic secretion, the basic mechanism has been largely conserved.

Various proteins play roles in the several stages of vesicular release and recycling (**FIGURE 13.15**). The first stage is vesicular *mobilization* or *targeting*. At any time, many of the vesicles in a presynaptic terminal are located away from active zones in the storage compartment. Before these vesicles can release their contents, they must be mobilized to move up to release sites at the active zones. At rest, the protein *synapsin* attaches vesicles to the actin cytoskeleton, and in mobilization this attachment is released when synapsin is phosphorylated, permitting the vesicle to migrate to an active zone. A vesicle may *attach* or *tether* reversibly to the terminal membrane, followed by *docking* (which is irreversible) (see Figure 13.15A).

Docking involves the interaction of proteins called *SNAREs*; vesicular v-SNAREs intertwine with terminal-membrane t-SNAREs to hold the docked vesicle at the active zone. A major v-SNARE is called *vesicle-associated membrane protein* (*VAMP*, also called *synaptobrevin*). It connects with the t-SNAREs *syntaxin* and *SNAP-25*. Docked vesicles are primed by proteins, including Munc18 and complexin, readying the vesicles for rapid release (see Figure 13.15B).

Depolarization triggers Ca^{2+} entry through voltage-gated Ca^{2+} channels, and the Ca^{2+} ions bind to the vesicle protein *synaptotagmin*, triggering formation of a fusion pore that may lead to complete exocytotic fusion of the docked vesicle and the presynaptic membrane (see Figure 13.15C). Another protein, *rab3*, is active in vesicular mobilization and recycling, and may inhibit excessive fusion and release. Cytoplasmic proteins may disassemble the SNARE complex after fusion. The protein dynamin promotes the pinching off of vesicles, through an ATP-dependent mechanism, and as noted above, vesicular endocytosis is associated with the protein clathrin.

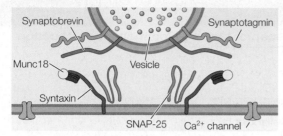

(A) Docked vesicle

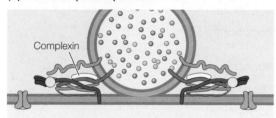

(B) Vesicle in primed position

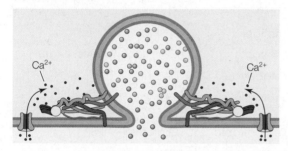

(C) Vesicle fusion

FIGURE 13.15 Vesicular docking and fusion release neurotransmitters Several proteins function in transmitter release from synaptic vesicles. SNARE proteins include synaptobrevin (a v-SNARE, vesicle-associated protein); and the t-SNAREs syntaxin and SNAP-25, which are attached to the terminal membrane. (A) A vesicle docks at the active zone, allowing its v-SNAREs to interact reversibly with t-SNAREs. Syntaxin is held in a folded, inactive conformation by Munc18. (B) The vesicle is primed by the formation of a SNARE complex. Syntaxin is open and the complex of v-SNAREs and t-SNAREs holds the vesicle at the release site. The cytoplasmic protein complexin stabilizes the complex. (C) The vesicle fuses. Ca^{2+} entry through voltage-gated Ca^{2+} channels triggers fusion by the binding of Ca^{2+} to the vesicular protein synaptotagmin. The Ca^{2+}–synaptotagmin complex binds to the SNARE complex, displacing complexin and inducing pore formation.

SUMMARY
Presynaptic Neurons Release Neurotransmitter Molecules in Quantal Packets

■ Small-molecule neurotransmitters are synthesized predominantly at axon terminals and are transported into synaptic vesicles.

■ Neurotransmitters are released by presynaptic depolarization, which opens voltage-gated Ca^{2+} channels at active zones. Calcium ions trigger neurotransmitter release.

■ Neurotransmitter is released in quantal packets, several thousand molecules at a time. Each quantum corresponds to a synaptic vesicle.

■ Synaptic vesicles fuse with the presynaptic membrane to release their transmitter contents by exocytosis. Vesicular membranes are retrieved, refilled with neurotransmitter, and recycled.

■ Specific proteins associated with synaptic vesicles play different roles in vesicular targeting, docking, fusion, and retrieval.

Neurotransmitters Are of Two General Kinds

Several dozen chemical compounds have been identified as possible neurotransmitters, and the list of such compounds continues to increase. There are two major kinds of neurotransmitters: *small-molecule neurotransmitters* (mostly amines and amino acids) and *neuropeptides* (**TABLE 13.2**). Some neurotransmitters are well known; others are still being discovered.

Our knowledge of different neurotransmitter systems bears little relation to their physiological prevalence. *Cholinergic* synapses (defined previously as synapses in which the neurotransmitter is acetylcholine) are best known because the neuromuscular junction is an easily studied model. Synapses in which the neurotransmitter is norepinephrine, or noradrenaline (termed *noradrenergic* or *adrenergic* synapses), are the next best known, because there are peripheral noradrenergic synapses in the sympathetic nervous system. It is estimated, however, that fewer than 10% of the synapses in the vertebrate CNS are cholinergic and fewer than 1% employ catecholamines such as norepinephrine. Neurotransmitter identity is not always known, but amino acids appear to be the most abundant and widespread CNS neurotransmitters. This section discusses selected aspects of neurotransmitter function; it is beyond our scope to survey the subject comprehensively. The references at the end of the chapter and the back of the book (see Additional References) provide more extensive information.

Neurons have one or more characteristic neurotransmitters

Neurons are metabolically specialized to synthesize and release a particular neurotransmitter or combination of neurotransmitters. A cholinergic neuron, then, expresses the genes for proteins that synthesize and transport acetylcholine. A neuron, however, may contain and release more than one kind of neurotransmitter; such multiple neurotransmitters released by single neurons are termed **cotransmitters**. Many neurons, for example, contain and apparently release a small-molecule neurotransmitter and one or more neuropeptides.

Even a single neurotransmitter may produce a variety of postsynaptic effects, as we will discuss shortly. Moreover, a postsynaptic neuron typically receives synapses from many kinds of presynaptic neurons, and each kind may release a different neurotransmitter. Thus although it is relatively sound to characterize a neuron in terms of the neurotransmitter it releases (e.g., as a cholinergic neuron), it is usually unsound to characterize it as responding to a particular

TABLE 13.2 Some neurotransmitters, key synthetic enzymes, and receptors of vertebrate central nervous systems[a]

Neurotransmitter	Enzyme	Receptor	Receptor type	Common mode of action
Amines				
Acetylcholine	Choline acetyltransferase	Nicotinic	Ionotropic	EPSP
		Muscarinic M_1–M_5	Metabotropic	G protein → IPSP
Dopamine	Tyrosine hydroxylase (TH)	d_1	Metabotropic	
		d_2	Metabotropic	
		d_3	Metabotropic	
Norepinephrine	TH and dopamine β-hydroxylase	$\alpha_{1,2,3}$	Metabotropic	
		$\beta_{1,2}$	Metabotropic	
Serotonin	Tryptophan hydroxylase	5-HT_1	Metabotropic	
		5-HT_2	Metabotropic	
		5-HT_3	Ionotropic	
Amino acids				
Glutamate	(General metabolism)	AMPA	Ionotropic	EPSP
		NMDA	Ionotropic	Ca^{2+} second messenger
		Metabotropic	Metabotropic	DAG/IP_3
GABA	Glutamic acid decarboxylase	$GABA_A$	Ionotropic	IPSP
		$GABA_B$	Metabotropic	G protein → IPSP
Glycine	(General metabolism)	GlyR	Ionotropic	IPSP
Peptides	(Protein synthesis)	Various	Metabotropic	G protein–coupled (some tyrosine kinase)

[a]This table is not exhaustive; there are more transmitters, and more receptors for each transmitter.

neurotransmitter (e.g., as a cholinoceptive neuron), because any neuron may respond to several different neurotransmitters.

An agent is identified as a neurotransmitter if it meets several criteria

There is general agreement on the experimental criteria for identifying the neurotransmitter at a synapse:

- The candidate neurotransmitter must be *present in the presynaptic terminal*, along with its synthetic machinery (enzymes, precursors, transporters).
- The candidate neurotransmitter is *released upon presynaptic stimulation*, in amounts sufficient to exert a postsynaptic action.
- When the candidate neurotransmitter is added to extracellular fluid in moderate concentrations, it should *mimic the effects of presynaptic stimulation*. For example, it should induce the same changes in permeability to ions as the synaptic action does.
- A *mechanism for removal* of the candidate neurotransmitter should exist. This removal mechanism can be by enzymatic

inactivation or by reuptake into cells. (We should point out, however, that many peptide neurotransmitters may have no specific removal mechanism.)

- The *effects of drugs* on transmission at a synapse may help identify its neurotransmitter and receptors. For example, curare is an ACh receptor antagonist that blocks neuromuscular transmission, and it also blocks the effect of ACh experimentally applied through a micropipette.

Note that these criteria were developed at a time when only fast, direct synaptic transmission was known. Some criteria are more essential than others, and some may be modified as our views of synaptic processes expand.

It is difficult to demonstrate all of these criteria at a particular synapse, especially in the CNS. For example, the demonstration of release of a candidate neurotransmitter requires that it be either assayed in situ or collected in sufficient quantity to be assayed. It further requires demonstration that the released neurotransmitter came from the presynaptic terminal. Such demonstration is rarely possible in the brain, where neurotransmitter identification is often inferred solely from chemical evidence of neurotransmitter presence. Rigorous tests employing the full spectrum of criteria are important where possible, to prevent uncritical acceptance of every synaptically active agent as a presumed neurotransmitter. Because these criteria are so difficult to satisfy experimentally, we have a long list of *possible* and *probable* neurotransmitters and a shorter list of cases in which the neurotransmitter at a particular synapse is convincingly demonstrated.

Vertebrate neurotransmitters have several general modes of action

The principal synaptic neurotransmitters of vertebrates are summarized in Table 13.2. The table could be much longer, because there are other small-molecule neurotransmitters and perhaps 90 neuroactive peptides. It is difficult to generalize about neurotransmitter functions because all neurotransmitters work in different ways and mediate different functional effects at different synapses. Nevertheless, here are some very broad generalizations about functional classes of neurotransmitters and synapses in the CNS of vertebrates:

- Most synapses (numerically) in the CNS use amino acid neurotransmitters. Most fast EPSPs result from glutamate; most fast IPSPs result from GABA or glycine.

- Biogenic amines (acetylcholine, norepinephrine, dopamine, serotonin) are found in relatively few neurons, but these neurons have widely projecting endings that appear to release transmitter over broader areas than discrete synapses (termed *volume transmission*). Many receptors for these neurotransmitters have slow actions that modulate neuronal activities, rather than mediating fast excitation or inhibition.

- Peptides are present in substantial numbers of CNS neurons. A neuroactive peptide may be co-released with one or more small-molecule neurotransmitters and may function as a cotransmitter with slow synaptic effects.

Rather than surveying the metabolism and action of specific neurotransmitters, let's discuss a few neurotransmitter-related concepts that are important for understanding general synaptic functions.

MULTIPLE RECEPTORS Many neurotransmitters can mediate different postsynaptic actions at different postsynaptic cells. For example, acetylcholine excites skeletal muscle via EPSPs but inhibits vertebrate heart muscle via hyperpolarizing IPSPs. These effects involve different permeability changes in the different postsynaptic cells (P_{Na} and P_K in skeletal muscle; P_K in heart muscle). Completely different postsynaptic receptors mediate these different effects: one a ligand-gated channel and one a *G protein–coupled receptor*. The two different kinds of acetylcholine receptors were first characterized many years ago by their pharmacology (i.e., the effects of drugs on them). The ACh receptor of skeletal muscle is stimulated by nicotine and hence is termed **nicotinic**. As noted on the previous page, a nicotinic ACh receptor is blocked by curare. The ACh receptor of heart muscle is stimulated by muscarine and hence is termed **muscarinic**; it is blocked by atropine. Muscarinic ACh receptors are the end-receptors of parasympathetic signals in the autonomic nervous system (see Chapter 15).

Most neurotransmitters affect more than one kind of neurotransmitter receptor, and these may be of different classes (e.g., ligand-gated channels and G protein–coupled receptors). The different receptor classes that respond to a particular neurotransmitter were originally characterized by pharmacology, as with nicotinic and muscarinic acetylcholine receptors. More recently, researchers have cloned the genes for neurotransmitter receptors, enabling them to associate dozens of receptors with a particular neurotransmitter. Some well-studied neurotransmitter receptors are listed in Table 13.2.

TERMINATION OF NEUROTRANSMITTER ACTION: ENZYMES AND REUPTAKE Neurotransmitters are generally active for only a short time. Neurotransmitter release, diffusion, and receptor binding occur within a few milliseconds. For normal synaptic function, neurotransmitter molecules must be cleared from the synaptic cleft. The temporal and spatial effects of neurotransmitter action are limited in two ways: by *enzymatic destruction* of the neurotransmitter molecules and by *reuptake* (using active transport to retrieve the neurotransmitter or its products).

At the neuromuscular junction, ACh is enzymatically digested by the enzyme acetylcholinesterase (AChE), located in the synaptic cleft (see Figure 13.11). AChE acts very rapidly; the entire sequence of ACh release, diffusion, binding to ACh receptors, and digestion requires about 5 ms. Some ACh molecules are destroyed even before they can bind to receptors. The products of ACh breakdown are choline and acetate. Choline is transported back into the presynaptic terminal by a specific high-affinity transporter in the terminal membrane. Thus the action of AChE both terminates the postsynaptic effects of ACh and (via the transporter) provides choline, the rate-limiting substrate for resynthesis of ACh in the presynaptic terminal.

The termination of neurotransmitter action by a localized enzyme is not a feature of all chemical synapses. For many neurotransmitters (catecholamines, amino acids), the termination and reuptake processes are combined. For example, the neurotransmitter norepinephrine (noradrenaline) is itself actively transported back into the presynaptic cell by a high-affinity transporter. Although enzymes that catabolize norepinephrine are present, they act more slowly than the reuptake system. Thus the synaptic action of norepinephrine is terminated by reuptake of the neurotransmitter rather than by enzymatic destruction. Glial cells can also actively take up neurotransmitters. Reuptake transporters are linked to Na^+, using Na^+ entry down its electrochemical gradient to drive transmitter uptake against a concentration gradient.

The "recycling" of neurotransmitter molecules—with or without enzymatic conversion—is a process distinct from the recycling of vesicular membranes discussed earlier. Whereas vesicular recycling involves endocytotic pinching off of *organelles* (multimolecular pieces of vesicular membrane), in neurotransmitter recycling a transporter actively transports *molecules* of neurotransmitter or neurotransmitter metabolite across the membrane against a concentration gradient. The two processes occur independently of each other.

PEPTIDE NEUROTRANSMITTERS DIFFER FROM SMALL-MOLECULE NEUROTRANSMITTERS IN SYNTHESIS, RELEASE, AND TERMINATION Peptide neurotransmitters are chains of amino acids, typically 3–55 amino acids long. Unlike small-molecule neurotransmitters, which are synthesized in the axon terminals, peptide neurotransmitters are synthesized in the cell body of a neuron and must be transported down the axon for release. All peptides are synthesized on ribosomes of the rough endoplasmic reticulum; the neuronal axon lacks ribosomes and performs no protein synthesis.

TABLE 13.3 Comparison of small-molecule and peptide neurotransmitters

	Small-molecule neurotransmitters	Neuropeptides
Synthesis site	Axonal terminal or varicosity	Nucleus/ER as propeptide
Vesicles	Small clear vesicles	Large dense-cored vesicles
Release	Low-frequency stimulation	High-frequency stimulation
Inactivation	Reuptake or specific enzymes	Extracellular peptidases

Peptide neurotransmitters are synthesized as part of a larger precursor polypeptide, called a *propeptide*. A propeptide typically contains several copies of the peptide neurotransmitter within its amino acid sequence. After synthesis, the propeptide molecules are packed into secretory vesicles called *large dense-cored vesicles*, which are distinct from the smaller vesicles of small-molecule transmitters. The large dense-cored vesicles are transported down the axon, and in the vesicle the propeptide is cleaved enzymatically into smaller pieces that may include several copies of the active peptide.

Release of peptide neurotransmitters is by processes similar to the release of small-molecule neurotransmitters, but with some differences. Exocytosis of large dense-cored vesicles is not at active zones, and thus is not as near to the sites of Ca^{2+} entry. Probably for this reason, the release of peptide neurotransmitters requires a higher frequency of presynaptic action potentials than does the release of small-molecule transmitters, presumably allowing more Ca^{2+} entry and buildup of Ca^{2+} concentration over a larger area.

Peptide neurotransmitters are not retrieved once they have been released; eventually they are digested by nonspecific extracellular peptidases. Peptide neurotransmitters may become depleted as a result of the "long supply chain" for their synthesis, as well as the absence of their recycling. In contrast, for small-molecule neurotransmitters, resupply can usually keep up with release because the neurotransmitter can be locally retrieved or rapidly resynthesized in the axon terminal.

TABLE 13.3 compares several key differences between small-molecule and peptide neurotransmitters.

Neurotransmitter systems have been conserved in evolution

Most neurotransmitters that act at vertebrate synapses are also present in the nervous systems of the major invertebrate groups. For example, evidence supports neurotransmitter roles for acetylcholine, GABA, glutamate, dopamine, and serotonin (also called *5-hydroxytryptamine*, or *5-HT*) among coelenterates, nematodes, annelids, arthropods, and molluscs. Genomic studies reinforce the evolutionary similarities of neurotransmitter systems in different phyla; for example, the genomes of the nematode worm *Caenorhabditis* and of *Drosophila* and other insects have been completely sequenced, so their entire collections of neurotransmitter enzymes and receptors are known. They have nicotinic ACh receptors, GABA receptors, excitatory glutamate receptors, and many G protein–coupled receptors related to their vertebrate counterparts.

The same neurotransmitters, however, may be employed in different roles in different phyla. For example, most evidence indicates that in arthropods, glutamate is the major excitatory neuromuscular transmitter and acetylcholine is the major sensory neurotransmitter, whereas their roles are reversed in the vertebrates.

These observations suggest that the evolution of neurotransmitter systems has been conservative; the same or related neurotransmitters and receptors are conserved across many phyla. Even the peptide neurotransmitters, which might be expected to show greater differences among phyla than the small-molecule amino acids and amines, are organized in protein families that indicate their evolutionary relationship across phyla.

SUMMARY
Neurotransmitters Are of Two General Kinds

- Neurotransmitters can be small molecules or peptides. Perhaps a dozen small-molecule neurotransmitters and several dozen peptide neurotransmitters have been identified.

- A neuron can be identified by its characteristic neurotransmitter, but a single neuron may produce and release more than one neurotransmitter.

- For any neurotransmitter there are several receptors. Different kinds of receptors for a transmitter may coexist in the same organism and the same neuron.

- Most fast synapses in CNSs employ glutamate for EPSPs and GABA or glycine for IPSPs.

- Many receptors for small-molecule neurotransmitters, and probably for all peptides, act metabotropically and mediate slow synaptic potentials and modulatory responses.

- Peptides are synthesized in the neuronal cell body and transported down the axon packed in vesicles, unlike small-molecule transmitters, which are synthesized locally in axon terminals.

- The synaptic action of small-molecule neurotransmitters is terminated by reuptake or by enzymatic destruction.

Postsynaptic Receptors for Fast Ionotropic Actions: Ligand-Gated Channels

As we noted earlier, physiologists classify the postsynaptic neurotransmitter receptors into two broad categories, depending on their mechanism of action. *Ionotropic receptors* produce their effects *directly*; the neurotransmitter binds to the active site of the receptor protein, which is a *ligand-gated channel* (see Figure 2.27A). The receptor–channel opens in response to the binding to allow ions to pass into and out of the postsynaptic cell. *Metabotropic receptors* are not channels; instead they produce their effects *indirectly*. Neurotransmitter molecules bind to and activate a metabotropic receptor, which in turn initiates a cascade of signal transduction messenger molecules that may eventually modulate an ion channel in the postsynaptic membrane. Because an ionotropic receptor is a single macromolecular unit, it works rapidly and produces fast PSPs. Because a metabotropic receptor produces a cascade of reactions of

TABLE 13.4 Ionotropic and metabotropic receptors: Structural, functional, and mechanistic differences

Characteristic	Ionotropic receptors	Metabotropic receptors
Receptor molecule	Ligand-gated channel receptor	G protein–coupled receptor
Molecular structure	Four or five subunits around an ion channel	Protein with seven transmembrane segments; no channel
Molecular action	Open ion channel	Activate G protein; metabolic cascade
Second messenger	No	Yes (usually)
Gating of ion channels	Direct	Indirect (or none)
Type of synaptic effect	Fast EPSP or IPSP	Slow PSPs; modulatory changes (in channel properties, cell metabolism, or gene expression)

separate proteins, its effects are slower and longer-lasting. **TABLE 13.4** compares some of the key structural, functional, and mechanistic differences between ionotropic and metabotropic receptors.

ACh receptors are ligand-gated channels that function as ionotropic receptors

The best-known examples of ionotropic receptors are the nicotinic acetylcholine (ACh) receptors (discussed earlier) that produce EPSPs at the vertebrate neuromuscular junction. Molecular biologists first isolated nicotinic ACh receptors in the electric organs of marine fish such as skates and rays (see Box 20.1), which consist almost entirely of modified neuromuscular junctions with dense concentrations of ACh receptors. These nicotinic ACh receptors are strikingly

similar to those of other vertebrate neuromuscular junctions. A nicotinic ACh receptor is a glycoprotein composed of five subunits, including two α subunits (**FIGURE 13.16A**). Each subunit has four helical, membrane-spanning hydrophobic segments: M1 through M4 (**FIGURE 13.16B,C**). The polar M2 domains face the interior of the protein to form the central ion channel. The different subunits of the ACh receptor share considerable sequence homology and presumably evolved from a common ancestor.

Each of the two α subunits has an ACh-binding site on its extracellular side (**FIGURE 13.16D,E**). When the receptor binds two ACh molecules, the five subunits change their conformations to open a central channel large enough for Na⁺ and K⁺ ions to pass through it. The channel binds ACh and remains open only for a

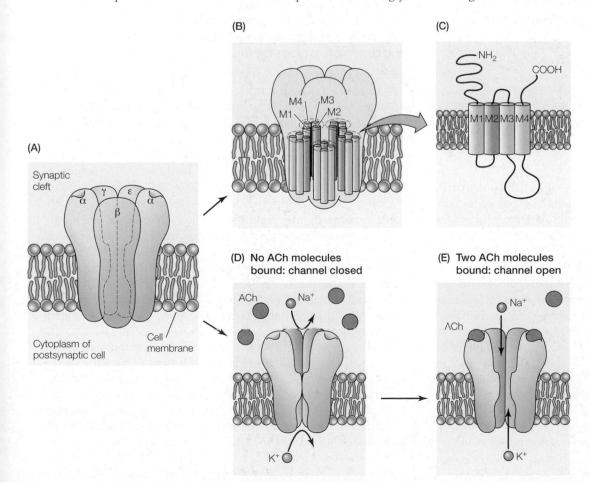

FIGURE 13.16 The molecular structure and function of a ligand-gated channel, the nicotinic acetylcholine receptor (A) The receptor has five protein subunits; the two α subunits each contain an ACh-binding site. (B) The five subunits surround a central ion channel, and each subunit has a similar structure, with four membrane-spanning segments (M1–M4). The M2 segments line the pore. (C) Each subunit is a single polypeptide chain. (D) When ACh is not bound to the receptor, its inner channel, which narrows in the region of the lipid bilayer, is closed. (E) The binding of an ACh molecule to each of the two α subunit–binding sites leads to a conformation change that opens the channel to a diameter of 0.6 nm, allowing Na⁺ and K⁺ ions to flow through.

(A) Patch clamp of ACh receptor–channels

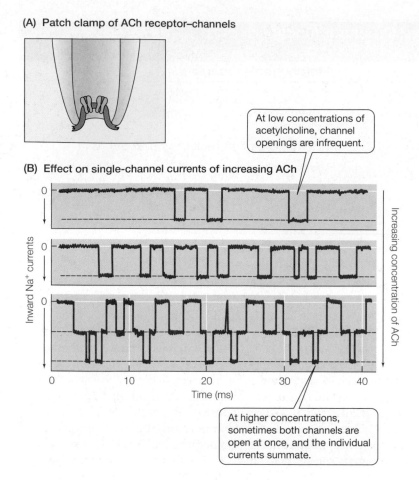

(B) Effect on single-channel currents of increasing ACh

At low concentrations of acetylcholine, channel openings are infrequent.

Increasing concentration of ACh

Inward Na⁺ currents

Time (ms)

At higher concentrations, sometimes both channels are open at once, and the individual currents summate.

FIGURE 13.17 Patch-clamp recordings of acetylcholine receptor–channel currents (A) A fine glass electrode is sealed to the membrane with suction, and the patch of membrane is pulled free of the rest of the cell. In this example, the patch contains two nicotinic acetylcholine receptors. Opening of a receptor–channel allows current to flow, as shown in the recordings in (B). (B) Each opening results in an inward current, shown as a downward deflection. The frequency of opening increases with increasing ACh concentration.

short time (about 1 ms). In fact, the channel may flicker open and closed during this brief interval. With prolonged exposure to ACh (e.g., if acetylcholinesterase is inhibited), the ACh receptor enters a third, *desensitized* state in which ACh is bound but the channel is closed. Desensitized receptors remain desensitized until the ACh molecules dissociate from the receptor.

The current flowing through a single ACh receptor, and the additive nature of the currents flowing through more than one channel, can be demonstrated by patch clamp. The patch-clamp technique (described in Chapter 12; page 323) records single-channel currents, as shown in **FIGURE 13.17**. By slowly increasing the ACh concentration inside the pipette (which in this inside-out patch is equivalent to the synaptic cleft), we control the probability that a channel will open. At a low ACh concentration, only one channel opens at a time (top recording in Figure 13.17B). Increasing the ACh concentration slightly causes each channel to open more frequently (center recording in Figure 13.17B), and increasing the ACh concen-

tration still further causes both channels to open simultaneously for intermittent periods (bottom recording in Figure 13.17B).

To summarize, the patch-clamp experiment demonstrates four characteristics of ligand-gated channel receptors and the ionic currents they produce once they are activated:

1. The opening of a ligand-gated channel is an all-or-none phenomenon.

2. The probability that a channel will open depends on the concentration of neurotransmitter at the receptor.

3. The net ionic current through the open channel provides that channel's contribution to a synaptic potential.

4. The currents through all open channels can be summated and constitute the synaptic current.

The properties of ACh receptors as ligand-gated channels can be compared to those of voltage-gated channels that underlie the action potential. The major difference is in the control of permeability changes of the molecules. In the voltage-gated Na⁺ channel, permeability depends directly on the membrane potential (see Figures 12.16 and 12.17). Permeability of the ACh receptor is essentially voltage-independent and depends instead on the binding of the neurotransmitter ACh. As noted previously, this difference in the control of permeability to ions of the two molecules results in the difference in properties of the potentials they produce: Action potentials are all-or-none and propagated; synaptic potentials are graded in amplitude and spread decrementally.

Many, but not all, ligand-gated channel receptors have evolved from a common ancestor

We have discussed nicotinic acetylcholine receptors in some detail, as examples of receptor proteins mediating direct, fast action of neurotransmitters. Studies have clarified the molecular structures of other neurotransmitter receptors that act as ligand-gated channels. Most of these structures are strikingly similar to the structure of the ACh receptors of neuromuscular junctions. The many ligand-gated channels that mediate PSPs in response to GABA, glycine, or serotonin (see Table 13.2) are composed of five subunits, each with four membrane-spanning segments, with considerable sequence homology to the ACh receptor. Therefore these receptors appear to have evolved from a common ancestor and are termed the **ligand-gated channel superfamily**, comparable to the voltage-gated channel superfamily discussed in Chapter 12 (see pages 326–327).

In contrast, ionotropic receptors for the excitatory amino acid glutamate appear evolutionarily unrelated to the ligand-gated channel superfamily. Ionotropic glutamate receptors each contain four relatively large subunits that have three membrane-spanning segments. The glutamate receptor subunits share little or no sequence homology with members of the ligand-gated channel superfamily, and doubtless evolved independently. Functions of glutamate receptors are discussed in the section "Synaptic Plasticity: Synapses Change Properties with Time and Activity" (see page 360).

SUMMARY

Postsynaptic Receptors for Fast Ionotropic Actions: Ligand-Gated Channels

■ The receptors that produce fast PSPs are ligand-gated channels. They are receptor–channels because the same molecule is both the receptor and the ion channel.

■ The nicotinic acetylcholine receptor of the neuromuscular junction is the model ligand-gated channel. It contains five homologous subunits that surround a central ion channel that opens to allow Na^+ and K^+ ions to flow across the membrane.

■ A ligand-gated channel opens briefly in response to binding two molecules of neurotransmitter, contributing to the synaptic current that produces a PSP.

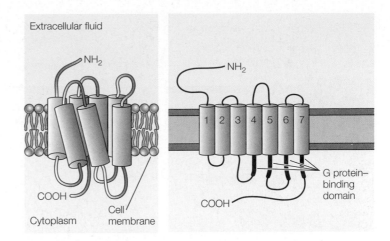

FIGURE 13.18 G protein–coupled receptors share a common structure A GPCR has seven membrane-spanning segments (TM1–7) (shown pictorially on the left and lined up on the right). Several of the intracellular loops can interact with G proteins. Neurotransmitter-binding sites are part of the membrane-spanning segments of the receptors for most small-molecule transmitters.

Postsynaptic Receptors for Slow, Metabotropic Actions: G Protein–Coupled Receptors

Our discussion of synaptic action so far has largely considered only fast synapses such as vertebrate neuromuscular junctions and synapses onto spinal motor neurons. In these synapses, neurotransmitter produces an EPSP or IPSP by means of a brief increase in permeabilities to ions. Until the 1980s such fast, direct synaptic transmission was thought to be *the* mechanism of synaptic action. Several lines of investigation led to a broader view of synaptic function that included much longer time courses and wider synaptic actions than those of fast PSPs. Researchers discovered (1) second messenger–mediated cell signaling (see Chapter 2; pages 65-68) in nonneural and neural cells, (2) slower synaptic actions such as synaptic potentials mediated by *decreased* permeability (to be discussed shortly), and (3) peptide neurotransmitters and cotransmitters that did not seem to produce classical fast PSPs.

Physiologists now understand that fast PSPs represent *one* kind of synaptic mechanism, not the only kind, and that synapses can also produce long-lasting metabolic effects. Metabotropic receptors can alter permeability to ions to change membrane potential indirectly (termed *indirect gating* of ion channels) but can also induce other metabolic changes that don't gate ion channels at all.

G protein–coupled receptors initiate signal transduction cascades

The major group of receptors that mediate metabotropic synaptic actions are **G protein–coupled receptors** (**GPCRs**), so called because they activate other membrane proteins termed G proteins. A GPCR is not an ion channel, and so is unlike a ligand-gated channel. All GPCRs have similar overall protein structures and have considerable sequence homology with each other. Thus GPCRs constitute an evolutionary superfamily, like the superfamilies of voltage-gated channels and ligand-gated channels. Because GPCRs have seven transmembrane segments (**FIGURE 13.18**), they are sometimes termed the *7-TM superfamily*. For many 7-TM neurotransmitter receptors, ligands bind near the extracellular ends of some transmembrane domains, whereas several cytoplasmic

domains (adjacent to TM5, TM6, TM7, and perhaps TM4) mediate G protein binding.

GPCRs are widespread initiators of signal transduction cascades, some activated by neurotransmitters and others by other signals. Receptors for peptide hormones (see Chapter 16) and some sensory receptor molecules (see Chapter 14) are also members of the GPCR superfamily. Clearly this mechanism of cellular response to extracellular signals arose early in evolutionary history and has been exploited for many functions.

Metabotropic receptors act via second messengers

Metabotropic receptors typically act to increase the concentration of an intracellular second messenger. Recall from Chapter 2 that a **second messenger** is an intracellular signaling molecule that carries the signal to the interior of the cell, altering some activity of the cell in response to activation of a surface membrane receptor (see Figure 2.30). The best-known second messenger is the cyclic nucleotide **3′-5′-cyclic adenosine monophosphate** (**cyclic AMP**, or **cAMP**).

Let's consider a specific example of how a neurotransmitter receptor acts via a second messenger to influence cellular metabolism (**FIGURE 13.19**). Consider a synapse at which the presynaptic neuron releases the neurotransmitter norepinephrine (noradrenaline). Norepinephrine acts on a GPCR, which, when activated by binding neurotransmitter, activates a membrane protein called a **G protein**. G proteins are so named because they bind guanosine nucleotides, which regulate their activity. Both the GPCR and the G protein can diffuse laterally in the fluid mosaic membrane of the postsynaptic neuron, so when the GPCR is activated, it can bump into the G protein and activate it (step ❶ in Figure 13.19). The G protein is bound to the inner leaflet of the cell membrane. It consists of α, β, and γ subunits. In their inactive form, the G-protein subunits are bound together, and the α subunit is bound to a molecule of *guanosine diphosphate* (GDP). When the G protein is activated by a

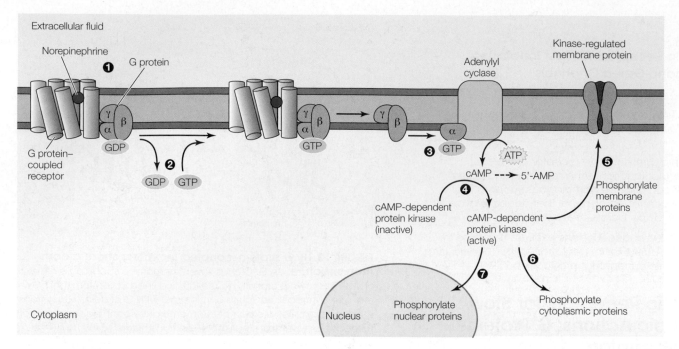

FIGURE 13.19 Metabotropic receptors: cAMP as a second messenger Some neurotransmitters act via GPCRs to alter cellular metabolism, rather than via ligand-gated channels. In this example, norepinephrine binds to a GPCR ❶ to activate a G protein ❷. The G protein has three subunits; when the protein is inactive, the α subunit binds GDP. The G protein is activated when it encounters an activated receptor; it exchanges the GDP for a GTP, and the α subunit dissociates from the joined β and γ subunits. The α subunit in turn activates the enzyme adenylyl cyclase ❸. Adenylyl cyclase catalyzes the conversion of ATP to the second messenger cAMP. Cyclic AMP activates a protein kinase ❹, which phosphorylates proteins to activate or inactivate them. Cyclic AMP–dependent protein kinase can phosphorylate membrane proteins ❺, which can indirectly gate them open or closed. The kinase can also phosphorylate cytoplasmic proteins (such as metabolic enzymes) ❻ and nuclear proteins that regulate gene expression ❼. Thus neurotransmitter binding can have widespread, long-lasting effects on postsynaptic cells.

GPCR, the α subunit releases its GDP and acquires a molecule of *guanosine triphosphate* (*GTP*) (step ❷); then the G protein dissociates from the GPCR, and the α subunit dissociates from the β and γ subunits, which remain joined to each other (step ❸).

Once it has been activated by the receptor, the G protein can activate another class of membrane protein termed an *intracellular effector*. In the case illustrated in Figure 13.19, the intracellular effector is an enzyme, adenylyl cyclase. Adenylyl cyclase (once activated by the active G protein) catalyzes the cytoplasmic conversion of ATP to cAMP. (Usually the dissociated α subunit is the active form of the G protein; the β and γ subunits remain together and play regulatory roles. In some cases, however, the linked β and γ subunits can activate an effector; see Figure 13.20.) The α subunit has GTPase activity, so eventually the GTP bound to it is degraded to GDP. When this occurs, the α subunit deactivates and reassociates with the regulatory β and γ subunits.

Intracellular second messengers such as cAMP can exert widespread metabolic effects, by activating a **protein kinase** (step ❹), an enzyme that phosphorylates proteins. An increase in intracellular cAMP concentration leads to increased activation of cAMP-dependent protein kinase. (cAMP action is eventually terminated by the enzyme phosphodiesterase, which converts the cAMP molecules to 5′-AMP.) Activated cAMP-dependent protein kinase phosphorylates proteins (steps ❺–❼), altering their structure and activity. Thus stimulation of a neurotransmitter-dependent adenylyl cyclase leads to protein phosphorylation.

In postsynaptic neurons, the proteins phosphorylated often include membrane ion channels; phosphorylation alters their gating

to change the membrane's permeability to ions and thereby change membrane potential (step ❺). This indirect gating of ion channels is one way in which metabotropic receptors function, but it is not the only way! Cyclic AMP–dependent protein kinase can also phosphorylate cytoplasmic proteins involved in control of cellular metabolism (step ❻), and even nuclear proteins that regulate gene expression (step ❼). Uncovering the fact that synaptic actions can control cellular metabolism and gene expression is one of the most important advances in our understanding of synaptic physiology in the last 30 years.

Cell signaling initiated by GPCRs underlies the action of many other controls of cellular activities, as discussed in Chapters 2 and 16. Both at metabotropic synapses and elsewhere, GPCRs can act via a variety of mechanisms in addition to the production of cAMP diagrammed in Figure 13.19. As one example, other GPCRs can activate a different G protein that *inhibits* adenylyl cyclase, decreasing rather than increasing the concentration of cAMP in the cell.

Other mechanisms of G protein–mediated activity

Some G proteins can activate ion channels directly without employing a second messenger. This direct G-protein action is present in cholinergic synapses on cardiac muscle tissue, which, unlike skeletal muscle, is inhibited by acetylcholine. The inhibitory action of ACh is mediated by muscarinic ACh receptors, which are GPCRs. In mammalian cardiac muscle fibers, the GTP-bound α subunit dissociates from the activated G protein, and the β and γ subunits bind directly to K⁺ channels, opening them to produce

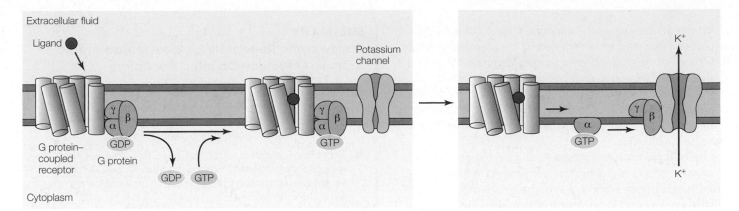

FIGURE 13.20 G proteins can themselves activate ion channels, without a second messenger In this example of acetylcholine-mediated inhibition of vertebrate heart muscle fibers, a muscarinic acetylcholine receptor (a GPCR) activates a G protein, which can itself gate a potassium (K⁺) channel open without involving a second messenger. In this case the joined β and γ subunits activate the channel, rather than the more common activation by an α subunit.

IPSPs (**FIGURE 13.20**). This mechanism demonstrates three significant features of synaptic function:

1. A particular neurotransmitter can mediate very different effects by activating different kinds of receptors.

2. G proteins can act on channels directly without an intervening second messenger (although this is not their usual mechanism).

3. Channels can be gated by signals other than voltage or direct neurotransmitter binding.

GPCRs can act via second-messenger systems other than cyclic AMP. The number of known systems is small, although different kinds of cell-signaling mechanisms are still being discovered. One other second messenger is *cyclic GMP* (*cyclic guanosine monophosphate*), the production of which is similar to that of its adenosine analog cAMP. That is, some neurotransmitter receptors act via a G protein to activate a guanylyl cyclase to produce cyclic GMP, which activates a cGMP-dependent protein kinase, leading to protein phosphorylation.

Another way GPCRs can act is via second-messenger systems involving products of membrane lipid metabolism and Ca²⁺ ions. Most lipids in membranes are phospholipids (see Chapter 2; page 37), in which one of the fatty acids in a neutral fat is replaced by a phosphate to which is bound a small organic residue. One common membrane phospholipid is phosphatidylinositol 4,5-bisphosphate (PIP₂). Neurotransmitters can act via GPCRs to lead to the production of second messengers from PIP₂ (**FIGURE 13.21**). For example, norepinephrine can activate a GPCR, activating a G protein so that its α subunit activates a membrane-bound enzyme, *phospholipase C*. Phospholipase C catalyzes the hydrolysis of PIP₂ into **inositol trisphosphate** (**IP₃**) and **diacylglycerol** (**DAG**).

Both IP₃ and DAG act as second messengers, via different paths that reflect their structures. IP₃, which had formed the polar head of the phospholipid, is a polar molecule that freely diffuses through the cytoplasm. Its major action is to release Ca²⁺ ions

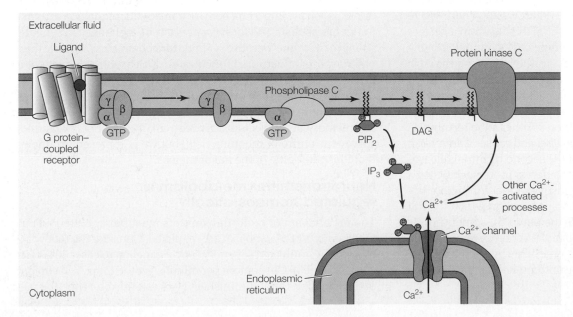

FIGURE 13.21 Diacylglycerol and inositol trisphosphate are other second messengers Another pathway of cellular regulation via GPCRs is activation (via a G protein) of the enzyme phospholipase C, which cleaves the membrane phospholipid phosphatidylinositol 4,5-bisphosphate (PIP₂) into two second messengers: diacylglycerol (DAG) and inositol trisphosphate (IP₃). DAG activates a protein kinase (kinase C), and IP₃ mobilizes Ca²⁺ ions from intracellular stores. Both pathways play important regulatory roles in cell metabolism.

from intracellular stores such as the endoplasmic reticulum. DAG, in contrast, is nonpolar and is constrained to the membrane lipid bilayer. It diffuses through the bilayer to activate *protein kinase C*. Thus DAG as a second messenger leads to protein phosphorylation, as does cAMP. The Ca^{2+} released by action of IP_3 can enhance the activation of protein kinase C, as well as activating **calmodulin** (**CaM**), which with Ca^{2+} activates *calcium/calmodulin-dependent protein kinase*, or *CaM kinase*.

The ability of neurotransmitters to stimulate the production of second messengers is important, in part because it provides one possible mechanism whereby synaptic transmission can mediate relatively slow and long-lasting effects. The direct action of neurotransmitters on ion channels in membranes has a time course of milliseconds. The time course of effects mediated by second messengers may be less than a second (e.g., in slow PSPs mediated by decreased permeabilities to ions, discussed later in this section). At the other extreme, second-messenger actions may underlie synaptic changes involved in learning and memory, with a time course of days or years, as we will see.

G protein–coupled receptors mediate permeability-decrease synaptic potentials and presynaptic inhibition

GPCRs mediate a variety of synaptic effects, as we have indicated. They may lead to indirect opening of ion channels, but they can also act to close ion channels or act on postsynaptic targets that are not ion channels. Here we consider two of these alternative possibilities: permeability-decrease synaptic potentials that are mediated by ion-channel closing, and presynaptic inhibition.

Permeability-decrease synaptic potentials were first studied at synapses in vertebrate sympathetic ganglia. It should be clear from the Goldman equation (Equation 12.6) that any changes in a postsynaptic cell's permeability to ions—decreases as well as increases—change the membrane potential. Thus a neurotransmitter receptor that leads to a *decrease* in permeability to an ion species also produces a synaptic potential. Postsynaptic neurons in bullfrog sympathetic ganglia characteristically have two distinct phases of synaptic response to stimulation of preganglionic axons: a fast EPSP lasting about 30 ms, and a slow EPSP lasting several seconds. The fast EPSP results from ACh acting at nicotinic receptors (ligand-gated channels) to increase permeability to ions; the slow EPSP results from a permeability *decrease*, due to the closing of a type of K^+ channel, moving the membrane potential away from E_K. In the latter case the ACh released by the presynaptic neurons acts on muscarinic GPCRs to indirectly gate the closing of the K^+ channels.

Presynaptic inhibition (PSI) is a specific inhibitory interaction in which one axon terminal ends on another axon terminal (a configuration called an *axo-axonal synapse*) and causes a *decrease* in the amount of neurotransmitter that the second terminal releases per action potential. Presynaptic inhibition is more specific than the more common postsynaptic (or just synaptic) inhibition, because it opposes only those excitatory synapses on which the PSI neuron ends. In many cases GPCRs mediate PSI, sometimes by decreasing the amplitude or duration of the action potential in the inhibited presynaptic terminal, and sometimes by decreasing the amount of Ca^{2+} entry and neurotransmitter release in response to the depolarization.

SUMMARY
Postsynaptic Receptors for Slow, Metabotropic Actions: G Protein-Coupled Receptors

- Many neurotransmitter receptors act via second messengers, triggering metabolic cascades in postsynaptic neurons. These metabotropic receptor effects are often slow and long-lasting.

- G protein–coupled receptors (GPCRs) are the major receptors of metabotropic synapses. All GPCRs have seven membrane-spanning segments, and all are evolutionarily related.

- GPCRs act via G proteins. A G protein has three subunits; normally the α subunit becomes activated when it dissociates from the regulatory β and γ subunits.

- An activated G protein activates an intracellular effector, usually to produce an intercellular second messenger.

- Second messengers of importance in metabotropic synapses include cyclic AMP, the membrane phospholipid derivatives DAG and IP_3, and Ca^{2+} ions.

- Most second messengers activate protein kinases, which phosphorylate proteins such as ion channels and change their activity.

- G proteins can activate ion channels directly.

- Metabotropic receptors play roles in slow synaptic potentials in which permeability to ions decreases, and in presynaptic inhibition.

Synaptic Plasticity: Synapses Change Properties with Time and Activity

The parameters of each step of synaptic transmission may change quantitatively over time. Presynaptically, neurotransmitter can be synthesized, stored, and released at different rates, and postsynaptic sensitivity to the neurotransmitter may be increased or decreased in different circumstances. Even electrically mediated transmission can be altered by changes in pH and in Ca^{2+} concentration that can close gap-junction channels. Changes in the parameters of synaptic transmission are important both for homeostatic regulation of transmitter metabolism and for *synaptic plasticity*—changes in synaptic strength over time. Because synaptic functions are more labile than other aspects of neuronal function, such as axonal conduction, it is widely supposed that synaptic plasticity is the mechanism that underlies how nervous system function changes over time. Thus the synaptic bases of nervous system development, and of learning and memory, are subjects of active current investigation. We cannot survey the entire burgeoning field, but we will present examples that illustrate some of the major themes.

Neurotransmitter metabolism is regulated homeostatically

The metabolism of neurotransmitters must be regulated just as other aspects of metabolism are regulated. Consider the following example: A stimulated superior cervical ganglion releases 10% of its acetylcholine (ACh) content per minute, yet the total ACh content of the ganglion is not diminished. How can this be, when the rate of stimulated release is 50 times greater than the resting release rate? Clearly, the rate of synthesis of ACh following stimulated release must also increase 50-fold if there is no depletion of ACh.

Although not all the mechanisms controlling the increased synthesis of ACh are known, the increased availability of free choline plays a major role, because choline is normally the rate-limiting substrate in ACh synthesis. When more ACh is released into the synaptic cleft, more choline is produced by acetylcholinesterase. Thus more choline is taken up by the choline transporter at the presynaptic terminal and is available for resynthesis of ACh by choline acetyltransferase (see Figure 13.11). At other synapses, control of the synthesis and action of neurotransmitters may be more complex than in this simple example.

Several important advances in the pharmacological treatment of psychiatric conditions have focused on manipulation of neurotransmitter synthesis or reuptake pathways. For instance, selective serotonin reuptake inhibitors (SSRIs) function—as the name would suggest—to reduce the amount of the neurotransmitter serotonin that is taken up after release into the synaptic cleft. The resulting increased availability of serotonin to postsynaptic receptors has proven to be a strikingly effective treatment for depression and other mood disorders.

Learning and memory may be based on synaptic plasticity

Synaptic potentials have time courses of milliseconds to seconds—long enough to have a transient effect on the excitability of postsynaptic cells. However, if synapses are involved in the long-term behavioral changes of learning and memory (an assertion for which there is increasing evidence), then neurons should demonstrate changes in synaptic strength—synaptic plasticity—that have a suitably long time course of minutes, days, or weeks. Synaptic strength is usually measured as the amplitude of a postsynaptic potential in response to a presynaptic action potential.

In many synapses, the amplitudes of individual postsynaptic potentials are not constant over time. **Synaptic facilitation** is an increase in amplitude of postsynaptic potentials in response to successive presynaptic impulses (**FIGURE 13.22A**). A decrease in amplitude of postsynaptic potentials with successive presynaptic impulses is termed **synaptic antifacilitation**, or *synaptic depression* (**FIGURE 13.22B**). *Both synaptic facilitation and antifacilitation result from changes in the amount of neurotransmitter liberated per presynaptic*

impulse. These changes are known to be calcium-dependent, but their mechanisms are otherwise incompletely understood.

Facilitation of synaptic transmission is often especially pronounced after *tetanic* stimulation of presynaptic neurons—that is, stimulation by trains of stimuli at a rate of about 10–100 per second for several seconds. The response to a single presynaptic impulse may be elevated severalfold after tetanic stimulation, and although the effect diminishes over time, it may persist for hours. This extended enhancement of synaptic response is termed **posttetanic potentiation**. Posttetanic potentiation indicates that synaptic efficacy can change with use, and these changes can be long-lasting. Particularly long-term potentiation changes have been reported in the hippocampus and cerebral cortex of the vertebrate brain, regions that are implicated in learning and memory functions. We will discuss these changes after considering the anatomically simpler example of the marine mollusc *Aplysia*.

Habituation and sensitization in *Aplysia*

Habituation and sensitization are two simple forms of behavior that occur in nearly all kinds of animals. **Habituation** is defined as the decrease in intensity of a reflex response to a stimulus when the stimulus is presented repeatedly. **Sensitization** is the prolonged enhancement of a reflex response to a stimulus, which results from the presentation of a second stimulus that is novel or noxious. Habituation and sensitization are considered simple forms of learning—that is, of modification of behavior with experience.

Reflexive gill withdrawal in *Aplysia* is a behavioral response that is subject to habituation and sensitization and is amenable to study of its synaptic basis. The gill of *Aplysia* withdraws in response to mechanical stimulation of the animal's siphon or mantle shelf (**FIGURE 13.23A**). The amplitude of gill withdrawal decreases with repeated low-frequency stimulation; that is, the response habituates. After a shock to the head, the response to siphon stimulation is again large; that is, it is *sensitized* by the head shock (**FIGURE 13.23B**).

Eric Kandel, a Nobel Prize–winning American neurobiologist, together with his colleagues, mapped the neural circuit of the gill-withdrawal reflex and determined the synaptic locus of the habituation and sensitization. Habituation of the gill-withdrawal response results from a waning of synaptic excitation of gill motor neurons by sensory neurons, and the time course of the decrease in sensory-to-motor EPSPs closely parallels the time course of behavioral habituation (**FIGURE 13.23C**). This decrease of EPSP amplitude results not from any postsynaptic change, but rather from a decrease in the number of quanta of neurotransmitter released by the sensory nerve endings. Thus the synaptic basis of habituation in *Aplysia* is *antifacilitation* of the sensory synaptic terminals.

Sensitization of the gill-withdrawal response by head shock also occurs at the sensory-to-motor synapses (see Figure 13.23C). In contrast to habituation, a sensitizing stimulus increases the amount of neurotransmitter released per impulse at the sensory-neuron terminal. This facilitation apparently results from activation of synaptic endings of sensitizing interneurons that end on the sensory terminals in axo-axonal synapses. Thus presynaptic *facilitation* is the synaptic basis of behavioral sensitization.

How is the amount of neurotransmitter release diminished during habituation and increased by sensitization? Evidence indicates that the Ca^{2+} current entering the presynaptic terminal during an

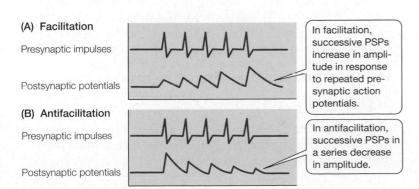

(A) **Facilitation**

Presynaptic impulses

Postsynaptic potentials

In facilitation, successive PSPs increase in amplitude in response to repeated presynaptic action potentials.

(B) **Antifacilitation**

Presynaptic impulses

Postsynaptic potentials

In antifacilitation, successive PSPs in a series decrease in amplitude.

FIGURE 13.22 Synaptic facilitation and antifacilitation

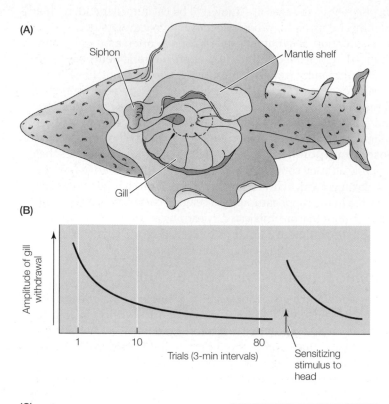

(A)

Siphon

Mantle shelf

Gill

(B)

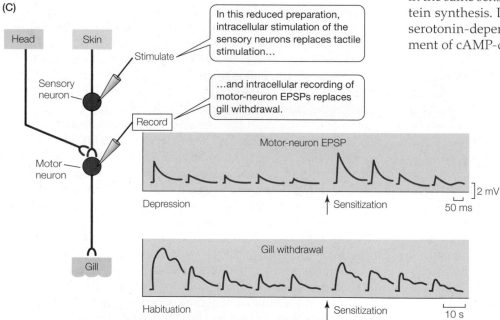

(C)

In this reduced preparation, intracellular stimulation of the sensory neurons replaces tactile stimulation…

…and intracellular recording of motor-neuron EPSPs replaces gill withdrawal.

FIGURE 13.23 Habituation and sensitization in *Aplysia* gill withdrawal (A) The gill-withdrawal reflex of *Aplysia* is a response to stimulation of the siphon or the mantle shelf. The animal retracts the gill. (B) The gill-withdrawal reflex habituates with repeated stimulation and is recovered following a sensitizing stimulus to the head. (C) Changes in synaptic activity of a reduced preparation (diagrammed) are comparable to habituation and sensitization responses of the whole animal. Stimulation of a nerve from the head replaces head shock. The decrement and enhancement of the motor-neuron EPSPs mirror the behavioral habituation and sensitization, respectively, of gill withdrawal. This result suggests that the sensory-to-motor neuron synapse is the primary site of the behavioral plasticity. (Note the longer time course of the behavioral responses.)

impulse is depressed during habituation. This finding suggests that there is a progressive, long-lasting inactivation of Ca^{2+} channels with habituation, allowing less Ca^{2+} to enter and to trigger neurotransmitter release. The presynaptic facilitation underlying sensitization is caused by an increased Ca^{2+} influx.

Some facilitating interneurons release 5-HT (serotonin), which acts to increase the amount of cAMP in the sensory terminals (**FIGURE 13.24**). The cAMP acts via a cAMP-dependent protein kinase to phosphorylate K^+ channels in the terminal, and thereby to decrease the K^+ current that normally terminates the action potential. This K^+ channel inactivation prolongs the action potential, leading to an increase in the Ca^{2+} influx and in resultant neurotransmitter release. Cyclic AMP–dependent protein kinase can also phosphorylate other proteins (e.g., to increase the mobilization of stored neurotransmitter to the release sites), and serotonin can also act at the synapse through other serotonin receptors and other second messengers, such as diacylglycerol (DAG).

The studies on *Aplysia* outlined here have determined the anatomical location of two forms of behavioral plasticity in specific, identifiable synapses and have made considerable progress in defining the synaptic mechanisms producing these changes. Short-term sensitization in *Aplysia* lasts about an hour, but with repeated training sessions sensitization may persist for more than 3 weeks. This long-term sensitization depends on changes in the same sensory-to-motor synapses, but it requires new protein synthesis. Long-term sensitization is driven by persistent serotonin-dependent elevation of cAMP, leading to the movement of cAMP-dependent protein kinase to the nucleus, where the kinase activates gene transcription via cAMP-dependent transcriptional regulatory proteins such as CREB (cAMP response element–binding protein).

Investigators have also demonstrated *classical conditioning* of the *Aplysia* gill-withdrawal response: a learned association between two stimuli. The synaptic basis of classical conditioning is more complex than that of sensitization. In part it depends on similar presynaptic mechanisms of enhanced neurotransmitter release, but there is also a postsynaptic component that involves NMDA-type glutamate receptors (see page 364).[2] The details of these mechanisms are complicated, but the major point is that synaptic mechanisms similar to those that underlie sensitization and habituation can also be employed to explain classical conditioning in *Aplysia*.

Long-term potentiation in the hippocampus

Synaptic changes in mammalian brains that are long-lasting have been explored extensively during the last 35 years, most notably in the hippocampus. The hippocampus is a deep forebrain structure that is strongly implicated in spatial learning (such as maze learning) and in memory formation. The circuitry of the hippocampus

[2] D. G. Glanzman and coworkers have shown that longer-term sensitization in *Aplysia* also has a postsynaptic component that depends on NMDA receptors.

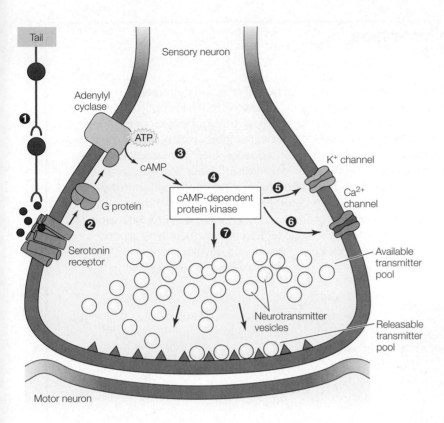

Tail

Sensory neuron

Adenylyl cyclase

ATP

❸

cAMP

❹

cAMP-dependent protein kinase

❺

K⁺ channel

Ca²⁺ channel

❻

G protein

❷

❼

Serotonin receptor

Available transmitter pool

Neurotransmitter vesicles

Releasable transmitter pool

Motor neuron

FIGURE 13.24 A model of *Aplysia* sensitization Transmitter release in the sensory terminals decreases with habituation and increases with sensitization. With habituation, fewer Ca²⁺ channels open in response to action potentials in the sensory terminal. The resulting decrease in Ca²⁺ influx decreases transmitter release. Serotonergic neurons promote sensitization ❶. Serotonin acts ❷ to activate an adenylyl cyclase in the terminals, which stimulates the synthesis of cAMP ❸. Cyclic AMP in turn acts on a cAMP-dependent protein kinase ❹ to phosphorylate proteins, including a K⁺ channel ❺; this leads to a decrease in the repolarizing K⁺ current and a broadening of the action potential. The increase in the duration of the action potential increases the time during which Ca²⁺ channels can open ❻, leading to a greater influx of Ca²⁺ and increased release. The kinase also acts on Ca²⁺ channels and on mobilization of neurotransmitter vesicles ❼. Other serotonin effects are not shown. (After Kandel et al. 1995.)

(A) Experimental setup

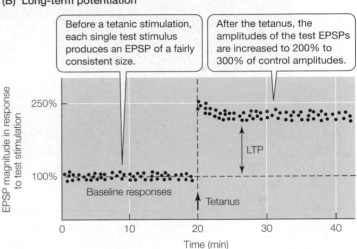

Hippocampus of rodent brain

Recording

Testing stimulus

Hippocampal slice

CA1 neuron

CA3 neuron

(B) Long-term potentiation

Before a tetanic stimulation, each single test stimulus produces an EPSP of a fairly consistent size.

After the tetanus, the amplitudes of the test EPSPs are increased to 200% to 300% of control amplitudes.

EPSP magnitude in response to test stimulation

250%

LTP

100%

Baseline responses

Tetanus

0   10   20   30   40

Time (min)

FIGURE 13.25 Long-term potentiation in the hippocampus Synapses in the hippocampus and in other brain areas exhibit long-term potentiation (LTP), a long-lasting increase in EPSP amplitude after a tetanic (strong, repeated train) stimulus. (A) The hippocampus of mammalian brains contains several synaptic pathways and types of neurons. Here we consider synapses (of CA3 neurons) onto CA1 neurons. For simplicity, only one neuron of each type is shown in the diagram. (B) Long-term potentiation may persist for many hours.

is well characterized, and because it is largely restricted to two dimensions, it can be studied in hippocampal slices that allow stable, long-term recordings from visible neurons (**FIGURE 13.25A**). Circuits in the hippocampus (and elsewhere in the brain) undergo prolonged changes when strongly stimulated, and some of these changes are associative; that is, they are specific to the pathways that are stimulated together.

We will examine associative long-term potentiation in a particular region (CA1) of the hippocampus. **Long-term potentiation** (**LTP**) is a long-lasting enhancement of synaptic transmission following intense stimulation (**FIGURE 13.25B**); it resembles posttetanic potentiation but is more prolonged. Associative LTP is specific to the activated pathway and requires the cooperative interaction of many presynaptic neurons to depolarize a postsynaptic CA1 cell adequately.

LONG-TERM POTENTIATION IN THE BRAIN INVOLVES CHANGES IN SYNAPSE STRENGTH As first pointed out by Donald Hebb, learning could depend on "successful" synapses getting stronger, with a "successful" synapse meaning one in which the presynaptic and postsynaptic neurons have been active or depolarized at the same time: "Neurons that fire together wire together." A synapse that undergoes a long-term change in strength as a result of coincident activation of the pre- and postsynaptic neurons is called a *Hebbian synapse*. The synapses that undergo LTP in the CA1 region of the hippocampus are Hebbian because potentiation depends on the associated activity of presynaptic and postsynaptic cells. A critical feature of LTP is that any synapse that is active while the

BOX 13.1 Synapse Formation: Competing Philosophies
MATTHEW S. KAYSER

How does the nervous system form synapses during development? Much of what we know comes from the model system of the vertebrate neuromuscular junction (NMJ). The neuromuscular synapse is a reliable relay, so that an action potential traveling in a motor axon will consistently result in contraction of the muscle fiber it innervates. Interestingly, during development, multiple motor axons can innervate a muscle fiber initially (see figure), but at a mature NMJ just one axon innervates one muscle fiber. How is it decided which motor axon will innervate a given muscle fiber? The outcome is more analogous to Darwinian natural selection than Calvinist predestination: Motor axons engage in a competition for space on a muscle fiber.

Single mature neurons in the CNS—in sharp contrast to single mature muscle fibers at the NMJ—receive thousands of inputs. That is, the dendrites of a single neuron may make synaptic contact with axons of thousands of other neurons. This raises the question of how the brain can possibly become wired properly. A leading hypothesis suggests molecular matchmaking occurs, wherein the appropriate axons and dendrites recognize cues on one another. **Box Extension 13.1** describes more about how synapses form during development in the brain and at neuromuscular junctions.

A neuromuscular junction with multiple innervation

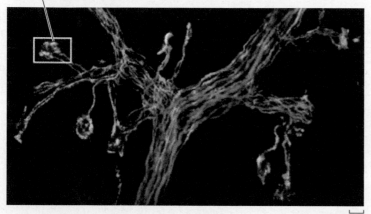

5 μm

Multiple axons can innervate each muscle fiber at immature NMJs in mice In this image (on day 8 of postnatal development) some neurons have been labeled blue and others green. The postsynaptic receptors are red. The muscle fibers themselves are not visible. (From Walsh and Lichtman 2003.)

postsynaptic cell is strongly depolarized will be potentiated, for a period that can last hours in hippocampal slices and weeks in intact animals. Interestingly, some of the mechanisms controlling synaptic plasticity in the mature nervous system may function similarly to those that control synapse formation in neural development (**BOX 13.1**).

Establishment of hippocampal LTP (**FIGURE 13.26**) depends on two related glutamate receptors: NMDA receptors (so named because they also are activated by the drug N-methyl-D-aspartate) and other ionotropic glutamate receptors, most notably AMPA receptors (activated by the drug α-amino-3-hydroxy-5-methylisoxazole-4-propionic acid). AMPA receptors produce fast EPSPs. NMDA receptors also produce EPSPs, but they have an important functional property: They work only when the postsynaptic cell is depolarized. As Figure 13.26A shows, an NMDA receptor is activated by glutamate, but at resting membrane potential its ion channel is blocked by a bound Mg^{2+} ion, so few or no ions flow through it. The EPSP produced under these conditions depends on AMPA receptors. In contrast, when the postsynaptic cell is substantially depolarized, the bound Mg^{2+} ion is released; with glutamate binding to NMDA receptors, a considerable number of Na^+ and Ca^{2+} ions can enter the cell through the unblocked channels (see Figure 13.26B). In

this way, an NMDA receptor functions as a molecular coincidence detector: Both postsynaptic depolarization and presynaptic stimulation must occur at the same time for NMDA receptor–mediated ion flux to occur.

The Ca^{2+} ions entering the postsynaptic cell through NMDA receptors act as second messengers, activating Ca^{2+}-dependent signaling molecules such as Ca^{2+}/calmodulin-dependent kinase II (CaMKII) and protein kinase C (see Figure 13.26B,C). Thus two features of NMDA receptors are important for the establishment of LTP: (1) Their full activation requires a Hebbian simultaneous depolarization of the presynaptic and postsynaptic cells; and (2) they have indirect, long-lasting effects mediated by second messengers.

Whereas long-lasting changes in synapse strength are clearly induced by postsynaptic activation of NMDA receptors, what mechanisms actually underlie the increased synaptic strength observed in LTP? For more than a decade, there was a hearty scientific debate on whether LTP is a reflection of changes in neurotransmitter release or neurotransmitter receptor properties (in other words, presynaptic or postsynaptic alterations). It now seems clear that a major mechanism for enduring changes in synapse strength involves modulation of the postsynaptic AMPA-type ionotropic glutamate receptors touched on above. Ca^{2+} flux through NMDA receptors triggers a series of

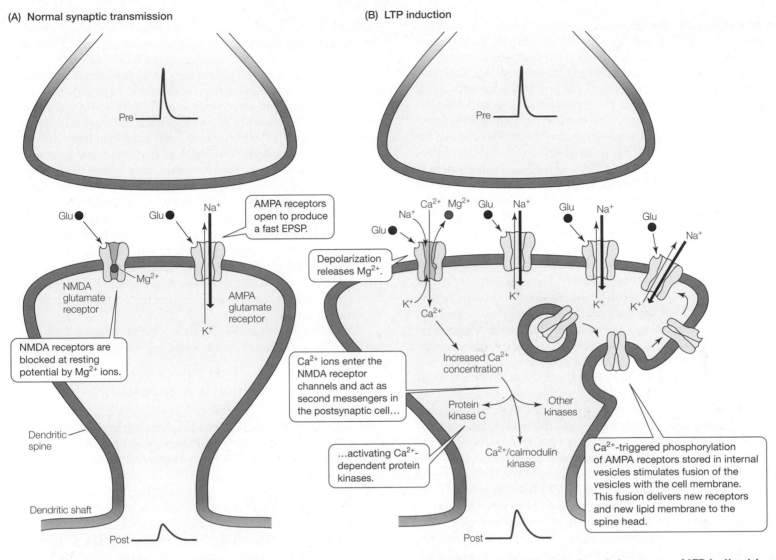

(A) Normal synaptic transmission

Pre

AMPA receptors open to produce a fast EPSP.

Glu

Glu

Na⁺

NMDA glutamate receptor

Mg²⁺

AMPA glutamate receptor

K⁺

NMDA receptors are blocked at resting potential by Mg²⁺ ions.

Dendritic spine

Dendritic shaft

Post

(B) LTP induction

Pre

Ca²⁺ Mg²⁺ Glu Na⁺

Na⁺

Glu

Glu Na⁺

Glu

Na⁺

Depolarization releases Mg²⁺.

K⁺

Ca²⁺

K⁺

K⁺ K⁺

Increased Ca²⁺ concentration

Ca²⁺ ions enter the NMDA receptor channels and act as second messengers in the postsynaptic cell…

Protein kinase C

Other kinases

…activating Ca²⁺-dependent protein kinases.

Ca²⁺/calmodulin kinase

Ca²⁺-triggered phosphorylation of AMPA receptors stored in internal vesicles stimulates fusion of the vesicles with the cell membrane. This fusion delivers new receptors and new lipid membrane to the spine head.

Post

(C)

Glutamate uncaging (6 ms, 0.5 Hz)

−12 s −4 s 4 s 12 s 20 s 28 s 36 s 44 s

52 s 60 s 68 s 76 s 84 s 92 s 100 s 108 s

2.0

1.6

Lifetime (ns)

3 mm

FIGURE 13.26 Induction and maintenance of LTP in the hippocampus Induction of long-term potentiation (LTP) is postsynaptic. (A) At the unpotentiated synapse, glutamate neurotransmitter is released and acts at various kinds of glutamate receptors, principally AMPA receptors. NMDA receptors are blocked by Mg²⁺. (B) Tetanic stimulation depolarizes the postsynaptic cell substantially, releasing the Mg²⁺ block and allowing glutamate to activate the NMDA receptors, leading to Ca²⁺ entry. The resultant phosphorylation events lead to an enhanced postsynaptic response via insertion of internal AMPA receptors into the postsynaptic membrane, extensive downstream signaling, and synaptic growth. (C) The fluorescence micrographs in this time series show the effect of increased glutamate on a single dendritic spine (the green spot to the left of the vertical green dendrite). Localized release of glutamate (at the white triangle) increases the CaMKII activity (red, peaking at 12 s) and then increases the volume of only the dendritic spine immediately adjacent to the glutamate release. (C from Lee et al. 2009.)

downstream events leading to phosphorylation and subsequent insertion of new AMPA receptors into the postsynaptic cell membrane (see Figure 13.26B). The increased number of postsynaptic AMPA receptors results in increased amplitude of synaptic responses. In addition, alteration in the biophysical properties of these channels contributes to the observed potentiation.

Of course, one might imagine that our brains would not work optimally if synaptic contacts were only able to get stronger. As it turns out, certain stimuli result in a *long-term depression* (*LTD*) of synaptic responses. At least some forms of LTD are also NMDA receptor–dependent, which raises the question of how both LTP

and LTD could rely on NMDA receptor activity and the resultant Ca²⁺ influx. Although the answers are not totally clear, it is thought that high postsynaptic levels of Ca²⁺ over a short period result in LTP, whereas lower levels of Ca²⁺ over a prolonged period result in LTD. Stimuli known to result in LTD set into motion a cascade of events leading to removal of AMPA receptors from the postsynaptic membrane and, in turn, reduced amplitude of postsynaptic responses. Because the insertion or removal (processes known as trafficking) of AMPA receptors is a key event in the modification of synaptic strength, one could ask what happens if the signals controlling AMPA receptor trafficking are perturbed. Researchers recently asked this very question using mutant mice with single amino acid changes to specific phosphorylation sites on their AMPA receptors. These mutations interfere with the normal phosphorylation and dephosphorylation—and thus the trafficking—of the AMPA receptor proteins. The researchers found that the mutant mice have abnormal LTP and LTD, as well as memory defects in spatial learning tasks. Thus LTP and LTD not only reflect the dynamic nature of synaptic contacts, but also highlight the fact that changes in synaptic strength—both increases and decreases—are involved in memory formation.

LONG-TERM MEMORIES CAN INVOLVE CHANGES TO THE PHYSICAL STRUCTURE OF NEURONS   The establishment of LTP (or LTD) requires NMDA receptor–mediated Ca^{2+} flux and AMPA receptor insertion (or removal), but some memories, as they say, last a lifetime. How are more permanent changes to synapses accomplished? The answer appears to have at least two components. First, the so-called late phase of LTP (persistent changes lasting hours to even weeks) requires the synthesis of new proteins as well as gene transcription. Inhibition of protein synthesis at the time an animal is being trained in a new task does not affect the animal's ability to learn, but it prevents the consolidation ("conversion") of the new memory into a long-term one. Some of the new required proteins can be made locally at a synapse. However, some signals ultimately activate processes in the nucleus. In the nucleus, transcription factors such as CREB are known to be of particular importance in turning on the transcription of genes encoding proteins that are critical in maintaining memories. What is the character of these proteins and how are they transported to the correct synapse where the signal began? These are significant questions for the next generation of neuroscientists!

The second manner in which changes to synaptic transmission are made more permanent involves structural alterations to synapses themselves. Excitatory synapses usually occur on the stereotyped morphologic protrusions known as dendritic spines (see Figure 13.5). Stimuli resulting in LTP, or even the experimental focal release of glutamate right next to a spine, have an interesting effect: The head of the affected spine gets larger (see Figure 13.26C). So in concert with the prolonged increase in response amplitude described above comes an increase in the volume of the dendritic spines on which the LTP-inducing stimuli occurred. Investigators hypothesize that the same vesicles that supply new AMPA receptors to the postsynaptic density also deliver fresh lipid materials to allow expansion of the postsynaptic membrane (see Figure 13.26B). The enlargement of individual spines also requires activation of NMDA receptors and CaMKII, indicating that spine growth likely is a structural basis of LTP. Moreover, changes in dendritic spine morphology are synapse-specific—another important property of LTP. Neighboring, unstimulated spines or portions of the dendrite are unaffected. In this way, only the "successful" synapses are strengthened, without nonspecifically reinforcing other, unpaired contacts.

So far we have covered the mechanisms by which individual synapses and spines respond to plasticity-inducing stimuli. But in addition to spine expansion, changes in experience can also result in the growth of new dendritic spines altogether. Karel Svoboda and colleagues have characterized how dendritic spines behave in functioning cortical circuits of awake animals. Svoboda's group made cranial viewing windows in the skulls of mice and repeatedly (even over the course of months) imaged portions of dendrites closest to the surface of the brain, coming from neurons that were made to fluoresce bright green. The type of neurons chosen were those from barrel cortex, the somatosensory area of the rodent brain receiving input from the animal's whiskers (**FIGURE 13.27**). By trimming the whiskers of a mouse in a particular pattern, Svoboda and colleagues were able to alter the animal's sensory experience, and found that the changing sensory input drove the formation of new dendritic spines and synaptic contacts, while eliminating others. In other words, alterations in external stimuli allow formation of new synapses and actually reshape the *physical* structure of dendrites in a mouse's brain.

Remember, however, the elegance of brain circuitry: Nothing lasts forever, as the next mind-altering experience is just around the bend.

Long-term potentiation is a necessary component of learning

LTP was originally studied as a long-lasting synaptic change that might underlie learning and memory, a molecular *correlate* of the changes needed for long-term behavioral changes to have a synaptic basis. The existence of LTP does not prove that it is involved in learning and memory—only that it could be. Is there direct evidence that LTP is a necessary component of learning? The answer is yes. Studies that knock out (block the expression of) particular genes show that the absence of NMDA receptors, of CaM kinase, or of other kinases disrupts both LTP and spatial learning of a mouse trying to negotiate a water maze (see page 510). As described above, the same holds true for manipulations that disrupt AMPA receptor insertion into the postsynaptic membrane.

One of the most compelling studies correlating NMDA receptors, LTP, and learning and memory is that of *Doogie* mice, introduced at the beginning of the chapter. Joe Tsien and coworkers genetically engineered mice to overexpress a particular subunit of their NMDA receptors. This subunit is normally abundant in developing mice but less so in adults. An important functional difference is that NMDA receptors containing the juvenile subunit remain open longer than NMDA receptors containing only adult subunit types. Mice of the *Doogie* strain (with more NMDA receptors containing the juvenile subunit) have more potent hippocampal LTP when their brains are tested, presumably because more Ca^{2+} enters through the longer-opening NMDA receptors during LTP induction. The mice also perform better than average mice on several sorts of learning tasks, and they remember novel stimuli longer. Researchers have now created or identified at least 33 mutant mouse strains that, like *Doogie*, have enhanced cognitive abilities and enhanced LTP. These studies reinforce the correlation between LTP and spatial learning and memory; moreover, they demonstrate that manipulation of the

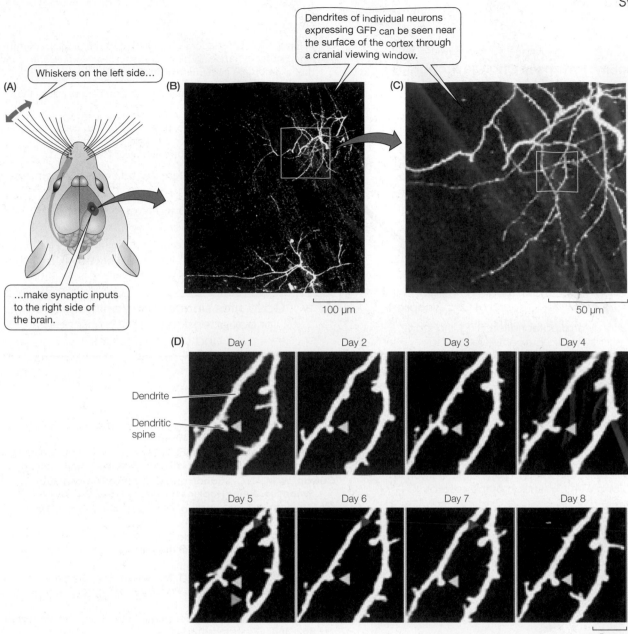

FIGURE 13.27 Repeated imaging of dendrites through cranial windows in living rodents (A) Sensory neurons from the left whiskers of a rodent relay to make synaptic inputs in somatosensory cortex on the right side of the animal's brain. Each whisker drives input onto neurons occupying only a discrete area of the cortex, known as a barrel (box in B). (C) Higher magnification of the boxed dendrites in (B). (D) Repeated long-term imaging through the cranial window of the branch of a dendrite (the area boxed in C) reveals transient (blue arrowhead), semistable (red arrowheads), and stable (yellow arrowheads) protrusions. (After Trachtenberg et al. 2002; courtesy of Karel Svoboda.)

molecular substrates of LTP can enhance learning and memory, as well as disrupt it.

Beyond these correlational studies, is there also direct evidence that learning induces LTP in the brain? Mark Bear and colleagues have finally confirmed this 40-year-old conjecture. While putting mice through a training protocol in which the mice learned to avoid a mild shock, the investigators used multielectrode arrays to record from multiple areas of hippocampus area CA1 at once, because it is not known where in CA1 LTP might occur with this particular task. They found in awake and behaving animals that the process of learning the task results in long-term potentiation of synaptic responses using the same signaling pathways as electrophysiologically induced LTP. This work and a growing number of subsequent studies strongly suggest that LTP really is a molecular basis for learning and memory.

The extent of synaptic plasticity even in adult nervous systems represents a staggering expansion of our understanding of the range of mechanisms and functions of synaptic physiology. Synapses transmit signals, by both rapid and prolonged mechanisms. But synapses also modulate properties—of synapses, of neurons, and of behavior.

SUMMARY

Synaptic Plasticity: Synapses Change Properties with Time and Activity

■ Neuronal stimulation that increases the rate of neurotransmitter release also increases rates of neurotransmitter resynthesis. The homeostatic mechanisms of this regulation involve both substrate availability and more complex mechanisms.

■ With a train of presynaptic action potentials, the amplitudes of the resultant postsynaptic potentials may increase (facilitation) or decrease (antifacilitation). Thus the synaptic transfer of information depends on its history.

■ The synaptic bases of behavioral habituation, sensitization, and classical conditioning in *Aplysia* depend on second messenger–mediated control of the amount of neurotransmitter released at CNS synapses.

■ Hippocampal long-term potentiation (LTP) is a long-lasting change in synaptic properties related to learning and memory. The induction of LTP depends on NMDA receptors that respond to both glutamate neurotransmitter and postsynaptic depolarization, to allow Ca^{2+} entry into the postsynaptic cell.

■ LTP is maintained by means of Ca^{2+}-dependent second-messenger pathways that make the postsynaptic cell more sensitive to glutamate neurotransmitter. Insertion of AMPA receptors into the postsynaptic membrane increases the amplitude of the postsynaptic response, and occurs along with expansion of the area of dendritic spines.

■ Studies that manipulate the expression of critical genes in the LTP metabolic pathway significantly affect learning and memory in mice.

STUDY QUESTIONS

1. What are the functional advantages and disadvantages of electrical synapses?

2. What adaptive advantage do synapses provide in nervous systems (i.e., why isn't the nervous system just a mesh of cells cytoplasmically connected through gap junctions)?

3. Why is it difficult to show vesicular release of neurotransmitter molecules by electron microscopy?

4. In his book *Ionic Channels of Excitable Membranes*, Bertil Hille characterized the importance of calcium ions: "Calcium channels … serve as the only link to transduce depolarization into all the nonelectrical activities controlled by excitation. Without Ca^{2+} channels our nervous system would have no outputs." Discuss this statement with reference to synaptic function.

5. What are the criteria for identification of the neurotransmitter at a particular synapse? Which two criteria are *sufficient* to conclude that a candidate is the neurotransmitter?

6. Why (in evolutionary terms) do you think there are so many kinds of neurotransmitters?

7. Why are there multiple receptor subtypes for each neurotransmitter? Why not a few receptors and lots of neurotransmitters acting on each receptor?

8. Discuss the functional differences between ionotropic and metabotropic receptors.

9. Would you classify NMDA receptors as ionotropic or metabotropic? Why?

10. Do you think there is one cellular mechanism of learning and memory or several? Cite studies in *Aplysia* and in the mammalian hippocampus to support your answer.

Go to **sites.sinauer.com/animalphys4e**
for box extensions, quizzes, flashcards,
and other resources.

REFERENCES

Allen, N. J. 2014. Astrocyte regulation of synaptic behavior. *Annu. Rev. Cell Dev. Biol.* 30: 439–463.

Bear, M. F., and B. W. Connors. 2015. *Neuroscience: Exploring the Brain*, 4th ed. Lippincott Williams & Wilkins, Baltimore, MD.

Cowan, W. M., T. C. Südhof, and C. F. Stevens (eds.) 2001. *Synapses*. Johns Hopkins University Press, Baltimore, MD.

Hille, B. 2001. *Ion Channels of Excitable Membranes*, 3rd ed. Sinauer, Sunderland, MA.

Holtmaat, A., and K. Svoboda 2009. Experience-dependent structural synaptic plasticity in the mammalian brain. *Nat. Rev. Neurosci.* 10: 647–658.

Kandel, E. R., J. H. Schwartz, T. M. Jessell, S. A. Siegelbaum, and A. J. Hudspeth (eds.). 2013. *Principles of Neural Science*, 5th ed. McGraw-Hill, New York.

Lehrer, J. 2009. Neuroscience: small, furry … and smart. *Nature* 461: 862–864. Genes for smart mice.

Leng, G., and M. Ludwig. 2008. Neurotransmitters and peptides: whispered secrets and public announcements. *J. Physiol. (London)* 586: 5625–5632.

Nicholls, J. G., A. R. Martin, P. A. Fuchs, and three additional authors. 2012. *From Neuron to Brain*, 5th ed. Sinauer, Sunderland, MA. One of the clearest and most experiment-driven textbook discussions of synaptic physiology.

Südhof, T. C. 2013. A molecular machinery for neurotransmitter release: synaptotagmin and complexin in the SNARE–SM cycle. *Nat. Med.* 19: 1227–1231. A Nobel Prizewinner explains vesicular release of neurotransmitters.

Tsien, J. Z. 2000. Building a brainier mouse. *Sci. Am.* 282(4): 62–68. A straightforward introduction to the genetically engineered *Doogie* mouse strain, and to experiments indicating the roles of synaptic long-term potentiation in learning and memory.

See also **Additional References** *and* **Figure and Table Citations**.

Sensory Processes

Sensory systems provide animals with essentially all of the information they have about their external environment—as well as most of the information they have about their internal environment. For example, consider the nocturnal encounter between a hunting bat and a moth in the photo above. The bat finds its prey by echolocation: It emits ultrasonic cries (sounds at frequencies above the limit of human hearing) and hears the echoes of its cries that bounce off objects in the environment, rather like human detection of submarines by sonar. Moths, however, have evolved auditory organs that are very sensitive to bat cries. A moth will fly away from faint bat cries (at distances beyond the range of the bat's echo detection), but in response to louder cries (when the bat is close enough to get clear echoes) the moth will either take erratic and apparently random evasive action or power-dive to the ground.

Clearly it is of selective advantage for bats to have evolved auditory systems that allow them to use echolocation to detect and capture insects in the dark, while also detecting and avoiding obstacles. Likewise it is of selective advantage for moths to detect and evade hunting bats. We will examine aspects of auditory sensory systems in both bats and moths later in this chapter, along with other sensory systems that provide animals with information about their environments.

A nocturnal encounter between a moth and a hunting bat Bats echolocate to orient and to capture insects, and moths have evolved auditory systems that help them evade bat predators. In this case a moth hears the bat's cries and is able to evade the bat by diving toward the ground. Sensory systems may be energetically expensive, but the information they provide is often crucial for survival.

Sensory systems of all kinds depend on specialized sensory receptor cells that respond to stimuli, either environmental stimuli or stimuli arising inside the body. Different sensory cells respond to different stimuli, and they vary greatly in **sensitivity** (the ability to distinguish among stimuli of different intensity) and **specificity** (the ability to distinguish among stimuli of different types). Auditory reception in bats and moths is both sensitive and specific, having been shaped by long periods of natural selection. But all cells are somewhat responsive to aspects of their environment and thus subserve some functions that can be considered sensory. For example, bacteria and protists respond to light and to chemical gradients. Cellular responses of this kind presumably preceded the evolution of specialized sensory neurons. A **sensory receptor cell** is a cell that is specialized to transform the energy of a stimulus into an electrical signal. The kind of stimulus that excites different receptor cells may be chemical, mechanical, or electromagnetic. A **stimulus** is a form of external energy (external to the cell) to which a sensory receptor cell can respond.

Humans have studied sensory functions since the ancient Greeks, and currently new experimental approaches are revolutionizing our understanding of how sensory systems work. Recording techniques such as patch-clamp recording (see Figure 12.17) have greatly increased the ease of registering the responses of sensory cells. Imaging methods have allowed the examination of activity of many neurons at once. Molecular techniques have also been important, both for identifying the genes and proteins responsible for sensory detection and for manipulating the expression of sensory genes to clarify their functions (see Chapter 3, pages 79–83).

Organization of Sensory Systems

Sensory receptor cells normally function as parts of a larger system, rather than in isolation. They are commonly clustered together in **sense organs**, anatomical structures that are specialized for the reception of particular kinds of stimuli. Usually a sense organ contains many similar receptor cells, as well as several kinds of nonneural tissues. For example, the vertebrate eye is a sense organ that contains photoreceptor cells, as well as nonneural tissues such as those that make up the cornea and iris. We can also speak of **sensory systems**, defined as sense organs and all of their associated central nervous system (CNS) processing areas. For example, the vertebrate visual system includes the eyes and the central areas in the brain that are primarily concerned with processing visual information.

The basic function of a sensory receptor cell is to convert stimulus energy into an electrical signal, a process known as **sensory transduction**. Sensory transduction requires specialized molecules called **sensory receptor molecules** (or simply **receptor molecules**), which are particularly sensitive to a sensory stimulus. Sensory receptor molecules initiate the transduction of the stimulus to produce an electrical response called a *receptor potential*. The receptor potential may lead to action potentials carried to the CNS, where sensory information is processed. Different kinds of sensory receptor cells have different receptor molecules that make them sensitive to different stimuli. The receptor molecules are all membrane proteins. Receptor cells often have modified cell membranes to increase surface area and thus receptor number and resultant sensitivity. The two common ways in which receptor cells increase their outer membrane surface area are via cilia and microvilli (microscopic fingerlike projections; see page 41).

Sensory receptor cells can be classified in four different ways

The oldest way of classifying sensory receptor cells is by **sensory modality**—the subjective nature of the sensory stimulus. Aristotle distinguished five primary senses: vision, hearing, touch, smell, and taste. As **TABLE 14.1** shows, however, animals perceive many other stimuli besides these classical five senses. Humans are aware of additional sensory modalities such as balance and temperature, and there are many other modalities that do not normally enter our consciousness (such as muscle length and blood oxygen partial pressure). In addition, many animals appear to possess receptor cells sensitive to modalities and qualities of stimuli not sensed by humans. These include electric and magnetic fields and ultraviolet radiation.

A second classification is based on the *form of stimulus energy* that excites sensory receptor cells at the receptor surface (see Table 14.1). Photoreceptors, electroreceptors, and magnetoreceptors all respond to different forms of electromagnetic energy; auditory receptors, mechanoreceptors, and vestibular receptors (for balance) are all excited by mechanical energy. Olfactory (smell) and taste receptors are chemoreceptor cells that respond to chemical energy: They have receptor proteins that bind specific chemicals, and the change in free energy associated with this binding triggers a conformation change in the protein to induce a response.[1]

A third classification of sensory receptor cells is according to their *mechanism of transduction*. Recall from Chapter 13 that the synaptic neurotransmitter receptor molecules are of two types (see Figure 13.6): *ionotropic receptors* or *ligand-gated channels*, in which the same molecule binds neurotransmitter and acts as the ion channel; and *metabotropic* or *G protein–coupled receptors* (*GPCRs*), in which binding of neurotransmitter activates a G protein, leading to a metabolic cascade that ultimately opens membrane channels. Sensory receptor molecules bear striking similarities to neurotransmitter receptors and divide into the same two classes of mechanism (see Table 14.1 and **FIGURE 14.1**). In **ionotropic transduction** (see Figure 14.1A), the sensory stimulus is received and then transduced into an electrical signal, a depolarization produced by increased membrane permeability to cations, resulting in cation (mostly Na^+) entry into the cell. The same receptor molecule produces both the reception and the transduction. All forms of mechanoreception, including hearing, seem to be ionotropic. Thermoreception, electroreception, and some forms of taste reception also use ionotropic transduction. Other sensory systems use **metabotropic transduction**, mean-

---

[1] The definition of a stimulus as a form of energy that stimulates a receptor cell can be difficult to grasp, particularly for chemoreceptors. Thinking of mechanical, photic, and thermal stimuli as energy is fairly intuitive, but the energy in a chemical stimulus is less clear. Chemoreceptor proteins bind chemical molecules in the same way that enzymes bind substrates, using weak bonds rather than covalent bonds (see Chapter 2). There is a change in free energy associated with this binding, and this change in free energy leads to a conformational change in the bound receptor protein that initiates the cell's response to the stimulus. This energy change is quite different from the energy transformations in a molecule (e.g., a glucose molecule) when it is metabolized and the energy from breaking its covalent bonds is released.

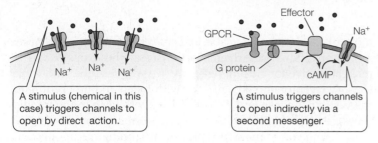

(A) Ionotropic transduction

Na+   Na+   Na+

A stimulus (chemical in this case) triggers channels to open by direct action.

(B) Metabotropic transduction

Effector

GPCR

Na+

G protein

cAMP

A stimulus triggers channels to open indirectly via a second messenger.

FIGURE 14.1 Two kinds of sensory transduction mechanisms
Sensory stimuli activate sensory receptor molecules in one of two ways. (A) Stimulus-gated ion channels open directly in response to an applied stimulus, constituting an ionotropic mechanism. (B) Stimulus energy activates a metabotropic G protein–coupled receptor (GPCR), triggering a metabolic cascade (see Figure 13.6D) that results in activation of the sensory cell. The examples shown are chemoreceptor proteins, but the two types of transduction apply to all kinds of sensory receptors.

ing the sensory receptor molecule acts like a neurotransmitter- or hormone-activated GPCR in activating a metabotropic cascade (see Figures 2.27, 13.19, and 14.1B). The receptor molecule activates a G protein, which in turn activates an effector molecule to alter the concentration of a second messenger that triggers ion permeability changes, leading to the cell's response. Vision, vertebrate olfaction, and some forms of taste reception use metabotropic transduction. Both ionotropic and metabotropic sensory receptor molecules are membrane proteins, as earlier noted.

Finally, receptor cells can be classified according to the *location* of the source of the stimulus energy relative to the body. **Exteroceptors** are sensory cells that respond to stimuli outside the body, such as light or sound. **Interoceptors** respond to internal stimuli, such as the pH or osmotic concentration of the blood.

Sensory receptor cells transduce and encode sensory information

Sensory receptor cells have two functional roles. First, as we noted above, a sensory receptor cell *transduces* some form of stimulus energy, converting it to an electrical signal termed a *receptor potential*. The receptor potential is usually a depolarization of the sensory cell, which (if it reaches threshold) can trigger action potentials that propagate to the CNS. (Some sensory receptor cells do not have axons, but instead synaptically excite separate sensory neurons to generate action potentials, as shown, for example, in Figures 14.7 and 14.15.) Second, a sensory receptor cell *encodes* information about a stimulus; this information is carried via trains of action potentials that are transmitted to the CNS. It is worth pointing out that the only information that the CNS receives to tell an animal what is happening in the external environment comes from the encoded signals of the animal's peripheral sensory receptor cells. Thus the encoding of sensory signals must allow the CNS to get all necessary information: What's out there? Where? How much? How is it changing over time?

A key part of the encoding of sensory information is "keeping the wiring straight," by maintaining within the CNS an orderly segregation of axons from different sensory receptor cells that project to different specific CNS locations. Visual afferent axons project to different areas of the brain than do auditory afferent axons, skin touch receptors, and so on. This observation leads to an important generalization: *The sensory modality or quality of sensation associated with a stimulus depends solely on which receptor cells are stimulated, rather than on how they are stimulated.* **FIGURE 14.2** illustrates this generalization, known as the principle of **labeled lines**. Most sensory receptor cells are quite specific in the kind of stimulus to which they respond, but any form of stimulus energy may excite a receptor cell if there is enough of it. For example, any stimulus that excites photoreceptors is perceived as light, whether the stimulus is actually light, a poke in the eye ("seeing stars"), or electrical stimulation of the optic nerve. Another key aspect of "keeping the wiring straight" is that for each kind of receptor cell, the afferent axons typically maintain a geometric arrangement within the CNS that mirrors the geometric arrangement of the receptor cells, so that the axons project onto orderly maps in the CNS. This pattern facilitates central decoding of where each stimulus comes from.

TABLE 14.1 Sensory receptor cells differ in the kind of stimulus that excites them and in their mechanism of transduction		
Receptor type	**Stimulus perceived**	**Mechanism of transduction**
Mechanoreceptors	Touch, pressure, proprioception	Ionotropic
Vestibular receptors	Balance; body position and movement	Ionotropic
Osmoreceptors	Osmotic pressure	Ionotropic
Auditory receptors	Sound	Ionotropic
Thermoreceptors	Heating and cooling	Ionotropic
Electroreceptors	Electric fields in water	Ionotropic
Some taste chemoreceptors	Salty and sour in vertebrates; insects?	Ionotropic
Some taste chemoreceptors	Sweet, bitter, and umami (proteinaceous) in vertebrates	Metabotropic
Olfactory chemoreceptors	Chemicals generally from a distance	Metabotropic in vertebrates; ionotropic or mixed (insects)
Photoreceptors	Light	Metabotropic
Magnetoreceptors	Position or change of magnetic field	Unknown

(A) The labeled-lines principle

In the PNS, different kinds of sensory receptor cells lead to generation of similar action potentials.

In the CNS, different sensory neurons go to different brain regions, where action potentials are interpreted as different kinds of sensory stimulation.

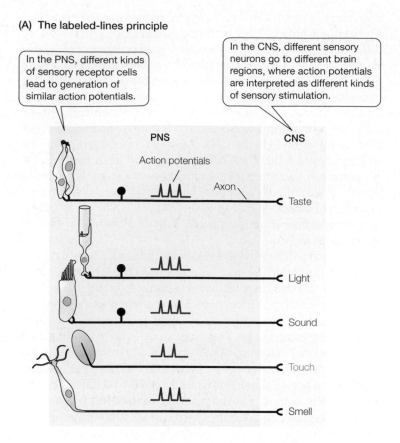

PNS

CNS

Action potentials

Axon

Taste

Light

Sound

Touch

Smell

(B) Regions of the cerebral cortex that process different sensory qualities

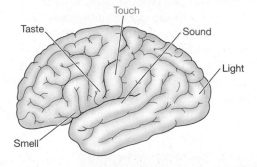

Touch

Taste

Sound

Light

Smell

FIGURE 14.2 The principle of labeled lines in sensory systems (A) In the peripheral nervous system (PNS), receptor cells sensitive to different kinds of stimuli send similar kinds of signals (action potentials) to the CNS. The CNS interpretation of the sensory modality depends on which lines (axons) convey the signals. Note that the sensory receptor cells for touch and smell are neurons with axons that enter the CNS, whereas the sensory cells that detect sound and taste have no axons but instead synaptically excite sensory neurons. (B) In the CNS, different sensory pathways (with intervening synaptic relays) ultimately project to different regions of the cerebral cortex, providing the anatomical basis for the principle of labeled lines.

The rest of this chapter will review our present understanding of several of the most important and well-studied sensory organs. We will consider both transduction and selected examples of sensory coding of information, emphasizing the five common senses that have been the most thoroughly investigated. We will choose our examples primarily from arthropods and mammals, because the

molecular techniques available for *Drosophila* (fruit flies) and mice, for example, have made the sensory organs in these organisms the most attractive for current research. We will describe ionotropic sensory receptors first, beginning with touch and proceeding to the vestibular sense and hearing. Then, with taste in the middle (because it exhibits transduction mechanisms of both kinds), we will continue on to metabotropic sensory receptors, for olfaction and finally for vision. Sensory reception of polarized light and of magnetic stimuli is discussed in Chapter 18.

SUMMARY
Organization of Sensory Systems

- Sensory receptor cells respond to stimulation by a form of energy. Most sensory cells are specialized to respond to one form of stimulus energy.

- Sense organs contain clusters of similar receptor cells as well as nonneural cells.

- Receptor cells transduce stimulus energy into an electrical response, usually a depolarizing receptor potential. The transduction depends on specific receptor molecules and can be ionotropic (directly opening ion channels) or metabotropic (triggering a metabolic cascade via a G protein–coupled receptor, or GPCR).

- The receptor potential in a sensory neuron can trigger action potentials that propagate to the CNS.

- Sensory receptor cells often have cilia or microvilli that increase the area of the membrane surface.

Mechanoreception and Touch

All cells are somewhat responsive to mechanical stimulation, but **mechanoreceptors** are specialized to respond to different types of mechanical stimuli. Various mechanoreceptors mediate the senses of touch, pressure, equilibrium, and hearing, as well as certain types of osmotic stimulation. Some of these are treated separately later in this chapter. Here we introduce general features of sensory receptor function, using an insect mechanoreceptor.

Insect bristle sensilla exemplify mechanoreceptor responses

Insects have a hard exoskeleton covered with sensory bristles or hairs, each of which is a miniature sense organ called a **sensillum** (plural *sensilla*). Many sensilla are mechanosensory, whereas others are predominantly chemosensory (but may also contain a mechanoreceptor). Mechanosensory bristles are hollow and contain sensory neuron endings. The structure of a bristle sensillum is shown in **FIGURE 14.3A**. A bristle shaft or hair extends from the body of the insect as a part of its exoskeleton. The very tip of the mechanoreceptor cell's dendrite is attached to the inside of the bristle shaft at its base. When the bristle is moved, it deforms the membrane at the distal tip of the mechanoreceptor dendrite and opens *stretch-activated channels*. These channels open in response to stretch and allow cations to flow through the channel. The net inward current carried by cations produces a **receptor potential**, defined as the primary electrical response of a sensory receptor cell to stimulation, the output of sensory transduction. In bristle

(A) **An insect mechanosensory bristle**

(B) **Mechanoreceptor responses to bristle displacement**

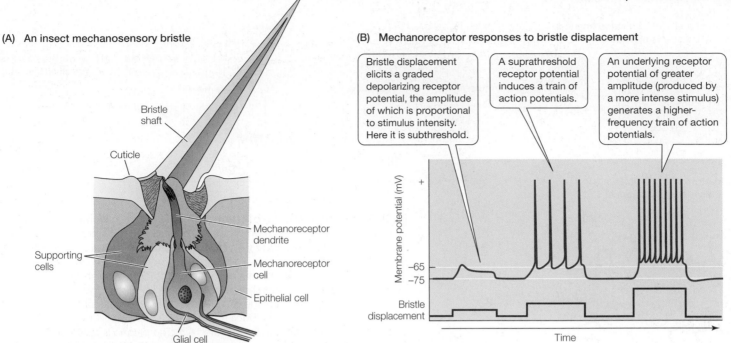

FIGURE 14.3 **Insect cuticular mechanoreception** (A) An insect mechanosensory bristle (bristle sensillum) contains a bipolar sensory neuron, the dendrite of which is distorted by movement of the bristle shaft, leading to activation of the neuron. (B) The general response of a receptor cell to stimulation, illustrated here with an insect sensillum, is to produce a graded, depolarizing receptor potential, the amplitude of which is proportional to the intensity of the stimulus. The receptor potential induces a train of action potentials, the frequency of which is a code for stimulus intensity. (A after Thurm 1964, Bullock and Horridge 1965, and Keil 1997.)

mechanoreceptor neurons, and in most sensory cells, the receptor potential is a depolarization (the membrane potential becoming less inside-negative, typically toward zero) (**FIGURE 14.3B**). The stretch-activated channels in the bristle mechanoreceptor membrane are nonselective cation channels, permeable to both Na^+ and K^+. Thus the ionic basis of the receptor potential is similar to the ionic basis of fast excitatory postsynaptic potentials (EPSPs), and like fast EPSPs, the receptor potential has a reversal potential near zero (see page 346).

The receptor potential, if suprathreshold, depolarizes the sensory neuron enough to generate action potentials, which then propagate to the CNS. The magnitude of the stimulus is *encoded* by the sensory receptor cell. Figure 14.3B shows that the stronger the stimulus deflecting the bristle, the greater the receptor potential depolarization and the higher the frequency of action potentials in the sensory neuron. The CNS, then, receives a coded message, which it decodes from the train of action potentials (as well as from the identity of the cell that is activated). The frequency of the action potentials denotes the strength of the stimulus, and the cell's identity denotes that the stimulus is a mechanical deflection of a bristle at a certain position on the insect's body.

What is the nature of the stretch-activated ion channels that respond to mechanical stimuli to generate a receptor potential? Ion channels sensitive to membrane stretch and pressure are actually rather widespread in cells, as was discovered after the invention of the patch-clamp technique. In the experiment depicted in **FIGURE**

14.4A, a patch-clamp electrode was sealed onto an embryonic skeletal muscle cell and then removed to form an isolated patch, much as we described earlier for voltage-gated Na^+ channels (see Figure 12.17). When pressure was applied directly to the membrane patch by sucking the patch farther into the electrode and stretching the membrane, a channel opened. The greater the negative pressure (indicated by millimeters of mercury), the more often and longer the channel opened. In a patch of membrane that has been pulled away from the cell, the opening of the channel must result directly from the stretch, rather than indirectly from a metabotropic cascade (because any second messenger would be washed away by the bathing solution).

Further information about stretch-activated channels underlying mechanoreception comes from molecular genetic studies. The ion channel responsible for the bristle mechanoreceptor response was identified by screening for mutations in *Drosophila*. These experiments identified a channel called NOMPC (for *no mechanoreceptor potential C*), the gene for which was then cloned. The channel protein's structure is shown in **FIGURE 14.4B**. The structure is somewhat similar to that of a voltage-gated K^+ channel or one of the domains of a voltage-gated Na^+ channel (see Figure 12.21) with six transmembrane sequences and a P loop that lines the channel pore. The NOMPC channel is not actually voltage-gated, however; it is a member of a different but related channel family, called the **TRP (transient receptor potential) channel** family (see Box Extension 12.1). As we will see, TRP channels are central to transduction in many kinds of sensory receptor cells, including receptors for touch, taste, smell, temperature, and some visual systems. **TABLE 14.2** summarizes our present knowledge of the major transduction ion channels in sensory receptor cells. The TRP channels in mechanoreceptors tend to have many *ankyrin repeats* (see Figure 14.4B), a protein structural motif used to link the protein to elements of the cytoskeleton. This linkage allows the channels to open in response to stretch, enabling cations to flow through the channels to produce a depolarizing receptor potential.

(A) Opening of a stretch-activated channel shown by patch-clamp recording

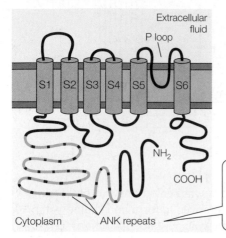

Pipette with patch of membrane

Stretch-activated channel

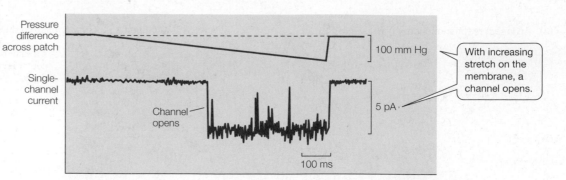

Pressure difference across patch

Single-channel current

Channel opens

100 mm Hg

5 pA

100 ms

With increasing stretch on the membrane, a channel opens.

(B) The structure of the NOMPC mechanosensory ion channel

Extracellular fluid

P loop

S1 S2 S3 S4 S5 S6

NH₂

COOH

Cytoplasm   ANK repeats

Ankyrin repeats attach the channel protein to cytoskeleton proteins, allowing stretch to open the channel.

FIGURE 14.4 Stretch-activated channels (A) Patch-clamp recording from a patch of membrane isolated from a mammalian muscle fiber (see Figure 12.17). An increasing pressure difference across the patch (shown as millimeters of mercury [mm Hg]) distorts the membrane patch and opens the stretch-activated channel (single-channel currents shown in picoamps [pA]; see Figure 12.17). (B) Protein subunit of the NOMPC ion channel of *Drosophila* (see page 373), which transduces mechanical deformation into an electrical receptor potential (see Figure 14.3). The protein is related to a voltage-gated K⁺ channel protein (see Figure 12.21) but without the voltage sensitivity; instead its ankyrin (ANK) repeats attach the protein to cytoskeletal components to provide mechanical sensitivity. Like the voltage-gated K⁺ channel protein, the NOMPC ion channel has six transmembrane α-helices (S1–S6) and a P loop. (Note: The structure of the mammalian channel shown in [A] may differ from that of the *Drosophila* channel shown in [B].) (A after Hamill 2006.)

Touch receptors in the skin of mammals have specialized endings

The organs mediating touch in the skin of mammals have been much studied, although the channels responsible for transduction have not yet been identified. Touch receptor cells in mammalian skin involve an association of epithelial cells with the distal endings of neurons that have their cell bodies in the dorsal root ganglia adjacent to the spinal cord (see Figure 15.3). These sensory neurons, called **dorsal root ganglion (DRG)** cells, send their distal processes into the skin and their central axons into the dorsal or sensory part of the spinal cord. The distal processes of the DRG cells form four kinds of specialized endings with epithelial cells, illustrated for the non-hairy

TABLE 14.2	Major transduction ion channels in sensory receptor cells			
Sense	**Type of animal**	**Kind of channel**	**Direct, ionotropic activation of channel by:**	**Indirect, G protein–coupled receptor activation of channel via:**
Touch	Insect	TRP[a]	Stretch	—
	Vertebrate	TRP	Stretch, temperature	—
Hearing	Insect	TRP	Stretch	—
	Vertebrate	?	Stretch (tip links)	—
Heat, cold	Vertebrate	TRP	Temperature	—
Taste	Insect	?	?	?
	Vertebrate	TRP (sweet, umami, bitter)	—	PLC → IP₃ → Ca²⁺
		ENaC (salt)	Na⁺	—
Smell	Insect	ORₓ/OR83b?	Odorant binding to ORₓ?	?
	Vertebrate	CNG (MOE)	—	cAMP
		TRP (VNO)	—	PLC → DAG/IP₃
Vision	Insect	TRP	—	PLC → DAG/IP₃
	Vertebrate	CNG	—	cGMP

[a]Abbreviations: cAMP, cyclic adenosine monophosphate; cGMP, cyclic guanosine monophosphate; CNG, cyclic nucleotide–gated channel; DAG, diacylglycerol; ENaC, epithelial sodium channel; IP₃, inositol trisphosphate; MOE, main olfactory epithelium; ORₓ/OR83b, different odorant receptor proteins may dimerize with OR83b to form or modulate a channel; PLC, phospholipase C; TRP, transient-receptor-potential channel; VNO, vomeronasal organ.

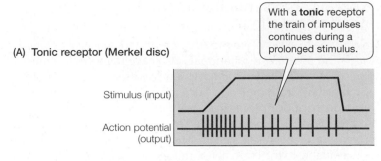

Meissner corpuscle (touch)

Merkel disc (touch and pressure)

Free nerve endings (pain, itch, temperature)

Ruffini ending (pressure)

Nerves

Pacinian corpuscle (vibration)

Sweat gland

Epidermis

Dermis

FIGURE 14.5 Mechanoreceptor cells in mammalian skin A small area of skin contains many mechanosensory endings of sensory neurons, the cell bodies of which are located in dorsal root ganglia (see Figure 15.3). The sensory endings have four kinds of specialized endings with epithelial cells. Merkel discs and Meissner corpuscles are superficial, just beneath the epidermis. Pacinian corpuscles and Ruffini endings are larger and more deeply located. All respond to mechanical stimulation. Free nerve endings respond to other stimuli.

skin of mammals in **FIGURE 14.5**. The most important of these specialized endings for the tactile sensing of form and texture is the *Merkel disc* just below the skin epidermis, formed by the association of a Merkel cell with a nerve ending. A single DRG cell sends its distal process into many fine branches, which end in association with several Merkel cells. The nerve endings respond directly or indirectly to indentation of the skin. It is not clear whether transduction occurs in the nerve endings or in the associated epithelial Merkel cells. The Merkel cells appear to contain and release neurotransmitters, and mutant mice that lack Merkel cells lack the sensitivity to light touch that is characteristic of the Merkel disc response. Adjacent to the Merkel discs are the *Meissner corpuscles*, formed from two to six sensory neuron endings associated with Schwann cells and collagen (not shown). Deeper within the skin are two more elaborate types of organs sensitive to mechanical stimuli, called *Ruffini endings* and *Pacinian corpuscles*. There are also sensory endings located around mammalian hair follicles, as well as free nerve endings that mediate pain and temperature sensation.

For most sensory receptors, the frequency of action potentials in response to a continuous and constant stimulation decreases over time, a process termed sensory **adaptation**. When recordings are made from the axons of DRG cells, two basic types of responses can be recorded, called **tonic** (slowly adapting) and **phasic** (rapidly adapting). Typical examples are given in **FIGURE 14.6**. Tonic responses decrease slowly in frequency and generally continue for as long as the stimulus is present. The DRG sensory endings associated with the Merkel discs and Ruffini endings give tonic responses. Phasic responses, because they adapt rapidly, generally signal *changes* in touch or pressure. Meissner corpuscles have phasic responses, giving a burst of action potentials as the stimulus is applied but ceasing during a maintained stimulus. Pacinian corpuscles deep within the skin adapt so rapidly that they normally give only a single action potential at the onset or offset of a prolonged stimulus. Thus they are sensitive only to sudden indentation of the skin or to vibration (which changes rapidly). The sensory ending of a Pacinian corpuscle is encased in a bulb formed by multiple layers of thin, concentric accessory cells called *lamellae*. These lamellae are mostly responsible for

(A) Tonic receptor (Merkel disc)

With a **tonic** receptor the train of impulses continues during a prolonged stimulus.

Stimulus (input)

Action potential (output)

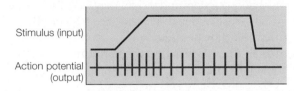

(B) Tonic receptor (Ruffini ending)

Stimulus (input)

Action potential (output)

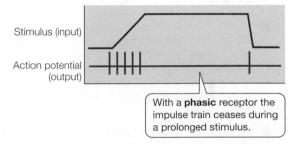

(C) Phasic receptor (Meissner corpuscle)

Stimulus (input)

Action potential (output)

With a **phasic** receptor the impulse train ceases during a prolonged stimulus.

(D) Extremely phasic receptor (Pacinian corpuscle)

Stimulus (input)

Action potential (output)

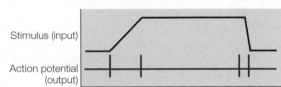

FIGURE 14.6 Tonic and phasic receptors (A) Tonic (slowly adapting) receptor cells have a slow and incomplete decrease in impulse frequency. A Merkel disc is a tonic receptor. (B) A Ruffini ending is also tonic in its response. (C) Phasic (rapidly adapting) receptor cells have a rapid, complete decrease in impulse frequency and thus convey information about *change* in stimulus intensity. A Meissner corpuscle is a phasic receptor. (D) A Pacinian corpuscle is extremely phasic, normally generating only a single action potential at the beginning and end of a sudden stimulus.

the phasic nature of the response, because when they are removed, the sensory ending of the Pacinian corpuscle responds more like a tonic receptor. The multiple layers of the lamellae apparently absorb the energy of the mechanical stimulus, so that only rapid changes in pressure are communicated deep within the lamellar wrappings to the sensory ending. Many sensory systems contain separate tonic and phasic receptor cells, but their mechanisms of sensory adaptation may differ. The Pacinian corpuscle depends on accessory structures to provide its phasic property, but other phasic sensory receptors do not have analogous accessory structures, and their phasic responses to a prolonged stimulus appear to depend on properties of the transduction channels or on other elements of the neuronal membrane.

Proprioceptors monitor internal mechanical stimuli

In addition to exteroceptive mechanoreceptors sensitive to external touch and pressure, most animals also have interoceptive mechanoreceptors, which monitor movement, position, mechanical stress, and tension within the body. Such internal mechanoreceptors are termed **proprioceptors** (from the Latin *proprius*, "one's own"). Strictly speaking, proprioceptors are mechanoreceptors associated with the musculoskeletal system. They provide most of the information available to the brain about muscle contraction, position, and movement of parts of the body, although other receptors, such as skin mechanoreceptors and vision, may make secondary contributions.

The best-understood proprioceptive organ in vertebrates is the **muscle spindle** organ, which monitors the length of a skeletal muscle. We discuss this sense organ in Chapter 19, along with consideration of its physiological roles.

SUMMARY
Mechanoreception and Touch

■ Mechanoreceptors have many sensory functions. In addition to surface mechanoreceptors that convey information about environmental touch and pressure, mechanoreceptors can serve as proprioceptors that monitor body and limb position and muscle length and force. (They can also serve as equilibrium and auditory receptors, as we will see in the next section.)

■ Mechanoreceptors have stretch-activated ion channels that mediate ionotropic transduction.

■ Many sensory receptors produce a response that diminishes over time and are said to adapt to sustained stimulation. Tonic (slowly adapting) receptors signal the intensity and duration of a stimulus, whereas phasic (rapidly adapting) receptors signal changes in stimulus intensity.

Vestibular Organs and Hearing

Most animals have mechanoreceptor organs for orientation to gravity and for sound detection. Even a simple animal like a jellyfish has an organ for orientation with respect to gravity, called a **statocyst**, some version of which is present in members of all animal phyla. A statocyst contains grains of sand or a secretion of calcium carbonate. This relatively dense mineral material sinks within the statocyst and stimulates receptor cells beneath it by bending their cilia. In this way an uncomplicated structure can provide reliable information about orientation relative to gravity, as well as movement and acceleration of the animal. Many kinds of animals also use mechanoreceptors to detect sound, by a variety of mechanisms.

Insects hear with tympanal organs

Sound consists of waves of compression of air or water, which propagate away from a vibrating source. Suppose we have a loudspeaker, whose cone or membrane is vibrating in and out. When the membrane pushes out, it compresses air molecules, momentarily increasing the air pressure. When the membrane vibrates back into the speaker, there is a momentary rarefaction of air molecules (the same number of molecules occupies more volume) and the pressure decreases. Therefore the vibrating speaker cone causes a repeating pressure wave in the air, the frequency of which is identical to the frequency of movement of the speaker cone. This wave of sound pressure propagates away from the speaker cone at the speed of sound, with a frequency that is a function of the rate of repetition of the pressure wave. Auditory organs are specialized to detect such waves of sound pressure and sometimes to measure their frequencies.

In insects the most common form of auditory organ is the **tympanal organ**, in which a thin cuticular tympanum (eardrum) is displaced by sound waves. Mechanosensory cells are attached to the tympanum and are stimulated by its movement, much like the receptor cell of the bristle sensillum. Tympanal organs may occur at any of several locations on an insect's body, including the thorax (noctuid moths), abdomen (locusts, cicadas), legs (crickets, katydids), or labial palps (sphingid moths). Auditory organs seem to have evolved repeatedly in different insect groups, at different locations on the body.

One of the simplest and best-studied tympanal organs is that of noctuid moths such as the moth at the beginning of this chapter. Each of the paired thoracic tympanal organs contains only two neurons that respond to sound. The frequency sensitivity ranges from 3 to 150 kilohertz (kHz), with maximum sensitivity at 50–70 kHz. Most of this range is *ultrasonic*—above the frequency range of sound audible to humans (20 Hz–20 kHz)—but it matches the frequency range of ultrasonic cries of echolocating bats (**BOX 14.1**). The two auditory cells of a moth's tympanic organ (called A_1 and A_2) respond similarly to ultrasonic pulses, but the threshold (lowest detectable sound intensity) of A_2 is 100 times higher than that of A_1. The cells convey no information about sound frequency. Sound intensity is coded by the impulse frequency in each receptor cell, by a shorter response latency with increasing intensity, and by recruitment of the high-threshold A_2 cell.

The behavioral significance and neurophysiology of auditory responses in moths have been well studied. Moth tympanal organs provide good directional information about the source of ultrasound. If a bat emits ultrasonic pulses to the left of a moth, the response of the moth's left "ear" (tympanal organ) will be greater. The left A_1 cell will respond with a shorter latency and a higher frequency than the right A_1 cell. Using this asymmetric response, the moth can turn away from the sound source and fly away. This response is effective for predator avoidance if the bat is distant, but because bats are stronger fliers, it is ineffective if the bat is close enough to detect echoes from the moth. A nearby bat emits an ultrasonic cry

---

BOX 14.1 Echolocation

Many kinds of bats have poor vision and yet fly well at night, avoiding obstacles and catching insects at rates as high as two per second. They orient by emitting ultrasonic pulses (i.e., sound at frequencies too high to be audible to humans) and detecting echoes reflected by objects around them. They are able to use the information in the auditory echoes to locate and discriminate prey insects, and catch them in the open and in wooded environments. **Box Extension 14.1** describes how echolocating bats detect and catch insects.

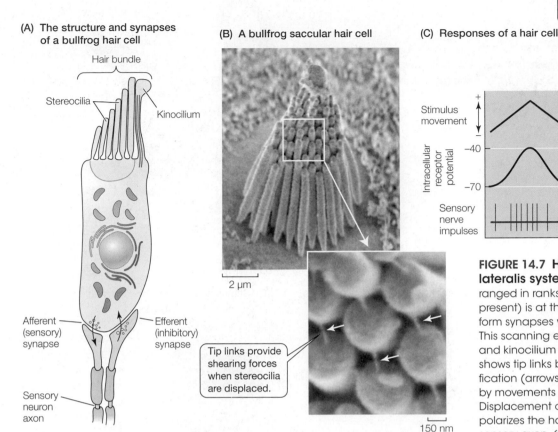

(A) **The structure and synapses of a bullfrog hair cell**

Hair bundle

Stereocilia

Kinocilium

Afferent (sensory) synapse

Efferent (inhibitory) synapse

Sensory neuron axon

(B) **A bullfrog saccular hair cell**

2 μm

Tip links provide shearing forces when stereocilia are displaced.

150 nm

(C) **Responses of a hair cell**

Movement of stereocilia toward the right stretches the tip links, opening ion channels.

Stimulus movement

Intracellular receptor potential

−40

−70

Sensory nerve impulses

FIGURE 14.7 Hair cells of the vertebrate acoustico-lateralis system (A) Hair cells contain stereocilia arranged in ranks from shortest to longest. A kinocilium (if present) is at the end with the longest stereocilia. Hair cells form synapses with afferent and efferent nerve axons. (B) This scanning electron micrograph shows the stereocilia and kinocilium of a bullfrog saccular hair cell. The inset shows tip links between adjacent stereocilia at high magnification (arrows). (C) Hair cells are depolarized and excited by movements of the stereocilia toward the kinocilium. Displacement of stereocilia away from the kinocilium hyperpolarizes the hair cell, decreasing impulse frequency in the sensory axon. (Micrographs courtesy of Peter Gillespie; from Strassmaier and Gillespie 2002.)

loud enough to stimulate one or both A_2 cells, and triggers a very different response from the moth. Instead of turning away, the moth flies erratically or dives to the ground—responses presumably more evolutionarily adaptive than an attempt to "outrun" the bat once detected. In addition, some moths such as the toxic dogbane tiger moth (*Cycnia tenera*) emit ultrasonic clicks, using a separate thoracic sound-producing tymbal organ, in response to attacking bat cries. These clicks may confuse or "jam" the bat's sonar system and may also serve to advertise the moth's unpalatability.

Most insect tympanal organs are similar to those of moths, in that they are sensitive detectors and encoders of sound intensity over a certain range of frequencies but are poor at detecting frequency differences. Some insects, however, have a limited ability to discriminate sound frequencies, and others have specialized tympanal organs with which they can localize relatively low-frequency sounds. There is some evidence in *Drosophila* that the ion channels responsible for hearing are TRP channels related to the NOMPC channels of bristle sensilla, but characterization of auditory transduction in insects is not yet complete.

Vertebrate hair cells are used in hearing and vestibular sense

Hair cells are the sensory mechanoreceptor cells in the vertebrate acoustico-lateralis system, which includes the vestibular organs (for balance and detection of acceleration), the lateral line system of surface receptors in fish and amphibians (which detect water flow as well as other stimuli), and the mammalian *cochlea*, an auditory organ that we discuss shortly. A vertebrate **hair cell** (**FIGURE 14.7**) is an epithelial cell, and like all epithelial cells, it has an api-

cal surface that faces an overlying lumen and a basal surface that faces underlying tissues. A hair cell has at its apical end a tuft of microvilli. These microvilli are collectively called a **hair bundle** because they resemble microscopic hairs. The individual microvilli have the unfortunate name of *stereocilia* (singular *stereocilium*)—unfortunate because they are most definitely not cilia but instead are true microvilli, containing numerous actin fibers that make them rigid. Hair cells do not possess axons and do not generate action potentials. Instead, they release neurotransmitter substance onto afferent neurons that conduct action potentials into the CNS.

The hair cells of the acoustico-lateralis system all work in the same basic way: The bending of the stereocilia is transduced into a receptor potential. The stereocilia of a hair bundle are arranged in order of increasing height (see Figure 14.7A). Hair cells of some species may also have a single true cilium, called a *kinocilium*, as shown in Figure 14.7A for the vestibular hair cells of a bullfrog. Mammalian auditory hair cells, however, do not have kinocilia, and even in the bullfrog the kinocilium can be removed with no apparent loss of function. The actin in a stereocilium makes it rigid, and the diameter of each stereocilium narrows at its junction with the rest of the hair cell, as shown in the scanning micrograph of Figure 14.7B. When pushed to the side, the stereocilia pivot at their bases. The pivoting of the displaced stereocilia produces a shearing force between neighboring stereocilia that is transduced into a change in membrane potential. Hair cells are directionally sensitive, as shown in Figure 14.7C. Displacement of the hair bundle toward the tallest of the stereocilia depolarizes the hair cell and increases the amount of neurotransmitter it releases (from a resting level). The increased release of neurotransmitter onto

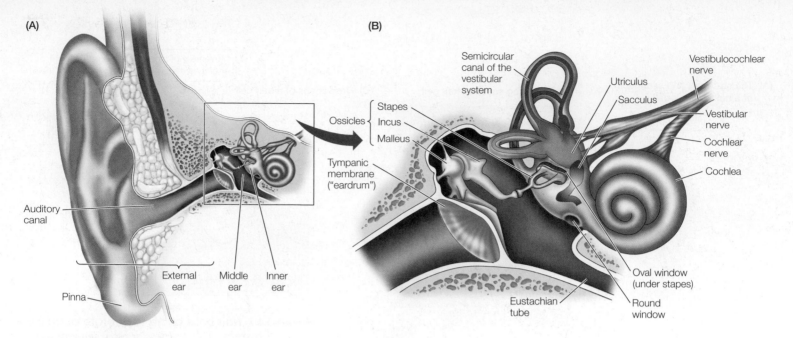

FIGURE 14.8 **Anatomy of the mammalian ear** (A) Structure of the human ear. (B) Components of the inner ear. The semicircular canal receptors are stimulated by head rotation. The utriculus and sacculus contain macular hair cells that are stimulated by linear motion of the head and by gravity. The cochlea contains auditory receptors. (C) The three semicircular canals of the inner ear are at approximately right angles to each other, so that any angular movement of the head stimulates at least one of them. (D) With rotation of the head, fluid movement in the canal stimulates the hair cells.

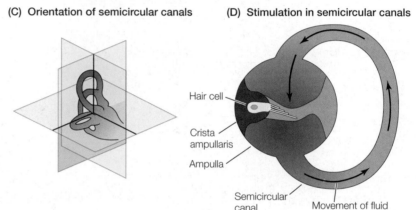

postsynaptic cranial sensory neurons increases the frequency of action potentials they produce. Displacement toward the shortest of the stereocilia hyperpolarizes the cell, decreases the transmitter released from the resting tonic level, and decreases cranial sensory neuron activity.

The stereocilia are joined to their neighbors by filamentous *tip links* (see arrows in Figure 14.7B inset), which are directly involved in producing the hair-cell response. When hair cells are exposed to a solution with a very low Ca^{2+} concentration (say 1 micromolar [μM]), the tip links break and the voltage response of the hair cell is abolished. Over a period of about 24 h in normal levels of Ca^{2+}, the tip links reform, and the hair-cell response returns with approximately the same time course. Movement of the stereocilia in one direction presumably stretches the tip links and opens cation channels near the outer ends of the stereocilia, permitting ion fluxes that depolarize the cell. Movement in the opposite direction causes the channels to close. The channels open within microseconds of displacement, permitting the receptor potentials of auditory hair cells to follow sound frequencies of several kilohertz. Because of this extraordinarily fast response, investigators believe that the forces exerted by the attached tip links directly gate the channel opening. The channels themselves are nonselective cation channels, and there appear to be only a small number in each hair cell—perhaps no more than one or two at the end of each tip link. The molecular identity of the channels is not yet known, in part because of the small number of channels per cell.

Vertebrate vestibular organs sense acceleration and gravity

Vertebrate **vestibular organs**, by definition, subserve sensory functions of acceleration and balance. The paired vestibular organs lie adjacent to the auditory organs in the *inner ear* and use similar hair cells. They are called vestibular because their hair cells project into fluid-filled chambers reminiscent of tiny rooms. Because of their complex geometry, the vestibular chambers and the neighboring chambers of the cochlea are collectively termed the *labyrinth*. **FIGURE 14.8A** shows the structure of the human ear. The *external ear* consists of the pinna and air-filled external auditory canal. The eardrum separates the external ear from the *middle ear*, which is also air-filled and contains the bones that transmit sound vibrations from the eardrum to the inner ear. We return later to these structures in the section on hearing. **FIGURE 14.8B** shows the inner ear, including the vestibular organs and cochlea, in greater detail. Each vestibular organ comprises three **semicircular canals**, which detect angular acceleration of the head and body, and two **otolith organs**, called the **sacculus** and **utriculus**, which detect linear movement and acceleration. The cochlea is involved in hearing. All of these chambers are filled with fluid, and many are continuous with each other.

The three semicircular canals on one side of the head are oriented at approximately right angles to one another, with one canal in each of the x, y, and z planes (**FIGURE 14.8C**). This conformation allows detection of angular movement in all three axes in space. At the base of each canal, a region called the *ampulla* contains a cluster of hair cells in a structure called the *crista ampullaris* (**FIGURE 14.8D**). Acceleration of the head causes fluid (endolymph) in the ampulla to slosh against the hair bundles of the hair cells, like water sloshing in a bowl when the bowl is suddenly moved. This movement of fluid pushes against the crista ampullaris and deflects the bundles of the hair cells, opening or closing mechanoreceptive channels as in the bullfrog hair cells. Because for each of the three major geometrical planes (*x, y, z*) there are two semicircular canals, one on each side of the head, rotational movements may produce sloshing of fluid in one direction on the right side of the head and in the other direction on the left side. As a result, the hair cells may depolarize on one side and hyperpolarize on the other. This information, carried into the CNS by axons of afferent neurons, makes possible a determination of the direction of head movement.

The utriculus and sacculus also contain hair cells. The hair cells in these structures are contained in a region called the **macula** (plural *maculae*). The macula is oriented approximately horizontally for the utriculus and vertically for the sacculus. The hair cells are covered by a gelatinous mass called the *otolithic membrane* into which the hair bundles protrude. A dense network of crystals of calcium carbonate lies on top of the otolithic membrane. The maculae are sensitive to orientation with respect to gravity as well as head movements. When the head moves, these crystals (collectively called an *otolith*) tend to lag behind as an inertial mass, causing the otolithic membrane to slide against the hair bundle. This deflects the hair bundle and produces a change in hair cell membrane potential.

Sound stimuli create movements in the vertebrate cochlea that excite auditory hair cells

The mammalian ear is adapted to receive and amplify sound-pressure waves, in order to detect both the amplitude and frequency of sound. High-frequency waves produce *high-pitch* sounds, and low-frequency waves produce *low-pitch* sounds. The *intensity* (loudness) of a sound depends on the amplitude of the sound waves, which correlates with how far a sound source, such as a speaker cone, pushes in and out.

The mammalian ear consists of three parts: an **external ear** distal to the eardrum, an air-filled **middle ear**, and a liquid-filled **inner ear**, which consists, in part, of the **cochlea** (see Figure 14.8A and B). Sound-pressure waves vibrate the eardrum (tympanic membrane), and this vibration is transmitted to the membranous *oval window* of the inner ear (see Figure 14.8B) by three middle-ear ossicles: the *mulleus, incus*, and *stapes*. In addition to the ossicles, the middle ear contains two muscles, the tensor tympani (associated with the eardrum) and the stapedius (associated with the stapes). These muscles can contract to damp the movements of the ossicles, protecting the auditory membranes from damage by loud sounds. The eustachian tube connects the middle ear with the pharynx, equalizing pressure in the middle ear with environmental pressure.

The major function of the middle-ear ossicles is to transfer sound energy from air to the liquid of the inner ear. Airborne sound striking a liquid surface is almost all reflected; only about 1/30th of the sound energy is transferred to the liquid. This is why you cannot hear your friends at poolside when you are swimming under water. The energy transfer is poor because liquids cannot be compressed, so they have a low volume of movement in response to sound pressure. These considerations mean that for significant transfer of energy from the eardrum (vibrating in air) to the liquid medium of the inner ear, the pressure developed at the inner ear must be augmented. The middle-ear ossicles achieve this increase in pressure by applying forces from a relatively large area (55 mm^2 for the human eardrum) onto a much smaller area (3 mm^2 for the foot plate of the stapes, which covers the oval window). The concentration of force on a small area provides the necessary increase in pressure, allowing efficient transfer of sound energy.

The cochlea is a coiled tube containing chambers filled with fluid. A multicellular, membranous structure termed the **basilar membrane** separates the cochlea into an upper chamber (scala vestubuli) and a lower chamber (scala tympani). As the oval window moves, it creates fluid movement in the cochlea. This movement vibrates the basilar membrane and stimulates the auditory hair cells that sit on it. The basilar membrane varies in width and thickness along its length: It is narrow, thick, and rigid at the base of the cochlea near the oval window, and wider, thinner, and more flexible at the apex, farthest from the oval window (**FIGURE 14.9A**).

(A) The basilar membrane (top view)

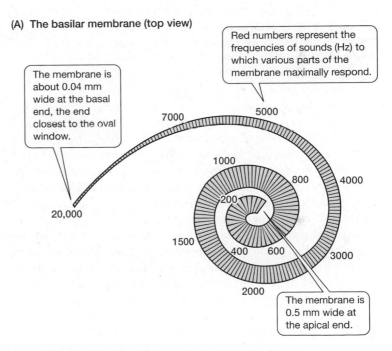

Red numbers represent the frequencies of sounds (Hz) to which various parts of the membrane maximally respond.

The membrane is about 0.04 mm wide at the basal end, the end closest to the oval window.

The membrane is 0.5 mm wide at the apical end.

(B) A diagram with the cochlea unwound (side view)

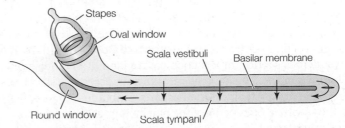

FIGURE 14.9 Anatomy of the cochlea (A) The surface of the basilar membrane, seen from above. The basilar membrane is narrower and stiffer at its basal end (near the oval window) than at its apical end. (B) A diagrammatic representation of how the inner ear would appear if the cochlea were unwound. The basilar membrane (seen in side view) separates the upper scala vestibuli from the lower scala tympani.

FIGURE 14.10 Amplitude of movement of the basilar membrane differs at different sound frequencies (A) Von Békésy's results with cadavers and loud sounds showed that the position of maximum movement is a function of sound frequency. (B) In living cochleas, amplification of basilar-membrane movement by the cochlear amplifier produces a sharper peak of membrane displacement. Furosemide blocks hair cell transduction, decreasing the amplitude of basilar-membrane displacement by disrupting the cochlear amplifier. In (B) the amplitudes of movement (shown in the y axis) are magnified relative to the position along the length of the basilar membrane (shown in the x axis).

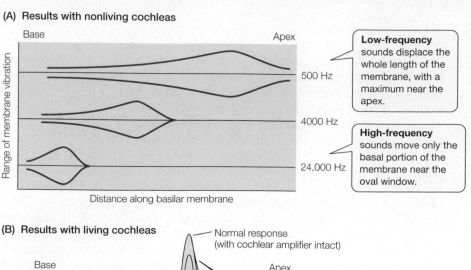

(A) Results with nonliving cochleas

Low-frequency sounds displace the whole length of the membrane, with a maximum near the apex.

High-frequency sounds move only the basal portion of the membrane near the oval window.

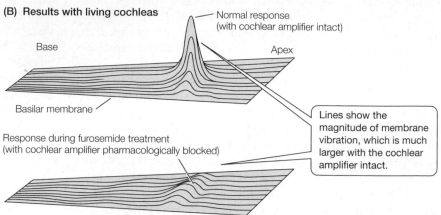

(B) Results with living cochleas

Normal response (with cochlear amplifier intact)

Basilar membrane

Response during furosemide treatment (with cochlear amplifier pharmacologically blocked)

Lines show the magnitude of membrane vibration, which is much larger with the cochlear amplifier intact.

The structure of the cochlea is more easily understood if we schematically uncoil it, as shown in **FIGURE 14.9B**. Sound waves pushing in at the oval window set up traveling pressure waves in the fluid-filled cochlea, and these produce minute movements of the basilar membrane. The differences in width and rigidity along its length give the basilar membrane a variable mechanical compliance, so that sound waves of different frequencies vibrate the basilar membrane maximally at different points along its length (see Figure 14.10). High-frequency pressure waves vibrate the stiffer, narrower part of the basilar membrane at the basal end near the oval window. Low-frequency sounds maximally vibrate the basilar membrane at its broader apical end.

The selective movement of the basilar membrane was first described by Georg von Békésy (1899–1972). Using cochleas of human cadavers, he demonstrated the traveling wave of basilar membrane movement and showed that for every frequency there is a different place of maximum amplitude of the traveling wave along the length of the basilar membrane (**FIGURE 14.10A**). Modern experiments on living cochleas have used emission of gamma rays to measure basilar-membrane movements. These experiments have confirmed von Békésy's observations in general, but they show that the basilar-membrane movements are more sharply localized than his measurements indicated (**FIGURE 14.10B**). The greater resolution of the living cochlea can be blocked by inhibiting the hair cells with a chemical called furosemide, which is known to block hair cell transduction (**FIGURE 14.10C**). This result shows that there must be some active component in the living cochlea that contributes to sound localization along the basilar membrane. This active component of the basilar membrane response to sound is called the **cochlear amplifier**, which we will discuss later.

When the basilar membrane moves, it stimulates the hair cells in a region of the cochlea called the **organ of Corti. FIGURE 14.11A** shows a cross section of the cochlea, divided by the horizontal basilar membrane. The hair cells and various accessory structures of the organ of Corti sit on the basilar membrane and vibrate up and down with it. Stereocilia of the hair cells project into a separate fluid compartment of the cochlea, called the *scala media*.

The hair cells of the organ of Corti are of two kinds: Typically there are three rows of **outer hair cells** and a single row of **inner hair cells** (**FIGURE 14.11B**). Auditory hair cells are similar to the bullfrog hair cells described earlier, but they lose their kinocilium during development and thus have only stereocilia. The hair cells are covered by a flap of tissue termed the *tectorial membrane*. The stereocilia of the hair cells are very close to or in contact with the tectorial membrane (**FIGURE 14.12A**). As the basilar membrane moves up and down, it causes the stereocilia to push up against the tectorial membrane, so that the hair bundle is displaced. Displacement in one direction depolarizes the hair cell membrane potential, and displacement in the opposite direction hyperpolarizes it. Intracellular responses of auditory hair cells have been recorded—a feat made difficult by the small size and inaccessibility of the cells. Both inner and outer hair cells depolarize in response to tones, the amount of depolarization depending on sound intensity and frequency. Auditory hair cells make synaptic contact with afferent neurons of the auditory (cochlear) nerve and also receive efferent synapses. Most (80%–95%) of the afferent neurons synapse with the *inner* hair cells, which in humans represent only about 20%–25% of the approximately 24,000 hair cells. Thus the inner hair cells are the major source of auditory input to the brain.

The outer hair cells also respond to sounds and appear to be largely responsible for the cochlear amplifier in mammals (see Figure 14.10B). Outer hair cells change length by as much as 4% in response to changes in their membrane potential (**FIGURE 14.12B**). These movements are extremely fast and can track frequencies of several kilohertz. Changes in length of outer hair cells in response to sound-induced receptor potentials probably amplify the local movement of the basilar membrane and thus amplify the forces acting on nearby inner hair cells. A motor protein in the outer hair cell membrane called *prestin* seems to be responsible for the cell's shortening and lengthening, because outer hair cells of knockout mice lacking prestin no longer move in response to changes in membrane potential. Mice lacking prestin show a dramatic loss in auditory sensitivity, indicating that this protein and the outer hair cells play an important role in the function of the cochlea. However, some vertebrate species lack outer hair cells or prestin but still appear

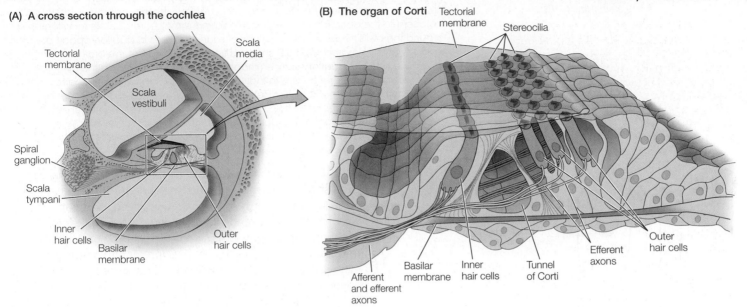

(A) A cross section through the cochlea

Tectorial membrane

Scala media

Scala vestibuli

Spiral ganglion

Scala tympani

Inner hair cells

Basilar membrane

Outer hair cells

(B) The organ of Corti

Tectorial membrane

Stereocilia

Outer hair cells

Afferent and efferent axons

Basilar membrane

Inner hair cells

Tunnel of Corti

Efferent axons

FIGURE 14.11 Internal structure of the mammalian cochlea
(A) A cross section through one turn of the cochlea. (B) Detail showing inner and outer hair cells of the organ of Corti, which sits on the basilar membrane. Both inner and outer hair cells synapse with afferent (sensory) and efferent (motor) axons, but most afferent axons receive synaptic input from inner hair cells, and most efferent axons end on outer hair cells, which are thought to magnify local movements and serve as a cochlear amplifier.

to have some cochlear amplification. Thus there may be additional mechanisms for a cochlear amplifier in some groups.

The organization of hair cells on the basilar membrane allows for specific populations of the hair cells to respond to specific sound frequencies, and for their respective afferent neurons to send action potentials to the CNS. In this way, coding for high- and low-frequency sounds (high and low pitch) is *spatially mapped* on the basilar membrane. This spatial coding of *frequency information* is maintained in the auditory pathways leading to the auditory cortex. As in other sensory systems, information about the intensity (loudness) of the sound stimulus is conveyed by the frequency of action potentials produced in afferent neurons (which in turn depends on the amount of neurotransmitter released by the hair cells).

The localization of sound is determined by analysis of auditory signals in the CNS

Vertebrates analyze sound in various ways. For example, human brains contain pathways that are most sensitive to tones, clicks, and speech sounds, and different sorts of information are extracted from these different kinds of sound stimuli. We will consider only one

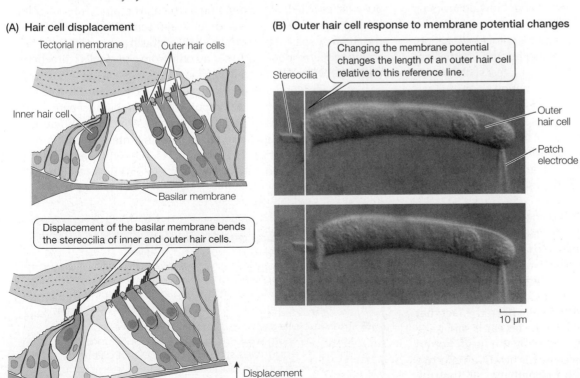

(A) Hair cell displacement

Tectorial membrane

Outer hair cells

Inner hair cell

Basilar membrane

Displacement of the basilar membrane bends the stereocilia of inner and outer hair cells.

Displacement of basilar membrane

(B) Outer hair cell response to membrane potential changes

Stereocilia

Changing the membrane potential changes the length of an outer hair cell relative to this reference line.

Outer hair cell

Patch electrode

10 µm

FIGURE 14.12 Movement of the basilar membrane stimulates auditory hair cells (A) Basilar membrane displacement bends the stereocilia of hair cells against the overlying tectorial membrane. This bending transduces sound vibration into electrical signals of the hair cells. (B) Outer hair cells change length in response to changes in their membrane potential. Depolarization of an isolated outer hair cell causes it to shorten (below), whereas hyperpolarization causes lengthening (above). These rapid movements provide positive feedback to amplify the movements of the basilar membrane and enhance hearing sensitivity. (Micrographs courtesy of Jonathan Ashmore; from Holley and Ashmore 1988.)

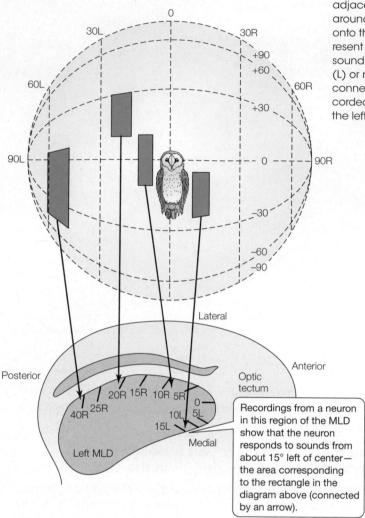

FIGURE 14.13 A map of auditory space in the brain of a barn owl Individual neurons in the midbrain auditory area (MLD, nucleus mesencephalicus lateralis dorsalis, adjacent to the optic tectum) respond to sound stimuli from discrete locations in space around the owl's head, so the locations of sound sources in auditory space are mapped onto the midbrain area. The MLD shown is from the left side of the brain. Rectangles represent the areas of auditory space (mapped onto a globe around the owl) from which sound sources evoke responses in individual space-specific neurons. Sounds to the left (L) or right (R) excite neurons in different positions along the MLD, as shown by the lines connecting the rectangles to the positions in the MLD in which neuronal activity was recorded. Only the horizontal component of the MLD map is shown; 10L means 10° to the left of center, 5R is 5° to the right, and so on. (After Knudsen and Konishi 1978.)

Recordings from a neuron in this region of the MLD show that the neuron responds to sounds from about 15° left of center—the area corresponding to the rectangle in the diagram above (connected by an arrow).

aspect of CNS auditory information processing: How do animals localize the source of a sound stimulus?

Humans are capable of reasonably accurate auditory localization, especially in the horizontal (left–right) plane, but our abilities are greatly overshadowed by those of other vertebrates, such as owls and bats. We will first examine the basic information an animal can use to localize sound, and then consider how this information is processed in the brain. We start with the observation that a single ear can provide no information about where a sound comes from; any auditory localization requires *comparison* of the responses of two ears. We can clarify this statement by examining the two sorts of information used in auditory localization: sound *time* difference and sound *intensity* difference.

First, any sound that is not straight ahead (or behind) will have a **time difference**, arriving at the two ears at slightly different times. A sound from a source that is offset to the left will reach the left ear first, with a difference in time of arrival that increases with increasing offset. Second, an offset sound will have an **intensity difference**: It will be louder in the ear that more directly faces the sound source. The intensity difference results from the fact that sounds do not go around corners well, so the far ear is said to be in a *sound shadow*, shielded from the sound by the head. Sound shadowing is a function of frequency because the head is an effective barrier to high-frequency sounds (with wavelengths smaller than the head) but little barrier to low-frequency sounds (with wavelengths

bigger than the head). Because high-frequency sounds have more sound shadowing, the difference in sound intensity between the ears is significant only for high-frequency sounds. For humans, both time differences and intensity differences provide information about horizontal (right–left) location of a sound source, but we have limited ability to localize the vertical (up–down) origin of sounds.

Owls, by contrast, have excellent abilities to localize sounds both horizontally and vertically, and using this ability, they can catch mice hidden under leaf litter. Like other vertebrates, a barn owl uses time differences to determine left–right origin of sounds. Both owls and echolocating bats (see Box 14.1) can use time differences as small as 10 microseconds (μs) in auditory localization! The owl's ability to localize the *vertical* component of a sound depends on a structural asymmetry of its two ears: An owl's right ear points more upward; its left ear, more downward. Barn owls have a ruff of sound-reflecting feathers around the face that act as parabolic reflectors, sharpening the directionality of hearing and enhancing the right–left asymmetry. Because of this asymmetry, at higher sound frequencies (with more sound shadowing), a sound from above will stimulate the right ear more than the left, so the intensity difference between ears will code the vertical component of a sound source.

How does the owl's brain put together the information about the horizontal axis of a sound source (coded primarily by time differences) and the vertical axis (coded by intensity differences)? Neurons in a region of the owl midbrain (the nucleus mesencephalicus lateralis dorsalis, or MLD, corresponding to the mammalian inferior colliculus) represent a map—an orderly representation in the brain of the auditory space around the owl's head. Each neuron is said to be *space-specific*, responding to sound only from a particular direction (**FIGURE 14.13**). Nearby cells respond to sound from nearby areas, so the outside world of sound sources is mapped in two dimensions onto the MLD. This map is computed from two parallel pathways in the brainstem and midbrain—one processing time-difference (left–right) information, and one processing intensity-difference (up–down) information. Presumably it is this computed auditory map that allows an owl to localize mice in darkness to within 1° of the sound source.

SUMMARY
Vestibular Organs and Hearing

■ Hair cells are sensitive and versatile vertebrate mechanoreceptors that transduce displacement of stereocilia into a receptor potential. They are the major receptors of vertebrate hearing and equilibrium sense.

■ The structure of the vertebrate ear effectively conveys sound-pressure waves into the inner ear. Sounds of different frequencies stimulate hair cells at different locations along the length of the basilar membrane of the cochlea.

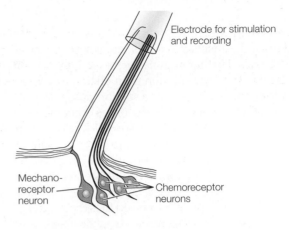

■ Central auditory pathways of vertebrates sort coded information about sounds in order to discriminate and map different sound frequencies and locations. The auditory systems of insects, although less complex, can nonetheless provide them with behaviorally important information.

Chemoreception and Taste

Chemoreception is the sensory response to a chemical stimulus. Chemoreception emerged very early in evolution. Even bacteria are able to detect attractive and repellent chemicals in their environment. The cell membrane of a bacterium contains receptor proteins for a variety of chemicals, including sugar, amino acids, and small peptides. The binding of an attractant to one of these receptors produces a change in the phosphorylation of a second messenger, which directly regulates the movement of the flagellum and directs the bacterium toward the food source. Animals detect chemicals with complex and sophisticated systems, which are the subject of this and the next section of this chapter. We divide chemoreceptive systems into two major categories: **taste** (the gustatory sense) and **olfaction** (the sense of smell).

In terrestrial animals, the distinction between taste and olfaction is relatively clear. Taste is mediated by specialized chemoreceptive organs generally located in or near the animal's mouth, and the stimuli are in liquid form. Olfaction also occurs in specialized structures, for example on antennae in insects and other arthropods and in the nose in vertebrates. The stimuli for olfaction are airborne. They must, however, dissolve in the liquid of the insect bristle or mucus of the vertebrate nasal passage before binding to the receptor molecules.

For animals that live in water, the distinction between taste and smell is less clear, because all chemical stimuli are dissolved in water in the aquatic environment. Nevertheless, the terms *taste* and *olfaction* remain useful because many aquatic animals have separate chemoreceptive organs that serve different functions and respond differently to chemical stimuli at lower concentrations and at higher concentrations. Lobsters, for example, orient and search in response to chemicals at low concentrations (presumably from distant, smell-like sources), which stimulate chemoreceptors on the antennules. High concentrations of chemicals (tastes) can trigger feeding movements by stimulating chemoreceptors on the mouthparts. Fish also have separate organs of taste and smell. They have taste buds structurally similar to mammalian taste buds, not only near the mouth but on the skin, which help them decide what to eat. Fish also have a nasal olfactory epithelium with receptor cells similar to our own, which can sense chemicals in dilute concentration (often from distant sources). It is with its nose that a female salmon finds its breeding ground.

Because the distinction between taste and olfaction is clearest among terrestrial animals, and the greatest progress has been made understanding chemoreception in terrestrial insects and mammals, we focus on these groups in our examination.[2]

---

[2] We should point out, however, that chemoreception includes more than just taste and olfaction. In mammalian nasal chemoreception alone there are multiple receptor fields and their pathways: the main olfactory organ; the accessory olfactory, or vomeronasal, organ (discussed later); and at least three other chemosensory organs. Mammals, other vertebrates, and invertebrates often have multiple types of chemoreception not readily described as taste or olfaction.

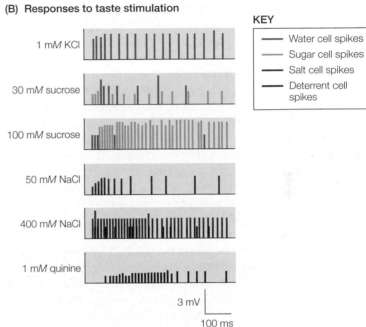

(B) Responses to taste stimulation

1 mM KCl

30 mM sucrose

100 mM sucrose

50 mM NaCl

400 mM NaCl

1 mM quinine

3 mV |

100 ms

KEY
— Water cell spikes
— Sugar cell spikes
— Salt cell spikes
— Deterrent cell spikes

FIGURE 14.14 *Drosophila* **taste sensillum and its responses to different stimulatory solutions** (A) A typical taste sensillum on the leg tarsus contains four chemoreceptor cells and a mechano-receptor cell. A pipette electrode both stimulates the sensillum and records action potentials from the sensory cells. (B) Each of the four chemoreceptor cells has spikes (extracellularly recorded action potentials) of a different size. Here the spike responses of different cells are rendered in different colors. The water cell responds best to water (with just enough salt to conduct charges), the sugar cell responds best to sugar solutions, and the salt cell responds best to increasing salt concentrations. The deterrent cell responds to quinine and other bitter alkaloids and also to high concentrations of salt.

Insect taste is localized at chemoreceptive sensilla

Insect taste receptor cells are located within bristles or sensilla, similar in their anatomy to the bristle mechanosensory sensillum of Figure 14.3A. Within the shaft of a taste sensillum there are dendrites of two to four chemoreceptor cells and often a single mechanoreceptor (**FIGURE 14.14A**). The shaft of the sensillum has one or more tiny holes at its tip, through which water and taste molecules can enter. In flies such as the blowfly, these sensilla are located on the *tarsus* (the terminal segment of the leg) and the *labellum* (the tip of the extensible proboscis or mouth used for feeding). A fly detects food in part by stepping on it. Stimulation of even a single tarsal sensory sensillum with a sugar solution elicits extension of the proboscis. Further stimulation of a labellar sensillum on the proboscis with sugar solution then elicits drinking behavior, in which the proboscis works as a suction pump.

FIGURE 14.14B shows the responses of chemoreceptor cells in a tarsal taste sensillum to different stimuli. The cells were stimulated by placing a pipette containing plain water or water with varying concentrations of sugar, salt, or quinine on top of the sensillum shaft, so that the stimulating fluid could enter the tiny holes at the shaft tip. The chemoreceptor cells produce action potentials, which can be recorded with an electrode. In flies there is generally one cell (termed the *sugar cell*) that responds most strongly to solutions of sugars. The frequency of action potentials in the sugar cell increases with increasing concentration of sugar and thus encodes information about sugar concentration. A second cell (the *salt cell*) responds preferentially to a range of salts, particularly to monovalent cations. A third cell (the *water cell*) responds best to plain water (although a very dilute salt concentration is necessary for recording). The response of the water cell diminishes with increasing concentration of any dissolved substances. Some sensilla have a fourth cell termed the *deterrent cell*, which responds to deterrent alkaloids as well as to very high salt concentrations. (The deterrent cell is sometimes called the *bitter cell*, by imperfect analogy with bitter taste in mammals.)

The molecular mechanisms of insect taste transduction are diverse and not completely clear. Responses to sugars and to bitter substances are probably mediated by receptor proteins with seven transmembrane segments. In *Drosophila* a large family of proteins has been described, the genes for which are specifically expressed in the taste bristles. This family of protein molecules (called *GR*, or gustatory *receptor*, proteins) comprises 68 proteins in several subfamilies, which are selectively expressed in genetically predetermined places such as on the labellum of the proboscis. A single gustatory receptor cell expresses the genes for several receptor proteins, but with genes for different groups of proteins expressed in different cells. In some cells, receptor proteins of one subfamily appear to mediate taste of sugars, and in different cells the proteins of another subfamily detect deterrent compounds, that is, compounds that signal repulsion from the stimulus. A third subfamily of receptor proteins is implicated in detecting attractive and inhibitory pheromones in courtship. (GR proteins are unrelated to G protein–coupled receptors [GPCRs], and their N-terminal and C-terminal ends are on opposite sides of the membrane from GPCRs.) Their mechanism of action is not clear but may be ionotropic. Mechanisms of detecting salt and water are probably ionotropic (as for mammals; see below) but are little understood.

Taste in mammals is mediated by receptor cells in taste buds

Vertebrate taste receptor cells are epithelial sensory cells that synapse onto terminals of cranial sensory neurons, which then carry the signals to taste centers in the brain. The taste cells are grouped together on the tongue and back of the mouth in **taste buds**. In mammals, the taste buds of the tongue are confined to small swellings known as **papillae** (**FIGURE 14.15A**), of which there are three kinds: the *fungiform papillae* near the front of the tongue, the *foliate papillae* on the lateral borders, and the large *cir-*

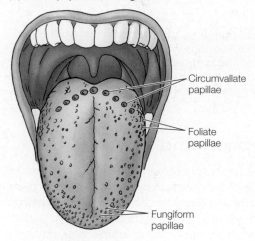

(A) Taste papillae on tongue

Circumvallate papillae

Foliate papillae

Fungiform papillae

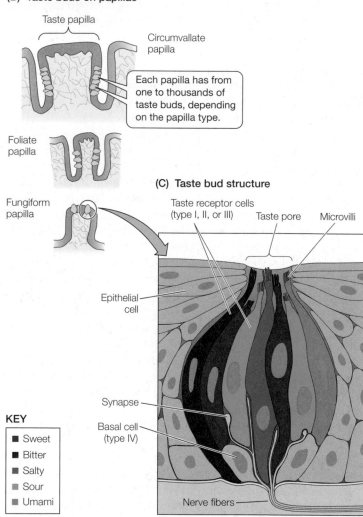

(B) Taste buds on papillae

Taste papilla

Circumvallate papilla

Each papilla has from one to thousands of taste buds, depending on the papilla type.

Foliate papilla

Fungiform papilla

(C) Taste bud structure

Taste receptor cells (type I, II, or III)

Taste pore

Microvilli

Epithelial cell

Synapse

Basal cell (type IV)

Nerve fibers

KEY
■ Sweet
■ Bitter
■ Salty
■ Sour
■ Umami

FIGURE 14.15 Mammalian taste buds (A) Taste buds are localized at taste papillae on the tongue. (B) Different kinds of taste papillae contain differing numbers of taste buds. (C) A taste bud contains many taste cells of different types; some form discrete synapses onto afferent sensory neurons. The taste cells of a taste bud extend microvilli through a taste pore to contact saliva. Receptor molecules in the microvillar membrane are exposed to taste stimuli at the surface of the tongue. Each taste bud cell contains receptor molecules for one taste quality (sweet, bitter, salty, sour, or umami). (C after Kandel et al. 2000.)

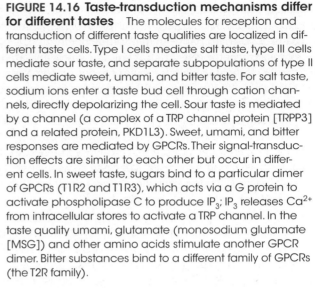

FIGURE 14.16 Taste-transduction mechanisms differ for different tastes The molecules for reception and transduction of different taste qualities are localized in different taste cells. Type I cells mediate salt taste, type III cells mediate sour taste, and separate subpopulations of type II cells mediate sweet, umami, and bitter taste. For salt taste, sodium ions enter a taste bud cell through cation channels, directly depolarizing the cell. Sour taste is mediated by a channel (a complex of a TRP channel protein [TRPP3] and a related protein, PKD1L3). Sweet, umami, and bitter responses are mediated by GPCRs. Their signal-transduction effects are similar to each other but occur in different cells. In sweet taste, sugars bind to a particular dimer of GPCRs (T1R2 and T1R3), which acts via a G protein to activate phospholipase C to produce IP_3; IP_3 releases Ca^{2+} from intracellular stores to activate a TRP channel. In the taste quality umami, glutamate (monosodium glutamate [MSG]) and other amino acids stimulate another GPCR dimer. Bitter substances bind to a different family of GPCRs (the T2R family).

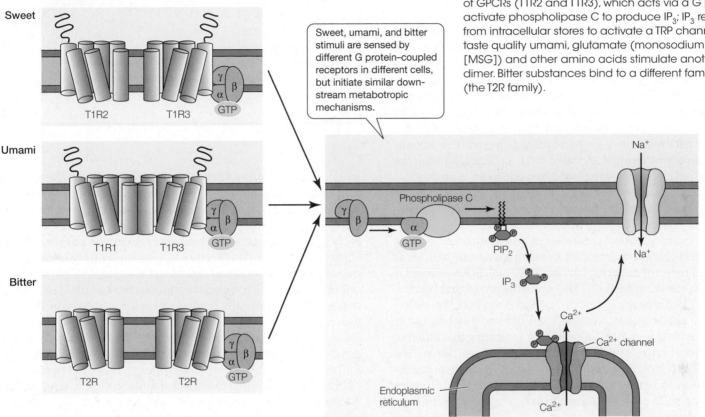

cumvallate papillae, which are at the very back and easily seen with the naked eye. The taste buds are located within the papillae, with only a few on a single fungiform papilla but with thousands on a single circumvallate papilla (**FIGURE 14.15B**).

All taste buds have a similar structure (**FIGURE 14.15C**). Each includes 50–150 slender, elongate cells of at least four types that differ in morphology and staining properties. Until recently investigators thought that only one of these cell types (type III) was a mature taste receptor, because it was the only type that made defined synapses with taste sensory neurons. Recent studies, however, show that types I, II, and III taste cells all mediate different taste qualities. Type IV cells (basal cells) are round, basally located, and serve as precursors to new taste receptor cells. Studies with markers of cell division such as bromodeoxyuridine (BrdU) have shown that individual taste receptor cells have lifetimes of only about 5–10 days and that they are constantly being replaced by new receptor cells derived from the basal cells.

Mechanisms of taste transduction differ for different taste qualities. In mammals, including humans, there are five distinct categories of tastes:[3] the familiar sour, salty, sweet, and bitter and the more recently characterized *umami* (from a Japanese word meaning "delicious"), which is the taste of monosodium glutamate and aspartate, common in proteinaceous food. These five tastes and their distinct transduction mechanisms appear to be separately localized in different taste receptor cells. Two of these mechanisms are ionotropic (for salty and sour), and the remaining three are metabotropic, mediated by GPCRs.

The simplest transduction scheme is for salty taste (**FIGURE 14.16**). Here the taste receptor cells have channels permeable to Na^+. When the Na^+ ion concentration within the mouth increases, the membrane potential of a salt receptor cell depolarizes. The

---

[3] It may turn out that there are more than five taste categories; for example, the taste category oleogustus (taste sensitivity to fatty acids) has recently been proposed.

reason can be seen from the Goldman equation (see page 318): Any increase in external concentration of permeable Na^+ ions will cause a positive change in membrane potential. The Na^+ channels of taste receptor cells are not voltage-gated and are unrelated to the voltage-gated Na^+ channels of axons. In some species they can be blocked by the compound amiloride and seem to be related to the voltage-independent Na^+ channels of the kidney. Sensation of salt by a direct, ionotropic permeability to Na^+ would seem an insensitive mechanism for salt sensation, because a fairly large increase in Na^+ ion concentration would be required to produce a depolarization. However, salt taste *is* insensitive: For a food to taste salty, the concentration of NaCl must be of the order of 50 millimolar (mM), a relatively high level.

The sour taste of acidity is also mediated by a channel (see Figure 14.16). This channel is probably formed from two related proteins (PKD1L3 and TRPP3, also known as PKD2L1), which are expressed together in a distinct subset of taste receptor cells. In mice, when the gene for TRPP3 is knocked out so that this protein is no longer expressed, the mice have no reaction to sour stimuli, even though responses to the other four classes of taste stimuli are unaffected. It is unlikely that the channels formed by interaction of these proteins function simply by being permeable to H^+ and depolarizing the membrane as described by the Goldman equation, because H^+ is normally too low in concentration to adequately depolarize the receptor cell by this means. Instead, H^+ may modulate the permeability of the channels to more numerous cations.

The other three tastes are all sensed by metabotropic GPCRs (see Figure 14.16). The receptor for sweet is apparently a dimer of two related receptor molecules called T1R2 and T1R3. Umami is sensed by a similar dimer, of T1R3 and another related protein, T1R1. The T1R3 receptor is therefore common to both the sweet- and umami-sensing cells, whereas the T1R2 and T1R1 receptors are expressed in different cell populations, indicating that different groups of taste bud cells mediate the sweet and umami flavors. The T1R2 and T1R3 proteins seem to be the only GPCRs required for sweet taste, as shown by knockout experiments. When the genes for T1R2 and T1R3 are knocked out in mice, the ability of the mice to detect sweet compounds disappears, and similar experiments show that T1R1 and T1R3 are required for the detection of amino acids (umami). Sweet taste reception in hummingbirds is a special case, discussed in **BOX 14.2**.

The family of GPCRs that sense bitter compounds is much larger and more disparate in amino acid sequence than those that sense sweet or umami, probably because we sense as bitter a much greater variety of chemical compounds with differing structures. The receptor molecules are nevertheless related to one another and form the T2R family of GPCRs, which has approximately 30 members. The T2R receptors are expressed in a unique population of taste bud cells that do not also express any of the T1R receptors.

The signal transduction cascades for sweet, umami, and bitter all seem to use similar G proteins, which activate a phospholipase C, producing the two second messengers IP_3 and diacylglycerol (DAG; see Figure 13.21). The IP_3 leads to release of Ca^{2+} from intracellular stores, which opens yet another TRP channel in the outer cell membrane (called TRPM5). In mice, knocking out the genes for either the phospholipase C or TRPM5 produces major deficits

---

BOX 14.2 | **Genomics and sweet taste in hummingbirds**

Genomic analyses provide critical information about the evolution of sensory systems. A fascinating example is the evolution of sweet taste in hummingbirds, the only vertebrate group known to use a nonstandard receptor for tasting sugars. **Box Extension 14.2** describes how hummingbirds are thought to have evolved this uniquely repurposed sensory mechanism.

---

in the sensing of sweet, umami, and bitter but leaves the sensing of salty and sour entirely intact. This shows that the metabotropic mechanisms of detection in the tongue are independent of the ionotropic mechanisms.

Because single taste bud cells seem to contain only one of the kinds of channels or GPCR combinations shown in Figure 14.16, the transduction of taste would appear to be segregated into separate cell populations distributed throughout the tongue. These cells make synapses onto the cranial sensory neurons, which carry their signals into the brain. Recent studies indicate that some of the synaptic interactions between taste cells and sensory neurons are diffuse and perhaps indirect, via other taste cells. Perhaps for this reason, recordings from single cranial axons and from single CNS taste neurons often reveal sensitivities to more than one taste. These observations have spawned many theories about how axons with multiple sensitivities can mediate our distinct sensations of salty, sour, sweet, bitter, and umami.

Recent experiments indicate that processing in the CNS somehow segregates the signals of the different taste responses despite their apparent mixing at the level of the cranial nerve fibers. It is possible to induce in a mouse the expression of the gene for a novel GPCR, which is activated by a compound that is normally tasteless. With the novel GPCR, the mouse can taste this compound; the taste is sensed as sweet if the novel receptor is directed to be expressed in the taste bud receptor cells that normally express T1R2 receptors (the "sweet" taste bud cells), or the taste is sensed as bitter if expressed in cells that normally express T2R receptors. Thus even if single cranial nerve axons transmit signals from more than one type of cell, the CNS sorts these signals to produce distinct sensations, which seem to depend entirely on which taste bud receptor cells have been stimulated. Some taste neurons in the brain respond to many taste qualities, whereas others are more selective to one taste, leading to ongoing questions about how the coding and decoding of taste information work.

SUMMARY
Chemoreception and Taste

■ Most animals possess two types of chemoreceptors for external stimuli: contact or taste chemoreceptors that respond to near-field chemicals at relatively high concentrations, and distance or olfactory chemoreceptors that respond to low concentrations of chemicals from sources over a larger area. This generalization is useful but oversimplifies a greater diversity of external chemical senses, as well as internal chemoreceptors involved in homeostatic regulation.

■ Taste chemoreceptors of mammals monitor five taste qualities: sweet, sour, salty, bitter, and umami. Insects have taste sensilla that provide at least analogous information.

■ Transduction mechanisms of chemoreceptors are diverse, both within an animal and across animal phyla. Taste sensory transduction in mammals may involve ionotropic activation of ion channels (salty, sour) or G protein–coupled receptors (sweet, bitter, umami).

Olfaction

Some insects, such as male moths, have elaborate olfactory systems (**FIGURE 14.17**). Each pinnate antenna of a male hawkmoth (*Manduca sexta*) contains an estimated 360,000 olfactory receptor cells, and each antenna of a male polyphemus moth (*Antheraea polyphemus*) (shown in the figure) has about 260,000. Such numbers represent a substantial fraction of the neurons in the insect's nervous system, and represent a large energy investment.

The sensilla on the antennae of insects that detect chemicals at a distance are similar in structure to the taste sensilla (see Figure 14.14A). Instead of one or a few small holes at the tip of the sensillum shaft, however, the entire shaft length is perforated by numerous tiny pores only 10 nanometers (nm) in diameter, which lead into tiny channels called *tubules* penetrating the exoskeleton. The receptor cells send thin dendritic processes into the shaft, where they are bathed in a fluid called the *sensillar lymph*. Odorant molecules enter the pores, travel down the tubules, and dissolve in the sensillar lymph. Terrestrial insects (and terrestrial mammals) have *odorant binding proteins* that help the volatile odorant molecules move into the aqueous environment surrounding the olfactory receptors. Ultimately the odorant molecules bind to olfactory receptor (OR) protein molecules on the dendrite's cell membrane. The *Drosophila* antennal OR proteins have been identified; they are not related to classical GPCRs but rather appear to form heteromers with an ion channel protein (OR83b) in which the OR provides the odorant binding specificity and leads to fast opening of the OR83b channel. Other studies suggest that the ORs may also have a slower G protein–linked action that reinforces and prolongs their ionotropic action.

Most varieties of insect olfactory receptor neurons are classified as *odor-generalist cells* and are used to detect a wide variety of odors. Insects, however, have another olfactory system on their antennae, specialized to detect only a very few chemicals with high sensitivity. The chemoreceptor cells used in this system are sometimes referred to as *odor-specialist cells* and are typically responsive to pheromones.

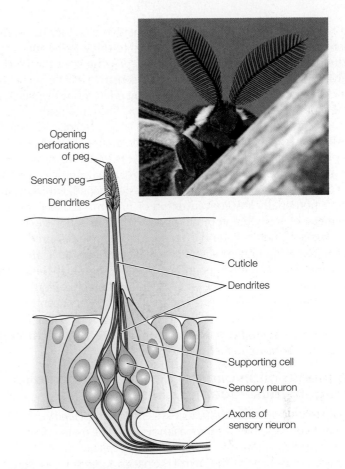

FIGURE 14.17 Insect olfactory receptors detect pheromones and other chemicals Each pinnate antenna of a male *Antheraea polyphemus* moth may have as many as 260,000 chemosensory neurons, many of which are specialized to detect a chemical signal (pheromone) released by a female moth.

As we will describe in Chapter 16 (see pages 456–457), a *pheromone* is a metabolite released into the outside world by an individual that is used for communicating with other individuals of the same species, including eliciting specific behavioral or systemic responses when detected by those individuals. Various insects have sex-attractant (and sex-deterrent) pheromones, oviposition-deterrent pheromones, trail-marking pheromones, alarm pheromones, and colony-recognition pheromones, and it is these chemicals that the odor-specialist cells are programmed to detect.

Chemoreception of pheromones has been most thoroughly studied for the sex attractants of moths. Female moths release pheromones that, when detected, induce the males to fly upwind to find the females. Males are extraordinarily sensitive to the attractant released by females of their species and can respond over large distances. A female gypsy moth produces 1 mg of disparlure, its sex attractant, an amount theoretically sufficient to attract a billion males.

In the silkworm moth (*Bombyx mori*), the first insect species for which a sex-attractant pheromone was characterized, the male has large pinnate or comblike antennae (like those of *Antheraea polyphemus* shown in Figure 14.17) with up to 50,000 sensilla responsive to the sex-attractant pheromone. Because of their pinnate branching and the density of their pheromone-sensitive olfactory hairs, the antennae of *Bombyx* males are estimated to catch up to one-third of the pheromone molecules in the air passing over them. The pheromone receptor cells are exquisitely sensitive and very specific in their responses. The major sex-attractant pheromone in *Bombyx* is bombykol, a 16-carbon unsaturated alcohol. The hair sensilla contain chemoreceptor cells specifically sensitive to bombykol, and recordings indicate that binding of one or two molecules of bombykol to a receptor cell elicits enough depolarization to generate an impulse. The threshold for the behavioral response of a male moth is reached with activation of about 200 receptor cells per second, which occurs at a phenomenally low concentration of 1000 molecules per cubic centimeter of air (which contains about 30 billion billion molecules per cubic centimeter at standard temperature and pressure)!

The sex-attractant detectors of a male moth are specialized to detect one or a very few compounds with exceedingly high sensitivity. This is quite different from the odor-generalist chemoreceptors, which detect a much broader range of chemicals at higher concentrations. The division of odor-detection cells into odor specialists and odor generalists is also found in many mammals, which have two major olfactory systems for detecting chemicals.

The mammalian olfactory epithelium contains odor-generalist receptor cells

All vertebrates have a main olfactory system, and most land vertebrates also have an accessory (vomeronasal) system. The main system's olfactory receptive surface, termed the *olfactory epithelium*, lines part of the internal nasal cavity. The area of the nasal mucosa that constitutes the olfactory epithelium varies greatly among species, being only 2–4 cm^2 in humans but 18 cm^2 in dogs and 21 cm^2 in cats. Humans have an estimated 10^7 olfactory receptor neurons, and dogs up to 4×10^9. In contrast to this large number of olfactory neurons, a mammalian auditory nerve contains the axons of only 3×10^4 neurons.

Each olfactory receptor cell is a bipolar neuron with a cell body in the olfactory epithelium (**FIGURE 14.18**). A single, narrow dendrite extends from the cell body to the mucus-covered epithelial surface and ends in a dendritic knob, which projects into the layer of mucus (secreted by supporting cells and glands). From this knob project 20–30 olfactory cilia, which extend and intermesh within the mucous layer. The membranes of these cilia are the sites of olfactory transduction. As with insect olfactory sensilla, odorant molecules first dissolve in the liquid of the mucus and interact with odorant-binding proteins before interacting with receptor molecules of the mammalian olfactory cilia.

The olfactory receptor cells have fine, unmyelinated axons that extend a short distance (through the cribriform plate of the ethmoid bone) to the olfactory bulb of the forebrain, where they synapse with second-order cells in the olfactory bulb. The receptor axons are typically only 0.2 micrometers (μm) in diameter, and are among the smallest axons in the nervous system. Olfactory receptor cells undergo continuous turnover throughout adult life, and they were the first mammalian neurons discovered to do so. The receptor cells differentiate from basal cells in the epithelium, develop dendrites and axons that connect with the second-order cells in the olfactory bulb, and have life spans of about 60 days before they degenerate.

The function of the olfactory receptor cells has been clarified by molecular genetic and physiological approaches. In the 1990s, Linda Buck and Richard Axel cloned the first of the olfactory recep-

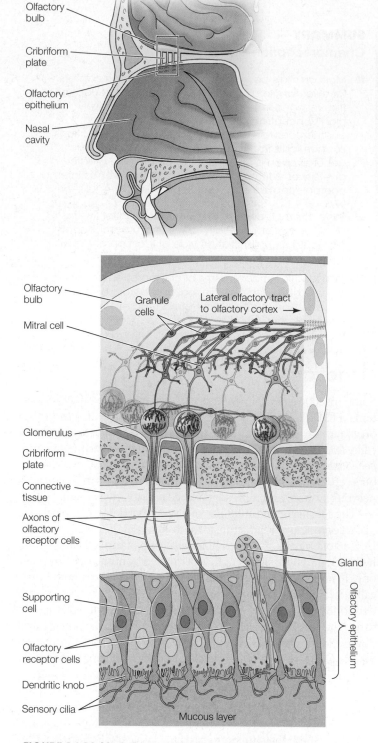

FIGURE 14.18 Vertebrate olfactory receptors Olfactory receptor cells are small bipolar neurons, the sensory cilia of which extend into the mucous layer of the nasal cavity. Their axons perforate the bone of the cribriform plate to end in glomeruli of the olfactory bulb of the brain. Mitral cells and granule cells integrate olfactory information, and the mitral cell axons carry the information to the olfactory cortex.

tor (OR) proteins. Inferring that the OR proteins would be GPCRs specifically expressed in the olfactory epithelium, they looked for olfactory proteins with sequences known to be widely shared by GPCR molecules. This led to the discovery of the first 18 members of a family of genes, now known in mice to contain nearly 1000 members and to constitute about 3% of the total number of genes

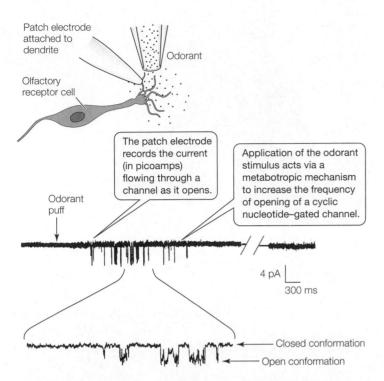

FIGURE 14.19 Odorant excitation of a vertebrate olfactory receptor cell opens individual ion channels An individual olfactory receptor cell was isolated from the olfactory epithelium of a salamander, and its response to odor stimulation was recorded with a patch electrode. The membrane patch remained attached to the cell. A brief puff of odorant led to opening of single channels in the patch, as shown by recorded single-channel currents. (After Firestein et al. 1991, in Fain 2003.)

in the mouse genome. For this groundbreaking work, Buck and Axel were awarded the Nobel Prize in 2004.

The OR proteins all have seven transmembrane domains like other GPCRs (see Figure 13.18), with considerable sequence variability in transmembrane regions 3, 4, and 5. These variable regions are thought to be the areas of the molecule that bind to a diverse array of odorants. Similar OR protein families are found in all vertebrates so far studied. The activation of olfactory GPCRs leads to the opening of cyclic nucleotide–gated channels to produce a receptor potential (**FIGURE 14.19**). Stuart Firestein and colleagues placed a patch pipette on the dendrite of an olfactory receptor cell from a salamander and stimulated the cell with odorant.[4] The odorant bound to receptor molecules lateral to the cell-attached patch, and yet channels opened inside the patch area, showing the involvement of a second messenger. The odorant receptor molecules initiated a G protein–mediated signal cascade, producing intracellular cAMP that diffused in the cytoplasm and opened cyclic nucleotide–gated channels in the patch.

FIGURE 14.20 diagrams the signal cascade in mammalian olfactory receptor cells. The binding of an odorant to an OR protein initiates the exchange of GTP in place of GDP on the α subunit of a G protein (G_{olf}) specifically expressed in olfactory receptor cells. The G_{olf} α–GTP then binds to and activates an adenylyl cyclase, which synthesizes cAMP. The cAMP attaches to the cytoplasmic surface of the channel, which closely resembles a voltage-gated K^+ channel in its structure but has a binding site for cyclic nucleotides near its –COOH terminus (see Box Extension 12.1). The opening of the channel causes Na^+ to flow into the cell and depolarize it.

The cyclic nucleotide–gated channels are also permeable to Ca^{2+}, and the flow of Ca^{2+} into the cell stimulates the opening of

[4] In this cell-attached patch recording the patch pipette was sealed to the dendrite membrane, but the patch remained attached to the rest of the cell. Odor stimuli were applied to the outer face of the membrane lateral to the patch, and the sealed electrode prevented receptor molecules from diffusing in the membrane from outside the patch area to within it.

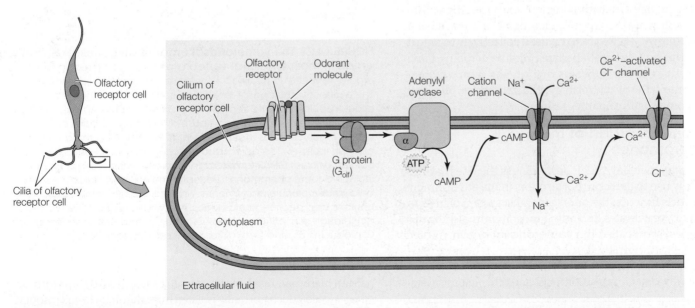

FIGURE 14.20 Olfactory transduction mechanisms in cilia membranes of olfactory receptor cells Many odorants act to increase cAMP. The odorant binds to an odorant receptor on the ciliary membrane; the receptor activates a G protein to activate adenylyl cyclase, producing cAMP. Cyclic AMP binds to and opens a cation channel, allowing entry of Na^+ and Ca^{2+} ions and depolarizing the cell. Ca^{2+} binds to Ca^{2+}-activated Cl^- channels, which permit Cl^- efflux that augments the depolarization.

a Ca^{2+}-activated Cl^- channel, causing negatively charged Cl^- to flow *out* of the cell, producing further depolarization. The outflow of Cl^- through the Ca^{2+}-activated Cl^- channels ensures that the olfactory cells will continue to respond even if the Na^+ concentration in the mucus is decreased, as can occur, for example, after immersion of the animal's head in water. When the mucus, which has high Na^+ and Cl^- concentrations relative to the receptor cell cytoplasm, is diluted by water, the concentration gradient for Cl^- favors its leaving the cell (whereas the gradient for Na^+ entry is reduced). The transduction cascade shown in Figure 14.20 is the primary cascade used by the olfactory receptor cells. Knocking out the genes for G_{olf}, adenylyl cyclase, or the cyclic nucleotide–gated channel produces mice with greatly reduced electrical responses to odorants and almost no sense of smell.

Each olfactory receptor cell expresses the gene for only one kind of GPCR. However, recordings from the receptor cells show that many cells have rather broad sensitivities to different odorant molecules. This means that many odorants can bind to one kind of receptor protein, and that any particular odorant is likely to bind to several different receptor proteins, probably with greater affinity (and at lower concentration) for some proteins than for others. Different odorants would then stimulate different *populations* of receptor cells expressing different proteins, and the combined signal from the different populations would constitute the combinatorial code for each different odorant stimulus. Such a system is likely to provide greater flexibility and adaptability than one in which each olfactory GPCR is absolutely specific for a single chemical structure. The broader binding affinities of olfactory GPCRs may also allow the olfactory system to respond to more odors than if each of the several hundred or even 1000 genes coded for proteins with narrow selectivity.

The olfactory receptor cells send their axons to the neighboring olfactory bulb in the CNS. The axon of each receptor cell terminates within a globular cluster called a *glomerulus* (see Figure 14.18). All of the receptor cells expressing a particular receptor molecule terminate in only one or two glomeruli, so that the many receptor types are kept segregated in distinct regions of processing within the bulb. This segregation seemed puzzling at first, because if responses from many GPCRs are to be used collectively to detect distinct odors, the signals must at some point be combined and compared. Recent evidence indicates that signals from different glomeruli converge at the next level in the olfactory cortex, allowing the necessary comparison and integration.

The vomeronasal organ of mammals detects pheromones

The olfactory system we have been describing is the one that vertebrates normally use to detect airborne odors in the environment and resembles in its function the odor-generalist system of insects. In addition, many vertebrates, including most mammals,[5] have a second olfactory system called the **vomeronasal organ**, located below the main olfactory epithelium on each side of the nose (**FIGURE 14.21A**). The vomeronasal organ appears to function in a more odor-specialist way, detecting pheromones and other chemical sig-

[5] In humans and apes the vomeronasal organ regresses during fetal development and is thought to be nonfunctional in adults. There is behavioral evidence for human detection of pheromones, but it may be via the main olfactory epithelium rather than the regressed vomeronasal organ.

(A) Vomeronasal anatomy

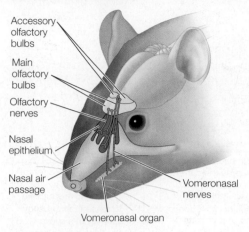

(B) Transduction in vomeronasal receptor cells

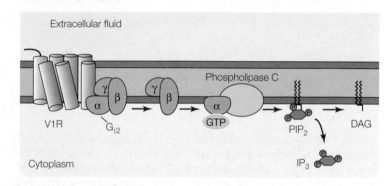

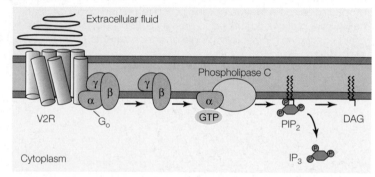

FIGURE 14.21 The vomeronasal organ is an accessory olfactory organ in most terrestrial vertebrates (A) The paired vomeronasal organs lie ventral to the main olfactory epithelium (left side shown here). The axons of vomeronasal sensory neurons project to the accessory olfactory bulb, separate from the main olfactory bulb. Apical and basal parts of the vomeronasal organ have different populations of vomeronasal sensory neurons that project to different parts of the accessory olfactory bulb. (B) The apical and basal vomeronasal sensory neurons contain different GPCR families and bind different classes of chemicals, including some pheromones: Receptor molecules in apical neurons bind small-molecule pheromones, whereas receptor molecules in basal neurons bind peptide pheromones. The two kinds of GPCRs (V1R, V2R) act through different G proteins (G_{i2} and G_o) to activate phospholipase C, producing IP_3 and DAG. (After Spehr and Munger 2009.)

nals. In mammals, each of the paired vomeronasal organs forms a self-enclosed pouch normally isolated from the air breathed through the nose. When an animal detects an unusual odor such as that associated with a pheromone, the organ pumps air into its lumen so that the air wafts over the vomeronasal receptor cells. In some reptiles, pheromones are delivered to the organ from the tongue.

A snake or lizard repeatedly flicks out and withdraws its tongue to sample the air and deposit molecules such as pheromones or prey odorants onto the surface of the vomeronasal organ. The forked tongue provides two-point sampling for the detection of gradients of the concentration of chemical stimuli.

The histology of the mammalian vomeronasal epithelium is similar to that of the main olfactory epithelium, but the receptor cells have microvilli instead of cilia at their locus of transduction. The vomeronasal receptor cells in mice express about 300 genes for receptors in two families, called V1R and V2R (**FIGURE 14.21B**). These receptors are GPCRs only distantly related to the OR proteins of the primary epithelium. The V1R and V2R receptors are expressed in separate cell types in different regions of the vomeronasal organ; the two receptor classes act via different G proteins to activate phospholipase C, producing IP_3 and DAG from PIP_2 (as in some synapses [see Figure 13.21] and in some taste receptors [see Figure 14.16]). Ultimately this cascade (which may also involve generation of arachidonic acid) leads to opening of a TRP channel to depolarize the membrane of the vomeronasal receptor cell.

Preliminary recordings from neurons in the vomeronasal organ seem to show that, like the odor-specialist cells of male moths, the vomeronasal receptor cells respond specifically to one or only a few compounds with high sensitivity. They predominantly mediate responses to pheromones, which play important roles in a variety of rodent behaviors, including pregnancy block, defense and recognition of the young, mating, and intraspecific aggression. Whether pheromones also play a role in human behavior is a question still widely debated. Although most pheromonal sensory responses in mammals are mediated by the vomeronasal organ, recent rodent studies show that the main olfactory epithelium is also involved in pheromonal communication, whereas the vomeronasal organ also processes some non-pheromonal odors.

SUMMARY
Olfaction

- Olfactory chemoreceptors of the main olfactory epithelium of vertebrates are neuronal receptor cells with cilia that contain intramembrane receptor proteins. Each receptor cell expresses the gene for one of these membrane receptor proteins, and all the receptor neurons that express that same protein synapse in the same glomerulus of the olfactory bulb. Insect olfactory neurons have broadly similar connection patterns but unrelated receptor proteins.

- Vertebrate olfactory receptor proteins are G protein–coupled receptors, which stimulate production of a second messenger, cAMP.

- The vomeronasal organ of vertebrates is an accessory olfactory organ that senses pheromonal and other stimuli. Vomeronasal sensory cells are microvillar rather than ciliary, and express GPCR proteins that stimulate production of IP_3 and DAG.

Photoreception

Visual systems—the systems that carry out photoreception and the processing of visual signals—have been more extensively studied and are better understood than any other sensory system. This is in part a reflection of the importance of vision in our lives, but it is also due to certain technical advantages. **Photoreceptors**—sensory receptor cells that are sensitive to light—are easier to identify, isolate, grind up, clone, and study with physiological techniques than other receptor cells. The photoreceptor protein *rhodopsin* was the first G protein–coupled receptor to be sequenced, cloned, and examined with X-ray crystallography, and the details of the G-protein cascade in vertebrate photoreceptors are more fully understood than any other G-protein cascade in nature. Visual processing in the brain, treated in the next section, has provided unique insight into the working of the visual cortex, which has illuminated not only the study of sensory processing but also basic mechanisms of synaptic integration in the CNS.

Photoreceptor cells and eyes of different groups have evolved similarities and differences

Photoreception—the response of a sensory cell to light—arose early in evolution. All organisms detect light by using a pigment, termed a **photopigment**, that absorbs the light. Although bacteria have several kinds of photopigments, all animals employ one dominant photopigment, rhodopsin, for photoreception. Because the photopigment is associated with cell membranes, all photoreceptor cells (light-responsive cells) have greatly increased membrane surface areas that increase their light sensitivity. Photoreceptor cells are subdivided into ciliary and rhabdomeric. In **ciliary photoreceptor** cells, modified cilia contain the rhodopsin molecules. **Rhabdomeric photoreceptor** cells, in contrast, have collections of microvilli that increase the membrane surface area. Vertebrate photoreceptor cells are ciliary, and arthropod photoreceptor cells are rhabdomeric, but many other phyla have examples of both photoreceptor types, sometimes in the same animal.

Most animal phyla contain examples of eyes (light-responsive sense organs) that differ in type and complexity. Eyes, as noted earlier, are sense organs that contain photoreceptors mediating light responses; eyes may also form images of the visual world that provide information to allow an animal to localize and identify visual stimuli. There are two major kinds of image-forming eyes. In a **camera eye** (see Figure 1.7), a lens forms an inverted image on an array of photoreceptors at the back of the eye. In a **compound eye**, many facets called *ommatidia* (singular *ommatidium*), each with its own lens, together produce what is called a *mosaic image*—each ommatidium conveys information about one part of the visual world, and the animal's nervous system constructs the image as a mosaic of "tiles" of individual ommatidial responses.

The evolution of eyes is a subject of considerable speculation, now informed by recent genomic studies. There is evidence that most eyes evolved from a common ancestor, despite many differences in the organization of different eyes. As stated above, all eyes employ the same kind of photopigment (rhodopsin, discussed further below). Moreover, disparate eyes share homologous genes that regulate eye development. A single gene, *PAX6*, can initiate eye development in diverse animals, and *PAX6* interacts similarly with other regulatory genes in the eye development of many phylogenetic groups of animals. It is likely that basic genetic plans for eye development and function evolved very early, but specific eye lineages (such as the camera eyes of cephalopods

FIGURE 14.22 A hypothesis showing the evolution of eyes A rhodopsin-based photo-receptor cell may have become a prototype eye by a developmental association with a screening pigment cell that provided directional selectivity, under control of *PAX6* or a related developmental gene. The prototype eye is thought to have evolved monophyletically into more complex eyes, including simple camera eyes (snail, scallop), complex camera eyes (cephalopod, vertebrate), and compound eyes (fly, ark clam). The hypothesis is based on homologies of genes for photoreception and for eye development in different groups. (After Gehring 2005.)

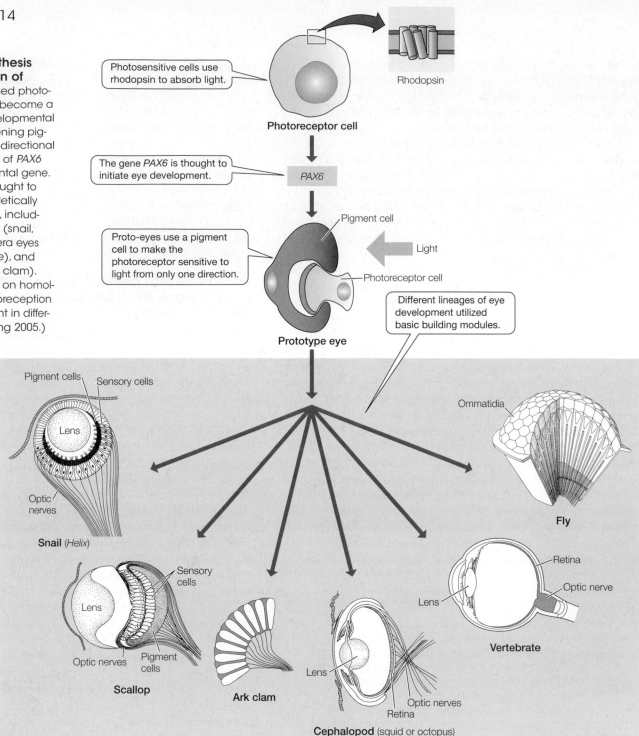

Photosensitive cells use rhodopsin to absorb light.

Rhodopsin

Photoreceptor cell

The gene *PAX6* is thought to initiate eye development.

PAX6

Pigment cell

Proto-eyes use a pigment cell to make the photoreceptor sensitive to light from only one direction.

Light

Photoreceptor cell

Prototype eye

Different lineages of eye development utilized basic building modules.

Pigment cells   Sensory cells

Lens

Optic nerves

Snail (*Helix*)

Sensory cells

Lens

Optic nerves   Pigment cells

Scallop

Ark clam

Lens

Retina

Optic nerves

Cephalopod (squid or octopus)

Ommatidia

Fly

Retina

Optic nerve

Lens

Vertebrate

[squids and octopuses] and of vertebrates; see Figure 1.7) evolved separately, using parallel expressions of these developmental and functional modules. **FIGURE 14.22** shows one hypothesis of a common evolution of photoreceptors and eyes. A rhodopsin-based photoreceptor cell could have become a directionally sensitive prototype eye by assembly with a pigment cell that screened the photoreceptor from light on one side; this assembly may have been under control of *PAX6*. Such a prototype eye may then have evolved into other simple eyes with multiple receptor cells, and into more complex eyes with lenses. The examples shown in Figure 14.22 include simple camera eyes (snail, scallop), more complex camera eyes (cephalopod, vertebrate), and compound eyes (ark clam, fly).

Rhodopsin consists of retinal conjugated to opsin, a G protein–coupled receptor

A photopigment consists of a protein containing an associated nonpeptide organic molecule called a *chromophore* (from the Greek meaning "color bearing"). The absorption of a photon of light by a chromophore produces a chemical reaction, and this triggers a transduction cascade.

The chromophore of animal photoreceptors is *retinal*, which is bound to the integral membrane protein *opsin*, to produce the light-sensitive pigment **rhodopsin. FIGURE 14.23A** shows the chemical structure of retinal, which is the aldehyde of vitamin A. The form of retinal used in vision in most animals, including all mammals,

(A) Skeleton structure of retinal

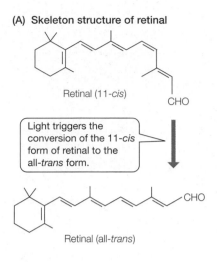

Retinal (11-*cis*)

CHO

Light triggers the conversion of the 11-*cis* form of retinal to the all-*trans* form.

CHO

Retinal (all-*trans*)

(B) Rhodopsin structure

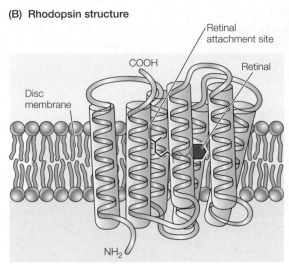

Disc membrane

COOH

Retinal attachment site

Retinal

NH₂

FIGURE 14.23 Rhodopsin is a photopigment composed of two parts: retinal and opsin (A) Skeleton chemical structure of retinal, which exists in two isomers (11-*cis* and all-*trans*). Light triggers a conformation change from 11-*cis* to all-*trans* retinal. (B) Three-dimensional structure of the protein (opsin) in vertebrate rhodopsin. Seven α-helical regions of the protein span the membrane; retinal is attached to an amino acid residue in the seventh helix. (The red retinal would actually be hidden behind the nearer helices but is shown as if it were visible through them.)

is 11-*cis* retinal.[6] Investigators have identified many mammalian rhodopsins with different light-absorption spectra. Because they all use the same chromophore, the differences among rhodopsins must result mostly from differences in the amino acid composition of the opsin moiety. In all species so far investigated, the aldehyde end of retinal is covalently bound to the amino group of a lysine of the opsin protein. This bond buries the chromophore deep within the middle of the rhodopsin molecule. The structure of rhodopsin in **FIGURE 14.23B** shows the seven transmembrane α-helices characteristic of all GPCRs, and the position of the bound retinal.

The absorption of light by 11-*cis* retinal produces a **photochemical reaction**, which twists the aldehyde tail of the chromophore around one of its double bonds and produces all-*trans* retinal (see Figure 14.23A). Because retinal is intimately associated with opsin, the photoconversion to all-*trans* retinal is followed by a series of spontaneous changes in conformation of opsin, producing activated rhodopsin (also called metarhodopsin II). This action induced by light is very rapid, producing activated rhodopsin in about 1 millisecond (ms).

Rhodopsin activates a G protein signal-transduction cascade, like other members of the GPCR superfamily (see Chapter 13, pages 357–360). The details of this cascade can differ in different kinds of animals and sometimes even in different photoreceptors of the same animal. We once again take as our representative examples *Drosophila* and vertebrates, including mammals.

Phototransduction in *Drosophila* leads to a depolarizing receptor potential

All insects and many other arthropods have compound eyes, which (as noted earlier) consist of many clustered cell groups called ommatidia, each ommatidium with its own lens and photoreceptors called **retinular cells** (**FIGURE 14.24**). The eight or more retinular cells of an ommatidium are arranged in a circle, like sections of an orange but more elongate. The transduction cascade of a retinular cell is localized to the membranes of its microvilli, which are arranged along one edge of the retinular cell in an array called a **rhabdomere**. The microvillar membranes of the rhabdomere con-

tain not only the rhodopsin photopigment, but also the G proteins and associated proteins, including the channels that produce the electrical response to light.

Phototransduction in insects is similar at the outset to that in vertebrates (described later) but involves a different second messenger. Figure 14.24C shows the phototransduction cascade in *Drosophila* photoreceptors. Absorption of a photon causes a change in the conformation of rhodopsin, leading to activation of a G protein by exchange of GTP for GDP on the G-protein α subunit. The activated G protein then activates a phospholipase C, which produces the second messengers IP₃ and diacylglycerol (DAG). DAG (or a lipid metabolite of it) is thought to directly open two cation channels, both of which are TRP channels. Opening of the cation-permeable channels produces a depolarization, which spreads along a short axon to the photoreceptor synaptic terminal and triggers the release of synaptic transmitter. *Drosophila* photoreceptors, like most arthropod photoreceptors, do not generate action potentials.

As shown in Figure 14.24C, several of the proteins involved in photoreceptor transduction are bound together by a cytoplasmic scaffolding protein. This tight organization may minimize the time required for the interactions of the proteins of the cascade and the production of the electrical signal, thereby increasing the speed of phototransduction.

The vertebrate eye focuses light onto retinal rods and cones

In contrast to the compound eye of insects, all vertebrates have camera eyes. In this type of eye, the cornea and lens focus an inverted image of the visual field on the **retina**, the photoreceptor-containing layer at the back of the eye (**FIGURE 14.25A**). Light is *refracted* (light rays are bent) at surfaces where materials differ in density (quantified as the refractive index). For *terrestrial* vertebrates the greatest amount of refraction occurs at the interface between the air and the cornea, which differ greatly in refractive index and thus produce greater bending. Refraction by the lens is secondary because the lens and the aqueous humor have more similar refractive indices. Lens refraction serves primarily to focus the image by changing the shape of the lens. In the eye of *aquatic* vertebrates, the cornea does little refraction, because its refractive index is similar

---

[6] Insects use a slightly different 3-hydroxy-11-*cis* retinal, and some vertebrate species living in freshwater use 3-dehydro-11-*cis* retinal, with two double bonds in the ring.

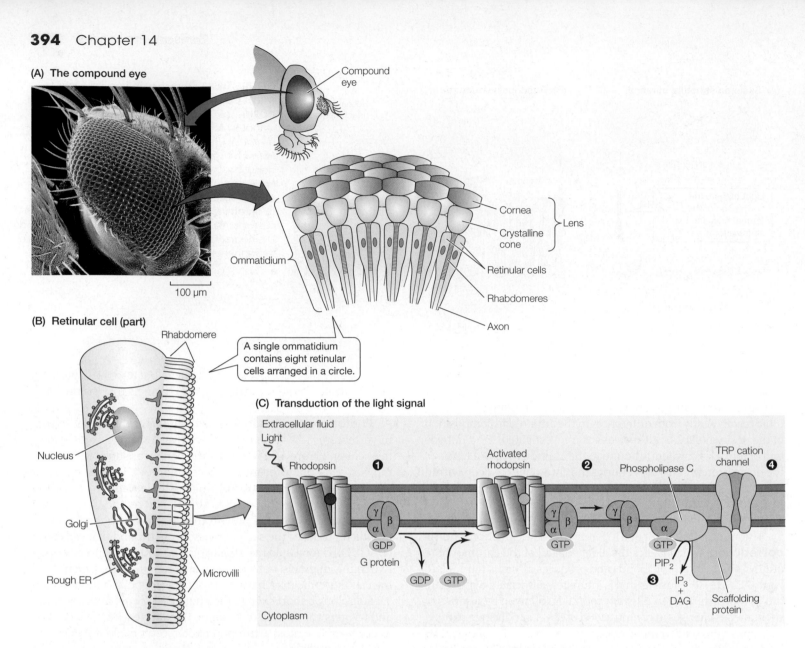

(A) The compound eye

Compound eye

100 µm

Ommatidium

Cornea ⎫
Crystalline cone ⎬ Lens
Retinular cells
Rhabdomeres
Axon

(B) Retinular cell (part)

Rhabdomere

A single ommatidium contains eight retinular cells arranged in a circle.

Nucleus

Golgi

Rough ER

Microvilli

(C) Transduction of the light signal

Extracellular fluid

Light

Rhodopsin ❶

Activated rhodopsin ❷

Phospholipase C

TRP cation channel ❹

γ β
α
GDP

G protein

GDP   GTP

γ β
α
GTP

γ β

α
GTP

PIP₂

❸ IP₃ + DAG

Scaffolding protein

Cytoplasm

FIGURE 14.24 Photoreceptor transduction in the compound eye of the fruit fly *Drosophila* (A) Scanning electron micrograph of a compound eye. Each ommatidium of the compound eye contains its own lens system and eight elongated photoreceptive retinular cells, only two of which are visible in each ommatidium here. (B) Part of one of the retinular cells; each retinular cell has a rhabdomere composed of microvilli facing the midline of the ommatidium. (C) The microvillar membrane contains rhodopsin and the other molecular constituents of the signal-transduction cascade, bound together by a scaffolding protein. Light-activated rhodopsin activates a G protein ❶ to activate phospholipase C ❷, catalyzing the formation of IP₃ and DAG ❸. These second messengers lead to opening of TRP cation channels ❹.

to that of water. Instead, a thick, nearly spherical lens accomplishes most of the refraction; the refractive index of the lens is greater at its center than at its edges, thereby minimizing distortion. A spherical fish lens cannot change shape, but instead focuses by moving anteriorly and posteriorly, like a camera lens.

The retina of the vertebrate eye is a developmental outgrowth of the brain. It contains rod and cone photoreceptor cells, as well as a network of neurons—*horizontal cells, bipolar cells, amacrine cells*, and *ganglion cells*—which perform the first stages of visual integration (**FIGURE 14.25B**). A pigmented epithelium lies at the back of the retina. It absorbs light not captured by the photoreceptors and performs many important metabolic functions, including the control of the ionic environment around the rods and cones and the synthesis of 11-*cis* retinal. The retina is said to be inverted,

with the photoreceptors in the outermost layer, farthest away from incoming light. This inverted structure is a consequence of the way in which the retina develops in the embryo, as the more distal layer of a two-layered optic cup.

Although light must pass through all the retinal layers to reach the outer segments of the rods and cones, the retinal layers do not degrade the image greatly because they are quite transparent. Some light scattering does occur, however, and many retinas have a central high-acuity region in which the intervening cell layers and blood vessels are displaced to the side. In humans this region is the **fovea**, a depression 1.5 mm in diameter (5° of visual angle) (see Figure 14.25A). The central 1° of the fovea contains tightly packed cones to the exclusion of other neurons. Rod photoreceptors are absent in this central part of the human fovea but greatly outnumber

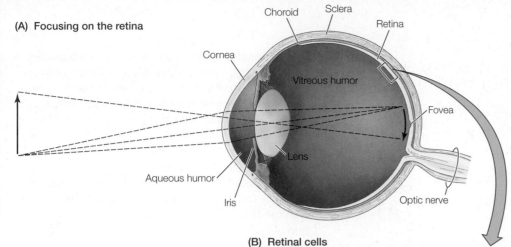

(A) Focusing on the retina

Choroid

Sclera

Retina

Cornea

Vitreous humor

Fovea

Aqueous humor

Lens

Iris

Optic nerve

FIGURE 14.25 Structure of the mammalian eye and retina (A) The cornea and the lens focus an inverted image on the retina at the back of the eye. (B) The photoreceptors (rods and cones) are in an outer nuclear layer at the back of the retina. They are connected via bipolar cells to ganglion cells, in a "straight-through" pathway. Horizontal cells form a lateral pathway in the outer plexiform layer, and amacrine cells form a lateral pathway in the inner plexiform layer. The ganglion cells are the output of the retina, their axons forming the optic nerve. (After Dowling 1979 in Kandel et al. 1995.)

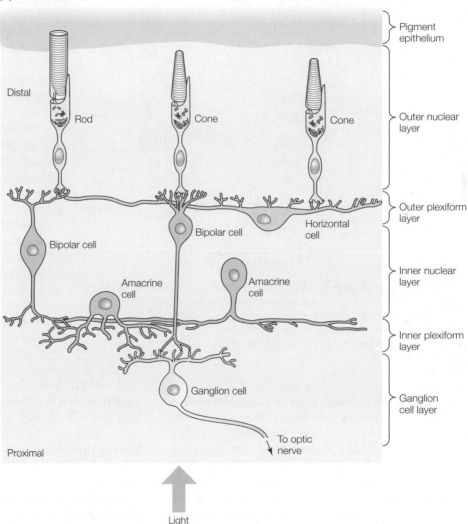

(B) Retinal cells

Pigment epithelium

Distal

Rod

Cone

Cone

Outer nuclear layer

Outer plexiform layer

Bipolar cell

Bipolar cell

Horizontal cell

Inner nuclear layer

Amacrine cell

Amacrine cell

Inner plexiform layer

Ganglion cell

Ganglion cell layer

To optic nerve

Proximal

Light

cones elsewhere in the retina. Primates and some birds have well-developed foveas, and many other vertebrates have a less elaborate and broader area of relatively high acuity called the *area centralis*.

Another consequence of the inverted retina is that the axons of retinal ganglion cells, which form the optic nerve, come off the inner side of the retina, facing the lens. The axons exit through the retina at the *optic disc*, producing a blind spot in the visual field. Humans are normally unaware of this blind spot because it falls in the binocular visual field, so that one eye supplies the information missing in the blind spot of the other eye. Moreover, we depend

on the fovea for much of our high-acuity vision, and we make unconscious rapid eye movements, further decreasing any visual deficit resulting from the blind spot.

Rods and cones of the retina transduce light into a hyperpolarizing receptor potential

FIGURE 14.26 shows the two kinds of photoreceptors in the retina, **rods** and **cones**. The more sensitive rods are used in dim light, and the cones are used in brighter light, for color vision, and for high-acuity vision in humans and other animals having a fovea. Nocturnal

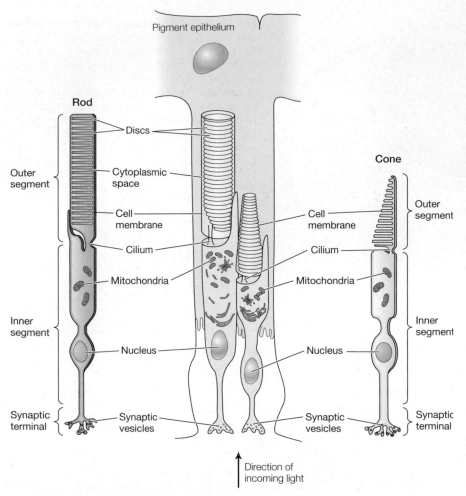

Pigment epithelium

Rod

Outer segment

Discs

Cytoplasmic space

Cell membrane

Cilium

Mitochondria

Inner segment

Nucleus

Synaptic terminal

Synaptic vesicles

Cone

Cell membrane

Outer segment

Cilium

Mitochondria

Inner segment

Nucleus

Synaptic vesicles

Outer segment

Inner segment

Synaptic terminal

Direction of incoming light

FIGURE 14.26 Vertebrate photoreceptors Rods and cones are shown both in situ and schematically. Both rods and cones have an inner segment that contains the nucleus and synaptic terminal, and an outer segment that contains ordered lamellae bearing photopigment molecules. In cones these lamellae are invaginations of the outer membrane, but in rods they are internalized discs that are discontinuous with the outer membrane.

animals tend to have retinas in which most or all photoreceptors are rods, whereas cones predominate in retinas of strongly diurnal animals. Both rods and cones have an **outer segment** containing the photosensitive membranes, and an **inner segment** containing the nucleus, mitochondria, other cell organelles, and the synaptic terminal. The inner and outer segments are connected by a short ciliary stalk, and the outer segment is derived from a modified cilium. The photoreceptor outer segments contain many flattened *lamellae* of membranes derived from the cell membrane. In the cones, these lamellae retain continuity with the outer cell membrane, so the lumen of each lamella is continuous with extracellular space. In rods, in contrast, the lamellae become separated from the outer membrane and form internalized flattened **discs** (see Figure 14.26). Several hundred to 1000 discs, stacked like pancakes, fill the rod outer segment. The membranes of rod discs and cone invaginated lamellae contain the photopigment rhodopsin. Because the disc membranes and the cell membrane are discontinuous in rods, the light-induced changes in rhodopsin at the disc membrane must somehow affect the ion permeability of the outer membrane. An electrical change in the disc membrane will not directly affect the

surface membrane because these two membranes are not in continuity. Instead, there is an *intracellular messenger* that conveys a change from the discs to the outer surface membrane. This intracellular messenger is **cyclic GMP** (**cyclic guanosine monophosphate**, or **cGMP**; see Figure 2.30).

The transduction of light into an electrical signal in rod and cone photoreceptors has four stages: First, light activates rhodopsin; second, activated rhodopsin stimulates a G protein to activate a phosphodiesterase enzyme; third, the enzyme decreases the concentration of cyclic GMP in the photoreceptor cytoplasm; and fourth, the decrease in cyclic GMP closes cyclic nucleotide–gated ion channels similar to the channels in olfactory receptor cells. Because the photoreceptor channels close, the Na^+ influx decreases and the photoreceptor *hyperpolarizes* rather than depolarizes. Most of our knowledge of phototransduction comes from rods (which are bigger than cones), but the mechanisms are qualitatively similar for both.

FIGURE 14.27 shows the mammalian transduction cascade in more detail. In the dark, the cGMP concentration within a rod outer segment is relatively high. The cGMP binds to cyclic nucleotide–gated channels, opening them and thereby keeping the permeability of the outer segment membrane high for cations, including Na^+. The cGMP-gated channels of vertebrate photoreceptors are not voltage-gated, although they are structurally and evolutionarily related to voltage-gated channels. Instead they open when they bind cGMP on the cytoplasmic surface (see Box Extension 12.1).

The light-induced change in rhodopsin activates a series of reactions at the disc membrane that result in an enzymatic degradation of cGMP. Figure 14.27 shows this process. Light-stimulated rhodopsin (metarhodopsin II) activates the G protein **transducin**, which is closely related to the G proteins that mediate metabotropic synaptic actions (see Figure 13.19) and chemoreception (see Figure 14.20). The activated G protein stimulates **cGMP phosphodiesterase** (**PDE**), an enzyme in the disc membrane that hydrolyzes cGMP to 5′-GMP. Activation of PDE involves dissociation of its catalytic subunit from a regulatory subunit that inhibits its activity. The activated PDE decreases the cytoplasmic concentration of cGMP, and the cation channels close. Thus the second messenger–mediated response to light in a rod outer segment is decreased Na^+ influx, producing hyperpolarization.

Because the channels are relatively open in the dark, cations flow into the cell and keep it depolarized (**FIGURE 14.28A**). Recall from the Goldman equation in Chapter 12 (Equation 12.6, page 318) that the membrane potential depends on the ratio of permeabilities to Na^+ and K^+. Because the resting Na^+ permeability of a rod or cone is higher than in a normal neuron, the membrane potential is less negative. In fact, the resting membrane potential of a rod or cone is about −30 mV, much more depolarized than the resting membrane potential of a typical neuron. This constant flow of Na^+ into the cell produces a **dark current** (**FIGURE 14.28B**), which must be

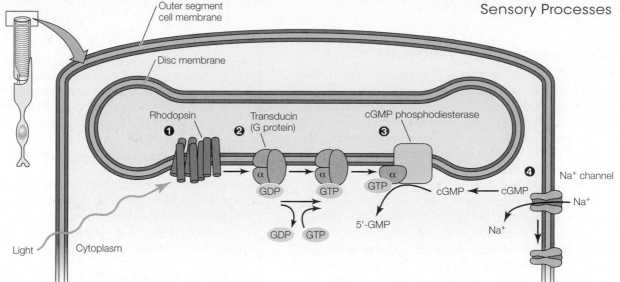

Outer segment
cell membrane

Disc membrane

Rhodopsin ❶

Transducin (G protein) ❷

cGMP phosphodiesterase ❸

α GDP

α GTP

α GTP

GDP GTP

5'-GMP

cGMP ← cGMP

❹ Na⁺ channel

Na⁺

Na⁺

Light

Cytoplasm

FIGURE 14.27 Phototransduction closes cation channels in a rod outer segment In the dark, the cation channels are kept open by intracellular cGMP; they conduct an inward current, carried largely by Na⁺. When light strikes the photoreceptor, these channels are closed by a G protein–coupled mechanism. ❶ Rhodopsin molecules in the disc membrane absorb light and are activated. ❷ The activated rhodopsin stimulates a G protein (transducin), which in turn activates cGMP phosphodiesterase. ❸ The phosphodiesterase catalyzes the breakdown of cGMP to 5'-GMP. ❹ As the cGMP concentration decreases, cGMP detaches from the cation channels, which close.

constantly counteracted by the expenditure of ATP to pump Na⁺ out of the cell with a very active Na⁺–K⁺-ATPase located in the inner segment. This large expenditure of ATP makes the metabolic rate of a rod in darkness perhaps the highest of any cell in the body.

The dark current keeps the rod relatively depolarized. The light-induced decrease in cytoplasmic concentration of cGMP shifts the equilibrium of cGMP binding to the channel protein so that cGMP dissociates from the channel and the channel closes. The response to light is thus a decrease in the dark current (**FIGURE 14.28C**), which causes the membrane potential to become more negative, producing a hyperpolarizing receptor potential that is graded according to the intensity of the light flash (see Figure 14.28A).

Vertebrate rods, like the photoreceptors of *Drosophila*, can produce a detectable change in dark current from the absorption of a single photon. Each photon absorbed by rhodopsin leads to approximately 10^6 Na⁺ ions *not* entering a rod outer segment. One functional attribute of an intracellular messenger is to provide the necessary amplification for this process. The activation of one rhodopsin molecule can lead to the activation of many transducin and phosphodiesterase molecules, causing the enzymatic degradation of a large number of cGMP molecules. This decreased cGMP concentration can close many Na⁺ channels, blocking the entry of enough Na⁺ ions to produce a change in membrane potential of the order of 1 mV. If phototransduction were to use an ionotropic mechanism, the absorption of a photon would only be able to open or close a single channel, and the sensitivity would be insufficient to achieve the detection of a single quantum of light.

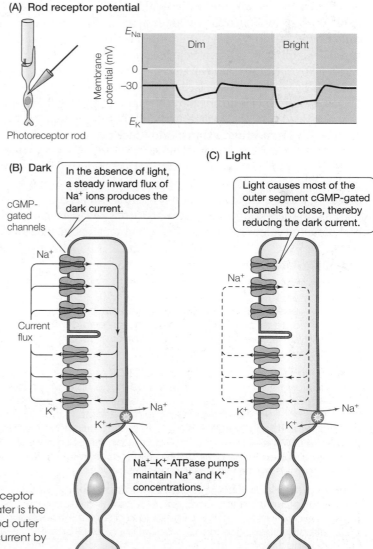

(A) Rod receptor potential

Photoreceptor rod

E_{Na}

Dim

Bright

Membrane potential (mV)

0

−30

E_K

(B) Dark

In the absence of light, a steady inward flux of Na⁺ ions produces the dark current.

cGMP-gated channels

Na⁺

Current flux

K⁺

Na⁺

K⁺

Na⁺–K⁺-ATPase pumps maintain Na⁺ and K⁺ concentrations.

(C) Light

Light causes most of the outer segment cGMP-gated channels to close, thereby reducing the dark current.

Na⁺

K⁺

Na⁺

K⁺

FIGURE 14.28 Light hyperpolarizes vertebrate photoreceptors (A) Retinal rods and cones are relatively depolarized in the dark, and their receptor potential in light is a graded hyperpolarization. The brighter the light, the greater is the hyperpolarization. (B) A dark current, carried largely by Na⁺ ions, enters the rod outer segment in the dark and depolarizes it. (C) Light acts to decrease the dark current by closing cGMP-gated Na⁺ channels, leading to hyperpolarization.

Enzymatic regeneration of rhodopsin is slow

If rods are stimulated by very bright light, the photoreceptors recover very slowly. This is because the bright light has converted 11-*cis* retinal to all-*trans* retinal in a large fraction of the pigment of the eye, a process called *bleaching*. After a bleach greater than about 1% of the photopigment, recovery of rod vision is prolonged; this slow adjustment to darkness is called **dark adaptation**.

Complete recovery of sensitivity after bleaching illumination requires the regeneration of the all-*trans* retinal back to 11-*cis* retinal. Regeneration can occur either photochemically or enzymatically. In insects, most regeneration occurs photochemically. The meta-rhodopsin with all-*trans* retinal can absorb a second photon and convert directly back to 11-*cis* rhodopsin without requiring metabolic energy. This has obvious advantages but one disadvantage: Because light illuminating the eye will generally stimulate both rhodopsin and metarhodopsin, the photoreceptors will usually contain some combination of both pigments. Many arthropods start afresh every day, breaking down and resynthesizing the light-sensitive microvilli and regenerating the rhodopsin on a circadian basis.

For vertebrate rods, most if not all regeneration normally occurs by an enzymatic mechanism. After bleaching of the 11-*cis* retinal to all-*trans* retinal, the all-*trans* retinal becomes unbound from the opsin protein. The all-*trans* retinal (aldehyde) is then converted to all-*trans* retinol (alcohol) by an enzyme in the photoreceptor called *retinol dehydrogenase*. The retinol is then transported by a specialized transport protein called *interphotoreceptor retinoid binding protein*, or *IRBP*, to the layer of cells behind the photoreceptors called the *pigment epithelium* (see Figures 14.22, 14.25B, and 14.26). The all-*trans* retinol is taken up by the pigment epithelial cells and re-isomerized back to 11-*cis* retinol and then converted to 11-*cis* retinal. The 11-*cis* retinal is transported back to the photoreceptors again by IRBP, and it then recombines with opsin by the reformation of the covalent bond between the retinal aldehyde and the amino group of the lysine in opsin. This procedure, although cumbersome and slow, assures the regeneration of the full complement of visual pigment in the rod, and visual sensitivity returns to its maximum dark-adapted value only after all of the pigment has been restored.

SUMMARY
Photoreception

- The vertebrate eye is a camera eye that focuses light onto retinal rod and cone photoreceptors. Rods and cones are unusual in that light produces a hyperpolarizing receptor potential.

- The photopigment rhodopsin is a GPCR molecule conjugated to retinal. It is contained in membranes of outer segments of vertebrate rods and cones. When rhodopsin absorbs light, it acts via a G protein to decrease the concentration of cGMP in the cytoplasm, leading to closing of cGMP-gated Na^+ channels that keep the photoreceptor depolarized in the dark. Light-induced closure of these channels hyperpolarizes the photoreceptors.

- In arthropods such as *Drosophila* the photopigment rhodopsin is similar to that of vertebrates and activates a similar G protein, but it is linked to a different intracellular effector and leads to the production of DAG and IP_3, opening ion channels and producing a depolarizing receptor potential.

- Rhodopsin is deactivated and ultimately regenerated to 11-*cis* rhodopsin after activation. In vertebrates, most regeneration is a slow enzymatic process, part of which occurs outside the photoreceptors in the adjacent pigment epithelium.

Visual Sensory Processing

Photoreceptors respond to light, but the vertebrate visual system responds to pattern: *contrasts*, or changes in light level and color over space and time. This conversion—from sensitivity to light into sensitivity to contrast—occurs partly in the retina and partly in higher visual-processing areas of the brain.

Retinal neurons respond to contrast

The aspects of visual stimuli that are most important for the behavior of animals (including humans) are *patterns* of light and darkness and color, rather than the overall light level. In behavioral terms, the significant features of the visual world are spatial patterns of visual stimuli that represent objects in the world, and temporal patterns that indicate movements. To a frog, for example, a small dark area in the visual field may have great behavioral importance (a fly for lunch)—particularly if it moves relative to the rest of the visual field. Changes in overall illumination may be less important, merely indicating, for example, that a cloud has passed in front of the sun.

The photoreceptors themselves respond primarily to the light level in one point in space. Therefore we would expect that visual systems might integrate signals from receptors in ways that abstract the behaviorally significant spatial patterns and movements of stimuli. Although considerable progress has been made in understanding how visual signals are processed and used behaviorally in insects, visual integration has been studied most extensively in the vertebrate visual system. In this section we consider how neural circuits in the retina perform the early stages of this visual integration.

The retina contains the rod and cone photoreceptors and four kinds of integrating neurons (see Figure 14.25B): bipolar cells, horizontal cells, amacrine cells, and ganglion cells. **Bipolar cells** receive input from photoreceptors at the *outer plexiform layer* near the photoreceptors, and they synapse on amacrine and ganglion cells in the *inner plexiform layer*. **Horizontal cells** extend tangentially, connecting different regions of the outer plexiform layer. **Amacrine cells** mediate similar tangential interconnections in the inner plexiform layer. Retinal **ganglion cells** are the output of the retina; their axons form the optic nerve that extends to the brain.

How is sensory input from the photoreceptors transformed by these neurons as it makes its way through the retina? We begin our examination of retinal integration by looking at the response properties of the retinal ganglion cells, which provide the sensory output from the retina. This will give us an overview of how the visual image is processed by the retinal cells. The ganglion cells, in contrast to photoreceptors, each respond to stimulation over a relatively large visual area—the ganglion cell's *receptive field*. For the vertebrate visual system, the **receptive field** of a neuron such as the ganglion cell is defined as the area of the retina (or the area of the visual field) within which the membrane potential or impulse activity of that neuron can be influenced by light. For a rod or cone, the receptive field corresponds for the most part to the retinal area

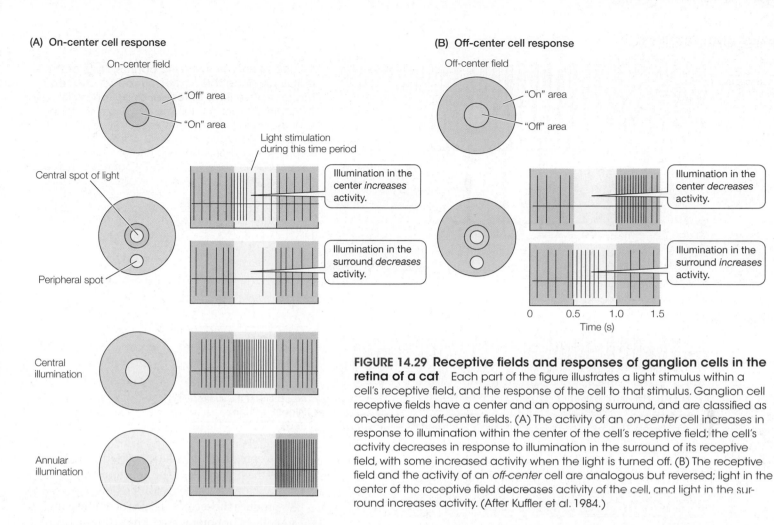

(A) On-center cell response

On-center field

"Off" area

"On" area

Central spot of light

Peripheral spot

Light stimulation during this time period

Illumination in the center *increases* activity.

Illumination in the surround *decreases* activity.

Central illumination

Annular illumination

Diffuse illumination

0   0.5   1.0   1.5
Time (s)

(B) Off-center cell response

Off-center field

"On" area

"Off" area

Illumination in the center *decreases* activity.

Illumination in the surround *increases* activity.

0   0.5   1.0   1.5
Time (s)

FIGURE 14.29 Receptive fields and responses of ganglion cells in the retina of a cat Each part of the figure illustrates a light stimulus within a cell's receptive field, and the response of the cell to that stimulus. Ganglion cell receptive fields have a center and an opposing surround, and are classified as on-center and off-center fields. (A) The activity of an *on-center* cell increases in response to illumination within the center of the cell's receptive field; the cell's activity decreases in response to illumination in the surround of its receptive field, with some increased activity when the light is turned off. (B) The receptive field and the activity of an *off-center* cell are analogous but reversed; light in the center of the receptive field decreases activity of the cell, and light in the surround increases activity. (After Kuffler et al. 1984.)

occupied by the receptor itself. For visual interneurons such as retinal ganglion cells, in contrast, the receptive field is typically much larger, embracing an area containing many photoreceptors, and can also include different regions giving responses of different polarity.

Let's begin by examining the receptive fields of two ganglion cells that are sensitive to contrast. The fields of these cells are typical of many of the ganglion cells in mammals and 90% of those in the human retina. We can record from ganglion cells with an extracellular microelectrode, etched to a fine tip and insulated with resin except for a few microns at the tip. Such a microelectrode records the impulses of neurons in the immediate vicinity. We can map the receptive fields of a neuron by recording its responses when small spots of light are shone either directly onto the retina or onto a screen in front of the eye. If light on a particular spot elicits a response, then that spot is considered to be part of the receptive field of the neuron being recorded.

FIGURE 14.29 shows the responses of the two retinal ganglion cells to different kinds of light stimulation. The cells have receptive fields divided into two areas: a *center* and a *surround*. The first cell

to the left, termed an **on-center cell**, increases its rate of impulse discharge when the center of its receptive field is illuminated by a spot of light (see Figure 14.29A). The same spot of light, however, suppresses activity when it is presented in the larger surrounding part of the receptive field. The on-center cell is maximally stimulated when the entire center of its receptive field, but none of its surround, is illuminated. The cell's activity is maximally inhibited or suppressed when the surround, but not the center, is illuminated by an annulus (ring) of light. Such suppression is followed by an increased discharge when the light is turned off (an "off response"). Diffuse light stimulation, covering the entire receptive field of the ganglion cell, has little effect on the cell's activity because the excitatory effect of light at the center and the inhibitory effect of light in the surround are antagonistic, canceling each other out.

The cell represented in Figure 14.29B, called an **off-center cell**, also has a receptive field with a concentric, antagonistic center and surround. The off-center cell, however, is inhibited by light in its center and excited by light in its surround. Its receptive field is thus the opposite of the field of the on-center cell.

The two retinal ganglion cells just described are typical of many of the ganglion cells in cat and monkey retinas, which contain roughly equal numbers of these two types. There are other classifying features of ganglion cell organization, which we discuss later (see Box 14.3). Furthermore, many mammals and other vertebrates have ganglion cells with properties that are more complex than those illustrated in Figure 14.29. Nevertheless, the

(A) Straight-through pathways

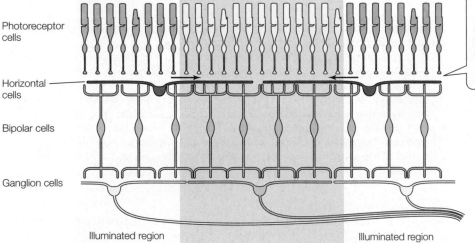

Photoreceptor cells

Bipolar cells

Ganglion cells

Illuminated region

(B) Lateral pathways

Photoreceptor cells

Horizontal cells

Bipolar cells

Ganglion cells

Illuminated region   Illuminated region

Light in the surround opposes the effect of light in the center because the synaptic actions of horizontal cells oppose the actions of the straight-through pathway.

FIGURE 14.30 The synaptic connections of the retina produce the center–surround, concentric receptive field of retinal ganglion cells (A) The radial, straight-through pathways from receptors to bipolar cells (in red) to the dendritic tree of a retinal ganglion cell produce the center of the ganglion cell's receptive field. (Horizontal cells and amacrine cells are present but not shown.) (B) Lateral pathways such as the horizontal-cell pathways shown (in red) produce the surround of the ganglion cell's receptive field. (Amacrine cells, not shown, also contribute to lateral pathways.) (After Masland 1986.)

preceding description of response properties of contrast-sensitive on-center and off-center ganglion cells is a sufficient starting place for an analysis of retinal integration.

In order to see how the response properties and receptive fields of contrast-sensitive ganglion cells are derived from neural circuits in the retina, we next examine the synaptic connections and properties of the other retinal neurons. **FIGURE 14.30** shows the basic circuit organization of the vertebrate retina. Rods and cones (the photoreceptor cells) synapse on bipolar cells and horizontal cells. Horizontal cells synapse onto rods and cones, and through them influence bipolar cells. Bipolar cells synapse on amacrine cells and ganglion cells. Amacrine cells (see Figure 14.25B) synapse back onto bipolar cells, onto other amacrine cells, and onto ganglion cells. For simplicity, amacrine cells are not shown in Figure 14.30.

We can distinguish between two sorts of retinal pathways: **straight-through pathways**, which project radially through the retina at right angles to its surface (photoreceptors → bipolar cells → ganglion cells) (see Figure 14.30A); and **lateral pathways**, which extend along the retinal sheet, via horizontal cells and amacrine cells (see Figure 14.30B). The straight-through pathways give rise to the properties of the *center* of a contrast-sensitive ganglion cell's

receptive field, and the lateral pathways give rise to the properties of the antagonistic *surround*.

How do these two antagonistic pathways produce the receptive fields of retinal ganglion cells? For cone vision, the straight-through pathway makes up the center of a ganglion cell's receptive field (**FIGURE 14.31A**). Light in the center of the receptive field excites cones, hyperpolarizing them. A cone synapses onto two kinds of retinal bipolar cells: *off-center* bipolar cells in which light in the center of the receptive field hyperpolarizes the cell, and *on-center* bipolar cells in which light in the center of the receptive field depolarizes the cell.

The light stimulus in Figure 14.31A thus leads to hyperpolarization of an off-center bipolar cell, which synapses onto an off-center ganglion cell and hyperpolarizes it. Light in the center hyperpolarizes cones to *depolarize* an on-center bipolar cell, which synapses onto an on-center ganglion cell and depolarizes it. The ganglion cells generate action potentials when they depolarize; however, rods and cones, bipolar cells, and horizontal cells do not normally generate action potentials.

Light in the surround of a ganglion cell's receptive field antagonizes these effects, in part by activating horizontal cells of the lateral pathway (**FIGURE 14.31B**). Stimulated cones in the surround area synapse onto horizontal cells, the lateral processes of which synapse back onto cone terminals to oppose the effects of light on the cones (i.e., horizontal cells depolarize cone terminals). This produces bipolar cells with receptive fields having an antagonistic center and surround. Light in the *surround* of an off-center bipolar cell depolarizes it (leading to depolarization of the surround of an off-center ganglion cell), and light in the surround of an on-center

(A) The straight-through pathway mediates the center of a receptive field

(B) The lateral pathway mediates the surround of a receptive field

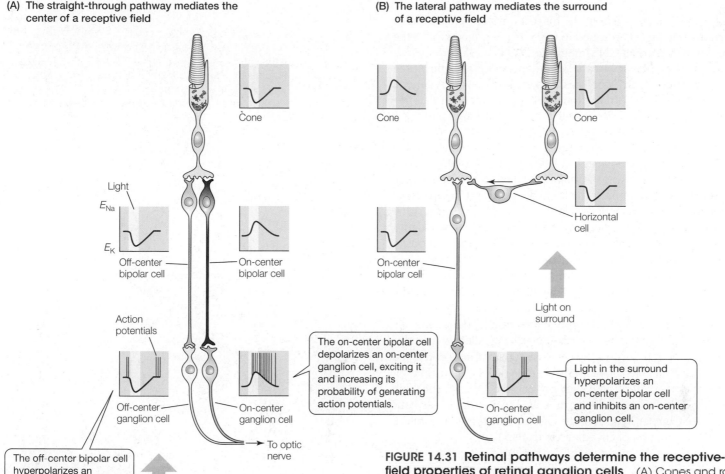

Light

E_{Na}

E_K

Cone

Off-center bipolar cell

On-center bipolar cell

Action potentials

Off-center ganglion cell

On-center ganglion cell

To optic nerve

The on-center bipolar cell depolarizes an on-center ganglion cell, exciting it and increasing its probability of generating action potentials.

The off-center bipolar cell hyperpolarizes an off-center ganglion cell, inhibiting any generation of action potentials.

Light on center

Cone

Cone

Horizontal cell

On-center bipolar cell

Light on surround

On-center ganglion cell

Light in the surround hyperpolarizes an on-center bipolar cell and inhibits an on-center ganglion cell.

FIGURE 14.31 Retinal pathways determine the receptive-field properties of retinal ganglion cells (A) Cones and rods contribute to the *center* of a ganglion cell's receptive field via the straight-through pathway. A cone, for example, that is excited by light hyperpolarizes, hyperpolarizing an off-center bipolar cell and depolarizing an on-center bipolar cell. (B) Cones and rods contribute to the *surround* of a ganglion cell's receptive field via the lateral pathway. An illuminated cone, for example, hyperpolarizes horizontal cells, which end on the synaptic endings of other cones and depolarize them (opposing the effects of light). (Horizontal-cell actions also oppose light effects on off-center bipolar cells, depolarizing them and exciting off-center ganglion cells; these interactions are not shown.)

bipolar cell hyperpolarizes it (leading to hyperpolarization of the surround of an off-center ganglion cell). This inhibition of neighboring cells in an array is termed *lateral inhibition*. Horizontal cells do not mediate *all* of the lateral inhibition that produces the surround of a ganglion cell's receptive field. Amacrine cells provide a second layer of lateral inhibitory interaction in the retina, enhancing the center–surround antagonism and in some cases imparting to ganglion cells a sensitivity to moving light stimuli.

Amacrine cell pathways can contribute to more complex receptive field properties of ganglion cells. In mammals (including primates) and most other vertebrates, there are ganglion cells that respond to movement and even to the direction of movement, firing rapidly when a stimulus moves in one direction and not at all for motion in the opposite direction. Amacrine cells are thought to play a key role in producing these receptive fields. In frogs, they probably also contribute to the receptive fields of ganglion cells that respond to quite specific features of a visual stimulus. One type of ganglion cell responds only to a small, dark convex edge that moves relative to the background. The optimal size of a dark stimulating object (1°) is about the size of a fly at striking distance, and stimuli that activate these ganglion cells tend to trigger the frog's feeding-strike movements. It is plausible to interpret ganglion

cells with these properties as "fly detectors" adapted to respond to a specific, behaviorally significant feature of the visual world. A small population of ganglion cells with similar response properties may also exist in mammals, where instead of catching flies they may help direct eye and head movements to bring moving objects to the center of gaze.

The vertebrate brain integrates visual information through parallel pathways

In the preceding discussion of the retina we have seen that the activities of ganglion cells (the retinal output neurons) convey information about visual contrast rather than overall level of illumination. Activity in an on-center ganglion cell signals that a region of the visual field (corresponding to the center of its receptive field) is brighter than the surround. Corresponding activity in an off-center ganglion cell indicates that the center of its receptive field is darker than the surround. This responsiveness to stimulus pattern or contrast is continued in CNS visual projections in the brain. We will examine CNS integration of visual information as an example from which we can derive some general principles of the way in which a brain processes sensory information. A general principle is that visual information is conveyed over several different CNS pathways in the vertebrate brain, the different pathways conveying information about different aspects of a complex visual stimulus, such as color, fine details of form, and stimulus movements that elicit responsive eye movement.

In fish and amphibians the major visual projection of the optic nerve is to the optic tectum of the midbrain. The optic nerves cross the brain midline at the optic chiasm and connect to neurons in the contralateral (opposite) tectum. In mammals the region homologous to the optic tectum is the superior colliculus. Visual projections to the superior colliculus from on- and off-center ganglion cells, and also from movement- and direction-selective cells, are thought to be important in many aspects of visual behavior, including the control of eye movements. Another sparse but important projection arises from a specialized type of intrinsically light-sensitive ganglion cell, which has its own photopigment (called *melanopsin*). These ganglion cells, unlike the contrast-sensitive cells presented in Figure 14.29, are sensitive to the absolute level of illumination regardless of its pattern, and they project to centers in the CNS controlling pupillary movements, as well as to the suprachiasmatic nucleus (see Chapter 15 and Figure 15.16), where they play a critical role in the regulation of circadian rhythms.

In mammals the major visual projection is the *geniculostriate system* (**FIGURE 14.32**). In this pathway the axons of retinal ganglion cells that form the optic nerve synapse in a region of the thalamus termed the **lateral geniculate nucleus** (**LGN**). Neurons of the LGN project to the **primary visual cortex** at the posterior end of the cerebrum. Unlike the optic projections of lower vertebrates, the projections of most mammals are only partially crossed at the optic chiasm. In cats and primates, the projections of the nasal (inner) half of the retina cross to the contralateral (opposite) side, whereas those of the temporal (outer) half go to the ipsilateral (same) side LGN. This mixing of input from the two eyes allows mammals with forward-facing eyes to merge binocular input for depth perception at the visual cortex.

David Hubel and Torsten Wiesel contributed greatly to our understanding of the CNS processing of visual information with a series of studies of the response properties of visual neurons in cats and monkeys, for which they won the Nobel Prize in 1981. They recorded from visual cells of anaesthetized animals while projecting patterned stimuli (light and dark bars, edges, and spots) onto a screen in front of the animal. They found that LGN neurons had response properties similar to those of retinal ganglion cells: They respond to stimuli to only one eye and have concentric receptive fields with an antagonistic center and surround, either on-center or off-center.

The receptive fields of neurons in the visual cortex are quite different from those of the retina and LGN. We can differentiate two major kinds of neurons in the visual cortex: *simple cells* and *complex cells*. Both types are binocular: They can respond to visual stimuli presented to either eye, although one eye may predominate (an effect called *ocular dominance*). Both kinds of cells have a preferred *axis of orientation* of the visual stimulus, a term best explained if

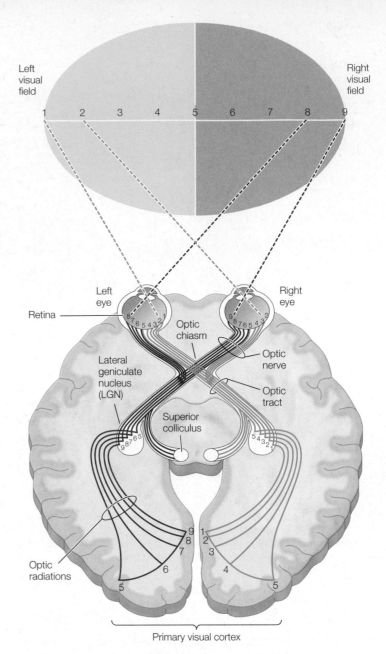

FIGURE 14.32 Central visual projections of a mammal As shown in this cutaway view of the brain (seen from above), only part of the optic tract crosses the midline at the optic chiasm. Therefore a stimulus in the *left* visual field (the left half of the visible world) projects to the *right* lateral geniculate nucleus (LGN) and visual cortex via both eyes (blue pathway). Conversely, a stimulus in the *right* visual field projects to the *left* LGN and visual cortex (red pathway). The spatial order of the visual field is mapped in the visual cortex, as indicated by numbers (1–9). (After Frisby 1980.)

we describe the receptive fields of a few simple cells in the visual cortex of a cat.

First, recall that a receptive field is the area of the retina or of the visual field in which light stimuli influence the activity of a particular neuron. **FIGURE 14.33** shows the receptive fields of three **simple cells**, as they might be mapped with small spots of light shone either directly onto the retina or onto a projection screen viewed by the animal. Plus signs (+) denote areas in which a light spot excites the simple cell, increasing the frequency of its discharge of action potentials. Minus signs (–) denote areas in which the frequency of action potentials decreases when the light spot is on and increases when the spot is turned off ("off response"). The

(A) Dark bar   **(B) Light bar**   **(C) Light–dark edge**

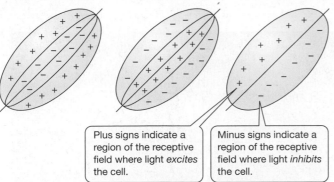

> Plus signs indicate a region of the receptive field where light *excites* the cell.

> Minus signs indicate a region of the receptive field where light *inhibits* the cell.

FIGURE 14.33 Receptive fields of simple cells in the visual cortex of a cat Each receptive field is an area of the retina that affects the activity of a single cortical cell. All three receptive fields shown here are of cells with the same axis of stimulus orientation; other cell fields have different axes of orientation. Cortical simple cells typically respond best to (A) a dark bar, (B) a light bar, or (C) a light–dark edge.

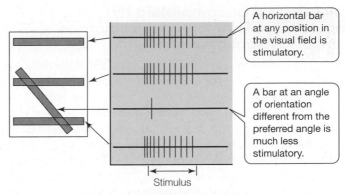

> A horizontal bar at any position in the visual field is stimulatory.

> A bar at an angle of orientation different from the preferred angle is much less stimulatory.

Stimulus

FIGURE 14.34 The receptive field of a complex cell in the visual cortex Complex cells respond best to a bar (or edge) at the correct angle of orientation, which in this case is horizontal. Unlike simple cells, however, the stimulatory bar may be anywhere within the receptive field.

positions of excitatory and inhibitory responses to spots of light reveal the extent of the receptive field of the cell and also the type of visual stimulus that most effectively elicits its response.

Each of the receptive fields in Figure 14.33 has a line drawn through it that defines its axis of orientation. To be most effective, a stimulus must be aligned with this axis of orientation. The optimal stimuli producing the largest number of action potentials for these three cells are as follows: a dark bar in the central band of the receptive field; a light bar in the central band of the field; and an edge on the center line that is light on the upper left and dark on the lower right. In all cases the stimulus pattern must be correctly aligned and oriented. In Figure 14.33A, for example, a vertical dark bar will be a weak or ineffective stimulus because it falls on both excitatory and inhibitory areas. For the same reason, visual cortical cells are insensitive to changes in overall illumination.

The three receptive fields shown in Figure 14.33 have the same axis of orientation, which is to say that the best edge or bar stimulus in the visual field is at the same angle for each cell. In fact, all of the cells in a small area of the primary visual cortex have a similar axis of orientation. All axes of orientation, however, are represented in the cortex, in different bands for different axes. If the recording electrode is advanced sideways through the cortex, it records from cells with progressively changing axes of orientation, a finding that suggests that the primary visual cortex is organized in columns or wedges with an orderly arrangement of axes. There is also an orderly arrangement of bands of ocular dominance, some areas containing neurons dominated by input from the ipsilateral eye and some by the contralateral eye. Overall, a pinwheel-shaped region of the cortex about 1 × 1 mm is subdivided into areas containing cells responding to all axes of orientation, as well as ocular-dominance bands from both eyes.

Interspersed in the cortex among the simple cells are **complex cells**, which are also responsive to bars and edges and have receptive fields with a preferred axis of orientation. Unlike simple cells, however, complex cells are insensitive to the *position* of a stimulatory bar or edge within the receptive field. As illustration of this defining feature of a complex cell, **FIGURE 14.34** shows the receptive field of a complex cell that responds to a horizontal bar anywhere within its sizeable receptive field. The same bar stimulus presented at an angle different from the preferred axis of orientation is much less

stimulatory. No part of the receptive field of a complex cell can be defined as excitatory or inhibitory; instead the stimulus *pattern* (a dark horizontal bar) is excitatory. Therefore complex cells have strict requirements about stimulus form and orientation, but less-strict requirements about stimulus position within the field.

It is not completely clear how the receptive field properties of cortical simple and complex cells result from synaptic input from other cells. Early experiments suggested a *hierarchical* organization in which many similar geniculate cells with receptive field centers aligned in a row converged onto a single simple cell in the cortex. Careful studies of cortical physiology indicate that this notion is probably true for simple cells. Complex cells were envisioned to receive convergent synaptic input from several simple cells, and in turn to synapse upon still higher-order cells. Aspects of this scheme are no doubt correct, but complex cells have been shown to receive some synaptic input directly from geniculate axons. These and other studies show that there is also a *parallel* organization of CNS visual projections. **BOX 14.3** discusses further these aspects of higher visual integration.

BOX 14.3 **What roles do individual neurons play in higher visual integration?**

Central integration of visual information in mammals and other vertebrates involves both a hierarchical organization and a parallel organization. That is, several parallel pathways each consist of a hierarchy of projections to higher visual centers through multiple neurons and synapses. Three classes of visual pathways project from the retina to the lateral geniculate nucleus (LGN), each consisting of different kinds of matched retinal ganglion cells and LGN cells that differ in size and in response properties. In primates these classes have been termed *magnocellular* (large cells), *parvocellular* (smaller, medium-sized cells), and *koniocellular* (still smaller cells). **Box Extension 14.3** describes how these parallel paths convey different sorts of visual information, and asks how individual neurons and ensembles of neurons process visual information in the brain.

Color vision is accomplished by populations of photoreceptors that contain different photopigments

The ability to distinguish color depends on the differential sensitivities of photopigments to different wavelengths of light. Although many animals are color-blind, many other animals with well-developed diurnal visual systems have evolved color vision. Examples include several orders of insects, crustaceans such as mantis shrimp, teleost fish, frogs, turtles, lizards, birds, and primates.

Theories of color vision are strongly based on human perceptual studies. In 1801, Thomas Young (1773–1829) proposed that human color perception was based on separate receptor classes sensitive to red, green, and blue light. This theory was supported by perceptual observations that any color could be duplicated by a mixture of three primary colors.

The physiological basis of this trichromaticity theory has been clarified in the last 40 years. Humans and other primates have three populations of cone photoreceptors, which are sensitive to different wavelengths of light. The three cone classes have slightly different photopigments of the rhodopsin family, each with a somewhat different opsin sequence that alters its sensitivity to light wavelengths. **FIGURE 14.35** illustrates the absorption spectra of the three classes of primate cones determined by measuring light absorption by each pigment at different wavelengths. Spectral characteristics can also be determined by testing the effectiveness of stimulation by light of different wavelengths to define an action spectrum. The three types of cones are usually called *red cones*, *green cones*, and *blue cones*, loosely approximating the color of light to which they are most sensitive. The spectral sensitivities of the three types of cones are rather broad and overlapping, so that the perception of color must be based on the ratio of excitation of different cone populations. For example, we might expect perception of long-wavelength red light to depend on an analyzer that is excited by red cones but inhibited by green cones. Just this sort of integration occurs in the retina.

Two kinds of *color-opponent* processes are present in the mammalian retina: red–green opponency and blue–yellow opponency. These opponent processes explain why we can perceive a color as bluish green or as reddish yellow (orange), but we do not ordinarily perceive a color as reddish green or as bluish yellow. Many of the ganglion cells of a primate retina that are stimulated by cones have color-opponent properties. Red–green opponent ganglion cells, for example, have concentric antagonistic receptive fields but are also color-opponent. One cell may be inhibited by red light in its receptive-field center and excited by green light in its surround; another may be excited by red light in the center and inhibited by green in the surround, and so on. Other classes of ganglion cells may be excited by blue light and inhibited by yellow (a sum of red and green cone input). Blue–yellow opponent cells lack center–surround antagonism.

Color-opponent retinal ganglion cells project to the LGN and are relayed to the primary visual cortex. Color information in the visual cortex appears to be integrated by clusters of cells in separate districts in the cortex, located separately from other cells that are not color-selective but instead process information about brightness contrast. This apparent segregation of color channels and so-called

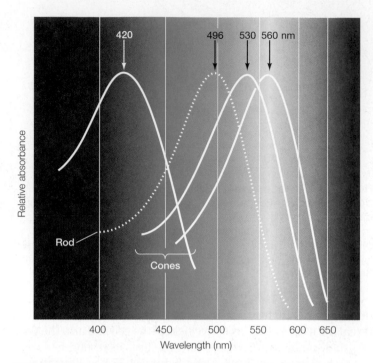

FIGURE 14.35 Spectral sensitivities of human retinal cones Spectral sensitivities were determined by measuring the absorption spectra of single cones. The three types of cones have photopigments with different absorption maximums: 420 nm (blue cones), 530 nm (green cones), and 560 nm (red cones). Note that the "green" and "red" cones have similar spectral sensitivities and are both most excited by yellowish light. Rods have an absorption peak of 496 nm.

achromatic channels is another example of parallel organization in the mammalian visual system.

Mechanisms for color vision in animals other than mammals are always based on several populations of receptor cells with different spectral sensitivities, although these receptors are not always cones. Fish such as carp have three populations of cones analogous to those of primates, with absorption maximums of 455 nm, 530 nm, and 625 nm, respectively. In contrast, frogs have two or more classes of rod photopigments; their color vision involves input from both rods and cones. Birds, mice, and some other vertebrates have ultraviolet visual sensitivity. Many insects have well-developed mechanisms for color vision, including receptors sensitive to ultraviolet radiation. The different retinular cells in an insect ommatidium often contain different photopigments, and these provide information for making color discriminations. The honeybee has photoreceptors with maximum sensitivities to 350 nm, 450 nm, and 550 nm, and readily distinguishes both ultraviolet radiation and colors of visible light. Many white flowers seem featureless to us but reveal striking patterns under ultraviolet illumination, and the bees use these patterns to identify pollen sources.

This last example is a reminder that many animals use sensory abilities foreign to us, and in fact we use many sensory faculties of which we are largely unaware, as well as the senses of which we are conscious. Several of the following chapters will consider additional roles and uses of sensory information.

SUMMARY
Visual Sensory Processing

■ Neural circuits of the vertebrate retina integrate the responses of retinal photoreceptors to excite and inhibit retinal ganglion cells. Ganglion cell receptive fields may be excited or inhibited by light at the center of the field, whereas light in the surround antagonizes the effect of light in the center.

■ Straight-through pathways (photoreceptor → bipolar cell → ganglion cell) produce the center (on- or off-center) of a ganglion cell's receptive field. Lateral pathways through horizontal cells and amacrine cells produce the antagonistic surround.

■ Axons of ganglion cells make up the optic nerve, relaying visual information to several brain areas. The geniculostriate pathway projects to the lateral geniculate nucleus (LGN) and from there to the primary visual cortex.

■ Simple and complex cells in the primary visual cortex respond to light or dark bars or edges oriented at particular angles.

■ Parallel pathways in the visual cortex convey information about different aspects of a visual stimulus, such as details of visual form, movement, color, and binocular determination of object distance.

■ Color vision depends on the ratio of activation of three classes of cone photoreceptors sensitive to different wavelengths of light. Retinal circuitry integrates color contrasts based on red–green and blue–yellow opponencies.

STUDY QUESTIONS

1. Suppose that a sensory neuron in the periphery generates a train of action potentials, and synaptically excites an interneuron in the CNS. What does this signal mean? What additional information does the interneuron need in order to decode the message?

2. Suppose that in a particular sensory receptor, the amplitude of the receptor potential increases linearly with the log of the stimulus intensity. The receptor potential depolarizes the cell to produce a train of action potentials, and the frequency of the action potentials increases linearly with increasing receptor-potential depolarization (above threshold). There is a maximum frequency of action potentials, beyond which additional increase in stimulus intensity does not further increase action-potential frequency. Graph these results as the amount of response as a function of stimulus intensity (or log intensity). If these results are typical of many receptors, what do they tell you about how sensory receptors encode stimulus intensity?

3. What is adaptation of a sensory receptor? At what stage(s) does it occur? What are its functions?

4. Where is rhodopsin localized in a vertebrate retinal rod? Where is the receptor potential generated? How does the transduction mechanism of the rod connect these two sites? How does it increase light sensitivity?

5. Diagram and describe how lateral inhibition via horizontal cells produces the surround of a retinal ganglion cell's receptive field.

6. To what features of the visual world do simple and complex cells of the mammalian visual cortex respond? How do these cells differ?

7. Insects and vertebrates have analogous proprioceptors that monitor muscle length and tension, joint positions, limb forces, and so on. Insects, however, do not have statocysts to monitor the direction of gravitational force. How could an insect detect gravity when at rest? What could be the possible adaptive advantage of the *absence* of statocysts for a flying insect?

8. Among vertebrates, acoustico-lateralis hair cells are used to detect water currents in lateral line organs of fish, gravity and acceleration in the labyrinth, and sound in the cochlea. How can similar receptor cells mediate all these responses?

9. Compare the mechanisms and accuracy of auditory localization of a sound source in a human, an owl, and a bat.

10. Mechanisms of chemoreceptor transduction are of two broad kinds, which parallel the two major mechanisms of synaptic transmission. Describe these, and suggest an evolutionary speculation for the similarities.

Go to **sites.sinauer.com/animalphys4e**
for box extensions, quizzes, flashcards,
and other resources.

REFERENCES

Bargmann, C. I. 2006. Comparative chemosensation from receptors to ecology. *Nature* 444: 295–301.

Conner, W. E., and A. J. Corcoran. 2011. Sound strategies: The 65 million-year-old battle between bats and insects. *Annu. Rev. Entomol.* 57: 21–39.

Cronin, T. W., S. Johnsen, N. J. Marshall, and E. J. Warrant. 2014. *Visual Ecology.* Princeton University Press, Princeton, NJ. A good recent monograph covering many aspects of visual function in diverse animals.

Delmas, P., J. Hao, and L. Rodat-Despoix. 2011. Molecular mechanisms of mechanotransduction in mammalian sensory neurons. *Nat. Rev. Neurosci.* 12: 139–153.

Fain, G. L. 2003. *Sensory Transduction.* Sinauer, Sunderland, MA. An excellent monograph covering how sensory receptors work.

Fettiplace, R., and K. X. Kim. 2014. The physiology of the mechano-electrical transduction channels for hearing. *Physiol. Rev.* 94: 951–986.

Frings, S. 2009. Primary processes in sensory cells: current advances. *J. Comp. Physiol. A* 195: 1–19.

Hughes, H. C. 1999. *Sensory Exotica: A World Beyond Human Experience.* Bradford Book/MIT Press, Cambridge, MA. A delightful, accessible treatment of animal senses not shared by humans.

Jiang, Y., and H. Matsunami. 2015. Mammalian odorant receptors: functional evolution and variation. *Curr. Opin. Neurobiol.* 34: 54–60.

Kaupp, U. B. 2010. Olfactory signaling in vertebrates and insects: differences and commonalities. *Nat. Rev. Neurosci.* 11: 188–200.

Nassi, J. J., and E. M. Callaway. 2009. Parallel processing strategies of the primate visual system. *Nat. Rev. Neurosci.* 10: 360–372.

Palkar, R., E. K. Lippoldt, and D. D. McKerny. 2015. The molecular and cellular basis of thermosensation in mammals. *Curr. Opin. Neurobiol.* 34: 14–19.

See also **Additional References** *and* **Figure and Table Citations**.

Nervous System Organization and Biological Clocks

15

Pictured here is a star-nosed mole (*Condylura cristata*). Star-nosed moles, like other moles, are *fossorial*: They live underground and dig for their food, which consists of small invertebrates. Moles have greatly reduced visual systems and are often considered blind, but some are sensitive at least to light levels. Star-nosed moles have evolved a stellate array of fleshy, fingerlike structures (shown in the photograph). The function of this "star" is tactile; it contacts the ground as many as 10–15 times per second and enables the mole to forage efficiently on small prey. The star-nosed mole may seem like an odd animal with which to introduce the integrative functions of nervous systems. But it illustrates two general points. First, the activities of the many cells in the mole's body must be controlled and coordinated in order for the mole to function and behave as a mole, rather than as just a mass of cells. In Chapter 2 we introduced controls *within* cells, mediated by cell-membrane receptors, second messengers, enzymes, transcription factors, and the like. Our focus here is control, coordination, and integration of activities *among* cells—that is, *inter*cellular control. Such control is the major functional role of the nervous system, and of the endocrine system. In the last three chapters we considered the cellular elements of nervous systems, the synaptic interactions of neurons, and the ways in which environmental information is acquired

How does this star-nosed mole control and integrate the functions of all the cells in its body?

by sensory processes. In this chapter we consider how entire nervous systems are organized for the specific and adaptive control of sets of cells. (Endocrine control is treated separately, in Chapter 16.)

The second point that the star-nosed mole illustrates is that control systems such as the nervous system are not only reactive, but also proactive. That is, although a nervous system functions to respond to changes in the outside world—stimuli in the animal's environment—it also coordinates its own activities without waiting for stimuli. Animals have *intrinsic* or *endogenous* functions such as daily rhythms of activity. Even if an animal cannot see day and night in its environment, its behavioral and physiological activities continue to cycle on a daily basis that anticipates the day–night cycle. This anticipatory activity is controlled by a *biological clock*—an endogenous, physiological timekeeping mechanism—that allows the animal to know when day will start, whether the animal sees light or not. Anticipation of this sort can be highly advantageous to an animal because it permits the animal to prepare physiologically and behaviorally for the new day, rather than merely waiting in a state of total ignorance about when the day will arrive. That this intrinsic rhythm persists without environmental cues doesn't diminish their importance, however. Environmental cues are necessary to reset the clock, *entraining* it to the outside world so that it does not drift earlier or later. In this chapter, in addition to discussing nervous systems in general, we will examine the biological clocks that control endogenous rhythms, as an example of the intrinsic, anticipatory functions of nervous systems.

The Organization and Evolution of Nervous Systems

The organization of neurons into functional nervous systems is what allows for the complexity of the neural control of animal physiology and behavior. We can define a **nervous system** as an organized constellation of cells (neurons and support cells) specialized for the repeated conduction of electrical signals within and between cells. These signals pass from sensory cells and neurons to other neurons and then to muscles, glands, or other organs that carry out actions. Nervous systems integrate the signals of converging neurons, generate new signals, and modify the properties of neurons based on their interactions. Nearly all animals have nervous systems. All nervous systems share similar characteristics, although they vary in the complexity of their organization and of their behavioral output.

Before we proceed to a closer look at the organization and evolution of nervous systems, we need to develop a framework of basic terms and organizational concepts that will facilitate our discussion. Some of these terms and concepts, although initially presented here, will be revisited and refined later in the chapter as well.

The nervous system in most types of animals consists of two major divisions: the *central nervous system* and the *peripheral nervous system*. The **central nervous system** (**CNS**) consists of relatively large structures such as the brain and spinal cord in which large numbers of neurons and support cells are anatomically juxtaposed and interact to achieve integrative functions. The CNS is rich with the cell bodies and processes (axons and dendrites) of neurons. Some neurons, called **interneurons**, are confined to the CNS. Other neurons are at least partially outside the CNS: Those that convey information to the CNS are **sensory neurons**, and those

that convey information out of the CNS to control muscles or other effectors are **motor neurons**. An **effector** is an organ, tissue, or cell that *acts*—that carries out functions such as motion or secretion—under the direction of the nervous system (or endocrine system). Muscles and glands are examples of effectors. The **peripheral nervous system** (**PNS**), then, consists of all the processes and cell bodies of sensory and motor neurons that are present outside the CNS (including autonomic ganglia and the enteric nervous system, considered later).

In the PNS, a **nerve** consists of the axons of multiple neurons bundled together into a structure resembling a cable of telephone wires. Although individual axons are too minute to be seen without a microscope, a nerve is macroscopically visible because it consists of many axons. If one transects a nerve and looks at either stump under a microscope, one sees many axons in cross section, just as one sees the cross sections of many wires when one cuts a wire cable. Axons of multiple neurons are often bundled together in the CNS as well as in the PNS. Such bundles in the CNS are not called *nerves*, however. Instead they are called *tracts*, *commissures*, or *connectives*, as we will discuss later. Another term of importance in describing the anatomy of nervous systems is *ganglion*. A ganglion, speaking macroscopically, is a swelling positioned along a nerve or connective. Because further details of ganglion structure and function differ in different groups of animals, we will postpone a closer look at them until later in this chapter.

Physiologists recognize two primary divisions of the vertebrate PNS: the *somatic* and the *autonomic nervous systems*. These same divisions are sometimes recognized in the PNSs of invertebrates. The **somatic nervous system** is the part of the PNS that controls the skeletal (striated) muscles that generally produce voluntary movements; skeletal muscles are thus called **somatic effectors**. Sensory reception of external stimuli and transmission of this sensory information are also functions of the somatic nervous system. The **autonomic nervous system**, by contrast, is the part of the PNS that controls **autonomic effectors** (or **internal effectors**), defined to include all neuron-controlled effectors other than the striated muscles, such as cardiac muscle, smooth (nonstriated) muscles, and glands. The autonomic nervous system also has sensory neurons that convey information to the CNS about the internal organs. The somatic nervous system controls most observable behavior and therefore is the part of the PNS with which we are most familiar. Autonomic effectors exert most of their effects on visceral organs, internally and invisibly.

Nervous systems consist of neurons organized into functional circuits

A good way to start a discussion of the cellular organization of nervous systems is to recall the example of the cockroach in Chapter 12 (page 309). As is typical of nervous systems in general, in the nervous system of the cockroach, cellular elements sense the environment, send signals to other cells (neurons) in the CNS, and ultimately control and coordinate cells of effectors to generate physiological or behavioral outputs. In the specific case we discussed (see Figure 12.3), wind-receptor sensory neurons were excited by an environmental mechanical stimulus, and those sensory neurons in turn synaptically stimulated interneurons that excited motor neurons to induce contraction of muscle cells, producing a reflexive behavioral response—a jump. **FIGURE 15.1** represents the cockroach's neural

(A) The startle response circuit of the cockroach

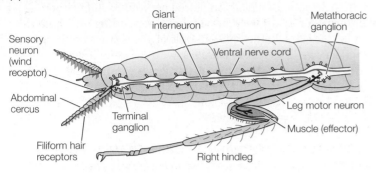

(B) Simplified diagram of the response circuit

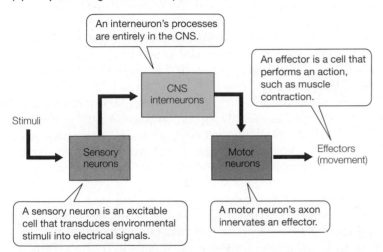

An interneuron's processes are entirely in the CNS.

An effector is a cell that performs an action, such as muscle contraction.

Stimuli

CNS interneurons

Sensory neurons

Motor neurons

Effectors (movement)

A sensory neuron is an excitable cell that transduces environmental stimuli into electrical signals.

A motor neuron's axon innervates an effector.

FIGURE 15.1 Neuronal elements in a nervous system: a neural circuit mediating the cockroach startle response (A) The startle response circuit (introduced in Figure 12.3) involves three neuronal elements: sensory neurons, interneurons, and motor neurons. (B) A simplified diagram of this circuit shows the basic functions of neural circuits in nervous systems. Sensory neurons convey signals about environmental stimuli to the CNS. These signals are integrated by CNS neurons and can trigger or modulate output signals of motor neurons to control effectors such as contracting muscle cells.

circuit as a block diagram, simplified to illustrate how this simple reflex in a cockroach exemplifies the general functional features of nervous systems:

- Neurons are organized in circuits in such a way that they can elicit a coordinated, adaptive response of effectors.
- Sensory receptor cells (which, like neurons, are excitable cells) transform environmental stimuli into electrical signals.
- CNS interneurons integrate signals from sensory receptors and other signals arising within the animal, generating an integrated pattern of impulses.
- Motor commands are sent out from the CNS to effectors.

Many types of animals have evolved complex nervous systems

We have little direct knowledge of the evolution of nervous systems, which are rarely preserved in the fossil record. Theories of nervous system evolution are based on interpretations of the anatomy and the molecular genetics of living groups, a risky proposition because all groups alive today are highly evolved and none can be

---

Did nervous systems evolve only once? All multicellular animals except sponges have neurons and nervous systems, but perhaps not all nervous systems are homologous. Genomic analyses have found many genes important for nervous system organization to be present in primitive, unicellular protists, and could have been employed in evolution of neurons on more than one occasion. **Box Extension 15.1** describes the organization and evolution of nervous systems in different animal groups.

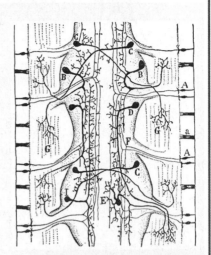

Diagram of the nervous system of a worm by S. Ramón y Cajal The cell labels are as follows: A, sensory cells of the skin; B–E, motor neurons; G, terminal ramifications of motor neurons in muscles; and I, interneurons. (From Ferreira et al. 2014.)

taken as representing a primitive condition. Comparative studies of living animals show that the *neurons* of the nervous systems of all animals, although diverse in form, are quite similar in their functional properties. For example, the neurons of all phyla have common molecular bases for their excitability and intercellular communication, with homologous voltage-gated channels and synaptic mechanisms. Moreover, the genetic controls of nervous system development show striking homologies in a wide range of phyla. The major changes in the evolutionary history of nervous systems appear to have involved changes in the *complexity of organization* of neurons into systems, rather than changes in the neurons themselves.

Two major trends characterize the evolution of nervous systems in the bilaterally symmetrical phyla of animals: centralization and cephalization. **Centralization** of nervous systems refers to a structural organization in which integrating neurons are collected into central integrating areas rather than being randomly dispersed. **Cephalization** is the concentration of nervous structures and functions at one end of the body, in the head. **BOX 15.1** further discusses the evolution of diverse nervous systems; here we consider two major types of relatively complex nervous systems, those of arthropods and of vertebrates.

ARTHROPOD CENTRAL NERVOUS SYSTEMS ARE ORGANIZED AS CHAINS OF SEGMENTAL GANGLIA Animals with relatively complex CNSs exhibit two different major forms of CNS organization: *ganglionic* nervous systems characteristic of protostomes, and *columnar* nervous systems characteristic of vertebrates and other deuterostomes (see the back endpapers for the protostome/deu-

(A) Dorsal view of the central nervous system

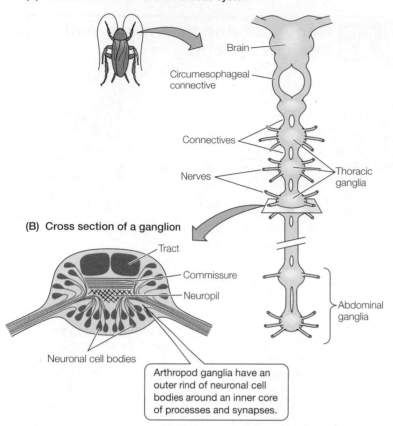

(B) Cross section of a ganglion

Arthropod ganglia have an outer rind of neuronal cell bodies around an inner core of processes and synapses.

FIGURE 15.2 The organization of an arthropod central nervous system (A) The CNS, shown here in a dorsal view, consists of a chain of segmental ganglia linked by connectives. The circumesophageal connectives link the anterior *brain* to the *ventral nerve cord*, which consists of the linked ganglia posterior to the circumesophageal connectives. (Some abdominal ganglia are omitted.) (B) A ganglion, shown in cross section, contains an outer rind of cell bodies and an inner core of synaptic neuropil and of axons (tracts and commissures).

terostome distinction). To see the features of ganglionic nervous system organization, we focus here on arthropods. Aspects of the organization of a ganglionic nervous system are also present in annelids and molluscs.

In arthropods, the CNS consists of a chain of segmental ganglia. **Ganglia** (singular *ganglion*) are swellings containing discrete aggregations of nerve cell bodies and processes. The chained ganglia are linked by paired bundles of axons called **connectives** (**FIGURE 15.2**). The CNS of an arthropod such as a cockroach consists of an anterior *brain* and a *ventral nerve cord* that is linked to the brain by connectives encircling the esophagus. The ventral nerve cord is a chain of ganglia linked by connectives—one ganglion for each thoracic and abdominal body segment. (Some arthropods show secondary fusion of some of these segmental ganglia.)

Each ganglion in the CNS of an arthropod consists of an outer *rind* and an inner *core*. The rind consists mostly of cell bodies of neurons and is devoid of axons and synapses. Indeed, nearly all neuronal cell bodies of arthropods are confined to the rinds of the CNS ganglia, the major exceptions being cell bodies of sensory neurons, many of which are located in the PNS. The inner core of each ganglion contains two regions: a region of synaptic contacts between axons and dendrites that is termed the **neuropil** (or *neuropile*) and a region of **tracts** (bundles) of axonal processes within the ganglion.

In arthropod or other ganglionic nervous systems, there are four terms for a bundle of nerve axons, depending on where the bundle is located. In the PNS a bundle of axons is a *nerve*, between ganglia in the CNS it is a *connective*, within a ganglion it is a *tract*, and between right and left sides of a bilaterally symmetrical ganglion it is a **commissure**. The terms *nerve*, *tract*, and *commissure* have the same meanings for vertebrate nervous systems, but vertebrate CNSs do not have connectives.

THE VERTEBRATE CENTRAL NERVOUS SYSTEM IS A CONTINUOUS COLUMN Vertebrate CNSs, in contrast to those of arthropods, are classified as columnar because they consist of a continuous column of neural tissue, with cell bodies and synaptic areas intermingled. The CNS of vertebrates consists of a brain and a spinal cord (**FIGURE 15.3**). It differs from the ganglionic CNSs of arthropods (and other protostomes) in several respects, some of which have already been mentioned. The vertebrate CNS is dorsal and hollow, and it develops from a neural tube that invaginates from the dorsal surface of the embryo. The nerve cords of arthropods, in contrast, are ventral and solid, do not arise by invagination, and have connectives between CNS ganglia. The vertebrate CNS, reflecting its origin as a continuous tube, is not clearly divided into ganglia and connectives, as is the arthropod CNS (compare Figure 15.3A with Figure 15.2A).

Despite the differences in organization between ganglionic and columnar CNSs, there is good evidence that the centralized organization of nervous systems in bilaterally symmetrical animals is an ancient characteristic that evolved once. In 1822, the French biologist Étienne Geoffroy Saint-Hilaire suggested that vertebrates were related to other Bilateria (animal phyla with bilateral symmetry) such as worms and arthropods, but that vertebrates were inverted, so that what had been a ventral nervous system in other groups became dorsal in the inversion. Anton Dohrn and others championed this idea in the later nineteenth century, but the idea was not taken seriously until it received strong support from recent studies of expression of homologous patterning genes in the development of nervous systems in different phyla. Similarities in patterns of gene expression and control between vertebrates and protostomes (such as fruit flies and annelid worms) strongly suggest a common origin of CNS organization, with an inversion of the body axis of vertebrates as Saint-Hilaire envisioned.

SUMMARY
The Organization and Evolution of Nervous Systems

■ Animals have evolved nervous systems with varying degrees of centralization and complexity. There are homologies between the nervous systems of different animal groups.

■ Most phyla of animals have bilateral symmetry and have evolved central nervous systems (CNSs) that centralize control functions. Sensory neurons convey information into the CNS, and motor neurons convey outward commands to effectors. CNSs usually have some degree of cephalization (concentration of neural structures into a clear anterior brain).

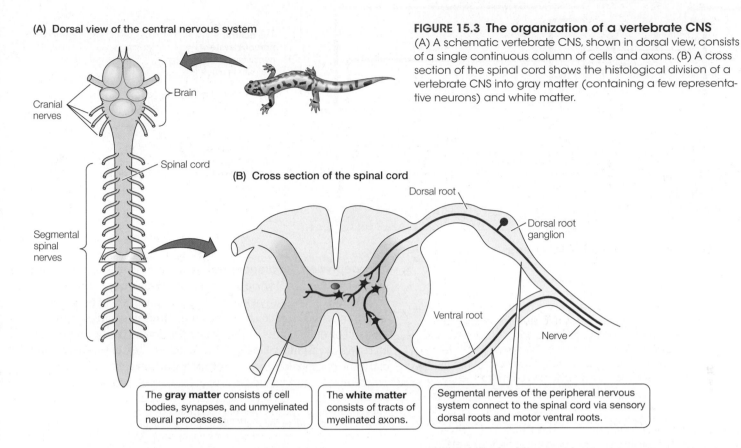

(A) Dorsal view of the central nervous system

Cranial nerves

Brain

Spinal cord

Segmental spinal nerves

FIGURE 15.3 The organization of a vertebrate CNS
(A) A schematic vertebrate CNS, shown in dorsal view, consists of a single continuous column of cells and axons. (B) A cross section of the spinal cord shows the histological division of a vertebrate CNS into gray matter (containing a few representative neurons) and white matter.

(B) Cross section of the spinal cord

Dorsal root

Dorsal root ganglion

Ventral root

Nerve

The **gray matter** consists of cell bodies, synapses, and unmyelinated neural processes.

The **white matter** consists of tracts of myelinated axons.

Segmental nerves of the peripheral nervous system connect to the spinal cord via sensory dorsal roots and motor ventral roots.

■ Arthropods have a ganglionic nervous system, one major form of nervous system organization. The arthropod CNS is a ventral ladderlike chain of segmental paired ganglia joined by connectives. A vertebrate CNS, in contrast, is a continuous column of cells and axons.

The Vertebrate Nervous System: A Guide to the General Organizational Features of Nervous Systems

The nervous systems of most animals tend to share common organizational features. Here we discuss these organizational features using vertebrate nervous systems as examples. Keep in mind, however, that many of these organizational features apply to nervous systems in general. The vertebrate nervous system is organized into different regions that are discrete in gross structure, although neurons and their functions may cross these boundaries.

Nervous systems have central and peripheral divisions

The division of nervous systems into central and peripheral divisions was stressed earlier but deserves reiteration because it is of such pivotal importance. For a vertebrate, the CNS consists of the brain and spinal cord, and the PNS consists of nerves that connect the CNS to various parts of the body (**FIGURE 15.4**). Peripheral nerves contain axons of **afferent** neurons—neurons that carry nerve impulses *toward* the CNS (e.g., sensory neurons)—and axons of **efferent** neurons that carry nerve impulses *away* from the CNS

(e.g., motor neurons). The vertebrate PNS also includes **peripheral ganglia**, which are collections of neuronal cell bodies associated with peripheral nerves. (These should not be confused with the *CNS* ganglia of arthropod nervous systems.) As in other animals, the vertebrate PNS conveys sensory input to the CNS, and it conveys motor output (to control muscles and other effectors) from the CNS to the periphery. Effector functions include contraction, secretion, emission of light and heat, electric organ discharge, and other actions.

The central nervous system controls physiology and behavior

The vertebrate CNS demonstrates the two general principles of organization of complex nervous systems: centralization and cephalization. Functionally, all neural activity is funneled into the CNS via afferent sensory neurons and sent out from the CNS by efferent motor neurons to effectors (see Figure 15.4). All significant integration and processing of neural activity occurs in the CNS: There are no sensory-to-motor synapses in the PNS, and therefore no peripheral integration takes place.[1]

Histologically, the vertebrate CNS consists of two types of tissue: gray matter and white matter. **Gray matter** is composed of intermingled neuronal cell bodies, processes, and synaptic contacts. **White matter**, by contrast, consists entirely of tracts of myelinated axons (see page 310); it is the myelin that imparts a distinctive white appearance to the tissue. In the spinal cord the white matter is external and the gray matter is internal (see Figure

---

[1] This generalization is not true for the enteric nervous system, the branch of the autonomic nervous system that controls the gut (discussed later in this section).

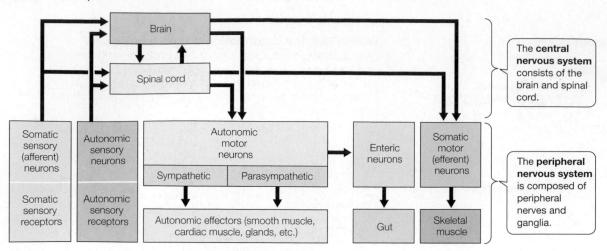

FIGURE 15.4 Divisions of the vertebrate nervous system are interconnected The most basic distinction is between the CNS and the PNS. The PNS has sensory and motor divisions. The somatic nervous system includes somatic receptors and afferent sensory neurons (these might be parts of the same sensory cell, or different cells), and efferent motor neurons controlling striated skeletal muscle. The autonomic nervous system includes autonomic sensory neurons and efferent neurons controlling internal autonomic effectors. The enteric division of the autonomic nervous system has some communication with the CNS but functions rather independently to control the gut.

15.3B). This arrangement does not hold in the brain, however, where gray matter is often external to white matter and the arrangement is considerably more complicated. For example, brains have a more layered gross organization: Often the outer surface layer is called a **cortex**.

Two types of circuits characterize the functional organization of the spinal cord: local and ascending/descending. Local circuits exist within single segments of the spinal cord. In a local circuit (such as the type of circuit that mediates a simple spinal reflex; see Chapter 19), sensory neurons entering a segment—and interneurons in the segment—control motor output of the same segment. This local control is thought to be primitive. In addition to this local control, sensory input is relayed to the brain and contributes to higher integration. Such transmission of information from the spinal cord to the brain is called *ascending*. In addition, *descending* transmission of information occurs from the brain to the spinal cord, permitting the brain to exert dominance over the rest of the CNS.

The nerves in the PNS that connect to the CNS are called **cranial nerves** if they connect to the brain and **spinal nerves** if they connect to the spinal cord. The brain receives sensory input from sensory neurons in its cranial nerves, and also via ascending pathways from the spinal cord. Brain motor neurons traveling within cranial nerves directly control effectors of the head. Much of the brain's output, however, serves to control or modulate the spinal cord. The spinal nerves are arranged segmentally, as we will discuss shortly. The spinal cord receives sensory input through the dorsal roots of its spinal nerves (see Figure 15.3B). It sends motor output to the periphery of the body via the ventral roots of the spinal nerves.

All vertebrate brains share a common structural organization, exhibiting three major regions: **forebrain, midbrain,** and **hindbrain** (**FIGURE 15.5**). Five subdivisions of these regions are recognized:

telencephalon and **diencephalon** (forebrain), **mesencephalon** (midbrain), and **metencephalon** and **myelencephalon** (hindbrain). Each of these divisions contains many tracts of nerve axons and many clusters of cell bodies termed **nuclei** (singular *nucleus*).[2] Two major outgrowths of the dorsal portion of the brain become increasingly prominent in higher vertebrates: the **cerebellar cortex** of the metencephalon and the **cerebral cortex** of the telencephalon. The cerebral cortex includes the olfactory cortex, hippocampus, and

[2] In vertebrates, clusters of neuronal cell bodies are usually called *nuclei* in the CNS and *ganglia* in the PNS. The basal ganglia of the brain constitute an exception to this terminology; they are actually CNS nuclei.

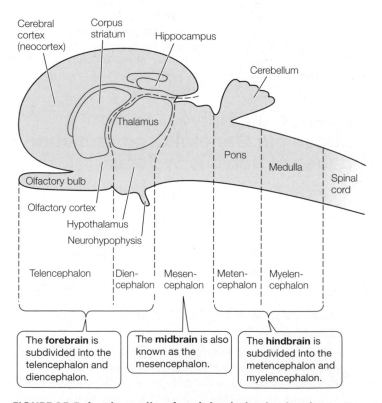

FIGURE 15.5 A schematic of vertebrate brain structure The brain has three major parts—forebrain, midbrain, and hindbrain—which are divided into five subdivisions, as shown. Some major structures of these five brain regions are labeled. In this side view, the anterior end is to the left.

TABLE 15.1 Some major areas of the mammalian brain and their general functions
Most functions involve several brain areas, and many brain areas and functions are omitted.

Major brain division	Brain subdivision	Area	Major functions
Forebrain	Telencephalon	Cerebral cortex	Higher sensory, motor, and integrative functions
		Hippocampus	Learning and memory
		Basal ganglia	Motor control
		Limbic system	Emotions
	Diencephalon	Thalamus	Major sensory relay
		Hypothalamus	Homeostatic and endocrine regulation; circadian clock
Midbrain	Mesencephalon	Superior colliculus	Visual integration
		Inferior colliculus	Auditory integration
Hindbrain	Metencephalon	Cerebellum	Motor coordination
		Pontine motor nuclei	Descending motor control
	Myelencephalon	Medulla oblongata	Autonomic and respiratory control

(in mammals) neocortex. **TABLE 15.1** lists some brain structures and their functions. We discuss several aspects of brain function in the next section, and we discuss several of the brain regions further in Chapters 14 and 19.

The CNS, along with the endocrine system, exerts control over the functions of an animal's organ systems (its physiology) and also over its behavior, including all movements and externally observable activity. We will take the simplifying view that behavior is the province of the skeletal muscles, controlled via the somatic nervous system. In contrast, the autonomic nervous system exerts physiological control of internal organ systems via other effectors. This distinction is not completely true, because (for example) reproductive behavior includes glandular secretion and smooth muscle contractions mediated by the autonomic nervous system, whereas breathing (lung ventilation) is a physiological function mediated by somatic control of skeletal muscles of the diaphragm and ribs.

Still, the distinction between somatic control of behavior through skeletal muscle and autonomic control of organ systems through other effectors, although imprecise, is a useful generalization.

Five principles of functional organization apply to all mammalian and most vertebrate brains

PRINCIPLE 1: BRAIN FUNCTION IS SOMEWHAT LOCALIZED
Neurons in different anatomical regions of the brain play different functional roles. That is, you can point to a part of the brain and reasonably ask what functional activities occur there. For example, **FIGURE 15.6** shows some major areas of the cerebral cortex that are involved in sensory processing and in speech. Several kinds of studies demonstrate localization of function: stimulation studies, lesion studies that destroy or isolate brain tissue in a region, strokes (which interfere with blood circulation to a particular part of the

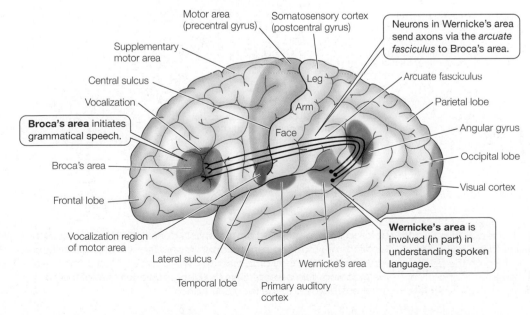

FIGURE 15.6 Localization of function in the human brain This surface view of the left cerebral hemisphere shows primary sensory areas (visual cortex, green; primary auditory cortex, blue; somatosensory cortex, yellow); motor areas (orange); and areas involved in language. Although the functions are rather localized, circuits interconnect several areas, as shown for spoken language. Note that recent imaging studies are refining this classic view of speech areas.

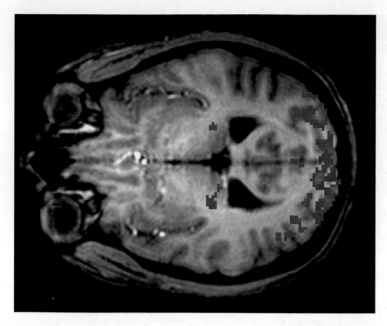

FIGURE 15.7 Functional neuroimaging demonstrates localization of function in the human cerebral cortex Functional magnetic resonance imaging (fMRI) shows increased activity of particular brain areas. This fMRI sectional image of a brain (seen from above; left is anterior) shows increased neural activity during a visual stimulus. The neural activity (measured as increased flow of oxygenated blood and shown in red) increases in the lateral geniculate nucleus of the thalamus (middle) and the primary visual cortex (posterior). (See Figure 14.32 for a diagrammatic view of this pathway and Box 15.2 for a discussion of fMRI as an imaging method.) (From Chen et al. 1999.)

brain, resulting in death of neurons and disruption of a particular sensory or motor process), and functional imaging studies.

Modern imaging studies are particularly important in clarifying location of function. Imaging methods such as functional magnetic resonance imaging (fMRI) (**FIGURE 15.7, BOX 15.2**) show localized increased metabolic activity that results from increases in neuronal electrical activity. These areas of increased activity show where particular functions are localized. Note that localization of function does *not* mean that a particular brain area is involved in only one function, or that any function involves only one discrete brain area. A function is often apt to involve a *circuit* in the brain (a network of synaptically interconnected and interacting neurons, which may be widespread) rather than a discrete *center*.

PRINCIPLE 2: BRAINS HAVE MAPS The brain maintains information about the body's anatomical organization in terms of topographic representations, or **maps**. The brain contains many mapped representations that record and recall the parameter of *where* a stimulus occurs or an effector is to be controlled. For example, the sensory surface of the body is mapped onto the primary somatosensory area of the cerebral cortex (**FIGURE 15.8B**) to form a **somatotopic map** (a map of the body projected to a brain area). An analogous but rougher somatotopic map of motor control exists on the surface of the primary motor cortex (**FIGURE 15.8A**); stimulation of a part of the motor cortex will elicit movement of the corresponding part of the body.

There are additional sensory and motor maps in other areas of the cerebral cortex, as well as in many noncerebral brain structures. The visual system, for example, exhibits a point-to-point correspondence between areas in the visual field in the outside world and areas in the retina, which we discussed in Chapter 14. This geometric orientation is preserved in visual pathways, so that

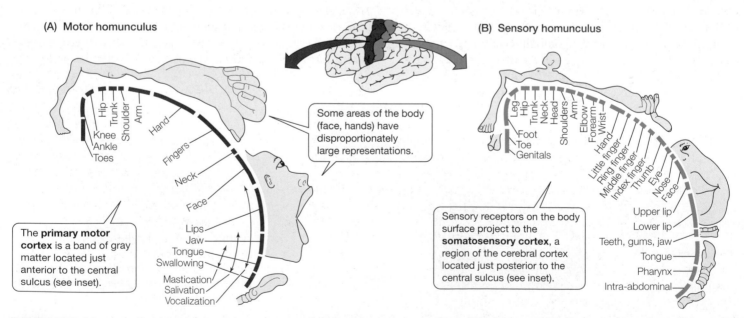

FIGURE 15.8 Maps in the human brain In what is called a somatotopic map, parts of the body are mapped onto the cerebral cortex in a way that preserves their anatomical position on the body. (A) Neurons in the primary motor cortex control movements of parts of the body, and are loosely organized as a somatotopic map that is less detailed than that of the somatosensory cortex. (B) Sensory receptor locations are mapped somatotopically onto the somatosensory cortex. A drawing of the body projected onto the brain is called a *homunculus,* meaning "little person." Note the disproportionate size of the cortical areas devoted to the face and hands.

BOX 15.2 Functional Magnetic Resonance Imaging
SCOTT A. HUETTEL

Functional magnetic resonance imaging (fMRI) is a method for detecting the functional activity of different areas of the brain. The use of fMRI has grown dramatically over the past 25 years and is now the dominant research method in cognitive neuroscience. Each year, several thousand articles are published that use fMRI to study topics as basic as the structure of memory and as complex as the foundations of moral cognition. What accounts for this remarkable growth?

What information does fMRI provide, and how does that information lead to inferences about brain function? What new directions will fMRI researchers pursue in the coming years, as scientists make new discoveries both about the technique itself and about the functional organization of the brain? **Box Extension 15.2** describes the development of the fMRI technique and its current and potential future uses.

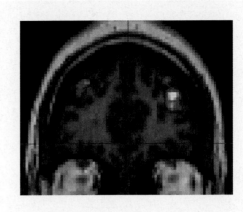

---

the retinal map of the world is projected to each of the major visual centers in the brain. However, maps in the brain are by no means universal; many areas of the mammalian brain lack topographic organization.

PRINCIPLE 3: SIZE MATTERS   In general, the bigger the brain area, the more neurons are present in it, and the more complex the integration that occurs there. Thus mammalian brains, with tens of billions of neurons, perform more complex integration than do insect brains with tens of thousands of neurons. By the same token, the massive enlargement of the cerebral cortex in the course of mammalian evolution (as a result of the developmental proliferation of greatly increased numbers of neurons) attests to major increases in processing capability of mammalian, and especially primate, brains. Primate brains are not "more evolved" than the brains of fish and amphibians, because fish and amphibians have had as long a time to evolve and can be considered to be as well adapted to their ecological niches as primates are. Nevertheless, the great expansion of cerebral cortex in primates has allowed types of neural function (e.g., language, culture) not found in brains with fewer neurons.

PRINCIPLE 4: VERTEBRATE BRAIN EVOLUTION HAS INVOLVED REPEATED EXPANSION OF FOREBRAIN AREAS   The hindbrains of fish and amphibians are similar in structure to the corresponding regions in mammalian brains, but birds and mammals have evolved more complex forebrain structures, most notably the mammalian cerebral cortex, as noted previously. According to most evolutionary neurobiologists, many different vertebrate lineages underwent evolutionary expansion of their brains, expanding and reorganizing preexisting components of the forebrain in the process. Mammals reorganized the dorsal pallium into a neocortex, which underwent expansion in several lineages, notably primates. Forebrain expansions in fish and birds have involved different homologies. The "new" forebrain structures are thus not completely new, but rather represent expansions and elaborations of preexisting structures. They may subsume some of the functions of the more posterior structures of the midbrain and hindbrain, but they

do not replace these "older" structures, which persist with less apparent change. The addition of forebrain structures leads to the popular description that humans have an anterior, telencephalic "mammalian brain" and an older "reptilian brain," consisting of the deeper structures of the midbrain and hindbrain.

PRINCIPLE 5: NEURAL CIRCUITS ARE PLASTIC   We tend to think of the synaptic connections between neurons as "hardwired"— rather fixed and unchanging—but there is abundant evidence that this is not the case. Instead, as we discussed in Chapter 13, synapses are *plastic*: They change with development, maturation, and experience such as learning. Noting a difference between short-term memory and long-term memory will clarify this point: *Short-term memory*, memory of events of the last minutes to hours, can be disrupted by a concussion or by electroconvulsive shock. *Long-term memories*, in contrast, are not lost after a concussion, because they are thought to be stored more permanently as changes in the "wiring"—changes in the strengths of synaptic interconnections of neurons. This storage of long-term memory may be an example of how neural connections are plastic, or subject to change in strength and effectiveness.

We know also that synaptic connections are plastic during development of the nervous system; they are made and broken as an animal's brain matures, and they are subject to competition between neuron endings for synaptic sites on a postsynaptic target cell (see Box 13.1). Scientists believe that plasticity in development may have mechanisms similar to the plasticity associated with learning and memory that we explored in Chapter 13.

Moreover, there is abundant evidence that new neurons develop throughout life in the nervous systems of both vertebrates and invertebrates. An ongoing proliferation of new neurons has long been accepted for invertebrates and lower vertebrates, and more recent studies clearly demonstrate neuronal proliferation in some parts of adult mammalian nervous systems. Evidence that new neurons proliferate and differentiate throughout life has fueled widespread interest in the use of transplants of undifferentiated stem cells in an attempt to counteract neuron losses in neurodegenerative diseases such as Parkinson's disease and Alzheimer's disease.

(A) African hedgehog (*A. albiventris*)

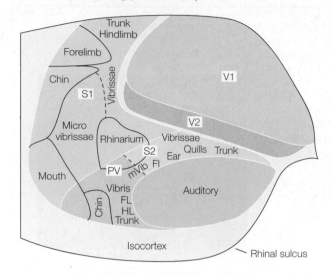

(B) Star-nosed mole (*C. cristata*)

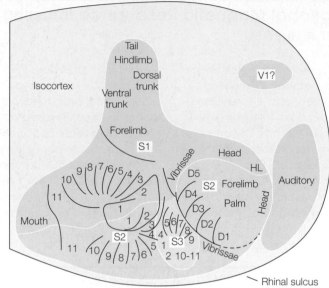

(C) Cortical magnification in star-nosed mole

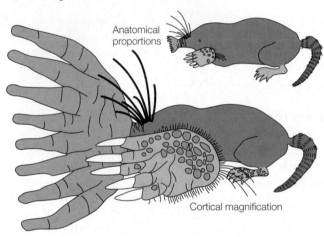

FIGURE 15.9 Maps in the brain differ for different animal groups
Side views of the cerebral cortex of an African hedgehog and a star-nosed mole (anterior is to left). (A) The hedgehog brain shows a localization of cerebral cortical function typical of small and relatively unspecialized mammals, with large areas devoted to visual (V1, V2) and auditory functions, a primary somatosensory area (S1), and two smaller somatosensory areas (S2, PV) below it. (B) The cerebral cortex of the star-nosed mole exhibits differences in area of cortical representation, reflecting the animal's unusual sensory capacities. The visual area (V1?) is greatly reduced, in accordance with the fossorial mole's greatly diminished visual function. The auditory area is displaced posteriorly, and the somatosensory areas are expanded, especially the ventral ones. The three sets of radiating lines numbered 1–11 indicate three areas that represent the star nose. (C) Depicted here is the disproportionally large representation of the star nose and of the digging forelimb in the star-nosed mole. (After Catania 1999 and 2005.)

STAR-NOSED MOLES EXEMPLIFY THESE PRINCIPLES OF FUNCTIONAL ORGANIZATION The star-nosed mole with which we opened this chapter provides a clear example of these organizational principles. **FIGURE 15.9A,B** shows side views of the cerebral cortex of an African hedgehog (*Atelerix albiventris*) and a star-nosed mole. The hedgehog brain (see Figure 15.9A) shows a localization of cerebral cortical function typical of small and relatively unspecialized mammals, with large areas devoted to visual and auditory functions, a primary somatosensory area (S1), and two smaller somatosensory areas (S2, PV) below it. The cerebral cortex of the star-nosed mole (see Figure 15.9B) has impressive differences in size of the cortical representations. Both brains demonstrate localization of function (as in Figure 15.7), with particular areas performing particular functions, and both brains show somatotopic maps, with particular areas of the body projecting to particular places in the brain (as shown in Figure 15.8). The size differences in the brain representations of the star-nosed mole reflect the unimportance of vision and the importance of somatosensory functions, particularly of the "star," which has three somatosensory representations that together occupy more area than that devoted to the rest of the body (**FIGURE 15.9C**). Finally, the star-nosed mole shows evidence of plasticity in development of the brain; a few moles develop with 12 rather than 11 rays in the star, and developing sensory neurons

from the "extra" ray induce a corresponding extra cortical representation to which they indirectly project!

The peripheral nervous system has somatic and autonomic divisions that control different parts of the body

As noted previously, physiologists recognize two primary divisions of the vertebrate PNS: the *somatic* and the *autonomic* nervous systems (see Figure 15.4). Here we focus on the properties of these two divisions. We start with a relatively brief discussion of the somatic nervous system. Thereafter, most of this section is devoted to the autonomic nervous system.

Because the somatic nervous system is the part of the PNS that controls the skeletal (striated) muscles, it controls the muscles of locomotion and other body movements, speech, and breathing. Thus it controls most observable behavior. In addition to its motor elements, the somatic nervous system also includes the somatic sensory receptors for touch, hearing, vision, taste, olfaction, and so forth. We have examined aspects of the somatic nervous system in earlier chapters and in the first pages of this chapter. Here we consider only its overall organization.

Somatic motor and sensory neurons exit and enter the CNS in the cranial and spinal nerves (as do autonomic neurons, discussed

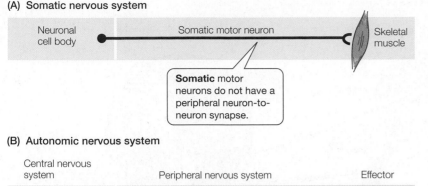

(A) Somatic nervous system

Neuronal cell body · Somatic motor neuron · Skeletal muscle

Somatic motor neurons do not have a peripheral neuron-to-neuron synapse.

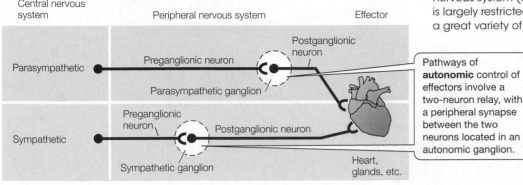

(B) Autonomic nervous system

Central nervous system · Peripheral nervous system · Effector

Parasympathetic · Preganglionic neuron · Postganglionic neuron · Parasympathetic ganglion

Sympathetic · Preganglionic neuron · Postganglionic neuron · Sympathetic ganglion

Pathways of **autonomic** control of effectors involve a two-neuron relay, with a peripheral synapse between the two neurons located in an autonomic ganglion.

Heart, glands, etc.

FIGURE 15.10 The organization of the mammalian somatic and autonomic nervous systems, including examples of effector organs (A) Skeletal muscle is under somatic control. Each motor neuron extends from the CNS to the effector it controls. (B) The heart is under autonomic control, via chains of two neurons with a synapse in a peripheral ganglion. Sympathetic ganglia are typically close to the spinal cord; parasympathetic ganglia are typically near target organs. Preganglionic autonomic neurons secrete the neurotransmitter acetylcholine (ACh); postganglionic sympathetic neurons secrete norepinephrine, and postganglionic parasympathetic neurons secrete ACh. The enteric nervous system (not shown) has some connection to the CNS but is largely restricted to the gut. The enteric nervous system employs a great variety of neurotransmitters.

next). Mammals have 12 pairs of *cranial nerves*, rather specialized in function, that are numbered 1 to 12 using Roman numerals. Some of the cranial nerves are associated with the major sense organs of the head (I, olfactory; II, optic; VIII, auditory). Other cranial nerves have principally motor functions, or have mixed sensory and motor functions. The vagus nerve (X) innervates the larynx and some other somatic components, while also being a major component of the autonomic nervous system.

The spinal nerves are arranged segmentally, with one pair of spinal nerves per vertebra along the spinal column. These nerves include both sensory and motor somatic neurons. The sensory neurons enter the spinal cord in the dorsal roots of the spinal nerves; the cell bodies of those neurons are located in **dorsal root ganglia**, enlargements of the dorsal roots outside the spinal cord (see Figure 15.3B). The ventral roots of the spinal nerves (see Figure 15.3B) contain axons of the somatic motor neurons that innervate the skeletal muscles. They also contain autonomic neurons that innervate autonomic ganglia. An important property of the somatic nervous system is that somatic motor neurons directly synapse on muscle fibers, without synapsing on other neurons after leaving the CNS (**FIGURE 15.10A**). (This is in basic contrast to motor neurons of the autonomic nervous system, as we discuss below.)

The autonomic nervous system has three divisions

The vertebrate autonomic nervous system is usually defined as if it were a motor system—the division of the PNS that controls the autonomic effectors. However, the autonomic nervous system also includes sensory neurons that convey afferent signals from internal organs to the CNS. Many invertebrate animals have nervous system divisions that control visceral functions. These divisions are sometimes described as *autonomic* by analogy to vertebrate autonomic nervous systems. The autonomic effectors that are controlled at least partly by the vertebrate autonomic nervous system include the following:

- Smooth muscles throughout the body, such as those in the gut wall, blood vessels, eyes (iris muscles), urinary bladder, hair follicles, spleen, airways of the lungs, and penis.

- Many exocrine glands, such as sweat glands, tear glands, and the exocrine portion of the pancreas. (**Exocrine glands** discharge secretions into the environment or into internal body cavities, in contrast to endocrine glands, which secrete into the blood or other body fluids.)

- A few endocrine glands, notably the adrenal medullary glands (called chromaffin tissue in some vertebrates) that secrete epinephrine (adrenaline).

- Acid-secreting cells of the stomach.

- The pacemaker region and other parts of the heart.

- The brown adipose tissue of mammals (a heat-producing tissue).

- The swim bladders and integumentary chromatophores (color-change cells) of fish.

The vertebrate autonomic nervous system is usually considered to consist of three divisions—*sympathetic, parasympathetic,* and *enteric*—first described by John Langley (1852–1925). Langley based his classification on anatomy, not function. Therefore his divisions are not always distinctly different in function. Today his scheme is considered valid for mammals. It is often used for nonmammalian vertebrates also, although its application to these other groups remains debatable.

The **sympathetic** and **parasympathetic divisions** in mammals, by definition, functionally link the CNS with autonomic effectors. The two divisions characteristically have opposing effects on the autonomic effectors, so that sympathetic actions are said to mediate "fight-or-flight" responses, whereas parasympathetic actions mediate "rest-and-digest" functions (see page 420). The **enteric division**, in contrast, is largely contained in the walls of the gut, although it has some connection with the CNS.

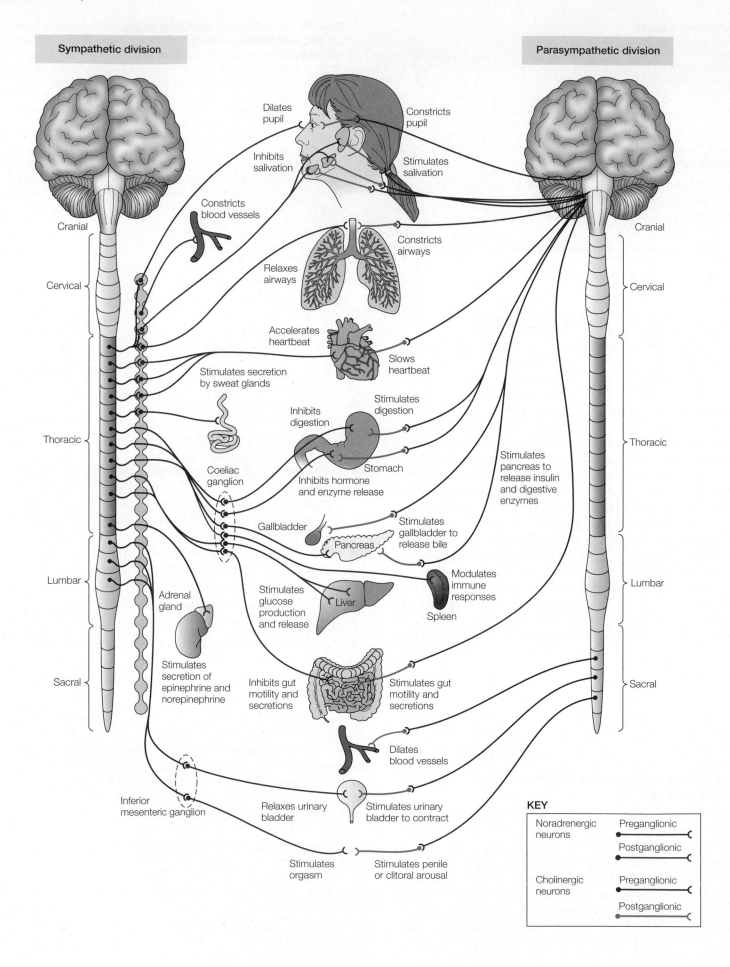

Sympathetic division

Parasympathetic division

Cranial

Cervical

Thoracic

Lumbar

Sacral

Cranial

Cervical

Thoracic

Lumbar

Sacral

Dilates pupil

Inhibits salivation

Constricts blood vessels

Relaxes airways

Accelerates heartbeat

Stimulates secretion by sweat glands

Inhibits digestion

Coeliac ganglion

Inhibits hormone and enzyme release

Gallbladder

Stimulates glucose production and release

Adrenal gland

Stimulates secretion of epinephrine and norepinephrine

Inhibits gut motility and secretions

Inferior mesenteric ganglion

Relaxes urinary bladder

Stimulates orgasm

Constricts pupil

Stimulates salivation

Constricts airways

Slows heartbeat

Stimulates digestion

Stomach

Stimulates pancreas to release insulin and digestive enzymes

Stimulates gallbladder to release bile

Pancreas

Modulates immune responses

Liver

Spleen

Stimulates gut motility and secretions

Dilates blood vessels

Stimulates urinary bladder to contract

Stimulates penile or clitoral arousal

KEY

Noradrenergic neurons	Preganglionic
	Postganglionic
Cholinergic neurons	Preganglionic
	Postganglionic

◀ **FIGURE 15.11 Parasympathetic and sympathetic divisions of the autonomic nervous system provide dual innervation of most visceral autonomic effectors** Parasympathetic nerves and ganglia are shown on the right and sympathetic nerves and ganglia on the left, but both have paired right and left sides.

ANATOMY OF THE SYMPATHETIC AND PARASYMPATHETIC DIVISIONS A key attribute of the sympathetic and parasympathetic divisions is that they have motor pathways in which signals exit the CNS and control autonomic effectors. These motor pathways are characterized by a *peripheral synapse*—that is, an "extra" synapse interposed between the CNS and the ultimate synaptic ending on effector tissue (**FIGURE 15.10B**). This peripheral synapse is a characteristic of the sympathetic and parasympathetic divisions of all vertebrates, so autonomic motor signals in both divisions traverse a two-neuron relay to reach their effectors from the CNS.

The peripheral synapses between the first and second neurons in the sympathetic and parasympathetic divisions are located within clusters of neuronal cell bodies called **autonomic ganglia**. The neurons that extend from the CNS to the ganglia are termed **preganglionic neurons**, whereas those extending from the ganglia to the effectors are termed **postganglionic neurons**. The mammalian sympathetic and parasympathetic divisions differ in the positions of the ganglia. In the parasympathetic division, the ganglia are located mostly at or near the effectors, so preganglionic parasympathetic neurons are long and postganglionic parasympathetic neurons are short (see Figure 15.10B). In contrast, sympathetic ganglia are located mostly near the spinal cord, so preganglionic sympathetic neurons are short, and postganglionic sympathetic neurons are long.

The mammalian sympathetic and parasympathetic divisions also differ in that motor neurons of the two divisions exit the CNS in nerves associated with different regions of the CNS (**FIGURE 15.11**). Parasympathetic preganglionic neurons exit the CNS from two regions: cranial and sacral. For this reason, the parasympathetic division is sometimes called the *craniosacral* division. The cranial group of nerves consists of four of the pairs of cranial nerves: the oculomotor (III), facial (VII), glossopharyngeal (IX), and vagus (X) nerves. The sacral group of nerves emerges from the posterior part of the spine.

The preganglionic neurons of the sympathetic division exit the CNS in nerves of the thoracic and lumbar regions of the spine (see Figure 15.11). For this reason, the sympathetic division is also called the *thoracolumbar* division. Most of the nerves terminate in sympathetic ganglia immediately lateral to the spine. These **paravertebral ganglia** occur segmentally (at regularly repeating intervals) along the length of the spine and are interconnected by peripheral longitudinal nerve connectives, forming a **sympa-**

thetic chain on each side of the vertebral column. In contrast to this general pattern, some preganglionic sympathetic neurons terminate in ganglia more distant from the spine, as in the coeliac ganglion (solar plexus). Some directly innervate the medullary tissue of the adrenal glands (see Chapter 16).

FUNCTION OF THE SYMPATHETIC AND PARASYMPATHETIC DIVISIONS The postganglionic neurons of the parasympathetic and sympathetic divisions release different chemical neurotransmitter substances at their synapses with effector cells. Parasympathetic postganglionic neurons typically release acetylcholine and thus are termed **cholinergic**. Most sympathetic postganglionic neurons release mainly catecholamines—chiefly norepinephrine (noradrenaline) in mammals—and thus are called **adrenergic** (or noradrenergic).

Autonomic effectors may be innervated by one or both of the sympathetic and parasympathetic divisions. In mammals many smooth muscles of blood vessels and the piloerector (hair-erecting) muscles of the hair follicles receive only sympathetic innervation. Many other effectors, in contrast, receive both sympathetic and parasympathetic innervation (see Figure 15.11). The responses elicited by the two divisions in such cases usually oppose each other. For instance, the pacemaker region of the heart (cells that initiate the heartbeat) is innervated by both divisions. Parasympathetic postganglionic neurons, which secrete acetylcholine as their neurotransmitter, decrease the heart rate. In contrast, sympathetic postganglionic neurons, which secrete norepinephrine, increase heart rate. (Receptors for these neurotransmitters are listed in Table 13.2.)

The parasympathetic and sympathetic divisions play functional roles that reflect their tendency to act in opposition, and demonstrate the integrative actions of autonomic control (**TABLE 15.2**). The parasympathetic division (especially its cranial part) tends to promote processes that restore body reserves of energy (e.g., stimulating digestion). In contrast, the sympathetic division promotes mobilizing body energy reserves (e.g., promoting blood flow to muscles) and inhibits some processes that restore reserves; it is particularly activated in the face of stress, and it readies the body

TABLE 15.2 Major actions of sympathetic and parasympathetic divisions in vertebrates

Process	Parasympathetic effect	Sympathetic effect
Digestion: gastrointestinal secretion and motility	Stimulates	Inhibits
Heartbeat	Slows	Increases rate and force
Blood vessels	Usually dilates (when present)	Constricts vessels to kidneys and gut; dilates vessels to skeletal muscles
Blood pressure	Decreases	Increases
Lung passages	Constricts	Dilates
Secretion of epinephrine and norepinephrine by adrenal medullary glands	—	Stimulates

to meet stress. Direct sympathetic innervation mediates most of the effects listed for the sympathetic division in Table 15.2, but the sympathetic nervous system also stimulates the medullary tissue of the adrenal glands to secrete epinephrine (adrenaline) and norepinephrine, which act hormonally to complement sympathetic action on effectors. Circumstances that might provoke strong, concerted activation of the sympathetic division include vigorous exertion, pain, threats to safety, and exposure to physical extremes such as severe heat or cold. It is through their integrative and opposed actions that the sympathetic division has been said to prepare an individual for "fight-or-flight" whereas the parasympathetic division is said to promote "rest-and-digest" functions.

THE ENTERIC NERVOUS SYSTEM CONTROLS THE GUT The enteric division of the autonomic nervous system consists of elaborate networks of neurons located entirely within the walls of the gut. It has been likened to the nerve nets of Cnidaria (see Box 15.1) with a relatively diffuse structural organization and pattern of synaptic interactions among gut sensory neurons, interneurons, and motor neurons. The enteric division controls peristalsis, segmentation, and other patterns of contraction of the smooth muscles of the gut wall that serve to move digested materials through the gut. The enteric nervous system's function is largely autonomous of CNS control, although the CNS can modulate enteric neurons and synapses via sympathetic and parasympathetic nerves.

The enteric nervous system contains large numbers of neurons: about 200–600 million in humans. This is far more neurons than in other peripheral organs and probably exceeds the number of neurons in the spinal cord! Recent evidence shows that about 90% of the visceral fibers in the parasympathetic vagus nerve carry information from the gut to the brain, rather than from brain to gut. This finding, as well as the large number of enteric neurons, suggests that the enteric nervous system plays other roles in addition to moving material through the gut.

SUMMARY
The Vertebrate Nervous System: A Guide to the General Organizational Features of Nervous Systems

- The CNS of vertebrates consists of the brain and spinal cord. Cranial and spinal nerves emanate from the CNS to form the PNS. The brain is divided into a forebrain, midbrain, and hindbrain; the forebrain is enlarged in birds and especially in mammals.

- Vertebrate brain functions are somewhat localized. However, brain functions are also somewhat distributed, involving circuits rather than centers.

- Many vertebrate brain regions preserve the orderly spatial arrangements of the corresponding external world, for example, as somatotopic maps of body sensory input and motor output.

- Brains change with development, experience, and learning and memory. Understanding the structural and synaptic bases of these changes is a major challenge to investigators.

- The PNS of vertebrates has a somatic division that controls skeletal muscle and an autonomic division that controls effectors associated with internal organs. The autonomic nervous system is divided into sympathetic and parasympathetic divisions, which usually have opposite physiological effects, and the enteric division, which controls gut contraction and other aspects of digestive tract physiology.

Biological Clocks

Animals (and other organisms) possess endogenous physiological timing mechanisms termed **biological clocks** that rhythmically modulate the functioning of cells, tissues, and organs. Biological clocks endow an animal with an intrinsic **temporal organization**, a timed pattern of change in physiology or behavior that is independent from a change in environment (see Chapter 1, pages 16–19, for discussion of temporal frameworks in animal physiology). Biological clocks are typically operations of the nervous system, controlling physiological and behavioral processes via nervous and neuroendocrine output. Characteristically, the physiological state of an animal is endogenously different at different times of day, or in different seasons of the year. One such change is the sleep–wake cycle, discussed in **BOX 15.3**.

Biological clocks orchestrate daily and seasonal changes, controlling and integrating the changes in physiological states. **FIGURE 15.12** shows the daily variation in several physiological functions in a 24-year-old man. These data are from one of the earliest studies designed to test whether humans have endogenous, physiological mechanisms for keeping track of time. The man was placed in living quarters that were entirely isolated from the outside world. He had no clocks in his environment and was unable to distinguish between day and night. Thus he slept, ate, and urinated without any environmental cues. Three dramatic conclusions can be reached by study of these results. First, the man continued to exhibit regular cycles in all the variables studied. Second, he tended to exhibit internal synchronization of his cycles: On any one day, he tended to have the highest rectal temperature and to excrete the most during the block of time when he was principally awake. Finally, however, he did not stay synchronized with the outside world: Relative to clock time in the outside world, his cycles became later and later as the days went by, so that—for example—after 13 days, he elected to be awake after midnight rather than before midnight (outside world time).

Put loosely, the man in Figure 15.12 was able to keep track of time endogenously, but his internal biological clock was not able to stay precisely synchronized with outside time, so that his rhythm was said to be *free-running* (see below). As we will see, these properties are very general among animals.

Organisms have endogenous rhythms

A *rhythm* is a regular, cyclical variation in function. Rhythms that continue in the absence of environmental information about time, such as those of the man in Figure 15.12, are termed **endogenous rhythms**. The first organisms in which endogenous rhythms were demonstrated were certain plants that raise their leaves during some

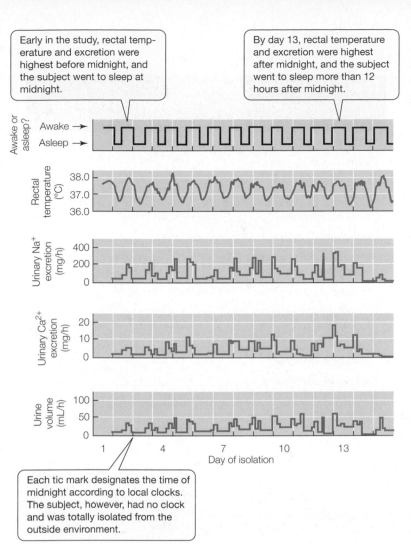

Early in the study, rectal temperature and excretion were highest before midnight, and the subject went to sleep at midnight.

By day 13, rectal temperature and excretion were highest after midnight, and the subject went to sleep more than 12 hours after midnight.

Each tic mark designates the time of midnight according to local clocks. The subject, however, had no clock and was totally isolated from the outside environment.

FIGURE 15.12 Daily rhythm of several physiological functions in a human A young man lived by himself in an apartment without clocks, telephones, windows, or other avenues whereby he could know the time of day in the outside world. He could turn the lights on and off, prepare food, go to bed, urinate, and engage in other processes of daily living, but his timing of those activities was based entirely on his endogenous, physiological sense of time. This graph shows his patterns of variation in five remotely monitored functions during his first 15 days without knowledge of external time. (After Wever 1979.)

times of day and lower them at others.[3] In 1729, the Frenchman M. de Mairan reported that these plants continue to raise and lower their leaves in approximately a daily rhythm even when they are kept in constant darkness and at constant temperature. That is, their rhythm of leaf movements is intrinsic or endogenous: It continues in more or less a daily pattern *even when the plants are denied environmental sources of information about the time of day*. Experiments of a similar nature have since been performed on many plant and animal systems, and daily rhythms in many types of function in many types of organisms have been shown to persist in constant laboratory environments.

The **period** of a rhythm is the amount of time between a particular part of the rhythm in one cycle (one day) and that same part in the next cycle. Typically the period is measured as the time between the start of one day's episode of activity and the start of the next day's. For example, in the case of Mairan's plants, the period could be measured as the time between the start of leaf raising on one day and the start of leaf raising on the next.

An endogenous rhythm that has a period of about a day is termed a **circadian rhythm** (*circa*, "about"; *dies*, "a day"). Not all daily rhythms prove to be endogenous when tested. Only those

---

[3] In our discussion of rhythms and biological clocks, we will consistently use the words *day* and *daily* to refer to the 24-h day, not just to hours of daylight.

<table>
<tr><td>

BOX 15.3

</td><td>

Sleep   DAVID S. GARBE

</td></tr>
</table>

Why animals sleep continues to be one of the most elusive and mysterious questions in biology. Sleep, nonetheless, is found widely among animals—from relatively simple phyla such as worms (e.g., *Caenorhabditis elegans*) to higher-order groups such as humans—suggesting that sleep is a required, evolutionarily conserved behavior. Yet from certain perspectives, sleep could be considered disadvantageous to an organism's overall survival and fitness because while animals are sleeping, they are not immediately able to eat, mate, or protect themselves from predation. Still we as humans sleep for approximately one-third of our entire lives. Most scientists conclude that there must be an overriding

benefit of sleep that compensates for the lack of interaction with the surrounding environment. In fact, sleep is crucial for survival. Long-term sleep deprivation is lethal to rodents and fruit flies (*Drosophila melanogaster*). Moreover, disturbances in sleep and its regulation are associated with several chronic human disease states, including insulin resistance and diabetes, fatal familial insomnia, shift-work disorder, certain neurodegenerative diseases, and major depressive disorder. **Box Extension 15.3** discusses functions of sleep and mechanisms of its regulation.

Leopard (*Panthera pardus*)

422 Chapter 15

that are endogenous—that can persist in the absence of environmental information about the time of day—are properly termed circadian. Circadian rhythms appear to occur in all eukaryotes and some prokaryotes. Thus a capacity for endogenous rhythmicity is believed to be an ancient feature of life. **TABLE 15.3** lists some examples of the known endogenous circadian rhythms.

Under normal conditions in a state of nature, circadian rhythms are tightly coupled to environmental cues such as a daily light–dark cycle. **FIGURE 15.13A** shows the locomotor activity and metabolic rate of a chaffinch (*Fringilla* sp.) initially kept on a normal light–dark cycle. In diurnal species such as a chaffinch, locomotor activity, metabolic rate, and many other physiological variables increase during the day, usually starting near dawn. Two rhythms are said to be **in phase** if they occur synchronously. Thus, for a chaffinch exposed to a normal light–dark cycle, as you can see from Figure 15.13A, both locomotor activity and metabolic rate are in phase with the light–dark cycle.

To test whether a daily rhythm is an endogenous circadian rhythm, an experimenter must remove the environmental timing information. In the case of the chaffinch, this means removing the light–dark cycle. When a chaffinch is exposed to constant dim light (**FIGURE 15.13B**), both its locomotor rhythm and rhythm of oxygen consumption persist—demonstrating that they are endogenous rhythms—but the rhythms fail to remain synchronized with the time dawn would have come each day. Instead, under these conditions, the period of each rhythm is a little shorter than a day—about 23 h—meaning that with each passing 24-h day, activity starts earlier and earlier relative to the time dawn would have come. When environmental cues are absent, the biological rhythm that persists is said to **free-run** or to be a **free-running rhythm**. Like the chaffinch and like the man in Figure 15.12, most organisms have rhythms with free-running periods that are circadian—close to but not exactly equal to 24 h.

The difference between a free-running rhythm and one that is synchronized to environmental cues is easier to see if the records of activity on successive days are stacked one below the last, to make a chart called an *actogram*. **FIGURE 15.14** shows the activity rhythms of two nocturnal flying squirrels (*Glaucomys volans*) for 23 consecutive days stacked in this way: One squirrel was studied in a normal light–dark cycle and the other squirrel was studied in constant darkness. The light–dark cycle (when present) synchronizes the activity rhythm of a flying squirrel, bringing it into phase, so that the onsets of activity periods are lined up at the same time each day, as seen in Figure 15.14A. However, when a squirrel is

TABLE 15.3	**Some processes that show circadian rhythmicity in animals and other eukaryotes**

Locomotor activity in many vertebrates and invertebrates

Sleep–wake cycles in many animals

Metabolic rate in many animals

Variations of body temperature (including torpor) in birds and mammals

Urine output and drinking in mammals

Adrenocortical hormone secretion and epidermal mitosis in mammals

Integumentary color change in fish and crabs

Oviposition, mating, and emergence of adults from pupae in insects

Female pheromone release and male pheromone sensitivity in insects

Mating in *Paramecium*

Bioluminescence and photosynthetic capacity in dinoflagellate algae

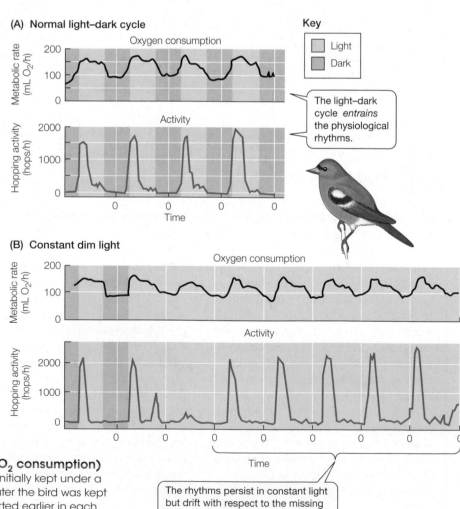

FIGURE 15.13 Circadian rhythm of metabolic rate (O₂ consumption) and motor activity for a chaffinch (A) The bird was initially kept under a normal light–dark cycle, which entrained the rhythm. (B) Later the bird was kept in constant dim light. Note that the free-running rhythm started earlier in each successive 24-h period. Zero on the *x* axis denotes midnight. (After Pohl 1970.)

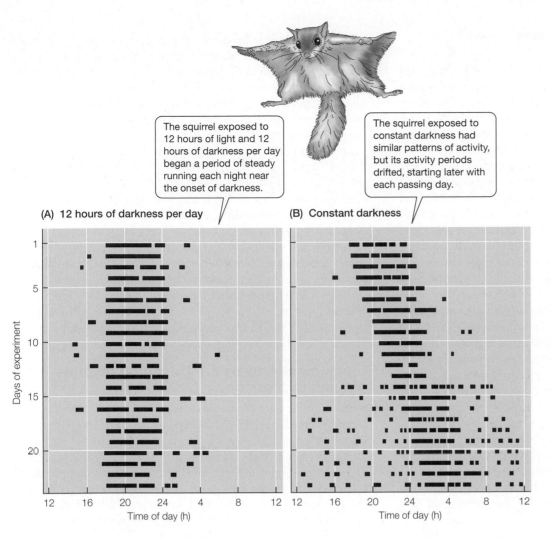

The squirrel exposed to 12 hours of light and 12 hours of darkness per day began a period of steady running each night near the onset of darkness.

The squirrel exposed to constant darkness had similar patterns of activity, but its activity periods drifted, starting later with each passing day.

(A) 12 hours of darkness per day

(B) Constant darkness

FIGURE 15.14 Activity rhythms of two nocturnal flying squirrels (*Glaucomys volans*) Activity was recorded on a running wheel over a period of 23 days at 20°C. Each horizontal line represents a 24-h day. Times on the *x* axis are expressed in the 24-h format; hour 12 is noon and hour 24 is midnight. Turning of the running wheel activated a pen to record a short vertical line for each rotation of the wheel; these vertical blips are usually so close together that they are fused and give the appearance of a heavy, continuous horizontal bar during periods of steady running. The activity pattern of the squirrel in (A) was entrained by the environmental light–dark cycle. The free-running rhythm of the squirrel in (B) was about 21 min longer than 24 h. (After P. J. DeCoursey in Campbell and Reece 2002.)

placed in constant darkness and has no environmental information about the time of day, the squirrel's endogenous rhythm of activity persists, but because the period of the endogenous circadian rhythm is not precisely 24 h, the free-running rhythm drifts in its timing (see Figure 15.14B). Specifically, in this case the period of the free-running rhythm is *more* than 24 h, and therefore the activity interval drifts to occur later and later each day.

The process by which a biological rhythm is brought into phase with an environmental rhythm is called **entrainment**. During this process, the environmental cues *entrain* the biological rhythm, as illustrated in Figure 15.14A. An environmental cue that is capable of entraining (setting the phase) of a biological rhythm is called a **phasing factor** or **zeitgeber** (a term adopted from German and meaning "time-giver"). In nature, the onset of darkness each night cues the activity of nocturnal flying squirrels—not so much directly as indirectly—by resetting the biological clock that generates the circadian rhythm (see below). The squirrels do not wait in total ignorance each day to see when darkness will arrive. Rather, they have an endogenous sense of the time of day, and the onset of darkness simply serves as a cue that maintains a *precise* 24-h rhythm in a system that, in itself, would maintain an *approximate* 24-h rhythm.

Several types of environmental stimuli serve as phasing factors for circadian rhythms. Daily cycles of light intensity entrain the great majority of rhythms. In addition, rhythms can be entrained by cycles of temperature, sound, food availability, social interaction, or other parameters.

Biological clocks generate endogenous rhythms

A *biological clock* is a physiological mechanism that times an endogenous rhythm. This statement does not explain the mechanism by which the clock works; it merely presents the logical necessity of the existence of such a mechanism. Most (but not all) biological clocks are located in the animal's nervous system, as might be expected for their control functions. In examples such as those discussed above, when we speak of entraining a circadian rhythm to a light–dark cycle, it is really the biological clock that is entrained. The endogenous rhythm is the output of the clock. In essence, the clock controls effectors that allow investigators to see what the clock is doing.

Often a biological clock is localized in a discrete region of the nervous system. For example, the biological clock controlling circadian rhythms in vertebrates is located in the suprachiasmatic nucleus of the brain (discussed later in this chapter). In insects and molluscs, the eyes—or structures closely associated with the eyes—often act as the principal circadian control centers or pacemakers. If the optic lobes of the brain are transplanted from one cockroach to another, for example, the recipient takes on the rhythms of the donor!

Control by biological clocks has adaptive advantages

The major adaptive advantage of biological clocks is that they are predictive: They enable an animal to anticipate and prepare for regular environmental changes. Biological clocks exert *feed-forward control* over effectors, in contrast to homeostatic feedback control (see Box 10.3). Feed-forward control, by definition, initiates changes in physiological systems, rather than correcting for changes after they happen. An animal that is strictly dependent on external cues must wait until the cues appear to trigger or stimulate a response. An animal with an internal clock, however, can anticipate when a physiological or behavioral action will be necessary and can initiate it unbidden.

Circadian clocks permit timing of processes during periods of the 24-h day when environmental cues about time are vague or unreliable. For example, consider the star-nosed mole with which we started this chapter. Moles rarely emerge above ground, and they have reduced visual systems and poor form vision. However, they have well-developed neural connections from their eyes to the suprachiasmic nuclei (the master circadian clock; see below). The activity rhythms of moles have been little studied, but robust activity rhythms occur in mole rats, an unrelated group of fossorial rodents that also live completely underground and have reduced vision. More work remains to be done on rhythmic activities in moles, but the available evidence suggests that even living in a completely subterranean habitat doesn't eliminate the adaptive value of circadian rhythms.

Circadian clocks also enable animals to measure changes in **photoperiod**, the number of hours of daylight in a 24-h day. Many animals depend on changes in photoperiod over the course of the year for timing annual events in their life cycles. For example, the long photoperiods of spring may be used as a cue for reproduction or migration.

Finally, circadian clocks enable some animals to use the sun to determine the compass direction, for example in migration. Consider the fact that, if you see the sun on the horizon, you know the direction is west only if you know the time is afternoon. Similarly, certain animals can determine compass directions from the position of the sun, but only if they know the time of day. In these animals, circadian clocks provide the time-of-day information necessary to use the sun as a compass. Homing pigeons orienting relative to the sun, for example, will orient at wrong compass directions if their circadian clocks have been abnormally shifted (see Chapter 18, Figure 18.5).

Endogenous clocks correlate with natural history and compensate for temperature

Free-running circadian rhythms of animals have periods that are longer or shorter than 24 h, as we have seen. For many animals the period is correlated with natural history. Whereas nocturnal animals often have periods of free-running rhythm that are longer than 24 h (and that thus drift later and later each day, as Figure 15.14B illustrates), many diurnal animals have periods shorter than 24 h. However, there are many exceptions. Moreover, experimenters have found that the period of an animal's free-running rhythm is also affected by the level of constant illumination under which the rhythm is measured. The way in which light pulses or light–dark cycles entrain a circadian rhythm also differs somewhat for diurnal and for nocturnal animals.

One remarkable feature of the clocks controlling circadian rhythms is that their timing is relatively insensitive to tissue temperature. Although some clocks can be entrained by temperature changes, the free-running period of the clock itself does not speed up or slow down much with changes in cellular temperature. As seen in Chapter 10 (pages 243–244), the rates of most metabolic processes are quite sensitive to body temperature; heart rate, breathing rate, and metabolic rate, for example, are likely to double or triple if the body temperature of an animal is raised by 10°C. In sharp contrast, the frequencies of free-running circadian rhythms typically increase by less than 5% when body temperature is elevated by 10°C.

A biological clock would obviously be of little use if it were highly sensitive to temperature; imagine the chaos if our wristwatches were to double their rate when warmed by 10°C! The low thermal sensitivity of biological clocks is therefore adaptive. Given, however, that the primary timing mechanisms of these clocks operate on a cellular level, how do the clocks manage to be so immune to the thermal effects that so strongly influence most metabolic processes? This is a major unsolved question in circadian physiology.

Clock mechanisms are based on rhythms of gene expression

Genetic and molecular studies have clarified the mechanisms of action of circadian clocks. Investigators have identified mutations that modify or disrupt clock function in the fruit fly *Drosophila melanogaster*, hamsters, mice, and other model organisms. For example, a mutation in golden hamsters (*Mesocricetus auratus*) causes the activity rhythm of the animals to exhibit an exceptionally short (20-h) free-running period. By determining the biochemical consequences of such mutations, investigators have identified many of the key components of clock mechanisms.

The timekeeping mechanism in a cell typically depends on a rhythmic alternation between enhanced and inhibited expression of key *clock genes* that are broadly homologous among phyla. As diagrammed in **FIGURE 15.15A**, an activator, such as the enhanced expression of a clock gene (increased transcription and translation), leads to increased levels of the protein encoded by the gene. The protein, however, acts as a repressor of expression of the gene. That is, the protein eventually suppresses the expression of its own gene, for example, by interfering with the action of transcription factors that promote the gene's expression. Provided that there are appropriate time delays, such a mechanism can cycle back and forth between two states of gene expression in much the same way that a pendulum swings between two extremes of position, permitting accurate timekeeping. The details of the timekeeping mechanism vary from one group of organisms to another and are proving often to be exceedingly complex. **FIGURE 15.15B** shows core elements of the timekeeping mechanism in neurons of insects and mammals. A mechanism of this sort is often called a **circadian oscillator** because timekeeping is achieved by oscillation between two states of gene expression.

The loci of biological clock functions vary among animals

Animals exhibit circadian organization throughout their bodies: Many tissues are capable of acting as circadian clocks. Typically,

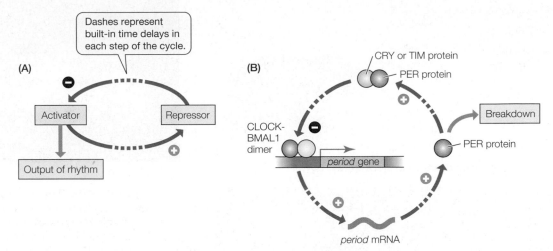

FIGURE 15.15 Cellular mechanisms of circadian timekeeping
(A) A universal model of a circadian timekeeping mechanism, thought to apply to all organisms. A negative feedback loop will produce a periodic rhythm if time delays occur between activation and self-repression. (B) The molecular basis of circadian rhythms in mammals and insects. CLOCK and BMAL1 protein dimers activate transcription of the *period* gene (*per*), producing per mRNA. PER proteins accumulate in the cytoplasm and eventually translocate into the nucleus, where they associate with Cryptochrome (CRY) proteins in mammals or Time-less (TIM) proteins in flies. The PER/CRY (or PER/TIM) dimers inhibit the activity of CLOCK/BMAL1, thus repressing synthesis of PER proteins. PER proteins break down gradually, which eventually relieves repression of the *per* gene and starts the cycle over again. Many details of the molecular clock are not shown; for example, another cycle controls periodic activation of the *bmal1* gene, and other gene products affect the translocation and the rate of degradation of PER protein. (After Herzog 2007.)

however, one tissue acts as a *master clock* that entrains, or imposes its rhythm on, all the other tissues. The entrainment ensures that arrays of tissues and organs ordinarily exhibit synchronous rhythms.

In mammals, the master circadian clock resides in the paired **suprachiasmatic nuclei** of the hypothalamic region of the diencephalon. Each *suprachiasmic nucleus* (SCN) is just dorsal to the optic nerve at the optic chiasm (**FIGURE 15.16A**). Neurons in the SCN express a rhythmic circadian activity of clock genes (**FIGURE 15.16B**). A stunning experiment demonstrated the primacy of the SCN. Researchers destroyed the paired SCNs in a group of genetically normal hamsters. Later they implanted in each hamster paired SCNs taken from a mutant hamster that exhibits an unusual, 20-h free-running activity period. Although the genetically normal hamsters exhibited normal free-running activity rhythms before destruction of their SCNs, they did not show circadian rhythms of activity when they lacked SCNs. **FIGURE 15.16C** illustrates this loss of circadian rhythmicity after SCN destruction. After the hamsters received replacement SCNs, they once again exhibited circadian rhythms, but the free-running period of the rhythms was the unusual, short period characteristic of the mutant donor hamsters.

Individual neurons in the SCN are independently rhythmic when maintained in tissue culture. Communication between neurons in the SCN, as well as between the SCN and the rest of the body, remains incompletely understood; ventral and dorsal SCN neurons differ in neurotransmitters, intrinsic rhythmicity, and connections with other brain areas. Neural connections from the eyes in mammals provide information to the SCN about the daily light–dark cycle in the environment. Interestingly, the light sensors responsible for this entrainment are specialized photosensitive ganglion cells that employ the photopigment *melanopsin*, rather than the (rhodopsin-based) rod and cone photoreceptors of the rest of the visual system.

The SCN is not the only anatomical location of circadian control in mammals, although it is the principal control center and the best understood. The retinas of the eyes are also endogenously rhythmic but do not seem to exert substantial direct control over other tissues. Sometimes certain circadian rhythms in addition to the retinal rhythms persist in mammals after SCN inactivation, pointing to additional clocks. For example, liver cells can maintain a circadian rhythm that can be entrained by feeding. Such peripheral clocks are probably controlled (entrained) by the SCN in normal circumstances.

One important output of the SCN clock controls the pineal gland. The **pineal gland** is a small, unpaired gland that forms embryologically as an evagination of the roof of the brain and is found in virtually all vertebrates. Its principal hormonal secretion is **melatonin**, a compound synthesized from the amino acid tryptophan. In mammals, according to current evidence, the pineal gland is not independently rhythmic; it secretes melatonin in a circadian rhythm because of circadian control from the SCN. Pineal melatonin is secreted at night in mammals (both diurnal and nocturnal) and in virtually all other vertebrates. Thus melatonin is sometimes called the *darkness hormone*. Pineal melatonin is also of great importance in controlling many seasonal rhythms, such as reproduction.

The pineal physiology of nonmammalian vertebrates often differs from that of mammals in two important ways. First, the nonmammalian pineal gland may be endogenously rhythmic and thus can act as a primary circadian control center. Second, the pineal gland is often light-sensitive and acts as a "third eye," providing extraocular information on the environmental day–night cycle (light may reach the pineal gland through the skull). Interaction between the SCN and the pineal gland in the control of circadian rhythms in nonmammalian vertebrates is complex, diverse, and

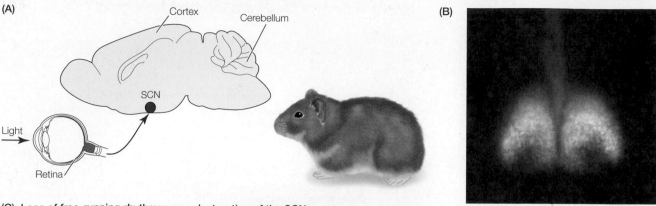

(A)

Cortex

Cerebellum

SCN

Light

Retina

(B)

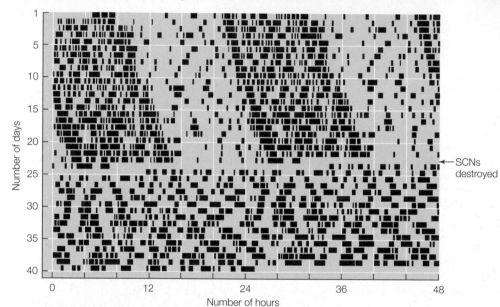

(C) Loss of free-running rhythms upon destruction of the SCNs

Number of days

Number of hours

← SCNs destroyed

FIGURE 15.16 The paired suprachiasmatic nuclei in the brain constitute the major circadian clock of mammals (A) The location of the paired SCNs in the ventral hypothalamus, dorsal to the optic chiasm (the crossing of the optic nerves). The sagittal section shows a side view near the midline (anterior to the left). (B) Slice of the paired SCNs showing PER expression visualized with a luciferase reporter. A video of the SCN slice over the course of 7 days shows that PER expression pulses with a circadian rhythm. (C) Actogram showing the loss of a free-running circadian activity rhythm following destruction of the SCNs in a golden hamster (*Mesocricetus auratus*). (B from Welsh et al. 2010.)

not well understood. The retinas in nonmammalian vertebrates are endogenously rhythmic, but as in mammals, the retinas seem not to serve as circadian control centers for the rest of the body.

Studies of transcription profiling in various tissues of different animals show that there are daily rhythms of transcription of hundreds of genes (see Chapter 3, pages 82–83). In many cases the circadian nature of these daily rhythms has not yet been demonstrated, but it is likely that they result from the output of the circadian clock in the SCN, either directly or via entrainment of other local circadian clocks.

Circannual and circatidal clocks: Some endogenous clocks time annual or tidal rhythms

Daily rhythms have been the most studied of all biological rhythms, and so we have emphasized them in our discussion of biological clocks. However, animals exhibit rhythmic physiological and behavioral variations that operate on other timescales as well. Annual rhythms of reproduction, migration, fat accumulation, dormancy, and so on are well-known examples. Animals living along the seashore often display rhythms synchronized with the tides, which usually rise and fall every 12.4 h (half a lunar day). For instance, fiddler crabs (*Uca* spp.) that scavenge for food on the sand or mud exposed by low tide become rhythmically more active at each time of low tide.

Some, but not all, annual and tidal rhythms are endogenous: They persist even when animals are placed in a laboratory environ-

ment where they are denied environmental information about the time of year or the time of the tidal cycle. Typically the periods of the free-running endogenous rhythms are only *approximately* a year or a tidal cycle in length. Thus the endogenous rhythms are termed **circannual** or **circatidal**.

Under natural conditions, of course, certain environmental parameters vary in phase with the annual or tidal cycle. The endogenous circannual and circatidal rhythms of animals become entrained, so in nature the biological rhythms are kept in phase with the actual seasons and tides. For instance, the annual cycle of photoperiod length (long days in summer, short days in winter) is the phasing factor for certain circannual rhythms; and features of ebbing and flowing tidal water, such as mechanical agitation, serve as phasing factors for some of the circatidal rhythms. It is not clear whether the endogenous timing mechanisms for circannual and circatidal rhythms depend on circadian oscillators. Arguments have been presented on both sides. As yet, the nature of these timing mechanisms remains unresolved.

Interval, or "hourglass," timers can time shorter intervals

In addition to circadian oscillators—which rhythmically cycle—animals appear to possess physiological timing mechanisms that permit timing of *parts* of days by functioning like stopwatches or hourglasses. These noncyclic timers are called **interval timers** or **"hourglass" timers**. Once activated on a given day, they measure the passage of time; but like stopwatches, they are noncyclic and must be restarted to operate again. Male pigeons, for example, seem to use an interval timer to determine how long they incubate eggs in a particular stint; they stay on the eggs for a relatively fixed length of time after they start, regardless of the time of day when they start. Recent research locates the neurophysiological sites of the interval timers of birds and mammals in different parts of the brain than the circadian oscillators.

We tend to think of nervous systems as devices that take in sensory information and transform it into motor output. Circadian rhythms and other endogenous rhythms remind us that behavioral output reflects intrinsic organization of the CNS, as well as responses to sensory input. The persistence of circadian clocks in fossorial animals such as moles suggests that these endogenous systems have selective importance, even in cases where this adaptive value is hard to see.

SUMMARY
Biological Clocks

■ A circadian rhythm has a period of about a day. It is an example of an endogenous rhythm, one that does not require sensory information for timing.

■ A circadian rhythm of an animal will drift, or free-run, in constant light or darkness, when there are no sensory timing cues. A light–dark cycle entrains the circadian rhythm to exactly 24 h.

■ A biological clock is the physiological basis of an animal's ability to time an endogenous rhythm. Biological clocks exert rhythmically changing control, modulating the outputs of the nervous and endocrine systems to prepare an animal for daily changes and seasonal changes. In mammals, the suprachiasmatic nucleus (SCN) of the brain is the principal biological clock for circadian rhythms.

■ Animals may possess other timing mechanisms for shorter rhythmic periods (such as circatidal rhythms) or longer periods (such as circannual rhythms) than those of circadian rhythms.

STUDY QUESTIONS

1. How does the physiological control exerted by the nervous system and endocrine system relate to the concept of homeostasis?

2. Compare and contrast the nervous system organization in arthropods and vertebrates. What are their functional similarities and differences?

3. How, in general, is the vertebrate autonomic nervous system organized? Is it redundant to have separate sympathetic and parasympathetic control of many organs?

4. The sympathetic and parasympathetic divisions of the autonomic nervous system employ the same neurotransmitter (acetylcholine) for preganglionic neurons, but different neurotransmitters for postganglionic neurons (norepinephrine for sympathetic, and acetylcholine for parasympathetic; see Figure 15.10). How would it affect autonomic function if the situation were reversed—that is, if the preganglionic neurotransmitters were different and the postganglionic neurotransmitters were the same?

5. What adaptive advantages might centralization and cephalization offer in the evolution of nervous system organization?

6. Mammals have brains that are more complex than those of fish and amphibians, particularly in terms of expansion of the cerebral cortex. Does this increased complexity make mammals more advanced and fish and amphibians more primitive? Why or why not?

7. The paired suprachiasmatic nuclei (SCNs) usually function as the master circadian clock in mammals. In some circumstances the circadian rhythms of animals may become split, with some effectors following one free-running rhythm and some following another. Give two hypotheses of how such a split might happen—one compatible with an SCN always being a master clock and one not.

8. One of the first genes that was determined to control circadian-clock timing is the *per* gene in *Drosophila*. Mutants of this gene have shorter or longer circadian free-running rhythms, as well as shorter or longer periods of a much faster rhythmic courtship song. What does this observation suggest about the relationship between circadian rhythms and shorter, hourglass-timing rhythms?

Go to **sites.sinauer.com/animalphys4e**
for box extensions, quizzes, flashcards,
and other resources.

REFERENCES

Bell-Pedersen, D., V. M. Cassone, D. J. Earnest, and four additional authors. 2005. Circadian rhythms from multiple oscillators: lessons from diverse organisms. *Nat. Rev. Genet.* 6: 544–556. A particularly insightful and readable account of circadian timekeeping throughout the living world.

Dunlap, J. C., J. J. Loros, and P. J. DeCoursey (eds.). 2004. *Chronobiology: Biological Timekeeping*. Sinauer Associates, Sunderland, MA. Probably the best single reference on circadian and other biological clocks.

Emes, R. D., and S. G. Grant. 2012. Evolution of synapse complexity and diversity. *Annu. Rev. Neurosci.* 35: 111–131.

Gershon, M. D. 1998. *The Second Brain*. HarperCollins, New York. An engaging popular book on the enteric nervous system and control of the gut. Good treatment of paths of discovery of neurotransmitters and autonomic control.

Greenspan, R. J. 2007. *An Introduction to Nervous Systems*. Cold Spring Harbor Laboratory Press, Cold Spring Harbor, NY. An excellent and accessible view of invertebrate nervous systems.

Hastings, M. H., M. Brancaccio, and E. S. Maywood. 2014. Circadian pacemaking in cells and circuits of the suprachiasmatic nucleus. *J. Neuroendocrinol.* 26: 2–10.

Kaas, J. H. (ed.). 2009. *Evolutionary Neuroscience*. Academic Press, San Diego, CA.

Linden, D. J. 2007. *The Accidental Mind*. Belknap Press of Harvard University Press, Cambridge, MA. A popular account of structural organization and evolution of vertebrate brains.

Striedter, G. 2006. *Principles of Brain Evolution*. Sinauer Associates, Sunderland, MA.

Welsh, D. K., J. S. Takahashi, and S. A. Kay. 2010. Suprachiasmatic nucleus: cell autonomy and network properties. *Annu. Rev. Physiol.* 72: 551–577.

See also **Additional References** *and* **Figure and Table Citations**.

Endocrine and Neuroendocrine Physiology

In the fall, bears prepare to spend up to 7 months hibernating in their dens. They eat voraciously, store fat, and become obese. They will use this stored fat during the following winter months of food deprivation. Insulin is one of several hormones that influence bears' nutritional status. It stimulates adipose and muscle tissues to take up glucose and fatty acids when these molecules are present in the blood. Recently investigators found that the sensitivity of fat cells, but not muscle cells, to insulin varies over the seasons in grizzly bears (*Ursus arctos horribilis*). In late fall the bears' adipose cells became exquisitely sensitive to insulin so that they avidly took up nutrient molecules to store as fat. A short time later, when the bears entered their dens, the adipose cells became *insulin-resistant*. Even though insulin was still present in the blood, adipose cells did not respond to it. In the absence of the insulin signal, the stored fat was broken down and used for ATP production. In the spring, when the bears emerged from their dens, the adipose cells again became responsive to insulin, although they were not as sensitive as they would become near the end of the summer. These results suggest that grizzly bears exploit seasonal variations in the sensitivity to insulin—specifically in adipose cells—to control the use of fat as an energy source.

The grizzly bear (*Ursus arctos horribilis*) Bears eat large amounts during the summer and store fat that they will draw on during winter months in their dens.

This example illustrates the role of a single endocrine mechanism in ensuring normal reversible obesity as a means for the grizzly bear to respond to annual changes in its environment. It also illustrates the potential value of understanding the physiology of an "extreme animal" in light of human medicine. In the case of insulin, for example, people with Type II diabetes are insulin-resistant and typically obese. Unraveling the molecular mechanisms in the grizzly bear's metabolism that intersect with those of humans diagnosed with Type II diabetes may provide clues to developing new therapies.

In this chapter we will use a few examples to illustrate the principles of endocrine functions. We will see that hormones (often together with neural signals) change both an animal's physiology and its behavior. Although physiologists' understanding of endocrine systems of many invertebrate groups is still incomplete, we will see that the basic principles of endocrine function apply to both vertebrates and invertebrates.

Introduction to Endocrine Principles

Both hormones and neurohormones are endocrine chemical signals that are carried in the blood. They are similar to neurotransmitters because they exert their effects by binding to receptor molecules expressed by target cells. The main difference, shown in **FIGURE 16.1**, is that a neurotransmitter crosses a narrow synaptic gap (see Figure 16.1A), whereas an endocrine signal circulates in the blood over large distances (see Figure 16.1B,C). The figure also illustrates the concept that chemical signals reach their targets over different distances, ranging from locally secreted autocrine and paracrine signals (see Figure 16.1D) to neurotransmitters at synapses to endocrines carried greater distances in the blood. Because hormone and neurohormone molecules are carried throughout the bloodstream, they can influence large populations of target cells, as long as the target cells express receptor molecules for the hormone. Therefore circulation of hormones and neurohormones permits widespread responses. These responses occur with a delay, relative to responses to neural signals.

Two main factors contribute to slowed responses to endocrine signals. First, hormones require travel time to reach target cells. And second, certain hormones control gene transcription and the synthesis of proteins by target cells, so the responses they initiate are exhibited only after a delay, when protein synthesis is accomplished. Responses to hormones may be brief or last as long as hours or days.

Defined specifically, a **hormone** is a chemical substance produced and released by nonneural endocrine cells or by neurons; it exerts regulatory influences on the function of other, distant cells reached via the blood; and it is effective at very low concentrations (as little as 10^{-12}

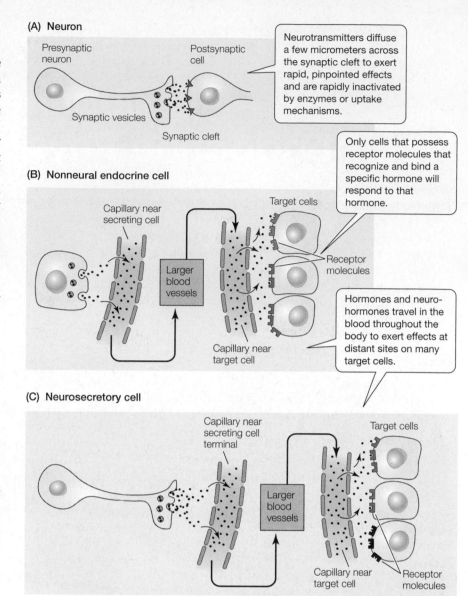

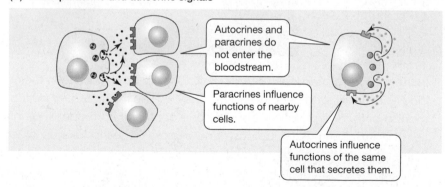

FIGURE 16.1 Chemical signals act over short and long distances within the body (A) A neuron releases neurotransmitter molecules that act on receptor molecules of the postsynaptic cell. (B) A nonneural endocrine cell secretes hormone molecules that enter a capillary (or hemolymph) and are carried throughout the bloodstream. Hormones enter and leave capillaries through spaces between the endothelial cells that make up the capillary wall. (C) A neurosecretory cell secretes hormone like a neuron releases neurotransmitter, and the hormone enters and leaves the blood in the same manner as a hormone from a nonneural endocrine cell. (D) Paracrine and autocrine signals diffuse locally to activate receptors on neighboring cells (paracrine) or on the same cell (autocrine). They do not enter the blood.

molar [*M*])[1]. Hormones released by neurons are often referred to as neurohormones, and the neurons as neuroendocrine or neurosecretory cells. The secretory cells that produce hormones secrete them into the surrounding extracellular fluid, from which they diffuse into capillaries.[2] A table of the major mammalian endocrine and neuroendocrine tissues, their secretions, and their main actions at target tissues can be found in Appendix K.

The secretory cells may be organized into discrete organs termed **endocrine glands** (also called *ductless glands* because they lack outflow ducts), or they may be isolated cells or clusters of cells distributed among the cells of other tissues. For example, the thyroid gland is a discrete structure that secretes thyroid hormones. By contrast, G cells, which secrete gastrin, are distributed in the gastric mucosa of the mammalian stomach (see Chapter 6). Interestingly, the same molecule may serve both as a hormone and as another type of chemical signal in the same organism. In mammals, for example, cholecystokinin (CCK) is not only a hormone secreted by cells in the intestine, but also functions as a neurotransmitter or neuromodulator in the central nervous system (CNS). Alternatively, a hormone can circulate widely in the general circulation and also serve locally as a paracrine. For example, testosterone exerts widespread effects as a hormone and also functions as a paracrine within the testis to support spermatogenesis.

Hormones bind to receptor molecules expressed by target cells

Although a hormone circulates past many cells, it influences the functions of only certain cells, called target cells, which express *receptor molecules* that specifically bind the hormone. Consider thyroid hormones, for example. These hormones, secreted by the thyroid gland, exert a wide range of metabolic, structural, and developmental effects on many different tissues (see Table 16.3 and Appendix K). They have such widespread effects because many different cells of the body possess receptor molecules that recognize thyroid hormones. Typical target cells express thousands of receptor molecules for a particular hormone. The *sensitivity* of a target cell to a hormone depends on the number of functional receptor molecules the target cell expresses for that hormone. The sensitivity of a target cell can change under different conditions because the number of receptor molecules that recognize that hormone can increase (by upregulation) or decrease (by downregulation). In addition, many target cells express separate populations of different types of receptor molecules, so they are capable of responding to more than one hormone. These variations in the types and numbers of receptor molecules expressed by target cells contribute to the immense versatility of hormonal regulation in animals. An addi-

tional consideration to keep in mind is that a target cell's response to a hormone at any moment depends not only on the number of receptor molecules it expresses for that hormone but also on the hormone's concentration in the blood.

Concentrations of hormones in the blood vary in response to varying conditions

Most endocrine cells synthesize and release some hormone all the time, but the rate of release is variable, depending on mechanisms of control. Often neurons or other hormones control the rates of synthesis and secretion of hormones. In general, the higher the rate at which a hormone is secreted, the higher its concentration in the blood, and the greater its effect on target cells. Circulating hormone molecules are enzymatically degraded at their targets or by organs (such as the liver and kidneys in vertebrates). Therefore the moment-to-moment blood concentration of a hormone represents a balance between the *rate of addition* of hormone to the blood (by secretion) and the *rate of removal* from the blood (by metabolic destruction and excretion). Hormone concentration depends primarily on the rate of addition to the blood, because the rate of removal is relatively constant. A hormone's half-life—the time required to reduce the concentration by one-half—indicates its rate of removal from the blood and thus the duration of its activity.

Some hormones may be converted to a more active form after secretion by a process termed **peripheral activation**. For example, thyroid hormone is secreted mainly as the four-iodine molecule tetraiodothyronine, or T_4. After T_4 is secreted, its target cells enzymatically remove one iodine to form triiodothyronine, or T_3, which is more physiologically active than T_4.

Most hormones fall into three chemical classes

TABLE 16.1 summarizes the characteristics of the following three chemical classes of hormones:

1. **Steroid hormones** are synthesized from cholesterol (**FIGURE 16.2**). In vertebrates, the gonads and the adrenal cortex secrete steroid hormones, as do the skin and, in pregnant mammals, the placenta. The molting hormones of arthropods (e.g., ecdysone) are also steroids. Steroid hormones are lipid-soluble, so they can pass through cell membranes to reach receptor molecules located inside their target cells. In some cells, lipid-soluble hormones (e.g., estrogen) are transported across the membrane. One transporter of these hormones is *megalin*, an integral protein receptor molecule of the target cell membrane that brings lipid-soluble hormones (which often form a complex with carrier molecules) into the cell by endocytosis. We will see that a few steroid hormones also exert their effects by binding to receptors expressed on the surface membrane of the target cells.

2. **Peptide** and **protein hormones** are structured from chains of amino acids (**FIGURE 16.3**). In vertebrates, they include antidiuretic hormones, insulin, and growth hormone. Examples of peptide and protein hormones in invertebrates include the gamete-shedding hormone of sea stars and the diuretic hormones of insects. Peptide and

---

[1] In the 1970s, radioimmunoassay (RIA) techniques became available, which made it possible to detect and measure blood levels of hormones with levels of accuracy that had not previously been possible. The technique markedly advanced both basic research and clinical applications. Rosalyn Yalow shared the Nobel Prize in Physiology or Medicine in 1977 in recognition of her development of RIA techniques, along with Roger Guillemin and Andrew V. Schally, who were recognized for their studies of the production of peptide hormones in the brain.

[2] In animals with open circulatory systems, the blood and extracellular fluid blend to form hemolymph (see page 695). In these animals, hormones are released directly into, and circulated in, the hemolymph.

FIGURE 16.2 Steroid hormones are derived from cholesterol Steroidogenesis begins with the formation of pregnenolone. Different steroid-secreting cells possess different enzymes that modify pregnenolone to produce specific steroid hormones. The skin cells that produce inactive vitamin D are an exception: They do not cleave the side chain to form pregnenolone.

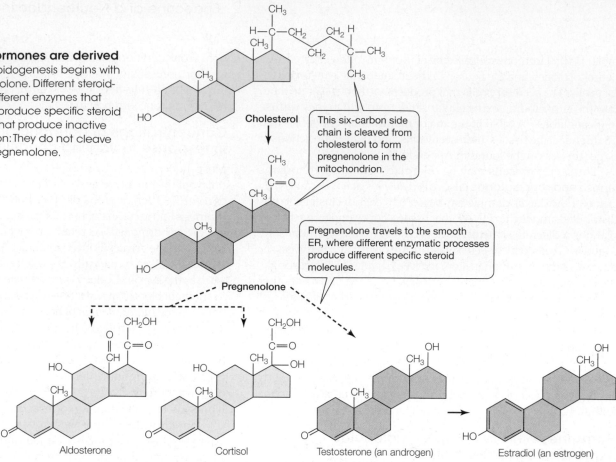

This six-carbon side chain is cleaved from cholesterol to form pregnenolone in the mitochondrion.

Pregnenolone travels to the smooth ER, where different enzymatic processes produce different specific steroid molecules.

Aldosterone

Cortisol

Testosterone (an androgen)

Estradiol (an estrogen)

protein hormones vary enormously in molecular size, from tripeptides (such as thyrotropin-releasing hormone, which consists of 3 amino acids) to proteins containing nearly 200 amino acids (such as growth hormone). Often hormones consisting of assemblages of amino acids are simply called *peptide hormones* (blurring the size distinction), and we will usually follow that practice. Peptide hormones are soluble in aqueous solutions.

3. **Amine hormones** are modified amino acids (**FIGURE 16.4**). **Melatonin**, secreted by the vertebrate pineal gland (see Chapter 15), is derived from tryptophan,

whereas the **catecholamines** and **iodothyronines** are derived from tyrosine. *Catecholamines* are found widely as synaptic transmitter substances in both invertebrates and vertebrates. However, three catecholamines also serve as hormones in vertebrates: epinephrine (also called adrenaline), norepinephrine (noradrenaline), and dopamine. *Iodothyronines*, the thyroid hormones, are found only in vertebrates. They are synthesized by the thyroid gland and have the unique property of being rich in iodine. Whereas melatonin and the catecholamines are soluble in water, the iodothyronines are soluble in lipids.

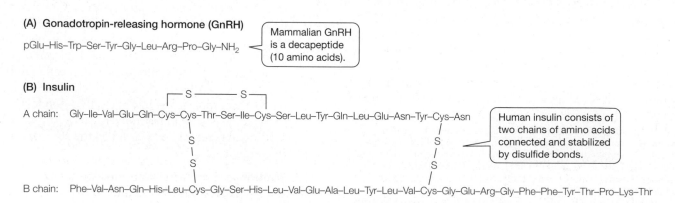

(A) Gonadotropin-releasing hormone (GnRH)

pGlu–His–Trp–Ser–Tyr–Gly–Leu–Arg–Pro–Gly–NH₂

Mammalian GnRH is a decapeptide (10 amino acids).

(B) Insulin

A chain: Gly–Ile–Val–Glu–Gln–Cys–Cys–Thr–Ser–Ile–Cys–Ser–Leu–Tyr–Gln–Leu–Glu–Asn–Tyr–Cys–Asn

B chain: Phe–Val–Asn–Gln–His–Leu–Cys–Gly–Ser–His–Leu–Val–Glu–Ala–Leu–Tyr–Leu–Val–Cys–Gly–Glu–Arg–Gly–Phe–Phe–Tyr–Thr–Pro–Lys–Thr

Human insulin consists of two chains of amino acids connected and stabilized by disulfide bonds.

FIGURE 16.3 Peptide and protein hormones consist of assemblages of amino acids (A) The amino acids on both ends of GnRH are modified by posttranslational modification. Therefore pGlu at the N terminus of the peptide is a modified form of glutamic acid, and glycine at the opposite end has undergone C-terminal amidation. (B) The A chain of human insulin consists of 21 amino acids, and the B chain consists of 30.

TABLE 16.1 Peptide, steroid, and amine hormones of vertebrates

Property	Peptides	Steroids	Amine hormones		
			Catecholamines	Thyroid hormones	Melatonin
Site of secretion	Most sites in Appendix K, except adrenal cortex and medulla, thyroid and pineal glands, and skin	Adrenal cortex, gonads, skin, and (in pregnant mammals) placenta	Adrenal medulla (epinephrine and norepinephrine), hypothalamus (dopamine)	Thyroid gland	Pineal gland
Structure	Chains of amino acids	Derived from cholesterol	Derived from tyrosine	Derived from tyrosine and iodine	Derived from tryptophan
Solubility	Water-soluble	Lipid-soluble	Water-soluble	Lipid-soluble	Water-soluble
Synthesis and storage	Synthesized at rough ER; processed in Golgi apparatus; stored in vesicles in advance of use	Synthesized on demand in intracellular compartments; not stored	Synthesized in cytoplasm and stored in vesicles	Made prior to use and stored in a colloid island within the gland	Synthesized in the cytoplasm and stored in vesicles
Secretion	Exocytosis	Simple diffusion through cell membrane	Exocytosis	Simple diffusion through cell membrane	Exocytosis
Transport	Dissolved in plasma; some bound to carrier proteins	Bound to carrier proteins	Dissolved in plasma	Bound to carrier proteins	Dissolved in plasma
Half-life	Minutes	Hours	Seconds to minutes	Days	Minutes
Location of receptor molecules	Surface of target cell membrane	Cytoplasm or nucleus (some steroids bind to cell-surface receptors)	Surface of target cell membrane	Nucleus	Surface of target cell membrane
Action at target cell	Activate second-messenger systems or alter membrane channels	Alter gene expression; activated genes initiate transcription and translation	Activate second-messenger systems	Alter gene expression; activated genes initiate transcription and translation	Activates second-messenger systems
Response of target cell	Change activity of preexisting proteins, some of which may induce new protein synthesis	Synthesize new proteins; some may change activity of preexisting proteins	Change activity of preexisting proteins	Synthesize new proteins	Changes activity of preexisting proteins

Source: After Gardner and Shoback 2007.

Hormone molecules exert their effects by producing biochemical changes in target cells

To initiate changes in target cells, hormones first bind to specific receptor molecules. With some exceptions, nonpolar (hydrophobic) hormones bind to intracellular receptors and polar hormones (hydrophilic) to cell-surface receptors. Three types of receptor molecules are important in mediating hormone actions: intracellular receptors, G protein–coupled membrane receptors, and enzyme-linked membrane receptors (see pages 62–65).

Lipid-soluble hormones (steroids and iodothyronines) bind to intracellular receptors. Because they are lipid-soluble, these hormones enter target cells by diffusing through the lipid bilayer of the cell membrane (see Figure 2.27) or are carried bound to a lipoprotein molecule. Their receptors are located either in the cyto-

plasm or in the nucleus. When the hormone molecule binds to the receptor molecule, it forms a hormone–receptor complex that acts as a transcription factor that interacts with the target cell's DNA to alter gene expression. By turning processes of transcription and translation on or off, the hormone directly influences the synthesis of proteins by the target cell. New proteins, which can be enzymes or structural proteins, carry out the target cell's physiological response. Because the production of new proteins requires time, there is a delay (ranging from many minutes to hours) between the hormone's binding to the intracellular receptor and the target cell's response. Once the proteins are synthesized, the response lasts until they are degraded. In addition, we now know that certain target cells of some steroid hormones, such as aldosterone, estrogen, and certain glucocorticoids (see Box 16.1), express *cell-surface* receptors for those hormones. When steroid hormones bind to receptors

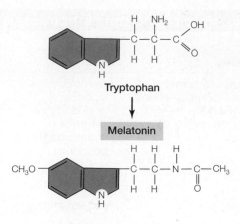

FIGURE 16.4 Amine hormones are derived from amino acids Catecholamine-secreting cells use biochemical pathways beginning with tyrosine to produce dopamine, norepinephrine, and epinephrine. The level of expression of specific enzymes in a catecholamine-secreting cell determines which catecholamine is produced in greatest abundance. In the thyroid gland, enzymatic reactions in the follicle cells and colloid start with tyrosine to produce iodothyronines (thyroid hormones). Similarly, tryptophan is modified biochemically in the pineal gland to produce melatonin.

on the cell membrane, they do not exert their effects by genomic means, but instead function like water-soluble hormones, exerting rapid *non-genomic* effects.

The water-soluble peptide and catecholamine hormones bind receptor molecules that are located in the cell membrane and have hormone-binding sites that face the extracellular fluid (see Figure 2.27B). These receptors mediate hormone actions by regulating ion-channel permeability or by activating an intracellular second-messenger system, which transduces the hormonal signal into a response of the target cell (see Figures 2.29 and 2.30). Some peptide hormones, such as insulin and growth hormone, bind to enzyme-linked membrane receptors (see Figure 2.27C). Peptide hormones exert their effects primarily by changing the activities of existing proteins, although some may also alter gene activities. Target cells can change their physiology quickly by using preexisting proteins, so their responses to these hormones can be measured within minutes.

Water-soluble carrier proteins in the blood transport lipid-soluble hormones and many water-soluble hormones

Carrier proteins (see Table 16.1, "Transport") bind to hormone molecules reversibly and noncovalently. Free and bound forms of the hormone are in equilibrium in the blood. Only the free form is physiologically active. In the vicinity of target cells, when some of the free hormone molecules bind to receptor molecules of a target tissue, some bound hormone molecules unbind from the carrier proteins, and the equilibrium in the blood is maintained. Thus carrier proteins provide a reservoir of bound hormone that can be drawn on. Furthermore, carrier proteins ensure that lipid-soluble hormones remain dissolved in the aqueous plasma. Finally, carrier proteins protect hormones from rapid inactivation and excretion and thus extend their half-lives. In humans, the half-life of the steroid hormone cortisol is 60–90 min, and that of thyroxine is several

days. In contrast, peptide hormones have half-lives in the range of minutes. To maintain a steady level of a peptide hormone in the blood for a period longer than a few minutes, the endocrine cells must continue to secrete it. This is also the case for the catecholamine epinephrine, which has a half-life of no more than 1–2 min.

SUMMARY
Introduction to Endocrine Principles

- Hormone molecules are synthesized, stored, and released by nonneural endocrine cells or neurons, circulate in the blood at low concentrations, exert their effects on target tissues, and are metabolically destroyed or directly excreted from the body.

- The magnitude of a hormone's effect on target cells depends on both the abundance of receptor molecules with which it can bind and its concentration in the blood. Blood concentration of a hormone depends on a balance between the rate of synthesis and the rate of degradation or excretion. The rate of synthesis and secretion of a particular hormone is often governed by another hormone; some endocrine cells also receive neural input.

- Hormones are categorized into three main classes: steroids, peptides and proteins, and amines. The same chemical messenger may function as a hormone in one context and as a different type of chemical messenger in another.

- The half-lives of hormones vary depending on their chemical class, ranging from seconds to hours or days. Carrier proteins in the blood transport all lipid-soluble, nonpolar hormones and many water-soluble, polar ones. Only free hormone molecules are able to bind to receptor molecules in or on target cells.

- Both lipid-soluble and water-soluble hormones initiate biochemical changes in their target cells by binding to receptor molecules. Measurable responses to water-soluble and lipid-soluble hormones that bind cell-surface receptor molecules occur with a shorter delay than do responses to lipid-soluble hormones that bind intracellular receptors to initiate genomic actions.

Synthesis, Storage, and Release of Hormones

In this section we compare the cellular mechanisms of synthesis, storage, and secretion of two chemical classes of hormones: peptides and steroids. We use insulin of vertebrates as our primary example of a peptide hormone. Insulin lowers the concentration of blood glucose by stimulating cells to take up glucose (as well as amino acids) from the blood. It also promotes synthesis of nutrient storage molecules. Insulin is synthesized in pancreatic endocrine β (beta) cells of the islets of Langerhans, which are embedded in the pancreatic exocrine tissue that secretes digestive enzymes. Insulin follows the pattern typical for the synthesis of most secreted peptides, in which a large precursor molecule, a *preprohormone*, is modified by *posttranslational processing* to a *prohormone* and finally a mature hormone.

Peptide hormones are synthesized at ribosomes, stored in vesicles, and secreted on demand

The insulin molecule consists of two peptide chains, designated A and B, connected by disulfide bonds (see Figure 16.3B). Although the amino acid sequences of the two chains vary somewhat among species, the general structure of the molecule, the amino acid sequences of certain regions, and the positions of the disulfide bonds are all highly conserved. The β cell DNA encodes the structure of the insulin molecule, which is transcribed into messenger RNA and translated into its initial amino acid sequence as a preprohormone at ribosomes of the rough endoplasmic reticulum (ER).

FIGURE 16.5 illustrates and explains the sequence of steps that change preproinsulin to proinsulin and then mature insulin. Mature insulin and the cleaved C-peptide are stored in vesicles ready for release. They are secreted by calcium-dependent exocytosis. When blood concentrations of glucose increase, β cells depolarize, causing voltage-gated Ca^{2+} channels to open and allow an influx of Ca^{2+} ions. The rate of insulin secretion is modulated by several factors, the most important of which are the levels of glucose and certain amino acids in the blood. As cells in the body respond to insulin's signal to take up glucose, the blood glucose concentration falls, and the β cells are no longer stimulated to secrete insulin. By this *negative feedback* mechanism, the blood glucose concentration

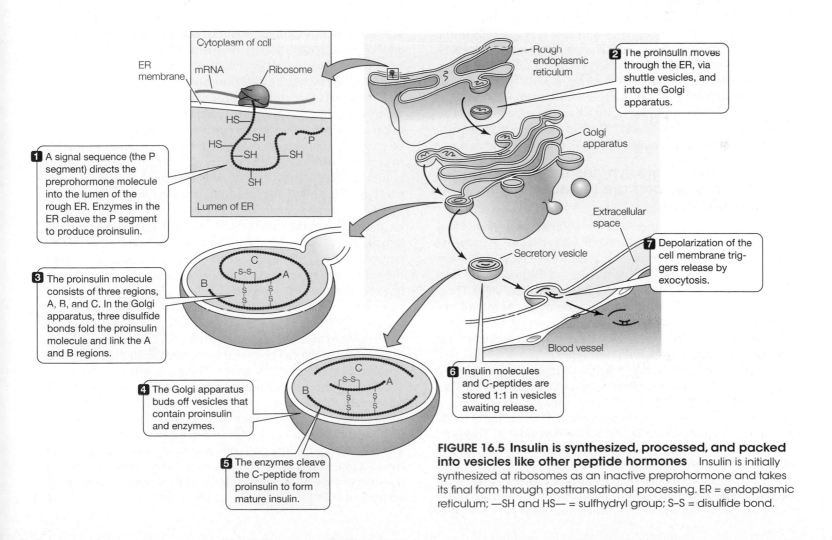

1 A signal sequence (the P segment) directs the preprohormone molecule into the lumen of the rough ER. Enzymes in the ER cleave the P segment to produce proinsulin.

2 The proinsulin moves through the ER, via shuttle vesicles, and into the Golgi apparatus.

3 The proinsulin molecule consists of three regions, A, B, and C. In the Golgi apparatus, three disulfide bonds fold the proinsulin molecule and link the A and B regions.

4 The Golgi apparatus buds off vesicles that contain proinsulin and enzymes.

5 The enzymes cleave the C-peptide from proinsulin to form mature insulin.

6 Insulin molecules and C-peptides are stored 1:1 in vesicles awaiting release.

7 Depolarization of the cell membrane triggers release by exocytosis.

FIGURE 16.5 Insulin is synthesized, processed, and packed into vesicles like other peptide hormones Insulin is initially synthesized at ribosomes as an inactive preprohormone and takes its final form through posttranslational processing. ER = endoplasmic reticulum; —SH and HS— = sulfhydryl group; S–S = disulfide bond.

is kept relatively constant. Other factors also stimulate insulin secretion; these include parasympathetic stimulation of the islets of Langerhans and gastrointestinal hormones secreted by the digestive tract in the presence of food. Both of these stimuli act before blood glucose levels actually increase and therefore function as *feed-forward* mechanisms. Sympathetic stimulation of the islets of Langerhans inhibits secretion of insulin. Without insulin promoting its uptake into cells, glucose remains available in the blood to provide a source of energy for the sympathetic "fight-or-flight" response.

During secretion, C-peptide is released one-to-one with insulin. Because the half-life of insulin is only 1–2 min, and that of C-peptide is about 30 min, clinicians measure the blood concentration of C-peptide as a reflection of insulin secretion by β cells. The function of C-peptide at its target cells is a topic of active investigation, and current studies suggest that one of its functions is to exert an anti-inflammatory effect on the endothelial cells that line blood vessels.

Whereas preproinsulin is produced and processed by a single type of cell, the pancreatic β cell, other preprohormones are synthesized by different types of cells and processed into different end products for secretion. An example is preproopiomelanocortin, the precursor of proopiomelanocortin (POMC). POMC is synthesized by several different cells, including cells in the anterior pituitary gland, brain, skin, and mammalian placenta. Each type of cell expresses different enzymes to carry out posttranslational processing of POMC and so produces different end products. Some types of cells co-secrete more than one end product. For example, adrenocorticotropic hormone (ACTH) and melanocyte-stimulating hormone (MSH) are produced by different types of cells in the anterior pituitary by enzymatically cleaving POMC at different sites. The enzymes of the ACTH-secreting cells cleave POMC in such a way that another hormone, β-endorphin, is also produced and co-secreted with ACTH. β-Endorphin plays an important role in pain control, which we will see when we consider the mammalian stress response (see Figure 16.12). Cells of the hypothalamus also process POMC to produce significant amounts of β-endorphin.

Steroid hormones are synthesized on demand prior to secretion, and are released into the blood by diffusion

Steroid hormones are synthesized from cholesterol. In vertebrates, some cholesterol is obtained from animal fats in the diet. *Steroidogenic* (steroid-producing) endocrine cells and liver cells also synthesize cholesterol. In vertebrates, endocrine cells of the gonads, adrenal cortex, skin, and (in mammals) the placenta produce steroid hormones. Once these cells synthesize or take in the cholesterol, enzymes cleave the six-carbon side chain to form pregnenolone (except in skin cells), which then enters one of several possible biochemical pathways (see Figure 16.2). Different types of steroidogenic cells have different sets of enzymes, so that each produces a different major end product. The enzymes necessary for steroid synthesis are contained within intracellular compartments: Those involved in converting cholesterol to pregnenolone are located in the mitochondria, and most of the others are in the smooth ER. Therefore pregnenolone molecules must travel from the mitochondria to the smooth ER for further enzymatic conversions.

Unlike peptide hormones, steroid hormones are not stockpiled in vesicles prior to secretion. Instead, when the cell is stimulated, they are made from precursors stored in lipid droplets and immediately secreted. Secretion is accomplished by diffusion through the cell membrane. Therefore, whereas the blood concentration of a peptide hormone is determined by the rate of release of preexisting stored hormone, the blood concentration of a steroid hormone is determined by the rate of synthesis—and immediate release—of the hormone.

SUMMARY
Synthesis, Storage, and Release of Hormones

- Peptide hormones are synthesized by transcription of DNA, translation, and posttranslational processing. They are stored in vesicles and secreted on demand by exocytosis.

- Steroid hormones are synthesized from cholesterol. Steroidogenic cells use different biochemical pathways and sets of enzymes to produce different steroid hormones. Steroid hormones are synthesized on demand and secreted by diffusion through the cell membrane.

Types of Endocrine Cells and Glands

Endocrine cells are commonly divided into two major classes: **epithelial** (also called **nonneural**) **endocrine cells** and **neurosecretory cells**. When cells of these types are organized into discrete glands, the glands are described by the same terminology: **epithelial** (**nonneural**) **glands** and **neurosecretory glands**. The endocrine secretions of either type of cell or gland are properly termed hormones, but those of neurosecretory structures are often distinguished by being called **neurohormones**.

The signals that stimulate secretion by nonneural endocrine cells are usually other hormones, although some, such as the islets of Langerhans in the vertebrate pancreas, also receive neural input. By contrast, neurosecretory endocrine cells are always signaled to secrete their neurohormones by synaptic input from typical neurons. Thus neurosecretory cells interface directly with the nervous system. Both neurons and neurosecretory cells typically generate action potentials and release their products by exocytosis: neurotransmitter into the synaptic cleft and neurohormones into the blood. The fundamental similarity between neurons and neurosecretory cells suggests evolutionary continuity between the neural and the endocrine control systems. We do not know, however, whether neurosecretory cells evolved from neurons, or vice versa, or whether both types of cells have a common ancestry.

The cell bodies of neurosecretory cells are located within the CNS, but their axons extend outside the CNS. Neurohormones are synthesized in the cell bodies, transported down the axons, and released at the ends of the axons. The axon terminals are often contained within a **neurohemal organ**, which is an anatomically distinct site for the release of neurohormones. It consists of one or more clusters of axon terminals and a rich supply of blood vessels. Neurohemal organs occur in both vertebrates and invertebrates. The pars nervosa of the posterior pituitary gland is a prominent neurohemal organ in vertebrates (see Figure 16.6). The corpus allatum of insects, discussed later in this chapter, is another prominent example.

Endocrine glands are categorized as *discrete glands* (such as the thyroid), *diffuse glands* (such as the gastrin-secreting cells of the mammalian stomach), or *intermediate glands* (such as the islets of Langerhans in the pancreas). Studies of endocrine glands in different animals suggest that discrete glands, such as the adrenal glands, may have evolved from more diffuse secretory cells of ancestral animals. This idea is supported by the anatomical differences among adrenal glands of amphibians, birds (avian reptiles), and mammals. The adrenal gland of mammals is a discrete gland that sits adjacent to the kidney. The gland consists of an inner *medulla* that secretes the catecholamines epinephrine and norepinephrine and an outer *cortex* that secretes several different steroid hormones (see Figure 16.9). In birds, the adrenal gland is also discrete, but the catecholamine-secreting cells are distributed among the steroid-secreting cells rather than being segregated into different regions. In amphibians, the adrenal gland isn't discrete at all. Instead its different cell types form patches on the kidney.

SUMMARY
Types of Endocrine Cells and Glands

- Epithelial (nonneural) endocrine cells are generally controlled by hormones. (Some, such as the β cells of the pancreas, also receive neural input.)

- Neurosecretory cells are always controlled by synaptic input from neurons. Neurons and neurosecretory cells are thought to be related evolutionarily, but their origins are not known.

- Endocrine glands may be discrete, diffuse, or intermediate. Certain discrete glands appear to have evolved from diffusely distributed cells.

Control of Endocrine Secretion: The Vertebrate Pituitary Gland

In this section we use the vertebrate pituitary gland as an example to illustrate two major controls of secretion: neural control of neurosecretory cells and neurosecretory control of endocrine cells. The principles of control described in this example also apply to other endocrine tissues in both vertebrates and invertebrates. This section will also set the stage for introducing the concept that neural and neurosecretory controls are also influenced by feedback mechanisms.

The pituitary gland lies immediately below the hypothalamus and consists of two parts: the **adenohypophysis**, commonly called the **anterior pituitary**, and the **neurohypophysis**, commonly called the **posterior pituitary**. In development, the anterior pituitary forms from a dorsal evagination (outpocketing) of the oral cavity called Rathke's pouch. This completely nonneural tissue pinches off from the oral cavity to associate closely with the posterior pituitary, which is an extension of the hypothalamus.

The posterior pituitary illustrates neural control of neurosecretory cells

The posterior pituitary (neurohypophysis) consists of bundles and terminations of axons that originate in the hypothalamus (**FIGURE 16.6**). Hypothalamic neurosecretory cells extend their axons through the *median eminence*, which forms part of the floor of the hypothala-

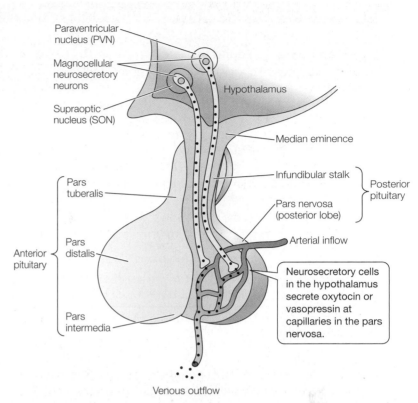

FIGURE 16.6 The hypothalamus and posterior pituitary All vertebrate pituitary glands have two parts, posterior and anterior, but they vary in specific morphology. This diagram and that in Figure 16.7 are based on the structure of the mammalian pituitary gland. The magnocellular cell bodies of neurosecretory cells located in nuclei in the hypothalamus extend their axons along the infundibular stalk to the pars nervosa of the posterior pituitary. These axon terminals secrete oxytocin and vasopressin by exocytosis into the general circulation.

mus, along the *infundibular stalk*, and into the *pars nervosa* ("nervous part"), where the axons terminate at a rich network of capillaries. (The pars nervousa is also called the neural lobe or posterior lobe.)

In most mammals, two peptide hormones are released into the blood in the pars nervosa: vasopressin (VP) and oxytocin. **Vasopressin**, also called **antidiuretic hormone** (**ADH**), limits the production of urine and also stimulates constriction of arterioles. The functions of **oxytocin** (which is produced in both males and females) include causing contractions of the uterus during birth and ejection of milk by the mammary glands during suckling. Endocrinologists originally believed that the hormones of the pars nervosa were synthesized there. However, starting in the 1930s, research revealed that these hormones are actually synthesized by neurosecretory cells that have their cell bodies within the hypothalamus. These cells are large and termed **magnocellular neurons**. In mammals, two clusters of cell bodies in the hypothalamus, the **paraventricular nucleus** and **supraoptic nucleus**, are the main sites of production of vasopressin and oxytocin (see Figure 16.6). When the neurosecretory cells are stimulated, they generate action potentials that propagate from the hypothalamus to their axon terminals in the pars nervosa. Here they release hormone by exocytosis into the extracellular fluid near capillaries, and the hormone diffuses into the blood.

TABLE 16.2 Posterior pituitary nonapeptides found in vertebrates

Common name	Found in	Amino acid site 1	2	3	4	5	6	7	8	9
Arginine vasopressin (AVP) (antidiuretic hormone, ADH)	Most mammals	Cys	Tyr	**Phe**	Gln	Asn	Cys	Pro	**Arg**	Gly(NH₂)
Lysine-vasopressin (LVP)	Pig, peccary, hippopotamus	Cys	Tyr	**Phe**	Gln	Asn	Cys	Pro	**Lys**	Gly(NH₂)
Arginine vasotocin (AVT)	Nonmammalian vertebrates	Cys	Tyr	**Ile**	Gln	Asn	Cys	Pro	**Arg**	Gly(NH₂)
Oxytocin	Most mammals	Cys	Tyr	**Ile**	Gln	Asn	Cys	Pro	**Leu**	Gly(NH₂)
Mesotocin	Reptiles (including birds), amphibians, lungfish	Cys	Tyr	**Ile**	Gln	Asn	Cys	Pro	**Ile**	Gly(NH₂)
Isotocin	Most bony fish	Cys	Tyr	**Ile**	Ser	Asn	Cys	Pro	**Ile**	Gly(NH₂)

Note: Red text indicates differences in amino acids among the vasopressins and oxytocin. Mesotocin and isotocin are shown for comparison.
Source: After Bentley 1998.

The neurosecretory cells that produce and secrete vasopressin and oxytocin receive and integrate synaptic input from a host of neurons, which provide *neural control*. Vasopressin cells, for example, receive input about blood volume and the osmotic concentration of body fluids. When they receive signals reporting high osmotic concentration and/or low blood volume, they secrete vasopressin, which triggers processes involved in retaining water. Likewise, oxytocin cells respond to signals from the mammary glands when suckling occurs or from the cervix of the uterus during labor and birth (see page 492).

Vasopressin and oxytocin, which are very similar in their amino acid sequences (**TABLE 16.2**), are considered members of a family of hormones that are probably descended from a single ancestral peptide. They are both nonapeptides (have nine amino acids). In most mammals they differ at only two amino acid sites, yet these structural differences underlie profound differences in function. Table 16.2 shows vasopressins and oxytocins in different vertebrate species that have additional variations in amino acid composition. Investigators analyze both chemical and genetic information about the peptides within families of hormones to learn about their common ancestral origins.

The anterior pituitary illustrates neurosecretory control of endocrine cells

The anterior pituitary (adenohypophysis) is nonneural endocrine tissue (**FIGURE 16.7**). It is subdivided into the *pars distalis*, *pars intermedia*, and *pars tuberalis* (see Figure 16.6). The exact positions and relative sizes of these parts vary greatly from one animal group to another, and in some groups not all parts are present. All the hormones of the anterior pituitary are synthesized and secreted by endocrine cells within its tissues. Different specific populations of cells secrete different hormones. All anterior pituitary hormones are peptides, proteins, or glycoproteins.

Anterior pituitary hormones are categorized into two main groups, tropic or direct acting, according to their target tissues. The **direct acting** hormones exert their principal effects on nonendocrine

tissues. **Growth hormone** (**GH**), for example, influences growth and nutrient metabolism in tissues such as fat and muscle. Other direct acting hormones are **prolactin** and **melanocyte-stimulating hormone** (**MSH**). By contrast, the **tropic hormones** control other endocrine tissues. Hormones that influence the functions of other endocrine tissues have the suffix *-tropic* in their names, or are called *tropins*. **Thyroid-stimulating hormone** (**TSH**), for example, is also

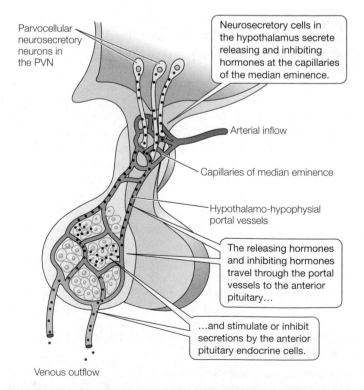

Parvocellular neurosecretory neurons in the PVN

Neurosecretory cells in the hypothalamus secrete releasing and inhibiting hormones at the capillaries of the median eminence.

Arterial inflow

Capillaries of median eminence

Hypothalamo-hypophysial portal vessels

The releasing hormones and inhibiting hormones travel through the portal vessels to the anterior pituitary…

…and stimulate or inhibit secretions by the anterior pituitary endocrine cells.

Venous outflow

FIGURE 16.7 The hypothalamus and anterior pituitary Parvocellular neurosecretory neurons in the hypothalamus secrete releasing and inhibiting hormones that control the secretions of nonneural anterior pituitary endocrine cells. PVN = paraventricular nucleus.

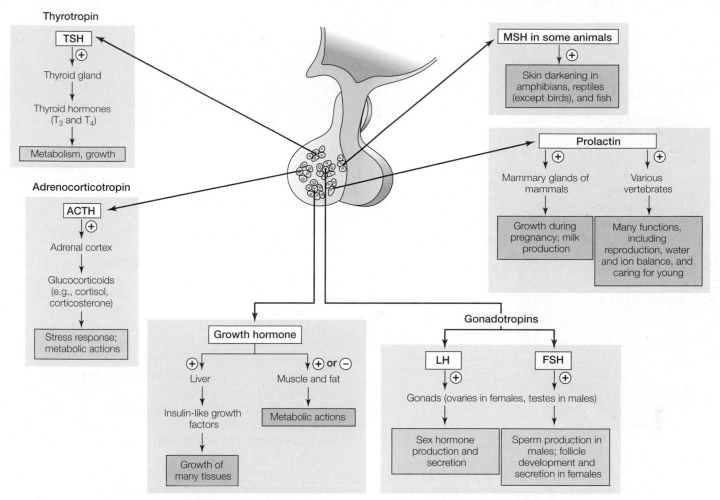

FIGURE 16.8 The anterior pituitary hormones   Separate populations of cells in the anterior pituitary secrete different hormones. (+) = stimulatory; (–) = inhibitory; TSH = thyroid-stimulating hormone; ACTH = adrenocorticotropic hormone; MSH = melanocyte-stimulating hormone; LH = luteinizing hormone; FSH = follicle-stimulating hormone.

called *thyrotropin*. This anterior pituitary hormone supports and maintains the tissues of the thyroid gland and also stimulates the gland to secrete thyroid hormones. If a target gland is deprived of input from its tropic hormone, the gland not only stops secreting hormone, but also shrivels in size. **Adrenocorticotropic hormone (ACTH)**, **luteinizing hormone (LH)**, and **follicle-stimulating hormone (FSH)** are also tropic hormones produced by the anterior pituitary. The general functions of the anterior pituitary hormones are listed in Appendix K and **FIGURE 16.8**.

Secretions of the anterior pituitary gland are controlled by neurohormones secreted by neurosecretory cells in the hypothalamus. These cells are small and called **parvocellular neurons**. The hypothalamus and anterior pituitary are connected by a specific, dedicated vascular pathway termed the **hypothalamo–hypophysial portal system** (see Figure 16.7). Capillaries of the median eminence coalesce into *portal vessels* that extend to the anterior pituitary, where they branch to form capillary beds around the endocrine cells of the anterior pituitary. The portal vessels are a short, direct path for neurohormones to travel quickly, without dilution, from the capillaries in the median eminence to the anterior pituitary capillaries.

The portal system carries parvocellular neurohormones that control the secretions of anterior pituitary cells (see Figure 16.7).

Some of these neurohormones stimulate secretion of hormones and are called *releasing hormones (RHs)*. Others, called *inhibiting hormones (IHs)*, inhibit secretion of anterior pituitary hormones. Each neurohormone is specific in its actions (see Appendix K). For example, *thyrotropin-releasing hormone (TRH)* from the hypothalamus stimulates the secretion of TSH in the anterior pituitary, and TSH travels through the general circulation to stimulate release of thyroid hormones from the thyroid gland. Like the anterior pituitary tropic hormones, the RHs are also tropic. They both stimulate secretion by anterior pituitary endocrine cells and maintain their vigor.

We can think of the hypothalamo–hypophysial portal system as providing an interface between the brain and much of the endocrine system. The posterior pituitary is an extension of the brain. The anterior pituitary—although not part of the brain—is under the control of the brain and regulates the functions of many other tissues, including several endocrine glands. The vertebrate pituitary gland illustrates how the integrative capabilities of the CNS influence endocrine function. The CNS receives input from a multitude of sensory receptors that monitor conditions both within the animal and in its outside environment. Neurons provide synaptic input to neurosecretory cells, which in turn influence the physiology of other cells, including endocrine cells. (In some other glands, such as the islets of Langerhans, neurons that are not part of the hypothalamo–hypophysial system influence endocrine cells directly). Neural control of the endocrine system is essential, but endocrine control of the nervous system is also important. In the next section, we consider examples of the mutual interactions of both systems.

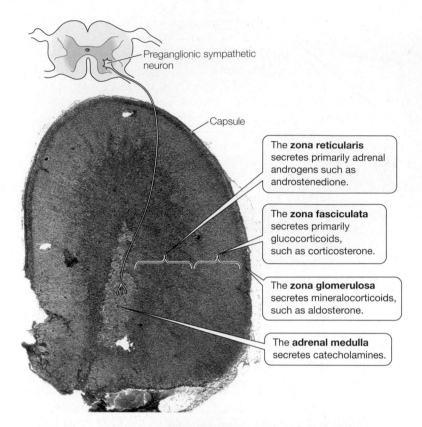

Preganglionic sympathetic neuron

Capsule

The **zona reticularis** secretes primarily adrenal androgens such as androstenedione.

The **zona fasciculata** secretes primarily glucocorticoids, such as corticosterone.

The **zona glomerulosa** secretes mineralocorticoids, such as aldosterone.

The **adrenal medulla** secretes catecholamines.

FIGURE 16.9 The adrenal gland consists of an inner medulla and an outer cortex This section of a mouse adrenal gland shows the different parts characteristic of mammalian adrenal glands. The medulla is homologous to a sympathetic ganglion, except that the postganglionic cells do not possess axons. The cortex surrounding the medulla has three distinct layers, each specialized to secrete a different category of steroid hormones. Because the cells of the zonae reticularis and fasciculata are slightly intermingled, the zona reticularus secretes some glucocorticoids (in addition to androgens) and the zona fasciculata secretes some androgens (in addition to glucocorticoids). The zona glomerulosa secretes only mineralocorticoids. (Courtesy of Judith Wopereis, Smith College.)

Hormones and neural input modulate endocrine control pathways

When the secretions of one endocrine gland act on another in a sequence, endocrinologists speak of the system as an **axis**. For example, the hypothalamus–anterior pituitary–thyroid axis represents a hormonal sequence from TRH to TSH to thyroid hormones. Similar axes are also found in invertebrates. In this section we use the hypothalamus–pituitary–adrenal cortex (HPA) axis to illustrate how rates of endocrine secretion can be *modulated*, or changed, by hormonal and neural influences.

The adrenal gland (**FIGURE 16.9**) secretes several hormones. Here we limit our consideration to the **glucocorticoids** (cortisone, cortisol, and corticosterone), a class of steroid hormones so named because (among other functions) they promote an increase in the blood concentration of glucose ("gluco") and because they are secreted by the adrenal cortex ("cortico"). The main glucocorticoid produced in primates and fish is cortisol, whereas in reptiles (including birds), amphibians, and rodents it is corticosterone (see Box 16.1 on page 456). Stressful or challenging conditions cause increased glucocorticoid secretion. (We consider the mammalian stress response later in this chapter.)

Glucocorticoid secretion is controlled by the HPA axis (**FIGURE 16.10**). Brain neural activity is integrated by neurons that secrete **corticotropin-releasing hormone** (**CRH**) into the capillaries of the median eminence. CRH is carried in the portal system to the anterior pituitary, where it stimulates adrenocorticotropic cells to secrete ACTH into capillaries leading to the general circulation. ACTH is carried to the adrenal cortex, where it stimulates glucocorticoid secretion. Glucocorticoids act at widespread target tissues to influence many physiological processes.

HORMONAL MODULATION Any stage in this control pathway can be modulated. *Negative feedback* (see Figure 16.10) is the most widespread type of **hormonal modulation**. In negative feedback, a hormone causes changes in its control pathway that tend to suppress its own secretion. In the HPA axis, high glucocorticoid levels tend to suppress secretion of CRH by the hypothalamus and ACTH by the anterior pituitary, and also to reduce the responsiveness of the ACTH cells to CRH. As a result, the pituitary secretes less ACTH, and the adrenal cortex receives less stimulus to secrete glucocorticoids. Mechanisms of negative feedback do not reduce hormone secretion to zero, but instead *stabilize* blood concentrations of hormones. Occasionally, hormonal modulation involves *positive feedback*, such as that resulting in the explosive increase in oxytocin secretion during the process of birth (see page 492).

In addition to feedback mechanisms, other types of hormonal modulation can affect endocrine control pathways. For example, hormones that are ancillary parts of a pathway can alter a target gland's response to a particular hormone. In the HPA pathway, vasopressin acts together with CRH to increase the secretion of ACTH from the anterior pituitary. Not all vasopressin-secreting neurosecretory cells in the hypothalamus are magnocellular neurons that extend their axons to the posterior pituitary. Some are parvocellular neurons that terminate their axons on the capillary bed of the median eminence. When released, vasopressin from these neurons circulates through the portal vessels to the anterior pituitary. By itself vasopressin has little effect on the ACTH-secreting cells. However, when these cells receive signals from both vasopressin and CRH, their secretion is greater than it would be under the influence of CRH alone. This sort of effect, in which one hormone can amplify the effect of another, is called **synergism**.

A hormone can influence the effects of another hormone at the same target tissue in three different ways: by synergism (producing an enhanced response such as we have seen with vasopressin and CRH), **permissiveness** (in which the presence of one hormone is *required* for the other to exert its full effect), or **antagonism** (in which one hormone opposes the action of another). One example of permissiveness is that of cortisol permitting the catecholamine norepinephrine to cause constriction (narrowing of the diameter) of blood vessels, a function necessary to maintain normal blood pressure. Because cortisol *must* be present for vasoconstriction to occur, basal levels of this glucocorticoid are necessary for homeostatic regulation of blood pressure. Without treatment, people with a condition called "adrenal insufficiency" (who are unable to secrete adequate amounts of hormones, including cortisol, from

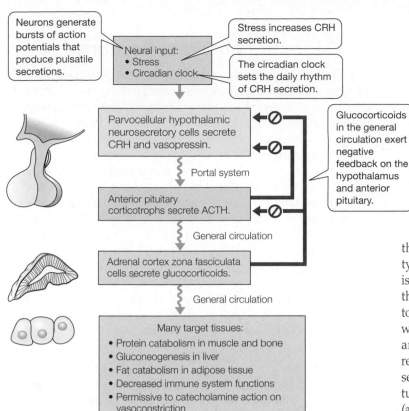

Neurons generate bursts of action potentials that produce pulsatile secretions.

Neural input:
• Stress
• Circadian clock

Stress increases CRH secretion.

The circadian clock sets the daily rhythm of CRH secretion.

Parvocellular hypothalamic neurosecretory cells secrete CRH and vasopressin.

Portal system

Anterior pituitary corticotrophs secrete ACTH.

General circulation

Adrenal cortex zona fasciculata cells secrete glucocorticoids.

General circulation

Glucocorticoids in the general circulation exert negative feedback on the hypothalamus and anterior pituitary.

Many target tissues:
• Protein catabolism in muscle and bone
• Gluconeogenesis in liver
• Fat catabolism in adipose tissue
• Decreased immune system functions
• Permissive to catecholamine action on vasoconstriction

FIGURE 16.10 Hormonal, neural, and feedback mechanisms modulate the action of the HPA axis Glucocorticoids secreted by the adrenal cortex zona fasciculata exert effects on target tissues and also provide negative feedback to the anterior pituitary and the hypothalamus. ACTH from the anterior pituitary exerts negative feedback on the hypothalamus. CRH = corticotropin-releasing hormone; ACTH = adrenocorticotropic hormone.

the adrenal cortex) are at risk of death if they experience a stress, such as a hemorrhage, that requires systemic vasoconstriction.

An example of antagonism is the interaction between insulin and glucagon. We know that insulin from β cells in the pancreas promotes uptake of glucose from the blood by many different tissues. **Glucagon** is a hormone secreted by α cells in the islets of Langerhans, and it functions to oppose the action of insulin: It stimulates the release of glucose and fatty acids into the blood. The balanced actions of these two hormones help maintain stable levels of glucose in the blood.[3] In situations, such as stress, in which higher blood concentrations of glucose are required to respond to a crisis, glucagon secretion increases and insulin secretion decreases. **FIGURE 16.11** shows blood glucose levels in dogs given insulin alone or in combination with glucagon and epinephrine. Epinephrine and glucagon are both antagonists of insulin, and they work synergistically to oppose insulin's action.

NEURAL MODULATION   Neural modulation also affects endocrine control pathways. In the HPA axis, for example, neurons reporting increases or decreases in stress provide synaptic input to CRH neurosecretory cells in the hypothalamus (see Figure 16.10).

[3] Antagonism, in the context of endocrinology, means that two hormones have opposing actions, but it does not indicate their mechanisms of action. For example, opposing hormones may trigger different biochemical pathways in the target cell, or they may induce downregulation of receptors for the opposing hormone in the target cell.

Furthermore, neuronal biological clocks influence hormone secretion. For example, blood levels of cortisol regularly rise and fall in a daily rhythm, being highest in the early morning and lowest during the evening. These cyclic changes in secretion are driven by a circadian clock in the brain that sends input to the hypothalamic CRH neurosecretory cells. Changes in lighting and feeding schedules can perturb these diurnal changes in glucocorticoid secretion, and stress can disrupt the diurnal rhythm.

Finally, synaptic input causes most hypothalamic and pituitary hormones to be secreted *in pulses,* so that there are intermittent brief periods of high hormone concentration in the blood. (When a hormone is continuously present, target cells typically downregulate their receptor molecules for that hormone; it is thought that one advantage of pulsatile release is that it prevents this *desensitization* of target cells.) Synaptic input to the neurosecretory cells governs the pulsatile release of hypothalamic hormones, which in turn causes pulsatile output of pituitary hormones. The amplitude of a pulse of hormone (i.e., the amount of hormone released during a pulse) can vary. For example, the CRH neurons secrete pulses of hormone two or three times per hour. The amplitude of each pulse increases in the early morning in diurnal species (and early evening in nocturnal species) according to input from the circadian clock. It also increases when neural activity signals the presence of one or more stressors. Ultimately, the *combination* of neural modulation and hormonal modulation determines the minute-to-minute blood levels of hormones.

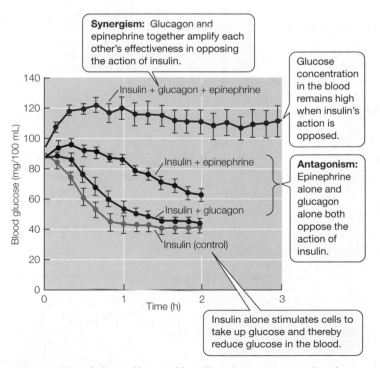

FIGURE 16.11 Interactions of insulin, glucagon, and epinephrine illustrate synergism and antagonism between hormones Blood glucose levels were measured in 29 dogs infused with insulin alone or in combination with glucagon and epinephrine. Glucagon and epinephrine both reduce insulin's stimulation of cells to take up glucose from the blood, and they exert a greater effect when they are present together. Error bars show the standard errors. (After Sacca et al. 1979.)

SUMMARY
Control of Endocrine Secretion: The Vertebrate Pituitary Gland

■ The vertebrate pituitary gland consists of the adenohypophysis (anterior pituitary) and the neurohypophysis (posterior pituitary). Posterior pituitary hormones (vasopressin and oxytocin) are secreted in response to neural activity. Secretions of anterior pituitary hormones are controlled by releasing hormones (RHs) and inhibiting hormones (IHs) from the hypothalamus, which are transported to the anterior pituitary through the hypothalamo–hypophysial portal system.

■ The rate of and pattern of hormone secretion are influenced by a combination of hormonal modulation (such as feedback mechanisms, synergism, permissiveness, and antagonism) and neural modulation (such as sensory input and clock mechanisms).

The Mammalian Stress Response

In this section we use the mammalian stress response (**FIGURE 16.12**) to illustrate and integrate several of the principles of endocrinology discussed in previous sections. The stress response is an adaptation that allows an animal to respond immediately in a generalized way to a threatening or challenging situation. Stressors experienced by animals include being wounded, being exposed to thermal extremes (birds and mammals) or other hostile environmental conditions, being forced to exercise vigorously, and experiencing troublesome social conditions or high levels of emotion. We tend to think of stressors as negative challenges to survival, but in some instances they may heighten experiences in a positive way. In humans, and perhaps other animals as well, stressors over which an individual perceives a sense of control can be rewarding. For example, seeking novel situations generates stress but also facilitates intellectual and emotional growth. Interestingly, feeding and sexual activity—both essential behaviors for biological success—also stimulate the stress system.

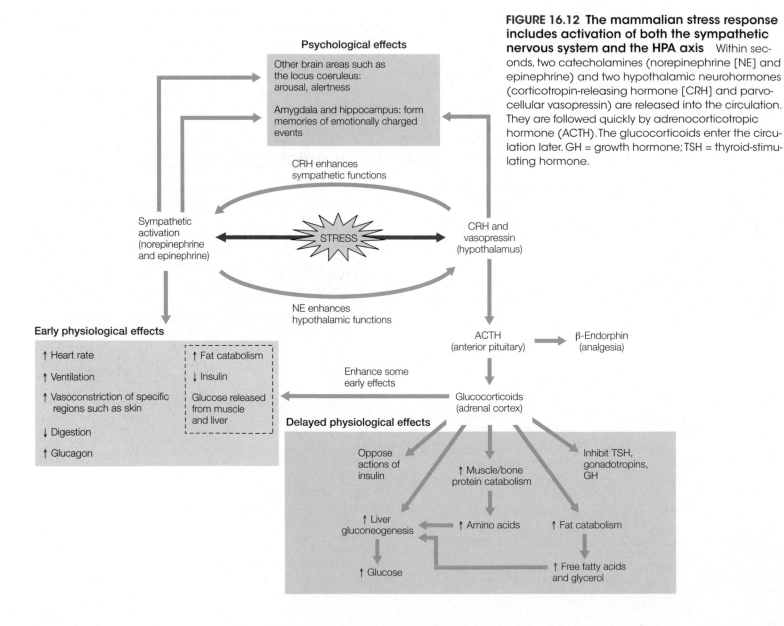

FIGURE 16.12 The mammalian stress response includes activation of both the sympathetic nervous system and the HPA axis Within seconds, two catecholamines (norepinephrine [NE] and epinephrine) and two hypothalamic neurohormones (corticotropin-releasing hormone [CRH] and parvocellular vasopressin) are released into the circulation. They are followed quickly by adrenocorticotropic hormone (ACTH). The glucocorticoids enter the circulation later. GH = growth hormone; TSH = thyroid-stimulating hormone.

During the stress response, heart and breathing rates increase, cognition and alertness are sharpened, metabolic processes release stored energy, oxygen and nutrients are directed to the CNS and to those sites in the body that are stressed the most, and feeding and reproduction are curtailed. All of these changes ensure survival in an acute crisis. Different stressors differentially turn on different components of the stress response; however, any one stressor that is sufficiently potent will turn on a generalized stress syndrome. When the stressor is no longer present, feedback mechanisms ensure that the stress response is turned off. In the classic example of a zebra chased by a lion, the threat is clear, the physiological response is swift, and the episode has a finite end. However, if a stressor (physical or emotional) persists for long periods, physiological responses that are adaptive in the short term become damaging in the long term.

The autonomic nervous system and HPA axis coordinate the stress response to an acute threat

The rat detects the cat and runs. Within seconds of the threat, the rat's sympathetic nervous system releases catecholamines (epinephrine and norepinephrine) from sympathetic nerve terminals and the adrenal medulla, and hypothalamic neurosecretory cells release CRH (and often vasopressin) into the hypothalamo–hypophysial portal system. A few seconds later, the anterior pituitary secretes ACTH and β-endorphin. Thus two output systems, the sympathetic "fight-or-flight" system and the HPA axis, together mount the response to a stressor. Their functions are not independent, but intermingled. For example, in addition to its role as a neurohormone that stimulates ACTH secretion, CRH also acts as a neurotransmitter in other areas of the brain, where it stimulates the sympathetic nervous system. Researchers uncovered this additional role of CRH by injecting it into the brain ventricles of dogs and rats whose pituitary glands had been removed. These animals secreted no ACTH, because the ACTH-secreting cells were gone. However, injected CRH caused increases in blood concentrations of catecholamines from sympathetic neurons and associated increases in blood pressure and heart rate. Experiments such as these reveal that one of the functions of CRH is to link the sympathetic and adrenocortical branches of the stress response. CRH also acts as a neurotransmitter or neuromodulator in the amygdala and hippocampus (which function together to form memories of emotionally charged events).

The two output branches of the stress response are also linked by norepinephrine. The CRH neurosecretory cells in the hypothalamus receive noradrenergic synaptic input from several different nuclei of the brain. Some of these nuclei are innervated by neurons using CRH as their neurotransmitter, so that reciprocal interactions are possible. Like CRH, norepinephrine also provides input to the amygdala and hippocampus. Although researchers understand many CRH and norepinephrine connections in the brain, they do not know what neurocircuitry upstream of these pathways actually turns on the stress response.

The physiological effects of the stress response occur in two phases (see Figure 16.12). The early effects are seen within less than 1 min, when the catecholamines (epinephrine and norepinephrine)[4]

trigger increases in heart and respiration rates, blood pressure, and other sympathetic responses. These changes provide increased blood flow to the skeletal muscles and heart, as well as increased air flow into and out of the lungs as the bronchial airways increase in diameter. Blood vessels to the skin constrict, diverting blood from sites of possible injury. Digestive functions are suppressed. Arousal of the CNS and alertness are promoted. Epinephrine stimulates the release of glucose into the blood by triggering the breakdown of glycogen stored in the liver (see Figure 2.29) and muscles, and it also stimulates the release of fatty acids from lipid stores. Epinephrine in the blood and norepinephrine from sympathetic nerve terminals both inhibit insulin secretion and stimulate glucagon secretion from the islets of Langerhans. Ordinarily, increased glucose in the blood would stimulate insulin secretion, which would promote the uptake of glucose from the blood by all tissues except brain and exercising skeletal muscle. By inhibiting insulin secretion and stimulating glucagon secretion, the catecholamines ensure plentiful levels of glucose in the blood to fuel physical exertion and maintain brain function.

Additional effects occur during the first phase of the stress response. For example, both vasopressin and epinephrine (in addition to CRH) synergistically stimulate the secretion of ACTH. Furthermore, a variety of experiments have shown that, in addition to stimulating glucocorticoid secretion, ACTH facilitates learning, which may contribute to an animal's preparedness in responding to a similar stressor in the future. Finally, ACTH is co-secreted with the endogenous opiate β-endorphin (both are cleavage products of POMC; see page 436). β-Endorphin is also produced by POMC cells in the hypothalamus and may contribute to *analgesia*; that is, it may decrease the animal's perception of pain.

In the second phase of the stress response (see Figure 16.12), glucocorticoids secreted by the adrenal cortex set into motion delayed physiological effects. The full effects of glucocorticoids on target tissues can be detected about 1 h after the stress response is initiated.[5] Glucocorticoids reinforce the actions of the sympathetic nervous system and have additional metabolic effects that facilitate the release of usable sources of energy into the bloodstream. They stimulate the catabolism of protein in muscle and (at high levels) bone, and they stimulate the liver to use the released amino acids to produce glucose in a process called *gluconeogenesis*. The liver cells release this newly formed glucose into the blood. Like epinephrine and norepinephrine, glucocorticoids oppose the action of insulin and ensure fuel availability. Glucocorticoids also stimulate catabolism of fats so that fatty acids can be used as an alternative energy source by all tissues except the brain (which uses only glucose in the short term but can use ketoacids made from fatty acids in the liver during starvation). The metabolic actions of glucocorticoids, coordinated with those of the catecholamines, ensure glucose availability to the brain in the face of required physical exertion and possible enforced fasting (for example, while hiding from a predator or recovering from a wound). The amino acids released by protein catabolism are also available for tissue repair.

In addition to their important metabolic effects, glucocorticoids increase their permissive effect on vasoconstriction stimulated by

---

[4] Catecholamines act quickly because they are stored in vesicles ready for release, and because, as we saw earlier in this chapter, their target cells respond through second-messenger systems that use preexisting proteins.

[5] The effects of glucocorticoids take longer to appear because these steroid molecules must be produced from precursors prior to secretion. Once secreted, they typically exert genomic effects in their target cells, which then synthesize new proteins.

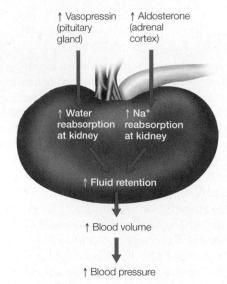

FIGURE 16.13 The regulation of blood volume after blood loss Vasopressin and aldosterone exert effects at the kidney to conserve fluid and thereby help maintain adequate blood volume and blood pressure.

the catecholamines, as we saw earlier. They also inhibit the secretion of gonadotropins (FSH and LH), thyrotropin (TSH), and growth hormone (GH) from the anterior pituitary. Assuming the chase is short, and, for example, the rat pops into its burrow before the cat seizes it, the inhibitory effects of glucocorticoids on reproduction and growth are minimal. Safe in its burrow, the rat experiences diminished sympathetic neural responses, and the glucocorticoid molecules in its general circulation feed back negatively on the CRH and ACTH cells of the HPA axis. Thus the glucocorticoids themselves modulate the stress response. With decreased ACTH in the circulation, the adrenal cortex secretes decreased amounts of glucocorticoids, and concentrations in the blood return to basal levels.

If an animal experiences a severe wound and loses blood, several mechanisms are called into play to correct the loss of blood volume (**FIGURE 16.13**). Blood volume directly affects blood pressure, and blood pressure is necessary to ensure adequate delivery of oxygen and nutrients to the brain and other essential organs. Catecholamines released by the already activated sympathetic nervous system stimulate the heart and blood vessels to maintain blood pressure. Vasopressin released from the posterior pituitary stimulates the nephrons in the kidney to retain water, and aldosterone secreted by cells in the zona glomerulosa of the adrenal cortex acts at the nephrons to retain sodium, which exerts an osmotic holding effect on water. The combined actions of these hormones are aimed at conserving fluid volume, which translates to adequate blood volume and blood pressure. These ideas are further explored later in this chapter (see pages 449–451 and Figures 16.16 and 16.17).

The HPA axis modulates the immune system

The *immune system* works to prevent the invasion of foreign pathogens and to search out and destroy those that sneak through natural barriers. It neutralizes toxins and disposes of dead, damaged, and abnormal cells. During the early phases of the stress response, the catecholamines and glucocorticoids (still at low concentrations) stimulate the immune system (**FIGURE 16.14**). Stimulating the

immune system ensures that a wounded animal that escaped a predator doesn't succumb to bacterial infection from the wound. The immune response often causes inflammation in response to infection or a wound. At higher concentrations (in later stages of the stress response and during recovery), glucocorticoids have anti-inflammatory effects and thus keep the immune system from overreacting and damaging healthy cells and tissues. Researchers now know that a web of chemical pathways allows communication among the nervous, endocrine, and immune systems. These three systems interact continuously to maintain homeostasis.

When certain cells of the immune system detect bacterial or viral pathogens or tumor cells, they release **cytokines**, which bind with specific receptor molecules on target cells to communicate specific excitatory or inhibitory messages. In the stress response, these cytokines travel in the blood to the hypothalamus, where they stimulate CRH neurosecretory cells. This chemical connection directly informs the CNS that the animal has detected an invading stressor. Using cytokines to turn on the stress response accomplishes two goals. First, the physiological responses of the HPA axis, such as the mobilization of energy stores, help the animal fight infection. Second, the glucocorticoids (at high concentrations) inhibit the production of agents that cause inflammation (such as prostaglandins). By muting inflammation, they modulate (keep in check) the immune response.

Studies on experimental animals have indicated, on a whole-organism scale, a relationship between the stress response and inflammatory disease. For example, if the HPA axis of normal rats is disrupted by removing their pituitary gland, they become susceptible to inflammation. Some genetic strains of rats have also shown this association. The Lewis strain of rat, for example, has an impaired HPA axis and secretes little CRH in response to stress; this strain of rat is highly susceptible to inflammatory and autoimmune disease. Injecting glucocorticoids into these animals improves their resistance to inflammation. Conversely, the Fischer strain of rat has an HPA axis that responds excessively to stress; it is highly resistant to inflammatory disease, presumably because it secretes high levels of glucocorticoids.

Communication among the nervous, endocrine, and immune systems is not limited to cytokines interacting with the CNS. Some cells of the immune system can also stimulate glucocorticoid secretion without involving the CNS. In the presence of pathogens, these cells synthesize the hormone ACTH and secrete it into the blood. This ACTH acts at the adrenal cortex in the same way it would have had it been secreted by cells in the anterior pituitary. This single example illustrates the interrelatedness of the three systems. We have seen that cells of the immune system secrete hormones, that cells in the nervous system have receptors for cytokine signals produced by immune cells, and that hormones affect the functions of both the nervous system and the immune system. The field of *neuroimmunomodulation* is ripe for future investigations, which have the potential to reveal the mechanisms underlying "mind–body interactions."

Chronic stress causes deleterious effects

Acute stressors are of short duration. All of the components of the stress response to an acute stressor contribute positively to the survival of a rat getting away unscathed from a cat, or a gazelle narrowly escaping with a bleeding gash inflicted by the jaws of a

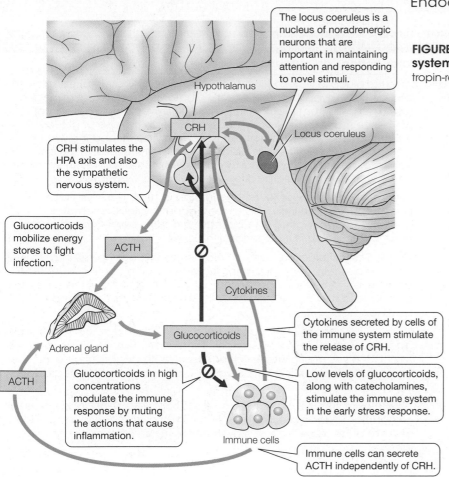

The locus coeruleus is a nucleus of noradrenergic neurons that are important in maintaining attention and responding to novel stimuli.

Hypothalamus

CRH

Locus coeruleus

CRH stimulates the HPA axis and also the sympathetic nervous system.

ACTH

Glucocorticoids mobilize energy stores to fight infection.

Cytokines

Adrenal gland

Glucocorticoids

Cytokines secreted by cells of the immune system stimulate the release of CRH.

ACTH

Glucocorticoids in high concentrations modulate the immune response by muting the actions that cause inflammation.

Low levels of glucocorticoids, along with catecholamines, stimulate the immune system in the early stress response.

Immune cells

Immune cells can secrete ACTH independently of CRH.

FIGURE 16.14 The central nervous system and the immune system interact during the stress response CRH = corticotropin-releasing hormone; ACTH = adrenocorticotropic hormone.

glucocorticoids vary normally according to a circadian rhythm, and studies reveal that many amphibians, mammals, and reptiles (including birds) also show seasonal changes in glucocorticoid concentrations. In many (but not all) of these animals, glucocorticoids are highest during the breeding season. Investigators have proposed that glucocorticoids may be secreted maximally during periods of the year when energy demands are highest, or when they are necessary to support physiological processes associated with season-specific behaviors, or when they are needed to prime vascular, immune, cognitive, and metabolic systems in anticipation of events that occur during specific seasons. (These events could include breeding, disease, predation, and severe weather.) The roles of glucocorticoids in different species and the effects of different stressors on the physiology of animals in their natural environments are topics of active inquiry. An understanding of animals' abilities to respond to the challenges of environmental changes and extremes, including those imposed by human factors, will be important in developing successful conservation, environmental, and animal husbandry programs.

hyena. However, the stress response can be maladaptive when it is induced in animals exposed to stressors for long periods, or exposed repeatedly to stressors such as those related to social rank in dominance hierarchies or to "psychological" stressors.

Researchers and clinicians suspect that continuously constricted blood vessels and retention of salt and fluid when no blood is lost contribute to hypertension (high blood pressure) and other cardiovascular maladies. Prolonged exposure to glucocorticoids also gives rise to other pathological conditions. The catabolic actions of glucocorticoids cause muscle wasting and bone thinning. Their suppression of immune-system functions causes susceptibility to infections and disease. Chronic activation of the HPA axis suppresses reproductive functions, a phenomenon seen in highly trained athletes (both male and female), ballet dancers, starving animals, and people with anorexia nervosa. High levels of glucocorticoids have also been implicated in causing atrophy of dendrites of neurons in the hippocampus, and even shrinkage of the hippocampus itself. Thus continuous or repeated bouts of stress can potentially damage a part of the brain that is especially important in forming memories of emotionally charged events. Because of their far-reaching effects, stress and anxiety are subjects of intense ongoing research.

Plasma glucocorticoid concentrations show seasonal variations

Much of our understanding of the stress response has been gained from experiments using laboratory animals. However, increasing numbers of investigators are now paying attention to the roles of the stress response, and glucocorticoids in particular, in wild animals in their natural environments. We know that blood concentrations of

SUMMARY
The Mammalian Stress Response

■ The stress response is a generalized constellation of physiological changes aimed at ensuring survival when an animal is exposed to real or perceived hostile or challenging conditions.

■ The major physiological actions in the stress response include mobilizing stored energy and inhibiting energy storage; enhancing cardiovascular and respiratory functions; increasing alertness and cognition; inhibiting feeding, digestion, and reproduction; and modulating immune function. If an animal experiences loss of blood, hormones are secreted that promote the retention of water and solutes.

■ The same effects of the stress response that are essential for survival of an animal exposed to an acute stressor can be deleterious during periods of prolonged stress.

■ Wild animals in their natural environments experience seasonal variations in blood concentrations of glucocorticoids.

Endocrine Control of Nutrient Metabolism in Mammals

This is the first of three sections that address endocrine controls of three mammalian physiological processes. We begin with endocrine controls of nutrient metabolism and follow with considerations of salt and water balance and calcium metabolism. Animals acquire

TABLE 16.3 Hormones involved in mammalian nutrient metabolism

Hormone (source)	Main stimulus for secretion	Major actions on nutrient metabolism
Insulin (β cells of the endocrine pancreas)	High blood glucose; high blood amino acids; gastrointestinal hormones; parasympathetic stimulation	Increases glucose uptake from the blood by resting skeletal muscle cells and fat cells (hypoglycemic effect); promotes formation of glycogen (glycogenesis) from glucose in muscle and liver; stimulates fat cells to synthesize triglycerides from glucose; promotes the use of glucose in ATP production; increases uptake of free fatty acids and triglyceride synthesis by fat cells; inhibits breakdown of triglycerides; increases uptake of amino acids by muscle and liver, promotes protein synthesis, and inhibits protein degradation; inhibits gluconeogenesis
Glucagon (α cells of the endocrine pancreas)	Low blood glucose; high blood amino acids; sympathetic stimulation	Increases blood glucose levels by stimulating glycogenolysis and gluconeogenesis in liver cells (hyperglycemic effect); promotes breakdown of triglycerides to increase blood levels of free fatty acids and glycerol; inhibits triglyceride synthesis
Epinephrine (adrenal medulla)	Sympathetic stimulation during stress and exercise	Increases blood glucose levels by promoting glycogenolysis in both liver and muscle cells; often promotes triglyceride degradation to increase blood levels of free fatty acids and glycerol; inhibits insulin release and antagonizes glucose uptake by cells; stimulates glucagon and ACTH secretion
Glucocorticoids (zona fasciculata of adrenal cortex)	Stress	Enhance the actions of glucagon and catecholamines; increase blood glucose levels by antagonizing cellular glucose uptake and promoting glycogenolysis and gluconeogenesis; promote degradation of triglycerides to form free fatty acids and glycerol; promote degradation of proteins to form free amino acids
Thyroid hormones (thyroid gland)	Secreted continuously; increased by TRH and TSH, which are influenced by exposure to cold in human newborns (but not adults) and other animals	Promote oxidation of nutrients (raise metabolic rate); enhance effects of catecholamines on metabolism; stimulate growth hormone secretion and enhance the effects of growth hormone on synthesis of new proteins
Growth hormone (anterior pituitary gland)	Secreted continuously with a circadian rhythm; increased by growth hormone releasing hormone and decreased by somatostatin (influenced by stress, exercise, hypoglycemia)	Promotes protein synthesis and growth; increases uptake of amino acids by liver and muscle; enhances breakdown of triglycerides to increase blood levels of free fatty acids; antagonizes glucose uptake by muscles
Androgens (gonads and adrenal cortex)	Secreted continuously from puberty onward	Promote protein synthesis and growth of muscle
Osteocalcin (osteoblasts of bone)	Bone turnover during growth and remodeling	Stimulates β cell proliferation and insulin secretion; stimulates adipose cells to secrete adiponectin, which increases target cells' sensitivity to insulin; these combined effects influence whole-body glucose metabolism

nutrients when they eat, but many animals do not eat continuously. Still, their cells need nutrients all the time. Moreover, their cells may require the three major classes of nutrients—carbohydrates, lipids, and proteins—in very different proportions than are found in digested foods. Thus, providing continuous and appropriate nutrients to all tissues requires controls not only of feeding but also of storage, mobilization, and molecular interconversions of nutrients. **TABLE 16.3** lists several hormones that influence nutrient metabolism. Insulin and glucagon, in particular, play major roles in nutrient metabolism.

Insulin regulates short-term changes in nutrient availability

Many mammals go through bouts of feeding separated by several hours of not feeding. To prevent alternations of feast and famine at the cellular level, mechanisms are set in motion that favor storage of nutrient molecules immediately after a meal and mobilization

of nutrients from storage depots as the hours pass until the next meal. Insulin is the most important hormone involved in managing short-term fluctuations of nutrient availability.

During digestion, several factors—including rising concentrations of glucose and amino acids in the blood, gastrointestinal hormones, and parasympathetic activity—stimulate the β cells in the pancreatic islets of Langerhans to increase their secretion of insulin. Insulin is the dominant hormone in the blood during the fed state. It favors the *storage* of all three major classes of nutrients. It promotes the uptake of glucose, fatty acids, and amino acids from the blood into tissues, especially muscle and fat. In the case of glucose, for example, insulin binds to its receptor molecule[6] on a target cell and triggers a series of signal transduction events that lead to the incorporation of specific glucose transporter molecules (called GLUT4 transporters) into the membrane of the target cell.

[6] Insulin binds to an enzyme-linked membrane receptor that triggers the action of tyrosine kinase.

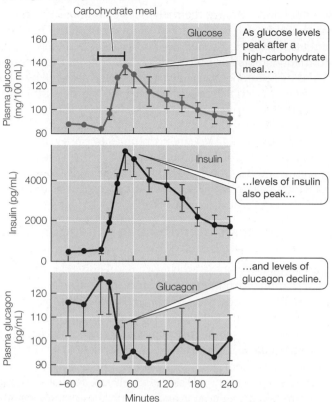

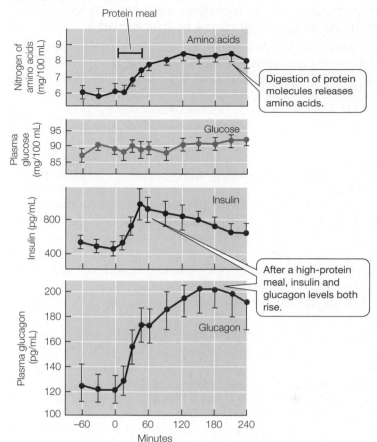

FIGURE 16.15 Hormone and nutrient levels in the blood of healthy human subjects before and after a meal (A) Average levels for 11 subjects who consumed a high-carbohydrate meal of white bread and boiled spaghetti, corn, rice, and potatoes. (B) Average levels for 14 subjects who consumed a high-protein meal of boiled lean beef. Error bars show the standard errors. The experimenters used different scales on the y axes to emphasize the changes in hormone levels. Thus the rise in insulin is notably smaller in (B) than in (A), and the rise of glucagon is more robust in (B) than its decline in (A). (After Müller et al. 1970.)

The GLUT4 transporters allow the target cell to take up glucose by facilitated diffusion. The brain and liver, however, do not depend on insulin to stimulate the uptake of glucose. Instead, they both have constitutive glucose transporter molecules that permit them to take up glucose continuously. In addition, skeletal muscles in the process of exercising do not depend on insulin for glucose uptake either. Although resting skeletal muscles require insulin to promote incorporation of GLUT4 transporters into their cell membranes, the contractions of exercising muscles stimulate *insulin-independent* incorporation of GLUT4 transporters into the membrane. Because insulin causes a decrease in blood glucose levels, it is said to exert a *hypoglycemic effect*.

Insulin also promotes synthesis of nutrient storage molecules: glycogen from glucose, triglycerides (lipids) from fatty acids or glucose, and proteins from amino acids. At the same time insulin promotes the formation of these large molecules, it inhibits the enzymes that break them down.

Insulin secretion decreases as digestion comes to an end, and the ebbing of insulin levels in the blood is often the only endocrine change necessary for a shift to net *mobilization* of nutrients from stores. As insulin declines, breakdown of stored glycogen and lipids begins, and glucose and fatty acids are released into the blood. The overall pattern of insulin secretion—high levels in the fed state and low levels in the unfed state—tends to *stabilize blood concentrations of nutrients*.

FIGURE 16.15A shows the average rise and decline in plasma levels of insulin for several people after a high-carbohydrate meal. With increased glucose in the blood, insulin secretion increases; as blood glucose levels decline, so do blood levels of insulin. The spike in blood glucose following the meal shows that blood nutrient concentrations are not completely stable. However, concentrations remain far more stable than they would without the negative feedback mediated by insulin. No other hormone in the body can lower blood glucose levels. This point is made dramatically clear by people with *diabetes mellitus*, who secrete abnormally low amounts of insulin or who have diminished tissue responsiveness to insulin. After a high-carbohydrate meal, individuals with untreated diabetes experience far higher blood glucose concentrations than those without diabetes. In fact, the blood glucose levels of diabetics become so high that their kidneys are unable to recover all the glucose filtered from the blood in the process of urine formation (see Chapter 29), and glucose is excreted in their urine and wasted. Chronic high levels of glucose cause damage to the eyes, kidneys, blood vessels, and nervous system.

Given its importance, it is not surprising that insulin secretion and the sensitivity of the body's cells to it are influenced by several other hormones over the long term. For example, adiponectin (secreted by adipose tissue) increases cell sensitivity to insulin, and osteocalcin (secreted by osteoblast cells of bone) promotes insulin secretion and also increases cells' sensitivity to it.

Glucagon works together with insulin to ensure stable levels of glucose in the blood

Glucagon is a peptide hormone secreted by the α cells of the pancreatic islets. The main stimuli for its release are low levels of glucose in the blood, sympathetic stimulation of the α cells, and high levels of amino acids in the blood. Glucagon's main effect is to increase the production of glucose and its release into the blood. Because it causes blood glucose levels to rise, glucagon is said to exert a *hyperglycemic effect*, the opposite of insulin's hypoglycemic effect. Glucagon stimulates cells in the liver to break down glycogen in a process called *glycogenolysis* and to release the resulting glucose into the blood. Glucagon also exerts effects opposite to those of insulin with regard to fats. It inhibits triglyceride (lipid) synthesis and stimulates adipose (fat) cells to break down triglycerides into fatty acids and glycerol and release these products into the blood. In addition, glucagon stimulates *gluconeogenesis* in liver cells. In gluconeogenesis, new glucose molecules are formed from noncarbohydrate molecules, mainly amino acids and glycerol obtained from the breakdown of triglyceride molecules. Proteins and fats are both mobilized from body tissues at low insulin levels. Thus amino acids from protein breakdown and glycerol from fat breakdown become available for gluconeogenesis in liver cells.

As blood glucose levels rise, glucagon secretion tends to decrease by negative feedback. Consequently, from meal to meal, both insulin and glucagon contribute to stable blood glucose levels. Under conditions of stress (and exercise), however, sympathetic stimulation causes secretion of epinephrine from the adrenal medulla as well as increased synaptic input to the α cells. Epinephrine has the dual effect of stimulating the α cells to secrete glucagon and inhibiting the β cells from secreting insulin. This arrangement ensures increased glucose availability without hindrance from insulin.

Glucagon is secreted when blood levels of glucose and fatty acids are low, a condition typical of the unfed state. However, the rate of glucagon secretion is not increased by low levels of amino acids. Instead, glucagon secretion increases when blood levels of amino acids are *high*. Therefore, although glucagon is the dominant hormone during the unfed state, it is often secreted during the fed state, depending on the nutrient composition of a meal. When a high-carbohydrate meal is consumed by healthy human subjects, blood levels of glucose rise, insulin secretion increases, and glucagon secretion decreases (see Figure 16.15A). Under these conditions, the low levels of glucagon reinforce the actions of insulin. After a high-protein meal, however, *both* insulin and glucagon rise (**FIGURE 16.15B**). The rise in insulin promotes the incorporation of absorbed amino acids into body proteins. The rise in glucagon under these circumstances has an adaptive advantage because a high-protein meal in itself supplies little glucose, yet the brain's preferred energy source is glucose. Increased glucagon ensures an output of glucose from liver glycogen stores even in the face of high insulin levels.

The interactions between insulin and glucagon in managing the appropriate use and storage of foodstuffs are key to maintaining nutrient homeostasis. We have seen that absorbed nutrients, gastrointestinal hormones, and sympathetic and parasympathetic inputs act at the α and β cells of the islets of Langerhans to influence secretion of glucagon and insulin. As a final consideration, we need to remember that the brain (especially the hypothalamus) continually integrates afferent information provided by secreted hormones and nutrients themselves. The brain receives information about short-term energy availability from the presence of nutrients such as glucose and free fatty acids in the blood. It receives information about long-term energy stores from the presence of circulating hormones such as leptin, which is secreted by adipose cells. Brain neural activity transduces these inputs into efferent signals that coordinate glucose production by the liver, insulin and glucagon secretion in the pancreas, and glucose uptake by muscle cells. Ongoing research continues to clarify and enhance our understanding of the broad, integrative framework involved in regulating body fat stores and blood glucose levels. These research efforts may reveal possibilities for treatment of obesity and diabetes, two major public-health concerns.

Other hormones contribute to the regulation of nutrient metabolism

Other hormones that exert their own regulatory roles in other systems also influence nutrient metabolism. For example, both growth hormone and glucocorticoids act *synergistically* with epinephrine to enhance epinephrine's effect on lipid breakdown. Other hormones play an essential *permissive* role in nutrient metabolism. Background levels of glucocorticoids, for example, are essential for preventing levels of blood glucose from plummeting during fasting and other stresses. Not only do glucocorticoids stimulate glucose formation, but they are also required for glucagon and epinephrine to exert their effects.

Several hormones play key roles in the growth of young animals. Growth hormone, thyroid hormones, and androgens are important because they promote the formation of proteins. For example, the presence of testicular androgens—principally testosterone—during puberty brings about the greater muscular development that occurs in boys as compared with girls. These hormones also work together synergistically to enhance each other's effects. Thyroxine, for example, enhances the effects of growth hormone in young animals.

Hormones also alter nutrient metabolism during exercise and fasting. Both circumstances require mobilization of metabolic fuels from stores. During both, insulin secretion declines. Without insulin's effects, glycogen breakdown, lipid breakdown, protein breakdown, and gluconeogenesis all occur at heightened rates. Glucagon secretion rises during both exercise and fasting and stimulates glycogenolysis and gluconeogenesis in liver cells. Other hormones may also play a role, but exercise and fasting do not necessarily elicit the same hormonal responses. For example, epinephrine secretion increases during exercise but does not consistently increase during fasting. Furthermore, prolonged fasting, but not exercise, often leads to a decline in thyroid hormone levels; this decline lowers metabolic demands and conserves fuels.

SUMMARY
Endocrine Control of Nutrient Metabolism in Mammals

- Insulin is secreted when nutrient molecules are abundant in the blood. It exerts a hypoglycemic effect by promoting uptake and storage of nutrients and inhibiting degradation of glycogen, lipids, and proteins. In the absence of insulin, nutrient molecules are mobilized to enter the blood from storage sites.

■ Glucagon is secreted when blood glucose levels are low. It exerts a hyperglycemic effect by stimulating the breakdown of glycogen (glycogenolysis), the breakdown of triglyceride molecules, and the formation of glucose from noncarbohydrate sources (gluconeogenesis).

■ Growth hormone, glucocorticoids, epinephrine, thyroid hormones, and androgens play permissive and synergistic roles in nutrient metabolism.

Endocrine Control of Salt and Water Balance in Vertebrates

We devote the last part of this book (Chapters 27–30) to the strategies animals use to maintain salt and water balance in a variety of environments. In this section we explore the endocrine controls employed by vertebrates to maintain appropriate body fluid volume and appropriate concentrations of salts. Salts dissociate into ions in solution and therefore are also referred to as electrolytes. Because dissolved particles exert osmotic pressure that holds water, they have a direct effect on fluid volume in the animal. Furthermore, fluid volume has a direct effect on arterial blood pressure, which is essential for ensuring delivery of respiratory gases and nutrients to cells and removal of metabolic by-products (see Chapter 25). Therefore the hormones involved in salt and water balance have far-reaching effects. The dominant salt in the extracellular fluid is NaCl, which dissociates into Na^+ and Cl^- ions. The main hormones involved in salt and water balance affect the movements of Na^+ and water. Cl^- typically follows Na^+ passively.

Antidiuretic hormones conserve water

Earlier in this chapter we discussed vasopressin (also called antidiuretic hormone, or ADH), which is produced by magnocellular neurosecretory cell bodies in the hypothalamus and released from their axon terminals in the posterior pituitary gland. This hormone acts to conserve water by preventing the production of a large volume of urine. All major classes of vertebrates produce hormones with antidiuretic action. Most mammals use *arginine vasopressin* (AVP), but some pigs and their relatives use *lysine vasopressin* (LVP), and nonmammalian vertebrates use the closely related *arginine vasotocin* (AVT) (see Table 16.2). In all vertebrates, the target tissue of these hormones is the nephron of the kidney. Antidiuretic hormones stimulate the reabsorption of water from the lumen of the nephron. This means that instead of being excreted in the urine, water is returned to the extracellular fluid.

The action of AVP has been studied extensively in mammals. Its effect is to stimulate the incorporation of specific **aquaporin (AQP)** (water channel) molecules (see Chapter 5; page 126) into the membranes of epithelial cells in the collecting duct of the nephron (**FIGURE 16.16**). Different types of aquaporins exist permanently

in various regions of the nephron. However, AQP-2 molecules are present in the apical membranes (those facing the lumen) of the cells of the collecting duct only when vasopressin is present. The epithelial cells of the tubules of the nephron are connected by *tight junctions*, which prevent movement of substances, including water, between cells. Thus water in the lumen of the tubule is destined for excretion unless it can pass through the epithelial cells back into the interstitial fluid and plasma. Receptor molecules for AVP are located on the basal side of the cells. When AVP is secreted from the posterior pituitary, it travels in the general blood circulation to the kidneys and binds to these receptors. Through a second-messenger system, AVP stimulates the movement of AQP-2 molecules from intracellular storage vesicles to the apical membrane facing the lumen. Experiments show that when exposed to AVP, the epithelial cells begin to increase their permeability to water within 1 min, and reach peak permeability in about 40 min. Water moves out of the lumen osmotically. It passes into the epithelial cell and out the basal end of the cell through a different type of AQP channel (AQP-3) that is always present and open.

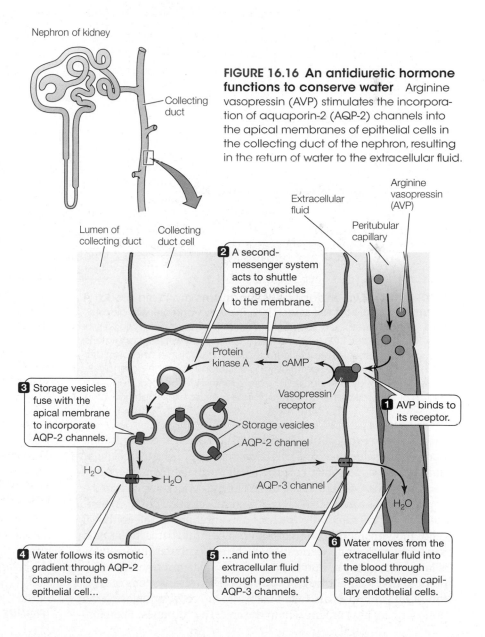

FIGURE 16.16 An antidiuretic hormone functions to conserve water Arginine vasopressin (AVP) stimulates the incorporation of aquaporin-2 (AQP-2) channels into the apical membranes of epithelial cells in the collecting duct of the nephron, resulting in the return of water to the extracellular fluid.

Nephron of kidney

Collecting duct

Lumen of collecting duct

Collecting duct cell

Extracellular fluid

Peritubular capillary

Arginine vasopressin (AVP)

2 A second-messenger system acts to shuttle storage vesicles to the membrane.

Protein kinase A ← cAMP

Vasopressin receptor

1 AVP binds to its receptor.

3 Storage vesicles fuse with the apical membrane to incorporate AQP-2 channels.

Storage vesicles

AQP-2 channel

H_2O → H_2O

AQP-3 channel

H_2O

4 Water follows its osmotic gradient through AQP-2 channels into the epithelial cell…

5 …and into the extracellular fluid through permanent AQP-3 channels.

6 Water moves from the extracellular fluid into the blood through spaces between capillary endothelial cells.

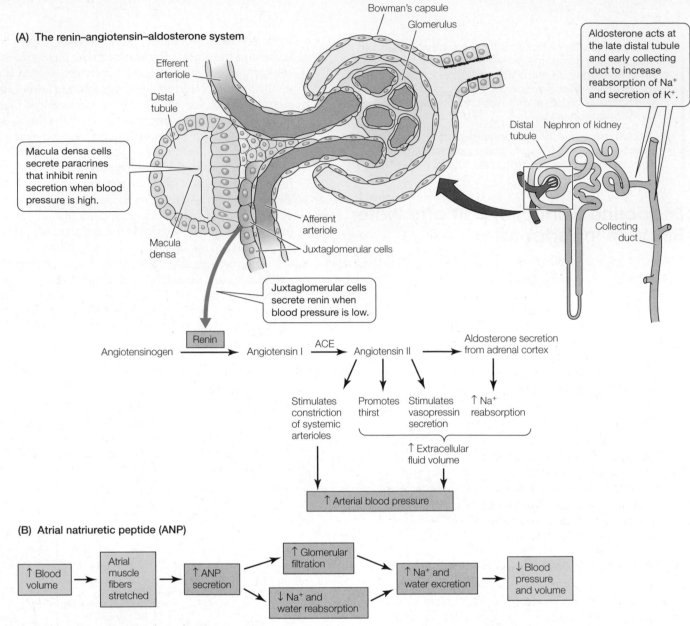

(A) The renin–angiotensin–aldosterone system

Bowman's capsule

Glomerulus

Efferent arteriole

Distal tubule

Macula densa cells secrete paracrines that inhibit renin secretion when blood pressure is high.

Macula densa

Afferent arteriole

Juxtaglomerular cells

Juxtaglomerular cells secrete renin when blood pressure is low.

Aldosterone acts at the late distal tubule and early collecting duct to increase reabsorption of Na⁺ and secretion of K⁺.

Distal tubule   Nephron of kidney

Collecting duct

Angiotensinogen → Renin → Angiotensin I → ACE → Angiotensin II → Aldosterone secretion from adrenal cortex

Stimulates constriction of systemic arterioles / Promotes thirst / Stimulates vasopressin secretion / ↑ Na⁺ reabsorption

↑ Extracellular fluid volume

↑ Arterial blood pressure

(B) Atrial natriuretic peptide (ANP)

↑ Blood volume → Atrial muscle fibers stretched → ↑ ANP secretion → ↑ Glomerular filtration / ↓ Na⁺ and water reabsorption → ↑ Na⁺ and water excretion → ↓ Blood pressure and volume

FIGURE 16.17 Two endocrine mechanisms maintain blood volume and blood pressure   (A) The renin–angiotensin–aldosterone system corrects low extracellular fluid volume and low blood pressure. Renin initiates the steps that lead to the production of angiotensin II, which exerts several effects to increase blood volume and pressure and also stimulates the secretion of aldosterone from the adrenal cortex. The juxtaglomerular cells secrete renin in response to reduced stretch of the wall of the afferent arteriole (which results from low blood pressure) and to sympathetic stimulation. ACE = angiotensin-converting enzyme. (B) Atrial natriuretic peptide (ANP) corrects high extracellular fluid volume and high blood pressure. Specialized muscle cells in the atria of the heart secrete ANP when they are stretched. ANP inhibits reabsorption of Na⁺ at the nephron and also inhibits secretion of vasopressin and renin to reduce extracellular fluid volume and blood pressure.

When the extracellular fluid has a high osmotic concentration (for example, due to excess salt consumption) or the extracellular fluid volume is low (such as due to hemorrhage from a wound), neurons in the CNS stimulate the AVP neuroendocrine cells in the hypothalamus to secrete AVP. In the presence of AVP, AQP-2 channels are incorporated into the apical membranes of the epithelial cells, allowing reabsorption of water. When the extracellular fluid has a low osmotic concentration or the extracellular fluid volume is large (for example, due to excess water consumption), the AVP neuroendocrine cells do not secrete AVP. In the absence of AVP, the AQP-2 channels are taken back into the cells' cytoplasm, and none (or very few) are present in the apical membranes. Therefore

water is not reabsorbed—no matter what the osmotic gradient is. The water is excreted because it cannot escape the lumen.

The renin–angiotensin–aldosterone system conserves sodium and excretes potassium

Aldosterone is a steroid hormone produced and secreted by the zona glomerulosa cells of the adrenal cortex (see Figures 16.2 and 16.9). It is called a *mineralocorticoid* because it contributes to the balance of minerals—mainly sodium and potassium—in the body. The main target tissues of aldosterone are epithelial cells of the late distal tubule and early collecting duct of the nephron (**FIGURE 16.17A**). Aldosterone stimulates these cells to secrete K⁺

into the lumen (for excretion) and to reabsorb Na⁺ (take it out of the lumen and return it to the interstitial fluid and plasma).[7] When aldosterone is secreted, it enters the target cells and binds to its receptor in the cytoplasm. The hormone–receptor complex enters the nucleus and initiates transcription of the DNA to synthesize new Na^+–K^+-ATPase pumps and Na⁺ and K⁺ channels for incorporation into the cell membrane. The new pumps and channels speed the reabsorption of Na⁺ and the secretion of K⁺. High plasma K⁺ stimulates aldosterone secretion, which stimulates excretion of K⁺. This function ensures low levels of K⁺ in the extracellular fluid, a condition essential to maintaining cell resting membrane potentials. Low arterial blood pressure is a very powerful stimulus for aldosterone secretion. One way to correct low blood pressure is to increase the extracellular fluid volume (or preserve it, in the case of blood loss). Na⁺ (and Cl⁻) within the extracellular fluid exerts an osmotic "hold" on water. This effect, combined with increased reabsorption of water, increases the volume of the extracellular fluid and therefore increases arterial blood pressure.

This mechanism is called into play during the stress response if an individual is wounded. When low blood pressure is detected, specialized cells in the kidney secrete a substance called **renin**. These cells, called **juxtaglomerular (JG) cells** or granular cells, are modified smooth muscle cells located in the walls of the afferent arterioles leading to the glomerulus of the nephron (see Figure 16.17A). The JG cells are sensitive to stretch. When they are stretched *less* (as in the case of low blood pressure), they increase their secretion of renin. The JG cells are also innervated by sympathetic nerve fibers and receive sympathetic stimulation to secrete renin when blood pressure is low. Finally, renin secretion is regulated by paracrines from the *macula densa*, a group of specialized cells that monitor fluid and solute flow through the distal tubule. When blood pressure is low, the macula densa cells reduce their paracrine-mediated inhibition of the JG cells. Mammals, reptiles (including birds), amphibians, and bony and cartilaginous fish all have JG cells that secrete renin. Its actions have been studied most thoroughly in mammals.

Renin is the first substance in a chain that leads to the secretion of aldosterone (see Figure 16.17A). Renin is secreted into the blood and, in mammals, has a half-life of about 20 min. It interacts enzymatically with a large protein called **angiotensinogen**, which is produced by the liver and is continuously present in the blood. Renin cleaves a peptide bond of angiotensinogen to produce a ten–amino acid peptide called **angiotensin I**. Angiotensin I itself does not have a physiological effect. Circulating in the blood, angiotensin I encounters a membrane-bound enzyme called **angiotensin-converting enzyme (ACE)**, which is produced by the endothelial cells of the blood vessels. ACE cleaves two amino acids from angiotensin I to make the eight–amino acid peptide hormone **angiotensin II**. Angiotensin II stimulates the secretion of aldosterone from the adrenal cortex and causes constriction of systemic arterioles. In addition, angiotensin II stimulates secretion of vasopressin and promotes thirst and drinking. These actions contribute to raising the extracellular fluid volume and therefore raising arterial blood pressure.

[7] Aldosterone also has other target tissues, including the urinary bladder (amphibians, reptiles other than birds, and mammals), sweat glands (mammals), salt glands (reptiles, including birds), salivary glands (mammals), and intestine (mammals, birds, and amphibians). At all of these target tissues, its effect is to retain Na⁺ in the body.

Interestingly, the toxin of the Brazilian pit viper *Bothrops jararaca* kills by causing a catastrophic drop in blood pressure. A component in the toxin specifically blocks the action of ACE and therefore halts production of angiotensin II and secretion of aldosterone. Knowing this physiology, researchers at the pharmaceutical firm Bristol-Myers Squibb studied the molecular structure of the most active component in the venom, which gave them an understanding of the structure of the active site of ACE. They then designed a drug that interacts specifically with the active site and prevents its enzymatic action. This drug, captopril, was the first "ACE-inhibitor" drug used to treat hypertension (high blood pressure).

Atrial natriuretic peptide promotes excretion of sodium and water

Atrial natriuretic peptide (ANP) is produced by specialized muscle cells in the atria of the heart and is secreted when the heart muscle cells are stretched. High arterial blood pressure and large amounts of sodium, which lead to expanded extracellular fluid volume, stimulate ANP secretion (**FIGURE 16.17B**). ANP has been identified in most vertebrates. It acts on the distal parts of nephrons to inhibit the reabsorption of Na⁺ and increase its excretion. ANP also inhibits the secretion of vasopressin, renin, and aldosterone, thereby counteracting the hormones that conserve water and Na⁺. Finally, ANP increases the rate of filtration of blood at the glomerulus, which causes more fluid to be excreted. Consequently, by its many effects, ANP reduces the extracellular fluid volume and corrects conditions of high arterial blood pressure. The mechanisms by which ANP exerts these actions are not yet understood and are being actively investigated. ANP has also been found to stimulate the excretion of Na⁺ from the gills of bony fish.

> **SUMMARY**
> **Endocrine Control of Salt and Water Balance in Vertebrates**
>
> ■ Hormones continuously regulate the balance of salt and water in vertebrates.
>
> ■ Vasopressins are peptide neurohormones that stimulate the conservation of water.
>
> ■ Aldosterone is a steroid hormone that stimulates the excretion of K⁺ and conservation of Na⁺. It is part of the renin–angiotensin–aldosterone system that is set in motion under conditions of low arterial blood pressure.
>
> ■ Atrial natriuretic peptide (ANP) exerts many different actions, all of which stimulate the excretion of Na⁺ and water.

Endocrine Control of Calcium Metabolism in Mammals

Calcium is finely regulated in both the extracellular and intracellular fluids. We know that the intracellular Ca^{2+} concentration is kept very low ($<10^{-7} M$) so that transient influxes of Ca^{2+} from intracellular organelles or the extracellular fluid will trigger cellular responses such as muscle contraction or exocytosis. The extracellular concentration of Ca^{2+} is higher than that in the cytoplasm (~1.2 millimolar [mM]), but it too is maintained within very narrow limits.

Low levels of Ca^{2+} cause hyperexcitability of nerve and muscle membranes, which can produce muscle twitches and sensations of tingling and numbness. High levels of Ca^{2+} cause reduced excitability that produces hyporeflexia (slower reflexes) and lethargy. Three hormones—*parathyroid hormone* (*PTH*), active vitamin D (*1,25-dihydroxycholecalciferol* or *1,25[OH]₂D₃*), and *calcitonin*—acting at bone, the intestine, and the kidney regulate the extracellular Ca^{2+} concentration (**FIGURE 16.18**). The total amount of calcium in the blood plasma includes that bound to proteins (mainly albumin), in complexes with anions such as phosphate, sulfate, and citrate, and free as ionized Ca^{2+}. Only free Ca^{2+} (about half the total calcium) is biologically active.

Parathyroid hormone increases Ca^{2+} in the blood

In mammals, PTH, an 84–amino acid peptide hormone, is secreted by *chief cells* of the parathyroid glands. The paired parathyroid glands are located on the dorsal surface of the thyroid gland. The chief cells produce PTH continuously and secrete it by exocytosis. Their rate of secretion depends on the concentration of Ca^{2+} in the extracellular fluid, increasing when extracellular Ca^{2+} is low and decreasing when it is high. The chief cells possess a G protein–coupled calcium-sensing receptor on their cell membrane that acts via the enzyme phospholipase C to *inhibit* the secretion of PTH. Thus less PTH is secreted when extracellular Ca^{2+} is high. When extracellular Ca^{2+} is low, the inhibition is lifted and more PTH is secreted.

PTH binds to several different target tissues to cause an increase in extracellular Ca^{2+}. It exerts its effect at all targets by binding to cell-surface G protein–coupled receptors associated with the enzyme adenylyl cyclase, thereby stimulating production of the second messenger cAMP (see Figure 2.30). In bone, PTH acts synergistically with active vitamin D to cause bone resorption (breakdown), a process that releases both Ca^{2+} and phosphate into the extracellular fluid. In the kidney, PTH stimulates certain cells of the nephron to reabsorb Ca^{2+}, and other cells to excrete phosphate anions. This combined effect ensures that reabsorbed Ca^{2+} does not form a complex with phosphate but instead remains free and biologically active, neatly sidestepping the potential problem raised by the release of both Ca^{2+} and phosphate during bone resorption. Finally, PTH stimulates certain cells of nephrons to convert inactive vitamin D to its active form. One action of active vitamin D is to promote Ca^{2+} absorption from the intestine.

Active vitamin D increases Ca^{2+} and phosphate in the blood

Vitamin D is a steroid hormone. The inactive form of vitamin D (cholecalciferol, also called vitamin D₃) is produced in skin cells by the action of UV light; it is also obtained from the diet and from vitamin supplements. Cholecalciferol circulates to the liver, where it is converted to 25-hydroxycholecalciferol, and then to the kidney, where enzymatic actions in cells of the nephron convert it to its active form, $1,25(OH)_2D_3$. All of these forms of vitamin D are circulated in the blood plasma bound to carrier proteins. PTH, low extracellular Ca^{2+}, and low phosphate all stimulate conversion of vitamin D₃ to its active form. Active vitamin D targets three main

types of cells: epithelial cells of the small intestine, osteoblasts of bone, and cells in the kidney nephron (see Figure 16.18). At the intestinal epithelium, $1,25(OH)_2D_3$ exerts its action through gene transcription that leads to the synthesis and activation of an intracellular calcium-binding protein, *calbindin*. Calbindin facilitates absorption of Ca^{2+} from the intestinal lumen into the blood. Another action of $1,25(OH)_2D_3$ is to act synergistically with PTH to release Ca^{2+} and phosphate from bone. Both hormones target bone-forming cells called *osteoblasts*, but the peptide PTH binds to a membrane receptor, and $1,25(OH)_2D_3$ binds to an intracellular receptor. Through separate actions, the two hormones synergistically lead to the production and release of cytokines from the osteoblasts. The cytokines are paracrine signals that trigger the differentiation and activation of *osteoclasts*. Active osteoclasts break down bone (resorb it) and, in the process, liberate Ca^{2+} and phosphate ions. Finally, $1,25(OH)_2D_3$ stimulates the reabsorption of both Ca^{2+} and phosphate ions at the kidney nephron. However, the actions of PTH on regulating these ions at the nephron appear to be much more robust than those of $1,25(OH)_2D_3$.

Calcitonin opposes bone resorption and decreases Ca^{2+} and phosphate in the blood

Calcitonin is a 32–amino acid peptide hormone secreted by *C* (*clear*) *cells* in the thyroid gland in mammals (see Figure 16.18B). C cells possess a G protein–coupled membrane calcium-sensing receptor similar to that in PTH-secreting cells. However, whereas high extracellular Ca^{2+} *inhibits* the release of PTH, it *stimulates* the release of calcitonin. Calcitonin acts directly at the osteoclasts to inhibit bone resorption and the release of Ca^{2+} and phosphate ions. Despite laboratory experiments that have shown these cellular effects, observations in humans strongly suggest that calcitonin does not affect in any measurable way the functions of osteoclasts in the skeletons of adult humans. For example, removing the thyroid gland (the source of calcitonin) while sparing the parathyroid glands has no effect on bone metabolism. However, in other mammals (and perhaps also in humans), calcitonin plays an important role in preventing bone resorption in pregnant and lactating females. It thereby protects the maternal skeleton and diverts dietary calcium to the fetus or lactating neonate. At the kidney nephron, calcitonin facilitates the excretion of both Ca^{2+} and phosphate.

Calcitonin is used clinically to treat diseases such as osteoporosis and Paget's disease (a disease characterized by overactive osteoclasts that weaken the bones by excessive resorption). Interestingly, the calcitonin used to treat humans comes from fish, salmon in particular. Fish also regulate Ca^{2+} in the extracellular fluid within narrow limits, and they use calcitonin to inhibit the resorption of Ca^{2+} from bone. Salmon calcitonin has a similar, but not identical, amino acid structure to that of the mammalian hormone. It is favored for therapeutic use because it binds to the same receptor molecule on target cells as mammalian calcitonin does, but it binds more tightly and is degraded more slowly (that is, it has a longer half-life). Currently bisphosphonates, which cause osteoclasts to undergo apoptosis, are also used as treatments to slow bone thinning.

FIGURE 16.18 Parathyroid hormone (PTH), active vitamin D, and calcitonin finely regulate Ca²⁺ in the extracellular fluid   Negative feedback mechanisms modulate these hormones' rates of secretion. The text explains that these hormones also regulate extracellular phosphate. (A) When extracellular Ca²⁺ is low, secretion of PTH and active vitamin D increases. These hormones act at target tissues to increase extracellular Ca²⁺. (B) When extracellular Ca²⁺ is high, calcitonin secretion increases, and secretion of PTH and active vitamin D decreases. Calcitonin acts at its target tissues to decrease extracellular Ca²⁺.

SUMMARY
Endocrine Control of Calcium Metabolism in Mammals

- Three hormones finely regulate the extracellular concentration of calcium ions in mammals: parathyroid hormone (PTH), active vitamin D (1,25-dihydroxycholecalciferol, or 1,25[OH]₂D₃), and calcitonin.

- The peptide PTH is secreted by chief cells of the parathyroid glands when extracellular Ca²⁺ is low. PTH stimulates cells in the nephron of the kidney to reabsorb Ca²⁺ and excrete phosphate and also to increase the conversion of inactive vitamin D to active vitamin D. In concert with active vitamin D, PTH stimulates bone resorption. These functions all contribute to increased Ca²⁺ in the extracellular fluid.

- Active vitamin D is formed when extracellular Ca²⁺ is low. It is a steroid and binds to intracellular receptors to influence gene transcription in its target tissues. It promotes absorption of dietary Ca²⁺ across the intestinal epithelium, the resorption of bone, and the reabsorption of both Ca²⁺ and phosphate in the nephron. Its actions increase Ca²⁺ and phosphate in the extracellular fluid.

- The peptide calcitonin targets osteoclasts in bone to inhibit bone resorption. At the kidney nephron, calcitonin increases the excretion of calcium and phosphate ions. These actions decrease calcium and phosphate in the extracellular fluid.

Endocrine Principles in Review

In the examples so far described in this chapter, we have seen the far-reaching and complex interrelationships between the endocrine system and other systems within the organism. In future chapters we will present comparable examples of hormonal regulation that are woven into the tapes-

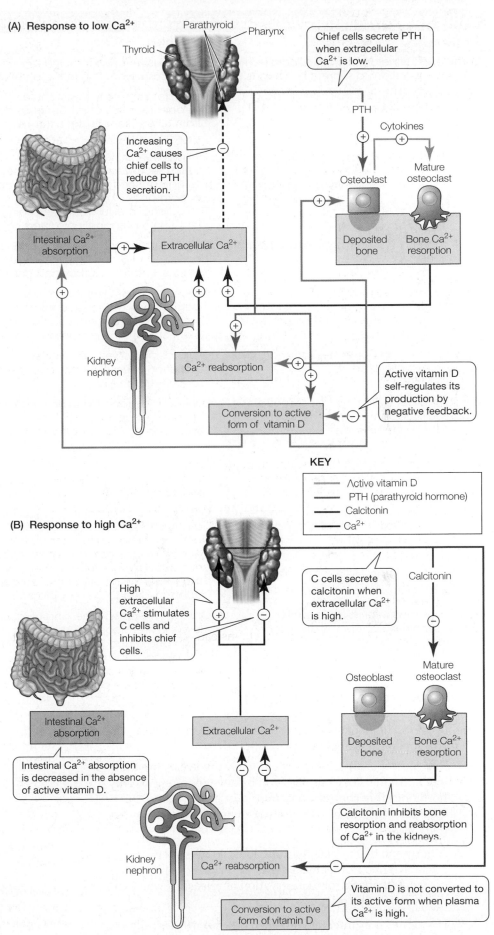

TABLE 16.4 Generalizations about the complexity of endocrine and neuroendocrine function

Feature	Examples
Some endocrine organs are specialized for a single function: to secrete hormones.	The exclusive function of the thyroid gland is to secrete hormones. The exclusive function of the anterior pituitary gland is to secrete hormones.
Some organs secrete hormones and also perform other functions.	The testes secrete testosterone and also produce sperm cells. The pancreas secretes hormones from specific endocrine cells (insulin from β, glucagon from α, and somatostatin from δ cells) and also secretes digestive enzymes from exocrine structures.
A single endocrine organ may secrete more than one hormone, and different control mechanisms may control their secretion.	The anterior pituitary gland consists of discrete populations of endocrine cells, each of which secretes a specific hormone. Secretion is controlled by hypothalamic RHs and IHs. The adrenal cortex secretes glucocorticoids and mineralocorticoids under the control of hormones. The adrenal medulla secretes epinephrine and norepinephrine in response to synaptic input from preganglionic fibers of the sympathetic nervous system.
More than one tissue may secrete the same hormone.	Both the pancreas and the hypothalamus secrete the hormone somatostatin (also called GHIH at the hypothalamus). Both the anterior pituitary and certain immune system cells secrete ACTH.
A single hormone may affect the functions of more than one target cell, and the actions of the target cells may produce different outcomes.	Glucocorticoids stimulate protein degradation in muscle, triglyceride degradation in adipose tissue, and gluconeogenesis in the liver. Oxytocin causes contraction of the uterine smooth muscle during birth and milk ejection from the mammary gland during suckling.
Secretion rates of hormones may vary over time in a cyclic pattern.	Glucocorticoid secretion varies on a diurnal cycle (influenced by the circadian clock). Secretion of estrogens and progesterone varies over the estrous or menstrual cycles of female mammals.
A single target cell may express receptor molecules for, and therefore respond to, more than one hormone.	Liver cells respond to insulin by taking up glucose and converting it into glycogen for storage (glycogenesis). The same liver cells respond to glucagon by breaking down glycogen (glycogenolysis), diminishing glycogen synthesis, and producing glucose from noncarbohydrate molecules (gluconeogenesis).
A single endocrine or neuroendocrine cell may secrete more than one hormone.	Because of the particular enzymes it uses in posttranslational processing, a POMC-expressing endocrine cell of the anterior pituitary secretes ACTH, β-endorphin, and other fragments.
The same molecule may be a hormone in some contexts and a different type of chemical signal in others.	CRH is a neurohormone released into the hypothalamo–hypophysial portal system and a neurotransmitter at other sites in the CNS. Norepinephrine is a hormone when released into the blood from the adrenal medulla and a neurotransmitter when released from postganglionic fibers of the sympathetic nervous system. Although the chemical nature of the signaling molecule remains the same, the nature of the response is determined by the mechanisms set in motion upon binding to the receptor molecule.
Hormones may interact with each other synergistically, permissively, or antagonistically.	Glucagon, epinephrine, and glucocorticoids interact synergistically to increase glucose levels in the blood. Insulin and glucagon exert antagonistic effects that influence glucose concentration in the blood: Insulin decreases glucose in the blood, whereas glucagon increases it. Glucocorticoids are necessary to permit norepinephrine to cause contraction of smooth muscle cells in the walls of small blood vessels.

try of overall homeostasis. Five prevailing patterns of hormonal control appear repeatedly:

1. The control of any single system likely involves more than one hormone. For example, fluid balance requires not only hormones that control water release or conservation, but also other hormones that control solute concentrations in body fluids.

2. A hormone that affects the functions of one system probably affects other systems as well. For example, a hormone that controls conservation or release of water will directly affect blood pressure.

3. Hormones may interact with each other synergistically, permissively, or antagonistically.

4. Most endocrine controls are inextricably associated with neural controls. We also know that the immune system interacts with the endocrine and nervous systems in some arenas; future research will probably reveal that all three systems are more intimately related than we now understand.

5. Many molecules that function as hormones in one context function as different types of chemical signals in different contexts.

This list of five patterns of hormonal control gives an overview, but it is possible to make further generalizations about the complexity of endocrine function. **TABLE 16.4** presents several specific ideas to keep in mind when considering endocrine function.

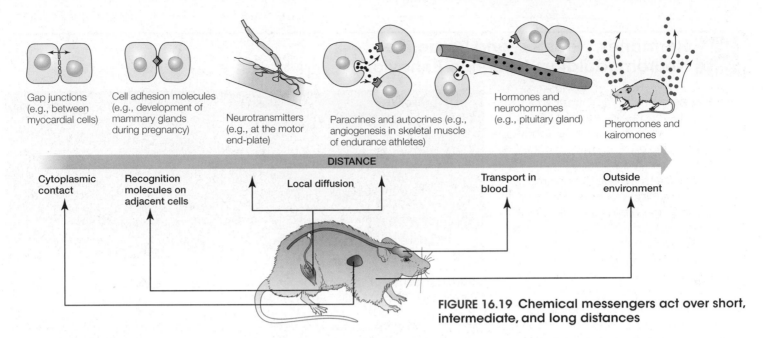

Gap junctions (e.g., between myocardial cells)

Cell adhesion molecules (e.g., development of mammary glands during pregnancy)

Neurotransmitters (e.g., at the motor end-plate)

Paracrines and autocrines (e.g., angiogenesis in skeletal muscle of endurance athletes)

Hormones and neurohormones (e.g., pituitary gland)

Pheromones and kairomones

DISTANCE

Cytoplasmic contact

Recognition molecules on adjacent cells

Local diffusion

Transport in blood

Outside environment

FIGURE 16.19 Chemical messengers act over short, intermediate, and long distances

Chemical Signals along a Distance Continuum

Cells communicate with one another by a variety of mechanisms that can be broadly categorized into six main groups based on the distances involved (**FIGURE 16.19**):

1. *Gap junctions* (see page 339) are formed by connexon protein channels between adjacent cells. When these channels are open, they allow ions and other small molecules to diffuse directly from one cell to the next.

2. *Cell adhesion molecules* (*CAMs*) on the external surface of cell membranes play important roles in signaling between adjacent cells involved in embryonic development, wound repair, and cellular growth and differentiation.

3. *Neurotransmitters* (see Chapter 13) are released by presynaptic neurons in response to electrical signals. They diffuse across a narrow synaptic gap to interact with receptor molecules on a postsynaptic cell, which may be a neuron, muscle cell, or endocrine cell.

4. *Paracrines* and *autocrines* diffuse relatively short distances to influence cells in the local environment—including themselves, in the case of autocrines.

5. *Hormones* and *neurohormones* are carried in the blood and specialized for long-distance communication within the animal. (From our study of the stress response, we know that *cytokines* also communicate across long distances; however, they are not always blood-borne, and they function locally as well.)

6. Some chemical signals act outside the animal. For example, animals of the same species communicate with **pheromones**, whereas animals detect **kairomones** to obtain chemical information about members of a different species. Chemical signals that act outside the body on members of the same or different species are also called *ectocrines*.

Having previously considered most of these classes of intercellular communication, we will now briefly turn our attention to examples of local chemical messengers—paracrines—and the external chemical messengers—pheromones and kairomones.

Paracrines are local chemical signals distributed by diffusion

Two examples of locally acting chemical messengers are *cytokines* and *neuromodulators*. *Cytokines* are peptides or proteins secreted by many different types of cells. They are made on demand when required to play a regulatory role. Acting locally as paracrines, cytokines control cell development and differentiation as well as the immune response. Certain cytokines control the process of *angiogenesis* (the development of new blood vessels). Angiogenesis is important in the growing animal and also in wound healing, in rebuilding the uterine lining following menstruation, and in increasing the vascular supply to cardiac and skeletal muscle during endurance training by athletes (see Figure 16.19 and Chapter 21). A particularly well known large group of cytokines is the *interleukins*. Interleukin 2, for example, is a cytokine that enhances the activity of cytotoxic (killer) T cells in the immune response. Cytokines involved in the immune response may also travel long distances in the blood, just like hormones do.

Neuromodulators may be released from synaptic nerve terminals along with neurotransmitters and—like transmitters—bind to specific receptor sites typically on postsynaptic cells. They tend to exert their effects more slowly and for a longer duration than do neurotransmitters. One way in which neuromodulators modify the postsynaptic response to a neurotransmitter is by changing the conductances of specific membrane channels. Like paracrines in general, neuromodulators can diffuse from their site of release and influence the responsiveness of not only their immediate postsynaptic cell but also nearby cells that have appropriate receptor sites. Because of their long-term influences and their ability to affect a population of neighboring cells, neuromodulators have the potential to modify the functions of entire neuronal circuits. This

BOX 16.1 Hormones, Neurons, and Paracrine Neuromodulators Influence Behavior

During the breeding season, mating pairs of the rough-skinned newt (*Taricha granulosa*) engage in clasping behavior called *amplexus*. The male mounts the female and grasps her with both hind- and forelimbs, and the pair maintains their embrace for hours. However, they disengage within minutes if they are exposed to an acute stressor, or are injected experimentally with corticosterone, the stress hormone in amphibians. The ability to uncouple quickly, become watchful, and escape if necessary provides an adaptive advantage to the newts. Researchers have shown that clasping behavior is controlled by specific neurons in the hindbrain. It is enhanced by arginine vasotocin (see Table 16.2) and diminished by corticosterone. Corticosterone stimulates these neurons to secrete endocannabinoids (eCBs, endogenously produced marijuana-like molecules) which serve as *paracrine neuromodulators* within the hindbrain. The eCBs are the mediators of clasp inhibition. As is the case in mammals (see Figures 16.10 and 16.12), a stress stimulus in amphibians turns on the HPA[*] axis and triggers the release of corticosterone into the circulation (acute stress → ↑CRH → ↑ACTH → ↑corticosterone). This steroid targets

Courtship behavior   The male rough-skinned newt (*Taricha granulosa*) clasps the female in a posture called amplexus.

the hindbrain clasp-controlling neurons. It doesn't bind with intracellular receptor molecules but instead binds to specific membrane G protein–coupled receptor molecules and exerts its action by nongenomic means, thereby ensuring a rapid response. Corticosterone stimulates the hindbrain neurons to release eCBs, which in turn inhibit "upstream" neural activity that ordinarily would stimulate the neurons that trigger clasping behavior. The eCBs thereby reduce the probability of action potential generation by the cell that produced them—the neuron that stimulates clasping behavior. Additional experiments showed that corticosterone reduced the enhancing effect of vasotocin on clasping behavior. In this case, corticosterone caused a reduction of vasotocin receptors expressed by the hindbrain neurons

that control clasping. In sum, a stressful stimulus reduces activity of the hindbrain clasping-behavior neurons by reducing both synaptic excitation (mediated by eCBs) and vasotocin's enhancing effect (by reducing receptors for vasotocin). The animals separate until a safer time to mate. This example of mating in newts illustrates the richness of interactions between endocrine, paracrine, and neural signals in controlling behaviors that ultimately determine the success of an animal. The rough-skinned newt also employs another strategy to ensure its survival: Its skin produces the potent neurotoxin tetrodotoxin, which inhibits the generation of nerve and muscle action potentials and provides a highly effective defense against predators.

[*] The adrenal tissue in amphibians is called the interrenal gland, so the axis is more accurately termed *HPI, hypothalamic–hypophysial–interrenal.*

effect on the crustacean stomatogastric ganglion is described in Box 19.2. Neuromodulators also act in concert with other chemical signals to influence behavior (**BOX 16.1**).

Pheromones and kairomones are used as chemical signals between animals

Pheromones are chemical signals that act between members of the same species. They are produced within the animal and then released into the environment. In many animals, they convey information that signals social status (for example, sex or dominance), sexual readiness, food trails, and alarm, and they elicit behaviors that are typically stereotyped and not modified by experience. Pursuing a potential mate, mating behavior, and aggressive behavior to protect a territory are often set in motion by pheromones. Physiological functions, such as the onset of puberty and estrous cycling, are also influenced by pheromones.

Many vertebrates detect pheromones with a sensory structure called the *vomeronasal organ* (*VNO*) (see Chapter 14). In several amphibians and reptiles, the VNO is the major olfactory organ—that is, the VNO detects both general odors and pheromones. For example, the Komodo dragon (*Varanus komodoensis*) specializes in eating carrion. This large lizard follows the scent of a recently dead or dying animal by repeatedly sampling the air with its tongue and then touching the tongue to the roof of its mouth, where there are openings to the VNO. Interestingly, the male rough-skinned newt described in Box 16.1 rubs his mental (chin) gland on the snout of the female during mating. It is proposed that the gland generates courtship pheromones that reach the female's VNO through the nasal passages and make her receptive to mating. Many mammals, such as rodents, use the VNO to detect pheromones exclusively. Early experiments showed that surgically ablating the VNO resulted in impaired behaviors: Males would neither defend their territories nor mount females; females would not adopt the lordosis position for mating, and they seldom became pregnant. Recent studies in mice using both surgical ablation experiments and knockout animals that lack functional VNO neurons suggest that both mate preference and sexual behavior are controlled by pheromones. Invertebrates also use pheromones, which they detect with a variety of chemosensory structures. The sensory apparatus used by male silkworm moths to detect *bombykol*, the powerful sex-attractant pheromone released by females, is described in Chapter 14.

The molecular structures of most pheromones are not known, but those that have been identified vary widely. Bombykol is a fat-derived alcohol; aphrodisin, the pheromone in the vaginal secretions of female hamsters, is a protein. The pheromone in the saliva of boars is a steroid; if a female pig detects this pheromone, she will stand motionless in the lordosis position.

Whereas pheromones released into the environment convey information between members of the *same* species, other categories of chemicals released by animals give information (unintentionally) to members of *different* species. *Kairomones* are chemicals produced by animals of one species that are released into the environment—not for the purpose of communicating, but in the process of accomplishing some other function. Members of a second, detecting species exploit these kairomones in a way that benefits them but may be a detriment to the members of the releasing species. Kairomones elicit both behavioral and physiological responses. For example, one species may release a kairomone that reveals its location to a potential predator and elicits foraging behavior in the predator. Or a predator may release a kairomone that informs potential prey of its presence. Such an enemy-avoidance kairomone allows the prey to respond defensively. For example, water fleas of the genus *Daphnia* detect kairomones released by predatory fly larvae in the surrounding water; they respond by developing spines and "helmets" that may protect them from the predators.

Additional interspecies chemical signals include *allomones* and *synomones*. Allomones are released by a species for its own advantage, such as a predator emitting a chemical that attracts prey. Synomones are released by members of one species that provide information advantageous to both its own species and also to the members of a different species; for example, both species could benefit from avoiding a toxic plant. As researchers learn more about the physiological processes triggered by intraspecific and interspecific informational chemicals, we can look forward to enlarging our understanding of physiology in the context of ensuring survival and reproductive success in the natural world.

SUMMARY
Chemical Signals along a Distance Continuum

- Chemical signals fall along a "distance spectrum" ranging from molecules that signal between individual cells and over short distances, to hormones and neurohormones that travel long distances in the blood, to chemical signals released by animals into the environment. Receptor molecules of target cells in tissues or sense organs detect these signals to trigger a functional response.

- Locally acting paracrines and autocrines include neuromodulators and cytokines.

- Pheromones are chemical signals released into the environment that convey information to animals of the same species. Kairomones, allomones, and synomones are chemicals released into the environment that convey information between members of different species.

Insect Metamorphosis

In this section we explore the hormones that control metamorphosis in insects. Hormones play as varied and extensive roles in invertebrates as they do in vertebrates. Color changes, spawning, metamorphosis, water balance, nutrient metabolism, growth, maturation, and death are just a few of the functions controlled by hormones in invertebrates. We have chosen insect metamorphosis as our example because it demonstrates that the same principles of endocrine function found in vertebrates also apply to invertebrates. In addition, because insects have great agricultural, environmental, medical, and forensic importance, their natural history and physiology have been studied extensively (**BOXES 16.2 AND 16.3**).

Insects have separate sexes. They engage in courtship behavior that leads to sperm being deposited and stored in the female reproductive tract. Fertilization does not take place at the time of mating. It occurs later, usually just before the female lays her eggs. After various periods, depending on the species and environmental conditions, the eggs hatch. From hatching onward, insects undergo a complex developmental journey to reach sexual maturity. The life cycles of most insects involve changes in the form of the animal at different stages, a process called **metamorphosis** (*meta*, "change"; *morph*, "structure or form"). Although insects and vertebrates are evolutionarily divergent, we will find that their endocrine systems have independently evolved many similar features.

Insect metamorphosis may be gradual or dramatic

There are two main types of metamorphosis (**FIGURE 16.20**). Insects such as bugs, grasshoppers, and cockroaches go through gradual metamorphosis and are referred to as hemimetabolous (*hemi*, "partial or gradual"; *metabolous*, "change"). Animals such as flies, beetles, butterflies, and moths go through dramatic metamorphosis and are referred to as holometabolous (*holo*, "complete").

In hemimetabolous development, the egg hatches into a nymph, which goes through several **molts** (see Figure 16.20A). With each molt, epidermal cells underlying the cuticle (exoskeleton) synthesize a new cuticle, and the old cuticle is shed along with the linings of the foregut, hindgut, and tracheae—a process called **ecdysis**. The new cuticle is expanded while it is still soft and pliable. To expand it, the animal takes air into the foregut and "puffs itself up." The swallowed air applies pressure on the hemolymph (blood) and forces it into narrow lanes. The increased pressure inside the body helps fill out or unfurl external structures. Once expanded, the new cuticle hardens. The larger cuticle provides room for internal structures to grow before the next molt. The periods between

FIGURE 16.20 Two types of metamorphosis
(A) Cockroaches go through gradual or partial (hemimetabolous) metamorphosis. (B) Moths go through complete (holometabolous) metamorphosis.

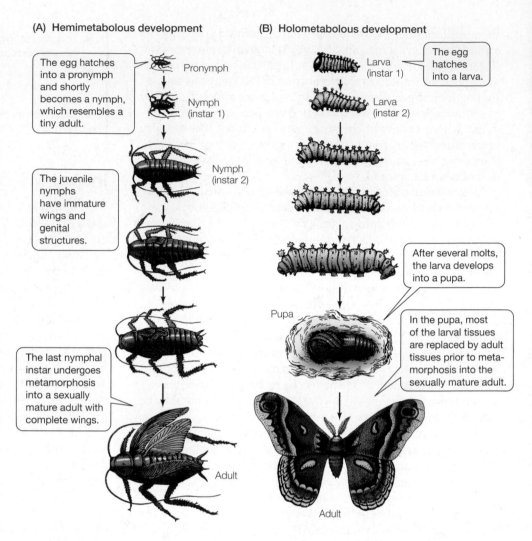

(A) Hemimetabolous development

The egg hatches into a pronymph and shortly becomes a nymph, which resembles a tiny adult.

Pronymph

Nymph (instar 1)

Nymph (instar 2)

The juvenile nymphs have immature wings and genital structures.

The last nymphal instar undergoes metamorphosis into a sexually mature adult with complete wings.

Adult

(B) Holometabolous development

The egg hatches into a larva.

Larva (instar 1)

Larva (instar 2)

After several molts, the larva develops into a pupa.

Pupa

In the pupa, most of the larval tissues are replaced by adult tissues prior to metamorphosis into the sexually mature adult.

Adult

molts are called **instars**. Hemimetabolous nymphs go through four to eight instars; each species has a characteristic number. The last nymphal instar undergoes metamorphosis into the adult. Adults have complete wings and are sexually mature. Adults do not grow or undergo additional molts.

In holometabolous development, the egg hatches into a *larva* (see Figure 16.20B). Depending on the species, larvae are referred to as grubs, caterpillars, or maggots. Like hemimetabolous nymphs, holometabolous larvae go through several molts and expand the new cuticle. With each molt, the animal increases in size. The larvae of holometabolous insects are the forms that usually cause crop damage. After several molts, holometabolous larvae enter a stage called the **pupa**, in which most of the larval tissues are destroyed and replaced by adult tissues. The pupa has a much thicker cuticle than the larva or adult. The pupa then metamorphoses into an adult. Adults are specialized for reproduction. In some species, such as silkworm moths, the adult may not even feed.

BOX 16.2 Juvenile Hormone (JH) in the Laboratory, Silk Industry, and Environment

In the laboratory, it is possible to remove the corpora allata surgically from experimental animals and thus remove the source of juvenile hormone (JH). When this procedure is carried out on early instars (beyond the second), instead of molting into another juvenile form, the experimental animals metamorphose into tiny, sterile adults. Or the converse experiment can be done: Last-instar larvae or nymphs can be treated with additional JH, either by implanting donor corpora allata from earlier-stage instars or by applying JH to the animals' surfaces. These experimental animals do not metamorphose into adults, but instead molt into giant extra-instar larvae or nymphs.

Silk growers have used this phenomenon to their advantage. The larvae of

the domesticated silkworm *Bombyx mori* spin silk filaments to make their cocoons. Spraying late-instar larvae with an analog of JH prevents them from pupating, so they molt instead into extra-instar larvae. When these larger larvae pupate, they spin more silk to make larger cocoons. Using this technique, silk growers can obtain a 10%–20% increase in usable silk filaments.

JH and its analogs are also used in pest control. Certain mosquito species known to be vectors of West Nile, chikungunya, and dengue viruses are often controlled by adding the JH analog methoprene to storm-water catch basins. Catch basins are parts of drainage systems that hold water for various periods of time and so provide areas for mosqui-

toes to breed. Methoprene added to the water mimics JH and prevents the emergence of adult animals. Preventing a large proportion of an insect population from reaching sexual maturity reduces its reproductive capability and overall impact. Methoprene has been proven effective in controlling insect pests in their adult stage, such as mosquitoes, fleas, and biting flies. However, because the larvae of many insects cause severe damage to agricultural crops, researchers would like to find a way to antagonize (instead of mimic) JH. A JH antagonist that induced larvae to metamorphose early into adults would curtail their voracious feeding stages. So far, however, effective JH antagonists for broadscale use have not been identified.

BOX 16.3 Insects in Forensics and Medicine

Certain insects, such as the blowflies, feed and lay their eggs only on *necrotic* (dead) tissue. This characteristic has been employed by forensic scientists as one method to determine the time of death of bodies found under suspicious circumstances. Studies of many different insects have revealed a progression of species that occupy a corpse. The blowflies (bluebottles and greenbottles) arrive first and lay their eggs in moist regions (see the figure). The eggs hatch, and maggots (larvae) develop and feed on the tissues. Flesh flies soon are drawn by olfactory cues to the decomposing flesh, and they too lay eggs that develop into maggots. Maggots of blowflies and flesh flies go through three larval instars. Depending on how long the cadaver is left to decompose, the maggots may pupate and metamorphose into adults. The presence of pupal cases gives clues to investigators about time elapsed since death.

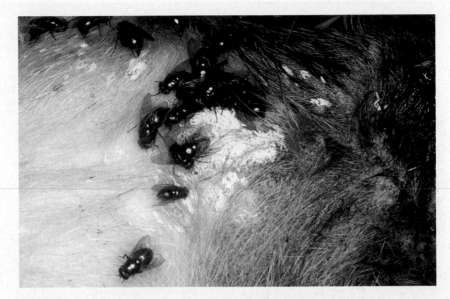

Blowflies lay their egg masses on the surface of a pig carcass The eggs will hatch into maggots, which feed on the decomposing tissue and go through successive developmental stages. Forensic entomologists use pigs for their studies because pig and human tissues are similar to each other.

As the maggots feed, beetles arrive to feed on both the corpse and the maggots. Wasps and other flies follow in progression. After several months, mites, along with skin and hide beetles, and finally clothes moths, explore the dried remains. Forensic entomologists collect pupal cases, excrement, and other evidence of the animals that roved the cadaver. They take into account environmental conditions (such as temperature and humidity), the species of insects on a corpse, and the details of the life cycle of each and "work backward" in time to estimate when death occurred. Insects can also give clues about the site of death. Because

some insects have a very narrow range of habitats, a species of insect found on a corpse outside the insect's range indicates that the body was moved.

The larvae of some of the same insects that are attracted to corpses also feed on the necrotic tissues of wounds of living people. During the American Civil War, surgeons recorded that wounds of soldiers with maggots on them were cleaner and healed more quickly than wounds not visited by maggots. These days, physicians usually clean wounds by *surgical debridement*, using instruments to scrape away dead tissue so that living tissue can grow in the healing process. However, they

also use maggots of blowflies to debride wounds that are especially resistant to healing. Once or twice a week, blowfly larvae are applied to the surface of wounds at a density of five to eight larvae per square centimeter. The larvae are voracious feeders, but they do not eat living tissue. They seek out every bit of necrotic tissue in the wound, so they succeed in doing a far better job of cleaning than surgical instruments can do. Larvae applied to wounds must be sterile. The eggs laid by adult blowflies are soaked in solutions of Lysol or Clorox and then kept in sterile culture dishes until they hatch into sterile larvae.

The holometabolous life cycle of the domesticated silkworm moth *Bombyx mori* is shown in **FIGURE 16.21**.

Hormones and neurohormones control insect metamorphosis

Three main hormones control insect metamorphosis: prothoracicotropic hormone (PTTH, a small protein), ecdysone (a steroid), and juvenile hormone (JH, a *terpene*, which is a fatty acid derivative). Gene knockout experiments in silkworms recently showed that JH exerts its effects only after the second instar, and investigators are currently pursuing studies of other chemical signals that control the early stages of development. The chemical structures of JH and

20-hydroxyecdysone (20E)—the active form of ecdysone—are shown in **FIGURE 16.22**. Although the specific mechanisms of metamorphosis differ from species to species, investigators have shown that PTTH, JH, and 20E typically function together. We will see that several additional hormones also play important roles. **TABLE 16.5** summarizes the endocrine and neuroendocrine secretions that achieve insect metamorphosis, and **FIGURE 16.23** illustrates the main neural and endocrine structures involved.

In both hemimetabolous and holometabolous insects, the molting process begins with increased activity of neuroendocrine cells in the brain that produce the neurohormone PTTH. Like neuroendocrine cells in vertebrates, these cells receive and inte-

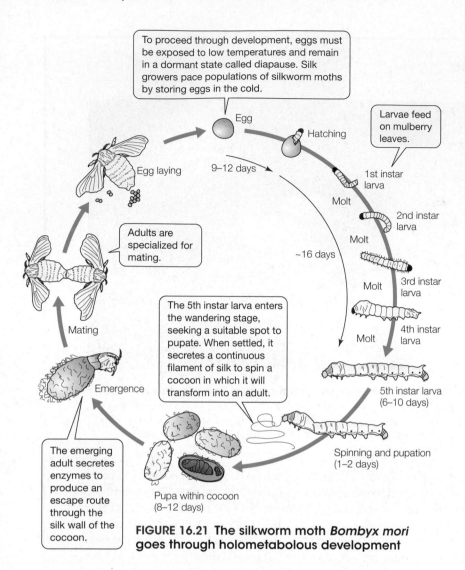

To proceed through development, eggs must be exposed to low temperatures and remain in a dormant state called diapause. Silk growers pace populations of silkworm moths by storing eggs in the cold.

Larvae feed on mulberry leaves.

Egg

Hatching

Egg laying

9–12 days

1st instar larva

Molt

2nd instar larva

Molt

~16 days

3rd instar larva

Molt

4th instar larva

Adults are specialized for mating.

Mating

The 5th instar larva enters the wandering stage, seeking a suitable spot to pupate. When settled, it secretes a continuous filament of silk to spin a cocoon in which it will transform into an adult.

Molt

5th instar larva (6–10 days)

Emergence

Spinning and pupation (1–2 days)

The emerging adult secretes enzymes to produce an escape route through the silk wall of the cocoon.

Pupa within cocoon (8–12 days)

FIGURE 16.21 The silkworm moth *Bombyx mori* goes through holometabolous development

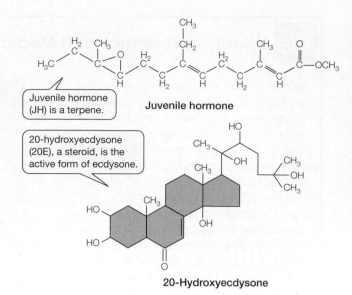

Juvenile hormone (JH) is a terpene.

Juvenile hormone

20-hydroxyecdysone (20E), a steroid, is the active form of ecdysone.

20-Hydroxyecdysone

FIGURE 16.22 Juvenile hormone (JH) and 20-hydroxy-ecdysone (20E) are both lipid-soluble hormones

JH maintains juvenile (immature) characteristics in the developing animal (exerting its effects after the second instar). This hormone is released from nonneural endocrine cells in the corpora allata. Thus each corpus allatum is both a site for neurosecretory release of PTTH (a neurohemal organ) *and* a nonneural endocrine gland that secretes JH. When 20E acts on the epidermis *and* the hemolymph level of JH is high, the insect will molt into another, larger juvenile form—a larger larva (in holometabolous insects) or nymph (in hemimetabolous insects). The corpora allata become inactive during the last larval or nymphal instar. Therefore, when ecdysone is secreted to initiate the next molt, there is little or no JH present. When 20E is secreted in the presence of very low hemolymph levels of JH, the epidermis will produce adult or pupal structures. The nymph of a hemimetabolous insect will develop into an adult, and the larva of a holometabolous insect will develop into a pupa. In holometabolous insects, ecdysone is secreted again (and converted to 20E) at the end of pupation. Because no JH is present, 20E triggers metamorphosis into the adult form.

Both 20E and JH exert their effects on target cells by genomic means. Both hormones are soluble in lipids, pass through the cell membrane, and bind with intracellular receptors in their target cells. The hormone–receptor complexes activate or inhibit the transcription of specific genes. As the insect goes through its life cycle, appropriate levels of 20E and JH control expression of suites of larval, pupal (in holometabolous forms), and adult genes. The proteins encoded by specific genes influence the functions of the target cells.

In addition to PTTH, 20E, and JH, additional hormones ensure the progress of an insect through its life cycle. Every time the insect molts (undergoes ecdysis), it must break out of its old cuticle. Contractions of the body-wall skeletal muscles are essential for ecdysis. Current data indicate that the Inka cells located on tracheae secrete pre-ecdysis triggering hormone (PETH) and ecdysis-triggering hormone (ETH). These peptides stimulate neuronal circuits in the

grate synaptic inputs, and generate action potentials that trigger hormone secretion by exocytosis. **FIGURE 16.24** shows PTTH neuroendocrine cells in the brain of a moth. Their axons extend to a pair of structures closely associated with the brain, the **corpora allata** (singular *corpus allatum*). In some insects, the PTTH axons may also terminate in the corpora cardiaca. The axon terminals secrete PTTH from these structures into the hemolymph. A variety of signals trigger molting, all mediated ultimately by the nervous system. Day length, temperature, crowding, and certain behaviors[8] are known to set the molting process in motion. Neurons that detect these factors send excitatory synaptic signals to the PTTH cells in the brain to stimulate secretion.

The PTTH is carried in the hemolymph to the thorax, where it stimulates the **prothoracic glands** to secrete ecdysone. Ecdysone is a prohormone that is converted by peripheral activation to **20-hydroxyecdysone** (**20E**) at its target tissues. A major target tissue of 20E is the epidermis, which lies just below the cuticle. 20E, also called *molting hormone*, stimulates the epidermis to secrete enzymes that digest away the old cuticle and synthesize a new one.

---

[8] A behavioral example is *Rhodnius prolixus*, the blood-sucking bug that transmits Chagas disease. It must engorge itself with a blood meal to trigger the hormonal sequence that mediates molting to the next instar.

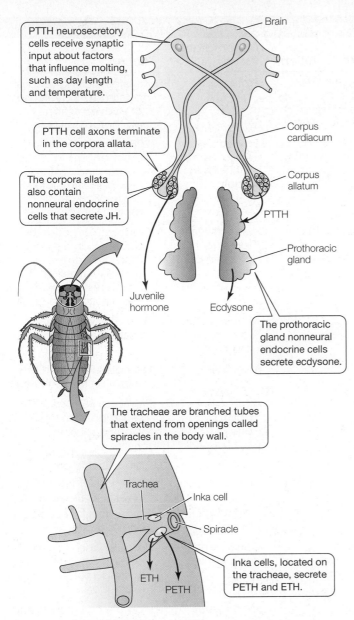

PTTH neurosecretory cells receive synaptic input about factors that influence molting, such as day length and temperature.

PTTH cell axons terminate in the corpora allata.

The corpora allata also contain nonneural endocrine cells that secrete JH.

Brain

Corpus cardiacum

Corpus allatum

PTTH

Prothoracic gland

Juvenile hormone

Ecdysone

The prothoracic gland nonneural endocrine cells secrete ecdysone.

The tracheae are branched tubes that extend from openings called spiracles in the body wall.

Trachea

Inka cell

Spiracle

ETH

PETH

Inka cells, located on the tracheae, secrete PETH and ETH.

FIGURE 16.23 Endocrine and neuroendocrine structures involved in the control of insect metamorphosis Ecdysone (molting hormone), JH, pre-ecdysis triggering hormone (PETH), and ecdysis-triggering hormone (ETH) are all secreted by nonneural endocrine cells. PTTH is secreted by neurons with cell bodies in the brain and secretory axon terminals in the corpora allata. In some insects, the PTTH axons may terminate in the corpora cardiaca.

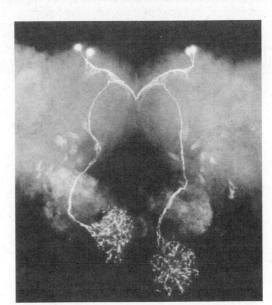

FIGURE 16.24 Neuroendocrine cells secrete PTTH Four PTTH-secreting cells in the brain of the moth *Manduca sexta* were stained with an immunofluorescent dye. The axons cross at the midline and extend to the corpora allata on the opposite side, where they form highly branched terminals. The width of the brain is about 1.5 mm. (Courtesy of Walter Bollenbacher; from O'Brien et al. 1988.)

required for ecdysis at the end of each instar in both holometabolous and hemimetabolous insects.

After metamorphosis, the adult is poised for reproduction. For reproduction to succeed, the corpora allata, which became inactive and stopped secreting JH before the final molt, must begin to secrete JH again in the adult. JH functions as a gonadotropin in adults to support the production of fertile eggs and sperm. (Recently, peptide molecules isolated from ovaries, oviducts, and brains of insects have also been shown to influence gonadal function in insects. These molecules are subjects of active inquiry.) In addition, JH stimulates the production of pheromones required for mating. Ecdysone is also produced in the adult—not by the prothoracic glands, but by the ovaries. JH, functioning now as a gonadotropin, causes its secretion. Ecdysone, converted to 20E, stimulates the fat body, an organ that performs functions similar to those of the vertebrate liver, to produce yolk proteins, which are transported to the ovary for incorporation into eggs. With adults now capable of producing gametes and releasing chemical signals that attract mates, the insect life cycle is ready to be repeated.

SUMMARY
Insect Metamorphosis

■ Insect metamorphosis illustrates the convergent evolution of endocrine and neuroendocrine functions between vertebrates and invertebrates.

■ Insects change form in the course of their life cycles. Hemimetabolous insects go through gradual metamorphosis, and holometabolous insects go through complete metamorphosis.

(Continued)

CNS that control motor programs for stereotyped, coordinated muscle contractions. PETH and ETH are released prior to every molt in both holometabolous and hemimetabolous insects.[9] Eclosion hormone (EH) and corazonin, both neuropeptide hormones secreted by the brain, signal the Inka cells to secrete PETH and ETH. Thus neural, neuroendocrine, and endocrine functions are all essential for coordinating the physiological processes leading to motor behaviors that cause shedding of the old cuticle. After ecdysis, the new cuticle expands, and the neurohormone bursicon promotes its hardening and darkening. Similar behavioral sequences are

[9] At each molt, ecdysone, through its genomic actions, stimulates PETH/ETH secretion by Inka cells and increased expression of ETH receptors on their target neurons in the brain.

TABLE 16.5 Major hormones and neurohormones that control insect metamorphosis

Hormone	Type of molecule	Type of signal	Site of secretion	Major target tissue	Action
Prothoracicotropic hormone (PTTH)	Protein (~5000 molecular weight)	Neuroendocrine	Brain, with axon terminals extending to corpora allata	Prothoracic glands	Initiates molting (ecdysis) by stimulating release of ecdysone from prothoracic glands
Ecdysone (peripherally activated to 20-hydroxyecdysone, 20E)	Steroid	Endocrine	Prothoracic glands in larva/nymph; ovary in adult	Epidermis in larva/nymph; fat body in adult	20E promotes cellular mechanisms to digest old cuticle and synthesize new one; stimulates production of yolk proteins in adult
Juvenile hormone (JH)	Terpene (fatty acid derivative)	Endocrine	Corpora allata	Epidermis in larva/nymph; ovary in adult	Opposes formation of adult structures and promotes formation of larval/nymphal structures; functions as a gonadotropin in adult
Crustacean cardioactive peptide (CCAP)	Peptide	Neuroendocrine	Abdominal ganglia	Motor neurons to body-wall muscles	Promotes initiation of ecdysis, wing expansion
Corazonin	Peptide	Neuroendocrine	Brain	Inka cells	Promotes PETH and ETH secretion from Inka cells
Eclosion hormone (EH)	Peptide	Neuroendocrine	Brain	Inka cells, possibly others	Promotes PETH and ETH secretion from Inka cells, wing expansion
Pre-ecdysis triggering hormone (PETH)	Peptide	Endocrine	Inka cells of tracheae	Neuronal circuits in brain	Coordinates motor programs to prepare for shedding the cuticle
Ecdysis triggering hormone (ETH)	Peptide	Endocrine	Inka cells of tracheae	Neuronal circuits in brain	Coordinates final motor programs for escaping from old cuticle
Bursicon	Heterodimer protein (~30 kDa molecular weight)	Neuroendocrine	Brain and abdominal ganglia	Cuticle and epidermis	Hardens and darkens new cuticle after each ecdysis; promotes wing expansion
Insulin-like peptides	Peptide	Neuroendocrine	Brain	Most tissues	Promote cell growth and proliferation

Sources: After Randall et al. 2002; Zitnan et al. 2003; Zitnan et al. 2007; Zitnan and Adams 2012; White and Ewer 2014.

- Environmental and behavioral signals mediated by the nervous system initiate molting by providing synaptic input to the PTTH neuroendocrine cells in the brain. These cells secrete PTTH, which stimulates secretion of ecdysone from the prothoracic glands. Ecdysone is converted to 20-hydroxyecdysone (20E) by peripheral activation.

- 20E stimulates the epidermis to secrete enzymes required for the molting process. At each molt, the epidermis lays down a new cuticle beneath the old one.

- Under the control of PETH and ETH, the insect performs stereotyped muscle contractions in order to shed the old cuticle.

- JH, secreted by nonneural endocrine cells in the corpora allata, prevents metamorphosis into the adult form. The relative amounts of JH and 20E in the hemolymph determine whether the epidermis will produce juvenile, pupal (in holometabolous forms), or adult structures.

- In adults, JH functions as a gonadotropin, stimulates the production of sex-attractant pheromones, and stimulates the secretion of ecdysone, which promotes incorporation of yolk into eggs.

STUDY QUESTIONS

1. Explain why the effects of steroid hormones are seen after a longer delay than the effects of peptide hormones.

2. Construct a table to compare the hormones involved in water and salt balance. Include the site of secretion and molecular structure of each hormone, the principal stimuli that cause its secretion, its effect on target cells, and other information you consider important.

3. Insulin secretion is essential in metabolism. What type of molecule is insulin? What structure secretes it? What factors control its secretion? What is/are its target tissue(s)? What effect(s) does it exert?

4. Compare and contrast the functions and modes of action of the following pairs of hormones: insulin and glucagon; active vitamin D and parathyroid hormone; adrenocorticotropic hormone (ACTH) and cortisol.

5. Compare and contrast the structure and functions of hypothalamic neurosecretory neurons that send their axons to the posterior pituitary gland with those of hypothalamic neurosecretory neurons that have axons terminating on the capillary network of the median eminence.

6. List the similarities and differences between prothoracicotropic hormone (PTTH) and ecdysone. Consider the structural and functional characteristics of both molecules.

7. In patients receiving cortisone drug therapy, adrenocorticotropic hormone (ACTH) secretion decreases, and the adrenal glands shrink. Using your knowledge of the HPA axis, feedback mechanisms, and the tropic functions of hormones, explain why patients experience these changes.

8. Using your knowledge of the stress response, explain why the physiological changes of the stress response are important to survival in the short term and deleterious when they persist over long periods.

9. Define/describe negative feedback in endocrine controls and explain its physiological significance.

10. Discuss antagonism, permissiveness, and synergism in the context of hormone interactions.

Go to **sites.sinauer.com/animalphys4e**
for box extensions, quizzes, flashcards,
and other resources.

REFERENCES

Bentley, P. J. 1998. *Comparative Vertebrate Endocrinology*, 3rd ed. **Cambridge University Press, Cambridge, UK.** A detailed text replete with examples illustrating the physiology of hormones and their evolution in vertebrate animals.

Daimon, T., M. Uchibori, H. Nakao, and two additional authors. 2015. Knockout silkworms reveal a dispensable role for juvenile hormones in holometabolous life cycle. *Proc. Natl. Acad. Sci. U.S.A.* 112: E4226–E4235. Gene knockout experiments showed that JH exerts its effects only after the second instar in a holometabolous insect. The authors propose that another yet-to-be-identified factor controls earlier embryonic development.

Davis, A., E. Abraham, E. McEvoy, and six additional authors. 2015. Corticosterone suppresses vasotocin-enhanced clasping behavior in male rough-skinned newts by novel mechanisms interfering with V1a receptor availability and receptor-mediated endocytosis. *Horm. Behav.* 69: 39–49. A study of the circuitry, receptors and chemical messengers involved in controlling amplexus in the rough-skinned newt.

Døving, K. B., and D. Trotier. 1998. Structure and function of the vomeronasal organ. *J. Exp. Biol.* 201: 2913–2925. A comprehensive description of the organ that detects and processes pheromonal signals.

Landry, C. S., M. D. Ruppe, and E. G. Grubbs. 2011. Vitamin D receptors and parathyroid glands. *Endocr. Pract.* 17 Suppl. 1: 63–68. A review of the functions of vitamin D and its role in calcium metabolism and parathyroid gland functions.

Melmed, S. (ed.). 2011. *The Pituitary*, 3rd ed. **Academic Press, New York.** A compendium of chapters that address the overall functions of the pituitary gland, its association with the hypothalamus, and the roles of hormones secreted by different populations of cells.

Nelson, O. L., H. T. Jansen, E. Galbreath, and nine additional authors. 2014. Grizzly bears exhibit augmented insulin sensitivity while obese prior to a reversible insulin resistance during hibernation. *Cell Metab.* 20: 376–382. A study of changes of the insulin receptor of adipose cells from being highly sensitive to insulin during the summer and becoming insulin-resistant during winter hibernation.

Tirindelli, R., M. Dibattista, S. Pifferi, and A. Menini. 2009. From pheromones to behavior. *Physiol. Rev.* 89: 921–956. A review of mammalian pheromones and the behaviors they elicit, their chemosensory receptors and systems, transduction mechanisms, and central projection pathways involved in integration.

Yosten, G. L. C., and G. R. Kolar. 2015. The physiology of proinsulin C-peptide: Unanswered questions and a proposed model. *Physiology* 30: 327–332. A review of the synthesis of C-peptide in the β cells of the islets of Langerhans, its interactions with insulin, and its effects on the functions of endothelial cells.

Zitnan, D. and M. E. Adams. 2012. Neuroendocrine regulation of ecdysis. In L. I. Gilbert (ed.), *Insect Endocrinology* pp. 253–309, **Academic Press, Boston.** This detailed study of ecdysis is Chapter 7 of a comprehensive collection of chapters on many insect hormones, including prothoracotropic hormone, insulin-like peptides, juvenile hormone, and bursicon.

See also **Additional References** *and* **Figure and Table Citations**.

"Mad as a March hare" is a 500-year-old expression—referenced often in literature and film—that refers to one of the most visible of all mammalian reproductive behaviors. The species of hare that likely gave rise to the expression (*Lepus europaeus*) was introduced to the British Isles from continental Europe at the time of the Roman Empire. In Britain, the animals begin each year in February or March to "box" with each other in broad daylight. Boxing hares stand straight upright, kicking and batting each other with their forelegs. They also dash and leap in other forms of "madness." *Harebrained* refers to people who behave like them. Several other species of hares and rabbits are known to box as well. At least in the two or three species that have been studied, boxing is primarily an interaction between females and males in which the females—being at a stage not receptive to mating—repel male advances. Boxing mirrors the reproductive biology of the hares and rabbits in two ways. First, being seasonal, it reflects the seasonality of reproduction. Second, it emphasizes that animals often need to be in a specialized state of receptivity for mating to take place.

Rabbits and hares are classified as **induced ovulators** because in them, **ovulation**—the release of mature eggs from the ovaries of a female—is induced by, and dependent on, stimuli generated by the actual act of copulation. Whenever animals of any sort engage in

Boxing is a captivating prelude to the mating season in some hares and rabbits
The antics of boxing hares in the month of March were the inspiration for Lewis Carroll's invention of the March Hare in *Alice in Wonderland*.

FIGURE 17.1 The physiology of induced ovulation in female rabbits (A) The surge in the concentration of luteinizing hormone (LH) in the blood plasma following copulation. Data are averages for nine animals. (B) Major steps leading to ovulation following copulation. A sequence of this sort is often described as a *neuroendocrine reflex* because both neural and endocrine steps are involved in bringing about a stereotyped outcome in response to a specific stimulus. In the hypothalamo–hypophysial portal vasculature, blood flows from the hypothalamus to the anterior pituitary gland; for greater detail on this system, see Figure 16.7. GnRH = gonadotropin-releasing hormone; LH = leuteinizing hormone. (A after Tsou et al. 1977.)

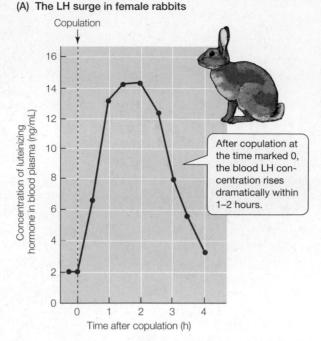

(A) The LH surge in female rabbits

After copulation at the time marked 0, the blood LH concentration rises dramatically within 1–2 hours.

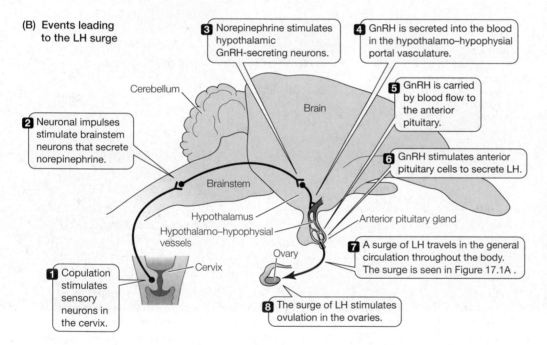

(B) Events leading to the LH surge

1 Copulation stimulates sensory neurons in the cervix.

2 Neuronal impulses stimulate brainstem neurons that secrete norepinephrine.

3 Norepinephrine stimulates hypothalamic GnRH-secreting neurons.

4 GnRH is secreted into the blood in the hypothalamo–hypophysial portal vasculature.

5 GnRH is carried by blood flow to the anterior pituitary.

6 GnRH stimulates anterior pituitary cells to secrete LH.

7 A surge of LH travels in the general circulation throughout the body. The surge is seen in Figure 17.1A.

8 The surge of LH stimulates ovulation in the ovaries.

Cerebellum
Brain
Brainstem
Hypothalamus
Hypothalamo–hypophysial vessels
Cervix
Ovary
Anterior pituitary gland

ovulators include certain cats, shrews, and camelids. Humans, most rodents,[1] dogs, and farm animals such as cows and sheep are spontaneous ovulators.

When physiologists study reproduction, they investigate the mechanisms by which induced and spontaneous ovulation occur. They also investigate many other aspects. Reproduction is a stunningly extended process when one considers all the steps required for one generation of adults to beget another generation of adults, and physiologists study many of those steps. To set the stage for identifying all the aspects of reproduction studied by physiologists, let's look further at our opening example. Of all the rabbits and hares, the best understood species is the European rabbit (*Oryctolagus cuniculus*), which not only exists in the wild but also has been domesticated as the "laboratory rabbit" used extensively in research. As we now look in more depth at the reproductive biology of rabbits and hares, we will use information on the European rabbit. Other rabbits and hares are believed in general to be similar.

The endocrine system typically plays a key role in the coordination of reproduction in species throughout the animal kingdom. Exemplifying this point, the endocrine system plays a pivotal role in coordinating copulation and ovulation in rabbits and other induced ovulators. Soon after a female rabbit copulates, her pituitary gland dramatically increases its secretion of the peptide hormone **luteinizing hormone** (**LH**) into her blood, resulting in a sudden, steep increase in the blood concentration of LH (**FIGURE 17.1A**). This **LH surge** is responsible for inducing the release of eggs from her ovaries.

Copulation initiates a *sequence* of neuronal, neuroendocrine, and endocrine events that culminate in the LH surge (**FIGURE 17.1B**). First (step ❶ in the figure), the act of copulation physically stimulates the cervix and provides other sensory inputs that excite sensory neurons in and around the cervix to produce nerve impulses. These signals are relayed by neuronal pathways to the brain, where the signals (step ❷) stimulate brainstem neurons that have their cell bodies in the brainstem but send neuronal processes to the hypothalamus. When activated, these brainstem neurons increase secretion of the neurotransmitter norepinephrine (NE) into the hypothalamus; genes involved in NE synthesis in the neurons have been shown by genomic studies to exhibit increased transcription (heightened mRNA synthesis) within 20–40 min after a female rabbit mates. Within an hour after copulation (steps ❸ and ❹), neurosecretory cells in the hypothalamus, stimulated

sexual reproduction, one of the key requirements for success is *synchronization* that ensures that competent eggs meet competent sperm. The achievement of this essential condition is particularly straightforward in the induced ovulators. In them, because copulation itself is the stimulus for mature eggs to be shed from the ovaries into the female reproductive tract, eggs remain in the ovaries until sperm are provided, and then the eggs move promptly from the ovaries to the reproductive tract to meet the sperm. One might expect fertilization of virtually 100% of ovulated eggs under such circumstances, and in rabbits that is the outcome. Despite these advantages, induced ovulation occurs in only a minority of mammal species. Most mammals—indeed most animals of all kinds—are classified as **spontaneous ovulators** because ovulation results from processes endogenous to the female, more or less independent of mating. In addition to rabbits and hares, the induced

---

[1] Rabbits and hares are not rodents; they are lagomorphs.

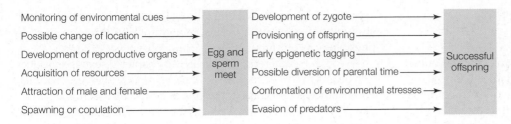

FIGURE 17.2 The complexity of reproduction Reproduction involves much more than the meeting of eggs and sperm. This diagram is meant to be suggestive rather than comprehensive. Exact processes vary from species to species, but the number of processes is always great.

by NE, initiate a massive increase in secretion of gonadotropin-releasing hormone (GnRH; see Figure 16.3) into the blood in the hypothalamo–hypophysial portal system (see Figure 16.7). The GnRH is carried by blood flow in the portal system (step ❺) to the anterior pituitary gland and stimulates cells in the anterior pituitary to secrete LH into the general circulation at a greatly increased rate (step ❻). As we have seen in Figure 17.1A, this leads to a dramatic rise in the concentration of LH in the blood: the LH surge (step ❼).

Eggs released from the ovaries in response to the LH surge (step ❽ in Figure 17.1B) arrive in the oviducts (the passageways to the uterus; see Figure 17.10A) about 10 h after copulation. Sperm have also arrived there by then, and fertilization occurs there.

As already suggested, the events we have described cannot occur at just any time of year in rabbits that are living in their natural environment. Instead, wild rabbits in places such as northern Europe exhibit a strongly seasonal cycle of reproductive competence, being ready to reproduce only during the warmer months of the year. In male rabbits, the testes regress each autumn; by winter, they cease sperm production and are only a third or a half of their summer size. Following the winter solstice in late December, however—in a context of ever-increasing hours of daylight per day as time passes—the testes regrow, under photoperiodic control, and ultimately return to full sperm production, a process termed **testicular recrudescence**. The rabbit version of "March madness" then begins. Female rabbits—as is true of females in most species of mammals (although not humans)—are attracted to males and will mate with them only at highly circumscribed times when the females are in a specialized state of reproductive receptivity termed **estrus**[2] or "heat" (see page 481). Periods of estrus cease to occur in the winter months. They resume, however, at the same time of year as the males regain testicular function.

Estrus—a change of behavior—is induced by effects of **estrogens** on the brain; estrogens are a class of steroid hormones (see Figure 16.2). The ovaries secrete estrogens, thereby stimulating estrus, at times when mature eggs are present: an elegant synchronization mechanism that ensures that a female rabbit enters estrus when—and only when—she has mature eggs. Then, when she mates, the act of copulation itself induces the release of the mature eggs so they can be fertilized.

A property of rabbits and hares that tends to seem strange to human eyes is that their nursing is very brief. A mother rabbit, after she gives birth, typically covers her young in a nest and departs except for a single 3-min nursing visit each day. Youngsters must obtain all their daily food in that brief interval. Natural historians believe, without firm evidence, that leaving the young alone for most of the time is a way to protect them from predators by making

them as inconspicuous as possible. The same strategy is employed by many African antelopes in the days immediately following birth.

With their mother visiting only briefly each day, nestling rabbits must orient quickly to their mother's teats when she arrives. Recent research has revealed that a pheromone emitted by the mother is crucial for this task. A **pheromone** is a chemical compound (or mix of compounds) emitted into the environment by one individual of a species that elicits specific behavioral responses from other individuals of the same species (see page 456). The pheromone employed in rabbit nursing is apparently secreted by each nipple onto the skin surface of the nipple, where it coats the skin and gets into the milk. Young are guided to the nipple by their attraction to the pheromone. Pheromones are also commonly employed in reproduction in other ways. In animals ranging from crabs and insects to mammals, for example, adult males of many species are attracted to adult females by pheromones the females emit (see page 456).

Animal reproduction—as studied by biologists today—is a process with many steps that collectively may extend over a long period of time. **FIGURE 17.2** summarizes certain of the major steps, some of which are illustrated by our opening discussion of rabbits and hares. Considered narrowly, sexual reproduction is achieved when sperm and egg meet and conception occurs. However, many processes must occur prior to the fusion of egg and sperm to ensure that they meet and that the new zygote has a chance of success. The reproductive organs of the two parents must develop to a state of readiness—illustrated by testicular recrudescence in male rabbits and hares. The parents must time their development of readiness correctly relative to the outside environment so that sperm and egg meet at a time that's promising for success—illustrated by spring mating in rabbits and hares. Male and female must develop attraction toward each other—illustrated by female estrus. Processes such as these are as integral to reproduction as the actual meeting of sperm and egg. After conception has taken place, many processes must occur for the zygote to develop into an independent offspring. The energy and nutrient needs of the developing zygote must be met, for example. In some types of animals this means that offspring must be able to feed themselves at an early date. In others the parents must provide food—illustrated by female rabbits and hares providing milk. When the parents provide food or protection, large amounts of parental time and energy may be invested in the offspring long after sperm and egg have met.

Mechanisms of coordination also deserve emphasis because they play major roles during all stages of the reproductive process. The nervous and endocrine systems are vital, as illustrated by the coordination of copulation and fertilization in induced ovulators (see Figure 17.1). Pheromones are also often vital agents of coordination.

[2] The noun is spelled *estrus*, but the adjective is spelled *estrous*. Thus it is correct to speak of "estrus" and of "estrous cycles."

The Two Worlds of Reproductive Physiology

In the study of physiology, there are always two principal questions to ask about a system, as we stressed in Chapter 1. One is the question of mechanism: How does the system work? The other is the question of adaptation and origins: Why did the system evolve? Chapter 1 emphasizes that these two questions usually need to be answered separately because answering one of the questions does not imply the answer to the other (see page 10).

The biological study of reproduction today is divided between these two questions in an especially vivid way. Part of the reason is that reproduction is staggeringly diverse. To see how great this diversity is, contrast for a moment two vertebrate systems: circulation and reproduction. All vertebrates are sufficiently similar in their circulatory systems that if you study the properties of the circulatory system in one group (e.g., mammals), you already know a great deal about the circulatory system in the other groups. In sharp contrast, when considering reproduction in the full sense diagrammed in Figure 17.2, there are dozens of different major permutations of the reproductive system. Some fish produce millions of offspring each time they reproduce, whereas others of the same body size, living in the same waters, produce fewer than 20. Some mammals breed in the spring, some in the fall. Mammals give birth to live young, birds lay eggs. Birds in the tropics lay systematically fewer eggs in a clutch than birds elsewhere. The offspring of some small mammals are highly developed at birth, whereas those of others are helpless at birth. Gender is generally a genetically determined, fixed trait in mammals. In contrast, gender is determined by environmental temperature in some turtles, and many fish can change gender during their adult lives. These are just a few of the ways in which the reproductive process varies among vertebrates, and variation in invertebrates is even greater.

For some physiologists who study reproduction, the question of adaptation and origins is paramount. The enormous diversity of reproduction inspires deep curiosity about why such diversity has evolved. Physiologists ask questions such as: How can we better understand the evolutionary diversification? How important is the environment in determining the modes of reproduction that have evolved? What are the potential advantages of particular reproductive traits, and can a trait that is advantageous in some contexts be disadvantageous in others? Inevitably, these physiologists take a strongly ecological and evolutionary approach to understanding reproduction.

For other physiologists who study reproduction, mechanism is the central question. As they focus on any particular reproductive system—such as the menstrual cycle in primates—they want to know: What cellular processes are involved? What are all the hormones and neuronal pathways that control the process? How is coordination achieved?

These two approaches to the study of reproduction are not mutually exclusive. They are in fact highly interactive. For instance, as mechanism-oriented biologists elucidate the fine details of mechanisms, their discoveries enhance the capacities of adaptation-oriented biologists to think realistically about evolutionary diversification.

As we now move forward in this chapter, we will first take a brief look at some of the major attributes of reproduction that physiologists study. Then we will devote several sections to the study of adaptive patterns, taking a strongly ecological–evolution-ary approach. Finally we will take a close look at the mechanistic study of reproductive systems, using mammalian reproductive physiology as our focus.

We stress *sexual* reproduction in this chapter. Asexual reproduction occurs in some animals and can be important where it occurs. For instance, a reef coral is a colony of genetically identical hydra-like polyps produced asexually by budding. *Parthenogenesis*—the development of an egg without fertilization—is another type of asexual reproduction that occurs in some animals, such as certain fish, crustaceans, and insects.

What Aspects of Reproduction Do Physiologists Study?

In seeking to understand reproduction throughout the animal kingdom, physiologists investigate all the attributes of animals that they discover are relevant. Thus no simple list can fully describe the topics that draw interest. Nonetheless, whether physiologists are most interested in adaptations and origins or in mechanisms, they focus on roughly a half-dozen aspects of reproduction, which we briefly describe here. **FIGURE 17.3** provides illustrations.

- *Mate association.* Potential mates need to find each other and communicate sexual readiness. Physiologists study the signaling and sensory processes that mediate these activities, including pheromones and behaviors such as estrus.

- *Control of the annual cycle of reproduction.* Most animal species reproduce at particular times of the year. Physiologists study the adaptive implications of seasonality and the mechanisms by which reproduction is coordinated with the annual environmental cycle. They also study seasonal changes in structure and function—and migrations animals undertake to be in suitable environments when they reproduce (e.g., see Chapters 9, 18, and 30).

- *Functions of reproductive organs and cells.* Physiologists study how eggs are made, how erections occur, how sperm cells are sustained and sometimes transformed after being discharged from a male, how milk is synthesized, how nutrients are exchanged in placentas, and all the other activities of reproductive organs and cells.

- *Coordination of the reproductive organs by neural, endocrine, and neuroendocrine processes.* Physiologists study how an animal's reproductive organs are coordinated and controlled during an episode of reproductive activity.

- *Delivery of resources to offspring by parents.* Parents sometimes transfer large quantities of resources to their offspring by prenatal processes (e.g., by synthesizing egg yolk) or postnatal ones (e.g., by bringing food items to young or nursing them with milk). **Offspring provisioning** is a general term that refers to these processes. Physiologists study the implications of provisioning for the energy and nutrient budgets of parents and offspring, and they study the mechanisms of offspring provisioning.

- *Physiology and environmental relations of the young.* Offspring face distinctive challenges because they are small, immature, and often live in environments different from those of their parents. Physiologists study the distinctive

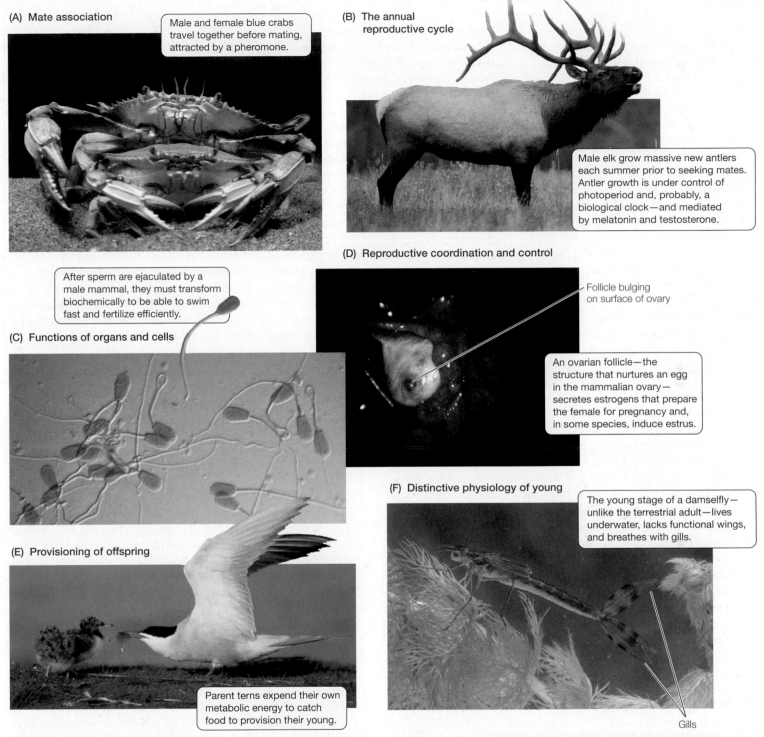

(A) Mate association

Male and female blue crabs travel together before mating, attracted by a pheromone.

(B) The annual reproductive cycle

Male elk grow massive new antlers each summer prior to seeking mates. Antler growth is under control of photoperiod and, probably, a biological clock—and mediated by melatonin and testosterone.

After sperm are ejaculated by a male mammal, they must transform biochemically to be able to swim fast and fertilize efficiently.

(C) Functions of organs and cells

(D) Reproductive coordination and control

Follicle bulging on surface of ovary

An ovarian follicle—the structure that nurtures an egg in the mammalian ovary—secretes estrogens that prepare the female for pregnancy and, in some species, induce estrus.

(F) Distinctive physiology of young

The young stage of a damselfly—unlike the terrestrial adult—lives underwater, lacks functional wings, and breathes with gills.

(E) Provisioning of offspring

Parent terns expend their own metabolic energy to catch food to provision their young.

Gills

FIGURE 17.3 The aspects of reproduction studied by physiologists (A) Female blue crabs (*Callinectes sapidus*) can mate only when they have just molted. Male and female crabs travel together for days prior to mating so the male will be present when the female molts. (B) In the Rocky Mountain elk (*Cervus elaphus*; also called red deer) shown, antler growth—a remarkable form of annual regeneration—is strongly controlled by photoperiod. Evidence from related species suggests that photoperiod entrains an endogenous quasi-annual (circannual) biological clock. (C) Human sperm (shown) and other mammalian sperm are poorly effective at fertilizing eggs when first ejaculated. They must undergo a process termed *capacitation* (discussed later in this chapter) in the female reproductive tract to be able to fertilize efficiently. Capacitation involves biochemical restructuring of sperm membranes and all other sperm parts. (D) In a mammalian ovary (such as the human ovary shown), each egg, as discussed later in this chapter, is nurtured by a follicle. Follicles enlarge, bulge out on the ovarian surface (seen here), and become endocrine structures as they mature. Through its hormonal secretions, a follicle helps prepare a female's body to nurture the egg that will be released when the follicle ruptures. (E) Terns, such as this common tern (*Sterna hirundo*), catch fish by diving from above into fish schools. They feed whole fish to their offspring. (F) Damselflies are insects that start life as aquatic nymphs and later metamorphose into terrestrial adults. Shown here is an aquatic nymph of the shy emerald damselfly (*Lestes barbarus*). The gills are external. When a nymph metamorphoses into a terrestrial adult, its wings emerge fully formed from the wing primordia, which are visible on the dorsal side of the nymph shown.

functional attributes and adaptations of the young at each stage of development—and the maturation process itself—emphasizing that reproduction is successful only if the young succeed in meeting each challenge they face in their environment, so that they survive to attain sexual maturity. In this book, the physiology of young, developing animals is a principal theme in Chapters 4 and 11 and also receives attention in many other chapters, including this one.

Lists such as this are important. Equally important, however, are efforts to understand how all the aspects of reproduction are integrated together in the life histories of animals. Although studies of integration can become very complex, physiologists and other biologists are now undertaking more and more research on integration.

One approach to understanding mechanisms of integration is to identify animals in which diverse reproductive traits occur together in sets and search for the mechanisms that link the traits in each set. Cases are known in which all the traits in a particular set are under the control of a single hormone. Genomic studies have recently revealed that astoundingly, at least in a few cases, all the traits of a set are genetically linked in the genome (**BOX 17.1**).

The Environment as a Player in Reproduction

As already mentioned, when physiologists study reproduction with an emphasis on adaptation and origins, they inevitably take a strongly ecological–evolutionary approach in which the environment is of central importance. The environment sometimes also must be considered even in the case of highly mechanistic studies. For example, as we discussed in Chapter 4, the environment can exert strong effects on the timing of the onset of puberty in human populations. Girls in subsistence cultures are often 2–3 years older at menarche (first menstruation) than girls in well-fed

cultures. Observations such as these emphasize that studies of the mechanisms of puberty cannot ignore the environment in which puberty unfolds.

Speaking in general, why is the environment important to consider? The most direct answer is that when an animal undertakes a reproductive effort, the environment can exert a dominant, controlling influence on the effort's success. Both the physical and the biological environment are important. If offspring are produced during seasons of severe cold or heat, they may die, even though they would have lived if produced when physical conditions were benign. Similarly, if they are produced during a season of low food productivity, they may die of starvation or nutritional deficiencies. Food shortages are legendary for sometimes decimating offspring, thereby nullifying the reproductive efforts of parents. For example, the eggs of most fish have so little yolk that when larvae emerge from the eggs, they soon run out of yolk and confront urgent needs for food. Many instances are known in which virtually 100% of fish larvae have starved because food organisms in their environment were not available in close synchrony with hatching.

For animals that live in environments with regular seasonal cycles, the reproductive cycle is nearly always synchronized with the seasonal cycle. In many places, for example, the spring of the year is marked by benign physical conditions and profuse production of food, such as growth of grasses on land or multiplication of algal populations in aquatic habitats. As we are all well aware, many species of animals have evolved life cycles in which their reproduction is synchronized with the spring season.

Temperature and photoperiod are often used as cues

For the synchronization just mentioned to take place, animals must have evolved suitable control mechanisms. Thus the nature of the control mechanisms becomes an important question for research on reproductive physiology. Among the environmental characteristics used as cues, two—temperature and photoperiod—are most important, and each deserves brief discussion.

In the spring of the year, a rising environmental temperature is often correlated with plant growth and therefore with an increased food supply. One reason for this correlation is that many plants themselves use warm temperatures as cues to grow or flower. These facts and others make it entirely logical that animal species often use a rising environmental temperature as a cue to reproduce. As we are all aware, however, temperature is not an infallible cue. Sometimes high temperatures occur in the middle of winter, or weeks of cold occur during spring. Temperature cueing, therefore, can go awry.

By contrast, **photoperiod**—the number of hours of daylight per 24-h day—varies in a reliable, mathematically exact way with the time of year. The photoperiod on June 1 is always the same, for example. Thus when animals use photoperiod as a cue for spring reproduction, they are using an environmental characteristic that provides unambiguous information.

Although the use of photoperiod as a cue for reproduction is conceptually simple, physiologists have found that when animals use photoperiodic cues, the diversity of *mechanisms* they employ borders on bewildering in its complexity. Several mechanisms have evolved, and the mechanisms used can differ not only among major animal groups but also among related species or even populations within a species. Here we will focus just on mammals.

In mammals two basic mechanisms exist for using photoperiod to synchronize reproduction with the seasons. Some species (e.g., sheep, sika deer, some ground squirrels, and European hamsters [*Cricetus cricetus*]) synchronize their annual reproductive cycles with the seasons by use of internal circannual clocks that endogenously time events for an entire year.[3] In these species, information on the prevailing photoperiod in the environment is used to entrain (adjust) the internal circannual clocks so they are synchronized with annual events in the outside world.

Most species of mammals, by contrast, do not use endogenous circannual clocks. Instead, they continually adjust their reproductive physiology to correspond to the immediately prevailing photoperiod in their environment. For example, many mice use long photoperiods as an indicator of spring and summer, and they remain in a state of reproductive readiness for as long as the photoperiod is long; thus they stay in a state of readiness indefinitely if they are artificially provided with unending long-photoperiod days. Conversely, if they are exposed to short-photoperiod days for an extended period, they stop being in a state of readiness.[4]

In mammals (and also in some other vertebrates), the hormone *melatonin*—secreted by the pineal gland—participates as an endocrine signal in the photoperiodic control of reproduction. Melatonin

is secreted at night. Melatonin secretion is thus inversely related to photoperiod, decreasing as the number of hours of light per day increases (see page 425). Melatonin, however, does not have a simple concentration-dependent effect on target tissues. Instead, patterns of melatonin secretion are best thought of as *codes* that tissues throughout the body can use to obtain information on the photoperiod in the environment.

Animals of temperate climates that breed in the spring often use both temperature and photoperiod to synchronize reproduction with the environment. A common pattern is for individuals to enter a state of reproductive readiness in synchrony with photoperiodic cues—long days—and then mate or spawn when they also experience warm temperatures. Temperature cues are used to "fine-tune" the timing determined by photoperiod.

Within ecological communities, species vary in the strength of their reliance on photoperiod and temperature. Sometimes, as a consequence, global warming can desynchronize relationships among species in a community, because warming (a change of temperature) has no effect on photoperiod. Suppose, to illustrate, that there are two species, animal A and plant B, and suppose that when species A produces young in the spring, the young principally eat the new spring growth of B. Moreover, suppose that A uses principally temperature cues to time its reproduction, whereas B uses principally photoperiodic cues. In a case like this, global warming can induce reproduction in A to occur earlier than in the past (because early spring days are warmer than they have been in the past), while B produces spring growth at the same time in the spring as it did in the past. The young of A, produced early, may then find too little food (see Box 10.1).

Latitudinal trends graphically illustrate the importance of the environment

The intimate relationship between environment and reproduction is vividly illustrated by latitudinal trends in reproduction. As an example, **FIGURE 17.4** depicts the timing of pregnancies at a wide range of latitudes in a set of closely related mouse species (principally the deer mouse, *Peromyscus maniculatus*). Although our knowledge of the cues used by these mice is not complete, both temperature and photoperiod are likely important. Extensive experiments have demonstrated the importance of photoperiod. Under short-photoperiod conditions (much as we saw earlier in rabbits and hares), females cease their cycles of estrus and ovulation (see next paragraph), and in males the testes regress to small size (by apoptosis), cease sperm production, and are withdrawn into the abdomen.[5] These changes are reversed when the mice experience longer photoperiods.

The mice are spontaneous ovulators. During their reproductive season, females that are not pregnant or lactating exhibit an endogenously timed **estrous cycle**, whereby they enter estrus (receptivity to males) and spontaneously ovulate every 6 days. Within the time span of the reproductive season, this cycling continues until the females get pregnant. As Figure 17.4 shows, however, the length of the reproductive season varies dramatically with latitude, being short at high latitudes (where favorable spring and summer conditions do not last long) and longer at mid-latitudes. Whatever

---

[3] *Circannual* means "about a year." The term is used to refer to biological clocks that time processes over an entire year, because such clocks do not keep perfect time and thus are accurate only to approximately a year.

[4] A common phenomenon seen in species of this type—a phenomenon that modifies their relation to photoperiod—is *refractoriness*, a type of noncyclic endogenous timing process. Refractoriness refers to an animal's becoming unresponsive to a photoperiod after being exposed to it for an extended length of time (e.g., several months). Several species of mammals, for example, if exposed continually to short-photoperiod days, eventually cease responding to the short photoperiod and begin to exhibit the reproductive readiness they display during long photoperiods.

[5] Individuals in a single population vary substantially in how rigidly they respond to short days by entering a nonreproductive state.

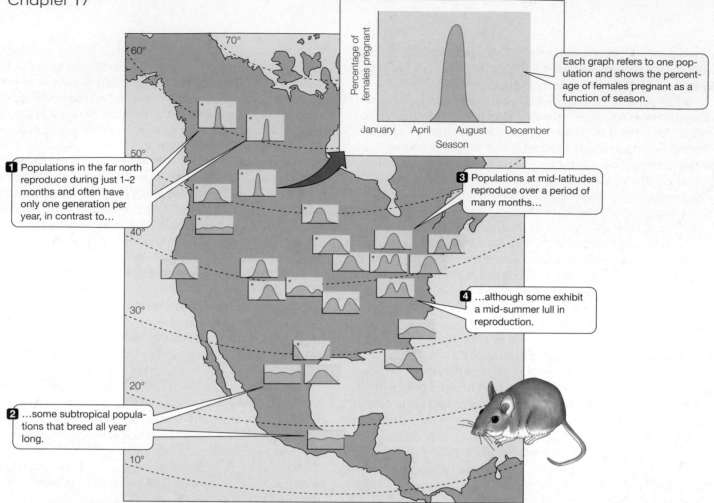

**FIGURE 17.4 Latitudinal trends in the seasonality of repro-
duction in white-footed mice, deer mice, and other related
mice** These mice (members of the genus *Peromyscus*) are often
the dominant mouse-sized mammals in North American ecosystems.
Graphs show the percentage of pregnant females as a function of
season. Each graph refers to one population and is plotted at the
locality where the population was studied. Most of the populations
shown are strongly seasonal. The details of their seasonality depend
on the latitude where they exist, however. All the plots marked with an
asterisk (*) are for populations of a single species, the deer mouse
(*Peromyscus maniculatus*). (After Bronson 1989.)

the length of the reproductive season, when a female becomes
pregnant, her estrous cycle stops. When she gives birth several
weeks later, however, she immediately goes into estrus and thus
can immediately become pregnant again.[6] If she does not become
pregnant in the estrus associated with birth, she later resumes
estrous cycling and can become pregnant in that way. Male mice
in the reproductive season produce sperm constantly.

An important consequence of the latitudinal variation in
length of the reproductive season in the mice is that the number
of generations per year varies with latitude. Newborns become
sexually mature in less than 2 months. Therefore, if the reproductive season lasts 6 months or more—as it does at southern latitudes
(see Figure 17.4)—youngsters can readily have offspring or even
grand-offspring in the same year as their birth. In contrast, only a
single generation per year is likely at high northern latitudes where
the reproductive season is short.

These latitudinal patterns in the reproductive output of *Peromyscus* mice dramatize the critical role the environment can have

in the present. The patterns also have implications for the evolution
of reproduction. When the mice at high latitudes shut down their
reproductive systems in early autumn each year, they do so because
of the reproductive control systems they have evolved. In places
such as northern Canada, young produced late in the year would
likely not survive because of the harsh winter environment, yet
the production of such young would impose costs on the parents.
With these realities in mind, we can hypothesize that the mice have
evolved control systems that ensure young will not be produced at
times later than the early autumn.

The major message of this section is that, for animals living in
their natural environments, reproduction cannot be understood
without taking the environment into account. Thus, to end the
section, it is appropriate to recognize explicitly that the environment
of an animal is, by definition, the animal's full set of surroundings. *Environment* does not simply mean physical conditions, such
as temperature or rainfall. It also includes other organisms that
are present. Organisms used for food can be vitally important
for reproduction, as we have already mentioned. Predators and
parasites can be important too. Even other individuals of the same

[6] This phenomenon is called *postpartum estrus* and is discussed later in this
chapter.

species can be important. For example, male elk (*Cervus elaphus*; see Figure 17.3B) in Scotland—where they are called red deer—rarely father offspring before they are 5 years old, even though they are *physiologically* able to do so by 2 years of age. The social challenges they face in relating to other males of their species account for the difference; a male must be big and experienced enough to dominate other males before he will gain access to females. Similarly, in some species of dragonflies and coral-reef fish, males need to control territories to have ready access to females; males that cannot control territories have relatively little opportunity to mate.

Animals living in distinctive habitats often use distinctive cues for reproduction

In deserts and some other environments, favorable conditions for reproduction occur unpredictably rather than in dependable cycles. To time their reproduction, animals in such environments may not use cyclic cues such as photoperiod but instead respond to the immediate presence of essential resources. Rainfall is often the principal cue to reproduce in desert animals (see Box 21.2).

Along seacoasts, ecologically important rhythms occur because of the lunar tides, which occur on a 24.8-h lunar-day cycle rather than the 24-h solar-day cycle. Animals living or breeding along the seacoasts are believed often to use tidal cues, or possibly direct lunar cues, to time their reproduction.

Reproduce Once or More Than Once?

In this and the next four sections we will continue to discuss animal reproduction from an ecological–evolutionary perspective, emphasizing diversity and adaptation. We have already mentioned some of the paramount questions that define this approach to the study of reproduction, such as: When animals differ in a reproductive trait, why have they diverged and what are the advantages and disadvantages in each case? What are the ecological and evolutionary implications of different types of reproduction?

As we approach reproduction from this perspective, a pivotal consideration is how many times an animal might be expected to reproduce. The answer can affect the way we think of just about everything else. For example, there are species in which each individual is programmed to reproduce at only a single time in its life. In cases like this—considered from a strictly reproductive point of view—there is little or no reason to protect the health or even the life of the parents following reproduction.

When each individual is physiologically programmed to reproduce at only a single time in its life, the condition is described as **semelparity**, and the species is said to be **semelparous**. The four diverse species featured in **FIGURE 17.5** represent some of the best-studied examples, and we will discuss them as a way of learning more about semelparity. The common marine worm

(A) Nereid worm (*Nereis virens*)

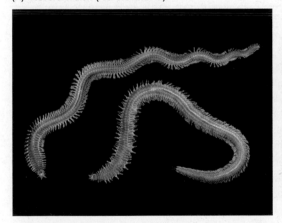

(B) Octopus

(C) Mayfly

(D) Sockeye salmon

FIGURE 17.5 These semelparous species are physiologically programmed to reproduce during only a single period of time in their lives, after which they die (A) The marine polychaete annelid worm *Nereis virens*. (B) The octopus *Octopus vulgaris*. (C) The mayfly *Ephemera danica*. Shown here is the adult stage. Mayflies spend most of their lives as aquatic, water-breathing nymphs. (D) A breeding male sockeye salmon (*Oncorhynchus nerka*). These fish do not feed after they start their upriver migration to spawning areas. They develop hook-shaped jaws that they can no longer close but that are useful in mating competition with other males. They also develop humped backs and brilliant red coloration.

Nereis virens (see Figure 17.5A) lives for several nonreproductive years, burrowing in bottom sediments, before it initiates reproduction.[7] At that point, an individual metamorphoses into a specialized reproductive form called an *epitoke*, which has a distinctive external appearance and accumulates large numbers of sperm or eggs in its coelomic cavity. Eventually, male epitokes swim up into the water and release all of their sperm at the entrances of burrows occupied by females, which release eggs. Both sexes then undergo programmed death (phenoptosis). A neurohormone secreted by the brain controls this life cycle. It is secreted during most of the life of an individual, inhibiting sexual maturation. When secretion stops, partly under photoperiodic control, the animal is doomed: It matures, produces gametes, spawns, and dies. Probably, endocrine or neuroendorine controls are paramount in orchestrating the single act of reproduction in other semelparous animals as well.

Semelparity is common in octopuses. Adult males of the species shown (see Figure 17.5B) place packets of sperm in the reproductive tracts of adult females during mating. Each female subsequently lays about 250,000 fertilized eggs over a period of a month. She then stops searching for food for herself and guards the eggs for 1–2 months, until they hatch. Guarding the eggs is her final act, because she dies soon after her eggs hatch. Males also die at about the same time.

Semelparity is also common in insects, among which the mayflies provide the most dramatic examples. Mayflies spend most of their lives in streams as aquatic, sexually immature nymphs. With the last molt of their exoskeleton—at a time when the individuals of some species are already more than a year old—the mayflies emerge from the water as flying adults (see Figure 17.5C), often in fabulously large swarms. These adults cannot feed for the very fundamental reason that they lack mouthparts! The adults live for only a few minutes in some species—or a few hours or at most a few days in others—during which they mate and lay eggs.

The species of Pacific salmon of North America (genus *Oncorhynchus*) also exhibit semelparity. For example, sockeye salmon (see Figure 17.5D), after growing to adulthood over a period of 1–4 years in the Pacific Ocean, migrate up rivers—sometimes hundreds of miles—before spawning. Both sexes stop feeding as soon as they enter the rivers from the sea. They then must obtain energy to power their upriver migration entirely by catabolizing their own body substance, particularly fat. After reaching their destination, they spawn and die.

Semelparity is exceedingly rare in mammals but not unknown. Males of certain antechinuses ("marsupial mice") of Australia die abruptly after just one reproductive opportunity (**BOX 17.2**).

As already suggested, an important feature of semelparity is that the vigor of parental individuals can be sacrificed to an unusually great extent to provide resources for offspring. In fact, during the stage when semelparous adults prepare for spawning or mating, they often catabolize large portions of their own tissues to provide energy and nutrients for gamete synthesis and, shortly afterward, breeding. By the time Pacific salmon spawn, for example, they are often capable of little else because they have profoundly depleted their body resources.

[7] Details of life history often differ among populations of a single species. Details mentioned in this paragraph and the next are for representative populations of each species.

BOX 17.2 Semelparity in a Mammal

Males of the brown antechinus (*Antechinus stuartii*)—a species of small Australian marsupial, sometimes loosely called a *marsupial mouse*—lead lives governed by pressures to reproduce. Only dominant males succeed in mating with females. Therefore, to be ensured of impregnating a female and passing his genes on to another generation, a male must achieve dominance.

Photoperiod, acting via the pineal gland and its hormone melatonin, causes a precise synchronization of the reproductive cycles of these animals. All the males in a particular population become sexually mature at the same time. For several weeks after becoming sexually mature, males engage in intense fights with other males in an effort to become dominant and succeed in mating.

In about 2 months, after all the females are pregnant, all the males, whether or not they have mated, become sterile and begin to die. Studies show that the immune systems of the males are suppressed, and microorganisms and parasites invade these already stressed animals. In the laboratory, male *Antechinus* live far longer than in the wild, but just like in the wild, they become sterile after the mating period. Their testes fill with connective tissue and no longer produce sperm or hormones.

Most species of animals are **iteroparous**, meaning that individuals are physiologically capable of two or more separate periods of reproductive activity during their lives (i.e., multiple *iterations* of reproduction). In many interoparous species, as we know, individuals reproduce at least once per year for as many years as they live. That is, individuals undergo reproductive *cycles* (which semelparous individuals do not). These cycles are known often to be principally under endocrine and neuroendocrine control, whether ovulation is induced or spontaneous. We have already discussed some of these cycles in our considerations of hares and mice, and later in this chapter we will look in greater detail at the endocrine and neuroendocrine controls.

The question of parental investment in offspring is far more complex in iteroparous animals than in semelparous ones. This is so because when iteroparous parental individuals produce offspring in any one period of reproductive activity, they remain able to produce more offspring in the future. Consider a species in which a female produces 10 offspring per year and has an expected reproductive life span of 10 years. After a female has reproduced for her first time, she has produced only 10 of her possible lifetime total of 100 offspring. To produce the other 90, her own life and health must be preserved. Lifetime reproductive output is one of the most crucial of all factors operative in evolution by natural selection. In iteroparous species, if we think of parents as reproducing early and late in their lives, we would expect the evolution of physiological control systems that—during early reproduction—allocate adequate resources to the parents so that they can reproduce later in life, rather than sacrificing everything for the early offspring.

Eggs, Provisioning, and Parental Care

As will become ever more clear during our discussion of reproduction from an ecological–evolutionary perspective, almost all the features of an animal's reproductive process interrelate with each other. We have already mentioned the question of resources given to offspring by parents as we have stressed other aspects. Now we will center our attention on that question.

The most common way that animals deliver resources to their offspring is by providing provisions in eggs. With this in mind, we quickly see that the norm among animals is to provide offspring with only very small amounts of resources. Small, lightly provisioned eggs that hatch into small, immature, and lightly provisioned offspring probably represent the primitive condition, and eggs of this sort are common in modern aquatic species. In many cases the numbers of these small eggs are enormous. A single female sea urchin, for example, may spew 1 million, or even 20 million, small eggs into the sea in one reproductive event. A single 1-m-long Atlantic codfish (*Gadus morhua*) releases about 3 million eggs each time she spawns, and if she can live long enough to reach a size of 1.3 m, she will release 9 million eggs at each spawning. One can reason that a female has a certain total amount of yolk that she is able to invest in her eggs in a reproductive cycle. Animals like the urchins and cod essentially divide this total into millions of tiny pieces. Millions of eggs can therefore be produced. However, each egg is provisioned with only a tiny bit of yolk, and offspring run out of provisions quickly. Offspring, therefore, must feed themselves soon after hatching, or die.

An important aspect of reproductive diversification is that many animals have evolved strategies of delivering increased quantities of provisions to each offspring. The deposition of more yolk in each egg is perhaps the simplest means of increasing provisions. Many sharks and rays illustrate this phenomenon dramatically, as seen in the species of dogfish shark in **FIGURE 17.6**, in which the amount of yolk per egg is enormous.[8] With so much yolk going to each egg, only a few eggs can be produced; a 1-m-long dogfish shark produces fewer than 20 eggs in a single reproductive event, in sharp contrast to a cod of similar length, which produces 3 million eggs. The advantage of increased provisioning is that yolk-rich eggs—compared with yolk-poor ones—tend to yield bigger, more mature offspring that are more likely to survive. Per generation, 1% of shark eggs can be expected to survive to adulthood, whereas survivorship of cod eggs is often 0.00001%. In addition to sharks and rays, other animals noted for producing eggs that contain relatively large amounts of yolk include birds, turtles, lizards, and other nonavian reptiles—and some invertebrates, such as squid.

Species that provision their offspring with egg yolk have faced a trade-off during their evolution. For any given total parental investment, a species can produce lots of yolk-poor eggs that have little individual chance of survival. Alternatively, the species can produce small numbers of yolk-rich eggs that have a far greater individual chance of survival. There are cases (e.g., bony fish vs. cartilaginous fish) in which related species have followed dramatically divergent evolutionary paths. Biologists as yet cannot offer a full explanation.

Besides yolk-rich eggs, there are other mechanisms of providing lots of provisions to each offspring. These additional provisioning

The *entire* egg of a codfish is about this size in relation to the young dogfish, even though the two species are of similiar adult sizes. Whereas an adult female codfish makes millions of eggs per reproductive event, a female dogfish makes fewer than 20.

2 cm

FIGURE 17.6 Abundant provisioning of offspring with yolk in a species of dogfish shark (*Squalus acanthias*) In this species, after eggs are fully formed, they are retained in the reproductive tract of the mother while the young undergo their initial development. During this developmental phase, the young live off the yolk provided to each egg when the egg was formed; they receive no further nutritional assistance from their mother. The youngsters pictured here have already used more than half their yolk to grow to the stage shown. Note that the sac containing the yolk of each individual is profusely vascularized by the individual's circulatory system; nutrient molecules are mobilized from the yolk by blood flow through the yolk-sac blood vessels. An egg of a codfish is shown at about the same scale for comparison.

mechanisms—which are sometimes combined with yolk provisioning—can be classified as *prenatal* or *postnatal*.

Speaking of *prenatal* provisioning, one mechanism is the transfer of nutrients from the bloodstream of the mother to the bloodstream of each developing offspring. This, of course, is what placental mammals do; the placenta is a structure specialized to facilitate such nutrient transfer, as discussed further later in this chapter. Such nutrient transfer also occurs in placenta-like structures that have evolved in some other vertebrates, including certain lizards, requiem sharks, and hammerhead sharks. A second mechanism of prenatal provisioning is for the mother to produce eggs that do not develop to maturity on their own but instead serve as food for the offspring that develop.[9] Some snails, such as the large whelks of the Atlantic seaboard of the United States, provide examples. These snails deposit numerous eggs in capsules; inside each capsule, only a few individuals develop fully, and as those individuals develop, they feed on the other eggs in the capsule. Similarly, within litters of great white, mako, and porbeagle sharks—which develop in their mother's reproductive tract—some individuals eat others. In a great white shark, the reproductive tract has two branches, and by the time offspring are fully developed, there is only a single survivor in each branch! This mode of provisioning is fundamentally similar

---

[8] Yolk is the only means of provisioning in the species shown.

[9] The eggs that are eaten are sometimes "programmed" to be used in this way, as by being infertile and therefore incapable of development, but in general the process of determining which eggs serve for food is not understood. In invertebrate examples, the eggs that are eaten are called *nurse eggs*.

to yolk provisioning, in that the mother provides the resources to produce the youngsters that end up being used as food.

Considering *postnatal* provisioning, lactation in mammals represents one example (see Figures 17.19 and 17.20). Postnatal provisioning also occurs when parents gather food for their young, as when birds feed fish or insects to their offspring (see Figure 17.3E). Similarly, certain wasps lay their eggs on prey that will serve as food for the offspring that hatch from the eggs.

From an ecological–evolutionary perspective, a central point regarding provisioning of all sorts deserves to be reemphasized: Whereas provisioning aids offspring, it imposes costs on parents. Reproducing parents must eat sufficiently, above and beyond their own needs, to provide the yolk, milk, or hapless siblings they donate to their offspring.

A dramatic point to recognize is that, of all animals, mammals and birds provide the greatest provisions per offspring to their young. Consequently, while young mammals and birds start life with particularly abundant resources, their parents face exceptional reproductive costs per offspring (see Figure 17.19).

External or Internal Fertilization?

External fertilization represents the primitive state in animals. Today it usually persists in the types of aquatic animals that make small, lightly provisioned eggs. Females release the eggs into the water, and males release sperm into the water in the same vicinity. Sometimes, as in certain fish, the males are quite meticulous in matching the spatial pattern of their sperm output to the spatial distribution of the eggs. Still, by the time fertilization occurs, the lightly provisioned eggs are outside the mother and on their own.

Internal fertilization, which has evolved multiple times independently, sets the stage for two phenomena that are also highly relevant to provisioning: (1) the production of encased, yolk-rich eggs and (2) internal development. Large eggs encased in shells or tough capsules—and provisioned with abundant yolk—are made by birds. They are also made by lizards, turtles, and skates and many other cartilaginous fish. Internal fertilization, which occurs in all these groups, is required for such eggs because fertilization must occur inside the female reproductive tract before the shell or capsule (which would block fertilization) is added. Substantial amounts of yolk can also be incorporated.

Fertilization must also be internal if offspring are to undergo their early development inside the reproductive tract of the mother. Placental or placenta-like provisioning can occur only if fertilization is internal.[10]

Internal fertilization occurs by multiple means in various types of animals. In mammals, a penis inserted into the female reproductive tract squirts a fluid containing sperm into the tract.

---

[10] When offspring develop in the reproductive tract of their mother, they exit their mother as small animals—a phenomenon called *viviparity*—instead of exiting as eggs (*oviparity*). Several types of provisioning can occur in cases of viviparity. Sometimes, as exemplified by the dogfish in Figure 17.6, offspring that develop in the reproductive tract of their mother are nourished entirely by egg yolk. This condition has been termed *yolk sac viviparity* or *ovoviviparity*. When, as happens in certain sharks, certain young provision themselves by eating other young while in their mother's reproductive tract, the condition is termed *cannibal viviparity*. Also, of course, nutrients may be transferred from the mother's blood to the offspring's blood across a placenta or placenta-like structure (occasionally called *placental viviparity*).

In most birds, there is no penis; instead, sperm are transferred when a male and female press their cloacal apertures together. In many invertebrates and in sharks and rays, sperm are transferred in physically discrete packets (*spermatophores*); for example, male sharks insert a modified pelvic fin (clasper) in the female genital orifice and wash sperm packets down a groove in the fin.

The Timing of Reproductive Cycles

As we have already stressed, for animals that live in environments with regular seasonal cycles, the reproductive cycle is nearly always timed to coordinate with the seasonal cycle. An important consideration, however, is that there are often several steps in the reproductive process, and potentially it would be adaptive for a species to coordinate some steps separately from others. This is the primary subject we will consider in this section.

In some species, individual steps in the reproductive process are rigidly linked to each other. Therefore, although reproduction is coordinated with the environment, the individual steps in the reproductive process cannot be separately coordinated. Typically in theses cases mating leads promptly to fertilization, fertilization leads promptly to embryonic development, and development adheres to a relatively rigid schedule, culminating in hatching or birth at a relatively fixed time after fertilization. Humans illustrate this sort of sequencing in an especially inflexible way.

Many animal species, however, have evolved mechanisms of decoupling successive steps in the reproductive process, so that the time that elapses between one step and the next is flexible. Such mechanisms increase options for certain steps to be coordinated with environmental conditions independently of other steps. Two common mechanisms are sperm storage and embryonic diapause.

Sperm storage permits flexible timing between copulation and fertilization

Sperm storage in the female reproductive tract is a common mechanism that provides for flexible timing between one major step of the reproductive cycle—copulation—and the next—fertilization. Female blue crabs (*Callinectes sapidus*; see Figure 17.3A) provide a vivid illustration. A female crab is typically able to copulate for only a few hours in her life: the hours that elapse between the time she sheds her next-to-last exoskeleton and the time her final exoskeleton starts to harden. After she obtains sperm in this brief period, however, she may make and fertilize new masses of eggs for a year or more! The mechanism that permits this phenomenon is long-term storage of viable sperm in her reproductive tract. A female crab that copulates in the autumn of one year, for example, may simply allow the harsh months of winter to pass by and then use the sperm from her autumn mating to produce a batch of fertilized eggs in the spring of the next year. Queen honeybees store sperm for years, and certain bats do so for over half a year. Sperm storage occurs also in certain other crustaceans and insects, and in certain sharks, salamanders, turtles, snakes, and birds.

Embryonic diapause permits flexible timing between fertilization and the completion of embryonic development

Embryonic diapause—a programmed state of arrested or profoundly slowed embryonic development—occurs in many sorts of animals and serves to decouple fertilization and the full develop-

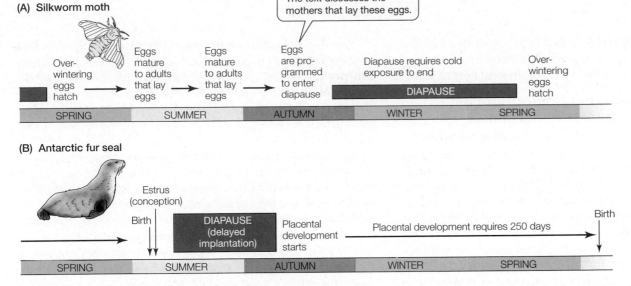

FIGURE 17.7 Embryonic diapause is employed by both of these animals to synchronize reproductive events advantageously with seasonal environmental conditions (A) Female silkworm moths (*Bombyx*) that lay eggs in the autumn produce diapausing eggs that arrest their development and stay arrested until they have had a long period of cold exposure, ensuring they do not hatch until spring. (B) In Antarctic fur seals (*Arctocephalus gazella*), placental development lasts only about 250 days. Diapause—here called delayed implantation—is employed to ensure that placental development does not start until an advantageous birthing season is only 250 days in the future. (There are many varieties of silkworm moths, which differ in details of their life cycles. Shown here is the life cycle of the type termed *multivoltine*. In the fur seal diagram, "placental development" is used as a shorthand to refer to the period between implantation and birth.)

ment of embryos. In cases of embryonic diapause, embryos start to develop but then stop for a while before continuing. The length of the period of arrested development is typically flexible; thus embryonic diapause provides flexibility in the timing between fertilization and completion of embryonic development. Mechanisms of embryonic diapause have evolved independently many times. Because of this, the mechanisms often differ greatly in morphological and biochemical detail from one animal group to another. Here we will discuss insects and mammals. Embryonic diapause is also known to occur among crustaceans and mites among invertebrates and among certain fish, turtles, and birds.

Embryonic diapause and other forms of diapause are common in insects

The eggs laid by silkworm moths (*Bombyx*) provide a particularly well understood example of embryonic diapause in insects. The exact nature of diapause in these eggs depends on the developmental background of the mother. Sometimes egg diapause is deterministic, and that's what we will discuss here. In this case, following copulation and fertilization, the mother lays eggs that are programmed to undergo a complete arrest of development (cessation of mitosis) when they reach the gastrula stage of their embryonic maturation. The eggs are programmed to develop this way while they are still forming in the mother's ovaries: A peptide neurohormone (*diapause hormone*) secreted by cells in the mother's subesophageal ganglion acts on the eggs to program them. These diapausing eggs are laid in the autumn. After they have entered their programmed developmental arrest, they require exposure to cold to emerge from their arrested state; specifically, they must be exposed for at least 2 months to a temperature of about 5°C or lower.[11] The low temperature terminates ("breaks") the dia-

pause; consequently, development resumes soon after the cold exposure. In natural habitats, these interacting processes—the programmed appearance of diapause and the need for extended cold exposure to terminate it—ensure that fertilized eggs laid in the autumn do not erroneously complete development and hatch prior to winter. Instead, the eggs wait until winter has passed to hatch (**FIGURE 17.7A**).

One of the well-studied migratory locusts, *Locusta migratoria*, also provides an instructive example of the use of embryonic diapause (migratory locusts are discussed further in Chapter 4). In a way that's very reminiscent of the deer mice and white-footed mice in Figure 17.4, these locusts exhibit marked latitudinal trends in their seasonality of reproduction. Whereas only one generation is completed per year at cold, high latitudes, two generations may be completed at mid-latitudes, and even three or four generations at still warmer latitudes. Each adult female lays multiple batches of eggs before she dies. The eggs are programmed to develop promptly into new individuals if the mother has experienced long days, but the eggs are programmed to enter embryonic diapause if the mother has experienced short days. The short days of autumn lead all the mothers in a population to lay diapausing eggs. Thus each population experiences a full hiatus in its reproductive output in autumn. Winter cold is required to terminate the diapause. Thus the population regains its reproductive potential only the next spring, when the diapausing eggs—which have overwintered—resume development and mature into the next year's first adults. Just as in the mice, reproduction is physiologically programmed to cease in winter.[12]

Insects as a group, in fact, make extremely common use of programmed quiescent stages to synchronize their reproductive cycles to the annual seasonal cycle. In many species, the larva, pupa, or adult—instead of the egg—is the stage that enters programmed

---

[11] If diapausing eggs are not exposed to cold, they eventually resume development, but not for a very long time—a year or longer.

[12] The details of the locust description given here are from studies of Japanese populations.

quiescence (the quiescence is then called *larval*, *pupal*, or *adult diapause*). Regardless of the stage, programmed quiescence typically acts to interrupt the progress of the life cycle at a stressful time of year.

Delayed implantation and postpartum estrus play important timing roles in mammals

Embryonic diapause is usually called **delayed implantation** in placental mammals because the arrest of embryonic development in these animals occurs when an embryo is at the blastocyst stage of its maturation, a stage that occurs just prior to implantation of the embryo in the uterine wall. In some species, delayed implantation is classified as *obligate* because it is a feature of all pregnancies. Its role in such cases is typically to match the length of the reproductive cycle to the length of the calendar year. Antarctic fur seals (*Arctocephalus gazella*) illustrate delayed implantation in this role (**FIGURE 17.7B**). A female fur seal goes into estrus and mates in the days immediately after she gives birth during the early summer. If she conceives a new offspring, she cannot afford to give birth to it either much earlier or much later than early summer in the following year. If she were to give birth earlier—in the Antarctic spring—her newborn could immediately experience overwhelming cold. If she were to give birth later, her newborn could have too short a period of summer to mature before its first winter. In fact, the average time that elapses between the day a female copulates and the day she gives birth is almost exactly 365 days! Consider, however, that an embryo's placental development in these seals requires only about 250 days. What explains, then, that the time from copulation to birth is 365 days? Delayed implantation. After an egg is fertilized and develops to the blastocyst stage, it goes into developmental arrest (in the uterus) for 3–4 months. It stays in this arrested state for long enough that when, finally, it resumes development and implants (under photoperiodic control), 250 days remain before the proper birthing moment.

Delayed implantation is classified as *facultative* in species of mammals in which it can either occur or fail to occur in any particular pregnancy. White-footed mice (*Peromyscus leucopus*), which are abundant in many North American forests and which we have discussed before (see Figure 17.4), exemplify facultative delayed implantation. Their time between conception and birth can be as short as 23 days or as long as 37 days (a >50% difference)! Facultative delayed implantation is responsible for this variation. In 23-day gestations, no delay of implantation occurs; 23 days, in other words, is the period required for development to be completed without interruptions. Delayed implantation can occur, however, and can be of variable length. A 5-day delay results in a 28-day gestation. A 10-day delay causes a 33-day gestation. Patterns of this sort are common in rodents. Detailed studies of laboratory mice have started to unravel the complex physiology and hormonal controls involved. In those mice, in the midst of a delay, not only is the embryo's development arrested, but also the uterine wall is unresponsive to the embryo's presence (although ultimately the uterus must accept implantation of the embryo). The arrested state is ended by a pulse of estrogen secreted by the mother's ovaries. The estrogen brings the uterus into a receptive state, and the uterus then secretes a uterine hormone, catecholestrogen, which activates the embryo. After this cross-talk between the uterus and embryo, implantation soon occurs, and the embryo completes intrauterine development.

Another reproductive timing mechanism that has evolved in some mammals is postpartum estrus. In species that exhibit **postpartum estrus**, a female becomes receptive to mating—and is capable of conceiving—in the immediate aftermath of giving birth (*postpartum*, "following parturition"). Postpartum estrus occurs in many true seals, sea lions, and fur seals, as we have seen. Horses and zebras, some kangaroos, and rabbits and hares also exhibit it.

In rodents such as we have been discussing, postpartum estrus is common, and it often functions in a coordinated way with delayed implantation to control the timing of the births of successive litters so as to optimize litter spacing relative to parental resources. If a female mouse or rat undergoes postpartum estrus and successfully mates during the estrus, she goes through a period of nurturing two litters of young at once: While one litter (just born) grows in the nest, a second simultaneously grows in her uterus! This process gives the animals great reproductive potential but can be very demanding on the mother. Often, as is true in the deer mice and white-footed mice we previously discussed (see Figure 17.4), the length of time required for an uninterrupted gestation (23 days in the white-footed mouse) is almost the same as that required for nestlings to mature from birth to weaning (21 days). Therefore the litter developing in the uterus can have grown to full prenatal size—placing its greatest prenatal demands on the mother—at the very time the litter in the nest is nearing weaning and consuming milk at a maximum rate. Facultative delayed implantation *is used to prevent this stressful coincidence* by offsetting the development of the two litters. When white-footed mice *without nestlings* conceive litters, they exhibit no delay of implantation; gestation is 23 days. However, when they conceive in a postpartum estrus and thus *have nestlings*, they often exhibit a delay of implantation of a week or longer. Consequently, the intrauterine young are still at a relatively early stage of development when the nestling young reach weaning age and are most demanding.

Some iteroparous animals reproduce only once a year

In certain ways, the most dramatic examples of reproductive seasonality are provided by iteroparous species in which individuals in theory could reproduce many times per year but in fact reproduce just once per year, often at a highly restricted and defined time. The palolo worms of the tropical South Pacific (*Palola viridis*) must rank as the most extraordinary of such animals. Individual palolo worms spend their lives burrowing within the coral reefs. Once a year, the posterior end of each individual transforms into a reproductive entity—a type of epitoke (see page 474)—filled with eggs or sperm. All of these epitokes simultaneously detach from the parent worms, swarm to the sea surface within a span of 30 min, and release their gametes at the time of the third quarter of the moon in October or November! This spectacular process is the single mating event of the year, and individuals do not reproduce again until a year later. The moon phase seems to function in this case as a synchronization factor; when all individuals spawn at the third quarter of the moon, they are assured of spawning together.[13]

Several species of foxes and wolves also breed a single time per year. In their own way, they seem as extraordinary as the palolo

[13] The details of the palolo description given here are for a population in Samoa that has been carefully studied.

FIGURE 17.8 An iteroparous, monestrous mammal The red fox (*Vulpes vulpes*) goes into estrus and ovulates only once a year.

worms. A female red fox (*Vulpes vulpes*) (**FIGURE 17.8**), for example, enters estrus for only a single 1- to 6-day interval—and ovulates just once—in a year. Even if she fails to conceive or loses her young, she next enters estrus and ovulates a full year later. Mammals of this sort are called *monestrous* ("one estrus"). Many species of birds that breed at temperate latitudes also reproduce just once a year; they produce a single clutch of eggs in the spring (usually employing photoperiod and temperature as timing cues), and even if that clutch fails, they do not produce a second.

Large-bodied iteroparous mammals, which have long gestation periods (see Figure 1.11), are also examples of animals that breed just once a year. They face distinctive and intriguing challenges to synchronize their reproductive cycles with the annual seasonal cycle because their lengths of gestation are often a large fraction of a year (see Figure 1.11). To see the challenge, consider a hypothetical mammal with an 8-month gestation. If two adults mate in the late spring or summer—a fine time for them—their offspring will be born in the harsh middle of winter unless a winter birth is prevented by some sort of reproductive specialization.

One option for large-bodied mammals to achieve synchronization of their reproductive cycle and the seasonal cycle is the one we have already discussed: delayed implantation (see Figure 17.7B). However, certain groups of mammals—such as sheep, goats, deer, and primates—have never evolved delayed implantation. The option that some such species in temperate latitudes employ is **short-day breeding**. These species resemble most temperate-zone mammals in that they have distinct periods of the year when they are physiologically ready and unready to reproduce. Unlike most temperate-zone mammals, however, they come into reproductive readiness when the photoperiod (hours of light per day) is short or decreasing, rather than long or increasing. Scottish sheep provide the classic example.[14] In the spring, the males have small testes and little blood testosterone. Days start to shorten following the summer solstice (about June 21), and as the photoperiod declines, male sheep begin by late summer to exhibit increased testis size

[14] As discussed earlier, sheep have an endogenous circannual rhythm that controls annual reproductive events. The rhythm is entrained to photoperiod in the prevailing environment in such a way that reproductive readiness is enhanced as photoperiod decreases.

and elevated testosterone concentration. Mating occurs between early October and mid-December. With a gestation of about 5 months, the young are born between March and May, when spring conditions prevail.

Besides delayed implantation and short-day breeding, a third option for synchronizing the reproductive cycle and seasonal cycle in large mammals is more difficult to document but may in fact be employed by zebras and horses. That option is the specialized evolution of a 12-month placental gestation. The plains zebra has a gestation (without delayed implantation) of almost exactly 12 months, and the wild (Przewalski) horse has a gestation of 11.5 months.

Some large mammals are distinctive among all iteroparous animals in that they are unable to reproduce every year when all the requirements of both prenatal and postnatal development are taken into account. African elephants provide an extreme example. They have a gestation of 22 months—the longest known—and nurse for 3 years. The average time between births is 6–7 years. Some of the large whales have gestations of 1–1.5 years and give birth only every 2–3 years.

SUMMARY
The Timing of Reproductive Cycles

■ In iteroparous animals that live in environments with regular seasonal cycles, the reproductive cycle is nearly always timed to coordinate with the seasonal cycle of environmental conditions, in at least certain ways. Photoperiod and environmental temperature are the most commonly used environmental cues employed to achieve this coordination. The most dramatic aspect of the coordination is that individuals or populations often become physiologically incapable of reproducing during the most unfavorable time of year (winter).

■ Sperm storage and embryonic diapause are commonly employed mechanisms that permit certain steps in the reproductive process to be coordinated relatively independently of other steps. Sperm storage uncouples the times of mating and fertilization. Embryonic diapause uncouples the times of fertilization and completion of embryonic development.

■ Embryonic diapause in placental mammals is called delayed implantation. It may be obligate or facultative. Obligate delayed implantation is typically employed to create down time so as to adjust the total length of the reproductive cycle to be about 365 days. Facultative delayed implantation in small mammals is typically used to create offsets in developmental timing between litters that a mother is simultaneously nursing and nurturing in her uterus.

■ Among small- and medium-sized iteroparous species, a minority reproduce just once per year at highly circumscribed times (e.g., palolo worms, red foxes, some birds). Large-bodied mammals with gestation periods that are lengthy, yet shorter than 12 months, often employ delayed implantation or short-day breeding to ensure that their young are born at a favorable time of year. The mammals of largest size, such as elephants and large whales, require more than a single year to complete a reproductive cycle.

Sex Change

As already mentioned, the mode of sex determination in mammals—rigid genetic determination during early development—is not universal. Some types of animals instead have modes of sex determination in which one individual can assume either sex, often with the individual's environment playing a major role. These phenomena have attracted a great deal of attention from physiologists with both ecological–evolutionary and mechanistic interests.

Here we discuss species in which individuals can change sex during their lifetimes. Teleost fish, gastropod molluscs, and marine annelid worms are the three groups in which this phenomenon is best known. Fertilization is typically external in the species involved; their sex change, therefore, does not require a wholesale alteration of sexual anatomy, such as exchanging a penis for a vagina, or vice versa. However, an individual that has been producing sperm starts to produce eggs instead, or one that has been producing eggs makes sperm. Often, behavior, body coloration, and other secondary sexual characteristics change as well (**FIGURE 17.9**).

The formal name for sex change is **sequential hermaphroditism**. A *hermaphrodite* is an individual that produces both eggs and sperm in its lifetime. A *sequential* hermaphrodite produces eggs at one time and sperm at a different time, rather than both together. A sex-changing species is **protandrous**—meaning "man first"—if individuals start as males and later turn into females (*proto*, "first in time"; *andros*, "man"). A species is **protogynous** if individuals start as females and later become males. In some species, an individual can switch back and forth between the two sexes more than once.

Biologists think that sex change has evolved in cases in which *lifetime reproductive output* is maximized by changing sex. Fish species in which males control harems of females provide an instructive illustration. To control a harem, a male must be large so he can ward off other males. If individuals are physiologically and morphologically able to adopt either sex, we can reason that an individual will do best in terms of leaving offspring if it starts its reproductive life as a female; when young and small, it would be unable to control a harem, and thus would be unable to breed in a consistent way if it were male. When the individual grows large enough to control a harem, however, it will do best in terms of leaving offspring if it is male because then it will be able to mate with many other individuals (an entire harem) rather than being just one female in a crowd of many females. These concepts are believed to explain why many harem-forming species of coral-reef fish have evolved to be protogynous sequential hermaphrodites—starting life as females and converting later to males—rather than adopting a single lifelong sex.

The endocrine basis of sex change, although in general not well understood, is starting to become clear in some of the protogynous harem-forming fish. Sex steroids, such as estrogens, and some other hormones, such as arginine vasotocin (AVT), are involved. In bluehead wrasse (see Figure 17.9), for example, when a female individual has the opportunity to become the dominant male, she undergoes a rapid fourfold increase in hypothalamic transcription of mRNA for AVT, and the consequent increase in AVT seems to account for a sharp increase in her expression of male sexual behavior.

Another type of fluid sex determination is seen in sea turtles and many other nonavian reptiles. In these egg-laying animals, nest temperature determines whether eggs develop into males or females—a phenomenon called **temperature-dependent sex determination**.

(A) Female

(B) Male

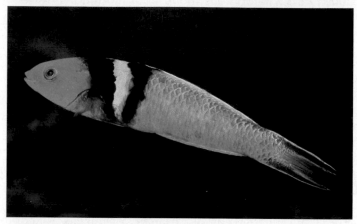

FIGURE 17.9 Sex change: A bluehead wrasse (*Thalassoma bifasciatum*) when female and, later, when male Individuals of this species of coral-reef fish are often protogynous sequential hermaphrodites. In this case, an individual starts life as a female (A) but later becomes male (B) if the opportunity arises. (A courtesy of Aaron M. Florn.)

Reproductive Endocrinology of Placental Mammals

Now we turn to a focus on the mechanistic study of reproduction. Just as we have included information on mechanisms in earlier sections of the chapter, we will mention ecological–evolutionary thinking at times in this section. The two viewpoints—mechanistic and ecological–evolutionary—are both crucial for a full study of reproductive physiology.

Endocrinology is the preeminent topic in the mechanistic study of reproduction. For exploring reproductive endocrinology in depth, humans and other mammals provide excellent illustrations for several reasons. First, mammalian reproductive endocrinology is the best understood in the animal kingdom and often serves as a model for studies of other animals. In addition, of course, we are intrinsically interested in our own reproductive endocrinology, and knowledge of it has important applications, as in contraception. As we discuss reproductive endocrinology, we will also integrate information on other related topics, such as morphology and gamete production. Most species of mammals are spontaneous ovulators, and in discussing endocrinology in this section, we assume spontaneous ovulation unless stated otherwise.

Females ovulate periodically and exhibit menstrual or estrous cycles

According to the prevailing view of most investigators, female mammals produce all the egg cells—oocytes—they will ever have during their fetal life. A female's total number of egg cells—a crucial factor for her fertility—is thus defined at birth. Recent studies on mice have suggested, to the contrary, that additional egg cells may be produced after birth and into adulthood from germline stem cells; this evidence is still being evaluated. In the fetal ovaries, diploid germ cells called **oogonia** divide repeatedly by mitosis to produce many oogonia. Each of these oogonia undergoes the very first stage of meiosis—during which its chromosomes replicate (see Appendix H)—during fetal life or shortly afterward. The cells are then called **primary oocytes**. Thereafter the cells remain as primary oocytes, without completing the first meiotic division, until they are ovulated following a female's attainment of reproductive maturity at puberty.

After a female reaches sexual maturity, one or a few of the primary oocytes in her ovaries mature and undergo ovulation—release from the ovaries—during each of her ovulation cycles. Because all the primary oocytes are formed during fetal life, some must remain in the ovaries, awaiting ovulation, for many years in long-lived mammals. Oocytes of elephants and some whales, for example, may remain viable for 60 years! An individual primary oocyte completes its first meiotic division at the time it is ovulated. It will complete its second meiotic division only if it is fertilized.

In most species of mammals that display spontaneous ovulation, females ovulate in cycles until they become pregnant (at which time the pregnancy interrupts the ovulation cycles). Primates such as humans, gorillas, and chimpanzees menstruate in each cycle that does not result in pregnancy; that is, after pregnancy has failed to occur, they shed the uterine lining as a blood-tinged discharge from the vagina. Their cycles are thus called *menstrual cycles*. In other species of mammals, females do not menstruate. Instead, the principal outward manifestation of their ovulation cycles is that they go into estrus (heat) in approximate synchrony with ovulation. As mentioned previously, their cycles are termed *estrous cycles*, and during estrus they use behaviors and other signals, such as pheromones, to indicate that they are sexually receptive. Estrus lasts, depending on the species, for a few hours to a few days during each cycle. Menstrual and estrous cycles, as indicated, are named for their outward manifestations: menstruation and estrus. An important fact to recognize is that these outward manifestations occur at dramatically different points in the underlying physiological cycles. Whereas estrus occurs at the same time as ovulation, menstruation occurs long after ovulation, when the uterine lining is being shed following a failure of pregnancy to be achieved.

One of the most important universal properties in mammalian reproduction is that ovulation always occurs in response to a surge of luteinizing hormone (LH) released from the anterior pituitary gland. You will recall that in induced ovulators, copulation itself sets in motion the chain of events that culminates in the LH surge (see Figure 17.1). In spontaneous ovulators, the LH surge is produced endogenously in the female (independent of copulation) by a series of interacting endocrine and neuroendocrine events over the course of the menstrual or estrous cycle.

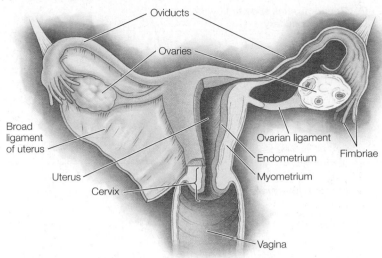

(A) Internal organs (frontal view)

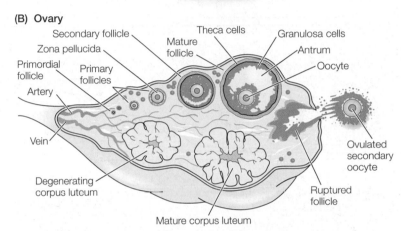

(B) Ovary

FIGURE 17.10 The human female reproductive system (A) The internal organs of the female reproductive system viewed from the front. The oviducts are sometimes called *fallopian tubes*. Whole organs are shown at the left, whereas the organs on the right are shown in sagittal section; moreover, the vagina and cervix are fully in sagittal section, and the broad ligament is omitted on the right. (B) An ovary, showing the developmental stages of follicles and corpora lutea. Each ovary is about 2–4 cm long. At any one time, it contains follicles and corpora lutea in various stages of development. In this diagram, the earliest stages are shown at the upper left, and successively more mature stages are depicted in order in a clockwise loop. This arrangement is for pedagogical purposes only: All the stages can in fact be found throughout the ovary.

The status of the uterus is modulated during each menstrual or estrous cycle. The inner lining of the uterus is prepared for pregnancy in coordination with ovulation. If pregnancy fails to occur, the lining regresses and then is prepared again in the next menstrual or estrous cycle.

We will use the human menstrual cycle as our common thread in describing the endocrine controls of these processes, while also noting parallel aspects of estrous cycles. In preparation for a full discussion of the human menstrual cycle, we first need to look at the cellular and morphological changes that occur in the ovaries and uterus. **FIGURE 17.10** shows the anatomy of the human female reproductive system. **FIGURE 17.11** displays many of the major morphological and endocrinological events in the human menstrual cycle. In Figure 17.11, all four parts (A–D) are plotted in relation to a single time axis, shown at the bottom. The beginning of the first

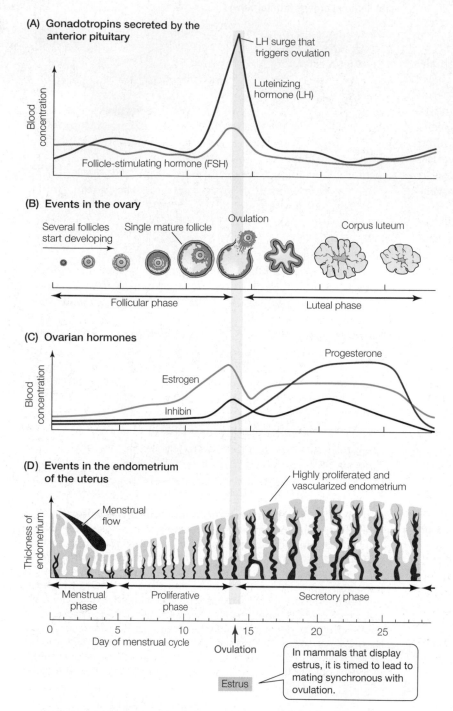

(A) Gonadotropins secreted by the anterior pituitary

LH surge that triggers ovulation

Luteinizing hormone (LH)

Blood concentration

Follicle-stimulating hormone (FSH)

(B) Events in the ovary

Several follicles start developing

Single mature follicle

Ovulation

Corpus luteum

Follicular phase

Luteal phase

(C) Ovarian hormones

Blood concentration

Progesterone

Estrogen

Inhibin

(D) Events in the endometrium of the uterus

Highly proliferated and vascularized endometrium

Thickness of endometrium

Menstrual flow

Menstrual phase

Proliferative phase

Secretory phase

0   5   10   15   20   25

Day of menstrual cycle

Ovulation

Estrus

In mammals that display estrus, it is timed to lead to mating synchronous with ovulation.

FIGURE 17.11 A synoptic view of events in the human female reproductive cycle The blood concentrations of (A) anterior pituitary hormones and (C) ovarian hormones vary in relation to cellular changes in (B) the ovaries and (D) the uterus in a highly orchestrated way. All four parts of this figure are plotted in relation to a single time axis (at the bottom), with the first day of menstruation considered to be "day 0." In mammals with estrous cycles, events are basically similar, although there are many differences of detail. In such animals, as shown at the bottom of the figure, estrus (heat) is coordinated to occur synchronously with ovulation; this coordination is achieved by effects of estrogen (see part C) and other hormones on brain function. Estrus lasts from a few hours to a few days, depending on the type of mammal, whereas ovulation occurs at a single point in time. Note that menstruation and estrus—the outward manifestations of menstrual and estrous cycles—occur at dramatically different times of the underlying physiological cycle; menstruation occurs roughly halfway between one ovulation and the next, whereas estrus occurs when ovulation occurs.

day of menstruation is arbitrarily considered to be "day 0" on that axis, and the axis covers 28 days, the average length of a menstrual cycle.[15] As we discuss the cellular and morphological changes in the ovaries and uterus, we will focus on parts (B) and (D) of the figure. Later we will consider all four parts of Figure 17.11 together as we integrate all of the events that occur in the menstrual cycle.

EVENTS IN THE OVARIES Soon after a female is born, each primary oocyte in her ovaries becomes enclosed by a single layer of somatic cells. An oocyte and its layer of somatic cells are then together called a **primordial follicle**. After a female has reached reproductive maturity, a subset of her primordial follicles is recruited to mature further during each menstrual or estrous cycle—a process termed **folliculogenesis**. The part of the menstrual or estrous cycle during which follicles mature is known as the **follicular phase** of the cycle.

Early in the follicular phase, the somatic cells enclosing each recruited primordial follicle become more cuboidal and come to be known as **granulosa cells**. This marks the transformation of the follicle to a **primary follicle**. As further maturation occurs, the primary oocyte increases in size and secretes around itself a noncellular layer of glycoproteins, the **zona pellucida**. In addition, the granulosa cells increase in number and form multiple layers. The next step is maturation into a **secondary follicle** (see Figure 17.10B). In this process, a fluid-filled cavity, the *antrum*, opens up within the layers of granulosa cells. Moreover, connective tissue at the outer margin of the granulosa cells differentiates into a layer of **theca cells**.[16]

Although many follicles begin to mature in the ovaries at the start of each follicular phase, only some of the follicles attain full maturity. In species that ordinarily give birth to a single offspring, such as humans, just one *dominant follicle* matures fully in most menstrual or estrous cycles. That follicle grows to be especially large by mitotic proliferation of its granulosa and theca cells and expansion of the fluid-filled antrum. Ultimately, when the follicular phase of the cycle ends with the LH surge, the oocyte of this dominant follicle will complete its first meiotic division (a division that began when the female was a fetus in her mother's uterus); one of the daughter cells produced by this division will receive almost all the cytoplasm and will become the **secondary oocyte** that will be fertilized if mating is successful (see Appendix H). When the follicle is fully developed (see Figure 17.10B), it is known as a **mature**, **Graafian**, or **preovulatory follicle**. It becomes positioned just under the outer epithelium of the ovary and bulges outward into the space surrounding the ovary (see Figure 17.3D).

---

[15] Normal human menstrual cycles are 21–42 days long, and normal menstrual periods last 2–8 days.

[16] Although maturation to the secondary follicle stage requires follicle-stimulating hormone (FSH) and LH, development to the primary follicle stage is not believed to require hormonal support and therefore can occur without FSH and LH.

In humans, whereas a late primary follicle is only about 0.02 cm in diameter, a fully mature follicle is 1.5–2.0 cm in diameter! All the other follicles that initiated maturation at the start of the follicular phase—that is, all those except the single dominant one—degenerate by a process of programmed cell death termed **atresia**.[17] In species of mammals that normally give birth to litters of multiple offspring, atresia also occurs, but multiple follicles develop to full maturity and undergo ovulation.

During the maturation of a follicle, two-way communication between the oocyte and the granulosa cells coordinates the follicle's development. The oocyte—surrounded by the zona pellucida—and the granulosa cells nearest it make cytoplasmic connections across the zona pellucida by means of gap junctions. Both the oocyte and granulosa cells also secrete paracrine agents. The granulosa cells provide signals and nutrients that support the oocyte's development and influence which oocyte genes are expressed. The oocyte, in turn, paces follicular development and sends signals that stimulate granulosa proliferation.

When the LH surge activates the process of ovulation, enzymes secreted by the fully mature follicle break down the thin layer of overlying ovarian epithelium and the juxtaposed follicular wall. Antral fluid pours out of the ruptured follicle into the space surrounding the ovary, carrying with it the oocyte surrounded by the zona pellucida and attached granulosa cells (see Figure 17.10B).

Fertilization of an ovulated oocyte, if it occurs, normally takes place in the oviducts (also called the fallopian tubes). Each ovary is positioned near the opening of an oviduct (see Figure 17.10A), and oocytes released from the ovary are swept into the opening, as by ciliary action.

In the ovary, ovulation marks the end of the follicular phase and the start of the **luteal phase**. The cells of the ruptured mature follicle within the ovary reorganize through proliferation, vascularization, and other processes to form a structure called the **corpus luteum** (plural *corpora lutea*) (see Figure 17.10B). If the oocyte in the oviduct is not fertilized, the corpus luteum—known as a **corpus luteum of the cycle**—secretes hormones for a finite time and then degenerates. In humans a corpus luteum of the cycle functions for about 10 days and then starts to regress; in the average cycle, the luteal phase ends on day 28, when the corpus luteum stops functioning entirely. If fertilization occurs, the corpus luteum—known as a **corpus luteum of pregnancy**—grows even further and continues to secrete hormones that are essential for establishing and maintaining pregnancy. In some types of mammals, such as many rodents, the luteal phase is highly attenuated because the corpus luteum of the cycle is very short-lived.

A key point to recognize is that both the follicles and corpora lutea are endocrine structures. In fact, when we speak of "ovarian hormones," we are really speaking of secretions of the follicles and corpora lutea.

EVENTS IN THE UTERUS The glandular, epithelial lining of the uterus, known as the **endometrium** (see Figure 17.10A), cyclically prepares for pregnancy, and then, if pregnancy fails to occur, reverts to an unprepared state (see Figure 17.11D), as already mentioned. In mammals with menstrual cycles—namely humans and some other primates—the endometrium sloughs off during menstruation and then regrows as the uterus prepares again for pregnancy. In species of mammals with estrous cycles, the endometrium is not sloughed off but is simply resorbed.

The ways in which the phases of the uterine cycle in humans relate to the phases of the ovarian cycle are seen in Figure 17.11 (compare parts B and D). The **menstrual phase** of the uterine cycle, lasting about 5 days, is the part when sloughing occurs. It is followed by the **proliferative phase**, during which the endometrium undergoes rapid thickening—from about 1 mm thick to 3–5 mm thick—and redevelops glands and circulatory vascularization. These uterine phases occur simultaneously with the follicular phase in the ovaries. The uterine proliferative phase is synchronized to reach completion approximately when ovulation occurs. The uterus then enters its **secretory phase**, which overlaps with the ovarian luteal phase. In the secretory phase, the endometrium matures to the point that it is fully ready to accept implantation of an embryo and provide the embryo with nutritive support during its early postimplantation development. Implantation occurs about midway through the secretory phase if the egg that is ovulated is fertilized.

ENDOCRINE CONTROL OF THE FOLLICULAR PHASE Menstrual and estrous cycles are orchestrated by neurohormones released from the brain and by hormones released from the anterior pituitary gland and ovaries. To understand the human menstrual cycle, we need to follow events occurring over time at four different sites in the body: the hypothalamus, anterior pituitary gland, ovaries, and uterus. Figure 17.11 synthesizes many of the major endocrinological events and shows how they synchronize with cellular and morphological events. For convenience, we divide our discussion into three sections, this one on the follicular phase (see Figure 17.11B) and the next two on ovulation and the luteal phase.

Two gonadotropic hormones (gonadotropins) secreted by the anterior pituitary gland play major roles in controlling menstrual and estrous cycles: **luteinizing hormone (LH)** and **follicle-stimulating hormone (FSH)**. The secretion of LH and FSH is controlled by **gonadotropin-releasing hormone (GnRH)**, a neurohormone released by GnRH-secreting neuroendocrine cells in the hypothalamus of the brain (see Appendix K), and by hormones—discussed later—secreted by the ovaries. The GnRH-secreting cells—which receive and integrate a variety of neural and neuroendocrine inputs—generate bursts of action potentials roughly every 1.5 h (with some variation depending on the phase of the menstrual or estrous cycle), and these bursts lead to synchronous release of GnRH by means of exocytosis. Accordingly, GnRH is secreted in pulses.[18] The pulses of GnRH are carried by blood flow in the hypothalamo–hypophysial portal system to the anterior pituitary gland (see Figure 16.7). There they stimulate the secretion of LH and FSH in pulses.

---

[17]A human female has 200,000–400,000 primary oocytes in her ovaries at the time of puberty. Over her lifetime, only about 400 of these oocytes will mature and be ovulated. The rest will be lost through the atresia that occurs in each menstrual cycle.

[18] The rate and amplitude of the pulsed release are important factors in determining the actions of GnRH. The pulsed pattern is in itself highly significant in that artificially induced steady release of GnRH leads, in many circumstances, to strong suppression of LH and FSH secretion.

FIGURE 17.12 Hormonal control of estrogen production and secretion by an ovarian follicle The top of this diagram shows the base of the brain and the pituitary gland (see Figure 16.7). The bottom shows a maturing follicle in the ovary. Estrogen (principally estradiol) is synthesized and secreted by each such follicle. Estrogen synthesis by a follicle requires both LH and FSH and entails cooperative function of the theca and granulosa cells.

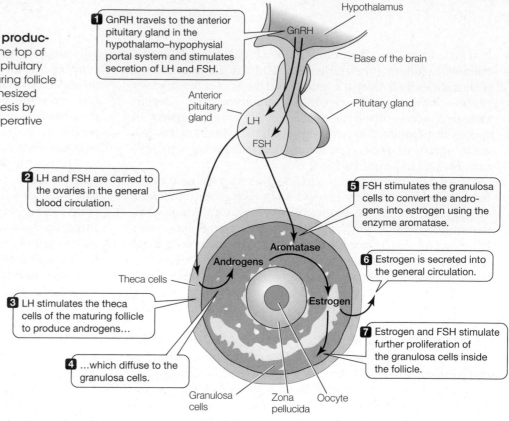

1 GnRH travels to the anterior pituitary gland in the hypothalamo–hypophysial portal system and stimulates secretion of LH and FSH.

2 LH and FSH are carried to the ovaries in the general blood circulation.

3 LH stimulates the theca cells of the maturing follicle to produce androgens…

4 …which diffuse to the granulosa cells.

5 FSH stimulates the granulosa cells to convert the androgens into estrogen using the enzyme aromatase.

6 Estrogen is secreted into the general circulation.

7 Estrogen and FSH stimulate further proliferation of the granulosa cells inside the follicle.

The GnRH-secreting neuroendocrine cells in the hypothalamus are influenced by many factors and in many ways serve as the final, integrative communication path for multiple inputs affecting reproductive physiology—inputs that converge on those cells. Malnutrition or stress, for example, can impair GnRH secretion (and therefore reproduction). In seasonal breeders, environmental signals such as photoperiod—integrated by the nervous system—often affect a female's reproductive status by influencing the pulsatile activity of the GnRH-secreting cells. **Kisspeptin neurons** in the forebrain—relatively recently discovered—play critical roles in the control of the GnRH-secreting cells. Because kisspeptin neurons have receptors for steroid hormones, they can respond to steroid (e.g., estrogen) signals, and they possibly represent the principal pathway by which the GnRH-secreting neurons are affected by steroid signals. Kisspeptin neurons act on the GnRH-secreting cells by secretion of **kisspeptin**, a peptide that acts—sometimes at exceedingly low concentrations—as an excitatory signal to the GnRH-secreting cells. Although we will discuss kisspeptin neurons chiefly in relation to the menstrual and estrous cycles, they are also critically involved in controlling the early stages of puberty. The recent discovery that kisspeptin neurons help control early puberty is exciting because the mechanism of puberty initiation remains unknown and represents one of the great remaining mysteries in the study of the life cycle.

During the follicular phase of the ovarian cycle, blood concentrations of LH and FSH remain relatively flat until a few days before ovulation (see Figure 17.11A). Then they increase dramatically, peaking at the time of ovulation. As the blood concentrations of LH and FSH rise, both hormones act on the follicles developing in the ovaries. The initial target of LH in each maturing follicle is the layer of theca cells that encloses the follicle. FSH acts on the granulosa cells. These two target tissues, working in concert, result in secretion of the steroid hormone **estrogen** by the follicles (we use the generic term *estrogen* for simplicity, although in fact two or more chemically specific estrogens are produced, the principal one being *estradiol*). As outlined in **FIGURE 17.12**, when LH binds to theca cell receptors, those cells are stimulated to produce androgens. The androgens diffuse through the theca cell membranes to the nearby granulosa cells. The binding of FSH to granulosa cell receptors stimulates the action of the enzyme **aromatase**, which converts the androgens into estrogen by a chemical process termed **aromatization** (see Figure 16.2). The granulosa and theca cells "need each other" to synthesize estrogen because only the theca cells can synthesize the androgens that are the precursors of estrogen, and only the granulosa cells can produce aromatase, which is required to convert androgens to estrogen.

Blood levels of estrogen increase slowly during the early stages of the follicular phase, then rise dramatically prior to ovulation (see Figure 17.11C). Estrogen itself plays a role in this increase by acting locally as a paracrine/autocrine agent within the ovary. In combination with FSH and growth factors, estrogen stimulates the proliferation of granulosa cells. This is a classic example of *positive feedback* (see Box 1.1) in that estrogen produced by the granulosa cells stimulates mitotic multiplication of the granulosa cells themselves (see Figure 17.12), and then as the cells increase in number, they produce increasing amounts of estrogen, a process that supports their continued proliferation.

The granulosa cells of the developing follicles change in their ability to respond to LH as they mature. Although they do not possess receptors for LH during the early part of follicular development, the granulosa cells of dominant follicles later produce LH receptors and incorporate them into their cell membranes. The LH receptors of the mature granulosa cells permit these cells to respond to the surge of LH that triggers ovulation.

The estrogen secreted by the follicles affects the anterior pituitary gland and probably the hypothalamus. The effects of estrogen on these structures depend on its concentration in the blood. During the early part of the follicular phase, when blood estrogen levels are low, estrogen exerts a negative feedback effect that keeps blood levels of LH and FSH low. During the late part of the follicular phase, however, estrogen (now at a high blood concentration; see Figure 17.11C) causes increased secretion of LH and FSH by the gonadotropin-secreting cells in the anterior pituitary, and it also—in most studied species—causes increased GnRH secretion by the GnRH-secreting neuroendocrine cells in the hypothalamus. This effect on GnRH secretion is mediated by kisspeptin neurons; current evidence indicates that high levels of estrogen stimulate kisspeptin neurons to deliver excitatory signals to the GnRH-secreting cells to increase their secretion of GnRH. The increased GnRH secretion,

TABLE 17.1 Endocrine and neuroendocrine cells and the major secretions involved in ovarian cycles and pregnancy in female mammals, emphasizing humans

Cells	Hormone	Action
Follicular phase		
Hypothalamic GnRH cells	GnRH	Stimulates secretion of FSH and LH from anterior pituitary cells
Anterior pituitary gonadotropin cells	LH	Stimulates theca cells to secrete androgens during the follicular phase Surge of LH triggers final maturation of the oocyte and ovulation Following ovulation, LH initiates transformation of follicle cells to the corpus luteum
	FSH	Stimulates aromatase action in granulosa cells to convert androgens to estrogen
Theca cells of ovarian follicle	Androgens	Diffuse from theca cells to granulosa cells for conversion to estrogen
Granulosa cells of ovarian follicle	Estrogen (principally estradiol)	Acting together with FSH, stimulates proliferation of granulosa cells At low concentrations, has a negative feedback effect on the anterior pituitary (and possibly the hypothalamus), keeping FSH and LH secretions low At high concentrations, has a positive feedback effect on the anterior pituitary and often the hypothalamus, promoting the LH surge Promotes estrous behavior in species that exhibit estrus Promotes growth of endometrium and development of endometrial progesterone receptors
	Inhibin	Inhibits FSH secretion
Luteal phase		
Corpus luteum cells	Progesterone (the dominant secretion)	Causes endometrium to become secretory; promotes relaxation of uterine and oviduct smooth muscles
	Estrogen	Acting together with progesterone, reduces secretion of FSH and LH from the anterior pituitary; thereby greatly suppresses folliculogenesis in primates and slows folliculogenesis in other mammals
	Inhibin	Inhibits FSH secretion
Pregnancy		
Placental cells	Chorionic gonadotropin	Secreted by embryonic placental cells in primates and horses, in which it rescues the corpus luteum and ensures the maintained function of the corpus luteum
	Progesterone	Opposes stimulatory effect of estrogen on uterine smooth muscle until late in pregnancy Stimulates the secretion of prolactin from the anterior pituitary Acting together with estrogen and prolactin, promotes growth of mammary glands
	Estrogen	Acting together with progesterone and prolactin, promotes growth and development of mammary glands Acting together with progesterone, prevents milk secretion by mammary glands Prepares the uterine smooth muscle for parturition by promoting production of oxytocin receptors and synthesis of connexins that form gap junctions between muscle cells Stimulates enzymatic breakdown of cervical collagen fibers, thereby softening the cervix
	Lactogen (chorionic somato-mammotropin)	Alters maternal glucose and fatty acid metabolism to shunt glucose and fatty acids to the fetus; may contribute to development of capacity for lactation

combined with the enhanced secretory capability of the pituitary gonadotropin-secreting cells that respond to GnRH, culminates in a huge output of LH—the LH surge—that triggers ovulation.

In addition to its effect on the anterior pituitary and hypothalamus, estrogen stimulates growth of the endometrium of the uterus by binding with specific receptors in the endometrial cells. Indeed, the rise in blood levels of estrogen during the late part of the follicular phase in the ovary (see Figure 17.11C) is responsible for the proliferative phase of the uterus (see Figure 17.11D). Among its many effects, estrogen stimulates the endometrial cells to produce receptor molecules for **progesterone**, another steroid hormone, in preparation for events that occur after ovulation.

During the final part of the follicular phase, the granulosa cells of dominant follicles increase their secretion of **inhibin**, a hormone that inhibits FSH secretion from the anterior pituitary. The first section of **TABLE 17.1** summarizes the cells and hormones that are active during the follicular phase.

ENDOCRINE CONTROL OF OVULATION When the anterior pituitary gland secretes LH in amounts sufficient to produce the surge in blood LH concentration (see Figure 17.11A), the LH exerts multiple effects on the granulosa cells of the dominant follicle (or follicles) that bring about dramatic events within hours. LH causes the granulosa cells to secrete chemical mediators that induce the oocyte to complete its first meiotic division (see Appendix H). LH also causes the granulosa cells to begin secreting progesterone, decrease their secretion of estrogen, and release enzymes and prostaglandins that lead to breakdown of the outer follicular membranes and overlying ovarian epithelium. As the membranes and epithelium rupture, antral fluid and the oocyte are released from the ovary. Finally, LH initiates morphological and biochemical changes in the remaining granulosa cells and theca cells, causing them to transform into a corpus luteum.[19]

ENDOCRINE CONTROL OF THE LUTEAL PHASE After the LH surge is over, the newly formed corpus luteum begins to function. At this time the anterior pituitary gland secretes a low level of LH, which maintains the corpus luteum. The corpus luteum secretes progesterone (which now becomes a major ovarian hormone), estrogen, and inhibin, and during the middle of the luteal phase, blood levels of these hormones increase, peaking in association with maximum endometrial thickness (see Figure 17.11C). Progesterone is the major hormone secreted by the corpus luteum and rises to a particularly high blood concentration in the luteal phase. Progesterone and estrogen, acting in concert, exert negative feedback on the anterior pituitary (and probably the hypothalamus) to keep GnRH, LH, and FSH secretions low. Inhibin also suppresses FSH secretion.

In primates, development of new follicles is greatly suppressed during the luteal phase, mediated in part by the low blood levels of LH and FSH that exist during that phase. In some other mammals, folliculogenesis can occur in all phases of the cycle but is reduced during the luteal phase.

The corpus luteum is essential for establishing conditions that permit implantation and pregnancy. Progesterone secreted by the corpus luteum in the luteal phase is especially important in preparing the uterus for implantation. The endometrial lining of the uterus is able to respond to progesterone at this time because of the process noted earlier, that estrogen secreted in the follicular phase stimulated the endometrium to synthesize progesterone receptors. Stimulated by progesterone, the exocrine glands of the endometrium secrete glycogen and enzymes, and additional blood vessels develop in the endometrium. The thickened and secretory endometrium is poised then to support implantation and nourishment of an embryo[20] if fertilization occurs. Progesterone also inhibits contractions of smooth muscle cells in the walls of the uterus and oviducts. The second part of Table 17.1 summarizes the hormones that are active during the luteal phase: the part of the cycle dominated by the corpus luteum.

In primates (although not all mammals), the cells of the corpus luteum have a limited life span in the absence of pregnancy. If pregnancy occurs, the embryo, in many mammal species, provides a hormonal or paracrine signal that rescues the corpus luteum by preventing its degeneration, as detailed in the discussion of pregnancy. However, if pregnancy does not occur, the luteal cells stop secreting hormones and degenerate. This process is completed within about 14 days after ovulation in humans, as mentioned earlier.

One important effect of the degeneration of the corpus luteum at the end of the luteal phase is that it sets the stage for a new cycle of folliculogenesis (i.e., the start of a new follicular phase). Without a functional corpus luteum, blood concentrations of progesterone, estrogen, and inhibin decrease to the low levels that characterize the end of the luteal phase and the start of the follicular phase (see Figure 17.11C). The hypothalamus and anterior pituitary gland are thereby freed of negative feedback, so that secretion of LH and FSH can rise again, initiating another round of follicular development in the ovaries.

Another important effect of the degeneration of the corpus luteum occurs in the uterus: The endometrium—because it is no longer supported by luteal hormones—is resorbed or, in species that menstruate, sloughed off. In mammals that menstruate, the endometrium responds to the loss of luteal hormones by secreting prostaglandins, which initiate deterioration of its superficial layers. The prostaglandins cause constriction of the uterine blood vessels and contraction of the myometrium, the smooth muscle in the uterine wall. The closing off of the blood supply prevents O_2 and nutrients from reaching the cells in the endometrium, and the cells begin to die. The superficial layers of cells are then lost in menstrual flow, leaving just the thin base of the endometrium to start the next cycle. The uterine blood vessels dilate following their initial constriction. The increased blood flow then causes the walls of weakened capillaries to break, and bleeding occurs. Menstrual flow is thus a mixture of sloughed-off endometrial cells and blood.

Males produce sperm continually during the reproductive season

The anatomy of the human male reproductive system (**FIGURE 17.13**) is typical of the general pattern of organization seen throughout the eutherian mammals. The penis contains the urethra, which allows the passage of urine during urination and semen during ejaculation.

THE TESTES AND PRODUCTION OF SPERM The paired **testes** (testicles), responsible for sperm and hormone production, are contained in the **scrotum**, a sac that suspends the testes outside the body cavity. Most mammals exhibit this condition. In human males, the testes are normally in the scrotum all the time. In many other species, such as certain mice, the testes are in the scrotum only in the reproductive season, being drawn inside the abdomen at other times. All tissues are susceptible to heat damage if their temperature goes high enough. The testes of most mammals are unusual in that they are damaged by exposure to ordinary abdominal temperature (37°C). The testes need to be about 2°C cooler than that for normal production of viable sperm, and being in the scrotum permits this. Why the testes require a relatively low temperature remains a mystery. The testes are internal in dolphins, whales, and seals but are believed to be kept cooler than most internal tissues in those species by specialized arrays of arteries and veins supplied with blood that has been cooled by flowing through vessels close to the body surface.[21] The testes are also internal in

---

[19] The transformation of the residual follicle into a corpus luteum was named *luteinization* in the early study of endocrinology. This explains why the hormone that promotes the transformation was named *luteinizing hormone*. *Lutein* refers to yellow pigment. A corpus luteum is yellow in some cases, explaining why its formation was called luteinization and why it was given the name "yellow body."

[20] A developing mammal starts to be termed an *embryo* at about the time it implants in the endometrium and is considered to be an embryo thereafter. In medical terminology, a human is an *embryo* from implantation to the end of the eighth week of pregnancy, and a *fetus* from the ninth week until birth.

[21] Countercurrent heat exchange (see pages 268–269) occurs in these arrays of arteries and veins. The venous blood flowing through the arrays comes from the body surface, where it is cooled. By virtue of the countercurrent exchange, arterial blood flowing toward the testes is efficiently cooled, prior to reaching the testes, by losing heat to the venous blood.

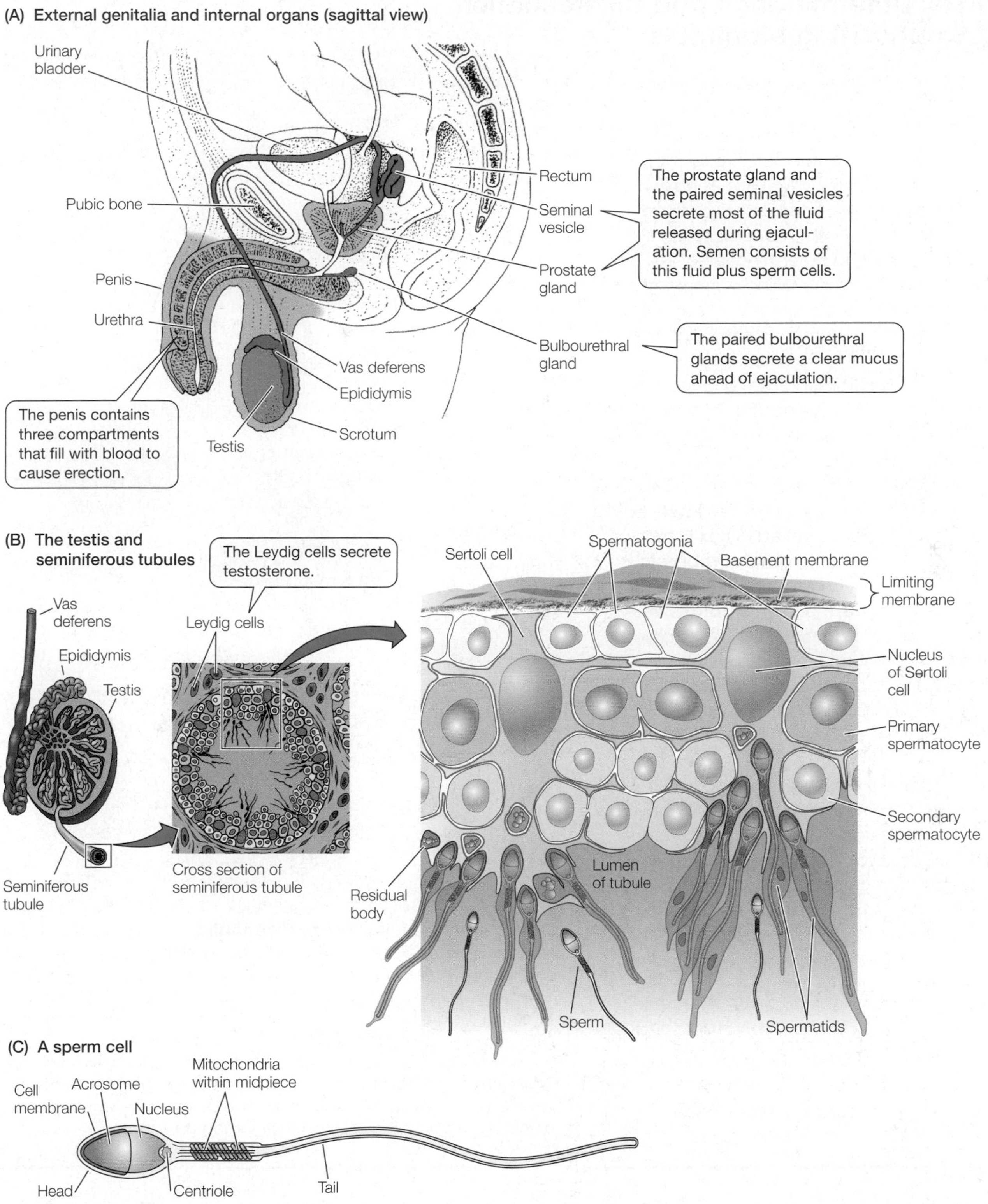

(A) External genitalia and internal organs (sagittal view)

Urinary bladder

Pubic bone

Penis

Urethra

Rectum

Seminal vesicle

Prostate gland

Bulbourethral gland

Vas deferens

Epididymis

Scrotum

Testis

The prostate gland and the paired seminal vesicles secrete most of the fluid released during ejaculation. Semen consists of this fluid plus sperm cells.

The paired bulbourethral glands secrete a clear mucus ahead of ejaculation.

The penis contains three compartments that fill with blood to cause erection.

(B) The testis and seminiferous tubules

The Leydig cells secrete testosterone.

Vas deferens

Epididymis

Testis

Leydig cells

Seminiferous tubule

Cross section of seminiferous tubule

Sertoli cell

Spermatogonia

Basement membrane

Limiting membrane

Nucleus of Sertoli cell

Primary spermatocyte

Secondary spermatocyte

Residual body

Lumen of tubule

Sperm

Spermatids

(C) A sperm cell

Cell membrane

Acrosome

Nucleus

Mitochondria within midpiece

Head

Centriole

Tail

FIGURE 17.13 The human male reproductive system (A) The male reproductive system, colored various shades of blue in this diagram, consists of the external genitalia (penis and scrotum), the testes, ducts such as the vas deferens on each side of the body, and the accessory glands (a single prostate gland, paired seminal vesicles, and paired bulbourethral glands). The urethra is colored white to symbolize that its function is only partly reproductive, inasmuch as it conveys both urine and semen to the outside. (B) Each testis is divided into 200–300 compartments containing coiled seminiferous tubules that are continuous with the epididymis and vas deferens. Each seminiferous tubule, which is bounded by a basement membrane, consists of a single layer of Sertoli cells interspersed with sperm-producing cells at various stages of development. Leydig cells occur in the connective tissue that fills the spaces between adjacent seminiferous tubules. In the development of sperm, spermatogonia develop into primary spermatocytes, then secondary spermatocytes, then spermatids, and finally sperm. (C) A mature sperm cell possesses a nucleus with a haploid set of chromosomes, a flagellum tail for motility, mitochondria that produce ATP, and an acrosome, which contains enzymes that help the sperm penetrate to the egg during the process leading to fertilization.

BOX 17.3 Sex Determination and Differentiation, Emphasizing Mammals

Animals use a variety of mechanisms to determine the sex of an individual. Sometimes the sex of an individual is not hardwired by genes, but instead, environmental factors play major roles in sex determination. We have seen two examples earlier in this chapter: sex change based on harem formation or other social factors in certain fish, and determination of sex by the body temperature prevailing during embryonic development in sea turtles and some other nonavian reptiles. In mammals, sex is determined genetically.

The external genital structures of a mammalian embryo are said to be **indifferent** during early development because they have the potential to differentiate into either male or female structures. The indifferent state is illustrated in a 4-week-old human embryo at the top of the figure. As the gonads develop (as discussed in **Box Extension 17.3**), if they are testes, the **genital tubercle** is hormonally signaled to develop into the glans ("head") of the penis, the **labioscrotal swellings** are signaled to develop into the scrotum, and the other primordia form male structures as well. If the gonads develop into ovaries, all these structures differentiate into female external genitalia. For example, the genital tubercle becomes the glans of the clitoris.

The gonads are also indifferent early in the development of a mammalian embryo: They can become either testes or ovaries. The indifferent gonads are positioned in the mid-abdomen. The gonads stay there throughout development in females. In males of most species of mammals, the gonads migrate away from their initial position as gestation proceeds, so that at birth they are either in the scrotum (as in humans) or poised to enter the scrotum.

Box Extension 17.3 discusses the sexual differentiation of internal structures at greater length. It also discusses the genetic and endocrine mechanisms that control sexual differentiation.

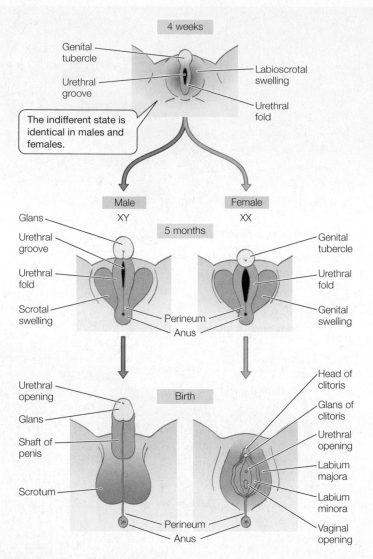

Differentiation of the external genitalia Colors are used to identify homologous tissues. For example, the tissue labeled yellow is of the same developmental origin in all five drawings. In early development—as typified by a human embryo at 4 weeks of developmental age (seen at the top)—the external genital structures are identical regardless of an individual's sex. In male embryos (XY), secretion of testosterone by the gonads when they differentiate causes the external genitalia to differentiate to become male in form, starting after the 7th week of gestation in humans. Without testosterone, the genitalia become female in form. The external genitalia can be definitively distinguished as male or female at about the 12th week of human gestation.

---

elephants, hyraxes, and elephant shrews, and in those species they might function at abdominal temperature.

Sperm are produced in coiled **seminiferous tubules** (see Figure 17.13B). The seminiferous tubules of a testis merge to join a coiled tube, the **epididymis**, located next to the testis within the scrotum. The epididymis becomes continuous with the **vas deferens**, which empties into the urethra. In cross section (see Figure 17.13B), one can see that the internal structure of each seminiferous tubule is composed of sperm-producing cells in various stages of develop-

ment, plus a single layer of critically important somatic cells, the **Sertoli cells**, which support and regulate the production of sperm. The seminiferous tubules account for about 85% of testis volume, and the Sertoli cells account for much of the tissue volume of the seminiferous tubules. Sertoli cells multiply during puberty, but their number in an adult is generally considered to be fixed. Strong correlations exist between the number of Sertoli cells and both testis size and the rate of sperm production. Connective tissue fills the spaces between the seminiferous tubules. **Leydig cells** (also

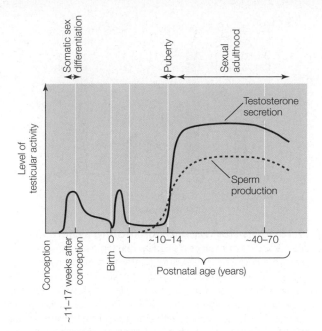

FIGURE 17.14 Testosterone secretion and sperm production during the life span of human males Note that the time scale is not even. (After Troen and Oshima 1981.)

called **interstitial cells**) located in the connective tissue secrete the androgen **testosterone**.

ENDOCRINE CONTROL IN MALES The Leydig cells exhibit a fascinating lifetime pattern of testosterone secretion, which we will discuss using the human male as an example (**FIGURE 17.14**). The Leydig cells secrete testosterone in abundance during the first trimester of embryonic life, when the hormone plays a key role in sexual differentiation of the penis, scrotum, seminal vesicles, and other male structures (**BOX 17.3**). By the 12th to 14th week after conception, the form of the genitalia is determined, and soon the Leydig cells become less active for the remainder of uterine life. They then secrete testosterone in abundance fol-

lowing birth. The function of this episode of secretion, which peaks at about 1 month after birth, is uncertain. The Leydig cells then become quiescent, in this case for more than a decade, until puberty starts and blood testosterone rises again. Thereafter testosterone secretion remains high throughout life, although it declines gradually after middle age.

In adult testes, the production of testosterone is controlled by pituitary gonadotropins, and the production of sperm is controlled by both the gonadotropins and testosterone. The brain and pituitary hormones involved are the same as those in females. GnRH is released from the hypothalamus (in pulses), and FSH and LH are secreted from the anterior pituitary gland. The Leydig cells are the target tissue of LH, which binds to receptors on the cell membranes and stimulates the cells to produce and secrete testosterone. The Sertoli cells are the target tissue of FSH and also are influenced by testosterone from the Leydig cells. Both testosterone and FSH, acting indirectly through effects on Sertoli cells, play roles in stimulating spermatogenesis (production of sperm), although the importance of FSH in spermatogenesis varies among species of mammals; whereas testosterone is essential for spermatogenesis, in humans FSH plays more of an augmenting role and is not absolutely required for sperm production (although one or more pituitary agents are required inasmuch as pituitary removal stops spermatogenesis). Besides paracrine substances that promote the proliferation and differentiation of sperm, the Sertoli cells secrete several additional substances, including (1) products that nourish the developing sperm; (2) inhibin, which inhibits FSH secretion from the anterior pituitary; and (3) fluid that fills the lumens of the seminiferous tubules. This fluid contains **androgen-binding protein** that binds testosterone and thereby keeps the hormone at a high concentration within the lumens of the tubules. **TABLE 17.2** summarizes the cells and hormones that are active in male reproduction.

Testosterone secreted into the general circulation is responsible for masculinizing many tissues. At puberty, for example, it promotes growth of the penis and testes, deepening of the voice, and growth

TABLE 17.2 Endocrine and neuroendocrine cells and secretions involved in male reproduction, emphasizing humans

Cells	Hormone	Action
Hypothalamic GnRH cells	GnRH	Stimulates secretion of FSH and LH from anterior pituitary cells
Anterior pituitary cells	LH	Stimulates Leydig cells to secrete testosterone
	FSH	Required for development of, and supports, Sertoli cells; stimulates Sertoli cells to support spermatogensis but in a secondary role to testosterone
Leydig cells	Testosterone	Required for mitosis and meiosis of spermatogenesis. Stimulates Sertoli cells to support and regulate spermatogenesis. Exerts negative feedback on the anterior pituitary and the hypothalamus Mediates secondary sexual characteristics such as growth of facial hair and muscular strength During early development, mediates sexual differentiation of reproductive organs (see Box 17.3) During early development, mediates sexual differentiation of certain aspects of brain neuroendocrine function and certain other aspects of brain fine structure and function (at least some of these effects are mediated by aromatization products of testosterone)
Sertoli cells[a]	Inhibin	Inhibits FSH secretion

[a]Sertoli cells also secrete *nonhormonal* products, notably substances that nourish the differentiating sperm, extracellular fluid (including androgen-binding protein) that fills the seminiferous tubules, and paracrine agents.

of facial hair. It contributes to an adolescent growth spurt in muscle mass in boys and to muscular strength throughout life.

Negative feedback effects of testosterone and inhibin on the anterior pituitary and hypothalamus keep FSH and LH secretions relatively low and steady from day to day in males. This pattern contrasts with the dramatic cycles of anterior pituitary (and ovarian) secretions seen in females (see Figure 17.11). The relatively constant levels of hormones in males sustain continuous production of sperm. Whereas this constancy of hormone concentrations and sperm production prevails throughout the year in adult human males, it prevails only in the reproductive season in seasonal breeders. In the many species of mammals that cease reproduction for part of the year, hormonal changes are instrumental in shutting down and later reactivating sperm production.

ERECTION AND EJACULATION The shaft of the human penis is filled almost entirely by three compartments of spongy tissue that can be expanded with blood. Erection entails inflating these compartments with blood until the penis is stiff. In males of several types of mammals, such as bats and carnivores, the penis contains a bone—the baculum—that helps provide stiffness for copulation.

Nitric oxide (NO), a messenger molecule, is the immediate mediator of erection. Erotic thoughts or physical stimuli lead to release of NO from parasympathetic nerve endings in the penis. The NO leads to dilation of blood vessels that permit blood to fill the spongy tissues. Investigators have recently discovered a positive feedback loop for NO. After NO from nerve endings initiates erection, the flux of moving blood causes further release of NO from endothelial cells in blood vessel walls. The positive feedback loop is important because NO is very short-lived and needs to be generated rapidly if its concentration is to be high. NO acts by mediating synthesis of cyclic GMP, a second messenger (see Figure 2.30). Current medications for erectile dysfunction inhibit an enzyme that breaks down cyclic GMP, thereby permitting the second messenger to act longer.

Semen, the fluid expelled from the penis during ejaculation, consists of sperm plus secretions of accessory sex glands. In humans, the most important accessory sex glands (see Figure 17.13A) are the prostate gland and seminal vesicles—which provide most of the fluid in semen—and the bulbourethral glands, which secrete a clear mucus ahead of ejaculation. After sperm mature in the testes, they are stored in the epididymis and vas deferens. They are mixed with the accessory-gland fluids as ejaculation occurs. The accessory-gland fluids provide fructose and other energy sources for the sperm; they contain acid–base buffers and other agents that create a suitable environment for the sperm; and they contain messenger compounds such as prostaglandins.

Pregnancy and birth are orchestrated by specialized endocrine controls

Fertilization usually occurs in the upper third of the oviduct, near the ovary (see Figure 17.10A). Sperm must travel there through the female reproductive tract to meet the secondary oocyte(s). Whereas millions of sperm enter the female reproductive tract in a single act of mating, very few actually arrive at the oocyte, and only one fuses with it to form the zygote. While sperm are in the female reproductive tract, they go through a process called **capacitation**,

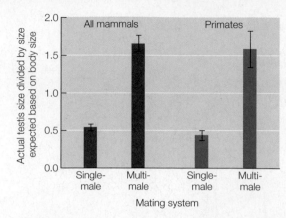

FIGURE 17.15 Testis size depends on mating system in mammals Species of mammals are placed in two categories: *single-male mating* in which the usual pattern is for one male to copulate with a female during a single mating period, and *multiple-male mating* in which the usual pattern is for multiple males to copulate with a female during a single mating period. Because testis size varies in a regular, allometric fashion with body size in all mammals analyzed together, an "expected" testis size can be calculated for each species simply by use of the species' body size. The y axis shows the actual testis size as a ratio of this expected size. Number of species analyzed is 54 for all mammals, 25 for primates. Error bars show ± 1 standard error. (After Kenagy and Trombulak 1986.)

which is essential for them to be capable of fertilizing the oocyte (see Figure 17.3C). Capacitation enhances the abilities of sperm to swim rapidly and to fuse with the cell membrane of the oocyte. If a female mammal mates with more than one male in a brief period of time, **sperm competition** is said to occur within the female reproductive tract as the sperm from the two sources vie to fertilize the oocyte. In species of mammals with a routine pattern of synchronous insemination by multiple males, the males tend to have dramatically large testes—which enable them to enter large numbers of sperm into the sperm competition (**FIGURE 17.15**).

In the period following ovulation, the oocyte is retained in the upper third of the oviduct because the lumen of the oviduct, owing to effects of estrogen, is constricted sufficiently to block travel toward the uterus. The lumen of the oviduct is not pinched off entirely, however, and therefore the tiny sperm can swim through to reach the oocyte.

The oocyte is enclosed in the zona pellucida and surrounded by granulosa cells when in the oviduct (**FIGURE 17.16A**). For fertilization to occur, sperm must traverse these structures to reach the cell membrane of the oocyte. The head of a sperm has enzymes on its outer surface that break down the extracellular matrix between granulosa cells, allowing the sperm to penetrate. To breach the zona pellucida (**FIGURE 17.16B**), a sperm releases enzymes by exocytosis from the acrosome (see Figure 17.13C) in its head. The exocytosis of these enzymes, called the **acrosomal reaction**, is triggered only when the sperm head binds to protein molecules, which are species-specific, on the zona pellucida. After its acrosomal reaction is initiated, a sperm cuts a channel through the zona pellucida. The head of the sperm then adheres to the cell membrane of the oocyte. Only a few sperm get this far. The cell membrane of just one of these sperm fuses with the cell membrane of the oocyte, and the cytoplasms of the two gametes become continuous. **Fertilization** refers to this specific event, which results in the formation of a **zygote**.

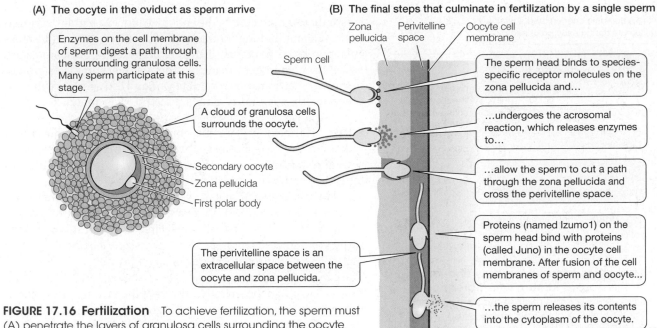

(A) The oocyte in the oviduct as sperm arrive

Enzymes on the cell membrane of sperm digest a path through the surrounding granulosa cells. Many sperm participate at this stage.

A cloud of granulosa cells surrounds the oocyte.

Secondary oocyte

Zona pellucida

First polar body

The perivitelline space is an extracellular space between the oocyte and zona pellucida.

(B) The final steps that culminate in fertilization by a single sperm

Zona pellucida

Perivitelline space

Oocyte cell membrane

Sperm cell

The sperm head binds to species-specific receptor molecules on the zona pellucida and…

…undergoes the acrosomal reaction, which releases enzymes to…

…allow the sperm to cut a path through the zona pellucida and cross the perivitelline space.

Proteins (named Izumo1) on the sperm head bind with proteins (called Juno) in the oocyte cell membrane. After fusion of the cell membranes of sperm and oocyte…

…the sperm releases its contents into the cytoplasm of the oocyte.

FIGURE 17.16 Fertilization To achieve fertilization, the sperm must (A) penetrate the layers of granulosa cells surrounding the oocyte and (B) both cut a channel through the zona pellucida and fuse with the oocyte cell membrane. In conjunction with the final step shown here, the sperm flagellum (tail) is also taken into the oocyte where it is broken down in the cytoplasm (not shown). (B after Primak-off and Myles 2002.)

Two crucial processes quickly follow fertilization. One is a reaction—termed the **cortical reaction**—that blocks more than one sperm from fertilizing the oocyte. The second is completion of meiosis by the oocyte. During the cortical reaction, organelles called **cortical granules** in the fertilized oocyte's peripheral cytoplasm release substances into the extracellular space around the oocyte that alter the cell membrane as well as the zona pellucida. These alterations prevent adhesion or fusion of more than one sperm, ensuring that only one haploid set of paternal chromosomes is admitted to the oocyte. Meanwhile, the oocyte completes the second meiotic division (see Appendix H), and thereafter only a single haploid set of maternal chromosomes remains in the oocyte. The two haploid sets of chromosomes join to make the zygote's diploid set of chromosomes.

IMPLANTATION, LUTEAL RESCUE, AND EARLY PREGNANCY

Implantation is the entry of the early embryo into the cellular matrix of the endometrium, the uterine epithelium. For this process to occur, the newly conceived individual must travel to the uterus. As it does so, mitotic cell divisions take place, and the zygote (the single cell formed by fertilization) matures into a hollow early embryo termed a blastocyst. The trip down the oviduct is relatively slow; in humans, the early embryo arrives in the uterus about 4 days after fertilization took place in the upper third of the oviduct. Travel down the oviduct becomes possible because the rising levels of progesterone secreted by the corpus luteum induce the smooth muscles of the oviduct to relax, so the lumen of the oviduct opens. Once in the uterus, the blastocyst "hatches" from the zona pellucida and remains free (for about 3 days in humans) before implanting.

The blastocyst consists of cells that will ultimately develop into the newborn individual, plus cells collectively termed the **trophoblast** that will not contribute to the new individual but instead help form the placenta. A portion of the outer surface of the blastocyst is specialized to make contact with the endometrium and begin the process of implantation. To enable the blastocyst to bury itself in the nutrient-rich endometrium, enzymes are secreted by trophoblast cells.

Progesterone is required for the endometrium to remain in a highly developed state, rather than being sloughed off or resorbed. The placenta is the principal source of the necessary progesterone after the placenta develops. However, in early pregnancy—the days or weeks immediately following implantation—the corpus luteum is the primary source of progesterone and must continue to secrete it at high levels. The corpus luteum must therefore be rescued so that it does not degenerate, as it would in the absence of pregnancy. In humans, the shift from dependence on the corpus luteum as the primary source of progesterone to dependence on the placenta happens after about 50–70 days (pregnancy lasts about 266 days). The corpus luteum must thus survive for at least 50–70 days rather than atrophying in 14 days, as it does in the absence of pregnancy.

To rescue the corpus luteum, different species use different mechanisms. In primates, an embryonic membrane that participates in the formation of the placenta—the chorion—secretes the hormone **chorionic gonadotropin** (**CG**), which rises in blood concentration following fertilization (**FIGURE 17.17**) and acts on the corpus luteum to extend the corpus luteum's life. CG is in the same glycoprotein chemical family as LH, and genomic evidence indicates it evolved relatively recently in geologic time—in the common ancestor of the anthropoid primates (monkeys, apes, and humans)—by duplication of a gene for LH and subsequent evolutionary modification of the gene. CG is excreted in the urine and serves as the basis for pregnancy tests; detection of CG in the urine indicates that an embryo has implanted and is developing. Uniquely among nonprimate mammals, horses independently evolved a gonadotropin secreted by the chorion that functions similarly to the primate CG to rescue the corpus luteum. In rodents, the hormone prolactin, released by the anterior pituitary gland, sustains the corpus luteum.

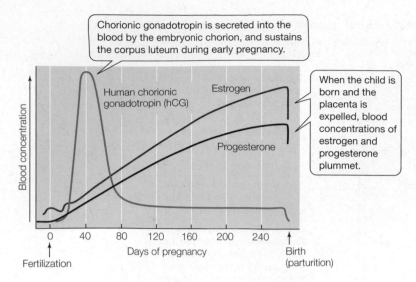

Chorionic gonadotropin is secreted into the blood by the embryonic chorion, and sustains the corpus luteum during early pregnancy.

When the child is born and the placenta is expelled, blood concentrations of estrogen and progesterone plummet.

Human chorionic gonadotropin (hCG)

Estrogen

Progesterone

Blood concentration

0   40   80   120   160   200   240

Days of pregnancy

Fertilization   Birth (parturition)

FIGURE 17.17 Humans and other primates employ chorionic gonadotropin to rescue the corpus luteum when pregnancy occurs Blood concentrations in a human mother are shown. During the first part of pregnancy, the hormone chorionic gonadotropin—secreted by the chorion, an embryonic membrane that contributes to formation of the placenta—sustains the life of the corpus luteum, which secretes estrogen and progesterone into the blood. In later pregnancy, the placenta becomes the principal source of estrogen and progesterone and also secretes them into the general circulation.

THE PLACENTA At first, the endometrium provides the embryo with sufficient nutrients and other forms of metabolic support. However, as time passes, the implanted embryo interacts with maternal uterine structures to form the **placenta**, a structure in which blood vessels of the maternal and embryonic circulatory systems are so closely juxtaposed that O_2 and other materials can move readily between the bloodstreams of the mother and embryo.[22] Four or five quite distinct structural types of placentas are known in various mammals. In all the types, the maternal and embryonic bloodstreams are separated by intervening cellular structures, such as the endothelia of blood capillaries (see Figure 2.6). Consequently, although many substances move readily between the two bloodstreams by diffusion and other transport processes (see Chapter 5), the bloods do not normally mix. The flow of maternal blood through the placenta supplies the embryo with O_2 and nutrients.[23] It also removes CO_2 and other metabolic by-products.

In addition to being a nutritive and waste-removal structure, the placenta is also an endocrine structure (see Table 17.1), as already mentioned. The mature placenta secretes estrogen (mostly estriol in humans and many other mammals), progesterone, and other hormones that support physiological functions during pregnancy and prepare the mother's body for birth and lactation. Progesterone

[22] Where we here speak of the "embryo" and the "embryonic" circulation and bloodstream, these would be called the "fetus" and the "fetal" circulation and bloodstream during the later stages of intrauterine development because as noted before, an embryo comes to be called a fetus as it matures.

[23] Substances such as drugs and alcohol in the maternal blood can also cross the placenta. Some are known to cause physical or behavioral birth defects by disrupting development of the embryo/fetus. When a mother consumes alcohol during pregnancy, for example, her child can suffer a suite of negative effects known as *fetal alcohol syndrome*.

is essential for maintaining the endometrium in a highly developed state and for inhibiting contractions of the myometrium (the muscular portion of the uterus). Estrogen (produced through aromatase action from androgens made in other tissues) stimulates growth of the myometrium. Both progesterone and estrogen contribute to preparing the mammary glands (breasts) for lactation.

BIRTH The factor(s) that initiate birth (parturition) remain largely unknown. We know, however, that during the period prior to birth, several physiological changes occur that prepare the mother and fetus (late embryo) for the birth process.

Blood estrogen, present at ever-increasing levels as pregnancy progresses (see Figure 17.17), prepares the myometrium for contractile activity in two important ways. First, estrogen stimulates the myometrial smooth muscle cells to synthesize *connexins*, the protein molecules that join together to make up *gap junctions* (see Figure 2.7). The gap junctions that are formed allow electrical activity to spread from cell to cell in the smooth muscles of the myometrium, so that the entire myometrium can produce highly coordinated contractions. Second, estrogen stimulates the myometrial smooth muscle cells to produce receptors for the hormone **oxytocin**. In some mammals, although probably not humans, **relaxin**—a peptide hormone secreted by corpora lutea—also plays an important role in preparing for birth. Secretion of relaxin during the period leading up to birth induces remodeling and softening of the cervix (see Figure 17.10A); it also facilitates the establishment of flexible, elastic connections between pubic bones. In these ways, relaxin demonstrably eases the exit of fetuses from the uterus, facilitating rapid delivery.

Oxytocin is the principal hormone controlling the forces that produce delivery. It is secreted by hypothalamic neurosecretory cells and released from the posterior pituitary gland at the time of birth, and it stimulates uterine smooth muscle cells to produce and secrete prostaglandins. Both the oxytocin and the prostaglandins strongly stimulate contractile activity by the smooth muscle cells. As birth becomes imminent, forceful contractions spread over the body of the uterus toward the cervix and vagina (see Figure 17.10A). The contractions force the fetus against the cervix, thereby stimulating mechanoreceptors there. These conditions set up the positive feedback loop shown in **FIGURE 17.18**: The mechanoreceptors in the cervix send action potentials to the hypothalamus, where they stimulate neurosecretory cells to secrete oxytocin. Having recently incorporated oxytocin receptors into their cell membranes, the myometrial smooth muscle cells respond to the oxytocin with contractions. With increased contractions forcing the fetus against the cervix, the cervical mechanoreceptors are stimulated even more, and more oxytocin is secreted. As myometrial contractions increase, the cervical opening dilates (widens) to permit passage of the fetus. These events reach a crescendo with delivery. After the fetus is born, the placenta (called the afterbirth) is also delivered. Without the placenta as an endocrine structure, the mother's blood levels of progesterone and estrogen then plummet toward pre-pregnancy levels (see Figure 17.17).

Lactation, a costly part of reproduction, is governed by neuroendocrine reflexes

A defining characteristic of mammals is that females have mammary glands that provide milk to their newborn young. **Lactation** is the process of producing and providing milk. In most species

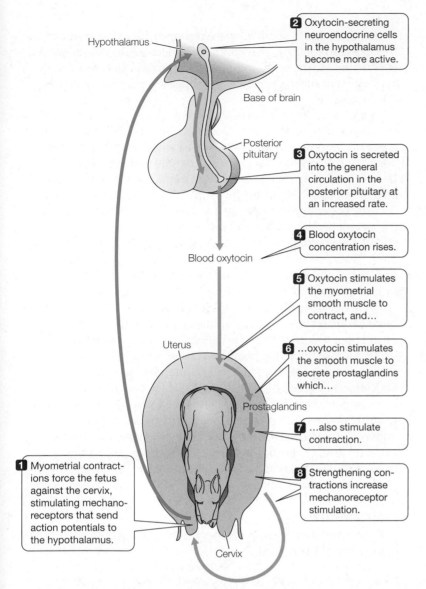

2 Oxytocin-secreting neuroendocrine cells in the hypothalamus become more active.

Hypothalamus

Base of brain

Posterior pituitary

3 Oxytocin is secreted into the general circulation in the posterior pituitary at an increased rate.

Blood oxytocin

4 Blood oxytocin concentration rises.

5 Oxytocin stimulates the myometrial smooth muscle to contract, and...

Uterus

6 ...oxytocin stimulates the smooth muscle to secrete prostaglandins which...

Prostaglandins

7 ...also stimulate contraction.

1 Myometrial contract-ions force the fetus against the cervix, stimulating mechano-receptors that send action potentials to the hypothalamus.

8 Strengthening con-tractions increase mechanoreceptor stimulation.

Cervix

FIGURE 17.18 Positive feedback during birth Oxytocin, prosta-glandins, and mechanical stimuli at the cervix participate in a posi-tive feedback loop that progressively forces the cervix to open wide enough to permit the fetus to be born.

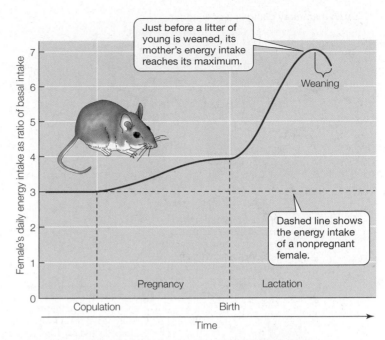

Just before a litter of young is weaned, its mother's energy intake reaches its maximum.

Weaning

Dashed line shows the energy intake of a nonpregnant female.

Pregnancy

Lactation

Copulation

Birth

Time

FIGURE 17.19 Costs of reproduction in small mammals Shown are average values determined from a meta-analysis of data on 17 species of small mammals, mostly mice and rats. Required rates of energy intake are expressed as ratios of the species-specific basal rate of energy intake (calculated from basal metabolic rate; see Chapter 7). For females living in the wild, the required rate of energy intake when they are not pregnant is about three times the basal rate. (After Thompson 1992.)

of mammals that have been studied, lactation is the most costly phase of reproduction for the mother. As **FIGURE 17.19** shows, for example, pregnancy is far less costly than lactation in small mammals such as mice and rats. When a litter of young reaches the end of its time in the nest (just prior to weaning) and is de-manding milk at its highest rate, the mother needs to take in more than twice as much energy per day as she would require if she were not reproducing.

Lactation consists of two functionally distinct steps. The first is **milk production** or secretion—in which milk is synthesized by cells of the mammary glands and secreted into cavities within the glands. The second step is **milk ejection** or **let-down**—in which milk is expelled from the cavities of the mammary glands into the mouth of the offspring. Milk—which varies widely among species in its detailed composition—is an essential source of food until the time of weaning for most newborns. Immediately after birth, the

mammary glands in many species—prior to secreting milk—secrete fluids termed *colostrums*, which are rich in antibodies and vitamins.

All mammary glands have a similar basic structure, although species vary widely in the size and number of their mammary glands. A mammary gland (**FIGURE 17.20A**) consists of many hollow, semi-spherical glandular structures called **alveoli** (singular *alveolus*) connected to milk ducts. The secretory epithelial cells of an alveolus synthesize milk and secrete it into the alveolar lumen. Contractile myoepithelial cells surround the alveolar epithelial cells. When the myoepithelial cells contract, they squeeze the alveolus and force milk from the alveolar lumen out via the ducts of the mammary gland.

Lactation is controlled principally by two hormones (**FIGURE 17.20B**). In fact, these two hormones control the two steps of lacta-tion. Prolactin secreted from the anterior pituitary gland stimulates *milk secretion* by the alveolar epithelial cells. Oxytocin secreted from the posterior pituitary gland stimulates *milk ejection*. Oxytocin thus has two major reproductive roles: It helps orchestrate uterine contraction during birth, and it stimulates milk ejection.

In females that are not lactating, dopamine—also called prolactin-inhibiting hormone (PIH)—inhibits secretion of pro-lactin. Dopamine is secreted from neuroendocrine cells in the hypothalamus and carried to the anterior pituitary gland in the hypothalamo–hypophysial portal system. During late pregnancy and lactation, dopamine secretion *decreases*, thereby permitting secretion of prolactin. In addition, hypothalamic thyrotropin-releasing hormone (TRH) stimulates secretion of prolactin.

(A) A mammary gland in a cow udder

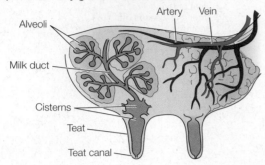

(B) Hormonal control of lactation

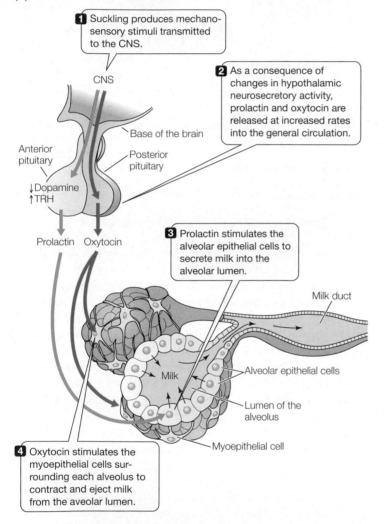

1 Suckling produces mechano-sensory stimuli transmitted to the CNS.

2 As a consequence of changes in hypothalamic neurosecretory activity, prolactin and oxytocin are released at increased rates into the general circulation.

3 Prolactin stimulates the alveolar epithelial cells to secrete milk into the alveolar lumen.

4 Oxytocin stimulates the myoepithelial cells surrounding each alveolus to contract and eject milk from the aveolar lumen.

CNS

Base of the brain

Anterior pituitary

Posterior pituitary

↓Dopamine ↑TRH

Prolactin   Oxytocin

Milk duct

Milk

Alveolar epithelial cells

Lumen of the alveolus

Myoepithelial cell

FIGURE 17.20 Mammary glands and lactation   (A) The structure of mammary glands as illustrated by the cow udder. Mammary alveoli (singular alveolus) synthesize milk, which exits by ducts. Cisterns, which are not found in humans or several other mammals, collect and store milk. (B) The base of the brain (top) and two alveoli in a mammary gland (bottom) diagrammed to show the hormonal control of lactation. Suckling stimulates the secretion of prolactin and oxytocin by the pituitary gland. Prolactin stimulates milk synthesis. Oxytocin stimulates milk ejection. Other neural signals, such as those activated by an infant's cry or a clanking milk bucket in a dairy barn, can stimulate oxytocin (but not prolactin) secretion and promote milk ejection. CNS = central nervous system, TRH = thyrotropin-releasing hormone.

Oxytocin acts by stimulating the myoepithelial cells surrounding the alveoli to contract. This contraction ejects milk. Milk ejection is essential for the offspring to obtain milk because the simple application of sucking forces to the nipples is inadequate to draw out enough milk.

Suckling by offspring is a potent stimulus for the secretion of both prolactin and oxytocin, as emphasized by Figure 17.20B. The nipples are richly innervated and very sensitive to mechanical stimulation. Suckling triggers mechanosensory signals that are transmitted by neurons to the central nervous system. These neural signals affect hypothalamic dopamine- and TRH-secreting neurons to decrease dopamine secretion and increase TRH secretion—thereby increasing prolactin secretion by the pituitary gland—and they stimulate hypothalamic neuroendocrine cells to secrete oxytocin. These control sequences are often described as being *neuroendocrine reflexes* because they are relatively automatic processes mediated by both neural and endocrine elements.

In many types of mammals, estrus and/or ovulation are suppressed during lactation in response to high levels of prolactin. Such suppression of estrus is called *lactation anestrus*, and the suppression of ovulation is called *lactation anovulation*. Prolactin-mediated suppression of GnRH secretion by the hypothalamic GnRH-secreting cells, leading to reduced FSH and LH secretion, is the primary basis for these phenomena. In human populations that do not use contraceptives, breast-feeding appears to play an important role in determining spacing between births. Studies of prolactin levels in both !Kung and American mothers show that frequent nursing maintains high blood levels of prolactin, which tend to suppress ovulation.

SUMMARY
Reproductive Endocrinology of Placental Mammals

- The ovaries produce oocytes and secrete hormones. Oocytes are shed from the ovaries during each estrous or menstrual cycle during the reproductive season. Each cycle has three main ovarian phases: development of follicles (follicular phase), ovulation, and function of the corpus luteum (luteal phase). The uterine endometrium grows thicker prior to ovulation and becomes secretory after ovulation.

- The gonadotropins known as luteinizing hormone (LH) and follicle-stimulating hormone (FSH)—released in response to gonadotropin-releasing hormone (GnRH)—stimulate granulosa cells in ovarian follicles to secrete estrogen (particularly estradiol). Estrogen acts as a paracrine/autocrine agent that stimulates proliferation of granulosa cells. It also acts as a blood-borne hormone that stimulates growth of the uterine endometrium, affects behavior in species with estrus, and feeds back on the anterior pituitary gland and hypothalamus.

■ Ovulation is induced by copulation in some species, but is spontaneous in most. A surge in secretion of LH triggers ovulation in either case. In spontaneous ovulators, the surge is a consequence of endogenous interactions of endocrine and neuroendocrine tissues, notably the hypothalamus, anterior pituitary gland, and ovarian follicles. A key part of the process is stimulation of kisspeptin neurons by estrogen, followed by kisspeptin-stimulated secretion of GnRH.

■ After ovulation, the cells of each ruptured ovarian follicle reorganize into a corpus luteum, which secretes progesterone, estrogen, and inhibin. These hormones inhibit or decrease folliculogenesis in the ovaries by reducing secretion of LH and FSH from the anterior pituitary. Progesterone supports the secretory state of the uterine endometrium and inhibits contraction of the smooth muscles of the myometrium and oviducts.

■ If fertilization does not occur, the corpus luteum degenerates, and uterine endometrial tissue is resorbed or discharged as menstrual flow. If fertilization does occur, the corpus luteum is rescued and continues to function. Its function is essential for maintenance of the uterus and placenta in early pregnancy.

■ Fertilization occurs in the oviduct when a secondary oocyte is present and sperm have undergone capacitation. Upon fusion of the two gametes' cell membranes, the oocyte completes meiosis and undergoes the cortical reaction, which blocks fertilization by any additional sperm. The zygote moves down the oviduct to the uterus, developing into a blastocyst, which ultimately implants (possibly following delayed implantation in some species).

■ After implantation, the embryonic trophoblast and maternal endometrium form the placenta, a structure in which embryonic and maternal blood vessels closely intermingle, permitting exchange of O_2, nutrients, and wastes between the two bloodstreams (although not mixing of blood). The placenta secretes hormones, such as progesterone, which are crucial for maintaining the placenta and sustaining pregnancy.

■ Birth is accomplished by coordinated contractions of the uterine myometrium (smooth muscle). Birth is facilitated by a positive feedback loop—mediated partly by hypothalamic oxytocin secretion—that causes the contractions to become more and more powerful.

■ Milk is produced by alveolar epithelial cells in the mammary glands. Prolactin stimulates milk secretion by the epithelial cells. Oxytocin causes milk ejection (milk let-down) by stimulating the contraction of myoepithelial cells surrounding the alveoli. Suckling increases secretion of both prolactin and oxytocin.

■ Males produce sperm continuously during the breeding season. Sperm are produced with the aid of Sertoli cells in the seminiferous tubules in the testes. Leydig cells embedded in connective tissue between the seminiferous tubules secrete testosterone. The functions of the seminiferous tubules and Leydig cells are controlled by continuous secretion of LH and FSH, released in response to GnRH.

■ In males, LH stimulates the Leydig cells. Testosterone from the Leydig cells promotes sperm production via effects on the Sertoli cells, sometimes with additive effects exerted by FSH on the Sertoli cells.

■ Erection of the penis results from blood flow into spongy tissue, controlled by nitric oxide (NO). The parasympathetic nervous system initiates NO production, but then further NO is released from blood vessel endothelial walls, forming a positive feedback loop. During ejaculation, sperm are mixed with secretions of male accessory glands (principally the prostate gland and paired seminal vesicles in humans) to produce the semen that is emitted.

STUDY QUESTIONS

1. "Breeding like rabbits" is a common expression for having large numbers of offspring. Rabbits are indeed noted for their great reproductive potential. Rabbits exhibit induced ovulation and postpartum estrus. Define these reproductive traits, and explain how they help endow rabbits with their considerable reproductive potential.

2. Suppose you are in charge of a captive breeding program for a species of zoo animal that is similar to humans in its reproductive endocrinology. By study of hormone signals, how could you tell when a female ovulates? If you permit her to mate at that time, how could you later tell from study of hormone signals whether she has conceived? Explain your answers.

3. Estimate the age of a primary oocyte within a dominant follicle in the ovary of a woman on her 18th birthday. Explain your reasoning.

4. Explain the advantages that both insects and mammals can realize from embryonic diapause.

5. Ovulation in mammals, whether it is induced or spontaneous, occurs in response to a surge of luteinizing hormone (LH). Compare and contrast the ways the LH surge is generated in induced and spontaneous ovulators.

6. In discussions of feedback in biological systems, negative feedback is typically emphasized, and positive feedback is often treated as an anomaly. In fact, positive feedback is common in reproductive physiology. List two or three examples of positive feedback. Explain how positive feedback functions to advantage in each case. If possible, discuss whether homeostasis is always ideal.

7. Consider a species, such as a lizard, in which females produce sets of eggs with large, heavy quantities of yolk. Suppose you hypothesize that one of the costs of reproduction is that females cannot run as fast or escape predators as well when they are in the process of producing such eggs. Describe a manipulative experiment you could carry out to test your hypothesis, and explain how you would interpret your results. (For an example, see the paper by Miles et al. in the References.)

8. What are the relative advantages and disadvantages of semelparity and iteroparity? In answering, consider iteroparous species that have both short and long expected life spans in their natural environments.

9. Knowing what you do about the feedback effects of testosterone on the anterior pituitary gland, explain the reason that male

athletes who take anabolic steroids become temporarily sterile (anabolic steroids are steroids that mimic some actions of testosterone).

10. Aromatase inhibitors are a new generation of drugs used to treat women who have estrogen-sensitive breast cancers (cancers that grow most rapidly when estrogen is present). Explain why aromatase inhibitors are useful in these cases.

11. Prolactin and oxytocin are both involved in lactation. Define, describe, and explain the functional significance of each of these hormones.

Go to **sites.sinauer.com/animalphys4e**
for box extensions, quizzes, flashcards,
and other resources.

REFERENCES

Bronson, F. H. 1989. *Mammalian Reproductive Biology.* University of Chicago Press, Chicago. For those interested in an ecological perspective on reproductive biology, this is probably the best book ever written. Even though the reference list is getting dated, the book remains a treasure trove of insights and research ideas.

Eppig, J. J., K. Wigglesworth, and F. L. Pendola. 2002. The mammalian oocyte orchestrates the rate of ovarian follicular development. *Proc. Natl. Acad. Sci. U.S.A.* 99: 2890–2894. A study of the nature of communication between the oocyte and granulosa cells in the ovarian follicle.

Goldman, B. D. 2001. Mammalian photoperiodic system: Formal properties and neuroendocrine mechanisms of photoperiodic time measurement. *J. Biol. Rhythms* 16: 283–301. This paper provides a brief, notably accessible explanation of mechanisms of photoperiodic time measurement in mammals.

Kauffman, A. S., D. K. Clifton, and R. A. Steiner. 2007. Emerging ideas about kisspeptin-GPR54 signaling in the neuroendocrine regulation of reproduction. *Trends Neurosci.* 30: 504–511.

Lopes, F. L., J. A. Desmarais, and B. D. Murphy. 2004. Embryonic diapause and its regulation. *Reproduction* 128: 669–678. A modern review of a process used by many types of animals to time the birth of offspring in ways that optimize the chances of survival of mother and young.

Miles, D. B., B. Sinervo, and W. A. Frankino. 2000. Reproductive burden, locomotor performance, and the cost of reproduction in free ranging lizards. *Evolution* 54: 1386–1395.

Munday, P. L., P. M. Buston, and R. R. Warner. 2006. Diversity and flexibility of sex-change strategies in animals. *Trends Ecol. Evol.* 21: 89–95.

Nelson, R. J. 2011. *An Introduction to Behavioral Endocrinology*, 4th ed. Sinauer, Sunderland, MA.

Peterson, C., and O. Söder. 2006. The Sertoli cell—a hormonal target and "super" nurse for germ cells that determines testicular size. *Horm. Res.* 66: 153–161.

Reece, W. O., H. H. Erickson, J. P. Goff, and E. E. Uemura (eds.). 2015. *Dukes' Physiology of Domestic Animals*, 13th ed. Wiley, New York. A classic reference on the physiology of domestic animals, including reproductive physiology.

Rommel, S. A., D. A. Pabst, and W. A. McLellan. 1998. Reproductive thermoregulation in marine mammals. *Am. Sci.* 86: 440–448. How do dolphins keep their testes cool?

Taylor, S., and L. Compagna. 2016. Avian supergenes. *Science* 351: 446–447. An easy-to-read introduction to the potential roles of supergenes in reproduction.

See also **Additional References** *and* **Figure and Table Citations**.

This newly hatched marine turtle crawling toward the sea is beginning an astounding migration during which it will spend up to three decades in the Pacific Ocean before returning to its birth site (Malaysia) to breed. In the South Atlantic Ocean, green turtles (*Chelonia mydas*) feed off the coast of Brazil and then migrate eastward across more than 2200 km of open ocean to nest at Ascension Island, an isolated speck of land in the middle of the South Atlantic. In the Northern Hemisphere, loggerhead turtles (*Caretta caretta*) hatch on Florida beaches, swim out to the Gulf Stream, and spend years circulating around the North Atlantic before returning to U.S. waters.

Because it may take an individual sea turtle 30 years to reach sexual maturity, researchers had been uncertain whether the turtles were returning to their *natal* beaches. Accumulated evidence suggests that they do. For example, no sea turtle tagged at Ascension Island has ever been found to nest elsewhere. Moreover, analysis of mitochondrial DNA indicates that green turtles nesting at different sites are genetically distinct, a fact suggesting that gene exchange between different colonies is relatively rare. The reproductive success of sea turtles is based on their finely tuned sense of place. As the breeding season arrives, sea turtles move unerringly toward a precise destination, irrespective of their individual positions at sea.

Sea turtles are masters of animal navigation. As physiologists use the term, **navigation** is the act of moving on a particular course, or toward a specific destination, by using sensory cues to determine direction and position. As sea turtles demonstrate, navigational abilities facilitate specialized locomotor behaviors such as migration and homing. **Migration** is the periodic movement of an animal from one region to another. Migratory periods may be prompted by seasonal changes, as is often the case with migratory birds. A migratory cycle, however, may extend over an animal's lifetime, as is true of Pacific salmon (*Oncorhynchus* spp.), which return from the sea to their natal lakes to spawn and die. Whereas migration may involve movements on a global scale, homing is often a more localized behavior.

Newly hatched sea turtles begin a lifetime of navigation

Homing is the ability of an animal to find its way repeatedly to a *specific point*, most often its nesting or dwelling place, although some animals, such as honeybees (*Apis mellifera*), use their homing abilities to pinpoint food sources too.

The dividing line between migration and homing is not always clear, and some navigating animals appear to engage both abilities selectively. Homing performance, however, usually declines as the distance from home or a target increases, and an animal's homing abilities may be disrupted completely if it is passively displaced (e.g., blown off course by a storm or deliberately moved by a researcher) from its path of travel.

It's not surprising that animal navigation has been a subject of scientific interest to researchers. Lacking maps, compasses, and other human cultural accoutrements, how do navigating animals gather and interpret cues about their position on the planet? Do they possess physiological (sensory) mechanisms unavailable to humans? Is the ability to navigate mainly a result of learning and memory of past experience? And what is the evolutionary importance for the animal of the ability to navigate?

The Adaptive Significance of Animal Navigation

Physiologists presume that navigational abilities evolved because they represent an adaptive advantage to the animals that possess these skills. Before examining how animals navigate, let's consider *why* it may be advantageous for them to do so.

Navigational abilities promote reproductive success

Most (but hardly all) animals reproduce at a time and place that presents the lowest degree of environmental stress to the parents and their offspring. For example, long-distance migrants such as humpback whales (*Megaptera novaeangliae*) spend the summer feeding in the food-rich polar seas of the Northern Hemisphere. In the winter, however, they migrate to more tropical waters to reproduce, often traveling as far as 3200 km. Surprisingly, humpback whales apparently do not eat during the winter breeding season; only 1 out of 2000 caught in one study had food in the stomach. The winter migration is *not* to areas of high food productivity, but rather to areas of warm water that engender less thermoregulatory stress, especially for the young.

Many animals exhibit *natal philopatry*, the tendency either to stay in the area of their origin or to return to it to reproduce. In many species of migrating birds, such as the wood thrush (*Hylocichla mustelina*), males return every spring to reestablish the same territories each year. In this case, learned familiarity may contribute to

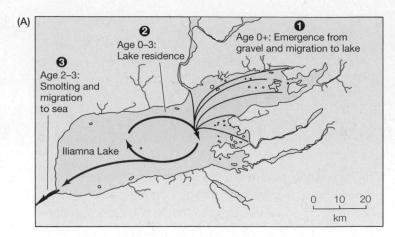

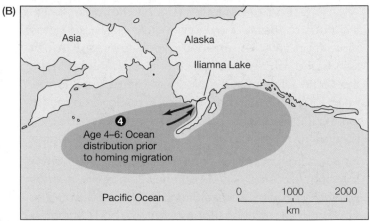

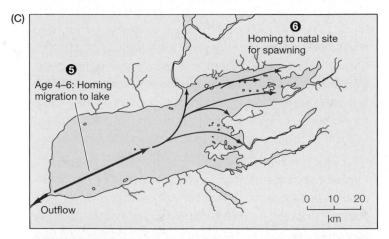

FIGURE 18.1 Salmon life-cycle migration and homing The life cycle of sockeye salmon (*Oncorhynchus nerka*) from Iliamna Lake in Alaska. (A) Newly hatched salmon emerge from gravel of streams and beaches and spend up to 3 years in the lake before transforming to smolts that migrate to the sea. (B) Ocean distribution of maturing salmon prior to their homing migration back to Iliamna Lake. (C) Homing migration back to the lake and to their natal site for spawning. (D) A male in the reproductive phase (right) has red sides, a deep body, and elongated, hooked jaws. (A–C after Dittman and Quinn 1996.)

the advantage of philopatry. Other cases, however, do not involve practice. Anadromous fish (i.e., fish that ascend rivers from the sea to breed), such as Pacific salmon, provide the most dramatic example of how animal navigational skills are employed to accomplish this reproductive aim (**FIGURE 18.1**).

Several species of salmon (*Oncorhynchus*) hatch in freshwater streams and lakeshores in the North Pacific rim. After spending 1–3 years in downstream waters and lakes as juveniles, they metamorphose into smolts and migrate downstream to the Pacific. The maturing salmon spend a few years in the Alaskan Gyre, a large circulating current that offers plentiful foraging opportunities. The adult salmon of North American species share the gyre with those of Asiatic species. Although the various species feed together, they do not breed in the gyre. Spawning occurs only after adult salmon migrate through the open ocean to their natal river and then, using the imprinted scent of their natal lake or stream, home accurately to it. Here they reproduce once and die.

Presuming that natal philopatry evolved by natural selection, what is its adaptive advantage that has led to such impressive navigational ability in salmon? Probably the most important advantage is that the ability to home to the natal stream returns locally adapted individuals to appropriate environments. Salmon populations are often strongly adapted to local breeding environments, and this adaptation reduces the fitness of strays that disperse instead of returning to the natal site.

However, *perfect* site fidelity would be expected to be disadvantageous because conditions change from year to year at a breeding habitat, so there should be some dispersal to adjust to these temporal changes. Salmon populations appear to maximize reproductive fitness by natural selection favoring a balance of a high percentage of return to natal sites and a low percentage of dispersal. By using navigational skills to feed globally but spawn parochially, salmon thus maintain distinct populations that are well adapted to their breeding sites.

Navigational abilities facilitate food acquisition

The examples cited of sea turtles, humpback whales, and Pacific salmon all suggest a common conclusion: The best place to feed isn't always the best place to breed. Using their navigational abilities, migratory animals can position themselves favorably at food sources, accumulating chemical energy until they reach breeding readiness. By spending their summers feeding in polar seas, humpback whales take advantage of mineral-rich upwellings from the ocean floor that support an abundant food chain, as well as long summer days that maximize the time these animals can spend foraging.

Animal navigators also employ homing skills to acquire food. For example, several species of birds store or cache seeds, hiding them and then retrieving them days or months later. Clark's nutcrackers (*Nucifraga columbiana*) and pinyon jays (*Gymnorhinus cyanocephalus*) dig holes in sandy ground and bury seeds. An individual of either species is able to locate thousands of cache sites and also to remember which sites it has already visited and emptied. A single Clark's nutcracker may hide 33,000 seeds in a season at approximately 6600 locations, and it can remember these locations and retrieve the seeds with a degree of success that is nothing less than astonishing.

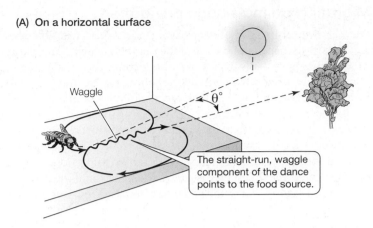

(A) On a horizontal surface

Waggle

θ°

The straight-run, waggle component of the dance points to the food source.

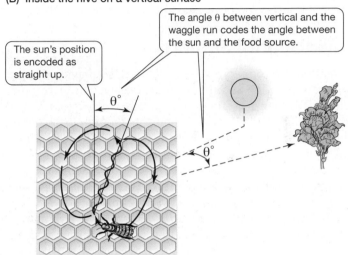

(B) Inside the hive on a vertical surface

The angle θ between vertical and the waggle run codes the angle between the sun and the food source.

The sun's position is encoded as straight up.

θ°

θ°

FIGURE 18.2 The waggle dance of honeybees conveys information about the direction and distance of a food source A returning forager that has discovered a nectar source dances either outside the hive on a horizontal surface (A) or inside the hive on a vertical surface (B). Distance is coded by the duration of the waggle run.

In another example of using homing skills to acquire food, foraging honeybees may take hours or days to first locate a new source of nectar, but then new workers will appear at the food source within minutes of the first bee's return to the hive. Karl von Frisch (1886–1982) demonstrated that a returning forager performs a "dance" that conveys information to other bees about the direction and distance of the food source (**FIGURE 18.2**). The dance consists of looping turns alternating with a straight-line portion in which the signaling bee waggles her abdomen. Typically the dance is performed in the hive, on the vertical surface of a honeycomb, out of view of the sun. The angle between the sun and the food source is translated into an angle between vertical and the waggle component of the dance, so that other workers leaving the hive know the correct direction of the food source. Food location by honeybees certainly requires navigation, but it is somewhat atypical of homing behavior because it involves social cooperation and communication among individuals.

Migrating animals need navigation

For migrating animals, the adaptive value of navigation is difficult to separate from the adaptive value of the migration itself. If migratory movements have evolved by natural selection, they must convey adaptive advantages that outweigh their energetic and informational costs. Sometimes the adaptive advantages of migration are obvious; for example, birds that feed on flying insects in temperate summers must go south in the winter (when the insects at temperate latitudes are dormant), unless they hibernate or find a different winter food. Navigation is clearly important in migration, and not only for knowing which way is south. Most seasonal migrants do not just move south in the fall and north in the spring; instead they travel along defined routes to restricted destinations (more restricted for some than others). Therefore detailed navigational ability of migrants presumably imparts its own selective advantage, one that is separable (at least in principle) from the migratory journey itself.

Navigational Strategies

Navigation is a complex instance of animal behavior, involving sensory integration and motor control, as well as learning and memory. Hence it is useful to approach this topic by studying the repertoire of behavioral mechanisms that animals exhibit as they navigate, rather than diving immediately into a search for underlying physiological mechanisms or cellular functions. Biologists who study animal behavior recognize five behavioral strategies that animals use to navigate (**TABLE 18.1**): *trail following, piloting, path integration, compass navigation, and map-and-compass navigation.*

It is noteworthy that the names of these strategies have analogies with human navigational techniques. But because some navigating species have sensory abilities and integrative capacities that humans lack, these named strategies may not actually encompass *all* the navigational techniques that animals employ. Moreover, terms such as *compass* and *map* are metaphors when applied to animals; we cannot know how animals actually experience the compasses and maps that their behavior suggests they possess.

Trail following is the most rudimentary form of animal navigation

Most people are familiar with following a trail through a forest or across a field in which each spot of trampled earth follows in a continuous series from start to finish. Similarly, **trail following** is a navigational strategy that is accomplished by detection of an interconnected series of local sensory cues (i.e., cues that are immediately proximate to an animal). Although humans are used to thinking about trails in visual terms, for many navigating species of arthropods, such as ants, a trail may consist solely of olfactory cues.

Trail-laying ants have an abdominal scent gland that is used to lay a *trail pheromone*, a chemical signal that marks a path to a food source. Each ant senses the olfactory profile of its path of travel and, in turn, deposits olfactory markers that it and other members of the colony can use to travel between a nest and the food source. (As we will discuss shortly, other kinds of ants employ different navigational strategies.)

Piloting animals follow a discontinuous series of learned cues

A person who knows that a hot-dog stand is located at the base of a tower, and who heads toward the tower to find the stand, is employing piloting (also called *beaconing*). Using this navigational strategy, **piloting** animals follow a discontinuous series of learned landmarks to determine where they are in their path of travel with respect to their destination. (A **landmark** is a discrete, sensed marker of position, usually but not necessarily visual.) The ability to use learned landmarks to navigate means that sensory cues need not be connected in a relatively unbroken series, as is the case with trail following. Moreover, because piloting is learned behavior, most animals that employ this strategy improve their navigational performance with age and experience. Although landmark-based learning such as piloting is considered relatively simple behavior, we will see that it is difficult to distinguish from more behaviorally sophisticated processes, such as map-and-compass navigation, a topic we cover later in the chapter (see page 506).

As our example of the hot-dog stand suggests, animals often use piloting as a strategy to find their way to a specific destination, a fact that Nikolaas Tinbergen (1907–1988) proved in a classic field experiment with digger wasps (*Philanthus triangulum*). Tinbergen placed a ring of pinecones around the entrance of a wasp's burrow. When the wasp emerged, she flew around the entrance for about 6 s before disappearing to hunt for food. Tinbergen established experimentally that the wasp used the pinecones as a local landmark for navigation by shifting the pinecone ring about 30 cm from the entrance while the wasp was away. On returning, the wasp flew to the center of the pinecone ring and was unable to find the nest. If the pinecone ring was moved farther than 1–2 m from the nest, the wasp even failed to locate the ring (and the nest). This latter observation suggests that, for the digger wasp, it is primarily the final moments in navigating to a destination that depend on piloting; the wasp uses other sensory cues to return to the correct neighborhood.

Birds that cache seeds also appear to use local landmarks as a primary mechanism in relocating their food. In one experiment, Clark's nutcrackers in an aviary hid pine seeds in an open arena strewn with landmark objects. After the birds had cached the seeds, the array of objects on one side of the arena was displaced by 20 cm, whereas the array on the other side was untouched. When the nutcrackers returned to search for their caches, they probed the ground at the "correct" places *with respect to the moved objects*, missing the caches by 20 cm on the manipulated side of the arena

Strategy	Description and source of information
Trail following, route learning	Continuous cues about trail, landmarks
Piloting	Landmarks (discontinuous)
Path integration	Integrated direction and distance information
Compass navigation	Celestial or magnetic cues, learned or genetic
Map-and-compass navigation	Compass cues, landmark- or gradient-based map

TABLE 18.1   A classification of strategies in animal navigation

Source: After Papi 1992.

and retrieving them accurately on the undisturbed side. In the middle of the arena, the birds probed about 10 cm away from the caches, indicating that they were using information from both displaced and undisplaced landmarks.

Path integration is a form of dead reckoning

As with piloting, animals that exhibit **path integration** behavior use this navigational strategy to accomplish homing, but unlike animals engaged in piloting, they do *not* refer to landmarks. In path integration, the navigating animal somehow keeps a running tally of its past directions and distances traveled, and it integrates this information in a manner that allows it to set a direct route back to its starting point at any time. Desert ants (*Cataglyphis fortis*) provide an excellent example of path integration behavior: They forage along tortuous, novel paths but can return from any point directly back to their hidden nesting site once they have located food (**FIGURE 18.3**).

Path integration in desert ants is equivalent to the method of dead reckoning often used by early mariners. *Dead reckoning* was a way of determining a ship's position at sea without using landmarks or star positions. The sailors kept a record (the "logbook") of the ship's direction and speed, and accumulated this information into an aggregate direction and distance. Mariners used the position of the sun to ascertain their direction, and they assessed speed by throwing overboard a piece of wood (the "log") attached to a long, knotted line that was cast off or played out as the ship sailed. The number of knots of line that were cast off over a period of time gave the speed and was entered in the logbook.

Ants navigate in a featureless desert rather than a featureless ocean, but their strategy is similar to sailors' technique of dead reckoning. Ants gather directional information by using the sun's position (or polarized light) as a compass. The ants measure distance by integrating proprioceptive information while walking, rather like counting steps for humans. Experimenters captured ants that had just found food and were ready to return home. They altered the length of the ant's legs, cutting them short in some ants and gluing longer "stilts" on others (**FIGURE 18.4**), and displaced them as in Figure 18.3B. The ants made systematic errors in their attempt to return home: The ants with stilts walked too far before beginning their random search of the expected home area, whereas the ants with stumps underestimated the distance and shifted to random search too early.

Other animals use different methods to estimate distance traveled. Bees assess the distance they travel by **optic flow**—the flow of visual images past them as they move. The closer the bees are to objects they pass, the greater the objects' apparent motion is, and the greater the bees' estimation of distance. To confirm that honeybees use optic flow, researchers manipulated the apparent distance of objects along their flight path. When bees were made

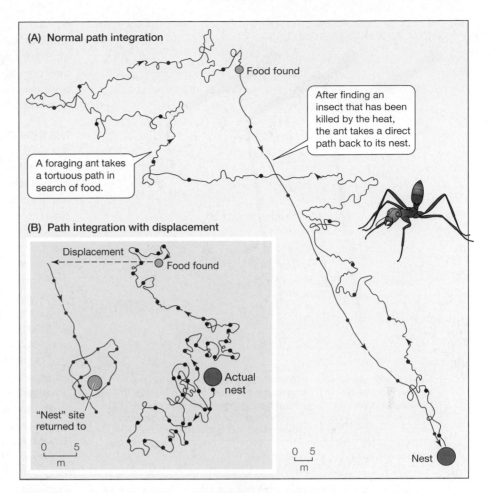

FIGURE 18.3 Path integration in desert ants   (A)After finding food via a circuitous path, an ant takes a direct path back to its nest that does not depend on landmarks, but rather on path integration—adding all the direction and distance vectors in the tortuous outward path—and using a sun compass or polarized-sky-light compass to determine the direction back. (B) If an ant is artificially displaced when it has found food, it returns to where home would have been without the displacement, using external compass cues and ignoring local landmarks (except at the presumed nest location). Points mark positions at 1-min intervals. (After Collett and Zeil 1998.)

to fly through a narrow tunnel with finely patterned walls, their optic flow increased and they overestimated their travel distances. Moreover, they communicated their overestimate to other bees in their waggle dance (see Figure 18.2), so that other bees (not flying through the tunnel) searched for the food source at too great a distance from the hive.

Animals can derive compass information from environmental cues

A **compass** is a mechanism that indicates geographical direction. Animals use compass information to *orient*—that is, to choose a direction in which to head. Orientation is a necessary component of compass navigation and map-and-compass navigation; no matter how you determine that you are north of your goal and need to go south, you need a mechanism to tell you which way is south. Animals derive compass information for navigation from a variety of environmental cues, such as the position of the sun and stars, the quality of polarized light, or the lines of force in Earth's magnetic field. It is useful to remember that we are using *compass* here as a

(A)

Stilts   Normal   Stumps

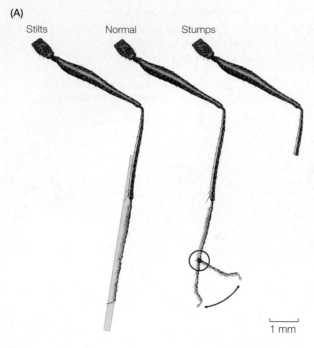

1 mm

(B)

The ants with shortened legs did not walk far enough, because their stride was shorter than normal.

The ants on stilts walked too far. They walked the number of steps that would have brought them to their nest with normal legs, but they overshot because their stride was longer than normal.

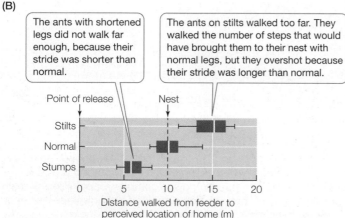

FIGURE 18.4 Desert ants estimate distance by monitoring steps Ants integrate proprioceptive information as a measure of distance while walking, analogous to humans counting steps. (A) Experimenters altered the length of legs of captured ants that had just found food and were ready to return home. They cut short the legs of some ants ("stumps") and glued extensions ("stilts") on others. Each ant was displaced as in Figure 18.3B. (B) The distance the ants walked homeward (before initiating random search) was a direct function of the new leg length. Ants with extended legs (stilts) walked too far, and ants with short legs (stumps) underestimated the distance to home, as if all the ants walked equivalent numbers of steps. (From Wittlinger et al. 2006.)

metaphorical designation: Investigators can only infer the presence of a compass on the basis of an animal's behavior.

To study compass-based navigational strategies, researchers usually manipulate the sensory basis of a suspected animal compass while they measure the animal's *initial orientation*, the direction in which the animal orients its body and movements at the start of an actual or intended journey. In some cases this measurement might consist of observing the flight of a bird to vanishing point—the point at which it moves beyond the observer's viewing horizon. In the case of migratory birds, measuring initial orientation may

entail quantifying the direction of *migratory restlessness*. Birds in migratory phase will orient in a cage, facing and hopping in the direction of intended migration. Researchers can alter the suspected compass cues and then observe whether the orientation direction of migratory restlessness changes.

THE SUN COMPASS Many animals use the position of the sun as a source of compass information. In an early example demonstrating the presence of a sun compass, Gustav Kramer (1910–1959) showed that caged songbirds in migratory phase changed the orientation of their migratory restlessness when he altered the apparent position of the sun by using mirrors. The other original demonstration of sun-compass navigation was for foraging bees. The bee's waggle dance (see Figure 18.2) conveys information about the direction of a nectar source as an angle relative to the sun's position. Other worker bees in the hive use this sun-compass information to fly directly to the nectar source.

The sun moves across the sky from east to west at about 15° per hour, so animals must know the time of day to determine a compass heading from the sun's position. (Details of the sun's path also depend on latitude and season.) An animal that navigates using a sun compass must integrate solar position and circadian time information in order to orient its path of travel. The caged birds and the bees described earlier could maintain their sense of direction at different times of the day using their circadian clocks to account for the sun's movement.

Researchers can disrupt a suspected sun compass by resetting the animal's circadian clock, usually by imposing an artificial light–dark cycle that is different from natural sunrise and sunset (see Chapter 15, pages 421–423). If an animal experiences a shifted light–dark cycle, one in which the lights come on at noon and go off at midnight (rather than at 6:00 AM and 6:00 PM, respectively), the new light–dark cycle will entrain the animal's circadian clock, causing the animal to misinterpret the sun's position.

Consider the following example: Guided by its normal circadian clock, a pigeon flying south at 9:00 AM will orient 45° to the right of the sun, which will be in the southeast (**FIGURE 18.5A**). If the pigeon flies south at 3:00 PM, it will orient 45° to the left of the sun, which will be in the southwest. A pigeon whose circadian clock is set ahead 6 h, however, will make predictable errors in sun-compass navigation (**FIGURE 18.5B**). If the pigeon thinks it is 3:00 PM when it is really 9:00 AM, the bird will fly due east, 90° from its intended southerly route, because it will fly 45° to the left of the sun. Experiments such as this one show that homing pigeons and most diurnal migrating birds use a sun compass as their primary mechanism of navigation.

THE POLARIZED-LIGHT COMPASS In addition to using a sun compass to navigate, many insects and birds appear to use polarized light as a source of compass information. This means of navigating is helpful on partly cloudy days, after sunset, or whenever the position of the sun is obscured from view.

Atmospheric particles polarize sky light. Recall from physics that white light from the sun is unpolarized; that is, it is a bundle of electromagnetic waves, each with an electrical vector (e-vector) that vibrates at right angles to the line of propagation of the light ray. In unpolarized light the e-vectors of different waves are at an

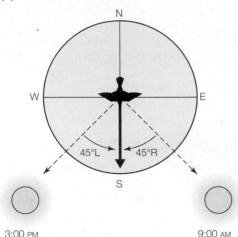

(A) Normal circadian clock

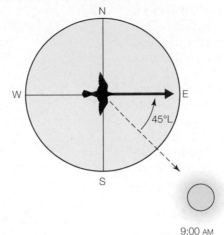

(B) Circadian clock set ahead 6 hours

3:00 PM   9:00 AM

9:00 AM

FIGURE 18.5 Homing pigeons use a sun compass on sunny days The direction in which individual pigeons vanished over the horizon from a release point north of home. (A) Control pigeons oriented in the homeward direction (south) when released at any time of day. They used the sun's position and their internal circadian clocks to determine which direction was south. (B) Pigeons whose circadian clocks had been shifted 6 h ahead misinterpreted the sun's position and departed approximately 90° to the left of the homeward direction. If released at 9:00 AM, they thought it was 3:00 PM and departed 45° to the left of the sun's position (appropriate for 3:00 PM).

infinite number of planes (**FIGURE 18.6A**). **Polarized light**, however, vibrates in only one plane with respect to its line of propagation. A fraction of the energy in a ray of sunlight will become polarized if it is reflected from a particle. Light reflecting from a surface at 90° from the incoming light ray is fully polarized; light reflecting at other angles is incompletely polarized.

Earth's atmosphere contains an abundance of suitably reflective particles—dust, water droplets, and ice crystals—that can polarize sunlight and provide the photic information that is the basis of an animal's polarized-light compass. Sky light in the vicinity of the sun reaches an observer's eye in a direct path, so it is unpolarized (see Figure 18.6A). Light reaching the eye from other parts of the sky, however, is reflected from atmospheric particles. As the reflection angle (from the sun, to the particles, to the eye) increases up to 90°, the proportion of light reaching the eye becomes increasingly polarized. Sky light, then, is maximally polarized in a band that is 90° away from the sun's position (**FIGURE 18.6B**).

Human eyes cannot detect useful differences in the polarization of sky light, but the eyes of many navigating insects (and birds) can. A bee, for example, has eyes that are sensitive to polarized ultraviolet light. As long as a patch of blue sky remains in view, a bee can detect the plane of light polarization and use this information to infer the position of the sun. Researchers cannot say with certainty *how* an animal experiences polarized light, but to extract compass information, it must be able to detect the angle of polarization, as well as gradients in the degree of light polarization (see Figure 18.6B).

The effect of light polarization detected by the bee may not be altogether different from what people experience when they

wear polarized sunglasses. The polarizing filter in the lenses absorbs the horizontally polarized components of sunlight and passes the vertically polarized components. By tilting the head, a wearer of polarized glasses will observe that the sky appears to be noticeably darker in a band that is at right angles to the sun. Light

(A) Polarization of reflected sunlight

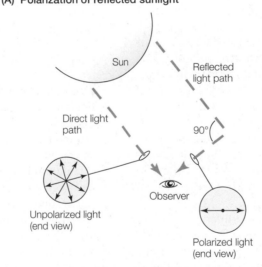

Sun

Reflected light path

Direct light path

90°

Observer

Unpolarized light (end view)

Polarized light (end view)

FIGURE 18.6 Polarization of sky light can aid in determining the sun's position (A) The blue sky results from reflected scatter of blue and ultraviolet sunlight by particles in the atmosphere. Sunlight is unpolarized; its electrical vector (e-vector) is at right angles to the direction of propagation of the light wave, but it can be at any direction. The insets show end views looking into the light; for unpolarized light, arrows show e-vectors at all orientations. In contrast, the reflected light is polarized, with its e-vector in only one direction (here shown as horizontal in the end view). (B) The pattern of polarized light at two solar positions: 25° (left) and 60° (right) above the horizon. The plane of polarization is at right angles to the plane of light scattering, and the degree of polarization (indicated by the thickness of the orange bars) is strongest at 90° from the sun. (B after Wehner 1997.)

(B) Polarization indicates the sun's position

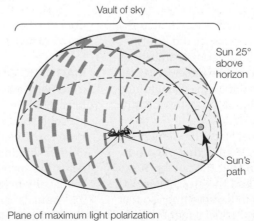

Vault of sky

Sun 25° above horizon

Sun's path

Plane of maximum light polarization

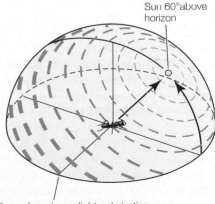

Sun 60° above horizon

Plane of maximum light polarization

reaching the eye from this part of the sky is maximally polarized, and the polarizing filter passes only the fraction of the light that is vertically polarized. The degree of darkness and the tilt angle of the wearer's head provide sensory cues from which the sun's position can be inferred.

How do the photoreceptor cells of insects respond to polarized light? First, an individual rhodopsin molecule must be differentially sensitive to the plane of light polarization. A rhodopsin molecule maximally absorbs light when the e-vector of the light is parallel to the long axis of the retinal chromophore (see Figure 14.23). Second, the many rhodopsin molecules in a photoreceptor must be aligned similarly, so that they all absorb light at the same plane of polarization.

In arthropod eyes, rhodopsin is localized in rhabdomeres, collections of microvilli arrayed at right angles to the long axis of the photoreceptor (retinular) cell (see Figure 14.24). The rhodopsin molecules are preferentially oriented so that the 11-*cis* retinal chromophore of each is parallel to the long axis of the microvillus. Therefore light with its e-vector parallel to the long axis of the microvilli is preferentially absorbed. Different photoreceptor cells have their microvillar arrays aligned at right angles to each other and thus are differentially responsive to light at different planes of polarization.

THE STAR COMPASS   Many species of birds that are normally diurnal migrate at night, perhaps to escape predation. They cannot effectively use the sun as a compass (although some nocturnally migrating birds do use the direction of the setting sun and sky-light polarization for initial nighttime orientation). The moon is an unreliable source of nocturnal directional information; it is visible at night only half the time, and its phases (resulting from a different periodicity than that of the sun) make lunar navigation a complex problem. The stars of the night sky provide more reliable information and are used by nocturnal migrants as a **star compass**. Earth's rotation makes the stars appear to sweep across the sky like the sun, but the region around Polaris (the North Star) provides a stable reference point for north (in the Northern Hemisphere; see Figure 4.3).

Experiments by the behaviorist Stephen Emlen convincingly demonstrated that some nocturnal migrants use star patterns in the night sky as a compass to determine their heading. Emlen raised indigo buntings (*Passerina cyanea*) so that they could see the night sky but not local landmarks. When the birds reached spring migratory phase, he measured their pattern of migratory restlessness. Caged buntings that could see the night sky exhibited the expected north-oriented pattern of restlessness, as did buntings that could see the identical Northern Hemispheric constellations projected in a planetarium (**FIGURE 18.7A**). When Emlen projected the same constellations so that they faced south in the planetarium, the buntings reversed their pattern of migratory restlessness by 180° to the south (**FIGURE 18.7B**).

To orient successfully, young birds must learn the elements of the night sky during a critical period prior to their first migration. Birds in the Northern Hemisphere learn that the northern sky rotates around Polaris (the North Star), and they learn star patterns within 35° of Polaris. Emlen raised indigo buntings so that they never saw the real night sky, but only artificial night skies in

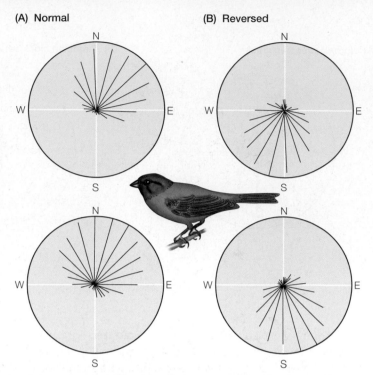

FIGURE 18.7 Planetarium experiments demonstrate that nocturnally migrating birds use star patterns for orientation Indigo buntings were raised so that they could see the night sky but not local landmarks. The orientation preferences of their migratory restlessness (indicated by the radiating black lines) were then tested in a planetarium with either normal star patterns (A) or star patterns reversed in direction (B). The results for two buntings shown here reveal that the star pattern is the dominant determinant of orientation direction. (After Emlen 1967.)

a planetarium. One group saw the normal pattern of stellar rotation around Polaris. The other group saw the normal star pattern, but instead of rotating around Polaris, the constellations rotated around Betelgeuse, a bright star in the constellation Orion. When the birds came into fall migratory condition, Emlen tested their orientation in a planetarium, under a stationary sky. The control birds oriented away from Polaris, that is, away from their stellar north. The experimental birds oriented away from Betelgeuse, indicating that they had learned a star map in which the stationary star (Betelgeuse) was considered north! Apparently rotation of the star pattern is necessary for learning the location of stellar north, but it is not necessary for using the pattern once that pattern has been learned.

Recent studies indicate that seals and dung beetles can also orient to the night sky; the dung beetles use the Milky Way rather than learned constellations.

MAGNETIC COMPASSES   Earth acts like a gigantic bar magnet; it has north and south poles that are connected by magnetic lines of force (**FIGURE 18.8A**). Earth's magnetic field, in principle, provides a reliable source of directional information: The magnetic poles are close to the geographical poles, and they never change during an animal's lifetime. Because humans are quite unaware of magnetic fields, it is surprising to find that many species of animals can detect Earth's magnetic field and can use it to navigate.

Animals can sense the polarity and the angular inclination of Earth's magnetic field—two qualities that provide the navigational

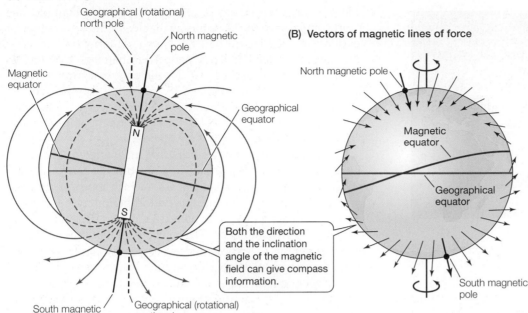

(A) Lines of magnetic force

Geographical (rotational) north pole

North magnetic pole

Magnetic equator

Geographical equator

N

S

Both the direction and the inclination angle of the magnetic field can give compass information.

South magnetic pole

Geographical (rotational) south pole

(B) Vectors of magnetic lines of force

North magnetic pole

Magnetic equator

Geographical equator

South magnetic pole

FIGURE 18.8 Earth's magnetic field can provide compass information (A) Lines of magnetic force leave the south magnetic pole vertically, curve around Earth's surface, and reenter vertically over the north magnetic pole. (B) Vectors of magnetic lines of force at different points on Earth's surface. Arthropods and some vertebrates use a directional compass (arrowheads point toward magnetic north), and many vertebrates use an inclination compass (into the ground = toward the nearer pole). (A after Goodenough et al. 2001; B after Wiltschko and Wiltschko 1996.)

cues that animals use as a **magnetic compass**. Magnetic lines of force have a *polar component*: their north and south polarity. They also have an *inclination component*: the "dip angle" of the lines of force relative to Earth's surface (**FIGURE 18.8B**). Magnetic lines of force exit Earth nearly vertically at the south magnetic pole, and their angle of inclination steadily decreases at lower latitudes until they are parallel to Earth's surface at the magnetic equator. From there, their angle of inclination steadily increases until the lines of force are again nearly vertical at the north magnetic pole.

Accordingly, animals may use either a *magnetic polarity compass* (one sensing directional polarity) or an *inclination compass* (one sensing the magnetic inclination angle) to navigate. Experimental evidence shows that arthropods such as bees sense magnetic polarity, whereas birds, turtles, and other reptiles sense the inclination angle. Using an inclination compass, the direction "downward into Earth" is always poleward, as close inspection of Figure 18.8B reveals. Thus the downward inclination is toward magnetic north in the Northern Hemisphere but toward magnetic south in the Southern Hemisphere. At the magnetic equator the lines of force are parallel to Earth's surface, and an animal's inclination compass cannot discriminate between north and south. For this reason, birds that obtain navigational cues from an inclination compass may become disoriented as they migrate past the magnetic equator if they don't have a backup source of navigational information.

Researchers infer the presence of a magnetic compass from animal behavior. To detect an animal's use of a magnetic compass, a researcher may experimentally alter the magnetic sensory cues that the animal receives. One way to do this is to place the test animal in a box made of a type of metal (mu metal) that shields it from Earth's magnetic field. Alternatively, researchers may alter the magnetic field in the animal's vicinity by using permanent magnets or by placing tiny Helmholtz coils around the suspected locations of the animal's magnetic sensory apparatus (**FIGURE 18.9A**). A Helmholtz coil consists of a pair of ring-shaped electromagnets. When an electric current is passed through the coils, a magnetic

field is established in the space between the coils that can be used to nullify or reverse the effect of Earth's magnetic field. By placing a Helmholtz coil to either side of a homing pigeon's head, for example, researchers can control the bird's perception of magnetic fields.

The ability of animals to sense magnetic fields may be a primitive, relatively widespread mechanism, because it is found in some bacteria as well as many animals. Magnetic orientation is often used as a backup for other forms of compass navigation, as exemplified in homing pigeons. Recall from our discussion earlier that pigeons whose circadian clocks had been shifted forward 6 h flew in the wrong direction when they attempted to navigate south by following their sun compass (see Figure 18.5). However, the clock-shifted birds had no trouble navigating in the desired direction on *cloudy* days, which suggests they have another navigational mechanism that is not based on solar position.

Evidence demonstrates that this second navigational system in homing pigeons is a magnetic compass. Pigeons with magnets attached to their backs or heads home normally on sunny days but are disoriented on overcast days; control pigeons with brass bars attached to their backs instead of magnets home normally on both sunny and overcast days. What's more, homing pigeons wearing Helmholtz coils (see Figure 18.9A) experience similar disruptions of their homing abilities when released on cloudy days. When the inclination angle of the induced field between the coils matches that of Earth's magnetic field, the pigeons orient normally (**FIGURE 18.9B**). But when the inclination angle of the induced field is reversed, the pigeons fly in the opposite direction from their intended course home (**FIGURE 18.9C**).

Another line of evidence for animals' use of a magnetic compass comes from studying the effects produced by natural magnetic anomalies (distortions of Earth's magnetic field by geological peculiarities) on a navigating animal's path of travel. On overcast days, pigeons homing in a northeasterly direction toward Boston sometimes become disoriented at a major magnetic anomaly called Iron Mountain, in Rhode Island. The mountain's metallic composition

(A) The experimental setup

(B) Magnet in normal orientation

(C) Magnet in reversed orientation

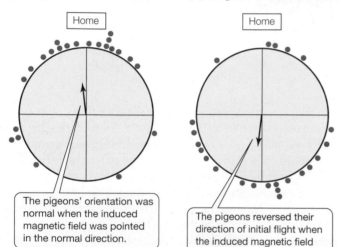

The pigeons' orientation was normal when the induced magnetic field was pointed in the normal direction.

The pigeons reversed their direction of initial flight when the induced magnetic field was reversed.

FIGURE 18.9 Changing the magnetic field changes the orientation of released pigeons (A) A small Helmholtz coil is shown attached to a pigeon's head, with a power pack on the pigeon's back. Reversing the direction of electric current flow through the coil reverses the direction of the magnetic field. (B,C) Pigeons with Helmholtz coils were released south of home on overcast days. They interpreted the direction in which magnetic lines dip into Earth as north. Each dot outside the circle represents the direction in which a released pigeon vanished over the horizon. The arrow at the center of each circle is the mean vanishing bearing for the group. (After Walcott and Green 1974.)

bends the magnetic-field lines in its vicinity, sending the hapless birds off course. In the ocean, magnetic-field lines cluster together to form high-intensity ridges or spread apart to form low-intensity troughs that can be mapped by satellite. Whales and dolphins are more likely to swim ashore and become stranded at locations where troughs in magnetic-field intensity meet the shore. This correlation suggests that whales migrate using a magnetic compass and perhaps follow the path of magnetic troughs.

Other organisms, including salamanders, turtles, salmon, and bacteria, have been shown to orient to magnetic fields. Such orientation is commonly weaker than that to solar or other celestial cues, supporting the idea that magnetic orientation is a primitive, widespread mechanism, often used when other cues are unavail-

able. Monarch butterflies (*Danaus plexippus*), for example, use a sun compass when the sun is available, but in the absence of celestial cues, they use magnetic-compass information to orient and navigate. Detection of magnetic fields in monarch butterflies and in *Drosophila* fruit flies may depend on cryptochrome-based light reception (see Boxes 18.1 and 18.2). Bees are also able to detect magnetic fields, and the orientation of their dances at the hive can be influenced by magnetic fields. However, magnetic information plays little role in ordinary bee navigation.

For any animal that navigates using magnetic-compass cues, the magnetic response is innate and presumably genetically determined. But unlike the star compass—which appears to be fixed once it has been learned during a critical period in the animal's life—an animal's magnetic compass seems to be subject to at least some degree of recalibration. The ability to recalibrate is important because migrants to Arctic latitudes must resolve conflicts between magnetic and celestial navigational cues. The north magnetic pole and the geographical North Pole are not identical (see Figure 18.8). Magnetic-field information points to a different location than celestial-compass information does, and the disparity between the two increases at higher and higher latitudes. For this reason, migrating Arctic birds must pause in their journey at intervals in order to allow their magnetic compasses to realign with celestial-compass cues such as polarized sky light. If prevented from pausing long enough to recalibrate, the birds may become disoriented.

The physiological mechanisms by which animals detect magnetic fields are discussed in **BOX 18.1**.

Some animals appear to possess a map sense

Most animals probably do not use simple compass navigation in migration or homing without also using landmark or map information. In cases of **map-and-compass navigation**, the animal possesses in its brain some sort of representation of its position, and the position of its goal. This is the *map,* and the animal's *compass* provides bearings relative to the map.

For most animals that appear to have a map sense, we do not know the map's basis, although it might be based on various sensory cues. A map could be based on geomagnetic information, because both the vertical angle and the strength of magnetic fields increase near the magnetic poles. The disorientation of homing pigeons and migrating whales in the vicinity of magnetic anomalies suggests that magnetic information is an important basis of their maps. Solar cues could also provide map information; the height of the sun above the horizon varies with the latitude, and the times of sunrise and sunset vary with longitude. For some migrating creatures, infrasounds (very low-frequency sound waves—for example, from waves crashing on a distant shore) may serve as the basis of the map. Even olfactory cues are suspected of forming the basis of the map sense, as appears to be the case with pigeons. Whatever its basis, map construction and use represent a sophisticated instance of animal learning.

DISPLACEMENT EXPERIMENTS TEST FOR A MAP SENSE Displacement experiments show the difference between path integration and compass navigation on the one hand, and complex map-and-compass navigation on the other. If a desert ant is experimentally moved while it is feeding, its return path makes no

BOX 18.1 Magnetoreceptors and Magnetoreception KENNETH J. LOHMANN

The idea that animals perceive Earth's magnetic field was once dismissed as impossible by physicists and biologists alike. Earth's field is much too weak for an organism to detect, the argument went, and there are no possible biological mechanisms capable of converting magnetic-field information into electrical signals used by the nervous system.

Over time, however, evidence accumulated that animals do indeed perceive magnetic fields, until even the most hardened skeptics came to accept the idea. Today it now seems clear that diverse animals, ranging from invertebrates such as molluscs and insects to vertebrates such as sea turtles and birds, exploit information in Earth's field to guide their movements over distances both large and small. What has remained mysterious is exactly how they do this.

Determining how the magnetic sense functions is an exciting frontier of sensory physiology. For sensory systems such as vision, hearing, and smell, the cells and structures involved in perceiving relevant sensory stimuli have been largely identified, and the basic way in which the sense operates is understood. In contrast, the cells that function as receptors for the magnetic sense have not been identified with certainty in any animal. Even the basic principles around which magnetic sensitivity is organized remain a matter of debate. This box continues on the web in **Box Extension 18.1**. There you will find more information on magnetoreception in animals and on mechanisms that may underlie it.

---

correction for this passive displacement (see Figure 18.3). Therefore the ant has no map sense that allows it to detect its displacement relative to its goal.

Birds, in contrast, are more likely to correct for a passive displacement, whether experimental or by natural means such as a storm. If a bird migrating southwest purely by compass navigation became displaced, say, 100 km to the southeast, it would continue to go southwest. If the bird were using map-and-compass navigation, it would realize it had been displaced (apparently without using local, familiar landmarks) and would correct its path to the west or northwest.

FIGURE 18.10 illustrates the distinction between path integration and compass navigation. European starlings (*Sturnus vulgaris*) migrate from Baltic breeding grounds to wintering grounds in France and England. Migrants were captured in the Netherlands, transported to Switzerland, and released. The juvenile starlings continued southwest (by compass navigation) and ended up in Spain, but the older, experienced birds that had developed a map sense could detect the displacement and reorient to a new northwestern path. Map-and-compass navigation is sometimes called *true navigation*, an unfortunate term because it (wrongly) suggests that other methods of an animal's finding its way are not really navigation. Because we find that compass orientation is relatively complex, we may expect the basis of the less-understood map sense to be complex as well.

PIGEONS MAY HAVE AN OLFACTORY MAP Some of the clearest studies of map sense have focused on pigeons, and the dominant hypothesis of map sense in pigeons is the *olfactory-map hypothesis*. Displaced pigeons can determine their position relative to home without using visual or proprioceptive information from the outward journey to the release site. Therefore investigators conclude that the pigeons must have some sort of map sense that, with the compass information previously discussed, allows them to return from more than 700 km away. According to the olfactory-map hypothesis, the pigeon's map sense is based on an odor profile somewhat analogous to the one used by homing salmon. The odor profile could consist of a mosaic of discrete odor sources or a few odor gradients.

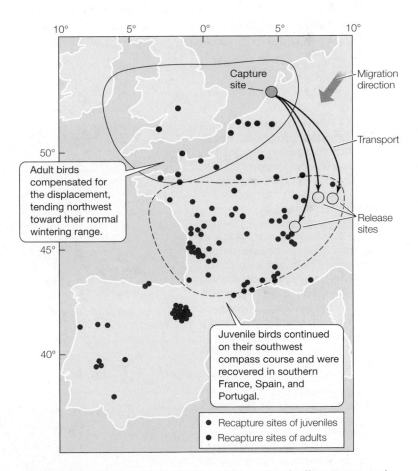

FIGURE 18.10 Migrating adult and juvenile starlings respond differently to displacement European starlings were captured in the Netherlands while in their autumn migration southwest from northeastern Europe and were transported to release sites in Switzerland. The normal wintering range is bounded by the solid line; a dashed line shows a similar area displaced to the same extent as the displacement of the released birds. The adults' compensation for displacement suggests the use of map-and-compass navigation. (After Perdeck 1958.)

Several lines of experimental evidence support the olfactory-map hypothesis. Pigeons are usually housed in somewhat open structures (lofts), often on the roofs of buildings. A pigeon that has its olfactory sense disrupted has difficulty homing to its loft. Manipulations of odors around the home loft can also affect homing performance, perhaps by affecting the learning of an olfactory map. In one example, pigeons were raised in a loft that was exposed to normal odors but also had an odor of benzaldehyde blown in by a fan from the north-northwest. Thus they presumably learned to associate north-northwest wind with benzaldehyde. At the end of the summer they were transported to various distant, unfamiliar release sites, exposed to benzaldehyde, and released. They flew primarily south, regardless of the release site. Control birds raised without benzaldehyde, but exposed to it just prior to release, oriented toward home normally from the release sites. The simplest interpretation of these findings is that the pigeons learned to associate benzaldehyde odor with north-northwest wind, so the experimental pigeons thought they had been released at a benzaldehyde source north-northwest of home, and flew south-southeast.

Other investigators have criticized the olfactory-map hypothesis on several grounds. Olfactory anesthesia or nerve cuts to make pigeons unable to smell may have indirect effects on homing performance that have been mistakenly interpreted as affecting an olfactory map. Some indirect effects can be ruled out; for example, the olfactory manipulations do not simply make the animal sick. Other effects, such as a possible indirect disruption of the pigeons' use of magnetic signals, are harder to rule out. In addition, some experiments have been hard to replicate, perhaps because investigators in different countries raise their birds in different styles of lofts that may affect olfactory-map learning. Overall, however, there is substantial evidence to infer roles of olfactory cues in pigeon homing, and so we can consider the olfactory-map hypothesis to be supported by evidence but not proven.

Sea turtles exemplify the degree of our understanding of navigation

Sea turtles illustrate what we know and don't know about the behavioral mechanisms of navigational control. Turtles use different cues in navigation at different stages in their life cycle, and some of these cues are much easier to analyze than others. New hatchlings starting off the beach appear to engage in a sequence of orienting mechanisms. Loggerhead sea turtles hatching on Florida beaches initially orient toward a lighter sky, which will be toward water (because the water reflects more sunlight or moonlight and makes the sky lighter). This light orientation and a tendency to move down a slope take the hatchlings to the water, where they

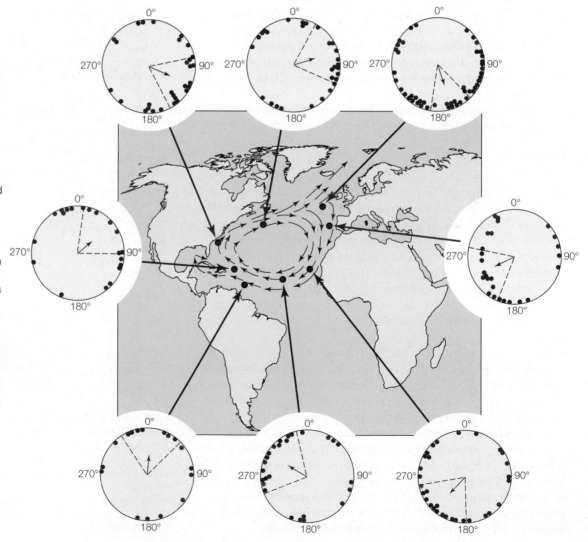

FIGURE 18.11 Hatchling turtles orient to artificial magnetic fields in ways that suggest a magnetic map Newly hatched loggerhead turtles were placed in artificial magnetic fields characteristic of eight locations (dots on map) along their prospective migratory route. Arrows on the map show major currents of the North Atlantic Subtropical Gyre, around which the turtles swim as they mature. Each dot in the orientation diagrams represents the mean angle of oriented swimming of one hatchling. The arrow in the center of an orientation diagram indicates the mean angle of orientation of each group; arrow length shows the statistical degree of orientation of the group. All eight groups showed statistically significant orientation to the artificial magnetic fields. The angle of orientation differed greatly for the groups tested with artificial magnetic fields corresponding to different locations. Moreover, the orientation at each simulated location was appropriate to keep turtles within the gyre, rather than (for example) straying into the fatally cold water of the North Atlantic and North Sea. Dashed lines represent the 95% confidence interval for the mean angle; data are plotted relative to magnetic north. (After Lohmann et al. 2012.)

swim out continuously for 24 h. The primary orientation cue for this frenzied outward swim is the waves; the turtles head into the waves, which come (with rare exceptions) from the open ocean. (During the exceptions, young turtles swim into the waves even if that orientation takes them back to shore.)

The young turtles can also detect Earth's magnetic field, and magnetic-compass orientation may replace wave orientation after their initial outward swim. This navigational sequence is thought to take the turtles out to the Gulf Stream, one of the currents flowing around the North Atlantic Subtropical Gyre (**FIGURE 18.11**). Adult turtles must use map-and-compass navigation to return to their natal beach to nest (in contrast to juveniles, which only have to reach the gyre). Tracking studies indicate that the adults navigate by direct, straight-line routes. Recent evidence indicates that even the younger turtles can use magnetic information as a map component that helps keep them in the gyre as they mature.

Experimental replication of the parameters of the magnetic field at different points in the turtles' migration route leads to different swimming orientations that match the turtles' swimming direction at that location (see Figure 18.11). For example, caged turtles in an artificial magnetic field that replicates the natural magnetic field off the northern coast of Spain swim southward, a direction that keeps them in the North Atlantic Subtropical Gyre rather than washing into northern waters off Scandinavia. Similar experiments with turtles as well as with salamanders, spiny lobsters, and birds all show that experimental replication of the magnetic-field vectors and strengths of a particular location can alter orientation behavior in predictable ways. These studies suggest that animals can use magnetic information as a map (as well as a compass), a finding that may begin to clear up the murkiest area in complex navigation.

Innate and Learned Components of Navigation

We have indicated that many elements of navigation, such as using landmarks, imprinting to a home stream, and constructing a map, are largely learned behaviors. Here we discuss the roles of innate ("hardwired") behavior and of learning in navigation, and we consider how vertebrates learn their spatial surroundings.

Some forms of navigation have strong innate aspects

Two examples demonstrate innate components of navigation that do not depend on prior learning. Monarch butterflies summer throughout the United States north of the Gulf states. Those butterflies east of the Rocky Mountains migrate south every fall to overwintering sites in forested mountains of central Mexico. In the spring the same individuals migrate

north to the U.S. Gulf Coast states, where females lay eggs on milkweed. A new generation continues the migration to summering grounds in the northern states and produces two or more short-lived generations in a summer. The autumn migrants that return to the same restricted overwintering sites in the following year are thus descendants, two to five generations removed, of ancestors that occupied the sites the previous winter! The navigational performance of the monarch butterfly therefore must have a strong innate, genetic component; the butterflies cannot have learned the overwintering location from previous experience. The sensory and genetic bases of navigation in monarch butterfly migration are discussed in **BOX 18.2**.

Some birds also have innate, apparently genetic components of their migratory navigation. European blackcaps (*Sylvia atricapilla*), for example, migrate from Europe to Africa around the Mediterranean Sea—those in western Europe going southwest via Gibraltar, and those in eastern Europe going southeast via Syria. When birds from the two sides of the divide are raised in isolation, they orient in cages in the correct direction for their area of origin, despite the lack of opportunity to learn their way. Cross-breeding experiments further demonstrate that the difference in orientation between eastern and western populations is genetic.

The hippocampus is a critical brain area for vertebrate spatial learning and memory

As we have discussed, an animal's ability to navigate often involves learning, such as the recognition of landmarks by some homing animals, or the imprinting of home-stream odors by salmon. Spatial

---

BOX 18.2 Migratory Navigation in Monarch Butterflies: Sensory and Genomic Information

We think of vertebrates such as fish, turtles, birds, and mammals as champion long-distance migrants, but insects such as butterflies and moths also migrate long distances, despite their smaller-sized and apparently flimsier bodies. As we have discussed, the navigation of monarch butterflies cannot be learned, because the annual cycle of northward spring migration and southward fall migration involves different generations. Fall migrants are born in the north and navigate to restricted overwintering sites that their great-grandparents and great-great-grandparents left the previous winter.

The physiological and genetic bases of monarch migration, and especially the sensory basis of the navigation that underlies it, have been clarified in an ongoing series of experiments. **Box Extension 18.2** shows that migrating monarch

Monarch butterfly (*Danaus plexippus*)

butterflies use a time-compensated sun compass and a magnetic compass in migratory navigation, as well as information about day length and temperature in controlling and modulating migratory behavior. Genomic analysis is beginning to show the genetic underpinnings of this migratory performance.

learning is crucial for many sorts of navigation, and for vertebrates a major locus of spatial learning is the *hippocampus* of the brain.

By surgically creating hippocampal lesions in test animals and monitoring their performance in mazes, researchers have shown the importance of the hippocampus in spatial learning. One standard test of spatial learning in laboratory rodents is the *Morris water maze*, which consists of a hidden, submerged escape platform in a pool of turbid water. The test animal must learn the location of the platform from local cues (landmarks) that are situated outside the pool. Mice or rats with hippocampal lesions do not remember the location of the platform, and they continue to swim about randomly, trial after trial. Another means of testing spatial learning is the *radial-arm maze*, which consists of several blind corridors that extend symmetrically from a central chamber. The researcher repeatedly baits the end of one arm (or shifts the bait from arm to arm in a recurring pattern); the test animal must remember which arm is baited to earn the reward. Normal mice and rats remember the arms they have visited and do not reenter them during a test session, but mice and rats with hippocampal lesions repeatedly enter the same arm and often bypass the arms they have not yet visited.

Does spatial learning in the hippocampus explain how food-caching birds such as Clark's nutcrackers store information about the locations of thousands of cache sites? In fact, birds with bilateral hippocampal lesions continue to hide food normally, but they cannot recover their caches. They can, however, remember simple, nonspatial associative tasks (such as pairing a sound with a reward). Thus the deficit appears to be specific for spatial learning and memory. Also suggesting the importance of this brain region in spatial learning is that researchers have noticed that the hippocampus is significantly larger in food-caching birds than in size-matched species that do not cache. Moreover, for species that cache, preventing caching behavior in immature birds appears to diminish hippocampal growth.

The hippocampus also appears to be involved in homing in pigeons. It is critical for learning two aspects of pigeon navigation: piloting and navigational maps. Inexperienced pigeons with hippocampal lesions are impaired in learning local landmarks for piloting, as might be expected from the studies with seed-caching birds and rodents. They also fail to orient correctly from distant, unfamiliar release sites, indicating that they have failed to learn a navigational map. Experienced pigeons that receive hippocampal lesions *after* learning a navigational map orient correctly from distant release sites, but they still have difficulty with local, landmark-based navigation. Therefore the hippocampus appears necessary throughout life for landmark-based navigation but not for using a navigational map once that map has been learned. Interestingly, the left hippocampus appears necessary for map learning in pigeons, but not the right.

How might the hippocampus (and related brain areas) function in spatial orientation? There are two aspects or strategies of spatial learning and memory, called *allocentric* (world-centered) and *egocentric* (body-centered) representations. These are comparable to two kinds of video games: the maplike bird's-eye-view strategy games and the "first-person action" games in which the screen shows the view ahead of the character. Either allocentric or egocentric representations can be used to orient in an environment. It is not exactly clear *how* the hippocampus stores and retrieves spatial memory, but it involves long-term potentiation (LTP) (see Chapter 13, page 363). Gene knockouts that disrupt LTP in mice disrupt spatial learning in tests such as the Morris water maze. Overexpression of genes that enhance LTP improves learning and memory of familiar objects, spatial tasks such as the Morris water maze, and other memory tasks.

Several classes of neurons have properties that appear to underlie navigation. For example, chronic electrical recordings from single neurons in the hippocampus of rats or mice show that some neurons generate action potentials only when an animal is in a particular part of a familiar area (see Figure 18.12B). These hippocampal neurons are termed **place cells** because their activity encodes the spatial position of the animal. The existence of place cells in the hippocampus suggests that this brain area creates some sort of a spatial map of the environment in the brain. The spatial representations of place cell activity are dynamic; as an animal is exposed to new environments, some place cells may change their specification to incorporate new areas.

Brain areas near the hippocampus (sometimes grouped with it in a region called the *hippocampal formation*) are also involved in processing spatial information used in navigation. The entorhinal cortex, located in the medial temporal lobe of the forebrain, is the main interface between the hippocampus and the neocortex. Neurons in one part of the entorhinal cortex are called **grid cells**, because they are active when the animal is in any of several grid-like locations in its environment (**FIGURE 18.12**). Grid cells do not provide locale information as place cells do; instead, they increase their activity when a rat walks through the vertices of an invisible grid of equilateral triangles that measure off distances within the environment. Different grid cells have somewhat different scales, representing 30-cm grids in the most dorsal part of the area and 70-cm grids in the ventral part. The finding of grid neurons is certainly consistent with the idea that animals form a kind of maplike representation of their environment in their brains.

At least two other classes of neurons are implicated in spatial representation. Investigators have recorded from neurons—termed **head direction cells**—that encode the direction the animal's head is pointing. These neurons occur in a circuit in the rodent forebrain that extends from the dorsal tegmental nucleus to the entorhinal cortex, and are activated by vestibular and other sensory information. Head direction cells are direction-specific but not location-specific, unlike place cells, which are location-specific but not direction-specific. Finally there are *boundary cells* (or border cells) that are activated near borders in the animal's environment.

Such studies of the hippocampus and other brain areas are starting to build a physiological basis of navigation. It is not clear just how animals use the neural information these studies describe, but the findings are consistent with what animals would need for complex spatial representation. One can imagine how animals might use place cells that encode landmark-based learning of familiar areas, head direction cells that encode compass-heading information (and receive input from various directional sensors), and grid cells that provide a distance scale. In 2014 the Nobel Foundation awarded Nobel Prizes to John O'Keefe for pioneering work on place cells, and to May-Britt Moser and Edvard Moser for characterization of grid cells. These exciting developments provide an important link between brain processes and the complex spatial behaviors of navigation, leading ultimately to the navigational basis of migration, homing, and other spatial behaviors.

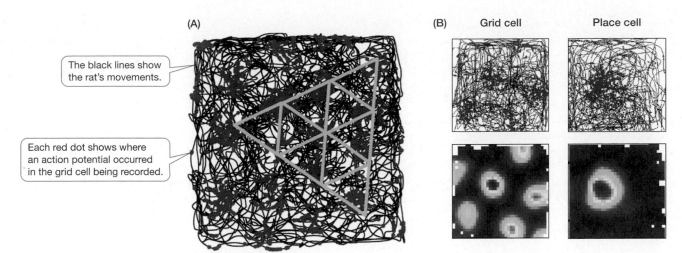

FIGURE 18.12 Grid cells and place cells suggest how the hippocampal formation of the brain plays roles in spatial learning (A) A grid cell recorded from the entorhinal cortex in the brain of a foraging rat. The black line shows the tortuous path of the rat's movement through a 1.5 m-wide square enclosure. Each red dot corresponds to an action potential (spike) in the recording, showing the rat's position at that time. Blue equilateral triangles superimposed over the red areas of grid cell activity show the regular hexagonal structure of the (invisible) grid pattern of areas of activity. (B) Com-

parison of activity of a grid cell (left) and a hippocampal place cell (right). Each cell is represented (above) as the animal's trajectory with spike locations as in (A), and (below) as a color-coded map of spike rate and location (warmer colors are higher spike rates). An individual hippocampal place cell is active when the rat is in a particular part of the arena. Place cell coding of location is rather stable from day to day and is learned during exploration of initially unfamiliar environments. (From Moser et al. 2015.)

STUDY QUESTIONS

1. Explain the effect on sun-compass orientation of a circadian-clock shift to a light schedule that is 4 h later (e.g., from lights-on at 6:00 AM to lights-on at 10:00 AM).

2. List the advantages and disadvantages to an animal of using information from magnetic, sun, star, and polarized-light compasses.

3. Experienced homing pigeons were thought not to need any information about the outward journey to a release site in order to home successfully. Much evidence for this conclusion came from depriving the birds of visual, magnetic, and vestibular information that could be used for path integration. What other sorts of studies would you want performed before accepting this conclusion?

4. Outline how a vertebrate might use place cells, head direction cells, and grid cells in different forms of navigation.

Go to **sites.sinauer.com/animalphys4e** for box extensions, quizzes, flashcards, and other resources.

REFERENCES

Burgess, N. 2014. The 2014 Nobel Prize in Physiology or Medicine: a spatial model for cognitive neuroscience. *Neuron* 84: 1120–1125.

Frost, B. J., and H. Mouritsen. 2006. The neural mechanisms of long-distance animal navigation. *Curr. Opin. Neurobiol.* 16: 481–488. Excellent overview of studies on the navigation mechanisms discussed in this chapter.

Geva-Sagiv, M. L. Las, Y. Yovel, and N. Ulanovsky. 2015. Spatial cognition in bats and rats: from sensory acquisition to multiscale maps and navigation. *Nat. Rev. Neurosci.* 16: 94–108. An attempt to integrate large-scale behavioral data on navigation with smaller-scale neurophysiological data.

Gould, J. L., and C. G. Gould. 2012. *Nature's Compass: The Mystery of Animal Navigation.* Princeton University Press, Princeton, NJ.

Guerra, P. A., and S. M. Reppert. 2015. Sensory basis of lepidopteran migration: focus on the monarch butterfly. *Curr. Opin. Neurobiol.* 34: 20–28.

Jeffery, K. J. (ed.). 2003. *The Neurobiology of Spatial Behaviour.* Oxford University Press, New York. This multiauthor review volume contains interesting and readable reviews of many issues in animal navigation.

Lohmann, K. J., C. M. Lohmann, and C. S. Endes. 2008. The sensory ecology of ocean navigation. *J. Exp. Biol.* 211: 1719–1728.

Lohmann, K. J., N. F. Putman, and C. M. F. Lohmann. 2012. The magnetic map of hatchling loggerhead sea turtles. *Curr. Opin. Neurobiol.* 22: 336–342.

Moser, E. I., E. Kropf, and M.-B. Moser. 2008. Place cells, grid cells, and the brain's spatial representation system. *Annu. Rev. Neurosci.* 31: 69–89.

Reppert, S. M., P. A. Guerra, and C. Merlin. 2016. Neurobiology of monarch butterfly migration. *Annu. Rev. Entomol.* 61: 25–42.

Srinivasan, M. V. 2015. Where paths meet and cross: navigation by path integration in the desert ant and honeybee. *J. Comp. Physiol., A* 215: 533–546.

Zhan, S., W. Zhang, K. Niitepõld, and seven additional authors. 2014. The genetics of monarch butterfly migration and warning coloration. *Nature* 514: 317–321.

See also **Additional References** *and* **Figure and Table Citations**.

PART IV
Movement and Muscle

Previous page
Henkel's leaf-tailed gecko (*Uroplatus henkeli*) uses powerful muscles that are finely controlled by neural input to leap between tree branches
This gecko lives in tree tops in the tropical rain forest of eastern Madagascar, and it descends to the ground only to lay eggs in leaf litter on the forest floor. It is nocturnal and hunts for insects and snails in its treetop habitat. A master of disguise, Henkel's leaf-tailed gecko exhibits mottled colors and a leaf-shaped tail that allow it to blend in with its surroundings. When attacked, it can shed the tail (which will regenerate) as a defense mechanism.

Control of Movement

The Motor Bases of Animal Behavior

19

A robot salamander shows how features of the neural control of locomotion may have eased the transition from swimming to walking in vertebrate evolution (Photograph of robot salamander from Ijspeert et al. 2007.)

When a salamander moves around, eats, or mates, its nervous system sends output to effectors that produce its behavior. All externally observable behavior is a direct result of the activation of effectors. That is, all behavior that can be observed consists of movements (usually resulting from muscle contractions), sounds (also from muscle contractions), gland secretions, color changes, and other outputs of effectors. Therefore the mechanisms by which animals generate patterns of behavior involve the coordinated control of muscles and other effectors.

In this chapter we consider the neural control and coordination of movement, particularly the generation of motor behaviors such as walking, swimming, and flying. The salamander shown above resembles the first land-based tetrapod vertebrates: It can swim by undulating its body from side to side, and can walk by stepping its legs in a coordinated sequence. The robot shown in the inset can do the same, because the investigators who built it modeled its locomotor control after that of the salamander. In doing so they have shown that the neural circuitry that controls and coordinates swimming behavior and walking behavior in salamanders may also provide a straightforward explanation for the transition from swimming to walking—a transition likely to have been important in vertebrate evolution.

Neural Control of Skeletal Muscle Is the Basis of Animal Behavior

A major function of an animal's nervous system is to generate its behavior. The ways in which the coordinated motor output of a nervous system produces behavior may be as simple as the withdrawal of a limb in response to a painful stimulus, or the tail-flip escape of a crayfish, or as complex as the mating behavior of a stickleback fish or of a songbird. In this chapter we concentrate on the neural circuits that produce relatively simple patterns of motor output to generate behavioral elements such as locomotion. (More elaborate behavioral performances are less readily understood in terms of the patterns of motor output of nervous systems to produce them, but physiologists consider them to represent elaborations of the simpler patterns discussed here.)

A major question in the analysis of the neural basis of behavior is, what are the neural circuits—the assemblies of neurons and the patterns of synaptic interconnections between them—that produce particular patterns of behavioral movements? In this chapter we will examine several neural circuits and relate their actions to the behaviors they mediate. We will compare neural circuitry of invertebrates and vertebrates, illustrate models of circuits that are thought to underlie rhythmic patterns of behavior such as walking or flying, and look closely at the control of movements of vertebrate animals.

Invertebrate neural circuits involve fewer neurons than vertebrate circuits

Investigators have elucidated the neural circuits underlying several behaviors in invertebrates (mostly arthropods and molluscs) and in vertebrates. The circuits in these two groups of animals are qualitatively similar, but vertebrate circuits always include many more neurons than do those in invertebrates. As we will see in Chapter 20, an arthropod muscle receives innervation from only a few motor neurons, whereas a vertebrate muscle receives innervation from hundreds of motor neurons.[1] Similar disparities in neuron number occur in the central nervous system (CNS), with vertebrates having 10^4–10^5 times as many neurons as arthropods or molluscs.

This contrast leads to two generalizations about invertebrate and vertebrate neural circuits. First, in many cases an invertebrate neuron may be a uniquely *identified neuron*—that is, a neuron whose structure, location, electrical activity, or other properties are sufficiently distinctive that the neuron can be recognized and studied in every individual of a species. In contrast, nearly all vertebrate neurons cannot be uniquely identified, but can be recognized only as members of a population. The second generalization is that single individual neurons play functional roles in invertebrate circuits, whereas many neurons participate in a particular function in vertebrate neural circuits. For example, a single arthropod neuron may act as a *command neuron*, a neuron whose activity is sufficient to command a particular element of behavior. For vertebrates, a larger number of neurons nearly always act together to provide such a command function.

---

[1] Note that a whole muscle of a vertebrate is composed of thousands of muscle fibers, and each muscle fiber receives a synapse from only one motor neuron (see Chapter 13).

We have already described two simple invertebrate neural circuits, each mediating a **reflex**, which we can define as a simple, graded response to a specific stimulus. One of these circuits mediates the startle escape response of cockroaches, discussed in Chapter 12 (see Figure 12.3); the other controls the gill withdrawal reflex in the marine mollusc *Aplysia*, discussed in Chapter 13 (see Figure 13.23). Each of these circuits depends on only tens of neurons in the circuit. In both of these examples, a modest number of mechanosensory neurons excite CNS interneurons to excite some motor neurons (while inhibiting others), leading to selective activation of muscles to produce a reflexive response. Several neural circuits of invertebrates, whether reflexive as in these examples or centrally programmed, are well understood because of the relatively small number of neurons they contain. The vertebrate circuits we examine next, in contrast, contain thousands of individual neurons, although the neurons are the same few types as for invertebrates.

Vertebrate spinal reflexes compensate for circumstances, as well as initiate movements

Spinal reflexes are mediated by the neural circuits of the vertebrate spinal cord. As a result of pioneering studies by Charles Sherrington (1857–1952) and Ivan Pavlov (1849–1936) on spinal reflexes at the beginning of the twentieth century, analysis of behavior in terms of spinal reflexes dominated studies of neural circuits through most of the twentieth century. These spinal reflexes are therefore perhaps the best-known vertebrate neural circuits.

In spinal reflexes, sensory input (from receptors of the skin, muscles, tendons, and joints) enters the spinal cord through the dorsal roots (**FIGURE 19.1**). This sensory input, via intervening synapses in the spinal cord, excites some motor neurons and inhibits others (see Figure 13.7), leading to movements by selectively activating muscular contraction. The sensory inputs from different populations of receptors have different connections in the spinal cord and thereby initiate different reflexes. We will examine two of the many reflexes of the mammalian hindlimb that have been studied extensively: the stretch reflex and the flexion reflex.

THE STRETCH REFLEX　The first spinal reflex we will consider is the **stretch reflex** (or *myotatic reflex*). The stretch reflex tends to oppose stretch of a muscle. A familiar example of the stretch reflex is the knee-jerk response to a tap on the patellar tendon at the knee joint, a test that is a staple of routine medical examinations (see Figure 19.1). Although the patellar reflex is familiar, it does not give a clear picture of the function of stretch reflexes, which are essential for the maintenance of posture and the coordination of movements. It is important to remember that all skeletal muscles are continuously involved in generating stretch reflexes. When the doctor taps you on the patellar tendon, the hit of the hammer stretches the knee extensor muscles in your thigh. This stretch stimulates *muscle spindles* (**BOX 19.1**) located in the muscle, which contain stretch-sensitive endings of sensory receptors wrapped spirally around the noncontractile portions of specialized *intrafusal* muscle fibers.

The sensory axons associated with muscle spindles are known as **1a afferent** fibers—*afferent* meaning "conducting toward the CNS," and *1a* because they are the largest and most rapidly con-

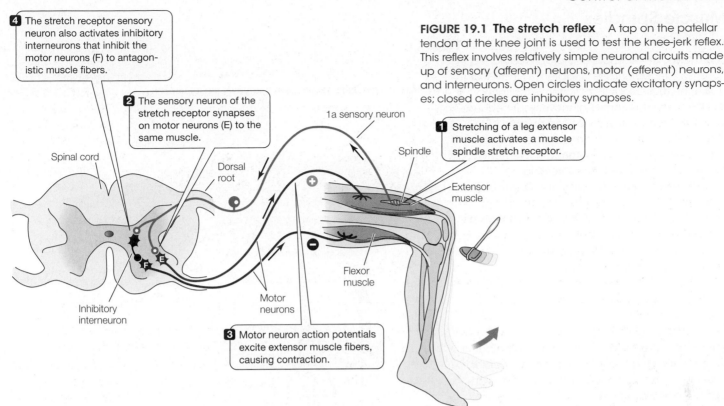

4 The stretch receptor sensory neuron also activates inhibitory interneurons that inhibit the motor neurons (F) to antagonistic muscle fibers.

2 The sensory neuron of the stretch receptor synapses on motor neurons (E) to the same muscle.

1 Stretching of a leg extensor muscle activates a muscle spindle stretch receptor.

1a sensory neuron

Spindle

Spinal cord

Dorsal root

Extensor muscle

Inhibitory interneuron

Motor neurons

Flexor muscle

3 Motor neuron action potentials excite extensor muscle fibers, causing contraction.

FIGURE 19.1 The stretch reflex A tap on the patellar tendon at the knee joint is used to test the knee-jerk reflex. This reflex involves relatively simple neuronal circuits made up of sensory (afferent) neurons, motor (efferent) neurons, and interneurons. Open circles indicate excitatory synapses; closed circles are inhibitory synapses.

ducting sensory fibers in the body. (Not all muscle-spindle sensory neurons are of the 1a class, but we will simplify the discussion by considering only the 1a sensory neurons here.)

The 1a axons from muscle spindles enter the spinal cord and make direct, excitatory synaptic contact with motor neurons to the same muscle (see Figure 19.1). This direct synaptic excitation is unusual; most vertebrate sensory neurons directly synapse only onto interneurons (intrinsic neurons that do not leave the CNS). The simplest manifestation of a stretch reflex, then, involves only two kinds of neurons: 1a sensory neurons and motor neurons.

When a muscle spindle is stretched, its 1a afferent neuron generates a train of nerve impulses. These impulses elicit excitatory postsynaptic potentials (EPSPs) in motor neurons, which—if their depolarizations exceed threshold—lead to motor neuron impulses and ultimately to contraction of the stretched muscle, and the leg kicks upward. Muscle spindles are said to be *in parallel* to the extrafusal fibers that generate the substantive contractile forces of a muscle because they act *beside* the extrafusal fibers (see Box 19.1). Because muscle spindles are in parallel to the force-producing extrafusal muscle fibers, they are sensitive to muscle *length*. They respond to increasing muscle length by increasing their activity.

Other aspects of the neural circuit of the stretch reflex make its function more complex. The 1a afferent axons also synapse on other neurons in addition to motor neurons. Some of these connections stimulate excitatory interneurons that excite the motor neurons to the same or to other muscles that work in parallel with the first muscle. Other synapses excite inhibitory interneurons that inhibit the motor neurons to the antagonist (opposing) muscle. Thus a tap on the patellar tendon not only excites motor neurons to the extensor muscle to produce the familiar knee jerk, but also inhibits motor neurons to the antagonist flexor muscle (see Figure 19.1).

These synaptic connections illustrate one of the most basic features of reflexes and of the organization of motor systems: the principle of **reciprocity**. Muscles (or groups of muscles) tend to be arranged in antagonist pairs that oppose each other, such as the flexor muscles that bend the knee and the extensor muscles that straighten it. The principle of reciprocity states that any signal that activates movements, whether it is the sensory input to a reflex or a command of the CNS, is coordinated to contract a set of muscles that work together (the *agonists*) while relaxing the opposite (*antagonist*) set. This reciprocal control of muscles ensures that two mutually antagonistic muscles do not usually counteract each other and suppress the movement. Thus in the stretch reflex, stretch of the extensor muscle activates extensor motor neurons to contract the extensor and relieve the stretch, while inhibiting flexor motor neurons to prevent co-contraction of the antagonist flexor muscle.

Another added complexity in the stretch reflex (and in motor circuits in general) is the number of neurons involved in even the simplest behavioral act. Figure 19.1 shows one 1a sensory neuron and one extensor motor neuron, but these only represent larger populations of these neuron types. A large leg muscle contains many muscle spindles, so its stretch activates many sensory neurons. Moreover, the muscle is supplied by at least 300 motor neurons. Unlike the situations in crayfish and *Aplysia*, none of these neurons are identified; you cannot find the same extensor motor neuron from animal to animal. Each 1a sensory neuron synapses with most, and probably all, of the motor neurons to the same muscle, as well as with many interneurons of different types. For example, in addition to the neurons that are part of the stretch reflex circuit, many interneurons convey 1a sensory information to higher brain centers.

The stretch reflex circuit illustrates the principle of **divergence** of CNS neural connections: Each presynaptic neuron usually contacts many postsynaptic neurons. The converse principle, **convergence**, also occurs, because each postsynaptic neuron is contacted by many presynaptic neurons. For example, each of the motor neurons receives

BOX
19.1 **Muscle Spindles**

A muscle spindle is an example of a *proprioceptor*—a mechanosensory receptor (see Chapter 14) that is associated with the musculoskeletal system. Proprioceptors are important for the control of movement because they provide an animal with information about where the parts of its body are positioned in space—information that is necessary both to program a movement and to monitor how the movement is progressing.

The *muscle spindle organ* monitors the *length* of a skeletal muscle. Each muscle spindle consists of eight to ten specialized muscle fibers, called **intrafusal fibers**, that are contained within a capsule of connective tissue (**Part 1** of figure). Intrafusal fibers are in contrast to **extrafusal fibers**, which are the normal "working" contractile fibers that bear loads and generate movements. When an intrafusal fiber is stretched by elongation of the whole muscle, the sensory neuron terminal associated with the central portion of the fiber is distorted; the distortion increases the cation permeability of the nerve terminal membrane and produces a graded receptor potential which, if threshold is reached, gives rise to action potentials (**Part 2** of figure).

The functional organization of a muscle spindle has several salient features:

1. The intrafusal fibers of muscle spindles are too few and weak to generate appreciable tension themselves. Because they are located *in parallel* to the tension-producing extrafusal fibers, muscle spindles do not sense muscle tension.

2. Muscle spindles provide information about muscle length and, in some animals, about changes in length. Their activity thus provides information about both muscle length (position) and rate of change in length (velocity of movement).

3. The intrafusal fibers of muscle spindles in vertebrates receive motor innervation. In birds and mammals, a separate class of small motor neurons termed *gamma (γ) motor neurons* makes synaptic contact on the contractile portions of the intrafusal fibers. Signals from the CNS that cause contraction of the ends of the intrafusal fibers regulate the degree of stretch on the central noncontractile region, and thereby influence the degree of distortion of the 1a afferent nerve terminal. The combined actions of alpha (α) motor neurons (to the extrafusal fibers), γ motor neurons (to the intrafusal fibers), and 1a afferent sensory information contribute to smooth, coordinated motions (see pages 519–521).

Vertebrates have additional types of proprioceptors. Two examples are *Golgi tendon organs*, which are embedded in tendons in series with working muscle fibers and monitor muscle tension, and *joint receptors*, which are embedded in connective tissue at a joint and respond over a certain range of joint positions or movements.

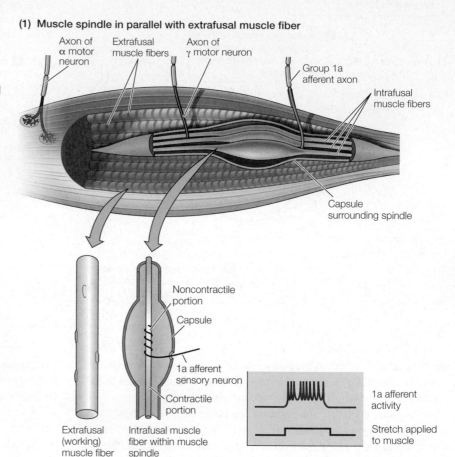

(1) Muscle spindle in parallel with extrafusal muscle fiber

Axon of α motor neuron
Extrafusal muscle fibers
Axon of γ motor neuron
Group 1a afferent axon
Intrafusal muscle fibers
Capsule surrounding spindle

Noncontractile portion
Capsule
1a afferent sensory neuron
Contractile portion

Extrafusal (working) muscle fiber
Intrafusal muscle fiber within muscle spindle

1a afferent activity
Stretch applied to muscle

(2) Muscle-spindle responses

1a afferent neuron

Less stretch than at rest opens fewer channels, and the frequency of action potentials decreases.

Distortion of the spiral nerve ending opens more channels and generates a greater frequency of action potentials.

Extrafusal fiber
Intrafusal fiber

Muscle length: At rest   Lengthened   Contracted and shortened

1a afferent activity

At resting length, the 1a afferent fiber generates tonic activity.
Stretch of the muscle increases 1a afferent activity.
Shortening of the muscle during contraction decreases 1a afferent activity.

Vertebrate muscle-spindle stretch receptors   (1) Skeletal muscles contain muscle spindles arranged in parallel with tension-producing extrafusal muscle fibers. The cell bodies of the 1a afferent sensory neurons are in the dorsal root ganglia. (2) Muscle spindles are sensitive to changes in muscle *length*.

input from about 10,000 synapses, representing many 1a sensory neurons and many more excitatory and inhibitory interneurons. Thus the cartoon view of the circuit for a stretch reflex in Figure 19.1 is a great oversimplification.

THE FLEXION REFLEX When you step on a tack, you reflexively withdraw your foot from the offending stimulus. Your foot is drawn upward by contraction of the flexor muscles of the thigh. The neural circuit mediating this flexion reflex is shown at the left in **FIGURE 19.2**. A diverse array of sensory neurons known as *flexion-reflex afferents* have endings in the skin, muscles, and joints; some of these are sensitive to painful and noxious stimuli. The flexion-reflex afferents make excitatory synaptic contacts on interneurons in the CNS that in turn excite motor neurons to the flexor muscles, as well as inhibitory interneurons that inhibit motor neurons to the extensor muscles. Thus, as in the stretch reflex (and in other spinal reflexes), synaptic interactions in the spinal cord maintain the reciprocity of action between antagonist pools of flexor and extensor motor neurons. Unlike the sensory neurons of the stretch reflex, however, flexion-reflex afferents make only indirect connections to motor neurons, via at least one layer of intervening interneurons.

The obvious function of the flexion reflex is protective; the offended limb is flexed, lifted, and withdrawn from a painful and potentially damaging stimulus. The reflex circuit is relatively short, local, and rapid. Of course, flexion-reflex afferents also connect to other interneurons that ascend the spinal column to the brain, so you become aware of the painful stimulus. This slower process occurs while the reflex flexion is taking place, so in most cases the foot is lifted (or the hand is withdrawn from the hot stove) before you are aware of the stimulus triggering the withdrawal. Note that many receptors other than pain receptors can trigger flexion reflexes, and that the main function of the flexion-reflex afferents is to provide proprioceptive and cutaneous information to the brain and spinal cord, not just to elicit flexion reflexes.

If you stepped on a tack with your left foot while your right foot was lifted off the ground, it would be a good idea to extend your right foot while flexing your left foot. In fact, one component of the flexion reflex ensures this. As Figure 19.2 shows, flexion-reflex afferents synapse onto interneurons that cross the midline of the spinal cord and indirectly excite extensor motor neurons of the contralateral ("opposite side") leg. Thus the right leg is extended (by exciting extensor motor neurons and inhibiting flexor motor neurons) while the stimulated left leg is flexed (by exciting flexor motor neurons and inhibiting extensor motor neurons). The reflex extension of the contralateral leg has been given a separate name—the *crossed extension reflex*—but functionally it is an integral part of the flexion reflex, a product of the synaptic connections "wired in" to the spinal cord.

This example illustrates that reflexes do not operate in a vacuum, influencing only a single antagonist pair of muscles. Instead, reflexes may have diverse and widespread effects, and they must interact with all other synaptic influences on motor neurons.

FUNCTIONAL ROLES OF REFLEXES We noted previously that flexion reflexes have a clear protective role, but the functions of the stretch reflex are more complicated. We can illustrate one aspect of the function of stretch reflexes with the following example: Sup-

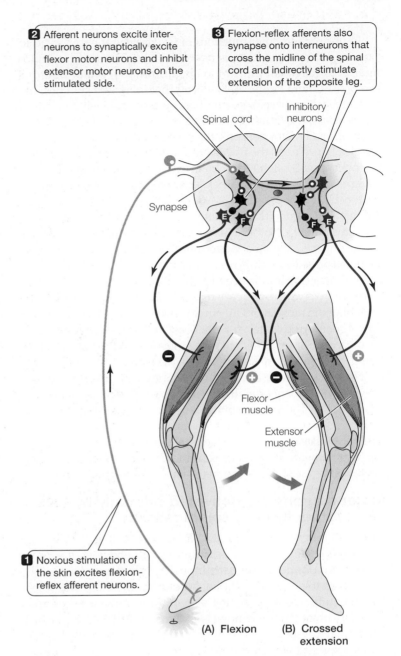

2 Afferent neurons excite interneurons to synaptically excite flexor motor neurons and inhibit extensor motor neurons on the stimulated side.

3 Flexion-reflex afferents also synapse onto interneurons that cross the midline of the spinal cord and indirectly stimulate extension of the opposite leg.

Inhibitory neurons

Spinal cord

Synapse

E F   F E

Flexor muscle

Extensor muscle

1 Noxious stimulation of the skin excites flexion-reflex afferent neurons.

(A) Flexion   (B) Crossed extension

FIGURE 19.2 The neural circuit of the flexion reflex and the crossed extension reflex The flexion reflex mediates protective withdrawal of a limb (A), whereas the crossed extension reflex extends the opposite limb for compensatory support (B). Smaller arrows indicate direction of nerve impulse propagation; plus signs (+) indicate increases in activity, and minus signs (–) indicate decreases. Open circles indicate excitatory synapses; closed circles are inhibitory synapses. F, flexor motor neuron; E, extensor motor neuron.

pose that while you are standing, a large monkey or a small person jumps on your back. The added weight will cause your knees to start to buckle, stretching the extensor muscles and activating the sensory neurons of muscle spindles. This sensory activity will reflexively excite motor neurons to the extensor muscles, generating more muscle force to counteract the increased load and maintain upright posture. This scenario illustrates a functional role of the stretch reflex in maintaining posture, by counteracting changes in load, muscle fatigue, or other factors.

To understand other functions of the stretch reflex, we need to consider another complexity in its organization. The stretch receptor organ receives motor innervation by **gamma (γ) motor neurons** (see Box 19.1 and Figure 19.3). Recall that in muscle spindles, the stretch-sensitive 1a sensory neurons are associated with intrafusal muscle fibers. The intrafusal muscle fibers are innervated by a population of small motor neurons, the γ motor neurons. The extrafusal muscle fibers (i.e., all the fibers that are not part of muscle spindles) are innervated by alpha (α) motor neurons. (When the term *motor neuron* is used without the Greek-letter prefix, it denotes an α motor neuron. Hence the previous discussions of vertebrate motor neurons in Chapter 13 and in this chapter refer to α motor neurons.)

Activation of γ motor neurons excites the 1a afferent neuron by contracting the contractile ends of intrafusal fibers, thereby stretching the noncontractile central sensory portion of the spindle, and consequently distorting the distal end of the 1a afferent neuron. Therefore there are two ways to increase muscle-spindle receptor activity: by stretching the muscle and by γ motor-neuron activity. Activation of γ motor neurons can "take up the slack" in intrafusal muscle fibers during muscle shortening, allowing muscle spindles to continue to signal length changes. This function is termed *gain control*.

Note that activity of γ motor neurons to intrafusal fibers and activity of α motor neurons to the surrounding extrafusal fibers have opposite effects on muscle-spindle sensory activity. Action potentials of the γ motor neurons that cause contraction of the intrafusal muscle fibers stimulate the 1a sensory neuron, whereas action potentials of the α motor neurons that cause contraction of the extrafusal muscle fibers relieve the stretch stimulus on the muscle spindle.

Motor neurons are activated primarily by CNS input rather than by spinal reflexes

Although reflexes play important roles in controlling the activity of motor neurons and coordinating motor behaviors, it is more accurate to consider the *primary* input to the motor circuitry of the spinal cord to be the descending input from the brain and spinal CNS circuits; the sensory fibers mediating spinal reflexes are a *secondary* input. This viewpoint is shown in **FIGURE 19.3**, in which the primary descending inputs enter from the top of the diagram and sensory inputs enter from the bottom. Such a view implies that a major role of the sensory input to the spinal cord is to supply *sensory feedback* that can modulate or correct the responses of motor neurons to CNS signals.

To illustrate how spinal reflex circuits provide sensory feedback, let's consider how the stretch reflex compensates for a resistance or load during the execution of a centrally generated, voluntary movement. To emphasize that similar neural circuits control arms and legs, we will use arm movements for this illustration. Suppose you decide to pick up a pamphlet from a table (**FIGURE 19.4**). Because this is a voluntary movement, the CNS must program the activation of motor neurons, rather than sensory input initiating the movement. Essentially the CNS estimates the amount of force necessary to pick up the pamphlet and sends a command to the motor neurons to generate that force. At the same time, the stretch reflex mediates **load compensation**, augmenting the contraction if there is extra weight or resistance added to the intended movement.

The CNS command for a voluntary movement excites both α and γ motor neurons, a process termed **α–γ coactivation**.

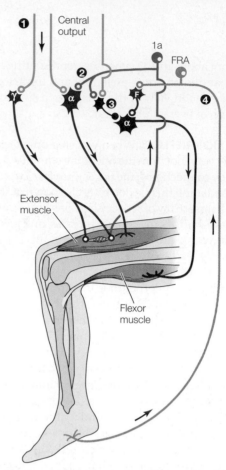

FIGURE 19.3 The basic circuit diagram of the ventral horn of the mammalian spinal cord The motor output neurons of the spinal cord are the alpha (α) motor neurons to flexor and extensor muscles, and gamma (γ) motor neurons to intrafusal fibers of muscle spindles. Input pathways to the motor neurons and interneurons of the ventral horn include CNS descending pathways from the brain (❶, shown in green); the 1a afferent (shown in blue) excitatory (❷) and inhibitory (❸) pathways from muscle spindles; and the flexion-reflex afferent (FRA) pathway (❹, shown in orange). Local interneurons shown are a 1a inhibitory interneuron (black) and an interneuron (F; magenta) in the flexion-reflex pathway. Arrows indicate direction of impulse propagation.

This coactivation has two functions. First, it ensures that the ongoing sensitivity of the muscle spindle is maintained during muscle shortening. In the absence of coactivation, a contraction that shortens the muscle would slacken the intrafusal muscle fiber and unload the muscle spindle, decreasing its sensitivity. Coactivation prevents this decrease. Second, coactivation allows the spindle to determine whether the muscle shortens during the intended movement. Suppose that the pamphlet is not very heavy, and the CNS has correctly estimated its weight. (We will call this the *no-load* condition, although it is more accurately described as light-load.) As shown in Figure 19.4A, the coactivation of α and γ motor neurons activates contraction of both the intrafusal fibers and the extrafusal fibers of the working muscle. In the absence of a substantial load, the extrafusal muscle fibers shorten to flex the arm, allowing the intrafusal fiber associated with a muscle spindle to shorten as it contracts. The shortening of the intrafusal fiber as the γ motor neurons activate it decreases its tension and lessens its activation of the stretch receptor. Because shortening unloads the muscle spindle, the 1a afferent neuron of the stretch receptor

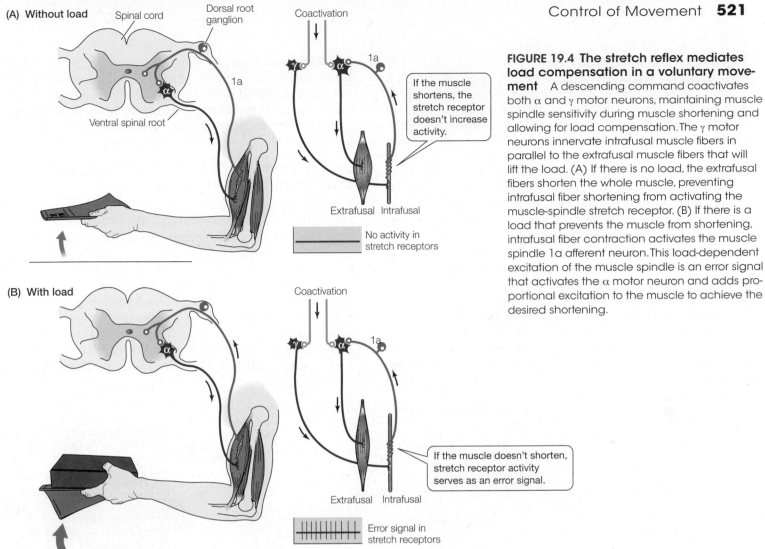

(A) Without load

Spinal cord

Dorsal root ganglion

Coactivation

Ventral spinal root

1a

If the muscle shortens, the stretch receptor doesn't increase activity.

Extrafusal   Intrafusal

No activity in stretch receptors

(B) With load

Coactivation

1a

If the muscle doesn't shorten, stretch receptor activity serves as an error signal.

Extrafusal   Intrafusal

Error signal in stretch receptors

FIGURE 19.4 The stretch reflex mediates load compensation in a voluntary movement A descending command coactivates both α and γ motor neurons, maintaining muscle spindle sensitivity during muscle shortening and allowing for load compensation. The γ motor neurons innervate intrafusal muscle fibers in parallel to the extrafusal muscle fibers that will lift the load. (A) If there is no load, the extrafusal fibers shorten the whole muscle, preventing intrafusal fiber shortening from activating the muscle-spindle stretch receptor. (B) If there is a load that prevents the muscle from shortening, intrafusal fiber contraction activates the muscle spindle 1a afferent neuron. This load-dependent excitation of the muscle spindle is an error signal that activates the α motor neuron and adds proportional excitation to the muscle to achieve the desired shortening.

generates few if any action potentials. That is, if the muscle shortens when it is signaled to shorten, no follow-up is needed.

Now suppose that the pamphlet is very heavy and the force estimate of the CNS is insufficient (see Figure 19.4B). (We will call this the *loaded* condition.) Coactivation excites the α and γ motor neurons as before, but now the extrafusal muscle fibers do not shorten, because of the unanticipated load. In the absence of shortening of the whole muscle, excitation of an intrafusal fiber by γ motor neurons will activate the stretch receptor, producing a train of action potentials. This stretch receptor activity constitutes an *error signal*, a measure of how much the muscle failed to shorten as commanded.

As Figure 19.4 shows, the stretch receptor's 1a axon makes excitatory synaptic contact with an α motor neuron that innervates the working extrafusal muscle fibers. Activity in the stretch receptor neuron (the error signal) excites proportional activity in the α motor neuron, generating additional tension in the working muscle to overcome the load. The neuron pair of the stretch receptor and α motor neuron is a reflex circuit that functions as a *load-compensating servo loop*, detecting an error (failure to shorten) and counteracting it ("more force, please") within a centrally commanded movement. Although recent studies show that α and γ motor neurons are not always coactivated, the role of coactivation in load compensation is an important aspect of γ motor neuronal function.

SUMMARY
Neural Control of Skeletal Muscle Is the Basis of Animal Behavior

- The pattern of motor output of a nervous system produces behavioral actions. The nervous system can generate motor patterns centrally or in response to discrete stimuli. Simple, stimulus-evoked responses are often reflexive: Stronger stimulation evokes stronger responses.

- Neural circuits that generate simple patterns of behavior in invertebrates typically involve relatively small numbers of identifiable neurons. The circuits for similar acts of vertebrates involve many more individual neurons, which are not uniquely identifiable.

- Vertebrate spinal reflexes (such as the stretch reflex and the flexion reflex) have the simplest and best-understood neural circuits for a vertebrate behavior. Even the simplest vertebrate reflexes, however, have large numbers of neurons in a circuit.

- The primary synaptic input of spinal motor neurons is from the central nervous system (CNS); sensory (reflex) input is secondary. Many reflexes mediate adjustments of centrally programmed movements, such as load compensation.

Neural Generation of Rhythmic Behavior

Most animal behavior consists not just of isolated single acts of the sorts that we have been discussing, but rather of action patterns: *sequences* of effector actions that result from sequences of motor output of the nervous system. These sequences of motor activity are patterned in space and time. For example, consider the activity of your nervous system that is required to pick up a pencil. First you extend your arm by contracting muscles at the shoulder and upper arm, then you flex your fingers to oppose your thumb, and then you elevate and flex the arm to lift the pencil. You may alter the activity of postural muscles in your trunk or legs to compensate for the arm movements. This motor performance may involve varying amounts of visual, tactile, and proprioceptive sensory input; moreover, the temporal and spatial pattern of the sequence of contractions may differ considerably from one time to the next. Variability of this sort heightens the difficulty of investigating action patterns.

As a way to approach the study of sequences of behavioral action patterns, neurophysiologists have concentrated on **rhythmic behavior**: stereotyped, repetitive sequences of movement such as walking, swimming, and flying in which the motor output is stable, repeatable, and predictable from cycle to cycle of the activity. We now examine several examples of neurophysiological analysis of rhythmic behavior, attempting to extract principles that may be of general importance in motor control systems.

Locust flight results from an interplay of CNS and peripheral control

Let's begin our exploration of the control of rhythmic behavior by asking: How does a locust fly? As **FIGURE 19.5A** shows, the movement of a single wing of a flying locust can be viewed as a simple up-and-down oscillation, generated by a set of elevator (or levator) and depressor muscles. The electrical activity of these muscles can be recorded from a tethered locust flying in a wind stream. This activity consists of alternating bursts of muscle potentials—the depressors being activated when the wings are up, and the levators being activated when the wings are down. Because each muscle depolarization results from an action potential in a motor neuron to that muscle, it is clear that flight results from the generation in

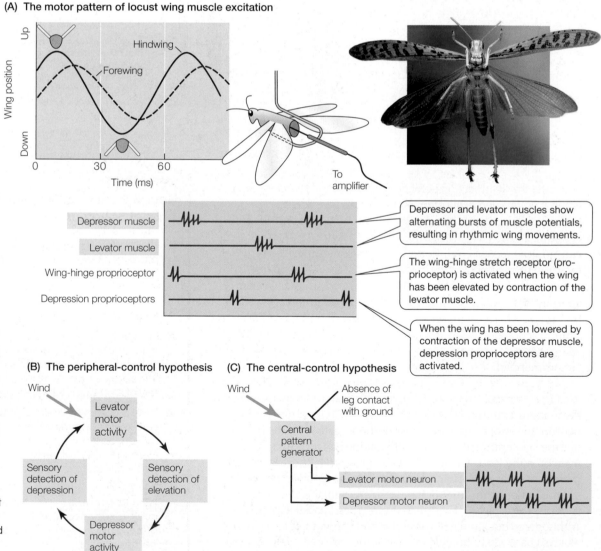

(A) The motor pattern of locust wing muscle excitation

Depressor and levator muscles show alternating bursts of muscle potentials, resulting in rhythmic wing movements.

The wing-hinge stretch receptor (proprioceptor) is activated when the wing has been elevated by contraction of the levator muscle.

When the wing has been lowered by contraction of the depressor muscle, depression proprioceptors are activated.

FIGURE 19.5 Control of flight in the locust (A) Cyclic wing movements and the associated temporal pattern of motor and sensory activity, recorded from a tethered locust. Two hypotheses could explain the generation of the motor pattern of wing muscle excitation: a peripheral-control hypothesis (B), in which sensory feedback resulting from a movement triggers the next movement; and a central-control hypothesis (C), in which a central pattern generator (CPG) produces the motor pattern without requiring moment-to-moment sensory timing. (The CPG is gated by tonic sensory input: wind on the head and the absence of leg contact with the ground.)

the CNS of alternating bursts of action potentials in levator and depressor motor neurons. This kind of pattern—alternating bursts of activity in motor neurons to antagonist muscles—underlies most forms of rhythmic behavior.

How are the motor neurons to antagonist muscles activated in alternation to produce a rhythmic movement such as that of a locust wing? Historically, two kinds of hypotheses have been advanced to explain the neural basis of rhythmic movements: peripheral control and CNS control. According to the hypothesis of **peripheral control**, each movement activates receptors that trigger the next movement in the sequence. The position of a locust wing is monitored by several proprioceptors (see Figure 19.5A): a single wing-hinge stretch receptor that generates a train of impulses when the wing is elevated, and several other receptors that are activated when the wing is depressed.

Locust flight could (in principle) operate by peripheral control, by having sensory feedback from wing sensory receptors activate the motor neurons for the next movement (**FIGURE 19.5B**). Thus elevation of the wings would excite the wing-hinge stretch receptor, which would synaptically excite depressor motor neurons, thereby lowering the wing. The lowered wing would terminate excitation of the wing-hinge stretch receptor and would excite the depression-sensitive receptors, which would synaptically excite levator motor neurons, elevating the wing and completing the cycle. The peripheral-control hypothesis is also called the *chained-reflex* hypothesis because each movement is a reflex response to sensory feedback resulting from the last movement.

According to the hypothesis of **central control**, locust flight is sustained by a **central pattern generator** (**CPG**)—a neural circuit in the CNS that can generate the sequential, patterned activation of motor neurons to antagonistic muscles that underlies a behavior pattern, without requiring sensory feedback to trigger the next movement. Thus, in central control of locust flight, the basic pattern of alternation of activation of levator and depressor motor neurons would result from an intrinsic CPG rather than from a chained reflex (**FIGURE 19.5C**).

How would one determine whether peripheral control or central control is responsible for the patterned motor activity underlying locust flight? The obvious answer is to remove the relevant sensory input, a process termed *deafferentation*. In the locust, most if not all wing sensory input can be removed by cutting the nerves to the wing-hinge area. Donald Wilson, who pioneered analysis of locust flight control, found that tethered locusts could maintain flight in the absence of wing sensory feedback. This result indicates that central control (not peripheral control) is responsible for locust flight. However, he observed that the flight frequency of the test animals was slower than normal. When he stimulated the cut sensory nerve stumps or the ventral nerve cord with temporally unpatterned stimulation, normal flight frequency was restored. This restoration suggested that the sensory stimulation provided general excitation to the CNS but was not necessary to supply *timing* information for pattern generation. These experiments demonstrated the existence of a CPG for locust flight.

Subsequent experiments in other animals have shown that many patterns of rhythmic behavior are under central control. These rhythmic activities include walking, swimming, feeding, and breathing or ventilation (see Chapter 23) in a variety of invertebrates and vertebrates. Thus the concept of central pattern generation is a generally important aspect of the control of coordinated behavior.

The hypotheses of central control and peripheral control may appear to be logical alternatives, but they are not mutually exclusive. Therefore the demonstration of a CPG for a behavior pattern such as locust flight does not mean that sensory input is unimportant. Sensory feedback can play significant roles in a centrally controlled behavior. This statement may seem paradoxical, but consider the fact that if you were walking down the sidewalk and suddenly lost all sensation to your legs, you would probably still be able to generate the motor output sequence of walking. Evidence from cats (see page 528) suggests that most mammals have a CPG for walking. Does that mean that sensory input is irrelevant? Of course not.

For example, sensory input may affect the quality of performance of walking and is essential for correcting the basic pattern, such as when one is walking over uneven terrain. In the locust, several functions of sensory feedback have been found. First, as already noted, sensory input has a generally stimulatory effect, speeding up the flight rhythm. Moreover, sensory feedback can provide specific timing information, adding an element of peripheral control to the system. Electrophysiological studies have shown that wing proprioceptors *do* have the synaptic effects diagrammed in Figure 19.5B: The stretch receptor monitoring wing elevation excites depressor motor neurons, and depression-sensitive receptors excite levator motor neurons. Thus the synaptic connections necessary for a chained reflex are present, and these reflexes operate with latencies appropriate to reinforce the flight rhythm—although they are *not necessary* for it.

Sensory feedback can *entrain* a CPG. In another experiment, one wing of a tethered, flying locust was moved up and down at a set frequency by a motor, the forced cyclic movement overriding normal flight-generated sensory feedback. When the forced movement of the one wing was at a rate close to the normal flight frequency, the flight frequency (recorded from muscles to all four wings) changed to match the driving frequency of the motor! Therefore sensory information from the driven wing can entrain the CPG to the driven frequency.

Our conclusions at this point are these: (1) There is a CPG for flight that can maintain the flight pattern in the absence of sensory timing information, and (2) sensory timing information (when present) can reset the CPG, entraining it to a slightly different driven frequency.[2] Thus the original hypotheses of central and peripheral (reflex) control are not mutually exclusive. The CPG is *sufficient* to maintain flight, but this sufficiency does not rule out contributions of peripheral control. Similar interactions of CPGs with sensory entrainment have been demonstrated in other activities and animals, such as swimming in dogfish sharks. The relative contributions of central control and peripheral control can be expected to differ in different cases. It is likely, however, that the two kinds of controls interact in most cases.

---

[2] The roles of the CPG and the sensory timing information in this case are analogous to those in a circadian (about 24-h) endogenous activity rhythm of an animal. As discussed in Chapter 15, many animals kept in constant light conditions will exhibit an activity rhythm with a period near (but not exactly) 24 h. If a light–dark cycle (such as 12 h light, 12 h dark) is added, it provides timing information that will entrain the endogenous circadian rhythm to an exactly 24-h period (see Figure 15.14). In the same manner, sensory timing information in locust flight can entrain the CPG.

There are different mechanisms of central pattern generation

How do neurons and networks of neurons in the CNS act as a CPG that determines the spatiotemporal patterns of motor output to generate a rhythmic behavioral pattern? Because many of the rhythmic behavior patterns studied are oscillatory, cycling in a roughly sine-wave fashion, the CPGs underlying them have been termed **oscillators**. Studies have clarified the neural basis of some of these oscillators.

OSCILLATOR THEORY   In theory, there are two logical categories of oscillators: cellular oscillators and network oscillators. The CPGs that have been studied appear to employ a mix of these two kinds of oscillatory mechanisms in differing degree—some with cellular oscillators playing a dominant role, and others dominated by network properties. We will introduce the theoretical types first and then describe their roles in a real example of a central-pattern-generating neural circuit.

A **cellular oscillator** is a neuron that generates temporally patterned activity by itself, without depending on synaptic interaction with other cells. Such cells may generate endogenous bursts of action potentials (**FIGURE 19.6A**), or they may show oscillations of membrane potential without generating any action potentials (**FIGURE 19.6B**). The underlying mechanisms of oscillation may be similar for both types because some cells that generate impulse bursts will continue to oscillate after impulse generation is blocked with tetrodotoxin (TTX) (see Chapter 12, page 326).

Cellular oscillators are thought to play a role in central pattern generation in several cases studied, including those controlling molluscan feeding, crustacean heartbeat, and crustacean scaphognathites (gill bailers) (see Figure 23.29).

A **network oscillator** is a network of neurons that interact in such a way that the output of the network is temporally patterned, although no neuron in the network functions as a cellular oscillator. Thus the oscillatory or pattern-generating property is said to be an *emergent property* of the network, resulting from cellular interactions in the network rather than from intrinsic cellular properties. The simplest model of an oscillatory network, termed a *half-center model*, is shown in **FIGURE 19.6C**. Two neurons (or pools of neurons) receive a common tonic excitatory ("command") input but synaptically inhibit each other. With command excitation, one neuron will have a lower threshold and fire first, inhibiting the other. When the first neuron stops generating impulses, the other is released from inhibition and generates a train of impulses, inhibiting the first.

The half-center model appears straightforward but is actually rather unstable in its simplest form. Unless an additional time-dependent property is added to allow the first half-cycle to run down, the first neuron to reach threshold will tend to remain active and perpetually inhibit the other. There are several possible mechanisms to "fatigue" a half-center, such as postinhibitory rebound, slow inactivation of the impulse-generating capability of the cells, or antifacilitation of the inhibitory synapses (see Figure 13.22).

Another model network oscillator, which is more stable, contains three or more neurons that inhibit each other in a cyclic inhibitory loop. **FIGURE 19.6D** shows a *closed-loop model* that would produce a stable pattern of bursts in the sequence 1–2–3–1–2–3… without any cell possessing endogenous oscillator properties. Because there are three neurons inhibiting each other rather than two, one neuron

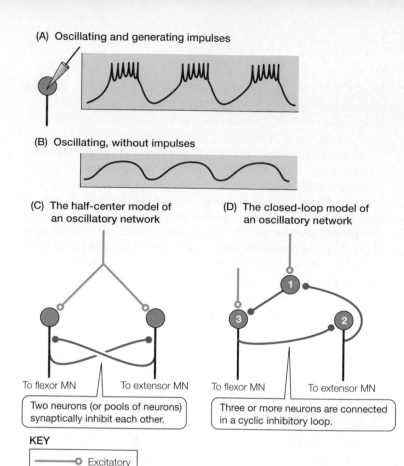

(A) Oscillating and generating impulses

(B) Oscillating, without impulses

(C) The half-center model of an oscillatory network

(D) The closed-loop model of an oscillatory network

To flexor MN   To extensor MN

Two neurons (or pools of neurons) synaptically inhibit each other.

To flexor MN   To extensor MN

Three or more neurons are connected in a cyclic inhibitory loop.

KEY

○ Excitatory

● Inhibitory

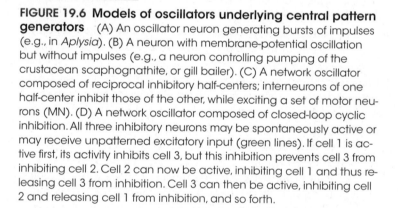

FIGURE 19.6 Models of oscillators underlying central pattern generators   (A) An oscillator neuron generating bursts of impulses (e.g., in *Aplysia*). (B) A neuron with membrane-potential oscillation but without impulses (e.g., a neuron controlling pumping of the crustacean scaphognathite, or gill bailer). (C) A network oscillator composed of reciprocal inhibitory half-centers; interneurons of one half-center inhibit those of the other, while exciting a set of motor neurons (MN). (D) A network oscillator composed of closed-loop cyclic inhibition. All three inhibitory neurons may be spontaneously active or may receive unpatterned excitatory input (green lines). If cell 1 is active first, its activity inhibits cell 3, but this inhibition prevents cell 3 from inhibiting cell 2. Cell 2 can now be active, inhibiting cell 1 and thus releasing cell 3 from inhibition. Cell 3 can then be active, inhibiting cell 2 and releasing cell 1 from inhibition, and so forth.

is always being released from inhibition and is thus able to fire and inhibit its follower.

Neural circuits acting as network oscillators are reported to underlie several sorts of rhythmic behaviors, including the neurogenic leech heartbeat, as well as swimming in leeches, molluscs (*Tritonia* and *Clione*), lampreys, and clawed toad (*Xenopus*) tadpoles. Each described circuit is different, but each displays some of the mutual inhibition that is characteristic of half-center models, as well as closed-loop inhibitory and excitatory elements.

In the nervous systems of animals, CPGs can combine the properties of both cellular oscillators and network oscillators. Such a **hybrid oscillator** might have one or more oscillatory cells acting within a network that stabilizes and reinforces the oscillation. Most CPGs will probably turn out to be hybrid oscillators with a mixture of cellular and network oscillatory properties. In **BOX 19.2** we describe one of the most completely studied CPG networks, the hybrid oscillatory network of the crustacean stomatogastric ganglion.

BOX 19.2 Circuits and Elaborations of Central Pattern Generators: The Stomatogastric Ganglion

Central pattern generators (CPGs) control many rhythmic behavior patterns, at least in large part. One of the most completely studied CPG networks is that of the crustacean stomatogastric ganglion. It is a hybrid oscillatory network, combining the properties of both cellular oscillators and network oscillators. This example introduces principles of general importance about oscillatory circuits and their control. We will then ask, how elaborate a behavioral performance might be programmed by a CPG?

The **stomatogastric ganglion** sits on the external surface of the stomach of a lobster, crayfish, crab, or other decapod crustacean. It contains about 30 neurons, most of which are motor neurons controlling stomach muscles. Why is studying the stomatogastric ganglion important? The main reason is that the stomatogastric ganglion generates two robust rhythms (and participates in two others) with a network of only 30 neurons! In addition, crustacean stomachs are ectodermal, chitin-lined, and controlled by striated muscle; thus their control is more like that of an appendage than what we would expect of a stomach. Crabs and lobsters swallow large pieces of food and chew the pieces with chitinous teeth in their stomachs, so the behavior mediated by the stomatogastric ganglion is more analogous to the control of vertebrate jaws than that of vertebrate stomachs.

As we discussed in Chapter 6 (see Box 6.3), the crustacean stomach consists of two chambers: an anterior cardiac chamber containing teeth that function as a *gastric mill* to grind and chew food, and a posterior pyloric chamber containing a sieve that serves to keep food particles from passing to the rest of the gut until the particles are small enough. **Figure A** shows a simplified neural circuit

(Continued)

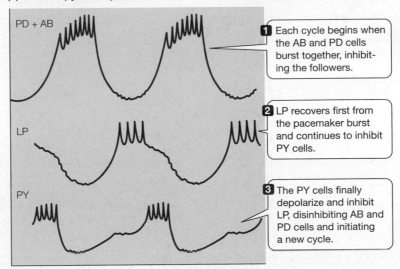

(1)

Brain — Stomatogastric ganglion — Motor nerve — Dorsal dilator muscle

Gastric mill

Cardiac sac

Pylorus

Constrictor muscles

Stomatogastric nerve

Esophageal ganglion — Esophagus

Commissural ganglion

Pyloric dilator muscle

(2) The pyloric circuit (simplified)

The AB cell generates endogenous bursts of action potentials.

AB/PD

PY   LP

(3) A basic pyloric rhythm

PD + AB

1 Each cycle begins when the AB and PD cells burst together, inhibiting the followers.

LP

2 LP recovers first from the pacemaker burst and continues to inhibit PY cells.

PY

3 The PY cells finally depolarize and inhibit LP, disinhibiting AB and PD cells and initiating a new cycle.

FIGURE A The network of neurons producing the pyloric rhythm of the crustacean stomatogastric ganglion   (1) The stomatogastric ganglion lies on the dorsal surface of the heart; the stomatogastric nerve connects it to the esophageal ganglion and the rest of the CNS. The 30 or so neurons of the stomatogastric ganglion receive modulatory input from extrinsic neurons. (2) The pyloric circuit of stomatogastric ganglion neurons (simplified). AB is a strong cellular oscillator that serves as the pacemaker for the rhythm. The AB and PD cells are strongly electrically coupled and represented by a single symbol. They are interconnected with other (follower) cells by inhibitory chemical synapses. (3) A basic pyloric rhythm.

BOX 19.2 **Circuits and Elaborations of Central Pattern Generators: The Stomatogastric Ganglion** (*Continued*)

and rhythmic output of one rhythm of the stomatogastric ganglion: the *pyloric rhythm* that controls the straining of food particles by the pyloric filter.

The pyloric circuit (see Figure A2) acts as a hybrid oscillator, containing an oscillator neuron (AB) that serves as the pacemaker for the rhythm. The oscillator cell is tightly electrically coupled to two pyloric dilator (PD) neurons so that these three generate bursts of action potentials together, inhibiting the other neurons in the network (see Figure A3). The oscillatory AB neuron and the coupled PD neurons burst first, inhibiting follower cells (LP, PY, and two others not shown). At the end of the AB/PD burst, the LP cell recovers from inhibition faster than the PY cells; therefore the LP cell bursts next and prolongs PY inhibition. PY neurons then burst and inhibit LP, until the next AB/PD burst starts a new cycle.

The pyloric circuit thus has both cellular oscillator and network oscillator properties. The generation of the pyloric rhythm depends primarily on the AB cel-

lular oscillatory neuron, but its triphasic cycle (AB/PD → LP → PY → ...) and timing depend on the strength and time course of inhibitory synapses and on intrinsic currents of the follower cells.

The rhythms and circuits of the stomatogastric ganglion exemplify another feature that may be of general importance: They are profoundly subject to *modulation*. The stomatogastric ganglion receives about 100 axons of neurons from other parts of the nervous system, many of which can secrete neuromodulators that act diffusely in the small ganglion to alter its motor output. At least 15 modulators are present, including the amines serotonin, dopamine, octopamine, and histamine; the classical neurotransmitters acetylcholine and GABA; and several peptides, including proctolin and FMRFamide-like and cholecystokinin (CCK)-like peptides.

The most common effect of a neuromodulator is to initiate and maintain rhythmic activity in a network. For example, adding serotonin, octopamine, or

dopamine to a previously quiescent isolated stomatogastric ganglion induces a pyloric rhythm (**Figure B**), although the rhythms induced by the three modulators differ in detail. In general, the stomatogastric ganglion requires permissive modulatory input from extrinsic neurons for the expression of its rhythms. Many of the CNS and sensory neurons that provide this modulatory input are well characterized.

The neuromodulators of the stomatogastric ganglion act in two ways: They alter the intrinsic membrane properties of individual stomatogastric neurons, and they alter the strengths and dynamics of synaptic connections of the neurons. Modulatory effects on intrinsic neuronal currents can induce cellular oscillation (many stomatogastric neurons are conditional oscillators), excite or inhibit particular neurons, or alter other excitable properties. Moreover, modulators can make individual synapses more or less potent, changing the functional circuit connections, as well as cellular activities.

(1) CNS attached

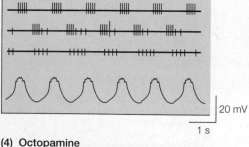

20 mV

1 s

(2) CNS detached by sucrose block

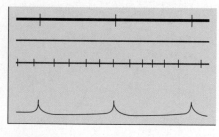

(3) Dopamine

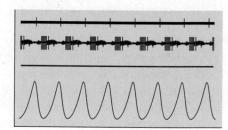

(4) Octopamine

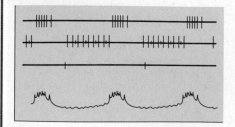

(5) Serotonin

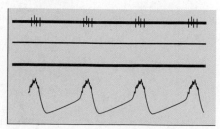

> The top three traces in each part are extracellular recordings (voltage versus time) from motor nerves containing axons of different neurons; the bottom trace is an intracellular recording from cell AB.

FIGURE B Neurotransmitters modulate the pyloric rhythm of the stomatogastric ganglion   (1) When the stomatogastric ganglion is attached to the rest of the central nervous system (CNS), endogenous neuromodulators maintain a normal pyloric rhythm. (2) Separating the ganglion from the rest of the CNS (by sucrose block or by cutting the stomatogastric nerve) often leads to cessation of bursts of rhythmic activity, as shown. (3–5) Addition of any of the indicated neuromodulators to the bathing solution reestablishes the rhythm in the isolated ganglion. Note that the pattern of the rhythm differs somewhat with different modulators. These effects result from specific, individualized actions of the modulators on both cellular ionic currents and synaptic strengths. (From Harris-Warrick and Flamm 1986.)

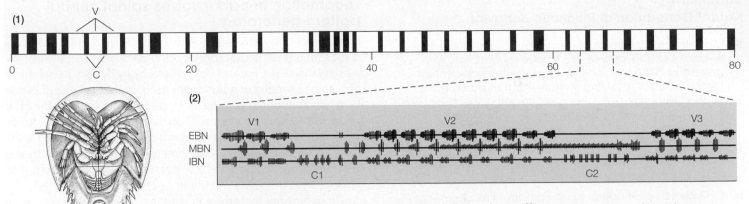

FIGURE C Central pattern generation of long-term behavior patterns   The pattern of long-term sequential alternation of gill-ventilation (V) and gill-cleaning (C) motor rhythms persists in an isolated abdominal nerve cord of the horseshoe crab *Limulus*. (1) An 80-min period of recorded activity during which gill-ventilation bouts (white) and gill-cleaning bouts (black) alternate fairly regularly. Such stable alternation is common in intact animals. (2) An approximately 140-s record of left branchial nerves of the first abdominal ganglion (EBN = external branchial nerve; MBN = medial branchial nerve; IBN = internal branchial nerve). Three bouts of gill-ventilation rhythm (V1–V3) alternate with two bouts of gill-cleaning pattern (C1–C2). See Figure 23.27D for a view of an exposed book gill. (After Wyse et al. 1980.)

It is at once exhilarating and sobering to realize that neuronal circuits such as those of the stomatogastric ganglion are not rigidly "hard-wired," but rather are plastic and malleable—exhilarating because the ability of neuromodulation to free a circuit from the "tyranny of the wiring" may underlie adaptive plasticity of neural control of behavior, but sobering because of the realization that a circuit diagram such as that in Figure A2 is descriptive of only one state of a dynamically shifting circuit. Neurons can even shift from one functional circuit to another, firing in "gastric time" or "pyloric time" under modulatory influence, and circuit elements can combine to form new patterns of output. Growing evidence suggests that these roles of neuromodulators are of widespread importance among CPGs.

Central pattern generators can underlie relatively complex behavior

How elaborate a behavioral performance can we expect to depend on mechanisms of central pattern generation? It is, after all, a long way from a short-term rhythmic activity such as locust flight to complex behavior patterns such as the ritualized mating behavior of many animals. Are complex patterns such as courtship rituals simply elaborations and chains of centrally pro-

grammed acts? For technical reasons, it has been difficult for scientists to explore the neurophysiological bases of increasingly complex behaviors, but some progress has been made.

Central-pattern-generating mechanisms have been shown to be sufficient for some behavior patterns that are significantly more complex and longer lasting than the simple cyclic patterns already described. One such long-term sequence involves gill movements of the horseshoe crab *Limulus*. Figure 23.27 shows the unusual *book gills* of these animals. These gills consist of many thin sheets of tissue termed *lamellae* (singular *lamella*) (the "pages" of the book) positioned under segmental flaps termed *gill plates*, shown in **Figure C**. Rhythmic beating of the gill plates ventilates the gill lamellae. This gill ventilation, however, is often intermittent: Periods of rhythmic ventilation a few minutes long may alternate with shorter periods of quiescence (apnea) or of gill cleaning. Gill cleaning is an intricate behavioral pattern in which the paired gill plates are brought across the midline and a part of one plate rhythmically cleans the lamellae of the opposite book gill. Horseshoe crabs roughly alternate between two mirror-image patterns of gill cleaning, termed *left-leading* and *right-leading*.

When the abdominal ventral nerve cord is dissected out of a *Limulus* and its

unstimulated motor activity is recorded in isolation, the motor output pattern underlying all of the behavior just described persists (see Figure C): Periods of a ventilatory motor output rhythm alternate with periods of a gill-cleaning motor pattern. Moreover, the rough alternation between left-leading and right-leading gill cleaning can also be expressed in isolation. Thus long and relatively elaborate sequences of behaviorally significant motor patterns can be expressed in isolated CNS tissue, without muscles, movement, or sensory feedback.

Behavioral patterns that are still more complex have been analyzed in insects, including molting activity (see Figure 16.20) of crickets and moths and reproductive behavior in leeches and in several insects. These stereotyped behaviors appear to have centrally patterned components but also to have stages at which appropriate sensory feedback is necessary to proceed to the next stage. The central motor programs could function as modules, like stored subroutines in a computer program; when "called" by the CNS, a module would generate a particular motor pattern, after which sensory input would determine how to proceed next. It is likely that increasingly complex behavior patterns will have increasingly elaborate interactions between sensory components and central motor programs.

SUMMARY
Neural Generation of Rhythmic Behavior

■ Most rhythmic patterns of animal behavior (walking, swimming, flying, and so on) involve a central pattern generator (CPG), which can produce the basic motor pattern without requiring sensory input at particular times in the cycle. The CPG interacts with sensory feedback from the cyclic movements, which can entrain the CPG.

■ CPG circuits may depend on cellular oscillators, network oscillators, or a combination of both, as in the crustacean stomatogastric ganglion.

■ CPG circuits are subject to neuromodulation, in which a neurotransmitter/neuromodulator can alter circuit function to generate or alter rhythmic output.

Control and Coordination of Vertebrate Movement

The principles of central pattern generation and the interaction of central and peripheral control of movement were first developed from invertebrate studies, principally with arthropods and molluscs. In this section we consider the degree to which these principles also apply to vertebrates. We can start with the question: How does a cat walk? For the moment, let's consider the cat nervous system as composed simply of three compartments that can influence movement: brain, spinal cord, and sensory input (**FIGURE 19.7**).

The immediate generators of walking movements in a cat are the spinal motor neurons that control the limb muscles. The spinal circuitry associated with these motor neurons was introduced earlier (see Figure 19.3). The motor neurons receive direct or indirect synaptic input from three sources: (1) descending input from the brain, (2) sensory input from proprioceptors and other receptors in the periphery, and (3) local input from intrinsic spinal circuits. If the spinal motor neurons are to be activated in the correct spatiotemporal pattern to produce walking, what are the roles of these three compartments in generating this pattern?

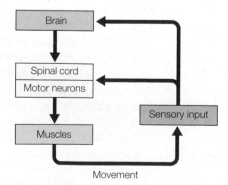

Movement

FIGURE 19.7 The major components of the control of movement in a vertebrate Motor neurons in the spinal cord activate muscles as a result of receiving three sorts of inputs: input from local spinal circuits, descending input from the brain, and sensory input. The movement produced by muscle contractions leads to further sensory input, termed sensory feedback.

Locomotion in cats involves spinal central pattern generators

In arthropod systems such as those discussed in Box 19.2, the compartments of neural control could be experimentally isolated with relative ease. For example, it is technically easy to isolate the abdominal ganglia of a horseshoe crab, and to ask the question: What do abdominal ganglia do by themselves, without the brain and without sensory feedback? Such questions are harder to investigate experimentally in vertebrates, in which an isolated spinal cord usually dies quickly. With refinements of technique in the last 35 years, however, it has become possible to perform experiments analogous to those in invertebrates.

To determine their roles in walking, it is possible to separate the functions of the brain, spinal cord, and sensory feedback experimentally in a vertebrate such as a cat, by lesions in the CNS or peripheral nerves. A cat is able to make fairly normal stepping movements on a treadmill after transection of the spinal cord to remove brain influence. First, in chronic (long-term) experiments, cats with the spinal cord transected 1–2 weeks after birth recover the ability to walk on a treadmill at a speed dependent on the treadmill speed. Second, in acute (short-term) experiments, spinally transected cats can walk on a treadmill if given the norepinephrine precursor L-dopa or the norepinephrine receptor stimulator clonidine.

These experiments show that the brain does not need to provide timing information for walking. Noradrenergic fibers descending from the brain in intact cats presumably command or enable the expression of the walking pattern by spinal circuits, but they are not necessary for timing the stepping cycle of a limb; certainly injected L-dopa does not provide timing information. In other experiments, cats with brain sections (at level 1 in **FIGURE 19.8**) can walk on a treadmill when given unpatterned electrical stimulation to a mesencephalic locomotor command region. With increasing strength of stimulation, the rate of locomotion increases and the gait changes to a trot and finally to a gallop. Thus the brain may initiate locomotion and modulate it subject to conditions, but the brain is not necessary for generating the locomotor pattern.

Sensory feedback from the hindlimbs is also unnecessary for hindlimb stepping movements, as can be shown by experiments similar to those just described. Cats with or without spinal transection (at level 2 in Figure 19.8) can make normally alternating stepping sequences following hindlimb deafferentation by cuts of the dorsal roots that contain the sensory afferent axons. (For the spinally transected cats, walking is initiated with L-dopa or clonidine.) These experiments indicate that the cat spinal cord contains a CPG for walking movements. Similar experiments indicate that fish, salamanders, toads, and turtles also have spinal locomotor CPGs.

Sensory feedback can still play important functional roles in locomotion of intact vertebrates. Spinal reflexes stabilize and modulate the effects of centrally patterned locomotor output, but spinal reflexes themselves may also be modulated by the CPG. For example, the effect of mechanical stimulation of the top of the foot of a walking cat depends on the position of the foot in the stepping cycle. If the foot is off the ground and swinging forward, it is lifted higher when stimulated ("exaggerated flexion"). If the foot is on the ground and bearing the cat's weight, the same stimulation produces a more forceful extension. This reversal of a spinal reflex

(A)

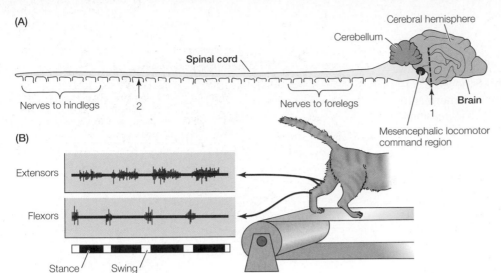

(B)

Extensors

Flexors

Stance   Swing

FIGURE 19.8 Spinal and brain control of mammalian locomotion (A) The spinal cord and lower brainstem of a cat isolated from cerebral hemispheres by transection at level 1. Electrical stimulation of the mesencephalic locomotor command region can produce locomotion in this preparation. Transection of the spinal cord at level 2 isolates the hindlimb segments of the cord. The hindlimbs are still able to walk on a treadmill after recovery from surgery. (B) Locomotion on a treadmill of a cat with a spinal transection at level 2. Reciprocal bursts of electrical activity are recorded from flexors during the swing phase of walking and from extensors during the stance phase. (After Kandel et al. 1995.)

(which is clearly adaptive for stable walking) shows that the CNS events of the stepping cycle can strongly modulate reflex function.

The experiments described in this section demonstrate that the mechanisms of control of rhythmic locomotor movements are fundamentally similar in many invertebrates and vertebrates. Although the cellular aspects may vary (e.g., different network mechanisms of central pattern generation), the functional roles of central and reflex aspects of control appear to be similar in many cases.

Central pattern generators are distributed and interacting

So far we have considered a single CPG controlling a single limb. What about coordination of multiple limbs in walking? In fact, there

are multiple CPGs to control different segments of the animal, and these CPGs must all interact to control the entire patterned movement. Let's return to the salamander with which we started the chapter. Salamanders swim by bending their bodies laterally in S curves, which progress posteriorly to propel the animal forward. This swimming pattern is similar to that of some fish and may be a primitive pattern for vertebrates. The traveling wave of contractions results from the activities of many segmental trunk CPGs that interact to coordinate the swimming movement. Salamanders also have leg-controlling CPGs for walking, and these coordinate with each other (so that opposite legs alternate, for example) and also interact with the segmental body CPGs (**FIGURE 19.9**). As in the cat, stimulation of a mesencephalic locomotor command center

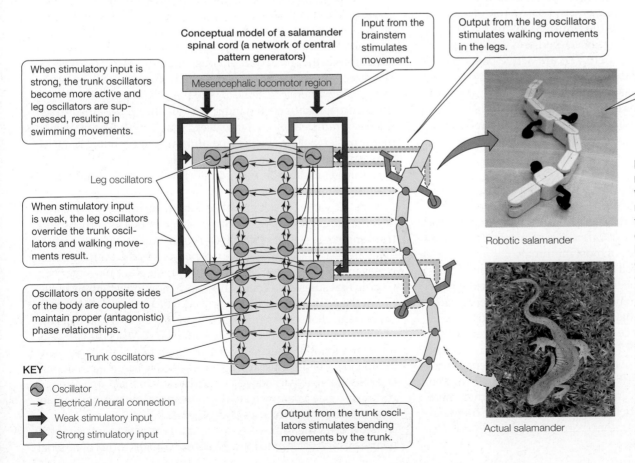

FIGURE 19.9 Design of a robot salamander that walks and swims The robot has segmental oscillators controlling body bending, and leg oscillators controlling leg stepping. The oscillators are stimulated by brainstem commands, diagrammed by blue and red arrows, and are coupled to maintain correct phase relationships (black arrows). During walking, the leg oscillators override the trunk oscillators. With stronger stimulatory input, leg oscillators saturate and stop oscillating, allowing trunk oscillators to generate the traveling waves of swimming. (After Ijspeert et al. 2007.)

can elicit walking; stronger stimulation elicits faster walking, and ultimately, swimming movements. Figure 19.9 shows how the network of interacting CPGs is modeled in the robotic salamander to produce this transition from walking to swimming with increasingly stronger command input. The investigators arranged the coupling between oscillators in the robot so that the leg oscillators overrode trunk oscillators, but the leg oscillators saturated (stopped oscillating) at high frequencies. With these interactions, investigators found that the robot could simulate effectively both salamander walking and swimming, with reasonably "natural" transitions between the two! This modeling study suggests that the known interactions of CPGs are sufficient to coordinate different gaits as well as their transitions.

The generation of movement involves several areas in the vertebrate brain

The vertebrate brain is profoundly important in the control of movement. We have discussed experiments showing that patterned locomotor movements can persist in spinally transected vertebrates and hence do not require the brain. This finding, however, does not contradict the importance of the brain in initiation, coordination, and regulation of normal movements. Let's now consider the ways in which brain areas interact with sensory input and spinal centers in movement control. These have been studied most extensively in mammals.

Until recently, the production and control of complex motor functions have been substantially attributed to brain structures such as the cerebral cortex, basal ganglia, and cerebellum. In such views, the spinal cord was assigned a subservient function in the production of movement, playing a largely passive role of relaying the commands dictated to it by the brain. Many recent studies (including the locomotion studies described previously) provide evidence that the spinal motor circuits are active participants in several aspects of the production of movement, contributing to functions that had been ascribed to "higher" brain regions. Moreover, the roles of various brain areas in motor control can be difficult to separate from sensory, motivational, and other aspects of brain function. Views on motor control are changing as a result of new data and interpretations.

CEREBRAL CORTEX We begin our examination of the execution of a voluntary movement with the motor areas of the cerebral cortex (**FIGURE 19.10**). The **primary motor cortex** (or simply *motor cortex*) lies just anterior to the central sulcus, a prominent valley in the convoluted cortical surface of most mammals.

Early studies demonstrated that electrical stimulation of areas of the primary motor cortex elicited movements of particular parts of the body, with a point-to-point correspondence between the area stimulated and the movements produced. Thus the body regions are represented on the surface of the primary motor cortex by a somatotopic map (**FIGURE 19.11**; see also Figure 15.8). The motor cortical somatotopic map was long viewed as a detailed representation of individual body parts (such as digits of the hand) or even of individual muscles, but recent studies support a rougher, more complex map of movement patterns, organized to promote coordination among muscles and joints rather than to control single muscles.

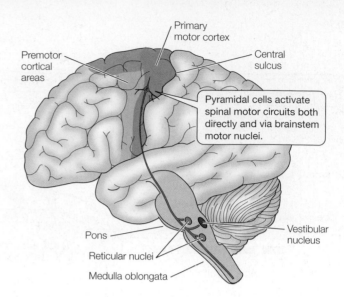

FIGURE 19.10 The location of major motor areas of the human cerebral cortex and brainstem The primary motor cortex is just anterior to the central sulcus (which separates it from the somatosensory cortex). Anterior to the primary motor cortex is a mosaic of areas collectively termed the premotor region. Neurons from the primary motor cortex descend to activate motor nuclei of the brainstem (via the corticobulbar tract) and circuits of the spinal cord (via the corticospinal tract).

The neurons of the primary motor cortex that mediate motor responses to stimulation are pyramidal cells (neurons with pyramid-shaped cell bodies), the axons of which synapse on brainstem motor nuclei and also continue down the spinal cord as major components of the corticospinal tract. (This tract is known as the *pyramidal tract* because the axons funnel through a pyramid-shaped structure on the ventral surface of the brainstem—not because the cells are pyramidal neurons.) The corticospinal axons end primarily on interneurons in the spinal cord, although in primates some axons also end directly on spinal motor neurons. The neurons of the primary motor cortex therefore activate spinal motor circuits directly via the corticospinal tract, and indirectly (via the corticobulbar tract) through brainstem motor nuclei (see Figure 19.10). The brainstem motor nuclei include the pontine and medullary reticular nuclei, the vestibular nucleus, and (in some species) the red nucleus. These nuclei are generally more important in involuntary postural control than in voluntary movements.

The activation of pyramidal cells in the primary motor cortex governs the expression of voluntary movements. The corticospinal tract appears essential for voluntary movement, but subcortical areas are also important. Activity of neurons in the primary motor cortex precedes and correlates well with voluntary movements, a finding that suggests a control function. Individual neuron activities in the primary motor cortex encode the force and direction of movements. For some neurons the amount of activity predicts the *amount* of force of a movement, for others it predicts the *change* of force, and for still others it predicts the *direction* of the movement independent of the muscles and forces used to achieve it. This last type of cellular activity suggests that neurons in the primary motor cortex can code a parameter of the movement that is more abstract than the forces and muscles that generate it.

How does the initiation of a voluntary movement trigger the neural events that lead to the activation of motor-cortex pyramidal

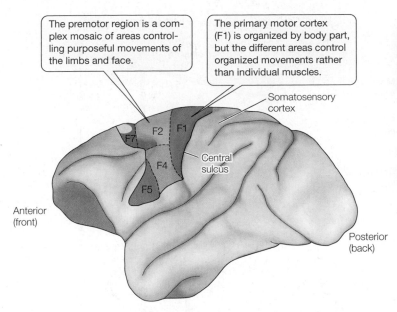

The premotor region is a complex mosaic of areas controlling purposeful movements of the limbs and face.

The primary motor cortex (F1) is organized by body part, but the different areas control organized movements rather than individual muscles.

Somatosensory cortex

Central sulcus

Anterior (front)

Posterior (back)

FIGURE 19.11 The primary motor cortex and premotor region of a rhesus monkey The primary motor cortex (F1) is just anterior to the central sulcus and is somatotopically organized: Neurons in different areas control movements of different parts of the body. A mosaic of premotor cortical areas (F2–F7) receives inputs from different regions of the cortex and is active in organizing and executing different kinds of movements. For example, F5 organizes grasping movements and has spatially overlapping projections to the hand and mouth. (F3 and F6 are on the medial surface and not visible in this view.) (After Rizzolatti and Lupino 2001.)

cells? How does the motor cortex interact with other brain areas, such as the cerebellum and basal ganglia, to produce smoothly coordinated, skilled movements without our having to expend continuous conscious effort?

For the first question, we know from recordings and from imaging studies that extensive areas of the brain are active before a decision to move. For example, if we place surface electroencephalogram (EEG) electrodes on the skull of a subject and ask her to move one finger whenever she wishes, we record a consistent pattern of activity from much of the entire cortical surface: a small, widespread rising wave of electrical activity that precedes the movement by about 800 milliseconds (ms). This electrical activity, termed a *readiness potential*, becomes localized to the relevant portion of the primary motor cortex only in the last 50–80 ms preceding the movement. The decision to initiate a voluntary movement appears to involve many areas of the cortex, including so-called association areas, and to be passed to a specific motor cortical site for initiation of the movement.

In addition to the primary motor cortex, other regions of the cerebral cortex participate in organizing movements. The frontal cortex anterior to the primary motor cortex (see Figure 19.11) consists of a mosaic of areas that participate in the planning, organization, and execution of purposeful movement. These areas had been thought to activate motor cortical regions in a hierarchical arrangement, but more recent primate studies show that some areas of the anterior mosaic connect directly to the spinal cord. Moreover, electrical stimulation of both motor and premotor cortices with stimulus trains longer than 100 ms leads to coordinated movements of different types. These results suggest that regions of both primary motor cortex and the mosaic of premotor areas control different categories of complex movement, rather than being hierarchically arranged.

One complication of the function of premotor areas is that the properties of neurons in these areas are not simply motor. An interesting example is that of **mirror neurons**, found in area F5 in monkeys (see Figure 19.11). Individual mirror neurons are activated when a monkey generates a particular movement such as reach-and-grasp. But they are also activated when the monkey sees another individual (another monkey or the experimenter)

make the corresponding movement. Mirror neurons, then, appear to code for the abstract concept of the movement rather than its execution. Mirror neurons may function in the understanding of actions, as well as in imitative learning.

The premotor areas may project directly to the primary motor cortex, but they (and other cerebral areas) also have important projections to subcortical areas. Many studies suggest that both the initial preprogramming of a movement and its modification once initiated involve interaction of the cerebral cortex with two subcortical areas that are important in voluntary movement. Next we will consider these subcortical areas: the cerebellum and the basal ganglia.

CEREBELLUM The **cerebellum** is a large, highly convoluted structure at the dorsal side of the hindbrain. It is present in all vertebrates. The cerebellum regulates movement indirectly, adjusting the descending motor output of other brain areas. The cerebellum is clearly involved in the coordination of movement, as demonstrated by the effects of cerebellar lesions in various animals, including humans. Voluntary movements are still possible following cerebellar lesions, but they are clumsy and disordered, lacking the smooth and effortless precision of normal movements. Movements are accompanied by tremor, and patients with cerebellar injuries report that they have to concentrate on each part of a movement, joint by joint. The cerebellum supports the smooth and coordinated execution of complex movements, by evaluating motor commands and sensory feedback to provide error correction signals for motor control during a movement.

The cerebellum contains two major parts: an outer **cerebellar cortex** and underlying **deep cerebellar nuclei**. The sole output of the cerebellar cortex is to the deep cerebellar nuclei. Three functional divisions of the cerebellar cortex receive inputs from and project (send outputs) to different parts of the brain via different deep cerebellar nuclei: the *vestibulocerebellum* (posterior, interacting

(A) The cellular structure of the cerebellar cortex

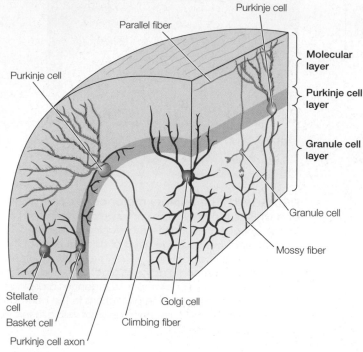

(B) Synaptic interactions of the cerebellar cortex

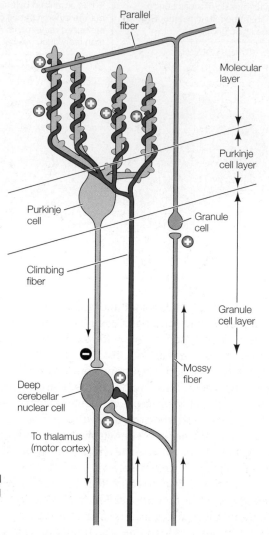

FIGURE 19.12 The neuronal organization of the mammalian cerebellar cortex (A) Cerebellar inputs are mossy fibers and climbing fibers. The output neuron is the Purkinje cell. Local interneurons are the granule cell, stellate cell, basket cell, and Golgi cell. Parallel fibers are granule cell axons that run along a folium (fold), at right angles to the planar dendritic trees of Purkinje cells. (B) Synaptic interactions in the cerebellar cortex may provide a model basis for motor learning. The two inputs to the cerebellum (mossy fibers and climbing fibers) converge on Purkinje cells. The learning of a motor task involves heterosynaptic interaction in which a climbing fiber depresses the synaptic actions of the parallel fibers on a Purkinje cell. Excitatory synapses are indicated with a plus sign (+) and inhibitory synapses with a minus sign (–).

with the vestibular system), the medial *spinocerebellum* (coordinating ongoing movement via its output directed toward the motor cortex and brainstem motor nuclei), and the lateral *cerebrocerebellum* (concerned with motor planning as well as with planning and sequencing of non-motor cognitive behavior).

The cellular architecture and synaptic interactions of the cerebellar cortex are elegantly precise and are as well known as those of any other area of the brain. As **FIGURE 19.12A** shows, the cerebellar cortex contains five types of neurons and two principal types of input fibers. The axons of Purkinje cells constitute the only output of the cerebellar cortex; these end in the deep cerebellar nuclei below the cortical surface. The major synaptic interactions of the cerebellar cortex are shown in **FIGURE 19.12B**. Climbing fibers make powerful excitatory 1:1 synaptic contacts with Purkinje cells. Mossy fibers, in contrast, provide divergent excitatory input to many granule cells. Axons of granule cells ascend to the surface layer of the cortex and branch in opposite directions to become the parallel fibers, which make excitatory synaptic contacts with the other types of cerebellar cortical cells. The roles of climbing-fiber and mossy-fiber input are not completely clear. Climbing fibers are thought to convey error signals (sensory feedback from errors in movements), whereas mossy fibers may convey broader information about the sensory context of a movement.

The synaptic interaction of parallel fibers and Purkinje cells is especially prominent. The parallel fibers pass through the flattened, planar dendrites of Purkinje cells at right angles (see Figure 19.12B). Each Purkinje cell receives excitatory synapses from about 100,000 parallel fibers (in addition to 1 climbing fiber). Thus the climbing-fiber and mossy-fiber inputs differ greatly in the degree of divergence and convergence of their synaptic effects.

With the exception of the granule cells, which exert excitatory synaptic effects, all of the other cell types of the cerebellar cortex are *inhibitory* in their effects. Basket cells, stellate cells, and Golgi cells mediate different sorts of inhibition within the cerebellar cortex. Moreover, the Purkinje axons that are the sole output of the cerebellar cortex are inhibitory in their effects on the deep cerebellar nuclei. The inhibitory output of the cerebellar cortex is balanced at the level of the deep cerebellar nuclei, because the climbing fibers and mossy fibers make excitatory synapses on the neurons of the deep nuclei. The inhibitory output of the Purkinje cells of the cerebellar cortex may then refine and sculpt movements by selectively opposing the excitatory effects of climbing- and mossy-fiber input at the deep cerebellar level.

Despite extensive studies of the circuitry of the cerebellum (only superficially described here), we still do not have a clear understanding of how it modulates or coordinates movements.

There are several models of cerebellar action—some stressing millisecond timing of motor discharges, others stressing corrective feedback or coordination of disparate body regions and muscle groups. One popular model is that the cerebellum learns motor tasks so that they can be performed unconsciously and automatically in the appropriate sensory context. This motor learning is thought to involve changes in synaptic strength, specifically long-term depression at synapses of parallel fibers onto Purkinje cells (see Figure 19.12B). This long-term synaptic depression and the related long-term potentiation are discussed in Chapter 13 (see pages 362–367).

BASAL GANGLIA   The **basal ganglia** are a set of nuclei (clusters of brain neurons) located in the forebrain and midbrain, under the cerebral hemispheres. The most important areas (in terms of motor control) are the *caudate nucleus*, the *putamen*, and the *globus pallidus*. The caudate nucleus and putamen are similar in origin and function, and together they are termed the *neostriatum* (or simply *striatum*). The caudate nucleus and putamen receive excitatory input from many parts of the cerebral cortex, both motor and association areas. The caudate nucleus and putamen send inhibitory neurons to the globus pallidus. The major output of the basal ganglia is inhibitory; neurons from the internal segment of the globus pallidus inhibit neurons in the thalamus that excite the cerebral motor cortex.

The basic circuit of the basal ganglia is considered an example of a **loop** circuit—one in which the output of the circuit loops back to the site of the circuit's input. (This designation is not strictly accurate, however, because the motor cortex affected by basal ganglia output is much more restricted than the broad areas of cerebral input to the basal ganglia.) Loop circuits are common in the vertebrate brain, and they appear to be important for many aspects of motor control, emotions, and other brain activities. The basal ganglia also form a second loop circuit, between the striatum and the substantia nigra. Dopaminergic neurons from the substantia nigra project to the striatum and receive inhibitory feedback from it. These dopamine neurons modulate the strength of basal ganglia pathways (see below).

The basal ganglia are important in selecting movements, suppressing competing or unwanted movements, and initiating the selected movement. Most of the neurons in the circuits of the basal ganglia are inhibitory, so these functions involve considerable inhibitory interaction. **FIGURE 19.13** shows how the basic synaptic connections of the basal ganglia function to disinhibit movement. The output of the globus pallidus pars interna (GPi) inhibits movements. For the initiation of a movement, this tonic inhibition is lifted, by *disinhibition*. The striatum (caudate nucleus and putamen) receives excitatory input from the cerebral cortex, and striatal neurons inhibit neurons in the globus pallidus via two pathways. In the *direct* pathway there are two inhibitory synapses: Striatal neurons inhibit GPi, and GPi neurons inhibit neurons of the thalamus. Thus activation of striatal

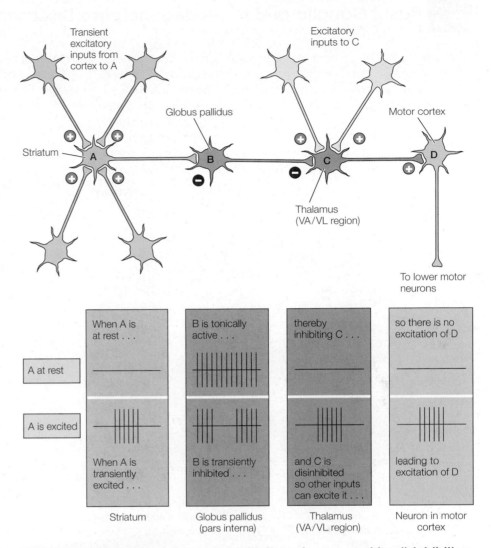

FIGURE 19.13 Basal ganglia aid the initiation of movement by disinhibition In the direct pathway through the basal ganglia, neurons from the cerebral cortex excite striatal neurons, which inhibit neurons in the globus pallidus pars interna. These globus pallidus neurons inhibit neurons in the thalamus that promote movement. Cerebral cortical activation of striatal neurons transiently inhibits the tonic inhibitory output of the globus pallidus, disinhibiting the thalamus and activating a movement. Excitatory synapses are indicated with a plus sign (+) and inhibitory synapses with a minus sign (–). VA/VL, ventral anterior and ventral lateral regions of the thalamus.

neurons inhibits GPi neurons and disinhibits the thalamus, thereby allowing a movement (see Figure 19.13). The *indirect* pathway, in contrast, involves a chain of three inhibitory neurons: Striatal neurons inhibit neurons of the globus pallidus pars externa (GPe), GPe neurons inhibit GPi neurons, and GPi neurons inhibit the thalamus. This triple inhibition means that striatal activity via the indirect path will suppress other activity in the thalamus, strengthening the tonic suppression of other unwanted movements and preventing them from competing with the movement selected.

The balanced roles of the direct and indirect pathways in the basal ganglia serve in the selection, suppression, and initiation of movements. These roles contrast with the cerebellum, which appears to "fine-tune" movement, coordinating and smoothly modifying the execution of a movement to match a command. Degenerative changes in neurons of the basal ganglia underlie human movement disorders such as Parkinson's disease and Huntington's disease (**BOX 19.3**).

Basal Ganglia and Neurodegenerative Diseases

In humans, two movement disorders result from neurodegenerative changes in basal ganglia function: Parkinson's disease and Huntington's disease (Huntington's chorea). Parkinson's disease is characterized by difficulty in initiating movements (*akinesia*), so a simple task such as climbing stairs or getting up from a chair becomes almost impossible to carry out. Akinesia is often accompanied by postural rigidity and by tremors in limbs at rest. Huntington's chorea represents the converse of parkinsonism: Movements

(Continued)

Normal and abnormal circuitry of the basal ganglia

(1) Location of the basal ganglia in a cross section of the human brain. The basal ganglia include the caudate nucleus and putamen (orange in the diagram; together called the striatum), the external and internal segments of the globus pallidus (red-violet), the subthalamic nuclei (green), and in the midbrain (not shown), the substantia nigra. The caudate nucleus and putamen, although anatomically separated from each other by white matter of the internal capsule, are functionally similar. (2) Basic circuit in the normal brain. Excitatory input (green) to the putamen (and caudate nucleus) activates neurons of the direct and indirect pathways. Neurons in the direct pathway inhibit (red arrow) neurons in the internal segment of the globus pallidus (GPi), and the GPi neurons in turn inhibit neurons in the thalamus and suppress movement generation by the motor cortex. In contrast, putamen neurons in the indirect pathway inhibit neurons in the external segment of the globus pallidus (GPe), and the GPe neurons in turn inhibit GPi neurons. Thus the direct and indirect pathways have opposing, but balancing, effects on the ease of movement generation. Modulating input from dopaminergic neurons in the substantia nigra pars compacta (SNc) enhances the direct pathways and inhibits the indirect pathway. Other circuitry through the subthalamic nucleus (STN) is not considered here. (3) In Parkinson's disease the dopaminergic neurons of the SNc degenerate, unbalancing their modulatory effects on the striatum (putamen and caudate nucleus). Direct pathway output of the striatum is weakened and indirect output is strengthened, leading to excessive inhibitory output of GPi, suppressing movement generation. (4) In Huntington's disease, striatal inhibitory neurons of the indirect pathway degenerate, leading to insufficient inhibitory output of GPi and excessive movement generation. (2–4 after Kandel et al. 2000.)

(1) Cerebrum

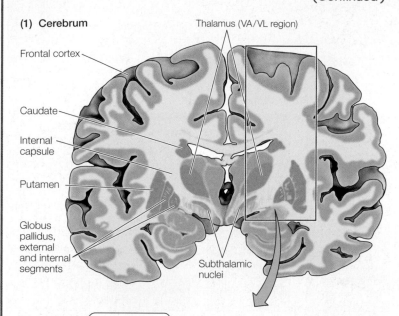

(2) Normal

Direct pathway facilitates movement.

Indirect pathway inhibits movement.

(3) Parkinson's disease

(4) Huntington's disease

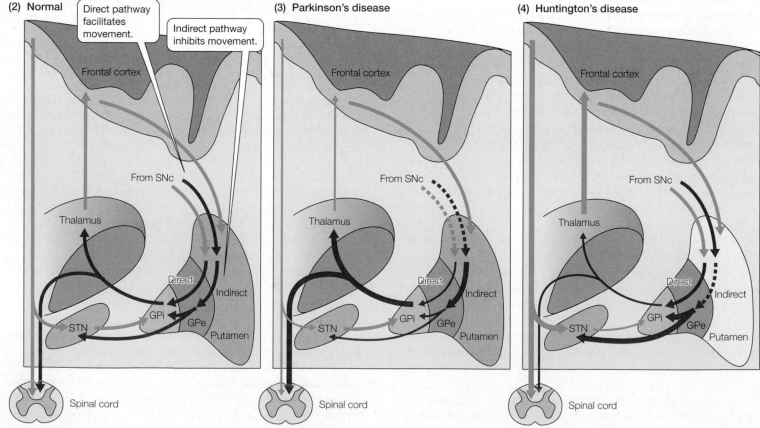

(Continued)

occur uncontrolledly and are difficult to stop. Both *chorea* (uncontrolled but coordinated jerky movements) and *athetosis* (slow writhing movements) are associated with damage to the striatum.

How can degeneration of neurons in the basal ganglia produce such opposite effects? First we need to examine the neuronal circuitry of the normal basal ganglia in more detail. Recall that there are two major circuits on the basal ganglia, the *direct pathway* and the *indirect pathway* (see page 533). Figure 19.13 shows how the direct pathway promotes the generation of a voluntary movement by disinhibition. The indirect pathway, as

we noted, opposes this action by adding another layer of inhibition. The figure shows the location and circuitry of the basal ganglia.

Both Parkinson's and Huntington's diseases involve neuronal degeneration and altered neurotransmitter activity in the basal ganglia. In Parkinson's disease, dopaminergic neurons in the substantia nigra degenerate. The synaptic endings of these dopamine neurons are in the striatum, so degeneration of the neurons deprives the striatum of dopaminergic input, which normally enhances the direct pathway through the globus pallidus and suppresses the indirect pathway. Parkinsonian symptoms, then, reflect an

increased inhibitory output, leading to an inability to select and change behavioral activities. Dopamine replacement therapy (through provision of the dopamine precursor L-dopa, which crosses the blood–brain barrier) alleviates many symptoms of Parkinson's disease, at least for a while. Implants of dopaminergic tissue into the striatal area may provide longer-lasting relief from Parkinson's symptoms. In Huntington's disease, in contrast, extensive loss of neurons in the striatum decreases striatal inhibition of the indirect pathway. Chorea then results from a decreased inhibitory output of the basal ganglia, decreasing the ability to suppress unwanted movements.

THE INTERACTION OF BRAIN AREAS IN MOVEMENT CONTROL   We will now attempt to integrate the hypothesized roles of the cerebral cortex, cerebellum, and basal ganglia in the control of voluntary movement. As **FIGURE 19.14** shows, the planning and programming of a movement can be viewed as separate from the execution of the movement. We can suppose that the decision to move starts in the association cortex (cortex that is not linked to any particular sensory or motor system), because the readiness potentials recorded prior to a movement are not localized to a specific cerebral area. Two loop circuits from the association cortex are thought to be involved in preprogramming a movement: one loop goes through the basal ganglia (selection and initiation) and another through the lateral cerebrocerebellum (initial programming). Both loops feed back to the motor cortex via the ventrolateral nucleus of the thalamus. The motor cortex then generates the appropriate pattern of activity to initiate the movement.

Information about the command is sent to the spinocerebellum, via several subcortical nuclei. This process, termed *command monitoring*, "informs" the cerebellum of the intended movement. The spinocerebellum also receives ascending information—both

sensory information about joint position and muscle tension, and CNS information from spinal and brainstem motor centers. The spinocerebellum may integrate this feedback information about the state of lower motor centers (internal feedback) and about the periphery (external feedback) with the monitored cerebral command. The cerebellar output can then modify and correct the command on a continuous basis as the movement evolves, using an integral of all relevant information (command, motor state, and sensory feedback). This continuous correction is presumably faster and smoother than, say, a correction system based on sensory feedback alone.

There remains considerable controversy over the roles of all these brain areas in the control of movement, and the areas are also implicated in other functions in addition to motor control. Even if the preceding description of brain area interaction in the execution of a voluntary movement is correct, it begs other questions (such as, how is a decision to move actually made?). Nevertheless, the relations shown in Figure 19.14 illustrate how brain areas may interact in planning, coordinating, and commanding movements in mammals.

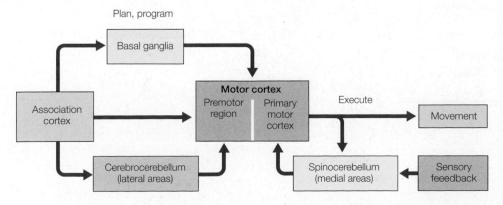

Plan, program

FIGURE 19.14 The interaction of brain areas in the planning, execution, and control of voluntary movement   The sensory association cortex (as well as other cortical areas) funnels activity to premotor cortical areas, including loops through the basal ganglia and the cerebrocerebellum, in the planning and programming of the movement. Activity passes to the primary motor cortex for executing the movement, with correction from a cerebellar loop through the spinocerebellum. Inputs to the cerebral cortex pass through the thalamus (not shown).

SUMMARY
Control and Coordination of Vertebrate Movement

■ The vertebrate brain and spinal cord interact in the generation of behavior patterns such as locomotion and voluntary movements.

■ Tetrapod vertebrates have a spinal CPG for stepping during walking. Descending commands from the brain can activate the CPG, and sensory feedback can modulate it.

■ Several brain areas are important in generating and coordinating movements in mammals. In the cerebral cortex, the primary motor cortex directly activates spinal motor centers to generate movements; premotor cortical areas are involved in planning and organizing movements.

■ The cerebellum and the basal ganglia are connected to the cerebral cortex in looping circuits. The cerebellum is active in coordinating movements and in motor learning. The basal ganglia are involved in the selection and initiation of movements by disinhibition. Parkinson's and Huntington's diseases stem from abnormalities in function of the basal ganglia.

STUDY QUESTIONS

1. Suppose that an arthropod such as a locust or crayfish has about 10^5 neurons, a fish 10^8, and a rat 10^{10}. Is the behavior of the fish 1000 times more complicated than that of the arthropod? Why or why not? What does the difference in number of neurons suggest about how these different animals generate patterns of behavior?

2. How do we find out whether a vertebrate such as a cat or a salamander has a central pattern generator (CPG) for walking?

3. A neural circuit for a CPG (see Figure 19.6) may include motor neurons or may consist entirely of interneurons. How do you think the inclusion of motor neurons might restrict the flexibility of a CPG circuit? For what sorts of behavior patterns do you think a CPG circuit might include motor neurons?

4. Why would an oscillator of the closed-loop model be expected to provide a more stable rhythmic output than one employing the half-center model?

5. Some movements (such as picking up a cup) are visually guided, or *steered*. Others (such as throwing a football) cannot be corrected once launched and are termed *ballistic*. How, and to what extent, might you expect the neural control of these movements to differ?

Go to **sites.sinauer.com/animalphys4e**
for box extensions, quizzes, flashcards,
and other resources.

REFERENCES

Büschges, A., H. Scholz, and A. El Manira. 2011. New moves in motor control. *Curr. Biol.* 21: R513–R524. An excellent review of motor control issues, including optogenetic approaches.

Carew, T. J. 2000. *Behavioral Neurobiology: The Cellular Organization of Natural Behavior.* Sinauer, Sunderland, MA. Excellent, accessible text on neuroethology, explicating the neural basis of animal behavior for a variety of well-studied cases.

Catela, C., M. M. Shin, and J. S. Dasen. 2015. Assembly and function of spinal circuits for motor control. *Annu. Rev. Cell Dev. Biol.* 31: 669–698.

Drew, T., and D. S. Marigold. 2015. Taking the next step: cortical contributions to the control of locomotion. *Curr. Opin. Neurobiol.* 33: 25–33.

Georgopoulos, A. P., and A. F. Carpenter. 2015. Coding of movements in the motor cortex. *Curr. Opin. Neurobiol.* 33: 34–39.

Goulding, M. 2009. Circuits controlling vertebrate locomotion: moving in a new direction. *Nat. Rev. Neurosci.* 10: 507–518.

Grillner, S. 2006. Biological pattern generation: the cellular and computational logic of networks in motion. *Neuron* 52: 751–766.

Katz, P. S. 2016. Evolution of central pattern generators and rhythmic behaviours. *Phil. Trans. R. Soc., B* 371(1685): 20150057. doi: 10.1098/rstb.2015.0057

Marder, E. 2012. Neuromodulation of neuronal circuits: back to the future. *Neuron* 76: 1–11.

Marder, E., and D. Bucher. 2001. Central pattern generators and the control of rhythmic movements. *Curr. Biol.* 11: R986–R996. An excellent review of neural mechanisms generating and modulating behaviorally important rhythmic motor patterns.

Nelson, A. B., and A. C. Kreitzer. 2014. Reassessing models of basal ganglia function and dysfunction. *Annu. Rev. Neurosci.* 37: 117–135.

Rybak, I. A., K. J. Dougherty, and N. A. Shevtsova. 2015. Organization of the mammalian locomotor CPG: review of computational model and circuit architectures based on genetically identified spinal interneurons. *eNeuro* 2(5): ENEURO.0069-15.2015.

See also **Additional References** *and* **Figure and Table Citations**.

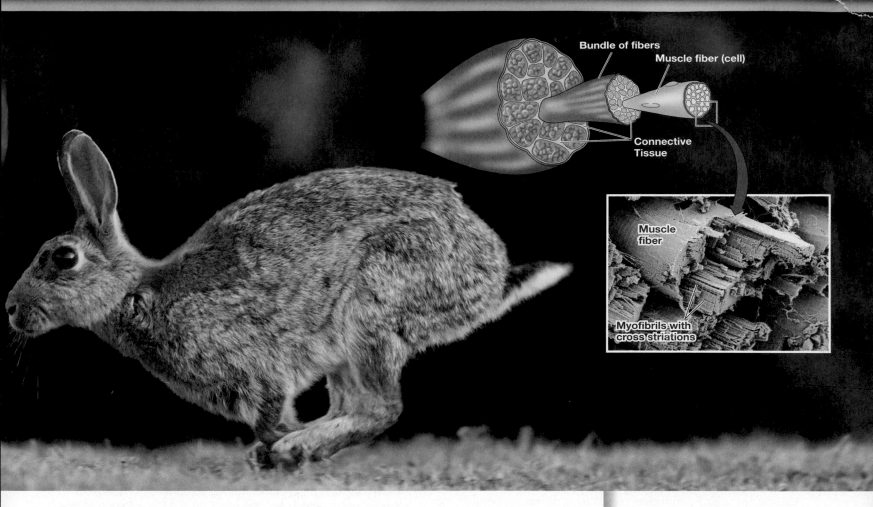

Bundle of fibers
Muscle fiber (cell)
Connective Tissue
Muscle fiber
Myofibrils with cross striations

Muscles are specialized for movement. All animals use muscles to generate movements that accomplish behaviors, such as bees collecting nectar from flowers or a rabbit dashing away from a predator. Animals also use muscles to accomplish physiological functions, such as generating heartbeats or mixing and propelling foods for digestion.

Muscle is a tissue that consists of contractile cells. To produce movements, muscle cells use the *molecular motor* **myosin**[1] to capture and convert the chemical energy of ATP into the mechanical energy of movement. Myosin is a large protein that interacts with another protein, **actin**, to generate force. Myosin and actin are referred to as *contractile proteins*.

All animal phyla have two categories of muscle cells: striated and smooth (or unstriated). **Striated muscle cells** have transverse bands, giving them a striped appearance. The pattern of bands reflects the organization of myosin and actin into regularly repeating units, called *sarcomeres*. In vertebrates, **skeletal** muscles (attached to bones) and **cardiac** (heart) muscles are both striated muscles. The figure shown here of a whole skeletal muscle illustrates the

Skeletal muscles consist of bundles of longitudinally arranged muscle fibers (cells) Connective tissue wraps around individual muscle fibers and bundles of fibers, and it weaves into the tendon which attaches the muscle to the skeleton. The scanning electron micrograph reveals individual fibers of the rabbit psoas muscle, a skeletal muscle that rotates and flexes the thigh. Each muscle fiber contains longitudinally arranged *myofibrils* with cross striations that delineate *sarcomeres* arranged end to end. (Micrograph courtesy of Richard Briggs, Smith College.)

---

[1] The myosin molecules that produce contractile force in smooth and striated muscles are classified as myosin II. They are part of the myosin superfamily, which consists of at least 18 different classes of myosins found in protozoans, fungi, plants, and animals.

longitudinal arrangement of muscle cells in this type of muscle, and the scanning electron micrograph reveals the internal components of a single striated muscle cell. Each cell (fiber) contains *myofibrils* arranged in parallel, and each myofibril has cross-striations, which delineate sarcomeres.

Smooth (**unstriated**) **muscle cells** also use actin and myosin to generate contractions, but these proteins are not organized into sarcomeres. Smooth muscles of vertebrates are found primarily in hollow or tubular organs such as the intestine, uterus, and blood vessels. Invertebrates also have striated and smooth muscles, but they are not always found in the same distribution as in vertebrates. In arthropods, for example, the skeletal (attached to the exoskeleton) and cardiac muscles are both striated, but so are muscles of the alimentary (digestive) tract.

In this chapter we consider muscles in light of the major themes of this book: mechanism and adaptation. We examine first the physiological and biochemical mechanisms that underlie muscle contraction, and then the adaptations of certain muscles specialized to perform different functions. Because vertebrate muscles have been the focus of in-depth experimental studies, we examine them in detail, and we complement our observations with examples of some well-studied invertebrate muscles.

Vertebrate Skeletal Muscle Cells

Skeletal muscles (see chapter opener figure) are composed of bundles of long, cylindrical **muscle fibers**, or **muscle cells** (the two terms are used interchangeably). Connective tissue surrounds individual muscle fibers, bundles of fibers, and the muscle itself. The connective tissue holds fibers together, provides a matrix for nerve fibers and blood vessels to gain access to the muscle cells, contributes elasticity to the whole muscle, and weaves itself into **tendons**. The tendons attach the muscles to bones and thereby transmit force generated by the muscle fibers to the skeleton. Whereas small muscles may contain only a few hundred muscle fibers, large limb muscles of mammals contain thousands of fibers. Single muscle fibers can be as long as 0.3 m (1 ft). Single fibers are typically 10 to 100 micrometers (μm) in diameter, although some (such as those in certain Antarctic fish) can reach several hundred μm in diameter. Skeletal muscle fibers are multinucleate (contain many nuclei) because they form developmentally by the fusion of individual uninucleate cells called **myoblasts**. A muscle fiber is surrounded by a cell membrane sometimes referred to as the **sarcolemma** (**FIGURE 20.1A**) (The prefixes *myo-* and *sarco-* both denote "muscle.")

Each muscle fiber contains hundreds of parallel, cylindrical **myofibrils** (**FIGURE 20.1B,C** and the opening scanning electron micrograph). The myofibrils are 1 to 2 μm in diameter and as long as the muscle fiber. Each myofibril has regularly repeating, transverse bands. The major bands are the dark **A bands** and the lighter **I bands**.[2] In the middle of each I band is a narrow, dense **Z disc**, or **Z line**. The portion of a myofibril between two Z discs is called a

---

[2] Microscopists gave the A band its name because they observed that it is *anisotropic* (strongly polarizes visible light). They named the I band *isotropic* because it does not polarize visible light. The names of other components of the sarcomere come from the German language. The Z disc that separates sarcomeres comes from *Zwischenscheibe* ("between line"); the bright H zone at the center of the A band from *hell* ("clear or bright"); and the M line down the middle of the H zone from *Mittelscheibe* ("middle line").

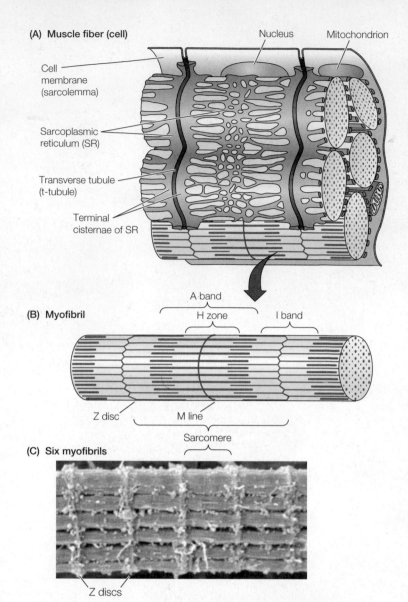

FIGURE 20.1 Each muscle cell contains many myofibrils (A) The myofibrils are enclosed by the sarcoplasmic reticulum, which forms intimate associations with transverse tubules that extend from the cell membrane (sarcolemma). The transverse tubules and sarcoplasmic reticulum couple excitation of the muscle fiber membrane with contraction. (B) Contraction is accomplished by interaction of the myosin and actin filaments of the sarcomeres, which are arranged end to end to form the myofibrils. (C) This enlarged segment of the scanning electron micrograph in the chapter opener figure illustrates a series of sarcomeres in six myofibrils. Their alignment in register gives a striated appearance.

sarcomere (see Figure 20.1B,C). Thus one myofibril consists of a longitudinal series of repeating sarcomeres. The Z discs of adjacent myofibrils are lined up in register with each other, so the pattern of alternating A bands and I bands appears continuous for all the myofibrils of a muscle fiber. This alignment of banding within a muscle fiber gives the fiber its striated appearance.

Higher-magnification electron micrographs show that the myofibrils contain two kinds of **myofilaments** (**FIGURE 20.2**). The **thick filaments** (see Figure 20.2C) are composed primarily of the protein *myosin* and are confined to the A band of each sarcomere. A single thick filament consists of 200 to 400 myosin molecules. The **thin filaments** (see Figure 20.2B) are composed primarily of *actin*. A single thin

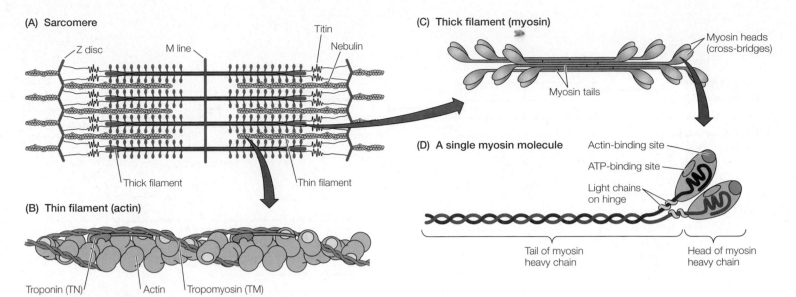

(A) Sarcomere

Z disc · M line · Titin · Nebulin

Thick filament · Thin filament

(B) Thin filament (actin)

Troponin (TN) · Actin · Tropomyosin (TM)

(C) Thick filament (myosin)

Myosin heads (cross-bridges)

Myosin tails

(D) A single myosin molecule

Actin-binding site

ATP-binding site

Light chains on hinge

Tail of myosin heavy chain · Head of myosin heavy chain

FIGURE 20.2 The sarcomere is the functional unit of striated muscle (A) Thick and thin myofilaments overlap and slide by each other to generate contractions. (B) Each thin filament is made of two chains of globular actin molecules arranged in a loose helix. The regulatory proteins tropomyosin (TM) and troponin (TN) are also components of the thin filament. (C) Myosin molecules form the thick filament. (D) Each myosin molecule contains two heavy chains of amino acids. The two chains coiled around each other form the tail. The amino-terminal end of each heavy chain forms one of the heads. The head region has a site for binding actin and a different site for binding and hydrolyzing ATP. A hinge region connects the head to the tail. The myosin molecule also includes two smaller light chains associated with each head. Thus each complete myosin molecule contains six polypeptide chains: two heavy and four light. The molecular composition of the heavy and light chains varies in different types of muscles. The different myosin isoforms of heavy chains and light chains confer variations of functional properties, such as the rate at which the myosin ATPase hydrolyzes ATP.

filament consists of two chains of globular actin molecules wrapped around each other in a loose helix. Thin filaments are anchored to proteins in the Z discs (see Figure 20.2A). They extend from the Z discs partway into the A bands of each flanking sarcomere, where they interdigitate with thick filaments. The central region of the A band, which contains only thick filaments and appears lighter than the rest of the A band, is called the **H zone**. A narrow dense region called the **M line** bisects the H zone.

In the M line, the thick filaments of the myofibril are webbed together with accessory proteins to maintain their regular spacing. The Z disc and M line ensure that neither the thin filaments nor the thick filaments float free. In vertebrates, the thick filaments are about 1.6 μm long and 12 to 15 nanometers (nm) across. The thin filaments are about 1.0 μm long and 7 to 8 nm across. Muscle fibers also contain *intermediate filaments*, so named because their diameters are about 10 nm, intermediate between those of thick and thin filaments. Intermediate filaments contribute to the architectural integrity of the muscle fiber. The protein *desmin*, for example, forms a scaffold around Z discs of adjacent myofibrils to hold them together. This scaffold extends to the cytoskeleton that lies beneath the cell membrane as well as to the nucleus and mitochondria. These connections help maintain the structural organization of the cell during contractile activity.

Cross sections of a myofibril show the relationship of thick and thin filaments in a sarcomere (**FIGURE 20.3**). A cross section through the I band shows only thin filaments. A section through the part of the A band in which the thick and thin filaments overlap shows each thick filament surrounded by six thin filaments. A section through the H zone shows only thick filaments.

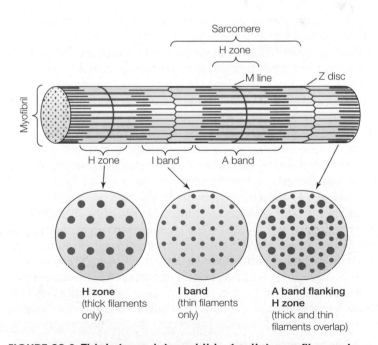

Sarcomere

H zone

M line · Z disc

Myofibril

H zone · I band · A band

H zone (thick filaments only)

I band (thin filaments only)

A band flanking H zone (thick and thin filaments overlap)

FIGURE 20.3 Thick (myosin) and thin (actin) myofilaments are arranged in parallel in a sarcomere Cross sections of a single myofibril illustrate the regions of overlap of the thick and thin myofilaments. Both the M line and the Z disc contain accessory proteins that anchor the thick and thin filaments.

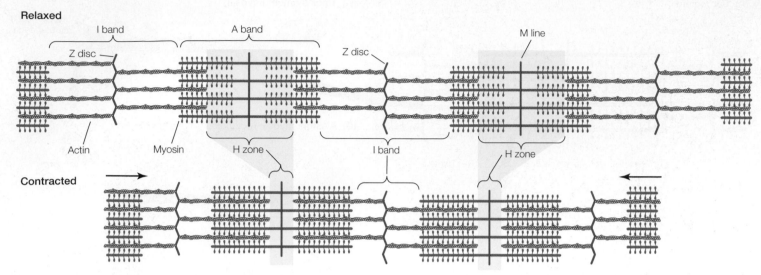

FIGURE 20.4 Thick and thin myofilaments slide by one another to produce contractions Cross-bridges of the thick filaments draw the thin filaments toward the center of each sarcomere. The myofilaments do not shorten, but the I band and H band of each sarcomere do. Because the sarcomeres are arranged in series in a myofibril, the entire myofibril shortens. Arrows indicate shortening of two adjacent sarcomeres during contraction.

The myosin molecules have radial projections on them called heads or **cross-bridges** (see Figure 20.2C,D). When the muscle cell is stimulated to contract, the myosin cross-bridges interact transiently with the overlapping actin thin filaments. The interactions of the myosin cross-bridges with actin molecules generate the force for muscle contraction. These contractile proteins function together with additional proteins in the sarcomere. For example, **titin** and **nebulin** (see Figure 20.2A) are giant proteins that help align actin and myosin. A single molecule of titin attaches to both the Z disc and the M line and spans the distance between them.[3] Titin's properties vary along its length: The region along the I band is highly folded and elastic; the region along the A band is integrated into the lattice of the thick filaments and is inelastic. This big titin molecule maintains the thick filament at the center of the sarcomere, and its elasticity over the region of the I band confers the ability of the sarcomere to spring back after the muscle fiber is stretched.

Nebulin is inelastic; it runs the length of a thin filament and stabilizes it. The nebulin molecule also specifies the length of the thin filament to optimize the overlap between thick and thin filaments. **Obscurin** is a third giant protein, which binds with proteins in the M line and links to the sarcoplasmic reticulum. Like titin and nebulin, obscurin plays an important role in maintaining the structural organization of the sarcomere. **Troponin** (**TN**) and **tropomyosin** (**TM**) are protein molecules associated with the actin chains of the thin filaments (see Figure 20.2B). They regulate the

process of contraction by controlling whether or not the myosin cross-bridges can interact with the thin filaments.[4]

Thick and thin filaments are polarized polymers of individual protein molecules

Individual myosin molecules are large proteins of about 500 kilodaltons (kDa), each consisting of two globular heads joined to a long rod, or tail. The heads are the cross-bridges, and the tail contributes to the backbone of the thick filament (see Figure 20.2C,D). During polymerization the myosin molecules orient themselves with their tails pointing toward the center of the thick filament and their heads toward the ends. As a result, the two halves of the thick filament become mirror images of each other with a short bare zone of only tails in the middle of the filament. The cross-bridges on either side of the bare zone point in opposite directions.

Each actin molecule is a globular protein (42 kDa) called *G-actin*. G-actin monomers form chains of *F-actin* (filamentous actin). The two chains of F-actin wind around each other in a loose helix (see Figure 20.2B). Like the myosin molecules in thick filaments, G-actin molecules in thin filaments are arranged so that those on one side of the Z disc have one orientation, and those on the other side have the opposite orientation. The consequence of the polarized organization of the thick and thin filaments is that the cross-bridges in contact with the thin filament act like oars to pull the thin filaments toward the center of the sarcomere.

Thus, when a muscle fiber contracts, the thick and thin filaments do not shorten but instead slide by one another. In 1954 two independent teams—A. F. Huxley and R. Niedergerke, and H. E. Huxley and J. Hanson—formulated the *sliding-filament theory* of muscle contraction, which has since been amply confirmed. It states that the force of contraction is generated by the cross-bridges of the thick filaments attaching to the thin filaments and actively pulling them toward the center of the sarcomere (**FIGURE 20.4**).

---

[3] Titin (also known as connectin) is the largest known protein. Composed of nearly 27,000 amino acids, it has a molecular weight of 3 million Daltons (Da). Whereas the thick and thin myofilaments of similar length are polymers made up of hundreds of myosin or actin molecules, a single molecule of titin extends from the Z disc to the center of the sarcomere!

[4] The characteristics of many proteins in striated muscle fibers continue to be studied and refined. Currently, investigators are using proteomics to characterize the profiles of many proteins in different muscles and to follow changes that occur within muscles over time, such as those in holometabolous insects that undergo complete metamorphosis (see Chapter 16) or those in disease states.

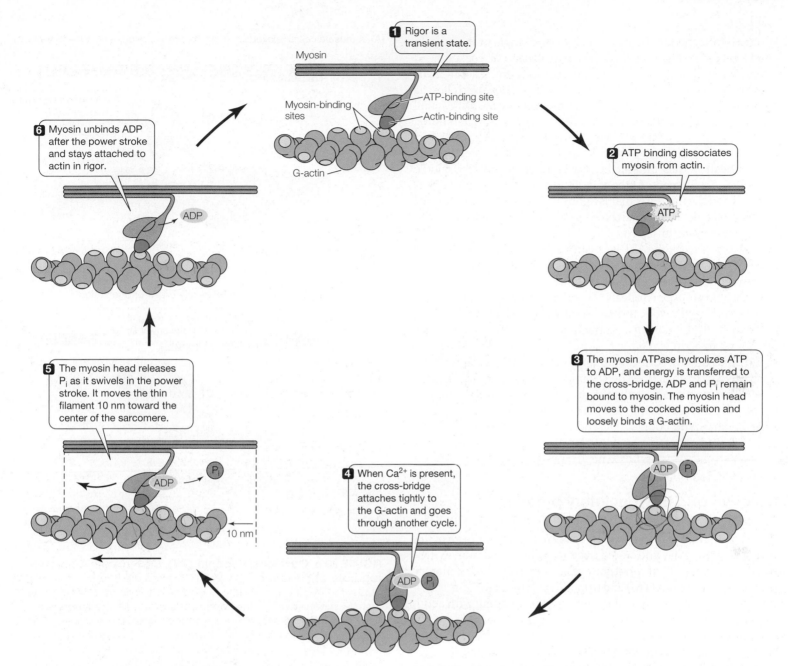

FIGURE 20.5 A single cross-bridge cycle uses one molecule of ATP and moves the actin filament about 10 nm Each cross-bridge goes through several cycles during a single contraction. The two myosin heads function independently, and only one binds to actin at a time. Structural studies suggest that no more than four myosin heads can attach over a span of seven G-actin monomers.

Muscles require ATP to contract

Myosin heads cyclically attach to actin molecules and then swivel to pull on the actin filament. Each myosin head has two binding sites: one for actin and the other for ATP. The binding site for ATP is an ATPase with enzymatic activity that splits inorganic phosphate from the ATP molecule and captures the released energy. The energy is used to power cross-bridge action.

The cycle of molecular interactions underlying contraction is shown in **FIGURE 20.5**. In step ❶ the myosin head is bound to actin at the conclusion of a power stroke. This is the *rigor* conformation, as in *rigor mortis,* in which muscles become stiff after death because of the absence of ATP. In rigor, the actin and myosin filaments are rigidly fixed in place until a new molecule of ATP binds to the myosin head and triggers its *unbinding* from actin (step ❷). In life, the rigor stage of each cross-bridge cycle is brief because the globular myosin head readily binds ATP and detaches from actin. It is important to understand that the detachment of myosin from actin requires the *binding* of ATP to change the conformation of myosin's actin-binding site, but it does not require the energy derived from the ATP.

Once released from actin, the myosin head hydrolyzes the ATP to ADP and inorganic phosphate (P_i) and the released energy is transferred to the cross-bridge and stored there. The ADP and

P_i remain attached to the head. A change in angle of the myosin head (termed *cocking*) accompanies hydrolysis. The cocked head interacts loosely with a G-actin monomer to form an actin–myosin–ADP–P_i complex (step ❸). To make another power stroke, the complex must bind tightly to the actin monomer (step ❹). We will see in the next paragraphs and Figure 20.6 that this tight binding requires the presence of Ca^{2+} ions. When the complex binds tightly to actin it triggers the power stroke and release of P_i (step ❺). The myosin head swivels, pulling the attached actin toward the middle of the myosin filament. At the end of the power stroke, the ADP is released and the myosin remains tightly bound to the actin (step ❻). A new molecule of ATP then binds to the myosin head, triggering its release from actin.

With each cycle, one ATP is consumed, and the myosin molecule moves the actin filament a short distance (about 10 nm). During a single contractile event (stimulated by a muscle fiber action potential), each cross-bridge repeats several binding/unbinding cycles. The cross-bridges work independently and asynchronously, so that at any instant during a contraction some of the cross-bridges are bound to actin while the rest are in other phases of the cycle. At no time during contraction do all cross-bridges simultaneously detach from actin. The summed effect of all the repeated cross-bridge cycles is to pull the thin filaments toward the middle of the sarcomere. In living, relaxed muscle, all cross-bridges have stored energy and bound ADP and P_i (as in Figure 20.5 step ❸), but most are temporarily unable to bind tightly to actin to generate a power stroke. As we will see next, cross-bridge cycling requires the presence of calcium.

Calcium and the regulatory proteins tropomyosin and troponin control contractions

In a resting muscle, each myosin head has detached from actin, hydrolyzed the ATP, and stored the energy obtained from hydrolysis. It is "primed" for another cycle. However, the regulatory proteins tropomyosin (TM) and troponin (TN) prevent contraction by inhibiting the myosin heads from tightly engaging with actin. TM is a protein dimer of two polypeptides that form an α-helical coiled-coil, which lies along the groove between the two actin chains of the thin filament (**FIGURE 20.6**).

A single TM molecule extends the length of seven globular actin molecules. Each TM molecule is associated with one TN molecule. In the resting state (see Figure 20.6A), the TM molecule lies over the myosin-binding sites of the adjacent actin molecules and prevents myosin cross-bridges from interacting with actin. TN is a golf club–shaped complex of three subunits. The "handle" is troponin T (TN-T), which binds to tropomyosin. The "club" includes troponin I (TN-I), which binds to actin, and troponin C (TN-C), which binds Ca^{2+} ions. For contraction to occur, Ca^{2+} must bind to TN.

The key physiological regulator of muscle contraction is calcium. When Ca^{2+} ions bind to TN-C, they trigger conformational changes that remove TM's steric blocking of the myosin-binding sites on actin (see Figure 20.6B). Once interaction between actin and myosin is possible, the primed myosin cross-bridges are permitted to go through repeated cross-bridge cycles as long as Ca^{2+} is present in the cytoplasm. The muscle will therefore contract only when Ca^{2+} ions are available to bind TN. In relaxed skeletal muscle fibers, the

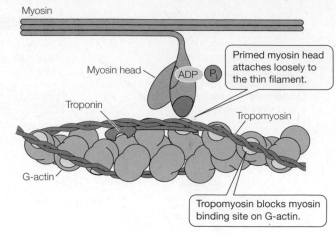

(A) A muscle cell is relaxed when no Ca^{2+} ions are present in the cytoplasm

Myosin

Myosin head

ADP P_i

Primed myosin head attaches loosely to the thin filament.

Troponin

Tropomyosin

G-actin

Tropomyosin blocks myosin binding site on G-actin.

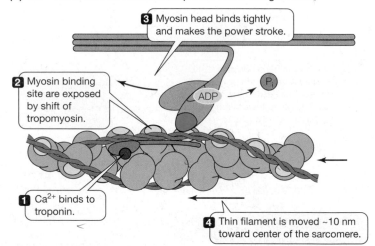

(B) Ca^{2+} ions released from the SR permit cross-bridge action

❸ Myosin head binds tightly and makes the power stroke.

❷ Myosin binding site are exposed by shift of tropomyosin.

ADP

P_i

❶ Ca^{2+} binds to troponin.

❹ Thin filament is moved ~10 nm toward center of the sarcomere.

FIGURE 20.6 Ca^{2+} ions, troponin (TN), and tropomyosin (TM) regulate contraction   (A) When Ca^{2+} ions are scarce in the cytoplasm, the TN-I subunit binds to two adjacent actin monomers, and the TN-T subunit binds to the tropomyosin molecule. These connections hold TM in a position that covers the myosin-binding sites on actin and inhibits cross-bridge action. (B) The TN-C subunit binds to Ca^{2+} ions when they are released from the sarcoplasmic reticulum. This binding causes conformational changes that detach TN-I from actin and allow TM to roll over the actin surface. The changed position of TM, as well as allosteric changes, permits cross-bridge action.

intracellular concentration of calcium is extremely low: less than $1 \times 10^{-7}\ M$, which is below the concentration that will induce (by mass action) calcium association with troponin.

SUMMARY
Vertebrate Skeletal Muscle Cells

■ Each whole muscle consists of bundles of longitudinally arrayed muscle fibers, which in turn consist of longitudinally arranged myofibrils made up of thick (myosin) and thin (actin) myofilaments organized into sarcomeres.

- A single myofibril consists of a longitudinal series of sarcomeres. The myofibrils are aligned in register in a muscle fiber so that they give the fiber a striped, or striated, appearance. Titin, nebulin, and obscurin are giant proteins that maintain the internal structural organization of the muscle fiber.

- The contractile proteins actin and myosin polymerize in a polarized fashion to form the thin and thick filaments. During contraction, the heads of individual myosin molecules bind to sites on individual actin molecules and draw the thin filaments toward the center of each sarcomere.

- Each myosin head also functions as an ATPase to provide the energy required to power cross-bridge motion. In relaxed muscle, each cross-bridge contains ADP, P_i, and stored energy obtained from the hydrolysis of ATP. The cross-bridge is oriented in the cocked position, loosely attached to actin, and is primed for its next power stroke.

- The regulatory proteins troponin (TN) and tropomyosin (TM), located on the thin filament, inhibit myosin cross-bridges from interacting tightly with actin, except when cytoplasmic Ca^{2+} is elevated. When Ca^{2+} binds to TN-C, it triggers conformational changes that allow myosin cross-bridges to bind tightly to myosin-binding sites on actin molecules and produce a power stroke.

Excitation–Contraction Coupling

Neural excitation triggers skeletal muscle contraction. Each skeletal muscle fiber is innervated by a motor neuron at the motor end-plate (**FIGURE 20.7**; see also Figure 13.8). An action potential conducted to the axon terminal of the motor neuron releases acetylcholine (ACh), which binds to postsynaptic ACh receptors in the end-plate and causes permeability changes that depolarize the muscle fiber membrane (sarcolemma) and produce an action potential. Depolarization of the muscle fiber is referred to as *excitation*. This excitation leads to rapid activation of the contractile machinery of the muscle fiber. The relationship between depolarization and contraction is called **excitation–contraction coupling**.

Excitation and contraction are coupled by two separate but intimately associated membrane systems (see Figures 20.1 and 20.7). The first of these is a system of tubules that is continuous with the sarcolemma: the **transverse tubules**, or **t-tubules**. Each t-tubule dips into the muscle fiber at an angle perpendicular to the sarcolemma, transverse to the long axis of the muscle fiber. The t-tubule invaginations occur at regular intervals along the length of the sarcolemma. The position of invagination varies among phyletic groups, usually at the level of the Z discs (e.g., amphibian muscle) or at the junction of the A and I bands (e.g., mammalian and reptilian muscles). Because the t-tubule membrane is a continuation of the outer sarcolemma, the tubule lumen is continuous with extracellular space. When the sarcolemma is depolarized, the t-tubules conduct this excitation deep into the interior of the muscle fiber. The t-tubules come into close association with the second membrane system required for excitation–contraction coupling, the **sarcoplasmic reticulum (SR)**.

The SR is a branching lacework of tubules contained entirely within the muscle fiber. Each myofibril is enveloped in SR. The SR membrane has Ca^{2+}-ATPase active-transport pumps that maintain a low concentration of Ca^{2+} ions in the cytoplasm and a high concentration ($\sim 1 \times 10^{-3}$ M) of Ca^{2+} ions within the SR.[5]

The SR between two t-tubules is called an SR *compartment*. Each compartment of the SR forms a sleeve of branching tubules around each myofibril (see Figure 20.1A). Enlarged sacs called *terminal cisternae* (singular *cisterna*) lie next to the t-tubules. In resting muscle, Ca^{2+} is largely confined to the terminal cisternae of the SR. Once an action potential conducted along the sarcolemma depolarizes the t-tubule, Ca^{2+} ions are released from the SR into the cytoplasm. How does depolarization of the t-tubule membrane produce Ca^{2+} release from the separate membrane system of the sarcoplasmic reticulum? In skeletal muscle, the two membrane systems are linked by two kinds of membrane proteins (see Figure 20.7): the *dihydropyridine receptors* (DHPRs) of the t-tubules and the *ryanodine receptors* (RyRs) of the SR. Both of these proteins were named for the drugs that bind to them specifically. In skeletal muscle, the t-tubular DHPRs and the SR RyR Ca^{2+} channels interact directly with each other in a one-on-one fashion.

The DHPRs are voltage-sensitive calcium channels in some tissues. However, in skeletal muscle, the DHPRs do not open to permit calcium flux from the extracellular fluid into the cytoplasm. Instead, their sensitivity to voltage changes plays a significant role in excitation–contraction coupling. When the t-tubule is depolarized, the DHPRs change conformation. Because the DHPRs are intimately associated with the RyRs, their altered form causes the RyRs to open. When the RyRs open, Ca^{2+} diffuses out of the SR into the cytoplasm.[6]

Figure 20.7 illustrates a motor neuron action potential triggering excitation (top panel) and contraction (middle panel) of a muscle fiber. When the RyR channels of the SR open, Ca^{2+} ions rapidly diffuse the short distance to the adjacent myofilaments, bind to troponin, and initiate processes that allow cross-bridge action. In vertebrate skeletal muscle, sufficient Ca^{2+} diffuses to the myofibrils so that every TN–TM complex moves to allow all cross-bridges to function. Indeed, the cytoplasmic concentration of Ca^{2+} increases from $<10^{-7}$ M to $>10^{-6}$ M. When the muscle fiber action potential ends (bottom panel), the t-tubules repolarize and the RyRs close. Ca^{2+} ions no longer leave the SR, and those that left when the channels were open are returned by the action of Ca^{2+}-ATPase pumps. As the cytoplasmic Ca^{2+} concentration decreases, Ca^{2+} ions unbind from TN, TM again blocks the myosin-binding sites on actin to prevent cross-bridge action, and relaxation occurs. In some fast-contracting muscles, *parvalbumin* (a low-molecular-weight protein in the cytoplasm) binds Ca^{2+}. The action of parvalbumin in concert with the SR Ca^{2+}-ATPase pumps enhances the rate of removal of Ca^{2+} from TN and hastens

---

[5] The sarcoplasmic reticulum is homologous to the smooth endoplasmic reticulum of other cells. Its Ca^{2+}-ATPase pumps are often referred to as SERCA (SarcoEndoplasmic Reticulum Calcium transport ATPase). Ca^{2+} is stored in the SR both free and bound to the protein calsequestrin.

[6] Cardiac muscle also has both DHPR and RyR proteins (in different isoforms), but their interaction is entirely indirect. The DHPR in cardiac muscle functions as a Ca^{2+} channel. It opens in response to depolarization and lets in extracellular Ca^{2+} from the t-tubular lumen. This Ca^{2+} from outside the cell opens the RyR calcium channel of the SR by a process called Ca^{2+}-induced Ca^{2+} release. The indirect coupling seen in cardiac muscle is probably more primitive and was replaced in skeletal muscle by evolution of the faster direct coupling of the two proteins.

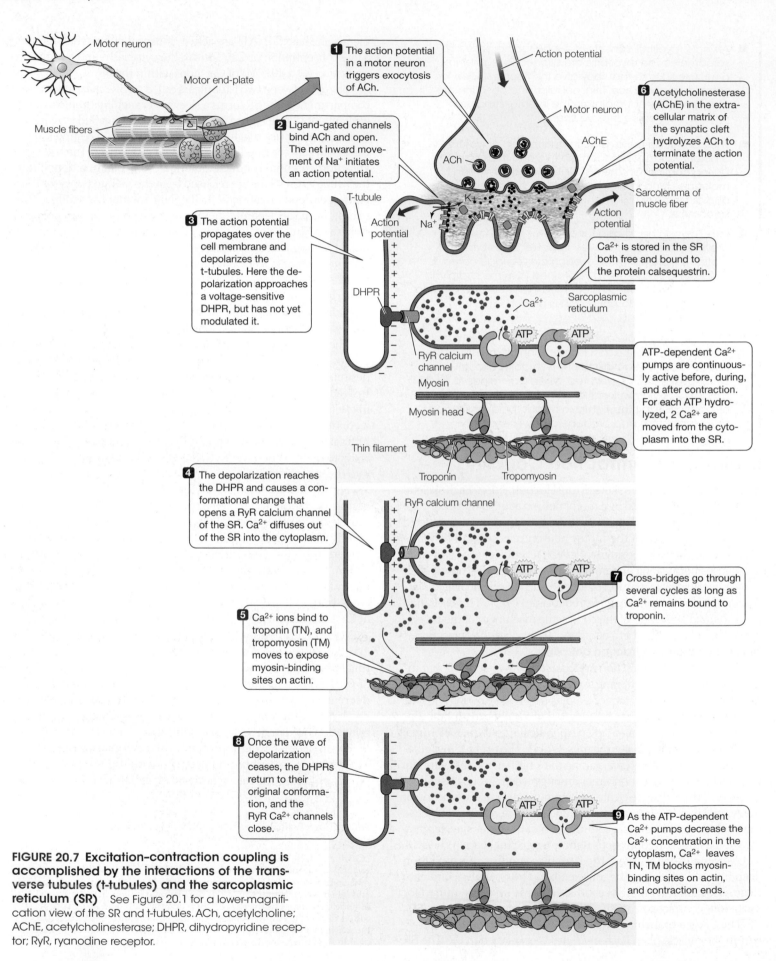

Motor neuron

Motor end-plate

Muscle fibers

1 The action potential in a motor neuron triggers exocytosis of ACh.

Action potential

Motor neuron

AChE

6 Acetylcholinesterase (AChE) in the extra-cellular matrix of the synaptic cleft hydrolyzes ACh to terminate the action potential.

2 Ligand-gated channels bind ACh and open. The net inward movement of Na⁺ initiates an action potential.

ACh

Sarcolemma of muscle fiber

T-tubule

K⁺

Action potential

Na⁺

Action potential

3 The action potential propagates over the cell membrane and depolarizes the t-tubules. Here the depolarization approaches a voltage-sensitive DHPR, but has not yet modulated it.

Ca^{2+} is stored in the SR both free and bound to the protein calsequestrin.

DHPR

Ca^{2+}

Sarcoplasmic reticulum

RyR calcium channel

ATP   ATP

Myosin

Myosin head

Thin filament

Troponin   Tropomyosin

ATP-dependent Ca^{2+} pumps are continuously active before, during, and after contraction. For each ATP hydrolyzed, 2 Ca^{2+} are moved from the cytoplasm into the SR.

4 The depolarization reaches the DHPR and causes a conformational change that opens a RyR calcium channel of the SR. Ca^{2+} diffuses out of the SR into the cytoplasm.

RyR calcium channel

ATP   ATP

7 Cross-bridges go through several cycles as long as Ca^{2+} remains bound to troponin.

5 Ca^{2+} ions bind to troponin (TN), and tropomyosin (TM) moves to expose myosin-binding sites on actin.

8 Once the wave of depolarization ceases, the DHPRs return to their original conformation, and the RyR Ca^{2+} channels close.

ATP   ATP

9 As the ATP-dependent Ca^{2+} pumps decrease the Ca^{2+} concentration in the cytoplasm, Ca^{2+} leaves TN, TM blocks myosin-binding sites on actin, and contraction ends.

FIGURE 20.7 Excitation–contraction coupling is accomplished by the interactions of the transverse tubules (t-tubules) and the sarcoplasmic reticulum (SR) See Figure 20.1 for a lower-magnification view of the SR and t-tubules. ACh, acetylcholine; AChE, acetylcholinesterase; DHPR, dihydropyridine receptor; RyR, ryanodine receptor.

relaxation. Quicker relaxation ensures sooner readiness for the next contraction. The Ca^{2+} ions bound to parvalbumin later unbind and are transported back into the SR.

SUMMARY
Excitation–Contraction Coupling

■ The sarcoplasmic reticulum (SR) sequesters Ca^{2+} ions to keep the cytoplasmic concentration of Ca^{2+} low. The terminal cisternae of the SR possess RyR calcium channels. Transverse tubules include voltage-sensitive DHPRs that come into intimate contact with the RyRs of the SR.

■ Each skeletal muscle contraction is initiated by an action potential in a motor neuron that releases acetylcholine, which in turn gives rise to a muscle fiber action potential.

■ The action potential propagates over the cell membrane of the muscle fiber and depolarizes the DHPRs in the t-tubules. The DHPRs cause the RyR calcium channels to open and allow Ca^{2+} ions to diffuse out of the terminal cisternae of the SR into the cytoplasm.

■ Ca^{2+} ions bind to TN and cause conformational changes of TN and TM that expose the myosin-binding sites of adjacent actin molecules. Previously primed myosin heads bind to the actin sites. Repeated cross-bridge cycles continue as long as sufficient Ca^{2+} is present. The cross-bridges move the thick and thin filaments relative to each other, pulling the thin filaments toward the center of the sarcomere.

■ Once the muscle fiber action potential is over, the RyR channels close. The Ca^{2+}-ATPase pumps of the SR sequester Ca^{2+} back into the SR. As the Ca^{2+} concentration in the cytoplasm decreases, Ca^{2+} dissociates from TN, and the TN–TM complex again prevents actin–myosin interactions. The muscle relaxes. Parvalbumin (prevalent in fast muscles) also binds cytoplasmic Ca^{2+} and thereby enhances the rate of relaxation.

Whole Skeletal Muscles

Many skeletal muscles in vertebrates work in **antagonistic pairs** arranged around joints. When one muscle shortens, its antagonist lengthens; and vice versa. Muscles generate force only by contraction; they lengthen passively. The antagonistic arrangement ensures relengthening of the member of a pair that shortened during contraction. For example, the quadriceps muscles on the front of the thigh and the hamstring muscles on the back of the thigh work together as an antagonistic pair of muscles. The hamstring muscles shorten to bend the knee joint. The quadriceps muscles shorten to straighten the knee joint. Often muscles work in combination with connective tissues that store elastic energy. For example, grazing animals such as camels and cows use muscles to pull their heads down to feed. Lowering the head stretches a ligament that attaches to the back of the head at one end and to the vertebral column at the other end. The stretched ligament stores energy like that in a stretched spring. This energy is expended as the ligament springs back to its original length, helping the muscles that lift the animal's head.

Muscle contraction is the force generated by a muscle during cross-bridge activity

Although the term *contraction* suggests that the muscle shortens during cross-bridge activity, this is not always the case. For example, you can "tighten up" your biceps without allowing your elbow joint to flex. Even though cross-bridge cycling occurs, the bones do not move, and the whole muscle stays the same length. This type of contraction is called **isometric** ("same length") **contraction**. The sarcomeres in the muscles shorten slightly during isometric contraction (the biceps "bulges") because they pull on elastic elements within the muscle. Elastic structures include not only the connective tissue surrounding the muscle fibers, which continue into tendons, but also components of the myofibrils such as titin and the cross-bridge links themselves. **FIGURE 20.8** illustrates the relationship of contractile and elastic components in a muscle.

Whole muscles can indeed shorten. For example, when your hand brings a heavy book toward your face, the biceps muscle shortens to decrease the angle at your elbow. This type of contraction, in which the whole muscle shortens, is called **concentric contraction** because the muscular action brings the hand closer to the center of the body. Cross-bridges can also be active when the muscle is lengthening. For example, if you hold a 10-pound weight in your hand with your elbow bent, and slowly extend your arm, the sarcomeres of your biceps are lengthening at the same time that the cross-bridges are generating force. Similarly, when you go through the motion of sitting down, or hike down an incline, the quadriceps muscles on the top of your thighs are actively contracting, but the muscles are actually longer than they are when the knee is not bent. The contractile activity in these cases is resisting stretch imposed by an external force. These contractions are called **lengthening**, or **eccentric**, **contractions**. Lengthening contractions are thought to produce minor damage to muscle fibers that lead to delayed soreness following exercise.

Concentric and eccentric contractions are both examples of **isotonic** ("same tension") **contractions** (see Figure 20.8B), which we explain below. Muscles hardly ever produce pure isotonic contractions. Most muscle activity involves dynamic combinations of both isometric and isotonic contractions. Physiologists separate these types of contractions experimentally in order to study particular properties of muscles.

A muscle exerts its force on a **load**. For example, when you lift an object with your hand, the load on which the biceps muscle exerts force includes the mass of the lower arm plus the mass of the object. The force of the muscle is opposed by the force of the load. If the force developed by a muscle is greater than the force exerted on it by a load, the muscle will change length.

Once the muscle begins to change length, the force it produces is constant and equal to the force of the load. Experimenters recording *isotonic* contractions measure changes in *length* of the muscle. If the force exerted by the load is greater than the muscle force (e.g., an extremely heavy weight), the muscle will produce an isometric contraction. Experimenters recording *isometric* contractions measure the *tension* developed by the muscle. **Tension** is the force exerted on a load by a unit of cross-sectional area of muscle.[7]

---

[7] Physiologists often use the terms *muscle tension* and *muscle force* interchangeably. Tension has the units of force/cross-sectional area. We know that cross-bridge action at the level of the sarcomere underlies the action produced by a whole muscle. The tension generated by a muscle fiber is directly proportional to the number of attached cross-bridges between the thick and thin filaments.

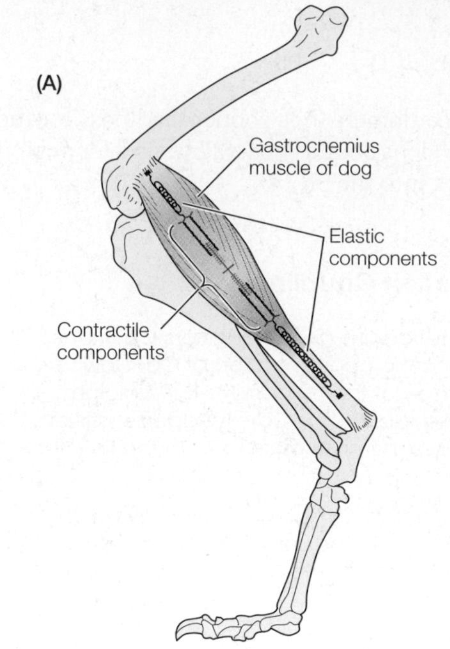

FIGURE 20.8 Contractile and elastic components interact during contraction The cross-bridge action within sarcomeres pulls on immediately adjacent sarcomeres and also on elastic elements within the muscle. (A) A single schematic sarcomere represents contractile components that are associated with intracellular and extracellular elastic elements. (B) Schematic representations show contractile and elastic elements at rest (when cross-bridges are not active) and during contractions (when cross-bridges cycle).

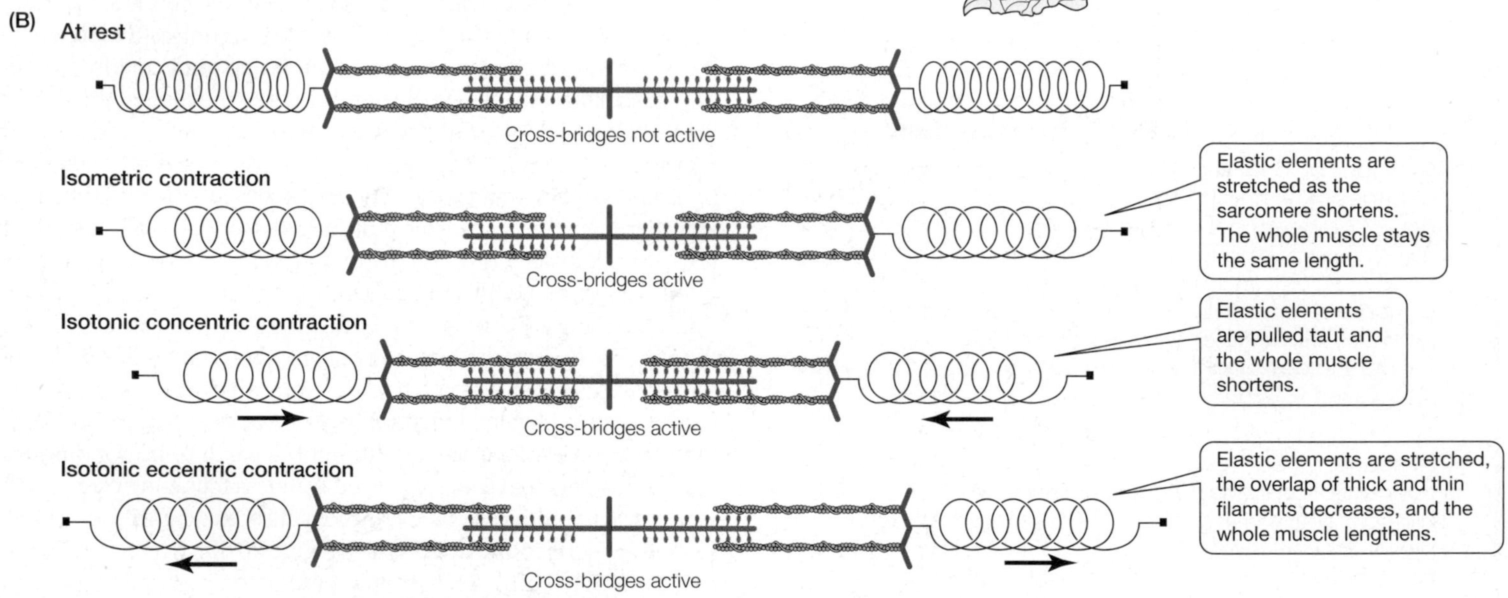

To record isometric contractions, experimenters usually attach the muscle to a very stiff force transducer that measures tension (force/cross-sectional area) while permitting only minuscule changes in length.

A twitch is the mechanical response of a muscle to a single stimulus

FIGURE 20.9 shows the twitch response of the same mammalian muscle recorded under isometric and isotonic conditions. Both twitches have three phases: a latent period, a contraction phase, and a relaxation phase. The isometric twitch has a brief latent period before any contractile tension is recorded (see Figure 20.9A); this latent period largely reflects the time required for excitation–contraction coupling to occur. The isotonic twitch has a longer latent period (see Figure 20.9B). Before the muscle can lift the load and shorten, excitation–contraction coupling must occur, and the cross-bridges must develop enough tension isometrically to overcome the force exerted by the load. If the load were greater, the latency would be longer because additional time would be required to develop tension to equal the heavier load.

The velocity of shortening decreases as the load increases

Isotonic recordings are ideal for revealing that the load directly influences the velocity (speed) at which a muscle shortens. You know from experience that you can lift a pencil faster than an unabridged dictionary, and you cannot lift your car at all. The fact that velocity of shortening decreases progressively with increasing loads is referred to as the **load–velocity relationship**. Current models suggest that greater loads somehow decrease the rate at which the myosin heads detach from actin, and therefore slow the speed of shortening. This relationship is also referred to as the **force–velocity relationship**. The words *load* and *force* can be used interchangeably because—in isotonic contractions—the force produced by the muscle equals the force of the load. Indeed, the force generated by the muscle decreases with velocity of shortening because there is decreased probability of cross-bridge action as a sarcomere's speed starts from zero to reach a finite value. The load–velocity (force–velocity) relationship is considered further in Chapter 21 (see page 567) and is illustrated in Figure 21.1, which also shows muscle *power* (power = force × velocity).

(A) Isometric recording (same length) measures change in tension

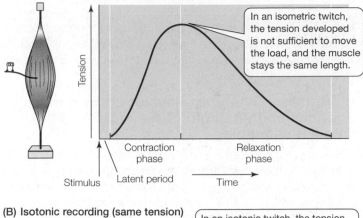

In an isometric twitch, the tension developed is not sufficient to move the load, and the muscle stays the same length.

Tension

Stimulus   Latent period   Contraction phase   Relaxation phase   Time

(B) Isotonic recording (same tension) measures change in length

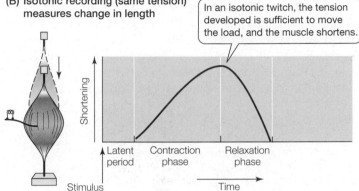

In an isotonic twitch, the tension developed is sufficient to move the load, and the muscle shortens.

Shortening

Stimulus   Latent period   Contraction phase   Relaxation phase   Time

FIGURE 20.9 Recordings reveal differences between isometric and isotonic contractions (twitches) (A) In the isometric experimental arrangement, the muscle contracts but is not allowed to move a load. (B) In the isotonic experimental arrangement, the muscle shortens and moves the load, once its contractile activity generates tension (force) that equals the force of the load.

successive twitches produced add to each other, so the overall response is greater than the twitch response to a single stimulus. Such addition is called **summation**. Summation can be recorded either isometrically or isotonically. **FIGURE 20.10** shows isometric records of summation. The electrical events triggering the contractions (the action potentials of the motor neurons and muscle fibers) are all-or-none and do not sum. However, because the action potentials last only 1 to 2 ms, and a muscle twitch lasts many milliseconds, the muscle can generate a second (or even multiple) action potential(s) before the end of the twitch produced by the first action potential.

The amplitude of the summed contractions depends on the interval of time between stimuli. Low frequencies of stimulation produce contractions that sum but are not fused. Higher-frequency stimulation produces a fused contraction called **tetanus**—the maximum contractile response the muscle can achieve. In mammalian muscle, the amplitude of the tetanus is usually three or four times the amplitude of a single twitch. In amphibian muscle, the tetanic response can exceed ten times the amplitude of a single twitch.

The frequency of action potentials determines the tension developed by a muscle

Depending on the muscle, twitches can vary from tens to hundreds of milliseconds (ms) in duration—much longer than the duration of the skeletal muscle action potential, which is about 2 ms. When a muscle is stimulated more than once within a brief period, the

A sustained high calcium concentration in the cytoplasm permits summation and tetanus

As we noted earlier, each action potential triggers the release of a sufficient number of Ca^{2+} ions from the SR to allow every TN–TM complex to move away from the myosin-binding sites on the actin thin filaments. Thus every cross-bridge in every stimulated skeletal muscle fiber is capable of interacting with actin and pulling the thin

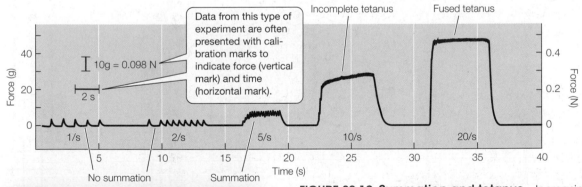

Data from this type of experiment are often presented with calibration marks to indicate force (vertical mark) and time (horizontal mark).

$10g = 0.098\ N$

2 s

1/s   2/s   5/s   10/s   20/s

Incomplete tetanus   Fused tetanus

No summation   Summation

Time (s)

FIGURE 20.10 Summation and tetanus Increasing the frequency of stimulation produces summation of twitches up to a maximum contractile response called *fused tetanus*. In this example, short trains of stimuli were applied to the sciatic nerve that innervates the gastrocnemius muscle of a frog. The muscle was allowed to rest briefly between trains of stimuli that were applied at different frequencies. Experimenters use known weights to calibrate the recording apparatus. Because weight is the magnitude of the force of gravity on an object, it is expressed as the product of the mass of the object (in kilograms, kg) times the strength of the gravitational field (9.8 N/kg). Therefore tension produced by the muscle is accurately expressed in units of newtons (N). (Published values of muscle tension are often expressed in units of grams instead of newtons.) *Inset*: Tetanus toxin prevents the exocytosis of neurotransmitter from inhibitory interneurons onto α motor neurons in the spinal cord. The motor neurons lack inhibitory synaptic input to balance excitatory inputs, and they become hyperexcitable. In severe cases, essentially all motor neurons generate action potentials and cause tetanic contractions of the skeletal muscles. (Painting by Sir Charles Bell, 1809.)

filaments toward the center of the sarcomere. If all cross-bridges are fully engaged, how is it possible to produce a tetanic force several times greater than the response to a single stimulus?

The answer is that the contractile apparatus requires *time* to pull on the various elastic components of the muscle. The elastic components include the connective tissue associated with the muscle fibers and the elastic components of the myofibrillar apparatus. These all lie in series with each other. Thus the elastic structures are referred to as *series elastic elements*. For maximum tension to be recorded at the ends of the muscle fiber, the elastic elements must be stretched taut.

During a single twitch produced by a single action potential, the Ca^{2+} released into the cytoplasm is pumped back into the SR before the cross-bridges can fully stretch out the elastic elements.[8] Successive action potentials, however, open the RyR channels with sufficient frequency to keep the cytoplasmic concentration of Ca^{2+}

---

[8] The condition of the muscle fiber during the time Ca^{2+} ions are available to permit cross-bridge action is often referred to as the *active state*.

high enough so that the actin-binding sites for myosin remain exposed over time. Thus cross-bridges can cycle repeatedly until the elastic elements are stretched taut and the full contractile potential of the muscle fiber is realized.

The amount of tension developed by a muscle depends on the length of the muscle at the time it is stimulated

Whole skeletal muscles, because of their attachments to bones (or to exoskeleton in invertebrates), do not change greatly in length. Nevertheless, muscles develop the most tension if they start contracting at an ideal initial length. Isometric recordings from isolated whole muscles illustrate this idea. **FIGURE 20.11A** shows the tension produced by a muscle when it was set at several different lengths prior to stimulation. Maximum tension was achieved when the muscle was set at lengths near its normal relaxed length in the animal. When the muscle was set at shorter lengths or stretched to longer lengths, the development of tension dropped off.

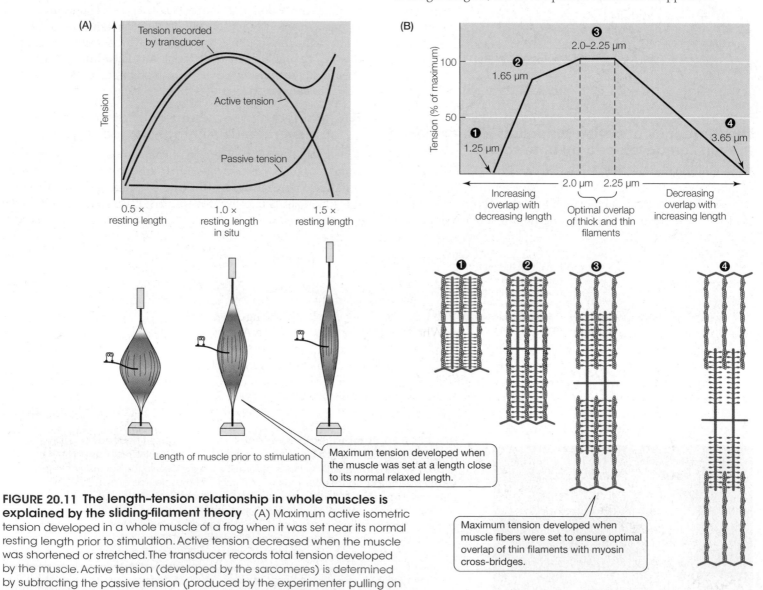

FIGURE 20.11 The length–tension relationship in whole muscles is explained by the sliding-filament theory (A) Maximum active isometric tension developed in a whole muscle of a frog when it was set near its normal resting length prior to stimulation. Active tension decreased when the muscle was shortened or stretched. The transducer records total tension developed by the muscle. Active tension (developed by the sarcomeres) is determined by subtracting the passive tension (produced by the experimenter pulling on the muscle to stretch it) from the total tension recorded during contraction. (B) Isometric tension recorded from single fibers is related to the sarcomere length set prior to stimulation. It is at its maximum when the actin filaments overlap the greatest number of myosin cross-bridges. (B from Gordon et al. 1966.)

This **length–tension relationship** is entirely explained by the sliding-filament model of muscle contraction. Elegant isometric recordings of tension developed by *single* frog skeletal muscle fibers unambiguously demonstrated the length–tension relationship at the level of the sarcomeres. **FIGURE 20.11B** shows the tension developed by single muscle fibers set at different lengths prior to stimulation. The set length of the muscle fiber affected the length of the sarcomeres within it and therefore the degree of overlap of the thick and thin filaments within each sarcomere. The experimenters plotted the amount of tension developed upon stimulation as a function of sarcomere length. Maximum tensions were recorded when the sarcomere lengths were set near those found in the intact animal.

The diagrams in Figure 20.11B illustrate that, at the lengths that yielded maximum tension, the overlap of thick and thin filaments permits optimal cross-bridge binding with actin (❸). Stretching or compressing the sarcomeres leads to less tension developed in response to stimulation. Sarcomeres set at longer-than-ideal lengths have less overlap of thick and thin filaments and therefore fewer available sites for myosin cross-bridges to bind (❹). At sarcomere lengths that are shorter than ideal, the thin filaments overlap each other, probably interfering with myosin cross-bridge action (❷), and ultimately push up against the Z disc (❶).

The striking agreement of the length–tension curve of single muscle fibers with the observed regions of filament overlap strongly implies that each cross-bridge contributes an independent and equal increment of tension, and provides compelling support for the sliding-filament theory of muscle contraction. Below we will see the powerful effects of sarcomeres arranged in series within myofibrils, and of myofibrils arranged in parallel.

In general, the amount of work a muscle can do depends on its volume

Work performed by a muscle can best be understood by the use of isotonic recording. **Work** is the product of force produced by the muscle and the distance that the muscle shortens. **FIGURE 20.12** shows the distance a muscle shortened when it was given increasingly greater loads to lift. The muscle shortened the greatest distance when it lifted no load. It did not shorten at all when the force of the load exceeded the maximum force it could develop. When the muscle lifted no load, although it shortened, it performed little work because it exerted negligible force. When the muscle attempted to lift a very heavy load, it exerted isometric force but performed no work because it did not move the load. At intermediate loads, the muscle did increasing amounts of work, up to about 40% of the maximum load, and then it did progressively decreasing amounts of work while lifting loads of increasing mass.

The *force* exerted by a muscle is proportional to the cross-sectional area (CSA) of its contractile elements. In all muscles examined that use actin and myosin as contractile proteins, the diameters of the thick and thin filaments are essentially the same. This means that a cross section through the contractile components of any muscle would reveal the same number of cross-bridges per unit of area. Because of this constant number of cross-bridges, most vertebrate skeletal muscle fibers (and many invertebrate muscle fibers) exert about the same amount of force per unit of area. A muscle that has a greater total CSA would therefore be able to produce a greater total force than a thinner muscle, because of the additive effect of myofibrils in parallel. Investigators refine this concept by taking into account

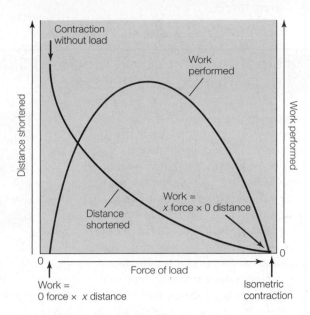

FIGURE 20.12 Work of contraction Isotonic recordings show that the muscle shortens the greatest distance when it lifts no load, and shortens progressively shorter distances with increasing loads. Multiplying the force developed (equal to the force of the load, in newtons, N) by the distance shortened for each load (in meters, m) gives a curve that represents work performed by the muscle (in joules, J). (After Schmidt-Nielsen 1997.)

the orientation of the muscle fibers within a given muscle. In some muscles, the individual muscle fibers are arranged in parallel and extend the full length of the muscle. However, other muscles may include shorter muscle fibers arranged at an angle to the long axis of the muscle. This arrangement adds complexity to understanding the amount of force generated by a particular type of muscle.

The *length* of a muscle fiber does not contribute to the force it generates. However, the length is important in determining how much work the muscle can do. The sarcomeres in most vertebrates are about 2.5 μm long. Thus, if each sarcomere contracts, for example, 10% of its length during a twitch, a muscle that has myofibrils consisting of 100 sarcomeres in series will shorten 25 μm. A muscle that has myofibrils consisting of 300 sarcomeres in series would shorten three times that distance. Assuming that both muscles had the same CSA, they would both exert the same force. However, the longer muscle would perform more work because work is the product of force times the distance shortened. A muscle that was both greater in CSA and longer would produce even more work because it would both exert greater force and shorten a greater distance. Thus muscles that have a greater volume of contractile machinery are generally capable of doing greater work.

The length of a muscle fiber also affects the velocity of shortening. Because the sarcomeres are arranged in series, velocities—like length changes—are also additive. For example, if each sarcomere shortened at a speed of 20 μm/s, then a 100-sarcomere myofibril would shorten at (20 μm/s) × 100 = 2 mm/s. A 300-sarcomere myofibril would shorten at a speed of 6 mm/s. Therefore, assuming sarcomeres of equivalent lengths, the longer a muscle fiber, the greater its velocity of shortening.

Interestingly, the muscles of some animals have been so drastically modified that they possess hardly any contractile machinery at all. For example, the electric organs of some fish do not do contractile work but instead produce electric shocks (**BOX 20.1**).

BOX 20.1 Electric Fish Exploit Modified Skeletal Muscles to Generate Electric Shocks

In addition to using skeletal muscles for locomotion, electric fish have incorporated highly modified skeletal muscle cells—called *electrocytes*—into **electric organs** (**EOs**), which they use for stunning prey, exploring the environment, and communicating. Researchers recently investigated the specific effects on prey of electric discharges produced by the electric eel *Electrophorus electricus* from freshwater rivers of South America (Figure A). They first showed that high-voltage electric discharges produced by the eel's EO specifically stimulated the prey's efferent motor neurons, causing them to release acetylcholine (ACh) at skeletal muscle end-plates. The researchers then demonstrated how the electric eels time the discharges so that they either immobilize free-swimming fish or reveal the locations of hidden fish. When an eel targets free-swimming prey, it emits high-frequency discharges (~400 Hz) that remotely stimulate the prey's motor neurons at such a high frequency that they cause the skeletal muscles to undergo tetanus and temporarily immobilize the prey. During this short period of immobility, the electric eel captures the prey. To flush out hidden prey, the electric eel emits two or three high-voltage discharges that stimulate the prey's motor neurons to produce full-body twitches that give away the animal's hiding spot. Within 20 to 40 ms of having revealed the prey's location—before the prey can escape—the eel attacks it with a high-frequency discharge and immobilizes it for capture.

Some strongly electric fish were known in ancient times. Aristotle described how the Mediterranean torpedo, which can deliver shocks of 50 to 60 V and several amperes of current, hides in wait to stun and then consume its prey. Francesco Redi (in 1671) and Stephano Lorenzini (in 1678) dissected torpedoes and con-

FIGURE A The electric eel *Electrophorus electricus*

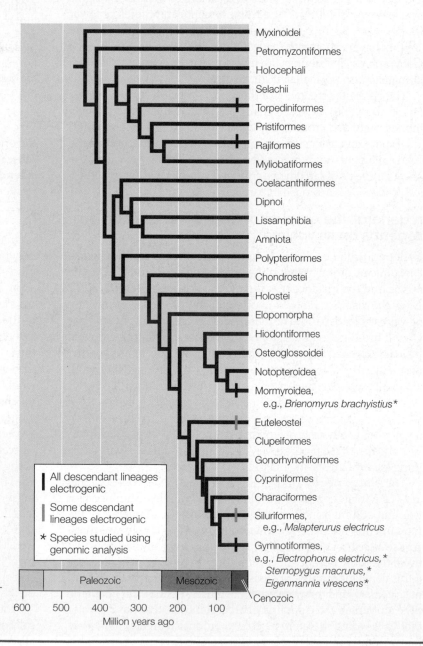

Myxinoidei
Petromyzontiformes
Holocephali
Selachii
Torpediniformes
Pristiformes
Rajiformes
Myliobatiformes
Coelacanthiformes
Dipnoi
Lissamphibia
Amniota
Polypteriformes
Chondrostei
Holostei
Elopomorpha
Hiodontiformes
Osteoglossoidei
Notopteroidea
Mormyroidea,
    e.g., *Brienomyrus brachyistius**
Euteleostei
Clupeiformes
Gonorhynchiformes
Cypriniformes
Characiformes
Siluriformes,
    e.g., *Malapterurus electricus*
Gymnotiformes,
e.g., *Electrophorus electricus,**
    *Sternopygus macrurus,**
    *Eigenmannia virescens**

| All descendant lineages electrogenic
| Some descendant lineages electrogenic
* Species studied using genomic analysis

Paleozoic   Mesozoic   Cenozoic
600   500   400   300   200   100
Million years ago

FIGURE B Phylogeny of electric fishes This phylogenetic tree shows that electric organs evolved independently in six different orders of vertebrates. (After Gallant et al. 2014.)

cluded that the EO was derived from muscle. Nearly 300 years later, in the 1950s, fish with EOs that generate only weak electrical pulses were discovered in Africa and South America.

Combined observations of EOs in many unrelated fish indicate that EOs have evolved independently at least six different times. Recently, researchers have studied the genomic basis of the convergent evolution of EOs. **Figure B** illustrates the phylogeny of known groups of electric fish and identifies those used in a study of gene expression in EO tissues. The cellular morphology of EOs varied greatly among the animals studied (marked with asterisks), yet these divergent animals showed striking convergence in patterns of upregulation and downregulation of genes and transcription factors expressed to produce the physiological features of EOs.

Whereas genes involved in differentiation of muscle structures and functions are downregulated in all EOs, those involved in EO function are upregulated. For example, the genes for certain ion channels and transporters are highly expressed, which would contribute to increased excitability of the electrocytes. Certain collagen genes are also upregulated, presumably to produce the insulating connective tissue that prevents dissipation of the current produced by the electrocytes. Finally, the electric fish from divergent taxa all upregulated genes for insulin-like growth factors (IGFs), which turn on cellular mechanisms that contribute to large cell size. Thus, despite the widely varied morphologies of EO tissues in divergent fish, selection pressures appear to have exploited convergent use of gene expression patterns that yield the features of functional EOs.

In the process of studying the physiology of EOs and learning how electric fish use them in their native habitats, investigators have also made discoveries about convergent evolution, animal behavior, and cellular differentiation. Furthermore, because EOs yield abundant amounts of nearly pure excitable membranes, they provide tissues for ever-more-sophisticated studies of channels and membrane receptors. Finally, the large quantities of AChE in many EOs provide a plentiful source of enzyme for studies of anticholinesterases, which are of interest both to environmentalists (who find synthetic anticholinesterases in toxic wastes) and investigators seeking ways to prolong ACh signals at specific synapses. The structure and diverse functions of EOs are explored in **Box Extension 20.1**.

SUMMARY
Whole Skeletal Muscles

- Cross-bridge activity within individual muscle fibers accounts for the force generated by a muscle. Force exerted by a muscle is proportional to the cross-sectional area of its contractile elements.

- The tension (force per cross-sectional area) generated by a whole muscle is directly related to the number of actively contracting muscle fibers.

- The amount of tension developed by each contracting fiber in a muscle is determined by the frequency of action potentials from its motor neuron (to produce summation of twitches and tetanus) and the length of the muscle fiber at the time it is stimulated (the length–tension relationship).

- The speed with which a muscle shortens decreases as the load it lifts increases (the load–velocity relationship).

- Work performed by a muscle is the product of force produced by the muscle and the distance it shortens.

Muscle Energetics

In this section we examine the sources of energy available to muscle fibers and the ways in which energy is used by different types of muscle fibers.

ATP is the immediate source of energy for powering muscle contraction

ATP performs at least three functions in the contraction–relaxation cycle:

1. ATP binding to the cross-bridge (but not hydrolysis) is necessary for detachment of myosin from actin.

2. Hydrolysis of ATP primes (cocks) the myosin cross-bridge in preparation for binding to actin and undergoing a power stroke.

3. Energy from the hydrolysis of ATP drives the calcium pump that transports Ca^{2+} ions into the sarcoplasmic reticulum.

Interestingly, muscle contains only enough ATP (2–4 mM) to sustain contraction for a few seconds. Thus nearly all forms of muscular work require the production of ATP while the muscle is active. The rate of work produced depends on the rate at which ATP is provided to the contractile apparatus.

In broad outline, vertebrate muscle fibers possess three biochemical mechanisms that produce ATP (see Chapter 7 for a detailed discussion of ATP resupply):

❶ *Use of the phosphagen creatine phosphate.* Phosphagens temporarily store high-energy phosphate bonds. The high-energy phosphate of creatine phosphate can be donated to ADP to produce ATP, as shown in Figure 8.7. Creatine phosphate is produced in resting muscle from creatine and ATP. The formation of ATP from creatine phosphate is driven by simple mass action. Whereas creatine phosphate is the phosphagen found in muscles of vertebrates, it and other phosphagens, such as arginine phosphate, are found in invertebrates (see Chapter 8, pages 198–199).

❷ *Anaerobic glycolysis.* This form of catabolism requires no oxygen (see Figure 8.5). It must have glucose or glycogen

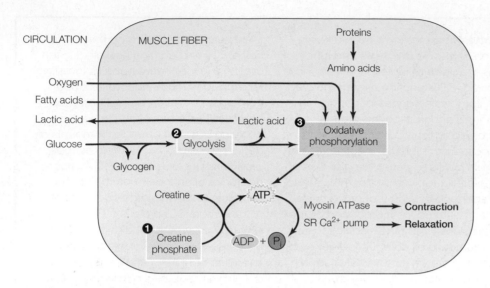

FIGURE 20.13 Three metabolic pathways supply the ATP for muscle contractile activity: ❶ Transfer of high-energy phosphate from creatine phosphate to ADP; ❷ anaerobic glycolysis; and ❸ aerobic catabolism involving oxidative phosphorylation.

as fuel. In addition to ATP, it produces *lactic acid*, which in vertebrates is always retained in the body and disposed of metabolically.

❸ *Aerobic catabolism.* This form of catabolism requires oxygen and can use all three major classes of foodstuff as fuel (see Figures 8.1–8.3). It produces ATP principally by *oxidative phosphorylation*. Its other major products are CO_2 and H_2O.

FIGURE 20.13 illustrates major elements in the production and use of ATP in a vertebrate muscle fiber. The three mechanisms of ATP production differ greatly in how fast they can make ATP when operating at peak output, how much ATP they can make, and how quickly they can accelerate their rate of ATP production. **TABLE 20.1** summarizes these attributes (see Table 8.1 for more details).

If a resting muscle is called upon suddenly to engage in all-out effort, creatine phosphate supplies much of the ATP in the first seconds because phosphagen-based ATP synthesis can be accelerated very rapidly. The rate of ATP supply is exceedingly high because of the intrinsic properties of the phosphagen mechanism. But because

the muscle soon runs out of creatine phosphate, this high rate of ATP supply is short-lived.

With continued contractile activity, anaerobic glycolysis takes over as the principal mechanism of ATP synthesis. Because the peak rate of ATP synthesis by anaerobic glycolysis is lower than that using creatine phosphate, the rate of ATP supply to the contractile apparatus decreases (although it is still very high). Anaerobic glycolysis can make somewhat more total ATP than phosphagen, but it, too, exhausts its ability to make ATP if a muscle stays in a state of all-out exertion.

At that point, aerobic catabolism becomes the sole source of ATP. The rate of ATP supply falls still further because aerobic catabolism exhibits the lowest rate of ATP synthesis. But the aerobic mechanism can make ATP on a sustained basis. These transitions in the biochemistry of ATP synthesis are the reason that the rate of work by a muscle declines with time during all-out exercise. Figure 8.12 illustrates this concept.

Vertebrate muscle fibers vary in their use of ATP

Tonic muscle fibers are relatively rare. They are found mainly in postural muscles of lower vertebrates.[9] Tonic muscle fibers do not generate action potentials, but they do undergo changes in membrane potential. They contract more slowly than any other types of vertebrate muscle fibers, and their slow cross-bridge cycling permits long-lasting contractions with low energetic costs. The most common types of muscle fibers are **twitch fibers**. These fibers generate action potentials, and each action potential gives rise to a muscle twitch (see Figure 20.9). The biochemical and metabolic features of twitch fibers vary, which gives them different contractile capabilities.

Twitch fibers are generally classified into three main categories based on differences in isoforms of the myosin ATPase and metabolic features of the cells: *slow oxidative (SO)*, *fast oxidative glycolytic (FOG)*, and *fast glycolytic (FG)* fibers (**TABLE 20.2**). In mammals, the myosin ATPases in FG and FOG fibers hydrolyze about 600 ATP molecules per second, whereas those in SO fibers split ATP about half as rapidly. Because the rate of ATP hydrolysis governs the rate of cross-bridge cycling, higher ATPase activity allows faster contraction.

It is important to remember that the amount of tension developed *per cross-bridge cycle* is the same in both fast and slow types of muscle, but the number of cycles accomplished per unit of time differs. Earlier we saw that the velocity of contraction of a muscle fiber depends on the load being moved (the *load–velocity*, or *force–velocity, relationship*).

	Use of phosphagen	Anaerobic glycolysis	Aerobic catabolism
TABLE 20.1 Characteristics of the three principal mechanisms of ATP regeneration in vertebrate muscle			
Peak rate of ATP synthesis	Very high	High	Moderate
Total possible yield of ATP in one episode of use	Small	Moderate	High (maintained indefinitely)
Rate of acceleration of ATP production	Fast	Fast	Slow

Note: See Table 8.1 for more detail.

[9] In mammals, tonic fibers occur only as intrafusal fibers of muscle spindles and in extraocular muscles. In many mammals, the extraocular muscles, which control complex motions of the eyes, also contain extremely fast-contracting muscle fibers.

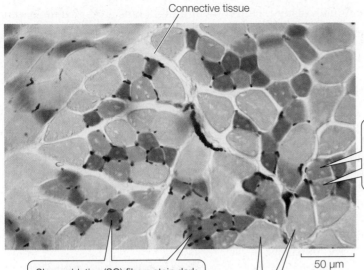

Connective tissue

FIGURE 20.14 Whole muscles often consist of mixtures of different types of fibers

Fast oxidative glycolytic (FOG) fibers have abundant mitochondria and are intermediate in diameter and abundance of capillaries.

50 μm

Slow oxidative (SO) fibers stain dark because they contain abundant mitochondria. They have small diameters and are surrounded by many capillaries (stained black).

Fast glycolytic (FG) fibers have few mitochondria, large diameters, and few adjacent capillaries.

This relationship applies to all muscle fibers but varies depending on the type. Therefore the velocity of contraction depends on *both* the type of muscle fiber and the load against which it exerts force.

The SO fibers are mitochondria-rich and poised to make ATP principally by aerobic catabolism; they have small diameters and are rich in myoglobin (an intracellular hemoglobin-like molecule that facilitates O_2 transport), red (because of the myoglobin), well supplied with blood capillaries, and slow to fatigue. The FG fibers have large diameters, are invested with fewer capillaries, have little myoglobin, and are *white*. They have few mitochondria and make ATP mainly by anaerobic glycolysis; they are rich in glycogen (the principal fuel of anaerobic glycolysis), quickly accumulate lactic acid, and fatigue rapidly. The FOG fibers are intermediate. Although the FOG isoform of myosin is different from that of FG fibers, the activity of myosin ATPase in FOG fibers is high, and they are therefore capable of rapid tension development. Unlike FG fibers, however, FOG fibers are relatively rich in mitochondria, and because they make ATP aerobically, they are relatively resistant to fatigue.

Whole muscles often include mixtures of different types of muscle fibers, and their performance reflects their composition. The light micrograph in **FIGURE 20.14** shows an example of SO, FOG and FG fibers intermingled within a muscle. Table 20.2 provides a comparative guide to the three fiber types. The distinctions, although useful, should not be viewed too rigidly, however, because the fiber types vary considerably in other characteristics. For example, fibers of a given type may differ from each other because they have different isoforms of troponin, tropomyosin, or other proteins. Furthermore, varying

TABLE 20.2 Characteristics of mammalian skeletal twitch muscle fibers			
	Slow oxidative (SO)	**Fast oxidative glycolytic (FOG)**	**Fast glycolytic (FG)**
Myosin ATPase activity	Slow	Fast	Fast
Speed to reach peak tension	Slow	Intermediate to fast	Fast
Duration of twitches	Long	Short	Short
Rate of Ca^{2+} uptake by sarcoplasmic reticulum	Slow to intermediate	High	High
Resistance to fatigue	High	Intermediate	Low
Number of mitochondria	Many	Many	Few
Myoglobin content	High	High	Low
Color	Red	Red	White
Diameter of fiber	Small	Intermediate	Large
Number of surrounding capillaries	Many	Many	Few
Levels of glycolytic enzymes	Low	Intermediate	High
Ability to produce ATP using oxidative phosphorylation	High	High	Low
Force developed per cross-sectional area of entire fiber	Low	Intermediate	High
Function in animal	Posture	Standing, walking, rapid repetitive movements	Jumping, bursts of high-speed locomotion
Frequency of use by animal	High	Intermediate to high	Low

Note: The names of different types of skeletal muscle fibers vary in the literature. Slow oxidative fibers are also called Type I; and fast oxidative fibers, Type IIa. Different types of fast glycolytic fibers are found in mammals, IIb in small mammals and IIx in large mammals.

conditions of use can cause one type of fiber to be converted into another (see Chapter 21).

Different fiber types are specifically adapted to serve different functions, which gives muscles a broad repertoire of contractile abilities. SO fibers (and tonic fibers when present) do not generate much tension, but they operate efficiently and without fatigue. They are adapted for isometric postural functions and for small, slow movements. The myosin isoform in SO fibers has low ATPase activity and therefore can produce tension economically. FOG fibers generate more tension and faster contractions, yet they are fatigue-resistant. They are adapted for repeated movements such as locomotion. FG fibers generate rapid contractions and large increments of tension, but they lack endurance. They are used for occasional, forceful, fast movements such as leaps or bursts of speed in escape or prey capture.

The ankle extensors in the cat hindlimb illustrate the functional roles of different fiber types. Three muscles—the *soleus*, *medial gastrocnemius*, and *lateral gastrocnemius*—compose the ankle extensors. They all insert on the Achilles tendon at the heel. The soleus contracts slowly and consists entirely of SO fibers. It is most active in postural standing. The medial and lateral gastrocnemii are faster muscles of mixed fiber composition. For example, the medial gastrocnemius contains approximately 45% FG, 25% FOG, and 25% SO fibers. (The remaining 5% of fibers are intermediate in their properties between those of FG and FOG fibers.)

Because the FG fibers have relatively greater diameters (see Table 20.2), the 45% that they contribute to the muscle fibers contributes 75% of the maximum total tension of the medial gastrocnemius. However, walking and most running use only about 25% of the maximum tension of the medial gastrocnemius. This is the amount of tension produced by the FOG and SO fibers without any contribution from the FG fibers. Thus these fatigue-resistant fibers are sufficient for most locomotion. The large force contributed by FG fibers is believed to be reserved for short bursts of contraction required in motions such as jumping.

Different animals employ different types of muscles that contribute to their achieving success

In fish, the trunk muscles of the body are divided into separate regions of red slow muscle and white fast muscle. The muscle fibers in the two regions bear many histological, biochemical, and physiological similarities to mammalian fiber types. SO-like fibers are found in the red muscle and FG-like fibers in the white muscle. The slow red muscle makes up less than 10% of the total trunk muscle in most fish species, and it never exceeds 25%. Yet only the slow red muscle is used at all speeds of steady cruising. The white fast muscle that constitutes the great bulk of the muscle mass is used only for bursts of high-speed swimming, and it fatigues rapidly. The shear size of the white muscle is a testimony to the extreme importance to the fish of being able to accelerate rapidly through its dense water environment when necessary to capture food or escape a predator.

Several animals possess exceptional muscles that are adapted for very rapid contractions. Certain vertebrates possess rapidly contracting muscles that consist of fibers that are oxidative and fatigue-resistant. Hummingbird flight muscle, for example, can contract and relax at frequencies approaching 80 times/s (hertz, Hz), so the contraction–relaxation cycle is completed in less than 15 ms. Sound-producing muscles of insects, fish, birds, and bats can be even faster.[10] In all of these cases there are extreme adaptations for rapid generation of tension, and also rapid relaxation.

Experimenters have shown that three main factors contribute to increased speeds of contraction: (1) myosin isoforms capable of rapid cross-bridge cycling, (2) troponin isoforms that have a low affinity for Ca^{2+} so that Ca^{2+} unbinds rapidly, and (3) increased density of Ca^{2+}-ATPase pumps in the SR and parvalbumin in the cytoplasm for rapid relaxation. Large amounts of ATP are required to support rapid cross-bridge cycling and pump functions, and not surprisingly, these muscles require a well-developed SR, many mitochondria, and a rich supply of capillaries to deliver O_2 and nutrients.

The benefit of rapid contraction brings with it a cost of limited ability to generate tension, because space in cells is limited. In most muscles used for locomotion, about 90% of the space is filled by myofibrils; mitochondria, glycogen, and SR fill the remaining 10% of the space. Consider the tail-shaker muscle of the rattlesnake, which can produce contractions at a frequency of up to 90 Hz (at optimum temperatures). Rattlesnakes make themselves conspicuous by rattling their tails continuously, sometimes for hours. The tail-shaker muscle fibers have high metabolic demands and require reserves of fuel, abundant mitochondria, and also extensive SR.

The space required for these "supporting" components necessarily limits the space available for contractile proteins (the tension-generating components). Indeed, in rattlesnake shaker muscle fibers, only about 30% of the space is occupied by myofibrils. The remaining space is filled by SR (26%), glycogen (17%), and mitochondria (26%). The diminished contractile machinery results in less ability to generate tension. These muscle fibers illustrate a general point: that space in cells can be at a premium, and thus trade-offs may be required among various cell components. In contrast, the asynchronous flight muscles of some insects produce extremely high frequencies of robust contractions, and their fibers contain a large volume of contractile elements with relatively few mitochondria and little SR. In a dramatic departure from all other known skeletal muscles, the asynchronous flight muscles of insects have evolved an excitation–contraction mechanism in which one action potential triggers several contraction–relaxation cycles. In *Drosophila melanogaster*, for example, the motor neurons to the asynchronous flight muscles generate action potentials at a rate of about 5 Hz, but the muscles' contraction frequency is approximately 200 Hz. The wing beats of these insects result from changes in the shape of the thorax produced by opposing sets of muscles that are alternately activated by stretch. The action potentials ensure that sufficient Ca^{2+} is present in the cytoplasm to permit actin–myosin cross-bridge action. In **BOX 20.2** we compare asynchronous and synchronous flight muscles of insects.

---

[10] The sound-producing muscles of the male toadfish swim bladder are the fastest known vertebrate muscles, contracting at frequencies of up to 200 Hz. In insects, the sound-producing muscle of the shrill-chirping male cicada can contract and relax at a frequency of 550 Hz!

BOX 20.2 Insect Flight

Humans have long admired and envied the ability of other animals to fly. Insects—from lazily looping butterflies to dive-bombing mosquitoes—have captivated our attention. Insect flight muscles possess the familiar features of striated skeletal muscle fibers found in other animals. The

Honeybee (*Apis mellifera*)

myofibrils are organized into sarcomeres; t-tubules and sarcoplasmic reticulum are present; and Ca^{2+} ions bind to the TN–TM complex to permit cross-bridge cycling that produces tension. **Box Extension 20.2** describes the special features of insect flight muscles that underlie insects' aerial feats.

SUMMARY
Muscle Energetics

- Contractile activity requires the hydrolysis of ATP to provide energy for cross-bridge power strokes and to support the Ca^{2+}-ATPase pumps of the sarcoplasmic reticulum.

- ATP is produced by three principal means: (1) transfer of the high-energy phosphate from creatine phosphate to ADP, (2) glycolysis, and (3) oxidative phosphorylation.

- Vertebrate muscle fibers are classified into different types on the basis of their biochemical and metabolic features, and each type is adapted to serve different functions. Muscles usually contain a mixture of different fiber types.

- Muscles adapted for extremely rapid contractions typically produce less tension than muscles that contract at slower rates. The presence of large numbers of mitochondria and abundant SR reduces the cross-sectional area of contractile machinery, and therefore the ability to generate tension.

Neural Control of Skeletal Muscle

Whole skeletal muscles in both vertebrates and invertebrates produce smooth, fluid movements that are generated by continuous and finely controlled neural input. Unlike smooth and cardiac muscles (which may generate contractions endogenously and may respond to hormonal as well as neural control), skeletal muscles contract only when stimulated by motor neurons. Two contrasting evolutionary approaches are known to provide gradation of tension

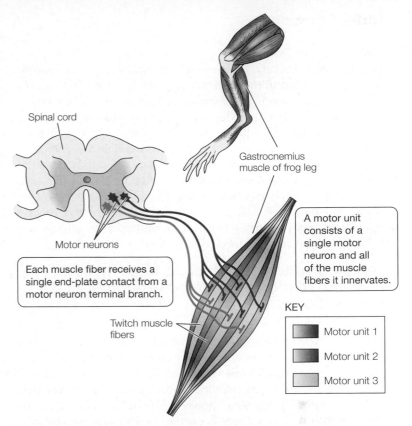

FIGURE 20.15 Vertebrate skeletal muscles consist of many different, independent motor units An action potential in a motor neuron stimulates an action potential and contraction in all of the muscle fibers it innervates. Varying the number of active motor units varies the amount of tension produced by the whole muscle.

in a muscle, one exemplified by vertebrates (the *vertebrate plan*) and the other by arthropods (the *arthropod plan*). In most of the well-studied invertebrate groups other than arthropods, muscle tension is controlled by variations on the arthropod plan.

The vertebrate plan is based on muscles organized into motor units

A vertebrate skeletal muscle is innervated typically by about 100 to 1000 motor neurons. The axon of each motor neuron branches to innervate several muscle fibers, and each muscle fiber receives synaptic input from only one motor neuron. A motor neuron and all the muscle fibers it innervates are collectively termed a **motor unit** (**FIGURE 20.15**). When the motor neuron generates an action potential, all of the muscle fibers in the motor unit generate action potentials and contract to produce a twitch. Trains of action potentials of increasing frequencies can produce summation of twitches up to fused tetanic contraction.[11] Thus, as in whole muscles, the amount of tension produced by a single motor unit can be varied by varying the frequency of action potentials generated by the motor neuron. Although the amount of tetanic tension varies in different animals, in many vertebrate muscles it is about two to five times greater than the twitch tension.

A more dramatic effect on the amount of tension developed by a whole muscle can be accomplished by varying the number

[11] In mammals, fused tetanus occurs at about 300 action potentials/s in slow-twitch, oxidative muscle fibers, and at about 100 action potentials/s in fast-twitch, glycolytic fibers.

of active motor units. Increasing the number of active motor units is called **recruitment of motor units**. Recruitment requires stimulating increasing numbers of motor neurons that innervate the muscle. For example, the tension in a muscle innervated by 100 motor neurons could be graded in 100 steps by recruitment. The amount of tension developed by the whole muscle increases as more motor units are activated (recruited). Recruitment is the dominant means used to control the amount of tension produced in vertebrate twitch muscles. Varying the number of active motor units, as well as the timing of their activation, ensures precise and smooth movements. The elastic properties of the muscle also contribute to the smoothness of movement.

The innervation of vertebrate tonic muscle is intermediate between the vertebrate and arthropod plans

Like twitch muscles, tonic muscles also are made up of motor units that consist of a single motor neuron and the muscle fibers it innervates. However, whereas each fiber of a twitch muscle has a single end-plate contact near the middle of the fiber, each muscle fiber of a tonic muscle receives many synaptic contacts distributed over its length. This pattern, shown in **FIGURE 20.16A**, is termed **multiterminal innervation**. An action potential generated by a motor neuron produces an excitatory postsynaptic potential (EPSP) at each of the distributed junctions on all fibers in the motor unit. Tonic muscle fibers have little or no ability to generate action potentials. Instead, each depolarizing EPSP spreads passively over a region of membrane and down the t-tubules in that area. Because an EPSP is produced at each of the many terminals along the entire length of the fiber, the depolarized t-tubules trigger opening of RyRs at many points, and contractile elements are activated along the entire fiber. The amount of tension generated depends directly on the amount of depolarization produced by the EPSPs.

The arthropod plan employs multiterminal and polyneuronal innervation

Although the fibers of arthropod skeletal muscles share many features of vertebrate skeletal muscle, including the organization of thick and thin filaments into sarcomeres and excitation–contraction coupling by way of t-tubules and SR, they show interesting differences in their patterns of innervation. A typical arthropod whole muscle is innervated by one to ten motor neurons, in contrast to the hundreds or thousands of motor neurons that innervate a whole vertebrate muscle. Most individual arthropod muscle *fibers* are innervated by more than one motor neuron, a pattern termed **polyneuronal innervation** (**FIGURE 20.16B**).

As in vertebrate tonic muscle, each neuron in arthropod skeletal muscle branches to provide multiterminal innervation to several muscle fibers. Arthropod muscle fibers typically do not generate all-or-none action potentials. (Insect flight muscles, which do generate action potentials, are an exception.) Because arthropod muscle fibers are innervated polyneuronally, the motor units of arthropods overlap; each muscle fiber is part of several motor units. Thus arthropods have only a few overlapping motor units per muscle, whereas vertebrates have many, nonoverlapping motor units per muscle.

Some arthropod muscles are innervated by both excitatory and inhibitory motor neurons. This feature—distinctly different from vertebrate muscles, which are innervated solely by excitatory neurons—allows *peripheral inhibition*. In arthropods, the excitatory transmitter is typically glutamate (not acetylcholine) and the inhibitory transmitter is gamma-aminobutyric acid (GABA). The algebraic summation of graded inhibitory postsynaptic potentials (IPSPs) and EPSPs *in the muscle fiber* determines the amount of tension developed. The greater the depolarization, the greater the amount of Ca^{2+} released from the SR, and the greater the tension developed. Thus the dominant mechanism for controlling tension in arthropod muscles is *controlling the degree of depolarization of muscle*

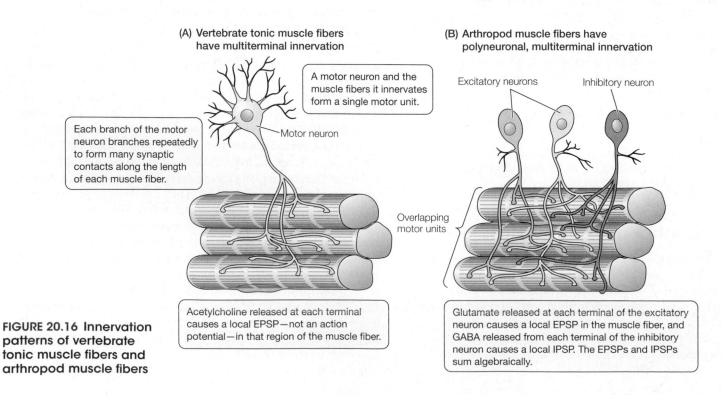

(A) Vertebrate tonic muscle fibers have multiterminal innervation

A motor neuron and the muscle fibers it innervates form a single motor unit.

Motor neuron

Each branch of the motor neuron branches repeatedly to form many synaptic contacts along the length of each muscle fiber.

Acetylcholine released at each terminal causes a local EPSP—not an action potential—in that region of the muscle fiber.

(B) Arthropod muscle fibers have polyneuronal, multiterminal innervation

Excitatory neurons

Inhibitory neuron

Overlapping motor units

Glutamate released at each terminal of the excitatory neuron causes a local EPSP in the muscle fiber, and GABA released from each terminal of the inhibitory neuron causes a local IPSP. The EPSPs and IPSPs sum algebraically.

FIGURE 20.16 Innervation patterns of vertebrate tonic muscle fibers and arthropod muscle fibers

fibers, which in turn depends on the frequency of action potentials in the excitatory and inhibitory motor neurons.

Arthropod muscle fibers have a range of speeds of contraction, but unlike in vertebrate fibers, the velocity of contraction of arthropod muscle fibers is associated with different sarcomere lengths: Short-sarcomere fibers contract quickly, and long-sarcomere fibers contract slowly. Most arthropod muscles contain a variety of fibers with different sarcomere lengths and contraction speeds. Some muscles, however, are composed of all long-sarcomere slow fibers or all short-sarcomere fast fibers. For example, the muscles of crayfish and lobsters that control flexion and extension of the abdomen are made up of either all fast or all slow fibers. Thus there are fast and slow flexor muscles and fast and slow extensor muscles. The slow flexor and extensor muscles each receive up to five excitatory motor neurons and one inhibitory neuron. Many of the fast muscles receive three excitatory axons and one inhibitory axon.

An additional pattern of innervation is found only in insects. The skeletal muscles in insects receive synaptic input not only from excitatory and inhibitory neurons but also from a third type of neuron that releases octopamine or tyramine. The octopamine/tyramine transmitters do not directly trigger or inhibit muscle contraction, but instead perform two different functions that affect muscle activity. First, they *modulate* neuromuscular activity elicited by input from glutamatergic excitatory motor neurons and GABA-ergic inhibitory neurons. For example, octopamine accelerates the relaxation rate of muscles by influencing the functions of chloride and potassium channels in the muscle membrane. Second, the octopamine/tyramine neurons to skeletal muscle fibers also promote glycolysis, and therefore ATP production from carbohydrates, during contractions. This direct neural control of metabolism plays an adaptive role in adjusting muscle ATP production to the energy demands of motor behaviors. Sometimes, however, it is not adaptive to use carbohydrates as metabolic fuel. Indeed, the flight muscles of certain insects that possess synchronous flight muscles (see Box Extension 20.2) switch from carbohydrate to lipid metabolism during long flights. For example, locusts have synchronous flight muscles and are well known for their ability to fly across oceans. In these animals, the octopamine neurons to the flight muscles stimulate glycolysis of carbohydrate stores during rest, but these neurons are inhibited during flight. In the absence of octopamine input, the flight muscles metabolize lipids instead of carbohydrates.

SUMMARY
Neural Control of Skeletal Muscle

- The neuromuscular organization of vertebrates is characterized by many nonoverlapping motor units, each controlled by a single motor neuron. Each muscle fiber within a motor unit generates an action potential that spreads rapidly over the entire cell membrane and triggers contraction.

- Vertebrate tonic fibers are organized into nonoverlapping motor units. They usually do not generate action potentials. Each fiber is innervated by a branch of a motor neuron that makes multiple synaptic contacts along its length. Excitatory postsynaptic potentials (EPSPs) produce local contractions near each synaptic contact.

- The neuromuscular organization of arthropods is characterized by few motor neurons, overlapping motor units, and in some cases, by peripheral inhibition. Each muscle fiber is typically innervated by more than one motor neuron, and each neuron makes multiple synaptic contacts on the fiber. Arthropod muscle fibers typically do not generate action potentials. Instead, the postsynaptic potentials produced at several points along the length of the fiber provide graded electrical signals that each trigger the contractile machinery in a small section of the fiber and control the degree of tension developed. Insect muscles are innervated not only by excitatory and inhibitory neurons but also by neurons that release octopamine or tyramine at synaptic contacts. These transmitters modulate neuromuscular activity and regulate energy metabolism.

Vertebrate Smooth (Unstriated) Muscle

Whereas the main function of skeletal muscle in vertebrates is to accomplish locomotion, and that of cardiac muscle is to pump blood through the heart, smooth muscles are important in the homeostatic functions of many different systems within vertebrate animals. Smooth muscles are found in the gastrointestinal, respiratory, reproductive, and urinary tracts and in the blood vessels. In addition, smooth muscles are in the eye (they control the size of the pupil and shape of the lens) and at the base of hairs or feathers (see Chapter 10, page 265). In hollow and tubular organs, smooth muscles permit a variety of functions, including changing size and volume (such as the bladder or stomach), propelling materials along a tube (such as chyme through the intestine or urine through the ureter), and maintaining tension for long periods of time (as in the walls of arterioles or sphincter muscles).

Compared with vertebrate skeletal muscles, which are relatively uniform in their structure and function, smooth muscles are richly varied in their architectural arrangement within organs, the types of stimuli that trigger their contraction, and the types of electrical activities they generate. Smooth muscles use the contractile proteins actin and myosin, but these proteins are not organized into sarcomeres, and so smooth muscle cells do not appear striated. Interdigitating myosin and actin filaments are organized into bundles around the periphery of the cell, and cross-bridge action causes them to slide by one another to accomplish contraction (**FIGURE 20.17A**). Smooth muscle cells have a greater proportion of actin relative to myosin than do striated muscles—a difference that is reflected in the larger ratio of thin to thick filaments in smooth muscles (about 12–15 thin filaments per thick filament) relative to striated muscles (about 2–4 thin filaments per thick filament). The actin filaments attach to **dense bodies** in the cytoplasm and on the inner surface of the cell membrane. Intermediate filaments also attach to the dense bodies to help form a stable cytoskeleton. The thick myosin filaments have cross-bridges along their entire length (unlike skeletal and cardiac thick filaments, which are "bald" in the middle). The myosin structure increases the probability that actin-binding sites will overlap cross-bridges even when the muscle is stretched.

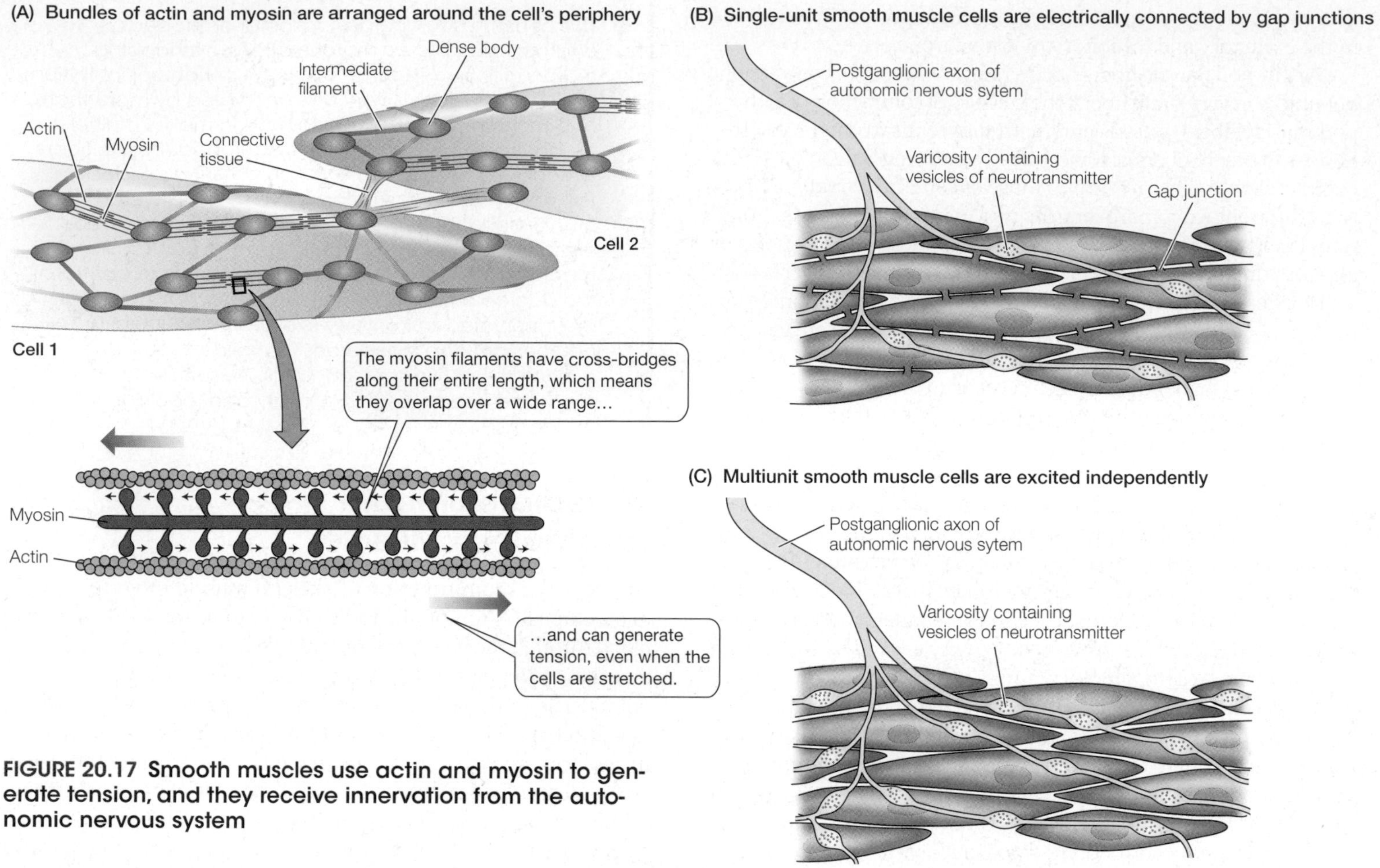

(A) Bundles of actin and myosin are arranged around the cell's periphery

The myosin filaments have cross-bridges along their entire length, which means they overlap over a wide range…

…and can generate tension, even when the cells are stretched.

(B) Single-unit smooth muscle cells are electrically connected by gap junctions

(C) Multiunit smooth muscle cells are excited independently

FIGURE 20.17 Smooth muscles use actin and myosin to generate tension, and they receive innervation from the autonomic nervous system

Smooth muscle cells range from 40 to 600 μm in length (shorter than most skeletal muscle fibers). They are widest in diameter around the single nucleus (2–10 μm, only somewhat wider than the 1–2 μm diameter of a single myofibril of a skeletal muscle fiber) and taper toward the ends. This cell shape is referred to as *spindle-shaped*. When a smooth muscle cell is stimulated to contract, the cross-bridge action of the peripherally arranged myofilaments causes the cell to shorten and plump up in the center. Smooth muscle cells are linked by connective tissue to each other and to surrounding connective tissue to ensure transmission of contractile force throughout a tissue or organ.

In addition to their small dimensions and single nucleus, smooth muscle cells are characterized by the absence of transverse tubules, troponin, and nebulin. They have a reduced sarcoplasmic reticulum (SR) but typically have *caveolae*, invaginations of the cell membrane that are thought to contribute to the rise of Ca^{2+} in the cytoplasm when the cell is activated. As in cardiac and skeletal muscles, the myosin ATPase of smooth muscle hydrolyzes ATP to power cross-bridge motions. However, the smooth muscle myosin ATPase hydrolyses ATP much more slowly than do the ATPases of different types of skeletal and cardiac myosins. Because of the slow rate of ATP hydrolysis, myosin cross-bridges cycle at a slower rate in smooth muscle, resulting in slower speed of contraction and longer contraction time. Many smooth muscles can maintain contractions for long periods using only a small portion of available cross-bridges and small expenditures of energy. The smooth muscle of the esophageal sphincter that guards the opening of the stomach is a good example. Except when food is swallowed, this muscle stays contracted continuously and prevents stomach acid and enzymes from entering the esophagus.

Smooth muscle cells are broadly classified

One useful classification scheme differentiates vertebrate smooth muscle into two main types: *single-unit* and *multiunit* smooth muscles. In **single-unit smooth muscle**, the muscle cells are electrically coupled by gap junctions (**FIGURE 20.17B**). Because of this coupling, groups of muscle cells are depolarized and contract together, functioning as a single unit. The smooth muscles of the gastrointestinal tract and small-diameter blood vessels are examples of the single-unit type. Single-unit smooth muscle is often spontaneously active, with electrical activity propagating from cell to cell via the gap junctions. This type of muscle can also be activated by stretch. Neural and hormonal controls may modulate the endogenous activity to varying degrees.

Multiunit smooth muscles have few if any gap junctions, so the muscle cells function as independent units (**FIGURE 20.17C**). They are innervated by autonomic nerves, and individual cells are under more direct neural control than are cells of single-unit smooth muscles. Multiunit smooth muscles may or may not generate action potentials, and they may be activated hormonally or by local chemical stimuli as well as neurally. They are not stretch-sensitive. Smooth muscles of the hair and feather erectors, eye, large arteries, and respiratory airways are examples of multiunit smooth muscles.

The smooth muscle of the mammalian uterus changes between multiunit and single-unit depending on circulating levels of reproductive steroid hormones. For example, during late pregnancy, when estrogen is present in the blood at high levels, the uterine smooth muscle cells form gap junctions that electrically couple adjacent cells. Thus the uterus is able to function as a single-unit smooth muscle to produce coordinated contractions during the birthing process.

A second classification used to distinguish different types of smooth muscle is based on contractile and electrophysiological properties. **Tonic smooth muscles**, such as those in the airways and certain sphincter muscles, maintain contractile force ("tone") for long periods and do not generate spontaneous contractions or action potentials (although they do undergo changes in membrane potential). **Phasic smooth muscles**, such as those in the stomach and small intestine, produce rhythmic or intermittent activity. They contract rapidly, produce spontaneous contractions, and generate action potentials that propagate through gap junctions from cell to cell. These gastrointestinal muscles can also be classified as single-unit smooth muscles. It is important to keep in mind that although classification schemes are useful when considering smooth muscles broadly, smooth muscles are hugely diverse, so not all of them fit neatly into specific categories.

Some smooth muscle cells undergo slow-wave changes in membrane potential in the absence of external stimulation. Slow waves recorded in these muscle cells may trigger action potentials if they exceed the voltage threshold of the fiber. In smooth muscle, inward Ca^{2+} current produces the rising phase of the action potential, so action potentials lead to a direct increase in intracellular Ca^{2+} concentration. However, because action potentials are not required to open voltage-gated Ca^{2+} channels, even subthreshold depolarizations will allow an influx of Ca^{2+} ions that may produce measurable tension in the muscle.

Ca²⁺ availability controls smooth muscle contraction by myosin-linked regulation

Like skeletal and cardiac muscles, smooth muscles maintain a low resting internal Ca^{2+} concentration (using pumps in both the SR and the cell membrane), and a rise in cytoplasmic Ca^{2+} initiates contraction. As we have seen, different smooth muscle cells respond to different types of stimuli, some respond to more than one type of stimulus, and some undergo spontaneous changes in membrane potential. The sum of inputs and membrane functions determines the moment-to-moment level of Ca^{2+} in the cytoplasm. Because the amount of available cytoplasmic Ca^{2+} determines the degree of force generated by the contractile proteins, smooth muscle cells produce *graded contractions*.

When a smooth muscle cell is stimulated to contract, Ca^{2+} enters the cytoplasm down its concentration gradient from both the SR and the extracellular space. The cells are small, so diffusion distance for Ca^{2+} is short, whether it enters from the SR or across the cell membrane. Unlike skeletal and cardiac muscles, smooth muscles do not use the thin filament regulatory proteins troponin (TN) and tropomyosin (TM) to regulate contraction. Instead, the proteins that regulate smooth muscle contraction are on the thick filament. **FIGURE 20.18** shows that Ca^{2+} activates smooth muscle predominantly by regulating the phosphorylation of myosin light chains (MLCs) (see Figure 20.2D). In this *myosin-linked regulation*,[12] entering Ca^{2+} ions combine with the Ca^{2+}-binding protein *calmodulin*, which is present in the cytoplasm. The Ca^{2+}–calmodulin complex activates the enzyme *myosin light-chain kinase (MLCK)*, which phosphorylates one of each pair of MLCs of individual myosin molecules. Phosphorylation of the MLCs enhances the ATPase activity of the myosin heads and triggers them to bind to actin filaments and generate cross-bridge motions. As in skeletal muscle, hydrolysis of one ATP molecule powers one cross-bridge cycle. As long as Ca^{2+} is present and the MLC remains phosphorylated, repeated cross-bridge cycles occur. The degree of force generated by the contractile proteins reflects the number of active cross-bridges. The number of active cross-bridges increases with increasing Ca^{2+} in the cytoplasm. This property of smooth muscle, in which the number of active cross-bridges is variable, is similar to cardiac muscle. However it is distinctly different from skeletal muscle, in which a muscle action potential triggers the release of sufficient Ca^{2+} from the SR to allow every cross-bridge to be active.

Relaxation is accomplished by the pumping of Ca^{2+} from the cytoplasm into the SR or out of the cell. As free Ca^{2+} decreases, Ca^{2+} unbinds from calmodulin, and the MLCK becomes inactive. Another enzyme, *myosin light-chain phosphatase (MLCP)*, dephosphorylates the light chains.

The processes shown in Figure 20.18 are often modulated by additional signaling molecules that influence the functions of MLCK and MLCP. For example, a smooth muscle cell activated by one type of signal may also receive an additional signal to turn on the G protein RhoA and its effector Rho kinase (ROK). The RhoA/ROK system inhibits MLCP. By preventing the dephosphorylation of MLCs, more myosin heads retain ATPase activity, so that—at any level of Ca^{2+} in the cytoplasm—more cross-bridges are active and generating force. This effect is referred to as *Ca²⁺ sensitization*. Interestingly, in tonic smooth muscles that produce especially prolonged contractions, such as the lower esophageal sphincter, the cross-bridges remain attached to actin in a **latch state** long after cytoplasmic Ca^{2+} is reduced. In this condition, ATP replaces the bound ADP of the myosin head ATPase extremely slowly, so the majority of cross-bridges "latched" to actin maintain tension without using ATP. The mechanisms responsible for maintaining the energy-saving latch state, and terminating it, are not fully clarified. It is possible, for example, that the relative activities of

---

[12] Additional mechanisms, including those that involve proteins on the thin filament, also play a role in regulating contraction in smooth muscle. Caldesmon, calponin, and tropomyosin are three such proteins thought to influence cross-bridge action.

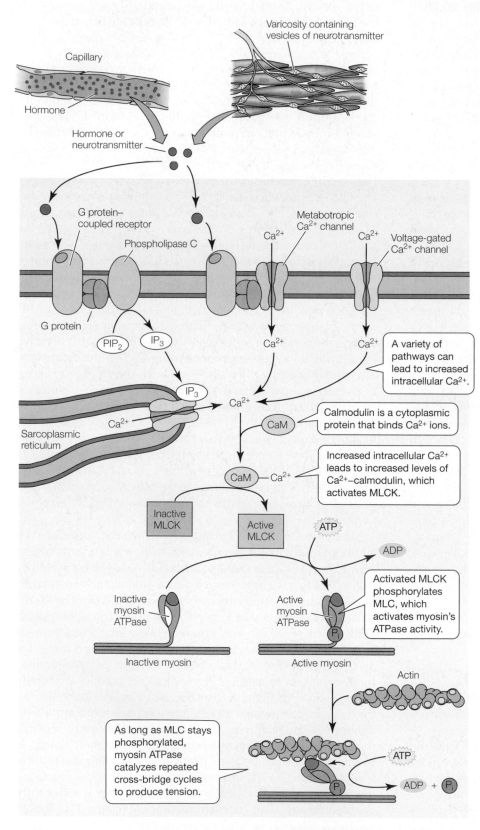

Capillary

Hormone

Varicosity containing vesicles of neurotransmitter

Hormone or neurotransmitter

G protein–coupled receptor

Phospholipase C

G protein

PIP₂

IP₃

Metabotropic Ca²⁺ channel

Ca²⁺

Voltage-gated Ca²⁺ channel

Ca²⁺

Ca²⁺

A variety of pathways can lead to increased intracellular Ca²⁺.

IP₃

Ca²⁺

Sarcoplasmic reticulum

Ca²⁺

CaM

Calmodulin is a cytoplasmic protein that binds Ca²⁺ ions.

CaM — Ca²⁺

Increased intracellular Ca²⁺ leads to increased levels of Ca²⁺–calmodulin, which activates MLCK.

Inactive MLCK

Active MLCK

ATP

ADP

Inactive myosin ATPase

Active myosin ATPase

Activated MLCK phosphorylates MLC, which activates myosin's ATPase activity.

Pᵢ

Inactive myosin

Active myosin

Actin

As long as MLC stays phosphorylated, myosin ATPase catalyzes repeated cross-bridge cycles to produce tension.

ATP

Pᵢ

ADP + Pᵢ

FIGURE 20.18 Myosin-linked regulation of smooth muscle contraction requires Ca²⁺ ions, calmodulin, and myosin light-chain kinase MLC, myosin light chain; MLCK, myosin-light chain kinase.

MLCK and MLCP are involved, or that proteins of the thin filament play a role, or that second messengers, such as those influenced by the paracrine nitric oxide (NO), have an effect.

Myosin-linked regulation of contraction (instead of troponin–tropomyosin–actin-linked regulation) also occurs in the muscles of molluscs and several other invertebrate groups. The muscles that hold shut the shells of bivalve molluscs (such as scallops) are known to remain contracted for hours or even days with very little O₂ consumption. These muscles actively contract, but relax extremely slowly, a condition termed the *catch state*. In this state, the muscles are stiff and resistant to stretch. Some investigators have suggested that an intermediate state of actin–myosin–ADP similar to the latch state of vertebrate smooth muscle could account for the economical maintenance of tension in molluscan muscle. An alternative idea is that the catch state is produced by the formation of a rigid network of connections between the myofilaments, a condition not dependent on myosin cross-bridges. These ideas are currently under investigation.

The autonomic nervous system (ANS) innervates smooth muscles

Postganglionic sympathetic and parasympathetic axons branch and ramify among the muscle cells (see Figure 20.17). The autonomic axons have repeated swellings, or varicosities, near their terminations, giving them a beaded appearance. The varicosities of the postganglionic axons function in a similar way to the presynaptic axon terminals of the somatic nervous system. Neurotransmitters are synthesized in the varicosities, stored in vesicles, and released by exocytosis. Neural activity triggers transmitter release, and the transmitter molecules diffuse over the surface of the muscle cells until they encounter receptor molecules. Unlike skeletal muscle fibers, smooth muscle cells lack distinct postsynaptic regions such as end-plates.

We conclude this section with two comparisons that illustrate the versatility of function that can be attained by interactions of the ANS and smooth muscles. In the first case, the smooth muscle of the urinary bladder wall is innervated by both divisions of the ANS. The parasympathetic transmitter acetylcholine stimulates the smooth muscle cells to contract and cause voiding of urine. By contrast, the sympathetic transmitter norepinephrine inhibits the smooth muscle cells from contracting so that urine is retained. Thus different behaviors of the same muscle can be achieved by activating different divisions of the ANS. A second comparison points out that the same neurotransmitter from the same autonomic division can cause either relaxation or contraction depending on the type of receptor expressed by the smooth muscles. Thus whereas norepinephrine causes relaxation of smooth muscle cells of the bladder wall, it causes contraction of the smooth muscle cells in most blood vessels. In sum, combining different presynaptic and postsynaptic elements permits rich versatility in controlling functions to achieve homeostasis. Chapter 15 provides a detailed description of the autonomic nervous system.

SUMMARY
Vertebrate Smooth (Unstriated) Muscle

■ Smooth muscles make up the walls of tubular and hollow organs, and are found in the eye and at the base of hairs and feathers. Smooth muscles contract slowly because their myosin ATPase isomers hydrolyze ATP very slowly. Some types of smooth muscles maintain contractions for protracted lengths of time using very little energy.

■ Smooth muscle cells are small, spindle-shaped, and uninucleate. They contain thin actin filaments and thick myosin filaments arranged around the periphery of the cell. Although the thick and thin filaments overlap with each other, they do not form sarcomeres, which accounts for the muscles' "smooth" appearance.

■ Smooth muscles receive innervation from the autonomic nervous system, and may be influenced by hormones, paracrines, and even stretch. Smooth muscles vary in the number of gap junctions present and in their contractile and electrophysiological properties. Cells in *single-unit* smooth muscles are connected by numerous gap junctions so that excitation spreads from cell to cell. Cells in *multiunit* smooth muscles have few gap junctions and function independently of each other. *Tonic* smooth muscles contract for long periods of time and typically generate only graded membrane depolarizations. *Phasic* smooth muscles produce rhythmic or intermittent contractions and generate action potentials.

■ Smooth muscle contraction is controlled by Ca^{2+}, which enters the cytoplasm from the extracellular space or SR and binds to calmodulin. The Ca^{2+}–calmodulin complex activates MLCK, which phosphorylates myosin light chains and thereby increases the ATPase activity of the myosin head. Because the number of active cross-bridges in a smooth muscle varies depending on the amount of Ca^{2+} present at any given time, smooth muscle cells are capable of producing graded contractions. Relaxation occurs when cytoplasmic Ca^{2+} decreases, Ca^{2+} unbinds from calmodulin, and MLCK is no longer activated. MLCP dephosphorylates the light chains. Other signaling pathways can influence MLCK and MLCP activity and thus modulate the Ca^{2+} sensitivity of smooth muscle cells.

Vertebrate Cardiac Muscle

Vertebrate cardiac muscle, the muscle that forms the walls of the heart and functions to propel blood through the vascular system,

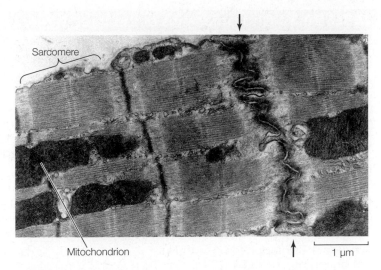

FIGURE 20.19 Cardiac muscle   Striated cardiac muscle fibers are connected by intercalated discs (arrows), which include electrical gap junctions and two types of mechanical connections called desmosomes and fasciae adherentes.

is discussed in Chapters 12 and 25. We note its main features here to provide a comparison with smooth and skeletal muscle. Cardiac muscle is classified as striated because its myofibrils are organized into sarcomeres, which possess the same structural and regulatory proteins that skeletal muscle sarcomeres have (see pages 538–539) (**FIGURE 20.19**). The cells are typically branched instead of straight like skeletal muscle fibers or spindle-shaped like smooth muscle cells. They are usually uninucleate. In mammals, the SR and t-tubules are well developed, but they are variable in other vertebrate animals.

Cardiac muscle fibers have functional properties that contribute to their effectiveness in pumping blood. First, they are characterized by the presence of **intercalated discs** between adjacent cells. Intercalated discs include gap junctions and localized mechanical adhesions called *desmosomes* and *fasciae adherentes* (singular *fascia adherens*). The adhesions provide mechanical strength so that the force of contraction generated by one cell can be transmitted to the next to ensure coordinated pumping. The electrical coupling at gap junctions ensures that all cells connected by gap junctions contract (beat) nearly synchronously. Gap junctions and desmosomes are illustrated in Figure 2.7.

A second property of cardiac muscle cells is that they are capable of generating endogenous action potentials at periodic intervals. Typically, specialized pacemaker cells with the fastest endogenous rate impose their rhythm on the contractile activity of the rest of the heart. Finally, the action potentials of vertebrate cardiac fibers have very long durations, typically 100 to 500 ms (see Figure 12.23). Their long durations ensure a prolonged contraction rather than a brief twitch. Indeed, the action potentials last as long as the contractions. Because the cardiac cells are refractory during the prolonged action potentials, contractions cannot sum; thus the coordinated pumping of blood is ensured.

TABLE 20.3 summarizes the properties of vertebrate skeletal, smooth, and cardiac muscles.

TABLE 20.3 Characteristics of the three major types of muscles in vertebrates

	Skeletal	Multiunit smooth	Single-unit smooth	Cardiac
Structure	Large, cylindrical, multinucleate fibers	Small, spindle-shaped, uninucleate cells	Small, spindle-shaped, uninucleate cells	Branched uninucleate fibers, shorter than skeletal muscle fibers
Visible striations	Yes	No	No	Yes
Mechanism of contraction	Thick myosin and thin actin filaments slide by each other	Thick myosin and thin actin filaments slide by each other	Thick myosin and thin actin filaments slide by each other	Thick myosin and thin actin filaments slide by each other
Cross-bridge action regulated by Ca^{2+} ions	Yes	Yes	Yes	Yes
Innervation	Somatic nervous system initiates contractions	Autonomic nervous system initiates contractions	Autonomic nervous system modulates contractions	Autonomic nervous system modulates contractions
Spontaneous production of action potentials by pacemakers	No	No	Yes	Yes
Hormones influence function	No	Yes	Yes	Yes
Gap junctions present	No	No (few)	Yes	Yes
Transverse tubules	Yes	No	No	Yes
Sarcoplasmic reticulum	Abundant	Sparse	Sparse	Moderate
Source of Ca^{2+} ions for regulation	Sarcoplasmic reticulum	Extracellular fluid and sarcoplasmic reticulum	Extracellular fluid and sarcoplasmic reticulum	Extracellular fluid and sarcoplasmic reticulum
Troponin and tropomyosin	Both present	Tropomyosin only	Tropomyosin only	Both present
Ca^{2+} regulation	Ca^{2+} and troponin; tropomyosin–troponin complex moves to expose myosin-binding sites on actin	Ca^{2+} and calmodulin; phosphorylation of myosin light chains	Ca^{2+} and calmodulin; phosphorylation of myosin light chains	Ca^{2+} and troponin; tropomyosin–troponin complex moves to expose myosin-binding sites on actin
Speed of contraction (reflecting myosin ATPase activity)	Varies from fast to slow depending on fiber type	Very slow	Very slow	Slow

STUDY QUESTIONS

1. Knowing the dimensions of a vertebrate skeletal muscle and the relationship between the SR and the myofilaments, estimate the approximate distance that a single Ca^{2+} ion would travel from a terminal cisterna of the SR to a TN-binding site.

2. Experimenters can separate F-actin thin myofilaments from myosin thick myofilaments. First they homogenize muscle cells in a blender (to break cell membranes); then they place the homogenate in a Ca^{2+}-free "relaxing solution" that contains ATP. Explain why ATP must be present and Ca^{2+} ions must not be present in order to isolate thick and thin myofilaments from each other.

3. List and describe the events that take place (and the structures involved) between excitation of the skeletal muscle cell membrane by an action potential and the initiation of cross-bridge action.

4. Combining your knowledge of rates of diffusion with your knowledge of muscle physiology, explain why it is advantageous for oxidative muscle fibers (which depend on aerobic metabolism to generate ATP) to have smaller diameters than glycolytic muscle fibers have.

5. What is the difference between a single cross-bridge power stroke and a single twitch of a skeletal muscle fiber?

6. (a) In skeletal muscle, if all cross-bridges are activated when a single action potential triggers Ca^{2+} release from the SR, why is the amount of tension produced by a train of action potentials greater than the amount of tension of a single twitch? (b) The maximum contractile response of a muscle stimulated by a train of stimuli is called tetanus. Tetanus toxin prevents inhibitory control of motor neurons, which as a consequence release ACh from their terminals unremittingly. Describe the effects of tetanus toxin on contractile activity of skeletal muscles in an individual infected by *Clostridium tetanii*, the bacterium that produces tetanus toxin.

7. Arthropod muscle fibers typically do not generate action potentials. Using your knowledge of their innervation, explain how their contractile elements are activated in a rapid and coordinated fashion.

8. Describe the organization of a motor unit in vertebrate skeletal muscle, and explain how recruitment of motor units influences the amount of tension produced by a whole muscle.

9. Two muscles have the same diameter, but one is twice as long as the other. Which muscle produces more work? Explain your answer.

10. Contraction in both skeletal and smooth muscles requires the influx of Ca^{2+} into the cytoplasm. Compare and contrast the locations and functions of the molecules to which Ca^{2+} binds in skeletal and smooth muscles, and explain the steps that lead to cross-bridge cycling in each type of muscle.

Go to **sites.sinauer.com/animalphys4e**
for box extensions, quizzes, flashcards,
and other resources.

REFERENCES

Brini, M., T. Cali, D. Ottolini, and E. Carafoli. 2013. The plasma membrane calcium pump in health and disease. *FEBS J.* 280: 5385–5397. A review highlighting the importance of active transport in maintaining low levels of cytoplasmic calcium.

Catania, K. 2014. The shocking predatory strike of the electric eel. *Science* 346: 1231–1234. A description of the electric eel's behaviors in hunting and attacking prey, and the neurophysiology underlying these behaviors.

Clark, K. A., A. S. McElhinny, M. C. Beckerle, and C. C. Gregorio. 2002. Striated muscle cytoarchitecture: an intricate web of form and function. *Annu. Rev. Cell Dev. Biol.* 18: 637–706. A description of many cytoskeletal protein molecules associated with the sarcomeres, and their contributions to contractile function.

Gallant, J. R., L. L. Traeger, J. D. Volkening, and 13 additional authors. 2014. Genomic basis for the convergent evolution of electric organs. *Science* 344: 1522–1525. A genomic study pointing out that electric organs have evolved independently at least six times, yet key transcription factors and developmental processes of electric organs show remarkable evidence of convergent evolution.

Gordon, A. M., A. F. Huxley, and F. J. Julian. 1966. The variation in isometric tension with sarcomere length in vertebrate muscle fibers. *J. Physiol. (London)* 184: 170–192. Evidence at the level of the sarcomere to account for the length–tension relationships observed in whole muscles.

Hooper, S. L., K. H. Hobbs, and J. B. Thuma. 2008. Invertebrate muscles: thin and thick filament structure; molecular basis of contraction and its regulation, catch and asynchronous muscle. *Prog. Neurobiol.* 86: 72–127. A review that compares general features of vertebrate muscles with muscles of several invertebrate phyla.

Keynes, R. D., and H. Martins-Ferreira. 1953. Membrane potentials in the electroplates of the electric eel. *J. Physiol. (London)* 119: 315–351. This classic paper describes the first use of intracellular microelectrodes to record the membrane potentials of the electroplates of *Electrophorus electricus* L.

Meyer, L. C., and N. T. Wright. 2013. Structure of giant muscle proteins. *Front. Physiol.* 4: 368. Published online, doi: 10.3389/fphys.2013.00368. A review of the molecular structures of titin, nebulin, and obscurin and their roles in maintaining the functional organization of sarcomeres.

Pflüger, H.-J., and C. Duch. 2011. Dynamic neural control of insect muscle metabolism related to motor behavior. *Physiology* 26: 293–303. A brief comparison of vertebrate and invertebrate muscles and an overview of motor innervations of insect skeletal muscles.

Rome, L. C. 2006. Design and function of superfast muscles: new insights into the physiology of skeletal muscle. *Annu. Rev. Physiol.* 68: 192–221. A review of the adaptations of rapidly contracting muscles and the trade-offs made to achieve high-speed contractions.

Siegman, M. J., S. Davidheiser, S. U. Mooers, and T. M. Butler. 2013. Structural limits on force production and shortening of smooth muscle. *J. Muscle Res. Cell Motil.* 34: 43–60. Comparison of two different smooth muscles using physiological and ultrastructural methods showed that both active and passive components of the muscles' structures influence functional behaviors.

Zalk, R., O. B. Clarke, A. des Georges, and six additional authors. 2015. Structure of a mammalian ryanodine receptor. *Nature* 517: 44–49. Ryanodine receptors rapidly release Ca^{2+} from intracellular stores. Knowledge of their structure aids in understanding their functions and therefore possible applications in developing therapies for disorders such as muscular dystrophy and heart failure.

See also **Additional References** *and* **Figure and Table Citations**.

The success of any living species is achieved by the ability of its individuals to pass their genes on to their offspring, and to ensure survival of sufficient numbers of the offspring so that they in turn reach reproductive maturity. To achieve success, animals need to obtain nourishment, grow, and reproduce—and also escape predators. Muscle tissues support all of these abilities. Indeed, in vertebrates, striated muscles constitute nearly half of the tissues of an animal. Muscle tissue continually synthesizes and degrades proteins, assembles and disassembles contractile elements, and maintains itself through repair. Imagine these processes occurring not only in actively contracting skeletal muscles but also in cardiac muscle, which (in mammals) contracts 30–700 times per minute! The dynamic nature of muscle tissue underlies its *phenotypic plasticity* (see Chapter 4, page 94).

Governed by a myriad of controls, muscle is capable of changing in both mass and cellular characteristics. The remarkable ability of muscle to change with use is especially interesting because muscle cells (muscle fibers) in adults are postmitotic; that is, once a muscle forms, the cells do not divide by mitosis, and their number cannot increase. For example, we know from Chapters 8 and 20 that muscles typically consist of mixtures of different types of muscle fibers that are characterized by different speeds of contraction and different metabolic properties.[1] Researchers have found that a muscle's activity causes interconversion between certain of these types of fibers, but not a change in the number of fibers. Furthermore, it is widely agreed that when a muscle increases in bulk, it does so by **hypertrophy**—that is, by adding structural proteins to individual cells—not by adding new cells by mitosis. When a muscle is not used, it becomes smaller because the individual muscle fibers lose actin and myosin components

---

[1] The three main types of fibers are slow-twitch fibers (*slow oxidative* [*SO*]) and two types of fast-twitch fibers (*fast oxidative gycolytic* [*FOG*] and *fast glycolytic* [*FG*]; see Chapter 20, Table 20.2).

Different activities produce changes in skeletal muscles
Here we see astronauts participating in extravehicular activity outside the International Space Station and basketball players fighting for possession of the ball. These activities place very different demands on skeletal muscles and influence their phenotype.

of myofibrils. This reduction of muscle mass is called **atrophy**. Atrophy of a muscle can also result from loss of cells, a phenomenon seen in some disease states and also in aging.

Other changes also occur in response to different types of activity. For example, a weight lifter does resistance exercises (such as lifting free weights or working against an external resistance such as a strength-training machine) that lead to hypertrophy of the exercised muscle fibers. These exercises stimulate the individual fibers to increase their synthesis of actin and myosin, which form more myofibrils. The exercised muscle fibers also add nuclei by fusing with satellite cells (mononuclear precursors of muscle cells) that lie close to a muscle fiber's cell membrane. The added nuclei support the overall functions of a larger-volume cell. Recent experiments in rodents showed that added nuclei persist even when a once-hypertrophied cell undergoes atrophy. This result led researchers to suggest that the nuclei of hypertrophied muscles may play a role in "muscle memory," the ability of previously trained muscles to regain their former strength with less retraining than was required to gain strength initially.

Different changes take place in the individual fibers in the leg muscles of a person training for long-distance running. These muscle fibers form more and larger mitochondria so that more ATP is produced by oxidative phosphorylation. Unlike the fibers in resistance-trained muscles, the fibers in the endurance-trained muscles of runners show relatively little hypertrophy. This morphology is in keeping with maintaining short diffusion distances between the blood in capillaries and the interior of the muscle fibers. As we will see, endurance exercise also stimulates more capillaries to form around the muscle fibers. Although the muscles of athletes provide extreme examples, similar but more subtle changes occur in any person's muscles over the course of day-to-day activities.

There is no shortage of examples demonstrating that muscles can—and do—change in response to functional demands. Researchers are actively studying the signals that trigger these changes. How does the nature of a muscle's mechanical activity cause individual muscle fibers to convert from one type to another? What conditions lead to atrophy? Indeed, the aim of many ongoing studies is to find ways to inhibit the muscle atrophy that occurs under conditions such as microgravity during space travel, in diseases such as muscular dystrophy, and in aging. What conditions lead to hypertrophy? Answers to questions such as these have potential use not only in health care but also in economic contexts. For example, research in the meat and livestock industries is aimed at defining conditions that maximize growth rates and also provide optimum quality of meats. Studies of humans and laboratory mammals have contributed a great deal to our current understanding of use-related changes in muscle. Studies of other animals provide complementary perspectives of muscle function. In this chapter we explore several studies on vertebrate striated muscles that highlight their remarkable phenotypic plasticity.

Muscle Phenotypes

The phenotype of muscle tissue depends on the type of actions it performs. In the following sections we will use endurance exercise and resistance

exercise to illustrate the ways in which muscle tissue changes to respond optimally to different actions. **Endurance exercise**, such as long-distance running, cycling, or swimming, is exercise that involves repetitive actions that generate relatively low forces. Slow-twitch fibers that have narrow diameters and depend on aerobic metabolism play a dominant role in endurance exercise. They are well designed for maintaining isometric force economically and for carrying out repetitive isotonic contractions. As we saw in Chapter 20, most movements involve both isometric and isotonic contractions. Isometric contractions do not shorten the whole muscle or move the limb; they typically occur at the beginning and at the end of an action. Isotonic contractions move the limb. **Resistance**, or strength, **exercise**, such as stair running or weight lifting, is exercise that involves fewer repetitions of movements that generate large forces. Fast-twitch fibers that depend more on anaerobic metabolism are important in resistance exercises, which are often referred to as *power pursuits*.

The different types of muscle fibers contain different *isoforms* (that is, different molecular forms) of two important molecules, the myosin heavy chain of the thick filament and the Ca^{2+}-ATPase pump of the sarcoplasmic reticulum (SR). The different isoforms of these two molecules strongly influence a muscle fiber's functional properties. The rate of ATP hydrolysis by the ATPase of the myosin head is directly related to the rate at which cross-bridges cycle, and therefore to the speed of contraction. The rate at which Ca^{2+} ions are taken back into the SR from the cytoplasm also affects the twitch. Different isoforms of the Ca^{2+}-ATPase pump have different kinetics and are expressed at different levels in muscle fibers. The faster the cytoplasm is cleared of Ca^{2+} ions by the pump, the faster a single twitch is completed. The different isoforms of myosin and the SR Ca^{2+}-ATPase explain several of the properties of the three main types of muscle fibers described in Chapter 20, Table 20.2—slow oxidative (SO), fast oxidative glycolytic (FOG), and fast glycolytic (FG).

TABLE 21.1 shows that SO (also called Type I) fibers have the slowest myosin isoform and thus the slowest rates of cross-bridge cycling. SO fibers also have a slow Ca^{2+}-ATPase isoform in the SR. They contain abundant mitochondria and tend to be fatigue-resistant. FOG (also called Type IIa) fibers have a myosin isoform that hydrolyzes ATP faster than does the myosin isoform of SO muscle fibers, and a fast Ca^{2+}-ATPase isoform in the SR; they are relatively resistant to fatigue. FG (in humans, also called Type IIx) fibers have the fastest myosin isoform and a fast SR Ca^{2+}-ATPase

TABLE 21.1 Muscle fiber types and molecular isoforms			
General terms	**Slow oxidative (SO)**	**Fast oxidative glycolytic (FOG)**	**Fast glycolytic (FG)**
Human terms	Type I	Type IIa	Type IIx[a]
Myosin heavy-chain isoform	Slow cross-bridge cycling	Rapid cross-bridge cycling	Rapid cross-bridge cycling
Sarcoplasmic reticulum Ca^{2+}-ATPase	Slow Ca^{2+} uptake by SR	Fast Ca^{2+} uptake by SR	Fast Ca^{2+} uptake by SR
Speed of contraction	Slow	Fast	Fast

[a]Early studies on humans referred to what are now known as IIx fibers as IIb fibers. In current nomenclature, IIb fibers are a type of fast muscle fiber found in small mammals. An additional type of fast fiber found in small mammals is IId/x.

isoform. FG fibers have the fastest speed of contraction. Indeed, in humans the FG (Type IIx) muscle fibers can contract up to ten times faster than SO (Type I) fibers. FG (Type IIx) fibers typically have the largest diameters of the three main fiber types (see Figures 8.13 and 20.14), contain relatively few mitochondria, and fatigue easily.

Motor units composed of different fiber types have different physiological characteristics. The muscle fibers of any single *motor unit* (a motor neuron and all of the muscle fibers it innervates; see Figure 20.15) are all of the same fiber type. The muscle fiber type is strongly influenced by the motor neuron. We know from early experiments that some motor units could be converted from one type to another by cross-innervation (cutting the original nerve fibers and allowing different ones to innervate the muscle fibers). Typically, motor units are recruited in a fixed order: first SO, then FOG, and finally FG. The two fast fiber types are recruited to produce more powerful isotonic movements than slow fibers produce, or to supplement isometric contractions produced by slow fibers. The pattern of recruitment and the diversity of fibers with different velocities of contraction and fatigability allow a muscle to shorten at varied, appropriate speeds as it generates the forces required for different types of movement.

Power output determines a muscle's contractile performance, and changes in response to use and disuse

We know from Chapter 20 (see page 549) that the greater the cross-sectional area devoted to the contractile elements in a muscle fiber, the greater the force it can generate. We also know that the speed of contraction is determined by the rate at which a given myosin isoform hydrolyzes ATP and produces cross-bridge actions. These qualities—*force* generated and *velocity* (speed) of shortening—determine the contractile performance of a muscle. Indeed, the force generated by a muscle multiplied by the velocity of shortening determines the mechanical power produced:

$$\text{Power} = \text{force} \times \text{shortening velocity} \qquad (21.1)$$

FIGURE 21.1 shows a *power curve* superimposed on the force–velocity relationship described in Chapter 20 (see page 546). The force–velocity (load–velocity) relationship reflects the interaction between the force a muscle generates against a load and the speed at which it shortens: The velocity of shortening decreases as the load increases.

Points on the power curve are determined from the instantaneous product of force and velocity. Power is zero (P_0) when the muscle shortens against no load (at maximum velocity, V_{max}) and also when the muscle contracts isometrically but does not shorten. Power output of most muscles is maximal when the muscle shortens at 20%–40% of V_{max} and at about 30%–40% of the load that prevents it from shortening. Power is measured in watts: 1 watt = 1 newton (N) × 1 m/s, or 1 watt = 1 joule (J)/s.

We know that in isotonic contractions the force generated by a muscle equals the force of the load it moves. Thus the *x* axis of the graph indicates the force exerted by the load, and it also reflects the force produced by the muscle. For a particular muscle, the cross-sectional area of the myofibrils in its muscle fibers determines the maximum force it can produce. The *y* axis of the graph, which indicates velocity of shortening, is related to the myosin isoforms expressed by the motor units of a particular muscle. Therefore

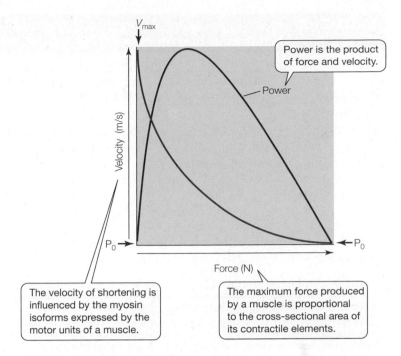

Power is the product of force and velocity.

Power

The velocity of shortening is influenced by the myosin isoforms expressed by the motor units of a muscle.

The maximum force produced by a muscle is proportional to the cross-sectional area of its contractile elements.

FIGURE 21.1 The power a muscle is capable of generating determines its functional capabilities The blue line represents the force–velocity relationship; the red line represents the power curve. Power is zero (P_0) when shortening velocity is at its maximum (no load; V_{max}) or at zero (isometric contraction).

the force–velocity relationship—and thus the power output—of a whole muscle will be determined by the cross-sectional area of individual muscle fibers (how many myofibrils they contain) and the different fiber types that make up the muscle. Because the myofibrils are structured similarly in all skeletal muscle fiber types, a unit of cross-sectional area will generate the same force in both slow and fast types of fibers. However, the *power* is much greater for fast-twitch fibers than for slow-twitch fibers because of their greater speed of contraction. When the individual fibers of muscles change in response to different kinds of use or to disuse, their cross-sectional area (determined by hypertrophy or atrophy) and speed of contraction (determined by fiber type) also change. The changes directly affect power output and therefore contractile performance.

Endurance training elicits changes in fiber type, increased capillary density, and increased mitochondrial density

In an average active person, the muscles used in locomotion typically have a mix of about half slow Type I fibers and half fast fibers; the majority of fast fibers are Type IIa.[2] Elite athletes show distinct differences from the average person and also from one another. These differences often reflect the aerobic demands of a particular sport. For example, the leg muscles of marathon runners who train for endurance tend to have a preponderance of small-diameter Type I fibers and relatively few large-diameter fast fibers, which are mainly Type IIa. By contrast, the leg muscles of sprinters who

[2] Different muscles in the body of a human or other animal differ in their composition of fast and slow fiber types depending on the muscle's function (see Chapter 20, pages 552–554).

TABLE 21.2 Fiber type distribution and capillaries around each fiber before and after endurance training in two studies

Fiber type	Before Training		After Training	
	Distribution (%) (mean ± S.E.M.)	Capillaries around each fiber (mean ± S.E.M.)	Distribution (%) (mean ± S.E.M.)	Capillaries around each fiber (mean ± S.E.M.)
7 women				
Type I	58.2 ± 2.8	4.11 ± 0.15	57.5 ± 2.9	5.04 ± 0.21**
Type IIa	24.9 ± 2.6	3.4 ± 0.16	31.6 ± 2.7*	4.15 ± 0.21**
Type IIx	11.8 ± 2.7	2.33 ± 0.19	2.7 ± 2.5*	2.68 ± 0.14**
Intermediate	5.2	6.9	7.9*	8.0**
5 men				
Type I	39 ± 2.1	3.9 ± 0.18	42 ± 2.2	5.4 ± 0.32*
Type IIa	36 ± 2.9	4.2 ± 0.2	42 ± 2.4*	5.5 ± 0.45*
Type IIx	20 ± 1.6	3.0 ± 0.22	13 ± 1.5*	4.2 ± 0.5*

Sources: After Ingjer 1979 (women) and Andersen and Henriksson 1977 (men).

Note: The data shown from the study on women were based on 168–265 muscle fibers analyzed per subject prior to training, and 137–197 per subject after training. Asterisks indicate significant differences from pre-training values: *, $P < 0.005$; **, $P < 0.01$. The intermediate category of fibers had histological properties intermediate between those of Type IIa and Type IIx fibers, suggesting that different myosin isoforms were coexpressed at the same time. In the study on men, 1035 ± 126 fibers were analyzed per subject prior to training, and 937 ± 270 fibers per subject after training. Asterisks indicate significant differences from pre-training values ($P < 0.05$).

follow resistance-training guidelines tend to have a preponderance of large-diameter fast fibers (a mix of Types IIa and IIx) and relatively few Type I fibers. Between these extremes, humans and other animals show considerable variation in the distributions of fiber types within their muscles. It appears likely that individuals vary genetically, with some programmed to have muscles more like those of a marathoner and others to have muscles more like those of a sprinter. Still, studies of humans and experimental animals have shown that exercise training can cause some degree of interconversion between these fiber types, in particular between Type IIa and Type IIx fibers.

TABLE 21.2 presents the results of two studies of endurance training in humans. These "classic" histological studies in both men and women illustrate in single experiments the now well-known association between endurance training and changes in muscle fiber-type composition and also sprouting of new capillaries to improve circulatory supply to the muscle. Subsequent and ongoing studies have shed light on the mechanisms underlying these changes by applying biochemical, molecular biological, genomic, proteomic, and imaging techniques. In one of these early studies, seven previously untrained women, 21–24 years old, did supervised cross-country running for 24 weeks, running 45 min/day, 3 days a week. In the other study, five previously untrained men, 20–23 years old, trained for a period of 8 weeks by pedaling a bicycle ergometer (a bicycle equipped to measure the work done by muscles) 40 min/day, 4 days a week. In both studies, samples of muscle tissue were taken before and after training from the *vastus lateralis* muscle of the quadriceps group (**FIGURE 21.2**). **FIGURE 21.3** illustrates the instruments used in a **needle biopsy**, the procedure used to obtain the samples. The *vastus lateralis* is a mixed muscle, with representation of all three fiber types. In an average active person, the myosin isoform profile is about 50% Type I, 40% Type IIa, and 10% Type IIx.

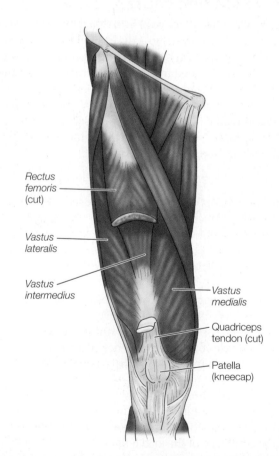

Rectus femoris (cut)

Vastus lateralis

Vastus intermedius

Vastus medialis

Quadriceps tendon (cut)

Patella (kneecap)

FIGURE 21.2 Quadriceps muscles of the anterior thigh The four muscles of the quadriceps group lie on the top of the thigh and insert on the quadriceps tendon. The *vastus lateralis*, frequently biopsied for studies on human muscle, is the most lateral of the four muscles.

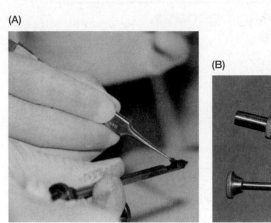

(A)

(B)

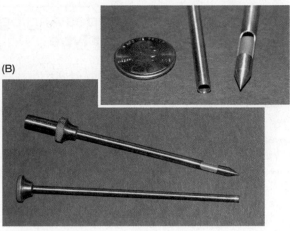

FIGURE 21.3 A needle biopsy is used to obtain samples of muscle tissue (A) The biopsy sample is a small plug of tissue removed from the muscle. (B) The biopsy apparatus consists of a pointed outer needle and an inner slider with a razor-sharp cutting edge. After application of local anesthetic, the apparatus is inserted into the muscle through an incision. A small amount of surrounding muscle tissue bulges into the opening on the side of the pointed needle. The inner slider is pressed forward to sever the protruding bit of muscle tissue. The slider is withdrawn and the tissue sample removed from the pointed needle using forceps. (A courtesy of S. P. Scordilis, Smith College.)

In both studies, the proportion of Type I fibers remained unchanged as a result of training. However, endurance training caused a significant change in the proportions of the two types of fast fibers in the *vastus lateralis* muscle. The proportion of Type IIa fibers increased and was accompanied by a decrease in the proportion of Type IIx fibers. The authors of the study on the women noted that the proportion of Type I fibers in the tested muscles of these subjects (~58%) was greater than usually found in untrained female subjects. Indeed, the proportion of Type I fibers in the *vastus lateralis* muscles of eight subjects who dropped out of this study after only a few weeks was approximately 47%. This observation raised the question of whether the anatomy or physiology of volunteers in long-term studies influences their interest in participating. The authors suggested that the seven women who completed the nearly half-year study may have chosen to participate in (and stick with) it because the fiber-type composition of their muscles (determined by their genetic make-up) contributed to their inherent ability to respond to endurance training.

Because the tissue samples showed no evidence of cells dying or new cells being formed, the results are interpreted to mean that Type IIx fibers changed into Type IIa fibers. Changes in gene expression must have occurred to accomplish the shift in proportions of the fast fiber types observed in both studies. Some of the original Type IIx muscle fibers must have repressed their genes for the faster IIx myosin isoform and switched on their genes for the IIa myosin isoform. The signals that trigger this change in gene expression—and subsequent changes in muscle structure—are topics of active inquiry. It is interesting that interconversions between the two types of fast fibers appear to occur readily, whereas transformations from Type II fibers to Type I fibers, and vice versa, do not. Researchers have observed conversions of fast fibers into slow fibers in rodent muscles treated with more aggressive techniques than possible in exercise-training programs. Whether such changes could be achieved under physiological conditions over longer periods than those so far tested in humans is still an open question. It is possible that remodeling of entire motor units, including the motor neurons, is necessary to achieve more extensive interconversions between slow- and fast-twitch muscle fibers.

The two studies also showed that endurance training led to an increase in the number of capillaries in contact with each muscle fiber. New capillaries grow by sprouting branches from existing vessels, a process called **angiogenesis**. Subsequent studies demonstrated that exercised muscles produce and release the cytokine **vascular endothelial growth factor** (**VEGF**), which stimulates angiogenesis (**BOX 21.1**).

FIGURE 21.4 shows the upregulation of VEGF in response to exercise in six untrained men. Their biopsied muscle samples were analyzed biochemically for VEGF content and histologically to determine changes in the abundance of capillaries. At the beginning of the study, each subject exercised only the left leg for 30 min. In this pre-training *knee-extensor exercise* bout, the subjects repeatedly

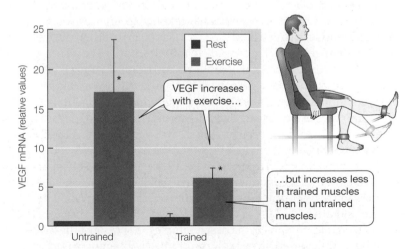

VEGF increases with exercise...

...but increases less in trained muscles than in untrained muscles.

FIGURE 21.4 VEGF responses to a single bout of endurance exercise Vascular endothelial growth factor (VEGF) mRNA increased in the *vastus lateralis* muscle after a single bout of knee-extensor exercise in both untrained and trained muscles, but the increase was attenuated in the trained muscles. Morphological studies of the muscle tissues showed a significant increase in the number of capillaries around each fiber after training, from about 3.6 capillaries per fiber in the untrained muscle to about 4.3 capillaries per fiber in the trained muscle. Bars are means ± 1 S.E.M., and * indicates a significant difference from untrained values ($P < 0.05$). Tissue samples were taken from six subjects. (After Richardson et al. 2000.)

| BOX 21.1 | In Endurance Training, VEGF Production Precedes Angiogenesis, and Angiogenesis Precedes Changes of Fiber Type |

Experiments using laboratory mice complement those of human subjects in studies of the effects of endurance training. Researchers followed the time course of changes in the *plantaris* muscle over a period of 28 days. (The plantaris lies beneath the *gastrocnemius* muscle of the lower leg.) They found that VEGF production occurred within 3 days of the start of running and was followed by increased numbers of capillaries after about 7 days. The conversion of the fastest muscle fibers to intermediate Type IIa fibers was complete after 14 days. Data from these experiments appear in **Box Extension 21.1**.

contracted the quadriceps muscle group against a load to extend the leg from a "sitting angle" to a straightened position. About 1 h after the exercise bout, biopsies were taken from the *vastus lateralis* muscle of the exercised leg (Untrained, Exercise sample) and from the same muscle of the rested right leg (Untrained, Rest sample). The subjects then trained the left quadriceps muscle (with varying exercises and loads to optimize training) for a period of 8 weeks. At the end of the training period, they performed the same exercise as in the pre-training bout (but at a higher load), and biopsies of the exercised *vastus lateralis* were taken about 1 h afterward (Trained, Exercise sample). Two days later, biopsies were taken from the same, now rested, muscle (Trained, Rest sample).

The biopsies showed low levels of VEGF mRNA at rest in both untrained and trained muscles. A single exercise bout initiated rapid upregulation of VEGF mRNA—within 1 h—in both untrained and trained muscles. Interestingly, the response in the trained muscles was less than that in the untrained muscles. Morphological studies of the same biopsied tissues showed that the training program induced significant angiogenesis, with an 18% increase in the number of capillaries around each muscle fiber. When chronic exercise training upregulates VEGF repeatedly, capillaries are stimulated to proliferate. Once angiogenesis has taken place in the exercise-adapted muscle, a single bout of exercise appears to stimulate less upregulation of VEGF. These studies strongly indicate that exercise causes mechanical and/or metabolic perturbations in muscle cells that trigger increased expression of VEGF. The VEGF released into the interstitial space acts as a paracrine to induce angiogenesis.

Endurance exercise increases not only the density of capillaries in muscles, but also the number and size of mitochondria. **FIGURE 21.5** shows data obtained in an experiment in which five women and five men (29 ± 5.1 years old)—all previously untrained—participated in a 6-week endurance regimen using bicycle ergometers. Tissue samples were taken by biopsy from the *vastus lateralis* muscle before and after the training period. Measurements of structures in electron micrographs of the tissue samples showed that the total volume of mitochondria per volume of muscle fiber increased by 40% and that the volume occupied by lipid droplets within the muscle fibers nearly doubled, increasing by 97%. In addition, the number of capillaries per muscle fiber increased by 26%.

Recent and ongoing studies of *mitochondrial biogenesis* have shown that exercise also stimulates changes in the structural and enzymatic proteins of the mitochondria themselves. These phenotypic changes contribute to greater resistance to fatigue. More mitochondria permit the cells to use more oxygen (supplied by capillaries). Furthermore, the newly generated mitochondria tend

to make more use of fatty acids as their primary substrate instead of glucose. Using fatty acids decreases lactic acid production, saves glycogen, and spares creatine phosphate (see Figure 20.13). The increase in lipids within the cells parallels this shift. The signals that transduce the effects of exercise into mitochondrial biogenesis appear within a few hours of an exercise bout. In sum, endurance exercise triggers multifaceted responses in skeletal muscles.

Three main sets of genes exert specific and independent effects that coincide to enhance the muscles' functions during endurance exercise: (1) contractile protein genes (for specific myosin isoforms), (2) angiogenesis genes (for growth factors such as VEGF that support angiogenesis), and (3) mitochondrial genes (for mitochondrial biogenesis). The fact that all of these phenotypic features occur in response to endurance training strongly suggests that a single factor regulates the transcriptional programs of all three sets of these genes. Indeed, investigators studying both human and rodent skeletal muscles have identified a transcriptional

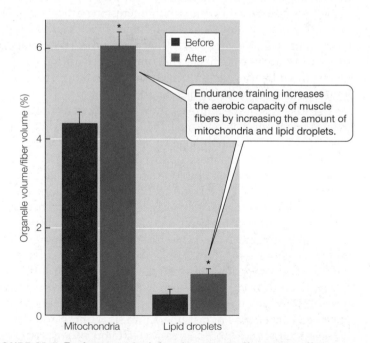

Endurance training increases the aerobic capacity of muscle fibers by increasing the amount of mitochondria and lipid droplets.

FIGURE 21.5 Endurance training increases the proportion of cell volume occupied by both mitochondria and lipid droplets Tissue samples of *vastus lateralis* muscle were taken from five women and five men who trained on a bicycle ergometer 30 min/day, 5 times each week for a period of 6 weeks. Bars represent means ± 1 S.D., and * indicates a significant difference ($2P < 0.05$ in a paired *t* test). (After Hoppeler et al. 1985.)

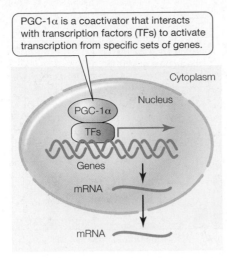

PGC-1α is a coactivator that interacts with transcription factors (TFs) to activate transcription from specific sets of genes.

Cytoplasm

Nucleus

PGC-1α

TFs

Genes

mRNA

mRNA

FIGURE 21.6 PGC-1α is a coactivator of muscle gene expression (After Carter et al. 2015.)

coactivator—PGC-1α—that ties together these effects. PGC-1α is a protein that coactivates many transcription factors (**FIGURE 21.6**). One of its isoforms is **PGC-1α1**, which has been shown to drive the transcriptional programs of mitochondrial biogenesis, conversion of muscle fibers to oxidative types, and angiogenesis. We describe here experiments that used wild-type and transgenic mice to demonstrate that exercise-induced angiogenesis requires the action of PGC-1α1.

Eight-week-old mice were housed individually in cages, half of which were equipped with an exercise wheel connected to a data-acquisition system, as shown in **FIGURE 21.7A**. Mice are nocturnal, and those in cages with exercise wheels ran voluntarily during several consecutive nights (**FIGURE 21.7B**). Mice housed in cages without exercise wheels remained sedentary. Quadriceps muscles of control (wild-type, WT) mice were compared with those of mice that were bred to lack PGC-1α1 specifically in skeletal muscles. Such mice are called *muscle-specific PGC-1α knockout (MKO)* mice. After 14 days of endurance training or being sedentary, the mice were killed and the quadriceps muscles cross-sectioned and stained with a specific marker for endothelial cells (the cells that make up capillary walls). **FIGURE 21.7C** shows that both WT and MKO mice that did not exercise showed no evidence of angiogenesis. Interestingly, although exercised WT and MKO mice both logged similar distances on the exercise wheel, only the quadriceps muscles of the WT mice had an increased density of capillaries. The exercised muscles of MKO mice did not undergo angiogenesis. These data provide strong evidence that PGC-1α1 is required to induce VEGF and angiogenesis. In further experiments, the investigators demonstrated a link between β-adrenergic sympathetic stimulation (which occurs in exercise) and induced PGC-1α1. **FIGURE 21.8A** summarizes the effects of the isomer PGC-1α1 that are triggered by endurance exercise.

Resistance training causes hypertrophy and changes in fiber type

Resistance training aims at increasing muscle size (hypertrophy) and strength. Resistance-exercise programs use repetitions of short, intensive bouts of shortening and/or lengthening contractions as well as isometric contractions. Key to resistance training is keeping exercise bouts as short bursts so that they stimulate hypertrophy of the muscle fibers but not the angiogenesis and mitochondrial biogenesis that are produced by longer-duration endurance exercise. Therefore resistance training usually does not increase the aerobic capacity of the muscle or produce an increase in capillary density. In addition to stimulating hypertrophy, resistance training produces changes in the muscle's fiber-type composition similar to those seen in endurance training. Biopsies of muscle tissue taken from previously untrained human subjects who partici-

(A) The experimental setup

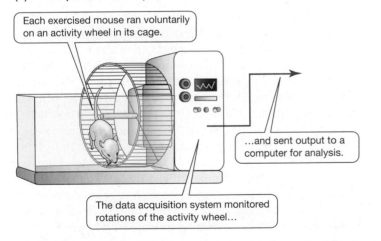

Each exercised mouse ran voluntarily on an activity wheel in its cage.

...and sent output to a computer for analysis.

The data acquisition system monitored rotations of the activity wheel...

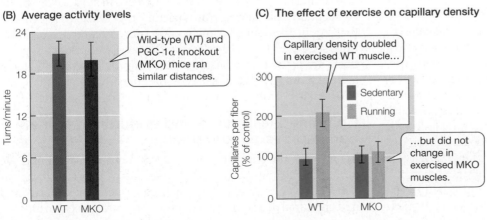

(B) Average activity levels

Wild-type (WT) and PGC-1α knockout (MKO) mice ran similar distances.

Turns/minute

24
18
12
6
0

WT   MKO

(C) The effect of exercise on capillary density

Capillary density doubled in exercised WT muscle...

Capillaries per fiber (% of control)

300
200
100
0

Sedentary
Running

...but did not change in exercised MKO muscles.

WT   MKO

FIGURE 21.7 Angiogenesis induced by endurance training requires the transcriptional coactivator PGC-1α1 (A) Exercised wild-type (WT) and PGC-1α1 knockout (MKO) mice—the latter bred not to express PGC-1α1 in their muscles—ran on activity wheels in their cages; sedentary mice did not have access to activity wheels. (B) Exercising WT and MKO mice logged similar numbers of revolutions (turns) of their activity wheels. (C) Exercised WT mice produced increased capillaries in their quadriceps muscles; however, even though exercised MKO mice ran a similar amount as the WT mice, they did not produce increased capillaries. Neither WT nor MKO sedentary mice produced increased capillaries. (B, C after Chinsomboon et al. 2009.)

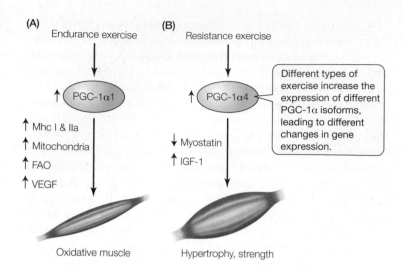

(A) Endurance exercise

↑ PGC-1α1

↑ Mhc I & IIa
↑ Mitochondria
↑ FAO
↑ VEGF

Oxidative muscle

(B) Resistance exercise

↑ PGC-1α4

↓ Myostatin
↑ IGF-1

Hypertrophy, strength

Different types of exercise increase the expression of different PGC-1α isoforms, leading to different changes in gene expression.

FIGURE 21.8 Endurance exercise and resistance exercise induce different isoforms of the transcriptional coactivator PGC-1α (A) Endurance exercise induces PGC-1α1, and (B) resistance exercise induces PGC-1α4. PGC-1α1 causes production of myosin heavy-chain (MhC) isoforms expressed preferentially in oxidative muscles (Mhc I and Mhc IIa), mitochondrial biogenesis, fatty acid oxidation (FAO), and angiogenesis through production of VEGF. PGC-1α4 causes downregulation of myostatin and upregulation of insulin-like growth factor-1 (IGF-1). (After Ruas et al. 2012.)

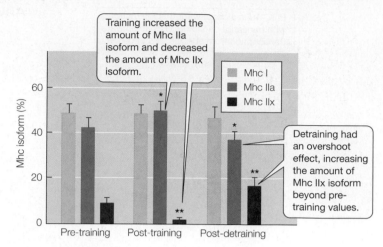

Training increased the amount of Mhc IIa isoform and decreased the amount of Mhc IIx isoform.

Detraining had an overshoot effect, increasing the amount of Mhc IIx isoform beyond pre-training values.

Mhc I
Mhc IIa
Mhc IIx

Mhc isoform (%)

Pre-training   Post-training   Post-detraining

FIGURE 21.9 Fast myosin heavy-chain isoforms change during resistance training and detraining Biopsied tissue samples were taken from the *vastus lateralis* of nine subjects immediately before training, immediately after training, and after the 90-day period of detraining. Tissue samples were analyzed to determine the composition of myosin heavy-chain (Mhc) isoforms, which correlated with fiber type. The bars represent means ± S.E.M., and * indicates a significant difference: *, $P < 0.05$, **, $P < 0.01$. (After Andersen and Aagaard 2000.)

pate in resistance-training programs over several weeks typically show little change in the proportion of Type I fibers in the trained muscle. Instead, resistance training causes the transformation of Type IIx fibers into Type IIa fibers. Investigators have proposed that mechanical deformations of the muscle fiber membrane and cytoskeleton could stimulate a stretch-activated signal that influences gene expression. Such a signal could promote expression of the Type IIa genes and repress expression of the Type IIx genes. In addition, a second isoform of PGC-1α, **PGC-1α4**, is produced in response to resistance exercise and stimulates hypertrophy of muscle cells (**FIGURE 21.8B**). This isoform causes downregulation of the protein myostatin and upregulation of insulin-like growth factor-1 (IGF-1). We will see that both of these molecules are involved in determining muscle mass by balancing protein synthesis and degradation (see Figure 21.17).

Both resistance-trained and endurance-trained muscles continue to remodel during taper

Athletes preparing for competitions often include a period of reduced intensity training, called *taper*, prior to the competition. Using this strategy can yield a 2%–4% increase in performance. The changes that occur during taper are influenced by the type of endurance or resistance demands on the muscles but always include changes in fast myosin isomers. **FIGURE 21.9** shows changes that occurred in muscles that underwent a period of resistance training followed by a period of detraining. This early study provides a foundation for training programs that incorporate periods of taper (reduced-intensity exercise) prior to competitions. Nine untrained men, 27 ± 3 years old, underwent supervised resistance training of the legs three times a week. After 90 days of training, the subjects returned

to their previous (nontraining) level of activity for the next 90 days (a period of detraining). Needle biopsies were taken from the *vastus lateralis* muscle immediately before training, at the end of the training period, and after 90 days of detraining.

Analysis of the tissue samples showed that neither training nor detraining had an effect on the amount of myosin heavy-chain (Mhc) I isoform present in the *vastus lateralis* muscle (see Figure 21.9). As expected, after 3 months of training, the amount of Mhc IIa isoform increased in all nine subjects, from about 42% to about 50%, and the amount of Mhc IIx isoform decreased, from about 9% to about 2%. Interestingly, additional changes occurred during the 90 days of detraining after resistance exercises were stopped. The Mhc IIx isoform increased to about 17%—twice its proportion in untrained muscles—during the detraining period. This increase was accompanied by a corresponding decrease in the Mhc IIa isoform, which declined to an average of 37%. Because large-diameter, glycolytic Type IIx fibers produce the fastest and most powerful contractions, they offer sprinters a strong competitive advantage. Thus, to increase the fastest fibers in their muscles, sprinters often follow a heavy regimen of resistance exercise with tapered training (not so drastic as detraining) for a few weeks in advance of a major competition. This training plan is useful for specialized competitions that require the exertion of high power over a very limited period of time, such as a 60- or 100-m sprint. However, because the Type IIx fibers are prone to fatigue, resistance-training programs that aim toward more hypertrophied Type IIa fibers in exercised muscles will benefit performance in a broader range of competitions.

Combined resistance and endurance training can improve performance

In exploring the effects of endurance and resistance training, we have seen that both produce similar changes in fiber-type composition. However, endurance training triggers increased capillary and mitochondrial density, but not hypertrophy, and resistance training produces hypertrophy without increased capillaries and mitochondria. Most studies on human subjects have considered the effects of endurance and resistance training separately. What would be the effects if athletes undertook both types of training together?

Recently, researchers found that combined resistance and endurance training improved the performance of elite male cyclists, who are recognized as endurance athletes. Fourteen men, 19.5 ± 0.8 years old, participated in 16 weeks of either endurance training alone or a combination of resistance and endurance training (seven men assigned to each training group). Both groups performed 10–18 h of endurance (cycling) training each week. In addition, the combined strength-and-endurance-training group followed a regimen of resistance exercises (knee extension, incline leg press, hamstring curls, and calf raises) with carefully timed rest periods. Before and after the training period, biopsies were taken from the *vastus lateralis* muscle, maximum force exerted by the quadriceps muscle was measured as an indicator of muscle strength, and endurance was assessed by all-out cycling trials on an ergometer. Compared with cyclists in the endurance-training group, those men who underwent combined endurance and resistance training showed greater endurance capacity, increased muscle strength, an increase in Type IIa fibers, and a reduction in Type IIx fibers. Significantly, the endurance phenotype of the muscle fibers was retained: They did not show hypertrophy, and there was no change in the number of capillaries around individual muscle fibers. The results of this study suggest that combined resistance and endurance training can improve the performance of endurance athletes—as long as careful attention is paid to the type and timing of resistance-training exercises.

Hypertrophy also occurs in cardiac muscles

Like skeletal muscle, mammalian cardiac muscle increases in size by hypertrophy (adding proteins to individual cells), not by hyperplasia (adding new cells). Normal physiological hypertrophy of the heart occurs during growth from birth to adulthood as well as in response to changing physiological conditions such as exercise training and pregnancy. Increased pressure or volume of blood in the heart chambers stimulates the heart cells (uninucleate *myocytes*) to add proteins to myofibrils, which increases cross-sectional area, and to add sarcomeres to the ends of myofibrils, which lengthens the cells. The increased cross-sectional area of the myocytes produces thicker heart walls, and the increased cell length increases the internal diameter of the heart chambers. Both of these changes occur mainly in the left ventricle. New capillaries are also added during hypertrophy (correlated with increased levels of VEGF). These morphological changes enhance cardiac function by allowing increased oxygen consumption by the myocytes, increasing the force and speed of contraction, and increasing the volume of blood pumped out of the heart with each heartbeat.

Different types of exercise have different effects on the heart muscle. Endurance training stimulates increased wall thickness and increased internal diameter of the left ventricle. By contrast, resistance training such as weight lifting stimulates increased wall thickness but produces little change in the internal diameter of the left ventricle. Sports such as cycling that involve both endurance and resistance training stimulate intermediate changes.

Hypertrophy in response to imposed physiological demands on the heart is reversible when those demands diminish. For example, during pregnancy a woman's blood volume increases by up to 50% and her circulatory system grows an entire new circuit of vessels through the placenta. Both of these changes profoundly affect the heart's function. Physiologists have equated a pregnant woman's cardiovascular demands to those of an endurance-trained runner. After birth, these demands are reduced, and the heart muscle atrophies to its pre-pregnancy state.

Similarly, the hearts of grizzly bears (*Ursus arctos horribilus*) undergo reversible changes in size. During summer months, the animals are active and their hearts beat at an average rate of 84 beats/min. During the 4–6 months of winter, when the animals hibernate, the heart rate slows to an average of 19 beats/min. Along with this reduced function, the mass of the ventricular tissue atrophies by about 26%. When a bear arouses from hibernation and resumes its summer activities, its heart regains its summer size through hypertrophy.[3]

In humans, heart tissue also becomes hypertrophic under pathological conditions such as chronic high blood pressure or following heart attacks. Researchers have shown that cardiac myocytes undergo different cellular changes during pathological hypertrophy than they do during normal, physiological hypertrophy. For example, pathological hypertrophy is associated with cell death, and the remaining cells switch from using mainly lipids as an energy source to using glucose. Because pathological hypertrophy is associated with increased mortality, researchers are interested in detailing the distinction between physiological and pathological hypertrophy and designing therapies to prevent pathological hypertrophy or ameliorate its effects. Recently they have turned to an unusual model, the Burmese python (*Python bivittatus*, also called *P. molurus*), to better understand normal hypertrophy.

The cardiac myocytes of the Burmese python show an extraordinary capacity to grow in response to stimuli. Metabolically speaking, a python leads a quiet life punctuated by occasional large meals that may be up to 50% of its body mass. With eating, rapid and dramatic changes occur. The alimentary tract elaborates structurally to support digestive processes (see pages 161–162), and other organs also increase in mass, including the liver, pancreas, kidneys, lungs, and heart. Indeed, within 48 h of feeding, the mass of the ventricle of the heart (pythons have just one ventricle; see Chapter 25) increases by 40% (**FIGURE 21.10**)!

The python's enlarged heart allows for increased arterial blood flow to support the increased metabolic demands of digesting and assimilating the meal. For example, the digestive tissues consume more oxygen as they produce enzymes and transport nutrient molecules. Other tissues increase metabolic activity to store assimilated nutrients or build structural and functional molecules. During digestion, the pythons in the study shown in Figure 21.10 increased their oxygen consumption per unit of mass of body tissue nearly sevenfold. Thus cardiac hypertrophy following a meal provides an exquisite adaptation to meet transiently increased metabolic demands. And the change is reversible. When digestion ceases, the python's heart diminishes in size until the next meal is consumed.

---

[3] Studies of heart functions in American black bears (*Ursus americanus*) and brown bears (*Ursus arctos*) have reported similar reductions in heart rate during hibernation (correlating with reduced metabolic rate), but not all of these investigations measured changes in muscle mass of the bears' hearts.

FIGURE 21.10 The heart of the Burmese python (*Python bivittatus*, also called *P. molurus*) undergoes hypertrophy within 48 h of a meal Like other carnivorous nonavian reptiles, *P. bivittatus* experiences periods of metabolic quiescence between bouts of feeding, and a rapid increase in metabolism shortly after a meal. Several tissues increase in mass, including the heart. The mass of the ventricle, cardiac myosin heavy-chain (Mhc) mRNA, and DNA per unit of mass of ventricular tissue all show significant changes within 2 days of consuming a meal. The changes are reversible. Mhc mRNA expression is shown relative to 18S ribosomal subunit mRNA expression, which is constant. Six animals were included in each experimental group. The bars represent mean values ± 1 S.E.M., and * indicates a significant difference ($P < 0.05$) from the fasting value, determined by a one-tail t test. (After Andersen et al. 2005.)

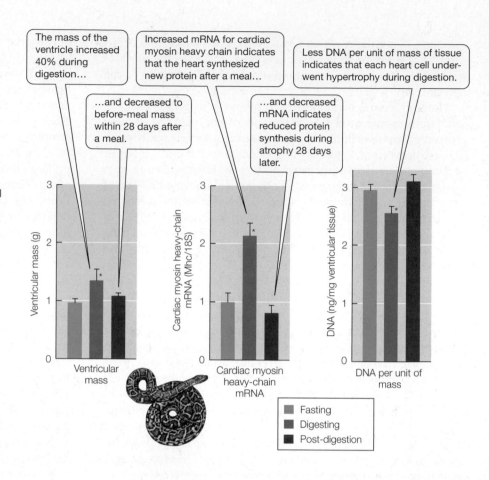

The study illustrated in Figure 21.10 also showed that the huge growth of the python's heart was accomplished by the synthesis of new contractile proteins in the cardiac muscle cells. The researchers studied several different parameters in pythons under three different conditions: after fasting for 28 days, 2 days after consuming a meal of rats equal to 25% of their body weight, and 28 days after the meal. First the researchers determined that fluid accumulation did not account for the increase in the mass of the python's ventricle. They measured both the wet mass of the ventricle and its dry mass (after removal of water). They found that the dry mass constituted the same proportion of the wet mass in all three conditions. Had added fluid contributed to the increase in size of the ventricle, the dry mass sampled 48 h after feeding would have constituted a *smaller* proportion of the wet mass. Thus fluid accumulation did not appear to contribute to the increased mass.

To determine if increased protein synthesis accounted for the increase in wet mass of the ventricle during digestion, the researchers measured the amount of mRNA for the cardiac myosin heavy chain, a major protein constituent of cardiac myocytes. They found that the amount of this mRNA more than doubled, indicating that myosin was being synthesized at a greater rate during digestion than before and after digestion (see Figure 21.10). The researchers also found that the mass of DNA per unit of mass of the ventricle decreased during digestion. This result means that the mass of the tissue increased without an accompanying increase in DNA, and it indicates that the uninucleate myocytes did not divide. Thus hyperplasia did not account for the increase in mass of the ventricle. Instead hypertrophy, the same process that accounts for increase of heart size in mammals, produced the increase in ventricular mass of the python heart.

Cardiac hypertrophy clearly occurs following a meal, but the signal(s) that trigger it remain elusive. Recently, researchers observed that the increased metabolic demands following a meal cause reduced blood O_2 levels, which lead to increased contractile work of the heart. They proposed that the increased mechanical demands on the ventricular myocytes are responsible for initiating molecular processes that result in their hyptertrophy. Another research group examined the molecular and cellular mechanisms that underlie cardiac hypertrophy in pythons and in mice (**FIGURE 21.11**). These researchers found that the heart of a fasted python grew in size when the animal was infused with blood plasma from a fed python for a period of 48 h (see Figure 21.11A). (They used a catheter inserted into the snake's hepatic vein to introduce the added plasma.) This result indicates that something in a fed python's plasma was able to induce hypertrophy in an unfed animal. The researchers used gas chromatography to analyze the constituents of python plasma, and they turned up three fatty acids that increased upon feeding: myristic acid, palmitic acid, and palmitoleic acid. When the researchers infused fasted pythons with a mixture of these fatty acids (in a molar ratio of 1 palmitoleic acid : 6 myristic acid : 16 palmitic acid), they observed an increase in the size of the heart that was similar to that seen in a fed python 3 days after eating (days post-feeding [DPF]; see Figure 21.11A). Amazingly, when mouse pups were infused with the same fatty acid mixture for 7 days, their ventricles grew by about 10% (see Figure 21.11B).

In different experiments, the researchers found that culture medium containing either plasma from a fed python or the fatty acid mixture stimulated hypertrophy of cultured rat ventricular cells (see Figure 21.11C). The molecular characteristics of the hypertrophic rat cells were those of physiological, not pathological, hypertrophy. This study's findings are encouraging to cardiologists in their quest for interventions that would steer cardiac myocytes away from pathological hypertrophy in human disease states. The finding that the same combination of molecules stimulates physiological hypertrophy in widely divergent species is also intriguing from an evolutionary perspective.

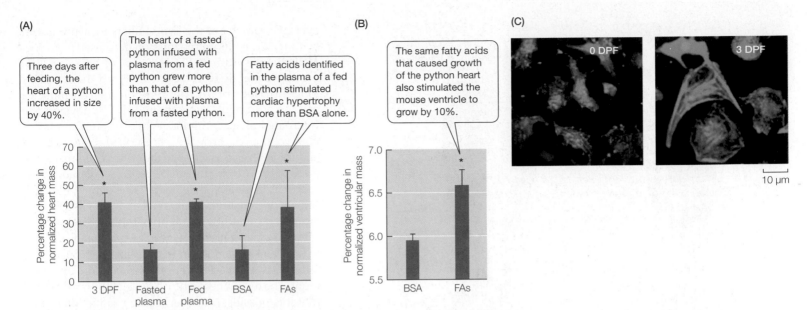

FIGURE 21.11 Fatty acids (FAs) in the plasma of fed Burmese pythons stimulate growth of the python heart, mouse heart, and rat cardiac cells   As described in the text, researchers identified three FAs that increased in the plasma of fed pythons. To determine whether these FAs were acting to induce cardiac hypertrophy, the researchers infused unfed pythons with an artificial mixture of the three FAs dissolved in a bovine serum albumin solution (BSA). (A) Three pythons were infused for 48 h with plasma from a fed or a fasted python, with BSA alone, or with the FAs in BSA and were then compared with pythons that had been fed 3 days earlier (3 days post-feeding [DPF]). The python heart mass was normalized by dividing heart weight by body weight. Error bars represent ± SE; * indicates $P < 0.05$ for tested python heart versus a fasted heart. (B) Six mice were infused for 7 days with either BSA alone or the FAs in BSA. The mouse heart mass was normalized by dividing left ventricular mass by tibia length. The error bars represent ± SE; * indicates $P < 0.05$ for the effect of FAs versus BSA on mouse ventricle. (C) Cultured rat cardiac myocytes were treated with plasma from fasted pythons (0 DPF) or from fed pythons (3 DPF). The fed plasma caused hypertrophy of the myocytes. Similar results were obtained by treating cultured cells with the FA mixture (not shown). In these fluorescence micrographs, the nuclei are blue and a cytoskeletal protein in the cytoplasm is green. (After Riquelme et al. 2011.)

Atrophy

Muscle that is not used will atrophy. When a person wears a cast on her leg, for example, that immobilizes the muscles of the leg, she can lose as much as 20% of the mass of the affected muscles in just a few weeks. The tissue wastes away, and for this reason atrophy is also referred to as *wasting*. Clinical forms of *disuse atrophy* arise not only from limb casting but also from bed rest, spinal cord injuries that cut off nerve input to skeletal muscles, and direct injuries to nerves that innervate muscles. Furthermore, disuse atrophy in sedentary elderly persons adds to and compounds the inevitable atrophy that occurs with aging. Because skeletal muscles play important roles in maintaining posture and producing body movements, impairment of their functions has a strong impact both on the quality of life of individuals and on broader dimensions of public health. Prolonged disuse can lead to changes in muscle structure and function that require lengthy rehabilitation programs and major commitments of health care resources. Atrophy also occurs during starvation, in chronic diseases such as HIV/AIDS, and in genetic diseases such as muscular dystrophy.

Because muscle is the body's main protein store, all muscles continuously synthesize and degrade structural proteins. Normally these two processes are kept in balance. When muscles atrophy, the muscle fibers decrease their uptake of amino acids from the circulation and reduce protein synthesis. However, the most important cause of atrophy appears to be increased enzymatic breakdown of proteins. With the loss of actin and myosin, the myofibrils are reduced, and the diameter of the muscle fibers decreases. Additional structural changes include a reduction in numbers of both mitochondria and nuclei. Biochemical changes reflect these structural changes. For example, atrophic muscles have decreased amounts of actin mRNA, cytochrome *c* mRNA, and oxidative enzymes. As we will see, the molecular signals that trigger these catabolic actions (in which protein degradation exceeds protein synthesis) are not simply the opposite of those that trigger anabolic actions (net protein synthesis) in hypertrophy.

Humans experience atrophy in microgravity

One cause of disuse atrophy is weightlessness during space travel. In one study, researchers used electron microscopy to examine the fine structure of muscle tissue samples taken from four astronauts who spent 17 days on a space shuttle mission. Forty-five days before launch, tissue samples were taken from the *soleus* muscle by needle biopsy. Within 3 h after the shuttle had landed, postflight biopsies were taken. In humans, the *soleus* consists of about 70% slow Type I fibers. It is an important weight-bearing ("antigravity") muscle that lies just beneath the *gastrocnemius* muscle on the posterior of the lower leg (**FIGURE 21.12**). Electron micrographs of longitudinal sections of two Type I muscle fibers (**FIGURE 21.13**) provide evidence of atrophy that occurred in microgravity. Compared with the myofibrils in the preflight tissue, the diameters of the postflight myofibrils were drastically reduced, indicating loss of both thick and thin myofilaments. Indeed, within 17 days, the

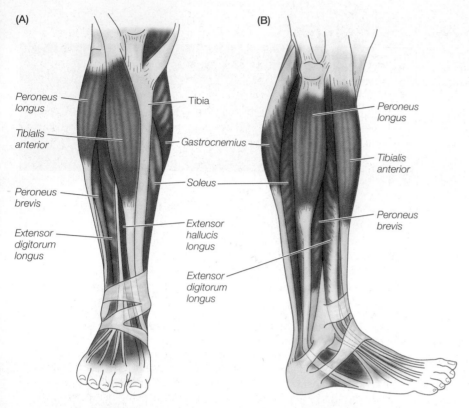

(A)

Peroneus longus

Tibialis anterior

Peroneus brevis

Extensor digitorum longus

Tibia

Gastrocnemius

Soleus

Extensor hallucis longus

Extensor digitorum longus

(B)

Peroneus longus

Tibialis anterior

Peroneus brevis

FIGURE 21.12 Major muscles of the human lower leg (A) The *tibialis anterior* muscle lies lateral to the tibia. (B) The *gastrocnemius* muscle, seen in profile, forms the calf. The *soleus* muscle is partially covered by the *gastrocnemius*.

average diameter of the Type I fibers measured from all four astronauts on the mission decreased from 96 ± 1 micrometers (μm) to 88 ± 1 μm.

Subsequently researchers analyzed the effects of longer periods of microgravity on the calf muscles of nine astronauts who spent 161–192 days on the International Space Station. While in space, the astronaut volunteers carried out prescribed resistance and endurance exercises. Despite the exercise regimens, the mass of the astronauts' calf muscles decreased, as did that of individual muscle fibers. For example, the Type I fibers of the *soleus* diminished an average of 20% in diameter. In addition, the fiber-type composition of the *gastrocnemius* and *soleus* muscles shifted from slow to fast. The reductions in muscle size correlated with reductions in muscle strength tested before and after flight. The researchers who carried out this study noted that it is possible that even greater reductions could have occurred had the astronauts not exercised while on the Space Station.

Because the production of maximum force is proportional to the cross-sectional area of a muscle, atrophy reduces both maximum force production and power output. Decreased contractile performance caused by atrophy could contribute to a variety of risks during space flights, including inability to perform emergency escape procedures and to carry out tasks specific to a mission. The studies of astronaut muscles emphasize the

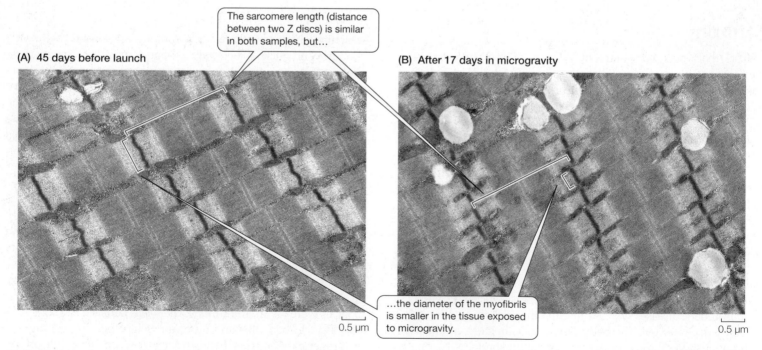

(A) 45 days before launch

The sarcomere length (distance between two Z discs) is similar in both samples, but...

(B) After 17 days in microgravity

...the diameter of the myofibrils is smaller in the tissue exposed to microgravity.

0.5 μm

0.5 μm

FIGURE 21.13 Disuse atrophy occurs during space travel These electron micrographs show longitudinal sections of Type I fibers from the *soleus* muscle of an astronaut (A) 45 days before launch and (B) within 3 h of landing after a 17-day space shuttle mission. To minimize any effects of locomotion on the muscle fibers, the astronauts used wheelchairs after landing until the biopsies were taken. The dramatic reduction in myofibrillar diameter is indicated by the smaller Z discs (brackets) in the postflight tissue. These slow Type I fibers have relatively little sarcoplasmic reticulum. The large lipid droplets indicate that these postflight muscles accumulated lipids. Interestingly, electron micrographs of muscle fibers of astronauts who spent 6 months on the International Space Station showed reduced lipid accumulation. (From Widrick et al. 1999.)

importance of understanding the changes of muscle in microgravity and developing exercise plans that not only stave off reduction of muscle mass but also ensure cardiovascular fitness. Until these gains are made, atrophy will remain a major obstacle to long-term interplanetary space travel for humans.

Disuse influences the fiber-type composition of muscles

Studies on the *soleus* muscles of small mammals under conditions of disuse have shown increases in Type II fibers and decreases in Type I fibers.[4] These results were observed using different experimental techniques, such as immobilizing the hindlimb with a cast to set the muscle in a shortened position (to eliminate the possibility of its exerting force on a load or receiving any mechanical stimulus) and denervation (to eliminate neural stimulation). Changes in fiber-type composition appear to occur very shortly after the onset of disuse. One study reported that rat *soleus* muscles immobilized in a shortened position for 5 days began to transcribe the fast Type IIb genes within 1 day of immobilization.

In humans, some short-term studies of disuse (such as imposed bed rest) have found changes in fiber-type composition, whereas others have found little evidence of change. However, we know that long periods of no activity at all produce profound changes in the paralyzed muscles of persons with spinal cord injuries. Muscles paralyzed for several years are severely atrophied and have hardly any slow Type I fibers. Instead, they consist of fast oxidative Type IIa and fast glycolytic Type IIx fibers, with a preponderance of Type IIx fibers. These muscles also express a fast SR Ca^{2+}-ATPase.

---

[4] As in humans, the *soleus* is normally composed of predominantly slow muscle fibers. The *soleus* muscle of rats consists of 89% Type I fibers and 11% Type IIa fibers, whereas the *soleus* muscle of cats consists of 99% Type I fibers and only 1% Type IIa fibers. The fast Type IIb is entirely absent from the *soleus* in both rats and cats.

Taking these observations into account, researchers have proposed that the fast Type IIx fiber is the "default" fiber type of muscles. If a muscle fiber does not produce contractions that generate force, or is not mechanically stretched, it expresses the Type IIx genes, produces the fastest myosin isoform, and attains a fast phenotype. Mechanical activity is thought to activate expression of the Type I and Type IIa genes, and probably to repress the Type IIx genes.

Muscles atrophy with age

Starting at about age 40, humans lose as much as 1%–2% of muscle mass each year, and this rate accelerates after age 65. The loss occurs in men and women and all ethnicities. It is reflected in declining contractile performance of even the most physically fit individuals. **FIGURE 21.14** shows the record speeds achieved by athletes in track and field events. No matter the event, peak speed declines with age. Peak performance is reached at different ages for events of different distances. Thus sprinters in the 200-m dash achieve their maximum performance at about 21 or 22 years of age, whereas record marathon runners peak in their late 20s. In addition, the slopes of decline in speed with age are steeper for the fast events, which depend more on power output, than for the marathon, which requires endurance. These observations suggest that power output (combined force and velocity of shortening) decreases more rapidly with age than does stamina.

What happens to muscles as they age that causes them to decrease in contractile performance? If disuse atrophy were the only change, then world-class athletes who maintain rigorous exercise programs would not show a decline. But they do. In fact, aging muscles show not only disuse atrophy of individual fibers, but also the loss of muscle fibers. Age-related loss of muscle mass, with its related loss of strength and function, is called **sarcopenia**. Initial studies on aging muscles suggested that the death of motor neurons in the spinal cord was the main cause of the loss of muscle fibers. All of the muscle fibers of a motor unit of a dead motor neuron would lose their innervation and also die (see Figure 20.15). However, only about 10%–15% of motor neurons die during normal aging. Current evidence suggests that some but not all of the distal branches of an aging motor neuron become abnormal and disconnect from the muscle fibers they innervate. The disconnected muscle fibers die. The remaining axon branches of the neuron retain functional contacts with the muscle fibers, but the muscle has lost a portion of the muscle fibers of that motor unit.

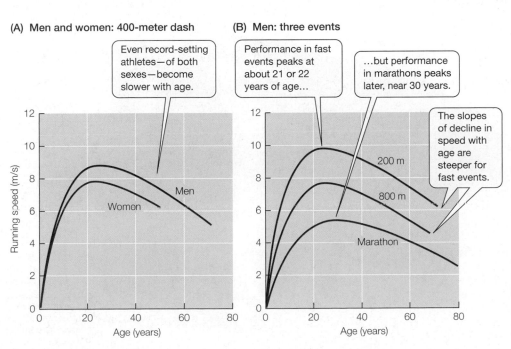

FIGURE 21.14 Record speeds achieved by athletes decrease with age (A) Record speeds for men and women running in the 400-m dash. (B) Record running speeds for men in events of three different lengths. The curves were fitted to data obtained from published records. (After Moore 1975.)

Other studies have shown that aging muscle fibers retain the molecular machinery for regeneration. Through normal daily activities, muscles of all ages undergo minor injuries and continual processes of repair. Studies of aging animals show that their muscles actively express molecules that trigger repair and regeneration (including proliferation of satellite cells; see pages 580–581). Despite this, regenerative processes of many fibers ultimately fail before completion, and atrophy results. Numerous growth factors and hormones influence the maintenance and repair of motor neurons, nerve–muscle synapses, and muscle cells. Studies on the effects of these chemical signals and the roles they play in the interactions between muscle fibers and motor neurons may yield information on treatments that could prevent or delay the loss of muscle mass.

The loss of muscle mass in an elderly person contributes to an inability to adjust posture and maintain balance, which leads to the potential for falls or inappropriate movements that lead to injury. However, encouraging studies strongly suggest that resistance training maintained for prolonged periods is feasible and increases not only muscle strength but also coordination. Additional studies have shown that endurance exercise diminishes the loss of mitochondria seen in aging muscles and therefore improves the cells' metabolic activity.[5] Thus caregivers encourage elderly persons to remain active. Conditioning exercises to prevent injuries and reduce atrophy from disuse can forestall the effects of muscle fiber loss (which at present is inevitable) and contribute to ensuring mobility and independence well into old age. Additional studies also recommend appropriate nutrition with adequate amounts of protein to

maintain muscle mass. It is generally believed that younger persons can also forestall some of the effects of aging by exercising regularly to maintain endurance capacity, muscle strength, and coordination.

Some animals experience little or no disuse atrophy

As we saw in Chapter 20, the muscles of all vertebrate animals are structurally and functionally very similar. Yet although humans and many experimental mammals experience notable losses of muscle after remarkably short periods of disuse, other animals can spend extended periods hibernating or estivating and have very little loss of muscle structure and function. **BOX 21.2** describes the amazing resistance to atrophy shown by an estivating Australian frog. Studies of hibernating bears (**FIGURE 21.15**) also show little disuse atrophy. The bears remain inactive in their winter dens for 5–7 months a year. Remote video recordings of hibernating brown bears show them lying in a curled position for about 98% of each day. Nevertheless, their muscles undergo little decline in contractile capability. Field biologists give heart-stopping accounts of accidentally disturbing overwintering bears and discovering that they show no loss of locomotor ability. Experiments performed to test the strength of the *tibialis anterior* (TA; see Figure 21.12) muscles of black bears in the fall, shortly after they entered their dens, and again in the spring, shortly before they emerged, revealed that the muscles lost only 25%–30% of their strength (measured as production of force) after 110–130 days of inactivity. A subsequent study of the TA muscles of bears in their dens focused on measurements of contractile parameters (such as contraction and relaxation times) and fatigue. The results showed only marginal reductions in contractile capabilities, further supporting the idea that bears resist atrophy of locomotor muscles despite inactivity and not eating.

---

[5] Interestingly, studies on aging mice engineered to overexpress PGC-1α (which is stimulated by endurance training) showed that the muscles maintained mitochondrial biogenesis and function. These data indicate that PGC-1α is essential to reap the effects of endurance training in aging muscle (see Figure 21.7).

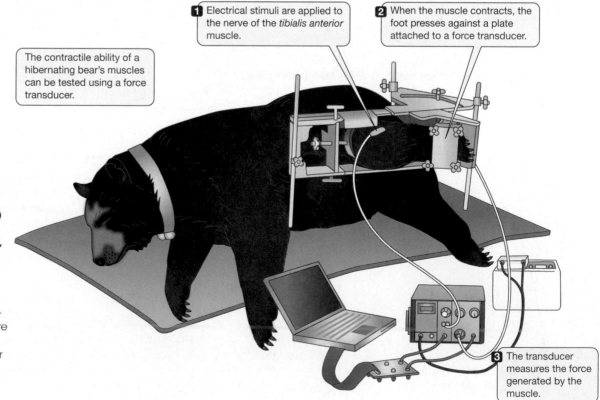

FIGURE 21.15 The American black bear (*Ursus americanus*) spends several months each winter in its den without eating, drinking, defecating or urinating Females give birth and suckle their cubs during this period. Hibernating bears were anesthetized in their dens during early and late winter and taken to the laboratory where they were fitted with an apparatus used to assess muscle strength. After testing, the bears were returned to their dens. (After Harlow et al. 2001.)

The contractile ability of a hibernating bear's muscles can be tested using a force transducer.

1 Electrical stimuli are applied to the nerve of the *tibialis anterior* muscle.

2 When the muscle contracts, the foot presses against a plate attached to a force transducer.

3 The transducer measures the force generated by the muscle.

No Time to Lose

The Australian green-striped burrowing frog (*Cyclorana alboguttata*) can undergo months or even years of immobility and starvation with little muscle atrophy. *C. alboguttata* estivates underground within a cocoon of shed skin and mucus and emerges only during periods of heavy summer rains, which do not necessarily happen every year. During the few weeks when water is available, these frogs must breed and feed before again becoming dormant. To be successful in accomplishing these goals in a limited amount of time, they require muscles that are immediately functional. Several studies have shown that frogs that estivated up to 9 months showed remarkable preservation of the *gastrocnemius*, a large muscle required for jumping. Other, non-jumping muscles decreased somewhat in diameter, and in vitro tests showed greater fatigue and somewhat slower rates of contraction. However, the power output of the muscles was still similar to that of controls. Thus, over long periods of disuse and anorexia, the muscles of *Cyclorana* show selective atrophy that allows them to retain contractile properties that ensure a frog's ability to emerge and breed.

Frogs typically have low metabolic rates, which decrease to extremely low levels during estivation. Still, over several months of anorexia, lipid reserves would be depleted and proteins would be catabolized for fuel to maintain metabolism. In *Cyclorana*, the proteins of non-jumping muscles appear to be catabolized preferentially in order to conserve the myofibrils of the jumping muscles for actions required at emergence.

Consistent with the low metabolic rates, biochemical studies of *Cyclorana* skeletal muscles showed significantly reduced mitochondrial production of reactive oxygen species (ROS), such as hydrogen peroxide. ROS are released from mitochondria during aerobic metabolism, and they are known to cause cellular damage (see Box 8.1). To protect against ROS, many cells produce antioxidants, such as superoxide dismutase, that scavenge and inactivate ROS. It is possible that the frog's low metabolic rate is one factor that slows disuse atrophy of its muscles. Low mitochondrial activity would yield

Green-striped burrowing frog (*Cyclorana alboguttata*)

very low levels of ROS and keep protein degradation and atrophy at a minimum. A possible additional protective measure in estivating *Cyclorana* is that the jumping muscles produce higher levels of antioxidants than do the non-jumping muscles.

Cyclorana's cardiac muscle cells respond differently to estivation. Unlike skeletal muscles cells, they continue mitochondrial production of ROS, reflecting continued production of ATP.

These and other studies have led researchers to propose that overwintering bears (which do not eat, drink, urinate, or defecate) may preserve skeletal muscle protein by recycling nitrogen from urea into amino acids that can be used in protein synthesis. Indeed, measurements of protein metabolism in biopsied samples of *vastus lateralis* muscles taken from bears suggest that maintained protein balance contributes to preserving the integrity of muscles. Researchers collected biopsies in summer and during early and late winter. Protein synthesis was greater than protein breakdown in the summer, but synthesis and breakdown occurred at equal rates throughout the winter; thus the muscles did not experience net protein loss. In dormancy, bears periodically increase the blood flow to their limbs and shiver (showing some similarity to the bouts of *arousal* seen in small mammalian hibernators; see Chapter 11, pages 296–297). They may use shivering as a form of isometric exercise to maintain a low level of muscle use. Studying the functions of animals with muscles that resist atrophy may provide insights into how to preserve structure and function in animals (including humans) whose muscles are susceptible to wasting.

Regulating Muscle Mass

Skeletal muscles move all parts of the skeleton, generate heat, and are involved in regulating metabolism. Maintaining muscle mass, therefore, is essential to homeostasis. It is not surprising that multiple (and overlapping) controls influence the balance between anabolic and catabolic processes in this crucial tissue. As we know, net protein synthesis takes place in hypertrophy, and net protein degradation occurs in atrophy. Earlier we saw that the coactivator PGC-1α4 promotes hypertrophy in response to resistance exercise (see Figure 21.8B). Here we describe two mechanisms that control hypertrophy and atrophy: myostatin and the PI3K–Akt1 pathway.

Myostatin

Currently researchers are avidly investigating the effects of **myostatin** on muscle mass. This growth factor (also called growth and differentiation factor-8, GDF-8) was discovered in 1997 and found to be expressed specifically in developing and adult skeletal muscles. Its function is to regulate muscle mass. Laboratory animals with

(A)

(B)

FIGURE 21.16 Whippets illustrate the role of myostatin as a negative growth regulator (A) This double-muscled dog is homozygous for a mutation of the myostatin gene that results in impaired production of myostatin. (B) This dog does not have the mutation; it expresses myostatin, which limits the growth of its skeletal muscles.

mutations of the myostatin gene that make the myostatin protein inactive show remarkably increased skeletal muscle mass. Thus myostatin is referred to as a negative growth regulator.[6]

Experiments using laboratory animals and animals with natural mutations indicate that without myostatin's regulatory effects, hypertrophy (increase in fiber size) occurs as a result of increased protein synthesis and satellite cell activation. Myostatin binds to a receptor on the muscle cell membrane to induce intracellular signaling sequences that control growth of the cell. The growth factor negatively regulates additional features of the whole muscle, including the amount of fat deposited between muscle fibers. Studies have also shown that the myostatin gene is highly conserved in vertebrate evolution. The coding sequence of the biologically active region is identical in mouse, rat, human, pig, chicken, and turkey myostatins. Myostatin also regulates muscle growth in zebrafish, but its gene sequence is not identical to that found in mammals and birds.

Genetic studies of animals such as racing dogs and the Belgian Blue breed of cattle—and now one human child—have uncovered natural mutations in the myostatin gene that render the myostatin nonfunctional. Animal breeds with the mutation are often referred to as "double-muscled" because of the muscles' extreme bulk. For example, **FIGURE 21.16** compares two whippets (greyhound-like racing dogs). The whippet in Figure 21.16A is homozygous for a mutation of the myostatin gene that causes reduced production of myostatin and a striking double-muscled phenotype. Breeders describe these dogs as "bully" whippets. The whippet in Figure 21.16B is homozygous for the normal myostatin gene. Whippets that are heterozygous for the myostatin gene show some increased muscle bulk and are swifter in races than dogs that have two copies

[6] Myostatin protein is synthesized by muscle cells and secreted into the extracellular space. It functions as an autocrine chemical signal by binding to its specific receptor on the membrane of the muscle cell that secreted it.

of the normal myostatin gene. Owners report that the homozygous double-muscled dogs experience cramping in the shoulder and thigh, and they are seldom used in racing.

Myostatin is of interest to researchers in human medicine, animal husbandry, and human performance. Medical therapies that reduced myostatin expression could be used to diminish muscle wasting in disease states and in aging. Ongoing research efforts in animal breeding are aimed at selecting, or developing through genetic engineering techniques, animals that produce high quantities of succulent meat. Such animals must also be cost-effective to breed (for example, the heavily muscled pelvis of a myostatin mutant female can impede calving, making such an animal cost-*in*effective to breed). Following the publication of a report in 2004 of a child possessing a myostatin mutation, coaches worldwide realized that reduced myostatin expression could have a positive effect on athletic ability. Indeed, the mother of the child was muscular, and other family members were known to be especially strong. Currently there is lively interest in assessing the extent of polymorphisms of the myostatin gene in the human population, and the possibility of identifying "natural" athletes who could be trained to achieve peak performance.

The PI3K–Akt1 pathway

Experiments using mouse models have revealed a pivotal protein—**Akt1**—in muscle cells that appears to regulate the balance between synthesis and degradation. Akt1 (also called protein kinase B [PKB]) is a signal-transduction molecule that regulates several molecules that promote both net protein synthesis and cell survival. Its function is strongly influenced by mechanical stimuli (that is, use and disuse) and endocrine signals. **FIGURE 21.17** shows that Akt1 simultaneously promotes protein synthesis and inhibits protein degradation. When a muscle exerts contractile force against a load, its cells secrete IGF-1 (insulin-like growth factor-1, also referred to as IGF-I). There are several different isoforms of IGF-1. Some circulate in the blood; for example, a major producer of IGF-1 is the liver in response to growth hormone. Other isoforms of IGF-1 act as paracrines or autocrines within the muscle. In the model shown in Figure 21.17, IGF-1 binds to its receptor on the muscle cell membrane and triggers signals that involve phosphoinositol 3-kinase (PI3K) and Akt1 molecules. When Akt1 is activated by phosphorylation, it sets in motion molecular processes that result in increased protein synthesis. At the same time, some of the phosphorylated Akt1 molecules enter the nucleus and prevent the transcription of genes that turn on pathways leading to protein degradation. Thus, in response to contractile activity, activated Akt1 ensures protein synthesis and limits protein degradation. The hormone insulin, binding to its receptor on the muscle cell membrane, also stimulates protein synthesis by the Akt1 pathway. The IGF-1–PI3K–Akt1 pathway also balances protein synthesis and degradation in cardiac muscle cells.

When skeletal muscle fibers (which are multinuclear) become hypertrophic, they typically have the same DNA-to-protein ratio as do cells with smaller cross-sectional areas. To direct the protein synthesis necessary for repair and hypertrophy, a skeletal muscle cell incorporates satellite cells that lie just outside the muscle cell membrane, and these provide additional nuclei (and DNA) to support its increased functions. Experiments suggest that an

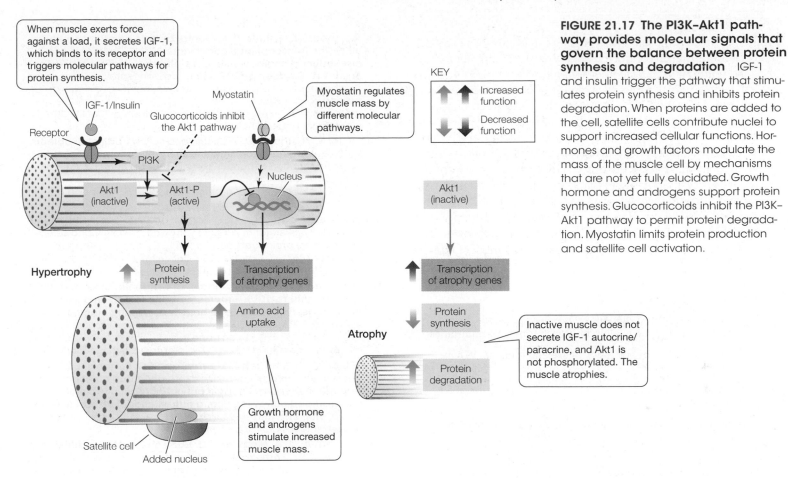

When muscle exerts force against a load, it secretes IGF-1, which binds to its receptor and triggers molecular pathways for protein synthesis.

Myostatin regulates muscle mass by different molecular pathways.

Glucocorticoids inhibit the Akt1 pathway

KEY

Increased function

Decreased function

Inactive muscle does not secrete IGF-1 autocrine/paracrine, and Akt1 is not phosphorylated. The muscle atrophies.

Growth hormone and androgens stimulate increased muscle mass.

FIGURE 21.17 The PI3K–Akt1 pathway provides molecular signals that govern the balance between protein synthesis and degradation IGF-1 and insulin trigger the pathway that stimulates protein synthesis and inhibits protein degradation. When proteins are added to the cell, satellite cells contribute nuclei to support increased cellular functions. Hormones and growth factors modulate the mass of the muscle cell by mechanisms that are not yet fully elucidated. Growth hormone and androgens support protein synthesis. Glucocorticoids inhibit the PI3K–Akt1 pathway to permit protein degradation. Myostatin limits protein production and satellite cell activation.

important function of locally acting IGF-1 is to stimulate satellite cells to proliferate and fuse with the muscle fibers. Thus net protein synthesis in the muscle fiber occurs in coordination with the activation of satellite cells.

The basic Akt1 pattern is intertwined with additional processes. First, the molecules downstream from Akt1 that are involved in protein synthesis are influenced by additional factors, including available amino acids and ATP production within the cell. Second, several genes direct atrophy, and Akt1 may not inhibit all of them. Finally, there are three isoforms of Akt—Akt1, Akt2, and Akt3—each encoded by a separate gene. Whereas Akt1 regulates growth, Akt2 regulates metabolism of muscle cells in response to insulin binding to its receptor on the cell membrane (see Chapter 16, page 446). Akt3 is not expressed in muscle. Interestingly, in muscle tissue, IGF-1 and insulin overlap in function. Both stimulate protein synthesis via Akt1, and IGF-1 appears able to regulate glucose metabolism to some extent. In the overall life cycle, researchers have found that IGF-1 is essential to ensure that protein synthesis exceeds degradation during growth of young animals. In adult animals, protein synthesis is needed mainly for maintenance and repair and IGF-1 plays a lesser role—unless its secretion is stimulated by increased loading (such as resistance training), in which case its action is stepped up to produce hypertrophy. Insulin is essential for skeletal muscle metabolism in both young and adult animals, but it is not so important in hypertrophy. Growth hormone (GH) and androgens such as testosterone also promote protein synthesis. Studies are ongoing to clarify the molecular mechanisms by which these hormones work.

Atrophy occurs in an inactive muscle (such as one in a cast that does not exert force against a load) because the PI3K–Akt1 pathway is not activated, or is activated to a lesser degree. Akt1 directs less protein synthesis, the atrophy genes it had kept in check are expressed, degradation of proteins exceeds synthesis, and there is a reduction in mass (see Figure 21.17). The Akt1 pathway is also inhibited by glucocorticoids. As we saw in Chapter 16 (see pages 443–444), glucocorticoids are secreted as part of the *stress response* to stimulate protein catabolism in muscles. The liver uses the released amino acids as a carbon source to produce glucose that is released into the blood as a fuel, especially for the brain. This response is very important to survival during starvation, for example. But corticosteroids such as prednisone, which are prescribed clinically to combat inflammation, produce muscle wasting as an unwanted side effect. Researchers are actively investigating possible interventions in the PI3K–Akt1 pathway to prevent atrophy.

Summary

In this chapter we have considered the remarkable malleability of striated muscle. We have seen that muscles respond to use or disuse, and that they respond in particular ways to specific kinds of use. Muscle fibers decrease in mass by atrophy and increase by hypertrophy. Depending on conditions of use, they express different isoforms of functional molecules, such as the myosin heavy chain of the thick filament, the Ca^{2+}-ATPase pump of the sarcoplasmic reticulum (SR), or mitochondrial enzymes. The functions of muscle fibers are also influenced by their associations with capillaries (which proliferate around muscle fibers engaged in endurance exercise), motor neurons (which control the fiber type of an entire motor unit and also maintain a muscle fiber's viability), and hormones and growth factors (which affect the balance of protein synthesis and degradation in the muscle fiber). Although muscle fibers change in response to specific activities, an individual's genetic make-up also influences muscle function, for example by directing the proportions of different fiber types within a skeletal muscle or the degree of hypertrophy that muscle fibers can achieve. Because research on muscle plasticity is motivated to a large extent by interest in human health and performance, humans and small mammals are often used as research subjects. However, other animal models (including amphibians, nonavian reptiles, and large mammals) provide useful perspectives for understanding the plastic potential of muscle.

STUDY QUESTIONS

1. List and describe three changes in muscles that occur during endurance training and explain how each improves endurance.

2. Define and describe hypertrophy of skeletal muscles. What conditions stimulate it?

3. Discuss the adaptive advantage of cardiac hypertrophy in a human and in a python.

4. Myostatin is termed a negative growth regulator. Explain the meaning of this term, and describe the consequences of gene mutations that cause myostatin to be nonfunctional.

5. Describe and discuss molecular mechanisms that contribute to the balance between protein synthesis and degradation in regulating muscle mass.

6. Speculate on the reasons dormant animals experience little disuse atrophy, despite inactivity and anorexia.

7. Why can't resistance exercises prevent decreased power output in aging muscles?

Go to **sites.sinauer.com/animalphys4e**
for box extensions, quizzes, flashcards,
and other resources.

REFERENCES

Andersen, J. L., and P. Aagaard. 2010. Effects of strength training on muscle fiber types and size; consequences for athletes training for high-intensity sport. *Scand. J. Med. Sci. Sports* 20 (Suppl. 2): 32–38. A review of the effects of strength training on muscle fiber-type composition, contractile properties, and hypertrophy with the aim of defining useful training programs.

Chinsomboon, J., J. Ruas, R. K. Gupta, and six additional authors. 2009. The transcriptional coactivator PGC-1α mediates exercise-induced angiogenesis in skeletal muscle. *Proc. Natl. Acad. Sci. U.S.A.* 106 (50): 21401–21406. Endurance training stimulated angiogenesis in skeletal muscle of wild-type mice, but similar training of knockout mice lacking PGC-1α1 did not stimulate angiogenesis. These experiments provide evidence that PGC-1α1 is required for angiogenesis in skeletal muscle.

Fitts, R. H., S. W. Trappe, D. L. Costill, and seven additional authors. 2010. Prolonged space flight-induced alterations in the structure and function of human skeletal muscle fibres. *J. Physiol.* 588: 3567–3592. A study of the characteristics of muscles of nine volunteer astronaut subjects who spent 6 months on the International Space Station.

Mantle, B. L., N. J. Hudson, G. S. Harper, and two additional authors. 2009. Skeletal muscle atrophy occurs slowly and selectively during prolonged aestivation in *Cyclorana alboguttata* (Gunther 1867). *J. Exp. Biol.* 212: 3664–3672. Jumping muscles were preferentially spared during aestivation.

McGee-Lawrence, J., P. Buckendahl, C. Carpenter, and three additional authors. 2015. Suppressed bone remodeling in black bears conserves energy and bone mass during hibernation. *J. Exp. Biol.* 218: 2067–2074. A study of serum markers of bone remodeling combined with bone biopsies clarifies how eucalcemia, bone mass, and strength are preserved during hibernation.

Reilly, B. D., A. J. R. Hickey, R. L. Cramp, and C. E. Franklin. 2014. Decreased hydrogen peroxide production and mitochondrial respiration in skeletal muscle but not cardiac muscle of the green-striped burrowing frog, a natural model of muscle disuse. *J. Exp. Biol.* 217: 1087–1093. A study of the production of reactive oxygen species during aestivation.

Riquelme, C. A., J. A. Magida, B. C. Harrison, and four additional authors. 2011. Fatty acids identified in the Burmese python promote beneficial cardiac growth. *Science* 334: 528–531. A study that identifies molecular components that underlie the hypertrophy of the python heart and shows that these same components also cause cellular and organ hypertrophy in rats.

Ruas, J. L, J. P. White, R. R. Rao, and 14 additional authors. 2012. A PGC-1α isoform induced by resistance training regulates skeletal muscle hypertrophy. *Cell* 151: 1319–1331. The PGC-1α isoform PGC-1α4 upregulates IGF-1 and downregulates myostatin. This study complements that of Chinsomboon et al. 2009, which showed that PGC-1α1 induces angiogenesis with endurance training.

Slay, C. E., S. Enok, J. W. Hicks, and T. Wang. 2014. Reduction of blood oxygen levels enhances postprandial cardiac hypertrophy in Burmese python (*Python bivittatus*). *J. Exp. Biol.* 217: 1784–1789. Experiments suggest that cardiac hypertrophy is stimulated when metabolic demands of systemic tissues outpace the level of oxygen delivery.

Yan, Z., M. Okutsu, Y. N. Akhtar, and V. A. Lira. 2011. Regulation of exercise-induced fiber type transformation, mitochondrial biogenesis, and angiogenesis in skeletal muscle. *J. Appl. Physiol.* 110: 264–274. A review of signal transduction pathways involved in adaptations to exercise. The results described are based largely on experiments that used genetically engineered animals.

See also **Additional References** *and* **Figure and Table Citations**.

PART V
Oxygen, Carbon Dioxide, and Internal Transport

Previous page
By making a tube in which it lives, this marine annelid worm (*Portula* sp.) is well poised to defend itself against predators by quickly withdrawing into the tube when danger approaches. However, supplies of oxygen (O_2) inside the tube are too meager to meet metabolic needs. Worms such as this get their O_2 from the open water. For O_2 uptake, the worms have evolved elaborate arrays of pinnately divided tentacles that they project into the open water, as seen here. The tentacles serve as gills. They collectively present a large surface area to the water for gas exchange, and blood containing a hemoglobin-like O_2-transport pigment circulates between them and the body of the worm in the tube. Worms such as these can regenerate the full array of tentacles and even their entire anterior end, including the brain, if a predator succeeds in nipping off exposed body parts.

Introduction to Oxygen and Carbon Dioxide Physiology

These aquatic animals are collecting oxygen (O_2) from the water in which they live. The O_2 is dissolved in the environmental water and must move from the water to each animal's cells for the cells to use it. As we discuss the movement of gases, we will follow the same convention we adopted in Chapter 5 and use the word **transport** as a general term to refer to all movements; thus when we speak of "gas transport," we will refer in an entirely general way to any and all movements of gases from place to place, regardless of mechanism. An important characteristic of animals in their relations to O_2 is that, insofar as we now know, *active* transport mechanisms for O_2 do not exist (see page 113). Thus, for O_2 to move from the environmental water to the cells of an animal, conditions must be favorable for *passive* transport toward the cells along the entire path that O_2 follows to reach the cells.[1] The coral-reef worms in the photograph —which have gills composed of spectacular pinetree-shaped arrays of tentacles—have an elaborate circulatory system, filled with blood that is rich in a hemoglobin-like O_2-transport

---

[1] See Chapter 5 for a rigorous discussion of the distinction between *active* and *passive* transport across cell membranes. A quick way to distinguish them here is to say that ATP is used in active transport but not passive transport. *Passive* transport is always in the direction of equilibrium (see page 112). *Active* transport can be opposite to that direction because during active transport, metabolic energy is employed. Because there are no active-transport mechanisms for O_2, the direction of O_2 movement must always be toward equilibrium.

Both of these kinds of aquatic animals take up dissolved O_2 from the water by breathing with gills The orange, pinetree-shaped structures are the gills of coral-reef worms (*Spirobranchus*). These worms (often called "Christmas tree worms") live in tubes embedded in the corals. They acquire O_2 by projecting their gills—consisting of highly vascularized tentacles—into the water. The tentacles also collect food. Diving beetles, which are common inhabitants of freshwater ponds, ostensibly breathe by acquiring bubbles of air from the atmosphere. However, a beetle's bubble also serves as a gill to acquire O_2 from the water while the beetle is underwater, and this gill action often accounts for most of a beetle's O_2 uptake during a dive.

pigment. As their blood flows through their gill tentacles, O_2 enters the blood from the environmental water by passive diffusion, provided conditions are favorable for passive transport in that direction; then the circulation of the blood delivers the O_2 to all cells. The bubble carried by the water beetle also acts as a gill. People often think that a water beetle gains O_2 solely by picking up a bubble of air from the atmosphere. In reality, a beetle gains far more O_2 by having the bubble serve as a gill while the beetle is underwater. When a beetle is submerged, O_2 moves steadily into its bubble from the water, provided conditions are favorable for passive diffusion in that direction. The O_2 is then distributed to all the beetle's cells by entering a system of gas-filled tubules (the tracheal system) that ramifies throughout the insect body.

How can we determine if conditions are favorable for O_2 to enter an animal from its environment, rather than moving in the opposite direction? This is one of the central questions we will address in this chapter. As we discuss O_2 transport, we will also discuss carbon dioxide (CO_2) transport, because O_2 and CO_2 are the principal gases that are consumed by and produced by cellular respiration (aerobic catabolism). Together, O_2 and CO_2 are called the **respiratory gases**.

Of all the exchanges of materials between an animal and its environment, the exchanges of the respiratory gases are usually the most urgent. A person, for example, dies within minutes if denied O_2 but can live for days without exchanging nutrients, nitrogenous wastes, or water. The urgency of the need for O_2 arises from the role that O_2 plays as the final electron acceptor in cellular respiration. As discussed in Chapter 8 (see Figure 8.3), energy cannot be transferred from bonds of food molecules to bonds of ATP by the aerobic catabolic apparatus of a cell unless O_2 is available in the cell's mitochondria to combine with electrons exiting the electron-transport chain. The need to void CO_2 is ordinarily not as urgent as the need to acquire O_2. Nonetheless, export of CO_2 is often a pressing concern because accumulation of CO_2 in the body can rapidly acidify the body fluids and exert other harmful effects.

The respiratory gases move from place to place principally by two mechanisms: *simple diffusion* and *convection (bulk flow)*. In the case of O_2, these in fact are the only mechanisms of transport because, as already stressed, active transport of O_2 is unknown. Carbon dioxide, however, is sometimes actively transported across cell membranes in the form of bicarbonate ions (HCO_3^-) formed by reaction with water.

An important first step in understanding respiratory gases and gas transport is to address the concept of *chemical potential* and how it is expressed in studies of gases. In Chapter 5, when we discussed the diffusion of glucose and other uncharged solutes in aqueous solutions, we concluded that (1) a solute such as glucose always diffuses from regions of a solution where it is relatively high in *concentration* to regions where it is relatively low in *concentration*, and that (2) the rate of diffusion of such a solute from region to region is directly proportional to its difference in concentration between regions (see Equation 5.1). Whereas these principles apply to the sorts of solutes and situations that have dominated our attention in earlier parts of this book, the principles are not in fact entirely general.

A truly general statement of the principles of diffusion is worded in terms of **chemical potential**. The abstract definition of chemical potential is a topic in physical chemistry; loosely, chemical potential is the strength of the tendency of a chemical substance to undergo a physical or chemical change. A practical definition of chemical potential for our purpose here—the study of chemical diffusion—is to say that chemical potential provides the basis for a *truly general law of diffusion*: In all cases of diffusion, without exception, materials tend to move in net fashion from regions where their *chemical potential* is high to regions where their *chemical potential* is low, and at a rate proportional to the difference in chemical potential.

In the study of diffusion *within a single aqueous solution*, the *concentration* of a solute (the amount of solute per unit of volume) is a useful surrogate for *chemical potential* because, for most purposes, if one uses diffusion equations expressed in terms of the concentrations of solutes, one obtains correct answers.

Concentration, however, is not a useful measure of chemical potential when analyzing the diffusion of materials *between gas mixtures and aqueous solutions*. Materials such as O_2 and CO_2 exist in both phases—both gas mixtures and aqueous solutions—and they diffuse between them. If one attempts to analyze their diffusion between phases on the basis of their concentrations, *grossly erroneous conclusions can be reached*. To analyze such diffusion accurately, a direct measure of chemical potential (not an indirect or surrogate measure) is needed. Physiologists use *partial pressure* to express chemical potential in the study of gases. Thus, as we proceed in our study of gases, we will often encounter the concept of partial pressure. We now turn to a discussion of partial pressure itself and of the relation of partial pressure to concentration.

The Properties of Gases in Gas Mixtures and Aqueous Solutions

In addition to existing in air and other types of gas mixtures, gases also *dissolve* in aqueous solutions, as already suggested. When a gas dissolves in an aqueous solution, molecules of the gas become distributed among the H_2O molecules in much the same way as glucose molecules or Na^+ ions are incorporated among H_2O molecules when solids dissolve. The molecules of a gas *disappear* when they go into solution. They do not appear as tiny bubbles any more than glucose in solution appears as tiny sugar grains; when we see bubbles of gas, they represent gas that is *not* in solution. In this section we first address gases in gas mixtures, then gases dissolved in aqueous solutions. We will sometimes refer to these two conditions as the *gas phase* and *liquid phase*, respectively.

Gases in the gas phase

The modern study of gases in the gas phase traces back to John Dalton (1766–1844), who articulated the *law of partial pressures*. According to this concept, the total pressure exerted by a mixture of gases (such as the atmosphere) is the sum of individual pressures exerted by each of the several component gases in the mixture (**FIGURE 22.1**). The individual pressure exerted by any particular gas in a gas mixture is termed the **partial pressure** of that gas. An important property of the partial pressure of each gas in a mixture

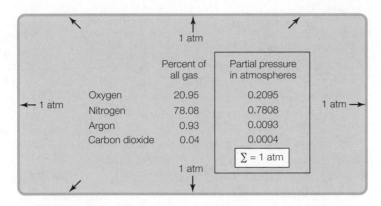

FIGURE 22.1 The total pressure exerted by a mixture of gases is the sum of the partial pressures exerted by the individual constituents of the mixture The diagram shows a container surrounding a body of dry atmospheric air at sea level. Data on the four most abundant constituents of dry air are shown. The air exerts a total pressure of 1 atmosphere (1 atm), which is the sum of the partial pressures. Each constituent would exert its same partial pressure even if the other constituents were absent.

is that it is independent of the other gases present. Moreover, in a volume of mixed gases, each component gas behaves in terms of its partial pressure as if it alone occupied the entire volume. Thus the partial pressure of each gas in a gas mixture can be calculated from the **universal gas law**,

$$PV = nRT \qquad (22.1)$$

where P represents pressure, V is volume, n is the number of moles (the quantity) of gas, R is the universal gas constant, and T is the absolute temperature.[2] To use the universal gas law to calculate the partial pressure of a particular gas in a gas mixture, one sets n equal to the molar quantity of the *particular gas* of interest and V equal to the volume occupied by the *gas mixture as a whole*.

A simple procedure exists to calculate the partial pressure of a gas in a mixture if one knows the proportions of the various gases in the mixture. The proportions of gases in a mixture are expressed as *fractional concentrations*. The **mole fractional concentration** of any particular gas in a mixture is the fraction of the total moles of gas present represented by the gas in question. To illustrate, in a volume of dry atmospheric air near sea level, the number of moles of O_2 is 20.95% of the total number of moles of all gases taken together; therefore the mole fractional concentration of O_2 in dry air is 0.2095. The **volume fractional concentration** of a particular gas in a mixture is the fraction of the total volume represented by that particular gas. Equal molar quantities of different gases occupy essentially equal volumes. Thus the volume fractional concentration of a gas in a mixture is essentially the same as its mole fractional concentration. This means, for example, that if we remove the

O₂ from a volume of dry atmospheric air at a given temperature and pressure and we then restore the remaining gas to the same temperature and pressure, the final volume will be 20.95% less than the original volume. It is easy to show, using the universal gas law (see Equation 22.1), that the partial pressure exerted by each gas in a mixture is "its fraction" of the total pressure, where "its fraction" means the mole fractional concentration or volume fractional concentration of the gas. Expressed algebraically, if P_{tot} is the total pressure of a gas mixture, P_x is the partial pressure of a particular gas (x) in the mixture, and F_x is the mole or volume fractional concentration of that gas, then

$$P_x = F_x P_{tot} \qquad (22.2)$$

To illustrate the application of Equation 22.2, let's calculate the partial pressures of O_2, nitrogen (N_2), and CO_2 in dry atmospheric air. Near sea level, dry air consists chemically of 20.95% O_2, 78.08% N_2, and 0.04% CO_2.[3] If the total pressure of the air is 1 atmosphere (atm), then the partial pressure of O_2 in the air is $F_x P_{tot} = (0.2095)$ (1 atm) = 0.2095 atm, and the partial pressures of N_2 and CO_2 are 0.7808 atm and 0.0004 atm (see Figure 22.1). We discuss other units for pressure later in this chapter.

What is the relation between the *partial pressure* of a gas in a gas mixture and the *concentration* (amount per unit of volume) of the gas in the mixture? If one rearranges the universal gas law (Equation 22.1), one gets $P = (n/V)RT$. If n is the molar amount of a particular gas in a gas mixture that occupies a total volume V, then n/V is the concentration of the gas, and P is its partial pressure. Moreover, RT is a constant if the temperature T is constant. Thus, in a gas phase, the *partial pressure* and the *concentration* of any particular gas are simply proportional to each other at any given T. For example, if there are two gas mixtures that are identical in temperature and if the concentration of O_2 is 5 millimole per liter (mmol/L) in one and 10 mmol/L in the other, the partial pressure of O_2 will be exactly twice as high in the latter.

Gases in aqueous solution

The partial pressure of a gas dissolved in an aqueous solution (or any other sort of solution) is defined to be equal to the partial pressure of that gas in a gas phase with which the solution is at equilibrium. To illustrate, consider what happens if O_2-free water is brought into contact with air containing O_2 at a partial pressure of 0.21 atm. Let's assume, specifically, that the volume of air is so great that as O_2 dissolves in the water, there is essentially no change in the O_2 concentration of the air, and thus the partial pressure of O_2 in the air remains 0.21 atm. Oxygen will dissolve in the water until equilibrium is reached with the air. Then the partial pressure of O_2 *in the aqueous solution* will be 0.21 atm. If this solution is later exposed to air that contains O_2 at a partial pressure of 0.19 atm, the solution will lose O_2 to the air until a new equilibrium is established. The partial pressure of O_2 in the solution will then be 0.19 atm. The term **tension** is sometimes used as a synonym for partial pressure when speaking of gases in aqueous solutions, as when the O_2 partial pressure is called the "O_2 tension." We do not use the tension terminology in this book.

---

[2] The partial pressure of a particular gas is symbolized by writing P with the molecular formula of the gas as a subscript. Thus, for example, the partial pressure of O_2 is symbolized P_{O_2}. In the SI, R is 8.314 J/mol•K (where K represents a Kelvin and is equivalent to 1 degree Celsius). Values for R in other systems of units are listed in standard reference books on chemistry and physics. When we say T is absolute temperature, we mean it is temperature on the Kelvin scale; in a room at 20°C, for example, T is 293 K.

[3] In the mid-twentieth century, the value for CO_2 was 0.03%. Burning of coal, petroleum, and forest wood has raised it to 0.04%.

The *partial pressure* and the *concentration* of a gas in an aqueous solution are proportional to each other, but the nature of this proportionality is more complicated than that in gas phases. **Henry's law** is the fundamental law that relates partial pressure and concentration in aqueous solutions. There are several ways in which this law is expressed, and the various expressions employ several different, but related, coefficients.[4] Our approach here is to use the **absorption coefficient**, defined to be the dissolved concentration of a gas when the partial pressure of the gas in solution is 1 atm. If P_x is the partial pressure of a particular gas (x) in solution, C_x is the dissolved concentration of the gas, and A is the absorption coefficient,

$$C_x = AP_x \qquad (22.3)$$

In gas phases, where C_x is also proportional to P_x, the equation relating C_x and P_x is essentially identical for all gases (because all adhere to the universal gas law). In aqueous solutions, however, the proportionality constant A varies a great deal, depending on the specific gas of interest, temperature, and salinity (values of A can be looked up in reference books on gas physical chemistry).

The coefficient A, the absorption coefficient, is in fact a measure of gas *solubility*. A high absorption coefficient signifies high solubility, meaning that a lot of gas will dissolve at any particular partial pressure. With this in mind, three important characteristics of gases dissolved in aqueous solutions come to light when absorption coefficients are examined:

- The solubilities of different gases are different. Specifically, CO_2 has a far higher solubility than O_2 or N_2. The absorption coefficients of CO_2, O_2, and N_2 in cold (0°C) distilled water are 77, 2.2, and 1.1 mmol/L, respectively. These absorption coefficients tell you that if each of these gases is brought to a partial pressure of 1 atm in cold distilled water, the amount dissolved in each liter of water will be 77 mmol of CO_2 but only 2.2 mmol of O_2 and 1.1 mmol of N_2.

- The solubilities of gases in aqueous solutions decrease strongly with increasing water temperature. This is true of all gases. To illustrate, the absorption coefficients of O_2 in distilled water at 0°C, 20°C, and 40°C are 2.2, 1.4, and 1.0 mmol/L, respectively. Thus, if the partial pressure of O_2 is 1 atm, a liter of distilled water dissolves 2.2 mmol of O_2 when at 0°C but less than half as much, 1.0 mmol, when at 40°C. Gases tend to come out of solution and form bubbles as water warms (**FIGURE 22.2A**).

- The solubilities of gases in aqueous solutions decrease with increasing salinity. For instance, the absorption coefficients of O_2 at a fixed temperature (0°C) in distilled water, 80% seawater, and full-strength seawater are 2.2, 1.8, and 1.7 mmol/L, respectively. Increasing the salinity of an aqueous solution tends to drive gases out of solution by decreasing the solubilities of the gases, a phenomenon called the *salting-out effect* (**FIGURE 22.2B**).

---

[4] During the history of the study of gases in aqueous solution, an uncommonly large number of coefficients were defined by various scientists working on the subject. Many of these have survived into the present time and today are referred to by names such as *Henry's law coefficients*, the *Bunsen coefficient*, and the *Ostwald coefficient*. Although differing in detail, all describe the same principles and can be interconverted using standard equations.

(A) A sealed sample bottle of cold creek water after sitting in a warm room

(B) A carbonated beverage to which grains of salt have been added

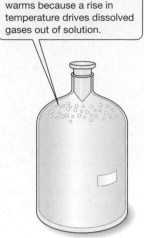

Bubbles form when water warms because a rise in temperature drives dissolved gases out of solution.

Each grain of salt increases salinity in its vicinity, driving dissolved CO_2 out of solution.

Grain of salt

FIGURE 22.2 Everyday illustrations of the effects of temperature and salinity on gas solubility

SUMMARY
The Properties of Gases in Gas Mixtures and Aqueous Solutions

- The total pressure of a gas mixture is the sum of the partial pressures exerted by the individual gases in the mixture. The partial pressure of each gas is independent of the other gases present.

- In gas mixtures, the concentration and the partial pressure of any given gas are simply proportional to each other. The proportionality coefficient is the same for all gases, at any given temperature, because it is a corollary of the universal gas law.

- When gases dissolve in aqueous solutions, they disappear into solution, just as sugars disappear when they dissolve. Bubbles of gas, no matter how tiny, are not in solution.

- The partial pressure of a gas dissolved in an aqueous solution is equal to the partial pressure of the same gas in a gas phase with which the solution is at equilibrium.

- In an aqueous solution, the concentration and the partial pressure of any given dissolved gas are proportional. However, the coefficient of proportionality, termed the absorption coefficient, varies greatly from gas to gas, and depends on temperature and salinity.

- Gas solubility decreases with increasing temperature and also with increasing salinity.

Diffusion of Gases

Simple diffusion, as earlier noted, is one of the two principal mechanisms of respiratory gas transport. When gases diffuse from place to place, they do so by the same fundamental mechanism by which solutes diffuse through solutions (see Chapter 5, page 105). Gas molecules move ceaselessly at random on an atomic-molecular scale. Merely by the operation of the laws of probability, when the chemical potential of a gas differs from place to place, these random movements carry more gas molecules away from regions of high chemical potential than into such regions. Macroscopically, therefore, net gas transport occurs.

The fundamental law of gas diffusion is that *gases diffuse in net fashion from areas of relatively high partial pressure to areas of relatively low partial pressure. This is true within gas mixtures, within aqueous solutions, and across gas–water interfaces.* In view of this law, the coral-reef worms we discussed at the start of this chapter can get O_2 from their environmental water only if the partial pressure of O_2 dissolved in the water is higher than the partial pressure of O_2 dissolved in their blood as the blood flows through their gill tentacles. Similarly, the water beetle can get O_2 from the water only if the partial pressure of O_2 dissolved in the water exceeds the partial pressure of O_2 in its bubble gas.

The diffusion of gases in the direction of the partial-pressure gradient does *not* necessarily mean that diffusion occurs in the direction of the concentration gradient. This is an enormously important point. To see it clearly, we need first to consider some elementary cases that are, in a sense, deceptively simple. Let's consider diffusion *within* a gas mixture of *uniform temperature*—or *within* an aqueous solution of *uniform temperature and salinity*. In these cases, if the partial pressure of a particular gas is greater in one region than in another, the concentration is also greater. Thus, *within* such gas mixtures and aqueous solutions, when gases diffuse from high to low partial pressure, they also diffuse from high to low concentration. These elementary cases are merely simple and special cases, however, and in that way they are deceptively simple.

Under more complex circumstances, situations commonly exist in which gases—while obeying the fundamental law of gas diffusion—diffuse from low concentration to high concentration! Water beetles often exemplify this scenario. If we assume a uniform temperature of 20°C, and if we assume a beetle's environmental water is at equilibrium with the atmosphere so that the O_2 partial pressure in the water is 0.21 atm, the O_2 concentration in the environmental water is about 0.3 mmol/L (calculated from Equation 22.3 using the value for *A* already mentioned). Air is dramatically richer in O_2 than water is under most circumstances, as we discuss later. Specifically, atmospheric air at 20°C, with an O_2 partial pressure of about 0.21 atm (see Figure 22.1), has an O_2 concentration of about 8.6 mmol/L. Let's consider a beetle that grabbed a fresh bubble from the atmosphere a while ago and has removed 50% of the O_2 from its bubble (**FIGURE 22.3**). The partial pressure of O_2 in this beetle's bubble gas will be about 0.1 atm, and the concentration of O_2 will be about 4.3 mmol/L (50% of the values in the atmospheric air). Diffusion will steadily transport O_2 from the environmental water into the beetle's bubble because the partial pressure of dissolved O_2 in the water, 0.21 atm, is far higher than the partial pressure in the bubble gas, 0.1 atm. The beetle's bubble, in other words, will act as a gill! This is true even though the *concentration* of O_2 in the beetle's bubble gas, 4.3 mmol/L, is far higher than the concentration in the water, 0.3 mmol/L. Oxygen will travel from high partial pressure to low partial pressure, as it always does, but in this situation involving both gas and liquid phases, it will move from low concentration to high concentration. Always, *equilibrium is attained with respect to any given gas when the partial pressure of the gas is uniform everywhere in a system.*

Although we will use the principles of gas diffusion in this book primarily to analyze O_2 and CO_2, these principles apply to

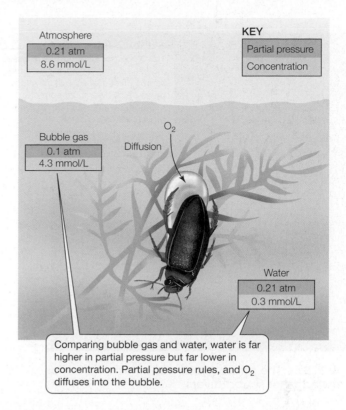

FIGURE 22.3 The function of a water beetle's bubble as a gill: O_2 levels in the atmosphere, water, and bubble gas— and O_2 diffusion The upper value in each box is the partial pressure of O_2, and the lower value is the O_2 concentration. The beetle acquires each bubble from the atmosphere. Here the beetle is assumed to have used 50% of the O_2 originally in its bubble. The volume of the bubble is assumed to stay constant as O_2 is withdrawn; this is approximately true because the principal gas in the bubble, N_2, tends to remain in the bubble even as O_2 is used. Temperature is 20°C everywhere. See the text for further explanation.

all substances that exist as gases under physiological conditions.[5] Thus, for example, partial pressures govern the diffusion of gaseous anesthetics or gaseous poisons into and out of an animal's body.

Partial pressures also govern the diffusion of N_2, a topic crucial for understanding the threat of decompression sickness in scuba divers (**FIGURE 22.4**) and dolphins (see Chapter 26). When a scuba diver is at any particular depth, his or her lungs are kept normally expanded by the operation of the scuba air regulator, which ensures that the air in the lungs is at a pressure equal to the ambient (environmental) water pressure at that depth. Suppose a diver is at a depth of 20 m, where the ambient water pressure is 3 atm. The total air pressure in the diver's lungs is then also 3 atm, and the partial pressure of N_2 in the lung air is 78% of the total, or 2.3 atm (assuming the diver's tanks are filled with ordinary air). If the diver remains at 20 m long enough for his or her blood to come to equilibrium with the lung air, the partial pressure of N_2 dissolved in the diver's blood will be 2.3 atm. This is not a problem as long as the diver stays at depth. However, the high N_2 partial pressure will force macroscopic bubbles to form in the diver's blood

---

[5] Water vapor (gaseous water) is a bit of a special case because, unlike other gases, it does not merely dissolve in aqueous solutions but can *become* liquid water or be generated from liquid water. It is addressed in Chapter 27.

The blood partial pressures of gases increase during diving because the air in a diver's lungs is at elevated pressure. A diver must correctly manage the increased blood partial pressures to dive safely.

FIGURE 22.4 The principles of gas diffusion are vital knowledge for scuba divers

if he or she suddenly comes to the surface. The reason can be seen by imagining that an extremely minute, microscopic bubble forms in the blood (some authorities think such miniscule bubbles are always present within liquids). Let's assume the bubble contains only N_2 (a best-case scenario). Then, if the diver has come to the surface and is under 1 atm of pressure, the N_2 partial pressure in the minute bubble (gas in the gas phase) will be 1 atm. The partial pressure of N_2 dissolved in the diver's blood, however, is 2.3 atm. Thus N_2 will diffuse rapidly from the blood into the bubble, and the bubble will grow to macroscopic size. You can get an excellent visual image of this process by watching the formation of CO_2 bubbles within a recently opened bottle of pop or beer, a situation that adheres to the same physical laws. Bubbles in the blood are extremely dangerous (see Chapter 26), and their formation must be prevented. Standard practice in scuba diving is to avoid staying too long at depth. Suppose, however, that a diver has been at 20 m for such a long time that his or her blood N_2 partial pressure is 2.3 atm. Before surfacing, the diver must have adequate diving skills to come partway to the surface and remain there until the blood N_2 partial pressure falls to a safe level.

The equation for the *rate* of gas diffusion is similar in form to the equation for the rate of solute diffusion presented in Chapter 5 (see page 106). Consider a fluid system (consisting of a gas phase, liquid phase, or both) in which a gas, such as O_2, is at a relatively high partial pressure, P_1, in one region and a relatively low partial pressure, P_2, in another. Think of an imaginary plane that is perpendicular to the direction of diffusion between the two regions. Let J be the rate of net movement of gas through the plane, per unit of cross-sectional area. Then

$$J = K\frac{P_1 - P_2}{X} \qquad (22.4)$$

where X is the distance separating P_1 and P_2 and K is a proportionality factor that is often called the **Krogh diffusion coefficient**. As

you will see, the rate of diffusion J is directly proportional to the difference in partial pressures but is inversely proportional to the distance separating the two partial pressures (see Table 5.1). The diffusion coefficient K depends on the particular diffusing gas, the temperature, and the ease with which the gas is able to pass through the particular material separating the regions of different partial pressure. If the two regions of interest are separated by a layer of tissue (e.g., a gill epithelium), K is termed the **gas permeability** of the tissue.

Gases diffuse far more readily through gas phases than through aqueous solutions

The ease with which gases diffuse is far greater when they are diffusing through air than through water. The Krogh diffusion coefficient (K) for O_2, for example, is about 200,000 times higher in air than in water at 20°C! One way to see the enormous implications of this difference is to consider O_2 diffusing between two regions that have a particular difference in O_2 partial pressure (i.e., a particular value for $[P_1 - P_2]$ in Equation 22.4). If the diffusion is occurring through air rather than water, the length of the diffusion path can be 200,000 times longer and still have the same rate of O_2 transport.

An interesting application of these principles is provided by analyzing the O_2 supply to mice in underground burrows, or the O_2 supply to eggs of sea turtles buried in beach sand. If the soil or sand is dry and porous, O_2 is often supplied (chiefly or entirely) by diffusion directly through the soil or sand. More specifically, O_2 is supplied by diffusion through the network of minute gas-filled spaces among the soil or sand particles. There are countless different paths—countless "angles of approach"—by which diffusion can occur from the atmosphere to the underground animals or eggs through the soil or sand. When transport along all these paths is summed, diffusion through the gas-filled spaces in soil or sand often proves adequate to meet O_2 needs. This is true, however, *only* if the spaces among the soil or sand particles are *gas-filled*. If the spaces become water-filled, the rate of O_2 diffusion drops by a factor of about 200,000—a circumstance that can have drastic consequences for the underground animals or eggs (**FIGURE 22.5**).

Diffusion through water—according to a commonly used rule of thumb—can meet the O_2 requirements of living tissues only if the distances to be covered are about *1 millimeter* or less (**BOX 22.1**)! This rule has many important applications. A dramatic application concerns the consequences of liquid accumulation in a person's lungs: Just a small accumulation of body fluids in the terminal air spaces of a person's lungs immediately creates a dire medical emergency because of the small diffusion distance that is tolerable with water present.

The difference between air and water in the ease of diffusion of CO_2 is less than that for O_2, but still substantial. The Krogh diffusion coefficient for CO_2 at 20°C is about 9000 times greater in air than in water.

Gas molecules that combine chemically with other molecules cease to contribute to the gas partial pressure

Only gas molecules that exist as free gas molecules contribute to the partial pressure of a gas. This may seem an odd point to stress until you realize that, especially in body fluids, gas molecules often undergo chemical reactions. Molecules of O_2 in the blood of a

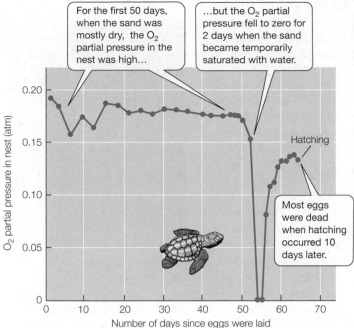

For the first 50 days, when the sand was mostly dry, the O_2 partial pressure in the nest was high…

…but the O_2 partial pressure fell to zero for 2 days when the sand became temporarily saturated with water.

Hatching

Most eggs were dead when hatching occurred 10 days later.

FIGURE 22.5 Replacement of air with water in the interstitial spaces of beach sand can cause anoxia in a sea turtle nest because diffusion is far slower through water than through air Female loggerhead sea turtles (*Caretta caretta*) come ashore to lay their eggs on ocean beaches. They bury the eggs about 40 cm deep in the sand, cover them, and leave. The eggs are dependent on O_2 diffusion through the sand to obtain O_2 during the long period (50–70 days) they require to become mature enough to hatch. The nest described here—which contained 117 eggs—was positioned unusually close to the surf's edge, and the sand overlaying the nest became submerged underwater when the eggs were about 54 days old. O_2 partial pressure in the nest was measured at regular time intervals with a fiber-optic probe permanently implanted in the nest. (Data kindly provided by Nathan A. Miller.)

person or coral-reef worm, for example, combine chemically with hemoglobin or hemoglobin-like pigments, and those molecules of O_2 then *do not* contribute to the partial pressure of O_2 in the blood. Only O_2 molecules that are free in solution contribute to the partial pressure. Similarly, CO_2 molecules added to blood—or

to an environmental liquid such as seawater—may react with H_2O to yield chemical forms such as HCO_3^- (bicarbonate ions). Only the molecules of CO_2 that are free in solution as unchanged CO_2 molecules contribute to the partial pressure of CO_2.

One noteworthy implication of these principles is that if you know the partial pressure of a gas in an aqueous solution and you use the absorption coefficient (*A*) to calculate the concentration of the gas in the solution from the partial pressure (see Equation 22.3), the concentration you obtain is that of gas free in solution as unchanged gas molecules. Any gas that is present in chemically combined form (or chemically altered form) is not included in the concentration calculated in this way.

A second noteworthy implication is that *only the gas molecules that are free in solution as unchanged gas molecules affect the direction*

BOX 22.1 Diffusion through Tissues Can Meet O_2 Requirements over Distances of Only 1 Millimeter or Less

August Krogh (1874–1949), one of the great names in respiratory physiology, was the first to quantify the distance over which the simple diffusion of O_2 through tissues might meet the O_2 requirements of life. As his model, he considered a spherical cell with a rate of O_2 consumption of 0.1 mL O_2/g•h (relatively low) and an O_2 partial pressure of about 0.21 atm at the cell surface (relatively high). He calculated that diffusion of O_2 through such a cell would meet the O_2 demands of all its parts only if the cell radius were no greater than 0.9 mm! Modern recalculations come up with similar results. Thus Krogh's major conclusion still stands: Diffusion from a high-partial-pressure source through the aqueous medium of tissue can be expected to supply the ordinary O_2 requirements of aerobic catabolism over only short distances—approximately 1 mm or less.

Over truly minute distances, such as the distance across a cell membrane, diffusion transport is very fast. However, the rate of diffusion falls as distance increases (see Table 5.1). Thus, in general, for O_2 transport to occur at a rate sufficient to supply the needs of tissues at distances greater than about 1 mm, convective transport—transport by a moving fluid, such as circulation of blood—needs to supplement diffusion.

The larvae of bony fish—some of which are the smallest of all vertebrate animals—present important examples of the application of these principles. The young of some species are so immature at hatching that they lack an effective circulatory system. Thus their interior tissues must receive O_2 from their body surfaces largely by diffusion. As a young larva grows, diffusion becomes less and less certain to be able to meet all its O_2 needs because its

A 3-week-old larva of the anchovy *Engraulis mordax*, common along the West Coast of the United States
The O_2 demand of the larva is met primarily by diffusion. The larva's average body radius is about 0.6 mm and approaching the maximum that calculations for the species indicate is compatible with uptake of O_2 by diffusion, yet the circulatory and gill-breathing systems remain immature. (Graham 1990 provides a more complete discussion; after Kramer and Ahlstrom 1968.)

deepest tissues come to be positioned farther from its body surface. In a 3-week-old larval anchovy (illustrated), gill development and circulatory development are still incomplete, yet the thickness of the body is close to the maximum over which diffusion can meet O_2 needs. As a larval fish develops through such a transition phase—a stage at which its convective-transport mechanisms are required but still immature—it is particularly vulnerable to dying if it finds itself in low-O_2 water.

and rate of gas diffusion because only those gas molecules affect the partial-pressure gradient. This point is exceedingly important in many situations, one of which is the dynamics of O_2 uptake by an animal's blood as the blood flows through the gills or lungs. Let's consider an animal with hemoglobin in its blood. When O_2 diffuses from the environment into the animal's blood, the O_2 molecules that combine with hemoglobin do not increase the blood's partial pressure of O_2. Thus the hemoglobin-combined O_2 molecules do not interfere with further O_2 diffusion into the blood. To illustrate with an extreme case, suppose that all the O_2 that diffuses into an animal's blood over a period of time combines with hemoglobin. In this case, the blood O_2 partial pressure stays constant, and if the environmental O_2 partial pressure also stays constant, the difference of partial pressure driving diffusion stays constant as well. The rate of O_2 diffusion will then remain undiminished even as more and more O_2 enters the blood.

SUMMARY
Diffusion of Gases

- Gases always diffuse from regions of high partial pressure to regions of low partial pressure and at a rate that is proportional to the difference in partial pressure. A gas is at equilibrium in a system when its partial pressure is uniform throughout.

- Diffusion occurs much more readily through air than water. Specifically, the Krogh diffusion coefficient is 200,000 times greater for O_2 and 9000 times greater for CO_2 in air than in water (at 20°C).

- When gas molecules undergo chemical combination, they cease to contribute to the partial pressure of the gas in question and thus no longer affect the direction or rate of diffusion of the gas.

Convective Transport of Gases:
Bulk Flow *carried by fluid*

Besides diffusion, the second major mechanism of respiratory gas transport is convection. **Convective gas transport**—also called **transport by bulk flow**—occurs when a gas mixture or an aqueous solution *flows* and gas molecules in the gas or liquid are carried from place to place by the fluid flow. The transport of O_2 by blood flow in an animal's circulatory system provides an example.

Convective gas transport is typically *far faster* than gas diffusion because convection (bulk flow) moves gas molecules in a deterministic, forced fashion rather than depending on random molecular movements. The two principal processes by which animals set fluids in motion to transport gases are breathing and the pumping of blood. Both cost metabolic energy. The reward an animal gets for making the metabolic investment is that it is able to speed the transport of O_2 and CO_2 from place to place.

Convection does not always require muscular effort from an animal. Ambient winds and water currents, in fact, often move O_2 and CO_2 from place to place in the environments of animals in ways that are helpful, as when global-scale water currents renew the O_2 supplies of the deep sea. Sometimes, ambient winds and water currents even induce internal flow within animals or animal-built structures (**BOX 22.2**).

The precise effects of fluid flow on gas transport depend in part on the geometry of the flow. Two geometries of importance are depicted in **FIGURE 22.6**. One is **unidirectional flow** through a tube, such as the flow of blood through a blood vessel. The second is **tidal flow** (back-and-forth flow) in and out of a blind-ended cavity, such as occurs in the lungs of mammals. To exemplify how the rate of convective gas transport can be calculated, let's use as a model the simple case of O_2 transport by the unidirectional flow of a fluid through a tube (see Figure 22.6A) when there is no exchange of gases across the walls. This sort of model applies, for example, to O_2 transport by blood flow through the arteries of animals because arteries basically act as conduits, neither adding O_2 to the blood passing through them nor removing O_2. In this case,

$$\text{Rate of convective gas transport} = C_T F \qquad (22.5)$$

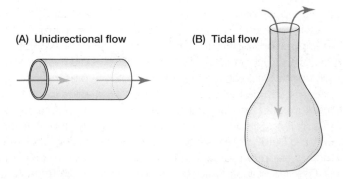

(A) Unidirectional flow   **(B) Tidal flow**

FIGURE 22.6 Two types of convective transport Arrows depict bulk flow of fluid.

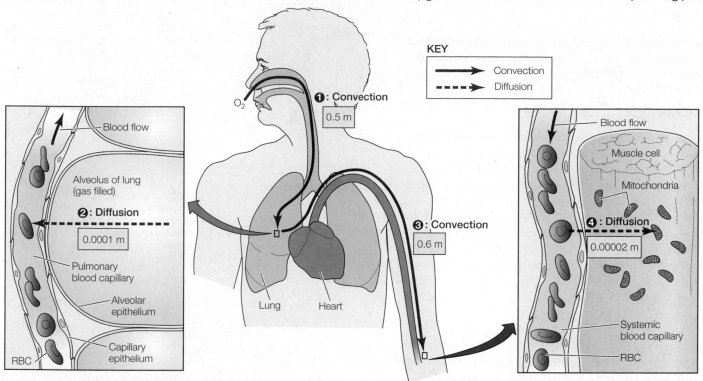

FIGURE 22.7 Convection (bulk flow) and diffusion alternate in transporting O₂ from the atmosphere to the mitochondria in a person Four steps are recognized, with the distance covered in each step shown in meters: ❶ Convective transport by flow of inhaled air from the atmosphere to the depths of the lung. ❷ Diffusion across a gas-filled alveolar end sac (including alveolus), then through the alveolar epithelium and the epithelium of a pulmonary blood capil-lary, and finally into a red blood cell (RBC), where O_2 combines with hemoglobin. Only the final 0.000006 m of this step requires diffusion through tissue or liquid; the initial 0.000094 m is covered by diffusion through gas. ❸ Convective transport by the circulation of the blood. ❹ Diffusion from an RBC in a systemic blood capillary to a mitochondrion in a muscle cell. Mitochondria are actually much smaller relative to the muscle cell than shown.

where C_T represents the *total* concentration of gas in the fluid (including both gas that is chemically combined [e.g., O_2 combined with hemoglobin] and gas that is free in solution) and F represents the rate of fluid flow.[6] As the equation makes clear, the rate of convective gas transport can be increased by increasing the concentration or the flow rate. Mammals and some other animals with very high demands for O_2 have evolved favorable modifications of *both* of these properties to enhance the rate of convective O_2 transport by the flow of blood through their arteries. The high concentration of hemoglobin in mammalian blood enables each volume of blood to carry 50 times more O_2 than it could carry in simple solution, and the intense work done by the mammalian heart propels blood through the arteries at very high rates (e.g., 5 L/min through the systemic aorta of an adult person during rest, and 35 L/min in trained athletes during high exertion).

Gas transport in animals often occurs by alternating convection and diffusion

Although in tiny animals O_2 may be able to move as fast as is needed from the environment to all parts of the body by diffusion (see Box 22.1), exclusive reliance on diffusion is not possible in animals that are larger than 2–3 g, or even in many animals that are smaller than that. The reason is *distance*. In most animal body plans, diffusion

of O_2 within the body occurs through liquid media—body fluids and tissues. Because diffusion through liquid media is inherently slow, it can meet transport needs only over very short distances (see Box 22.1). For aquatic animals, even O_2 transport to the gills from the environment must occur through a liquid medium, and thus again, diffusion can suffice only if the distances to be covered are very short. Animals that face these diffusion limitations typically employ convective transport (bulk flow) to solve the problem.

In animals that employ convective transport, diffusion and convection typically alternate as O_2 makes its way from the environment to the mitochondria. During this alternation, diffusion meets the needs for gas transport over short distances, whereas convective transport is employed for the long hauls. An apt example is provided by O_2 transport from the atmosphere to the mitochondria in a person (**FIGURE 22.7**). The first step in this transport is that O_2 must travel a distance of about 0.5 m from the atmosphere to minute air passageways deep in the lungs. For this transport to occur at an adequate rate, convection must be employed. Thus we breathe, moving large masses of air rapidly (by bulk flow) over long distances (step ❶ in Figure 22.7). Oxygen then moves by diffusion over the next 0.0001 m of distance; in this step (step ❷), the O_2 crosses the minute alveolar end sacs deep in the lungs and the two epithelia (alveolar and capillary) that separate the lung air from the blood. The next long haul is the movement of O_2 from the lungs to the systemic tissues (step ❸). Transport to a muscle in the forearm, 0.6 m from the lungs, is an example. This transport must be by convection—in this case, the circulation of the blood. Finally,

---

[6] F is used here as a general symbol for the rate of fluid flow. According to established convention in circulatory and respiratory physiology, the fluid flow rate would be symbolized by Q if the fluid were a liquid (e.g., blood) or V if it were a gas (e.g., air).

O_2 must travel 0.00002 m from blood within systemic blood capillaries into mitochondria within tissue cells (step ❹). This last step occurs by diffusion.

The Oxygen Cascade

The concept of the **oxygen cascade** is an especially informative way to summarize the transport of O_2 from the environment to the mitochondria of an animal. In everyday language, a *cascade* is a series of steep waterfalls along a stream. Suppose that a waterwheel (or turbine) is positioned at the base of the cascade shown in **FIGURE 22.8A**. The force available to turn the waterwheel is that available when the water drops from height 4 to height 5. The water starts at height 1 as it enters the cascade, but gradually, as it flows through the cascade, it drops from height 1 to 2, and then from height 2 to 3, and so forth, so that the final fall of the water—which provides the actual force to turn the waterwheel—is only from height 4 to height 5. The *oxygen cascade* is analogous to such a water cascade. To construct an oxygen cascade, one plots the O_2 partial pressure at successive steps along the path between an animal's environmental source of O_2 and its mitochondria, as shown in **FIGURE 22.8B**. Because O_2 is not subject to active transport, the partial pressure always drops with each step, and thus the plot resembles the hydraulic analog we have discussed, a cascading mountain stream.

Why is the *partial pressure* plotted in an oxygen cascade? The answer to this question is an essential concept for having a full understanding of the physiology of O_2 transport. The mitochondria of an animal are where O_2 is ultimately used, and—as seen in step ❹ of Figure 22.7—*O_2 enters the mitochondria by diffusion at a rate that depends on the difference in O_2 partial pressure between the blood in systemic capillaries and the mitochondria themselves.* Just as a waterwheel on a stream requires water of a certain height immediately upstream to function, the mitochondria in each cell require a certain partial pressure of O_2 at the cell surface (i.e., in the capillary blood at the cell surface) if they are to receive O_2 fast enough by diffusion to meet their O_2 needs. *From this perspective, the entire point of the O_2-transport system in an animal is to maintain an adequately high O_2 partial pressure at cell surfaces throughout the body.* With each step along the oxygen cascade, the O_2 partial pressure drops. The partial pressure at the cell surfaces must nonetheless be kept high enough for O_2 to diffuse to the mitochondria at an adequate rate.

Let's now review in greater detail the particular oxygen cascade shown in Figure 22.8B: the cascade in healthy, resting people at sea level. The first step, the transport of O_2 from the ambient air to the gas in the alveolar end sacs in the lungs, occurs by convection, as we have seen. When people breathe in their normal way, the drop in O_2 partial pressure during this step is from about 0.2 atm in the ambient air[7] to about 0.13 atm in alveolar gas. The second step of the O_2 cascade, the transport of O_2 across the alveolar sacs and

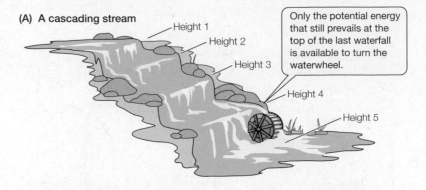

(A) A cascading stream

Height 1
Height 2
Height 3
Height 4
Height 5

Only the potential energy that still prevails at the top of the last waterfall is available to turn the waterwheel.

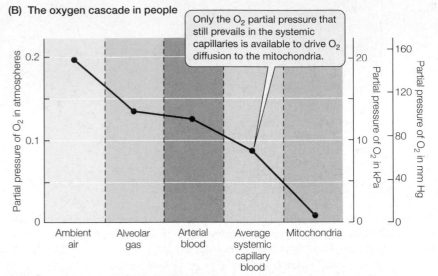

(B) The oxygen cascade in people

Only the O_2 partial pressure that still prevails in the systemic capillaries is available to drive O_2 diffusion to the mitochondria.

Partial pressure of O_2 in atmospheres
Partial pressure of O_2 in kPa
Partial pressure of O_2 in mm Hg

Ambient air — Alveolar gas — Arterial blood — Average systemic capillary blood — Mitochondria

FIGURE 22.8 The concept of the physiological oxygen cascade is based on an analogy with a cascade along a mountain stream (A) In a cascading stream, water loses potential energy each time it falls lower. (B) In a person, the O_2 partial pressure drops with each step in oxygen transport. In (A), the height numbers refer to sequentially lower heights along the stream's path; kinetic energy is ignored but would have to be taken into account for a full physical analysis. In (B), the numerical values are representative values for a healthy, resting individual at sea level; the partial pressure depicted for the mitochondria is estimated rather than measured and (as would be true during rest at sea level) is higher than the minimum required for unimpaired aerobic catabolism. The scales at the right show O_2 partial pressure in kilopascals (kPa) and millimeters of mercury (mm Hg).

through the epithelia separating the alveolar gas from the blood, occurs by diffusion. Its rate therefore depends on the difference in O_2 partial pressure between the alveolar gas and the blood. In healthy lungs, a partial-pressure difference of about 0.007 atm is sufficient to cause O_2 to diffuse at the rate required. Lungs damaged from smoking or disease often require a larger partial-pressure difference, which might mean that the blood partial pressure is lower than normal. The third step of the cascade is the convective step between arterial blood and the blood in systemic blood capillaries. Whereas the arterial blood has an O_2 partial pressure of about 0.12 atm, the partial pressure in the capillaries averages about 0.09 atm.[8]

---

[7] We say "about 0.2 atm" here because in the real world, air has a variable content of water vapor (see Chapter 27). In the open atmosphere, the gas other than water vapor (the "dry air") is always 20.95% O_2. However, this gas is diluted when water vapor is present, lowering the overall O_2 percentage and O_2 partial pressure.

[8] The average partial pressure in the capillaries is dynamically determined by the rate at which blood flow brings O_2 to the capillaries and the rate at which diffusion into tissues removes O_2 from them. Although the average capillary partial pressure is a useful concept, an important factor in reality, as discussed in Chapter 24, is that the blood O_2 partial pressure in capillaries is actually not a single number. Instead, because O_2 diffuses out of capillaries all along their lengths, the blood O_2 partial pressure drops from a relatively high value at the arterial ends of capillaries to a lower value at the venous ends.

The final step of the oxygen cascade is the "payoff" of O_2 transport: the diffusion-mediated transport of O_2 from the blood in systemic capillaries to the mitochondria in surrounding cells, driven by the difference in O_2 partial pressure between the blood and the mitochondria. The reason the O_2 partial pressure in the mitochondria is lower than that in the capillary blood is that the mitochondria constantly draw down the partial pressure in their vicinity by chemically consuming O_2, converting it to water. For aerobic catabolism to be unimpaired, the *mitochondrial* partial pressure cannot be allowed to fall below about 0.001 atm. The *capillary* O_2 partial pressure must therefore be kept sufficiently higher than 0.001 atm for diffusion to occur at a rate equal to the rate of mitochondrial O_2 consumption. Keeping the capillary O_2 partial pressure from slipping below this required value, as already stressed, is the key task of all the O_2-transport processes. In healthy people resting at sea level (see Figure 22.8B), the average capillary O_2 partial pressure is well above the minimum value required, providing a wide margin of safety. However, during strenuous exercise, as we shall see in upcoming chapters, all the systems responsible for O_2 delivery may need to operate near their limits if they are to keep the capillary partial pressure above the minimum value required.

Expressing the Amounts and Partial Pressures of Gases in Other Units

Up to now, to keep units of measure simple while we have developed the principles of gas transport, we have always expressed the amounts of gases in moles and the partial pressures in atmospheres. Most of the alternative units in use—such as expressing amounts in grams or pressures in kilopascals—are very straightforward (see Appendix A).[9]

What can be confusing is that amounts of gases are often expressed as *volumes of gases*. The amount of a gas in moles is simply proportional to the volume of the gas *provided the volume is expressed at* **standard conditions of temperature and pressure** (**STP**), meaning a temperature of 0°C (273 K) and a pressure of 1 atm (101 kPa; 760 mm Hg). This relation is so simple that it does not even vary among gases; for essentially all gases, 1 mole occupies 22.4 L at STP. The reason that volumetric expressions can get confusing is that the volume occupied by a given molar amount of gas depends on temperature and pressure (see Equation 22.1). Because of this dependency, if we are thinking about a gas that is at a temperature and pressure different from 0°C and 1 atm, the gas has *two* volumes of potential interest. One is the volume the gas actually occupies under the conditions where it exists. The other is its volume at

STP. Suppose, for example, that we are studying the lung gas of a person who is standing at a location where the barometric pressure is 740 mm Hg and we are concerned about 0.0446 mole of O_2 in the lung gas. In the lungs, where this gas is at a temperature of 37°C and a pressure of 740 mm Hg, the O_2 occupies a volume of 1170 mL. However, this same quantity of O_2 occupies only 1000 mL at STP. Appendix C provides more detail on these calculations. When physiologists use gas volume as a way to express the absolute amount of gas present (the molar quantity of the gas), the volume at STP is the volume used, because it is this volume that bears an unvarying, one-to-one relation to the number of moles.

Another aspect of volumes that can be confusing arises from the common practice, among biologists, of expressing the amount of gas *dissolved in an aqueous solution* as a volume of gas at STP. Let's illustrate. Suppose that a liter of an aqueous solution has dissolved in it an amount of O_2 that, in gaseous form, would occupy 2 mL at STP. Biologists then often say that the concentration of O_2 in the solution is "2 mL O_2/L." Except for the units used, this expression is no different from saying that there is a certain number of moles dissolved per liter. What can be confusing is that the gas occupies the stated volume only if it is *removed* from solution and placed at STP. When the gas is *actually in* solution, it—in essence—does not occupy any volume at all.

The Contrasting Physical Properties of Air and Water

Air and other gas phases differ dramatically from water and aqueous solutions in many physical properties that are of critical importance for the physiology of the respiratory gases. We have already stressed that gases diffuse much more readily through air than through water. Other properties—such as density and viscosity—also differ dramatically.

Water is much more dense and viscous than air. At 20°C, for example, the density of water (about 1 g/mL) is more than 800 times higher than the density of sea-level air (about 0.0012 g/mL). The viscosity of water is 35 times higher than that of air at 40°C and more than 100 times higher at 0°C. Because of water's greater density and viscosity, water-breathing animals must generally expend more energy than air-breathing ones to move a given volume of fluid through their respiratory passages.

Dramatic differences also typically exist between air and water in natural environments in the amounts of O_2 they contain per unit of volume (**TABLE 22.1**). Ordinary atmospheric air is about 21% O_2, as we have seen. This means that if air is at sea level and at 0°C, it contains 210 mL of O_2, measured at STP, per liter. Warming the air lowers its absolute concentration of O_2 a little because gases expand and become more rarefied when they are heated, but even at 24°C, atmospheric air at sea level contains 192 mL of O_2 (measured at STP) per liter.[10] The amounts of O_2 per liter in water are dramatically lower because they depend on the *solubility* of O_2 in water, and O_2 is not particularly soluble. If freshwater at 0°C is equilibrated with atmospheric air at sea level so that the partial pressure of O_2 in the

---

[9] The units of measure for pressure, although fundamentally straightforward, can present problems at present because several disparate systems of units exist, and different branches of physiology have progressed to different degrees in adopting SI units. Moreover, in the United States (although not elsewhere), there has been great resistance in everyday life to abandoning old-fashioned units of measure such as millimeters of mercury (mm Hg) and pounds per square inch (psi). The basic SI unit of pressure is the pascal (Pa)—equal to a newton per square meter (N/m^2). This is such a tiny pressure that nearly always in physiology, the kilopascal (kPa) is the SI unit used. The millimeter of mercury (mm Hg), although now being left behind, has seen extensive use in physiology for many decades; it is the pressure exerted by a column of mercury 1 mm high under standard gravitational acceleration. Another old unit used often in physiology for many decades is the torr (Torr), which is essentially identical to a millimeter of mercury. 1 kPa = ~7.5 mm Hg. In addition, 1 kPa = ~0.01 atm. An atmosphere (atm) is 760 mm Hg or about 101 kPa.

[10] Air at 24°C contains 210 mL of O_2 per liter when the volume of O_2 is measured at 24°C. However, the air is less dense than air at 0°C and thus contains less mass of O_2 per liter than air at 0°C. This effect of temperature is removed by correcting to STP. The O_2 that occupies 210 mL at 24°C will occupy only 192 mL at STP.

TABLE 22.1 The usual maximum concentration of O_2 in air, freshwater, and seawater at three temperatures

The concentrations listed are for air at sea level and for fully aerated freshwater and seawater equilibrated with such air. In other words, the O_2 partial pressure is 0.21 atm in all cases. For the most part, actual O_2 concentrations in natural environments are either as high as shown or lower (because of O_2 depletion by organisms).

	Concentration of O_2 (mL O_2 at STP/L) at specified temperature		
	0°C	12°C	24°C
Air	210	200	192
Freshwater	10.2	7.7	6.2
Seawater[a]	8.0	6.1	4.9

[a] The values given are for full-strength seawater having a salinity of 36 g/kg.

water is identical to that in the atmosphere (0.21 atm), the water dissolves 10.2 mL of O_2 per liter. Seawater, because of its salinity, dissolves less O_2 under the same conditions (see Table 22.1). Both freshwater and seawater dissolve less O_2 as they are warmed because the solubility of O_2 decreases as temperature increases. Overall, as is dramatically evident in Table 22.1, *the maximum O_2 concentration likely to occur in bodies of water is 5% or less of the concentration in air at sea level.* One way to appreciate the quantitative significance of these numbers is to consider a hypothetical terrestrial or aquatic animal that is trying to obtain a liter of O_2 by completely extracting the O_2 from a volume of its environmental medium. At 0°C, this animal would need to process 4.8 L of air, 98 L of freshwater, or 125 L of seawater!

Water actually presents animals with a *combination* of properties that together make water a far more difficult place to acquire O_2 than air. First, each liter of water is typically more costly than a liter of air to pump during breathing because of the relatively high density and viscosity of water. Second, each liter of water has a lower O_2 reward to provide (see Table 22.1). Together, these considerations mean that a water-breathing animal must often work much harder than an air-breathing one to obtain a given quantity of O_2, which means that in the water breather, a greater fraction of the O_2 taken up must be dedicated to obtaining more O_2. Although in resting people the cost of ventilating the lungs with air is 1%–2% of the whole-body metabolic rate, in resting fish the cost of ventilating the gills with water is probably near 10% of the whole-body metabolic rate.[11] The dramatic differences in the physical properties of air and water also undoubtedly help explain why the highest absolute metabolic rates in the animal kingdom are found in air breathers: insects, mammals, and birds. A high absolute rate of O_2 uptake is far more feasible in air than in water.

Respiratory Environments

When we consider the respiratory environments of animals in nature, biotic processes are as important as physical effects in determining the concentrations and partial pressures of O_2 and CO_2. The processes at work in a portion of a terrestrial or aquatic environment

---

[11] Measurement of the value for fish has proved challenging, and a wide range of values has been reported in the experimental literature. The value of 10% is often cited as a "best guess" of the average.

are diagrammed in **FIGURE 22.9**. The animals and plants living there exert strong influences on the local concentrations and partial pressures of O_2 and CO_2. During the day, with adequate sunlight, photosynthetic organisms add O_2 to the air or water and extract CO_2. Animals, bacteria, and fungi remove O_2 and add CO_2, and at night the photosynthetic organisms do the same. Any portion of the environment, as the figure shows, exchanges O_2 and CO_2 with neighboring regions by way of diffusion and convection (breezes or water currents). Diffusion always tends to equalize partial pressures across boundaries, and convection usually does so as well (by physical mixing). Thus, in any portion of the environment, it

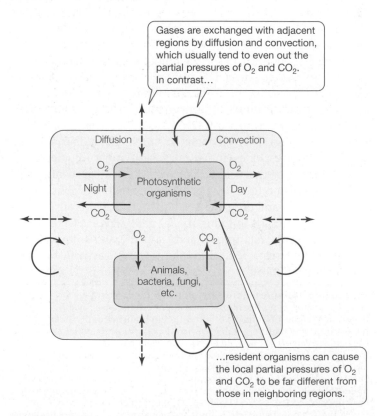

Gases are exchanged with adjacent regions by diffusion and convection, which usually tend to even out the partial pressures of O_2 and CO_2. In contrast…

…resident organisms can cause the local partial pressures of O_2 and CO_2 to be far different from those in neighboring regions.

FIGURE 22.9 The processes that affect the partial pressures of O_2 and CO_2 within a portion of a terrestrial or aquatic environment The outer box (shaded blue) symbolizes a portion of the natural world.

is common to find that the resident organisms collectively raise or lower the O_2 partial pressure relative to that in neighboring regions, whereas diffusion and convection simultaneously tend to even out the partial pressure from place to place. The relative strengths of these processes determine whether the O_2 partial pressure in the portion of interest becomes different from, or remains similar to, the partial pressures in neighboring regions.

In open environments on land, because breezes are ubiquitous and diffusion occurs relatively rapidly in air, the fractional composition of dry air is virtually uniform from place to place, both at sea level and over the altitudinal range occupied by animals. Oxygen, for example, represents 20.95% of the volume of dry air just about everywhere. When the concentrations and partial pressures of gases vary from place to place in open-air environments, they do so principally because of altitude (see Figure 1.15); at 4500 m (14,800 ft)—one of the highest altitudes where human settlements occur—the concentration and the partial pressure of O_2 are only about 60% as high as at sea level because the air (while still 21% O_2) is under less pressure. In contrast to open terrestrial environments, in secluded ones, such as underground burrows, the metabolic activities of resident organisms often draw down the local concentration and partial pressure of O_2 and increase the level of CO_2. These local changes are possible in secluded places because convective mixing with the open atmosphere is restricted in such places.

Regional differences in the concentrations and partial pressures of O_2 and CO_2 are far more common in aquatic environments than in terrestrial ones because in water the processes that act to even out gas levels are weaker than in air. With the evening-out processes less effective, local biotic processes are readily able to modify local gas levels. Thus, even in a completely unobstructed body of water such as a lake, the concentrations and partial pressures of gases often vary considerably from place to place (see Figure 1.16). In secluded aquatic environments populated by organisms having substantial O_2 demands, O_2 partial pressures approaching zero can readily occur.

STUDY QUESTIONS

1. Explain in your own words why the oxygen cascade is presented as a cascade of partial pressures.

2. Plot the rate of O_2 consumption of a fish as a function of water temperature (see Figure 10.9 if you need help). Then, based on information from this chapter, plot the O_2 concentration of fully aerated water as a function of water temperature. Why do physiologists sometimes say that high water temperatures create a "respiratory trap" for fish, a "trap" in which the fish are caught between two conflicting trends?

3. A team of investigators is out on a boat on a lake on a marvelous, sunny summer day, and they are taking water samples from various depths in the lake so as to construct a plot of dissolved O_2 concentration versus depth, similar to the plot in Figure 1.16. When the investigators bring up a water sample from a depth of 20 m, why is it imperative that they measure its dissolved O_2 concentration immediately, rather than letting it sit on the deck for 5 minutes? (Hint: Look at Figure 22.2A.)

4. In the sediments at the bottoms of ponds and lakes, there is often absolutely no O_2 at depths of 0.5 cm or more, even if the water above is rich in O_2. How is this possible?

5. The absorption coefficient of O_2 in seawater at 0°C is 1.7 mmol/L. What is the concentration of O_2 if the O_2 partial pressure is 0.1 atm? Express your answer first in mmol/L, then in mL/L.

6. Archimedes' principle states that when an object is immersed in a fluid, it is buoyed up by a force equal to the weight of the fluid displaced. Assume (slightly incorrectly) that tissue has the same density as water. Using Archimedes' principle, explain why the gill filaments of aquatic animals typically flop into a drooping mass in air even though they float near neutral buoyancy in water. How do these differences between air and water affect the ability of gills to function in gas exchange?

7. For a biologist, what are the three or four most important things to know about gas solubility?

8. Compare and contrast diffusion and convection. In what way do they "alternate" in the O_2 transport system of a mammal?

9. The hemoglobin in mammalian blood is usually thought of simply as increasing the amount of O_2 that can be carried by each liter of blood. However, in a lecture on hemoglobin, a respiratory physiologist made the following statement: "The presence of hemoglobin in the blood also makes possible the rapid uptake of O_2 by the blood as it flows through the lungs." Explain the lecturer's point.

10. When a water beetle is placed in a laboratory situation where the atmosphere is pure O_2 and the water that the beetle is in is equilibrated with the atmosphere, after the beetle obtains a new bubble from the atmosphere, it cannot stay underwater for nearly as long as it can when the atmosphere is ordinary air. This is true because the bubble does not operate effectively as a gill. Explain. Assume (almost accurately) that CO_2 added to the bubble diffuses quickly out of the bubble into the surrounding water. (Hint: What is the partial pressure of O_2 in the water and in the bubble?)

11. Helicopters have been used to move young salmon from lake to lake for purposes of aquaculture in remote areas such as Alaska. From the viewpoint of the behavior of dissolved gases, why is it important for helicopters used in this way to stay at very low altitudes? (Hint: After fish were transported at altitudes where the total ambient pressure was 0.9 atm, high percentages suffered disease or death because of gas bubble formation in various tissues.)

Go to **sites.sinauer.com/animalphys4e**
for box extensions, quizzes, flashcards,
and other resources.

REFERENCES

Boutilier, R. G. 1990. Respiratory gas tensions in the environment. *Adv. Comp. Environ. Physiol.* 6: 1–13.

Brubakk, A. O., and T. S. Neuman (eds.). 2003. *Bennett and Elliott's Physiology and Medicine of Diving*, 5th ed. Saunders, New York.

Denny, M. W. 1993. *Air and Water: The Biology and Physics of Life's Media*. Princeton University Press, Princeton, NJ.

Farhi, L. E., and S. M. Tenney (eds.). 1987. *The Respiratory System*, vol. 4, *Gas Exchange*. (Handbook of Physiology [Bethesda, MD], section 3). Oxford University Press, New York. Includes articles on the physics of gases.

Graham, J. B. 1990. Ecological, evolutionary, and physical factors influencing aquatic animal respiration. *Am. Zool.* 30: 137–146.

Hlastala, M. P., and A. J. Berger. 2001. *Physiology of Respiration*, 2nd ed. Oxford University Press, New York.

Severinghaus, J. W. 2003. Fire-air and dephlogistication. Revisionisms of oxygen's discovery. In R. C. Roach, P. D. Wagner, and P. H. Hackett (eds.), *Hypoxia: Through the Lifecycle (Advances in Experimental Medicine and Biology*, vol. 543), pp. 7–19. Kluwer Academic, New York. Written by a leader in the study of respiratory physiology, this fascinating paper discusses the history of one of the greatest of all scientific discoveries, the discovery of oxygen.

Shams, I., A. Avivi, and E. Nevo. 2005. Oxygen and carbon dioxide fluctuations in burrows of subterranean blind mole rats indicate tolerance to hypoxic-hypercapnic stresses. *Comp. Biochem. Physiol., A* 142: 376–382.

Vogel, S. 1994. *Life in Moving Fluids: The Physical Biology of Flow*, 2nd ed. Princeton University Press, Princeton, NJ.

Weibel, E. R. 1984. *The Pathway for Oxygen*. Harvard University Press, Cambridge, MA.

Weiss, T. F. 1996. *Cellular Biophysics*, vol. 1. MIT Press, Cambridge, MA.

See also **Additional References** *and* **Figure and Table Citations**.

External Respiration

The Physiology of Breathing

Tunas represent one of the pinnacles of water breathing. If they lived on land, where they could be readily observed, they would be classified metaphorically with wolves, African hunting dogs, and other strong, mobile predators. Judging by the length of time spent in motion, tunas are actually more mobile than any terrestrial predator. Using their aerobic, red swimming muscles, they swim *continuously*, day and night, at speeds of one to two body lengths per second; some species cover more than 100 km per day during migrations.[1] Tunas thus rank with the elite endurance athletes among fish, the others being salmon, mackerel, billfish, and certain sharks. To meet the relentless O_2 demands of their vigorous lifestyle, tunas require a respiratory system that can take up O_2 rapidly from the sea and a circulatory system that can deliver O_2 rapidly from the gills to tissues throughout the body. They are, in fisheries ecologist John Magnuson's memorable phrase, "astounding bundles of adaptations for efficient and rapid swimming."

Tunas breathe with gills, as do most fish. Their gills are hardly average, however. Instead, tuna gills are exceptionally specialized for O_2 uptake, illustrating a general principle in the

Throughout the animal kingdom, species that depend on vigorous endurance exercise for survival—such as tunas— must be able to acquire oxygen rapidly and steadily Although water is not a particularly rich source of O_2, tunas have gills and breathing processes that enable them to live as highly active predators. This is a southern bluefin tuna (*Thunnus maccoyii*).

---

[1] Burst speeds are 12–15 body lengths per second, but they are not relevant to our discussion here because they are powered relatively anaerobically (by the white swimming muscles).

study of respiration: that a single type of breathing system may exhibit a wide range of evolutionary refinements in various species. Based on studies of yellowfin tunas and skipjack tunas, the gills of a tuna have about eight times more surface area than the gills of a rainbow trout—a relatively average fish—of equal body size. If the gas-exchange membranes of the gills of a small, 1-kg (2.2-pound) tuna were flattened, they would form a square measuring 1.3 m on each side. The gill membranes of tunas are also exceptionally thin: Whereas the average distance between blood and water is about 5 μm in a rainbow trout's gills, it is 0.6 μm in the gills of a skipjack or yellowfin tuna. Compared with average fish, tunas have evolved gills that present an extraordinarily large surface of extraordinarily thin membrane to the water for gas exchange.

Most fish, including rainbow trout, drive water across their gills by a pumping cycle that is powered by their buccal and opercular muscles (discussed later in this chapter). Some species are adept at alternating between this mechanism of ventilating their gills and another mechanism termed **ram ventilation**. During ram ventilation, a fish simply holds its mouth open while it swims powerfully forward, thereby "ramming" water into its buccal cavity and across its gills; in this way, the swimming muscles assume responsibility for powering the flow of water across the gills. During their evolution, tunas completely abandoned the buccal–opercular pumping mechanism and became *obligate ram ventilators*, a distinction they share with just a few other sorts of fish. As obligate ram ventilators, they have no choice regarding how much of their time they spend swimming. They must swim continuously forward, or they suffocate! Physiologists debate whether ram ventilation is intrinsically superior as a way of moving water across the gills. Less debatable is the fact that tunas achieve extraordinarily high rates of water flow using ram ventilation. During routine cruising, a small, 1-kg skipjack or yellowfin tuna drives about 3.6 L of water across its gills per minute; this is seven times the resting flow rate in a 1-kg rainbow trout, and twice the maximum rate of the trout.

Fundamental Concepts of External Respiration

In all animals, the systems used to exchange respiratory gases with the environment can be diagrammed as in **FIGURE 23.1**. A **gas-exchange membrane** or **respiratory exchange membrane**—a thin layer of tissue consisting typically of one or two simple epithelia—separates the internal tissues of the animal from the environmental medium (air or water).[2] **External respiration** or **breathing**—the topic of this chapter—is the process by which O_2 is transported to the gas-exchange membrane from the environmental medium and by which CO_2 is transported away from the membrane into the environmental medium. **Ventilation** is bulk flow (convection) of air or water to and from the gas-exchange membrane during breathing. Not all animals employ ventilation; in some, breathing can occur by diffusion rather than convection.

Oxygen always crosses the gas-exchange membrane by diffusion, as stressed in Chapter 22. This means that for O_2 to enter an animal from the environment, the partial pressure of O_2 on the

[2] *Membrane* is used here in an entirely different way than when speaking of a cell membrane or intracellular membrane. The gas-exchange membrane is a tissue formed by one or more layers of cells.

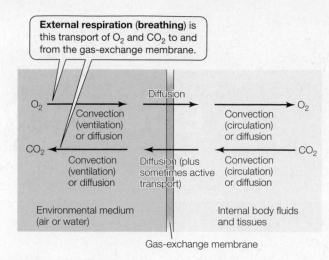

FIGURE 23.1 Generalized features of animal gas exchange O_2 and CO_2 move between the environmental medium and the internal tissues of an animal across a gas-exchange membrane.

inside of the gas-exchange membrane *must* be lower than that on the outside. The fact that O_2 enters animals by diffusion explains why the area and thickness of the gas-exchange membrane play critical roles in O_2 acquisition. The rate of diffusion across a membrane increases in proportion to the area of the membrane. Furthermore, according to the fundamental diffusion equation (see Equation 22.4), the rate of diffusion across a membrane increases as the thickness of the membrane decreases. These physical laws explain why an expansive, thin gas-exchange membrane is a great asset for animals such as tunas that must acquire O_2 at high rates. As for CO_2, diffusion is the exclusive mechanism by which it crosses the gas-exchange membrane in some animals, including humans; in others, however, although diffusion is the principal CO_2 transport mechanism, active transport also occurs. Active transport of CO_2 (as HCO_3^-) is best documented in freshwater animals (see Figure 5.15).

Usually, just a single part of an animal is identified by name as a "breathing organ." Other parts may participate in gas exchange with the environment, however. Let's briefly consider some examples that illustrate the range of possibilities. In mammals and tunas, most of the body surface—the skin—is of very low permeability to gases. Therefore just the identified breathing organs—the lungs and gills—take up almost all O_2 and void almost all CO_2. In a typical adult frog, by contrast, the skin is permeable to gases; O_2 and CO_2 are exchanged to a substantial extent across the skin as well as the lungs, and therefore the identified breathing organs—the lungs—are not the *only* breathing organs. Similarly, some fish breathe with their stomachs as well as their gills, sea stars breathe with their tube feet as well as their gills (branchial papulae), and some salamanders, lacking lungs, breathe only with their skin.

In organs that are specialized for external respiration, the gas-exchange membrane is typically thrown into extensive patterns of invagination or evagination, which greatly increase the membrane surface area. For physiologists, *gills* and *lungs* are generic labels that refer to two such patterns (**FIGURE 23.2**). **Gills** are respiratory structures that are evaginated from the body and surrounded by the environmental medium. **Lungs**, by contrast, are respiratory structures that are invaginated into the body and contain the environmental medium. The adjective **branchial** refers to structures or processes associated with gills, whereas **pulmonary** refers to those associated with lungs.

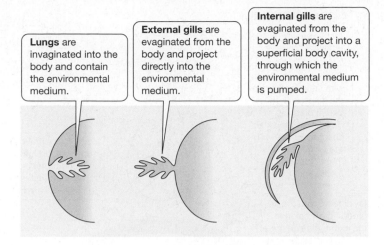

Lungs are invaginated into the body and contain the environmental medium.

External gills are evaginated from the body and project directly into the environmental medium.

Internal gills are evaginated from the body and project into a superficial body cavity, through which the environmental medium is pumped.

FIGURE 23.2 Three types of specialized breathing structures

Although exceptions occur, water breathing is usually by gills, whereas air breathing is usually by lungs. The comparative method strongly indicates that lungs are adaptive for terrestrial life (see Figure 1.18). One advantage of lungs on land is that their finely divided elements receive structural support by being embedded in the body. The finely divided elements of gills project into the environmental medium. This is not a problem in water because there, the fine, evaginated gill processes are supported by the water's substantial buoyant effect.

Gills can be *external* or *internal* (see Figure 23.2). **External gills** are located on an exposed body surface and project directly into the surrounding environmental medium. **Internal gills** are enclosed within a superficial body cavity. Whereas external placement permits ambient water currents to flow over the gills, internal placement usually requires an animal to use metabolic energy to ventilate them. Internal placement has its advantages nonetheless. When the gills are internal, the enclosing structures physically protect them and may help canalize the flow of water across the gills in ways that enhance the efficiency or control of breathing.

Ventilation of lungs, gills, or other gas-exchange membranes may be *active* or *passive*. Ventilation is **active** if the animal creates the ventilatory currents of air or water that flow to and from the gas-exchange membrane, using forces of suction or positive pressure that it generates by use of metabolic energy (as by contracting muscles or beating cilia). Ventilation is **passive** if *environmental* air or water currents directly or indirectly induce flow to and from the gas-exchange membrane (see Box 22.2). Active ventilation, although it uses an animal's energy resources, is potentially more reliable, controllable, and vigorous than passive ventilation. Active ventilation may be **unidirectional, tidal (bidirectional)**, or **nondirectional**. It is unidirectional if air or water is pumped over the gas-exchange membrane in a one-way path. It is tidal if air or water alternately flows to and from the gas-exchange membrane via the same passages (see Figure 22.6); mammalian lungs illustrate tidal ventilation. Ventilation is nondirectional if air or water flows across the gas-exchange membrane in many directions; animals with external gills that they wave back and forth in the water exemplify nondirectional ventilation.

In air-breathing animals with lungs, the lungs are usually ventilated. Some lungs, however, exchange gases with the environment entirely by diffusion and are termed **diffusion lungs**. Some insects and spiders, for example, are believed to breathe with diffusion lungs. Within a diffusion lung, the air is still, and O_2 and CO_2 travel the full length of the lung passages by diffusion.

A **dual breather**, or **bimodal breather**, is an animal that can breathe from either air or water. Dual breathers often have at least two distinct respiratory structures, which they employ when breathing from the two media. Examples include certain air-breathing fish (with both lung- and gill-breathing) and amphibians (with both lung- and skin-breathing).

SUMMARY
Fundamental Concepts of External Respiration

■ Oxygen always crosses the gas-exchange membrane by diffusion. This means that O_2 enters an animal only if the O_2 partial pressure on the outside of the gas-exchange membrane is higher than that on the inside.

■ Breathing organs are categorized as gills if they are evaginated structures that project into the environmental medium. They are lungs if they are invaginated structures that contain the medium.

■ Ventilation is the forced flow (convection) of the environmental medium to and from the gas-exchange membrane. It is categorized as active if an animal generates the forces for flow using metabolic energy. Ventilation may be unidirectional, tidal, or nondirectional.

Principles of Gas Exchange by Active Ventilation

Active ventilation is very common, and its analysis involves several specialized concepts that apply to a variety of animals. Here we discuss the principles of active ventilation.

When an animal ventilates its breathing organ (e.g., gills or lungs) directionally—either unidirectionally or tidally—a discrete current of air or water flows to and from the gas-exchange membrane. The rate of O_2 uptake by the breathing organ then depends on (1) the volume flow of air or water per unit of time and (2) the amount of O_2 removed from each unit of volume:

$$\text{Rate of } O_2 \text{ uptake (mL } O_2/\text{min)} = V_{\text{medium}} (C_I - C_E) \quad \textbf{(23.1)}$$

where V_{medium} is the rate of flow (L/min) of the air or water through the breathing organ, C_I is the O_2 concentration of the inhaled (inspired) medium (mL O_2/L medium), and C_E is the O_2 concentration of the exhaled (expired) medium. The difference $(C_I - C_E)$ represents the amount of O_2 removed from each unit of volume of the ventilated medium. The *percentage* of the O_2 available in the inhaled medium that is removed is $100(C_I - C_E)/C_I$. This ratio—known as the **oxygen utilization coefficient**, **oxygen extraction coefficient**, or **oxygen extraction efficiency**—expresses how thoroughly an animal is able to use the O_2 in the air or water it pumps through its lungs or gills.

To illustrate these calculations, consider a fish for which the water entering the mouth contains 6 mL O_2/L, the water exiting the gills contains 4 mL O_2/L, and the rate of ventilation is 0.5 L/min. According to Equation 23.1, the fish's rate of O_2 uptake is 0.5 L/min × 2 mL O_2/L = 1 mL O_2/min. Moreover, its oxygen utilization coefficient is 33%—meaning that the fish is removing 33% of the

O_2 from each volume of water it pumps and is allowing the other 67% to flow out with the exhaled water.

Tunas are especially efficient in using the O_2 in the water they drive over their gills. Whereas rainbow trout use 33%, yellowfin and skipjack tunas use 50%–60%.

The O_2 partial pressure in blood leaving a breathing organ depends on the spatial relation between the flow of the blood and the flow of the air or water

The O_2 partial pressure in the blood leaving a breathing organ can in many ways be considered the best single measure of the breathing organ's effectiveness. This is clear from the oxygen cascade concept (see Figure 22.8B): An animal with a high O_2 partial pressure in the blood leaving its breathing organ is particularly well poised to maintain an O_2 partial pressure in its mitochondria that is sufficiently high for aerobic catabolism to proceed without being O_2-limited.

Breathing organs with different designs exhibit inherent differences in the blood O_2 partial pressure they can maintain. These differences depend on the spatial relation between the flow of blood and the flow of the air or water. The differences are not absolute, because in all designs, the blood O_2 partial pressure is affected by additional factors besides spatial flow relations. Nonetheless, the flow relations between the blood and the air or water have great importance.

To explore the implications of various designs, let's start by considering tidally ventilated breathing organs, such as the lungs of mammals (**FIGURE 23.3**). Tidally ventilated breathing organs are distinguished by the fact that the medium (air in this case) next to the gas-exchange membrane is never fully fresh. The explanation is that such breathing organs are never entirely emptied between breaths. Consequently, when an animal breathes in, the fresh medium inhaled mixes—inside the breathing organ—with stale medium left behind by the previous breathing cycle, and because of this mixing, the O_2 partial pressure of the medium next to the gas-exchange membrane is lower—often much lower—than the partial pressure in the outside environment. The O_2 partial pressure in the blood leaving the breathing organ is lower yet (see Figure 23.3), because a partial-pressure gradient must exist between the medium next to the gas-exchange membrane and the blood for O_2 to diffuse from the medium into the blood. Characteristically, in tidally ventilated structures, the O_2 partial pressure of the blood leaving the breathing organ is *below* the O_2 partial pressure of the *exhaled* medium.

When ventilation is unidirectional rather than tidal, the two most obvious relations that can exist between the flow of the medium and the flow of the blood are *cocurrent* and *countercurrent*. In the cocurrent arrangement (**FIGURE 23.4A**), the medium flows along the gas-exchange membrane in the same direction as the blood, resulting in **cocurrent gas exchange**. In the countercurrent arrangement (**FIGURE 23.4B**), the medium and blood flow in opposite directions, and **countercurrent gas exchange** occurs. *Concurrent* is sometimes used as a synonym of *cocurrent*. Thus *cocurrent gas exchange* may also be called *concurrent gas exchange*.

In an organ that exhibits cocurrent gas exchange, when O_2-depleted afferent[3] blood first reaches the gas-exchange membrane, it

[3] *Afferent* means "flowing toward"; in this case it refers to blood flowing toward the gas-exchange membrane. *Efferent* means "flowing away" and refers to blood flowing away from the gas-exchange membrane.

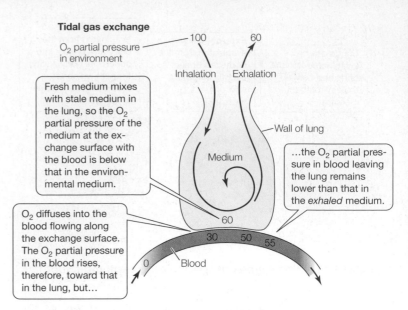

FIGURE 23.3 Tidal gas exchange: O_2 transfer from the environmental medium to the blood in a tidally ventilated lung Because a tidally ventilated lung is never fully emptied, fresh medium mixes in the lung with stale medium. Numbers are O_2 partial pressures in arbitrary units: The blood arriving at the breathing organ is arbitrarily assigned a value of 0, whereas the atmosphere is arbitrarily assigned a value of 100.

meets fresh, incoming medium, as shown at the *left* of Figure 23.4A. Then, as the blood and medium flow along the exchange membrane in the same direction, they gradually approach equilibrium with each other at an O_2 partial pressure that is intermediate between their respective starting partial pressures. When the blood reaches the place where it leaves the exchange membrane, its final exchange of O_2 is with medium that has a partial pressure considerably below that of the environmental medium. Cocurrent gas exchange therefore resembles tidal exchange, in that the O_2 partial pressure of blood leaving the breathing organ cannot ordinarily rise above the partial pressure of *exhaled* medium.

In an organ that exhibits countercurrent gas exchange, when O_2-depleted afferent blood first reaches the gas-exchange membrane, it initially meets medium that has already been substantially deoxygenated, as shown at the *right* of Figure 23.4B. However, as the blood flows along the exchange surface in the direction opposite to the flow of medium, it steadily encounters medium of higher and higher O_2 partial pressure. Thus, even as the blood picks up O_2 and its partial pressure rises, a partial-pressure gradient favoring further uptake of O_2 is maintained. The final exchange of the blood is with fresh, incoming medium of high O_2 partial pressure. Countercurrent exchange is thus an intrinsically more effective mode of exchange than either tidal or cocurrent exchange. One way to see this clearly is to note that the blood O_2 partial pressure created by countercurrent exchange is characteristically *much higher than the partial pressure in exhaled medium*; in principle, the O_2 partial pressure of the blood leaving the breathing organ might even approach equality with the O_2 partial pressure in *inhaled* medium. Moreover, if you compare Figures 23.4A and B, you will see that the O_2 partial pressure of the medium falls more—and that of the blood rises more—in 23.4B: Countercurrent exchange achieves a more complete transfer of O_2 from the medium to the blood than cocurrent (or tidal) exchange under comparable conditions.

(A) Cocurrent (concurrent) gas exchange

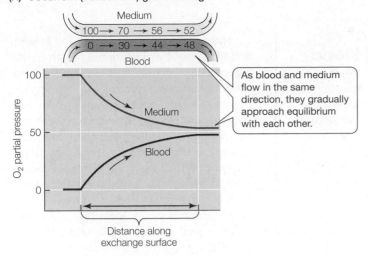

As blood and medium flow in the same direction, they gradually approach equilibrium with each other.

(B) Countercurrent gas exchange

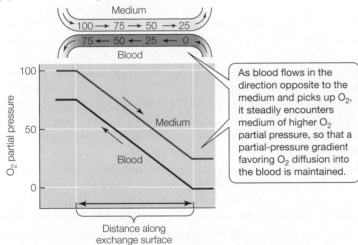

As blood flows in the direction opposite to the medium and picks up O_2, it steadily encounters medium of higher O_2 partial pressure, so that a partial-pressure gradient favoring O_2 diffusion into the blood is maintained.

FIGURE 23.4 Cocurrent and countercurrent gas exchange: Two modes of O_2 transfer from the environmental medium to the blood when ventilation is unidirectional The upper diagram in each case depicts the flow of medium and blood along the gas-exchange membrane. Numbers are O_2 partial pressures in arbitrary units, as specified in the caption of Figure 23.3. The blood reaches a higher O_2 partial pressure when the exchange is countercurrent because in that case the blood exchanges with fresh medium just before leaving the gas-exchange membrane.

When ventilation is unidirectional, a third possibility is **crosscurrent gas exchange**. In this type of exchange, the blood flow breaks up into multiple streams, each of which undergoes exchange with the medium along just part of the path followed by the medium (**FIGURE 23.5**). Some blood therefore undergoes gas exchange exclusively with O_2-rich medium (although other blood exchanges with O_2-poor medium). Cross-current exchange permits the O_2 partial pressure of the mixed blood leaving the breathing organ to be higher than that of exhaled medium, but it does not permit as high a blood O_2 partial pressure as countercurrent exchange under comparable circumstances.

The modes of exchange between the blood and the medium can be ranked in terms of their intrinsic ability to create a high O_2 partial pressure in the blood leaving the breathing organ (corresponding to their intrinsic efficiency in transferring O_2 from the medium to the blood): *Countercurrent exchange is superior to cross-current exchange, and cross-current exchange is superior to cocurrent or tidal*

Cross-current gas exchange

FIGURE 23.5 Cross-current gas exchange: A third mode of O_2 transfer from the environmental medium to the blood when ventilation is unidirectional Numbers are O_2 partial pressures in arbitrary units, as specified in the caption of Figure 23.3. The afferent blood vessel breaks up into many vessels that "cross" the path followed by the medium; each of these vessels makes exchange contact with just a limited part of the structure through which the medium is flowing. These vessels then coalesce to form a single efferent vessel. Cross-current exchange is intermediate between cocurrent and countercurrent exchange in its intrinsic gas-transfer efficiency.

exchange. As already noted, however, this ranking is not absolute because additional factors affect the ways in which real breathing organs function in real animals.

Arterial CO_2 partial pressures are much lower in water breathers than air breathers

Thus far we have discussed only O_2 in the blood leaving the gas-exchange surface. What about CO_2? The most important pattern in blood CO_2 partial pressure is a distinction between water and air breathers. *In water breathers, the partial pressure of CO_2 in the blood leaving the breathing organs is always similar to the CO_2 partial pressure in the ambient water*, regardless of whether gas exchange is tidal, cocurrent, countercurrent, or cross-current. If the ambient water is near equilibrium with the atmosphere and therefore near zero in its CO_2 partial pressure (see Figure 22.1), the blood is also near zero in CO_2 partial pressure. For example, in fish in aerated water, the partial pressure of CO_2 in blood leaving the gills is typically about 0.3 kilopascals (kPa) (2 millimeters of mercury [mm Hg]). In sharp contrast, *in air breathers the partial pressure of CO_2 in the blood leaving the breathing organs is usually far above the CO_2 partial pressure in the atmosphere.* In mammals and birds, for example, the partial pressure of CO_2 in blood leaving the lungs is more than ten times the corresponding value in fish; in humans it is 5.3 kPa (40 mm Hg)! In vertebrates, systemic arterial blood pumped by the heart to supply O_2 to the body is derived from the blood leaving the breathing organs. Thus, from what we have said, if we survey species in which arterial CO_2 partial pressure has been measured, we can understand why the arterial CO_2 partial pressure is typically only 0.1–0.6 kPa (1–4 mm Hg) in water-breathing vertebrates, whereas it is far higher, 4.0–5.3 kPa (30–40 mm Hg), in air-breathing vertebrates. *When vertebrates and other animals emerged onto land in the course of evolution, the level of CO_2 in their blood and other body fluids shifted dramatically upward.*

These patterns are explained principally by the differing physical and chemical properties of water and air. More specifically, they are explained by the analysis of capacitance coefficients, as explained in **BOX 23.1**.

BOX 23.1

Capacitance Coefficients Explain the Relative Partial Pressures of O_2 and CO_2 in Arterial Blood

During breathing, the extent to which the CO_2 partial pressure of the respired medium changes relative to its change in O_2 partial pressure tends to be sharply different between water breathers and air breathers because of a property of water and air called the **capacitance coefficient**. For any particular gas (O_2 or CO_2) and any particular respired medium (water or air), the capacitance coefficient is defined to be the change in total gas *concentration* in the respired medium per unit of change in gas *partial pressure* in the respired medium. **Box Extension 23.1** explains by use of a figure how capacitance coefficients place constraints on animal physiology.

SUMMARY
Principles of Gas Exchange by Active Ventilation

- The oxygen utilization coefficient during breathing is the percentage of the O_2 in inhaled medium that an animal removes before exhaling the medium.

- The four major types of gas exchange that can occur during directional ventilation can be ranked in terms of their inherent ability to establish a high O_2 partial pressure in blood exiting the breathing organ. Countercurrent gas exchange ranks highest. Cross-current gas exchange ranks second. Cocurrent and tidal gas exchange rank third.

- Air breathers tend to have much higher CO_2 partial pressures in their systemic arterial blood than water breathers. This difference arises during breathing and is principally a consequence of the differing physical and chemical properties of air and water.

Low O_2: Detection and Response

Most animals are fundamentally aerobic. Thus when the O_2 levels in an animal's tissues become unusually low, the low O_2 levels typically mean trouble. Animals detect low O_2 levels—and respond to them—at two different scales. The mechanisms involved in detection and response are elaborate. Their elaborateness reflects the extreme importance of O_2.

One scale of response to low O_2 is that of the whole body: the scale of organ systems. In many types of animals, the O_2 partial pressure in the circulating blood is monitored continuously, and if the blood O_2 partial pressure falls, the function of the breathing system is modulated to reverse this trend. Water breathers, for example, commonly respond to a decreased blood O_2 partial pressure by increasing their rate of gill ventilation. Air breathers may increase their rate of lung ventilation by breathing at a higher rate. Responses such as these—which we will discuss often in this chapter—help protect an animal's cells from experiencing hypoxia.[4]

The second scale of response is intracellular, in cells throughout the body. The intracellular responses are elicited if the rate at which the breathing and circulatory systems supply O_2 fails to keep pace with the rate of mitochondrial O_2 utilization—resulting in cellular hypoxia.

---

[4] *Hypoxia*, as mentioned in Chapter 8, refers to an especially low level of O_2 in tissues or cells.

Hypoxia-inducible factors 1 and **2 (HIF-1** and **HIF-2)** are the most central players in the intracellular responses. They are closely related transcription factors, which bind with specific *hypoxia-response elements* (nucleotide base sequences) of DNA to activate gene transcription. In the last two decades there has been an explosion of knowledge regarding these transcription factors, which are implicated in many disease states as well as playing crucial roles in the ordinary lives of animals. HIF-1 is an ancient molecule, identified in invertebrates as well as in all groups of vertebrates.

HIF-1 and HIF-2 increase in concentration in a cell when the cell experiences hypoxia, and they affect the transcription of dozens or hundreds of target genes, which have been identified through transcription profiling (transcriptomics) (see Chapter 3). HIF-1 and HIF-2 are different in their effectiveness in activating various genes in a tissue. Moreover, their target genes differ to some extent from tissue to tissue within a single species, and they also differ among species. Overall, HIF-1 and HIF-2 have extremely broad ranges of potential action, and many of their actions aid cells that are confronting hypoxia. In vertebrates they cause certain types of cells to increase secretion of *erythropoietin*, which enhances production of red blood cells (see Box 24.2). They also increase intracellular synthesis of several types of molecules, including enzymes of anaerobic glycolysis and a form of mitochondrial cytochrome oxidase that is particularly efficient in using O_2. Besides these effects and many others, they also promote *angiogenesis*, the development of new blood capillaries. The increase in capillaries shortens the average distance between tissue cells and capillaries, thereby aiding O_2 diffusion into the cells.

The basic manner in which the intracellular concentrations of these transcription factors are regulated is exemplified in **FIGURE 23.6** using HIF-1. A complete HIF-1 molecule consists of α and β subunits. The β subunits are present regardless of cellular O_2 level. Moreover, the α subunits are constantly being not only synthesized but also tagged for breakdown by the ubiquitin–proteasome system (see Figure 2.23). When the O_2 level in a cell decreases, breakdown of α subunits is inhibited. Therefore the concentration of α subunits

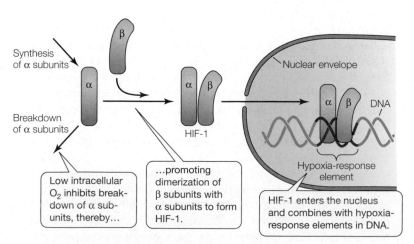

Synthesis of α subunits

Breakdown of α subunits

α $\quad$ β

α β

HIF-1

Nuclear envelope

α β DNA

Hypoxia-response element

Low intracellular O_2 inhibits breakdown of α subunits, thereby...

...promoting dimerization of β subunits with α subunits to form HIF-1.

HIF-1 enters the nucleus and combines with hypoxia-response elements in DNA.

FIGURE 23.6 Formation and action of HIF-1

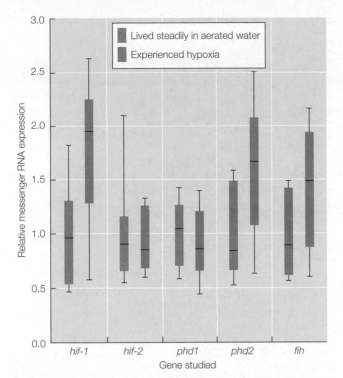

FIGURE 23.7 A genomic assessment of acclimation in the HIF signaling system Expression of five genes that code for compounds in the HIF signaling system was measured in the heart in two groups of epaulette sharks (*Hemiscyllium ocellatum*). One group lived continuously in aerated water. The other experienced eight brief (2-h) episodes of hypoxic water over a period of 4 days. Expression is measured on a relative scale. For each group, the upper and lower box limits correspond to the 75th and 25th percentiles of the data, the line in the box marks the median, and the bars extending from the box show the data range. The differences between the two groups were statistically significant for the *hif-1* gene (which codes for HIF-1) and the *phd2* gene. (After Rytkönen et al. 2012.)

rises, and more HIF-1 molecules are formed by dimerization of α and β subunits. The HIF-1 molecules enter the nucleus, where they bind with specific DNA sequences (the hypoxia-response elements), which in turn control gene transcription.

Some species of animals are well adapted to living in environments where the ambient O_2 level is prone to fall to low levels. For example, the epaulette shark (*Hemiscyllium ocellatum*) lives in places where it experiences periods of very low environmental O_2. In a recent genomic study, researchers asked whether the HIF signaling system undergoes acclimation when fish that have been living in high-O_2 water begin to experience low O_2. One can reason that if a fish that's been living in high-O_2 water is about to confront a period of low environmental O_2, the fish might profit by upregulating its HIF-1 signaling system, recognizing that it will depend on the HIF signaling system to help its tissues cope with the hypoxia they may soon experience. Using genomic methods, researchers measured the expression in the heart muscle of five genes that code for HIF-1 and other compounds in the HIF signaling system. Sharks that were exposed to a series of eight brief episodes of low O_2 in their environment showed upregulation of expression of some of these genes, compared with sharks that lived in high-O_2 water all the time (**FIGURE 23.7**).

Some human cultural groups live at high altitudes, where they steadily experience low atmospheric O_2 concentrations. As we will discuss later in this chapter (see Box 23.2), recent evidence points

to a major role for the HIF signaling system in allowing these people to succeed.

The two scales of response to low O_2 that we noted at the start of this section in fact interact with each other. As already mentioned, for example, in vertebrates HIF-1 and HIF-2 play central roles in controlling the production of red blood cells. In this way, HIF-1 and HIF-2 help control the O_2-carrying ability of each unit of volume of blood, which is a major factor at the scale of organ systems.

Introduction to Vertebrate Breathing

The vertebrates living today are usually thought of, in a rough way, as representing an evolutionary sequence. In actuality, of course, today's fish were not the progenitors of today's amphibians, and today's lizards (or other nonavian reptiles) were not the progenitors of today's mammals and birds. Thus when we think of the sequence from fish to mammals and birds, caution is always called for in thinking of it as an evolutionary sequence. Comparisons among today's animals nonetheless often provide revealing insights into trends that occurred during evolution. Before we start our study of breathing in the various vertebrate groups, an overview of general trends will help place the individual vertebrate groups in a larger context.

One property of breathing organs that is of enormous importance is the total surface area of the gas-exchange membrane. This area is typically an allometric function of body size among species within any one group of phylogenetically related vertebrates. The allometric relation differs, however, from one group to another, as seen in **FIGURE 23.8A**. To explore Figure 23.8A, let's first focus on the lines for amphibians and reptiles other than birds. These lines are similar. Moreover, the lines for most groups of fish (e.g., toadfish and dogfish) are also similar. These similarities tell us two things. First, the total lung area of a nonavian reptile is roughly similar to the total lung area of an amphibian of the same body size. Second, the total gill area of a fish of particular body size is roughly similar to the total lung area of an amphibian or reptile of that size—suggesting that when vertebrates emerged onto land, there was not immediately much of a change in the area of the gas-exchange surface in their breathing organs. If we now turn our attention to the lines for mammals and birds, we notice that those groups, although independently evolved, are similar to each other. Moreover, they exhibit a dramatic step upward in the area of gas-exchange surface in their lungs by comparison with amphibians or nonavian reptiles. In brief, the mammals and birds independently evolved lungs with markedly enhanced gas-exchange surface areas—probably in association with their evolution of homeothermy. As we saw in Chapter 10, homeothermy increases an animal's metabolic rate by a factor of at least four to ten. Thus lungs with an enhanced ability to take up O_2 and void CO_2 are required by homeothermic animals. Remarkably, tunas—noteworthy for being extremely active fish, as discussed at the opening of this chapter—have gill surface areas that approximate the lung surface areas of equal-sized mammals and birds!

The fact that mammals and birds have exceptionally large gas-exchange surface areas does not mean that they have large lungs compared with reptiles or amphibians. In fact, the opposite is often true. For example, if a rodent is compared with a like-size lizard, snake, or turtle, the lung volume of the reptile is likely to be

(A) Area of the gas-exchange membrane vs. body size

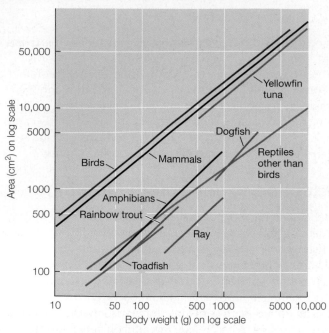

(B) Thickness of the gas-exchange membrane vs. body size

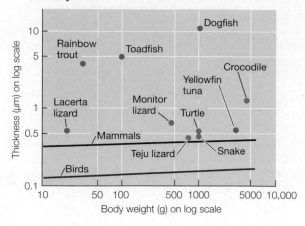

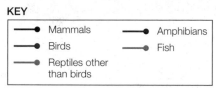

KEY

Mammals	Amphibians
Birds	Fish
Reptiles other than birds	

FIGURE 23.8 Total area and thickness of the gas-exchange membrane in the gills or lungs of vertebrates as functions of body size The lines for mammals and birds in both (A) and (B) and those for amphibians and reptiles other than birds in (A) are for many species (e.g., about 40 species of mammals, ranging in size from shrews to horses). The lines for fish in (A) are for various-sized individuals of single species. The thickness in (B) is the average distance between the blood and the water or air. (After Perry 1990.)

at least five times greater than that of the rodent. Yet the surface area of the gas-exchange membrane is likely to be ten times greater in the rodent than in the reptile! The explanation for the high gas-exchange surface area in the lungs of a mammal or bird is that the lungs of these animals are extraordinarily densely filled with branching and rebranching airways, as we will see in this chapter. In nonavian reptiles and amphibians, in contrast, the lungs typically have parts that are simply like balloons—little more than a sheet of tissue surrounding an open central cavity—and even the parts that are subdivided are much less elaborately subdivided than mammalian or avian lungs. Thus, whereas the lungs are large in a reptile compared with a rodent, they provide the reptile with a comparatively small area of gas-exchange membrane.

The thickness of the barrier between the blood and the environmental medium also shows significant evolutionary trends in the major vertebrate groups. In most fish, the sheet of gill tissue between blood and water is roughly 5–10 μm thick. Vertebrate lungs uniformly have a much reduced barrier between blood and air (**FIGURE 23.8B**). Mammals tend to have a thinner sheet of tissue between blood and air than lizards, crocodilians, or other nonavian reptiles. To a dramatic extent, the tissue in birds is thinner yet, roughly 0.2 μm. Tunas, which are highly specialized for O_2 uptake compared with most fish, have a blood–water barrier in their gills that is similar in thickness to the blood–air barrier in mammals.

The skin varies widely in its role as a gas-exchange site in vertebrates (**FIGURE 23.9**). Some fish and most reptiles have evolved skins that permit little O_2 or CO_2 exchange. In contrast, the skin can be responsible for 25% or more of gas exchange in other fish, in certain snakes and turtles, and in many amphibians; in the lungless salamanders, which not only lack lungs but have an epidermis that

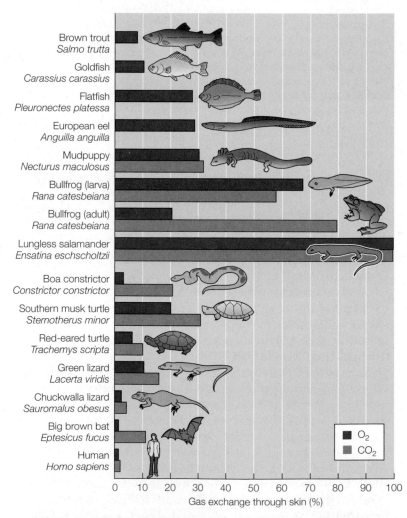

FIGURE 23.9 The percentage of O_2 and CO_2 exchange that occurs across the skin in vertebrates The exact extent of skin breathing within a species often depends on environmental conditions (e.g., temperature). Some authorities place the bullfrog in the genus *Lithobates*. (After Feder and Burggren 1985.)

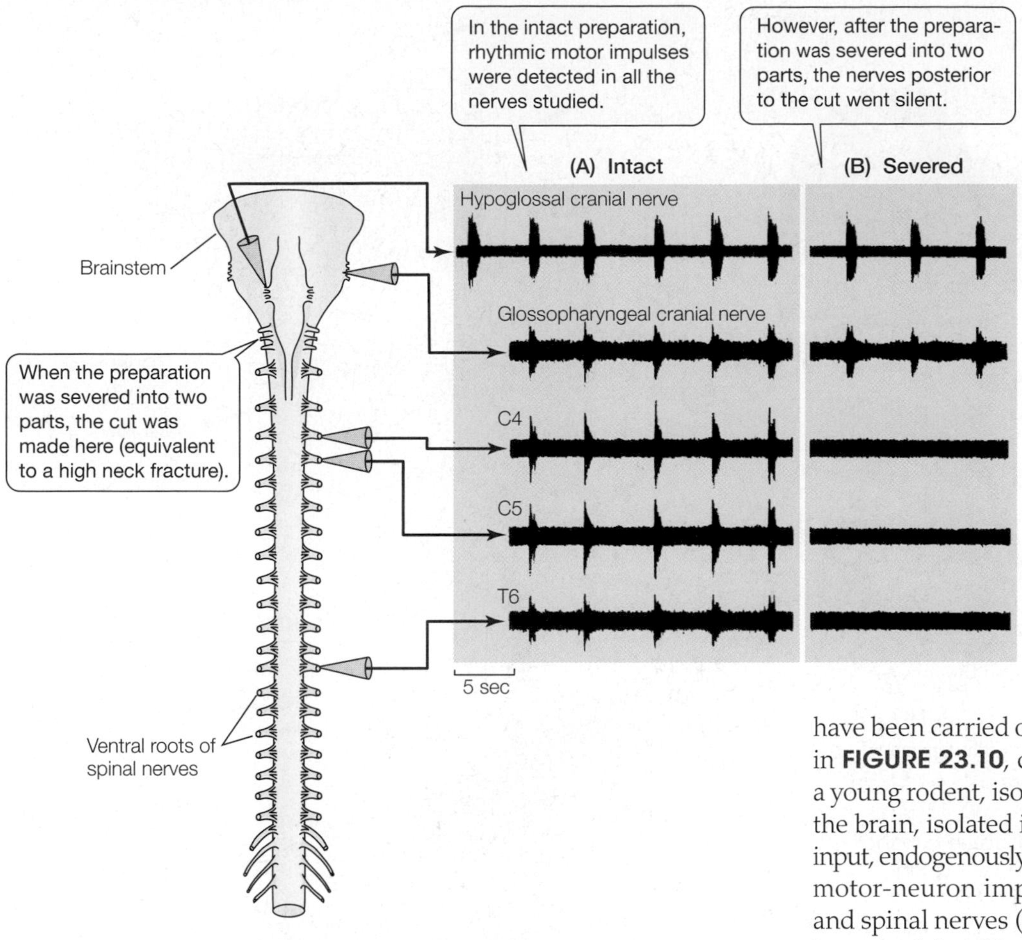

In the intact preparation, rhythmic motor impulses were detected in all the nerves studied.

However, after the preparation was severed into two parts, the nerves posterior to the cut went silent.

When the preparation was severed into two parts, the cut was made here (equivalent to a high neck fracture).

Brainstem

Ventral roots of spinal nerves

(A) Intact

(B) Severed

Hypoglossal cranial nerve

Glossopharyngeal cranial nerve

C4

C5

T6

5 sec

FIGURE 23.10 A central pattern generator in the brainstem originates motor-neuron impulses that produce the breathing rhythm An *isolated* portion of the central nervous system of a neonatal rat, cut free from all connections with the rest of the animal, was used for this study. It included the brainstem and spinal cord. Recording electrodes were attached to two cranial nerves (hypoglossal and glossopharyngeal) and the ventral roots of three spinal nerves: C4 and C5 (C = cervical) and T6 (T = thoracic). Records show electrical activity as a function of time. (From Feldman et al. 1988.)

is vascularized (highly unusual), 100% of gas exchange occurs across the skin. Most mammals and birds resemble humans (see Figure 23.9) in relying almost entirely on their lungs. Among terrestrial vertebrates, the role of the skin in breathing is related to the skin's desiccation resistance. Groups of terrestrial vertebrates that are well defended against water loss through their skin, such as mammals and birds, tend to exhibit little exchange of O_2 or CO_2 through their skin, whereas groups, such as frogs and salamanders, that are poorly defended against cutaneous water loss tend to engage in significant cutaneous breathing. Modifications that render the skin poorly permeable to water (see page 771) also make it poorly permeable to O_2 and CO_2.

The control of the active ventilation of the gills or lungs is the final subject that deserves mention in introducing vertebrate breathing. Of central importance is the control of the rhythmic muscle contractions that produce breathing movements, such as our own inhalation and exhalation movements. The muscles responsible for breathing movements in all vertebrates are skeletal muscles. They therefore require stimulation by motor neurons for *each and every contraction* they undergo. The breathing rhythm originates in sets of neurons that form central pattern generators (see Chapter 19); in a process called **rhythmogenesis**, these sets of neurons initiate rhythmic outputs of nerve impulses (action potentials) that travel to the breathing muscles and stimulate them into rhythmic patterns of contraction. The central pattern generators for breathing in all vertebrates are believed to be located in the brainstem: in the medulla, and sometimes other associated parts, of the brain. Although outside influences routinely play roles in controlling breathing, these central pattern generators are absolutely essential for creating the breathing rhythm. Accordingly, if a vertebrate's spinal cord is severed just posterior to the brainstem, breathing

stops instantly, because the neuronal outputs from the brainstem are unable to travel to the breathing muscles by way of spinal nerves. Experiments that illustrate these vital points have been carried out on the type of preparation shown at the left in **FIGURE 23.10**, consisting of the brainstem and spinal cord of a young rodent, isolated from the rest of the body. The medulla of the brain, isolated in this preparation from any sensory neuronal input, endogenously generates rhythmic bursts of ventilation-driving motor-neuron impulses that are detectable in multiple cranial and spinal nerves (see Figure 23.10A). However, if the spinal cord is severed (see Figure 23.10B), the spinal nerves posterior to the injury go silent, paralyzing the ventilatory muscles they service.[5]

Humans and most other mammals exhibit **continuous breathing**, meaning that each breath is followed promptly by another breath in a regular, uninterrupted rhythm. Birds and most fish also usually display continuous breathing. Lizards, snakes, turtles, crocodilians, amphibians, and air-breathing fish, in contrast, usually exhibit **intermittent breathing**, or **periodic breathing**, defined to be breathing in which breaths or sets of breaths are regularly interrupted by extended periods of **apnea**—that is, periods of no breathing. During intermittent breathing in these vertebrates, each period of apnea follows an inspiration. The lungs, therefore, are inflated during apnea. The glottis—the opening of the airways into the buccal cavity—is closed during apnea in the groups of vertebrates that display intermittent breathing. Because of this glottal closure, the inspiratory muscles can relax during the apnea without causing air to be expelled from the lungs.

SUMMARY
Introduction to Vertebrate Breathing

■ The gill surface area of most fish of a given body size is similar to the lung surface area of amphibians and nonavian reptiles of the same size. Compared with the latter groups, mammals and birds have much more lung surface area—helping to meet their far higher needs for gas exchange. The barrier between the blood and the air or water in the breathing organs is notably thin in mammals and thinnest in birds.

(Continued)

[5] More information on rhythmogenesis is presented later in this chapter in the discussion of mammals.

FIGURE 23.11 The branchial breathing system in teleost (bony) fish (A) The lateral view shows the orientation of the gill arches under the operculum. (B and C) Consecutive enlargements show the structure of the gill array. (D) An enlarged view of a filament and three of the secondary lamellae on its upper surface (those on the lower surface not shown), showing that blood flow within the secondary lamellae is countercurrent to water flow across them. Blood flows in a sheet through each secondary lamella, as seen in the section of the foremost one. (After Hill and Wyse 1989.)

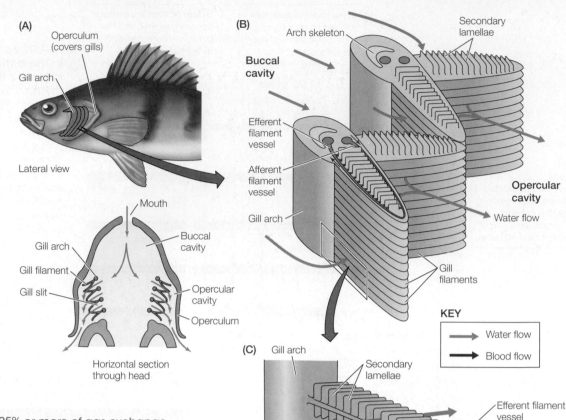

■ The skin can account for 25% or more of gas exchange in some fish, turtles, and other nonavian reptiles, and up to 100% in some amphibians. The skin is a minor contributor to gas exchange, however, in mammals and birds.

■ The breathing muscles of vertebrates are skeletal muscles activated by motor-neuron impulses. The breathing rhythm originates in a central pattern generator in the brainstem.

Breathing by Fish

Many fish start life as tiny larvae that breathe only by diffusion of gases across their general body surfaces (see Box 22.1). Because fish larvae lack specialized breathing organs, it can be easy to disregard these early stages in the study of breathing. However, gas-exchange insufficiencies are sometimes responsible for mass deaths of larvae. Thus the physiology of early diffusion–respiration stages can be of great importance ecologically and evolutionarily. As a young fish grows, its body becomes too thick for diffusion to suffice (see Box 22.1). At the same time, its gills develop and mature—as does its circulatory system, which is required for O_2 from the gills to reach the rest of its body.

For our study of gill breathing, we focus here on teleosts, the principal group of bony fish. The buccal cavity of a teleost communicates with the environment not only by way of the mouth but also by way of lateral pharyngeal openings, the **gill slits**. The gills are arrayed across these lateral openings, and a protective external flap, the **operculum**, covers the set of gills on each side of the head. Looking in detail at the structure of the gill apparatus, there are four **gill arches** that run dorsoventrally between the gill slits on each side of the head (**FIGURE 23.11A**); the arches, which are reinforced with skeletal elements, provide strong supports for the gills proper. Each gill arch bears two rows of **gill filaments** splayed out laterally in a V-shaped arrangement (**FIGURE 23.11B**);

collectively, the rows of filaments form a continuous array shaped like a series of Vs (VVVV) separating the buccal cavity on the inside from the opercular cavity on the outside. Each gill filament bears a series of folds, called **secondary lamellae**, on its upper and lower surfaces (**FIGURE 23.11C**); the lamellae run perpendicular to the long axis of the filament and number 10–40 per *millimeter* of filament length on each side. The secondary lamellae—which are richly perfused with blood and are thin-walled—are the principal sites of gas exchange.

Water flows along the surfaces of the secondary lamellae from the buccal side to the opercular side. Conversely, blood within the secondary lamellae flows in the opposite direction, as seen in **FIGURE 23.11D**. Countercurrent gas exchange, therefore, can occur along the gas-exchange membranes. The percentage of O_2 extracted from ventilated water by resting teleosts has been reported to range from very high values of 80%–85% down to 30% or less; the latter values raise questions about the exact nature of gas exchange in the fish involved, because the values are lower than would be expected from effective countercurrent exchange.

Gill ventilation is usually driven by buccal–opercular pumping

In general, water flow across the gills of a teleost fish is maintained almost without interruption by the synchronization of two pumps: a **buccal pressure pump**, which develops positive pressure in the buccal cavity and thus forces water from the buccal cavity through the gill array into the opercular cavity, and an **opercular suction pump**, which develops negative pressure in the opercular cavity and thus sucks water from the buccal cavity into the opercular cavity. The relative dominance of the two pumps varies from species to species; here we take a generalized view. We look first at the action of each pump separately and then at the integration of the pumps over the breathing cycle. It will be important to remember throughout that water flows from regions of relatively high pressure to ones of relatively low pressure.

THE BUCCAL PRESSURE PUMP The stage is set for the buccal pressure pump to operate when a fish fills its buccal cavity with water by depressing the floor of the cavity while holding its mouth open. The lowering of the buccal floor increases the volume of the buccal cavity, thereby decreasing buccal pressure below ambient pressure and causing an influx of water. The mouth is then closed, and the buccal pump enters its positive-pressure phase. The fish raises the floor of the buccal cavity during this phase. This action increases the buccal pressure above ambient pressure and drives water from the buccal cavity through the gills into the opercular cavities. Thin flaps of tissue, which act as passive valves, project across the inside of the mouth opening from the upper and lower jaws. During the refilling phase of the buccal cycle, when buccal pressure is below ambient, these flap valves are pushed inward and open by the influx of water through the mouth. During the positive-pressure phase, however, the flap valves are forced against the mouth opening on the inside and help to prevent water from exiting the buccal cavity through the mouth.

THE OPERCULAR SUCTION PUMP A teleost fish is able to expand and contract its opercular cavities by lateral movements of its opercular flaps and other muscular actions. Running around the rim of each operculum is a thin sheet of tissue that acts as a passive valve, capable of sealing the slitlike opening between the opercular cavity on the inside and the ambient water on the outside. The negative-pressure phase—suction phase—of the opercular pump occurs when the opercular cavity is expanded. At this time, the pressure in the cavity falls below the pressures in the buccal cavity and the ambient water. The negative pressure in the opercular cavity sucks water from the buccal cavity into the opercular cav-

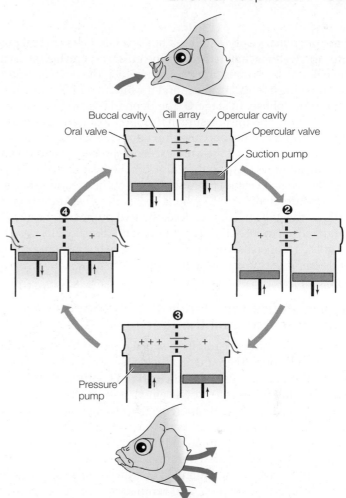

FIGURE 23.12 The breathing cycle in teleost fish Plus (+) and minus (–) symbols indicate pressures *relative to ambient pressure*. The buccal and opercular pumps are represented by pistons. Blue arrows represent water flow. Arrows through the gill array indicate water flow over the gills. Stages **❷** and **❹** are transitional and short in duration. (After Hughes 1961, Moyle 1993.)

ity through the gill array. Water would also be sucked in from the environment were it not for the action of the opercular rim valve, which is pushed medially against the fish's body wall by the higher ambient pressure, sealing the opercular opening and preventing influx of ambient water. After its sucking phase, the opercular pump enters its discharge phase. The cavity is contracted, raising the pressure inside to be higher than ambient pressure; this pressure difference forces the rim valve open and discharges water through the opercular opening.

INTEGRATION OF THE TWO PUMPS PRODUCES NEARLY CONTINUOUS, UNIDIRECTIONAL FLOW The temporal integration of the buccal and opercular pumps is diagrammed in **FIGURE 23.12**. In stage **❶**, the buccal cavity is being refilled. Expansion of the buccal cavity produces a pressure below ambient; therefore, if the buccal pump were the only pump, flow of water through the gills from the buccal side would not occur at this time, and in fact, there would be backflow through the gills into the buccal cavity because of the lowered buccal pressure. It is during stage **❶**, however, that the opercular pump is in its sucking phase. The pressure in the opercular cavity is reduced to a level far below buccal pressure, and water is drawn through the gills from the buccal cavity. Stage ❷ is a short transition period in which the opercular pump is completing its sucking phase and the buccal pump is beginning its pressure phase. In stage ❸, the opercular pump is in its discharge phase, and the pressure in the opercular cavity is elevated. However, because the buccal pump is simultaneously in its pressure phase, the buccal pressure exceeds opercular pressure, and water again flows through the gills from the buccal cavity. Only in stage ❹, which occupies just a short part of the breathing cycle, is the pressure gradient in the direction favoring backflow of water through the gills. In all, therefore, the two pumps are beautifully integrated to produce almost continuous, unidirectional flow across the gills: The opercular pump sucks while the buccal pump is being refilled, and the buccal pump develops positive pressure while the opercular pump is being emptied.

Many fish use ram ventilation on occasion, and some use it all the time

When a swimming fish attains a speed of 50–80 cm/s or greater, its motion through the water, if it holds its mouth open, can itself elevate the buccal pressure sufficiently to ventilate the gills at a rate adequate to meet O_2 requirements. Many fast-swimming fish, in fact, cease buccal–opercular pumping when they reach such speeds and employ *ram ventilation*, which in theory lowers the metabolic cost of ventilation (because ventilation is achieved simply as a by-product of swimming). As mentioned at the start of this chapter, in ram ventilation the gills are ventilated by a swimming fish's forward motion, and ventilation is therefore powered by the swimming muscles.

Tunas, mackerel, dolphinfish, bonitos, billfish (e.g., marlins), and lamnid sharks swim continuously and use ram ventilation all the time. In the tunas, at least, ram ventilation is obligatory, as already mentioned, because the buccal–opercular pumping mechanisms have become incapable of producing a sufficiently vigorous ventilatory stream.

Fish physiologists have long wondered whether the delicate structures of the gills might be thrown into disarray by the forces of high-speed water movement over the gills when powerful swimmers such as tunas employ ram ventilation. Recently progress has been made in better understanding gill structure in these fish. The gills of tunas and some other powerful swimmers, it turns out, have evolved structural reinforcements (**FIGURE 23.13**). These structural specializations are believed to stabilize the geometry of the filaments and secondary lamellae when the gills are subjected to high-velocity water flow.

Decreased O_2 and exercise are the major stimuli for increased ventilation in fish

Fish are capable of as much as 30-fold changes in their rates of gill ventilation by buccal–opercular pumping. One potent stimulus to increase the rate of ventilation is a decrease in the partial pressure of O_2 in the environment or blood, detected by chemoreceptor cells in the gills. Exercise also is a potent stimulus of ventilation in fish. Carbon dioxide, however, tends to be a relatively weak ventilatory stimulus in fish—in sharp contrast to mammals, in which ventilation is extremely sensitive to elevated CO_2. The

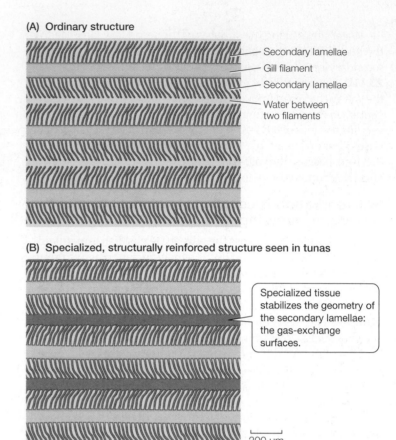

(A) Ordinary structure

— Secondary lamellae
— Gill filament
— Secondary lamellae
— Water between two filaments

(B) Specialized, structurally reinforced structure seen in tunas

Specialized tissue stabilizes the geometry of the secondary lamellae: the gas-exchange surfaces.

200 μm

FIGURE 23.13 Specialized gill reinforcements in tunas Each part shows a longitudinal section through three vertically adjacent gill filaments on a single gill arch. For orientation, compare with Figure 23.11C. (A) The ordinary structure of the gills in teleost fish. (B) The specialized structure seen in parts of the gills of many species of tunas, in which tissue not only connects the outer ends of all the secondary lamellae on each filament but also bridges the lamellae on one filament to those on the next adjacent filament. (After Wegner et al. 2013.)

question arises as to why CO_2 plays such different roles in the two groups. The answer probably lies in the high solubility of CO_2 in water (see Box 23.1 for a more rigorous treatment). Put loosely, for a water breather, although O_2 can be difficult to acquire, CO_2 is easy to excrete. As discussed earlier, water breathers never increase the CO_2 partial pressure by much in the water passing over their gills; therefore CO_2 is unlikely to be a sensitive indicator of their ventilatory status.

Changes in the rate of ventilation are not the only means employed by fish to adjust gill O_2 exchange. Some fish, for example, exhibit *lamellar recruitment*, by which they adjust the proportion of the secondary lamellae that are actively perfused with blood. Whereas only about 60% of lamellae may be perfused at rest, 100% may be perfused during exercise or exposure to reduced O_2.

Several hundred species of bony fish are able to breathe air

Many species of bony fish have evolved mechanisms for tapping the rich O_2 resources of the air. Nearly 400 species of air-breathing fish are known, especially in freshwater. Most retain functional gills and are dual breathers, acquiring O_2 from both water and air. The extent to which these fish rely on the atmosphere depends on several factors. They typically increase their use of air as the level of dissolved O_2 in their aquatic habitat falls. They also tend to resort increasingly to air breathing as the temperature rises, because high temperatures elevate their O_2 needs. Notably, air-breathing fish typically void most of their CO_2 into the water—across their gills or skin—even when relying on the atmosphere for most of their O_2. What is the adaptive value of air breathing? The usual hypothesis is that air breathing arose in groups of fish that were living over evolutionary scales of time in O_2-poor waters (see page 23), as a means of solving the problem of O_2 shortage in the water (see also page 687).

Some air-breathing fish lack marked anatomical specializations for exploiting the air. American eels (*Anguilla rostrata*) are examples. They sometimes come out onto land in moist situations, and they then meet about 60% of their O_2 requirement by uptake across their skin and 40% by buccal air gulping. Their gills, which are quite ordinary, are probably the primary site of O_2 uptake from the air they gulp.

In most air-breathing fish, some part or branch of the alimentary canal has become specialized as an air-breathing organ. The specialized region varies greatly among species—reflecting the fact that air breathing has evolved independently more than 20 times. The specialized region is always highly vascularized, and its walls may be thrown into extensive patterns of evagination or invagination. In some species the buccal cavity is specialized for air breathing—as is true in electric eels (*Electrophorus electricus*), which have vast numbers of vascularized papillae on the walls of their buccal cavity and pharynx. The opercular cavities form air-breathing organs in some species; mudskippers (*Periophthalmus*), for example, breathe air using expanded gill chambers lined with vascularized, folded membranes. Many air-breathing species employ *suprabranchial chambers*, situated in the dorsal head above the gills. Quite a few air-breathing fish employ vascularized portions of the stomach to breathe. Others, notably species of armored catfish (family Callichthyidae), employ the intestine; in these, half or more of the intestinal length is highly vascularized and devoted to breathing from air that is swallowed and later expelled via the anus. The swim bladder (gas bladder) is used as an air-breathing organ by many fish. The "tinkering" aspect of evolution (see page 10) is nowhere better illustrated than in the fantastic diversity of body parts that fish have diverted from old functions to the task of getting O_2 from the atmosphere.

The air-breathing organs of fish are most often inflated by buccal pumping. A fish takes air into its buccal cavity, then closes its mouth and compresses the buccal cavity. In this way, air is driven into its stomach, swim bladder, or other air-breathing structure.

A potential problem for air-breathing fish is that O_2 taken up from the atmosphere may be lost to O_2-poor water across their gills! This possibility probably helps explain why the gills of these animals are often reduced in comparison with those of other fish. In extreme cases, the gills are so atrophied that air breathing is obligatory—as is true of electric eels, which drown if they cannot obtain air. Many air-breathing fish have also evolved specialized circulatory shunts by which oxygenated blood can bypass the gas-exchange surfaces of their gills, thereby limiting O_2 loss across the gills.

Of all air-breathing fish, the six species of **lungfish** (**dipnoans**) (**FIGURE 23.14A**) have received the most attention because they

(A) An African lungfish in the genus *Protopterus*

(B) The inner wall of a lungfish lung

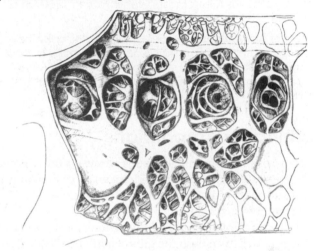

FIGURE 23.14 Lungfish and their lungs This figure features African lungfish (genus *Protopterus*); two other genera of lungfish occur in South America and Australia. (A) *Protopterus dolloi*. (B) The inner wall of part of a lung of *Protopterus aethiopicus*. The respiratory surface area of the lung is greatly enhanced by a complex pattern of vascularized folds. The side compartments in the wall of the lung open to a central cavity that runs the length of the lung. The cavity communicates anteriorly with a short pulmonary canal leading to the esophagus. (B from Poll 1962.)

are believed to be the modern fish that most closely resemble the ancestral fish that gave rise to terrestrial vertebrates. The walls of the lungs of lungfish (**FIGURE 23.14B**) are thrown into complex arrays of interconnected folds—with the folds arranged roughly in tiers from low folds to high folds—resembling the walls of many amphibian lungs. Why are the air-breathing organs of lungfish called *lungs*, and why are the fish themselves named *lung*fish? These names arose decades ago because the air-breathing organs were viewed as being "particularly homologous" to the lungs of terrestrial vertebrates. Many morphologists now conclude, however, that swim bladders could just as justifiably be called lungs by this standard. To a physiologist, all invaginated breathing organs are lungs (see Figure 23.2), and thus all of the air-breathing organs of fish are lungs.

SUMMARY
Breathing by Fish

- The secondary lamellae are the principal sites of gas exchange in fish gills. Countercurrent gas exchange occurs in the lamellae.

- Water flow across the gills is essentially unidirectional. It is driven by a buccal pressure pump and an opercular suction pump that act in an integrated rhythm, so that the buccal pump drives water across the gills when the opercular pump is being emptied of water and the opercular pump sucks water across the gills while the buccal pump is being refilled with water.

- Some fish turn to ram ventilation when swimming fast enough. Others, such as tunas, are obligate ram ventilators and must swim all the time to avoid suffocation.

- A lowered O_2 partial pressure in the blood is a more potent stimulus for increased ventilation in fish than an elevated CO_2 partial pressure.

- Most of the 400 or so species of air-breathing fish have an air-breathing organ that is derived from the buccal cavity, opercular cavity, stomach, or intestines—or one that originates as an outpocketing of the foregut (e.g., swim bladder).

Breathing by Amphibians

Amphibians, of all the vertebrate groups, mix water and air breathing to the greatest extent. Many move from an aquatic environment to a terrestrial one during their individual development, and many are dual breathers as adults.

The gills of aquatic amphibian larvae (tadpoles) are of different origin and structure than the gills of adult fish. They develop as outgrowths of the integument of the pharyngeal region and project into the water from the body wall (**FIGURE 23.15A**). The gills are external in all young amphibian larvae; in salamander larvae they remain so, but in the larvae of frogs and toads (anurans), an outgrowth of the integument, termed the *operculum* (different from the bony operculum of a fish), soon encloses the gills in a chamber that opens to the outside posteriorly. Ventilation of the gills enclosed in the opercular cavity is accomplished by buccal pumping. The gills of amphibians are generally lost at metamorphosis, but external gills remain throughout life in certain aquatic salamanders, such as mudpuppies (*Necturus*). Mudpuppies ventilate their gills by waving them back and forth in the water and increase the frequency of these movements in response to decreased O_2 or increased temperature.

In most amphibians, paired lungs develop from the ventral wall of the pharynx near the time of metamorphosis. Each lung is classified as **unicameral** (*uni*, "one"; *cameral*, "chamber") because it is a single sac with an open, undivided central cavity that provides access to any side compartments that may be formed by the folding of the walls. The lungs of many adult amphibians are simple, well-vascularized sacs; their internal surface area is increased little, if at all, by folding, and in this respect they are less well developed than the lungs of lungfish. The inner walls of the lungs of frogs and toads are often more elaborate; they may be thrown into complex

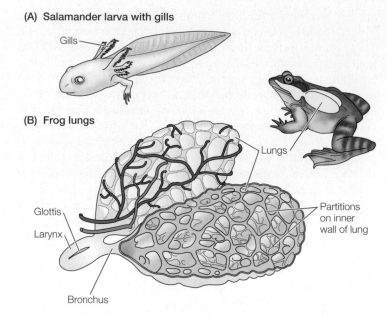

FIGURE 23.15 Breathing organs of amphibians (A) The external gills of a 3-week-old salamander larva (*Ambystoma maculatum*). (B) The lungs of a frog (*Rana temporaria*). The dorsal half of one lung has been cut off to reveal the compartmentalization of the inner wall by interconnected folds in multiple tiers. Modern amphibians lack a trachea; their lungs connect almost directly to the pharynx. (B after Poll 1962.)

patterns of interconnected folds in multiple tiers, giving them a honeycombed appearance (**FIGURE 23.15B**).

Lunged amphibians fill their lungs by buccopharyngeal pressure. This basic mechanism is presumably carried over from their piscine ancestors and, as mentioned earlier, is often employed by amphibian larvae to ventilate their gills. In frogs, although several patterns differing in detail have been reported, the essentials of the buccopharyngeal pressure pump are uniform. Air is taken into the buccal cavity through the nares or mouth when the pressure in the cavity is reduced by lowering its floor. Then, when the floor of the buccal cavity is raised with the mouth closed and the nares at least partially sealed by valves, the increase in pressure forces air into the lungs. The inflation of the lungs elevates the air pressure within them. Thus the lungs would discharge upon opening of the mouth or nares were it not for the glottis (see Figure 23.15B), which is closed by muscular contraction after inhalation. A period of apnea (no breathing), with the lungs inflated, then follows (i.e., breathing is intermittent). The nares are opened during the apneic period, and a frog often pumps air in and out of its buccal cavity through its nares at that time by raising and lowering the floor of the buccal cavity, termed *buccopharyngeal pumping*. Then the glottis is opened, and air from the lungs is exhaled.

Exhalation in a frog results in part from elastic recoil of the expanded lungs and may also be promoted by contraction of muscles in the walls of the lungs and body wall. Exhalation is described as having both **passive** and **active** components. In the study of the forces that drive lung volume changes, *passive* means "not involving contraction of muscles" and refers to forces developed by simple elastic rebound. *Active*, by contrast, refers to forces developed by muscular contraction.

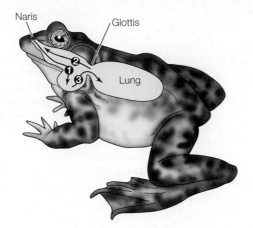

FIGURE 23.16 **The three major steps in the ventilatory cycle of an adult bullfrog (*Rana catesbeiana*)** Some authorities name this species *Lithobates catesbeianus*. (After Gans 1970.)

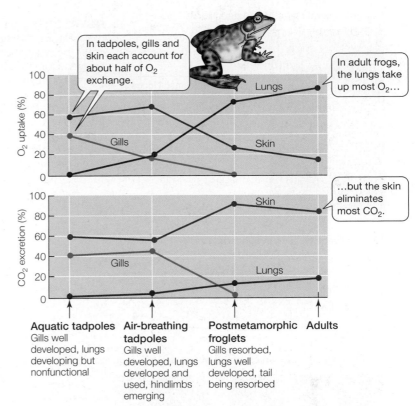

FIGURE 23.17 **The development of external respiration in the bullfrog (*Rana catesbeiana*)** Shown are the percentages of O_2 uptake and CO_2 excretion that occur across the gills, lungs, and skin of bullfrogs as they develop from tadpoles to adults. The animals were studied at 20°C and had free access to well-aerated water and air. Some authorities name this species *Lithobates catesbeianus*. (After Burggren and West 1982.)

The bullfrog (*Rana catesbeiana*) provides a well-studied specific example of the pulmonary breathing cycle. A bullfrog fills its buccal cavity in preparation for inflation of its lungs *before* it empties its lungs; to do so, with its glottis closed, it inhales air, which mostly comes to lie in a posterior depression of the buccal floor (step ❶, **FIGURE 23.16**). Next (step ❷), the glottis is opened, and pulmonary exhalant air passes in a coherent stream across the dorsal part of the buccopharyngeal cavity to exit through the nares. The fresh air in the depression of the buccal floor is then driven into the lungs when the buccal floor is raised with the nares closed (step ❸). An important effect of the buccopharyngeal pumping between breaths is that it washes residual pulmonary exhalant air out of the buccal cavity, so that when the next pulmonary ventilatory cycle begins, the buccal cavity is filled with a relatively fresh mixture.

Gills, lungs, and skin are used in various combinations to achieve gas exchange

A central question in amphibian respiratory physiology is how the total gas exchange of an individual is partitioned among the available gas-exchange sites: the gills, lungs, and skin. Bullfrogs, to continue with them as an example, start their lives without lungs, and when they are living at 20°C as aquatic tadpoles, their gills and skin each account for about half of their O_2 and CO_2 exchange (**FIGURE 23.17**). As bullfrog tadpoles mature and their lungs become functional, their lungs gradually assume primary responsibility for their O_2 uptake. In adulthood, the lungs take up most O_2. The lungs, however, do not play a large role in CO_2 exchange at any age; instead, when the gills are lost at metamorphosis, the skin increases its role in eliminating CO_2 (see Figure 23.17). The pattern seen in adult bullfrogs at 20°C—that the lungs are primarily responsible for O_2 uptake, whereas the skin eliminates most CO_2—is common in amphibian adults at such temperatures.

Some species of frogs in temperate regions of the world hibernate at the bottoms of ponds and lakes during winter. All their O_2 and CO_2 exchange is then across their skin. One reason the skin can suffice under these circumstances is that the hibernating animals have very low needs for O_2 and CO_2 exchange because of their low body temperatures and seasonal metabolic depression. The relatively high permeability of the skin is also important: Whereas it increases rates of dehydration on land, it makes underwater breathing by adults possible.

Breathing by Reptiles Other than Birds

When we consider the lizards, snakes, turtles, and crocodilians, an important first point to make is that in most of them, the lungs take up essentially all O_2 and eliminate essentially all CO_2. The skin of reptiles is generally much less permeable than amphibian skin, meaning that it protects far better against evaporative dehydration, but it does not readily allow the respiratory gases to pass through.

The simplest type of reptilian lung—seen in most lizards and snakes—is unicameral (single-chambered); it is a saclike structure with an open central cavity. Unicameral lungs are sometimes well perfused with blood throughout, but sometimes they are well perfused only at the anterior end and are balloonlike at the posterior end (**FIGURE 23.18A**). In the well-perfused parts where O_2 and CO_2 are principally exchanged with the blood, the walls are thrown into a honeycomb-like pattern of vascularized folds, increasing their surface area (**FIGURE 23.18B,C**). Air flows in and out of the central cavity during breathing, but gas exchange between the central cavity and the depths of the honeycomb-like cells on the walls is probably largely by diffusion.

A major evolutionary advance observed in several groups of nonavian reptiles is that, in each lung, the main lung cavity has become subdivided by major septa into numerous smaller parts, forming a **multicameral**—multiple-chambered—lung, as seen in **FIGURE 23.18D**. A multicameral lung can provide a great deal more

(A) Unicameral lungs of a plated lizard

Some unicameral lungs are perfused with blood principally at just the cranial (anterior) end.

FIGURE 23.18 Lizard lungs (A) Freshly dissected unicameral lungs from an African plated lizard (*Gerrhosaurus* sp.). Blood trapped within the tissue is responsible for the red color. The highly localized presence of blood at the cranial (anterior) ends of the lungs reflects the fact that virtually all perfusion takes place within the cranial portions of the lungs in some species with unicameral lungs. (B) A drawing of the gross internal structure of the unicameral lung of another type of lizard, the green lizard (*Lacerta viridis*). (C) Scanning electron micrograph of the lung wall of a tegu lizard (*Tupinambis nigropunctatus*), showing the honeycomb-like pattern of vascularized partitions. Magnification: 25×. (D) Gross internal structure of the highly developed multicameral lung of a monitor lizard (*Varanus exanthematicus*). (A courtesy of Tobias Wang; B and D courtesy of Hans-Rainer Duncker, reprinted from Duncker 1978; C courtesy of Daniel Luchtel and Michael Hlastala.)

(B) A unicameral lung in a lacertid lizard

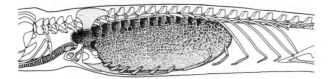

(C) Scanning electron micrograph of the wall of a tegu lizard lung

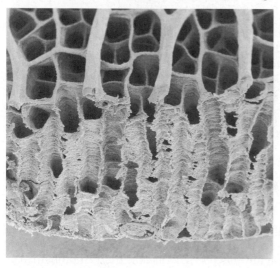

(D) The elaborate, multicameral lung of a monitor lizard

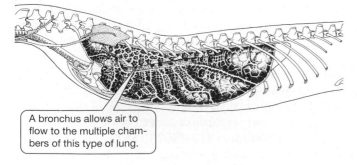

A bronchus allows air to flow to the multiple chambers of this type of lung.

and turtles. These groups all resemble each other in their septation, in that—when viewed in three dimensions—each lung is divided into three rows of chambers, with four or more chambers in each row. A noteworthy evolutionary development in the multicameral lung is the appearance of a cartilage-reinforced tube (*bronchus*) that runs lengthwise through the lung (see Figure 23.18D). This tube allows air to flow to all of the multiple chambers in the lung.

During ventilation the lungs in nonavian reptiles are filled principally or exclusively by *suction* (also termed *aspiration*) rather than by buccal pressure: Air is drawn into the lungs by an expansion of the lung volume, which creates a subatmospheric pressure within the lung chambers. This mode of ventilation represents a major evolutionary transition from the earlier buccal-pressure–filled lungs of air-breathing fish and amphibians. It is a transition that is carried forward to the mammals and birds, which also employ suction to fill their lungs. Because suction is created by the action of thoracic and abdominal muscles, not by buccal muscles, the evolution of suction ventilation liberated the buccal cavity from one of its ancient functions, allowing it to evolve in new directions without ventilatory constraints.

Suction is developed in the lungs of reptiles in two different ways. In one sort of breathing cycle, seen in at least some snakes, thoracic and abdominal expiratory muscles compress the lungs to a volume smaller than their passive relaxation volume during the exhalation of air; suction for inhalation is then developed when the lungs later rebound elastically to larger size. In the other sort of breathing cycle, seen in lizards and some crocodilians, inspiratory muscles actively create suction in the lungs during inhalation by expanding the lungs to a size larger than their passive relaxation volume; then, when the muscular activity stops, the lungs rebound elastically, becoming smaller, and the elastic rebound contributes to exhalation.

Lizards provide an instructive example to study in more detail. Unlike modern amphibians, lizards have well-developed ribs. Running over and between the ribs on each side of the body are **intercostal muscles** (*costa*, "rib") that—by means of their contractions—can expand or contract the volume enclosed by the rib cage. When a resting lizard inhales, certain of its intercostal muscles are activated and expand the rib cage, a mechanism sometimes called the **costal suction pump**. After the inflation of the lungs has occurred, the glottis is closed and the inspiratory muscles relax. The inhaled air is then held in the lungs for several seconds to several minutes of apnea, while often the buccal cavity is ventilated by buccopharyngeal pumping, thought generally to aid olfaction. Exhalation then occurs, followed quickly by another inhalation.

Recent research has demonstrated that when lizards are walking or running, some of their intercostal muscles help produce the back-and-forth flexions of the body that are so characteristic of lizard

surface area of gas-exchange membrane per unit of lung volume than a unicameral lung because the septa between lung chambers, not just the outer walls, can develop elaborate, highly folded gas-exchange surfaces. The multicameral type of lung occurs in monitor lizards (see Figure 23.18D), reptiles noted for their dramatically active ways of life and relatively high aerobic competence. It is also found in crocodilians locomotion. This involvement of the intercostals in locomotion can interfere with their ability to develop ventilation forces. Some, but not all, species of lizards overcome this problem by using buccal pressure to help fill their lungs while they are walking or running.

Sea turtles and some crocodilians exhibit the most structurally elaborate lungs seen in the nonavian reptiles.

Breathing by Mammals

Mammals and birds possess the most elaborate lungs of all animals. Their lungs are independently evolved and built on very different principles. The extreme intricacy of mammalian lungs is illustrated by the plastic cast of a person's airways shown in **FIGURE 23.19**.

The lung system of an adult human consists of 23 levels of airway branching. The *trachea* first branches to form two major airways, the *primary bronchi* (singular *bronchus*), which enter the two lungs (see Figure 23.19). Each primary bronchus then branches and rebranches dendritically (as a tree branches) within the lung, giving rise to *secondary* and *higher-order bronchi* of smaller and smaller diameter, and then to ever-smaller fine tubes known as *bronchioles*. At the outer limits of this branching tree of airways, 23 branches away from the trachea, the final bronchioles end

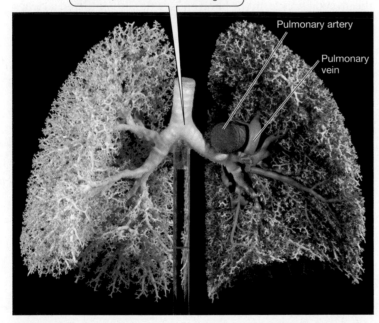

FIGURE 23.19 Human lungs The lung on the left shows only the airways, whereas the lung on the right shows the airways and blood vessels. Note the large sizes of the major blood vessels, reflecting the high rates at which blood must flow through the lungs. To visualize the airways, the lung passages were injected with a yellow plastic. Similarly, the arteries and veins were injected with red and blue plastic, respectively. After the plastic hardened, the tissue was removed, leaving just the plastic to mark the airways and blood vessels. Only the airways and vessels of relatively large diameter are preserved with this technique: The airway system and the arterial and venous systems branch far more finely than seen here. Note that the arteries and veins tend to branch in parallel with the airways. (Courtesy of Ewald R. Weibel, Institute of Anatomy, University of Berne, Switzerland.)

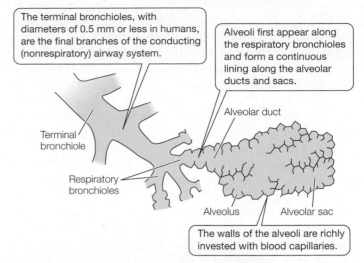

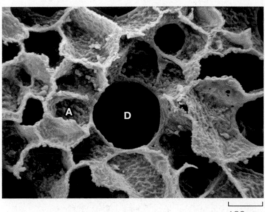

FIGURE 23.20 Respiratory airways of the mammalian lung (A) Longitudinal section of the final branches of the airways in a mammalian lung, showing respiratory bronchioles, alveolar ducts, alveolar sacs, and alveoli. (B) Scanning electron micrograph of a 0.4-mm^2 area of a human lung, showing alveolar ducts (D) and alveoli (A). (A after Hildebrandt and Young 1960; B courtesy of Ewald Weibel.)

blindly in **alveolar ducts** and **alveolar sacs**, the walls of which are composed of numerous semispherical outpocketings, each called an **alveolus** (*alveolus*, "hollow cavity") and collectively termed **alveoli** (**FIGURE 23.20**). The structure is so elaborate that describing it quantitatively is a great challenge, and estimates still undergo revision. According to the latest values, as summarized by Ewald Weibel, there are almost 500 million alveoli in the two lungs (taken together) of a human adult; they vary in size but average about 0.25 mm in diameter. The alveoli make up a total area of gas-exchange membrane averaging about 130 m^2! The floor of an 80-student classroom is likely to have a similar area. Thus, by virtue of elaborate branching, a highly vascularized interface between air and blood that is the size of a large classroom floor is fit within the compact volume occupied by our lungs.

The trachea and bronchi and all but the last few branches of bronchioles in a mammal's lungs are not much involved in gas exchange; they are thus known as the **conducting airways** and are said to constitute the **anatomical dead space** of the lungs.

They are lined with a relatively thick epithelium and do not receive a particularly rich vascular supply. Gas exchange between air and blood occurs in the **respiratory airways** of the lungs (see Figure 23.20A), which consist of *respiratory bronchioles* (the last two or three branches of bronchioles), *alveolar ducts* (through-passages lined with alveoli), and *alveolar sacs* (end sacs lined with alveoli). The pulmonary walls of the respiratory airways are composed of a single layer of thin, highly flattened epithelial cells and are richly supplied with blood capillaries. The alveoli constitute most of the gas-exchange surface. In them, blood and air are separated by just two thin epithelia (the pulmonary alveolar epithelium and the capillary epithelium) and a basement membrane in between. The total average diffusive thickness of these structures is only 0.3–0.6 μm. Recent research indicates that type IV collagen in the basement membrane is critical for imparting mechanical strength to these very thin structures: strength adequate to prevent them from rupturing, despite the fact that they are exposed to considerable physical stresses during breathing.

The total lung volume is employed in different ways in different sorts of breathing

When mammals breathe at rest, they do not come close to inflating their lungs fully when they inhale, and they do not come close to deflating their lungs fully when they exhale. Thus a wide margin exists for increasing the use of total lung volume. The **tidal volume** is the volume of air inhaled and exhaled per breath. In resting young men, the volume of the lungs at the end of inhalation is about 2900 mL, whereas that at the end of exhalation is about 2400 mL. Thus the *resting tidal volume* is about 500 mL (**FIGURE 23.21**). The maximum volume of air that an individual can expel beyond the resting expiratory level is termed the resting **expiratory reserve volume**. In healthy young men, it is about 1200 mL; that is, of the 2400 mL of air left in the lungs at the end of a resting exhalation, 1200 mL can be exhaled by maximum expiratory effort, but 1200 mL (termed the *residual volume*) cannot be exhaled at all (see Figure 23.21). The maximum volume of air that can be inhaled beyond the resting inspiratory level is the resting **inspiratory reserve volume**. In healthy young men, it is about 3100 mL; thus the total lung volume at the end of a maximum inspiratory effort is about 6000 mL (see Figure 23.21). Using the terminology developed here, when mammals increase their tidal volume above the resting level, they do so by using parts of their inspiratory and expiratory reserve volumes. The maximum possible tidal volume, sometimes termed **vital capacity**, is attained by fully using both reserves and thus is the sum of the resting tidal volume and the resting inspiratory and expiratory reserve volumes: about 4800 mL in young men. The vital capacity of humans tends to be increased by physical training, but advancing age and some diseases tend to decrease it.

The gas in the final airways differs from atmospheric air in composition and is motionless

The gas in the alveoli, which is the gas that undergoes exchange with the blood, differs dramatically in composition from atmospheric air in all mammals. The most fundamental reason is that the alveolar sacs form the blind ends of tidally ventilated airways that are never fully emptied. To see the implications of this anatomical fact, consider a resting person. At the end of a resting exhalation, the lungs contain 2400 mL of stale air, about 170 mL of which is in the anatomical dead space (the conducting airways). When the person then inhales, the stale air in the anatomical dead space moves deeper into the lungs, into the respiratory airways, meaning that the entire 2400 mL of stale air is in the respiratory airways after inhalation. Of the 500 mL of fresh atmospheric air inhaled during a resting breath, about 330 mL passes through the anatomical dead space and enters the respiratory airways; the other 170 mL—the last of the air to be inhaled—simply fills the anatomical dead space and is later exhaled, unused. All things considered, at the end of a resting inhalation, the gas in the respiratory airways consists of a mix of 2400 mL of stale air and 330 mL of fresh atmospheric air. Accordingly, the O_2 partial pressure in the alveoli is bound to be far below the atmospheric O_2 partial pressure, and the CO_2 partial pressure in the alveoli is far above the atmospheric partial pressure. Another significant property that arises from these quantitative realities is that the gas partial pressures in the alveoli do not change much between inhalation and exhalation. Using the values for resting humans, only about 12% of the air in the respiratory airways at the end of an inhalation is fresh, whereas 88% is carried over from previous breaths. The large carry-over from breath to breath helps impart stability to the gas composition deep in the lungs.

The exact partial pressures of gases that prevail in the alveoli depend *dynamically* on the *rate* at which fresh air is brought into the depths of the lungs and the *rate* of gas exchange with the blood. Ventilatory control systems (to be discussed later) ordinarily adjust the rate of ventilation relative to the rate of gas exchange with the

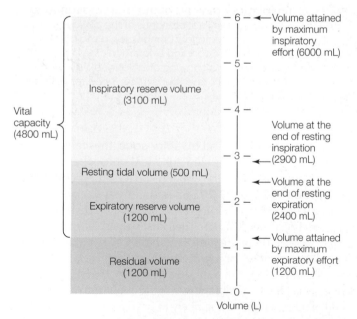

FIGURE 23.21 Dynamic lung volumes in healthy young adult men The lung volumes shown include the anatomical dead space as well as the respiratory airways. The inspiratory and expiratory reserve volumes shown are the reserves available during resting breathing. The residual volume is the volume remaining in the lungs after maximum expiratory effort. Values shown are averages for 70-kg men.

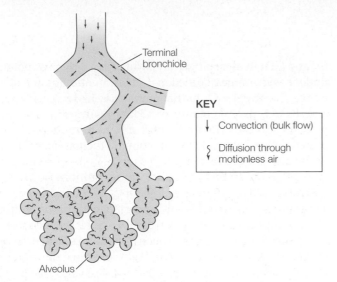

Terminal
bronchiole

KEY

↓ Convection (bulk flow)

↕ Diffusion through
motionless air

Alveolus

FIGURE 23.22 Mechanisms of gas transport in the final branches of mammalian lungs during inhalation Although gases are drawn by convection (bulk flow) into the finest bronchioles, the gases in the alveolar sacs and alveoli are motionless. Therefore O_2 must travel by diffusion across the final, very short distance it must cover to reach the gas-exchange membrane, and CO_2 must diffuse in the opposite direction. (After Weibel 1984.)

blood so that certain set-point partial pressures of O_2 and CO_2 are maintained in the alveolar gas. In humans near sea level, the partial pressure of O_2 in alveolar gas is nearly always about 13.3 kPa (100 mm Hg), and that of CO_2 is about 5.3 kPa (40 mm Hg).[6]

The gas occupying the final respiratory airways of a mammal is essentially motionless (**FIGURE 23.22**). This unexpected property is still another consequence of the anatomical fact that the alveolar sacs are the blind ends of a tidally ventilated airway system. During inhalation, convective air movement carries air rapidly down the trachea and through the various bronchi. As the air flows ever deeper into the lungs, however, it slows, because the collective volume of the airways rapidly increases as the airways branch into greater and greater numbers. Convective airflow ceases before the air reaches the alveolar sacs.

Because the gas in the blind ends of the airways is motionless, O_2 and CO_2 transport through that gas must occur by diffusion (see Figure 23.22). The layer of motionless gas is very thin (recall, for instance, that human alveoli average only 0.25 mm in diameter). Therefore diffusion across the motionless layer of gas can readily occur fast enough for rates of O_2 and CO_2 exchange between the lung air and the blood to be adequate (see Table 5.1). This happy situation is contingent, however, on the alveoli being filled with gas, not water. If the alveoli become filled with body fluid, rates of diffusion sharply plummet. That is why any disease that causes even a minute chronic accumulation of body fluid within the airways is a mortal threat.

---

[6] There are some data that indicate that these values vary with body size among species of mammals, and a partial explanation for this pattern would be the regular relation between breathing rate and body size. Small-bodied mammals take many more breaths per minute than large-bodied ones (see Chapter 7). Correlated with this potentially greater influx of fresh air relative to lung volume, mammals smaller than humans may have consistently higher values for the alveolar partial pressure of O_2 than humans (e.g., 14.5 kPa measured in rats). Conversely, mammals larger than humans may have lower values than humans (e.g., 10 kPa in horses).

The forces for ventilation are developed by the diaphragm and the intercostal and abdominal muscles

Unlike other vertebrates, mammals have a true **diaphragm**: a sheet of muscular and connective tissue that completely separates the thoracic and abdominal cavities (see Figure 1.18). The diaphragm is dome-shaped, projecting farther into the thorax at its center than at its edges. Contraction of the diaphragm tends to flatten it, pulling the center away from the thorax toward the abdomen. This movement increases the volume of the thoracic cavity, resulting in expansion of the lungs and inflow of air by suction.

The **external** and **internal intercostal muscles** that run obliquely between each pair of adjacent ribs are also important in ventilation. Contraction of the external intercostals rotates the ribs anteriorly and outward, expanding the thoracic cavity. The contractile filaments of the internal intercostals run (roughly speaking) perpendicular to those of the externals, and in general their contraction rotates the ribs posteriorly and inward, decreasing the volume of the thoracic cavity. You can easily demonstrate the action of these muscles on yourself by consciously expanding and contracting your rib cage while monitoring your ribs with your fingers.

To fully understand ventilation, it is essential to recognize that the lungs and thoracic wall together form an elastic system. Much like a hollow rubber ball with a hole on one side, they assume a certain equilibrium volume, known as their **relaxation volume**, if they are free of any external forces. For the volume of the lungs to deviate from the relaxation volume, muscular effort must be exerted. In adult human males, the relaxation volume of the lungs, measured as the volume of gas they hold at relaxation, is about 2400 mL. This volume is the same as that of the lungs after a resting exhalation (see Figure 23.21): The lungs and thorax are in their relaxed state after a resting exhalation.

Inhalation in humans at rest is active (meaning that it entails muscular effort). During inhalation, the lungs are *expanded to greater than their relaxation volume* by contraction of the diaphragm, external intercostal muscles, and anterior internal intercostals. Mammals do not close the glottis during ordinary breathing. Thus inhalation continues only as long as the inspiratory muscles contract. Exhalation in resting humans is largely or completely passive (meaning that it does not entail muscular effort). When the inspiratory muscles cease to contract at the end of inhalation, lung volume *returns elastically to its passive equilibrium state: the relaxation volume.*

When mammals exercise, they typically increase not only their tidal volume but also their breathing frequency. Additional muscular activity is required, therefore, both to amplify changes in lung volume during each breath and to hasten the inspiratory and expiratory processes. In people, the external intercostal muscles assume a greater role in inhalation during exercise than they do during rest; whereas expansion of the rib cage by these muscles during quiet breathing is of relatively minor importance compared with the action of the diaphragm, expansion of the rib cage by the external intercostals during heavy exertion accounts for about half of the inspiratory increase in lung volume. Active forces also contribute to exhalation in people during exercise. The most important muscles in exhalation are the internal intercostals, which actively contract the rib cage, plus the muscles of the abdominal wall, which contract the abdominal cavity, forcing the diaphragm upward into the thoracic cavity. These muscles hasten exhalation during exercise and may also compress the lungs beyond their relaxation volume, thereby enhancing tidal volume by use of some of the expiratory reserve volume.

Although all mammals tend to use the same basic groups of muscles for ventilation, the relative importance of the muscle groups varies. In large quadrupeds, for example, movements of the rib cage tend to be constrained, and the diaphragm bears especially great responsibility for ventilation.

The control of ventilation

We can think of ventilation as consisting of a rhythm—inhalation alternating with exhalation—that is modulated from time to time to meet changing demands for O_2 and CO_2 exchange.

A KEY SITE OF ORIGIN OF THE VENTILATORY RHYTHM IS THE PRE-BÖTZINGER COMPLEX The location of the specific neurons that are most important for rhythmogenesis—generation of the breathing rhythm—is now well established. These neurons are found in a bilaterally arrayed pair of neuron clusters, called the **pre-Bötzinger complex**, within the ventrolateral medulla of the brainstem. Thin medullary tissue slices containing the pre-Bötzinger complex—slices that are entirely isolated from the rest of the body—endogenously produce neural outputs that control the routine breathing rhythm and that evidently also play roles in controlling sighs (**FIGURE 23.23**). A question that remains is whether the central pattern generator in the pre-Bötzinger complex can, in itself, produce the fully formed rhythmogensis observed in intact animals. Current evidence indicates it must interact with one or more additional, nearby central pattern generators to achieve this.

VENTILATION IS MODULATED BY CHEMOSENSATION OF CO_2, H^+, AND O_2 As briefly mentioned earlier, the partial pressures of

O_2 and CO_2 in alveolar gas are ordinarily held at set-point levels under a wide range of functional states. In humans at rest, for example, the alveolar O_2 partial pressure is 13.3 kPa (100 mm Hg), and the CO_2 partial pressure is 5.3 kPa (40 mm Hg). These partial pressures remain the same during light to moderately intense exercise. Only during heavy exertion do the gas partial pressures in the alveoli deviate more than slightly from the resting values. To explain the stability of alveolar gas composition, ventilatory controls based on chemosensation of CO_2, H^+, and O_2 are critical, although they are not the only important controls of ventilation.

Controls based on sensation of CO_2 and H^+ are the most potent chemosensory controls of ventilation in mammals. When the concentration of CO_2 in the blood or other body fluids rises, the concentration of H^+ typically increases in parallel, and vice versa (see page 659).[7] Thus the blood concentrations of CO_2 and H^+ vary together. The blood concentrations of *both* CO_2 and H^+ are independently sensed by chemosensitive neural zones near the ventral surface of the medulla (the retrotrapezoid nucleus being of particular importance). A deviation of either concentration from its normal level *potently* influences breathing, and because both concentrations tend to vary together, they often exert synergistic effects. Ventilation increases or decreases so as to bring the concentrations of CO_2 and H^+ back toward normal—a negative feedback system. For example, if the CO_2 concentration of the blood is elevated, ventilation is increased, resulting in a greater rate of CO_2 exhalation. The potency of these effects is illustrated by the fact that an increase of just 0.5 kPa (4 mm Hg) in a person's arterial partial pressure of CO_2—from 5.3 to 5.8 kPa (40 to 44 mm Hg)—will cause the volume of air ventilated per minute to approximately *double*. During high-intensity exercise, as discussed in Chapter 8, production of lactic acid by anaerobic glycolysis

---

[7] CO_2 has been aptly termed a "gaseous acid" because it reacts with H_2O to form carbonic acid (H_2CO_3), which then dissociates to form H^+ and HCO_3^-. The changes in H^+ concentration are sometimes expressed as changes in pH. When H^+ concentration increases, pH decreases, and vice versa.

FIGURE 23.23 The likely fountainhead of breathing A 1-mm-thick slice of the medulla of the brain of a nestling mouse produces electrical signals that, judging from their patterns, control the rhythm of ordinary breathing and sighs. Each heavy black line is a recording of electrical activity as a function of time. The recordings presented here have been integrated to remove random noise, explaining why they are smoother than those in Figure 23.10. (After Lieske et al. 2000.)

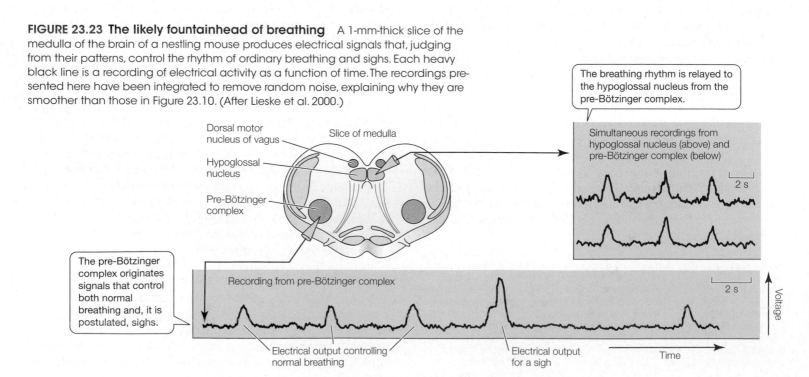

The breathing rhythm is relayed to the hypoglossal nucleus from the pre-Bötzinger complex.

Simultaneous recordings from hypoglossal nucleus (above) and pre-Bötzinger complex (below)

Dorsal motor nucleus of vagus

Slice of medulla

Hypoglossal nucleus

Pre-Bötzinger complex

2 s

The pre-Bötzinger complex originates signals that control both normal breathing and, it is postulated, sighs.

Recording from pre-Bötzinger complex

2 s

Voltage

Electrical output controlling normal breathing

Electrical output for a sigh

Time

often increases the H^+ concentration of the blood, in addition to the effects of CO_2 on H^+. Ventilatory drive caused by this acidification sometimes is so potent that it gives athletes a sense of profound discomfort because of the intensity of their breathing.

Let's now turn to the role of O_2. Blood hypoxia in mammals is detected principally by chemoreceptive bodies outside the central nervous system: the **carotid bodies** and **aortic bodies**. There are two carotid bodies (each measuring about 0.5 cm in humans), positioned along the two common carotid arteries, near where each branches to form internal and external carotids. The carotid bodies receive blood flow from the carotid arteries and relay sensory information on the blood O_2 partial pressure to the brainstem via the glossopharyngeal nerves. Although humans have only O_2-sensing carotid bodies, dogs, cats, and many other mammals also have O_2-sensing aortic bodies, located along the aortic arch.

Both the carotid and aortic bodies are richly perfused with blood from major arteries, and because of this they are in excellent positions to monitor arterial O_2 partial pressure. Ordinarily the arterial O_2 partial pressure must fall far below normal before ventilation is reliably stimulated. Sensation of CO_2 and H^+ by the medulla of the brain is therefore paramount in regulating ventilation under usual resting conditions. In humans, if the arterial O_2 partial pressure falls below 7–8 kPa (50–60 mm Hg) (as compared with a normal arterial value of ~12.7 kPa [~95 mm Hg]), marked stimulation of ventilation occurs. Such a large drop in arterial O_2 partial pressure is unusual. Thus sensation of O_2 partial pressure is of key importance in ventilatory regulation only under specialized circumstances. One of these is exposure to high altitude (**BOX 23.2**). The carotid and aortic bodies become more sensitive to lowered O_2 when the blood concentration of CO_2 or H^+ is elevated.

BOX 23.2 Mammals at High Altitude, With Notes on High-Flying Birds

The environment at high montane altitudes is challenging in many respects. It can be cold, windy, dehydrating, and high in ultraviolet radiation. The most immediate challenge for a mammal or bird at high altitude, however, is to meet the O_2 demands of its cells, because the source of O_2—the atmosphere—is rarefied. This low level of O_2 cannot be escaped. People can escape cold temperatures by building dwellings, and animals can do so by building nests, but the low level of O_2 in the ambient air at high altitude exists everywhere, including in a dwelling or nest. Here we discuss several dimensions of this challenge; other aspects are discussed in Boxes 8.3 ($\dot{V}_{O_2max}$) and 24.5 (blood and circulation).

Permanent human settlements occur at 3500–4500 m (11,500–15,000 ft) on the Andean and Tibetan Plateaus. The people in these settlements have ordinary resting and maximum rates of O_2 consumption compared with members of the general population at sea level, and they lead active lives—despite the fact that the atmospheric O_2 partial pressure is only 60%–65% as high as at sea level.

Evolutionary adaptation, based on natural selection and gene-frequency changes, has probably occurred in the high-altitude populations in the Andes and Tibet. These populations, which began independently, have existed now for

10,000–20,000 years—representing hundreds of generations. Recent research has established beyond doubt that the Andean and Tibetan populations differ in striking ways in their physiological attributes at high altitude.

When a person born and reared at low altitude moves to the high mountains, his or her functional traits change over time as acclimatization occurs. However, such a person may never acquire a close physiological resemblance to people born and reared at high altitude.

From what we have said thus far, you can see that for analyzing people at high altitude, we need to distinguish three groups: newly arrived lowlanders, acclimatized lowlanders, and native highlanders. Moreover, we need to recognize distinct populations of native highlanders. Among other mammals, many species resemble humans in being predominantly of lowland distribution, and they therefore—at least in principle—present the same complexities. By contrast, some species, such as llamas, are endemic to high altitudes (found only there in a wild state).

In the study of mammals that have principally lowland distributions—including humans—a major theme is that exposure to high altitudes may sometimes trigger responses that evolved for reasons

other than adaptation to high altitude. Consider a population living at low altitude. Tissue hypoxia—a state of too little O_2 in the tissues—will arise inevitably in individuals of such a population from time to time. Tissue hypoxia will occur, for example, in people with anemic diseases or in individuals who suffer severe blood loss in accidents or wars. Accordingly, defenses against tissue hypoxia will evolve. Such defenses are well known in modern lowland populations, as we discuss later in this box and in Box 24.5. If individuals from such a lowland population go to high altitude, their whole body will be subjected to hypoxia because of the new environment. This hypoxia caused by high altitude may then trigger responses that evolved to defend against hypoxia occurring at low altitudes (caused, for example, by anemia or blood loss). Such responses may be detrimental (maladaptive) at high altitude. In brief, we must be

(Continued)

alert to the possibility that some of the responses seen at high altitude may be misplaced responses!

Let's now look at some of the information available on high-altitude physiology. The figure depicts the oxygen cascades of native lowland Peruvians living at sea level and of native Peruvian highlanders at 4500 m in the Andes. You'll notice that despite the large drop in *ambient* O_2 partial pressure at 4500 m, the *venous* partial pressure of the highlanders is reduced only a little. Comparing the two populations, the venous O_2 partial pressure is kept similar: It is *conserved*. To understand why this occurs, physiologists study all the steps in the oxygen cascade (see page 594). From inspection of the figure, you can see that the conservation of venous O_2 partial pressure in the Andean highlanders results from significant reductions in two of the partial pressure drops (steps) of the oxygen cascade. The drop in partial pressure between ambient air and alveolar gas is about 4.3 kPa (32 mm Hg) at high altitude and therefore is much smaller than the drop at sea level, 5.7 kPa (43 mm Hg); and the drop between arterial blood and mixed venous blood is about 1.5 kPa (11 mm Hg) at high altitude, versus 7.3 kPa (55 mm Hg) at sea level. The arterial-to-venous drop is a topic in blood gas transport and is discussed in Box 24.5. Here we examine *lung function* and *systemic tissue physiology*.

One of the most important defenses that lowland people marshal at high altitude is **hyperventilation**, defined to be an increase in the rate of lung ventilation associated with any given rate of O_2 consumption. When lowlanders first move to high altitude, a prompt (acute) increase in their rate of ventilation occurs; this increase is probably activated principally by the reduction in their arterial O_2 partial pressure, sensed by the carotid bodies. As lowlanders pass their first days at high altitude, their rate of ventilation becomes even higher, evidently because of an increasing physiological sensitivity of the breathing control mechanisms to hypoxic stimulation. The hyperventilation

of newly arrived lowlanders accelerates the flux of fresh air to their lungs and clearly helps them maintain a relatively high O_2 partial pressure in their alveolar gas despite the fact that they are breathing rarefied air. Nonetheless, based on the information available, hyperventilation gradually subsides if lowlanders spend extended lengths of time at altitude. Among native highlanders, Tibetans and Andeans differ strikingly. Tibetan highlanders exhibit marked chronic hyperventilation; at a given O_2 demand, their ventilation rate is roughly twice that of people residing at sea level. For them, hyperventilation is permanent! Andean highlanders exhibit less of a hyperventilation response. Most species of nonhuman mammals at high altitude display some degree of hyperventilation.

Although processes such as hyperventilation (and others discussed in Box 24.5) help keep O_2 partial pressures in the systemic blood capillaries from falling excessively at high altitude, capillary O_2 partial pressures do in fact decline. In the people at sea level in the figure, blood enters the systemic capillaries at an arterial O_2 partial pressure of about 12.5 kPa (94 mm Hg) and exits at a mixed venous O_2 partial pressure of about 5.2 kPa (39 mm Hg). In the people at 4500 m, blood enters at a much lower partial pressure, 5.9 kPa (44 mm Hg), and exits at a modestly lower one, 4.4 kPa (33 mm Hg). Thus the O_2 partial pressure in the capillaries—which drives O_2 diffusion to the mitochondria in cells—is, on average, reduced at high altitude, a common circumstance in mammals. Great interest is focused at present on how the tissues of mammals accommodate to this condition. Investigations of various species indicate that *tissue-level adjustments* that offset the effects of chronic

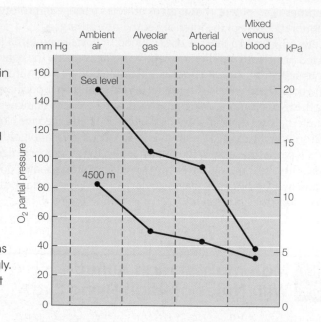

Oxygen cascades of people at sea level and high altitude Two groups of native male Peruvians were studied at their altitudes of residence. (Data from Torrance et al. 1970.)

tissue hypoxia are common, either as a consequence of acclimatization or as a result of adaptive evolution.

Genomic scientists are trying to identify genes that have been subject to natural selection in native highland populations. These studies have recently hit pay dirt in finding that genes in the HIF-2 signaling pathway (see page 604) have been subject to strong positive natural selection in Tibetan highlanders during the approximately 20,000 years since the Tibetan Plateau was colonized with permanent human settlements. These genes may prove to affect HIF-2 signaling in ways that aid life at high altitude, such as by controlling red blood cell production in advantageous ways (see Box 24.5). **Box Extension 23.2** discusses elevated pulmonary blood pressure in humans at high altitude, specific tissue-level adjustments, HIF involvement, and llamas as examples of native highland mammals. It also addresses high-flying birds, especially the bar-headed goose (see page 22).

VENTILATION IS ALSO MODULATED BY CONSCIOUS CONTROL, LUNG MECHANOSENSORS, AND DIRECT EFFECTS OF EXERCISE The most obvious type of modulation of ventilation in humans is conscious control. We can temporarily stop breathing by choosing to stop. There are, in addition, other types of control besides the chemosensory ones.

One well-understood set of controls is based on mechanoreceptors in the lungs, which sense stretch or tension in the airways. Information from these receptors is relayed via sensory neurons to the brainstem, where signals for inhalation tend to be inhibited by lung expansion and excited by lung compression. Certain of these mechanosensory responses are known as the *Hering-Breuer reflexes*.

It is now well established that during exercise there are controls operating in addition to chemosensory ones. Whereas these other controls are important, they are not well understood. As already stressed, the arterial partial pressures of O_2 and CO_2 remain little changed from resting values during light to moderate exercise. This stability results *because* of adjustments in the ventilation rate, which is increased in tandem with the metabolic rate. However, arterial gas partial pressures are far too stable during light to moderate exercise to *account* for observed increases in ventilation on the basis of the simple chemosensory negative feedback systems we have discussed up to here; for example, whereas an increase of about 0.5 kPa (4 mm Hg) in the arterial CO_2 partial pressure is required to bring about a doubling of ventilation rate, the measured CO_2 partial pressure during exercise may not be elevated to that extent even when ventilation has reached 10–15 times the resting rate! In addition to the controls mediated by gas partial pressures, there is increasing evidence for the existence of controls that are initiated in direct association with the muscular movements of exercise. These controls are postulated to take two forms: First, parts of the brain that initiate motor signals to the exercising muscles might simultaneously initiate stimulatory signals to the breathing centers. Second, sensors of movement or pressure in the limbs might stimulate the breathing centers based on the vigor of the limb activity they detect. One persuasive piece of evidence for these sorts of controls is that when people suddenly begin to exercise at a moderate level, their ventilation rate undergoes a marked increase within just *one or two breaths*; this response is far too rapid to be mediated by changes in the chemical composition of the body fluids. Another piece of evidence is that in many species of mammals, breathing movements and limb movements are synchronized during running.

BOTH TIDAL VOLUME AND BREATHING FREQUENCY ARE MODULATED BY CONTROL SYSTEMS The overall rate of lung ventilation depends on two properties: the tidal volume, V_T, and the frequency of breaths, f, usually expressed as the number of breaths per minute. The product of these is the **respiratory minute volume**:

$$\underset{\text{(mL/min)}}{\text{Respiratory minute volume}} = \underset{\text{(mL/breath)}}{V_T} \times \underset{\text{(breaths/min)}}{f} \qquad \textbf{(23.2)}$$

To illustrate, resting humans have a tidal volume of about 500 mL and breathe about 12 times per minute. Thus their respiratory minute volume is about 6 L/min.

Both tidal volume and breathing frequency are increased during exercise and during other states that increase the rate of metabolism. Humans and other mammals, however, cannot maximize both of these variables simultaneously because the time needed for one breathing cycle tends to increase as the tidal volume increases. Nonetheless, during vigorous exercise, trained athletes are able to maintain a tidal volume of at least 3 L while breathing at least 30 times per minute. In this way, their respiratory minute volume can reach greater than 100 L/min—more than 15 times the resting value.

The **alveolar ventilation rate**, typically expressed as the **alveolar minute volume**, is defined to be the rate at which new air is brought into the alveoli and other *respiratory* airways. This rate is important because the air that reaches the respiratory airways is the air that can undergo gas exchange with the blood. The alveolar minute volume is calculated by subtracting the volume of the anatomical dead space, V_D, from the tidal volume and multiplying by the breathing frequency:

$$\text{Alveolar minute volume} = (V_T - V_D) \times f \qquad \textbf{(23.3)}$$

Another property of importance relating to the respiratory airways is the fraction of all inhaled air that reaches them. This fraction—calculated by dividing the alveolar minute volume (Equation 23.3) by the total minute volume (Equation 23.2)—is $(V_T - V_D)/V_T$.

From the expression just described, you can see that—with V_D assumed to be constant—the fraction of air reaching the respiratory airways increases as the tidal volume increases. This fact helps resolve a paradox. When the overall ventilation rate is increased, the oxygen utilization coefficient increases. Although humans, for example, use about 20% of the O_2 in the air they breathe when their tidal volume is 500 mL, they use about 30% when their tidal volume is 2000 mL. How is this possible if, as we have often emphasized, the control systems typically keep the alveolar O_2 partial pressure constant? The paradox is resolved by recognizing two aspects of gas exchange. First, air that reaches the respiratory airways always gives up about the same fraction of its O_2 (accounting for the constancy of alveolar O_2 partial pressure). Second, however, a greater proportion of all the air that is breathed actually enters the respiratory airways as the tidal volume increases.

In species of different sizes, lung volume tends to be a constant proportion of body size, but breathing frequency varies allometrically

If we look at the full range of mammals, ranging in size from shrews to whales, there is a strong inverse (and allometric) relation between breathing frequency and body size. Humans breathe about 12 times per minute at rest. A mouse breathes 100 times per minute! This dramatic effect is a logical consequence of several facts. First, lung volume tends, on average, to be a relatively constant fraction of total body volume: Lung volume in liters averages about 6% of body weight in kilograms. Second, resting tidal volume tends consistently to be about one-tenth of lung volume or 0.6% of body weight. Third, related to these points, when mammals of all sizes are at rest, the amount of O_2 they obtain *per breath* is approximately a constant proportion of their body weight. However, as emphasized in Chapter 7, the resting weight-specific rate at which mammals metabolically consume O_2 increases allometrically as body size decreases. If the O_2 demand per unit of body weight in a small mammal is greater than that in a large mammal, and yet the small animal obtains about the same amount of O_2 per breath per unit of weight, then the small animal must breathe more frequently.

Pulmonary surfactant keeps the alveoli from collapsing

The alveoli may be thought of as aqueous bubbles because their gas-exchange surfaces are coated with an exceedingly thin water layer. If the alveoli were composed only of water, they would follow the physical laws of simple aqueous bubbles. One such law is that the tendency of a bubble to collapse shut increases as its radius decreases.[8] Thus, during exhalation, there is a risk that as the radius of an alveolus decreases, the alveolus might collapse shut by emptying entirely into the airways of the lung. This possibility may sound like a remote conjecture from a physics book. In fact, however, until this application of physics to biology was appreciated, thousands of human babies died every year because of bubble physics, as we discuss soon.

The alveoli in normal lungs do not behave as simple aqueous bubbles because of the presence of a complex mixture of metabolically produced lipids and proteins called **pulmonary surfactant** (*surfactant*, "surface active agent"). About 90% of pulmonary surfactant is lipids, mostly phospholipids (amphipathic molecules, as seen in Chapter 2). Pulmonary surfactant is synthesized by specialized pulmonary epithelial cells, which secrete the lipoprotein complex as vesicles. After secretion, phospholipids from the vesicles associate with the surface of the thin water layer that lines each alveolus, where they radically alter the surface-tension properties of the water. Their overall effect is to reduce the surface tension below that of pure water. This effect in itself helps prevent alveoli from collapsing shut because the tendency of bubbles to collapse decreases as their surface tension decreases (explaining why soap bubbles linger longer than bubbles of pure water). The most profound effect of pulmonary surfactant, however, is that it gives the alveoli a *dynamically variable surface tension*. With surfactant present, as an alveolus enlarges, its surface tension increases, an effect that impedes further enlargement and helps prevent the alveolus from expanding without limit. Conversely, as an alveolus decreases in size, its surface tension decreases, helping to prevent any further decrease in size. Surfactant, therefore, helps keep all alveoli similar in size.

Infants born prematurely sometimes lack adequate amounts of pulmonary surfactant. Their alveoli therefore lack the protections of surfactant, and many alveoli collapse shut during each exhalation. An affected infant then must inhale with sufficient force to reopen the alveoli. The process is tiring and damages the alveoli. Death rates were very high until therapies based on knowledge of pulmonary surfactant were introduced.

The use of knockout mice and other genomic methods is leading to rapid advances in understanding of the surfactant proteins. One of them, *protein B*, is now known to be essential for life, evidently because it plays indispensable roles in controlling the distribution of surfactant lipids.

Pulmonary surfactants that share basic chemical similarities have been reported from the lungs of all groups of terrestrial vertebrates, lungfish, and some other air-breathing fish. Thus the pulmonary surfactants have a long evolutionary history, dating back at least

---

[8] This is an implication of Laplace's Law, which—applied to bubbles—states that $\Delta P = 2T/r$, where r is the radius of a bubble, ΔP is the pressure difference between the inside and outside of the bubble, and T is the tension in the walls of the bubble. If everything is considered constant except ΔP and r, the pressure difference required to keep a bubble open is seen to increase as the radius r decreases.

---

to the origins of vertebrate air breathing. The roles of pulmonary surfactants in animals other than mammals are incompletely known but are gradually being better understood.

SUMMARY
Breathing by Mammals

■ The lungs of mammals consist of dendritically branching airways that end blindly in small, thin-walled, well-vascularized outpocketings, the alveoli. The airways exhibit 23 levels of branching in the human adult lung, giving rise to 500 million alveoli. The airways in a mammalian lung are categorized as conducting airways, where little gas exchange with the blood occurs, and respiratory airways, where most gas exchange with the blood takes place.

■ Because of the blind-ended structure of the mammalian lung, the gas in the alveoli always has a substantially lower O_2 partial pressure and higher CO_2 partial pressure than atmospheric air.

■ Contraction of the diaphragm is a principal force for inhalation in mammals, especially large quadrupeds. External intercostal muscles may contribute to inhalation; internal intercostal muscles and abdominal muscles may contribute to exhalation. Inhalation occurs by suction as the lungs are expanded by contraction of inspiratory muscles. At rest, exhalation occurs passively by elastic rebound of the lungs to their relaxation volume when the inspiratory muscles relax.

■ The breathing rhythm in mammals originates in a central pattern generator in the pre-Bötzinger complex in the medulla of the brainstem.

■ The most potent chemosensory stimulus for increased ventilation in mammals is a rise in blood CO_2 partial pressure and/or H^+ concentration, sensed in the medulla. The blood O_2 partial pressure, ordinarily a less influential factor in controlling ventilation, is sensed by the carotid bodies along the carotid arteries (humans) or by carotid and aortic bodies (certain other mammals). The control of ventilation during exercise involves stimuli generated in association with limb movement as well as chemosensory controls.

■ Pulmonary surfactant, a mix of lipids and proteins that affects surface tension, makes a critical contribution to maintaining the proper microscopic conformation of the lungs in all air-breathing vertebrates.

Breathing by Birds

The lungs of birds, although they are logical derivatives of the types of lungs thought to exist in the common ancestors of birds and mammals, differ in fundamental structural features from the lungs of mammals and all other modern vertebrates except certain crocodilian reptiles. The structural difference between avian and mammalian lungs inevitably invites comparison. Are avian lungs functionally superior to mammalian lungs? Some authorities conclude that the designs of the lungs in birds and mammals are "different but equal"—that is, equal in their gas-exchange ability. Other authorities conclude that the lungs of birds are in fact superior organs of gas exchange. As evidence, they point to the fact that bird

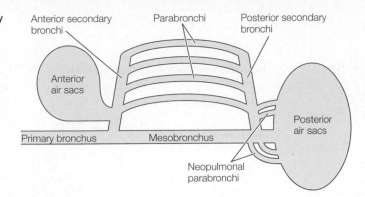

(A) **Anatomy**

Anterior secondary bronchi

Parabronchi

Posterior secondary bronchi

Anterior air sacs

Posterior air sacs

Primary bronchus

Mesobronchus

Neopulmonal parabronchi

FIGURE 23.24 Airflow in the lungs and air sacs of birds (A) Basic anatomy of the avian lung and its connections with the air sacs. In this presentation, the anterior and posterior groups of secondary bronchi are each represented as a single passageway. The tubes labeled *parabronchi* are those of the dominant paleopulmonal system. (B, C) Airflow during (B) inhalation (when the air sacs undergo expansion) and (C) exhalation (when the air sacs undergo compression).

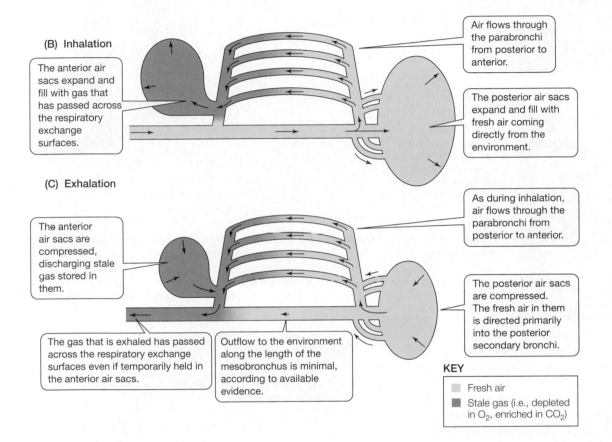

(B) **Inhalation**

Air flows through the parabronchi from posterior to anterior.

The anterior air sacs expand and fill with gas that has passed across the respiratory exchange surfaces.

The posterior air sacs expand and fill with fresh air coming directly from the environment.

(C) **Exhalation**

As during inhalation, air flows through the parabronchi from posterior to anterior.

The anterior air sacs are compressed, discharging stale gas stored in them.

The gas that is exhaled has passed across the respiratory exchange surfaces even if temporarily held in the anterior air sacs.

Outflow to the environment along the length of the mesobronchus is minimal, according to available evidence.

The posterior air sacs are compressed. The fresh air in them is directed primarily into the posterior secondary bronchi.

KEY

Fresh air

Stale gas (i.e., depleted in O_2, enriched in CO_2)

lungs have relatively large surface areas for gas exchange and thin gas-exchange membranes (see Figure 23.8). They also point to the fact that some birds, such as certain geese, cranes, and vultures, can *fly*—not just mope around and survive—near or above the altitude of Mt. Everest. The design of the bird lung, in comparison with the mammalian lung, may be an advantage at high altitude, in part because (as we will see) cross-current gas exchange prevails in the bird lung instead of tidal exchange. High-flying birds are discussed in Box 23.2.

A bird's trachea bifurcates to give rise to two primary bronchi, which enter the lungs. Here the similarity to mammals ends. The primary bronchus that enters each lung *passes through* the lung, being known as the **mesobronchus** within the lung. Two groups of branching **secondary bronchi** arise from the mesobronchus. One group, which arises at the anterior end of the mesobronchus, spreads over the ventral surface of the lung. The other group originates toward the posterior end of the mesobronchus and spreads over the dorsolateral lung surface. For simplicity, we call these the *anterior* and *posterior* groups of secondary bronchi, although

they are formally termed the *medioventral* and *mediodorsal* groups, respectively. Also for simplicity, each group is represented as just a single passageway in **FIGURE 23.24A**.

The anterior and posterior secondary bronchi are connected by a great many small tubes, 0.5–2.0 mm in internal diameter, termed **tertiary bronchi** or **parabronchi** (four are shown in Figure 23.24A). As depicted in **FIGURE 23.25**, the central lumen of each parabronchus gives off radially along its length an immense number of finely branching **air capillaries**. The air capillaries are profusely surrounded by blood capillaries and are the sites of gas exchange. They are only 3–14 μm in diameter (large birds tending to have diameters greater than those of small birds), and collectively they form an enormous gas-exchange surface amounting to 200–300 mm^2/mm^3 of tissue in the parabronchial walls. Air flows through the central lumen of each parabronchus, but exchange between the central lumen and the surfaces of its air capillaries is probably largely by diffusion. The parabronchi, air capillaries, and associated vasculature constitute the bulk of the lung tissue of a bird.

(A) Scanning electron micrograph of parabronchi in longitudinal section

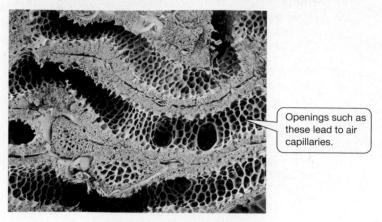

Openings such as these lead to air capillaries.

(B) Scanning electron micrograph of a parabronchus in cross section

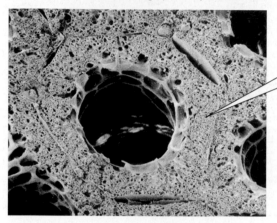

This tissue consists of intermingled air capillaries and blood capillaries.

(C) A parabronchus and associated vasculature

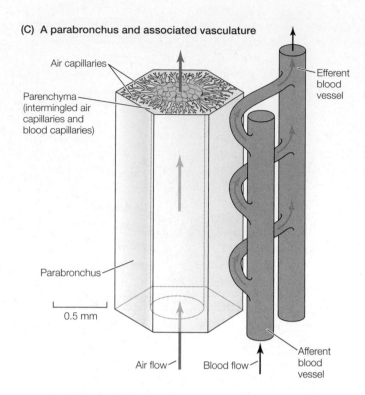

Air capillaries

Parenchyma (intermingled air capillaries and blood capillaries)

Efferent blood vessel

Parabronchus

0.5 mm

Air flow   Blood flow   Afferent blood vessel

FIGURE 23.25 Parabronchi and air capillaries: The gas-exchange sites in avian lungs (A) Scanning electron micrograph of the lung of a chicken (*Gallus*). Magnification: 12× (B) Scanning electron micrograph of a single parabronchus of a chicken lung in cross section. Magnification: 43× (C) Diagram of the structure of a parabronchus and how blood flow relates to the parabronchus. (A and B courtesy of Dave Hinds and Walter S. Tyler.)

A bird's **air sacs**, which are part of the breathing system, are located *outside* the lungs and occupy a considerable portion of the thoracic and abdominal body cavities (**FIGURE 23.26**). Usually there are nine air sacs, divisible into two groups. The *anterior air sacs* (cervical, anterior thoracic, and interclavicular) connect to various anterior secondary bronchi. The *posterior air sacs* (abdominal and posterior thoracic) connect to the posterior portions of the mesobronchi. (Each mesobronchus terminates at its connection with an abdominal air sac.) The air sacs are thin-walled, poorly vascularized structures that play little role in gas exchange between the air and blood. Nonetheless, as we will see, they are essential for breathing.

The structures of the lung described thus far are present in all birds, and their connections with the air sacs are similar in all birds. These lung structures are collectively termed the **paleopulmonal system**, or simply **paleopulmo**. Most birds, in addition, have a more or less extensively developed system of respiratory parabronchial tubes—termed the **neopulmonal system**—running directly between the posterior air sacs and the posterior parts of the mesobronchi and posterior secondary bronchi (see Figure 23.24A). The neopulmonal system is especially well developed in songbirds. The paleopulmonal system, nonetheless, is always dominant.

Ventilation is by bellows action

Avian lungs are compact, rigid structures. Unlike mammalian lungs, they undergo little change in volume over the course of

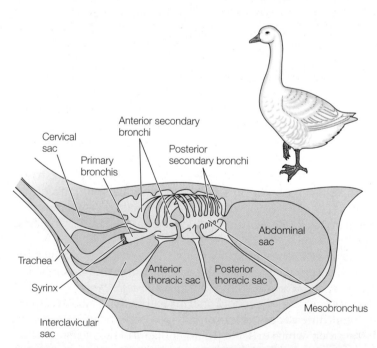

Cervical sac

Primary bronchis

Anterior secondary bronchi

Posterior secondary bronchi

Abdominal sac

Trachea

Syrinx

Anterior thoracic sac

Posterior thoracic sac

Mesobronchus

Interclavicular sac

FIGURE 23.26 The air sacs of a goose and their connections to the lungs Air sacs are blue, lungs are light orange. All the air sacs are paired, except the single interclavicular sac. (After Brackenbury 1981.)

each breathing cycle. The air sacs, by contrast, expand and contract substantially and, like bellows, suck and push gases through the relatively rigid airways of the lungs. To a dramatic extent relative to mammalian ventilation, this avian process is an energetically inexpensive way to move air.

The part of the rib cage surrounding the lungs themselves is relatively rigid. During inhalation, other parts of the rib cage (especially those posterior to the lungs) are expanded by contraction of internal intercostal muscles and certain other thoracic muscles, and the sternum swings downward and forward. These movements enlarge all the air sacs by expanding the thoracoabdominal cavity. Some of the external intercostals and abdominal muscles compress the thoracoabdominal cavity and air sacs during exhalation. Resting birds typically breathe at only about one-half or one-third the frequency of resting mammals of equivalent body size, but the birds have greater tidal volumes.

Air flows unidirectionally through the parabronchi

Air flows unidirectionally through the parabronchi of the paleopulmonal system. To see how this occurs, we must describe the movement of air during both inhalation and exhalation. During inhalation, both the anterior and posterior sets of air sacs expand. Suction, therefore, is developed in *both* sets of air sacs, and *both* receive gas. As depicted in **FIGURE 23.24B**, air inhaled from the atmosphere flows through the mesobronchus of each lung to enter the posterior air sacs and posterior secondary bronchi. Simultaneously, the air entering the posterior secondary bronchi is drawn anteriorly through the parabronchi by suction developed in the expanding anterior air sacs. Three aspects of the events during inhalation deserve emphasis. First, the posterior air sacs are filled with relatively fresh air coming directly from the environment. Second, the anterior air sacs are filled for the most part with stale gas that has passed across the respiratory exchange surfaces in the parabronchi. Finally, the direction of ventilation of the parabronchi in the paleopulmonal system is from posterior to anterior.

During exhalation, both sets of air sacs are compressed and discharge gas. As shown in **FIGURE 23.24C**, air exiting the posterior air sacs predominantly enters the posterior secondary bronchi to pass anteriorly through the parabronchi. This air is relatively fresh, having entered the posterior sacs more or less directly from the environment during inhalation. Gas exiting the parabronchi anteriorly, combined with gas exiting the anterior air sacs, is directed into the mesobronchus via the anterior secondary bronchi and exhaled. Recall that the anterior air sacs were filled with stale gas from the parabronchi during inhalation. Thus the exhaled gas is mostly gas that has passed across the respiratory exchange surfaces. Three aspects of the expiratory events deserve emphasis: First, the relatively fresh air of the posterior air sacs is directed mostly to the parabronchi. Second, most of the gas that is exhaled from the lungs has passed across the respiratory exchange surfaces. Finally, air flows through the parabronchi of the paleopulmonal system from posterior to anterior, just as it does during inhalation.

One of the greatest remaining questions in the study of avian lungs is *how* air is directed along its elaborate (and in some ways counterintuitive) paths through the paleopulmonal system and air sacs. Passive, flaplike valves appear to be entirely absent. Active,

---

BOX 23.3 **Bird Development: Filling the Lungs with Air Before Hatching**

The lungs of both mammals and birds initially develop in a fluid-filled condition. Young animals of both groups therefore face the problem of filling their lungs with air so as to be able to breathe when they are born or hatched. When mammals are born, they are able to fill their lungs sufficiently to survive by inflating them suddenly with their first breath from the atmosphere (all nonavian reptiles do likewise). Birds, however, cannot inflate their lungs in this way: The air capillaries in their lungs cannot be inflated suddenly out of a collapsed state because the lungs are relatively rigid and the air capillaries have extremely small diameters. Another obstacle to a sudden-inflation strategy for birds is that avian lungs probably will not work correctly unless *every* critical airway becomes gas-filled, because the pattern of airflow through the lungs is determined by complex aerodynamic interactions among the airways. Birds have thus evolved a way to *fill their lungs with air gradually* before the lungs become essential for breathing.

During most of a bird's development inside an egg, its breathing organ is a highly vascular chorio-allantoic membrane pressed against the eggshell on the inside. Oxygen and CO_2 pass between the atmosphere and the membrane by diffusion through gas-filled pores in the eggshell.

As an egg develops, it dehydrates by controlled loss of water vapor outward through the eggshell pores, a process that leads to the formation of a gas-filled space, the **air cell**, inside the egg at its blunt end. About 1–2 days before a young bird hatches, it starts to breathe from the air cell, inhaling and exhaling gas. During the ensuing hours until it hatches, the bird makes a *gradual* transition from gas exchange across its chorio-allantoic membrane to full-fledged pulmonary breathing. The air capillaries, in fact, undergo most of their prehatching development during this period. The airways and air capillaries in the lungs fill with gas and thus are already gas-filled by the time hatching begins and the chorio-allantoic membrane is left behind.

---

muscular valves could be present, but evidence for their existence is at best circumstantial. Most present evidence suggests that the complex architecture of the lung passages creates aerodynamic conditions that direct air along the inspiratory and expiratory paths without need of either passive or active valves.

Ventilation of the neopulmonal system is incompletely understood. Probably, however, airflow through many of the neopulmonal parabronchi is bidirectional (see Figure 23.24B,C).

As discussed in **BOX 23.3**, birds face unique challenges at hatching because of their lungs, which at that point must take over full responsibility for gas exchange.

The gas-exchange system is cross-current

When the unidirectional flow of air through the paleopulmonal parabronchi in the lungs of birds was first discovered, countercurrent exchange between the blood and air was quickly hypothesized. Soon, however, this hypothesis was disproved by clever experiments, which showed that the efficiency of gas exchange between air and blood is not diminished if the direction of airflow in the parabronchi is artificially reversed. Morphological and functional studies have now shown convincingly that blood flow in the respiratory exchange vessels of the circulatory system occurs in a cross-current pattern relative to the flow of air through the parabronchi (see Figures 23.5 and 23.25C).

SUMMARY
Breathing by Birds

■ The lungs of birds are relatively compact, rigid structures consisting mostly of numerous tubes, running in parallel, termed parabronchi. Fine air capillaries, extending radially from the lumen of each parabronchus, are the principal sites of gas exchange. Air sacs, which are nonrespiratory, are integral parts of the breathing system.

■ The lungs are ventilated by a bellows action generated by expansion and compression of the air sacs.

■ Airflow through the parabronchi of the paleopulmonal system (the major part of the lungs) is posterior to anterior during both inhalation and exhalation. Cross-current gas exchange occurs.

Breathing by Aquatic Invertebrates and Allied Groups

Many small aquatic invertebrates, and some large ones, have no specialized breathing organs. They exchange gases across general body surfaces, which sometimes are ventilated by swimming motions or by cilia- or flagella-generated water currents. Many larvae and some adults also lack a circulatory system. Thus gases move within their bodies by diffusion or by the squishing of body fluids from place to place. To the human eye, these sorts of gas-exchange systems are confining. They suffice only if the animals are tiny (see Box 22.1) or have specialized body plans, such as those of flatworms and jellyfish. Most cells of a flatworm are near a body surface because the worm's body is so thin. Most cells of a jellyfish are near a body surface because a jellyfish's body is organized with its active tissues on the outside and primarily low-metabolism, gelatinous tissue deep within.

Adults of the relatively advanced phyla of aquatic invertebrates typically have gills of some sort. The gills of the various major phyletic groups are often independently evolved. Thus, whereas they all are evaginated and project into the water (meeting the definition of gills), they vary widely in their structures and in how they are ventilated (**FIGURE 23.27**).

Molluscs exemplify an exceptional diversity of breathing organs built on a common plan

The phylum Mollusca nicely illustrates that within a phyletic group, a single basic sort of breathing apparatus can undergo wide diversification. In molluscs, outfolding of the dorsal body wall produces a sheet of tissue, the **mantle** (responsible for secreting the shell), that overhangs or surrounds all or part of the rest of the body, thereby enclosing an external body cavity, the **mantle cavity**. The gills of molluscs typically are suspended in the mantle cavity and thus are internal gills (see Figure 23.2). Certain of the aquatic snails provide a straightforward example. In them (**FIGURE 23.28A**), a series of modest-sized *gill leaflets* hangs in the mantle cavity and is ventilated unidirectionally by ciliary currents. Blood flow through the leaflets, in at least some cases, is opposite to the direction of water flow. Thus countercurrent gas exchange occurs.

One major modification of the gills in molluscs is the evolution of extensive *sheetlike gills* in the clams, mussels, oysters, and other lamellibranch ("sheet-gilled") groups. In these groups, four gill sheets, or lamellae, composed of fused or semifused filaments, hang within the mantle cavity (**FIGURE 23.28B**). Cilia on the gill sheets drive incoming water through pores on the gill surfaces into water channels that run *within* the gill sheets; the water channels then convey the water to exhalant passages. The direction of water flow within the water channels is opposite to the direction of blood flow in the major gill blood vessels, meaning that countercurrent gas exchange can again occur. The specialized sheetlike gills of these molluscs represent, in part, an adaptation for feeding: As the abundant flow of incoming water passes through the arrays of pores leading to the interior water channels of the gill sheets, food particles suspended in the water are captured for delivery to the mouth (a type of suspension feeding; see page 145). In some (not all) molluscs with sheetlike gills, the food-collection function has become paramount: Respiratory gas exchange across general body surfaces suffices to meet metabolic needs. Thus the "gills" have become primarily feeding organs.

In the cephalopod molluscs—the squids, cuttlefish, and octopuses—it is not so much the gills that are specialized, but the mechanism of ventilation. The gills are feathery structures that follow the usual molluscan plan of being positioned in the mantle cavity (**FIGURE 23.28C**). They are ventilated, however, by muscular contraction rather than beating of cilia. Cephalopods swim by using muscular contractions of the mantle; they alternately suck water into the mantle cavity via incurrent openings and then drive it forcibly outward through a ventral funnel by mantle contraction, producing a jet-propulsive force. The gills are ventilated (in countercurrent fashion) by the vigorous flow of water used for propulsion. Some species move so much water for propulsion that they use only a small fraction of the O_2 in the water: 5%–10%.

A final specialization worthy of note in molluscs is the evolution of the mantle cavity into a lung in the dominant group of snails and slugs that live on land, a group known aptly as the *pulmonates* (**FIGURE 23.28D**). In the terrestrial pulmonates, gills have disappeared, and the walls of the mantle cavity have become highly vascularized and well suited for gas exchange. Some species are thought to employ the mantle cavity as a diffusion lung, but others ventilate it by raising and lowering the floor of the cavity.

Decapod crustaceans include many important water breathers and some air breathers

In the decapod crustaceans—which include many ecologically and commercially important crabs, shrimps, lobsters, and crayfish—the head and thorax are covered with a continuous sheet of exoskeleton, the *carapace*, that overhangs the thorax laterally, fitting more or less closely around the bases of the thoracic legs. The carapace encloses two lateral external body cavities—the **branchial chambers**—in which the gills lie (**FIGURE 23.29A**).

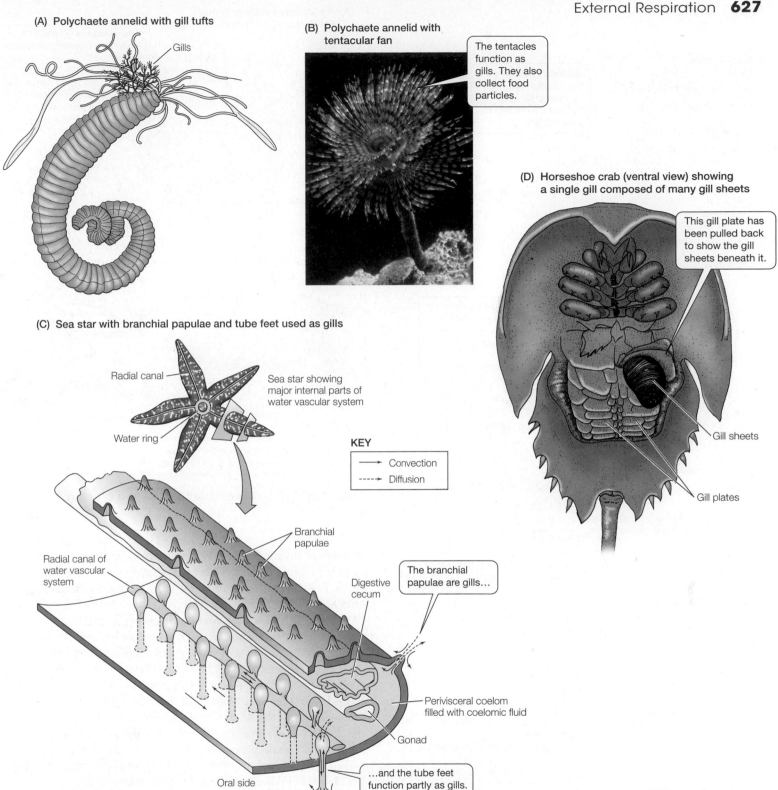

(A) Polychaete annelid with gill tufts

Gills

(B) Polychaete annelid with tentacular fan

The tentacles function as gills. They also collect food particles.

(D) Horseshoe crab (ventral view) showing a single gill composed of many gill sheets

This gill plate has been pulled back to show the gill sheets beneath it.

Gill sheets

Gill plates

(C) Sea star with branchial papulae and tube feet used as gills

Radial canal

Sea star showing major internal parts of water vascular system

Water ring

KEY

→ Convection

--→ Diffusion

Branchial papulae

Radial canal of water vascular system

Digestive cecum

The branchial papulae are gills...

Perivisceral coelom filled with coelomic fluid

Gonad

...and the tube feet function partly as gills.

Oral side

FIGURE 23.27 A diversity of gills in aquatic invertebrates
(A) This terebellid worm (*Amphitrite*), a type of marine annelid, lives inside a tube it constructs and can pump water in and out of the tube. (B) This fanworm, another type of marine annelid, also lives in a tube, but when undisturbed, it projects its well-developed array of pinnately divided tentacles into the ambient water. The tentacles are used for both feeding and respiratory gas exchange; they are ventilated by the action of cilia on the tentacles. (C) Sea stars bear many thin-walled, fingerlike projections from their coelomic cavity, termed *branchial papulae* ("gill processes"), on their upper body surfaces; respiratory gases pass between the coelomic fluid and ambient water by diffusion through the walls of the papulae. Similarly, gases diffuse between the coelomic fluid and ambient water through the tube feet and associated parts of the water vascular system. Cilia accelerate these processes by circulating fluids over the inner and outer surfaces of the papulae and tube feet. (D) Horseshoe crabs (*Limulus*) have unique *book gills*, consisting of many thin gill sheets arranged like pages of a book. The book gills are protected under thick *gill plates*, which undergo rhythmic flapping motions that ventilate the gills. (D after a drawing by Ralph Russell, Jr.)

(A) Aquatic snail

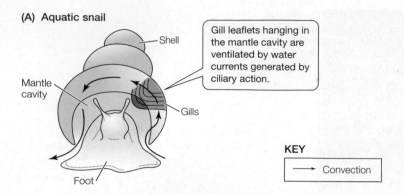

Shell

Mantle cavity

Gills

Foot

Gill leaflets hanging in the mantle cavity are ventilated by water currents generated by ciliary action.

KEY

→ Convection

(B) Clam (a lamellibranch mollusc)

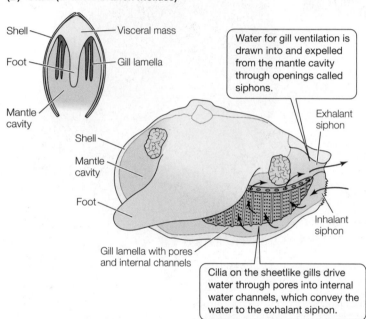

Shell

Visceral mass

Foot

Gill lamella

Mantle cavity

Shell

Mantle cavity

Foot

Gill lamella with pores and internal channels

Water for gill ventilation is drawn into and expelled from the mantle cavity through openings called siphons.

Exhalant siphon

Inhalant siphon

Cilia on the sheetlike gills drive water through pores into internal water channels, which convey the water to the exhalant siphon.

(C) Squid (a cephalopod)

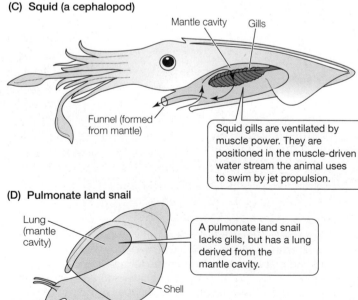

Mantle cavity

Gills

Funnel (formed from mantle)

Squid gills are ventilated by muscle power. They are positioned in the muscle-driven water stream the animal uses to swim by jet propulsion.

(D) Pulmonate land snail

Lung (mantle cavity)

Shell

A pulmonate land snail lacks gills, but has a lung derived from the mantle cavity.

FIGURE 23.28 The diversification of the breathing system in molluscs

(A) A transverse section through the thorax of a crayfish

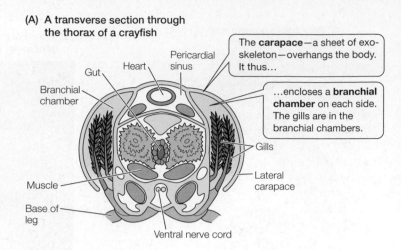

Gut

Heart

Pericardial sinus

Branchial chamber

The **carapace**—a sheet of exoskeleton—overhangs the body. It thus…

…encloses a **branchial chamber** on each side. The gills are in the branchial chambers.

Gills

Lateral carapace

Muscle

Base of leg

Ventral nerve cord

(B) A lateral view showing the gills under the carapace

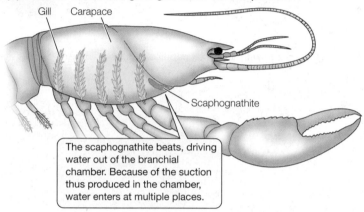

Gill

Carapace

Scaphognathite

The scaphognathite beats, driving water out of the branchial chamber. Because of the suction thus produced in the chamber, water enters at multiple places.

FIGURE 23.29 The gills and ventilation in a crayfish

The gills arise from near the bases of the thoracic legs. Each gill consists of a central axis to which are attached many richly vascularized lamellar plates, filaments, or dendritically branching tufts. The gill surfaces—like all the external body surfaces of crustaceans—are covered with a chitinous cuticle. The cuticle on the gills is thin, however, and permeable to gases.

Ventilation in crustaceans is always accomplished by muscular contraction, because the body surfaces of crustaceans lack cilia. In decapod crustaceans, each branchial chamber is ventilated by a specialized appendage—the *scaphognathite* or *gill bailer*—located toward its anterior end (**FIGURE 23.29B**). Beating of this appendage, under control of nerve impulses from a central pattern generator in the central nervous system, generally drives water outward through an anterior exhalant opening. Negative pressure is thus created within the branchial chamber, drawing water in at various openings. Ventilation is unidirectional, and countercurrent exchange may occur.

Some crabs and crayfish have invaded the land, especially in the tropics. All the semiterrestrial and terrestrial species retain gills, which are supported to some degree by their cuticular covering. One trend observed in semiterrestrial and terrestrial crabs is that the gills tend to be reduced in size and number by comparison with aquatic crabs. A second trend is that the branchial chambers tend to be enlarged ("ballooned out"), and the tissue[9] that lines the chambers tends to be specialized by being well vascularized, thin, and thrown into folds that increase its surface area. These trends—the reduction of the gills

[9] This tissue is called the *branchiostegites*.

and the development of lunglike branchial chambers—are strikingly parallel to those seen earlier in air-breathing fish and air-breathing (pulmonate) snails. The branchial chambers of crabs on land are typically ventilated with air by beating of the scaphognathites. In the species that are semiterrestrial (amphibious), the gills are kept wet by regular trips to bodies of water. Recent evidence suggests that in these crabs, O_2 is chiefly taken up by the branchial-chamber epithelium, whereas CO_2 is chiefly voided across the gills. In fully terrestrial species of crabs, water is not carried in the branchial chambers, and the branchial-chamber epithelium may bear chief responsibility for exchange of both O_2 and CO_2.

SUMMARY
Breathing by Aquatic Invertebrates and Allied Groups

■ The gills of various groups of aquatic invertebrates are often independently evolved. Wide variation thus exists in both gill morphology and the mode of gill ventilation.

■ A single basic sort of breathing apparatus can undergo wide diversification within a single phyletic group. This general principle is illustrated by the molluscs, the great majority of which have breathing organs associated with the mantle and located in the mantle cavity. Whereas both aquatic snails and lamellibranchs employ ciliary ventilation, the gills are modest-sized leaflets in snails, but expansive sheets (used partly for feeding) in the lamellibranchs. Cephalopods, such as squids, ventilate their gills by muscular contraction. Most land snails lack gills and breathe with a lung derived from the mantle cavity.

Breathing by Insects and Other Tracheate Arthropods

The insects (**FIGURE 23.30**) have evolved a remarkable strategy for breathing that is entirely different from that of most metabolically active animals. Their breathing system brings the gas-exchange surface itself close to every cell in the body. Thus, with some thought-provoking exceptions, the cells of insects get their O_2 directly from the breathing system, and the circulatory system plays little or no role in O_2 transport. Insect blood, in fact, usually lacks any O_2-transport pigment such as hemoglobin.

The body of an insect is thoroughly invested with a system of gas-filled tubes termed **tracheae** (**FIGURE 23.31A,B**). This system opens to the atmosphere by way of pores, termed **spiracles**, located at the body surface along the lateral body wall. Tracheae penetrate into the body from each spiracle and branch repeatedly, collectively reaching all parts of the animal (only major branches are seen in Figure 23.31). The tracheal trees arising from different spiracles typically join via large *longitudinal* and *transverse connectives* to form a fully interconnected tracheal system. The spiracles, which number from 1 to 11 pairs, are segmentally arranged and may occur on the thorax, abdomen, or both (but not the head). Usually they can be closed by spiracular muscles. Although tracheal breathing is best understood in insects, there are other tracheate arthropods: Most notably, certain groups of spiders and ticks have tracheal systems. Some spiders and other arachnids have *book lungs*, unique breathing

FIGURE 23.30 A praying mantis, one of the largest existing insects To look at a praying mantis, one could imagine it breathing through its mouth. Nothing could be further from the truth.

organs that sometimes function in parallel with tracheal systems and sometimes are the sole breathing organs (**BOX 23.4**).

The tracheae of an insect develop as invaginations of the epidermis and thus are lined with a thin cuticle. Typically the cuticle is thrown into spiral folds, providing resistance against collapse. The tracheae become finer with increasing distance from the spiracles and finally give rise to very fine, thin-walled end-tubules termed **tracheoles**, believed to be the principal sites of O_2 and CO_2 exchange with the tissues. Tracheoles are perhaps 200–350 μm long and are believed to end blindly. They generally taper from a lumen diameter approximating 1 μm at their origin to 0.05–0.20 μm at the end. The walls of the tracheoles and the finest tracheae are about 0.02–0.2 μm thick—exceedingly thin by any standard (see Figure 23.8B).

Although the layout of the tracheal system varies immensely among various species of insects, the usual result is that all tissues are thoroughly invested with fine tracheae and tracheoles. The degree of tracheation of various organs and tissues tends to vary directly with their metabolic requirements. For the most part, the tracheoles run between cells. However, in the flight muscles of many species, the tracheoles penetrate the muscle cells, indenting the cell membranes inward, and run among the individual myofibrils, in close proximity to the arrays of mitochondria. The average distance between adjacent tracheoles within the flight muscles of strong fliers is often only about 3 μm. The nervous system, rectal glands, and other active tissues—including muscles besides the flight muscles—also tend to be richly supplied by the tracheal system, although intracellular penetration is not nearly as common as in flight muscles. In the epidermis of the bug *Rhodnius*, which has been carefully studied, tracheoles are much less densely distributed than in active flight muscles, but nonetheless, cells are usually within

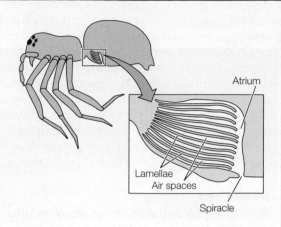

BOX 23.4 The Book Lungs of Arachnids

Some arachnids possess a novel type of respiratory structure, the **book lung**. Scorpions have only book lungs. Many species of spiders also have book lungs, but they may have systems of tracheae as well. The number of book lungs in an individual arachnid varies from a single pair (as in certain spiders) to four pairs (in scorpions). Book lungs are invaginations of the ventral abdomen, lined with a thin chitinous cuticle. Each book lung consists of a chamber, the atrium, which opens to the outside through a closable ventral pore, the spiracle (see figure). The dorsal or anterior surface of the atrium is thrown into many lamellar folds: the "pages of the book." Blood streams through the lamellae, whereas the spaces between adjacent lamellae are filled with gas. The lamellae commonly number into the hundreds, and the blood-to-gas distance across their walls is often less than 1 μm. Some book lungs may function as diffusion lungs, whereas others are clearly ventilated by pumping motions. They oxygenate the blood, which then carries O_2 throughout the body.

A book lung The section shows the internal structure of a book lung in a two-lunged spider. (After Comstock 1912.)

30 μm of a tracheole. In other words, no cell is separated from a branch of the tracheal system by more than two or three other cells!

The terminal ends of the tracheoles are sometimes filled with liquid when insects are at rest. During exercise, or when the insects are exposed to O_2-deficient environments, the amount of liquid decreases and gas penetrates farther into the tracheoles. This process facilitates the exchange of O_2 and CO_2 because of the greater ease of diffusion in gas than liquid.

Distensible enlargements of the tracheal system called **air sacs** are a common feature of insect breathing systems (**FIGURE 23.31C**) and may occur in the head, thorax, or abdomen. Some air sacs are swellings along tracheae, whereas others form blind endings of tracheae. Air sacs tend to be particularly well developed in active insects, in which they may occupy a considerable fraction of the body volume.

Diffusion is a key mechanism of gas transport through the tracheal system

The traditional dogma has been that the tracheal system of most insects functions as a diffusion lung, meaning that gas transport through the system occurs solely by diffusion. This dogma is presently undergoing profound revision. Diffusion, nonetheless, seems likely to be an important gas-transport mechanism in subparts of the tracheal system in most or all insects and may be the sole transport process in some. Diffusion can occur fast enough to play this role because the tracheae are gas-filled.

Because of the importance of diffusion in insect breathing and because diffusion transport tends to be slow when distances are great, physiologists have long wondered whether insect body size is limited by the nature of the insect breathing system. A recent study using a new X-ray technique to visualize the tracheal system (see Figure 23.31B) revealed that in beetles of different body sizes, the volume of the tracheal system is disproportional to body size. In stark contrast to mammals, a greater proportion of body space is devoted to the breathing system in big beetles than in small ones. This trend could represent an evolutionary compensation for diffusion limitations in large-bodied insects. If so, the trend would suggest that a tracheal breathing system poses obstacles to the evolution of large body size.

When considering an individual insect, a question that arises is how the *rate* of diffusion can be varied to correspond to the insect's needs for O_2. Although diffusion may sound like a process that is purely physical and therefore independent of animal needs, in fact its rate responds to an insect's metabolic needs because the animal's metabolism alters the partial pressures of gases. Suppose that an insect breathing by diffusion has an adequate rate of O_2 transport when (1) the atmospheric O_2 partial pressure is at the level marked ❶ in **FIGURE 23.32A** and (2) the O_2 partial pressure at the inner end of its tracheal system is at the level marked ❷. If the insect suddenly increases its rate of O_2 consumption, its end-tracheal O_2 partial pressure will fall because of the increased rate of O_2 removal from the tracheae. This decline in the end-tracheal partial pressure will increase the difference in partial pressure between the two ends of the tracheal system and thus accelerate O_2 diffusion. Suppose that the difference in partial pressure between level ❶ and level ❸ in Figure 23.32A is sufficient for O_2 to diffuse fast enough to meet the insect's new (higher) O_2 need. The end-tracheal partial pressure will then fall to level ❸ and stabilize. In this way, the rate of diffusion will automatically increase to meet the insect's heightened O_2 need.

Of course, there are limits to the ability of the process just described to increase the rate of O_2 diffusion. The end-tracheal O_2 partial pressure must itself remain sufficiently high for O_2 to diffuse from the ends of the tracheae to the mitochondria in cells. If an insect's oxygen cascade follows line ❶-❷-❹ in **FIGURE 23.32B** when the insect has a low rate of cellular O_2 use, it might follow line ❶-❸-❺ when the insect's rate of O_2 use is raised. The mitochondrial O_2 partial pressure would then be very low, and a further increase in the rate of O_2 diffusion might not be possible while keeping the mitochondrial O_2 level adequate.

(A) Major parts of the tracheal system in a flea

Thoracic spiracle 1

Abdominal spiracles 1–7

Abdominal spiracle 8

Tracheae

0.5 mm

(C) Air sacs in the abdomen of a worker honeybee

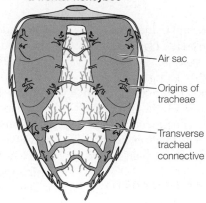

Air sac

Origins of tracheae

Transverse tracheal connective

FIGURE 23.31 Insects breathe using a tracheal system of gas-filled tubes that—branching and rebranching—reach all tissues from the body surface   (A) The principal tracheae in a flea in the genus *Xenopsylla*—an insect that has ten pairs of spiracles (two thoracic pairs and eight abdominal). (B) The tracheae in a tiny living adult carabid beetle in the genus *Notiophilus*, visualized by a cutting-edge technique, synchrotron X-ray phase-contrast imaging. (C) Air sacs and associated tracheae in the abdomen of a worker honeybee (*Apis*). Additional air sacs occur in the head and thorax. (A after Wigglesworth 1935; B from Socha et al. 2010; image courtesy of Jake Socha.)

(B) Tracheae in a carabid beetle

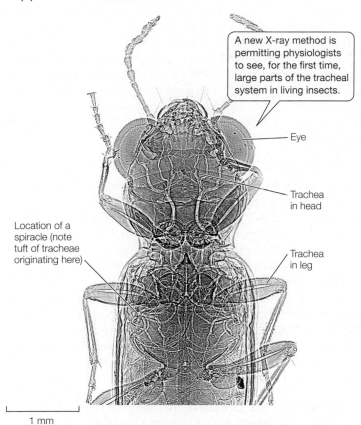

A new X-ray method is permitting physiologists to see, for the first time, large parts of the tracheal system in living insects.

Eye

Trachea in head

Location of a spiracle (note tuft of tracheae originating here)

Trachea in leg

1 mm

Some insects employ conspicuous ventilation

Conspicuous (macroscopic) ventilation of the tracheal system occurs in some large species of insects at rest and is common among active insects. Grasshoppers and locusts, for example, are easily seen to pump their abdomens, and abdominal pumping occurs also in bumblebees, ants, and some other insects. The abdominal pumping motions alternately expand and compress certain of the tracheal airways, either causing tidal ventilation or causing air to be sucked in via certain spiracles and expelled via others, flowing unidirectionally through parts of the tracheal system in between. Air sacs, when present (as they are in grasshoppers and bumblebees, for example), commonly act as bellows during such muscular pumping movements; they may be compressed to only 25%–50% of their full size during each cycle of compression. A mechanism of

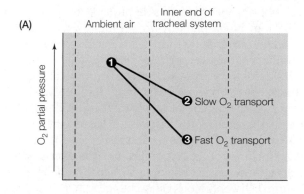

(A)

Ambient air

Inner end of tracheal system

O_2 partial pressure

❶

❷ Slow O_2 transport

❸ Fast O_2 transport

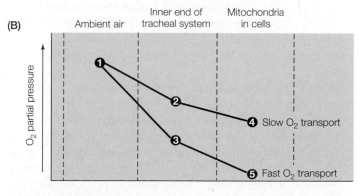

(B)

Ambient air

Inner end of tracheal system

Mitochondria in cells

O_2 partial pressure

❶

❷

❸

❹ Slow O_2 transport

❺ Fast O_2 transport

FIGURE 23.32 Insect oxygen cascades assuming oxygen transport by diffusion   These diagrams outline simplified thought exercises. (A) A drop in the O_2 partial pressure at the inner end of the tracheal system from level ❷ to level ❸ will speed diffusion through the tracheal system by increasing the difference in partial pressure from one end to the other. (B) For the rate of diffusion to the mitochondria to be accelerated, the difference in partial pressure between the inner end of the tracheal system and the mitochondria must be increased—as well as the difference between the ambient air and the inner tracheae. Eventually no further increase in the rate of diffusion will be possible because the mitochondrial partial pressure will become too low for mitochondrial function.

conspicuous tracheal ventilation that is important in many insects during flight is **autoventilation**: ventilation of the tracheae supplying the flight muscles *driven by flight movements*.

Physiologists have generally hypothesized that during conspicuous ventilation, air is forced to flow only in major tracheae, with diffusion being the principal mode of gas transport through the rest of the tracheal system. According to this hypothesis, the function of conspicuous ventilation is essentially to *reduce the path length for diffusion* by moving air convectively to a certain depth in the tracheal system.

Microscopic ventilation is far more common than believed even 15 years ago

A revolution is underway in the understanding of microscopic ventilation: forced airflow that occurs on such fine scales that it is impossible to detect without the use of technology. Probably the most dramatic recent discovery is that when microscopic X-ray videos are made of various insects—such as beetles, crickets, and ants—the major tracheae in the head and thorax are observed to undergo cycles of partial compression and relaxation, a process named *rhythmic tracheal compression*. These pulsations occur every 1–2 s and are substantial: Each compression reduces the volumes of the tracheae to 50%–70% of their relaxed volumes. These microscopic cycles of tracheal compression probably move gases convectively. They may be particularly important for O_2 delivery to the head and brain, recognizing that tracheae to the head connect to the atmosphere at thoracic spiracles and thus may be long.

Similar X-ray studies have revealed that in some insects under some conditions, tracheae undergo massive rhythmic collapsing movements during which they completely empty and refill. This is observed, for example, in some moth caterpillars exposed to hypoxia.

One of the first sorts of evidence for microscopic ventilation was the discovery and analysis of **discontinuous gas exchange** in diapausing pupae[10] of moths some decades ago. The hallmark and defining feature of discontinuous gas exchange is that an insect releases CO_2 to the atmosphere in dramatic, intermittent bursts, whereas the insect takes up O_2 from the atmosphere at a relatively steady rate. The observed pattern of CO_2 release arises in large part from spiracular control. In the periods between one burst of CO_2 release and the next, the spiracles are closed or partly closed, and CO_2 produced by metabolism *accumulates in body fluids* by dissolving and reacting to form bicarbonate (HCO_3^-). Because O_2 is removed from the tracheal airways by metabolism during these periods, but the CO_2 produced by metabolism temporarily accumulates in the body fluids rather than in the airways, a partial vacuum—a negative pressure—can develop in the airways. When such a negative pressure develops, it sets the stage for microscopic ventilation because atmospheric air can be sucked into the tracheal airways convectively on an inconspicuous, microscopic scale when the insect partly or fully opens its spiracles. Investigators have directly assessed in several species whether inward suction of air actually occurs, and it does in some (not all) of the species tested. No one knows the depth to which air is drawn, but it travels by convection at least through the spiracles and into the major tracheae. Discontinuous gas exchange is known today to occur widely in quiescent or resting insects, plus certain ticks, mites, and spiders.

In addition to the forms of microscopic ventilation we have already discussed, several other types have been reported during the last 25 years. These include processes named *miniature ventilation pulses* in grasshoppers and tiny *Prague cycles* of CO_2 release in beetles.

According to the old dogma, insects that were not conspicuously pumping their abdomens or ventilating in other conspicuous ways were breathing entirely by diffusion. The evidence is now overwhelming that convective phenomena are widely employed by visibly motionless insects, but many mysteries remain regarding the exact interplay of convection and diffusion in the tracheal system.

Control of breathing

A vulnerability of the insect respiratory system is that it can permit rapid evaporative loss of body water. The gas in the tracheal airways is humid—ordinarily saturated with water vapor—and when the spiracles are open, only a minute distance separates the humid tracheal gas from the atmosphere. Outward diffusion of water vapor can accordingly be rapid. Insects commonly solve this problem by keeping their spiracles partly closed—or by periodically opening and closing them—whenever compatible with their needs for O_2 and CO_2 exchange. If the spiracles of resting insects are experimentally forced to remain fully open all the time, the rate of evaporative water loss increases 2- to 12-fold, demonstrating the importance of keeping them partly closed.

In insects using diffusion transport, it is common for the spiracles to be opened more fully or frequently as the insects become more active. The greater opening of the spiracles facilitates O_2 transport to the tissues, although it also tends to increase evaporative water loss. Insects that ventilate their tracheal systems by abdominal pumping or other conspicuous mechanisms are well known to increase their rates of ventilation as they become more active.

What is the chemosensory basis for spiracular control? The most potent stimulus for opening of the spiracles in insects is an increase in the CO_2 partial pressure and/or H^+ concentration of the body fluids. A decrease in the O_2 partial pressure in the body fluids may also stimulate spiracular opening but typically offers far less potent stimulation. In these respects, the control of the spiracles in insects resembles the control of pulmonary ventilation in mammals.

Aquatic insects breathe sometimes from the water, sometimes from the atmosphere, and sometimes from both

Many insect species live underwater in streams, rivers, and ponds during parts of their life cycles. The aquatic life stages of some of these species lack functional spiracular openings and obtain O_2 by taking up dissolved O_2 from the water using superficial arrays of fine tracheae. These insects often have dense proliferations of fine tracheae under their general integument. Many have **tracheal gills**: evaginations of the body surface that are densely supplied with tracheae and covered with just a thin cuticle—a remarkable parallel with the evolution of ordinary gills in numerous other groups of aquatic animals. Tracheal gills may be positioned on the outer body surface or in the rectum. In aquatic insects that breathe with tracheal gills or other superficial tracheae, the tracheal system remains gas-filled. Oxygen diffuses into the tracheal airways from the water across the gills or other superficial tracheae. Thereafter

---

[10] *Diapause* is a programmed resting stage in the life cycle.

the gas-filled tracheal system serves as the path of least resistance for the O_2 to move throughout the body.

Other aquatic insects retain functional spiracles and have evolved alternative ways of interfacing their tracheal breathing systems with the ambient water or air. The aquatic insects with functional spiracles breathe from external gas spaces. There are three distinctive ways in which they do so. The simplest to understand is the system used by insects such as mosquito larvae, which hang at the water's surface and have their functional spiracles localized to the body region that contacts the atmosphere. Such insects breathe from the atmosphere, much in the way that terrestrial insects do.

A second strategy employed by aquatic insects with functional spiracles is to carry a conspicuous bubble of gas captured from the atmosphere. Many water beetles, for example, carry a conspicuous bubble either under their wings or at the tip of their abdomen (see opening photo of Chapter 22). Their functional spiracles open into the bubble and exchange O_2 and CO_2 with the gas in the bubble. As explained previously (see Figure 22.3), a remarkable attribute of a bubble like this is that it *acts as a gill*: Dissolved O_2 from the water diffuses into the bubble. Thus the insect is able to remove much more O_2 from the bubble than simply the amount captured from the atmosphere. A conspicuous bubble gradually shrinks, a process that decreases its surface area and impairs O_2 diffusion into the bubble because the rate of diffusion depends on bubble surface area. A bubble of this sort must therefore be periodically renewed with air from the atmosphere.

The third strategy employed by aquatic insects with functional spiracles is certainly the most unexpected. It is also a type of bubble breathing, but a very different type. In some aquatic insects, parts of the body surface are covered extremely densely with fine water-repelling hairs; the bug *Aphelocheirus aestivalis*, for example, has 2–2.5 million of these hairs per square *millimeter*! Such densely distributed water-repelling hairs on the body surface trap among themselves a thin, almost invisible film of gas *that cannot be displaced*. This film of gas, known as a **plastron**, is incompressible and permanent. Thus its surface area remains constant, and it can serve as a gill (an air space into which O_2 diffuses from the water) for an indefinite period. Some plastron-breathing aquatic insects remain submerged continuously for months!

SUMMARY
Breathing by Insects and Other Tracheate Arthropods

- Insects and many arachnids breathe using a tracheal system that connects to the atmosphere by way of spiracles on the body surface and ramifies throughout the body so that gas-filled tubes bring O_2 close to all cells.

- The modes of gas exchange through the tracheal system include diffusion, conspicuous ventilation (such as abdominal pumping and autoventilation), and several forms of microscopic ventilation.

- Aquatic insects may lack functional spiracles and breathe using superficial tracheal beds. Alternatively, they may have functional spiracles and breathe from the atmosphere, large bubbles, or plastrons.

STUDY QUESTIONS

1. Lungs ventilated with water occur in some animals—most notably sea cucumbers—but are rare. Why would water lungs be unlikely to be favored by natural selection? Give as many reasons as possible.

2. In Chapter 1 we discussed François Jacob's question of whether evolution is more like tinkering or engineering. Jacob's view is that tinkering is a far better analogy than engineering. How could the evolution of air breathing in fish be used to argue for the validity of the tinkering analogy?

3. Outline the differences among the three most sophisticated lungs found in modern animals: the mammalian lung, the avian lung, and the insect tracheal system.

4. Suppose a mammal's tidal volume is 2 L, its tracheal volume is 80 mL, its anatomical dead space volume is 350 mL, and its breathing frequency is 9 breaths per minute. What is its alveolar minute volume?

5. When researchers first discovered that airflow through a bird's paleopulmonal parabronchi is unidirectional, the question arose as to whether gas exchange is countercurrent, cocurrent, or cross-current. Some ingenious investigators carried out experiments in which they measured the efficiency of gas exchange between air and blood in duck lungs when parabronchial airflow was in its normal direction and when the direction of parabronchial airflow was artificially reversed. The efficiency did not change. How is this evidence *against* countercurrent and cocurrent gas exchange? How is this evidence *for* cross-current exchange?

6. What is the evidence that the breathing rhythm in mammals originates in the brainstem?

7. Explain mechanistically how hyperventilation alters the oxygen cascade of an animal.

8. Arthropods are distinguished by having an exoskeleton and lacking external cilia. How do these traits of the phylum affect the breathing structures and ventilation modes of arthropods?

9. Why do fish suffocate when taken out of the water and placed in air, whereas mammals suffocate if the air in their alveoli is replaced with water?

10. In your own words, explain how a person's pulmonary oxygen utilization coefficient can increase as the person's rate of ventilation increases even though the composition of alveolar gas is held constant by negative feedback controls.

11. Insects of giant size compared with today's insects are found in the fossil record. There were times in the past when the atmospheric concentration of O_2 was 1.5 times today's concentration. Some investigators hypothesize that insect gigantism was permitted by high O_2 levels. Explain why this hypothesis is plausible.

Go to **sites.sinauer.com/animalphys4e**
for box extensions, quizzes, flashcards, and other resources.

REFERENCES

Beall, C. M. 2007. Two routes to functional adaptation: Tibetan and Andean high-altitude natives. *Proc. Natl. Acad. Sci. U.S.A.* 104: 8655–8660.

Daniels, C. B., and S. Orgeig. 2003. Pulmonary surfactant: The key to the evolution of air breathing. *News Physiol. Sci.* 18: 151–157.

Duncker, H.-R. 2004. Vertebrate lungs: structure, topography and mechanics. A comparative perspective of the progressive integration of respiratory system, locomotor apparatus and ontogenetic development. *Respir. Physiol. Neurobiol.* 144: 111–124. An occasionally opinionated, but consistently intellectually rewarding, survey of the evolution of vertebrate lungs. A strength of this paper is that it emphasizes how lung evolution has interacted with and sometimes constrained the evolution of other structural and physiological attributes of developing and adult vertebrates.

Graham, J. B. 1997. *Air-Breathing Fishes*. Academic Press, New York.

Guyenet, P. G., and D. A. Bayliss. 2015. Neural control of breathing and CO_2 homeostasis. *Neuron* 87: 946–961. An accessible article on one of the central contemporary research areas in mammalian respiratory physiology, including up-to-date material on effects of exercise.

Kaiser, A., C. J. Klok, J. J. Socha, and three additional authors. 2007. Increase in tracheal investment with beetle size supports hypothesis of oxygen limitation on insect gigantism. *Proc. Natl. Acad. Sci. U.S.A.* 104: 13198–13203.

Maina, J. N. 2000. What it takes to fly: The structural and functional respiratory refinements in birds and bats. *J. Exp. Biol.* 203: 3045–3064. An intriguing, comparative inquiry into the question of how the respiratory systems of bats and birds are adapted to meet the high gas-exchange requirements of flight.

Scheid, P., and J. Piiper. 1997. Vertebrate respiratory gas exchange. In W. H. Dantzler (ed.), *Comparative Physiology*, vol. 1 (Handbook of Physiology [Bethesda, MD], section 13), pp. 309–356. Oxford University Press, New York. An overview of gas-exchange principles and the comparative physiology of breathing in vertebrates, written by two long-term leaders in these fields. An excellent source to master the quantitative analysis of tidal, cocurrent, countercurrent, and cross-current gas exchange.

Scott, G. R. 2011. Elevated performance: the unique physiology of birds that fly at high altitudes. *J. Exp. Biol.* 214: 2455–2462.

Semenza, G. L. 2007. Life with oxygen. *Science* 318: 62–64. A short but stimulating introduction to the molecular biology of oxygen in animals. Use it as a launching point into the literature it cites.

Storz, J. F., G. R. Scott, and Z. A. Cheviron. 2010. Phenotypic plasticity and genetic adaptation to high-altitude hypoxia in vertebrates. *J. Exp. Biol.* 213: 4125–4136. In terms of achieving clarity within an extremely complex area of research, probably the best paper ever written on high-altitude physiology of animals; emphasizes mammals and birds.

Weibel, E. R. 1984. *The Pathway for Oxygen*. Harvard University Press, Cambridge, MA. One of the all-time great treatments of its subject, the pathway that oxygen follows from atmosphere to mitochondria, written by a founder of the field of morphometric lung analysis. Becoming dated, unfortunately.

Weibel, E. R. 2009. What makes a good lung? The morphometric basis of lung function. *Swiss Medical Weekly* 139: 375–386. A stimulating and easy-to-read review of the latest information on human lung structure and function, including numerous illustrations.

West, J. B., and A. M. Luks. 2015. *West's Respiratory Physiology: The Essentials*, 10th ed. Lippincott Williams & Wilkins, Philadelphia. An articulate, compact treatment of medical respiratory physiology.

See also **Additional References** *and* **Figure and Table Citations**.

Transport of Oxygen and Carbon Dioxide in Body Fluids

(with an Introduction to Acid–Base Physiology)

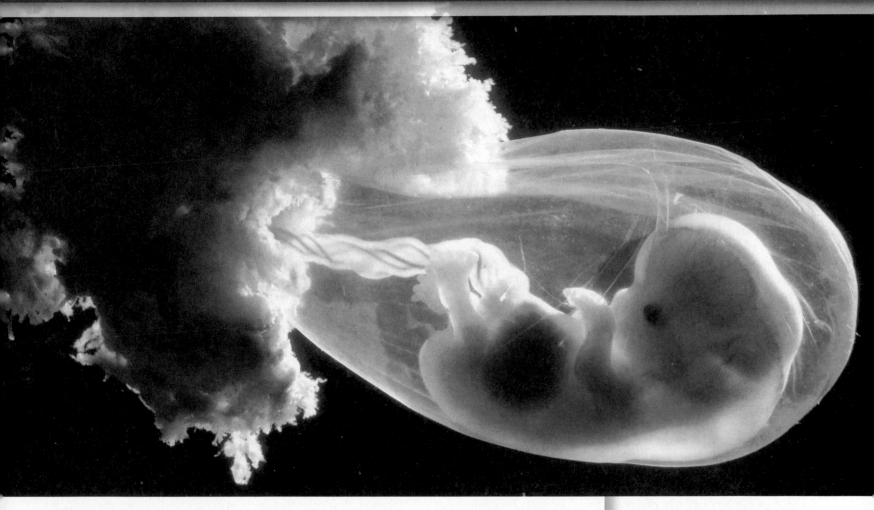

A developing mammalian fetus receives oxygen and voids carbon dioxide by means of a close juxtaposition of its own blood vessels with those of its mother's circulatory system in the placenta. The fetus's heart pumps blood through the umbilical cord to the placenta, where that blood picks up O_2 that its mother's blood has brought there from her lungs. The now-oxygenated fetal blood returns to the fetus through the umbilical cord and is circulated to all the parts of the body of the fetus. By 10 weeks after conception, a human fetus already has hemoglobin-rich blood. At that age, the amount of hemoglobin per unit of volume in the fetus's blood has reached 50% of the adult concentration and is increasing rapidly, so that it will be about 80% of the adult value at 20 weeks of age.

In the placenta, O_2 must cross from the mother's blood to the fetus's blood by diffusion through tissues separating the two circulatory systems. The detailed way in which this occurs remains a topic of active research. The basic options for the mode of gas transfer are countercurrent gas exchange, cross-current gas exchange, and cocurrent (concurrent) gas exchange—the same options we discussed in Chapter 23 (pages 602–603) for the transfer of O_2 between fluid streams. The mode of gas transfer in the human placenta remains uncertain for two reasons. First, experiments cannot be done on human fetuses. Second, other

A human fetus obtains oxygen (O_2) by pumping blood along its umbilical cord to the placenta, where its blood picks up O_2 from its mother's blood Hemoglobin plays a major role in the acquisition and transport of O_2 by the fetus. In the mother, O_2 taken up in her lungs combines with hemoglobin in her red blood cells and is carried by blood flow to the placenta in that form. In the placenta, the fetus's hemoglobin combines with O_2 that is released from the mother's hemoglobin. The O_2 combined with fetal hemoglobin is then carried, by the circulation of the fetus's blood, from the placenta into the body of the fetus, where the O_2 is used.

species of placental mammals exhibit such wide diversity in the morphology and physiology of their placentas that researchers are not certain which animal model would best reveal how the human placenta works. Enough is known about placental physiology in several mammalian species to make clear, however, that—contrary to expectation—substantial impediments to O_2 transfer from maternal to fetal blood often exist in mammalian placentas.[1] Placentas are emphatically not like lungs, in which high rates of air and blood flow and minutely thin intervening membranes result readily in dramatic blood oxygenation.

An important reason a human fetus can in fact obtain enough O_2 from its placenta is that the fetus produces a different molecular form of hemoglobin from the one its mother produces. This is also true in many other species of placental mammals that have been studied. For reasons we discuss principally later in this chapter, fetal hemoglobin has a higher affinity for O_2 than adult hemoglobin does. This greater affinity has two important, interrelated consequences. First, the *difference* in affinity between the maternal and fetal hemoglobins means that O_2 has a chemical tendency to leave the lower-affinity hemoglobin of the mother to bind with the higher-affinity hemoglobin of the fetus. Second, the *high absolute affinity* of fetal hemoglobin means that it can become well oxygenated even if the O_2 partial pressure in the fetal blood remains relatively low, as it typically does.

The hemoglobins are one of several types of **respiratory pigments** or **oxygen-transport pigments** that animals have evolved. The defining property of the respiratory pigments is that they undergo *reversible combination with molecular oxygen* (O_2). Thus they can pick up O_2 in one place, such as the lungs of an adult or the placenta of a fetus, and release the O_2 in another place, such as the systemic tissues[2] of the adult or fetus. All the types of respiratory pigments are *metalloproteins*: proteins that contain metal atoms, exemplified by the iron in hemoglobin. In addition, all are strongly colored at least some of the time, explaining why they are called *pigments*.

The most straightforward function of the respiratory pigments is to increase the amount of O_2 that can be carried by a unit of volume of blood. Although O_2 dissolves in the blood plasma[3] just as it dissolves in any aqueous solution (see Chapter 22), the solubility of O_2 in aqueous solutions is relatively low, meaning that the amount of O_2 that can be carried in dissolved form per unit of volume is not high. When a respiratory pigment is present in the blood, however, the blood can carry O_2 in two ways: in chemical combination with the pigment as well as in simple solution. Therefore a respiratory pigment increases the **oxygen-carrying capacity** of blood, meaning the total amount of O_2 that can be carried by each unit of volume. In some cases, the increase is very large. For example, when the blood of an adult person leaves the lungs, it contains

almost 200 mL of O_2 per liter of blood in chemical combination with hemoglobin and about 4 mL of O_2 per liter in solution. Thus the blood's concentration of O_2 is increased about 50-fold by the presence of hemoglobin. This means, among other things, that the heart can work far less intensely; roughly calculated, the circulation of 1 L of actual human blood delivers the same amount of O_2 as would the circulation of 50 L of blood without hemoglobin.

Multiple molecular forms of hemoglobin occur, as already exemplified by the contrast between fetal and maternal hemoglobins. Not only may one species have multiple molecular forms, but different species have different forms. Thus the word *hemoglobin* refers to a family of many compounds, not just a single compound. To emphasize this fact, we refer to these compounds as *hemoglobins* (plural) rather than just *hemoglobin* (singular). All the hemoglobins—plus a great diversity of other globin proteins—are coded by genes of a single ancient gene family. Natural selection and other processes have modified the genes in this gene family over evolutionary time, giving rise to the great diversity of hemoglobins and other globin proteins in modern organisms.

Hemoglobins have several functions; that is, their functions are not limited just to increasing the blood's oxygen-carrying capacity. Blood hemoglobins, for example, play important roles as *buffers* and participate in blood CO_2 *transport* as well as O_2 transport. Moreover, specialized hemoglobins are found *within* muscle cells or nerve cells (neurons), where they often *facilitate diffusion of O_2* into the cells and potentially serve as *intracellular storage depots for O_2*. Fast-breaking research indicates also that hemoglobins within some muscle cells serve in intricate ways both to *synthesize and break down intracellular nitric oxide* (*NO*), a compound that potently *controls mitochondrial respiration* (mitochondrial O_2 consumption and ATP production) in the muscle cells. This chapter emphasizes the role of hemoglobins in blood O_2 transport but touches on the other functions as well.

One could aptly say that a revolution is currently underway in the study of the respiratory pigments. The driving forces in this revolution are molecular sequencing, genomics, applications of advanced chemical analysis, and phylogenetic reconstruction. For instance, because of the availability of relatively cheap molecular sequencing tools, it is becoming routine—as it has not been before—to know the entire amino acid sequences of respiratory-pigment molecules that are being compared. Genomics facilitates the widespread search for respiratory-pigment molecules and has led to the discovery of new ones.

A final introductory point worth noting is that when hemoglobins or other respiratory pigments combine with O_2, they are said to be **oxygenated**, and when they release O_2, they are **deoxygenated**. They are not said to be *oxidized* and *reduced*. The reason for these distinctions is that the process by which a respiratory pigment combines with O_2 is not chemically equivalent to oxidation. During the oxygenation of a hemoglobin molecule, for example, although electrons are partially transferred from iron atoms in the hemoglobin molecule to the O_2, the transfer is not complete, as it would be in full-fledged oxidation. In fact, if a hemoglobin molecule accidentally becomes truly oxidized (so that its iron atoms are converted from their ordinary ferrous state to the ferric state), the molecule (now called *methemoglobin*) loses its ability to combine with O_2! The prefixes **oxy-** and **deoxy-** are used to specify the oxygenated and deoxygenated states of respiratory-pigment molecules. Hemoglobin,

---

[1] The placental O_2 partial pressure is strikingly low during the first trimester of human intrauterine development, for example. To explain this unexpected state, one hypothesis is that the low partial pressure helps limit formation of reactive oxygen species (see Box 8.1), which might be particularly damaging to the early developmental stages.

[2] The **systemic tissues** are all tissues other than the tissues of the breathing organs.

[3] The **plasma** of the blood is the aqueous solution in which the cells are suspended. Operationally, plasma is obtained by removing all cells from blood (e.g., by centrifugation).

BOX 24.1 Absorption Spectra of Respiratory Pigments

The hemoglobins and other respiratory pigments—like all pigments—differentially absorb various wavelengths of light. The pattern of absorption by a pigment when expressed as a function of wavelength, is known as an **absorption spectrum** (plural *spectra*). The absorption spectrum of a specific respiratory pigment (e.g., human hemoglobin) changes with the oxygenation or deoxygenation of the pigment, as shown in the accompanying figure. These changes are qualitatively evident to our eyes: We know, for example, that oxygenated hemoglobin (bright red) differs in color from deoxygenated hemoglobin (purple-red). By using quantitative light-absorption measurements, the percentage of heme groups that are oxygenated in blood can be determined. This is the principle behind the finger probes—known as *pulse oximeters*—that are used to monitor arterial blood oxygenation in

hospital patients. **Box Extension 24.1** explains how a pulse oximeter measures the percentage of oxygenated heme groups in arterial blood and why it is called a "pulse" oximeter.

Absorption spectra for fully oxygenated and fully deoxygenated human hemoglobin To measure absorption, light of each wavelength is passed through a hemoglobin solution of defined concentration and optical path length (in the case shown here, the concentration was 1 mM, and the light path through the solution was 1 cm long). The fraction of the incoming photon energy that fails to pass through the solution is measured. From the data, one calculates the *extinction coefficient*, which is a measure of the absorption of the light by the hemoglobin: A high extinction coefficient signifies high absorption. (After Waterman 1978.)

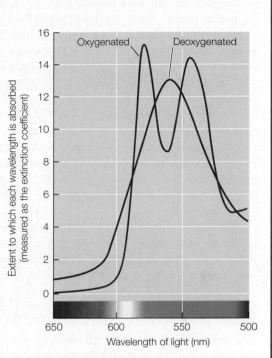

---

for example, is called **oxyhemoglobin** when it is combined with O_2 (oxygenated) and **deoxyhemoglobin** when it is not combined with O_2 (deoxygenated). Respiratory pigments change color when they are oxygenated and deoxygenated, and measures of these color changes can be used to monitor the oxygenation and deoxygenation of blood (**BOX 24.1**).

The Chemical Properties and Distributions of the Respiratory Pigments

Four chemical categories of respiratory pigments are recognized: **hemoglobins**, **hemocyanins**, **hemerythrins**, and **chlorocruorins**. The prefix *hemo-* is from the Greek for "blood," explaining its use in the names of three of the pigment categories. Like the hemoglobins, the other categories are groups of related compounds, not single chemical structures.

Many of the important chemical properties of the respiratory pigments resemble the properties of the enzyme proteins we studied in Chapter 2. The parallels are so great, in fact, that biochemists have occasionally dubbed the respiratory pigments "honorary enzymes." The point of mentioning these parallels is not to suggest that respiratory pigments are enzymes; in terms of their principal functions, they are *not*. The point, instead, is to highlight that, based on your knowledge of enzyme proteins, you will find that you already know a great deal about the molecular features of respiratory pigments.

When a hemoglobin molecule, for example, combines with O_2, it does so at *defined binding sites*, resembling the way in which

enzymes combine with their substrates at defined binding sites. Moreover, the combination of the O_2-binding sites with O_2 is *highly specific* and occurs by *noncovalent, weak bonding* (see Box 2.1), just as enzyme–substrate binding is specific and noncovalent. Accordingly, O_2 is a *ligand* of hemoglobin, based on the definition of "ligand" we developed in Chapter 2. When a hemoglobin molecule combines with O_2, it undergoes a change in its molecular conformation (shape) that is analogous to the conformational change an enzyme molecule undergoes when it combines with its substrate; the ability of a hemoglobin molecule to *flex* in this way is an essential attribute of its function, just as molecular flexibility is critical for enzyme function. One of a hemoglobin molecule's most important properties is its *affinity* for O_2, meaning the ease with which it binds with the O_2 molecules it encounters; thus a hemoglobin molecule (like an enzyme) is characterized in part by *how readily it binds* with its primary ligand.

A hemoglobin molecule also has specific sites at which it combines with *ligands other than O_2*. Using the same terminology we used in Chapter 2 in connection with enzymes, such ligands (e.g., H^+ and CO_2) are *allosteric ligands* or *allosteric modulators*, because when they bind with their specific sites on a hemoglobin molecule, they affect the ability of the hemoglobin to bind with its primary ligand, O_2. Allosteric ligands, for example, can potently affect a hemoglobin molecule's affinity for O_2. Within a hemoglobin molecule, just as in an enzyme molecule, allosteric ligands exert their effects *at a distance*; that is, the binding sites for allosteric ligands on a hemoglobin molecule are separate from the O_2-binding sites, and when allosteric ligands affect O_2 binding, they do so by modifying the conformation and flexibility of the molecule as a whole.

Hemogloblin molecules are usually *multisubunit proteins*; that is, each whole molecule consists of two or more proteins bonded together by noncovalent bonds. Multisubunit hemoglobins have an O_2-binding site on each subunit and thus have multiple O_2-binding sites. In common with multisubunit enzymes that exhibit cooperativity among substrate-binding sites, multisubunit hemoglobin molecules exhibit *cooperativity among their O_2-binding sites*, meaning that binding of O_2 to any one site on a molecule affects how readily the other sites bind O_2.[4] These interactions among O_2-binding sites themselves occur at a distance; the various O_2-binding sites on a multisubunit hemoglobin are separate and distinct, and they influence each other by effects that are relayed through the structure of the whole multisubunit molecule, rather than by direct site-to-site effects.

The points we have made using hemoglobin as an example apply to the other categories of respiratory pigments as well. Thus, in the study of all respiratory pigments, it is helpful to keep these points in mind.

Despite sharing many key properties with enzymes, the respiratory pigments differ from enzymes in a major way: *They do not modify their primary ligand*. After they combine with O_2, they later release the same molecule, O_2.

Hemoglobins contain heme and are the most widespread respiratory pigments

The chemical structures of all hemoglobin molecules share two features. First, all hemoglobins contain **heme** (**FIGURE 24.1A**), which is a particular metalloporphyrin containing iron in the ferrous state (ferrous protoporphyrin IX). Second, the heme is noncovalently bonded to a protein known as a **globin** (**FIGURE**

---

[4] In the terminology developed in Chapter 2, this is *homotropic* cooperativity. See page 49 for more on cooperativity within multisubunit proteins.

24.1B). The combination of heme with globin accounts for the name *hemoglobin*. Oxygen binds at the heme site at a ratio of one O_2 molecule per heme. In all hemoglobin molecules, the heme is identical. The multiple molecular forms of hemoglobin differ in their protein (globin) structures (and in the numbers of unit molecules of hemoglobin that are linked together).

Biochemical studies reveal that small changes in the *protein structure* of a hemoglobin molecule can cause highly significant alterations in the *functional properties* of the molecule. There are, to illustrate, more than 100 known mutant forms of human hemoglobin. Each human globin protein consists of more than 140 amino acids, and most of the mutant forms differ from the normal form in just one of those amino acids. Nonetheless, many mutant forms differ markedly from the normal form in their functional properties; they may differ in their affinity for O_2 or in other key properties, such as their solubility or structural stability.

The blood hemoglobins of vertebrates are almost always four-unit (tetrameric) molecules (**FIGURE 24.1C**) that can bind a total of four O_2 molecules. The molecular weight of each unit molecule is typically about 16,000–17,000 daltons (Da). Thus the four-unit blood hemoglobins have molecular weights of approximately 64,000–68,000 Da. Two types of globins, termed α and β, are found in adult blood hemoglobins. Most biochemists agree that the ancestral genes for the two types originated by gene duplication about 500 million years ago. Each molecule of adult blood hemoglobin consists of two α units and two β units. The human α-globin contains 141 amino acids, whereas the human β-globin contains 146 amino acids. Although other vertebrate species are also described as having α- and β-globins, the particular chemical structures of those globins vary from species to species. Relatively huge hemoglobin molecules are found in the blood of some invertebrates, as we will soon see.

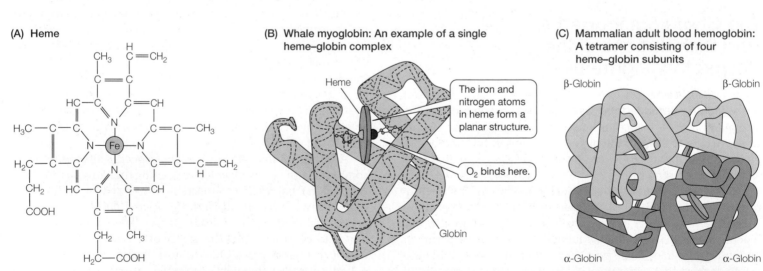

(A) Heme

(B) Whale myoglobin: An example of a single heme–globin complex

Heme

The iron and nitrogen atoms in heme form a planar structure.

O_2 binds here.

Globin

(C) Mammalian adult blood hemoglobin: A tetramer consisting of four heme–globin subunits

β-Globin   β-Globin

α-Globin   α-Globin

FIGURE 24.1 The chemical structure of hemoglobin (A) The structure of heme: Ferrous iron forms a complex with protoporphyrin. The positions assigned to double and single bonds in the porphyrin ring are arbitrary because resonance occurs. (B) A single heme–globin complex. The specific molecule shown is myoglobin (muscle hemoglobin) taken from the muscle of a whale. The structure of the globin protein includes eight segments in which the amino acid backbone of the protein (seen spiraling inside the cylindrical outline) forms a helix. The outer, cylindrical part of the drawing shows the major contours of the globin structure. (C) A tetrameric hemoglobin molecule of the sort found in mammalian red blood cells. In adults, each tetramer consists of two α-globins, two β-globins, and a total of four heme groups. (B and C after Dickerson and Geis 1983.)

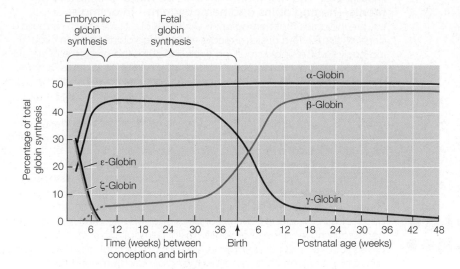

FIGURE 24.2 Human developmental changes in the types of globins synthesized for incorporation into blood hemoglobins Each blood-hemoglobin molecule consists of four globin units plus four O_2-binding heme groups (see Figure 24.1C). In early embryos, the principal globins synthesized are α-globin, ε-globin (epsilon-globin), and ζ-globin (zeta-globin); *embryonic hemoglobins* are made up principally of these globins (e.g., one common form consists of two α-globin and two ε-globin units). By about 8 weeks after conception, α-globin and γ-globin (gamma-globin) are the principal globins synthesized, and most hemoglobin molecules are *fetal hemoglobin* molecules consisting of two α-globin and two γ-globin units. Following birth, synthesis of γ-globin ebbs, whereas that of β-globin increases, so that by 20 weeks of postnatal age, the blood hemoglobin is predominantly *adult hemoglobin* consisting of α- and β-globins. The dashed part of the β-globin curve is postulated rather than empirical. (After Wood 1976.)

Animals sometimes have hemoglobins inside muscle cells or inside the cells of other tissues besides blood. Such hemoglobins typically differ from blood hemoglobins in their chemical structure. The muscle hemoglobins, termed **myoglobins** (*myo-*, "muscle"), of vertebrates provide apt examples. Located in the cytoplasm of muscle fibers (muscle cells), they tend to be especially abundant in cardiac muscle fibers and in the slow oxidative (SO) class of skeletal muscle fibers (see page 202). When present at high concentrations, they impart a reddish color to the tissue; "red" muscles are red because of myoglobins. Unlike blood hemoglobins, vertebrate myoglobins appear always to be single-unit (monomeric) molecules (see Figure 24.1B). They also have distinctive globins. In adult humans, for example, the globin of myoglobin is of different structure than the α- or β-globins.

In addition to varying spatially—from tissue to tissue—within an animal, the chemical nature of hemoglobin often also changes temporally over the life cycle. For example, as already mentioned, the blood hemoglobin of fetal mammals is often different from that of the adults of their species. In humans, fetuses synthesize α-globin (as adults do) and γ-globin (gamma-globin), which differs from the β-globin synthesized by adults (**FIGURE 24.2**); each fetal blood-hemoglobin molecule consists of two α-globin and two γ-globin subunits. Earlier in development, as Figure 24.2 shows, still different globins are expressed.

Although the adult blood hemoglobin of many animals (e.g., humans and most other mammals) is of essentially uniform composition, in many species of poikilothermic vertebrates and invertebrates, the blood of adults normally consists of mixes of two, three, or even ten or more chemically different forms of hemoglobin. A relatively simple example is provided by the blood hemoglobin of the sucker fish *Catostomus clarkii*, which consists of about 80% of one major type of hemoglobin and 20% of another. When multiple chemical forms of hemoglobin occur in a species, the forms sometimes differ substantially in their O_2-binding characteristics. Possession of multiple blood hemoglobins may thus permit a species to maintain adequate O_2 transport over a broader range of conditions than would be possible with only a single hemoglobin type.

THE DISTRIBUTION OF HEMOGLOBINS Hemoglobins are the most widely distributed of the respiratory pigments, being found in at least nine phyla of animals (**FIGURE 24.3**) and even in some protists and plants. They are the only respiratory pigments found in vertebrates, and with a few interesting exceptions (see Chapter 3), all vertebrates have hemoglobin in their blood. The blood hemoglobins of vertebrates are always contained in specialized cells, the **red blood cells** (**erythrocytes**), discussed in **BOX 24.2**.

Among the invertebrates, the distribution of hemoglobins is not only wide but sporadic. Hemoglobins may occur within certain subgroups of a phylum but not others, and even within certain species but not other closely related species. Sometimes, among all the members of a large assemblage of related species, only an isolated few possess hemoglobins. The evolution of the wide but sporadic distribution of hemoglobins certainly provokes curiosity. According to the prevailing view at present, the hemoglobin gene family originated even before animals did, and therefore genes of the family are potentially present in all evolutionary lines of animals. The genes are sometimes fully functional and expressed in modern animals, and sometimes not—accounting for the hemoglobin distribution observed.

The circulating hemoglobins of invertebrates may be found in blood, or they may occur in other moving fluids, such as coelomic fluids. Sometimes, as in vertebrates, these hemoglobins are contained within cells and thus categorized as *intracellular*. The intracellular hemoglobins of invertebrates are always of relatively low molecular weight (~14,000–70,000 Da); structurally, they are generally one-, two-, or four-unit molecules. By contrast, the blood hemoglobins of some invertebrates are dissolved in the blood plasma and thus categorized as *extracellular*. Earthworms (*Lumbricus*), for example, have hemoglobin dissolved in their blood plasma, which when held to the light is wine red and clear—quite unlike vertebrate bloods, which are opaque because of their high concentrations of red blood cells. The extracellular, dissolved hemoglobins of invertebrates are—almost always—relatively huge, multiunit molecules, having molecular weights of 0.2–12 million Da. There are 144 O_2-binding sites *in each molecule* of earthworm (*Lumbricus*) hemoglobin!

The concentration of blood hemoglobin in some invertebrates changes so dramatically from one environment to another that the animals change color. Water fleas (*Daphnia*), for example, have little hemoglobin and are pale when they have been living in O_2-rich waters. However, if they are placed in O_2-poor waters, they increase their levels of hemoglobin within days and become bright red (see Figure 24.20).

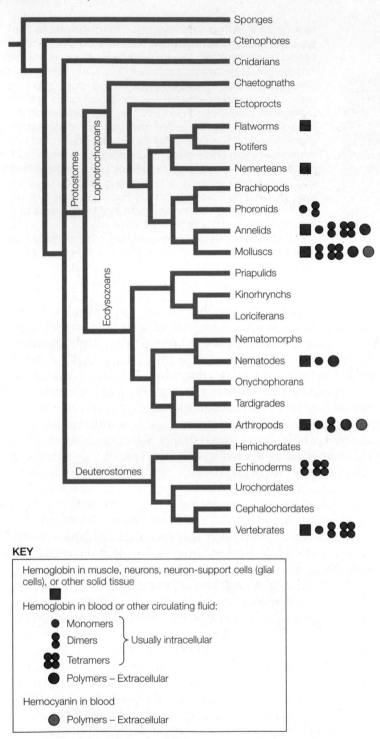

KEY

Hemoglobin in muscle, neurons, neuron-support cells (glial cells), or other solid tissue

■

Hemoglobin in blood or other circulating fluid:

● Monomers
⦁ Dimers ⎫ Usually intracellular
✖ Tetramers ⎭

● Polymers – Extracellular

Hemocyanin in blood

● Polymers – Extracellular

FIGURE 24.3 The distribution of the two major respiratory pigments—hemoglobins and hemocyanins—in animals A red square indicates that hemoglobins occur in solid tissues such as muscle or nerve tissues. Red circles indicate that hemoglobins occur in circulating body fluids. A single small red circle indicates the presence in circulating body fluids of hemoglobins consisting of one unit molecule of heme plus globin: hemoglobin monomers. A pair or foursome of small red circles symbolizes hemoglobin dimers or tetramers, respectively. A large red circle indicates polymeric hemoglobins of high molecular weight, consisting of many joined unit molecules. The polymeric hemoglobins are always extracellular—dissolved in the circulating fluid. The monomeric, dimeric, and tetrameric forms of hemoglobin, with few exceptions, are intracellular—contained within circulating cells such as erythrocytes. A large blue circle indicates polymeric hemocyanins of high molecular weight dissolved in the blood; this is the only circumstance in which hemocyanins occur. In each group labeled here as having hemoglobin or hemocyanin, not all species in the group have it; nor does each species with hemoglobin necessarily have all the chemical forms of hemoglobin shown. In vertebrates, for example, although most have blood hemoglobin, icefish do not (see Chapter 3); and the monomeric and dimeric forms of blood hemoglobin occur only in cyclostome fish, with tetrameric forms being found in all other vertebrates that have blood hemoglobin. Similarly, whereas some arthropods and molluscs have hemoglobins as symbolized here, the majority lack them. This summary is not exhaustive, and it assumes echiurids are annelids. (Hemoglobin data from Terwilliger 1980; cladogram after Sadava et al. 2008.)

Copper-based hemocyanins occur in many arthropods and molluscs

Hemocyanins are found in just two phyla—the arthropods and the molluscs (see Figure 24.3)—but clearly rank as the second most common class of respiratory pigments. In turning to the hemocyanins, we encounter a minor problem that they share with the chlorocruorins and hemerythrins: The names given to these compounds provide no clue to their chemical structures. Hemocyanins do not contain heme, iron, or porphyrin structures. The metal they contain is copper, bound directly to the protein. The arthropod and mollusc hemocyanins exhibit consistent structural differences and are clearly of separate evolutionary origin. Thus they are distinguished as *arthropod hemocyanins* and *mollusc hemocyanins*. Each O_2-binding site of a hemocyanin contains two copper atoms; thus the binding ratio is one O_2 molecule per two Cu. In both phyla, hemocyanins are invariably found dissolved in the blood plasma, not in cells, and are typically large molecules (4–9 million Da in molluscs, 0.5–3 million Da in arthropods) that have numerous O_2-binding sites. The number of binding sites per molecule is as high as 160 in some cases. Although hemocyanins are colorless when deoxygenated, they turn bright blue when oxygenated. Species that have high concentrations of hemocyanins are dramatically blue-blooded!

The molluscs that possess hemocyanins include the squids and octopuses (cephalopods), many chitons and gastropods (snails and slugs), and a relatively small subset of bivalves. Hemocyanins are not present in most bivalve molluscs (clams, scallops, and the like); indeed, bivalves usually lack circulating respiratory pigments of any kind. Groups of arthropods in which hemocyanins are important include the decapod crustaceans (crabs, lobsters, shrimps, and crayfish), the horseshoe crabs, and the spiders and scorpions. Even some (relatively primitive) insects have recently been discovered to have hemocyanins. Hemocyanins are never found within muscle or other solid tissues. Certain molluscs that have blood hemocyanins have hemoglobins in their muscles, neurons, or gills.

Hemoglobins—usually single-unit molecules—are found widely in solid tissues of invertebrates, not only in muscles but also in certain other tissues. Both muscle and nerve hemoglobins occur, for example, in a wide variety of molluscs and annelids; the nerve hemoglobins may be present in neurons per se or in support cells (glial cells), but either way, they sometimes impart a striking pinkish or red color to the ganglia or nerves. Although insects usually lack circulating respiratory pigments, large numbers of insect species have hemoglobins in the fat body or parts of the tracheal system. Such hemoglobins in some backswimmer bugs, for example, store O_2 for release to the tracheae during diving.

BOX 24.2   Blood Cells and Their Production

The red blood cells (erythrocytes, RBCs) of vertebrates vary in size, shape, and other properties. Mammals have relatively small RBCs, usually 4–10 μm in diameter; human RBCs, for example, average 7.4 μm. Some other groups of vertebrates have distinctly larger RBCs; the oval RBCs of frogs and toads, for example, average 23 × 14 μm in their major dimensions. A significant difference between the RBCs of mammals and those of other vertebrates is that the mature RBCs of mammals are essentially devoid of cell organelles; they have no nucleus, mitochondria, or ribosomes. The RBCs of all other vertebrates, and all the respiratory pigment–containing blood cells of invertebrates that have been studied, are nucleated, emphasizing how distinctive mammalian RBCs are in this regard. One way of expressing the blood's content of RBCs is as the **hematocrit**, defined to be the percentage of total blood volume occupied by the RBCs. The normal hematocrit for people at sea level is 42%–45%.

The process by which the body makes RBCs is called **erythropoiesis**. Because RBCs have relatively short life spans, they are continually being replaced. The rate of turnover seems to be particularly rapid in mammals, possibly because mature mammalian RBCs—lacking a nucleus and ribosomes—lack any ability to repair proteins. The average human RBC lasts 4 months. We replace almost 1% of our RBCs every day, meaning that we make about 2 million new RBCs per second! The principal site of erythropoiesis in adult mammals is the soft interior of the bones, the **bone marrow**.

Erythropoiesis is under endocrine control. In mammals, the principal control is exerted by a glycoprotein hormone, **erythropoietin**. When low-O_2 conditions are detected in the body, erythropoietin is secreted in increased amounts and accelerates erythropoiesis. The kidneys are the principal site of erythropoietin secretion in adult mammals. Molecular probes for erythropoietin messenger RNA (mRNA) have now established that

erythropoietin is synthesized by secretory cells in *interstitial tissue* located between adjacent nephron tubules in the cortex of the kidneys (see Figure 29.6). Hypoxia—a low level of O_2—in the kidneys causes increased transcription of the erythropoietin gene in the secretory cells, thereby increasing formation of mRNA for synthesis of erythropoietin. The erythropoietin then travels in the blood to the bone marrow, where it stimulates production of RBCs.

Upregulation of the erythropoietin gene by hypoxia is mediated by the transcription factor *hypoxia-inducible factor 1* (*HIF-1*), discussed in Chapter 23 (see Figure 23.6). This transcription factor, now known to be enormously significant in responses to hypoxia throughout the animal kingdom, in fact was discovered through studies of the regulation of the erythropoietin gene. HIF-1 plays multiple regulatory roles, including the upregulation of pathways that aid delivery of iron to the bone marrow when erythropoiesis is stimulated.

Chlorocruorins resemble hemoglobins and occur in certain annelids

Chlorocruorins, also sometimes called *green hemoglobins*, occur in just four families of marine annelid worms, including the fan worms and feather-duster worms that are so popular with aquarists. Chlorocruorins are always found extracellularly, dissolved in the blood plasma. They have close chemical similarities to the extracellular hemoglobins found dissolved in the blood plasma of many other annelids. Like the extracellular hemoglobins, they are large molecules, with molecular weights of close to 3 million Da, composed of unit molecules consisting of iron-porphyrin groups conjugated with protein. They bind one O_2 per iron-porphyrin group. The chlorocruorins differ from hemoglobins in the type of iron porphyrin they contain.[5] This difference gives the chlorocruorins a distinctive and dramatic color. In dilute solution, they are greenish. In more concentrated solution, they are deep red by transmitted light but greenish by reflected light.

Iron-based hemerythrins do not contain heme and occur in three or four phyla

Hemerythrins have a distribution that is puzzling because it is both limited and far-flung, encompassing three or four phyla. Cir-

culating hemerythrins occur in a single family of marine annelid worms (the magelonids), in the sipunculid worms (which have been a separate phylum but might be annelids), in many brachiopods (lamp shells), and in some species of the small phylum Priapulida. Despite their name, hemerythrins do not contain heme. They do contain iron (ferrous when deoxygenated), bound directly to the protein. Each O_2-binding site contains two iron atoms, and there is one such site per 13,000–14,000 Da of molecular weight. In some instances (including, for example, some annelids that lack circulating hemerythrin), single-unit hemerythrins, known as myohemerythrins, occur within muscle cells. Better known are the circulating hemerythrins, which are always located intracellularly, in blood or coelomic cells, and typically have molecular weights of 40,000–110,000 Da; many are octomers, having eight O_2-binding sites per molecule. Hemerythrins are colorless when deoxygenated but turn reddish violet when oxygenated.

SUMMARY
The Chemical Properties and Distributions of the Respiratory Pigments

■ The four chemical classes of respiratory pigments are all metalloproteins. They bind reversibly with O_2 at specific O_2-binding sites associated with the metal atoms in their molecular structures.

(Continued)

---

[5] The porphyrin differs from heme in that one of the vinyl chains (—CH=CH$_2$) on the periphery of the protoporphyrin ring in heme (see Figure 24.1A) is replaced with a formyl group (—CHO).

- In hemoglobins, the unit molecule consists of heme bonded with protein (globin). The heme structure—an iron (ferrous) porphyrin—is identical in all hemoglobins. The globin, however, varies widely among species and among different molecular forms of hemoglobin within any single species.

- Hemoglobins are the most common and widespread respiratory pigments, occurring in at least nine phyla. Virtually all vertebrates have blood hemoglobin. The blood-hemoglobin molecules of vertebrates are usually tetramers consisting (in adults) of two α-globin and two β-globin unit molecules; they always occur in red blood cells. Although many invertebrates also have hemoglobins in blood cells, some invertebrates have hemoglobins dissolved in their blood plasma.

- Hemocyanins are the second most common of the respiratory pigments in animals. They contain copper and turn bright blue when oxygenated. There are two types of hemocyanins, which are of separate evolutionary origin: arthropod hemocyanins (occurring in crabs, lobsters, crayfish, horseshoe crabs, spiders, and some other arthropods) and mollusc hemocyanins (occurring in squids, octopuses, many snails, and some other molluscs). Hemocyanins are always dissolved in the blood plasma.

- Chlorocruorins, which are similar to hemoglobins, occur in only four families of marine annelid worms, and are always dissolved in the blood plasma.

- Hemerythrins are non-heme, iron-containing respiratory pigments that have a limited and scattered distribution, occurring in three or four different invertebrate phyla.

The O₂-Binding Characteristics of Respiratory Pigments

A key tool for understanding the function of a respiratory pigment is the *oxygen equilibrium curve*. In a body fluid containing a respiratory pigment, there is a large population of O_2-binding sites. Human blood, for example, contains about 5.4×10^{20} heme groups per 100 mL. The combination of O_2 with each individual O_2-binding site is *stoichiometric*: One and only one O_2 molecule can bind, for example, with each heme group of a hemoglobin or with each pair of copper atoms in a hemocyanin. However, in blood, where there are great numbers of O_2-binding sites, all sites do not simultaneously bind with O_2 or release O_2. Instead, the fraction of the O_2-binding sites that are oxygenated varies in a *graded* manner with the O_2 partial pressure. The **oxygen equilibrium curve**, also sometimes called the **oxygen dissociation curve**, shows the functional relation between the percentage of binding sites that are oxygenated and the O_2 partial pressure (**FIGURE 24.4A**).

The respiratory pigment in the blood of an animal is said to be **saturated** if the O_2 partial pressure is high enough for all O_2-binding sites to be oxygenated. Accordingly, the percentage of binding sites that are oxygenated is often termed the **percent saturation**. The blood's *oxygen-carrying capacity*, an important property mentioned earlier, is the amount of O_2 carried per unit of volume at saturation. Ordinary human blood, for example, has an oxygen-carrying capacity of about 20 mL O_2 per 100 mL of blood

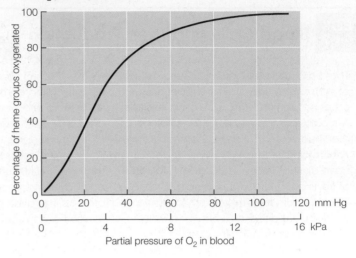

(A) Percentage of heme groups oxygenated as a function of O_2 partial pressure

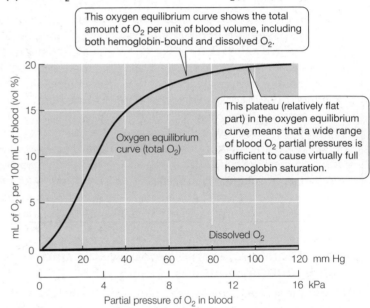

(B) Blood O_2 concentration as a function of O_2 partial pressure

This oxygen equilibrium curve shows the total amount of O_2 per unit of blood volume, including both hemoglobin-bound and dissolved O_2.

This plateau (relatively flat part) in the oxygen equilibrium curve means that a wide range of blood O_2 partial pressures is sufficient to cause virtually full hemoglobin saturation.

Oxygen equilibrium curve (total O_2)

Dissolved O_2

FIGURE 24.4 A typical oxygen equilibrium curve for human arterial blood presented in two different ways (A) The percentage of heme groups oxygenated as a function of the O_2 partial pressure. (B) The total blood O_2 concentration—including both hemoglobin-bound and dissolved O_2—as a function of the O_2 partial pressure; the portion of the total O_2 present as dissolved O_2 is plotted at the bottom. Normal arterial values of CO_2 partial pressure, pH, and temperature are assumed. In humans, as in other animals, significant individual variation occurs. Partial pressure is shown in two systems of units, the SI system (kPa) and a traditional system (mm Hg)—as will often be the case in this chapter. (After Roughton 1964; B assumes an O_2 concentration of 20 vol % at 16 kPa.)

and actually has that concentration of O_2 when saturated.[6] The volume of gas carried per 100 volumes of blood is often termed the **volumes percent (vol %)** of gas. In that system of units, the oxygen-carrying capacity of ordinary human blood is 20 vol %.

---

[6] Gas volumes are always expressed at standard conditions of temperature and pressure (see Appendix C) unless otherwise stated. Such volumes are proportional to molar quantities, as discussed in Chapter 22.

The oxygen equilibrium curve can be presented in two ways. Figure 24.4A—showing the *percentage of oxygenated binding sites* (the *percent saturation*) as a function of O_2 partial pressure—exemplifies one of these. The alternative presentation, seen in **FIGURE 24.4B**, shows the *blood O_2 concentration* as a function of the O_2 partial pressure. To calculate this alternative form of the curve from the first form, one needs merely to convert the percentage of oxygenated binding sites at each partial pressure into the corresponding blood O_2 concentration. For most purposes, this conversion can be carried out by use of the oxygen-carrying capacity: The O_2 concentration at each partial pressure is the oxygen-carrying capacity multiplied by the percentage of oxygenated binding sites.[7]

Because O_2 *dissolves* in the blood plasma, blood in fact contains O_2 in two forms: dissolved and bound to the respiratory pigment. The amount of dissolved O_2 per unit of blood volume simply follows the principles of gas solution (see Equation 22.3). Therefore it is proportional to the O_2 partial pressure, producing a straight-line relation as seen at the bottom in Figure 24.4B. In humans and most other vertebrates, dissolved O_2 accounts for just a small fraction of all O_2 in the blood (see Figure 24.4B).

Later in this chapter, we will see that the O_2-binding properties of respiratory pigments are often affected by temperature, pH, and other properties of the blood chemical environment. We will also discuss the reason for the sigmoid shape of the oxygen equilibrium curve (see Figure 24.4). Before we consider those factors, however, it is important to understand the basic principles of respiratory-pigment function in living animals and to appreciate the interpretive value of oxygen equilibrium curves. To these ends, and recognizing that refinements will later be needed, let's look at the fundamentals of O_2 transport by our own blood.

Human O_2 transport provides an instructive case study

To understand the uptake of O_2 by the blood in a person's lungs, it is important to recall from Chapter 23 that breathing maintains the O_2 partial pressure in the alveolar gases of our lungs at about 13.3 kPa (100 mm Hg). Blood arriving at the alveoli has a lower O_2 partial pressure. Thus O_2 diffuses into the blood from the alveolar gas, raising the blood O_2 partial pressure as the blood passes through the lungs. As the blood O_2 partial pressure rises, at each partial pressure the hemoglobin in the blood takes up the amount of O_2 that is dictated by its oxygen equilibrium curve (see Figure 24.4).[8]

If, in the lungs, the O_2 partial pressure of the blood were to rise to the alveolar partial pressure, 13.3 kPa (100 mm Hg), we can see from Figure 24.4 that the hemoglobin in the blood would become virtually saturated with O_2. In fact, mixed blood leaving the lungs is at a somewhat lower O_2 partial pressure: 12.0–12.7 kPa (90–95 mm Hg) in a person at rest. This lower partial pressure hardly affects the blood O_2 content, however, because as shown by the oxygen equilibrium curve, there is a plateau in the relation between the blood O_2 concentration and the O_2 partial pressure at these high

partial pressures; provided the blood O_2 partial pressure is high enough to be in the plateau region, hemoglobin will be almost saturated with O_2 regardless of the partial pressure. The alveolar O_2 partial pressure could even vary a bit, and still, because of the plateau—a property of the hemoglobin—the blood leaving the lungs would remain almost entirely saturated. The close "matching" of the saturation partial pressure of hemoglobin and the alveolar partial pressure represents a striking evolutionary coadaptation: *The hemoglobin molecule has evolved O_2-binding properties that suit it to oxygenate well at the O_2 partial pressures maintained in the lungs by the breathing system.*

After leaving the lungs, blood flows to the left side of the heart and is pumped to the systemic tissues. To understand the events in the systemic tissues, it is crucial to recall that in the mitochondria, O_2 is continually being combined with electrons and protons to form H_2O. By this process, O_2 molecules are removed from solution, and the O_2 partial pressure in and around the mitochondria is lowered. Blood arriving in capillaries of systemic tissues from the lungs has a high O_2 partial pressure; O_2 thus diffuses from the blood to the mitochondria (see Figure 22.7). During this diffusion, dissolved O_2 leaves the blood, and the O_2 partial pressure of the blood falls. As this occurs, hemoglobin releases (unloads) O_2, thereby making hemoglobin-bound O_2 available to diffuse to the mitochondria. The oxygen equilibrium curve (see Figure 24.4) is a key to understanding the unloading of O_2 from hemoglobin: As the blood O_2 partial pressure falls, the amount of O_2 released from hemoglobin at each O_2 partial pressure is dictated by the curve.

Knowing that hemoglobin leaves the lungs in a virtually saturated condition, we can calculate its yield of O_2 to the systemic tissues by obtaining a measure of its degree of saturation after it has passed through the systemic tissues. The simplest way to obtain this measure is to determine the degree of saturation in blood drawn from the great veins leading back to the heart; such blood is termed **mixed venous blood** because it represents a mixture of the venous blood coming from all parts of the body. In people at rest, the O_2 partial pressure of mixed venous blood is about 5.3 kPa (40 mm Hg). From the oxygen equilibrium curve (see Figure 24.4B), we can see that blood at this partial pressure contains about 15 mL of O_2/100 mL. Recalling that arterial blood contains about 20 mL of O_2/100 mL, we see that the O_2 content of the blood falls by about 5 mL of O_2/100 mL when the blood circulates through the systemic tissues in humans at rest. In other words, as shown by the "Rest" arrow in **FIGURE 24.5**, about 5 mL of O_2 is released from each 100 mL of blood. The release of O_2 from the blood is often expressed as the **blood oxygen utilization coefficient**, defined to be the *percentage* of arterial O_2 that is released to the systemic tissues. In people at rest, recognizing that arterial blood contains about 20 mL of O_2/100 mL and that about 5 mL of O_2/100 mL is released to the tissues, the oxygen utilization coefficient is about 25%. That is, only one-fourth of the O_2 brought to the systemic tissues in the arterial blood is actually used at rest.

THE SIGNIFICANCE OF MIXED VENOUS O_2 PARTIAL PRESSURE The O_2 partial pressure of mixed venous blood represents an average of the O_2 partial pressures of blood leaving the various systemic tissues. It thus allows us to gauge the blood's *overall* drop

---

[7] For exacting work, the dissolved O_2, discussed in the next paragraph, has to be calculated separately from the pigment-bound O_2 and the two amounts added.

[8] Hemoglobin also plays an important role in *speeding* the uptake of O_2 by the blood, as discussed in Chapter 22 (see page 592).

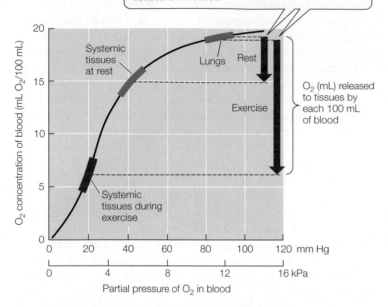

Arrows show the drop in blood O_2 concentration as blood from the lungs flows through the systemic tissues. Each 100 mL of blood yields much more O_2 during vigorous exercise (right arrow) than during rest (left arrow), because oxygenation in the lungs remains similar but deoxygenation in the systemic tissues is increased.

FIGURE 24.5 Oxygen delivery by human blood at rest and during vigorous exercise The oxygen equilibrium curve shown is that for human arterial blood (see Figure 24.4B). The thickened, shaded areas on the curve show representative ranges of blood O_2 concentration and O_2 partial pressure in the lungs (blue), the systemic tissues during rest (green), and the systemic tissues during vigorous exercise (red). The vertical purple arrows to the right show how much O_2 is delivered to the tissues by each 100 mL of blood during rest and exercise. All values are semi-quantitative; the intent of this diagram is conceptual rather than literal. Tissue values are mixed venous blood values. Effects of pH and other variables of the blood-hemoglobin milieu are not included.

in O_2 partial pressure during circulation through all tissues combined. It does not necessarily reflect, however, the drop in partial pressure as the blood flows through any particular tissue; blood entering a particular tissue at a partial pressure of 12.7 kPa (95 mm Hg) might exit at a partial pressure that is either higher or lower than the mixed venous partial pressure. The mixed venous partial pressure is, in fact, a *weighted* average of the O_2 partial pressures of blood leaving the various tissues. It is weighted according to the rate of blood flow through each tissue; tissues with high rates of blood flow influence the mixed venous partial pressure more than those with low rates of flow do.

THE DETERMINANTS OF A TISSUE'S VENOUS O_2 PARTIAL PRESSURE The O_2 partial pressure to which the blood falls in its passage through a particular tissue is not a static property of that tissue. Instead, it is a dynamic and changing property. It depends on the rate of blood flow through the tissue, the arterial O_2 partial pressure, the amount of hemoglobin per unit of blood volume, and the tissue's rate of O_2 consumption. To illustrate, if the rate of blood flow through a tissue decreases while all the other relevant factors remain unchanged, each unit of blood volume will have

to yield more O_2 in its passage through the tissue, and therefore the venous partial pressure will decline. Similarly, an increase in the rate of O_2 consumption by a tissue will cause a decrease in the tissue's venous partial pressure. The venous partial pressures normally seen in people at rest result from the set of conditions that ordinarily prevail at rest.

DELIVERY OF O_2 DURING EXERCISE As discussed in Chapter 23, controls on breathing tend to keep the alveolar O_2 partial pressure stable, near 13.3 kPa (100 mm Hg), as people exercise more and more intensely. During vigorous exercise, however, the *blood* O_2 partial pressure reached in the lungs tends to decline, compared with the resting blood partial pressure; this decline occurs in part because blood passes through the pulmonary circulation faster as the intensity of exercise increases, thus decreasing the time available for equilibration between the alveolar gas and blood. The relative flatness of the oxygen equilibrium curve at high O_2 partial pressures again comes to the rescue (see Figure 24.4). Even if the blood passing through the lungs reaches a partial pressure of only 11 kPa (80 mm Hg)—which is often the case during intense exercise—the O_2 *content* of the arterial blood is hardly affected. For simplicity, we treat the O_2 content of the arterial blood as a constant as we discuss exercise in more detail.

The modest utilization of blood O_2 at rest leaves a large margin to increase utilization during exercise. As we have seen, only about 25% of the O_2 carried by the systemic arterial blood is used when people are at rest. The remaining amount, the amount of O_2 in mixed venous blood, is called the **venous reserve**. During exercise, more O_2 is withdrawn from each unit of blood volume as the blood passes through the systemic tissues, and the venous reserve becomes smaller.

A highly significant attribute of blood O_2 transport during rest is that the resting mixed venous O_2 partial pressure, averaging 5.3 kPa (40 mm Hg), is low enough to be *below the plateau* displayed by the oxygen equilibrium curve at high partial pressures (see the green-shaded part of the curve in Figure 24.5). During exercise, therefore, when the venous O_2 partial pressure declines below its resting value, it does so on the *steep* part of the oxygen equilibrium curve (**FIGURE 24.6**). Consequently, relatively small decreases in the venous O_2 partial pressure result in relatively large increases in the yield of O_2 from the blood. This point is illustrated in Figure 24.6. For discussing the figure, recall first that at rest, a drop in partial pressure from an arterial value of 12.0–12.7 kPa (90–95 mm Hg) to the resting venous value of 5.3 kPa (40 mm Hg)—a total drop of 6.7–7.4 kPa (50–55 mm Hg)—causes release of about 5 mL of O_2 from each 100 mL of blood. As the figure shows, a further drop of just 2 kPa (15 mm Hg) to a venous partial pressure of 3.3 kPa (25 mm Hg) causes the blood to release another 5 mL of O_2 from each 100 mL of blood, thus *doubling* the O_2 yield. Moreover, a still further drop of just 1.3 kPa (10 mm Hg) to a venous partial pressure of 2 kPa (15 mm Hg) *triples* the yield of O_2 from the blood! This steep release of O_2 is a consequence of the binding characteristics of the hemoglobin molecule.

How great is the actual O_2 delivery during exercise in mammals? Over a wide range of exercise states, the O_2 partial pressure of blood leaving the working skeletal muscles is about 2.7 kPa (20 mm Hg) in humans and also in several other species on which measurements have been made. At this value, the oxygen utiliza-

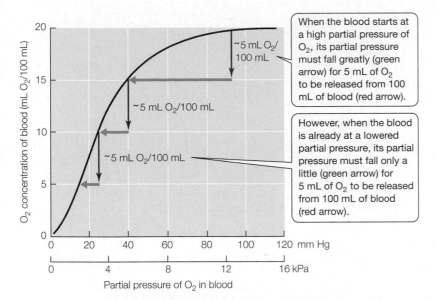

FIGURE 24.6 As the O$_2$ partial pressure of blood falls, less and less of a drop in partial pressure is required to cause unloading of 5 mL of O$_2$ from each 100 mL of blood   Each green horizontal arrow depicts the drop in O$_2$ partial pressure required to cause the unloading depicted by the red vertical arrow to its right. The oxygen equilibrium curve shown is that for human arterial blood (see Figure 24.4B).

tion coefficient for blood flowing through the working skeletal muscles is about 65% (as compared with 25% at rest). We have said that a partial pressure of about 2.7 kPa (20 mm Hg) prevails in the blood leaving the muscles over a wide *range* of exercise states; that is, even as the muscles work harder and demand more O$_2$, in this range little change occurs in their venous partial pressure or in the amount of O$_2$ they obtain from each unit of blood volume. This stability of venous partial pressure occurs because the *rate of blood flow* to the muscles is adjusted: As the O$_2$ demand of the muscles rises, their rate of blood flow increases in parallel, enabling them to draw O$_2$ from an enhanced volume of blood per unit of time. Of course, the rate of blood flow cannot increase indefinitely. Once it is maximized, further increases in a muscle's intensity of work result in further decreases in the venous O$_2$ partial pressure. Indeed, during extreme exertion, the O$_2$ partial pressure of blood leaving some muscles may fall close to zero, signifying virtually complete deoxygenation of the blood, corresponding to 100% O$_2$ utilization.

As the O$_2$ partial pressure of blood in the systemic capillaries declines, there is a risk that the rate of O$_2$ diffusion from the blood to the mitochondria will become too low to support aerobic catabolism. The venous O$_2$ partial pressure below which aerobic catabolism becomes impaired is known as the **critical venous O$_2$ partial pressure**. It is approximately 1.3 kPa (10 mm Hg) in mammalian muscles. As we have seen, the rate of blood flow through muscles is usually increased sufficiently to maintain the venous O$_2$ partial pressure above this critical level over a wide range of exercise states. Human hemoglobin yields about 90% of its O$_2$ before the venous partial pressure falls below the critical level (see Figure 24.4A). In this respect we see once more that the O$_2$-binding properties of hemoglobin are closely integrated with other physiological features.

Let's now look briefly at *whole-body* O$_2$ utilization and O$_2$ delivery during exercise. Although blood draining active muscles may be rather thoroughly deoxygenated during heavy exercise, the partial pressure of *mixed* venous blood generally does not fall below 2.1–2.7

kPa (16–20 mm Hg) in humans, even during strenuous work, because blood from the exercising muscles mixes in the great veins with blood from other parts of the body in which O$_2$ utilization is not so great. The whole-body oxygen utilization coefficient therefore rises to a peak of about 60%–75% during exercise—indicating that 2.5–3.0 times more O$_2$ is extracted from each volume of blood than is extracted at rest, as illustrated in Figure 24.5. In average young people, the rate of blood circulation can be increased to 4–4.5 times the resting level. These values, taken together, show that the total rate of O$_2$ delivery by the circulatory system can increase to 10–13 times the resting rate. Trained athletes often achieve still higher O$_2$ delivery rates, principally because endurance training increases the rate at which a person's heart can pump blood.

THE "MOLECULAR DESIGN" OF HUMAN HEMOGLOBIN
We have seen in this section that (1) human hemoglobin is nearly saturated at the O$_2$ partial pressures that are maintained in the lungs by breathing; (2) the oxygen equilibrium curve of hemoglobin is nearly flat at pulmonary O$_2$ partial pressures, so that high oxygenation is ensured regardless of variation in pulmonary function; and (3) the oxygen equilibrium curve is shaped in such a way that 90% of the O$_2$ bound to hemoglobin can be released for use at blood partial pressures that are compatible with full mitochondrial function. These functional properties of human hemoglobin are consequences of its chemical structure, and its normal chemical structure is but one of thousands of possible structures. Many physiologists have concluded that the human hemoglobin molecule provides a particularly convincing example of "evolutionary molecular design." Natural selection has produced a molecule with functional properties that are integrated in strikingly harmonious ways with the attributes of the organs that provide O$_2$ to the blood and draw O$_2$ from the blood.

A set of general principles helps elucidate O$_2$ transport by respiratory pigments

From our study of hemoglobin function in people, we can state four key principles that are useful for understanding the function of blood respiratory pigments in general:

1. To determine the extent of pigment oxygenation, ask first: What are the blood O$_2$ partial pressures established in the breathing organs? Then examine the oxygen equilibrium curve to determine the extent of pigment oxygenation at those partial pressures.

2. To determine the extent of pigment deoxygenation in systemic tissues, start by acquiring information on blood O$_2$ partial pressures in those tissues. The mixed venous O$_2$ partial pressure is a useful and easily measured indicator, although one must remember that it does not necessarily provide information on O$_2$ release in any *particular* tissue. After the O$_2$ partial pressure in the systemic tissues has been measured or estimated, examine the oxygen equilibrium curve to determine the extent of pigment deoxygenation in the systemic tissues.

3. To compute circulatory O$_2$ delivery, the rate of blood flow is as important as the yield of O$_2$ per unit of blood volume,

because O_2 delivery is the product of flow rate and O_2 yield per unit of volume. Complexity is introduced by the fact that these two factors are not independent: The rate of blood flow helps to determine the venous O_2 partial pressure and thus the yield of O_2 per unit of blood volume.

4. The function of the O_2-transport system is strongly affected by exercise. Full understanding of the function of an O_2-transport system requires that animals be studied over a range of physiological conditions.

The shape of the oxygen equilibrium curve depends on O_2-binding site cooperativity

What determines the shape of the oxygen equilibrium curve? As we explore this question, vivid parallels to principles we addressed in the study of enzymes will again be evident. In Chapter 2 (see page 46), we saw that when the catalytic sites of a particular enzyme function independently of each other, a hyperbolic relation exists between enzyme activity and substrate concentration; when the sites exhibit cooperativity, however, a sigmoid relation occurs. Similarly, when the O_2-binding sites of a respiratory pigment function independently, the oxygen equilibrium curve is hyperbolic, but when they exhibit cooperativity, a sigmoid curve results.

Hyperbolic oxygen equilibrium curves are exemplified by the vertebrate myoglobins (**FIGURE 24.7A**). The vertebrate (and most invertebrate) myoglobins contain just one O_2-binding site (heme) per molecule. Thus their O_2-binding sites function independently of each other, and the chemical reaction between a myoglobin and O_2 can be written simply as

$$Mb + O_2 \rightleftharpoons MbO_2 \qquad \text{(24.1)}$$

where Mb is a molecule of deoxymyoglobin and MbO_2 is one of oxymyoglobin. According to the principles of mass action (see page 50), increasing the partial pressure (and thus the chemical potential) of O_2 will shift this reaction to the right, increasing myoglobin oxygenation. Mass-action principles applied to such a simple chemical reaction also predict that the fraction of myoglobin molecules oxygenated will increase as a *hyperbolic* function of the O_2 partial pressure, as is observed (see Figure 24.7A).

Because vertebrate blood hemoglobins have four O_2-binding sites within each molecule, the opportunity exists for cooperativity. In fact, positive cooperativity occurs in these hemoglobins: Binding of O_2 at one or two of the O_2-binding sites on a molecule of blood hemoglobin alters the conformation of the molecule in ways that enhance the affinity of the remaining sites for O_2, meaning that a partially oxygenated molecule is more likely than an entirely deoxygenated one to bind additional O_2.[9] The consequence is a sigmoid oxygen equilibrium curve, exhibiting a particularly steep relation between O_2 binding and O_2 partial pressure in the mid-range of O_2 partial pressures. **FIGURE 24.8** presents oxygen equilibrium curves for the blood of 11 animal species, including 6 vertebrates that have four-unit hemoglobins and 5 invertebrates that have high-molecular-weight hemoglobins or hemocyanins

[9] Because the four O_2-binding sites are located within the four different protein subunits of the hemoglobin tetramer, the cooperativity displayed by the tetramer is often termed *subunit interaction*. It used to be termed *heme–heme interaction*, but this term has been dropped because the interaction between the O_2-binding sites is indirect, not directly between one heme and another.

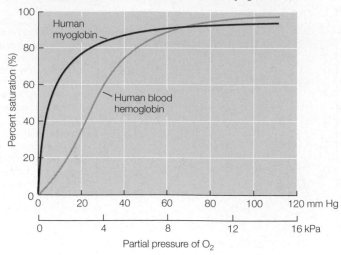

(A) The oxygen equilibrium curve for human myoglobin

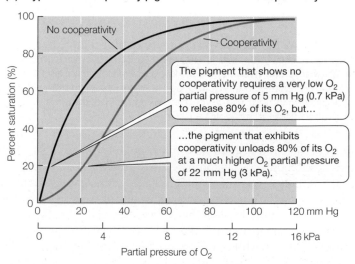

(B) Hypothetical respiratory pigments that differ in cooperativity

The pigment that shows no cooperativity requires a very low O_2 partial pressure of 5 mm Hg (0.7 kPa) to release 80% of its O_2, but…

…the pigment that exhibits cooperativity unloads 80% of its O_2 at a much higher O_2 partial pressure of 22 mm Hg (3 kPa).

FIGURE 24.7 Respiratory pigments display hyperbolic or sigmoid oxygen equilibrium curves depending on whether they exhibit cooperativity in O_2 binding (A) The hyperbolic oxygen equilibrium curve of human myoglobin—a pigment that exhibits no cooperativity—compared with the sigmoid curve of human blood hemoglobin—a pigment that displays cooperativity. Both curves were determined under similar conditions: 38°C, pH 7.40. (B) Comparison of oxygen equilibrium curves for two hypothetical pigments that reach saturation at about the same O_2 partial pressure, but differ in whether they exhibit cooperativity. (A after Roughton 1964.)

with numerous O_2-binding sites. All the curves are sigmoid to some degree, indicating that intramolecular cooperativity occurs in all cases. The extent of cooperativity, which varies from one respiratory pigment to another, is usually expressed using a mathematical index called the **Hill coefficient** (n), named after A. V. Hill (1886–1977), a Nobel laureate. The coefficient is 1.0 for pigments that show no cooperativity (e.g., myoglobins) and reaches 6 or more in some high-molecular-weight pigments with very high cooperativity.[10]

Although in the last paragraph we emphasized the effect of cooperativity on the oxygenation (loading) of respiratory pigments, cooperativity also affects deoxygenation (unloading). During deoxygenation of a molecule that exhibits cooperativity, removal of O_2 from some of the O_2-binding sites tends to decrease the affinity of the remaining sites for O_2, thereby promoting even further deoxygenation.

[10] Mammalian hemoglobins exhibit values of 2.4–3.0.

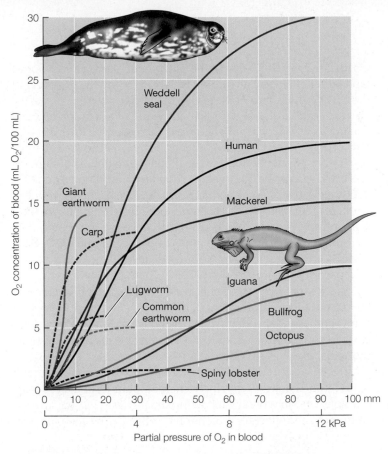

FIGURE 24.8 A diversity of blood oxygen equilibrium curves The blood oxygen equilibrium curves of 11 animal species vary in two ways. First, they vary in shape, a property that reflects the different molecular forms of the respiratory pigments in different species. Second, they vary in height, a property that reflects how much respiratory pigment is present per unit of blood volume (oxygen-carrying capacity). Species: bullfrog, *Rana catesbeiana* (sometimes called *Lithobates catesbeianus*); carp, *Cyprinus carpio*; common earthworm, the nightcrawler *Lumbricus terrestris*; giant earthworm, the 1-m-long South American earthworm *Glossoscolex giganteus*; iguana, *Iguana iguana*; lugworm, the seacoast annelid *Arenicola* sp.; mackerel, *Scomber scombrus*; octopus, the giant octopus *Enteroctopus dofleini* of the North American Pacific coast; spiny lobster, *Panulirus interruptus*; Weddell seal, *Leptonychotes weddelli*. (After Hill and Wyse 1989.)

FIGURE 24.7B shows the consequences,[11] using two hypothetical pigments that are similar in the O_2 partial pressure at which they become saturated, but differ in that one exhibits cooperativity whereas the other does not. If we assume that both pigments are initially fully oxygenated and ask how they behave during deoxygenation, it is clear that in the mid-range of O_2 partial pressures, the pigment showing cooperativity deoxygenates more readily, giving up more of its O_2 at any given O_2 partial pressure. In a sentence, whether a molecule of a respiratory pigment is loading or unloading, cooperativity enhances the *responsiveness* of the process to changes in the O_2 partial pressure within the mid-range of partial pressures.

Respiratory pigments exhibit a wide range of affinities for O_2

The respiratory pigments of various animals vary widely in how readily they combine with O_2, a property known as their **affinity** for O_2. Pigments that require relatively high O_2 partial pressures

---

[11] Figure 24.6 also does so.

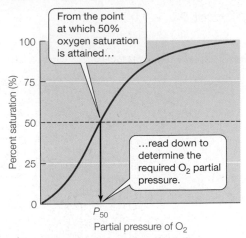

FIGURE 24.9 How to measure P_{50}

for full loading and that conversely unload substantial amounts of O_2 at relatively high partial pressures are said to have a relatively **low affinity** for O_2. Pigments that load fully at low partial pressures and consequently also require low partial pressures for substantial unloading are said to have a relatively **high affinity** for O_2. Affinity for O_2 is an *inverse* function of the O_2 partial pressure required for loading: The *higher* the O_2 partial pressure required to load a pigment, the *lower* is the pigment's affinity for O_2. The hemoglobins of humans and carp (see Figure 24.8) provide examples of pigments that differ in their affinity for O_2. Human hemoglobin requires a far higher O_2 partial pressure to become saturated than carp hemoglobin, indicating that the human hemoglobin combines less readily with O_2 and has a lower affinity.

A convenient index of O_2 affinity is P_{50} (pronounced "P fifty"), defined to be the partial pressure of O_2 at which a pigment is 50% saturated. **FIGURE 24.9** shows how P_{50} is measured. With Figure 24.9 in mind, a glance at Figure 24.8 reveals that human hemoglobin has a much higher P_{50} ($\cong 3.5$ kPa in arterial blood) than carp hemoglobin ($\cong 0.7$ kPa). *Affinity and P_{50} are inversely related: As P_{50} increases, O_2 affinity decreases.*

In the jargon of respiratory-pigment physiology, lowering the O_2 affinity is said to "shift the oxygen equilibrium curve to the right." To explain, **FIGURE 24.10** shows that a rightward shift (a shift from the blue to the red curve) reflects a higher P_{50} and therefore a lower O_2 affinity. Raising the O_2 affinity (decreasing the P_{50})—as would occur by shifting from the red to the blue curve—is said to "shift the curve to the left."

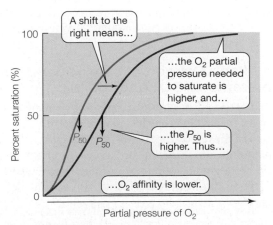

FIGURE 24.10 A "shift to the right" Such a shift reflects decreased O_2 affinity.

(A) Human hemoglobin at various pH levels

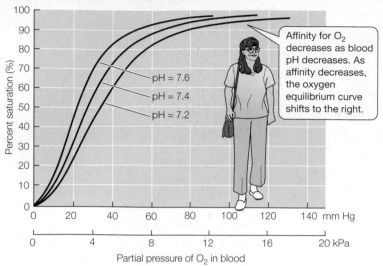

Affinity for O_2 decreases as blood pH decreases. As affinity decreases, the oxygen equilibrium curve shifts to the right.

pH = 7.6
pH = 7.4
pH = 7.2

(B) Dog hemoglobin at various CO_2 partial pressures

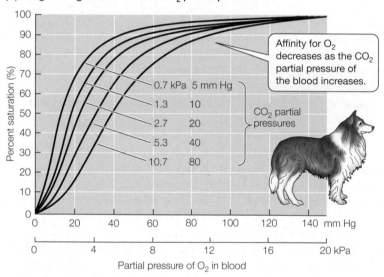

Affinity for O_2 decreases as the CO_2 partial pressure of the blood increases.

CO_2 partial pressures

0.7 kPa 5 mm Hg
1.3   10
2.7   20
5.3   40
10.7   80

FIGURE 24.11 The Bohr effect: Affinity for O_2 decreases as pH decreases or CO_2 partial pressure increases (A) Oxygen equilibrium curves of human hemoglobin at three different pHs at 38°C. In resting humans, the normal pH of arterial blood is about 7.4, whereas that of mixed venous blood is about 0.04 unit less. (B) Oxygen equilibrium curves of dog hemoglobin at five different CO_2 partial pressures at 38°C. The data in part (B) are from the original work of Bohr and his coworkers. (After Roughton 1964.)

The Bohr effect: Oxygen affinity depends on the partial pressure of CO_2 and the pH

In a body fluid or tissue containing a respiratory pigment, a decrease in the pH or an increase in the CO_2 partial pressure often causes the O_2 affinity of the respiratory pigment to decrease, thus shifting the oxygen equilibrium curve to the right. This effect, illustrated for the blood hemoglobins of humans and dogs in **FIGURE 24.11**, is known as the **Bohr effect** or **Bohr shift**,[12] in commemoration of

Christian Bohr (1855–1911), the prominent Danish physiologist (and father of Nobel laureate Niels Bohr) who led the discovery of the effect in 1904. Part of the reason that an increase in CO_2 partial pressure causes such a shift is that the pH of a solution tends to decline as its CO_2 partial pressure is increased.[13] However, CO_2 also exerts a direct negative effect on the O_2 affinities of some respiratory pigments, such as the blood hemoglobins of humans and other mammals. Recognizing that protons (H+ ions) and CO_2 itself can exert independent affinity-lowering effects, modern workers often distinguish two types of Bohr effects: a **fixed-acid Bohr effect**—which results from influences of the proton (H+) concentration on respiratory-pigment molecules—and a **CO_2 Bohr effect**—which results from the immediate influences of increased CO_2 partial pressure.

Species that show these effects vary widely in the *magnitudes* of the effects. One reason is that Bohr effects have probably evolved several times independently and thus have a different molecular basis in some animals than others. Even species with the same molecular mechanism often vary widely in details.

Protons exert their effects on O_2 affinity by combining with pigment molecules. Referring to hemoglobin (Hb) as a specific example, we can write the following *strictly conceptual* equation to summarize the effects of protons on O_2 affinity (the equation does not reflect the true stoichiometry of the reaction):

$$HbO_2 + H^+ \rightleftharpoons HbH^+ + O_2 \qquad (24.2)$$

Increasing the H+ concentration tends to increase the combination of Hb with H+, thus shifting the chemical reaction in Equation 24.2 to the right and favoring dissociation of O_2. The H+ ions bind at sites on the hemoglobin molecules (e.g., at histidine residues) different from the O_2-binding sites. Thus H+ acts as an *allosteric* modulator of O_2 binding. CO_2 also combines chemically with pigment molecules and functions as an allosteric modulator in cases in which it exerts direct effects on affinity.

The Bohr effect often has adaptive consequences for O_2 delivery. The CO_2 partial pressure is generally higher, and the pH is generally lower, in the systemic tissues than in the lungs or gills. Because of this, a respiratory pigment that displays a Bohr effect shifts to lower O_2 affinity each time the blood enters the systemic tissues and reverts back to higher O_2 affinity each time the blood returns to the breathing organs. The shift to lower affinity in the systemic tissues promotes release of O_2 because it facilitates deoxygenation. Conversely, the shift back to higher affinity in the breathing organs promotes uptake of O_2 by facilitating oxygenation. **FIGURE 24.12** illustrates the net effect of this shifting back and forth between two oxygen equilibrium curves as the blood flows between the breathing organs and systemic tissues. At any given O_2 partial pressures in the arterial and venous blood, more O_2 is delivered to the systemic tissues than would be if the pigment followed just one or the other equilibrium curve alone.

During exercise, the CO_2 partial pressure in the systemic tissues often rises above that prevailing during rest because of the increased production of CO_2. Furthermore, the pH in the systemic tissues often falls below the resting pH, not only because of the elevated

[12] In unusual cases, such as some species of molluscs and spiders, Bohr effects opposite to the usual direction, termed *reverse Bohr effects*, are observed.

[13] As already noted in Chapter 23, CO_2 has been aptly termed a "gaseous acid" because it reacts with H_2O to produce H+. The chemistry of these reactions is presented at length later in this chapter.

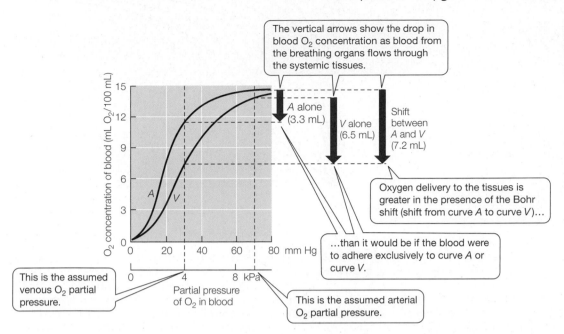

The vertical arrows show the drop in blood O_2 concentration as blood from the breathing organs flows through the systemic tissues.

A alone (3.3 mL)

V alone (6.5 mL)

Shift between *A* and *V* (7.2 mL)

Oxygen delivery to the tissues is greater in the presence of the Bohr shift (shift from curve *A* to curve *V*)...

...than it would be if the blood were to adhere exclusively to curve *A* or curve *V*.

This is the assumed venous O_2 partial pressure.

This is the assumed arterial O_2 partial pressure.

FIGURE 24.12 The Bohr effect typically enhances O_2 delivery in an animal The diagram shows oxygen equilibrium curves for arterial blood (*A*) and venous blood (*V*) in a hypothetical animal. The venous blood displays a reduced O_2 affinity because its CO_2 partial pressure and H^+ concentration are higher than those in arterial blood. The magnitude of this Bohr shift is exaggerated for clarity. The three bold arrows show unloading under three different assumptions. The top of each arrow is the O_2 concentration of blood as it leaves the breathing organs; the bottom is the O_2 concentration of the blood as it leaves the systemic tissues. The numeric values next to the arrows are the volumes of O_2 delivered per 100 mL of blood.

CO_2 partial pressure but also because acid metabolites—such as lactic acid—often accumulate during exercise. These changes often augment the Bohr shift during exercise, thereby enhancing O_2 delivery to the active tissues.

Now it will be clear why we indicated earlier in this chapter that refinements would ultimately be needed to our initial analysis of O_2 delivery in humans. We based our earlier analysis on the arterial oxygen equilibrium curve alone (see Figure 24.4), whereas in reality, Bohr shifts occur as blood flows between the lungs and systemic tissues. In humans at rest, venous blood is slightly more acidic (pH 7.36) than arterial blood (pH 7.40). Moreover, the CO_2 partial pressure is higher in mixed venous blood than in arterial blood: about 6.1 kPa (46 mm Hg) in venous blood and 5.3 kPa (40 mm Hg) in arterial. Looking at Figure 24.11, you can see that these differences in pH and CO_2 partial pressure are sufficient to cause small but significant Bohr shifts of the oxygen equilibrium curve as blood flows between the lungs and systemic tissues.

To fully understand respiratory-pigment function, it is important that, before closing this section, we consider not only how pH can affect oxygenation, but also how oxygenation can affect pH. Let's return to the conceptual equation, Equation 24.2, that describes the reaction of H^+ ions with respiratory pigments (assuming that a fixed-acid Bohr effect exists). Earlier we stressed one perspective on this equation; namely, that an increase in H^+ concentration will push the chemical reaction to the right, decreasing the tendency of pigment molecules to bind to O_2. Now we also stress that the equation has a complementary and equally important property: Removal of O_2 from pigment molecules will pull the chemical reaction to the right, causing the pigment molecules to take up H^+ from their surroundings. When blood passes through the systemic tissues, metabolism tends to increase the H^+ concentration of the blood solution. Simultaneously, however, because of the diffusion of O_2 out of the blood, respiratory-pigment molecules unload O_2 and thus bind with H^+. This removal of free H^+ from the blood, induced by the deoxygenation of the pigment molecules, limits the increase in the blood concentration of H^+—and the decrease in blood pH—caused by the metabolic addition of CO_2 and H^+.

The Root effect: In unusual cases, CO_2 and pH dramatically affect the oxygen-carrying capacity of the respiratory pigment

In some types of animals, because of distinctive properties of their respiratory pigments, an increase in the CO_2 partial pressure or a decrease in the pH of the blood not only causes a Bohr effect, but also reduces the amount of O_2 the respiratory pigment binds when saturated.[14] The reduction in the amount of O_2 bound to the pigment at saturation (**FIGURE 24.13**) is termed the **Root effect**, after its discoverer. Root effects of sizable magnitude are not common. Among vertebrates, they are observed only in fish, principally teleost fish. Some molluscs also show either normal or reversed Root effects.

Root effects provide a mechanism by which the O_2 *partial pressure* of even well-oxygenated blood can be dramatically increased under the control of blood pH. To see this, consider the hemoglobin in the blood of eels when it is fully loaded with O_2 (see Figure 24.13).[15] At a pH of 7.54, the hemoglobin is chemically combined with about 12.6 mL of O_2 per 100 mL of blood. Acidification to a pH of 7.35 lowers the O_2-binding capacity of the hemoglobin because of the Root effect, so that the hemoglobin can chemically combine with only about 9.4 mL O_2/100 mL. In this way O_2 *is forced off the hemoglobin by the change of pH*. The acidification from pH 7.54 to 7.35 forces the eel hemoglobin to unload 3.2 mL of O_2 into each 100 mL of blood! The O_2 released goes into blood solution; it has no other immediate place to go. *By dissolving, it dramatically elevates the blood O_2 partial pressure.*

The Root effect is employed in various species of teleost fish to help create high O_2 partial pressures in two regions of the body: the swim bladder and the eyes. In both types of organs, the pH of well-oxygenated blood is lowered by a tissue-specific addition of lactic acid, which induces a rise in the blood O_2 partial pressure

---

[14] Some modern authorities view the Root effect as an exaggerated Bohr effect.

[15] Although Figure 24.13 serves as a useful visual guide, the insight it provides into hemoglobin function is qualitative, not quantitative, because the O_2 concentrations shown include dissolved O_2.

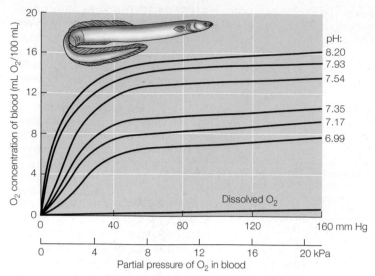

FIGURE 24.13 The Root effect in eels: Acidification lowers the oxygen-carrying capacity of hemoglobin Oxygen equilibrium curves are shown for the whole blood of eels (*Anguilla vulgaris*) at six different pH levels (at 14°C). Saturation of hemoglobin is indicated when the slope of an oxygen equilibrium curve parallels the slope of the dissolved O_2 line. Experiments on some fish have shown that O_2 binding by hemoglobin is reduced at low pH even when the hemoglobin is exposed to an O_2 partial pressure of 140 atmospheres (atm) (14,000 kPa)! (After Steen 1963.)

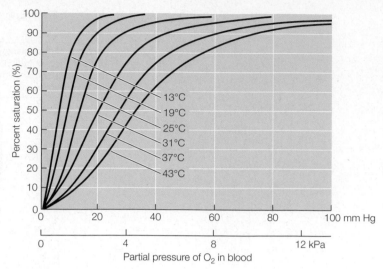

FIGURE 24.14 An increase in temperature typically causes a decrease in O_2 affinity Oxygen equilibrium curves are shown for human blood at six different temperatures, with pH held constant at 7.4. These results show the pure effect of changes in temperature because of the constancy maintained in pH. The results, however, tend to understate the effects of temperature in many real-life situations because when the pH is not artificially controlled, a rise in blood temperature typically induces a decrease in blood pH, as discussed later in this chapter (see Figure 24.24), meaning that the immediate effects of temperature are often reinforced by thermally induced fixed-acid Bohr effects. The CO_2 partial pressure was held constant during the studies shown. (After Reeves 1980.)

because of the Root effect. Moreover, in both types of organs, this rise in the blood O_2 partial pressure is *amplified* by a countercurrent vascular arrangement (a rete mirabile) that favors multiplication of the initial effect.[16] The creation of a high O_2 partial pressure helps inflate the swim bladder in many fish (swim-bladder gas is often principally O_2). The retinas of some fish are so poorly vascularized that they require a high surrounding O_2 partial pressure to acquire enough O_2 to function properly. Recently, a convincing case has been made that, in the course of evolution, the first role of the Root effect in fish was oxygenation of the retina. Later, at least four different lines of fish independently evolved the use of the Root effect in O_2 secretion to inflate the swim bladder.

Thermal effects: Oxygen affinity depends on tissue temperature

The O_2 affinity of respiratory pigments is often inversely dependent on temperature (**FIGURE 24.14**). Increases in temperature decrease affinity, whereas decreases in temperature increase affinity (changes in temperature only rarely affect the O_2 content of blood at saturation, however). When humans or other mammals exercise, if the blood temperature in their exercising muscles exceeds the temperature in their lungs, thermal shifts in affinity will enhance O_2 delivery to the muscles in a manner much like that already described for the Bohr effect (see Figure 24.12). In total, therefore, unloading of O_2 to the exercising muscles will be promoted in a concerted manner by both a temperature effect and a Bohr effect, both of which independently tend to decrease the O_2 affinity of the respiratory pigment (and thereby facilitate O_2 unloading) when the blood passes through the muscles. Conversely to this happy state,

temperature effects may become a problem in the hypothermic limbs of mammals in Arctic climates—a matter addressed in **BOX 24.3** in relation to recent studies of a resurrected ancient protein, the hemoglobin of the woolly mammoth.

Organic modulators often exert chronic effects on oxygen affinity

Organic compounds synthesized by metabolism often play major roles as allosteric modulators of the function of respiratory pigments. In vertebrates, the principal compounds acting in this role are organophosphate compounds within the red blood cells, which affect hemoglobin O_2 affinity. The organophosphate of chief importance in most mammals, including humans, is **2,3-bisphosphoglycerate**, which is synthesized in red blood cells from intermediates of glycolysis. This compound is sometimes called **BPG** or **2,3-BPG**, but more commonly, for historical reasons, the abbreviated name **2,3-DPG** (standing for 2,3-diphosphoglycerate) is used. The effect of 2,3-DPG is to reduce the O_2 affinity (raise the P_{50}) of the hemoglobin molecules with which it binds. As shown by the black line in **FIGURE 24.15**, hemoglobin O_2 affinity is therefore a function of the 2,3-DPG concentration. The hemoglobin of humans and most other mammals is continuously exposed to and modulated by 2,3-DPG within the red blood cells. Thus, as stressed in Figure 24.15, the "normal" O_2 affinity of human hemoglobin in the red blood cells is in part a consequence of modulation by a "normal" 2,3-DPG concentration within the cells.[17]

---

[16] The process of countercurrent multiplication is explained, in a different context, in Chapter 29 (see pages 792–793).

[17] Some mammals have hemoglobins that are not modulated by organophosphate compounds under ordinary physiological conditions. Included are some ruminants, cats, civets, and related species. Their hemoglobins, within the red blood cells, display functionally appropriate O_2 affinities without 2,3-DPG, and the red blood cells of adult animals of these types usually contain little 2,3-DPG.

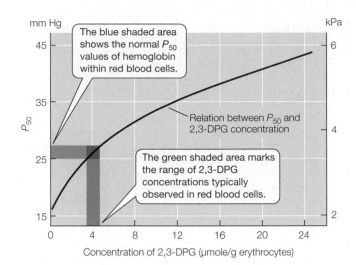

mm Hg

The blue shaded area shows the normal P_{50} values of hemoglobin within red blood cells.

Relation between P_{50} and 2,3-DPG concentration

The green shaded area marks the range of 2,3-DPG concentrations typically observed in red blood cells.

Concentration of 2,3-DPG (μmole/g erythrocytes)

FIGURE 24.15 The normal P_{50} of human hemoglobin within red blood cells depends on a normal intracellular concentration of 2,3-DPG There is usually about one 2,3-DPG molecule per hemoglobin molecule in human red blood cells. The temperature (37°C), CO_2 partial pressure (5.3 kPa), and extracellular pH (7.4) were held constant during the measurements presented here. (After Duhm 1971.)

In nonmammalian vertebrates, modulation of hemoglobin O_2 affinity by red blood cell organophosphates is also very common, although the specific phosphate compounds that bind with and allosterically affect hemoglobin vary from one taxonomic group to another and usually do not include 2,3-DPG. ATP and guanosine triphosphate (GTP) are generally the principal organophosphate modulators in fish. In birds, inositol pentaphosphate (IPP) and ATP are especially important. As in mammals, when organophosphates act as modulators in these other groups, their effect is to lower O_2 affinity.

Chronic changes in the concentration of organophosphate modulators in red blood cells serve as *mechanisms of acclimation or acclimatization* in many vertebrates. People suffering from anemia, to mention one example, often exhibit a chronic increase in the concentration of 2,3-DPG in their red blood cells; the O_2 affinity of their hemoglobin is thereby lowered by comparison with the usual affinity. The resulting shift to the right in their oxygen equilibrium curve is not great enough to cause any substantial impairment of O_2 loading in their lungs, but it significantly facilitates O_2 unloading in their systemic tissues (**FIGURE 24.16**). Thus each molecule of hemoglobin, on average, delivers more O_2 from the lungs to the systemic tissues during each passage through the circulatory system. In anemic people, this effect helps offset the disadvantage of having a reduced amount of hemoglobin per unit of blood volume.

BOX 24.3 **Resurrection of the Blood Hemoglobin of the Extinct Woolly Mammoth: Evidence for an Ancient Adaptation to the Challenges of Regional Hypothermia**

Using genomic methods, researchers recently resurrected the hemoglobin of the extinct woolly mammoth (*Mammuthus primigenius*) so they could study the hemoglobin directly. The mammoth was an abundant resident of Arctic and sub-Arctic environments—in sharp contrast to its extant relatives, the African and Asian elephants. Arctic mammals often permit tissue temperatures in their appendages to fall far below the temperature of the body core (see Figure 10.32), raising the possibility that the O_2 affinity of hemoglobin might be raised to such a high level by low tissue temperatures that the appendage tissues are subjected to impaired O_2 offloading. Was this a problem for woolly mammoths? By study of the resurrected mammoth hemoglobin, the researchers concluded that, in fact, the blood hemoglobin of the woolly mammoth had adaptive specializations that made it relatively insensitive to low temperatures.

The researchers extracted DNA from a 43,000-year-old, permafrost-preserved mammoth femur collected in Siberia. They then amplified and sequenced the genes in the DNA that coded for the α- and β-globin chains of hemoglobin (they located these genes based on homology with the known genes in today's African and Asian elephants). Using the ancient DNA nucleotide sequences to predict the amino acid sequences in the ancient α- and β-globin chains, they discovered that the mammoth's α- and β-globin proteins differed from those of the Asian elephant at one and three amino acid positions, respectively. The researchers then, in essence, introduced the genes for Asian elephant globins into the bacterium *Escherichia coli*, which faithfully synthesized elephant hemoglobin. Finally, by use of site-directed mutagenesis, the researchers modified the in-

troduced elephant genes at the positions that needed to be changed for the *E. coli*-synthesized globin proteins to match those once circulating in the blood of the extinct woolly mammoth. Thereafter the *E. coli* produced authentic woolly mammoth hemoglobin, which was purified and studied to determine its thermal and O_2-transport properties. **Box Extension 24.3** discusses additional details and provides references.

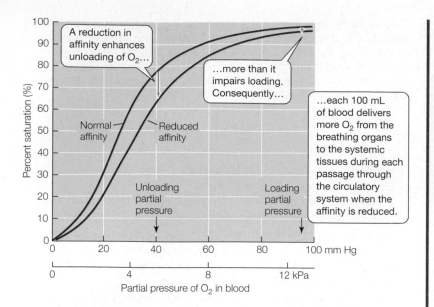

FIGURE 24.16 A decrease in the O_2 affinity of hemoglobin can aid O_2 delivery to the systemic tissues when the O_2 partial pressure in the breathing organs remains high Two human oxygen equilibrium curves, representing normal and reduced O_2 affinities, are shown. The loading O_2 partial pressure in the lungs is assumed to be 12.7 kPa (95 mm Hg), and the unloading O_2 partial pressure in the systemic tissues is assumed to be 5.3 kPa (40 mm Hg). The green vertical arrows show the changes in percent saturation at these two partial pressures caused by the shift from normal affinity to reduced affinity (for simplicity and clarity, other effects on affinity, such as Bohr effects, are ignored, and the reduction in affinity is exaggerated). The principles elucidated here apply to gill breathers as well as lung breathers.

The arthropod hemocyanins are well known to be modulated by organic compounds such as lactate ions, dopamine, and trimethylamine. In many crustaceans, for example, O_2 affinity is elevated by increasing plasma concentrations of lactate ions (specifically L-lactate ions), which exert their effects by binding to specific allosteric sites on the hemocyanin molecules. When animals such as blue crabs (*Callinectes sapidus*) engage in exercise that produces lactic acid (see Chapter 8), the affinity-increasing effect of the lactate ions offsets the large affinity-decreasing effect of the acidification of their blood (Bohr effect), helping to ensure that their hemocyanin remains capable of fully loading with O_2 in the gills.

Inorganic ions may also act as modulators of respiratory pigments

Concentrations of specific inorganic ions in blood cells or blood plasma sometimes allosterically modulate the O_2 affinity or other attributes of respiratory pigments. Recent research, for example, has revealed that in many ruminant mammals and certain bears, the concentration of Cl^- in the red blood cells is a critical allosteric modulator of hemoglobin function. The divalent ions Ca^{2+} and Mg^{2+} are important allosteric modulators of hemocyanin in crustaceans. Blue crabs exposed to O_2-poor waters, for example, increase their blood Ca^{2+} concentration, which raises the O_2 affinity of their hemocyanin.

SUMMARY
The O_2-Binding Characteristics of Respiratory Pigments

■ The oxygen equilibrium curve of a respiratory pigment, which shows the relation between the extent of O_2 binding by the pigment and the O_2 partial pressure, is a key tool for interpreting respiratory-pigment function. The shape of the oxygen equilibrium curve depends on the degree of cooperativity among O_2-binding sites on respiratory-pigment molecules. When there is no cooperativity—as is the case when each molecule has only a single O_2-binding site—the oxygen equilibrium curve is hyperbolic. The curve is sigmoid when molecules have multiple O_2-binding sites that exhibit positive cooperativity. Hyperbolic curves are the norm for myoglobins; sigmoid curves are the norm for blood pigments.

■ The Bohr effect is a reduction in O_2 affinity caused by a decrease in pH and/or an increase in CO_2 partial pressure. The Bohr effect typically enhances O_2 delivery because it promotes O_2 unloading in systemic tissues while promoting loading in the breathing organs.

■ The Root effect, which occurs only rarely, is a substantial reduction of the oxygen-carrying capacity of a respiratory pigment caused by a decrease in pH and/or an increase in CO_2 partial pressure. In teleost fish it helps inflate the swim bladder and oxygenate the retina.

■ Elevated blood temperatures often decrease the O_2 affinity of respiratory pigments.

■ Organic molecules and inorganic ions frequently serve as allosteric modulators of respiratory-pigment function. 2,3-DPG (2,3-BPG) in the red blood cells of mammals, for example, chronically decreases the O_2 affinity of the hemoglobin in the cells.

The Functions of Respiratory Pigments in Animals

It would be hard to exaggerate the diversity of functional properties found among animal respiratory pigments. The *oxygen affinity* (P_{50}) of respiratory pigments varies from less than 0.2 kPa to more than 7 kPa. *Cooperativity* (the Hill coefficient, n) varies from 1 to more than 6. The *concentration* of the respiratory pigment in an animal's blood may be so low that the pigment merely doubles the oxygen-carrying capacity of the blood in comparison with the dissolved O_2 concentration; alternatively, a pigment may be so concentrated that it allows blood to carry 80 times more O_2 than can be dissolved. A respiratory pigment may or may not exhibit a *Bohr effect* or *temperature effect*. One pigment may be modulated by 2,3-DPG, another by ATP. With this diversity of properties, even when the respiratory pigments of various animals carry out a single function, they do so in a diversity of detailed ways.

Respiratory pigments, moreover, are presently known to carry out at least eight different functions—meaning that, overall, they have a very wide range of action. The functions are not mutually exclusive; often a single respiratory pigment carries out two or more functions simultaneously. Although we will cover only a few functions in any detail, all eight deserve recognition:

1. Respiratory pigments in blood (or other circulating body fluids) typically aid the *routine transport of O_2 from the breathing organs to the systemic tissues*. This is the function to which we have devoted most of our attention up to this point in the chapter.

2. Respiratory pigments in the blood of some invertebrates probably function primarily as *O_2 stores*, rather than participating in routine O_2 transport. The pigments that fit this description have very high O_2 affinities. Consequently, they hold so tightly to O_2 that they probably do not unload under routine conditions. Instead, they seem to release their O_2 when animals face severe O_2 shortages. In certain species of tube-dwelling marine worms, for example, O_2 bound to a high-affinity blood hemoglobin is believed to be unloaded primarily during periods when the worms do not breathe, when their tissue O_2 partial pressures fall very low.

3. Blood respiratory pigments often serve as major *buffers of blood pH* and thereby play key roles in blood acid–base regulation. As already mentioned (see Equation 24.2), this buffering is often of an "active" sort, in the sense that the affinity of the respiratory pigments for H^+ changes as they unload and load O_2. The pigments tend to remove H^+ from solution as they become deoxygenated and release H^+ into solution as they become oxygenated. We return to this topic later in this chapter (e.g., see Figure 24.23).

4. Blood respiratory pigments often play critical roles in *CO_2 transport*, as we will also see later in the chapter.

5. Hemoglobins in the cytoplasm of muscle cells (myoglobins), or in the cells of other solid tissues, play two principal respiratory roles. First, they *increase the rate of O_2 diffusion through the cytoplasm of the cells*, a phenomenon that in muscle cells is called **myoglobin-facilitated O_2 diffusion**: At any given difference in O_2 partial pressure between the blood capillaries and the mitochondria of the cells, O_2 diffuses through the cytoplasm to the mitochondria faster if myoglobin is present. The second role played by hemoglobins within solid tissues is *O_2 storage for the tissues*. The myoglobin-bound O_2 store in skeletal muscles, for example, can be called upon at the start of sudden, vigorous muscular work to help sustain aerobic ATP production while circulatory O_2 delivery is still being accelerated to meet the heightened O_2 need.[18]

6. Sometimes, respiratory pigments *act as enzymes*, not in carrying out their roles in O_2 transport, but in other contexts. At least in mammals, for example, deoxymyoglobin and deoxyhemoglobin catalyze the local formation of the critically important signaling compound nitric oxide (NO) from nitrite in certain settings.

7. Respiratory pigments occasionally play *nonrespiratory transport roles*. In at least some species of worms that have symbiotic sulfur-oxidizing bacteria (see Figure 6.16), for example, the blood hemoglobin has sulfide-binding sites,

and it transports S^{2-} as well as O_2 from the gills to the organ in which the bacteria live.

8. Finally, the fastest-breaking story in the contemporary study of respiratory pigments is the increasing recognition that at least in mammals, myoglobins are sometimes intimately involved in several *tissue functions other than O_2 supply*, especially the regulation of mitochondrial respiration, as addressed in **BOX 24.4**.

Patterns of circulatory O_2 transport: The mammalian model is common but not universal

Circulatory O_2 transport in most animals qualitatively follows the pattern we described earlier for mammals (see Figure 24.5). This pattern has several major features, which, for example, can be seen in the O_2 transport physiology of rainbow trout (**FIGURE 24.17**). First, the blood respiratory pigment reaches near-saturation in the lungs or gills when the animals are living in well-aerated environments. Second, the respiratory pigment yields just a modest fraction of its O_2 to the systemic tissues during circulation at rest, meaning that venous blood in resting individuals is far from being fully deoxygenated. Third, the large resting venous O_2 reserve is used (i.e., venous blood becomes more deoxygenated) during exercise or other states of heightened metabolism. Thus increased tissue O_2 demands are met by increasing the amount of O_2 delivered per unit of blood volume, as well as by increasing the rate of blood flow.

Squids and octopuses are important examples of animals that follow a different pattern of circulatory O_2 transport, and thus

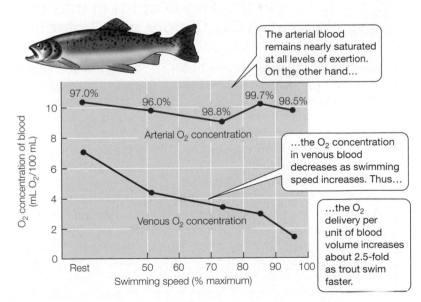

FIGURE 24.17 Blood O_2 transport in rainbow trout in relation to exercise The lines show the average O_2 concentration of arterial and venous blood in trout (*Oncorhynchus mykiss*) at rest and swimming at various speeds in well-aerated water. The numbers above the arterial points show the average arterial percent saturation of the particular fish studied at each speed. As fish increase their speed from rest to maximum, they increase O_2 delivery per unit of blood volume about 2.5-fold. Trout also increase their rate of circulation about 3-fold. Thus the trout increase the total rate of O_2 delivery to their tissues about 7-fold. (After Jones and Randall 1978.)

---

[18] This role is discussed at length in Chapters 8 and 20.

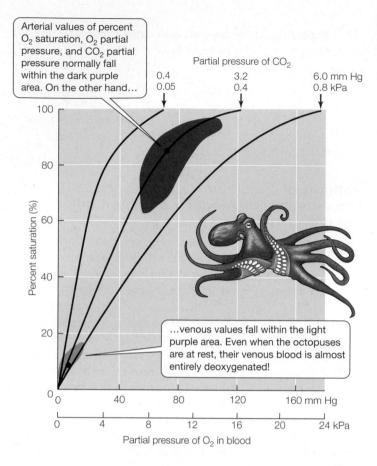

Arterial values of percent O_2 saturation, O_2 partial pressure, and CO_2 partial pressure normally fall within the dark purple area. On the other hand...

Partial pressure of CO_2

| 0.4 | 3.2 | 6.0 mm Hg |
| 0.05 | 0.4 | 0.8 kPa |

...venous values fall within the light purple area. Even when the octopuses are at rest, their venous blood is almost entirely deoxygenated!

Percent saturation (%)

Partial pressure of O_2 in blood

FIGURE 24.18 Blood O_2 delivery in an octopus: Even at rest, octopuses have almost no venous reserve The octopuses (*Enteroctopus dofleini*) studied were resting or only moderately active in well-aerated water. The three oxygen equilibrium curves correspond to three different blood CO_2 partial pressures (i.e., the hemocyanin exhibits a Bohr effect). All the data obtained on arterial blood fall within the dark purple area, whereas the data on venous blood fall within the light purple area. The two dots represent the approximate means for arterial and venous blood. (After Johansen and Lenfant 1966.)

illustrate that the pattern observed in mammals and fish is not universal. The squids and octopuses that have been studied have only a very small venous O_2 reserve when they are at rest: *Inactive* individuals use 80%–90% of the O_2 available in their arterial blood (**FIGURE 24.18**). Thus, when the animals exercise, they have little room to increase unloading of O_2 from their hemocyanin, and they must meet their heightened O_2 demands almost entirely by increasing their circulatory rates. This pattern places high demands on their hearts and constrains their ability to exercise, as we will see in more detail in Chapter 25 (see Box 25.3). The inherently small venous O_2 reserve of the squids and octopuses also limits their ability to live in poorly aerated waters. If a squid or octopus ventures into O_2-poor waters and consequently can't oxygenate its arterial blood fully, it can't compensate to any great degree (as a fish can) by enhancing the deoxygenation of its venous blood, because the venous blood is already highly deoxygenated even in aerated waters. Squids and octopuses are notoriously intolerant of low-O_2 environments.

BOX 24.4 Heme-Containing Globins in Intracellular Function: Myoglobin Regulatory and Protective Roles, Neuroglobins, and Cytoglobins

A revolution is underway in the understanding of the roles of globins in intracellular function. New roles of myoglobin are being documented or hypothesized. In addition, new intracellular globins—not known in the twentieth century—have been discovered.

Based on research using myoglobin knockout mice and other methods, researchers now hypothesize that in cardiac muscle and possibly other types of muscle, myoglobin plays a key role in the regulation of mitochondrial respiration, serves as a defense against reactive oxygen species (see Box 8.1), and helps control mitochondrial substrates. We say more here about just the first of these roles. Nitric oxide (NO) potently inhibits cytochrome oxidase (see Figure 8.3) and in this way serves as a key regulator of the rate of mitochondrial O_2 consumption and ATP synthesis in at least certain muscles (notably cardiac). When O_2 is

relatively abundant in a cell, myoglobin becomes oxygenated, forming oxymyoglobin (oxyMb). OxyMb breaks down NO, a process that prevents NO inhibition of cytochrome oxidase, thereby permitting the mitochondria to use O_2 to synthesize ATP when O_2 is available. Conversely, when O_2 is low in abundance in a cell, deoxymyoglobin (deoxyMb) forms. DeoxyMb acts as an enzyme that catalyzes NO synthesis; the NO inhibits cytochrome oxidase and thereby inhibits mitochondrial use of O_2 and ATP synthesis. In these ways, myoglobin is a principal player in regulating mitochondrial function to match the availability of O_2.

In 2000, a heme-containing globin expressed in the brain of humans and mice was discovered (based on genomics research) and named **neuroglobin (Ngb)**. Neuroglobins are now known to occur in most (possibly all) vertebrates. They are intracellular (in cytoplasm)

and have been observed (usually at low concentration) in most brain neurons, peripheral neurons, the retina, some endocrine glands (e.g., adrenal), and the sperm-producing tissues of the testicles. The functions of neuroglobins are gradually being elucidated. They bind O_2 reversibly with high affinity (like myoglobins). Their chief function may be to act as O_2 stores for the central nervous system and retina. Animals genetically engineered to overexpress neuroglobins recover from strokes better than controls do, suggesting that the neuroglobin O_2 store helps protect neurons when their external O_2 supply is cut off. Neuroglobins might also function in antioxidant defense (see Box 8.1) or as sensors of metabolic stress. **Box Extension 24.4** discusses neuroglobin structure and another recently discovered set of cytoplasmic globins, the cytoglobins.

Regardless of the exact pattern of circulatory O_2 delivery an animal displays, the oxygen-carrying capacity of its blood—which depends on the amount of respiratory pigment per unit of blood volume—is a key determinant of how much O_2 can be delivered to its tissues. As already seen in Figure 24.8, animals display a wide range of oxygen-carrying capacities. The range of known values in animals that have blood respiratory pigments is from about 30–40 mL O_2/100 mL of blood in some diving mammals to just 1–2 mL O_2/100 mL in many crustaceans and molluscs. Among vertebrates, a rough correlation exists between metabolic intensity and the oxygen-carrying capacity of the blood; mammals and birds usually have carrying capacities of 15–20 mL O_2/100 mL, whereas fish, amphibians, and nonavian reptiles usually have less hemoglobin per unit of volume and have carrying capacities of 5–15 mL O_2/100 mL. Active species of fish such as tunas and lamnid sharks tend to have higher oxygen-carrying capacities than do related sluggish species.

Animals with hemocyanin tend to have low oxygen-carrying capacities. Squids and octopuses exhibit the highest carrying capacities known for hemocyanin-containing bloods, and their carrying capacities are just 2–5 mL O_2/100 mL (at or below the lower end of the range for fish). Animals with hemocyanin—which is always dissolved in the blood plasma, not contained in blood cells—probably cannot have much higher carrying capacities because the hemocyanin concentrations needed for higher capacities would make their blood too viscous to pump.

Individual animals can vary their oxygen-carrying capacity by raising or lowering the amount of respiratory pigment per unit of blood volume. The most common responses of this sort are long-term, occurring during acclimation or acclimatization to changed environments (to be discussed shortly). Some vertebrates, however, can acutely change their carrying capacity because they can remove red blood cells from their blood, store the cells, and quickly release them back into the blood. Horses, dogs, and some seals are well known to store massive quantities of red blood cells in their spleen when at rest. When the cells are needed during exercise, they are quickly released back into the blood under control of the sympathetic nervous system. Foxhounds, for example, can promptly increase their oxygen-carrying capacity from 16 to 23 mL O_2/100 mL in this way.

Respiratory pigments within a single individual often display differences in O_2 affinity that aid successful O_2 transport

Two respiratory pigments often exist within one animal and pass O_2 from one to the other. Most commonly, this occurs in animals that have myoglobins. In these animals, the blood respiratory pigment (hemoglobin or hemocyanin) and the myoglobin act as a sort of "O_2 bucket brigade": The blood pigment carries O_2 from the lungs or gills to the muscles, and then passes the O_2 to the myoglobin in the muscle cells. This process is typically aided by differences in O_2 affinity. Specifically, the myoglobin typically has a higher O_2 affinity—a lower P_{50}—than the blood pigment; one can see in Figure 24.7A, for example, that the P_{50} of human myoglobin (about 0.8 kPa, 6 mm Hg) is far lower than that of human blood hemoglobin (about 3.5 kPa, 27 mm Hg). The higher O_2 affinity of the myoglobin means that it tends to load with O_2

at the expense of unloading of the blood hemoglobin. Thus the difference in affinity promotes transfer of O_2 from the blood to the muscle cells.

Affinity relations also promote the transfer of O_2 from mother to fetus across the placenta in placental mammals. Generalizing across species, the P_{50} of fetal blood is typically less than the P_{50} of maternal blood by 0.4–2.3 kPa (3–17 mm Hg). Because the fetal blood has a higher O_2 affinity, it tends to oxygenate by drawing O_2 from the maternal blood. The relatively high affinity of the fetal blood also means that it is able to become relatively well oxygenated even if the O_2 partial pressure in the placenta is relatively low. Several specific mechanisms account for the differences in O_2 affinity between fetal and maternal bloods in various species. In humans and other primates, the difference occurs because the chemical structures of the fetal and maternal hemoglobins are different, as mentioned at the beginning of this chapter. One key effect of these structural differences is that the fetal hemoglobin is less sensitive to 2,3-DPG; because 2,3-DPG lowers affinity, the diminished sensitivity of fetal hemoglobin to 2,3-DPG raises its O_2 affinity. In some other species, such as dogs and rabbits, the hemoglobins in the fetus and mother are chemically the same; the reason the fetal affinity is higher is that fetal red blood cells have lower intracellular concentrations of 2,3-DPG than maternal red blood cells. In still other species of mammals, additional mechanisms of raising the fetal O_2 affinity are observed; ruminants, for example, have fetal forms of hemoglobin that are intrinsically higher in affinity than maternal hemoglobin, without 2,3-DPG modulation.

The relatively high O_2 affinity of fetal hemoglobin is not necessarily the only factor that promotes O_2 transfer from the mother's blood to the blood of the fetus. An extremely interesting additional factor is that often the loss of CO_2 from the fetal blood to the maternal blood induces a synchronous rise in fetal O_2 affinity and fall in maternal O_2 affinity because of Bohr effects in the two hemoglobins: Bohr effects that have these opposing but reinforcing consequences!

Evolutionary adaptation: Respiratory pigments are molecules positioned directly at the interface between animal and environment

A dramatic property of the respiratory pigments is that they are molecules that, in a way, actually form part of the interface between an animal and its environment: They pick up O_2 from the environment and deliver it to cells deep within tissues. Moreover, evolution has produced hundreds of different molecular forms of the respiratory pigments. Because of these considerations, the respiratory pigments have long been regarded as prime subjects for the study of evolutionary molecular adaptation.

Such studies have revealed that often species that have long histories of existence in low-O_2 environments have evolved respiratory pigments with higher O_2 affinities than related species living in high-O_2 environments. This common pattern is well illustrated by the fish in Figure 24.8 and by other fish: Carp and catfish, which often inhabit waters low in O_2, have average P_{50} values of 0.1–0.7 kPa (1–5 mm Hg)—meaning their hemoglobins load particularly well at low O_2 partial pressures—whereas mackerel and rainbow trout, which live in well-aerated waters, have far higher P_{50} values

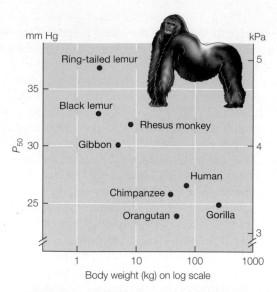

FIGURE 24.19 The O_2 affinity of the hemoglobin in the whole blood of primates is a regular function of body size Small-bodied species tend to exhibit lower O_2 affinity—and thus higher P_{50}—than large-bodied ones. (After Dhindsa et al. 1972.)

of 2.1–2.4 kPa (16–18 mm Hg).[19] One reason goldfish survive the tender loving care of kindergarteners is that these members of the carp family have high-affinity hemoglobins that can load well in O_2-poor water. Mammal species that live underground typically have evolved higher O_2 affinities than aboveground species of the same body size. Similarly, some species native to high altitudes have evolved higher O_2 affinities than lowland species (see Box 24.5).

Another thought-provoking evolutionary pattern that has been discovered is the relation between O_2 affinity and body size in groups of related species. In mammals and some other vertebrate groups, the O_2 affinity of blood hemoglobin tends to decrease as body size decreases: Small species have relatively high P_{50} values and therefore relatively low O_2 affinities (**FIGURE 24.19**). Natural selection is hypothesized to have favored this pattern because of the inverse relation between weight-specific metabolic rate and body size (see Chapter 7). Arterial blood oxygenates similarly in all species of aboveground mammals near sea level because the O_2 partial pressure in the lungs is high enough in all such species to be on the plateaus of their oxygen equilibrium curves (where differences in affinity have little effect; see Figure 24.16). The lower-affinity hemoglobins in the smaller species unload O_2 to the tissues more readily, however. In this way, the lower affinity in the small species is hypothesized to help them meet their higher weight-specific O_2 needs.

Of course, it is exciting to find trends that make sense, but sometimes when physiologists have compared the O_2 affinities of related species, they have found no clear patterns, or even trends opposite to those expected. At present, a comprehensive *predictive* theory of affinity adaptation does not exist. An important reason is that when affinity is modified in the course of evolution, the changes can potentially affect both loading and unloading. Although a decrease in affinity, for example, could aid O_2 delivery by promoting O_2

unloading in the systemic tissues, it could potentially also diminish O_2 delivery by interfering with O_2 loading in the breathing organs. For sorting out these complexities, a crucial question is whether the O_2 partial pressure in the breathing organs is high enough to cause full oxygenation. To explain, consider a case in which arterial O_2 partial pressures are consistently high enough for respiratory pigments to be well oxygenated regardless of O_2 affinity. In this case, the principal effect of low affinity is to promote unloading of O_2 in the systemic tissues, which augments O_2 delivery.[20] You will recognize this argument. It is exactly why researchers think that small-bodied mammals living aboveground near sea level can benefit by evolving relatively low affinities (see Figure 24.19).

The respiratory-pigment physiology of individuals undergoes acclimation and acclimatization

When individual animals are exposed chronically to reduced O_2 availability in their environments, they often respond with chronic alterations of their respiratory-pigment physiology. The most common response of this sort in both vertebrates and invertebrates is for the concentration of the respiratory pigment in the blood to be increased. Fish, for example, often increase the concentration of red blood cells in their blood when they live in poorly oxygenated waters.

In addition to the "quantitative strategy" of increasing the amount of respiratory pigment per unit of blood volume, animals also often modify the O_2-binding properties of the pigments. Sometimes this is achieved by synthesizing different molecular forms. A dramatic example is provided by the water flea, *Daphnia*, a small, hemoglobin-synthesizing crustacean common in freshwater ponds. When *Daphnia* that have been living in O_2-rich water are transferred into O_2-poor water, hypoxia-inducible transcription factors (HIFs) are released and affect DNA transcription by modulating hypoxia-response elements in the promoter regions for the globin genes (see page 604). Multiple globin types can be synthesized, and after the transfer to O_2-poor water, the mix of globins is modified. In fact, new mRNAs can appear *within minutes*, and new hemoglobin molecules—composed of different proportions of globin subunits than the preexisting molecules—can appear within 18 h. The new molecular forms of hemoglobin have a higher O_2 affinity than the preexisting ones. Thus, over the first 11 days in O_2-poor water, the O_2 affinity rises (P_{50} falls) as the concentration of hemoglobin also rises (**FIGURE 24.20**)! Together, these changes give the *Daphnia* a greatly enhanced capability to acquire O_2 from their environment. The *Daphnia* also, as mentioned earlier, change color. Pale at the start, they turn brilliant red (see Figure 24.20).

For modifying the O_2-binding properties of respiratory pigments, perhaps a more common strategy is not to alter the molecular forms of pigments synthesized but to modulate preexisting types in advantageous ways. When fish, for example, are transferred from well-aerated to poorly aerated waters, they do not typically alter their hemoglobin types, but they often decrease the concentrations of ATP and GTP within their red blood cells over time. These chronic changes in the intracellular modulators of hemoglobin raise its O_2 affinity. Blue crabs, as noted earlier, chronically raise the concentration of Ca^{2+} in their blood when exposed to O_2-poor

---

[19] These measurements were made at approximately the same CO_2 partial pressures and temperatures.

[20] Figure 24.16, although it applies to changes in O_2 affinity within a species, illustrates this effect.

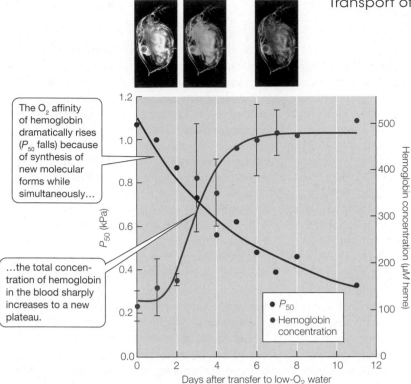

The O_2 affinity of hemoglobin dramatically rises (P_{50} falls) because of synthesis of new molecular forms while simultaneously...

...the total concentration of hemoglobin in the blood sharply increases to a new plateau.

- P_{50}
- Hemoglobin concentration

FIGURE 24.20 When water fleas are transferred to O_2-poor water, their O_2-transport system undergoes rapid acclimation because of altered gene expression The water fleas (*Daphnia magna*) had been living in well-aerated water and were transferred at time 0 to water in which the O_2 partial pressure (3 kPa) was only 15% as high as in well-aerated water. The composition and concentration of their blood hemoglobin were then monitored for 11 days. The animals change color, as shown by the photographs. Symbols are means; error bars delimit ± 1 standard deviation. (After Paul et al. 2004; photos courtesy of Shinichi Tokishita.)

waters, apparently by mobilizing Ca^{2+} from the exoskeleton; the effect is to raise the O_2 affinity of their hemocyanin. **BOX 24.5** discusses how mammals respond to the low atmospheric O_2 partial pressures of high altitudes.

Icefish live without hemoglobin

We now end our discussion of the transport of O_2 by respiratory pigments by recalling a group of unusual and puzzling vertebrates, the Antarctic icefish: animals that, although reasonably large, have no O_2 transport by blood hemoglobin because they have no hemoglobin in their blood. Of all vertebrates, the icefish—which we discussed at length in Chapter 3—are the only ones that lack blood hemoglobin as adults. As stressed earlier, their habitat is undoubtedly critical in permitting them to live without blood hemoglobin. The Antarctic seas tend to be consistently well aerated, and the temperature of the water is typically frigid (near –1.9°C) year-round. Because of the low temperature, the solubility of O_2 is relatively high, not only in the ambient water but also in the fish's blood. Despite the advantages of high O_2 solubility, the blood oxygen-carrying capacities of icefish (about 0.7 mL O_2/100 mL) are only about one-tenth as high as those of related red-blooded Antarctic fish (6–7 mL O_2/100 mL). The icefish circulate their blood exceptionally rapidly, evidently to compensate for the fact that each unit of blood volume carries relatively little O_2. They have evolved hearts that are dramatically larger than those of most fish of their body size; with each heartbeat, they therefore pump at least four to ten times more blood than is typical. In the microcirculatory beds of their tissues, they also have blood vessels that are of exceptionally large diameter. These large vessels allow rapid blood flow to occur with exceptionally low vascular resistance.

BOX 24.5 **Blood and Circulation in Mammals at High Altitude**

The study of blood O_2 transport in humans and other mammals at high altitude is, in its own particular way, one of the most intriguing chapters in the annals of evolutionary physiology. This is true because in the past 30 years, the blood responses of lowland people at high altitude have morphed from being touted as exceptional examples of *adaptation* to being cited as defining examples of *maladaptation*. A key reason for the change of perspective has been a gradual recognition of the important point discussed in Box 23.2 that responses of predominantly lowland species—such as humans—at high altitude may sometimes represent misplaced expressions of responses that evolved in lowland populations to meet lowland challenges. For example, when the low atmospheric O_2 partial pressure at

altitude induces tissue hypoxia, the hypoxia might trigger responses—not necessarily advantageous at high altitude—that evolved to help with lowland anemia. As we discuss blood and circulation at high altitude, keep in mind a critical point emphasized in Box 23.2: *Lowland* people and *lowland* species spending time at high altitude need to be distinguished from *native highland* groups.*

If you think back to the oxygen cascade for people in the high Andes in Box 23.2, you will recall that when people—whether native lowlanders or highlanders—are exposed to the reduced atmospheric partial pressure of O_2 at high

---
* Altitude physiology is discussed also in Boxes 8.3 and 23.2 (which includes information on high-flying birds).

altitude, they do not experience an equal reduction in their venous O_2 partial pressure. The venous partial pressure, in fact, is reduced *far* less than the atmospheric partial pressure. A key reason for this conservation of venous O_2 partial pressure is blood O_2 transport. The drop in O_2 partial pressure between arterial and venous blood is much smaller at high altitude than at sea level (see Box 23.2). This smaller drop in O_2 partial pressure is

(Continued)

BOX 24.5 **Blood and Circulation in Mammals at High Altitude** (*Continued*)

important because it helps keep the O_2 partial pressure in the systemic tissues from falling too low.

The principal explanation for the reduced arteriovenous (a-v) drop in O_2 partial pressure at high altitude does not entail any special adaptations. Instead, the reduced a-v drop is simply a consequence of the shape of the mammalian oxygen equilibrium curve. Living at high altitude lowers the arterial O_2 partial pressure. Figure 24.6 illustrates the consequence: When the arterial O_2 partial pressure is moved off the plateau of the equilibrium curve, there is a sharp reduction in the a-v drop in partial pressure required for the blood to yield any particular quantity of O_2.

In the search for special high-altitude adaptations, three aspects of blood and circulation have been studied: (1) the oxygen-carrying capacity of the blood, (2) the hemoglobin O_2 affinity, and (3) the rate of blood circulation.

Regarding the oxygen-carrying capacity, when *lowland* people and some other species of *lowland* mammals go to high altitude, their oxygen-carrying capacities typically rise to well above sea-level values. Secretion of erythropoietin (see Box 24.2) is increased, causing an increase in the number of red blood cells (RBCs) per unit of blood volume: a state known as **polycythemia** ("many cells in the blood"). This change can be dramatic. For example, if lowland people go from sea level to 4000–5000 m, their oxygen-carrying capacity may increase from 20 to 28 mL O_2 per 100 mL of blood. This sort of response was long touted as a vivid illustration of adaptative phenotypic plasticity. By now, however, sufficient comparative data have accumulated that we can make the following statement with good confi-

dence: Species of mammals (and birds) that are *native* to high altitudes do *not* have unusually high RBC concentrations or oxygen-carrying capacities. Moreover, among people, some native highland peoples—notably the Tibetan highlanders—do not exhibit the strong erythropoietin response shown by lowlanders and have oxygen-carrying capacities near those of lowlanders at sea level. Why is an elevated RBC concentration in general *not* favored at high altitude? Researchers now have evidence that an elevated RBC concentration can make the blood too viscous, placing a greater workload on the heart and sometimes interfering with regional blood flow. In an effort to carry out a direct test, researchers have medically removed RBCs from lowland people displaying high RBC concentrations at high altitude; some (but not all) studies of this sort have found that the subjects experienced either no change or an improvement in their ability to function. Overall, careful comparative studies have shown that evolution favors little or no increase in RBC concentration at high altitude. The response of lowland people probably evolved as a mechanism for lowlanders near sea level to compensate for anemia (caused by disease or blood loss) and is a misplaced response—triggered erroneously—at high altitude.

With regard to O_2 affinity, *lowland* humans and some other *lowland* species undergo an increase in the concentration of 2,3-DPG in their RBCs at high altitude. When this change was first discovered, it was claimed to help prevent tissue hypoxia by lowering the O_2 affinity of hemoglobin and thus promoting O_2 unloading into the systemic tissues. By now we realize that this claim might not be even theoretically correct because it

is myopically focused on just one part of the oxygen equilibrium curve and fails to consider effects on loading as well as unloading. More to the point, the collection of comparative data on many additional species now permits confidence in the following conclusion: Species of mammals (and birds) *native* to high altitudes typically have either ordinary O_2 affinities or particularly *high*—sometimes dramatically high—O_2 affinities (which help hemoglobin take up O_2 in the lungs). Thus, if lowland humans at high altitude have a reduced affinity, we must be wary of interpreting it as being beneficial. This topic is discussed further in **Box Extension 24.5**.

Regarding the rate of circulation, although an increase might at first seem logically to be expected at high altitude, cardiac output is not systematically elevated in humans or other mammals, either at rest or at any given level of exercise. An increase in circulatory rate is not a general attribute of high-altitude animals, and theoretical analyses discussed in Box Extension 24.5 clarify why. This said, researchers recently found that in the special case of Tibetan highlanders, circulatory rate is unusually high and a key to limiting tissue hypoxia.

In all, the study of blood and circulation at high altitude has a complex history, which we can see in retrospect got off on the wrong foot because researchers sometimes assumed uncritically that the responses of lowland humans must be beneficial. Taking a broad view, hemoglobin O_2 affinity is often particularly high in native highland mammals and birds, and this is the most convincing generality now known in the study of blood and circulation.

SUMMARY
The Functions of Respiratory Pigments in Animals

■ Respiratory pigments are diverse in their functional properties. The functions they can potentially perform include O_2 transport, facilitation of CO_2 transport, transport of substances other than respiratory gases, blood buffering, facilitation of O_2 diffusion through the cells of solid tissues such as muscle, and O_2 storage in blood or solid tissues.

■ Blood respiratory pigments typically become well oxygenated in the breathing organs, and when animals are at rest, the respiratory pigments typically release only a modest fraction of their O_2 to the systemic tissues (25% in humans). During exercise, O_2 delivery is enhanced by increases in both the extent of pigment unloading and the rate of blood flow.

- The O_2 affinities of respiratory pigments are often critical for pigment function. When O_2 is transferred from one respiratory pigment to another in an individual animal—as when blood hemoglobin donates O_2 to myoglobin—it is usual for the pigment receiving the O_2 to have a higher O_2 affinity. Comparing related species, those with long evolutionary histories in O_2-poor environments often have evolved blood respiratory pigments with particularly high O_2 affinities.

- Respiratory-pigment physiology undergoes acclimation, as by changes in pigment amounts, synthesis of new molecular forms, or modulation of preexisting forms.

Carbon Dioxide Transport

Carbon dioxide dissolves in blood as CO_2 molecules, but usually only a small fraction of the carbon dioxide in blood is present in this chemical form (about 5% in human arterial blood). Thus the first step in understanding carbon dioxide transport is to discuss the other chemical forms in which carbon dioxide exists in blood. Because carbon dioxide can be present in multiple chemical forms, not just CO_2, we must distinguish the material from its exact chemical forms. We do this by speaking of "carbon dioxide" when we refer to the sum total of the material in all its chemical forms and by specifying the chemical form (e.g., CO_2) when we refer to a particular form.

When carbon dioxide dissolves in aqueous solutions, it undergoes a series of reactions. The first is hydration to form *carbonic acid* (H_2CO_3):

$$CO_2 + H_2O \rightleftharpoons H_2CO_3 \qquad \textbf{(24.3)}$$

The second is dissociation of the carbonic acid to yield *bicarbonate* (HCO_3^-) and a proton:

$$H_2CO_3 \rightleftharpoons H^+ + HCO_3^- \qquad \textbf{(24.4)}$$

Bicarbonate can then dissociate further to yield *carbonate* (CO_3^{2-}) and an additional proton. This final dissociation, however, occurs to only a small extent in the body fluids of most animals. Moreover, although carbonic acid is an important intermediate compound, it never accumulates to more than very slight concentrations. For most purposes, therefore, the reaction of CO_2 with water can be viewed simply as yielding HCO_3^- and protons:

$$CO_2 + H_2O \rightleftharpoons HCO_3^- + H^+ \qquad \textbf{(24.5)}$$

Equation 24.5 emphasizes that carbon dioxide *acts as an acid* in aqueous systems because it reacts to produce H^+; as mentioned earlier, it has been aptly termed a "gaseous acid."

The extent of bicarbonate formation depends on blood buffers

Almost no bicarbonate is generated when CO_2 is dissolved in distilled water or a simple salt (NaCl) solution. However, bicarbonate is typically the dominant form in which carbon dioxide exists in the bloods of animals. How can we explain these two, seemingly contradictory, statements? The answer lies in the factors that affect bicarbonate formation, which we now examine.

Suppose that we bring a liter of an aqueous solution—initially devoid of carbon dioxide—into contact with a gas that acts as a source of CO_2, and that this gas remains at a constant CO_2 partial pressure regardless of how much CO_2 it donates to the solution. From Chapter 22, we know that after the solution comes to equilibrium with the gas, the concentration of carbon dioxide in solution in the form of CO_2 will be simply proportional to the CO_2 partial pressure. Thus the amount of CO_2 taken up in *dissolved* form by our liter of solution will depend simply on the principles of gas solubility. In contrast, the extent of bicarbonate formation is governed, not by the principles of solubility, but by the action of compounds that act as *buffers of pH*. In blood, these are the blood buffers. For our immediate purposes, the function of the blood buffers that deserves emphasis is that, under conditions when the concentration of H^+ is being driven upward, they are able to restrain the rise in concentration by removing free H^+ ions from solution (we'll return to a fuller description of buffer function shortly).

How do blood buffers determine the amount of HCO_3^- formation? A straightforward way to see the answer is to return to the analysis of the solution mentioned in the last paragraph and apply the principles of mass action (see page 50) to Equation 24.5. According to the principles of mass action, the following equation holds true at equilibrium:

$$\frac{\left[HCO_3^-\right]\left[H^+\right]}{\left[CO_2\right]} = K \qquad \textbf{(24.6)}$$

where the square brackets signify the concentrations of the various chemical entities, and K is a constant. Because $[CO_2]$ is a constant at equilibrium in our solution at a given CO_2 partial pressure, and because K is also a constant, Equation 24.6 reveals that the amount of HCO_3^- formed per unit volume of solution depends inversely on the H^+ concentration. If $[H^+]$ is kept relatively low, $[HCO_3^-]$ at equilibrium will be relatively high, meaning that a lot of HCO_3^- will be formed as the system approaches equilibrium. However, if $[H^+]$ is allowed to rise to high levels, $[HCO_3^-]$ at equilibrium will be low, meaning little HCO_3^- will be formed. When carbon dioxide enters our solution from the gas and undergoes the reaction in Equation 24.5, the degree to which the H^+ made by the reaction is allowed to accumulate, driving $[H^+]$ up, is determined by the buffers in the solution. If the buffers are ineffective, the H^+ produced by the reaction will simply accumulate as free H^+ in the solution; thus $[H^+]$ will rise rapidly to a high level, and the entire reaction will quickly reach an end point with little uptake of carbon dioxide and little formation of HCO_3^-. However, if the buffers are highly effective, so that most H^+ is removed from solution as it is formed, $[H^+]$ will stay low, and a great deal of carbon dioxide will be able to undergo reaction, causing a large buildup of HCO_3^-.

Let's now speak about buffers in more detail. Buffer reactions are represented by the general equation

$$HX \rightleftharpoons H^+ + X^- \qquad \textbf{(24.7)}$$

where X^- is a chemical group or compound that can combine reversibly with H^+. When H^+ is added to a buffered solution, the buffer reaction is shifted to the left, removing some of the H^+ from free solution (as already stressed). However, if H^+ is extracted from a buffered solution, the reaction shifts to the right, releasing free H^+ from compound HX. In brief, a buffer reaction acts to stabilize $[H^+]$. Together, HX and X^- are termed a **buffer pair**. According to the principles of mass action, the following equation describes a buffer reaction at equilibrium:

$$\frac{[H^+][X^-]}{[HX]} = K' \qquad \text{(24.8)}$$

where K' is a constant that depends on the particular buffer reaction and the prevailing conditions, notably temperature. The negative of the common logarithm of K' is symbolized **pK'**, just as the negative of the logarithm of [H$^+$] is called pH. The *effectiveness* with which a particular buffer reaction (a particular buffer pair) is able to stabilize [H$^+$] is greatest when half of the X$^-$ groups are combined with H$^+$ and half are not; that is, the change in pH caused by the addition or removal of H$^+$ is minimized when [HX] = [X$^-$]. From Equation 24.8, it is clear that for [HX] and [X$^-$] to be equal, [H$^+$] must equal K'; that is, pH must equal pK'. Therefore the buffering effectiveness of any given buffer reaction is greatest when the prevailing pH matches the pK' of the reaction. Applying this principle to the blood of an animal (it also applies to other solutions), we can say that the blood may contain an enormous variety of potential buffer pairs, but typically *the buffer reactions that are important will be those with pK' values within one pH unit of the pH prevailing in the blood.*

The blood of mammals and most other vertebrates is highly effective in buffering the H$^+$ generated from CO$_2$ because the blood has a high concentration of effective buffer groups. These groups are found mostly on blood protein molecules, *especially hemoglobin!* Two types of chemical groups are particularly noteworthy as buffer groups because they are abundant and have appropriate pK' values: the terminal amino groups of protein chains and the imidazole groups found wherever the amino acid histidine occurs in protein structure. *The imidazole groups are the dominant buffering groups.* The buffering of human blood is so effective that when CO$_2$ undergoes the reaction in Equation 24.5, forming HCO$_3^-$ and H$^+$, the buffer groups remove more than 99.999% of the H$^+$ produced from free solution! This buffering permits a great deal of HCO$_3^-$ to be formed. Thus the blood can take up a great deal of carbon dioxide.

Carbon dioxide transport is interpreted by use of carbon dioxide equilibrium curves

Blood equilibrium curves for carbon dioxide have interpretive value similar to that of oxygen equilibrium curves. To understand the use of carbon dioxide equilibrium curves, we must first establish the meaning of the **total carbon dioxide concentration** of the blood. Suppose that some blood is brought to equilibrium with an atmosphere containing no CO$_2$, so that the CO$_2$ partial pressure of the blood is zero. Suppose that the blood is then exposed to an atmosphere containing CO$_2$ at some fixed, positive partial pressure. And suppose that as the blood comes to equilibrium with the new atmosphere, we measure the total quantity of CO$_2$ it takes up, regardless of the chemical form assumed by the CO$_2$ in the blood. This quantity—the total amount of CO$_2$ that must enter each unit of blood volume to raise the blood CO$_2$ partial pressure from zero to any particular positive CO$_2$ partial pressure—is termed the blood's *total carbon dioxide concentration* at that partial pressure. A plot of the total carbon dioxide concentration as a function of CO$_2$ partial pressure is known as a **carbon dioxide equilibrium curve** or **carbon dioxide dissociation curve (FIGURE 24.21A)**.

What determines the *shape* of the carbon dioxide equilibrium curve? In mammals, carbon dioxide exists in blood in three principal

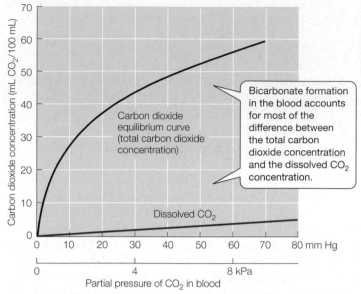

(A) Human arterial blood

Carbon dioxide equilibrium curve (total carbon dioxide concentration)

Dissolved CO$_2$

Bicarbonate formation in the blood accounts for most of the difference between the total carbon dioxide concentration and the dissolved CO$_2$ concentration.

Partial pressure of CO$_2$ in blood

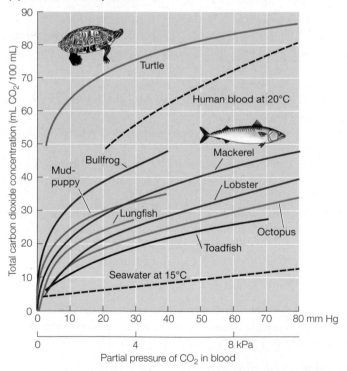

(B) Blood of nine species

Turtle

Human blood at 20°C

Mackerel

Bullfrog

Mud-puppy

Lobster

Lungfish

Octopus

Toadfish

Seawater at 15°C

Partial pressure of CO$_2$ in blood

FIGURE 24.21 Carbon dioxide equilibrium curves (A) The carbon dioxide equilibrium curve of fully oxygenated human blood at normal body temperature. The portion of the total carbon dioxide concentration attributable to dissolved CO$_2$ is shown at the bottom. (B) Carbon dioxide equilibrium curves for oxygenated blood of nine species at 15°–25°C. Because all curves were not determined at exactly the same temperature, some of the differences among curves may arise from temperature effects. Species: bullfrog, *Rana catesbeiana* (sometimes called *Lithobates catesbeianus*); lobster, *Panulirus vulgaris*; lungfish, *Neoceratodus forsteri*; mackerel, *Scomber scombrus*; mudpuppy, *Necturus maculosus*; octopus, *Octopus macropus*; toadfish, *Opsanus tau*; turtle, *Pseudemys floridana*. (After Hill and Wyse 1989.)

chemical forms, and thus the total carbon dioxide concentration has three components. Two, as we have already discussed, are dissolved CO_2 and HCO_3^-. The third is carbon dioxide that is directly chemically combined (in a reversible manner) with amino groups on hemoglobin and other blood proteins, forming **carbamate groups** (—NH—COO$^-$) (also called *carbamino groups*). Typically, in both mammals and other types of animals, the great preponderance of blood carbon dioxide is in the form of HCO_3^-; 90% of the carbon dioxide in human blood, for example, is in that form. The shapes of the carbon dioxide equilibrium curves of animals are thus determined largely by the kinetics of HCO_3^- formation in their bloods. This means that the shapes depend on the blood buffer systems: the concentrations of buffer groups, their pK' values, and the extent to which their capacities for H^+ uptake are being called upon to buffer acids other than CO_2.

A diversity of carbon dioxide equilibrium curves is found in the animal kingdom (**FIGURE 24.21B**). If we compare air-breathing and water-breathing animals, we find that they typically operate on substantially different parts of their carbon dioxide equilibrium curves. The reason, as discussed in Chapter 23 (see Box 23.1), is that air breathers typically have far higher arterial CO_2 partial pressures than water breathers do. For example, the systemic arterial CO_2 partial pressure in resting mammals and birds breathing atmospheric air— being at least 3.3 kPa (25 mm Hg)—is far higher than that commonly observed in gill-breathing fish in well-aerated waters, 0.1–0.4 kPa (1–3 mm Hg). In air breathers, the CO_2 partial pressure of blood rises from a high arterial value to a still higher venous value as the blood circulates through the systemic tissues, meaning that the part of the carbon dioxide equilibrium curve that is used is the part at relatively high CO_2 partial pressures. In water breathers, by contrast, both the arterial and venous CO_2 partial pressures are relatively low; the part of the equilibrium curve that is used by water breathers is therefore the steep part at relatively low CO_2 partial pressures.

The Haldane effect: The carbon dioxide equilibrium curve depends on blood oxygenation

The carbon dioxide equilibrium curve of an animal's blood commonly changes with the state of oxygenation of the respiratory pigment (the O_2-transport pigment) in the blood, a phenomenon named the **Haldane effect** after one of its discoverers. When a Haldane effect is present, *deoxygenation promotes CO_2 uptake by the blood, whereas oxygenation promotes CO_2 unloading.* Thus the total carbon dioxide concentration at any given CO_2 partial pressure is greater when the blood is deoxygenated than when it is oxygenated (**FIGURE 24.22**). The reason for the Haldane effect is that the buffering function of the respiratory pigments—which play major buffer roles—depends on their degree of oxygenation. Deoxygenation of a respiratory pigment alters its buffering function in such a way that it tends to take up more H^+ and lower the blood concentration of H^+. According to Equation 24.6, this means that when a respiratory pigment becomes deoxygenated, more HCO_3^- can form, and the blood therefore reaches a higher total carbon dioxide concentration. This phenomenon is the necessary converse of the Bohr effect, as noted earlier (see page 648).

The functional significance of the Haldane effect is illustrated in the inset of Figure 24.22 using CO_2 transport in resting humans as an example. Point A shows the total carbon dioxide concentration and CO_2 partial pressure in arterial blood, whereas point V shows the values in venous blood. The arrows between A and V represent the *functional relation between total carbon dioxide concentration and CO_2 partial pressure in the body*, where oxygenation changes simultaneously with the uptake and release of CO_2. Note that the slope of this functional relation is steeper than the slope of any of the equilibrium curves in Figure 24.22 for blood at a fixed level of oxygenation (red and purple lines). Thus, when the CO_2 partial pressure shifts back and forth between its values in arterial and venous blood (A and V), the blood takes up and releases more

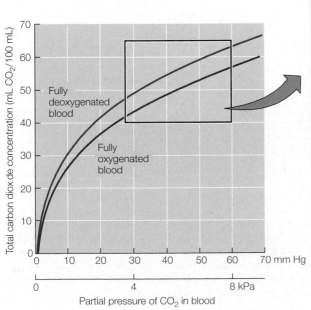

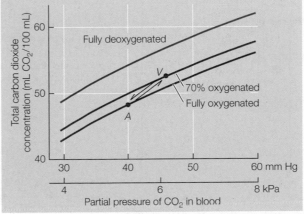

FIGURE 24.22 The Haldane effect and its implications for human carbon dioxide transport The principal graph (left) shows carbon dioxide equilibrium curves for fully oxygenated and essentially fully deoxygenated human blood, illustrating the Haldane effect. The inset (above) summarizes carbon dioxide *transport* in humans at rest. Point A represents arterial blood, which is fully oxygenated and has a CO_2 partial pressure of about 5.3 kPa (40 mm Hg). Point V represents mixed venous blood, which is about 70% oxygenated and has a CO_2 partial pressure of about 6.1 kPa (46 mm Hg). The arrows show the functional relation between total carbon dioxide concentration and CO_2 partial pressure as blood circulates through the body, becoming alternately arterial (A) and venous (V).

CO_2 than would be possible without the Haldane effect. In this way, hemoglobin function *simultaneously* aids CO_2 transport and O_2 transport!

Critical details of vertebrate CO_2 transport depend on carbonic anhydrase and anion transporters

An important attribute of the hydration of CO_2 to form bicarbonate (Equation 24.5) is that it occurs relatively slowly in the absence of catalysis (requiring a minute or so to reach equilibrium). The native slowness of this reaction presents a potential bottleneck in the blood's ability to take up CO_2 as bicarbonate in the systemic tissues and release CO_2 from bicarbonate in the lungs. The enzyme **carbonic anhydrase (CA)** greatly accelerates the interconversion of CO_2 and HCO_3^-, thereby preventing this reaction from acting as a bottleneck.[21] The reaction is the only one in CO_2 transport known to be catalyzed.

The morphological location of CA has important consequences for CO_2 transport. In vertebrates, CA is found within the red blood cells but almost never free in the blood plasma. Sometimes CA is

[21] When CO_2 is hydrated to form HCO_3^- by carbonic anhydrase catalysis, H_2CO_3 is not formed as an intermediate. Instead, the reaction proceeds by a pathway not involving H_2CO_3 formation.

also found associated with the inner endothelial walls of blood capillaries, such as lung or skeletal muscle capillaries. A key point is that *CA is both essential and localized*.

With this in mind, let's discuss what happens when CO_2 from metabolism enters the blood in a systemic capillary (**FIGURE 24.23**). CO_2 diffuses readily into the red blood cells (possibly mediated in part by aquaporin AQP-1). There it encounters CA and is quickly converted to HCO_3^- and H^+. In fact, if there is no membrane-bound CA in the capillary walls or other CA outside the red blood cells, virtually all the reaction of CO_2 to form HCO_3^- and H^+ occurs inside the cells. Hemoglobin—the most important blood buffer—is immediately available inside the red blood cells to take up H^+ and thus play its critical role in promoting HCO_3^- formation. In fact, because hemoglobin is undergoing deoxygenation as CO_2 is added to the blood, hemoglobin develops a greater affinity for H^+ just as it is needed. The red blood cell membranes of nearly all vertebrates are well endowed with a transporter protein—a *rapid anion exchange protein* (often termed the **band 3 protein**)—that facilitates diffusion of HCO_3^- and Cl^- across the membranes in a 1:1 ratio. The HCO_3^- that is formed and buffered inside red cells thus tends to diffuse out into the plasma, so that the plasma ultimately carries most of the HCO_3^- added to the blood in the systemic capillaries. As HCO_3^- diffuses

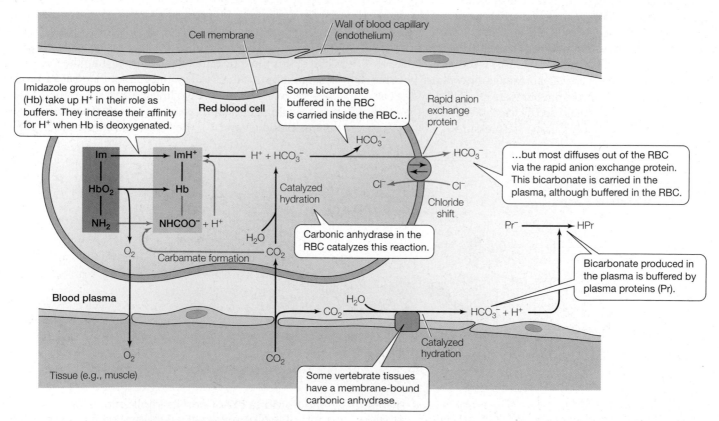

FIGURE 24.23 The major processes of CO_2 uptake by the blood in a systemic blood capillary of a vertebrate The molecules boxed in red and purple (HbO_2 and Hb) represent oxygenated and deoxygenated hemoglobin, respectively. Three elements of molecular structure are highlighted in each hemoglobin molecule: the O_2-binding site (symbolized Hb), an imidazole buffering group (Im), and an amino group (NH_2) that can participate in carbamate forma-

tion. The processes shown in this figure occur in reverse in the lungs or gills. Where the labels refer to the site of buffering, they are specifying where the H^+ generated during bicarbonate production is taken up by buffering compounds. Carbamate formation occurs to a significant extent in mammals, but not necessarily in other vertebrates. Hb, hemoglobin; RBC, red blood cell; Pr, plasma proteins.

out of the red blood cells into the plasma, Cl⁻ diffuses into the cells from the plasma—a process called the **chloride shift**. In tissues that have a membrane-bound CA associated with the blood-capillary endothelium, such as the skeletal muscles of at least certain vertebrates, some rapid formation of HCO_3^- and H^+ occurs in the plasma, where the HCO_3^- must be buffered by plasma proteins. All these events occur in reverse when the blood flows through the lungs or gills.

The operations of these kinetic details govern the exact ways in which CO_2 is transported under any set of conditions. For example, from recent research, we know that the operations of the kinetic details differ during exercise and rest—a difference that may ultimately prove to be critical for a full understanding of exercise physiology.

SUMMARY
Carbon Dioxide Transport

■ The carbon dioxide equilibrium curve, which shows the relation between the total carbon dioxide concentration of blood and the CO_2 partial pressure, is a key tool for analyzing carbon dioxide transport. In water breathers, the CO_2 partial pressures of both systemic arterial blood and systemic venous blood are typically low and on the steep portion of the carbon dioxide equilibrium curve. In air breathers, blood CO_2 partial pressures tend to be far higher and therefore on the flatter portion of the carbon dioxide equilibrium curve.

■ Most carbon dioxide carried in blood is typically in the form of bicarbonate, HCO_3^-. The extent of HCO_3^- formation depends on blood buffers and determines the shape of the carbon dioxide equilibrium curve. Because respiratory pigments are major blood buffers, they play major roles in carbon dioxide transport.

■ The Haldane effect, which is in part the necessary converse of the Bohr effect, is an increase in the total carbon dioxide concentration of the blood caused by deoxygenation of the respiratory pigment. The Haldane effect aids carbon dioxide transport by promoting CO_2 uptake by the blood in the systemic tissues and CO_2 loss from the blood in the breathing organs.

■ Rapid uptake of CO_2 by the blood or loss of CO_2 from the blood requires the action of carbonic anhydrase, an enzyme localized to certain places (e.g., red blood cells).

Acid–Base Physiology

The pH of the body fluids cannot vary far from normal levels without serious functional consequences. In humans, for instance, the normal pH of arterial blood at 37°C is about 7.4, and a person will lie near death if his or her pH rises to just 7.7 or falls to 6.8! Abnormal H^+ concentrations inflict their adverse effects to a large extent by influencing the function of proteins. As buffer groups on a protein molecule take up or lose H^+, the electrical charge of the whole molecule is rendered more positive or negative; beyond certain limits, these changes result in changes in molecular conformation or other properties that interfere with proper protein function. For the electrical-charge and ionization status of a protein molecule

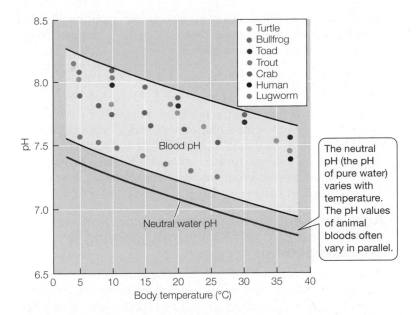

FIGURE 24.24 Normal blood pH is a temperature-dependent variable As the neutral pH varies with body temperature, blood pH—which is more alkaline than neutral—often varies in parallel; in species that follow this pattern, the blood pH tends to remain alkaline to a fixed extent. Data are shown for six poikilothermic species and for humans. Species differ in how alkaline their blood is relative to the neutral pH. (After Dejours 1981.)

to remain within limits compatible with protein function, the pH must remain within parallel limits.

The neutral pH is defined to be the pH of pure water. By this definition, as shown in **FIGURE 24.24**, the neutral pH varies with temperature, being higher at low temperatures than at high ones. In poikilotherms, the normal blood pH often varies with body temperature in parallel with the neutral pH. Specifically, a common pattern is that, within a species, the blood pH is displaced by a relatively fixed amount to the alkaline side of the neutral pH, rising and falling with body temperature to maintain this fixed displacement (see Figure 24.24). A species that follows this pattern is said to maintain a *constant relative alkalinity* of blood pH. In a species of this sort, the pH inside cells, the intracellular pH, also parallels neutral pH (although intracellular pH and blood pH are different from each other).

In the early days of the study of acid–base physiology, studies of humans and other large mammals gave rise to the notion that the pH in any particular region of the body is always regulated at a single, invariant level (e.g., 7.4 in human arterial blood). We now realize that this type of pH regulation is a special case that occurs only in animals that maintain a constant deep-body temperature. Considering all animals taken together, the pH that is maintained by acid–base regulatory mechanisms—whether in the blood or inside cells—is more commonly a temperature-dependent variable.

What advantage might animals gain by increasing their pH as their body temperature falls? According to the leading hypothesis, the **alphastat hypothesis**, the *changes* in pH are a means of maintaining a *constant* state of electrical charge on protein molecules.

The reason that changes in pH are required is that changes in temperature alter the chemical behavior of the buffer groups on protein molecules. Most importantly, as temperature falls, the pK' values of imidazole groups increase. This means that at reduced temperatures, imidazole groups increase their inherent tendency to combine with H^+. If this change in chemical behavior were unopposed, more of the imidazole groups on proteins would be combined with H^+ at low temperatures than at high ones. Decreasing the H^+ concentration (raising the pH) at low temperatures serves to oppose the heightened tendency of the imidazole groups to take up H^+. Accordingly, it helps prevent the proportion of positively charged groups on proteins from changing.

Acid–base regulation involves excretion or retention of chemical forms affecting H^+ concentration

When a process occurs that tends to cause a protracted increase in the amount of acid in the body of an animal, maintenance of the animal's temperature-dependent normal pH requires that other processes be set in motion that will either export acid from the body or increase the body's content of base. Conversely, if a disturbance occurs that decreases body acid, acid–base regulation requires a compensating uptake of acid or export of base. There are two competing "worldviews" of acid–base regulation.[22] In our brief overview here, we adopt the simpler of the two, which focuses on adjustments in CO_2, H^+, and HCO_3^-.

The concentration of CO_2 in the body fluids of an animal can be raised or lowered to assist acid–base regulation. This is especially true in terrestrial animals, which (in contrast to aquatic ones) normally have relatively high blood partial pressures of CO_2. Suppose that a person's blood becomes too acidic. One possible compensatory response is for the person to increase lung ventilation, thereby lowering the CO_2 partial pressure in the blood and other body fluids. Lowering the blood CO_2 partial pressure will pull Equation 24.5 to the left and thus lower the blood H^+ concentration. Slowing of lung ventilation, by contrast, can assist with acid–base regulation if the body fluids become too alkaline. The slowing of ventilation will promote accumulation of CO_2 in the body fluids and cause Equation 24.5 to be shifted to the right, providing more H^+.

Animals often have the ability to exchange H^+ itself with the environment, and this ability also can be used for acid–base regulation. Because H^+ is not a gas, it must be transported in liquid solution. In terrestrial animals, responsibility for the export of H^+ from the body rests with the kidneys. Humans, for example, are routinely confronted with an excess of H^+ from their diet, and they void the excess principally in their urine; this urinary elimination of H^+ can be curtailed entirely, however, when appropriate. In aquatic animals, including both fish and crustaceans, H^+ is exchanged with the environment by the gill epithelium (see Box 5.2, for example).

Bicarbonate ions are also exchanged with the environment to assist acid–base regulation. The HCO_3^- exchanges are mediated principally by the kidneys in terrestrial animals but, it appears, principally by the gill epithelium in fish and crabs (see Box 5.2). Bicarbonate functions as a base. If retention of HCO_3^- in the body is increased, Equation 24.5 is shifted to the left, tending to remove H^+ from solution in the body fluids, making the body fluids more alkaline. Conversely, increased elimination of HCO_3^- tends to raise the H^+ concentration of the body fluids.[23]

Disturbances of acid–base regulation fall into respiratory and metabolic categories

Disturbances of the pH of the body fluids are categorized as **acidosis** or **alkalosis**. Acidosis occurs when the pH of the body fluids is shifted to the acid side of an animal's normal pH at a given body temperature. Alkalosis is a shift in pH to the alkaline side of an animal's normal pH. Disturbances of pH are also classified as **respiratory** or **metabolic** according to their primary cause.

The *respiratory* disturbances of pH are ones that are brought about by an abnormal rate of CO_2 elimination by the lungs or gills. **Respiratory alkalosis** arises when the exhalation of CO_2 is abnormally increased relative to CO_2 production, causing the CO_2 partial pressure in the body fluids to be driven below the level needed to maintain a normal pH. Panting by mammals, for example, sometimes causes respiratory alkalosis (see page 272). **Respiratory acidosis** occurs when the exhalation of CO_2 is impaired and metabolically produced CO_2 therefore accumulates excessively in the body. Prolonged breath-holding, for example, can cause respiratory acidosis.

Whereas the blood property that is *initially* altered in *respiratory* disturbances of pH is the CO_2 partial pressure, *metabolic* disturbances of pH—by definition—initially alter the blood *bicarbonate* concentration. **Metabolic alkalosis** and **metabolic acidosis** both have numerous possible causes. Metabolic acidosis, for example, can result from excessive loss of HCO_3^- in gastrointestinal fluids during chronic diarrhea. Metabolic acidosis can also result from excessive addition of H^+ to the body fluids, as when lactic acid is accumulated during vigorous exercise; the added H^+ from lactic acid reacts with the pool of HCO_3^- in the body fluids, lowering the concentration of HCO_3^-.

Animals typically respond to disturbances of pH by marshaling their acid–base regulatory mechanisms. Lung ventilation by human athletes performing work of ever-increasing intensity provides a striking and interesting example. When athletes are not accumulating lactic acid, they simply increase their rate of lung ventilation in parallel with their rate of CO_2 production. However, when athletes work intensely enough that they accumulate lactic acid, they increase their rate of lung ventilation more than their rate of CO_2 production. This disproportional increase in ventilation, an example of *hyperventilation*, causes CO_2 to be exhaled from the body faster than it is being produced. The CO_2 partial pressure in the blood and body fluids is thereby lowered, helping to limit the degree of acidosis caused by the accumulation of lactic acid.

In the study of global climate change, a concern that has recently become a primary focus is ocean acidification (**BOX 24.6**).

---

[22] The books by Davenport and Stewart in the References and Additional Readings, respectively, provide readable introductions to these two worldviews. For those who become interested in the strong ion difference approach, not covered here, the reference by Johnson et al. in the Additional Readings is also worthwhile.

[23] One way to view this effect of HCO_3^- elimination is to recognize that HCO_3^- originates from H_2CO_3; when HCO_3^- is eliminated, just the H^+ of H_2CO_3 remains in the body fluids, acidifying them.

BOX 24.6 Acidification of Aquatic Habitats

Compared with the bloods of animals, the waters of most natural aquatic habitats, such as the oceans, are not buffered in a way that would significantly impede a rise in H+ concentration caused by addition of acidic materials. *Acid rain* has been a recognized problem in bodies of freshwater—streams, ponds, and lakes—for many decades. It is often caused by sulfur and nitrogen oxides liberated into the atmosphere by the combustion of fossil fuels. The oxides react with water to form acids, such as sulfuric acid.

Acidification caused by atmospheric CO_2 is a more recently identified challenge. The CO_2 concentration of Earth's atmosphere has risen from about 300 parts per million (ppm) to about 400 ppm in the past century because of the burning of wood, coal, and petroleum. Because of the principles we have discussed, the CO_2 concentration in the waters of the ocean has increased (see Chapter 22), and this has made the oceans more acidic by about 0.1 pH unit by driving the reaction shown in Equation 24.5 to the right. Animals are not always able to regulate processes that are affected by the environmental acidification. For example, the acidification tends to erode certain types of animal skeletons that are composed of calcium carbonate (e.g., skeletons of reef corals) or to interfere with production of such skeletons. It can also interfere with sensory and developmental processes. **Box Extension 24.6** provides references for further reading.

Rising atmospheric CO_2 acidifies oceans, leading to chemical reactions that can dissolve calcium carbonate skeletons

$$CO_2$$

$$CO_2 + H_2O \rightarrow HCO_3^- + H^+$$

$$H^+ + CO_3^{2-} \rightarrow HCO_3^-$$

$$CaCO_3 \rightarrow Ca^{2+} + CO_3^{2-}$$
(calcified skeletons and other structures)

SUMMARY
Acid–Base Physiology

■ The neutral pH varies with temperature, being higher at low temperatures than at high ones. In animals with variable body temperatures, the normal blood pH often varies in parallel with the neutral pH, being displaced in the alkaline direction to a constant extent (constant relative alkalinity).

■ Acidosis and alkalosis are categories of acid–base disturbance. They occur, respectively, when the blood pH is to the acid or alkaline side of an animal's normal pH for the prevailing body temperature. Either sort of disturbance can be respiratory (originating because of changes in CO_2 loss by breathing) or metabolic (originating because of changes in the blood bicarbonate concentration).

■ Within their range of acid–base regulation, animals correct chronic acid–base disturbances by modulating the elimination of CO_2, H+, and HCO_3^- in regulatory ways.

STUDY QUESTIONS

1. While touring a saltwater aquarium, suppose you see a striped bass, a hammerhead shark, an octopus, a feather-duster worm, and a lobster. What type of respiratory pigment would you expect to find in the blood of each?

2. One could say that a respiratory pigment with relatively low O_2 affinity is potentially disadvantageous for loading, but advantageous for unloading. Explain both parts of this statement.

3. In most species of mammals, the O_2 affinity of a fetus's blood hemoglobin is greater than that of its mother's blood hemoglobin. However, mammal species are not all the same in the *mechanism* that causes the affinities to be different. Specify three distinct mechanisms for the difference in affinity between fetal and maternal blood hemoglobin. Recall from Chapter 1 that François Jacob argued that evolution is analogous to tinkering rather than engineering. Considering the mechanism of the fetal–maternal difference in O_2 affinity, would you say that the evolution of the mechanism provides evidence for Jacob's argument? Explain.

4. Outline the ways in which mammalian hemoglobin *simultaneously* plays important roles in O_2 transport, CO_2 transport, and control of blood pH.

5. A fish swims from a body of cool water into a body of warm water. As its body temperature rises, its rate of O_2 consumption increases. The warm water, however, is likely to have a lower concentration of dissolved O_2 than the cool water because the solubility of O_2 in water decreases as temperature increases. These two factors taken together—an increase in the fish's rate of O_2 consumption and a decrease in the dissolved O_2 concentration of its environmental water—can make it difficult for the fish to obtain enough O_2 to meet its needs. The two factors can act as a two-pronged trap. Actually, however, the fish may face a three-pronged trap. How is the increase in temperature likely to affect the O_2 affinity of the fish's hemoglobin, and how could the effect on hemoglobin add even further to the challenge the fish faces? Does global warming pose concerns of this sort?

6. When fishing boats pull trawling nets through the water, many fish avoid being caught by vigorously swimming away. Others, after vigorous escape swimming, get caught, but later are released because they are not of legal size or are not the species desired. Fish that escape or are released sometimes die anyway. The accumulation of lactic acid from anaerobic work in such fish seems in certain cases to be a key factor in their deaths. How could a large lactic acid accumulation in a fish interfere with its ability to obtain enough O_2 to survive?

7. To study the chemical properties of the blood hemoglobin of a vertebrate, it might seem convenient to remove the hemoglobin from the red blood cells so that the hemoglobin is in simple aqueous solution. However, removing the hemoglobin from red blood cells often promptly alters its O_2-binding characteristics. Why?

8. Give an example of respiratory alkalosis and one of metabolic acidosis. In each case, explain how your example illustrates that type of acid–base disturbance.

9. As noted in this chapter, respiratory pigments that are dissolved in the blood plasma usually have very high molecular weights. The statement has been made that in animals with dissolved respiratory pigments, "the polymerization of unit respiratory-pigment molecules into high-molecular-weight polymers allows the blood solution to have a high oxygen-carrying capacity without having its osmotic pressure boosted to high levels by the presence of the respiratory pigment." Explain. (Hint: Review in Chapter 5 how dissolved entities affect the colligative properties of solutions.)

10. Among related species (e.g., vertebrates), there is often a positive correlation between the *oxygen*-carrying capacity of blood and the height of the *carbon dioxide* equilibrium curve (the total carbon dioxide concentration at high CO_2 partial pressures). Why?

11. Studies have shown that reindeer and muskox, two Arctic mammals (see Chapter 11), have evolved hemoglobins that are unusually low in their sensitivity to temperature compared with the hemoglobins of most large mammals. The researchers who made this discovery hypothesized that a particularly low thermal sensitivity is required for hemoglobin to unload O_2 to an adequate extent in the distal parts of these animals' legs, where tissue temperatures may be 25°C cooler than in the thorax (see Figure 10.32). Explain the rationale for this hypothesis. In answering, discuss how the temperature of the hemoglobin molecules changes as blood flows between the lungs and the legs in cold weather. If possible, design studies to test the hypothesis.

Go to **sites.sinauer.com/animalphys4e**
for box extensions, quizzes, flashcards,
and other resources.

REFERENCES

Berenbrink, M. 2007. Historical reconstructions of evolving physiological complexity: O_2 secretion in the eye and swimbladder of fishes. *J. Exp. Biol.* 209: 1641–1652.

Cameron, J. N. 1989. *The Respiratory Physiology of Animals*. Oxford University Press, New York. Extensive discussion of blood gas transport, including a set of thorough case studies of animals such as blue crabs, trout, and humans.

Carter, A. M. 2009. Evolution of factors affecting placental oxygen transfer. *Placenta* 23: S19–S25.

Davenport, H. W. 1974. *The ABC of Acid–Base Chemistry*, 6th ed. University of Chicago Press, Chicago. A peerless text for learning the concepts of the traditional theory of acid–base physiology, in which CO_2, HCO_3^-, and H^+ are the major players.

Dickerson, R. E., and I. Geis. 1983. *Hemoglobin: Structure, Function, Evolution, and Pathology*. Benjamin-Cummings, Menlo Park, CA. A book one will never forget. The combination of science and art is masterful.

Doney, S. C., V. J. Fabry, R. A. Feely, and J. A. Kleypas. 2009. Ocean acidification: the other CO_2 problem. *Annu. Rev. Mar. Sci.* 1: 169–192.

Jensen, F. B. 2004. Red blood cell pH, the Bohr effect, and other oxygenation-linked phenomena in blood O_2 and CO_2 transport. *Acta Physiol. Scand.* 182: 215–227. This is one of several commendable reviews that appeared together on the occasion of the 100th anniversary of the original publication of the Bohr effect. Although it has a fairly narrow focus, the paper is accessible, and reading it provides a good experience of seeing how systems interact in the transport of the respiratory gases.

Paul, R. J., B. Zeis, T. Lamkemeyer, and two additional authors. 2004. Control of oxygen transport in the microcrustacean *Daphnia*: regulation of haemoglobin expression as central mechanism of adaptation to different oxygen and temperature conditions. *Acta Physiol. Scand.* 182: 259–275.

Schneider, H. 2011. Oxygenation of the placental-fetal unit in humans. *Respir. Physiol. Neurobiol.* 178: 51–58.

Storz, J. F., G. R. Scott, and Z. A. Cheviron. 2010. Phenotypic plasticity and genetic adaptation to high-altitude hypoxia in vertebrates. *J. Exp. Biol.* 213: 4125–4136.

Waser, W., and N. Heisler. 2005. Oxygen delivery to the fish eye: Root effect as crucial factor for elevated retinal P_{O_2}. *J. Exp. Biol.* 208: 4035–4047.

Weber, R. E. 2007. High-altitude adaptations in vertebrate hemoglobins. *Resp. Physiol. Neurobiol.* 158: 132–142. A superlative treatment of vertebrate hemoglobins at altitude.

West, J. B., R. B. Shoene, A. M. Luks, and J. S. Milledge. 2012. *High Altitude Medicine and Physiology*, 5th ed. CRC Press, Boca Raton, FL. J. B. West has been a leader in the medical study of the physiology humans at high altitude, and this book provides a first-rate introduction to the field in detail.

See also **Additional References** *and* **Figure and Table Citations**.

When we look at the dynamism of an athlete's muscles during intense activity, we can imagine that the muscles are a force unto themselves. The muscles, however, can be only as effective as their circulation permits them to be. For each cell in a muscle, the streaming of blood in nearby capillaries is a lifeline for resupply with O_2 and other necessities. Although a muscle may have a highly developed contractile apparatus, it is able to contract only with as much endurance and power as its rate of ATP production permits, and during sustained exertion, the capacity of a muscle to produce ATP is determined by the rate at which O_2 is brought to it by the circulation of the blood (see Chapter 9).

In ancient human societies, food was obtained and battles were fought with human muscles. Thus the rate of O_2 supply to the muscles by the circulation helped determine whether people and their families thrived or perished. In industrialized societies, the rate of O_2 supply to the muscles by the circulation helps determine whether a person can perform a job requiring hard manual labor—or be a successful basketball player, long-distance runner, or tennis champion.

Besides transporting commodities such as O_2, the circulatory system also performs other vital functions. Blood pressure is used to initiate the formation of urine. Blood pressure is

The heart circulates blood to every tissue and organ, sustaining all bodily strength   For most of human history, people did their work and fought their wars with human muscle—at intensities dependent on blood circulation. For the most part today, we realize the potency of muscle action only in more limited contexts, such as participation in sports or outdoor recreations. Circulation is always the indispensible companion, nonetheless. The muscles can work no harder than their circulation permits during most forms of exertion.

also used to stiffen erectile tissues in both sexes during sexual intercourse. In men, the circulatory system of the testes plays a thermoregulatory role, helping to keep the testes cooler than the abdominal organs. Vascular heat exchangers in the arms of both women and men—and in the appendages of various other species of mammals—reduce energy costs by limiting heat losses from the body in chilly environments.

Ancient people undoubtedly were as aware as we are that when the body is cut, it oozes blood regardless of where the cut is made. This observation was the first sign that blood at high pressure streams through every region of tissue. The idea that the blood makes a round-trip through the body—the concept that it circulates—was first put forth by William Harvey in 1628. However, Harvey and his contemporaries could not possibly have understood the circulation as we do today, because blood capillaries were unknown in his time, and O_2 was not identified as a defined chemical element for another 150 years. From Harvey's time to the present, scientists have learned steadily more about the tasks that are accomplished by the circulation of the blood.

The very word *circulation* has taken on a progressively different meaning as knowledge has expanded. For Harvey, *circulation* meant round-trip blood flow in major blood vessels. A generation after Harvey, early microscopists discovered that minute blood vessels just barely wider than red blood cells—the capillaries—weave among the cells of every tissue. *Circulation* then included tissue blood flow, and bleeding from every cut could finally be understood. In the nineteenth century, scientists demonstrated that the blood brings O_2 to cells (see Box 7.1). In the twentieth century, hormones, antibodies, and immune cells were discovered, and the concept of *circulation* was expanded to include the transport of these agents from one region of the body to another.

To define **circulation** today in a way that is relevant to all kinds of animals, two perspectives can be taken. From the perspective of mechanics, circulation is the pressure-driven bulk flow of a body fluid called **blood** through a system of tubular vessels or other passages that brings the fluid to all parts of the body. The system of vessels or other blood passages—plus the blood itself—is called the **circulatory system**. When we think of circulation, however, we usually do not think of it in only these mechanical terms. From a second perspective, circulation is defined by what it accomplishes. Thus, for us today, circulation is a pressure-driven bulk flow of fluid that rapidly transports O_2, CO_2, nutrients, organic wastes, hormones, agents of the immune system, heat, and other commodities throughout the body and that often provides a source of hydraulic pressure for organ function.

The *speed* of transport by the circulation is one of its most central and defining attributes. As we first saw in discussing Table 5.1, diffusion through aqueous solutions is too slow to transport commodities at biologically significant rates over distances exceeding 1 mm or so. Only very small animals, therefore, can depend on diffusion as their sole means of internal transport (see Box 22.1). Convective transport—transport by bulk flow of body fluids—is intrinsically far faster than diffusion. Consequently, as stressed in Chapter 22 (see Figure 22.7), animals that are larger than 1 mm or so generally require blood circulation (or some other form of bulk flow of body fluids) to move commodities from place to place in their bodies at adequate rates.

As we study the circulation, both of the defining perspectives we have identified will be important. Looking back at the person in our opening photograph, for example, one of our two key questions must be how his circulatory system itself works: How do his heart and vascular system function to bring blood to and from the cells in his head, hands, skeletal muscles, and all other parts of his body at the rate required, and how are these processes regulated? Our second key question must focus on consequences: What functions are accomplished by his circulation?

In humans and in most other types of animals, the transport of O_2 is by far the most pressing and urgent function performed by the circulation.[1] That is, of all the commodities that tissues require to be brought to them by the circulation of blood, O_2 is the one that, by far, they can least afford to have brought more slowly. This observation has both evolutionary and contemporary implications. In most types of modern animals, *metabolic intensity* and the *peak capacity of the circulatory system to transport* O_2 are strongly correlated. These properties have likely coevolved over the course of evolutionary time, neither getting far ahead of the other because there would be few selective factors favoring a mismatch. Similarly, when we look at minute-by-minute variations in the circulatory function of individual animals, we find usually that tissue O_2 needs drive changes in blood-flow rates: The rate of blood flow rises and falls as the metabolic need for O_2 increases and decreases. These principles explain why O_2 transport often receives paramount attention in the study of circulatory systems, even though circulation is essential for a great many functions.

Hearts

A logical starting point for study of the circulation is the hearts of animals. A **heart** is a discrete, localized pumping structure. Some animals that have a circulatory system lack a heart; in many annelid worms, for example, the blood is propelled through the circulatory system entirely by peristaltic contractions of blood vessels. Hearts are very common in circulatory systems, however, and often assume principal responsibility for driving the flow of blood through the blood vessels.

In some types of animals, such as arthropods, the heart is **single-chambered**, consisting of a single muscular tube or sac. In others, such as vertebrates, the heart is composed of two or more compartments through which blood passes in sequence, and thus is **multichambered**. Many types of animals, in addition to their principal heart, possess other hearts that assist with the pumping of blood through localized parts of the body. Such secondary or local hearts are called **accessory hearts** or **auxiliary hearts**.

The muscle tissue of a heart, composed of *cardiac muscle*, is known as the **myocardium** (*myo*, "muscle"; *cardium*, "heart"). **Cardiac muscle**, one of the major types of muscle, typically has distinctive structural and physiological properties in comparison with other types of muscle (e.g., skeletal muscle). **BOX 25.1** highlights the distinctive properties of vertebrate cardiac muscle. Some of those properties are discussed principally in other chapters, and in those cases, Box 25.1 provides cross-references to the relevant text.

To study the morphology of a heart in detail, there is no more appropriate example than our own (**FIGURE 25.1**). The hearts of

---

[1] Insects and other tracheate arthropods are dramatic exceptions to this statement, as discussed later in the chapter.

<table>
<tr><td>**BOX 25.1**</td><td>**The Structure and Function of Vertebrate Cardiac Muscle**</td></tr>
</table>

Vertebrate cardiac muscle has distinctive traits that suit it for its specialized role. Some of these traits are discussed fully in this chapter. Others are discussed in detail in Chapters 12, 13, and 20. Here we briefly note the properties that receive detailed treatment in the other chapters.

Structurally, cardiac muscle is a type of striated muscle, and in this respect it resembles skeletal muscle (see Table 20.3). Cardiac muscle is structurally very different from skeletal muscle, however, in that adjacent muscle cells (muscle

FIGURE A Cardiac muscle cells with intercalated disc

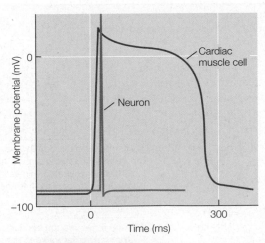

FIGURE B Action potentials

fibers) are joined by specialized *intercalated discs* (**Figure A**; see also page 561) that impart important mechanical and electrical properties. Where two cells meet at an intercalated disc, they are joined together by strong mechanical adhesions, so that mechanical forces developed in one cell are transmitted to the other, helping to achieve mutually reinforcing force generation. The two cells that meet at an intercalated disc are also joined to each other there by gap junctions (see Figures 2.7 and 13.2), meaning that the cytoplasm of each cell is continuous with that of the other cell. At the gap junctions, an action potential (wave of cell-membrane depolarization) in one cell is transmitted electrically—and therefore very rapidly—to the other cell (see pages 338–339). Thus when one cardiac muscle cell generates an action

potential and contracts, adjacent cells quickly generate action potentials and contract almost synchronously. Because of the control of contraction in this way, all the cells in the wall of a heart chamber contract together.

Individual action potentials in cardiac muscle cells are long, drawn-out events by comparison with those in neurons (Figure B). During an action potential in a cardiac muscle cell, depolarization of the cell membrane—which is the immediately effective stimulus for contraction—lasts about 100–500 milliseconds (ms), whereas depolarization in neurons typically lasts less than 1 ms. The mechanism of the long depolarization is discussed in Chapter 12 (see Figure 12.23). Its function is to ensure that contraction is prolonged, rather than being just a brief twitch. Cardiac muscle cells must contract for about 100–500 ms during

each heartbeat for the heart to pump blood effectively.

The cellular process that initiates each heartbeat is also a property of the cardiac muscle. The pacemaker, which initiates each beat, is composed of specialized muscle cells. Although we say much about this topic in this chapter, the electrical properties of the cells that compose the pacemaker are discussed in Chapter 12. As explained there (see pages 329–330), the membrane potential across the cell membrane in these cells does not stay at a stable resting value between beats. Instead, after a heartbeat, the membrane potential spontaneously drifts in the direction of ever-increasing depolarization. Because of this drift, the membrane potential eventually becomes depolarized enough to initiate a new action potential, which triggers the next heartbeat.

other mammals and of birds are similar. The left side of the human heart, which consists of two chambers—a weakly muscular **atrium** and a strongly muscular **ventricle**—receives freshly oxygenated blood from the lungs and pumps it to the systemic tissues of the body.[2] Blood arrives in the **left atrium** via the **pulmonary veins** that drain the lungs.[3] It leaves the **left ventricle** via a single massive artery, the **systemic aorta**, which branches to send arterial

vessels to the head, arms, abdomen, and all other body regions, even the myocardium itself. Passive valves, consisting of flaps of connective tissue covered with endothelial tissue, are positioned between the atrium and ventricle (the **left atrioventricular valve**) and between the ventricle and aorta (the **aortic valve**); these valves allow blood to flow freely in the correct direction but prevent it from flowing backward. After blood leaves the systemic aorta, it passes through the **systemic circuit**—the blood vessels that take blood to and from the systemic tissues—and ultimately returns in the great collecting veins (**venae cavae**; singular *vena cava*) to the heart, where it enters the **right atrium** and then the **right ventricle**. The function of the right side of the heart is to pump

---

[2] The **systemic tissues** are all the tissues other than the tissues of the breathing organs.

[3] By definition, **veins** are vessels that carry blood toward the heart, and **arteries** are vessels that carry blood away from the heart.

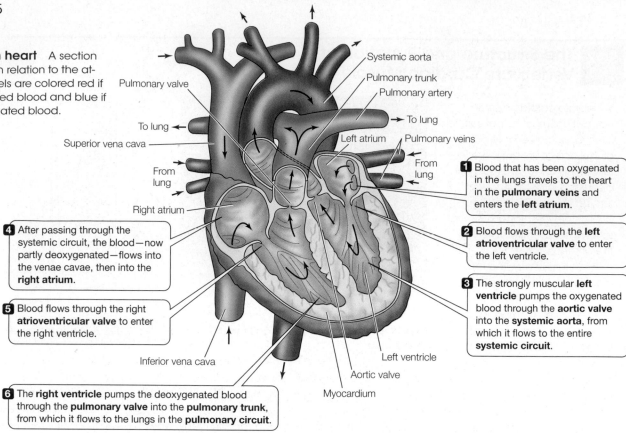

FIGURE 25.1 The human heart A section through the heart, shown in relation to the attached blood vessels. Vessels are colored red if they carry freshly oxygenated blood and blue if they carry partly deoxygenated blood.

Pulmonary valve

Systemic aorta

Pulmonary trunk

Pulmonary artery

To lung

To lung

Superior vena cava

Left atrium   Pulmonary veins

From lung

From lung

Right atrium

1 Blood that has been oxygenated in the lungs travels to the heart in the **pulmonary veins** and enters the **left atrium**.

4 After passing through the systemic circuit, the blood—now partly deoxygenated—flows into the venae cavae, then into the **right atrium**.

2 Blood flows through the **left atrioventricular valve** to enter the left ventricle.

3 The strongly muscular **left ventricle** pumps the oxygenated blood through the **aortic valve** into the **systemic aorta**, from which it flows to the entire **systemic circuit**.

5 Blood flows through the right **atrioventricular valve** to enter the right ventricle.

Inferior vena cava

Left ventricle

Aortic valve

Myocardium

6 The **right ventricle** pumps the deoxygenated blood through the **pulmonary valve** into the **pulmonary trunk**, from which it flows to the lungs in the **pulmonary circuit**.

blood through the **pulmonary circuit**—the blood vessels that take blood to and from the lungs. The right ventricle propels blood into a large vessel, the **pulmonary trunk**, which divides to form the **pulmonary arteries** to the lungs. As in the left heart, passive flap valves prevent backward flow in the right heart; these valves are positioned between the atrium and ventricle (the **right atrioventricular valve**) and between the ventricle and pulmonary trunk (the **pulmonary valve**). After blood has been oxygenated in the lungs, it returns to the left atrium.

The heart as a pump: The action of a heart can be analyzed in terms of the physics of pumping

During the beating cycle of any type of heart, the period of contraction is called **systole** (pronounced with a long *e*: *sis-tuh-lee*), and the period of relaxation is termed **diastole** (*dy-as-tuh-lee*). The heart is a pump, and we can understand its workings as a pump by analyzing pressure, flow, and volume during these periods. Here, as an example, we analyze the workings of the human left heart (left atrium and ventricle) shown in **FIGURE 25.2**.

At the time marked by the arrow at the bottom of Figure 25.2, ventricular systole begins. Whereas the pressure inside the ventricle was lower than that inside the atrium during the time just before the arrow, as soon as the ventricle starts to contract (marked by the arrow), the ventricular pressure rises abruptly to exceed the atrial pressure, causing the atrioventricular valve between the chambers to flip shut. For a brief interval of time (about 0.05 s), however, the ventricular pressure remains below the pressure in the systemic aorta, meaning that the aortic valve is not forced open. During this interval, therefore, both the inflow and outflow valves of the ventricle are shut. The volume of blood in the ventricle during this time is thus constant, and the interval is called the phase of **isovolumetric contraction** ("contraction with unchanging volume") or **isometric**

contraction. The contraction of the ventricle on the fixed volume of blood within causes the blood pressure inside the ventricle to rise rapidly. As soon as the ventricular pressure rises high enough to exceed the aortic pressure, the aortic valve flips open, and the blood in the ventricle accelerates extremely rapidly, gushing out into the aorta (thus increasing aortic pressure). The opening of the aortic valve marks the start of the phase of **ventricular ejection**. Toward the end of this phase, the aortic pressure comes to exceed the ventricular pressure slightly, but ejection of blood into the aorta continues for a while—at a rapidly falling rate—because of blood momentum. Ultimately, the ventricle starts to relax. The ventricular pressure then falls rapidly away from the aortic pressure, and the aortic valve shuts. A period of **isovolumetric relaxation** follows, as ventricular pressure falls with both the inflow and outflow valves shut. When the ventricular pressure drops below the atrial pressure, the atrioventricular valve opens inward to the ventricle, and **ventricular filling** begins. Most filling of the ventricle occurs *before* atrial systole—that is, before the atrial muscle contracts; the motive force for this filling is the pressure built up by *accumulation* of pulmonary venous blood in the atrium. When atrial systole occurs, it forces some additional blood into the ventricle just before the next ventricular systole.

In thinking of any heart as a pump, its most important attribute is the volume of blood it pumps per unit of time, known as the **cardiac output**. (In the case of the mammalian or avian heart, the term *cardiac output* refers specifically to the output of the left ventricle into the systemic aorta unless stated otherwise.) The cardiac output is the product of the heart rate and the **stroke volume**, the volume of blood pumped per heart cycle:

$$\text{Cardiac output} = \text{heart rate} \times \text{stroke volume}$$
$$\text{(mL/minute)} \quad \text{(beats/minute)} \quad \text{(mL/beat)} \quad (25.1)$$

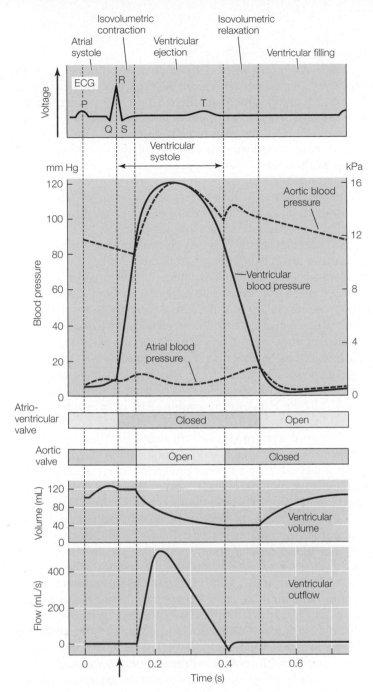

FIGURE 25.2 The heart as a pump: The dynamics of the left side of the human heart The heart cycle is divided into five phases, labeled at the top and demarcated by the vertical lines that run through the diagram. The diagram shows the synchronous changes that occur in left ventricular blood pressure, systemic aortic blood pressure, left atrial blood pressure, ventricular volume, the rate of blood flow out of the ventricle, and the closing and opening of the atrioventricular and aortic valves in humans at rest. The arrow at the bottom marks the start of ventricular systole. The ECG (see top panel) is the electrocardiogram, discussed later in this chapter.

The circulation must deliver O₂ to the myocardium

The myocardium of any heart (the "heart muscle") performs sustained, vigorous work, and its cells therefore are especially dependent on a steady O_2 supply. In most vertebrates, the ventricular myocardium is second only to the brain in its reliance on aerobic catabolism and in the urgency with which it requires O_2.

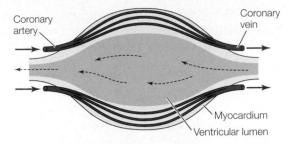

(A) Compact myocardium with coronary arteries and veins

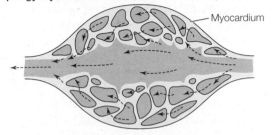

(B) Spongy myocardium with little or no development of coronary vessels

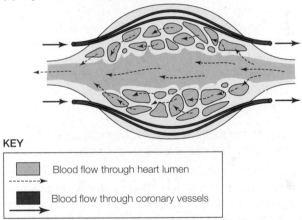

(C) Myocardium composed of outer compact tissue and inner spongy tissue

KEY

Blood flow through heart lumen

Blood flow through coronary vessels

FIGURE 25.3 Three systems evolved by animals to supply O₂ to the myocardium The myocardium is sometimes supplied with O_2 by blood flowing through a coronary circulatory system, whereas sometimes it is supplied by blood flowing through the heart lumen. A major reason to be aware of this distinction is that, although blood pumped into coronary arteries is always well oxygenated, the blood flowing through the heart lumen may not be. Each drawing shows a stylized heart. (A) The compact myocardium of mammals and birds is supplied by coronary arteries and veins. (B) A fully spongy myocardium (e.g., characteristic of most teleost fish) is oxygenated mostly by luminal blood. (C) Sometimes the ventricular myocardium consists of an outer, compact layer with a coronary circulation and an inner spongy layer; this arrangement is seen in salmonid fish, tunas, and sharks.

In mammals and birds, the ventricular myocardium is classified as **compact** because its muscle cells are packed closely together, much as cells are in other sorts of muscles. Blood passing through the *ventricular lumen*—the open central cavity of the right or left ventricle—cannot flow directly among the myocardial muscle cells because of their close packing. The myocardium, therefore, is not oxygenated by the blood flowing through the heart lumen. Instead, the ventricular myocardium in mammals and birds is supplied with tissue blood flow and O_2 by a system of blood vessels called the *coronary circulation* (**FIGURE 25.3A**). **Coronary arteries** branch from the systemic aorta at its very beginning and carry freshly oxygenated blood to capillary beds throughout the myocardium; the blood then flows into **coronary veins**, which carry it out of the myocardium and into the right atrium. If a coronary artery becomes blocked, the part of the myocardium it supplies quickly deteriorates because of O_2 deprivation, explaining why occlusions in the coronary arteries are extremely dangerous.

The original, evolutionarily primitive arrangement for oxygenation of the myocardium in vertebrates[4] is believed to be that shown in **FIGURE 25.3B**. This arrangement is common today in teleost (bony) fish, amphibians, and nonavian reptiles. The ventricular myocardium is classified as **spongy** because the muscle tissue is thoroughly permeated by an anastomosing (branching and rejoining) network of open spaces. The blood that is passing through the lumen of the ventricle—ventricular *luminal* blood—flows among the spaces of the spongy tissue, and the myocardial cells obtain their O_2 from the luminal blood. A shortcoming of this arrangement, as we will see later in this chapter, is that luminal blood is not always well oxygenated.

In some fish—including salmonids, tunas, and sharks—as well as some amphibians and nonavian reptiles, the ventricular myocardium consists of an outer compact muscle layer and an inner spongy layer (**FIGURE 25.3C**). The compact layer is supplied by coronary blood vessels that receive freshly oxygenated blood. The spongy layer obtains O_2 from the luminal blood, although it may also receive branches of the coronary circulation. Recent studies, discussed later, have revealed that in hearts of this sort, the proportions of compact and spongy myocardium sometimes vary dramatically within a single species.

The electrical impulses for heart contraction may originate in muscle cells or neurons

The rhythmic contraction of a heart reflects a rhythmic depolarization of the cell membranes of its constituent muscle cells. As described in Chapter 20, the cell membranes of muscle cells are polarized electrically; when a muscle cell is at rest, its cell membrane is polarized with the inside negative and the outside positive, but a wave of depolarization—an action potential—spreads along the cell membrane during contraction. In fact, *depolarization of the cell membrane of a muscle cell is the immediate stimulus for the cell to contract* (see Chapter 20).[5] A key question about any heart is: Where does the impetus for the rhythmic depolarization of the muscle cells originate? Do the muscle cells themselves spontaneously depolarize in a rhythmic manner? Or are they induced to depolarize by electrical impulses arriving from other cells? In either case, which are the cells that originate depolarization? A heart's **pacemaker** is the cell or set of cells that spontaneously *initiates* the rhythm of depolarization in the heart.[6]

---

[4] In general, relatively little is known about myocardial O_2 supply in invertebrates.

[5] To refresh your memory regarding the meaning of depolarization, see page 313 in Chapter 12. Cardiac muscle cells often exhibit specialized depolarization processes by comparison with neurons. The specialized depolarization of vertebrate cardiac muscle cells is described in Box 25.1. Depolarization refers to a decrease in the absolute value of the voltage difference across the cell membrane, such that the inside becomes less negative.

[6] See Box 25.1 for discussion of the cellular physiology of pacemaker depolarization.

In some animals, each electrical impulse to contract originates in muscle cells or modified muscle cells; the hearts of such animals are described as **myogenic** ("beginning in muscle"). In other animals, each impulse to contract originates in neurons (nerve cells), and the heart is termed **neurogenic** ("beginning in neurons"). Let's explore some of the attributes of myogenic and neurogenic hearts by looking at a classic example of each.

MYOGENIC HEARTS The hearts of vertebrates are myogenic. In almost all cases, they are innervated, but they continue to beat even if all nervous connections are stripped away.

An important feature of vertebrate heart muscle is that adjacent muscle cells are *electrically coupled*. As discussed in Box 25.1, electrical coupling occurs at gap junctions, which in humans and other mammals occur primarily at specialized regions of intercellular contact known as *intercalated discs*. Because adjacent cells are electrically coupled, depolarization of any one cell in the myocardium *directly* and *quickly* causes depolarization of neighboring cells. In turn, those cells induce their neighbors to depolarize, and so on. Thus, within large regions of the heart muscle, once depolarization is initiated at any point, it rapidly spreads—directly from muscle cell to muscle cell—to all cells in the region, leading all to contract together, as a unit (illustrated in Box 7.5).

Most or all of the muscle cells in a vertebrate heart possess an inherent ability to undergo spontaneous rhythmic depolarization and contraction. Thus pieces of muscle cut from any part of the heart will beat. In the intact heart, of course, individual bits of heart muscle do not depolarize and contract on their own—at their own rhythms. Instead, all the cells in the myocardium are controlled by a particular group of specialized muscle cells: the pacemaker. In fish, amphibians, and nonavian reptiles, the pacemaker is located in the wall of the sinus venosus, the first heart chamber (see Figure 25.14B), or at the junction of the sinus venosus and atrium. In birds and mammals, in which the sinus venosus has become incorporated into the atrium, the pacemaker is located in the wall of the right atrium (**FIGURE 25.4**) and is known as the **sinoatrial (S-A) node** (**sinus node**). The cells of the pacemaker are modified in comparison with most heart muscle cells; for example, they have a relatively poorly developed contractile apparatus. They are fundamentally muscle cells, however, meaning that the heart is myogenic. A critical attribute of the pacemaker cells is that they exhibit the highest frequency of spontaneous depolarization of all cells in the heart,[7] and therefore are normally the first to depolarize at each heartbeat. By thus *initiating* a wave of depolarization that spreads throughout the heart (in the form of action potentials), they *impose* their rhythm of depolarization on the heart as a whole.

The process by which depolarization spreads through the vertebrate heart or any other myogenic heart is known as **conduction**. Critical details of conduction in the mammalian heart depend on key structural features of the heart. The myocardium of the two atria of the heart is separated, for the most part, from the myocardium of the two ventricles by a layer of fibrous connective tissue across which myocardial cells are not electrically coupled by gap junctions and through which depolarization therefore cannot pass. In the mammalian heart, the one "electrical window" through this

---

[7] The concept of spontaneous depolarization is discussed in Box 25.1 and in Chapter 12 (see p. 329).

(A) The conducting system and sinoatrial node

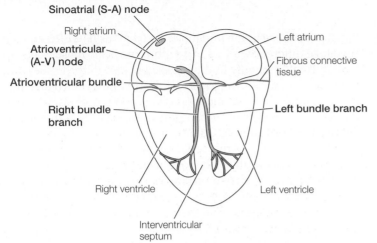

Sinoatrial (S-A) node

Right atrium

Atrioventricular (A-V) node

Atrioventricular bundle

Right bundle branch

Left atrium

Fibrous connective tissue

Left bundle branch

Right ventricle

Left ventricle

Interventricular septum

FIGURE 25.4 The conducting system and the process of conduction in the mammalian heart (A) The morphological arrangement of the conducting system and the position of the sino-atrial node. The branches of the right and left bundle branches are in fact more elaborate than shown; traveling along the inner surfaces of the ventricles and across the ventricular cavities, they run to much of the inner wall of each ventricle. (B) The initiation and conduction of depolarization during a heartbeat. Box 7.5 shows actual images of the spread of depolarization in the surface layers of the ventricles. (A after Scher and Spach 1979; B after Rushmer 1976.)

(B) The initiation and spread of depolarization during a heartbeat

KEY

■ Depolarized

□ Not depolarized

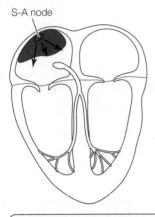
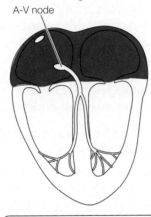
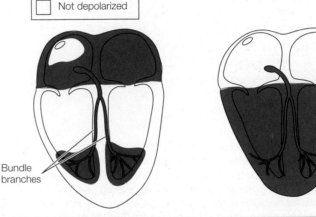

S-A node

A-V node

Bundle branches

1 Depolarization begins in the S-A node and spreads outward through atrial muscle.

2 Although depolarization spreads rapidly throughout the atrial muscle, its spread into the A-V node is delayed. The depolarized atria start to contract.

3 Once the A-V node becomes depolarized, the depolarization spreads very rapidly into the ventricles along the conducting system. Atrial muscle starts to repolarize.

4 The nearly simultaneous depolarization of cells throughout the ventricular myocardium leads to forceful ventricular contraction.

fibrous layer is provided by a **conducting system** composed of specialized muscle cells. As shown in Figure 25.4A, the conducting system starts with a group of cells in the right atrial wall known as the **atrioventricular (A-V) node**. Emanating from this node is a bundle of cells called the **atrioventricular bundle** (*common bundle, bundle of His*), which penetrates the fibrous layer and enters the *interventricular septum*—the wall of tissue that separates the right and left ventricles. Once in the septum, the atrioventricular bundle divides into right and left portions, the **bundle branches**, which travel along the right and left surfaces of the septum and connect with systems of large, distinctive muscle cells, the **Purkinje fibers**, that branch into the ventricular myocardium on each side.

The conducting system of the mammalian heart has two key *functional* properties: (1) Depolarization enters and traverses the A-V node relatively slowly, and (2) depolarization spreads down the atrioventricular bundle, bundle branches, and systems of Purkinje fibers much more rapidly than it could travel through ordinary ventricular muscle. The implications of these properties become apparent when we consider the sequence of events during a heartbeat, shown in Figure 25.4B. Steps **1** and **2** show that once the sinoatrial (S-A) node initiates a heartbeat by depolarizing spontaneously, the depolarization spreads rapidly throughout the muscle of both atria, leading to atrial contraction. Spread into the

ventricular muscle does not occur as rapidly, however, because it is dependent on activation of the conducting system, and the spread of depolarization into and through the initial part of the conducting system—the A-V node—is relatively slow (step **2**). This slowness of depolarization of the A-V node is responsible for the sequencing of contraction: atrial contraction distinctly first, ventricular contraction distinctly second. Once the A-V node is activated, depolarization sweeps rapidly down the conducting system into the ventricles (step **3**), precipitating wholesale ventricular depolarization and contraction (step **4**). The rapid delivery of the depolarizing wave to far-flung parts of the ventricular tissue by the conducting system ensures that all parts of the ventricular myocardium contract approximately together.

NEUROGENIC HEARTS The defining feature of neurogenic hearts is that the rhythmic depolarization responsible for initiating the heartbeats originates in nervous tissue. The hearts of lobsters are well-documented examples. Each muscle cell in a lobster heart is innervated and typically contracts when and only when stimulated to do so by nerve impulses (neuronal action potentials). As shown in **FIGURE 25.5**, a **cardiac ganglion** consisting of nine neurons is attached to the inside of the dorsal wall of the heart. The axonal processes of the five most anterior neurons (numbered 1–5)

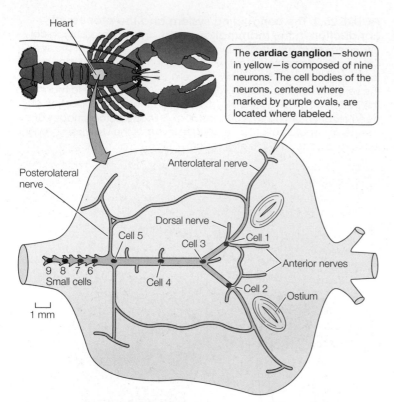

Heart

The **cardiac ganglion**—shown in yellow—is composed of nine neurons. The cell bodies of the neurons, centered where marked by purple ovals, are located where labeled.

Posterolateral nerve

Anterolateral nerve

Dorsal nerve

Cell 5
Cell 3
Cell 1

9 8 7 6
Small cells

Cell 4

Cell 2

Anterior nerves

Ostium

1 mm

FIGURE 25.5 The neurogenic heart of a lobster and the cardiac ganglion that initiates and controls its contractions A dorsal view of the heart of the American lobster (*Homarus americanus*), showing the cardiac ganglion, which is positioned on the inside of the dorsal heart wall. Neuronal processes go out from each of the nine cell bodies in the ganglion and together form the structure of the ganglion. The posterior four neurons (numbers 6–9) are small, whereas the anterior five are large. Neuronal processes exit the ganglion to innervate the cells of the heart muscle. Regulatory neurons from the central nervous system enter the ganglion in the dorsal nerve. The ostia (slitlike openings through the heart wall) are discussed later in this chapter. (After Hartline 1967.)

innervate the heart muscle. Those of the four posterior neurons (numbered 6–9) are confined to the ganglion and make synaptic contact with the five anterior neurons. One of the posterior neurons ordinarily assumes the role of pacemaker. This neuron functions as a cellular oscillator and central pattern generator (see Chapter 19): Periodically and spontaneously, it produces a train of impulses, which excite the other posterior neurons. The impulses from the posterior neurons activate the five anterior neurons, which in turn send trains of impulses to the muscle cells of the heart, causing the latter to contract approximately in unison. If the ganglion and heart muscle are dissected apart, the ganglion continues to produce bursts of impulses periodically, but the muscle ordinarily stops contracting! Other animals known or believed to have neurogenic hearts include other decapod crustaceans (e.g., crabs, shrimps, and crayfish), horseshoe crabs (*Limulus*), and spiders and scorpions.

A heart produces an electrical signature, the electrocardiogram

When a mass of heart muscle is *in the process* of being depolarized, such that some regions of cells are depolarized already and others await depolarization, a difference in electrical potential exists between the extracellular fluids in the depolarized regions of the

muscle and those in the undepolarized regions (**FIGURE 25.6A**). A voltage difference of this sort within the heart muscle sets up ionic currents, not only in the muscle but also in the tissues and body fluids surrounding the heart. In this way, the voltage difference within the heart induces voltage differences elsewhere in the body, even between various parts of the external body surface. **Electrocardiograms** (**ECGs**, **EKGs**) are measurements over time of voltage differences of this sort. They are recorded using extracellular electrodes, usually placed on the body surface. To record the elementary ECG[8] of a person, a physician or nurse places electrodes on the skin of the person's two arms and left leg. The electrodes detect voltage differences on the skin surface that are induced by voltage differences within the heart muscle.

The ECGs of two species, human and octopus, are shown in **FIGURE 25.6B**. The waveforms in the human ECG are named with letters (**FIGURE 25.6C**). The **P wave** is produced by the depolarization of the myocardium of the two atria (= atrial contraction). The Q, R, and S waves, together known as the **QRS complex**, arise from the depolarization of the myocardium of the two ventricles (= ventricular contraction). Repolarization of the ventricles generates the **T wave**.[9] Figure 25.2 shows the relation of the ECG waveforms to mechanical events during the heart pumping cycle.

Heart action is modulated by hormonal, nervous, and intrinsic controls

Heart action is subject to hormonal, nervous, and intrinsic controls. The controls we are typically most aware of are hormonal. When we are frightened and our heart pounds and races, the hormones epinephrine and norepinephrine, secreted by the adrenal medullary glands, are in part responsible for the heart stimulation we experience.

Nearly all hearts—whether myogenic or neurogenic—are innervated by neurons coming from the central nervous system, termed *regulatory neurons*. Some of these neurons stimulate increased heart action, whereas others are inhibitory. In the mammalian heart, both the sinoatrial node—the pacemaker—and the muscle cells of the myocardium are profusely innervated by the sympathetic and parasympathetic divisions of the autonomic nervous system. Sympathetic impulses delivered to the S-A node increase the frequency of spontaneous depolarization by the pacemaker cells (by affecting ion channel proteins) and thus raise the heart rate, whereas parasympathetic impulses exert opposite effects. Sympathetic impulses delivered to the cells of the myocardium markedly enhance the force and speed of their contraction, whereas parasympathetic impulses reduce the force and speed of contraction. When people exercise, sympathetic stimulation of the heart is increased. Most vertebrates are similar to mammals in that their hearts receive both sympathetic excitatory innervation and parasympathetic inhibitory innervation.[10] In lobsters, the cardiac ganglion is innervated by both excitatory and inhibitory regulatory neurons; these neurons modulate both the frequency and intensity

[8] More advanced electrocardiograms used for detailed diagnostic purposes require the attachment of electrodes at numerous additional positions on the chest.

[9] The waveform produced by repolarization of the atria is typically not seen because it is obscured by the QRS complex.

[10] Some teleost (bony) fish are exceptions in that they have only parasympathetic inhibitory innervation. In hagfish (primitive jawless fish), the heart seems to lack any innervation.

(A) Relative charges in myocardial extracellular fluids as the human ventricular myocardium depolarizes

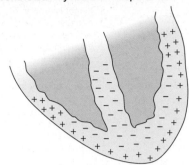

(B) Electrocardiograms of human and octopus

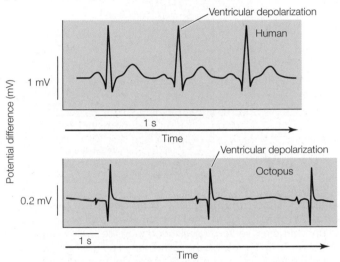

(C) Waveforms in the normal human electrocardiogram

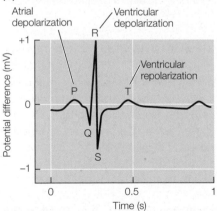

FIGURE 25.6 Electrocardiography (A) Relative electrical charges in the extracellular fluids of the human ventricular myocardium at an instant during passage of a wave of depolarization. The part of the ventricular myocardium lying nearest the ventricular chambers depolarizes first because it is the part supplied immediately by the branches of the conducting system (see Figure 25.4). (B) Electrocardiograms of a human and an octopus (*Eledone cirrhosa*) during three heartbeats. The human ECG was obtained using electrodes placed on the skin surface of the right arm and left leg. Electrodes attached to the surface of the ventricle (main body) of the systemic heart were used to record the octopus ECG. (C) A human ECG during one heartbeat with waveforms identified. (Octopus ECG after Smith 1981.)

of the bursts of impulses generated by the ganglion and thus affect the heart rate and the force of heart contraction.

Intrinsic controls of heart action are controls that occur without the mediation of hormones or extrinsic neurons. The **Frank–Starling mechanism** refers to a vitally important intrinsic control of the vertebrate heart; namely, that stretching of the cardiac muscle tends to increase the force of its contraction by an effect at the cellular level. This mechanism plays an important role in enabling the heart to match its output of blood to its input. Consider, for example, what happens when the rate of blood flow into a heart chamber is increased. Because the heart chamber then tends to take in more blood in the time between beats, it becomes more stretched (distended) between beats. Because of the Frank–Starling mechanism, the heart muscle then intrinsically contracts more forcefully, which enhances ejection of blood—a response that tends to match heart output to the increased blood received. The Frank–Starling mechanism is important in all vertebrate hearts studied. It is particularly dramatic in its effects in fish hearts. Lobster hearts function similarly to vertebrate hearts in that they intrinsically increase both the rate and force of their contraction as they are stretched. The mechanism of this response in lobsters is, at least in part, quite different from that in vertebrates, however, because the cardiac ganglion is involved. Stretch induces the ganglion to fire more frequently and intensely.

SUMMARY
Hearts

- The output of a heart, known as the cardiac output, depends on the heart rate and stroke volume.

- The cells in the heart muscle, the myocardium, must have means of receiving O_2. In some hearts the myocardium is spongy, and blood flowing through the heart chambers flows through the spongy spaces, supplying O_2 to the cells. In other hearts, including those of mammals, the myocardium is compact and is supplied with blood and O_2 by means of coronary blood vessels.

- A heart is myogenic if the depolarization impulses required for heartbeats originate in muscle cells or modified muscle cells. A heart is neurogenic if the impulses originate in neurons. Vertebrate hearts are myogenic. Hearts of adult decapod crustaceans are neurogenic.

- In the mammalian heart, the sinoatrial node in the wall of the right atrium acts as pacemaker, initiating waves of depolarization. Conduction of a wave of depolarization from the atria to the ventricles occurs through the conducting system, which ensures both that the ventricles contract later than the atria and that the entire ventricular myocardium contracts approximately at once.

- When a part of the myocardium is in the process of contracting, voltage differences in the extracellular fluids develop between regions of muscle cells that have already undergone depolarization and regions that have not. These differences can be detected on the body surface. The electrocardiogram is a recording of such differences as a function of time.

- The rate and force of heart contraction are governed by nervous, endocrine, and intrinsic controls.

Principles of Pressure, Resistance, and Flow in Vascular Systems

Having discussed the fundamental features of the hearts of animals, we now need to turn our attention to the perfusion of the vascular system. **Perfusion** refers to the forced flow of blood through blood vessels.

The **blood pressure** produced by the heart—or, in some animals, by other muscular activity—is the principal factor that causes blood to flow through the vascular system. What we mean by *blood pressure* is the amount by which the pressure of the blood exceeds the ambient pressure. Blood pressure is often expressed in *kilopascals* (*kPa*) by physiologists, but usually in *millimeters of mercury* (*mm Hg*) in medicine and related disciplines.[11] When we say that the blood pressure in a vessel in an animal is 10 kPa (= 75 mm Hg), this means that the pressure there is 10 kPa, or 75 mm Hg, higher than the pressure simultaneously present in the animal's surrounding environment.

In arteries, the blood pressure rises and falls over the heart cycle. The highest pressure attained at the time of cardiac contraction is termed the **systolic pressure**, whereas the lowest pressure reached during cardiac relaxation is the **diastolic pressure**. In young adult humans at rest, the systolic pressure in the systemic aorta is usually about 16 kPa or 120 mm Hg, and the aortic diastolic pressure is about 10 kPa or 75 mm Hg. When these pressures are measured for clinical reasons, the results are often expressed as a pseudo-ratio—for example, 120/75 ("120 over 75"). The **mean pressure** in an artery is obtained by averaging the pressure over the entire cardiac cycle; it usually does not equal the average of the two extreme pressures, systolic and diastolic, because the systolic and diastolic phases are not the same in duration. In resting young adults, the mean pressure in the systemic aorta is ordinarily about 12.7 kPa or 95 mm Hg.

In addition to the pressures produced dynamically by the beating of the heart, pressures resulting from *fluid-column effects* can also be important in circulatory systems. Any unobstructed vertical column of fluid exerts a pressure—termed a *hydrostatic pressure*—that increases as its height increases (**FIGURE 25.7A**). Because blood in the vessels of an animal forms fluid columns, fluid-column pressures are present in circulatory systems. The pressure produced by the beating of the heart is added (in an algebraic sense discussed in the next paragraph) to the fluid-column pressures that are present in arteries to determine the total arterial blood pressure. These relations are relevant for the clinical measurement of blood pressure. The pressure of interest during clinical measurement is that produced by the heart, unconfounded by fluid-column effects. The measurement device should therefore be in the same horizontal plane as the heart.

In a vertical column of blood, each 13 cm of height exerts about 10 mm Hg of pressure. At levels of the body below the heart, this hydrostatic fluid-column pressure adds to the pressure contributed by the heart; thus, in a person standing up, the blood pressure in arteries in the legs is far above that in the systemic aorta (**FIGURE 25.7B**). At levels of the body above the heart, some of the pressure developed by the heart is lost in simply supporting the fluid column of blood. Thus the blood pressure in the arteries of the neck and

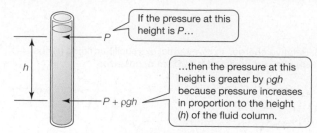

(A) The physics of fluid-column effects in an unobstructed vertical tube

If the pressure at this height is *P*...

...then the pressure at this height is greater by ρ*gh* because pressure increases in proportion to the height (*h*) of the fluid column.

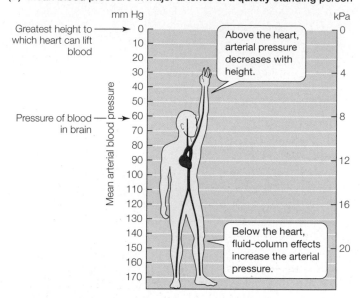

(B) Mean blood pressure in major arteries of a quietly standing person

mm Hg   kPa

Greatest height to which heart can lift blood

Above the heart, arterial pressure decreases with height.

Pressure of blood in brain

Mean arterial blood pressure

Below the heart, fluid-column effects increase the arterial pressure.

FIGURE 25.7 Fluid-column effects on blood pressure in the arterial vascular system (A) The physics of fluid-column effects in an unobstructed, vertically positioned tube filled with a nonmoving fluid. The symbol *h* represents the difference in height between two points within the fluid column, ρ is the mass density of the fluid, and *g* is acceleration due to gravity. (B) Fluid-column effects on arterial blood pressure in a person standing quietly. (B after Rushmer 1976.)

head decreases by approximately 10 mm Hg for every 13 cm of height above the heart.[12]

Considering all the factors that could influence blood flow, how can we predict the *direction* of flow? Throughout most of this chapter, we will make simplifying assumptions so that we will be able to analyze blood flow based just on the pressures produced by the beating of the heart. However, it is important to be aware of a more general concept—the **total fluid energy** of the blood—that allows one to analyze blood flow in any situation, without simplifying assumptions.

Blood can possess three forms of energy that affect its flow, and the *total fluid energy* of the blood is the sum of these three forms, as shown in **FIGURE 25.8**. The first form of energy (labeled ❶ in Figure 25.8) is the pressure the blood is under because of the beating of the heart; this is a form of potential energy, which means that it can produce motion, but is not motion itself. The second form of energy (❷) is the blood's energy of motion—its kinetic energy. The third form (❸) is the potential energy the blood possesses because of its position in Earth's gravitational

---

[11] Appendix A and footnote 9 on page 595 of Chapter 22 discuss the relations among these units.

[12] These same relations do not hold in any sort of simple fashion in veins because the veins, instead of being unobstructed, are obstructed by venous valves.

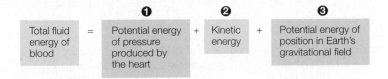

FIGURE 25.8 Total fluid energy: The true driving force for blood flow Kinetic energy is energy of motion. Potential energy is not motion but can produce motion.

field. *Blood always flows from where its total fluid energy is higher to where its total fluid energy is lower.*

Why be concerned with this complex concept? Actually, we have already encountered two situations in which the total fluid energy must be taken into account to explain the flow of blood. First, we saw in Figure 25.2 that toward the end of ventricular systole in the human heart, blood briefly continues to flow out of the left ventricle into the systemic aorta, even though the blood pressure in the ventricle (factor ❶ in the total fluid energy) has fallen below the pressure in the systemic aorta. If flow were governed only by simple pressure differences, blood would not flow out of the ventricle when the pressure in the ventricle is lower than that in the aorta. Blood in fact flows out of the ventricle at the end of ventricular systole, however, because at that time, the blood on the ventricular side of the aortic valve has a greater forward momentum—a greater kinetic energy (factor ❷)—than that on the aortic side. Thus the *total fluid energy*—the true driving force for blood flow—is higher in the ventricle than in the aorta.

The second case we have encountered that cannot be understood without taking account of total fluid energy is the flow of blood in arteries into our legs when we are standing up. Looking at Figure 25.7, you can see that at any given time, if one uses a simple pressure-measurement device to measure the existing pressure in the systemic aorta of a person standing up and the simultaneous pressure in the arteries in the lower legs, the pressure in the lower legs is perhaps 70–80 mm Hg *higher*.[13] Thus, if pressure alone governed blood flow, blood would flow from the legs into the aorta, not the other way around. Here again, to understand the flow that actually occurs, one must analyze the total fluid energy. A simple pressure-measurement device confounds and confuses two distinct factors in the total fluid energy: the pressure produced by the beating of the heart (factor ❶) and the blood's potential energy of position in Earth's gravitational field (factor ❸). These factors become confounded by a simple pressure-measurement device because in parts of the body below the heart, high potential energy of position in the blood near the heart is recorded as high pressure in the blood below the heart. To analyze blood flow accurately, one must go back to the equation for total fluid energy and analyze its three components in an unconfounded way. The pressure produced by the heart (factor ❶) is slightly higher in the aorta than in the major leg arteries, and furthermore, the blood in the aorta has a greater potential energy of position (factor ❸) than that in the leg arteries because it is at a greater elevation (put loosely, the blood in the aorta tends to "fall" into the legs). Considering all three factors, the *total fluid energy* is higher in the aorta, and therefore blood flows into the legs.

For many purposes, the analysis of blood flow can be usefully simplified by making two assumptions that in fact are often reasonably realistic: (1) Assume that the kinetic energy of the blood

(factor ❷) does not vary from place to place within the system being analyzed. This assumption is often quite realistic. For example, in human arteries, kinetic energy accounts for a very small fraction of the total fluid energy (only 1%–3% in the systemic aorta), and therefore little accuracy is lost by disregarding it. (2) Assume that the animal under study is in a horizontal posture. For a person or other animal lying on a horizontal surface, all blood is roughly in one horizontal plane, and one can therefore reasonably assume that potential energy of position (factor ❸) is equal everywhere. The horizontal posture is in fact the posture usually used for experiments or analysis, and one can demonstrate that its use usually does not detract from the generality of conclusions reached.

With these simplifying assumptions made, blood flow can be analyzed using the pressure developed by the heart as the sole driving force.[14] This is the approach we will use except in special cases.

The rate of blood flow depends on differences in blood pressure and on vascular resistance

Already in the nineteenth century, physiologists were seeking to understand the perfusion of blood vessels by analyzing the steady, nonturbulent flow of a simple liquid such as water through a horizontal, rigid-walled tube (**FIGURE 25.9A**). Their analyses led to insights that are still considered important today. The factors that determine the rate of flow (mL/minute) from one end of a tube to the other, they discovered, are the pressure at the entry to the tube (P_{in}), the pressure at the exit (P_{out}), the radius of the lumen of the

---

[14] Under these simplifying assumptions, the total fluid energy is equal to the pressure developed by the heart—factor ❶ in Figure 25.8—because factors ❷ and ❸ do not apply. Thus blood flows from where the pressure developed by the heart is high to where it is low.

(A) Pressures and dimensions that affect the rate of flow

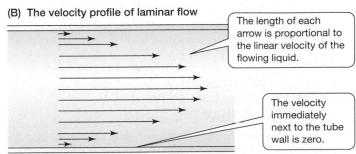

(B) The velocity profile of laminar flow

The length of each arrow is proportional to the linear velocity of the flowing liquid.

The velocity immediately next to the tube wall is zero.

FIGURE 25.9 The physics of flow through tubes (A) Critical factors for understanding the rate of flow using the Poiseuille equation. (B) Linear velocity as a function of distance from the tube wall. A microscopically thin layer of liquid touching the tube wall does not move at all. The velocity profile seen here applies when a simple liquid such as water flows in a laminar fashion through a tube; the velocity profile for blood differs from that for a simple liquid.

---

[13] The true value depends on how tall the person is.

tube (r), the tube length (l), and the viscosity of the liquid (η).[15] The formula relating these quantities is named the **Poiseuille equation** or **Hagen–Poiseuille equation**, after Jean Poiseuille (1797–1869) and Gotthilf Hagen (1797–1884), who derived it:

$$\text{Flow rate} = (P_{in} - P_{out})\left(\frac{\pi}{8}\right)\left(\frac{1}{\eta}\right)\left(\frac{r^4}{l}\right) \qquad (25.2)$$

According to the Poiseuille equation, increasing the difference in pressure between the ends of a tube increases the rate of flow through the tube. Raising the viscosity of the fluid diminishes the rate of flow. The final term in the equation is an important geometric term: It shows that the rate of flow through a tube is a direct function of *the fourth power* of the radius of the lumen. Because of this relation, the rate of flow is extraordinarily sensitive to changes in the radius of the lumen. If water is flowing through a tube at a certain rate and the radius of the lumen is reduced to half of the starting radius, the rate of flow falls to 1/16 of the original rate!

The Poiseuille equation, as mentioned, was derived to apply to simple liquids such as water flowing through unbranched, rigid-walled tubes. Blood is not a simple liquid because it contains suspended cells, and blood vessels are not unbranched or rigid-walled. Nonetheless, the Poiseuille equation often proves to be a useful approximate model for understanding the flow of blood through blood vessels. From the Poiseuille equation, we see that *when muscles in the walls of a blood vessel change the radius of the vessel by contracting or relaxing, they exert profound (fourth power) control over the rate of flow through the vessel.*

Another equation that is useful for understanding the rate of steady blood flow through a horizontal system of blood vessels is

$$\text{Flow rate} = \frac{\Delta P}{R} \qquad (25.3)$$

where ΔP is the difference in blood pressure between the entry vessels of the vascular system and the exit vessels, and R is the resistance to flow through the system, termed the **vascular resistance**. This equation is analogous to Ohm's law in electrical circuits. It simply says that the rate of flow increases when the difference in pressure increases, but the rate of flow decreases when the vascular resistance increases. If we consider a simple tubular vessel, an easy relation exists between Equation 25.3 and the Poiseuille equation. Because ΔP and ($P_{in} - P_{out}$) represent the same quantity in this case, the resistance R is equal to $8\eta l/\pi r^4$. One can see that *resistance is inversely proportional to the fourth power of the vessel radius.* Halving the radius of a vessel increases the resistance to flow through the vessel by a factor of 16.

The dissipation of energy: Pressure and flow turn to heat during circulation of the blood

When blood flows through a horizontal blood vessel or system of vessels in an animal, the heart maintains a high pressure at the entry end. This pressure drives the blood through the vessel or system of vessels. However, as the blood passes through, the pressure becomes diminished. Why?

To answer this question accurately, we need to look briefly in more detail at the nature of flow through a tube or blood vessel. The

linear velocity of a bit of liquid in a stream flowing through a tube is defined to be the length of the tube traveled per unit of time. According to ideal flow theory, when the flow of a liquid through a tube is steady and nonturbulent, the liquid moves in a series of infinitesimally thin, concentric layers (laminae) that differ in their linear velocities. This type of flow, called **laminar flow**, is illustrated in **FIGURE 25.9B**. The outermost of the concentric layers of liquid—the layer immediately next to the wall of the tube—does not move at all. Layers closer and closer to the center move faster and faster.

A crucial aspect of this sort of flow is that the adjacent layers of the liquid that are moving at different linear velocities do not slip effortlessly past each other. Instead, there is a sort of friction that must be overcome to make them move relative to each other. The total magnitude of this *internal friction* in a moving liquid depends in part on the dimensions of the tube. In addition, it depends on a property intrinsic to the particular liquid, namely the **viscosity** (specifically, *dynamic viscosity*) of the liquid. Viscosity refers to *a lack of intrinsic slipperiness between liquid layers moving at different linear velocities*; liquids that are particularly low in internal slipperiness have high viscosities and exhibit syruplike properties. The internal friction within a moving liquid is very real and, like other friction, results in the degradation of kinetic energy (energy of motion) into heat. Thus, as a liquid flows through a tube, some of its kinetic energy is steadily degraded.[16]

Understanding that energy of motion is degraded to heat, we can now see the broad outlines of the energetics of blood flow through a horizontal system of blood vessels. The pressure provided at the entry end by the heart is a form of potential energy. Some of this potential energy is converted to kinetic energy: energy of motion of the blood. Then, along each millimeter of the tubular system through which the blood flows, some of the kinetic energy is lost as heat in overcoming internal friction. *Ultimately, therefore, pressure is converted to heat.* During horizontal flow, the drop in pressure from one point in a tubular system to another point downstream is in fact a good index of the heat produced in overcoming opposing viscous forces. *Thus the drop in blood pressure from place to place can be used as a measure of the energy cost of blood flow.*

To apply these concepts, let's consider blood flow through arteries and capillaries in our circulatory system. From one end of an artery to the other, the mean blood pressure caused by pumping of the heart changes by only a small amount; for example, the pressure in an artery in our wrist—0.7 m from our heart—is only about 0.4 kPa (3 mm Hg) lower than the pressure in our systemic aorta. This small drop in pressure tells us that the cost of driving blood through the arteries is low. Shortly we will see that in blood capillaries the blood pressure drops precipitously as blood flows through. This steep pressure drop signifies that capillaries, unlike the arteries, are very costly to perfuse.

---

[15] We will describe viscosity shortly. For the moment, it refers to how syruplike a liquid is.

[16] See Chapter 7 for a discussion of energy and energy degradation. Sometimes people say that pumping the blood through vessels involves overcoming friction between the blood and the vessel walls. This is not correct. In fact, there is no frictional force to be overcome between a laminarly flowing liquid and the walls of a tube because no relative motion occurs between the walls and the liquid layer next to them (that layer is still). The frictional resistance to the flow of a liquid is *entirely internal to the liquid.*

(A) The circulatory plan

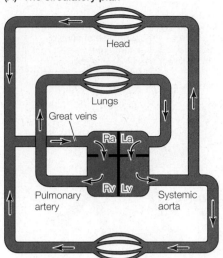

Head

Lungs

Great veins

Ra La

Pulmonary artery

Rv Lv

Systemic aorta

Limbs, body wall, abdomen, systemic circulation of thorax

(B) A schematic of the circulatory plan emphasizing that the systemic and pulmonary circuits are connected in series

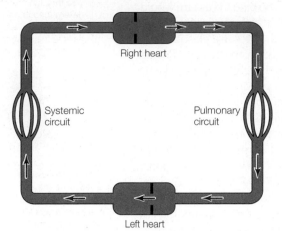

Right heart

Systemic circuit

Pulmonary circuit

Left heart

FIGURE 25.10 The circulatory plan in mammals and birds (A) The circulatory plan as it exists geometrically in the body. (B) The same plan, redrawn as a schematic to emphasize the arrangement of the pulmonary and systemic circuits in series with each other. Red and blue portions carry relatively oxygenated and deoxygenated blood, respectively. Ra, right atrium of the heart; La, left atrium; Rv, right ventricle; Lv, left ventricle.

SUMMARY
Principles of Pressure, Resistance, and Flow in Vascular Systems

■ Blood pressure is measured relative to environmental pressure; it is the extent to which the pressure in the blood exceeds that in the environment.

■ During steady flow of blood through horizontal vessels or systems of vessels, the rate of blood flow is directly proportional to the difference of pressure between the inlet and outlet. It is also inversely proportional to vascular resistance. Thus the equation describing blood flow (Equation 25.3) is analogous to Ohm's law. According to the Poiseuille equation (Equation 25.2), vascular resistance varies inversely with the fourth power of vessel radius.

■ Blood pressure declines during the flow of blood through vessels because the potential energy represented by the pressure is converted to kinetic energy, which then is converted to heat in overcoming viscous resistance to flow. During steady flow through a horizontal system, this drop in blood pressure is a measure of the energy cost of perfusion.

Circulation in Mammals and Birds

We will talk about the circulatory systems of mammals and birds together because both groups have four-chambered hearts, and they have essentially identical circulatory plans. Mammals and birds evolved these properties independently, however. As shown by the circulatory plan in **FIGURE 25.10A**, O_2-depleted blood returning from the systemic tissues enters the right heart via the great veins and is pumped by the right ventricle to the lungs, where O_2 is taken up and CO_2 is released. The blood oxygenated in the lungs then travels to the left heart and is pumped by the left ventricle to the systemic aorta, which divides to supply all the systemic tissues. A key feature of this circulatory plan is that it places the lungs in series with the systemic tissues. This series arrangement, emphasized in **FIGURE 25.10B**, maximizes the efficiency of O_2 delivery to the systemic tissues: All the blood pumped to the systemic tissues by the heart is freshly oxygenated, and the tissues receive blood that is at the full level of oxygenation achieved in the lungs. These features are important ways in which the circulatory systems of mammals and birds are able to meet the high O_2 demands of these animals.

The circulatory system is closed

Circulatory systems are classified as **open** or **closed**, depending on whether the entire circulatory path is enclosed in discrete vessels. In an *open* system, the blood leaves discrete vessels and bathes at least some nonvascular tissues directly. In a *closed* system, there is always at least a thin vessel wall separating the blood from the other tissues. The distinction between open and closed systems is relative because there are many intergradations. Mammals and birds, as well as other vertebrates, have essentially closed circulatory systems.

Each part of the systemic vascular system has distinctive anatomical and functional features

The blood vessels at various points in the systemic vascular system differ anatomically and functionally in important ways. In this section we discuss the major types of vessels in the order in which blood passes through them. A significant point to mention at the start is that in vertebrates, *all* types of vessels—and the chambers of the heart—are lined on the inside with a single-layered epithelium termed (for historical reasons) the **vascular endothelium**. The cells of the endothelium are exceedingly important: They perform many functions, which are only gradually being understood. For example, some endothelial cells secrete agents into the blood— such as nitric oxide or prostacyclin (prostaglandin I_2)—that affect the contraction and relaxation of vascular smooth muscle or help control clotting. Endothelial cells also sometimes synthesize active hormones from hormone precursors in the blood or terminate hormone action (e.g., by degrading hormones). Endothelial cells also participate in immune responses.

ARTERIES The great arteries have thick walls that are heavily invested with smooth muscle and with elastic and collagenous connective tissue. Thus they are equipped to convey blood under considerable pressure from the heart to the peripheral parts of the circulatory system. The elasticity of the great arteries enables them to perform important hydrodynamic functions. If the heart were to discharge blood into rigid, inelastic tubes, the blood pressure would oscillate violently upward and downward with each contraction and relaxation of the heart. Instead, the arteries are elastic. They stretch when they receive blood discharged from the heart. Some of the energy of each heart contraction is thereby stored as elastic potential energy in the artery walls, and consequently the increase in arterial pressure during systole is limited to some extent. The energy stored elastically at the time of systole is released as the arteries rebound to their unstretched dimensions during diastole. In this way, some of the energy of heart contraction is used to maintain the pressure in the great arteries between contractions. The end result is that arterial elasticity has two effects: Variations in arterial pressure over the cardiac cycle are reduced—termed the *pressure-damping effect*—and a substantial pressure is maintained in the arteries even when the heart is at rest between beats—termed the *pressure-reservoir effect*.

The arteries become smaller as they branch outward toward the periphery of the circulatory system. The walls of the arteries simultaneously become thinner, a fact that at first appears paradoxical when we recall that the mean blood pressure diminishes hardly at all in the arteries. The paradox is resolved in good measure by a principle identified by the great mathematician Pierre Simon, the Marquis de Laplace (1749–1827), and now known as **Laplace's law**.[17] This law deals with the relation between the pressure in the lumen and the wall tension in hollow structures. As applied to tubes, it says that when the pressure in the lumen of a tube exceeds that outside the tube by any given amount, the circumferential tension (stretch) developed within the walls of the tube is directly proportional to the tube radius:

$$T = r\Delta P \qquad (25.4)$$

where T is wall tension, r is the radius of the lumen, and ΔP here represents the pressure difference across the walls. Because of this relation, even though a small artery may be exposed to the same blood pressure as a large one (and therefore have the same ΔP), the tension developed within its walls is lower than that developed within the walls of the large artery. Accordingly, the walls of small arteries need not be as well fortified as those of large arteries to resist overexpansion. The same principle explains why blood capillaries can be exceedingly thin-walled and yet resist substantial pressures.

MICROCIRCULATORY BEDS The systemic arteries ultimately deliver blood to networks of microscopically tiny blood vessels in all the systemic organs and tissues. These **microcirculatory beds**, diagrammed in **FIGURE 25.11**, consist of three types of vessels: *arterioles*, *capillaries*, and *venules*.

Arterioles, although minute, have the same basic structure as arteries: Their walls consist of smooth muscle and connective tissue. The mean diameter of the lumen of the arterioles of humans is about 30 micrometers (µm). Notably, the walls, which average about 20 µm in thickness, are so invested with muscle and connective tissue that they are almost as thick as the lumen is wide. The smooth muscles in the walls of the arterioles are exceedingly

---

[17] Laplace's law applies quantitatively only to simple elastic materials such as rubber. It is a useful principle for analyzing complex elastic structures like arterial walls, but the detailed study of arteries involves additional considerations.

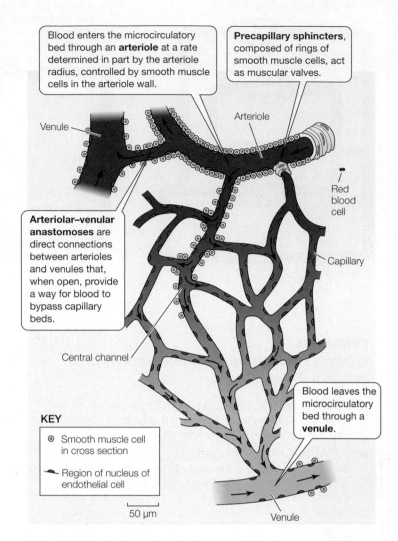

FIGURE 25.11 A microcirculatory bed In this microcirculatory bed of a mammal, capillaries form an anastomosing network between an arteriole at the top and a venule at the bottom. The endothelial cells that line the arterioles and form the walls of the capillaries are thin and flat, except where each cell is thickened in the region of its nucleus. Precapillary sphincters and arteriolar–venular anastomoses are opened and closed by smooth muscle cells. (After Copenhaver et al. 1978.)

important because they are responsible for the *vasomotor control of blood distribution*. The adjective **vasomotor** refers to changes in the radius of the lumen of blood vessels. A decrease in the radius of the lumen is called **vasoconstriction**, whereas an increase is called **vasodilation**. In our earlier discussion of the Poiseuille equation (see Equation 25.2), we noted that the rate of flow through a tubular vessel is extremely sensitive to the radius of the lumen of the vessel. By contracting and relaxing, the smooth muscles in the wall of an arteriole control the radius of the lumen of the vessel and thereby profoundly affect the rate of blood flow into the capillary beds that the arteriole supplies. Control of the arteriolar muscles is mediated by the sympathetic division of the autonomic nervous system, by circulating hormones, and by chemical mediators at the local tissue level, such as nitric oxide (NO), a vasodilator that is released from the vascular endothelium. In addition to the arterioles, both precapillary sphincters (see Figure 25.11) and small terminal arteries also participate in controlling blood flow to capillary beds.

An especially familiar example of the control of tissue perfusion by arterioles is provided by the responses of skin blood flow to warm and cold environments. As we discussed in Chapter 10 (see pages 267–268), humans and other mammals often maintain relatively vigorous blood flow to their skin surface in warm environments, but reduce blood flow to the skin surface in cold environments. When flow is reduced, the effect occurs because arterioles supplying blood to the superficial microcirculatory beds of the skin are vasoconstricted by the action of the sympathetic nervous system and local vasoconstrictive agents. Modulation of arteriolar–venular anastomoses (see Figure 25.11) also plays a role. Changes in the perfusion of skeletal muscles during exercise provide another important example of arteriolar control: Blood flow through a skeletal muscle can be increased by a factor of ten or more in a person during exercise. Part of the reason for the increased blood flow is arteriolar vasodilation, which appears to be mediated principally by local effects of metabolites produced in exercising muscles. A final familiar example of the control of tissue perfusion by arterioles and small arteries is the erection of the penis. During sexual arousal, parasympathetic neurons and cells in the local vascular endothelium of the penis release nitric oxide, which acts as a potent signal for dilation of the arterioles and small arteries that supply blood to the penis's spongy erectile tissues (see page 490 for more detail).

Arteriolar control of microcirculatory beds is one of the premier attributes of closed circulatory systems. The heart produces pressure that is transmitted to all the microcirculatory beds in the body by way of the arteries. This driving force is always available at the entry to each microcirculatory bed, ensuring that vasodilation or vasoconstriction of arterioles will cause immediate changes in tissue perfusion. Each microcirculatory bed has its own arterioles—which determine the rate of flow into the bed—and thus is readily controlled independently of other microcirculatory beds. These features permit *highly sensitive temporal and spatial control of blood distribution*.

From the arterioles, the blood typically enters the **capillaries**. The walls of capillaries consist of *only the vascular endothelium*—a single layer of highly flattened cells resting on an outer basement membrane (see Figure 2.6C). The walls of capillaries are very thin (<1 μm) and often fenestrated to some extent—meaning that physical openings through the walls are present, such as gaps between neighboring endothelial cells (see page 725). Moreover, in most capillaries studied (including pulmonary ones), aquaporins—water channel proteins—are abundant in the cell membranes of the endothelial cells. Because of these features, the capillaries are the *preeminent sites of exchange of O_2, water, and other materials between the blood and the tissues*. The lumens of the capillaries are often barely wide enough to allow red blood cells to pass through in single file. A **capillary bed** (see Figure 25.11 and Figure 27.2) consists of many capillaries that branch and anastomose among the cells of a tissue.

The aquaporins in the endothelial cell membranes of capillaries clearly facilitate osmosis between the blood inside and the tissue fluids outside the capillaries, although often the capillary endothelium seems to be adequately permeable to water without them, making their roles unclear. In the brain, aquaporins in astrocytes—the nonneural brain cells that, in intimate juxtaposition with capillaries, help form the blood–brain barrier—seem clearly to be involved in regulating cerebrospinal fluid volume.

The density of capillaries is a remarkable subject. Tissues differ in how many capillaries they have per unit of tissue volume; capillary density is particularly high in skeletal muscles, the myocardium, and the brain. In such tissues—rich in capillaries—the exchange surface provided by the capillary beds is nothing short of amazing. A cubic centimeter of skeletal or cardiac muscle may well contain 10–20 *meters* of capillaries! In cross sections of mammalian gastrocnemius (calf) muscles, one can see from 300 to more than 600 capillaries per *square millimeter*. Within the capillary beds of skeletal muscles, only some of the capillaries are open at rest, but all may open during exercise. It is also possible for new capillaries to develop, or for old ones to disappear, during acclimation to changed environmental conditions, aging, or disease. Modern theories hold that a pathological loss of capillaries may be an invisible but major contributing factor to several chronic debilitating diseases. In a state of health, **angiogenesis**—the process of forming new capillaries and other microcirculatory elements—is an important process, controlled by *angiogenic activators* and *inhibitors* such as paracrine agents (see pages 604–605) for the roles of hypoxia-inducible factors). New capillaries are often formed in response to wounding, athletic training, or exposure to low O_2 partial pressures. The capillary beds drain into **venules**, which are small vessels with thin walls (2–5 μm in humans) containing connective tissue and muscle cells.

VEINS The blood flows from the microcirculatory beds back to the heart through a series of veins of increasing diameter. As we will see, the blood pressure has declined precipitously by the time the blood leaves the microcirculatory beds. Thus the walls of the veins need not be capable of resisting high tensions and are thin compared with those of the arteries. To a far greater extent than the arterial system, the venous system contains passive one-way valves. These valves are sheets of tissue shaped so that they permit flow toward the heart, but pinch shut if flow reverses. The valves help ensure that blood will move consistently toward the heart in the veins even though the driving pressures are low and variable. The pressure developed by the heart is not entirely absent in the veins and aids the return of blood to the heart. However, the contraction of *skeletal muscles* is also important; when the muscles contract, they squeeze the blood in nearby veins, and because the valves in the veins permit only one-way flow, the squeezing moves blood toward the heart. Veins are sometimes important because of their *capacitive* properties: By distending, they can increase their capacity to hold blood and thus accommodate blood volume that cannot be housed elsewhere in the circulatory system.

Mammals and birds have a high-pressure systemic circuit

The blood pressures maintained in the great systemic arteries of mammals and birds are the highest blood pressures found in any animals (**TABLE 25.1**). The need for these high pressures arises from the particular flow requirements of mammals and birds and from their systemic vascular resistance: a classic application of Equation 25.3. Mammals and birds require high rates of blood flow because of their high O_2-transport demands. The resistance to flow through their systemic circuits is high, however. To maintain high rates of flow in the face of this high resistance, mammals and birds require exceptionally high pressures in their systemic arteries. Why is the

TABLE 25.1	Systolic and diastolic blood pressures in the arteries leaving the heart and the cardiac outputs of some resting vertebrates[a]		
	Blood pressure (mm Hg)[b]		Cardiac output (mL/kg·min)
Species	Systolic	Diastolic	
Human (young adult male)	120	75	80–90
Bottlenose dolphin	150	121	47–105
Horse	171	103	150
Ground squirrel	139	99	313
Laboratory rat	130	91	209
Bobwhite quail	147	132	—
Pekin duck	181	134	—
Sparrow	180	140	—
Turtle	25	10	57
Iguana	48	37	58
Leopard frog	32	21	20–30
Rainbow trout	45	33	18–37
Catfish	40	30	11
Dogfish shark	30	24	25

Sources: After Farrell 1991; Sturkie 1976.

[a]Where both systemic and pulmonary arteries leave the heart, the data are for systemic arteries.

[b]To obtain blood pressures in units of kilopascals, divide the values given by 7.5.

resistance to flow so high in the systemic circuit? The Poiseuille equation (see Equation 25.2) tells us that vessels of tiny radius tend to make particularly great contributions to vascular resistance. In the closed circulatory system of a vertebrate, every tissue is densely invested with microscopically minute vessels through which the blood must be forced to flow. Collectively, these minute vessels give the systemic circuit its high resistance.

As blood enters the terminal arteries (the final branches of the arterial vasculature) and then the microcirculatory beds, the total cross-sectional area of the vasculature increases markedly, and simultaneously, the average linear velocity of the blood decreases dramatically (**FIGURE 25.12A**). Although individual capillaries are minute in cross section, they are so incredibly numerous that their collective cross-sectional area greatly exceeds that of the systemic aorta. The blood slows—that is, its linear velocity drops—where it has a greater cross-sectional area to pass through, in much the same way as a woodland stream slows where it widens. Capillaries are so short (<1 mm) that the blood must have a very low linear velocity when it passes through them for it to remain long enough to exchange with the tissues.

Even though the blood slows dramatically as it passes through the terminal arteries, arterioles, and capillaries, these are the vessels where the blood loses most of its pressure (**FIGURE 25.12B**). In humans at rest, the mean blood pressure falls from about 12 kPa to 8

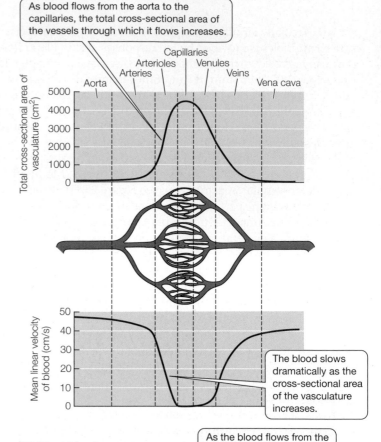

(A) Vascular cross-sectional area and blood linear velocity

As blood flows from the aorta to the capillaries, the total cross-sectional area of the vessels through which it flows increases.

The blood slows dramatically as the cross-sectional area of the vasculature increases.

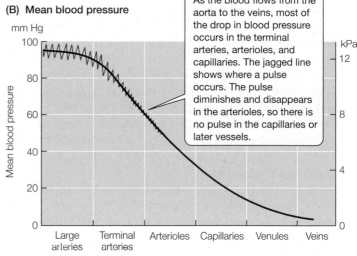

(B) Mean blood pressure

As the blood flows from the aorta to the veins, most of the drop in blood pressure occurs in the terminal arteries, arterioles, and capillaries. The jagged line shows where a pulse occurs. The pulse diminishes and disappears in the arterioles, so there is no pulse in the capillaries or later vessels.

FIGURE 25.12 Blood flow in the human systemic vasculature (A) Cross-sectional area and linear velocity of blood flow in the various parts of the systemic vasculature. Although the cross-sectional area of individual capillaries is minute, the capillaries are so numerous that their collective cross-sectional area greatly exceeds that of the systemic aorta. (B) Mean blood pressure in the various parts of the systemic vasculature. The jagged line does not represent the actual pulse (variation of blood pressure between systole and diastole), but symbolizes where the pulse occurs. (A from Feigl 1974.)

kPa (90 mm Hg to 60 mm Hg) as blood flows through the terminal arteries. Then the mean pressure falls to about 4.4 kPa (33 mm Hg) as the blood flows through the arterioles, and to about 2 kPa (15 mm Hg) as it flows through the capillaries and initial venules. A pressure drop of only about 0.4 kPa (3 mm Hg) is sufficient to move blood at a high linear velocity from the systemic aorta to the

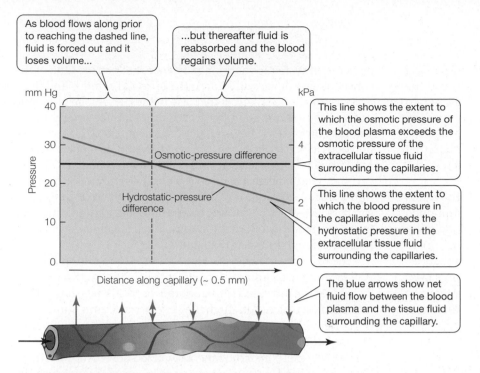

As blood flows along prior to reaching the dashed line, fluid is forced out and it loses volume...

...but thereafter fluid is reabsorbed and the blood regains volume.

This line shows the extent to which the osmotic pressure of the blood plasma exceeds the osmotic pressure of the extracellular tissue fluid surrounding the capillaries.

This line shows the extent to which the blood pressure in the capillaries exceeds the hydrostatic pressure in the extracellular tissue fluid surrounding the capillaries.

The blue arrows show net fluid flow between the blood plasma and the tissue fluid surrounding the capillary.

Osmotic-pressure difference

Hydrostatic-pressure difference

Distance along capillary (~ 0.5 mm)

FIGURE 25.13 Fluid exchange across mammalian systemic capillary walls: The Starling–Landis hypothesis The walls of blood capillaries are *fenestrated* in most tissues, meaning that they are densely perforated by minute physical pores such as gaps between adjacent cells (see page 725). Ultrafiltration (pressure-driven bulk flow) occurs readily through these fenestrations. Osmosis of water occurs readily because of the thinness of the endothelial cells, the fenestrations, and the presence of aquaporins in the cell membranes. On the graph, the osmotic-pressure difference can be represented as a horizontal line because the losses and gains of fluid volume are not great enough to change the relative osmotic pressures substantially. Values are approximate.

wrist in the major arteries of the arm. However, a 15-fold greater pressure drop, of about 6 kPa (45 mm Hg), is required to move blood at a low linear velocity over a distance of *a few millimeters* from the beginnings of arterioles to the initial venules in a microcirculatory bed! These quantitative measurements demonstrate emphatically that a closed circulatory system is energetically costly.

The venous part of the systemic circuit is a low-pressure, low-resistance system. The pressure developed by the left ventricle is essentially entirely degraded to heat by the time the blood reaches the entry of the great veins into the right atrium (see Figure 25.12B).

Fluid undergoes complex patterns of exchange across the walls of systemic capillaries

A capillary has fluid on both sides of its walls: blood plasma on the inside and extracellular tissue fluid (tissue interstitial fluid) on the outside (see Figure 27.2). The mean blood pressure in systemic capillaries typically exceeds the hydrostatic pressure in the surrounding extracellular tissue fluid. This difference in *hydrostatic* pressure favors pressure-driven oozing (bulk flow) of fluid out of the blood plasma across the capillary walls: a process termed **ultra-filtration** (see page 780). However, the blood plasma has a higher *osmotic* pressure than the extracellular tissue fluid. This difference in osmotic pressure favors the osmosis of water into the blood plasma across the capillary walls. The interplay of the processes favoring efflux of fluid from capillaries and those favoring influx has attracted considerable interest, in part because imbalances can give rise to serious disease states. Yet fluid exchanges in the capillaries are not yet thoroughly understood.

Studies by E. H. Starling (1866–1927) and E. M. Landis (1901–1987) in the early twentieth century produced a model of fluid exchange across capillary walls (**FIGURE 25.13**) that today is still believed to summarize the essentials in many tissues. The osmotic pressure of the blood plasma in most mammals exceeds that of the extracellular tissue fluid by about 3.3 kPa (25 mm Hg) along the entire length of the capillaries. This difference, called the **colloid osmotic pressure** of the blood plasma, arises because the plasma is richer than the tissue fluid in dissolved proteins (e.g., albumins)—large molecules that do not pass freely through capillary walls (see page 127). At the arterial ends of the capillaries, the hydrostatic pressure of the blood (the blood pressure) exceeds that of the tissue fluid by more than 4 kPa (30 mm Hg); because the hydrostatic-pressure difference favoring efflux of fluid from the capillaries is thus greater than the osmotic-pressure difference favoring influx, fluid is forced out of the capillaries in net fashion. However, as the blood flows through the capillaries, its hydrostatic pressure falls. At the venous ends of the capillaries, the hydrostatic pressure of the blood exceeds that of the tissue fluid by only about 2 kPa (15 mm Hg). Therefore, at the venous ends, the osmotic-pressure difference favoring influx of fluid into the capillaries exceeds the hydrostatic-pressure difference favoring efflux, and a net influx of fluid into the capillaries occurs.

The **Starling–Landis hypothesis**, stated briefly, is that the blood plasma loses volume in the initial segments of systemic blood capillaries but regains volume in the final segments. *The overall effect is often a net loss*: Fluid from the blood plasma tends to be transferred to the extracellular tissue fluid in a net fashion. The fluid added to the tissue fluid is picked up by the *lymphatic system*, which ultimately returns it to the blood. Tissues swell (a state termed *tissue edema*) if this fluid is not picked up.

The pulmonary circuit is a comparatively low-pressure system that helps keep the lungs "dry"

Discussion of the pulmonary circulation has been put off to this point because an understanding of the forces affecting fluid movement across capillary walls is essential for grasping a significant problem that could arise in perfusing the lungs. Rapid net loss of fluid from the blood plasma across the walls of the pulmonary capillaries could flood the terminal air spaces in the lungs—the alveoli of mammals or the air capillaries of birds—with liquid. This, in turn, would impair O_2 and CO_2 exchange.[18] Indeed, when a persistent state of alveolar flooding—called *pulmonary edema*—occurs during disease states, it is a life-threatening emergency. Persistent flooding of the alveoli or air capillaries does not normally occur, however, because the pulmonary circulation is a comparatively low-resistance, low-pressure system.

The vascular resistance of the pulmonary circuit is far lower than that of the systemic circuit because the pulmonary circulatory path is much shorter than the systemic path, lacks arterioles, and

---

[18] The rate of diffusion of O_2 and CO_2 is dramatically lower through water than through air, as emphasized in Chapter 22 (see page 590).

exhibits other distinctive properties. The volume of blood passing through the pulmonary circuit per unit of time is the same as the volume passing through the systemic circuit because the two are connected in series (see Figure 25.10). However, because the pulmonary vascular resistance is low, the arterial pressure required to drive this flow in the pulmonary circuit is far lower than that in the systemic circuit (see Equation 25.3). In humans at rest, the mean pressure maintained in the pulmonary arteries by the right ventricle is about 1.9 kPa (14 mm Hg), in contrast to the much higher mean pressure, about 12.7 kPa (95 mm Hg), maintained in the systemic aorta by the left ventricle. The mean blood pressure in the pulmonary capillaries is about 1 kPa (8 mm Hg), in contrast to about 3.3 kPa (25 mm Hg) in the systemic capillaries. The low pressure in the pulmonary capillaries largely precludes any net loss of liquid from the plasma across the capillary walls by ultrafiltration. Thus the air spaces of the lungs remain air-filled, not liquid-filled.[19]

During exercise, blood flow is increased by orchestrated changes in cardiac output and vascular resistance

The circulatory system of mammals and birds operates at a relatively leisurely pace when individuals are at rest. Exercise heightens demands on the circulatory system and thus brings out its full capabilities. A useful equation for understanding O_2 delivery in the systemic circuit is

$$\text{Rate of } O_2 \text{ delivery} = \text{cardiac output} \times (\text{arterial } O_2 \text{ concentration} - \text{venous } O_2 \text{ concentration}) \quad \textbf{(25.5)}$$

During exercise, if we assume that arterial blood is always fully oxygenated, the rate of O_2 delivery can be increased by increasing the rate of blood flow (i.e., increasing cardiac output) or by extracting more O_2 from each unit of volume of blood that circulates (i.e., decreasing venous O_2 concentration). *Both* of these strategies of increasing O_2 delivery are in fact simultaneously employed in mammals and birds. During vigorous exercise in humans, for example, venous blood becomes much more deoxygenated than at rest, as seen in Figure 24.5. Simultaneously, in average young adults the cardiac output can increase by at least a factor of four over the resting level through increases in both heart rate and stroke volume. Athletes who are successful in endurance events tend to have relatively high maximum cardiac outputs; some can sustain cardiac outputs that are six to seven times their resting values.

A substantial decline in the total resistance of the systemic vasculature occurs during exercise. Were it not for this drop in systemic vascular resistance, the aortic blood pressure would have to increase dramatically to drive blood through the vasculature at rates that are several times faster than at rest (see Equation 25.3). In fact, during whole-body exercise, the mean blood pressure in the human systemic aorta increases by only about 2.7 kPa (20 mm Hg), even when cardiac output far exceeds its resting level.

Vasodilation in the vascular beds of the active muscles (including the breathing muscles) is responsible for much of the

decrease in systemic vascular resistance that occurs. One of the consequences of vasodilation in the terminal arteries and arterioles is that there is a great increase in the percentage of muscle capillaries that are open and carrying blood (as briefly noted before). This increase helps reduce the resistance to through-flow of blood. It also reduces the average diffusion distance between capillaries and muscle cells, and it permits the capillary beds to carry a greatly increased flow of blood without major changes in the linear velocity (or residence time) of blood in individual capillaries. Endurance training tends to produce a chronic increase in the number of capillaries per muscle cell, an effect that clearly helps meet high muscle O_2 demand.

The response of the systemic vascular system during exercise is in fact highly coordinated and adaptive, in that it preferentially distributes the increased cardiac output to the tissues that require increased perfusion. Although the skeletal muscles in humans receive about 20% of the cardiac output at rest, they receive 70%–80% or more of the much increased cardiac output during vigorous whole-body exertion. Simultaneously, blood flow to the intestines, kidneys, liver, and inactive muscles may be substantially reduced by vasoconstriction. The brain receives about the same flow of blood regardless of exercise state.[20]

Species have evolved differences in their circulatory physiology

When mammal species of various body sizes are compared, there is a very apparent allometric relation between resting cardiac output and body size: Cardiac output per unit of body weight tends to increase as body size decreases. Thus small mammals meet their relatively high weight-specific demands for O_2 transport in part by maintaining relatively high weight-specific rates of blood flow. In relation to body size, small mammal species typically do not have hearts larger than those of large species. Instead, as we discussed in Chapter 7 (see Table 7.4), the principal way in which small species maintain their relatively high weight-specific cardiac outputs is by having relatively high heart rates (e.g., 600 beats/min in mice!). Birds tend to have larger hearts, lower heart rates, and higher aortic blood pressures (see Table 25.1) than mammals of equivalent body size.

Many species face unique cardiovascular challenges. Among them, giraffes present one of the most thought-provoking cases. Full-grown giraffes can be more than 5 m tall. Because of their long necks, their brains can be 1.6 m above their hearts, a distance that is more than four times greater than in humans. Moreover, because of their tall columns of blood, standing giraffes have exceptionally high blood pressures in their legs and feet. A few aspects of giraffe circulatory physiology are agreed upon by all authorities who have studied them. One is that giraffes have an exceptionally well developed left ventricle and maintain unusually high systemic aortic blood pressures. Their mean aortic pressure when they are standing and at rest is about 29 kPa (220 mm Hg), in contrast to about 13 kPa (100 mm Hg) in most other mammals.

The question of how blood is circulated to the brain in a standing giraffe is a topic of contentious debate. Some physiologists argue that the arteries and veins of the head and neck function

---

[19] Current theories of gas exchange postulate that exceedingly thin and highly transient layers of liquid may accumulate under some normal physiological conditions (e.g., vigorous exercise) and affect the details of the gas-exchange process.

[20] Responses of the circulation to high altitude are discussed in Box 24.5.

analogously to a siphon system, in which the height of the brain would pose no special challenges. The majority of physiologists who have studied the question, however, have concluded that the siphon analogy is flawed. These physiologists conclude that giraffes require their exceptionally high aortic blood pressures to be able to lift blood high enough against gravity to reach their brains and still have enough residual pressure at head height to perfuse their brains. According to this line of argument, if giraffes had mean aortic pressures like those in humans, blood would not even reach their brains! The mean human aortic pressure (12.7 kPa or 95 mm Hg) can lift blood to a height of only 1.2 m (see Figure 25.7B).

Vascular countercurrent exchangers play important physiological roles

Animals have often evolved structures in which blood vessels carrying blood in opposite directions are in contact with each other. Typically, arterial vessels are closely juxtaposed to venous ones. Usually the walls of these vessels are too thick to permit O_2, nutrients, or other macromolecules to pass through. However, heat can pass through, and the vascular structures are called *countercurrent heat exchangers* in reference to blood flowing in opposite directions. As explained in Chapter 10, heat moves especially efficiently from vessels carrying relatively warm blood into ones carrying cool blood when countercurrent heat exchange prevails.

One important function of countercurrent heat exchange in male mammals is helping to keep the testes in the scrotum cooler than the rest of the body. The arteries that take blood to the testes and the veins that return blood from them are arranged into a countercurrent heat exchanger. Heat is thereby short-circuited (see Figure 10.36) out of the arterial blood into the cooler venous blood before the arterial blood reaches the testes, helping to keep them at the diminished temperature they require for sperm production (see page 486).

Countercurrent heat exchangers also help minimize the amount of heat lost from the appendages of mammals and birds in cold environments (see Figure 10.35), and they play a key role in brain cooling (see Figure 10.37). In the small subset of fish that are warm-bodied, countercurrent heat exchangers are essential for keeping heat in the tissues (see Figure 10.46). Moreover, countercurrent blood flow in tiny vessels with O_2-permeable walls is vital for inflating the swim bladder with gas in many types of fish.

SUMMARY
Circulation in Mammals and Birds

- Mammals and birds, like virtually all other vertebrates, have closed circulatory systems, meaning that the blood always remains within blood vessels lined with vascular endothelium.

- The pulmonary and systemic circuits are connected in series. The left ventricle develops high pressures to force blood through the high-resistance systemic circuit. The right ventricle develops lower pressures to force blood through the low-resistance pulmonary circuit.

- In the systemic circuit, arteries convey blood over relatively long distances with little loss of blood pressure; arteries also perform pressure-damping and pressure-reservoir functions because of their elasticity. Arterioles in the systemic microcirculatory beds exert fine spatial and temporal control over blood flow by contraction and relaxation of the smooth muscles in their walls (vasomotor controls). The capillaries are the principal sites of exchange between the blood and systemic tissues because their walls consist of just a single layer of fenestrated endothelial cells rich in aquaporins, and because they are densely distributed.

- As blood flows through systemic capillaries, blood pressure tends to force fluid to pass outward through the capillary walls by ultrafiltration. The colloid osmotic pressure of the blood plasma tends to cause fluid movement into the blood. The net effect of this interplay is a loss of fluid, which is picked up by the lymphatic system. The lower blood pressures in the pulmonary circuit help to prevent pulmonary flooding (edema).

- During exercise, cardiac output is augmented by increases in both heart rate and stroke volume. Arterial blood pressure does not rise excessively because vascular resistance is decreased, mainly by vasodilation in active muscles.

Circulation in Fish

Fish in general resemble the vast majority of other vertebrates in having closed circulatory systems. The circulatory plan of most fish is illustrated in **FIGURE 25.14A**. Blood is pumped anteriorly by the heart into the **ventral aorta**, which distributes it to the afferent gill vessels.[21] The blood then passes through the blood channels of the gills and is brought by the efferent gill vessels to the **dorsal aorta**, a large dorsal artery that distributes the blood to the systemic tissues. After perfusing the systemic capillaries, the blood returns in the veins to the heart. As in mammals and birds, the circulatory plan places the breathing organs in series with the systemic tissues, thus ensuring efficient O_2 transport.

There are two ways in which the circulatory plan of fish poses potential problems that do not exist in the mammalian and avian plan. First, in fish there is no heart between the breathing-organ circulation and the systemic circulation to impart fresh energy to oxygenated blood as it leaves for the systemic tissues. When the heart pumps blood into the ventral aorta, the pressure it provides must be sufficient to drive the blood through the resistances of *both* the gill vasculature *and* the systemic vasculature. Second, oxygenation of the myocardium depends fully or partly on O_2 gained from the blood flowing through the lumen of the heart in fish (see Figure 25.3B,C), and this blood is *relatively deoxygenated* because it has just passed through the systemic circuit.

The fish heart (**FIGURE 25.14B**) consists of four chambers arranged in series: a **sinus venosus** into which the great veins empty, an **atrium**, a **ventricle**, and a bulbous segment that empties into the ventral aorta. The main propulsive force is developed by the ventricle. In elasmobranch fish (sharks, skates, and rays) and lungfish (dipnoans),

---

[21] *Afferent* means "going toward," and *efferent* means "going away."

(A) The circulatory plan

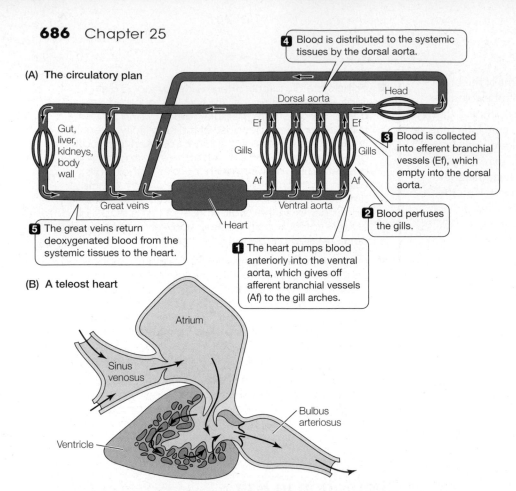

4 Blood is distributed to the systemic tissues by the dorsal aorta.

3 Blood is collected into efferent branchial vessels (Ef), which empty into the dorsal aorta.

2 Blood perfuses the gills.

1 The heart pumps blood anteriorly into the ventral aorta, which gives off afferent branchial vessels (Af) to the gill arches.

5 The great veins return deoxygenated blood from the systemic tissues to the heart.

(B) A teleost heart

FIGURE 25.14 The circulatory plan in gill-breathing fish (A) The overall circulatory plan of fish. Red and blue portions carry relatively oxygenated and deoxygenated blood, respectively. (B) The heart of a teleost fish. As discussed earlier (see Figure 25.3), the myocardium in teleosts is usually, at least in part, spongy and oxygenated by blood flow through the ventricular lumen, as shown. In both teleost fish and elasmobranch fish (sharks, skates, and rays), the heart consists of four sequential chambers. (B after Randall 1968.)

the bulbous segment—known as the **conus arteriosus**—includes cardiac muscle and contracts in sequence with the ventricle, helping to pump the blood. In teleost fish,[22] the bulbous segment—called the **bulbus arteriosus**—consists of vascular smooth muscle and elastic tissue and does not contract in sequence with the other heart chambers; it seems to act primarily as an elastic chamber that smoothes pressure oscillations and serves as a pressure reservoir between heart contractions (see pages 675–676).

Fish generally have far smaller hearts and far lower cardiac outputs than mammals or birds of similar body size (see Table 25.1). Their lower cardiac outputs correlate with their lower O_2 demands: Fish have far lower metabolic rates than mammals (see Chapter 10) and thus can satisfy their O_2 needs with lower rates of blood flow. Fish differ from mammals in that the heart empties almost completely by the end of each systole. Fish also maintain lower arterial pressures: The mean pressure in the ventral aorta is typically between 3 kPa (20 mm Hg) and 12 kPa (90 mm Hg)—usually toward the lower end of this range. The blood pressure drops significantly as the blood passes through the gills. Because of this drop, the mean pressure in the dorsal aorta—the pressure available to perfuse the systemic circuit—is generally only 60%–80% as high as the pressure produced by the heart.

In many fish the ventricular myocardium is entirely spongy—consisting of a mesh of relatively narrow, long sheets of myocardial cells. The sheets are called *trabeculae* and, attached at both ends to the ventricular walls, crisscross parts of the ventricular lumen, being bathed on either side by the luminal blood (usually systemic venous blood) that is passing through (and being pumped by) the ventricle. Some fish have, in addition, a layer of compact myocardium that surrounds the spongy myocardium (see Figure 25.3B,C). The compact myocardium is vascularized by coronary blood vessels; the blood pumped into these vessels is *fully oxygenated*. Among all fish that have been studied, the percentage of the myocardium that is compact varies from 0% (i.e., entirely spongy myocardium) to about 60%.

Different species of fish exhibit a wide and interesting range in the size and performance of the heart, correlated with how physically active the animals are. Species that are relatively inactive and sluggish tend to have relatively small hearts, little development of compact myocardium, and low cardiac outputs for their body sizes, whereas athletic species (e.g., salmon and tunas) tend to have large hearts, great development of compact myocardium, and high cardiac outputs. Tunas, the supremely active fish we featured at the start of Chapter 23, stand out among all fish in the performance of their circulatory systems. Yellowfin and skipjack tunas have weight-specific cardiac outputs at rest that are twice as high as the highest known in other fish; their ventral aortic blood pressures are also twice as high, accounting for the ability of the blood to flow through the vascular system at a high rate. The tunas are also among the fish with the highest amounts of hemoglobin per unit of blood volume. Thus tunas not only employ the most sophisticated breathing systems found among fish to obtain O_2, but also exploit both major factors in Equation 25.5 to sustain exceptionally high rates of O_2 delivery to their tissues: They pump blood rapidly, and each unit of blood volume is able to deliver a large amount of O_2.

The plasticity and adaptability of the fish heart are remarkable. Because of phenotypic plasticity, the size of the ventricle in an individual can vary up to threefold as a function of the individual's environment or other circumstances. Recently, various populations of sockeye salmon (*Oncorhynchus nerka*) in the Fraser River system of British Columbia have been studied. Because individuals in each population return reliably to that particular population's spawning area, the populations are distinct. When adults undertake their physically demanding upriver migration prior to spawning, those in some populations face a far more strenuous task that those in others; for example, some must swim more than 1000 km to their spawning areas, whereas others swim only about 100 km. Among populations of this single species, the percentage of compact myocardium tends to be significantly higher in those that must exert high migration effort than in those with easy migrations (**FIGURE 25.15**). Compact myocardium is believed to be advantageous for fish that engage in demanding exercise because it reliably receives fully oxygenated blood owing to its coronary circulation.

[22] Teleost fish are the principal group of fish with bony skeletons; the great majority of bony fish are teleosts.

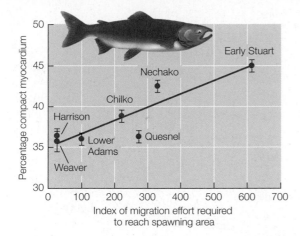

FIGURE 25.15 In seven populations of sockeye salmon, the percentage of the ventricular myocardium that is compact tends to increase as required migration effort increases The data are for females in seven populations of sockeye salmon (*Oncorhynchus nerka*) that migrate for spawning into the Fraser River system in British Columbia, Canada. Each symbol corresponds to one population and is labeled with the name of the tributary where the population spawns. The index of migration effort was calculated numerically by multiplying the river distance a population must travel by the speed of the opposing river current the fish must swim against. Whereas the Harrison and Weaver populations swim upstream only about 120 km, the Early Stuart population swims about 1080 km. Error bars represent ± 1 standard error. (After Eliason et al. 2011.)

Exercise in fish is only beginning to be well understood. During exercise, some species emphasize increases in stroke volume to enhance cardiac output, whereas others emphasize increases in heart rate. Rainbow trout, often a model species for exercise studies, can—on average—increase their cardiac output about threefold as they swim faster, mostly by increasing stroke volume. They also deoxygenate their blood during exercise more than they do at rest (see Figure 24.17), thereby increasing the transport of O_2 per unit of blood volume by as much as twofold. The combination of the two effects permits rainbow trout to attain a sixfold increase in their rate of circulatory O_2 delivery.

Enhanced blood deoxygenation during exercise (see Figure 24.17) has potentially important implications for myocardial O_2 supply in fish because of the fact that the spongy part of the myocardium obtains its O_2 from the venous blood flowing through the heart lumen (see Figure 25.3B). As the intensity of exercise increases, the myocardium requires O_2 at a greater rate, yet the partial pressure of O_2 decreases in the blood flowing through the heart lumen because of the increased unloading of O_2 to the systemic tissues. The diminishing O_2 supply to the spongy myocardium could place limits on exercise performance, especially in O_2-poor environments.

The circulatory plans of fish with air-breathing organs (ABOs) pose unresolved questions

A few hundred species of fish are able to breathe air, as discussed in Chapter 23 (see pages 610–611). In most cases, the air-breathing organs (ABOs) of these species are derived from structures—such as the mouth membranes, gut, or swim bladder—that are primitively served by the *systemic* circulation. Accordingly, the *oxygenated* venous blood leaving the air-breathing organ typically flows into the *systemic* venous vasculature—not the systemic arterial vasculature!

Consider, for illustration, the electric eel (*Electrophorus electricus*), an obligate air breather. Most of its O_2 uptake occurs in its mouth cavity and pharynx; the lining of the mouth cavity is thoroughly

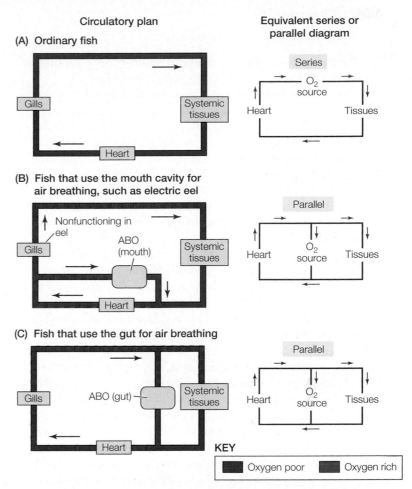

FIGURE 25.16 The circulatory plans of some air-breathing fish The proportions of red and blue in all of the circulatory paths represent the approximate proportions of oxygenated and deoxygenated blood carried in the vessels. (A) The circulatory plan in ordinary, water-breathing fish, shown as a base of comparison. (B) The circulatory plan in the electric eel (*Electrophorus electricus*) and certain other fish that have an air-breathing organ (ABO) derived from the mouth cavity. The diagram applies specifically to the electric eel, in which the gills are not effective in taking up O_2. Afferent vessels to the ABO are derived from the afferent gill vessels; efferent vessels from the ABO connect to systemic veins. The circulatory plan in fish that use the swim bladder as an air-breathing organ is similar. In some fish species that have ABOs of these sorts, the gills remain partly functional, so in aerated water, there is at least some increase in oxygenation when blood passes through the gills. (C) The circulatory plan in certain catfish (e.g., *Hoplosternum, Plecostomus*) and other fish in which part of the stomach or intestine serves as an ABO. Afferent vessels to the ABO are derived from the dorsal aorta; efferent vessels enter systemic veins. (After Johansen 1970.)

covered with highly vascularized tufts of tissue, so much so that it resembles a red cauliflower. The circulatory plan of the electric eel is shown in **FIGURE 25.16B**. The afferent blood vessels to its air-breathing organ (its mouth cavity) arise from vessels leading toward the gills. The efferent vessels from the air-breathing organ, carrying oxygenated blood, empty into the systemic venous vasculature. Thus, in contrast to the usual fish pattern (**FIGURE 25.16A**) and to the pattern in mammals and birds, the circulatory plan places the O_2 source in *parallel* with the circulation of the systemic tissues. That is, as shown by the schematic diagram of the parallel arrangement at the right of Figure 25.16B, one bloodstream splits to send blood to both the O_2 source and systemic tissues, and blood leaving these two sites combines.

The implications of the parallel arrangement are profound. As shown in Figure 25.16B, oxygenated blood from the air-breathing organ freely *mixes* with deoxygenated blood from the systemic tissues in the systemic veins and heart. Thus the heart pumps a mix of deoxygenated and oxygenated blood to both the air-breathing organ and the systemic arteries. The degree of O_2 saturation of hemoglobin in the blood pumped by the heart of the electric eel never exceeds 60%–65%! This is true even though the blood leaving the air-breathing organ itself may be more than 90% saturated. Not just mouth breathers, but also the fish species that have a gastrointestinal air-breathing organ (**FIGURE 25.16C**)—and those that employ the swim bladder as an air-breathing organ—typically have circulatory plans that place the air-breathing organ in parallel with the systemic tissues.

In considering the implications of parallel circulatory plans, biologists have traditionally emphasized that the mixing of the oxygenated and deoxygenated bloods *reduces the efficiency of O_2 transport*. Blood that has been deoxygenated in the systemic tissues is in part recycled directly back to the systemic tissues; and blood oxygenated in the air-breathing organ is in part recycled directly back to the air-breathing organ. According to the traditional interpretation of most biologists, these consequences are, quite simply, shortcomings. When these fish evolved air breathing, according to this line of thinking, they would have benefited if the organs they adopted for air breathing had evolved a modified venous vasculature—a vasculature that delivered the oxygenated blood to systemic *arteries*. Instead, they retained a disadvantageous, primitive venous vasculature that empties oxygenated blood into systemic *veins* carrying deoxygenated blood.

An entirely new perspective has been brought to this matter in the past 15 years. This perspective emphasizes that the spongy myocardium gets its O_2 from the blood passing through the heart chambers. When a fish obtains its O_2 from the air, mixing the oxygenated blood from its air-breathing organ with its systemic venous blood could be an advantage because such mixing ensures that the blood flowing through the heart chambers contains increased O_2 for the heart muscle. This hypothesis is still being evaluated.

Lungfish have specializations to promote separation of oxygenated and deoxygenated blood

The lungfish (dipnoans), which are usually considered the modern fish that most closely resemble the progenitors of terrestrial vertebrates (see page 611), display circulatory plans that differ from those of nearly all other air-breathing fish. Here we consider just the genus *Protopterus*, the African lungfish. These lungfish (one of which is pictured in Box 25.2) have evolved a circulatory system that can maintain considerable separation between the blood oxygenated in their lungs and other, deoxygenated blood.

The lungs of *Protopterus* have a venous vasculature that is dramatically different from that of most air-breathing organs in fish: The veins from the lungs lead directly into the left side of the atrium of the heart, rather than connecting with the systemic venous vasculature. This means that blood from the lungs is kept anatomically separate from systemic venous blood until both streams of blood have entered the atrium. The sinus venosus, which receives only systemic venous blood in *Protopterus*, connects to the right side of the atrium.

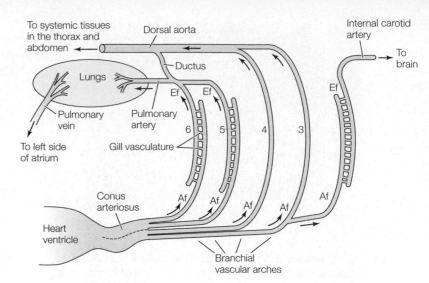

FIGURE 25.17 The branchial vascular arches of a lungfish and their relation to heart, lungs, and systemic tissues In this diagram of one side of *Protopterus aethiopicus*, the branchial vascular arches are numbered according to the aortic arches they represent (other numbering systems are sometimes used). Branchial arches 3 and 4 go to gill arches that lack gill lamellae; those branchial vascular arches do not break up into capillaries, but instead form direct through-connections to the dorsal aorta. Gills of a rudimentary sort are found on the gill arches supplied by branchial vascular arches 5 and 6. Within those gill arches, the afferent branchial arteries (Af) break up to supply the gill lamellae, and blood is collected from the lamellae by efferent branchial vessels (Ef). A similar arrangement is found in the most anterior gill arch, as shown. The efferent vessels of arches 5 and 6 form the pulmonary artery. The vessel called the ductus connects the pulmonary artery and dorsal aorta; flow through it is not well understood, although it is generally believed to help blood bypass the lungs during periods when ventilation of the lungs is suspended. Arrows show directions of blood flow. (After Laurent et al. 1978.)

The heart of *Protopterus* is very different from that of most fish in three respects. First, the atrium and ventricle are partly (but not completely) divided into right and left halves by septa. Second, the conus arteriosus, a sharply twisted tube, possesses two longitudinal ridges that project toward each other from opposite sides of its lumen, partially dividing the lumen into two channels. Third, the four pairs of afferent branchial arteries arise immediately from the anterior end of the conus arteriosus (rather than having a ventral aorta in between), as do the homologous vessels of amphibians.

As shown in **FIGURE 25.17**, two of the four pairs of afferent branchial arteries—constituting aortic arches 3 and 4—arise from the ventral channel of the conus arteriosus and travel to anterior gill arches that lack gill lamellae; these arteries do not break up into capillaries, but instead form direct through-connections to the dorsal aorta. The other two pairs of afferent branchial arteries—constituting aortic arches 5 and 6—lead from the dorsal channel of the conus arteriosus to two pairs of posterior gill arches that retain rudimentary gills; these arteries break up to supply the gills, but the gill capillaries are of large diameter and can be bypassed to at least some extent by way of vascular shunts, meaning that blood often encounters only a relatively low resistance in flowing from the afferent to the efferent vessels of these gill arches. The efferent vessels give rise to the arteries that carry blood to the lungs—arteries homologous to the pulmonary arteries of tetrapod vertebrates.

To understand blood flow through the central circulation[23] of lungfish, *functional* studies are obviously needed, because the anatomy by itself does not mandate particular patterns of flow.

[23] The term **central circulation** refers to the heart and the veins and arteries that immediately connect to the heart.

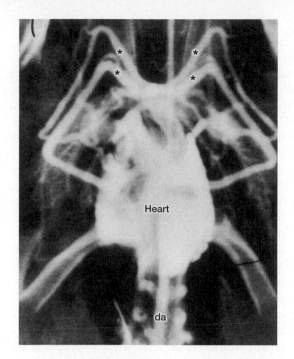

FIGURE 25.18 Oxygenated blood from the lungs of a lungfish is pumped selectively by the heart into certain branchial vascular arches The fish under study (*Protopterus aethiopicus*) was viewed from below using X-rays. A fluid opaque to X-rays was injected into the pulmonary vein, flowed into the heart, and was being pumped by the heart when this image was made. White-colored parts of the image show the location of the fluid at the moment the image was made. Note that of the four pairs of branchial vascular arches that arise directly from the heart, only two pairs—those labeled with asterisks—received substantial amounts of the X-ray–opaque fluid. The arches receiving the fluid were numbers 3 and 4 (compare Figure 25.17). This image shows that blood from the lungs is pumped by the heart into these arches preferentially and thus flows primarily to the dorsal aorta (da). (Courtesy of Kjell Johansen; from Johansen and Hol 1968.)

The functional studies that have been done, although they vary in results, demonstrate that oxygenated pulmonary venous blood and deoxygenated systemic venous blood in fact follow substantially different paths through the central circulation.

One method of studying blood-flow patterns is to inject fluids that are opaque to X-rays into selected vessels and, using X-rays, monitor where these fluids are carried by the blood.[24] Studies of this sort show that oxygenated venous blood from the lungs tends to follow a course through the left parts of the atrium and ventricle, and is delivered preferentially to the ventral channel of the conus arteriosus. Then (**FIGURE 25.18**) the blood is pumped into vascular arches 3 and 4, which carry the blood directly to the dorsal aorta (see Figure 25.17). In this way, the blood oxygenated in the lungs is directed preferentially into arteries of the systemic circuit. Systemic venous blood tends to pass through the right parts of the atrium and ventricle. That blood then appears in X-ray images either to be distributed about evenly to all four pairs of vascular arches (those numbered 3–6) or to be delivered preferentially to the dorsal channel of the conus arteriosus and into vascular arches 5 and 6, which in part supply blood to the pulmonary arteries.

Another way to study blood-flow patterns is to monitor blood O_2 levels. In one study of *Protopterus*, when the O_2 partial pressure in systemic venous blood averaged 2 mm Hg and that in pulmonary venous blood was 46 mm Hg, the blood pumped into vascular arches 3 and 4 for direct passage to the dorsal aorta had an average O_2 partial pressure of 38 mm Hg, showing that it consisted mostly of oxygenated blood from the lungs. Simultaneously, the blood pumped to vascular arches 5 and 6 had a much lower O_2 partial pressure.

Physiologists have long pondered whether there might be advantages to having an incompletely divided heart. Considering lungfish, one line of thinking focuses on the fact that they are intermittent breathers. Because they alternate between periods of breathing and breath-holding, their lungs alternate between periods when their O_2 concentration is high and low. Their incompletely divided heart permits them to *redistribute* the total outflow of blood from the heart in a way that birds and mammals cannot (**BOX 25.2**)—they can pump a large fraction of their cardiac output to their lungs when the lung air is rich in O_2 but pump a small fraction of their cardiac output to the lungs when the lung air is O_2-depleted. This redistribution indisputably occurs at times and may be a specific evolutionary reason that the central circulation has remained incompletely divided.

---

[24] The same sort of technique is used to visualize blood flow to organs such as the brain and myocardium in medical practice.

BOX 25.2 An Incompletely Divided Central Circulation Can Be an Advantage for Intermittent Breathers

African lungfish (*Protopterus annectens*)

Intermittent breathing is common in lungfish, amphibians, lizards, snakes, and turtles—groups in which the heart is incompletely divided. It is also common in crocodilians, in which the central circulation outside the heart is incompletely divided. An intermittent breather, having ventilated its lungs with air, holds its breath for a substantial time before ventilating again (see page 607). Intermittent breathing presents both opportunities and challenges because the lungs vary from time to time in how effectively they are able to oxygenate the blood. Immediately after an animal has taken a series of breaths, the air in the lungs is rich in O_2, and blood flowing through the lungs can become well oxygenated. However, after a long interval of apnea (cessation of breathing), the air in the lungs may be depleted of O_2, and little opportunity may exist for blood flowing through the lungs to gain O_2.

A great deal of research has convincingly demonstrated that most or all air-breathing vertebrates with incompletely divided central circulations in fact modulate blood flow to their lungs independently of flow to the rest of the body—at least to some degree and under certain circumstances. They increase blood flow to the lungs immediately after each period of breathing, whereas they decrease it toward the end of each period of apnea. In doing so, they achieve **ventilation–perfusion matching** to some degree. That is, they achieve functionally efficient matching of air flow and blood flow to the lungs. These topics are discussed further in **Box Extension 25.2**.

SUMMARY
Circulation in Fish

■ In most fish, the heart pumps blood to the gills, after which the blood passes through the systemic circuit before returning to the heart. The gill and systemic circuits are arranged in series.

■ In air-breathing fish, blood leaving the air-breathing organ (ABO) usually mixes with systemic venous blood. Thus the circulation of the ABO is in parallel with the systemic circuit. This arrangement decreases the efficiency of O_2 transport but may help oxygenate the myocardium.

■ Lungfish have a modified central circulation in which blood from the lungs enters the left side of the atrium, and the atrium, ventricle, and conus arteriosus are partly divided. Deoxygenated and oxygenated blood can be kept relatively separate and pumped selectively to the lungs and systemic circuit. Redistribution of cardiac output is possible, and occurs in synchrony with the intermittent breathing cycle.

Circulation in Amphibians and in Reptiles Other than Birds

Three interrelated themes have dominated research on circulation in amphibians and nonavian reptiles. One is *selective distribution*: To what extent is oxygenated blood sent to the systemic tissues, and deoxygenated blood to the lungs? The second is *maintenance of different blood pressures in the systemic and pulmonary circuits*: Can high pressures be maintained in the systemic circuit to produce high flow rates in that circuit while low pressures are maintained in the pulmonary circuit to prevent flooding the airways with fluid? The third is *redistribution of cardiac output*: To what extent is flow to the lungs adjusted separately from that to the rest of the body? These phenomena interact in complicated ways, but we will discuss them as distinct processes.

In the lung-breathing amphibians and nonavian reptiles, two completely separate atrial chambers are present. Oxygenated blood from the lungs enters the left atrium, whereas systemic venous blood enters the right atrium via the sinus venosus. Separation of oxygenated and deoxygenated blood is anatomically guaranteed until the blood enters the ventricle, which in these groups consists of an outer compact myocardium with a coronary circulation and an inner spongy myocardium oxygenated by blood flowing through the heart lumen (see Figure 25.3C). After blood leaves the anatomically separate atria and enters the ventricle, the central circulation is incompletely divided in one way or another.

In amphibians, the ventricle entirely lacks a septum. The ventricular lumen, however, is not a wide-open cavity because it is crisscrossed with ribbons and cords of the spongy myocardium. The ventricle discharges into a contractile conus arteriosus, which in turn typically discharges into paired carotid, systemic, and pulmonary arteries. The lumen of the conus in most amphibians is incompletely divided by a complexly twisted, longitudinal ridge of tissue (the *spiral fold*).

Although amphibians are diverse, it is clear that many—including certain frogs, toads, and salamanders—are capable of substantial selective distribution of oxygenated and deoxygenated blood (**FIGURE 25.19**). American bullfrogs (*Lithobates catesbeianus*, once called *Rana catesbeiana*) provide a dramatic example. In one

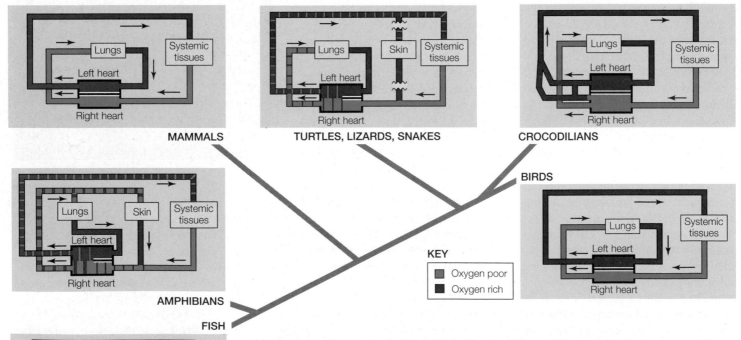

FIGURE 25.19 The typical circulatory plans of the major vertebrate groups, plotted on a phylogenetic tree of the groups The proportions of red and blue in all of the circulatory paths represent the approximate proportions of oxygenated and deoxygenated blood carried in the vessels. In the groups with incompletely divided hearts, average or ordinary performance is shown here; performance during long periods of breath-holding or diving might be quite different. Among the "turtles, lizards, snakes," only some of the turtles absorb O_2 across their skin in significant amounts, explaining why the skin circulation is drawn as indefinite. (Based on a concept by Tobias Wang, with thanks.)

study, 91% of the pulmonary venous blood arriving in the heart was channeled into the systemic arteries, whereas 84% of the systemic venous blood was directed into the pulmonary arteries. Such a high degree of selective distribution functionally places the pulmonary and systemic circuits in series, promoting efficient O_2 transport. When amphibians achieve such feats, they do so with a ventricle that entirely lacks a septum and with a conus arteriosus that is only incompletely divided! Physiologists do not know exactly how.

The skin of amphibians is an important site of O_2 uptake (see page 613). Some of the major arteries to the skin arise from the pulmonary arteries (also known as *pulmocutaneous arteries*). Thus, if deoxygenated blood is preferentially pumped to the lungs, it is also preferentially pumped to the skin. The veins draining the skin connect with the general systemic venous vasculature—a property traditionally viewed as disadvantageous because it allows blood oxygenated in the skin to mix with deoxygenated systemic venous blood rather than staying separate. However, from the newly hypothesized viewpoint that emphasizes adequate O_2 for the myocardium, the arrangement of the cutaneous veins could be neutral or beneficial. The amphibian myocardium gets much of its O_2 from the blood flowing through the heart lumen, and cutaneous flow adds O_2 to the venous blood entering the right heart (see Figure 25.19).

In nonavian reptiles, just as in mammals and birds, the principal arteries arise directly from the ventricle of the heart; there is no intervening conus arteriosus. In *turtles, snakes, and lizards*, the ventricle is incompletely divided into three chambers by muscular ridges and partial septa. Although pulmonary and systemic venous bloods are potentially able to mix in the ventricle in these groups, numerous studies have demonstrated that dramatic selective blood distribution occurs (see Figure 25.19). The pulmonary arteries receive mostly deoxygenated blood under many circumstances, for example. The mechanism of selective distribution varies. Two of the best-understood cases are pythons and the strikingly active varanid lizards. In them, prominent ridges of tissue that incompletely divide the ventricular chambers are pressed so tightly against opposing structures during heart contraction as to create complete anatomical blocks to blood flow from one chamber to another. Thus, although the ventricular chambers that eject blood into the pulmonary and systemic circuits are not separated by a complete septum, they are temporarily separate during systole. Other temporary physical barriers within the heart also play roles.

Regarding blood pressures in the systemic and pulmonary circuits, the pythons and varanid lizards are unique among the reptiles we are here discussing in that they exhibit dramatically different blood pressures in different ventricular chambers during systole—a phenomenon that requires complete physical separation between the chambers. Specifically, the pressure in the chamber pumping to the systemic arteries is high (much higher than that observed in most reptiles), while simultaneously the pressure in the chamber pumping to the lungs remains low—low enough to prevent ultrafiltration of fluid into the lung air spaces. The pythons and varanids, in other words, resemble mammals and birds in being able to employ high pressures to drive a high rate of flow in their systemic circuit while not endangering lung function. In other noncrocodilian reptiles that have been studied and in amphibians, no means exist to achieve complete physical separation between parts of the ventricular lumen during systole. Thus the systemic

and pulmonary circuits receive blood at the same pressure, and that pressure has to be low enough so as not to endanger lung function.

Regarding redistribution of cardiac output, lizards, snakes, and turtles—and also amphibians—frequently display intermittent breathing, and most species that have been studied redistribute their cardiac output during periods of intermittent breathing, so that flow to the lungs is high after breaths have been taken but low after periods of apnea (see Box 25.2). Certain diving turtles and snakes provide especially dramatic examples. When their lung air becomes O_2-depleted during a dive, they continue to circulate blood round and round in their systemic circuit while sending hardly any blood to their lungs![25] The control of the distribution of cardiac output is becoming better understood. In amphibians, which entirely lack a ventricular septum, blood is distributed into the pulmonary and systemic circuits as dictated by peripheral vascular resistance; the lungs get a lot of flow when their vascular resistance is low relative to the systemic circuit's vascular resistance, and vice versa. This same phenomenon is important in lizards, snakes, and turtles, but in addition in these groups, the regulation of temporary anatomical barriers in the heart plays an important role in determining the paths that blood follows when ejected.

Burmese pythons (*Python bivittatus,* also called *P. molurus*) have been used in many studies as model organisms for understanding the implications of a severe feast-and-famine life history (see pages 161–162). They often go for long periods without eating but can eat huge meals when the opportunity arises. Between times of feast and famine, a python extensively remodels its body. The heart is included in this remodeling, as discussed in Chapter 21 (see Figure 21.10). Recent genomic studies have established that when a python that has been fasting succeeds in catching an animal to eat, dozens of myocardial genes undergo expression changes in just the first 1–3 days, as the heart greatly increases its size and synthesizes enhanced levels of contractile proteins.

In the *crocodilian reptiles*, the ventricle is completely divided into two chambers by a septum. This does not mean, however, that the crocodilians resemble mammals and birds—far from it. Crocodilians have two systemic aortas, which arise from the left and right ventricles (**FIGURE 25.20A**). The two aortas are connected shortly after their exit from the ventricles by an aperture called the **foramen of Panizza**, and the vascular beds perfused by the two aortas are also connected more distally. The pulmonary artery arises from the right ventricle. Functional studies have demonstrated that the crocodilian heart can achieve virtually perfect selective distribution of deoxygenated blood to the lungs and oxygenated blood to the systemic circuit. The way this happens is that deoxygenated systemic venous blood in the right ventricle can be pumped exclusively into the pulmonary artery because valves and pressure relationships prevent it from entering the systemic aorta that leaves the right ventricle (**FIGURE 25.20B**; see also Figure 25.19). This flow pattern is not guaranteed by anatomy, however, and flow may be redistributed during diving or breath-holding (**FIGURE 25.20C**).

[25] In the professional literature on the cardiovascular physiology of animals with incompletely divided central circulations, the variable distribution of total cardiac output to the pulmonary and systemic circuits is often called *shunting. Right-to-left* (R-L) *shunting* is diversion of systemic venous blood directly back into the systemic circuit, bypassing the lungs. This is the sort of shunting displayed by the diving turtles. *Left-to-right* (L-R) *shunting* is diversion of pulmonary venous blood directly back to the lungs.

FIGURE 25.20 Blood flow in the heart ventricles and the systemic and pulmonary arteries of crocodilian reptiles (A) The basic plan of the ventricles and vessels. (B) During steady air breathing, blood flows readily from the right ventricle into the pulmonary artery. Consequently, the systolic pressure in the right ventricle does not rise particularly high, and in fact remains lower than the pressure in the left systemic aorta, so the passive flap valve between the right ventricle and left systemic aorta is not forced open. Blood from the left ventricle is pumped mostly into the right systemic aorta; some may enter the left systemic aorta through the foramen of Panizza, although this does not always occur. (C) During diving or prolonged breath-holding, resistance to flow into the pulmonary artery is raised. Blood therefore cannot flow out of the right ventricle as easily, the right ventricular pressure rises, and the flap valve into the left systemic aorta is pushed open during ventricular systole, permitting some of the systemic venous blood pumped by the right ventricle to bypass the lungs. In these ways, the incompletely divided central circulation permits pulmonary blood flow to be reduced during times when lung air is O_2-depleted (see Box 25.2).

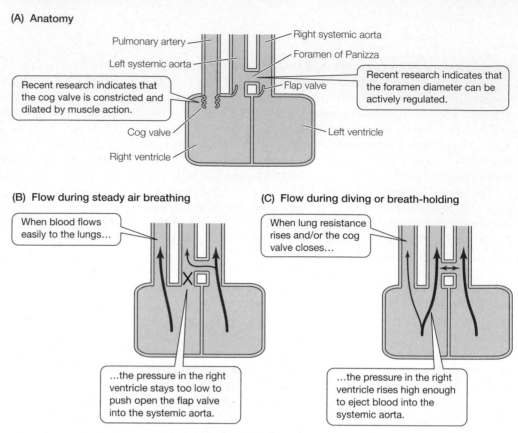

(A) Anatomy

Pulmonary artery
Left systemic aorta
Right systemic aorta
Foramen of Panizza
Flap valve
Cog valve
Right ventricle
Left ventricle

Recent research indicates that the cog valve is constricted and dilated by muscle action.

Recent research indicates that the foramen diameter can be actively regulated.

(B) Flow during steady air breathing

When blood flows easily to the lungs…

…the pressure in the right ventricle stays too low to push open the flap valve into the systemic aorta.

(C) Flow during diving or breath-holding

When lung resistance rises and/or the cog valve closes…

…the pressure in the right ventricle rises high enough to eject blood into the systemic aorta.

Concluding Comments on Vertebrates

In studying the circulatory systems of fish, amphibians, and reptiles other than birds, there is perhaps a yearning to find a linear advance toward the condition of mammals and birds. The facts, however, do not at all reflect such a linear progression (see Figure 25.19). The groups of animals alive today are not in the business of striving to be mammals and birds (how could they be?).

Among air-breathing vertebrates, logic suggests that efficiency of O_2 transport is promoted by having the pulmonary and systemic circuits connected in series during periods of time when the lungs are being ventilated and the lung air is rich in O_2. In mammals and birds, which breathe continuously, the lung air is always rich in O_2. Thus efficiency is achieved by having an anatomically deterministic connection of the pulmonary and systemic circuits in series.

The other air-breathing vertebrates besides mammals and birds fill their lungs with air only intermittently. In these vertebrates, as explained in Box 25.2, an *incomplete* anatomical division of the central circulation can potentially be superior to an anatomically rigid separation in its ability to promote efficiency of O_2 transport. Recognizing this point, it is clear that the circulatory plans of the "lower" vertebrates may have their own virtues and may have been subject to positive natural selection, rather than being evolutionary relics. There are at least a half dozen other rationales, proposed by biologists, for reaching the same conclusion. Empirical evidence for or against these evolutionary hypotheses is needed, and gathering such evidence is perhaps the central challenge at present for research in the comparative physiology of vertebrate circulation.

Invertebrates with Closed Circulatory Systems

Closed circulatory systems occur in only a few groups of invertebrates, notably some annelid worms and the cephalopod molluscs (the squids and octopuses and their relatives). The cephalopods, which we emphasize here, are of great interest because they are extraordinarily active animals compared with most members of their phylum (e.g., clams and snails). Some squids, in fact, jet around at such high speeds that they rank with fish as being among the most active of all aquatic animals. The giant squids are the largest invertebrates on Earth. The circulatory system of a squid or octopus consists of extensive networks of arteries and veins—both of which are muscular—joined in large part by capillary beds.

A squid or octopus has a principal heart, called its **systemic heart**, and two weaker, auxiliary **branchial hearts**. The systemic heart—which is myogenic—consists of a powerful muscular chamber valved at its inflows and outflows. **FIGURE 25.21** shows the circulatory plan. Blood enters the systemic heart from the gills and is pumped from the systemic heart into major arteries (aortae) that take it to the systemic tissues. As blood returns to the vicinity of the systemic heart in the major veins (which, unlike vertebrate veins, often display contractions that move the blood), it is split into two symmetrical paths to be directed to the gills. Near the base of each gill is a bulbous branchial heart, which pumps blood into an afferent branchial vessel. The blood then passes through capillaries in the gill to arrive in the efferent branchial vessel and return to the systemic heart. This circulatory plan resembles that of mammals, birds, and most fish in that it places the breathing organs and the systemic tissues strictly in series. The arrangement of hearts may seem a bit odd at first, but is actually "close to home." As seen

(A) The circulatory plan

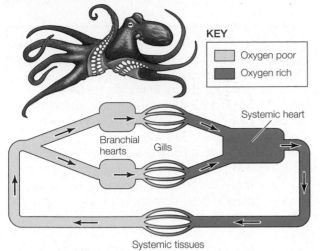

KEY

Oxygen poor

Oxygen rich

Systemic heart

Branchial hearts

Gills

Systemic tissues

(B) The hearts and gills of an octopus

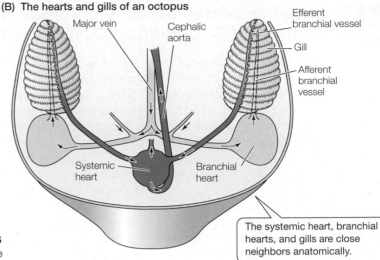

Major vein

Cephalic aorta

Efferent branchial vessel

Gill

Afferent branchial vessel

Systemic heart

Branchial heart

The systemic heart, branchial hearts, and gills are close neighbors anatomically.

FIGURE 25.21 The circulatory plan of squids and octopuses (A) The arrangement of the hearts, gills, and systemic tissues relative to one another. Hemocyanin, the respiratory pigment of squids and octopuses, turns blue when well oxygenated but is clear or nearly clear when deoxygenated. (B) A more realistic drawing of the central circulatory system of an octopus. (B after Johansen and Lenfant 1966.)

by comparing Figures 25.10B and 25.21A, the arrangement of hearts in a squid or octopus is identical to that in a mammal, except that the respiratory pumps are anatomically separate from the systemic pump.

Physiologically, the circulatory system of squids and octopuses resembles that of vertebrates far more than it does that of the other molluscan groups (which have low-pressure, open circulatory systems). In the octopus *Enteroctopus dofleini* and the highly active squid *Loligo pealeii*, the mean blood pressure produced by the systemic heart is 4–5 kPa (30–37 mm Hg), and by the time the blood has passed through the systemic tissues and reached the great systemic veins, its mean pressure has dropped nearly to zero. The

systemic pressure gradient in cephalopods thus resembles that in fish (see Table 25.1). The cardiac output of resting or mildly active *E. dofleini* is typically about 10–20 mL/kg•min—also in the range seen in fish (see Table 25.1). When the vascular resistance of the systemic circuit in cephalopods is calculated (see Equation 25.3), it is relatively high, as would be expected for a closed circulatory system. Overall, therefore, the cephalopods are similar to fish in that they maintain reasonably rapid rates of blood flow through a systemic circuit of relatively high resistance by maintaining relatively high pressures in the systemic arteries.

Exercise in cephalopods presents intriguing questions, which we have already started to discuss in earlier chapters. **BOX 25.3**, in addition to summarizing points made in those other chapters, explains the special burden borne by the circulatory system during cephalopod exercise.

BOX 25.3 **Bearing the Burden of Athleticism, Sort of: A Synthesis of Cephalopod O$_2$ Transport**

Both the vertebrate and the invertebrate worlds have produced high-speed, swimming predators. The principal vertebrates in this ecological role are fish. As improbable as it may seem when gazing at a clam or snail, the molluscs have produced the invertebrate entry: the squids. The fossil record indicates that swimming cephalopods were abundant in the oceans before the appearance of marine fish and dominated in playing the role of fast-swimming predators. The arrival of fish then produced competition, which the fish won. Some species of squids alive today have been as successful as fish. However, the squids might have dominated the oceans had fish not

appeared, and today it is the fish that dominate.

To generate swimming power, the cephalopods became committed to jet propulsion early in their evolution and have never turned back. Compared with propulsion of the sort employed by fish, jet propulsion is inherently inefficient. For example, a squid might well use twice the O$_2$ of a like-sized fish to swim at half the speed of the fish. Thus the systems that supply O$_2$ to the tissues of squids are asked to assume a large burden.

The gills of cephalopods are positioned in the mantle cavity, where they have always been throughout the evolution of molluscs (see Figure 23.28). Being

in the mantle cavity, the gills are exposed to extreme rates of water flow during jet propulsion and have no difficulty obtaining O$_2$, as we saw in Chapter 23 (see page 626).

The circulatory system is where limits develop. For one thing, molluscs are committed to hemocyanin dissolved in the blood plasma as their respiratory pigment. Dissolved hemocyanin makes the blood ever more viscous as its concentration rises. Squids and octopuses have

(Continued)

Bearing the Burden of Athleticism, Sort of: A Synthesis of Cephalopod O$_2$ Transport *(Continued)*

evolved the highest concentrations of hemocyanin in any animals: They seem to have raised blood viscosity as high as it can go. Even thus maximally concentrated, hemocyanin allows the blood leaving the gills to carry only about 5 mL O$_2$ per 100 mL of blood, as we saw in Chapter 24. Hemoglobin in red blood cells allows many fish to have oxygen-carrying capacities more than twice as high. As also seen in Chapter 24 (see Figure 24.18), squids and octopuses, even when *approximately at rest*, are believed to have almost no venous O$_2$ reserve. Therefore, when they start jetting about, their sole available option to increase O$_2$ transport is to increase their rate of blood flow by increasing cardiac work.

If a squid and a fish were both to increase cardiac output fourfold during exercise, the squid would increase its rate of circulatory O$_2$ delivery about fourfold, but the fish might increase its rate tenfold by simultaneously tapping its venous O$_2$ reserve to increase the amount of O$_2$ delivered per unit of blood volume (see Figure 24.17). This difference illustrates the implications of the resting venous O$_2$ reserve: If there is little resting venous reserve, O$_2$ delivery can be increased only by increasing the rate of blood flow.

Viewed metaphorically, the sophisticated circulatory system of the cephalopods is doomed to bearing an almost impossible burden by a series of irrevocable evolutionary commitments in

other functional attributes, such as the commitment to jetting as the source of propulsive power and the commitment to hemocyanin. When a squid powers up, its O$_2$ cost of swimming is high compared with that of a fish, and—to make things worse—because the blood oxygen-carrying capacity and venous O$_2$ reserve of a squid are comparatively low, the squid's hearts must bear almost the entire responsibility of increasing O$_2$ delivery. Although some squids fortunately prosper nonetheless, the "O$_2$ equation" is probably a key reason that the squids as a group were outcompeted by fish over the eons in the oceans.

Invertebrates with Open Circulatory Systems

The animals with open circulatory systems—which include all arthropods and most molluscs—commonly have a well-developed central heart. The heart may discharge into a network of arteries, and the arteries may lead to capillary beds of discrete, minute vessels lined only with a single layer of endothelial cells. Ultimately, however, in an open circulatory system, the blood exits discrete vessels (**FIGURE 25.22**), and thereafter it flows through lacunae and sinuses. **Lacunae** are small spaces among cells of nonvascular organs and tissues. A tissue with lacunae is often thoroughly permeated by an anastomosing network of lacunar spaces—a network that brings blood close to all the cells in the tissue. Lacunar networks are like capillary networks in this respect. However, lacunar channels are characteristically irregular in shape and are not in the form of discrete vessels lined

(A) Shortly before arrival of dye

(B) 1 second later

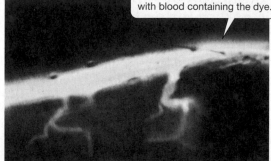

At this moment, a major leg artery and some side branches become filled with blood containing the dye.

(C) Another moment later

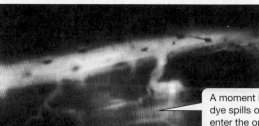

A moment later, blood containing the dye spills out of discrete vessels to enter the open part of the circulation.

FIGURE 25.22 The transition of blood flow from the closed part of the circulation to the open part A fluorescent dye was injected into arteries of a spider (*Pholcus*). As the dye traveled with the flowing blood, it was imaged using fluorescent videomicroscopy. (A) A portion of a leg just before the arrival of blood containing dye. (B) The same spot 1 s later. (C) The same spot a moment later, showing blood and dye flowing out of discrete vessels. (Photos courtesy of R. J. Paul; from Paul et al. 1994.)

with endothelium. **Sinuses** are larger blood spaces, commonly representing thoroughfare channels for the blood. Lacunae and sinuses in animals with open circulatory systems are sometimes bounded by a membrane of some type. However, they are believed by most physiologists to be sometimes bounded simply by cells of ordinary, nonvascular tissues. This latter point is of key importance because it means that the blood directly bathes the cells of the nonvascular tissues, and *there is in fact no clear distinction between blood and extracellular tissue fluid (interstitial fluid)*. In recognition of this fact, some authors call the blood **hemolymph** ("blood lymph").

As already suggested, experts debate the exact nature of the lining of lacunae and sinuses in some cases. Regardless of the ultimate outcome of these debates, it is clear that the design of the circulatory system in animals with open circulatory systems is significantly different from that in vertebrates or cephalopods. In the latter groups, not only is blood *carried to* the depths of each tissue by discrete vessels, but it remains in vessels as it *passes through* each tissue, and it is *collected from* each tissue by discrete vessels. In animals with open circulatory systems, after blood leaves the arteries (or sometimes capillaries), it is left to follow paths through lacunae and sinuses—paths that in many cases seem ill-defined—before it is again channeled into discrete vessels.

The crustacean circulatory system provides an example of an open system

Some immature, small, or sessile crustaceans lack a heart and blood vessels. Their circulation is exclusively through sinuses and lacunae, and the propulsive force is provided by ordinary body movements. In some other crustaceans, a heart is present, but empties through holes rather than arteries. In still others, arteries are present, but they end abruptly after extending only a short distance from the heart. Here we emphasize the decapod crustaceans—crabs, crayfish, lobsters, and their relatives—which have especially elaborate open circulatory systems.

In adult decapod crustaceans, the heart is a single-chambered saclike structure—positioned in the dorsal thorax—the beat of which is initiated neurogenically (see Figure 25.5).[26] *All the vessels connected to the heart are arteries* (as is typical of crustaceans in general). The arteries are valved at their origins and leave the heart in several directions (**FIGURE 25.23A**). Blood enters the heart not through vessels, but through slits in the heart wall, called **ostia** (singular *ostium*), of which decapods usually have three pairs. The heart is suspended by elastic **suspensory ligaments** within a bounded sinus, the **pericardial sinus** (**FIGURE 25.23B,C**). The only blood channels that enter the pericardial sinus are "veins" that bring blood from the gills or other respiratory surfaces.[27]

When the heart contracts at systole, the ostia are closed by muscular tension, flap valves, or both; blood within the heart thus cannot spew out through the ostia into the pericardial sinus, and is driven into the arteries. Contraction of the heart stretches the suspensory ligaments attached to the heart wall. During diastole,

---

[26] The heart is myogenic in the embryos and juveniles of at least some decapod crustaceans and is partly or wholly myogenic in adults of some other types of crustaceans.

[27] We term these channels "veins," in quotation marks, because they are not typical venous vessels of the sort seen in vertebrates or cephalopods, but are usually called *veins* nonetheless.

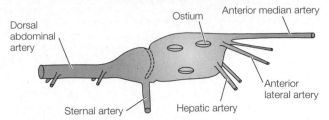

(A) The heart and the arteries emanating from the heart

Dorsal abdominal artery

Ostium

Anterior median artery

Anterior lateral artery

Hepatic artery

Sternal artery

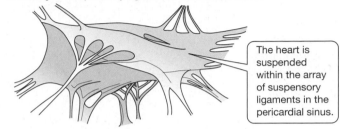

(B) The array of suspensory ligaments around the heart

The heart is suspended within the array of suspensory ligaments in the pericardial sinus.

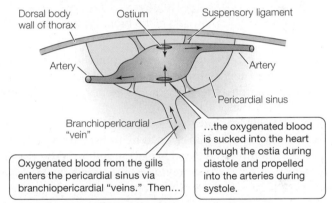

(C) Flow of blood through the central circulation

Dorsal body wall of thorax

Ostium

Suspensory ligament

Artery

Artery

Pericardial sinus

Branchiopericardial "vein"

Oxygenated blood from the gills enters the pericardial sinus via branchiopericardial "veins." Then…

…the oxygenated blood is sucked into the heart through the ostia during diastole and propelled into the arteries during systole.

FIGURE 25.23 The heart of a decapod crustacean (A) The heart and the arteries leaving it. All the vessels connected to the heart in a crustacean are arteries. (B) The heart is surrounded by an array of elastic suspensory ligaments that run between the heart wall and the wall of the surrounding pericardial sinus, suspending the heart in the pericardial sinus. (C) The position of the heart in the pericardial sinus and the pattern of blood flow through the central circulation. (A after Wilkens 1999; B after Plateau 1880.)

elastic rebound of these ligaments expands the heart back to its presystolic volume. This elastic rebound is the primary force for refilling. As the heart is stretched open, the pressure within is reduced below that in the pericardial sinus, and blood is sucked inward from the pericardial sinus through the ostia.

The walls of the arteries of crustaceans, although elastic, are typically nonmuscular. Thus the arteries are neither pulsatile (as in cephalopod molluscs) nor capable of vasomotor control (as in vertebrates)—raising the question of whether they can exert any control over the regional distribution of cardiac output. By now, however, it is clear that blood distribution is under control of highly localized muscular valves in the arteries. The best studied are the **cardioarterial valves**, positioned at the points where the various arteries leave the heart. Each valve contains innervated muscle. Excitatory nerve impulses, by stimulating contraction of a

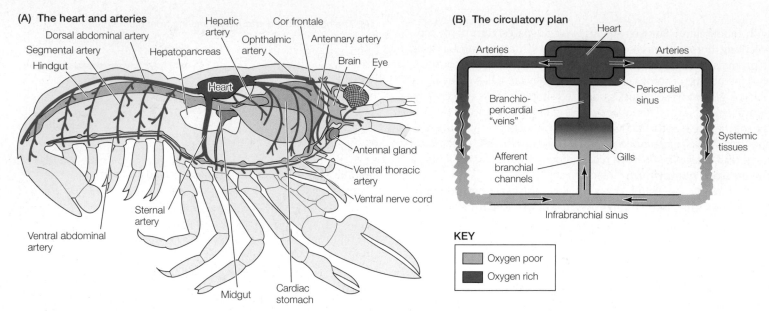

(A) The heart and arteries

Dorsal abdominal artery
Segmental artery
Hindgut
Hepatopancreas
Hepatic artery
Ophthalmic artery
Cor frontale
Antennary artery
Brain Eye
Heart
Antennal gland
Ventral thoracic artery
Ventral nerve cord
Sternal artery
Ventral abdominal artery
Midgut
Cardiac stomach

(B) The circulatory plan

Heart
Arteries Arteries
Branchio-pericardial "veins"
Pericardial sinus
Afferent branchial channels
Gills
Systemic tissues
Infrabranchial sinus

KEY

	Oxygen poor
	Oxygen rich

FIGURE 25.24 Circulation through the body of a crayfish or lobster (A) The heart and major arteries. The *cor frontale* is an accessory heart (powered by skeletal muscles) positioned upstream from the extensive vascular networks of the brain; it is believed to help ensure that adequate perfusion pressures are maintained in the brain. (B) The circulatory plan. Because the respiratory pigment of decapod crustaceans is hemocyanin, the blood turns bluish when oxygenated but is clear when fully deoxygenated. (A after McLaughlin 1980.)

valve, can limit blood flow into the artery guarded by that valve; simultaneously, inhibitory impulses sent to other valves might enhance blood flow into other arteries. Hormones also help control the cardioarterial valves.

Decapod crustaceans have extensive arterial systems (**FIGURE 25.24A**). Branching arteries lead blood from the heart to most regions of the body. Sometimes the blood is discharged from the arteries directly to lacunar networks. Sometimes the arteries lead to capillary beds, which then discharge into lacunar networks. Prominent capillary beds are present, for example, in the brain and the ganglia of the ventral nerve cords. In all, the arterial systems of decapods

are often impressively elaborate, especially in comparison with some of the other groups of arthropods such as insects (**BOX 25.4**).

In a decapod crustacean, after blood leaves the arteries or capillaries and enters lacunar networks throughout the body, it ultimately drains into a system of sinus thoroughfare channels located ventrally along the length of the animal. In these ventral sinuses, blood from the posterior regions of the body flows anteriorly, and that from the anterior regions of the body flows posteriorly. As shown in **FIGURE 25.24B**, both flows converge on a ventral thoracic sinus termed the **infrabranchial sinus** (*infrabranchial*, "below the gills"). Afferent branchial channels that carry blood into the gills arise from this

---

BOX 25.4 Circulation and O₂: Lessons from the Insect World

Like all arthropods, insects have an open circulatory system. A dorsal vessel that runs along most of the body is divisible into a posterior heart (usually restricted to the abdomen) and an anterior dorsal aorta that runs forward into the thorax and head, as shown in the figure. The heart has ostia and refills much as a crustacean heart does. It contracts in a peristaltic wave, forcing blood into the dorsal aorta, which often continues the peristaltic wave of contraction. The dorsal aorta is the only blood vessel in many species of insects. In others, segmentally arranged lateral arteries branch off from the dorsal vessel. The major blood vessels in either case end abruptly with little branching, discharging blood directly to

the lacunar circulation. The blood-vessel system is far less extensive than that in decapod crustaceans, and there are no capillary beds.

The modest development of the circulatory system in insects seems paradoxical at first. Many insects are very active animals with high O₂ demands; during sustained flight, some insects attain weight-specific rates of O₂ consumption that not only are many times higher than any reported for crustaceans, but are among the highest in the entire animal kingdom. In phyla other than the arthropods (most notably chordates and molluscs), groups with a capacity for relatively intense aerobic catabolism show unambiguous circulatory refinements in

comparison with related groups with a lower capacity for aerobic catabolism: Mammals have far faster rates of blood flow than fish, for example. The circulatory system of insects, however, is not more advanced than that of crustaceans or other arthropods, despite the extraordinary metabolic intensity of many insects.

The paradox of the modest circulatory system of insects finds its likely answer in the tracheal breathing system. In an insect, O₂ is brought from the atmosphere to all cells by a system of gas-filled tracheal tubes. Because this system functions *independently* of the circulatory system (see Figure 23.31), the latter plays little or no role in O₂ transport. The insects have a lesson to teach: Namely, sinus. Blood flows into these channels, traverses systems of small sinuses or lacunae in the gill filaments or lamellae, and then exits the gills by way of efferent branchial channels. Importantly, these channels discharge into passageways—the branchiopericardial "veins"—that lead directly into the pericardial sinus. Therefore, after a period of seemingly poorly controlled flow following its discharge from the arterial system, the blood is channeled in a highly significant way prior to its return to the heart. The gills are placed in series with the rest of the circulation, and oxygenation of the blood pumped by the heart is ensured.[28]

Like fish, decapod crustaceans perfuse the systemic tissues and the gills in sequence, without repumping the blood between the two sites. In decapods, however—unlike in fish—blood leaving the heart passes through the systemic circuit first and the gill circuit second. Thus blood pumped to the systemic circuit is at the pressure developed by the heart. The magnitude of the drop of blood pressure in the systemic circuit is known in only a few species, but in general ranks as being moderately large; the difference between arterial pressure and infrabranchial-sinus pressure is 1.3–2 kPa (10–15 mm Hg) in spiny lobsters (*Panulirus*) and about 1 kPa (6–7 mm Hg) in American lobsters (*Homarus*). The gill circulation typically poses a low resistance and requires only a very small pressure difference to be perfused (about 0.2 kPa, or 1–2 mm Hg).

Open systems are functionally different from closed systems but may be equal in critical ways

In animals with open circulatory systems, the pressures in the blood spaces are often affected substantially by body movements—a phenomenon that must be considered for the function of these systems to be analyzed. When body movements create pressure

differences from one part of an animal's body to another, they aid blood flow; in American lobsters, for example, flexion of the abdomen—an important swimming movement (see page 189)—raises pressures in the abdomen above those in the thorax and can cause blood flow toward the thorax to speed up by a factor of ten or more. Body movements and changes in posture, however, also sometimes affect the *overall* level of pressure *throughout* the body; a change in posture, for instance, might compress all the blood spaces in the body and pressurize the blood everywhere. If pressure is changed to the same extent everywhere, blood flow is not immediately affected. Only *differences* in pressure cause flow.

When we speak of circulatory systems as being "high-pressure" or "low-pressure" systems, the *pressure gradient required to perfuse them*—not the overall pressure that exists everywhere—is the pertinent characteristic. If, for example, a pressure gradient of 0.5 kPa (4 mm Hg) is sufficient to drive blood through an animal's circulatory system, the system is a "low-pressure" system. This remains true even if the animal is in a posture that pressurizes its whole body to a high level.[29]

Having clarified the meaning of terms, we can now make the important generalization that open circulatory systems are typically low-pressure or moderate-pressure systems. In bivalve molluscs and nonpulmonate snails, arterial pressures typically exceed venous pressures by just 0.1–0.5 kPa (1–4 mm Hg) or even less. The pressure gradient in the spiny lobsters mentioned in the previous section (1.3–2 kPa, or 10–15 mm Hg) ranks with the highest known to occur with regularity in animals having open circulatory systems.

The relatively low pressure gradients in open circulatory systems have often been interpreted to mean that blood flow through open systems is sluggish. Low pressure gradients do not necessarily

---

[28] A small fraction of the blood returning to the heart bypasses the gills by flowing through the tissue lining the gill chambers. This circulatory path is greatly elaborated in air-breathing crabs that use the lining of the gill chambers as a breathing surface.

[29] For instance, if the arterial pressure in a crustacean were 40 mm Hg and the infrabranchial-sinus pressure were 36 mm Hg, the circulatory system would rank as a "low-pressure" system, even though the two pressures are themselves relatively high, because the pressure gradient required for perfusion would be just 4 mm Hg.

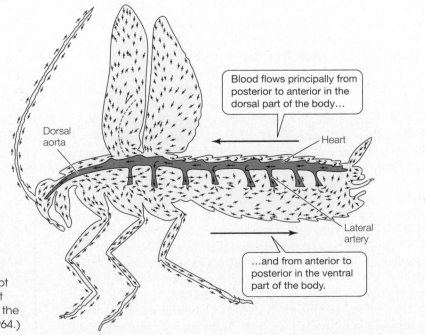

the circulatory system can remain quite simple in even highly active animals if it need not assume the burden of O$_2$ transport. The tasks the circulatory system must generally perform in insects (e.g., nutrient and hormone transport) are far less urgent than O$_2$ transport. Thus the insect circulatory system can be quite modest. The decapod crustaceans, in contrast, depend on their circulatory system to supply their tissues with O$_2$. Thus the decapods require a relatively sophisticated version of the arthropod circulatory system.

Dorsal aorta

Blood flows principally from posterior to anterior in the dorsal part of the body...

Heart

Lateral artery

...and from anterior to posterior in the ventral part of the body.

Blood flow through the tissues of an insect is principally through lacunae and sinuses Heart ostia are not shown. Insects often have accessory pulsatile structures that aid flow through their legs, wings, and antennae. In general, the blood of insects lacks a respiratory pigment. (From Jones 1964.)

TABLE 25.2 Systemic circulatory function: Decapod crustaceans compared with fish

Characteristics of circulatory function	Spiny lobster	Rock crab	Starry flounder	Rainbow trout
Principal features of circulatory function				
Rate of O_2 delivery to tissues (mL O_2/kg·min)[a]	0.80	0.60	0.46	0.65
Rate of blood flow through systemic circuit (mL blood/kg·min)[b]	128–148	125	39	18
Pressure change to perfuse systemic circuit (mm Hg)[c]	14	3	16	22
Systemic resistance (pressure change divided by flow rate)[d]	0.1	0.03	0.4	1.2
Secondary information				
Heart rate (beats/min)	65	101	35	63
Stroke volume (mL/kg·stroke)	2.1	1.2	1.2	0.3
Blood pressure in major systemic arteries (mm Hg)[c]	35	10	18	26
Blood pressure in major systemic veins or venous sinuses (mm Hg)[c]	21	7	2	4
Blood oxygen-carrying capacity (mL O_2/100 mL blood)	2.0	1.3	5.7	7.8

Source: Hill and Wyse 1989; based on a compilation of data from several sources. Species included are *Panulirus interruptus* (spiny lobster), *Cancer productus* (rock crab), *Platichthys stellatus* (starry flounder), and *Oncorhynchus mykiss* (rainbow trout). Body weights of animals studied were 200–700 g. Temperature during studies was 8°C–16°C. The rock crab periodically ceases heart action; values given are averages measured 5 min or more after a cardiac pause.

[a]Measured as the animals' rates of O_2 consumption.

[b]Equals cardiac output in steady state.

[c]Pressure change is the difference between blood pressures in arteries and in veins (or venous sinuses); both of these pressures are listed in the "secondary information" below. The arterial pressures were measured in systemic arteries, except for the crab pressure, which was measured as systolic ventricular pressure and thus probably overestimates arterial pressure. The venous pressures were measured in the systemic veins of the fish and in the infrabranchial sinus of the crustaceans.

[d]The values of systemic resistance are in units of mm Hg·kg·min/mL blood.

mean sluggish flow, however, because the rate of flow depends on resistance as well as pressure (see Equation 25.3). The first four rows of **TABLE 25.2** compare four critical functional attributes of the open circulatory systems in a lobster and crab with those of the closed circulatory systems of fish of similar body sizes. As the first row shows, all the species have *similar* rates of circulatory O_2 delivery under the conditions of study. The crustaceans (see the second row) circulate their blood *at least three times faster* than the fish, however. The crustaceans, that is, are the exact opposite of being sluggish! Although the crustaceans (third row) maintain *lower pressure gradients* than the fish, circulation is fast in the crustaceans because (fourth row) the *resistance to flow* through the crustacean circulatory systems is much lower than the resistance to flow in the fish.

Not all animals with open circulatory systems have high rates of blood flow. In the decapod crustaceans, however, the overall picture is that pressure gradients—although low—are adequate to circulate blood rapidly because the resistance to flow is low. The resistance is low precisely because the circulatory systems of the crustaceans are open; blood flows through lacunae and sinuses during much of its travel, rather than through minutely narrow capillary vessels. Why do the crustaceans *need* their high blood-flow rates? One reason, certainly, is that their blood oxygen-carrying capacities are low (1–2 mL O_2/100 mL of blood, compared with 6–8 mL O_2/100 mL in fish), meaning that each unit of volume of blood they circulate is limited in the O_2 it can deliver.

Inevitably, the question arises of whether open or closed circulatory systems are "better." Data such as those in Table 25.2 suggest that the systems are "different but equal" in key respects—at least when sophisticated crustaceans are compared with fish. The greatest void that remains in physiological knowledge of open circulatory systems is the question of how well the flow of blood can be directed and controlled. Physiologists have tended to assume that little direction or control is possible in open systems. Increasing evidence exists, however, that open circulatory systems can direct blood flow far more exactly than physiologists have assumed—by using cardioarterial valves or other (mostly unknown) mechanisms. The novel study in **FIGURE 25.25**, for example, shows that in the leg muscles of a spider, *lacunar* blood flow tends to be most vigorous near the particular sets of muscle cells that are most O_2-dependent. Closed circulatory systems are justly touted for their ability to exert fine spatial and temporal control over blood flow. Although open systems seem unlikely to be the equal of closed systems in this respect, the difference between the two is probably smaller than traditionally believed.

SUMMARY
Invertebrates with Open Circulatory Systems

■ Most invertebrates have open circulatory systems in which blood leaves discrete vessels and flows through systems of lacunae and sinuses, where it comes into contact with ordinary tissue cells (accounting for the fact the blood is sometimes called hemolymph).

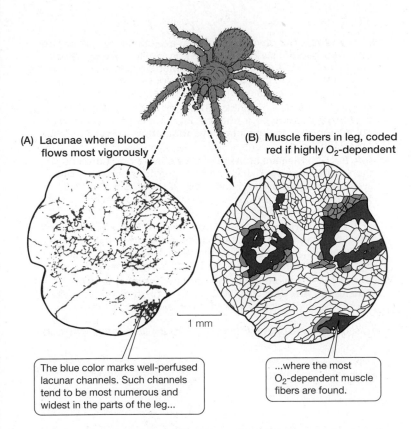

(A) Lacunae where blood flows most vigorously

(B) Muscle fibers in leg, coded red if highly O_2-dependent

1 mm

The blue color marks well-perfused lacunar channels. Such channels tend to be most numerous and widest in the parts of the leg...

...where the most O_2-dependent muscle fibers are found.

FIGURE 25.25 The microanatomy of blood flow in a lacunar system: Blood flows most vigorously where it is needed most Adjacent cross sections of a tarantula's leg are shown. The tissue seen inside the cross sections is mostly muscle, that is, the tissue of the walking muscles. (A) Locations where blood flows particularly vigorously in the lacunar networks of the muscles. In this cross section, hemocyanin was stained immunohistochemically using monoclonal antibodies (antibodies specific to hemocyanin). The stain was then visualized. (B) Locations of muscle fibers that are particularly O_2-dependent. In this cross section, fibers were assessed using their profiles of catabolic enzymes (similar to Figure 8.13). The fibers colored dark red are the most poised to produce ATP aerobically and are the most O_2-dependent. Those colored light red are also aerobically poised, but less so. (From Paul et al. 1994.)

- Animals with open circulatory systems typically have a heart, and they may have extensive systems of blood vessels, even including capillary beds. The blood ultimately leaves the vessels, however.

- Open circulatory systems tend to be characterized by relatively small changes of blood pressure across the systemic circuit compared with those in closed circulatory systems. Resistance is low in open systems because blood is not forced through capillary beds in most or all tissues. Thus the rate of blood flow may be high despite the small pressure changes.

- Little is known about the spatial and temporal control of blood flow in animals having open circulatory systems. Some control is known to be possible, however (as by cardioarterial valves in crustaceans), and control is probably more sophisticated than generally assumed in the past.

STUDY QUESTIONS

1. Imagine that the digestion of a meal has just provided a new supply of glucose and amino acids in your midgut (small intestine). Then imagine all the steps that will take place as your body uses the glucose and amino acids and disposes of the waste products that are produced. List all the individual and specific functions your circulatory system will perform during this sequence of events. Arrange your list in the correct temporal order.

2. A current working hypothesis is that human diseases characterized by incurable weakness and lethargy may sometimes arise in part from loss of blood capillaries. Basing your answer on the laws of diffusion, explain how transport to and from a tissue would be affected if 20% of its capillaries were lost.

3. Thomas Kuhn, a great historian of science, said that "big ideas" in science are not discovered at fixed moments in history, but are discovered over extended periods of time by multiple scientists. We often hear it said that William Harvey "discovered the circulation of the blood." Kuhn would say that this is not true. He would say that when we speak of today's meaning of the "circulation of the blood," the concept of "circulation" required centuries to be discovered. Pretend you are Kuhn, and argue for his points of view. Then, speaking for yourself, explain why you agree or disagree with Kuhn.

4. Figure 25.12B shows mean blood pressures at the entrances and exits of the large arteries, the terminal arteries, the arterioles, and the capillaries in the systemic circuit of a human at rest. Using those pressures and assuming a cardiac output of 6 L/min, calculate the collective vascular resistance of the large arteries, the terminal arteries, the arterioles, and the capillaries.

5. Looking ahead to Chapter 29, the process that initiates urine formation in most types of vertebrates is ultrafiltration of fluid from the blood plasma into kidney tubules under the force of blood pressure. What properties of the blood plasma in the urine-forming structures of the kidney are likely to affect the *rate* of ultrafiltration into the kidney tubules? Why is failure of urine production a potential side effect of heart disease?

6. During exercise—as compared with rest—what are the aspects of blood O_2 transport that can be modified to increase O_2 delivery to the exercising muscles? Based on the information discussed in this chapter, how do squids differ from vertebrates in the specific ways in which they increase O_2 delivery?

7. List the major types of blood vessels in the systemic circuit of a mammal, and outline the *functions* of each type of vessel.

8. What are the pros and cons of closed and open circulatory systems? Consider not only the type of information in Table 25.2 but also the other relevant information discussed in this chapter.

9. Mechanistically, why is pulmonary edema a threat to life? Thinking as creatively as possible, list three distinctly different potential causes of pulmonary edema.

10. Many objects designed by engineers have "circulatory systems" of one type or another. Describe at least three or four examples. Why do engineers sometimes include systems of flowing fluids in objects they design? We often hear it said that animal systems can teach us how to design engineered systems. If you were a biologist teaching a group of engineers about the lessons of animal circulatory systems, what principles would you emphasize?

Go to **sites.sinauer.com/animalphys4e**
for box extensions, quizzes, flashcards,
and other resources.

REFERENCES

Burggren, W., A. Farrell, and H. Lillywhite. 1997. Vertebrate cardiovascular systems. In W. H. Dantzler (ed.), *Comparative Physiology*, vol. 1 (Handbook of Physiology [Bethesda, MD], section 13), pp. 215–308. Oxford University Press, New York. A thorough and articulate review of most aspects of vertebrate circulatory function.

Burggren, W. W., B. B. Keller, and C. Weinstein (eds.). 1997. *Development of Cardiovascular Systems: Molecules to Organisms*. Cambridge University Press, New York.

Claireaux, G., D. J. McKenzie, A. G. Genge, and three additional authors. 2005. Linking swimming performance, cardiac pumping ability and cardiac anatomy in rainbow trout. *J. Exp. Biol.* 208: 1775–1784.

Farmer, C. G. 1999. Evolution of the vertebrate cardiopulmonary system. *Annu. Rev. Physiol.* 61: 573–592.

Farrell, A. P., E. J. Eliason, E. Sandblom, and T. D. Clark. 2009. Fish cardiorespiratory physiology in an era of climate change. *Can. J. Zool.* 87: 835–851. The best short review of cardiorespiratory function in fish.

Glass, M. L., and S. C. Wood (eds.). 2009. *Cardio-respiratory Control in Vertebrates*. Springer-Verlag, New York.

Hall, J. E. 2011. *Guyton and Hall Textbook of Medical Physiology*, 12th ed. Saunders, Philadelphia. A human physiology text with a tradition of excellence in its treatment of circulatory physiology.

Hicks, J. W. 2002. The physiological and evolutionary significance of cardiovascular shunting patterns in reptiles. *News Physiol. Sci.* 17: 241–245. A particularly useful modern summary of a complex issue, in which the potential advantages of reptilian heart morphology and function are thoroughly elucidated.

Hillman, S. S., and M. S. Hedrick. 2015. A meta-analysis of *in vivo* vertebrate cardiac performance: implications for cardiovascular support in the evolution of endothermy. *J. Exp. Biol.* 218: 1143–1150.

Jensen, B., J. M. Nielsen, M. Axelsson, and three additional authors. 2010. How the python heart separates pulmonary and systemic blood pressures and blood flows. *J. Exp. Biol.* 213: 1611–1617. Dramatic and well-presented research on the function of a nonmammalian heart.

O'Dor, R., H. O. Pörtner, and R. E. Shadwick. 1990. Squid as elite athletes: Locomotory, respiratory, and circulatory integration. In D. L. Gilbert, W. J. Adelman, Jr., and J. M. Arnold (eds.), *Squid as Experimental Animals*, pp. 481–503. Plenum, New York.

Wilkens, J. L. 1999. Evolution of the cardiovascular system in Crustacea. *Am. Zool.* 39: 199–214. An analysis of the pros and cons of open circulatory systems in an evolutionary context.

See also **Additional References** *and* **Figure and Table Citations**.

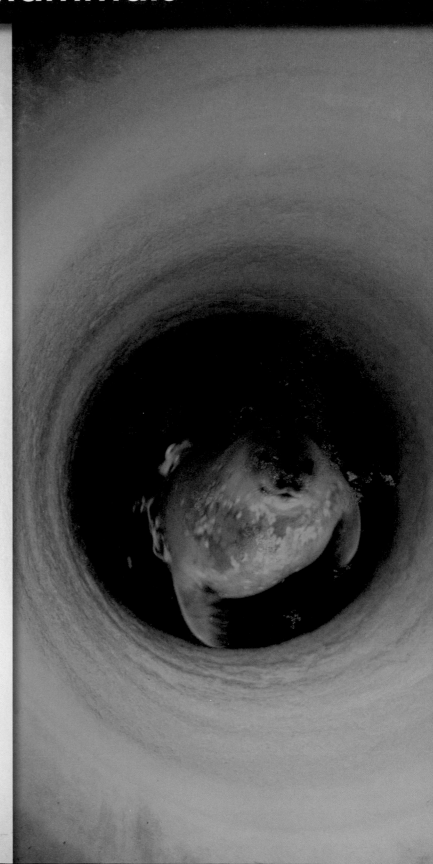

The Weddell seal (*Leptonychotes weddellii*) captures all its food underwater, yet depends entirely on the atmosphere for oxygen. It seems, when first considered, to be one of the most physiologically improbable creatures one could imagine. Not only is it incapable of breathing where it hunts, it is large (400–500 kg), homeothermic, entirely predatory, and confronted by cold throughout its life. A Weddell seal lives through all seasons in the frigid Antarctic, either hauled out on ice sheets or diving in seawater at –2°C to capture fish, crustaceans, and squids. Despite the seeming improbability of such animals, they are ecologically successful. Early Antarctic explorers, such as the Scottish sea captain James Weddell, brought home news of stupendous populations of diving mammals. Weddell seals today number near 1 million.

Worldwide, there are more than 80 species of whales and dolphins, plus more than 30 species of seals and sea lions. They are descended from terrestrial ancestors, probably lured to the sea by a bounty of food. As they evolved methods of feeding on the ocean's riches, however, marine mammals did not evolve ways of gleaning O_2 from the sea, and thus they remain tied to their ancestors' mode of breathing.

Modern technology is enabling us to realize more fully than ever the enormous role of diving in the lives of these animals. **FIGURE 26.1**, for example, shows the dives of three seal species along a relatively short stretch of coastline in Antarctica. Each vertical yellow trace represents a dive. The seals dive often as they travel horizontally from place to place.

Diving by marine mammals first came under serious study by physiologists about 1935. From the beginning, physiological studies have focused on two basic questions:

A Weddell seal diving through a hole in the Antarctic ice sheet   These large seals, found only in Antarctica, collect all their food by diving. They often cut and maintain holes through the ice sheet so as to be able to access their dual habitats: water for feeding and air for breathing.

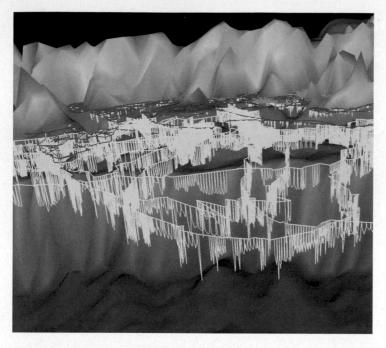

FIGURE 26.1 Natural dives of three seal species in the wild
The image is a three-dimensional representation of diving and surface movement by seals in Antarctica. Each dive (regardless of species) is represented by a vertical yellow path, the length of which represents dive depth. The lines at the water surface are the movement tracks of individual seals, color-coded to designate species: yellow, southern elephant seal (*Mirounga leonina*); green, Weddell seal (*Leptonychotes weddellii*); red, crabeater seal (*Lobodon carcinophagus*). The average depth of elephant seal dives was 360 m—a fact that provides insight into the depth scale. Because crabeater seals predominantly eat krill swarming in the upper 100 m of water, they do not dive to particularly great depths in comparison with Weddell and elephant seals, which seek fish and squids. (Courtesy of Daniel Costa.)

First, how do animals that are strictly dependent on the atmosphere for O_2 meet their metabolic energy demands during long periods underwater? Second, how do diving mammals cope with the high pressures they encounter at depth? This chapter focuses on the first question, but considers the second briefly at the end. The individual development of diving competence, as animals mature from birth to full adulthood, is a critical topic in the study of diving: As a youngster matures, it must meet its metabolic needs for energy at each stage of its life even though its development of full diving competence may require many months. We discussed the development of diving in Chapter 4. Here our focus is on adults.

Diving Feats and Behavior

The diving feats of seals and whales have been appreciated in a limited way for centuries. During the heroic era of whaling (chronicled most famously in Herman Melville's *Moby-Dick*), for example, the whalers were amazed by the depths to which wounded whales could dive. Sometimes a sperm whale had to be cut loose after "sounding" so deep as to draw out two lengths of harpoon line, each more than 370 m long. When physiological research on animal diving began, few techniques existed for the study of free-ranging animals in the wild, and thus the first investigations were mostly carried out in bathtubs and swimming pools. By 1970, however,

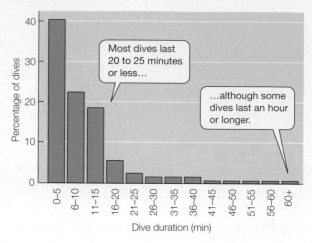

FIGURE 26.2 Durations of dives by wild Weddell seals These data represent more than 1000 dives by six free-living seals. Each vertical bar shows the percentage of dives falling within a particular duration category. Most dives are far shorter than the maximum duration displayed by the species. (After Kooyman et al. 1980.)

innovative scientists were starting to take advantage of the revolution in technology to monitor the diving behavior of free-living animals. They discovered that fact is indeed sometimes stranger than fiction. There are mammals that sometimes electively stay underwater—holding their breath—for 2 h, and there are ones that go so deep that they leave the atmosphere—their source of O_2—more than a mile behind.

The Weddell seal proved to be a perfect subject for early studies of diving behavior in the wild because the only devices initially available to obtain data on the durations and depths of dives were innovative but primitive instruments that—after being attached to an animal for a recording period—had to be reclaimed for the data to be acquired. Weddell seals living on ice sheets gain access to the water for feeding by cutting and maintaining holes through the ice. After a seal dives through an ice hole, it must ultimately return to that or another hole to breathe. Gerald Kooyman and his colleagues capitalized on this trait. They would attach a data recorder to a wild Weddell seal and then permit the seal to live in its ordinary way for several days, whereupon they would find the seal at one of its breathing holes to remove the device. Using this approach, they were able—for the first time with any species—to describe the durations and depths of thousands of voluntary dives.

The most fantastic revelation of the early studies was that Weddell seals sometimes stay submerged voluntarily for more than 1 h (**FIGURE 26.2**); we know today that their dives occasionally last as long as 80 min! Equally important, however, was the revelation that the vast majority of dives by Weddell seals are considerably shorter than the maximum durations observed; most dives last 20–25 min or less. The pattern of diving depths resembles that of diving durations in that, during most dives, seals go to depths that are considerably shallower than the maximum the species is capable of reaching (**FIGURE 26.3**). On rare occasions, a Weddell seal descends to nearly 600 m (0.37 mile), but few dives are deeper than 400 m. While diving, a seal is sometimes subject to stupendous physical pressures. A useful rule of thumb is that water pressure increases by about 1 atmosphere (atm) (101 kPa) for every 10 m of depth. Thus a seal diving to 400 m voluntarily subjects itself to about 40 atm of hydrostatic pressure.

Today, technology has advanced to the point that data can be radioed from free-living marine mammals and picked up on

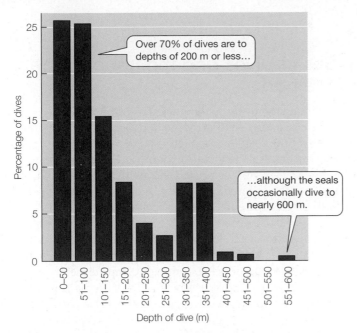

Over 70% of dives are to depths of 200 m or less...

...although the seals occasionally dive to nearly 600 m.

FIGURE 26.3 Depths of dives by wild Weddell seals These data represent more than 380 dives by 27 free-living individuals. Each vertical bar shows the percentage of dives falling within a particular depth category. Most dives are substantially shallower than the maximum depth displayed by the species. (After Kooyman 1966.)

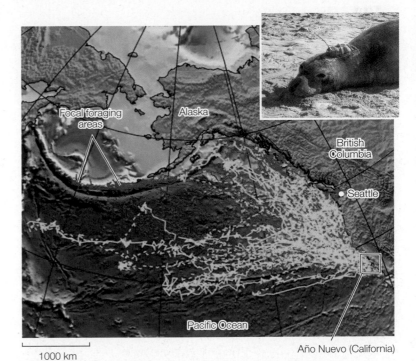

1000 km

Año Nuevo (California)

FIGURE 26.4 Migrations of northern elephant seals The elephant seals (*Mirounga angustirostris*) that breed at Año Nuevo in California get their food at distant places in the Pacific Ocean. Shown here are the migration routes of more than 30 animals from the Año Nuevo rookery. They were equipped with radio transmitters that signaled their location (see inset) and followed by satellite. The "focal foraging area" of an animal, which usually looks like a tight cluster of data in these records, is the part of the ocean where the seal lingered for many weeks of diving and feeding after its outward journey. When animals leave Año Nuevo, they tend to travel without pause to their focal foraging areas. During successive migratory trips away from Año Nuevo, individuals often return to their specific focal foraging areas using much the same outbound and inbound travel routes on each trip. Red represents males; yellow-green, females. (Courtesy of Dan Costa; after Le Boeuf et al. 2000.)

a global scale by satellite receivers. One dramatic study using this technology was carried out on the northern elephant seals (*Mirounga angustirostris*) that breed in a famous colony at Año Nuevo in California, south of San Francisco. Both sexes spend most of their lives at sea. They visit land on two occasions each year to mate, molt, and (in the case of females) rear young. Between these visits to land, the seals undertake long oceanic migrations, often across half the breadth of the Pacific Ocean, during which they feed intensively. The longitudes and latitudes of individual animals can be followed throughout their migrations using radio and satellite technology (**FIGURE 26.4**). For monitoring diving depths and durations, instruments that log the data rather than transmit them are still sometimes used. Modern digital data loggers, however, have prodigious capabilities compared with the original data loggers. Therefore a data logger attached to a seal at the start of its migration can record all the animal's dives throughout its entire migratory trip, and the diving data can later be correlated with the seal's satellite-recorded longitudes and latitudes. **FIGURE 26.5** shows the depths of all the dives undertaken by Moo, a male northern elephant seal, while he migrated between Año Nuevo and his principal foraging area near the Aleutian Islands. Moo dove at all hours, day and night. His deepest dive was to about 750 m (0.47 mile). Another northern elephant seal was once observed to dive to almost 1600 m (essentially 1 mile). Most dives by northern elephant seals, however, are to depths far less than these maxima, as exemplified by Moo's diving record (see Figure 26.5).

The marine mammals are far from uniform in their diving capabilities. The major groups—the true seals, the fur seals, and the whales—have different phylogenetic histories, and within each group, species have diversified. Weddell seals, elephant seals, hooded seals, and ribbon seals—all true (phocid) seals—are among the most proficient divers. One of the longest voluntary dives ever recorded, 2 h, was observed in a southern elephant seal. Elephant seals also attain astounding depths; a male southern elephant seal

was recorded recently to dive to over 2000 m (1.2 miles) on more than 160 different occasions, reaching a maximum depth of 2150 m (1.33 miles)—probably the deepest dive ever observed. Sperm whales and certain of the beaked whales also rank with the most proficient divers. Dives to 1900–2000 m have been recorded in sperm whales (*Physeter macrocephalus*) and Cuvier's beaked whales (*Ziphius cavirostris*). Sperm whales dive routinely for 40–50 min and occasionally dive for longer than 2 h. The opposite end of the spectrum of diving proficiency is exemplified by certain of the fur seals, such as northern fur seals (*Callorhinus ursinus*).[1] Based on records of more than 3000 dives, northern fur seals do not dive longer than 8 min or deeper than 260 m. Despite wide variation in the extreme dives of which species are capable, all species of diving mammals (and birds) seem to adhere to the two important generalizations that we have already illustrated: (1) The durations of most dives are substantially shorter than the maximum duration of which each species is capable, and (2) most dives are to depths substantially more shallow than the species-specific maximum depth. Dives of record duration and depth, although they are awe-inspiring and require physiological explanation, are uncommon.

---

[1] The fur seals and sea lions—which compose the group known as otariid seals—tend as a group to be less proficient divers than the true (phocid) seals.

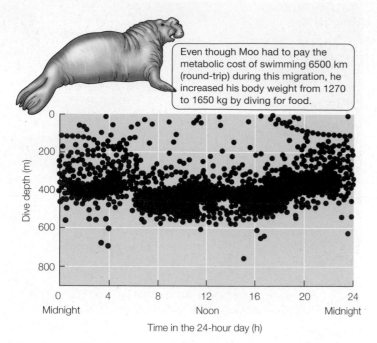

Even though Moo had to pay the metabolic cost of swimming 6500 km (round-trip) during this migration, he increased his body weight from 1270 to 1650 kg by diving for food.

FIGURE 26.5 Diving by Moo, a male northern elephant seal, as he traveled to and from his principal foraging area The maximum depth reached by Moo during each of his dives is plotted against the time of day of the dive. Each symbol represents a single dive. This plot includes all dives that Moo undertook during several weeks while he migrated from Año Nuevo to his focal foraging area in Alaskan waters, and back. Moo's dives were recorded by a time–depth logger with prodigious memory that was attached to him before his departure from Año Nuevo and recovered after his return. (After Le Boeuf et al. 2000.)

To put the performances of marine mammals into perspective, it is instructive to look at the capabilities of a representative terrestrial mammal, *Homo sapiens*. Outside of athletic competition, trained human breath-hold divers are generally limited to about 3 min of submergence when at rest and about 90 s when swimming. For comparing general populations of marine mammals with general populations of humans, people who dive in traditional occupations have been of particular interest. The *ama* of Korea and Japan, for example, earn their living by diving for shellfish and edible seaweeds. Their abilities are typical of those exhibited by other people, such as pearl and sponge divers, who employ breath-hold diving as a career. The ama who dive in deep waters receive assistance during descent and ascent so that their time at the bottom can be maximized. They carry weights to aid descent and are pulled back to the water's surface on a rope by an assistant in a boat. They routinely remain submerged for 60–80 s and reach depths of 15–25 m. They rest only about 1 min between dives and thus average about 30 dives per hour. The extremes recorded for human performance during breath-hold diving are attained by competitive divers and are difficult to summarize because several distinct types of competitive diving exist and have their own rules. When assistance can be employed in both descent and ascent, depths of about 200 m in dives lasting more than 4 min are achieved. Dives of such depths and durations can be carried out only by *exceptional* people in the entire global human population and are dangerous even for them. However, such dives seem trivial and commonplace for *average* individuals of many species of marine mammals.

Types of Dives and the Importance of Method

An important theme throughout science—dramatized in the study of diving physiology—is that the methods employed in experiments often affect the results and, therefore, our perception of reality. During the first decades of the modern study of diving physiology (roughly 1935–1970), scientists believed that they should "bring diving into the laboratory" where they could establish experimental controls and employ sophisticated instrumentation. In the lab studies that were carried out, animals were often strapped to a movable platform and lowered under water whenever an investigator wished to elicit diving responses. By the 1960s, research of this sort had produced such a seemingly complete picture of diving physiology that most biologists believed that diving was a "solved problem." Then some radical investigators began to study animals such as Weddell seals in the wild. The methods they had available to use were initially primitive compared with those available in the lab. Nonetheless, their early studies showed that the physiology of *voluntary* diving in the wild often differs from that of *forced* diving in the lab. Perceptions thus changed. The initial view had been that diving physiology shows little variation from one dive to another because it is controlled by highly stereotyped physiological reflexes. The newer view—now abundantly confirmed—was that marine mammals use a spectrum of physiological strategies during different sorts of dives. Today, even as we recognize that field studies are essential and advances in technology continue to expand the range of physiological mechanisms that can be studied in the field, lab studies continue also to be essential because certain mechanisms can be studied only in the lab.

Recognizing that lab and field studies of diving may produce different results, one of the first questions to ask about any set of data is whether it was obtained by forcing animals under water or by studying free-living animals diving voluntarily. Another important question is whether the dives studied were long or short *relative to the species-specific maximum dive length*, because we know now that qualitatively different suites of responses are often marshaled in long and in short dives. Still another key question is whether the studied animals were active or quiet during diving. These, then, are some of the important organizing distinctions among dives in the study of diving physiology:

- Forced or voluntary?
- Short or long (relative to the maximum length for the species)?
- Quiet or active?

Physiology: The Big Picture

When diving mammals break contact with the atmosphere, they carry O_2 with them in three major internal stores: O_2 bound to blood hemoglobin, O_2 bound to muscle myoglobin, and O_2 contained in air in their lungs. These stores permit aerobic catabolism to continue to some extent during a dive. The internal stores of O_2 are adequate in principle to permit all the tissues of a diver's body to function aerobically throughout a relatively short dive. When we look at the physiology of actual dives, we find, in fact, that voluntary dives of relatively short duration are mostly or completely aerobic in a variety of diving species, according to available evidence.

However, internal O_2 stores are utterly inadequate to permit fully aerobic function throughout a diver's body during protracted dives. How, then, are some species able to survive for 30 min, 60 min, or even longer without breathing? Our basic concept of how energy needs are met during protracted dives was first proposed by Laurence Irving (1895–1979) in 1934. He recognized that certain tissues—notably the central nervous system and heart—are predominantly or exclusively dependent on aerobic catabolism for production of ATP; they need O_2 on a steady basis and are quickly damaged by O_2 insufficiency. However, other tissues—such as skeletal muscle—have a well-developed ability to meet their ATP demands anaerobically and thus are relatively tolerant of O_2 deprivation. Irving then reasoned that dives can be prolonged if animals "reserve" a portion of their O_2 supplies for the tissues that are O_2-dependent. During a dive, if all tissues have equal access to an animal's entire O_2 store, then the concentration of O_2 throughout the body will fall quickly to such a low level that the O_2-dependent tissues are impaired. Under such circumstances, a seal or whale will need to surface, even though many of its tissues could continue to function—anaerobically—for a longer time. However, if some O_2 is reserved for use by the O_2-dependent tissues during a dive, then those tissues can continue to have adequate O_2 even while other parts of the body exhaust their O_2 supplies and turn to anaerobic catabolism, thereby extending the time the animal can remain submerged.

The preferential delivery of some O_2 to the O_2-dependent tissues is achieved, as Irving predicted, by adjustments of circulatory function. These adjustments were first elucidated by Irving, working with Per Scholander (1905–1980), in lab experiments on seals forced under water. During forced dives, blood flow is curtailed to many body regions—such as the skeletal muscles of the appendages and torso, the skin, the gut, and the kidneys—by vasoconstriction of the arterial vessels that supply those regions. The tissues that are deprived of active blood flow can make use of the hemoglobin-bound O_2 in the small volume of blood that passes through their capillaries, and they can use their myoglobin-bound O_2. However, as those limited and local O_2 stores are depleted, the circulation-deprived tissues turn to anaerobic catabolism, and lactic acid accumulates in them. *With the circulation to many parts of the body curtailed, the heart pumps blood primarily between itself and the lungs and head.* The O_2 stores of the circulating blood are thereby reserved primarily for the O_2-dependent tissues—the brain and heart—and whatever O_2 is extracted from the air in the lungs is likewise delivered preferentially to those tissues. Consequently, adequate O_2 partial pressures can be maintained in the O_2-dependent tissues for a long period. Lactic acid produced by the skeletal muscles and other circulation-deprived tissues tends to remain sequestered in those tissues during a dive, precisely because the tissues receive little or no blood flow. When the animal surfaces, however, circulation to such tissues is restored, and there is a sudden rise of lactic acid in the circulating blood. *The observation that circulating lactic acid increases principally after a forced dive was one of the earliest pieces of evidence that circulatory function is sometimes radically altered during diving.*

When animals are forcibly submerged, the adjustments in the pattern of blood flow just described tend to occur rapidly, consistently, and to a profound extent. Accordingly, during the era prior to 1970, when diving was studied mainly in labs, these responses were labeled a *diving reflex*. Today, however, based on studies of voluntarily diving animals, physiologists recognize that the responses of the circulatory system are not nearly as inflexible and stereotyped as once thought. Indeed, as suggested earlier, relatively little redistribution of blood flow is believed now to occur during voluntary dives that are short enough for all energy needs to be met by aerobic catabolism using an animal's O_2 stores. Seals and whales in the wild undergo a profound redistribution of blood flow when they dive voluntarily for long periods. Animals that are forced under water probably exhibit stereotypic and "reflexive" circulatory responses because they sense that they have no control over the length of time they will be submerged, and thus they consistently marshal the responses they employ for prolonged diving in the wild.

From this overview, you can see that the O_2 stores, circulatory physiology, and metabolic physiology of marine mammals all play critical roles in diving physiology. The next three sections discuss these three elements in more detail.

The Oxygen Stores of Divers

The size of a diving mammal's total O_2 store is obviously a key determinant of how long the animal can stay submerged. Among other things, a dive can last only as long as the brain is supplied with O_2, and the size of the O_2 store helps determine how long the supply of O_2 to the brain can be sustained.

The blood O_2 store tends to be large in diving mammals

The amount of O_2 stored in the blood depends on three features: (1) the oxygen-carrying capacity of the blood, (2) the total volume of blood, and (3) the degree to which the blood is fully loaded (saturated) with O_2 at the time of submergence.

Regarding the oxygen-carrying capacity, although some species of diving mammals have values that are well within the ordinary range for nondiving, terrestrial mammals, some other diving species have exceptionally high values. Bottlenose dolphins, northern fur seals, and Steller (northern) sea lions illustrate the first group; they have oxygen-carrying capacities of 17–22 volumes percent (vol %)[2]—quite ordinary for mammals. In contrast, species of diving mammals known to have especially high oxygen-carrying capacities include the harbor seal (26–29 vol %), sperm whale (31 vol %), Weddell seal (29–36 vol %), and ribbon seal (34 vol %). Among the seals as a group, there is a trend for species that undergo long dives—such as the three true seals just mentioned—to have higher oxygen-carrying capacities than species that perform shorter dives.

Blood volumes in some representative terrestrial species—humans, dogs, horses, and rabbits—average 60–110 mL per kilogram of body weight. Although some diving species have blood volumes in the same range, some of the accomplished divers such as harbor and ribbon seals have blood volumes of 130–140 mL/kg, and the blood volumes of Weddell seals, elephant seals, and sperm whales are 200–250 mL/kg—two to three times as high as is typical of terrestrial mammals.

---

[2] As applied to blood (see Chapter 24), O_2 *volumes %* is the amount of O_2—expressed as volume in milliliters at standard conditions of temperature and pressure—that is present per 100 mL of blood (including O_2 present both in combination with hemoglobin and in solution).

(A) A bottlelnose dolphin observed at two depths

(1) Near the surface

(2) At a depth of 300 m

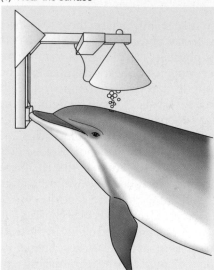

A dolphin's flexible thoracic wall is pushed inward by the high water pressure at depth, reducing the volume of the thoracic cavity and the volume of air in the lungs.

(B) Lungs of a harbor porpoise visualized by computed tomography at two depths

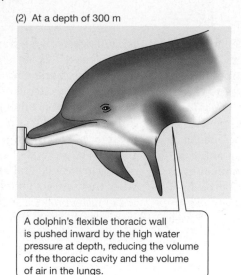

(1) At the surface (0 m)

Lungs

100 mm

(2) At depth of 100 m

FIGURE 26.6 Because the thorax is highly compressible in marine mammals, lung volume decreases dramatically as ambient pressure increases   (A) Portrayed here is a bottlenose dolphin (*Tursiops truncatus*) that was trained to push a button to signal its presence as it traveled between the surface of the water and a depth of 300 m. When it exhaled near the surface of the water (1), its exhalant air was collected for analysis. When it pushed the button at 300 m (2), a camera was activated to take its photograph. The ambient pressure at 300 m was about 30 atm. The photographs of the dolphin at that depth revealed that its thoracic wall was pushed far inward by the elevated ambient pressure. (B) The lungs inside the thorax of a harbor porpoise (*Phocoena phocoena*) at two different ambient pressures. Although the porpoise had died of an accident, it was freshly dead and undamaged. As the ambient water pressure was raised to simulate depths of 0 and 100 m, the porpoise's lungs were visualized using a CT (computed tomography) scanner, with scaling identical in both images. Note that *both* the thorax and lungs decreased in volume as ambient pressure increased. (A after Ridgway et al. 1969; B after Moore et al. 2011.)

An animal's maximum possible blood store of O_2 is calculated by multiplying the oxygen-carrying capacity of its blood by its blood volume. This figure, although only indirectly relevant to normal physiology (because the entire volume of blood is never fully oxygenated), is useful for comparing species. The maximum possible blood O_2 store of humans and horses is about 14–15 mL O_2 per kilogram of body weight. In contrast, far higher O_2 storage capacities are found in the species of diving mammals that combine the advantages of both a high oxygen-carrying capacity and a high blood volume. The maximum blood O_2 store in Weddell seals, elephant seals, and sperm whales is 60–85 mL O_2/kg—four to six times higher than the stores of humans and horses.

Diving mammals have high myoglobin concentrations and large myoglobin-bound O_2 stores

The amount of O_2 stored as oxymyoglobin at the time of submergence depends on how much myoglobin is present in each unit of muscle tissue. *One of the most consistent features of diving species of mammals is that, relative to terrestrial species, they have dramatically high myoglobin concentrations in their skeletal muscles.* The skeletal muscles of some proficient divers are so rich in myoglobin that they are almost black. In humans and horses, skeletal muscles contain about 4–9 mg of myoglobin per gram of wet weight. By contrast, sperm whales and harbor, Weddell, elephant, and ribbon seals have 55–70 mg of myoglobin per gram of wet weight, and hooded and harp seals have 85–95 mg/g!

Oxymyoglobin represents an essentially private store of O_2 for the muscles. As discussed in Chapter 24 (see page 655), myoglobin has such a high affinity for O_2 that it typically draws O_2 from blood hemoglobin rather than donating O_2 to the blood. Thus, even if muscles receive blood circulation during a dive, oxymyoglobin within the muscles does not yield much O_2 to the blood for use elsewhere in the body. Instead, the O_2 remains bound to the myoglobin until the O_2 partial pressure in the muscles falls to a low level; then it is donated to the muscle mitochondria to permit continued aerobic ATP production in the muscles. An interesting point is that the O_2 partial pressure in a muscle *must* be low for the muscle's myoglobin to unload. One role of the severe reduction of blood flow to the skeletal muscles during prolonged dives is probably to ensure that the O_2 partial pressure in the skeletal muscles falls promptly to a level that will permit myoglobin to be tapped for O_2.

Diving mammals vary in their use of the lungs as an O_2 store

To understand lung function in diving mammals, one must first recognize the unusual structural flexibility of the thorax in these animals. In the deep-diving marine mammals that have been studied, the thorax is structured in ways that permit the thoracic walls to be shoved freely inward as the outside water pressure increases (**FIGURE 26.6A**). Thus, over extensive ranges of depth and pressure, the *thoracic cavity and lungs are freely compressible* (**FIGURE 26.6B**). If the pressure applied to the thorax is increased tenfold, for example, the volume of air in the lungs is reduced to roughly one-tenth of what it previously was.[3] Humans and other terrestrial mammals are different in that their thoracic walls are structured less flexibly and resist compression.

One might at first think that a large air store in the lungs would be of unquestionable advantage for a diving mammal. Three considerations, however, argue against this conclusion. First, a large amount of air in the lungs can strongly buoy a diving animal upward, forcing it to work hard to remain submerged. Second, the alveoli are believed to be typically the first parts of the lungs to collapse as the lung air compresses at depth. Hence, at depth, the lung air comes to be contained mostly in the *conducting* airways—the trachea, bronchi, and non-respiratory bronchioles—where the O_2 in the air becomes unavailable because little O_2 transfer to the blood can occur in the conducting airways. As a consequence of this process, whereas diving mammals can make effective use of their pulmonary O_2 stores during shallow diving, they are not necessarily able to do so at depth. A final consideration is that a large pulmonary air store means not only a large store of O_2 in the lungs, but also a large store of N_2 (air is 78% N_2). A large N_2 store can increase the likelihood of decompression sickness, as discussed later in this chapter.

The size of the air store in a marine mammal's lungs at the start of a dive depends on two factors: the *volumetric capacity* of the lungs (the amount of air they can hold when fully inflated) and the degree to which the animal inflates its lungs before diving. Regarding the first factor, marine mammals do not as a rule have exceptionally large lungs; their volumetric capacities per unit of body weight are generally similar to those of terrestrial mammals, or just modestly larger. Moreover, many of the deep-diving seals and whales have volumetric capacities that are relatively low by comparison with those of other species of seals or whales of similar size that dive more shallowly. Regarding the second factor—the degree of lung inflation—some marine mammals dive after a vigorous inhalation and thus may carry an amount of lung air that approaches their volumetric capacity. This appears to be true, for example, of whales (including dolphins). By contrast, many deep-diving species of true (phocid) seals dive following *exhalation*; their lungs are filled to just 20%–60% of their volumetric capacity when they submerge. All things considered, little premium is placed on having an exceptionally large pulmonary air (and O_2) store in marine mammals.

[3] Too few actual measurements are available to determine if lung volume responds to changes of outside pressure identically as would be predicted from a simple application of the universal gas law (see Equation 22.1).

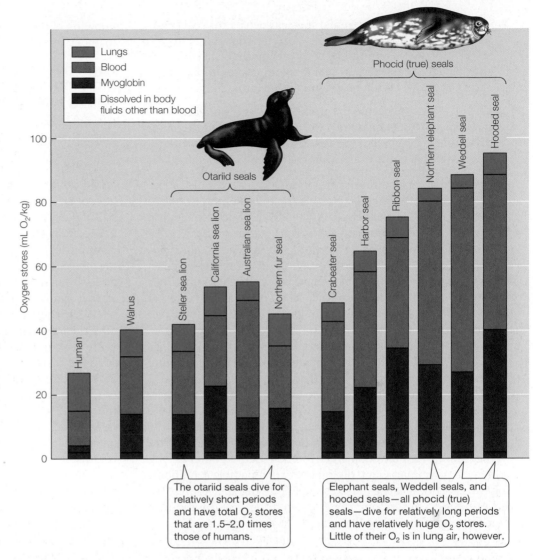

The otariid seals dive for relatively short periods and have total O_2 stores that are 1.5–2.0 times those of humans.

Elephant seals, Weddell seals, and hooded seals—all phocid (true) seals—dive for relatively long periods and have relatively huge O_2 stores. Little of their O_2 is in lung air, however.

FIGURE 26.7 A comparison of the total O_2 stores of adults in 11 species of marine mammals and humans Included are members of three seal families: six phocids (true seals), four otariids (sea lions and fur seals), and the walrus (classified in its own family). The total O_2 available per kilogram of body weight for each species is estimated by adding the O_2 dissolved in body fluids other than blood throughout the body, the O_2 bound to myoglobin in muscles, the O_2 in blood (mostly bound to hemoglobin), and the O_2 in lung air. The lungs of humans are assumed to be fully inflated, but those of the other species are assumed to be only partly inflated because many seals dive after exhaling. The data for seals were calculated by one research team using a standardized set of procedures and assumptions (Burns et al. 2007); the data for humans, although comparable, are from other sources.

Total O_2 stores never permit dives of maximum duration to be fully aerobic

Three major points emerge from the study of the O_2 stores of marine mammals:

1. Some species of diving mammals have much greater total O_2 stores per unit of body weight than terrestrial mammals because they have high blood oxygen-carrying capacities and high blood volumes—giving them high blood O_2 stores—and they have high concentrations of myoglobin in their muscles—giving them high myoglobin-bound O_2 stores (**FIGURE 26.7**).

2. Comparing diving mammals, species such as fur seals and sea lions that dive for relatively short periods tend to have smaller O_2 stores per unit of body weight than species such as harbor and ribbon seals that are more proficient as divers and dive for longer periods (see Figure 26.7).

3. The O_2 stores of marine mammals of all kinds are utterly inadequate to sustain a rate of O_2 consumption during long dives that is equivalent to the rate of O_2 consumption of these animals while they are at rest and breathing air.

The third point is sufficiently important to deserve illustration. Paul Ponganis and his colleagues estimated the total available O_2 store of a 450-kg Weddell seal to be 38.8 L.[4] Such a seal has a resting rate of O_2 consumption, when it is breathing air, of 1.9–2.3 L O_2/min. Thus the seal could sustain its resting, aerial rate of O_2 consumption for 17–20 min during diving if it completely used its available O_2 store. Actually, however, Weddell seals sometimes dive for 60–80 min (see Figure 26.2). This sort of result is typical of diving species; dives of maximum length always last from two to several times longer than would be predicted if an animal were to function aerobically at the rate seen during rest in air. Another way to see that O_2 stores are inadequate to account fully for diving performance is to recognize that the diving capabilities of long-duration divers are disproportional to their O_2 stores. For example, consider harbor seals and humans, two species of roughly the same body size. Although the weight-specific O_2 stores of harbor seals are 2–2.5 times higher than those of humans (see Figure 26.7), an average seal can remain submerged for more than 12 times as long as an average human when diving under comparable conditions. Regardless of how large the O_2 stores of diving species may be, these stores do not in themselves explain the dive durations of which the animals are capable.

Circulatory Adjustments during Dives

The circulation holds a special place in the chronicles of diving physiology because the very first physiological observations on diving were measures of heart rates. Starting in 1870, the French physiologist Paul Bert (1833–1886) studied ducks and found that their heart rates decreased from 100 to 14 beats/min when he forced them under water. A decrease in heart rate during diving is called **diving bradycardia** (*brady*, "slow"). Bert and others soon demonstrated that the phenomenon is a consistent feature of forced submergence in diving mammals and birds. The universality of diving bradycardia quickly persuaded physiologists that the slowing of the heart is important in permitting animals to stay submerged for extended periods. But how is it important?

Irving and Scholander took the next crucial step. In the 1930s, they postulated that in addition to the heart, other parts of the cardiovascular system also respond during diving. One of their earliest tests of their idea was to study bleeding from the paw skin of seals. Small cuts that bled freely when the seals were breathing air stopped bleeding suddenly and completely when the seals were pushed under water! From simple observations like this, the modern revolution in diving physiology began, and after seven decades of

knowing of the existence of diving bradycardia, scientists began to understand its true significance.

Regional vasoconstriction: Much of a diving mammal's body is cut off from blood flow during forced or protracted dives

After physiologists realized that a marine mammal's pattern of blood flow might change during dives, they employed several techniques to examine how the vascular system functions during diving. One such technique involves the use of a *contrast fluid* that is impenetrable by X-rays. When a contrast fluid is injected into an animal's bloodstream, its flow can be observed on X-ray images. In one of the watershed studies in the history of diving physiology, a group of diagnostic radiologists headed by Klaus Bron injected a contrast fluid into the systemic aorta of a harbor seal, then took X-ray images to observe subsequent blood flow in the seal's visceral arterial system while the animal was breathing air and while it was forcibly submerged (**FIGURE 26.8**).

When the seal was breathing air, an X-ray image taken 0.5 s after the injection of the contrast fluid showed that blood flowed vigorously from the aorta into abdominal arteries that branch off from the aorta, such as the coeliac and renal arteries (see Figure 26.8A). Within 2 s after injection, the left kidney was illuminated in the X-ray image, showing that the elaborate arterial system within the kidney had filled with blood that contained the contrast fluid (see Figure 26.8B). After only 6 s, the coeliac and renal arteries had already faded from view (see Figure 26.8C), demonstrating that the contrast fluid had flowed through them and exited.

The sequence of X-ray images taken while the seal was submerged revealed a strikingly different circulatory pattern. At 1 s after injection, the X-ray image showed that the blood containing the contrast fluid had penetrated only a short distance from the aorta into the coeliac and renal arteries and stopped, leaving the branches of those arteries unfilled (see Figure 26.8D; compare with Figure 26.8A). Even after 4 s, the left kidney remained dark, indicating that most of its arterial vessels had not received blood containing contrast fluid (see Figure 26.8E). As long as 14 s after injection, the bases of the coeliac and renal arteries remained filled with contrast fluid (see Figure 26.8F), indicating that the blood in those arteries had stagnated and was unable to flow through the vessels.

Bron's pioneering radiological study confirmed that blood flow to major parts of a marine mammal's body is profoundly curtailed during a forced dive. The coeliac and renal arteries supply blood to the stomach, spleen, liver, pancreas, and kidneys. During a dive, blood is unable to flow freely (if at all) into any of those visceral organs.

Vasoconstriction—under control of the sympathetic nervous system—is responsible for cutting off the blood flow. Strikingly, although arterioles are the typical sites of vasomotor control in mammals (see page 680), pronounced vasoconstriction occurs in *sizable arteries* in at least some diving species. Arteries constrict shut, and the organs they supply are denied blood flow (see Figure 26.8D–F). The parts of a diving mammal's body that receive little or no blood flow during forced dives commonly include the animal's limbs, the skeletal muscles of its torso, its pectoral muscles, its skin and body wall, and many visceral organs as already discussed.

Simultaneously, because vasoconstriction occurs selectively, blood flows freely—or relatively freely—to a diving animal's brain, lungs, and myocardium. A pundit once said that a seal becomes

---

[4] To calculate this amount from Figure 26.7, multiply the size of the O_2 store per kilogram, 86 mL O_2/kg, by the seal's body weight, 450 kg.

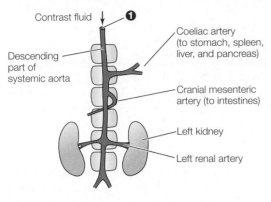

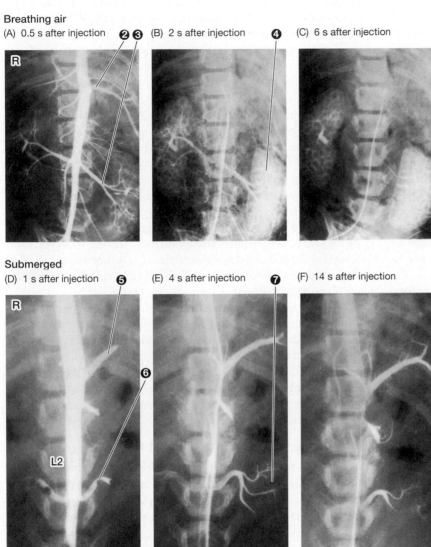

Breathing air
(A) 0.5 s after injection ❷ ❸ (B) 2 s after injection ❹ (C) 6 s after injection

Submerged
(D) 1 s after injection ❺ (E) 4 s after injection ❼ (F) 14 s after injection

FIGURE 26.8 Patterns of blood flow in the circulatory system are radically changed during forced or prolonged submergence X-ray images of the visceral cavity of a harbor seal (*Phoca vitulina*), viewed from its ventral side, after injection of contrast fluid during air breathing (A–C) and during forced submergence (D–F). The images show that although the blood labeled with contrast fluid flowed into all major branches of the aorta within seconds during air breathing, the blood was unable to flow freely out of the aorta when the seal was submerged. ❶ Contrast fluid was injected into the descending portion of the systemic aorta. Within 0.5 s after the contrast fluid was injected when the seal was breathing air, the coeliac artery ❷ and left renal artery ❸ were filled with blood containing contrast fluid; and within 2 s after injection, the entire left kidney ❹ was filled with blood containing contrast fluid. When the seal was submerged, however, the coeliac artery ❺ and renal artery ❻ did not readily fill with blood containing contrast fluid, and after 4 s, the kidney ❼ remained dark—showing that it had not received blood with contrast fluid. L2 = second lumbar vertebra; R = right side of seal. (Photographs courtesy of Klaus Bron; from Bron et al. 1966.)

a "heart-lung-brain machine" during forced dives (**FIGURE 26.9**). Blood is pumped by the right heart to the lungs, travels from the lungs to the left heart, then is pumped by the left heart to the head, and finally returns to the right heart to be pumped again to the lungs. The rate of blood flow to the brain during diving tends to remain similar to the rate during air breathing. Meanwhile, as the blood travels round and round between heart and head, flow to the parts of the body posterior to the heart is cut off or severely restricted.

Diving bradycardia matches cardiac output to the circulatory task

After experiments had revealed that blood flow to many regions of a diving mammal's body is curtailed during forced or protracted dives, diving bradycardia could at last be understood for what it is: a single part of an integrated, body-wide reorganization of cardiovascular function. During a forced or protracted dive, vasoconstriction greatly reduces the dimensions of the active circulatory system. Accordingly, less output of blood from the heart is required. An analogy is provided by a faucet that sends a flow of water to six hoses; if five of the hoses are pinched off, the faucet can maintain unaltered flow to the sixth hose with only a sixth of its preexisting output. Bradycardia is a mechanism that matches the heart's output (the cardiac output) to the dimensions of the vascular system being perfused. In marine mammals, the stroke volume of the heart changes to only a modest extent, if at all, during diving, according to studies of several species. Thus cardiac output declines during a dive roughly in proportion to the decline in heart rate (see Equation 25.1).

Studies of blood pressure demonstrate in a particularly graphic way that the drop in cardiac output during a dive is matched to the reduction in the dimensions of the active circulatory system. As discussed in Chapter 25 (see Equation 25.3), the systemic arterial blood pressure depends on two factors: the rate of blood flow out of the heart into the vascular system and the resistance to blood flow posed by the vascular system. A change in either factor without a compensatory adjustment in the other can severely disturb blood pressure. The vasoconstriction that occurs during a dive increases the overall resistance to blood flow posed by the vascular system. However, the blood pressure in the great systemic arteries remains unaltered or changes only modestly. These two facts demonstrate that cardiac output is reduced during dives in a highly integrated way that closely matches the increase in peripheral vascular resistance.

Cardiovascular responses are graded in freely diving animals

Cardiovascular function in free-living marine mammals undergoing voluntary dives has proved particularly challenging to study but is gradually becoming understood. Today, for example, methods exist to

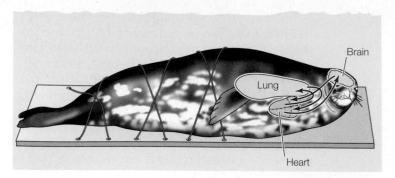

FIGURE 26.9 The forcibly submerged seal as a "heart-lung-brain machine" With blood flow to much of the body curtailed by vasoconstriction, active circulation is principally from the right heart to the lungs, the lungs to the left heart, the left heart to the brain, and the brain back to the right heart.

monitor the electrocardiogram continuously in unfettered animals at sea. From use of such methods, we know that the heart rates of freely diving mammals typically decrease in a *graded* manner as the animals increase the durations of their dives (**FIGURE 26.10**). During forced dives, heart rate responses are more of an "on–off" sort; the heart rate of a forcibly submerged harbor seal, for instance, drops to less than 10% of its pre-dive level within 10 s every time the animal is submerged. An important distinction between voluntary and forced dives, therefore, is the graded versus stereotyped nature of the heart rate response during the two sorts of dives.

Ideal methods still do not exist to study vasoconstriction directly in freely diving animals. For two major reasons, however,

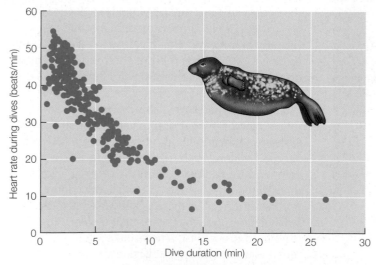

FIGURE 26.10 Diving heart rate varies with dive duration in a graded manner in freely diving seals Grey seals (*Halichoerus grypus*) living freely in the ocean near Scotland were monitored using radio and sonic transmitters. When not diving, their average heart rate was 119 beats/min. During their voluntary dives, the seals always exhibited bradycardia, but they adjusted the degree of bradycardia depending on how long they spent underwater. This graded bradycardia is in contrast to the stereotyped bradycardia of forcibly submerged seals, which typically exhibit a maximum or near-maximum drop in heart rate whenever forced under water. (After Thompson and Fedak 1993.)

scientists believe that the vasoconstrictor response is also graded during voluntary dives. First, as we have discussed, heart rate and vasoconstriction are believed to be integrated in a way that matches cardiac output to the dimensions of the vascular system requiring blood flow. If this integrated response occurs consistently, then little vasoconstriction would occur in animals that exhibit just a small reduction of heart rate, and progressively greater vasoconstriction would develop as bradycardia becomes more profound. A second line of evidence that corroborates this view is provided by studies of systemic organ function. Although studies of organ function do not tell a completely consistent story, they often indicate that vasoconstriction is graded in voluntary dives. Urine formation by the kidneys, for instance, requires blood flow. Thus one can learn about blood flow to the kidneys during dives by use of chemical markers that permit measurement of urine formation. In free-living Weddell seals, urine formation seems to continue during relatively short dives (indicating continued blood flow) but seems to stop during long dives (indicating cessation of blood flow).

The current working hypothesis of most physiologists who study diving is that the entire suite of cardiovascular responses to voluntary diving occurs in a graded manner. During protracted voluntary dives, free-living animals probably function much like forcibly submerged ones: They undergo profound vasoconstriction accompanied by a profound drop in heart rate, and large parts of the body are cut off from active blood flow. During relatively short voluntary dives, however, vasoconstriction is probably modest, so that only a small reduction in cardiac output is warranted, and most (or all) parts of the body continue to receive blood flow. When all parts of the body receive blood flow, all can share the blood O_2. This means that a dive cannot be of extreme length, but it also means that the animal avoids the stresses of anaerobic catabolism (discussed in Chapter 8 and later in this chapter).

One must wonder how the elaborate cardiovascular responses seen in diving mammals evolved. These responses in fact seem to be specializations of phylogenetically ancient responses to asphyxic conditions, responses that occur widely in vertebrates (**BOX 26.1**).

Red blood cells are removed from the blood between dive sequences in some seals

Species of seals with large blood O_2 stores, such as Weddell seals and ribbon seals (see Figure 26.7), typically have exceptionally large concentrations of red blood cells in their blood *when they are diving*. Although an elevated red blood cell concentration enables the blood to store a large amount of O_2, it has a downside: It increases the viscosity of the blood, forcing the heart to work harder to pump blood. In at least some of the true (phocid) seals, red blood cells are partly removed from the circulating blood and stored in the spleen when the animals are resting at the water's surface or on land. Then the cells are returned to the blood during diving.[5] The blood of a Weddell seal, for instance, although 38% red blood cells (by volume) when the animal is resting in air, may become enriched to 52% red blood cells during diving. The removal of red blood cells from the blood when a seal is not diving means that during such rest periods the heart does not have to work exceptionally hard to pump blood. Red blood cells require at least 10–20 min to move between the spleen and blood. Thus the cells do not move in and out of the

---

[5] Certain terrestrial mammals also vary the red blood cell concentration of their blood by sequestering cells in the spleen, as noted on page 655.

BOX 26.1 The Evolution of Vertebrate Cardiac and Vascular Responses to Asphyxia

In the decades following Scholander and Irving's seminal discoveries regarding the diving physiology of marine mammals, Scholander and others sought to learn when the diving responses originated in the evolution of vertebrates. They were intrigued to find similar responses in fish taken out of water! Two sorts of fish that they studied were flying fish and grunion, both of which breathe with gills but occasionally emerge into air voluntarily. Flying fish sometimes spend tens of seconds out of water as they skitter across the sea surface, and grunion sometimes spend minutes high on beaches where they slither out of the water to mate and lay eggs. The researchers found that both species exhibit profound bradycardia (as the figure shows), plus evidence of peripheral vasoconstriction, when they are out of water. The common denominator between diving marine mammals and these fish is that both are asphyxic (unable to breathe) when they undergo bradycardia and peripheral vasoconstriction. Scholander thus argued that bradycardia and peripheral vasoconstriction first evolved in fish ancestral to today's fish as defenses against asphyxia. By now there is a large body of evidence indicating that bradycardia is a typical response of both cartilaginous and bony fish when they are exposed to O_2-poor (hypoxic) water, reinforcing the view that bradycardia is an ancient response of vertebrates to O_2 insufficiency.

If, in fact, mammals inherited the rudiments of their cardiovascular responses to asphyxia from piscine ancestors, one would guess that those responses would be observed in many kinds of mammals, not just diving ones. In fact, the species of mammals that habitually dive are not alone in undergoing bradycardia and peripheral vasoconstriction when they are unable to obtain O_2 by breathing. Scholander himself reported evidence for bradycardia and peripheral vasoconstriction in full-term fetal humans and other mammals when they pass through the birth canal, and he observed bradycardia in human pearl divers. Abundant evidence exists today that adult humans and adults of at least some other terrestrial mammals routinely display bradycardia and redistribution of blood flow (e.g., restriction of blood flow to skeletal muscles) when their whole bodies are submerged, or even if they simply immerse their faces in a bowl of water. Humans, however, are not "just like"

marine mammals: Human physiological responses to being submerged are not as profound as those seen in marine mammals and are not coordinated in the same way (e.g., arterial blood pressure often soars in humans during long breath-hold dives because the output of blood from the heart and the resistance to blood flow through the vascular system are mismatched).

The marine mammals—viewed from the perspectives discussed here—seem in a sense to have "perfected" responses that all or most mammals share and that the mammals may well have inherited from fish. An important objective for future research is to understand better what exactly occurred during this "perfecting" process in the course of evolution and how exactly the control mechanisms in marine mammals differ from those in terrestrial ones.

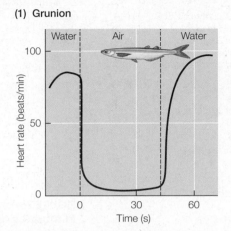 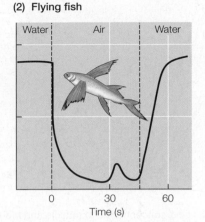

(1) Grunion   **(2) Flying fish**

The heart rates of fish removed from water   The graphs show Scholander's original data for (1) grunion and (2) flying fish. (After Scholander 1964.)

---

blood with each dive, but enter the blood during sequences of dives and are withdrawn during extended periods of rest.

Metabolism during Dives

The stage is now set to discuss metabolism during dives. Is it aerobic or anaerobic, or is it both? In terrestrial mammals that are resting or just modestly active, blood flows freely throughout the body—making O_2 available to all tissues—and all tissues make ATP aerobically. A similar picture seems to exist during short, voluntary dives by many marine mammals, even though they are not breathing; the circulatory system remains open to most or all

body regions, and the tissues receive enough O_2 from the body's O_2 stores to function aerobically. A dramatically different picture develops during forced or prolonged dives.

The body becomes metabolically subdivided during forced or protracted dives

The intense peripheral vasoconstrictor response that occurs during forced or protracted dives effectively divides a diving animal's body into two metabolically distinct parts. This subdivision is illustrated in **FIGURE 26.11** by the classic data that Scholander and Irving gathered on forcibly submerged harbor seals. The tissues that are denied blood flow during submergence, such as the skeletal muscles

(A) Oxygen concentration in muscle and blood

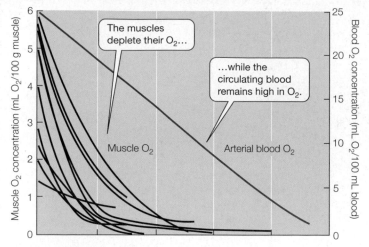

(B) Lactic acid concentration in muscle and blood

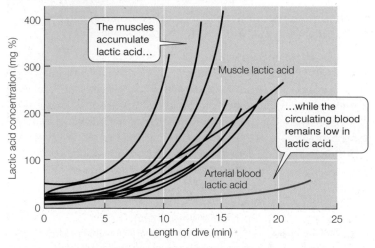

FIGURE 26.11 Metabolic subdivision of the body in seals during forced submergence Levels of O_2 and lactic acid in the dorsal torso muscles and circulating arterial blood are shown for forcibly submerged harbor seals (*Phoca vitulina*) as functions of their time underwater. (A) Muscle O_2 concentration in each of ten seals and the average O_2 concentration in circulating arterial blood. (B) Muscle lactic acid concentration in each of ten seals and the average lactic acid concentration in circulating arterial blood. (From Scholander et al. 1942.)

of the torso, initially continue to metabolize aerobically, using local O_2 stores such as O_2 bound to myoglobin. However, as shown by the muscle data in Figure 26.11, the tissues denied blood flow reduce their O_2 supplies to nearly zero long before a dive is over (see Figure 26.11A), and simultaneously they start to accumulate lactic acid (see Figure 26.11B) as they turn to anaerobic glycolysis to synthesize ATP. The O_2 concentration of the circulating arterial blood in a submerged seal falls much more slowly than the O_2 concentration of the skeletal muscles (see Figure 26.11A); therefore, long after the skeletal muscles have exhausted their O_2, the brain, myocardium, and other perfused tissues receive substantial O_2 supplies from the blood. The perfused tissues accordingly remain aerobic and do not produce lactic acid for most of the duration of a dive, as shown by the fact that little lactic acid accumulates in the circulating arterial blood during submergence (see Figure 26.11B).

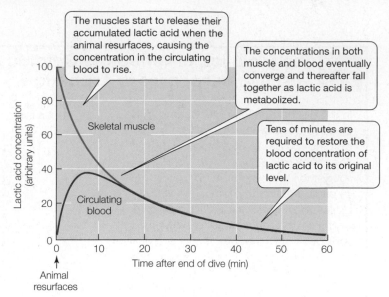

FIGURE 26.12 The aftermath of a prolonged dive: Lactic acid in muscles and blood The graph shows concentrations of lactic acid in skeletal muscles (blue line) and in circulating blood (red line) in the hour following a prolonged dive (a 43-min dive by an adult Weddell seal is assumed). Lactic acid concentrations are expressed relative to the concentration in muscle at dive's end, arbitrarily set equal to 100 units. Blood values are empirical, whereas muscle values are based on the assumption of exponential kinetics. (After Butler and Jones 1997.)

All things considered, you can see that well before a long dive is over, a seal's body becomes divided into two regions, one of which remains aerobic while the other becomes O_2-depleted and dependent on anaerobic ATP production.

An important point to note is that lactic acid remains sequestered in the skeletal muscles and other vasoconstricted tissues while a dive is in progress, rather than entering the circulating blood (see Figure 26.11B). However, when a dive ends and the diving animal starts to breathe again, blood flow is promptly restored to the vasoconstricted tissues, and the lactic acid accumulated in them is washed out. The washout results in a relatively rapid rise in the lactic acid concentration of the circulating blood, followed by a gradual decline. **FIGURE 26.12** shows the prevailing view of the processes at work after breathing is resumed. The concentration of lactic acid in the skeletal muscles falls exponentially. After blood lactic acid has risen for several minutes, it also starts to fall because the metabolic processes that clear lactic acid from the blood (see Figure 8.6) eventually outpace the entry of lactic acid into the blood from the muscles. The time scale is highly significant. Because the metabolism of lactic acid is slow (as emphasized in Chapter 8; see pages 200–201), the concentration of lactic acid in muscles and blood may not return to baseline levels *for many tens of minutes after a dive*.

Metabolic limits on dive duration are determined by O_2 supplies, by rates of metabolic O_2 use and lactic acid production, and by tissue tolerances

A diving mammal usually elects to end a dive before its metabolic limits are reached. However, a highly protracted dive could be terminated by exhaustion of O_2 in the part of the body receiving active circulation, by excessive accumulation of lactic acid in anaerobic tissues, or by other metabolic limitations.

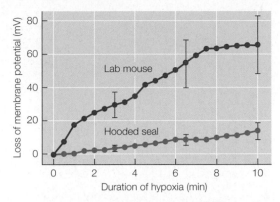

FIGURE 26.13 Loss of normal membrane potential in brain neurons of hooded seals and lab mice during exposure to tissue hypoxia The normal membrane potential in neurons of both species is about 64 mV (inside negative). The y axis shows how much of this potential is lost as a function of time when, starting at time 0, neurons are continuously exposed to severe hypoxia. A value of 40 mV on the y axis, for example, indicates that membrane potential has fallen by 40 mV, from 64 mV to 24 mV. Error bars show ± 1 standard error. (After Folkow et al. 2008.)

Three factors determine the limits of endurance of the O_2-dependent tissues that receive active blood circulation during dives: (1) the magnitude of the O_2 store available to those parts of the body, (2) the rate of use of the O_2 store, and (3) the extent to which the partial pressure of O_2 can fall before impairing function. A scattering of interesting insights are available regarding each of these considerations. For example, even though the myocardium is believed to have a continuous requirement for O_2, it may nonetheless start to employ anaerobic glycolysis to some extent after several minutes of diving. Both this partial recourse to anaerobic catabolism and the drop in cardiac work associated with diving bradycardia reduce the heart's O_2 needs and thus help postpone the time when O_2 supplies become inadequate to sustain myocardial function.

The brain, in contrast to the heart, is believed to be entirely aerobic in diving mammals, just as it is in terrestrial mammals. However, several types of evidence indicate that the brain remains functional at lower O_2 partial pressures in at least some seals than in terrestrial mammals. Recent studies, for example, have been done on the membrane potentials of individual neurons in brain slices taken from a highly defined brain region of hooded seals (*Cystophora cristata*) and lab mice. As discussed in Chapter 12, a relatively normal membrane potential is required for neurons to produce action potentials (impulses), which are essential for brain function. Seal neurons are dramatically more resistant to loss of membrane potential when placed in low-O_2 conditions than are mouse neurons (**FIGURE 26.13**). This probably helps explain observations that indicate that seals—of at least some species—are able to remain conscious when blood O_2 has fallen to levels that would cause blackout in most mammals. Neuronal mechanisms that permit the brain cells to maintain conscious function at very low O_2 levels permit the seals to make almost complete use of their O_2 stores and thus dive especially long. Research is under way to better understand hypoxia resistance in the neurons of seal brains. Recent research points in an utterly unexpected direction, namely that neuroglobin-containing glial (non-neuronal) cells in the brain may play a major role in the defense of the neurons.

METABOLIC RATE DURING DIVING An animal's metabolic rate during diving is one of the most important factors in determin-

ing metabolic limits on dive duration. A low metabolic rate could slow both the rate of O_2 depletion and the rate of lactic acid accumulation, thereby lengthening a dive regardless of whether O_2 or lactic acid sets the limits. Physiologists have sought evidence for depressed metabolism in diving mammals from the dawn of the modern study of diving. Rigorous measurements of metabolic rates during diving are not easy to obtain, however, and knowledge of this key subject remains incomplete.

The preponderance of available evidence indicates that submergence commonly brings about a depression of metabolism. Good insight is provided, for example, by studies of free-ranging Weddell seals that were trained to breathe consistently from a monitored source of air for hours at a time as they engaged in repeated voluntary dives (most of which were short enough to be fully aerobic). Remarkably, the average rate of O_2 consumption of these *active* animals was equal to or lower than that of *resting* but nondiving seals!

Depression of metabolism during diving seems paradoxical because swimming ought to increase an animal's metabolic rate. This paradox is far from fully resolved. Hypothermia seems to be a common energy-sparing mechanism among seals: They often let their tissues cool during diving. We will return soon to specific mechanisms by which the diving metabolic rate can be kept low.

ADAPTATIONS TO THE ACCUMULATION OF METABOLIC END PRODUCTS Both the aerobic and anaerobic catabolic pathways produce end products that accumulate in a diving mammal: CO_2 and lactic acid, respectively. The ways in which a diver responds to the buildup of these compounds are important determinants of maximum dive duration.

Acidification is one concern. Both CO_2 and lactic acid tend to cause the pH of tissues and body fluids to decline. How is the pH kept from falling so low during a dive that it forces the dive to end? Part of the answer is that diving species are noted for having particularly high blood buffering capacities.

Stimulation of the urge to breathe is another concern. Anyone who has ever held his or her breath as long as possible underwater knows that eventually the urge to breathe becomes impossible to resist. The underlying causes of this phenomenon are the buildup of blood CO_2 and the drop of blood pH: In terrestrial mammals, both factors are potent stimuli for pulmonary ventilation (see page 618). Diving species exhibit blunted (i.e., reduced) ventilatory sensitivity to changes in blood CO_2 and pH compared with terrestrial species. An example is provided by free-ranging harbor seals that were exposed to elevated concentrations of CO_2 in their breathing air. As the CO_2 concentration was raised, the seals increased their ventilation rates, but only about half as much as humans would. A blunted drive to breathe helps a diving mammal stay submerged for long periods.

The Aerobic Dive Limit: One of Physiology's Key Benchmarks for Understanding Diving Behavior

In protracted dives, the accumulation of lactic acid has major behavioral consequences, for three reasons. First, ridding the body of lactic acid requires a lot of time (see Figure 26.12). Second, O_2 is required for the process (see page 197), meaning that a diving mammal typically must stay at the water's surface where it can

TABLE 26.1	Average surface times required for adult Weddell seals to dissipate accumulations of lactic acid	
Lactic acid accumulation (mg lactic acid/100 mL blood)[a]	**Time required to return to resting level (min)[b]**	
20	11	
40	27	
80	70	
120	105	
145	120	

Source: After Kooyman et al. 1980.

[a]Lactic acid accumulation refers to the peak blood concentration measured during dive recovery. This concentration provides a reasonable estimate of the total-body accumulation.

[b]The times listed are for seals that remain continuously at the water's surface until the blood concentration of lactic acid is restored to resting. The resting level of lactic acid is about 5 mg/100 mL blood.

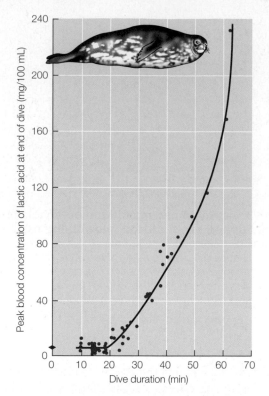

FIGURE 26.14 Peak concentration of lactic acid in arterial blood of freely diving, adult Weddell seals following dives of various durations Because the circulatory system is fully open after a dive, lactic acid produced anywhere in the body during a dive, even if temporarily sequestered while a seal was submerged, appears in the arterial blood after a dive. The peak blood concentration after a dive therefore provides an estimate of the total-body net production of lactic acid during the dive. The diamond on the *y* axis marks the average resting concentration of lactic acid in arterial blood. (After Kooyman et al. 1980.)

breathe, or return often to the surface, while it metabolizes lactic acid. Third, if a diving mammal has a lactic acid burden to metabolize following a protracted dive, it cannot engage immediately in a second highly protracted dive. This is so because the lactic acid of a second dive simply adds to the preexisting lactic acid, and there is a cap on the total lactic acid that can be accumulated and tolerated (see Chapter 8).

The lengths of time required to rid the body of lactic acid are impressively long, as already stressed. **TABLE 26.1**, for example, lists the lengths of time Weddell seals must stay at the water's surface to fully metabolize various accumulations of lactic acid—a requirement to regain their full range of behavioral options.

Accumulation of lactic acid can be avoided if dives are kept short. This is illustrated in **FIGURE 26.14**, which shows one of the most famous and important sets of results ever reported in the study of diving physiology. To gather the data shown, Gerald Kooyman and his colleagues drew blood from wild Weddell seals when the seals returned to ice holes to breathe following voluntary dives. Recall that although lactic acid may be sequestered within skeletal muscles *during* a dive, it is released into the general blood circulation immediately *after* a dive (see Figure 26.12). The investigators simply waited long enough after each dive for the release process to drive the blood concentration to its peak level before they drew a seal's blood. Then they could estimate the seal's total-body accumulation of lactic acid from the blood concentration. Put another way, the total-body accumulation and the peak blood concentration are highly correlated. Figure 26.14 shows how the peak blood concentration of lactic acid varies with dive duration. Weddell seals accumulate lactic acid when they dive for 25 min or longer. In fact, they accumulate very high levels during lengthy dives. However, they do not accumulate lactic acid above the resting, baseline level if they dive for 20 min or less.

Based on the data shown in Figure 26.14, Kooyman and his colleagues defined the concept of the **aerobic dive limit (ADL)**, which is the longest dive that can be undertaken without a net accumulation of lactic acid above the resting level. In adult Weddell seals, the ADL is 20–25 min. Unfortunately, measuring lactic acid levels in freely diving animals is a challenge, and plots like the one in Figure 26.14 are available for few species. Investigators therefore sought alternative methods to estimate the ADL. The most popular alternative is to assume that a dive will not cause accumulation of lactic acid if the O_2 cost to meet all the metabolic demands of the dive is less than the total available O_2 store, and to then calculate the ADL from the O_2 store and diving metabolic rate. Estimates based on the two methods are not always similar. To distinguish them, investigators usually use the expressions *ADL* and *cADL*; in this system, *ADL* symbolizes a value based on direct measures of lactic acid, whereas *cADL* symbolizes a value calculated from O_2 stores.[6] Here, for simplicity, we will simply use *ADL* to refer to the length of the longest dive possible without net accumulation of lactic acid, regardless of how it has been estimated.

Just as species of marine mammals vary widely in their diving competence, they also vary widely in their ADLs. Compared with the ADL of Weddell seals, the ADLs of fur seals and California sea lions (both otariids), for example, are far shorter, about 4–6 min. In contrast, ADLs in some species are far longer. Some fairly crude estimates indicate that sperm whales and male southern elephant seals have ADLs of 40–50 min!

---

[6] There is now a push to distinguish the two methods more emphatically, by using a new expression, the *diving lactate threshold* (*DLT*), to refer to estimates from lactic acid.

Why does the ADL matter? *A central hypothesis of modern diving physiology is that it is adaptive for diving mammals to keep most of their dives shorter than their species-specific ADL.* By keeping their dives fully aerobic, they avoid lengthy recovery times to metabolize lactic acid. Accordingly, they can minimize their time at the surface and maximize their time underwater.

How exactly do fully aerobic diving patterns translate into increased underwater time? When a dive is fully aerobic, replenishing O_2 stores is the only task an animal must carry out to recover. Vigorous breathing can replenish body O_2 stores rapidly. In fact, based on observations on several species, marine mammals require only 1–4 min to recover at the water surface—between one dive and the next—*when their dives are shorter than the ADL*. Short surface intervals, in turn, provide lots of underwater time for foraging. Consider, for instance, an adult Weddell seal that undertakes six 15-min dives. Because these dives are shorter than its ADL and fully aerobic, the seal will need to be at the surface for only about 4 min between dives. Thus, counting dive time and recovery time, the six dives will take 114 min, of which 90 min will be spent underwater. The seal will therefore be able to spend 80% of its time underwater, foraging and feeding. Now let's look, in contrast, at the seal's use of time if it dives just once during 114 min for as long as it can while still being fully recovered when the total time is over. A 44-min dive will cause accumulation of lactic acid: specifically about 80 mg per 100 mL of blood (see Figure 26.14). This accumulation of lactic acid, to be metabolized, will require about 70 min of recovery at the surface (see Table 26.1). A 44-min dive and recovery will therefore take the entire 114 min. This total time is the same as required for the six 15-min dives. However, the seal diving for 44 min will spend 40%—not 80%—of its time underwater! In brief, because of the long recovery times necessitated by anaerobic metabolism, a diver can typically spend a greater fraction of its time foraging and feeding if it makes many fully aerobic dives, each shorter than its ADL, than if it makes just a few lengthy dives.

Dives shorter than the ADL are also postulated to be adaptive because they permit homeostasis to be maintained throughout the body with little or no interruption. As we discussed earlier, investigators think that most or all organs receive a continuing blood flow and O_2 supply during dives that are relatively short.[7] Short dives, therefore, permit most or all organs to continue functioning in an approximately normal way. Protracted dives, by contrast, can force many organs away from homeostasis; for instance, in a protracted dive, the kidneys may be forced to stop urine production, and enzymes throughout the body may be forced to function at highly altered pH.

If dives shorter than the ADL are, in fact, adaptive as hypothesized, then diving mammals are expected to elect short dive lengths during their natural diving behavior. With this thought, we come full circle. We stressed at the start of this chapter that dives of maximum duration, although important and amazing, tend to be uncommon: Most dives are far shorter than the species-specific maximum. Now we can ask more specifically whether most dives are shorter than the species-specific ADL, as expected. Often (although not always), the answer is yes. For example, about 95% of the voluntary dives of wild, adult Weddell seals are shorter than the 20- to 25-min ADL of adult Weddell seals. Adult grey seals have a shorter ADL, about 10 min, and more than 90% of their dives are shorter than 10 min. The ADL of adult bottlenose dolphins is about 4 min, and more than 90% of their dives are shorter than 4 min.

The aerobic dive limit—a *physiological* feature of marine mammals—is therefore an important *behavioral* benchmark. Animals may dive for far longer than their ADL when faced with extraordinary behavioral challenges, such as avoiding danger or searching for new foraging areas. However, dives that are longer than the ADL require long recovery times and force many organs away from homeostasis. Therefore dives are usually kept shorter than the ADL. The significance of large body O_2 stores becomes clearer in this light: Large O_2 stores give a species a high ADL, thereby permitting a greater range of diving options while animals adhere to the "rule" that the ADL is the upper length limit for the majority of dives.

Marine mammals exploit multiple means of reducing their metabolic costs while underwater

Intense study of ADL in recent years has led to a renewed focus on the metabolic rates of diving mammals while they are submerged. This is true because the underwater metabolic rate is one of the principal determinants of the ADL. To illustrate, consider a seal that has a usable O_2 store of 3000 mL. If its metabolism is fully aerobic and its metabolic rate during diving is 300 mL O_2/min, its ADL is 10 min. If its metabolic rate could be halved, its ADL would be 20 min.

How might animals reduce their metabolic costs while diving? Here we mention some results of recent research on three different aspects of diving physiology:

- *Tissue cooling.* Earlier we noted that seals often let their body temperatures decline during diving. In hooded seals (*Cystophora cristata*), at least, investigators have recently shown that tissue cooling is actually promoted during diving because shivering is inhibited. Faced with a particular cold stress, the seals shiver when on land to keep warm, but they do not shiver underwater. Avoiding the cost of shivering while diving gives the seals a longer ADL, because if shivering occurred, it would consume a portion of their O_2 stores.

- *Delay in food processing.* A provocative recent report indicates that grey seals, when capturing prey underwater, postpone processing their prey until they are breathing air. In this way, they do not need to meet the O_2 costs of specific dynamic action (see Figure 7.5) while they are diving.

- *Limiting the costs of swimming.* Swimming costs are currently the best understood of the topics discussed here. By employing miniature video cameras and other high-technology devices, investigators have established in the last 20 years that seals and dolphins often limit their costs of underwater travel by employing gliding (**FIGURE 26.15A,B**) and other high-efficiency modes of locomotion. A principle taught to all scuba divers is that as soon as a diver becomes negatively buoyant, sinking becomes self-reinforcing; a little

---

[7] This is not to say that blood flow is unaltered. Flow is likely to be redistributed during short dives much as it is redistributed in terrestrial animals during exercise, with some organs receiving relatively more of the total flow than they do during rest and some receiving less. Nonetheless, the current working hypothesis is that blood flow (at one rate or another) is maintained to most or all organs during short dives.

FIGURE 26.15 Energy-sparing behaviors of freely diving Weddell seals

Behaviors that spare energy also spare O_2. (A) A Weddell seal outfitted with data and video recorders. Recorders like these are used to gather the types of data presented in parts B and C. Seals tolerate the presence of the recorders and seem to dive normally despite their presence. (B) A three-dimensional record of a dive by a Weddell seal, color-coded to record the type of locomotion at each stage. Data were gathered by use of recorders like those in A and remote-sensing devices such as accelerometers. (C) Simultaneous records from remote detectors of forward speed and hind-flipper activity, showing the type of evidence that is interpreted as indicating stroke-and-glide locomotion. (B and C after data of Randall Davis, Lee Fuiman, Markus Horning, and Terrie Williams, with gratitude.)

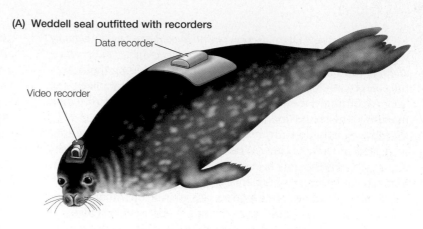

(A) Weddell seal outfitted with recorders

Data recorder

Video recorder

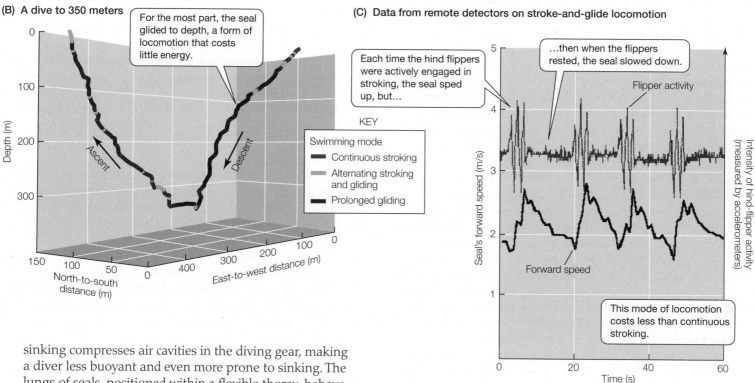

(B) A dive to 350 meters

For the most part, the seal glided to depth, a form of locomotion that costs little energy.

Ascent

Descent

KEY

Swimming mode
- Continuous stroking
- Alternating stroking and gliding
- Prolonged gliding

(C) Data from remote detectors on stroke-and-glide locomotion

Each time the hind flippers were actively engaged in stroking, the seal sped up, but...

...then when the flippers rested, the seal slowed down.

Flipper activity

Forward speed

This mode of locomotion costs less than continuous stroking.

sinking compresses air cavities in the diving gear, making a diver less buoyant and even more prone to sinking. The lungs of seals, positioned within a flexible thorax, behave as compressible air cavities, and seals seem often to employ the self-reinforcing nature of negative buoyancy to sink, holding their bodies almost motionless as they glide downward. Animals that glide to depth (see Figure 26.15B) avoid the metabolic costs (e.g., O_2 costs) of using muscle power to descend. Research has also shown that the cost of swimming is lower if seals alternately propel themselves and glide, rather than propelling themselves by muscle power continually. Telemetry data indicate that seals in fact often swim by the energy-sparing mechanism of alternately stroking with their flippers and gliding (**FIGURE 26.15C**).

Decompression Sickness

Many "diver's diseases" have been described in the annals of human diving, and physiologists have long been concerned about whether and how these diseases are avoided by marine mammals. Although nitrogen narcosis (altered cognitive function attributed to high N_2 partial pressure) is starting to receive attention in studies of marine mammals, the disease that has been studied the most is

decompression sickness, also called the *bends*. An informative starting point is to examine the etiology of this disease in human divers.

Human decompression sickness is usually caused by N_2 absorption from a compressed-air source

When diving humans unambiguously develop decompression sickness, they are almost always diving with a source of compressed air (such as a scuba tank) that steadily resupplies their lungs with air. During a dive, the compressed-air source maintains the air pressure in the lungs at a level high enough to equal the ambient water pressure at all depths. This arrangement prevents the lungs from collapsing under the force of the ambient pressure and allows continued breathing. It also, however, means that an unusually elevated N_2 partial pressure is maintained continuously in the lung air. Assuming that a diver is breathing pressurized ordinary air, 78% of the air is N_2. Thus, if the total pressure at depth is 3 atm, the partial pressure of N_2 in lung air is 2.3 atm; and if the total pressure is 5 atm, the partial pressure of N_2 is almost 4 atm. Because these elevated N_2 partial pressures are maintained continuously in the lungs during dives with a compressed-air source, all the body tissues gradually come to equilibrium with them by dissolving more N_2 than ordinary. If a dive is long enough, all tissues will in fact dissolve enough extra N_2 to have a N_2 partial pressure equal to that in the lung air.

Decompression sickness may then occur if the person suddenly surfaces ("decompresses"). Under such circumstances, as explained in Chapter 22 (see pages 589–590), dissolved N_2 can rapidly come out of solution and cause macroscopic bubbles to form within the blood and other tissues, much as bubbles form within a bottle of soda water when the cap is removed and the contents are decompressed. Bubbles formed in this way are generally believed to be the primary agents of decompression sickness. The most common symptom is throbbing pain in the joints and muscles of the arms and legs (the "bends"). In addition, an afflicted person may have neurological symptoms, such as paralysis, and severe breathing problems (the "chokes"). Exactly how the bubbles cause these symptoms remains a topic of ongoing research. Bubbles can block blood flow, press on nerve endings, and even disturb the structures of proteins because of electrical phenomena at gas–water interfaces.

The factors that determine whether *clinical* decompression sickness occurs are not completely known. One consideration is that, after a person surfaces, the N_2 partial pressure in his or her lung air is restored to its ordinary value, and the N_2-charged blood and other tissues start to lose N_2 into the lung air. In this way, excess dissolved gas is steadily eliminated by way of the lungs. If the N_2 overload is not too great, this elimination may lower the N_2 partial pressure in the blood and other tissues rapidly enough that even if macroscopic bubbles start to form, their growth and proliferation are halted before clinical symptoms occur. As a very rough rule of thumb, humans can surface immediately without fear of the bends if their blood and tissue N_2 partial pressure is less than 2 atm. Otherwise they must alter their behavior and surface gradually to avoid illness (which in severe cases can be fatal).

Breath-hold dives must be repeated many times to cause decompression sickness in humans

When we consider humans undergoing *breath-hold* diving, a crucial difference from diving with compressed air is immediately apparent. A breath-hold diver descends with only a small quantity of N_2 in his or her lung air (the quantity contained in the lung air upon submergence); his or her pulmonary N_2 supply is not steadily renewed. During descent to depth, the lungs of a breath-hold diver are compressed under the force of the increasing ambient pressure, and the N_2 partial pressure in the lungs increases initially to high levels, just as in diving with compressed air. This process creates a partial-pressure gradient favoring the transfer of N_2 from the lungs to the blood and other tissues. However, because the *quantity* of N_2 in the lungs is limited, only a limited quantity can be transferred. In humans, the amount of N_2 that is transferred to the blood and other tissues during a single breath-hold dive is far too small to cause decompression sickness.

What happens, however, if a person undergoes *many repeated* breath-hold dives? If the time between successive dives is insufficient for the tissues to release accumulated dissolved N_2 after each

dive, the tissue N_2 partial pressure can conceivably be elevated *in a series of upward steps* to a threatening level. This possibility seems to be far from theoretical. Several reports exist of people who have developed symptoms of decompression sickness after sequences of many breath-hold dives. In a classic instance described by Poul-Erik Paulev (1935–), for example, an individual complained of decompression symptoms after diving about 60 times to depths of 15–20 m over a period of 5 h.

Marine mammals have been thought—perhaps erroneously—to avoid decompression sickness during deep dives by alveolar collapse

Marine mammals are breath-hold divers. We must wonder, then, if they dive in ways whereby the N_2 transfer to their tissues by breath-hold dives might additively become great enough to cause decompression sickness. In fact, calculations and experiments indicate that if N_2 is presumed to be able to move freely from the lungs into the rest of the body, N_2 partial pressures in the blood and other tissues of diving mammals may often rise to threatening levels following many repeated dives.

Evidence indicates, however, that during deep dives, decompression sickness is in fact generally prevented in marine mammals. How? For 70 years, *alveolar collapse* has been considered the primary mechanism. Because of the compressibility of the thorax in marine mammals, the volume of the air in their lungs decreases as the animals descend during diving (see Figure 26.6). Moreover, because of distinctive patterns of structural reinforcement in their lungs, the *respiratory* parts of the lungs (see Figure 23.20) are far more prone to collapsing shut than any of the other lung parts. These factors taken together have been hypothesized to mean that as the animals descend to great depths, their lung air comes to be contained entirely in the *conducting* (nonrespiratory) airways of their lungs (**FIGURE 26.16**). In this way, below a certain depth, N_2 transfer from the lungs into the blood and other tissues cannot occur, because the N_2 in the lung air is safely sequestered in the parts of the lungs where gas exchange between the lung air and blood is not possible. Oxygen, as we stressed earlier, is also sequestered within the conducting airways. The sequestration of O_2 might simply

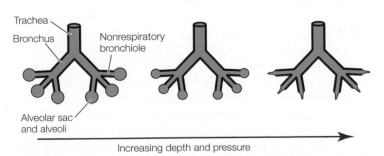

FIGURE 26.16 The hypothesis of preferential collapse of the alveoli and alveolar sacs at depth As the lung air is compressed to a smaller volume at depth, the alveoli and alveolar sacs collapse preferentially, and all air moves into the trachea, bronchi, and nonrespiratory bronchioles—the conducting airways (anatomical dead space)—where gas exchange with the blood is negligible.

be a price an animal must pay to avoid decompression sickness, or possibly—as discussed in the next section of this chapter—it may hold advantages of its own.

Without doubt, alveolar collapse helps prevent N_2 in the lungs from dissolving in the blood and other body fluids by sequestering the N_2 in the lungs. But how complete is this sequestration? Until about 15 years ago, physiologists believed that alveolar collapse occurs at relatively low ambient pressures and thus occurs quickly as an animal descends into deep waters (e.g., by the time 50 m is reached).

From that viewpoint, only shallow dives could present a problem. Certain marine mammals (e.g., some dolphins, fur seals, and sea lions) dive repeatedly to depths that, although sufficient to elevate the pulmonary N_2 pressure to levels of questionable safety, are clearly too shallow to induce full alveolar collapse. During repeated shallow dives of this sort, N_2 might invade the blood and other tissues sufficiently to pose a risk of decompression sickness. Experiments on trained bottlenose dolphins undergoing repeated dives to modest depths suggest that this worry is realistic. After an hour of shallow diving, the animals developed muscle N_2 partial pressures—1.7 to 2.1 atm—that in a human would be only marginally safe.

Deep dives are now a new concern. Under the postulated scenario in Figure 26.16, deep dives would not be a worry: The alveoli would collapse shut early in each dive, sequestering N_2 in the lungs. However, recent empirical evidence indicates that full collapse may not reliably occur at as shallow depths as previously thought during deep dives. Ultrasound scans for bubbles, for example, reveal that some deep-diving species have bubbles in their tissues when they finish sequences of multiple deep dives.

If a purely *physical* mechanism (alveolar collapse) turns out not to be fully adequate to prevent decompression sickness, *behavioral* prevention may be required. Physiologists have started to wonder if diving marine mammals need to manage the threat of decompression sickness behaviorally just as human divers must. If a person accumulates excess dissolved N_2, he or she slows ascent so N_2 can outgas from the tissues and body fluids without formation of clinically consequential bubbles. Perhaps that is what diving mammals do.

Decompression sickness is an unresolved phenomenon

Decompression sickness is in the spotlight now because, among other things, certain beaked whales seem to develop widespread bubble formation in their tissues after deep diving while being exposed to high-intensity sonar signals of human origin. Such bubble formation strongly suggests that the whales accumulate excess dissolved N_2 in their tissues during diving. The sonar could conceivably destabilize the dissolved state. Alternatively, it might disorient the animals so much that their normal behaviors are disrupted and they ascend too rapidly. Efforts to understand these phenomena have accentuated how little is confidently known about diving diseases in marine mammals.

A Possible Advantage for Pulmonary O_2 Sequestration in Deep Dives

A recent hypothesis regarding the relation between alveolar collapse and lung O_2 provides a thought-provoking note on which to end this chapter. From the beginning of the scientific study of marine mammals, investigators have tended to perceive the lungs as analogous to scuba tanks. Only a defective scuba tank would fail to deliver all its O_2. Thus, when lungs fail to deliver all their O_2, that failure is perceived as a flaw. Alveolar collapse in deep-diving marine mammals, in this view, is simultaneously advantageous and flawed. It is advantageous because even if not perfect, it helps prevent much of the lung N_2 from dissolving in the blood. It is flawed because it denies an animal the use of some of its O_2 store. Translated into evolutionary terms, this view presupposes that alveolar collapse was favored by natural selection as a means of preventing decompression sickness, with the sequestration of O_2 being simply a negative side effect. The new hypothesis postulates that in deep divers, O_2 sequestration by alveolar collapse is itself an advantage.

As a diving mammal ascends to the water's surface at the end of a deep dive, the total pressure in its lungs declines as its lungs expand. The partial pressure of O_2 in the lung air accordingly decreases precipitously as the mammal ascends. With these thoughts in mind, it is clear that the *amount* of O_2 in the lung air needs to be great enough that the O_2 partial pressure does not fall to a dangerously low level during ascent. To illustrate, suppose that during a dive to 400 m, a seal starts with 0.1 mol of O_2 in its lungs but uses 95% of it, leaving only 5%. At 400 m, the O_2 partial pressure in its lung air will still be higher than the normal value at the sea surface because of the compression of the lung air at depth. As the seal nears the surface at the end of its dive, however, the O_2 partial pressure in the expanding lung air will fall to be only 5% of normal, a level low enough to cause blackout (unconsciousness), even in some seals. Moreover, with the lung air so dilute in O_2, O_2 might easily diffuse *from the blood into the lung air*! Thus, even if the blood is at a high enough O_2 partial pressure to prevent blackout when ascent begins, the blood O_2 partial pressure might fall to blackout-inducing levels before ascent is completed. These worries suggest that it is important for the lungs to retain a sizable portion of their O_2 throughout a deep dive. Thus, in addition to impeding transfer of N_2 to the seal's tissues and body fluids, alveolar collapse may in fact be a mechanism for preventing the risk of blackout during ascent by ensuring that enough O_2 remains in the lungs to keep the O_2 partial pressure acceptably high in the ever-expanding pulmonary air at the dive's end.

One reason to end on this novel note is that these thoughts emphasize the highly *interactive* nature of the challenges faced by the lungs, blood, and other tissues during all stages of diving. A second reason is to stress the importance of striving to see physiological challenges from the point of view of the animals studied. Lungs cease to look like mere scuba tanks when viewed from the perspective of a deep-diving marine mammal.

STUDY QUESTIONS

1. Comparative physiology is sometimes defined as being the identification and use of "ideal" species for the study of each phenomenon of interest. The Weddell seal is the best known of all diving mammals. In what ways has it been the "ideal" species for the study of voluntary diving in the wild?

2. Outline the pros and cons of carrying lots of lung air during a dive.

3. Based on the study of O_2 needs and stores, the aerobic dive limit (ADL) for young Weddell seals weighing 140 kg is calculated to be 10 min, whereas that for fully grown 400-kg Weddell seals is calculated to be about 20 min. Why might small individuals in general be expected to have shorter ADLs than large individuals? To carry out an empirical study of the changes in ADL during postnatal development in a species, what experiments and measurements would you plan?

4. There is evidence that marine mammals practice unihemispheric sleep: sleep that occurs in only one brain hemisphere at a time, so that while one hemisphere sleeps, the other is awake. Such sleep is essentially unheard of in terrestrial mammals (although common in birds). What might be the advantages of unihemispheric sleep for a marine mammal?

Go to **sites.sinauer.com/animalphys4e**
for box extensions, quizzes, flashcards,
and other resources.

REFERENCES

Burns, J. M., K. C. Lestyk, L. P. Folkow, and two additional authors. 2007. Size and distribution of oxygen stores in harp and hooded seals from birth to maturity. *J. Comp. Physiol., B* 177: 687–700. A wide-ranging and thought-provoking overview of comparative diving physiology in seals, with valuable attention to the whole life cycle.

Clark, C. A., J. M. Burns, J. F. Schreer, and M. O. Hammill. 2007. A longitudinal and cross-sectional analysis of total body oxygen store development in nursing harbor seals (*Phoca vitulina*). *J. Comp. Physiol., B* 177: 217–227.

Costa, D. P., N. J. Gales, and M. E. Goebel. 2001. Aerobic dive limit: How often does it occur in nature? *Comp. Biochem. Physiol., A* 129: 771–783.

Folkow, L. P., J.-M. Ramirez, S. Ludvigsen, and two additional authors. 2008. Remarkable neuronal hypoxia tolerance in the deep-diving adult hooded seal (*Cystophora cristata*). *Neurosci. Lett.* 446: 147–150. A short paper that highlights challenges still to be met and the tantalizing future of highly interdisciplinary research.

Kooyman, G. L. 2006. Mysteries of adaptation to hypoxia and pressure in marine mammals. *Mar. Mamm. Sci.* 22: 507–526. A fascinating retrospective on the modern history of diving studies. Also a frank assessment of some of the key topics that remain enigmatic today, such as the physiology of decompression sickness.

Kooyman, G. L., E. A. Wahrenbrock, M. A. Castellini, and two additional authors. 1980. Aerobic and anaerobic metabolism during voluntary diving in Weddell seals: Evidence of preferred pathways from blood chemistry and behavior. *J. Comp. Physiol., B* 138: 335–346. A groundbreaking study.

Le Boeuf, B. J., D. E. Crocker, D. P. Costa, and three additional authors. 2000. Foraging ecology of northern elephant seals. *Ecol. Monogr.* 70: 353–382. A tour de force on the migrations and foraging behavior of elephant seals as studied with modern technology. For the curious mind, this is a report that generously stimulates the development of hypotheses for future studies of diving.

Lindholm, P., and C. E. G. Lundgren. 2009. The physiology and pathophysiology of human breath-hold diving. *J. Appl. Physiol.* 106: 284–292.

McIntyre, T., P. J. N. de Bruyn, I. J. Ansorge, and four additional authors. 2010. A lifetime at depth: vertical distribution of southern elephant seals in the water column. *Polar Biol.* 33: 1037–1048.

Ponganis, P. J., J. U. Meir, and C. L. Williams. 2011. In pursuit of Irving and Scholander: a review of oxygen store management in seals and penguins. *J. Exp. Biol.* 214: 3325–3339. A thought-provoking and thoroughly up-to-date review of O_2 stores and their management during diving. Very rewarding.

Sparling, C. E., M. A. Fedak, and D. Thompson. 2006. Eat now, pay later? Evidence of deferred food-processing costs in diving seals. *Biol. Lett.* 3: 94–98.

Tyack, P. L., M. Johnson, N. A. Soto, and two additional authors. 2006. Extreme diving of beaked whales. *J. Exp. Biol.* 209: 4238–4253.

See also **Additional References** *and* **Figure and Table Citations**.

PART VI
Water, Salts, and Excretion

Previous page
Drinking is so mundane in modern human life that we can forget its challenges and life-or-death significance for many animals in a state of nature. Giraffes, although they do not live in Earth's driest places, occupy dry savannas where drinking water may at times be scarce or salty. Physiologists study the principles that determine how much drinking water is needed and how salty the water can be and still suffice. This giraffe—along with zebras, warthogs, and Eland antelopes—was photographed in Selenkay, Amboseli Porini, Kenya.

Crustaceans like this blue crab must molt to grow. Between molts, the body of a crustacean is enclosed in a tough, calcified exoskeleton. The exoskeleton has the advantage that it acts like armor, protecting the soft tissues inside from predators. It has the disadvantage, however, that it must be shed for the animal to increase in size. A blue crab must molt more than 25 times in the course of its 2-year life. The blue crab shown is pulling itself out of its old exoskeleton as it undergoes one of its final molts; the principal claw-bearing legs are usually the last to come out, and the crab is at the stage where it is working to free those legs.

Dramatic changes in water physiology play major roles in the molting process. Prior to the actual time that a crab molts, seams along the posterior edge of its dorsal exoskeleton are weakened under hormonal control. Then as molting begins—also under hormonal control—the crab's body takes on a great deal of excess water from the surrounding environment, becoming grossly swollen (see inset). This swelling can increase the weight of the animal's tissues by 50%, or sometimes even 100%. The swelling of the body plays two major roles in the molting process. First, the swelling cracks open the old exoskeleton along its seams, as the inset shows. The crab can then exit its old exoskeleton by pulling all its body parts out backward. After the crab has exited, the swelling plays a second major role: It enables the crab to

Blue crabs dramatize the dynamism of animal water and salt relations when they shed their exoskeleton to grow
When a blue crab (*Callinectes sapidus*) sheds its exoskeleton, it takes on extra water in a carefully orchestrated way and swells. The swelling cracks open the old exoskeleton (inset) so the crab can pull itself out. The living crab is at the left in the main photograph, whereas the structure at the right is nothing but its old exoskeleton, now abandoned.

start *very quickly* to make a new, larger exoskeleton. Because of its swelling with water, a crab's body is bigger as soon as the animal crawls out of its old exoskeleton. Synthesis of the new, bigger exoskeleton can therefore begin immediately. Later, the crab will grow into its new exoskeleton, voiding excess water as it does so. In this way, it is protected inside its new exoskeleton while most of its growth takes place.

The swelling that blue crabs and other aquatic crustaceans routinely undergo serves as a visible, dramatic reminder that watery solutions play crucial roles in the lives of animals. It also helps emphasize the *dynamism* of water exchanges in all animals: Body water comes and goes, often turning over very rapidly. Mammals the size of mice commonly turn over 25% of their water every day in their natural habitats. People, being larger, turn over only about 7%. Even a turnover rate of 7% per day, however, means that a person's body weight could, in principle, increase or decrease by 5 kg (11 pounds) in a single day merely because of an imbalance in water gains and losses.

Despite the dynamism of water turnover, most animals, when healthy, maintain a relatively steady body-water content by means of highly responsive mechanisms that match water gains and water losses. The routine swelling of crustaceans during molting is really just a variation on this theme of water balance. To swell, the crustaceans unbalance their water gains and losses in a controlled way, but between molts they regulate their body-water content.

Two principal types of body fluids are recognized in animals: the **intracellular fluids** inside cells and the **extracellular fluids** outside cells. The extracellular fluids are divided into two subparts: the **interstitial fluids** found between cells in ordinary tissues[1] and the **blood plasma**, the part of the blood other than the blood cells.[2] Each body fluid is an aqueous solution in which are dissolved a variety of inorganic ions—such as Na^+, Cl^-, and K^+—plus organic compounds such as the plasma proteins. The intracellular fluids, interstitial fluids, and blood plasma are often described as the **fluid compartments** of the body.

In this chapter and the three that follow, we discuss the nature and significance of the body fluids, the challenges animals face to maintain the composition and volume of their three body-fluid compartments within viable limits, and the mechanisms of body-fluid regulation. The present chapter focuses on concepts and mechanisms that pertain to animals in general or to large subsets of animals. Later chapters discuss the specifics of particular animal groups, environments, and organs involved in water–salt physiology.

The Importance of Animal Body Fluids

One sign of the importance of body fluids is simply their abundance. Collectively they account for more than half of an animal's body weight in most cases. They represent about 60% of body weight in adult humans and in many other animals (**FIGURE 27.1**). In some types of animals, such as certain of the gelatinous invertebrates, they account for as much as 95%.

---

[1] For simplicity, in earlier chapters the *interstitial fluids* have been referred to by their more general name *extracellular fluids*.

[2] In animals with open circulatory systems, the interstitial fluids and blood plasma are identical. See Chapter 25.

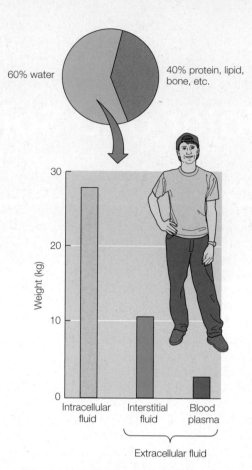

FIGURE 27.1 Body fluids account for 60% of the body weight of young adults The bar chart shows the normal weights of the intracellular fluid and the two categories of extracellular fluids—interstitial fluid and blood plasma—in a 70-kg person. (See Figure 6.2 for the composition of the 40% of the body that is not body fluids.)

For the most part, the cells, the subcellular structures, and the proteins and other molecules in an animal's body are bathed with body fluids or in contact with body fluids. The body fluids, therefore, constitute the immediate environment for cells and molecules in the body (see page 13). The *compositions* of the body fluids constitute the context in which cells, organelles, and molecules function.

The particular *inorganic ions*—often called *salts* or *electrolytes*—dissolved in the body fluids, and the concentrations of those ions, are important aspects of body-fluid composition for several reasons. One is that the ionic composition of body fluids affects the crucial three-dimensional molecular conformations of enzymes and other proteins. Ion concentrations in the intracellular and interstitial fluids are also important in maintaining correct electrical gradients across cell membranes, and they play key roles in nerve-impulse transmission and muscle excitation (see Chapters 12 and 20). Too little intracellular potassium (K^+), for example, can alter heart rhythm by affecting key ion gradients across the cell membranes of the pacemaker cells.

The *water* in body fluids is also important for several reasons. The conceptually simplest reason is that water is the matrix in which the ions are dissolved; the amount of water is as instrumental in determining ion concentrations as are the ions themselves. Water is also important because it affects the *volumes* of cells and tissues. In addition, osmotic movements of water affect the hydrostatic pressures that prevail in the body.

Another important aspect of solutes and water is that we now recognize that a solution has a complex *structure* which arises from interactions among the ions, water molecules, and organic molecules in the solution. Individual ions in an aqueous solution, for example, induce the water molecules immediately surrounding them to orient around the ions, forming shells of hydration. Molecules of water also associate in structured ways with proteins and other macromolecules (constituting "water of hydration"). In regions of free water, the H_2O molecules structure themselves by mutual hydrogen bonding into arrays. The physiological implications of the structured nature of solutions are just starting to be explored, and there can be little doubt that new insights into the roles of body fluids will emerge as the structured nature of solutions is taken into full account.

All animals have evolved specific, *controlled* relations of their body-fluid composition to the composition of the external environment in which they live. In some cases, the control exerted by animals over their body-fluid composition is dramatic. Freshwater animals, for example, expend energy to maintain ionic concentrations that are far higher in their body fluids than in freshwater. By contrast, the water–salt regulatory processes in some other animals are subtle. Many invertebrates that live in the open ocean, for example, maintain just relatively small differences in salt composition between their blood plasma and seawater. Whether the differences between an animal's body fluids and its external environment are dramatic or subtle, actively maintained differences always exist. This observation suggests that for all animals, the effects of salts and water on protein function and other functional properties have led to fine-tuning of body-fluid composition during the course of evolution and to the evolution of mechanisms that can control the water–salt composition of the body fluids.

The Relationships among Body Fluids

The three major body-fluid compartments of an animal—the intracellular fluids, interstitial fluids, and blood plasma—interact and affect one another. **FIGURE 27.2** emphasizes that the three compartments are closely juxtaposed within an animal's tissues. The intracellular fluids are separated from the interstitial fluids just by the cell membranes of the tissue cells. The interstitial fluids are separated from the blood plasma only by the endothelium of the blood capillaries (a single cell layer).

Brisk exchange of water and ions often occurs between the intracellular fluids and the interstitial fluids across cell membranes. As discussed in Chapter 5, this is true in part because cell membranes are permeable to water and ions; water can cross cell membranes by osmosis, and ions can cross by diffusion. Moreover, ions are often transported between the intracellular and interstitial fluids by facilitated diffusion or active transport, mediated by transporter proteins in the cell membranes.

Brisk exchange of water and ions also often occurs between the interstitial fluids and the blood plasma. One reason is that the blood-capillary endothelium that separates the interstitial fluids and blood plasma is in most cases perforated relatively densely by minute, physical pores. Some of these pores are formed by tiny gaps (about 4 nanometers [nm] in diameter) *between* cells in the endothelium; other pores are formed in some tissues by holes

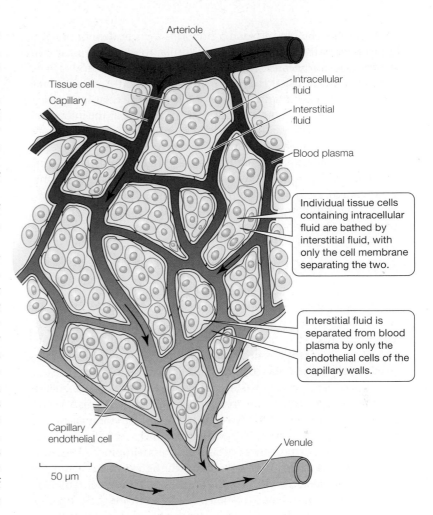

FIGURE 27.2 The three major types of body fluids are closely juxtaposed This diagram of a microcirculatory bed in a mammal shows the close juxtaposition of the three major types of body fluids. For simplicity, red blood cells and other blood cells are not shown. (See Figure 25.11 for more on the structure of microcirculatory beds.)

through endothelial cells.[3] Water and ions freely cross the capillary endothelium through these pores by osmosis and diffusion, bringing the fluids on each side close to equilibrium with each other. In addition, as discussed in Chapter 25 (see Figure 25.13), the elevated blood pressure inside capillaries forces a bulk flow of fluid—containing both water and ions—out of the blood plasma and into the interstitial fluids by ultrafiltration.

When physiologists study the water–salt physiology of whole animals, they nearly always sample just the blood plasma to gain insight into the status of the animals, rather than trying to sample all fluid compartments. They take this approach because blood can be sampled easily by simply inserting a syringe needle. What, however, can we deduce about the composition of the other body fluids from measures on just the blood plasma?

---

[3] The capillaries in various tissues of an animal's body vary in their types and densities of pores. In the human body, for example, liver capillaries are very porous. Brain capillaries, however, virtually lack pores. The capillaries in most tissues are intermediate. The pores are often called **fenestrations**.

Actually, the blood plasma provides a great deal of insight. The interstitial fluid is typically very similar to the blood plasma in *osmotic pressure* because of the free exchange of water and ions by osmosis and diffusion that occurs through blood-capillary pores and by other mechanisms.[4] Moreover, the intracellular fluid inside cells ordinarily has about the same *osmotic pressure* as the interstitial fluid and blood plasma. Thus, if you measure the osmotic pressure of the blood plasma, you usually know the osmotic pressures of the interstitial and intracellular fluids. What about *ions*? The interstitial fluid is typically very similar to the blood plasma in ionic composition. However, the intracellular fluid differs dramatically in ionic composition from the interstitial fluid (and the blood plasma) because of the action of ion pumps (see Figure 5.10).

Types of Regulation and Conformity

Three types of regulation of the composition of the blood plasma are possible: *osmotic regulation, ionic regulation*, and *volume regulation*. Some animals—such as humans and other mammals—regulate in all three ways. Other animals, however, are conformers in certain aspects of their water–salt physiology.[5]

Osmotic regulation (osmoregulation) is the maintenance of a constant or nearly constant osmotic pressure in the blood plasma. To illustrate osmotic regulation, consider an aquatic animal that is placed in environmental water of various osmotic pressures. If the animal exhibits perfect osmotic regulation, its blood osmotic pressure will remain the same regardless of the osmotic pressure of the environmental water, as seen in **FIGURE 27.3A**. The dashed line in the figure is a line of equality between the blood osmotic pressure and the ambient (environmental) osmotic pressure and thus is called the **isosmotic line** (*iso-*, "equal"). A perfect osmoregulator exhibits no tendency to follow the isosmotic line. The opposite of osmotic regulation is **osmotic conformity**. In an osmotic conformer, as seen in **FIGURE 27.3B**, the blood osmotic pressure always equals the osmotic pressure of the environmental water and thus falls on the isosmotic line.

Actual animals often display gradations or mixes of osmotic regulation and osmotic conformity. **FIGURE 27.3C**, for example, shows the relation between blood osmotic pressure and ambient osmotic pressure in three marine invertebrates. The shrimp is an almost perfect osmoregulator; its blood osmotic pressure varies just a little with the osmotic pressure of its external environment. The crab exhibits impressive but imperfect osmotic regulation in waters

---

[4] Blood plasma is typically more concentrated than interstitial extracellular fluid in dissolved proteins that cannot pass through the capillary endothelium. This disparity produces a difference in osmotic pressure between the two fluids that can be of critical importance for understanding certain phenomena, such as bulk-fluid exchanges across capillary walls (see Figure 25.13 and the discussion of the Starling–Landis hypothesis in Chapter 25). The *absolute* osmotic pressures of the blood plasma and interstitial fluid are quite similar nonetheless. In human systemic tissues (see Chapter 25), the osmotic pressure of the blood plasma is about 3.3 kilopascals (kPa) (25 millimeters of mercury [mm Hg]) higher than the osmotic pressure of the interstitial fluid. In the osmolarity system of units, 3.3 kPa is equivalent to about 1.5 milliosmolar (mOsm). Thus, when the blood plasma has an osmotic pressure of 300 mOsm (as is typical of human blood), the interstitial fluid has an osmotic pressure of 298.5 mOsm. For most purposes, therefore, the blood plasma and interstitial fluid can be considered to be virtually identical in osmotic pressure.

[5] See page 14 and Figure 1.6 for the distinction between *regulation* and *conformity*.

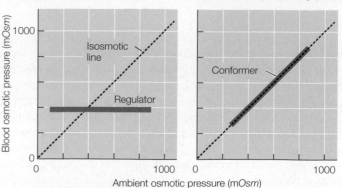

(A) Osmotic regulation (idealized) (B) Osmotic conformity (idealized)

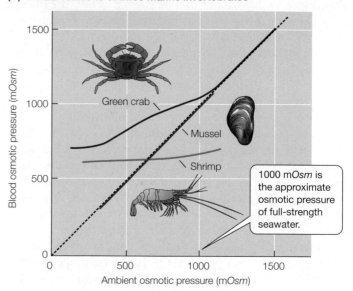

(C) Actual relations of three marine invertebrates

1000 mOsm is the approximate osmotic pressure of full-strength seawater.

FIGURE 27.3 Osmotic regulation and conformity In each graph, each solid line shows the osmotic pressure of blood plasma as a function of the ambient (environmental) osmotic pressure; the dashed line is a line of equality between blood osmotic pressure and ambient osmotic pressure (an isosmotic line). Osmotic pressures are expressed in units of milliosmolarity (mOsm). (A, B) The osmotic pressure of the blood plasma as a function of the osmotic pressure of the ambient water in (A) a perfect osmotic regulator and (B) a perfect osmotic conformer. (C) The osmotic pressure of the blood plasma as a function of the ambient osmotic pressure is shown for three species of marine invertebrates: *Mytilus edulis*, the blue mussel; *Carcinus maenas*, the green crab; and *Palaemonetes varians*, a species of grass shrimp. The mussel is a strict osmotic conformer. The crab regulates in waters more dilute than seawater but is an osmotic conformer at higher ambient osmotic pressures. The shrimp regulates over a wide range of ambient osmotic pressures. (C after Hill and Wyse 1989.)

more dilute than 1000 milliosmolar (m*Osm*)—seawater concentration—but it is an osmotic conformer in waters more concentrated than 1000 m*Osm*. The mussel is an osmotic conformer at all the concentrations studied.

The concepts of osmotic regulation and conformity apply to terrestrial animals as well. Humans, for example, maintain a remarkably stable blood-plasma osmotic pressure of about 300 m*Osm* whether they drink a lot of water per day or little per day. Humans, therefore, are excellent osmotic regulators.

Ionic regulation of the blood plasma is the maintenance of a constant or nearly constant concentration of an inorganic ion in the blood plasma. There are as many potential types of ionic regulation as there are ions because each ion is subject to ion-specific physiological controls. The opposite of ionic regulation is **ionic conformity**; if an animal is an ionic conformer for a particular ion, it allows the concentration of the ion in its blood plasma to match the concentration in its external environment.

Volume regulation is the regulation of the total *amount* of water in a body fluid. Volume regulation of the blood plasma, for example, is regulation of the amount of water in the plasma. Animals generally regulate their blood volume. We, for example, maintain an approximately constant amount of water in our blood plasma regardless of how much water we drink per day. Blue crabs illustrate both the presence and absence of volume regulation. At the specific times in their lives when they swell and molt, volume regulation is suspended; their blood takes on water, and their bodies swell. At other times, blue crabs are volume regulators, as already stressed.

The concept of volume regulation applies to the interstitial and intracellular fluids as well as to the blood plasma. For example, if cells maintain a constant amount of intracellular water, they exhibit cell-volume regulation.

True **volume conformity** refers to completely passive changes of body-fluid volume (e.g., driven by osmosis). It occurs only rarely in animals.

The integrated study of osmotic regulation, ionic regulation, and volume regulation can seem confusing at first because although the three types of regulation are in fact distinct, they are often intimately related. To illustrate, consider a freshwater fish that takes on a quantity of H_2O from its dilute environment by osmosis. This influx of H_2O into its body will tend to have three *simultaneous* effects:

- It will tend to lower the osmotic pressure of the fish's blood plasma (a challenge to osmotic regulation).
- It will tend to dilute ions in the fish's blood plasma (a challenge to ionic regulation).
- It will tend to increase the volume of water in the fish's blood plasma and other body fluids (a challenge to volume regulation).

Simultaneous effects such as these can create the illusion that osmotic regulation, ionic regulation, and volume regulation are the same thing. They are in fact different, however, as we will see by example later in this chapter. *An exceedingly important tool for understanding animal water–salt physiology is to maintain the conceptual distinction between the three types of regulation and analyze problems in water–salt physiology from all three perspectives.*

Both regulation and conformity have pros and cons. If an animal exhibits osmotic and ionic regulation of its blood plasma, the composition of its interstitial fluid is also regulated, and therefore its *cells*—which are bathed by interstitial fluid—experience a constant osmotic–ionic environment. This constancy can be of advantage to the cells, but it costs energy: Animals must invest energy to maintain constancy. If animals exhibit osmotic–ionic conformity in their blood plasma and interstitial fluid, they avoid such energy costs, but their cells must then cope with changing osmotic–ionic conditions in the interstitial fluid that bathes them.[6]

[6] See Chapter 1 (page 16) for a more complete discussion of the pros and cons of regulation and conformity (i.e., homeostasis and lack of homeostasis).

SUMMARY
Types of Regulation and Conformity

- Osmotic regulation is the maintenance of a steady osmotic pressure in the blood plasma.
- Ionic regulation of any particular inorganic ion is the maintenance of a steady concentration of that ion in the blood plasma.
- The concepts of osmotic and ionic regulation can also be applied to the interstitial and intracellular fluids.
- Volume regulation is the maintenance of a steady volume (amount) of water in the body as a whole or in a particular fluid compartment. Cell-volume regulation is regulation of the amount of intracellular water in a cell.

Natural Aquatic Environments

Some aquatic animals live in environments that are approximately uniform and stable in their water–salt composition. The animals that live in the open ocean are in this category. Full-strength seawater is approximately the same in its concentrations of major ions (**TABLE 27.1**) everywhere in the open ocean. The **salinity** of a body of water—defined to be the number of grams of dissolved inorganic matter per kilogram of water—is a measure of the total concentration of all salts taken together. Seawater in the open ocean has a uniform salinity, 34–36 g/kg.[7] An important and easily remembered feature of full-strength seawater is that its osmotic pressure is almost exactly 1000 m*Osm*, as already noted.

Freshwater is usually defined as water having a salinity less than 0.5 g/kg (the worldwide average for lakes and rivers is 0.1–0.2 g/kg). Freshwater is essentially a uniform environment in terms of its osmotic pressure, which is always very low (0.5–15 m*Osm*). Ion concentrations in freshwater present a more complex picture: Although ions are always dilute in freshwater (see Table 27.1),

[7] Several alternative ways exist to write this unit of measure. To refer to grams per kilogram (g/kg), some scientists use the symbol ‰; others write *parts per thousand* (*ppt*) or *parts per mille*. Salinity can also be expressed using *practical salinity units*; values in this system are dimensionless (have no units) but are numerically very similar to values expressed in g/kg.

TABLE 27.1 Concentrations of major ions in seawater and freshwater

Ion	Concentration (m*M*) Seawater	Freshwater[a]
Sodium (Na$^+$)	470	0.35
Chloride (Cl$^-$)	548	0.23
Magnesium (Mg^{2+})	54	0.21
Sulfate (SO$_4{}^{2-}$)	28	0.19
Calcium (Ca^{2+})	10	0.75
Potassium (K$^+$)	10	0.08
Bicarbonate (HCO$_3{}^-$)	2	1.72

Sources: Data for seawater from Barnes 1954; data for freshwater from Bayly and Williams 1973.

[a]Freshwater values are worldwide averages for rivers.

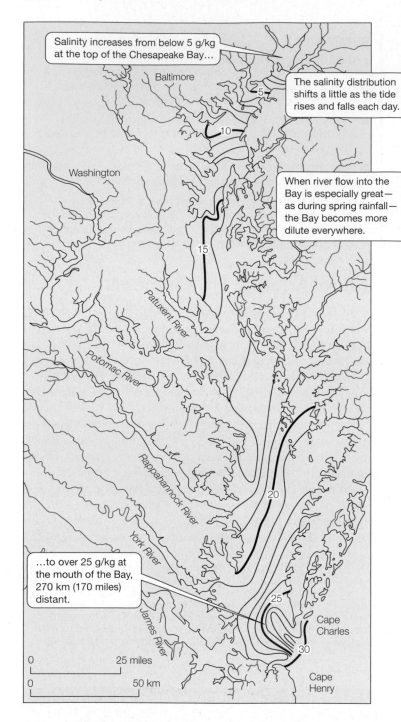

Salinity increases from below 5 g/kg at the top of the Chesapeake Bay...

The salinity distribution shifts a little as the tide rises and falls each day.

When river flow into the Bay is especially great—as during spring rainfall—the Bay becomes more dilute everywhere.

...to over 25 g/kg at the mouth of the Bay, 270 km (170 miles) distant.

A typical estuarine shoreline (Chesapeake Bay)

FIGURE 27.4 Salinity trends in an estuary Salinity varies greatly along the length of the Chesapeake Bay, one of the world's most thoroughly studied estuaries. The map shows a typical pattern of salinity in the water near the surface. Numbers are salinities in grams per kilogram. Each line connects all the surface waters of a particular salinity; therefore, for example, the line for 20 g/kg connects all the places where the salinity of the surface water is 20 g/kg. (Salinity distribution after McHugh 1967.)

Many of the places where brackish waters occur are classified as **estuaries**. An estuary is any body of water that is partially surrounded by land and that has inflows of both freshwater and seawater. Estuaries are of great importance in human affairs. They also are among the most interesting aquatic habitats physiologically because of the dramatic variability they often display in their salinity spatially and temporally—presenting animals with unusual challenges because of the great variation in environmental osmotic pressure and ion concentrations. Brackish waters are usually defined to have salinities between 0.5 and 30 g/kg—corresponding to osmotic pressures of 15 to 850 mOsm. As illustrated by the Chesapeake Bay (**FIGURE 27.4**), a single estuary may exhibit almost this entire range of salinities within a distance of 270 km (170 miles). Estuaries are the characteristic habitat of blue crabs, which—when not molting—are effective osmoregulators at brackish salinities. As a blue crab travels into waters of various salinities in an estuary, its blood osmotic pressure remains almost constant everywhere it goes (**FIGURE 27.5**).

Natural Terrestrial Environments

Animals on land are surrounded by a fluid—air—that contains water only in the gaseous state and, of course, is essentially free of salts. One might guess that terrestrial animals would commonly be able to gain water from the air when the humidity of the air is high. That is not the case, however. For most terrestrial animals most of the time, the atmosphere is a sink for water: Animals lose water to it by evaporation. Because living in air is inherently dehydrating, the study of the water–salt physiology of terrestrial animals is dominated by the study of water.

For scientists interested in the water relations of terrestrial animals, the world's deserts are particularly important and intriguing habitats because they present animals with the extremes of terrestrial water stress. Deserts cover substantial areas of Earth's landmasses (**FIGURE 27.6**) and on average are expand-

certain ions vary in concentration from one body of freshwater to another in ways that are biologically consequential. Calcium (Ca^{2+}) is particularly noteworthy. Although always dilute, it is distinctly higher in concentration in some bodies of freshwater—termed "hard"—than in others—termed "soft." These variations in Ca^{2+} concentration can exert substantial effects on the water–salt physiology of freshwater animals by affecting their membrane permeabilities and sometimes other functional properties.

Where ocean water mixes with freshwater along coastlines, waters of intermediate salinity, termed **brackish waters**, are formed. In these places, both osmotic pressure and ion concentrations in the water are often highly variable not only from place to place but also from time to time.

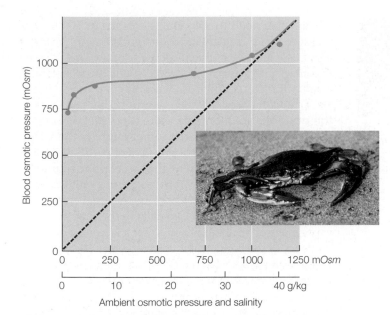

FIGURE 27.5 The responses of a resident osmotic regulator to variations in salinity in an estuary Blue crabs (*Callinectes sapidus*) are abundant throughout the Chesapeake Bay and other similar estuaries along the Atlantic and Gulf coasts of the United States. The graph shows the average osmotic pressure of the blood plasma of blue crabs as a function of the osmotic pressure and salinity of the ambient water during nonmolting periods of their lives, when they osmoregulate. The dashed line is the isosmotic line. (After Kirschner 1991.)

ing. Although deserts are often hot, they are not necessarily so, because they may occur at high altitudes or polar latitudes. Deserts are defined by their dryness, not their temperature. A simple definition of a desert is that it receives less than about 25 cm (10 inches) of rain or other precipitation per year; we discuss more refined definitions in Chapter 30.

To understand the relations of animals to atmospheric water in deserts or other terrestrial habitats, one must study the principles of **evaporation**, the change of water from a liquid to a gas. Evaporation is a special case of gas diffusion, which (as discussed in Chapter 22) is analyzed using the *partial pressures* of gases.[8]

Gases always diffuse in net fashion from regions of high partial pressure to regions of low partial pressure. Thus, if a body fluid, or any other aqueous solution, is in contact with the atmosphere, net evaporation occurs if the partial pressure of water in the solution exceeds that in the atmosphere, and the rate of evaporation increases as the difference in partial pressure increases.

What do we mean by the partial pressure of water in a solution and in the atmosphere? Let's start with the latter. Gaseous water, called **water vapor**, is simply a gas like any other gas. Thus it is a constituent of the atmosphere in the same way that other gases, such as O_2, are. The partial pressure of water vapor—often called the **water vapor pressure**—in the atmosphere is simply the portion of the total atmospheric pressure that is exerted by the water vapor

---

[8] If you are unfamiliar with the concept of *partial pressure*, you should review pages 586–588 and Figures 22.1 and 22.3.

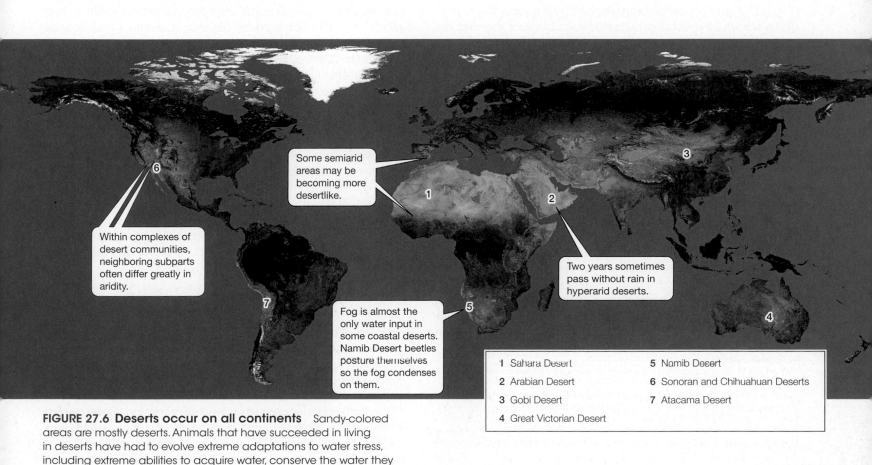

Some semiarid areas may be becoming more desertlike.

Within complexes of desert communities, neighboring subparts often differ greatly in aridity.

Two years sometimes pass without rain in hyperarid deserts.

Fog is almost the only water input in some coastal deserts. Namib Desert beetles posture themselves so the fog condenses on them.

1 Sahara Desert	5 Namib Desert
2 Arabian Desert	6 Sonoran and Chihuahuan Deserts
3 Gobi Desert	7 Atacama Desert
4 Great Victorian Desert	

FIGURE 27.6 Deserts occur on all continents Sandy-colored areas are mostly deserts. Animals that have succeeded in living in deserts have had to evolve extreme adaptations to water stress, including extreme abilities to acquire water, conserve the water they have, and tolerate dehydration.

TABLE 27.2 The saturation water vapor pressure at selected temperatures

The saturation water vapor pressure is independent of other gases; it does not depend on the composition of the air. This table also shows the mass of water per unit of volume when the saturation water vapor pressure prevails.

Temperature (°C)	Saturation water vapor pressure (mm Hg)	Saturation water vapor pressure (kPa)	Mass of water per unit of air volume at saturation (mg H_2O/L)
0	4.6	0.61	4.9
10	9.2	1.23	9.4
20	17.5	2.33	17.3
30	31.8	4.24	30.4
37[a]	47.1	6.28	43.9
40	55.3	7.37	51.1

[a]Data for 37°C are included because 37°C is the usual deep-body temperature of humans and other placental mammals.

present. It can be calculated from the universal gas law (see Equation 22.1) and is independent of the partial pressures of the other gases in the atmosphere. **Humidity** is an informal term referring loosely to the water content of air. Although several measures of humidity are in common use, the *water vapor pressure* is the most useful for physiological analysis.[9]

Unlike other gases in the atmosphere, water vapor displays an upper limit on its partial pressure: The water vapor pressure can rise only to a certain maximum in air of a particular temperature. The limit on water vapor pressure is a direct consequence of the fact that (unlike the other atmospheric gases) water can exist as a liquid (not just a gas) under ordinary atmospheric conditions. Air that has reached its maximum water vapor pressure is said to be **saturated**, and its water vapor pressure is termed the **saturation water vapor pressure**. If a body of air has reached saturation with water vapor, it cannot hold any more water in the gaseous state. Thus, if water vapor is added to such air from an outside source, the excess water vapor promptly condenses out in the form of liquid water droplets (e.g., fog forms in this manner). The saturation water vapor pressure increases dramatically with the temperature of air, as **TABLE 27.2** shows.

To illustrate the immediate significance of these concepts, consider that we humans exhale air at a temperature greater than 30°C, whereas a toad with a body temperature of 20°C breathes out air at 20°C. The air exhaled is essentially saturated with water vapor in both cases. Warmer air holds more water vapor when saturated, however (see Table 27.2), and therefore humans lose more water than toads lose with each liter of air they exhale.

[9] In addition to being expressed simply as the water vapor pressure, the humidity of air is often expressed *relative* to the air's temperature-specific *saturation water vapor pressure*, discussed in the next paragraph. One expression of this sort is the **saturation deficit**, which is the *difference* between the actual, prevailing water vapor pressure and the saturation water vapor pressure. Another such expression of humidity is the **relative humidity**, defined to be the *ratio* of the actual water vapor pressure over the saturation water vapor pressure.

Now we need to ask what is meant by the partial pressure of water in an aqueous solution. Any particular aqueous solution (e.g., freshwater or seawater), if it is placed in contact with air in a closed system, will tend to establish a characteristic, equilibrium water vapor pressure in the air. That vapor pressure is the **water vapor pressure of the aqueous solution**. The water vapor pressure of *pure* liquid water depends on the temperature of the water. Specifically, the water vapor pressure of pure liquid water at a particular temperature is the same as the saturation water vapor pressure of air at the same temperature; Table 27.2, therefore, can be used to look up the water vapor pressure of pure liquid water. One way to think of the water vapor pressure of liquid water is that it is a measure of the tendency of the liquid water to inject water vapor into air. Liquid water at 30°C has a much greater tendency to inject water vapor into air than liquid water at 10°C.

In addition to depending on temperature, the water vapor pressure of an aqueous solution depends also on its solute concentration. The water vapor pressure is in fact a *colligative* property of a solution (see page 122). Raising the concentration of dissolved entities in a solution lowers the water vapor pressure of the solution. This effect is relatively small at the concentrations of most animal body fluids; even a 1-*Osm* solution has a water vapor pressure that is 98% as high as that of pure water. Sometimes, however, the effect of solutes is physiologically important. For instance, if salt left behind from evaporated sweat is allowed to accumulate indefinitely on a person's skin so that newly secreted sweat becomes highly concentrated by dissolving accumulated salt, the water vapor pressure of the sweat can be reduced to be only 75% of that of pure water—an effect that can significantly impair vaporization of the sweat.

Having now discussed the partial pressure of water both in air and in aqueous solutions, let's now turn to the physical laws that govern the evaporation of water from terrestrial animals. As mentioned earlier, water always diffuses from regions of relatively high water vapor pressure to regions of lower water vapor pressure. Thus water evaporates from an aqueous solution (e.g., a body fluid) if the water vapor pressure of the solution exceeds the water vapor pressure of the air next to the solution. The *rate* of evaporation depends on the difference in water vapor pressure. Specifically, if J is the net rate of evaporation per unit of solution surface area, WVP_s is the water vapor pressure of the solution, and WVP_a is the water vapor pressure in the air, then

$$J = K\frac{WVP_s - WVP_a}{X} \tag{27.1}$$

where X is the distance separating WVP_s and WVP_a, and K is a proportionality factor.

From Equation 27.1, we see that the rate at which an animal loses water by evaporation in a terrestrial environment depends partly on the environmental humidity, expressed as WVP_a. As we know from everyday experience, lowering the water vapor pressure of the air (lowering the humidity) speeds evaporation if all other factors are held constant. The rate of evaporation also depends on the water vapor pressure of the body fluid from which evaporation is occurring, WVP_s. Thus, for example, evaporation occurs faster from body fluids that are warm than from ones that are cooler. If a mammal or bird—or a lizard basking in the sun—is exposed to the open air in a hot desert, the animal may face exceptional risks of dehydration by evaporation because its body fluids are warm (meaning WVP_s is high; see Table 27.2) while simultaneously the desert air is dry (WVP_a is low).

A factor that is not immediately evident in Equation 27.1 is that the rate of evaporation also depends on the rate of air movement. If an animal is standing in still air, evaporation from the animal itself tends to humidify the air immediately next to its skin, creating a boundary layer of elevated WVP_a near the skin, very similar to the boundary layer depicted in Figure 5.3. The operative value of WVP_a in Equation 27.1 is therefore raised, slowing evaporation. A wind, however, blows water-vapor-laden air away from the skin surface, replacing it with drier air from the open atmosphere, thereby decreasing the WVP_a next to the skin and speeding evaporation. A wind in a desert can be extremely dehydrating.

In an animal that is losing water by evaporation through its integument (skin), K in Equation 27.1 is the *permeability of the integument to water*. A low value of K represents the chief physiological defense that animals can marshal to protect themselves from high rates of evaporative desiccation. A low value of K—signifying a low integumentary permeability to water—slows the loss of body water by evaporation.

To complete our discussion of the diffusion of water vapor in terrestrial animals, let's return to a point made at the start: Animals tend to lose water to the atmosphere, not gain water from it. Put another way, animals rarely gain water by condensation.

When is condensation possible? For most types of animals, condensation can occur only when the body surface is cooler than the air. Under such circumstances—which are uncommon—animals can function like glasses of iced tea. We are all familiar with the fact that gaseous water from the atmosphere condenses into water droplets on the outside of a cold glass of iced tea on a humid summer day. This process depends in part on the physics of the *initiation* of water-droplet formation, an advanced topic in physical chemistry. However, after minute (invisibly tiny) water droplets have been initiated, their *growth* follows the principles of water diffusion we are discussing. Water will diffuse from the water vapor in the atmosphere into a droplet of liquid water—causing the droplet to grow—only if the water vapor pressure in the atmosphere exceeds the water vapor pressure of the liquid water in the droplet. A *cold* water droplet has a relatively low water vapor pressure (see Table 27.2). If the water vapor pressure in the air is simultaneously high, water will move from its gaseous state in the air into the liquid state in the droplet. In this way, minute droplets will grow into big, visible droplets.

For the most part, the body surfaces of animals are not cooler than the air, explaining why animals usually lose water by evaporation rather than gaining it by condensation. Occasionally, however, animals in special circumstances have cool body surfaces. For example, when lizards that have spent the night in chilly underground burrows emerge into the open air in the morning, their skin can be cooler than the air; when the air is humid, water droplets can then form on their skin and be ingested. Such condensation is transitory,

however, because the lizards warm as time passes. As soon as they are as warm as the air, they start to lose water by evaporation.[10]

SUMMARY
Natural Terrestrial Environments

■ The water vapor pressure of air is the partial pressure of water vapor in the air and is the most useful expression of humidity for analysis of evaporation and condensation.

■ The water vapor pressure of an aqueous solution is the equilibrium water vapor pressure the solution tends to create in juxtaposed air if the solution and air are sealed in a closed system.

■ Water vapor diffuses from regions of high water vapor pressure to regions of low water vapor pressure. Thus evaporation occurs if the water vapor pressure of an aqueous solution exceeds that of the surrounding air. Evaporation takes place at a rate proportional to the difference in vapor pressure.

Organs of Blood Regulation

Animals living in their natural environments routinely experience conditions that tend to change their blood composition. A mammal exposed to the dryness of desert air, for example, loses water by evaporation, and its loss of water tends to raise the osmotic pressure of its blood, concentrate ions in its blood, and decrease the volume of its blood (challenges to osmotic regulation, ionic regulation, and volume regulation). Another example would be a fish that migrates from the lower part of the Chesapeake Bay (where the salinity is high) into the upper part (where the salinity is near that of freshwater) (see Figure 27.4). In the dilute water, the fish will take on water at an accelerated rate by osmosis, and the influx of water will tend to lower the osmotic pressure of its blood, lower its blood ion concentrations, and expand its blood volume, as we saw earlier. Animals such as mammals and fish are regulators of blood composition and respond to such challenges in negative feedback fashion (see Box 1.1). Certain of their organs act to reverse changes in blood composition, keeping their blood characteristics stable despite the environmental challenges they encounter.

Among the organs involved, the kidneys play particularly important roles, not only in mammals and fish but also in most other types of animals. Indeed, the most fundamental function of

---

[10] Animals in *saturated air* could, in principle, gain water at a low rate while at the *same* temperature as the air, because the solutes in body fluids lower the body-fluid water vapor pressure slightly. However, in reality, when animals are in saturated air, their metabolic heat production raises their temperature to be at least slightly above air temperature, forcing the water vapor pressure gradient to favor evaporation. Thus, without a cooling mechanism, water is not gained by diffusion from the air. Of course, evaporation of body fluids can cool an animal's body surfaces. However, thermodynamics dictates that evaporative cooling cannot be sufficient to cause simultaneous condensation (one process cannot cause evaporation and condensation simultaneously). Thus, for water to be obtained by diffusion from air because of body cooling, the cooling must be caused by some process other than simultaneous evaporation. Some insects, ticks, and other terrestrial arthropods do not follow the physical rules we are discussing here and can gain water steadily from atmospheric water vapor; these cases are discussed in Chapter 28 (see page 772).

kidneys is to *regulate the composition of the blood plasma by removing water, salts, and other solutes from the plasma in controlled ways*. Other organs also play major roles in the regulation of blood composition. For example, the gills of aquatic animals are typically important organs of blood regulation, and salt glands (discussed in Chapter 28) are important in certain birds, lizards, turtles, and other reptiles.

In this chapter we emphasize the kidneys, and more specifically we emphasize a conceptual (rather than a mechanistic) understanding of their function.[11] Two reasons for taking a conceptual approach are paramount. First, some of the most important concepts of kidney function apply almost universally—providing insights that pertain regardless of the specifics of various kidney types. Second, the concepts that apply to kidney function can often be applied as well to other organs (e.g., salt glands) that regulate blood composition.

Kidneys are fluid-processing organs: They start with blood plasma and produce urine. Many of the effects of kidney function on blood composition can be analyzed by comparing the output of the kidneys with their input—that is, by comparing the urine to the blood plasma. This sort of comparison is usually carried out by use of **U/P ratios**: ratios of urine (U) composition over plasma (P) composition.

The osmotic U/P ratio is an index of the action of the kidneys in osmotic regulation

The **osmotic U/P ratio** is the osmotic pressure of the urine divided by the osmotic pressure of the blood plasma. For example, if an animal's urine osmotic pressure is 150 m*Osm* and its plasma osmotic pressure is 300 m*Osm*, its osmotic U/P ratio is 0.5. Urine may, in principle, be *isosmotic*, *hyperosmotic*, or *hyposmotic* to the blood plasma.[12] The osmotic U/P ratio reflects this **relative osmoticity** of the urine. If U/P = 1, the urine is isosmotic to the plasma. If U/P

---

[11] The mechanisms of operation and the anatomy of kidneys are discussed in Chapter 29.

[12] If A and B are two solutions and A has a higher osmotic pressure than B, then A is *hyperosmotic* to B, whereas B is *hyposmotic* to A, as explained in Chapter 5. If two solutions have the same osmotic pressure, they are *isosmotic* to each other.

< 1, the urine is hyposmotic to the plasma. If U/P > 1, the urine is hyperosmotic to the plasma. The kidneys of an animal typically have control over the U/P ratio and can adjust it within a species-specific range. Humans, for example, can have an osmotic U/P ratio as high as 4 or as low as 0.1.

To explore the interpretive value of the U/P ratio, let's start by considering a freshwater fish. As discussed in Chapter 5 (see Figure 5.19), the body fluids of a freshwater fish have an osmotic pressure far higher than the osmotic pressure of freshwater. That is, the blood plasma is strongly hyperosmotic to freshwater. Suppose that a fish takes a quantity of pure water—H_2O—into its body fluids by osmosis from the pond or stream in which it lives. This water will dilute the fish's blood and reduce its plasma osmotic pressure. Can the fish restore its original plasma osmotic pressure (i.e., can it osmoregulate) by producing urine? A bit of reflection will reveal that the answer is yes *only if the fish is able to produce urine that is more dilute than its plasma—that is, hyposmotic urine (U/P < 1)*.

A urine that is hyposmotic to the blood plasma *preferentially voids water*. By this we mean that, relative to the blood plasma, urine of this sort is richer in water and poorer in dissolved solutes. Therefore, when the urine is excreted, it disproportionally depletes the blood plasma of water. Because of this preferential removal of water from the plasma—and the converse, the preferential retention of solutes in the plasma—voiding the urine acts to elevate the osmotic pressure of the plasma toward its original level. One way to see this point is to contrast this outcome with what would happen if a fish's urine were always isosmotic to its blood plasma. A fish with a U/P ratio of 1 merely excretes water and solutes in the same ratio at which they exist in its blood plasma. Thus, if the fish's plasma were too dilute, it would remain too dilute regardless of how much isosmotic urine (U/P = 1) the fish might excrete.

From our analysis of the freshwater fish, we arrive at two general principles of kidney function (**FIGURE 27.7**):

■ The production of urine isosmotic to blood plasma (U/P = 1) cannot serve directly to change the osmotic pressure of the plasma or bring about osmotic regulation of the plasma.

U/P ratio	Implications for excretion		
	Effects on water excretion	Effects on solute excretion	Effects on composition of blood plasma
U/P = 1 (isosmotic urine)	Water is excreted in the same relation to solutes as prevails in the blood plasma.	Solutes are excreted in the same relation to water as prevails in the blood plasma.	The formation of urine leaves the ratio of solutes to water in the blood plasma unchanged, thus does not alter the plasma osmotic pressure.
U/P < 1 (hyposmotic urine)	Water is preferentially excreted. Urine contains more water relative to solutes than plasma.	Solutes are preferentially held back from excretion. Urine contains less solutes relative to water than plasma.	The ratio of solutes to water in the plasma is shifted upward. The osmotic pressure of the plasma is raised.
U/P > 1 (hyperosmotic urine)	Water is preferentially held back from excretion. Urine contains less water relative to solutes than plasma.	Solutes are preferentially excreted. Urine contains more solutes relative to water than plasma.	The ratio of solutes to water in the plasma is shifted downward. The osmotic pressure of the plasma is lowered.

FIGURE 27.7 The interpretive significance of the osmotic U/P ratio The terms *solute* and *solutes* refer to total numbers of osmotically effective dissolved entities.

■ The production of hyposmotic urine (U/P < 1) aids osmotic regulation of the blood plasma if an animal's plasma has become too dilute and the plasma osmotic pressure therefore needs to be raised.

As one might expect, most freshwater animals have evolved kidneys that have the capacity to make urine that is hyposmotic to their plasma. We humans, as well as most other terrestrial animals, also have that capacity, which serves us well after an evening of too much iced tea or beer.

If the plasma osmotic pressure of an animal has been raised to abnormally *high* levels, urine that is more concentrated than the plasma must be produced to correct the problem. Such urine preferentially voids solutes (and preferentially retains water), thereby lowering the ratio of solutes to water in the plasma. From this analysis, we arrive at a third principle of kidney function (see Figure 27.7):

■ The production of hyperosmotic urine (U/P > 1) aids osmotic regulation of the blood plasma if an animal's plasma has become too concentrated and the plasma osmotic pressure therefore needs to be lowered.

The ability to produce urine that is hyperosmotic to the blood plasma is not nearly as widespread as the ability to produce hyposmotic urine. The greatest capacities to concentrate the urine are found in mammals, birds, and insects—all primarily terrestrial groups that frequently face risks of dehydration.

The effects of kidney function on volume regulation depend on the amount of urine produced

The kidneys help regulate the *quantity of water* in an animal's body—that is, they aid volume regulation—by voiding greater or lesser *amounts* of water as required. We ourselves provide a familiar example: We make a lot of urine after drinking a lot of water, but we make little urine if we are short of water.

The kidneys, in fact, can play a critical role in *volume regulation* even when not playing any direct role in *osmotic regulation*. In this respect, kidney function illustrates that volume regulation and osmotic regulation are *distinct processes*, as stressed previously. To illustrate these points, let's consider freshwater crabs. These are species of true crabs that live in rivers and lakes, mostly in tropical and subtropical parts of the world. Freshwater crabs provide striking examples of animals in which the kidneys participate in volume regulation but not osmotic regulation. The crabs are dramatically hyperosmotic to the freshwater in which they live and thus experience a steady osmotic flux of water into their body fluids. To meet this challenge to volume regulation, the crabs produce a substantial flow of urine; each day, their kidneys excrete the same amount of water as they gain by osmosis. However, at least in the species that have been investigated, the kidneys of freshwater crabs are unable to produce urine that is more dilute than the blood plasma. Their urine is always isosmotic to the plasma (U/P = 1). Consequently, the production of urine by the crabs does not alter their plasma osmotic pressure. Although the kidneys of freshwater crabs help them dispose of their excess *volume* of water, *other* organs must maintain the high *osmotic pressure* of their blood.

The effects of kidney function on ionic regulation depend on ionic U/P ratios

The action of an animal's kidneys in ionic regulation can be analyzed in ways closely analogous to the analysis of osmotic regulation (see Figure 27.7). For each ion, an ionic U/P ratio can be computed; it is the concentration of that ion in the urine divided by the concentration of the ion in the blood plasma. The sodium U/P ratio, for example, is the urine Na^+ concentration divided by the plasma Na^+ concentration. To see the interpretive value of an ionic U/P ratio, let's continue with Na^+. If the sodium U/P ratio is greater than 1, the urine contains more Na^+ per unit of water volume than the plasma; thus the excretion of urine preferentially voids Na^+ and lowers the plasma Na^+ concentration. Conversely, if the sodium U/P ratio is less than 1, the excretion of urine acts to retain Na^+ preferentially in the body and raise the plasma Na^+ concentration.

The kidneys can play a role in *ionic regulation* even when not playing any direct role in *osmotic regulation*. In this way, the kidneys illustrate that ionic regulation is a distinct concept from osmotic regulation. Marine teleost (bony) fish are good examples of animals in which the kidneys participate in ionic regulation but not osmotic regulation. These fish are hyposmotic to the seawater in which they live. Therefore they lose water osmotically to their environment while simultaneously they gain ions by diffusion from the seawater. Both of these processes tend to raise the osmotic pressure and the ion concentrations of their blood plasma. The marine teleost fish produce a urine that is isosmotic to their plasma (osmotic U/P = 1); their urine, therefore, can play no direct role in solving their *osmotic* regulatory problem. However, their urine differs dramatically from their blood plasma in its *solute composition*. In particular, the U/P ratios for Mg^{2+}, SO_4^{2-}, and Ca^{2+} are far greater than 1. The excretion of urine by these fish therefore serves the important *ionic* regulatory role of keeping down the internal concentrations of these ions, which the fish tend to gain from the seawater.

SUMMARY
Organs of Blood Regulation

■ The effects of kidney function on the composition of the blood plasma are analyzed using osmotic and ionic U/P ratios. Figure 27.7 summarizes the interpretation of U/P ratios.

■ Osmotic regulation, volume regulation, and ionic regulation are separable kidney functions in the sense that the kidneys can participate in volume regulation while simultaneously not aiding osmotic regulation, or they can carry out ionic regulation independently of osmotic regulation.

Food and Drinking Water

The specific composition of food and drinking water often has major implications for the water–salt physiology of animals living in their natural environments—illustrating once again that physiology and ecology are intimately related. To start our discussion of this topic, let's focus on the relative osmoticities of predators and their prey.

When one animal captures and eats another, the water–salt composition of the prey animal—not just its nutrient content—may be significant for the predator. Consider, for example, predator–prey relations in the ocean. Marine mammals and teleost fish are dramatically hyposmotic to seawater. However, most marine invertebrates are approximately isosmotic to seawater. Recognizing these properties, when a mammal or fish consumes a meal of invertebrates, the body fluids of its prey are markedly more concentrated in salts than its own body fluids are. The predator must therefore eliminate excess salts to maintain its normal body-fluid composition. In contrast, consider a mammal or fish that consumes a meal of fish. In this case, the body fluids of the prey are similar in salt concentration to those of the predator. Thus the fish-eating predator incurs little or no excess salt load when it eats, in contrast to the large salt load incurred by the invertebrate-eating predator. A fish-eating predator benefits from the work that its prey performed to maintain body fluids more dilute than seawater—an intriguing lesson in ecological energetics.

Salty drinking water may not provide H$_2$O

When animals drink water rich in salts, the water may not serve as a useful source of H$_2$O. Whether an animal can gain H$_2$O by drinking salty water (e.g., seawater) depends on whether the animal can eliminate the salts from the salty water using less H$_2$O than was ingested with them. This principle, which applies to all animals, is a critical consideration when people suffering from dehydration are presented with the option of drinking salty water.

We have all heard Coleridge's famous line from *The Rime of the Ancient Mariner*, "Water, water, everywhere, nor any drop to drink." Sailors desperate for water discovered long ago that drinking ocean water was worse than drinking no water at all: Drinking the seawater paradoxically dehydrated them. We now know that a key consideration in understanding this paradox is that the maximum Cl$^-$ concentration that the human kidney can produce in the urine is lower than the concentration of Cl$^-$ in seawater. Therefore, if people drink seawater, the Cl$^-$ they ingest can be excreted only by voiding more H$_2$O than was taken in with the Cl$^-$. That is, such people not only must use all the H$_2$O ingested with the seawater to excrete the Cl$^-$; they must also draw on other bodily reserves of H$_2$O, thereby dehydrating their tissues. Some animals are able to excrete salts at higher concentrations than humans can and thus are able to gain H$_2$O by drinking salty solutions such as seawater (by excreting the salts in less H$_2$O than was ingested with them).

Plants and algae with salty tissue fluids pose challenges for herbivores

Some plants in terrestrial environments—particularly ones native to deserts—have very salty tissue fluids. If herbivores eat such plants, they receive a substantial salt load along with the food value of the plants.

The soils in some desert regions are very saline. One reason for this condition is that salts tend to accumulate over eons of time in the places where rain settles in deserts; the evaporation of rain water leaves the salts it contains behind in the soil, and each rainfall adds to the salts left by preceding rainfalls. Plants called **halophytes** ("salt plants") root in these saline soils and often have high salt concentrations in their tissue fluids. Such plants form a major part of the diet of desert sand rats (*Psammomys obesus*) and are consumed in large quantities at times by dromedary camels. The total salt concentration in some halophytes exceeds that of seawater by as much as 50%. Many of the halophytes are succulent plants with juicy leaves. Animals that eat them obtain considerable water, but they obtain a large salt load as well.

Analytically, the salt levels of salty plants pose much the same problems for animals as the problem posed by salty drinking water. Suppose, for example, that the Na$^+$ concentration in a halophyte's tissues is five times that in mammalian blood plasma. A mammal would then require kidneys that can produce a sodium U/P ratio greater than 5 to be able to excrete the Na$^+$ and obtain a net gain of H$_2$O from the plants.[13] Most mammals cannot achieve such a high sodium U/P ratio. Sand rats, however, have kidneys with legendary concentrating abilities. Therefore, after scraping off and discarding the saltiest parts of the leaves, sand rats are able to eat halophytes without ill effects from the salt they ingest. Accordingly, they can eat foods that other desert rodents must avoid.

Air-dried foods contain water

Many terrestrial animals consume air-dried seeds or other dry plant matter. These air-dried foods contain moisture, even though they are ostensibly dry. The moisture they contain is significant, particularly for animals that live where drinking water is difficult to find.

Air-dried foods equilibrate with air moisture. Accordingly, they vary in their water content as the humidity varies. Whereas "dry" barley grain, for example, contains almost 4 g of water per 100 g dry weight at 10% relative humidity, its water content is five times higher at 76% relative humidity. When air-dried plant material is exposed to an altered air humidity, its moisture content changes within hours. Two humidity patterns are of importance to animals in this regard. First, the relative humidity of the air tends to rise at night, and second, it tends to be higher belowground than aboveground. Animals that get water from air-dried food can often increase their water intake by feeding at night or by storing the food in burrows prior to ingesting it.

Protein-rich foods can be dehydrating for terrestrial animals

Because carbohydrates and lipids consist primarily of carbon, hydrogen, and oxygen, their oxidation during metabolism results mostly in formation of CO$_2$ and H$_2$O. The CO$_2$ is exhaled into the atmosphere, and the H$_2$O contributes to an animal's water resources. Proteins, by contrast, contain large amounts of nitrogen, and their catabolism results in nitrogenous wastes.

The products of protein catabolism can affect a terrestrial animal's water balance when they must be excreted in solution in the urine. In mammals, for example, the principal nitrogenous waste is urea, a highly soluble compound voided in the urine. The amount of urinary water required to void urea depends on the urea-concentrating ability of an animal's kidneys. When a mammal is producing urine with as high a urea concentration as it can, a high-protein meal often forces the animal to void more water (to get rid of the urea) than a low-protein meal.

---

[13] This is just a rough calculation in the case of the plants because the organic constituents of plants must also be considered.

TABLE 27.3 Average gross amount of metabolic water formed in the oxidation of pure foodstuffs

The values in this table apply to the oxidation of materials that have been absorbed from a meal and to the oxidation of materials stored in the body. To emphasize this, the materials are called foodstuffs rather than foods. The gross amount of metabolic water formed is, by definition, simply the amount made by the oxidation reactions.

Foodstuff	Grams of H₂O formed per gram of foodstuff
Carbohydrate[a]	0.56
Lipid	1.07
Protein with urea production[b]	0.40
Protein with uric acid production[b]	0.50

Source: After Schmidt-Nielsen 1964.

[a]Starch is assumed for the specific value listed.

[b]Water yield in protein catabolism depends on the nitrogenous end product.

Metabolic Water

When organic food molecules are aerobically catabolized, water is formed, as illustrated by the equation for glucose oxidation:

$$C_6H_{12}O_6 + 6\,O_2 \rightarrow 6\,CO_2 + 6\,H_2O \qquad (27.2)$$

The water produced by catabolic reactions such as this is known as **metabolic water (oxidation water)**, in contrast to **preformed water**, which is water taken in as H₂O from the environment. Metabolic water is produced by all animals. **TABLE 27.3** lists the amount of water formed per gram of foodstuff oxidized.

The simple fact that aerobic catabolism produces water does not by itself mean that animals gain water in *net* fashion from catabolism. This is true because the catabolism of foodstuffs not only produces water, but also obligates an animal to certain water losses. The **obligatory water losses** of catabolism are the losses that must take place for catabolism to occur. To assess the *net* impact of catabolism on the water balance of an animal, the obligatory water losses must be subtracted from the gains of metabolic water. This important principle will be exemplified after we briefly discuss the *respiratory, urinary,* and *fecal* components of obligatory water loss. Detailed aspects of this topic can vary, depending on the type of animal and on peculiarities of diet. For simplicity, we assume here a terrestrial mammal (or similar animal) that is eating foods of ordinary salt content (i.e., not highly salty foods).

The **obligatory respiratory water loss** is defined to be the loss of water that is necessary to obtain O₂ for catabolism. Aerobic catabolism requires O₂ (see Equation 27.2), and when animals breathe to obtain O₂, they lose water by evaporation. The aerobic catabolism of all types of food molecules causes obligatory respiratory water loss. The magnitude of the loss depends on a species' physiology of breathing and on the humidity of the ambient air.

The **obligatory urinary water loss** is the loss of urine water that is mandated by the ingestion or catabolism of food molecules. Protein catabolism is the usual cause of obligatory urinary water loss. As we have seen, protein catabolism produces nitrogenous wastes (urea

in mammals) that demand urine excretion. The excretion of such wastes obligates water excretion. The catabolism of carbohydrates and lipids does not yield products that must be excreted in urine, and thus is not a cause of obligatory urinary water loss.

The **obligatory fecal water loss** is the loss of water that must occur in feces for food catabolism to take place. Obligatory fecal water loss occurs only when *ingested* foods are catabolized because the catabolism of organic materials stored in the body, such as lipid stores, produces no fecal waste. Ingested foods usually contain *preformed* water. If an animal must lose more water in its feces than it took in as preformed water with its ingested food, it incurs a net fecal water loss that is required for it to catabolize the food. This net loss is the obligatory fecal water loss.[14]

The use of these concepts is exemplified in **BOX 27.1**, which focuses on kangaroo rats—animals that thrive in some of the driest places in North America's southwestern deserts. The kangaroo rats in the studies discussed were given no water to drink and fed nothing but air-dried barley grain. Remarkably, they did not suffer from dehydration, because they were able to gain metabolic water in net fashion by catabolizing the barley. Note that their net water gain is computed in two steps. The first step is to determine the immediate (gross) yield of metabolic water from the oxidation of the food molecules absorbed from a gram of ingested barley. The second step is to subtract the obligatory respiratory, urinary, and fecal water losses required to catabolize a gram of ingested barley.

Metabolic water matters most in animals that conserve water effectively

At moderate temperatures and even at low relative humidities (as low as about 20%), kangaroo rats and some other desert rodents can live indefinitely on air-dried seeds without drinking water, meeting most of their water needs with metabolic water and the remainder with the small amounts of preformed water in the seeds. Under similar conditions, most other mammals would quickly die. These striking contrasts have given rise to a myth that desert rodents produce especially large amounts of metabolic water.

In fact, *the amount of metabolic water produced per gram of food oxidized is fixed by chemistry*; it depends simply on the stoichiometry of the aerobic catabolic pathways. Thus, for a given type of food, all animals produce the same amount of metabolic water per gram of food oxidized.[15]

Why, then, do some animals depend more on metabolic water than others? The answer lies in *water conservation*: Some animals *conserve* water more effectively than others. If two animals have the same metabolic rate and oxidize the same foods, both will produce the same amount of metabolic water. If one conserves water poorly and thus has a high overall rate of water turnover—lots of water entering and leaving its body each day—its production of metabolic water will be small relative to its total water intake and loss; thus its

[14] If the feces contain less water than was taken in with ingested food, the animal realizes a net gain of *preformed* water, which is not a factor in metabolic water calculations.

[15] The only exception is that when protein is catabolized, the amount of metabolic water produced depends to some extent on the particular nitrogenous end product made (see Table 27.3) because the chemistry of the reactions depends on the particular end product. The nitrogenous end products are discussed in detail in Chapter 29.

BOX 27.1 Net Metabolic Water Gain in Kangaroo Rats

Desert kangaroo rats (*Dipodomys*) were studied at an air temperature of 25°C and a relative humidity of 33%. They were fed air-dried barley grain and given no drinking water. This box shows how obligatory water losses are taken into account to calculate the kangaroo rats' *net* gain of metabolic water.

Gross Metabolic Water Production (0.54 g of H_2O per gram of barley)

The food molecules absorbed by a kangaroo rat from a gram of ingested barley yield about 0.54 g of metabolic water during cellular oxidation. This value is calculated by first estimating the amounts of carbohydrate, lipid, and protein absorbed and catabolized during the processing of a gram of barley. The water yields of oxidizing the three components are then calculated from Table 27.3 and summed.

Obligatory Water Losses (total: 0.47 g of H_2O per gram of barley)

1. *Respiratory* (0.33 g of H_2O per gram of barley): Oxidation of the food molecules absorbed from a gram of barley requires consumption of about 810 mL of O_2. To acquire this amount of O_2 by breathing, a kangaroo rat loses about 0.33 g of water by pulmonary evaporation under the conditions of the experiments.

2. *Urinary* (0.14 g of H_2O per gram of barley): The protein absorbed from a gram of ingested barley yields about 0.03 g of urea. Kangaroo rats can concentrate urea sufficiently in their urine that this amount of urea can be excreted in about 0.14 g of water.

3. *Fecal* (0 g of H_2O per gram of barley): About 0.1 g of preformed water is obtained with each gram of barley ingested. However, in a kangaroo rat, the feces resulting from the digestion of a gram of barley contain only about 0.03 g of water. Because the water lost in feces is less that that gained with the food, there is no net fecal water loss, and therefore the obligatory fecal water loss is 0. A kangaroo rat, in fact, *gains preformed water* by digesting barley because it loses less water in its feces than it takes in with the barley.

A kangaroo rat This species, *Dipodomys merriami*, is one of several species of kangaroo rats noted for their success in the deserts and semideserts of the American Southwest.

Net Gain of Metabolic Water (0.07 g of H_2O per gram of barley)

The total of all the obligatory water losses during the catabolism of a gram of ingested barley is 0.47 g. To calculate the net gain of metabolic water, this total must be subtracted from the gross metabolic water production per gram of barley, 0.54 g. The net gain is therefore 0.07 g of water per gram of ingested barley; in other words the rats have increased their body H_2O content by means of metabolic production of H_2O. This gain of metabolic water can be used to offset other water losses, or if it represents an excess, it can be excreted.

---

metabolic water will represent just a small part of its water budget. If the other animal conserves water well and thus has a low overall rate of water turnover, its production of metabolic water will be large relative to its total water intake and loss—and will represent a large part of its water budget.

To illustrate, suppose we place kangaroo rats and laboratory rats at an air temperature of 25°C and a relative humidity of 33% and we feed both species the same food: air-dried barley grain. Let's assume (as is approximately true) that individuals of the two species have similar rates of metabolism. If we provide no drinking water, the kangaroo rats will thrive—living largely on metabolic water—but the laboratory rats will deteriorate and ultimately require drinking water to save their lives. Chemical stoichiometry dictates that the two species produce similar amounts of metabolic water. How, then, can we explain why only one species survives without drinking water? The answer is seen in **TABLE 27.4**, which presents a *full* accounting of the effects of catabolism on water balance. The kangaroo rats conserve water more effectively than the laboratory rats; they have lower urinary water losses because they can concentrate urea to a greater degree in their urine, and they have lower fecal water losses, in part because they produce drier feces. When the obligatory water losses of the two species are subtracted from the gross amount of

TABLE 27.4 Approximate catabolic gains and losses of water in kangaroo rats (*Dipodomys*) and laboratory rats (*Rattus*) when eating air-dried barley grain and denied drinking water

Study animals were caged at 25°C and 33% relative humidity. The values given are grams of H_2O per gram (dry weight) of barley ingested. Values for kangaroo rats are from Box 27.1.

Category of water gain or loss	Kangaroo rats	Laboratory rats
Gross metabolic water produced	0.54 g/g	0.54 g/g
Obligatory water losses		
Respiratory	0.33	0.33
Urinary	0.14	0.24
Fecal	0.00	0.03
Total obligatory water losses	0.47	0.60
Net gain of metabolic water	+ 0.07	−0.06

metabolic water produced, the kangaroo rats—because they conserve water so well—enjoy a *net gain* of metabolic water. The laboratory rats, however, suffer a *net loss*. Note that the critical difference between them is not in how abundantly they *produce* metabolic water. Exceptional *water conservation* in the kangaroo rats is what permits them to live on metabolic water.

SUMMARY
Metabolic Water

■ Metabolic water is produced by all animals in amounts determined by the chemical stoichiometry of the oxidation of organic molecules.

■ To determine the *net* effect of catabolism on water balance, obligatory respiratory, urinary, and fecal water losses must be subtracted from gross metabolic water production.

■ The role played by metabolic water in the overall water budget of an animal depends on the animal's capacity to *conserve* body water.

Cell-Volume Regulation

One of the most important attributes of cells in the body is that they have particular *volumes*. As noted earlier, cell membranes are generally sufficiently permeable to water that the intracellular fluids remain isosmotic to the extracellular fluids bathing the cells (the interstitial fluids) simply by osmosis. This condition may at first sound benign, but it in fact means that any change in the osmotic pressure of the extracellular fluids is a threat to the maintenance of a constant cell volume. If the osmotic pressure of the extracellular fluids decreases, cells tend to take on water by osmosis and swell. If the extracellular osmotic pressure increases, cells tend to lose water and shrink. How, then, can cells display **cell-volume regulation**—that is, maintain a constant cell volume—when the osmotic pressure of the extracellular fluids changes?

For cells to exhibit volume regulation, they must alter their total content of osmotically effective dissolved entities. To see this, consider the model cell in **FIGURE 27.8**. In ❶, the cell contains ten osmotically effective dissolved entities and is at its normal size. If the cell is transferred into a solution that has an osmotic pressure only half as great as the starting solution, its initial response ❷ is to take on water by osmosis and swell to twice its normal volume, thereby reducing the concentration of dissolved entities inside by half. How can the cell restore its normal volume if it remains in the dilute solution? The answer is that the cell must reduce its content of dissolved entities. As seen in ❸, if the cell halves the number of dissolved entities in its intracellular fluid, it will return to its original volume while remaining isosmotic to the dilute solution. Suppose now that we return to the original cell ❶ and we transfer it into a solution that has an osmotic pressure double that of the starting solution. The cell's initial response ❹ will then be to lose water osmotically and shrink to half its normal volume, doubling the concentration of dissolved entities inside. To restore its normal volume in the concentrated ambient solution, the cell must increase its content of dissolved entities. As seen in ❺, if the cell doubles

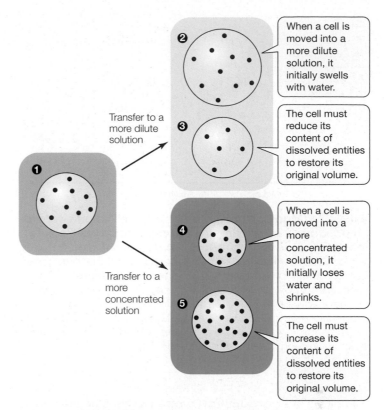

FIGURE 27.8 The fundamental principles of cell-volume regulation Dots represent dissolved entities in the intracellular fluid of a model cell. Shading around cells represents the osmotic pressure of the solution bathing the cells (darker blue symbolizes higher osmotic pressure). The solutions surrounding the cells at the right have osmotic pressures that are half as great (top) and twice as great (bottom) as the osmotic pressure of the solution at the left.

When a cell is moved into a more dilute solution, it initially swells with water.

The cell must reduce its content of dissolved entities to restore its original volume.

When a cell is moved into a more concentrated solution, it initially loses water and shrinks.

The cell must increase its content of dissolved entities to restore its original volume.

Transfer to a more dilute solution

Transfer to a more concentrated solution

the number of dissolved entities in its intracellular fluid, it will return to its original volume while remaining isosmotic to the concentrated solution.

We see from this exercise that *if the extracellular fluids of an animal become diluted, its cells must reduce their intracellular numbers of osmotically effective dissolved entities to retain their original volumes.* Conversely, *if the animal's extracellular fluids become more concentrated, cell-volume regulation requires that cells increase their intracellular numbers of osmotically effective dissolved entities.* These insights help clarify the potential advantages of osmoregulation of the extracellular fluids: If organs such as the kidneys can keep the extracellular fluids at a constant osmotic pressure, the cells are freed from any requirement to adjust their content of solutes for cell-volume regulation.

When circumstances require cells to undergo cell-volume regulation, certain intracellular solutes are increased or decreased in amount more than others: Cells are selective in which solutes they modulate. As an introduction to this complex subject, let's focus on the modulation of organic solutes. Invertebrates—meaning most animals—employ organic solutes as their principal agents of cell-volume regulation. Certain mammalian tissues also emphasize organic solutes, notably the brain and the renal medulla, the interior tissue of each kidney (see page 798).[16]

---

[16] However, many types of mammalian cells have evolved a different type of strategy.

KEY

- Inorganic ion
- Organic solute molecule

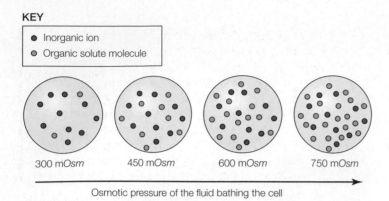

300 mOsm   450 mOsm   600 mOsm   750 mOsm

Osmotic pressure of the fluid bathing the cell

FIGURE 27.9 Many animal cells achieve cell-volume regulation principally by altering their content of organic solute molecules Here we assume for simplicity that *only* organic solutes are modulated. As the model cell is exposed to a higher and higher external osmotic pressure (from left to right), it adds organic solute molecules to its intracellular fluids. Consequently, it maintains *both* a constant cell volume *and* a constant total intracellular concentration of inorganic ions. Real cells typically adjust inorganic ions to some extent while emphasizing modulation of organic solutes. The principal intracellular inorganic ions are K^+, Mg^{2+}, Na^+, and Cl^-.

Looking at Figure 27.3C, you can see that blue mussels are osmoconformers. When a mussel is placed in a more dilute or more concentrated environment, its blood osmotic pressure changes to match the new environmental osmotic pressure. Cells throughout its body therefore confront volume-regulation challenges. The cells of osmoregulators also face such challenges (although not as dramatically) because osmoregulating animals are rarely perfect regulators. For example, blue crabs (see Figure 27.5) exhibit small changes in their blood osmotic pressure if they move into waters with a substantially altered salinity.

To carry out cell-volume regulation—as we have said—the cells of mussels, blue crabs, and other invertebrates principally modulate their content of organic solutes. **FIGURE 27.9** illustrates this mechanism in action. By modulating its content of organic solutes, a cell can regulate its volume while also keeping its intracellular concentrations of inorganic ions relatively constant. That is, the cells achieve cell-volume regulation and ionic regulation simultaneously!

What are the mechanisms by which cells modify the numbers of dissolved organic molecules in their intracellular fluid? Here we discuss free amino acids, which are major agents of intracellular volume regulation in many sorts of animals, including mussels and blue crabs. When animals are transferred into more saline environmental water, multiple processes are employed to raise intracellular quantities of free amino acids, including decelerated amino acid catabolism, accelerated synthesis of new amino acids, accelerated breakdown of intracellular proteins to release amino acids, and accelerated active transport of amino acids into the cells. When animals are transferred into more dilute environmental water, the processes employed to decrease intracellular quantities of free amino acids are in many ways the opposite. For example, catabolism of amino acids is accelerated. An increase in excretion of ammonia (NH_3)—a product of amino acid catabolism (see Figure 6.3B)—is often easily detected when animals are transferred into dilute waters!

Intracellular organic solutes have been of extreme importance in the *evolution* of the composition of the intracellular fluids. Most vertebrates and freshwater invertebrates have evolved intracellular fluids that—speaking very roughly—are one-third as concentrated (in total osmotic pressure) as the intracellular fluids evolved by most invertebrates in the open ocean. Does that mean that the intracellular fluids are three times *saltier* in the ocean invertebrates? No: All these animals are similar in their intracellular inorganic ion concentrations! They differ in their concentrations of intracellular organic solutes. In much the same way as illustrated in Figure 27.9, as animals evolved different *total* intracellular concentrations, most of the difference was accounted for by evolving different concentrations of organic solutes rather than inorganic ions.

SUMMARY
Cell-Volume Regulation

■ For a cell to maintain a constant volume, it must reduce the amounts of osmotically effective dissolved entities in its intracellular fluid when the osmotic pressure of the surrounding extracellular fluid falls, and it must increase its content of dissolved entities when the osmotic pressure of the extracellular fluid rises.

■ Organic molecules such as free amino acids are the principal intracellular solutes employed for cell-volume regulation by cells of invertebrates. They are also the principal intracellular solutes employed in certain of the tissues (e.g., brain and renal medulla) of vertebrates.

■ The use of organic molecules as principal agents of cell-volume regulation permits simultaneous ionic and volume regulation of the intracellular fluid.

From Osmolytes to Compatible Solutes: Terms and Concepts

Any solute that exerts a sufficiently large effect on the osmotic pressure of a body fluid to be of consequence for understanding water–salt physiology is termed an **osmolyte** (or **osmotic agent**), regardless of what other functions it may perform. An osmolyte that an individual animal or individual cell increases or decreases in amount to achieve osmotic regulation or cell-volume regulation is called an **osmotic effector**.

Animals, as we have just seen, are selective in which solutes they employ as intracellular osmotic effectors: They generally use organic solutes, rather than inorganic ions, in this role. Why animals employ organic solutes is an important contemporary question in the study of water–salt physiology. The hypothesis favored by most physiologists is that the solutes animals preferentially employ as osmotic effectors are ones that have minimal effects on proteins and other macromolecules. Such solutes, termed *compatible solutes*, can be increased and decreased in a cell without greatly altering the cell's enzyme kinetics or other aspects of cell protein function.

The term *compatible solute* was first used more than 35 years ago to refer to a solute that, at high concentrations, does not interfere with the ability of enzymes to function effectively. Modern definitions are less specific and more rough-and-ready, but they retain the spirit of the original. Today, a **compatible solute** is a solute that,

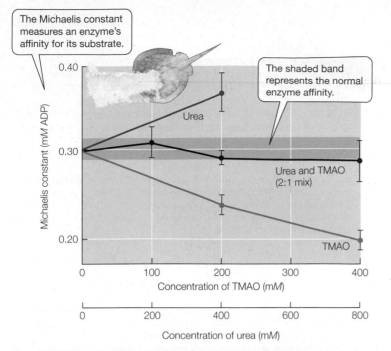

The Michaelis constant measures an enzyme's affinity for its substrate.

The shaded band represents the normal enzyme affinity.

FIGURE 27.10 Counteracting solutes in a stingray The enzyme pyruvate kinase was extracted from round stingrays (*Urolophis halleri*) for study. The affinity of the enzyme for one of its substrates (ADP, adenosine diphosphate) was measured. When the enzyme was independently exposed to an increasing concentration of urea (red line) or an increasing concentration of trimethylamine oxide (TMAO; green line), its affinity was strongly affected and driven out of the normal range (shaded area) in opposite directions. However, when the enzyme was exposed to a mix of urea and TMAO (black line), it exhibited approximately its normal affinity regardless of how high the concentrations of the two organic solutes were raised. Affinity decreases as the Michaelis constant increases (see Chapter 2). (After Yancey et al. 1982.)

when concentrated enough to contribute significantly to the osmotic pressure of a body fluid, has little or no effect on the structure and function of macromolecules in the body fluid. Whereas inorganic ions generally perturb enzymes or other macromolecules when their concentrations are substantially altered, some types of organic solutes are outstanding examples of compatible solutes. Particularly noteworthy compatible solutes include: (1) certain of the free standard amino acids,[17] (2) betaines, which are fully methylated forms of amino acids, (3) taurine, and (4) glycerol.

The concept of *counteracting solutes* is also significant for understanding the roles of solutes in water–salt physiology. Unlike compatible solutes, individual counteracting solutes may have strong effects on macromolecules. However, the effects of one counteracting solute are "opposite" to those of others, meaning that the solutes can offset each other's effects. **Counteracting solutes** (also called *compensatory solutes* or *chemical chaperones*) are osmolytes that act *in teams* of two or more to modify the osmotic pressures of body fluids without greatly perturbing macromolecules; as teams, they are relatively innocuous because the individual solutes of a team have mutually offsetting effects on the macromolecules.

The defining example of the concept of compatible solutes is provided by the interplay of *urea* and *methylamines* in the body fluids

[17] Among the standard amino acids (those used in protein synthesis), glycine, arginine, proline, and serine—when present as free amino acids—have particularly small effects on macromolecules and thus stand out as superior compatible solutes.

of sharks, skates, rays, and a few other groups of marine fish. Urea tends to have strong destabilizing and inhibiting effects on enzymes and other macromolecules. Conversely, certain methylamine compounds—such as trimethylamine oxide (TMAO), glycine betaine, and sarcosine—tend to stabilize and activate enzymes, and thus can counteract the effects of urea. In animals that employ urea as an osmolyte, one or more methylamines are usually also present in quantities that more or less exactly "titrate away" the effects of the urea (**FIGURE 27.10**). This is the case in the sharks, skates, and rays, which we discuss fully in Chapter 28.

STUDY QUESTIONS

1. In your own words, explain why an animal with excessively high blood osmotic pressure must be able to produce urine hyperosmotic to its blood plasma if its kidneys are to help correct the problem. In your answer, explain why neither urine that is isosmotic to the blood plasma nor urine that is hyposmotic to the plasma would help.

2. Based on the physical laws of evaporation, explain why the high body temperatures of mammals and birds make them prone to having higher rates of evaporative water loss than poikilothermic terrestrial vertebrates with lower body temperatures.

3. People who live and work in heated homes and office buildings in places with cold winters are well aware that the air inside becomes very dry during the winter season. Explain why the air inside a heated building is often low in humidity when the air outside is cold. Keep in mind that outside air is circulated inside by incidental or forced flow. (Hint: Consult Table 27.2.)

4. When animals oxidize stored fat, they produce metabolic water. Even though the production of metabolic water from stored fat follows principles of chemical stoichiometry, the *net* gain of water that animals realize from the oxidation of stored fat depends on the humidity of the atmosphere. Explain why. Does the net gain increase or decrease as the atmosphere becomes more humid?

5. When blue crabs living in full-strength seawater swell during molting, they take on the H_2O that bloats their bodies from the seawater in which they live. They obtain some of the H_2O by drinking. In addition, data show that the activity of Na^+–K^+-ATPase in their gills increases as they start to swell during molting. This rise in ATPase activity suggests that the gills increase active transport of ions from the surrounding seawater into the blood of the crabs. How could this process help account for uptake of H_2O?

6. Some species of animals gain physiological advantages by exploiting the specialized physiological capabilities of other species. Explain how marine mammals that prey on teleost fish rather than invertebrates illustrate this principle. Also explain how carnivorous mammals that prey on herbivorous mammals in deserts illustrate the same principle.

7. The kidney function of freshwater crabs illustrates that volume regulation can occur independently of osmotic regulation, and the kidney function of marine teleost fish illustrates that ionic regulation can occur independently of osmotic regulation. Explain both points.

8. People dying of dehydration because of lack of freshwater to drink inevitably ponder drinking their own urine. Analyzing the matter by use of the principles developed in this chapter, is a dehydrated person likely to gain H_2O by drinking his or her urine?

9. In many parts of the world, if one goes out at daybreak in autumn, it is a common sight to see a layer of fog formed just above the surface of the water of ponds and lakes. A fog consists of minute droplets of liquid water suspended in the atmosphere. When a layer of fog forms above a pond, the pond itself is the source of the water in the atmospheric water droplets. Inevitably, the temperature of the water in a pond is higher than the temperature of the air on mornings when fog layers form. Using the quantitative information in Table 27.2, explain why these fog layers form.

10. How are compatible intracellular solutes employed in cell-volume regulation, and why do they have advantages as intracellular osmotic effectors?

11. Consider a set of related terrestrial animals of various body sizes. Body surface area (SA) is an allometric function of body weight (W): $SA \propto W^{0.67}$. From Chapter 7, the metabolic rate (MR) is also an allometric function of body weight; at rest, for example, a likely relation would be $MR \propto W^{0.7}$. Predict the relation between rate of evaporative water loss (EWL) and body size, *taking account of both integumentary and respiratory EWL*. Explain your logic and assumptions.

Go to **sites.sinauer.com/animalphys4e**
for box extensions, quizzes, flashcards,
and other resources.

REFERENCES

Bennion, B. J., and V. Daggett. 2004. Counteraction of urea-induced protein denaturation by trimethylamine N-oxide: A chemical chaperone at atomic resolution. *Proc. Natl. Acad. Sci. U.S.A.* 101: 6433–6438.

Bradley, T. J. 2009. *Animal Osmoregulation.* Oxford University Press, Oxford, UK. A short textbook on the water–salt relations of animals.

Campbell, G. S., and J. M. Norman. 1988. *An Introduction to Environmental Biophysics*, 2nd ed. Springer-Verlag, New York. This book discusses all aspects of the biophysics of water in terrestrial systems.

Frank, C. L. 1988. Diet selection by a heteromyid rodent: Role of net metabolic water production. *Ecology* 69: 1943–1951. Do desert animals select foods based on the promise of the foods to provide a net yield of metabolic water? This interesting study at the interface of physiology and behavior addresses this question.

Gilles, R., and E. Delpire. 1997. Variations in salinity, osmolarity, and water availability: Vertebrates and invertebrates. In W. H. Dantzler (ed.), *Comparative Physiology*, vol. 2 (Handbook of Physiology [Bethesda, MD], section 13), pp. 1523–1586. Oxford University Press, New York. A detailed but well-integrated review that includes advanced information on most of the topics covered in this chapter.

Hochachka, P. W., and G. N. Somero. 2002. *Biochemical Adaptation*, 2nd ed. Oxford University Press, New York. The single best modern discussion of the biochemistry of water and solutes in the animal kingdom. Essential reading for anyone interested in the evolution of cell-volume regulation and the aspects of physiology related to such regulation.

Mangum, C. 1992. Physiological aspects of molting in the blue crab *Callinectes sapidus*. *Am. Zool.* 32: 459–469.

Somero, G. N. 1992. Adapting to water stress: Convergence on common solutions. In G. N. Somero, C. B. Osmond, and C. L. Bolis (eds.), *Water and Life*, pp. 3–18. Springer-Verlag, New York. Although becoming dated, this paper is still noteworthy as a compact, easy-to-read overview of the challenges of cell-volume stress and the responses of cells to such stress, emphasizing compatible solutes.

Taylor, J. R. A., and W. M. Kier. 2003. Switching skeletons: Hydrostatic support in molting crabs. *Science* 301: 209–210.

Yancey, P. H. 2005. Organic osmolytes as compatible, metabolic and counteracting cytoprotectants in high osmolarity and other stresses. *J. Exp. Biol.* 208: 2819–2830. An up-to-date review of organic osmolytes, emphasizing their unexpected complexity of action and the need to move beyond labels that oversimplify their roles.

See also **Additional References** *and* **Figure and Table Citations**.

Water and Salt Physiology of Animals in Their Environments

The body fluids of marine teleost (bony) fish—such as these coral reef fish—are far more dilute than the seawater in which they are swimming. Such fish have blood osmotic pressures of about 300–500 milliosmolar (m*Osm*), whereas seawater has an osmotic pressure of approximately 1000 m*Osm*. From the viewpoint of fluid concentration, marine teleost fish are packets of low-salinity fluids cruising about within a high-salinity environment!

Because the body fluids of a marine teleost fish are not at equilibrium with the seawater surrounding the fish, passive processes occur that tend to alter the composition of the body fluids. Water tends to move out of the body fluids into the surrounding seawater by osmosis because the body fluids have a lower osmotic pressure than seawater. Conversely, several inorganic ions diffuse inward from the concentrated seawater. Both the outward osmosis of water and the inward diffusion of ions tend to concentrate the fish's body fluids. Recognizing these processes, we see that marine teleost fish must steadily expend energy to maintain their body fluids out of equilibrium with the seawater in which they swim.

When we reflect on the questions that are raised by the body-fluid composition of marine teleost fish, we quickly recognize that the questions of *mechanism* and *origin* stressed in Chapter 1

These ocean fish (powder blue surgeonfish, *Acanthurus leucosternum*) expend energy to keep their body fluids more dilute than seawater Major questions raised are why they do so and what mechanisms they employ.

are both important. *How* do the fish keep their body fluids more dilute than seawater, and *why* do they do so? These same two questions—which are the central questions of mechanistic physiology and evolutionary physiology—arise in the study of the water–salt physiology of all animals.

In terms of their water–salt physiology, animals have been versatile in adapting to an astounding range of environments on Earth—not just seawater and ordinary freshwater, but also salt lakes far more concentrated than seawater, glacial ponds almost as dilute as distilled water, estuaries with highly variable salinity, and terrestrial environments ranging from rainforests to extreme deserts. Each type of habitat poses distinct challenges, and animals of diverse types that live in a particular habitat often have converged on similar mechanisms for meeting the challenges. For these reasons, the detailed study of animal water–salt physiology is logically organized around habitats. We take that approach in this chapter. We start by focusing on freshwater habitats, in part because we have already emphasized them in Chapter 5, and in part because the study of freshwater fish will help set the stage for a better understanding of ocean fish.

Animals in Freshwater

The animals living today in freshwater are descended from ocean-living ancestors: The major animal phyla originated in the oceans and later invaded all other habitats. Seawater was probably somewhat different in its total salinity and salt composition in the early eras of animal evolution than it is today. Nonetheless, when the animal phyla invaded freshwater from the oceans, there can be no doubt that they encountered a drastic reduction in the concentration of their surroundings. The osmotic pressure of freshwater is typically less than 1% as high as that of seawater today, and the major ions in freshwater are very dilute compared with their concentrations in seawater (see Table 27.1).

All freshwater animals regulate their blood[1] osmotic pressures at levels hyperosmotic to freshwater and are therefore classified as

hyperosmotic regulators. As **TABLE 28.1** shows, the blood osmotic pressures of various types of freshwater animals span an order of magnitude, but even freshwater mussels, which are among the most dilute animals on Earth, have blood that is substantially more concentrated than freshwater; body fluids as dilute as freshwater seem to be incompatible with life. The solutes in the blood plasma of freshwater animals are mainly inorganic ions; Na^+ and Cl^- dominate. As Table 28.1 shows, each of the individual inorganic ions in the blood plasma of freshwater animals is—in almost all cases—substantially more concentrated in the blood than in freshwater.

Passive water and ion exchanges: Freshwater animals tend to gain water by osmosis and lose major ions by diffusion

Being hyperosmotic to their surroundings, freshwater animals tend to gain water continuously by osmosis, and this water gain tends to dilute their body fluids. The relatively high concentrations of ions in their blood suggest that the net diffusion of ions tends to be from their blood into the ambient water. The analysis of ion diffusion is actually more complex, however, because—as discussed in Chapter 5—ion diffusion depends on electrical gradients as well as concentration gradients. When all the complexity is taken into account, nevertheless (see pages 110–111), the direction of diffusion of the major ions—such as Na^+ and Cl^-—is as the concentration gradients suggest: from the blood into the environmental water. This loss of major ions by diffusion tends, like the osmotic water gain, to dilute the body fluids of a freshwater animal as shown at the top in **FIGURE 28.1**.

In a broad sense, we expect a freshwater animal's energy costs for osmotic and ionic regulation to depend directly on the animal's rates of passive water gain and passive ion loss. The more rapidly water is taken up by osmosis, and the more rapidly ions are lost by diffusion, the more rapidly an animal will need to expend energy to counteract these processes so as to maintain a normal blood composition. Three factors determine the rates of passive exchange of water and ions: (1) the magnitudes of the osmotic and ionic gradients between the blood and ambient water, (2) the permeability of an animal's outer body covering to water and ions, and (3) the surface area across which exchange is occurring. The first two of these factors deserve further discussion.

---

[1] In this book, we always use the term *blood* to refer to the fluid that is circulated within the circulatory system, although some authors use the term *hemolymph* to refer to that fluid in invertebrates that have open circulatory systems (see page 695).

TABLE 28.1 The composition of blood plasma in some freshwater animals
The worldwide average composition of river water is shown for comparison.

Animal	Osmotic pressure (milliosmole per kg of H_2O)	Ion concentrations (millimole per liter)					
		Na^+	K^+	Ca^{2+}	Mg^{2+}	Cl^-	HCO_3^-
Freshwater mussel (*Anodonta cygnaea*)	44	16	0.5	6	0.2	12	12
Snail (*Viviparus viviparus*)	76	34	1.2	5.7	< 0.5	31	11
Crayfish (*Astacus fluviatilis*)	436	212	4.1	16	1.5	199	15
Mosquito larva (*Aedes aegypti*)	266	100	4.2	—	—	51	—
Brown trout (*Salmo trutta*)	326	161	5.3	6.3	0.9	119	—
Frog (*Rana esculenta*)[a]	237	109	2.6	2.1	1.3	78	27
River water	0.5–10	0.4	0.1	0.8	0.2	0.2	1.7

Source: Hill and Wyse 1989; river water data from Table 27.1.
[a] The frog has recently been renamed *Pelophylax esculentus*.

(A) Problems of passive water and salt exchange faced by freshwater animals

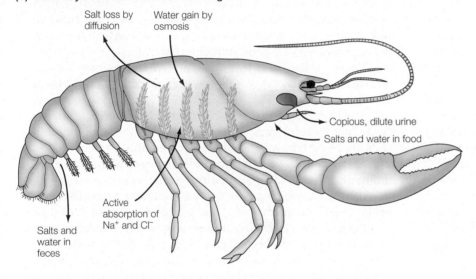

Salt loss by diffusion

Water gain by osmosis

Carapace

Antennal gland (green gland), responsible for urine formation

Blood
~ 400 mOsm
~ 200 mM Na⁺

Ambient water
~3 mOsm
~0.4 mM Na⁺

The gills, which are covered by the carapace and not visible externally, project into the branchial chambers, through which water is pumped.

A freshwater animal tends to gain water and lose salts, especially across its gills.

(B) Summary of all water and salt exchanges

Salt loss by diffusion

Water gain by osmosis

Copious, dilute urine

Salts and water in food

Active absorption of Na⁺ and Cl⁻

Salts and water in feces

FIGURE 28.1 Water–salt relations in a freshwater animal   (A) A freshwater animal, such as a crayfish, faces challenges because of passive water and salt exchange. The numbers are generalized, approximate values for the osmotic pressure and Na⁺ concentration found in the blood of a crayfish and the ambient water. Values for a specific species of crayfish under specific study conditions are given in Table 28.1. (B) A summary of all the major processes of water and salt exchange, including the energy-requiring processes the animal uses to maintain water–salt balance. The antennal glands, or green glands, which function as the kidneys of a crayfish, open at the bases of the second antennae.

OSMOTIC AND IONIC GRADIENTS   Most types of freshwater animals have *far* less concentrated body fluids than their marine relatives. The decapod crustaceans (e.g., crayfish, crabs, and lobsters) illustrate this general pattern. Although most marine decapods are essentially isosmotic to seawater (about 1000 mOsm), most freshwater decapods have blood osmotic pressures of 500 mOsm or less (e.g., ~440 mOsm in the crayfish in Table 28.1). Similarly, although marine molluscs are approximately isosmotic to seawater, freshwater molluscs have far lower blood osmotic pressures (e.g., ~40–80 mOsm in those in Table 28.1). The lower blood concentrations seen in freshwater animals result in smaller osmotic and ionic gradients between their blood and the freshwater environment than would otherwise be the case. For example, if decapod crustaceans and molluscs had retained their ancient blood concentrations when they invaded freshwater, the osmotic difference between their blood and freshwater would be almost 1000 mOsm. The actual osmotic difference between the blood and the surrounding water in freshwater decapod crustaceans and molluscs is far lower be

cause their blood is less concentrated than that of their marine progenitors. In the crayfish in Table 28.1, for example, the osmotic difference between the blood and the surrounding water is about 440 mOsm.

The evolution of more-dilute blood when animals invaded freshwater was probably an adaptation to reduce the energy costs of living in freshwater. More-dilute blood means smaller osmotic and ionic gradients between the blood and freshwater, and thus lower rates of water gain and ion loss by osmosis and diffusion.

PERMEABILITIES   The permeability of the integument[2] of a freshwater animal to water and ions is in general relatively low. Freshwater crayfish, for example, are no more than 10% as permeable to water, Na⁺, and Cl⁻ as marine decapod crustaceans of similar body size. The low permeabilities evolved by freshwater animals are important in reducing their rates of passive water and ion exchange and thus in reducing their energy costs of maintaining a normal blood composition. For a freshwater animal (or any other animal that maintains a difference in composition between its blood and the ambient water), a low integumentary permeability is analogous to an insulatory pelage in an Arctic mammal; the low permeability slows the processes that tend to bring the blood and ambient water to equilibrium, just as pelage insulation slows heat losses that tend to cool an Arctic mammal to ambient temperature.

If freshwater animals did not need to breathe, they might cover themselves entirely in an integument of extremely low permeability to water and ions. However, they do need to breathe, and there seems to be no way to make gills that are both highly permeable to O_2 and poorly permeable to H_2O and inorganic ions. Thus, just as the gills of freshwater animals provide a "window" for O_2 to enter the body, they provide a window for water to enter by osmosis and for ions to leave by diffusion. In fact, the very attributes of gills that are virtues

---

[2] *Integument* is a general term for the outer body covering. For example, the integument of a vertebrate is its skin, and the integument of an arthropod is its exoskeleton (cuticle) or shell.

TABLE 28.2 Rates of urine production—and osmotic and Na⁺ U/P ratios—in some freshwater animals

Animal	Rate of urine production (mL/100 g body wt·day)	Osmotic U/P ratio[a]	Na⁺ U/P ratio[b]
Snail (*Viviparus viviparus*)	36–131	0.20	0.28
Crayfish (*Astacus fluviatilis*)	8	0.10	0.006–0.06
Mosquito larva (*Aedes aegypti*)	≤20	0.12	0.05
Frog (*Rana clamitans*)	32	—	—
Clawed toad (*Xenopus laevis*)	58	0.16	0.10
Goldfish (*Carassius auratus*)	33	0.14	0.10

Source: Hill and Wyse 1989.

[a]The osmotic U/P ratio is the osmotic pressure of the urine divided by the osmotic pressure of the blood plasma.

[b]The Na⁺ U/P ratio is the urine Na⁺ concentration divided by the plasma Na⁺ concentration.

for O_2 uptake—high permeability and large surface area—are negatives for water–salt balance. A common pattern in freshwater animals is for little osmosis and diffusion to occur across the general integument—because the integument is poorly permeable to water and ions—and for most osmosis and diffusion to occur across the gills (plus possibly a few other localized body surfaces[3]).

The importance of the gills as windows for passive water and ion exchange has an interesting and significant implication: Differences in whole-body permeability to water and salts among related freshwater animals are sometimes secondary effects of differences in their metabolic intensities and demands for O_2. Species with high O_2 demands often have gill systems that are particularly well suited to rapid inward rates of O_2 diffusion. As a corollary, their gill systems also permit particularly rapid rates of water uptake by osmosis and ion loss by diffusion. In these cases, high rates of water–salt exchange are consequences of the evolution of high metabolic intensity.

Most types of freshwater animals share similar regulatory mechanisms

Most types of freshwater animals share a fundamentally similar suite of mechanisms for osmotic–ionic regulation. This suite of mechanisms is found in such phylogenetically diverse groups as freshwater teleost fish, lampreys, frogs, toads, soft-shelled turtles, freshwater mussels, crayfish, earthworms, leeches, and mosquito larvae.

URINE As we have seen, freshwater animals are faced with a continuous influx of excess water by osmosis. They void this excess water by making a copious (abundant) urine. A goldfish or frog, for example, might excrete urine equivalent to one-third of its body weight per day (**TABLE 28.2**). Because urine production balances osmotic water gain, the rate of urinary water excretion provides

a measure of the rate of osmotic water influx. The daily osmotic water influx of a goldfish or frog is therefore equal to one-third of its body weight![4]

The urine of freshwater animals, in addition to being produced in abundance, is typically markedly hyposmotic to their blood plasma and contains much lower concentrations of Na⁺ and Cl⁻ than the plasma. That is, the U/P ratios (urine:plasma ratios) for osmotic pressure, Na⁺, and Cl⁻ are far less than 1 in these animals (see Table 28.2). Recall from Figure 27.7 that when the osmotic U/P ratio is less than 1, urine production tends to raise the plasma osmotic pressure. Similarly, when the U/P ratio for an ion is less than 1, urine production tends to raise the plasma concentration of that ion. Typically, therefore, the kidneys of a freshwater animal not only solve the animal's volume-regulation problem by voiding the animal's excess volume of water, but also aid osmotic and ionic regulation by helping to maintain a high osmotic pressure and high ion concentrations in the blood.

Whereas the urine of freshwater animals is generally copious and dilute, an important concept to keep in mind is that kidneys are *regulatory* organs: They characteristically *adjust their function in ways that help to maintain stability of volume and composition in the body fluids*. Thus the exact volume and composition of the urine vary with circumstances. For example, if a freshwater animal experiences an increase in the rate at which it takes in water by osmosis, its kidneys ordinarily increase their rate of urine production.

Although freshwater animals typically limit the concentrations of Na⁺ and Cl⁻ in their urine to low levels, some loss of these ions in the urine is inevitable. This urinary loss of ions can pose a threat to the integrity of the body fluids when Na⁺ and Cl⁻ are in short supply. The rate of loss of ions in the urine depends in part on the rate of urine production, and therefore on the rate of osmotic water flux into an animal. Any factor that increases an animal's rate of osmotic water influx tends to increase the animal's rate of ion loss. We see, therefore, that *volume regulation and ionic regulation are basically at conflict with each other in freshwater animals*.

ACTIVE ION UPTAKE IN FRESHWATER ANIMALS IN GENERAL An important way that freshwater animals replace lost Na⁺ and Cl⁻ is that they actively transport both ions into their blood directly from the pond or river water in which they live. We have just seen that freshwater animals lose ions in their urine. Earlier we saw that they also lose ions by direct outward diffusion across their permeable body surfaces. The ions lost in these two ways need to be replaced. Freshwater animals do this by taking up Na⁺ and Cl⁻ by active transport from their ambient water. Other ions may also be transported inward in this way. Freshwater fish, for example, take up Ca^{2+} from the ambient water by active transport.

---

[3] The membranes of the buccal and opercular cavities in fish, for example, are relatively permeable compared with most body surfaces and are important sites of passive water–salt exchange. Little water or ion exchange occurs across an adult fish's outer skin.

[4] If a 70-kg person had a similar weight-specific rate of water uptake, he or she would gain 23 L (6 gallons) of water per day and thus would need to excrete 23 L of urine per day.

The capacities of most freshwater animals for active uptake of Na^+ and Cl^- are remarkable. For example, some crayfish, fish, and frogs—which have Na^+ and Cl^- concentrations of 100–200 mM in their blood plasma—can actively take up Na^+ and Cl^- in net fashion from ambient waters as dilute as 0.01 mM (four orders of magnitude more dilute than their blood)! The site of active ion uptake is usually the gills or the general integument. In teleost (bony) fish and decapod crustaceans (e.g., crayfish), the site of uptake is the gill epithelium. In frogs, active ion uptake occurs across the gills when the animals are tadpoles but across the skin when they are adults.[5] Active ion uptake also occurs across the general integument in leeches and aquatic oligochaete worms.

The cellular-molecular mechanisms of active Na^+ and Cl^- uptake by freshwater animals are featured as one of the focal examples in Chapter 5. Although details may vary, the following key points are believed to apply to all or most groups of freshwater animals:

- The active uptake of ions from the ambient water requires ATP. Thus active ion uptake places demands on an animal's energy resources.

- The mechanisms that pump Na^+ and Cl^- from the ambient water into the blood are typically different and independent from each other.

- The Cl^- pump typically exchanges bicarbonate ions (HCO_3^-) for Cl^- ions, in this way remaining electroneutral (**FIGURE 28.2**).

- The Na^+ pump typically exchanges protons (H^+) for Na^+ ions (or possibly exchanges ammonium ions, NH_4^+, for Na^+ in some groups of animals), thereby remaining electroneutral (see Figure 28.2).

- The HCO_3^- and H^+ pumped from the blood into the ambient water by the Cl^- and Na^+ pumps are produced by aerobic catabolism, being formed by the reaction of metabolically produced CO_2 with H_2O (see Figure 28.2). Thus the Na^+ and Cl^- pumps participate in removal of metabolic wastes.

- Because HCO_3^- and H^+ are principal players in acid–base regulation (see page 664), the Na^+ and Cl^- pumps sometimes play critical roles in the acid–base physiology of freshwater animals.

ACTIVE ION UPTAKE IN FRESHWATER FISH: THE GILLS AS ION-REGULATORY ORGANS The gills of teleost fish—although often discussed simply as breathing organs—in fact carry out two major functions that serve homeostasis: In both freshwater and seawater, the gills function both as *ion-regulatory organs* and as *gas-exchange organs*. The gills are the principal sites where Na^+ and Cl^- are taken up by active transport from freshwater, as we have already said. During the early development of freshwater fish, the gills—despite being thought of usually as breathing organs—assume their ion-uptake function first. To be more specific, when larvae first hatch out of the egg, both functions are carried out by the general integument, according to studies of rainbow trout

[5] Recent research has revealed that an important, deadly fungal pathogen of amphibians, *Batrachochytrium dendrobatidis*, severely disrupts ion uptake across the skin, and this may be the chief way the fungus kills.

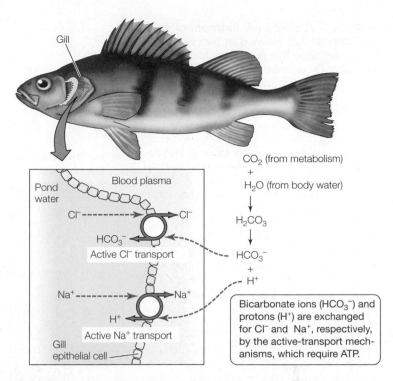

FIGURE 28.2 Ion exchanges mediated by active Na^+ and Cl^- transport in the gill epithelium of freshwater teleost fish The mechanisms of active transport exist within single epithelial cells. The view here is a whole-epithelium view and therefore, as discussed in Chapter 5 (see Figure 5.14), does not specify the cell-membrane mechanisms involved. The cell-membrane mechanisms are discussed in Box 5.2.

(*Oncorhynchus mykiss*) living in freshwater. Responsibility for Na^+ uptake from the ambient water shifts to the gills about 15–16 days after hatching, whereas responsibility for O_2 uptake doesn't shift to the gills until 23–28 days.

The two images of a single gill filament in **FIGURE 28.3** help emphasize the two major functions of the gills in adult teleosts. Figure 28.3A shows the microscopic structure of the filament. It consists of a thin, principal part—shaped somewhat like the blade of a feather—that bears many thin folds, the secondary lamellae, on its upper and lower surfaces. Blood flows through all these parts. The secondary lamellae greatly increase the surface area across which O_2 can diffuse inward from the ambient water into the blood (see Figure 23.11). Figure 28.3B shows the same filament visualized in a way that reveals the presence and location of the gill epithelial proteins that are instrumental in ion transport between the ambient water and blood; in this specific case (although not always), the epithelial cells expressing the ion-transport proteins are located in parts of the filament other than the secondary lamellae.

The method used to obtain Figure 28.3B—immunocytochemistry—is worth brief mention before we go further because it is the principal method used at present to study gill ion-transport functions throughout the animal kingdom. Two fluor-labeled antibodies were used: one antibody against Na^+–K^+-ATPase and another antibody against a cotransporter protein (not an ATPase) termed NKCC-1, which transports Na^+, K^+, and Cl^- ions in fixed ratios during each transport cycle (see page 117). When the filament

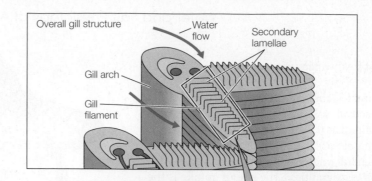

FIGURE 28.3 A single gill filament of an adult teleost fish viewed in two ways that emphasize the two principal gill functions: O₂ uptake and ion pumping The color drawing shows the overall gill structure. The gill filament seen in (A) and (B) is from a euryhaline species that occurs over wide ranges of salinity in salt marshes, the killifish *Fundulus heteroclitus*. This specimen was captured in nearly freshwater (salinity 4‰) in coastal Virginia. The images were acquired by confocal microscopy. (A) The structure of the filament. (B) The locations of two ion-transport proteins in a particular plane (optical section) of the filament: Na⁺–K⁺-ATPase (red) and NKCC-1 (green). Cell nuclei are labeled dark blue. Yellow represents places where both transport proteins occurred together (red + green = yellow). See Figure 23.11 for more on gill structure. (A and B courtesy of Aaron M. Florn.)

(A) The O₂ uptake function of a gill filament:
The expanded surface area for gas exchange

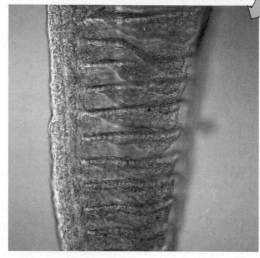

(B) The ion-transport function of a gill filament:
Ion-transport proteins labeled red and green

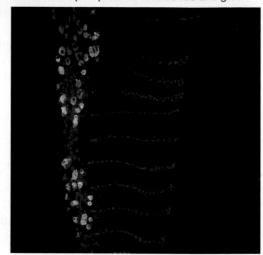

was exposed to the antibodies, they bound where Na⁺–K⁺-ATPase and NKCC-1 occurred, respectively. Then, when the filament was scanned with lasers that excited the fluors, the antibody against Na⁺–K⁺-ATPase glowed red, whereas that against NKCC-1 glowed green. Thus red shows where Na⁺–K⁺-ATPase was located, green shows where NKCC-1 was located, and yellow (the combination of red and green) shows where both gill epithelial proteins occurred in approximately the same location.

Let's now focus briefly on the cellular level of organization. The gill epithelium in fish consists principally of two types of cells: (1) *mitochondria-rich cells* (MRCs), also called *chloride cells*, and (2) *pavement cells*. The MRCs are considered to be the principal (although not exclusive) sites of active ion transport in the gills.[6] They are a central focus of research on ion transport in both freshwater and marine fish, and **BOX 28.1** discusses their properties and diversity. **FIGURE 28.4** shows a portion of the gill epithelium of a freshwater fish visualized by scanning electron microscopy (the magnification is far greater than in Figure 28.3). In this electron micrograph, we see a single freshwater-type MRC surrounded by pavement cells. Regarding the two major functions of the gills, O₂ uptake is believed to occur principally across pavement cells, which usually occupy more than 90% of the gill epithelium and are thinner than MRCs. As for ion uptake, physiologists would like to know precisely where Cl⁻ uptake and Na⁺ uptake occur, but the problem of deducing the detailed molecular mechanisms of ion transport and localizing them to particular cells is exceedingly challenging, and probably the localization of pumps to cell types will require considerably more research. Current models of Na⁺ and Cl⁻ uptake in freshwater

fish are discussed in Box 5.2, and modulation of gill function when migratory fish swim between freshwater and seawater is discussed later in this chapter (see page 759).

The number of MRCs in the gill epithelium of a fish living in freshwater is variable and under adaptive (partly hormonal) control. One condition that increases the number of MRCs is alkalosis: excess blood bicarbonate (HCO₃⁻) (see pages 644). During alkalosis,

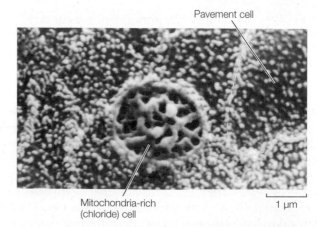

Pavement cell

Mitochondria-rich (chloride) cell

1 μm

FIGURE 28.4 A mitochondria-rich cell surrounded by pavement cells in the gill epithelium of a freshwater teleost fish The image is a scanning electron micrograph of the outer surface of the gill epithelium of a brown bullhead (*Ictalurus nebulosus*) that had been living in ordinary freshwater. One pavement cell is outlined (yellow dashed line) to show its limits. (See Box 28.1 for detail on mitochondria-rich cells.) (Photograph courtesy of Greg Goss and Steve Perry; from Goss et al. 1998.)

[6] Based on the immunocytochemical evidence, the labeled cells in Figure 28.3B are MRCs.

BOX 28.1 Fish Mitochondria-Rich Cells and Their Diversity

Mitochondria-rich cells (MRCs)—also called **mitochon-drion-rich cells**, **chloride cells**, or **ionocytes**—have two distinctive morphological features, both indicative of high metabolic activity: They contain large numbers of mito-chondria and an elaborate system of intracellular mem-branes (this system is continuous with the basolateral cell membrane). MRCs are typically also strikingly rich in Na^+–K^+-ATPase by comparison with most cells—another sign of high metabolic activity. Certain MRCs contain more than 100 million molecules of Na^+–K^+-ATPase per cell, one of the highest abundances known. MRCs are in general believed to be the principal sites of active ion transport in the gills of teleost fish.

A discovery of great significance—which has emerged with full clarity in just the last 20 years—is that there are *multiple types of MRCs*. For example, largely owing to the revolution in immunocytochemistry, researchers now recognize types of MRCs that differ biochemically: These MRCs can differ in their *quantities* of key ion-transport proteins and in their *molecular forms* of the proteins. MRCs with different molecular forms of Na^+–K^+-ATPase occur.

Based on the latest evidence, a fish capable of living in both freshwater and seawater typically has different types of MRCs in its gills—dubbed *freshwater and seawater types*—in the two environments. When the fish is transferred from one environment to the other, it switches types by re-placing or transforming its MRCs. Moreover, a fish may have two types of MRCs present in its gills in one environment. For example, rainbow trout (*Oncorhynchus mykiss*) living in freshwater have at least two types.

As yet researchers have not created a standardized nomenclature for the types of MRCs. Reading the research literature published prior to about 1995 can be confusing because, at the time, physiologists tended to think of MRCs as being relatively homogeneous and in general spoke of them as if they are all the same.

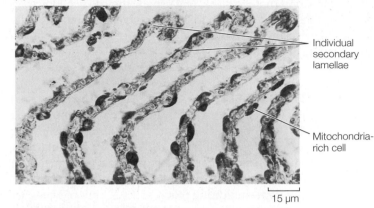

(A) Trout living in ordinary freshwater

Individual secondary lamellae

Mitochondria-rich cell

15 μm

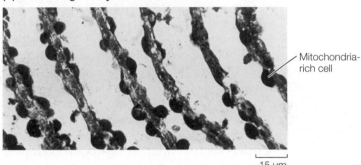

(B) Trout living in very "soft" freshwater

Mitochondria-rich cell

15 μm

FIGURE 28.5 Cellular acclimation to living in two types of water in the gill epithelium of freshwater fish   Seen here are sections of the secondary lamellae in the gills of rainbow trout (*Oncorhynchus mykiss*), viewed using light microscopy and stained to show mitochondria-rich cells. (A) Tissue section from a fish that had been living in ordinary freshwater with a Ca^{2+} concentration of 0.4 millimolar (mmol/L). (B) Tissue section from a fish that had been living for 2 weeks in very "soft" freshwater with a Ca^{2+} concen-tration of 0.05 mmol/L. (Photographs courtesy of Steve Perry; from Perry 1998.)

the MRCs, besides becoming more numerous, also modify their cell proteins—upregulating a key Cl^-/HCO_3^- countertransport protein that exports HCO_3^- from the body fluids in exchange for Cl^- (see Figure B in Box Extension 5.2).

A second, and fascinating, condition that leads to increased numbers of MRCs in freshwater fish is life in very "soft" water: water of exceptionally low Ca^{2+} concentration (**FIGURE 28.5**). Freshwater fish acquire most of their Ca^{2+} from the water in which they live, rather than from their food. The MRCs (or a subset of them) are the sites of active Ca^{2+} uptake. When fish are living in Ca^{2+}-poor waters, an increase in the number of MRCs is believed to help them acquire sufficient Ca^{2+}. However, increasing the number of MRCs can also interfere with uptake of O_2! Recent research on several species has shown that in fish living in very soft water, the replacement of pavement cells by MRCs in the

secondary lamellae can double the average diffusion distance between blood and water in the gills, because MRCs are thick (see Figure 28.5)—thicker than the pavement cells they replace. This doubling of the diffusion distance measurably interferes with O_2 uptake. *Thus freshwater fish exhibit a trade-off between their ability to take up Ca^{2+} and their ability to take up O_2; increasing one ability decreases the other.*

The concept of trade-offs is a major theme in modern ecology and evolutionary biology. The situation in freshwater fish just described is one of the physiological trade-offs that, considering all of animal physiology, is best understood at a cellular level.

FOOD AND DRINKING WATER   Freshwater animals of all types—fish, crayfish, and so forth—gain ions from their food, in addition to acquiring them by active uptake from the ambient wa-ter. The role of food in meeting ion needs is not well understood, although inputs of ions by active transport are generally thought to exceed those from food. In addition to eating food, freshwater animals also have the opportunity to drink water. But do they? Freshwater animals typically must produce urine at a very high rate just to deal with their passive osmotic water influxes. Thus one would not expect them to drink, and usually they do not. However, recent studies of teleost fish in freshwater have revealed that some species—especially when they are larvae—do drink sufficiently to raise their total water influx by 5%–50% more than their osmotic influx alone. The reasons for this drinking remain obscure.

QUANTITATIVE EXAMPLE AND COST ESTIMATES   The usual pattern of water–salt balance in freshwater animals is summarized in Figure 28.1B (showing a crayfish) and in Figure 28.8A (showing a fish). To review this pattern in words, let's look quantitatively at the gains and losses of water and Na^+ in a freshwater crayfish (*Astacus*). When fasting at 20°C, a 29-g crayfish excretes about 2.4 mL of urine per day—indicating that it gains 2.4 mL of water per day by osmosis, principally across its gills. The crayfish's urine is very dilute in Na^+ (1 m*M* Na^+). Therefore only about 2–3 micromoles (μmole) of Na^+ is lost per day in its urine. The animal's loss of Na^+ by direct diffusion into the surrounding water is much greater, approximately 240 μmole/day. The Na^+ lost by excretion and diffusion is replaced by active Na^+ uptake across the gills at a rate near 240 μmole/day.

According to a relatively recent study using modern methods, the energy cost of osmotic–ionic regulation in freshwater animals is about 3%–7% of their resting metabolic rate. The study—which focused on rainbow trout and mudpuppies (aquatic amphibians)—was based on measurements of ion-pumping rates in the gills, skin, and kidneys, plus information on the ATP demands of the ion pumps.

A few types of freshwater animals exhibit exceptional patterns of regulation

The typical pattern of osmotic–ionic regulation we've described is not observed in all freshwater animals. As usual, a look at exceptions can be as conceptually revealing as a look at the rule. Here we will look just at those exceptional freshwater animals that fail to produce a dilute urine.

The freshwater crabs are outstanding examples. These animals are unfamiliar to most North American readers because historically they have not occurred in North American waters.[7] Nonetheless, freshwater crabs are common on most other continents, usually at tropical or subtropical latitudes. Two species that have been studied are *Potamon niloticus*, an African crab, and *Eriocheir sinensis*, found in Asia and Europe. Freshwater crabs typically maintain high blood osmotic pressures in comparison with other freshwater animals: about 500–650 m*Osm* in *Potamon* and *Eriocheir*, for example (compare Table 28.1). As usual, the major blood solutes are Na^+ and Cl^-. The feature that makes freshwater crabs distinctive is that their urine is virtually isosmotic to their blood plasma. Its ionic composition is also very similar to that of blood plasma. Consequently the loss of Na^+ and Cl^- *per unit of volume* of urine in the freshwater crabs is very high by comparison with that in most freshwater animals. How do the crabs compensate? One well-known part of the answer is that the bodies of freshwater crabs exhibit extraordinarily low permeability to water. Because

of this low permeability, the crabs experience relatively low rates of osmotic water influx, and therefore the rates at which they must excrete their high-concentration urine are low. *Potamon* and *Eriocheir*, for instance, excrete water equivalent to only about 0.6%–3.6% of their body weight per day (compare Table 28.2). The unusually low urine output of the freshwater crabs—which results from their unusually low permeability to water—helps limit the rate at which they lose ions. The total quantities of Na^+ and Cl^- that the crabs lose—by both diffusion and urine excretion—although large by comparison with the quantities lost by freshwater crayfish, are "manageable," in the sense that the ions can be replaced by active uptake (across the gills) from the ambient water.

A few exceptional teleost fish living in freshwater also excrete urine that is nearly isosmotic to their blood plasma. One well-studied example is the toadfish *Opsanus tau*, a primarily marine fish that enters freshwater creeks. The freshwater crabs and the toadfish seem likely to be relatively recent immigrants to freshwater. Their short evolutionary history in freshwater helps explain why their kidneys have not evolved the ability to make dilute urine.

Why do most freshwater animals make dilute urine?

The freshwater crabs and the toadfish prove that production of dilute urine is not a necessity for life in freshwater; the production of relatively concentrated urine increases ion losses per unit of volume, but if total urinary losses can be restrained enough that ions can be replaced, existence in freshwater is possible. Why, then, have the vast majority of freshwater animals evolved the capacity to make urine that is dramatically hyposmotic to their blood plasma?

The answer is probably energy savings. To produce a dilute urine, the kidneys start with a fluid that is as concentrated as blood plasma and actively extract NaCl from it. Every Na^+ or Cl^- ion removed from the urine prior to excretion is an ion that does not have to be replaced by active uptake from the ambient water. In the urine—as it is formed in the kidneys—the concentrations of Na^+ and Cl^- are initially as high as in the blood plasma; only gradually—as the ions are reabsorbed—do urine ion concentrations fall to low levels. In contrast, the concentrations of Na^+ and Cl^- in the ambient water are *always very low*. Two lines of argument—one based on thermodynamic principles and the other on the molecular details of transport mechanisms—indicate that active uptake of Na^+ or Cl^- from a relatively concentrated source costs less energy *per ion* than uptake from a dilute source. Thus removing ions from urine prior to excretion is less costly than replacing the same ions from the ambient water.

SUMMARY
Animals in Freshwater

■ All freshwater animals are hyperosmotic to the water in which they live. They tend to gain water by osmosis and lose ions by diffusion, especially across their permeable gill membranes. These passive fluxes of water and ions tend to dilute their body fluids.

■ To void their excess of water, freshwater animals produce a copious urine.

---

[7] The freshwater crab *Eriocheir* has recently been introduced into some river systems along the West Coast of the United States and is now established there as an alien species.

- In nearly all freshwater animals, the urine is dilute compared with the blood plasma. The dilute condition of the urine helps to maintain not only the blood osmotic pressure but also blood concentrations of major ions at levels higher than those in the environment.

- To replace ions lost by direct diffusion into the environment and excretion in urine, freshwater animals take up Na+, Cl−, and some other ions by active transport. The gill epithelium is the principal site of active ion uptake in adult teleost fish and crayfish. Foods also help to replenish ions.

Animals in the Ocean

Animal life in the oceans is far more phylogenetically diverse than that in the other major habitats on Earth, probably in good part because animals originated in the oceans. Today, all phyla and most classes of animals have marine representatives.

Many phylogenetic groups of animals moved from the oceans to freshwater and the land over the course of evolutionary time. In turn, many freshwater and terrestrial groups reinvaded the oceans. Consequently, whereas some modern marine animals have a continuously marine ancestry, others trace their history to forms that occupied other major habitats.

Such a phylogenetic history has sometimes left major imprints on the water–salt physiology of modern marine animals. For example, although the octopus and the marine teleost fish seen swimming in seawater in **FIGURE 28.6** may look like they would have similar blood osmotic concentrations, their concentrations actually are very different, as we will see in the following sections. History is believed to be the explanation. Whereas octopuses probably trace a continuously marine ancestry, marine teleost fish are likely descended from freshwater ancestors.

Most marine invertebrates are isosmotic to seawater

Most marine invertebrates are isosmotic, or nearly so, to seawater. Included are the marine molluscs—exemplified by the octopus in Figure 28.6A—and such other marine animals as sponges, coelenterates, annelids, echinoderms, and most arthropods. For the most part, these animals are products of lines of evolution that never left the sea. They have always lived in seawater, and this probably explains why they have the simplest possible osmotic relation to seawater. The osmotic pressure of seawater is about 1000 m*Osm*, and the osmotic pressure of marine invertebrates' body fluids is about the same. Being essentially isosmotic to their environment, marine invertebrates do not tend to gain or lose water by osmosis to any great extent: They do not face problems of osmotic regulation.

The solutes in the blood plasma of marine invertebrates are mostly inorganic ions, and the ionic composition of their blood plasma tends to be grossly similar to that of seawater. Despite this similarity, as exemplified in **TABLE 28.3**, the ionic composition of the blood plasma seems universally to differ in detail from the ionic composition of seawater. A particular ion often proves to be relatively concentrated in some animal species but relatively dilute in others; for example, Mg^{2+} is relatively high in concentration in the blood plasma of the squid *Loligo* but low in that of the crab *Carcinus*. The adaptive significance of such differences in blood ionic composition is generally unknown.

Isosmotic marine invertebrates maintain the differences in ionic composition between their blood plasma and seawater by ionic regulatory processes. These animals are typically relatively permeable to ions (and water). Ions therefore tend to diffuse between their blood and seawater with ease, following their electrochemical gradients. One process these animals commonly use to maintain their blood ionic composition is active uptake of ions from seawater at the body surface or from ingested seawater

FIGURE 28.6 Two ocean animals with different blood osmotic pressures, an invertebrate with blood isosmotic to seawater and a teleost fish with blood dramatically hyposmotic to seawater (A) The octopus (*Octopus cyanea*) belongs to the group of marine molluscs known as cephalopods, which also includes squids and cuttlefish. (B) The teleost fish at the right is *Zebrasoma veliferum*. Both species shown here are found in Hawaii.

TABLE 28.3 The composition of the blood plasma or other extracellular body fluids in some marine invertebrates and hagfish

All these animals are isosmotic to seawater. The ion concentrations listed are for animals living in seawater of the composition specified in the last row of the table.

Animal and body fluid	Ion concentration (mmol/kg H₂O)					
	Na⁺	K⁺	Ca²⁺	Mg²⁺	Cl⁻	SO₄²⁻
Mussel (*Mytilus*), blood plasma	474	12.0	11.9	52.6	553	28.9
Squid (*Loligo*), blood plasma	456	22.2	10.6	55.4	578	8.1
Crab (*Carcinus*), blood plasma	531	12.3	13.3	19.5	557	16.5
Sea urchin (*Echinus*), coelomic fluid	474	10.1	10.6	53.5	557	28.7
Jellyfish (*Aurelia*), mesogleal fluid	474	10.7	10.0	53.0	580	15.8
Hagfish (*Myxine*), blood plasma	537	9.1	5.9	18.0	542	6.3
Seawater	478	10.1	10.5	54.5	558	28.8

Source: After Potts and Parry 1964.

in the gut. A second common process is kidney regulation of blood composition. In crustaceans, molluscs, and some other groups, although the excretory organs make a urine that is approximately isosmotic to the blood plasma, they alter the urine's ionic composition, thereby contributing to ionic regulation. For example, in most marine decapod crustaceans, the urine is richer in Mg^{2+} and SO_4^{2-} than the blood plasma (ionic U/P = 1.1–4.2 in several species), which helps keep plasma concentrations of these ions lower than seawater concentrations (see Figure 27.7).

Hagfish are the only vertebrates with blood inorganic ion concentrations that make them isosmotic to seawater

The hagfish, an exclusively marine group of jawless primitive vertebrates, resemble the great majority of marine invertebrates in two key respects: (1) Their blood is approximately isosmotic to seawater, and (2) their blood solutes are principally Na^+, Cl^-, and other inorganic ions (see Table 28.3). The ionic regulatory processes of hagfish are similar to those of osmoconforming marine invertebrates. Hagfish appear to be the only modern vertebrates that trace a continuously marine ancestry (see Box 28.2).

The marine teleost fish are markedly hyposmotic to seawater

As stressed at the start of this chapter, the marine teleost (bony) fish are **hyposmotic regulators**: Their blood osmotic pressures are far lower than the osmotic pressure of the seawater in which they swim. As in freshwater teleosts, Na^+, Cl^-, and other inorganic ions constitute most of the solutes in the blood plasma. One of the most intriguing questions about marine teleosts concerns the *origin*

of their hyposmotic state: *Why* is their blood plasma dramatically more dilute in ions and lower in osmotic pressure than seawater? Most biologists conclude, as discussed in **BOX 28.2**, that the dilute body fluids of marine teleosts are an evolutionary vestige: These fish are generally believed to be descended from ancient ancestors that lived in freshwater.

The blood osmotic pressures of marine teleost fish are typically 300–500 m*Osm*—higher than those of freshwater teleosts (about 250–350 m*Osm*), but not exceptionally so. Evidently, when the teleost fish invaded the oceans from freshwater in the course of their evolution, they evolved modest increases in the concentration of total blood solutes. This change served to reduce the difference in concentration between their blood plasma and the environmental water (the seawater) in their new habitat.

Despite such a change, *today's marine teleost fish—because of their profoundly dilute state relative to seawater—face a difference between their blood plasma and their environmental water that is far greater than that faced by freshwater teleosts*. In freshwater teleosts, blood osmotic pressure averages about 300 m*Osm* higher than the osmotic pressure of freshwater. In marine teleosts, however, the difference between the blood and environmental osmotic pressures is about 600 m*Osm*: approximately twice as great![8] This consideration in itself would tend to saddle the marine fish with a relatively high rate of osmotic water flux. Marine teleosts, however, are typically less permeable to water than freshwater teleosts, so in fact the osmotic fluxes experienced by the two groups are roughly similar in magnitude (for a given body size).

Of course, although the osmotic fluxes of marine and freshwater teleosts may be similar in magnitude, they are totally opposite in direction. Because a marine teleost is hyposmotic to seawater, water tends to *leave* its body osmotically, rather than entering as it does in a freshwater teleost. *For a hyposmotic animal, the ocean is a desiccating environment.*

The concentrations of Na^+, Cl^-, Mg^{2+}, SO_4^{2-}, and some other inorganic ions are far lower in the blood plasma of marine teleosts than in seawater, suggesting that marine teleost fish also face problems of inward diffusion of multiple ions. Moreover, the concentration gradients between the blood plasma and the environment for the two major plasma ions, Na^+ and Cl^-, are large by comparison with the (oppositely directed) gradients seen in freshwater teleosts. Nonetheless, actual rates of ion diffusion depend on *electrical gradients* and gill *permeability*, not just on ion *concentration gradients*. When all these factors are taken into account, inward Na^+ diffusion turns out not to be much of a problem for marine teleosts (or may not even occur in some species), because the gill epithelium is positively charged on the inside, repelling Na^+. In contrast, some other ions—most notably Cl^-—tend to diffuse into the blood plasma of marine teleosts from seawater at substantial rates, tending to concentrate the body fluids of the fish.

---

[8] Average blood osmotic pressure in marine teleosts is about 400 m*Osm*, and the osmotic pressure of seawater is about 1000 m*Osm*—a difference of 600 m*Osm*.

Where Were Vertebrates at Their Start?

The traditional hypothesis of most biologists is that vertebrates originated as jawless animals in the oceans about 500 million years ago. The lineage leading to modern hagfish (which are themselves jawless) then never left the oceans. Today's hagfish, in this view, are the one group of modern vertebrates or vertebrate-like animals (craniates) that have lived in the oceans throughout their evolutionary history. According to the traditional hypothesis, early jawless vertebrates entered freshwater, and jaws then originated in freshwater. Thus, according to the traditional hypothesis, all jawed vertebrates (plus modern lampreys) are descended from freshwater ancestors. Specifically, the jawed fish living in the oceans today arose through reinvasion of the oceans from freshwater.

The water–salt physiology of modern vertebrates provides one of the principal arguments for this traditional hypothesis.

The total blood-plasma salt concentrations of all jawed vertebrates alive today are monotonously similar, and the concentrations seen in vertebrates are more like those in freshwater invertebrates than in marine invertebrates. These patterns suggest that all the living jawed vertebrates had ancient ancestors that lived in freshwater and that the blood ion concentrations of vertebrates became relatively fixed at that time in evolution. According to this view, modern marine teleost fish have blood salt concentrations far lower than those in seawater because they are descended from freshwater ancestors.

You will notice the circularity of reasoning here. The features of the blood composition of modern animals are used as evidence for a freshwater origin of jawed vertebrates. The purported freshwater origin of jawed vertebrates is then used to explain the blood composition of modern animals. This unsettling state of affairs arises in part because the fossil record is too sketchy to provide firm independent confirmation or refutation of the freshwater-origins hypothesis for jawed vertebrates.

Some evolutionary biologists believe that the available circumstantial evidence supports a scenario different from the traditional one. One alternative view, for example, is that the earliest vertebrates were closely associated with the continental margins and experienced both marine and freshwater environments. All scenarios for the early evolution of vertebrates concur in postulating that the blood composition of modern marine jawed vertebrates is a consequence of early evolutionary experience with freshwater. No other interpretation seems plausible.

REPLACEMENT OF WATER LOSSES   Marine teleost fish lose water by osmosis and, to a lesser extent, by urine production. To replace the water they lose (and thereby volume regulate), these fish *drink seawater*. Although some drink an amount of seawater that is less than 1% of their body weight per day, others drink more than 50% per day, and the average is probably 10%–20% per day.

At first sight, drinking seawater seems to be a straightforward way to obtain water. Consider, however, that when seawater is first taken into the gut of a marine teleost, it is strongly *hyperosmotic* to the fish's blood plasma. Consequently, H_2O is predicted to travel by osmosis *out of the blood plasma into the ingested seawater in the gut*, not vice versa—and that is exactly what happens. Studies of several species show that as ingested seawater travels through the esophagus, stomach, and (in at least some instances) anterior intestine, not only do Na^+ and Cl^- diffuse into the blood across the gut wall, but also H_2O *enters the gut fluids* by osmosis. Gradually, therefore, the ingested seawater in the gut expands in volume and is diluted. Water uptake from the gut fluids eventually occurs, nonetheless. This is true because in later parts of the intestine, Na^+ and Cl^- are *actively transported out of the gut contents into the blood*. This ATP-requiring, active uptake of Na^+ and Cl^- into the blood creates conditions that favor the *osmotic uptake of water*. In the simplest cases to understand, the active uptake of the ions from the gut fluids renders the gut fluids hyposmotic to the blood. Often, however, a process called *near-isosmotic fluid transport* occurs, in which the gut fluids and blood plasma remain approximately isosmotic as water moves briskly by osmosis into the blood; in this case, highly localized osmotic gradients *within the epithelium* of the intestines are involved in translating ion uptake into water uptake (termed *local osmosis*). Recent evidence, discussed more on page 760, indicates that aquaporins in the intestinal epithelium are instrumental in facilitating water uptake from the gut in fish in seawater.

By the time ingested seawater is completely processed, about 50%–85% of the H_2O in the seawater is absorbed into the blood. However, a much greater proportion of the NaCl in the ingested seawater—often more than 97%—is absorbed. This is true because NaCl absorption is required to drive the absorption of H_2O. The influx of NaCl into the blood aggravates the problems of Na^+ and Cl^- regulation that the fish face. Accordingly, *in marine teleosts—as in freshwater teleosts—the process of volume regulation worsens the problems of ionic regulation.*[9]

Divalent ions[10] in ingested seawater are handled very differently from the monovalent ions, Na^+ and Cl^-. The gut epithelium is poorly permeable to the major divalent ions, Mg^{2+} and SO_4^{2-}. Consequently, although the divalent ions diffuse into the blood to a small extent as seawater passes through the gut, for the most part they remain in the gut and are expelled in the feces.

---

[9] Volume regulation and ionic regulation are inextricably linked in the marine and freshwater teleosts because the major solutes of the body fluids in both of these groups of fish are inorganic ions.

[10] *Divalent* ions have two charges per ion, either two positive charges or two negative charges.

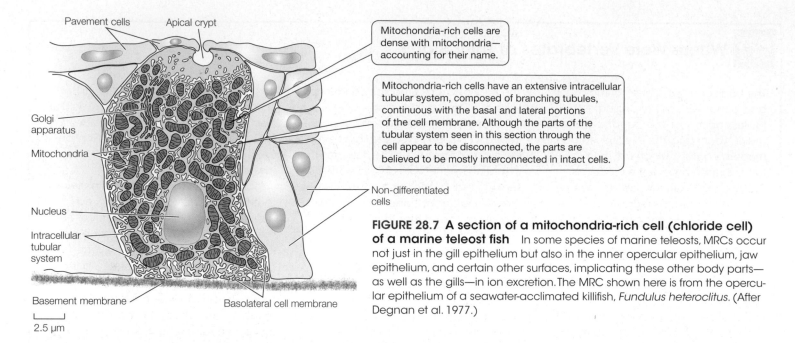

Pavement cells   Apical crypt

Mitochondria-rich cells are dense with mitochondria—accounting for their name.

Mitochondria-rich cells have an extensive intracellular tubular system, composed of branching tubules, continuous with the basal and lateral portions of the cell membrane. Although the parts of the tubular system seen in this section through the cell appear to be disconnected, the parts are believed to be mostly interconnected in intact cells.

Golgi apparatus

Mitochondria

Non-differentiated cells

Nucleus

Intracellular tubular system

Basement membrane

Basolateral cell membrane

2.5 μm

FIGURE 28.7 A section of a mitochondria-rich cell (chloride cell) of a marine teleost fish   In some species of marine teleosts, MRCs occur not just in the gill epithelium but also in the inner opercular epithelium, jaw epithelium, and certain other surfaces, implicating these other body parts—as well as the gills—in ion excretion. The MRC shown here is from the opercular epithelium of a seawater-acclimated killifish, *Fundulus heteroclitus*. (After Degnan et al. 1977.)

URINE   We now turn to the question of how marine teleosts eliminate the excess ions that enter their body fluids from the gut, or that diffuse into their body fluids from seawater across their gills or other external body surfaces. For the most part, excess *divalent* ions in the body fluids are removed by excretion in the urine, whereas excess *monovalent* ions are excreted by the gills.

The kidneys of marine teleosts typically excrete urine that is about as concentrated as they can possibly produce, approximately isosmotic to the blood plasma.[11] The fact that the osmotic pressure of the urine matches that of the plasma (osmotic U/P $\cong$ 1) means that the excretion of urine cannot help the fish with their osmotic regulatory problem (see Figure 27.7). However, the *ionic composition* of the urine differs dramatically from that of the plasma, and the kidneys are the principal organs that carry out ionic regulation of Mg^{2+}, SO_4^{2-}, and Ca^{2+}. Whereas U/P ratios for Na^+, Cl^-, and K^+ are below 1, those for Mg^{2+} and SO_4^{2-} are *far* greater than 1. The kidneys thereby void the major divalent ions preferentially in relation to water and keep plasma concentrations of those ions from increasing.

For every milliliter of water that is first ingested and absorbed and then excreted as urine, a marine teleost is left with an excess of solutes because, although the water enters its body hyperosmotic to its body fluids, the water leaves its body isosmotic to its body fluids. From the viewpoint of osmoregulation, therefore, production of urine by marine teleost fish is an outright liability, and we would expect the fish to limit their volume of urine to the minimum necessary for excretion of solutes that are not excreted by other routes. Nitrogenous wastes and the principal ions, Na^+ and Cl^-, are voided across the gills. Thus the role of the kidneys in marine teleosts is largely limited to excretion of divalent ions, and the rate of urine production can be low. The urine volumes of several species have been measured to be just 0.5%–3.5% of body weight per day (compare Table 28.2).

EXTRARENAL NaCl EXCRETION BY THE GILLS   The gills of an adult marine teleost assume primary responsibility for excreting excesses of the major ions, Na^+ and Cl^-, from the blood plasma

into the surrounding ocean. The excretion of Cl^- is active and is carried out by seawater-type mitochondria-rich cells (MRCs) in the gill epithelium (**FIGURE 28.7**). These cells are often called *chloride cells* in the study of marine teleosts because of their well-established excretion of Cl^-. In fish soon after hatching, as discussed in Chapter 4 (see Figure 4.6), the MRCs are principally found in the general integument, but soon the cells become localized to the gills. **BOX 28.3** outlines the mechanism these cells employ to pump Cl^-. Although the gill epithelium is believed always to transport Na^+ as well as Cl^- out of the blood into the seawater, the excretion of Na^+ occurs by mixed mechanisms; Na^+ excretion is probably active in about half the species that have been studied, but passive in the others (in which Na^+ *diffuses* outward, attracted by an outside-negative electrical gradient generated by active Cl^- excretion). The elimination of Cl^- and Na^+ by the gills of marine teleost fish provides our first example of **extrarenal salt excretion**: excretion of inorganic ions by structures other than the kidneys.

Present evidence indicates that excretion of NaCl by the gills in many teleosts is accomplished without concomitant excretion of water; the material excreted is purely ions. Thus, in addition to voiding NaCl from the blood (ionic regulation), the process produces a fluid that is essentially infinitely higher in osmotic pressure than the blood plasma. The process therefore tends to lower the osmotic pressure of the plasma (see Figure 27.7) and maintains the blood osmotic pressure at a level below the ambient osmotic pressure. *The gills are in fact the sites where osmotic regulation is principally accomplished.*

QUANTITATIVE EXAMPLE AND COST ESTIMATES   In **FIGURE 28.8**, the pattern of water–salt regulation in marine teleost fish is summarized (see Figure 28.8B) and contrasted with the pattern in freshwater teleost fish (see Figure 28.8A). Let's review the pattern in marine teleost fish by making use of quantitative data for one particular species that has been thoroughly studied, the southern flounder (*Paralichthys lethostigma*). An individual flounder that weighs about 1 kg loses water equivalent to about 7.9% of its body

[11] The kidneys of fish are incapable of producing urine that is hyperosmotic to blood plasma.

(A) Freshwater teleost

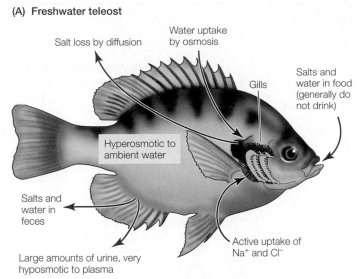

(B) Marine teleost

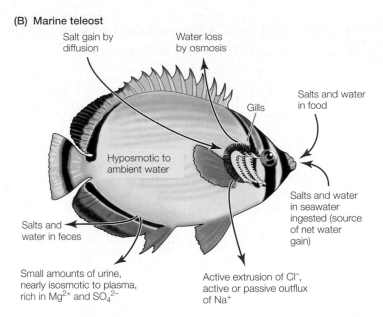

FIGURE 28.8 Contrasting water–salt relations in freshwater and marine teleost fish

weight per day because of osmosis from its body fluids into the surrounding seawater. To replace this water, and also to replace urinary water losses of 0.4% of body weight per day and fecal losses of 2.7% per day, the fish drinks seawater equivalent to 11% of its body weight per day. From the seawater it ingests, the fish absorbs 76% of the H_2O, but in doing so it absorbs much higher percentages of the Na^+ (99%) and Cl^- (96%). The flounder produces a scanty urine having a U/P ratio for Mg^{2+} of about 100 and a U/P ratio for SO_4^{2-} of 330; the urine removes excesses of both of these divalent ions from the body fluids. The gills of the flounder excrete virtually all of the excess monovalent ions, Na^+ and Cl^-.

The energy cost of Na^+ regulation, Cl^- regulation, and osmotic regulation in marine teleosts has been estimated in several species (including tunas) to be 8%–17% of the resting metabolic rates of the fish, based on measured ion-pumping rates and the known ATP costs of pumping. Rainbow trout—which can live in freshwater or seawater—are estimated to devote 8% of their resting energy use to Na^+, Cl^-, and osmotic regulation when living in seawater, but 3% when living in freshwater. The higher cost in seawater reflects the fact, earlier stressed, that for a teleost fish, the difference in concentration between the blood and the ambient water is about twice as great in seawater as in freshwater.

BOX 28.3 Epithelial NaCl Secretion in Gills, Salt Glands, and Rectal Glands

NaCl secretion by the gill epithelium of marine teleost fish is believed to occur by the mechanism shown in the figure (see box continuation, next page). The same model of the secretory mechanism is believed to apply to additional NaCl-secreting structures that we will soon discuss in this chapter: the cranial salt glands of marine birds and sea turtles, and the rectal salt glands of marine sharks, skates, and rays.

To understand the model, let's focus first on active Cl^- secretion by the mitochondria-rich cell (MRC) in the figure. The Cl^--transporting protein shown in the basolateral membrane of the cell, often called NKCC, is *not* an ATPase. Thus Cl^- transport is by *secondary* rather than primary active transport (see page 115). NKCC is an electroneutral cotransporter that moves one Na^+ ion, one K^+ ion, and two Cl^- ions into the cell across the

basolateral membrane on each transport cycle; in the jargon of the study of transporter proteins, it is a *Na–K–2Cl co-transporter*. The energy for Cl^- transport is supplied from ATP indirectly by the action of $Na^+–K^+$-ATPase (see page 117), also located in the basolateral membrane. The $Na^+–K^+$-ATPase uses ATP-bond energy to pump Na^+ out of the cell, thereby creating a strong electrochemical gradient favoring diffusion of Na^+ from the blood into the cell. Following its electrochemical gradient, Na^+ diffuses into the cell bound to NKCC, and this process brings Cl^- from the blood into the cell. The entry of Cl^- into the cell creates an electrochemical gradient favoring the diffusion of Cl^- out of the cell, and Cl^- exits the cell by way of Cl^- channel proteins in the apical cell membrane. In the case of the gill epithelium of marine teleosts, the specific type

of channel protein that is dominant is known as *CFTR* (*cystic fibrosis transmembrane conductance regulator*), and Cl^- leaving the cell enters the ambient water. Potassium (K^+) ions accumulated in the cell by the actions of $Na^+–K^+$-ATPase and NKCC simply diffuse back to the blood by way of K^+ channels.

The Cl^- and K^+ channels are gated channels in at least some cases. Secretion is controlled in part by regulatory mechanisms that open and close the channels. In the cells of avian salt glands, the Cl^- and K^+ channels are known to be Ca^{2+}-activated; thus secretion is under immediate control of a Ca^{2+}-based second-messenger system (see Figure 2.30).

Let's now turn to the question of how Na^+ is secreted across the epithelium. The transfer of Cl^- across the epithelium

(Continued)

BOX 28.3 Epithelial NaCl Secretion in Gills, Salt Glands, and Rectal Glands *(Continued)*

by the mechanism we have described is electrogenic, and as shown in the figure, it renders the apical side of the epithelium electrically negative relative to the blood. This difference in electrical potential across the epithelium can be large enough to create an electrochemical gradient that favors *diffusion* of Na⁺ from the blood outward across the epithelium (Na⁺ is repelled from the positively charged inner epithelial surface and attracted to the negatively charged outer surface). Current evidence indicates that when Na⁺ travels outward across the epithelium by diffusion, it follows paracellular (between-cell) pathways. In some cases, diffusion is believed to be the sole mechanism of Na⁺ secretion, but in other cases there is evidence for secretion of Na⁺ by active transport.

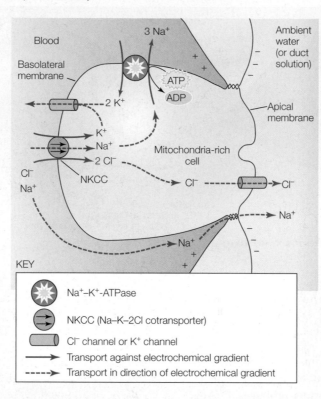

KEY

🟊 Na⁺–K⁺-ATPase

⬤ NKCC (Na–K–2Cl cotransporter)

▭ Cl⁻ channel or K⁺ channel

→ Transport against electrochemical gradient

- - → Transport in direction of electrochemical gradient

The probable mechanism of epithelial NaCl secretion The diagram shows a mitochondria-rich cell (chloride cell) flanked on either side of its apical surface by two other cells (which may or may not themselves be MRCs) in a secretory epithelium. In the gills of marine teleosts, NaCl is secreted directly into the ambient water; in the cranial salt glands of birds and sea turtles and the rectal salt glands of elasmobranch fish, NaCl is secreted into ducts, and the salty solution thus formed flows out of the animal by way of a duct system. NKCC, Na⁺–K⁺-ATPase, and the K⁺ channel are proteins in the basolateral cell membrane. The Cl⁻ channel is a protein in the apical membrane.

Some arthropods of saline waters are hyposmotic regulators

Quite a few arthropods that live in the ocean or more-saline waters, such as salt lakes, maintain their blood osmotic pressure at a level hyposmotic to the water in their environment. These animals include brine shrimp ("sea monkeys"), the insects that live (usually as larvae) in salty waters, and some marine crabs and shrimps. Biologists generally believe that the ancestors of most of these animals lived in more-dilute habitats, and their body fluids bear an imprint of that earlier time. Their mechanisms of hyposmotic regulation have been well studied in a few cases and usually parallel those of marine teleost fish.

Marine reptiles (including birds) and mammals are also hyposmotic regulators

The sea turtles, sea snakes, penguins, gulls, whales, seals, and other marine reptiles and mammals—like marine teleost fish—are markedly hyposmotic to seawater. All are descended from terrestrial ancestors, and their blood compositions are clearly carryovers from their ancestors. The blood osmotic pressures of all these marine vertebrates tend to be about 400 m*Osm*: just modestly higher than the values seen in modern-day terrestrial and freshwater vertebrates.

Because the marine nonavian reptiles, birds, and mammals are air breathers, they do not expose permeable respiratory membranes to seawater. Another advantage of their terrestrial heritage is that they have inherited integuments that were originally adapted to

limiting water losses in the dehydrating terrestrial environment, so they tend to exhibit low integumentary permeabilities.

These animals nonetheless confront problems of water loss and salt loading. They lose water, for example, by pulmonary evaporation during breathing; they also lose water to some extent across their skin, not only when they are immersed in seawater, but also when they are exposed to the air. These animals often gain excess salts from the foods they eat; for example, when they prey on marine plants or invertebrates that are isosmotic to seawater, they ingest body fluids that have far higher salt concentrations than their own. In addition, they probably often take in quantities of seawater with the foods they eat, although, for the most part, they are thought not to drink seawater.

MARINE REPTILES (INCLUDING BIRDS) The marine turtles and lizards—in common with other nonavian reptiles—are generally not able to produce urine that is more concentrated in total solutes than their blood plasma. The urine-concentrating capabilities of marine birds are incompletely understood, but for most species the maximum urine concentration appears to be isosmotic to the blood plasma or only modestly hyperosmotic (U/P ≤ 2). Because of these meager concentrating abilities, the kidneys of marine turtles, lizards, and birds are in general not able—by themselves—to maintain the blood of the animals hyposmotic to seawater (see Figure 27.7).

Obviously, then, these animals require a way to excrete salts in a more concentrated state than their kidneys can provide. Virtually

(A) Herring gull

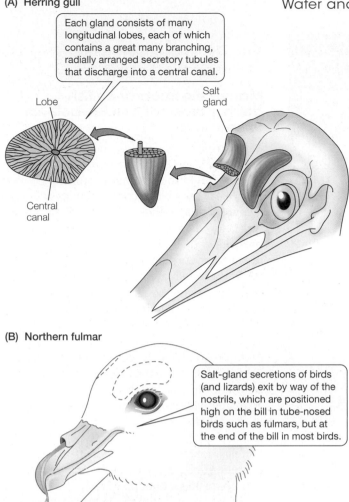

Each gland consists of many longitudinal lobes, each of which contains a great many branching, radially arranged secretory tubules that discharge into a central canal.

Lobe

Salt gland

Central canal

(B) Northern fulmar

Salt-gland secretions of birds (and lizards) exit by way of the nostrils, which are positioned high on the bill in tube-nosed birds such as fulmars, but at the end of the bill in most birds.

FIGURE 28.9 Avian salt glands The salt glands of birds are located above the eyes. Ducts carry the secretions of the salt glands to the nasal passages, and the secretions drip out from the external nares (nostrils). (A) The structure of the salt glands of a herring gull. Each salt gland lies in a shallow depression in the skull above the eye. (B) A northern fulmar—a type of oceanic bird—showing the outlines of the salt glands above the eyes and the dripping of salt-gland secretions from the nostrils. (A after Schmidt-Nielsen 1960; B after Goldstein 2002.)

TABLE 28.4 Concentration of Na⁺ in the salt-gland secretions of marine birds, turtles, and lizards	

Cl^- concentrations are typically about the same as Na^+ concentrations. All data are for adults except those for the green sea turtles.

Animal	Na⁺ concentration (mM)
Marine iguana (*Amblyrhynchus cristatus*)	840
Loggerhead sea turtle (*Caretta caretta*)	730–880
Newly hatched green sea turtle (*Chelonia mydas*)	460–830
Herring gull (*Larus argentatus*)	600–800
Brown pelican (*Pelecanus occidentalis*)	600–750
Humboldt penguin (*Spheniscus humboldti*)	725–850
Leach's storm-petrel (*Oceanodroma leucorhoa*)	900–1100
Standard seawater	470

Sources: Schmidt-Nielsen and Fange 1958; Schmidt-Nielsen 1960; Marshall and Cooper 1988.

all of them have **salt glands**, organs of *extrarenal salt excretion* that make up for the meager concentrating abilities of their kidneys.[12]

These glands are located in the head, as illustrated for birds in **FIGURE 28.9**. They put out secretions that are dramatically hyperosmotic to the blood (by a factor of four to five in many species). Moreover, these secretions (as shown in **TABLE 28.4**) contain concentrations of Na^+ and Cl^- (and K^+ as well) that exceed those in seawater. Thus birds, lizards, and turtles with salt glands are, in principle, able to extract pure H_2O from seawater; they could drink seawater and void the major monovalent ions in less H_2O than they ingested, retaining the excess H_2O in their bodies. The salt gland secretions are discharged into the nasal passages in birds (see Figure 28.9B) and lizards. In sea turtles the secretions are emitted like tears. The cellular mechanism of salt secretion by

the salt glands conforms to the model in Box 28.3, at least in birds and sea turtles.

The ingestion of a salt load by an animal with salt glands is promptly followed by an increase in the rate of secretion by the glands. Control of this response, at least in birds, is mediated principally by the parasympathetic division of the autonomic nervous system. When osmoreceptors located in or near the heart and brain detect high blood osmotic pressures, the parasympathetic nervous system releases acetylcholine in the salt glands; this chemical message induces gated Cl^- and K^+ channels in the mitochondria-rich cells (chloride cells) of the salt glands (see figure in Box 28.3) to open, activating secretion.

In addition to these acute responses, salt glands also undergo chronic responses (acclimatization). For example, if an individual bird experiences a chronic increase in salt ingestion—as it would after migrating from a freshwater habitat to an ocean habitat—its salt glands typically increase in size, concentrating ability, and peak secretory rate. These changes reverse if the bird returns to freshwater.

The tears observed flowing down the faces of sea turtles when they emerge onto beaches to lay eggs are of some renown. We now understand that they are secretions of salt glands, not tears of emotion. If you watch a fulmar or gull standing by the ocean, you will see—emerging from its nostrils—droplets of salt-gland secretions (see Figure 28.9B), often flicked away by a shake of its head.

Sea snakes, it has recently been discovered, differ in their water–salt relations from other marine reptiles, even though they have salt glands (which empty into the mouth). Based on study of

---

[12] Salt glands have been reported in marine lizards (e.g., the Galápagos iguana), sea turtles, marine snakes, and 14 orders of marine birds. However, they have not been reported in the passerine ("perching") birds.

TABLE 28.5 Average composition of blood plasma and excretory fluids in two sharks and the coelacanth

For the dogfish shark and coelacanth, which were living in seawater, the composition of the seawater during study is given. The bull sharks were acclimated to seawater or freshwater for only a week; thus their plasma composition may not have been entirely stabilized. TMAO = trimethylamine oxide.

	Osmotic pressure (mOsm)	Solute concentration (mM)			
		Na⁺	Cl⁻	Urea	TMAO
Dogfish shark (*Squalus acanthias*) living in seawater					
Blood plasma	1018	286	246	351	71
Urine	780	337	203	72	6
Rectal-gland secretion	1018	540	533	15	—
Seawater	930	440	496	0	0
Bull shark (*Carcharhinus leucas*)					
Blood plasma when living in seawater	940	304	315	293	47
Blood plasma when living in freshwater	595	221	220	151	19
Coelacanth (*Latimeria chalumnae*) living in seawater					
Blood plasma	931	197	187	377	122
Urine	961	184	15	388	94
Seawater	1035	470	548	0	0

Sources: Hill and Wyse 1989; Pillans et al. 2005.

one common species (*Hydrophis platurus*), the salt glands are not effective enough to prevent dehydration. The snakes spend long periods in a relatively dehydrated state, and to rehydrate, they need to drink freshwater (or other dilute water).

MARINE MAMMALS Mammals, as a group, are capable of producing the most concentrated urine of all vertebrates. This ability is believed to be the key for marine mammals such as seals and whales to display hyposmotic regulation. Salt glands or other mechanisms of extrarenal salt excretion are not known in mammals.

As important as the kidneys are in marine mammals, the urine-concentrating abilities of these animals are not dramatic; their concentrating abilities are not particularly high in comparison with those of nonmarine mammals of similar body size. Partly for this reason, the overall patterns of water and salt balance in seals and whales remain open to debate. Existence on a diet of teleost fish poses no great challenges. To date, however, the data available leave unclear whether or how most species could exist while chronically eating only invertebrates, which often have substantially saltier body fluids than fish.

Another area of uncertainty is drinking. Although seals and whales are thought generally not to drink seawater, research in the last 25 years has revealed that some species of both groups do drink under certain circumstances; some fur seals, for instance, drink seawater when hauled out on land for weeks in hot climates during their breeding season. Physiologists are debating the potential advantages of drinking for animals that in general cannot concentrate salts in

their urine to levels higher than the concentrations seen in seawater (see page 733).

Marine elasmobranch fish are hyperosmotic but hypoionic to seawater

The marine sharks, skates, and rays—collectively known as the *elasmobranch fish*—have evolved a novel solution to the osmotic problems of living in the sea. Their blood concentrations of *inorganic ions* are similar to those of marine teleost fish and well below those in seawater. However, the osmotic pressure of their blood is slightly *higher* than that of seawater. As illustrated by the sharks in **TABLE 28.5**, these fish are able to be hyperosmotic to seawater—even though their blood has far lower concentrations of inorganic ions than seawater—because their blood plasma and other body fluids have high concentrations of two *organic solutes*: urea and, to a lesser extent, trimethylamine oxide (TMAO).[13] Because their blood is hyperosmotic to seawater, the marine elasmobranchs experience a small *osmotic influx* of water, in sharp contrast to the marine teleosts, which confront relentless osmotic desiccation. The hyperosmoticity of the elasmobranchs—caused by their high blood concentrations of urea and TMAO—is, in effect, a mechanism for obtaining water.

Elasmobranchs are specialized to produce and retain urea. In sharp contrast to teleosts, elasmobranchs typically synthesize urea as their principal nitrogenous product of protein catabolism (**BOX 28.4**). Of all the thousands of species of teleosts, fewer than ten are known to employ urea in this role; the others employ ammonia. In contrast, the use of urea as the principal nitrogenous product is universal in marine elasmobranchs. Urea accumulates in the body fluids of marine elasmobranchs because of specializations of their kidneys and gills. Elasmobranchs reabsorb urea from their urine as the urine forms in their kidneys, possibly by use of active urea transport. Moreover, the gills of marine elasmobranchs also retain urea because they have a dramatically low permeability to urea and, according to recent discoveries, they actively return outgoing urea to the blood plasma.

Urea in high concentrations can alter the structures of proteins, and the concentration of urea is kept low in most vertebrates (about 2–7 mM in human plasma). Plasma concentrations of urea in marine elasmobranchs—usually 300–400 mM—are "out of sight" by comparison. Some enzymes and other macromolecules in elasmobranchs have evolved exceptional resistance to urea's denaturing effects. Some elasmobranch organs, such as the heart, have in fact become dependent on urea for proper function.

Recent studies have revealed, however, that many elasmobranch proteins are just as sensitive to urea's denaturing effects as homologous proteins in other vertebrates. How can this be? A key part of the answer is that TMAO serves as a *counteracting solute*. In the amounts present, TMAO offsets the effects of urea (see Figure

---

[13] For the chemical structures of urea and TMAO, see Figure 29.24.

| BOX 28.4 | The Evolution of Urea Synthesis in Vertebrates |

Because proteins are 16% nitrogen by weight, the disposition of nitrogen is a significant matter when proteins are catabolized. Simple deamination of amino acids during protein breakdown leads to formation of ammonia (NH_3) as the nitrogen-containing end product of catabolism. Urea (see the structure to the right) is one of the major alternative nitrogenous end products. The synthesis of urea costs energy: Each urea molecule requires the energy from four or five ATP molecules for its synthesis. This cost is an "extra cost" that is avoided if ammonia is made instead of urea.

The biochemical pathway by which urea is synthesized from protein nitrogen in vertebrates is known as the *ornithine–urea cycle*. In the last 30 years, a consensus has emerged that the ornithine–urea cycle existed in the earliest vertebrates. That is, the earliest vertebrates are believed to have had genes coding for all the ornithine–urea cycle enzymes.

Despite its antiquity, urea synthesis is observed today in only a minority of modern vertebrates, which have a scattered distribution in the vertebrate phylogenetic tree (see figure). These include the elasmobranch fish, coelacanth fish, mammals, most amphibians, and some others. Two principal advantages of urea synthesis seem to account for the cases in which vertebrates invest extra energy to make urea rather than ammonia from their waste protein nitrogen. First, urea is sometimes employed as an osmolyte to raise the osmotic pressure of the blood; it is used in this way by some marine fish—most notably elasmobranchs—to render the blood hyperosmotic to seawater. Second, urea is sometimes employed as a detoxification compound for waste

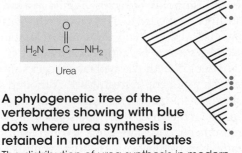

A phylogenetic tree of the vertebrates showing with blue dots where urea synthesis is retained in modern vertebrates
The distribution of urea synthesis in modern vertebrates is distinctly scattered. The tree is presented in detail in Box Extension 28.4.

nitrogen. Urea is far less toxic than ammonia and therefore is far better suited to being accumulated in the body than ammonia is. **Box Extension 28.4** presents the phylogenetic tree in detail and discusses the evolution of urea synthesis more thoroughly.

27.10), evidently by opposing effects of urea on deleterious interactions of proteins with solvent water, interactions that if unopposed cause protein unfolding.

In most aquatic animals, the blood osmotic pressure is attributable primarily to inorganic ions dissolved in the blood plasma. Because of this, problems of osmotic and ionic regulation are related in particular ways: If an animal tends to gain water by osmosis, it tends to lose ions by diffusion, and vice versa.

These relations are uncoupled in the marine elasmobranch fish because about 40% of the blood osmotic pressure is attributable to urea and TMAO rather than inorganic ions. Because the elasmobranchs are slightly hyperosmotic to seawater, they tend to gain water by osmosis, but because their blood ion concentrations are below those in seawater, they also tend to gain excess ions by diffusion from seawater. As a consequence of the fact that water enters elasmobranchs osmotically, they need not drink to obtain water, and therefore—unlike teleosts—do not incur the NaCl load caused by drinking seawater (see page 751).

Excess salts are removed from the body fluids of elasmobranchs by the kidneys and, extrarenally, by **rectal salt glands**. The salt glands, consisting of thousands of secretory tubules, void into the rectum a secretion (see the data for the dogfish shark in Table 28.5) that is isosmotic to the blood, but contains only traces of urea and approximates or exceeds seawater in its concentrations of Na^+ and Cl^-. The mechanism of NaCl secretion is as described in Box 28.3. Whether active ion excretion occurs across the gills is an unresolved question.

Ever since the "elasmobranch strategy" of water–salt regulation in the sea (**FIGURE 28.10**) was discovered, biologists have speculated

about its possible advantages over the "teleost strategy."[14] Until recently, the usual conclusion was that the elasmobranch strategy costs less energy because marine elasmobranchs are able to obtain H_2O by "cost-free" osmosis, whereas marine teleosts must drink seawater and pump NaCl out of it to get H_2O. The error in this view, we now recognize, is that the osmosis of water into a marine elasmobranch is *not* "cost-free." To keep its blood hyperosmotic

---

[14] The assumption is that marine elasmobranchs and teleosts both inherited low blood salt concentrations from freshwater ancestors, but they have diverged in how they manage the consequences.

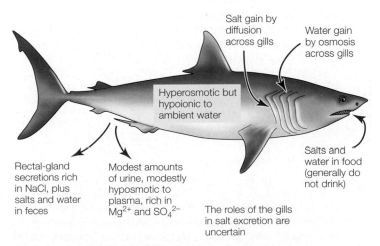

FIGURE 28.10 Water–salt relations in a marine shark
Protein-rich foods are required for adequate urea synthesis.

to seawater, an elasmobranch must synthesize urea, which costs more ATP (see Box 28.4) than merely making ammonia from waste nitrogen (as most teleost fish do). The elasmobranch might also need to pay ATP costs to recover urea from its urine and intercept urea diffusing outward across its gills. In a careful analysis, Leonard Kirschner concluded that the costs of the elasmobranch and teleost strategies are essentially the same. At least in terms of energy, the strategies seem to be "different but equal."

About 15% of the elasmobranch species alive today occur in dilute brackish waters or in freshwater. Although some are permanent residents of freshwater, most also occur in the ocean. A well-studied example of the latter is the bull shark (*Carcharhinus leucas*), famed for its rare but devastating attacks on coastal swimmers. When the elasmobranchs that live in the ocean venture into dilute waters, they lower their blood urea concentrations somewhat by decreasing urea synthesis and retention. Nonetheless, they retain elevated blood urea concentrations, as illustrated by bull sharks (see Table 28.5), even though doing so promotes osmotic uptake of excess water in dilute environments.

In addition to marine elasmobranchs, two other types of marine fish maintain high blood concentrations of urea and TMAO: the coelacanth and the holocephalans (chimaeras). The coelacanth (see Table 28.5) is a particularly interesting case because it is the only living example of the crossopterygian fish, the presumed ancestors of the terrestrial vertebrates.

SUMMARY
Animals in the Ocean

■ Most marine invertebrates are approximately isosmotic to seawater, but their blood differs from seawater in ionic composition. They exhibit ionic regulation but have little or no need for osmotic regulation. Hagfish display the same pattern.

■ Marine teleost fish are hyposmotic to seawater, apparently because they are descended from freshwater or coastal ancestors.

■ Because they are hyposmotic to seawater, marine teleosts tend to lose water by osmosis and gain ions by diffusion. To replace water, they drink; however, to absorb H_2O from the seawater in their gut, they must actively take up NaCl, increasing their problem of salt loading. Their kidneys make urine that is approximately isosmotic to their blood plasma but rich in divalent ions, thereby assuming chief responsibility for divalent ion regulation. Monovalent ions are excreted across their gills; although Cl⁻ is secreted actively into the ambient water by mitochondria-rich (chloride) cells, Na⁺ secretion is often secondary to Cl⁻ secretion and passive.

■ Marine birds, turtles, and lizards have cranial salt glands that permit them to excrete ions at higher concentrations than possible in their urine.

■ Marine mammals lack salt glands but have kidneys that can produce more-concentrated urine than reptiles (including birds). Their urine-concentrating abilities are not exceptional compared with those of other mammals, however, and their water–salt balance is not entirely understood.

■ Marine elasmobranch fish, although they have blood ion concentrations far lower than those of seawater, are slightly hyperosmotic to seawater because of high concentrations of two counteracting organic solutes, urea and trimethylamine oxide (TMAO). Unlike teleosts, therefore, elasmobranchs need not drink and need not incur an extra NaCl load to gain H_2O from ingested seawater.

Animals That Face Changes in Salinity

Many aquatic animals face large changes in the salinity of the waters they occupy during their lifetimes. These include (1) animals such as salmon and eels that migrate long distances between rivers and the open ocean and (2) animals that live near the margins of the continents. Along coastlines, waters of intermediate salinity—brackish waters—occur in estuaries, salt marshes, and other settings (see Chapter 27). Ocean animals that venture into brackish coastal waters encounter lower salinities than they experience when living in the open ocean. Freshwater animals face elevated salinities when they enter brackish waters. Some species live principally *within* estuaries; they face changes in salinity as they move from place to place (see Figure 27.4) or as tides or other water movements shift the waters around them.

In their relations to changing salinities, animals are often categorized as *stenohaline* or *euryhaline*. **Stenohaline** species are able to survive within only narrow ranges of ambient salinity. **Euryhaline** species, in contrast, can survive within broad ranges of salinity.

Animals are also classified as *osmoconformers* or *osmoregulators* (see Figure 27.3). **Osmoconformers**—sometimes described as **poikilosmotic**—permit their blood osmotic pressure to match the ambient osmotic pressure. **Osmoregulators**—sometimes called **homeosmotic**—maintain a relatively constant blood osmotic pressure even as the ambient osmotic pressure rises and falls.

Most species of invertebrates that occur in the open ocean are stenohaline osmoconformers; when they are placed in brackish waters, their blood osmotic pressure falls, and because they cannot tolerate blood osmotic pressures much lower than those they have in seawater, they do not prosper or may die. Certain marine osmoconformers are exceptional, however, in that they are euryhaline. Oysters and mussels provide outstanding examples; despite being osmoconformers, some species thrive over wide ranges of salinity, from seawater itself to waters less than 20% as concentrated as seawater. For osmoconformers to be so euryhaline, their cells must have remarkable abilities to function over wide ranges of blood osmotic pressure. The cells of euryhaline osmoconformers are noted for having dramatic powers of cell-volume regulation (see Figure 27.8).

An intriguing and commercially important illustration of how water–salt physiology can feature in the lives of euryhaline osmoconformers is provided by the story of MSX, a debilitating protistan parasite of the commercial oyster (*Crassostrea virginica*) of the Atlantic seaboard of the United States. These oysters live in estuaries, where the ambient salinity varies from place to place. Because their blood osmotic pressure matches the ambient osmotic pressure, their blood osmotic pressure also varies from place to place. The MSX parasite cannot survive in an oyster if its blood

(A) Hyper-isosmotic regulators

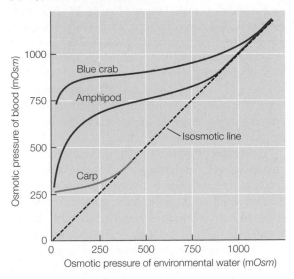

FIGURE 28.11 Types of osmotic regulation For each animal shown, blood osmotic pressure is plotted as a function of the osmotic pressure of the ambient water. Each dashed line is a line of equality between blood osmotic pressure and ambient osmotic pressure (an isosmotic line). (A) Three species of hyper-isosmotic regulators. Such regulation is typical of freshwater animals that enter brackish waters (e.g., the carp); it occurs also in many crabs of shores or estuaries (e.g., the blue crab, *Callinectes sapidus*, shown) and in some euryhaline annelids and amphipods (e.g., the amphipod *Gammarus oceanicus*, shown). (B) Four species of hyper-hyposmotic regulators. Such regulation occurs in many shore crabs (e.g., the fiddler crab, *Uca pugilator*, and the lined shore crab, *Pachygrapsus crassipes*), semiterrestrial crabs (e.g., the ghost crab, *Ocypode cursor*), coastal shrimps, and animals adapted to inland saline environments (e.g., the brine shrimp, *Artemia salina*), as well as euryhaline and migratory fish. (After D'Orazio and Holliday 1985; Greenaway 1988; Hill and Wyse 1989; Kirschner 1991.)

(B) Hyper-hyposmotic regulators

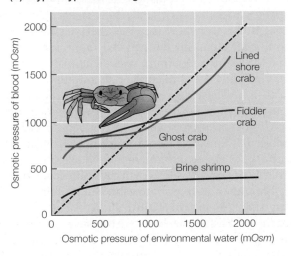

is more dilute than about 400 m*Osm*. For the oysters, therefore, ambient waters with osmotic pressures lower than 400 m*Osm* are safe havens from the parasite. In the Chesapeake Bay (see Figure 27.4), serious spread of MSX occurs during droughts. When there is little rain, rivers bring less freshwater into the Bay, and the salinity rises above 400 m*Osm* in places where it is ordinarily lower. The oysters living in such places experience a rise in blood osmotic pressure and become vulnerable to the parasite.

Among animals that are osmoregulators, regulation is often limited to certain ranges of ambient osmotic pressure. Thus different categories of regulators are recognized. In one common pattern, called **hyper-isosmotic regulation**, a species keeps its blood more concentrated than the environmental water at low environmental salinities, but allows its blood osmotic pressure to match the ambient osmotic pressure at higher salinities. Species that are predominantly freshwater animals but venture into brackish waters typically show this pattern, as do many coastal marine invertebrates (**FIGURE 28.11A**). Animals exhibit hyper-isosmotic regulation when they possess mechanisms of hyperosmotic regulation but lack mechanisms of hyposmotic regulation.

A second major category of regulators consists of those that keep their blood more concentrated than the environmental water at low environmental salinities but more dilute than the environmental water at high environmental salinities. This pattern is called **hyper-hyposmotic regulation** and requires mechanisms of both hyperosmotic and hyposmotic regulation. It is observed in salmon, eels, and other migratory fish and in a variety of crustaceans (**FIGURE 28.11B**).

Both osmoconforming and osmoregulating species occur among the crustaceans that live in the oceans and also in the marine annelids and some other related sets of marine animals. In these groups, euryhalinity and osmoregulatory ability tend to be correlated: The most euryhaline species are typically those that osmoregulate to a comparatively strong extent. Success in dilute waters in marine crustaceans, annelids, and other such groups, therefore, has been achieved by protecting the cells of the body from exposure to low blood osmotic pressures, in contrast to the oysters and mussels earlier discussed.

Migratory fish and other euryhaline fish are dramatic and scientifically important examples of hyper-hyposmotic regulators

The fish that migrate between freshwater and the oceans typically breed in one habitat and undergo much of their growth and maturation in the other. Some species—termed **anadromous** ("running upward")—ascend rivers and streams from the oceans to breed; these fish include salmon and certain smelts, shad, and lampreys. Other species—termed **catadromous** ("running downward")—grow in freshwater and descend to the oceans for breeding; they include the freshwater eels (genus *Anguilla*) of North America, Europe, and East Asia.

The migratory fish are superb osmoregulators. They function as hyperosmotic regulators when in freshwater and as hyposmotic regulators when in seawater, and they are so effective in both habitats that their blood osmotic pressure generally changes only a little between the two. Chinook salmon (*Oncorhynchus tshawytscha*), for example, have a plasma osmotic pressure averaging about 410 m*Osm* when in the ocean and about 360 m*Osm* when at their freshwater spawning grounds.

The mechanisms of regulation employed by migratory fish—and other euryhaline teleosts—in seawater and in freshwater are the same as those we earlier discussed for marine and freshwater tele-

(A) Responses of gill proteins to transitions between freshwater and seawater

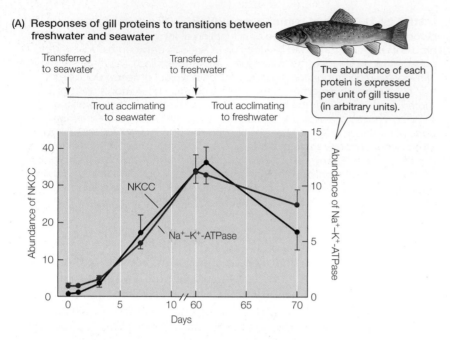

The abundance of each protein is expressed per unit of gill tissue (in arbitrary units).

(B) Gill tissue in which NKCC is stained for identification

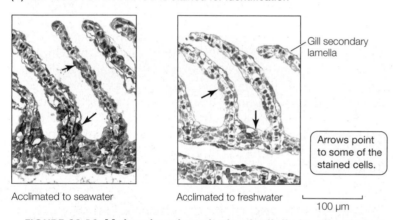

Gill secondary lamella

Arrows point to some of the stained cells.

Acclimated to seawater

Acclimated to freshwater

100 µm

FIGURE 28.12 Molecular phenotypic plasticity in gills of trout transferred between freshwater and seawater (A) Brown trout (*Salmo trutta*) that had been living in freshwater were transferred to seawater on day 0. After living in seawater for 60 days (note the break in the *x* axis), they were returned to freshwater. Abundances of Na⁺–K⁺-ATPase (α subunit) and NKCC (Na–K–2Cl cotransporter) were quantified by immunocytochemistry. (B) Images of gill secondary lamellae, from fish acclimated more than 60 days to seawater or freshwater, in which NKCC is visualized immunocytochemically by use of a monoclonal antibody specific to the protein, resulting in a red color (nuclei are stained blue). Note that staining is more intense in the seawater-acclimated gill. (A after Tipsmark et al. 2002; B courtesy of Christian Tipsmark, from Tipsmark et al. 2002.)

osts (see Figure 28.8). Thus, when the fish move from freshwater to seawater, they reverse the direction of active NaCl transport across their gills (inward transport in freshwater, outward in seawater); they greatly increase the rate at which they drink; they decrease their rate of urine production; and they switch from producing urine that is markedly hyposmotic to their blood plasma to producing urine that is approximately isosmotic to the plasma. In their intestinal epithelium, the activity of the NaCl-uptake mechanisms and the abundance of aquaporins also increase when they enter seawater.

Some crustaceans that exhibit hyper-hyposmotic regulation, such as fiddler crabs (*Uca*; see Figure 28.11B), are known to display similar and equally dramatic shifts in their regulatory mechanisms as they move between salinities.

The migratory fish—and other euryhaline teleosts—have been and continue to be the most important of all fish for studies of the physiological regulation of water–salt relations. They are studied intensely because their regulatory systems meet dramatic challenges and thus provide vivid insight into regulation in action.

One major objective of modern research on fish water–salt physiology is to understand the molecular mechanisms of successful transitions between freshwater and seawater. Studies of gill function provide a good illustration. In recent years, researchers have established that the gills of an individual fish undergo extensive molecular remodeling during transitions between freshwater and seawater—remodeling that leads to distinctive *freshwater and seawater gill phenotypes*. These phenotypic adjustments include critical changes in the cell morphology and the suites of ion-transport proteins in the mitochondria-rich cells (MRCs) that are so important for gill ion transport (see Box 28.1). Using monoclonal antibodies to assay defined cell-membrane proteins of the MRCs by use of immunocytochemistry, for example, researchers have studied the concentrations and types of Na⁺–K⁺-ATPase and NKCC (a Na–K–2Cl cotransporter) during freshwater-to-seawater transitions. Both of these ion-transport proteins are predicted—from knowledge of molecular transport mechanisms in teleost fish (see Boxes 5.2 and 28.3)—to increase in individuals transferred from freshwater to seawater. In studies of brown trout (*Salmo trutta*), quantitative changes in the proteins follow this prediction, as seen in **FIGURE 28.12**: The proteins increase in the gill MRCs when trout are transferred to seawater and decrease when the fish are returned to freshwater. Studies of several other species confirm these results. Moreover, research on Atlantic salmon (*Salmo salar*) and some other species reveals that the *molecular form* of Na⁺–K⁺-ATPase also changes between freshwater and seawater, implying that the detailed function of the ATPase is modulated. Aquaporins constitute another area of molecular research. Recent studies (e.g., on eels, *Anguilla japonica*) indicate that aquaporins are upregulated in the intestinal epithelium following transfer to seawater—a response predicted to facilitate uptake of H₂O from ingested seawater.

A second major objective of research today is to clarify the complex endocrine controls of water–salt physiology. Years ago, investigators discovered that hypophysectimized fish died when transferred to freshwater, but they could be rescued by the specific adenohypophysial hormone prolactin. Those experiments established that hormonal controls are of vital importance in water–salt physiology; prolactin, in particular, plays a key role in freshwater, not only in reducing the permeability of the gills to Na⁺ but also in augmenting urine flow. By now, several other hormones are known to be of central importance in water–salt physiology, including cortisol, growth hormone, insulin-like growth factor, and thyroid hormone.

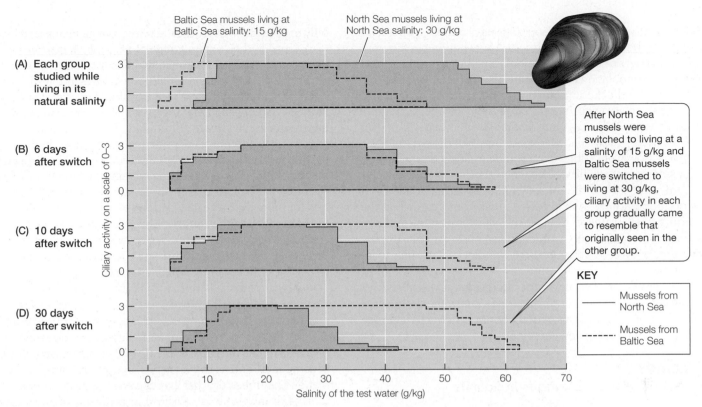

FIGURE 28.13 Acclimation of mussels to changed salinity
Blue mussels (*Mytilus edulis*) were collected from the North Sea, where the ordinary ambient salinity is about 30 g/kg, and from the Baltic Sea, where the ordinary ambient salinity is about 15 g/kg (for reference, open-ocean seawater has a salinity of about 35 g/kg). After initial testing (A), each group was switched to living in water of the opposite salinity and periodically retested (B–D). The aspect of their physiology studied was the activity of their gill cilia—important for pumping water through the body so food and O_2 can be collected. Ciliary activity (*y* axis) was scored on a scale of 0 (low) to 3 (high) as a function of salinity (*x* axis). Blue mussels are of great importance in natural ecosystems and aquaculture. (After Theede 1965.)

A memorable and informative experiment was done recently by simply adding NaCl to the diets of rainbow trout kept in freshwater. One might conclude that the trout were fooled by the dietary salt because they responded by modifying their gill phenotype to that of seawater-acclimated fish! For example, they upregulated the genes coding for $Na^+–K^+$-ATPase and NKCC. The experiment indicates that salt exposure is a trigger for phenotypic remodeling.

Genomic studies point to greater gene-expression changes in crustaceans than fish

Over the past 10 years many studies have used newly available genomic methods to measure altered expression of key ion-transport genes in the gills after animals are switched from one environment to another, such as from seawater to freshwater. Alan E. Wilson and colleagues recently completed a meta-analysis of almost 60 such studies. The goal of meta-analysis is to use statistical methods to analyze numerous *sets* of data simultaneously, to identify patterns that the data sets collectively indicate are present.

The meta-analysis by Wilson and his colleagues, based on data gathered in many labs on many species exposed to a variety of salinity changes, did identify a general pattern: Crustaceans tend to exhibit about threefold greater changes in expression of the genes coding for $Na^+–K^+$-ATPase and NKCC than teleost fish do. This genomic insight will need now to be assessed to determine the causes and consequences of this pattern of gene expression at higher levels of organization, such as in tissues and organs (see Figure 3.8).

Animals undergo change in all time frames in their relations to ambient salinity

The relations of animals to salinity can change in all the time frames we highlighted in Chapter 1 (see Table 1.2). Besides acute responses (the responses that individuals undergo soon after the salinity of their environment becomes altered), another time frame in which individuals respond is the chronic time frame (i.e., acclimation or acclimatization). The responses of trout shown in Figure 28.12 provide one example of acclimation. Another instructive example comes from studies of groups of blue mussels (*Mytilus edulis*) collected from the North Sea—where the salinity of the ambient water was 30 g/kg—and from the brackish Baltic Sea—where the salinity was 15 g/kg. **FIGURE 28.13A** shows the ranges of salinities at which the two groups of mussels were able to maintain ciliary activity (rated on a scale of 0–3 on the *y* axis) at the time of collection. The groups were then switched in the salinities at which they lived. As each group acclimated to its new salinity (**FIGURE 28.13B–D**), the range of salinities over which it could maintain ciliary activity gradually shifted. After 30 days, the North Sea animals living at a salinity of 15 g/kg displayed normal ciliary activity over approximately the same salinity range as originally seen in the Baltic Sea animals; the Baltic Sea animals living at a salinity of 30 g/kg also acclimated, coming to resemble the original North Sea animals. These results reveal acclimation in action and suggest that the original difference between the North Sea and Baltic Sea mussels was largely a consequence of acclimation (individual phenotypic plasticity).

Evidence also suggests that populations of a species can *evolve* differences in their water–salt physiology when living in different environments over multiple generations. An intriguing illustration is provided by populations of lampreys (*Petromyzon marinus*) that have become landlocked in North American freshwater lakes. Lampreys, which are anadromous like salmon, ordinarily migrate to the ocean when they are young adults. The populations that are landlocked in lakes have had no experience with the sea for many generations, however. Adults from some landlocked populations exhibit osmoregulatory difficulties when they are placed in water that is only half the salinity of seawater. However, adults from migratory populations—when tested—can osmoregulate at the full salinity of seawater *even before* they have migrated and had actual experience with salty waters. These observations strongly suggest genetic divergence between the landlocked and migratory populations.

Many animals are also known to undergo *developmental changes* in their water–salt physiology. Recall, for example, the changes we have discussed in the locations of mitochondria-rich cells (chloride cells) in developing fish (see Figure 4.6).

SUMMARY
Animals That Face Changes in Salinity

■ Some groups of marine invertebrates, such as molluscs, are uniformly osmoconformers. The euryhaline species in these groups are tolerant of wide ranges of blood osmotic pressure.

■ Other groups of marine invertebrates, such as crustaceans, include osmoconforming and osmoregulating species. In general in these groups, there is a correlation between osmoregulation and euryhalinity: The euryhaline species are osmoregulators.

■ Animals that are hyper-isosmotic regulators have mechanisms for hyperosmotic regulation but not hyposmotic regulation. Hyper-hyposmotic regulators have mechanisms for both types of regulation.

■ Euryhaline fish, such as species that migrate between seawater and freshwater, are excellent hyper-hyposmotic regulators. When they transition between freshwater and seawater, they undergo many changes in gill, kidney, and intestinal function—including molecular remodeling—under control of prolactin, cortisol, and other hormones.

Responses to Drying of the Habitat in Aquatic Animals

Residents of puddles, small ponds, intermittent streams, and the like are often confronted with drying of their habitat. The lungfish (dipnoan fish), which have lungs and fleshy fins, are of particular interest in this regard because of their relatively close relation to the fish that gave rise to amphibians. All species of lungfish occur in transient bodies of freshwater. If the habitat dries out, an African lungfish (*Protopterus aethiopicus*) digs a chamber in the bed of the lake or stream where it has been living; in the chamber, the fish curls up and secretes mucus that hardens into a cocoon opening only to its mouth. The cocoon acts as a barrier to evaporative water loss. The fish then enters a state of *metabolic depression* (see page 208): Its metabolic rate ultimately drops to about 10% of the ordinary resting level. This *hypometabolism* reduces its rate of respiratory water loss, and also its rate of use of stored energy. The lungfish's kidneys virtually stop making urine—keeping water in the body but compelling wastes to accumulate. The lungfish switches from producing ammonia as its principal nitrogenous end product to producing urea, a far less toxic compound (see Box 28.4), and urea may accumulate in its blood to levels approaching those of marine elasmobranch fish. The lungfish can survive in this dormant condition for more than a year!

Many other freshwater animals burrow into the substrate—often encasing themselves in mucous coverings—and enter a resting condition during times of drought. Included are some leeches, snails, water mites, and amphibians.

Anhydrobiosis—"remaining alive without water"—refers to survival while dried as fully as possible by desiccation in air. It represents the extreme in animal desiccation tolerance (**BOX 28.5**). A wide diversity of small animals from freshwater, saline, and terrestrial habitats are capable of anhydrobiosis, during which they become inert and as dry as paper or any other air-dried organic matter: They are so dry that they become like dust. In anhydrobiosis these animals are often tolerant of a variety of environmental extremes, not just extreme dryness, and often they can endure the air-dried state for many years. They frequently blow about in the wind, springing magically back to an animated life when they land in water. Biologists have long recognized two advantages of anhydrobiosis: It permits survival without water and can aid dispersal. A striking experiment on anhydrobiotic bdelloid rotifers has recently revealed a new potential advantage: escape from pathogens. Populations of rotifers exposed to a potentially lethal fungus were subjected to desiccation for various lengths of time. If desiccation continued for 4–5 weeks, 80%–90% of the populations were fungus-free after rehydration and the rotifers lived—because the rotifers tolerated desiccation longer than their fungal pathogen (**FIGURE 28.14**).

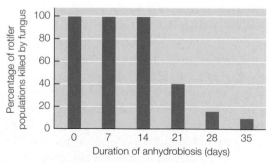

FIGURE 28.14 A long period of anhydrobiosis enhances survival of bdelloid rotifer populations because, during desiccation, the rotifers outlast their fungal pathogen Populations of the bdelloid rotifer *Habrotrocha elusa* were seeded with conidia of the fungal parasite *Rotiferophthora angustispora* and desiccated. The graph shows the percentage of rotifer populations killed by fungal growth following rehydration, as a function of the time until the populations were rehydrated. (After Wilson and Sherman 2010.)

BOX 28.5 Anhydrobiosis: Life as Nothing More than a Morphological State

Anhydrobiosis occurs particularly in a great variety of small freshwater animals. Some of these, such as many nematodes and bdelloid rotifers—and freshwater as well as terrestrial tardigrades (see photo)—are able to enter anhydrobiosis at any stage of their life cycle. In some other types of animals, anhydrobiosis is possible only during certain life stages, such as the eggs of some water fleas (cladocerans) and other crustaceans, the larvae of some chironomid flies, the embryonic cysts of coelenterates, and the gemmules of sponges. The encysted embryos (gastrulae) of brine shrimp (*Artemia*; often marketed as aquarium food under the strange name "sea monkeys") provide some of the best-studied examples of anhydrobiosis.

Being dried in air does not mean that every molecule of water is gone. Proteins and other materials tend to hold on to adhesively and cohesively bound water (more so when the humidity of the air is high than when it is low). Nevertheless, anhydrobiotic forms are very dry, typically having less than 2% as much water as

they have when hydrated. Some require certain minimum air humidities to survive. Some, however, can survive drying to the ultimate possible extent: Brine shrimp embryos, for example, can survive drying in a vacuum, which reduces their water content to 0.007 g per gram of dry weight, and recently anhydrobiotic tardigrades were shown to survive exposure to the vacuum of space!

Many anhydrobiotic forms, when dried to their maximum tolerable extent, cease to exhibit any of the usual signs of metabolism; for example, they neither consume O_2 nor accumulate wastes. Thus, as stated by one authority, they become "nothing more than a morphological state." They are described as *ametabolic*. More specifically, they are **reversibly ametabolic** because, although their metabolism stops, it can be restarted. These ametabolic forms raise interesting questions about how animal life is to be defined, inasmuch as metabolism is often included in the list of properties that distinguish living systems from nonliving ones.

0.1 mm

A tardigrade or "water bear" These tiny animals, frequently found in moss, can dry completely and survive. Shown is *Macrobiotus* sp., not in anhydrobiosis, imaged by scanning electron microscopy.

The disaccharide trehalose often accumulates in animal forms entering a state of anhydrobiosis. Trehalose prevents the structures of macromolecules, cell membranes, and intracellular membranes from being permanently destabilized by the loss of water. The way it does so remains debated.

Animals on Land: Fundamental Physiological Principles

As we now turn to animal life on land, we return to considering animals in their active, alert states, going about their daily lives. They will be our focus except for occasional brief discussions of dormancy.

As emphasized already, animal life originated and spent much of its early evolutionary history in water. The earliest animals that ventured to spend time on land, to consume terrestrial organisms as food, and ultimately, to develop on land were able to escape competitors and predators in their primordial aquatic habitat. For this reason, positive selective pressure for terrestriality must have been great. However, early animal life was adapted to living in an abundance of water. Evaporative losses of water on land posed a physiological threat of enormous importance for all stages of the life cycle.

We will focus on *water* in our discussion of animals on land. Although terrestrial animals sometimes face problems of salt balance, water balance usually represents their most pressing challenge in the realm of water–salt physiology.

The distinction between *humidic* and *xeric* animals provides a useful organizing principle for the study of water relations in terrestrial animals. The **humidic animals** are those that, although

they live on land, are restricted to humid, water-rich microenvironments. The **xeric animals** are those that are capable of living in dry, water-poor environments.[15]

The humidic animals include earthworms, slugs, centipedes, most amphibians, and most terrestrial crabs. Some live underground. Others live in leaf litter or under logs or rocks. The majority of frogs and toads stay in or near bodies of water, and when they venture away from water, they remain in protected microenvironments, such as the tall grass frequented by leopard frogs. Some humidic animals, such as most amphibians and all terrestrial crabs, still resemble their aquatic progenitors in that they require standing water to breed.

The major groups of xeric animals are the mammals, birds, reptiles other than birds, insects, and arachnids (e.g., spiders and ticks). Although xeric animals often seek protected, humid microenvironments, they are not stringently tied to such environments, as humidic animals are. The xeric animals can live successfully in the open air, and many of them—as they go about their daily lives—routinely expose themselves to the full drying power of

---

[15] The term *xeric* has a standardized meaning and is widely used. However, there is no standardized term to describe the animals restricted to moist habitats; although we use *humidic*, alternative terms are used in other books and articles. The term *mesic* is sometimes applied to animals intermediate between those that are xeric and those that are humidic.

the terrestrial environment. Some thrive in deserts and other equally dry environments, such as grain stores.

In many ways, the physiological difference between humidic and xeric animals is a distinction in how *rapidly* they get into trouble by loss of water in desiccating environments. The humidic animals dehydrate rapidly in dry environments. They therefore cannot remain long in such places, as xeric animals can.

A low integumentary permeability to water is a key to reducing evaporative water loss on land

Evaporation is one of the chief modes by which terrestrial animals lose water. In this section we begin our analysis of the physiological principles of living on land by discussing evaporation across the integument of the body. As discussed in Chapter 27 (see Equation 27.1), the rate of evaporation through an animal's integument depends on the *difference in water vapor pressure between the animal's body fluids and the air,* and it depends on the *permeability of the integument to water* (K in Equation 27.1).

A high integumentary permeability to water ranks as one of the most important specific characteristics that restrict humidic animals to their protective microhabitats. The skin of an earthworm, the skin of most amphibians, and most of the fleshy surfaces of a snail or slug, for example, have high permeabilities and provide little barrier to water loss: These animals often lose water *through their integuments* at rates that are 50%–100% as great as rates of evaporation from *open dishes of water* of equivalent surface area! With such a high integumentary permeability, a humidic animal can restrict its integumentary rate of evaporation only by limiting the difference in water vapor pressure that exists across its integument. From the viewpoint of physics, this explains why humidic animals are tied to humid habitats, where the air has a high water vapor pressure.

The xeric animals have integuments with a low permeability to water. Indeed, the evolution of a low integumentary permeability to water is one of the most important steps toward a xeric existence. In all the major xeric groups—vertebrate and invertebrate—*microscopically thin layers of lipids* are responsible for low integumentary water permeability. In mammals, birds, and nonavian reptiles (e.g., lizards and snakes), the lipid layers are structurally heterogeneous, lamellar complexes of lipids and keratin, less than 10 micrometers (μm) thick, located in the stratum corneum, the outermost layer of the epidermis of the skin. The principal lipids present are ceramides, cholesterol, and free fatty acids. Mammals, birds, and nonavian reptiles differ in histological details of the lipid layers, and they evidently evolved their lipid layers independently. In insects and arachnids, the lipids responsible for low integumentary permeability—such as long-chain hydrocarbons and wax esters—are contained in the outermost layer of the exoskeleton. This layer, termed the *epicuticle,* is only 1–2 μm thick.

Because virtually all the resistance to water loss across the integument of xeric animals resides in microscopically thin lipid layers, the physical toughness of the integument is not an index of resistance to water loss. A common misconception, for example, is that the scales of lizards and snakes block water loss; the real block is the lipid layer, a microscopic property of just the stratum corneum. Many millipedes and centipedes have sturdy exoskeletons that seem just as tough as those of insects, yet their exoskeletons are far more permeable to water than those of insects. Such millipedes

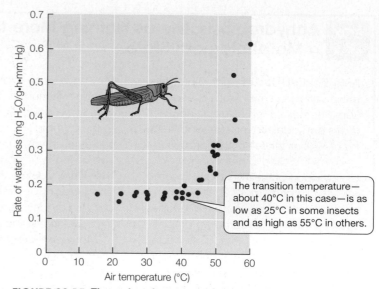

FIGURE 28.15 The rate of evaporative water loss of insects often starts to increase abruptly at a transition temperature   The graph shows how the rate of evaporative water loss of dead African migratory locusts (*Locusta migratoria*) increases as temperature increases. The rate of water loss is expressed "per mm Hg," referring to the difference (expressed in millimeters of mercury) between the actual water vapor pressure of the air during the measurements and the saturation vapor pressure at the various temperatures. Measured and expressed in this way, changes in the rate of evaporative water loss reflect changes in the water permeability of the integument. (After Loveridge 1968.)

The transition temperature—about 40°C in this case—is as low as 25°C in some insects and as high as 55°C in others.

and centipedes either lack the micrometer-thin epicuticular lipid layer or possess lipids of different types than insects. They are far more humidic than insects because of these microscopic differences.

Lipids reorganize and undergo phase alterations as their temperature changes, as evidenced by observing kitchen lipids such as butter. Researchers have long known that if the temperature of an insect (or arachnid) is gradually raised, water permeability increases just slightly up to a certain temperature—called the **transition temperature**—and then increases dramatically (**FIGURE 28.15**). The marked increase in permeability at the transition temperature is a consequence of lipid melting. Although the transition temperature is often so high that it would not be experienced by individuals in nature, this is not always the case. The cockroach *Periplaneta americana* (a common household pest), for example, experiences a marked increase in permeability starting at 25°C–30°C and thus might naturally encounter temperatures high enough to disrupt its protection against water loss. Temperature effects within the skin of vertebrates are far more complex because of changes in blood flow and other processes, but careful studies reveal that in at least some cases, a rise in epidermal tissue temperature decreases the effectiveness of the cutaneous lipid layer as a water barrier.

Within sets of related xeric animals, the chemical composition of the lipid layer can vary widely. Because of these lipid composition differences and also sometimes because of structural differences or differences in amounts of lipids, the lipid layer can be far more effective as a water barrier in some species than in other related species. **TABLE 28.6** shows, for example, that the resistance of skin to water loss—measured in a standardized way—varies widely *within* groups of xeric vertebrates as well as between such groups.

TABLE 28.6 Resistance of the skin to evaporative water loss in vertebrates

Values are for a standardized area of skin. Where a range is listed, the species-specific average resistance was measured in a variety of species (e.g., a variety of bird species), and the range listed is the range of these species-specific averages.

Group of animals	Resistance (s/cm)[a]
Ranid frogs and bufonid toads	0–3
Colubrid snakes (e.g., racers)	150–890
Viperid snakes (e.g., vipers)	790–1690
Iguanid lizards	110–1360
Birds	30–200
Human	380
House mouse	160

Source: Lillywhite 2006.

[a]Resistance is calculated as the inverse of conductance. Conductance is the rate of water loss across the skin—expressed as grams H_2O per cm^2 of skin per second—divided by the driving force, which in this case is the difference in water activity across the skin—expressed in water density units of grams H_2O per cm^3 of air. For a units analysis, one divides $g/cm^2 \cdot s$ by g/cm^3. Thus the units of conductance are cm/s, and the units of resistance, the inverse, are s/cm.

The lipid composition of the water barrier may even differ among populations of a single species and give rise to significant differences between populations in their physiology of water balance (**FIGURE 28.16**). At still another level of organization, cases are known—as in certain desert larks—in which the lipid composition of individuals changes as a consequence of acclimation to different environments.

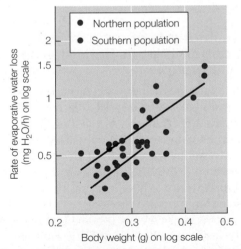

FIGURE 28.16 Differentiation between populations of one species in lipid-mediated protection against evaporative water loss The graph shows the rate of evaporative water loss at 25°C as a function of body weight in live grasshoppers of a single species (*Melanoplus sanguinipes*) from two geographically separate populations in northern and southern California. According to current evidence, these populations exhibit genetically controlled differences in the lipid composition of their epicuticular layer. These differences contribute to the greater resistance of the southern animals to water loss. In each population, the rate of water loss is an allometric function of body weight, so the relation plots as a straight line on log–log coordinates (see Appendix F). (After Gibbs 1998.)

Respiratory evaporative water loss depends on the function of the breathing organs and the rate of metabolism

Some groups of humidic animals have respiratory surfaces that are directly exposed to the air. Earthworms, some isopods, and some amphibians, for instance, breathe substantially or entirely across their general integument (skin or exoskeleton). This arrangement is a disadvantage from the viewpoint of evaporative water loss because movement of air across exposed respiratory surfaces can greatly exceed that necessary for exchange of O_2 and CO_2, so water loss can be much greater than the minimum required for respiratory gas exchange.

Most terrestrial animals have evolved invaginated respiratory structures (see Figure 1.20). In the xeric groups, breathing is carried out exclusively by such invaginated structures, and the general integument is virtually impermeable to O_2 and CO_2. The enormous advantage of this arrangement is that access of air to the thin, moist respiratory membranes can be closely controlled and thus limited to the levels required for exchange of O_2 and CO_2. The mammals, birds, and other reptiles control access of air to their lungs by regulating their breathing movements. Insects close and open the spiracles of their tracheal system (see Figure 23.31).

THE EFFECT OF TEMPERATURE ON THE WATER VAPOR CONTENT OF AIR A physical law of great consequence for warm-bodied air breathers is that when air is saturated with water vapor, its content of water per unit of volume approximately *doubles* with every 11°C increase in temperature (see Table 27.2).[16] Changes of air temperature can thus exert a strong influence on the amount of water carried by air movement.

When a mammal or bird inhales air into its lungs, the temperature of the air is raised to deep-body temperature, and the air becomes saturated with water vapor at its new, elevated temperature. As a consequence, depending on conditions, a substantial amount of water may be added to the inhaled air. For example, consider a mammal that inhales *saturated* air at 20°C. Such air contains about 17 mg H_2O/L (see Table 27.2). By the time the air reaches the lungs, it is saturated at 37°C, and it therefore contains 44 mg H_2O/L (see Table 27.2). Thus, even though the air is saturated to begin with, it contains 27 mg/L more water once it is in the lungs, all of this added water being drawn from the animal's body. If the air is then exhaled without modification, it carries all the added water away into the environment.

WATER CONSERVATION BY COOLING OF EXHALANT AIR When a mammal or bird exhales, the exhaled air usually is saturated with water vapor. However, in many mammals and birds, if the air is exhaled by way of the nasal passages, the *temperature* of the air is reduced before the air leaves the nostrils, a process that lowers the saturation water vapor pressure and therefore reduces the amount of water the air carries away.[17] Suppose the mammal we discussed in the preceding paragraph were to reduce the temperature of air from its lungs to 25°C before exhaling the air. On leaving the body, the air would then contain 23 mg H_2O/L

[16] As seen in Table 27.2, the saturation water vapor pressure of air increases with temperature, and the amount of water vapor per unit of volume increases in parallel.

[17] Cooling of nasal exhalant air also takes place in some lizards when they are maintaining high, behaviorally regulated body temperatures.

(see Table 27.2). The air would still carry away some body water (it entered the body with 17 mg H_2O/L). However, the reduction in the temperature of the exhalant air would cause 78% of the water added during inhalation to be recovered before exhalation.

In those mammals and birds that reduce the temperature of air before it is exhaled, the air is cooled by a countercurrent mechanism in the nasal passages. To understand the process, let's first look in more detail at what happens during inhalation, using our example of a mammal breathing 20°C air. As inhaled ambient air travels up the nasal passages, it is progressively warmed to about 37°C, and it takes up water vapor as its temperature is elevated. The heat that warms the air and the latent heat of vaporization for the added water vapor are drawn from the walls of the nasal passages.[18] Thus the walls of the nasal passages are cooled by the process of inhalation. The outer ends of the nasal passages are typically cooled most, and the inner ends least. During the ensuing exhalation, air coming up from the lungs arrives at the interior ends of the nasal passages at a temperature of 37°C and saturated. However, as the air moves down through the nasal passages toward the nostrils, it encounters the increasingly cooler surfaces created by the previous inhalation. Thus the air being exhaled is cooled as it travels toward the nostrils. This cooling lowers the saturation water vapor content of the air (see Table 27.2), causing water to condense out of the air onto the nasal-passage walls. The overall process is considered a *countercurrent* process because it depends on flow of air in opposite directions.

If the cooling of nasal exhalant air seems unfamiliar, it may be because only a small degree of cooling occurs in humans. In contrast, the cooling of nasal exhalant air in small mammals is dramatic, as illustrated by the data plotted as black dots in **FIGURE 28.17**; a small mammal that inhales air at 20°C and warms it to 37°C in its lungs might well exhale the air at 22°C–23°C.

A USEFUL MODEL OF RESPIRATORY EVAPORATIVE WATER LOSS

An insight-promoting way to think about the rate of respiratory evaporative water loss is to recognize that it depends on (1) an animal's rate of O_2 consumption and (2) the amount of water lost per unit of O_2 the animal consumes:[19]

$$\text{Rate of water loss} = \text{rate of } O_2 \text{ consumption} \times \text{water loss per unit of } O_2 \text{ consumed} \quad (28.1)$$

The principal insight to gain from this equation is that an animal's rate of metabolism is a major determinant of its rate of evaporative water loss. Mammals and birds, as we stressed in Chapters 7 and 10 (see Figure 7.9), typically have rates of O_2 consumption that are far higher than those of lizards, snakes, or other nonavian reptiles of similar body size. Mammals and birds therefore tend to have high rates of respiratory evaporative water loss by comparison with lizards, snakes, and other nonavian reptiles.[20]

---

[18] The nasal passages are not simple tubes. Their walls often consist of elaborate surface-enhancing structures (*turbinates*) over which air flows, as can be seen in the insets in Figure 10.37.

[19] If the rate of O_2 consumption is measured in milliliters of O_2 per hour (mL O_2/h), and if the water loss per unit of O_2 consumed is measured as mg H_2O/mL O_2, when these two factors are multiplied, the units of the result will be mg H_2O/h. That is, the result will be a measure of the rate of water loss in milligrams per hour.

[20] A compensation is that animals with relatively high rates of O_2 consumption also have relatively high rates of metabolic water production (e.g., see Figure 28.18).

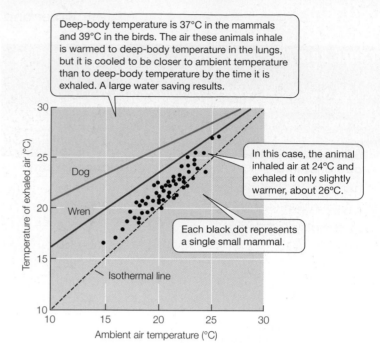

FIGURE 28.17 The temperature of air exhaled from the nostrils as a function of ambient air temperature in mammals and birds The black dots are data for individual small mammals (e.g., shrews, bats, mice, squirrels, and rabbits); 18 species are included. The solid lines are average results for mongrel domestic dogs (blue) and cactus wrens (*Campylorhynchus brunneicapillus*) (red). The dashed line is a line of equality between the temperature of exhaled air and the ambient air temperature (an isothermal line). (From Hill and Wyse 1989.)

The amount of water lost per unit of O_2 consumed is affected by several factors. One is the temperature of the exhaled air we just discussed (lower exhalant temperatures mean lower water loss). Another is the efficiency of the breathing organs in removing O_2 from inhaled air.

An animal's total rate of evaporative water loss depends on its body size and phylogenetic group

If we sum an animal's rates of integumentary and respiratory water loss, we get the animal's *total rate of evaporative water loss* (EWL). Within sets of phylogenetically related species, the broad statistical trend is for the total rate of EWL measured under particular conditions to vary allometrically with body size. This important pattern is illustrated in **FIGURE 28.18** using birds as examples. Small-bodied species tend to have higher weight-specific rates of EWL than related large-bodied species.

There are two reasons why small-bodied species tend to have relatively high weight-specific rates of EWL. First, small animals tend to have a greater body surface area per unit of weight than related large animals (see Equation 7.6); therefore they tend to have relatively high weight-specific rates of *integumentary* water loss. Second, small animals tend to have a higher metabolic rate per unit of body weight than related large animals (see Figure 7.9); therefore they tend to have relatively high weight-specific rates of *respiratory* water loss (see Equation 28.1).

There are also consistent differences *among phylogenetic groups* in their total rates of EWL, as we have stressed. Animals in humidic phylogenetic groups, such as amphibians, have highly permeable integuments and other properties that give them high total rates

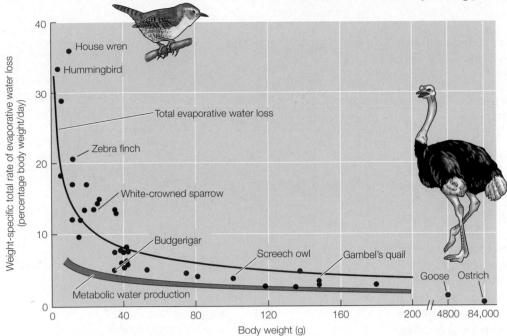

FIGURE 28.18 Within a phylogenetic group, the total rate of evaporative water loss is an allometric function of body size This trend is illustrated here (black line and black dots) using data for birds resting at 23°C–25°C in relatively dry air. Evaporative water loss is expressed in weight-specific units (the expression *percentage of body weight per day* is equivalent to *mg H_2O lost per 100 mg of body weight per day*). The black line is fitted statistically to the individual data points (black dots). The red area delimits the gross rates at which birds of various body sizes are expected to produce metabolic water (expressed as percentage of body weight per day). (From Hill and Wyse 1989.)

of EWL in comparison with similar-sized animals in xeric groups. Some of these differences in total rates of EWL are brought to light in **FIGURE 28.19**. All the vertebrates in the figure are about the same in body size; therefore, comparing them brings out differences among the phylogenetic groups. Lizards (and other nonavian reptiles) have very low total rates of EWL compared with amphibians because the lizards have low-permeability integuments (see Table 28.6) and enclosed breathing systems. Mammals and birds share these basic properties of lizards, but they have higher total rates of EWL than

lizards (and other nonavian reptiles). Why? The principal reason is that they have far higher metabolic rates than lizards. The two species of semiterrestrial crabs in Figure 28.19 are similar in size to the vertebrates; semiterrestrial crabs, as can be seen, are similar to amphibians in their rates of EWL and are humidic. The isopods and insects in the figure are tiny compared with the other animals shown. The isopods combine the disadvantages of small size and poor defenses against EWL. Their total weight-specific rates of EWL are staggering. One can only marvel at the insects, especially the

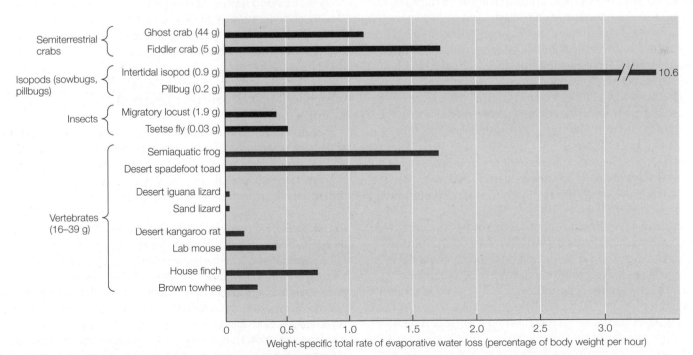

FIGURE 28.19 The total rate of evaporative water loss varies greatly among different types of vertebrates and arthropods All the animals shown were studied in dry or relatively dry air at 25°C–32°C. The vertebrates (red) are all similar in body size (16–39 g). Body weights are given for the individual arthropod species (blue). Evaporative water loss is expressed in weight-specific units (the expression *percentage of body weight per hour* is equivalent to *mg H_2O*

lost per 100 mg of body weight per hour). Species, listed from top to bottom: semiterrestrial crabs, *Ocypode quadrata* and *Uca annulipes*; isopods, *Ligia oceanica* and *Armadillidium vulgare*; insects, *Locusta migratoria* and *Glossina morsitans*; amphibians, *Rana temporaria* and *Scaphiopus couchii*; lizards, *Dipsosaurus dorsalis* and *Uma notata*; mammals, *Dipodomys merriami* and *Mus domesticus*; birds, *Carpodacus mexicanus* and *Pipilo fuscus*. (After Hill and Wyse 1989.)

tsetse flies (weighing 0.03 g), which have such excellent defenses against water loss that they have relatively low total rates of EWL per unit of weight, despite being very small.

Excretory water loss depends on the concentrating ability of the excretory organs and the amount of solute that needs to be excreted

In addition to evaporative water loss, excretion in urine is a second major way that terrestrial animals lose water. Like aquatic animals, terrestrial ones modulate the concentration, composition, and volume of their urine to serve changing requirements for osmotic, ionic, and volume regulation. We humans, for example, when dehydrated, produce a limited volume of urine that is hyperosmotic to our blood plasma (our osmotic U/P ratio can reach 4, meaning that our urine osmotic pressure can be four times the osmotic pressure of our blood plasma). However, after we have consumed large amounts of water, we void a copious urine that is hyposmotic to our blood plasma (our osmotic U/P ratio can be as low as 0.1–0.2). While recognizing the fundamental *regulatory* role of urine excretion, a key question in the study of water balance in terrestrial animals is how effectively the animals can *minimize* their urinary water losses. There are two basic ways to reduce the amount of water lost in urine. One is to concentrate the urine, thereby decreasing the amount of water required to excrete a given amount of solute. The second is to reduce the amount of solute excreted in the urine.

URINE-CONCENTRATING ABILITY   Most of the humidic terrestrial animals, such as earthworms and amphibians, are unable to raise the osmotic pressure of their urine above that of their blood plasma. Lizards, snakes, and turtles, although xeric, are also generally incapable of making urine that is hyperosmotic to their blood plasma. By contrast, three of the major xeric groups—insects, mammals, and birds—have evolved the ability to make hyperosmotic urine: urine more concentrated than their blood plasma. This unusual capability evolved independently in the three groups—a fact that emphasizes the selective advantages of being able to make concentrated urine in animals confronted with desiccation stress. Let's look at the three in more detail.

In terrestrial insects, maximum osmotic U/P ratios of 2–4 have been observed in certain blowflies, desert locusts, and stick insects. Mealworms (*Tenebrio*), which live in dry grain stores, can produce urine with a U/P ratio of 8.

Mammals hold the world records for urine-concentrating ability. Although mammals display an enormous range of urine-concentrating abilities, the U/P ratios achieved by many species are well above those seen in any other animals. To illustrate both of these points, the maximum reported osmotic U/P ratio is about 3 for muskrats, 4 for humans, 8 for dromedary camels, 9 for laboratory rats, 14 for Merriam's kangaroo rats and Mongolian gerbils (*Meriones*), and—at the highest extreme—about 26 for certain species of Australian desert hopping mice (e.g., *Notomys alexis*)! A significant and unexpected feature of mammalian kidney function is that the maximum concentrating ability tends to decrease with increasing body size, as seen in **FIGURE 28.20**. Much of the scatter in Figure 28.20 correlates with habitat; the highest values at a given

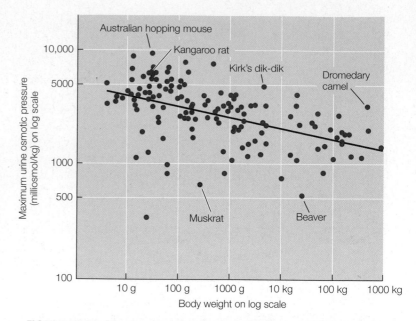

FIGURE 28.20 Maximum urine-concentrating ability in mammals: The maximum concentration is in part a function of body size   Each dot represents one species. Data for 146 species of mammals are included. The species chosen for labeling were deliberately selected to represent extremes. Because the overall relation is allometric, it plots as a straight line on log–log coordinates (see Appendix F). Because all mammals have approximately the same plasma osmotic pressure, a plot of the maximum osmotic U/P ratio as a function of body size would resemble this plot. (After Beuchat 1990.)

body size tend to represent species that confront relatively severe threats of desiccation or high dietary salt loads.

In birds, the most evident point to stress is that, as a group, birds are far inferior to mammals in their abilities to concentrate their urine. Although osmotic U/P ratios approaching 6 have been claimed in a few species of birds, the maximum U/P ratio ranges from 1.5 to 2.5 in most species, including many that live in arid regions.

Some terrestrial lizards and birds (e.g., ostriches and roadrunners) have cranial salt glands, which assist with secreting Na^+, K^+, and Cl^-. The secretions of the salt glands have higher ion concentrations than the urine in these animals. Thus the salt glands play an important role in decreasing the water expended to void excess ions.

REDUCING THE AMOUNT OF DISSOLVED MATTER EXCRETED IN THE URINE   Waste nitrogen from the catabolism of proteins is usually excreted in the urine by terrestrial animals.[21] One way to reduce the water demands of excretion is to incorporate the waste nitrogen into chemical compounds that are virtually insoluble—or poorly soluble—in water, thereby reducing the amount of material voided in solution. It is a testimony to the advantages of this type of nitrogen excretion that it has evolved independently many times: Insects, arachnids, some terrestrial snails, a few xeric frogs, birds, and other reptiles all produce poorly soluble nitrogenous wastes. The principal poorly soluble compounds employed are uric acid,

[21] See Chapter 29 (page 809) for a more thorough discussion of nitrogen excretion.

urate salts, allantoin, and guanine. Urate salts such as sodium and potassium urate—excreted by birds, other reptiles, and some additional groups—have the advantage that they carry away not only nitrogen but also inorganic cations in precipitated form.

By no means do all terrestrial animals exclude waste nitrogen from being excreted in solution, however. Many of the humidic animals—including earthworms, isopods, and most amphibians—excrete nitrogen principally as urea or ammonia, both of which are highly soluble. Mammals, paradoxically, also produce mostly urea (see Boxes 28.4 and 29.4).

If highly soluble nitrogenous wastes are voided in the urine, they demand water for their excretion. However, some of the animal groups that produce highly soluble wastes have evolved means of reducing the water demands. Mammals, for example, possess world-record abilities to concentrate urea in their urine (see Box 29.4). Some isopods, snails, and land crabs void ammonia as a gas.

Terrestrial animals sometimes enter dormancy or tolerate wide departures from homeostasis to cope with water stress

Many terrestrial animals, both humidic and xeric, enter dormancy—often called estivation—in response to immediate or predictable water stress. Mammals and birds that enter water-related dormancy—including desert ground squirrels and poorwills—often undergo metabolic depression and become hypothermic (see page 274). Metabolism is depressed during dormancy in poikilotherms as well.[22] Metabolic depression has several advantages. First, an animal in metabolic depression can live on body fat or other stored foods for a long time, and thus it can remain continuously in a protective microenvironment. Second, the animal's requirement for O_2 is reduced, thereby reducing its respiratory evaporation. Finally, nitrogenous and other wastes are produced at a low rate, which may be vital in permitting protracted existence with little or no urine output.

In addition to dormancy, another "change of status" that animals—particularly poikilotherms—commonly undergo during water stress is to permit large changes to occur in their body-fluid volume or composition while they continue to be active. Species that live in places where they are prone to dehydration are often especially tolerant to such changes. Tortoises in the Mohave Desert, for example, sometimes drop in body weight by 40%—and beetles in East African savannas sometimes lose 80% of their body water—because of dehydration during droughts or dry seasons. When Claude Bernard spoke of the *internal environment* in his groundbreaking studies that led to the concept of homeostasis, he was referring to the body fluids (see page 13). The ability to remain active and functional despite profound alteration of the body fluids has been termed **anhomeostasis** and can be a key to existence during water stress.

The total rates of water turnover of free-living terrestrial animals follow allometric patterns

A logical way to conclude our introductory discussion of animals on land is to focus on the total rates of *water turnover* (water lost and

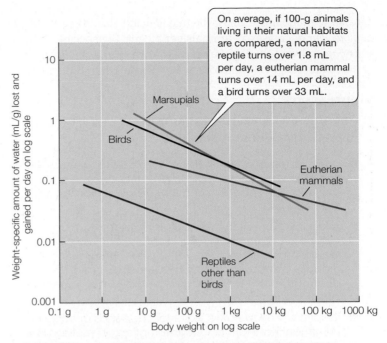

FIGURE 28.21 Terrestrial vertebrates living freely in their natural habitats: their total daily rates of water turnover in relation to body size The water turnover is the amount of water lost and gained per day when animals are in water balance. It is plotted here in weight-specific terms (mL/g). The weight-specific version of Equation 28.2 is $T/W = aW^{(b-1)}$. Each line is a plot of this equation. Because the relation for each group is allometric, it plots as a straight line on log–log coordinates. (After Nagy and Peterson 1988.)

gained per day) of terrestrial animals when they are living freely in their natural habitats. The rate of water turnover is typically measured by use of isotopically labeled water (e.g., heavy water). An animal living in the wild is trapped, injected with a known amount of labeled water, and turned loose. Days later, it is recaptured, and from the amount of labeled water remaining in its body, its rate of water turnover while free can be calculated. For an animal in water balance, water lost by evaporation, urination, and other processes each day is replaced by gains of water. The rate of water turnover measures the rates of these mutually balancing processes. A high rate of water turnover means that an animal loses and replaces a lot of water per day. Life can be precarious for such an animal because if an imbalance develops, it can lead rapidly to a crisis.

FIGURE 28.21 shows the general water-turnover patterns of vertebrates living in the wild. Note that mammals and birds of any given body size turn over far more water per day than lizards, snakes, or other nonavian reptiles of the same size. *Within* any one phylogenetic group, as might be guessed from what we have said before in this chapter, the total rate of water turnover, *T*, tends to be an allometric function[23] of body weight (*W*):

$$T = aW^b \qquad (28.2)$$

The exponent *b* is typically 0.6–0.8. This means that if we consider the relation between the *weight-specific* rate of water turnover (*T/W*) and weight, the exponent (*b* – 1) is negative: –0.2 to –0.4. There-

---

[22] *Suspension* of metabolism during anhydrobiosis is discussed in Box 28.5.

[23] See Appendix F for a discussion of allometric functions.

fore, as seen in Figure 28.21, the rate of water turnover *per gram of body weight* decreases as animals get bigger: Big species within a particular phylogenetic group tend to turn over a *smaller fraction* (lower percentage) of their total water per day than little ones.

SUMMARY
Animals on Land: Fundamental Physiological Principles

■ Humidic terrestrial animals are restricted to humid, water-rich microenvironments. Xeric terrestrial animals are those that are capable of a fully exposed existence in the open air.

■ A low integumentary permeability to water—which reduces integumentary evaporative water loss—is required for animals to be xeric. All the major xeric groups—insects, arachnids, birds, nonavian reptiles, and mammals—have low permeabilities because of integumentary lipids.

■ Respiratory evaporative water loss depends directly on (1) an animal's rate of O_2 consumption (its metabolic rate) and (2) the amount of H_2O lost per unit of O_2 consumed. One way to reduce the latter in mammals and birds is countercurrent cooling of nasal exhalant air.

■ The animals with the lowest total rates of evaporative water loss (EWL) are those, such as lizards, that combine the advantages of low integumentary permeability to water, tightly controlled access of air to breathing organs, and low metabolic rates.

■ Water loss in urine can be reduced by producing concentrated urine (which reduces the amount of water needed to void soluble wastes) or by producing poorly soluble nitrogenous end products such as uric acid (which remove waste nitrogen from solution). Only three groups of animals can make urine hyperosmotic to their blood plasma: insects, birds, and mammals.

■ Within groups of related species, water dynamism tends to vary allometrically with body size. Weight-specific EWL and weight-specific total water turnover tend to decrease as size increases.

Animals on Land: Case Studies

Now let's look at the water relations of some particular groups of animals on land. Doing so will provide an opportunity to integrate the points made in the previous section and discuss some new features of interest and importance. Chapter 30 continues this discussion of case studies with a focus on large-bodied mammals in deserts and dry savannas, such as camels and oryxes.

Amphibians occupy diverse habitats despite their meager physiological abilities to limit water losses

The terrestrial amphibians provide an instructive case study because they have invaded an impressive variety of habitats, from the shores of ponds to, quite literally, deserts. Yet despite this diversity of habitats, most species are humidic animals that, regardless of where they live, are remarkably similar to one another in their physiological water-balance characteristics. Their diversification into a wide range

of habitats has depended to a substantial extent on the evolution of *protective behaviors* and *advantageous patterns of seasonality*.

Most species of amphibians, including most that live in deserts, share several attributes that significantly limit their physiological capacity to restrain water losses. First and foremost, they have an integument that poses little barrier to evaporative water loss. Second, they incorporate waste nitrogen mostly into urea, a highly soluble compound requiring considerable amounts of water for its excretion. Moreover, although amphibians are notably adept at simply shutting off urine outflow when faced with dehydration, they are unable, when they do excrete urine, to produce a urine any more concentrated in total solutes than their blood plasma.

Amphibians have the same basic sources of water as most other animals: preformed water in food, preformed water taken in as "drink," and metabolic water. For the most part, adult amphibians are carnivores. Their food is therefore succulent, but it yields a lot of urea, which they cannot excrete in concentrated form. Significantly, if an amphibian is eating insects, the total amount of water it gets from its food (preformed and metabolic) is likely to be no more than about 15% of the amount it needs just to excrete the urea it produces from its food. This calculation emphasizes the overwhelming importance of "drink" as a water source for the majority of amphibians.

Most amphibians do not in fact drink, but instead absorb water across their skin. This absorption does not necessarily require immersion in water. Many species can gain water at substantial rates merely by pressing their ventral skin against moist soil, moss, or other substrates.[24] A region of the ventral skin at the posterior of the abdomen and extending onto the thighs—called the *pelvic patch* or *seat patch*—is often specialized for rapid water uptake. Its water permeability is modulated by insertion and retrieval of aquaporins in the cell membranes;[25] when aquaporins are inserted, the water permeability of the patch cells is increased. A medium-sized and well-hydrated leopard frog (*Rana pipiens*) sitting on wet soil in its native habitat might well absorb 6–10 g of water *per hour* from the soil across its ventral surfaces, while it simultaneously loses a like amount into the air by evaporation across its dorsal surfaces—a dramatic display of dynamism!

When they are away from sources of water, most terrestrial amphibians are able to ward off dehydration for a time by using their bladder as a canteen. If an animal starts to dehydrate, the cells in the walls of its bladder are rendered permeable to water by aquaporin insertion in the cell membranes. NaCl is actively transported out of the bladder, thereby removing solute from the bladder contents and promoting osmotic outflux of water. The capacity of the bladder to hold fluid in terrestrial frogs and toads is remarkable: The water contained in the filled bladder is equal to 20%–50% of an animal's bladder-empty weight. By contrast, in strictly aquatic amphibians, the bladder is usually tiny.

HORMONAL CONTROL OF RESPONSES TO DEHYDRATION In terrestrial amphibians, the neurohypophysial hormone arginine

---

[24] Amphibians are far from being the only animals that take up water from moist substrates. The phenomenon has been documented in certain isopods, millipedes, insects, spiders, scorpions, land crabs, and snails.

[25] Such aquaporins are regulated acutely in a cell by trafficking between intracellular locations where they are nonfunctional and the cell membrane where they enhance the permeability of the cell to transcellular osmosis.

vasotocin (see Table 16.2), called *antidiuretic hormone* (ADH), activates a suite of coordinated responses that collectively retard or reverse the process of dehydration. Release of ADH is stimulated if the volume of the body fluids (e.g., blood plasma) is decreased or if their osmotic pressure is increased. An amphibian's overall reaction to ADH has appropriately been called the amphibian *water-balance response*. In its complete form (not shown by all species), this response involves changes at three sites in the body: the kidneys, bladder, and skin. First, ADH causes the kidneys to reduce their rate of urine production and elevate the urine concentration toward isosmoticity with the blood by mechanisms discussed in Chapter 29 (see Figure 29.5). Second, ADH stimulates the bladder cells to increase their rate of NaCl reabsorption and their permeability to water by aquaporin insertion in the cell membranes, responses that augment return of water from the bladder contents to the blood. Finally, ADH causes the ventral skin through which water absorption occurs to increase its capacity for water influx—facilitating rehydration—by stimulating increased blood flow and aquaporin insertion in the cell membranes.

ADH is by no means the only hormone active in water–salt physiology. For example, *hydrins* synergize with ADH in some contexts. *Angiotensin II* (see page 451) has been shown to be a principal controller of "cutaneous drinking" in frogs, stimulating the animals to press their ventral skin against moist substrates.

HOW DO DEHYDRATION-PRONE AMPHIBIANS LIVE IN DESERTS?

As already mentioned, some species of frogs and toads, such as *Bufo cognatus* and the spadefoot *Scaphiopus couchii* in North America, have skin that provides no more protection against evaporative water loss than that of semiaquatic frogs such as leopard frogs (see Figure 28.19) yet live successfully in deserts or other arid habitats. Such desert species are in fact remarkably similar to the majority of terrestrial amphibians in all physiological respects, although some species show modest quantitative improvements over amphibians that live in moist habitats, such as by having a larger bladder, a somewhat greater tolerance of dehydration, or an accelerated pace of rehydration.

Behavior and seasonal dormancy are critical keys to the success of these desert amphibians. Dehydration can kill them *in an hour—or just a few hours*—if they are exposed in the desert. Stringent behavioral control of water loss is therefore a requirement of life. These desert amphibians spend much of their time in protective microhabitats, especially in burrows underground, and are largely nocturnal. They also employ seasonal dormancy to simply "retire from the scene" and protect their water status during dry seasons. Spadefoot toads (*S. couchii*), for example, spend many months of each year in dormancy. Overall, these desert amphibians are reclusive animals, holed up in secluded places during much of their lives. For some, dormancy dominates their lives more than activity. Their reward is that they are able to survive in deserts despite the high permeability of their skin and other vulnerabilities.

"RADICAL" PHYSIOLOGICAL SPECIALIZATIONS OCCUR IN SOME ARBOREAL, ARID-ZONE FROGS

For an amphibian to exist in arid places without being restricted to a secluded life, it must have evolved superior physiological mechanisms to cope with the challenges of dehydration stress. Biologists are gradually learning more about such mechanisms in several types of unusual

FIGURE 28.22 Arboreal frogs of the genus *Phyllomedusa* spread protective lipids secreted by integumentary glands over their skin surface The lipids sharply reduce the rate of evaporative water loss across the frogs' skin. The spreading is carried out by a series of stereotyped limb movements, as shown. (After Blaylock et al. 1976.)

arboreal frogs that live exposed lives in arid or semiarid habitats. Frogs of this sort in two genera—*Phyllomedusa* of South and Central America and *Chiromantis* of Africa—have been studied for several decades and are known to have physiological abilities to conserve water that are extraordinarily different from those of most amphibians. One distinctive trait of these frogs is that their integumentary permeability to water is exceptionally low; their rates of evaporative water loss are consequently little different from those of some lizards of similar size.[26] Cutaneous lipids are responsible for this low skin permeability in both genera. The lipids are spread on the outside of the skin rather than being incorporated within the skin tissue. *Phyllomedusa*, for example, secretes lipids (mainly waxy esters) from skin glands and spreads them on its skin surface (**FIGURE 28.22**). A second highly distinctive trait of these genera is that they excrete much of their nitrogenous waste as poorly soluble uric acid or urates (80% in *P. sauvagei*, for example).

The tiny, dramatically colorful reed frogs (*Hyperolius*) of the African savannas represent another remarkable group of arboreal frogs—a group that is only now starting to be well understood. Some species have exceptionally low skin permeabilities. They do not, however, routinely produce uric acid or other related compounds as do the frogs just discussed. During the dry season, reed frogs remain in exposed locations on the branches of bushes and trees even as they undergo profound dehydration. They stop producing

---

[26] Expressed in the same units as used in Table 28.6, the resistance to evaporative water loss of the skin in species of *Phyllomedusa* and *Chiromantis* is generally 200–400 s/cm.

urine at such times, and much of their waste nitrogen accumulates in their body fluids as urea. However, as they dehydrate, they start to synthesize guanine from the waste nitrogen retained in their bodies; in *H. viridiflavus*, 25% of waste nitrogen becomes guanine. Guanine is a low-solubility purine like uric acid (see Figure 29.24). Remarkably, the frogs deposit much of the guanine they synthesize in skin cells (iridophores), and the little animals turn bright white from its presence. The formation of guanine lengthens the time the frogs can store waste nitrogen (because it keeps the nitrogen out of solution), and it reduces solar heating by increasing the reflectance of the skin to incoming radiation!

Xeric invertebrates: Because of exquisite water conservation, some insects and arachnids have only small water needs

Certain insects are among the most successful of all animals in severe desert conditions. Being successful, they provide succulent food for other, less-adept desert dwellers.

Certainly one of the most intriguing phenomena in the living world is presented by desert ants that feed on other desert insects killed by heat. Life for diurnal insects in severe deserts is so tenuous that, every day, some individuals accidentally die of overheating despite extraordinary adaptations for desert existence. Desert ants of several species scavenge the bodies of such heat-killed insects. The deaths often occur in the heat of the day. Thus, to get moisture from their prey, the ants must venture forth from benign underground burrows in the heat of the day to gather the dead bodies before the sun quickly bakes the bodies dry. Species of such ants in deserts around the world have independently evolved unusually long legs (**FIGURE 28.23**). This is believed to be related to the fact than air temperature declines extremely steeply with altitude above the sun-heated sand. The stiltlike legs—although they elevate an ant's body just millimeters higher than it would otherwise be—can reduce the air temperature to which the body is exposed by 10°C. The ants also exploit the steep temperature gradient at times by taking breaks from desert foraging to climb up on pebbles or other high points in the desert terrain to reach even lower air temperatures than their stiltlike legs permit. These ants can tolerate tissue temperatures of 52°C–55°C (126°F–131°F). They

FIGURE 28.23 Diurnal desert ants that collect heat-killed insects are noted for long, stiltlike legs that keep them above the intensely hot sand surface Shown is *Cataglyphis diehli*. (Photograph by Rüdiger Wehner.)

thus rank with the most heat-tolerant of all animals. Nonetheless, they heat up promptly under intense sun and can easily suffer heat death themselves if they fail to act quickly as they exit their burrows, seek out recently heat-killed prey, and return to underground safety. Accurate *navigation* is crucial: After ants have traveled hundreds of meters to find prey in featureless deserts, they must find their burrows again. Ants of the genus *Cataglyphis* in the Sahara Desert have been shown to have evolved one of the most remarkable of animal navigation systems despite having a brain that weighs 1/10 of a milligram (see Figures 18.3 and 18.4).

Deserts are just the driest of the habitats in which insects and arachnids (e.g., spiders and scorpions) live. Some species prosper in a variety of other arid places. The suite of characteristics that permits many species to prosper in semiarid and arid habitats includes several physiological attributes that promote highly effective water conservation: high integumentary resistance to water loss (provided chiefly by epicuticular lipids); stringent limitation of respiratory water loss by control of the opening of the spiracles; excretion of waste nitrogen in poorly soluble forms; and an ability (at least in many insects) to produce concentrated urine (maximum osmotic U/P = 2–8).

Some flightless insects, ticks, and mites, in addition, have a way to obtain water that is *unique* in the animal kingdom: They are able to gain water from the gaseous water vapor in the air in a *steady* manner while they are at the same temperature as the air. For example, the desert cockroach *Arenivaga investigata* can gain water from the air steadily even when the ambient relative humidity is as low as 79%–83%. The mealworms (*Tenebrio*) that people often use as food for pets can gain water down to 88% relative humidity, and firebrats (*Thermobia*) can do so down to 45%. A water gain of about 10% of body weight per day is the rule when these insects are dehydrated and living in humid air. The mechanism of water uptake in many cases remains subject to debate. The site of uptake in the desert cockroach and some ticks is the mouth. Mealworms and firebrats, by contrast, absorb water via the rectum. Some investigators believe that true, primary active transport of H_2O sometimes occurs in these arthropods, although most disagree. A mechanism that is known to operate in some species is the production—at the mouth or rectum— of *localized*, superficial pockets of body fluids with such high solute concentrations that they have water vapor pressures below ambient water vapor pressure;[27] water vapor diffuses from the atmosphere into such body fluids, following the vapor pressure gradient.

Few water budgets have been worked out for insects or arachnids, but it is clear that certain species are so effective in limiting water losses that they can maintain water balance at moderate temperatures and low humidities while having *no drinking water* and *eating only air-dried foods*. Common examples include grain beetles and clothes moths. Besides the small amounts of preformed water in air-dried foods, the only sources of water for such insects are metabolic water and whatever water they may gain from atmospheric water vapor.

Because of their short generation times, insects are among the most useful animals for studies of physiological evolution using laboratory populations. Experiments using insects demonstrate that the resistance of animals to water stress can respond dramatically to evolutionary selective pressures (**BOX 28.6**).

---

[27] Recall from Chapter 5 that the water vapor pressure of a solution is a colligative property and decreases as solute concentration increases.

BOX
28.6 **The Study of Physiological Evolution by Artificial Selection**

One way to learn how the physiological features of organisms respond to evolutionary selective pressures is to expose animals to artificial (human-mediated) selection. For example, the evolutionary physiology of desiccation resistance can be studied by selecting—generation after generation—animals that are particularly resistant to desiccation and examining the traits of their descendants. In one extensive series of studies of fruit flies (*Drosophila melanogaster*), flies in successive generations of desiccation-selected populations—here termed SEL populations—were subjected to a desiccation treatment to select the parents of the next generation. Specifically, some of the flies in each generation were exposed to dry air, shortly after their emergence as adults, until 80% died; only the 20% that survived were bred to create the next generation. At each generation, flies in the SEL populations that were not subjected to desiccation treatment were used to test the populations' characteristics. The flies of control populations—CONT populations—never experienced desiccation stress.

Flies taken from the SEL and CONT populations were subjected to short-term tests of their desiccation tolerance at each

generation. In these tests, the flies in SEL populations lived longer during exposure to desiccation than those in CONT populations. **Part 1** of the figure shows the average number of hours that SEL flies lived longer as a function of the number of generations of selection. The SEL flies survived more than 30 h longer after 37 generations. Two physiological mechanisms have been identified that explain the greater capacity of SEL flies to survive desiccation. First, because of evolution driven by artificial selection, flies in SEL populations have more body water than CONT flies (**Part 2** of the figure). Second, the flies in SEL populations exhibit greater resistance to water loss.

Differences between flies in desiccation-selected (SEL) populations and control (CONT) populations in relation to the number of generations of selection Flies used to test the population characteristics shown were chosen at random from the two types of populations. Five separate populations of each type were studied at each generation of selection. Error bars show standard errors. (After Archer et al. 2007.)

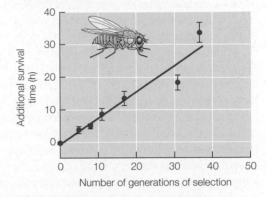

(1) Additional number of hours that flies in SEL populations survive a desiccation test, relative to CONT flies

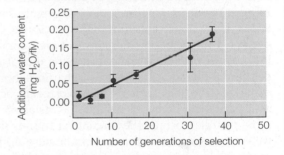

(2) Additional water content (mg H_2O/fly) of flies in SEL populations, relative to CONT flies

Xeric vertebrates: Studies of lizards and small mammals help clarify the complexities of desert existence

When one treks to the driest deserts, one sees only a few sorts of animals leading active lives. Lizards and small mammals[28] are two groups that especially stand out, in addition to insects and arachnids. A fact that draws interest to both the lizards and small mammals is that they often have no chance of finding drinking water for months on end. Because they cannot travel far, they have drinking water only when local rains provide it, and rains come only sporadically. Desert lizards characteristically eat insects or living plant tissues, foods that supply significant amounts of preformed water. Many species of small mammals in deserts also acquire substantial quantities of preformed water in their food because they eat primarily insects or plants; in North America, these animals include pack rats, grasshopper mice, and ground squirrels. In all the major deserts of the world, however, there have evolved extraordinary species of small mammals that live away from drinking water and eat principally air-dried seeds or other air-dried plant matter. The kangaroo rats (*Dipodomys*) of North America's southwestern

deserts (see the opening photo of Chapter 29) provide classic examples. Others include the kangaroo mice and pocket mice of North America, some gerbils and jerboas of Old World deserts, and the hopping mice of Australian deserts.

A significant, intriguing trait of the lizards that prosper away from drinking water is that, like other lizards, they are diurnal. The key traits that permit the existence of lizards as diurnal animals in the driest places on Earth include, first, their relatively low metabolic rates (see Figure 7.9). A low metabolic rate not only can greatly reduce water losses, it also reduces food needs—an asset in habitats where populations of food organisms are themselves stressed and relatively unproductive. Second, the desert lizards excrete their nitrogenous wastes as water-sparing uric acid or urates. They also use behavior to avoid stresses that are avoidable, as by moving into underground burrows or shadows during the heat of the day. Some species employ salt glands. Finally, desert lizards tend to be remarkably tolerant of large shifts in their body-fluid composition, such as high blood solute concentrations, during dehydration. They can survive perturbations of their body fluids—often for long periods—that would kill a mammal or bird.

The small mammals of deserts operate on a different, higher scale of metabolic intensity than the lizards. High metabolic intensity is in

[28] See Chapter 30 for a discussion of camels, oryxes, and other large mammals in deserts and dry savannas.

itself a liability in deserts; it raises the rate of respiratory evaporative water loss, as we have seen, and it can contribute enough endogenously produced heat to add significantly to heat stress. Small mammals that live in deserts have, in general, evolved lower basal metabolic rates than nondesert mammals of the same body size. Some species, moreover, undergo daily torpor or estivation when they are short of food or dehydrated. Still, when small mammals are active, their metabolic rates are far higher than those of lizards.

The species of small desert mammals that, by far, have attracted the most curiosity are the ones that eat primarily air-dried plant matter. In classic studies 60 years ago that contributed to the genesis of modern animal physiology, researchers produced the water-balance summary for kangaroo rats (*D. merriami*) in **FIGURE 28.24**. As we discuss this information, it will be important to keep in mind that the animals were studied at 25°C (77°F). They had no drinking water and were fed only barley grain. Each of the five lines in the figure shows how a key attribute of their water physiology varied with the humidity of the air. The red lines show the animals' *minimum* water losses by evaporation and elimination of urine and feces. Kangaroo rats have several specializations for conserving water. They exhibit exceptionally low cutaneous permeability to water; they cool their exhalant air by nasal countercurrent exchange; they can produce very concentrated urine (osmotic U/P = 14); and they can restrict their fecal water losses exceptionally. Their minimum evaporative, urinary, and fecal water losses are *stacked on top of each other* in the figure so that the heavy red line at the top represents their *total* water losses. Evaporative losses decrease with increasing humidity, but minimum urinary and fecal losses are independent of humidity, so the animals' total water losses decrease as humidity increases. The blue lines show the water inputs of the kangaroo rats. Their production of metabolic water is the same at all humidities because it depends on metabolic rate, which is the same regardless of humidity. In contrast, the rats' input of preformed water increases as humidity increases because air-dried plant matter, such as barley grain, comes to equilibrium with the water vapor in the air and contains more water when the humidity is high. The animals' inputs of metabolic and preformed water are stacked so that the heavy blue line represents their total water intake.

Kangaroo rats can be in water balance if their total water inputs equal or exceed their total, minimum water losses. Based on Figure 28.24, therefore, the animals can be in water balance while eating air-dried grain and drinking nothing if the relative humidity is above about 10%. Most of their water input is metabolic water. As stressed in Chapter 27 (see page 736), this is not because they produce exceptional amounts of metabolic water. It is because they conserve water so well that metabolic water can meet most of their needs.

Now we need to recall that the studies in Figure 28.24 were done at 25°C. The researchers who carried out the studies obtained information on soil temperatures and nocturnal air temperatures in the parts of the southwestern deserts of the United States where they worked.[29] They concluded that kangaroo rats in those deserts are usually not exposed to temperatures higher than 25°C. Being nocturnal, the kangaroo rats live in their cool burrows during the heat of the day and emerge onto the desert surface only in the cool of the night. Thus the researchers concluded—based on Figure 28.24—that kangaroo rats are able to stay in water balance in the desert while eating air-dried seeds or other air-dried plant material and drinking nothing.

[29] For soil temperatures they used the information in Figure 1.18.

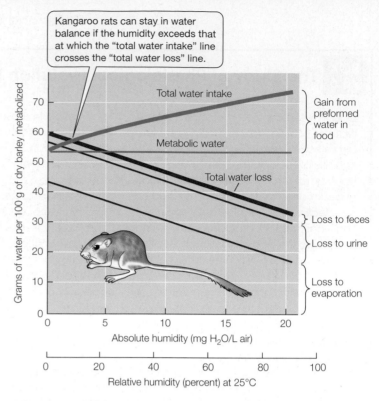

Kangaroo rats can stay in water balance if the humidity exceeds that at which the "total water intake" line crosses the "total water loss" line.

FIGURE 28.24 A kangaroo rat water budget For this study, carried out at 25°C, kangaroo rats (*Dipodomys merriami*) were fed husked barley grain at equilibrium with atmospheric moisture and provided no drinking water (they usually do not drink even if water is available). The water losses shown (red lines) are the minimum possible water losses; in actuality, if water intake exceeds minimum losses, the animals increase their losses (as by excreting more-dilute urine) so that losses match inputs (blue lines). The amounts of water graphed on the y axis are those gained or lost each time 100 g (dry weight) of barley is consumed. Under the conditions of study, the animals normally consume 100 g of barley in about a month. (After Schmidt-Nielson and Schmidt-Nielsen 1951.)

How is the water physiology of kangaroo rats affected if the ambient temperature is different from 25°C? The *dominant* modes of water gain and loss are metabolic water production (MWP) and evaporative water loss (EWL) (see Figure 28.24). A straightforward (although not complete) way to gain insight into the effect of temperature on water balance is to examine how MWP and EWL relate to each other as temperature varies. The two are often expressed as a ratio of gain over loss—**MWP/EWL**—for this purpose. The numerator, MWP, tends to *increase* as ambient temperature decreases because MWP varies with metabolic rate, which increases as temperature decreases below the thermoneutral zone (see Figure 10.28). The denominator, EWL, by contrast, typically tends to *decrease* as ambient temperature decreases in small mammals and birds. The MWP/EWL ratio (water gain over water loss) therefore becomes dramatically more favorable as the ambient temperature falls (**FIGURE 28.25**).

Based on this analysis of the MWP/EWL ratio, we would expect kangaroo rats in the wild to be under far less water stress during the cool seasons of the year than during the warm seasons. In fact, studies of kangaroo rats (*D. merriami*) in the wild match this expectation. In one population, the urinary osmotic pressure of the rats averaged about 1000 mOsm (U/P = 3) in midwinter but rose to about 4000 mOsm (U/P = 11) in midsummer. Low winter temperatures apparently placed the rats in such a favorable situa-

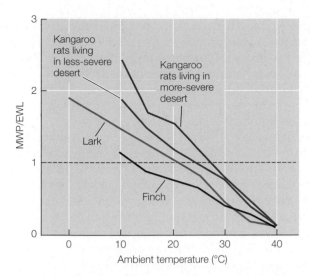

FIGURE 28.25 An index of water balance: metabolic water production (MWP) as a ratio of total evaporative water loss (EWL) This ratio provides a useful (although incomplete) index of ability to stay in water balance for species in which metabolic water production and evaporation are the principal processes of water gain and loss. Shown are data for two populations of kangaroo rats, *Dipodomys merriami*, and two species of desert birds, the dune lark (*Mirafra erythrochlamys*) and zebra finch (*Poephilia guttata*). (Data from Tracy and Walsberg 2001; Williams 2001.)

Xeric vertebrates: Some desert birds have specialized physiological properties

The fundamental conflict between heat balance and water balance in hot deserts is emphasized by the study of birds because most desert birds are diurnal and therefore do not evade the heat of the day. Despite the fact that birds often confront the stresses of deserts head-on, the species of desert birds first studied seemed to exhibit remarkably few specializations for desert existence. Birds as a group have higher body temperatures than mammals, are especially tolerant of hyperthermia, synthesize uric acid, and can fly to distant watering places. Such traits, shared by birds as a group, seemed initially to be sufficient for successful desert existence. A dogma developed that the presence of birds in deserts is largely a consequence of a happy marriage between the standard features of all birds and the requirements of desert life.

Physiologists have recently come to realize that this dogma is not correct. For example, when data on large numbers of species are analyzed statistically, desert birds turn out, on average, to be systemically different from other birds. One difference is that desert birds tend to exhibit relatively high resistance to evaporative water loss.

A particularly instructive comparative study is being carried out on a group of closely related birds, the Old World larks, which occupy an extreme diversity of habitats. Whereas certain species of Old World larks occur in moist habitats, others occur in semiarid places, and some live in hot, hyperarid deserts. The hoopoe lark (*Alaemon alaudipes*), seen in the inset in **FIGURE 28.26**, exemplifies the latter. When hoopoe larks are living in hyperarid deserts such as those of the Arabian Peninsula, they eat insects and other arthropods and thereby get preformed water from their food, but

tion that they had a water surplus and didn't need to concentrate their urine maximally.

Although most populations of kangaroo rats live in places where they experience the conditions we have discussed up to now, some populations live in far hotter places (in the Sonoran Desert) where the animals probably encounter temperatures as high as 35°C (95°F) both in their burrows and on the desert surface at night. These animals cannot achieve water balance while eating air-dried foods, and they can't make up the difference by drinking because no drinking water is present. Fortunately insects are able to exist and accumulate body fluids in this severe environment. The kangaroo rats probably achieve water balance by adding insects and green-plant parts to their usual diet of air-dried seeds: a striking example of one species taking advantage of adaptations of other species.

FIGURE 28.26 Lark species along a gradient of water availability Water availability in a species' habitat is quantified by an index that can vary from 1.5 (very arid) to 3.5 (moist). In each graph, each symbol represents a different species of lark. Metabolic rates and rates of water loss are expressed using indices designed to be independent of body weight. Metabolic rates shown are basal metabolic rates (red) and average free-living metabolic rates (black). Rates of water loss shown are rates of total evaporative water loss (red) and free-living water turnover rates (black). Activity is percentage of daylight hours spent active. Shown in the photograph is a desert-dwelling hoopoe lark (*Alaemon alaudipes*), one of the species found in the most arid habitats. (After Tieleman 2005.)

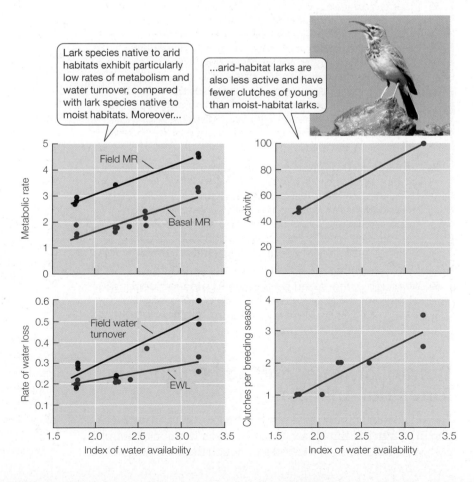

they never drink—water is almost never within flight distance. As Figure 28.26 illustrates, when diverse species of larks are arrayed along an axis of water availability—with a water-availability index of 1.5 representing a very arid habitat and an index of 3.5 representing a moist habitat—the species native to dry habitats display distinctive features. Included are physiological specializations: They exhibit low metabolic heat production and low water turnover. The low water turnover of the dry-habitat species is accounted for, in part, by cutaneous lipids that, in comparison with those of other larks, are unusually protective, and this difference in water loss between desert birds and other birds correlates with differences in the composition of their cutaneous lipids.

These new revelations in the study of birds help emphasize that although molecular biology is one of today's most important cutting edges, the comparative study of related species (the products of evolution) in divergent habitats remains a powerful source of insight for understanding life on Earth.

SUMMARY
Animals on Land: Case Studies

■ Most terrestrial amphibians have meager physiological abilities to limit water loss because their skin is highly permeable to water and they cannot make urine that is hyperosmotic to their body fluids. Stringent behavioral control of water balance and seasonal dormancy are essential for their success in arid places. A few types of arboreal amphibians that live in arid areas have unusual adaptations such as cutaneous lipids that protect against rapid evaporative water loss.

■ Insects and lizards are among the animals that are most physiologically capable of living in the driest places on Earth. Their key traits for existence in extreme places include very low integumentary permeability to water, relatively low metabolic rates, excretion of poorly soluble nitrogenous wastes, and tolerance of profound changes in body-fluid composition. Insects can produce hyperosmotic urine and sometimes gain water from atmospheric water vapor, but the fact that they are small is in itself a physiological (although not behavioral) disadvantage.

■ Some small mammals that eat predominantly air-dried foods (e.g., seeds) live in deserts without needing to drink. In addition to having highly evolved physiological mechanisms of water conservation, they depend on behavioral selection of relatively benign microhabitats to maintain water balance. In the hottest places they live, they probably must supplement their diet with water-rich foods such as insects.

■ Although some desert birds seem to succeed because of general avian properties that are of advantage under desert conditions, others exhibit dramatic specializations for desert existence.

Control of Water and Salt Balance in Terrestrial Animals

The kidneys and other organs responsible for water and salt regulation in terrestrial animals are generally under hormonal control in both vertebrates and invertebrates, although the nervous system is

sometimes a principal player, as in the control of avian salt glands. Some of the most important hormones involved are the *antidiuretic* and *diuretic* hormones. **Diuresis** is the production of an abundant (and usually dilute) urine. A **diuretic** hormone promotes diuresis. An **antidiuretic** hormone opposes diuresis or, in other words, modulates the excretory organs so that a relatively low volume of (usually concentrated) urine is produced.

Vertebrates are considered to produce only antidiuretic hormones. In insects, however, diuretic as well as antidiuretic hormones occur. Certain blood-sucking insects, for example, have a diuretic hormone that is secreted after a blood meal, promoting rapid excretion of much of the water in the blood and thereby concentrating the nutritious part of the meal (e.g., proteins) in the gut.

In vertebrates, which will be our focus in this brief discussion, three hormones or types of hormones play particularly important roles in the regulation of water–salt physiology:

1. **Antidiuretic hormone** (**ADH**), produced by the neurohypophysis (see pages 449–451 and Table 16.2)

2. Mineralocorticoids, most notably **aldosterone**, produced by the adrenal cortex or homologous interrenal tissue (see page 440)

3. **Natriuretic** hormones (see page 451)

Focusing on ADH first, its principal effect in mammals—and its principal effect on the kidneys in other terrestrial vertebrates—is to control the renal excretion of *pure water* (*osmotically free water*) relatively independently of solute excretion. To see this point, let's use a mammal as an example. Consider that a mammal has a certain *quantity* of urea, salts, and other solutes that it must excrete per day. If the solutes collectively are excreted at the maximum concentration the individual can achieve, the accompanying water loss can be considered to be strictly obligated by solute excretion. However, if the solutes are excreted at less than maximum concentration, then the urine contains additional water that is not obligated by solute excretion; in essence, the urine is diluted by the addition of pure water above and beyond the amount needed to void solutes, and the additional water represents a specific excretion of water itself. The urine can therefore be considered to consist of two components: (1) the solutes and their associated water and (2) a quantity of additional pure water. ADH controls the magnitude of the latter component. If a person, for instance, has a constant daily solute output, but consumes little water on one day and a lot of water the next day, ADH is secreted on the first day, and—because of its antidiuretic effect—restricts the amount of water excreted with the solutes. However, ADH secretion is reduced on the second day; this permits diuresis and thus the excretion of a great deal of water with the solutes.[30]

The principal effect of aldosterone is to cause the kidneys to hold back Na^+ from excretion while promoting the excretion of K^+ in the urine. These actions of aldosterone most obviously affect the quantities of Na^+ and K^+ in the body fluids. Less obviously, the action of aldosterone is one of the most important factors in the routine regulation of extracellular-fluid volume (including blood volume). To understand this latter role of aldosterone, consider that Na^+ is largely excluded from intracellular fluids by being actively transported out of cells (see Figure 5.10). Increases or decreases in the amount of

---

[30] The action of ADH is discussed in more detail in Chapters 16 and 29.

Na$^+$ in the body therefore lead to quantitatively similar increases and decreases in the amount of Na$^+$ (and accompanying anions, notably Cl$^-$) in the extracellular fluids. When Na$^+$ is retained in the body by being held back from the urine—and Na$^+$ thus accumulates in the extracellular fluids—the systems controlling the osmotic pressure of the extracellular fluids cause water to be retained as well, so as to maintain a normal extracellular-fluid osmotic pressure. In that way, the extracellular-fluid volume is expanded. Conversely, if the body's Na$^+$ content is reduced, water from the extracellular fluids is excreted to maintain a stable extracellular-fluid osmotic pressure. Actually, although aldosterone has its major effects on the kidneys, it often functions as an all-purpose Na$^+$-retention hormone. In mammals, for example, it stimulates the salivary glands, sweat glands, and intestines to increase reabsorption of Na$^+$. Moreover, aldosterone stimulates salt appetite.

The natriuretic hormones—often termed *natriuretic peptides* because they are types of peptides—promote addition of Na$^+$ to the urine, as their name indicates (*natri*, "sodium"; *uretic*, "having to do with excretion in the urine"). Despite an explosion of knowledge since 1990, many uncertainties remain about the functioning of the natriuretic hormones because there are multiple chemical forms—each of which potentially has multiple effects—which can differ from one set of vertebrates to another. The heart and certain brain regions (e.g., hypothalamus) are the chief sites of secretion in mammals. The atrial natriuretic peptide (ANP) of mammals is the best understood of these hormones. It is called *atrial natriuretic peptide* because it is produced principally by the atria of the heart (see page 451). Mammalian ANP has actions that in many ways are opposite to those of aldosterone. It inhibits aldosterone secretion and directly modulates the kidneys to promote Na$^+$ excretion by increasing both urine volume and urine Na$^+$ concentration.

The mechanisms of control of the secretion of ADH, aldosterone, and ANP are complex and incompletely understood. All of these hormones participate in negative feedback systems—now discussed—that act to stabilize the volume and osmotic pressure of the extracellular body fluids.

For the volume of the extracellular body fluids to be regulated, either the volume itself or reliable correlates of volume must be *sensed*, so that the regulatory systems will "know" whether to promote an increase or a decrease in volume at any particular moment. Probably volume itself is not sensed. However, good evidence exists that certain correlates of volume *are* sensed; for example, both the blood pressure and the extent to which blood-vessel walls are stretched are functions of blood volume, and pressure and stretch receptors that participate in volume regulation are known (e.g., in and around the heart). Similarly, if the osmotic pressure of the extracellular body fluids is to be regulated, either it or close correlates must be sensed. No doubt exists that receptors for osmoregulation are present (e.g., in the hypothalamus), but whether they respond to osmotic pressure itself, Na$^+$ concentration, or other correlated properties remains debated.

Secretion of ADH is controlled in part by changes in blood volume; pressure sensors and other sensors of volume affect ADH secretion by way of nervous inputs to the hypothalamus and also via the renin–angiotensin–aldosterone system (see next paragraph). Decreases in blood pressure activate ADH secretion, a response favoring fluid retention. Secretion of ADH is also under the control of osmoreceptors or other detectors of the concentration of the body fluids. Increases in the osmotic pressure of the body fluids induce increased ADH secretion; the ADH then favors the specific retention of water by the renal tubules, thereby tending to lower the osmotic pressure of the body fluids.

Aldosterone secretion is controlled to a major extent by another hormonal system, the *renin–angiotensin system* (see Figure 16.17), which itself is partly under the control of blood-pressure receptors and other detectors of blood volume. A decrease in blood pressure, signifying a reduction in blood volume, activates secretion by the kidneys of the hormone *renin* (pronounced "ree-nin"), which in turn causes formation in the blood of *angiotensin II*. The angiotensin stimulates the adrenal glands to secrete aldosterone, which induces increased Na$^+$ reabsorption from the urine, tending to expand extracellular-fluid volume and raise blood pressure.

Secretion of ANP is stimulated by expansion of extracellular-fluid volume, which is detected by stretching of the walls of the atria in the heart. ANP then promotes loss of extracellular fluid. One of ANP's principal overall effects is a decrease in blood pressure.

SUMMARY
Control of Water and Salt Balance in Terrestrial Animals

■ The control of body-fluid volume, composition, and osmotic pressure is mediated mostly by hormones that are secreted under control of negative feedback systems. Stretch or pressure receptors provide information on blood volume, and osmoreceptors provide information on blood osmotic pressure.

■ In vertebrates, antidiuretic hormone (ADH) regulates the amount of pure, osmotically free water that is excreted by the kidneys; it does so by controlling whether a more-than-minimum amount of water is excreted with solutes.

■ Aldosterone and natriuretic hormones in vertebrates act to promote Na$^+$ retention or Na$^+$ excretion, respectively. The control of body Na$^+$ content by these hormones helps to control extracellular-fluid volume because body Na$^+$ is present mostly in the extracellular fluids.

STUDY QUESTIONS

1. It has been said that in terms of water balance, behavior can compensate for physiology, or physiology can compensate for behavior. Do you agree? Give examples to support your answer.

2. When a salmon or other teleost fish migrates from seawater into freshwater, what are all the changes that take place or are likely to take place in its patterns of water–salt physiology?

3. Theory predicts that when both terrestrial and freshwater animals evolve higher metabolic rates, they can be expected to evolve greater challenges to maintaining water balance. Explain why this is so for both terrestrial and freshwater animals.

4. We noted in Chapter 10 that dogs are believed to benefit when they pant by breathing at a fixed resonant frequency. Although dogs inhale and exhale exclusively through the nose when not under heat stress, they exhale orally to some (variable) extent when panting. Air exhaled by way of the mouth remains nearly at deep-body temperature. Explain how a panting dog could *vary its rate of evaporative cooling*—even while *breathing at a fixed frequency*—by modulating how much it exhales by way of its nose or mouth.

5. Animals often face trade-offs, and one of the goals of modern physiology is to understand mechanistically why improvement of performance in one way may degrade performance in other ways. In freshwater fish, explain why O_2 uptake and Ca^{2+} uptake probably cannot be simultaneously maximized. In other words, why is there a trade-off between the ability to take up O_2 and the ability to take up Ca^{2+}?

6. Being as specific as possible, discuss how global warming can be expected to alter the water physiology of animals and what the ecological consequences might be. Figure 28.25 might provide a useful starting point, but do not limit yourself to it. Parks set aside to protect animals have defined geographical boundaries. How might the relation between global warming and water physiology affect the effectiveness of the park program?

7. Walter Cannon, who coined the term *homeostasis*, argued that lizards and amphibians are less highly evolved than mammals and birds because they are not as fully homeostatic as are mammals and birds. Considering water–salt relations and any other aspects of physiology you find to be pertinent, explain in detail why you agree or disagree.

8. Green crabs (*Carcinus maenas*) are hyperosmotic regulators in brackish waters. When water salinity is lowered, a green crab responds by increasing its rate of urine production; its urine output, on average, rises from 4% to 30% of its body weight per day when the ambient salinity is reduced from 35 g/kg to 14 g/kg. Explain the value of this response. If one arranges to keep a green crab in full-strength seawater and bathe just its antennules with an alternative water source, urine production increases as the salinity of the water bathing the antennules is lowered. What can one conclude from this result?

9. Related species of terrestrial animals typically display allometric relations between body-water dynamism and body size. For example, the weight-specific rate of evaporative water loss tends to decrease allometrically as body size increases. What are the mechanistic reasons for these relationships?

10. We discussed the fact in Chapter 23 (see Figure 23.11) that the efficiency of O_2 exchange across the gills of teleost fish is enhanced by countercurrent exchange between the water pumped over the gills and the blood flowing through the secondary lamellae of the gills. A recent review article makes the point that the countercurrent arrangement of blood flow and water flow—which has positive consequences for O_2 exchange—has disadvantageous side effects because it enhances osmotic water uptake by freshwater fish, osmotic water loss by marine teleosts, loss of NaCl by diffusion in freshwater fish, and NaCl gain by diffusion in seawater teleosts. Do you agree or disagree? Explain.

11. Probably the most complete account of water and salt balance in a marine mammal is not for a seal or whale but for a remarkable bat, *Pizonyx vivesi*, that lives on desert islands in the Gulf of California, where it subsists on fish and crustaceans. It gets much of its water as preformed and metabolic water from its food. However, it needs more water than its food provides, and the amount of additional water it needs increases as it flies more, because flight greatly increases its rate of evaporative water loss. The bat can concentrate NaCl to about 620 mM in its urine. Based on this information and the composition of seawater (see Table 27.1), could it maintain water balance by drinking increasing amounts of seawater as it flies more? What additional information would you need to be certain?

Go to **sites.sinauer.com/animalphys4e** for box extensions, quizzes, flashcards, and other resources.

REFERENCES

Bradley, T. J. 2009. *Animal Osmoregulation.* **Oxford University Press, Oxford, UK.** A short textbook on the water–salt physiology of animals.

Bradley, T. J., A. E. Williams, and M. R. Rose. 1999. Physiological responses to selection for desiccation resistance in *Drosophila melanogaster. Am. Zool.* **39: 337–345.**

Evans, D. H. 2008. Teleost fish osmoregulation: what we have learned since August Krogh, Homer Smith, and Ancel Keys. *Am. J. Physiol.* **295: R704–R713.**

Evans, D. H. (ed.). 2009. *Osmotic and Ionic Regulation. Cells and Animals.* **CRC Press, Boca Raton, FL.** An advanced treatise with chapters written by experts on most of the major groups of animals.

Fu, C., J. M. Wilson, P. J. Rombough, and C. J. Brauner. 2010. Ions first: Na+ uptake shifts from the skin to the gills before O_2 uptake in developing rainbow trout, *Oncorhynchus mykiss. Proc. R. Soc. London, Ser. B* **277: 1553–1560.**

Hammerschlag, N. 2006. Osmoregulation in elasmobranchs: a review for fish biologists, behaviourists and ecologists. *Mar. Freshw. Behav. Physiol.* **39: 209–228.**

Jönsson, K. I., E. Rabbow, R. O. Shill, and two additional authors. 2008. Tardigrades survive exposure to space in low Earth orbit. *Curr. Biol.* **18: R729–R731.**

Kirschner, L. B. 2004. The mechanism of sodium chloride uptake in hyperregulating aquatic animals. *J. Exp. Biol.* **207: 1439–1452.**

Perry, S. F., L. Rivero-Lopez, B. McNeill, and J. Wilson. 2006. Fooling a freshwater fish: how dietary salt transforms the rainbow trout gill into a seawater gill phenotype. *J. Exp. Biol.* **209: 4591–4596.**

Peterson, C. C. 1996. Anhomeostasis: Seasonal water and solute relations in two populations of the desert tortoise (*Gopherus agassizii***) during chronic drought.** *Physiol. Zool.* **69: 1324–1358.** A report of a study in which the concept of homeostasis did not seem to apply.

Walsberg, G. E. 2000. Small mammals in hot deserts: Some generalizations revisited. *BioScience* **50: 109–120.** Although this article focuses exclusively on small mammals in hot deserts, it is worthwhile reading for anyone interested in the physiology of free-living animals because of its commentaries on the use and abuse of model systems.

Williams, J. B., and B. I. Tieleman. 2001. Physiological ecology and behavior of desert birds. In V. Nolan, Jr., and C. F. Thompson (eds.), *Current Ornithology,* **vol. 16, pp. 299–353. Springer, New York.** An excellent example of a probing exploration of accumulated information to ask the question: Do the available data support the old paradigm, or do they actually support a new paradigm?

See also **Additional References** *and* **Figure and Table Citations**.

An animal's body fluids are dynamic—continuously gaining and losing water, inorganic ions, and organic solutes. When a kangaroo rat in the desert is successfully maintaining water balance in moderate summer weather, about 15% of the water in its body fluids is lost and replaced each day. This rate of turnover is low for a mammal of its body size; the mice and rats of moist temperate habitats often turn over 35% or more of their water per day. Humans, being larger, have lower percentage turnover rates, but still when when people go about their ordinary daily lives, about 7% of their body water is lost and replaced every day. Inorganic ions and organic solutes in the body fluids also undergo incessant turnover.

The great dynamism of the body fluids means that their composition is continuously in danger of being shifted away from normal. As water, for example, leaves the body fluids and is replaced each day, any mismatch between the rates of loss and replacement can quickly cause the body fluids to become too dilute or too concentrated.

For the blood and other body fluids to be maintained at a normal composition, an animal requires organs that are capable of correcting any departures from normal that develop. The kidneys bear primary responsibility for this task in terrestrial animals such as kanagaroo rats. In aquatic animals such as fish and crayfish, the kidneys and gills are the organs primarily responsible.

Species of small mammals that have evolved in deserts are, among all animals, the extreme performers in their ability to concentrate their urine Excretion of dissolved wastes in as little water as possible is a key to maintaining water balance in water-poor environments. Around the world, each desert has its own assemblage of small mammals. Some of these unique animals—such as the kangaroo rats (main photo) of the American southwest—have been carefully studied. Others—such as the jerboas (inset) of Middle East deserts—have been little studied by physiologists. Shown here are Merriam's kangaroo rat (*Dipodomys merriami*) from the American southwest and the four-toed jerboa (*Allactaga tetradactyla*) from the northern Sahara in Egypt and Libya.

Kangaroo rats and other desert rodents that have been studied have a status in the physiology of kidney function not unlike the status of cheetahs in the physiology of running. The kidneys of these desert rodents represent the ultimate product of evolution in their ability to concentrate urine. Most types of animals cannot concentrate their urine at all: That is, most cannot produce urine with an osmotic pressure higher than the osmotic pressure of their blood plasma. Some insects can make urine that is 6–8 times higher in osmotic pressure than their blood plasma. Birds also can concentrate their urine, typically to an osmotic pressure that is 3 times their blood osmotic pressure, or lower. Many desert rodents, however, achieve far higher urine concentrations than any of these other groups of animals. Certain species of kangaroo rats can make urine that is 14 times higher in osmotic pressure than their blood plasma, and some of the desert hopping mice of Australia can make urine more than 20 times higher.[1]

The advantage of the high concentrating ability of the kidneys of desert rodents is made clear when we recognize the challenges these animals face in their severe environments. During the dynamic daily flux of water and solutes in and out of the body, the blood plasma of desert rodents tends often to be shifted toward concentrations higher than normal; dehydration in the desert, for example, raises the plasma concentration. When an animal's kidneys can produce urine with a higher osmotic pressure than the blood plasma, they can dilute the blood.[2] The kidneys of desert rodents, with their unique abilities to produce urine that is very hyperosmotic to the blood plasma, are exceptionally suited to lowering the osmotic pressure of the blood plasma when the blood has become too concentrated.

Considering animals in general, what are **kidneys**? We will see in this chapter that the kidneys of various types of animals are very diverse in morphology as well as in details of their physiology. All kidneys, however, have three features in common. First, they all consist of tubular elements that discharge directly or indirectly to the outside world. Second, they all produce and eliminate aqueous solutions derived from the blood plasma or other extracellular body fluids. Third, *their function is the regulation of the composition and volume of the blood plasma and other extracellular body fluids by means of controlled excretion of solutes and water.*

Urine, the product of the kidneys, is typically a complex solution containing multiple inorganic and organic solutes. All the constituents of the urine—including the water—are drawn from the blood plasma, and the urine concentration of each affects the blood concentration according to the principles we discussed in Chapter 27 (see page 733 and Figure 27.7). The urine often contains nitrogenous wastes, but the role of the urine is much more far-reaching than merely excreting waste nitrogen. The urine of a mammal, for example—although it contains urea (the nitrogenous end product)—also contains Na^+, Cl^-, K^+, PO_4^{3-}, SO_4^{2-}, creatinine, and numerous other components. The kidneys excrete each of these in greater or lesser amounts day by day, closely regulating the concentration of each in the blood plasma. The kidneys also excrete greater or lesser amounts of H^+ in the urine, thereby helping to maintain a steady blood pH. Moreover, the kidneys regulate the osmotic pressure of the blood by means of the controlled excretion of water relative to total solutes. It seems almost impossible—but is true—that the kidneys perform all these functions *simultaneously* by structuring the composition of a *single* fluid output: the urine.

Basic Mechanisms of Kidney Function

Urine formation can usually be conceptualized as occurring in two steps, although these "steps" may sometimes be partly contemporaneous. First, an aqueous solution, called **primary urine**, is introduced into the kidney tubules. Second, this solution is modified as it moves through the kidney tubules and other excretory passages, ultimately becoming the **definitive urine** that is eliminated.

Primary urine is introduced into kidney tubules by ultrafiltration or secretion

One widespread mechanism by which fluid is introduced into kidney tubules is *ultrafiltration*. This is the mechanism used in most vertebrates and in many invertebrates, such as molluscs and decapod crustaceans (e.g., crayfish and crabs). Ultrafiltration into a kidney tubule occurs when the hydrostatic pressure is higher outside the tubule than inside the tubule lumen[3] at a place where the tubule wall is structured in a specialized, minutely porous way that permits fluid to pass through the wall. Under these circumstances, the difference in hydrostatic pressure—provided the difference is great enough[4]—forces fluid to enter the tubule lumen through the wall by means of pressure-driven *bulk flow*, or *streaming*. This flow is termed **ultrafiltration**. The process is literally a form of *filtration* because solutes of large molecular size typically are unable to pass through the wall of the tubule. Thus the fluid introduced into the tubule lumen—which is termed a **filtrate** or **ultrafiltrate**—consists only of water and the subset of solutes that are able to stream through with the water. The blood plasma is the source of the water and solutes that stream through. Although there are exceptions in some groups of invertebrates, the *blood pressure produced by the heart* is typically the pressure that drives ultrafiltration, explaining why heart disease can interfere with urine formation.

To understand the formation of primary urine by ultrafiltration more fully, let's examine the process in the vertebrate kidney. Each kidney consists of many tubules, called **nephrons**, the walls of which consist of a single layer of epithelial cells (see Figure 2.6B). As diagrammed in **FIGURE 29.1A**, each nephron *begins blindly* with its walls thrown into a hemispherical, invaginated structure termed a **Bowman's capsule**, named after William Bowman (1816–1892), who first described it. Tucked inside each Bowman's capsule is an anastomosing cluster of blood capillaries, termed a **glomerulus** (**FIGURE 29.1B**), which is supplied with blood at relatively high pressure by branches of the renal artery. A Bowman's capsule and its glomerulus together constitute a

---

[1] Urine this concentrated is almost rich enough in salts for the dissolved salts to crystallize out of solution.

[2] This important point is explained in Chapter 27 (see Figure 27.7).

[3] The *lumen* of a hollow structure such as a kidney tubule is the open central cavity.

[4] We discuss this important topic shortly.

(A) The general form of a vertebrate nephron at the end where primary urine is formed

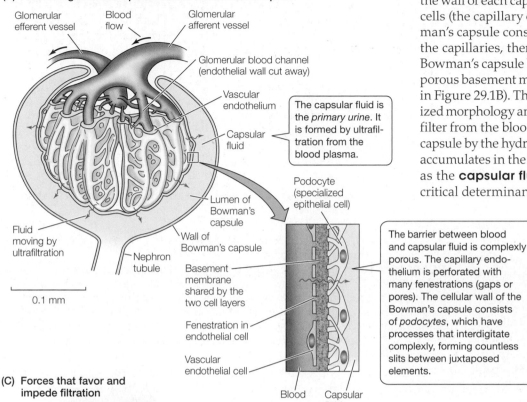

Bowman's capsule

(B) A human glomerulus positioned in a Bowman's capsule

Glomerular efferent vessel

Blood flow

Glomerular afferent vessel

Glomerular blood channel (endothelial wall cut away)

Vascular endothelium

> The capsular fluid is the *primary urine*. It is formed by ultrafiltration from the blood plasma.

Capsular fluid

Podocyte (specialized epithelial cell)

Fluid moving by ultrafiltration

Lumen of Bowman's capsule

Wall of Bowman's capsule

Nephron tubule

Basement membrane shared by the two cell layers

> The barrier between blood and capsular fluid is complexly porous. The capillary endothelium is perforated with many fenestrations (gaps or pores). The cellular wall of the Bowman's capsule consists of *podocytes*, which have processes that interdigitate complexly, forming countless slits between juxtaposed elements.

Fenestration in endothelial cell

Vascular endothelial cell

0.1 mm

Blood   Capsular fluid

(C) Forces that favor and impede filtration

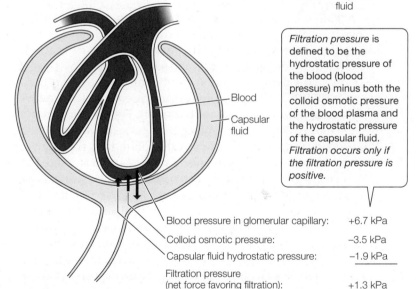

Blood

Capsular fluid

> *Filtration pressure* is defined to be the hydrostatic pressure of the blood (blood pressure) minus both the colloid osmotic pressure of the blood plasma and the hydrostatic pressure of the capsular fluid. *Filtration occurs only if the filtration pressure is positive.*

Blood pressure in glomerular capillary:	+6.7 kPa
Colloid osmotic pressure:	−3.5 kPa
Capsular fluid hydrostatic pressure:	−1.9 kPa
Filtration pressure (net force favoring filtration):	+1.3 kPa

FIGURE 29.1 The structural and functional basis for formation of primary urine by ultrafiltration in the vertebrate kidney (A) The blind end of a vertebrate nephron, where ultrafiltration occurs. (B) A human renal corpuscle, consisting of glomerulus and Bowman's capsule. The vascular endothelium that forms the walls of the glomerular capillaries has been cut away at the top of the drawing, so that only the blood channels are shown there. The Bowman's capsule is drawn diagrammatically; the inner membrane of the capsule actually interdigitates with the sheets of vascular endothelium so that there is intimate juxtaposition of all blood capillaries and the capsular membrane. (C) The forces of hydrostatic pressure and colloid osmotic pressure that affect the rate of filtration: The relative lengths of the black arrows symbolize the relative magnitudes of these forces. (B after Elias et al. 1960.)

renal corpuscle.[5] The glomerular capillaries are intimately juxtaposed to the inner wall of the Bowman's capsule. Moreover, the wall of each capillary consists of a single layer of epithelial cells (the capillary endothelium), just as the wall of the Bowman's capsule consists of a single layer of cells. The lumen of the capillaries, therefore, is separated from the lumen of the Bowman's capsule by only two layers of cells plus a porous basement membrane between them (as shown in Figure 29.1B). These intervening structures have specialized morphology and act as a filter. Fluid is forced through this filter from the blood plasma into the lumen of the Bowman's capsule by the hydrostatic pressure of the blood. The fluid that accumulates in the lumen of the Bowman's capsule is known as the **capsular fluid** and is the *primary urine*. Although a critical determinant of whether a solute will pass through is its molecular size, molecular charge and shape can also be significant; thus the filter has complex features, which include, but are not limited to, simple physical pores or slits.

Current theories regarding the function of the filter place particular importance on the cellular wall of the Bowman's capsule, which is composed of specialized cells called **podocytes** (drawn highly diagrammatically in Figure 29.1B). The podocytes have processes, and the processes of neighboring podocytes interdigitate in geometrically intricate ways, creating countless narrow slits between the processes. The assembly of processes and slits is called the *slit diaphragm*, believed to be the most critical part of the filter.

Inorganic ions and small organic molecules such as glucose, urea, and amino acids move freely with filtered fluid as it passes from the blood plasma into the lumen of a Bowman's capsule. Thus the concentrations of these solutes are virtually the same in the capsular fluid—the primary urine—as in the blood plasma. In contrast, solutes with molecular weights of about 10,000 daltons or more—such as albumins and other plasma proteins—are essentially unable to pass through the structures that separate the blood plasma and the capsular fluid. The primary urine, therefore, closely resembles the blood plasma in its composition of inorganic ions and low-molecular-weight organic solutes, but differs from the plasma in being almost devoid of high-molecular-weight organic solutes such as proteins.

Because proteins remain more concentrated in the blood plasma than in the capsular fluid, the osmotic pressure of the blood plasma is higher than the osmotic pressure of the capsular fluid (see page 126). This difference in osmotic pressure is called the *colloid osmotic pressure* of the blood. Taking the colloid osmotic pressure into account, there are two processes that tend to cause water (H_2O) to move between the blood plasma and capsular fluid. The first is the *difference in osmotic pressure*, which tends to cause osmosis of water from the capsular fluid into the blood

---

[5] Another name for *renal corpuscle* is *Malpighian corpuscle*. Sometimes the entire renal corpuscle is called a *glomerulus*.

...asma. The second is the *difference in hydrostatic pressure*, ...hich tends to cause bulk flow of water from the blood ...asma into the capsular fluid. *Net filtration* of fluid into ...e capsular lumen will occur only if the difference in ...ydrostatic pressure is greater than the difference in ...smotic pressure. In the renal corpuscles of the species ...mammals that have been used as model systems for ...search, the blood pressure (hydrostatic pressure of the ...ood) is about 6.7 kPa, and the opposing hydrostatic ...essure in the capsular fluid is about 1.9 kPa, meaning ...at the difference in hydrostatic pressure is about 4.8 ...Pa. The colloid osmotic pressure averages about 3.5 ...Pa. Thus the net force favoring filtration—termed the *filtration pressure*—is about 1.3 kPa, as shown in **FIGURE 29.1C**. The blood pressure in the glomerular capillaries is significantly higher than the blood pressure in most capillaries in mammals, helping to promote filtration and formation of primary urine. Part of the reason for the high capillary blood pressure is that the arterioles leading to the glomeruli are relatively large in diameter and thus offer a relatively low resistance as blood flows to the glomeruli.

The rate of primary-urine formation by all of an animal's kidney tubules taken together is called the **filtration rate**. In vertebrates, it is termed specifically the **glomerular filtration rate**, or **GFR**. Adult humans, for example, have a GFR of about 120 mL/min. At this rate, *the equivalent of all the plasma water in a person's body is filtered about every 30 minutes!* This fact points to an important property of vertebrates, namely that the GFR *greatly* exceeds the rate of excretion of definitive urine. Most of the filtered water is ultimately reabsorbed back into the blood, rather than being excreted. The sheer magnitude of the rate of filtration means, however, that the nephrons have very intimate access to the blood plasma to carry out their function of regulating plasma composition.

The rate of production of definitive urine by an individual vertebrate animal can, in principle, be controlled in part by regulating the GFR. This mode of controlling urine flow is employed to some degree by mammals. It is employed to a greater extent by other types of vertebrates. There are two principal ways to adjust the GFR. One is to vary the rate of filtration into all the nephrons of the kidneys collectively. The second is to increase or decrease the numbers of nephrons that are actually functioning as filtration units at any given time. The latter strategy is the norm in nonmammalian vertebrates. The rate of filtration into an individual nephron depends on the nephron's glomerular blood pressure, which is modulated by vasomotor changes in the diameter (and hence flow resistance) of the glomerular afferent vessel. Vasomotor changes of this sort are under the control of the autonomic nervous system and circulating hormones. Variation in the GFR is not the only way in which the rate of production of definitive urine can be controlled. Animals can also modulate the rate at which the nephrons reabsorb filtered fluid prior to excretion; this, in fact, is the preeminent process of urine volume control in mammals, as discussed later.

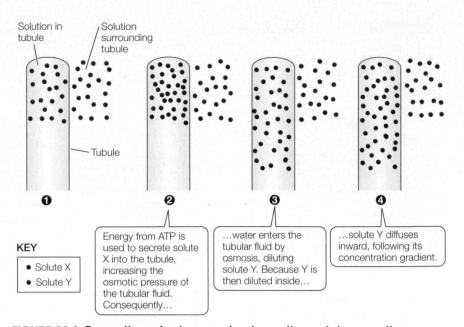

KEY
- Solute X
- Solute Y

❷ Energy from ATP is used to secrete solute X into the tubule, increasing the osmotic pressure of the tubular fluid. Consequently...

❸ ...water enters the tubular fluid by osmosis, diluting solute Y. Because Y is then diluted inside...

❹ ...solute Y diffuses inward, following its concentration gradient.

FIGURE 29.2 Formation of primary urine by active solute secretion In this model system, there are two uncharged solutes. Although the renal tubule is completely surrounded by the outside solution, only a small sample of the outside solution is shown at the upper right of the tubule in each step. For simplicity, the outside solution is assumed to stay constant in volume and composition. Movement of water into the tubule is represented by an increase in the length of the tubule filled with solution.

In addition to ultrafiltration, **active solute secretion** is the second mechanism by which water and solutes can be moved into kidney tubules to form the primary urine. This is the mechanism employed, for example, by insects and some marine fish.

To see how urine formation can be initiated by secretion, consider **FIGURE 29.2**, which presents the essentials of a secretory system in a conceptual, stepwise fashion. For simplicity, only two uncharged solutes are assumed to be present. Moreover, the fluid outside the tubule is assumed to be abundant, so that over short periods of time, movements of solutes and water into the tubule do not greatly modify its composition. At the start, which is labeled step ❶, the osmotic pressure and the concentrations of both solutes are equal on the inside and outside of the kidney tubule. In step ❷, an active-transport pump uses energy from ATP to secrete a quantity of solute X (symbolized with blue dots) into the lumen of the kidney tubule, increasing the inside concentration of X and also increasing the inside osmotic pressure. In step ❸, water moves inward by osmosis—following the osmotic gradient that was set up by secretion of solute X—and the volume of fluid in the tubule increases. Because of this increase in volume, the inside concentration of solute Y, initially the same as the outside concentration, is reduced so that it is now lower than the outside concentration. In step ❹, solute Y diffuses inward following its concentration gradient. Although simplified and artificial, this model system demonstrates that *active secretion of even just a single solute into a kidney tubule can lead to passive influx of water and other solutes.* Thus a complex solution of many solutes can be introduced into the lumen of a kidney tubule from the body fluids bathing the tubule by a secretory mechanism. During the operation of a secretory system, the epithelium of the kidney tubule acts as something of a filter. The permeability of the epithelium to the various solutes that *might* passively diffuse into the tubular lumen determines which solutes *do*, in fact, enter.

Whether the process of primary-urine formation is ultrafiltration or secretion, energy is required. In ultrafiltration systems, energy is expended in maintaining a suitably high blood pressure to cause net filtration. In secretory systems, energy is expended by the active-transport pump responsible for solute secretion.

The predominant regulatory processes in kidney function: After primary urine forms, solutes and water are recovered from it for return to the blood, and some solutes are added from the blood

As the fluid introduced into a kidney tubule moves down the tubule and through other parts of an animal's excretory system, it is typically altered extensively in volume and composition before it is eliminated as definitive urine. Most of the water in the primary urine is usually reabsorbed and returned to the blood plasma. Solutes may be reabsorbed and returned to the blood plasma—lowering the amounts excreted—or they can be added from the blood.

These processes that occur *after* primary-urine formation are the *predominant regulatory processes* in kidney function. That is, they are the predominant processes by which the formation of urine ultimately regulates the composition and volume of the blood plasma and other body fluids. This last statement is of central importance. *Regulating the composition and volume of the blood plasma and other body fluids is the function of the kidneys.* Their function is not to regulate the urine composition and volume. Instead, the formation of urine is a means to an end. It is a means to the specific end of regulating the body fluids.

As urine flows through a kidney tubule, it is separated from blood capillaries or blood spaces by the epithelial wall of the tubule, a single layer of cells. This epithelium is typically differentiated into distinct regions along the length of the tubule. Within each of these regions, the epithelial cells express distinctive membrane proteins, such as ion channels, transporters, and aquaporins; and the cells may have a distinctive structure. These properties give each region of the tubule distinctive abilities to reabsorb water and solutes from the tubular fluid—returning them to the blood plasma—and to secrete solutes from the blood plasma into the tubular fluid. The processes carried out by each region of a kidney tubule, and the permeability properties of each region, are commonly under endocrine control and hormonally modulated in regulatory ways.

In mammals, the regulatory exchange of solutes and water between urine and blood plasma is complete (and the composition of the definitive urine is fixed) when the urine leaves the kidneys. This is not always the case, however. In many types of animals other than mammals, solutes and/or water are further exchanged between the urine and blood plasma in the urinary bladder, cloaca, or other postrenal ("after kidney") structures, before the urine is finally excreted from the body.

SUMMARY
Basic Mechanisms of Kidney Function

- Primary urine is formed by ultrafiltration or by active solute secretion.

- During ultrafiltration, fluid is driven by elevated hydrostatic pressure from the blood plasma into the kidney tubules through intervening epithelia and basement membranes that act as a filter. The filtrate, which is the primary urine, is almost identical to blood plasma in its composition, except that it lacks high-molecular-weight solutes such as plasma proteins.

- In cases in which primary urine is formed by active solute secretion, the process that initiates and drives primary-urine formation is the active transport of one or more solutes into the kidney tubules. Water then follows by osmosis, and other solutes enter by diffusion, following electrochemical gradients set up by the active solute transport and osmosis.

- As primary urine flows through the kidney tubules, it undergoes exchange with the blood plasma by active or passive transport of solutes and by osmosis of water across the epithelial walls of the tubules. These processes are the *predominant regulatory processes in the kidney tubules*: They determine the ways in which the production of urine ultimately alters the composition and volume of the blood plasma. The urine produced by the kidneys is sometimes (as in mammals) the definitive urine, but in many animals, further regulatory exchange between urine and blood plasma occurs by postrenal processing.

Urine Formation in Amphibians

The amphibians provide an excellent starting point for the study of vertebrate nephron function. Much is known about the amphibian nephron because of practical considerations that make the nephrons of amphibians relatively easy to study. Furthermore, the amphibian nephron can reasonably be considered a "generalized" vertebrate nephron. Our purpose in this section is not only to describe how amphibians form urine, but also to bring out many additional general principles of vertebrate kidney function by example.

Each nephron of an amphibian (**FIGURE 29.3A,B**) consists of (1) a Bowman's capsule; (2) a convoluted segment known as the **proximal convoluted tubule**; (3) a short, relatively straight segment of small diameter, the **intermediate segment**; (4) a second convoluted segment known as the **distal convoluted tubule**; and (5) a relatively straight segment, the **collecting tubule**.[6] The nephrons are microscopic in diameter but macroscopic in length; in an average-sized toad, for example, each might be 1 cm long. Hundreds or thousands of nephrons are found in each kidney, and the nephrons constitute much of the bulk of the kidney tissue. In each nephron, the structure and function of the nephron epithelium change along the length of the nephron, from one segment to the next of the major nephron segments we have described. Structure

---

[6] The names of the nephron segments are not standardized. For example, the *collecting tubule* is sometimes called the *initial collecting duct*.

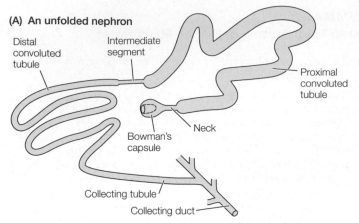

(A) An unfolded nephron

Distal convoluted tubule

Intermediate segment

Proximal convoluted tubule

Neck

Bowman's capsule

Collecting tubule

Collecting duct

FIGURE 29.3 Amphibian nephrons and their connections to collecting ducts (A) An unfolded amphibian nephron. (B) A nephron of the toad *Bufo bufo*, shown realistically in its natural configuration. Symbols help to trace the nephron along the course of its intricate geometry. Two of the segments—part of the proximal convoluted tubule and part of the late distal convoluted tubule—are shown in cross section at greater magnification than the main drawing. (C) A single collecting duct, showing the connections of the collecting tubules of many nephrons. (B after Møbjerg et al. 1998; C after Huber 1932.)

and function also often change *within* each major nephron segment (e.g., along the length of the distal convoluted tubule). In each kidney, the collecting tubules of all the nephrons feed into **collecting ducts** (**FIGURE 29.3C**), and all the collecting ducts connect to a single **ureter**, which carries fluid from the kidney to the bladder.

The proximal convoluted tubule reabsorbs much of the filtrate—returning it to the blood plasma—without changing the osmotic pressure of the tubular fluid

In an amphibian's kidneys, the amount of water filtered each day typically far exceeds the amount that needs to be excreted. The same can be said of Na^+ and Cl^-, which are the principal solutes in the blood plasma and therefore also in the filtrate: Na^+ and Cl^- enter the Bowman's capsules as briskly as water does, yet amphibians often need ultimately to conserve Na^+ and Cl^- to the maximum possible extent (see page 744). A high rate of filtration ensures that the nephrons have intimate access to the blood plasma to perform regulatory functions, as noted earlier. However, a high filtration rate also necessitates reabsorption of much of the water and NaCl filtered.

The reabsorption begins in the proximal convoluted tubule. Na^+ is actively reabsorbed across the walls of the proximal tubule. Cl^- may also be reabsorbed actively in some species, but in general its reabsorption is passive, induced by the electrical gradient set up by active Na^+ reabsorption. Although the quantities of Na^+ and Cl^- reabsorbed in the proximal tubule are substantial, the osmotic pressure of the tubular fluid does not fall in the proximal tubule. Instead, the tubular fluid—which is isosmotic to the blood plasma when introduced into the Bowman's capsule by ultrafiltration[7]—remains isosmotic to the plasma as it flows through the proximal tubule. Its osmotic pressure remains unchanged because as NaCl is reabsorbed, a proportional reabsorption of water from the tubular fluid occurs simultaneously. The epithelial walls of the proximal tubule are freely permeable to water.[8] Water therefore moves out of the tubular fluid by osmosis rapidly enough that the active reabsorption of NaCl does not produce a lower osmotic pressure in the tubular fluid than in the blood; the water is said to undergo *near-isosmolar transport* driven by the NaCl reabsorption. In those species of amphibians that have been studied, 20%–40% of the filtered NaCl and water are reabsorbed in the proximal tubule. Even as these large *amounts* of NaCl and water are reabsorbed,

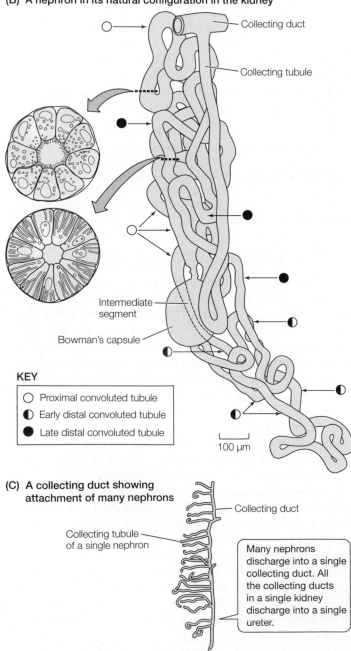

(B) A nephron in its natural configuration in the kidney

Collecting duct

Collecting tubule

Intermediate segment

Bowman's capsule

KEY

○ Proximal convoluted tubule

◐ Early distal convoluted tubule

● Late distal convoluted tubule

100 μm

(C) A collecting duct showing attachment of many nephrons

Collecting duct

Collecting tubule of a single nephron

Many nephrons discharge into a single collecting duct. All the collecting ducts in a single kidney discharge into a single ureter.

To ureter

---

[7] The difference in osmotic pressure that actually exists between the filtrate and the blood plasma—the colloid osmotic pressure—is large enough to affect filtration, as discussed earlier (see Figure 29.1C). However, not only in amphibians but also in other vertebrates, the difference in osmotic pressure is less than 1% of the *absolute* osmotic pressure of either the filtrate or plasma. Thus for most purposes, the filtrate and the plasma can be considered isosmotic.

[8] This high permeability is presumably a consequence of abundant constitutive (i.e., chronically present) aquaporins in the cell membranes of the epithelial cells.

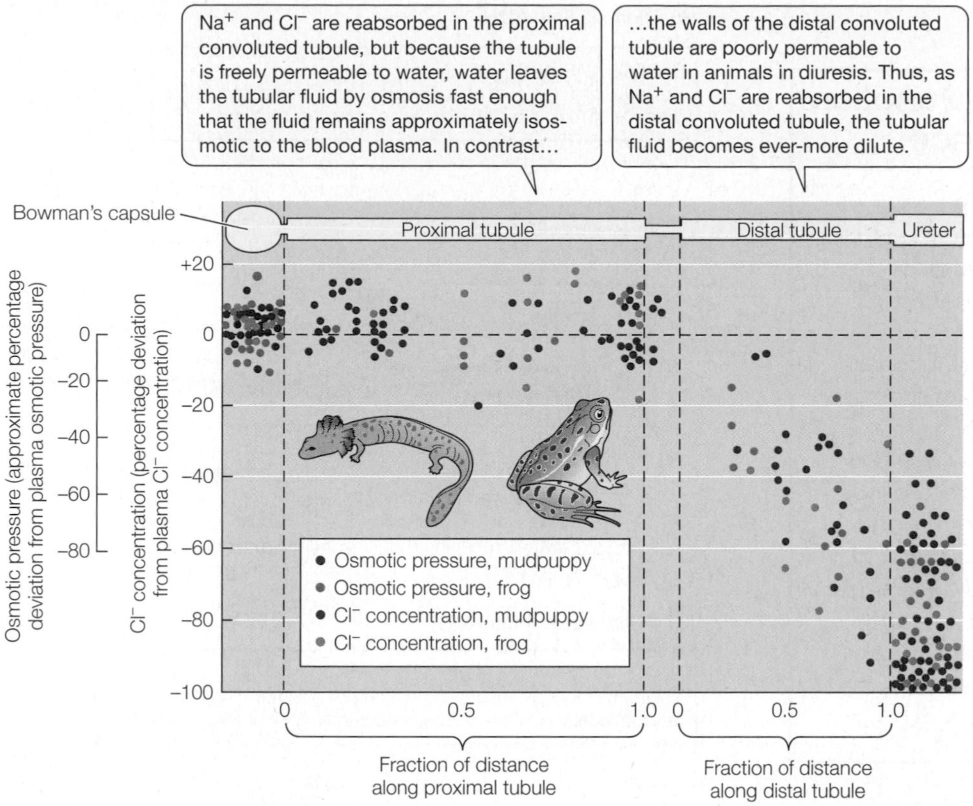

Na+ and Cl– are reabsorbed in the proximal convoluted tubule, but because the tubule is freely permeable to water, water leaves the tubular fluid by osmosis fast enough that the fluid remains approximately isosmotic to the blood plasma. In contrast…

…the walls of the distal convoluted tubule are poorly permeable to water in animals in diuresis. Thus, as Na+ and Cl– are reabsorbed in the distal convoluted tubule, the tubular fluid becomes ever-more dilute.

FIGURE 29.4 Urine formation in amphibians during diuresis (rapid urine flow) The osmotic pressure and Cl– concentration of urine as it flows through the nephrons of two species of amphibians—the semiterrestrial leopard frog (*Rana pipiens*) and the aquatic mudpuppy (*Necturus maculosus*). Fluid was sampled for analysis by use of minute pipettes inserted into the nephrons (see Box 29.2). Concentrations are expressed as percentage deviations from plasma concentrations; for example, a value of –40 indicates that the concentration in the tubular fluid was below the plasma concentration by an amount equal to 40% of the plasma concentration. (After Walker et al. 1937.)

the *concentrations* of ions and water in the tubular fluid—as shown in **FIGURE 29.4**—remain unaltered because ions and water are removed in proportion to each other (**BOX 29.1**).

Another important process that takes place in the proximal tubule is the reabsorption of glucose. Glucose in the blood plasma is a valuable metabolite that—because of its small molecular size—cannot be withheld from the primary urine during ultrafiltration. However, in amphibians and other vertebrates, glucose is promptly reclaimed and returned to the blood. Glucose is reabsorbed into the cells of the proximal tubule (and then passed to the blood plasma) by *secondary* active transport driven by the *primary* active transport of Na+—a mechanism similar to that diagrammed in Figure 5.12. Amino acids are also valuable organic molecules that are freely carried into the Bowman's capsules by ultrafiltration because of their small size. Their reabsorption begins in the proximal tubule. **BOX 29.2** discusses some of the methods used to study kidney function: methods that have played important roles in creating the knowledge discussed here and throughout the chapter.

The distal convoluted tubule can differentially reabsorb water and solutes, thereby regulating the ratio of water to solutes in the body fluids

Active reabsorption of NaCl from the tubular fluid continues in the distal convoluted tubule. In this way, the quantity of NaCl destined for excretion from the body—that is, removal from the body fluids—is gradually lowered toward the level that is appropriate for maintenance of internal NaCl balance.

A major function of the distal convoluted tubule in many amphibians—a function sometimes shared by the collecting ducts and urinary bladder—is control of the excretion of *pure* water, often termed **osmotically free water**. By controlling the excretion of free water, the distal tubule helps to control both the amount of water in the body fluids and the concentrations of solutes in body fluids, including concentrations in the blood plasma. Recall from Chapter 28

BOX 29.1 Quantity versus Concentration

When analyzing kidney function, it is important to maintain a clear distinction between measures of *quantity* (or *mass*) and measures of *concentration*. The importance of this distinction is illustrated nicely by the events in the proximal tubule of amphibians. As shown in Figure 29.4, the *concentrations* of Na+, Cl–, and water in the tubular fluid remain, on average, unchanged. Yet the *quantities* of these substances exiting the proximal tubule are much lower than those entering.

Measures of quantity and concentration are each informative, although in different ways. Quantity is an absolute measure, whereas concentration is a relative measure (quantity of solute relative to quantity of water). As a general principle, measures of *quantity* provide the most direct insight into questions of salt and water *balance*. For instance, to determine whether an animal is in Na+ balance, you would measure the quantity of Na+ gained per day and the quantity lost per day (including the quantity lost in urine) and compare them. Although urine *concentrations* are not directly useful for balance calculations, *concentrations* provide the most direct insight into the effects of urine production on blood-plasma composition. For instance—as explained in Figure 27.7—if you wanted to know whether the kidneys are lowering the Na+ concentration of the blood plasma, you would examine the urine Na+ concentration relative to the plasma Na+ concentration. Urine production is lowering the plasma Na+ concentration if the urine Na+ concentration is greater than the plasma Na+ concentration (meaning that Na+ U/P > 1.0).

<table>
<tr><td>**BOX 29.2**</td><td>**Methods of Study of Kidney Function: Micropuncture and Clearance**</td></tr>
</table>

Some of the methods used to study kidney function, although technically difficult, are intuitively easy to understand. A technique of this sort that has revolutionized renal physiology is **micropuncture**. Fine micropipettes are inserted into individual nephrons at identified points, permitting samples of tubular fluid to be withdrawn for analysis of composition. Such samples from amphibian nephrons reveal, for example, that the glucose concentration falls virtually to zero by the end of the proximal convoluted tubule. This is how we know that the proximal tubule is the site of glucose reabsorption.

A method that is not so intuitively simple to understand—but important in both physiological research and medical practice—is the study of **renal clearance**. Clearance studies are used to measure the glomerular filtration rate and can be used to quantify the reabsorption or secretion of solutes in the renal tubules. **Box Extension 29.2** explains the principles and uses of renal clearance studies.

(see page 776) that the water in urine may be considered to consist of two parts: (1) water that is *required* to accompany excreted solutes and (2) additional water that may be excreted but is not required for solute excretion. The second component may be considered to represent an excretion of pure, or "free," water—a regulated removal of water from the body fluids—precisely because it is not required for solute excretion.

In amphibians, the amount of water that is *required* to be excreted with solutes is determined by the fact that the maximum possible osmotic U/P ratio is 1.0: The urine osmotic pressure cannot exceed the plasma osmotic pressure.[9] This means that *at least* enough water must be excreted with solutes in the urine to create a solution that is isosmotic to the blood plasma. If the urine osmotic pressure of an amphibian in fact equals the animal's plasma osmotic pressure, the urine contains *only* water that is *required* for solute excretion. That is, the urine contains no water of the second kind: no pure, osmotically free water. However, if the urine osmotic pressure of an amphibian is less than the animal's plasma osmotic pressure (osmotic U/P < 1), the urine carries an "extra" quantity of water, an amount not strictly required by solute excretion. This extra quantity represents an excretion of pure water—removal of water from the body fluids. This excretion of pure water can be varied: The more dilute the urine, the more free water it contains. Thus *an animal can control its excretion of water independently of its excretion of solutes by varying the osmotic pressure of its urine*.

The extent of pure-water excretion is controlled in the distal convoluted tubule by varying the degree to which osmotic water reabsorption keeps pace with solute reabsorption there. The extent of water reabsorption is controlled by modulating the permeability of the walls of the tubule to water. This control of permeability is exercised at least partly by **antidiuretic hormone (ADH)** secreted

(A) Diuresis (low ADH)

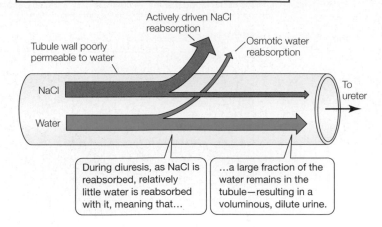

BLOOD-PLASMA EFFECTS:
Relatively high removal of pure water from plasma, tending to produce or maintain high plasma solute concentrations.

During diuresis, as NaCl is reabsorbed, relatively little water is reabsorbed with it, meaning that…

…a large fraction of the water remains in the tubule—resulting in a voluminous, dilute urine.

(B) Antidiuresis (high ADH)

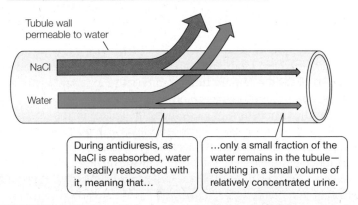

BLOOD-PLASMA EFFECTS:
Relatively little removal of pure water from plasma, tending to leave plasma solute concentrations unchanged.

During antidiuresis, as NaCl is reabsorbed, water is readily reabsorbed with it, meaning that…

…only a small fraction of the water remains in the tubule—resulting in a small volume of relatively concentrated urine.

FIGURE 29.5 Major solute and water fluxes in the late distal convoluted tubule during diuresis and antidiuresis The pathways followed by NaCl and water are symbolized by the relative sizes of the arrows. (A) In diuresis, low permeability of the tubule walls to water impedes osmotic water reabsorption; thus a relatively large fraction of the water remains in the tubule, and the ratio of solute to water (the osmotic pressure) in the tubular fluid is dramatically reduced. (B) In antidiuresis, the tubule walls are more permeable to water; thus a larger fraction of the water is reabsorbed, and the ratio of solute to water in the tubular fluid is affected relatively little. This conceptual diagram is simplified in two ways: First, it assumes that NaCl and water reabsorption occur in the same parts of the tubule, and second, it ignores solutes other than NaCl (e.g., urea).

by the neurohypophysis (posterior pituitary gland).[10] As we now explain these points, refer to **FIGURE 29.5** for a visual summary.

When ADH levels are low (see Figure 29.5A), the wall of the distal convoluted tubule has a low permeability to water. Consequently, NaCl reabsorption and water reabsorption from the tubular fluid are significantly uncoupled. The active reabsorption of NaCl tends to dilute the tubular fluid and thus create an osmotic gradient that favors water reabsorption by outward osmosis. However, the low permeability of the tubule wall to water impedes osmosis. This limitation of water

---

[9] Recall from Chapter 27 that the osmotic U/P ratio is the ratio of urine osmotic pressure to plasma osmotic pressure.

[10] The antidiuretic hormone of amphibians, birds, and nonavian reptiles is *arginine vasotocin*; see Table 16.2 on page 438.

reabsorption has three important and complementary consequences. First, relatively little water is returned to the body fluids. Second, NaCl reabsorption makes the tubular fluid more dilute than the blood plasma, both in osmotic pressure and in ion concentrations. This dilution is progressive: As fluid flows through the distal tubule, the fluid becomes ever-more dilute as ever-more ions are reabsorbed from it (see Figure 29.4). The active reabsorption of solutes from the urine across tubule walls that are poorly permeable to water is, in its fundamentals, the *universal mechanism by which animals make urine hyposmotic to the blood plasma*, and here we see that mechanism in action. The third principal consequence of low permeability to water in the amphibian distal convoluted tubule is that a high proportion of the water that enters the distal tubule passes through to be excreted in the urine. Considering the second and third consequences together, one can see that in the presence of low levels of ADH, the urine is *dilute* and *voluminous*. It carries away far more water than is necessary just to excrete solutes, and thus has a high content of pure, osmotically free water. The most important points to understand are the effects on the body fluids, because the function of producing urine is to regulate the volume and composition of the blood plasma and other body fluids. Under the circumstances we are discussing, the urine extracts relatively large amounts of pure water from the body fluids, tending to reduce the volume of the body fluids and raise the concentrations of solutes in them (see Figure 27.7).

When ADH levels are high (see Figure 29.5B), the presence of ADH induces the wall of the late distal convoluted tubule (the half or so of the distal tubule closest to the collecting tubule) to become relatively permeable to water, and the distal tubule then functions more like the proximal tubule. Osmotic water reabsorption is promoted. Thus, as NaCl is reabsorbed, more water is reabsorbed than when ADH levels are low. Focusing on immediate consequences, there are again three. First, a relatively large amount of water is returned to the body fluids. Second, the tubular fluid stays more nearly isosmotic to the blood plasma than when ADH levels are low. Third, a smaller proportion of the water that enters the distal tubule passes through to be excreted. In the presence of high levels of ADH, therefore, the urine is relatively *concentrated* and *scanty*.[11] It contains relatively small amounts of water extracted from the body fluids and carries away little or no pure, osmotically free water. As before, the most important points to understand are the effects on the body fluids, because the function of producing urine is to regulate the volume and composition of the body fluids. Under the circumstances we are now discussing, the urine extracts little or no pure water from the body fluids, tending to leave the concentrations of solutes in the body fluids unaltered (see Figure 27.7).

You can see now why ADH has the name it does. Recall from Chapter 28 that **diuresis** is production of abundant urine. High levels of ADH promote the opposite: **antidiuresis**.

ADH is believed to control the water permeability of the amphibian distal convoluted tubule by controlling the insertion and retrieval of aquaporin proteins (see page 800) in cell membranes in parts of the tubular epithelium. When the level of ADH is high, aquaporins are inserted into the cell membranes, and—with the water channels therefore in place in the cell membranes—water can pass through the epithelium relatively readily by osmosis. When the level of ADH is low, aquaporins are retrieved from the cell membranes (i.e., returned

to intracellular locations where they are nonfunctional), and osmosis through the epithelium is impeded. More will be said of aquaporin function later in this chapter.

Active H^+ secretion into the tubular fluid is an additional function that is known to occur in the distal convoluted tubule. The amount of H^+ added is adjusted to maintain a normal pH in the body fluids.

ADH exerts an elaborate pattern of control over nephron function

In amphibians—and also in birds, lizards, and other reptiles—ADH not only increases the permeability of parts of the distal convoluted tubule to water, but also decreases the glomerular filtration rate. Specifically, ADH reduces the GFR in these vertebrate groups by reducing the numbers of actively filtering nephrons, an effect mediated by inducing vasoconstriction in glomerular afferent blood vessels. The decrease in GFR tends to reduce urine flow and promote water retention in the body, thereby complementing the increase in water reabsorption induced by ADH in the distal tubules.

ADH has also been shown in some frogs and toads to increase the rate of active NaCl reabsorption from the renal tubules. This effect, like the others mentioned, also tends to reduce urine volume and promote water retention because it enhances solute-driven water reabsorption and decreases the solute load of the urine.

Clearly, ADH mediates a multifaceted *pattern* of control over nephron function. If an amphibian experiences excess water influx—as can occur during hours of immersion in freshwater—secretion of ADH is reduced. Then the GFR is relatively high, distal-tubule reabsorption of water is relatively low, and a voluminous, dilute urine results. If dehydration sets in, ADH is secreted from the neurohypophysis, apparently under the control of osmoreceptors (which detect an increase in body-fluid osmolarity) and of pressure or stretch receptors (which signal a decrease in blood volume). The ADH induces a reduction in GFR, an increase in distal-tubule water reabsorption, and an increase in NaCl reabsorption, thereby promoting water retention and production of a scanty, concentrated urine. The renal responses to ADH are not as well developed in some amphibian species from consistently moist or wet habitats as they are in species that are more terrestrial and thus more likely to experience dehydration (see pages 770–771).

The bladder functions in urine formation in amphibians

In many species of amphibians, the bladder not only stores urine but also plays a substantial role in adjusting the volume and composition of the urine. In these species, the function of the bladder can be described very much in the way we have described that of the distal convoluted tubules. The bladder wall is poorly permeable to water when ADH levels are low but becomes quite permeable to water when ADH levels are high; the participation of aquaporins in these changes of permeability in the amphibian bladder has been directly demonstrated. NaCl is actively reabsorbed across the bladder wall, and this reabsorption is stimulated by ADH.

The amphibian excretory system has mechanisms to promote excretion of urea

Urea is the principal compound used to excrete waste nitrogen in most adult amphibians. The nephrons, bladder, and other excretory passages of adults seem generally to be poorly permeable to urea.

---

[11] Recall that in amphibians, the maximum urine concentration is isosmotic to the plasma.

Thus urea introduced into the tubular fluid tends to be retained in it and removed from the body by excretion; moreover, as water is reabsorbed from the tubular fluid, urea in the fluid tends to be concentrated. Filtration is one process by which urea enters the nephrons, and in many amphibians it is probably the sole process. However, in at least some ranid frogs (e.g., bullfrogs), urea is also actively secreted into the tubular fluid across the nephron walls.

SUMMARY
Urine Formation in Amphibians

- A primary function of the proximal convoluted tubule of the amphibian nephron is the return of both water and solutes to the body fluids by the isosmotic reduction of urine volume. NaCl is actively reabsorbed from the tubular fluid. Because the epithelial wall of the proximal tubule is permeable to water, water exits the tubular fluid by osmosis, keeping the tubular fluid isosmotic to the blood plasma.

- Glucose and amino acids are actively reabsorbed from the tubular fluid in the proximal tubule, returning them to the body fluids.

- The distal convoluted tubule *differentially* returns water and solutes to the body fluids; in the process it helps regulate plasma composition and determines the volume and osmotic concentration of the definitive urine produced by the kidney. An important mechanism by which control of distal-tubule function is exercised is that the epithelial wall of the distal convoluted tubule can have high or low permeability to water, depending on blood levels of antidiuretic hormone (ADH) secreted by the neurohypophysis (posterior pituitary).

- When ADH levels are low, the distal-tubule epithelium is poorly permeable to water. Active reabsorption of NaCl returns NaCl to the body fluids and dilutes the tubular fluid. However, relatively little water is returned to the body fluids because water cannot readily move out of the tubular fluid by osmosis. The volume of the tubular fluid remains high, and both the osmotic pressure and the NaCl concentration of the fluid become progressively lower as the tubular fluid flows through the tubule.

- When ADH levels are high, aquaporins are believed to be inserted into cell membranes in the distal-tubule epithelium, causing the water permeability of the epithelium to become high. As active reabsorption of NaCl takes place, osmosis carries water out of the tubular fluid. Thus relatively high amounts of water are returned to the body fluids. The volume of the tubular fluid is reduced, and the fluid remains similar to the blood plasma in its osmotic pressure and NaCl concentration.

Urine Formation in Mammals

The nephrons of amphibians, as noted earlier, may reasonably be considered to represent the generalized vertebrate condition. The nephrons of lizards, snakes, turtles, and crocodilians resemble them. Mammalian nephrons differ, however. Compared with an amphibian nephron, each nephron of a mammal has an added, long segment of tubule, positioned between the proximal and distal convoluted tubules. This added segment is arranged in the shape of a hairpin loop and—having first been described by Jacob Henle in the 1860s—is called the **loop of Henle** (pronounced *Hen-lee*). An additional "innovative" feature of the mammalian kidney is that the loops of Henle of the various nephrons in a kidney, along with the collecting ducts, are arranged in parallel arrays, giving the kidney a pronounced macroscopic structure not seen in the kidneys of amphibians or reptiles. The loops of Henle and their parallel arrangement provide the anatomical basis for the production of urine that is more osmotically concentrated than blood plasma: *hyperosmotic urine*. Amphibians, lizards, snakes, turtles, and crocodilians—lacking these anatomical attributes—cannot produce hyperosmotic urine.

We saw in Chapter 28 that the ability of mammals to concentrate their urine is one of their most dramatic and important adaptations for life on land. Now, as we study the mammalian kidney, we will examine the mechanism by which their urine is concentrated. Mechanisms for producing urine that is hyperosmotic to the blood plasma might seem simple to evolve. The history of life offers a very different verdict, however. In the entire animal kingdom, only three major groups have mastered the task: mammals, birds, and insects. In each case, the ability to concentrate the urine has opened up new habitats and ways of life—such as by aiding certain small mammals, like the kangaroo rats described at the start of this chapter, to survive as seed eaters in deserts. Thus, as we examine the mammalian mechanism of concentrating urine, we focus on a physiological attribute of enormous ecological and evolutionary significance.

The nephrons, singly and collectively, give the mammalian kidney a distinctive structure

The loop of Henle in a mammalian nephron consists of two long and parallel tubes, termed *limbs*, connected by a hairpin bend: The **descending limb** leads from the proximal convoluted tubule to the bend, and the **ascending limb** runs from the bend to the distal convoluted tubule (**FIGURE 29.6B**). The descending limb begins with a segment of relatively large diameter termed a **thick segment**, and the ascending limb terminates with a thick segment. Interposed between these thick segments, at various positions and for various lengths, is a segment of very small diameter, the **thin segment**. The epithelium of the thin segment differs cytologically from that of the intermediate segment discussed earlier and occurs only in mammals and birds. The loop of Henle varies considerably in length among species of mammals and among the nephrons within the kidneys of any one species.

As can be seen in Figure 29.6B, the Bowman's capsules and convoluted tubules of the nephrons in each kidney of a mammal are aggregated toward the outer surface of the kidney, whereas the loops of Henle and collecting ducts project inward, toward the **renal pelvis**, a tubular structure that represents the expanded inner end of the ureter that drains the kidney (**FIGURE 29.6A**). Because of this highly ordered arrangement of the renal tubules, histologically distinct layers are evident in the gross structure of the kidney tissue. In sagittal section, the tissue of each kidney consists of an outer layer, the **cortex**, which surrounds an inner body of tissue, the **medulla** (see Figure 29.6A). The cortex (see Figure 29.6B) consists of Bowman's capsules, convoluted tubules, the beginnings of collecting ducts, and associated vasculature. The medulla consists of loops of Henle and collecting ducts, as well as their associated vasculature. Within the medulla, the loops of Henle and collecting ducts run in parallel to one another.

(A) Kidney in sagittal section

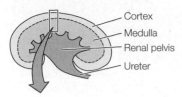

- Cortex
- Medulla
- Renal pelvis
- Ureter

FIGURE 29.6 The human kidney, emphasizing the nephrons and their blood supply (A) A sketch of a kidney in sagittal section, showing the basic features of kidney structure. (B) A more detailed look at nephrons and their blood supply. The structures of the nephrons and vasculature seen here—and the subdivision of the kidney into cortex and medulla—are similar in all mammals, although the gross shape of the kidney varies. (B after Smith 1951.)

(B) Detail of nephrons and blood supply

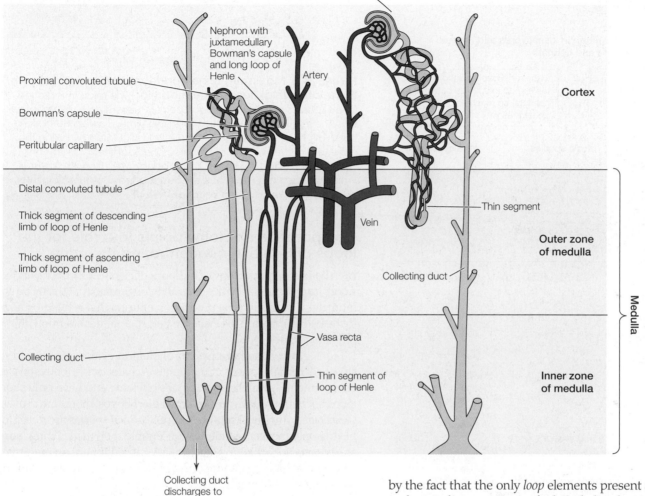

Nephron with outer cortical Bowman's capsule and short loop of Henle

Nephron with juxtamedullary Bowman's capsule and long loop of Henle

Artery

Proximal convoluted tubule

Bowman's capsule

Peritubular capillary

Distal convoluted tubule

Thick segment of descending limb of loop of Henle

Thick segment of ascending limb of loop of Henle

Vein

Cortex

Thin segment

Collecting duct

Outer zone of medulla

Medulla

Collecting duct

Vasa recta

Thin segment of loop of Henle

Inner zone of medulla

Collecting duct discharges to renal pelvis

To get oriented to fluid-flow patterns in the mammalian kidney, let's now trace the path of fluid through a nephron, focusing on the nephron to the left in Figure 29.6B. After filtration into the Bowman's capsule, fluid moves first through the proximal convoluted tubule and then descends into the medulla in the loop of Henle. After rounding the bend of the loop, the fluid returns to the cortex, passes through the distal convoluted tubule, and leaves the nephron to enter a collecting duct. The fluid then again passes through the medulla, this time in the collecting duct. After the fluid is discharged from the collecting duct into the renal pelvis, it flows into the ureter and to the bladder to be excreted. A convention worthy of note is that when fluid flows from the cortex toward the medulla, it is said to move *deeper* into the kidney.

We have already mentioned that the various nephrons in the kidney of a species may have loops of Henle of different lengths. Nephrons differing in this regard are positioned differently within the kidney, a fact that contributes to gross kidney structure. As can be seen at the left side of Figure 29.6B, there is a region deep in the medulla—termed the **inner zone** of the medulla—that is defined

by the fact that the only *loop* elements present are *thin* descending and ascending segments of *relatively long* loops of Henle. The surrounding, more superficial layer of the medulla is the **outer zone**. Loops of Henle that project into the inner zone are termed **long loops**. Loops that turn back within the outer zone of the medulla or within the cortex are called **short loops**. The thin segments of long and short loops of Henle differ cytologically. Bowman's capsules may be positioned near the outer cortical surface, at mid-depth in the cortex, or within the cortical tissue next to the medulla; the last location is termed the *juxtamedullary* ("near the medulla") position. As depicted in Figure 29.6B, nephrons with short loops tend to have their Bowman's capsules positioned toward the outer cortex, whereas those having long loops tend to have midcortical or juxtamedullary capsules. Laboratory rats have about 30,000 nephrons (of all types combined) in each kidney. Domestic dogs have about 400,000, and humans have 0.4–1.2 million.

A final morphological feature of importance is that the thick ascending segment of each nephron, near its outer (upper) end, passes immediately next to the Bowman's capsule of the very same nephron.[12] At the point where this occurs, the wall of the thick

---

[12] Figure 29.6B is drawn to emphasize other features and does not show this.

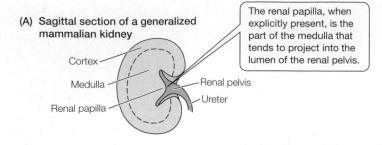

(A) Sagittal section of a generalized mammalian kidney

Cortex
Medulla
Renal papilla
Renal pelvis
Ureter

The renal papilla, when explicitly present, is the part of the medulla that tends to project into the lumen of the renal pelvis.

(B) Comparative kidney structure in insectivores and rodents from aquatic, mesic, and arid habitats

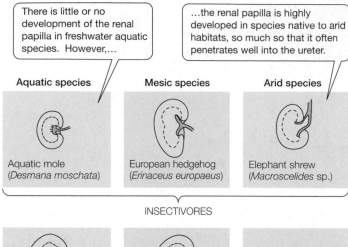

There is little or no development of the renal papilla in freshwater aquatic species. However,…

…the renal papilla is highly developed in species native to arid habitats, so much so that it often penetrates well into the ureter.

| Aquatic species | Mesic species | Arid species |

Aquatic mole (*Desmana moschata*)
European hedgehog (*Erinaceus europaeus*)
Elephant shrew (*Macroscelides* sp.)

INSECTIVORES

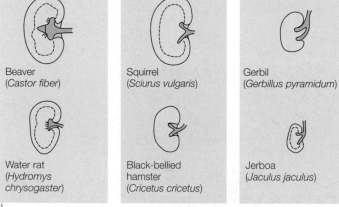

Beaver (*Castor fiber*)
Squirrel (*Sciurus vulgaris*)
Gerbil (*Gerbillus pyramidum*)

Water rat (*Hydromys chrysogaster*)
Black-bellied hamster (*Cricetus cricetus*)
Jerboa (*Jaculus jaculus*)

RODENTS

FIGURE 29.7 Evolutionary development of the renal papilla in mammals native to different habitats (A) Sagittal section of a generalized mammalian kidney, showing the location of the renal papilla. (B) Kidney structures of insectivores (e.g., shrews and moles) and rodents (e.g., rats and squirrels) from aquatic, mesic, and arid habitats. (After Sperber 1944.)

ascending segment is modified, forming a set of specialized cells, the **macula densa**. The macula densa and other associated cells form a structure called the **juxtaglomerular apparatus**.[13] Specialized vascular endothelial cells in this apparatus are responsible for secreting the key hormone renin (pronounced *ree-nin*), which controls secretion of another hormone, aldosterone, which is a major controller of renal ion excretion (see pages 450–451 and Figure 16.17).

---

[13] Be certain not to confuse the juxtamedullary capsules and the juxtaglomerular apparatus.

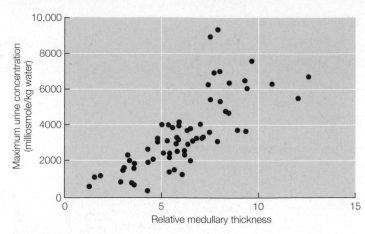

FIGURE 29.8 Maximum urine concentration correlates with the relative thickness of the medulla The relation is plotted for 68 species of mammals, each point representing a different species. The relative medullary thickness is a dimensionless number. To calculate it, an index of kidney size is first calculated by taking the cube root of the product of the three principal linear dimensions (length, width, and thickness) of the kidney. Medullary thickness is then expressed as a ratio of the index of kidney size to obtain relative medullary thickness. (After Beuchat 1990.)

Comparative anatomy points to a role for the loops of Henle in concentrating the urine

Even before the physiology of the loops of Henle began to be understood, morphological evidence strongly suggested that the loops are intimately involved in the production of urine that is hyperosmotic to the blood plasma. This evidence helped to center attention on the physiology of the loops.

One type of comparative morphological evidence comes from studies of certain species of mammals—characteristic of freshwater environments—that lack long loops of Henle and have only short loops, so that they have no inner medulla. Hippos, mountain beavers (*Aplodontia*), and muskrats are examples. Such species are noted for having only meager abilities to concentrate their urine. Long loops are essential for achieving high urinary concentrations; in mammals that achieve high concentrations, at least 15%–20% of the nephrons have long loops of Henle.

Another type of comparative morphological evidence comes from studies of the **renal papilla** (**FIGURE 29.7A**). Not all mammals have a grossly apparent renal papilla. Commonly, however, the renal medulla has a roughly pyramidal shape and forms a projection into the lumen of the renal pelvis. This projection, the renal papilla, is composed in major part of long loops of Henle. Thus the prominence of the renal papilla provides an indication of the number and length of long loops in a mammal's kidney. In 1944, Ivar Sperber (1914–2006) reported seminal observations on the papilla in about 140 species of mammals from diverse habitats. He found that the papilla was uniformly poorly developed in species inhabiting freshwater habitats. The papilla was more evident in species from mesic (moderately moist) habitats and was most developed in species from arid habitats (**FIGURE 29.7B**). Insofar as habitat may be taken as an indicator of demand for urinary concentration, Sperber's results indicated that there is a greater evolutionary development of the long loops of Henle in species that produce relatively concentrated urine.

Inspired by Sperber's work, comparative studies have since been conducted on **medullary thickness**. The thickness of the medulla provides a measure of the lengths of the longest loops of Henle. A problem that needs to be addressed in such comparative studies is

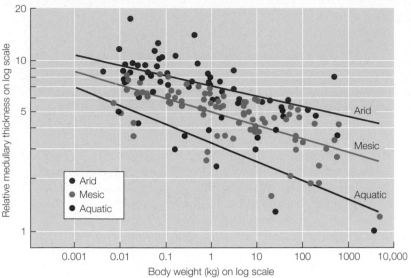

FIGURE 29.9 The relation between relative medullary thickness and body size depends on whether mammals are native to arid, mesic, or aquatic habitats Each point represents a different mammal species. The three lines are fitted statistically through the points for the arid, mesic, and freshwater aquatic species. The straight lines on this log–log plot indicate that the relations are allometric (see Appendix F). To interpret this plot, keep in mind that logarithmic scales tend to cause visual compression of data. The differences in relative medullary thickness among the arid, mesic, and aquatic mammals of a particular body size are substantial; for example, the medullary thickness of a representative 1-kg arid species is more than twice that of a 1-kg aquatic species. See the legend of Figure 29.8 for a description of how relative medullary thickness is calculated. (After Beuchat 1996.)

that medullary thickness depends on the body size of mammals; humans, for example, have a very thick medulla compared with all species of mice, merely because humans are more than 1000 times larger than mice. To remove the effects of absolute kidney size, medullary thickness is expressed as a ratio of kidney size. This ratio is called **relative medullary thickness**. A high relative medullary thickness means that the longest loops of Henle are long relative to the overall dimensions of the kidney. By now, data are available on many species, and as **FIGURE 29.8** shows, urinary concentrating ability is strongly correlated with relative medullary

thickness: Species with high relative medullary thickness tend to be able to produce especially concentrated urine.

The latest incarnation of Sperber's work is shown in **FIGURE 29.9**, where relative medullary thickness is plotted as a function of body weight (size) for mammals from three types of habitats. This modern analysis reveals that the relative thickness of the medulla tends to decrease allometrically with body size (just as concentrating ability tends to decrease with body size, as seen in Figure 28.20). Habitat, however, is a significant factor. At any given body size, mammals from arid habitats tend to have the thickest medullas and longest loops of Henle, whereas those from freshwater aquatic habitats have the thinnest and shortest, and those from intermediate mesic habitats are in between.

A dramatic morphological comparison of the kidneys of three species of rodents of roughly similar body size is seen in **FIGURE 29.10**. Two of the species, the Mongolian gerbil and sand rat, evolved

(A) Laboratory rat (*Rattus norvegicus*)

(B) Mongolian gerbil (*Meriones shawii*)

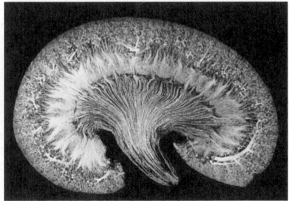

(C) Sand rat (*Psammomys obesus*)

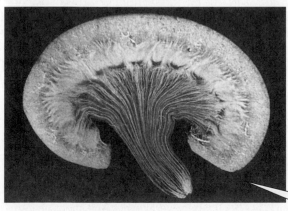

(D) A long-looped nephron in a sand rat kidney

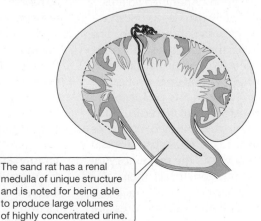

The sand rat has a renal medulla of unique structure and is noted for being able to produce large volumes of highly concentrated urine.

FIGURE 29.10 Kidney structure visualized by injection of the microvasculature (A–C) Mid-sagittal sections of the kidneys of three species of rodents of similar adult body size, in which the microscopic blood vessels of the kidneys have been injected with rubber for visualization. (D) A drawing of a sand rat kidney showing a nephron with a long loop of Henle. (Photographs in A–C courtesy of Lise Bankir [see Bankir and de Rouffignac 1985]; D after Kaissling et al. 1975.)

in deserts. Both have far more-prominent renal papillae (singular *papilla*) and thicker medullas than the laboratory rat (see Figure 29.10A–C). Moreover, the sand rat has a thicker, longer papilla than the gerbil. Detailed studies of the sand rat reveal that its renal medulla is particularly elaborately organized; in comparison with most mammals, an especially high proportion of the long loops of Henle in the sand rat extend far into the papilla (see Figure 29.10D), rather than turning back only a fraction of the way toward the tip. The sand rat, when living in its natural habitat, experiences far higher dietary salt loads than the gerbil because it subsists largely on succulent plants of very high salt content (see page 734), whereas the gerbil is a seed eater. The sand rat can produce a slightly more concentrated urine than the gerbil (6300 milliosmolar [m*Osm*] versus 5000 m*Osm*, respectively). What is more striking, however, is that the sand rat produces far greater *volumes* of highly concentrated urine than the gerbil. The longer, thicker papilla in the sand rat kidney correlates with the species' ability to produce an abundance of concentrated urine.

Countercurrent multiplication is the key to producing concentrated urine

When renal physiologists finally figured out how mammals make urine hyperosmotic to their blood plasma, they were guided to the loops of Henle by Sperber's studies of comparative kidney morphology. We will soon return to the loops, but first we need to distinguish *urea* and *nonurea solutes* and discuss the immediate concentrating process for the latter. The nonurea solutes are simply the solutes other than urea. They consist mostly of inorganic ions such as Na^+, K^+, Cl^-, and SO_4^{2-}. An important operational parameter is the osmotically effective concentration of all the nonurea solutes taken together, termed the **total concentration of nonurea solutes**.

THE IMMEDIATE CONCENTRATING PROCESS FOR NONUREA SOLUTES The immediate concentrating process for the nonurea solutes is the removal of water from the urine as it flows through the collecting ducts to leave the kidney. Recall that on its way out of the kidney, urine is discharged from the nephrons into the collecting ducts, and then flows down the collecting ducts—passing first through the renal cortex and then the medulla—prior to being discharged into the renal pelvis and ureter (see Figure 29.6B). At the point where urine enters the collecting ducts, the total concentration of nonurea solutes in the urine is *lower* than that in the blood plasma. However, when a mammal is in a state of antidiuresis, as the urine passes in the collecting ducts through deeper and deeper layers of the medulla, its total concentration of nonurea solutes is progressively elevated: The urine ultimately reaches a concentration of nonurea solutes far above the plasma concentration. The immediate mechanism that concentrates the nonurea solutes during this process is movement of water out of the urinary fluid by osmosis. Nonurea solutes are largely trapped within the collecting ducts because the collecting-duct walls are poorly permeable to such solutes. Thus, as water passes by osmosis out of the urine, the nonurea solutes in the urine become more concentrated. Why does water undergo osmosis out of the urine? The fluids that *surround* the collecting ducts in the medulla, known as the **medullary interstitial fluids**, have a high NaCl concentration. In fact, their NaCl concentration rises steadily with increasing depth in the medulla, so that in the deepest parts of the medulla the osmotic pressure attributable to NaCl is *far* above plasma osmotic pressure. During

antidiuresis, the cells of the collecting-duct walls are freely permeable to water. As urine inside the collecting ducts flows deeper into the medulla and encounters ever-more-concentrated medullary interstitial fluids just on the other side of the collecting-duct epithelium, water progressively moves by osmosis out of the urine into the medullary interstitial fluids.

An important attribute of these processes is that *a high NaCl concentration on the outside of the collecting ducts serves to concentrate not only NaCl, but also many other nonurea solutes, on the inside.* This happens because the solutes involved cannot readily cross the walls of the collecting ducts, yet the cells in the duct walls (the duct epithelium) are freely permeable to water. Because of this difference between permeability to solutes and to water, when high interstitial NaCl concentrations are encountered deep in the medulla, the primary process of equilibration between the urine and the medullary interstitial fluid is osmosis. As this osmosis occurs, nonurea solutes in the urine are concentrated *indiscriminately* until their total osmotically effective concentration matches the total osmotically effective concentration of NaCl in the medullary interstitial fluids.

A SINGLE EFFECT BASED ON ACTIVE NaCl TRANSPORT Now we must consider how the gradient of NaCl concentration in the medullary interstitial fluids is created. The loops of Henle are responsible. The first step in understanding how the loops of Henle produce the NaCl gradient is to study a phenomenon, termed the *single effect*, that is well documented in the outer zone of the medulla, where the thick segments of the ascending limbs of the loops of Henle occur.

The cells in the walls of the ascending thick segment of a loop of Henle actively transport NaCl from the tubular fluid inside the loop into the adjacent medullary interstitial fluid. The *consequences* of this NaCl transport, illustrated in **FIGURE 29.11**, depend on the permeability characteristics of the ascending limb and the adjacent descending limb of the loop of Henle. The walls of the ascending limb are essentially impermeable to water. Thus the active transport of NaCl out of the tubular fluid inside the ascending limb creates a difference in osmotic pressure between that fluid and the adjacent interstitial fluid, in addition to decreasing the NaCl concentration of the fluid inside the ascending limb and increasing the NaCl concentration of the interstitial fluid. The permeability characteristics of the descending limb appear to vary from species to species. Nonetheless, by passive processes of one sort or another, the fluid inside the descending limb readily approaches equilibrium or near-equilibrium with the interstitial fluid in terms of osmotic pressure and ion concentrations.

In a few words, the active transport of NaCl out of the ascending limb lowers the NaCl concentration and osmotic pressure of the ascending-limb fluid and raises the NaCl concentration and osmotic pressure of both the adjacent interstitial fluid and adjacent descending-limb fluid. These differences between the ascending-limb fluid and the *adjacent* interstitial and descending-limb fluid represent the **single effect** of the active-transport mechanism.

COUNTERCURRENT MULTIPLICATION The major hurdle in understanding how mammals produce concentrated urine was crossed in the 1940s and 1950s when Werner Kuhn (1899–1968), Heinrich Wirz (1914–1993), and several other investigators demonstrated that the concept of **countercurrent multiplication** ap-

(A) Initial condition

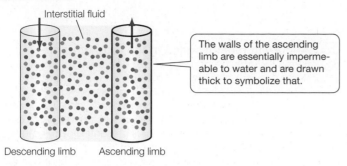

Interstitial fluid

The walls of the ascending limb are essentially impermeable to water and are drawn thick to symbolize that.

Descending limb   Ascending limb

(B) Processes that generate the single effect

Active transport of NaCl out of the fluid in the ascending limb dilutes that fluid and concentrates the interstitial fluid. The walls of the descending limb are permeable to water, so the fluid inside becomes concentrated by losing water osmotically to the interstitial fluid and sometimes by gaining Na^+ and Cl^- by diffusion.

(C) The single effect

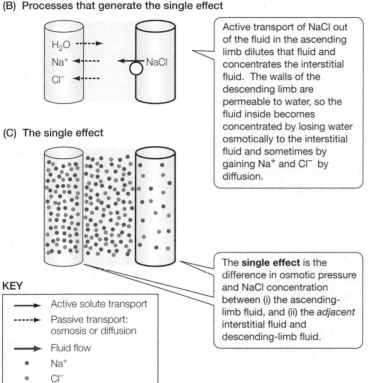

The **single effect** is the difference in osmotic pressure and NaCl concentration between (i) the ascending-limb fluid, and (ii) the *adjacent* interstitial and descending-limb fluid.

KEY

→ Active solute transport

----▸ Passive transport: osmosis or diffusion

→ Fluid flow

• Na^+

• Cl^-

FIGURE 29.11 Generation of the single effect in the loop of Henle Shown here are the ascending limb of the loop of Henle and adjacent descending limb in the outer zone of the medulla, where the thick segments of the ascending limbs occur. As a thought exercise, the diagrams show how the single effect can be generated from scratch. (A) In the initial condition, all the fluids are identical in their osmotic pressures and ion concentrations. (B) The processes that generate the single effect. (C) The single effect that is produced. The osmotic pressure and the concentrations of ions in the ascending-limb fluid are lowered from their original levels, whereas the osmotic pressure and the concentrations of ions in the interstitial and descending-limb fluids are raised.

plies to the loops of Henle. In the classic model of countercurrent multiplication generated by their work, it was assumed that all parts of each ascending limb actively transport NaCl in the manner just described. Here we develop that classic model. Later we discuss complexities introduced by more recent research.

The hairpin shape of a loop of Henle sets up two fluid streams that are oppositely directed (countercurrent), intimately juxtaposed, and connected. These properties are all requirements for a countercurrent multiplier system to operate. Such a system also requires a metabolic energy investment within the system (**BOX 29.3**). The energy investment in the loop of Henle is provided by the active NaCl transport we have already discussed, which—at an ATP cost—creates a difference in osmotic pressure and ion concentration between adjacent parts of the oppositely directed fluid streams—the single effect (see Figure 29.11C).

The countercurrent multiplier system multiplies the single effect. To be more specific, the single effect amounts to a difference of roughly 200 m*Osm* oriented *from side to side* in the loop of Henle. The countercurrent multiplier system multiplies this difference into a much larger difference in concentration *from end to end* in the loop (**FIGURE 29.12A**). An end-to-end difference of 600 m*Osm*

BOX 29.3 **Countercurrent Multipliers versus Countercurrent Exchangers**

When oppositely directed fluid streams are closely juxtaposed and commodities are actively or passively exchanged between them, the effect of the countercurrent arrangement is to preserve or magnify differences in the levels of those commodities from *end to end* along the axis of fluid flow. The countercurrent arrangement has this effect because it *impedes end-to-end flux* of commodities that are actively or passively exchanged between the fluid streams.

Two functional types of countercurrent systems are recognized: *active* and *passive*. The active systems are **countercurrent multipliers**, exemplified by the loops of Henle. The passive systems are called **countercurrent exchangers** (or *countercurrent diffusion exchangers*)

and are exemplified by the heat exchangers in the appendages of mammals (see Figure 10.35B).

In an active system, metabolic energy is used *within the countercurrent system itself* to induce flux of commodities into or out of the fluid streams; within the loop of Henle, for example, ATP energy is used to transport NaCl out of the ascending limb. In a passive system, fluxes of commodities into or out of the fluid streams occur without expenditure of metabolic energy *in the countercurrent system itself*. In the heat exchanger in Figure 10.35B, for example, heat does not move out of one blood vessel and into another because of any metabolic energy expenditure *within* the countercurrent system; instead, heat follows temperature gradi-

ents that exist because energy expenditure *elsewhere* in the body has caused the body core to be warmer than the environment.

Active countercurrent systems produce—they *create*—differences in levels of commodities from end to end along their axis of flow. Note, for instance, that if energy expenditure (ion pumping) in the loops of Henle were turned off, the gradient of osmotic pressure and NaCl concentration from the outer to the inner end of the loops would disappear. Passive systems, by contrast, do not *create* end-to-end differences; instead they *preserve* or *accentuate* end-to-end differences that already exist for other reasons.

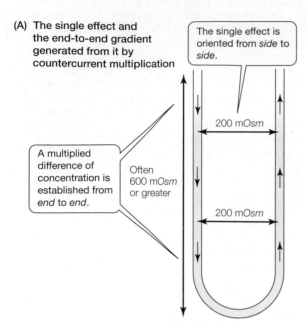

(A) The single effect and the end-to-end gradient generated from it by countercurrent multiplication

The single effect is oriented from *side* to *side*.

A multiplied difference of concentration is established from *end* to *end*.

Often 600 m*Osm* or greater

200 m*Osm*

200 m*Osm*

FIGURE 29.12 Countercurrent multiplication in the loop of Henle
(A) The distinction between the side-to-side (transverse) difference in osmotic pressure and the end-to-end (axial) difference in the loop of Henle. The side-to-side difference is the single effect. The end-to-end difference is generated from the side-to-side difference by countercurrent multiplication. (B) The *process* by which countercurrent multiplication occurs. The numbers are osmotic pressures in units of milliosmolarity (m*Osm*). The operation of the multiplier is presented conceptually as a series of alternating steps. In ❶, the entire system is at 300 m*Osm*. In ❷, a single-effect osmotic gradient of 200 m*Osm* is created all along the loop, and in ❸, fluid flows through the loop. These steps are repeated in ❹ through ❽. The amount of fluid movement through the loop decreases progressively in ❸, ❺, and ❼. Fluid entering the descending limb is always at 300 m*Osm*, creating a tendency for the osmotic pressure at the cortical end of the descending limb and interstitial space to remain near 300 m*Osm*. Although both (A) and (B) are presented in terms of osmotic pressures, the differences in osmotic pressure are paralleled by differences in NaCl concentration. The brilliant pedagogical scheme in (B) was conceived by Robert F. Pitts (1908–1977). (B after Pitts 1974.)

(B) The process of countercurrent multiplication

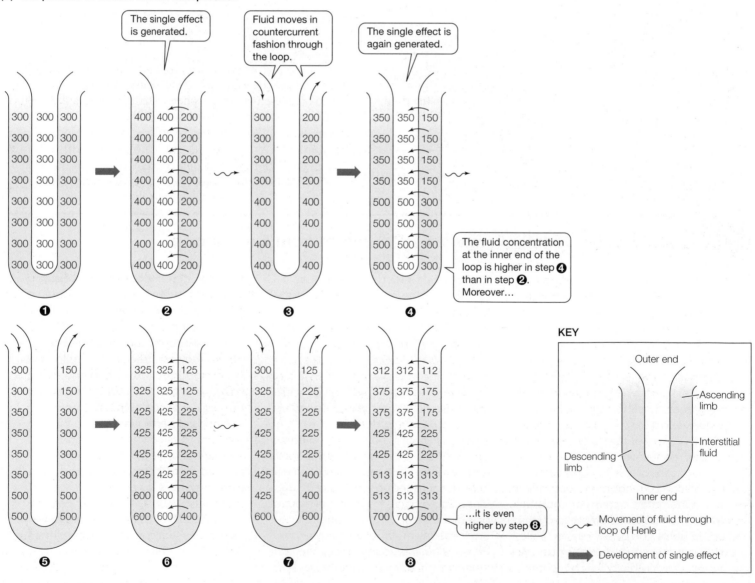

would not be unusual. Many mammals can create an end-to-end difference that is much greater.

The mechanism of countercurrent multiplication is diagrammed in **FIGURE 29.12B**. Although osmotic pressures are shown in the figure and the following discussion is phrased in those terms, it will be important to remember that differences in osmotic pressure in the loop of Henle are paralleled by differences in NaCl concentration. In step ❶ of Figure 29.12B, the entire loop of Henle and the interstitial space are filled with fluid of the same osmotic pressure as that exiting the proximal convoluted tubule—approximately isosmotic to the blood plasma (300 m*Osm*). In step ❷, active transport establishes a single-effect osmotic gradient of 200 m*Osm* all along the loop. In step ❸, fluid moves through the loop in countercurrent fashion. Fluid that was concentrated in the descending limb during step ❷ is thus brought around into the ascending limb and now lies opposite to the descending limb, so that both limbs and the interstitial space are filled with concentrated fluid at the inner end of the loop. Now when, in step ❹, the single-effect osmotic gradient is again established, the interstitial fluid is elevated to 500 m*Osm* at the inner end, rather than the 400 m*Osm* developed in step ❷, and the fluid in the descending limb also reaches this higher osmotic concentration of 500 m*Osm*. Steps ❺ and ❻, and steps ❼ and ❽, repeat this process and should be studied in order to see how the countercurrent multiplier works. *Fluid concentrated in the descending limb moves around into the ascending limb, setting the stage for the single effect to produce an ever-increasing osmotic concentration in the interstitial fluid and descending limb at the inner end of the loop.* Meanwhile, the steady influx of 300-m*Osm* fluid into the beginning of the descending limb, and the dilution of the ascending-limb fluid as it flows from deep in the medulla to the top of the ascending limb, combine to keep the osmotic pressure of the interstitial fluid at the outer (cortical) end of the loop near 300 m*Osm*. Thus the difference in osmotic pressure between the two ends of the loop becomes greater and greater, so much so that it greatly exceeds the single effect (see Figure 29.12A).

As noted previously, during the early years when the countercurrent multiplication concept was initially applied to understanding mammalian kidney function, the single effect was postulated to be created along the entire length of a loop of Henle by active NaCl transport out of the ascending limb. However, by 1970, research had established that the thin segment of the ascending limb deep in the medulla is unlikely to be carrying out such active transport. That discovery started a saga that continues unended today. Active transport of NaCl out of the tubular fluid occurs in the thick segment of the ascending limb, and a consensus exists that the single effect is created according to the classic model (see Figure 29.11) in the outer region of a loop of Henle where the thick ascending segment occurs. However, the single effect is now assumed to be created by some other mechanism in the inner region of a loop of Henle where the ascending limb is thin.

All this said, it is important to return to the big picture: Countercurrent multiplication of a single effect along much or all of the length of a loop of Henle creates a large gradient of osmotic pressure from one end of the loop to the other. In the medulla, there are thousands of loops of Henle, all aligned in parallel. We would expect that all these loops, by their combined action, would create in the medullary tissue as a whole a dramatic gradient of increasing osmotic pressure from the outer side of the tissue (next to the cortex) to the inner side of the tissue (farthest from the cortex). The classic data that originally confirmed this expectation are shown in **FIGURE 29.13**.

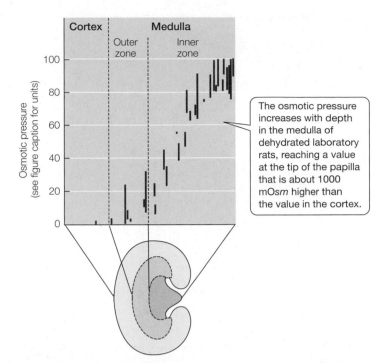

The osmotic pressure increases with depth in the medulla of dehydrated laboratory rats, reaching a value at the tip of the papilla that is about 1000 m*Osm* higher than the value in the cortex.

FIGURE 29.13 Osmotic pressure increases with depth in the medulla   Each vertical red line shows the range of osmotic pressures measured at a particular depth in the cortex or medulla of kidneys taken from five dehydrated laboratory rats. All the rats, in addition to being sampled at various places, were deliberately sampled at the tip of the papilla, accounting for the cluster of data there. On the *y* axis, 0 represents an osmotic pressure equal to that of the blood plasma in the general circulation, whereas 100 represents the highest osmotic pressure measured (about 1000 m*Osm* greater). Intermediate osmotic pressures are scaled relative to the two extremes; specifically, any particular measured osmotic pressure (OP) is expressed as 100 × (measured OP – plasma OP)/(maximum OP – plasma OP). Throughout the cortex, the osmotic pressure is equivalent to the osmotic pressure of plasma in the general circulation. The increase in osmotic pressure with depth in the medulla is attributable both to an increase in NaCl concentration and to an increase in urea concentration. (After Wirz et al. 1951.)

CONCLUDING POINTS ON THE MECHANISM OF CONCENTRATING NONUREA SOLUTES   **FIGURE 29.14A** summarizes the changes in the total concentration of nonurea solutes in the tubular fluid of nephrons and collecting ducts when the kidney of a mammal is producing concentrated urine. As fluid in a nephron travels down the descending limb of the loop of Henle, its concentration of nonurea solutes rises, reaching a high level at the hairpin bend of the loop. Thereafter, as the fluid comes back out of the medulla in the ascending limb of the loop of Henle, its concentration of nonurea solutes falls, so that by the time the fluid exits the loop, it is actually *more dilute* than when it started and more dilute than the blood plasma. Then, however, the fluid makes a final, crucial pass through the medulla, traveling down a collecting duct to be discharged into the renal pelvis. On this pass, final concentration of the nonurea solutes occurs.

The *total concentration of nonurea solutes* in the definitive urine depends on the *NaCl concentration* of the interstitial fluids of the innermost medulla, because in a kidney producing concentrated urine, the urine osmotically equilibrates with those interstitial fluids just before leaving the kidney (see Figure 29.14A). In turn, the inner-medullary NaCl concentration itself depends on the

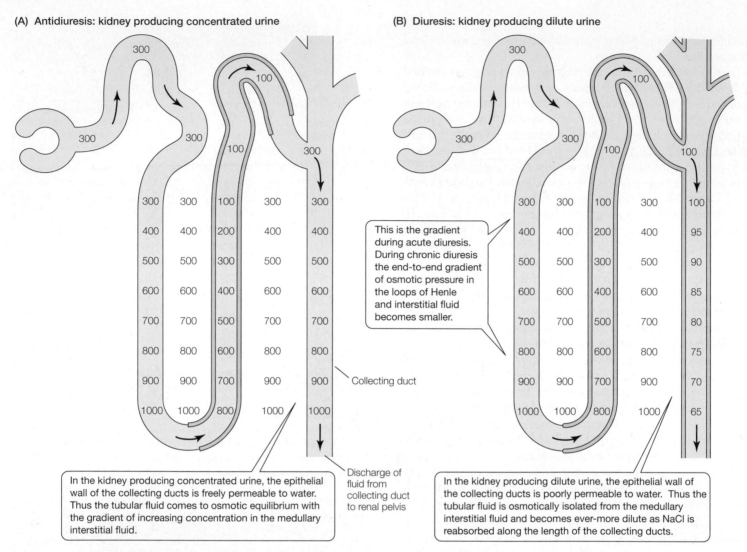

FIGURE 29.14 Osmotic pressures attributable to nonurea solutes in the nephrons and collecting ducts during antidiuresis and diuresis Thick yellow borders symbolize tubules that are poorly permeable to water. Tubules without yellow borders are permeable to water. The change in the water permeability of the collecting ducts between antidiuresis (A) and diuresis (B) is mediated by insertion and removal of aquaporins in apical cell membranes of the collecting-duct epithelium, as discussed later. The interstitial fluids (white areas) exhibit similar gradients of osmotic pressure throughout the medulla. The numbers, expressed in units of milliosmolarity, are approximate and intended only to illustrate general trends.

properties of the countercurrent multiplier system, including the size of the single effect, the rate of fluid flow through the loops of Henle, and the lengths of the loops. Lengthening of the loops tends to increase the end-to-end gradient of NaCl concentration that can be maintained by the loops and thus tends to raise the inner-medullary NaCl concentration. This explains why, among related species of similar body size, the species with a relatively thick medulla and prominent renal papilla tend to be capable of producing relatively concentrated urine (see Figures 29.7 and 29.8).[14]

CONCENTRATION OF UREA The mechanisms that concentrate urea differ from those that concentrate the nonurea solutes.

[14] When comparing species that cover a wide range of body sizes, *relative* loop length—estimated as relative medullary thickness—is a far better predictor of concentrating ability than *absolute* loop length (see Figure 29.8). Factors other than absolute length thus clearly play major roles in kidney concentrating function, but these additional factors are not yet understood.

Whereas the walls of the collecting ducts block most nonurea solutes in the urine and medullary interstitial fluid from diffusing to electrochemical equilibrium, they permit urea to do so. Urea is present at high concentration in the medullary interstitial fluid, and when mammals are in an antidiuretic state, the walls of the collecting ducts *in the inner medulla* permit urea to diffuse freely between the urine inside the ducts and the inner-medullary interstitial fluid (this diffusion is mediated by a facilitated-diffusion *urea transporter [UT] protein* that is dramatically upregulated by ADH). Basically, therefore, high urea concentrations in the urine reflect the diffusion of urea to concentration equilibrium across the walls of the inner-medullary collecting ducts.

How does urea come to be present at high concentration in the medulla? Put simply, much more urea is filtered than is excreted, and some of the urea reabsorbed along the nephrons accumulates in the medulla. The thick ascending segment of the loop of Henle, the distal convoluted tubule, and the cortical and outer-medullary parts of the collecting duct are poorly permeable to urea. During antidiuresis, at least in the cortical and outer-medullary collecting duct, water is drawn out of the tubular fluid by osmosis.[15] Because the permeability to urea in these tubular regions is low, urea—trapped inside—becomes concentrated in the tubular fluid as water is lost. The important net result is that the tubular fluid has a high urea concentration by the time it flows into the inner-medullary collecting duct, which is highly permeable to urea during antidiuresis. Urea therefore diffuses from the tubular fluid into the interstitial fluid in the inner medulla, a process that charges the inner-medullary interstitial fluid with urea. According to present thinking, this entire sequence of events is *self-reinforcing* because urea enters the tubular fluid in the loop of Henle. That is, as tubular fluid flows through the loop, it picks up urea from the inner-medullary interstitial fluid, and soon afterward that tubular fluid flows into the inner-medullary collecting duct. Because of this process (sometimes called *urea recycling*) the urea concentration of tubular fluid arriving in the inner-medullary collecting duct tends automatically to rise in parallel with the urea concentration of the interstitial fluid. Thus, with a steady influx of new urea from filtration, a gradient favorable for diffusion of urea *into* the interstitial fluid from the inner-medullary collecting duct is maintained, even though the interstitial-fluid concentration may rise to a high level. A high concentration of urea in the interstitial fluid promotes a high urinary concentration of urea because diffusive outflux of urea from the collecting-duct fluid continues only to the point of concentration equilibrium with the medullary interstitial fluid.

How does the process of urea concentration relate to the process by which nonurea solutes are concentrated? Although the answer is intricate, two important global points should be made. First, because urea and nonurea solutes are concentrated by rather separate mechanisms, a high urea concentration in the urine does not in any simple mathematical fashion displace nonurea solutes or reduce the concentration of nonurea solutes that is possible. The urine of a mammal can *simultaneously* contain high concentrations of both urea and nonurea solutes. The second point to be made is that the osmotic reabsorption of *water* from the urine in the inner medulla is controlled by the processing of the *nonurea* solutes. Because urea diffuses to concentration equilibrium across the walls of the inner-medullary collecting ducts, it does not (except transiently) make a direct contribution to the *difference* in osmotic pressure between the collecting-duct fluid and the interstitial fluid. The *difference* in osmotic pressure—which is responsible for concentrating nonurea solutes in the urine by governing the osmotic reabsorption of water—is a consequence of different concentrations of the nonurea solutes.

THE BLOOD SUPPLY OF THE MEDULLA: THE VASA RECTA   The blood capillaries of the medulla form hairpin loops—known as **vasa recta**—that parallel the loops of Henle. This arrangement, diagrammed in Figure 29.6B, is vividly evident in Figure 29.10A–C, in which the structures visualized are the blood vessels.

The looped shape of the vasa recta prevents the circulation of blood to the medulla from destroying the concentration gradients of NaCl and urea in the medullary interstitium. To see this, consider what would happen if blood, after flowing into the medulla from

the cortex, simply exited the medulla on the inner (i.e., deep, pelvic) side. The walls of blood capillaries are freely permeable to water and small solutes. Thus, as blood flowed from the cortex, deeper and deeper into the medulla—encountering ever-more-concentrated interstitial fluids—it would lose water to the interstitial fluids by osmosis and take up NaCl and urea by diffusion. Exiting on the inner (deep) side of the medulla, the blood would leave all that water behind and take the solutes away, diluting the medullary interstitial fluids in both ways. Instead, after flowing from the cortex to the inner medulla, the blood reverses direction and flows back to the cortex. On its way out, as it encounters ever-more-dilute interstitial fluids, it reabsorbs water and yields NaCl and urea, reversing the processes that occurred on the way in. The familiar tendency of countercurrent flow to preserve gradients oriented parallel to the axis of flow is once again evident. The vasa recta act as *countercurrent diffusion exchangers* (see Box 29.3).

An important function of blood flow through the vasa recta is to remove water from the medullary interstitial fluids. The final process of concentrating the urine, as we have seen, entails osmotic movement of water from the collecting ducts into the inner-medullary interstitial fluid. This water, if allowed to accumulate, would itself dilute the inner-medullary fluid and thereby diminish the medullary concentration gradient. The flow of blood through the vasa recta carries the water away. Evidently, as the blood dynamically loses water during its passage into the medulla and regains water during its passage out, the colloid osmotic pressure resulting from the blood proteins introduces a bias for the gains of water by the blood to exceed losses.

CELL-VOLUME REGULATION, COMPATIBLE SOLUTES, AND COUNTERACTING SOLUTES IN THE MEDULLA   The cells in the medulla of the kidney—such as those in the walls of the loops of Henle and vasa recta—are unique among the cells in a mammal's body in that they must tolerate exposure to very high solute concentrations and osmotic pressures in the interstitial fluids that bathe them. The medullary cells must have high levels of intracellular solutes to maintain normal cell volumes rather than being shriveled by osmotic water loss, as explained in Figures 27.8 and 27.9. In comparison with all other cells in a mammal's body, the renal medullary cells are noted for having exceptionally high intracellular concentrations of organic osmolytes of metabolic origin, notably polyhydric alcohols and methylamines (**FIGURE 29.15**). These organic compounds serve as *compatible solutes* (see page 738): They balance the high extracellular NaCl concentration while having relatively small effects on cell macromolecules.[16]

The regulatory roles of the kidney tubules in overview: The concentrating and diluting kidney and the control of transitions

Thus far we have focused on how the mammalian kidney can produce urine more concentrated than the blood plasma. The mammalian kidney resembles other kidneys, however, in that it

---

[15] NaCl is actively transported out of the tubular fluid in these parts of the collecting duct, causing water to move out by osmosis.

[16] The high concentrations of urea in the renal medulla are themselves a challenge to the function of medullary cells because urea can perturb enzymes and other proteins. There is some evidence that the methylamines accumulated in medullary cells (see Figure 29.15) help to offset the perturbing effects of urea. That is, the methylamines act as counteracting solutes (see page 739).

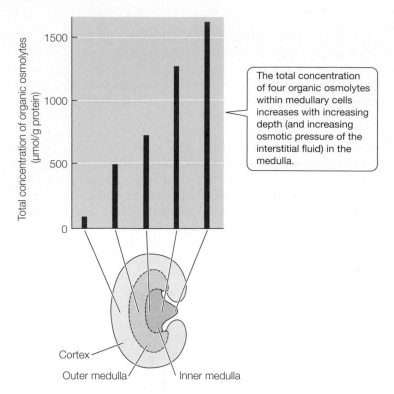

The total concentration of four organic osmolytes within medullary cells increases with increasing depth (and increasing osmotic pressure of the interstitial fluid) in the medulla.

FIGURE 29.15 Cell-volume regulation by organic osmolytes in the medulla of the kidney The cells in the renal medulla produce high intracellular concentrations of organic osmolytes as a means of regulating cell volume in the face of the high osmotic pressures in the interstitial fluids bathing them. The data shown are for normally hydrated laboratory rats. Each bar represents, at the designated anatomical location, the sum of the four principal organic osmolytes. The four osmolytes are two polyhydric alcohols (sorbitol and *myo*-inositol) and two trimethylamines (glycine betaine and glycerophosphorylcholine). Concentration is expressed as total micromoles of osmolytes per gram of tissue protein. (After Beck et al. 1998.)

carries out many processes simultaneously as it performs its overall function of regulating the composition and volume of the blood plasma and other body fluids. In this section we take more of an overview of nephron and collecting-duct function in mammals. A useful way to approach this task is, first, to discuss how multiple solutes and water are processed when the kidney is producing a concentrated urine, and then discuss—in a synthetic way—how the kidney functions when producing dilute urine and how the switch between concentration and dilution is regulated.

AN OVERVIEW OF EVENTS IN THE CONCENTRATING KIDNEY
Glomerular filtration is, of course, the first step in forming urine. In comparison with other vertebrates, mammals—with some known exceptions (e.g., dromedary camels)—tend to maintain relatively stable GFRs and adjust their rate of urine production principally by adjusting the fraction of filtered fluid that they ultimately reabsorb and return to the blood prior to excretion. The fluid introduced into the Bowman's capsule of a nephron by filtration is approximately isosmotic to the blood plasma and contains similar concentrations of inorganic ions, glucose, and amino acids. A major function of the proximal convoluted tubule is net reabsorption of NaCl and water—net return of NaCl and water to the body fluids. In fact, 60%–80% of the filtered amounts of NaCl and water are reabsorbed by the time the tubular fluid reaches the beginning of the loop of Henle. The

cells of the epithelial walls of the proximal tubule are freely permeable to water because of aquaporins (discussed shortly), so water exits osmotically as NaCl and other solutes are reabsorbed, and the tubular fluid stays isosmotic to the blood plasma. Glucose, many amino acids, and HCO_3^- (bicarbonate ion) are almost completely reabsorbed and returned to the blood plasma in the proximal tubule.

A major contemporary area for research is establishing the molecular basis for the function of the proximal convoluted tubule and all other segments of the kidney tubules. The ultimate goal of this research is to understand every aspect of reabsorption and secretion along all parts of the renal tubules in terms of the specific transporter proteins, channel proteins, and other molecules that mediate the processes. **FIGURE 29.16A** summarizes the major molecular ion-transport mechanisms in the epithelial cells of the wall of the early proximal convoluted tubule. Na^+ reabsorption from the urine is driven by primary active transport carried out by Na^+–K^+-ATPase (see page 115) in the basolateral membrane, which creates a Na^+ electrochemical gradient across the apical membrane favoring Na^+ uptake from the tubular fluid. The reabsorption of glucose and amino acids occurs by secondary active transport (see page 115).

Aquaporins, as earlier mentioned, provide the molecular basis for the high water permeability of the proximal tubule epithelium. The aquaporins in cell membranes of the epithelial cells of the proximal tubule are classified as *constitutive* because they are always present in the cell membranes; their levels are not much affected by external agents.

After fluid leaves the proximal convoluted tubule, its next step is to travel through the loop of Henle. Although the tubular fluid enters the loop isosmotic to plasma (~300 m*Osm*), it exits the loop hyposmotic to plasma (perhaps 100–150 m*Osm*), as we have seen (see Figure 29.14A). In the ascending thick segment of the loop of Henle, active NaCl transport out of the tubular fluid is a key process that both creates the single effect for countercurrent multiplication, and accounts for the hyposmotic state of the tubular fluid as it leaves the loop of Henle. **FIGURE 29.16B** presents a current model of the molecular biology of the active NaCl transport out of the tubular fluid in the ascending thick segment. **Loop diuretics**—medications employed to treat hypertension (high blood pressure)—are targeted at the Na–K–2Cl cotransporter. These medications inhibit NaCl transport out of the tubular fluid by inhibiting the cotransporter, resulting in increased Na^+ excretion and water excretion, which tend to decrease the volume of the blood plasma (see page 777).

After exiting the loop of Henle, the tubular fluid passes through the distal convoluted tubule. The epithelial walls of much or all of the distal convoluted tubule are poorly permeable to water and actively transport NaCl out. Thus the tubular fluid remains strongly hyposmotic to the blood plasma (see Figure 29.14A). Potassium (K^+) is added to the tubular fluid (partly passively, partly actively) in the distal convoluted tubule and cortical collecting duct. This addition of K^+ controls the amount of K^+ that is removed from the body fluids and eliminated in the urine because most K^+ from filtration was reabsorbed from the tubular fluid in earlier parts of the nephron.

Perhaps 5% or less of the originally filtered volume reaches the collecting duct. In the concentrating kidney, the collecting duct is permeable to water because of an aquaporin-based mechanism discussed in the next section.[17] Thus dilute tubular fluid arriving in the collecting duct promptly comes to isosmoticity with the cortical

---

[17] The terminal distal tubule may also be permeable to water.

(A) Early proximal convoluted tubule

In all three parts of the kidney tubule, Na⁺ diffuses into the epithelial cells from the tubular fluid because there is an electrochemical gradient favoring such diffusion.

In the early proximal tubule, the tubular fluid is rich in glucose and amino acids, and much of the Na⁺ entry into a cell occurs by means of cotransporters that bring about the secondary active transport of glucose and amino acids into the cell. Only the Na–glucose cotransporter is shown here.

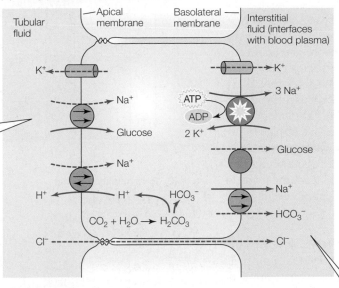

KEY

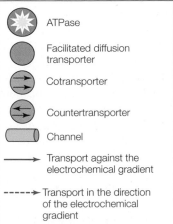

ATPase	
Facilitated diffusion transporter	
Cotransporter	
Countertransporter	
Channel	
——→	Transport against the electrochemical gradient
- - -→	Transport in the direction of the electrochemical gradient

In all three parts of the kidney tubule, energy for Na⁺ reabsorption comes from ATP used for primary active transport by Na⁺–K⁺-ATPase. The ATPase removes Na⁺ from each type of epithelial cell across the basolateral cell membrane. Na⁺ enters each cell across the apical cell membrane by diffusion down the electrochemical gradient generated by Na⁺–K⁺-ATPase. The membrane proteins involved in Na⁺ entry are different in all three types of epithelial cells, however.

(B) Thick ascending limb of loop of Henle

In the thick ascending limb, much of the Na⁺ entry into a cell occurs by means of a Na–K–2Cl cotransporter that carries K⁺ and Cl⁻ inward by secondary active transport. **Loop diuretics** inhibit this cotransporter.

(C) Collecting duct Na⁺-reabsorbing cell (principal cell)

In the collecting duct, Na⁺ enters the principal cells by a channel. The collecting-duct principal cells are **the** *main targets of aldosterone*, which promotes Na⁺ reabsorption by increasing synthesis of the Na⁺ channel protein and the Na⁺–K⁺-ATPase protein, as well as other actions. **Diuretic drugs such as amiloride** block or inhibit the Na⁺ channel.

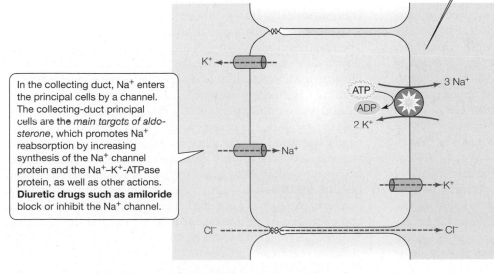

FIGURE 29.16 Major molecular mechanisms of NaCl reabsorption and associated processes in three parts of the mammalian kidney tubule   Each drawing shows a representative epithelial cell in the epithelial wall of the tubule. (A) The early proximal convoluted tubule. Cl⁻ is removed from the tubular fluid at this site by simple diffusion, largely via paracellular pathways, following an electrochemical gradient created by Na⁺–K⁺-ATPase. The Na–H countertransporter in the apical membrane moves H⁺ into the tubular fluid, where the H⁺ combines with tubular bicarbonate (HCO₃⁻), forming CO₂, which enters the cell, supplying CO₂ to the intracellular reaction shown. (B) The thick segment of the ascending limb of the loop of Henle. Here, Cl⁻ is reabsorbed through the cells, rather than by the paracellular route, in a process mediated by a Na–K–2Cl cotransporter in the apical membrane. (C) In the collecting ducts, different cells carry out Na⁺ and Cl⁻ reabsorption. The drawing shows a Na⁺-reabsorbing cell, known as a *principal cell*. In all three segments, most K⁺ brought into cells by Na⁺–K⁺-ATPase diffuses out via channels.

interstitial fluid (~300 m*Osm*) by osmotic outflux of water (see Figure 29.14A). The tubular fluid then descends deeper and deeper into the medulla, from the cortical to the pelvic end of the collecting duct. As it does so, it encounters an ever-higher interstitial NaCl concentration and attains higher concentrations of urea and nonurea solutes by the mechanisms we have discussed. Water is reabsorbed osmotically and returned to the blood plasma in the vasa recta. Especially in the cortical part of the collecting duct, but also in the inner-medullary part, NaCl is actively reabsorbed (**FIGURE 29.16C**). This reabsorption of NaCl in the collecting duct determines the final amount of NaCl that is removed from the body fluids and excreted. It also plays a key role in controlling urine volume—the amount of water removed from the body fluids—because by reducing the amount of nonurea solute in the urine, it enhances osmotic return of water from the urine to the blood. In the end, mammals in antidiuresis typically excrete only 1% or less of the filtered NaCl and water.

THE DILUTING KIDNEY AND THE REGULATION OF SWITCHES BE-TWEEN CONCENTRATION AND DILUTION Individual mammals are typically capable of adjusting the concentration and volume of their urine over broad ranges, thereby modulating the effects of urine production on the concentration and volume of the blood plasma and other body fluids. A person in antidiuresis might produce urine that is as concentrated as about 1200 m*Osm* (U/P = 4) and limited in volume to less than 1% of the filtered amount. In diuresis, by contrast, that person might produce urine as dilute as about 50 m*Osm* (U/P = 0.2) and increase the volume to about 15% of the filtered amount. The effects on the body fluids vary commensurately. *In antidiuresis the high osmotic concentration of the urine tends to dilute the body fluids (see Figure 27.7), and the low urine volume has the synergistic effect of conserving water. Conversely, in diuresis, the low urine osmotic concentration tends to concentrate the body fluids (see Figure 27.7), and the high urine volume voids water.*

The principal agent of control of switches between antidiuresis and diureses is antidiuretic hormone (ADH). The ADH of most mammals is *arginine vasopressin*; therefore ADH is often called **vasopressin** in books on mammalian physiology and medicine. As knowledge advances, the known effects of ADH become more extensive and complex. The action of ADH that is of most central importance in mammals is that it modulates the permeability of the collecting ducts to water.

The effect of ADH on the permeability of the collecting-duct epithelium is mediated by a specific molecular form of aquaporin, AQP-2, that is inserted into and retrieved from the apical cell membranes of the collecting-duct epithelial cells; Figure 16.16 shows this process in detail. The presence of ADH causes insertion of aquaporin molecules into the apical cell membranes and an increase in epithelial permeability to water (**FIGURE 29.17**). When ADH levels fall, the aquaporin molecules are retrieved from the apical cell membranes and epithelial permeability to water decreases.

In the concentrating kidney, although we have not said so heretofore, the high permeability of the collecting ducts to water is elicited by high blood levels of ADH. This high water permeability permits water to leave the collecting ducts by following the osmotic gradient between the collecting-duct fluid and medullary interstitial fluid. The osmotic exit of water, as previously stressed, accounts for both the *low volume* and *high concentration* of the urine produced during antidiuresis.

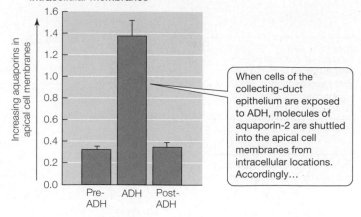

(A) Number of aquaporin molecules in apical cell membranes as a ratio of number in intracellular membranes

> When cells of the collecting-duct epithelium are exposed to ADH, molecules of aquaporin-2 are shuttled into the apical cell membranes from intracellular locations. Accordingly…

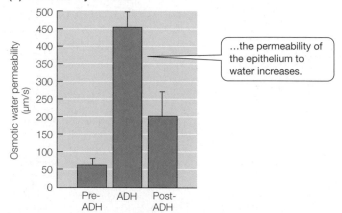

(B) Permeability to water

> …the permeability of the epithelium to water increases.

FIGURE 29.17 The collecting-duct epithelium: Cellular position of molecules of aquaporin-2 (AQP-2), and permeability to water, when ADH is present or absent Studies were carried out on collecting ducts from the inner medulla. The distribution of AQP-2 molecules was determined by visualizing and directly counting the molecules by means of immunological labels and electron microscopy. According to the shuttle hypothesis, the AQP-2 molecules in each epithelial cell are shuttled back and forth between the apical cell membrane and intracellular vesicular membranes. (A) The number of AQP-2 molecules in apical cell membranes as a ratio of the number in intracellular vesicular membranes. (B) Permeability of the collecting-duct epithelium to water. (After Knepper et al. 1996.)

When blood levels of ADH are low and aquaporins are retrieved from the apical cell membranes of the collecting-duct epithelium, the collecting ducts are poorly permeable to water. The distal convoluted tubules are also poorly permeable. Thus, during diuresis, from the time the tubular fluid exits the loops of Henle to the time it is discharged into the renal pelvis, it is blocked from coming freely to osmotic equilibrium with the surrounding cortical and medullary interstitial fluids. Recall that the tubular fluid is hyposmotic to plasma when it exits the loops; therefore it would lose water osmotically if it could. However, the low water permeability of the walls of the distal tubules and collecting ducts in diuresis—when ADH levels are low—impedes such water loss. **FIGURE 29.14B** shows a further consequence of the low water permeability: As NaCl is actively reabsorbed in the distal tubules and collecting ducts during diuresis, the tubular fluid becomes ever-more hyposmotic to the plasma. The urine produced is both *abundant* (because of little water reabsorption) and *dilute* (because of the diluting process just described).

Notice how fundamentally similar the action of ADH is in mammals and in amphibians. In both groups—and indeed, in all groups of tetrapod vertebrates—the primary effect of ADH on the renal tubules is to increase the permeability to water of tubular epithelia that otherwise are poorly permeable.[18] This increase in permeability to water has the important consequence that it allows the tubular fluids to come to osmotic equilibrium with the fluids surrounding the tubules. In amphibians, the fluids surrounding the distal tubules and collecting ducts are osmotically similar to the blood; thus the presence of ADH causes production of urine that approaches isosmoticity with the blood. In mammals, the collecting ducts are surrounded by fluids that are hyperosmotic to the blood. Consequently, ADH causes production of hyperosmotic urine. In both mammals and amphibians, ADH principally controls the excretion of *water* and thus controls the removal of *water* from the body fluids.[19] Although the amount of each nonurea *solute* excreted is adjusted by solute-specific tubular mechanisms (e.g., active reabsorption or secretion), the concentration of ADH determines the amount of water that is extracted from the body fluids and excreted with the solutes.

When an individual mammal switches between chronic antidiuresis and chronic diuresis, an additional change besides the permeability adjustments occurs: The magnitude of the osmotic gradient in the medullary interstitial fluids—the gradient between the cortex and inner medulla—diminishes. For example, in a dog shifted from chronic antidiuresis to chronic diuresis, the osmotic pressure of the inner-medullary interstitial fluid might change from 2400 mOsm (about 2100 mOsm higher than the cortical osmotic pressure) to 500 mOsm (about 200 mOsm higher). Periods of hours or days are required for such changes to be fully realized.

ADH is not the only hormone that controls kidney function. As discussed in Chapter 28 (see page 777), aldosterone and natriuretic hormones help control the reabsorption and secretion of Na^+ and K^+. In addition, calcitonin affects renal function, and the kidneys themselves employ paracrines, such as eicosanoids and kinins, as *local* chemical messengers.

Modern molecular and genomic methods create new frontiers in the study of kidney function

Modern molecular and genomic methods are enabling kidney researchers to study subjects that seemed utterly beyond reach 25 years ago. A stunning example is provided by recent studies of the fine structure of the mammalian renal medulla. Different parts of the tubules in the medulla often differ categorically in one or more cell-membrane proteins. When this is the case, elements can be distinguished by immunological labels. For the immunocytochemical study in **FIGURE 29.18A**, fluorescent antibodies were prepared against three distinguishing proteins. A blue-fluorescing antibody was prepared against aquaporin-2, which is found only in collecting ducts. Red- and green-fluorescing antibodies were prepared against aquaporin-1 and a urea transporter found, respectively,

---

[18] This is postulated to occur in all cases by aquaporin insertion.

[19] See Chapter 28 (page 776) for a full explanation of this point and Figure 29.5 for a diagram of how changes in tubular permeability to water can alter the amount of water excreted in the urine.

(A) Direct fluorescence image of a longitudinal section

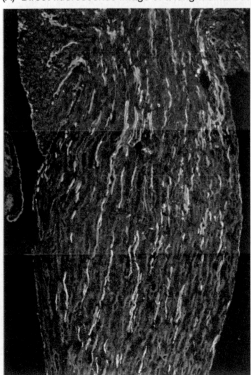

0.5 mm

(B) Reconstruction of the tip of the medulla

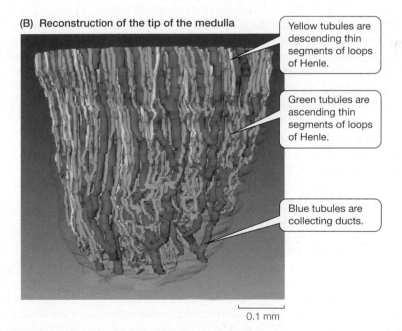

Yellow tubules are descending thin segments of loops of Henle.

Green tubules are ascending thin segments of loops of Henle.

Blue tubules are collecting ducts.

0.1 mm

FIGURE 29.18 Studies of the fine structure of the papilla of the medulla in laboratory rats The two parts use different color codes. (A) A portion of a single nal section through the medulla, orientated so the tip is down. Blue-, red-, and g ing antibodies were prepared against three different proteins that distinguish ducts (blue), descending thin segments of the loops of Henle (red), and d recta (green). (B) A three-dimensional reconstruction of the loops of He ducts in the tip of the papilla. The outer epithelium of the papilla is sho courtesy of Thomas Pannabecker; from Pannabecker and Dantzler

TABLE 29.1	Results of a comprehensive transcriptomic study of the inner-medullary collecting duct: Ten highly expressed genes and the proteins for which they code[a]

Gene	Protein
Actb	Beta actin (a form of nonmuscle actin)
Actg1	Gamma actin (a form of nonmuscle actin)
Aqp2	Aquaporin-2 water channel
Aqp3	Aquaporin-3 water channel
Aqp4	Aquaporin-4 water channel
Atp1b1	Na^+–K^+-ATPase β1 polypeptide
Atp1a1	Na^+–K^+-ATPase α1 polypeptide
Clic1	Chloride channel
Sgk1	Serum/glucocorticoid regulated protein kinase
Slc14a2	Solute carrier family 14 urea transporter

Source: After Uawithya et al. 2007.

[a]Animals studied were pathogen-free laboratory rats (*Rattus norvegicus*).

in the descending thin segments of the loops of Henle and the descending vasa recta. Figure 29.18A shows just a single section through the kidney. By synthesizing information from many sections, the three-dimensional fine structure can be reconstructed. **FIGURE 29.18B** shows the collecting ducts and the ascending and descending thin segments of loops of Henle in just the very tip—*only 0.5 mm long!*—of the medullary papilla. One discovery from this detailed study of structure is that there is far more contact among the loops of Henle and the collecting ducts in the medullary tip than would be predicted from a random arrangement of the two types of tubules relative to each other.

Genomic studies are being used to develop comprehensive lists of all the proteins present in specific parts of the kidneys. For example, a recent transcriptomic study of the inner-medullary collecting duct identified almost 8000 proteins. The genes for some are highly expressed (**TABLE 29.1**). These genes encode proteins that, for the most part, are already known to play important roles in kidney function. For example, the genes for aquaporins, Na^+–K^+-ATPase, _____ are among the most highly expressed. The _____ ighly expressed gene, *Sgk1*, is _____ e control (see Figure 29.16C). _____ ome highly expressed genes do _____ the inner-medullary collecting _____ le 29.1), for example, seem likely _____ I signals to control aquaporins, _____ mented in the inner-medullary _____ 0 proteins identified, there are _____ any transcription factors—about _____ Identifying the proteins present _____ comprehensive genomic survey _____ ctional research.

Young ... e longitudi- ... een-fluoresc- ... the collecting ... scending vasa ... e and collecting ... wn in gray. (Images ... 007.)

SUMMARY
Urine Formation in Mammals

■ The loops of Henle, collecting ducts, and vasa recta form parallel arrays in the medulla of the mammalian kidney, creating the structural basis for the ability to form urine hyperosmotic to the blood plasma. Among species of mammals of a particular body size, the species with long loops of Henle tend to be able to produce more concentrated urine than those with shorter loops.

■ The proximal convoluted tubule reabsorbs—and returns to the body fluids—much of the NaCl and water from the filtrate by processes that do not alter the osmotic pressure of the tubular fluid. It also fully reabsorbs glucose and amino acids, returning them to the body fluids.

■ After the tubular fluid passes through the loop of Henle, it is less concentrated than when it entered. Nonetheless, processes in the loop of Henle create the gradients of osmotic pressure and NaCl concentration in the medullary interstitial fluid that are responsible for the ultimate concentration of the urine. In the part of the loop where the ascending limb is thick, active NaCl transport creates a single-effect difference in osmotic pressure and NaCl concentration between adjacent parts of the ascending and descending limbs. By acting as a countercurrent multiplication system, the loop generates a difference in osmotic pressure and NaCl concentration from end to end that is much larger than the single effect.

■ During antidiuresis, as tubular fluid makes its last pass through the medulla in the collecting ducts, nonurea solutes are concentrated because the collecting-duct walls are freely permeable to water, permitting osmotic equilibration between the tubular fluid and the medullary interstitial fluid. The high permeability of the collecting-duct epithelial walls to water results from insertion of aquaporin-2 molecules into cell membranes in response to ADH (vasopressin).

■ During diuresis, the collecting-duct walls are poorly permeable to water, so tubular fluid is osmotically isolated from the medullary interstitial fluid and can be diluted by solute reabsorption.

Urine Formation in Other Vertebrates

Freshwater and marine teleost fish differ in nephron structure and function

The sort of nephrons we described in amphibians apparently evolved in their freshwater progenitors, because the nephrons of nearly all freshwater teleost (bony) fish are structurally similar to those of amphibians. In freshwater fish, as in amphibians, the distal convoluted tubule plays a key role in diluting the urine. The walls of the tubule are nearly impermeable to water. Thus, as NaCl is reabsorbed and returned to the body fluids, water remains behind in the tubule, and a dilute urine is produced. The effect is to help keep the blood osmotic pressure high (see Figure 27.7) despite the water overload that occurs in freshwater fish because of inward osmosis from the environment.

Marine teleost fish commonly lack the distal convoluted tubule. If they are descended from freshwater ancestors, as is usually thought (see Box 28.2), the absence of the distal tubule probably represents a secondary loss rather than a primitive condition. The reason for the loss seems straightforward: Marine teleosts are hyposmotic to their seawater environment and thereby face continuous osmotic desiccation. They have no need of a nephron segment specialized for the production of a voluminous, dilute urine rich in osmotically free water.

In addition to differing in the presence or absence of the distal convoluted tubule, freshwater and marine teleosts differ in other ways. Freshwater teleosts typically have relatively large numbers of nephrons and well-developed glomeruli. Their GFRs are relatively high, as suits animals that have excesses of water that must be voided in urine. In contrast, marine teleosts tend to have relatively few nephrons and small glomeruli. They have low GFRs, a condition that seems adaptive for animals that face desiccation and produce relatively little urine.

Many marine teleosts—according to present evidence—do not form their primary urine entirely by ultrafiltration. Instead, they form their primary urine partly by *secretion* into the proximal tubules. The mechanism of secretion is that ions—including Na^+, Cl^-, Mg^{2+}, and SO_4^{2-}—are actively transported into the proximal tubules, and water and other solutes follow (see Figure 29.2).

In about 30 known species of marine teleosts—described as **aglomerular**—the trend toward small glomeruli in the marine environment is carried to its logical extreme, and the nephrons lack glomeruli. These aglomerular species form their primary urine entirely by secretion. Aglomerularism has evolved on three independent occasions, suggesting it is adaptive under some circumstances. Some seahorses and pipefish, some Antarctic fish, and the oyster toadfish (*Opsanus tau*) are aglomerular.

Some of the most interesting fish from the viewpoint of kidney function are the euryhaline teleost species that can live in either freshwater or seawater, such as salmon and migratory eels. The control of kidney function in teleosts has in fact been most thoroughly studied in some of these species. When a euryhaline fish is transferred from seawater to freshwater, it typically undergoes a large increase in GFR, mediated for the most part by an increase in the number of filtering nephrons. Active secretion of Mg^{2+} and SO_4^{2-} into the urine, which is vigorous in seawater, is curtailed in freshwater (where ambient Mg^{2+} and SO_4^{2-} concentrations are vastly lower). Moreover, when fish are transferred from seawater to freshwater, the nephrons—and sometimes other excretory structures (e.g., the bladder)—undergo decreases in their overall permeability to water, a change that favors water excretion. Prolactin, arginine vasotocin (the "ADH" of fish; see Table 16.2), and angiotensin II are implicated in controlling these changes, but the controls are not well understood.

The reptiles other than birds have nephrons like those of amphibians, but birds have some mammalian-type nephrons

The nephrons of lizards, snakes, turtles, and crocodilians are broadly similar to those of amphibians. Birds, by contrast, have a range of nephron forms, which are usually categorized into two major types (**FIGURE 29.19**). Some of the nephrons of birds have short, uncomplicated proximal and distal tubules, and they lack loops of Henle. These nephrons superficially resemble the nephrons of nonavian reptiles in structure, and they are called **loopless**

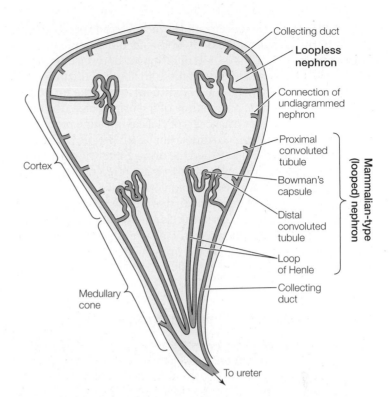

FIGURE 29.19 One lobe of a bird's kidney in cross section (After Willoughby and Peaker 1979.)

nephrons. Other avian nephrons have a loop of Henle interposed between the proximal and distal convoluted tubules and are called **looped nephrons** or **mammalian-type nephrons**. These nephrons have relatively large glomeruli and elaborate proximal tubules. Approximately 10%–30% of the nephrons in a bird's kidney are typically of the mammalian type; the remainder are of the loopless type. The mammalian-type nephrons are organized into sets. Among the nephrons of a set, the Bowman's capsules and proximal and distal convoluted tubules are all positioned near the same part of the kidney that houses the loopless nephrons, but the loops of Henle all project in a compact parallel array toward the direction of the ureter. Each parallel array of loops of Henle is called a medullary cone (see Figure 29.19). Each kidney includes many cones. Collecting ducts carrying the outflow from both the loopless and the mammalian-type nephrons run through the medullary cones on their way to the ureter.

Neither the nephrons in the kidneys of nonavian reptiles nor the loopless nephrons in a bird's kidneys can produce urine that is hyperosmotic to blood. However, the loops of Henle of the mammalian-type nephrons in a bird's kidney carry out countercurrent multiplication and can raise the urine osmotic pressure above the blood osmotic pressure. Details of the countercurrent mechanism in birds are probably different from those in mammals. For most species of birds, the maximum urine osmotic pressure is no more than about 2.5 times blood osmotic pressure.

Uric acid, the principal nitrogenous end product of birds and most other reptiles, is introduced into the nephrons by filtration and secretion. It is actively secreted into the urine as the urine flows through the nephrons, and this secretion accounts for the greater part of the excreted amount.

In both birds and nonavian reptiles, the ureters do not discharge directly to the outside of the body but instead discharge into the cloaca. From the cloaca, the urine is often moved by reverse peristalsis into the lower intestine. Because of these attributes, both the cloaca and the intestine may reclaim constituents from the urine and modify the composition and volume of the urine before the urine is excreted. At least four categories of cloaca–intestine processing are recognized, correlated to some extent with the life histories of the birds and other reptiles. When the urine enters the cloaca from the ureters, the uric acid and urates in it are often present largely in the form of supersaturated colloidal suspensions stabilized by specific proteins. The uric acid and urates are then precipitated into the form of tiny solid particles in the cloaca–intestine, forming the thick white paste that is often seen mixed with the feces of a bird or other reptile. Precipitation *after* the urine has left the ureters helps prevent clogging of the renal tubules with the precipitate.

SUMMARY
Urine Formation in Other Vertebrates

- Freshwater teleost fish have nephrons structurally similar to amphibian nephrons. Marine teleost fish, however, usually lack the distal convoluted tubule and have a relatively poorly developed glomerular filtration apparatus that seems often to be supplemented by active solute secretion. A few marine fish are aglomerular and depend entirely on secretion.

- Birds and other reptiles have nephrons structurally similar to amphibian nephrons. Birds, in addition, have mammalian-type nephrons (with loops of Henle) organized into parallel arrays—the medullary cones—in which urine hyperosmotic to blood plasma can be made.

Urine Formation in Decapod Crustaceans

An adult crayfish, crab, lobster, or other decapod crustacean has two renal organs, known as **antennal glands** or **green glands**, which are located in its head and open to the outside independently near the bases of its second antennae (**FIGURE 29.20A**). Each antennal gland is basically a single tube, sometimes loosely described as resembling "a single giant nephron." In a freshwater crayfish (**FIGURE 29.20B**), each antennal gland begins with a closed **end sac** or **coelomosac** lying to the side of the esophagus. Following the end sac is the **labyrinth** (or green body), a sheet of spongy tissue consisting of a channel that branches and anastomoses extensively along its length. The **nephridial canal**, which also has a spongy internal morphology, leads from the labyrinth to the expanded **bladder**, and the bladder empties to the outside. The nephridial canal is found only in certain freshwater decapod crustaceans.

The walls of the end sac are thin, and arteries from the heart supply a network of small vessels or lacunae on the outer surface of the end sac (see Figure 25.23). This morphological evidence has long suggested that fluid enters the end sac by filtration under the force of blood pressure. Additional morphological evidence for this concept is provided by the presence of cells resembling podocytes

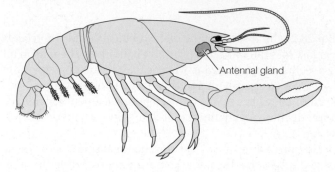

(A) Position of the antennal gland (green gland)

Antennal gland

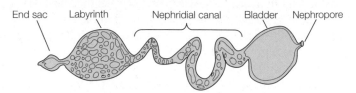

(B) Antennal gland unfolded, with urine properties plotted below corresponding anatomical locations

End sac   Labyrinth   Nephridial canal   Bladder   Nephropore

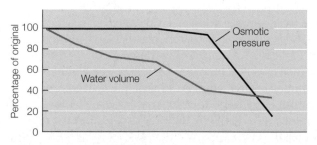

FIGURE 29.20 The antennal gland and urine formation in a freshwater crayfish (A) The position of the antennal gland on the right side of a crayfish's body. (B) Although the nephridial canal is in fact tightly convoluted and partly enveloped by the sheetlike labyrinth, the antennal gland can be stretched out to reveal its parts. The graph shows measured changes in the osmotic pressure and water content of the urine as it passes through the parts of the antennal gland in crayfish (*Austropotamobius pallipes* and *Orconectes virilis*) living in freshwater. Values are plotted immediately below the anatomical locations where they were measured and are expressed as percentages of the values in the end sac. (After Riegel 1977.)

(see Figure 29.1) in the end-sac epithelium. The physiological evidence that is available supports the hypothesis that primary urine is formed in the end sac by filtration from the blood. The composition of the urine is modified in all the structures through which it flows as it passes through the antennal gland. The labyrinth of American lobsters, for example, is known to reclaim glucose, and probably reclaims amino acids, from the filtrate. In marine crabs, the bladder is an important site where glucose is reabsorbed from the urine and Mg^{2+} is secreted into the urine. Unfortunately, a truly synthetic understanding of the handling of solutes by all parts of an antennal gland is not yet available for any species.

Modifications of the osmotic pressure of the urine have been a major focus of study in freshwater decapods. The labyrinth is by all accounts incapable of rendering the urine hyposmotic to the blood. There is a much-emphasized correlation between the presence of a nephridial canal and the ability to produce urine hyposmotic to the blood. Freshwater crayfish, which can make dilute urine, have a nephridial canal, which is often hypothesized to function in ways analogous to the vertebrate distal convoluted tubule. The nephridial canal, however, is absent in ocean decapods, which produce urine isosmotic to their blood plasma; and it is also absent in freshwater crabs, such as *Eriocheir sinensis* (see page 748), which are recent immigrants to freshwater and unable to make dilute urine. Some studies on crayfish indicate that the bladder helps produce dilute urine (see Figure 29.20B). Excretion of dilute urine by a crayfish helps keep the osmotic pressure of its body fluids high (see Figure 27.7). Moreover, because crayfish actively reabsorb NaCl from the primary urine when they are producing a dilute definitive urine, the process helps retain NaCl in the body fluids.

Urine Formation in Molluscs

The renal organs of molluscs are tubular or saccular structures, called **nephridia**, or **kidneys**, that empty into the mantle cavity or directly to the outside. Bivalves, most cephalopods (octopuses and squids), and some gastropods have two kidneys, but most gastropods have only one. The kidneys are closely associated with the heart or hearts. In fact, the pericardial fluid in the pericardial cavity that surrounds the heart is the primary urine in most molluscs! The pericardial fluid flows into a canal called the **renopericardial canal**, which is part of the kidney.

The most thoroughly understood molluscan kidney is that of the giant octopus *Enteroctopus dofleini*, an ocean mollusc—isosmotic to seawater—found along the Pacific coast of North America. The kidneys of octopuses and squids are associated with the branchial (gill) hearts rather than the systemic heart (**FIGURE 29.21**). Each branchial heart bears a thin-walled protuberance, the *branchial heart*

appendage, which communicates with the lumen of the heart. In *Enteroctopus*, the pericardial cavity of each branchial heart encloses only the side of the heart bearing the heart appendage, as seen in Figure 29.21. A kidney connects to each pericardial cavity. Compelling evidence exists in *Enteroctopus* and certain other cephalopods that the pericardial fluid is an ultrafiltrate of the blood, forced into the pericardial cavity across the branchial heart appendage under the force of pressure developed in the heart. In each kidney, this filtrate flows through a long renopericardial canal and then an enlarged **renal sac** before being discharged into the mantle cavity. Studies have shown that the renopericardial canal alters the composition of the urine. Glucose and amino acids are promptly reabsorbed there and returned to the body fluids, for example—a process reminiscent of their prompt reclamation in vertebrate nephrons.

Urine Formation in Insects

The formation of urine has been much more thoroughly studied in insects than in any other group of invertebrates. Most insects possess Malpighian tubules,[20] and these tubules are often called the *excretory tubules*. A point to be stressed from the outset, however, is that the hindgut is as important as the Malpighian tubules in the formation of urine.

The **Malpighian tubules** are long, slender, blind-ended structures that typically arise from the junction of the midgut and hindgut (**FIGURE 29.22**). They number from 2 to more than 200, depending on the species. Projecting into the hemocoel, they are bathed by the blood (hemolymph). The walls of the tubules consist of a single layer of epithelial cells, surrounded on the outside by

---

[20] The tubules are named after Marcello Malpighi (1628–1694), one of the great early microscopists, who was the first to describe the blood capillaries and renal corpuscles of vertebrates as well as the Malpighian tubules and tracheae of insects.

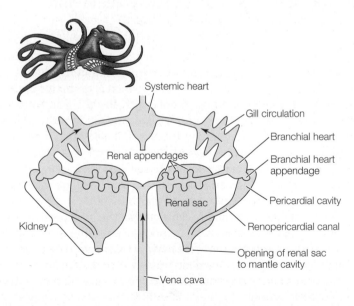

FIGURE 29.21 The kidneys of an octopus (*Enteroctopus*) and their relations to the circulatory system   The kidneys are yellow, and the circulatory system is blue. After the principal vein returning blood from the systemic tissues, the vena cava, branches, each branch passes by one of the renal sacs, and there it bears many glandular diverticula, called *renal appendages*, which are closely juxtaposed to the walls of the sac. In *Enteroctopus*, ammonia is believed to be secreted into the renal sacs across the renal appendages. Blood then travels to the branchial hearts, where ultrafiltration occurs. (After Martin and Harrison 1966.)

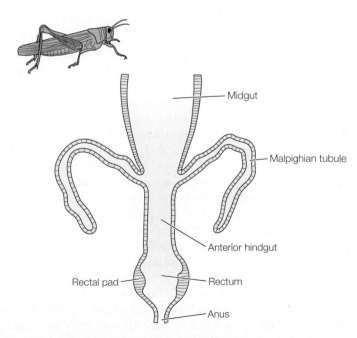

FIGURE 29.22 The posterior gut and Malpighian tubules of an insect   The Malpighian tubules empty into the gut at the junction of the midgut and hindgut.

a thin basement membrane. Although the tubules exhibit little histological differentiation along their length in some species, they are differentiated into two to six (or possibly more) distinct regions in numerous others. In many species, the various tubules within an individual are morphologically similar, but in others, two or more types of tubules are present.

The hindgut, which is lined with cuticle (see Box 6.3), typically consists of a relatively small-diameter **anterior hindgut** (*ileum* or *intestine*) and an expanded posterior part, the **rectum** (see Figure 29.22). The walls of the anterior hindgut usually consist of a single layer of cuboidal or squamous epithelial cells. The walls of the rectum are similar to those of the anterior hindgut in some insects, but in many other species, parts of the rectal wall consist of thick columnar epithelial cells, sometimes associated with secondary cell layers. These thickened parts of the rectal wall are termed **rectal pads** or **rectal papillae** (see Figure 29.22), depending on their gross morphology.

The Malpighian tubules form and sometimes modify the primary urine

The function shared by the Malpighian tubules of all insects studied is the formation of the primary urine. The tubules are not supplied with blood vessels, and filtration does not occur. Instead, the primary urine is formed by a secretory mechanism in insects (see Figure 29.2). In the most common scenario, potassium chloride (KCl) is secreted vigorously by the Malpighian tubule epithelium into the lumen of the tubule from the blood bathing the tubule—so vigorously that the K^+ concentration in the tubular fluid is 6–30 times higher than the blood K^+ concentration. According to theories that have rapidly matured in the last 30 years, the K^+ secretion occurs by secondary active transport; a H^+-ATPase uses ATP bond energy to create electrochemical gradients that drive the secondary active transport of K^+. The secretion of K^+ into a Malpighian tubule is electrogenic, and Cl^- accompanies the K^+ passively by following the electrical gradient (inside positive) set up by K^+ secretion. The result is that KCl is secreted at a cost of ATP. The flux of KCl into the lumen of a Malpighian tubule drives osmotic entry of water, which typically occurs briskly enough that the tubular fluid remains approximately isosmotic to the blood (a case of near-isosmotic water transfer). Many additional solutes then enter the tubular fluid passively, as by following solute concentration gradients set up by the osmotic influx of water; these solutes include amino acids, sugars, diverse organic wastes and toxins, and several inorganic ions. Proteins are largely excluded from the tubular fluid because of their molecular size. Certain organic compounds—notably some detoxification products plus secondary compounds synthesized by plants to deter herbivory (e.g., alkaloids)—are actively secreted into the primary urine by some insects.

Although KCl is most commonly the principal salt secreted to initiate primary-urine formation, NaCl plays this role in some species. There are also species in which KCl predominates under some conditions, whereas NaCl does under other conditions.

As primary urine flows through the Malpighian tubules toward the gut, the tubular epithelium may reabsorb salts, water, or other molecules such as glucose—returning them to the blood. In the end, the fluid that enters the hindgut from the Malpighian tubules is approximately isosmotic to the blood and contains numerous solutes. Its solute composition is quite unlike that of the blood, however. In particular, the fluid that enters the hindgut is typically far richer in KCl than the blood is—a consequence of the secretory mechanism of primary-urine formation.

The rate at which primary urine is formed can be strikingly high relative to the total volume of the insect's body fluids. Current estimates, for example, indicate that in a female yellow-fever mosquito (*Aedes aegypti*), during an ordinary 24-h day when she has *not* taken a blood meal, the Malpighian tubules produce primary urine equivalent to 12 times her total body volume of extracellular fluid! If this sounds strange, it is really not; recall the very high rate at which human kidneys produce primary urine—also equivalent to about 12 times the entire extracellular fluid volume every day. In insects, as in vertebrates, most primary urine is reabsorbed (mostly by the hindgut) rather than being excreted. The overall process—a high rate of primary-urine formation followed by a high rate of reabsorption—gives the excretory system intimate access to the blood to carry out its regulatory functions, as already stressed.

A noteworthy aspect of the reabsorption process in insects is the reclamation of KCl. KCl must be secreted into the Malpghian tubules at a high rate to drive the production of primary urine, but it could not be lost from the body fluids at that rate. Instead, most KCl is reabsorbed back into the blood and recycled to produce more primary urine.

The reabsorption of KCl and water from the primary urine sometimes starts in the lower parts of the Malpighian tubules. However, the reabsorption occurs predominantly in the hindgut, especially in the rectum.

The hindgut modulates urine volume and composition in regulatory ways

After urine is discharged from the Malpighian tubules, it flows with the feces through the hindgut, where its composition, concentration, and volume are modified, resulting in the definitive urine, which is excreted. The rectum (which is far better understood than the anterior hindgut) not only reabsorbs—and returns to the blood—most of the water, K^+, Na^+, and Cl^- introduced into the hindgut by the Malpighian tubules, but also often reabsorbs amino acids, acetate, and phosphate. The rectum also has some secretory functions. For example, H^+ is secreted from the blood into the urine in the rectum, and the resulting acidification contributes to the precipitation of uric acid and urates there.

Research has increasingly clarified that the insect rectum has impressive *regulatory* abilities. It can modify the volume, composition, and osmotic pressure of the urine in ways that serve to regulate the volume, composition, and osmotic pressure of the blood. The rectum adjusts the osmotic pressure of the urine by varying the relative rates of reabsorption of water and total solutes. It also adjusts the ionic composition of the urine. In one set of experiments (**TABLE 29.2**), for example, fasting locusts were permitted to drink either tap water or a saline solution containing K^+, Na^+, Cl^-, and other ions. The rectum in the water-fed locusts reclaimed ions, returning them to the blood: It *lowered* ion concentrations in the rectal fluid (soon to be excreted as urine), compared with concentrations in the anterior-hindgut fluid. However, the rectum in the saline-fed locusts *raised* ion concentrations in the rectal fluid, compared with the anterior-hindgut fluid. The saline-fed locusts also accumulated

TABLE 29.2 Average composition of the rectal fluid and other body fluids in locusts (*Schistocerca gregaria*) that were deprived of food and provided with either tap water or a saline solution to drink

Experimental treatment	Fluid	Osmotic pressure (m*Osm*)	Ion concentration (m*M*)		
			Cl⁻	Na⁺	K⁺
Water-fed	Rectal fluid	820[a]	5	1	22
	Anterior-hindgut fluid	420	93	20	139
	Blood	400	115	108	11
Saline-fed	Rectal fluid	1870	569	405	241
	Anterior-hindgut fluid	—	192	67	186
	Blood	520	163	158	19

Source: After Phillips 1964.
[a]The high osmotic pressure in the scanty rectal fluid of water-fed animals is presumed to be caused by organic solutes.

greater volumes of urine in the rectum, so that overall, the quantities of ions excreted in their urine were hundreds of times greater than those excreted by the water-fed animals. In this way the rectum played a major role in helping to regulate blood ion concentrations.

The study of the hormonal control of urine production in insects is a burgeoning field at present, in part because of the expectation that the next generation of controls for insect pests might include procedures that defeat vital control mechanisms. Numerous diuretic and antidiuretic neurohormones—which affect both Malpighian-tubule and rectal function—have been identified in various species.

PRODUCTION OF URINE HYPEROSMOTIC TO THE BLOOD Insects are one of the three major groups of animals that can produce urine that is hyperosmotic to their blood plasma (mammals and birds are the other two groups). When insects produce hyperosmotic urine (see page 768), the process of concentration usually occurs in the rectum. At least three different mechanisms of concentrating the urine have evolved.

In insects that have rectal pads or rectal papillae—such as cockroaches (*Periplaneta*), desert locusts (*Schistocerca*), and blowflies (*Calliphora*)—the urine in the rectal lumen is concentrated by *water reabsorption in excess of solute reabsorption*. This water reabsorption is highly intriguing because it can continue even when the osmotic pressure of the rectal contents has risen to be two or more times higher than the osmotic pressure of the blood bathing the rectum! The existence of this seemingly paradoxical process has been demonstrated in several ways. Perhaps the most compelling evidence comes from experiments in which the rectum was filled with a *pure* solution of a solute (e.g., trehalose) that is neither reabsorbed nor secreted across the rectal wall. The *amount* of such a solute in the rectum is constant during the course of an experiment. When locusts (*Schistocerca*) were treated in this way, they reabsorbed water from their rectal contents and the rectal-fluid osmotic pressure rose. After the rectal-fluid osmotic pressure had become twice as high as blood osmotic pressure, the rectal-fluid osmotic pressure *continued to rise* and reached nearly three times the blood osmotic pressure. Results of this sort show that the rectal wall can move water *against* a large, opposing osmotic gradient between the rectal fluid on the inside and blood on the outside. The results show, moreover, that

in the short term, this water reabsorption can occur even in the absence of simultaneous solute reabsorption.

The mechanism of such water reabsorption was mystifying for many years. Now, however, a consensus exists that this water reabsorption is a case of osmosis on a microscopic scale: *local osmosis*. The mechanism depends in part on a complex microarchitecture in the rectal pads or papillae. The details of structure and possibly of function vary from species to species. Here we focus on the blowfly (*Calliphora*) as an example.

In the rectal papilla of a blowfly (**FIGURE 29.23A**), adjacent cells of the columnar epithelium are tightly joined on the side facing the rectal lumen and on the opposite (basal) side, but in between, the cells are separated by an elaborate network of minute channels and spaces, here termed the **intercellular spaces** (*intercellular*, "between cells"). The network of intercellular spaces communicates at the apex of the papilla with subepithelial spaces—here called **infundibular channels**—that are positioned under the basal side of the epithelial cell layer and connect with general blood spaces. Researchers hypothesize that the epithelial cells actively secrete solutes into the intercellular spaces, thereby rendering the fluid in the intercellular spaces strongly hyperosmotic to both the blood and the fluid in the rectal lumen (**FIGURE 29.23B**). Osmosis then carries water out of the rectal lumen into the intercellular spaces; that is, because of the *locally* high osmotic pressure in the intercellular spaces, water is osmotically withdrawn from the rectal fluid, even though the latter is thereby made increasingly hyperosmotic to the blood. Entry of water into the intercellular spaces adds volume to the fluid in the spaces and thereby causes fluid to flow in streams through the intercellular spaces toward the apex of the papilla and then through the infundibular channels toward the main blood cavity of the body (hemocoel). The fluid exiting the intercellular spaces is highly concentrated, but as it flows under the epithelial cells in the infundibular channels, solutes are believed to be actively or passively reabsorbed from the fluid into the cells across membranes poorly permeable to water, with two highly significant consequences. First, the fluid flowing through the infundibular channels is diluted, so that *in the end a fluid rich in water, rather than in solutes, is returned to the blood*—helping, for instance, to keep the osmotic pressure of the blood of a dehydrated insect from rising too high. Second,

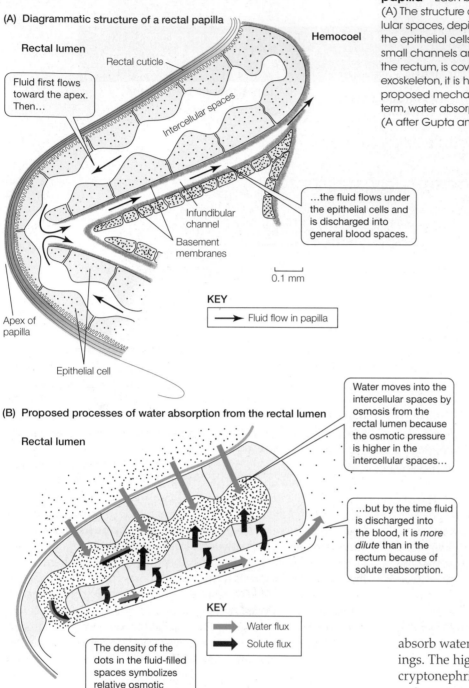

(A) Diagrammatic structure of a rectal papilla

Rectal lumen

Hemocoel

Rectal cuticle

Fluid first flows toward the apex. Then...

Intercellular spaces

...the fluid flows under the epithelial cells and is discharged into general blood spaces.

Infundibular channel

Basement membranes

0.1 mm

Apex of papilla

KEY

→ Fluid flow in papilla

Epithelial cell

(B) Proposed processes of water absorption from the rectal lumen

Rectal lumen

Water moves into the intercellular spaces by osmosis from the rectal lumen because the osmotic pressure is higher in the intercellular spaces...

...but by the time fluid is discharged into the blood, it is *more dilute* than in the rectum because of solute reabsorption.

KEY

→ Water flux

→ Solute flux

The density of the dots in the fluid-filled spaces symbolizes relative osmotic pressures.

FIGURE 29.23 The structure and function of the blowfly rectal papilla Each blowfly (*Calliphora erythrocephala*) has four rectal papillae. (A) The structure of a papilla, shown highly diagrammatically. The intercellular spaces, depicted for simplicity as a single broad cavity running through the epithelial cells, actually consist of a complex, interconnecting network of small channels and spaces *between* the epithelial cells. The papilla, being in the rectum, is covered with a thin cuticle; although this material is part of the exoskeleton, it is highly permeable to water and to solutes of small size. (B) The proposed mechanism of water absorption from the rectal lumen. In the short term, water absorption can occur without solute absorption from the rectum. (A after Gupta and Berridge 1966.)

to the excrement of other insects, and ostensibly dry. The structural basis for concentrating and drying the excrement is a specialized association between the Malpighian tubules and rectum: the **cryptonephridial complex**. In this complex, the distal parts of the Malpighian tubules (the parts nearest the blind ends)—which float freely in the hemocoel of most insects—are closely associated with the outer rectal wall, and these parts of the tubules and the rectum are together enclosed by a **perinephric membrane**, which separates them from the hemocoel. KCl and NaCl are actively transported from the blood into the lumen of each of these cryptonephric Malpighian tubules. In a marked departure from the usual condition in insects, however, water is prevented from entering the tubular fluid from the blood (probably by water-impermeability of the perinephric membrane). Because the fluid in the cryptonephric Malpighian tubules is formed by the inward secretion of ions without water, it has a dramatically high osmotic pressure (far higher than blood osmotic pressure), creating a gradient that favors osmotic reabsorption of water from the closely juxtaposed rectal lumen. One advantage for the insects is that (probably) this arrangement allows them to take up water vapor from the air: an extremely unusual capability for animals. Mealworms and some of the other insects with a cryptonephridial complex are known to absorb water vapor from the atmosphere across their rectal linings. The high concentrations of salts in the tubular fluids of the cryptonephric Malpighian tubules—which give the fluids a low water vapor pressure—appear to be responsible (see page 732).

A third mechanism by which insects produce concentrated urine is known in the small subset of insects that live in saline waters. Some, at least, produce concentrated urine by secreting ions into their rectal fluid. Living in salty water, they face the challenge of keeping their body fluids from becoming too concentrated in ions. They respond by secreting ions out of their blood in abundance.

solutes are returned to the epithelial cells and thus can again be secreted into the intercellular spaces, permitting continued osmotic water absorption from the rectal fluid without great need for new solutes from any source. The nature of the solutes involved is not fully resolved, although Na^+, K^+, and Cl^- are strongly implicated; some organic solutes also play roles.

A second type of concentrating mechanism has been described in insects that have a *cryptonephridial complex*. These insects include mealworms (larval *Tenebrio molitor*), certain larval and adult coleopterans (beetles), and certain larval lepidopterans (butterflies and moths). Mealworms can produce pellets of excrement (feces and urine combined) that are particularly concentrated relative

SUMMARY
Urine Formation in Insects

■ Primary urine is introduced into the Malpighian tubules by a secretory process usually based on active transport of KCl into the tubular fluid. As the primary urine flows down the Malpighian tubules, it may be modified by reabsorption or secretion, but typically remains isosmotic to the blood.

- The Malpighian tubules empty into the hindgut at the junction of the midgut and hindgut.

- The rectum modifies the volume, composition, and osmotic pressure of the urine in ways that help regulate the volume, composition, and osmotic pressure of the blood. The production of hyposmotic urine occurs by reabsorption of solutes in excess of water. Two of the known mechanisms of producing hyperosmotic urine, however, enable insects to reabsorb water in excess of solutes. Some of the insects that produce hyperosmotic urine in this way do so by local osmosis and solute recycling in rectal pads or papillae; others do so with a cryptonephridial complex. Saline-water insects may form hyperosmotic urine by secretion of solutes into the rectum.

Nitrogen Disposition and Excretion

When animals catabolize organic molecules to release chemical energy, the atoms of the molecules appear in a variety of catabolic end products. During aerobic catabolism, the three most abundant atoms—carbon, hydrogen, and oxygen—appear in CO_2 and H_2O. The CO_2 is typically voided promptly into the environment across lungs, gills, or skin. The H_2O (metabolic water) simply becomes part of an animal's body water resources. The fourth most abundant atom is nitrogen, which is a characteristic constituent of proteins and nucleic acids. The disposition of nitrogen atoms from catabolism is not as simple as that of carbon, hydrogen, and oxygen.

Some of the compounds into which animals incorporate nitrogen during catabolism are shown in **FIGURE 29.24**. Each of these nitrogenous end products has advantages and disadvantages for the animals that synthesize it. For example, some of the compounds are relatively cheap to make, whereas others are low in toxicity. There is no single end product that is ideal in all ways. Moreover, whereas end products often represent wastes to be excreted, they sometimes can be used in adaptive ways and have value. With all these considerations in mind, we cannot be surprised that animals have evolved a variety of strategies for dealing with the nitrogen atoms released from organic molecules by catabolism.

The relation between nitrogen excretion and kidney function varies from one group of animals to another. Mammals, birds, and nonavian reptiles exemplify one end of the spectrum: They excrete nitrogenous end products entirely in their urine. At the other end of the spectrum, there are many aquatic animals in which nitrogenous end products are excreted mainly across the gills or skin, and the kidneys play little or no role.

Animals often produce two or more nitrogenous end products. There are several reasons for this. One is that nitrogen is a major constituent of nucleic acids as well as proteins, and often the catabolic pathways involved in breaking down nucleic acids produce a different nitrogenous end product than those responsible for breaking down proteins or amino acids. Humans and other primates, for example, synthesize uric acid from the nitrogen of nucleic acid purines, but they synthesize principally urea from protein nitrogen.[21] Protein catabolism dominates as a source of nitrogen. About 95% of waste nitrogen is from protein catabolism, and the products of protein catabolism therefore dominate.

[21] The affliction known as gout results from abnormal uric acid metabolism.

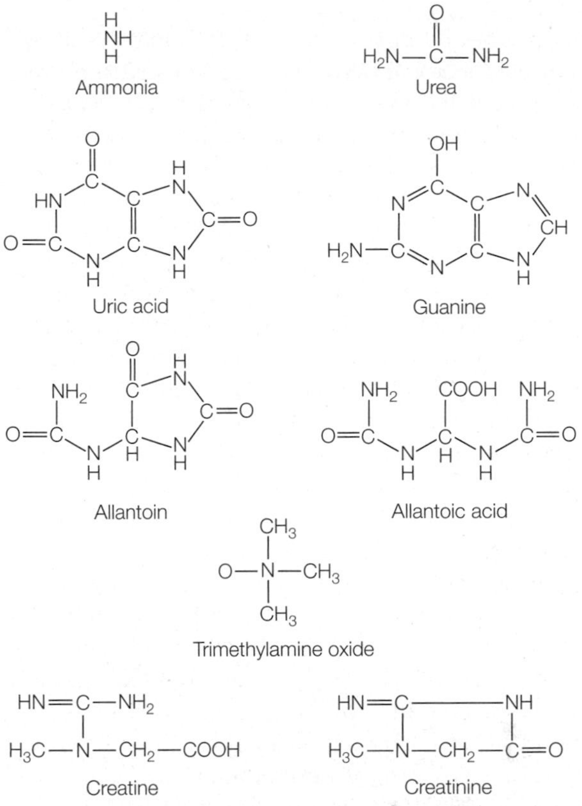

FIGURE 29.24 Some nitrogenous compounds excreted by animals Uric acid and guanine are purines. Allantoin and allantoic acid are poorly soluble breakdown products of uric acid. Trimethylamine oxide and its precursor, trimethylamine, are found in a variety of marine animals but do not occur in freshwater animals; both are highly soluble. Creatine and its internal anhydride, creatinine, occur as relatively minor excretory compounds in many vertebrates and some invertebrates. Some animals, mostly invertebrates, also lose significant amounts of amino acids to the environment.

When ammonia (NH_3) or the ammonium ion (NH_4^+) is the principal nitrogenous end product of an animal, the animal is described as **ammonotelic**. If urea is the principal nitrogenous end product, an animal is termed **ureotelic**. If uric acid is the principal end product, an animal is **uricotelic**.

Ammonotelism is the primitive state

Ammonia (NH_3) reacts with hydrogen ions to form the ammonium ion (NH_4^+). At the ordinary pH values of animal body fluids and tissues, this reaction is shifted strongly toward the formation of NH_4^+. For simplicity, we use the word *ammonia* here to refer to either NH_3 or NH_4^+. Ammonia is clearly the primitive nitrogenous end product of animals. One piece of strong evidence for this view is the fact that the great majority of today's ocean invertebrates are ammonotelic.

Ammonia is highly toxic. For example, even at low concentrations, it can disrupt neuron function, the integrity of the blood–brain barrier, and gill permeability. Blood concentrations are ordinarily kept low: usually under 0.3 millimolar (mM) in vertebrates, for example.

Because of its toxicity, ammonia cannot ordinarily be allowed to accumulate in an animal's body. Thus, for an animal to be ammonotelic, it must have a means of unfailingly voiding ammonia as rapidly as it is formed by catabolism. Aquatic animals can meet this challenge because of the abundance of water in which they live; often, much of the ammonia they produce is voided directly into the ambient water across their gills or other external body surfaces. Not only most ocean invertebrates, but also most other water-breathing aquatic animals, are ammonotelic. Both freshwater and marine teleost fish are typically ammonotelic, and both the aquatic tadpoles of amphibians and adult aquatic amphibians (e.g., mudpuppies) are ammonotelic.

Ammonotelism, however, is unusual on land. A terrestrial animal is likely to depend on excretion in urine to rid itself of nitrogenous wastes. Urinary excretion of ammonia requires the excretion of a great deal of water.[22] Because of this and because an animal often faces an urgent need to prevent ammonia accumulation in its body, excretion of ammonia in the urine might sometimes require a terrestrial animal to void large amounts of water when water itself is in short supply.

Although ammonotelism is unusual on land, some *humidic* terrestrial animals are either ammonotelic or at least produce substantial quantities of ammonia. Some of the earliest studies on this phenomenon were carried out on terrestrial isopod crustaceans (e.g., pillbugs), which are ammonotelic. They void much of their ammonia into the atmosphere as NH_3 gas! In this way, they avoid use of water to get rid of ammonia. Many terrestrial snails, although not ammonotelic, void substantial NH_3 gas as well. In both isopods and snails, the fundamental reason for ammonia production may be that ammonia is useful to them because it plays a role in the deposition of calcium carbonate in their exoskeleton or shell. Production of NH_3 gas is also known to occur in some land crabs and a few species of amphibious fish.

Ammonia is the cheapest nitrogenous end product to produce. This probably explains why ammonotelism is so common among aquatic animals, which can easily evade ammonia's toxicity by excreting it freely into the water they occupy. Ammonia is generally formed during the catabolism of proteins by way of reactions that have no ATP cost. Some of these, termed *transamination* reactions, move amino groups to particular amino acids for which deamination enzymes exist; then the latter amino acids are *deaminated* (see Figure 6.3B).

Urea is more costly to synthesize but less toxic than ammonia

Urea is highly soluble and generally diffuses readily across membranes. Although hardly benign in its effects on macromolecules, it is far less toxic than ammonia. In humans, blood concentrations are normally in the range of 3–7 mM, and much higher concentrations, although abnormal, can be tolerated. As discussed in Chapter 28, very high blood urea concentrations occur in marine elasmobranch fish (≥ 300 mM) and some other animals.

If urea is less toxic than ammonia, why are so many animals ammonotelic? There are probably several reasons. One, certainly, is that urea is more costly to make than ammonia. The synthesis of each urea molecule requires the energy from four or five high-energy phosphate bonds (equivalent to that released by converting four or five ATP molecules to ADP).[23] As is so often the case in biology, animals face trade-offs. Ammonia is toxic but cheap; urea is less toxic but more costly.

[22] This topic is discussed at greater length in the next section.

[23] Authorities differ in whether they estimate the cost to be four or five per molecule.

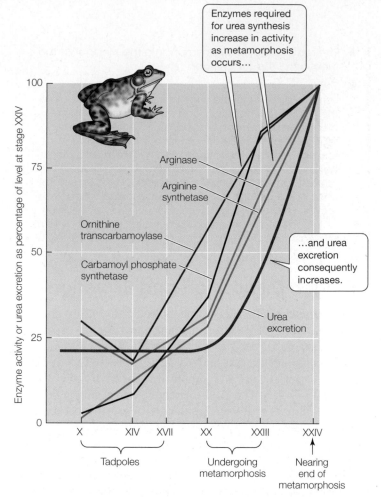

FIGURE 29.25 Bullfrogs shift from ammonotelism to ureotelism as they undergo metamorphosis American bullfrogs (*Lithobates catesbeianus*, until recently named *Rana catesbeiana*), which start life as aquatic tadpoles, metamorphose into semiterrestrial adults. Developmental biologists identify the successive stages of development using the stage numbers on the x axis. The activities in the liver of four enzymes of the metabolic pathway that synthesizes urea (the ornithine–urea cycle) are shown as functions of stage. The excretion of urea is also shown (red line). ("Arginine synthetase" is now recognized to represent the activity of two enzymes that together produce arginine.) (After Brown and Cohen 1958.)

Most of the animals that routinely employ urea as their principal nitrogenous end product are terrestrial vertebrates. Adult terrestrial amphibians are predominantly ureotelic. All mammals are ureotelic, as are some turtles. Terrestrial invertebrates have more often evolved uricotelism; only some flatworms and earthworms are at times ureotelic. For reasons that remain baffling in some cases, certain ocean aquatic animals are ureotelic, including some small, planktonic crustaceans and some larval fish.

There is persuasive evidence that vertebrates adopted ureotelism when they emerged onto land. One line of supportive evidence is the fact that modern terrestrial amphibians are ureotelic, in contrast to freshwater fish, which are nearly always ammonotelic. Another line of evidence is that, although the tadpoles of amphibians are usually ammonotelic, they express the enzymes for urea synthesis increasingly as they go through metamorphosis, eventually becoming ureotelic adults (**FIGURE 29.25**).

For terrestrial animals that excrete nitrogenous wastes in their urine, the advantage of ureotelism is that urea excretion requires less water than ammonia excretion. This is a direct consequence of the lower toxicity of urea. Because urea is less toxic than ammonia, steady-state blood concentrations of urea in ureotelic vertebrates—expressed in molar terms—are typically *at least* 20 times higher than steady-state blood concentrations of ammonia in ammonotelic vertebrates. Thus, if a urea-producing and an ammonia-producing species were both to excrete identical quantities of nitrogen in their urine at identical urine-to-plasma ratios, the water cost for the urea-producing species would be no more than 1/40th of that for the ammonia-producing species.[24] These considerations have no significance for aquatic animals that void nitrogen across general body surfaces. The considerations have great significance, however, for animals on land that excrete nitrogen in urine.

Some animals employ ureotelism for functions other than the simple, routine excretion of nitrogen, meaning that the evolution of ureotelism cannot always be interpreted strictly in terms of waste processing. Three cases in which ureotelism has an adaptive advantage are worthy of note:

1. Some ocean fish use urea as an osmolyte to aid osmoregulation in seawater—notably marine elasmobranchs, holocephalans, and the coelacanth *Latimeria*. This use of urea—which entails perpetually high urea concentrations in the body fluids—is discussed at length in Chapter 28 (see page 756).

2. Some aquatic, ammonotelic vertebrates switch to producing urea when they face crises that prevent them from being able to get rid of ammonia. For example, lungfish (see page 762) and some freshwater teleost species from stillwater environments switch from being ammonotelic to being ureotelic when they are confronted with drying of their habitat or other stresses. They then can stop voiding urine and allow their nitrogenous waste to accumulate in their body fluids, a strategy that would be impossible if they were synthesizing highly toxic ammonia.

3. Recently it was discovered that when toadfish living in the ocean excrete urea, the urea has value because it interferes with the ability of predators to find the toadfish by chemosensory processes.

The most thoroughly known biochemical mechanism for the synthesis of urea from protein nitrogen—the mechanism used by all the vertebrates that make urea and some invertebrates—is the **ornithine–urea cycle**, a set of biochemical reactions requiring five enzymes.[25] In the ornithine–urea cycle, one of the nitrogen atoms incorporated into urea originates from free ammonia derived from deamination reactions, especially deamination of glutamic acid. The second nitrogen incorporated into urea comes from the amino group of aspartic acid. Amino groups from most amino acids can

be transferred to glutamic acid or aspartic acid by transamination reactions. In vertebrates that synthesize urea, the liver is the one tissue that expresses the full suite of enzymes of the ornithine–urea cycle.

The earliest vertebrates probably had all the genes required for the ornithine–urea cycle (see Box 28.4). A current working hypothesis, therefore, is that all living vertebrates have the genes, and that the existence of a functioning ornithine–urea cycle in a species depends on whether the genes are *expressed* in the species. If the ability to synthesize urea is such a "readily available" option, it would help explain why ureotelism occurs in a notably scattered and wide variety of vertebrates (see the cladogram in Box 28.4).

Uric acid and related compounds remove nitrogen from solution

An animal is classified as uricotelic if its primary nitrogenous end product is uric acid, the dihydrate of uric acid, urate salts, or a mix of these compounds, all of which are purines. These compounds all have low toxicities and solubilities. Uric acid itself is poorly soluble in water (0.4 millimoles—65 mg—dissolves in a liter at 37°C). The urate salts, although more soluble than uric acid, also have very low solubilities in comparison with urea or ammonia. Because of their low solubilities, uric acid and urates remove nitrogen from solution, reducing the water costs of excretion; they can be excreted as semisolid pastes or even as dry pellets or powders.

Experiments have shown that a variety of cations—including Na^+, K^+, NH_4^+, Ca^{2+}, and Mg^{2+}—can be incorporated into uric acid excrement in a poorly soluble state. Uncertainty exists over the chemical form assumed by these ions; they can be present as urate salts, but apparently they also can be bound in some manner to undissolved uric acid. Regardless, the cations are removed from solution, and this state can appreciably reduce the water demands of cation excretion. Calculations indicate, for example, that desert iguanas (*Dipsosaurus*) can void as much as 5000 milliequivalents of undissolved K^+ per liter of water when the K^+ is combined with uric acid—an effective concentration that is well above the highest K^+ concentration achieved by reptilian salt glands!

Focusing again on nitrogen, uric acid and urates not only permit nitrogen to be excreted with little water, but also have great advantages in times of water crisis when urine production is curtailed or stopped. If a ureotelic animal in water crisis stops producing urine, the urea concentration in its body fluids steadily rises because urea is so soluble that its solubility limits are never reached. This buildup cannot continue indefinitely because at high concentrations, urea becomes toxic. By contrast, if a uricotelic animal stops producing urine, uric acid and urates are deposited as precipitates within its body. Because the solubilities of these compounds are low, their concentrations in the body fluids cannot increase above low levels regardless of the amounts stored. Uric acid and urates, therefore, are suited to indefinite storage.

In addition to uric acid and urates, other purines or compounds derived from purines are sometimes employed as nitrogenous end products. The purine *guanine*—which is even less soluble than uric acid—is a primary nitrogenous end product in some spiders and other animals. *Allantoin* and *allantoic acid*—compounds formed by the partial breakdown of uric acid—may also be primary nitrogenous end products; although more soluble than uric acid, they have low solubilities.

---

[24] This is true because the urinary molar concentration of urea would be at least 20 times higher than that of ammonia, and each molecule of urea contains two atoms of nitrogen, rather than just one.

[25] Additional biochemical mechanisms that produce urea are known in some animals. The reactions of the ornithine–urea cycle are presented in any biochemistry text.

BOX 29.4 Why Are Mammals Not Uricotelic?

If birds, lizards, and snakes are uricotelic, why aren't mammals? An informed biologist knowing the patterns of nitrogen excretion in other animal groups, but knowing nothing about nitrogen excretion in mammals, would surely predict uricotelism in mammals, at least in desert species. Yet without known exception, all mammals are ureotelic. We do not know the answer to this riddle. However, a point worth stressing is that mammals are able to do things with urea that are unique.

The mammalian kidney is in a class by itself in its ability to concentrate urea. The maximum U/P ratio for urea in mammalian urine is typically higher than that for any other solute and can greatly exceed the maximum osmotic U/P ratio—signifying that urea can be concentrated to a much greater extent than solutes as a whole. In humans, for example, the maximum osmotic U/P ratio is about 4.2, and the maximum Cl⁻ U/P ratio is about 3.5, but the maximum urea U/P ratio is about 170. Many desert rodents can achieve urinary urea concentrations of 2.5–5.0 molar (*M*), corresponding to 70–140 g of nitrogen per liter! Because the urinary urea concentrations achieved by mammals are much greater than those attained by other animals, the water losses obligated by nitrogen excretion in mammals are exceptionally low in comparison with other ureotelic groups. In fact, urinary nitrogen-to-water ratios attained by desert rodents equal or exceed those observed in some of the uricotelic vertebrates that void their uric acid in a relatively fluid mix (e.g., certain birds). However, some birds and other reptiles that void uric acid in the form of relatively dry pellets achieve nitrogen-to-water ratios that are several times higher than the highest mammalian values.

Some researchers have argued that mammals have remained ureotelic because the elaborately developed countercurrent multiplication system of the mammalian kidney provides such great potential for concentrating urea that the selective advantage of uricotelism has been blunted. The reverse argument is that mammals for some reason were unable to evolve the biochemical and physiological attributes required for uricotelism. Being tied to ureotelism, so this argument goes, the mammals experienced great selective pressures to evolve exceptional urea-concentrating abilities.

Uricotelism—or the production of other purines as principal nitrogenous end products—is the most common state in terrestrial animals. This striking generalization is true even though uric acid probably requires considerably more energy per nitrogen atom for its synthesis than urea.[26] Birds, lizards, and snakes are uricotelic (**BOX 29.4**). (The white matter in bird droppings is uric acid.) As already mentioned, turtles that inhabit dry terrestrial habitats tend toward uricotelism. Most terrestrial invertebrates that live in the open air employ purines or purine derivatives as their primary nitrogenous end products. Most terrestrial insects, for example, employ uric acid, allantoin, or allantoic acid as their principal nitrogenous excretion. Spiders, scorpions, and certain ticks excrete mostly guanine. Temporary or permanent storage of purines in the body has been observed in many insects and snails and in certain land crabs. Stepping back from all these details, an important point to recognize is that several terrestrial *phyla* have converged on uricotelism, and this convergent evolution testifies to the advantages of poorly soluble nitrogenous end products for terrestrial existence.

The biochemical pathways employed for the synthesis of uric acid or related compounds from protein nitrogen are complicated.[27] However, they in fact are only relatively small modifications of very ancient and universal pathways for the synthesis of the purine constituents of DNA. This origin of the pathways helps explain how uricotelism (or "purinotelism") could have evolved independently in several phyla on land.

SUMMARY
Nitrogen Disposition and Excretion

■ Animals that synthesize ammonia or urea as their primary nitrogenous end product are termed, respectively, ammonotelic or ureotelic. Animals that synthesize mainly uric acid or urates are uricotelic.

■ Ammonotelism is the primitive condition and is seen in most water-breathing aquatic animals. Ammonia has the advantage of costing no extra ATP to produce. It is toxic, however. Thus, for an animal to be ammonotelic, the animal must have a means to void ammonia reliably as fast as it is produced so that blood levels are kept low. Aquatic animals void ammonia into the ambient water across their gills or general body surfaces.

■ Ureotelism is more costly than ammonotelism because producing urea has an ATP cost. Urea is far less toxic than ammonia, however. Ureotelism has evolved principally in certain groups of vertebrates, in which it usually serves one or more of three possible functions: reducing the water requirement of routine nitrogen excretion (e.g., terrestrial amphibians and mammals), adjusting the blood osmotic pressure in advantageous ways (e.g., elasmobranch fish), and detoxification of waste nitrogen during periods when water-stressed animals cease urine production.

---

[26] Although some biochemists calculate that the synthesis of uric acid from protein nitrogen costs about the same amount of ATP-bond energy per nitrogen atom as urea synthesis (2–2.5 high-energy phosphate bonds per atom), others calculate that each nitrogen atom costs as much as 6 high-energy phosphate bonds to be incorporated into uric acid. Regardless of that consideration, uric acid probably has a higher overall cost than urea in at least some animals because of extra processes that must be carried out to prevent it from precipitating prematurely.

[27] These pathways are reviewed in biochemistry texts.

■ Although uricotelism is even more costly per nitrogen atom than ureotelism, uric acid and related compounds have the advantage that they are so poorly soluble that they are low in toxicity, can be excreted in little water, and can be accumulated in the body indefinitely. Most groups of terrestrial animals, including invertebrates (e.g., insects) and vertebrates (e.g., birds, lizards, and snakes), are uricotelic or primarily produce other purines (e.g., guanine) or purine derivatives.

STUDY QUESTIONS

1. We have seen that animals often encounter problems in dealing with waste nitrogen. Considering ammonotelic, ureotelic, and uricotelic types of animals, explain why these problems are often greatest in carnivores (as compared with omnivores or herbivores).

2. Considering the distal convoluted tubule of the amphibian nephron, explain how changes in the permeability of the tubule wall to water affect the amount of pure, osmotically free water excreted in the urine. Define what is meant by pure, osmotically free water.

3. Outline how the orientation of nephrons relative to each other imparts gross structure to the kidneys of mammals and birds.

4. If you were attempting to tell whether an animal produces its primary urine by ultrafiltration or secretion, what measurements would you make *on the primary urine*? If your measurements indicated that ultrafiltration might be occurring, what other types of measurements would you make to determine whether physical and physicochemical conditions favorable to ultrafiltration exist? Explain.

5. When researchers first proposed the countercurrent multiplication hypothesis for concentration of urine in the mammalian kidney, there was great resistance to its acceptance in certain quarters. The anatomist Ivar Sperber, whose comparative morphological studies originally helped draw attention to the loops of Henle, pointed out that there were certain rodents in which the anatomy of the kidney should make it relatively simple to sample blood from the hairpin bends of the vasa recta deep in the medulla. Samples of such blood were obtained, and the osmotic pressure of this blood proved to be far higher than the osmotic pressure of blood in the general circulation. This research convinced doubters of the validity of the countercurrent multiplication process. Why does blood at the hairpin bends of the vasa recta have a high osmotic pressure, and why would knowing its osmotic pressure in the cases described provide strong support for the countercurrent multiplication hypothesis?

6. Production of any sort of nitrogenous waste other than ammonia costs energy. Name at least three distinctly different advantages an animal might gain by investing in production of urea or uric acid.

7. Explain how primary urine is introduced into the Malpighian tubules of an insect.

8. The immediate effect of ADH on the renal tubules of frogs and mice is the same, yet when ADH is secreted, frogs produce urine that is approximately isosmotic to their blood plasma, whereas mice produce urine far more concentrated than their blood plasma. Explain this difference in terms of the factors affecting osmosis in the kidneys of frogs and mice.

9. Drugs that increase urine flow (diuretic drugs) are often employed in the treatment of hypertension (high blood pressure) or other disease states. Three physiological categories of such drugs are ones that (i) function as loop diuretics, (ii) inhibit the action of aldosterone, and (iii) block Na^+ channels in the collecting ducts. Explain why each of these categories would be expected to increase Na^+ excretion and urine flow. (Hint: Rereading the section on hormones at the end of Chapter 28 might prove helpful.)

10. In mammals, the kidneys are the only organs that regulate routine excretion of water, salts, and nitrogenous wastes from the blood. As logical as this may sound to us, it is unique among vertebrates. For each of the other groups of vertebrates, describe the functions of as many organs as you can—in addition to the kidneys—that participate in these processes. Consider the discussions of these groups of animals in Chapter 28 as well as in this chapter.

11. Whenever the concentrating ability of mammalian kidneys has been studied in relation to the lengths of the loops of Henle in various species, a clear correlation between the two has been found—indicating that loop length matters—but in addition, there has been a great deal of scatter in the data (e.g., see Figure 29.8). Fifty years from now, physiologists will probably understand the mechanistic reasons *why* loop length is not a perfect predictor of concentrating ability. Suppose a government agency has decided to give you all the resources you need to study whatever you desire. As a brainstorming exercise, what specific aspects of mammalian kidney function other than loop length would you investigate to try to account better for differences among species in concentrating ability?

Go to **sites.sinauer.com/animalphys4e** for box extensions, quizzes, flashcards, and other resources.

REFERENCES

Beuchat, C. A. 1996. **Structure and concentrating ability of the mammalian kidney: Correlations with habitat.** *Am. J. Physiol.* **271: R157–R179.** A good example of modern comparative biology that takes full advantage of the massive amounts of data available in our "information era" to review what is known and generate hypotheses for future research.

Beyenbach, K. W. 2003. **Transport mechanisms of diuresis in Malpighian tubules of insects.** *J. Exp. Biol.* **206: 3845–3856.** An accessible, well-illustrated, and definitive review of Malpighian tubule function in insects, written by one of the leaders in the field. Emphasizes molecular mechanisms but includes an enlightening overview of organ function. A more advanced review by the same author and colleagues is included in the Additional References.

Beyenbach, K. W. 2004. **Kidneys sans glomeruli.** *Am. J. Physiol.* **286: F811–F827.**

Braun, E. J., and W. H. Dantzler. 1997. **Vertebrate renal system. In** W. H. Dantzler (ed.), *Comparative Physiology*, vol. 1 (Handbook of Physiology [Bethesda, MD], section 13), pp. 481–576. Oxford University Press, New York. A modern review of comparative kidney function in the vertebrates, emphasizing cellular and molecular physiology.

Eaton, D. C., and J. P. Pooler. 2013. *Vander's Renal Physiology.* McGraw-Hill, New York. A compact summary of medical renal physiology justly respected for its excellent pedagogy.

Evans, D. H. (ed.). 2009. *Osmotic and Ionic Regulation. Cells and Animals.* CRC Press, Boca Raton, FL.

Ip, Y. K., S. F. Chew, and D. J. Randall. 2004. Five tropical air-breathing fishes, six different strategies to defend against ammonia toxicity on land. *Physiol. Biochem. Zool.* 77: 768–782.

Knepper, M. A., and G. Gamba. 2004. Urine concentration and dilution. In B. M. Brenner (ed.), *Brenner and Rector's The Kidney*, 7th ed., vol. 1, pp. 599–636. Saunders, Philadelphia. An unusually articulate and clear review of concentration and dilution in the mammalian kidney, including a thorough discussion of models for the single effect in the inner medulla. Many enlightening diagrams.

LeMoine, C. M. R., and P. J. Walsh. 2015. Evolution of urea transporters in vertebrates: adaptation to urea's multiple roles and metabolic sources. *J. Exp. Biol.* 218: 1936–1945. A concise, current overview of the roles played by urea, focused on the urea transporter proteins that often are essential.

O'Donnell, M. J. 1997. Mechanisms of excretion and ion transport in invertebrates. In W. H. Dantzler (ed.), *Comparative Physiology*, vol. 2 (Handbook of Physiology [Bethesda, MD], section 13), pp. 1207–1289. Oxford University Press, New York.

Pannabecker, T. L. 2014. Comparative physiology and architecture associated with the mammalian urine concentrating mechanism: role of inner medullary water and urea transport pathways in the rodent medulla. *Am. J. Physiol.* 304: R488–R503.

Wright, P. A. 1995. Nitrogen excretion: three end products, many physiological roles. *J. Exp. Biol.* 198: 273–281. A succinct modern treatment of the end products of nitrogen catabolism.

See also **Additional References** *and* **Figure and Table Citations**

In deserts, large herbivores such as the oryx in this photograph are usually the equivalent of nomads, moving about to find moisture to survive. Similarly, in the dry savannas—the grassland plains—large herbivores often function as nomads or migrants. Conditions vary so much from time to time and from place to place in these arid zones that living in an invariant locale is often not a viable strategy. The same can be said for humans in a state of nature; most of the traditional cultures of the deserts and dry savannas were nomadic.

Our goal in this chapter is to explore how the oryxes and other species of sizable mammals—5 kg or larger in body weight—can live in hot deserts or dry-savanna ecosystems. One way to understand these mammals is to compare them with small species, so we will sometimes mention the kangaroo rats, gerbils, and other small mammals that coexist in arid habitats with the large species. However, the small mammals were discussed in Chapter 28, and we will not return to them here in any detail. The mammals of Africa will receive greatest emphasis because they form the most diverse and abundant set of large, arid-land mammals in the world today, and their diversity has drawn the interest of physiologists for decades.

Desert and Dry-Savanna Environments

Deserts and semideserts by most accountings cover about a third of the land on Earth (see Figure 27.6). There is no uncomplicated way to define a desert. One of the simplest definitions is that a desert receives less than about 25 cm (10 inches) of precipitation per year, but this is an imperfect standard because there are regions

Oryxes are extreme examples of large mammals that can survive indefinitely in hot deserts without drinking water Physiologists are still learning how these animals orchestrate their water losses and gains to stay in water balance in such water-poor and thermally stressful places. The animal shown is a gemsbok oryx (*Oryx gazella*), often called simply gemsbok.

that receive much more than that but are undoubtedly deserts. Following the lead of Imanuel Noy-Meir, most biologists prefer a definition that emphasizes two attributes of deserts that are of extreme importance for plants and animals. First, a **desert** is a place where precipitation is so low that *availability of water exerts a dominant controlling effect on biological processes*. Second, *when precipitation occurs in deserts, it comes in infrequent, largely unpredictable events*. Rains in deserts are highly irregular and unreliable in both time and space. One year may bring 5 times (or even 20 times) the rain of another. One 30-km² area may be drenched during a storm while a nearby area of similar size receives nothing, because rains in deserts are usually produced by isolated storms rather than broad fronts. Large herbivores must often conduct their lives in ways that—more than anything—are opportunistic, taking advantage of rain or moisture whenever and wherever it occurs.

What are the features of the dry savannas? The classic look of the **dry savannas** is of endless plains of grass with trees dotted here and there. Taking a big-picture view of Earth, dry savannas are in many cases neighboring environments to deserts. Gradients of moisture often occur on continental scales of space, with moist forests in some regions, deserts in others, and dry savannas between the two—in intermediate regions where there is more moisture than in the deserts, but far less than in the forests. In dry savannas—as in deserts—water is a dominant controlling factor for biological processes. This is true in a somewhat different way than in deserts, however. Savannas usually have discrete rainy seasons interrupted by discrete rainless seasons. Because of the rainless season each year, although on an annual basis savannas receive more rain than deserts do and receive it more predictably, the plants and animals living in savannas must often endure profound drought for long lengths of time. The Serengeti plains of East Africa are examples of this second sort of water-controlled ecosystem. Every year, during what's called the "long dry season," no rain falls for 4–6 months; thus streams dry up, and soils become so parched that the grasses turn crisp and brown.

Deserts and dry savannas exist for several reasons. The single most important cause of desert and semidesert conditions on a planetary scale is the global pattern of air movements, whereby air warmed at the equator rises to high altitudes and displaces air at those altitudes in such a way that high-altitude air descends to Earth's surface somewhere else. Air at high altitudes, partly because it is cold (see Table 27.2), contains little moisture, even when saturated with water vapor. Thus, in regions where high-altitude air tends consistently to fall to low altitude, the land can become parched. The present global pattern is for high-altitude air to descend in two bands encircling the globe at latitudes roughly 30° north and 30° south of the equator. Most of the world's great deserts (see Figure 27.6)—including the largest, the Sahara—are products of this process. Another common but more localized cause of desert or savanna conditions is rain-shadowing caused by highlands. If a region's only reliable source of moisture is winds blowing in from the ocean, and if intervening highlands force the winds to rise—so that the air cools and its moisture condenses to form rain or snow—the winds may have little moisture left by the time they blow into the region of interest. Near Los Angeles, for example, the coastal mountains force the prevailing winds blowing east from the Pacific Ocean to rise, causing mountain rain. Seen from above,

the land still farther east, the Mojave Desert, looks much like a dry, sandy-colored shadow cast by the mountains.

The existence and extent of deserts and dry savannas, we see, often depend on nothing more permanent than the gossamer movements of air. Accordingly, over the long reaches of geological time, deserts and savannas have come and gone as patterns of air movement have changed. The modern Sahara Desert, for example, has existed for less than 6000 years, although desert conditions have come and gone in North Africa for at least 7 million years. Currently the Sahara Desert is tending to expand north and south. For example, a recent consensus report by climate scientists concluded that the southern reaches of Europe (e.g., Italy and southern Spain) are likely to become far drier over the upcoming decades.

Despite the variability just discussed, deserts and savannas have probably always existed somewhere. Thus they have long presented plants and animals with special challenges. They have also long presented opportunities if a species is able to make do with little water and tap the meager water resources available.

The Relations of Animals to Water

Large mammals have both disadvantages and advantages compared with small mammals in hot deserts and dry savannas. Many of the most apparent disadvantages of large size are behavioral. Large mammals cannot take shelter underground by digging burrows, as small rodents do, and they are less likely than small species to find adequate shade or other protective microhabitats on the surface of the ground. In hot environments, large mammals therefore confront the heat more directly than small mammals do, and the heat stress they experience can place demands on their water resources. In contrast to the *behavioral* limitations they face, however, large mammals are often in a distinctly more favorable *physiological* position than small mammals are.

Large body size is a physiological advantage in terms of water costs

To explore the physiological implications of body size, let's compare mammals of large and small size while making two simplifying assumptions.[1] First, let's assume that the animals *thermoregulate physiologically by use of evaporative cooling*. Second, let's assume that the animals are actually confronted with stressfully hot conditions (rather than evading them). For an animal to thermoregulate under such circumstances, it must evaporate water to void from its body not only the *exogenous* heat that enters its body from the hot environment but also the *endogenous* heat produced by its own metabolism. Large body size is an advantage from both of these perspectives.

Speaking of exogenous heat inputs, large mammals enjoy an advantage because they have less body-surface area per unit of weight than small mammals do (their surface-to-volume ratios are lower; see Equation 7.6). When the air temperature is higher than body temperature and the sun is beating down, environmental heat enters an animal's body across its body surfaces. Because large mammals have relatively little body-surface area per unit of weight, they tend to experience less heat entry per unit of weight

---

[1] These assumptions are not realistic in all cases but help bring out important principles.

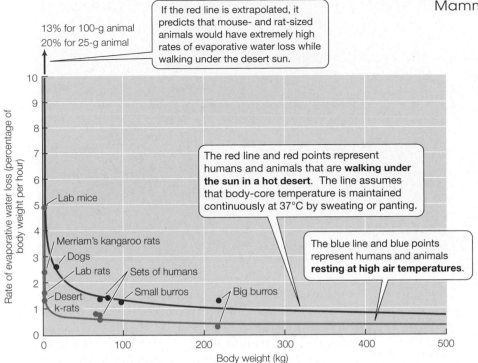

13% for 100-g animal
20% for 25-g animal

If the red line is extrapolated, it predicts that mouse- and rat-sized animals would have extremely high rates of evaporative water loss while walking under the desert sun.

The red line and red points represent humans and animals that are **walking under the sun in a hot desert**. The line assumes that body-core temperature is maintained continuously at 37°C by sweating or panting.

The blue line and blue points represent humans and animals **resting at high air temperatures**.

Lab mice

Merriam's kangaroo rats

Dogs

Lab rats · Sets of humans

Desert k-rats · Small burros · Big burros

FIGURE 30.1 The rate of evaporative water loss experienced by mammals exposed to heat stress depends strongly on their body size The graph shows weight-specific rates of evaporative water loss for exercising animals (red) and resting animals (blue) directly exposed to hot conditions. The resting humans and burros were studied while resting in the desert; the resting mice and rats were studied in a laboratory chamber maintained at an air temperature of 40°C (104°F). The equation for the red line is $E = 6.03W^{-0.33}$, where E is rate of evaporative water loss in percentage of body weight per hour and W is body weight in kilograms. The equation for the blue line is $E = 1.38W^{-0.21}$. The two species of kangaroo rats included in the resting studies are Merriam's kangaroo rats (*Dipodomys merriami*) and desert kangaroo rats (abbreviated "Desert k-rats"; *D. deserti*), weighing 34 g and 114 g, respectively. (After Schmidt-Nielsen 1954; Soholt et al. 1977.)

than small mammals do over any specified period of time. Thus their water costs to void the incoming heat are lower per unit of weight.

Speaking of endogenous heat inputs, large mammals are in a more favorable position than small ones because of the allometric relation between metabolic rate and body size. Large mammals tend to have far lower weight-specific metabolic rates than small mammals (see Figure 7.9). Thus their rates of endogenous heat production per unit of body weight are relatively low, and their weight-specific water costs to void endogenously produced heat are also low.

Quantitatively, the physiological advantages of large size are dramatic. In the 1930s, D. B. Dill (1891–1986)—a pioneer in the study of exercise and heat—undertook famous "walks in the desert" that provided the very first quantitative information on the water costs of mammalian exercise under hot conditions. He himself was one of the human subjects, and he took along with him both dogs and burros. Together, man and beast trekked under the searing sun in the Nevada desert, sweating or panting to thermoregulate; and as they walked, Dill measured their rates of dehydration.

As we now discuss Dill's results and other related findings, we will focus on the red and blue lines in **FIGURE 30.1**. Both lines show the rates at which animals must lose water by evaporation to thermoregulate under hot conditions as a function of body size. To interpret each line, it is important to recognize that the rate of evaporation is plotted in *weight-specific* terms on the *y* axis. Specifically, the rate of evaporation is expressed as grams of water lost from the body per 100 g of body weight during an hour. That is, the *y* axis shows *percentage of body weight lost per hour*.

The red line was constructed by the application of fundamental principles to Dill's original data, assuming that mammals maintain a body temperature of about 37°C as they exercise.[2] Later, Dill and collaborators gathered more data of a similar sort on burros and high-school students in the Colorado desert, accounting (along with Dill's original data) for the five data points plotted in red on the figure.

The red line shows that there is a very strong relation between body size and the water cost of thermoregulation when mammals are active in the daytime desert. As you can see, the water cost of thermoregulation is predicted to *soar to extraordinarily high levels* at small body sizes. According to the red line, if a 100-g rat were to walk along under the desert sun and thermoregulate by evaporation of water, its water cost would be almost 13% of its body weight *per hour*. If a 25-g mouse were to do so, its water cost would be about 20% per hour. Dehydration ordinarily becomes lethal under hot conditions when 10%–15% of body weight has been lost. Thus the rat walking along under the sun would die of dehydration after about an hour, and the mouse would die after a half hour! In other words, from what we know, it would be *impossible* for rat- and mouse-sized mammals to lead fully exposed, active lives in the daytime desert. Of course, rats and mice have never been actually tested to see if these predictions hold true for animals walking under the desert sun. This explains why the blue line (see Figure 30.1) is also plotted. The blue line and blue symbols show actual results for four species of rats and mice—as well as burros and humans—that were studied while resting under milder heat-stress conditions. These results verify that water costs rise steeply at body sizes as small as those of rats and mice. From this perspective, we see that small mammals are very fortunate that they can readily escape the full intensity of the daytime desert heat by burrowing and other *behavioral* means. *Physiologically*, small mammals have very little ability to face the full brunt of desert conditions.

Facing the full brunt of desert conditions, however, is exactly what large desert mammals must often do. And fortunately, as Figure 30.1 shows, large mammals are in a far better physiological position to do so than small mammals. Based on the red line, a 100-kg mammal, such as a small adult burro, can thermoregulate while exercising in the heat of the desert day at a water cost of just 1.3% of its body weight per hour, and a 400-kg mammal, such as an adult dromedary camel, can do so at a water cost of only 0.8% per hour. The water costs of large mammals in the desert are lower

[2] The humans, dogs, and burros used to construct the line in fact thermoregulate in approximately this way, although not all species do.

FIGURE 30.2 Common wildebeests are drinking-water-dependent antelopes that seek shade Also called blue wildebeests, common wildebeests (*Connochaetes taurinus*) weigh about 200 kg as adults. They seek shade when it is available, as under the acacia tree in the photograph on the right.

yet when the animals are not exercising, as shown by the blue line. Large size, in brief, is a great advantage physiologically when animals directly confront hot conditions. This is not to say that all large mammals can solve the problems they face in deserts by the simple evaporation of water; for animals that live where water is almost impossible to find, even a loss as low as 0.8% per hour during midday hours would become life-threatening within a few days. We will see later in this chapter that animals such as oryxes and camels have evolved specific adaptations that enable them to reduce water costs to levels far below the ordinary costs associated with their body sizes.

Coexisting species are diverse in their relations to drinking water

One might imagine that all large species of mammals that have long evolutionary histories in arid environments would have evolved minimal water requirements. That is not the case, however. Among the species that are common in dry, hot places, some are classified as **drinking-water-independent** because they are able to remain healthy for many days or weeks without access to drinking water. Although most of these animals must drink on occasion, some species may be literally independent of drinking water and *never* need to drink. However, in contrast to the drinking-water-independent animals, other coexisting species must drink each day, or at least every other day. These are classified as **drinking-water-dependent**.

Whether a particular species falls into the drinking-water-independent or the drinking-water-dependent category depends in part on the severity of conditions. A species that is independent of drinking water in the cool seasons of the year might, for example, become dependent in the hot seasons. Despite such ambiguities, the distinction between drinking-water-independent and drinking-water-dependent species is a useful organizing principle for discussing large desert and dry-savanna herbivores.

An outstanding example of coexisting drinking-water-independent and drinking-water-dependent species is provided by the fabled herds of herbivores in the Serengeti ecosystem (a dry-savanna system) and other similar ecosystems in East Africa and South Africa. Two of the most prominent drinking-water-dependent species in the Serengeti are wildebeests and zebras.[3] During the hot seasons of the year, they must drink every day or every other day. This means, in effect, that they cannot wander more than about 25 km (15 miles) from standing water. Species that often coexist with the wildebeests and zebras, but are drinking-water-independent, include Grant's gazelle, the common eland, and the dik-dik (dwarf antelope). These animals are able to travel far from standing water (because they need not return to it every day), and in that way they gain access to food resources that are unavailable to the drinking-water-dependent species. On even just modestly hot days in areas where trees are found, wildebeests seek shade in the heat of the day (**FIGURE 30.2**), whereas Grant's gazelles on such days are indifferent to sun or shade (**FIGURE 30.3**). This contrast in shade-seeking behavior correlates with the water physiology of the two species: The species that dehydrates more readily in the absence of drinking water also exploits shade more readily when shade is available.

The most dramatic manifestation of the difference between drinking-water-independent and drinking-water-dependent species in the Serengeti ecosystem is the differential participation of the two categories of animals in the annual migration. From the viewpoint of water physiology, the annual migration in the Serengeti is probably the most dramatic illustration in today's world of the way water can exert a dominant controlling effect on biological processes in an ecosystem.[4]

In the Serengeti, there is a gradient of decreasing moisture availability (increasing aridity) from the northwest to the southeast because of the positions of mountains and highlands, the directions of winds, and the courses followed by rivers; total annual rainfall in the northwest (~110 cm/year) is about twice that in the southeast (~50 cm/year). In the rainy season, enough rain falls everywhere

---

[3] The species discussed in this paragraph are the common or blue wildebeest (*Connochaetes taurinus*), the plains or Burchell's zebra (*Equus burchelli*), the common eland (*Tragelaphus oryx*; also called *Taurotragus oryx*), Grant's gazelle (*Gazella granti*), and Kirk's dik-dik (*Madoqua kirkii*).

[4] Recall Imanuel Noy-Meir's famous definition of a desert, mentioned at the start of this chapter.

FIGURE 30.3 Grant's gazelles are drinking-water-independent antelopes noted for their indifference to sun or shade These gazelles (*Gazella granti*) weigh about 50 kg as adults. Grant's gazelles are one of the species often reputed by naturalists to be able to live indefinitely without drinking water in the dry savannas and deserts.

for lush growth of grasses. However, during the long dry season between June and November, the southeastern plains typically receive almost no rain. As the dry season progresses, streams and rivers in the southeast first stop flowing, and they become nothing but a series of puddles and ponds. Then even the puddles and ponds dry up. The soil becomes parched, and the grasses turn brown and brittle. These seasonal patterns set the stage for the annual migration.

Not all species of large herbivores participate in the migration. In fact, most of the drinking-water-independent species do not. Grant's gazelles, elands, dik-diks, and most other drinking-water-independent species simply "stay put," even as the thronging herds of migrating animals move through.

The most abundant migratory species is the wildebeest, which in recent times has numbered about 1 million. Another highly visible migratory species is the zebra. Both the wildebeest and zebra are drinking-water-dependent. They spend the dry season in the far northwest, the wettest part of the ecosystem (**FIGURE 30.4A**). Then, as the rainy season starts in November or December, they move on a broad front from the northwest toward the southeastern plains (**FIGURE 30.4B**). While the rains fall in the southeast (**FIGURE 30.4C**), they find water and lush grass there. However, after the rains stop in the southeast during the following May or June, they must get out of the area. First they move west (**FIGURE 30.4D**), and then—as the landscape becomes ever drier—they move in thundering herds back to the far northwest (**FIGURE 30.4E**), where they have the best chance of finding the water they require during the season of little rain. As we discussed in Chapter 6 (see page 140), scientists have only hypotheses—not certain knowledge—about why wildebeests and zebras move from the northwest to the southeast when the rainy season arrives. However, there can be no doubt why these animals leave the southeast when the rains stop: They need to drink each day, and doing so in the southeast becomes impossible. Meanwhile, drinking-water-independent species are free to stay put or make just relatively subtle adjustments to the strong seasonality of water availability.

Although the Serengeti migration seems almost unique today, it was not unique primordially. Migratory herds evolved in other settings as well. For example, wildebeests in the large Kalahari ecosystem at the border of Botswana and South Africa used to migrate in much the same pattern as those in the Serengeti. Most of these other migrations have been ended by human intervention. The Kalahari migration, for instance, was disrupted by fences and the installation of wells, which the wildebeests congregate around rather than migrating to seek water.

Carnivorous mammals such as lions and cheetahs are generally assumed (on the basis of little evidence) to achieve water balance without having to drink much (if at all) because of the high water content of their foods. Some predators in the Serengeti are indifferent to the migration, whereas others follow the prey animals.

Water conflicts threaten animals and people

Water conflicts seem to be almost inevitable, as well as intrinsically ominous, in ecosystems where water exerts dominant controlling effects on biological processes. The conflicts that presently exist are worth noting because they threaten the future existence of the species and systems of life we are discussing.

One sort of water conflict is that between the water needs of desert and dry-savanna ecosystems and the water claims of human enterprises outside the ecosystems. In the United States, burgeoning cities in arid parts of the West and equally burgeoning demands for agricultural irrigation water have led to the draining and diversion of many of the rivers that once flowed through the western deserts and savanna-like prairies. A dramatic example is provided by the Colorado River, which is tapped for so many human uses as it flows from the Rocky Mountains to the Sea of Cortez (Gulf of California) that in many years it now runs out of water before reaching its ancient destination and dries up in the desert. Near the northern boundary of the Serengeti ecosystem, several actual or proposed human ventures pose threats of water conflict with the wildlife. The drinking-water-dependent species in the ecosystem, as we have seen, rely on a dependable supply of drinking water in the northwest to survive the dry season. The sufficiency of this supply is already less than fully adequate; during drought years, 20%–30% of the wildebeest have been known to die. The principal reliable water source in the northwest is the Mara River, which brings water into the Serengeti ecosystem from forested highlands farther north. Deforestation of the highlands, diversion of

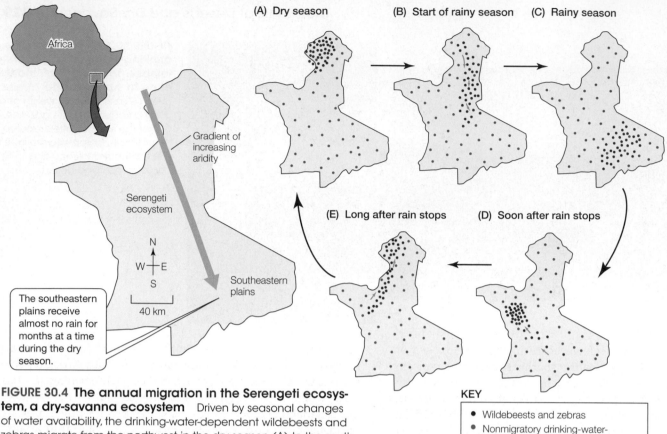

(A) Dry season　　(B) Start of rainy season　　(C) Rainy season

(E) Long after rain stops　　(D) Soon after rain stops

Gradient of increasing aridity

Serengeti ecosystem

Southeastern plains

The southeastern plains receive almost no rain for months at a time during the dry season.

40 km

KEY

- Wildebeests and zebras
- Nonmigratory drinking-water-independent species of herbivores

FIGURE 30.4 The annual migration in the Serengeti ecosystem, a dry-savanna ecosystem Driven by seasonal changes of water availability, the drinking-water-dependent wildebeests and zebras migrate from the northwest in the dry season (A) to the southeast in the rainy season (B,C) and back again (D,E), whereas most drinking-water-independent species adjust only in much more subtle ways to the seasonal rhythm of dry and rainy seasons. The Serengeti ecosystem as mapped here includes not just Serengeti National Park but also other surrounding conservation areas. The general trend is for the land to receive less and less rain per year as one moves from the northwest to the southeast.

river flow for irrigation uses, and diversion for hydroelectric power production have been proposed. All could create water conflicts with the wild animals in their dry-season refuge, potentially disrupting the migration and the entire Serengeti ecosystem.

Another major sort of water conflict is intimately linked with the traditional ways of life of indigenous peoples who live within desert and dry-savanna ecosystems. Throughout the world, nomadism was the traditional way of life for hundreds of such cultural groups. The indigenous people survived on very small amounts of water compared with modern urban water usage (**TABLE 30.1**), and they obtained the water they needed by moving with the rains. As notions of private land ownership spread around the world in

the last few centuries, such ways of life became threatened or were eliminated because private land ownership is not readily compatible with nomadism; a nomad today would be encroaching on one piece of private property after another. Thus, as we look at a modern-day Maasai (**FIGURE 30.5**) or modern-day representative of any other nomadic group, we see two things. First we are reminded of the peoples—not just the wild animals—that lived for millennia in the deserts and dry savannas, integrated with these places in their natural state. Second, we see the troubled ending of a way of life as modern governments impose changed relationships to the land and, importantly, to water. Many nomadic peoples have been settled on their own private parcels of land, but in the deserts and dry savannas, a single parcel is unlikely to provide adequate water through all years and all seasons; the people used to be nomads precisely because movement was necessary.

A debate that today has reached its time of ultimate resolution is whether some nomadic cultures should be given a chance to

TABLE 30.1　Water use by people in four sorts of communities in Arabia

People in indigenous desert settlements use one-tenth the water of people in modern towns. The figures are for all domestic water use, including drinking, washing, bathing, and other water demands.

Type of community	Domestic water use per person (L/day)
Modern Arabian town without major industry[a]	240
Traditional agricultural village	120
Small desert settlement with supply by government water truck	80
Small desert settlement with traditional water supply	28

Source: After Goudie and Wilkinson 1977.
[a]New York City has a similar usage rate.

FIGURE 30.5 A Maasai youth of college age He is a member of the current generation of a people who until recent times were nomads in the dry savannas of East Africa.

continue by protecting large tracts of land for nomadic use or even by assimilating the people into parks now reserved exclusively for wild animals. The challenges are particularly acute for the herding cultures, whose herds of cattle or goats traditionally provided them with a way to glean the essentials of life from the arid land, but today greatly magnify water conflicts. Protected parks for wild animals in deserts and dry savannas have nearly always had their borders set to include the most reliable water sources. When the herds kept by nomadic peoples are banned from the parks, the herds are excluded from the most dependable places to drink, but when the herds are let into the parks, they compete with the wild animals for water.

All species of large herbivores require considerable amounts of preformed water

Let's now take a more quantitative approach to understanding the water physiology of oryxes, gazelles, wildebeests, and the other large wild herbivores of deserts and dry savannas. Recall that under many conditions, small desert mammals such as kangaroo rats acquire well over half their water as metabolic water; at 25°C, for example, up to 90% of a kangaroo rat's water needs are met by metabolic water (see Figure 28.24). In sharp contrast, *all the species*

of large herbivores of deserts and dry savannas acquire most of their water as preformed water.

To see this more clearly, let's consider the water budgets of two very different mammals of similar body size: the highly drinking-water-dependent common wildebeest and the highly drinking-water-independent beisa oryx. These animals were studied under two sets of conditions. The first set of conditions was designed to be about as mild as the animals would ever experience in nature; the environmental temperature was kept constantly at about 22°C. The second set of conditions was intended to simulate more closely the high water stresses the animals sometimes encounter in the wild; during each 24-h day, the temperature was raised to 40°C (104°F) (simulating daytime heat) for 12 h and lowered to 22°C (simulating night) for the other 12 h. Under both sets of conditions, the animals were gradually given less and less preformed water per day until they reached the minimum they required to maintain health. They were then maintained on that minimum for 2 weeks so that their water budgets could be studied.

The results (**FIGURE 30.6**) reveal that the drinking-water-dependent wildebeest has a far greater total water need, and a far greater need for preformed water, than the drinking-water-independent oryx when both species are studied side by side under identical conditions. Physiologically, the wildebeest and oryx are dramatically different.

A second major insight revealed by the results, however, is that even the oryx—which is famous for being one of the most drinking-water-independent of all large mammals—requires considerable quantities of preformed water to stay in water balance. Metabolic water met only about 20% of the oryx's total water need under both sets of conditions (see Figure 30.6B). Preformed water, therefore, had to meet about 80% of its need. How is it possible for a species to be drinking-water-independent yet require preformed water to meet much of its daily water requirement? The answer is that the oryx must be able to find *foods that will provide sufficient preformed water* to meet much of its daily water need.

In their water budgets, the wildebeest and oryx are quite typical of the large herbivores that live in deserts and dry savannas. Keeping the data on the wildebeest and oryx in mind (see Figure 30.6),

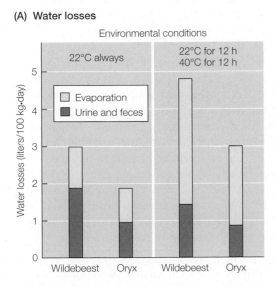

(A) Water losses

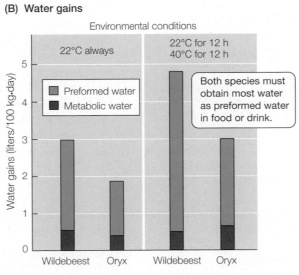

(B) Water gains

FIGURE 30.6 Water budgets of wildebeests and oryxes The wildebeests (*Connochaetes taurinus*) that were studied averaged 160 kg in body weight; the oryxes (*Oryx beisa*) were about the same weight, averaging 100 kg. The two species were studied in identical chambers that permitted their water losses (A) and gains (B) to be quantified. During some experiments (left in each part) the air temperature was a constant 22°C. During others (right) the air temperature was 22°C for half of each 24-h day and 40°C for the other half. The animals had been acclimated to water restriction before the start of these measurements and were maintained on water restriction throughout the study. (After Taylor 1968, 1970.)

we can make the following key points about the large herbivores as a group:

- *All* the species of large herbivores that live in deserts and dry savannas depend more on preformed water than on metabolic water to meet their water needs.

- Drinking-water-dependent species cannot get all the preformed water they need from their foods, either because they need more preformed water than available foods could ever provide or because they are ineffective in selecting and processing foods to maximize their acquisition of preformed water. Therefore they have to drink each day for their total intake of preformed water to be adequate to keep them healthy.

- Drinking-water-independent species, in contrast, can get all the preformed water they need to stay healthy—for many days in a row—from their foods. This is why they are classified as drinking-water-independent; it is why they do not require drinking water very often. This said, it is important to note that obtaining adequate preformed water from foods is far from simple, as we will soon see, because the foods available in deserts and dry savannas are often parched.

- Most drinking-water-independent species in fact need to drink occasionally during the hot–dry seasons of the year. Although the preformed water they get from their foods is *almost* enough to meet their total need for preformed water, it is not *fully* adequate. Thus, when they do not drink on a given day, they become a bit dehydrated. They become a bit more dehydrated the next day, and so forth, until ultimately—after a week or two (or longer)—they need to find drinking water.

- A few drinking-water-independent species are believed to be able to meet their entire need for preformed water from the preformed water in their foods even in the hot–dry seasons of the year, meaning that they need never drink. The oryx and eland are often cited as examples. Field biologists report that these species, when living in water-poor regions, are never observed to drink. Moreover, physiological data (which we will soon discuss) make it plausible to believe that these animals acquire enough preformed water from their foods that they never require drinking water.

Water and food resources in the deserts and dry savannas are often complex

Now that we have addressed the needs of large herbivores for preformed water in drink or food, a next logical step is to consider the nature of the drinking-water and food resources available to them. These resources are often far more complex in their properties than meets the eye.

One complexity in both deserts and dry savannas is that *when standing water is found, it is often salty water.* Recent measurements have revealed, for example, that in the central and southeastern Serengeti ecosystem (see Figure 30.4), salinities of 5–15 g/kg are common in the headwaters of major rivers, and salinities of 20–30 g/kg are common in stagnant pools or landlocked lakes (for comparison, as noted in Chapter 27, the salinity of full-strength seawater is 35 g/kg). Not all Serengeti waters are this salty; some have salinities lower than 1 g/kg. Nonetheless, one cannot help but be impressed with how salty the waters can be in this benign-looking grassland ecosystem. Some investigators postulate, in fact, that the immediate stimulus for the migrating herds to leave the southeast Serengeti at the end of the rainy season is the increasing salinity of the drinking water as the landscape dries.

Why are waters in deserts and dry savannas so often salty? Briefly speaking, soils in arid regions are commonly salty, and the waters become salty by dissolving salts out of the soils. A key reason the soils are salty is that they get so little rain that they are rarely thoroughly flushed with water. Typically, each rain brings a trace of salt with it.[5] When the rain hits the ground in arid regions, it soaks in to only a shallow depth and evaporates. Rain after rain causes salts to build up incrementally in the upper layers of the soil. These salts then dissolve into any pools of water that form on top of the soil, making the water salty until it itself evaporates, leaving the salts behind once more. This process is accentuated in low places where rain tends to settle in pools or flow in temporary streams. In such places, particularly large amounts of water accumulate after each rain and leave particularly large quantities of salts behind after evaporating.

When animals find salty water to drink, their kidneys must be able to excrete each ion in the water at a higher concentration than in the water itself if the animals are to gain H_2O by drinking the water (see page 734). Thus the evolution of kidneys with exceptional concentrating abilities has been important for desert and savanna mammals for two reasons: A high concentrating ability allows ordinary soluble wastes to be excreted with relatively little water, and a high concentrating ability permits animals to gain H_2O from saltier water sources than would otherwise be acceptable. Despite kidney specializations, when drinking waters approach half the concentration of seawater, they cease being useful H_2O sources for some of the species of mammals that are native to the deserts and dry savannas.

A second important complexity of water and food resources in deserts and dry savannas is that *in dry seasons, the leaves of plants often undergo large and rapid swings in their content of preformed water because of equilibration with the humidity of the air.* When this phenomenon was first reported about 50 years ago, it seemed that it might be merely a footnote in the biology of a few species. By now, however, desert biologists recognize the phenomenon as being often of extreme significance in the water biology of large herbivores. **FIGURE 30.7** shows how a perennial grass that is a favored food of herbivores varies in its moisture content because of equilibration with changing atmospheric humidity between midday, evening, and the dark hours before dawn in the Namib Desert. The measurements were made in the dry season, when the blades of the grass were dead and brown. Although ostensibly "dry" all the time, dead grass blades quickly equilibrate with changes in atmospheric humidity and have a far higher water content at the end of the night than at midday or evening. Accordingly, animals can increase their input of preformed water by eating preferentially in the hours around dawn, and research has shown that many species do.

A third aspect of the complexity of water and food resources has both obvious and less obvious aspects. In deserts and dry savannas, *both the growth and the nutritional composition of plants tend to be highly correlated with changes in water availability.* The

---

[5] Aerosolized seawater can be detected in tiny amounts in the atmosphere hundreds of miles inland from the oceans, for example. The salts dissolve in cloud droplets and are carried to Earth in rain.

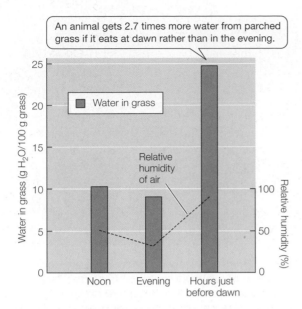

An animal gets 2.7 times more water from parched grass if it eats at dawn rather than in the evening.

FIGURE 30.7 The moisture content of "dry" grass varies with time of day or night Data were gathered on dead, brown blades of the perennial desert grass *Stipagrostis uniplumus* in the Namib Desert. The water content of the grass is highest near dawn, when the relative humidity is highest. (After Louw 1972.)

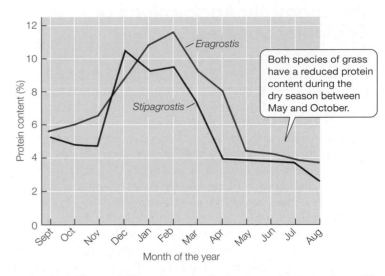

Both species of grass have a reduced protein content during the dry season between May and October.

FIGURE 30.8 The nutritional content of some plants varies with rainy and dry seasons The percentage of protein in two species of perennial grasses (*Eragrostis lehmanniana* and *Stipagrostis obtusa*) eaten by wildebeests and oryxes was measured throughout the year in the southern Kalahari Desert ecosystem. (After Lovegrove 1993.)

effects of rainfall on plant growth are obvious: In deserts, where rainfall occurs irregularly and unpredictably, plant growth is likewise irregular and unpredictable. In savannas, plant growth waxes in the wet season and wanes in the dry season. What is not obvious is that some plants vary markedly in their content of nutritionally important materials as rains come and go. Certain of the important perennial grasses in the Kalahari Desert ecosystem, for example, vary threefold or more in their protein content between the wet and dry seasons (**FIGURE 30.8**).[6] A recent study of grasses in a hyperarid Arabian desert showed that protein content (computed relative to plant dry weight) falls exponentially during long periods of drought, from about 9% in the early weeks of drought (just after the cessation of rain) to 5% a year later. After long periods of drought, the concentrations of phosphorus, copper, and other minerals are sometimes too low in desert plants to meet the nutritional needs of herbivores.

A final aspect of the complexity of water and food resources is *the formation of fog and the deposition of fog water on plants and other objects*. Fog and fog water are potentially significant sources of water in deserts or savannas located within about 80 km of seacoasts, particularly coasts where—because of upwelling or other oceanographic processes—there is a narrow band of cold seawater along the shore. Fogs often recur so frequently in such places that they can make a major contribution to the ecosystem water budget. Although several mechanisms of fog formation are likely, the most common is that warm moist air blowing landward from the open ocean is cooled just before making landfall by passing across the cold coastal band of water, and this cooling of the air causes its saturation vapor pressure to decline (see Table 27.2), forcing water out of the vapor state to form fog droplets. In hyperarid coastal regions that have frequent fogs—such as parts of the deserts in Namibia, Chile, and Oman—there is so little rainfall and so much fog that plants may

receive several times more water by deposition of fog water on their surfaces than they receive from rain! Some desert arthropods, such as certain beetles and scorpions, are specialists in gathering the fog water (**FIGURE 30.9**). For the large desert herbivores that are our focus in this chapter, the effects of fog are more indirect: Fogs provide water to the plants that these herbivores eat. This input of water not only increases the water content of the plants but helps maintain plant health, so that—for example—plant protein stays higher than it otherwise would, making the plants more nutritious. Certain of the grasses (*Stipagrostis*) in the Namib Desert have recently been shown to harvest fog droplets by having grooves in their grass blades, down which droplets flow, ultimately dripping off to strike the soil where their roots grow.

FIGURE 30.9 The desert beetle *Onymacris unguicularis* collects water from fog in the Namib Desert When fog rolls in from the sea, as it does often because of local conditions, the beetle goes to the crests of sand dunes and positions itself, as shown, in a "fog-basking" stance, in which its broad dorsal surface faces into the fog-laden seabreeze. As the fog water accumulates on the beetle's body, the water runs down to the beetle's mouth.

---

[6] The leaves and seed pods of Kalahari trees, however, do not show such strong seasonal changes.

The Dramatic Adaptations of Particular Species

For the most part, the physiologists who have made the effort to study the large mammals of deserts and dry savannas have been water-balance specialists. It's a commentary on human nature, rather than animal nature, that scientists—like all people—have personal proclivities that lead them to attach greater importance to some things they observe than to others. Water-balance physiologists tend to attach greatest importance to adaptations that permit animals to live with as little water as possible. With water-balance physiologists organizing most physiological research in the deserts and dry savannas, wildebeests—noted for their slavish dependence on sources of drinking water—have often been dismissed as uninteresting.

Before we turn to the "water-balance champions"—species famed for extreme degrees of independence from drinking water—we would do well to pause a moment to recognize how remarkably successful wildebeests are. Here is a stunning fact: In the Serengeti, wildebeests outnumber all the other species of large mammals combined! This is true even though they are noted more for needing water than for conserving it. Faced with life in an environment that dries out for months at a time, instead of evolving exceptional modes of surviving without drinking water, wildebeests have evolved ways of getting themselves reliably to places where they can find the drinking water they need. If wildebeests have extraordinary adaptations, their special abilities probably reside in their brains and sense organs, which physiologists other than water-balance physiologists will ultimately study. Clearly wildebeests are able to *navigate* during their long migratory treks each year. They probably have other exceptional nervous system abilities as well. Naturalists report, for example, that wildebeests in the northwestern Serengeti often start moving southeast at the start of the rainy season before rain is actually falling in the northwest. Do they hear distant thunder or smell distant rain and respond by moving out across a mental map? The answers cannot help but be fascinating.

Let's now discuss four case studies of the sort that intrigue water-balance physiologists: the cases of the oryxes, the Thomson's and Grant's gazelles, the sand gazelle, and the dromedary (one-humped) camel. These are all large animals that, because of their size, have limited opportunities to find protective microenvironments. Consequently, when they live in hot, dry places where they are exposed to environmental heat stress, they actually experience heat stress during all the hours it prevails, and they therefore confront directly the fundamental conflict between heat balance and water balance: Although use of water for evaporative cooling provides a *physically straightforward* way to rid the body of excess heat, it is not *ecologically straightforward* when environmental water is in short supply. We saw earlier that large body size is a physiological advantage in hot, dry environments (see Figure 30.1). Large size in itself, however, does not permit an animal to be completely independent of drinking water in deserts and dry savannas. All of these animals have evolved exceptional species-specific adaptations for maintaining water balance in places where water exerts a dominant controlling effect on life.

Oryxes represent the pinnacle of desert survival

Among the truly wild large mammals, oryxes (**FIGURE 30.10**) represent the pinnacle of evolution in their ability to survive in deserts. Dromedaries might be their equal physiologically, but wild dromedaries went extinct before historical times, and today's

FIGURE 30.10 Oryxes in deserts often exist on dead, dry grasses and the leaves of water-stressed bushes and trees Oryxes are able to stay in water balance with their only preformed water coming from such sources because they conserve water exceptionally well by mechanisms that are only partly understood. The animal shown is a gemsbok (*Oryx gazella*).

dromedaries are either domesticated or escaped from domestication. Four types of oryxes are recognized, although taxonomists debate whether some are species or subspecies. The three types that have featured in physiological research are the gemsbok oryx (*Oryx gazella*) and beisa oryx (*O. beisa*)—which closely resemble each other—and the Arabian oryx (*O. leucoryx*). The Arabian oryx, a species that weighs 70–100 kg, was driven to extinction in the wild for a time but has been reestablished. Oryxes can live in some of the most inhospitable places on Earth—including hyperarid deserts where summer air temperatures sometimes exceed 45°C (113°F), the sun shines inexorably, there is no drinking water except immediately after rains, and the rains themselves are so infrequent that average precipitation is less than 5 cm (2 inches) per year.

In one of the groundbreaking studies of oryxes, physiologists measured the total water turnover rates of *free-living* Arabian oryxes by use of isotopically labeled water (see page 769). They found the water turnover rates of the oryxes to be only one-quarter to one-half as high as would be expected for free-living mammals of their body size. The same team also measured evaporative water loss in *caged* Arabian oryxes. They found that as oryxes are acclimated to water restriction, they conserve water better: They reduce their metabolic rate (lowering respiratory evaporation), lower their total rate of evaporative water loss so it is only about half as high as in nonacclimated oryxes, and achieve a total rate of evaporative water loss that is only 40% as high as expected for ungulates of their body size studied under caged conditions. These relatively recent results confirm the conclusion reached by the earlier laboratory studies we discussed (see Figure 30.6) that oryxes have evolved unusually low water needs. Oryxes are far more effective than average mammals of their body size in limiting their rates of water loss and, therefore, the rates at which they must resupply themselves with water.

The *mechanisms* by which oryxes achieve exceptional water conservation and low water-turnover rates are only starting to be understood. The kidneys of beisa and Arabian oryxes can concentrate urine to an osmotic urine-to-plasma (U/P) ratio of about 8 (**TABLE 30.2**)—a high value for animals of their size (see Figure 28.20). Free-living Arabian oryxes lower their metabolic rates in summer

Species	Urine osmotic pressure (mosmol/kg H_2O)	Osmotic U/P ratio[b]	Fecal water content (g H_2O/100 g)
African buffalo (*Syncerus caffer*)	1120	4	
Hereford cow (*Bos taurus*)	1160	4	75
Zebu cow (*Bos indicus*)	1300	4	
Somali donkey (*Equus asinus*)	1680	5	61
Common wildebeest (*Connochaetes taurinus*)	1830	6	
Thomson's gazelle (*Gazella thomsoni*)	2640	7	
Arabian oryx (*Oryx leucoryx*)	2500	8	43
Beisa oryx (*Oryx beisa*)	3100	8	
Grant's gazelle (*Gazella granti*)	2790	8	
Dromedary camel (*Camelus dromedarius*)	3200	8	44
Dik-dik (*Madoqua kirkii*)	4760	~12	44

Sources: After Maloiy et al. 1979; Ostrowski et al. 2006; dik-dik urine data updated from Beuchat 1996.
[a]All values are from dehydrated animals.
[b]U/P ratios depend in part on plasma values, explaining why they do not always correlate exactly with urine osmotic pressures.

to be only half as high as in winter, based on doubly labeled water studies (see page 217). Although oryxes are famous for being indifferent to shade under many circumstances, they methodically remain inactive in shade—if they can find it—during the heat of the day when under water stress. At least two species sometimes dig shallow depressions in which they settle down; this behavior is thought, without evidence, to shield them from the sun or allow them to lose heat to the cool subsoil they expose. Pregnancy and lactation place substantial water demands on females, and a recent study indicates that in Arabian oryxes, reproduction is curtailed during drought; both mating behaviors and conceptions were only about 20% as high during a dry year as during a moist year.

The water-conserving mechanism that is now best understood in oryxes—and thought generally to be widely used by other desert and dry-savanna species—is modulation of body temperature in ways that minimize the use of water for thermoregulation (see pages 270–271). A person placed in a hot desert maintains a stable body-core temperature of about 37°C, but at the price of profuse sweating, which entails rapid water loss (up to 2L/h). Oryxes and other species of large mammals with long evolutionary histories in hot–dry environments are hypothesized in general to exploit the advantages of a *variable* body-core temperature when necessary to save water. The firmest supporting evidence for this hypothesis available today in any species is in fact provided by two recent studies of Arabian oryxes living wild in the deserts of Saudi Arabia. The body temperatures of the animals were monitored remotely for hundreds or thousands of hours by use of radiotelemetry in one study and by data loggers[7] in the other. The results of one study are presented in **FIGURE 30.11**. The average body-core temperature

[7] Data loggers are devices that, as they collect data, log the data into digital memory. They are later recovered and the data downloaded.

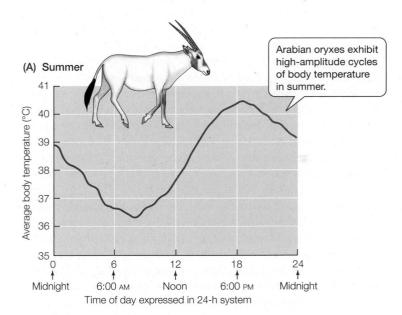

(A) Summer

Arabian oryxes exhibit high-amplitude cycles of body temperature in summer.

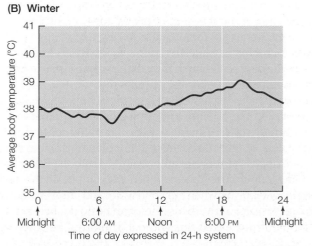

(B) Winter

FIGURE 30.11 Average body-core temperatures of free-living Arabian oryxes over the 24-h day The six oryxes (*Oryx leucoryx*) studied were living in their natural environment in Saudi Arabia and monitored using implanted radiotelemeters during (A) summer and (B) winter. (After Ostrowski et al. 2003.)

rose and fell to a small extent over the 24-h day during winter (see Figure 30.11B), when the ambient air temperature varied from an average low of 13°C at night to an average high of 27°C during the day. In summer, however, when the ambient air temperature varied from a nighttime low of 29°C to a daytime high of 44°C (111°F), the amplitude of the oryxes' cycle of body-core temperature became much larger (see Figure 30.11A). On average, the oryxes in summer let their body temperatures fall to 36.5°C overnight, but then rise to 40.5°C during the day: an average amplitude of about 4°C. The second, completely independent study on Arabian oryxes obtained similar results; the average amplitude in summer was 5°C. In the two studies taken together, three individuals were observed at times to allow their body-core temperature to cycle by 7.5°C–7.7°C between night and day—the highest amplitude daily cycle ever observed in any large mammal.

The oryxes' high-amplitude cycles of body temperature are believed to save water in two ways (see pages 270–271). First, when the animals let their body temperatures rise during the heat of the day, they are in effect *storing* heat instead of panting or sweating to get rid of it. They then exploit the coolness of the nighttime environment to lose the heat by nonevaporative means (convection and thermal radiation). The water savings per day from allowing the body temperature to cycle—rather than holding it constant by panting or sweating—is about 0.3–0.4 L for an oryx of average size. The elevation of body temperature during the day also has a second advantage: It *reduces the difference in temperature between an oryx's tissues and the hot environment during the heat of the day*, thereby slowing the rate at which heat enters its body.

Remote-monitoring techniques—such as radiotelemetry and automated data collection by data loggers—have only recently started to be used to study the physiology of desert and dry-savanna mammals, and only three studies have been completed on the body temperatures of free-living oryxes (none has been completed on free-living camels or gazelles). Two of the studies are those we have just discussed. The third study did not find large cycles of body temperature in the oryxes monitored (*O. gazella*). This may mean only that the animals were not experiencing a tendency to dehydrate. Many laboratory investigations have demonstrated that desert and dry-savanna mammals are in general far more likely to allow large swings in body temperature when they are tending to dehydrate than otherwise. Moreover, one of the studies on free-living Arabian oryxes found a correlation between the amplitude of body-temperature cycles and environmental moisture: The cycles became greater as the environment became drier. A goal for physiologists as they make greater use of remote-monitoring techniques will be to find answers to this key question: How and why do free-living animals vary the degree to which they make maximum use of the water-conserving mechanisms they possess?

Up to now we have discussed just one side of the water-balance equation in oryxes: their ability to limit water losses. What about water gains? Studies on free-living oryxes in the wild confirm that, as laboratory data suggest (see Figure 30.6), they must meet most of their water needs with preformed water. Metabolic water can account for only one-quarter or less of their water inputs in the wild.

Although oryxes drink when they find water and are even known at times to dig for water in riverbeds, there can be no doubt that they often live for months or even years without drinking. They then get the preformed water they need from their foods—an astounding proposition, because the foods available to them often seem to the eye to be distinctly unpromising. Oryxes eat the leaves of desert trees, such as acacia trees. These leaves, although they are firm and not ostensibly juicy, turn out, when analyzed, to be at least 50%–60% water, even in the midst of severe droughts. Oryxes also eat dried grasses and other dead, dry plants (see Figure 30.10), and when they do so, they time their feeding to occur principally in the predawn hours, when the plants are richest in water (see Figure 30.7). In dry times, oryxes become nomads in the deserts, seeking out places where fog, dew, or rain has allowed plants to build up more water than average. According to some orthodox scientific evidence plus reports of the San people, who have lived in African deserts and dry savannas for millennia, oryxes have uncanny abilities to find and excavate the underground dormancy organs of desert plants, structures that can be 1 m deep but contain 50%–70% water. All these strategies used by oryxes to obtain preformed water from food seem so marginal that they might not work to sustain life, and in fact there is no quantitative proof that they do work. However, they apparently must work, because the oryxes survive. One reason life is possible on such meager water resources is water conservation. Surviving on the water in the leaves of water-stressed or dead plants would be impossible without mechanisms that can profoundly limit the amount of water an animal needs.

Oryxes have been recorded to survive rainless droughts in the desert for at least 34 months. Sometimes their body condition declines under such extreme duress, and some die. Even after 2–3 years without drinking, however, the problem that undermines oryxes seems most likely *not* to be dehydration, based on studies of recently dead animals. Instead, they seem most likely to die of severe protein deficiency (or other nutritional deficiency) caused by the deteriorating nutritional quality of the plants in their environment.

Grant's and Thomson's gazelles differ in their relations to water

Grant's gazelle (*Gazella granti*) and Thomson's gazelle (*G. thomsoni*) are two look-alike antelopes, weighing 10–50 kg, that often coexist in abundance in the dry savannas of East Africa (see Figure 30.3). Both are drinking-water-independent[8] and often stand in the open throughout the day without seeking shade. For five or six decades the contrasts between these two species have highlighted the fact that straightforward studies of water input and water output do not always tell the entire story about the water relations of savanna animals.

When Grant's and Thomson's gazelles are subjected to standard water-balance analyses in laboratory settings, they prove to be similar. If anything, Thomson's gazelle seems more adept at conserving water during such tests. However, in their natural environment, the species clearly have different relations to water, and of the two, Thomson's gazelle paradoxically seems to be the inferior one in coping with water stress. In the Serengeti, Thomson's gazelles participate in the annual migration; despite being drinking-water-independent, they trek along in great numbers with the drinking-water-dependent wildebeests and zebras. Where savannas give way to deserts, although Grant's

---

[8] Recall that "drinking-water-independent" is a category of drinking behavior. Drinking-water-independent species do not need to drink each day; instead they can skip drinking for many days or weeks in a row. They may, however, need to drink on occasion.

gazelles occupy the deserts as well as the savannas, Thomson's gazelles stay largely in the moister savannas.

The study of physiology has yet to explain the divergence of the water relations of these two gazelle species in the wild. Some insight is probably provided by their different responses to extreme ambient temperatures in laboratory tests. Exposed to an air temperature of 45°C, Grant's gazelle allows its body temperature to rise to 45°C or higher; some individuals reach a rectal temperature of 46.5°C (116°F), among the highest ever recorded in a vertebrate. Thomson's gazelle, however, keeps its body temperature below 43°C, a process that means more panting and more use of water for thermoregulation. The two species may also differ in food selection (the plants they select to eat and the times when they eat them), behavioral thermoregulation, or other traits that only careful field studies will bring fully to light.

The sand gazelle is drinking-water-independent in hyperarid deserts

The sand gazelle (*Gazella subgutturosa*), which weighs about 20 kg and therefore is similar in size to Grant's and Thomson's gazelles, dramatically illustrates the physiological diversification that can occur within sets of closely related animals (in this case three species in the single genus *Gazella*). The sand gazelle lives in full-fledged deserts, including hyperarid deserts, and therefore—averaged over the course of an entire year—tends to face greater problems of water availability than savanna-dwelling Grant's and Thomson's gazelles. Recent research has revealed that the sand gazelle has evolved a strikingly low rate of evaporative water loss (EWL). Its total EWL is only about 20% as high as the EWL of Grant's or Thomson's gazelles—and also only about 20% as high as expected for an average ungulate of its body size. The mechanisms by which EWL is so profoundly reduced remain to be fully understood. Two mechanisms documented thus far are cycling of body temperature (summer amplitude: 2.6°C) and a reduction of metabolic rate (i.e., a reduction of internal heat production) by as much as 45% when acclimated to food and water restriction.

The dromedary camel does not store water, but conserves it and tolerates profound dehydration

The modern era of physiological research on large desert mammals began with studies on dromedary camels (**FIGURE 30.12**) by Knut and Bodil Schmidt-Nielsen in the mid-twentieth century. More has steadily been learned since then. Dromedaries drink, in part because they are domesticated and their human handlers water them. Nonetheless, they are capable of extreme performance in some of the most inhospitable places on Earth. During the cool seasons of the year, dromedaries are able to travel for several weeks in the desert and cover hundreds of miles—serving as beasts of burden—without drinking at all from start to finish.

From the time of ancient Rome until the Schmidt-Nielsens' pioneering studies, the reigning theory about the camels' low requirement for drinking water was that they had a canteen onboard. That is, people assumed that camels have a greater amount of body water than ordinary mammals when they start a trek. For centuries, debate therefore focused on the location of the canteen, with most commentators supposing that the rumen or parts of the rumen

FIGURE 30.12 Dromedary camels—the "ships of the desert"— do not carry extra water in their bodies, as legend has often held Instead, they depend on water conservation, an ability to eat diverse desert foods, and extreme tolerance of dehydration to be able to go for days or weeks without drinking. They have a long history of use as beasts of burden in much of northern Africa and southern Asia and have been imported into Australia. They are the largest of the mammals adapted to desert life, sometimes reaching adult weights of 500–600 kg.

played this role. After early biochemists established that water is made by metabolism, even the hump became a proposed site of canteen function. Not that the hump contains water: Scientists have long recognized that the hump is filled with fat. However, fat yields more than its own weight in H_2O when it is oxidized (see Table 27.3), and thus it seemed only logical that the fat in the hump was a lightweight way to carry H_2O. The Schmidt-Nielsens and later investigators systematically demolished all these myths, some of which had "provided the answer to the camel question" for 2000 years. A camel's rumen contains no more water than any other ruminant's rumen. When the total amount of water in a camel's body is measured, it is no more than the average for all mammals. The hump fat actually *costs* H_2O to oxidize because the H_2O lost in breathing to obtain O_2 for the oxidation of the fat exceeds the metabolic water produced (see page 735). In the end, researchers established beyond doubt that camels drink only to make up for prior water losses, never to store water in their bodies in anticipation of water needs during an upcoming trek.

There are two principal keys to the extraordinary ability of camels to survive in deserts for long periods without drinking. One of these is by now a familiar theme: Camels conserve water extraordinarily well. Based on studies of penned animals, we know that dromedaries dramatically exploit the water-conservation benefits of large daily changes in body temperature. When dehydrated, they sometimes allow their body temperature to rise as much as 6°C during the day and fall to the same extent in the cool of night. One reason they can permit their body temperature to rise as high as it does is that they keep their brain cooler than the rest of their body (see page 270)—a trait also reported in oryxes and gazelles. Dromedaries also employ several other water conservation strategies: (1) they can produce dry feces and concentrated urine (see Table 30.2); (2) they curtail their urine production relatively rapidly and profoundly when faced with dehydration; (3) they have thick, sometimes glossy, fur that acts as a heat shield, helping to slow influx of heat from the fur surface and reflect or reradiate incoming solar radiation (see page 270); and (4) they minimize heat influx behaviorally, as by steadily facing the sun when allowed to rest during the day, a practice by which they consistently present a narrow body profile to the sun's direct rays.

The second known key to the ability of camels to survive for weeks without drinking water is their unusual tolerance of dehydration. Species of mammals without long evolutionary histories in deserts, such as dogs or horses, are in danger of death if they dehydrate by 10%–15% of their body weight while under heat stress. Dromedaries, in contrast, can tolerate at least twice this extent of dehydration. There are known cases of dromedaries being alert and functional in the desert heat while dehydrated by 30%–40% of their body weight.

Thus, when a dromedary sets off on a desert trek, it has the same amount of body water, in relation to its body size, as you or I. However, it dehydrates by a smaller percentage of its body weight per day because it conserves water exceptionally well, and it can tolerate a level of total dehydration at least twice what we can tolerate. Along the way it munches desert plants, ranging from green leaves to seemingly impossible foods like dried-out thorn bushes and saltbushes, thereby replacing some of its water losses using preformed water from its food. Thus it can forestall its need to drink for many days in the summer and many weeks in the cool seasons.

STUDY QUESTIONS

1. What are the pros and cons of studying water physiology in the laboratory and in the field? Could an animal's water physiology be fully understood by field studies alone? By laboratory studies alone? Explain.

2. In Chapter 1 we noted that an animal's body size is one of its most important properties. Considering all the mammals that live in deserts, explain why their body sizes matter. Include discussion of physiological mechanisms where appropriate.

3. The effects of water loss in milk on the water budgets of lactating female mammals in arid environments are not well understood. Propose three specific hypotheses regarding mechanisms by which females might prevent water losses in their milk from forcing them into negative water balance (which could lead to fatal dehydration). For each hypothesis, design a study to test or evaluate the hypothesis.

4. Contrast the body-temperature responses of humans and Arabian oryxes during exposure to a hot–arid desert. In each case, state implications for water balance. Include both day and night in your analysis.

Go to **sites.sinauer.com/animalphys4e**
for box extensions, quizzes, flashcards,
and other resources.

REFERENCES

Armstrong, L. E., C. M. Maresh, C. V. Gabaree, and five additional authors. 1997. Thermal and circulatory responses during exercise: effects of hypohydration, dehydration, and water intake. *J. Appl. Physiol.* 82: 2028–2035.

Durant, S. M., T. Wacher, S. Bashir, and 39 additional authors. 2014. Fiddling in biodiversity hotspots while deserts burn? Collapse of the Sahara's megafauna. *Diversity and Distributions* 20: 114–122.

Hetem, R. S., W. M. Strauss, L. G. Fick, and five additional authors. 2010. Variation in the daily rhythm of body temperature of free-living Arabian oryx (*Oryx leucoryx*): does water limitation drive heterothermy? *J. Comp. Physiol., B* 180: 1111–1119.

Louw, G., and M. Seely. 1982. *Ecology of Desert Organisms.* Longman, New York. A fine introduction to the ecology of the world's deserts, written by two eminent desert biologists.

Louw, G. N. 1993. *Physiological Animal Ecology.* Longman Scientific & Technical, Harlow, England. The aging of this excellent book is sad. It remains an exceptional source—compact and lucid—on the basic biology of African deserts.

Lovegrove, B. 1993. *The Living Deserts of Southern Africa.* Fernwood Press, Vlaeberg, South Africa. It's hard to imagine a more delightful and astute book on deserts and desert organisms than this one, written for the general public by a physiologist and illustrated with a profusion of outstanding color photographs. Although limited to southern Africa in its coverage, the book provides insight into desert processes in general.

Monbiot, G. 1994. *No Man's Land.* Macmillan, London. An exploration into the fate of the nomadic peoples of the East African arid lands.

Oldfield, S. 2004. *Deserts. The Living Drylands.* MIT Press, Cambridge, MA. This book, although not focused on physiology and less rich with science than that by Lovegrove, covers all the deserts of the world and is bound to excite interest and curiosity about desert animals. Densely illustrated with outstanding photographs.

Ostrowski, S., J. B. Williams, and K. Ismael. 2003. Heterothermy and the water economy of free-living Arabian oryx (*Oryx leucoryx*). *J. Exp. Biol.* 206: 1471–1478. This paper not only provides an important new set of data, but offers a good review of the controversy swirling around the concept of adaptive heterothermy in desert and dry-savanna mammals. See also Ostrowski, S., and J. B. Williams. 2006. Heterothermy of free-living Arabian sand gazelles (*Gazella subgutturosa marica*) in a desert environment. *J. Exp. Biol.* 209: 1421–1429.

Ostrowski, S., J. B. Williams, P. Mésochina, and H. Sauerwein. 2006. Physiological acclimation of a desert antelope, Arabian oryx (*Oryx leucoryx*), to long-term food and water restriction. *J. Comp. Physiol., B* 176: 191–201.

Reisner, M. 1993. *Cadillac Desert: The American West and Its Disappearing Water,* rev. ed. Penguin Books, New York.

Schmidt-Nielsen, K. 1964. *Desert Animals.* Oxford University Press, London. One of the classic monographs in animal physiology. The book still inspires fascination about the relations of desert animals to water and stands as a model of clear thinking about its subject, although much of its detailed content is dated.

See also **Additional References** *and* **Figure and Table Citations**.

A

Appendices

The Système International and Other Units of Measure

Physiology is in transition to full use of a system of units of measure called the **Système International** (**SI**). Some subdisciplines of physiology today use almost entirely SI units, whereas other subdisciplines continue to use older systems of measure along with SI units. The SI recognizes seven base units of measure, listed in the first section of the accompanying table: the meter, kilogram, degree Celsius, second, ampere, mole, and candela. All other units in the SI are to be derived from these seven; thus, for example, the SI unit for velocity (distance per unit of time) is the meter/second. Many derived SI units are listed in the second section of the table. Some derived units are given special names; for example, the unit of force (mass × acceleration) is technically a (kilogram × meter)/second^2, which is more commonly known as a newton. The unit for energy (force × distance) is a newton × meter, called a joule. Note that when names of people are used for units of measure, the names are written entirely in lowercase (although *abbreviations* are uppercase; e.g., 1 newton is abbreviated 1 N).

Prefixes indicating orders of magnitude greater or smaller are acceptable in the SI. For instance, the prefix kilo- can be used to indicate "three orders of magnitude greater" (a kilojoule [kJ] is 1000 J). See Appendix B for definitions of prefixes. Note that in scientific writing, abbreviations for symbols are not followed by a period (unless by coincidence, as at the end of a sentence). For example, "kg"—not "kg."—is the correct abbreviation for "kilogram."

Two sorts of notation are used for derived units: *numerator/denominator* notation and *exponential* notation. In numerator/denominator notation the unit of velocity, for example, is the meter/second or m/s; in exponential notation the unit is written m•s^{-1}. In this text we use numerator/denominator notation because we consider it more intuitive and easier to understand, but many students are also familiar with exponential notation and find it easier for complex calculations. In this appendix we list units in both numerator/denominator notation and exponential notation.

Relations between SI units and selected traditional units are shown in the table, in the column labeled "Relations among units." Some relations between different sets of traditional units are also listed. To obtain the relations converse to those shown, divide both sides of the relevant equation by the number to the right of the equals sign. For instance, the table states, "1 m = 3.28 ft." Dividing both sides by 3.28 yields, "1 ft = 0.305 m."

More information on the SI and traditional units of measure can be obtained in *Handbook of Chemistry and Physics* (CRC Press, Boca Raton, FL) or *Lange's Handbook of Chemistry* (McGraw-Hill, New York).

Quantity	SI Unit		Relations among units
Base SI Units			
Length	meter (m)		1 m = 3.28 feet (ft)
			1 inch (in) = 25.4 millimeter (mm)
			1 statute mile (mi) = 1609.3 m
Mass	kilogram (kg)		1 kg = 2.20 pound, avoirdupois (lb)
			1 ounce, avoirdupois (oz) = 28.3 gram (g)
Temperature	1 degree Celsius (°C) = 1 kelvin (K)		A difference of 1°C = a difference of 1.8 degree Fahrenheit (°F)
Time	second (s)		—
Electric current	ampere (A)		—
Amount of substance	mole (mol)		—
Luminous intensity	candela (cd)		1 candela ≅ 1 candle (pentane)

Quantity	SI Unit		
	Numerator/ denominator notation	Exponential notation	
Derived SI Units			
Area	m^2	m^2	1 m^2 = 10,000 square centimeters (cm^2)
			1 m^2 = 10.8 ft^2
Volume	m^3	m^3	1 m^3 = 1 × 10^6 cm^3
			1 m^3 = 1000 liter (L)
			1 cm^3 = 1 milliliter (mL)
			1 U.S. gallon = 3.785 L
			1 U.S. fluid ounce = 29.6 mL
Density	kg/m^3	$kg \cdot m^{-3}$	1 kg/m^3 = 0.001 g/mL
Velocity	m/s	$m \cdot s^{-1}$	1 m/s = 3.28 ft/s
			1 m/s = 2.24 statute mile/hour
Acceleration	m/s^2	$m \cdot s^{-2}$	1 m/s^2 = 3.28 ft/s^2
Force	$kg \cdot m/s^2$ = 1 newton (N)	$kg \cdot m \cdot s^{-2}$ = 1 N	1 N = 0.102 kilogram of force
			1 N = 0.225 pound of force
			1 N = 1 × 10^5 dyne
Energy, work	$kg \cdot m^2/s^2$ = 1 N·m = 1 joule (J)	$kg \cdot m^2 \cdot s^{-2}$ = 1 N·m = 1 J	1 J = 0.239 calorie (cal)
			1 J = 1 × 10^7 erg
			1 J = 0.000948 British thermal unit
			1 J = 0.738 foot-pound
Power	$kg \cdot m^2/s^3$ = 1 J/s = 1 watt (W)	$kg \cdot m^2 \cdot s^{-3}$ = 1 J·s^{-1} = 1 W	1 W = 0.239 cal/s
			1 W = 0.0013 horsepower
			1 W = 3.41 British thermal unit/h
Pressure	$kg/(s^2 \cdot m)$ = 1 N/m^2 = 1 pascal (Pa)	$kg \cdot s^{-2} \cdot m^{-1}$ = 1 N·m^{-2} = 1 Pa	1 Pa = 0.0075 mm of mercury (mm Hg)
			1 kilopascal (kPa) = 1000 Pa
			1 atmosphere = 101.3 kPa
			1 atmosphere = 760 mm Hg
			1 mm Hg = 1 torr
			1 lb/in^2 = 6.89 kPa
Frequency	1/s = 1 hertz (Hz)	s^{-1} = 1 Hz	1 Hz = 1 cycle/s
Electric potential	$kg \cdot m^2/(s^3 \cdot A)$ = 1 W/A = 1 volt (V)	$kg \cdot m^2 \cdot s^{-3} \cdot A^{-1}$ = 1 W·A^{-1} = 1 V	—
Electric resistance	$kg \cdot m^2/(s^3 \cdot A^2)$ = 1 V/A = 1 ohm (Ω)	$kg \cdot m^2 \cdot s^{-3} \cdot A^{-2}$ = 1 V·A^{-1} = 1 Ω	—
Electric charge	s·A = 1 coulomb (C)	s·A = 1 C	1 C = 0.00028 ampere-hour

B

Prefixes Indicating Orders of Magnitude

You undoubtedly know that kilo- means "three orders of magnitude greater." This and a series of less-familiar prefixes are summarized for convenience in the following table.

Prefix	Abbreviation (placed before unit abbreviation)	Meaning in words	Meaning in powers of 10
tera-	T	12 orders of magnitude greater	$\times 10^{12}$
giga-	G	9 orders of magnitude greater	$\times 10^{9}$
mega-	M	6 orders of magnitude greater	$\times 10^{6}$
kilo-	k	3 orders of magnitude greater	$\times 10^{3}$
deci-	d	1 order of magnitude smaller	$\times 10^{-1}$
centi-	c	2 orders of magnitude smaller	$\times 10^{-2}$
milli-	m	3 orders of magnitude smaller	$\times 10^{-3}$
micro-	μ	6 orders of magnitude smaller	$\times 10^{-6}$
nano-	n	9 orders of magnitude smaller	$\times 10^{-9}$
pico-	p	12 orders of magnitude smaller	$\times 10^{-12}$
femto-	f	15 orders of magnitude smaller	$\times 10^{-15}$

Gases at Standard Temperature and Pressure

Temperature and pressure exert such great effects on gas volume that they must be specified for a volume to have meaning. Physiologists usually convert all measured gas volumes to volumes at **standard temperature and pressure (STP)**—0°C and 1 atmosphere—so that the volumes can be compared. Reported values are assumed to be at STP unless otherwise stated. The equation for making the conversion is

$$\text{volume at STP} = \text{volume measured} \times \frac{P_{meas}}{1 \text{ atm}} \times \frac{273 \text{ K}}{T_{meas}} \qquad \textbf{(C.1)}$$

where P_{meas} is the pressure prevailing during measurement, in atmospheres, and T_{meas} is the temperature prevailing during measurement, in absolute (Kelvin) degrees.

Suppose, to illustrate, that you study a bird in a chamber where the pressure is 0.92 atm and the temperature is 20°C. On the Kelvin scale, the temperature is 293 K (273 plus its value on the Celsius scale). Suppose the animal uses 50 mL of O_2, measured under the *prevailing* conditions, in an hour. Expressed at STP, the volume used is

$$\text{volume at STP} = 50 \text{ mL} \times \frac{0.92 \text{ atm}}{1 \text{ atm}} \times \frac{273 \text{ K}}{293 \text{ K}} = 42.9 \text{ mL} \qquad \textbf{(C.2)}$$

Thus the animal's rate of O_2 consumption *at STP*, a measure of its metabolic rate, is 42.9 mL O_2/hour.

A significant property of gases is that identical molar amounts of different gases occupy almost identical volumes when at the same pressure and temperature. A mole of gas occupies about 22.4 L at STP, whether the gas is O_2, N_2, or some other gas.

Fitting Lines to Data

Suppose you are interested in how two traits of an animal are related to each other. The traits might, for example, be resting heart rate and age, but here we will call them Y and X for generality. If you make measurements on ten individuals, you will have ten pairs of values that you can plot as in **FIGURE A1** to display visually the relation between the traits. Each point is for one of the studied individuals and represents that particular individual's measures for traits Y and X.

The points plotted seem to fall along a straight line. You could simply sketch a line through them, but that would be subjective. A better approach is to use a statistical procedure to determine which line best fits the data according to objective criteria. The procedure most commonly used to do this is called **linear least squares regression**. This procedure starts by assuming that a linear equation applies. That is, it assumes that the equation for the line of best fit will take the form $Y = a + bX$, where a and b are constants. An investigator must decide first whether that assumption is reasonable. If one doubts the suitability of the linear model, there are advanced statistical procedures that will help one to decide. Once the decision is made to use the equation $Y = a + bX$, the procedure of linear least squares regression provides an objective way to find the values for a (intercept) and b (slope) that position the line most appropriately relative to the data.

A proposed line relating Y to X is drawn in blue in **FIGURE A2**. The red lines in the figure show the vertical distance between each data point and the proposed line. The central proposition (axiom) of least squares regression, approximately stated, is that the best line relating Y to X is the one that minimizes the sum of the lengths of the red lines. Exactly stated, the procedure squares the length of each red line, adds the squares of all the red lines, and determines the best blue line as the one that minimizes this sum of squares (thus "least squares"). You might guess from appearances that the blue line in Figure A2 is not the best line relating Y to X, by this standard. It isn't, but you need not try one blue line after another to find the best. Instead, the calculus of minimization (i.e., minimization of the sum of squares) provides algorithms for carrying out the procedure, and these algorithms are simple enough to be programmed into even cheap calculators today. **FIGURE A3** shows the line that best fits the data by the least squares standard; the line is plotted, and its equation is given in the lower right.

An important reason for understanding the procedure as described is that you will see that the best line is strictly a function of the data (Y, X values) used for the computation. Different data invariably lead to different values for a and b. If one investigator measures Y and X on ten individuals and another measures them on ten different individuals, the investigators will obtain different values for a and b, and different "best" lines. Even the addition of one individual to a data set will change a and b; in **FIGURE A4**, a data point for an eleventh individual is added to the ten data points in Figure A3, resulting in a new line of best fit by the least squares procedure. These facts do not mean that the statistically fitted lines are "just guesses." The facts merely reflect the reality that a fitted line can be no more certain than the data on which the line is based. Ten data points typically leave much room for uncertainty. If you had 100 data points, the uncertainty would be diminished, and a and b would probably be altered only slightly by adding one more data point. Statistical procedures permit one to calculate a numerical measure of the level of uncertainty of a fitted line. Mastery of statistics is in fact crucial for the choice of line-fitting procedures, because there are several nuances involved in applying the procedures correctly.

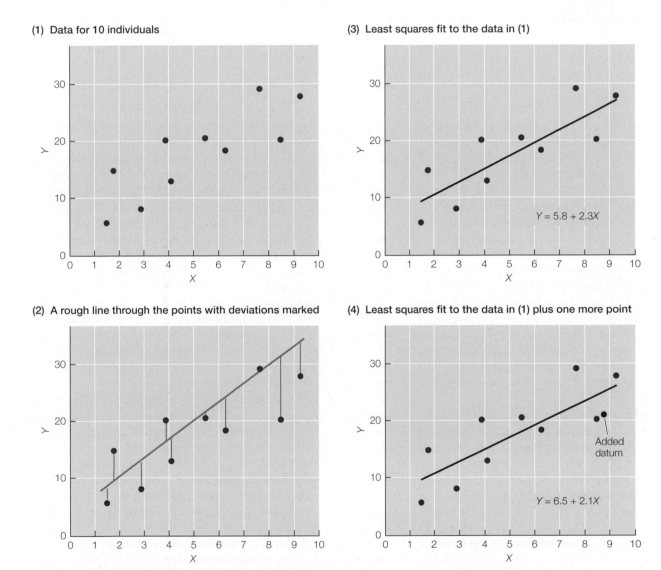

(1) Data for 10 individuals

(2) A rough line through the points with deviations marked

(3) Least squares fit to the data in (1)

$Y = 5.8 + 2.3X$

(4) Least squares fit to the data in (1) plus one more point

Added datum

$Y = 6.5 + 2.1X$

FIGURE A Challenges and procedures of statistical line fitting   (1) Ten X, Y data points measured on ten individuals. (2) A subjectively placed, proposed line (blue) to fit the data points. The red lines show the vertical deviations between the data points and line. (3) The line of best statistical fit to the ten points, based on the linear least squares model. (4) An additional data point has been added, and the line of best statistical fit has been recalculated to take account of the information provided by all 11 data points.

Logarithms

The ruler in **FIGURE A1** is calibrated in a linear scale of numbers of the sort we use in everyday life; each major marking is 1 unit greater than the preceding one. The ruler in **FIGURE A2** is calibrated in a nonlinear, *logarithmic* scale; each marking on the scale is an *order of magnitude* greater than the preceding one. Note that the logarithmic ruler covers a far wider range. That ruler runs between 0.01 and 100,000; for the linear ruler to do the same, it would need to be extremely long, or the markings would need to be extremely close. One reason logarithmic scales are sometimes used by biologists is their wide range. For instance, a physiologist comparing metabolic rates in all mammals would find a logarithmic scale useful because the range of mammalian adult body sizes—from 2 grams to over 4,000,000 grams—is far too wide to fit on any single linear scale. One type of ruler is as good as another in the abstract. One can calibrate a ruler in any way one pleases, provided account is taken of the particular calibration when measurements are interpreted and analyzed. Thus, if a logarithmic ruler suits one's purposes, one is free to use it.

The **common logarithm** of a number N is defined to be the value of L for which $10^L = N$. That is, common logarithms are exponents of 10. The ruler in **FIGURE A3** is the same as that in Figure A2 except that the numbers printed on the ruler are logarithmic values. Where the ruler in Figure A2 says "1000," that in Figure A3 says "3" because 3 is the common logarithm of 1000 ($10^3 = 1000$). Similarly, where the ruler in Figure A2 says "0.1," that in Figure A3 says "–1" because $10^{-1} = 0.1$. The scales (markings) on the rulers in Figure A2 and Figure A3 are logarithmic, but whereas the numbers on the ruler in Figure A2 are ordinary numbers, those on the ruler in Figure A3 are logarithms. Biologists usually use the sort of "hybrid" format in Figure A2—a logarithmic scale labeled with nonlogarithmic, ordinary numbers—rather than an "all-logarithmic" presentation like Figure A3.

The ruler in **FIGURE A4** shows a range of two orders of magnitude in greater detail than in Figure A2. The presentation is hybrid; that is, whereas the markings are logarithmic, the numbers are ordinary. Note that numbers that would be evenly spaced on a linear ruler—such as the numbers 2, 3, 4, and 5—are not evenly spaced on the ruler in Figure A4 because of the logarithmic scaling. Very commonly, logarithmic scales are printed as on the ruler in **FIGURE A5**, with the major markings labeled but the intermediate markings unlabeled. When you see this format, you can determine the meaning of the intermediate markings in the following way: As you move toward higher values between two major markings, the first intermediate mark is 2 times the next lower major mark, the second is 3 times the next lower major mark, and so forth.

Common logarithms are the principal ones used in physiology. However, powers of any number can in principle be used as logarithms. **Natural logarithms**, which are powers of the irrational number e (2.71828…), are second in importance in physiology. The number that is raised to a power to derive logarithms is called the **base** of the logarithms; thus common logarithms are base 10, whereas natural logarithms are base e.

The value x in the expression $\log_x$ is the base of a logarithm; by convention, the expression "log" (written without a subscript) means "$\log_{10}$" and "ln" means "$\log_e$." Logarithms in two different base systems are simply proportional to each other (if m and p are any two bases and c is a constant equal to $\log_p m$, $\log_p N = c \times \log_m N$). Thus *the mathematical nature of a relationship in logarithmic coordinates is not altered by the choice of base*; for example, if a relation is linear when expressed in common logarithms, it remains linear in natural logarithms.

Before the invention of the cheap electronic calculator around 1970, logarithms were essential as computational tools. Today, logarithms are rarely used that way and have two principal applications: graphical and mathematical/statistical, both mentioned in Appendix F.

(1) An ordinary ruler (linear scale)

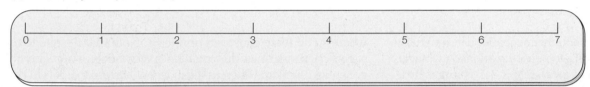

(2) A ruler scaled logarithmically

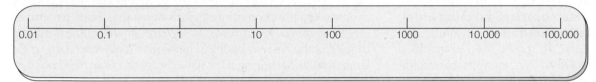

(3) The same ruler as in (2) with values expressed in logarithmic units

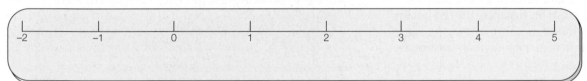

(4) The section of (2) between 1 and 100 with minor marks added

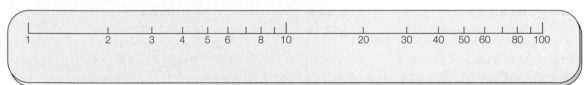

(5) Five orders of magnitude on a logarithmic ruler with only major marks labeled, showing interpretation of three minor marks

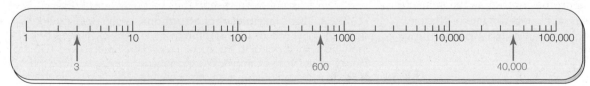

FIGURE A Five rulers illustrating properties and styles of labeling of linear and logarithmic scales of measurement

APPENDIX F

Exponential and Allometric Equations

The two most common types of nonlinear equations in the study of physiology are exponential and allometric equations. A variable Y is an **exponential function** of a variable X if Y changes by a fixed *multiplicative factor* every time X changes by a fixed *additive increment*. An example of exponential increase would be a rate that doubles every time the temperature increases 10°C. In this instance, the multiplicative factor would be 2, and the additive increment would be 10°C. If the actual rate at 0°C is 6 (in arbitrary units), the rate would be 12 at 10°C, 24 at 20°C, 48 at 30°C, and 96 at 40°C, as shown by the blue line in **FIGURE A**. The multiplicative factor need not be an integer; the green line in Figure A, for instance, is constructed from the same assumptions as the blue, except that the multiplicative factor is 1.73. Exponential decrease occurs when the multiplicative factor is less than 1; for instance, the red line in Figure A results when the factor is 0.5.

The general mathematical form of an exponential function is

$$Y = m \times 10^{n \times X} \qquad \text{(F.1)}$$

where m and n are constants. The number 10 is the *base* in this writing of the equation; any base, such as e, could be freely substituted for 10 in the general formulation, although in an actual numerical example, the value of n changes if the base is changed. The exponential equation is commonly applied to cases in which Y is time-dependent and X represents time. In such cases, since $Y = m$ when $X = 0$, m is simply the value of Y at the start (time zero). If n is positive, exponential increase occurs, whereas if n is negative, exponential decrease occurs. Whether increase or decrease is occurring, the rate of change becomes greater as the absolute value of n becomes greater; thus n represents the rate of change.

The **allometric function**—also often called a **power function**—is defined as follows:

$$Y = a \times X^b \qquad \text{(F.2)}$$

where a and b are constants and $b \neq 1$. Note that despite their superficial similarities, this equation is very different from the exponential function. Whereas the independent variable X is in the exponent in the exponential equation, it is raised to a power (thus *power* function) in the allometric equation. The allometric equation first became prominent in biology as a mathematical means of describing nonproportional relative growth of body structures. One of countless examples is the relative growth of the head and the rest of the body in people. The head of a newborn baby is far larger relative to the rest of the newborn's body than the head of an adult person is to the rest of the adult body. Thus, as people grow

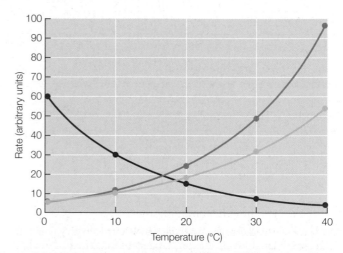

FIGURE A Three exponential equations Each point plotted along the blue line, at 10°C steps from left to right, is 2 times higher than the point preceding. Each point along the green line is 1.73 times higher. These examples of exponential increase illustrate the self-reinforcing nature of exponential change. Each point along the red line is 0.5 times the height of the preceding point, illustrating exponential decrease.

from birth to adulthood, an allometric, rather than proportional, equation describes the relation between head size and body size. Today, the allometric equation is extremely important in physiology, developmental biology, paleontology, and other fields that focus on the *relative scaling of animal traits*. If b is negative in Equation F.2, trait Y decreases as trait X increases (e.g., green line in **FIGURE B**). If b is positive but less than 1, trait Y increases as X increases, but not as fast (e.g., blue line in Figure B). A value of b greater than 1 signifies that Y increases disproportionately rapidly as X increases (e.g., red line in Figure B).

Logarithmic scales are very useful in the study of both exponential and allometric relations. If one takes the common logarithm of both sides of Equation F.1, one gets the *logarithmic form of the exponential equation*:

$$\log Y = \log m + n \cdot X \qquad \text{(F.3)}$$

This equation tells you that for an exponential relation, if you plot the logarithm of Y on one axis of a graph and X itself on the other axis—a plot called *semilogarithmic* because one axis is logarithmic and the other is not—you will get a straight line with slope n and intercept $\log m$.

The *logarithmic form of the allometric equation* is obtained by taking the logarithm of both sides of Equation F.2:

$$\log Y = \log a + b \cdot \log X \qquad \text{(F.4)}$$

This equation reveals that for an allometric relation, plotting the logarithm of Y on one axis and the logarithm of X on the other—a graph termed *log–log* because both axes are logarithmic—yields a straight line with b as slope and $\log a$ as intercept.

The constants in the exponential and allometric equations (n, m and a, b) are predicted from theory in some applications. More usually, however, they are determined empirically: Actual data are gathered on the relation between Y and X, and the data are used to estimate the values of the constants. For this approach, the usual practice is to plot the raw data (X,Y values) on the sort of graph that produces a straight line (semilogarithmic if exponential, log–log if allometric). Linear least squares regression (see Appendix D) is then used to fit a line through the data. Finally, the slope and intercept of the line are used to estimate the equation parameters. R. H. Peters (*The Ecological Implications of Body Size*, Cambridge University Press, Cambridge, 1983) thoughtfully discusses the theoretical details and limitations of this approach.

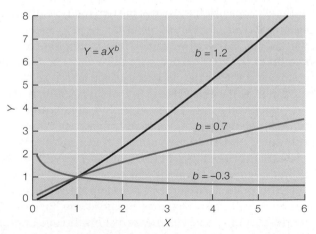

FIGURE B Three allometric equations The parameter a in the allometric equation (Equation F.2) is set equal to 1 in all three lines shown. The three lines differ in the value of b, the exponent in the allometric equation

Phylogenetically Independent Contrasts

The goal of the study of **phylogenetically independent contrasts**, stated simply, is to take phylogenetic information properly into account in the statistical analysis of comparative data. Suppose data have been gathered on 1000 species. The traditional approach to analyzing such data is to treat them as 1000 independent data points. However, when a family tree is available for the set of species, we know that the 1000 data points are not in fact equally independent. Some of the species will be more closely related than others; this means that some of the relations among data will be more likely to be affected by common ancestry than others. An extreme but quick-to-grasp example would be to think of eight friends who happen, actually, to be four pairs of identical twins. You could analyze information on them, such as eye color or muscle mass, as if they are eight independent people, or you could analyze the information with full recognition that some are much more closely related than others. Obviously, you might reach different conclusions by use of the two methods. The expression *phylogenetically independent contrasts* is used to refer both to (1) the general study of how phylogenetic information should be taken into account in the analysis of comparative data and (2) a specific method for such studies developed by Joseph Felsenstein. Here we use the term in its general sense.

Why, specifically, might the use of phylogenetically independent contrasts be important? For many purposes (but not all), the simplest way to understand the importance of this approach is to appreciate that the approach helps to minimize *pseudoreplication artifacts* in the statistical analysis of data.

To see this point, let's start by discussing the problem of *pseudoreplicates*: literally, *false replicates*. Suppose you have two types of animals, M and P, and you measure a physiological property on a single individual of each type. Suppose, moreover, that the individual of type M displays a greater value of the property than the individual of type P. It's obvious that—with only two data points—you could not conclude that *type M* has a greater value than *type P* with a statistically high level of certainty; perhaps the result for the single individual of M is merely a fluke—a statistical oddity. Suppose, however, that you write down the value you measured on the M individual 12 times, and you write down the value for the P individual 12 times, so that now you have 24

values. All 12 of the values for type M will be higher than all 12 of the values for type P, and if you simply enter the 24 values into a statistical test, the test will almost inevitably tell you that there is a statistically significant difference between types M and P. In the terminology of statistics, you have increased your total *sample size* from 2 to 24, and because large sample sizes tend to permit enhanced statistical certainty, this increase in sample size raises the likelihood of reaching a statistically significant conclusion. Of course, we realize in this blatant case that the analysis based on 24 values is totally bogus. You have measures on only two animals. You have not really increased your sample size: The number of useful values cannot be increased merely by writing the numbers down over and over! Most of your replicate values for each type are, in fact, *false* replicates, pseudoreplicates. A **pseudoreplicate** is a "measurement" that masquerades as an independent measure but, in fact, is a duplicate (or partial duplicate) of other measures—it is not fully independent of the other measures.

Pseudoreplication can be *far* more difficult to spot than it is in this blatant example. Sometimes biologists—despite trying to be straightforward and meticulous—are misled by complexities and subtleties into believing they have more independent measures than they actually do. Thus deceived, they use inflated sample sizes in statistical analyses, and they may reach conclusions that, in fact, their data do not support. For good reason, therefore, statistics books and courses warn biologists to be on the alert for pseudoreplication. (Sometimes these warnings are phrased in terms of undesirable "correlation" because pseudoreplicates are correlated with each other in a way—nonindependence—that can confound correct analysis.)

To see how knowledge of family trees is relevant, let's once again imagine that we have 24 pieces of data, but in this case they are actual data gathered by direct measurements on 24 different modern species. For each species, we have measures of a physiological trait T and body weight W. Thus the data plot as 24 points on a graph of T versus W (**FIGURE A**).

If we are interested in the relation between T and W, Figure A gives the impression that we have 24 independent data points relevant to our interest. Suppose, however, that an analysis of the family tree reveals that the 24 species are not in fact independent, but

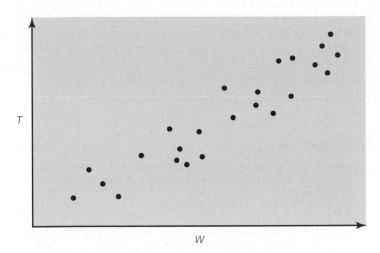

FIGURE A Measured data for 24 modern-day species of a single major type (e.g., 24 birds or 24 beetles) *T* is a physiological trait. *W* is body weight.

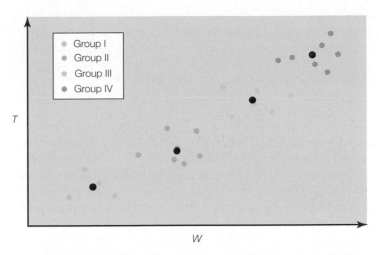

FIGURE C Estimated values (black symbols) for *T* and *W* in the four ancestors of groups I, II, III, and IV The values are reconstructed from analysis of the modern species in each group (colored symbols).

instead—much like the four pairs of twins mentioned before—they fall into four distinct groups based on their evolutionary relationships. Suppose, specifically, that all four species placed in group I in **FIGURE B** are *closely related* descendants of a single common ancestor. Suppose the same can be said of the seven species in group II, the six in group III, and the seven in group IV. Moreover, suppose that the four ancestral species—the common ancestors of groups I to IV—were themselves *distantly related*.

Recognition of this phylogenetic information alters the way in which we might view the raw data in Figure A. Initially we perceived the data as consisting of 24 independent data points. A more valid view might be that the four ancestral species diverged dramatically in traits *T* and *W*, and later—when the modern species evolved from the ancestral species—groups I, II, III, and IV simply retained the properties of their four distinct ancestors, with a relatively small amount of added variation. In this case, if we are trying to understand the fundamental relation between *T* and *W*,

the most accurate view might be that the number of *independent* data points relevant to our interest is 4 (not 24), as seen in **FIGURE C**. From a statistical viewpoint, the four blue data points in Figure B might best be viewed as pseudoreplicates (with a bit of added variation) of the one ancestral value for group I. Similarly, the seven orange data points might best be considered pseudoreplicates of the ancestral value for group II, and so forth.

Why would this matter? **FIGURE D** shows lines put through the two sets of data, the 24 data points in Figure A and the 4 data points (black symbols) in Figure C. Although one of these two lines does not recognize the family tree and the other does, they are almost identical. This is actually a common (although not universal) finding: Regressions and correlations are often similar whether or not phylogenetic information is taken into account. However, the purported *statistical certainty* of the two lines in Figure D is radically different because the sample size for one line is purported to be 24, whereas the sample size for the other is 4.

One goal in the study of phylogenetically independent contrasts is to articulate procedures for statistical analysis that identify and

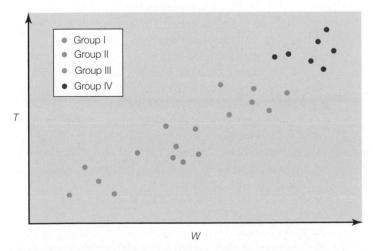

FIGURE B Related groups of species within the set of 24, identified by study of the family tree Each color identifies a distinct group. Each group of modern species is descended from a common ancestor that is only distantly related to the common ancestors of the other three groups.

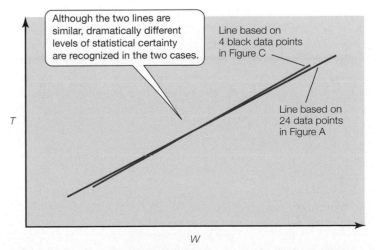

FIGURE D Lines of best statistical fit to the 24 data points in Figure A and the 4 data points (black symbols) in Figure C

use a *valid sample size*. Sample size is important because confidence in the outcome of an analysis depends on it. In the simple example we have used here, the true sample size—the number of *independent* pieces of information—is clearly far less than 24, and it might be as low as 4! This shift in the recognized sample size could cause us to downgrade considerably our confidence in the exact relation we have identified between *T* and *W*.

Methods based on phylogenetically independent contrasts are not without problems. One problem is that each such method makes its own assumptions about the way in which evolution proceeds. A common assumption, for example, is that evolution proceeds along paths (random walks) analogous to those exhibited by molecules undergoing Brownian motion—an assumption that probably applies to certain types of evolutionary change but certainly does not apply to others. A second problem is that the family tree of a group of animals is almost never known with anything approaching 100% certainty. An analysis based on a family tree is thus vulnerable to being rendered second-rate, or even worthless, because at a later date the family tree may well be revised.

Problems such as these help explain why *ordinary least squares regression applied to all available data points* is still commonly used (rather than having been displaced by methods based on phylogenetically independent contrasts). Some thoughtful investigators have decided that—at the present time in the maturation of evolutionary biology—the most prudent course of action is to analyze data by both ordinary least squares regression applied to all data and by methods based on phylogenetically independent contrasts. The two types of methods often agree relatively closely in the trends they identify. If they do not agree, then at least physiologists are put on notice that ambiguity exists.

REFERENCES

Garland, T., Jr., A. F. Bennett, and E. L. Rezende. 2005. Phylogenetic approaches in comparative physiology. *J. Exp. Biol.* 208: 3015–3035. This paper is listed first because it is a good starting point if you want to take the next step toward mastering this demanding topic.

Duncan, R. P., D. M. Forsyth, and J. Hone. 2007. Testing the metabolic theory of ecology: allometric scaling exponents in mammals. *Ecology* 88: 324–333.

Felsenstein, J. 2004. *Inferring Phylogenies.* Sinauer Associates, Sunderland, MA.

McKechnie, A. E., and B. O. Wolf. 2004. The allometry of avian basal metabolic rate: good predictions need good data. *Physiol. Biochem. Zool.* 77: 502–521.

Rohlf, F. J. 2006. A comment on phylogenetic correction. *Evolution* 60: 1509–1515.

Williams, J. B., and B. I. Tieleman. 2001. Physiological ecology and behavior of desert birds. *Curr. Ornithol.* 16: 299–353.

Mitosis and Meiosis

All cells use the information encoded in the genes of their DNA to orchestrate their functions. When cells divide, it is critical that the DNA be duplicated and distributed properly to the daughter cells so that they receive a complete set of operational instructions. To accomplish this function, DNA forms chromosomes that are inherited when the cell replicates itself in cell division. One chromosome includes one complete molecule of DNA, which incorporates many genes. The majority of cells in the body, the **somatic cells**, contain two copies of each DNA molecule, one inherited from each parent. Such cells are referred to as *diploid*, which means having two identical sets of chromosomes. A set of two copies of the same chromosome is called a *homologous pair*. By contrast, **germ cells**—sperm or ova—have only one copy of each DNA molecule, so they are *haploid*, which means having a single set of chromosomes. When a sperm cell and an ovum unite in fertilization, each contributes one member of each homologous pair, and the resulting zygote is diploid.

Animals use two types of cell division. A diploid somatic cell uses **mitosis** to produce two diploid daughter cells. A specialized diploid cell in the testis or ovary uses **meiosis** to produce haploid daughter cells that develop into germ cells. The life cycle of a cell is separated into two main segments: the time during which it is not dividing, which is called **interphase**, and the time during which it is actively dividing. Some cells, once differentiated, don't divide at all. A motor neuron in the vertebrate spinal cord is a good example of such a cell. It remains in interphase continuously. Other cells, such as the epithelial cells lining the gut, divide frequently by mitosis. The figures in this appendix outline the events that occur during mitosis (**FIGURE A**) and meiosis (**FIGURE B**) (see next pages for figures). In both types of cell division, the DNA is present as *chromatin* during interphase. It is incorporated into visible chromosomes only when the cell is dividing. In both types of division,

the DNA is replicated (doubled) while the cell is still in interphase (near the end of interphase), and duplicated chromosomes form in the first phase of cell division. The duplicates of each chromosome, called *sister chromatids*, are joined by a *centromere*.

The main difference between mitosis and meiosis is that a cell undergoing mitosis divides once after doubling its DNA, whereas a cell undergoing meiosis divides twice after doubling its DNA. In mitosis, at the conclusion of telophase (see Figure A), each daughter cell receives the same number of chromosomes as the original cell had prior to division; that is, each cell receives two copies of each chromosome—homologous pairs. At the conclusion of meiosis, each daughter cell receives only one copy of each chromosome (see Figure B), and thus has half the number of chromosomes the original cell had prior to the initiation of meiosis.

A further difference between meiosis and mitosis is that meiosis permits exchange of genetic material between chromosomes. When the duplicated homologous chromosomes align themselves in *tetrads* during late prophase I of meiosis (see Figure B), their close apposition allows parts of the nonsister chromatids to swap with each other. This physical exchange of genetic material, called *crossing-over*, produces chromosomes that have a completely new, unique mix of genes different from both of the original chromosomes. If these chromosomes contribute to the formation of a zygote in fertilization, the new individual will be unique, not only because it contains chromosomes from two different parents, but also because the genetic material of the contributing parents' chromosomes was reshuffled in the process of crossing-over. Biologists have observed that genetic diversity among the members of a species appears to contribute to the endurance of that species over long evolutionary periods. The reasons underlying this adaptive advantage of genetic diversity are being actively investigated.

FIGURE A Mitosis

MITOSIS

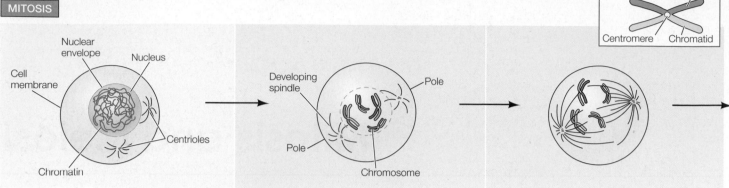

KEY

Chromatid

Centromere   Chromatid

Interphase:
The "unwound" DNA is present as chromatin, not as condensed chromosomes. When the cell prepares to divide, it replicates its DNA.

Prophase:
The nuclear envelope breaks down, the chromatin condenses into chromosomes, and a spindle of microtubules forms at the centromeres at each pole of the cell. For simplicity, we show just four chromosomes—each consisting of two sister chromatids (formed by DNA replication during the lead-up to prophase).

Prometaphase:
The chromosomes attach to spindle fibers at their centromeres.

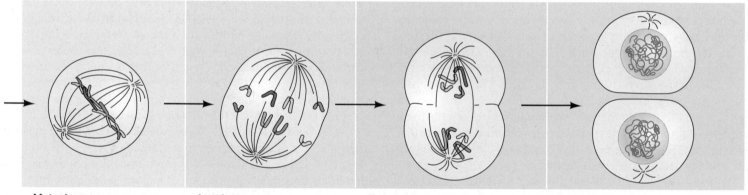

Metaphase:
The chromosomes align at the equator of the cell.

Anaphase:
The sister chromatids are pulled apart into daughter chromosomes.

Telophase:
The chromosomes approach the poles of the cell, and a contractile ring of actin causes the cell membrane to pinch in. Each daughter cell receives a daughter chromosome from each original chromosome, maintaining a full, diploid set of chromosomes in each cell.

Interphase:
The DNA reverts to the form of chromatin rather than discrete chromosomes.

FIGURE B Meiosis

MEIOSIS I

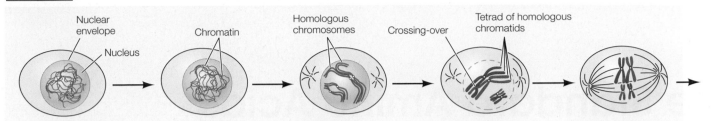

Interphase:
The DNA is present as chromatin rather than condensed chromosomes. Red and blue represent homologous DNA. DNA replicates.

Early prophase I

Mid prophase I:
For simplicity, we show only two pairs of homologous chromosomes. To distinguish the pairs, one is short, whereas the other is long. The two (homologous) chromosomes in a pair are distinguished by coloring one red and the other blue. Because of prior DNA replication, each chromosome consists of two sister chromatids, connected by a centromere (see Key in Figure A). The two members of a homologous pair come together and align with each other.

Late prophase I:
The nuclear envelope breaks down. The four chromatids of each homologous pair represent a tetrad in which physical exchange of chromosomal material can take place between nonsister chromatids. The exchange is called crossing-over.

Metaphase I:
The chromosomes, arrayed in homologous pairs, line up midway between the poles.

MEIOSIS II

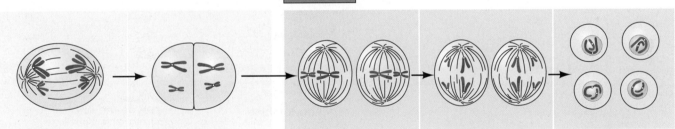

Anaphase I:
One chromosome of each pair is pulled toward each of the poles.

Telophase I:
Each daughter cell receives one already-replicated chromosome from each homologous pair. No further replication occurs.

Metaphase II:
This step is similar to Metaphase I. However, each cell has only one of each type of chromosome.

Anaphase II:
The sister chromatids of each chromosome are pulled toward opposite poles.

Telophase II:
Each of the four final daughter cells is haploid: Each has only one chromosome of each type.

The Standard Amino Acids

The standard amino acids are those employed in the synthesis of most proteins and polypeptides. The 20 most common standard amino acids are listed here, along with their usual abbreviations. There are two systems of abbreviation. In the three-letter system, each amino acid is represented by a three-letter code (the first letter of which is uppercase). In the one-letter system, each is represented by a single uppercase letter. Figure 6.3 shows the chemical structures of five of the standard amino acids.

In addition to the standard amino acids used in protein synthesis, many other amino acids are known and play various roles in organisms.

Names and abbreviations of the 20 most common standard amino acids		
Amino acid	Three-letter code	One-letter code
Alanine	Ala	A
Arginine	Arg	R
Asparagine	Asn	N
Aspartic acid	Asp	D
Cysteine	Cys	C
Glutamic acid	Glu	E
Glutamine	Gln	Q
Glycine	Gly	G
Histidine	His	H
Isoleucine	Ile	I
Leucine	Leu	L
Lysine	Lys	K
Methionine	Met	M
Phenylalanine	Phe	F
Proline	Pro	P
Serine	Ser	S
Threonine	Thr	T
Tryptophan	Trp	W
Tyrosine	Tyr	Y
Valine	Val	V

Basic Physics Terms

This appendix defines and interrelates basic terms in mechanics, the physics of energy, and the physics of electricity. It is deliberately selective rather than comprehensive. Symbols for *units* are roman (upright) and symbols for *variables* are italic (slanted); for example, m is meters and *m* is mass. Note that Appendix A treats the Système International (SI) and other units of measure, and Appendix C discusses gases at standard temperature and pressure. Physical principles not treated here are discussed elsewhere in the text; for example, the physics of gas transfer is discussed in Chapter 22.

Mechanics

Acceleration (*a*) Rate of change of velocity; $a = dv/dt$. Informally, rate of change of speed.

Force (*F*) A push or pull on a body; what it takes to make mass accelerate. $F = ma$ and is measured in *newtons* (N); see Appendix A for the derivation of the newton.

Mass (*m*) The amount of material in an object. The basic SI unit is the *kilogram* (kg), which in the SI is a unit of mass, not of weight. See also *Weight*.

Pressure Force per unit of area. The SI unit of pressure is the *pascal* (Pa; $1 \text{ Pa} = 1 \text{ N/m}^2$). Pressure differences determine direction when materials flow and affect the rate of flow, as discussed in detail in Chapter 25.

Speed See *Velocity*.

Velocity (*v*) Rate of movement; $v = dx/dt$, where *x* is distance and *t* is time. Velocity and speed are similar, except that velocity is a vector quantity that specifies direction as well as speed; speed is scalar and is independent of direction.

Weight (*w*) The force exerted on a mass by Earth's gravitational acceleration; $w = mg$, where *g* is the acceleration due to gravity (see also *Force*). However, as stated in the U.S. National Institute of Standards and Technology *Guide to the SI*, in everyday use and common parlance in the United States, *weight* is used as a synonym for *mass*. When *weight* is used in this way, it is expressed in kilograms.

Energy and related concepts

Energy The capacity to do work. In mechanics, this is divided into kinetic energy and potential energy. Energy is measured in *joules*.

Heat Heat (unlike temperature) is a form of energy: the energy that a substance possesses by virtue of the random motions of its atomic–molecular constituents. The amount of heat in a substance depends on both the number and kinds of atoms and molecules in the substance and on the speed of each. A large copper block with many copper atoms moving at a given average speed contains proportionally more heat energy than a small copper block with fewer atoms moving at the same speed, although they are at the same temperature. Chapter 10 discusses the physics of heat transfer, and of temperature and heat in general. See also *Temperature*.

Kinetic energy (*K*) Energy inherent in the motion of mass; it increases with increasing mass and velocity; $K = 1/2mv^2$.

Light Electromagnetic radiation with a wavelength of about 400–700 nanometers (nm); wavelengths shorter than about 400 nm are ultraviolet radiation, and wavelengths longer than about 700 nm are infrared radiation.

Potential energy Stored energy that can be released to do work. An object that gravity can move downhill, a coiled spring, a battery, and the chemical bond energy in molecules all represent examples of potential energy.

Power Work per unit of time. The standard unit of power is the *watt* ($1 \text{ W} = 1 \text{ J/s}$).

Sound Waves of compression of air or water, which propagate away from a vibrating source.

Temperature The atoms and molecules within any substance undergo constant random motions on an atomic–molecular scale. The temperature of a substance is a measure of the speed—or intensity—of these random motions, and is independent of the macroscopic mass of the substance (although it is dependent on molecular mass).

Work (*W*) The mechanical definition of work is the product of force and distance (*d*); $W = Fd$, measured in *joules*. The mechanical definition of work does not apply in all contexts of work in physiology; physiological work is discussed in Chapter 7, pages 167–170.

Electricity and related concepts

The flow of electricity through electrical circuits is analogous to the hydraulic flow of water or any other liquid through pipes.

Similar relationships and equations describe both systems, and the hydraulic analogy helps to develop an intuitive understanding of electrical concepts. The section in Chapter 12 titled "The Ionic Basis of Membrane Potentials" (starting on page 312) discusses electrical concepts in a physiological perspective.

Capacitance (*C*)   A measure of the ability of a nonconductor to store charge. Capacitance is the amount of charge stored per unit of voltage ($C = Q/E$) and is measured in *farads* (F). A capacitor has two conducting plates separated by an insulator; capacitance increases with increased size of the plates and with decreased distance between them. When a capacitor is first connected to a battery, positive charges flow onto one plate of the capacitor and negative charges onto the other. Although no charges cross from one plate to the other, a transient capacitive current flows because the charges on one plate will repel like charges and attract unlike charges on the other plate. See Figures 12.8 and 12.24.

Charge, electrical (*Q*)   All matter is made up of charged particles (protons and electrons). Atoms or molecules that have unequal numbers of positive and negative charges are called *ions*. Ions of unlike charge attract each other, and ions of like charge repel each other. Charge can be defined as the source of this electrical force of attraction/repulsion. The standard unit of charge is the *coulomb* (C), which is the charge carried by 6.24×10^{18} protons or by a like number of electrons. Charge relates to mass through Faraday's constant (96,500 C/mol of electrons).

Conductance, electrical   The inverse of resistance; the ease of current flow for a given difference of electrical potential or voltage, measured in *siemens* (S).

Current (*I*)   Electric current is the flow of charge, measured in *amperes* (1 A = 1 C/s). In biological systems, currents are carried by ions moving in solution. See also *Charge, electrical*.

Ohm's law $E = IR$. The voltage (difference of electrical potential) is the product of the current and the resistance in a circuit, or by rearrangement, the current equals the voltage over the resistance ($I = E/R$).

Potential difference   See *Voltage*.

Resistance (*R*)   The property that hinders the flow of current. In an analogous hydraulic system, for a given pressure difference, more water will flow through a pipe that presents relatively low resistance to current flow than through a pipe that presents relatively high resistance to current flow. Similarly, in an electrical circuit, the lower the resistance of an element of a circuit, the more current will flow through the element for a given difference of electrical potential or voltage. The unit of resistance is the *ohm* (Ω).

Voltage (*V* or *E*)   Potential difference; the electromotive force or electrical potential. (In common usage these terms mean the same thing.) The potential difference in an electrical circuit is analogous to the pressure difference in a hydraulic circuit. No current will flow between two points at the same pressure, or the same voltage. In a hydraulic circuit, a pump or a difference in height supplies the pressure difference (the latter because water flows downhill due to the force of gravity). In an electrical circuit, a battery separates positive and negative charges to provide the potential difference (acting as a voltage source). Potential difference is measured in *volts* (V); 1 V is the energy required to move 1 C of charge a distance of 1 m against a force of 1 N. Chapter 12, pages 315–319, describes how membrane permeability to ions generates a membrane potential difference.

Summary of Major Bloodborne Hormones in Mammals

We list here the major hormones and neurohormones secreted by endocrine and neuroendocrine tissues in mammals, along with their major functions. Given ongoing advances in technologies to measure blood-borne chemical signals and define their functions, it is likely that researchers will continue to identify hormonal roles for molecules not previously known to be hormones or neurohormones, and discover new functions of known hormones and neurohormones.

REFERENCES

Gardner, D. G., and D. Shoback. 2007. *Greenspan's Basic and Clinical Endocrinology*. McGraw-Hill Medical, New York.

Goodman, H. M. 2009. *Basic Medical Endocrinology*, 4th ed. Academic Press, New York.

Hadley, M. E., and J. E. Levine. 2007. *Endocrinology*, 6th ed. Benjamin Cummings, San Francisco.

Melmed, S. (ed.). 2011. *The Pituitary*, 3rd ed. Academic Press, New York.

Nelson, R. J. 2011. *An Introduction to Behavioral Endocrinology*, 4th ed. Sinauer, Sunderland, MA.

Secreting tissue	Hormone	Class of molecule	Main functions
Adrenal cortex	Aldosterone (mineralocorticoid)	All steroids	Stimulates Na^+ reabsorption and K^+ secretion in kidney.
	Androgens		Act on bone to cause growth spurt at puberty; increase sex drive in females by action on brain.
	Glucocorticoids (cortisol, cortisone, corticosterone)		Part of stress response; affect metabolism of many tissues to increase blood glucose and cause protein and fat catabolism.
Adrenal medulla	Epinephrine and norepinephrine	Catecholamines	Part of stress response; reinforce sympathetic nervous system; influence cardiovascular function and organic metabolism of many tissues.
Anterior pituitary gland (adenohypophysis)	Prolactin (PRL)	All peptides	Promotes development of mammary gland during pregnancy; stimulates milk synthesis and secretion during lactation; promotes caring for young by both males and females in many species of fish, birds, and mammals.
	Growth hormone (GH, somatotropin)		Stimulates growth and metabolism of bone and soft tissues; promotes protein synthesis, glucose conservation, and fat mobilization.
			Stimulates secretion of insulin-like growth factors (IGFs, also called somatomedins) by liver.
	Melanocyte-stimulating hormone (MSH)		In mammals, also produced in hypothalamus; reduces appetite, suppresses immune system. In amphibians, nonavian reptiles, and fish, causes skin darkening by stimulating dispersal of granules containing the pigment melanin.
	Adrenocorticotropic hormone (ACTH, corticotropin)		Stimulates glucocorticoid secretion by adrenal cortex and supports adrenal cortical tissue (tropic action).

(Continued)

Secreting tissue	Hormone	Class of molecule	Main functions
	Thyroid-stimulating hormone (TSH, thyrotropin)		Stimulates hormone synthesis and secretion by thyroid gland; is tropic.
	Follicle-stimulating hormone (FSH)		Stimulates sperm production in the testis, follicular growth in the ovary, and sex hormone production in male and female gonads; is tropic.
	Luteinizing hormone (LH)		Stimulates sex hormone production in male and female gonads, ovulation, and development of corpus luteum; is tropic.
	γ-Lipotropin and β-endorphin		Possibly decrease perception of pain.
Bone osteoblasts	Osteocalcin	Peptide	Regulates metabolism by increasing secretion of insulin, reducing fat storage by adipose cells, and increasing sensitivity of body's cells to insulin (demonstrated in mice). In combination with luteinizing hormone (LH), stimulates testosterone synthesis by Leydig cells in the testis (demonstrated in mice).
Fat (white adipose) tissue	Leptin	Peptide	Influences food intake (promotes weight loss), metabolic rate, and reproductive functions; regulates bone remodeling.
	Adipokines (e.g., adiponectin, apelin, vaspin)		Influence metabolism, inflammation, cardiovascular functions.
Gastrointestinal (GI) tract	Stomach: gastrin; ghrelin ("hunger hormone") Small intestine: secretin, cholecystokinin, glucagon-like peptide-1 (GLP-1), glucagon-like peptide-2 (GLP-2), glucose-dependent insulinotropic peptide; motilin, somatostatin	Peptides	Listed gastrointestinal peptides aid in digestion and absorption of nutrients by various actions on GI tract, liver, pancreas, and gallbladder. In addition, some serve to stimulate secretion of other hormones (e.g., GLP-1 stimulates insulin secretion).
Gonad: ovary and testis	Estrogens	Steroids	Promote maturation of follicle, ovulation, and secondary sex characteristics; prepare uterus for pregnancy by actions during estrous and menstrual cycles; promote closure of epiphyseal growth plate of bone.
	Progesterone	Steroid	Prepares uterus for pregnancy.
	Inhibin	Peptide	Inhibits secretion of FSH from anterior pituitary in males and females.
	Activins	Peptide	Stimulate secretion of FSH; promote spermatogenesis and follicle development.
	Relaxin	Peptide	May make cervix and pelvic ligaments more pliable for parturition; increases sperm motility in semen.
	Anti-Müllerian hormone	Peptide	Promotes differentiation of the bipotential gonad into testis in early embryonic development.
	Androgens	Steroids	Promote sperm production and secondary sexual characteristics; promote sex drive in males; enhance growth at puberty; promote closure of epiphyseal plate of bone.
Heart: cells in atria	Atrial natriuretic peptide (ANP)	Peptide	Promotes Na+ and water excretion in kidney.
Hypothalamus	Releasing and inhibiting hormones	All peptides (except DA)	Stimulate or inhibit secretion of anterior pituitary hormones; releasing hormones are tropic; secreted by parvocellular neurosecretory cells and carried in hypothalamo–hypophysial portal system.
	Corticotropin-releasing hormone (CRH)		Stimulates secretion of ACTH.
	Oxytocin		Stimulates secretion of ACTH.
	Vasopressins (ADH)		Stimulate secretion of ACTH.
	Thyrotropin-releasing hormone (TRH)		Stimulates secretion of TSH and prolactin.

Secreting tissue	Hormone	Class of molecule	Main functions
	Growth hormone–releasing hormone (GHRH)		Stimulates secretion of GH.
	Somatostatin (SS, GHIH)		Inhibits secretion of GH.
	Gonadotropin-releasing hormone (GnRH)		Stimulates secretion of FSH and LH.
	Gonadotropin-inhibiting hormone (GnIH)		Inhibits secretion of FSH and LH.
	Dopamine (DA, prolactin-inhibiting hormone, catecholamine)		Inhibits secretion of prolactin.
	Melanocyte-stimulating hormone–inhibiting hormone (MSH-IH)		Inhibits secretion of MSH.
	Ghrelin		Stimulates GH secretion.
	Kisspeptins		Stimulate GnRH neurons and serve as the main relay signal for sex steroid feedback to the GnRH cells.
	Orexins		Promote wakefulness; modulate reward pathways; stimulate food intake.
Kidney	Renin	Peptide	Cleaves angiotensinogen to form angiotensin I.
	Erythropoietin (EPO)	Peptide	Stimulates red blood cell production in bone marrow.
	Calcitriol (active form of vitamin D, $1,25[OH]_2D_3$)	Steroid	Increases body calcium and phosphate.
Liver	Angiotensinogen	Peptide	Converted in blood to angiotensin II, which stimulates aldosterone secretion, promotes thirst, and causes vasoconstriction and secretion of vasopressin.
	Insulin-like growth factors (IGFs, also called somatomedins)	Peptides	Promote cell division and growth of many tissues (IGFs are also secreted by other tissues, such as muscle).
Pancreas (endocrine cells)	Insulin (β cells)	Peptide	Promotes uptake and storage of nutrients by most cells.
	Glucagon (α cells)		Maintains blood levels of nutrients after a meal and during stress.
	Somatostatin (δ cells)		Inhibits digestion and absorption of nutrients by the gastrointestinal tract.
Parathyroid gland	Parathyroid hormone	Peptide	Increases Ca^{2+} and decreases phosphate levels in the blood by action on kidney and bone; stimulates vitamin D activation by action at kidney nephron.
Pineal gland	Melatonin	Amine	Controls circadian rhythms; seasonal breeding, migration, hibernation; possibly sexual maturity; causes skin lightening in amphibians and lampreys.
Placenta of pregnant female mammal	Estrogens and progesterone	Steroids	Support pregnancy and fetal and maternal development.
	Chorionic gonadotropin	Peptide	Extends functional life of corpus luteum in ovary (horses, primates).
	Relaxin	Peptide	May make cervix and pelvic ligaments more pliable for parturition.
	Placental lactogen	Peptide	May support mammary gland development during pregnancy; changes maternal metabolism to support fetus.
Posterior pituitary gland (neurohypophysis)	Vasopressins (ADH)	Peptides	Regulate water reabsorption at kidney; vasoconstriction; produced by magnocellular neurons in hypothalamus and transported to posterior pituitary.

(Continued)

Secreting tissue	Hormone	Class of molecule	Main functions
	Oxytocin		Stimulates uterine contraction during parturition and milk ejection from mammary glands during suckling.
Skin	Vitamin D	Steroid	Increases body calcium when activated.
Thymus gland	Thymosin, thymopoietin	Peptides	Stimulate T-lymphocyte development and proliferation.
	Calcitonin	Peptide	Decreases blood Ca^{2+} levels in some animals by inhibiting bone resorption (secreted by C cells).
Thyroid gland	Thyroid hormones (iodothyronines): Tetraiodothyronine (T_4, thyroxine) and triiodothyronine (T_3)	Iodinated amines	Increase metabolism of many tissues; necessary for normal growth and development of the nervous system.
	Calcitonin	Peptide	Decreases blood Ca^{2+} levels in some animals by inhibiting bone resorption (secreted by C cells).

Glossary

A

1a Afferent fiber An afferent fiber is the axon of an afferent neuron. 1a afferent fibers are the largest sensory axons in vertebrate peripheral nerves.

A band In striated muscle, the region of a sarcomere that spans the length of the thick filaments. It includes the H zone and regions on both ends where thick and thin filaments overlap. It appears dark in muscle sections prepared for microscopy. It takes its name from the fact that it is described as *anisotropic* because its refractive index in polarized light changes with the plane of polarization.

Absolute refractory period In, e.g., a neuron, the time during and immediately after an action potential in which the voltage threshold is infinite. Thus, no depolarization can exceed threshold and no new action potentials can be initiated.

Absorbed chemical energy See *absorbed energy*.

Absorbed energy In the study of nutritional physiology, the chemical-bond energy of the compounds that an animal absorbs (assimilates) from its digestive tract. Also called *assimilated energy*.

Absorption In the study of nutritional physiology, the entry of organic molecules into the living tissues of an animal from outside those tissues. Absorption includes the entry of materials from the lumen of the gut inasmuch as the lumen is continuous with the outside environment and therefore outside the animal. Also called *assimilation*.

Absorption coefficient The dissolved concentration of a gas when the partial pressure of that gas in solution is 1 atm. See also *Henry's law*.

Absorption efficiency In the study of nutritional physiology, the fraction of molecules or the fraction of chemical-bond energy absorbed from the digestive tract expressed in relation to the amount ingested. Also called *assimilation efficiency*.

Absorption spectrum The absorption of electromagnetic energy by a molecule or other object as a function of the wavelength of the energy.

Accessory heart In an animal with two or more hearts, a heart other than the primary heart. Also called *auxiliary heart*.

Acclimation A chronic response of an individual to a changed environment in cases in which the old and new environments differ in just one or two highly defined ways. A form of *phenotypic plasticity*. Acclimation is a laboratory phenomenon.

Acclimatization A chronic response of an individual to a changed environment in cases in which the new and old environments are natural environments that can differ in numerous ways, such as winter and summer environments, or low and high altitudes. A form of *phenotypic plasticity*.

Acidosis A state in which the pH of the body fluids is excessively acid.

Acoustico-lateralis system A vertebrate sensory system in which the sensory receptors are hair cells and their derivatives. Includes auditory receptors, organs of balance and gravity detection, and the lateral line system of fish and amphibians.

Acrosomal reaction The release of enzymes from the acrosome in the head of a sperm; occurs when the cell membrane of the sperm head binds to species-specific receptor sites on the zona pellucida surrounding the oocyte.

Actin One of the contractile proteins of muscle cells. Globular G-actin monomers polymerize to form the filamentous F-actin of the thin myofilaments. Actin also contributes to motility in many other kinds of cells.

Action potential A brief electrical signal of about 100 mV across the cell membrane of a neuron or other excitable cell. It is initiated by a depolarization above threshold and is propagated to the end of the axon or cell. Also called a *nerve impulse*.

Activation energy The minimal amount of energy a molecule must gain to enter its transition state during a chemical reaction. Without entering its transition state, a molecule cannot react. Enzymes speed reactions by lowering the activation energy, making achievement of the transition state more likely.

Active (exhalation) Exhalation driven by muscle action.

Active (ventilation) See *active change in lung volume*.

Active change in lung volume In the study of ventilation, a change in lung volume driven by muscle action. Contrast with *passive change in lung volume*.

Active evaporative cooling Evaporative cooling that occurs because evaporation is accelerated by some physiological process other than the simple production of heat, such as panting or sweating. See also *evaporative cooling*.

Active site A specific region of an enzyme molecule, at or near the surface of the molecule, where the enzyme binds with its substrate and where the conversion of substrate to product is catalyzed. Also called a *substrate-binding site*.

Active solute secretion Solute secretion by active transport. See *active transport*.

Active transport The transport of a solute across a membrane by a mechanism that is capable of using metabolic energy to cause solute molecules to move across the membrane. An active-transport mechanism is capable of making a solute move away from equilibrium (against its electrochemical gradient). Also called *uphill transport*.

Active zone The region of a presynaptic ending where neurotransmitter molecules are released.

Acute response A response exhibited by an animal during the first minutes or hours after an environmental change.

Adaptation (1) In evolution, a genetically controlled trait that, through the process of natural selection, has come to be present at high frequency in a population because it confers a greater probability of survival and successful reproduction in the prevailing environment than available alternative states. (2) In a sensory receptor, a decrease in the frequency of action potentials in response to a stimulus, during prolonged exposure to the stimulus, even though the stimulus is maintained at a constant level. *Tonic receptors* adapt slowly, and *phasic receptors* adapt rapidly.

Adenohypophysis The nonneural endocrine portion of the vertebrate pituitary gland. It is commonly called the anterior pituitary and includes three parts with variable representation in different species: the pars tuberalis, pars intermedia, and pars distalis.

Adequate stimulus   The kind of stimulus energy to which a receptor is most sensitive, or the kind of stimulus to which it normally responds.

Adipocyte   An animal cell specialized for the storage of fats or oils.

Adrenergic neuron   A neuron that synthesizes and releases norepinenphrine (noradrenaline) or epinephrine (adrenaline) as a neurotransmitter.

Adrenocorticotropic hormone (ACTH)   A tropic peptide hormone secreted by cells in the anterior pituitary gland (adenohypophysis); circulates in the general circulation and controls secretions of all three layers of the adrenal cortex.

Aerobic   Requiring O_2.

Aerobic capacity   A synonym for $\dot{V}_{O_2max}$.

Aerobic catabolic pathways   Catabolic pathways requiring O_2. Usually refers to the Krebs cycle and electron transport.

Aerobic dive limit (ADL)   In diving mammals and birds, the length of the longest dive that can be undertaken without net accumulation of lactic acid above the resting level.

Aerobic expansibility   An animal's peak rate of O_2 consumption during locomotion expressed as a ratio of its resting rate of O_2 consumption. Sometimes used to refer to the same calculation as *aerobic scope*.

Aerobic scope for activity   The differ-ence between an animal's peak rate of O_2 consumption during locomotion and its resting rate of O_2 consumption. Sometimes used to refer to the same calculation as *aerobic expansibility*.

Afferent   Going toward. Thus, for example, an afferent blood vessel carries blood toward an organ of interest. Contrast with *efferent*.

Afferent neuron   A neuron, normally sensory, that conducts signals from the periphery into the central nervous system.

Affinity   See *enzyme–substrate affinity, oxygen affinity*.

Aglomerular   Lacking a glomerulus.

Air capillaries   Minute, gas-filled, blind-ended channels that branch off from the principal respiratory tubules, the parabronchi, within the lungs of a bird and that act as the primary sites of respiratory gas exchange between the air and blood.

Air sacs   Expanded, gas-filled chambers connected to the breathing system in birds or insects. In birds, the air sacs (of which there are usually nine) act as bellows for ventilating the lungs.

Akt1   A signal transduction molecule that stimulates cells to survive and grow; activated by phosphatidylinositol 3-kinase (PI-3 K) when PI-3 K is activated by an extracellular signal. Also called *protein kinase B*.

Aldosterone   A hormone of vertebrates, secreted by the adrenal cortex or homologous interrenal tissue, that increases Na^+ retention by the kidneys and other organs.

Alkalosis   A state in which the pH of the body fluids is excessively alkaline.

All-or-none   Occurring fully or not at all; not graded.

Allometric equation   Two variables, X and Y, are related in an allometric manner when $Y = aX^b$ (b 1). See Appendix F. Also called the *power equation*.

Allosteric activation   Activation of an enzyme by allosteric modulation. See *allosteric modulation*.

Allosteric inhibition   Inhibition of an enzyme by allosteric modulation. See *allosteric modulation*.

Allosteric modulation   (1) In relation to enzymes, modulation of the catalytic properties of an enzyme by the binding of nonsubstrate ligands to specific nonsubstrate-binding sites, called regulatory sites or allosteric sites; a type of *cooperativity*, commonly important in the regulation of enzyme-catalyzed reactions and pathways. (2) In relation to respiratory pigments, modulation of O_2 binding by the binding of ligands other than O_2. See also *cooperativity*.

Allosteric modulator   (1) In relation to enzymes, a nonsubstrate ligand of an enzyme that modulates the catalytic activity of the enzyme by binding to a specific regulatory site on the enzyme molecule. (2) In relation to respiratory pigments, a ligand other than O_2 that modulates binding with O_2.

Allosteric site   See *allosteric modulation*.

Alpha (α) motor neuron   A relatively large motor neuron in a vertebrate spinal cord that innervates extrafusal muscle fibers.

α-γ coactivation   Synaptic activation of α and γ motor neurons together, so that a centrally commanded contraction of a muscle does not inactivate muscle spindle receptors.

Alphastat hypothesis   The hypothesis that in poikilotherms the pH of body fluids is typically regulated at a relatively fixed difference from neutral pH—even as the body temperature changes and the neutral pH therefore also changes—as a way of maintaining a relatively constant state of electrical charge on proteins (particularly on protein imidazole groups).

Alveolar sacs   In the lungs of a mammal, the finest airways, which are lined with alveoli and end blindly.

Alveolar ventilation rate   In the lungs of a mammal, the rate at which air is brought into the alveoli and other respiratory airways (collectively) of the lungs during breathing.

Alveoli (singular alveolus)   (1) In the lungs of vertebrates, the blind-ended terminations of the branchings of the respiratory tract that form the surface for exchange of gases between the air and the blood. In mammalian lungs, each alveolus is a semispherical outpocketing measuring less than 0.5 mm in diameter. (2) In the mammary glands of mammals, the hollow glandular structures that secrete milk.

Amacrine cell   A retinal neuron that mediates lateral antagonistic effects and is sensitive to visual movements.

Ambient   Relating to the surroundings of an organism. Synonymous with *environmental*; for instance, the ambient temperature is the environmental temperature.

Amine hormones   Chemical signals derived from the amino acids tyrosine (catecholamines and iodothyronines) or tryptophan (melatonin).

Ammonotelic   Incorporating most nitrogen from the catabolism of nitrogenous compounds into ammonia. "Most" is defined differently by different authorities; a common approach is to categorize an animal as ammonotelic if 50% or more of the nitrogen released by catabolism is incorporated into ammonia.

Amniotic egg   A vertebrate egg in which the developing embryo is enclosed in the amnion, a sac (extra-embryonic membrane) that contains a bathing amniotic fluid. The amnion evolved and develops in conjunction with other extra-embryonic membranes that support the embryo, notably the chorion, allantois, and yolk sac. The internally developing eggs of mammals are amniotic eggs, as are the cleidoic eggs of birds and other reptiles. See also *cleidoic egg*.

Amphipathic molecule   A molecule that consists of a polar subpart and a nonpolar subpart.

Amplification   In a sequence of enzyme-catalyzed biochemical reactions, an increase in the number of reacting molecules resulting from the fact that one enzyme molecule can catalyze the formation of more than one product molecule.

Amylase   A digestive enzyme that breaks down starch or glycogen.

Anabolism   Constructive metabolism; the set of metabolic processes that build relatively large molecules from smaller molecular building blocks using energy.

Anadromous   Relating to an aquatic animal that undergoes most of its growth in seawater but enters freshwater to breed.

Anaerobe   An anaerobic organism. See *anaerobic*.

Anaerobic Capable of functioning without O_2.

Anaerobic glycolysis The reactions that convert glucose to lactic acid.

Anastomose To form an interconnected network of tubules, vessels, or similar structures by patterns of branching, reconnection, and rebranching.

Anatomical dead space In the lungs of an animal, the sum total of the conducting airways; that is, that portion of the lungs that does not participate in the exchange of respiratory gases between air and blood.

Androgen A masculinizing hormone such as testosterone.

Angiogenesis The formation of new blood vessels (e.g., new capillaries) by sprouting of branches from existing vessels.

Angiotensin converting enzyme (ACE) A membrane-bound enzyme produced by vascular endothelial cells that converts inactive angiotensin I to angiotensin II.

Angiotensin I Formed by the action of renin, which catalyzes the cleavage of amino acids from angiotensinogen; an inactive component of the renin–angiotensin–aldosterone system.

Angiotensin II Formed by the action of angiotensin-converting enzyme; an active component of the renin–angiotensin–aldosterone system; stimulates secretion of aldosterone from the adrenal cortex and vasopressin from the posterior pituitary, and also promotes thirst and systemic vasoconstriction

Angiotensinogen A protein secreted by liver cells that is continuously present in the blood; the first molecule in the renin–angiotensin–aldosterone system.

Anhydrobiosis Survival while dried as fully as possible by desiccation in air.

Animal physiology The study of the functional properties of animals; the study of "how animals work."

Anion A negatively charged ion.

Annotation In the study of genomics, the process of adding interpretive information to gene identities. For example, if a genomic analysis indicates the likely presence of a particular gene that is known in some organisms to code for a detoxification enzyme, stating that the gene is "involved in detoxification" would be an act of annotation.

Anoxia In relation to the tissues of an animal, the state of being devoid of O_2.

Antagonism A type of influence of one substance (such as a hormone) in relation to another in which one opposes the action of the other on a target tissue.

Antagonist An opponent. In biochemical reactions, a substance that opposes the action of another substance. See also *antagonistic muscle pair*.

Antagonistic muscle pair Two muscles, or groups of muscles, that perform coordinated, opposing actions. When one muscle contracts to close the angle of a joint, its antagonist relaxes. To open the joint, the formerly relaxed muscle contracts and the formerly contracted muscle relaxes.

Antennal gland The urine-producing structure of a decapod crustacean. Also called a *green gland*.

Anterior pituitary The nonneural endocrine portion of the vertebrate pituitary gland. It includes three parts with variable representation in different species: the pars tuberalis, pars intermedia, and pars distalis. Also called the *adenohypophysis*.

Anti-Müllerian hormone See *Müllerian inhibitory substance*.

Antidiuresis The production of urine at a low rate. Such urine is usually concentrated as well as scanty.

Antidiuretic Opposing diuresis; promoting antidiuresis.

Antidiuretic hormone (ADH) A hormone that promotes antidiuresis. In mammals, also called *vasopressin*.

Antifreeze compound A metabolically synthesized compound, added to a body fluid, that has the principal function of lowering the freezing point of that body fluid.

Antifreeze protein A protein or glycoprotein that acts as an antifreeze compound. See *antifreeze compound*.

Antioxidants A collective term referring to molecules that prevent or delay damaging oxidation of macromolecules (or other molecules) involved in normal function by reactive oxygen species. See also *reactive oxygen species*.

Antiporter See *countertransporter*.

Aortic body In mammals, a small organ juxtaposed to the systemic aortic arch that senses O_2 in the arterial blood. Not found in humans.

Apical membrane In an epithelial cell, the part of the cell membrane that faces toward the lumen or open space lined by the epithelium of which the cell is a part.

Apical surface In an epithelium, the surface that faces toward the lumen or open space lined by the epithelium. Also called the *mucosal surface*.

Apnea The absence of breathing.

Aquaporin A chemically defined type of water channel, often highly specific for water as opposed to solutes. See also *water channel*.

Arachnid A spider, tick, scorpion, or mite.

Archimedes' principle The principle that states that a body immersed in a fluid is buoyed up by a force equal to the weight of the fluid displaced.

Aromatase The enzyme that converts androgens such as testosterone to estrogens such as estradiol.

Aromatization The process of converting androgens such as testosterone to estrogens such as estradiol, catalyzed by aromatase.

Arousal (1) In the study of hibernation, the emergence of an animal from hibernation, involving the rewarming of tissues to ordinary homeothermic temperatures. (2) In the study of sleep, the transition from sleep to wakefulness.

Arousal threshold In the study of sleep, a measure of the difficulty of waking a sleeping individual at a particular time; a low threshold signifies that sleep terminates relatively easily.

Arteriole A microscopically fine blood vessel with muscular walls that carries blood from arteries to capillaries in a microcirculatory bed of a vertebrate. Contraction and relaxation of the muscular walls controls the rate of blood flow to the capillaries supplied by the arteriole.

Artery A macroscopic blood vessel that carries blood away from the heart.

Asexual reproduction The formation of new individuals without the union of gametes from two different parents. The offspring are genetically identical to their parent.

Assimilated chemical energy In the study of nutritional physiology, the chemical-bond energy of the compounds that an animal assimilates (absorbs) from its digestive tract. Also called *absorbed energy*.

Assimilation In the study of nutritional physiology, synonymous with *absorption*.

Astrocyte A type of glial cell (non-neuronal cell) of the vertebrate central nervous system that regulates extracellular ion concentrations and metabolically supports neurons. Important in the *blood–brain barrier*.

Atmosphere A unit of measure for pressure. An *atmosphere* or, more technically speaking, a *normal atmosphere* is the average pressure exerted by Earth's atmosphere at sea level. It is quantitatively specified by international agreement to be 1.01325×10^5 pascal at 15°C. It is also specified to be equal to the pressure exerted by a column of mercury 760 mm high under standard gravitational acceleration.

ATP synthase A mitochondrial enzyme that synthesizes ATP using energy drawn from the mitochondrial proton electrochemical gradient.

ATPase An enzyme capable of hydrolyzing adenosine triphosphate (ATP) and thereby releasing energy from ATP.

Atresia In the ovary, the process by which an ovum or follicle stops developing and degenerates.

Atrial natriuretic peptide A member of a family of peptide hormones secreted by specialized cells in the atria of the heart that stimulate the excretion of water and sodium.

Atrioventricular (A-V) node In the mammalian heart, a cluster of modified atrial muscle cells that initiate the relay of depolarization from the atria to the ventricles.

Atrioventricular bundle A subpart of the conducting system in which depolarization spreads throughout the vertebrate heart. See *conducting system*.

Atrium In the study of hearts, a heart chamber; in a mammalian heart, one of the two weakly muscular chambers into which blood flows as it enters the heart. In the book lungs of arachnids, the chamber into which the multiple page-like lamellae project.

Atrophy A decrease in the mass of a tissue or organ by the loss of cells or of intracellular components of cells. Also called *wasting*.

Audition Hearing; sound detection and perception.

Autocrine A locally acting chemical signal that binds to receptors and exerts a regulatory effect on the same cell that secreted it.

Autonomic effector An effector other than skeletal muscle; includes smooth and cardiac muscles and tissues of the viscera and exocrine glands.

Autonomic ganglia A collection of peripheral (postganglionic) neurons in the autonomic nervous system; preganglionic neurons synapse onto these postganglionic neurons, which control autonomic effectors.

Autonomic nervous system (ANS) The division of the nervous system that innervates and controls autonomic effectors and conveys sensory information from internal organs.

Autoreceptor In synaptic processes, a receptor protein on a presynaptic terminal that is stimulated by neurotransmitter released by the same neuron.

Autotroph An organism that can obtain the energy it needs to stay alive from sources other than the chemical bonds of organic compounds. Some autotrophs (*photoautotrophs*) use photon energy; others (*chemoautotrophs*) use energy from inorganic chemical reactions. Contrast with *heterotroph*. See also *primary production*.

Autoventilation In an insect, ventilation of the treachae (breathing tubules) of the flight muscles by the flight-muscle contractions.

Auxiliary heart See *accessory heart*.

Average daily metabolic rate (ADMR) An animal's rate of metabolism averaged over all hours of the 24-hour day.

Avogadro's number The number of molecules in a mole; also, the number of independent dissolved entities in an osmole. Equal to 6.022×10^{23}.

Axis In the study of endocrinology, two or more hormone-secreting tissues that together form an hierarchical control system.

Axodendritic synapse A synapse of an axon onto the dendrite of another neuron.

Axon A process of a neuron specialized for conveying action potentials (usually) away from the cell body. An axon may be myelinated (ensheathed in *myelin*) or unmyelinated.

Axon hillock The region of a neuronal cell body from which an axon originates; often the site of impulse initiation.

Axon initial segment The region of an axon close to the cell body: often the site if impulse initiation.

Axonal transport The transport of materials within the cytoplasm of an axon; may be either anterograde (away from the cell body) or retrograde (toward the cell body).

Axosomatic synapse A synapse of an axon onto the soma of another neuron.

B

Baculum A rodlike structure of cartilage or bone that gives stiffness to the shaft of the penis of some species of mammals.

Basal ganglia In the vertebrate brain, a collection of cell groups in the ventral forebrain that organize motor behavior. The *caudate nucleus*, *putamen*, and *globus pallidus* are major components of the basal ganglia; the *subthalamic nucleus* and *substantia nigra* (midbrain) are usually included.

Basal lamina See *basement membrane*.

Basal membrane In an epithelial cell, the part of the cell membrane that faces toward the underlying tissue on which the epithelium rests or to which the epithelium is attached.

Basal metabolic rate (BMR) The metabolic rate of a homeothermic animal when it is in its thermoneutral zone, resting and fasting.

Basal surface In an epithelium, the surface that faces toward the underlying tissue on which the epithelium rests or to which the epithelium is attached. Also called the *serosal surface*.

Basement membrane A thin, permeable, noncellular, and nonliving sheet of matrix material on which an epithelium rests, or that surrounds a cell such as a muscle or fat cell. It is composed of glycoproteins and particular types of collagen and is secreted by cells; for example, the cells of an epithelium help secrete the basement membrane on which the epithelium rests. Also called a *basal lamina*.

Basilar membrane A membranous tissue within the cochlea of the vertebrate ear that contains the auditory sensory hair cells and is vibrated by sound waves.

Basolateral membrane In an epithelial cell, the basal cell membrane plus the lateral cell membranes between the basal membrane and the ring of tight junctions, in contradistinction to the *apical membrane*. See *basal membrane*.

Behavioral fever In a poikilotherm, a modification of behavioral thermoregulation caused by bacterial infection, such that the behaviorally regulated body temperature is higher than usual.

Behavioral thermoregulation (1) In poikilotherms, the maintenance of a relatively constant body temperature by behavioral means. (2) In homeotherms, the use of behaviors to assist in the maintenance of a relatively constant body temperature.

Bends See *decompression sickness*.

Bile salt A type of emulsifying compound, produced by the biliary system in the liver of a vertebrate, that plays a key role in lipid digestion and absorption because of its ability to emulsify lipids within the gut lumen.

Biliary system The system in the liver of a vertebrate that produces bile-containing bile salts.

Bimodal breather See *dual breather*.

Biological clock A physiological mechanism that gives an organism an endogenous capability to keep track of the passage of time.

Bioluminescence Biochemical production of light by cells.

Bipolar cell A type of neuron in the vertebrate retina that mediates the direct (straight-through) pathway connecting rods and cones to ganglion cells.

Bivalve mollusc A clam, mussel, scallop, oyster, or related animal.

Black-body temperature See *radiant temperature*.

Bladder In the study of urine production, an expanded chamber where urine is temporarily stored prior to being excreted. The bladder may exert no effect on the urine other than storing it (e.g., mammals) or may modify urine composition and volume (e.g., amphibians).

Blastocyst A hollow sphere of cells formed in early mammalian development that implants in the uterus. Its inner cell mass will form the embryo and extra-embryonic structures, and its outer trophoblast cells will contribute to formation of the placenta.

Blood The fluid (including suspended cells) that is circulated within the circulatory system.

Blood plasma The part of the blood that remains after blood cells are removed; the part of the blood other than cells.

Blood pressure The extent to which the pressure of the blood exceeds the ambient pressure.

Blood–brain barrier Phenomenologically, a tendency for many substances to exchange much less readily between the blood plasma and the extracellular tissue fluids in the brain than such exchanges in most other tissues. Structurally, a specialized morphology of blood capillaries and glial cells in the brain that interferes with passive transport between blood and extracellular tissue fluid.

Bohr effect A decrease in the O_2 affinity of a respiratory pigment (a shift of the oxygen equilibrium curve to the right) caused by a decrease in pH or an increase in CO_2 partial pressure.

Bohr shift Synonym for *Bohr effect*.

Bomb calorimeter A device used to measure the energy content of organic materials.

Bone marrow In vertebrates, spongy tissue within certain bones, responsible for producing red blood cells.

Book gills Unusual gill structures consisting of many sheets of tissue arrayed like pages of a book; found in horseshoe crabs.

Book lungs Lungs within which sheets of gas-exchange tissue alternate with sheetlike air spaces, like pages of a book slightly separated by air. Found in spiders, scorpions, and some other arachnids.

Bowman's capsule In a vertebrate kidney, the invaginated, blind end of a nephron; the site of formation of primary urine.

Brackish water Water that is intermediate in salinity between seawater and freshwater. One criterion sometimes used is that water is considered brackish if its salinity is between 0.5 and 30 g/kg.

Bradycardia A heart rate that is unusually low.

Brain The anterior enlargement of the central nervous system in an animal with a cephalized nervous system.

Branchial Relating to gills.

Breathing The process by which an animal gains O_2 from its environment and voids CO_2 into its environment. The environment may be air (air breathing) or water (water breathing).

Brown adipose tissue (BAT) A metabolically active form of adipose tissue, known only in mammals, that serves as the primary site of mammalian nonshivering thermogenesis and expresses a distinctive mitochondrial protein, *uncoupling protein 1* (UCP1). Also called *brown fat*. Contrast with *white adipose tissue*.

Brown fat Synonym for *brown adipose tissue*.

Brush border An apical epithelial surface bearing microvilli.

Buccal pressure pump The development of positive pressure within the buccal cavity of a vertebrate, used to force air into lungs or water across gills.

Buffer reaction A chemical reaction in a solution that tends to stabilize the pH of the solution by removing H^+ from the solution when H^+ is added by some external process and by adding H^+ to the solution when H^+ is removed by an external process.

Bulk flow Gross flow of a fluid (gas or liquid), such as occurs in a wind or water current.

Bulk solution Solution that is not immediately next to a membrane or other surface.

Bundle branches Subparts of the conducting system in which depolarization spreads throughout the vertebrate heart. See *conducting system*.

Burst exercise Sudden, intense exercise.

C

Ca^{2+}-ATPase An ATPase enzyme that pumps Ca^{2+} across a membrane. See *ATPase*.

Caching behavior The storage of food or other material for later use.

Caisson disease See *decompression sickness*.

Calmodulin (CaM) An intracellular protein that binds calcium ions to form a Ca–CaM complex, which can then act as an intracellular messenger to affect other proteins.

Calorie (cal) A unit of energy equal to the amount of heat required to raise the temperature of 1 g of water by 1°C, from 14.5°C to 15.5°C. Some people outside science distinguish the calorie from the kilocalorie (kcal; 1000 calories) by writing "calorie" (small "c") to represent the calorie and "Calorie" (large "c") to represent the kilocalorie. Sometimes, in fact, in writing outside science, "calorie" (small "c") is used to refer to the kilocalorie; thus one must be cautious in interpreting uses of "calorie" outside science.

Calorigenic effect of ingested food Synonym for *specific dynamic action*.

Camera eye An eye that optically resembles a camera, with a lens that focuses an image on a retina of light-sensitive cells.

cAMP See *cyclic adenosine monophosphate*.

Capacitance An electrical term meaning the ability of a capacitor or a capacitor-like structure, such as a cell membrane, to store electrical charges. A cell membrane acts like a capacitor because of its electrically insulating properties. Capacitance (C, in farads) is a measure of the amount of charge stored per unit of voltage. See *capacitor*.

Capacitance coefficient (β) In the study of respiratory gas exchange, the change in total gas *concentration* per unit of change in gas *partial pressure* in air, water, or a body fluid like blood. The capacitance coefficient sometimes differs from simple gas-solubility coefficients such as the absorption coefficient because in the case of the capacitance coefficient, the gas concentration includes bound gases, such as O_2 bound to hemoglobin.

Capacitation The final maturation of sperm that takes place in the female reproductive tract to make the sperm capable of rapid forward swimming and fertilization.

Capacitor Two conducting plates separated by an insulating layer. If the plates are close enough to each other, charges on one plate can electrostatically attract or repel charges on the other plate, even if charges cannot cross the insulation. A capacitor can store electrical charge.

Capillary A microscopically fine blood vessel, the wall of which consists of only a single layer of epithelial cells. Capillaries are the principal sites of exchange between blood and other tissues in a closed circulatory system.

Carbamate groups Chemical structures formed when CO_2 reversibly combines with amino groups of hemoglobin or other proteins.

Carbon dioxide dissociation curve See *carbon dioxide equilibrium curve*.

Carbon dioxide equilibrium curve In relation to the CO_2-carrying properties of blood, a graph of the *total carbon dioxide concentration* as a function of the CO_2 partial pressure of the blood. Also called the *carbon dioxide dissociation curve*.

Carbonic anhydrase (CA) An enzyme that accelerates the conversion of CO_2 and H_2O into HCO_3^- and H^+, or the reverse reaction.

Cardiac ganglion In arthropods, a ganglion, located in the heart, that is composed of the pacemaker neurons and other neurons that stimulate rhythmic contractions of the heart muscle cells.

Cardiac muscle Muscle that forms the wall of the heart. In vertebrates, cardiac muscle consists of branched, generally uninucleate, striated muscle cells that are connected by intercalated discs. The contractions of the cells are initiated by endogenously generated myogenic action potentials and may be modified by neural and hormonal factors. The cardiac muscle of invertebrates may consist of striated or nonstriated cells that may be endogenously active (myogenic) or controlled by neural input (neurogenic).

Cardiac output The volume of blood pumped by a heart per unit of time, calculated as the stroke volume multiplied by the number of beats per unit of time (heart rate). In the case of the mammalian or avian heart, the cardiac output is specifically the output of the left ventricle into the systemic aorta unless stated otherwise.

Carotid bodies In mammals, two small organs juxtaposed to the carotid arteries that sense O_2 in the arterial blood.

Carrier See *transporter*.

Carrier-mediated transport Solute transport across a membrane that requires the reversible, noncovalent binding of solute molecules with a protein (called a *transporter* or *carrier*) in the membrane. There are two kinds of carrier-mediated transport: *active transport* (which employs metabolic energy) and *facilitated diffusion* (which does not).

Catabolism Destructive metabolism; the set of metabolic processes by which complex chemical compounds are broken down to release energy, create smaller chemical building blocks, or prepare chemical constituents for elimination.

Catadromous Relating to an aquatic animal that undergoes most of its growth in freshwater but enters seawater to breed.

Catalyst A compound that facilitates a chemical reaction—in which covalent bonds are made or broken—without, in the end, being modified by the reaction. *Enzymes* are catalysts.

Catalytic rate constant (k_{cat}) The number of substrate molecules a unit of enzyme is capable of converting to product per unit of time when the enzyme is saturated.

Catalytic vacuole The part of an enzyme molecule that provides a suitable setting for catalysis during the conversion of substrate to product; includes, but is not limited to, the *active site*.

Catecholamines Chemical signals derived from the amino acid tyrosine that serve as neurotransmitters and also as three amine hormones (epinephrine, norepinephrine, and dopamine).

Cation A positively charged ion.

Cell body The portion of a neuron that contains the cell nucleus; also called the *soma* or *perikaryon*.

Cell membrane The membrane that encloses an animal cell and forms the outer boundary of the cell. Also called the *plasma membrane*.

Cell signal transduction The processes by which cells alter their intracellular activities in response to extracellular signals.

Cell theory The theory, developed in the nineteenth century, that organisms are composed of cells, which act as structural, functional, and developmental units of organization.

Cell-volume regulation The process by which the volume or size of a cell is kept stable.

Cellular oscillator A neuron that can generate a rhythmic change in membrane potential or activity without needing synaptic input to do so; one possible basis of a central pattern generator.

Central circulation The heart and the veins and arteries that connect immediately to the heart.

Central control Control of motor output to produce rhythmic movements, for which the pattern of activity depends on the CNS, without requiring temporally patterned sensory input. Contrast with *peripheral control*.

Central nervous system (CNS) The consolidated integrative part of an animal's nervous system; in vertebrates, consists of the brain and spinal cord.

Central pattern generator (CPG) A neural circuit (or a single neuron) that generates a behaviorally significant pattern of motor output in space and time without requiring temporally patterned sensory input.

Centralization Over the course of evolution, the tendency of animal groups to concentrate integrative neural functions into a central nervous system.

Cephalization The concentration of structures of the nervous system toward the anterior end of an animal, a trend underlying the evolution of anterior brains in many animal groups.

Cephalopod mollusc An octopus, squid, cuttlefish, or related animal.

Cerebellar cortex The outer layer of the cerebellum of the vertebrate hindbrain; involved in motor coordination and learning.

Cerebellum A prominent structure of the vertebrate hindbrain, concerned with motor coordination, posture, and balance. Consists of the *cerebellar cortex* and *deep cerebellar nuclei*.

Cerebral cortex The outer part of the cerebral hemispheres of the vertebrate forebrain; greatly enlarged in birds and mammals.

cGMP phosphodiesterase (PDE) An enzyme that hydrolizes cyclic GMP to 5′-GMP.

cGMP See *cyclic guanosine monophosphate*.

Channel A membrane protein that aids the passive transport of a solute across a membrane without undergoing any sort of chemical binding with that solute. Channels participate particularly in passive transport of inorganic ions across membranes. See also *water channel*.

Channel-mediated water transport Water transport by osmosis through a water channel. See *water channel*.

Chaperones See *molecular chaperones* or *chemical chaperones*.

Chemical chaperones Synonym for *counteracting solutes*.

Chemical energy (chemical-bond energy) Energy that is liberated or required when atoms are rearranged into new configurations. Animals obtain the energy they need to stay alive by reconfiguring atoms in food molecules, thereby liberating chemical energy.

Chemical potential Qualitatively speaking, the strength of the tendency of a chemical substance to undergo a physical or chemical change. Measures of chemical potential are useful for predicting the direction of change because chemical substances tend to go from high chemical potential to low chemical potential.

Chemical synapse A synapse that for signal transmission employs a chemical neurotransmitter that is released presynaptically and acts on postsynaptic neurotransmitter receptors.

Chemical-bond energy Energy that is liberated or required when chemical bonds are broken or made. Animals obtain the energy they need to stay alive by breaking chemical bonds in food molecules, thereby liberating chemical-bond energy.

Chemiosmotic hypothesis A hypothesis concerning the biochemical relationship between electron transport and oxidative phosphorylation in mitochondria. According to the hypothesis, electron transport pumps protons into the mitochondrial intermembrane space, and the back-diffusion of the protons through ATP synthase results in ATP formation.

Chemoautotroph An organism that can obtain the energy it needs to stay alive from energy-yielding inorganic chemical reactions.

Chemoreception A sensory response to a chemical stimulus. Chemoreception includes *taste* (also termed the *gustatory sense*) and *olfaction* (the sense of smell), as well as other chemical sensitivities.

Chemosynthetic autotroph Synonym for *chemoautotroph*.

Chloride cell See *mitochondria-rich cell*.

Chloride shift In vertebrate blood, diffusion of Cl^- from the blood plasma into the red blood cells in exchange for diffusion of HCO_3^- out of the cells, or the reverse process. Mediated by a membrane countertransporter often called the *band 3 protein*.

Chlorocruorin A type of respiratory pigment found in certain marine annelid worms, formed by the combination of hemelike structures with protein. Undergoes reversible combination with O_2 at the hemelike loci.

Cholinergic neuron   A neuron that synthesizes and releases acetylcholine as a neurotransmitter.

Chorionic gonadotropin (CG)   A hormone secreted by the embryonic placenta in horses and primates that maintains the function of the corpus luteum beyond the time when it would degenerate if pregnancy did not occur.

Chromatophore   (1) In most animals, a type of cell containing pigment granules that can undergo changes in dispersion, thereby altering their influence on the color of the skin or other structure; a color-change cell. (2) In cephalopod molluscs (e.g., squids), the same term is used to refer to a very different color-change structure, namely an organ composed of a pigment cell and many muscle cells that can change the size of the pigment cell.

Chronic response   A response expressed by an animal following days, weeks, or other prolonged exposure to new environmental conditions.

Ciliary photoreceptor   A photoreceptor cell in which the light-sensitive part is a modified cilium; characteristic of vertebrates. Contrast with *rhabdomeric photoreceptor*.

Circadian oscillator   The biological clock that times a circadian rhythm.

Circadian rhythm   An endogenous rhythm with a period of about a day (ca. 24 hours).

Circannual rhythm   An endogenous rhythm with a period of about a year.

Circatidal rhythm   An endogenous rhythm with a period approximating a tidal cycle (ca. 12.4 hours).

Circulation   The pressure-driven mass flow of blood through a system of tubular vessels or other passages that brings the blood to all parts of the body.

Circulatory system   The blood and the system of vessels or other passages through which it circulates.

Classical conditioning   A form of associative learning in which an animal learns the association between two stimuli.

Clearance   See *plasma clearance*.

Cleidoic egg   Metaphorically, a cleidoic egg is a "locked box" (*cleido*, "key") (in contrast to an egg that exchanges nutrients with its surroundings throughout development). In a cleidoic egg—exemplified by the eggs of birds and other reptiles—everything needed by an embryo to develop from a single cell to a hatchling is present within the shell of the egg. The egg includes all nutrients required by the embryo, a method of gas exchange, and mechanisms to store or eliminate metabolic wastes. See also *amniotic egg*.

Clone   A new individual, produced by asexual reproduction, that is genetically identical to its parent, or a group of such individuals. Used as a verb, *to clone* means to make an exact genetic copy.

Closed circulatory system   A circulatory system in which the blood is enclosed within blood vessels throughout and is therefore distinct from the interstitial fluids. Contrast with *open circulatory system*.

Cochlea   A part of the inner ear of many vertebrates, coiled in mammals, that contains the auditory sensory hair cells.

Cochlear amplifier   An active physiological process that amplifies the movement of the basilar membrane in the cochlea in response to sound.

Cocurrent   Referring to two fluids flowing in the same direction. Also called *concurrent*.

Collecting duct   In a vertebrate kidney, a duct that receives fluid from multiple nephrons and carries it to where the fluid will exit the kidney. Activities of the collecting duct epithelium typically modify fluid composition and volume along the way.

Colligative properties   The properties of an aqueous solution that depend simply on the number of dissolved entities per unit of volume, rather than on the chemical nature of the dissolved entities. The three principal colligative properties in animal physiology are osmotic pressure, freezing point, and water vapor pressure.

Colloid osmotic pressure   The difference in osmotic pressure that arises between two solutions on either side of a cell membrane or epithelium because the two solutions differ in their concentrations of nonpermeating protein solutes. Because the solutes responsible for colloid osmotic pressure cannot cross the membrane or epithelium, the colloid osmotic pressure represents a "fixed" difference in osmotic pressure between the solutions. Also called *oncotic pressure*.

Commissure   A bundle of axons that connects the two sides of a bilaterally symmetrical central ganglion or bilateral regions of a central nervous system.

Compact myocardium   Myocardium in which the muscle cells are so closely packed together that the tissue is not perfused by the blood flowing through the lumens of the heart chambers, and thus the tissue must be perfused by a coronary circulation. The myocardium of a mammal is compact.

Comparative method   A method of analysis that seeks to identify adaptive traits or adaptive evolutionary trends by comparing how a particular function is carried out by related and unrelated species in similar and dissimilar environments.

Compass   A mechanism that indicates geographical direction on the Earth's surface; most human-made compasses are magnetic.

Compass direction   North, south, east, west, or a combination of these; the direction of a movement or orientation as it could be described using a compass.

Compatible solute   A solute that, when concentrated enough to contribute significantly to the osmotic pressure of a fluid, has little or no effect on the structure and function of macromolecules with which the fluid is in contact.

Compensation   In the study of acclimation or acclimatization, the return of a physiological property toward its value that existed prior to an environmental change even though the animal remains in the changed environment. Compensation is *partial* if the physiological property returns only partly toward its preceding value. It is *complete* if the physiological property returns to the value that existed prior to the environmental change.

Complete compensation   See *compensation*.

Complex cell   A neuron found in the mammalian primary visual cortex that has an orientation-selective receptive field but lacks distinct subparts excited or inhibited by reception of light. A complex cell responds to a bar or edge of a certain size and orientation, anywhere in the cell's receptive field.

Compound eye   A multifaceted eye characteristic of arthropods, composed of many individual optical units called *ommatidia*.

Concentration gradient   Technically, the difference in the concentration of a solute between two places divided by the distance separating those two places. Often used more loosely to refer simply to a difference in concentration.

Concentric muscle contraction   An isotonic muscle contraction in which the muscle shortens while generating force.

Concurrent   See *cocurrent*.

Condensation   A change in the physical state of a compound from a gas to a liquid.

Conductance   A measure of how easily electrical current will flow through a conductive pathway. Contrast with electrical *resistance*. See also *thermal conductance*.

Conducting airways   In the lungs of an animal, the airways that do not participate in the exchange of respiratory gases between air and blood, but rather simply conduct air from one place to another.

Conducting system   A system of specialized muscle cells by which depolarization spreads throughout the vertebrate heart.

Conduction   (1) In the study of heat, the transfer of heat by intermolecular collisions through a material that is macroscopically motionless. (2) In the study of heart physiology, the process by which depolarization spreads through the vertebrate heart or any other myogenic heart. (3) In the study of electrical phenomena, the transmission of electrical currents or signals through a conductive pathway.

Cone　A type of photoreceptor in the vertebrate retina. Cones are smaller and less light sensitive than rods and are used for diurnal vision and color vision.

Conformity　A state in which an animal's internal conditions match the external environmental conditions.

Connective　A bundle of neuronal axons in the central nervous system that connects central ganglia; found in the ganglionic nervous systems of arthropods, annelids, and molluscs.

Connexin　The protein that makes up connexons.

Connexon　The protein channel of a gap junction at which cytoplasmic continuity is established between two adjacent cells. Two connexons, positioned respectively in the cell membranes of the two cells, form the channel. Channels of this sort electrically couple cells (permitting current flow between them) and permit small molecules to move between cells.

Constitutive enzyme　An enzyme (or other protein) that is always expressed in a tissue. Contrast with *inducible enzyme*.

Consumed energy　Energy that was initially in the form of chemical energy or another form of high-grade energy but that has been converted to heat.

Consumption, of energy　Conversion of chemical-bond energy to heat or external work.

Contact chemoreceptor　A chemosensory cell (of a terrestrial animal) that is normally stimulated by chemicals that contact it in a liquid. Contrast with *distance chemoreceptor*.

Continuous breathing　In a terrestrial vertebrate, inhaling and exhaling continuously (without significant interruptions).

Contraction　The condition in which a muscle is activated to produce force. In a skeletal muscle, contraction can be shortening (isotonic), isometric (remaining the same length), or lengthening (isotonic).

Control system　A system that sets the level of a particular variable that is being controlled. To do so, it uses information from sensors to determine signals it sends to effectors that can modify the controlled variable. Control systems often (but not always) operate on negative feedback and are stabilizing.

Controlled variable　In speaking of a control system (e.g., home thermostat), the variable that the system controls (room temperature in the case of a home thermostat).

Convection　Always refers to mass flow, but has different specific meanings depend-ing on context. (1) In the study of heat transfer, *convection* is the transfer of heat by the mass flow of a material substance (e.g., wind). (2) In the study of material transport, *convection* refers to (i) the flow of a fluid from place to place and (ii) the transport of molecules in the fluid from place to place by the fluid flow. Blood flow, for example, is a type of convection, and when the blood carries a material such as O_2 from place to place, the material is said to undergo convective transport.

Convective gas transport　Transport of a gas by bulk flow (e.g., by a wind).

Convergence　Coming together. In neurophysiology, a pattern in which signals from many presynaptic neurons come together to affect a particular postsynaptic neuron. Contrast with *divergence*.

Cooperativity　A type of chemical behavior that occurs in protein molecules that have multiple ligand-binding sites, in which the binding of any one site to its ligand may facilitate or inhibit the binding of other sites on the same molecule to their ligands. The binding sites do not interact directly; instead, binding at one site induces protein conformational shifts that affect other sites at a distance.

Coronary artery　An artery that carries blood into the myocardium of a heart.

Coronary vein　A vein that carries blood out of the myocardium of a heart.

Corpus allatum (plural corpora allata)　One of a bilateral pair of organs in insects that serve both as neurohemal organs, where prothoracicotropic hormone (PTTH) is released from axon terminals of neurosecretory cells in the brain, and as nonneural endocrine tissue, which secretes juvenile hormone.

Corpus luteum (plural corpora lutea)　An endocrine structure in the vertebrate ovary formed by reorganization of the cells of an ovarian follicle that has undergone ovulation. In mammals, it secretes progesterone, estrogen, and inhibin. If fertilization occurs, it remains active during pregnancy; if not, it degenerates.

Corpus luteum of pregnancy　A corpus luteum that remains active for an extended period of time because pregnancy prevents it from degenerating. See *corpus luteum*.

Corpus luteum of the cycle　A corpus luteum that undergoes spontaneous degeneration after a brief period of being active. See *corpus luteum*.

Cortex　The outer layer of something (such as the cerebrum or the adrenal gland), as opposed to the medulla, or inner portion.

Corticotropin-releasing hormone (CRH)　A tropic peptide chemical signal secreted by hypothalamic parvocellular neurosecretory neurons; travels in hypothalamo–hypophysial portal vessels to the anterior pituitary gland and stimulates secretion of adrenocorticotropic hormone.

Cost of transport　For an animal that is undergoing directional locomotion, the energy cost of covering a unit of distance.

Costal suction pump　The development of negative pressure within the thorax of an air-breathing vertebrate by action of the costal muscles between the ribs, used to suck air into the lungs.

Costamere　In muscle fibers, a complex of protein molecules that serves as a region of attachment between the myofibrils, sarcolemma, and extracellular matrix. Costameres are organized into many parallel bands that circumscribe each muscle fiber at regular intervals coincident with the Z discs.

Cotransmitter　In neurons that synthesize and release more than one kind of neurotransmitter molecule, the second kind of neurotransmitter.

Cotransporter　A transporter protein that obligatorily carries two different solutes in the same direction simultaneously. Through the participation of a cotransporter, the diffusion of one solute in the direction of its electrochemical gradient can cause a second solute to move away from equilibrium. Also called a *symporter*.

Counteracting solutes　Osmolytes that act in teams of two or more to modify the osmotic pressures of body fluids without greatly perturbing macromolecules because the individual members of the teams have mutually offsetting effects on the macromolecules.

Countercurrent　Referring to two fluids flowing in opposite directions.

Countercurrent exchange　Exchange of heat, O_2, or other substances by passive-transport processes (e.g., diffusion) between two closely juxtaposed fluid streams flowing in opposite directions (e.g., blood flowing in opposite directions in two blood vessels).

Countercurrent multiplication　A process that occurs in a system consisting of two juxtaposed fluid streams flowing in opposite directions, in which metabolic processes produce a difference (termed the *single effect*) between adjacent parts of the two streams. Because of the dynamics of such a system, the difference produced between adjacent parts of the two streams is multiplied to create a much larger difference between the two ends of the system.

Countertransporter　A transporter protein that obligatorily carries two different solutes in opposite directions simultaneously. Through the participation of a counter-transporter, the diffusion of one solute in the direction of its electrochemical gradient can cause a second solute to move away from equilibrium. Also called an *antiporter*.

Coupling   In the study of cell energy transduction, the use of energy released in the electron-transport chain to synthesize adenosine triphosphate (ATP). It can be graded; i.e., the two processes—electron transport and ATP synthesis—can be tightly or loosely coupled.

Covalent bond   A chemical bond in which atoms fully share electrons. Covalent bonds are strong and, for the most part in biological systems, require enzyme catalysis to be made or broken.

Covalent modulation   Modulation of the catalytic properties of an enzyme, or the functional properties of another type of protein, by chemical reactions that make or break covalent bonds between a modulator and the enzyme or other modulated protein. The most common mode of covalent modulation is phosphorylation and dephosphorylation catalyzed by protein kinases and protein phosphatases. Also called *covalent modification*.

Cranial nerves   Peripheral nerves that connect to the brain.

Critical temperature   (1) In the study of poikilotherms, a body temperature (high or low) at which animals have little or no ability to increase their rate of O_2 consumption above their resting rate—making them incapable of much physical activity. (2) In the study of homeotherms, an ambient temperature that defines the upper or lower limit of the thermoneutral zone; see also *lower-critical temperature* and *upper-critical temperature*.

Cross-bridge   The head of a myosin molecule interacting with actin molecules to produce muscle contraction by repeated oarlike power strokes. Each myosin molecule has two heads, and each head has an actin-binding site and an enzymatic site that binds and hydrolyzes ATP to liberate energy to fuel each power stroke.

Cross-current exchange   A type of exchange between two fluid streams, most commonly found in breathing organs in which O_2 and CO_2 are exchanged between air and blood. During cross-current exchange of this sort, the total blood flow is broken up into multiple separate streams, each of which "crosses" the flow of air at a particular point, thereby exchanging respiratory gases with the air at just one point along the path of airflow. The streams then coalesce to reestablish a unified total blood flow.

Cryptobiosis   Latent life. A resting state in which there is little or no metabolism, so that there are few, if any, signs that the organism is still alive.

Cryptonephridial complex   A specialized, close association between the Malpighian tubules and rectum that occurs in certain types of insects (e.g., mealworms) and plays a role in the production of concentrated urine and the absorption of water vapor from the atmosphere.

Current, electric   The flow of electrical charge.

Cutaneous   Relating to the skin.

Cuticle   The exoskeleton of an arthropod, or the material of which the exoskeleton is composed.

Cyclic adenosine monophosphate (cyclic AMP, cAMP)   A second messenger produced intracellularly in response to several neurotransmitters and hormones.

Cyclic guanosine monophosphate (cyclic GMP, cGMP)   An intracellular messenger in rod and cone photoreceptors and a second messenger in some neurotransmitter actions.

Cytochrome P450 enzymes   A set of detoxification enzymes that are upregulated when an animal is exposed to toxins.

Cytokines   Peptide regulatory molecules that are involved in cell development and differentiation and in immune responses.

Cytoskeleton   Intracellular structural support elements (e.g., microtubules, intermediate filaments, actin microfilaments) composed primarily of fibrous protein polymers.

Cytosol   See *intracellular fluids*.

D

$D_2{}^{18}O$ method   See *doubly labeled water method*.

Daily torpor   In mammals and birds, a form of controlled hypothermia in which the body temperature is able to approximate ambient temperature for part (but only part) of each 24-hour day, generally on many consecutive days.

Dale's principle   The idea that a neuron releases the same kind of neurotransmitter at all its endings. With the discovery of cotransmitters, Dale's principle may be revised to say "the same kind(s) of neurotransmitters."

Dalton (Da)   A unit of measure for atomic and molecular mass, equal to 1/12 the mass of an atom of the most abundant carbon isotope, ^{12}C. Thus the mass of a ^{12}C atom is 12 daltons.

Dark adaptation   Increase in light sensitivity of a photoreceptor or visual system after a period in the dark.

Dark current   The ionic current, carried mainly by Na^+ ions, that flows into the outer segments of vertebrate photoreceptors in the dark. Light absorption leads to closing of the Na^+ channels, turning off the dark current and hyperpolarizing the photoreceptor.

Data logger   A small microcomputer with a large amount of memory that can be placed in or on an animal to collect and store time-labeled information on physiological or behavioral variables. After the data logger is recovered from the animal, the information it has stored is off-loaded to a computer.

Deamination   Removal of nitrogen-containing amino groups from amino acids or proteins.

Decapod crustacean   A lobster, crayfish, crab, shrimp, or related animal.

Decompression sickness   A pathological state that arises after diving when bubbles are formed within body fluids because the reduction in pressure during surfacing allows gases (especially N_2) present at high dissolved partial pressures to come out of solution. Also called *the bends, caisson disease*.

Decremental spread   Spread of a signal in such a way that signal strength decreases with distance. In electrophysiology, passive voltage changes (synaptic potentials, receptor potentials, etc.) spread decrementally, in contrast to propagated action potentials. Also called *electrotonic conduction, passive spread*.

Deep cerebellar nuclei   Collections of neurons under the cerebellar cortex and receiving output from it.

Definitive urine   The final urine that is excreted by an animal. Contrast with *primary urine*.

Degradation of energy   Conversion of energy from a high-grade form to heat.

Degrade   In the study of energy, the process of converting chemical-bond energy or other forms of high-grade energy to heat.

Delayed implantation   Embryonic diapause in placental mammals. A programmed state of arrested embryonic development that occurs after an embryo has arrived in the uterus but before it implants in the uterine wall.

Denaturation   A change in the tertiary (three-dimensional) structure of an intact protein that renders the protein nonfunctional. May be reversible or irreversible. The primary structure is not altered during denaturation.

Dendrite   The receptive element of most neurons, which receives synaptic input from other neurons. Most neurons have many, multiply branching dendrites, in contrast to one sparsely branching axon.

Dendritic   Branching in a way that resembles the branching of a tree. Also: Of or having to do with a dendrite.

Dendritic spine   A knoblike protrusion on a dendrite of a neuron that is typically the postsynaptic site of a single synapse. Common in mammalian central neurons.

Dense bodies   Structures in the cytoplasm and on the inner cell membrane of smooth muscle cells to which actin filaments attach.

Deoxy- A prefix referring to a molecule that is capable of reversible combination with O_2 that is in a state of not being so combined (e.g., deoxyhemoglobin).

Deoxygenation The release of O_2 from a combined state with a respiratory pigment such as hemoglobin. Not equivalent to *reduction*.

Depolarization A decrease in amplitude of the inside-negative electrical potential of a cell membrane toward zero. More generally, any increase in the inside positivity of a cell membrane, even if it exceeds zero.

Desmosome A "spot weld" type of junction between two adjacent cells in which protein filaments intermingle across the space between the cells, thereby strengthening and stabilizing the morphological arrangement of the cells.

Developmental physiology The study of functional properties in successive stages of the development of an individual animal. It includes studies of function in early developmental stages as well as in adults.

Developmental polyphenism See *polyphenic development*.

Diacylglycerol (DAG) A molecule derived from membrane phospholipids that acts as a second messenger in cell signaling.

Diapause A programmed state of suspended development or suspended animation in the life history of an animal. See also *embryonic diapause, delayed implantation*.

Diaphragm A sheet of muscular and connective tissue that completely separates the thoracic and abdominal cavities, found only in mammals.

Diastole The period of relaxation during each beating cycle of a heart.

Diastolic pressure During the beating cycle of the heart, the blood pressure prevailing during periods of heart relaxation.

Diencephalon Part of the vertebrate forebrain that contains the thalamus and hypothalamus.

Diet-induced thermogenesis (DIT) A chronic increase in metabolic rate induced in certain types of animals by chronic overeating.

Diffusion See *simple diffusion*.

Diffusion lung A lung within which the air is still, so that O_2 and CO_2 must travel the full length of the lung passages by diffusion.

Digestion The process of splitting up ingested food molecules into smaller chemical components that an animal is capable of distributing to the tissues of its body. Most commonly carried out by hydrolytic enzymes.

Dihydropyridine receptor (DHPR) The voltage-sensitive molecule in the transverse tubules of vertebrate striated muscle fibers that links excitation to the release of Ca^{2+} ions from the sarcoplasmic reticulum. It serves as a Ca^{2+} channel in cardiac muscle and as a voltage sensor in skeletal muscle.

Dipnoan A lungfish. A member of the sarcopterygian class of bony fish, believed to be among the closest living relatives of the ancestors of amphibians. There are three genera of dipnoans: *Protopterus* in Africa, *Neoceratodus* in Australia, and *Lepidosiren* in South America.

Direct acting hormones A category of anterior pituitary chemical signals that exert their effects on nonendocrine target cells; not tropic. Contrast with *tropic hormones*.

Direct calorimetry Measurement of metabolic rates by quantifying heat and external work.

Direct measurement A measurement procedure that quantifies a property by measuring exactly what the *definition* of the property specifies. Contrast with *indirect measurement*.

Disaccharidase A digestive enzyme that breaks a disaccharide sugar molecule into two single sugar molecules.

Disc A flat structure. Retinal rods have membrane-bound discs that are the site of phototransduction of rhodopsin.

Distance chemoreceptor A chemosensory cell (of a terrestrial animal) that is normally stimulated by chemicals that are airborne over a considerable distance. Distance chemoreceptors are usually more sensitive than contact chemoreceptors. Contrast with *contact chemoreceptor*.

Diuresis The production of urine at a high rate. Such urine is usually dilute as well as abundant.

Diuretic Promoting diuresis.

Diurnal Active in the daytime.

Divergence Spreading apart. In neurophysiology, a pattern in which signals from a particular presynaptic neuron synaptically excite or inhibit many postsynaptic neurons. Contrast with *convergence*.

Diving bradycardia Slowing of the heart rate during diving.

DNA microarray A grid of numerous, diverse DNA spots that, by hybridizing with messenger RNA molecules (mRNAs) in a mix of mRNAs, reveals which mRNAs are present in cells or tissues—thereby revealing which genes are being transcribed. A major tool in transcription profiling. Also called a *DNA microchip* or *gene chip*.

Donnan equilibrium A complex multi-ionic equilibrium state that tends to be reached by the interacting diffusion of multiple permeating ions and water across a cell membrane or epithelium when there is a set of nonpermeating ions (ions that cannot cross the membrane or epithelium) that are more abundant on one side than on the other. Because of their content of nonpermeating anionic proteins and nucleic acids, animal cells would approach Donnan equilibrium if it were not for the fact that living processes hold them away from any sort of equilibrium.

Dorsal aorta In the circulatory system of a fish, the major vessel that conveys blood from the gills to the systemic tissues.

Dorsal root ganglion A type of peripheral ganglion found at the dorsal root of a spinal nerve, containing cell bodies of the sensory neurons in that nerve.

Doubly labeled water method A method used to measure the metabolic rate of a free-living animal in which water labeled with unusual isotopes of hydrogen and oxygen is injected into the animal. It estimates the animal's rate of CO_2 production. Also called the $D_2{}^{18}O$ *method*.

Downregulation A downward shift in the catalytic activity of an enzyme, the rate of functioning of a biochemical pathway, or the rate of some other similar process brought about in a controlled manner by a regulatory system.

Dry heat transfer Heat transfer by conduction, convection, or thermal radiation; heat transfer that does not involve evaporation or condensation of water.

Dual breather An animal that simultaneously possesses the ability to breathe from air and from water. Also called *bimodal breather*.

Dynamic viscosity See *viscosity*.

Dystrophin A cytoskeletal protein in muscle fibers that connects actin filaments of the cytoskeleton to a complex of proteins in the sarcolemma.

E

Eccentric muscle contraction An isotonic muscle contraction in which the muscle lengthens as it exerts force while resisting stretch. Also called a *lengthening contraction*.

Ecdysis The process of shedding the outer body covering. In arthropods, the shedding of the old cuticle (exoskeleton) from one instar as the animal makes the transition to the next instar. Also called *molting*.

Ecdysone The steroid prohormone secreted by the thoracic glands of arthropods that stimulates ecdysis, or molting. Also called *molting hormone*.

Ectotherm See *poikilotherm*.

Effector A tissue, organ, or cell that carries out functions under the direction of the nervous system or another physiological control system (e.g., the endocrine system).

Efferent   Going away. Thus, for example, an efferent blood vessel carries blood away from an organ of interest. Contrast with *afferent*.

Efferent neuron   A neuron that conveys signals from the central nervous system to the periphery, usually exerting motor control.

Efficiency of energy transformation   In any process that transforms high-grade energy from one form to another, the output of high-grade energy expressed as a ratio of the input of high-grade energy. See also *absorption efficiency, gross growth efficiency, net growth efficiency.*

Elasmobranch fish   Sharks, skates, and rays.

Electric current   The flow of electric charges. In animals, electric current is carried by ions, unlike the flow of electrons in man-made circuits.

Electrical energy   Energy that a system possesses by virtue of the separation of positive and negative electrical charges.

Electrical gradient   Technically, the difference in electrical potential (voltage) between two places divided by the distance separating those two places. Often used more loosely to refer simply to a difference in electrical potential.

Electrical synapse   A synapse at which current spreads directly from cell to cell through a low-resistance gap junction.

Electrocardiogram (EKG, ECG)   A recording as a function of time of differences in electrical potential set up in extracellular body fluids by the depolarization and repolarization of the myocardium during heart contraction and relaxation. These differences can be detected on the surface of the body and recorded from there.

Electrochemical equilibrium   A term that is synonymous with *equilibrium*, but emphasizes that the equilibrium state for an ion or other charged solute depends on both electrical and chemical effects. A charged solute is at electrochemical equilibrium across a membrane when the effect of concentration on its diffusion and the effect of the electrical potential difference on its diffusion sum to zero.

Electroencephalogram (EEG)   A record of gross electrical activity in the brain, usually recorded using multiple electrodes placed on the skin of the head or on the brain surface. In sleep studies, it is used to measure the stages of sleep during each sleep period.

Electrogenic pump   An active-transport process that pumps net charge across a membrane, acting to generate an electric current across the membrane and to produce a voltage difference across the membrane. Contrast with *electroneutral pump*.

Electrolyte   An inorganic ion in a body fluid.

Electron-transport chain   A series of compounds in the mitochondria (each compound capable of reversible oxidation and reduction) that passes electrons removed from food molecules to O_2, capturing released energy for use in producing ATP.

Electroneutral pump   An active-transport process that pumps charges across a membrane such that no difference of charge is created across the membrane; it therefore is not a current source. Contrast with *electrogenic pump*.

Electroreceptor   A sensory receptor cell that responds to environmental electrical stimuli (e.g., in weakly electric fish).

Electrotonic conduction   See *decremental spread*.

Embryonic diapause   A programmed state of arrested or profoundly slowed embryonic development. See also *delayed implantation*.

Empirical   Based on data rather than merely reasoning.

Endocrine cell   A nonneural epithelial cell or a neuron that secretes a hormone or neurohormone.

Endocrine gland   A gland or tissue without ducts that secretes a hormone into the blood. Nonneural endocrine cells may form glands that are discrete (with all cells grouped together), diffuse (with cells scattered within other tissues), or intermediate between discrete and diffuse. Contrast with *exocrine gland*.

Endogenous rhythm   A rhythmic pattern of physiological or behavioral activity, the rhythmicity of which arises as an intrinsic property of an animal's cells (e.g., in the nervous system) without need of external timing information.

Endometrium   In the mammalian uterus, the inner tissue layer that during pregnancy helps form the placenta.

Endopeptidase   A type of protein-digesting enzyme.

Endothelium   The epithelium that lines the heart and the lumen of blood vessels in vertebrates.

Endotherm   An animal in which the body temperature is elevated by metabolically produced heat.

Endothermy   Warming of the body of an animal by its metabolic heat production.

Endurance exercise   Exercise that consists of many repetitions of relatively low-intensity muscular actions over long periods of time. Exercise that emphasizes aerobic catabolism as the source of ATP. Also termed *endurance training*. Contrast with *resistance exercise*.

Energy   The ability to maintain or increase order in a system.

Energy degradation   See *degradation of energy*.

Energy metabolism   The set of processes by which energy is acquired, transformed, channeled into useful functions, and dissipated by cells or organisms.

Enteric division   One of three divisions of the autonomic nervous system; exerts largely autonomous control over the gut.

Entrainment   The process of synchronizing an endogenous rhythm to an environmental rhythm.

Environment   An organism's surroundings, including other organisms as well as the abiotic conditions that prevail in the surroundings, such as temperature. Sometimes called the *external environment* to distinguish it from the *internal environment*. See also *internal environment*.

Environmental   See *ambient*.

Enzyme   A molecule—usually a protein—that catalyzes a chemical reaction in which covalent bonds are made or broken.

Enzyme–substrate affinity   The proclivity of an enzyme to form a complex with its substrate when the enzyme and substrate meet.

Epicuticle   The outermost layer of the exoskeleton of an arthropod, where lipids that protect against desiccation are deposited in insects and arachnids.

Epididymis   A tubular structure, juxtaposed to the testis (testicle), where sperm are stored in anticipation of ejaculation.

Epigenetic mark   A change in a gene or its immediate biochemical environment *other than a change in the DNA sequence*—such as methylation of cytosine residues—that modifies expression of the gene and that, when the gene replicates, also replicates so that the resulting gene copies also have their expression modified in the same way. The gene is said to be *marked* or *tagged*.

Epigenetics   The study of modifications of gene expression that are transmitted when genes replicate despite there being *no change in the DNA sequence*. See also *epigenetic mark*.

Epigenome   The sum total of all epigenetic marks in an animal. See *epigenetic mark*s.

Epithelial endocrine cells   synthesize and secrete hormones; also called nonneural endocrine cells

Epithelial glands   endocrine structures made of epithelial endocrine cells; also called nonneural glands;

Epithelium   A sheet of cells that lines a cavity or covers an organ or body surface, thereby forming a boundary between functionally different regions of the body or between an animal and its external environment.

Epitoke   In polychaete annelid worms, a specialized reproductive stage that is formed by either transformation of an ordinary individual or budding from an ordinary individual.

Equilibrium The state toward which an *isolated* system changes; that is, the state toward which a system moves—internally—when it has no inputs or outputs of energy or matter. A system is *at equilibrium* when internal changes have brought it to an internally stable state from which further net change is impossible without system inputs or outputs. The state of equilibrium is a state of minimal capacity to do work under locally prevailing conditions. See also *electrochemical equilibrium*.

Equilibrium potential The membrane potential at which an ion species is at electrochemical equilibrium, with concentration-diffusion forces offset by electrical forces so that there is no net flux of that ion species across the membrane.

Erythrocyte See *red blood cell*.

Erythropoiesis Production of red blood cells.

Erythropoietin A hormone that stimulates production of red blood cells.

Essential amino acid A standard amino acid that an animal cannot synthesize and that thus must be obtained from food, microbial symbionts, or other sources besides biosynthesis by the animal.

Essential fatty acid A type of fatty acid that an animal cannot synthesize from scratch and that necessitates acquisition of precursors from food, microbial symbionts, or other sources besides biosynthesis by the animal. The omega-3 and omega-6 fatty acids are the major types.

Essential In the study of nutrition, a required material—such as an amino acid, fatty acid, vitamin, or mineral element—that an animal must obtain from outside sources, i.e., sources other than biosynthesis by the animal.

Estivation (1) In mammals and birds, a form of controlled hypothermia in which the body temperature is able to approximate ambient temperature continuously for two or more consecutive days during summer. (2) In other animals, a nonspecific term referring to a resting or dormant condition during the summer or during drought.

Estrogen A feminizing hormone such as estradiol. Estrogens, which may be secreted by the ovary, placenta, testis, and possibly the adrenal cortex, are essential for female secondary sexual characteristics and reproduction.

Estrous cycle A cycle of behavioral readiness to copulate (correlated with ovulation) in most female mammals.

Estrus A stage of the estrous cycle around the time of ovulation during which a female uses behaviors to indicate that she is ready to conceive offspring. Also called *heat*.

Estuary A body of water along a seacoast that is partially enclosed by land and that receives inputs of both freshwater and seawater; it is intermediate in salinity between freshwater and seawater.

Euryhaline Referring to aquatic animals able to live over a wide range of environmental salinities. Contrast with *stenohaline*.

Eurythermal Referring to poikilotherms, able to live over a broad range of body temperatures. Contrast with *stenothermal*.

Euthermia The state of having a usual or normal body temperature. Specifically, in the study of mammals or birds that undergo controlled hypothermia, the state of having a fully homeothermic body temperature (e.g., about 37°C in a placental mammal).

Evaporation A change in the physical state of a compound from a liquid to a gas; most commonly used to refer to water. Evaporation can in principle occur at any temperature.

Evaporative cooling Removal of heat by the evaporation of water. The evaporative cooling caused by the evaporation of a gram of water equals the *latent heat of vaporization* of water per gram.

Evolution A change in gene frequencies over time in a population. Evolution can result in *adaptation*, or it can be *nonadaptive*.

Excess postexercise oxygen consumption (EPOC) An elevation of the actual O_2 uptake by breathing above the theoretical O_2 requirement of rest when an animal is resting immediately after exercise; "breathing hard" after exercise. Also called *oxygen debt*.

Excitable cells Cells that can generate action potentials because their cell membranes contain voltage-gated channels, notably neurons and muscle cells.

Excitation Process tending to produce action potentials.

Excitation–contraction coupling In a muscle cell, linkage of the electrical excitation of the cell membrane with contractile activity by facilitation of the availability of Ca^{2+} in the cytoplasm. Key events occur at the transverse tubules and sarcoplasmic reticulum. Also called *E-C coupling*.

Excitatory postsynaptic potential (EPSP) A voltage change in a postsynaptic cell—normally a depolarization—that tends to excite the cell.

Excitatory Resulting in an increase in activity or probability of activity. In a neuron, depolarization is excitatory because it increases the likelihood of generation of action potentials. Contrast with *inhibitory*.

Exocrine gland A gland with ducts in which secretions exit the gland by way of the ducts, rather than being secreted into the blood. Examples include salivary glands and sweat glands. Contrast with *endocrine gland*.

Exopeptidase A type of protein-digesting enzyme.

Expiratory reserve volume The volume of air that remains in the lungs following a resting exhalation, not including the portion that cannot be exhaled under any conditions.

Exponential Referring to a type of relationship between two variables in which the dependent variable goes up in multiplicative steps as the independent variable goes up in additive steps. See Appendix F.

Expression Synthesis of the protein (or other functional product) encoded by a gene.

Expression profiling Sometimes used (but not in this book) as a synonym for *transcription profiling*. See *transcription profiling*.

External ear The portion of the ear external to the eardrum or tympanic membrane.

External environment See *environment*.

External respiration Breathing.

External work Mechanical work by an animal that involves applying forces to objects outside the animal's body. Locomotion is the principal example; other examples would be a squirrel chewing into a nut, or a mole pushing soil aside.

Exteroceptor A sensory receptor cell that is activated by stimuli from outside the body.

Extracellular digestion Digestion that takes place outside of cells, such as in the lumen of the digestive tract.

Extracellular fluids The aqueous solutions (body fluids) outside cells. In animals with closed circulatory systems, subdivided into *blood plasma* and *interstitial fluids*.

Extrafusal muscle fiber In vertebrate skeletal muscle, an "ordinary" muscle fiber that is not associated with a muscle-spindle stretch receptor.

Extrarenal salt excretion Excretion of inorganic ions by structures other than the kidneys, such as the gills in marine teleost fish and salt glands in marine birds.

F

Facilitated diffusion Passive transport of a solute across a membrane mediated by the noncovalent and reversible binding of solute molecules to a solute-specific transporter (carrier) protein in the membrane. Facilitated diffusion is the principal mode of passive transport of polar organic solutes, such as glucose and amino acids, across membranes.

Fast glycolytic (FG) muscle fibers Muscle fibers that are poised to make ATP principally by anaerobic catabolism, develop contractile tension rapidly, have relatively high peak power outputs, and fatigue relatively rapidly.

Fast oxidative glycolytic (FOG) muscle fibers Muscle fibers that produce ATP principally by aerobic catabolism, develop contractile tension at a rate intermediate between the rates of slow oxidative (SO) and fast glycolytic (FG) fibers, have intermediate peak power outputs, and exhibit intermediate resistance to fatigue.

Fasting (1) In life histories, failure to eat, often for prolonged periods of time, because of features intrinsic to an animal's life history. Often distinguished from *starvation*, which refers to extrinsically imposed food deprivation. (2) In metabolism studies, not eating for a sufficient period of time so as to end the specific dynamic action of the last meal. See *specific dynamic action*.

Feedback See *negative feedback* and *positive feedback*.

Feed-forward A concept in control theory that is in certain respects the opposite of *feedback*. During feedback, deviations of a controlled property from a set-point level are detected, and the control system responds by either diminishing the deviations (negative feedback) or enhancing them (positive feedback). During feed-forward control, however, a system is driven to change by an input external to itself, not by responses to deviations of its own performance from a set point. Contrast with *negative feedback, positive feedback*.

Fenestrations Minute, physical openings—pores—in the walls of blood capillaries. Some fenestrations are formed by tiny gaps (about 4 nm in diameter) *between* cells in the capillary endothelium; others are formed in some tissues by holes *through* the capillary endothelial cells. Water and ions can freely cross the capillary endothelium through fenestrations by osmosis and diffusion.

Fermentation Enzyme-catalyzed reactions that occur without O_2, such as reactions that accomplish the anaerobic breakdown of compounds to liberate energy for metabolic use.

Fermenter An animal that maintains a symbiosis with major populations of anaerobic, fermenting microbes (e.g., rumen microbes) that perform nutritional functions for the animal.

Fermenting microbes Anaerobic microbes that carry out fermentation. See *fermentation*.

Fever A regulated elevation of body temperature to a level higher than usual.

Fiber See *muscle fiber*.

Fick equation An equation used to predict the rate of diffusion. See Equation 5.1.

Field metabolic rate (FMR) The *average daily metabolic rate* of an animal when living free in its natural environment.

Filter feeding Feeding on objects suspended in water that are very small by comparison to the feeding animal, when the mechanism of food collection is some sort of sieving. If the mechanism is of another sort or is unknown, the feeding should be termed *suspension feeding*.

Filtrate See *ultrafiltrate*.

Filtration rate See *glomerular filtration rate*.

Fixed act A simple all-or-none behavioral response to a stimulus. Contrast with *reflex*.

Flexion reflex A reflex response that flexes or withdraws a limb from a painful or noxious stimulus.

Fluid A gas or liquid.

Fluid compartment A defined subpart of the body fluids, often distributed throughout the body rather than occupying a discrete physical location. The simplest subdivision of the body fluids recognizes three fluid compartments: blood plasma, interstitial fluids, and intracellular fluids.

Fluid mosaic model A theory of the nature of cell membranes. According to this theory, a cell membrane consists of a mosaic of protein and lipid molecules, all of which move about in directions parallel to the membrane faces because of the fluid state of the lipid matrix.

Fluidity In reference to the phospholipids in a cell membrane, the ability of individual phospholipid molecules to diffuse through the population of all such molecules in a membrane leaflet because the molecules are not covalently bonded to one another. Fluidity is quantitatively variable. Diffusion is relatively rapid in some membranes, which are said to exhibit high fluidity.

Fluorescence A phenomenon associated with bioluminescence in which preexisting light (e.g., from a bioluminescent mechanism) is absorbed and re-emitted at longer wavelengths. Fluorescence does not produce light de novo.

Follicle (1) In secretory tissues such as the thyroid gland, a globe-shaped, hollow structure enclosed by an epithelium. (2) In the ovary, an oocyte and the layers of somatic cells surrounding it.

Follicle-stimulating hormone (FSH) A tropic peptide chemical signal secreted by cells in the anterior pituitary gland (adenohypophysis); travels in the general circulation to the gonads; stimulates sperm production in males and follicle development and secretions in females; a gonadotropin.

Follicular phase In a female mammal, the phase of the menstrual or estrous cycle prior to ovulation, during which primary oocytes and associated somatic follicle cells develop, mature, and secrete hormones.

Foodstuff A material in the body, such as carbohydrate or lipid, that an animal uses as a source of energy or chemical building blocks. Foodstuffs are often derived directly from foods but differ from foods in being already in the body.

Force–velocity relationship Also called load–velocity relationship, shows that the speed at which a muscle contracts decreases as the load against which it exerts force increases.

Forebrain The anterior portion of the brain, consisting in vertebrates of the telencephalon and diencephalon.

Foregut In a vertebrate, the esophagus and stomach.

Foregut fermenter An animal that has a specialized foregut chamber housing communities of fermenting microbes that assist with the breakdown of food materials and that often provide biosynthetic capabilities the animal lacks. The microbial communities commonly include bacteria, protists, yeasts, and fungi.

Fossorial Living underground. Used usually to distinguish species that live underground but belong to phylogenetic groups that generally live aboveground.

Fourier's law of heat flow Synonym for the *linear heat transfer equation*.

Fovea A central region in a vertebrate retina specialized for high-resolution processing of visual information.

Frank–Starling mechanism An important intrinsic control mechanism of the vertebrate heart in which stretching of the cardiac muscle tends to increase the force of its contraction by an effect exerted at the level of individual muscle cells.

Free-running rhythm An endogenous rhythm that is not entrained by an environmental rhythm.

Freezing-intolerant An animal exposed to freezing temperatures that dies if it freezes.

Freezing point The highest temperature at which a liquid can turn to a solid and freeze. A colligative property.

Freezing-point depression (ΔFP) The *difference* (sign ignored) between the freezing point of a solution and the freezing point of pure water. For example, if the freezing point of a solution is –1.6°C, its freezing-point depression is 1.6°C. A colligative property.

Freezing-tolerant An animal that can survive freezing of its extracellular body fluids.

G

G protein A protein involved in signal transduction that is activated by binding with guanosine triphosphate (GTP). Some G proteins occur in cell membranes and are typically trimers; others occur intracellularly and are typically monomers.

G protein–coupled receptor   A membrane receptor protein that, when it binds to its specific extracellular signal ligand, relays a signal into the cell by activating G proteins in the cell membrane.

Gamete   A reproductive cell, also called a germ cell; an egg (ovum) produced by an ovary or a sperm produced by a testis.

Gametogenesis   The formation of haploid eggs or sperm (gametes) through the process of meiosis. It is called specifically *oogenesis* when referring to processes in the ovary and *spermatogenesis* in the testis.

Gamma (γ) motor neuron   A small motor neuron in a vertebrate spinal cord that innervates an intrafusal muscle fiber.

Ganglion (plural ganglia)   A discrete collection of neuronal cell bodies. In arthropod nervous systems, most ganglia are segmental components of the central nervous system; in vertebrates, ganglia are components of the peripheral nervous system.

Ganglion cell   An output cell of the vertebrate retina, with an axon extending in the optic nerve to visual processing areas of the brain.

Gap junction   A region where the cell membranes of adjacent cells are unusually close to each other and share channels (formed by adjoining connexons in vertebrates) that establish cytoplasmic continuity between the cells.

Gas tension   See *partial pressure*.

Gated channel   A channel that "opens" and "closes" to facilitate or inhibit solute passage. Some gated channels "open" and "close" in response to changes in ligand binding; others do so in response to voltage changes or other changes.

Gene deletion   Synonym for *gene knockout*. See *knockout animal*.

Gene expression profiling   See *transcription profiling*.

Gene family   A group of genes that are evolutionarily related; genes related by common descent from ancestral genes.

Gene knockout   See *knockout animal*.

Gene lineage   In the study of evolution, a set of genes that are related by descent from a common ancestral gene; also a representation of the family tree of such a set of genes.

Genetic drift   Changes in gene frequencies within a population over time resulting primarily from chance.

Genome   The full set of genetic material of an organism.

Genome-wide association study (GWAS)   A study that compares the complete DNA of individuals with a particular condition to the DNA of individuals without the condition, with the objective of identifying the genes that play roles in causing the condition.

Genomic imprinting   An epigenetic phenomenon in which, in a given individual, an allele inherited from the individual's father is expressed exclusively (or predominantly) relative to the allele of the same gene inherited from the individual's mother, or vice versa—because of *epigenetic marks*. At present, genomic imprinting is known to occur only in mammals, insects, and flowering plants.

Genomics   The study of the genomes (the full sets of genetic material) of organisms.

Germ cell   See *gamete*.

Ghrelin   A peptide hormone secreted by cells in the stomach and upper intestine that stimulates hunger.

Giant axons   Axons that are much bigger than other axons around them. Diameters vary from 1mm in squid to 4 μm in fruit-flies.

Gills   In the most general sense used by physiologists, structures specialized for external respiration that project from the body into the ambient medium and are thereby surrounded by the environmental medium. Contrast with *lungs*.

Glial cells   Cells in an animal's neural tissue (e.g., brain) other than neurons. Glial cells are considered support cells, ensheathing neuronal processes or regulating the metabolism of neurons. They may play secondary roles in signaling and integration. Also called *neuroglia*.

Globins   A family of structurally similar proteins believed to be evolutionarily related by common descent from ancestral protein forms. Proteins of this family occur, for example, in hemoglobin O_2-transport pigments.

Glomerular filtration rate (GFR)   The rate at which all the nephrons in the kidneys of a vertebrate collectively produce primary urine by ultrafiltration.

Glomerulus   A minute anastomosing cluster of blood capillaries associated with a nephron in the kidney of a vertebrate, serving as the site of formation of primary urine by ultrafiltration. The term is also sometimes used to refer not only to such a cluster of capillaries, but also to the Bowman's capsule with which it is associated.

Glucagon   A hormone secreted by alpha cells of the endocrine pancreas in response to low plasma levels of glucose and/or high plasma levels of amino acids; increases plasma glucose by stimulating glycogenolysis, gluconeogenesis, and breakdown of triglyceride molecules.

Glucocorticoids   Steroid hormones, such as cortisol and corticosterone, that are released from the cortex of the adrenal gland and regulate carbohydrate, lipid, and protein metabolism.

Gluconeogenesis   Production of glucose or glycogen from noncarbohydrate molecules such as lactic acid or amino acids.

Glycolysis   The reactions that convert glucose to pyruvic acid.

Goldman equation   An equation that describes membrane potential in terms of the concentrations of and membrane permeabilities to more than one ion species.

Gonadotropin   A hormone that stimulates the gonads (ovaries or testes) to produce gametes and secrete hormones, and also supports and maintains the gonadal tissue.

Gonadotropin-releasing hormone (GnRH)   In a vertebrate, a hypothalamic hormone that controls gonadotropin secretion from the anterior pituitary gland.

Graded potential   A voltage change that is variable in amplitude—that is, not all-or-none like an action potential. Examples include synaptic potentials and receptor potentials.

Granular cells   See *juxtaglomerular cells*.

Granulosa cells   Somatic cells surrounding the primary oocyte in an ovarian follicle.

Gray matter   A histological region of a vertebrate central nervous system that contains neuronal cell bodies, dendrites, and synapses as well as axons. See also *white matter*.

Green gland   See *antennal gland*.

Grid cell   A neuron, commonly in the entorhinal cortex of the brain, that generates action potentials when an animal is at one of several gridlike locations in its environment. The loci at which grid cells are active form an invisible regular hexagonal grid that can provide a distance calibration for a map representation, in the brain, of the animal's environment.

Gross growth efficiency   In a growing animal, the chemical-bond energy of new biomass added by growth expressed as a ratio of the animal's ingested energy over the same time period.

Growth hormone (GH)   A direct acting hormone secreted by cells of the anterior pituitary gland (adenohypophysis); acts at many target tissues to promote lean body mass by stimulating protein synthesis and breakdown of triglyceride molecules.

Gular fluttering   Rapid up-and-down oscillation of the floor of the mouth cavity of a bird or other reptile to enhance the rate of evaporative cooling by increasing air flow over moist membranes in the mouth.

Gustatory   Having to do with *taste*.

Gut microbiome   Populations of microbes, consisting of many species of bacteria and other heterotrophic microbes, living in the gut lumen of an animal.

Gut motility   See *motility*.

H

20-Hydroxyecdysone (20E)   The physiologically active form of the molting hormone ecdysone; stimulates the epidermis to replace the cuticle in successive molts; a gonadotropin in adults. See *ecdysone*.

H zone   In striated muscle, a region at the center of a sarcomere that contains only thick filaments; shortens during contraction. Also called *H band*.

Habituation   A simple, nonassociative form of learning characterized by a learned decrease in a behavioral response with repeated presentations of a nonthreatening stimulus.

Hagen-Poiseuille equation   See *Poiseuille equation*.

Hair bundle   The tuft of microvilli (stereocilia) at the apical end of a hair cell. It responds to mechanical stimulation.

Hair cell   A sensory epithelial cell in a vertebrate acoustico-lateralis system that transduces displacement of its apical stereocilia into an electrical signal.

Haldane effect   A shift of the carbon dioxide equilibrium curve of the blood caused by the oxygenation and deoxygenation of the blood respiratory pigment. In blood that exhibits a Haldane effect, the total carbon dioxide concentration is higher when the blood is deoxygenated than when it is oxygenated.

Half-life   The time required to reduce something by one-half; applies, for example, to the concentrations of substances such as hormones, stored voltages, and radioactivity.

Half-saturation constant (K_m)   See *Michaelis constant*.

Halophyte   A plant that roots in saline soils, often characterized by high salt concentrations in its tissue fluids.

Head direction cell   A neuron in the brain that generates action potentials when the animal's head is pointing in a particular direction. Head direction cells provide spatial information about the animal's bearings in its environment. They are presumably used with place cells and grid cells in spatial navigation. See *grid cell* and *place cell*.

Heart   A discrete, localized structure specialized for pumping blood. It may be neurogenic or myogenic and associated with an open or a closed circulatory system.

Heat   The energy that matter possesses by virtue of the ceaseless, random motions that all of the atoms and molecules of which it is composed undergo on an atomic-molecular scale of distance. Also called *molecular kinetic energy*.

Heat increment of feeding   Synonym for *specific dynamic action*.

Heat-shock proteins   An evolutionarily related group of proteins that are principally inducible—being expressed in the aftermath of heat stress or other stress—and that function as molecular chaperones. Heat-shock proteins use ATP to assist in the repair of stress-damaged proteins by preventing the damaged proteins from aggregating with one another and by promoting molecular folding patterns that restore them to correct three-dimensional conformations.

Hematocrit   In blood, the sum total volume of all cells expressed as a percentage of blood volume.

Heme   A particular metalloporphyrin containing iron in the ferrous state: ferrous protoporphyrin IX. It is the O_2-binding site in all hemoglobins.

Hemerythrin   A type of respiratory pigment found in scattered groups of animals in three or four phyla, consisting of an iron-based metalloprotein. Undergoes reversible combination with O_2 at iron-containing loci.

Hemocyanin   A type of respiratory pigment found in arthropods and molluscs, consisting of a copper-based metalloprotein. Undergoes reversible combination with O_2 at copper-containing loci.

Hemoglobin   A type of respiratory pigment formed by the combination of heme (ferrous protoporphyrin IX) with a globin protein. Undergoes reversible combination with O_2 at the heme loci.

Hemolymph   A synonym for *blood* in an animal that has an open circulatory system. The term emphasizes that the blood in such animals includes all extracellular fluids, and thus that there is no distinction between the fluid that is in the blood vessels at any one time and the interstitial fluid between tissue cells.

Henry's law   A law that relates partial pressure and concentration for gases dissolved in an aqueous solution (or in another liquid solution). One way to express this law is that $C = AP$, where C is dissolved concentration, P is partial pressure in atmospheres, and A is the *absorption coefficient* (dissolved concentration when partial pressure is 1 atm).

Hepatopancreas   An organ in crustaceans that connects with the stomach and functions in secretion of digestive enzymes; absorption of nutrient molecules; storage of lipids, glycogen, and Ca^{2+}; and sequestration of foreign compounds.

Hermaphrodite   An individual that possesses both ovaries and testes. An animal may be a *simultaneous* hermaphrodite, in which both types of gonads produce gametes at the same time, or a *sequential* hermaphrodite, in which only one type of gonad produces gametes at a time.

Heterothermy   The property of exhibiting different thermal relations from time to time or place to place. There are two types of heterothermy: (1) In *regional heterothermy*, some regions of an individual animal's body exhibit different thermal relations than other regions at the same time. (2) In *temporal heterothermy*, an individual exhibits one type of thermal relation at certain times and another type of thermal relation at other times (e.g., hibernation at some times and homeothermy at others).

Heterotroph   An organism that obtains the energy it needs to stay alive by breaking up organic compounds that it obtains from other organisms, thereby releasing the chemical-bond energy of those organic compounds. Contrast with *autotroph*.

Hibernaculum (plural hibernacula)   The place where an animal resides while in hibernation.

Hibernation   (1) In mammals and birds, usually refers to a form of controlled hypothermia in which the body temperature is able to approximate ambient temperature continuously for two or more consecutive days during winter. (2) In other animals, a nonspecific term referring to a resting or dormant condition during winter.

High-grade energy   Energy in a form that can do physiological work. Chemical, electrical, and mechanical energy are forms of high-grade energy.

High-throughput method   An analytical method that is carried out by computer programs and robots without much direct human attention, and thus can process samples at a relatively high rate.

Hill coefficient   A measure of the degree of cooperativity among O_2-binding sites in a molecule of a respiratory pigment.

Hindbrain   Posterior portion of a vertebrate brain, consisting of the cerebellum, pons, and medulla.

Hindgut   The posterior part of the gut. In a person, the large intestines and rectum. In an insect, the part of the gut posterior to the connections of the Malpighian tubules.

Hindgut fermenter   An animal that has a specialized hindgut chamber housing communities of fermenting microbes that assist with the breakdown of food materials and that often provide biosynthetic capabilities the animal lacks.

Histones   Basic proteins with which DNA (acidic) is complexed in the cell nucleus.

Hodgkin cycle   The cycle that explains the rising phase of an action potential: Depolarization opens voltage-gated Na^+ channels, increasing membrane permeability to Na^+. The resulting inflow of Na^+ further depolarizes the membrane, opening more Na^+ channels.

Homeosmotic animal Synonym for *osmo-regulator*.

Homeostasis Internal constancy and the physiological regulatory systems that automatically make adjustments to maintain it. In the words of Walter Cannon, who coined the term, "the coordinated physiological processes which maintain most of the [constant] states in the organism."

Homeotherm An animal that thermoregulates by physiological means (rather than simply by behavior).

Homeoviscous adaptation The maintenance of a relatively constant lipid fluidity regardless of tissue temperature.

Homing The ability of an animal to return to its home site after being displaced.

Homologous Relating to features of organisms that are similar because of common evolutionary descent.

Horizontal cell A neuron in the vertebrate retina that is part of the lateral pathway, mediating center–surround antagonistic effects in retinal neuron receptive fields.

Hormonal axis A hormonal sequence in which one hormone stimulates the secretion of a second hormone that may in turn stimulate secretion of a third hormone. Known hormonal axes consist of two or three hormones in sequence.

Hormonal modulation The influence of endocrine signals on the rate of hormone secretion by means of negative and positive feedback, synergism, antagonism, and permissiveness

Hormone A chemical substance, released by nonneural endocrine cells or by neurons, that is carried in the blood to distant target cells, where it exerts regulatory influences on their function. There are three main chemical classes of hormones: steroids, peptides or proteins, and amines.

Hourglass timer See *interval timer*.

Humidic Restricted to humid, water-rich terrestrial microenvironments; unable to live steadily in the open air.

Humidity A general term referring loosely to the concentration of gaseous water in a gas. See also *water vapor pressure*, *relative humidity*, and *saturation deficit*—all of which represent ways to express humidity.

Hybrid oscillator A neural circuit of central pattern generation that contains both cellular oscillator and network oscillator elements.

Hydrolytic enzyme An enzyme that catalyzes the break-up of large organic molecules into smaller parts by bond-splitting reactions with H_2O.

Hydrophilic Dissolving readily in water ("water loving").

Hydrophobic Not dissolving readily in water ("water hating"); typically lipid-soluble.

Hydrostatic pressure The sort of pressure that is developed in a fluid (gas or liquid) when forces are applied that tend to increase the amount of matter per unit of volume. Hydrostatic pressure is what is meant by everyday scientific uses of the word *pressure*. A tire pump, for instance, produces a hydrostatic pressure in a tire.

Hyper-hyposmotic regulator An aquatic animal that maintains a blood osmotic pressure that is (1) higher than the osmotic pressure of the water in which it lives when the ambient osmotic pressure is low, but (2) lower than the osmotic pressure of the water in which it lives when the ambient osmotic pressure is high. Such an animal may have a stable blood osmotic pressure over a wide range of ambient osmotic pressures.

Hyper-isosmotic regulator An aquatic animal that maintains a blood osmotic pressure that is (1) higher than the osmotic pressure of the water in which it lives when the ambient osmotic pressure is low, but (2) the same as the osmotic pressure of the water in which it lives when the ambient osmotic pressure is high.

Hyperbolic kinetics A type of saturation kinetics in which the velocity of a chemical reaction increases in a smooth, strictly asymptotic way toward its maximum.

Hyperosmotic Having a higher osmotic pressure. Said of a solution in comparison to another, specific solution. *Hyperosmotic* is a relative term that is meaningless unless the comparison solution is specified.

Hyperosmotic regulator An aquatic animal that maintains a blood osmotic pressure higher than the osmotic pressure of the water in which it lives.

Hyperpolarization A voltage change that makes a cell membrane potential more inside-negative (normally moves it further from zero).

Hyperthermia The state of having a body temperature that is higher than the temperature considered to be normal or usual.

Hypertrophy The addition of structural components to cells in a way that increases the size of a tissue or organ. In muscle, the addition of contractile proteins to skeletal muscle fibers.

Hyposmotic Having a lower osmotic pressure. Said of a solution in comparison to another, specific solution. *Hyposmotic* is a relative term that is meaningless unless the comparison solution is specified. The word is a contraction of *hypo-osmotic*.

Hyposmotic regulator An aquatic animal that maintains a blood osmotic pressure lower than the osmotic pressure of the water in which it lives.

Hypothalamo–hypophysial portal system A system of blood vessels in a vertebrate that connects capillaries in the hypothalamus to capillaries in the anterior pituitary; provides a direct pathway by which hypothalamic hormones can reach specific populations of cells in the adenohypophysis (anterior pituitary).

Hypothermia The state of having a body temperature that is lower than the temperature considered to be normal or usual.

Hypoxia Referring to the tissues of an animal, the state of having an unusually low level of O_2.

Hypoxia-inducible factor (HIF) A transcription factor that increases in concentration in the cytosol of a cell when the cell experiences hypoxia and that enters the nucleus, where it activates genes coding for hypoxia responses.

I

I band In striated muscle, a region of two sequentially adjacent sarcomeres that includes only thin filaments and is bisected by the Z disc; shortens during contraction. It appears light in muscle sections prepared for microscopy; takes its name from the fact that it is described as *isotropic* because its refractive index in polarized light changes only minimally with the plane of polarization.

Ice-nucleating agent A dissolved or undissolved substance that promotes freezing (i.e., limits supercooling).

Ideal gas law See *universal gas law*.

Imidazole group A type of chemical group, found on the amino acid histidine and some other compounds, that is often extremely important in buffering body fluids because of its particular chemical buffering properties.

Implantation During pregnancy in a female placental mammal, the entry of an early embryo into the cellular matrix of the inner uterine epithelium (*endometrium*).

Imprinting (in molecular genetics) See *genomic imprinting*.

In phase Occurring at approximately the same phase of a cycle as something else. Two events that occur at approximately the same phase of a cycle are said to be in phase.

Inactivation (in ion transport) The closing of an ion channel in response to a stimulus such as membrane depolarization. This occurs in a time-dependent manner.

Indirect calorimetry Measurement of metabolic rates by quantifying respiratory gas exchange or some other property besides heat and external work.

Indirect measurement A measurement procedure that quantifies a property by measuring something other than what the definition of the property specifies. Contrast with *direct measurement*.

Induced ovulation   Ovulation (release of an egg from the ovaries of a female) that results from, and is dependent on, stimuli generated by the actual act of copulation. Contrast with *spontaneous ovulation*.

Induced ovulator   A species in which a female ovulates as a direct response to copulation.

Inducible enzyme   An enzyme (or other protein) that is expressed only when "induced" by the presence of a molecule or condition that serves as an inducing agent. An inducible enzyme disappears from a cell or tissue when its inducing agent is absent, but is expressed when the cell or tissue is exposed to its inducing agent. Contrast with *constitutive enzyme*.

Infrabranchial sinus   In a crustacean, a ventral sinus where blood collects after perfusing the systemic tissues, just prior to return to the gills.

Ingested chemical energy   Synonym for *ingested energy*.

Ingested energy   The energy present in the chemical bonds of an animal's food.

Inhibition   Decreasing the probability or frequency of action potentials.

Inhibitory postsynaptic potential (IPSP)   A voltage change in a postsynaptic cell—normally a hyperpolarization—that tends to inhibit the cell.

Inhibitory   Resulting in a decrease in activity or probability of activity. In a neuron, hyperpolarization is inhibitory because it decreases the likelihood of generation of action potentials. Contrast with *excitatory*.

Inka cells   Endocrine cells associated with the tracheae of insects that secrete two peptides: pre-ecdysis triggering hormone and ecdysis triggering hormone.

Innate behavior   Behavior that has a strong genetic basis or results from genetic pre-programming.

Inner ear   In a vertebrate, the cochlea and the semicircular canals of the vestibular organ.

Inner hair cells   In the organ of Corti of the cochlea, a single row of hair cells that transduce sound vibrations into voltage changes that excite auditory sensory neurons.

Inner segment   The portion of a retinal rod or cone that does not contain specialized light-sensitive membranes.

Innervate   To provide neural input.

Inorganic ion   A non-carbon atom (e.g., Na^+) or group of atoms (e.g., SO_4^{2-}) that bears a net negative or positive charge.

Inositol trisphosphate (IP₃)   A molecule derived from membrane phospholipids that acts as a second messenger in cell signaling.

Insensible water loss   In humans, water loss across the skin without sweating; a synonym for *transpirational water loss* in people.

Insertion   The movement of channel or transporter proteins from inactive intracellular locations into the membrane where they are active.

Inspiratory reserve volume   The volume of air that can be taken into the lungs following a resting inhalation, up to the maximum lung inflation that can be achieved.

Instar   A period between ecdyses, or molts, in the arthropod life cycle.

Insulation   The resistance to dry heat transfer through a material or between an animal and its environment. See also *resistance* (meaning 2).

Integral membrane protein   A protein that is part of a cell membrane and cannot be removed without extraction procedures that take the membrane apart.

Integration   The coordination of input signals, as by summing, to provide a harmonious control of output. *Cellular integration* refers to the integration of signals within a cell, and *physiological integration* refers to the integration of sensory, central nervous system, and endocrine signals for harmonious control of effectors in the body.

Integument   The outer body covering of an animal, such as the skin of a vertebrate or the exoskeleton of an arthropod.

Intensity difference   In sensory processes, intensity refers to the amount of a stimulus, as opposed to the kind of a stimulus (modality or quality).

Intercalated disc   An intercellular contact between adjacent cardiac muscle fibers of vertebrates that contains desmosomes and gap junctions.

Intercellular fluids   See *interstitial fluids*.

Intercellular   Between cells.

Intercostal muscles   Sheetlike muscles that run between adjacent ribs, the contraction of which expands or contracts the volume of the rib cage.

Intermittent breathing   Breathing in which breaths or sets of breaths are regularly interrupted by extended periods of apnea.

Internal effectors   Effectors other than skeletal muscle, such as glands, cardiac muscle, and smooth muscle, which are controlled by the autonomic nervous system.

Internal environment   The environment of the cells within an animal's body. The set of conditions—temperature, pH, Na^+ concentration, and so forth—existing in the body fluids of an animal and therefore experienced by cells within the animal's body. Also called the *milieu intérieur*.

Internal work   Mechanical work performed inside the body, such as the beating of the heart or peristalsis in the gut.

Interneuron   A neuron that is confined to the central nervous system and is therefore neither a sensory neuron nor a motor neuron.

Internode   The region of a myelinated axon that lies between two nodes of Ranvier and is covered by a myelin sheath.

Interoceptor   A sensory receptor cell that is activated by stimuli within the body and thus monitors some aspect of the internal state.

Interspecific enzyme homologs   Different molecular forms of a single enzyme occurring in two or more related species; called *homologs* because they are assumed to be related by evolutionary descent.

Interstitial cells   See *Leydig cells*.

Interstitial fluids   The fluids between cells in tissues. More specifically, in animals with closed circulatory systems, the fluids between cells in tissues other than blood; that is, the interstitial fluids are the extracellular fluids other than the blood plasma. Sometimes called *intercellular fluids* or *tissue fluids*.

Interval timer   A biological clock that times an interval shorter than a day but appears to be noncyclic, having to be restarted each time it operates, like a kitchen timer or an hourglass. Also called an "hourglass" timer.

Intracellular digestion   Digestion that takes place within cells.

Intracellular fluids   The aqueous solutions inside cells. Also called the *cytosol*.

Intracellular membrane   A membrane, such as the endoplasmic reticulum, found inside a cell. Typically has a phospholipid-bilayer structure similar to that of the outer cell (plasma) membrane. Also called a *subcellular membrane*.

Intracrine   A peptide signaling molecule that regulates intracellular functions and is known to act as a hormone, paracrine, or autocrine in different contexts. It may be retained within the cell that synthesized it or internalized from the extracellular space.

Intrafusal muscle fiber   A specialized muscle fiber associated with a vertebrate muscle-spindle stretch receptor. Muscle stretching or activation of the intrafusal muscle fiber by a gamma motor neuron can activate the stretch receptor.

Intraluminal digestive enzyme   A digestive enzyme that is secreted into the lumen of the gut or another body cavity, where it mixes with food materials and digests them.

Inulin   A polysaccharide widely used in studies of plasma clearance. See *plasma clearance*.

Iodothyronines The amine hormones thyroxine and triiodothyronine, derived from the amino acid tyrosine and synthesized in the thyroid gland.

Ion An atom or group of atoms that bears a net negative or positive charge.

Ionic conformity A state in which the concentration of an inorganic ion in the blood plasma matches, and varies with, the concentration of that ion in the external environment.

Ionic hypothesis The organizing principle that the distributions of ions across the membrane of a cell, and the permeability of the membrane to ions, determine membrane potentials and cellular electrical events.

Ionic regulation The maintenance of a constant or nearly constant concentration of an inorganic ion in the blood plasma regardless of the concentration of that ion in the external environment.

Ionotropic receptor (in synaptic function) A neurotransmitter receptor molecule that changes the membrane permeability of the postsynaptic cell to particular ions when it binds neurotransmitter molecules; usually a *ligand-gated channel*.

Ionotropic transduction (in sensory function) A kind of sensory transduction in which a sensory receptor molecule is itself an ion channel, changing ion flow into the cell in direct response to a sensory stimulus. It is analogous to ionotropic synaptic action, and stands in contrast to metabotropic (sensory) transduction. See also *metabotropic transduction*.

Islets of Langerhans Clusters of endocrine cells distributed among the exocrine-gland tissue of the pancreas. They contain specific endocrine cells that secrete specific hormones: β (or B) cells secrete insulin, α (or A) cells glucagon, and δ (or D) cells somatostatin.

Isoform In relation to a protein that exists in multiple molecular forms, any one of those molecular forms.

Isolated system A defined part of the material universe that (at least as a thought exercise) cannot exchange either matter or energy with its surroundings.

Isometric contraction A contraction in which a muscle does not shorten significantly as it exerts force (tension) against a load it cannot move.

Isosmotic Having the same osmotic pressure. Said of a solution in comparison to another, specific solution. *Isosmotic* is a relative term that is meaningless unless the comparison solution is specified.

Isosmotic line A line, entered on a graph of internal versus external osmotic pressure, that connects all the points where internal

osmotic pressure equals external osmotic pressure.

Isotonic contraction A contraction in which a muscle changes its length as it exerts force (tension) against a load. The tension during contraction remains constant and equal to the force exerted by the load.

Isozymes Various molecular forms of a single kind of enzyme synthesized by one species. Also called *isoenzymes*.

Iteroparity A type of reproductive life history in which individuals are physiologically capable of two or more separate bouts of reproduction during their lives. Contrast with *semelparity*.

Iteroparous Characterized by *iteroparity*.

J

Juvenile hormone In insects, a lipid-soluble hormone secreted by the nonneural endocrine cells of the corpora allata that stimulates its target tissues to maintain immature characteristics.

Juxtaglomerular apparatus An assembly of cells in a vertebrate nephron where the juxtaglomerular cells are found. See *juxtaglomerular cells*.

Juxtaglomerular cells Specialized smooth muscle cells in the arterioles associated with the glomerulus of a vertebrate nephron that secrete renin in response to low blood pressure. Also called *granular cells*.

K

Kairomone A chemical signal released by a member of one species in the course of its activities that is detected and exploited by a member of another species. Its detection confers a benefit on the detecting species that is detrimental to the releasing species.

k_{cat} See *catalytic rate constant*.

Kidney An organ that regulates the composition and volume of the blood and other extracellular body fluids by producing and eliminating from the body an aqueous solution (*urine*) derived from the blood or other extracellular fluids.

K_m See *Michaelis constant*.

Knockout animal An animal that has been genetically engineered to lack functional copies of a gene of interest.

L

Labeled lines The principle that sensory neurons encode the modality or quality of a sensory stimulus by having different sensory receptor cells respond to different kinds of stimuli, so that the CNS can decode the stimulus by monitoring which axons ("lines") deliver action potentials.

Lactase persistence In human biology, continued synthesis (expression) in adulthood of the digestive enzyme lactase,

permitting milk sugar—lactose—to be digested in adulthood.

Lactation In mammals, the process of producing milk and providing the milk to nursing offspring.

Lacunae In an open circulatory system, minute spaces between nonvascular cells that serve as passageways for blood flow.

Laminar flow Flow of a fluid without turbulence. When a fluid flows without turbulence over a surface or through a tube, the flow can be envisioned as occurring in a series of thin layers (laminae) of fluid at progressively greater distances from the surface or tube walls. The layer juxtaposed to the surface or tube wall does not flow at all, and layers at increasing distances from it flow faster and faster.

Landmark A recognizable natural or artificial feature used for navigation, a feature that stands out from its near environment and may be visible from a long distance.

Laplace's law The tension developed within the walls of a hollow structure exposed to a particular difference in pressure between inside and outside is directly related to the radius of the structure.

Larva A free-living developmental stage that is very different in appearance from the adult stage of the same species.

Latch state A characteristic state of smooth muscle in which dephosphorylated myosin heads remain attached to actin and maintain tension for long periods of time.

Latent heat of vaporization The heat that must be provided to convert a material from a liquid to a gas at constant temperature (called *latent* because although heat is provided, the temperature of the material does not rise). It is expressed per unit of mass of the material under study.

Lateral geniculate nucleus (LGN) A region of the thalamus in the diencephalon of the vertebrate brain that receives axons of retinal ganglion cells and relays visual information to the primary visual cortex.

Lateral pathways In the retina, neural pathways that extend parallel to the retinal surface, rather than from the external to the internal side. Contrast with *straight-through pathways*.

Leaflet One of the two phospholipid layers within a cell membrane.

Leak channel A channel in a cell membrane that is normally open and mediates the membrane permeability (primarily to K^+ ions) that underlies the resting membrane potential.

Left-to-right shunting In the study of blood flow through the central circulation of an air-breathing fish, amphibian, or nonavian reptile, diversion of pulmonary venous blood directly back to the lungs, bypassing the systemic circuit.

Length constant (λ)   The distance along a cell over which a change in passively spreading electrical potential decays in amplitude to 37% of its amplitude at the origin.

Length–tension relationship   The relationship between the length of a whole muscle or a sarcomere prior to stimulation and the tension developed during isometric contraction. Maximum tension develops when there is optimal overlap of thick and thin filaments to allow effective cross-bridge action.

Lengthening contraction   An isotonic contraction in which a muscle lengthens as it is activated and exerts force. Also called an *eccentric contraction*.

Leptin   A peptide hormone secreted by white adipose tissue (white fat) that in some animals tends to inhibit feeding.

Leydig cells   Testosterone-secreting cells located in the connective tissue between the seminiferous tubules of the vertebrate testis; also called *interstitial cells*.

LH surge   In mammalian reproduction, a sharp increase in a female's blood concentration of luteinizing hormone, serving to cause ovulation.

Ligand   Any molecule that selectively binds noncovalently to a structurally and chemically complementary site on a specific protein. The substrate of an enzyme, for example, is a ligand of that enzyme.

Ligand-gated channel   A gated channel that opens to allow diffusion of a solute as a result of binding by a neurotransmitter (or other specific signaling molecule) to a receptor site on the channel protein.

Linear heat-transfer equation   An equation that relates heat loss from an animal to the difference between body temperature and ambient temperature by means of a proportionality coefficient. See Equation 10.9.

Linear velocity   In a moving fluid, the linear distance covered per unit of time.

Lipase   A type of lipid-digesting enzyme.

Load compensation   The ability to control the shortening of a muscle somewhat independently of the load against which it shortens.

Load   The force against which a contracting muscle exerts an opposing force, the latter being referred to as muscle tension.

Load–velocity relationship   The principle that the velocity of shortening of a muscle during isotonic contraction decreases as load increases.

Loading   Oxygenation of a respiratory pigment.

Long-term potentiation (LTP)   A stable, long-lasting increase in the amplitude of the response of a neuron after it has been stimulated repeatedly by presynaptic input at a high frequency.

Loop (circuit)   A loop circuit is one that feeds back onto itself; e.g., locus A sends a signal to B, which sends a signal back to A.

Loop of Henle   A portion of a nephron tubule shaped like a hairpin, found in the nephrons of mammals and some of the nephrons of birds, that is the site of countercurrent multiplication, the process responsible for production of urine hyperosmotic to the blood plasma.

Low-grade energy   Energy in a form that cannot do any physiological work (i.e., heat).

Lower-critical temperature   In a homeotherm, the ambient temperature that represents the lower limit of the thermoneutral zone. See also *thermoneutral zone*.

Luciferase   An enzyme that catalyzes oxidation of a luciferin, resulting in light production, during bioluminescence. Many different chemical forms of luciferase are known.

Luciferin   A compound capable of light emission during bioluminescence. Light emission occurs when luciferin is oxidized by an enzyme luciferase. Many different chemical forms of luciferin are known.

Lumen   The open central cavity or core of a "hollow" organ or tissue; for example, the open central core of a blood vessel.

Luminal   Related to the lumen.

Lungfish   See *dipnoan*.

Lungs   In the most general sense used by physiologists, structures specialized for external respiration that are invaginated into the body and thereby contain the environmental medium. Contrast with *gills*.

Luteal phase   In a female mammal, the ovarian phase of the menstrual or estrous cycle that follows ovulation, during which the corpus luteum forms, is functional, and then degenerates.

Luteinizing hormone (LH)   A tropic peptide chemical signal secreted by cells in the anterior pituitary gland (adenohypophysis); travels in the general circulation to the gonads; promotes secretion of sex hormones.

Lymph   In an animal with a closed circulatory system, the interstitial fluids. *Lymph* is often used in a more restrictive sense to refer specifically to excess volumes of interstitial fluid that are collected from the tissues and returned to the blood by the lymphatic vascular system.

Lymphatic vascular system   An elaborate system of vessels that approximately parallels the blood vascular system in vertebrates and serves to remove excess interstitial fluids (lymph) from tissues throughout the body, returning the fluids ultimately to the blood plasma.

M

M line   In muscle fibers, a web of accessory proteins at the center of a sarcomere that anchors the thick filaments and titin.

Macula   In the vertebrate ear, a sensory area in the vestibular organs containing hair cells that monitor tilt and acceleration of the head.

Magnetic compass   A mechanism by which an animal uses Earth's magnetic field to determine compass direction in navigation.

Magnetoreception   A mechanism by which specialized animal cells are able to detect Earth's magnetic field by direct or indirect means.

Magnocellular neurons   Large neurosecretory neurons with cell bodies in the supraoptic and paraventricular nuclei of the mammalian hypothalamus; synthesize oxytocin and vasopressins and secrete them into the general circulation from axon terminals in the pars nervosa

Maintenance   The set of processes that maintain the status quo in an animal's body, including circulation, breathing, and tissue repair.

Malpighian tubules   Fine tubules that initiate urine formation in insects. They empty their product into the gut at the junction of the midgut and hindgut.

Mantle   In a mollusc, a sheet of tissue that secretes the shell (in those molluscs with a shell) and that encloses an external body cavity, the *mantle cavity*. In squids, the mantle is muscular and pressurizes fluid for jet propulsion.

Mantle cavity   In molluscs, an external body cavity formed where a sheetlike outfolding of the dorsal body wall, the *mantle*, overhangs or surrounds all or part of the rest of the body. The gills typically are suspended in the mantle cavity.

Map-and-compass navigation   Navigation in which an animal has information about where it is in the world (the "map") and about the direction it should take to reach a destination (the "compass").

Maps   A map is a geometrical representation of space and spatial relationships. In animal navigation a map is an internally stored, geographical representation of a part of the world around an animal. A map may also refer to the topological representation in the brain of the body surface (somatotopic map) or the outside world (retinotopic map, auditory map).

Mark   See *epigenetic mark*.

Mass   An object's mass is a measure of the *amount of matter* in the object. In physics, this is defined and measured in terms of inertia relative to a standard.

Mass action   The inherent tendency for the reactants and products of a chemical reaction to shift in their concentrations, by way of the reaction, until an equilibrium state—defined by particular concentration ratios—is achieved. Mass action impels reactions toward equilibrium, although other processes may operate against equilibrium and therefore prevent an equilibrium from actually being established.

Maximal aerobic power   A synonym for $\dot{V}_{O_2max}$.

Maximal exercise   Exercise that requires an individual's maximal rate of O_2 consumption.

Maximum reaction velocity (V_{max})   The greatest rate at which an enzyme-catalyzed reaction can convert substrate to product with the number of active enzyme molecules that are present.

Mean pressure   During the beating cycle of the heart, the average blood pressure (averaged over the full beating cycle).

Mechanical energy   Energy of *organized* motion in which many molecules move simultaneously in the same direction. The energy of motion of a moving arm provides an example. Mechanical energy and heat are the two forms of kinetic energy.

Mechano growth factor   A locally acting peptide, produced by muscle fibers when they are stimulated to produce mechanical activity, that stimulates muscle satellite cells to proliferate and become incorporated into the muscle fibers.

Mechanoreceptor   A sensory receptor cell specialized to respond to mechanical stimulation.

Medulla   The interior tissue of a kidney. The interior tissue of the adrenal gland.

Melanocyte-stimulating hormone (MSH)   A direct acting hormone secreted by cells of the anterior pituitary gland (adenohypophysis); derived from proopiomelanocortin; promotes skin darkening in amphibians, reptiles (except birds), and fish.

Melatonin   An amine hormone derived from the amino acid tryptophan; synthesized in and secreted from the pineal gland; influences circadian and seasonal rhythms; promotes sleep.

Membrane fluidity   See *fluidity*.

Membrane potential   The potential difference (voltage) across a cell membrane or other selectively permeable membrane.

Membrane resistance (R_m)   The electrical resistance of a membrane per unit of area. Many cell membranes have an R_m of about 1000 ohm × cm^2 (1000 Ω × cm^2). (The unit is ohm × cm^2 rather than ohm/cm^2 because membrane resistance *decreases* as area increases, as more resistances are added in parallel.)

Membrane-associated digestive enzyme   A digestive enzyme that is positioned in the apical membranes of epithelial cells lining the gut or another body cavity. Food materials in the gut lumen must make physical contact with the gut epithelium to be digested by this sort of enzyme.

Menarche   First menstruation.

Menstrual cycle   The cycle in which oocytes mature and are ovulated periodically in females of some primate species; one phase of each cycle is characterized by menstruation, the shedding of the uterine lining in a blood-tinged discharge from the vagina.

Mesencephalon   The midbrain region of the vertebrate brain.

Mesic   Moderately moist. "Mesic animals" are terrestrial animals that live in moderately moist environments.

Metabolic acidosis or alkalosis   Acidosis or alkalosis that is caused by abnormal excretion or retention of bicarbonate (HCO_3^-) in the body fluids. Contrast with *respiratory acidosis* or *alkalosis*.

Metabolic depression   A reduction in the ATP needs of an animal (or a specific tissue) to below the level ordinarily associated with rest in a way that does not present an immediate physiological threat to life.

Metabolic rate   An animal's rate of energy consumption; the rate at which it converts chemical-bond energy to heat and external work.

Metabolic scaling   The regular allometric or quasi-allometric relationship that typically exists between metabolic rate and body weight within sets of related species. For example, small-bodied mammals have higher rates of metabolism per unit of body weight than large-bodied species do. This relationship, which is approximately allometric, represents a case of metabolic scaling.

Metabolic water   Water that is formed by chemical reaction within the body. For example, when glucose is oxidized, one of the products is H_2O that did not previously exist. Also called *oxidation water*. Contrast with *preformed water*. See also *net metabolic water production*.

Metabolism   The set of processes by which cells and organisms acquire, rearrange, and void commodities (e.g., elements or energy) in ways that sustain life.

Metabolism–size relation   See *metabolic scaling*.

Metabolite   An organic molecule of relatively low molecular weight (e.g., glucose, an amino acid, or lactic acid) that is currently being processed by metabolism.

Metabolomics   The study of all the organic compounds in cells or tissues other than macromolecules coded by the genome. The molecules encompassed by metabolomics are generally of relatively low molecular weight (roughly <1500 daltons).

Metabotropic receptor (in synaptic function)   A neurotransmitter receptor that acts via signal transduction to alter a metabolic function of the postsynaptic cell, often by stimulating production of a second messenger.

Metabotropic transduction (in sensory function)   Sensory transduction by means of a signal transduction cascade, rather than by direct ionotropic action. The sensory receptor molecule is a G protein that activates a second messenger, ultimately producing a receptor potential. See also *ionotropic transduction*.

Metalloprotein   A protein that includes one or more metal atoms in its structure.

Metamorphosis   The process of changing from one body form to another, such as changing from a larva to an adult.

Metencephalon   The anterior part of the vertebrate hindbrain, containing the cerebellum and pons.

Methylation   Covalent bonding of a methyl group (—CH_3) to another molecule.

Methylome   In the study of epigenetics, the sum total of an animal's methylated DNA sites.

Michaelis constant (K_m)   The half-saturation constant of an enzyme-catalyzed reaction that exhibits hyperbolic kinetics—that is, the concentration of substrate at which the reaction velocity is half of the maximal velocity. This constant is a measure of enzyme–substrate affinity.

Michaelis–Menten equation   An equation that describes the relation between reaction velocity and substrate concentration in an enzyme-catalyzed reaction that exhibits hyperbolic kinetics. See Equation 2.2.

Microarray   See *DNA microarray*.

Microbiome   See *gut microbiome*.

Microcirculatory bed   In a closed circulatory system, one of the systems of microscopically fine blood vessels that connect arteries and veins, weaving among ordinary tissue cells. In vertebrate systemic tissues, microcirculatory beds consist of arterioles, capillaries, and venules.

Microclimate   The set of climatic conditions (temperature, humidity, wind speed, and so forth) prevailing in a subpart of a larger environmental system.

Microenvironment   A place within a larger environment in which the physical and chemical conditions differ significantly from the average conditions characterizing the larger environment.

Microvilli   Microscopic finger-shaped projections from the apical membranes of certain types of epithelial cells. Microvilli occur most commonly in epithelia that are active in secretion or reabsorption.

Midbrain Middle region of the vertebrate brain, between the forebrain and the hindbrain.

Middle ear The portion of the vertebrate ear between the tympanic membrane and the cochlea.

Midgut In a person, the small intestines. More generally, the initial part of the intestines.

Migration A seasonal or other periodic movement of animals from one geographic region to another.

Milieu intérieur See *internal environment*.

Milk ejection (or let-down) In a lactating female mammal, release of milk from the mammary glands under positive pressure.

Miniature EPSP (mEPSP) A small excitatory postsynaptic potential at a neuromuscular junction or postsynaptic neuron produced by presynaptic release of a single quantal packet of neurotransmitter.

Minute volume The amount of air, water, or blood pumped per minute in a breathing system or circulatory system.

Mirror neuron A neuron in the primate brain that is activated when the animal performs a particular movement, or watches another animal perform the same movement.

Mitochondria-rich cell A type of epithelial cell specialized for ion transport, found particularly in the gill epithelia of fish but also in some other epithelia of aquatic animals. Characterized by an abundance of mitochondria and other signs of secretory or absorptive activity. Also called a *chloride cell*.

Mixed venous blood The blood in the great veins leading back to the heart, formed by mixing of the venous blood coming from the various regions and organs of the body.

Modality The subjective sensation of a particular sense, such as sight, taste, or hearing.

Molal A unit of measure of chemical concentration; specifically, the molal concentration is the number of moles mixed with a kilogram of water to make a solution.

Molar A unit of measure of chemical concentration; specifically, the molar concentration is the number of moles dissolved in a liter of solution.

Mole (mol) A set of 6.022×10^{23} items (i.e., an Avogadro's number of items). For example, a mole of a chemical compound contains 6.022×10^{23} molecules of the compound. The mass in grams of a mole of a chemical compound is identical to the formula mass of the compound. Thus, if you have 18 g of water, a compound that has a formula mass of 18, you have one mole of water, or 6.022×10^{23} water molecules. See also *Avogadro's number, osmole*.

Molecular chaperones Proteins that use energy from ATP to guide the folding of other proteins into correct three-dimensional configurations. Molecular chaperones ensure that target proteins assume correct tertiary structures during initial synthesis, and they sometimes are able to guide reversibly denatured proteins back to functional conformations, thereby preventing permanent denaturation.

Molecular kinetic energy See *heat*.

Molting hormone See *ecdysone*.

Molting See *ecdysis*.

Molts Processes that cast off body coverings, such as feathers, fur, and the cuticle (by ecdysis in hemimetabolous and holometabolous insects).

Monounsaturated fatty acid See *unsaturated fatty acid*.

Motility Any sort of muscular activity by the gut, such as peristalsis or segmentation.

Motor cortex The region of the mammalian cerebral cortex that lies anterior to the central sulcus and is concerned with motor behavior; includes the *primary motor cortex* and associated motor areas in the frontal lobe.

Motor neuron A neuron that conveys motor signals from the central nervous system to the periphery to control an effector such as skeletal muscle.

Motor unit A motor neuron and all the muscle fibers it innervates.

Mucosal surface See *apical surface*.

Müllerian inhibitory substance A glycoprotein hormone secreted by Sertoli cells in the developing testes of genetic males that causes degeneration of the Müllerian ducts, which would otherwise give rise to the uterus and oviducts. Also called *anti-Müllerian hormone, Müllerian inhibitory hormone*.

Multiterminal innervation A pattern of innervation in which a single axon branches near its end to make many synaptic contacts along the length of a muscle fiber.

Multiunit smooth muscles A muscle made up of unstriated muscle cells that are not connected to neighboring cells by gap junctions; receive input from the autonomic nervous system, as well as endocrine and paracrine stimuli. Contrast with *single-unit smooth muscle*.

Muscarinic Action of acetylcholine that is mimicked by muscarine, or receptors mediating the action. Contrast with *nicotinic*.

Muscle A group of muscle cells (muscle fibers) and associated tissues. It may be smooth, skeletal, or cardiac.

Muscle cells contractile cells that constitute muscle tissues.

Muscle fiber A term typically used to refer to skeletal or cardiac muscle cells, but sometimes also applied to smooth muscle cells. This term, now used broadly, was initially introduced to emphasize the long, fibrous appearance of certain muscle cells, such as vertebrate skeletal muscle cells (fibers), that are multinucleate because they are formed during development by the fusion of two or more embryonic myoblasts.

Muscle spindle A stretch receptor that is arranged in parallel with the tension-producing fibers of a muscle and sends action potentials to the central nervous system when the muscle is stretched.

Myelencephalon Posterior part of the vertebrate hindbrain.

Myelin An insulating sheath around an axon, composed of multiple wrappings of glial cell membranes, that increases the velocity of propagation of action potentials.

Myoblasts uninucleate embryonic muscle cells that fuse to form multinucleate muscle fibers

Myocardium The muscle tissue of a heart.

Myofibril A longitudinal component of a striated muscle cell that consists of a series of sarcomeres and extends the length of the cell. In cross section, a muscle cell consists of multiple myofibrils, each surrounded by a sleeve of sarcoplasmic reticulum.

Myofilament Either of the two types of longitudinal components of sarcomeres. Thick myofilaments consist of polymerized myosin molecules, and thin myofilaments consist of polymerized actin molecules.

Myogenic heart A heart in which the electrical impulse to contract during each beating cycle originates in muscle cells or modified muscle cells. Contrast with *neurogenic heart*.

Myoglobin Any type of hemoglobin found in the cytoplasm of muscle.

Myometrium In the mammalian uterus, the outer tissue layer, consisting of smooth muscle.

Myosin A molecular motor found in many types of cells that converts chemical energy of ATP into mechanical energy of motion. In muscle cells, it functions as a contractile protein. Myosin monomers polymerize to form thick myofilaments.

Myostatin An inhibitory growth factor that regulates the growth of skeletal muscles by tempering hyperplasia in developing muscles and hypertrophy of adult muscle fibers.

N

Na⁺–K⁺-ATPase A ubiquitous and extremely important transporter protein that directly cleaves ATP molecules to release energy (it is an ATPase) and uses the energy to transport Na^+ and K^+ ions in a 3:2 ratio (a case of primary active transport). The ion gradients created are often used as energy sources for secondary active transport.

Na⁺–K⁺ pump Active transport of Na⁺ out of a cell and K⁺ into a cell by Na⁺–K⁺-ATPase.

Natriuretic Promoting loss of sodium in the urine.

Navigation The act of moving on a particular course or toward a specific destination using sensory cues to determine direction and position.

Nebulin In a sarcomere, a large inelastic protein that extends along the thin filament from the Z disc to the margin of the H zone.

Needle biopsy A procedure that uses a specially constructed needle to insert into a tissue and extract a small sample for study.

Negative feedback A process by which the deviations of a property from a specific set-point level are opposed, thereby tending to keep the property at the set-point level.

Nephridium (plural nephridia) A term used to refer to a kidney in certain types of animals such as molluscs.

Nephron One of the tubules that forms urine in the kidney of a vertebrate.

Nernst equation An equation used to determine the equilibrium electrical potential for a particular ion, given the ion concentrations on both sides of a membrane.

Nerve A collection of axons running together in the peripheral nervous system.

Nerve impulse See *action potential*.

Nerve net A simple, uncentralized, and unpolarized network of neurons, found in cnidarians and locally in many other groups and considered to be a primitive stage in the evolution of nervous systems.

Nervous system An organized constellation of neurons and glial cells specialized for repeated conduction of electrical signals (action potentials) within and between cells. These signals pass from sensory receptors and neurons to other neurons and effectors. Nervous systems integrate the signals of convergent neurons, generate new signals, and modify the properties of neurons based on their interactions.

Net growth efficiency In a growing animal, the chemical-bond energy of new biomass added by growth expressed as a ratio of the animal's absorbed energy over the same time period.

Net metabolic water production The production of metabolic water by a process minus the losses of water that are obligatory for that process to take place.

Network oscillator A neural circuit that acts as a central pattern generator as a result of the synaptic interaction of its constituent neurons, rather than as a result of oscillator activity in single cells.

Neurogenic heart A heart in which the electrical impulse to contract during each beating cycle originates in neurons. Contrast with *myogenic heart*.

Neuroglia See *glial cells*.

Neurohemal organ An organ made up of axon terminals of neurosecretory cells in association with a well-developed bed of capillaries or other circulatory specializations, in which the axon terminals store neurohormones and secrete them into the blood.

Neurohormone A hormone secreted by a neuron (also called a neuroendocrine cell or neurosecretory cell) into the blood.

Neurohypophysis The neuroendocrine portion of the vertebrate pituitary gland in which neurohormones are released from axon terminals. It is an extension of the brain, commonly called the posterior pituitary. See *posterior pituitary*.

Neuromuscular junction The synaptic junction of a motor neuron and a muscle fiber.

Neuron A nerve cell; the fundamental signaling unit of the nervous system, composed of a cell body and elongated processes—dendrites and axon—that carry electrical signals.

Neuron doctrine The theory that the nervous system, like other organ systems, is composed of discrete cellular elements (neurons) that are its fundamental signaling elements.

Neuronal integration The process by which a postsynaptic neuron sums the inputs from several presynaptic neurons to control its generation of action potentials.

Neuropil A CNS region of small neural processes and synapses that is relatively devoid of neural cell bodies.

Neurosecretory cell A neuron that synthesizes and releases hormones.

Neurosecretory glands Endocrine structures made of neurosecretory cells.

Neurotransmitter A molecule that is used as a chemical signal in synaptic transmission.

Neutral pH The pH of pure water. The neutral pH varies with temperature; at any given temperature, pH values less that the neutral pH represent an acid solution, whereas pH values above the neutral pH represent an alkaline solution.

Newton's law of cooling A common, although not necessarily historically defensible, synonym for the *linear heat transfer equation*.

Nicotinic Action of acetylcholine that is mimicked by nicotine, or receptors mediating the action. Contrast with *muscarinic*.

Nocturnal Active in the nighttime.

Node of Ranvier In the myelin sheath surrounding an axon, spaces between adjacent glial cells. These interruptions in the sheath allow propagation of action potentials by saltatory conduction.

Nonadaptive evolution Evolution that occurs by processes other than natural selection and that therefore can produce traits that are not adaptations. See also *evolution, genetic drift, pleiotropy*.

Noncovalent bond A chemical bond that does not involve covalent bonding. Because they are not covalent, noncovalent bonds are flexible rather than rigid, and they can be made and broken with relative ease, without enzyme catalysis. There are four types: hydrogen bonds, ionic bonds, van der Waals interactions, and hydrophobic bonds. Also called *weak bonds*.

Nonelectrogenic See *electroneutral pump*.

Nonessential amino acid A standard amino acid that an animal can synthesize for itself.

Nonneural endocrine cells Cells that synthesize and secrete hormones; also called epithelial endocrine cells

Nonneural glands Endocrine structures made of nonneural endocrine cells; also called epithelial glands

Nonpolar molecule A molecule in which electrons are evenly distributed, the various regions of which are all therefore similar in charge.

Nonshivering thermogenesis (NST) In mammals and some birds, elevation of heat production for thermoregulation by means other than shivering. The same mechanisms potentially function in body weight regulation by serving to get rid of the energy value of excess organic food molecules in the form of heat.

Nonspiking neuron A neuron that transmits information without generating action potentials.

Norm of reaction In phenotypic plasticity, the specific relations between environments and phenotypes. If environment A causes phenotype M to be expressed, and environment B causes phenotype N to be expressed, the relations A-leads-to-M and B-leads-to-N constitute the norm of reaction.

Nucleating agent See *ice-nucleating agent*.

Nucleus (of nervous system) A cluster of functionally related neuronal cell bodies in a vertebrate central nervous system.

Nutrition The study of the chemical components of animal bodies and how animals are able to synthesize those chemical components from the chemical materials they collect from their environments.

O

Obscurin A giant protein in skeletal muscle fibers that links the M-line to the sarcoplasmic reticulum; contributes to maintaining the structural organization of the muscle fiber.

Off-center cell A neuron in the retina that is hyperpolarized in response to light in the center of its receptive field. Contrast with *on-center cell*.

Ohm's law The relationship of voltage (V), current (I), and resistance (R): $V = IR$.

Olfaction The sense of smell; chemoreception of molecules released at a distance away from the animal. Among chemoreceptors, olfactory receptors are typically more sensitive than taste receptors and respond to distant or dilute chemical stimuli (odorants) that are usually airborne in terrestrial animals.

Oligodendrocyte A type of ensheathing glial cell (non-neuron cell) in the vertebrate central nervous system.

Omega 3 and omega 6 fatty acids Fatty acids characterized by double bonds at particular points in their molecular structures. If the first double bond encountered when scanning a fatty acid molecule from its methyl (—CH_3) end is the third bond in the carbon-chain backbone of the molecule, the fatty acid is an omega 3 fatty acid. If the first double bond occurs at the sixth position, the fatty acid is of the omega 6 type.

On-center cell A neuron in the retina that is depolarized in response to light in the center of its receptive field. Contrast with *off-center cell*.

Oncotic pressure Synonym for *colloid osmotic pressure*.

Ontogeny Individual development.

Oogenesis The formation of haploid eggs through the process of meiosis in the ovary.

Oogonia Diploid cells in the ovary that proliferate by mitosis, then go through meiosis to give rise to female gametes (ova). In elasmobranchs, birds, and most mammals, oogonia are generally thought to proliferate only during embryonic or fetal life and remain suspended in an early stage of meiosis until the female becomes reproductively mature.

Open circulatory system A circulatory system in which the blood leaves discrete vessels and bathes at least some nonvascular tissues directly, meaning that blood and interstitial fluid are the same. Contrast with *closed circulatory system*.

Open system A defined part of the material universe that is not isolated, meaning that it is capable of exchanging matter, energy, or both with its surroundings.

Operculum In a fish, the external flap on each side of the head that covers the gills.

Opsin The protein part of the photopigment rhodopsin, which is a G protein–coupled receptor molecule.

Optic flow The pattern of apparent motion of objects and images in a visual field that is caused by the movement of an animal relative to its surroundings.

Optogenetics A type of manipulative method for study of neuron function (or function of other excitable cells) in which light-gated (light-modulated) ion channels—obtained from microorganisms—are expressed (through genetic engineering) in neuronal membranes of animals. Light signals can then be used to modulate the membrane potential of the neuronal membranes—for example, by inducing depolarization on very short time scales.

Ordinary least squares regression See Appendix D.

Organ of Corti A region of the cochlea in the vertebrate ear containing the inner and outer *hair cells* that transduce sound vibrations into electrical signals.

Orientation The way an organism positions itself in relation to environmental cues.

Ornithine–urea cycle A cyclic metabolic pathway that produces the nitrogenous end product urea.

Oscillator Something that oscillates—moves rhythmically around a central point. A pendulum is an oscillator. Theoretical descriptions of central pattern generators refer to them as oscillators.

Osmoconformer An aquatic animal that allows its blood osmotic pressure to match and vary with the environmental osmotic pressure. Also called a *poikilosmotic animal*.

Osmoconformity See *osmotic conformity*.

Osmolar (Osm) A unit of measure of osmotic pressure. A 1-osmolar solution is defined to be a solution that behaves osmotically as if it has one Avogadro's number of independent dissolved entities per liter.

Osmolarity The osmotic pressure of a solution expressed in osmolar units. See *osmolar*.

Osmole An Avogadro's number of osmotically effective dissolved entities; a set of 6.022×10^{23} osmotically effective dissolved entities.

Osmolyte Any solute that exerts a sufficiently large effect on the osmotic pressure of a body fluid to be of consequence for water–salt physiology.

Osmoregulation See *osmotic regulation*.

Osmoregulator An animal that maintains an approximately constant blood osmotic pressure even as the osmotic pressure of its environment varies.

Osmosis The passive transport of water across a membrane.

Osmotic conformity A state in which the osmotic pressure of the body fluids matches, and varies with, the osmotic pressure in the external environment.

Osmotic effector An osmolyte that an individual animal or individual cell increases or decreases in amount to achieve the regulation of the osmotic pressure of a body fluid or the regulation of cell volume. Also called an *osmoticum*.

Osmotic permeability Permeability to water moving by osmosis. See *permeability*.

Osmotic pressure The property of a solution that allows one to predict whether the solution will gain or lose water by osmosis when it undergoes exchange with another solution; osmosis is always from lower osmotic pressure to higher osmotic pressure. An alternative, measurement-oriented definition is that the osmotic pressure of a solution is the difference in hydrostatic pressure that must be created between the solution and pure water to prevent any net osmotic movement of water when the solution and the pure water are separated by a semipermeable membrane. A colligative property.

Osmotic regulation The maintenance of a constant or nearly constant osmotic pressure in body fluids regardless of the osmotic pressure in the external environment.

Osmotically free water Water that is excreted in the urine above and beyond whatever water is absolutely required for solute excretion.

Osmoticum See *osmotic effector*.

Ossicles The ossicles of the middle ear (maleus, incus, stapes) are three bones, attached to each other in a series, that convey sound pressure waves from the tympanic membrane (eardrum) to the liquid-filled cochlea of the inner ear.

Ostia In the heart of an arthropod, valved passages through the heart muscle through which blood enters the heart chamber.

Otolith organs An otolith is a hard structure in the ear of a vertebrate (oto = ear, lith = stone) that stimulates hair cells to produce the sense of gravity and linear acceleration. The saccule and utricule of the inner ear are the otolith organs that mediate these senses.

Outer hair cells In the organ of Corti of the cochlea, the three rows of hair cells that amplify the sound-produced local movements that stimulate the inner hair cells to activate cochlear sensory neurons.

Outer segment The part of a retinal rod or cone that contains stacked membranes of photopigment and transduces light into an electrical signal.

Overexpression A genetic manipulation whereby cells or tissues produce unusually large quantities of one or more messenger RNAs of interest.

Ovulation The process of releasing an egg (ovum) from the ovary.

Ovum A haploid gamete (egg) produced in the ovary.

Oxidation Removal of electrons or hydrogen atoms from a molecule.

Oxidation water Synonym for *metabolic water*.

Oxidative phosphorylation The formation of ATP using energy released by the transport of electrons through the electron-transport chain.

Oxidative stress See *oxygen stress*.

Oxy- A prefix referring to a molecule that is capable of reversible combination with O_2 that is in a state of being so combined (e.g., oxyhemoglobin).

Oxygen affinity The readiness with which a respiratory pigment such as hemoglobin combines with O_2. When oxygen affinity is high, a low O_2 partial pressure is sufficient to cause extensive oxygenation.

Oxygen cascade The sequential drop in the partial pressure of O_2 from one step to the next in the series of steps by which O_2 is transported from the environment of an animal to the animal's mitochondria.

Oxygen conformity A response in which an animal exposed to a decreasing O_2 concentration in its environment allows its rate of O_2 consumption to decrease in parallel.

Oxygen debt An older term for *excess postexercise oxygen consumption*.

Oxygen deficit A difference between actual O_2 uptake by breathing and the theoretical O_2 requirement of exercise during the first minutes at the start of exercise. During the oxygen deficit phase, aerobic catabolism based on O_2 uptake by breathing is unable to meet fully the ATP requirement of exercise, and other ATP-producing mechanisms (such as phosphagen use) must contribute ATP.

Oxygen dissociation curve See *oxygen equilibrium curve*.

Oxygen equilibrium curve Referring to the O_2-carrying properties of blood, a graph of the amount of O_2 per unit of blood volume as a function of the O_2 partial pressure of the blood. Also called the *oxygen dissociation curve*.

Oxygen regulation A response in which an animal exposed to a decreasing O_2 concentration in its environment maintains a stable rate of O_2 consumption.

Oxygen stress Collectively speaking, the destructive effects of reactive oxygen species on functional or structural properties of cells or tissues. See *reactive oxygen species*.

Oxygen utilization coefficient (1) In breathing, the fraction (or percentage) of the total O_2 in respired air or water that is removed by the breathing process. (2)

In circulation, the fraction of the total O_2 carried by blood that is removed from the blood as it passes around the body.

Oxygen-carrying capacity In a body fluid (e.g., blood) that contains a respiratory pigment, the amount of O_2 per unit of fluid volume when the respiratory pigment is saturated with O_2. More generally, the maximal amount of O_2 per unit of volume in a fluid under ordinary physiological conditions.

Oxygen-transport pigment See *respiratory pigment*.

Oxygenation (1) Referring to water, the dissolution of O_2 in that water. (2) Referring to respiratory pigments, the combination of those pigments with O_2; oxygenation of this sort is reversible and not equivalent to oxidation.

Oxyhemoglobin Hemoglobin combined with O_2.

Oxytocin A vertebrate neurohormone synthesized by neurosecretory cells that have their cell bodies in the hypothalamus and their axon terminals in the pars nervosa of the posterior pituitary. Its major functions are to stimulate contraction of myoepithelial cells of the mammary glands to cause milk ejection and contraction of the myometrium during parturition.

P

P-type ATPases One of the major families of ATPases (transporter proteins that directly cleave ATP to obtain energy for active transport).

P/O ratio During electron transport and oxidative phosphorylation, the number of ATP molecules produced per oxygen atom reduced to H_2O. A measure of the degree of coupling that exists between electron transport and oxidative phosphorylation.

P450 enzymes A set of inducible enzymes that play roles in in the detoxification of foreign compounds. See *cytochrome P450 enzymes*.

P_{50} The partial pressure of O_2 that causes a blood or respiratory pigment to become 50% saturated with O_2. A measure of O_2 affinity; a high P_{50} signifies a low O_2 affinity.

Pacemaker The cell or set of cells that spontaneously initiates a rhythm. In a heart, for example, the pacemaker is the cell or set of cells that spontaneously initiates the rhythmic contractions of the heart muscle.

Pacemaker potential An intrinsic depolarization of a neuron or other excitable cell, leading to an action potential and thus making the cell spontaneously active.

Panting An increase in the rate of breathing; often serves to increase the rate of evaporative cooling by increasing air flow over

moist surfaces of the airways of the breathing system.

Papilla A nipple-shaped structure. Mammalian taste buds are collected in lingual papillae on the tongue surface.

Parabronchi The smallest-diameter tubes in the lungs of a bird. They are numerous and collectively constitute most of the lung tissue. Air capillaries—the sites of O_2 and CO_2 exchange—connect to the lumens of the parabronchi.

Paracellular Referring to transport across an epithelium, transport that occurs between cells (rather than through cells).

Paracrine A locally acting chemical signal that binds to receptors and exerts a regulatory effect on cells in the neighborhood of the cell that released it.

Parallel In reference to the arrangement of parts in an electrical circuit, vascular system, or other analogous system in which substances flow from place to place, the parts of the system are in *parallel* if one path of flow branches to give rise to two or more paths in which the parts reside, so that only a fraction of the total flow passes through any one of the parts. Contrast with *series*.

Parasympathetic division A division of the vertebrate autonomic nervous system that is connected to the CNS via cranial and sacral nerves; the parasympathetic and sympathetic divisions tend to exert opposing controls on autonomic effectors.

Paraventricular nucleus A cluster of neurons in the hypothalamus; contains magnocellular neuroendocrine cell bodies that extend their axon terminals to the pars nervosa, where they secrete oxytocin and vasopressin into the general circulation.

Paravertebral ganglia Ganglia of the sympathetic nervous system that are next to vertebrae.

Parthenogenesis A form of asexual reproduction in which eggs produced by females develop into genetically identical clones of their female parent.

Partial compensation See *compensation*.

Partial pressure (1) The pressure exerted by a particular gas within a mixture of gases. In terms of the pressure it exerts (its partial pressure), each gas in a mixture behaves as if it alone occupies the entire volume occupied by the mixture; thus each gas's partial pressure can be calculated from the universal gas law. (2) The concept of partial pressure is also applied to gases dissolved in aqueous solution: The partial pressure of a gas in aqueous solution is equal to the partial pressure of the same gas in a gas phase with which the solution is at equilibrium. *Gas tension* is synonymous with *partial pressure* for a gas in solution. See also *universal gas law*.

Parvocellular neurons Neurons characterized by small cell bodies. In the hypothalamus, they synthesize and secrete releasing and inhibiting hormones that are carried in the hypothalamo–hypophysial portal vessels to the anterior pituitary gland (adenohypophysis).

Passive change in lung volume In the study of ventilation, a change in lung volume that is a consequence of elastic rebound, not driven by muscle action. Contrast with *active change in lung volume*.

Passive electrical properties The electrical properties of a cell that do not involve a change in membrane ion permeability, and thus involve no change in electrical resistance.

Passive exhalation Exhalation that occurs without muscular contraction, usually driven by elastic rebound of structures stretched during inhalation.

Passive spread See *decremental spread*.

Passive transport The transport of a material by a mechanism that is capable of carrying the material only in the direction of equilibrium.

Patch-clamp recording A method of measuring single-channel currents by sealing a glass capillary microelectrode to a patch of cell membrane. Other conformations of patch clamping can measure whole-cell current or voltage. See also *single-channel current recording*.

Path integration An animal's summation of the distances and directions of its past movements, so that the animal knows where it is (relative to home) without using landmarks.

Pay-as-you-go phase A phase during exercise when aerobic catabolism using O_2 taken up by breathing is meeting the full ATP requirement of the exercise.

Pejus temperatures In the study of poikilotherms, a range of body temperatures in which animal performance deteriorates as body temperature is gradually raised (upper pejus range) or lowered (lower pejus range). Pejus means *turning worse*. Contrast with *critical temperature*.

Peptide hormones Also called protein hormones; made of assemblages of amino acids and soluble in water (polar).

Perfusion The forced flow of blood through blood vessels.

Pericardial sinus A fluid-filled space or cavity surrounding the heart.

Perikaryon See *cell body*.

Period The length (in time or space) of a single cycle of a rhythmic activity.

Periodic breathing See *intermittent breathing*.

Peripheral activation Conversion of a hormone after secretion to a more physiologically potent form.

Peripheral control Control of motor output to produce rhythmic movements, for which peripheral sensory receptors are needed to provide timing information for each stage of the sequence. Contrast with *central control*.

Peripheral ganglion A collection of neuronal cell bodies outside the CNS.

Peripheral membrane protein A protein that is associated with a cell membrane, typically on just one side or the other, but that can be removed without destroying the membrane.

Peripheral nervous system (PNS) The portion of a nervous system outside of the central nervous system, consisting of afferent and efferent nerves that connect the central nervous system to various parts of the body.

Peristalsis One of the gut's principal modes of muscular activity, in which constriction of the gut at one point along its length initiates constriction at a neighboring point farther along the gut, producing a "wave" of constriction that moves progressively along the gut, propelling food material before it.

Permeability In reference to a cell membrane or epithelium, the ease with which a particular solute can move through it by diffusion, or the ease with which water can move through it by osmosis.

Permeating In reference to a solute, able to pass through a cell membrane or epithelium.

Permissiveness A type of relationship between hormones in which one hormone must be present to allow another hormone to exert its effect.

PGC-1α1 An isomer of the transcriptional coactivator PGC-1α that drives the transcription of genes involved in producing the oxidative phenotype of muscle fibers, including angiogenesis, mitochondrial biogenesis, and the expression of myosin heavy chains found in oxidative muscle fibers.

PGC-1α4 An isomer of the transcriptional coactivator PGC-1α that drives the transcription of genes involved in downregulating myostatin and upregulating IGF-1, a combination that leads to hypertrophy and contractile strength.

pH The negative of the common logarithm of the concentration of H^+. The pH is inversely related to the concentration of H^+ (acidity). A low pH signifies acid conditions, whereas a high pH signifies alkaline conditions. See also *neutral pH*.

Phasic receptor See *rapidly adapting receptor*.

Phasic smooth muscles Smooth muscles that produce rapid, intermittent contractions; the cells produce spontaneous action potentials and are electrically connected by gap junctions.

Phasing factor An environmental cue that can entrain a biological clock, synchronizing it to environmental changes (such as the daily light-dark cycle). Also called a *zeitgeber*.

Phenotypic plasticity The ability of an individual animal to express two or more genetically controlled phenotypes.

Pheromone A chemical signal that conveys information between two or more individuals that are members of the same species. It typically signals the sexual readiness or social status of the releasing animal and triggers stereotyped behaviors or physiological changes in the detecting animal.

Phosphagens Compounds that can donate high-energy phosphate bonds to ADP to make ATP; thus, compounds that act as stores of high-energy phosphate bonds. The two most common are creatine phosphate and arginine phosphate.

Phospholipid A lipid compound in which a phosphate group or groups occur. A typical membrane phospholipid consists of two hydrocarbon tails linked by a phosphate group to a compound such as choline.

Phosphorylation Covalent attachment of an orthophosphate group to an organic compound.

Photoautotroph An organism that can obtain the energy it needs to stay alive from photons; an organism capable of photosynthesis.

Photochemical reaction A chemical reaction triggered by light absorption.

Photoperiod Day length; the number of hours of daylight in a 24-hour day.

Photopigment An unstable pigment molecule that undergoes a chemical change when it absorbs light, e.g., rhodopsin.

Photoprotein A type of chemical complex instrumental in bioluminescence in some marine animals. The complex consists of luciferin, O_2, and an inactive form of catalyzing protein. Initiation of light production is often dependent on exposure of the photoprotein to Ca^{2+} or Mg^{2+}.

Photoreception Response of a sensory cell to light stimulation. Photoreceptor cells contain a photopigment that absorbs light and triggers a response.

Photoreceptor See *Photoreception*.

Photosynthetic autotroph Synonym for *photoautotroph*.

Phototransduction Generation of an electrical response in a photoreceptor cell in response to a light stimulus.

Phyletic See *phylogenetic*.

Phylogenetic Having to do with the evolutionary relationships of organisms; the patterns of relationship that organisms exhibit by virtue of common evolutionary descent. Also called *phyletic*.

Phylogenetic reconstruction A reconstruction of the family tree (the ancestry) of groups of related species, often using molecular genetic data.

Phylogenetically independent contrasts See Appendix G.

Physiological work Any process carried out by an animal that increases order and requires energy. For example, an animal does physiological work when it synthesizes proteins, generates electrical or chemical gradients by actively transporting solutes across cell membranes, or contracts muscles to move materials inside or outside its body.

Physiology The study of function; the study of "how organisms work."

Pilomotor Related to the erection or compression of the hairs in the fur (pelage) of a mammal under the control of muscles attached to the bases of the hairs.

Piloting The act of navigation by visual landmarks.

Pineal gland A small endocrine gland in the vertebrate brain that produces melatonin.

Place cell A neuron in the hippocampus that generates action potentials when an animal is in a particular part of its familiar environment. Place cells are so named because their activity encodes the spatial position of the animal.

Place theory The theory that the pitch or frequency of a sound is encoded by the place of maximal vibration and maximal hair cell stimulation along the length of the basilar membrane of the cochlea of the ear.

Placenta An organ formed jointly by the embryo and the mother in marsupial and eutherian mammals. It allows intimate juxtaposition (but not mixing) of the maternal bloodstream and the embryonic or fetal bloodstream for the exchange of materials.

Plasma clearance The volume of blood plasma that would have to be completely cleared of a solute to obtain the amount of that solute excreted in urine over a specified period of time.

Plasma membrane See *cell membrane*.

Plasma See *blood plasma*.

Pleiotropy The control of two or more distinct and seemingly unrelated traits by an allele of a single gene.

Podocytes Distinctive cells with numerous processes that occur in structures where ultrafiltration occurs, such as the vertebrate renal glomerulus. Intricate geometric arrays of the cell processes of multiple podocytes

are believed to form a critical part of the filter that determines which solutes can and cannot pass through during ultrafiltration.

Poikilosmotic animal See *osmoconformer*.

Poikilotherm An animal in which the body temperature is determined by equilibration of the body with the thermal conditions in the environment. Also called an *ectotherm*.

Poiseuille equation An equation that describes the quantitative relation between the rate of flow of fluid through a horizontal tube and factors such as pressure, luminal radius, and length. The equation states that flow rate depends directly on the fourth power of the luminal radius. See Equation 25.2. Also called the *Hagen-Poiseuille equation*.

Polar molecule A molecule in which electrons are unevenly distributed, so that some regions of the molecule are relatively negative while others are relatively positive.

Polarized light In a light beam, polarization state refers to the electric field, which in general vibrates orthogonally (at right angles) to the direction in which a light wave moves. In polarized light, the vibration direction is restricted (e.g., to a plane in plane polarized light); whereas, in unpolarized light, the vibration direction is unrestricted (i.e., vibrates in all directions orthogonal to the beam). See Figure 18.6.

Polycythemia A state of having an unusually high concentration of red blood cells in the blood; seen, for example, in humans and some other lowland mammals when they acclimatize to high altitudes.

Polyneuronal innervation A pattern of innervation in which a single muscle fiber receives synaptic contacts from more than one motor neuron.

Polyphenic development A developmental phenomenon in which one individual— or a set of *genetically identical* individuals— can assume two or more discrete, highly distinct body forms, induced by differences in the developmental environment. Particularly common in insects.

Polyphenism See *polyphenic development* and *seasonal polyphenism*.

Polysaccharide A carbohydrate molecule that consists of many simple-sugar (monosaccharide) molecules polymerized together. Starch, cellulose, and chitin are important examples.

Polyunsaturated fatty acid See *unsaturated fatty acid*.

Positive feedback A process by which deviations of a property from a specific set-point level are reinforced, thereby tending to cause escalating changes in the property.

Postabsorptive Referring to an animal that has not eaten for a sufficiently long time that the specific dynamic action of its last meal is over. See *specific dynamic action*.

Posterior pituitary The neuroendocrine portion of the vertebrate pituitary gland in which neurohormones are released from axon terminals. The posterior pituitary is an extension of the brain. Also called the *neurohypophysis*.

Postganglionic neuron In the autonomic nervous system, a neuron that extends from an autonomic ganglion to an effector.

Postgenomic Referring to the study of an animal species or other organism after its full genome is known.

Postpartum estrus In a female mammal, estrus that occurs soon after she has given birth.

Postsynaptic cell A neuron or effector cell that receives a signal (chemical or electrical) from a presynaptic cell at a synapse.

Postsynaptic density An accumulation of dense material beneath the membrane of a postsynaptic cell, visible in electron micrographs. It corresponds to the accumulation of proteins associated with the concentration of neurotransmitter receptors.

Postsynaptic potential See *synaptic potential*.

Posttetanic potentiation The augmentation of a postsynaptic potential following a period of rapidly repeated (tetanic) stimulation.

Power curve In a muscle, the relationship between the velocity of shortening and the force exerted against a load.

Power equation See *allometric equation*.

Pre-Bötzinger complex A bilaterally arrayed pair of neuron clusters within the medulla of the brainstem of a mammal, believed to be the source of the breathing rhythm.

Preferred body temperature The body temperature that is maintained in a poikilotherm by behavioral thermoregulation.

Preflight warm-up In an insect, warming of the flight muscles prior to flight. Preflight warm-up is often essential for flight because the flight muscles must be warm to generate enough power for flight. It can occur by behavioral means (e.g., sun basking) or by physiological means (shivering).

Preformed water Water that enters the body in the form of H_2O. Contrast with *metabolic water*.

Preganglionic neuron In the autonomic nervous system, a neuron that extends from the central nervous system to an autonomic ganglion.

Pressure The force a fluid (liquid or gas) exerts in a perpendicular direction on solid surfaces with which it is in contact. Expressed per unit of surface area.

Presynaptic cell A neuron or other cell that transmits a signal to a postsynaptic cell at a synapse.

Presynaptic inhibition (PSI)   Inhibition of a neuron by decreasing the amount of neurotransmitter released by an excitatory presynaptic neuron.

Presynaptic terminal   The ending of a neuronal axon at the presynaptic side of a synapse.

Primary active transport   Active transport driven by a mechanism that draws energy directly from ATP. Contrast with *secondary active transport*.

Primary follicle   In the vertebrate ovary, a primary oocyte surrounded by a single layer of somatic cells—called *granulosa cells*—that are of different histological appearance than the somatic cells of a primordial follicle. A primary follicle has started its final maturation and, unless interrupted, will gradually develop more than a single layer of somatic cells and undergo other changes, becoming a secondary follicle and then progressing further.

Primary motor cortex   A region of the cerebral cortex (anterior to the central gyrus) that exerts descending control on spinal motor neurons to control movements.

Primary oocyte   A cell in the vertebrate ovary that has initiated meiosis during the early development of a female and is destined to produce an ovum but that remains in arrest until the female becomes reproductively mature.

Primary production   The production of organic matter from inorganic chemical precursors. The most common type of primary production is photosynthesis.

Primary protein structure   The sequence of amino acids in a protein molecule.

Primary urine   The fluid initially introduced into the tubules of a kidney. It is processed as it flows through the kidney tubules, ultimately becoming the *definitive urine* that is excreted from the body.

Primary visual cortex   Posterior region of the mammalian cerebral cortex that receives input from the lateral geniculate nucleus; serves as the first stage of visual information processing in the cerebral cortex.

Primordial follicle   In a vertebrate ovary, a *primary oocyte* surrounded by a single layer of flattened somatic cells. This is the earliest stage of follicle development. See also *primary follicle*.

Proenzyme   An inactive form of a digestive enzyme, activated after it arrives at its site of action.

Profiling   A research strategy in which investigators look as comprehensively as possible at a class of compounds of interest, such as messenger RNAs (*transcription profiling*) or proteins (*protein profiling, proteomics*). Also called *screening*.

Progesterone   A sex steroid hormone secreted by the corpus luteum of most vertebrates and the placenta of eutherian mammals.

Prolactin   A hormone secreted by cells in the anterior pituitary that stimulates the production of milk in mammals and performs a variety of other regulatory functions in vertebrates related to reproduction, water and mineral balance, and caring for the young.

Propagation   Spatial transmission of a signal such as an action potential without any decrease in amplitude with distance.

Proprioceptor   A sensory receptor that provides an animal with information about the relative position or movement of parts of its body.

Protandrous   In a sequential hermaphrodite, starting life as a male and later switching to female. See *hermaphrodite*.

Protogynous   In a sequential hermaphrodite, starting life as a female and later switching to male. See *hermaphrodite*.

Proteasome   See *ubiquitin-proteasome system*.

Protein hormones   Also called peptide hormones; made of assemblages of amino acids and soluble in water (polar).

Protein kinase   A regulatory enzyme that covalently bonds a phosphate group to a protein using ATP as the phosphate donor.

Proteomics   The simultaneous detection and measurement of large suites of proteins being synthesized by cells or tissues. Protein profiling.

Prothoracic glands   The paired glands located in the thorax of insects that synthesize and secrete ecdysone in response to prothoracotropic hormone.

Prothoracotropic hormone   A tropic protein neurohormone of insects, secreted by neuroendocrine cells with cell bodies in the brain and axon terminals in the neurohemal region of the corpora allata; stimulates the prothoracic glands to secrete ecdysone.

Protogynous   In a sequential hermaphrodite, starting life as a female and later switching to male. See *hermaphrodite*.

Proton pump   In the stomach, the cellular mechanism that secretes acid into the stomach lumen during digestion.

Ptilomotor   Related to the erection or compression of the feathers in the plumage of a bird under the control of muscles attached to the bases of the feathers.

Pulmonary   Related to the lungs.

Pulmonary arteries   In a vertebrate, blood vessels that carry blood from the heart to the lungs. This blood is partially deoxygenated.

Pulmonary circuit   In the circulatory system of an air-breathing animal, the blood vessels that take blood to and from the lungs.

Pump   In the study of cellular-molecular biology, a mechanism of *active transport*.

Pupa   The stage of development in holometabolous insects in which the larval tissues are destroyed and replaced by adult tissues. The pupa metamorphoses into the adult.

Purinotelic   Incorporating most nitrogen from the catabolism of nitrogenous compounds into purines, such as uric acid, guanine, and xanthine.

Purkinje fibers   Modified muscle cells that form part of the conducting system in which depolarization spreads throughout the vertebrate heart. See *conducting system*.

Q

Q_{10}   See *temperature coefficient*.

QRS complex   A part of the electrocardiogram that occurs simultaneously with ventricular contraction.

Quanta   See *Quantal release*.

Quantal release   The release of neurotransmitter molecules in multimolecular packets (quanta) corresponding to exocytosis of synaptic vesicles.

Quantitative trait locus (QTL) analysis   A statistical method that links two types of information—phenotypic data (observable traits) and genotypic data (usually chromosomal markers)—in an attempt to explain the genetic basis of variation in complex traits. QTL analysis allows researchers to link complex phenotypes to specific regions of chromosomes.

Quaternary protein structure   In a protein molecule that is composed of two or more separate proteins, the three-dimensional arrangement of the protein subunits relative to one another.

R

Radiant heat transfer   Transfer of heat by electromagnetic radiation between two surfaces that are not in contact. Also called *thermal radiation heat transfer*.

Radiant temperature   The surface temperature of an object as judged by its rate of emission of electromagnetic energy. The radiant temperature is the surface temperature that would have to prevail for the object to emit energy at the rate it actually does if its emissivity were exactly 1 (i.e., if it were a true black body). Radiant temperature is calculated by measuring the intensity of radiation and entering with it in the Stefan-Boltzmann equation with the emissivity set equal to 1. Also called the *blackbody temperature*.

Radiotelemetry   The use of a radio transmitter placed in or on an animal to transmit data on physiological or behavioral variables.

Radular apparatus   A feeding apparatus found in snails, slugs, chitons, squids, and some other molluscs. The radula itself is a band of connective tissue, studded with teeth, that is pulled back and forth to create grinding or scraping action.

Ram ventilation A type of gill ventilation observed in certain types of fish (e.g., tunas) in which the fish holds its mouth open as it swims forward, thereby using its swimming motions to drive water over its gills.

Rapidly adapting receptor A sensory receptor cell that exhibits a rapidly decreasing response to a maintained stimulus. Also called a *phasic receptor*.

Rate-limiting reaction In a sequence of chemical reactions, the one that proceeds slowest, thereby setting the rate of the entire sequence.

Reaction velocity In an enzyme-catalyzed reaction, the rate at which product molecules are made from substrate molecules.

Reactive oxygen species (ROS) A term that refers collectively to metabolically produced molecules that have a great potential to act as oxidizing agents. Included are free radicals (e.g., superoxide)—which are molecules that have an odd number of electrons—and other oxidizing agents (e.g., hydrogen peroxide and ozone). Although many reactive oxygen species play roles in normal metabolism, they can also be destructive.

Receptive field In sensory systems, the region of a sensory surface within which stimulation changes the activity of a particular neuron.

Receptor A protein that binds noncovalently with specific molecules and, as a consequence of this binding, initiates a change in membrane permeability or cell metabolism. Receptors mediate the response of a cell to chemical messages (signals) arriving from outside the cell. Although most receptors are in the cell membrane, some are intracellular. See also *sensory receptor*.

Receptor adaptation See *adaptation* (meaning 2).

Receptor molecule See *sensory receptor molecule*.

Receptor potential The graded change in membrane potential that occurs in a sensory receptor cell when it is stimulated.

Receptor, sensory See *sensory receptor*.

Reciprocity A principle of motor control in which signals that activate motor neurons and muscles also inhibit activation of antagonist motor neurons and muscles.

Recruitment of motor units A mechanism by which the force of contraction of a whole muscle can be varied depending on the number of motor units stimulated to contract. See *motor unit*.

Rectal pads Pad-shaped structures, composed of thickened epithelial cells, in the rectum of some insects.

Rectal salt gland A salt-secreting gland found in the rectum of elasmobranch fish (e.g., sharks).

Red blood cell A hemoglobin-containing cell in the blood of an animal. Also called an *erythrocyte*.

Red muscle In general, a loose term referring to a vertebrate muscle that is rich in myoglobin and thus reddish in color. In fish, a large mass of muscle that consists almost entirely of myoglobin-rich muscle fibers that make ATP mostly by aerobic catabolism; routine cruising by fish is powered by red muscle.

Redox balance (reduction–oxidation balance) A state in which a cell has the capability to remove electrons from a compound that undergoes reversible reduction and oxidation as fast as electrons are added to the compound.

Reduction Addition of electrons or hydrogen atoms to a molecule.

Reflex A simple, relatively stereotyped, but graded behavioral response to a specific stimulus.

Regional heterothermy See *heterothermy*.

Regulation The maintenance of internal conditions at an approximately constant level while external conditions vary.

Relative humidity In air of a particular temperature, the existing water vapor pressure divided by the saturation water vapor pressure characteristic of that temperature.

Relative refractory period In a neuron or other excitable cell, the brief period following an action potential during the generation of another action potential is relatively difficult because voltage conditions are further than usual from threshold.

Relaxation volume The volume that a structure (e.g., a lung or the thorax) assumes when there are no muscular forces tending to expand or contract it.

Release-inhibiting hormone A hormone secreted by neuroendocrine cells in the hypothalamus of a vertebrate that travels to the anterior pituitary through the hypothalamo-hypophysial portal system and inhibits the secretion of a hormone by a specific population of anterior pituitary endocrine cells. All release-inhibiting hormones are peptides except the catecholamine dopamine, which is known to inhibit the secretion of prolactin.

Releasing hormone A peptide hormone secreted by neuroendocrine cells in the hypothalamus of a vertebrate that travels to the anterior pituitary through the hypothalamo–hypophysial portal system and stimulates the secretion of a hormone by a specific population of anterior pituitary endocrine cells; may also exert a tropic action to maintain and support those anterior pituitary cells.

Renal Related to kidneys.

Renal corpuscle In the kidney of a vertebrate, a glomerulus and its associated Bowman's capsule. See also *glomerulus*.

Renal pelvis In a vertebrate kidney, an expanded tubular structure that connects to the ureter. Urine from the nephrons enters the renal pelvis and then flows into the ureter.

Renin A substance secreted into the blood by juxtaglomerular cells of the vertebrate nephron in response to low blood pressure; converts angiotensinogen to angiotensin I.

Renin–angiotensin–aldosterone system A hormonal complex that ensures adequate arterial blood pressure. Set in motion by renin, it produces angiotensin II, which stimulates secretion of aldosterone (important in renal conservation of sodium) and vasopressin, promotes thirst, and stimulates vasoconstriction.

Residue In biochemistry, the modified form assumed by a small molecule when it is incorporated by covalent bonding into a larger molecule. For example, when an amino acid is incorporated into a protein, it is technically an amino acid *residue* because its structure is no longer the full structure of the free amino acid.

Resistance (1) In electrical circuits, the property that hinders the flow of electric current (charge movement) through a material, measured in ohms (Ω). It is the inverse of electrical conductance, the ease of current flow, measured in siemens. See also *membrane resistance*. (2) In heat transfer, the property that hinders dry heat transfer either through a material or between an animal and its environment. Speaking of the latter case, resistance to heat transfer is defined to be the difference between body temperature and ambient temperature divided by the rate of dry heat transfer between the animal and the environment. Contrast with *thermal conductance*. (3) See *vascular resistance*.

Resistance exercise Exercise that consists of relatively short periods of high-intensity muscular actions against a large load, often repeated with intervening interruptions. The periods of high-intensity activity are usually characterized by significant dependence on anaerobic mechanisms of ATP production. Also called *resistance training*. Contrast with *endurance exercise*.

Resonant frequency In any elastic system, the frequency at which the system oscillates when left alone following activation by a pulse of energy; the natural frequency. The energy cost of energy-driven oscillation tends to be lowest if the oscillation is at the resonant frequency of the system involved.

Respiratory acidosis or alkalosis Acidosis or alkalosis that is caused by an abnormally rapid or slow rate of removal of CO_2 from the body fluids by breathing. Contrast with *metabolic acidosis* or *alkalosis*.

Respiratory airways In the lung of a vertebrate, the airways in which O_2 and CO_2 are exchanged between the lung air and blood (i.e., excluding airways that are too poorly vascularized or too thick-walled for exchange to occur).

Respiratory chain Synonym for *electron transport chain*.

Respiratory gases Oxygen (O_2) and carbon dioxide (CO_2).

Respiratory pigments Any of the metalloprotein pigments that undergo reversible combination with O_2 and thus are able to pick up O_2 in certain places in an animal's body (e.g., the breathing organs) and release it in other places (e.g., systemic tissues). Respiratory pigments include hemoglobin, hemocyanin, hemerythrin, and chlorocruorin. Also called O_2-transport pigments.

Respiratory quotient (RQ) The moles of CO_2 produced by a cell expressed as a ratio of the moles of O_2 simultaneously consumed. Because the *RQ* changes with the type of foodstuff being oxidized, it can be used to assess what foodstuffs a cell is using in aerobic catabolism.

Resting membrane potential (V_m) The normal electrical potential across the cell membrane of a cell at rest.

Rete Shorthand for rete mirabile. See *rete mirabile*.

Rete mirabile A Latin expression, meaning literally "wonderful net." A morphological term referring to any intricately complex vascular system composed of closely juxtaposed, small-diameter arterial and venous blood vessels. The term *rete* is a shorthand synonym.

Reticular theory Theory in the nineteenth century, now discarded, that cells in the CNS were cytoplasmically connected to each other in a syncytial reticulum. Contrast *neuron doctrine*.

Retina The layer of photoreceptor cells and other neurons that line the inside of a vertebrate eye.

Retinal The aldehyde of vitamin A, one of two components of the photopigment rhodopsin.

Retinular cell Photoreceptor cell in the ommatidium of a compound eye.

Retrieval The movement of channel or transporter protein molecules out of the membrane where they are active to intracellular locations where they are inactive.

Retrograde messenger In a synapse, a chemical signal thought to be released by a postsynaptic cell that alters the synaptic properties (such as neurotransmitter release) of the presynaptic cell.

Reversal potential The membrane potential at which the amplitude of a voltage response (such as a postsynaptic potential or receptor potential) is zero because there is no net driving force for ion flow. For example, E_{EPSP} is the reversal potential of an EPSP.

Rhabdomere Collection of microvilli from the retinular cells of an ommatidium of a compound eye.

Rhabdomeric photoreceptor A photoreceptor cell in which the photopigment rhodopsin is collected in microvillar membranes. Contrast with *ciliary photoreceptor*.

Rhodopsin The light-absorbing pigment of photoreceptors that initiates the visual response to light; composed of retinal and the protein opsin.

Rhythmic behavior Stereotyped, repetitive sequences of movement such as walking, swimming, flying, and breathing.

Rhythmogenesis Generation of a rhythm. Usually refers to rhythm generation by neurons or sets of neurons, such as the sets of neurons that rhythmically originate nerve impulses that stimulate the breathing muscles to contract.

Right-to-left shunting In the study of blood flow through the central circulation of an air-breathing fish, amphibian, or nonavian reptile, diversion of systemic venous blood directly back into the systemic circuit, bypassing the lungs.

RNA interference (RNAi) A cellular process that destroys specific mRNA molecules (produced naturally in cells) when specific double-stranded RNA molecules are introduced into cells. The pathway effectively silences the genes that produced the targeted mRNAs.

Rod A type of photoreceptor in the vertebrate retina. Rods are larger than cones, respond at lower light levels, and are used for nocturnal vision.

Root effect A decrease in the amount of O_2 a respiratory pigment can bind at saturation—and thus a decrease in the oxygen-carrying capacity of blood—caused by a decrease in pH or an increase in CO_2 partial pressure. Unusual; observed only in certain fish and molluscs.

ROS See *reactive oxygen species*.

Rumen An expanded nonacidic stomach chamber in a ruminant mammal, housing a mixed microbial community of symbiotic microbes that perform functions important for the mammal's nutrition.

Ryanodine receptor The calcium channel of the sarcoplasmic reticulum in striated muscles.

S

Sacculus Otolithic organ in the vertebrate inner ear containing hair cells that are stimulated by vertical movements and forces (including gravity).

Salinity The sum total concentration of inorganic dissolved matter in water, usually expressed as grams of dissolved matter per kilogram of water.

Salt A synonym for *inorganic ion*, or a compound formed by inorganic ions.

Salt glands Organs other than kidneys that excrete concentrated solutions of inorganic ions. Examples include the cranial salt glands of marine birds and the rectal salt glands of marine elasmobranch fish.

Saltatory conduction Propagation of action potentials in a spatially discontinuous manner along a myelinated axon by jumping from one node of Ranvier to another.

Sarcolemma The cell membrane of a muscle fiber.

Sarcomere The contractile unit of striated muscle that consists of contractile, regulatory, and cytoskeletal proteins. Many sarcomeres in series, delineated by Z discs, constitute a myofibril.

Sarcopenia The loss of skeletal muscle mass as a result of aging. It involves both the loss of contractile proteins from individual muscle fibers and the loss of complete fibers by cell death.

Sarcoplasmic reticulum (SR) A system of internal compartments in a muscle cell that envelops myofibrils and stores Ca^{2+} ions. The sarcoplasmic reticulum has calcium ATPase pumps that transport Ca^{2+} ions into its lumen and calcium channels that open in response to excitation along the associated transverse tubules.

Satellite cells Muscle stem cells that lie immediately outside the sarcolemma.

Saturated air Air in which a temperature-specific maximum partial pressure of water vapor prevails. Fully humidified air.

Saturated enzyme An enzyme that is catalyzing its reaction at a maximum rate because substrate molecules are abundant enough that they bind with the enzyme at as great a number per unit of time as is possible.

Saturated fatty acid A fatty acid in which all the bonds between carbon atoms in the carbon-chain backbone of the molecule are single bonds. Also called a *SFA* or a *saturate*.

Saturated respiratory pigment A respiratory pigment that has combined with as much O_2 as it can possibly hold.

Saturated transporter A transporter protein that is catalyzing transport at a maximum rate because the molecules being transported are abundant enough that they bind with the transporter at as great a number per unit of time as is possible.

Saturation deficit In air of a particular temperature, the difference between the saturation water vapor pressure characteristic of that temperature and the existing water vapor pressure.

Saturation kinetics The kinetics characteristic of a chemical reaction or other chemical process that is limited to a maximum velocity by a limited supply of some type of molecule with which other molecules must reversibly combine for the reaction or process to take place.

Saturation water vapor pressure The maximum possible water vapor pressure that can stably exist in a gas of a particular temperature. If the water vapor pressure rises above the saturation water vapor pressure, condensation occurs. The saturation water vapor pressure varies strongly with temperature.

Scaffolding (or scaffold) proteins Proteins that function to bring other proteins together in a relatively stable complex. Presynaptic terminals have scaffolding protein complexes involved in neurotransmitter release, and postsynaptic cells have scaffolding complexes mediating neurotransmitter receptor trafficking and action.

Scaling The study of the relations between physiological (or morphological) features and body size within sets of phylogenetically related species, e.g., the study of metabolism-weight relations.

Schwann cell A type of ensheathing non-neuronal glial cell found in the vertebrate peripheral nervous system. Schwann cells form, for example, the myelin sheath of myelinated axons.

Screening See *profiling*.

Scrotum In a male mammal, an external sac that houses the testes (testicles).

Seasonal polyphenism A phenomenon seen in some insects that go through two or more generations per year, in which *genetically identical* individuals can assume two or more discrete, highly distinct body forms depending on the season during which they develop.

Second law of thermodynamics A law, believed to apply without exception in the biosphere, stating that if an isolated system undergoes internal change, the net effect of the change in the system as a whole is always to increase disorder (entropy).

Second messenger An intracellular signaling molecule that is produced inside a cell in response to the binding of a chemically different extracellular signaling molecule to specific cell-membrane receptors.

Secondary active transport Active transport driven by a mechanism that does not draw energy directly from ATP, but instead obtains it from the potential energy inherent in an electrochemical gradient. During secondary active transport, an electro-chemical gradient of a solute (e.g., Na^+) is created using ATP-bond energy, and the active-transport mechanism obtains its energy from that gradient. Contrast with *primary active transport*.

Secondary compounds Compounds of diverse chemical types (e.g., alkaloids, tannins, or terpenes) that are found in tissues and that have as their major function the deterrence of tissue consumption by a predator.

Secondary lamellae In fish gills, the microscopically fine folds of tissue on the surfaces of the gill filaments that serve as the primary sites of exchange of respiratory gases between the ambient water and blood.

Secondary oocyte The cell formed at the time of ovulation by the primary oocyte when it concludes its first meiotic division and extrudes the first polar body.

Secondary protein structure The arrangement of the amino acids within subregions of a protein molecule into highly regular geometric shapes. The two most common types of such highly ordered arrays of amino acids are the α-helix and the β-sheet (pleated sheet).

Segmentation One of the gut's principal modes of muscular activity, in which circular muscles contract and relax in patterns that push the gut contents back and forth.

Selective permeability The state of having a high permeability to some solutes but a low permeability to others. See also *permeability*.

Semelparity A type of reproductive life history in which individuals are physiologically capable of only one bout of reproduction during their lives. In semelparous species, individuals are often programmed to die after reproducing once. Contrast with *iteroparity*.

Semelparous Characterized by *semelparity*.

Semen The sperm-containing fluid emitted by a male mammal during ejaculation. Most of the fluid volume is composed of secretions of accessory sexual glands, (notably the prostate gland and seminal vesicles in humans).

Semicircular canal A component of the *vestibular organ* of the vertebrate inner ear containing receptors that respond to head rotation.

Seminiferous tubules Small tubules in the testes (testicles) of a vertebrate, responsible for sperm production.

Semipermeable membrane A membrane that is permeable only to water. All true semipermeable membranes are human-made, as there are no natural biological membranes that are strictly semipermeable.

Sense organ A complex multicellular structure specialized to detect a particular type of sensory stimulus.

Sensillum A sensory hair of arthropods; not related to vertebrate hair, but rather a hollow chitinous projection of the exoskeleton that is associated with sensory receptor neurons.

Sensitivity The ability of a sensory cell to distinguish stimuli of different intensity. Receptor sensitivity may also refer to the adequate stimulus, the kind of stimulus to which the receptor is responsive.

Sensitization Enhancement of a learned behavioral response to a harmless stimulus after exposure to a strong or harmful stimulus.

Sensory adaptation See *adaptation* (meaning 2).

Sensory modality The class of stimulus that evokes a sensory response. The classical sensory modalities are vision (light is the stimulus that evokes a response), hearing (sound), touch, smell, and taste.

Sensory neuron A sensory receptor that is a neuron, or a peripheral neuron that is excited by a non-neuronal sensory receptor cell.

Sensory receptor cell A sensory cell that is specialized to respond to a particular kind of environmental stimulus.

Sensory receptor molecule A molecule in a sensory receptor cell that is particularly sensitive to a kind of sensory stimulus, and that participates in transducing a stimulus into a cellular response.

Sensory system The sense organs (or other sensory receptors) for a particular sensory modality and all of the central processing areas and pathways associated with those organs (or other receptors).

Sensory transduction The process by which the energy of a physical stimulus is converted into an electrical signal in a sensory receptor cell.

Septate junction A type of junction between epithelial cells that differs in fine structure from a tight junction but otherwise has similar properties. Found in invertebrates.

Sequential hermaphroditism See *hermaphroditism*.

Series In relation to the arrangement of parts in an electrical circuit, vascular system, or other analogous system in which substances flow from place to place, the parts of the system are in *series* if they occur sequentially along a single path of flow, so that all flow must occur sequentially through all parts. Contrast with *parallel*.

Serosal surface See *basal surface*.

Sertoli cells   Somatic cells in the seminiferous tubules of the testes of vertebrates that function to support spermatogenesis.

Sexual reproduction   The formation of a new, genetically unique individual from the union of male and female gametes.

Shivering   (1) In a mammal or bird, the unsynchronized contraction and relaxation of motor units in skeletal muscles in high-frequency rhythms, producing heat rather than organized motion as the primary product. (2) In an insect, contraction of the flight muscles in a nonflying mode to generate heat rather than flight.

Short-chain fatty acids (SCFA)   Fatty acids that consist of relatively few carbon atoms, including acetic acid (2 carbons), propionic acid (3), and butyric acid (4). They are produced, for example, by symbiotic fermenting microbes and are readily absorbed and metabolized by animals. Also called *volatile fatty acids*.

Sigmoid kinetics   In an enzyme-catalyzed reaction, an s-shaped relation between reaction rate and substrate concentration. In the study of hemoglobin or other respiratory pigments, an s-shaped relation between the percentage of oxygenated O_2-binding sites and the O_2 partial pressure.

Signal transduction   In the study of cell signaling, the translation of a signal from one chemical form (e.g., a hormone) into another chemical form (e.g., a second messenger inside a cell).

Simple cell   A neuron found in the mammalian primary visual cortex that has an elongated orientation-selective receptive field, so that it responds most to a bar or edge at a particular angle of orientation. The receptive fields of simple cells have distinct excitatory and inhibitory subregions.

Simple diffusion   Transport of solutes, water, gases, or other materials that arises from the molecular agitation that exists in all systems above absolute zero and from the simple statistical tendency for such agitation to carry more molecules out of regions of relatively high concentration than into such regions.

Simple epithelium   An epithelium consisting of a single cell layer.

Single effect   The difference produced by use of metabolic energy between adjacent parts of the two oppositely flowing fluid streams in a countercurrent multiplier system. See also *countercurrent multiplication*.

Single-channel current recording   The ability to record the ionic currents flowing though a single membrane channel by attaching a microelectrode to the membrane surrounding the channel. See also *patch-clamp recording*.

Single-unit smooth muscle   A muscle made up of unstriated muscle cells electrically coupled to neighboring cells by gap junctions; when stimulated, all of the cells are depolarized and contract together. Contrast with *multiunit smooth muscle*.

Sinoatrial (S-A) node (sinus node)   In the heart of a mammal or bird, a cluster of modified muscle cells in the right atrial wall where the pacemaker cells for the heart are located, responsible for initiating heart beats.

Sinuses   In an open circulatory system, large spaces, between nonvascular structures, that serve as passageways for blood flow.

Skeletal muscle   Muscle that produces locomotory movements or other external movements of the body. In vertebrates, skeletal muscle is attached to the endoskeleton. In most types of invertebrates, it is attached to the exoskeleton, shell, or other external covering. Vertebrate skeletal muscle consists of large, cylindrical, multinucleate striated cells (*muscle fibers*). Invertebrate skeletal muscle may consist of striated or smooth muscle cells, depending on the phylogenetic group under consideration.

Sliding-filament theory   The well-documented theory that muscle contraction results from active interaction between thick and thin myofilaments, which causes them to slide past each other.

Slow oxidative (SO) muscle fibers   Muscle fibers that are poised to make ATP principally by aerobic catabolism, develop contractile tension slowly, have relatively low peak power outputs, and are relatively resistant to fatigue.

Slowly adapting receptor   A sensory receptor cell that responds to a maintained stimulus in a way that decreases slowly and incompletely. Also called a *tonic receptor*.

Smooth muscle   Muscle that consists of small, spindle-shaped, uninucleate cells without striations. Thin actin filaments and thick myosin filaments are present, but not organized into sarcomeres. In vertebrates, smooth muscle is found in hollow and tubular internal organs, such as certain blood vessels and the gut.

Sociobiology   The study of the social relations and behaviors of animals from an evolutionary perspective.

Solute   A substance that is in solution (i.e., dissolved in solvent).

Solvent drag   Movement of solutes (dissolved entities) through a membrane driven by simultaneous osmotic movement of water (solvent).

Soma (plural somata)   See *cell body*.

Somatic effectors   Skeletal (striated) muscle, as opposed to *internal effectors*.

Somatic nervous system   The part of the vertebrate peripheral nervous system that controls skeletal muscles and provides afferent information from sensory receptors not associated with internal organs.

Somatotopic map   The topographic representation or mapping of the body surface onto a region of the brain—for example, in the vertebrate somatosensory cortex.

Spatial summation   The summation of postsynaptic potentials that result from presynaptic action potentials at different synapses onto a single postsynaptic cell.

Specific dynamic action (SDA)   A rise in the metabolic rate of an animal caused by the processing of ingested food.

Specificity   Sensory specificity is the ability to distinguish among different stimulus types (modalities or qualities).

Spermatogenesis   The formation of haploid sperm through the process of meiosis in the testis.

Spermatozoa   Haploid gametes, produced by spermatogenesis in the testis. Also called *sperm cells* or *sperm*.

Sphincter   In a vertebrate, a circular muscle, located between two chambers, that can contract tightly and steadily (tonically) for long periods of time, thus preventing exchange between the chambers.

Spinal nerves   In a vertebrate, segmental nerves of the peripheral nervous system that attach to the spinal cord.

Spinal reflex   A reflex mediated by neural circuits of the vertebrate spinal cord.

Spiracle   In a terrestrial arthropod (e.g., insect), a porelike aperture on the surface of the body that opens into the breathing system (e.g., tracheal system).

Spongy myocardium   Myocardium in which the muscle cells are sufficiently widely spaced that the tissue is perfused by the blood flowing through the lumens of the heart chambers. The myocardium of many fish is spongy.

Spontaneous ovulation   Ovulation (release of an egg from the ovaries of a female) that results from processes endogenous to the female, more or less independent of the actual act of mating. Contrast with *induced ovulation*.

Standard amino acid   One of the 20–22 amino acids that are employed by organisms to synthesize proteins.

Standard metabolic rate (SMR)   The metabolic rate of a poikilothermic (ectothermic) animal when it is resting and fasting. The SMR is specific to the body temperature prevailing during measurement.

Standard temperature and pressure (STP)   A temperature of 0°C and a pressure of 1 atm (101 kPa; 760 mm Hg). Gas volumes are often expressed under these conditions as a way of standardizing the effects of temperature and pressure on volume.

Star compass A mechanism by which an animal can use the positions of stars and constellations (and an internal clock) to determine compass direction in nocturnal navigation.

Starling–Landis hypothesis The concept that the ventricle of the heart tends to contract with greater force when it is stretched to a great extent at the start of contraction rather than stretched to a small extent.

Statocyst A sense organ that can detect acceleration and the direction of gravitational force.

Statolith A stony mineral concretion in a statocyst that is denser than the medium in which it sits, so that it is pulled downward by gravity and stimulates mechanoreceptors.

Stefan-Boltzmann equation An equation that relates the temperature of a surface and the rate at which the surface emits electromagnetic energy.

Stenohaline Referring to aquatic animals, able to live only within a narrow range of environmental salinities. Contrast with *euryhaline*.

Stenothermal Referring to poikilotherms, able to live only within a narrow range of body temperatures. Contrast with *eurythermal*.

Steroid hormones Nonpolar hormones synthesized on demand from cholesterol, secreted by diffusion through the cell membrane and circulated in the blood or hemolymph bound to carrier molecules; typically bind to intracellular receptors of the target cells and exert actions through genomic means. Some bind to receptors on the target cell membrane and exert nongenomic actions.

Stimulus (1) At a cellular level, a form of energy (sometimes called *stimulus energy*) that excites sensory receptor cells; the form of energy is specific for each type of receptor cell. (2) At a whole-animal level, a change in the external environment or in internal conditions that an animal can detect and respond to. (2) A change in the external environment or in internal conditions that can be detected by an animal.

Stoichiometry The existence of fixed ratios in chemical reactions. For instance, the fact that hydrogen and oxygen atoms react in a 2-to-1 ratio to form water is an example of stoichiometry.

Stomatogastric ganglion A small ganglion containing 30 or more neurons that controls rhythmic activity of stomach muscles in crustaceans. The stomatogastric ganglion is a model system for study of central pattern generators.

Straight-through pathways In the retina, neural pathways that extend through the retina from external photoreceptors to internal ganglion cells. Contrast with *lateral pathways*.

Stress response The response of an animal to a threatening situation. In vertebrates, it typically involves functions of the autonomic nervous system and the hypothalamus–pituitary–adrenal cortex axis.

Stretch reflex A vertebrate spinal reflex in which muscle stretching activates a muscle spindle stretch receptor, generating nerve impulses in 1a afferent axons that excite motor neurons innervating the same muscle to oppose the stretch.

Stretch-gated channel A channel protein that opens and closes in response to stretching or pulling forces that alter the physical tension on a membrane.

Striated muscle cells Also called striated muscle fibers; in vertebrates, they make up skeletal and cardiac muscles.

Striated muscle Muscle that consists of cells in which the thick myosin and thin actin filaments are arranged in sarcomeres. The sarcomeres in a cell are aligned in register to form stripes or striations at right angles to the long axis of the cell. In vertebrates, skeletal and cardiac muscles are striated.

Stroke volume In reference to a heart or other organ that pumps fluid by rhythmic cycles of contraction, the volume of fluid pumped per cycle.

Subcellular membrane See *intracellular membrane*.

Submaximal exercise Exercise that requires less than an individual's maximal rate of O_2 consumption.

Substrate One of the initial reactants of an enzyme-catalyzed reaction.

Substrate-binding site See *active site*.

Substrate-level phosphorylations ATP-producing reactions other than those associated with the electron-transport chain.

Sulfur-oxidizing bacteria Bacteria that employ oxidation of reduced sulfur compounds (e.g., H_2S) as a source of energy so they can function as chemoautotrophs. See *chemoautotroph*.

Summation In excitable cells, the addition of graded subthreshold potentials (electrical events). In muscle cells, the addition of twitches (mechanical events) produced by high frequencies of action potentials.

Sun compass A mechanism by which an animal can use the sun's position and an internal clock to determine compass direction in navigation.

Supercooling Cooling of a solution to below its freezing point without freezing.

Supercooling point The highest temperature at which freezing is almost certain to occur promptly in a supercooled solution. (Freezing of a supercooled solution is probabilistic. At some temperatures freezing is unlikely. The temperature needs to be lowered to the supercooling point for prompt freezing to be likely.)

Suprachiasmatic nucleus The part of the hypothalamus of the brain that acts as the master circadian clock in mammals.

Supramaximal exercise Exercise that requires ATP at a greater rate than it can be made aerobically even when an individual's rate of O_2 consumption is maximized.

Supraoptic nucleus A cluster of neurons in the hypothalamus; contains magnocellular neuroendocrine cell bodies that extend their axon terminals to the pars nervosa where they secrete oxytocin and vasopressin into the general circulation.

Surface-to-volume ratio The ratio of the total area of the outer surface of a three-dimensional object over the volume of that object.

Suspension feeding Feeding on objects suspended in water that are very small by comparison to the feeding animal. See also *filter feeding*.

Sweating The secretion onto the skin surface of a low-salinity aqueous solution (*sweat*) by specialized *sweat glands* to increase the rate of evaporative cooling. Occurs only in some groups of mammals.

Symmorphosis A hypothesis about the evolution of multiple organ systems in a species, which posits that the performance limits of all systems remain roughly matched because it would make no sense for any one system to have evolved capabilities that could never be used because of more-restrictive limits in other systems.

Sympathetic chain The chain of segmental sympathetic ganglia in a vertebrate nervous system.

Sympathetic division A division of the vertebrate autonomic nervous system that is connected to the CNS via thoracic and lumbar spinal nerves; the sympathetic and parasympathetic divisions tend to exert opposing controls over autonomic effectors.

Symporter See *cotransporter*.

Synapse A specialized site of communication between two neurons, between a neuron and an effector, or between a non-neuronal sensory cell and a neuron.

Synaptic antifacilitation A decrease in the amplitude of postsynaptic responses to repeated presynaptic action potentials. Also called *synaptic depression*.

Synaptic cleft The extracellular gap between presynaptic and postsynaptic cells at a synapse, typically 20–40 nm wide.

Synaptic current The current that flows through the postsynaptic membrane in synaptic transmission and produces a postsynaptic potential.

Synaptic facilitation An increase in the amplitude of postsynaptic responses that occurs after repeated presynaptic action potentials.

Synaptic homeostasis A form of *synaptic plasticity* that provides a means for neurons and neuron circuits to maintain stable function in the face of perturbations (such as developmental or activity-dependent changes in synapse number or strength). Sleep may play roles in synaptic homeostasis.

Synaptic plasticity Change in properties of synapses or strength of synaptic interactions with time or circumstance. Changes of nervous system function during development or learning are thought to reflect synaptic plasticity.

Synaptic potential A graded change in a postsynaptic cell's membrane potential produced by synaptic input. Also called a *postsynaptic potential*.

Synaptic transmission The process whereby one neuron influences the excitability of another neuron or effector. Synaptic transmission can be either chemical or electrical.

Synaptic vesicles Membrane-bound vesicles in a presynaptic terminal, into which neurotransmitter molecules are concentrated.

Synergism Interactions between two or more agents (e.g., hormones) whereby they have a greater effect acting together than the simple sum of their individual effects.

Systemic circuit In a circulatory system, the blood vessels that take blood to and from the systemic tissues.

Systemic tissues All tissues other than those of the breathing organs.

Systole The period of contraction during each beating cycle of a heart.

Systolic pressure During the beating cycle of the heart, the blood pressure prevailing during periods of heart contraction.

T

Tachycardia A heart rate that is unusually high.

Tagged See *epigenetic mark*.

Target cell A cell that responds to a chemical signaling molecule such as a paracrine or hormone because it expresses specific receptors for that molecule. It may express receptors for more than one signaling molecule. Its sensitivity depends on the number of receptors present, and can be changed by upregulation or downregulation of the receptors.

Taste Chemoreception of stimuli that are dissolved or suspended in liquids, typically requiring higher stimulus concentrations than olfaction. Taste chemoreceptors are often, but not always, localized around the mouth.

Taste bud A collection of epithelial taste receptor cells and support cells on the tongue or, in fish, on the skin surface.

Telemetry See *radiotelemetry*.

Telencephalon The anterior part of the vertebrate forebrain.

Teleost fish The principal group of fish having bony skeletons.

Temperature A measure of the speed or intensity of the ceaseless random motions that all the atoms and molecules of any substance undergo on an atomic-molecular scale. More exactly, the temperature of a substance is proportional to the product of the mean square speed of random molecular motions and the molecular mass.

Temperature coefficient The ratio of the rate of a process at one body temperature over the rate of the same process at a body temperature 10°C lower. Symbolized Q_{10}.

Temperature-dependent sex determination The phenomenon, seen in some nonavian reptiles, whereby the sex of offspring is determined by their body temperature during early development.

Temperature gradient Technically, the difference in temperature between two places divided by the distance separating those two places. Often used more loosely to refer simply to a difference in temperature. Also called a *thermal gradient*.

Temporal heterothermy See *heterothermy*.

Temporal organization The tendency for a physiological or behavioral system to be organized in time.

Temporal summation The summation of synaptic potentials in response to repeated presynaptic action potentials at the same synapse.

Tendons Connective tissue attachments of skeletal muscles to bones.

Tension (gas) A synonym of partial pressure. See *partial pressure*.

Tension (muscular) The force produced by cross-bridge action in a contracting muscle.

Tertiary protein structure The natural arrangement of an entire protein molecule in three dimensions, including its secondary structure and the other patterns of folding that give the molecule its particular conformation. Tertiary structure is flexible because it is stabilized by noncovalent bonds.

Testicular recrudescence The regrowth of the testes—and their return to producing sperm—after a period when they have been reduced in size and nonfunctional.

Testosterone A sex steroid hormone, produced by the Leydig cells of the testes, that is essential for male secondary sexual characteristics and reproduction. An androgen.

Tetanus Summed twitches of skeletal muscles produced by trains of motor action potentials. *Fused tetanus* is a smooth rise in tension produced by a high-frequency train of action potentials. *Unfused tetanus* is produced by a lower-frequency train of action

potentials and shows some relaxation of each twitch between action potentials.

Theca cells Somatic connective tissue cells that form the outer layer of a developing ovarian follicle.

Thermal conductance A measure of the ease of dry heat transfer between an animal and its environment. Contrast with *resistance* (meaning 2).

Thermal gradient See *temperature gradient*.

Thermal hysteresis The phenomenon of the freezing point of a solution being different from its melting point.

Thermal hysteresis protein (THP) A protein or glycoprotein that acts as an antifreeze by chemically interfering with the formation or growth of ice crystals. Thermal hysteresis proteins lower the freezing point more than the melting point, thus their name.

Thermal radiation heat transfer See *radiant heat transfer*.

Thermogenesis In the context of thermal relations, production of heat for the specific function of warming tissues.

Thermogenic tissue or process In the context of thermal relations, a tissue or process that specifically increases production of heat when activated.

Thermogenin An older term referring to uncoupling protein 1. See *UCP1*.

Thermoneutral zone (TNZ) In a homeotherm, the range of ambient temperatures over which the metabolic rate is constant regardless of ambient temperature.

Thermoregulation The maintenance of a relatively constant body temperature.

Thick filaments Polymers of myosin molecules in muscle cells. See *myosin*.

Thin filaments Polymers of G-actin monomers in muscle cells. See *actin*.

Thyroid-stimulating hormone (TSH) A peptide tropic hormone secreted by the anterior pituitary gland; travels in the general circulation, binds to its receptors on cells of the thyroid gland and stimulates synthesis and secretion of thyroid hormones.

Tidal flow Flow that occurs alternately in and out through a single set of passageways.

Tidal volume In an animal that exhibits tidal breathing, the amount of air inhaled and exhaled per breath.

Tight junction A place where the cell membranes of adjacent cells in an epithelium are tightly joined so that there is no extracellular space between the cells.

Time constant (τ) The time required for an exponential process to reach 63% of completion. In neurophysiology, it is a measure of the time needed to change membrane potential and is proportional to the product of resistance and capacitance.

Time difference In sensory physiology, a time difference between responses of different receptors can localize a stimulus; for example, sound that reaches the left ear first comes from the left.

Time-energy budget A method used to estimate an animal's average daily metabolic rate (ADMR). In this method, the time per day spent in each type of activity is measured and multiplied by an estimate of the energy cost of the activity per unit of time to get the total daily energy cost of the activity. The total costs of all activities are then added to get the ADMR. See also *average daily metabolic rate*.

Tissue A group of similar cells organized into a functional unit.

Tissue fluids See *interstitial fluids*.

Titin A giant elastic protein molecule that in a striated muscle spans an entire half-sarcomere from Z disc to M line.

Tonic muscle fibers Smooth muscle cells that constitute tonic smooth muscles; they maintain contractile force (tone) for long periods but do not generate spontaneous contractions or action potentials.

Tonic receptor See *slowly adapting receptor*.

Tonic smooth muscles A category of smooth muscles that maintain contractile force (tone) for long periods of time, e.g., sphincter muscles.

Total carbon dioxide concentration The amount of CO_2 a solution or body fluid takes up per unit of volume to reach a particular CO_2 partial pressure, regardless of the chemical form the CO_2 assumes when in the solution or body fluid.

Total fluid energy An important concept in understanding blood flow, defined in Figure 25.8.

Totipotent Capable of all things. For animals, chemical energy is the only form of *totipotent* energy.

Trabeculae Strands of muscle tissue that run through the open central cavity of a heart chamber, crisscrossing from one part of the chamber wall to another.

Trachea A principal tube in the breathing system of a terrestrial animal, typically with reinforced walls. (1) In vertebrates, the trachea is the initial, large airway that carries air from the buccal cavity to the lungs. (2) In insects, a trachea is any of multitudinous airways that ramify throughout the body to form the tracheal respiratory system. See also *tracheole*.

Tracheal gills In an aquatic insect, evaginated breathing structures characterized by a high density of small tracheae, into which O_2 diffuses from the water.

Tracheole A very fine, thin-walled end-tubule at the innermost reaches of the tracheal breathing system of an insect or other tracheate arthropod. The tracheoles are believed to be the principal sites of O_2 and CO_2 exchange with the tissues.

Tract A bundle of axons within a vertebrate central nervous system.

Trafficking The movement of protein molecules between active and inactive locations in a cell, thereby controlling the functional activity of the protein molecules. A common scenario for trafficking is for transport-protein molecules to be active in transport when in the cell membrane but inactive when in an intracellular membrane. See *insertion* and *retrieval*.

Trail following A trail is a marked path or course; trail following is navigation by simply following an interconnected series of local sensory cues. These cues can be visual, olfactory, or any other modality.

Transcellular In the study of transport through an epithelium, transport that occurs through the cells rather than by passageways between cells.

Transcription factors Factors that control transcription of DNA.

Transcription profiling A research strategy in which investigators detect and measure large sets of messenger RNA molecules simultaneously, as a way of assessing patterns of gene transcription in cells or tissues. Also called *transcriptomics*.

Transcriptomics See *transcription profiling*.

Transducin A G protein that is activated by rhodopsin in photoreceptors, leading to a receptor potential.

Transduction (1) In the study of energy, the transformation of one form of energy into another. (2) In the study of cell signaling, the translation of a signal from one chemical form to another chemical form. (3) In neurophysiology, the conversion of stimulus energy into an electrical signal in sensory receptor cells; the electrical signal is usually a receptor potential.

Transient receptor potential (TRP) channel A kind of ion channel found in membranes of many sensory cells; it opens in response to stimuli and produces a receptor potential.

Transpirational water loss Water loss across the integument of an animal that occurs without sweating or any other active mechanism of transporting water across the integument; passive water loss across the integument.

Transport An entirely general term referring to any and all movements of solutes, water, gases, or other materials from place to place, regardless of the mechanisms of movement.

Transporter A membrane protein that mediates the transport of solute molecules

across a membrane and must undergo reversible, noncovalent bonding with the solute molecules in order to do so. Transporters participate in *facilitated diffusion* and *active transport*. Also sometimes called a *carrier*.

Transverse tubules (t-tubules) Finger-like indentations of the cell membrane at regular intervals over the entire surface of a muscle cell. Transverse tubules conduct electrical excitation into the interior of the cell and are intimately associated with the sarcoplasmic reticulum.

Triacylglycerol A fat or oil; a lipid composed of glycerol esterified with three fatty acids. Also called a *triglyceride*.

Triglyceride See *triacylglycerol*.

Trophoblast In a mammal, the outer layer of cells of a blastocyst that will form the fetal portion of the placenta.

Tropic action An action performed by a hormone acting on a target endocrine gland to stimulate secretion of hormone by the target gland and also to maintain the structure and function of the target gland.

Tropic hormones Chemical messengers that bind to receptors of endocrine target cells to both stimulate secretion of hormone and also maintain the structure and functional capabilities of the target cells. Contrast with *direct acting hormones*.

Tropomyosin (TM) In muscles, a coiled protein molecule that spans seven actin monomers on a thin filament and is associated with one troponin molecule.

Troponin (TN) In muscles, a protein consisting of three subunits that is associated with actin and tropomyosin on the thin filaments.

Twitch A single contraction and relaxation of a skeletal muscle fiber produced by an action potential that triggers release of Ca^{2+} ions from the sarcoplasmic reticulum. A single twitch is produced by many repeated cross-bridge power strokes that draw the thin filaments toward the center of each sarcomere.

Twitch fibers In vertebrates, skeletal muscle fibers that generate an action potential when stimulated by a motor neuron action potential and produce a contractile event called a twitch. See *twitch*.

Tympanal organ An organ of hearing in which sound vibrates a tympanal membrane ("eardrum") to activate auditory receptor cells. The term is usually used for insect hearing organs, although the vertebrate ear is also a tympanal organ.

U

U/P ratio The concentration of urine expressed as a ratio of the concentration of blood plasma. The *osmotic* U/P ratio is the ratio of urine osmotic pressure over plasma osmotic pressure. The U/P ratio for a particular *ion* is the ratio of the urine concentration of that ion over the plasma concentration of that ion.

Ubiquitination Tagging of proteins for destruction. Target proteins are linked with one or more molecules of ubiquitin, a small tagging protein. Proteins thus tagged are later broken up by proteasomes.

Ubiquitin–proteasome system A cellular mechanism that destroys proteins by breaking them up into their amino acids.

UCP1 An uncoupling protein that is expressed in the mitochondria of *brown adipose tissue* in mammals. UCP1 permits protons from the space between the inner and outer mitochondrial membranes to move to the matrix (core) of a mitochondrion without driving the production of ATP; thus it promotes uncoupling of oxidative phosphorylation and production of heat from the energy released by electron transport.

Ultrafiltrate An aqueous solution produced by ultrafiltration. Also called *filtrate*.

Ultrafiltration Pressure-driven bulk flow (oozing, streaming) of fluid out of the blood plasma across the walls of blood capillaries (or sometimes through other barriers), considered a form of filtration because solutes of high molecular weight are left behind while ones of low molecular weight travel with the fluid. In ordinary tissues, fluid that leaves the blood plasma by ultrafiltration enters the tissue interstitial fluids. In kidneys that form primary urine by ultrafiltration, the fluid that leaves the blood plasma enters the kidney tubules (e.g., nephrons).

Umami One of five taste qualities in mammals, stimulated by glutamate. It is the savory taste of proteinaceous foods.

Uncoupling of oxidative phosphorylation The state of making little or no ATP from the energy that is released by the transport of electrons through the electron-transport chain.

Uncoupling protein (UCP) Unless otherwise stated, refers to UCP1. See *UCP1*.

Universal gas law In the study of gases, an equation that relates pressure (P), volume (V), molar quantity (n), and absolute temperature (T): $PV = nRT$, where R is the universal gas constant (8.314 J/mol•K, where K is a Kelvin and equivalent to one degree Celsius). Also called the *ideal gas law*.

Unloading Deoxygenation of a respiratory pigment.

Unsaturated In lipid chemistry, characterized by one or more double bonds between carbon atoms in a carbon chain.

Unsaturated fatty acid A fatty acid in which one or more of the bonds between carbon atoms in the carbon-chain back-bone of the molecule are double bonds; also called a *UFA* or an *unsaturate*. *Mono-unsaturated fatty acids* (*MUFAs*) have a single double bond in their structure. *Poly-unsaturated fatty acids* (*PUFAs*) have two or more double bonds in their structure.

Unstriated muscle cells Contractile cells in which the actin and myosin myofilaments are not arranged in sarcomeres. See *smooth muscle*.

Uphill transport See *active transport*.

Upper-critical temperature In a homeotherm, the ambient temperature that represents the upper limit of the thermoneutral zone. See also *thermoneutral zone*.

Upregulation An upward shift in the catalytic activity of an enzyme, the rate of functioning of a biochemical pathway, or the rate of some other similar process brought about in a controlled manner by a regulatory system.

Ureotelic Incorporating most nitrogen from the catabolism of nitrogenous compounds into urea. "Most" is defined differently by different authorities; a common approach is to categorize an animal as *ureotelic* if 50% or more of the nitrogen released by catabolism is incorporated into urea.

Ureter The tube that carries urine out of a kidney, often discharging it into the bladder.

Uricotelic Incorporating most nitrogen from the catabolism of nitrogenous compounds into uric acid or closely similar compounds such as urate salts. "Most" is defined differently by different authorities; a common approach is to categorize an animal as *uricotelic* if 50% or more of the nitrogen released by catabolism is incorporated into uric acid.

Urine The fluid excreted by a kidney.

Utriculus Otolithic organ in the vertebrate inner ear containing hair cells that are stimulated by horizontal movements and accelerations.

V

$\dot{V}_{O_2max}$ An animal's maximal rate of O_2 consumption. It is usually measured by inducing an animal to exercise at its peak sustained exercise intensity. Also called *aerobic capacity*.

Vaporization A change in the physical state of a material from a liquid to a gas.

Vasa recta Blood vessels of minute diameter that are arranged in long hairpin shapes and that constitute the principal vasculature of the medulla of the mammalian kidney. They function as countercurrent diffusion exchangers because blood flow is in opposite directions in the two limbs of each hairpin.

Vascular endothelial growth factor (VEGF) A locally acting cytokine that stimulates angiogenesis.

Vascular endothelium See *endothelium*.

Vascular resistance The resistance to blood flow through a blood vessel or system of blood vessels, calculated as the pressure drop divided by the flow rate.

Vasoconstriction Reduction in the inside (luminal) diameter of a blood vessel by contraction of smooth muscles in the vessel wall.

Vasodilation An increase in the inside (luminal) diameter of a blood vessel permitted by relaxation of smooth muscles in the vessel wall.

Vasomotor Related to changes in the inside (luminal) diameters of blood vessels mediated by contraction and relaxation of smooth muscles in the blood vessel walls.

Vasopressin A neurohormone of mammals, consisting of nine amino acids, that stimulates the reabsorption of water by the collecting ducts in the kidneys. Also called *antidiuretic hormone*. Its molecular structure is similar to that of the neurohormone vasotocin that controls water conservation in many nonmammalian vertebrates. Both are synthesized by neurosecretory cells that have their cell bodies in the hypothalamus and their axon terminals in the pars nervosa of the posterior pituitary.

Vein A macroscopic blood vessel that carries blood toward the heart.

Venae cavae Large veins that collect venous blood from many parts of the body for return to the heart.

Venous reserve In the study of O_2 transport, the amount of O_2 remaining in venous blood after the blood has passed through the systemic tissues.

Ventilation Forced flow (convection) of air or water into and out of structures used for external respiration or over body surfaces used for external respiration.

Ventilation-perfusion matching In a breathing organ, matching of the rate of blood flow and the rate of ventilation so that efficiency is promoted in the transfer of O_2 from air or water to the blood.

Ventral aorta In fish, the vessel that carries blood from the heart to the gills.

Ventricle A relatively muscular heart chamber.

Vesicle A small, membrane-bound, spherical organelle in the cytoplasm of a cell.

Vestibular organ A vertebrate sense organ consisting of statocysts (maculae) and semicircular canals, which together detect gravity and acceleration.

Viscosity Internal friction in a moving fluid; a lack of intrinsic slipperiness between fluid layers that are moving at different linear velocities. Fluids that are particularly high in internal friction—low in internal slipperiness—have high viscosities and exhibit syruplike properties. Viscosity can also be thought of as resistance to shear forces within a moving fluid.

Vital capacity In the study of lung function, the difference between lung volume after maximal inhalation and lung volume after maximal exhalation.

Vitamin An organic compound that an animal must obtain from food, symbiotic microbes, or another source other than the animal's biosynthesis because the animal cannot synthesize it, yet requires it in small amounts.

Volatile fatty acids See *short-chain fatty acids*.

Voltage A measure of the potential energy present because of charge separation. Separated electrical charges exert electrostatic force on each other. These forces can cause charges to move, and when charges move, work is done. Voltage, also called *electrical potential* or *potential difference*, provides a measure of the rate of charge movement that can be achieved.

Voltage clamp An experimental method to measure ionic current flow by imposing a selected membrane potential on a cell and monitoring the exogenously applied current necessary to maintain that potential by bucking the ionic current.

Voltage threshold The critical value of membrane depolarization that is just enough to trigger an action potential.

Voltage-gated channel An ion channel that opens in response to membrane depolarization.

Voltage-gated channel superfamily The evolutionarily related set of known voltage-gated cation channels.

Volume regulation The maintenance of a constant or nearly constant volume (amount) of body fluid. The term can be applied to cells, in which case it refers to a constant volume of intracellular fluid. It can also be applied to whole animals, in which case it refers to a constant volume of all body fluids.

Volumes percent (vol %) Milliliters (at STP) of gas dissolved or chemically combined within a fluid per 100 milliliters of the fluid.

Vomeronasal organ An accessory olfactory organ of vertebrates that mediates many (but not all) sensory responses to pheromones.

W

Wasting See *atrophy*.

Water channel A membrane protein that aids passive water transport—osmosis—through a cell membrane.

Water vapor Water in the gaseous state.

Water vapor pressure (1) In reference to a gas phase, the prevailing partial pressure of gaseous water. (2) In reference to an aqueous solution, the partial pressure of gaseous water that the solution will create by evaporation in a gas phase with which it is in contact if permitted to come to equilibrium with the gas phase. A colligative property.

Water vapor pressure of the aqueous solution See *water vapor pressure*, definition (2).

Water-vapor-pressure depression The difference between the water vapor pressure of a solution and that of pure water under the same conditions. A colligative property.

Watt (W) A unit of power (rate of energy use) equal to one joule per second.

Weak bond See *noncovalent bond*.

Weight In biology, commonly used as a synonym for mass (as stated in the U.S. National Institute of Standards and Technology *Guide to the SI*, in common parlance in the United States, *weight* is used as a synonym for *mass*). When *weight* is used in this way, it is expressed in kilograms. In physics, the force exerted on a mass by Earth's gravitational acceleration, expressed in Newtons. For physics calculations, *weight* and *mass* are dramatically different in meaning; see any physics text.

Weight-specific metabolic rate An animal's metabolic rate divided by its weight: metabolic rate per unit of weight.

White adipose tissue The ordinary fat-storage tissue of vertebrates, exemplified by the "fat" we speak of in poultry or beef prepared for food. Also called *white fat*. Contrast with *brown adipose tissue*.

White matter A histological region of the vertebrate central nervous system that consists largely of tracts of neuronal axons. The abundance of myelin imparts a glistening white sheen to the tissue.

White muscle In general, a loose term referring to a vertebrate muscle that is poor in myoglobin and thus whitish (rather than reddish) in color. In fish, a large mass of muscle that consists almost entirely of myoglobin-poor muscle fibers that make ATP mostly by anaerobic catabolism; white muscle powers burst exercise in fish.

Work See *external work, internal work, physiological work*.

X

Xeric Able to live steadily in the open air and thus face the full drying power of the terrestrial environment.

Z

Z disc In striated muscle, a web of accessory proteins at each end of a sarcomere that anchors the proteins titin and nebulin and the actin thin filaments. Also called *Z line, Z band*.

Z line See *Z disc*.

Zeitgeber See *phasing factor*.

Zona pellucida An extracellular layer of glycoproteins secreted by a primary oocyte.

Zooxanthellae Symbiotic algae living in the tissues of animals, such as reef-building corals.

Zygote A genetically unique, diploid cell produced by the union of an egg (ovum) and a sperm in sexual reproduction.

Zymogen An inactive form of a digestive enzyme, activated after it arrives at its site of action.

Photograph Credits

Chapter 20 *Opener:* © Arterra Picture Library/Alamy Stock Photo. **20.1C:** Courtesy of Richard Briggs, Smith College. **20.14:** © Biophoto Associates/Science Source. **20.19:** © Dr. Richard Kessel/Visuals Unlimited, Inc. **Box 20.1A:** © Patrice Ceisel/Visuals Unlimited, Inc. **Box 20.2:** © iStock.com/Michael Meyer.

Chapter 21 *Opener:* © Chuck Myers/ZUMA Wire/Alamy Live News. *Opener Inset:* Courtesy of NASA. **21.3B:** David McIntyre. **21.13:** Courtesy of Danny A. Riley and James L. W. Bain, Medical College of Wisconsin, Milwaukee, WI. **21.16A:** © Bruce Stotesbury/PostMedia News/Zuma Press. **21.16B:** © kostudio/Shutterstock. **Box 21.1:** Courtesy of Ant Backer.

Chapter 22 *Opener, Worms:* © WaterFrame/Alamy Stock Photo. *Beetle:* © Robert Henno/Alamy Stock Photo. **22.4:** © Chris A Crumley/Alamy Stock Photo. **Box 22.2:** David McIntyre.

Chapter 23 *Opener:* © Nature Picture Library/Alamy Stock Photo. **23.14A:** © John G. Shedd Aquarium/Visuals Unlimited, Inc. **23.27B:** © blickwinkel/Alamy Stock Photo. **23.30:** © blickwinkel/Alamy Stock Photo. **Box 23.2:** © iStock.com/Dmitry Pichugin.

Chapter 24 *Opener:* © Leemage/Corbis. **Box 24.3:** © Science Photo Library/Alamy Stock Photo. **Box 24.5:** © iStock.com/Dmitry Pichugin.

Chapter 25 *Opener:* © Scott Barbour/Getty Images. **Box 25.1A:** © Dr. Richard Kessel/Visuals Unlimited, Inc. **Box 25.2:** © Joel Sartore/Getty Images.

Chapter 26 *Opener:* © Galen Rowell/Corbis. **26.4** *Inset:* © Daniel Costa.

Chapter 27 *Opener:* © Kevin Fleming/Corbis *Opener Inset:* Courtesy of Richard Hill. **27.4, 27.5:** Courtesy of Richard Hill. **27.6:** NASA Goddard Space Flight Center Image by Reto Stöckli; enhancements by Robert Simmon.

Chapter 28 *Opener:* © Georgette Douwma/Getty Images. **28.6A,B:** © Larry Jon Friesen. **28.22:** David McIntyre. **28.26:** © Ran Schols/ Buiten-beeld/Getty Images. **Box 28.5:** © Dr. Diane R. Nelson/Getty Images.

Chapter 29 *Opener, Kangaroo rat:* © Robert J. Erwin/Science Source. *Jerboa:* © reptiles4all/Getty Images.

Chapter 30 *Opener:* © Pete Mcbride/Getty Images. **30.2** *Left:* © iStock.com/Peter Miller. **30.2** *Right:* Courtesy of Richard Hill. **30.3** *Left:* © Photodisc Blue/Getty Images. **30.3** *Right:* © mason01/Getty Images. **30.5:** Courtesy of Richard Hill. **30.9:** © Michael & Patricia Fogden/Minden Pictures/Getty Images. **30.10:** © blickwinkel/Alamy Stock Photo. **30.12:** © Peter Adams/Getty Images.

Figure and Table Citations

Chapter 1

Coulianos, C.-C., and A. G. Johnels. 1963. Note on the subnivean environment of small mammals. *Ark. Zool.* 15: 363–370.

Crossin, G. T., S. G. Hinch, A. P. Farrell, D. A. Higgs, A. G. Lotto, J. D. Oakes, and M. C. Healey. 2004. Energetics and morphology of sockeye salmon: effects of upriver migratory distance and elevation. *J. Fish Biol.* 65: 788–810.

Dyrstad, S. M., A. Aandstad, and J. Hallén. 2005. Aerobic fitness in young Norwegian men: a comparison between 1980 and 2002. *Scand. J. Med. Sci. Sports* 15: 298–303.

Hendry, A. P., V. Castric, M. T. Kinnison, and T. P. Quinn. 2004. The evolution of phylopatry and dispersal. Homing versus straying in salmon. In A. P. Hendry and S. C. Stearns (eds.), *Evolution Illuminated. Salmon and Their Relatives*, pp. 52–91. Oxford University Press, New York.

Lloyd, J. E. 1966. *Studies on the Flash Communication System in* Photinus *Fireflies*. Miscellaneous Publications (University of Michigan. Museum of Zoology), no. 130. Museum of Zoology, University of Michigan, Ann Arbor.

Lynch, C. B. 1992. Clinal variation in cold adaptation in *Mus domesticus*: Verification of predictions from laboratory populations. *Am. Nat.* 139: 1219–1236.

Misonne, X. 1959. Analyse Zoogéographique des Mammifères de l'Iran. *Memoires, 2me Série, no. 59 Institut Royal Sciences Naturelles de Belgique*, Brussels.

Owen-Smith, R. N. 1988. *Megaherbivores*. Cambridge University Press, Cambridge.

Pandolf, K. B., and A. J. Young. 1992. Environmental extremes and endurance performance. In R. J. Shephard and P.-O. Åstrand (eds.), *Endurance in Sport*, pp. 270–282. Blackwell, London.

Scriber, J. M. 1973. Latitudinal gradients in larval feeding specialization of the world Papilionidae (Lepidoptera). *Psyche* 80: 355–373.

Tsai, Y.-H., C.-W. Li, T.-M. Hong, J.-Z. Ho, E.-C. Yang, W.-Y. Wu, G. Margaritondo, S.-T. Hsu, E. B. L. Ong, and Y. Hwu. 2014. Firefly light flashing: oxygen supply mechanism. *Physical Rev. Lett.* 113: 258103.

Videler, J. J. 1993. *Fish Swimming*. Chapman & Hall, New York.

Walls, G. L. 1942. *The Vertebrate Eye and Its Adaptive Radiation*. Cranbrook Institute of Science, Bloomfield Hills, MI.

Wells, M. J. 1966. Cephalopod sense organs. In K. M. Wilbur and C. M. Yonge (eds.), *Physiology of Mollusca*, vol. 2, pp. 523–545. Academic Press, New York.

Young, J. Z. 1971. *The Anatomy of the Nervous System of* Octopus vulgaris. Oxford University Press, London.

Chapter 2

Bozler, E. 1928. Über die Tätigkeit der einzelnen glatten Muskelfaser bei der Kontraktion. II Mitteilung: Die Chromatophorenmuskeln der Cephalopoden. *Zeit. vergl. Physiol.* 7: 379–406.

Logue, J. A., A. L. DeVries, E. Fodor, and A. R. Cossins. 2000. Lipid compositional correlates of temperature-adaptive interspecific differences in membrane physical structure. *J. Exp. Biol.* 203: 2105–2115.

Nilsson, H., and M. Wallin. 1998. Microtubule aster formation by dynein-dependent organelle transport. *Cell Motil. Cytoskeleton* 41: 254–263.

Powers, D. A., M. Smith, I. Gonzalez-Villasenor, L. DiMichele, D. Crawford, G. Bernardi, and T. Lauerman. 1993. A multidisciplinary approach to the selectionist/neutralist controversy using the model teleost, *Fundulus heteroclitus*. *Oxf. Surv. Evol. Biol.* 9: 43–107.

Stock, D. W., J. M. Quattro, G. S. Whitt, and D. A. Powers. 1997. Lactate dehydrogenase (LDH) gene duplication during chordate evolution: The cDNA sequence of the LDH of the tunicate *Styela licata*. *Mol. Biol. Evol.* 14: 1273–1284.

Chapter 3

Biron, D. G., L. Marché, F. Ponton, H. D. Loxdale, N. Galéotti, L. Renault, C. Joly, and F. Thomas. 2005. Behavioural manipulation in a grasshopper harbouring hairworm: A proteomics approach. *Proc. R. Soc. London, Ser. B* 272: 2117–2126.

Booth, F. W., and P. D. Neufer. 2005. Exercise controls gene expression. *Am. Sci.* 93: 28–35.

Cheng, C.-H. C., L. Chen, T. J. Near, and Y. Jin. 2003. Functional antifreeze glycoprotein genes in temperate-zone New Zealand Nototheniid fish infer an Antarctic evolutionary origin. *Mol. Biol. Evol.* 20: 1897–1908.

Gelfi, C., and 10 additional authors. 2004. New aspects of altitude adaptation in Tibetans: a proteomic approach. *FASEB J.* 18: 612–614.

Kalujnaia, S., I. S. McWilliam, V. A. Zaguinaiko, A. L. Feilen, J. Nicholson, N. Hazon, C. P. Cutler, and G. Cramb. 2007. Transcriptomic approach to the study of osmoregulation in the European eel *Anguilla anguillu*. *Physiol. Genomics* 31: 385–401.

Mahoney, D. J., G. Parise, S. Melov, A. Safdar, and S. A. Tarnopolsky. 2005. Analysis of global mRNA expression in human skeletal muscle during recovery from endurance exercise. *FASEB J.* 19: 1498–1500.

Malmendal, A., J. Overgaard, J. G. Bundy, J. G. Sørensen, N. C. Nielsen, V. Loeschcke, and M. Holmstrup. 2006. Metabolic profiling of heat stress: hardening and recovery of homeostasis in *Drosophila*. *Am. J. Physiol.* 291: R205–R212.

Moylan, T. J., and B. D. Sidell. 2000. Concentrations of myoglobin and myoglobin mRNA in heart ventricles from Antarctic fishes. *J. Exp. Biol.* 203: 1277–1286.

Near, T. J., S. K. Parker, and H. W. Detrich III. 2006. A genomic fossil reveals key steps in hemoglobin loss by the Antarctic icefishes. *Mol. Biol. Evol.* 23: 2008–2016.

Near, T. J., J. J. Pesavento, and C.-H. C. Cheng. 2004. Phylogenetic investigations of Antarctic notothenioid fishes (Perciformes: Notothenioidei) using complete gene sequences of the mitochondrial encoded 16S rRNA. *Mol. Phylogenet. Evol.* 32: 881–891.

Palstra, A. P., D. Crespo, G. E. E. J. M. van den Thillart, and J. V. Planas. 2010. Saving energy to fuel exercise: swimming suppresses ooctye development and downregulates ovarian transcriptomic response of rainbow trout *Oncorhynchus mykiss*. *Am. J. Physiol.* 299: R486–R499.

Podrabsky, J. E., and G. N. Somero. 2004. Changes in gene expression associated with acclimation to constant temperatures and fluctuating daily temperatures in an annual killifish *Austrofundulus limnaeus*. *J. Exp. Biol.* 207: 2237–2254.

Rund, S. S. C., T. Y. Hou, S. M. Ward, F. H. Collins, and G. E. Duffield. 2011. Genome-wide profiling of diel and circadian gene expression in the malaria vector *Anopheles gambiae*. *Proc. Natl. Acad. Sci. U.S.A.* 108: E421–E430.

Sea Urchin Genome Sequencing Consortium. 2006. The genome of the sea urchin *Strongylocentrotus purpuratus*. *Science* 314: 941–952.

Sidell, B. D., and K. M. O'Brien. 2006. When bad things happen to good fish: The loss of hemoglobin and myoglobin expression in Antarctic icefishes. *J. Exp. Biol.* 209: 1791–1802.

Storey, K. B. 2004. Strategies for exploration of freeze responsive gene expression: Advances in vertebrate freeze tolerance. *Cryobiology* 48: 134–145.

Todgham, A. E., and G. E. Hofmann. 2009. Transcriptomic response of sea urchin larvae *Strongylocentrotus purpuratus* to CO_2-driven seawater acidification. *J. Exp. Biol.* 212: 2579–2594.

Whitfield, C. W., Y. Ben-Shahar, C. Brillet, I. Leoncini, D. Crauser, Y. LeConte, S. Rodriguez-Zas, and G. E. Robinson. 2006. Genomic dissection of behavioral maturation in the honey bee. *Proc. Natl. Acad. Sci. U.S.A.* 103: 16068–16075.

Chapter 4

Bogin, B. 1999. *Patterns of Human Growth*, 2nd ed. Cambridge University Press, Cambridge.

Burns, J. M., K. C. Lestyk, L. P. Folkow, M. O. Hammill, and A. S. Blix. 2007. Size and distribution of oxygen stores in harp and hooded seals from birth to maturity. *J. Comp. Physiol. B* 177: 687–700.

Folkow, L. P., and A. S. Blix. 1999. Diving behaviour of hooded seals (*Cystophora cristata*) in the Greenland and Norwegian Seas. *Polar Biol.* 22: 61–74.

Folkow, L. P., E. S. Nordøy, and A. S. Blix. 2010. Remarkable development of diving performance and migrations of hooded seals (*Cystophora cristata*) during their first year of life. *Polar Biol.* 33: 433–441.

Heijmans, B. T., E. W. Tobi, A. D. Stein, H. Putter, G. J. Blauw, E. S. Susser, P. E. Slagboom, and L. H. Lumey. 2008. Persistent epigenetic differences associated with prenatal exposure to famine in humans. *Proc. Natl. Acad. Sci. U.S.A.* 105: 17046–17049.

Katoh, F., A. Shimizu, K. Uchida, and T. Kaneko. 2000. Shift of chloride cell distribution during early life stages in seawater-adapted killifish, *Fundulus heteroclitus*. *Zool. Sci.* 17: 11–18.

Mays, S. A. 1999. Linear and appositional long bone growth in earlier human populations: a case study from Mediaeval England. In R. D. Hoppa and C. M. Fitzgerald (eds.), *Human Growth in the Past: Studies from Bones and Teeth*, pp. 290–312. Cambridge University Press, Cambridge.

Simola, D. F., R. J. Graham, C. M. Brady, B. L. Enzmann, C. Desplan, A. Ray, L. J. Zwiebel, R. Bonasio, D. Reinberg, J. Liebig, and S. L. Berger. 2016. Epigenetic (re)programming of caste-specific behavior in the ant *Camponotus floridanus*. *Science* 351: 42.

Walker, R. 1983. *The Molecular Biology of Enzyme Synthesis*. Wiley, New York.

Wen, X., S. Fuhrman, G. S. Michaels, D. B. Carr, S. Smith, J. L. Barker, and R. Somogyi. 1998. Large-scale temporal gene expression mapping of central nervous system development. *Proc. Natl. Acad. Sci. U.S.A.* 95: 334–339.

Chapter 5

Bublitz, M., J. P. Morth, and P. Nissen. 2011. P-type ATPases at a glance. *J. Cell Sci.* 124: 2515–2519.

Karasov, W. H., and J. M. Diamond. 1988. Interplay between physiology and ecology in digestion. *BioScience* 38: 602–611.

Weiss, T. F. 1996. *Cellular Biophysics*. MIT Press, Cambridge, MA.

Chapter 6

Andrew, A. L., D. C. Card, R. P. Ruggiero, D. R. Schield, R. H. Adams, D. D. Pollack, S. M. Secor, and T. A. Castoe. 2015. Rapid changes in gene expression direct rapid shifts in intestinal form and function in the Burmese python after feeding. *Physiol. Genomics* 47: 147–157.

Arp, A. J., J. J. Childress, and C. R. Fisher, Jr. 1985. Blood gas transportation in *Riftia pachyptila*. *Bull. Biol. Soc. Wash.* no. 6: 289–300.

Brainerd, E. L. 2001. Caught in the crossflow. *Nature* 412: 387–388.

Buller, H. A., and R. J. Grand. 1990. Lactose intolerance. *Annu. Rev. Med.* 41: 141–148.

Burton, B. T., and W. R. Foster. 1988. *Human Nutrition*, 4th ed. McGraw-Hill, New York.

Hofmann, R. R. 1989. Evolutionary steps of ecophysiological adaptation and diversification of ruminants: A comparative view of their digestive system. *Oecologia* 78: 443–457.

Hyman, L. H. 1967. *The Invertebrates: Mollusca I*. McGraw-Hill, New York.

Mayntz, D., D. Raubenheimer, M. Salomon, S. Toft, and S. J. Simpson. 2005. Nutrient-specific foraging in invertebrate predators. *Science* 307: 111–113.

Morris, J. G. 1991. Nutrition. In C. L. Prosser (ed.), *Environmental and Metabolic Animal Physiology. Comparative Animal Physiology*, 4th ed., pp. 231–276. Wiley-Liss, New York.

Pengelley, E. T., and S. J. Asmundson. 1969. Free-running periods of endogenous circannian rhythms in the golden-mantled ground squirrel, *Citellus lateralis*. *Comp. Biochem. Physiol.* 30: 177–183.

Slijper, E. J. 1979. *Whales*. Cornell University Press, Ithaca, NY.

Stevens, C. E. 1977. Comparative physiology of the digestive system. In M. J. Swenson (ed.), *Duke's Physiology of Domestic Animals*, 9th ed., pp. 216–232. Cornell University Press, Ithaca, NY.

Chapter 7

Brown, A. C., and G. Brengelmann. 1965. Energy metabolism. In T. C. Ruch and H. D. Patton (eds.), *Physiology and Biophysics*, 19th ed., pp. 1030–1049. Saunders, Philadelphia.

Golley, F. B. 1960. Energy dynamics of a food chain of an old-field community. *Ecol. Monogr.* 30: 187–206.

Hayssen, V., and R. C. Lacy. 1985. Basal metabolic rates in mammals: Taxonomic differences in the allometry of BMR and body mass. *Comp. Biochem. Physiol., A* 81: 741–754.

Jobling, M. 1993. Bioenergetics: Feed intake and energy partitioning. In J. C. Rankin and F. B. Jensen (eds.), *Fish Ecophysiology*, pp. 1–44. Chapman & Hall, New York.

Kleiber, M. 1975. *The Fire of Life*, 2nd ed. Krieger, Huntington, NY.

Lasker, R. 1970. Utilization of zooplankton energy by a Pacific sardine population in the California current. In J. H. Steele (ed.), *Marine Food Chains*, pp. 265–284. University of California Press, Berkeley.

Lavoisier, A. L. 1862. *Oeuvres de Lavoisier*, vol. 2. Imprimerie Impériale, Paris.

Mandelbrot, B. B. 1983. *The Fractal Geometry of Nature*. W. H. Freeman, San Francisco.

McKechnie, A. E., and B. O. Wolf. 2004. The allometry of avian basal metabolic rate: good predictions need good data. *Physiol. Biochem. Zool.* 77: 502–521.

McNab, B. K. 1986. The influence of food habits on the energetics of eutherian mammals. *Ecol. Monogr.* 56: 1–19.

Noujaim, S. F., O. Berenfeld, J. Kalifa, M. Cerrone, K. Nanthakumar, F. Atienza, J. Moreno, S. Mironov, and J. Jalife. 2007. Universal scaling law of electrical turbulence in the mammalian heart. *Proc. Natl. Acad. U.S.A.* 104: 20985–20989.

Noujaim, S. F., E. Lucca, V. Muñoz, D. Persaud, O. Berenfeld, F. L. Meijler, and J. Jalife. 2004. From mouse to whale: a universal scaling relation for the PR interval of the electrocardiogram of mammals. *Circulation* 110: 2802–2808.

Owen-Smith, R. N. 1988. *Megaherbivores: The Influence of Very Large Body Size on Ecology*. Cambridge University Press, New York.

Roberts, J. L. 1957. Thermal acclimation of metabolism in the crab *Pachygrapsus crassipes* Randall. I. The influence of body size, starvation, and molting. *Physiol. Zool.* 30: 232–242.

Saunders, J. B. de C. M., and C. D. O'Malley. 1950. *The Illustrations from the Works of Andreas Vesalius of Brussels*. World, Cleveland, OH.

Seymour, R. S., and A. J. Blaylock. 2000. The principle of Laplace and scaling of ventricular wall stress and blood pressure in mammals and birds. *Physiol. Biochem. Zool.* 73: 389–405.

Templeton, J. R. 1970. Reptiles. In G. C. Whittow (ed.), *Comparative Physiology of Thermoregulation*, vol. 1, pp. 167–221. Academic Press, New York.

Whitford, W. G. 1973. The effects of temperature on respiration in the Amphibia. *Am. Zool.* 13: 505–512.

Chapter 8

Allen, D. G., G. D. Lamb, and H. Westerblad. 2008. Skeletal muscle fatigue: cellular mechanisms. *Physiol. Rev.* 88: 287–332.

Åstrand, P.-O., and K. Rodahl. 1986. *Textbook of Work Physiology: Physiological Bases of Exercise*, 3rd ed. McGraw-Hill, New York.

Bickler, P. E., and L. T. Buck. 2007. Hypoxia tolerance in reptiles, amphibians, and fishes: life with variable oxygen availability. *Annu. Rev. Physiol.* 69: 145–170.

Billeter, R., and H. Hoppeler. 1992. Muscular basis of strength. In P. V. Komi (ed.), *Strength and Power in Sport*, pp. 39–63. Blackwell, Oxford, UK.

Coyle, E. F. 1991. Carbohydrate metabolism and fatigue. In G. Atlan, L. Beliveau, and P. Bouissou (eds.), *Muscle Fatigue: Biochemical and Physiological Aspects*, pp. 153–164. Masson, Paris.

Erlandsen, H., E. E. Abola, and R. C. Stevens. 2000 Combining structural genomics and enzymology: completing the picture in metabolic pathways and enzyme active sites. *Curr. Opin. Struct. Biol.* 10: 719–730.

Fulco, C. S., P. B. Rock, and A. Cymerman. 1998. Maximal and submaximal exercise performance at altitude. *Aviat. Space Environ. Med.* 69: 793–801.

Hand, S. C., and E. Gnaiger. 1988. Anaerobic dormancy quantified in *Artemia* embryos: A calorimetric test of the control mechanism. *Science* 239: 1425–1427.

Hochachka, P. W., and G. N. Somero. 2002. *Biochemical Adaptation*. Oxford University Press, New York.

Husak, J. F., S. F. Fox, M. B. Lovern, and R. A. van den Bussche. 2006. Faster lizards sire more offspring: sexual selection on whole-animal performance. *Evolution* 60: 2122–2130.

Newsholme, E. A., E. Blomstrand, N. McAndrew, and M. Parry-Billings. 1992. Biochemical causes of fatigue and overtraining. In R. J. Shephard and P.-O. Åstrand (eds.), *Endurance in Sport*, pp. 351–364. Blackwell, Oxford, UK.

Nilsson, G. E., and P. L. Lutz. 2004. Anoxia tolerant brains. *J. Cereb. Blood Flow Metab.* 24: 475–486.

Saraste, M. 1999. Oxidative phosphorylation at the *fin de siècle*. *Science* 283: 1488–1493.

Ultsch, G. R., H. Borschung, and M. J. Ross. 1978. Metabolism, critical oxygen tension, and habitat selection in darters (*Etheostoma*). *Ecology* 59: 99–107.

van Waversveld, J., A. D. F. Addink, and G. van den Thillart. 1989. Simultaneous direct and indirect calorimetry on normoxic and anoxic goldfish. *J. Exp. Biol.* 142: 325–335.

Chapter 9

Alerstam, T., M. Hake, and N. Kjellén. 2006. Temporal and spatial patterns of repeated migratory journeys by ospreys. *Anim. Behav.* 71: 555–566.

Alerstam, T., and A. Hedenström. 1998. The development of bird migration theory. *J. Avian Biol.* 29: 343–369.

Åstrand. P.-O., and K. Rodahl. 1986. *Textbook of Work Physiology*, 3rd ed. McGraw-Hill, New York.

Bastien, G. J., B. Schepens, P. A. Willems, and N. C. Heglund. 2005. Energetics of load carrying in Nepalese porters. *Science* 308: 1755.

Brett, J. R. 1964. The respiratory metabolism and swimming performance of young sockeye salmon. *J. Fish. Res. Board Can.* 21: 1183–1226.

Carter, S. L., C. D. Rennie, S. J. Hamilton, and M. A. Tarnopolsky. 2001. Changes in skeletal muscle in males and females following endurance training. *Can. J. Physiol. Pharmacol.* 79: 386–392.

Dial, K. P., A. A. Biewener, B. W. Tobalske, and D. R. Warrick. 1997. Mechanical power output of bird flight. *Nature* 390: 67–70.

Gill, R. E., Jr., T. L. Tibbits, D. C. Douglas, C. M. Handel, D. M. Mulcahy, J. C. Gottschalck, N. Warnock, B. J. McCaffery, P. F. Battley, and T. Piersma. 2009. Extreme endurance flights by landbirds crossing the Pacific Ocean: ecological corridor rather than barrier? *Proc. R. Soc. London, Ser. B* 276: 447–457.

Hammond, K. A., and J. Diamond. 1997. Maximal sustained energy budgets in humans and animals. *Nature* 386: 457–462.

Hill, R. W., and G. A. Wyse. 1989. *Animal Physiology*, 2nd ed. HarperCollins, New York.

Koch, L. G., and S. L. Britton. 2005. Divergent selection for aerobic capacity in rats as a model for complex disease. *Integr. Comp. Biol.* 45: 405–415.

Lindstedt, S. L., J. F. Hokanson, D. J. Wells, S. D. Swain, H. Hoppeler, and V. Navarro. 1991. Running energetics in the pronghorn antelope. *Nature* 353: 748–750.

Mellish, J.-A. E., S. J. Iverson, and W. D. Bowen. 2000. Metabolic compensation during high energy output in fasting, lactating grey seals (*Halichoerus grypus*): metabolic ceilings revisited. *Proc. R. Soc. London, Ser. B* 267: 1245–1251.

Nagy, K. A., W. R. Siegfried, and R. P. Wilson. 1984. Energy utilization by free-ranging jackass penguins, *Spheniscus demersus*. *Ecology* 65: 1648–1655.

Piersma, T. 2011. Why marathon migrants get away with high metabolic ceilings: towards an ecology of physiological restraint. *J. Exp. Biol.* 214: 295–302.

Taylor, C. R., K. Schmidt-Nielsen, and J. L. Raab. 1970. Scaling of energetic cost of running to body size in mammals. *Am. J. Physiol.* 219: 1104–1107.

Tucker, V. A. 1968. Respiratory exchange and evaporative water loss in the flying budgerigar. *J. Exp. Biol.* 48: 67–87.

Tucker, V. A. 1969. The energetics of bird flight. *Sci. Am.* 220(5): 70–78.

Tucker, V. A. 1975. The energetic cost of moving about. *Am. Sci.* 63: 413–419.

van Ginneken, V., E. Antonissen, U. K. Müller, R. Booms, E. Eding, J. Verreth, and G. van den Thillart. 2005. Eel migration to the Sargasso: remarkably high swimming efficiency and low energy costs. *J. Exp. Biol.* 208: 1329–1335.

Videler, J. J. 1993. *Fish Swimming*. Chapman & Hall, London.

Weimerskirch, H., M. Le Corre, and C. A. Bost. 2008. Foraging strategy of masked boobies from the largest colony in the world: relationship to environmental conditions and fisheries. *Mar. Ecol. Progr. Ser.* 362: 291–302.

Chapter 10

Armitage, K. B., B. C. Woods, and C. M. Salsbury. 2000. Energetics of hibernation in woodchucks (*Marmota monax*). In G. Heldmaier and M. Klingenspor (eds.), *Life in the Cold*, pp. 73–80. Springer, New York.

Bauwens, D., P. E. Hertz, and A. M. Castilla. 1996. Thermoregulation in a lacertid lizard: The relative contributions of distinct behavioral mechanisms. *Ecology* 77: 1818–1830.

Block, B. A., and J. R. Finnerty. 1994. Endothermy in fishes: A phylogenetic analysis of constraints, predispositions, and selection pressures. *Environ. Biol. Fishes* 40: 283–302.

Brown, J. H., and G. A. Bartholomew. 1969. Periodicity and energetics of torpor in the kangaroo mouse, *Microdipodops pallidus*. *Ecology* 50: 705–709.

Bullock, T. H. 1955. Compensation for temperature in the metabolism and activity of poikilotherms. *Biol. Rev. Cambridge* 30: 311–342.

Carey, F. G., and J. M. Teal. 1969. Regulation of body temperature by the bluefin tuna. *Comp. Biochem. Physiol.* 28: 205–213.

Dawkins, M. J. R., and D. Hull. 1964. Brown adipose tissue and the response of new-born rabbits to cold. *J. Physiol. (London)* 172: 216–238.

Dawson, W. R., and G. A. Bartholomew. 1956. Relation of oxygen consumption to body weight, temperature, and temperature acclimation in lizards *Uca stansburiana* and *Sceloporus occidentalis*. *Physiol. Zool.* 29: 40–51.

Dizon, A. E., and R. W. Brill. 1979. Thermoregulation in tunas. *Am. Zool.* 19: 249–265.

Egginton, S., and B. D. Sidell. 1989. Thermal acclimation induces adaptive changes in subcellular structure of fish muscle. *Am. J. Physiol.* 256: R1–R9.

Eliason, E. J., T. D. Clark, M. J. Hague, L. M. Hanson, Z. S. Gallagher, K. M. Jeffries, M. K. Gale, D. A. Patterson, S. G. Hinch, and A. P. Farrell. 2011. Differences in thermal tolerance among sockeye salmon populations. *Science* 332: 109–111.

Fletcher, G. L., S. V. Goddard, P. L. Davies, Z. Gong, K. V. Ewart, and C. L. Hew. 1998. New insights into fish antifreeze proteins: Physiological significance and molecular regulation. In H. O. Pörtner and R. C. Playle (eds.), *Cold Ocean Physiology*, pp. 239–265. Cambridge University Press, New York.

Hart, J. S. 1957. Climatic and temperature induced changes in the energetics of homeotherms. *Rev. Can. Biol.* 16: 133–174.

Heinrich, B. 1971. Temperature regulation of the sphinx moth, *Manduca sexta*. I. Flight energetics and body temperature during free and tethered flight. *J. Exp. Biol.* 54: 141–152.

Heinrich, B. 1974. Thermoregulation in bumblebees. I. Brood incubation by *Bombus vosnesenskii* queens. *J. Comp. Physiol.* 88: 129–140.

Hill, R. W. 1975. Daily torpor in *Peromyscus leucopus* on an adequate diet. *Comp. Biochem. Physiol., A* 51: 413–423.

Hill, R. W., D. P. Christian, and J. H. Veghte. 1980. Pinna temperature in exercising jackrabbits, *Lepus californicus*. *J. Mamm.* 61: 30–38.

Hill, R. W., T. E. Muhich, and M. M. Humphries. 2013. City-scale expansion of human thermoregulatory costs. *PLoS ONE* 8(10): e76238.

Hochachka, P. W., and G. N. Somero. 2002. *Biochemical Adaptation: Mechanism and Process in Physiological Evolution*. Oxford University Press, New York.

Huey, R. B., and J. G. Kingsolver. 1993. Evolution of resistance to high temperature in ectotherms. *Am. Nat.* 142: S21–S46.

Humphries, M. M., D. W. Thomas, and J. R. Speakman. 2002. Climate-mediated energetic constraints on the distribution of hibernating mammals. *Nature* 418: 313–316.

Irving, L., and J. Krog. 1955. Temperature of skin in the arctic as a regulator of heat. *J. Appl. Physiol.* 7: 355–364.

Johnsen, H. K. 1988. Nasal heat exchange. An experimental study of effector mechanisms associated with respiratory heat loss in Norwegian reindeer (*Rangifer tarandus tarandus*). Dr. Philos. Thesis, University of Tromsø, Norway.

Johnson, R. E. 1968. Temperature regulation in the white-tailed ptarmigan, *Lagopus leucurus*. *Comp. Biochem. Physiol.* 24: 1003–1014.

Kiss, A. J., A. Y. Mirarefi, S. Ramakrishnan, C. F. Zukoski, A. L. DeVries, and C.-H. C. Cheng. 2004. Cold-stable eye lens crystallins of the Antarctic nototheniid toothfish *Dissostichus mawsoni* Norman. *J. Exp. Biol.* 207: 4633–4649.

Kraffe, E., Y. Marty, and H. Guderley. 2007. Changes in mitochondrial oxidative capacities during thermal acclimation of rainbow trout *Oncorhynchus mykiss*: roles of membrane proteins, phospholipids and their fatty acid compositions. *J. Exp. Biol.* 210: 149–165.

Maloney, S. K., and G. Mitchell. 1997. Selective brain cooling: Role of angularis oculi vein and nasal thermoreception. *Am. J. Physiol.* 273: R1108–R1116.

Nilsson, G. E., N. Crawley, I. G. Lunde, and P. L. Munday. 2009. Elevated temperature reduces the respiratory scope of coral reef fishes. *Glob. Change Biol.* 15: 1405–1412.

Perry, A. L., P. J. Low, J. R. Ellis, and J. D. Reynolds. 2005. Climate change and distribution shifts in marine fishes. *Science* 308: 1912–1915.

Pörtner, H. O., and R. Knust. 2007. Climate change affects marine fishes through oxygen limitation of thermal tolerance. *Science* 315: 95–97.

Root, T. 1988. Energy constraints on avian distributions and abundances. *Ecology* 69: 330–339.

Rubinsky, B., S. T. S. Wong, J.-S. Hong, J. Gilbert, M. Roos, and K. B. Storey. 1994. ^{1}H magnetic resonance imaging of freezing and thawing in freeze-tolerant frogs. *Am. J. Physiol.* 266: R1771–R1777.

Scholander, P. F., W. Flagg, V. Walters, and L. Irving. 1953. Climatic adaptation in arctic and tropical poikilotherms. *Physiol. Zool.* 26: 67–92.

Scholander, P. F., R. Hock, V. Walters, F. Johnson, and L. Irving. 1950. Heat regulation in some arctic and tropical mammals and birds. *Biol. Bull.* 99: 237–258.

Somero, G. N. 1997. Temperature relationships: From molecules to biogeography. In W. H. Dantzler (ed.), *Comparative Physiology*, vol. 2 (Handbook of Physiology [Bethesda, MD], section 13), pp. 1391–1444. Oxford University Press, New York.

Visser, M. E., A. J. van Noordwijk, J. M. Tinbergen, and C. M. Lessells. 1998. Warmer springs lead to mistimed reproduction in great tits (*Parus major*). *Proc. R. Soc. London, Ser. B* 265: 1867–1870.

Woods, W. A., Jr., B. Heinrich, and R. D. Stevenson. 2005. Honeybee flight metabolic rate: does it depend upon air temperature? *J. Exp. Biol.* 208: 1161–1173.

Chapter 11

Arnold, W. 1988. Social thermoregulation during hibernation in alpine marmots (*Marmota marmota*). *J. Comp. Physiol., B* 158: 151–156.

Barnes, B. M. 1996. Relationships between hibernation and reproduction in male ground squirrels. In F. Geizer, A. J. Hulbert, and S. C. Nicol (eds.). *Adaptations to the Cold*, pp. 71–81. University of New England Press, Armidale, Australia.

Barnes, B. M., and C. L. Buck. 2000. Hibernation in the extreme: Burrow and body temperatures, metabolism, and limits to torpor bout length in arctic ground squirrels. In G. Heldmaier and M. Klingenspor (eds.), *Life in the Cold*, pp. 65–72. Springer, New York.

Boyer, B. B., and B. M. Barnes. 1999. Molecular and metabolic aspects of mammalian hibernation. *BioScience* 49: 713–724.

Geiser, F., and G. J. Kenagy. 1987. Polyunsaturated lipid diet lengthens torpor and reduces body temperature in a hibernator. *Am. J. Physiol.* 252: R897–R901.

Heldmaier, G. 1993. Seasonal acclimatization of small mammals. *Verh. Dtsch. Zool. Ges.* 86(2): 67–77.

Heldmaier, G., S. Ortmann, and R. Elvert. 2004. Natural hypometabolism during hibernation and daily torpor in mammals. *Respir. Physiol. Neurobiol.* 141: 317–329.

Hull, D. 1973. Thermoregulation in young mammals. In G. C. Whittow (ed.), *Comparative Physiology of Thermoregulation*, vol. 3, pp. 167–200. Academic Press, New York.

Johnsen, H. K., A. Rognmo, K. J. Nilssen, and A. S. Blix. 1985. Seasonal changes in the relative importance of different avenues of heat loss in resting and running reindeer. *Acta Physiol. Scand.* 123: 73–79.

Markussen, K. A., A. Rognmo, and A. S. Blix. 1985. Some aspects of thermoregulation in newborn reindeer calves (*Rangifer tarandus tarandus*). *Acta Physiol. Scand.* 123: 215–220.

Meeks, N. D., and C. R. Cartwright. 2005. Caribou and seal hair: examination by scanning electron microscopy. In J. C. H. King, B. Pauksztat, and R. Storrie (eds.), *Arctic Clothing*, pp. 42–44. The British Museum Press, London.

Meng, M., G. C. West, and L. Irving. 1969. Fatty acid composition of caribou bone marrow. *Comp. Biochem. Physiol.* 30: 187–191.

Newmaster, S. G., I. D. Thompson, R. A. D. Steeves, A. R. Rodgers, A. J. Fazekas, J. R. Maloles, R. T. McMullin, and J. M. Fryxell. 2013. Examination of two new technologies to assess the diet of woodland caribou: video recorders attached to collars and DNA barcoding. *Can. J. For. Res.* 43: 897–900.

Nieminen, M. 1980. Nutritional and seasonal effects on the haematology and blood chemistry in reindeer (*Rangifer tarandus tarandus* L.). *Comp. Biochem. Physiol., A* 66: 399–413.

Nilssen, K. J., J. A. Sundsfjord, and A. S. Blix. 1984. Regulation of metabolic rate in Svalbard and Norwegian reindeer. *Am. J. Physiol.* 247: R837–R841.

Orpin, C. G., S. D. Mathiesen, Y. Greenwood, and A. S. Blix. 1985. Seasonal changes in the ruminal microflora of the high-arctic Svalbard reindeer (*Rangifer tarandus platyrhynchus*). *Appl. Environ. Microbiol.* 50: 144–151.

Power, G. G., T. R. Gunn, B. M. Johnston, and P. D. Gluckman. 1987. Oxygen supply and the placenta limit thermogenic responses in fetal sheep. *J. Appl. Physiol.* 63: 1896–1901.

Ruf, T., and W. Arnold. 2000. Mechanisms of social thermoregulation in hibernating alpine marmots (*Marmota marmota*). In G. Heldmaier and M. Klingenspor (eds.), *Life in the Cold*, pp. 81–94. Springer, New York.

Ruf, T., and W. Arnold. 2008. Effects of polyunsaturated fatty acids on hibernation and torpor: a review and hypothesis. *Am. J. Physiol.* 294: R1044–R1052.

Soppela, P., M. Nieminen, S. Saarela, and R. Hissa. 1986. The influence of ambient temperature on metabolism and body temperature of newborn and growing reindeer calves (*Rangifer tarandus tarandus* L.). *Comp. Biochem. Physiol., A* 83: 371–386.

Chapter 12

Bear, M. F., B. W. Conners, and M. A. Paradiso. 2001. *Neuroscience: Exploring the Brain*, 2nd ed. Lippincott Williams & Wilkins, Baltimore, MD.

Bullock, T. H., and G. A. Horridge. 1965. *Structure and Function in the Nervous Systems of Invertebrates*. W. H. Freeman, San Francisco.

Camhi, J. M. 1984. *Neuroethology: Nerve Cells and the Natural Behavior of Animals*. Sinauer, Sunderland, MA.

Long, S. B., E. B. Campbell, and R. MacKinnon. 2005. Crystal structure of a mammalian voltage-dependent *Shaker* family K+ channel. *Science* 309: 897–903.

Rasband, M. 2010. The axon initial segment and the maintenance of neuronal polarity. *Nat. Rev. Neurosci.* 11: 552–562.

Schmidt, R. F. (ed.). 1985. *Fundamentals of Neurophysiology*, 3rd ed. Springer, New York.

Chapter 13

Antonsen, B. L., and D. H. Edwards. 2003. Differential dye coupling reveals lateral giant escape circuit in crayfish. *J. Comp. Neurol.* 466: 1–13.

Kandel, E. R., J. H. Schwartz, and T. M. Jessell (eds.). 1995. *Essentials of Neural Science and Behavior*. Appleton & Lange, Norwalk, CT.

Kandel, E. R., J. H. Schwartz, and T. M. Jessell (eds.). 2000. *Principles of Neural Science*, 4th ed. McGraw-Hill, New York.

Lee, S.-J. R., Y. Escobedo-Lozoya, E. M. Szatmari, and R. Yasuda. 2009. Activation of CaMKII in single dendritic spines during long-term potentiation. *Nature* 458: 299–304.

Sanes, J. R., and J. W. Lichtman. 2001. Induction, assembly, maturation and maintenance of a postsynaptic apparatus. *Nat. Rev. Neurosci.* 2: 791–805.

Trachtenberg, J. T., B. E. Chen, G. W. Knott, G. Feng, J. R. Sanes, E. Welker, and K. Svoboda. 2002. Long-term in vivo imaging of experience-dependent synaptic plasticity in adult cortex. *Nature* 420: 788–794.

Waites, C. L., A. M. Craig, and C. C. Garner. 2005. Mechanisms of vertebrate synaptogenesis. *Annu. Rev. Neurosci.* 28: 251–274.

Walsh, M. K., and J. W. Lichtman. 2003. In vivo time-lapse imaging of synaptic takeover associated with naturally occurring synapse elimination. *Neuron* 37: 67–73.

Chapter 14

Bullock, T. H., and G. A. Horridge. 1965. *Structure and Function in the Nervous System of Invertebrates*, vol. 2. W. H. Freeman, San Francisco.

Dowling, J. E. 1979. Information processing by local circuits: The vertebrate retina as a model system. In F. O. Schmitt and F. G. Worden (eds.), *The Neurosciences: Fourth Study Program*, pp. 163–181. MIT Press, Cambridge, MA.

Fain, G. L. 2003. *Sensory Transduction*. Sinauer, Sunderland, MA.

Firestein, S., F. Zufall, and G. M. Shepherd. 1991. Single odor-sensitive channels in olfactory receptor neurons are also gated by cyclic nucleotides. *J. Neurosci.* 11: 3565–3572.

Frisby, J. P. 1980. *Seeing: Illusion, Brain and Mind*. Oxford University Press, Oxford, UK.

Gehring, W. J. 2005. New perspectives on eye development and the evolution of eyes and photoreceptors. *J. Hered.* 96: 171–184.

Hamill O. P., 2006. Twenty odd years of stretch-sensitive channels. *Pflügers Arch—Eur. J. Physiol.* 453: 333–351.

Holley M. C. and J. F. Ashmore. 1988. On the mechanism of a high-frequency force generator in outer hair cells isolated from the guinea pig cochlea. *Proc. R. Soc. London, Ser. B.* 232: 413–429.

Kandel, E. R., J. H. Schwartz, and T. M. Jessell (eds.). 1995. *Essentials of Neural Science and Behavior*. Appleton & Lange, Stamford, CT.

Kandel, E. R., J. H. Schwartz, and T. M. Jessell (eds.). 2000. *Principles of Neural Science*, 4th ed. McGraw-Hill, New York.

Keil, T. A. 1997. Functional morphology of insect mechanoreceptors. *Microsc. Res. Tech.* 39: 506–531.

Knudsen, E. I., and M. Konishi. 1978. Center-surround organization of auditory receptive fields in the owl. *Science* 202: 778–780.

Kuffler, S. W., J. G. Nicholls, and A. R. Martin, 1984. *From Neuron to Brain*, 2nd ed. Sinauer, Sunderland, MA.

Masland, R. H. 1986. The functional architecture of the retina. *Sci. Am.* 255(6): 102–111.

Spehr, M., and S. D. Munger. 2009. Olfactory receptors: G protein-coupled receptors and beyond. *J. Neurochem.* 109: 1570–1583

Strassmaier, M., and P. G. Gillespie. 2002. The hair cell's transduction channel. *Curr. Opin. Neurobiol.* 12: 380–386.

Thurm, U. 1964. Mechanoreceptors in the cuticle of the honey bee: Fine structure and stimulus mechanism. *Science* 145: 1063–1065.

Chapter 15

Campbell, N. A., and J. B. Reece. 2002. *Biology*, 6th ed. Benjamin Cummings, San Francisco.

Catania, K. C. 1999. A nose that looks like a hand and acts like an eye: the unusual mechanosensory system of the star-nosed mole. *J. Comp. Physiol., A* 185: 367–372.

Catania, K. C. 2005. Evolution of sensory specializations in insectivores. *Anat. Rec. A Discov. Mol. Cell Evol. Biol.* 297: 1038–1050.

Chen, W., X. H. Zhu, K. R. Thulborn, and K. Ugurbil. 1999. Retinotopic mapping of lateral geniculate nucleus in humans using functional magnetic resonance imaging. *Proc. Natl. Acad. Sci. U.S.A.* 96: 2430–2434.

Ferreira, F. R. M., M. I. Nogueira, and J. DeFelipe. 2014. The influence of James and Darwin on Cajal and his research into the neuron theory and evolution of the nervous system. *Front. Neuroanat.* 8: 1. doi: 10.3389/fnana.2014.00001

Herzog, E. D. 2007. Neurons and networks in daily rhythms. *Nat. Rev. Neurosci.* 8: 790–802.

Pohl, H. 1970. On the effect of light upon the circadian rhythms of metabolism and activity of the chaffinches *Fringilla coelebs*. *Z. vergl. Physiol.* 66: 141–163.

Welsh, D. K., J. S. Takahashi, and S. A. Kay. 2010. Suprachiasmatic nucleus: cell autonomy and network properties. *Annu. Rev. Physiol.* 72: 551–577.

Wever, R. A. 1979. *The Circadian System of Man. Results of Experiments under Temporal Isolation.* Springer-Verlag, New York.

Chapter 16

Bentley, P. J. 1998. *Comparative Vertebrate Endocrinology*, 3rd ed. Cambridge University Press, Cambridge.

Gardner, D. G., and D. Shoback. 2007. *Greenspan's Basic and Clinical Endocrinology*, 8th ed. McGraw-Hill Medical, New York.

Müller, W. A., G. R. Faloona, E. Aguilar-Parada, and R. H. Unger. 1970. Abnormal alpha-cell function in diabetes: Response to carbohydrate and protein ingestion. *New Eng. J. Med.* 283: 109–115.

O'Brien, M. A., E. J. Katahira, T. R. Flanagan, L. W. Arnold, G. Haughton, and W. E. Bollenbacher. 1988. A monoclonal antibody to the insect prothoracicotropic hormone. *J. Neurosci.* 8: 3247–3257.

Randall, D., W. Burggren, and K. French. 2002. *Eckert Animal Physiology: Mechanisms and Adaptations*, 5th ed. W. H. Freeman, New York.

Sacca, L., N. Eigler, P. E. Cryer, and R. S. Sherwin. 1979. Insulin antagonistic effects of epinephrine and glucagon in the dog. *Am. J. Physiol.* 237: E487–E492.

White, B. A., and J. Ewer. 2014. Neural and hormonal control of postecdysial behaviors in insects. *Annu. Rev. Entomol.* 59: 363–381.

Zitnan, D. and M.E. Adams. 2012. Neuroendocrine regulation of ecdysis. In L.I. Gilbert (ed.), *Insect Endocrinology*, pp. 253–309. Academic Press, Boston.

Zitnan, D., I. Zitnanová, I. Spalovská, P. Takác, Y. Park, and M. E. Adams. 2003. Conservation of ecdysis-triggering hormone signaling in insects. *J. Exp. Biol.* 206: 1275–1289.

Zitnan, D., Y.-J. Kim, I. Zitnanová, L. Roller, and M. E. Adams. 2007. Complex steroid-peptide-receptor cascade controls insect ecdysis. *Gen. Comp. Endocrinol.* 153: 88–96.

Chapter 17

Bronson, F. H. 1989. *Mammalian Reproductive Biology.* University of Chicago Press, Chicago.

Horton, B. M., W. H. Hudson, E. A. Ortlund, S. Shirk, J. W. Thomas, E. R. Young, W. M. Zinzow-Kramer, and D. L. Maney. 2014. Estrogen receptor α polymorphism in a species with alternative behavioral phenotypes. *Proc. Natl. Acad. Sci. USA* 111: 1443–1448.

Kenagy, G. J., and S. C. Trombulak. 1986. Size and function of mammalian testes in relation to body size. *J. Mamm.* 67: 1–22.

Primakoff, P., and D. G. Myles. 2002. Penetration, adhesion, and fusion in mammalian sperm–egg interaction. *Science* 296: 2138–2185.

Thompson, S. D. 1992. Gestation and lactation in small mammals: Basal metabolic rate and the limits of energy use. In T. E. Tomasi and T. H. Horton (eds.), *Mammalian Energetics*, pp. 213–259. Cornell University Press, Ithaca, NY.

Troen, P., and H. Oshima. 1981. The testis. In P. Felig, J. D. Baxter, A. E. Broadus, and L. A. Frohman (eds.), *Endocrinology and Metabolism*, pp. 627–668. McGraw-Hill, New York.

Tsou, R. C., R. A. Dailey, C. S. McLanahan, A. D. Parent, G. T. Tindall, and J. D. Neill. 1977. Luteinizing-hormone releasing hormone (LHRH) levels in pituitary stalk plasma during preovulatory gonadotropin surge of rabbits. *Endocrinology* 101: 534–539.

Chapter 18

Collett, T. S., and J. Zeil. 1998. Places and landmarks: An arthropod perspective. In S. Healy (ed.), *Spatial Representation in Animals*, pp. 18–53. Oxford University Press, Oxford, UK.

Dittman, A. H., and T. P. Quinn. 1996. Homing in Pacific salmon: Mechanisms and ecological basis. *J. Exp. Biol.* 199: 83–91.

Emlen, S. T. 1967. Migratory orientation in the indigo bunting, *Passerina cyanea*: Part I: Evidence for use of celestial cues. *The Auk* 84: 309–342.

Goodenough, J., B. McGuire, and R. A. Wallace. 2001. *Perspectives on Animal Behavior*, 2nd ed. Wiley, New York.

Lohmann, K. J., N. F. Putman, and C. M. F. Lohmann. 2012. The magnetic map of hatchling loggerhead sea turtles. *Curr. Opin. Neurobiol.* 22: 336–342.

Moser, M-B, D. C. Rowland, and E. I. Moser. 2015. Place cells, grid cells, and memory. *Cold Spring Harb. Perspect. Biol.* 7: a021808. doi: 10.1101/cshperspect.a021808

Papi, F. (ed.). 1992. *Animal Homing.* Chapman & Hall, London.

Perdeck, A. C. 1958. Two types of orientation in migrating starlings, *Sturmus vulgaris* L. and chaffinches, *Fringilla coelebs* L., as revealed by displacement experiments. *Ardea* 46: 1–37.

Walcott, C., and R. P. Green. 1974. Orientation of homing pigeons altered by a change in the direction of an applied magnetic field. *Science* 184: 180–182.

Wehner, R. 1997. The ant's celestial compass system: Spectral and polarization channels. In Lehrer M. (ed.), *Orientation and Communication in Arthropods*, pp. 145–185. Birkhäuser, Basel, Switzerland.

Wiltschko, W., and R. Wiltschko. 1996. Magnetic orientation in birds. *J. Exp. Biol.* 199: 29–38.

Wittlinger, M., R. Wehner, and H. Wolf. 2006. The ant odometer: Stepping on stilts and stumps. *Science* 312: 1965–1967.

Chapter 19

Harris-Warrick, R. M., and R. E. Flamm. 1986. Chemical modulation of a small central pattern generator circuit. *Trends Neurosci.* 9: 432–437.

Ijspeert, A. J., A. Crespi, D. Ryczko, and J.-M. Cabelguen. 2007. From swimming to walking with a salamander robot driven by a spinal cord model. *Science* 315: 1416–1420.

Kandel, E. R., J. A. Schwartz, and T. M. Jessell (eds.). 1995. *Essentials of Neural Science and Behavior.* Appleton & Lange, Norwalk, CT.

Kandel, E. R., J. H. Schwartz, and T. M. Jessell (eds.). 2000. *Principles of Neural Science*, 4th ed. McGraw-Hill, New York.

Rizzolatti, G., and G. Luppino. 2001. The cortical motor system. *Neuron* 31: 889–901.

Wyse, G. A., D. H. Sanes, and W. H. Watson. 1980. Central neural motor programs underlying short- and long-term patterns of *Limulus* respiratory activity. *J. Comp. Physiol.* 141: 87–92.

Chapter 20

Gallant, J. R., L. L. Traeger, J. D. Volkening, H. Moffett, P. H. Chen, C. D. Novina, G. N. Phillips, Jr., R. Anand, G. B. Wells, M. Pinch, R. Güth, G. A. Unguez, J. S. Albert, H. H. Zakon, M. P. Samanta, and M. R. Sussman. 2014. Genomic basis for the convergent evolution of electric organs. *Science* 344: 1522–1525.

Chapter 18

Gordon, A. M., A. F. Huxley, and F. J. Julian. 1966. The variation in isometric tension with sarcomere length in vertebrate muscle fibers. *J. Physiol. (London)* 184: 170–192.

Schmidt-Nielsen, K. 1997. *Animal Physiology: Adaptation and Environment*, 5th ed. Cambridge University Press, New York.

Chapter 21

Andersen, J. B., B. C. Rourke, V. J. Caiozzo, A. F. Bennett, and J. W. Hicks. 2005. Postprandial cardiac hypertrophy in pythons. *Nature* 434: 37–38.

Andersen, J. L., and P. Aagaard. 2000. Myosin heavy chain IIX overshoot in human skeletal muscle. *Muscle Nerve* 23: 1095–1104.

Andersen, P., and J. Henriksson. 1977. Capillary supply of the quadriceps femoris muscle of man: Adaptive response to exercise. *J. Physiol.* 270: 677–690.

Carter, H. N., C. C. W. Chen, and D. A. Hood. 2015. Mitochondria, muscle health, and exercise with advancing age. *Physiology* 30: 208–223.

Chinsomboon, J., J. Ruas, R. K. Gupta, R. Thom, J. Shoag, G. C. Rowe, N. Sawada, S. Raghuram, and Z. Arany. 2009. The transcriptional coactivator PGC-1α mediates exercise-induced angiogenesis in skeletal muscle. *Protc. Natl. Acad. Sci. U.S.A.* 106 (50): 21401–21406.

Harlow, H. J., T. Lohuis, T. D. I. Beck, and P. A. Iaizzo. 2001. Muscle strength in overwintering bears. *Nature* 409: 997.

Hoppeler, H., H. Howald, K. Conley, S. L. Lindstedt, H. Claassen, P. Vock, and E. R. Weibel. 1985. Endurance training in humans: Aerobic capacity and structure of skeletal muscle. *J. Appl. Physiol.* 59: 320–327.

Ingjer, F. 1979. Effects of endurance training on muscle fibre ATP-ase activity, capillary supply and mitochondrial content in man. *J. Physiol.* 294: 419–432.

Moore, D. H., II. 1975. A study of age group track and field records to relate age and running speed. *Nature* 253: 264–265.

Richardson, R. S., H. Wagner, S. R. D. Mudaliar, E. Saucedo, R. Henry, and P. D. Wagner. 2000. Exercise adaptation attenuates VEGF gene expression in human skeletal muscle. *Am. J. Physiol.* 279: H772–H778.

Riquelme, C. A., J. A. Magida, B. C. Harrison, C. E. Wall, T. G. Marr, S. M. Secor, and L. A. Leinwand. 2011. Fatty acids identified in the Burmese python promote beneficial cardiac growth. *Science* 334: 528–531.

Ruas, J. L, J. P. White, R. R. Rao, S. Kleiner, K. T. Brannan, B. C. Harrison, N. P. Greene, J. Wu, J. L. Estall, B. A. Irving, I. R. Lanza, K. A. Rasbach, M. Okutsu, K. S. Nair, Z. Yan, L. A. Leinwand, and B. M. Spiegelman. 2012. A PGC-1α isoform induced by resistance training regulates skeletal muscle hypertrophy. *Cell* 151: 1319–1331

Waters, R. E., S. Rotevatn, P. Li, B. H. Annex, and Z. Yan. 2004. Voluntary running induces fiber type-specific angiogenesis in mouse skeletal muscle. *Am. J. Physiol.* 287: C1342–C1348.

Widrick, J. J., S. T. Knuth, K. M. Norenberg, J. G. Romatowski, J. L. Bain, D. A. Riley, M. Karhanek, S. W. Trappe, T. A. Trappe, D. L. Costill, and R. H. Fitts. 1999. Effect of a 17 day spaceflight on contractile properties of human soleus muscle fibres. *J. Physiol.* 516: 915–930.

Chapter 22

Graham, J. B. 1990. Ecological, evolutionary, and physical factors influencing aquatic animal respiration. *Am. Zool.* 30: 137–146.

Kramer, D., and E. H. Ahlstrom. 1968. Distributional atlas of fish larvae in the California Current Region: Northern Anchovy, *Engraulis mordax*. Girard, 1951–1965. CalCOFI Atlas NO. 9. State of California Marine Research Committee.

Chapter 23

Brackenbury, J. H. 1981. Airflow and respired gases within the lung-air-sac system of birds. *Comp. Biochem. Physiol., A* 68: 1–8.

Burggren, W. W., and N. H. West. 1982. Changing respiratory importance of gills, lungs and skin during metamorphosis in the bullfrog *Rana catesbeiana*. *Respir. Physiol.* 47: 151–164.

Comstock, J. H. 1912. *The Spider Book*. Doubleday, Garden City, NJ.

Duncker, H. R. 1978. General morphological principles of amniotic lungs. In J. Piiper (ed.), *Respiratory Function in Birds, Adult and Embryonic*, pp. 2–15. Springer-Verlag, New York.

Feder, M. E., and W. W. Burggren. 1985. Skin breathing in vertebrates. *Sci. Am.* 253(5): 126–142.

Feldman, J. L., J. C. Smith, D. R. McCrimmon, H. H. Ellenberger, and D. F. Speck. 1988. Generation of respiratory pattern in mammals. In A. H. Cohen, S. Rossignol, and S. Grillner (eds.), *Neural Control of Rhythmic Movements in Vertebrates*, pp. 73–100. Wiley, New York.

Gans, C. 1970. Respiration in early tetrapods—The frog is a red herring. *Evolution* 24: 723–734.

Hildebrandt, J., and A. C. Young. 1960. Anatomy and physics of respiration. In T. C. Ruch and H. D. Patton (eds.), *Physiology and Biophysics*, 19th ed., pp. 733–760. Saunders, Philadelphia.

Hill, R. W., and G. A. Wyse. 1989. *Animal Physiology*, 2nd ed. Harper & Row, New York.

Hughes, G. M. 1961. How a fish extracts oxygen from water. *New Sci.* 11: 346–348.

Lieske, S. P., M. Thoby-Brisson, P. Telgkamp, and J. M. Ramirez. 2000. Reconfiguration of the neural network controlling multiple breathing patterns: Eupnea, sighs, and gasps. *Nat. Neurosci.* 3: 600–608.

Moyle, P. B. 1993. *Fish: An Enthusiast's Guide*. University of California Press, Berkeley, CA.

Perry, S. F. 1990. Recent advances and trends in the comparative morphometry of vertebrate gas exchange organs. *Adv. Comp. Environ. Physiol.* 6: 45–71.

Poll, M. 1962. Étude sur la structure adulte et la formation de sacs pulmonaires des Protoptères. *Ann. Mus. R. Afr. Centr. (Ser. 8)* 108: 129–172.

Rytkönen, K. T., G. M. C. Renshaw, P. P. Vainio, K. J. Ashton, G. Williams-Pritchard, E. H. Leder, and M. Nikinmaa. 2012. Transcriptional responses to hypoxia are enhanced by recurrent hypoxia (hypoxia preconditioning) in the epaulette shark. *Physiol. Genomics* 44: 1090–1097.

Socha, J. J., T. D. Förster, and K. J. Greenlee. 2010. Issues of convection in insect respiration: Insights from synchroton X-ray imaging and beyond. *Resp. Physiol. Neurobiol.* 173S: S65–S73.

Torrance, J. D., C. Lenfant, J. Cruz, and E. Marticorena. 1970. Oxygen transport mechanisms in residents at high altitude. *Respir. Physiol.* 11: 1–15.

Wegner, N. C., C. A. Sepulveda, S. A. Aalbers, and J. B. Graham. 2013. Structural adaptations for ram ventilation: Gill fusions in scombrids and billfishes. *J. Morphol.* 274: 108–120.

Weibel, E. R. 1984. *The Pathway for Oxygen*. Harvard University Press, Cambridge, MA.

Wigglesworth, V. B. 1935. The regulation of respiration in the flea, *Xenopsylla cheopis*, Roths. (Pulicidae). *Proc. R. Soc. London, Ser. B* 118: 397–419.

Chapter 24

Dejours, P. 1981. *Principles of Comparative Respiratory Physiology*, 2nd ed. Elsevier/North-Holland, New York.

Dhindsa, D. S., J. Metcalfe, and A. S. Hoversland. 1972. Comparative studies of the respiratory functions of mammalian blood. IX. Ring-tailed lemur (*Lemur catta*) and black lemur (*Lemur macaco*). *Respir. Physiol.* 15: 331–342.

Dickerson, R. E., and I. Geis. 1983. *Hemoglobin: Structure, Function, Evolution, and Pathology*. Benjamin Cummings, Menlo Park, CA.

Duhm, J. 1971. Effects of 2,3-diphosphoglycerate and other organic phosphate compounds on oxygen affinity and intracellular pH of human erythrocytes. *Pflügers Arch.* 326: 341–356.

Hill, R. W., and G. A. Wyse. 1989. *Animal Physiology*, 2nd ed. Harper & Row, New York.

Johansen, K., and C. Lenfant. 1966. Gas exchange in the cephalopod, *Octopus dofleini*. *Am. J. Physiol.* 210: 910–918.

Jones, D. R., and D. J. Randall. 1978. The respiratory and circulatory systems during exercise. In W. S. Hoar and D. J. Randall (eds.), *Fish Physiology*, vol. 7, pp. 425–501. Academic Press, New York.

Paul, R. J., B. Zeis, T. Lamkemeyer, M. Seidle, and R. Pirow. 2004. Control of oxygen transport in the microcrustacean *Daphnia*: regulation of haemoglobin expression as central mechanism of adaptation to different oxygen and temperature conditions. *Acta Physiol. Scand.* 182: 259–275.

Reeves, R. B. 1980. The effect of temperature on the oxygen equilibrium curve of human blood. *Respir. Physiol.* 42: 317–328.

Roughton, F. J. W. 1964. Transport of oxygen and carbon dioxide. In W. O. Fenn and H. Rahn (eds.), *Respiration*, vol. 1 (Handbook of Physiology [Bethesda, MD], section 3), pp. 767–825. American Physiological Society, Washington, D.C.

Sadava, D., H. C. Heller, G. H. Orians, W. K. Purves, and D. Hillis. 2008. *Life: The Science of Biology*, 8th ed. Sinauer Associates, Sunderland, MA.

Steen, J. B. 1963. The physiology of the swimbladder of the eel *Angullia vulgaris*. I. The solubility of the gases and the buffer capacity of the blood. *Acta Physiol. Scand.* 58: 124–137.

Terwilliger, R. C. 1980. Structures of invertebrate hemoglobins. *Am. Zool.* 20: 53–67.

Waterman, M. R. 1978. Spectral characterization of human hemoglobin and its derivatives. In S. Fleischer and L. Packer (eds.), *Methods in Enzymology*, vol. 52, pp. 456–463. Academic Press, New York.

Wood, W. G. 1976. Haemoglobin synthesis during human fetal development. *Brit. Med. Bull.* 32: 282–287.

Chapter 25

Copenhaver, W. M., D. E. Kelly, and R. L. Wood. 1978. *Bailey's Textbook of Histology*, 17th ed. Williams & Wilkins, Baltimore.

Eliason, E. J., T. D. Clark, M. J. Hague, L. M. Hanson, Z. S. Gallagher, K. M. Jeffries, M. K. Gale, D. A. Patterson, S. G. Hinch, and A. P. Farrell. 2011. Differences in thermal tolerance among sockeye salmon populations. *Science* 332: 109–122.

Farrell, A. P. 1991. Circulation of body fluids. In C. L. Prosser (ed.), *Environmental and Metabolic Animal Physiology* (*Comparative Animal Physiology*, 4th ed.), pp. 509–558. Wiley-Liss, New York.

Feigl, E. O. 1974. Physics of the cardiovascular system. In T. C. Ruch and H. D. Patton (eds.), *Physiology and Biophysics*, 20th ed., vol. 2, pp. 10–22. Saunders, Philadelphia.

Hartline, D. K. 1967. Impulse identification and axon mapping of the nine neurons in the cardiac ganglion of the lobster, *Homarus americanus*. *J. Exp. Biol.* 47: 327–346.

Hill, R. W., and G. A. Wyse. 1989. *Animal Physiology*, 2nd ed. Harper & Row, New York.

Johansen, K. 1970. Air breathing in fishes. In W. S. Hoar and D. J. Randall (eds.), *Fish Physiology*, vol. 4, pp. 361–411. Academic Press, New York.

Johansen, K., and R. Hol. 1968. A radiological study of the central circulation in the lungfish, *Protopterus aethiopicus*. *J. Morphol.* 126: 333–348.

Johansen, K., and C. Lenfant. 1966. Gas exchange in the cephalopod, *Octopus dofleini*. *Am. J. Physiol.* 210: 910–918.

Jones, J. C. 1964. The circulatory system of insects. In M. Rockstein (ed.), *The Physiology of Insecta*, vol. 3, pp. 1–107. Academic Press, New York.

Laurent, P., R. G. DeLaney, and A. P. Fishman. 1978. The vasculature of the gills in the aquatic and aestivating lungfish (*Protopterus aethiopicus*). *J. Morphol.* 156: 173–208.

McLaughlin, P. A. 1980. *Comparative Morphology of Recent Crustacea*. W. H. Freeman, San Francisco.

Paul, R. J., S. Bihlmayer, M. Colmorgen, and S. Zahler. 1994. The open circulatory system of spiders (*Eurypelma californicum, Pholcus phalangioides*): A survey of functional morphology and physiology. *Physiol. Zool.* 67: 1360–1382.

Plateau, F. 1880. Le coeur des crustacés décapodes. *Arch. Biol.* 1: 595–695.

Randall, D. J. 1968. Functional morphology of the heart in fishes. *Am. Zool.* 8: 179–189.

Rushmer, R. F. 1976. *Cardiovascular Dynamics*, 4th ed. Saunders, Philadelphia.

Scher, A. M., and M. S. Spach. 1979. Cardiac depolarization and repolarization and the electrocardiogram. In R. M. Berne (ed.), *The Cardiovascular System*, vol. 1 (Handbook of Physiology [Bethesda, MD], section 2), pp. 357–392. Oxford University Press, New York.

Smith, P. J. S. 1981. The octopod ventricular cardiogram. *Comp. Biochem. Physiol., A* 70: 103–105.

Sturkie, P. D. (ed.). 1976. *Avian Physiology*, 3rd ed. Springer-Verlag, New York.

Wilkens, J. L. 1999. Evolution of the cardiovascular system in Crustacea. *Am. Zool.* 39: 199–214.

Chapter 26

Bron, K. M., H. V. Murdaugh, Jr., J. E. Millen, R. Lenthall, P. Raskin, and E. D. Robin. 1966. Arterial constrictor response in a diving mammal. *Science* 152: 540–543.

Burns, J. M., K. C. Lestyk, L. P. Folkow, M. O. Hammill, and A. S. Blix. 2007. Size and distribution of oxygen stores in harp and hooded seals from birth to maturity. *J. Comp. Physiol. B* 177: 687–700.

Butler, P. J., and D. R. Jones. 1997. Physiology of diving of birds and mammals. *Physiol. Rev.* 77: 837–899.

Folkow, L. P., J.-M. Ramirez, S. Ludvigsen, N. Ramirez, and A. S. Blix. 2008. Remarkable neuronal hypoxia tolerance in the deep-diving adult hooded seal (*Cystophora cristata*). *Neurosci. Lett.* 446: 147–150.

Kooyman, G. L. 1966. Maximum diving capabilities of the Weddell seal, *Leptonychotes weddelli*. *Science* 151: 1553–1554.

Kooyman, G. L., E. A. Wahrenbrock, M. A. Castellini, R. W. Davis, and E. E. Sinnett. 1980. Aerobic and anaerobic metabolism during voluntary diving in Weddell seals: Evidence of preferred pathways from blood chemistry and behavior. *J. Comp. Physiol., B* 138: 335–346.

Le Boeuf, B. J., D. E. Crocker, D. P. Costa, S. B. Blackwell, P. M. Webb, and D. S. Houser. 2000. Foraging ecology of northern elephant seals. *Ecol. Monogr.* 70: 353–382.

Moore, M. J., T. Hammar, J. Arruda, S. Cramer, S. Dennison, E. Montie, and A. Fahlman. 2011. Hyperbaric computed tomographic measurement of lung compression in seals and dolphins. *J. Exp. Biol.* 214: 2390–2397.

Ridgway, S. H., B. L. Scronce, and J. Kanwisher. 1969. Respiration and deep diving in the bottlenose porpoise. *Science* 166: 1651–1654.

Scholander, P. F. 1964. Animals in aquatic environments: Diving mammals and birds. In D. B. Dill (ed.), *Adaptation to the Environment* (Handbook of Physiology [Bethesda, MD], section 4), pp. 729–739. Oxford University Press, New York.

Scholander, P. F., L. Irving, and S. W. Grinnell. 1942. Aerobic and anaerobic changes in seal muscles during diving. *J. Biol. Chem.* 142: 431–440.

Thompson, D., and M. A. Fedak. 1993. Cardiac responses of grey seals during diving at sea. *J. Exp. Biol.* 174: 139–164.

Chapter 27

Barnes, H. 1954. Some tables for the ionic composition of seawater. *J. Exp. Biol.* 31: 582–588.

Bayly, I. A. E., and W. D. Williams. 1973. *Inland Waters and Their Ecology*. Longman, Camberwell, Australia.

Hill, R. W., and G. A. Wyse. 1989. *Animal Physiology*, 2nd ed. HarperCollins, New York.

Kirschner, L. B. 1991. Water and Ions. In C. L., Prosser (ed.), *Environmental and Metabolic Animal Physiology* (Comparative Animal Physiology, 4th ed.), pp. 13–107. Wiley-Liss, New York.

McHugh, J. L. 1967. Estuarine nekton. In G. H. Lauff (ed.), *Estuaries*, pp. 581–620. American Association for the Advancement of Science, Washington, D.C.

Schmidt-Nielsen, K. 1964. Terrestrial animals in dry heat: Desert rodents. In D. B. Dill (ed.), *Adaptation to the Environment* (Handbook of Physiology [Bethesda, MD], section 4), pp. 493–507. Oxford University Press, New York.

Yancey, P. H., M. E. Clark, S. C. Hand, R. D. Bowlus, and G. N. Somero. 1982. Living with water stress: Evolution of osmolyte systems. *Science* 217: 1214–1222.

Chapter 28

Archer, M. A., T. J. Bradley, L. D. Mueller, and M. R. Rose. 2007. Using experimental evolution to study the physiological mechanisms of desiccation resistance in *Drosophila melanogaster*. *Physiol. Biochem. Zool.* 80: 386–398.

Beuchat, C. A. 1990. Body size, medullary thickness, and urine concentrating ability in mammals. *Am. J. Physiol.* 258: R298–R308.

Blaylock, L. A., R. Ruibal, and K. Platt-Aloia. 1976. Skin structure and wiping behavior of phyllomedusine frogs. *Copeia* 1976: 283–295.

Bradley, T. J., A. E. Williams, and M. R. Rose. 1999. Physiological responses to selection for desiccation resistance in *Drosophila melanogaster*. *Am. Zool.* 39: 337–345.

Degnan, K. J., K. J. Karnaky, Jr., and J. A. Zadunaisky. 1977. Active chloride transport in the in vitro opercular skin of a teleost (*Fundulus heteroclitus*), a gill-like epithelium rich in chloride cells. *J. Physiol.* 271: 155–191.

D'Orazio, S. E., and C. W. Holliday. 1985. Gill Na, K-ATPase and osmoregulation in the sand fiddler crab, *Uca pugilator*. *Physiol. Zool.* 58: 364–373.

Gibbs, A. G. 1998. Water-proofing properties of cuticular lipids. *Am. Zool.* 38: 471–482.

Goldstein, D. L. 2002. Water and salt balance in seabirds. In E. A. Schreiber and J. Burger (eds.), *Biology of Marine Birds*, pp. 467–483. CRC Press, Boca Raton, FL.

Goss, G. G., S. F. Perry, J. N. Fryer, and P. Laurent. 1998. Gill morphology and acid–base regulation in freshwater fishes. *Comp. Biochem. Physiol., A* 119: 107–115.

Greenaway, P. 1988. Ion and water balance. In W. W. Burggren and B. R. McMahon (eds.), *Biology of the Land Crabs*, pp. 211–248. Cambridge University Press, New York.

Hill, R. W., and G. A. Wyse. 1989. *Animal Physiology*, 2nd ed. Harper & Row, New York.

Kirschner, L. B. 1991. Water and ions. In C. L. Prosser (ed.), *Environmental and Metabolic Animal Physiology* (Comparative Animal Physiology, 4th ed.), pp. 13–107. Wiley-Liss, New York.

Lillywhite, H. B. 2006. Water relations of tetrapod integument. *J. Exp. Biol.* 209: 202–226.

Loveridge, J. P. 1968. The control of water loss in *Locusta migratoria migratorioides* R & F. I: Cuticular water loss. *J. Exp. Biol.* 49: 1–13.

Marshall, A. T., and P. D. Cooper. 1988. Secretory capacity of the lachrymal salt gland of hatchling sea turtles, *Chelonia mydas*. *J. Comp. Physiol., B* 157: 821–827.

Nagy, K. A., and C. C. Peterson. 1988. *Scaling of Water Flux Rate in Animals*. University of California Publications in Zoology, vol. 120. University of California Press, Berkeley.

Perry, S. F. 1998. Relationships between branchial chloride cells and gas transfer in freshwater fish. *Comp. Biochem. Physiol., A* 119: 9–16.

Pillans, R. D., J. P. Good, W. G. Anderson, N. Hazon, and C. E. Franklin. 2005. Freshwater to seawater acclimation of juvenile bull sharks (*Carcharhinus leucas*): plasma osmolytes and Na+/K+ ATPase activity in gill, rectal gland, kidney and intestine. *J. Comp. Physiol., B* 175: 37–44.

Potts, W. T. W., and G. Parry. 1964. *Osmotic and Ionic Regulation in Animals*. Pergamon, Oxford.

Schmidt-Nielsen, B., and K. Schmidt-Nielsen. 1951. A complete account of the water metabolism in kangaroo rats and an experimental verification. *J. Cell. Comp. Physiol.* 38: 165–181.

Schmidt-Nielsen, K. 1960. The salt-secreting gland of marine birds. *Circulation* 21: 955–967.

Schmidt-Nielsen, K., and R. Fange. 1958. Salt glands in marine reptiles. *Nature* 182: 783–785.

Theede, H. 1965. Vergleichende experimentelle Untersuchungen über die zelluläre Gefrierresistenz mariner Muscheln. *Kieler Meeresforschungen* 21: 153–166.

Tieleman, B. I. 2005. Physiological, behavioral, and life history adaptations of larks along an aridity gradient: a review. In G. Bota, M. B. Morales, S. Mañosa, and J. Camprodon (eds.), *Ecology and Conservation of Steppe-land Birds*. Lynx Edicions, Barcelona, Spain.

Tipsmark, C. K., S. S. Madsen, M. Seidelin, A. S. Christensen, C. P. Cutler, and G. Cramb. 2002. Dynamics of Na+, K+, 2Cl− cotransporter and Na+, K+-ATPase expression in the branchial epithelium of brown trout (*Salmo trutta*) and Atlantic salmon (*Salmo salar*). *J. Exp. Zool.* 293: 106–118.

Tracy, R. L., and G. E. Walsberg. 2001. Intraspecific variation in water loss in a desert rodent, *Dipodomys merriami*. *Ecology* 82: 1130–1137.

Williams, J. B. 2001. Physiological ecology and behavior of desert birds. In V. Nolan, Jr., and C. F. Thompson (eds.), *Current Ornithology*, vol. 16, pp. 299–353. Springer, New York.

Wilson, C. G., and P. W. Sherman. 2010. Anciently asexual bdelloid rotifers escape lethal fungal parasites by drying up and blowing away. *Science* 327: 574–576.

Chapter 29

Bankir, L., and C. de Rouffignac. 1985. Urinary concentrating ability: Insights from comparative anatomy. *Am. J. Physiol.* 249: R643–R666.

Beck, F.-X., W. G. Guder, and M. Schmolke. 1998. Cellular osmoregulation in kidney medulla. In F. Lang (ed.), *Cell Volume Regulation*, pp. 169–184. Karger, Basel, Switzerland.

Beuchat, C. A. 1990. Body size, medullary thickness, and urine concentrating ability in mammals. *Am. J. Physiol.* 258: R298–R308.

Beuchat, C. A. 1996. Structure and concentrating ability of the mammalian kidney: Correlations with habitat. *Am. J. Physiol.* 271: R157–R179.

Brown, G. W., Jr., and P. P. Cohen. 1958. Biosynthesis of urea in metamorphosing tadpoles. In W. D. McElroy and B. Glass (eds.), *A Symposium on the Chemical Basis of Development*, pp. 495–513. Johns Hopkins Press, Baltimore, MD.

Elias, H., A. Hossman, I. B. Barth, and A. Solmor. 1960. Blood flow in the renal glomerulus. *J. Urol.* 83: 790–798.

Gupta, B. L., and M. J. Berridge. 1966. Fine structural organization of the rectum in the blowfly, *Calliphora erythrocephala* (Meig.) with special reference to connective tissue, tracheae and neurosecretory innervation in the rectal papillae. *J. Morphol.* 120: 23–82.

Huber, G. C. 1932. Renal tubules. In E. V. Cowdry (ed.), *Special Cytology*, 2nd ed., vol. 2, pp. 933–977. Paul B. Hoeber, New York.

Kaissling, B., C. de Rouffignac, J. M. Barrett, and W. Kriz. 1975. The structural organization of the kidney of the desert rodent *Psammomys obesus*. *Anat. Embryol.* 148: 121–143.

Knepper, M. A., J. B. Wade, J. Terris, C. A. Ecelbarger, D. Marples, B. Mandon, C.-L. Chou, B. K. Kishore, and S. Nielsen. 1996. Renal aquaporins. *Kidney Int.* 49: 1712–1717.

Martin, A. W., and F. M. Harrison. 1966. Excretion. In K. M. Wilbur and C. M. Yonge (eds.), *Physiology of Mollusca*, vol. 2, pp. 275–308. Academic Press, New York.

Møbjerg, N., E. H. Larsen, and Å. Jespersen. 1998. Morphology of the nephron in the mesonephros of *Bufo bufo* (Amphibia, Anura, Bufonidae). *Acta Zool.* 79: 31–50.

Pannabecker, T. L., and W. H. Dantzler. 2007. Three-dimensional architecture of collecting ducts, loops of Henle, and blood vessels in the renal papilla. *Am. J. Physiol.* 292: F696–F704.

Phillips, J. E. 1964. Rectal absorption in the desert locust, *Schistocerca gregaria* Forskål. III. The nature of the excretory process. *J. Exp. Biol.* 41: 69–80.

Pitts, R. F. 1974. *Physiology of the Kidney and Body Fluids*, 3rd ed. Year Book Medical, Chicago.

Riegel, J. A. 1977. Fluid movement through the crayfish antennal gland. In B. L. Gupta, R. B. Moreton, J. L. Oschman, and B. J. Wall (eds.), *Transport of Ions and Water in Animals*, pp. 613–631. Academic Press, New York.

Smith, H. W. 1951. The Kidney: Structure and Function in Health and Disease. Oxford University Press, New York.

Sperber, I. 1944. Studies on the mammalian kidney. *Zool. Bidr. Uppsala* 22: 249–432.

Uawithya, P., T. Pisitkun, B. E. Ruttenberg, and M. A. Knepper. 2007. Transcriptional profiling of native inner medullary collecting duct cells from rat kidney. *Physiol. Genomics* 32: 229–253.

Walker, A. M., C. L. Hudson, T. Findley, Jr., and A. N. Richards. 1937. The total molecular concentration and the chloride concentration of fluid from different segments of the renal tubule of Amphibia: The site of chloride reabsorption. *Am. J. Physiol.* 118: 121–129.

Willoughby, E. J., and M. Peaker. 1979. Birds. In G. M. O. Maloiy (ed.), *Comparative Physiology of Osmoregulation in Animals*, vol. 2, pp. 1–55. Academic Press, New York.

Wirz, H., B. Hargitay, and W. Kuhn. 1951. Lokalisation des Konzentrierungsprozesses in der Niere durch direkte Kryoskopie. *Helv. Physiol. Pharmacol. Acta* 9: 196–207.

Chapter 30

Beuchat, C. A. 1996. Structure and concentrating ability of the mammalian kidney: Correlations with habitat. *Am. J. Physiol.* 271: R157–R179.

Goudie, A., and J. Wilkinson. 1977. *The Warm Desert Environment*. Cambridge University Press, New York.

Louw, G. N. 1972. The role of advective fog in the water economy of certain Namib desert animals. In *Symposia of the Zoological Society of London*, no. 31, pp. 297–314. Academic Press, New York.

Lovegrove, B. 1993. *The Living Deserts of Southern Africa*. Fernwood Press, Vlaeberg, South Africa.

Maloiy, G. M. O., W. V. MacFarlane, and A. Shkolnik. 1979. Mammalian herbivores. In G. M. O. Maloiy (ed.), *Comparative Physiology of Osmoregulation in Animals*, vol. 2, pp. 185–209. Academic Press, New York.

Ostrowski, S., J. B. Williams, and K. Ismael. 2003. Heterothermy and the water economy of free-living Arabian oryx (*Oryx leucoryx*). *J. Exp. Biol.* 206: 1471–1478.

Ostrowski, S., J. B. Williams, P. Mésochina, and H. Sauerwein. 2006. Physiological acclimation of a desert antelope, Arabian oryx (*Oryx leucoryx*), to long-term food and water restriction. *J. Comp. Physiol., B* 176: 191–201.

Schmidt-Nielsen, K. 1954. Heat regulation in small and large desert animals. In J. L. Cloudsley-Thompson (ed.), *Biology of Deserts*, pp. 182–187. Institute of Biology, London.

Soholt, L. F., D. B. Dill, and I. Oddershede. 1977. Evaporative cooling in desert heat: Sun and shade; rest and exercise. *Comp. Biochem. Physiol., A* 57: 369–371.

Taylor, C. R. 1968. The minimum water requirements of some East African bovids. In *Symposia of the Zoological Society of London*, no. 21, pp. 195–206. Academic Press, London.

Taylor, C. R. 1970. Strategies of temperature regulation: Effect on evaporation in East African ungulates. *Am. J. Physiol.* 219: 1131–1135.

Additional References

In this section, we provide extensive lists of references, in which you will find additional information.

We start with books that each cover a wide variety of physiological systems. These books are categorized into "General Works," "Works Dealing with Specific Groups of Invertebrates," and "Works Dealing with Specific Groups of Vertebrates." Many of these books are excellent sources for both information and added references on specific topics. They are not generally listed with individual chapters, however, because, in many cases, they would need to be listed with virtually every chapter, such is their breadth of coverage. We include some relatively old titles that have not yet been superseded in their coverage by more-recent publications.

Following the three lists of books of relatively broad interest, you will find additional lists of references specific to each chapter. They are listed by chapter number.

General Works

Barrington, E. J. W. 1979. *Invertebrate Structure and Function*, 2nd ed. Wiley, New York.

Dantzler, W. H. (ed.). 1997. *Comparative Physiology* (Handbook of Physiology [Bethesda, MD], section 13). 2 vols. Oxford University Press, New York.

Florkin, M., and B. T. Scheer (eds.). 1967–1979. *Chemical Zoology*. 11 vols. Academic Press, New York.

Fregly, M. J., and C. M. Blatteis (eds.). 1996. *Environmental Physiology* (Handbook of Physiology [Bethesda, MD], section 4). 2 vols. Oxford University Press, New York.

Fretter, V., and A. Graham. 1976. *A Functional Anatomy of Invertebrates*. Academic Press, New York.

Hochachka, P. W., and G. N. Somero. 2002. *Biochemical Adaptation. Mechanism and Process in Physiological Evolution*. Oxford University Press, New York.

McNab, B. K. 2002. *The Physiological Ecology of Vertebrates: A View from Energetics*. Cornell University Press, Ithaca, NY.

Michal, G., and D. Schomburg (eds.). 2012. *Biochemical Pathways: An Atlas of Biochemistry and Molecular Biology*, 2nd ed. Wiley, New York.

Prosser, C. L. (ed.). 1991. *Environmental and Metabolic Animal Physiology* (*Comparative Animal Physiology*, 4th ed.). Wiley-Liss, New York.

Prosser, C. L. (ed.). 1991. *Neural and Integrative Animal Physiology* (*Comparative Animal Physiology*, 4th ed.). Wiley-Liss, New York.

Seymour, R. S. (ed.). 1983. *Respiration and Metabolism of Embryonic Vertebrates*. Junk, Dordrecht.

Storey, K. B. (ed.). 2004. *Functional Metabolism. Regulation and Adaptation*. Wiley-Liss, Hoboken, NJ.

Urich, K. 1994. *Comparative Animal Biochemistry*. Springer-Verlag, New York.

Willmer, P., G. Stone, and I. Johnston. 2005. *Environmental Physiology of Animals*, 2nd ed. Blackwell, Malden, MA.

Withers, P. C. 1992. *Comparative Animal Physiology*. Saunders, Philadelphia.

Works Dealing with Specific Groups of Invertebrates

Anderson, O. R. 1988. *Comparative Protozoology: Ecology, Physiology, Life History*. Springer-Verlag, New York.

Bergquist, P. R. 1978. *Sponges*. University of California Press, Berkeley.

Bliss, D. E. (ed.). 1982–1985. *The Biology of Crustacea*. 10 vols. Academic Press, New York.

Burggren, W. W., and B. R. McMahon (eds.). 1988. *Biology of the Land Crabs*. Cambridge University Press, New York.

Chown, S. L., and S. W. Nicolson. 2004. *Insect Physiological Ecology*. Oxford University Press, New York.

Gilbert, L. I. 2011. *Insect Molecular Biology and Biochemistry*. Academic Press, New York.

Holdich, D. M. (ed.). 2002. *Biology of Freshwater Crayfish*. Blackwell Science, Oxford, UK.

Hughes, R. N. 1986. *A Functional Biology of Marine Gastropods*. Johns Hopkins University Press, Baltimore, MD.

Iatrou, K., and S. S. Gill (eds.). 2005. *Comprehensive Molecular Insect Science*. 7 vols. Elsevier, Amsterdam.

Klowden, M. J. 2007. *Physiological Systems in Insects*, 2nd ed. Academic Press, New York.

Lawrence, J. 1987. *A Functional Biology of Echinoderms*. Johns Hopkins University Press, Baltimore, MD.

Nentwig, W. (ed.). 1987. *Ecophysiology of Spiders*. Springer-Verlag, New York.

Sawyer, R. T. 1986. *Leech Biology and Behaviour*. 3 vols. Clarendon, Oxford.

Schram, F. R. 1986. *Crustacea*. Oxford University Press, New York.

Sonenshine, D. E. 1991–1993. *Biology of Ticks*. 2 vols. Oxford University Press, New York.

Sutton, S. L., and D. M. Holdich (eds.). 1984. *The Biology of Terrestrial Isopods*. Symposia of the Zoological Society of London, no. 53. Clarendon, Oxford.

Wilbur, K. M. (ed.). 1983–1988. *The Mollusca*. 12 vols. Academic Press, New York.

Works Dealing with Specific Groups of Vertebrates

Bemis, W. E., W. W. Burggren, and N. E. Kemp (eds.). 1987. *The Biology and Evolution of Lungfishes*. Liss, New York.

Biology of the Reptilia. 1969–2010. 22 vols. [A continuing series, with changing editors and publishers.]

Carrier, J. C., J. A. Musick, and M. R. Heithaus (eds.). 2010. *Sharks and Their Relatives II: Biodiversity, Adaptive Physiology, and Conservation*. CRC Press, Boca Raton, FL.

Carrier, J. C., J. A. Musick, and M. R. Heithaus (eds.). 2012. *Biology of Sharks and Their Relatives*, 2nd ed. CRC Press, Boca Raton, FL.

Evans, D. H., and J. B. Claiborne (eds.). 2005. *The Physiology of Fishes*, 3rd ed. CRC Press, Boca Raton, FL.

Farner, D. S., J. R. King, and K. C. Parkes (eds.). 1982–1985. *Avian Biology*, vols. 6–8. Academic Press, New York.

Feder, M. E., and W. W. Burggren (eds.). 1992. *Environmental Physiology of the Amphibians*. University of Chicago Press, Chicago, IL.

Finn, R. N., and B. G. Kapoor (eds.). 2008. *Fish Larval Physiology*. Science Publishers, Enfield, NH.

Fish Physiology. 1969–2012. 31 vols. Published by Academic Press; Elsevier. [A continuing series, with changing editors and publishers. Each volume has a specific subject and, in recent cases, a volume title.]

Foreman, R. E., A. Gorbman, J. M. Dodd, and R. Olsson (eds.). 1985. *Evolutionary Biology of Primitive Fishes*. Plenum, New York.

Hall, J. E. 2011. *Guyton and Hall Textbook of Medical Physiology*, 12th ed. Saunders, Philadelphia.

Hamlett, W. C. (ed.). 1999. *Sharks, Skates, and Rays. The Biology of Elasmobranch Fishes*. Johns Hopkins University Press, Baltimore, MD.

Handbook of Physiology. 1960–2001. American Physiological Society, Bethesda, MD. [A comprehensive mammalian physiology; numerous volumes and editors.]

Hochachka, P. W., and T. P. Mommsen (eds.). 1991–1995. *Biochemistry and Molecular Biology of Fishes*. 5 vols. Elsevier Science, New York.

King, A. S., and J. McLelland (eds.). 1979–1989. *Form and Function in Birds*. 4 vols. Academic Press, New York.

Rankin, J. C., and F. B. Jensen (eds.). 1993. *Fish Ecophysiology*. Chapman & Hall, New York.

Stevens, L. 2004. *Avian Biochemistry and Molecular Biology*. Cambridge University Press, New York.

Whittow, G. C. (ed.). 2000. *Sturkie's Avian Physiology*, 5th ed. Academic Press, New York.

Widmaier, E. P., H. Raff, and K. T. Strang. 2006. *Vander's Human Physiology: The Mechanisms of Body Function*, 10th ed. McGraw-Hill, Boston.

Chapter 1

Aprille, J. R., C. J. Lagace, J. Modica-Napolitano, and B. A. Trimmer. 2004. Role of nitric oxide and mitochondria in control of firefly flash. *Integr. Comp. Biol.* 44: 213–219.

Bartholomew, G. A. 1986. The role of natural history in contemporary biology. *BioScience* 36: 324–329.

Brett, J. R. 1983. Life energetics of sockeye salmon, *Oncorhynchus nerka*. In W. P. Aspey and S. I. Lustick (eds.), *Behavioral Energetics: The Cost of Survival in Vertebrates*, pp. 29–63. Ohio State University Press, Columbus.

Cabanac, M. 2006. Adjustable set point: to honor Harold T. Hammel. *J. Appl. Physiol.* 100: 1338–1346.

Calder, W. A., III. 1984. *Size, Function, and Life History.* Harvard University Press, Cambridge, MA.

Crespi, B. J. 2000. The evolution of maladaptation. *Heredity* 84: 623–629.

Crossin, G. T., S. G. Hinch, A. P. Farrell, D. A. Higgs, and M. C. Healey. 2004. Somatic energy of sockeye salmon *Oncorhynchus nerka* at the onset of upriver migration: a comparison among ocean climate regimes. *Fish. Oceanogr.* 13: 345–349.

Crossin, G. T., S. G. Hinch, A. P. Farrell, D. A. Higgs, A. G. Lotto, J. D. Oakes, and M. C. Healey. 2004. Energetics and morphology of sockeye salmon: effects of upriver migratory distance and elevation. *J. Fish Biol.* 65: 788–810.

Endler, J. A. 1986. *Natural Selection in the Wild.* Princeton University Press, Princeton, NJ.

Feder, M. E., A. F. Bennett, W. W. Burggren, and R. B. Huey (eds.). 1987. *New Directions in Ecological Physiology.* Cambridge University Press, New York.

Folk, D. G., and T. J. Bradley. 2005. Adaptive evolution in the lab: unique phenotypes in fruit flies comprise a fertile field of study. *Integr. Comp. Biol.* 45: 492–499.

Futuyma, D. J. 2009. *Evolution*, 2nd ed. Sinauer, Sunderland, MA.

Garland, T., Jr., and S. C. Adolph. 1991. Physiological differentiation of vertebrate populations. *Annu. Rev. Ecol. Syst.* 22: 193–228.

Garland, T., Jr., and P. A. Carter. 1994. Evolutionary physiology. *Annu. Rev. Physiol.* 56: 579–621.

Gotthard, K., and S. Nylin. 1995. Adaptive plasticity and plasticity as an adaptation: A selective review of plasticity in animal morphology and life history. *Oikos* 74: 3–17.

Gould, S. J., and E. S. Vrba. 1982. Exaptation: A missing term in the science of form. *Paleobiology* 8: 4–15.

Hendry, A. P., and S. C. Stearns (eds.). 2004. *Evolution Illuminated. Salmon and Their Relatives.* Oxford University Press, New York.

Hoffmann, A. A., and P. A. Parsons. 1997. *Extreme Environmental Change and Evolution.* Cambridge University Press, Cambridge.

Kingsolver, J. G., and R. B. Huey. 1998. Evolutionary analyses of morphological and physiological plasticity in thermally variable environments. *Am. Zool.* 38: 545–560.

Kvist, A., A. Lindström, M. Green, T. Piersma, and G. H. Visser. 2001. Carrying large fuel loads during sustained bird flight is cheaper than expected. *Nature* 413: 730–732.

Levins, R., and R. Lewontin. 1985. The organism as the subject and object of evolution. In R. Levins and R. Lewontin, *The Dialectical Biologist*, pp. 85–106. Harvard University Press, Cambridge, MA.

Lewontin, R. C. 1978. Adaptation. *Sci. Am.* 239(3): 212–230.

Liao, J. L., D. N. Beal, G. V. Lauder, and M. S. Triantafyllou. 2003. Fish exploiting vortices decrease muscle activity. *Science* 302: 1566–1569.

Lomolino, M. V., B. R. Riddle, R. J. Whittaker, and J. H. Brown. 2010. *Biogeography*, 4th ed. Sinauer, Sunderland, MA.

Madigan, M. T., and B. L. Marrs. 1997. Extremophiles. *Sci. Am.* 276(4): 82–87.

Mangum, C., and D. Towle. 1977. Physiological adaptation to unstable environments. *Am. Sci.* 65: 67–75.

Mayr, E. 1983. How to carry out the adaptationist program? *Am. Nat.* 121: 324–334.

Mazumder, A., W. D. Taylor, D. J. McQueen, and D. R. S. Lean. 1990. Effects of fish and plankton on lake temperature and mixing depth. *Science* 247: 312–315.

McNab, B. K. 2002. *The Physiological Ecology of Vertebrates.* Cornell University Press, Ithaca, NY.

Montgomery, J., and K. Clements. 2000. Disaptation and recovery in the evolution of Antarctic fishes. *Trends Ecol. Evol.* 15: 267–271.

Palstra, A. P., D. Crespo, G. E. E. J. M. van den Thillart, and J. V. Planas. 2010. Saving energy to fuel exercise: swimming suppresses oocyte development and downregulates ovarian transcriptomic response of rainbow trout *Oncorhynchus mykiss. Am. J. Physiol.* 299: R486–R499.

Parsons, P. A. 1992. Evolutionary adaptation and stress: The fitness gradient. In M. K. Hecht, B. Wallace, and R. J. MacIntyre (eds.), *Evolutionary Biology*, vol. 26, pp. 191–223. Plenum, New York.

Parsons, P. A. 1996. Stress, resources, energy balances, and evolutionary change. In M. K. Hecht, R. J. MacIntyre, and M. T. Clegg (eds.), *Evolutionary Biology*, vol. 29, pp. 39–72. Plenum, New York.

Peterson, C. C. 1996. Anhomeostasis: Seasonal water and solute relations in two populations of the desert tortoise (*Gopherus agassizii*) during chronic drought. *Physiol. Zool.* 69: 1324–1358.

Piersma, T., G. A. Gudmundsson, and K. Lilliendahl. 1999. Rapid changes in the size of different functional organ and muscle groups during refueling in a long-distance migrating shorebird. *Physiol. Biochem. Zool.* 72: 405–415.

Réale, D., D. Garant, M. M. Humphries, P. Bergeron, V. Careau, and P.-O. Montiglio. 2010. Personality and the emergence of the pace-of-life syndrome concept at the population level. *Phil. Trans. Roy. Soc. B* 365: 4051–4063.

Rose, M. R., and G. V. Lauder (eds.). 1996. *Adaptation.* Academic Press, New York.

Scharloo, W. 1989. Developmental and physiological aspects of reaction norms. *BioScience* 39: 465–471.

Schoenheimer, R. 1942. *The Dynamic State of Body Constituents.* Harvard University Press, Cambridge, MA.

Shadwick, R. E. 2005. How tunas and lamnid sharks swim: an evolutionary convergence. *Am. Sci.* 93: 524–531.

Shrimpton, J. M., D. A. Patterson, J. G. Richards, S. J. Cooke, P. M. Schulte, S. G. Hinch, and A. P. Farrell. 2005. Ionoregulatory changes in different populations of maturing sockeye salmon *Oncorhynchus nerka* during ocean and river migration. *J. Exp. Biol.* 208: 4069–4078.

Spicer, J. I., and K. J. Gaston (eds.). 1999. *Physiological Diversity and Its Ecological Implications.* Blackwell, Malden, MA.

Stanley, S. M. 1989. *Earth and Life through Time*, 2nd ed. W. H. Freeman, New York.

Travis, J. 1994. Evaluating the adaptive role of morphological plasticity. In P. C. Wainwright and S. M. Reilly (eds.), *Ecological Morphology*, pp. 99–122. University of Chicago Press, Chicago.

Tsai, Y.-H., C.-W. Li, T.-M. Hong, J.-Z. Ho, E.-C. Yang, W.-Y. Wu, G. Margaritondo, S.-T. Hsu, E. B. L. Ong, and Y. Hwu. 2014. Firefly light flashing: oxygen supply mechanism. *Physical Review Letters* 113: 258103.

Umina, P. A., A. R. Weeks, M. R. Kearney, S. W. McKecknie, and A. A. Hoffmann. 2005. A rapid shift in a classic clinal pattern in *Drosophila* reflecting climate change. *Science* 308: 691–693.

Waterlow, J. C., P. J. Garlick, and D. J. Millward. 1978. *Protein Turnover in Mammalian Tissues and in the Whole Body.* North-Holland, New York.

Waterman, T. H. 1999. The evolutionary challenges of extreme environments (part 1). *J. Exp. Zool.* 285: 326–359.

Waterman, T. H. 2001. Evolutionary challenges of extreme environments (part 2). *J. Exp. Zool.* 291: 130–168.

Weibel, E. R. 2000. *Symmorphosis. On Form and Function in Shaping Life.* Harvard University Press, Cambridge, MA.

Weibel, E. R., C. R. Taylor, and H. Hoppeler. 1991. The concept of symmorphosis: A testable hypothesis of structure-function relationship. *Proc. Natl. Acad. Sci. U.S.A.* 88: 10357–10361.

Williams, G. C. 1966. *Adaptation and Natural Selection.* Princeton University Press, Princeton, NJ.

Wilson, T., and J. W. Hastings. 1998. Bioluminescence. *Annu. Rev. Cell Dev. Biol.* 14: 197–230.

Chapter 2

Arnold, W., T. Ruf, F. Frey-Roos, and U. Bruns. 2011. Diet-independent remodeling of cellular membranes precedes seasonally changing body temperature in a hibernator. *PLoS One* 6: e18641.

Bennett, A. F. 1974. Enzymatic correlates of activity metabolism in anuran amphibians. *Am. J. Physiol.* 226: 1149–1151.

Cooper, G. M., and R. E. Hausman. 2009. *The Cell. A Molecular Approach*, 5th ed. Sinauer, Sunderland, MA.

Eisenmesser, E. Z., D. A. Bosco, M. Akke, and D. Kern. 2002. Enzyme dynamics during catalysis. *Science* 295: 1520–1523.

Fields, P. A., and G. N. Somero. 1998. Hot spots in cold adaptation: Localized increases in conformational flexibility in lactate dehydrogenase A_4 orthologs of Antarctic notothenioid fishes. *Proc. Natl. Acad. Sci. U.S.A.* 95: 11476–11481.

Figueroa, X. F., B. E. Isakson, and B. R. Duling. 2004. Connexins: gaps in our knowledge of vascular function. *Physiology* 19: 277–284.

Fujii, R. 2000. The regulation of motile activity in fish chromatophores. *Pigment Cell Res.* 13: 300–319.

Gibbs, A. G. 1998. The role of lipid physical properties in lipid barriers. *Am. Zool.* 38: 268–279.

Goodsell, D. S. 1992. A look inside the living cell. *Am. Sci.* 80: 457–465.

Gouaux, E., and R. MacKinnon. 2005. Principles of selective ion transport in channels and pumps. *Science* 310: 1461–1465.

Haddock, S. H. D., M. A. Moline, and J. F. Case. 2010. Bioluminescence in the sea. *Annu. Rev. Mar. Sci.* 2: 443–493.

Han, X., H. Cheng, D. J. Mancuso, and R. W. Gross. 2004. Caloric restriction results in phospholipid depletion, membrane remodeling, and triacylglycerol accumulation in murine myocardium. *Biochemistry* 43: 15584–15594.

Hartl, F. U., and M. Hayer-Hartl. 2002. Molecular chaperones in the cytosol: From nascent chain to folded protein. *Science* 295: 1852–1858.

Hawkins, J. D. 1996. *Gene Structure and Expression*, 3rd ed. Cambridge University Press, New York.

Jensen, M. Ø., V. Jogini, D. W. Borhani, A. E. Leffler, R. O. Dror, and D. E. Shaw. 2012. Mechanism of voltage gating in potassium channels. *Science* 336: 229–233.

Jones, C. T. (ed.). 1982. *The Biochemical Development of the Fetus and Neonate*. Elsevier Biomedical Press, New York.

Korkegian, A., M. E. Black, D. Baker, and B. L. Stoddard. 2005. Computational thermostabilization of an enzyme. *Science* 308: 857–860.

Kurland, C. G., L. J. Collins, and D. Penny. 2006. Genomics and the irreducible nature of eukaryote cells. *Science* 312: 1011–1014.

Livingstone, D. R., and J. J. Stegeman (eds.). 1998. Forms and functions of cytochrome P450. *Comp. Biochem. Physiol., C* 121: 1–412. [Numerous individual articles.]

Logue, J. A., A. L. DeVries, E. Fodor, and A. R. Cossins. 2000. Lipid compositional correlates of temperature-adaptive interspecific differences in membrane physical structure. *J. Exp. Biol.* 203: 2105–2115.

Maier, T., S. Jenni, and N. Ban. 2006. Architecture of a mammalian fatty acid synthase at 4.5 Å resolution. *Science* 311: 1258–1262.

Mäthger, L. M., and R. T. Hanlon. 2007. Malleable skin coloration in cephalopods: selective reflectance, transmission and absorbance of light by chromatophores and iridophores. *Cell Tissue Res.* 329: 179–186.

McIntosh, J. M., A. D. Santos, and B. M. Olivera. 1999. *Conus* peptides targeted to specific nicotinic acetylcholine receptor subtypes. *Annu. Rev. Biochem.* 68: 59–88.

Michal, G. (ed.). 1999. *Biochemical Pathways: An Atlas of Biochemistry and Molecular Biology*. Wiley, New York.

Mitic, L. L., and J. M. Anderson. 1998. Molecular architecture of tight junctions. *Annu. Rev. Physiol.* 60: 121–142.

Niwa, H., S. Inouye, T. Hirando, T. Matsuno, S. Kojima, M. Kubota, M. Ohashi, and F. I. Tsuji. 1996. Chemical nature of the light emitter of the *Aequorea* green fluorescent protein. *Proc. Natl. Acad. Sci. U.S.A.* 93: 13617–13622.

Palmer, T. 1995. *Understanding Enzymes*, 4th ed. Prentice Hall/Ellis Horwood, London.

Pierce, V. A., and D. L. Crawford. 1997. Phylogenetic analysis of glycolytic enzyme expression. *Science* 276: 256–259.

Price, N. C., and L. Stevens. 1999. *Fundamentals of Enzymology: The Cell and Molecular Biology of Catalytic Proteins*, 3rd ed. Oxford University Press, New York.

Ramachandran, V. S., C. W. Tyler, R. L. Gregory, D. Rogers-Ramachandran, S. Duensing, C. Pillsbury, and C. Ramachandran. 1996. Rapid adaptive camouflage in tropical flounders. *Nature* 379: 815–818.

Rasmussen, H. 1989. The cycling of calcium as an intracellular messenger. *Sci. Am.* 261(4): 66–73.

Razeghi, P., and H. Taegtmeyer. 2005. Cardiac remodeling. UPS lost in transit. *Circ. Res.* 97: 964–966. [UPS is a play on words and here stands for ubiquitin–proteasome system.]

Salisbury, S. M., G. G. Martin, W. M. Kier, and J. R. Schulz. 2010. Venom kinematics during prey capture in *Conus*: the biomechanics of a rapid injection system. *J. Exp. Biol.* 213: 673–682.

Sotaniemi, E. A., and R. O. Pelkonen (eds.). 1987. *Enzyme Induction in Man*. Taylor & Francis, New York.

Spiegel, A. M., T. L. Z. Jones, W. F. Simonds, and L. S. Weinstein. 1994. *G Proteins*. Landes, Austin, TX.

Stock, D. W., J. M. Quattro, G. S. Whitt, and D. A. Powers. 1997. Lactate dehydrogenase (LDH) gene duplication during chordate evolution: The cDNA sequence of the LDH of the tunicate *Styela licata*. *Mol. Biol. Evol.* 14: 1273–1284.

Storey, K. B. (ed.). 2004. *Functional Metabolism: Regulation and Adaptation*. Wiley, Hoboken, NJ.

Suraweera, A., C. Munch, A. Hanssum, and A. Bertolotti. 2012. Failure of amino acid homeostasis causes cell death following proteasome inhibition. *Mol. Cell* 48: 242–253.

Urich, K. 1994. *Comparative Animal Biochemistry*. Springer, New York.

Weiss, T. F. 1996. *Cellular Biophysics*. MIT Press, Cambridge, MA.

Wen, X., S. Fuhrman, G. S. Michaels, D. B. Carr, S. Smith, J. L. Barker, and R. Somogyi. 1998. Large-scale temporal gene expression mapping of central nervous system development. *Proc. Natl. Acad. Sci. U.S.A.* 95: 334–339.

Wickner, W., and R. Schekman. 2005. Protein translocation across biological membranes. *Science* 310: 1452–1456.

Williams, E. E., B. S. Stewart, C. A. Beuchat, G. N. Somero, and J. R. Hazel. 2001. Hydrostatic-pressure and temperature effects on the molecular order of erythrocyte membranes from deep-, shallow-, and non-diving mammals. *Can. J. Zool.* 79: 888–894.

Chapter 3

Berg, F., U. Gustafson, and L. Andersson. 2006. The uncoupling protein 1 gene (*UCP1*) is disrupted in the pig lineage: A genetic explanation for poor thermoregulation in piglets. *PLoS Genet.* 2: e129 doi: 10.1371/journal.pgen.0020129.

Burnett, K. G., and 25 additional authors. 2007. *Fundulus* as the premier teleost model in environmental biology: opportunities for new insights using genomics. *Comp. Biochem. Physiol., D* 2: 257–286.

Castillo-Davis, C. I., and D. L. Hartl. 2003. Conservation, relocation and duplication in genome evolution. *Trends Genet.* 19: 593–597.

Chang, P. L., J. P. Dunham, S. V. Nuzhdin, and M. N. Arbeitman. 2011. Somatic sex-specific transcriptome differences in *Drosophila* revealed by whole transcriptome sequencing. *BMC Genomics* 12: 364.

Cheng, C.-H. C., L. Chen, T. J. Near, and Y. Jin. 2003. Functional antifreeze glycoprotein genes in temperate-zone New Zealand Notothenid fish infer an Antarctic evolutionary origin. *Mol. Biol. Evol.* 20: 1897–1908.

Connor, K. M., and A. Y. Gracey. 2011. Circadian cycles are the dominant transcriptional rhythm in the intertidal mussel *Mytilus californianus*. *Proc. Natl. Acad. Sci. U.S.A.* 108: 16110–16115.

Epperson, L. E., J. C. Rose, R. L. Russell, M. P. Nikrad, H. V. Carey, and S. L. Martin. 2010. Seasonal protein changes support rapid energy production in hibernator brainstem. *J. Comp. Physiol., B* 180: 599–617.

Farrell, A. P., and J. F. Steffensen (eds.). 2005. *The Physiology of Polar Fishes*. Elsevier, Boston.

Flück, M., and H. Hoppeler. 2003. Molecular basis of skeletal muscle plasticity—From gene to form and function. *Rev. Physiol. Biochem. Pharmacol.* 146: 159–216.

Hollywood, K., D. R. Brison, and R. Goodacre. 2006. Metabolomics: Current technologies and future trends. *Proteomics* 6: 4716–4723.

Jarvis, E. D., S. Ribeiro, M. L. da Silva, D. Ventura, J. Vielliard, and C. V. Mello. 2000. Behaviourally driven gene expression reveals song nuclei in hummingbird brain. *Nature* 406: 628–632.

Kalujnaia, S., I. S. McWilliam, V. A. Zaguinaiko, A. L. Feilen, J. Nicholson, N. Hazon, C. P. Cutler, and G. Cramb. 2007. Transcriptomic approach to the study of osmoregulation in the European eel *Anguilla anguilla*. *Physiol. Genomics* 31: 385–401.

Kock, K.-H. 2005. Antarctic icefishes (Channichthyidae): A unique family of fishes. A review, part I. *Polar Biol.* 28: 862–895. [See also 28: 897–909.]

Kurland, C. G., L. J. Collins, and D. Penny. 2006. Genomics and the irreducible nature of eukaryote cells. *Science* 312: 1011–1014.

Leung, Y. F., and D. Cavalieri. 2003. Fundamentals of cDNA microarray data analysis. *Trends Genet.* 19: 649–659.

Lockhart, D. J., and E. A. Winzeler. 2000. Genomics, gene expression and DNA arrays. *Nature* 405: 827–836.

Mahoney, D. J., G. Parise, S. Melov, A. Safdar, and S. A. Tarnopolsky. 2005. Analysis of global mRNA expression in human skeletal muscle during recovery from endurance exercise. *FASEB J.* 10.1096/fj.04-3149fje.

Mehta, T. S., S. O. Zakharkin, G. L. Gadbury, and D. B. Allison. 2006. Epistemological issues in omics and high-dimensional biology: Give the people what they want. *Physiol. Genomics* Sept. 2006; doi: 10.1152/physiolgenomics.00095.2006.

Near, T. J., S. K. Parker, and H. W. Detrich III. 2006. A genomic fossil reveals key steps in hemoglobin loss by the Antarctic icefishes. *Mol. Biol. Evol.* 23: 2008–2016.

Nikinmaa, M. and D. Schlenk. 2009. Editorial. *Aquat. Toxicol.* 92: 113.

O'Brien, K. M., and I. A. Mueller. 2010. The unique mitochondrial form and function of Antarctic channichthyid icefishes. *Integr. Comp. Biol.* 50: 993–1008.

Pilegaard, H., G. A. Ordway, B. Saltin, and P. Neufer. 2000. Transcriptional regulation of ... expression in human skeletal muscle during recovery from exercise. *Am. J. Physiol.* 279: E806–E814.

Podrabsky, J. E., and G. N. Somero. 2004. Changes in gene expression associated with acclimation to constant temperatures and fluctuating daily temperatures in an annual killifish *Austrofundulus limnaeus*. *J. Exp. Biol.* 207: 2237–2254.

Price, D. P., V. Nagarajan, A. Churbanov, P. Houde, B. Milligan, L. L. Drake, J. E. Gustafson, and I. A. Hansen. 2011. The fat body transcriptomes of the yellow fever mosquito *Aedes aegypti*, pre- and post-blood meal. *PLoS One* 6: e22573.

Rund, S. S. C., T. Y. Hou, S. M. Ward, F. H. Collins, and G. E. Duffield. 2011. Genome-wide profiling of diel and circadian gene expression in the malaria vector *Anopheles gambiae*. *Proc. Natl. Acad. Sci. U.S.A.* 108: E421–E430.

Ruvinsky, A., and J. A. M. Graves (eds.). 2005. *Mammalian Genomics*. CABI Publishing, Cambridge, MA.

Sanoudou, D., E. Vafiadaki, D. A. Arvanitis, E. Kranias, and A. Kontrogianni-Konstantopoulos. 2005. Array lessons from the heart: focus on the genome and transcriptome of cardiomyopathies. *Physiol. Genomics* 21: 131–143.

Sea Urchin Genome Sequencing Consortium. 2006. The genome of the sea urchin *Strongylocentrotus purpuratus*. *Science* 314: 941–952.

Sidell, B. D., M. E. Vayda, D. J. Small, T. J. Moylan, R. L. Londraville, M.-L. Yuan, K. J. Rodnick, Z. A. Eppley, and L. Costello. 1997. Variable expression of myoglobin among the hemoglobinless Antarctic icefishes. *Proc. Natl. Acad. Sci. U.S.A.* 94: 3420–3424.

Simonson, T. S., and 11 additional authors. 2010. Genetic evidence for high-altitude adaptation in Tibet. *Science* 329: 72–75.

Stillman, J. H., and 11 additional authors. 2008. Recent advances in crustacean genomics. *Integr. Comp. Biol.* 48: 852–868.

Storey, K. B. 2004. Strategies for exploration of freeze responsive gene expression: advances in vertebrate freeze tolerance. *Cryobiology* 48: 134–145.

Sukumaran, S., W. J. Jusko, D. C. DuBois, and R. R. Almon. 2011. Light-dark oscillations in the lung transcriptome: implications for lung homeostasis, repair, metabolism, disease, and drug action. *J. Appl. Physiol.* 110: 1732–1747.

Todgham, A. E., and G. E. Hofmann. 2009. Transcriptomic response of sea urchin larvae *Strongylocentrotus purpuratus* to CO_2-driven seawater acidification. *J. Exp. Biol.* 212: 2579–2594.

Van der Weyden, L., D. J. Adams, and A. Bradley. 2002. Tools for targeted manipulation of the mouse genome. *Physiol. Genomics* 11: 133–164.

Whitfield, C. W., Y. Ben-Shahar, C. Brillet, I. Leoncini, D. Crauser, Y. LeConte, S. Rodriguez-Zas, and G. E. Robinson. 2006. Genomic dissection of behavioral maturation in the honey bee. *Proc. Natl. Acad. Sci. U.S.A.* 103: 16068–16075.

Williams, D. R., L. E. Epperson, W. Li, M. A. Hughes, R. Taylor, J. Rogers, S. L. Martin, A. R. Cossins, and A. Y. Gracey. 2005. Seasonally hibernating phenotype assessed through transcript screening. *Physiol. Genomics* 24: 13–22.

Yi, X., and 69 additional authors. 2010. Sequencing of 50 human exomes reveals adaptation to high altitude. *Science* 329: 75–78.

Chapter 4

Bartolomei, M. S., and A. C. Ferguson-Smith. 2011. Mammalian genomic imprinting. *Cold Spring Harb. Perspect. Biol.* 3: a002592. doi: 10.1101/cshperspect.a002592.

Beldade, P., and P. M. Brakefield. 2002. The genetics and evo-devo of butterfly wing patterns. *Nat. Rev. Genet.* 3: 442–452.

Bonasio, R., S. Tu, and D. Reinberg. 2010. Molecular signals of epigenetic states. *Science* 330: 612–616.

Burggren, W. W., and K. S. Reyna. 2011. Developmental trajectories, critical windows and phenotypic alteration during cardio-respiratory development. *Resp. Physiol. Neurobiol.* 178: 13–21.

DeWitt, T. J., A. Sih, and D. S. Wilson. 1998. Costs and limits of phenotypic plasticity. *Trends Ecol. Evol.* 13: 77–81.

Fowler, S. L., D. P. Costa, J. P. Y. Arnould, N. J. Gales, and C. E. Kuhn. 2006. Ontogeny of diving behaviour in the Australian sea lion: trials of adolescence in a late bloomer. *J. Anim. Ecol.* 75: 358–367.

Fowler, S. L., D. P. Costa, J. P. Y. Arnould, N. J. Gales, and J. M. Burns. 2007. Ontogeny of oxygen stores and physiological diving capability in Australian sea lions. *Funct. Ecol.* 21: 922–935.

Freeman, A. S., and J. E. Byers. 2006. Divergent induced responses to an invasive predator in marine mussel populations. *Science* 313: 831–833.

Fu, C., J. M. Wilson, P. J. Rombough, and C. J. Brauner. 2010. Ions first: Na^+ uptake shifts from the skin to the gills before O_2 uptake in developing rainbow trout, *Oncorhynchus mykiss*. *Proc. R. Soc. London B* 277: 1553–1560.

Gluckman, P. D., M. A. Hanson, T. Buklijas, F. M. Low, and A. S. Beedle. 2009. Epigenetic mechanisms that underpin metabolic and cardiovascular diseases. *Nat. Rev. Endocrinol.* 5: 401–408.

Gregg, C. 2010. Parental control over the brain. *Science* 330: 770–771.

Hazel, W. N. 2002. The environmental and genetic control of seasonal polyphenism in larval color and its adaptive significance in a swallowtail butterfly. *Evolution* 56: 342–348.

Heijmans, B. T., E. W. Tobi, A. D. Stein, H. Putter, G. J. Blauw, E. S. Susser, P. E. Slagboom, and L. H. Lumey. 2008. Persistent epigenetic differences associated with prenatal exposure to famine in humans. *Proc. Natl. Acad. Sci. U.S.A.* 105: 17046–17049.

Johnston, I. A., N. J. Cole, V. L. A. Vieira, and I. Davidson. 1997. Temperature and developmental plasticity of muscle phenotype in herring larvae. *J. Exp. Biol.* 200: 849–868.

Katoh, F., A. Shimizu, K. Uchida, and T. Kaneko. 2000. Shift of chloride cell distribution during early life stages in seawater-adapted killifish, *Fundulus heteroclitus*. *Zool. Sci.* 17: 11–18.

Leonard, W. R., and M. L. Robertson. 1994. Evolutionary perspectives on human nutrition: The influence of brain and body size on diet and metabolism. *Am. J. Human Biol.* 6: 77–88.

Leonard, W. R., J. J. Snodgrass, and M. L. Robertson. 2007. Effects of brain evolution on human nutrition and metabolism. *Annu. Rev. Nutr.* 27: 311–327.

Ma, Z., W. Guo, X. Guo, X. Wang, and L. Kang. 2011. Modulation of behavioral phase changes of the migratory locust by the catecholamine metabolic pathway. *Proc. Natl. Acad. Sci. U.S.A.* 108: 3882–3887.

Macqueen, D. J., D. H. F. Robb, T. Olsen, L. Melstveit, C. G. M. Paxton, and I. A. Johnston. 2008. Temperature until the "eyed stage" of embryogenesis programmes the growth trajectory and muscle phenotype of adult Atlantic salmon. *Biol. Lett.* 4: 294–298.

Mays, S. A. 1999. Linear and appositional long bone growth in earlier human populations: a case study from Mediaeval England. In R. D. Hoppa and C. M. Fitzgerald (eds.), *Human Growth in the Past: Studies from Bones and Teeth*, pp. 290–312. Cambridge University Press, Cambridge.

Murgatroyd, C., A. V. Patchev, Y. Wu, V. Micale, Y. Bockmühl, D. Fischer, F. Holsboer, C. T. Wotjak, O. F. Almeida, and D. Spengler. 2009. Dynamic DNA methylation programs persistent adverse effects of early-life stress. *Nature Neurosci.* 12: 1559–1566.

Noren, S. R., S. J. Iverson, and D. J. Boness. 2005. Development of the blood and muscle oxygen stores in gray seals (*Halichoerus grypus*): Implications for juvenile diving capacity and the necessity of a terrestrial postweaning fast. *Physiol. Biochem. Zool.* 78: 482–490.

Ong, K. K., M. L. Ahmed, and D. B. Dunger. 2006. Lessons from large population studies on timing and tempo of puberty (secular trends and relation to body size): The European trend. *Mol. Cell. Endocrinol.* 254/255: 8–12.

Ranade, S. C., A. Rose, M. Rao, J. Gallego, P. Gressens, and S. Mani. 2008. Different types of nutritional deficiencies affect different domains of spatial memory function checked in a radial arm maze. *Neuroscience* 152: 859–866.

Richmond, J. P., J. M. Burns, and L. D. Rea. 2006. Ontogeny of total body oxygen stores and aerobic dive potential in Steller sea lions (*Eumetopias jubatus*). *J. Comp. Physiol. B* 176: 535–545.

Weaver, I. C. G., N. Cervoni, F. A. Champagne, A. C. D'Alessio, S. Sharma, J. R. Seckl, S. Dymov, M. Szyf, and M. J. Meaney. 2004. Epigenetic programming by maternal behavior. *Nat. Neurosci.* 7: 847–854.

Wiltschko, W., P. Daum, A. Fergenbauer-Kimmel, and R. Wiltschko. 1987. The development of the star compass in garden warblers, *Sylvia borin*. *Ethology* 74: 285–292.

Chapter 5

Beyembach, K. W., and H. Wieczorek. 2005. The V-type H^+ ATPase: molecular structure and function, physiological roles and regulation. *J. Exp. Biol.* 209: 577–589.

Bublitz, M., H. Poulsen, J. P. Morth, and P. Nissen. 2010. In and out of the cation pumps: P-type ATPase structure revisited. *Curr. Opin. Struct. Biol.* 20: 431–439.

de Groot, B. L., and H. Grubmüller. 2001. Water permeation across biological membranes: Mechanism and dynamics of aquaporin-1 and GlpF. *Science* 294: 2353–2357.

Evans, D. H., P. M. Piermarini, and K. P. Choe. 2003. The multifunctional fish gill: dominant site of gas exchange, osmoregulation, acid-base regulation, and excretion of nitrogenous wastes. *Physiol. Rev.* 85: 97–177.

Fabra, M., D. Raldúa, D. M. Power, P. M. T. Deen, and J. Cerdà. 2005. Marine fish egg hydration is aquaporin-mediated. *Science* 307: 545.

Harvey, B., I. Lacoste, and J. Ehrenfeld. 1991. Common channels for water and protons at apical and basolateral cell membranes of frog skin and urinary bladder epithelia. *J. Gen. Physiol.* 97: 749–776.

Horisberger, J.-D. 2004. Recent insights into the structure and mechanism of the sodium pump. *Physiology* 19: 377–387.

Jensen, M. Ø., V. Jogini, D. W. Borhani, A. E. Leffler, R. O. Dror, and D. E. Shaw. 2012. Mechanism of voltage gating in potassium channels. *Science* 336: 229–233.

Kinne, R. K. H. (ed.). 1990. *Basic Principles in Transport*. Karger, Basel, Switzerland.

Kinne, R. K. H. (ed.). 1990. *Comparative Aspects of Sodium Cotransport Systems*. Karger, Basel, Switzerland.

Kirschner, L. B. 1991. Water and ions. In C. L. Prosser (ed.), *Environmental and Metabolic Animal Physiology*, pp. 13–107. Wiley-Liss, New York.

Lodish, H., A. Berk, P. Matsudaira, C. A. Kaiser, M. Krieger, M. P. Scott, L. Zipursky, and J. Darnell. 2004. *Molecular Cell Biology*, 5th ed. W. H. Freeman, San Francisco.

Maloney, P. C., and T. H. Wilson. 1985. The evolution of ion pumps. *BioScience* 35: 43–48.

Marshall, W. S. 2002. Na^+, Cl^-, Ca^{2+} and Zn^{2+} transport by fish gills: retrospective review and prospective synthesis. *J. Exp. Zool.* 293: 264–283.

Perry, S. F., A. Shahsavarani, T. Georgalis, M. Bayaa, M. Furimsky, and S. L. Y. Thomas. 2003. Channels, pumps, and exchangers in the gill and kidney of freshwater fishes: their role in ionic and acid-base regulation. *J. Exp. Zool.* 300A: 53–62.

Sasaki, S., K. Ishibashi, and F. Marumo. 1998. Aquaporin-2 and -3: Representatives of two subgroups of the aquaporin family colocalized in the kidney collecting duct. *Annu. Rev. Physiol.* 60: 199–220.

Scholander, P. F., and H. T. Hammel. 1976. *Osmosis and Tensile Solvent*. Springer, New York.

Sidell, B. D., and J. R. Hazel. 1987. Temperature affects the diffusion of small molecules through cytosol of fish muscle. *J. Exp. Biol.* 129: 191–203.

Somero, G. N., C. B. Osmond, and C. L. Bolis (eds.). 1992. *Water and Life*. Springer, New York.

Spring, K. R. 1998. Routes and mechanism of fluid transport by epithelia. *Annu. Rev. Physiol.* 60: 105–119.

Stein, W. D. 1990. *Channels, Carriers, and Pumps. An Introduction to Membrane Transport*. Academic Press, New York.

Tipsmark, C. K., K. J. Sorensen, and S. S. Madsen. 2010. Aquaporin expression dynamics in osmoregulatory tissues of Atlantic salmon during smoltification and seawater acclimation. *J. Exp. Biol.* 213: 368–379.

Toyoshima, C., H. Sasabe, and D. L. Stokes. 1993. Three-dimensional cryo-electron microscopy of the calcium ion pump in the sarcoplasmic reticulum membrane. *Nature* 362: 469–471.

Verkman, A. S. 1993. *Water Channels*. Landes, Austin, TX.

Chapter 6

Andrew, A. L., D. C. Card, R. P. Ruggiero, D. R. Schield, R. H. Adams, D. D. Pollack, S. M. Secor, and T. A. Castoe. 2015. Rapid changes in gene expression direct rapid shifts in intestinal form and function in the Burmese python after feeding. *Physiol. Genomics* 47: 147–157.

Arumugam, M., and 41 additional authors. 2011. Enterotypes of the human gut microbiome. *Nature* 473: 174–180.

Badman, M. K., and J. S. Flier. 2005. The gut and energy balance: visceral allies in the obesity wars. *Science* 307: 1909–1914.

Bairlein, F. 2002. How to get fat: nutritional mechanisms of seasonal fat accumulation in migratory songbirds. *Naturwissenschaften* 89: 1–10.

Baumann, P., N. A. Moran, and L. Baumann. 1997. The evolution and genetics of aphid endosymbionts. *BioScience* 47: 12–20.

Becerra, J. X. 1997. Insects on plants: Macroevolutionary chemical trends in host use. *Science* 276: 253–256.

Bels, V. L., M. Chardon, and P. Vandewalle (eds.). 1994. Biomechanics of feeding in vertebrates. *Adv. Comp. Environ. Physiol.* 18: 1–374.

Blaxter, K., and I. MacDonald (eds.). 1988. *Comparative Nutrition*. Libbey, London.

Boyd, P. W., and 22 additional authors. 2007. Mesoscale iron enrichment experiments 1993–2005: synthesis and future directions. *Science* 315: 612–617.

Burger, J., M. Kirchner, B. Bramanti, W. Haak, and M. G. Thomas. 2007. Absence of the lactase-persistence-associated allele in early Neolithic Europeans. *Proc. Natl. Acad. Sci. U.S.A.* 104: 3736–3741.

Burton, B. T., and W. R. Foster. 1988. *Human Nutrition*, 4th ed. McGraw-Hill, New York.

Canfield, D. E., A. N. Glazer, and P. G. Falkowski. 2010. The evolution and future of Earth's nitrogen cycle. *Science* 330: 192–196.

Carey, C. (ed.). 1996. *Avian Energetics and Nutritional Ecology*. Chapman & Hall, New York.

Castellini, M. A., and L. D. Rea. 1992. The biochemistry of natural fasting at its limits. *Experientia* 48: 575–582.

Chang, E. B., M. D. Sitrin, and D. D. Black. 1996. *Gastrointestinal, Hepatobiliary, and Nutritional Physiology*. Lippincott-Raven, Philadelphia.

Chivers, D. J., and P. Langer (eds.). 1994. *The Digestive System in Mammals: Food, Form and Function*. Cambridge University Press, New York.

Choat, J. H., and K. D. Clements. 1998. Vertebrate herbivores in marine and terrestrial environments: A nutritional ecology perspective. *Annu. Rev. Ecol. Syst.* 29: 375–403.

Coale, K. H., and 18 additional authors. 1996. A massive phytoplankton bloom induced by an ecosystem-scale iron fertilization experiment in the equatorial Pacific Ocean. *Nature* 383: 495–501.

Dissanayake, C. B., and R. Chandrajith. 1999. Medical geochemistry of tropical environments. *Earth-Sci. Rev.* 47: 219–258.

Douglas, A. E. 1995. The ecology of symbiotic microorganisms. *Adv. Ecol. Res.* 26: 69–103.

Eisner, T., and J. Meinwald (eds.). 1995. *Chemical Ecology: The Chemistry of Biotic Interaction*. National Academy Press, Washington, D.C.

Ferraris, R. P., and J. Diamond. 1997. Regulation of intestinal sugar transport. *Physiol. Rev.* 77: 257–302.

Fertin, A., and J. Casas. 2006. Efficiency of antlion trap construction. *J. Exp. Biol.* 209: 3510–3515.

Finney, L. A., and T. V. O'Halloran. 2003. Transition metal speciation in the cell: Insights from the chemistry of metal ion receptors. *Science* 300: 931–936.

Foley, W. J., and C. McArthur. 1994. The effects and costs of allelochemicals for mammalian herbivores: An ecological perspective. In D. J. Chivers and P. Langer (eds.), *The Digestive System in Mammals: Food, Form and Function*, pp. 370–391. Cambridge University Press, New York.

Fryxell, J. M., J. F. Wilmshurst, and A. R. E. Sinclair. 2004. Predictive models of movement by Serengeti grazers. *Ecology* 85: 2429–2435.

Gee, J. H. R. 1991. Specialist aquatic feeding mechanisms. In R. S. K. Barnes and K. H. Mann (eds.), *Fundamentals of Aquatic Ecology*, pp. 186–209. Blackwell, London.

Glendinning, J. I. 2007. How do predators cope with chemically defended foods? *Biol. Bull.* 213: 253–266.

Grieshaber, M. K., and S. Völkel. 1998. Animal adaptations for tolerance and exploitation of poisonous sulfide. *Annu. Rev. Physiol.* 60: 33–53.

Grill, H. J. 2010. Leptin and the systems neuroscience of meal size control. *Front. Neuroendocrinol.* 31: 61–78.

Hofmann, R. R. 1989. Evolutionary steps of ecophysiological adaptation and diversification of ruminants: A comparative view of their digestive system. *Oecologia* 78: 443–457.

Hooper, L. V., T. Midtvedt, and J. I. Gordon. 2002. How host-microbial interactions shape the nutrient environment of the mammalian intestine. *Annu. Rev. Nutr.* 22: 283–307.

Howe, H. F., and L. C. Westley. 1988. *Ecological Relationships of Plants and Animals*. Oxford University Press, New York.

Itan, Y., B. L. Jones, C. J. E. Ingram, D. M. Swallow, and M. G. Thomas. 2010. A worldwide correlation of lactase persistence phenotype and genotypes. *BMC Evol. Biol.* 10: 36.

Jørgensen, C. B. 1990. *Bivalve Filter Feeding*. Olsen & Olsen, Fredensborg, Denmark.

Kaput, J., and R. L. Rodriguez. 2004. Nutritional genomics: the next frontier in the postgenomic era. *Physiol. Genomics* 16: 166–177.

Karasov, W. H., and J. M. Diamond. 1985. Digestive adaptations for fueling the cost of endothermy. *Science* 228: 202–204.

Karasov, W. H., and J. M. Diamond. 1988. Interplay between physiology and ecology in digestion. *BioScience* 38: 602–611.

Kunze, W. A. A., and J. B. Furness. 1999. The enteric nervous system and regulation of intestinal motility. *Annu. Rev. Physiol.* 61: 117–142.

Karasov, W. H., and I. D. Hume. 1997. Vertebrate gastrointestinal system. In W. H. Dantzler (ed.), *Comparative Physiology*, vol. 1 (Handbook of Physiology [Bethesda, MD], section 13), pp. 409–480. Oxford University Press, New York.

Lechene, C. P., Y. Luyten, G. McMahon, and D. L. Distel. 2007. Quantitative imaging of nitrogen fixation by individual bacteria within animal cells. *Science* 317: 1563–1566.

Martin, R. E., and K. Kirk. 2007. Transport of the essential nutrient isoleucine in human erythrocytes infected with the malaria parasite *Plasmodium falciparum*. *Blood* 109: 2217–2224.

Martinez del Rio, C. 1994. Nutritional ecology of fruit-eating and flower-visiting birds and bats. In D. J. Chivers and P. Langer (eds.), *The Digestive System in Mammals: Food, Form and Function*, pp. 103–127. Cambridge University Press, New York.

Mayntz, D., D. Raubenheimer, M. Salomon, S. Toft, and S. J. Simpson. 2005. Nutrient-specific foraging in invertebrate predators. *Science* 307: 111–113.

McCue, M. D. 2007. Snakes survive starvation by employing supply- and demand-side economic strategies. *Zoology* 110: 318–327.

McCue, M. D. 2010. Starvation physiology: Reviewing the different strategies animals use to survive a common challenge. *Comp. Biochem. Physiol., A* 156: 1–18.

McDowell, L. R. 2003. *Minerals in Animal and Human Nutrition*, 2nd ed. Elsevier, Amsterdam.

McNaughton, S. J. 1988. Mineral nutrition and spatial concentrations of African ungulates. *Nature* 334: 343–345.

McNaughton, S. J. 1990. Mineral nutrition and seasonal movements of African migratory ungulates. *Nature* 345: 613–615.

Mellinger, J. (ed.). 1990. *Animal Nutrition and Transport Processes. 1. Nutrition in Wild and Domestic Animals*. Karger, Basel, Switzerland.

Mertz, W. 1981. The essential trace elements. *Science* 213: 1332–1338.

Michal, G. (ed.). 1999. *Biochemical Pathways. An Atlas of Biochemistry and Molecular Biology.* Wiley, New York.

Morris, J. G. 1991. Nutrition. In C. L. Prosser (ed.), *Environmental and Metabolic Animal Physiology. Comparative Animal Physiology,* 4th ed., pp. 231–276. Wiley-Liss, New York.

Morton, B. 1983. Feeding and digestion in Bivalvia. In A. S. M. Saleuddin and K. M. Wilbur (eds.), *The Mollusca,* vol. 5, pp. 65–147. Academic Press, New York.

Motta, P. J., and C. D. Wilga. 2001. Advances in the study of feeding behaviors, mechanisms, and mechanics of sharks. *Environ. Biol. Fishes* 60: 131–156.

Murray, M. G. 1995. Specific nutrient requirements and migration of wildebeest. In A. R. E. Sinclair and P. Arcese (eds.), *Serengeti II. Dynamics, Management, and Conservation of an Ecosystem,* pp. 231–256. University of Chicago Press, Chicago.

Nagy, K. A. 2001. Food requirements of wild animals: Predictive equations for free-living mammals, reptiles, and birds. *Nutr. Abstr. Rev., Ser. B* 71: 1R–12R.

O'Brien, D. M., C. L. Boggs, and M. L. Fogel. 2005. The amino acids used in reproduction by butterflies: a comparative study of dietary sources using compound-specific stable isotope analysis. *Physiol. Biochem. Zool.* 78: 819–827.

Olivera, B. M., J. Rivier, C. Clark, C. A. Ramilo, G. P. Corpuz, F. C. Abogadie, E. E. Mena, S. R. Woodward, D. R. Hillyard, and L. J. Cruz. 1990. Diversity of *Conus* neuropeptides. *Science* 249: 257–263.

Pandian, T. J., and F. J. Vernberg (eds.). 1987. *Animal Energetics,* 2 vols. Academic Press, New York.

Parker, K. L., P. S. Barboza, and M. P. Gillingham. 2009. Nutrition integrates environmental responses of ungulates. *Func. Ecol.* 23: 57–69.

Penry, D. L., and P. A. Jumars. 1987. Modeling animal guts as chemical reactors. *Am. Nat.* 129: 69–96.

Reilly, S. M., and G. V. Lauder. 1992. Morphology, behavior, and evolution: Comparative kinematics of aquatic feeding in salamanders. *Brain Behav. Evol.* 40: 182–196.

Riquelme, C. A., J. A. Magida, B. C. Harrison, C. E. Wall, T. G. Marr, S. M. Secor, and L. A. Leinwand. 2011. Fatty acids identified in the Burmese python promote beneficial cardiac growth. *Science* 334: 528–531.

Saffo, M. B. 1992. Invertebrates in endosymbiotic associations. *Am. Zool.* 32: 557–565.

Schondube, J. E., L. G. Herrera-M., and C. M. del Rio. 2001. Diet and the evolution of digestion and renal function in phyllostomid bats. *Zoology* 104: 59–73.

Schultz, S. G. (ed.). 1991. *The Gastrointestinal System,* 4 vols. (Handbook of Physiology [Bethesda, MD], section 6). American Physiological Society, Bethesda, MD.

Schwenk, K. (ed.). 2000. *Feeding: Form, Function, and Evolution in Tetrapod Vertebrates.* Academic Press, New York.

Secor, S. M., and J. Diamond. 1997. Effects of meal size on postprandial responses in juvenile Burmese pythons (*Python molurus*). *Am. J. Physiol.* 41: R902–R912.

Simon, M., M. Johnson, P. Tyack, and P. T. Madsen. 2009. Behaviour and kinematics of continuous ram filtration in bowhead whales (*Balaena mysticetus*). *Proc. Roy. Soc. London, Ser. B* 276: 3819–3828.

Singer, M. A. 2003. Do mammals, birds, reptiles and fish have similar nitrogen conserving systems? *Comp. Biochem. Physiol., B* 134: 543–558.

Spallholz, J. E. 1989. *Nutrition: Chemistry and Biology.* Prentice Hall, Englewood Cliffs, NJ.

Spor, A., O. Koren, and R. Ley. 2011. Unravelling the effects of the environment and host genotype on the gut microbiome. *Nat. Rev. Microbiol.* 9: 279–290.

Starck, J. M., and T. Wang (eds.). 2005. *Physiological and Ecological Adaptations to Feeding in Vertebrates.* Science Publishers, Enfield, NH.

Tunnicliffe, V. 1992. Hydrothermal-vent communities of the deep sea. *Am. Sci.* 80: 336–349.

Urich, K. 1994. *Comparative Animal Biochemistry.* Springer, New York.

Vitousek, P. M., S. Hattenschwiler, L. Olander, and S. Allison. 2002. Nitrogen and nature. *Ambio* 31: 97–101.

Vinagre, A. S., and R. S. M. Da Silva. 2002. Effects of fasting and refeeding on metabolic processes in the crab *Chasmagnathus granulata* (Dana, 1851). *Can. J. Zool.* 80: 1413–1421.

Vonk, H. J., and J. R. H. Western. 1984. *Comparative Biochemistry and Physiology of Enzymatic Digestion.* Academic Press, New York.

Wainwright, P. C., and S. M. Reilly (eds.). 1994. *Ecological Morphology.* University of Chicago Press, Chicago.

Ward, J. E., L. P. Sanford, R. I. E. Newell, and B. A. MacDonald. 1998. A new explanation of particle capture in suspension-feeding bivalve molluscs. *Limnol. Oceanogr.* 43: 741–752.

Weindruch, R. 1996. Caloric restriction and aging. *Sci. Am.* 274(1): 46–52.

Weiss, S. L., E. A. Lee, and J. Diamond. 1998. Evolutionary matches of enzyme and transporter capacities to dietary substrate loads in the intestinal brush border. *Proc. Natl. Acad. Sci. U.S.A.* 95: 2117–2121.

Werth, A. J. 2001. How do mysticetes remove prey trapped in baleen? *Bull. Mus. Comp. Zool.* 156: 189–203.

Wildish, D., and D. Kristmanson. 1997. *Benthic Suspension Feeders and Flow.* Cambridge University Press, Cambridge.

Woods, S. C., R. J. Seeley, D. Porte, Jr., and M. W. Schwartz. 1998. Signals that regulate food intake and energy homeostasis. *Science* 280: 1378–1382.

Wright, P. A., and P. M. Anderson (eds.). 2001. *Nitrogen Excretion.* Academic Press, New York.

Wright, S. H., and G. A. Ahearn. 1997. Nutrient absorption in invertebrates. In W. H. Dantzler (ed.), *Comparative Physiology,* vol. 2 (Handbook of Physiology [Bethesda, MD], section 13), pp. 1137–1205. Oxford University Press, New York.

Chapter 7

Alexander, R. M. 1999. *Energy for Animal Life.* Oxford University Press, New York.

Andrade, D. V., A. S. Abe, A. P. Cruz-Neto, and T. Wang. 2005. Specific dynamic action in ectothermic vertebrates: a general review on the determinants of the metabolic responses to digestion in fish, amphibians and reptiles. In J. M. Starck and T. Wang (eds.), *Physiological and Ecological Adaptations to Feeding in Vertebrates,* pp. 305–324. Science Publishers, Enfield, NH.

Beamish, F. W. H., and E. A. Trippel. 1990. Heat increment: A static or dynamic dimension in bioenergetic models? *Trans. Am. Fish. Soc.* 119: 649–661.

Bishop, C. M. 1999. The maximum oxygen consumption and aerobic scope of birds and mammals: Getting to the heart of the matter. *Proc. R. Soc. London, Ser. B* 266: 2275–2281.

Bridges, C. R., and P. J. Butler (eds.). 1989. *Techniques in Comparative Respiratory Physiology: An Experimental Approach.* Cambridge University Press, New York.

Brown, A. C., and G. Brengelmann. 1965. Energy metabolism. In T. C. Ruch and H. D. Patton (eds.), *Physiology and Biophysics,* 19th ed., pp. 1030–1049. Saunders, Philadelphia. [A classic.]

Brown, C. R., and J. N. Cameron. 1991. The induction of specific dynamic action in channel catfish by infusion of essential amino acids. *Physiol. Zool.* 64: 276–297.

Brown, J. H., and G. B. West (eds.). 2000. *Scaling in Biology.* Oxford University Press, New York.

Calder, W. A., III. 1984. *Size, Function, and Life History.* Harvard University Press, Cambridge, MA.

Calder, W. A., III. 1987. Scaling energetics of homeothermic vertebrates: An operational allometry. *Annu. Rev. Physiol.* 49: 107–120.

Capellini, I., C. Venditti, and R. A. Barton. 2010. Phylogeny and metabolic scaling in mammals. *Ecology* 91: 2783–2793.

Clarke, A., P. Rothery, and N. J. B. Isaac. 2010. Scaling of basal metabolic rate with body mass and temperature in mammals. *J. Anim. Ecol.* 79: 610–619.

Darveau, C.-A., R. K. Suarez, R. D. Andrews, and P. W. Hochachka. 2002. Allometric cascade as a unifying principle of body mass effects on metabolism. *Nature* 417: 166–170.

Dobson, G. P., and J. P. Headrick. 1995. Bioenergetic scaling: Metabolic design and body-size constraints in mammals. *Proc. Natl. Acad. Sci. U.S.A.* 92: 7317–7321.

Duncan, R. P., D. M. Forsyth, and J. Hone. 2007. Testing the metabolic theory of ecology: allometric scaling exponents in mammals. *Ecology* 88: 324–333.

Dyson, F. J. 1954. What is heat? *Sci. Am.* 191(3): 58–63.

Giampietro, M., and D. Pimentel. 1991. Energy efficiency: Assessing the interaction between humans and their environment. *Ecol. Econom.* 4: 117–144.

Glazier, D. 2009. A unifying explanation for diverse metabolic scaling in animals and plants. *Biol. Rev.* 85: 111–138.

Heglund, N. C., and G. A. Cavagna. 1985. Efficiency of vertebrate locomotory muscles. *J. Exp. Biol.* 115: 283–292.

Heusner, A. A. 1991. Size and power in mammals. *J. Exp. Biol.* 160: 25–54.

Houlihan, D. F., C. P. Waring, E. Mathers, and C. Gray. 1990. Protein synthesis and oxygen consumption of the shore crab *Carcinus maenas* after a meal. *Physiol. Zool.* 63: 735–756.

Hulbert, A. J., and P. L. Else. 2000. Mechanisms underlying the cost of living in animals. *Annu. Rev. Physiol.* 62: 207–235.

Jobling, M. 1993. Bioenergetics: Feed intake and energy partitioning. In J. C. Rankin and F. B. Jensen (eds.), *Fish Ecophysiology,* pp. 1–44. Chapman and Hall, London.

Kaiyala, K. J., and D. S. Ramsay. 2011. Direct animal calorimetry, the underused gold standard for quantifying the fire of life. *Comp. Biochem. Physiol., A* 158: 252–264.

Killen, S. S., D. Atkinson, and D. S. Glazier. 2010. The intraspecific scaling of metabolic rate with body mass in fishes depends on lifestyle and temperature. *Ecol. Lett.* 13: 184–193.

Kleiber, M. 1975. *The Fire of Life*, 2nd ed. Krieger, Huntington, NY. [Considered a classic.]

Kooijman, S. A. L. M. 1993. *Dynamic Energy Budgets in Biological Systems*. Cambridge University Press, New York.

LaBarbera, M. 1989. Analyzing body size as a factor in ecology and evolution. *Annu. Rev. Ecol. Syst.* 20: 97–117.

Levine, J. A., L. M. Lanningham-Foster, S. K. McCrady, A. C. Krizan, L. R. Olson, P. H. Kane, M. D. Jensen, and M. M. Clark. 2005. Interindividual variation in posture allocation: possible role in human obesity. *Science* 307: 584–586. [See also pp. 530–531.]

Lindstedt, S. L., and W. A. Calder III. 1981. Body size, physiological time, and longevity in homeothermic animals. *Q. Rev. Biol.* 56: 1–16.

Lomax, E. 1979. Historical development of concepts of thermoregulation. In P. Lomax and E. Schönbaum (eds.), *Body Temperature*, pp. 1–23. Dekker, New York.

Lovegrove, B. G. 2000. The zoogeography of mammalian basal metabolic rate. *Am. Nat.* 156: 201–219.

Mandelbrot, B. B. 1983. *The Fractal Geometry of Nature*. W. H. Freeman, San Francisco.

McCarthy, I. D., D. F. Houlihan, and C. G. Carter. 1994. Individual variation in protein turnover and growth efficiency in rainbow trout, *Oncorhynchus mykiss* (Walbaum). *Proc. R. Soc. London, Ser. B* 257: 141–147.

McKechnie, A. E., and B. O. Wolf. 2004. The allometry of avian basal metabolic rate: good predictions need good data. *Physiol. Biochem. Zool.* 77: 502–521.

McNab, B. K. 2002. *The Physiological Ecology of Vertebrates*. Cornell University Press, Ithaca, NY.

Mendelsohn, E. 1964. *Heat and Life: The Development of the Theory of Animal Heat*. Harvard University Press, Cambridge, MA.

Miller, P. J. (ed.). 1996. *Miniature Vertebrates: The Implications of Small Body Size*. Symposia of the Zoological Society of London, no. 69. Clarendon, Oxford, UK.

Pennycuick, C. J. 1992. *Newton Rules Biology: A Physical Approach to Biological Problems*. Oxford University Press, New York.

Perrin, N., and R. M. Sibly. 1993. Dynamic models of energy allocation and investment. *Annu. Rev. Ecol. Syst.* 24: 379–410.

Peters, R. H. 1983. *The Ecological Implications of Body Size*. Cambridge University Press, New York.

Prusiner, S., and M. Poe. 1970. Thermodynamic considerations of mammalian heat production. In O. Lindberg (ed.), *Brown Adipose Tissue*, pp. 263–282. American Elsevier, New York.

Rolfe, D. F. S., and G. C. Brown. 1997. Cellular energy utilization and molecular origin of standard metabolic rate in mammals. *Physiol. Rev.* 77: 731–758.

Rome, L. C., L. Flynn, E. M. Goldman, and T. D. Yoo. 2005. Generating electricity while walking with loads. *Science* 309: 1725–1728.

Secor, S. M., and M. Boehm. 2006. Specific dynamic action of ambystomatid salamanders and the effects of meal size, meal type, and body temperature. *Physiol. Biochem. Zool.* 79: 720–735.

Schmidt-Nielsen, K. 1984. *Scaling: Why Is Animal Size So Important?* Cambridge University Press, New York.

Sparling, C. E., M. A. Fedak, and D. Thompson. 2007. Eat now, pay later? Evidence of deferred food-processing costs in diving seals. *Biol. Lett.* 3: 94–98.

Walsberg, G. E., and T. C. M. Hoffman. 2006. Using direct calorimetry to test the accuracy of indirect calorimetry in an ectotherm. *Physiol. Biochem. Zool.* 79: 830–835.

Weibel, E. R. 1991. Fractal geometry: A design principle for living organisms. *Am. J. Physiol.* 261: L361–L369.

West, G. B., J. H. Brown, and B. J. Enquist. 1997. A general model for the origin of allometric scaling laws in biology. *Science* 276: 122–126.

West, G. B., J. H. Brown, and B. J. Enquist. 1999. The fourth dimension of life: Fractal geometry and allometric scaling of organisms. *Science* 284: 1677–1679.

White, C. R., and M. R. Kearney. 2013. Determinants of inter-specific variation in basal metabolic rate. *J. Comp. Physiol. B* 183: 1–26.

White, C. R., and R. S. Seymour. 2003. Mammalian basal metabolic rate is proportional to body mass$^{2/3}$. *Proc. Natl. Acad. Sci. U.S.A.* 100: 4046–4049.

White, C. R., T. M. Blackburn, and R. S. Seymour. 2009. Phylogenetically informed analysis of the allometry of mammalian basal metabolic rate supports neither geometric nor quarter-power scaling. *Evolution* 63: 2658–2667.

Wieser, W. 1984. A distinction must be made between the ontogeny and the phylogeny of metabolism in order to understand the mass exponent of energy metabolism. *Respir. Physiol.* 55: 1–9.

Chapter 8

Adamczewska, Z. M., and S. Morris. 2001. Metabolic status and respiratory physiology of *Gecarcoidea natalis*, the Christmas Island red crab, during the annual breeding migration. *Biol. Bull.* 200: 321–335.

Allen, D. G. 2004. Skeletal muscle function: role of ionic changes in fatigue, damage and disease. *Clin. Exp. Pharmacol. Physiol.* 31: 485–493.

Allen, D. G., G. D. Lamb, and H. Westerblad. 2008. Skeletal muscle fatigue: cellular mechanisms. *Physiol. Rev.* 88: 287–332.

Bass, J., and J. S. Takahashi. 2010. Circadian integration of metabolism and energetics. *Science* 330: 1349–1354.

Bellinger, A. M., M. Mongillo, and A. R. Marks. 2008. Stressed out: the skeletal muscle ryanodine receptor as a target of stress. *J. Clin. Inves.* 118: 445–453.

Bennett, A. F. 1972. The effect of activity on oxygen consumption, oxygen debt, and heart rate in lizards *Varanus gouldii* and *Sauromalus hispidus*. *J. Comp. Physiol.* 79: 259–280.

Bennett, A. F. 1991. The evolution of activity capacity. *J. Exp. Biol.* 160: 1–23.

Bickler, P. E., and L. T. Buck. 2007. Hypoxia tolerance in reptiles, amphibians, and fishes: life with variable oxygen availability. *Annu. Rev. Physiol.* 69: 145–170.

Billeter, R., and H. Hoppeler. 1992. Muscular basis of strength. In P. V. Komi (ed.), *Strength and Power in Sport*, pp. 39–63. Blackwell, Oxford, UK.

Boutilier, R. G., and J. St-Pierre. 2000. Surviving hypoxia without really dying. *Comp. Biochem. Physiol., A* 126: 481–490.

Chih, C. R., and W. R. Ellington. 1983. Energy metabolism during contractile activity and environmental hypoxia in the phasic adductor muscle of the bay scallop *Argopecten irradians concentricus*. *Physiol. Zool.* 56: 623–631.

Clanton, T. L., and P. F. Klawitter. 2001. Physiological and genomic consequences of intermittent hypoxia—Invited review: Adaptive responses of skeletal muscle to intermittent hypoxia: The known and the unknown. *J. Appl. Physiol.* 90: 2476–2487.

Cooke, R. 2007. Modulation of the actomyosin interaction during fatigue of skeletal muscle. *Muscle Nerve* 36: 756–777.

Coughlin, D. J., and L. C. Rome. 1996. The roles of pink and red muscle in powering steady swimming in scup, *Stenotomus chrysops*. *Am. Zool.* 36: 666–677.

DeZwann, A., and V. Putzer. 1985. Metabolic adaptations of intertidal invertebrates to environmental hypoxia (a comparison of environmental anoxia to exercise anoxia). *Symp. Soc. Exp. Biol.* 39: 33–62.

Dickson, K. A. 1996. Locomotor muscle of high-performance fishes: What do comparisons of tunas with ectothermic sister taxa reveal? *Comp. Biochem. Physiol., A* 113: 39–49.

Dohm, M. R., J. P. Hayes, and T. Garland, Jr. 1996. Quantitative genetics of sprint running speed and swimming endurance in the laboratory mouse (*Mus domesticus*). *Evolution* 50: 1688–1701.

Ellington, W. R. 2001. Evolution and physiological roles of phosphagen systems. *Annu. Rev. Physiol.* 63: 289–325.

England, W. R., and J. Baldwin. 1983. Anaerobic energy metabolism in the tail musculature of the Australian yabby *Cherax destructor* (Crustacea, Decapoda, Parastacidae): Role of phosphagens and anaerobic glycolysis during escape behavior. *Physiol. Zool.* 56: 614–622.

Famme, P., and J. Knudsen. 1984. Total heat balance study of anaerobiosis in *Tubifex tubifex* (Müller). *J. Comp. Physiol., B* 154: 587–591.

Gleeson, T. T. 1996. Post-exercise lactate metabolism: A comparative review of sites, pathways, and regulation. *Annu. Rev. Physiol.* 58: 565–581.

Guppy, M., and P. Withers. 1999. Metabolic depression in animals: physiological perspectives and biochemical generalizations. *Biol. Rev.* 74: 1–40.

Hand, S. C. 1991. Metabolic dormancy in aquatic invertebrates. *Adv. Comp. Environ. Physiol.* 8: 1–50.

Henry, R. P., C. E. Booth, F. H. Lallier, and P. J. Walsh. 1994. Post-exercise lactate production and metabolism in three species of aquatic and terrestrial decapod crustaceans. *J. Exp. Biol.* 186: 215–234.

Herbert, C. V., and D. C. Jackson. 1985. Temperature effects on the responses to prolonged submergence in the turtle *Chrysemys picta bellii*. *Physiol. Zool.* 58: 655–681.

Hochachka, P. W. 1998. Mechanism and evolution of hypoxia-tolerance in humans. *J. Exp. Biol.* 201: 1243–1254.

Hochachka, P. W., and M. Guppy. 1987. *Metabolic Arrest and the Control of Biological Time*. Harvard University Press, Cambridge, MA.

Hochachka, P. W., P. L. Lutz, T. J. Sick, and M. Rosenthal. 1993. *Surviving Hypoxia: Mechanisms of Control and Adaptation*. CRC Press, Boca Raton, FL.

Hoppeler, H., M. Vogt, E. R. Weibel, and M. Flück. 2003. Response of skeletal muscle mitochondria to hypoxia. *Exp. Physiol.* 88: 109–119.

Husak, J. F., S. F. Fox, M. B. Lovern, and R. A. van den Bussche. 2006. Faster lizards sire more offspring: sexual selection on whole-animal performance. *Evolution* 60: 2122–2130.

Husak, J. F., S. F. Fox, and R. A. van den Bussche. 2008. Faster male lizards are better defenders not sneakers. *Anim. Behav.* 75: 1725–1730.

Hylland, P., S. Milton, M. Pek, G. E. Nilsson, and P. L. Lutz. 1997. Brain Na$^+$/K$^+$-ATPase activity in two anoxia tolerant vertebrates: Crucian carp and freshwater turtle. *Neurosci. Lett.* 235: 89–92.

Jackson, D. C. 2000. Living without oxygen: Lessons from the freshwater turtle. *Comp. Biochem. Physiol., A* 125: 299–315.

Kesaraju, S., G. Nayak, H. M. Prentice, and S. L. Milton. 2014. Upregulation of Hsp72 mediates anoxia/reoxygenation neuroprotection in the freshwater turtle via modulation of ROS. *Brain Res.* 1582: 247–256.

Klingenberg, M., and K. S. Echtay. 2001. Uncoupling proteins: The issues from a biochemist point of view. *Biochim. Biophys. Acta Bioenerg.* 1504: 128–143. [This entire issue of the journal is devoted to papers on mitochondrial energetics and uncoupling proteins.]

Kohn, T. A., J. W. Curry, and T. D. Noakes. 2011. Black wildebeest skeletal muscle exhibits high oxidative capacity and a high proportion of type IIx fibres. *J. Exp. Biol.* 214: 4041–4047.

Kooyman, G. L. 1989. *Diverse Divers.* Springer, New York.

Lamb, G. D., D. G. Stephenson, J. Bangsbo, and C. Juel. 2006. Point:counterpoint: Lactic acid accumulation is an advantage/disadvantage during muscle activity. *J. Appl. Physiol.* 100: 1410–1414.

Le Galliard, J.-F., J. Clobert, and R. Ferrière. 2004. Physical performance and darwinian fitness in lizards. *Nature* 432: 502–505.

Lehninger, A. L., D. L. Nelson, and M. M. Cox. 2005. *Lehninger Principles of Biochemistry*, 4th ed. W. H. Freeman, San Francisco.

Lurman, G. J., C. H. Bock, and H.-O. Pörtner. 2007. An examination of the metabolic processes underpinning critical swimming in Atlantic cod (*Gadus morhua* L.) using *in vivo* ^{31}P-NMR spectroscopy. *J. Exp. Biol.* 210: 3749–3756.

Lutz, P. L., and G. E. Nilsson. 1997. Contrasting strategies for anoxic brain survival—Glycolysis up or down. *J. Exp. Biol.* 200: 411–419.

Lutz, P. L., and K. B. Storey. 1997. Adaptations to variations in oxygen tension by vertebrates and invertebrates. In W. H. Dantzler (ed.), *Comparative Physiology*, vol. 2 (Handbook of Physiology [Bethesda, MD], section 13), pp. 1479–1522. Oxford University Press, New York.

McArdle, W. D., F. I. Katch, and V. L. Katch. 2007. *Exercise Physiology: Energy, Nutrition, and Human Performance*, 7th ed. Lippincott Williams & Wilkins, Philadelphia.

Melzner, F., C. Bock, and H.-O. Pörtner. 2006. Critical temperatures in the cephalopod *Sepia officinalis* investigated using *in vivo* 31P NMR spectroscopy. *J. Exp. Biol.* 209: 891–906.

Nedergaard, J., V. Golozoubova, A. Matthias, A. Asadi, A. Jacobsson, and B. Cannon. 2001. UCP1: The only protein able to mediate adaptive non-shivering thermogenesis and metabolic inefficiency. *Biochim. Biophys. Acta Bioenerg.* 1504: 82–106. [This entire issue of the journal is devoted to papers on mitochondrial energetics and uncoupling proteins.]

Nilsson, G. E., and S. Östlund-Nilsson. 2004. Hypoxia in paradise: widespread hypoxia tolerance in coral reef fishes. *Proc. R. Soc. London., Ser. B* 271: S30–S33.

Officer, C. B., R. B. Biggs, J. L. Taft, L. E. Cronin, M. A. Tyler, and W. R. Boynton. 1984. Chesapeake Bay anoxia: Origin, development, and significance. *Science* 223: 22–27.

Pierce, V. A., and D. L. Crawford. 1997. Phylogenetic analysis of glycolytic enzyme expression. *Science* 276: 256–259.

Podrabsky, J. E., J. P. Lopez, T. W. M. Fan, R. Higashi, and G. N. Somero. 2007. Extreme anoxia tolerance in embryos of the annual killifish *Austrofundulus limnaeus*: insights from a metabolomic analysis. *J. Exp. Biol.* 210: 2253–2266.

Pörtner, H. O., D. M. Webber, R. K. O'Dor, and R. G. Boutilier. 1993. Metabolism and energetics in squid (*Illex illecebrosus, Loligo pealei*) during muscular fatigue and recovery. *Am. J. Physiol.* 265: R157–R165.

Raymond, J., and D. Segrè. 2006. The effect of oxygen on biochemical networks and the evolution of complex life. *Science* 311: 1764–1767.

Rolfe, D. F. S., and G. C. Brown. 1997. Cellular energy utilization and molecular origin of standard metabolic rate in mammals. *Physiol. Rev.* 77: 731–758.

Rome, L. C., P. T. Loughna, and G. Goldspink. 1985. Temperature acclimation: Improved sustained swimming performance in carp at low temperatures. *Science* 228: 194–196.

Rothwell, N. J., and M. J. Stock. 1979. A role for brown adipose tissue in diet-induced thermogenesis. *Nature* 281: 31–35.

Rowell, L. B., and J. T. Shepherd (eds.). 1996. *Exercise: Regulation and Integration of Multiple Systems.* (Handbook of Physiology [Bethesda, MD], section 12.) Oxford University Press, New York.

Saraste, M. 1999. Oxidative phosphorylation at the *fin de siècle. Science* 283: 1488–1493.

Seebacher, F., M. D. Brand, P. L. Else, H. Guderley, A. J. Hulbert, and C. D. Moyes. 2010. Plasticity of oxidative metabolism in variable climates: molecular mechanisms. *Physiol. Biochem. Zool.* 83: 721–732.

Seymour, R. S. 1997. Plants that warm themselves. *Sci. Am.* 273(3): 104–109.

Shick, J. M., J. Widdows, and E. Gnaiger. 1988. Calorimetric studies of behavior, metabolism and energetics of sessile intertidal animals. *Am. Zool.* 28: 161–181.

Sinervo, B., and R. B. Huey. 1990. Allometric engineering: An experimental test of the causes of interpopulational differences in performance. *Science* 248: 1106–1109.

Somero, G. N., and J. J. Childress. 1980. A violation of the metabolism-size scaling paradigm: Activities of glycolytic enzymes in muscle increase in larger-size fish. *Physiol. Zool.* 53: 322–337.

Storey, K. B. (ed.). 2004. *Functional Metabolism. Regulation and Adaptation.* Wiley-Liss, Hoboken, NJ.

Suarez, R. K. 1996. Upper limits to mass-specific metabolic rates. *Annu. Rev. Physiol.* 58: 583–605.

Taigen, T. L., and F. H. Pough. 1985. Metabolic correlates of anuran behavior. *Am. Zool.* 25: 987–997.

van Deursen, J., W. Ruitenbeek, A. Heerschap, P. Jap, H. ter Laak, and B. Wieringa. 1994. Creatine kinase (CK) in skeletal muscle energy metabolism: A study of mouse mutants with graded reduction in muscle CK expression. *Proc. Natl. Acad. Sci. U.S.A.* 91: 9091–9095.

van Waversveld, J., A. D. F. Addink, and G. van den Thillart. 1989. Simultaneous direct and indirect calorimetry on normoxic and anoxic goldfish. *J. Exp. Biol.* 142: 325–335.

Warren, D. E., S. A. Reese, and D. C. Jackson. 2006. Tissue glycogen and extracellular buffering limit the survival of red-eared slider turtles during anoxic submergence at 3°C. *Physiol. Zool. Biochem.* 79: 736–744.

West, J. B. 1996. Physiology of extreme altitude. In M. J. Fregly and C. M. Blatteis (eds.), *Environmental Physiology*, vol. 2 (Handbook of Physiology [Bethesda, MD], section 4), pp. 1307–1325. Oxford University Press, New York.

Chapter 9

Aarestrup, K., and 10 additional authors. 2009. Oceanic spawning migration of the European eel (*Anguilla anguilla*). *Science* 325: 1660.

Alerstam, T., M. Hake, and N. Kjellén. 2006. Temporal and spatial patterns of repeated migratory journeys by ospreys. *Anim. Behav.* 71: 555–566.

Alerstam, T., and A. Hedenström. 1998. The development of bird migration theory. *J. Avian Biol.* 29: 343–369.

Alexander, R. M. 1984. Walking and running. *Am. Sci.* 72: 348–354.

Armstrong, L. E., C. M. Maresh, C. V. Gabaree, J. R. Hoffman, S. A. Kavouras, R. W. Kenefick, J. W. Castellani, and L. E. Ahlquist. 1997. Thermal and circulatory responses during exercise: effects of hypohydration, dehydration, and water intake. *J. Appl. Physiol.* 82: 2028–2035.

Autumn, K., D. Jindrich, D. DeNardo, and R. Mueller. 1999. Locomotor performance at low temperature and the evolution of nocturnality in geckos. *Evolution* 53: 580–599.

Battley, P. F., T. Piersma, M. W. Dietz, S. Tang, A. Dekinga, and K. Hulsman. 2000. Empirical evidence for differential organ reductions during trans-oceanic bird flight. *Proc. R. Soc. London, Ser. B* 267: 191–195.

Bennett, A. F. 1991. The evolution of activity capacity. *J. Exp. Biol.* 160: 1–23.

Bishop, C. M. 1999. The maximum oxygen consumption and aerobic scope of birds and mammals: getting to the heart of the matter. *Proc. R. Soc. London, Ser. B* 266: 2275–2281.

Booth, F. W., and P. D. Neufer. 2005. Exercise controls gene expression. *Am. Sci.* 93: 28–35.

Brooke, M. de L. 2004. The food consumption of the world's seabirds. *Proc. R. Soc. London, Ser. B* (Suppl.) 271: S246–S248.

Brooks, G. A., T. D. Fahey, and K. M. Baldwin. 2005. *Exercise Physiology: Human Bioenergetics and Its Applications*, 4th ed. McGraw-Hill, Boston.

Bundle, M. W., K. S. Hansen, and K. P. Dial. 2007. Does the metabolic rate–flight speed relationship vary among geometrically similar birds of different mass? *J. Exp. Biol.* 210: 1075–1083.

Careau, V., D. Thomas, M. M. Humphries, and D. Réale. 2008. Energy metabolism and animal personality. *Oikos* 117: 641–653.

Carey, C. (ed.). 1996. *Avian Energetics and Nutritional Ecology.* Chapman & Hall, London.

Chai, P., and R. Dudley. 1995. Limits to vertebrate locomotor energetics suggested by hummingbirds hovering in heliox. *Nature* 377: 722–725.

Chassin, P. S., C. R. Taylor, N. C. Heglund, and H. J. Seeherman. 1976. Locomotion in lions: Energetic cost and maximum aerobic capacity. *Physiol. Zool.* 49: 1–10.

Davis, R. W., T. M. Williams, and G. L. Kooyman. 1985. Swimming metabolism of yearling and adult harbor seals *Phoca vitulina. Physiol. Zool.* 58: 590–596.

Dial, K. P., A. A. Biewener, B. W. Tobalske, and D. R. Warrick. 1997. Mechanical power output of bird flight. *Nature* 390: 67–70.

Dickinson, M. H., F.-O. Lehmann, and S. P. Sane. 1999. Wing rotation and the aerodynamic basis of insect flight. *Science* 284: 1954–1960.

Dohm, M. R., J. P. Hayes, and T. Garland. 1996. Quantitative genetics of sprint running speed and swimming endurance in laboratory house mice (*Mus domesticus*). *Evolution* 50: 1688–1701.

Egevang, C., I. J. Stenhouse, R. A. Phillips, A. Petersen, J. W. Fox, and J. R. D. Silk. 2010. Tracking of Arctic terns *Sterna paradisaea* reveals longest animal migration. *Proc. Natl. Acad. Sci. U.S.A.* 107: 2078–2081.

Ellington, C. P. 1991. Limitations on animal flight performance. *J. Exp. Biol.* 160: 71–91.

Engel, S., M. S. Bowlin, and A. Hedenström. 2010. The role of wind-tunnel studies in integrative research on migration biology. *Integr. Comp. Biol.* 50: 323–335.

Fedak, M. A., and H. J. Seeherman. 1979. Reappraisal of energetics of locomotion shows identical cost in bipeds and quadrupeds including ostrich and horse. *Nature* 282: 713–716.

Full, R. J. 1991. The concepts of efficiency and economy in land locomotion. In R. W. Blake (ed.), *Efficiency and Economy in Animal Physiology*, pp. 97–131. Cambridge University Press, Cambridge.

Full, R. J. 1997. Invertebrate locomotor systems. In W. H. Dantzler (ed.), *Comparative Physiology*, vol. 2 (Handbook of Physiology [Bethesda, MD], section 13), pp. 853–930. Oxford University Press, New York.

Gans, C., A. S. Gaunt, and P. W. Webb. 1997. Vertebrate locomotion. In W. H. Dantzler (ed.), *Comparative Physiology*, vol. 1 (Handbook of Physiology [Bethesda, MD], section 13), pp. 55–213. Oxford University Press, New York.

Giampietro, M., and D. Pimentel. 1991. Energy efficiency: Assessing the interaction between humans and their environment. *Ecol. Econom.* 4: 117–144.

Guglielmo, C. G. 2010. Move that fatty acid: fuel selection and transport in migratory birds and bats. *Integr. Comp. Biol.* 50: 336–345.

Harrison, J. F., and S. P. Roberts. 2000. Flight respiration and energetics. *Annu. Rev. Physiol.* 62: 179–205.

Hays, G. C., and R. Scott. 2013. Global patterns for upper ceilings on migration distance in sea turtles and comparisons with fish, birds and mammals. *Func. Ecol.* 27: 748–756.

Hopker, J., L. Passfield, D. Coleman, S. Jobson, L. Edwards, and H. Carter. 2009. The effects of training on gross efficiency in cycling: a review. *Int. J. Sports Med.* 30: 845–850.

Jones, J. H., and S. L. Lindstedt. 1993. Limits to maximal performance. *Annu. Rev. Physiol.* 55: 547–569.

Killen, S. S., I. Costa, J. A. Brown, and A. K. Gamperl. 2007. Little left in the tank: metabolic scaling in marine teleosts and its implications for aerobic scope. *Proc. R. Soc. London, Ser. B* 274: 431–438.

Koch, L. G., and S. L. Britton. 2005. Divergent selection for aerobic capacity in rats as a model for complex disease. *Integr. Comp. Biol.* 45: 405–415.

Komi, P. V. (ed.). 1992. *Strength and Power in Sport*. Blackwell, Oxford, UK.

Kvist, A., A. Lindström, M. Green, T. Piersma, and G. H. Visser. 2001. Carrying large fuel loads during sustained bird flight is cheaper than expected. *Nature* 413: 730–732.

Le Galliard, J.-F., J. Clobert, and R. Ferrière. 2004. Physical performance and darwinian fitness in lizards. *Nature* 432: 502–505.

McArdle, W. D., F. I. Katch, and V. L. Katch. 2007. *Exercise Physiology: Energy, Nutrition, and Human Performance*, 7th ed. Lippincott Williams & Wilkins, Philadelphia, PA.

Moreno, J., A. Carlson, and R. V. Alatalo. 1988. Winter energetics of coniferous forest tits Paridae in the north: The implications of body size. *Funct. Ecol.* 2: 163–170.

Mullen, R. K., and R. M. Chew. 1973. Estimating the energy metabolism of free-living *Perognathus formosus*. *Ecology* 54: 633–637.

Müller, U. K., E. J. Stamhuis, and J. J. Videler. 2000. Hydrodynamics of unsteady fish swimming and the effects of body size: Comparing the flow fields of fish larvae and adults. *J. Exp. Biol.* 203: 193–206.

Norberg, U. M. 1990. *Vertebrate Flight*. Springer, New York.

O'Dor, R. K., and D. M. Webber. 1991. Invertebrate athletes: Trade-offs between transport efficiency and power density in cephalopod evolution. *J. Exp. Biol.* 160: 93–112.

Parker, K. L., C. T. Robbins, and T. A. Hanley. 1984. Energy expenditures for locomotion by mule deer and elk. *J. Wildl. Manage.* 48: 474–488.

Peterson, C. C., K. A. Nagy, and J. Diamond. 1990. Sustained metabolic scope. *Proc. Natl. Acad. Sci. U.S.A.* 87: 2324–2328.

Piersma, T., and A. Lindström. 1997. Rapid reversible changes in organ size as a component of adaptive behaviour. *Trends Ecol. Evol.* 12: 134–138.

Pontzer, H. 2007. Effective limb length and the scaling of locomotor cost in terrestrial animals. *J. Exp. Biol.* 210: 1752–1761.

Rothe, H.-J., and W. Nachtigall. 1987a. Pigeon flight in a wind tunnel. I. Aspects of wind tunnel design, training methods and flight behaviour of different pigeon races. *J. Comp. Physiol., B* 157: 91–98.

Rothe, H.-J., and W. Nachtigall. 1987b. Pigeon flight in a wind tunnel. II. Gas exchange and power requirements. *J. Comp. Physiol., B* 157: 99–109.

Rowell, L. B., and J. T. Shepherd. 1996. *Exercise: Regulation and Integration of Multiple Systems*. (Handbook of Physiology [Bethesda, MD], section 12.) Oxford University Press, New York.

Schreiber, E. A., and J. Burger (eds.). 2002. *Biology of Marine Birds*. CRC Press, Boca Raton, FL.

Secor, S. M., and J. Diamond. 1997. Determinants of the postfeeding metabolic response in Burmese pythons, *Python molurus*. *Physiol. Zool.* 70: 202–212.

Shephard, R. J., and P.-O. Åstrand (eds.). 2000. *Endurance in Sport*, 2nd ed. Blackwell, Oxford, UK.

Speakman, J. R. 1997. *Doubly Labelled Water: Theory and Practice*. Chapman and Hall, London.

Suarez, R. K. 1996. Upper limits to mass-specific metabolic rates. *Annu. Rev. Physiol.* 58: 583–605.

Taylor, C. R. 1985. Force development during sustained locomotion: A determinant of gait, speed and metabolic power. *J. Exp. Biol.* 115: 253–262.

Taylor, C. R. 1987. Structural and functional limits to oxidative metabolism: Insights from scaling. *Annu. Rev. Physiol.* 49: 135–146.

Taylor, C. R., S. L. Caldwell, and V. J. Rowntree. 1972. Running up and down hills: Some consequences of size. *Science* 178: 1096–1097.

Taylor, C. R., G. M. O. Maloiy, E. R. Weibel, V. A. Langman, J. M. Z. Kamau, H. J. Seeherman, and N. C. Heglund. 1980. Design of the mammalian respiratory system. III. Scaling maximal aerobic capacity to body mass: Wild and domestic mammals. *Respir. Physiol.* 44: 25–37.

Taylor, C. R., A. Shkolnik, R. Dmi'el, D. Baharav, and A. Borut. 1974. Running in cheetahs, gazelles, and goats: Energy cost and limb configuration. *Am. J. Physiol.* 227: 848–850.

Townsend, C. R., and P. Calow (eds.). 1981. *Physiological Ecology: An Evolutionary Approach to Resource Use*. Sinauer, Sunderland, MA.

Tucker, V. A. 1975. The energetic cost of moving about. *Am. Sci.* 63: 413–419.

Tucker, V. A. 2000. The deep fovea, sideways vision and spiral flight paths in raptors. *J. Exp. Biol.* 203: 3745–3754.

van Ginneken, V., E. Antonissen, U. K. Müller, R. Booms, E. Eding, J. Verreth, and G. van den Thillart. 2005. Eel migration to the Sargasso: remarkably high swimming efficiency and low energy costs. *J. Exp. Biol.* 208: 1329–1335.

Videler, J. J. 1993. *Fish Swimming*. Chapman & Hall, London.

Videler, J. J. 2005. *Avian Flight*. Oxford University Press, New York.

Wagner, P. D. 1996. Determinants of maximal oxygen transport and utilization. *Annu. Rev. Physiol.* 58: 21–50.

Wainwright, P. C., and S. M. Reilly (eds.). 1994. *Ecological Morphology: Integrative Organismal Biology*. University of Chicago Press, Chicago.

Weibel, E. R., C. R. Taylor, and L. Bolis. 1998. *Principles of Animal Design: The Optimization and Symmorphosis Debate*. Cambridge University Press, New York.

Weimerskirch, H., M. Le Corre, and C. A. Bost. 2008. Foraging strategy of masked boobies from the largest colony in the world: relationship to environmental conditions and fisheries. *Mar. Ecol. Progr. Ser.* 362: 291–302.

Westerterp, K. R. 2001. Limits to sustainable human metabolic rate. *J. Exp. Biol.* 204: 3183–3187.

Wickler, S. J., D. F. Hoyt, E. A. Cogger, and M. H. Hirschbein. 2000. Preferred speed and cost of transport: The effect of incline. *J. Exp. Biol.* 203: 2195–2200.

Wisløff, U., S. M. Najjar, O. Ellingsen, P. M. Haram, S. Swoap, Q. Al-Share, M. Fernström, K. Rezaei, S. J. Lee, L. G. Koch, and S. L. Britton. 2005. Cardiovascular risk factors emerge after artificial selection for low aerobic capacity. *Science* 307: 418–420.

Woakes, A. J., and W. A. Foster (eds.). 1991. *The Comparative Physiology of Exercise. Journal of Experimental Biology*, vol. 160. Company of Biologists, Cambridge.

Wolf, T. J., P. Schmid-Hempel, C. P. Ellington, and R. D. Stevenson. 1989. Physiological correlates of foraging efforts in honey-bees: Oxygen consumption and nectar load. *Funct. Ecol.* 3: 417–424.

Young, J., S. M. Walker, R. J. Bomphrey, G. K. Taylor, and A. L. R. Thomas. 2009. Details of insect wing design and deformation enhance aerodynamic function and flight efficiency. *Science* 325: 1549–1552.

Yousef, M. K., M. E. D. Webster, and O. M. Yousef. 1989. Energy costs of walking in camels, *Camelus dromedarius*. *Physiol. Zool.* 62: 1080–1088.

Chapter 10

Alexander, R. M. (ed.). 1992. *Mechanics of Animal Locomotion*. Springer, New York.

Altringham, J. D., and B. A. Block. 1997. Why do tuna maintain elevated slow muscle temperatures? Power output of muscle isolated from endothermic and ectothermic fish. *J. Exp. Biol.* 200: 2617–2627.

Andrews, M. T. 2007. Advances in molecular biology of hibernation in mammals. *BioEssays* 29: 431–440.

Angilletta, M. J., Jr. 2009. *Thermal Adaptation: A Theoretical and Empirical Synthesis*. Oxford University Press, Oxford, UK.

Angilletta, M. J., Jr., A. F. Bennett, H. Guderley, C. A. Navas, F. Seebacher, and R. S. Wilson. 2006. Coadaptation: a unifying principle in evolutionary thermal biology. *Physiol. Biochem. Zool.* 79: 282–294.

Angilletta, M. J., Jr., P. H. Niewiarowski, and C. A. Navas. 2002. The evolution of thermal physiology in ectotherms. *J. Therm. Biol.* 27: 249–268.

Arnold, W., T. Ruf, F. Frey-Roos, and U. Bruns. 2011. Diet-independent remodeling of cellular membranes precedes seasonally changing body temperature in a hibernator. *PLoS One* 6: e18641.

Aschoff, J. 1981. Thermal conductance in mammals and birds: Its dependence on body size and circadian phase. *Comp. Biochem. Physiol., A* 69: 611–619.

Bakken, G. S., W. R. Santee, and D. J. Erskine. 1985. Operative and standard operative temperature: Tools for thermal energetics studies. *Am. Zool.* 25: 933–943.

Bale, J. S. 2002. Insects and low temperatures: from molecular biology to distributions and abundance. *Philos. Trans. R. Soc. London, Ser. B* 357: 849–861.

Beers, J. M., and N. Jayasundara. 2015. Antarctic notothenioid fish: what are the future consequences of 'losses' and 'gains' acquired during long-term evolution at cold and stable temperatures? *J. Exp. Biol.* 218: 1834–1845.

Bernal, D., K. A. Dickson, R. E. Shadwick, and J. B. Graham. 2001. Review: Analysis of the evolutionary convergence for high performance swimming in lamnid sharks and tunas. *Comp. Biochem. Physiol., A* 129: 695–726.

Blix, A. S. 2005. *Arctic Animals and Their Adaptations to Life on the Edge*. Tapir Academic Press, Trondheim, Norway.

Blix, A. S., L. Walløe, E. B. Messelt, and L. P. Falkow. 2010. Selective brain cooling and its vascular basis in diving seals. *J. Exp. Biol.* 213: 2610–2616.

Block, B. A., and J. R. Finnerty. 1994. Endothermy in fishes: A phylogenetic analysis of constraints, predispositions, and selection pressures. *Environ. Biol. Fishes* 40: 283–302.

Boyles, J. G., F. Seebacher, B. Smit, and A. E. McKechnie. 2011. Adaptive thermoregulation in endotherms may alter responses to climate change. *Integr. Comp. Biol.* 51: 676–690.

Brück, K. 1986. Basic mechanisms in thermal long-term and short-term adaptation. *J. Therm. Biol.* 11: 73–77.

Bruinzeel, L. W., and T. Piersma. 1998. Cost reduction in the cold: heat generated by terrestrial locomotion partly substitutes for thermoregulation costs in Knot *Calidris canutus*. *Ibis* 140: 323–328.

Burggren, W. W., and B. R. McMahon. 1981. Oxygen uptake during environmental temperature change in hermit crabs: Adaptation to subtidal, intertidal, and supratidal habitats. *Physiol. Zool.* 54: 325–333.

Cabanac, M. 2006. Adjustable set point: to honor Harold T. Hammel. *J. Appl. Physiol.* 100: 1338–1346.

Campbell, G. S., and J. M. Norman. 1998. *An Introduction to Environmental Biophysics*, 2nd ed. Springer, New York.

Cannon, B., and J. Nedergaard. 2012. Neither Brown nor white. *Nature* 488: 286–287.

Carey, F. G. 1982. A brain heater in the swordfish. *Science* 216: 1327–1329.

Carey, H. V., M. T. Andrews, and S. L. Martin. 2003. Mammalian hibernation: Cellular and molecular responses to depressed metabolism and low temperature. *Physiol. Rev.* 83: 1153–1181.

Chappell, M. A. 1980. Thermal energetics and thermoregulatory costs of small arctic mammals. *J. Mamm.* 61: 278–291.

Chen, L., A. L. DeVries, and C.-H. C. Cheng. 1997. Convergent evolution of antifreeze glycoproteins in Antarctic notothenoid fish and Arctic cod. *Proc. Natl. Acad. Sci. U.S.A.* 94: 3817–3822. [See also 94: 3811–3816.]

Cheng, C.-H. C., and H. W. Detrich III. 2007. Molecular ecophysiology of Antarctic notothenioid fishes. *Philos. Trans. R. Soc. London, Ser. B* 362: 2215–2232.

Christian, K. A., and B. W. Weavers. 1996. Thermoregulation of monitor lizards in Australia: An evaluation of methods in thermal biology. *Ecol. Monogr.* 66: 139–157.

Costanzo, J. P., and R. E. Lee. 2005. Cryoprotection by urea in a terrestrially hibernating frog. *J. Exp. Biol.* 208: 4079–4089.

Costanzo, J. P., R. E. Lee, Jr., and M. F. Wright. 1991. Glucose loading prevents freezing injury in rapidly cooled wood frogs. *Am. J. Physiol.* 261: R1549–R1553.

Costanzo, J. P., J. D. Litzgus, J. B. Iverson, and R. E. Lee, Jr. 2000. Seasonal changes in physiology and development of cold hardiness in the hatchling painted turtle *Chrysemys picta*. *J. Exp. Biol.* 203: 3459–3470.

Cziko, P. A., C. W. Evans, C.-H. C. Cheng, and A. L. Devries. 2006. Freezing resistance of antifreeze-deficient larval Antarctic fish. *J. Exp. Biol.* 209: 407–420.

Danks, H. V. 2004. Seasonal adaptations in Arctic insects. *Integr. Comp. Biol.* 44: 85–94.

Denlinger, D. L., and R. E. Lee. 2010. *Low Temperature Biology of Insects*. Cambridge University Press, Cambridge, UK.

Derocher, A. E., N. J. Lunn, and I. Stirling. 2004. Polar bears in a warming climate. *Integr. Comp. Biol.* 44: 163–176.

Duman, J. G. 2001. Antifreeze and ice nucleator proteins in terrestrial arthropods. *Annu. Rev. Physiol.* 63: 327–357.

Duman, J. G. 2015. Animal ice-binding (antifreeze) proteins and glycolipids: an overview with emphasis on physiological function. *J. Exp. Biol.* 218: 1846–1855.

Farmer, C. G. 2000. Parental care: The key to understanding endothermy and other convergent features in birds and mammals. *Am. Nat.* 155: 326–334.

Feder, M. E., and G. E. Hofmann. 1999. Heat-shock proteins, molecular chaperones, and the stress response: Evolutionary and ecological physiology. *Annu. Rev. Physiol.* 61: 243–282.

Fletcher, G. L., C. L. Hew, and P. L. Davies. 2001. Antifreeze proteins of teleost fishes. *Annu. Rev. Physiol.* 63: 359–390.

Florant, G. L. 1998. Lipid metabolism in hibernators: The importance of essential fatty acids. *Am. Zool.* 38: 331–340.

Fregly, M. J., and C. M. Blatteis. 1996. *Environmental Physiology*, vol. 1. (Handbook of Physiology [Bethesda, MD], section 4.) Oxford University Press, New York. [Many articles on thermal relations by recognized authorities.]

Fudge, D. S., E. D. Stevens, and J. S. Ballantyne. 1997. Enzyme adaptation along a heterothermic tissue: The visceral retia mirabilia of the bluefin tuna. *Am. J. Physiol.* 272: R1834–R1840.

Garland, T., Jr., and S. C. Adolph. 1991. Physiological differentiation of vertebrate populations. *Annu. Rev. Ecol. Syst.* 22: 193–228.

Gates, D. M. 1980. *Biophysical Ecology*. Springer, New York.

Geiser, F., and T. Ruf. 1995. Hibernation versus daily torpor in mammals and birds: Physiological variables and classification of torpor patterns. *Physiol. Zool.* 68: 935–966.

Geiser, F., W. Westman, and B. M. McAllan. 2006. Development of thermoregulation and torpor in a marsupial: energetic and evolutionary implications. *J. Comp. Physiol., B* 176: 107–116.

Goldman, K. J. 1997. Regulation of body temperature in the white shark, *Carcharodon carcharias*. *J. Comp. Physiol., B* 167: 423–429.

Graether, S. P., M. J. Kuiper, S. M. Gagné, V. K. Walker, Z. Jia, B. D. Sykes, and P. L. Davies. 2000. β-Helix structure and ice-binding properties of a hyperactive antifreeze protein from insects. *Nature* 406: 325–328.

Graves, J. E., and G. N. Somero. 1982. Electrophoretic and functional enzymic evolution in four species of eastern Pacific barracudas from different thermal environments. *Evolution* 36: 97–106.

Harrison, J. F., and J. H. Fewell. 2002. Environmental and genetic influences on flight metabolic rate in the honey bee, *Apis mellifera*. *Comp. Biochem. Physiol., A* 133: 323–333. [See Woods et al. 2005 for a sort of point–counterpoint.]

Heinrich, B., and H. Esch. 1994. Thermoregulation in bees. *Am. Sci.* 82: 164–170.

Hill, R. W., and D. L. Beaver. 1982. Inertial thermostability and thermoregulation in broods of redwing blackbirds. *Physiol. Zool.* 55: 250–266.

Hofmann, G. E. 2005. Patterns of Hsp gene expression in ectothermic marine organisms on small to large biogeographic scales. *Integr. Comp. Biol.* 45: 247–255.

Holland, L. Z., M. McFall-Ngai, and G. N. Somero. 1997. Evolution of lactate dehydrogenase-A homologs of barracuda fishes (genus *Sphyraena*) from different thermal environments: Differences in kinetic properties and thermal stability are due to amino acid substitutions outside the active site. *Biochemistry* 36: 3207–3215.

Holmstrup, M., M. Bayley, and H. Ramløv. 2002. Supercool or dehydrate? An experimental analysis of overwintering strategies in small permeable arctic invertebrates. *Proc. Natl. Acad. Sci. U.S.A.* 99: 5716–5720.

Huey, R. B., and J. G. Kingsolver. 1993. Evolution of resistance to high temperature in ectotherms. *Am. Nat.* 142: S21–S46.

Humphries, M. M., D. W. Thomas, and J. R. Speakman. 2002. Climate-mediated energetic constraints on the distribution of hibernating mammals. *Nature* 418: 313–316.

Irving, L. 1972. *Arctic Life of Birds and Mammals.* Springer, New York.

Jastroch, M., J. A. Buckingham, M. Helwig, M. Klingenspor, and M. D. Brand. 2007. Functional characterization of UCP1 in the common carp: uncoupling activity in the liver mitochondria and cold-induced expression in the brain. *J. Comp. Physiol., B* 177: 743–752.

Jessen, C. 2001. *Temperature Regulation in Humans and Other Mammals.* Springer, New York.

Johnston, I. A., and A. F. Bennett (eds.). 1996. *Animals and Temperature: Phenotypic and Evolutionary Adaptation.* Cambridge University Press, New York.

Johnston, I. A., and P. Harrison. 1985. Contractile and metabolic characteristics of muscle fibers from Antarctic fish. *J. Exp. Biol.* 116: 223–236.

Jones, J. C., M. R. Myerscough, S. Graham, and B. P. Oldroyd. 2004. Honey bee nest thermoregulation: diversity promotes stability. *Science* 305: 402–404.

Karasov, W. H., and J. M. Diamond. 1985. Digestive adaptations for fueling the cost of endothermy. *Science* 228: 202–204.

Kingsolver, J. G., H. A. Woods, L. B. Buckley, K. A. Potter, H. J. MacLean, and J. K. Higgins. 2011. Complex life cycles and the responses of insects to climate change. *Integr. Comp. Biol.* 51: 719–732.

Kluger, M. J. 1979. *Fever: Its Biology, Evolution, and Function.* Princeton University Press, Princeton, NJ.

Kluger, M. J. 1991. Fever: Role of pyrogens and cryogens. *Physiol. Rev.* 71: 93–127.

Kvadsheim, P. H., L. P. Folkow, and A. S. Blix. 2005. Inhibition of shivering in hypothermic seals during diving. *Am. J. Physiol.* 289: R326–R331.

Lannig, G., C. Bock, F. J. Sartoris, and H. O. Pörtner. 2004. Oxygen limitation of thermal tolerance in cod, *Gadus morhua* L., studied by magnetic resonance imaging and on-line venous oxygen monitoring. *Am. J. Physiol.* 287: R902–R910.

Layne, J. R., Jr., and M. G. Stapleton. 2009. Annual variation in glycerol mobilization and effect of freeze rigor on post-thaw locomotion in the freeze-tolerant frog *Hyla versicolor. J. Comp. Physiol., B* 179: 215–221.

Lee, R. E., Jr., and J. P. Costanzo. 1998. Biological ice nucleation and ice distribution in cold-hardy ectothermic animals. *Annu. Rev. Physiol.* 60: 55–72.

Linton, T. K., I. J. Morgan, P. J. Walsh, and C. M. Wood. 1998. Chronic exposure of rainbow trout (*Oncorhynchus mykiss*) to simulated climate warming and sublethal ammonia: A year-long study of their appetite, growth, and metabolism. *Can. J. Fish. Aquat. Sci.* 55: 576–586.

Logue, J. A., A. L. DeVries, E. Fodor, and A. R. Cossins. 2000. Lipid compositional correlates of temperature-adaptive interspecific differences in membrane physical structure. *J. Exp. Biol.* 203: 2105–2115.

Loomis, S. H. 1995. Freezing tolerance of marine invertebrates. *Oceanogr. Mar. Biol. Annu. Rev.* 33: 337–350.

Lovegrove, B. G. 2003. The influence of climate on the basal metabolic rate of small mammals: a slow-fast metabolic continuum. *J. Comp. Physiol. B* 173: 87–112.

Lynch, C. B. 1992. Clinal variation in cold adaptation in *Mus domesticus*: Verification of predictions from laboratory populations. *Am. Nat.* 139: 1219–1236.

Manis, M. L., and D. L. Claussen. 1986. Environmental and genetic influences on the thermal physiology of *Rana sylvatica. J. Therm. Biol.* 11: 31–36.

McKechnie, A. E., and B. G. Lovegrove. 2002. Avian facultative hypothermic responses: A review. *Condor* 104: 705–724.

McKechnie, A. E., and N. Mzilikazi. 2011. Heterothermy in Afrotropical mammals and birds: A review. *Integr. Comp. Biol.* 51: 349–363.

McNab, B. K. 2002. *The Physiological Ecology of Vertebrates: A View from Energetics.* Cornell University Press, Ithaca, NY.

Mitchell, G., and A. Lust. 2008. The carotid rete and artiodactyl success. *Biol. Lett.* 4: 415–418.

Moerland, T. S. 1995. Temperature: Enzyme and organelle. In P. W. Hochachka and T. P. Mommsen (eds.), *Environmental and Ecological Biochemistry,* pp. 57–71. Elsevier, New York.

Morgan, I. J., D. G. McDonald, and C. M. Wood. 2001. The cost of living for freshwater fish in a warmer, more polluted world. *Glob. Change Biol.* 7: 345–355.

Overgaard, J., J. G. Sørensen, S. O. Petersen, V. Loeschke, and M. Holmstrup. 2005. Changes in membrane lipid composition following rapid cold hardening in *Drosophila melanogaster. J. Insect Physiol.* 51: 1173–1182.

Parmesan, C. 2006. Ecological and evolutionary responses to recent climate change. *Annu. Rev. Ecol. Syst.* 37: 637–669.

Pörtner, H. O., and R. Knust. 2007. Climate change affects marine fishes through oxygen limitation of thermal tolerance. *Science* 315: 95–97.

Pörtner, H. O., and R. C. Playle (eds.). 1998. *Cold Ocean Physiology.* Cambridge University Press, New York.

Pough, F. H., and R. M. Andrews. 1984. Individual and sibling-group variation in metabolism of lizards: The aerobic capacity model for the origin of endothermy. *Comp. Biochem. Physiol., A* 79: 415–419.

Powers, D. A., M. Smith, I. Gonzalez-Villasenor, L. DiMichele, D. Crawford, G. Bernardi, and T. Lauerman. 1993. A multidisciplinary approach to the selectionist/neutralist controversy using the model teleost, *Fundulus heteroclitus. Oxf. Surv. Evol. Biol.* 9: 43–107.

Prinzinger, R., A. Pressmar, and E. Schleucher. 1991. Body temperature in birds. *Comp. Biochem. Physiol., A* 99: 499–506.

Rahmstorf, S., A. Cazenave, J. A. Church, J. E. Hansen, R. F. Keeling, D. E. Parker, and R. C. J. Somerville. 2007. Recent climate observations compared to projections. *Science* 316: 709.

Regehr, E. V., N. J. Lunn, S. C. Amstrup, and I. Stirling. 2007. Effects of earlier sea ice breakup on survival and population size of polar bears in western Hudson Bay. *J. Wildl. Manage.* 71: 2673–2683.

Roberts, S. P., and J. F. Harrison. 1998. Mechanisms of thermoregulation in flying bees. *Am. Zool.* 38: 492–502.

Robertshaw, D. 2006. Mechanisms for the control of respiratory evaporative heat loss in panting animals. *J. Appl. Physiol.* 101: 664–668.

Rolfe, D. F. S., and G. C. Brown. 1997. Cellular energy utilization and molecular origin of standard metabolic rate in mammals. *Physiol. Rev.* 77: 731–758.

Root, T. 1988. Energy constraints on avian distributions and abundances. *Ecology* 69: 330–339.

Root, T. L., J. T. Price, K. R. Hall, S. H. Schneider, C. Rosensweig, and J. A. Pounds. 2003. Fingerprints of global warming on wild animals and plants. *Nature* 421: 57–60.

Rubinsky, B., S. T. S. Wong, J.-S. Hong, J. Gilbert, M. Roos, and K. B. Storey. 1994. 1H magnetic resonance imaging of freezing and thawing in freeze-tolerant frogs. *Am. J. Physiol.* 266: R1771–R1777.

Schulte, P. M., M. Gomez-Chiarri, and D. A. Powers. 1997. Structural and functional differences in the promoter and 5′ flanking region of *Ldh-B* within and between populations of the teleost *Fundulus heteroclitus. Genetics* 145: 759–769.

Seebacher, F., M. D. Brand, P. L. Else, H. Guderley, A. J. Hulbert, and C. D. Moyes. 2010. Plasticity of oxidative metabolism in variable climates: molecular mechanisms. *Physiol. Biochem. Zool.* 83: 721–732.

Seymour, R. S. 1997. Plants that warm themselves. *Sci. Am.* 276(3): 104–109.

Seymour, R. S. 2010. Scaling of heat production by thermogenic flowers: limits to floral size and maximum rate of respiration. *Plant Cell Environ.* 33: 1474–1485.

Somero, G. N. 2002. Thermal physiology and vertical zonation of intertidal animals: Optima, limits, and costs of living. *Integr. Comp. Biol.* 42: 780–789.

Somero, G. N. 2004. Adaptation of enzymes to temperature: searching for basic "strategies." *Comp. Biochem. Physiol., B* 139: 321–333.

Southwick, E. E., and G. Heldmaier. 1987. Temperature control in honey bee colonies. *BioScience* 37: 395–399.

Speakman, J. R., and E. Król. 2010. Maximal heat dissipation capacity and hyperthermia risk: neglected key factors in the ecology of endotherms. *J. Anim. Ecol.* 79: 726–746.

Steudel, K., W. P. Porter, and D. Sher. 1994. The biophysics of Bergmann's rule: A comparison of the effects of pelage and body size variation on metabolic rate. *Can. J. Zool.* 72: 70–77.

Stirling, I., and C. L. Parkinson. 2006. Possible effects of climate warming on selected populations of polar bears (*Ursus maritimus*) in the Canadian Arctic. *Arctic* 59: 261–275.

Storey, K. B. (ed.). 2004. *Functional Metabolism. Regulation and Adaptation.* Wiley-Liss, Hoboken, NJ.

Storey, K. B., and J. M. Storey. 1996. Natural freezing survival in animals. *Annu. Rev. Ecol. Syst.* 27: 365–386.

Strathdee, A. T., and J. S. Bale. 1998. Life on the edge: insect ecology in Arctic environments. *Annu. Rev. Entomol.* 43: 85–106.

Tanaka, H., Y. Takagi, and Y. Naito. 2000. Behavioural thermoregulation of chum salmon during homing migration in coastal waters. *J. Exp. Biol.* 203: 1825–1833.

Tewsbury, J. J., R. B. Huey, and C. A. Deutsch. 2008. Putting heat on tropical animals. *Science* 320: 1296–1297.

Tieleman, B. I., and J. B. Williams. 1999. The role of hyperthermia in the water economy of desert birds. *Physiol. Biochem. Zool.* 72: 87–100.

Tomanek, L., and G. N. Somero. 1999. Evolutionary and acclimation-induced variation in the heat-shock responses of congeneric marine snails (genus *Tegula*) from different thermal habitats: Implications for limits of thermotolerance and biogeography. *J. Exp. Biol.* 202: 2925–2936.

Ultsch, G. R. 1989. Ecology and physiology of hibernation and overwintering in freshwater fishes, turtles and snakes. *Biol. Rev. Cambridge* 64: 435–516.

Van Breukelen, F., and S. L. Martin. 2002. Molecular biology of thermoregulation—Invited review: Molecular adaptations in mammalian hibernators:

unique adaptations or generalized responses? *J. Appl. Physiol.* 92: 2640–2647.

Vogt, F. D. 1986. Thermoregulation in bumblebee colonies. I. Thermoregulatory versus brood-maintenance behaviors during acute changes in ambient temperature. *Physiol. Zool.* 59: 55–59.

Walsberg, G. E. 1983. Coat color and solar heat gain in animals. *BioScience* 33: 88–91.

Watabe, S. 2002. Temperature plasticity of contractile proteins in fish muscle. *J. Exp. Biol.* 205: 2231–2236.

Weathers, W. W., and K. A. Sullivan. 1993. Seasonal patterns of time and energy allocation by birds. *Physiol. Zool.* 66: 511–536.

Wegner, N. C., O. E. Snodgrass, H. Dewar, and J. R. Hyde. 2015. Whole-body endothermy in a mesopelagic fish, the opah, *Lampris guttatus*. *Science* 348: 786–788.

Wiersma, P., A. Muñoz-Garcia, A. Walker, and J. B. Williams. 2007. Tropical birds have a slow pace of life. *Proc. Natl. Acad. Sci. U.S.A.* 104: 9340–9345.

Williams, D. R., L. E. Epperson, W. Li, M. A. Hughes, R. Taylor, J. Rogers, S. L. Martin, A. R. Cossins, and A. Y. Gracey. 2005. Seasonally hibernating phenotype assessed through transcript screening. *Physiol. Genomics* 24: 13–22.

Williams, J. B., and B. I. Tielman. 2002. Ecological and evolutionary physiology of desert birds: A progress report. *Integr. Comp. Biol.* 42: 68–75.

Wolf, B. O., and G. E. Walsberg. 1996. Thermal effects of radiation and wind on a small bird and implications for microsite selection. *Ecology* 77: 2228–2236.

Woods, W. A., Jr., B. Heinrich, and R. D. Stevenson. 2005. Honeybee flight metabolic rate: does it depend upon air temperature? *J. Exp. Biol.* 208: 1161–1173. [See Harrison and Fewell 2002 for a sort of point–counterpoint discussion.]

Zachariassen, K. E., and E. Kristiansen. 2000. Ice nucleation and antinucleation in nature. *Cryobiology* 41: 257–279.

Chapter 11

Arnold, W., T. Ruf, S. Reimoser, F. Tataruch, K. Onderscheka, and F. Schober. 2004. Nocturnal hypometabolism as an overwintering strategy in red deer (*Cervus elaphus*). *Am. J. Physiol.* 286: R174–R181.

Arnold, W., T. Ruf, F. Frey-Roos, and U. Bruns. 2011. Diet-independent remodeling of cellular membranes precedes seasonally changing body temperature in a hibernator. *PLoS ONE* 6(4): e18641.

Barboza, P. S., and K. L. Parker. 2008. Allocating protein to reproduction in Arctic reindeer and caribou. *Physiol. Biochem. Zool.* 81: 835–855.

Barnes, B. M., and H. V. Carey (eds.). 2004. *Life in the Cold. Evolution, Mechanisms, Adaptation, and Application.* Biological Papers of the University of Alaska, Institute of Arctic Biology, Fairbanks, AK.

Berg, F., U. Gustafson, and L. Andersson. 2006. The uncoupling protein 1 gene (*UCP1*) is disrupted in the pig lineage: A genetic explanation for poor thermoregulation in piglets. *PLoS Genetics* 2: 1178–1181.

Blix, A. S., H. J. Grav, K. A. Markussen, and R. G. White. 1984. Modes of thermal protection in newborn muskoxen (*Ovibos moschatus*). *Acta Physiol. Scand.* 122: 443–453.

Blix, A. S., L. Walløe, and L. P. Folkow. 2011. Regulation of brain temperature in winter-acclimatized reindeer under heat stress. *J. Exp. Biol.* 214: 3850–3856.

Boyer, B. B., and B. M. Barnes. 1999. Molecular and metabolic aspects of mammalian hibernation. *BioScience* 49: 713–724.

Brück, K. 1970. Nonshivering thermogenesis and brown adipose tissue in relation to age and their integration in the thermoregulatory system. In O. Lindberg (ed.), *Brown Adipose Tissue*, pp. 117–154. American Elsevier, New York. [This older reference is included because it is an English-language summary of a series of definitive studies, carried out by the author, mainly on development of thermoregulation in guinea pigs.]

Dark, J. 2005. Annual lipid cycles in hibernators: integration of physiology and behavior. *Annu. Rev. Physiol.* 25: 469–497.

Frank, C. L., S. Karpovich, and B. M. Barnes. 2008. Dietary fatty acid composition and the hibernation patterns in free-ranging arctic ground squirrels. *Physiol. Biochem. Zool.* 81: 486–495.

Frerichs, K. U., C. B. Smith, M. Brenner, D. J. DeGracia, G. S. Krause, L. Marrone, T. E. Dever, and J. M. Hallenbeck. 1998. Suppression of protein synthesis in brain during hibernation involves inhibition of protein initiation and elongation. *Proc. Natl. Acad. Sci. U.S.A.* 95: 14511–14516.

Gjøstein, H., Ø. Holand, and R. B. Weladji. 2004. Milk production and composition in reindeer (*Rangifer tarandus*): effect of lactational stage. *Comp. Biochem. Physiol., A* 137: 649–656.

Hansen, B. B., R. Aanes, I. Herfindal, J. Kohler, and B. E. Saether. 2011. Climate, icing, and wild arctic reindeer: past relationships and future prospects. *Ecology* 92: 1917–1923.

Heldmaier, G., and M. Klingenspor (eds.). 2000. *Life in the Cold*. Springer, New York.

Heller, H. C., and N. F. Ruby. 2004. Sleep and circadian rhythms in mammalian torpor. *Annu. Rev. Physiol.* 66: 275–289.

Irving, L. 1972. *Arctic Life of Birds and Mammals.* Springer-Verlag, New York. [A classic that, despite its age, still provides insight because of the author's intimate knowledge of Arctic animals.]

Johnsen, H. K., A. Rognmo, K. J. Nilssen, and A. S. Blix. 1985. Seasonal changes in the relative importance of different avenues of heat loss in resting and running reindeer. *Acta Physiol. Scand.* 123: 73–79.

Knott, K. K., P. S. Barboza, and R. T. Bowyer. 2005. Growth in arctic ungulates: postnatal development and organ maturation in *Rangifer tarandus* and *Ovibos moschatus*. *J. Mamm.* 86: 121–130.

Krupnik, I. 2011. 'How many Eskimo words for ice?' Collecting Inuit sea ice terminologies in the International Polar Year 2007–2008. *Canadian Geographer* 55: 56–68.

Markussen, K. A., A. Rognmo, and A. S. Blix. 1985. Some aspects of thermoregulation in newborn reindeer calves (*Rangifer tarandus tarandus*). *Acta Physiol. Scand.* 123: 215–220.

Mathiesen, S. D., C. G. Orpin, Y. Greenwood, and A. S. Blix. 1987. Seasonal changes in the cecal microflora of the high-Arctic Svalbard reindeer (*Rangifer tarandus platyrhynchus*). *Appl. Environ. Microbiol.* 53: 114–118.

Nedergaard, J., E. Connolly, and B. Cannon. 1986. Brown adipose tissue in the mammalian neonate. In P. Trayhurn and D. G. Nicholls (eds.), *Brown Adipose Tissue*, pp. 152–213. Edward Arnold, London.

Nedergaard, J., V. Golozoubova, A. Matthias, I. Shabalina, K. I. Ohba, K. Ohlson, A. Jacobsson, and B. Cannon. 2001. Life without UCP1: Mitochon-

drial, cellular and organismal characteristics of the UCP1-ablated mice. *Biochem. Soc. Trans.* 29: 756–763.

Nilssen, K. J., J. A. Sundsfjord, and A. S. Blix. 1984. Regulation of metabolic rate in Svalbard and Norwegian reindeer. *Am. J. Physiol.* 247: R837–R841.

Orpin, C. G., S. D. Mathiesen, Y. Greenwood, and A. S. Blix. 1985. Seasonal changes in the ruminal microflora of the high-Arctic Svalbard reindeer (*Rangifer tarandus platyrhynchus*). *Appl. Environ. Microbiol.* 50: 144–151.

Post, E., and M. C. Forchhammer. 2008. Climate change reduces reproductive success of an Arctic herbivore through trophic mismatch. *Philos. Trans. R. Soc., B* 363: 2369–2375.

Rothwell, N. J., and M. J. Stock. 1985. Biological distribution and significance of brown adipose tissue. *Comp. Biochem. Physiol., A* 82: 745–751.

Ruf, T., and W. Arnold. 2000. Mechanisms of social thermoregulation in hibernating alpine marmots (*Marmota marmota*). In G. Heldmaier and M. Klingenspor (eds.), *Life in the Cold*, pp. 81–94. Springer, New York.

Ruf, T., and W. Arnold. 2008. Effects of polyunsaturated fatty acids (PUFAs) on hibernation and torpor: A review and hypothesis. *Am. J. Physiol.* 294: R1044–R1052.

Sørmo, W., Ø. E. Haga, E. Gaare, R. Langvatn, and S. D. Mathiesen. 1999. Forage chemistry and fermentation chambers in Svalbard reindeer (*Rangifer tarandus platyrhynchus*). *J. Zool.* 247: 247–256.

Soppela, P., M. Nieminen, S. Saarela, J. S. Keith, J. N. Morrison, F. MacFarlane, and P. Trayhurn. 1991. Brown fat-specific mitochondrial uncoupling protein in adipose tissues of newborn reindeer. *Am. J. Physiol.* 260: R1229–R1234.

Symonds, M. E., A. Mostyn, S. Pearce, H. Budge, and T. Stephenson. 2003. Endocrine and nutritional regulation of fetal adipose tissue development. *J. Endocrinol.* 179: 293–299.

Sundset, M. A., P. S. Barboza, T. K. Green, L. P. Folkow, A. S. Blix, and S. D. Mathiesen. 2010. Microbial degradation of usnic acid in the reindeer rumen. *Naturwissenschaften* 97: 273–278.

Sundset, M. A., K. E. Praesteng, I. K. O. Cann, S. D. Mathiesen, and R. I. Mackie. 2007. Novel rumen bacterial diversity in two geographically separated sub-species of reindeer. *Microb. Ecol.* 54: 424–438.

Syroechkovskii, E. E. 1995. *Wild Reindeer*. Smithsonian Institution Libraries, Washington, D.C.

Tyler, N. J. C., G. Jeffery, C. R. Hogg, and K.-A. Stokkan. 2014. Ultraviolet vision may enhance the ability of reindeer to discriminate plants in snow. *Arctic* 67: 159–166.

Wang, L. C. H., and T. F. Lee. 1996. Torpor and hibernation in mammals: Metabolic, physiological, and biochemical adaptations. In M. J. Fregly and C. M. Blatteis (eds.), *Environmental Physiology*, vol. 1 (Handbook of Physiology [Bethesda, MD], section 4), pp. 507–532. Oxford University Press, New York.

Williams, C. T., A. V. Goropashnaya, C. L. Buck, V. B. Fedorov, F. Kohl, T. N. Lee, and B. M. Barnes. 2011. Hibernating above the permafrost: effects of ambient temperature and season on expression of metabolic genes in liver and brown adipose tissue of arctic ground squirrels. *J. Exper. Biol.* 214: 1300–1306.

Willis, C. K. R., R. M. Brigham, and F. Geiser. 2006. Deep, prolonged torpor by pregnant, free-ranging bats. *Naturwissenschaften* 93: 80–83.

Willis, J. S. 1982. The mystery of periodic arousal. In C. P. Lyman, J. S. Willis, A. Malan, and L. C. H. Wang (eds.), *Hibernation and Torpor in Mammals and Birds*, pp. 92–101. Academic Press, New York.

Zingaretti, M. C., F. Crosta, A. Vitali, M. Guerrieri, A. Frontini, B. Cannon, J. Nedergaard, and S. Cinti. 2009. The presence of UCP1 demonstrates that metabolically active adipose tissue in the neck of adult humans truly represents brown adipose tissue. *FASEB J.* 23: 3113–3120.

Chapter 12

Aidley, D. J. 1998. *The Physiology of Excitable Cells*, 4th ed. Cambridge University Press, New York.

Alle, H., and J. R. P. Geiger. 2008. Analog signaling in mammalian cortical axons. *Curr. Opin. Neurobiol.* 18: 314–320.

Arbas, E. A., R. B. Levine, and N. J. Strausfeld. 1997. Invertebrate nervous systems. In W. H. Dantzler (ed.), *Comparative Physiology*, vol. 2 (Handbook of Physiology [Bethesda, MD], section 13), pp. 751–852. Oxford University Press, New York.

Armstrong, C. M. 2007. Life among the axons. *Annu. Rev. Physiol.* 69: 1–18.

Armstrong, C. M., and B. Hille. 1998. Voltage-gated ion channels and electrical excitability. *Neuron* 20: 371–380.

Baranauskas, B. 2007. Ionic channel function in action potential generation: current perspective. *Mol. Neurobiol.* 35: 129–150.

Barker, A. J., and E. M. Ullian. 2010. Astrocytes and synaptic plasticity. *Neuroscientist* 16: 40–50.

Catterall, W. A. 1995. Structure and function of voltage-gated ion channels. *Annu. Rev. Biochem.* 64: 493–531.

Catterall, W. A. 2000. From ionic currents to molecular mechanisms: The structure and function of voltage-gated sodium channels. *Neuron* 26: 13–25.

Catterall, W. A. 2001. A one-domain voltage-gated sodium channel from bacteria. *Science* 294: 2306–2308.

Catterall, W. A. 2012. Voltage-gated sodium channels at 60: structure, function, and pathophysiology. *J. Physiol.* 590: 2577–2589.

Catterall, W. A. 2014. Structure and function of voltage-gated sodium channels at atomic resolution. *Exp. Physiol.* 99: 35–51.

Catterall, W. A., S. Cestele, V. Yarov-Yarovoy, F. H. Yu, K. Konoki, and T. Scheuer. 2007. Voltage-gated ion channels and gating modifier toxins. *Toxicon* 49: 124–141.

Clark, B. D., E. M. Goldberg, and B. Rudy. 2009. Electrogenic tuning of the axon initial segment. *Neuroscientist* 15: 651–668.

Deisseroth, K. 2014. Circuit dynamics of adaptive and maladaptive behaviour. *Nature* 505: 309–317.

Di Resta, C., and A. Becchetti. 2010. Introduction to ion channels. *Adv. Exp. Med. Biol.* 674: 9–21.

Dorsett, D. A. 1980. Design and function of giant fiber systems. *Trends Neurosci.* 3: 205–208.

Elinder, F., J. Nilsson, and P. Arhem. 2007. On the opening of voltage-gated ion channels. *Physiol. Behav.* 92: 1–7.

England, S., and M. J. de Groot. 2009. Subtype-selective targeting of voltage-dependent sodium channels. *Brit. J. Pharmacol.* 158: 1413–1425.

Enyedi, P., and G. Czirják. 2010. Molecular background of leak K^+ currents: two-pore domain potassium channels. *Physiol. Rev.* 90: 559–605.

Gao, S., and M. Zhen. 2011. Action potentials drive body wall muscle contractions in *Caenorhabditis elegans*. *Proc. Natl. Acad. Sci. U.S.A.* 108: 2557–2562.

Halassa, M. M., and P. G. Haydon. 2010. Integrated brain circuits: astrocytic networks modulate neuronal activity and behavior. *Annu. Rev. Physiol.* 72: 335–355.

Hartline, D. K., and D. R. Colman. 2007. Rapid conduction and the evolution of giant axons and myelinated fibers. *Curr. Biol.* 17: R29–R35.

Hille, B. 2001. *Ionic Channels of Excitable Membranes*, 3rd ed. Sinauer, Sunderland, MA.

Hodgkin, A. L. 1964. *The Conduction of the Nervous Impulse*. Liverpool University Press, Liverpool, UK.

Jan, L. Y. and Y. N. Jan. 2012. Voltage-gated potassium channels and the diversity of electrical signaling. *J. Physiol.* 590: 2591–2599.

Johnston, D., and S. M.-S. Wu. 1995. *Foundations of Cellular Neurophysiology*. MIT Press, Cambridge, MA.

Karmazinova, M., and L. Lacinova. 2010. Measurement of cellular excitability by whole cell patch clamp technique. *Physiol. Res.* 59 (suppl. 1): S1–S7.

Katz, B. 1966. *Nerve, Muscle, and Synapse*. McGraw-Hill, New York.

Kress, G. J., and S. Mennerick. 2009. Action potential initiation and propagation: upstream influences on neurotransmission. *Neuroscience* 158: 211–222.

Lai, H. C., and L. Y. Jan. 2006. The distribution and targeting of neuronal voltage-gated ion channels. *Nat. Rev. Neurosci.* 7: 548–562.

Laming, P. R., E. Syková, A. Reichenbach, G. I. Hatton, and H. Bauer (eds.). 1998. *Glial Cells: Their Roles in Behaviour*. Cambridge University Press, Cambridge.

Liebeskind, B. J., D. M. Hillis, and H. H. Zakon. 2015. Convergence of ion channel genome content in early animal evolution. *Proc. Natl. Acad. Sci. U.S.A.* 112: E846–E861.

Lipscombe, D., J. Q. Pan, and S. Schorge. 2015. Tracks through the genome to physiological events. *Exp. Physiol.* doi: 10.1113/EP085129. [Epub ahead of print]

Luján, R. 2010. Organisation of potassium channels on the neuronal surface. *J. Chem. Neuroanat.* 40: 1–20.

Maday, S., A. E. Twelvetrees, A. J. Moughamian, and E. L. F. Holzbaur, 2014. Axonal transport: cargo-specific mechanisms of motility and regulation. *Neuron* 84: 292–309.

Misonou, H. 2010. Homeostatic regulation of neuronal excitability by K^+ channels in normal and diseased brains. *Neuroscientist* 16: 51–65.

Morais-Cabral, J. H., Y. Zhou, and R. MacKinnon. 2001. Energetic optimization of ion conduction rate by the K^+ selectivity filter. *Nature* 414: 37–42.

Neher, E., and B. Sakmann. 1992. The patch clamp technique. *Sci. Am.* 266(3): 28–35.

Ogawa, Y., and M. N. Rasband. 2008. The functional organization and assembly of the axon initial segment. *Curr. Opin. Neurobiol.* 18: 307–313.

Paixao, S., and R. Klein. 2010. Neuron-astrocyte communication and synaptic plasticity. *Curr. Opin. Neurobiol.* 20: 466–473.

Pathak, M. M., V. Yarov-Yarovoy, G. Agarwal, B. Roux, P. Barth, S. Kohout, F. Tombola, and E. Y. Isacoff. 2007. Closing in on the resting state of the Shaker K^+ channel. *Neuron* 56: 124–140.

Polleux, F., and W. Snider. 2010. Initiating and growing an axon. *Cold Spring Harb. Perspect. Biol.* 2010; 2: a001925.

Rasband, M. 2010. The axon initial segment and the maintenance of neuronal polarity. *Nat. Rev. Neurosci.* 11: 552–562.

Salzer, J. L., P. J. Brophy, and E. Peles. 2008. Molecular domains of myelinated axons in the peripheral nervous system. *Glia* 56: 1532–1540.

Sands, Z., A. Grottesi, and M. S. P. Sansom. 2005. Voltage-gated ion channels. *Curr. Biol.* 15: R44–R47.

Sasaki, T., N. Matsuki, and Y. Ikegaya. 2011. Action-potential modulation during axonal conduction. *Science* 331: 559–601.

Thaxton, C., and M. A. Bhat. 2009. Myelination and regional domain differentiation in the axon. *Results Probl. Cell. Differ.* 48: 1–28.

Travis, J. 1994. Glia: The brain's other cells. *Science* 266: 970–972.

Vacher, H., D. P. Mohapatra, and J. S. Trimmer. 2008. Localization and targeting of voltage-dependent ion channels in mammalian central neurons. *Physiol. Rev.* 88: 1407–1447.

Varshney, A., and M. K. Matthew. 2004. Architecture of a membrane protein: The voltage-gated K^+ channel. *Curr. Sci.* 87: 166–174.

Waxman, S. G. 1983. Action potential propagation and conduction velocity. *Trends Neurosci.* 6: 157–161.

Yoshimura, T., and M. N. Rasband. 2014. Axon initial segments: diverse and dynamic neuronal compartments. *Curr. Opin. Neurobiol.* 27: 96–102.

Yu, F. H., V. Yarov-Yarovoy, G. A. Gutman, and W. A. Catterall. 2005. Overview of molecular relationships in the voltage-gated channel superfamily. *Pharmacol. Rev.* 57: 387–395.

Chapter 13

Aidley, D., and P. Stanfield. 1996. *Ion Channels: Molecules in Action*. Cambridge University Press, Cambridge.

Alberini, C. M. 2009. Transcription factors in long-term potentiation and synaptic plasticity. *Physiol. Rev.* 89: 121–145.

Albuquerque, E. X., E. F. Pereira, M. Alkandon, and S. W. Rogers. 2009. Mammalian nicotinoic acetylcholine receptors: from structure to function. *Physiol. Rev.* 89: 73–120.

Alvarez, V. A., and B. L. Sabatini. 2007. Anatomical and physiological plasticity of dendritic spines. *Annu. Rev. Neurosci.* 30: 79–97.

Antonsen, B. L., and D. H. Edwards. 2003. Differential dye coupling reveals lateral giant escape circuit in crayfish. *J. Comp. Neurol.* 466: 1–13.

Bailey, C. H., E. R. Kandel, and K. M. Harris. 2015. Structural components of synaptic plasticity and memory consolidation. *Cold Spring Harb. Perspect. Biol.* 7: a021758.

Balice-Gordon, R. J., and J. W. Lichtman. 1994. Long-term synapse loss induced by focal blockade of postsynaptic receptors. *Nature* 372: 519–524.

Becerer, U., and J. Rettig. 2006. Vesicle pools, docking, priming, and release. *Cell and Tissue Res.* 326: 393–407.

Bennett, M. V., and R. S. Zukin. 2004. Electrical coupling and neuronal synchronization in the mammalian brain. *Neuron* 41: 495–511.

Bliss, T., G. Collingridge, and R. Morris. 2004. *Long-term Potentiation: Enhancing Neuroscience for 30 Years*. Oxford: Oxford University Press.

Buffelli, M., R. W. Burgess, G. Feng, C. G. Lobe, J. W. Lichtman, and J. R. Sanes. 2003. Genetic evidence that relative synaptic efficacy biases the outcome of synaptic competition. *Nature* 424: 430–434.

Collingridge, G. L., S. Peineau, J. G. Howland, and Y. T. Wang. 2010. Long-term depression in the CNS. *Nat. Rev. Neurosci.*11: 459–473.

Colquin, D., and B. Sakmann. 1998. From muscle endplate to brain synapses: A short history of synapses and agonist-activated channels. *Neuron* 20: 381–387.

Connors, B., and M. A. Long. 2004. Electrical synapses in the mammalian brain. *Annu. Rev. Neurosci.* 27: 393–418.

Cooper, J. R., F. Bloom, and R. E. Roth. 2002. *The Biochemical Basis of Neuropharmacology*, 8th ed. Oxford University Press, New York.

Dalva, M. B., A. C. McClelland, and M. S. Kayser. 2007. Cell adhesion molecules: signaling functions at the synapse. *Nat. Rev. Neurosci.* 8: 206–220.

Eccles, J. C. 1982. The synapse: From electrical to chemical transmission. *Annu. Rev. Neurosci.* 5: 325–339.

Fu, M., and Y. Zuo. 2011. Experience-dependent structural synaptic plasticity in the cortex. *Trends Neurosci.* 34: 177–187.

Glanzman, D. G. 2007. Simple minds: the neurobiology of invertebrate learning and memory. In G. North and R. J. Greenspan (eds.), *Invertebrate Neurobiology*, pp. 347–380. Cold Spring Harbor Laboratory Press, Cold Spring Harbor, NY.

Hammond, C. 1996. *Cellular and Molecular Neurobiology*. Academic Press, San Diego, CA.

Han, X., M. Chen, F. Wang, M. Windrem, S. Wang, S. Shanz, Q. Xu, N. A. Oberheim, L. Bekar, S. Betstadt, A. J. Silva, T. Takano, S. A. Goldman, and M. Nedergaard. 2013. Forebrain engraftment by human glial progenitor cells enhances synaptic plasticity and learning in adult mice. *Cell Stem Cell* 12: 343–353.

Haydon, P. G., and M. Nedergaard. 2015. How do astrocytes participate in neural plasticity? *Cold Spring Harb. Perspect. Biol.* 7: a020438.

He, L., and L. G. Wu. 2007. The debate on the kiss-and-run fusion at synapses. *Trends Neurosci.* 30: 447–455.

Heuser, J. E., and T. S. Reese. 1979. Synaptic-vesicle exocytosis captured by quick-freezing. In F. O. Schmitt and F. G. Worden (eds.), *The Neurosciences: Fourth Study Program*, pp. 573–600. MIT Press, Cambridge, MA.

Hille, B. 2001. *Ion Channels of Excitable Membranes*, 3rd ed. Sinauer, Sunderland, MA.

Holtz, R. W., and S. K. Fisher. 1999. Synaptic transmission and cellular signaling: An overview. In G. Siegel, B. Agranoff, R. W. Albers, S. K. Fisher, and M. D. Uhler (eds.), *Basic Neurochemistry*, 6th ed. Lippincott Williams & Wilkins, Philadelphia.

Jessell, T. M., and E. R. Kandel. 1993. Synaptic transmission: A bidirectional and self-modifiable form of cell-cell communication. *Cell 72/Neuron 10* (Suppl.): 1–30.

Kennedy, M. B., H. C. Beale, H. J. Carlisle, and L. R. Washburn. 2005. Integration of biochemical signaling in spines. *Nat. Rev. Neurosci.* 6: 423–434.

Kononenko, N. L., and V. Haucke. 2015. Molecular mechanisms of synaptic membrane retrieval and synaptic vesicle reformation. *Neuron* 85: 484–496.

Kuno, M. 1995. *The Synapse: Function, Plasticity, and Neurotrophism*. Oxford University Press, Oxford, UK.

Lee, H. K., and 12 additional authors. 2003. Phosphorylation of the AMPA receptor GluR1 subunit is required for synaptic plasticity and retention of spatial memory. *Cell* 112: 631–643.

Lee, S. E., and 15 others. 2010. RGS14 is a natural suppressor of both synaptic plasticity in CA2 neurons and hippocampal-based learning and memory. *Proc. Natl. Acad. Sci. U.S.A.* 107: 16994–16998.

Lee, S.-J. R., Y. Escobedo-Lozoya, E. M. Szatmari, and R. Yasuda. 2009. Activation of CaMKII in single dendritic spines during long-term potentiation. *Nature* 458: 299–304.

Lendvai, B., E. A. Stern, B. Chen, and K. Svoboda. 2000. Experience-dependent plasticity of dendritic spines in the developing rat barrel cortex in vivo. *Nature* 404: 876–881.

Levitan, I. B., and L. K. Kaczmarek. 2002. *The Neuron: Cell and Molecular Biology*, 3rd ed. Oxford University Press, New York.

Lisman, J. E., S. Raghavachari, and R. W. Tsien. 2007. The sequence of events that underlie quantal transmission at central glutamatergic synapses. *Nat. Rev. Neurosci.* 8: 597–609.

Lüscher, C., and R. C. Malenka. 2012. NMDA receptor-dependent long-term potentiation and long-term depression (LTP/LTD). *Cold Spring Harb. Perspect. Biol.* 4: a005710.

Malenka, R. C., and M. F. Bear. 2004. LTP and LTD: an embarrassment of riches. *Neuron* 44: 5–21.

Marder, E. 1998. From biophysics to models of network function. *Annu. Rev. Neurosci.* 21: 25–45.

Marder, E. 2009. Electrical synapses: rectification demystified. *Curr. Biol.* 19: R34–R35.

Matsuzaki, M., N. Honkura, G. C. Ellis-Davies, and H. Kasai. 2004. Structural basis of long-term potentiation in single dendritic spines. *Nature* 429: 761–766.

Mayford, M., S. A. Siegelbaum, and E. R. Kandel. 2012. Synapses and memory storage. *Cold Spring Harb. Perspect. Biol.* 4: a005751.

McAllister, A. K. 2007. Dynamic aspects of CNS synapse formation. *Annu. Rev. Neurosci.* 30: 425–450.

Meyer, J. S., and L. F. Quenzer. 2005. *Psychopharmacology: Drugs, the Brain, and Behavior*. Sinauer Associates: Sunderland, MA.

Neves, G., S. F. Cooke, and T. V. P. Bliss. 2008. Synaptic plasticity, memory and the hippocampus: A neural network approach to causality. *Nat. Rev. Neurosci.* 9: 65–75.

Park, M., E. C. Penick, J. G. Edwards, J. A. Kauer, and M. D. Ehlers. 2004. Recycling endosomes supply AMPA receptors for LTP. *Science* 305: 1972–1975.

Park, M., J. M. Salgado, L. Ostroff, T. D. Helton, C. G. Robinson, K. M. Harris, and M. D. Ehlers. 2006. Plasticity-induced growth of dendritic spines by exocytic trafficking from recycling endosomes. *Neuron* 52: 817–830.

Powis, D. A., and S. J. Bunn (eds.). 1995. *Neurotransmitter Release and Its Modulation*. Cambridge University Press, Cambridge.

Pozo, K., and Y. Goda. 2010. Unraveling mechanisms of homeostatic synaptic plasticity. *Neuron* 66: 337–351.

Scemes, E., D. C. Spray, and P. Meda. 2009. Connexins, pannexins, innexins: novel roles of "hemichannels." *Pflugers Arch.* 457: 1207–1226.

Shepherd, G. M., and S. D. Erulkar. 1997. Centenary of the synapse: From Sherrington to the molecular biology of the synapse and beyond. *Trends Neurosci.* 20: 385–392.

Shepherd J. D., and R. L. Huganir. 2007. The cell biology of synaptic plasticity: AMPA receptor trafficking. *Annu. Rev. Cell Dev. Biol.* 23: 613–643.

Siddiqui, T. J., and A. M. Craig. 2011. Synaptic organizing complexes. *Curr. Opin. Neurobiol.* 21: 132–143.

Smart, T. G., and P. Paoletti. 2012. Synaptic neurotransmitter-gated receptors. *Cold Spring Harb. Perspect. Biol.* 4: a009662

Smith, S. M., R. Renden, and H. von Gersdorff. 2008. Synaptic vesicle endocytosis fast and slow modes of membrane retrieval. *Trends Neurosci.* 31: 559–568.

Söhl, G., S. Maxeiner, and K. Willecke. 2005. Expression and functions of neuronal gap junctions. *Nat. Rev. Neurosci.* 6: 191–200.

Squire, L. S., D. Berg, F. E. Bloom, S. du Lac, A. Ghosh, and N. C. Spitzer (eds.). 2012. *Fundamental Neuroscience*, 4th ed. Academic Press, San Diego, CA.

Stuart, G., N. Spruston, and M. Häusser (eds.). 2008. *Dendrites*. Oxford University Press, Oxford and New York.

Südhof, T. C. 2004. The synaptic vesicle cycle. *Annu. Rev. Neurosci.* 27: 509–547.

Tang, J., A. Maximov, O.-H. Shin, H. Dai, J. Rizo, and T. C. Südhof. 2006. A complexin/synaptotagmin 1 switch controls fast synaptic vesicle exocytosis. *Cell* 126: 1175–1187.

Thompson, S. M., and H. A. Mattison. 2009. Secret of synapse specificity. *Nature* 458: 296–297.

Trachtenberg, J. T., B. E. Chen, G. W. Knott, G. Feng, J. R. Sanes, E. Welker, and K. Svoboda. 2002. Long-term in vivo imaging of experience-dependent synaptic plasticity in adult cortex. *Nature* 420: 788–794.

Valenstein, E. 2005. *The War of the Soups and the Sparks: The Discovery of Neurotransmitters and the Dispute Over How Nerves Communicate*. Columbia University Press, New York.

Venkatakrishnan, A. J., X. Deupi, G. Lebon, C. G. Tate, G. F. Schertler, and M. M. Babu. 2013. Molecular signatures of G-protein-coupled receptors. *Nature* 494: 185–194.

Waites, C. L., A. M. Craig, and C. C. Garner. 2005. Mechanisms of vertebrate synaptogenesis. *Annu. Rev. Neurosci.* 28: 251–274.

Walsh, M. K., and J. W. Lichtman. 2003. In vivo time-lapse imaging of synaptic takeover associated with naturally occurring synapse elimination. *Neuron* 37: 67–73.

Whitlock, J. R., A. J. Heynen, M. G. Shuler, and M. F. Bear. 2006. Learning induces long-term potentiation in the hippocampus. *Science* 313: 1093–1097.

Yuste, R. 2010. *Dendritic Spines*. MIT Press, Cambridge, MA.

Chapter 14

Axel, R. 1995. The molecular logic of smell. *Sci. Am.* 273(4): 154–159.

Baccus, S. A. 2007. Timing and computation in inner retinal circuitry. *Annu. Rev. Physiol.* 69: 271–290.

Bandell, M., L. J. Macpherson, and A. Patapoutian. 2007. From chills to chilis: mechanisms for thermosensation and chemesthesis via thermoTRPs. *Curr. Opin. Neurobiol.* 17: 490–497.

Barth, F. G., and A. Schmid (eds.). 2001. *Ecology of Sensing*. Springer, Berlin.

Baxi, K. N., K. M Dorries, and H. L Eisthen. 2006. Is the vomeronasal system really specialized for detecting pheromones? *Trends Neurosci.* 29: 1–7.

Bell, J., S. Bolinowski, and M. H. Holmes. 1994. The structure and function of Pacinian corpuscles: A review. *Prog. Neurobiol.* 42: 79–128.

Benton, R. 2008. Chemical sensing in *Drosophila*. *Curr. Opin. Neurobiol.* 18: 357–363.

Berson, D. M. 2007. Phototransduction in ganglion-cell photoreceptors. *Pflüger's Arch.* 454: 849–855.

Bradbury, J. W., and S. L. Vehrencamp. 2011. *Principles of Animal Communication*, 2nd ed. Sinauer, Sunderland, MA.

Buck, L. B. 1996. Information coding in the vertebrate olfactory system. *Annu. Rev. Neurosci.* 19: 517–544.

Carew, T. J. 2000. *Behavioral Neurobiology: The Cellular Organization of Natural Behavior*. Sinauer, Sunderland, MA.

Catania, K. C. 2002. The nose takes a starring role. *Sci. Am.* 287(1): 54–59.

Chalfie, M. 2009. Neurosensory mechanotransduction. *Nat. Rev. Molec. Cell Biol.* 10: 44–52.

Chandrashekar, J., M. A. Hoon, N. J. Ryba, and C. S. Zuker. 2006. The receptors and cells for mammalian taste. *Nature* 444: 288–294.

Christensen, A. P., and D. P. Corey 2007. TRP channels in mechanosensation: direct or indirect activation? *Nat. Rev. Neurosci.* 8: 510–521.

Conway, B. R. 2009. Color vision, cones, and color-coding in the cortex. *Neuroscientist* 15: 274–290.

Corcoran, A. J., J. R. Barber, N. I. Hristov, and W. E. Conner. 2011. How do tiger moths jam bat sonar? *J. Exp. Biol.* 214: 2416–2425.

Corey, D. P. 2001. Transduction and adaptation in vertebrate hair cells. In C. I. Berlin and R. P. Bobbin (eds.), *Hair Cells: Micromechanics and Hearing*, pp. 1–25. Singular Thompson Learning, San Diego, CA.

Corey, D. P., and S. D. Roper (eds.). 1992. *Sensory Transduction: Society of General Physiologists, 45th Annual Symposium, Marine Biological Laboratory, Woods Hole, Massachusetts, 5–8 Sept. 1991*. Rockefeller University Press, New York.

Coway, B. R., S. Chatterjee, G. D. Field, G. D. Horwitz, E. N. Johnson, K. Koida, and K. Mancuso. 2010. Advances in color science: from retina to behavior. *J. Neurosci.* 30: 14955–14963.

Damann, N., T. Voets, and B. Nilius. 2008. TRPs in our senses. *Curr. Biol.* 18: R880–R889.

Delcomyn, F. 1998. *Foundations of Neurobiology*. Freeman, New York.

Devor, M. 1996. Pain mechanisms. *Neuroscientist* 2: 233–244.

Dhaka, A., V. Viswanath, and A. Patapoutian. 2006. TRP ion channels and temperature sensation. *Annu. Rev. Neurosci.* 29: 135–161.

Døving, K. B., and D. Trotier. 1998. Structure and function of the vomeronasal organ. *J. Exp. Biol.* 201: 2913–2925.

Dowling, J. 1987. *The Retina: An Approachable Part of the Brain*. Belknap Press, Cambridge, MA.

Dulac, C. 1997. How does the brain smell? *Neuron* 19: 477–480.

Dulac, C. 2000. Sensory coding of pheromone signals in mammals. *Curr. Opin. Neurobiol.* 10: 511–518.

Dusenbery, D. B. 1992. *Sensory Ecology: How Organisms Acquire and Respond to Information*. Freeman, New York.

Eatock, R. A., R. R. Fay, and A. N. Popper (eds.). 2006. *Vertebrate Hair Cells*. Springer Handbook of Auditory Research, vol. 27. Springer, New York.

Fernald, R. D. 2006. Casting a genetic light on the evolution of eyes. *Science* 313: 1914–1918.

Field, G. D., and E. J. Chichilnisky. 2007. Information processing in the primate retina. *Annu. Rev. Neurosci.* 30: 1–30.

Finger, T. E., W. L. Silver, and D. Restrepo (eds.). 2000. *The Neurobiology of Taste and Smell*, 2nd ed. Wiley-Liss, New York.

Frisby, J. P. 1980. *Seeing: Illusion, Brain and Mind*. Oxford University Press, Oxford, UK.

Fu, Y., and K. W. Yau. 2007. Phototransduction in mouse rods and cones. *Pflüger's Arch.* 454: 805–819.

Gilbertson, T. A., S. Damak, and R. F. Margolskee. 2000. The molecular physiology of taste transduction. *Curr. Opin. Neurobiol.* 10: 519–527.

Grant, L., and P. A. Fuchs. 2007. Auditory transduction in the mouse. *Pflüger's Arch.* 454: 793–804.

Hallem, E. A., A. Dahanukar, and J. R. Carlson. 2006. Insect odor and taste receptors. *Annu. Rev. Entomol.* 51: 113–135.

Halliday, T. 1998. *The Senses and Communication*. Springer, New York.

Hansson, B. S. (ed.). 1999. *Insect Olfaction*. Springer, Berlin.

Hansson, B. S. and M. C. Stensmyr. 2011. Evolution of insect olfaction. *Neuron* 72: 698–711.

Hara, T. J., and B. S. Zielinski (eds.). 2007. Sensory Systems Neuroscience (Fish Physiology V. 25), Elsevier, Amsterdam.

Hardie, R. C. 2007. TRP channels and lipids: From *Drosophila* to mammalian physiology. *J. Physiol. (London)* 578: 9–24.

Hubel, D. H., and T. N. Wiesel. 2005. *Brain and Visual Perception: The Story of a 25-Year Collaboration*. Oxford University Press, Oxford.

Hudspeth, A. J. 2005. How the ear's works work: mechanoelectical transduction and amplification by hair cells. *C. R. Biol.* 328: 155–162.

Hudspeth, A. J. 2008. Making an effort to listen: mechanical amplification in the ear. *Neuron* 59: 530–545.

Hughes, H. C. 1999. *Sensory Exotica: A World beyond Human Experience*. MIT Press, Cambridge, MA.

Isogai, Y., S. Si, L. Pont-Lezica, T. Tan, V. Kapoor, V. N. Murthy, and C. Dulac. 2011. Molecular organization of vomeronasal chemoreception. *Nature* 478: 241–245.

Jefferis, G. S. 2005. Insect olfaction: a map of smell in the brain. *Curr. Biol.* 15: R668–R670.

Johnson, K. O. 2001. The roles and functions of cutaneous mechanoreceptors. *Curr. Opin. Neurobiol.* 11: 455–461.

Jones, G. 2005. Echolocation. *Curr. Biol.* 15: R484–R488.

Kay, L. M., and M. Stopfer. 2006. Information processing in the olfactory systems of insects and vertebrates. *Semin. Cell. Dev. Biol.* 17: 433–442.

Kleene, S. J. 2008. The electrochemical basis of odor transduction in vertebrate olfactory cilia. *Chem. Senses* 33: 839–859.

Knudsen, E. 1981. The hearing of the barn owl. *Sci. Am.* 245(6): 113–125.

Knudsen, E. I., and Konishi, M. 1978. Center-surround organization of auditory receptive fields in the owl. *Science* 202: 778–780.

Kolb, H. 2003. How the retina works. *Am. Sci.* 91: 28–35.

Konishi, M. 2003. Coding of auditory space. *Annu. Rev. Neurosci.* 26: 31–55.

Kruger, L. 1996. *Pain and Touch*. Academic Press, San Diego, CA.

Lamb, T. D., S. P. Collin, and E. N. Pugh, Jr. 2007. Evolution of the vertebrate eye: opsins, photoreceptors, retina and eyecup. *Nat. Rev. Neurosci.* 8: 960–976.

Land, M. F. 2005. The optical structures of animal eyes. *Curr. Biol.* 15: R319–R323.

LeMasurier, M., and P. G. Gillespie. 2005. Hair-cell mechanotransduction and cochlear amplification. *Neuron* 48: 403–415.

Lindemann, B. 2001. Receptors and transduction in taste. *Nature* 413: 219–225.

Liu, C. and C. Montell. 2015. Forcing open TRP channels: mechanical gating as a unifying activation mechanism. *Biochem. Biophys. Res. Commun.* 460: 22–25.

Lohmann, K. J., and S. Johnsen. 2000. The neurobiology of magnetoreception in vertebrate animals. *Trends Neurosci.* 23: 153–159.

Lumpkin, E. A., and M. J. Collins 2007. Mechanisms of sensory transduction in the skin. *Nature* 445: 858–865.

Ma, M. 2007. Encoding olfactory signals via multiple chemosensory systems. *Crit. Rev. Biochem. Mol. Biol.* 42(6): 463–480.

Martin, K. 1994. A brief history of the "feature detector." *Cereb. Cortex* 4: 1–7.

McIlwain, T. T. 1996. *An Introduction to the Biology of Vision*. Cambridge University Press, New York.

Møller, A. R. 2003. *Sensory Systems: Anatomy and Physiology*. Academic Press, San Diego, CA.

Mombaerts, P. 1999. Molecular biology of odorant receptors in vertebrates. *Annu. Rev. Neurosci.* 22: 487–509.

Montell, C. 2009. A taste of *Drosophila* gustatory receptors. *Curr. Opin. Neurobiol.* 19: 1–9.

Moss, C. F., and S. R. Sinha 2003. Neurobiology of echolocation in bats. *Curr. Opin. Neurobiol.* 6: 751–758.

Mountcastle, V. B. 2005. *The Sensory Hand: Neural Mechanisms of Somatic Sensation*. Harvard University Press, Cambridge, MA.

Nakagawa, T., and L. B. Vosshall. 2009. Controversy and consensus: noncanonical signaling mechanisms in the insect olfactory system. *Curr. Opin. Neurobiol.* 19: 284–292.

Niven, J. E., and S. B. Laughlin. 2008. Energy limitation as a selective pressure on the evolution of sensory systems. *J. Exp. Biol.* 211: 1792–1804.

Nobili, R., F. Mammano, and J. Ashmore. 1998. How well do we understand the cochlea? *Trends Neurosci.* 21: 159–167.

Palmer, S. E. 1999. *Vision Science: Photons to Phenomenology*. MIT Press, Cambridge, MA.

Perl, E. R. 2007. Ideas about pain, a historical view. *Nat. Rev. Neurosci.* 8: 71–80.

Pickles, J. O. 1993. Early events in auditory processing. *Curr. Opin. Neurobiol.* 3: 558–562.

Priebe, N. J., and D. Ferster. 2008. Inhibition, spike threshold, and stimulus selectivity in primary visual cortex. *Neuron* 57: 482–497.

Ramsey, I. S., M. Delling, and D. E. Clapham. 2006. An introduction to TRP channels. *Annu. Rev. Physiol.* 68: 619–647.

Ressler, K. J., S. L. Sullivan, and L. B. Buck. 1994. A molecular dissection of spatial patterning in the olfactory system. *Curr. Opin. Neurobiol.* 4: 588–596.

Rodieck, R. W. 1998. *The First Steps in Seeing*. Sinauer, Sunderland, MA.

Roper, S. D. 2007. Signal transduction and information processing in mammalian taste buds. *Pflüger's Arch.* 454: 759–776.

Schild, D., and D. Restrepo. 1998. Transduction mechanisms in vertebrate olfactory receptor cells. *Physiol. Rev.* 78: 429–466.

Simmons, P. J., and D. Young. 1999. *Nerve Cells and Animal Behavior*, 2nd ed. Cambridge University Press, Cambridge.

Smith, D. V., and R. F. Margolskee. 2001. Making sense of taste. *Sci. Am.* 284(3): 32–39.

Solomon, S. G., and P. Lennie. 2007. The machinery of color vision. *Nat. Rev. Neurosci.* 8: 276–286.

Solomon, S. G., and P. Lennie. 2007. The machinery of colour vision. *Nat. Rev. Neurosci.* 8: 276–286.

Spehr, M., and S. D. Munger. 2009. Olfactory receptors: G protein-coupled receptors and beyond. *J. Neurochem.* 109: 1570–1583.

Tirindelli, R., M. Dibattista, S. Pifferi, and A. Menini. 2009. From pheromones to behavior. *Physiol. Rev.* 89: 921–956.

Torre, V., J. F. Ashmore, T. D. Lamb, and A. Menini. 1995. Transduction and adaptation in sensory receptor cells. *J. Neurosci.* 15: 7757–7768.

Vay, L., C. Gu, and P. A. McNaughton. 2012. The thermo-TRP ion channel family: properties and therapeutic implications. *Br. J. Pharmacol.* 165: 787–801.

Vollrath, M. A., K. Y. Kwan, and D. P. Corey. 2007. The micromachinery of mechanotransduction in hair cells. *Annu. Rev. Neurosci.* 30: 339–365.

Vosshall, L. B., and R. F. Stocker. 2007. Molecular architecture of smell and taste in *Drosophila*. *Annu. Rev. Neurosci.* 30: 505–533.

Walker, R. G., A. T. Willingham, and C. S. Zuker. 2000. A *Drosophila* mechanosensory transduction channel. *Science* 287: 2229–2234.

Wilson, R. I. 2013. Early olfactory processing in *Drosophila*: mechanisms and principles. *Annu. Rev. Neurosci.* 36: 217–241.

Wilson, R. I., and Z. F. Mainen. 2006. Early events in olfactory processing. *Annu. Rev. Neurosci.* 29: 163–201.

Wolken, J. J. 1995. *Light Detectors, Photoreceptors, and Imaging Systems in Nature.* Oxford University Press, New York.

Woolf, C. J., and Q. Ma. 2007. Nociceptors–noxious stimulus detectors. *Neuron* 55: 353–364.

Yarmolinsky, D. A., C. S. Zuker, and N. J. P. Ryba. 2009. Common sense about taste: from mammals to insects. *Cell* 139: 234–244.

Zarzo, M. 2007. The sense of smell: molecular basis of odorant recognition. *Biol. Rev.* 82: 455–479.

Zeki, S. 1993. *A Vision of the Brain.* Blackwell, London.

Zimmermann, K., A. Leffler, A. Babes, C. M. Cendan, R. W. Carr, J. Kobayashi, C. Nau, J. N. Wood, and P. W. Reeh. 2007. Sensory neuron sodium channel $Na_v1.8$ is essential for pain at low temperatures. *Nature* 447: 856–859.

Zufall, F., and T. Leinders-Zufall. 2007. Mammalian pheromone sensing. *Curr. Opin. Neurobiol.* 17: 483–489.

Chapter 15

Albrecht, U., and G. Eichele 2003. The mammalian circadian clock. *Curr. Opin. Genet. Dev.* 13: 271–277.

Allman, J. M. 1999. *Evolving Brains.* Scientific American Library, New York.

Arbas, E. A., I. A. Meinertzhagen, and S. R. Shaw. 1991. Evolution in nervous systems. *Annu. Rev. Neurosci.* 14: 9–38.

Arbas, E. A., R. B. Levine, and N. J. Strausfeld. 1997. Invertebrate nervous systems. In W. H. Dantzler (ed.), *Comparative Physiology*, vol. 2 (Handbook of Physiology [Bethesda, MD], section 13), pp. 751–852. Oxford University Press, New York.

Arendt, D., A. S. Denes, G. Jékely, and K. Tessmer-Raible. 2008. The evolution of nervous system centralization. *Philos. Trans. R. Soc. London, B* 363: 1523–1528.

Arendt, J. 1995. *Melatonin and the Mammalian Pineal Gland.* Chapman & Hall, New York.

Barinaga, M. 2002. How the brain's clock gets daily enlightenment. *Science* 295: 955–957.

Bass, J., and J. S. Takahashi. 2010. Circadian integration of metabolism and energetics. *Science* 330: 1349–1354.

Brancaccio, M., R. Enoki, C. N. Mazuskj. J. Jones, J. A. Evans, and A. Azzi. 2014. Network-mediated encoding of circadian time: the suprachiasmatic nucleus (SCN) from genes to neurons to circuits, and back. *J. Neurosci.* 34: 15192–15199.

Butler, A. B. 2000. Chordate evolution and the origin of craniates: an old brain in a new head. *Anat. Rec.* 261: 111–125.

Butler, A. B., and W. Hodos, 1996. *Comparative Vertebrate Neuroanatomy.* Wiley-Liss, New York.

Camhi, J. 1980. The escape system of the cockroach. *Sci. Am.* 243(6): 158–172.

Carmona, F. D., M. Glösmann, J. Ou, R. Jiménez, and J. M. Collison. 2010. Retinal development and function in a "blind" mole. *Proc. R. Soc. London, Ser. B* 277: 1513–1522.

Carter, R. 1998. *Mapping the Mind.* University of California Press, Berkeley.

Catania, K. C. 2005. Evolution of sensory specializations in insectivores. *Anat. Rec. A Discov. Mol. Cell Evol. Biol.* 297: 1038–1050.

Catania, K. C. 2005. Star-nosed moles. *Curr. Biol.* 15: R863-R864.

Colwell. C. S. 2011. Linking neural activity and molecular oscillations in the SCN. *Nat. Rev. Neurosci.* 12: 553–569.

Costa, R., F. Saldrelli, and C. P. Kyriacou. 2007. Evolution of behavioral genes. In G. North and R. J. Greenspan (eds.), *Invertebrate Neurobiology*, pp. 617–646. Cold Spring Harbor Laboratory Press, Cold Spring Harbor, NY.

Dibner, C., U. Schibler, and U. Albrecht 2010. The mammalian circadian timing system: organization and coordination of central and peripheral clocks. *Annu. Rev. Physiol.* 72: 517–549.

Do, M. T., and K.-W. Yau. 2010. Intrinsically photosensitive retinal ganglion cells. *Physiol. Rev.* 90: 1547–1581.

Donald, J. A. 1998. Autonomic nervous system. In D. H. Evans (ed.), *The Physiology of Fishes*, 2nd ed., pp. 407–439. CRC Press, Boca Raton, FL.

Emes, R. D., A. J. Pocklington, C. N. Anderson, A. Bayes, M. O. Collins, C. A. Vickers, M. D. Croning, B. R. Malik, J. S. Choudhary, J. D. Armstrong, and S. G. Grant. 2008. Evolutionary expansion and anatomical specialization of synapse proteome complexity. *Nat. Neurosci.* 11: 799–806. DOI: 10.1038/nn.2135

Foster, R. G., and L. Kreitzman. 2004. *Rhythms of Life: The Biological Clocks That Control the Daily Lives of Every Living Thing.* Yale University Press, New Haven.

Furness, J. B. 2006. *The Enteric Nervous System.* Blackwell, Oxford.

Furness, J. B. 2006. The organisation of the autonomic nervous system: peripheral connections. *Autonom. Neurosci.* 130: 1–5.

Golombek, D. A., and R. E. Rosenstein. 2010. Physiology of circadian entrainment. *Physiol. Rev.* 90: 1063–1102.

Hastings, J. W., B. Rusak, and Z. Boulos. 1991. Circadian rhythms: The physiology of biological timing. In C. L. Prosser (ed.), *Neural and Integrative Animal Physiology* (Comparative Animal Physiology, 4th ed.), pp. 435–546. Wiley-Liss, New York.

Hastings, M. H., A. B. Reddy, and E. S. Maywood. 2003. A clockwork web: circadian timing in brain and periphery, in health and disease. *Nat. Rev. Neurosci.* 4: 649–661.

Herzog, E. D. 2007. Neurons and networks in daily rhythms. *Nat. Rev. Neurosci.* 8: 790–802.

Jansen, A. S. P., X. V. Nguyen, V. Karpitskiy, T. C. Mettenleiner, and A. D. Loewy. 1995. Central command neurons of the sympathetic nervous system: Basis of the fight-or-flight response. *Science* 270: 644–646.

Kaas, J. H. (ed.-in-chief) 2007. *Evolution of Nervous Systems: A Comprehensive Reference.* (4 vols.) Academic Press/Elsevier, Amsterdam; Boston.

Ko, C. K., and J. S. Takahashi. 2006. Molecular components of the mammalian circadian clock. *Hum. Mol. Genet.* 15: R271–R277.

Koukhari, W. L, and R. B. Sothern. 2006. *Introducing Biological Rhythms: A Primer on the Temporal Organization of Life, with Implications for Health, Society, Reproduction and the Natural Environment.* Springer, New York.

Lehman, M. N., R. Silver, W. R. Gladstone, R. M. Kahn, M. Gibson, and E. L. Bittman. 1987. Circadian rhythmicity restored by neural transplant. Immunocytochemical characterization of the graft and its integration with the host brain. *J. Neurosci.* 7: 1626–1638.

Liebeskind, B. J., D. M. Hills, and H. H. Zakon. 2015. Convergence of ion channel genome content in early animal evolution. *Proc. Natl. Acad. Sci. U.S.A.* 112(8): E846–E851. doi: 10.1073/pnas.1501195112

Lowrey, P. L., and J. S. Takahashi. 2004. Mammalian circadian biology: elucidating genome-wide levels of temporal organization. *Annu. Rev. Genomics Hum. Genet.* 5: 407–441.

Marder, E. 2007. Searching for insight: Using invertebrate nervous systems to illuminate fundamental principles in neuroscience. In G. North and R. J. Greenspan (eds.), *Invertebrate Neurobiology*, pp. 1–18. Cold Spring Harbor Laboratory Press, Cold Spring Harbor, NY.

Meech, R. W., and G. O. Mackie. 2007. Evolution of excitability in lower metazoans. In G. North and R. J. Greenspan (eds.), *Invertebrate Neurobiology*, pp. 581–615. Cold Spring Harbor Laboratory Press, Cold Spring Harbor, NY.

Menaker, M. 2003. Circadian photoreception. *Science* 299: 213–214.

Moore, R. Y. 1995. Organization of the mammalian circadian system. In D. J. Chadwick and K. Ackrill (eds.), *Circadian Clocks and Their Adjustment* (Ciba Foundation Symposium, 183), pp. 88–106. Wiley, Chichester, UK.

Moroz, L. L. 2015. Convergent evolution of neural systems in ctenophores. *J. Exp. Biol.* 218: 598–611.

Nauta, W. J. H., and M. Feirtag. 1979. The organization of the brain. *Sci. Am.* 241(3): 88–111.

Nilsson, S. 1983. *Autonomic Nerve Function in the Vertebrates.* Springer, New York.

Nilsson, S., and S. Holmgren (eds.). 1994. *Comparative Physiology and Evolution of the Autonomic Nervous System.* Harwood, Langhorne, PA.

Northrop, R. B. 2000. *Endogenous and Exogenous Regulation and Control of Physiological Systems.* Chapman & Hall/CRC, Boca Raton, FL.

Palmer, J. D. 1995. *The Biological Rhythms and Clocks of Intertidal Animals.* Oxford University Press, New York.

Pittendrigh, C. 1993. Temporal organization: Reflections of a Darwinian clock-watcher. *Annu. Rev. Physiol.* 55: 16–54.

Ralph, M. R., and M. W. Hurd 1995. Circadian pacemakers in vertebrates. In D. J. Chadwick and K. Ackrill (eds.), *Circadian Clocks and Their Adjustment* (Ciba Foundation Symposium, 183), pp. 67–87. Wiley, Chichester, UK.

Ralph, M. R., R. G. Foster, F. C. Davis, and M. Menaker. 1990. Transplanted suprachiasmatic nucleus determines circadian period. *Science* 247: 975–978.

Ramkisoensing, A., and J. H. Meijer. 2015. Synchronization of biological clock neurons by light and peripheral feedback systems promotes circadian rhythms and health. *Front. Neurol.* 6: 128. doi: 10.3389/fneur.2015.00128

Refinetti, R. 2006. *Circadian Physiology,* 2nd ed. CRC Press/Taylor & Francis, Boca Raton, FL.

Reppert, S. M. 2006. A colorful model of the circadian clock. *Cell* 124: 233–236.

Reppert, S. M., and D. R. Weaver 2001. Molecular analysis of mammalian circadian rhythms. *Annu. Rev. Physiol.* 63: 647–676.

Reppert, S. M., and D. R. Weaver 2002. Coordination of circadian timing in mammals. *Nature* 418: 935–941.

Reppert, S. M., R. J. Gegear, and C. Merlin. 2010. Navigational mechanisms of migrating monarch butterflies. *Trends Neurosci.* 33: 399–406.

Ryan, T. J., and S. G. Grant, 2009. The origin and evolution of synapses. *Nat. Rev. Neurosci.* 10: 701–712.

Sehgal, A. (ed.). 2004. *Molecular Biology of Circadian Rhythms.* Wiley-Liss, Hoboken, NJ.

Strausfeld, N. J. 2009. Brain organization and the origin of insects: an assessment. *Proc. Biol. Sci.* 276: 1929–1937.

Strausfeld, N. J. 2012. *Arthropod Brains: Evolution, Functional Elegance, and Historical Significance.* Belknap Press of Harvard University Press, Cambridge, MA.

Swanson, L. W. 2003. *Brain Architecture: Understanding the Basic Plan.* Oxford University Press, New York.

Takahashi, J. S., F. W. Turek, and R. Y. Moore (eds.). 2001. *Circadian Clocks.* Handbook of Behavioral Neurobiology, vol. 12. Kluwer Academic/Plenum, New York.

Telford, M. J. 2007. A single origin of the central nervous system? *Cell* 129: 237–239.

Turek, F. W., and O. Van Reeth. 1996. Circadian rhythms. In M. J. Fregley and C. M. Blatteis (eds.), *Environmental Physiology,* vol. 2 (Handbook of Physiology [Bethesda, MD], section 4), pp. 1329–1359. Oxford University Press, New York.

Ulinski, P. S. 1997. Vertebrate nervous system. In W. H. Dantzler (ed.), *Comparative Physiology,* vol. 1 (Handbook of Physiology [Bethesda, MD], section 13), pp. 17–53. Oxford University Press, New York.

Underwood, H. A., G. T. Wassmer, and T. L. Page. 1997. Daily and seasonal rhythms. In W. H. Dantzler

(ed.), *Comparative Physiology,* vol. 2 (Handbook of Physiology [Bethesda, MD], section 13), pp. 1653–1763. Oxford University Press, New York.

Vogel, G. 2011. Telling time without turning on genes. *Science* 331: 391.

Wever, R. A. 1979. *The Circadian System of Man. Results of Experiments under Temporal Isolation.* Springer-Verlag, New York.

Yates, F. E. 1982. Outline of a physical theory of physiological systems. *Can. J. Physiol. Pharmacol.* 60: 217–248.

Young, M. W. 2005. *Circadian Rhythms.* (Methods in Enzymology, vol. 393.) Elsevier Academic Press, San Diego.

Zhang, E. R., and S. A. Kay. 2010. Clocks not winding down: unraveling circadian networks. *Nature Rev. Molec. Cell Biol.* 11: 764–776.

Chapter 16

Baker, J. D., and J. W. Truman. 2002. Mutations in the *Drosophila* glycoprotein hormone receptor, rickets, eliminate neuropeptide-induced tanning and selectively block a stereotyped behavioral program. *J. Exp. Biol.* 205: 2555–2565.

Bentley, G. E., K. Tsutsui, and L. J. Kriegsfeld. 2010. Recent studies of gonadotropin-inhibitory hormone (GnIH) in the mammalian hypothalamus, pituitary and gonads. *Brain Res.* 1364: 62–71.

Boerjan, B., P. Verleyen, J. Huybrechts, L. Schoofs, and A. De Loof. 2010. In search for a common denominator for the diverse functions of arthropod corazonin: A role in the physiology of stress? *Gen. Comp. Endocrinol.* 166: 222–233.

Bradshaw, S. D., and F. J. Bradshaw. 2002. Arginine vasotocin: Site and mode of action in the reptilian kidney. *Gen. Comp. Endocrinol.* 126: 7–13.

Butler, M., H. S. Ginsberg, R. A. LeBrun, and A. Gettman. 2010. Evaluation of nontarget effects of methoprene applied to catch basins for mosquito control. *J. Vector Ecol.* 35: 372–384.

Byrd, J. H., and J. L. Castner (eds.). 2001. *Forensic Entomology: The Utility of Arthropods in Legal Investigations.* CRC Press, Boca Raton, FL.

Chestnut, C. H., III, M. Azria, S. Silverman, M. Engelhardt, M. Olson, and L. Mindeholm. 2008. Salmon calcitonin: a review of current and future therapeutic indications. *Osteoporos. Int.* 19: 479–491.

Coddington, E., C. Lewis, J. D. Rose, and F. L. Moore. 2007. Endocannabinoids mediate the effects of acute stress and corticosterone on sex behavior. *Endocrinology* 148: 493–500.

Coddington, E., C. Lewis, J. D. Rose, and F. L. Moore. 2007. Endocannabinoids mediate the effects of acute stress and corticosterone on sex behavior. *Endocrinology* 148: 493–500.

d'Anglemont de Tassigny, X., and W. H. Colledge. 2010. The role of kisspeptin signaling in reproduction. *Physiology* 25: 207–217.

Denver, R. J. 1997. Environmental stress as a developmental cue: Corticotropin-releasing hormone is a proximate mediator of adaptive phenotypic plasticity in amphibian metamorphosis. *Horm. Behav.* 31: 169–179.

Denver, R. J. 2007. Endocannabinoids link rapid, membrane-mediated corticosteroid actions to behavior. *Endocrinology* 148: 490–492.

Freeman, M. E., B. Kanyicska, A. Lerant, and G. Nagy. 2000. Prolactin: Structure, function, and regulation of secretion. *Physiol. Rev.* 80: 1523–1631.

Gilbert, S. F. 2001. Ecological developmental biology: Developmental biology meets the real world. *Dev. Biol.* 233: 1–12.

Goffin, V., N. Binart, P. Touraine, and P. A. Kelly. 2002. Prolactin: The new biology of an old hormone. *Annu. Rev. Physiol.* 64: 47–67.

Hammes, A., and 10 additional authors. 2005. Role of endocytosis in cellular uptake of sex steroids. *Cell* 122: 751–762.

Henry, H. S., and A. W. Norman (eds.). 2003. *Encyclopedia of Hormones.* Academic Press, Amsterdam.

Insel, T. R., and L. J. Young. 2001. The neurobiology of attachment. *Nat. Rev. Neurosci.* 2: 129–136.

Jindra, M., S. R. Palli, and L. M. Riddiford. 2013. The juvenile hormone signaling pathway in insect development. *Annu. Rev. Entomol.* 58: 181–204.

Kimchi, T., J. Xu, and C. Dulac. 2007. A functional circuit underlying male sexual behaviour in the female mouse brain. *Nature* 448: 1009–1014.

Koshimizu, T., K. Nakamura, N. Egashira, M. Hiroyama, H. Nonoguchi, and A. Tanoue. 2012. Vasopressin V1a and V1b receptors: From molecules to physiological systems. *Physiol. Rev.* 92: 1813–1864.

Kostron, B., K. Marquardt, U. Kaltenhauser, and H. W. Honegger. 1995. Bursicon, the cuticle sclerotizing hormone: Comparison of its molecular mass in different insects. *J. Insect Physiol.* 41: 1045–1053.

Lee, D., I. Orchard, and A. B. Lange. 2013. Evidence for a conserved CCAP-signaling pathway controlling ecdysis in a hemimetabolous insect, *Rhodnius prolixus. Front. Neurosci.* 7: 207. doi: 10.3389/fnins.2013.00207

Lin, B. C., and T. S. Scanlan. 2005. Few things in life are "free": Cellular uptake of steroid hormones by an active transport mechanism. *Mol. Interv.* 5: 338–339.

McEwen, B. S. 2002. *The End of Stress as We Know It.* Joseph Henry Press, Washington, D.C.

Morton, G. J., and M. W. Schwartz. 2011. Leptin and the central nervous system control of glucose metabolism. *Physiol. Rev.* 91: 389–411.

Nijhout, H. F., and M. C. Reed. 2008. A mathematical model for the regulation of juvenile hormone titers. *J. Insect Physiol.* 54: 255–264.

O'Brien, M. A., E. J. Katahira, T. R. Flanagan, L. W. Arnold, G. Haughton, and W. E. Bollenbacher. 1988. A monoclonal antibody to the insect prothoracicotropic hormone. *J. Neurosci.* 8: 3247–3257.

Pagotto, U., M. Giovanni, D. Cota, B. Lutz, and R. Pasquali. 2006. The emerging role of the endocannabinoid system in endocrine regulation and energy balance. *Endocrinol. Rev.* 27: 73–100.

Pryce, C. R., D. Rüedi-Bettschen, A. C. Dettling, and J. Feldon. 2002. Early life stress: Long-term physiological impact in rodents and primates. *News Physiol. Sci.* 17: 150–155.

Riddiford, L. 2011. When is weight critical? *J. Exp. Biol.* 214: 1613–1615. [A discussion of papers leading to the concept that an insect must gain a critical weight before undergoing metamorphosis.]

Romero, L. M. 2002. Seasonal changes in plasma glucocorticoid concentrations in free-living vertebrates. *Gen. Comp. Endocrinol.* 128: 1–24.

Romero, L. M. 2006. Seasonal changes in hypothalamic-pituitary-adrenal axis sensitivity in free-living house sparrows (*Passer domesticus*). *Endocrinology* 149: 66–71.

Romero, L. M., and M. Wikelski. 2001. Corticosterone levels predict survival probabilities of Galapagos marine iguanas during El Niño events. *Proc. Natl. Acad. Sci. U.S.A.* 98: 7366–7370.

Ruther, J., T. Meiners, and J. L. M. Steidle. 2002. Rich in phenomena—lacking in terms: A classification of kairomones. *Chemoecology* 12: 161–167.

Sapolsky, R. M. 2005. The influence of social hierarchy on primate health. *Science* 308: 648–652.

Sapolsky, R. M., L. M. Romero, and L. U. Munck. 2000. How do glucocorticoids influence stress responses? Integrating permissive, suppressive, stimulatory, and preparative actions. *Endocrinol. Rev.* 21: 55–89.

Sawyer, T. K., V. J. Hruby, M. E. Hadley, and M. H. Engel. 1983. α–Melanocyte stimulating hormone: Chemical nature and mechanism of action. *Am. Zool.* 23: 529–540.

Sbarbati, A., and F. Osculati. 2006. Allelochemical communication in vertebrates: Kairomones, allomones and synomones. *Cells Tissues Organs* 183: 206–219.

Sehnal, F., I. Hansen, and K. Scheller. 2002. The cDNA-structure of the prothoracicotropic hormone (PTTH) of the silkmoth *Hyalophora cecropia*. *Insect Biochem. Mol. Biol.* 32: 233–237.

Sherman, R. A., M. J. R. Hall, and S. Thomas. 2000. Medicinal maggots: An ancient remedy for some contemporary afflictions. *Annu. Rev. Entomol.* 45: 55–81.

Simoncini, T., and A. R. Genazzani. 2003. Non-genomic actions of sex steroid hormones. *Eur. J. Endocrinol.* 148: 281–292.

Sisneros, J. A. 2009. Seasonal plasticity of auditory saccular sensitivity in the vocal plainfin midshipman fish, *Porichthys notatus*. *J. Neurophysiol.* 102: 1121–1131.

Sisneros, J. A., P. M. Forlano, D. L. Deitcher, and A. H. Bass. 2004. Steroid-dependent auditory plasticity leads to adaptive coupling of sender and receiver. *Science* 305: 404–407.

Sisneros, J. A., P. M. Forlano, R. Knapp, and A. H. Bass. 2004. Seasonal variation of steroid hormone levels in an intertidal-nesting fish, the vocal plainfin midshipman. *Gen. Comp. Endocrinol.* 136: 101–116.

Sorrells, S. F., and R. M. Sapolsky. 2007. An inflammatory review of glucocorticoid actions. *Brain Behav. Immun.* 21: 259–272.

Sorrells, S. F., J. R. Caso, C. D. Munhoz, and R. M. Sapolsky. 2009. The stressed CNS: When glucocorticoids aggravate inflammation. *Neuron* 64: 33–39.

Sternberg, E. M., and P. W. Gold. 2002. The mind-body interaction in disease. *Sci. Am.* (special edition) 12(1): 82–89.

Takei, Y., and S. Hirose. 2002. The natriuretic peptide system in eels: A key endocrine system for euryhalinity? *Am. J. Physiol.* 282: R940–R951.

Watson, R. D., E. Spaziani, and W. E. Bollenbacher. 1989. Regulation of ecdysone biosynthesis in insects and crustaceans: a comparison. In J. Koolman (ed.), *Ecdysone: From Chemistry to Mode of Action*, pp. 188–203. Thieme, Stuttgart.

White, B. A., and J. Ewer. 2014. Neural and hormonal control of postecdysial behaviors in insects. *Annu. Rev. Entomol.* 59: 363–381.

Wicher, D. 2004. Metabolic regulation and behavior: How hunger produces arousal—An insect study. *Endocr. Metab. Immune Disord. Drug Targets* 7: 304–310.

Wikelski, M., V. Wong, B. Chevalier, N. Rattenborg, and H. L. Snell. 2002. Marine iguanas die from trace oil pollution. *Nature* 417: 607–608.

Winslow, J. T., N. Hastings, C. S. Carter, C. R. Harbaugh, and T. R. Insel. 1993. A role for central vasopressin in pair bonding in monogamous prairie voles. *Nature* 365: 545–548.

Young, L. J., M. M. Lim, B. Gingrich, and T. R. Insel. 2001. Cellular mechanisms of social attachment. *Horm. Behav.* 40: 133–138.

Zitnan, D., I. Zitnanová, I. Spalovská, P. Takác Y. Park, and M. E. Adams. 2003. Conservation of ecdysis-triggering hormone signaling in insects. *J. Exp. Biol.* 206: 1275–1289.

Zitnan, D., T. G. Kingan, J. L. Hermesman, and M. E. Adams. 1996. Identification of ecdysis-triggering hormone from an epitracheal endocrine system. *Science* 271: 88–91.

Chapter 17

Adams, V. L., R. L. Goodman, A. K. Salm, L. M. Coolen, F. J. Karsch, and M. N. Lehman. 2006. Morphological plasticity in the neural circuitry responsible for seasonal breeding in the ewe. *Endocrinology* 147: 4843–4851.

Allen, W. R. 2001. Fetomaternal interactions and influences during equine pregnancy. *Reproduction* 121: 513–527.

Asa, C. S., and C. Valdespino. 1998. Canid reproductive biology: An integration of proximate mechanisms and ultimate causes. *Am. Zool.* 38: 251–259.

Bahat, A., I. Tur-Kaspa, A. Gakamsky, L. C. Giojalas, H. Breitbart, and M. Eisenbach. 2003. Thermotaxis of mammalian sperm cells: A potential navigation mechanism in the female genital tract. *Nat. Med.* 9: 149–150.

Bakker, J., and M. J. Baum. 2000. Neuroendocrine regulation of GnRH release in induced ovulators. *Front. Neuroendocrinol.* 21: 220–262.

Berenbaum, S. A. 1999. Effects of early androgens on sex-typed activities and interests in adolescents with congenital adrenal hyperplasia. *Horm. Behav.* 35: 102–110.

Berglund, A. 1990. Sequential hermaphroditism and the size-advantage hypothesis: an experimental test. *Anim. Behav.* 39: 426–433.

Caba, M., J. Bao, K.-Y. F. Pau, and H. G. Spies. 2000. Molecular activation of noradrenergic neurons in the rabbit brainstem after coitus. *Mol. Brain Res.* 77: 222–231.

Cooke, B. M., G. Tabibnia, and S. M. Breedlove. 1999. A brain sexual dimorphism controlled by adult circulating androgens. *Proc. Natl. Acad. Sci. U.S.A.* 96: 7538–7540.

Creel, S., D. Christianson, S. Liley, and J. A. Winnie, Jr. 2007. Predation risk affects reproductive physiology and demography of elk. *Science* 315: 960.

DeLamirande, E., P. Leclerc, and C. Gagnon. 1997. Capacitation as a regulatory event that primes spermatozoa for the acrosome reaction and fertilization. *Mol. Hum. Reprod.* 3: 175–194.

Dudas, B., and I. Merchenthaler. 2006. Three-dimensional representation of the neurotransmitter systems of the human hypothalamus: inputs of the gonadotropin hormone-releasing hormone neuronal system. *J. Neuroendocrinol.* 18: 79–95.

Dungan, H. M., D. K. Clifton, and R. A. Steiner. 2006. Minireview: Kisspeptin neurons as central processors in the regulation of gonadotropin-releasing hormone secretion. *Endocrinology* 147: 1154–1158.

Evans, J. P., and H. M. Florman. 2002. The state of the union: The cell biology of fertilization. *Nat. Cell Biol.* 4 Suppl.: S57–S63.

Frisch, R. E. 2002. *Female Fertility and the Body Fat Connection*. University of Chicago Press, Chicago.

Gilbert, S. F. 2010. *Developmental Biology*, 9th ed. Sinauer, Sunderland, MA.

Godwin, J., R. Sawby, R. R. Warner, D. Crews, and M. S. Grober. 2000. Hypothalamic arginine vasotocin mRNA abundance variation across sexes and with sex change in a coral reef fish. *Brain Behav. Evol.* 55: 77–84.

Goldstein, I., and the Working Group for the Study of Central Mechanisms in Erectile Dysfunction. 2000. Male sexual circuitry. *Sci. Am.* 283(2): 70–75.

Hayssen, V., A. van Tienhoven, and A. van Tienhoven. 1993. *Asdell's Patterns of Mammalian Reproduction. A Compendium of Species-specific Data*. Comstock Publishing, Ithaca, NY.

Herbison, A. E. 1998. Multimodal influence of estrogen upon gonadotropin-releasing hormone neurons. *Endocrinol. Rev.* 19: 302–330.

Hübner, K., G. Fuhrmann, L. K. Christenson, J. Kehler, R. Reinbold, R. De La Fuente, J. Wood, J. F. Strauss III, M. Boiani, and H. R. Schöler. 2003. Derivation of oocytes from mouse embryonic stem cells. *Science* 300: 1251–1256.

Hunt, P. A., and T. J. Hassold. 2002. Sex matters in meiosis. *Science* 296: 2181–2183.

Ikonomidou, C., and 11 additional authors. 2000. Ethanol-induced apoptotic neurodegeneration and fetal alcohol syndrome. *Science* 287: 1056–1060.

Johnson, J., J. Canning, T. Kaneko, J. K. Pru, and J. L. Tilly. 2004. Germline stem cells and follicular renewal in the postnatal mammalian ovary. *Nature* 428: 145–150.

Karsch, F. J., J. E. Robinson, C. J. I. Woodfill, and M. B. Brown. 1989. Circannual cycles of luteinizing hormone and prolactic secretion in ewes during prolonged exposure to a fixed photoperiod: evidence for an endogenous reproductive rhythm. *Biol. Reprod.* 41: 1034–1046.

Lombardi, J. 1998. *Comparative Vertebrate Reproduction*. Kluwer Academic Publishers, Boston.

Martin, K. L. M., R. C. Van Winkle, J. E. Drais, and H. Lakisic. 2004. Beach-spawning fishes, terrestrial eggs, and air breathing. *Physiol. Biochem. Zool.* 77: 750–759.

Maston, G. A., and M. Ruvolo. 2002. Chorionic gonadotropin has a recent origin within primates and an evolutionary history of selection. *Mol. Biol. Evol.* 19: 320–335.

Matzuk, M. M., K. H. Burns, M. M. Viveiros, and J. J. Eppig. 2002. Intercellular communication in the mammalian ovary: Oocytes carry the conversation. *Science* 296: 2178–2180.

McCracken, J. A., E. E. Custer, and J. C. Lamsa. 1999. Luteolysis: A neuroendocrine-mediated event. *Physiol. Rev.* 79: 263–323.

McGlothlin, J. W., and E. D. Ketterson. 2008. Hormone-mediated suites as adaptations and evolutionary constraints. *Phil. Trans. Roy. Soc. London B* 363: 1611–1620.

Naylor, R., S. J. Richardson, and B. M. McAllan. 2008. Boom and bust: a review of the physiology of the marsupial genus *Antechinus*. *J. Comp. Physiol., B* 178: 545–562.

Niswender, G. D., J. L. Juengel, P. J. Silva, M. K. Rollyson, and E. W. McIntush. 2000. Mechanisms controlling the function and life span of the corpus luteum. *Physiol. Rev.* 80: 1–29.

Paria, B. C., J. Reese, S. K. Das, and S. K. Dey. 2002. Deciphering the cross-talk of implantation: Advances and challenges. *Science* 296: 2185–2188.

Pepling, M. E. 2006. From primordial germ cell to primordial follicle: Mammalian female germ cell development. *Genesis* 44: 622–632.

Plant, T. M., and G. R. Marshall. 2001. The functional significance of FSH in spermatogenesis and the control of its secretion in male primates. *Endocr. Rev.* 22: 764–786.

Primakoff, P., and D. G. Myles. 2002. Penetration, adhesion, and fusion in mammalian sperm–egg interaction. *Science* 296: 2183–2185.

Ren, D., B. Navarro, G. Perez, A. C. Jackson, S. Hsu, Q. Shi, J. L. Tilly, and D. E. Clapham. 2001. A sperm ion channel required for sperm motility and male fertility. *Nature* 413: 603–609.

Renfree, M. B., and G. Shaw. 2000. Diapause. *Annu. Rev. Physiol.* 62: 353–375.

Rocha, F., A. Guerra, and A. F. González. 2001. A review of reproductive strategies in cephalopods. *Bio. Rev.* 76: 291–304.

Ross, R. M. 1990. The evolution of sex-change mechanisms in fishes. *Environ. Biol. Fish.* 29: 81–93.

Rothchild, I. 2003. The yolkless egg and the evolution of eutherian viviparity. *Biol. Reprod.* 68: 337–357.

Savage, D. D., M. Becher, A. J. de la Torre, and R. J. Sutherland. 2002. Dose-dependent effects of prenatal ethanol exposure on synaptic plasticity and learning in mature offspring. *Alcohol Clin. Exp. Res.* 26: 1752–1758.

Setchell, B. P. 1998. The Parkes lecture—heat and the testis. *J. Reprod. Fertil.* 114: 179–194.

Sherwood, O. D. 2004. Relaxin's physiological roles and other diverse actions. *Endocr. Rev.* 25: 205–234.

Short, R. V. 1998. Difference between a testis and an ovary. *J. Exp. Zool.* 281: 359–361.

Sisneros, J. A., P. M. Forlano, D. L. Deitcher, and A. H. Bass. 2004. Steroid-dependent auditory plasticity leads to adaptive coupling of sender and receiver. *Science* 305: 404–407.

Spehr, M., G. Gisselmann, A. Poplawski, J. A. Riffell, C. H. Wetzel, R. K. Zimmer, and H. Hatt. 2003. Identification of a testicular odorant receptor mediating human sperm chemotaxis. *Science* 299: 2054–2058.

Stern, J. M., M. Konner, T. N. Herman, and S. Reichlin. 1986. Nursing behaviour, prolactin and postpartum amenorrhoea during prolonged lactation in American and !Kung mothers. *Clin. Endocrinol.* 25: 247–258.

Stewart, F., and W. R. Allen. 1995. Comparative aspects of the evolution and function of the chorionic gonadotrophins. *Reprod. Domest. Anim.* 30: 231–239.

Strassmann, B. I. 1996. The evolution of endometrial cycles and menstruation. *Q. Rev. Biol.* 71: 181–220.

Suter, K. J., W. J. Song, T. L. Sampson, J.-P. Wuarin, J. T. Saunders, F. E. Dudek, and S. M. Moenter. 2000. Genetic targeting of green fluorescent protein to gonadotropin-releasing hormone neurons: Characterization of whole-cell electrophysiological properties and morphology. *Endocrinology* 141: 412–419.

Taylor, J. A., M.-L. Goubillon, K. D. Broad, and J. E. Robinson. 2007. Steroid control of gonadotropin-releasing hormone secretion: Associated changes in pro-opiomelanocortin and preproenkephalin messenger RNA expression in the ovine hypothalamus. *Biol. Reprod.* 76: 524–531.

Tyndale-Biscoe, H., and M. Renfree. 1987. *Reproductive Physiology of Marsupials.* Cambridge University Press, Cambridge.

Vandenbergh, J. G. 2003. Prenatal hormone exposure and sexual variation. *Am. Sci.* 91: 218–225.

Wassarman, P. M., L. Jovine, and E. S. Litscher. 2001. A profile of fertilization in mammals. *Nat. Cell Biol.* 3: E59–E64.

Western, P. S., and A. H. Sinclair. 2001. Sex, genes, and heat: Triggers of diversity. *J. Exp. Zool.* 290: 624–631.

Wilkening, R. B., and G. Meschia. 1992. Current topic: Comparative physiology of placental oxygen transport. *Placenta* 13: 1–15.

Wilson, J. D., F. W. George, and M. B. Renfree. 1995. The endocrine role in mammalian sexual differentiation. *Recent Prog. Horm. Res.* 50: 349–364.

Zakar, T., and F. Hertelendy. 2007. Progesterone withdrawal: key to parturition. *Am. J. Obstet. Gynecol.* 196: 289–296.

Chapter 18

Avens, L., and K. J. Lohmann. 2004. Navigation and seasonal migratory orientation in juvenile sea turtles. *J. Exp. Biol.* 207: 1771–1778.

Berthold, P. 1996. *Control of Bird Migration.* Chapman & Hall, London.

Bingman, V. P., G. E. Hough, M, C. Kahn, and J. J. Siegel. 2003. The homing pigeon hippocampus and space: in search of adaptive specialization. *Brain Behav. Evol.* 62: 117–127.

Buzsaki, G. 2005. Neurons and navigation. *Nature* 436: 781–782.

Cheng, K., A. Narendra, S. Sommer, and R. Wehner. 2009. Traveling in clutter: navigation in the Central Australian desert ant *Melophorus bagoti. Behav. Processes* 80: 261–268.

Cochran, W. W., H. Mouritsen, and Wikelski, 2004. Migrating songbirds recalibrate their magnetic compass daily from twilight cues. *Science* 304: 405–408.

Collett, M., and T. S. Collett. 2006. Insect navigation: measuring travel distance across ground and through air. *Curr. Biol.* 16: R887–R890.

Collett, T. S., and M. Collett. 2000. Path integration in insects. *Curr. Opin. Neurobiol.* 10: 757–762.

DeBose, J. L., and G. A. Nevitt. 2008. The use of odors at different spatial scales: comparing birds with fish. *J. Chem. Ecol.* 34: 867–881.

Emlen, S. T. 1970. Celestial rotation: Its importance in the development of migratory orientation. *Science* 170: 1198–1201.

Emlen, S. T. 1975. The stellar orientation system of a migratory bird. *Sci. Am.* 233(2): 102–111.

Esch, H. E., S. Zhang, M. V. Srinivasan, and J. Tautz. 2001. Honeybee dances communicate distances measured by optic flow. *Nature* 411: 581–583.

Etheredge, J. A., S. M. Perez, O. R. Taylor, and R. Jander. 1999. Monarch butterflies (*Danaus plexippus* L.) use a magnetic compass for navigation. *Proc. Natl. Acad. Sci. U.S.A.* 96: 1345–1346.

Freake, M. J., R. Muheim, and J. B. Phillips. 2006. Magnetic maps in animals: a theory comes of age? *Q. Rev. Biol.* 81: 327–347.

Giurfa, M., and E. A. Capaldi. 1999. Vectors, routes, and maps: New discoveries about navigation in insects. *Trends Neurosci.* 22: 237–242.

Harris, M. 1957. *Something about a Soldier.* Macmillan, New York.

Healy, S. (ed.). 1998. *Spatial Representation in Animals.* Oxford University Press, Oxford, UK.

Heinrich, B. 2014. *The Homing Instinct: Meaning and Mystery in Animal Migration.* Houghton Mifflin Harcourt.

Heyers, D., M. Manns, H. Luksch, O. Güntürkün, and H. Mouritsen. 2007. A visual pathway links brain structures active during magnetic compass orientation in migratory birds. *PloS One* 2(9): e930. doi 10.1371/journal.pone.0000937

Homberg, U. 2004. In search of the sky compass in the insect brain. *Naturwissenschaften* 91: 199–208.

Homberg, U., S. Heize, K. Pfeiffer, M. Kinoshita, and B. el Jundi. 2011. Central neural coding of sky polarization in insects. *Phil. Trans. R. Soc. B* 366: 680–687.

Iglói, K., C. F. Doeller, A. Berthoz, L. Rondi-Reig, and N. Burgess. 2010. Lateralized human hippocampal activity predicts navigation based on sequence or place memory. *Proc. Natl. Acad. Sci. U.S.A.* 107: 14466–14471.

Jacobs, L. F. 2003. The evolution of the cognitive map. *Brain Behav. Evol.* 62: 128–139.

Keeton, W. T. 1971. Magnets interfere with pigeon homing. *Proc. Natl. Acad. Sci. U.S.A.* 68: 102–106.

Keeton, W. T. 1974. The mystery of pigeon homing. *Sci. Am.* 231(6): 96–107. (Reprinted in Eisner, T., and E. O. Wilson. 1975. *Animal Behavior: Readings from Scientific American.* Freeman, San Francisco.)

Kravitz D. J., K. S. Saleem, C. I. Baker, and M. Mishkin. 2011. A new neural framework for visuospatial processing. *Nat. Rev. Neurosci.* 12217–12230.

Labhart, T., and E. P. Meyer. 2002. Neural mechanisms in insect navigation: Polarization compass and odometer. *Curr. Opin. Neurobiol.* 12: 707–714.

Lehrer, M. (ed.). 1997. *Orientation and Communication in Arthropods.* Birkhäuser, Basel, Switzerland.

Lew, A. R. 2011. Looking beyond the boundaries: time to put landmarks back on the cognitive map? *Psychol. Bull.* 137: 484–507.

Lohmann, K. J., C. M. Lohmann, and N. F. Putnam. 2007. Magnetic maps in animals: nature's GPS. *J. Exp. Biol.* 210: 3697–3705.

Lohmann, K. J., C. M. Lohmann, L. M. Ehrhart, D. A. Bagley, and T. Swing. 2004. Animal behaviour: Geomagnetic map use in sea-turtle navigation. *Nature* 428: 909–910.

Menzel R., R. J. De Marco, and U. Greggers. 2006. Spatial memory, navigation and dance behaviour in *Apis mellifera. J. Comp. Physiol., A* 192: 889–903.

Mizumori S. J., C. B. Puryear, and A. K. Martig. 2009. Basal ganglia contributions to adaptive navigation. *Behav. Brain Res.* 199: 32–42.

Mouritsen, H. 2003. Spatiotemporal orientation strategies in long-distance migrants. In P. Berthold, E. Gwinner, and E. Sonnenschein (eds.), *Avian Migration*, pp. 493–513. Springer, Berlin.

Muheim, R. 2011. Behavioural and physiological mechanisms of polarized light sensitivity in birds. *Philos. Trans. R. Soc. Lond. B.* 366: 763–771.

Muheim, R., J. B. Phillips, and S. Åkesson 2006. Polarized light cues underlie compass calibration in migratory songbirds. *Science* 313: 837–839.

O'Keefe, J., and L. Nadel. 1978. *The Hippocampus as a Cognitive Map.* Oxford University Press, Oxford, UK.

Papi, F. (ed.). 1992. *Animal Homing.* Chapman & Hall, London.

Pennisi, E. 2003. Monarchs check clock to chart migration route. *Science* 300: 1216–1217.

Poucet, B., P. P. Lenck-Santini, V. Paz-Villagran, and E. Save. 2003. Place cells, neocortex and spatial navigation: a short review. *J. Physiol. Paris* 97: 537–546.

Redish, A. D. 1999. *Beyond the Cognitive Map: From Place Cells to Episodic Memory*. MIT Press, Cambridge, MA.

Rochefort, C., A. Arabo, M. André, B. Poucet, E. Save, and L. Rondi-Reig. 2011. Cerebellum shapes Hippocampal spatial code. *Science* 334: 385–389.

von Frisch, K. 1967. *The Dance Language and Orientation of Bees*. Harvard University Press, Cambridge, MA.

Wallraff, H. G. 2004. Avian olfactory navigation: its empirical foundation and conceptual state. *Anim. Behav.* 67: 189–204.

Warrant, E., and M. Dacke. 2011. Vision and visual navigation in nocturnal insects. *Annu. Rev. Entomol.* 56: 239–254.

Wiltschko, W., and R. Wiltschko. 2005. Magnetic orientation and magnetoreception in birds and other animals. *J. Comp. Physiol. A* 191: 675–693.

Wittlinger, M., R. Wehner, and H. Wolf. 2007. The desert ant odometer: a stride integrator that accounts for stride length and walking speed. *J. Exp. Biol.* 210: 198–207.

Wolbers, T., and M. Hegarty. 2010. What determines our navigational abilities? *Trends Cogn. Sci.* 14: 138–146.

Chapter 19

Alford, S., E. Schwartz, and G. V. DiPrisco. 2003. The pharmacology of vertebrate spinal central pattern generators. *Neuroscientist* 9: 217–228.

Anstey, M. L., S. M. Rogers. S. R. Ott, M. Burrows, and S. J. Simpson. 2009. Serotonin mediates behavioral gregarization underlying swarm formation in desert locusts. *Science* 323: 627–630.

Apps, R., and M. Garwicz. 2005. Anatomical and physiological foundations of cerebellar information processing. *Nat. Rev. Neurosci.* 6: 297–311.

Bizzi, E., M. C. Tresch, P. Saltiel, and A. d'Avella. 2000. New perspectives on spinal motor systems. *Nat. Rev. Neurosci.* 1: 101–108.

Borgmann, A. and A. Büschges. 2015. Insect motor control: methodological advances, descending control, and inter-leg coordination on the move. *Curr. Opin. Neurobiol.* 33: 8–15.

Briggman, K. L., and W. B. Kristan, Jr. 2008. Multifunctional pattern-generating circuits. *Annu. Rev. Neurosci.* 31: 271–294.

Burrows, M. 1996. *The Neurobiology of an Insect Brain*. Oxford University Press, Oxford, UK.

Büschges, A., T. Akay, J. P. Gabriel, and J. Schmidt. 2008. Organizing network action for locomotion: insights from studying insect walking. *Brain Res. Rev.* 57: 162–171.

Camhi, J. M. 1984. *Neuroethology*. Sinauer, Sunderland, MA.

Cattaneo, L., and G. Rizzolatti. 2009. The mirror neuron system. *Arch. Neurol.* 66: 557–560.

Chesselet, M.-F., and J. M. Delfs. 1996. Basal ganglia and movement disorders: An update. *Trends Neurosci.* 19: 417–422.

Clarac, F., D. Cattaert, and D. Le Ray. 2000. Central control components of a "simple" stretch reflex. *Trends Neurosci.* 23: 199–208.

Cohen, A. H., S. Rossignol, and S. Grillner (eds.). 1988. *Neural Control of Rhythmic Movements in Vertebrates*. Wiley, New York.

Crapse, T. B., and M. A. Sommer. 2008. Corollary discharge across the animal kingdom. *Nat. Rev. Neurosci.* 9: 587–600.

Cruz-Bermudez, N. D., and E. Marder. 2007. Multiple modulators act on the cardiac ganglion of the crab, *Cancer borealis*. *J. Exp. Biol.* 210: 2873–2884.

Delcomyn, F. 1998. *Foundations of Neurobiology*. W. H. Freeman, New York.

Dickinson, M. H., C. T. Farley, R. J. Full, M. A. R. Koehl, R. Kram, and S. Lehman. 2000. How animals move: An integrative view. *Science* 288: 100–106.

Dickinson, P. S. 2006. Neuromodulation of central pattern generators in invertebrates and vertebrates. *Curr. Opin. Neurobiol.* 16: 604–614.

Dinstein, I., C. Thomas, M. Behrmann, and D. J. Heeger. 2008. A mirror up to nature. *Curr. Biol.* 18: R13–R18.

Dominici, N., Y. P. Ivanenko, G. Cappellini, A. d'Avella, V. Mondì, M. Cicchese, A. Fabiano, T. Silei, A. Di Paolo, C. Giannini, R. E. Poppele, and F. Lacquaniti. 2011. Locomotor primitives in newborn babies and their development. *Science* 334: 997–999.

Doyon, J., P. Bellec, R. Amsel, V. Penhune, O. Monchi, J. Carrier, S. Lehéricy, and H. Benali. 2009. Contributions of the basal ganglia and functionally related brain structures to motor learning. *Behav. Brain Res.* 199: 61–75.

Edwards, D. H., W. J. Heitler, and F. B. Krasne. 1999. Fifty years of a command neuron: The neurobiology of escape behavior in the crayfish. *Trends Neurosci.* 22: 153–161.

Ferrell, W. R., and U. Proske (eds.). 1995. *Neural Control of Movement*. Plenum, New York.

Georgopoulos, A. P. 1991. Higher motor control. *Annu. Rev. Neurosci.* 14: 361–377.

Georgopoulos, A. P. 1997. Neural networks and motor control. *Neuroscientist* 3: 52–60.

Getting, P. A. 1989. Emerging principles governing the operation of neural networks. *Annu. Rev. Neurosci.* 12: 185–204.

Gillette, R. and J. W. Brown. 2015. The sea-slug *Pleurobranchaea californica*: a signpost species in the evolution of complex nervous systems and behavior. *Integr. Comp. Biol.* 55: 1058–1069.

Graziano, M. 2006. The organization of behavioral repertoire in motor cortex. *Annu. Rev. Neurosci.* 29: 105–134.

Grillner, S., J. Hellgren, A. Ménard, K. Saitoh, and M. A. Wikström. 2005. Mechanisms for selection of basic motor programs—roles for the striatum and pallidum. *Trends Neurosci.* 28: 364–370.

Grillner, S., B. Robertson, and M. Stephenson-Jones. 2013. The evolutionary origin of the vertebrate basal ganglia and its role in action selection. *J. Physiol.* 591: 5425–5431.

Grillner, S., P. Wallén, K. Siatoh, A. Kozlov, and B. Robertson. 2008. Neural bases of goal-directed locomotion in vertebrates—an overview. *Brain Res. Rev.* 57: 2–12.

Harris-Warrick, R. M., and E. Marder. 1991. Modulation of neural networks for behavior. *Annu. Rev. Neurosci.* 14: 39–57.

Heckman, C. J., C. Mottram, K. Quinlan, R. Theiss, and J. Schuster. 2009. Motoneuron excitability: the importance of neuromodulatory inputs. *Clin. Neurophysiol.* 120: 2040–2054.

Hodson-Tole, E. F., and J. M. Wakeling. 2009. Motor unit recruitment for dynamic tasks: current understanding and future directions. *J. Comp Physiol., B* 179: 57–66.

Holstege, G. 1991. Descending motor pathways and the spinal motor system: limbic and non-limbic components. *Prog. Brain Res.* 87: 307–421.

Hooper, S. L., and R. A. DiCaprio. 2004. Crustacean motor pattern generator networks. *Neurosignals* 13: 50–69.

Ijspeert, A. J. 2008. Central pattern generators for locomotion control in animals and robots: a review. *Neural Netw.* 21: 642–653.

Ijspeert, A. J., A. Crespi, D. Ryczko, and J.-M. Cabelguen. 2007. From swimming to walking with a salamander robot driven by a spinal cord model. *Science* 315: 1416–1420.

Ito, M. 2006. Cerebellar circuitry as a neuronal machine. *Prog. Neurobiol.* 78: 272–302.

Jin, X. and R. M. Costa. 2015. Shaping action sequences in basal ganglia circuits. *Curr. Opin. Neurobiol.* 33: 188–196.

Johnson, B. R., and S. L. Hooper. 1992. Overview of the stomatogastric nervous system. In R. M. Harris-Warrick, W. Marder, A. I. Selverston, and M. Moulins (eds.), *Dynamic Biological Networks*, pp. 1–30. MIT Press, Cambridge, MA.

Kandel, E. R. 1976. *Cellular Basis of Behavior*. W. H. Freeman, San Francisco.

Kandel, E. R., J. H. Schwartz, and T. M. Jessell (eds.). 2000. *Principles of Neural Science*, 4th ed. McGraw-Hill, New York.

Kiehn, O. 2006. Locomotor circuits in the mammalian spinal cord. *Annu. Rev. Neurosci.* 29: 279–306.

Kreitzer, A. C. 2009. Physiology and pharmacology of striatal neurons. *Annu. Rev. Neurosci.* 32: 127–147.

Kristan, W. B., Jr., R. L. Calabrese, and W. O. Friesen. 2005. Neuronal control of leech behavior. *Prog. Neurobiol.* 76: 279–327.

Kullander, K. 2005. Genetics moving to neuronal networks. *Trends Neurosci.* 28: 239–247.

Latash, M. L., and F. Lestienne (eds.). 2006. *Motor Control and Learning*. Springer, New York.

Marder, E., and D. Bucher. 2001. Central pattern generators and the control of rhythmic movements. *Curr. Biol.* 11: R986–R996.

Marder, E., and R. L. Calabrese. 1996. Principles of rhythmic motor pattern generation. *Physiol. Rev.* 76: 687–717.

McCrea, D. A., and I. A. Rybak. 2008. Organization of mammalian locomotor rhythm and pattern generation. *Brain Res. Rev.* 57: 134–146.

Mullins, O. J., J. T. Hackett, J. T. Buchanan, and W. O. Friesen. 2011. Neuronal control of swimming behavior: comparison of vertebrate and invertebrate model systems. *Prog. Neurobiol.* 93: 244–269.

Mulloney, B., and C. Smarandache. 2010. Fifty years of CPGs: two neuroethological papers that shaped the course of neuroscience. *Front. Behav. Neurosci.* 4: 45. doi: 10.3389/fnbeh.2010.00045.

Ölveczky, B. P. 2011. Motoring ahead with rodents. *Curr. Opin. Neurobiol.* 21: 571–578.

Orlovsky, G. N., T. G. Deliagina, and S. Grillner. 1999. *Neuronal Control of Locomotion: From Mollusc to Man*. Oxford University Press, Oxford, UK.

Pearson, K. G. 1993. Common features of motor control in vertebrates and invertebrates. *Annu. Rev. Neurosci.* 16: 265–297.

Poulet, J. F. A., and B. Hedwig. 2007. New insights into corollary discharges mediated by identified neural pathways. *Trends Neurosci.* 30: 14–21.

Prochazka, A. and P. Ellaway. 2012. Sensory systems in the control of movement. *Compr. Physiol.* 2: 2615–2627.

Ritzmann, R. E., and A. Büschges. 2007. Adaptive motor behavior in insects. *Curr. Opin. Neurobiol.* 17: 629–636.

Rizzolatti, G., and G. Luppino. 2001. The cortical motor system. *Neuron* 31: 889–901.

Rossignol S., R. Dubuc, and J. P. Gossard. 2006. Dynamic sensorimotor interactions in locomotion. *Physiol. Rev.* 86: 89–154.

Rothwell, J. 1994. *Control of Human Voluntary Movement*, 2nd ed. Chapman and Hall, London.

Simmons, P. J., and D. Young. 1999. *Nerve Cells and Animal Behavior*, 2nd ed. Cambridge University Press, Cambridge.

Sommer, M. A., and R. H. Wurtz. 2008. Brain circuits for the internal monitoring of movements. *Annu. Rev. Neurosci.* 31: 317–338.

Squire, L. R., F. E. Bloom, and N. C. Spitzer (eds.). 2008. *Fundamental Neuroscience*, 3rd ed. Academic/Elsevier, San Diego, CA.

Stein, P. S. G., S. Grillner, A. I. Selverston, and D. G. Stuart (eds.). 1997. *Neurons, Networks, and Motor Behavior*. MIT Press, Cambridge, MA.

Tanji, J. 2001. Sequential organization of multiple movements: Involvement of cortical motor areas. *Annu. Rev. Neurosci.* 24: 631–651.

Turner, R. S., and M. Desmurget. 2010. Basal ganglia contributions to motor control: a vigorous tutor. *Curr. Opin. Neurobiol.* 20: 704–716.

Wagenaar, D. A., M. S. Hamilton, T. Huang, W. B. Kristan, and K. A. French. 2010. A hormone-activated central pattern generator for courtship. *Curr. Biol.* 20: 487–495.

Wyse, G. A. 2010. Central pattern generation underlying *Limulus* rhythmic behavior patterns. *Curr. Zool.* 56: 537–549.

Zill, S. N., and B. R. Keller. 2009. Neurobiology: reconstructing the neural control of leg coordination. *Curr. Biol.* 19: R371–R373.

Chapter 20

Arif, S. H. 2009. A Ca^{2+}-binding protein with numerous roles and uses: parvalbumin in molecular biology and physiology. *BioEssays* 31: 410–421.

Assad, C., B. Rasnow, and P. K. Stoddard. 1999. Electric organ discharges and electric images during electrolocation. *J. Exp. Biol.* 202: 1185–1193.

Berchtold, M. W., H. Brinkmeier, and M. Müntener. 2000. Calcium ion in skeletal muscle: Its crucial role for muscle function, plasticity, and disease. *Physiol. Rev.* 80: 1215–1265.

Bernal, D., K. A. Dickson, R. E. Shadwick, and J. B. Graham. 2001. Review: Analysis of the evolutionary convergence for high performance swimming in lamnid sharks and tunas. *Comp. Biochem. Physiol. A* 129: 695–726.

Biserova, N. M., and H.-J. Pflüger. 2004. The ultrastructure of locust pleuroaxillary "steering" muscles in comparison to other skeletal muscles. *Zoology* 107: 229–242.

Borovikov, Y. S., N. S. Sheludʹko, and S. V. Avrova. 2010. Molluscan twitching can control actin-myosin interaction during ATPase cycle. *Arch. Biochem. Biophys.* 495: 122–128.

Bossé, Y., A. Sobieszek, P. D. Pané, and C. Y. Seow. 2008. Length adaptation of airway smooth muscle. *Proc. Am. Thorac. Soc.* 5: 62–67.

Brini, M., and E. Carafoli. 2009. Calcium pumps in health and disease. *Physiol. Rev.* 1341–1378.

Callister, R. J., P. A. Pierce, J. C. McDonagh, and D. G. Stuart. 2005. Slow-tonic muscle fibers and their potential innervation in the turtle, *Pseudemys (Trachemys) scripta elegans. J. Morphol.* 264: 62–74.

Conley, K. E., and S. L. Lindstedt. 2002. Energy-saving mechanisms in muscle: The minimization strategy. *J. Exp. Biol.* 205: 2175–2181.

Di Biase, V., and C. Franzini-Armstrong. 2005. Evolution of skeletal type e-c coupling: a novel means of controlling calcium delivery. *J. Cell Biol.* 171: 695–704.

Dudley, R. 2000. *The Biomechanics of Insect Flight.* Princeton University Press, Princeton, NJ.

Edman, K. A. P., and R. K Josephson. 2007. Determinants of force rise time during isometric contraction of frog muscle fibres. *J. Physiol.* 580: 1007–1019.

Fernández, D. A., and J. Calvo. 2009. Fish muscle: the exceptional case of notothenioids. *Fish Physiol. Biochem.* 35: 43–52.

Few, W. P., and H. H. Zakon. 2007. Sex differences in and hormonal regulation of Kv1 potassium channel gene expression in the electric organ: molecular control of a social signal. *Develop. Neurobiol.* 67: 535–549.

Franzini-Armstrong, C., and F. Protasi. 1997. Ryanodine receptors of striated muscles: A complex channel capable of multiple interactions. *Physiol. Rev.* 77: 699–729.

Funabara, D., S. Kanoh, M. J. Siegman, T. M. Butler, D. J. Hartshorne, and S. Watabe. 2005. Twitchin as a regulator of catch contraction in molluscan smooth muscle. *J. Muscle Res. Cell Motil.* 26: 455–460.

Galler, S. 2008. Molecular basis of the catch state in molluscan smooth muscles: a catchy challenge. *J. Muscle Res. Cell Motil.* 29: 73–99.

Galler, S., J. Litzlbauer, M. Kröss, and H. Grassberger. 2010. The highly efficient holding function of the mollusk "catch" muscle is not based on decelerated myosin head cross-bridge cycles. *Proc. R. Soc. London, Ser. B* 277: 803–808.

Gordon, A. M., E. Homsher, and M. Regnier. 2000. Regulation of contraction in striated muscle. *Physiol. Rev.* 80: 853–924.

Gordon, S., and M. H. Dickinson. 2006. Role of calcium in the regulation of mechanical power in insect flight. *Proc. Nat. Acad. Sci. U.S.A.* 103: 4311–4315.

Gotter, A. L., M. A. Kaetzel, and J. R. Dedman. 1998. *Electrophorus electricus* as a model system for the study of membrane excitability. *Comp. Biochem. Physiol., A* 119: 225–241.

Granzier, H., J. Radke, J. Royal, Y. Wu, T. C. Irving, M. Gotthardt, and S. Labeit. 2007. Functional genomics of chicken, mouse, and human titin supports splice diversity as an important mechanism for regulating biomechanics of striated muscle. *Am. J. Physiol. Regul. Integr. Comp. Physiol.* 293: R557–R567.

Gregorio, C. C., H. Granzier, H. Sorimachi, and S. Labeit. 1999. Muscle assembly: A titanic achievement? *Curr. Opin. Cell Biol.* 11: 18–25.

Heuser, J. E. 1983. Structure of the myosin crossbridge lattice in insect flight muscle. *J. Mol. Biol.* 169: 123–154.

Heuser, J. E., and R. Cooke. 1983. Actin-myosin interactions visualized by the quick-freeze, deep-etch replica technique. *J. Mol. Biol.* 169: 97–122.

Hoh, J. F. Y. 2002. "Superfast" or masticatory myosin and the evolution of jaw-closing muscles of vertebrates. *J. Exp. Biol.* 205: 2203–2210.

Hooper, S. L., and J. B. Thuma. 2005. Invertebrate muscles: Muscle specific genes and proteins. *Physiol. Rev.* 85: 1001–1060. [A review of numerous proteins found in several invertebrate phyla and a comprehensive lists of references.]

Huxley, H. E. 1969. The mechanism of muscular contraction. *Science* 164: 1356–1366.

Johnston, I. A., D. A. Fernandez, J. Calvo, V. L. A. Vieira, A. W. North, M. Abercromby, and T. Garland, Jr. 2003. Reduction in muscle fibre number during the adaptive radiation of notothenioid fishes: A phylogenetic perspective. *J. Exp. Biol.* 206: 2595–2609.

Josephson, R. K., J. G. Malamud, and D. R. Stokes. 2000. Asynchronous muscle: A primer. *J. Exp. Biol.* 203: 2713–2722.

Katz, P. S. 2006. Comparative neurophysiology: An electric convergence in fish. *Curr. Biol.* 16: R327–R330.

Kessel, R. G., and R. H. Kardon. 1979. *Tissues and Organs. A Text-Atlas of Scanning Electron Microscopy.* Freeman, San Francisco.

Lanner, J. T., D. K. Georgiou, A. D. Joshi, and S. L. Hamilton. 2010. Ryanodine receptors: Structure, expression, molecular details, and function in calcium release. *Cold Spring Harb. Perspect. Biol.* 2(11): a003996.

Lieber, R. L., and S. R. Ward. 2011. Skeletal muscle design to meet functional demands. *Phil. Trans. R. Soc. B* 366: 1466–1467.

LeWinter, M. M., Y. Wu, S. Labeit, and H. Granzier. 2007. Cardiac titin: Structure, functions and role in disease. *Clin. Chim. Acta* 375: 1–9.

Lindstedt, S. L., P. C. LaStayo, and T. E. Reich. 2001. When active muscles lengthen: Properties and consequences of eccentric contractions. *News Physiol. Sci.* 16: 256–261.

Lindstedt, S. L., T. E. Reich, P. Keim, and P. C. LaStayo. 2002. Do muscles function as adaptable locomotor springs? *J. Exp. Biol.* 205: 2211–2216.

Marden, J. H. 2000. Variability in the size, composition, and function of insect flight muscles. *Annu. Rev. Physiol.* 62: 157–178.

Moller, P. 1995. *Electric Fishes: History and Behavior.* Chapman & Hall, London.

Murphy, R. A., and C. M. Rembold. 2005. The latch-bridge hypothesis of smooth muscle contraction. *Can. J. Physiol. Pharmacol.* 83: 857–864.

Noguchi, H., S. Takemori, J. Kajiwara, M. Kimura, K. Maruyama, and S. Kimura. 2007. Chicken breast muscle connectin: Passive tension and I-band region primary structure. *J. Mol. Biol.* 370: 213–219.

Notsu, E., and A. Matsuno. 2004. An ultrastructural and biochemical study of foot structure in "catch" smooth muscle cells of a clam. *Cell Struct. Funct.* 29: 43–48.

O'Connell, B., R. Blazev, and G. M. M. Stephenson. 2006. Electrophoretic and functional identification of two troponin C isoforms in toad skeletal muscle fibers. *Am. J. Physiol.* 290: C515–C523.

Periasamy, M., and A. Kalyanasundaram. 2007. SERCA pump isoforms: Their role in calcium transport and disease. *Muscle Nerve* 35: 430–442.

Perry, S. V. 2001. Vertebrate tropomyosin: distribution, properties and function. *J. Muscle Res. Cell Motil.* 22: 5–49.

Purslow, P. P. 2002. The structure and functional significance of variations in the connective tissue within muscle. *Comp. Biochem. Physiol., A* 133: 947–966.

Rall, J. A. 2014. Mechanism of Muscular Contraction. Perspectives in Physiology published on behalf of the American Physiological Society by Springer, New York. http://www.springer.com/series/11779

Rome, L. C., and A. A. Klimov. 2000. Superfast contractions without superfast energetics: ATP usage by SR-Ca²⁺ pumps and crossbridges in toadfish swimbladder muscle. *J. Physiol. (London)* 526: 279–286.

Rome, L. C., C. Cook, D. A. Syme, M. A. Connaughton, M. Ashley-Ross, A. Klimov, B. Tikunov, and Y. E. Goldman. 1999. Trading force for speed: Why superfast crossbridge kinetics leads to superlow forces. *Proc. Natl. Acad. Sci. U.S.A.* 96: 5826–5831.

Rome, L. C., D. A. Syme, S. Hollingworth, S. L. Lindstedt, and S. M. Baylor. 1996. The whistle and the rattle: The design of sound producing muscles. *Proc. Natl. Acad. Sci. U.S.A.* 93: 8095–8100.

Rossi, A. E., and R. T. Dirksen. 2006. Sarcoplasmic reticulum: The dynamic calcium governor of muscle. *Muscle Nerve* 33: 715–731.

Ruan, Y. C., W. Zhou, and H. C. Chan. 2011. Regulation of smooth muscle contraction by epithelium: role of prostaglandins. *Physiology* 26: 156–170.

Schmitz, H., M. C. Reedy, M. K. Reedy, R. T. Tregear, and K. A. Taylor. 1997. Tomographic three-dimensional reconstruction of insect flight muscle partially relaxed by AMPPNP and ethylene glycol. *J. Cell Biol.* 139: 695–707.

Shadwick, R. E. 2005. How tunas and lamnid sharks swim: an evolutionary convergence. *Am. Sci.* 93: 524–531.

Somlyo, A. P., and A. V. Somlyo. 2003. Ca²⁺ sensitivity of smooth muscle and nonmuscle myosin II: Modulated by G proteins, kinases, and myosin phosphatase. *Physiol. Rev.* 83: 1325–1358.

Stokes, D. R., and R. K. Josephson. 2004. Power and control muscles of cicada song: structural and contractile heterogeneity. *J. Comp. Physiol. A* 190: 279–290.

Strohm, E., and W. Daniels. 2003. Ultrastructure meets reproductive success: Performance of a spheid wasp is correlated with the fine structure of the flight-muscle mitochondria. *Proc. R. Soc. London, Ser. B* 270: 749–754.

Sutlive, T. G., J. R. McClung, and S. J. Goldberg. 1999. Whole-muscle and motor-unit contractile properties of the styloglossus muscle in rat. *J. Neurophysiol.* 82: 584–592.

Swartz, D. R., Z. Yang, A. Sen, S. B. Tikunova, and J. P. Davis. 2006. Myofibrillar troponin exists in three states and there is signal transduction along skeletal myofibrillar thin filaments. *J. Mol. Biol.* 361: 420–435.

Unguez, G. A., and H. H. Zakon. 2002. Skeletal muscle transformation into electric organ in *S. macrurus* depends on innervation. *J. Neurobiol.* 53: 391–402.

Vale, R. D., and R. A. Milligan. 2000. The way things move: Looking under the hood of molecular motor proteins. *Science* 288: 88–95.

Vigoreaux, J. O. 2001. Genetics of the *Drosophila* flight muscle myofibril: A window into the biology of complex systems. *BioEssays* 23: 1047–1063.

Vogel, S. 2001. *Prime Mover: A Natural History of Muscle*. W. W. Norton, New York.

Warrick, D. R., B. W. Tobalske, and D. R. Powers. 2005. Aerodynamics of the hovering hummingbird. *Nature* 435: 1094–1097.

Wu, C. H. 1984. Electric fish and the discovery of animal electricity. *Am. Sci.* 72: 598–607.

Xu, C., R. Craig, L. Tobacman, R. Horowitz, and W. Lehman. 1999. Tropomyosin positions in regulated thin filaments revealed by cryoelectron microscopy. *Biophys. J.* 77: 985–992.

Young, I. S., C. L. Harwood, and L. C. Rome. 2003. Cross-bridge blocker BTS permits direct measurement of SR Ca²⁺ pump ATP utilization in toadfish swimbladder muscle fibers. *Am. J. Physiol.* 285: C781–C787.

Young, R. C. 2007. Myocytes, myometrium, and uterine contractions. *Ann. N.Y. Acad. Sci.* 1101: 72–84.

Zakon, H. H., Y. Lu, D. J. Swickl, and D. M. Hillis. 2006. Sodium channel genes and the evolution of diversity in communication signals of electric fishes: convergent molecular evolution. *Proc. Natl. Acad. Sci. U.S.A.* 103: 3675–3680.

Zhang, W.-C., and 15 additional authors. 2009. Myosin light chain kinase is necessary for tonic airway smooth muscle contraction. *J. Biol. Chem.* 285: 5522–5531.

Zhang, W., and S. J. Gunst. 2008. Interactions of airway smooth muscle cells with their tissue matrix. *Proc. Am. Thorac. Soc.* 5: 32–39.

Chapter 21

Aagaard, P. J. L. Andersen, M. Bennekou, B. Larsson, J. L. Olesen, R. Crameri, S. P. Magnusson, and M. Kjaer. 2011. Effects of resistance training on endurance capacity and muscle fiber composition in young top-level cyclists. *Scand. J. Med. Sci. Sports* 21: e298–e307.

Aagaard, P., and J. L. Andersen. 2010. Effects of strength training on endurance capacity in top-level endurance athletes. *Scand. J. Med. Sci. Sports* 20 (Suppl. 2): 39–47.

Andersen, J. B., B. C. Rourke, V. J. Caiozzo, A. F. Bennett, and J. W. Hicks. 2005. Postprandial cardiac hypertrophy in pythons. *Nature* 434: 37–38.

Audet, G. N., T. H. Meek, T. Garland Jr and I. M. Olfert. 2011. Expression of angiogenic regulators and skeletal muscle capillarity in selectively bred high aerobic capacity mice. *Exp. Physiol.* 96: 1138-1150.

Bellinge, R. H. S., D. A. Liberles, S. P. A. Inschi, P. A. O'Brien, and G. K. Tay. 2005. Myostatin and its implications on animal breeding: a review. *Anim. Genet.* 36: 1–6.

Booth, F. W., and P. D. Neufer. 2005. Exercise controls gene expression. *Am. Sci.* 93: 28–35.

Brooks, N. E., K. H. Myburgh and K. B. Storey. 2011. Myostatin levels in skeletal muscle of hibernating ground squirrels. *J. Exp. Biol.* 214: 2522-2527.

Bruusgaard, J. C., I. B. Johansen, I. M. Egner, Z. A. Rana, and K. Gundersen. 2010. Myonuclei acquired by overload exercise precede hypertrophy and are not lost on detraining. *Proc. Natl. Acad. Sci. U.S.A.* 107: 15111–15116.

Carter, H.N., C.C.W. Chen, and D.A. Hood. 2015. Mitochondria, muscle health, and exercise with advancing age. *Physiology* 30: 208–223.

Catalucci, D., M. V. G. Latronico, O. Ellingsen, and G. Condorelli. 2008. Physiological myocardial hypertrophy: how and why? *Front. Biosci.* 13: 312–324.

Donahue, S. W., S. A. Galley, M. R. Vaughan, P. Patterson-Buckendahl, L. M. Demers, J. L. Vance, and M. E. McGee. 2006. Parathyroid hormone may maintain bone formation in hibernating black bears (*Ursus americanus*) to prevent disuse osteoporosis. *J. Exp. Biol.* 209: 1630–1638.

Edström, E., M. Altun, E. Bergman, H. Johnson, S. Kullberg, V. Ramírez-León, and B. Ulfhake. 2007. Factors contributing to neuromuscular impairment and sarcopenia during aging. *Physiol. Behav.* 92: 129-135.

Emery, A. E. H. 2002. The muscular dystrophies. *Lancet* 359: 687–695.

Goldspink, G. 2005. Mechanical signals, IGF-I gene splicing, and muscle adaptation. *Physiology* 20: 232–238.

Goldspink, G. 2006. Impairment of IGF-I gene splicing and MGF expression associated with muscle wasting. *Int. J. Biochem. Cell Biol.* 38: 481–489.

Gustafsson, T., T. Osterlund, J. N. Flanagan, F. von Waldén, T. A. Trappe, R. M. Linnehau and P. A. Tesch. 2010. Effects of 3 days unloading on molecular regulators of muscle size in humans. *J. Appl. Physiol.* 109: 721–727.

Haizlip, K. M., B. C. Harrison, and L. A. Leinwand. 2015. Sex-based differences in skeletal muscle kinetics and fiber-type composition. *Physiology* 30: 30–39.

Harlow, H. J., T. Lohuis, T. D. I. Beck, and P. A. Iaizzo. 2001. Muscle strength in overwintering bears. *Nature* 409: 997.

Hoffman, E. P., and G. A. Nader. 2004. Balancing muscle hypertrophy and atrophy. *Nat. Med.* 10: 584–585.

Hood, D. A., I. Irrcher, V. Ljubicic, and A.-M. Joseph. 2006. Coordination of metabolic plasticity in skeletal muscle. *J. Exp. Biol.* 209: 2265–2275.

Hoppeler, H., and M. Flück. 2002. Normal mammalian skeletal muscle and its phenotypic plasticity. *J. Exp. Biol.* 205: 2143–2152.

Hoppeler, H., H. Howald, K. Conley, S. L. Lindstedt, H. Claassen, P. Vock, and E. R. Weibel. 1985. Endurance training in humans: Aerobic capacity and structure of skeletal muscle. *J. Appl. Physiol.* 59: 320–327.

Hudson, N. J., and C. E. Franklin. 2002. Effect of aestivation on muscle characteristics and locomotor performance in the green-striped burrowing frog, *Cyclorana alboguttata*. *J. Comp. Physiol., B* 172: 177–182.

Hudson, N. J., and C. E. Franklin. 2002. Maintaining muscle mass during extended disuse: Aestivating frogs as a model species. *J. Exp. Biol.* 205: 2297–2303.

Hudson, N. J., M. B. Bennett, and C. E. Franklin. 2004. Effect of aestivation on long bone mechanical properties in the green-striped burrowing frog, *Cyclorana alboguttata*. *J. Exp. Biol.* 207: 475–482.

Hudson, N. J., S. A. Lehnert, A. B. Ingham, B. Symonds, C. E. Franklin, and G. S. Harper. 2006. Lessons from an estivating frog: sparing muscle protein despite starvation and disuse. *Am. J. Physiol.* 290: R836–R843.

Ingjer, F. 1979. Effects of endurance training on muscle fibre ATP-ase activity, capillary supply and mitochondrial content in man. *J. Physiol.* 294: 419–432.

Kayes, S. M., R. L. Cramp, N. J. Hudson, and C. E. Franklin. 2009. Surviving the drought: burrowing frogs save energy by increasing mitochondrial coupling. *J. Exp. Biol.* 212: 2248–2253.

Lin, D. C., J. D. Hershey, J. S. Mattoon, and C. T. Robbins. 2012. Skeletal muscles of hibernating brown bears are unusually resistant to effects of denervation. *J. Exp. Biol.* 215: 2081–2087.

Lin, J., H. Wu, P. T. Tarr, C. Y. Zhang, Z. Wu, O. Boss, L. F. Michael, P. Puigserver, E. Isotani, E. N. Olson, B. B. Lowell, R. Bassel-Duby, and B. M. Spiegelman. 2002. Transcriptional co-activator PGC-1α drives the formation of slow-twitch muscle fibres. *Nature* 418: 797–801.

Lohuis, T. D., H. J. Harlow, and T. D. Beck. 2007. Hibernating black bears (*Ursus americanus*) experience skeletal muscle protein balance during winter anorexia. *Comp. Biochem. Physiol., B* 147: 20–28.

Lohuis, T. D., H. J. Harlow, T. D. Beck, and P. A. Iaizzo. 2007. Hibernating bears conserve muscle strength and maintain fatigue resistance. *Physiol. Biochem. Zool.* 80: 257–269.

Mantle, B. L., H. Guderley, N. J. Hudson, and C. E. Franklin. 2010. Enzyme activity in the aestivating Green-striped burrowing frog (*Cyclorana alboguttata*). *J. Comp. Physiol., B* 180: 1033–1043.

McPherron, A. C., A. M. Lawler, and S.-J. Lee. 1997. Regulation of skeletal muscle mass in mice by a new TGF-β superfamily member. *Nature* 387: 83–90.

Moore, D. H. II. 1975. A study of age group track and field records to relate age and running speed. *Nature* 253: 264–265.

Mosher, D. S., P. Quignon, C. D. Bustamante, N. B. Sutter, C. S. Mellersh, H. G. Parker, and E. A. Ostrander. 2007. A mutation in the myostatin gene increases muscle mass and enhances racing performance in heterozygote dogs. *PLoS Genet.* 3(5): e79. doi: 10.1371/journal.pgen.0030079

Narici, M. V. and M. D. de Boer. 2011. Disuse of the musculo-skeletal system in space and on earth. *Eur. J. Appl. Physiol.* 111: 403–420.

Nelson, O. L., C. T. Robbins, Y. Wu, and H. Granzier. 2008. Titin isoform switching is a major cardiac adaptive response in hibernating grizzly bears. *Am. J. Physiol. Heart Circ. Physiol.* 295: H366–H371.

Proudfoot, C., D. F. Carlson, R. Huddart, C. R. Long, et al. 2015. Genome edited sheep and cattle. *Transgenic Res.* 24: 147–153.

Ratamess, N. A., B. A. Alvar, T. K. Evetoch, T. J. Housh, W. B. Kibler, W. J. Kraemer and N. T. Triplett. 2009. American College of Sports Medicine position stand. Progression models in resistance training for healthy adults. *Med. Sci. Sports. Exerc.* 41: 687–708.

Richardson, R. S., H. Wagner, S. R. D. Mudaliar, E. Saucedo, R. Henry, and P. D. Wagner. 2000. Exercise adaptation attenuates VEGF gene expression in human skeletal muscle. *Am. J. Physiol.* 279: H772–H778.

Schiaffino, S., M. Sandri, and M. Murgia. 2007. Activity-dependent signaling pathways controlling muscle diversity and plasticity. *Physiology* 22: 269–278.

Schuelke, M., K. R. Wagner, L. E. Stolz, C. Hübner, T. Riebel, W. Kömen, T. Braun, J. F. Tobin, and S.-J. Lee. 2004. Myostatin mutation associated with gross muscle hypertrophy in a child. *N. Engl. J. Med.* 350: 2682–2688.

Shavlakadze, T., and M. Grounds. 2006. Of bears, frogs, meat, mice and men: complexity of factors affecting skeletal muscle mass and fat. *BioEssays* 28: 994–1009.

Shields, R. K., and S. Dudley-Javoroski. 2006. Musculoskeletal plasticity after acute spinal cord injury: Effects of long-term neuromuscular electrical stimulation training. *J. Neurophysiol.* 95: 2380–2390.

Smith, K., K. Winegard, A. L. Hicks and N. McCartney. 2003. Two years of resistance training in older men and women: The effects of three years of detraining on the retention of dynamic strength. *Can. J. Appl. Physiol.* 28: 462–474.

Solomon, A. M., and P. M. G. Bouloux. 2006. Modifying muscle mass—the endocrine perspective. *J. Endocrinol.* 191: 349–360.

Spangenburg, E. E., D. Le Roith, C. W. Ward, and S. C. Bodine. 2008. A functional insulin-like growth factor receptor is not necessary for load-induced skeletal muscle hypertrophy. *J. Physiol.* 586: 283–291.

Symonds, B. L., R. S. James, and C. E. Franklin. 2007. Getting the jump on skeletal muscle disuse atrophy: preservation of contractile performance in aestivating *Cyclorana alboguttata* (Gunther 1867). *J. Exp. Biol.* 210: 825–835.

Takeda, S., F. Elefteriou, R. Levasseur, X. Liu, L. Zhao, K. L. Parker, D. Armstrong, P. Ducy, and G. Karsenty. 2002. Leptin regulates bone formation via the sympathetic nervous system. *Cell* 111: 305–317.

Thomas, D. R. 2007. Loss of skeletal muscle mass in aging: Examining the relationship of starvation, sarcopenia and cachexia. *Clin. Nutr.* 26: 389–399.

Tinker, D. B., H. J. Harlow, and T. D. I. Beck. 1998. Protein use and muscle-fiber changes in free-ranging, hibernating black bears. *Physiol. Zool.* 71: 414–424.

Trappe, S., D. Costill, P. Gallagher, A. Creer, J. R. Peters, H. Evans, D. A. Riley, and R. H. Fitts. 2009. Exercise in space: human skeletal muscle after 6 months aboard the International Space Station. *J. Appl. Physiol.* 106: 1159–1168.

Vestergaard, P., O.-G. Støen, J. E. Swenson, L. Moselkilde, L. Heickendorff and O. Fröbert. 2011. Vitamin D status and bone and connective tissue turnover in brown bears (*Ursus arctos*) during hibernation and the active state. *PLoS One* 6 (6): e21483. doi:10.1371/journal.pone.0021483

Walsh, F. S., and A. J. Celeste. 2005. Myostatin: a modulator of skeletal-muscle stem cells. *Biochem. Soc. Trans.* 33 (6): 1513–1517.

Waters, R. E., S. Rotevatn, P. Li, B. H. Annex, and Z. Yan. 2004. Voluntary running induces fiber type-specific angiogenesis in mouse skeletal muscle. *Am. J. Physiol.* 287: C1342–C1348.

Widrick, J. J., and 10 additional authors. 1999. Effect of a 17 day spaceflight on contractile properties of human soleus muscle fibres. *J. Physiol.* 516: 915–930.

Chapter 22

Clever, H. L., and R. Battino. 1975. The solubility of gases in liquids. In M. R. J. Dack (ed.), *Techniques of Chemistry*, vol. 8, part 1, pp. 379–441. Wiley, New York.

Dejours, P. 1988. *Respiration in Water and Air*. Elsevier, New York.

Fenn, W. O., and H. Rahn (eds.). 1964. *Respiration*, vol. 1 (Handbook of Physiology [Bethesda, MD], section 3). Oxford University Press, New York.

Knott, N. A., A. R. Davis, and W. A. Buttemer. 2004. Passive flow through an unstalked intertidal ascidian: orientation and morphology enhance suspension feeding in *Pyura stolonifera*. *Biol. Bull.* 207: 217–224.

Maclean, G. S. 1981. Factors influencing the composition of respiratory gases in mammal burrows. *Comp. Biochem. Physiol., A* 69: 373–380.

Vogel, S. 1988. How organisms use flow-induced pressures. *Am. Sci.* 76: 28–34.

Weibel, E. R. 2000. *Symmorphosis. On Form and Function in Shaping Life*. Harvard University Press, Cambridge, MA.

Chapter 23

Ainsworth, D. M., C. A. Smith, S. W. Eicker, N. G. Ducharme, K. S. Henderson, K. Snedden, and J. A. Dempsey. 1997. Pulmonary-locomotory interactions in exercising dogs and horses. *Respir. Physiol.* 110: 287–294.

Bernhard, W., P. L. Haslam, and J. Floros. 2004. From birds to humans: New concepts on airways relative to alveolar surfactant. *Am. J. Respir. Cell Mol. Biol.* 30: 6–11.

Bishop, C. M., R. J. Spivey, L. A. Hawkes, N. Batbayar, B. Chua, P. B. Frappell, W. K. Milsom, T. Natsagdorj, S. H. Newman, G. R. Scott, J. Y. Takekawa, M. Wikelski, and P. J. Butler. 2015. The roller coaster flight strategy of bar-headed geese conserves energy during Himalayan migrations. *Science* 347: 250–254.

Block, B. A., and E. D. Stevens (eds.). 2001. *Tuna: Physiology, Ecology, and Evolution*. Academic Press, New York.

Boutilier, R. G. (ed.). 1990. *Vertebrate gas exchange: From environment to cell. Adv. Comp. Environ. Physiol.* 6: 1–411.

Brainerd, E. L. 1994. The evolution of lung-gill bimodal breathing and the homology of vertebrate respiratory pumps. *Am. Zool.* 34: 289–299.

Brainerd, E. L., and T. Owerkowicz. 2006. Functional morphology and evolution of aspiration breathing in tetrapods. *Respir. Physiol. Neurobiol.* 154: 73–88.

Bramble, D. M., and F. A. Jenkins, Jr. 1993. Mammalian locomotor-respiratory integration: Implications for diaphragmatic and pulmonary design. *Science* 262: 235–240.

Burggren, W. W., and A. W. Pinder. 1991. Ontogeny of cardiovascular and respiratory physiology in lower vertebrates. *Annu. Rev. Physiol.* 53: 107–135.

Burggren, W. W., K. Johansen, and B. McMahon. 1985. Respiration in phyletically ancient fishes. In R. E. Forman, A. Gorbman, J. M. Dodd, and R. Olsson (eds.), *Evolutionary Biology of Primitive Fishes*, pp. 217–252. Plenum, New York.

Burnett, L. E., J. D. Holman, D. D. Jorgensen, J. L. Ikerd, and K. G. Burnett. 2006. Immune defense reduces respiratory fitness in *Callinectes sapidus*, the Atlantic blue crab. *Biol. Bull. Woods Hole* 211: 50–57.

Cameron, J. N. 1989. *The Respiratory Physiology of Animals*. Oxford University Press, New York.

Deeming, D. C., and M. W. J. Ferguson (eds.). 1991. *Egg Incubation: Its Effects on Embryonic Development in Birds and Reptiles*. Cambridge University Press, New York.

Dejours, P. 1988. *Respiration in Water and Air*. Elsevier, New York.

Dempsey, J. A., and A. I. Pack (eds.). 1995. *Regulation of Breathing*, 2nd ed. Dekker, New York.

Denny, M. W. 1993. *Air and Water: The Biology and Physics of Life's Media*. Princeton University Press, Princeton, NJ.

Dillon, M. E., M. R. Frazier, and R. Dudley. 2006. Into thin air: physiology and evolution of alpine insects. *Integr. Comp. Biol.* 46: 49–61.

Duncker, H.-R. 1974. Structure of the avian respiratory tract. *Respir. Physiol.* 22: 1–19.

Duncker, H.-R. 2001. The emergence of macroscopic complexity: An outline of the history of the respiratory apparatus of vertebrates from diffusion to language production. *Zoology* 103: 240–259.

Evans, D. H., P. M. Piermarini, and K. P. Choe. 2005. The multifunctional fish gill: Dominant site of gas exchange, osmoregulation, acid-base regulation, and excretion of nitrogenous waste. *Physiol. Rev.* 85: 97–177.

Farmer, C. G. 1999. Evolution of the vertebrate cardiopulmonary system. *Annu. Rev. Physiol.* 61: 573–592.

Farmer, C. G., and K. Sanders. 2010. Unidirectional airflow in the lungs of alligators. *Science* 327: 338–340.

Fedde, M. R., J. A. Orr, H. Shams, and P. Scheid. 1989. Cardiopulmonary function in exercising bar-headed geese during normoxia and hypoxia. *Respir. Physiol.* 77: 239–262.

Feldman, J. L., G. S. Mitchell, and E. E. Nattie. 2003. Breathing: Rhythmicity, plasticity, chemosensitivity. *Annu. Rev. Neurosci.* 26: 239–266.

Fincke, T., and R. Paul. 1989. Book lung function in arachnids. III. The function and control of the spiracles. *J. Comp. Physiol., B* 159: 433–441.

Floros, J., and P. Kala. 1998. Surfactant proteins: Molecular genetics of neonatal pulmonary diseases. *Annu. Rev. Physiol.* 60: 365–384.

Fregly, M. J., and C. M. Blatteis (eds.). 1996. *Environmental Physiology*, vol. 2. (Handbook of Physiology [Bethesda, MD], section 4.), Oxford University Press, New York. [See Part V: *The Terrestrial Altitude Environment*.]

Greenaway, P., and C. Farrelly. 1990. Vasculature of the gas-exchange organs in air-breathing brachyurans. *Physiol. Zool.* 63: 117–139.

Hawkes, L. A., and 12 additional authors. 2011. The trans-Himalayan flights of bar-headed geese (*Anser indicus*). *Proc. Natl. Acad. Sci. U.S.A.* 108: 9516–9519.

Henry, R. P. 1994. Morphological, behavioral, and physiological characterization of bimodal breathing crustaceans. *Am. Zool.* 34: 205–215.

Hetz, S. K., and T. J. Bradley. 2005. Insects breathe discontinuously to avoid oxygen toxicity. *Nature* 433: 516–519.

Hicks, J. W., T. Wang, and A. F. Bennett, 2000. Patterns of cardiovascular and ventilatory response to elevated metabolic states in the lizard *Varanus exanthematicus*. *J. Exp. Biol.* 203: 2437–2445.

Hlastala, M. P., and A. J. Berger. 2001. *Physiology of Respiration*, 2nd ed. Oxford University Press, New York.

Hochachka, P. W. 1998. Mechanism and evolution of hypoxia tolerance in humans. *J. Exp. Biol.* 201: 1243–1254.

Howe, S., and D. L. Kilgore, Jr. 1987. Convective and diffusive gas exchange in nest cavities of the northern flicker (*Colaptes auratus*). *Physiol. Zool.* 60: 707–712.

Innes, A. J., and E. W. Taylor. 1986. The evolution of air-breathing in crustaceans: A functional analysis of branchial, cutaneous and pulmonary gas exchange. *Comp. Biochem. Physiol., A* 85: 621–637.

King, A. S., and T. McLelland (eds.). 1989. *Form and Function in Birds*, vol. 4. Academic Press, New York.

LaBarbera, M. 1990. Principles of design of fluid transport systems in zoology. *Science* 249: 992–1000.

Lahiri, S., A. Roy, S. M. Baby, T. Hoshi, G. L. Semenza, and N. R. Prabhakar. 2006. Oxygen sensing in the body. *Prog. Biophys. Mol. Biol.* 91: 249–286.

Lieske, S. P., M. Thoby-Brisson, P. Telgkamp, and J. M. Ramirez. 2000. Reconfiguration of the neural network controlling multiple breathing patterns: Eupnea, sighs, and gasps. *Nat. Neurosci.* 3: 600–608.

Lighton, J. R. B. 1996. Discontinuous gas exchange in insects. *Annu. Rev. Entomol.* 41: 309–324.

Lighton, J. R. B. 1998. Notes from underground: Towards ultimate hypotheses of cyclic, discontinuous gas-exchange in tracheate arthropods. *Am. Zool.* 38: 483–491.

Maina, J. N. 1994. Comparative pulmonary morphology and morphometry: The functional design of respiratory systems. *Adv. Comp. Environ. Physiol.* 20: 111–232.

Maina, J. N., and J. B. West. 2004. Thin and strong! The bioengineering dilemma in the structural and functional design of the blood-gas barrier. *Physiol. Rev.* 85: 811–844.

Mill, P. J. 1997. Invertebrate respiratory systems. In W. H. Dantzler (ed.), *Comparative Physiology*, vol. 2 (Handbook of Physiology [Bethesda, MD], section 13), pp. 1009–1096. Oxford University Press, New York.

Milsom, W. K. 1991. Intermittent breathing in vertebrates. *Annu. Rev. Physiol.* 53: 87–105.

Monge, C., and F. León-Velarde. 1991. Physiological adaptation to high altitude: Oxygen transport in mammals and birds. *Physiol. Rev.* 71: 1135–1172.

Moore, L. G., F. Armaza V., M. Villena, and E. Vargas. 2000. Comparative aspects of high-altitude adaptation in human populations. In S. Lahiri, N. R. Prabhakar, and R. E. Forster II (eds.), *Oxygen Sensing: Molecule to Man*, pp. 45–62. Kluwer/Plenum, New York.

Mortola, J. P. 2001. *Respiratory Physiology of Newborn Mammals. A Comparative Perspective*. Johns Hopkins University Press, Baltimore, MD.

Owerkowicz, T., C. G. Farmer, J. W. Hicks, and E. L. Brainerd. 1999. Contribution of gular pumping to lung ventilation in monitor lizards. *Science* 284: 1661–1663.

Parent, R. A. (ed.) 2015. *Comparative Biology of the Normal Lung*, 2nd ed. Academic Press, San Diego.

Park, K., W. Kim, and H.-Y. Kim. 2014. Optimal lamellar arrangement in fish gills. *Proc. Natl. Acad. Sci. USA* 111: 8067–8070.

Paul, R., T. Finke, and B. Linzen. 1987. Respiration in the tarantula *Eurypelma californicum*: Evidence for diffusion lungs. *J. Comp. Physiol., B* 157: 209–217.

Perry, S. F. 1989. Mainstreams in the evolution of vertebrate respiratory structures. In A. S. King and T. McLelland (eds.), *Form and Function in Birds*, vol. 4, pp. 1–67. Academic Press, New York.

Perry, S. F. 1990. Recent advances and trends in the comparative morphometry of vertebrate gas exchange organs. *Adv. Comp. Environ. Physiol.* 6: 45–71.

Perry, S. F., R. J. A. Wilson, C. Straus, M. B. Harris, and J. E. Remmers. 2001. Which came first, the lung or the breath? *Comp. Biochem. Physiol., A* 129: 37–47.

Piiper, J. (ed.). 1978. *Respiratory Function in Birds, Adult and Embryonic*. Springer-Verlag, New York.

Powell, F. L., and S. R. Hopkins. 2004. Comparative physiology of lung complexity: Implications for gas exchange. *News Physiol. Sci.* 19: 55–60.

Rankin, J. C., and F. B. Jensen (eds.). 1993. *Fish Ecophysiology*. Chapman & Hall, New York.

Ruben, J. A., T. D. Jones, N. R. Geist, and W. J. Hillenius. 1997. Lung structure and ventilation in

theropod dinosaurs and early birds. *Science* 278: 1267–1270.

Scheid, P., and J. Piiper. 1972. Cross-current gas exchange in avian lungs: Effects of reversed parabronchial air flow in ducks. *Respir. Physiol.* 16: 304–312.

Schmitz, A., and S. F. Perry. 1999. Stereological determination of tracheal volume and diffusing capacity of the tracheal walls in the stick insect *Carausius morosus* (Phasmatodea, Lonchodidae). *Physiol. Biochem. Zool.* 72: 205–218.

Scott, G. R., and W. K. Milsom. 2007. Control of breathing and adaptation to high altitude in the bar-headed goose. *Am. J. Physiol.* 293: R379–R391.

Shadwick, R. W., R. K. O'Dor, and J. M. Gosline. 1990. Respiratory and cardiac function during exercise in squid. *Can. J. Zool.* 68: 792–798.

Sláma, K. 1999. Active regulation of insect respiration. *Ann. Entomol. Soc. Am.* 92: 916–929.

Smatresk, N. J. 1994. Respiratory control in the transition from water to air breathing in vertebrates. *Am. Zool.* 34: 264–279.

Smith, J. C., A. P. L. Abdala, I. A. Rybak, and J. F. R. Paton. 2009. Structural and functional architecture of respiratory networks in the mammalian brainstem. *Phil. Trans. R. Soc., B* 364: 2577–2587.

Snelling, E. P., R. S. Seymour, S. Runciman, P. G. D. Matthews, and C. R. White. 2011. Symmorphosis and the insect respiratory system: allometric variation. *J. Exp. Biol.* 214: 3225–3237.

Socha, J. J., T. D. Förster, and K. J. Greenlee. 2010. Issues of convection in insect respiration: Insights from synchroton X-ray imaging and beyond. *Resp. Physiol. Neurobiol.* 173S: S65–S73.

Sollid, J., and G. E. Nilsson. 2006. Plasticity of respiratory structures—Adaptive remodeling of fish gills induced by ambient oxygen and temperature. *Respir. Physiol. Neurobiol.* 154: 241–251.

Storz, J. F. 2010. Genes for high altitudes. *Science* 329: 40–41.

Sundin, L., M. L. Burleson, A. P. Sanchez, J. Amin-Naves, R. Kinkead, L. H. Gargaglioni, L. K. Hartzler, M. Weimann, P. Kumar, and M. L. Glass. 2007. Respiratory chemoreceptor function in vertebrates—comparative and evolutionary aspects. *Integr. Comp. Biol.* 47: 592–600.

Taylor, E. W., D. Jordan, and J. H. Coote. 1999. Central control of the cardiovascular and respiratory systems and their interactions in vertebrates. *Physiol. Rev.* 79: 855–916.

Wagner, P. D. 1996. Determinants of maximal oxygen transport and utilization. *Annu. Rev. Physiol.* 58: 21–50.

Ward, M. P., J. S. Milledge, and J. B. West (eds.). 2000. *High Altitude Medicine and Physiology*, 3rd ed. Arnold, London.

Weibel, E. R. 2000. *Symmorphosis*. Harvard University Press, Cambridge, MA.

West, J. B. 2006. Human responses to extreme altitudes. *Integr. Comp. Biol.* 46: 25–34.

West, J. B., R. R. Watson, and Z. Fu. 2007. The human lung: Did evolution get it wrong? *Eur. Respir. J.* 29: 11–17.

Westneat, M. W., J. J. Socha, and W.-K. Lee. 2008. Advances in biological structure, function, and physiology using synchrotron X-ray imaging. *Annu. Rev. Physiol.* 70: 119–142.

Westneat, M. W., O. Betz, R. W. Blob, K. Fezzaa, W. J. Cooper, and W.-K. Lee. 2003. Tracheal respiration in insects visualized with synchrotron X-ray imaging. *Science* 299: 558–560.

Wood, S. C. (ed.). 1989. *Comparative Pulmonary Physiology: Current Concepts.* Lung Biology in Health and Disease, vol. 39. Dekker, New York.

Chapter 24

Ackers, G. K., M. L. Doyle, D. Myers, and M. A. Daugherty. 1992. Molecular code for cooperativity in hemoglobin. *Science* 255: 54–63.

Adamczewska, A. M., and S. Morris. 1998. The functioning of the haemocyanin of the terrestrial Christmas Island red crab *Gecarcoidea natalis* and roles for organic modulators. *J. Exp. Biol.* 201: 3233–3244.

Angersbach, D. 1978. Oxygen transport in the blood of the tarantula *Eurypelma californicum*: pO_2 and pH during rest, activity, and recovery. *J. Comp. Physiol.* 123: 113–125.

Arp, A. J., J. J. Childress, and R. D. Vetter. 1987. The sulphide-binding protein in the blood of the vestimentiferan tube-worm, *Riftia pachyptila*, is the extracellular haemoglobin. *J. Exp. Biol.* 128: 139–158.

Avivi, A., F. Gerlach, A. Joel, S. Reuss, T. Burmester, E. Nevo, and T. Hankeln. 2010. Neuroglobin, cytoglobin, and myoglobin contribute to hypoxia adaptation of the subterranean mole rat *Spalax*. *Proc. Natl. Acad. Sci. USA* 107: 21570–21575.

Bailey, X., S. Vanin, C. Chabasse, K. Mizuguchi, and S. N. Vinogradov. 2008. A phylogenetic profile of hemerythrins, the nonheme diiron binding respiratory proteins. *BMC Evol. Biol.* 8: 244.

Bauer, C., G. Gros, and H. Bartels (eds.). 1980. *Biophysics and Physiology of Carbon Dioxide.* Springer-Verlag, New York.

Berg, J. M., J. L. Tymoczko, and L. Stryer. 2007. *Biochemistry*, 6th ed. Freeman, New York.

Bernal, D., K. A. Dickson, R. E. Shadwick, and J. B. Graham. 2001. Review: Analysis of the evolutionary convergence for high performance swimming in lamnid sharks and tunas. *Comp. Biochem. Physiol., A* 129: 695–726.

Booth, C. E., B. R. McMahon, and A. W. Pinder. 1982. Oxygen uptake and the potentiating effects of increased hemolymph lactate on oxygen transport during exercise in the blue crab, *Callinectes sapidus*. *J. Comp. Physiol.* 148: 111–121.

Bridges, C. R. 2001. Modulation of hemocyanin oxygen affinity: Properties and physiological implications in a changing world. *J. Exp. Biol.* 204: 1021–1032.

Brix, O., A. Bårdgard, S. Mathisen, N. Tyler, M. Nuutinen, S. G. Condo, and B. Giardina. 1990. Oxygen transport in the blood of arctic mammals: Adaptation to local heterothermia. *J. Comp. Physiol., B* 159: 655–660.

Brown, A. C., and N. B. Terwilliger. 1998. Ontogeny of hemocyanin function in the Dungeness crab *Cancer magister*: Hemolymph modulation of hemocyanin oxygen-binding. *J. Exp. Biol.* 201: 819–826.

Brunori, M., and D. Vallone. 2006. A globin for the brain. *FASEB J.* 20: 2192–2197.

Burggren, W., B. McMahon, and D. Powers. 1991. Respiratory functions of blood. In C. L. Prosser (ed.), *Environmental and Metabolic Animal Physiology* (*Comparative Animal Physiology*, 4th ed.), pp. 437–508. Wiley-Liss, New York.

Burmester, T., and T. Hankeln. 2007. The respiratory proteins of insects. *J. Insect Physiol.* 53: 285–294.

Campbell, K. L., and 14 additional authors. 2010. Substitutions in woolly mammoth hemoglobin confer biochemical properties adaptive for cold tolerance. *Nat. Genet.* 42: 536–540.

Clark, T. D., E. Sandblom, G. K. Cox, S. G. Hinch, and A. P. Farrell. 2008. Circulatory limits to oxygen supply during an acute temperature increase in the Chinook salmon (*Oncorhynchus tshawytscha*). *Am. J. Physiol.* 295: R1631–R1639.

Cosby, K., and 14 additional authors. 2003. Nitrite reduction to nitric oxide by deoxyhemoglobin vasodilates the human circulation. *Nat. Med.* 9: 1498–1505.

Decker, H., N. Hellmann, E. Jaenicke, B. Lieb, U. Meissner, and J. Markl. 2007. Minireview: Recent progress in hemocyanin research. *Integr. Comp. Biol.* 47: 631–644.

Edsall, J. T., and J. Wyman. 1958. *Biophysical Chemistry*, vol. 1. Academic Press, New York. [A classic reference on carbon dioxide chemistry.]

Egginton, S. 1997. A comparison of the response to induced exercise in red- and white-blooded antarctic fishes. *J. Comp. Physiol., B* 167: 129–134.

Erzurum, S. C., S. Ghosh, A. J. Janocha, W. Xu, S. Bauer, N. S. Bryan, J. Tejero, C. Hemann, R. Hille, D. J. Stuehr, M. Feelisch, and C. M. Beall. 2007. Higher blood flow and circulating NO products offset high-altitude hypoxia among Tibetans. *Proc. Natl. Acad. Sci. U.S.A.* 104: 17593–17598.

Farrell, A. P., and S. M. Clutterham. 2003. On-line venous oxygen tensions in rainbow trout during graded exercise at two acclimation temperatures. *J. Exp. Biol.* 206: 487–496.

Farrell, A. P., E. J. Eliason, E. Sandblom, and T. D. Clark. 2009. Fish cardiorespiratory physiology in an era of climate change. *Can. J. Zool.* 87: 835–851.

Flögel, U., A. Fago, and T. Rassaf. 2010. Keeping the heart in balance: the functional interactions of myoglobin with nitrogen oxides. *J. Exp. Biol.* 213: 2726–2733.

Geers, C., and G. Gros. 2000. Carbon dioxide transport and carbonic anhydrase in blood and muscle. *Physiol. Rev.* 80: 681–715.

Gros, G., B. A. Wittenberg, and T. Jue. 2010. Myoglobin's old and new clothes: from molecular structure to function in living cells. *J. Exp. Biol.* 213: 2713–2725.

Hankeln, T., and 22 additional authors. 2005. Neuroglobin and cytoglobin in search of their role in the vertebrate globin family. *J. Inorg. Biochem.* 99: 110–119.

Hardison, R. 1999. The evolution of hemoglobin. *Am. Sci.* 87: 126–137.

Hebbel, R. P., J. W. Eaton, R. S. Kronenberg, E. D. Zanjani, L. G. Moore, and E. M. Berger. 1978. Human llamas. *J. Clin. Inves.* 62: 593–600.

Heisler, N. (ed.). 1986. *Acid–Base Regulation in Animals.* Elsevier, New York.

Heisler, N. 1990. Acid–base regulation: Interrelationships between gaseous and ionic regulation. *Adv. Comp. Environ. Physiol.* 6: 211–251.

Hendgen-Cotta, U. B., and 11 additional authors. 2008. Nitrite reductase activity of myoglobin regulates respiration and cellular viability in myocardial ischemia-reperfusion injury. *Proc. Natl. Acad. Sci. U.S.A.* 105: 10256–10261.

Höpfl, G., O. Ogunshola, and M. Gassmann. 2003. Hypoxia and high altitude. The molecular response. In R. C. Roach, P. D. Wagner, and P. H. Hackett (eds.), *Hypoxia. Through the Lifecycle* (*Advances in Experimental Medicine and Biology*, vol. 543), pp. 89–115. Kluwer Academic, New York.

Houlihan, D. F., G. Duthie, P. J. Smith, M. J. Wells, and J. Wells. 1986. Ventilation and circulation during exercise in *Octopus vulgaris*. *J. Comp. Physiol., B* 156: 683–689.

Ingermann, R. L. 1997. Vertebrate hemoglobins. In W. H. Dantzler (ed.), *Comparative Physiology*, vol. 1 (Handbook of Physiology [Bethesda, MD], section 13), pp. 357–408. Oxford University Press, New York.

Johansen, K., and C. Lenfant. 1966. Gas exchange in the cephalopod, *Octopus dofleini*. *Am. J. Physiol.* 210: 910–918.

Johansen, K., G. Lykkeboe, R. E. Weber, and G. M. O. Maloiy. 1976. Respiratory properties of blood in awake and estivating lungfish, *Protopterus amphibius*. *Respir. Physiol.* 27: 335–345.

Johnson, R. L., Jr., G. J. F. Heigenhauser, C. C. W. Hsia, N. L. Jones, and P. D. Wagner. 1996. Determinants of gas exchange and acid–base balance during exercise. In L. B. Rowell and J. T. Shepherd (eds.), *Exercise: Regulation and Integration of Multiple Systems* (Handbook of Physiology [Bethesda, MD], section 12), pp. 515–584. Oxford University Press, New York.

Kanatous, S. B., and P. P. A. Mammen. 2010. Regulation of myoglobin expression. *J. Exp. Biol.* 213: 2741–2747.

Kiceniuk, J. W., and D. R. Jones. 1977. The oxygen transport system in trout (*Salmo gairdneri*) during sustained exercise. *J. Exp. Biol.* 69: 247–260.

Kraus, D. W., and J. M. Colacino. 1986. Extended oxygen delivery from the nerve hemoglobin of *Tellina alternata* (Bivalvia). *Science* 232: 90–92.

Lieb, B., K. Dimitrova, H.-S. Kang, S. Braun, W. Gebauer, A. Martin, B. Hanelt, S. A. Saenz, C. M. Adema, and J. Markl. 2006. Red blood with blue-blooded ancestry: Intriguing structure of a snail hemoglobin. *Proc. Natl. Acad. Sci. U.S.A.* 103: 12011–12016.

Linzen, B. (ed.). 1986. *Invertebrate Oxygen Carriers.* Springer-Verlag, New York.

Lurman, G. J., N. Koschnick, H.-O. Pörtner, and M. Lucassen. 2007. Molecular characterization and expression of Atlantic cod (*Gadus morhua*) myoglobin from two populations held at two different acclimation temperatures. *Comp. Biochem. Physiol., A* 148: 681–689.

Lykkeboe, G., and K. Johansen. 1982. A cephalopod approach to rethinking about the importance of the Bohr and Haldane effects. *Pacific Sci.* 36: 305–313.

Mangum, C. P. (ed.). 1992. *Blood and Tissue Oxygen Carriers. Adv. Comp. Environ. Physiol.* 13: 1–477.

Mangum, C. P. 1997. Invertebrate blood oxygen carriers. In W. H. Dantzler (ed.), *Comparative Physiology*, vol. 2 (Handbook of Physiology [Bethesda, MD], section 13), pp. 1097–1135. Oxford University Press, New York.

Mangum, C., and D. Towle. 1977. Physiological adaptation to unstable environments. *Am. Sci.* 65: 67–75. [For an update, see R. P. Mason, C. P. Mangum, and G. Godette. 1983. *Biol. Bull.* 164: 104–123.]

McMahon, B. R. 2001. Respiratory and circulatory compensation to hypoxia in crustaceans. *Respir. Physiol.* 128: 349–364.

McMahon, B. R., D. G. McDonald, and C. M. Wood. 1979. Ventilation, oxygen uptake and haemolymph oxygen transport, following enforced exhausting activity in the Dungeness crab *Cancer magister*. *J. Exp. Biol.* 80: 271–285.

Miller, K. I. 1994. Cephalopod haemocyanins: A review of structure and function. *Mar. Freshw. Behav. Physiol.* 25: 101–120.

Morris, S., and C. R. Bridges. 1994. Properties of respiratory pigments in bimodal breathing animals: Air and water breathing by fish and crustaceans. *Am. Zool.* 34: 216–228.

Nikinmaa, M. 1990. *Vertebrate Red Blood Cells*. Springer-Verlag, New York.

Pelster, B., and D. Randall. 1998. The physiology of the Root effect. In S. F. Perry and B. L. Tufts (eds.), *Fish Respiration*, pp. 113–139. Academic Press, New York.

Perry, S. F., and B. L. Tufts (eds.). 1998. *Fish Respiration*. Academic Press, New York.

Perry, S. F., and G. M. Gilmour. 2006. Acid-base balance and CO_2 excretion in fish: Unanswered questions and emerging models. *Respir. Physiol. Neurobiol.* 154: 199–215.

Perutz, M. F. 1978. Hemoglobin structure and respiratory transport. *Sci. Am.* 239(6): 92–125.

Perutz, M. F. 1990. Mechanisms regulating the reactions of human hemoglobin with oxygen and carbon monoxide. *Annu. Rev. Physiol.* 52: 1–25.

Ratcliffe, P. J., K.-U. Eckardt, and C. Bauer. 1996. Hypoxia, erythropoietin gene expression, and erythropoiesis. In M. J. Fregly and C. M. Blatteis (eds.), *Environmental Physiology*, vol. 2 (Handbook of Physiology [Bethesda, MD], section 4), pp. 1125–1153. Oxford University Press, New York.

Reeves, R. B., and H. Rahn. 1979. Patterns in vertebrate acid–base regulation. In S. C. Wood and C. Lenfant (eds.), *Evolution of Respiratory Processes*, pp. 225–252. Dekker, New York.

Riggs, A. F. 1998. Self-association, cooperativity and supercooperativity of oxygen binding by hemoglobins. *J. Exp. Biol.* 201: 1073–1084.

Rummer, J. L., D. J. McKenzie, A. Innocenti, C. T. Supuran, and C. J. Brauner. 2013. Root effect hemoglobin may have evolved to enhance general tissue oxygen delivery. *Science* 340: 1327–1329.

Schmidt-Nielsen, K., and C. R. Taylor. 1968. Red blood cells: Why or why not? *Science* 162: 274–275.

Seibel, B. A., and P. J. Walsh. 2003. Biological impacts of deep-sea carbon dioxide injection inferred from indices of physiological performance. *J. Exp. Biol.* 206: 641–650.

Snyder, G. K., and B. A. Sheafor. 1999. Red blood cells: Centerpiece in the evolution of the vertebrate circulatory system. *Am. Zool.* 39: 189–198.

Snyder, L. R. G. 1985. Low P_{50} in deer mice native to high altitude. *J. Appl. Physiol.* 58: 193–199.

Snyder, L. R. G., J. P. Hayes, and M. A. Chappell. 1988. Alpha-chain hemoglobin polymorphisms are correlated with altitude in the deer mouse, *Peromyscus maniculatus*. *Evolution* 42: 689–697. [See also pp. 681–688.]

Sonveaux, P., I. I. Lobysheva, O. Feron, and T. J. McMahon. 2007. Transport and peripheral bioactivities of nitrogen oxides carried by red blood cell hemoglobin: role in oxygen delivery. *Physiology* 22: 97–112.

Stewart, P. A. 1978. Independent and dependent variables of acid–base control. *Respir. Physiol.* 33: 9–26.

Stewart, P. A. 1981. *How to Understand Acid–Base: A Quantitative Acid–Base Primer for Biology and Medicine*. Elsevier, New York.

Storz, J. F. 2007. Hemoglobin function and physiological adaptation to hypoxia in high-altitude mammals. *J. Mamm.* 88: 24–31.

Storz, J. F. 2010. Genes for high altitudes. *Science* 329: 40–41.

Tetens, V., and N. J. Christensen. 1987. Beta-adrenergic control of blood oxygen affinity in acutely hypoxia exposed rainbow trout. *J. Comp. Physiol., B* 157: 667–675.

Tremper, K. K., and S. J. Barker. 1989. Pulse oximetry. *Anesthesiology* 70: 98–108.

Truchot, J. P. 1987. *Comparative Aspects of Extracellular Acid–Base Balance*. Springer-Verlag, New York.

Urich, K. 1994. *Comparative Animal Biochemistry*. Springer-Verlag, New York.

Wagner, P. D. 1996. Determinants of maximal oxygen transport and utilization. *Annu. Rev. Physiol.* 58: 21–50.

Wang, T., H. Busk, and J. Overgaard. 2001. The respiratory consequences of feeding in amphibians. *Comp. Biochem. Physiol., A* 128: 535–549.

Weber, R. E., and K. L. Campbell. 2011. Temperature dependence of haemoglobin-oxygen affinity in heterothermic vertebrates: mechanisms and biological significance. *Acta Physiol.* 202: 549–562.

Weber, R. E., T.-H. Jessen, H. Malte, and J. Tame. 1993. Mutant hemoglobins (α^{119}-Ala and β^{55}-Ser): functions related to high-altitude respiration in geese. *J. Appl. Physiol.* 75: 2646–2655.

Wells, R. M. G. 1990. Hemoglobin physiology in vertebrate animals: A cautionary approach to adaptationist thinking. *Adv. Comp. Environ. Physiol.* 6: 143–161.

West, J. B. 1984. Human physiology at extreme altitudes on Mount Everest. *Science* 223: 784–788.

West, J. B. 2006. Human responses to extreme altitudes. *Integr. Comp. Biol.* 46: 25–34.

Westen, E. A., and H. D. Prange. 2003. A reexamination of the mechanisms underlying the arteriovenous chloride shift. *Physiol. Biochem. Zool.* 76: 603–614.

Wilkening, R. B., and G. Meschia. 1992. Current topic: Comparative physiology of placental oxygen transport. *Placenta* 13: 1–15.

Wittenberg, J. B., and B. A. Wittenberg. 2003. Myoglobin function reassessed. *J. Exp. Biol.* 206: 2011–2020.

Chapter 25

Agnisola, C., and D. F. Houlihan. 1994. Some aspects of cardiac dynamics in *Octopus vulgaris* (LAM). *Mar. Freshw. Behav. Physiol.* 25: 87–100.

Badeer, H. S. 1997. Is the flow in the giraffe's jugular vein a "free" fall? *Comp. Biochem. Physiol., A* 118: 573–576. [See the article by Seymour et al. for a contrasting point of view.]

Bourne, G. B., and B. R. McMahon (eds.). 1990. The Physiology of Invertebrate Circulatory Systems [a symposium]. *Physiol. Zool.* 63: 1–189.

Boutilier, R. G., M. L. Glass, and N. Heisler. 1986. The relative distribution of pulmocutaneous blood flow in *Rana catesbeiana*: Effects of pulmonary or cutaneous hypoxia. *J. Exp. Biol.* 126: 33–39.

Brill, R. W., and P. G. Bushnell. 2001. The cardiovascular system of tunas. In B. A. Block and E. D. Stevens (eds.), *Tuna: Physiology, Ecology, and Evolution*, pp. 79–120. Academic Press, New York.

Burggren, W. W. 1987. Form and function in reptilian circulations. *Am. Zool.* 27: 5–19.

Burggren, W. W. 1995. Central cardiovascular function in amphibians: Qualitative influences of phylogeny, ontogeny, and season. *Adv. Comp. Environ. Physiol.* 21: 175–197.

Burggren, W. W., and A. W. Pinder. 1991. Ontogeny of cardiovascular and respiratory physiology in lower vertebrates. *Annu. Rev. Physiol.* 53: 107–135.

Burggren, W. W., and K. Johansen. 1986. Circulation and respiration in lungfishes (Dipnoi). *J. Morphol.* Suppl. 1: 217–236.

Bushnell, P. G., and D. R. Jones. 1994. Cardiovascular and respiratory physiology of tuna: Adaptations for support of exceptionally high metabolic rates. *Environ. Biol. Fishes* 40: 303–318.

Cameron, J. N. 1989. *The Respiratory Physiology of Animals*. Oxford University Press, New York.

Clark, T. D., E. Sandblom, G. K. Cox, S. G. Hinch, and A. P. Farrell. 2008. Circulatory limits to oxygen supply during an acute temperature increase in the Chinook salmon (*Oncorhynchus tshawytscha*). *Am. J. Physiol.* 295: R1631–R1639.

Cooke, I. M. 2002. Reliable, responsive pacemaking and pattern generation with minimal cell numbers: The crustacean cardiac ganglion. *Biol. Bull.* 202: 108–136.

Eliason, E. J., T. D. Clark, M. J. Hague, L. M. Hanson, Z. S. Gallagher, K. M. Jeffries, M. K. Gale, D. A. Patterson, S. G. Hinch, and A. P. Farrell. 2011. Differences in thermal tolerance among sockeye salmon populations. *Science* 332: 109–122.

Farmer, C. G. 1997. Did lungs and the intracardiac shunt evolve to oxygenate the heart in vertebrates? *Paleobiology* 23: 358–372.

Farrell, A. P. 1991. Circulation of body fluids. In C. L. Prosser (ed.), *Environmental and Metabolic Animal Physiology* (*Comparative Animal Physiology*, 4th ed.), pp. 509–558. Wiley-Liss, New York.

Farrell, A. P. 1991. From hagfish to tuna: A perspective on cardiac function in fish. *Physiol. Zool.* 64: 1137–1164.

Farrell, A. P., and S. M. Clutterham. 2003. On-line venous oxygen tensions in rainbow trout during graded exercise at two acclimation temperatures. *J. Exp. Biol.* 206: 487–496.

Farrell, A. P., D. L. Simonot, R. S. Seymour, and T. D. Clark. 2007. A novel technique for estimating the compact myocardium in fishes reveals surprising results for an athletic air-breathing fish, the Pacific tarpon. *J. Fish Biol.* 71: 389–398.

Forouhar, A. S., M. Liebling, A. Hickerson, A. Nasiraei-Moghaddam, H.-J. Tsai, J. R. Hove, S. E. Fraser, M. E. Dickinson, and M. Gharib. 2006. The embryonic vertebrate heart tube is a dynamic suction pump. *Science* 312: 751–754.

Fritsche, R., M. Axelsson, C. E. Franklin, G. G. Grigg, S. Holmgren, and S. Nilsson. 1993. Respiratory and cardiovascular responses to hypoxia in the Australian lungfish. *Respir. Physiol.* 94: 173–187.

Gamperl, A. K., and A. P. Farrell. 2004. Cardiac plasticity in fishes: environmental influences and intraspecific differences. *J. Exp. Biol.* 207: 2539–2550.

Grigg, G. C., and K. Johansen. 1987. Cardiovascular dynamics in *Crocodylus porosus* breathing air and during voluntary aerobic dives. *J. Comp. Physiol., B* 157: 381–392.

Hagensen, M. K., A. S. Abe, E. Falk, and T. Wang. 2008. Physiological importance of the coronary arterial blood supply to the rattlesnake heart. *J. Exp. Biol.* 211: 3588–3593.

Hamlett, W. C., and R. Muñoz-Chápuli (eds.). 1996. Comparative Cardiovascular Biology of Lower Vertebrates [a symposium]. *J. Exp. Zool.* 275: 71–251.

Hammel, H. T. 1999. Evolving ideas about osmosis and capillary fluid exchange. *FASEB J.* 13: 213–231.

Hargens, A. R., R. W. Millard, K. Pettersson, and K. Johansen. 1987. Gravitational haemodynamics and oedema prevention in the giraffe. *Nature* 329: 59–60. [A useful commentary is provided on pp. 13–14 of the same journal issue.]

Hartline, D. K. 1979. Integrative neurophysiology of the lobster cardiac ganglion. *Am. Zool.* 19: 53–65.

Hicks, J. W., A. Ishimatsu, S. Molloi, A. Erskin, and N. Heisler. 1996. The mechanism of cardiac shunting in reptiles: A new synthesis. *J. Exp. Biol.* 199: 1435–1446.

Hicks, J. W., and T. Wang. 1998. Cardiovascular regulation during anoxia in the turtle: An in vivo study. *Physiol. Zool.* 71: 1–14.

Hill, R. B. (ed.). 1992. Control of Circulation in Invertebrates [a symposium]. *Experientia* 48: 797–858.

Hoppeler, H., O. Mathieu, E. R. Weibel, R. Krauer, S. L. Lindstedt, and C. R. Taylor. 1981. Design of the mammalian respiratory system. VII. Capillaries in skeletal muscles. *Respir. Physiol.* 44: 129–150.

Huang, P. L., Z. H. Huang, H. Mashimo, K. D. Bloch, M. A. Moskowitz, J. A. Bevan, and M. C. Fishman. 1995. Hypertension in mice lacking the gene for endothelial nitric-oxide synthase. *Nature* 377: 239–242.

Huckstorf, K., and C. S. Wirkner. 2011. Comparative morphology of the hemolymph vascular system in krill (Euphausiacea; Crustacea). *Arthropod Struct. Dev.* 40: 39–53.

Huckstorf, K., G. Kosok, E.-A. Seyfarth, and C. S. Wirkner. 2013. The hemolymph vascular system in *Cupiennius salei* (Araneae: Ctenidae). *Zool. Anzeiger* 252: 76–87.

Johansen, K., and W. Burggren. 1980. Cardiovascular functions in the lower vertebrates. In G. H. Bourne (ed.), *Hearts and Heart-like Organs*, vol. 1, pp. 61–117. Academic Press, New York.

Johansen, K., and W. W. Burggren (eds.). 1985. *Cardiovascular Shunts*. Munksgaard, Copenhagen.

Jones, H. D., and D. Peggs. 1983. Hydrostatic and osmotic pressures in the heart and pericardium of *Mya arenaria* and *Anodonta cygnea*. *Comp. Biochem. Physiol., A* 76: 381–385.

Kiceniuk, J. W., and D. R. Jones. 1977. The oxygen transport system in trout (*Salmo gairdneri*) during sustained exercise. *J. Exp. Biol.* 69: 247–260.

King, A. J., S. M. Henderson, M. H. Schmidt, A. G. Cole, and S. A. Adamo. 2005. Using ultrasound to understand vascular and mantle contributions to venous return in the cephalopod *Sepia officinalis* L. *J. Exp. Biol.* 208: 2071–2082.

LaBarbera, M. 1990. Principles of design of fluid transport systems in zoology. *Science* 249: 992–1000.

Laughlin, M. H. 1987. Skeletal muscle blood flow capacity: Role of muscle pump in exercise hyperemia. *Am. J. Physiol.* 253: H993–H1004.

Lillywhite, H. B. 1996. Gravity, blood circulation, and the adaptation of form and function in lower vertebrates. *J. Exp. Zool.* 275: 217–225.

Martin, A. W. 1980. Some invertebrate myogenic hearts: The hearts of worms and molluscs. In G. H. Bourne (ed.), *Hearts and Heart-like Organs*, vol. 1, pp. 1–39. Academic Press, New York.

McGaw, I. J., and C. L. Reiber. 2002. Cardiovascular system of the blue crab *Callinectes sapidus*. *J. Morphol.* 251: 1–21.

McGaw, I. J., and S. D. Duff. 2008. Cardiovascular system of anomuran crabs, genus *Lopholithodes*. *J. Morphol.* 269: 1295–1307.

McMahon, B. R., J. L. Wilkens, and P. J. S. Smith. 1997. Invertebrate circulatory systems. In W. H. Dantzler (ed.), *Comparative Physiology*, vol. 2 (Handbook of Physiology [Bethesda, MD], section 13), pp. 931–1008. Oxford University Press, New York.

Nilsson, S. 1995. Central cardiovascular dynamics in reptiles. *Adv. Comp. Environ. Physiol.* 21: 159–173.

Paul, R. J., S. Bihlmayer, M. Colmorgen, and S. Zahler. 1994. The open circulatory system of spiders (*Eurypelma californicum, Pholcus phalangioides*): A survey of functional morphology and physiology. *Physiol. Zool.* 67: 1360–1382.

Pawloski, J. R., D. T. Hess, and J. S. Stamler. 2001. Export by red blood cells of nitric oxide bioactivity. *Nature* 409: 622–626.

Randall, D. J., and C. Daxboeck. 1982. Cardiovascular changes in the rainbow trout (*Salmo gairdneri* Richardson) during exercise. *Can. J. Zool.* 60: 1135–1140.

Randall, D. J., W. W. Burggren, A. P. Farrell, and M. S. Haswell. 1981. *The Evolution of Air Breathing in Vertebrates*. Cambridge University Press, New York.

Rose, R. A., J. L. Wilkens, and R. L. Walker. 1998. The effects of walking on heart rate, ventilation rate and acid–base status in the lobster *Homarus americanus*. *J. Exp. Biol.* 201: 2601–2608.

Satchell, G. H. 1991. *Physiology and Form of Fish Circulation*. Cambridge University Press, London.

Seymour, R. S., A. R. Hargens, and T. J. Pedley. 1993. The heart works against gravity. *Am. J. Physiol.* 265: R715–R720. [See the article by Badeer for a contrasting point of view.]

Shadwick, R. E. 1998. Elasticity in arteries. *Am. Sci.* 86: 535–541.

Shadwick, R. E., R. K. O'Dor, and J. M. Gosline. 1990. Respiratory and cardiac function during exercise in squid. *Can. J. Zool.* 68: 792–798.

Snyder, G. K., and B. A. Sheafor. 1999. Red blood cells: Centerpiece in the evolution of the vertebrate circulatory system. *Am. Zool.* 39: 189–198.

Syme, D. A., K. Gamperl, and D. R. Jones. 2002. Delayed depolarization of the cog-wheel valve and pulmonary-to-systemic shunting in alligators. *J. Exp. Biol.* 205: 1843–1851.

Taylor, E. W., D. Jordan, and J. H. Coote. 1999. Central control of the cardiovascular and respiratory systems and their interactions in vertebrates. *Physiol. Rev.* 79: 855–916.

Trueman, E. R., and A. C. Brown. 1985. Dynamics of burrowing and pedal extension in *Donax serra* (Mollusca: Bivalvia). *J. Zool.* 207A: 345–355.

Verkman, A. S. 2002. Aquaporin water channels and endothelial cell function. *J. Anat.* 200: 617–627.

Wagner, P. D. 1996. Determinants of maximal oxygen transport and utilization. *Annu. Rev. Physiol.* 58: 21–50.

Wall, C. E., S. Cozza, C. A. Riquelme, W. R. McCombie, J. K. Heimiller, T. G. Marr, and L. A. Leinwand. 2010. Whole transcriptome analysis of the fasting and fed Burmese python heart: insights into extreme physiological cardiac adaptation. *Physiol. Genomics* 43: 69–76.

Wang, T. 2011. Gas exchange in frogs and turtles: how ectothermic vertebrates contributed to solving the controversy of pulmonary oxygen secretion. *Acta Physiol.* 202: 593–600.

Wang, T., J. Altimiras, W. Klein, and M. Axelsson. 2003. Ventricular haemodynamics in *Python molurus*: separation of pulmonary and systemic pressures. *J. Exp. Biol.* 206: 4241–4245.

Weibel, E. R. 1984. *The Pathway for Oxygen*. Harvard University Press, Cambridge, MA. Dramatic and well-presented research on the function of a non-mammalian heart.

Wells, M. J. 1992. The cephalopod heart: The evolution of a high-performance invertebrate pump. *Experientia* 48: 800–808.

Wells, M. J., and P. J. S. Smith. 1987. The performance of the octopus circulatory system: A triumph of engineering over design. *Experientia* 43: 487–499.

Wells, M. J., G. G. Duthie, D. F. Houlihan, P. J. S. Smith, and J. Wells. 1987. Blood flow and pressure changes in exercising octopuses (*Octopus vulgaris*). *J. Exp. Biol.* 131: 175–187.

West, N. H., and D. R. Jones (eds.). 1994. The Form and Function of Open and Closed Circulations [a symposium]. *Physiol. Zool.* 67: 1257–1425.

Wujcik, J. M., G. Wang, J. T. Eastman, and B. D. Sidell. 2007. Morphometry of retinal vasculature in Antarctic fishes is dependent upon the level of hemoglobin in circulation. *J. Exp. Biol.* 210: 815–824.

Chapter 26

Bennett, K. A., B. J. McConnell, and M. A. Fedak. 2001. Diurnal and seasonal variations in the duration and depth of the longest dives in southern elephant seals (*Mirounga leonina*): Possible physiological and behavioral constraints. *J. Exp. Biol.* 204: 649–662.

Blix, A. S. 2011. The venous system of seals, with new ideas on the significance of the extradural intravertebral vein. *J. Exp. Biol.* 214: 3507–3510.

Boutilier, R. G., J. Z. Reed, and M. A. Fedak. 2001. Unsteady-state gas exchange and storage in diving marine mammals: The harbor porpoise and gray seal. *Am. J. Physiol.* 281: R490–R494.

Boyd, I. L., J. P. Y. Arnould, T. Barton, and J. P. Croxall. 1994. Foraging behaviour of Antarctic fur seals during periods of contrasting prey abundance. *J. Anim. Ecol.* 63: 703–713.

Brischoux, F., X. Bonnet, T. R. Cook, and R. Shine. 2008. Allometry of diving capacities: ectothermy vs. endothermy. *J. Evol. Biol.* 21: 324–329.

Bron, K. M., H. V. Murdaugh, Jr., J. E. Millen, R. Lenthall, P. Raskin, and E. D. Robin. 1966. Arterial constrictor response in a diving mammal. *Science* 152: 540–543. [A classic.]

Brubakk, A. O., and T. S. Neuman (eds.). 2003. *Bennett and Elliott's Physiology and Medicine of Diving*. 5th ed. Saunders, Philadelphia, PA. [Includes an up-to-date chapter on comparative diving physiology and extensive treatment of diving diseases.]

Burns, J. M. 1999. The development of diving behavior in juvenile Weddell seals: Pushing physiological limits in order to survive. *Can. J. Zool.* 77: 737–747.

Butler, P. J. 2006. Aerobic dive limit. What is it and is it always used appropriately? *Comp. Biochem. Physiol., A* 145: 1–6.

Butler, P. J., and D. R. Jones. 1997. Physiology of diving of birds and mammals. *Physiol. Rev.* 77: 837–899.

Davis, R. W. 2014. A review of the multi-level adaptations for maximizing aerobic dive duration in marine mammals: from biochemistry to behavior. *J. Comp. Physiol. B* 184: 23–53.

Davis, R. W., and S. B. Kanatous. 1999. Convective oxygen transport and tissue oxygen consumption in Weddell seals during aerobic dives. *J. Exp. Biol.* 202: 1091–1113.

Falke, K. J., R. D. Hill, J. Qvist, R. C. Schneider, M. Guppy, G. C. Liggins, P. W. Hochachka, R. E. El-liott, and W. M. Zapol. 1985. Seal lungs collapse during free diving: Evidence from arterial nitrogen tensions. *Science* 229: 556–558.

Ferretti, G. 2001. Extreme human breath-hold diving. *Eur. J. Appl. Physiol.* 84: 254–271.

Fuiman, L. A., K. M. Madden, T. M. Williams, and R. W. Davis. 2007. Structure of foraging dives by Weddell seals at an offshore isolated hole in the Antarctic fast-ice environment. *Deep-Sea Res. II* 54: 270–289.

Guppy, M., R. D. Hill, R. C. Schneider, J. Qvist, G. C. Liggins, W. M. Zapol, and P. W. Hochachka. 1986. Microcomputer-assisted metabolic studies of vol-untary diving of Weddell seals. *Am. J. Physiol.* 250: R175–R187.

Guyton, G. P., K. S. Stanek, R. C. Schneider, P. W. Ho-chachka, W. E. Hurford, D. G. Zapol, G. C. Lig-gins, and W. M. Zapol. 1995. Myoglobin saturation in free-diving Weddell seals. *J. Appl. Physiol.* 79: 1148–1155.

Halsey, L. G., P. J. Butler, and T. M. Blackburn. 2006. A phylogenetic analysis of the allometry of diving. *Am. Nat.* 167: 276–287.

Halsey, L. G., T. M. Blackburn, and P. J. Butler. 2006. A comparative analysis of the diving behaviour of birds and mammals. *Funct. Ecol.* 20: 889–899.

Hindle, A. G., J.-A. E. Mellish, and M. Horning. 2011. Aerobic dive limit does not decline in an aging pin-niped. *J. Exp. Biol.* 315: 544–552.

Hochachka, P. W. 2000. Pinniped diving response mechanism and evolution: A window on the para-digm of comparative biochemistry and physiology. *Comp. Biochem. Physiol., A* 126: 435–458.

Hong, S. K. 1996. The hyperbaric environment. In M. J. Fregly and C. M. Blatteis (eds.), *Environmen-tal Physiology*, vol. 2 (Handbook of Physiology [Bethesda, MD], section 4), pp. 975–1056. Oxford University Press, New York.

Hong, S. K., and H. Rahn. 1967. The diving women of Korea and Japan. *Sci. Am.* 216(5): 34–43. [Although published in 1967, still informative because of the detail included on traditional ama practices.]

Hooker, S. K. and 27 coauthors. 2012. Deadly diving? Physiological and behavioural management of de-compression stress in diving mammals. *Proc. Royal Soc. B* 279: 1041–1050.

Hooker, S. K., R. W. Baird, and A. Fahlman. 2009. Could beaked whales get the bends? Effect of div-ing behaviour and physiology on modelled gas exchange for three species: *Ziphius cavirostris, Me-soplodon densirostris,* and *Hyperoodon ampullatus. Resp. Physiol. Neurobiol.* 167: 235–246.

Hurley, J. A., and D. P. Costa. 2001. Standard metabolic rate at the surface and during trained submersions in adult California sea lions (*Zalophus california-nus*). *J. Exp. Biol.* 204: 3273–3281.

Joulia, F, F. Lemaitre, P. Fontanari, M. L. Mille, and P. Barthelemy. 2009. Circulatory effects of apnoea in elite breath-hold divers. *Acta Physiologica* 197: 75–82.

Kjekshus, J. K., A. S. Blix, R. Elsner, R. Hol, and E. Amundsen. 1982. Myocardial blood flow and me-tabolism in the diving seal. *Am. J. Physiol.* 242: R97–R104.

Kooyman, G. L. 1981. *Weddell Seal, Consummate Diver.* Cambridge University Press, New York.

Kooyman, G. L. 1989. *Diverse Divers: Physiology and Behavior.* Springer-Verlag, New York.

Kooyman, G. L., and P. J. Ponganis. 1998. The physi-ological basis of diving to depth: Birds and mam-mals. *Annu. Rev. Physiol.* 60: 19–32.

Kvadsheim, P. H., L. P. Folkow, and A. S. Blix. 2005. In-hibition of shivering in hypothermic seals during diving. *Am. J. Physiol.* 289: R326–R331.

Meir, J. U., C. D. Champagne, D. P. Costa, C. L. Wil-liams, and P. J. Ponganis. 2009. Extreme hypoxemic tolerance and blood oxygen depletion in diving el-ephant seals. *Am. J. Physiol.* 297: R927–R939.

Meir, J. U., C. D. Champagne, D. P. Costa, C. L. Wil-liams, and P. J. Ponganis. 2009. Extreme hypoxemic tolerance and blood oxygen depletion in diving el-ephant seals. *Am. J. Physiol.* 297: R927–R939.

Moore, M. J., T. Hammar, J. Arruda, S. Cramer, S. Den-nison, E. Montie, and A. Fahlman. 2011. Hyper-baric computed tomographic measurement of lung compression in seals and dolphins. *J. Exp. Biol.* 214: 2390–2397.

Noren, S. R., and T. M. Williams. 2000. Body size and skeletal muscle myoglobin of cetaceans: Adapta-tions for maximizing dive duration. *Comp. Bio-chem. Physiol., A* 126: 181–191.

Noren, S. R., T. M. Williams, D. A. Pabst, W. A. McLel-lan, and J. L. Dearolf. 2001. The development of diving in marine endotherms: Preparing the skel-etal muscles of dolphins, penguins, and seals for activity during submergence. *J. Comp. Physiol., B* 171: 127–134.

Ponganis, P. J., G. L. Kooyman, and M. A. Castellini. 1993. Determinants of the aerobic dive limit of Weddell seals: Analysis of diving metabolic rates, postdive end tidal P_{O_2}'s, and blood and muscle oxygen stores. *Physiol. Zool.* 66: 732–749.

Ridgway, S. H., and R. Howard. 1979. Dolphin lung collapse and intramuscular circulation during free diving: Evidence from nitrogen washout. *Science* 206: 1182–1183.

Ridgway, S. H., B. L. Scronce, and J. Kanwisher. 1969. Respiration and deep diving in the bottlenose por-poise. *Science* 166: 1651–1654. [A classic.]

Rommel, S. A., D. A. Pabst, and W. A. McLellan. 1998. Reproductive thermoregulation in marine mam-mals. *Am. Sci.* 86: 440–448.

Schreer, J. F., K. M. Kovacs, and R. J. O. Hines. 2001. Comparative diving patterns of pinnipeds and sea-birds. *Ecol. Monogr.* 71: 137–162.

Sparling, C. E., and M. A. Fedak. 2004. Metabolic rates of captive grey seals during voluntary diving. *J. Exp. Biol.* 207: 1615–1624.

Thompson, D., and M. A. Fedak. 1993. Cardiac re-sponses of grey seals during diving at sea. *J. Exp. Biol.* 174: 139–164.

Thornton, S. J., D. M. Spielman, N. J. Pelc, W. F. Block, D. E. Crocker, D. P. Costa, B. J. Le Boeuf, and P. W. Hochachka. 2001. Effects of forced diving on the spleen and hepatic sinus in northern elephant seal pups. *Proc. Natl. Acad. Sci. U.S.A.* 98: 9413–9418.

Watkins, W. A., M. A. Daher, K. M. Fristrup, T. J. Howald, and G. N. DiSciara. 1993. Sperm whales tagged with transponders and tracked underwater by sonar. *Mar. Mamm. Sci.* 9: 55–67.

Wienecke, B., G. Robertson, R. Kirkwood, and K. Law-ton. 2007. Extreme dives by free-ranging emperor penguins. *Polar Biol.* 30: 133–142.

Williams, E. E., B. S. Stewart, C. A. Beuchat, G. N. Somero, and J. R. Hazel. 2001. Hydrostatic-pres-sure and temperature effects on the molecular

order of erythrocyte membranes from deep-, shal-low-, and non-diving mammals. *Can. J. Zool.* 79: 888–894.

Williams, T. M., L. A. Fuiman, M. Horning, and R. W. Davis. 2004. The cost of foraging by a marine predator, the Weddell seal *Leptonychotes weddellii*: pricing by the stroke. *J. Exp. Biol.* 207: 973–982.

Williams, T. M., R. W. Davis, L. A. Fuiman, J. Francis, B. L. Le Boeuf, M. Horning, J. Calambokidis, and D. A. Croll. 2000. Sink or swim: Strategies for cost-efficient diving by marine mammals. *Science* 288: 133–136.

Chapter 27

Beyenbach, K. W. (ed.). 1990. *Cell Volume Regulation.* Karger, Basel, Switzerland.

Borowitzka, L. J. 1981. Solute accumulation and regu-lation of cell water activity. In L. G. Paleg and D. Aspinall (eds.), *The Physiology and Biochemistry of Drought Resistance in Plants*, pp. 97–130. Academic Press, New York.

Denny, M. W. 1993. *Air and Water: The Biology and Physics of Life's Media.* Princeton University Press, Princeton, NJ.

Evans, D. H. (ed.). 2009. *Osmotic and Ionic Regulation. Cells and Animals.* CRC Press, Boca Raton, FL. The first two chapters provide advanced treatments of water and solute transport, cell-volume regulation, and osmosensing.

Galcheva-Gargova, Z., B. Dérijard, I.-H. Wu, and R. J. Davis. 1994. An osmosensing signal transduc-tion pathway in mammalian cells. *Science* 265: 806–808.

Haond, C., L. Bonnal, R. Sandeaux, G. Charman-tier, and J.-P. Trilles. 1999. Ontogeny of intracel-lular isosmotic regulation in the European lobster *Homarus gammarus* (L.). *Physiol. Biochem. Zool.* 72: 534–544.

Jeziorski, A., and 14 additional authors. 2009. The widespread threat of calcium decline in fresh wa-ters. *Science* 322: 1374–1377.

Kirschner, L. B. 1991. Water and ions. In C. L. Prosser (ed.), *Environmental and Metabolic Animal Physiol-ogy (Comparative Animal Physiology*, 4th ed.), pp. 13–107. Wiley-Liss, New York.

Lang, F. (ed.). 1998. *Cell Volume Regulation.* Karger, Basel, Switzerland. [The one defect in this other-wise admirable book is that it does not notify read-ers that its coverage is mammals, not animals in general.]

Neufeld, D. S., and J. N. Cameron. 1994. Mechanism of the net uptake of water in moulting blue crabs (*Callinectes sapidus*) acclimated to high and low sa-linities. *J. Exp. Biol.* 188: 11–23.

Oschman, J. L. 1980. Water transport, cell junctions, and "structured water." In E. E. Bittar (ed.), *Mem-brane Structure and Function*, vol. 2, pp. 141–170. Wiley, New York.

Phlippen, M. K., S. G. Webster, J. S. Chung, and H. Dircksen. 2000. Ecdysis of decapod crustaceans is associated with a dramatic release of crustacean cardioactive peptide into the haemolymph. *J. Exp. Biol.* 203: 521–536.

Spring, K. R. 1998. Routes and mechanism of fluid transport by epithelia. *Annu. Rev. Physiol.* 60: 105–119.

Strange, K. (ed.). 1994. *Cellular and Molecular Physiol-ogy of Cell Volume Regulation.* CRC Press, Boca Ra-ton, FL.

Wheatly, M. G., and A. T. Gannon. 1995. Ion regulation in crayfish: Freshwater adaptations and the problem of molting. *Am. Zool.* 35: 49–59.

Williams, E. E., M. J. Anderson, T. J. Miller, and S. D. Smith. 2004. The lipid composition of hypodermal membranes from the blue crab (*Callinectes sapidus*) changes during the molt cycle and alters hypodermal calcium permeability. *Comp. Biochem. Physiol., B* 137: 235–245.

Chapter 28

Ahearn, G. A., J. M. Duerr, Z. Zhuang, R. J. Brown, A. Aslamkhan, and D. A. Killebrew. 1999. Ion transport processes of crustacean epithelial cells. *Physiol. Biochem. Zool.* 72: 1–18.

Antunes-Rodrigues, J., M. de Castro, L. L. K. Elias, M. M. Valença, and S. M. McCann. 2004. Neuroendocrine control of body fluid metabolism. *Physiol. Rev.* 84: 169–208.

Ar, A., and H. Rahn. 1980. Water in the avian egg: Overall budget of incubation. *Am. Zool.* 20: 373–384.

Archer, M. A., T. J. Bradley, L. D. Mueller, and M. R. Rose. 2007. Using experimental evolution to study the physiological mechanisms of desiccation resistance in *Drosophila melanogaster*. *Physiol. Biochem. Zool.* 80: 386–398.

Beuchat, C. 1990. Body size, medullary thickness, and urine concentrating ability in mammals. *Am. J. Physiol.* 258: R298–R308.

Bradley, T. J. 1987. Physiology of osmoregulation in mosquitoes. *Annu. Rev. Entomol.* 32: 439–462.

Bradshaw, S. D. 1997. *Homeostasis in Desert Reptiles.* Springer, New York.

Brill, R., Y. Swimmer, C. Taxboel, K. Cousins, and T. Lowe. 2001. Gill and intestinal Na$^+$–K$^+$ ATPase activity, and estimated maximal osmoregulatory costs, in three high energy-demand teleosts: yellowfin tuna (*Thunnus albacares*), skipjack tuna (*Katsuwonus pelamis*), and dolphin fish (*Coryphaena hippurus*). *Mar. Biol.* 138: 935–944.

Calder, W. A., III, and E. J. Braun. 1983. Scaling of osmotic regulation in mammals and birds. *Am. J. Physiol.* 244: R601–R606.

Carey, C. 1986. Tolerance of variation in eggshell conductance, water loss, and water content by red-winged blackbird embryos. *Physiol. Zool.* 59: 109–122.

Carpenter, R. E. 1968. Salt and water metabolism in the marine fish-eating bat, *Pizonyx vivesi*. *Comp. Biochem. Physiol.* 24: 951–964.

Champagne, A. M., A. Muñoz-Garcia, T. Shtayyeh, B. I. Tieleman, A. Hegemann, M. E. Clement, and J. B. William. 2012. Lipid composition of the stratum corneum and cutaneous water loss in birds along an aridity gradient. *J. Exp. Biol.* 215: 4299–4307.

Charmantier, G., C. Haond, J.-H. Lignot, and M. Charmantier-Daures. 2001. Ecophysiological adaptation to salinity throughout a life cycle: A review in homarid lobsters. *J. Exp. Biol.* 204: 967–977.

Coast, G. M. 1998. Insect diuretic peptides: Structures, evolution, and actions. *Am. Zool.* 38: 442–449.

Crowe, J. H., A. E. Oliver, and F. Tablin. 2002. Is there a single biochemical adaptation to anhydrobiosis? *Integr. Comp. Biol.* 42: 497–503.

Crowe, J. H., F. A. Hoekstra, and L. M. Crowe. 1992. Anhydrobiosis. *Annu. Rev. Physiol.* 54: 579–599.

Currie, S., and S. L. Edwards. 2010. The curious case of the chemical composition of hagfish tissues—50 years on. *Comp. Biochem. Physiol., A* 157: 111–115.

Danulat, E. 1995. Biochemical-physiological adaptations of teleosts to highly alkaline, saline lakes. In P. W. Hochachka and T. P. Mommsen (eds.), *Biochemistry and Molecular Biology of Fishes*, vol. 5, *Environmental and Ecological Biochemistry*, pp. 229–249. Elsevier, New York.

Diehl, W. J. 1986. Osmoregulation in echinoderms. *Comp. Biochem. Physiol., A* 84: 199–205.

Donini, A., and M. J. O'Donnell. 2005. Analysis of Na$^+$, Cl$^-$, K$^+$, H$^+$, and NH$_4^+$ concentration gradients adjacent to the surface of anal papillae of the mosquito *Aedes aegypti*: application of self-referencing ion-selective microelectrodes. *J. Exp. Biol.* 208: 603–610.

Evans, D. H. 2011. Freshwater fish gill ion transport: August Krogh to morpholinos and microprobes. *Acta Physiol.* 202: 349–359.

Evans, D. H., and J. B. Claiborne (eds.). 2006. *The Physiology of Fishes*, 3rd ed. CRC Press, Boca Raton, FL.

Evans, D. H., P. M. Piermarini, and K. P. Choe. 2003. The multifunctional fish gill: dominant site of gas exchange, osmoregulation, acid-base regulation, and excretion of nitrogenous wastes. *Physiol. Rev.* 85: 97–177.

Evans, T. G., and G. N. Somero. 2008. A microarray-based transcriptomic time-course of hyper- and hypo-osmotic stress signaling events in the euryhaline fish *Gillichthys mirabilis*: osmosensors to effectors. *J. Exp. Biol.* 211: 3636–3649.

Feder, M. E., and W. W. Burggren (eds.). 1992. *Environmental Physiology of the Amphibians*. University of Chicago Press, Chicago.

Frank, C. L. 1988. Diet selection by a heteromyid rodent: Role of net metabolic water production. *Ecology* 69: 1943–1951.

Gerstberger, R., and D. A. Gray. 1993. Fine structure, innervation, and functional control of avian salt glands. *Int. Rev. Cytol.* 144: 129–215.

Gibbs, A. G. 1998. Water-proofing properties of cuticular lipids. *Am. Zool.* 38: 471–482.

Gilles, R., and E. Delpire. 1997. Variations in salinity, osmolarity, and water availability: Vertebrates and invertebrates. In W. H. Dantzler (ed.), *Comparative Physiology*, vol. 2 (Handbook of Physiology [Bethesda, MD], section 13), pp. 1523–1586. Oxford University Press, New York.

Goldstein, D. L., J. B. Williams, and E. J. Braun. 1990. Osmoregulation in the field by salt-marsh savannah sparrows, *Passerculus sandwichensis beldingi*. *Physiol. Zool.* 63: 669–682.

Goss, G. G., S. F. Perry, J. N. Fryer, and P. Laurent. 1998. Gill morphology and acid–base regulation in freshwater fishes. *Comp. Biochem. Physiol., A* 119: 107–115.

Greenaway, P. 1988. Ion and water balance. In W. W. Burggren and B. R. McMahon (eds.), *Biology of the Land Crabs*, pp. 211–248. Cambridge University Press, New York.

Griffith, R. W. 1980. Chemistry of the body fluids of the coelacanth, *Latimeria chalumnae*. *Proc. R. Soc. London, Ser. B* 208: 329–347.

Griffith, R. W. 1994. The life of the first vertebrates. *BioScience* 44: 408–417.

Gutierrez, J. S., J. A. Masero, J. M. Abad-Gomez, A. Villegas, and J. M. Sanchez-Guzman. 2011. Understanding the energetic costs of living in saline environments: effects of salinity on basal metabolic rate, body mass and daily energy consumption of a long-distance migratory shorebird. *J. Exp. Biol.* 214: 829–835.

Hadley, N. F. 1994. *Water Relations of Terrestrial Arthropods*. Academic Press, New York.

Hand, S. C. 1992. Water content and metabolic organization in anhydrobiotic animals. In G. N. Somero, C. B. Osmond, and C. L Bolis (eds.), *Water and Life*, pp. 104–127. Springer, New York.

Hand, S. C., and I. Hardewig. 1996. Downregulation of cellular metabolism during environmental stress: Mechanisms and implications. *Annu. Rev. Physiol.* 58: 539–563.

Havird, J. C., R. P. Henry, and A. E. Wilson. 2013. Altered expression of Na$^+$/K$^+$-ATPase and other osmoregulatory genes in the gills of euryhaline animals in response to salinity transfer: A meta-analysis of 59 quantitative PCR studies over 10 years. *Comp. Biochem. Physiol. D* 8: 131–140.

Hazon, N., F. B. Eddy, and G. Flik (eds.). 1997. *Ionic Regulation in Animals: A Tribute to Professor W. T. W. Potts*. Springer, New York.

Heisler, N (ed.). 1995. *Mechanisms of Systemic Regulation: Acid–Base Regulation, Ion-Transfer and Metabolism. Adv. Comp. Environ. Physiol.* 22: 1–265.

Hillyard, S. D. 1999. Behavioral, molecular and integrative mechanisms of amphibian osmoregulation. *J. Exp. Zool.* 283: 662–674.

Hochachka, P. W., and T. P. Mommsen (eds.). 1995. *Biochemistry and Molecular Biology of Fishes*, vol. 5, *Environmental and Ecological Biochemistry*. Elsevier, New York.

Hwang, P.-P., and T.-H. Lee. 2007. New insights into fish ion regulation and mitochondrion-rich cells. *Comp. Biochem. Physiol., A* 148: 479–497.

Jönsson, K. I. 2003. Causes and consequences of excess resistance in cryptobiotic metazoans. *Physiol. Biochem. Zool.* 76: 429–435.

Kaneko, T., and F. Katoh. 2004. Functional morphology of chloride cells in killifish *Fundulus heteroclitus*, a euryhaline teleost with seawater preference. *Fish. Sci.* 70: 723–733.

Kaneko, T., S. Hasegawa, Y. Takagi, M. Tagawa, and T. Hirano. 1995. Hypoosmoregulatory ability of eyed stage embryos of chum salmon. *Mar. Biol.* 122: 165–170.

Kim, Y. K., S. Watanabe, T. Kaneko, M. D. Huh, and S. I. Park. 2010. Expression of aquaporins 3, 8 and 10 in the intestines of freshwater- and seawater-acclimated Japanese eels *Anguilla japonica*. *Fish. Sci.* 76: 695–702.

Kinne, R. K. H. (ed.). 1990. *Basic Principles in Transport*. Karger, Basel.

Kirschner, L. B. 1991. Water and ions. In C. L. Prosser (ed.), *Environmental and Metabolic Animal Physiology* (*Comparative Animal Physiology*, 4th ed.), pp. 13–107. Wiley-Liss, New York.

Kirschner, L. B. 1993. The energetics of osmotic regulation in ureotelic and hypoosmotic fishes. *J. Exp. Zool.* 267: 19–26.

Kirschner, L. B. 1995. Energetics of osmoregulation in fresh water vertebrates. *J. Exp. Zool.* 271: 243–252.

Kirschner, L. B. 1997. Extrarenal mechanisms in hydromineral and acid–base regulation of aquatic vertebrates. In W. H. Dantzler (ed.), *Comparative Physiology*, vol. 1 (Handbook of Physiology [Bethesda, MD], section 13), pp. 577–622. Oxford University Press, New York.

Koehn, R. K. 1983. Biochemical genetics and adaptation in molluscs. In K. M. Wilbur (ed.), *The Mollusca*, vol. 2, pp. 305–330. Academic Press, New York.

Koehn, R. K., and T. J. Hilbish. 1987. The adaptive importance of genetic variation. *Am. Sci.* 75: 134–141.

Land, S. C., and N. Bernier. 1995. Estivation: Mechanisms and control of metabolic suppression. In P. W. Hochachka and T. P. Mommsen (eds.), *Biochemistry and Molecular Biology of Fishes*, vol. 5, *Environmental and Ecological Biochemistry*, pp. 381–412. Elsevier, New York.

Lillywhite, H. B. 2006. Water relations of tetrapod integument. *J. Exp. Biol.* 209: 202–226.

Lillywhite, H. B., C. M. Sheehy III, F. Brischoux, and A. Grech. 2014. Pelagic sea snakes dehydrate at sea. *Proc. Roy. Soc. London B* 281: 20140119.

Louw, G. N. 1993. *Physiological Animal Ecology*. Longman Scientific & Technical, Harlow, England.

Lovett, D. L., M. P. Verzi, J. E. Burgents, C. A. Tanner, K. Glomski, J. J. Lee, and D. W. Towle. 2006. Expression profiles of Na$^+$, K$^+$-ATPase during acute and chronic hypo-osmotic stress in the blue crab *Callinectes sapidus*. *Biol. Bull.* 211: 58–65.

MacMillen, R. E., and D. S. Hinds. 1983. Water regulatory efficiency in heteromyid rodents: A model and its applications. *Ecology* 64: 152–164.

Marshall, A. T., and P. D. Cooper. 1988. Secretory capacity of the lachrymal salt gland of hatchling sea turtles, *Chelonia mydas*. *J. Comp. Physiol., B* 157: 821–827.

Marshall, W. S. 2002. Na$^+$, Cl$^-$, Ca^{2+} and Zn^{2+} transport by fish gills: retrospective review and prospective synthesis. *J. Exp. Zool.* 293: 264–283.

McClanahan, L., Jr. 1972. Changes in body fluids of burrowed spadefoot toads as a function of soil water potential. *Copeia* 1972: 209–216.

McCormick, S. D. 2001. Endocrine control of osmoregulation in teleost fish. *Am. Zool.* 41: 781–794.

McCormick, S. D., A. M. Regish, and A. K. Christensen. 2009. Distinct freshwater and seawater isoforms of Na$^+$/K$^+$-ATPase in gill chloride cells of Atlantic salmon. *J. Exp. Biol.* 212: 3994–4001.

Mitrovic, D., and S. F. Perry. 2009. The effects of thermally induced gill remodeling on ionocyte distribution and branchial chloride fluxes in goldfish (*Carassius auratus*). *J. Exp. Biol.* 212: 843–852.

Mommsen, T. P., and P. J. Walsh. 1989. Evolution of urea synthesis in vertebrates: The piscine connection. *Science* 243: 72–75.

Nagy, K. A., and C. C. Peterson. 1988. *Scaling of Water Flux Rate in Animals*. University of California Publications in Zoology, vol. 120. University of California Press, Berkeley.

O'Donnell, M. J. 1997. Mechanisms of excretion and ion transport in invertebrates. In W. H. Dantzler (ed.), *Comparative Physiology*, vol. 2 (Handbook of Physiology [Bethesda, MD], section 13), pp. 1207–1289. Oxford University Press, New York.

Onken, H., and S. Riestenpatt. 1998. NaCl absorption across split gill lamellae of hyperregulating crabs: Transport mechanisms and their regulation. *Comp. Biochem. Physiol., A* 119: 883–893.

Ortiz, R. M. 2001. Osmoregulation in marine mammals. *J. Exp. Biol.* 204: 1831–1844.

Perry, S. F. 1997. The chloride cell: Structure and function in the gills of freshwater fishes. *Annu. Rev. Physiol.* 59: 325–347.

Perry, S. F. 1998. Relationships between branchial chloride cells and gas transfer in freshwater fish. *Comp. Biochem. Physiol., A* 119: 9–16.

Perry, S. F., A. Shahsavarani, T. Georgalis, M. Bayaa, M. Furimsky, and S. L. Y. Thomas. 2003. Channels, pumps, and exchangers in the gill and kidney of freshwater fishes: Their role in ionic and acid–base regulation. *J. Exp. Zool. A* 300: 53–62.

Pillans, R. D., J. P. Good, W. G. Anderson, N. Hazon, and C. E. Franklin. 2005. Freshwater to seawater acclimation of juvenile bull sharks (*Carcharhinus leucas*): plasma osmolytes and Na$^+$/K$^+$-ATPase activity in gill, rectal gland, kidney and intestine. *J. Comp. Physiol., B* 175: 37–44.

Podrabsky, J. E., and S. C. Hand. 1999. The bioenergetics of embryonic diapause in an annual killifish, *Austrofundulus limnaeus*. *J. Exp. Biol.* 202: 2567–2580.

Rahn, H., C. V. Paganelli, I. C. T. Nisbet, and G. C. Whittow. 1976. Regulation of incubation water loss in eggs of seven species of terns. *Physiol. Zool.* 49: 245–259.

Remane, A., and C. Schlieper. 1971. Biology of brackish water. *Die Binnengewässer* 25: 1–372.

Schmidt-Nielsen, K. 1964. *Desert Animals*. Oxford University Press, London.

Schmuck, R., and K. E. Linsenmair. 1988. Adaptations of the reed frog *Hyperolius viridiflavus* (Amphibia, Anura, Hyperoliidae) to its arid environment. III. Aspects of nitrogen metabolism and osmoregulation in the reed frog, *Hyperolius viridiflavus taeniatus*, with special reference to the role of iridophores. *Oecologia* 75: 354–361.

Schmuck, R., and K. E. Linsenmair. 1997. Regulation of body water balance in reedfrogs (superspecies *Hyperolius viridiflavus* and *Hyperolius marmoratus*: Amphibia, Anura, Hyperoliidae) living in unpredictably varying savannah environments. *Comp. Biochem. Physiol., A* 118: 1335–1352.

Schroter, R. C., and N. V. Watkins. 1989. Respiratory heat exchange in mammals. *Respir. Physiol.* 78: 357–368.

Schwimmer, H., and A. Haim. 2009. Physiological adaptations of small mammals in desert ecosystems. *Integr. Zool.* 4: 357–366.

Shuttleworth, T. J., and J.-P. Hildebrandt. 1999. Vertebrate salt glands: Short- and long-term regulation of function. *J. Exp. Zool.* 283: 689–701.

Somero, G. N., C. B. Osmond, and C. L Bolis (eds.). 1992. *Water and Life*. Springer, New York.

Suzuki, M., and S. Tanaka. 2010. Molecular diversity of vasotocin-dependent aquaporins closely associated with water adaptation strategy in anuran amphibians. *J. Neuroendocrinol.* 22: 407–412.

Tieleman, B. I. 2005. Physiological, behavioral, and life history adaptations of larks along an aridity gradient: a review. In G. Bota, M. B. Morales, S. Mañosa, and J. Camprodon (eds.), *Ecology and Conservation of Steppe-land Birds*. Lynx Edicions, Barcelona, Spain.

Tracy, C. R. 1976. A model of the dynamic exchanges of water and energy between a terrestrial amphibian and its environment. *Ecol. Monogr.* 46: 293–326.

Tracy, R. L., and G. E. Walsberg. 2002. Kangaroo rats revisited: Re-evaluating a classic case of desert survival. *Oecologia* 133: 449–457.

Uchiyama, M., and N. Konno. 2006. Hormonal regulation of ion and water transport in anuran amphibians. *Gen. Comp. Endocrinol.* 147: 54–61.

Varsamos, S., C. Nebel, and G. Charmantier. 2005. Ontogeny of osmoregulation in postembryonic fish: A review. *Comp. Biochem. Physiol., A* 141: 401–429.

Voyles, J., and 10 additional authors. 2009. Pathogenesis of Chytridiomycosis, a cause of catastrophic amphibian declines. *Science* 326: 582–585.

Warburg, M. R. 1997. *Ecophysiology of Amphibians Inhabiting Xeric Environments*. Springer, New York.

Wehner, R. 2003. Desert ant navigation: how miniature brains solve complex tasks. *J. Comp. Physiol., A* 189: 579–588.

Wehner, R., and S. Wehner. 2011. Parallel evolution of thermophilia: daily and seasonal foraging patterns of heat-adapted desert ants: *Cataglyphis* and *Ocymyrmex* species. *Physiol. Entomol.* 36: 271–281.

Westin, L., and A. Nissling. 1991. Effects of salinity on spermatozoa motility, percentage of fertilized eggs and egg development of Baltic cod (*Gadus morhua*), and implications for cod stock fluctuations in the Baltic. *Mar. Biol.* 108: 5–9.

Wheatly, M. 1993. Physiological adaptations in decapodan crustaceans for life in fresh water. *Adv. Comp. Environ. Physiol.* 15: 77–132.

Wheatly, M. G., and A. T. Gannon. 1995. Ion regulation in crayfish: Freshwater adaptations and the problem of molting. *Am. Zool.* 35: 49–59.

Williams, J. B., and B. I. Tieleman. 2001. Physiological ecology and behavior of desert birds. In V. Nolan, Jr., and C. F. Thompson (eds.), *Current Ornithology*, vol. 16, pp. 299–353. Springer, New York.

Withers, P. C., and M. Guppy. 1996. Do Australian desert frogs co-accumulate counteracting solutes with urea during aestivation? *J. Exp. Biol.* 199: 1809–1816.

Wright, J. L., P. Westh, and H. Ramløv. 1992. Cryptobiosis in Tardigrada. *Biol. Rev.* 67: 1–29.

Wright, P. A., and M. D. Land. 1998. Urea production and transport in teleost fishes. *Comp. Biochem. Physiol., A* 119: 47–54.

Wright, P. A., and P. M. Anderson. 2001. *Nitrogen Excretion. Fish Physiology*, vol. 20. Academic Press, New York.

Yancey, P. H., M. E. Gerringer, J. C. Drazen, A. A. Rowden, and A. Jamieson. 2014. Marine fish may be biochemically constrained from inhabiting the deepest ocean depths. *Proc. Natl. Acad. Sci. USA* 111:4461–4465.

Zachariassen, K. E., and S. A. Pedersen. 2002. Volume regulation during dehydration in desert beetles. *Comp. Biochem. Physiol., A* 133: 805–811.

Chapter 29

Abramson, R. G., and M. S. Lipkowitz. 1990. Evolution of the uric acid transport mechanisms in vertebrate kidney. In R. K. H. Kinne (ed.), *Basic Principles in Transport*, pp. 115–153. Karger, Basel, Switzerland.

Agre, P. 2004. Aquaporin water channels (Nobel Lecture). *Angew. Chem.* 43: 4278–4290.

Atkinson, D. E. 1992. Functional roles of urea synthesis in vertebrates. *Physiol. Zool.* 65: 243–267. [Emphasizes a novel perspective on the evolution of ureo- and uricotelism.]

Bankir, L., and C. de Rouffignac. 1985. Urinary concentrating ability: Insights from comparative anatomy. *Am. J. Physiol.* 249: R643–R666.

Beuchat, C. A. 1990. Body size, medullary thickness, and urine concentrating ability in mammals. *Am. J. Physiol.* 258: R298–R308.

Beyenbach, K. W., H. Skaer, and J. A. T. Dow. 2010. The developmental, molecular, and transport biology of Malpighian tubules. *Annu. Rev. Entomol.* 55: 351–374.

Bradley, T. J. 1987. Physiology of osmoregulation in mosquitoes. *Annu. Rev. Entomol.* 32: 439–462.

Braun, E. J. 1999. Integration of organ systems in avian osmoregulation. *J. Exp. Zool.* 283: 702–707.

Brenner, B. M. (ed.). 2004. *Brenner and Rector's The Kidney*, 7th ed. Saunders, Philadelphia.

Brown, J. A., R. J. Balment, and J. C. Rankin (ed.). 1993. *New Insights in Vertebrate Kidney Function.* Cambridge University Press, New York.

Campbell, J. W. 1991. Excretory nitrogen metabolism. In C. L. Prosser (ed.), *Environmental and Metabolic Animal Physiology (Comparative Animal Physiology,* 4th ed.), pp. 277–324. Saunders, Philadelphia.

Campbell, J. W., J. E. Vorhaben, and D. D. Smith, Jr. 1987. Uricoteley: Its nature and origin during the evolution of tetrapod vertebrates. *J. Exp. Zool.* 243: 349–363.

Casotti, G., K. C. Richardson, and J. S. Bradley. 1993. Ecomorphological constraints imposed by the kidney component measurements in honeyeater birds inhabiting different environments. *J. Zool. (London)* 231: 611–625.

Christiansen, J. S., R. A. Dalmo, and K. Ingebrigtsen. 1996. Xenobiotic excretion in fish with aglomerular kidneys. *Mar. Ecol. Progr. Ser.* 136: 303–304.

Coast, G. M., I. Orchard, J. E. Phillips, and D. A. Schooley. 2002. Insect diuretic and antidiuretic hormones. *Adv. Insect Physiol.* 29: 279–409.

Dantzler, W. H. 1985. Comparative aspects of renal function. In D. W. Seldon and G. Giebisch (eds.), *The Kidney: Physiology and Pathophysiology,* pp. 333–364. Raven, New York.

Danzler, W. H. 2003. Regulation of renal proximal and distal tubule transport: sodium, chloride and organic anions. *Comp. Biochem. Physiol., A* 136: 453–478.

Danzler, W. H., T. L. Pannabecker, A. T. Layton, and H. E. Layton. 2011. Urine concentrating mechanism in the inner medulla of the mammalian kidney: role of three-dimensional architecture. *Acta Physiol.* 202: 361–378.

Dickenson, H., D. W. Walker, L. Cullen-McEwen, E. M. Wintour, and K. Moritz. 2005. The spiny mouse (*Acomys cahirinus*) completes nephrogenesis before birth. *Am. J. Physiol.* 289: F272–F279.

Dow, J. A. T. 2009. Insights into the Malpighian tubule from functional genomics. *J. Exp. Biol.* 212: 435–445.

Greenwald, L. 1989. The significance of renal relative medullary thickness. *Physiol. Zool.* 62: 1005–1014.

Hadley, N. F. 1994. *Water Relations of Terrestrial Arthropods.* Academic Press, New York. [Includes a good discussion of excretory structure and function.]

Heisler, N. 1995. Ammonia vs. ammonium: Elimination pathways of nitrogenous wastes in ammonotelic fishes. *Adv. Comp. Environ. Physiol.* 22: 63–87.

Hevert, F. 1984. Urine formation in the lamellibranchs: Evidence for ultrafiltration and quantitative description. *J. Exp. Biol.* 111: 1–12.

Jones, H. D., and D. Peggs. 1983. Hydrostatic and osmotic pressures in the heart and pericardium of *Mya arenaria* and *Anodonta cygnea. Comp. Biochem. Physiol., A* 76: 381–385.

Kinne, R. K. H. (ed.). 1989. *Structure and Function of the Kidney.* Karger, Basel, Switzerland.

Kinne, R. K. H. (ed.). 1990. *Basic Principles in Transport.* Karger, Basel, Switzerland.

Kinne, R. K. H. (ed.). 1990. *Urinary Concentrating Mechanisms.* Karger, Basel, Switzerland.

Lang, F. (ed.). 1998. *Cell Volume Regulation.* Karger, Basel, Switzerland. [Includes background chapters and a specific chapter on the kidney.]

Laverty, G., and E. Skadhauge. 2008. Adaptive strategies for post-renal handling of urine in birds. *Comp. Biochem. Physiol., A* 149: 246–254.

Little, C. 1965. The formation of urine by the prosobranch gastropod mollusc *Viviparus viviparus* Linn. *J. Exp. Biol.* 43: 39–54.

Long, W. S. 1973. Renal handling of urea in *Rana catesbeiana. Am. J. Physiol.* 224: 482–490.

Marshall, W. S., and M. Grosell. 2006. Ion transport, osmoregulation, and acid-base balance. In D. H. Evans and J. B. Claiborne (eds.), *The Physiology of Fishes,* 3rd ed., pp. 177–230. CRC Press, Boca Raton, FL.

Neuhofer, W., and F.-X. Beck. 2006. Survival in hostile environments: strategies of renal medullary cells. *Physiology* 21: 171–180.

Pannabecker, T. L., and W. H. Dantzler. 2006. Three-dimensional architecture of inner medullary vasa recta. *Am. J. Physiol.* 290: F1355–F1366.

Pannabecker, T. L., and W. H. Dantzler. 2007. Three-dimensional architecture of collecting ducts, loops of Henle, and blood vessels in the renal papilla. *Am. J. Physiol.* 292: F696–F704.

Regnault, M. 1987. Nitrogen excretion in marine and fresh-water Crustacea. *Biol. Rev.* 62: 1–24.

Reilly, R. F., and D. H. Ellison. 2000. Mammalian distal tubule: Physiology, pathophysiology, and molecular anatomy. *Physiol. Rev.* 80: 277–313.

Robinson, W. E., and M. P. Morse. 1994. Biochemical constituents of the blood plasma and pericardial fluid of several marine bivalve molluscs: Implications for ultrafiltration. *Comp. Biochem. Physiol., B* 107: 117–123.

Rupert, E. E. 1994. Evolutionary origins of the vertebrate nephron. *Am. Zool.* 34: 542–553.

Sasaki, S., K. Ishibashi, and F. Marumo. 1998. Aquaporin-2 and -3: Representatives of two subgroups of the aquaporin family colocalized in the kidney collecting duct. *Annu. Rev. Physiol.* 60: 199–220.

Schipp, R., and F. Hevert. 1981. Ultrafiltration in the branchial heart appendage of dibranchiate cephalopods: A comparative ultrastructural and physiological study. *J. Exp. Biol.* 92: 23–35.

Schnermann, J. B., and S. I. Sayegh. 1998. *Kidney Physiology.* Lippincott-Raven, Philadelphia.

Schondube, J. E., L. Gerardo-Herrera-M., and C. Martínez del Rio. 2001. Diet and the evolution of digestion and renal function in phyllostomid bats. *Zoology* 104: 59–73.

Singer, M. A. 2001. Of mice and men and elephants: Metabolic rate sets glomerular filtration rate. *Am. J. Kidney Dis.* 37: 164–178.

Singer, M. A. 2003. Do mammals, birds, reptiles and fish have similar nitrogen conserving systems? *Comp. Biochem. Physiol., B* 134: 543–558.

Skorecki, K., G. M. Chertow, P. A. Marsden, M. W. Taal, and A. S. L. Yu (eds.). 2016. *Brenner and Rector's The Kidney.* 10th ed. Elsevier, Philadelphia.

Suzuki, M., and S. Tanaka. 2010. Molecular diversity of vasotocin-dependent aquaporins closely associated with water adaptation strategy in anuran amphibians. *J. Neuroendocrinol.* 22: 407–412.

Tryggvason, K., and J. Wartiovaara. 2005. How does the kidney filter plasma? *Physiology* 20: 96–101.

Tyler-Jones, R., and E. W. Taylor. 1986. Urine flow and the role of the antennal glands in water balance during aerial exposure in the crayfish, *Austropotamobius pallipes* (Lereboullet). *J. Comp. Physiol., B* 156: 529–535.

Urich, K. 1994. *Comparative Animal Biochemistry.* Springer Verlag, New York.

Walsh, P. J. 1997. Evolution and regulation of urea synthesis and ureotely in (Batrachoidid) fishes. *Annu. Rev. Physiol.* 59: 299–323.

Walsh, P. J., and P. Wright (eds.). 1995. *Nitrogen Metabolism and Excretion.* CRC Press, Boca Raton, FL.

Weihrauch, D., S. Morris, and D. W. Towle. 2004. Ammonia excretion in aquatic and terrestrial crabs. *J. Exp. Biol.* 207: 4491–4504.

Wenning, A. 1996. Managing high salt loads: From neuron to urine in the leech. *Physiol. Zool.* 69: 719–745.

Wessing, A. (ed.). 1975. Excretion. *Fortschr. Zool.* 23: 1–362. [Includes articles on many animal groups.]

Withers, P. C. 1998. Urea: diverse functions of a 'waste' product. *Clin. Exp. Pharmacol. Physiol.* 25: 722–727.

Wright, P. A., and P. M. Anderson (eds.). 2001. *Nitrogen Excretion. Fish Physiology,* vol. 20. Academic Press, New York.

Yancey, P. H. 1992. Compatible and counteracting aspects of organic osmolytes in mammalian kidney cells in vivo and in vitro. In G. N. Somero, C. B. Osmond, and C. L Bolis (eds.), *Water and Life,* pp. 19–32. Springer-Verlag, New York.

Chapter 30

Evenari, M. 1985. The desert environment. In M. Evenari, I. Noy-Meir, and D. W. Goodall (eds.), *Ecosystems of the World,* vol. 12A, *Hot Deserts and Arid Shrublands,* pp. 1–22. Elsevier, New York.

Gauthier-Pilters, H., and A. I. Dagg. 1981. *The Camel: Its Evolution, Ecology, Behavior, and Relationship to Man.* University of Chicago Press, Chicago.

Gereta, E., E. Wolanski, M. Borner, and S. Serneels. 2002. Use of an ecohydrology model to predict the impact on the Serengeti ecosystem of deforestation, irrigation and the proposed Amala Weir Water Diversion Project in Kenya. *Ecohydrol. Hydrobiol.* 2: 135–142.

Goudie, A., and J. Wilkinson. 1977. *The Warm Desert Environment.* Cambridge University Press, New York.

Henley, S. R., D. Ward, and I. Schmidt. 2007. Habitat selection by two desert-adapted ungulates. *J. Arid Environ.* 70: 39–48.

Ismail, K., K. Kamal, M. Plath, and T. Wronski. 2011. Effects of an exceptional drought on daily activity patterns, reproductive behaviour, and reproductive success of reintroduced Arabian oryx (*Oryx leucoryx*). *J. Arid Environ.* 75: 125–131.

Kröpelin, S., and 14 additional authors. 2008. Climate-driven ecosystem succession in the Sahara: the past 6000 years. *Science* 320: 765–768.

Maloiy, G. M. O. (ed.). 1972. *Comparative Physiology of Desert Animals.* Symposia of the Zoological Society of London, no. 31. Academic Press, New York.

Maloney, S. K., A. Fuller, G. Mitchell, and D. Mitchell. 2002. Brain and arterial blood temperatures of free-ranging oryx (*Oryx gazella*). *Pflügers Arch.* 443: 437–445.

Mitchell, D., S. K. Maloney, C. Jessen, H. P. Laburn, P. R. Kamerman, G. Mitchell, and A. Fuller. 2002. Adaptive heterothermy and selective brain cooling in arid-zone mammals. *Comp. Biochem. Physiol., B* 131: 571–585. [A disputatious challenge to much of conventional wisdom about the ways that large arid-zone mammals reduce water losses.]

Nix, H. A. 1983. Climate of tropical savannas. In F. Bourlière (ed.), *Ecosystems of the World,* vol. 13, *Tropical Savannas,* pp. 37–62. Elsevier, New York.

Ostrowski, S., P. Mésochina, and J. B. Williams. 2006. Physiological adjustments of sand gazelles (*Gazella subgutturosa*) to a boom-or-bust economy: standard fasting metabolic rate, total evaporative water loss, and changes in the sizes of organs during food and water restriction. *Physiol. Biochem. Zool.* 79: 810–819.

Ostrowski, S., and J. B. Williams. 2006. Heterothermy of free-living Arabian sand gazelles (*Gazella subgutturosa marica*) in a desert environment. *J. Exp. Biol.* 209: 1421–1429.

Ruckstuhl, K. E., and P. Neuhaus. 2009. Activity budgets and sociality in a monomorphic ungulate: the African oryx (*Oryx gazella*). *Can. J. Zool.* 87: 165–174.

Schroter, R. C., D. Robertshaw, and R. Z. Filadi. 1989. Brain cooling and respiratory heat exchange in camels during rest and exercise. *Respir. Physiol.* 78: 95–105.

Spalton, J. A. 1999. The food supply of the Arabian oryx (*Oryx leucoryx*) in the desert of Oman. *J. Zool., London* 248: 433–441.

Taylor, C. R. 1968. The minimum water requirements of some East African bovids. *Symposia of the Zoological Society of London*, no. 21, pp. 195–206. Academic Press, London.

Taylor, C. R. 1970. Dehydration and heat: Effects on temperature regulation of East African ungulates. *Am. J. Physiol.* 219: 1136–1139.

Taylor, C. R. 1970. Strategies of temperature regulation: Effect on evaporation in East African ungulates. *Am. J. Physiol.* 219: 1131–1135.

Williams, J. B., S. Ostrowski, E. Bedin, and K. Ismail. 2001. Seasonal variation in energy expenditure, water flux and food consumption of Arabian oryx *Oryx leucoryx*. *J. Exp. Biol.* 204: 2301–2311.

Williamson, D. T. 1987. Plant underground storage organs as a source of moisture for Kalahari wildlife. *Afr. J. Ecol.* 25: 63–64.

Wilson, R. T. 1989. *Ecophysiology of the Camelidae and Desert Ruminants*. Springer-Verlag, New York.

Wolanski, E., and E. Gereta. 2001. Water quantity and quality as the factors driving the Serengeti ecosystem, Tanzania. *Hydrobiologia* 458: 169–180.

Yagil, R. 1985. *The Desert Camel: Comparative Physiological Adaptation*. Karger, New York.

Index

Page numbers in *italic* type indicate that the information will be found in an illustration. Page numbers in **bold** indicate where an entry will be found as a key term. Page numbers followed by an "n" and a number indicate that the information is in a numbered footnote.

A

A bands, **538,** 539, *540*
Abdominal ganglia, *410*
Abomasum, *150*
Absolute refractory period, **320,** 332
Absorbed chemical energy, **168–169**
Absorbed energy, **168–169**
Absorption, **152**
  arthropods and molluscs, 154
  hindgut, 153
  interrelationship with feeding, nutrition, and digestion, 132–133
  milk digestion in humans, 131–132, 162–163
  vertebrates, 152–153, 156–158
Absorption coefficient, **588,** 591
Absorption spectra, of respiratory pigments, **637**
Acacia trees, 826
Acanthurus leucosternum (powder blue surgeonfish), *741*
Acceleration perception, 379
Accessory olfactory bulbs, *390*
Accessory sex glands, 490
Acclimation, **17–18**
  changes in enzyme levels and, 53
  changes in myosin isoforms, 253–254
  modulation of the oxygen affinity of respiratory pigments and, 651
  in poikilotherms, 244–247, 253–254
  in respiratory physiology of individuals, 656–657
Acclimatization, **17–18**
  changes in enzyme levels and, 53
  changes in myosin isoforms, 253–254
  insulatory, **273**
  of metabolic endurance, **273**
  modulation of the oxygen affinity of respiratory pigments and, 651
  of peak metabolic rate, **272–273**
  in poikilotherms, 247, 253–254
  in respiratory physiology of individuals, 656–657
  seasonal, in homeotherms, 272–274
"ACE-inhibitors," 451
Acetate, 353
Acetylcholine (ACh)
  breakdown by acetylcholinesterase, 345
  cholinergic neurons, 419
  cholinergic synapses (*see* Cholinergic synapses)

contraction and relaxation of smooth muscles, 560, 561
effect on pacemaker potentials, 329
electric organs and, 550
excitation–contraction coupling, 543, *544*
homeostatic regulation of the metabolism of, 360–361
ligand-gated channels and, 64
quantal release by the presynaptic neuron, 347–351
release in the neuromuscular junction, 345
synthesis and storage in the presynaptic terminal, 348
synthetic enzyme and receptor, *352*
termination and reuptake, 353
Acetylcholine (ACh) receptor
  blocker, 352
  ligand-gated channels that function as ionotropic receptors, 355–356
  in the neuromuscular junction, 345
  nicotinic and muscarinic, *352*, 353
Acetylcholinesterase (AChE), 253, 345, 353, 361
Acetyl coenzyme A, *191*, 192, 348
Acid–base physiology
  acidification of aquatic habitats, 665
  ion exchange in freshwater animals and, 745
  overview and the alphastat hypothesis, 663–664
  processes of acid–base regulation, 664
  properties of buffers, 659–660
  respiratory and metabolic categories of disturbances, 664
  respiratory pigments as buffers, 653
  See also Blood buffers
Acidification
  of aquatic habitats, 665
  in diving marine mammals, 713
Acidosis, **664**
Acid rain, 665
Acids, secreted by the stomach, 153
Acid (sour) taste, *385*, 386
Acoutisco-lateralis system, 377–378
Acrosomal reaction, **490**
Acrosome, *487*
Actb gene, *802*
Actg1 gene, *802*
Actin
  atrophy and, 575
  as a contractile protein, **537**
  polymerization, 540
  See also Thin filaments
Action potential propagation
  characteristics of, 320
  factors affecting conduction velocity, 332–335
  local circuits of current, 331–332
  membrane refractory periods prevent bidirectional propagation, 332
  overview, 330–331

Action potentials, 306, **320**
  cardiac muscle, 329–330, 561, 669
  change in membrane permeability to ions, 321–326
  cockroach startle response, 309
  excitable cells and, 320
  excitation–contraction coupling, 543, *544*
  frequency of determines tension developed by a muscle, 547
  ion movements do not affect bulk ion concentrations, 326
  molecular structure of voltage-gated ion channels, 326–328
  phases of, 320, *321*
  positive feedback, 15
  smooth muscle, 559
  summation of, 343
  as voltage-dependent, all-or-none electrical signals, 320–321, 331
Activation energy, **47**
Active evaporative cooling, **264–265,** 270, 271–272
Active exhalation, **612**
Active site, of enzymes, **49**
Active solute secretion, **782–783**
Active transport
  in digestive absorption, 156–158
  generation of the single effect in the loop of Henle, 792, *793*
  ionic hypothesis of membrane potential, 319
  of ions, 318
  ion uptake in freshwater animals, 744–747
Active-transport mechanisms (active-transport pumps), **105, 113**
  basic properties of, 113–114
  as carrier-mediated transport, 113
  cell-membrane and whole-epithelium perspectives, 119–120
  control of passive water transport and, 127
  energy coupling via the potential energy of electrochemical gradients, 118
  epithelial ion-pumping mechanisms in freshwater fish, 120
  examples, 114–115
  overview and summary, 113, 121
  primary and secondary, 115–117, 119
  in a typical animal cell, *114*
Active ventilation, **601**
  mammals, 617–619, 621
  principles of gas exchange, 601–604
  vertebrates, control of, 607
Active vitamin D, 452, *453*
Active zones, **340**
Actograms, 422–423
Acute physiological responses, **16–17**
  to eating, 159–160
  enzymes and, 58
  poikilotherms, 243–244
Adaptation, **9, 27**

A Simplified Phylogenetic Tree of the Animals and Descriptions of Major Phyla

This tree and the accompanying descriptions emphasize the animals that receive greatest attention in this book. The descriptions highlight properties of importance in the study of physiology and are not intended as formal definitions. The numbers of species are estimates of numbers of living species currently described. The protostomes and deuterostomes differ in features of embryonic development, e.g., whether the mouth develops at or near the embryonic blastopore (protostomes) or far from it (deuterostomes).

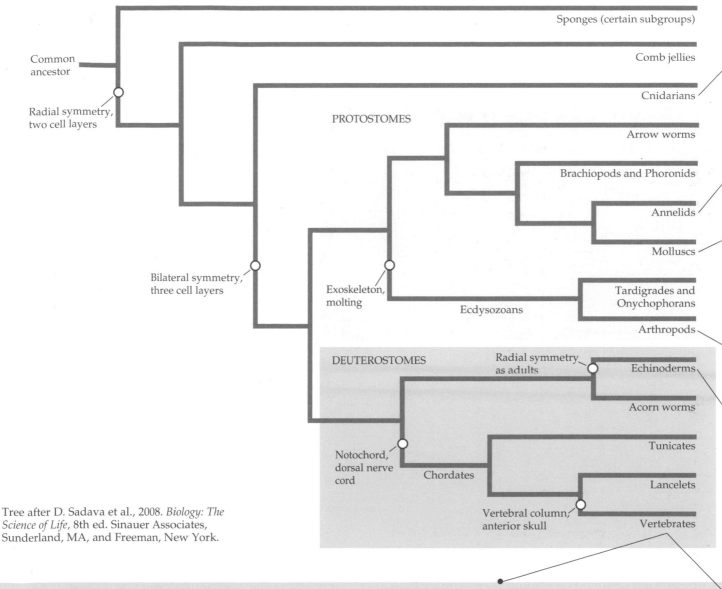

Common ancestor

Radial symmetry, two cell layers

Sponges (certain subgroups)

Comb jellies

Cnidarians

PROTOSTOMES

Arrow worms

Brachiopods and Phoronids

Annelids

Molluscs

Bilateral symmetry, three cell layers

Exoskeleton, molting

Ecdysozoans

Tardigrades and Onychophorans

Arthropods

DEUTEROSTOMES

Radial symmetry as adults

Echinoderms

Acorn worms

Notochord, dorsal nerve cord

Chordates

Tunicates

Lancelets

Vertebral column, anterior skull

Vertebrates

Tree after D. Sadava et al., 2008. *Biology: The Science of Life*, 8th ed. Sinauer Associates, Sunderland, MA, and Freeman, New York.

Vertebrates (e.g., fish, mammals), although not a phylum in themselves, are the principal subphylum of chordates. Have an internal skeleton (calcified in all but cartilaginous fish) with dorsal vertebral column. Major longitudinal nerve cord (spinal cord) dorsal. Head with brain, eyes, and other sense organs well developed. Circulatory system elaborate and closed, blood rich with hemoglobin. 65,000 species.

Fish (which include several distinct subgroups) are by far the most diverse vertebrates in number of species. Some, notably *elasmobranchs* (e.g., sharks, rays), have cartilaginous skeletons. Most, including *teleosts* (the principal bony fish), have calcified skeletons. Breathe with gills, but ~400 species can breathe air. Most ectothermic even if very large; some (notably tunas, lamnid sharks) endothermic. Some with (weak or strong) electric organs evolved from muscle. 33,000 species.

Amphibians are *tetrapods* (four-legged, or derived from four-legged, animals) that typically have water- and gas-permeable skin rich with secretory (e.g., mucus) glands. With exceptions, adults are physiologically poorly defended against desiccation and must stay near moisture to minimize dehydration and rehydrate. Over half are tied to water for reproduction, start life as freshwater larvae. 7500 species.

Reptiles other than birds—i.e., lizards, snakes, turtles, and crocodilians—are, for the most part, tetrapod ectotherms well defended against desiccation. Skin is less permeable than that of amphibians, sometimes of very low permeability. Usually terrestrial. Unlike amphibians, have amniotic eggs (as do birds and mammals). 9000 species.